Interest Boxes

Max Karl Ernst Ludwig Planck (1.2)

Diamond, Graphite, and Buckminsterfullerene: Substances Containing Only Carbon Atoms (1.8)

Water: A Unique Compound (1.11)

Blood: A Buffered Solution (1.20)

Highly Strained Hydrocarbons (2.11)

Baeyer and Barbituric Acid (2.11)

Cis-Trans Interconversion in Vision (3.4)

The Enantiomers of Thalidomide (4.13)

Chiral Drugs (4.13)

Lindlar's Catalyst (5.8)

Ethyne Chemistry or the Forward Pass? (5.12)

Kekulé's Dream (6.1)

Fossil Fuels: A Problematic Energy Source (8.0)

Decaffeinated Coffee and the Cancer Scare (8.8)

Food Preservatives (8.8)

The Concorde and Ozone Depletion (8.9)

Alkyl Halides as Survival Compounds (9.0)

Why Carbon Instead of Silicon? (9.4)

Solvation Effects (9.10)

S-Adenosylmethionine: A Natural Antidepressant? (9.11)

The Lucas Test (11.1)

Grain Alcohol and Wood Alcohol (11.1)

Biological Dehydrations (11.4)

Anesthetics (11.5)

Benzo(a)pyrene and Cancer (11.7)

Chimney Sweeps and Cancer (11.7)

An Inophorous Antibiotic (11.9)

Mustard Gas (11.10)

Antidote to a War Gas (11.10)

A Useful, Bad-Tasting Compound (11.11)

The Originator of Hooke's Law (12.10)

Ultraviolet Light and Sunscreens (12.16)

Anthocyanins: A Colorful Class of Compounds (12.19)

Nikola Tesla (13.1)

Buckyballs and AIDS (14.2)

Dioxin (15.10)

Peyote Cults (15.0)

A Few Words of Warning (15.12)

Nitrosamines and Cancer (15.12)

The Discovery of Penicillin (16.5)

Dalmations: Don't Try to Fool Mother Nature (16.5)

Aspirin (16.9)

Nature's Sleeping Pill (16.13)

Penicillin and Drug Resistance (16.14)

Penicillins in Clinical Use (16.14)

Making Soap (16.17)

Using ATP for Biosynthesis (16.18)

Nerve Impulses, Paralysis, and Insecticides (16.18)

Butanedione: An Unpleasant Compound (17.1)

Nonspectrophotometric Identification of Aldehydes and Ketones (17.7)

Preserving Biological Specimens (17.8)

β-Carotene (17.11)

Enzyme-Catalyzed Carbonyl Additions (17.12)

Enzyme-Catalyzed Cis-Trans Interconversion (17.16)

Blood Alcohol Content (18.2)

Treating Alcoholics with Antabuse (18.10)

An Unusual Antidote (18.10)

Fetal Alcohol Syndrome (18.10)

The Chemistry of Photography (18.11)

The Synthesis of Aspirin (19.9)

Measuring the Blood Glucose Levels of Diabetics (20.6)

Glucose/Dextrose (20.9)

Lactose Intolerance (20.17)

Galactosemia (20.17)

Why the Dentist is Right (20.18)

Controlling Fleas (20.18)

Synthetic Fibers (20.18)

Heparin (20.19)

Vitamin C (20.19)

The Wonder of Discovery (20.21)

Amino Acids and Disease (21.2)

Water Softeners: Examples of Cation-Exchange Chromatography (21.5)

Hair: Straight or Curly? (21.7)

Primary Structure and Evolution (21.11)

β-Peptides: An Attempt to Improve on Nature (21.13)

Vitamin B$_1$ (23.0)

Niacin Deficiency (23.2)

Heart Attacks: Assessing the Damage (23.6)

Phenylketonuria: An Inborn Error of Metabolism (23.6)

The First Antibacterial Drugs (23.9)

Too Much Broccoli (23.9)

Olestra: Nonfat with Flavor (24.3)

A Whale's Echolocation (24.3)

Multiple Sclerosis and the Myelin Sheath (24.4)

Is Chocolate a Health Food? (24.4)

Cholesterol and Heart Disease (24.9)

Clinical Treatment of High Cholesterol (24.9)

The Structure of DNA: Watson, Crick, Franklin, and Wilkens (25.0)

Sickle Cell Anemia (25.13)

Antibiotics that Act by Inhibiting Translation (25.13)

DNA Fingerprinting (25.15)

Designing a Polymer (26.7)

Porphyrin, Bilirubin, and Jaundice (27.4)

Cold Light (28.4)

Semisynthetic Drugs (29.0)

Orphan Drugs (30.13)

Student Aids

Derivation of the Henderson-Hasselbalch Equation (1.20)

A Few Words About Curved Arrows (3.6)

Rate Constants (3.7)

The Difference Between $\Delta G^{\dagger}$ and E_a (3.7)

Calculating Kinetic Parameters (3.7)

Borane and Diborane (3.17)

Configuration and Cycloalkenes (4.17)

Sodium Amide and Sodium (5.9)

Incipient Primary Carbocations (14.13)

Proton Transfer Steps (16.6)

The Role of Hydrates in the Oxidation of Primary Alcohols (18.2)

The Hunsdiecker Reaction (19.17)

Organic Chemistry

THIRD EDITION

Organic Chemistry

Paula Yurkanis Bruice

University of California, Santa Barbara

Prentice
Hall

PRENTICE HALL
Upper Saddle River, New Jersey 07458

Library of Congress Cataloging-in-Publication Data

Bruice, Paula Yurkanis
  Organic chemistry / Paula Yurkanis Bruice.—3rd ed.
    p. cm.
  Includes index.
  ISBN 0-13-017858-6
   1. Chemistry, Organic. I. Title.
  QD251.2.B784 2001
  547—dc21
                           00-044108
                                  CIP

Executive Editor: *John Challice*
Development Editor: *John Murdzek*
Executive Managing Editor: *Kathleen Schiaparelli*
Senior Marketing Manager: *Steve Sartori*
Editorial Assistant: *Gillian Buonanno*
Associate Editor: *Kristen Kaiser*
Manufacturing Manager: *Michael Bell*
Art Editor: *Karen Branson*
Art Director/Cover Designer: *Joseph Sengotta*
Photo Researcher: *Mary Teresa Grancoli*
Editor in Chief, Development: *Carol Trueheart*
Assistant Managing Editor, Science Media: *Alison Lorber*
Media Editor: *Paul Draper*
Assistant Vice President of Production and Manufacturing: *David W. Riccardi*
Associate Creative Director: *Carole Anson*
Art Manager: *Gus Vibal*
Interior Designer: *Judith A. Matz-Coniglio*
Front Cover Photo: *Soft coral display grainy yellow tentacles; Stephen Frink / The Stock Market*
Back Cover Photo: *Orange encrusting coral; courtesy of Phillip Crews, Department of Chemistry, University of California at Santa Cruz*
Photo Research Administrator: *Melinda Reo*
Art Studio: *Academy ArtWorks, Inc.*
Copyediting, Text Composition, and Electronic Page Makeup: *WestWords, Inc.*

Spectra reproduced by permission of Aldrich Chemical Co.

© 2001, 1998, 1995 by Prentice-Hall, Inc.
Pearson Education
Upper Saddle River, New Jersey 07458

Printed in the United States of America
10 9 8 7 6 5 4 3 2

ISBN 0-13-017858-6

Prentice-Hall International (UK) Limited, *London*
Prentice-Hall of Australia Pty. Limited, *Sydney*
Prentice-Hall Canada Inc., *Toronto*
Prentice-Hall Hispanoamericana, S.A., *Mexico*
Prentice-Hall of India Private Limited, *New Delhi*
Prentice-Hall of Japan, Inc., *Tokyo*
Pearson Education Asia Pte. Ltd., *Singapore*
Editora Prentice-Hall do Brasil, Ltda., *Rio de Janeiro*

To Meghan, Kenton, and Alec
with love and immense respect
and to Tom, my best friend

Brief Contents

Contents

PART I

An Introduction to the Study of Organic Chemistry

1   Electronic Structures and Bonding. Acids and Bases   2

HF

H_2O

NH_3

CH_4

2   An Introduction to Organic Compounds: Nomenclature, Physical Properties, and Representation of Structure   60

PART II

Hydrocarbons, Stereochemistry, and Resonance   109

3   Reactions of Alkenes. Thermodynamics and Kinetics   111

4   Stereochemistry: The Arrangement of Atoms in Space; The Stereochemistry of Addition Reactions   180

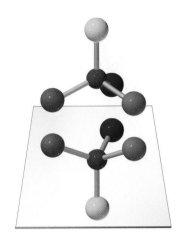

5   Reactions of Alkynes. Introduction to Multistep Synthesis   239

6   Electron Delocalization and Resonance   266

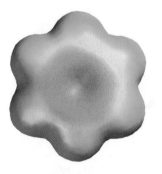

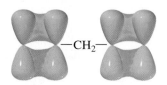

PART III

Substitution and Elimination Reactions 353

10 Reactions at an *sp³* Hybridized Carbon II:
Elimination Reactions of Alkyl Halides; Competition Between Substitution and Elimination **395**

11 Reactions at *sp³* Hybridized Carbon III:
Substitution and Elimination Reactions of Compounds with Leaving Groups Other Than Halogen. Organometallic Compounds **429**

CH_3OH

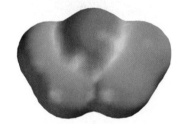

CH_3OCH_3

PART IV

Identification of Organic Compounds **479**

12 Mass Spectrometry, Infrared Spectroscopy, and Ultraviolet/Visible Spectroscopy **480**

13 NMR Spectroscopy 534

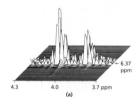

PART V

Aromatic Compounds 601

14 Aromaticity. Reactions of Benzene 602

15 Reactions of Substituted Benzenes 627

PART VI

Carbonyl Compounds 677

16 Carbonyl Compounds I:
Reactions of Carboxylic Acids and Their Derivatives with Oxygen And Nitrogen Nucleophiles 678

17 Carbonyl Compounds II:
Reactions of Carbonyl Compounds with Carbon and Hydrogen Nucleophiles;
Reactions of Aldehydes and Ketones with Oxygen and Nitrogen Nucleophiles;
Reactions of α,β-Unsaturated Carbonyl Compounds 735

formaldehyde

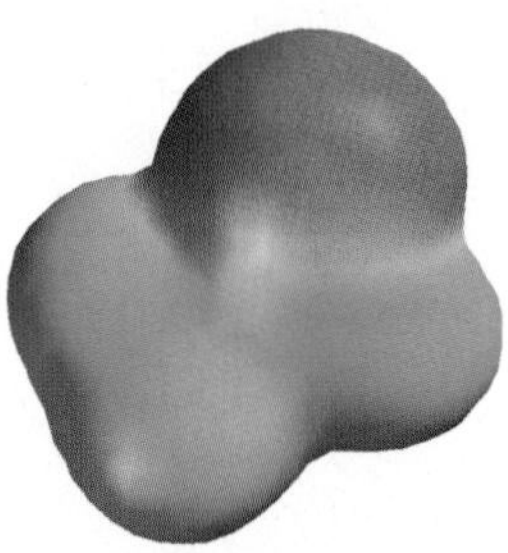

acetaldehyde

acetone

18 More About Oxidation–Reduction Reactions   789

19 Carbonyl Compounds III:
Reactions at the α-Carbon   827

acetyl-CoA

PART VII

Bioorganic Compounds   877

20 Carbohydrates   878

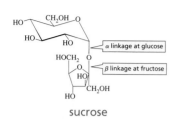

sucrose

21   Amino Acids, Peptides, and Proteins   916

22   Catalysis   957

23   The Organic Mechanisms of the Coenzymes. Metabolism   990

Dorothy Crowfoot Hodgkin

24   Lipids   1033

25   Nucleosides, Nucleotides, and Nucleic Acids   1063

PART VIII

Special Topics in Organic Chemistry 1103

26 Synthetic Polymers 1104

27 Heterocyclic Compounds 1133

pyrrole

28 Pericyclic Reactions 1164

furan

29 More About Multistep Organic Synthesis 1195

thiophene

30 The Organic Chemistry of Drugs:
Discovery and Design 1226

Paul Ehrlich

Appendices A-1

Preface

TO THE INSTRUCTOR

The guiding principle in writing this book was to create a text for students—a text that presents the material in a way that encourages students to think about what they have already learned and then apply this knowledge in a new setting. I want students to reason their way to a solution rather than memorize a multitude of facts, hoping they don't run out of memory before the course ends. I am convinced that the *what* of organic chemistry is much easier to grasp and retain if the *why* is understood.

Comments I have received from both colleagues and students on the second edition indicate the book is working in the way I had hoped. As much as I cherish having faculty tell me that their students are scoring higher than ever before on tests, nothing is more gratifying than hearing from the students themselves—a wonderful advantage of being the author of a textbook during the e-mail age. (Even my dog has received e-mail from students.) Many students have generously accredited their success in organic chemistry to the book—not giving themselves nearly enough credit for how hard they studied to achieve that success. And they always seem surprised that they have come to love "orgo" (on the East Coast) or "o-chem" (on the West Coast). I also hear from many pre-meds who say that the book gave them such a permanent understanding of organic chemistry that they found it to be the easiest part of that dreaded test, the MCAT.

As I strove to make this book even more useful to the student, I have relied on constructive critical comments from many of you. For these I am very grateful. I also kept a journal of questions students had when they came to my office hours. These questions let me know what sections in the book needed clarifying and what answers in the Study Guide/Solutions Manual needed more in-depth explanations. Most importantly, this analysis showed me where new problems needed to be created so that these same questions had less of a chance of being asked by students using the third edition. Because I teach a class of more than 400 students, I clearly have a vested interest in minimizing their confusion! In this edition, many sections have been rewritten to optimize readability and comprehension, and there are many new in-chapter problems, new end-of-chapter problems, and more solved problems to help students master organic chemistry by solving problems. There are also new interest boxes to show students the applications of organic chemistry, and additional margin notes to remind students of important concepts and principles.

I hope you find the third edition even more appealing to your students than the second. As always, I am eager to hear your comments—positive comments are the most fun, but critical comments are the most useful.

I would also like to draw your attention to the following features of the text:

A Functional Group Approach with a Mechanistic Organization That Ties Together Synthesis and Reactivity

This book is organized to discourage the student from rote memorization. The functional groups have been organized around mechanistic similarities—electrophilic additions, radical substitutions, nucleophilic substitutions, eliminations, electrophilic aromatic substitutions, nucleophilic acyl substitutions, and nucleophilic additions. This organization allows a lot of material to be understood based on unifying principles of reactivity.

In addition, instead of discussing the synthesis of a functional group when its reactivity is discussed—reactions that generally have little to do with each other—I discuss the synthesis of the compounds that are formed as a result of the functional group's reactivity. In the alkene chapter, for example, students learn about the reactions of alkenes but they do *not* learn at this point about the synthesis of alkenes. Instead, they learn about the synthesis of alkyl halides, alcohols, ethers, and alkanes—the compounds formed when alkenes react. Because alkenes are synthesized from the reactions of alkyl halides and alcohols, the synthesis of alkenes is covered when the reactions of alkyl halides and alcohols are discussed. Tying together the reactivity of a functional group and the synthesis of compounds resulting from its reactivity prevents the student from having to memorize lists of unrelated reactions. Understanding different ways functional groups can be prepared is useful in designing syntheses, however, so the different reactions that yield a particular functional group are compiled in Appendix IV. As students learn how to design syntheses, they appreciate the importance of reactions that change the carbon skeleton of a molecule; these reactions are compiled in Appendix V.

A Detailed Look at the Organization of the Text

The book has been divided into eight sections. Each section starts with a one-page overview so students can understand where they are "going." Chapter 1 provides a summary of the material that students need to recall from General Chemistry. The sections on acids and bases were rewritten to emphasize the relationship between acidity and stability of the conjugate base. (Acids and bases are covered even more extensively in the Study Guide/Solutions Manual.) In Chapter 2, students learn how to name five classes of organic compounds—those that will be the products of the reactions in the chapters that immediately follow. Chapter 2 also covers topics that are necessary before the study of reactions can begin—structures, conformations, and physical properties of organic compounds. Chapter 3 discusses the reactions of alkenes. In this chapter, students are first introduced to the concept of "curved arrows." (The Study Guide/Solutions Manual contains an extensive exercise on "electron pushing." I have found this exercise to be very successful in making my students comfortable with a topic that should be easy, but somehow perplexes even the best of them unless they have sufficient practice.) Chapter 3 also contains a discussion of thermodynamics and kinetics. This section was rewritten to make it even easier for students to understand. Rate equations are derived in an appendix for those who wish to give their students a more mathematical treatment, and the Study Guide/Solutions Manual contains a section on calculating kinetic parameters.

I chose to lead off with the reactions of alkenes because of their simplicity. Chapter 3 covers a wide variety of reactions but they all have similar mechanisms—an electrophile adds to the least substituted sp^2 carbon and a nucleophile adds to the other sp^2 carbon. The many reactions differ only in the nature of the electrophile and the nucleophile. Because organic chemistry is all about the interactions of electrophiles and nucleophiles, starting the study of organic reactions in a way that familiarizes students with a wide variety of electrophiles and nucleophiles makes sense. The reactions in Chapter 3 are discussed without regard to stereochemistry because I have found that students do well as long as only one new concept is introduced at a time. Chapter 4 reviews the isomers introduced in Chapters 2 and 3 (conformers and cis–trans isomers) and then discusses isomers that result from having a chirality center (the most recent IUPAC-approved term for a carbon bonded to four different groups). There is an extensive model-building exercise in the Study Guide/Solutions Manual to help the students get used to building molecular models, and the book's Companion Website gives students the opportunity to manipulate many molecules in three dimensions. Now that the students are comfortable with both isomers and electrophilic addition reactions, the two topics are considered together at the end of Chapter 4, where the stereochemistry of the addition reactions covered in Chapter 3 is presented. Chapter 5 covers alkynes. This chapter should build the students' confidence because of the similarity of the material to what was presented in Chapter 3.

Understanding electron delocalization is vitally important in organic chemistry, so it is covered in its own chapter (Chapter 6). This chapter is a continuation of the introduction to this topic in Chapter 1. Chapter 7 covers the reactions of dienes, allowing students to apply the concept of electron delocalization just studied to the electrophilic addition reactions that were mastered in Chapters 3 and 5. Chapter 8 covers the reactions of alkanes. The lack of a functional group in alkanes further emphasizes the importance of a functional group to chemical reactivity.

The next three chapters deal with substitution and elimination reactions at an sp^3 hybridized carbon. Chapter 9 covers substitution reactions of alkyl halides. Chapter 10 covers elimination reactions of alkyl halides and then goes on to consider competition between substitution and elimination. Chapter 11 covers substitution and elimination reactions at an sp^3 hybridized carbon when a group other than a halogen is the leaving group—reactions of alcohols, ethers, epoxides, thiols, sulfides, and quaternary ammonium ions.

Next are two spectroscopy chapters (mass spectrometry, IR spectroscopy, and UV/Vis spectroscopy in Chapter 12, followed by NMR spectroscopy in Chapter 13). Each spectral technique is written as a "stand-alone topic" so it can be covered independently at any time during the course. Chapter 12 opens with a table of functional groups for those who want to cover spectroscopy before students have been introduced to all the functional groups.

Chapter 14 covers aromaticity and the reactions of benzene. Chapter 15 discusses the reactions of substituted benzenes. Because not all organic courses cover the same amount of material in a semester, Chapters 12–15 have been strategically placed to come near the end of the first semester. Ending a semester before Chapter 12, Chapter 13, or Chapter 14, or after Chapter 15 won't interfere with the flow of information.

The order of the topics dealing with the chemistry of carbonyl compounds in Chapters 16 and 17, where carboxylic acid derivatives are discussed before aldehydes and ketones, was initially met with skepticism by many considering using the book. However, when the opinions of those who had used the two editions of the book were sought, there was strong agreement that this is a preferred order of treatment. Chapter 16 discusses the reactions of carboxylic acids and their derivatives with oxygen and nitrogen nucleophiles. Thus, the students are introduced to carbonyl chemistry by learning how tetrahedral intermediates partition. Chapter 17 starts by discussing the reactions of aldehydes, ketones, and carboxylic acid derivatives with strong (carbon and hydrogen) nucleophiles. By studying these compounds as a group, students see how the reactions of aldehydes and ketones differ from those of carboxylic acid derivatives. Then when they move on to study the formation and hydrolysis of imines, enamines, and acetals, they can easily understand these mechanisms because they are well versed in how tetrahedral intermediates partition. Over the years I have experimented with my classes and I believe this to be the most effective and easiest way to teach carbonyl chemistry. That being said, these chapters can easily be switched so long as Section 17.6 is skipped and then covered with Chapter 16.

Chapter 18 revisits reduction reactions, and discusses oxidation reactions, which my students tend to find troubling. I have found it helps to cover them all together as a unit. Because the reduction reactions appear in earlier chapters, this is an opportunity for students to extend their knowledge within the framework of what they already know. Chapter 19 deals with reactions at the α-carbon of carbonyl compounds.

One special note about amines. There is no question that amines are exceedingly important in organic chemistry. (In Chapter 30, The Organic Chemistry of Drugs, Table 30.1 lists the names, structures, and uses of the most widely prescribed drugs, and a majority of them are amines.) So then, why is there no separate chapter on amines?

My aim again is to organize organic chemistry in a way that allows students to understand and predict rather than memorize. This book covers the functional groups from the point of how they react rather than what they are called. Amines don't undergo additions, substitutions, or eliminations. Their reactivity lies in how they react with other organic compounds. Thus, all the chemistry that would be found in a typical chapter called Amines is in this book, but it is found when that reactivity with other compounds is discussed (for example, acid/base properties of aliphatic and aromatic amines—Chapters 1, 6, 15, and 27; amines as nucleophiles in substitution reactions—Chapters 9 and 16; amines as nucleophiles in addition reactions—Chapter 17; and elimination reactions of quarternary ammonium ions and amine oxides—Chapter 11). Amine synthesis is covered in those sections where amines are the products of the reactions being discussed (for example, the Gabriel synthesis of amines results from imide hydrolysis; amines are synthesized as the result of S_N2 reactions; and anilines are synthesized as a result of electrophilic aromatic substitution followed by reduction or as a result of nucleophilic aromatic substitution).

A Modular Organization in the Last Third of the Book

I anticipate that the first 21 chapters will be covered by most instructors during a year-long course. An instructor can then choose among the remaining chapters depending on his or her preference and the interests of the majority of the students enrolled in the course. Those teaching students whose interests are primarily in the biological and health sciences might be more inclined to cover Chapter 22 (Catalysis), Chapter 23 (The Organic Mechanisms of the Coenzymes. Metabolism), Chapter 24 (Lipids), and Chapter 25 (Nucleosides, Nucleotides, and Nucleic Acids). Those teaching courses designed for chemistry or engineering majors may decide to include Chapter 26 (Synthetic Polymers), Chapter 27 (Heterocyclic Compounds), Chapter 28 (Pericyclic Reactions), and Chapter 29 (More About Multistep Organic Synthesis). The book ends with a chapter on drug discovery and design—a topic that in my experience interests students enough that they will choose to read it on their own, even if it is not covered in the course.

Changes to This Edition—Organization

In response to reviewer comments, electron delocalization and resonance are now introduced in Chapter 1. The number and kinds of Kekulé structures students see at the beginning of their study has been increased. Chapter 2 now covers the conformations of both mono- and disubstituted cyclohexanes. Material on aromaticity that appeared in Chapter 6 of the second edition has been expanded and moved to a new Chapter 14. In Chapter 12, the treatment of mass spectrum fragmentation patterns has been reworked to emphasize the common behavior of a variety of compounds when they fragment. The treatment of IR spectroscopy has been reorganized and expanded and includes many additional FT-IR spectra. Because the NMR spectra in the book are FT-NMR spectra and because FT-NMR spectra are what most students see in their laboratory courses, the discussion of the theory of NMR spectroscopy in Chapter 13 describes Fourier transform spectrometers, a theory easier for students to understand than the traditional description of continuous wave spectrometers. In response to user feedback, more problems that involve interpreting spectra have been added to this edition. To help students analyze these specta, more signals have been enlarged. As in the last edition, most of the ^{1}H NMR spectra shown are 300 MHz FT-NMR spectra, which are more like the NMR spectra students are likely to see in their undergraduate laboratories and after they graduate. More spectroscopy problems have also been added to subsequent chapters to enhance the integration of spectroscopy throughout the course. In Chapter 19, the sections on enolate chemistry have been expanded and clarified.

Organic Chemistry with a Bioorganic Flavor

Bioorganic material is introduced throughout this text to encourage students to recognize that organic chemistry and biochemistry are not separate entities, but two parts of a continuum of knowledge. For example, students learn not just how carboxylic acids are activated for reaction in the laboratory, but also how they are activated for reaction in biological systems and why the mode of activation differs in the two situations. Once students learn how such things as electron delocalization, leaving-group tendency, electrophilicity, and nucleophilicity affect the reactions of simple organic compounds, they can appreciate how these same factors are involved in the reactions of more complicated organic molecules such as enzymes, nucleic acids, and vitamins.

In the first part of the book, the bioorganic material is limited primarily to interest boxes and the last sections of the chapters, so the material is available to the curious student but the instructor is not compelled to introduce bioorganic topics into the course. Later in the book there are several chapters that focus heavily on bioorganic topics, including the mechanisms of coenzymes, the chemistry of drug design, and the chemistry of biological macromolecules. Each instructor may choose which, if any, of these chapters to cover.

The chapters on bioorganic chemistry (Chapters 20–25) emphasize the chemistry associated with the topic. They contain more chemistry than one would expect to find in a biochemistry text. In the lipids chapter, for example, the mechanisms of prostaglandin formation (allowing students to understand how aspirin works), fat breakdown (allowing students to understand the odor associated with rancidity), and terpene biosynthesis are covered. The chapter on coenzymes emphasizes the role of vitamin B_1 as an electron delocalizer, vitamin K as a strong base, vitamin B_{12} as a radical initiator, biotin as a compound that can transfer a carboxyl group, and how the many different reactions of vitamin B_6 are controlled by the overlap of p orbitals. The chapter on nucleic acids explains mechanistically such things as the function of ATP, why DNA contains thymine instead of uracil, and how DNA strands are synthesized in the laboratory. The chapter on catalysis explains the various modes of catalysis that occur in organic reactions and then shows that these are identical to the modes of catalysis that occur in enzymatic reactions—all presented in a way that allows students to understand the lightening-fast rates of enzymatic reactions. Thus, these chapters are not a repetition of what will be covered in a biochemistry course, but are designed to serve as a bridge between the two disciplines.

An Early and Consistent Emphasis on Organic Synthesis

Students are introduced to synthetic chemistry and retrosynthetic analysis early in this book (Chapters 3 and 5, respectively), so they can use this technique throughout the course as they design multistep syntheses. In addition, there are six special sections spread throughout the book on synthesis design, each with a different focus. For example, one emphasizes the proper choice of reagents and reaction conditions to maximize the yield of the target molecule (Chapter 10), and another focuses on the synthesis of cyclic compounds (Chapter 17). Moreover, Chapter 29 (More About Multistep Organic Synthesis) discusses such things as protecting groups, control of stereochemistry during synthesis, and retrosynthetic analysis at a more advanced level. The use of combinatorial methods in organic synthesis is described in Chapter 30, which focuses on drug design.

Changes to This Edition—Synthesis

In this edition I have added more synthetic problems, in particular new multistep (road-map) problems. I also added more problems that involve the synthesis of compounds that students will recognize (like ketoprofen, Novocain, and Valium).

Margin Notes and Boxed Material to Engage the Student

Margin notes and biographical sketches appear throughout the text. The margin notes remind students of important principles, and the biographical sketches give students some appreciation of the history of chemistry and the people who contributed to that history. Interest Boxes have been included to bring life to the topic being discussed (e.g., Alkyl Halides as Survival Compounds, Chimney Sweeps and Cancer, Ultraviolet Light and Sunscreens, and Penicillin and Drug Resistance) or to give the student extra help (e.g., Calculating Kinetic Parameters, A Few Words About Curved Arrows, and Incipient Primary Carbocations).

Changes to This Edition—Interest Boxes

There are a dozen new boxes in this edition, including one on the pericyclic reaction that makes fireflies glow and one on how the popular new OTC antidepressant SAMe methylates neurotransmitters. I also extensively updated the boxes that remain from previous editions (for example, I mention the recent synthesis of octanitrocubane in the box on strained carbon compounds in Chapter 1).

My students really liked the margin notes in the second edition, so I added more of these in this edition. These emphasize core points and offer pedagogical advice.

Problems, Solved Problems, and Problem-Solving Strategies

The book contains more than 1500 problems. The problems within each chapter are primarily drill problems. They are designed so students can test themselves on the material just covered before they go on to the next section. Solutions to selected problems are shown to give the student an understanding of how to solve a problem, and short answers are provided in the Appendix for problems marked with a diamond so students can quickly test their understanding. Most chapters also contain at least one Problem-Solving Strategy that teaches students how to approach certain kinds of problems. For example, the Problem-Solving Strategy in Chapter 9 teaches students how to determine whether a reaction will be more apt to take place by an S_N1 or an S_N2 pathway. Each Problem-Solving Strategy is followed by an exercise that gives the student an opportunity to use the problem-solving skill just learned.

The end-of-chapter problems are of varying difficulty. The initial problems are drill problems that integrate material from the entire chapter. This provides a greater challenge by requiring the student to think about all the material in the chapter rather than individual sections. The end-of-chapter problems are purposefully not labeled to indicate the sections the material comes from. The problems become more challenging as the student proceeds through them, and material from more than one chapter is often integrated into later problems. I suspect the problems will not appear more difficult to the student, however, because as the student works through the problems, his or her ability increases and so does his or her confidence (that is why the more difficult problems are not marked as such). I made a concerted effort to make none of the problems so difficult that the student loses confidence. A few problems contain references to the journal in which the original work was published, enabling the curious student to find out more about the work that led to the conclusion shown in the problem.

Changes to This Edition—Problems

Much of the focus of this revision was on the problems, both in-chapter and end-of-chapter. There are 10% more problems in this edition, and there are more Solved Problems and more Problem-Solving Strategy essays. As mentioned previously, many of the new problems are multistep synthetic problems, designed to help students practice their synthetic reasoning skills.

Rich in Three-Dimensional, Computer-Generated Structures

The second edition was widely praised for its effective and useful art program. This edition continues to present energy-minimized, three-dimensional structures throughout the text to give the students an appreciation of the three-dimensional shapes of organic molecules.

Pedagogical, Efficient Use of Color

Organic chemistry is sufficiently challenging without forcing students to use a road map to understand the use of color in a textbook. Therefore, this book does not have such things as blue incoming groups and green leaving groups. Students will not find themselves asking "Why did the nucleophile change its color?" Color is used to make the book visually more appealing and to make learning easier by highlighting points of focus. An attempt has been made to make the color consistent (mechanism arrows are always red, for example), but there is no need in this text for a student to memorize a color palette.

Changes to This Edition—Art Program

Electrostatic potential maps have been added throughout this edition to allow students to visualize molecules in ways not possible before. With them, students can now *see* things like electron delocalization, how a formal positive charge is not always on the region in a molecule with the least electron density, the effect of electronegativity on pK_a, which atom of a cyclic bromonium ion is most susceptible to nucleophilic attack, the difference in the reactivity of substituted benzenes, the electrophilicity of the methyl group of an alkyl halide, and the nucleophilicity of the methyl group of an organometallic compound. There are hundreds of these new illustrations throughout the book.

INSTRUCTIONAL AIDS

The publisher offers a number of products for students and faculty that are designed to complete this learning package:

FOR THE STUDENT

Study Guide/Solutions Manual (ISBN 0-13-017859-4) The Study Guide/ Solutions Manual contains not just answers to the problems, but complete explanations of how the answers were obtained. The Study Guide/Solutions Manual also contains:

- a section on acid/base chemistry at a more advanced level than what is covered in the text, with a set of problems
- an 18-page exercise on "pushing electrons"
- an exercise on building molecular models
- an exercise on calculating kinetic parameters
- 21 practice tests

For the sake of making the path to mastery of the subject as smooth as possible, I thought it was important that I also write the Study Guide/Solutions Manual.

Bruice 3/e Companion Website (http://www.prenhall.com/bruice) This Companion Website for students features about 100 additional interactive, computer-graded problems for each chapter, regular updates of articles in the popular and lay scientific press that relate to organic chemistry, and a gallery of organic molecules that students can manipulate in real-time on their computers. And for students who are not familiar with the world wide web, we have:

Organic Media Companion for CW (ISBN 0-13-017863-2) This supplement comes FREE with each new copy of the text and is also available for separate purchase for students who buy used textbooks. It includes the web-enabled Organic Chemestry Student CD that accompanies the text as well as a brief, easy-to-understand book that describes the CD, how to use it, and how to use the Internet, world wide web, and Companion Website (CW) for the text. The CD is populated with multimedia elements and interactive exercises and offers students a one-click link to the Companion Website for the text.

Prentice Hall Molecular Model Kit (ISBN 0-205-508136-3) This best-selling model kit allows students to build space-filling and ball-and-stick models of common organic molecules. It allows accurate depiction of double and triple bonds, including heteroatomic molecules (which some model kits cannot handle well).

Prentice Hall Framework Molecular Model Kit (ISBN 0-13-330076-5) This model kit allows students to build scale models that show the mutual relations of atoms in organic molecules, including precise interatomic distances and bond angles. This is the most accurate model kit available.

Chemistry: Themes of the Times. This mini-newspaper is produced by *The New York Times* for students of chemistry. It features selected articles related to chemistry from the pages of *The New York Times* and is available free to adopters of the text.

If you wish, students' copies of the text may be accompanied by one of two molecular modeling tools:

The Molecular Modeling Workbook, featuring **SpartanView** and **SpartanBuild** software (ISBN 0-13-032026-9) This workbook includes a software tutorial and numerous challenging exercises students can tackle to solve problems involving structure building and analysis using the tools included in the two pieces of Spartan software. Available free when packaged with the text; please ask your Prentice Hall representative for details or e-mail *chemistryservice@prenhall.com*.

ChemOffice Ltd software, including ChemDraw LTD and Chem3D LTD. A free workbook is available on the Companion Website that includes a software tutorial and numerous exercises for each chapter of the text. The software is available at a substantial discount if purchased with the textbook; please ask your Prentice Hall representative for details or e-mail *chemistryservice@prenhall.com*.

FOR THE INSTRUCTOR

Organic Matter: A Presentational CD-ROM (ISBN 0-13-017861-6) This Instructor CD includes almost all images from the book, hundreds of 3-D images of molecules, animations of selected organic reactions, and videos of selected organic laboratory demonstrations. It also includes an easy-to-use interface that allows instructors to review material from the CD. Finally, it includes prebuilt Microsoft Power-Point slides for those instructors who prefer to use that software for their presentations. Works with both Windows and Macintosh.

Transparency package (ISBN 0-13-017850-0) This set features 250 full-color images from the text, a 50% increase over the last edition.

Test Item File (ISBN 0-13-031012-3) Written by Gary Hollis of Roanoke College, this test bank features over 2000 questions. For instructors who prefer to customize their own tests, we also feature the complete test item file electronically along with the new, easy-to-use TestManager software. This software also includes tools for course management and offering tests over a local area network. Available for Windows (ISBN 0-13-017865-9) and Macintosh (0-13-017866-7).

BlackBoard and WebCT Course Management The BlackBoard and WebCT course management systems equip faculty members with easy-to-use tools for creating sophisticated web-based educational programs. Prentice Hall provides robust content that is tailored specifically to the Bruice text. This content includes over 6000 prebuilt test questions, animations, 3-D (Chime) models, web links, and more. Using either of these course management systems, you can enhance an on-campus course or construct an entirely online course for distributed learning. Instructors with little or no technical experience can use a point-and-click navigation system to design their own on-line course components, including setting up course calendars, quizzes, assignments, lectures, and self-paced study help. Prebuilt courses in both BlackBoard and WebCT are available. These courses are available for a nominal fee over the price of the text (this fee includes software license fees). Please ask your Prentice Hall representative for details or e-mail *chemistryservice@prenhall.com.*

ACKNOWLEDGMENTS

It gives me great pleasure to acknowledge the dedicated efforts of many good friends that made this book a reality—John Murdzek, the development editor, who checked every word and every drawing that went into this book—he was an invaluable sounding board and his talent of making sense out of some of pretty muddled passages was at times miraculous; the comments and suggestions of E. Paul Papadopoulos of the University of New Mexico, Ronald Starkey of the University of Wisconsin, Green Bay, and Francis Klein of Creighton University have made an enormous difference; Terri O'Quin and Melvin Rueppel, who checked every inch of the book for accuracy; Edward Skibo of Arizona State University, who read and critiqued each chapter of the first edition as it was developed and whose many contributions persist in this edition; Alfred Burger, for the historical perspective he supplied for Chapter 30; Wayne Schnatter of Yeshiva University, for his contribution to Chapter 29; David L. Yerzley, M.D., for his assistance with the section on MRI; Andrei Blasko of Roche Bioscience, for his contribution to the discussion of 2-D NMR; John Perona of the University of California, Santa Barbara, who provided several of the molecular images used in this edition; Russell Kerr of Florida of Atlantic University and Phillip Crews of the University of California, Santa Cruz, who both helped with suggestions for the cover of this edition; Warren Hehre of Wavefunction, Inc., and Richard Johnson of the University of New Hampshire, for their advice on the electrostatic potential maps that appear in this book; and Chuck Huber, Chemistry Librarian at UCSB, who was always able to find what I needed. I also want to acknowledge the efforts of the many users of previous editions who generously gave their time to help me make this a better book. My particular thanks go to Barry Coddens and Terry Sheppard, Northwestern University; Barbara Schowen, University of Kansas; Jack Kirsch, University of California, Berkeley; Jim Long, University of Oregon; Tom Tidwell, University of Toronto; Peter Wagner, Michigan State University; Paige Phillips, Virginia Polytechnic Institute and State University; Barbara Mayer, California State University, Fresno; George Clemans, Bowling Green State University; Tom Nalli, Winona State University; Vincent Spaziano, St. Louis University; Paul Kropp, University of North Carolina, Chapel Hill; and Stanley Kudzin, State University of New York, New Paltz. I am also very grateful to my students, who pointed out sections that needed clarification, worked the problems, and searched for errors. Special thanks go to Bharat Pancholy and Benjamin Kay.

I am greatly indebted to the following colleagues whose suggestions and encouragement were invaluable:

Mahamed Asgar Ali, *Howard University*
Shelby R. Anderson, *Trinity College*
John Barbas, *Valdosta State University*
Rick Bolesta, *Mt. Hood Community College*
Joyce C. Brockwell, *Northwestern University*
Thomas A. Bryson, *University of South Carolina*
Paul Buonora, *University of Scranton*
George B. Clemans, *Bowling Green State University*
Barry A. Coddens, *Northwestern University*
John Cullen, *Monroe Community College*
Mark R. DeCamp, *University of Michigan, Dearborn*

Michael R. Detty, *State University of New York, Buffalo*
John DiCesare, *University of Tulsa*
Veljko Dragojlovic, *Nova Southeastern University*
Jeffrey Elbert, *South Dakota State University*
Jan M. Fleischer, *Earlham College*
Warren P. Giering, *Boston University*
Michael M. Haley, *University of Oregon*
James E. Hanson, *Seton Hall University*
David Harpp, *McGill University*
John Isidor, *Montclair State University*
Richard Johnson, *University of New Hampshire*
Dennis Lehman, *Harold Washington College*
William Loffredo, *East Stroudsberg University*
James W. Long, *University of Oregon*
Jerry Manion, *University of Central Arkansas*
Amanda Martin-Esker, *Northwestern University*
John N. Marx, *Texas Tech University*
Anthony Masulaitis, *Jersey City State College*
Barbara J. Mayer, *California State University, Fresno*
Robert McClelland, *University of Toronto*
Gary W. Morrow, *University of Dayton*
Thomas W. Nalli, *Winona State University*
Abby Parrill, *University of Memphis*
Lawrence Principe, *Johns Hopkins University*
Michael Rathke, *Michigan State University*
J. Ty Redd, *Southern Utah University*
Todd Richmond, *The Claremont Colleges*
Anthony Sky, *Lawrence Technical University*
Homer A. Smith, Jr., *University of Richmond*
Andrew B. Turner, *Saint Vincent College*
T. K. Vinod, *Western Illinois University*
Maria Vogt, *Bloomfield College*

I am deeply grateful to my editor, John Challice, who guided this book, and whose perceptiveness and astute editorial suggestions made all the difference. I am lucky to have his wisdom and friendship. I also want to thank the other talented and dedicated people at Prentice Hall who made this book a reality. Much thanks goes to Pat McCutcheon who kept everybody on track with infinite patience and made deadlines that weren't supposed to be possible.

I particularly want to thank the many wonderful and talented students I have had over the years who taught me how to be a teacher. And I want to thank my children, from whom I may have learned the most. Two special people, Summer Solstice and River Rose, were born while I wrote this edition. I look forward to the day when they can recognize their names in print.

I want to make this text as user friendly as possible, and I would appreciate any comments that will help me achieve this goal in future editions. If you find sections that could be clarified or expanded, or examples that could be added, please let me know. Finally, this edition has been painstakingly combed for typographical errors. Any that remain are my responsibility; if you find any, please let me know so they can be corrected in future printings.

Paula Yurkanis Bruice
University of California, Santa Barbara
pybruice@bioorganic.ucsb.edu

To The Student

Welcome to organic chemistry. This book has been written for you, one who is encountering the subject for the first time. The first thing you should do is familiarize yourself with the book. The material on the inside of the front and back covers is information that you may want to refer to many times during the course. The Key Terms and Summaries of Reactions at the end of the chapters, and the Glossary at the end of the book, can be useful study aids. Also look at the Appendices to see what kind of information is provided there. The electrostatic potential maps and the molecular models throughout the book are there to give you an appreciation of what molecules look like in three dimensions and how electronic charge is distributed within the molecule. Look at the margin notes as you read a chapter—they emphasize important points.

Be sure to work all the problems within each chapter. These are drill problems that will enable you to check whether or not you have mastered the material. Some of them are solved for you in the text. Others—those marked with a diamond—have short answers provided in the Appendix. You will also find Problem-Solving Strategies sprinkled throughout the text. These features provide detailed suggestions for how best to approach important problem types. Please read these Problem-Solving Strategies carefully, and don't be afraid to refer back to them when you're working the end-of-chapter problems.

Work as many end-of-chapter problems as you can. The more problems you work, the more comfortable you will be with the subject and the more prepared you will be for the material in subsequent chapters. Do not let any problem frustrate you. If you cannot figure out the answer in a reasonable amount of time, turn to the Study Guide/Solutions Manual to learn how you should have approached the problem. Later on go back and try to work the problem again without consulting the Study Guide/Solutions Manual.

The most important thing for you to remember in organic chemistry is **DO NOT GET BEHIND.** Organic chemistry consists of a lot of simple steps, so it is not a difficult subject. But the subject can become overwhelming if you don't keep up.

Before many of the theories and mechanisms were worked out, organic chemistry was a discipline that could be mastered only through memorization. Fortunately, that is no longer true. You will find many common threads that will allow you to use what you have learned in one situation to predict what will happen in other situations. So, as you read the book and study your notes, always try to understand *why* each thing happens. If the reasons behind the behavior are understood, most reactions can be predicted. If you approach the class with the misconception that you must memorize hundreds of unrelated reactions, it could be your downfall. There is simply too much material to memorize, and if all your knowledge is based on memorization, you won't have the necessary foundation on which to lay subsequent material. From time to time, some memorization will be required. Some fundamental rules have to be memorized, and you will have to memorize the common names of some organic compounds. But the latter should not be a problem. After all, your friends have common names and you've been able to learn them.

Students who take organic chemistry to gain entrance into medical school sometimes wonder why medical schools pay so much attention to how they do in organic chemistry. The importance of organic chemistry is not in the subject matter alone. Mastering organic chemistry requires a thorough understanding of fundamentals and the ability to use these fundamentals to analyze, classify, and predict. This parallels the study of medicine; a physician uses an understanding of fundamentals to analyze, classify, and diagnose.

Good luck in your study. I hope you will enjoy your course in organic chemistry and learn to appreciate the logic of the discipline. If you have any comments about the book or any suggestions about how it can be improved for the students who will follow you, I would love to hear from you. Positive comments are the most fun, but critical comments are the most useful.

Paula Yurkanis Bruice
pybruice@bioorganic.ucsb.edu

A Guide to Using This Text

The following pages walk you through some of the key features of this text. This book was designed with you, the student, in mind—to help you succeed in organic chemistry. My main goal in writing this text was to make the material you study as accesible and appealing as possible. I hope you enjoy your study of organic chemistry.

—Paula Bruice

STUDY AIDS

There are a number of features in the chapters themselves designed to help you learn the materials.

CHIRAL DRUGS

In 1992, the Federal Drug Administration (FDA) issued a policy statement that encouraged drug companies to use recent advances in synthetic and separation techniques to develop single enantiomer drugs. Such drugs are becoming more common. (S)-$(+)$-Ketamine is four times more potent an anesthetic than (R)-$(-)$-ketamine, and even more important, the disturbing side effects appear to be associated only with the (R)-$(-)$-enantiomer. The active ingredient of ibuprofen, the popular analgesic marketed as Advil, Nuprin, and Motrin, resides primarily in the (S)-$(+)$-enantiomer. The FDA has some concern about approving the drug as a single enantiomer because of potential drug overdoses. Can people who are used to taking two pills be convinced to take only one? It has recently been found that heroin addicts can be maintained with $(-)$-α-acetylmethadol for a 72-hour period compared to 24 hours with racemic methadone. This means less frequent visits to the clinic, and a single dose can get an addict through an entire weekend. The FDA may require drug companies that develop a drug as a single enantiomer rather than as the already approved racemic mixture to show that the missing enantiomer does not provide some type of beneficial effect.

▲ Boxes for Student Aid

Boxed materials bring life to the topic being discussed and provide extra motivation for you. For example, *Chiral Drugs* shows you how the stereochemistry you're learning is relevant in the real world of pharmaceuticals.

So far in our study of organic chemistry, we have paid attention to *why* a reaction occurs, *how* it occurs, and the *products* that are formed. A good way to review these reactions is to design syntheses because in so doing, you must be able to recall many of the reactions that you have learned.

Synthetic chemists consider time, cost, and yield in designing syntheses. In the interest of time, a well-designed synthesis should require as few steps (sequential reactions) as possible, and those steps should each involve a reaction that is easy to carry out. If two chemists in a pharmaceutical company were each asked to prepare a new drug, and one synthesized the drug in three simple steps while the other used 20 difficult steps, which chemist would be out looking for a new job? In addition, each step in the synthesis should result in the greatest possible yield of the desired product. The more reactant needed to synthesize one gram of product, the more expensive it is to make. Sometimes it is preferable to design a synthesis involving several steps if the starting materials are inexpensive, the reactions are easy to carry out, and the yield of each step is high. This would be better than designing a synthesis with fewer steps, that require expensive starting materials, and reactions that are more difficult or give lower yields.

The following examples will give you an idea of the type of thinking required for

**5.11
DESIGNING A
SYNTHESIS I: AN
INTRODUCTION TO
MULTISTEP
SYNTHESIS**

▲ Designing a Synthesis

Retrosynthetic analysis is introduced in Chapter 5 so you can use this technique throughout the course to design multistep syntheses. In addition, six detailed discussions of synthesis design are spread throughout the book, each with a different focus. There is also a chapter (Chapter 29) on more advanced multistep syntheses.

▲ Key Terms

There are a lot of new terms you'll need to learn as you study organic chemistry. To help you, you will find a list of important terms at the end of every chapter that you should know before proceeding. If you need to refresh your memory, you can use the handy page references to find the terms within the chapter.

It will be helpful to do the exercise on drawing curved arrows in the Study Guide/Solution Manual (Special Topic II).

have been able to predict that the product of the reaction of 2-butene and HBr is 2-bromobutane. Overall, the reaction involves the addition of 1 mole of HBr to 1 mole of the alkene. The reaction, therefore, is called an **addition reaction.** Because the first step of the reaction involves the addition of an electrophile (H⁺) to the alkene, the reaction is more precisely called an **electrophilic addition reaction.** Electrophilic addition reactions are characteristic reactions of alkenes.

At this point you may think that it would be easier just to memorize that

of the plane of the paper toward the viewer, and vertical lines represent the bonds that project back from the plane of the paper away from the viewer. The carbon chain is drawn vertically with C-1 at the top of the chain. This method of representing the arrangement of groups bonded to a chirality center is known as a **Fischer projection.**

original Fischer projections   Fischer projections used now

To draw enantiomers using a Fischer projection, draw the first enantiomer by arranging the four groups (or atoms) bonded to the chirality center in any order. Draw the second enantiomer by interchanging two of the groups (or atoms). It does not make any difference which two groups you interchange. (It is a good idea to make models to convince yourself that this is true.) Many organic chemists prefer to interchange the two horizontal groups because the enantiomers then look like mirror images on paper.

Emil Fischer (1852–1919) was born in a village near Cologne, Germany. He became a chemist against the wishes of his father, a successful merchant, who wanted him to enter

▲ Margin Notes

Margin notes highlight and remind you of important principles, and biographical sketches help you appreciate the history of chemistry and the diverse people who contributed to it.

SUMMARY OF REACTIONS

1. Electrophilic addition reactions

  a. Addition of hydrogen halides: ionic (Markovnikov orientation) (Section 3.9)

$$RCH{=}CH_2 \ + \ HX \ \longrightarrow \ RCHCH_3$$
$$\overset{|}{X}$$

$$HX = HF, HCl, HBr, HI$$

  b. Addition of hydrogen bromide: radical (apparent anti-Markovnikov orientation) (Section 3.18)

$$RCH{=}CH_2 \ + \ HBr \ \xrightarrow{\text{peroxide}} \ RCH_2CH_2Br$$

  c. Addition of halogen (Section 3.15)

$$RCH{=}CH_2 \ + \ Cl_2 \ \xrightarrow{CH_2Cl_2} \ RCHCH_2Cl$$
$$\overset{|}{Cl}$$

◄ Reaction Summaries

At the end of every chapter that discusses organic reactions is a summary of these reactions. Review these summaries and make sure that you can explain how each reaction occurs.

One of the greatest challenges facing organic chemistry students lies in the often abstract nature of the subject. To help you better visualize important concepts, we have developed striking artwork on paper and on the CD-ROM that accompanies this book.

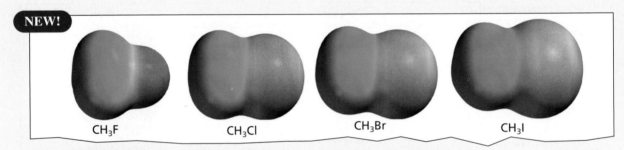

▲ Accurate and Complete Mechanisms
This text includes hundreds of complete mechanisms integrated into the text prose to promote true understanding, not just memorization. Make sure you understand and can predict each step.

NEW!

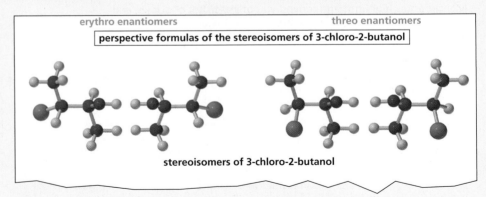

▲ Electrostatic Potential Maps
Electrostatic potential maps—now found throughout the textbook—help you visualize the electronic structure of molecules and atoms and give you a greater understanding of *why* and *how* reactions occur. Use these to better understand why certain molecules and ions behave the way they do.

erythro enantiomers   threo enantiomers

perspective formulas of the stereoisomers of 3-chloro-2-butanol

stereoisomers of 3-chloro-2-butanol

▲ Molecular Art
Energy-minimized, three-dimensional structures appear throughout the text to give you a more accurate appreciation for the "shapes" of organic molecules. Spend some time looking at these—remember, a picture is worth a thousand words!

> **PROBLEM-SOLVING STRATEGY**
>
> **a.** Which of the following compounds (**1–3**) will form hydrogen bonds between its molecules?
>
> **b.** Which will form hydrogen bonds with a solvent such as ethanol?
>
>   **1.** $CH_3CH_2CH_2OH$ **2.** $CH_3CH_2CH_2SH$ **3.** $CH_3OCH_2CH_3$
>
> In solving this type of question, start by defining the kind of compound that will do what is being asked.
>
> **a.** A hydrogen bond forms when a hydrogen that is attached to an O, N, or F of one molecule interacts with a lone pair of electrons on an O, N, or F of another molecule. Therefore, a compound that will form hydrogen bonds with itself must have a hydrogen bonded to an O, N, or F. Only compound **1** will be able to form hydrogen bonds with itself.
>
> **b.** Ethanol has an H bonded to an O, so it will be able to form hydrogen bonds with a compound that has a lone pair of electrons on an O, N, or F. Compounds **1** and **3** will be able to form hydrogen bonds with ethanol.
>
> Now continue on to Problem 18.

Developing effective problem-solving skills is essential to success in organic chemistry. Over 1500 problems throughout the text—and thousands more on the text's media components—teach you the skills you need to dissect and solve any problem.

▲ Problem-Solving Strategies

The **Problem-Solving Strategies** teach you how to approach certain kinds of problems, organize your thoughts, and improve your problem-solving abilities. Every strategy is followed by an exercise that allows you to immediately practice the strategy just discussed.

> **PROBLEM 8 ◆**
>
> Indicate whether each of the following structures has the *R* configuration or the *S* configuration:
>

> **PROBLEM 9 / SOLVED ◆**
>
> Do the following structures represent identical molecules or a pair of enantiomers?
>

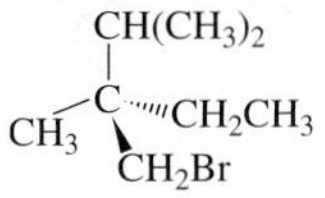

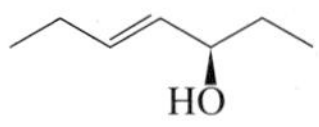

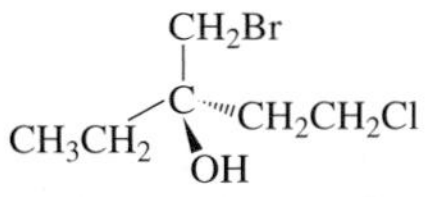

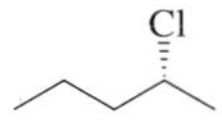

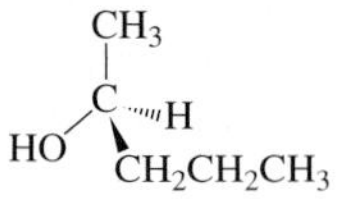

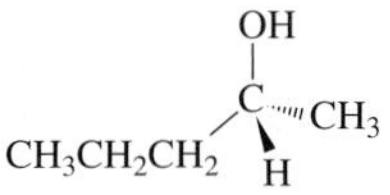

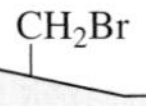

▲ Solved Problems and Practice Exercises

Throughout the text, solved problems walk you carefully through the detailed steps involved in solving a particular type of problem. You can then apply the process you just learned with the practice exercises that follow. A diamond beside a problem number tells you that the answer to the problem is in the text's appendix.

To give you as much help as possible, we've extended the text's message through an integrated teaching and learning program. Its media components use 3-D media to convey information in ways a textbook alone cannot. They also offer a wealth of problems, animations, course management tools, and much more.

Companion Website (CW)
www.prenhall.com/bruice

This innovative on-line resource center is designed specifically to support and enhance Bruice's text. CW offers:

- **A Problem-Solving Center,** which gives you access to 2000 new exercises to test your understanding of the content. Each question includes a hint, a cross-reference back to the text, and detailed feedback.

- **A Visualization Center,** where you can access 3-D (Chime) renderings of most of the important molecules in the text. You can click on a molecule, rotate it in three dimensions, and explore its structure in detail.

- **Applications,** in the Current Topics section, that provide you with current news articles that enrich your understanding of organic chemistry.

- **Dynamic Web Links,** which are a valuable source of supplemental information. These can also be used for on-line research projects.

- **Chemistry Drawing Problems,** where you use an on-line drawing tool to draw chemical structures. Each question provides feedback for the submitted drawings.

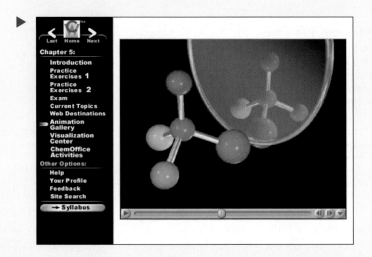

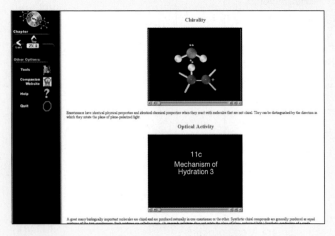

◀ **Organic Chemistry Student CD-ROM**

Populated with multimedia elements and interactive exercises throughout, this text-specific **Organic Chemistry Student CD-ROM**—*available FREE with the textbook*—arms you with powerful, easy-to-use technology. Book-specific, chapter-based, and media-rich, the Student CD-ROM places special emphasis on visualizing mechanisms through hands-on, animated presentations. It also offers videos, 3-D manipulable molecules, self-assessment exercises, and more. The Student CD-ROM is part of an integrated environment with the Companion Website; students with an Internet connection can link seamlessly from one to the other.

◀ Watch for this icon in this book; it tells you there is an interactive exercise on the CD-ROM relevant to what the text is discussing.

About the Author

Paula Bruice with Justus (formerly her lawyer-son's dog) and Marigold

Paula Yurkanis Bruice was raised primarily in Massachusetts, Germany, and Switzerland and was graduated from the Girls' Latin School in Boston. She received an A.B. from Mount Holyoke College and a Ph.D. in chemistry from the University of Virginia. She received an NIH postdoctoral fellowship for postdoctoral study in biochemistry at the University of Virginia Medical School, and she held a postdoctoral appointment in the Department of Pharmacology at Yale Medical School.

She has been a member of the faculty at the University of California, Santa Barbara since 1972, where she has received the Associated Students Teacher of the Year Award, the Academic Senate Distinguished Teaching Award, and two Mortar Board Professor of the Year Awards. Her research interests concern the mechanism and catalysis of organic reactions, particularly those of biological significance. Paula has a daughter and a son who are physicians and a son who is a lawyer. Her main hobbies are reading mystery/suspense novels and her pets (two dogs, two cats, and a parrot).

An Introduction to the Study of Organic Chemistry

The first two chapters cover a variety of topics that are designed to lay the foundation for your study of organic reactions.

Chapter 1 reviews the topics from general chemistry that will be important in your study of organic chemistry. The chapter starts with a description of the structure of atoms and then proceeds to a description of the structure of molecules. Acid-base chemistry is central to an understanding of organic reactions, so the concepts of acid-base chemistry are reviewed. You will see how the structure of a molecule affects its acidity, and how the acidity of a solution affects molecular structure.

To discuss organic compounds, you must be able to name them and to visualize their structures when you read or hear their names. In **Chapter 2,** you will learn how to name five different classes of organic compounds and begin to understand the rules used in naming compounds. Because the compounds examined in this chapter are the products of the reactions presented in the following 10 chapters, you will have the opportunity to review the nomenclature of these compounds as you proceed. You will also compare and contrast the structures and physical properties of these compounds, which makes learning about them a little easier than if each compound were presented separately. Because organic chemistry is a study of compounds that contain carbon, the last part of Chapter 2 discusses the spatial arrangement of the atoms in chains of carbon atoms and in rings of carbon atoms.

1

Electronic Structure and Bonding. Acids and Bases

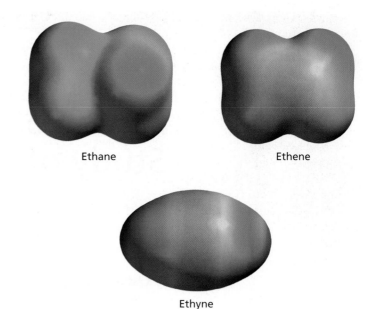

Ethane

Ethene

Ethyne

To stay alive, early humans must have been able to tell the difference between two kinds of material in their world. "You can live on roots and berries," they might have said, "but you can't live on dirt. You can stay warm by burning branches, but you can't burn rocks."

By the eighteenth century, scientists thought they had grasped the nature of that difference, and in 1807, Jöns Jakob Berzelius gave names to the two kinds of materials. Compounds derived from living organisms were believed to contain an unmeasurable vital force, the essence of life. These he called "organic." Compounds derived from minerals—those lacking that vital force—were "inorganic."

Because chemists could not create life in the laboratory, they assumed they could not create compounds with a vital force. With this mind-set, you can imagine how surprised chemists were in 1828 when Friedrich Wöhler produced an organic compound called urea by heating ammonium cyanate, a compound that did not come from a living organism.

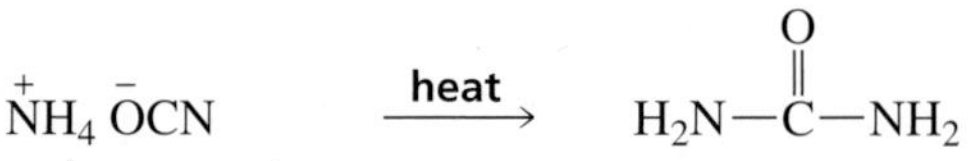

$$\overset{+}{N}H_4 \ \overset{-}{O}CN \quad \xrightarrow{\textbf{heat}} \quad H_2N-\overset{\overset{\displaystyle O}{\|}}{C}-NH_2$$

ammonium cyanate   urea

The urea Wöhler prepared was identical to the urea synthesized by mammals, even though he had prepared it by heating an inorganic salt. For the first time, an "organic" compound had been obtained from something other than a living system and certainly without the aid of any kind of vital force. Clearly, chemists needed a new definition for "organic compounds." **Organic compounds** are now defined as compounds that contain carbon.

Why is an entire branch of chemistry devoted to the study of carbon-containing compounds? First of all, the molecules that make life possible—proteins, enzymes, vitamins, lipids, carbohydrates, and nucleic acids—all contain carbon, so the chemical reactions that take place in living systems, including our own bodies, are organic reactions. Second, we depend on organic compounds that occur in nature for our food, for many of our medicines, for our clothing (cotton, wool, silk), and for energy (natural gas, petroleum). In addition to naturally occurring organic compounds, chemists have learned to synthesize millions of organic compounds never found in nature, including synthetic fabrics, plastics, synthetic rubber, medicines, and preservatives. Many of these synthetic compounds prevent shortages of naturally occurring products. For example, it has been estimated that if synthetic materials were not available for clothing, all of the arable land in the United States would have to be used for the production of cotton and wool just to provide enough material for clothing.

What makes carbon so special? Why are there so many carbon-containing compounds? The answer lies in carbon's position in the periodic table. Carbon is in the center of the second row of elements. The atoms to the left of carbon have a tendency to give up electrons, whereas the atoms to the right have a tendency to accept electrons.

the second row of the periodic table

Because carbon is in the middle, it neither readily gives up nor readily accepts electrons. Instead, it shares electrons. Not only can it share electrons with several different kinds of atoms, it can also share electrons with other carbon atoms, allowing the formation of a wide variety of chain-like molecules. Because carbon can share electrons with atoms other than carbon adds to the range of stable carbon-containing compounds. Consequently, carbon is able to form millions of stable compounds with a wide range of chemical properties simply by sharing electrons.

When we study organic chemistry, we study how organic compounds react. When an organic compound reacts, old bonds break and new bonds form. Bonds form when two atoms share electrons, and bonds break when two atoms no longer share electrons. How readily a bond forms and how easily it breaks depend on the particular electrons that are shared, which, in turn, depends on the atoms to which the electrons belong. So if we are going to start our study of organic chemistry at the beginning, we must start with an understanding of the structure of an atom and how the electrons in that atom are distributed.

Jöns Jakob Berzelius (1779–1848) not only coined the terms "organic" and "inorganic," but also invented the system of chemical symbols still used today. He published the first list of accurate atomic weights and proposed the idea that atoms carry an electric charge. He purified or discovered the elements cerium, lithium, silicon, thorium, titanium, and zirconium.

German chemist Friedrich Wöhler (1800–1882) began his professional life as a physician and later became a professor of chemistry at the University of Göttingen. Wöhler codiscovered the fact that two different chemicals could have the same molecular formula. He also developed methods of purifying aluminum—at the time, the most expensive metal on Earth—and beryllium.

1.1
THE STRUCTURE OF AN ATOM

An atom consists of a dense nucleus surrounded by electrons. The nucleus contains positively charged protons and neutral neutrons, so the nucleus is positively charged. The electrons are negatively charged. The amount of positive charge on a proton equals the amount of negative charge on an electron, so a neutral atom has an equal number of protons and electrons. Atoms can become charged if they gain or lose electrons. However, the number of protons in an atom does not change.

Protons and neutrons have approximately the same mass and are about 1800 times as massive as electrons. This means that most of the mass of an atom is in the nucleus. Most of the volume of an atom is occupied by its electrons, however, and it is the electrons that form chemical bonds.

The **atomic number** of an atom equals the number of protons in its nucleus. It is also the number of electrons that surround the nucleus of a neutral atom. For example, the atomic number of carbon is 6, which means that a neutral carbon atom has six protons and six electrons. Because the number of protons in an atom does not change, the atomic number of a particular element is always the same; all carbon atoms have an atomic number of 6.

The **mass number** of an atom is the sum of the number of its protons and the number of its neutrons. Not all carbon atoms have the same mass number because not all of them have the same number of neutrons. For example, 98.89% of naturally occurring carbon atoms have six neutrons (a mass number of 12), and 1.11% have seven neutrons (a mass number of 13). These two different kinds of carbon atoms (^{12}C and ^{13}C) are called isotopes. **Isotopes** have the same atomic number (i.e., the same number of protons) but different mass numbers because they have different numbers of neutrons. The chemical properties of isotopes of a given element are nearly identical.

Naturally occurring carbon also contains a trace amount of ^{14}C which has six protons and eight neutrons. This isotope of carbon is radioactive, decaying with a half-life of 5730 years. (The half-life is the time it takes for one-half of the nuclei to decay.) As long as a plant or animal is alive, it takes in as much ^{14}C as it excretes or exhales. When it dies, it no longer takes in ^{14}C, so the ^{14}C in the organism slowly decreases. Therefore, the age of an organic material can be determined by its ^{14}C content.

The **atomic weight** of an element is the average mass of the atoms in the naturally occurring element. Because an atomic mass unit (amu) is defined as exactly 1/12 of the mass of ^{12}C, the atomic weight of ^{12}C is 12.0000 amu; the atomic weight of ^{13}C is 13.0034 amu. Therefore, the atomic weight of carbon is 12.011 amu ($0.9889 \times 12.0000 + 0.0111 \times 13.0034 = 12.011$).

PROBLEM 1 ◆

Oxygen has three isotopes with mass numbers of 16, 17, and 18. The atomic number of oxygen is eight. How many protons and neutrons does each of the isotopes have?

1.2
DISTRIBUTION OF ELECTRONS IN AN ATOM

Electrons are continuously moving. Like anything that moves, electrons have kinetic energy, and this energy is what defies the attractive force of the positively charged protons that would otherwise pull the negatively charged electrons into the nucleus. For a long time, electrons were thought of as particles—infinitesimal planets orbiting the nucleus. In 1924, however, a French physicist named Louis de Broglie showed that electrons also have wave-like properties. He did this by combining a formula de-

MAX KARL ERNST LUDWIG PLANCK

Max Planck (1858–1947) was born in Germany, the son of a professor of civil law. He was a professor at the Universities of Munich (1880–1889) and Berlin (1889–1926). Two of his daughters died in childbirth and during World War I, one of his sons was killed in action. In 1918, Planck received the Nobel Prize in physics for his development of quantum theory. He became president of the Kaiser Wilhelm Society of Berlin (later renamed the Max Planck Society) in 1930. Planck felt it was his duty to remain in Germany during the Nazi era, but never supported the Nazi regime. He unsuccessfully interceded with Hitler on behalf of his Jewish colleagues and, as a consequence, was forced to resign from the presidency of the Kaiser Wilhelm Society in 1937. A second son was accused of taking part in the plot to kill Hitler and was executed. Planck lost his home to Allied bombings. He was rescued by Allied forces during the final days of the war.

veloped by Einstein that relates mass and energy with a formula developed by Planck relating frequency and energy. This startling idea spurred physicists to propose a mathematical concept known as quantum mechanics.

Quantum mechanics uses the same mathematical equations that describe the wave motion of a guitar string to characterize the motion of an electron around a nucleus. The version of quantum mechanics most useful to chemists was proposed by Erwin Schrödinger in 1926. According to Schrödinger, the behavior of each electron in an atom or a molecule can be described by a **wave equation.** A wave equation has a series of solutions that are known as **wave functions.** Solving the wave equation for a given electron tells us the volume of space around the nucleus where the electron is most likely to be found. This volume of space is called an **orbital.**

According to quantum mechanics, the electrons in an atom can be thought of as occupying a set of concentric shells that surround the nucleus. The first shell is the one closest to the nucleus. The second shell lies farther from the nucleus, and even farther out lie the third and higher-numbered shells. Each shell contains subshells known as **atomic orbitals.** Each atomic orbital has a characteristic shape and energy and occupies a characteristic volume of space, which is predicted by the Schrödinger equation. The closer the orbital is to the nucleus, the lower its energy.

The first shell contains only an *s* atomic orbital; the second shell contains *s* and *p* atomic orbitals; the third shell contains *s, p,* and *d* atomic orbitals; and the fourth and higher shells contain *s, p, d,* and *f* atomic orbitals (Table 1.1).

There is only one *s* atomic orbital in each shell. The second and higher shells each contain three degenerate *p* atomic orbitals. **Degenerate orbitals** are orbitals that have the same energy. The third and higher shells also contain five degenerate *d* atomic orbitals, and the fourth and higher shells contain seven degenerate *f* atomic orbitals. Because a maximum of two electrons can coexist in an atomic orbital (see the Pauli exclusion principle, below), the first shell, with only one atomic orbital, can contain no more than two electrons. The second shell, with four atomic orbitals—one *s* and three *p*—can have a total of eight electrons. Eighteen electrons can occupy the nine atomic orbitals—one *s*, three *p*, and five *d*—of the third shell, and 32 electrons can occupy the 16 atomic orbitals of the fourth shell. In studying organic chemistry, we will be concerned primarily with atoms that have electrons only in the first and second shells.

The **ground-state electronic configuration** of an atom describes the orbitals occupied by the atom's electrons when they are all in the available orbitals with the lowest energy. If energy is applied to an atom in the ground state, one or more electrons can jump into a higher energy orbital. The atom then would be in an **excited-state electronic configuration.** The ground-state electronic configurations of the 11 smallest atoms are shown in Table 1.2 (each arrow—whether pointing up or down—represents one electron). The following principles are used to determine which orbitals electrons occupy.

1. The **aufbau principle** (*aufbau* is German for "building up") tells you the first thing you need to know to be able to assign electrons to the various atomic

Prince Louis Victor Pierre Raymond de Broglie (1892–1987) studied history at the Sorbonne. During World War I he was stationed in the Eiffel Tower as a radio engineer. Intrigued by his exposure to radio communications, he returned to school after the war and earned a Ph.D. in physics. He received the Nobel Prize in physics in 1929, five years after obtaining his degree, for his work that showed electrons to have properties of both particles and waves.

The closer the orbital is to the nucleus, the lower its energy.

Erwin Schrödinger (1887–1961) was teaching physics at the University of Berlin when Hitler rose to power. Although not Jewish, Schrödinger left Germany to return to his native Austria, only to see it later taken over by the Nazis. He moved to the School for Advanced Studies in Dublin and then to Oxford University. In 1933 he shared the Nobel Prize in physics with Paul Dirac, a professor of physics at Cambridge University, for mathematical work on quantum mechanics.

TABLE 1.1 Distribution of Electrons in the First Four Shells That Surround the Nucleus				
	First shell	**Second shell**	**Third shell**	**Fourth shell**
Atomic orbitals	*s*	*s, p*	*s, p, d*	*s, p, d, f*
Number of atomic orbitals	1	1, 3	1, 3, 5	1, 3, 5, 7
Maximum number of electrons	2	8	18	32

Albert Einstein (1879–1955)
was born in Germany. When he was in high school his father's business failed and his family moved to Milan. Einstein had to stay behind because German law required compulsory military service after finishing high school. Einstein wanted to join his family in Italy. His high school math teacher wrote a letter saying that Einstein could have a nervous breakdown without his family and also that there was nothing left to teach him. Eventually Einstein was asked to leave the school because of his disruptive behavior. Popular folklore says he left because of poor grades in Latin and Greek, but his grades in those subjects were fine. Einstein was visiting the United States when Hitler came to power, so he accepted a position at the Institute for Advanced Study in Princeton, becoming a U. S. citizen in 1940. Although a lifelong pacifist, he wrote a letter to President Roosevelt warning of ominous advances in German nuclear research. This led to the creation of the Manhattan Project, which developed the atomic bomb and tested it in New Mexico in 1945.

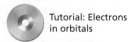
Tutorial: Electrons in orbitals

TABLE 1.2   The Ground-State Electronic Configurations of the Smallest Atoms

Atom	Name of element	Atomic number	$1s$	$2s$	$2p_x$	$2p_y$	$2p_z$	$3s$
H	Hydrogen	1	↑					
He	Helium	2	↑↓					
Li	Lithium	3	↑↓	↑				
Be	Beryllium	4	↑↓	↑↓				
B	Boron	5	↑↓	↑↓	↑			
C	Carbon	6	↑↓	↑↓	↑	↑		
N	Nitrogen	7	↑↓	↑↓	↑	↑	↑	
O	Oxygen	8	↑↓	↑↓	↑↓	↑	↑	
F	Fluorine	9	↑↓	↑↓	↑↓	↑↓	↑	
Ne	Neon	10	↑↓	↑↓	↑↓	↑↓	↑↓	
Na	Sodium	11	↑↓	↑↓	↑↓	↑↓	↑↓	↑

orbitals. According to this principle, an electron always goes into the available orbital with the lowest energy. The relative energies of the atomic orbitals are:

$$1s < 2s < 2p < 3s < 3p < 4s < 3d < 4p < 5s < 4d < 5p < 6s < 4f < 5d < 6p < 7s < 5f$$

Because a $1s$ atomic orbital is closer to the nucleus, it is lower in energy than a $2s$ atomic orbital, which is lower in energy (and is closer to the nucleus) than a $3s$ atomic orbital. Comparing atomic orbitals in the same shell, an s atomic orbital is lower in energy than a p atomic orbital, and a p atomic orbital is lower in energy than a d atomic orbital.

2. The **Pauli exclusion principle** states that (a) no more than two electrons can occupy each atomic orbital, and (b) the two electrons must be of opposite spin $(+1/2$ or $-1/2)$.[1] This is called an "exclusion" principle because it states that only so many electrons can occupy any particular shell. Notice in Table 1.2 that the two arrows in any one filled orbital point in opposite directions to indicate the opposite directions of electron spin.

From these first two rules, we can assign electrons to atomic orbitals for atoms that contain one, two, three, four, or five electrons. The single electron of a hydrogen atom occupies a $1s$ atomic orbital; the second electron of a helium atom fills the $1s$ atomic orbital; the third electron of a lithium atom occupies a $2s$ atomic orbital; the fourth electron of a beryllium atom fills the $2s$ atomic orbital; the fifth electron of a boron atom occupies one of the $2p$ atomic orbitals. Before we can continue to larger atoms (those containing six or more electrons), however, we need Hund's rule.

3. **Hund's rule** states that when there are degenerate orbitals—two or more orbitals with the same energy—an electron will occupy an empty orbital before it will pair up with another electron. This occurs because it takes energy to pair up electrons. The sixth electron of a carbon atom, therefore, goes into an empty $2p$ atomic orbital rather than pairing up with the electron already occupying a $2p$ atomic orbital.

Using these three rules, the location of the electrons in the remaining elements can be assigned.

[1] A helpful conceptual picture of electron spin (although not strictly correct because an electron has both particle-like and wave-like properties) is to imagine an electron as a spinning, charged sphere for which $+1/2$ and $-1/2$ indicate the two opposite directions in which it can rotate.

PROBLEM 2 ◆

Potassium has an atomic number of 19 and one unpaired electron. What orbital does the unpaired electron occupy?

PROBLEM 3 ◆

Write electronic configurations for chlorine (atomic number 17), bromine (atomic number 35), and iodine (atomic number 53).

In trying to explain why atoms form bonds, G. N. Lewis proposed that *an atom is most stable if it has a filled shell or an outer shell of eight electrons and no electrons of higher energy.* According to Lewis' theory, an atom will give up, accept, or share electrons in order to achieve a filled shell or an outer shell that contains eight electrons. This theory has come to be called the **octet rule.**

Lithium (Li) has a single electron in its 2s atomic orbital. If it loses this electron, the lithium atom ends up with a filled first shell and no other electrons—a stable configuration. Removing an electron from an atom takes energy. The energy required to remove an electron from an atom is called the **ionization energy** of that atom. Lithium has a relatively low ionization energy because it loses an electron relatively easily. Sodium (Na) has a single electron in its 3s atomic orbital. Consequently, sodium easily loses one electron, which leaves the sodium with an outer shell of eight electrons. Elements (such as lithium and sodium) that have low ionization energies are said to be **electropositive** (they readily lose an electron and thereby become positively charged). The elements in the first column of the periodic table are electropositive—each readily loses an electron because each has a single electron in an s atomic orbital.

Electrons in filled shells are called **core electrons.** They do not participate in chemical bonding. Electrons in shells that are not filled are called **valence electrons.** Lithium and sodium, for example, each have one valence electron. Because elements in the same column of the periodic table have the same number of valence electrons, and because the number of valence electrons is the major factor determining an element's chemical properties, elements in the same column of the periodic table have similar chemical properties. Thus, the chemical behavior of an element depends on its electronic configuration.

When we draw the electrons around an atom, as in the following equations, the electrons in any completely filled shells are not shown; only the valence electrons are shown. Each valence electron is shown as a dot. When the single valence electron of lithium or sodium is removed, the resulting atom—now called an ion—carries a positive charge.

$$Li \cdot \longrightarrow Li^+ + e^-$$

$$Na \cdot \longrightarrow Na^+ + e^-$$

Fluorine has seven valence electrons (Table 1.2). Consequently, fluorine readily acquires an electron in order to have an outer shell of eight electrons. When an atom such as fluorine or chlorine acquires an electron, energy is released. The energy released is called the **electron affinity.** The other elements in the same column as fluorine and chlorine (i.e., bromine and iodine) also need only one electron to have an outer shell of eight electrons, so they also readily acquire an electron. Elements that readily acquire an electron are said to be **electronegative** (they easily acquire an electron and thereby become negatively charged).

$$:\!\overset{..}{F}\!\cdot + e^- \longrightarrow :\!\overset{..}{F}\!:^-$$

$$:\!\overset{..}{C}l\!\cdot + e^- \longrightarrow :\!\overset{..}{C}l\!:^-$$

1.3
IONIC, COVALENT, AND POLAR BONDS

As a teenager, Austrian **Wolfgang Pauli (1900–1958)** *wrote articles on relativity that caught the attention of Albert Einstein. Pauli went on to teach physics at the University of Hamburg and at the Zurich Institute of Technology. When World War II broke out, he emigrated to the United States where he joined the Institute for Advanced Study at Princeton.*

Friedrich Hermann Hund (1896–1997) *was born in Germany. He was a professor of physics at several German universities, the last one being the University of Göttingen. He spent a year as a visiting professor at Harvard University. In February 1996, the University of Göttingen held a symposium to honor Hund on his 100th birthday.*

Ionic Bonds

Because sodium easily gives up an electron and chlorine readily acquires an electron, will sodium donate an electron to chlorine to form sodium chloride when sodium metal and chlorine gas are combined? Because the energy given off when chlorine accepts an electron (83.3 kcal/mol; 349 kJ/mol)[2] is less than the energy required to remove an electron from sodium (118 kcal/mol; 494 kJ/mol), you might predict that the reaction will not happen. However, when sodium metal and chlorine gas are mixed, crystalline sodium chloride (table salt) forms, so we know the reaction occurs. Where does the extra energy come from? The sodium ions and chloride ions in sodium chloride are held together because the opposite charges attract each other (Figure 1.1). These forces of attraction are what provide the energy required to transfer the electrons. Attractive forces between opposite charges are called **electrostatic attractions.** A bond is an attractive force between two atoms. A bond that is the result of only electrostatic attractions is called an ionic bond. Thus an **ionic bond** is formed when there is a *transfer of electrons.*

3-D Molecule: Sodium chloride lattice

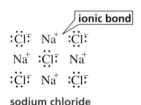

sodium chloride

Sodium chloride is an example of an ionic compound. **Ionic compounds** are formed when an element on the left side of the periodic table (an electropositive element) transfers one or more electrons to an element on the right side of the periodic table (an electronegative element).

Covalent Bonds

Instead of giving up or acquiring electrons, an atom can also achieve a filled outer shell of electrons by sharing electrons. For example, two fluorine atoms can each obtain a filled shell of eight electrons by sharing their unpaired valence electrons. A bond formed as a result of *sharing electrons* is called a **covalent bond.**

$$:\!\ddot{F}\!\cdot \; + \; \cdot\!\ddot{F}\!: \; \longrightarrow \; :\!\ddot{F}\!:\!\ddot{F}\!:$$

a covalent bond

Two hydrogen atoms can also react to form a covalent bond. By sharing electrons, each hydrogen acquires a completely filled first shell (two electrons).

$$H\!\cdot \; + \; \cdot\!H \; \longrightarrow \; H\!:\!H$$

Figure 1.1 ▶
(a) Crystalline sodium chloride. (b) The electron-rich chloride ions are red and the electron-poor sodium ions are blue.

[2]1 kcal = 4.184 kJ. Joules are the Système International (SI) units for energy, although many chemists use calories. We will use both in this book.

Similarly, hydrogen and chlorine can form a covalent bond by sharing electrons. In doing so, hydrogen fills its only shell and chlorine achieves an outer shell of eight electrons.

$$\text{H·} \;+\; \text{·C̈l:} \;\longrightarrow\; \text{H:C̈l:}$$

A hydrogen atom can also achieve a completely empty shell by losing an electron, thereby forming a positively charged hydrogen, which is called a **hydrogen ion.** A hydrogen ion is also known as a **proton** because when a hydrogen atom loses its valence electron, only the hydrogen nucleus—which consists of a single proton—remains. A hydrogen atom can achieve a completely filled first shell by gaining an electron, thereby forming a negatively charged hydrogen ion. A negatively charged hydrogen ion is called a **hydride ion.**

Bronze sculpture of Einstein on the grounds of the National Academy of Sciences in Washington, D.C.

$$\text{H·} \;\xrightarrow{\;-e^{-}\;}\; \text{H}^{+}$$
a hydrogen ion
a proton

$$\text{H·} \;\xrightarrow{\;+e^{-}\;}\; \text{H:}^{-}$$
a hydride ion

Oxygen needs to form two covalent bonds in order to fill its second shell. Nitrogen needs to form three covalent bonds and carbon must form four covalent bonds. Notice that all of the atoms in water, ammonia, and methane have filled electron shells.

$$2\,\text{H·} \;+\; \text{·Ö:} \;\longrightarrow\; \underset{\displaystyle \text{H}}{\text{H:Ö:}}$$
water

$$3\,\text{H·} \;+\; \text{·N̈·} \;\longrightarrow\; \underset{\displaystyle \text{H}}{\text{H:N̈:H}}$$
ammonia

$$4\,\text{H·} \;+\; \text{·C̈·} \;\longrightarrow\; \overset{\displaystyle \text{H}}{\underset{\displaystyle \text{H}}{\text{H:C:H}}}$$
methane

Polar Covalent Bonds

In the fluorine–fluorine and hydrogen–hydrogen covalent bonds shown previously, the atoms that share the bonding electrons are identical. Therefore, they share the electrons equally; that is, each electron spends as much time in the vicinity of one atom as in the other. An even, or nonpolar, distribution of charge results. Such a bond is called a **nonpolar covalent bond.**

In contrast, the bonding electrons in hydrogen chloride, water, and ammonia are more attracted to one atom than another because the atoms that share the electrons in these compounds are different and have different electronegativities. **Electronegativity** is the tendency of an atom to pull bonding electrons toward itself. The bonding electrons in these molecules are more attracted to the atom with the greater electronegativity, which results in a polar distribution of charge. A **polar covalent**

bond is a covalent bond between atoms of different electronegativities. The elec-tronegativities of some of the elements are shown in Table 1.3. Notice that elec-tronegativity increases as you go left-to-right across a row of the periodic table and increases as you go up any of the columns. (The column number of the periodic table tells you how many valence electrons an element has: lithium with one valence elec-tron is in column IA; fluorine with seven valence electrons is in column VIIA).

A polar covalent bond has a slight positive charge on one end and a slight nega-tive charge on the other. Polarity in a covalent bond is indicated by the symbols $\delta+$ and $\delta-$, which indicate partial positive and partial negative charges, respectively. The negative end of the bond is the end that has the more electronegative atom. The greater the electronegativity difference between the bonded atoms, the more polar the bond will be.

$$\overset{\delta+\ \ \delta-}{H-Cl} \qquad \overset{\delta+\ \ \delta-}{H-O} \qquad \overset{\delta+\ \ \delta-\ \ \delta+}{H-N-H}$$

The direction of bond polarity can be indicated using an arrow. By convention, the arrow indicates the direction the electrons are pulled, so the head of the arrow is at the negative end of the bond; a short, perpendicular line is drawn near the tail of the arrow to indicate the positive end of the bond.

$$H-Cl$$

You can think of ionic bonds and nonpolar covalent bonds as being at the oppo-site ends of a continuum of bond types. An ionic bond involves no sharing of elec-trons. A nonpolar covalent bond involves equal sharing. Polar covalent bonds fall

Tutorial: Electronegativity differences and bond types

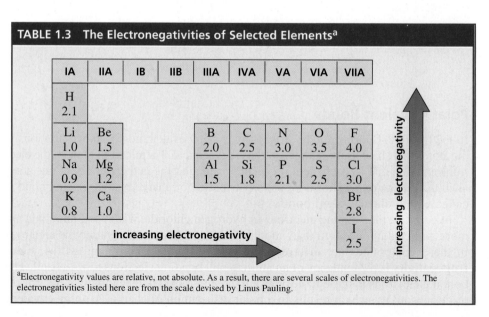

TABLE 1.3   The Electronegativities of Selected Elements[a]

IA	IIA	IB	IIB	IIIA	IVA	VA	VIA	VIIA
H 2.1								
Li 1.0	Be 1.5			B 2.0	C 2.5	N 3.0	O 3.5	F 4.0
Na 0.9	Mg 1.2			Al 1.5	Si 1.8	P 2.1	S 2.5	Cl 3.0
K 0.8	Ca 1.0							Br 2.8
								I 2.5

increasing electronegativity

[a]Electronegativity values are relative, not absolute. As a result, there are several scales of electronegativities. The electronegativities listed here are from the scale devised by Linus Pauling.

somewhere in between, and the greater the electronegativity difference between the atoms forming the bond, the closer the bond is to the ionic end of the continuum. Carbon-hydrogen bonds are relatively nonpolar because carbon and hydrogen have similar electronegativities (electronegativity difference = 0.4). Nitrogen-hydrogen bonds are relatively polar (electronegativity difference = 0.9), but not as polar as oxygen-hydrogen bonds (electronegativity difference = 1.4). The bond between sodium and chloride ions is closer to the ionic end of the continuum (electronegativity difference = 2.1), but sodium chloride is not as ionic as potassium fluoride (electronegativity difference = 3.2).

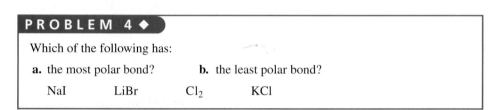

continuum of bond types

| ionic | polar | nonpolar |
| bond | covalent bond | covalent bond |

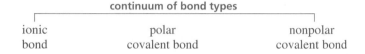

PROBLEM 4 ◆

Which of the following has:

a. the most polar bond?   **b.** the least polar bond?

    NaI   LiBr Cl_2 KCl

Understanding bond polarity is critical to understanding how organic reactions occur, because the central rule that governs the reactivity of organic compounds is that *electron-rich atoms or molecules are attracted to electron-deficient atoms or molecules.* You will find **electrostatic potential maps** throughout this book but we will frequently shorten the name to **potential map.** These maps allow you to see how charge is distributed in molecules. The electrostatic potential maps for LiH, H_2, and HF are shown below.

 3-D Molecules: LiH; H_2; HF

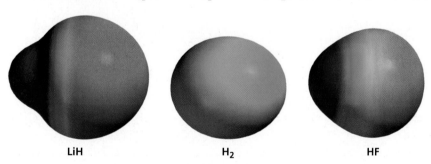

LiH H_2 HF

An electrostatic potential map shows the size and shape of a molecule. The colors indicate the charge distribution: red is used for electron rich regions (most negative electrostatic potential), and blue is used for electron poor regions (most positive electrostatic potential). Other colors indicate intermediate values of electron density.

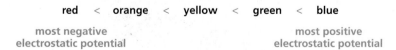

red < **orange** < **yellow** < **green** < **blue**

most negative
electrostatic potential   most positive
electrostatic potential

The electrostatic potential maps for LiH, H_2, and HF illustrate an important point. The size of an atom is not determined by its nucleus. Size is determined by the number of electrons that surround the nucleus. The negatively charged hydrogen in LiH is bigger than the neutral hydrogen in H_2 which, in turn, is bigger than the positively charged hydrogen in HF.

> ## PROBLEM 5 ◆
>
> After examining the electrostatic potential maps for LiH, HF, and H_2, answer the following questions:
>
> **a.** Which compound has the most positively charged hydrogen?
>
> **b.** Which compound has the most negatively charged hydrogen?
>
> **c.** Which compound has a nonpolar covalent bond?

The negative and positive ends of a polar bond give it a **dipole.** The size of the dipole is indicated by the dipole moment (μ). The **dipole moment** of a bond is defined as the magnitude of the charge (e) on the atom (either the partial positive charge or the partial negative charge because they have the same magnitude) times the distance between the two charges (d). A dipole moment is measured in a unit called the debye (D) (pronounced de-bye).

$$\text{dipole moment} = \mu = e \times d$$

Because the charge on an electron is 4.80×10^{-10} electrostatic units (esu) and the distance between charges in a polar bond is on the order of 10^{-8} cm, the product of charge and distance is on the order of 10^{-18} esu·cm. A dipole moment of 1.5×10^{-18} esu·cm is stated as 1.5 D. The dipole moments of some bonds commonly found in organic compounds are listed in Table 1.4.

TABLE 1.4   The Dipole Moments of Some Commonly Encountered Bonds

Bond	Dipole moment (D)	Bond	Dipole moment (D)
H—C	0.4	C—C	0
H—N	1.3	C—N	0.2
H—O	1.5	C—O	0.7
H—F	1.7	C—F	1.6
H—Cl	1.1	C—Cl	1.5
H—Br	0.8	C—Br	1.4
H—I	0.4	C—I	1.2

Peter Debye (1884–1966) was born in the Netherlands. He taught at the universities of Zürich (succeeding Einstein), Leipzig, and Berlin, but returned to his homeland in 1939 when the Nazis ordered him to become a German citizen. Upon visiting Cornell to give a lecture, he decided to stay and became a U.S. citizen in 1946. He received the Nobel Prize in chemistry in 1936 for his work on dipole moments and the properties of solutions.

In a molecule with only one covalent bond, the dipole moment of the molecule is identical to the dipole moment of the bond. For example, the dipole moment of hydrogen chloride (HCl) is 1.1 D because the dipole moment of the single hydrogen–chlorine bond is 1.1 D. The dipole moment of a molecule with more than one covalent bond depends on the geometry of the molecule as well as on the dipole moments of all the bonds in the molecule. We will examine the dipole moments of molecules with more than one covalent bond in Section 1.15 after you learn about the geometry of molecules.

> ## PROBLEM 6 / SOLVED
>
> Determine the partial charge on the oxygen atom in a C=O bond. The bond length is 1.22 $\overset{\circ}{\text{A}}^{*}$ and the bond dipole moment is 2.30 D.
>
> **SOLUTION**   If there were a full negative charge on the oxygen atom, the dipole moment would be:

$$(4.80 \times 10^{-10} \text{ esu})(1.22 \times 10^{-8} \text{ cm}) = 5.86 \times 10^{-18} \text{ esu} \cdot \text{cm} = 5.86 \text{ D}$$

Knowing that the dipole moment is 2.30 D allows us to calculate the partial charge on the oxygen atom.

$$\frac{2.30}{5.86} = 0.39$$

We find that the oxygen atom has an excess of about 0.4 electron and the carbon atom has a deficiency of about 0.4 electron.

PROBLEM 7

Use the symbols $\delta+$ and $\delta-$ to show the direction of polarity of the indicated bond in each of the following compounds (for example, $\overset{\delta-}{\text{HO}}-\overset{\delta+}{\text{H}}$):

a. H_3C-Cl **c.** H_3C-NH_2 **e.** $HO-Br$ **g.** $I-Cl$

b. $F-Br$ **d.** H_3C-OH **f.** $H_3C-MgBr$ **h.** H_2N-OH

The chemical symbols we have been using, in which the valence electrons in molecules are represented as dots, are called **Lewis structures.** The Lewis structures for H_2O, H_3O^+, HO^-, and H_2O_2 are shown below. When you draw a Lewis structure you must make sure that hydrogen atoms are surrounded by no more than two electrons, and C, O, N, and halogen (F, Cl, Br, I) atoms are surrounded by no more than eight electrons. In other words, they must obey the octet rule. Valence electrons not used in bonding are called **nonbonding electrons** or **lone-pair electrons.** Lewis structures are useful because they show us what atoms are bonded together and tell us whether any atoms possess lone-pair electrons or have a formal charge.

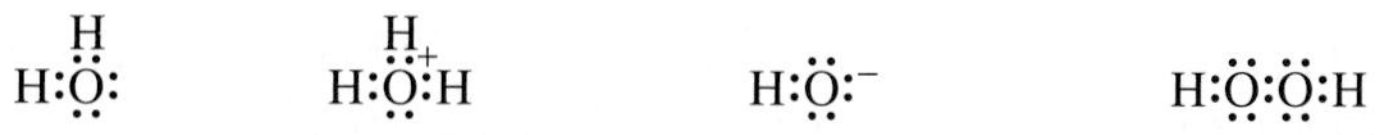

**1.4
LEWIS STRUCTURES**

water   hydronium ion   hydroxide ion   hydrogen peroxide

Once the atoms and the electrons are in place, you must determine whether a charge should be assigned to any of the atoms. A positive or a negative charge assigned to an atom is called a formal charge; the oxygen atom in the hydronium ion has a formal charge of $+1$, and the oxygen atom in the hydroxide ion has a formal charge of -1. A **formal charge** is the difference between the number of valence electrons an atom has when it is not bonded to any other atoms and the number of electrons it actually "owns" when it is bonded. An atom "owns" all of its nonbonding electrons and half of its bonding electrons.

formal charge = number of valence electrons −
          (number of nonbonding electrons + 1/2 number of bonding electrons)

For example, an oxygen atom has six valence electrons (Table 1.2). In water (H_2O), oxygen "owns" six electrons (four nonbonding electrons and half of the four bond-

---

*The angstrom (Å) is not a Système International unit. Those who opt to adhere strictly to SI units can convert it into picometers: 1 picometer (pm) $= 10^{-12}$ m; 1Å $= 10^{-10}$ m $= 100$ pm. Because the angstrom continues to be used by many organic chemists, we will use angstroms in this book.

*American chemist **Gilbert Newton Lewis (1875–1946)** was born in Weymouth, Mass. and received a Ph.D. from Harvard in 1899. He was the first to prepare "heavy water," which has deuterium atoms in place of the usual hydrogen atoms (D_2O versus H_2O). Because heavy water can be used as a moderator of neutrons, it became important for the development of the atomic bomb. Lewis started his career as a professor at the Massachusetts Institute of Technology and joined the faculty at the University of California, Berkeley, in 1912.*

ing electrons). Because the number of electrons it "owns" is equal to the number of its valence electrons $(6 - 6 = 0)$, the oxygen atom in water has no formal charge. The oxygen atom in the hydronium ion (H_3O^+) "owns" five electrons: two non-bonding plus three (half of six) bonding. Because the number of electrons it "owns" is one fewer than the number of its valence electrons $(6 - 5 = 1)$, its formal charge is $+1$. The oxygen atom in hydroxide ion (HO^-) "owns" seven electrons: six non-bonding plus one (half of two) bonding. Because it "owns" one more electron than the number of its valence electrons $(6 - 7 = -1)$, its formal charge is -1.

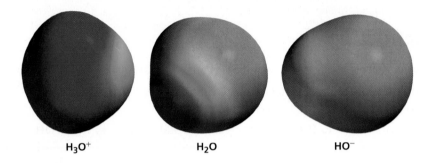

H_3O^+ H_2O HO^-

PROBLEM 8 ◆

A formal charge is a bookkeeping device. It does not necessarily indicate which atom possesses the greatest or least electron density. You can see this by examining the potential maps for H_2O, H_3O^+ and HO^-.

a. Which atom bears the formal negative charge in the hydroxide ion?

b. Which atom is the most negative in the hydroxide ion?

c. Which atom bears the formal positive charge in the hydronium ion?

d. Which atom is the most positive in the hydronium ion?

Knowing that nitrogen has five valence electrons (Table 1.2), convince yourself that the appropriate formal charges have been assigned to the nitrogen atoms in the following molecules.

ammonia   ammonium ion   amide anion   hydrazine

Carbon has four valence electrons, so take a moment to confirm why the carbon atoms in the following compounds have the indicated formal charges. A species containing a positively charged carbon atom is called a **carbocation,** and a species containing a negatively charged carbon atom is called a **carbanion.** (Recall that a *cation* is a positively charged ion, and an *anion* is a negatively charged ion.) Carbocations were formerly called carbonium ions, so you will see this term in the older literature. A species containing an atom with a single unpaired electron is called a **radical** (or a **free radical**).

methane   methyl cation   methyl anion   methyl radical   ethane
             a carbocation   a carbanion

Hydrogen has one valence electron, and the halogens have seven valence electrons each, so the following species have the indicated formal charges.

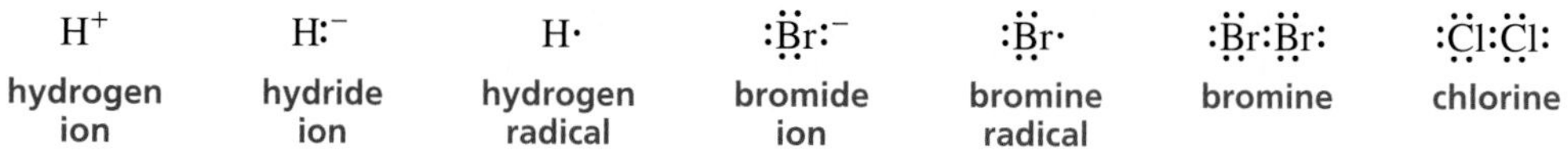

H⁺   H:⁻   H·   :Br:⁻   :Br·   :Br:Br: :Cl:Cl:

hydrogen   hydride   hydrogen   bromide   bromine   bromine   chlorine
ion   ion   radical   ion   radical

In studying the molecules in this section, notice that oxygen has *two* covalent bonds when it is neutral, nitrogen has *three* covalent bonds when it is neutral, and carbon has *four* covalent bonds when it is neutral. Neutral hydrogen and neutral halogens each have *one* covalent bond. If the atoms have more bonds or fewer bonds than the number required for a neutral molecule, they will have either a formal charge or an unpaired electron. These numbers are very important to remember when you are first drawing structures of organic compounds because they provide a quick way to recognize when you have made a mistake.

Movie: Formal charge

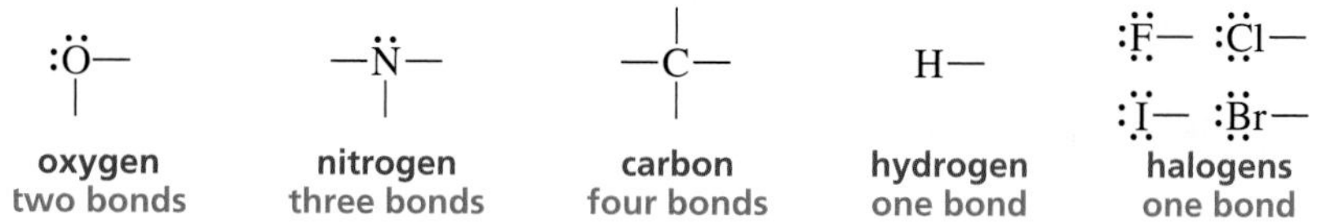

:Ö—   —Ṅ—   —C̣—   H—   :F̈— :C̈l—

                                                      :Ï— :B̈r—

oxygen   **nitrogen**   **carbon**   **hydrogen**   **halogens**
two bonds   three bonds   four bonds   one bond   one bond

In the Lewis structures for CH_2O_2, HNO_3, CH_2O, CO_3^{2-}, and N_2 shown next, notice that each atom has a complete octet (except hydrogen, which has a filled first shell) and that each atom has the appropriate formal charge. (In drawing a compound that has two or more oxygen atoms, avoid oxygen–oxygen single bonds. These are weak bonds, and few compounds have them.)

Lewis Structures

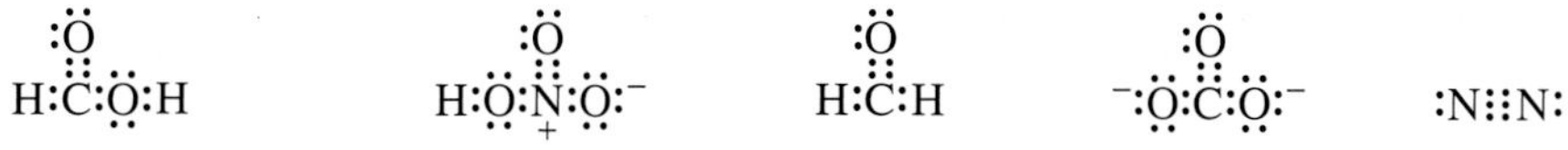

:Ö: :Ö: :Ö: :Ö:
H:C:Ö:H   H:Ö:N:Ö:⁻   H:C:H   ⁻:Ö:C:Ö:⁻   :N::N:
                      +

A pair of shared electrons can also be shown as a line between two atoms. (Compare the structures above with those below.)

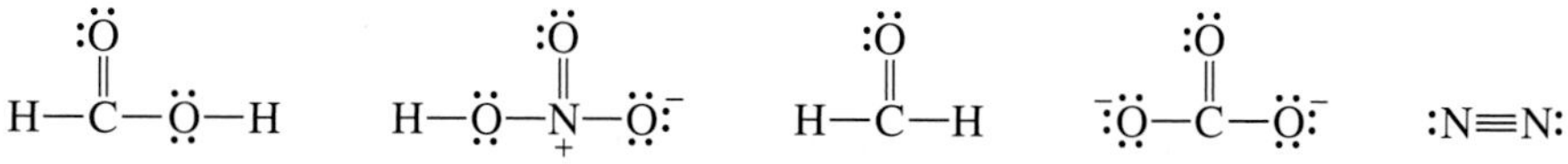

:Ö: :Ö: :Ö: :Ö:
‖   ‖   ‖   ‖
H—C—Ö—H   H—Ö—N—Ö:⁻   H—C—H   ⁻:Ö—C—Ö:⁻   :N≡N:
                      +

Suppose you are asked to draw the Lewis structure for HNO_2. First determine the total number of valence electrons (1 for H, 5 for N, 6 for each O $= 1 + 5 + 12 = 18$). Then use that number of electrons to form bonds and to fill octets with nonbonding electrons. If, after all the electrons have been assigned, any atom (other than hydrogen) does not have a complete octet, use a pair of nonbonding electrons to form a double bond.

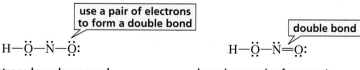

| use a pair of electrons to form a double bond |

H—Ö—Ṅ—Ö: | double bond |

                                        H—Ö—Ṅ=Ö:

18 electrons have been used   **by using a pair of oxygen's**
(6 bonding and 12 nonbonding)   **nonbonding electrons to form a**
N does not have a complete octet   **double bond, N gets a complete octet**

In **Kekulé structures,** the bonding electrons are drawn as lines and the nonbonding electrons are left out entirely unless they are needed to draw attention to some chemical property of the molecule. (However, even though they may not be shown, you should remember that neutral oxygen and nitrogen atoms always have nonbonding electrons: two pairs in the case of oxygen; one pair in the case of nitrogen.)

Kekulé Structures

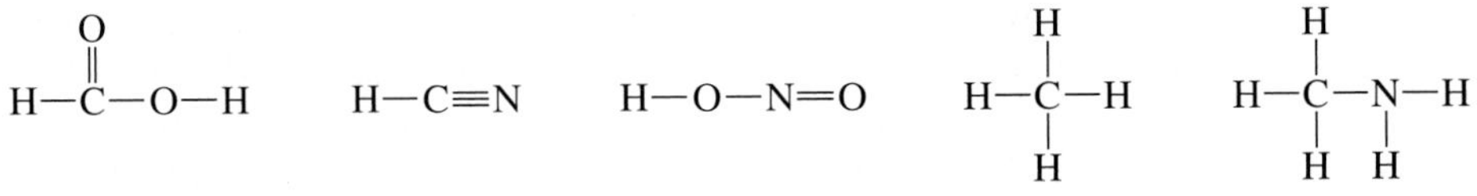

$$H-\overset{\overset{\displaystyle O}{\|}}{C}-O-H \qquad H-C\equiv N \qquad H-O-N=O \qquad H-\overset{\overset{\displaystyle H}{|}}{\underset{\underset{\displaystyle H}{|}}{C}}-H \qquad H-\overset{\overset{\displaystyle H}{|}}{\underset{\underset{\displaystyle H}{|}}{C}}-\overset{}{\underset{\underset{\displaystyle H}{}}{N}}-H$$

Frequently, structures are simplified by omitting some (or all) of the covalent bonds and listing atoms bonded to a particular carbon (or nitrogen or oxygen) next to it with a subscript to indicate the number of such atoms. These kinds of structures are called **condensed structures.**

Condensed Structures

$$HCO_2H \qquad HCN \qquad HNO_2 \qquad CH_4 \qquad CH_3NH_2$$

You can find more examples of condensed structures, and the conventions commonly used to create them, in Table 1.5.

TABLE 1.5 Kekulé and Condensed Structures

Kekulé Structure	Condensed Structures

Atoms bonded to a carbon are shown to the right of the carbon. Atoms other than H can be shown hanging from the carbon.

$$H-\overset{H}{\underset{H}{C}}-\overset{H}{\underset{Br}{C}}-\overset{H}{\underset{H}{C}}-\overset{H}{\underset{H}{C}}-\overset{H}{\underset{Cl}{C}}-\overset{H}{\underset{H}{C}}-H \qquad CH_3CHBrCH_2CH_2CHClCH_3 \quad \text{or} \quad CH_3\underset{Br}{CH}CH_2CH_2\underset{Cl}{CH}CH_3$$

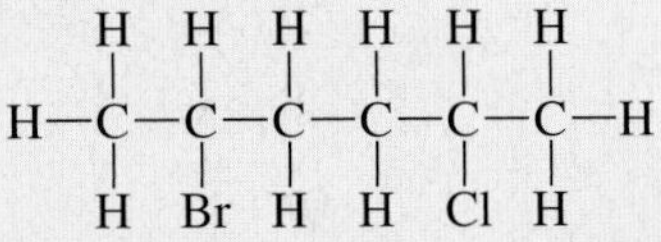

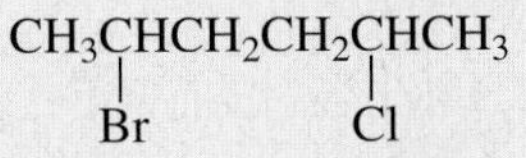

Repeating CH_2 groups can be shown in parentheses.

$$H-\overset{H}{\underset{H}{C}}-\overset{H}{\underset{H}{C}}-\overset{H}{\underset{H}{C}}-\overset{H}{\underset{H}{C}}-\overset{H}{\underset{H}{C}}-\overset{H}{\underset{H}{C}}-H \qquad CH_3CH_2CH_2CH_2CH_2CH_3 \quad \text{or} \quad CH_3(CH_2)_4CH_3$$

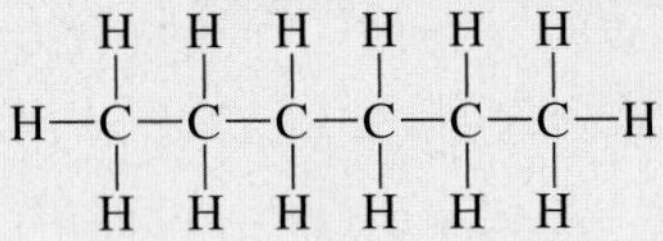

Groups bonded to a carbon can be shown (in parentheses) to the right of the carbon, or hanging from the carbon.

$$H-\overset{H}{\underset{H}{C}}-\overset{H}{\underset{H}{C}}-\overset{H}{\underset{CH_3}{C}}-\overset{H}{\underset{H}{C}}-\overset{H}{\underset{OH}{C}}-\overset{H}{\underset{H}{C}}-H \qquad CH_3CH_2CH(CH_3)CH_2CH(OH)CH_3 \quad \text{or} \quad CH_3CH_2\underset{CH_3}{CH}CH_2\underset{OH}{CH}CH_3$$

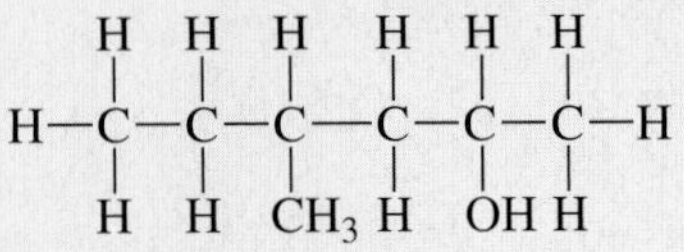

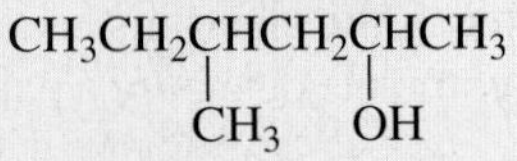

Groups bonded to the far right carbon are not put in parentheses.

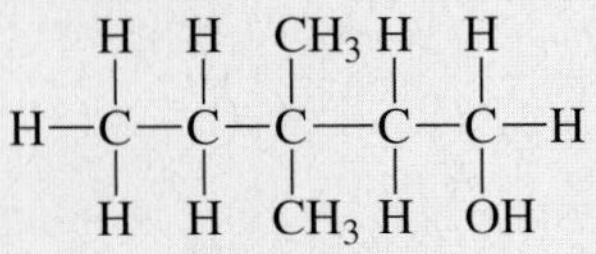

$$H-\overset{H}{\underset{H}{C}}-\overset{H}{\underset{H}{C}}-\overset{CH_3}{\underset{CH_3}{C}}-\overset{H}{\underset{H}{C}}-\overset{H}{\underset{OH}{C}}-H \qquad CH_3CH_2C(CH_3)_2CH_2CH_2OH \quad \text{or} \quad CH_3CH_2\overset{CH_3}{\underset{CH_3}{C}}CH_2CH_2OH$$

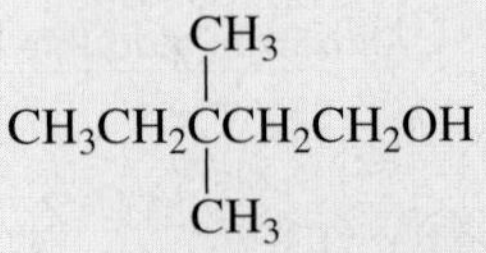

TABLE 1.5 (Continued)

| Kekulé Structure | Condensed Structures |

Two or more identical groups considered bonded to the "first" atom on the left can be shown (in parentheses) to the left of that atom, or hanging from the atom.

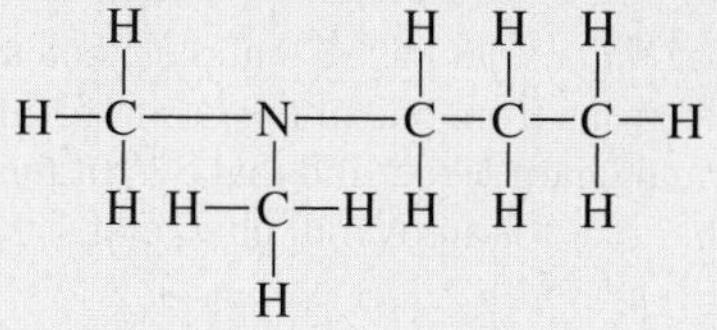

$(CH_3)_2NCH_2CH_2CH_3$ or $CH_3NCH_2CH_2CH_3$
$\qquad\qquad\qquad\qquad\qquad\qquad CH_3$

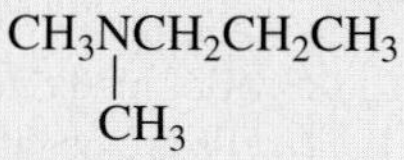

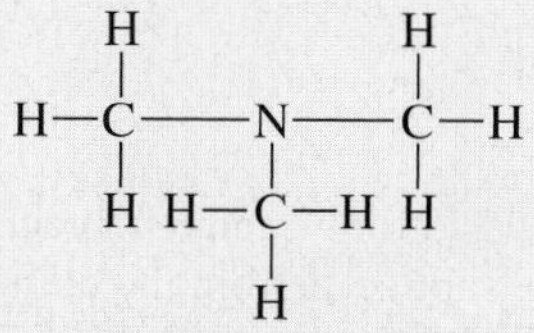

$(CH_3)_3N$ or CH_3NCH_3
$\qquad\qquad\qquad CH_3$

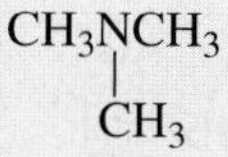

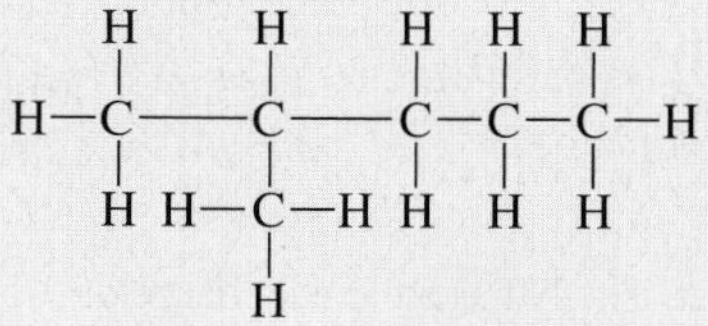

$(CH_3)_2CHCH_2CH_2CH_3$ or $CH_3CHCH_2CH_2CH_3$
$\qquad\qquad\qquad\qquad\qquad CH_3$

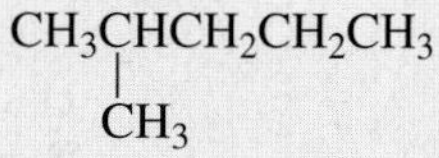

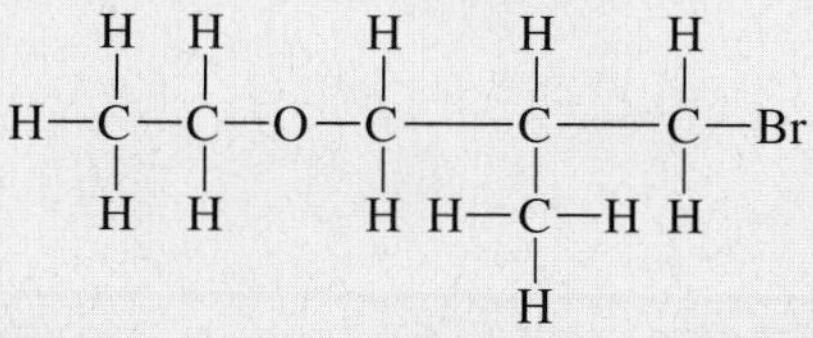

$CH_3CH_2OCH_2CH(CH_3)CH_2Br$ or $CH_3CH_2OCH_2CHCH_2Br$
$\qquad\qquad\qquad\qquad\qquad\qquad\qquad CH_3$

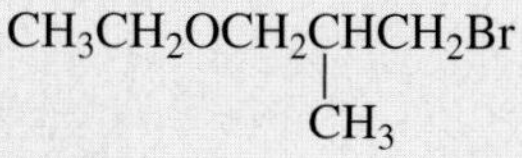

An oxygen doubly bonded to a carbon can be shown hanging off the carbon, or to the right of the carbon.

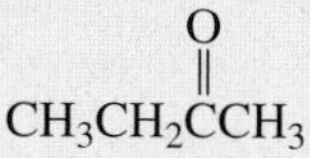

$CH_3CH_2\overset{\displaystyle O}{\overset{\|}{C}}CH_3$ or $CH_3CH_2COCH_3$ or $CH_3CH_2C(=O)CH_3$

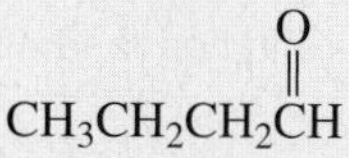

$CH_3CH_2CH_2\overset{\displaystyle O}{\overset{\|}{C}}H$ or $CH_3CH_2CH_2CHO$ or $CH_3CH_2CH_2CH=O$

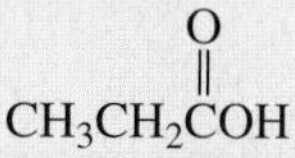

$CH_3CH_2\overset{\displaystyle O}{\overset{\|}{C}}OH$ or $CH_3CH_2CO_2H$ or CH_3CH_2COOH

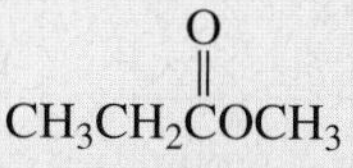

$CH_3CH_2\overset{\displaystyle O}{\overset{\|}{C}}OCH_3$ or $CH_3CH_2CO_2CH_3$ or $CH_3CH_2COOCH_3$

PROBLEM 9 / SOLVED

Draw the Lewis structure for each of the following:

a. NO_3^-	**d.** CO_2	**g.** $CH_3NH_3^+$	**j.** NaOH
b. NO_2^+	**e.** HCO_3^-	**h.** $^+C_2H_5$	**k.** NH_4Cl
c. NO_2^-	**f.** N_2	**i.** $^-CH_3$	**l.** Na_2CO_3

Solution to 9a   The only way you can arrange one N and three O's and avoid O—O bonds is to place the three O's around the N. The total number of valence electrons is 23 (5 for N, and 6 for each of the three O's). Because the species has one negative charge, add 1 to the number of valence electrons, for a total of 24. Use the 24 electrons to form bonds and to fill octets with nonbonding electrons.

$$:\ddot{O}:$$
$$|$$
$$:\ddot{O}—N—\ddot{O}:$$

When all 24 electrons have been assigned, you see that N does not have a complete octet. Complete N's octet by using a pair of nonbonding electrons from one of the O's to form a double bond. Check each atom to see if it has a formal charge. You find that two of the O's are negatively charged and the N is positively charged, for an overall charge of -1.

$$:\ddot{O}$$
$$\|$$
$$^-:\ddot{O}—\overset{+}{N}—\ddot{O}:^-$$

Solution to 9b The total number of valence electrons is 17 (5 for N, and 6 for each of the two O's). Because the species has one positive charge, subtract 1 from the number of valence electrons, for a total of 16. Use the 16 electrons to form bonds and to fill octets with nonbonding electrons.

$$:\ddot{O}—N—\ddot{O}:$$

Two double bonds are necessary to complete N's octet. The N has a formal charge of $+1$.

$$:O\!=\!\overset{+}{N}\!=\!O:$$

PROBLEM 10

a. Draw two Lewis structures for C_2H_6O. **b.** Draw three Lewis structures for C_3H_8O. (*Hint:* The two Lewis structures in **a** are constitutional isomers; they have the same atoms but differ in the way the atoms are connected. The three Lewis structures in **b** are also constitutional isomers.)

PROBLEM 11

Expand the following condensed structures to show the covalent bonds and nonbonding pairs of electrons:

a. $CH_3NHCH_2CH_3$ **c.** $(CH_3)_2CHCHO$

b. $(CH_3)_2CHCl$ **d.** $(CH_3)_3C(CH_2)_3CH(CH_3)_2$

1.5
ATOMIC ORBITALS

We have seen that electrons are distributed into different atomic orbitals (Table 1.2). An orbital is a three-dimensional region around the nucleus where there is a high probability of finding an electron. But what does an orbital look like? Mathematical calculations indicate that the s atomic orbital is a sphere with the nucleus at its center, and experimental evidence supports this theory. When we say that an electron occupies a $1s$ atomic orbital, we mean that there is a greater than 90% probability that the electron is in the space defined by the sphere. The **Heisenberg uncertainty principle** states that both the precise location and the momentum of an atomic particle cannot be simultaneously determined. This means we can never say precisely where an electron is—we can only describe its probable location. Because the average distance from the nucleus is greater for an electron in a $2s$ atomic orbital than for an electron in a $1s$ atomic orbital, a $2s$ atomic orbital is represented by a larger sphere. Consequently, the average electron density in a $2s$ atomic orbital is less than the average electron density in a $1s$ atomic orbital.

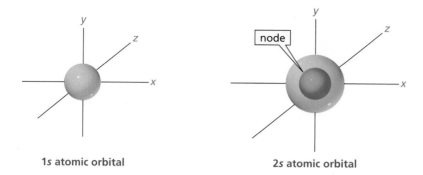

1s atomic orbital   2s atomic orbital

While an electron in a 1s atomic orbital can be anywhere within the 1s sphere, a 2s atomic orbital has a region where the probability of finding an electron falls to zero. This is called a **node.** To understand why nodes occur, remember that electrons have both particle-like and wave-like properties. A node results from the wave-like properties of an electron. To further understand, consider the two types of waves—traveling waves and standing waves. Traveling waves move through space; light is an example of a traveling wave. A standing wave is confined to a limited space; a vibrating string of a perfectly plucked guitar is an example of a standing wave—the string moves up and down, but the wave itself does not travel through space. A node is simply a region where a standing wave has an amplitude of zero; the guitar string has no transverse displacement. An electron is also a standing wave, but—unlike the wave created by a vibrating guitar string—it is three-dimensional. This means that the node of a 2s atomic orbital is actually a surface; it is a spherical surface within the 2s atomic orbital. Because the electron wave has zero amplitude at the node, there is zero probability of finding an electron at the node.

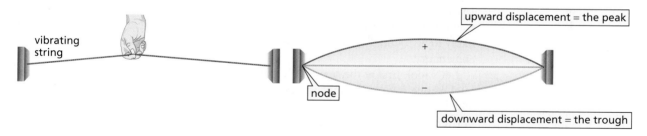

vibrating string

upward displacement = the peak

node

downward displacement = the trough

While 1s and 2s atomic orbitals resemble spheres, p atomic orbitals have two lobes. Generally, the lobes are depicted as having the shape of teardrops, but a computer-generated representation shows that they look more like doorknobs. The lobes are of opposite phase, which can be designated by a plus $(+)$ and a minus $(-)$ or by two different colors. (Notice that in this context $+$ and $-$ do not indicate charge; they indicate the phase of the orbital.) A nodal plane passes through the center of the nucleus, bisecting the two lobes of the p atomic orbital. Because the standing wave has zero amplitude at the node, there is zero probability of finding an electron in the nodal plane of the p atomic orbital.

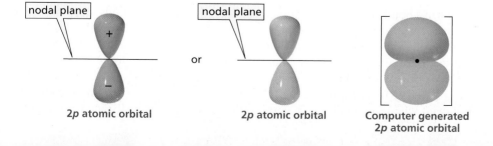

nodal plane   nodal plane

or

2p atomic orbital   2p atomic orbital   Computer generated
2p atomic orbital

Degenerate orbitals are orbitals that have the same energy.

We have seen that there are three degenerate p atomic orbitals (Section 1.2): The p_x orbital is symmetrical about the x-axis, the p_y orbital is symmetrical about the y-axis, and the p_z orbital is symmetrical about the z-axis. This means that each p orbital is perpendicular to the other two p orbitals. The energy of a $2p$ atomic orbital is slightly greater than that of a $2s$ atomic orbital because the average location of an electron in a $2p$ atomic orbital is farther away from the nucleus.

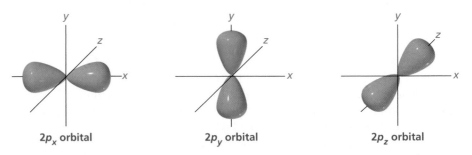

2p_x orbital   2p_y orbital   2p_z orbital

1.6 MOLECULAR ORBITALS AND BONDING

How do atoms form covalent bonds in order to form molecules? The Lewis model, in which atoms attain a complete octet simply by sharing electrons, represents only an approximation of what really happens when a covalent bond forms. As chemists learn more, theories become more sophisticated and more closely fit all the observations of real atoms and molecules. One drawback of the Lewis model is that it treats electrons like particles and does not take into account their wave-like properties.

Molecular orbital (MO) theory combines the tendency of atoms to fill their octets by sharing electrons (the Lewis model) with their wave-like properties. According to MO theory, covalent bonds result from the combination of atomic orbitals to form **molecular orbitals**—orbitals that belong to a whole molecule rather than being restricted to a single atom. Like an atomic orbital that describes the volume of space around the nucleus of an atom where an electron is likely to be found, a molecular orbital describes the volume of space about a molecule where an electron is likely to be found. Like atomic orbitals, molecular orbitals have a specific size, shape, and energy.

Let's look first at the bonding in a hydrogen molecule (H_2). As the $1s$ atomic orbital of one hydrogen atom approaches the $1s$ atomic orbital of a second hydrogen atom, they begin to overlap. As the atomic orbitals move closer together, the amount of overlap increases until the orbitals combine to form a molecular orbital. The bond that is formed when the two s atomic orbitals overlap is called a **sigma** (σ) **bond.** A sigma bond is cylindrically symmetrical—the electrons in a σ bond are symmetrically distributed about the internuclear axis (an imaginary line between the two nuclei).

H·   ·H   H : H   H : H
1s atomic   1s atomic   bonding molecular orbital
orbital   orbital

During bond formation, energy is released as the two orbitals start to overlap because the electron in each atom becomes attracted to the positively charged nucleus of the other atom as well as to its own nucleus (Figure 1.2). The more the orbitals overlap, the more the energy decreases until the atoms approach each other so closely that their positively charged nuclei start to repel each other. This causes a large increase in energy. Thus maximum stability (minimum energy) is achieved when the nuclei are a certain distance apart. This distance corresponds

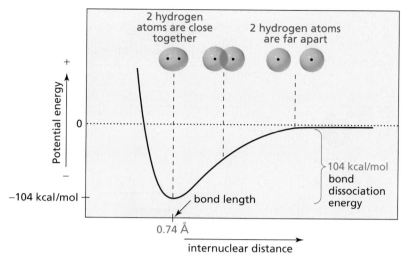

◀ **Figure 1.2**
The change in energy that occurs as two 1s atomic orbitals approach each other. The internuclear distance at minimum energy is the length of the hydrogen–hydrogen covalent bond.

Movie: H_2 bond formation

to the **bond length** of the new covalent bond. The bond length of the hydrogen–hydrogen bond is 0.74 Å.

As Figure 1.2 shows, there is a net release of energy when a covalent bond forms. When the hydrogen–hydrogen bond forms, 104 kcal/mol (435 kJ/mol) of energy is released. Breaking the bond requires precisely the same amount of energy. Thus, the **bond dissociation energy** is the energy required to break a bond, or the energy released when a bond is formed. Every covalent bond has a characteristic bond length and bond dissociation energy.

Maximum stability corresponds to minimum energy.

Orbitals are conserved—the number of atomic orbitals combined must equal the number of molecular orbitals formed. In discussing the formation of a hydrogen–hydrogen bond, however, we combined two atomic orbitals but discussed only one molecular orbital. Where is the other molecular orbital? It is there, but it contains no electrons.

The 1s atomic orbitals can combine in one of two ways. They can combine in a way that enhances each other, similar to two light waves or two sound waves that reinforce each other (Figure 1.3). This is called a σ **(sigma) bonding molecular**

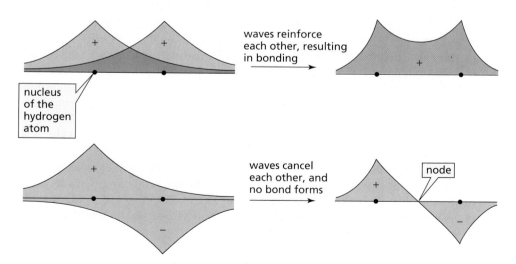

▲ **Figure 1.3**
The wave functions of two hydrogen atoms can interact to reinforce, or enhance, each other (top) or can interact to cancel each other (bottom).

Figure 1.4 ▶
Atomic orbitals of H· and molecular orbitals of H_2. Before covalent bond formation, each electron is in an atomic orbital. After covalent bond formation both electrons are in the bonding molecular orbital. The antibonding molecular orbital is empty.

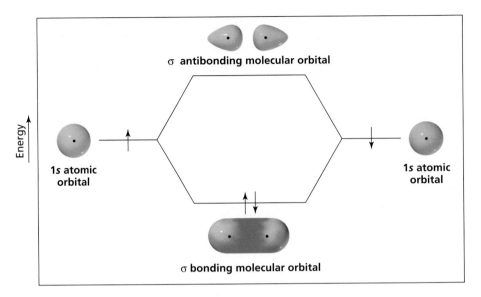

orbital. The overlapping orbitals have formed a σ bond. In a bonding molecular orbital, there is increased electron density between the nuclei which helps to bond the atoms together (Figure 1.4). The atomic orbitals can also combine in a way that cancels each other, producing a node between the nuclei (Figure 1.4). This is called a **σ antibonding molecular orbital.** An antibonding orbital can be indicated by an asterisk (σ^*). The cancellation is similar to the darkness that occurs when two light waves cancel each other or to the silence that occurs when two sound waves cancel each other because any electrons that occupy this orbital detract from, rather than aid, formation of a bond between the atoms. (Figure 1.3).

The aufbau principle and the Pauli exclusion principle, which are followed when electrons occupy atomic orbitals, are also followed when electrons occupy molecular orbitals—electrons occupy available orbitals with the lowest energy, and no more than two electrons can occupy a molecular orbital. So the two electrons of the covalent bond occupy the lower-energy σ bonding molecular orbital (Figure 1.4).

The electrons in the bonding molecular orbital are attracted to both positively charged nuclei. It is this increased electrostatic attraction that gives a covalent bond its strength. The greater the overlap of the atomic orbitals, the stronger the covalent bond. The more nuclei an electron "sees," the more stable it is. Consequently, the bonding molecular orbital is more stable (of lower energy) than the individual atomic orbitals (Figure 1.4). The antibonding molecular orbital, with no electrons between the nuclei, is less stable (of higher energy) than the atomic orbitals.

Using molecular orbital theory (and Figure 1.4), we can readily see that H_2^+ is not as stable as H_2 because H_2^+ has only one electron in the bonding orbital. We can also predict that He_2 does not exist. Because He_2 has four electrons, two electrons will fill the lower-energy bonding molecular orbital and the remaining two will fill the higher-energy antibonding molecular orbital. The two electrons in the antibonding molecular orbital will cancel the advantage to bonding gained by the two electrons in the bonding molecular orbital.

PROBLEM 12 ◆

Predict whether or not He_2^+ exists.

The covalent bond in F_2 results from end-on overlap of a $2p$ atomic orbital of one fluorine atom with a $2p$ atomic orbital of another fluorine atom.

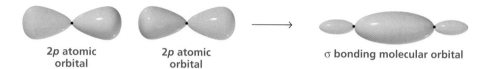

| 2p atomic orbital | 2p atomic orbital | σ bonding molecular orbital |

Because a $2p$ atomic orbital encompasses a greater volume of space than a $1s$ atomic orbital, the average electron density in a $2p$ atomic orbital is less than the average electron density in a $1s$ atomic orbital. So, when the two $2p$ orbitals of F_2 overlap, the average electron density between the nuclei in the resulting bonding molecular orbital is less than the average electron density between the nuclei in the bonding molecular orbital of H_2. Therefore, the electrons cannot pull the electrostatically repulsed fluorine nuclei as close together. Consequently, the F_2 bond is longer and weaker than the H_2 bond.

> The greater the electron density in the region of orbital overlap, the stronger the bond.

	Bond length	**Bond strength**
F—F	1.42 Å	38 kcal/mol or 159 kJ/mol
H—H	0.74 Å	104 kcal/mol or 435 kJ/mol

When a p orbital of one atom overlaps end-on with a p orbital of another atom, as it does in the fluorine–fluorine bond, a σ bond is formed (Figure 1.5). If the overlapping lobes of the p orbitals are in-phase (are the same color), a σ bonding molecular orbital is formed; if the lobes are out-of-phase (are different colors), a σ

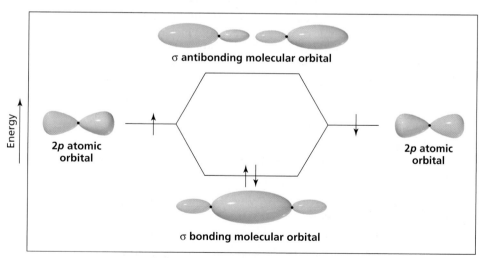

σ antibonding molecular orbital

Energy

2p atomic orbital

2p atomic orbital

σ bonding molecular orbital

◀ **Figure 1.5**
End-on overlap of two p orbitals to form a σ bonding molecular orbital and a σ antibonding molecular orbital.

antibonding molecular orbital is formed. The electron density of the σ bonding molecular orbital is concentrated between the nuclei, which causes the back lobes of the molecular orbital to be quite small.

In addition to overlapping end-on, two p atomic orbitals can also overlap side-to-side (Figure 1.6). A bond formed from the side-to-side overlap of p atomic orbitals is called a **pi (π) bond.** Side-to-side overlap of two in-phase p atomic orbitals forms a π bonding molecular orbital, whereas side-to-side overlap of two out-of-phase p orbitals forms a π antibonding molecular orbital (π^*). The maximum electron density in the π bonding orbital is on either side of the internuclear axis.

Figure 1.6 ▶
Side-to-side overlap of two
parallel *p* orbitals to form a π
bonding molecular orbital
and a π antibonding molecular orbital.

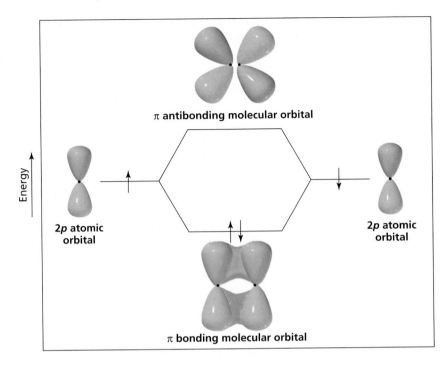

π antibonding molecular orbital

Energy

2*p* atomic
orbital

2*p* atomic
orbital

π bonding molecular orbital

**Computer generated
π antibonding
molecular orbital**

**Computer generated
π bonding
molecular orbital**

Organic chemists find that the information obtained from MO theory, where valence electrons occupy bonding and antibonding molecular orbitals, does not always
yield the most useful information about a covalent bond. The **valence-shell electron-
pair repulsion (VSEPR) model** combines the concept of atomic orbitals with the
Lewis concept of shared electron pairs and nonbonding electrons and adds a third
principle—minimization of electron repulsion. In this model, atoms share electrons
by overlapping their atomic orbitals and, because electron pairs repel each other, the
pairs of bonding electrons and pairs of nonbonding electrons around an atom are positioned as far apart as possible.

Because organic chemists generally think of chemical reactions in terms of the
changes that occur in the bonds of the reacting molecules, the VSEPR model often
provides the easiest way to visualize chemical change. However, the model is inadequate for molecules in excited states because it does not allow for antibonding orbitals. We will use both the MO model and the VSEPR model in this book. Our
choice will depend on which model provides the best description of the molecule
under discussion. (We will be using the VSEPR model in Sections 1.7–1.13.)

PROBLEM 13 ◆

a. Predict the relative lengths and strengths of the bonds in Cl_2 and Br_2.

b. Predict the relative lengths and strengths of the bonds in HF, HCl, and HBr.

**1.7
BONDING IN
METHANE AND
ETHANE.
SINGLE BONDS**

We will start the discussion of bonding in organic compounds by looking at the bonding in methane, a compound with only one carbon atom. We will then examine the
bonding in ethane (a compound with a carbon–carbon single bond), ethene (a compound
with a carbon–carbon double bond), and ethyne (a compound with a carbon–carbon
triple bond).

Next, we will look at bonds formed by atoms other than carbon that are commonly found in organic compounds—bonds formed by oxygen, nitrogen, and the halogens. Because the distribution of electrons in the atoms that make up a molecule determines the orbitals used in bond formation and because the orbitals used in bond formation determine the bond angles in a molecule, you will see that if you know the distribution of electrons in the atoms and the bond angles in the molecule, you can determine the orbitals that are involved in bond formation.

Bonding in Methane

Methane (CH_4) has four covalent carbon–hydrogen bonds. Because all four C—H bonds have the same length and all the bond angles are the same (109.5°), we can conclude that the four covalent bonds in methane are identical.

Four different ways to represent a methane molecule are shown here. In a perspective formula, bonds in the plane of the paper are drawn as solid lines, bonds protruding out of the plane of the paper toward the viewer are drawn as solid wedges, and those protruding back from the plane of the paper away from the viewer are drawn as hatched wedges. The electrostatic potential map shows that carbon and hydrogen share their bonding electrons relatively equally—there are no red areas representing high electron density or deep blue areas representing low electron density. In other words, there are no partial charges on any of the atoms in methane because the electronegativities of carbon and hydrogen are similar. Methane is a **nonpolar molecule.**

3-D Molecule:
Methane

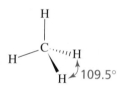

| perspective formula of methane | ball-and-stick model of methane | space-filling model of methane | electrostatic potential map for methane |

Initially it may seem surprising that carbon forms four covalent bonds, because carbon has only two unpaired electrons in its ground-state electronic configuration (Table 1.2). Therefore, one would expect it to form only two covalent bonds. However, if carbon formed only two covalent bonds, it would not complete its octet. In order to form four covalent bonds and thereby complete its octet, carbon must promote an electron from the $2s$ atomic orbital into the empty $2p$ atomic orbital. After promotion, there are four unpaired electrons in the new electronic configuration, so four covalent bonds can be formed.

$$\uparrow\downarrow \quad \underset{p_x}{\uparrow} \quad \underset{p_y}{\uparrow} \quad \underset{p_z}{\quad} \qquad \xrightarrow{\text{promotion}} \qquad \uparrow \quad \underset{p_x}{\uparrow} \quad \underset{p_y}{\uparrow} \quad \underset{p_z}{\uparrow}$$

$$\underset{s}{} \qquad\qquad\qquad\qquad\qquad\qquad \underset{s}{}$$

before promotion   after promotion

Because a p orbital is higher in energy than an s orbital, promotion of an electron from an s orbital to a p orbital requires energy. The amount of energy required is 96 kcal/mol (402 kJ/mol). Because the dissociation energy of a carbon–hydrogen bond is 105 kcal/mol (439 kJ/mol), the formation of four carbon–hydrogen bonds releases 420 kcal/mol of energy. If promotion did not occur, carbon could form only two covalent bonds for a net energy change of 210 kcal/mol. So, by spending 96 kcal/mol to promote an electron, an extra 114 kcal/mol (210 − 96) is released as a result of the formation of two additional covalent bonds (Figure 1.7).

Figure 1.7 ▶
As a result of electron promotion, carbon forms four covalent bonds with a release of 420 kcal/mol of energy. Without promotion, carbon would form two covalent bonds with a release of 210 kcal/mol of energy. Because it requires 96 kcal/mol to promote an electron, the overall energy advantage of promotion is 114 kcal/mol.

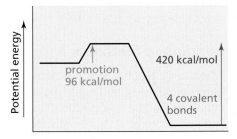

The four carbon–hydrogen bonds in methane are identical. Each has a bond length of 1.10 Å and breaking any one of the bonds requires 105 kcal/mol (439 kJ/mol). If carbon used an *s* orbital and three *p* orbitals to form these four covalent bonds, the bond formed with the *s* orbital would differ from the three bonds formed with *p* orbitals. How can carbon form four identical bonds using one *s* and three *p* orbitals?

Because all four bonds are identical, we know that carbon must have used hybrid orbitals. **Hybrid orbitals** are mixed orbitals. The concept of mixing orbitals, called **orbital hybridization,** was first proposed by Linus Pauling in 1931. If the one *s* and three *p* orbitals of the second shell are mixed together and then divided into four equal orbitals, each of the four resulting orbitals will be 1/4 *s* and 3/4 *p*. Such mixed orbitals are called sp^3 orbitals. Each sp^3 orbital has 25% *s* character and 75% *p* character. (The superscript 3 means that three *p* orbitals were mixed with one *s* orbital to form the hybrid orbitals.) The four sp^3 orbitals are degenerate. In other words, they have the same energy.

Linus Carl Pauling (1901–1994) *was born in Portland, Ore. A friend's home chemistry laboratory sparked Pauling's early interest in science. He received a Ph.D. from the California Institute of Technology and remained there for most of his academic career. He received the Nobel Prize in chemistry in 1954 for his work on molecular structure. Like Einstein, Pauling was a pacifist, winning the 1962 Nobel Peace Prize for his work on behalf of nuclear disarmament.*

Electron pairs spread themselves into space as far from each other as possible.

An sp^3 orbital has two lobes like a *p* orbital. The lobes differ in size, however, because the positive lobe of the *p* orbital adds to the *s* orbital, whereas the negative lobe of the *p* orbital subtracts from the *s* orbital (Figure 1.8). The larger lobe of the sp^3 orbital is used in covalent bond formation.

The four sp^3 orbitals arrange themselves in space in a way that allows them to get as far away from each other as possible (Figure 1.9a). This occurs because electrons repel each other and getting as far from each other as possible minimizes this repulsion (Section 1.6). When four orbitals spread themselves into space as far from each other as possible, they point toward the corners of a regular tetrahedron (a pyramid with four faces, each an equilateral triangle). Each of the four carbon–hydrogen bonds in methane is formed from overlap of the *s* orbital of a hydrogen with an sp^3 orbital of carbon (Figure 1.9b). Now you can understand why the four carbon–hydrogen bonds are identical.

The angle formed between any two bonds of methane is 109.5°. This bond angle is called the **tetrahedral bond angle.** A carbon such as the one in methane, which

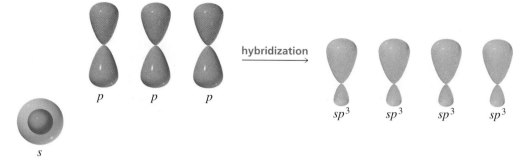

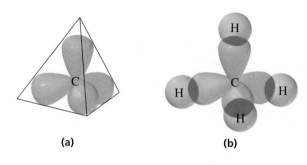

▲ **Figure 1.8**
An *s* orbital and three *p* orbitals hybridize to form four *sp*³ orbitals.

◄ **Figure 1.9**
(a) The four *sp*³ orbitals are directed toward the corners of a tetrahedron, causing each bond angle to be 109.5°. (b) An orbital picture of methane, showing the overlap of the *sp*³ orbitals of carbon with the *s* orbital of a hydrogen. (For clarity, the smaller lobes of the *sp*³ orbitals are not shown.)

forms covalent bonds using four equivalent *sp*³ hybrid orbitals, is called a **tetrahedral carbon.**

The postulation of hybrid orbitals may seem to be a theory contrived just to make things fit. And that is exactly what it is. But it is a theory that gives us a very good picture of how chemical compounds behave.

Note to the student

It is important to understand what molecules look like in three dimensions. Therefore, as you study a chapter you should go to the Web site and spend some time looking at the three-dimensional representations of molecules that can be found in the Visualization Center that accompanies each chapter (www.prenhall.com/bruice).

Bonding in Ethane

The two carbon atoms in ethane are tetrahedral. Each carbon uses four *sp*³ atomic orbitals to form four covalent bonds.

$$\begin{array}{c} \quad\ \ \text{H}\quad\text{H} \\ \quad\ \ | \quad\ | \\ \text{H}\!-\!\text{C}\!-\!\text{C}\!-\!\text{H} \\ \quad\ \ | \quad\ | \\ \quad\ \ \text{H}\quad\text{H} \end{array}$$

ethane

One *sp*³ orbital of one carbon overlaps an *sp*³ orbital of the other carbon to form the carbon–carbon bond. Each of the remaining three *sp*³ orbitals of each carbon overlaps the *s* orbital of a hydrogen to form a carbon–hydrogen bond. Thus, the

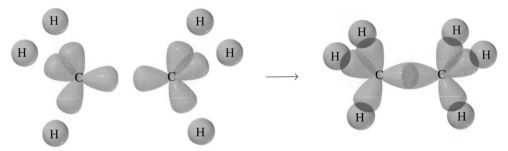

▲ **Figure 1.10**
An orbital picture of ethane. The carbon–carbon bond is formed by sp^3–sp^3 overlap, and each carbon–hydrogen bond is formed by sp^3–s overlap. (The smaller lobes of the sp^3 orbitals are not shown.)

carbon–carbon bond is formed by sp^3–sp^3 overlap and each carbon–hydrogen bond is formed by sp^3–s overlap (Figure 1.10). Each of the bond angles in ethane is nearly the tetrahedral bond angle of 109.5°, and the length of the carbon–carbon bond is 1.54 Å. Ethane, like methane, is a nonpolar molecule.

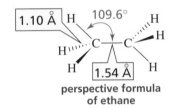

perspective formula
of ethane

3-D Molecule:
Ethane

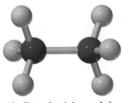

ball-and-stick model
of ethane

space-filling model
of ethane

electrostatic potential
map for ethane

All the bonds in methane and ethane are sigma (σ) bonds because they are all formed by end-on overlap of atomic orbitals. (All **single bonds** found in organic compounds are sigma bonds.)

PROBLEM 14

What orbitals are used to form the covalent bonds in propane ($CH_3CH_2CH_3$)?

1.8
BONDING IN ETHENE. DOUBLE BONDS

Each of the carbon atoms in ethene forms four bonds, but each is bonded to only three atoms.

$$\underset{H}{\overset{H}{\diagdown}}C=C\underset{H}{\overset{H}{\diagup}}$$

ethene
ethylene

To bond to three atoms, each carbon hybridizes three atomic orbitals. Because three orbitals are hybridized (an *s* orbital and two of the *p* orbitals), three hybrid orbitals are obtained. These are called *sp²* orbitals. After hybridization, each carbon atom has three degenerate *sp²* orbitals and one *p* orbital.

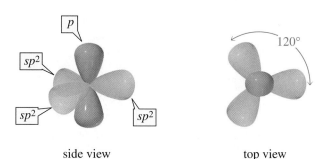

In order to get as far from each other as possible, the axes of the three *sp²* orbitals lie in a plane, directed to the corners of an equilateral triangle with the carbon nucleus in the center. This means that the bond angles are all close to 120°. Because the *sp²* hybridized carbon atom is bonded to three atoms that define a plane, it is called a **trigonal planar carbon.** The unhybridized *p* orbital is perpendicular to the triangular plane defined by the axes of the *sp²* orbitals (Figure 1.11).

◀ **Figure 1.11**
An *sp²* hybridized carbon. The three degenerate *sp²* orbitals lie in a plane. The unhybridized *p* orbital is perpendicular to the plane. (The smaller lobes of the *sp²* orbitals are not shown.)

side view   top view

The carbons in ethene form two bonds with each other. This is called a **double bond.** The two carbon–carbon bonds in the double bond are not identical. One of the carbon–carbon bonds in ethene results from the overlap of an *sp²* orbital of one carbon with an *sp²* orbital of the other carbon; this is a sigma (σ) bond because it is formed by end-on overlap (Figure 1.12a). Each carbon uses its other two *sp²* orbitals to overlap the *s* orbital of a hydrogen to form the carbon–hydrogen bonds. The second carbon–carbon bond results from the overlap of the two unhybridized *p* orbitals. They overlap in a side-to-side manner (Figure 1.12b) and therefore form a pi (π)

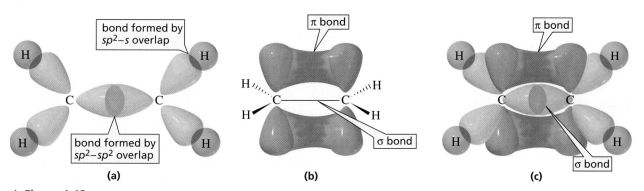

▲ **Figure 1.12**
(a) The carbon–carbon σ bond in ethene is formed by *sp²—sp²* overlap, and the carbon–hydrogen bonds are formed by *sp²—s* overlap.
(b) The carbon–carbon π bond is formed by side-to-side overlap of a *p* orbital of one carbon with a *p* orbital of the other carbon.
(c) The double bond has an electron-dense region above and below the plane containing the two carbons and four hydrogens.

bond. Thus, one of the bonds in a double bond is a σ bond and the other is a π bond. The carbon–hydrogen bonds are all σ bonds.

The two p orbitals that overlap to form the π bond must be parallel to each other for maximum overlap to occur. This forces the triangle formed by one carbon and two hydrogens to lie in the same plane as the triangle formed by the other carbon and two hydrogens. This means that all six atoms in ethene lie in the same plane, and the electrons in the p orbitals occupy a volume of space above and below this plane (Figure 1.12c). The electrostatic potential map for ethene clearly shows the electron-dense region above the plane defined by the two carbons and four hydrogens.

H   121.7°   H

1.08 Å   C=C

H   1.33 Å   H

**a double bond consists of
one σ bond and one π bond**

3-D Molecule:
Ethene

**ball-and-stick model
of ethene**   **space-filling model
of ethene**   **electrostatic potential map
for ethene**

Both bonds in the double bond of ethene contribute to its strength. In a carbon–carbon double bond, four electrons hold the carbons together; in a carbon–carbon single bond, only two electrons bind the atoms. This means that the carbon–carbon double bond in ethene is stronger (152 kcal/mol, 636 kJ/mol) and shorter (1.33 Å) than the carbon–carbon single bond in ethane (88 kcal/mol, 368 kJ/mol, and 1.54 Å).

PROBLEM 15 ◆

Which of the bonds of a carbon–carbon double bond has more effective orbital-orbital overlap, the σ bond or the π bond?

DIAMOND, GRAPHITE, AND BUCKMINSTERFULLERENE: SUBSTANCES CONTAINING ONLY CARBON ATOMS

Diamond is the hardest of all substances. Graphite, in contrast, is a slippery, soft solid most familiar to us as the "lead" in pencils. Both materials, in spite of their very different physical properties, contain only carbon atoms. The two substances differ only in the nature of the carbon–carbon bonds holding them together. Diamond consists of a rigid three-dimensional network of atoms, with each carbon bonded to four other carbons *via* sp^3 orbitals. The carbon atoms in graphite, on the other hand, are sp^2 hybridized, so each bonds to only three other carbon atoms. This trigonal planar arrangement causes atoms in graphite to lie in flat layered sheets that can shear off of neighboring sheets. As you rub a pencil across a page, sheets of carbon atoms shear off, leaving a thin trail of graphite. There is a third substance found in nature that contains only carbon atoms: buckminsterfullerene. Like graphite, buckminsterfullerene contains only sp^2 hybridized carbons, but instead of forming planar sheets, the sp^2 carbons in buckminsterfullerene form spherical structures. (Buckminsterfullerene is discussed in more detail in Section 14.2.)

The carbon atoms in ethyne are each bonded to only two atoms—a hydrogen and another carbon.

$$H—C≡C—H$$
ethyne
acetylene

Because each carbon forms covalent bonds with only two atoms, two orbitals (an *s* and a *p*) are hybridized. Two degenerate *sp* hybridized orbitals result. Each carbon atom in ethyne, therefore, has two *sp* hybridized orbitals and two unhybridized *p* orbitals (Figure 1.13).

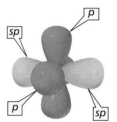

▲ **Figure 1.13**
An *sp* hybridized carbon. The two *sp* hybrid orbitals are oriented 180° away from each other, perpendicular to the two unhybridized *p* orbitals. (The smaller lobes of the *sp* orbitals are not shown.)

One of the *sp* orbitals of one carbon in ethyne overlaps an *sp* orbital of the other carbon to form a carbon–carbon σ bond. The other *sp* orbital of each carbon overlaps the *s* orbital of a hydrogen to form a carbon–hydrogen σ bond (Figure 1.14a). In order to minimize electron repulsions, the two *sp* hybrid orbitals point in opposite directions. Consequently, the bond angles are 180°. The two unhybridized *p* orbitals are perpendicular to each other and both are perpendicular to the *sp* orbitals.

Each of the unhybridized *p* orbitals engages in side-to-side overlap with a parallel *p* orbital on the other carbon, with the result that two π bonds are formed (Figure 1.14b). The overall result is a triple bond. A **triple bond** consists of one σ bond and two π bonds. Because the two unhybridized *p* orbitals on each carbon are perpendicular to each other, there is a region of high electron density above and below *and* in front of and in back of the internuclear axis of the molecule (Figure 1.14c). The electrostatic potential map for ethyne shows that there is a cylinder of electron density that wraps around the egg-shaped molecule.

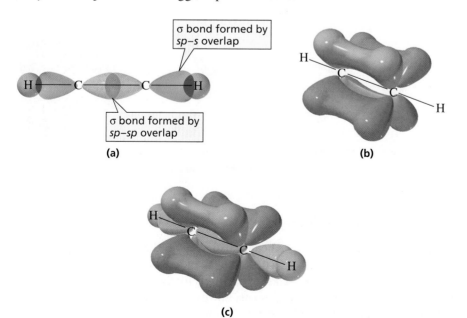

(a)

(b)

(c)

◄ **Figure 1.14**
(a) The carbon–carbon σ bond in ethyne is formed by *sp—sp* overlap, and the carbon–hydrogen σ bonds are formed by *sp—s* overlap. The carbon atoms in a triple bond and the atoms bonded to them are in a straight line.
(b) The two carbon–carbon π bonds are formed by side-to-side overlap of the *p* orbitals of one carbon with the *p* orbitals of the other carbon.
(c) The triple bond has an electron-dense region above and below *and* in front of and in back of the internuclear axis of the molecule.

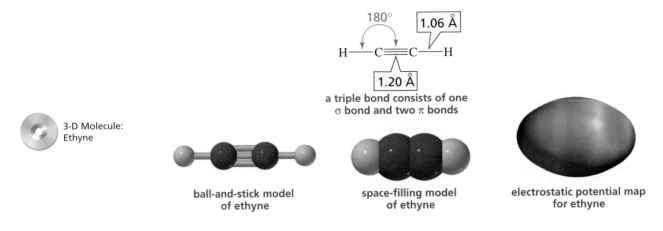

3-D Molecule:
Ethyne

a triple bond consists of one
σ bond and two π bonds

ball-and-stick model
of ethyne

space-filling model
of ethyne

electrostatic potential map
for ethyne

Because the two carbon atoms in a triple bond are held together by six electrons, a triple bond is stronger (200 kcal/mol, 837 kJ/mol) and shorter (1.20 Å) than a double bond.

1.10 BONDING IN THE METHYL CATION, THE METHYL RADICAL, AND THE METHYL ANION

Not all carbon atoms form four bonds. A carbon with a positive charge, a negative charge, or an unpaired electron forms only three bonds. We will now take a look at the bonding in carbon atoms that form three bonds.

The Methyl Cation ($^+CH_3$)

A positively charged carbon forms three covalent bonds, so it hybridizes three orbitals—an s orbital and two p orbitals. Therefore, it forms its three covalent bonds using three sp^2 orbitals. Its unhybridized p orbital remains empty. Because the three sp^2 orbitals lie in a plane, a carbocation is flat. The p orbital stands perpendicular to the plane defined by the three atoms covalently bonded to the positively charged carbon.

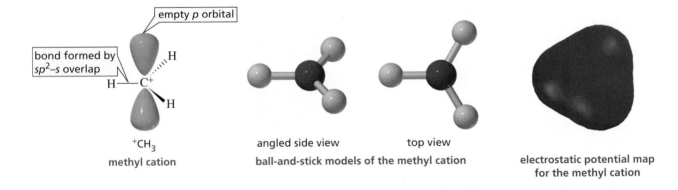

$^+CH_3$
methyl cation

angled side view   top view

ball-and-stick models of the methyl cation

electrostatic potential map
for the methyl cation

The Methyl Radical (·CH_3)

The carbon atom in the methyl radical is also sp^2 hybridized. The methyl radical differs by one unpaired electron from the methyl cation. That electron is in the p orbital. Notice the similarity in the ball-and-stick models for the methyl cation and the methyl radical. The electrostatic potential maps, however, are quite different because of the additional electron in the methyl radical.

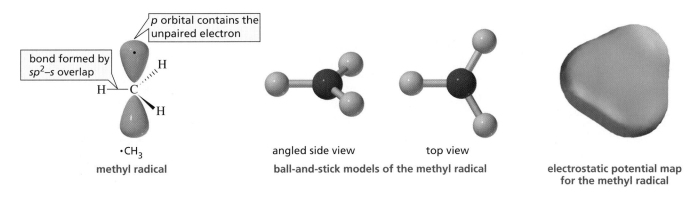

p orbital contains the unpaired electron

bond formed by sp^2-s overlap

$\cdot CH_3$

methyl radical

angled side view

top view

ball-and-stick models of the methyl radical

electrostatic potential map for the methyl radical

The Methyl Anion ($\bar{:}CH_3$)

A negatively charged carbon has three pairs of bonding electrons and one pair of nonbonding electrons. Because pairs of electrons are positioned as far apart as possible (Section 1.6), the four orbitals containing the bonding and nonbonding electrons point toward the corners of a tetrahedron. In other words, a negatively charged carbon is sp^3 hybridized. In the methyl anion, three of carbon's sp^3 orbitals each overlap the s orbital of a hydrogen and the fourth sp^3 orbital holds the nonbonding pair of electrons.

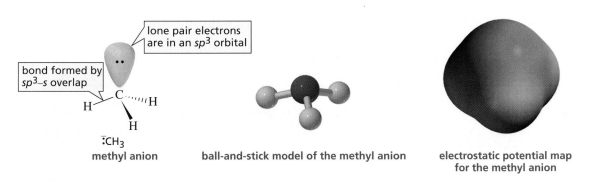

lone pair electrons are in an sp^3 orbital

bond formed by sp^3-s overlap

$\bar{:}CH_3$
methyl anion

ball-and-stick model of the methyl anion

electrostatic potential map for the methyl anion

Take a moment to compare the electrostatic potential maps for the methyl cation, the methyl radical, and the methyl anion.

The oxygen atom in water (H_2O) forms two covalent bonds. Because oxygen has two unpaired electrons in its ground-state electronic configuration (Table 1.2), it does not need to promote an electron to form the required number of covalent bonds. But this simple picture presents a problem. Oxygen has two nonbonding pairs of electrons. These two nonbonding pairs of electrons are chemically identical, which means that they must be in identical orbitals. However, the ground-state electronic configuration of oxygen suggests that one of the nonbonding pairs of electrons occupies an s orbital and the other pair occupies a p orbital, which would make them chemically nonidentical. Furthermore, if oxygen uses p orbitals to form the two covalent bonds as predicted by the ground-state electronic configuration, the oxygen–hydrogen bonds should have a bond angle of about 90° because the two p orbitals are at right angles to each other. However, the experimentally observed bond angle is 104.5°. In order to explain the observed bond angle and the fact that the two pairs of nonbonding electrons are identical, oxygen must use hybrid orbitals to form covalent bonds (just

1.11
BONDING IN WATER

as carbon does). The *s* orbital and the three *p* orbitals must hybridize to produce four sp^3 orbitals.

**second-shell electrons
of oxygen**

4 orbitals are hybridized

hybrid orbitals

Each of the two oxygen–hydrogen bonds is formed by the overlap of an sp^3 orbital of oxygen with the *s* orbital of a hydrogen. A nonbonding pair of electrons occupies each of the two remaining sp^3 orbitals.

The bond angle in water is a little smaller (104.5°) than the tetrahedral bond angle (109.5°) in methane—presumably because each nonbonding electron pair "sees" only one nucleus, which makes the nonbonding pair more diffuse than the bonding pair that "sees" two nuclei and is therefore relatively confined between the two nuclei. These more diffuse orbitals take up more room, causing the oxygen–hydrogen bonds to squeeze closer together, thereby decreasing the bond angle.

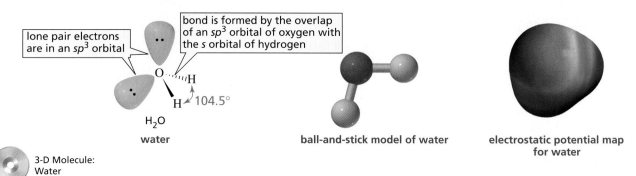

lone pair electrons are in an sp^3 orbital

bond is formed by the overlap of an sp^3 orbital of oxygen with the *s* orbital of hydrogen

H_2O
water

ball-and-stick model of water

electrostatic potential map for water

3-D Molecule:
Water

Compare the electrostatic potential map for water with that for methane. Water is a polar molecule. Methane is nonpolar.

PROBLEM 16 ◆

The bond angles in H_3O^+ are greater than _____ and less than _____.

WATER—A UNIQUE COMPOUND

Water is the most abundant compound found in biological organisms. It has unique properties that allowed life to originate and evolve. Its high heat of fusion (the heat required to convert a solid to a liquid) protects organisms from freezing at low temperatures. The high heat capacity (the heat required to raise the temperature of a substance a given amount) of water minimizes temperature changes in organisms, and its high heat of vaporization (the heat required to convert a liquid to a gas) allows animals to cool themselves with a minimal loss of body fluid. Because liquid water is more dense than ice, ice formed on the surface of water floats and insulates the water below. This is why oceans and lakes don't freeze from the bottom up. It's also why plants and aquatic animals can survive when the ocean or lake they live in freezes.

The experimentally observed bond angles in NH_3 are 107.3°. The bond angles indicate that nitrogen also uses hybrid orbitals when it forms covalent bonds. Like carbon and oxygen, the one s and three p orbitals of the second shell of nitrogen hybridize to form four degenerate sp^3 orbitals.

1.12 BONDING IN AMMONIA AND IN THE AMMONIUM ION

The nitrogen–hydrogen covalent bonds in NH_3 are formed from the overlap of an sp^3 orbital of nitrogen with the s orbital of a hydrogen. The single nonbonding pair of electrons occupies an sp^3 orbital. The bond angle (107.3°) is smaller than the tetrahedral bond angle (109.5°) because the nonbonding pair of electrons takes up more space than bonding electron pairs. Notice that the NH_3 bond angle is larger than the H_2O bond angle (104.5°) because nitrogen has only one nonbonding pair of electrons while oxygen has two nonbonding pairs.

3-D Molecule: Ammonia

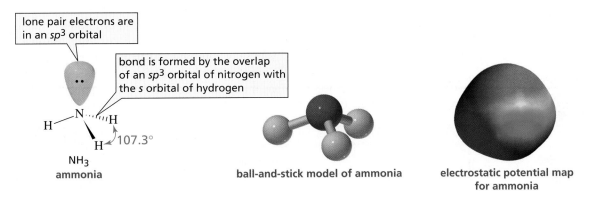

lone pair electrons are in an sp^3 orbital

bond is formed by the overlap of an sp^3 orbital of nitrogen with the s orbital of hydrogen

107.3°

NH_3
ammonia

ball-and-stick model of ammonia

electrostatic potential map for ammonia

Because the ammonium ion ($^+NH_4$) has four identical nitrogen–hydrogen bonds and has no nonbonding pairs of electrons, all the bond angles are 109.5°.

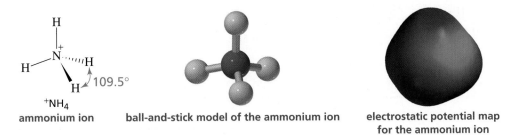

109.5°

$^+NH_4$
ammonium ion

ball-and-stick model of the ammonium ion

electrostatic potential map for the ammonium ion

PROBLEM 17 ◆

Which atom in the ammonium ion has the least electron density?

PROBLEM 18 ◆

Compare the electrostatic potential maps for methane, ammonia, and water. Which is the most polar molecule? Which is the least polar?

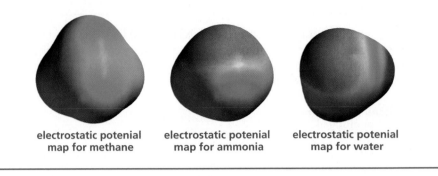

electrostatic potenial
map for methane

electrostatic potenial
map for ammonia

electrostatic potenial
map for water

1.13 BONDING IN THE HYDROGEN HALIDES

Fluorine, chlorine, bromine, and iodine are collectively known as the halogens. HF, HCl, HBr, and HI are called hydrogen halides. The hydrogen–halogen bond in the hydrogen halides is formed by the overlap of a *p* orbital of the halogen with the *s* orbital of hydrogen.

$$H-\ddot{\ddot{F}}:$$

hydrogen fluoride

In the case of fluorine, the *p* orbital used in bond formation belongs to the second shell of electrons. Chlorine uses a *p* orbital that belongs to the third shell of electrons. Because the average distance from the nucleus is greater for an electron in the third shell than for an electron in the second shell, the average electron density is less in a 3*p* orbital than in a 2*p* orbital. This means the electron density in the region where the *s* and *p* orbitals overlap decreases as the size of the halogen increases. Therefore, the hydrogen–halogen bond becomes longer and weaker as the atomic weight of the halogen increases (Table 1.6).

halogen halides

TABLE 1.6   Hydrogen–Halogen Bond Lengths and Bond Strengths

Hydrogen halide		Bond length (Å)	Bond strength $(DH°)^{a}$	
			kcal/mol	kJ/mol
H—F	H :F	0.917	136	571
H—Cl	H :Cl	1.2746	103	432
H—Br	H :Br	1.4145	87	366
H—I	H :I	1.6090	71	298

$^{a}DH°$ is the bond dissociation energy.

1.14 SUMMARY OF ORBITAL HYBRIDIZATION, BOND LENGTHS, BOND STRENGTHS, AND BOND ANGLES

All single bonds are σ bonds. All double bonds are composed of one σ bond and one π bond. All triple bonds are composed of one σ bond and two π bonds. *The hybridization of a carbon, oxygen, or nitrogen atom can be determined by the number of π bonds it forms: If it forms no π bonds, it is sp^3 hybridized; if it forms one π bond, it is sp^2 hybridized; if it forms two π bonds, it is sp hybridized.* The exceptions are carbocations and carbon radicals, which are sp^2 hybridized—not because they form a π bond, but because they have an empty or half-filled *p* orbital (Section 1.10).

$$CH_3-NH_2 \qquad \begin{matrix}CH_3\\ \diagdown\\ \diagup\\ CH_3\end{matrix}C=N-NH_2 \qquad CH_3-C\equiv N \qquad CH_3-OH \qquad CH_3-\overset{\overset{O}{\parallel}}{C}-OH \qquad O=C=O$$

$$\begin{matrix}\uparrow & \uparrow\\ sp^3 & sp^3\end{matrix} \qquad \begin{matrix}\uparrow & \uparrow & \uparrow & \uparrow\\ sp^3 & sp^2 & sp^2 & sp^3\end{matrix} \qquad \begin{matrix}\uparrow & \uparrow & \uparrow\\ sp^3 & sp & sp\end{matrix} \qquad \begin{matrix}\uparrow & \uparrow\\ sp^3 & sp^3\end{matrix} \qquad \begin{matrix}\uparrow & \uparrow & \uparrow\\ sp^3 & sp^2 & sp^3\end{matrix} \qquad \begin{matrix}\uparrow & \uparrow & \uparrow\\ sp^2 & sp & sp^2\end{matrix}$$

(The O above CH_3-C-OH is labeled $\longleftarrow sp^2$)

When comparing the lengths and strengths of carbon–carbon single, double, and triple bonds, we see that the more bonds holding two carbon atoms together, the shorter and stronger the carbon–carbon bond (Table 1.7). Triple bonds are shorter and stronger than double bonds, which are shorter and stronger than single bonds.

Because a double bond (a σ bond plus a π bond) is less than twice as strong as a single bond (a σ bond), we can conclude that a π bond is weaker than a σ bond. This stands to reason because the end-on overlap that forms σ bonds is better than the side-to-side overlap that forms π bonds. If we subtract the σ bond energy from the total bond energy of the double bond, we find that the π bond has a dissociation energy of 64 kcal/mol ($152 - 88 = 64$) or 268 kJ /mol. Because a σ bond formed by sp^2-sp^2 overlap is stronger (91 kcal/mol, 381 kJ/mol) than one formed by sp^3-sp^3 overlap, a better value for the strength of the π bond of ethene is 61 kcal/mol ($152 - 91 = 61$) or 255 kJ/mol.

The data in Table 1.7 indicate that a carbon–hydrogen σ bond is shorter and stronger than a carbon–carbon σ bond. The s orbital of hydrogen is closer to the nucleus than is the sp^3 orbital of carbon, so the carbon and hydrogen nuclei are closer together in sp^3-s overlap than are the carbon nuclei in sp^3-sp^3 overlap, causing the C—H bond to be shorter than a C—C bond. The carbon–hydrogen bond is stronger because there is greater electron density in the region of overlap of an sp^3 orbital with the s orbital (of hydrogen) than in the region of overlap of an sp^3 orbital with an sp^3 orbital (of carbon).

The length and strength of a C—H bond depends on the hybridization of the carbon atom to which the hydrogen is attached. The more s character in the orbital used by carbon to form the bond, the shorter and stronger the bond—again, because an s orbital is closer to the nucleus than a p orbital. So a carbon–hydrogen bond formed

The hybridization of a C, O, or N is $sp^{(3-\text{the number of }\pi\text{ bonds})}$.

A π bond is weaker than a σ bond.

TABLE 1.7   Comparison of the Bond Angles and the Lengths and Strengths of the Carbon—Carbon and Carbon—Hydrogen Bonds in Ethane, Ethene, and Ethyne.

Molecule	Hybridization of Carbon	Bond angles	Length of C—C bond (Å)	Strength of C—C bond (kcal/mol)	(kJ/mol)	Length of C—H bond (Å)	Strength of C—H bond (kcal/mol)	(kJ/mol)
ethane	sp^3	109.5°	1.54	88	368	1.10	101	423
ethene	sp^2	120°	1.33	152	636	1.08	107	448
ethyne	sp	180°	1.20	200	837	1.06	131	548

The more s character, the shorter and stronger the bond.

The more s character, the larger the bond angle.

by an sp hybridized carbon (50% s) is shorter and stronger than a carbon–hydrogen bond formed by an sp^2 hybridized carbon (33.3% s), which in turn is shorter and stronger than a carbon–hydrogen bond formed by an sp^3 hybridized carbon (25% s).

The bond angle also depends on the orbital used by carbon to form the bond. The greater the amount of s character in the orbital, the larger the bond angle. For example, sp hybridized carbons have bond angles of 180°, sp^2 hybridized carbons have bond angles of 120°, and sp^3 hybridized carbons have bond angles of 109.5°.

You may wonder how an electron knows what orbital it should go into. In fact, electrons know nothing about orbitals. They simply arrange themselves around atoms in the most stable manner possible. It is chemists who use the concept of orbitals to explain this arrangement.

PROBLEM 19 ◆

Why would you expect a carbon–carbon bond formed by sp^2—sp^2 overlap to be stronger than one formed by sp^3—sp^3 overlap?

PROBLEM 20

a. What is the hybridization of each of the carbon atoms in the following compound?

$$CH_3CHCH=CHCH_2C\equiv CCH_3$$
$$\overset{|}{CH_3}$$

b. What is the hybridization of each of the carbon, oxygen, and nitrogen atoms in the following compounds?

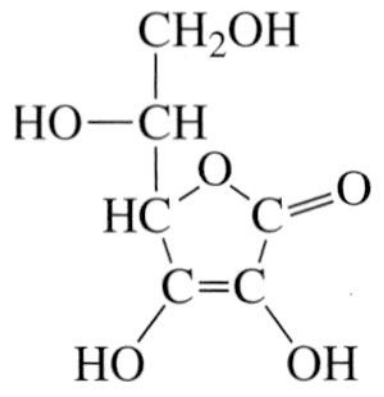

vitamin C 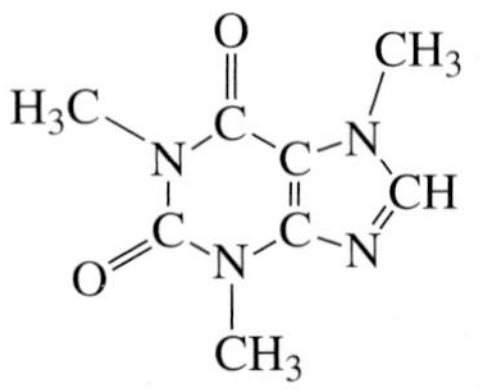   caffeine

PROBLEM 21

Describe the orbitals used in bonding and the bond angles in the following compounds. (*Hint:* see Table 1.7).

a. BeH_2 **b.** BH_3 **c.** CCl_4 **d.** CO_2 **e.** $HCOOH$ **f.** N_2

1.15 DIPOLE MOMENTS OF MOLECULES

We looked at the dipole moments of some commonly encountered covalent bonds in Section 1.3. For molecules that contain more than one covalent bond, the geometry of the molecule must be taken into account because both the *magnitude* and the *direction* of the individual bond dipole moments (the vector sum) are used to determine the over-all dipole moment of the molecule. For example, the carbon atom in carbon dioxide is bonded to two atoms, so it uses sp hybrid orbitals to form the carbon–oxygen σ bonds. The remaining two p orbitals on carbon form the two carbon–oxygen π bonds. Because *sp* hybrid orbitals form a bond angle of 180°, the individual carbon–oxygen bond dipole moments cancel each other, giving carbon dioxide a dipole moment of zero D.

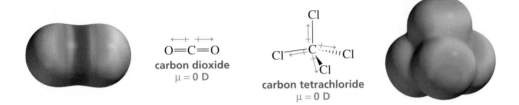

$$O=C=O$$
carbon dioxide
$\mu = 0$ D

carbon tetrachloride
$\mu = 0$ D

The carbon atom in carbon tetrachloride uses sp^3 hybrid orbitals. Because the four atoms bonded to carbon are identical and project symmetrically out from the central atom, the four bond dipole moments cancel and carbon tetrachloride has no dipole moment. Methane also has no dipole moment.

The dipole moment of chloromethane (CH_3Cl) is greater (1.87 D) than the dipole moment of the carbon–chlorine bond (1.5 D). This is because the direction of the C—H dipoles is such that they reinforce the dipole of the C—Cl bond—they are all in the same relative direction. The dipole moment of water (1.85 D) is greater than the dipole moment of a single oxygen–hydrogen bond (1.5 D) because the dipoles of the two O—H bonds of water reinforce each other. The nonbonding electrons also contribute to the dipole moment. Similarly, the dipole moment of ammonia (1.47 D) is greater than the dipole moment of a single nitrogen–hydrogen bond (1.3 D).

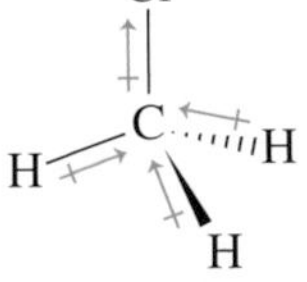

chloromethane
$\mu = $ **1.87 D**

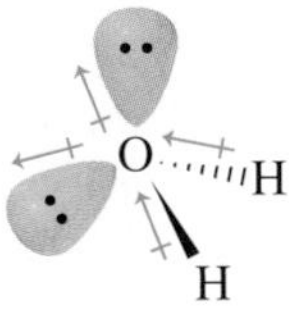

water
$\mu = $ **1.85 D**

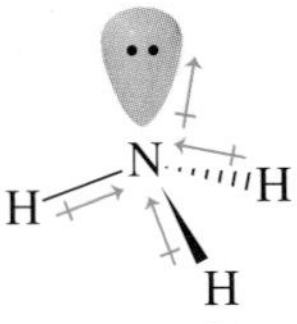

ammonia
$\mu = $ **1.47 D**

The potential maps of carbon dioxide, carbon tetrachloride, and methane show their even charge distributions. Chloromethane has an orange area because of the greater electron density on the chlorine atom. By comparing the potential maps of water and ammonia, you can see that water has the greater dipole moment—it has a more pronounced blue area.

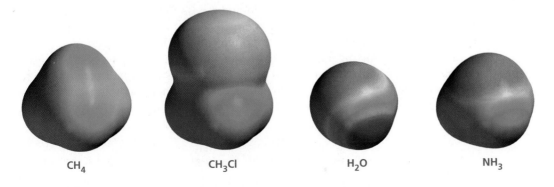

CH_4 CH_3Cl H_2O NH_3

PROBLEM 22

Account for the difference in the shape and color of the potential maps for ammonia and the ammonium ion in Section 1.12.

PROBLEM 23 ◆

Which of the following molecules would you expect to have a dipole moment of zero?

a. CH_3CH_3 **c.** CH_2Cl_2 **e.** $H_2C=CH_2$ **g.** NH_3

b. $H_2C=O$ **d.** $BeCl_2$ **f.** $H_2C=CHBr$ **h.** BF_3

1.16 AN INTRODUCTION TO ACIDS AND BASES

*Swedish chemist **Svante August Arrhenius (1859–1927)** received a Ph.D. from the University of Uppsala. Threatened with a low passing grade on his dissertation because his examiners did not understand his thesis on ionic dissociation, he sent his dissertation to several scientists who defended it. This dissertation earned Arrhenius the 1903 Nobel Prize in chemistry. He was the first to describe the "greenhouse" effect, predicting that as concentrations of atmospheric carbon dioxide (CO_2) increase, so will Earth's surface temperature (Section 12.10).*

Early chemists called any compound that tasted sour an acid (from *acidus,* Latin for "sour"). Some familiar acids are citric acid (found in lemons and other citrus fruits), acetic acid (found in vinegar), and hydrochloric acid (found in stomach acid—the sour taste associated with vomiting). Compounds that neutralize acids, such as wood ashes and other plant ashes, were called bases, or alkaline compounds ("ash" in Arabic is *al kalai*). Glass cleaners and solutions designed to unclog drains are alkaline solutions.

The terms "acid" and "base" were more precisely defined by Svante Arrhenius. He defined an acid as a species that ionizes in solution to release a hydrogen ion (H^+), and a base as a species that ionizes in solution to release a hydroxide ion (HO^-). His definition of an acid was sufficiently general, but his definition of a base applied to only one kind of base. Broader definitions were provided by Brønsted and Lowry in 1923.

In the Brønsted–Lowry definitions, an **acid** is a species that donates a proton, and a **base** is a species that accepts a proton. (Remember that hydrogen ions are also called protons.) In the following reaction, hydrochloric acid (HCl) meets the Brønsted-Lowry definition of an acid because it donates a proton to water. Water meets the definition of a base because it accepts a proton from HCl. Water can accept a proton because it has two lone pairs of electrons. Either lone pair can form a covalent bond with a proton. In the reverse reaction, H_3O^+ is an acid because it donates a proton to Cl^-, and Cl^- is a base because it accepts a proton from H_3O^+. Such proton-donating and proton-accepting species are commonly called **Brønsted acids** and **Brønsted bases,** respectively.

$$H\ddot{\underset{..}{C}l:\,\,+\,\,H_2\ddot{\underset{..}{O}}: \,\rightleftharpoons\, :\ddot{\underset{..}{C}}l:^- \,+\, H_3\ddot{O}^+}$$

    an acid   a base   a base   an acid

According to the Brønsted–Lowry definitions, any species that contains a hydrogen can potentially act as an acid, and any compound that contains a lone pair of electrons can act as a base. Both an acid and a base must be present in a proton-transfer reaction because an acid cannot donate a proton unless a base is present to accept it. **Proton-transfer reactions** are often called **acid-base reactions.**

*Born in Denmark, **Johannes Nicolaus Brønsted (1879–1947)** studied engineering before he switched to chemistry. He was a professor of chemistry at the University of Copenhagen. During World War II, he became known for his anti-Nazi position, and in 1947 was elected to the Danish parliament. He died before he could take his seat.*

In a reaction involving ammonia and water, ammonia (NH_3) is a base because it accepts a proton, and water is an acid because it donates a proton. In the reverse reaction, ammonium ion ($^+NH_4$) is an acid because it donates a proton, and hydroxide ion (HO^-) is a base because it accepts a proton.

$$\ddot{N}H_3 \,+\, H_2\ddot{\underset{..}{O}}: \,\rightleftharpoons\, {}^+NH_4 \,+\, H\ddot{\underset{..}{O}}:^-$$

    a base   an acid   an acid   a base

Notice that water can behave as an acid or as a base. It can behave as an acid because it has a proton that it can donate, but it can also behave as a base because it has a lone pair of electrons that can accept a proton.

When a compound loses a proton, the resulting species is called its **conjugate base.** Thus, Cl^- is the conjugate base of HCl, HO^- is the conjugate base of H_2O, and H_2O is the conjugate base of H_3O^+. When a compound accepts a proton, the resulting species is called its **conjugate acid.** Thus, $^+NH_4$ is the conjugate acid of NH_3, HCl is the conjugate acid of Cl^-, and H_2O is the conjugate acid of HO^-.

Acidity is a measure of how easily a compound gives up a proton. **Basicity** is a measure of how well a compound shares its electrons with a proton. A strong acid is one that gives up its proton easily. This means that its conjugate base must be weak because it has little affinity for the proton. A weak acid gives up its proton with difficulty, indicating that its conjugate base is strong because it has a high affinity for the proton. Thus, the following important relationship exists between an acid and its conjugate base: *The stronger the acid, the weaker its conjugate base.* For example, HBr is a stronger acid than HCl. This means that Br^- is a weaker base than Cl^-.

The stronger the acid, the weaker its conjugate base.

Thomas M. Lowry (1874–1936) was born in England, the son of an army chaplain. He earned a Ph.D. at Central Technical College, London (now Imperial College). He was head of chemistry at Westminister Training College and later at Guy's Hospital in London. In 1920 he became a professor of chemistry at Cambridge University.

PROBLEM 24 ◆

a. Draw the conjugate acid of each of the following:

   1. NH_3 2. Cl^- 3. HO^- 4. H_2O

b. Draw the conjugate base of each of the following:

   1. NH_3 2. HBr   3. HNO_3 4. H_2O

PROBLEM 25

a. Write an equation showing CH_3OH reacting as an acid and an equation showing it reacting as a base.

b. Write an equation showing NH_3 reacting as an acid and an equation showing it reacting as a base.

1.17 ORGANIC ACIDS AND BASES; pK_a AND pH

...hen a strong acid such as hydrochloric acid is dissolved in water, it dissociates almost completely, which means that products are favored at equilibrium. When a much weaker acid such as acetic acid is dissolved in water, it dissociates only to a small extent, so reactants are favored at equilibrium. Two half-headed arrows are used to designate equilibrium reactions. A longer arrow is drawn toward the species favored at equilibrium.

$$H\ddot{C}l: \ + \ H_2\ddot{O}: \ \rightleftharpoons \ H_3\ddot{O}^+ \ + \ :\ddot{\underset{..}{C}}l:^-$$
hydrochloric
acid

$$CH_3-\overset{\overset{\textstyle \ddot{O}:}{\|}}{C}-\ddot{O}H \ + \ H_2\ddot{O}: \ \rightleftharpoons \ H_3\ddot{O}^+ \ + \ CH_3-\overset{\overset{\textstyle \ddot{O}:}{\|}}{C}-\ddot{\underset{..}{O}}:^-$$
acetic acid

Whether a reversible reaction favors reactants or products at equilibrium is indicated by the **equilibrium constant** of the reaction (K_{eq}). Remember that brackets are used to indicate concentration in moles/liter $=$ molarity (M).

$$HA + H_2O \rightleftharpoons H_3O^+ + A^-$$

$$K_{eq} = \frac{[H_3O^+][A^-]}{[H_2O][HA]}$$

The degree to which an acid (HA) dissociates is described by its **acid dissociation constant (K_a)**. The acid dissociation constant is obtained by multiplying the equilibrium constant (K_{eq}) by the concentration of the solvent in which the reaction takes place.

$$K_a = K_{eq}[H_2O] = \frac{[H_3O^+][A^-]}{[HA]}$$

The stronger the acid, the smaller its pK_a.

The larger the acid dissociation constant, the stronger the acid (the more readily the compound gives up a proton). Hydrochloric acid has an acid dissociation constant of 10^7, whereas acetic acid has an acid dissociation constant of only 1.74×10^{-5}. For convenience, the strength of an acid is generally indicated by its **pK_a** value rather than its K_a value.

$$pK_a = -\log K_a$$

The pK_a of hydrochloric acid is -7 and the pK_a of acetic acid, a much weaker acid, is 4.76. Notice that the smaller the pK_a, the stronger the acid.

very strong acids	$pK_a < 1$
moderately strong acids	$pK_a = 1\text{–}5$
weak acids	$pK_a = 5\text{–}15$
extremely weak acids	$pK_a > 15$

Unless otherwise stated, the pK_a values given in this text indicate the strength of the acid *in water.* Later (in Section 9.10) you will see how the pK_a of an acid is affected when the solvent is changed.

The **pH** of a solution indicates the concentration of hydrogen ions in the solution. The concentration of hydrogen ions can be indicated as $[H^+]$ or, because a hydrogen ion in water is solvated, as $[H_3O^+]$. The lower the pH, the more acidic the solution.

$$pH = -\log[H_3O^+]$$

The pH of a solution can be changed simply by adding acid or base to the solution. Do not confuse pH and pK_a. The pH scale is used to describe the acidity of a *solution.* The pK_a is characteristic of a particular *compound,* much like a melting point or a boiling point—it tells how readily the compound gives up a proton.

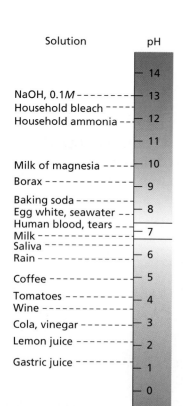

Solution	pH
	14
NaOH, 0.1M	13
Household bleach	
Household ammonia	12
	11
Milk of magnesia	10
Borax	9
Baking soda	
Egg white, seawater	8
Human blood, tears	
Milk	7
Saliva	
Rain	6
Coffee	5
Tomatoes	4
Wine	
Cola, vinegar	3
Lemon juice	2
Gastric juice	1
	0

PROBLEM 26 ◆

a. Which is a stronger acid, one with a pK_a of 5.2 or one with a pK_a of 5.8?

b. Which is a stronger acid, one with an acid dissociation constant of 3.4×10^{-3} or one with an acid dissociation constant of 2.1×10^{-4}?

PROBLEM 27 ◆

An acid has a K_a of 4.53×10^{-6}. What is its K_{eq}? ($[H_2O] = 55.5$ M)

Organic acids and bases will be very important when we discuss how and why organic compounds react. The most common organic acids are carboxylic acids. Acetic acid and formic acid are examples of carboxylic acids. Carboxylic acids have pK_a values ranging from about 3 to 5 (they are moderately strong acids). The pK_a values of a wide variety of organic compounds are given in Appendix II.

$$CH_3\overset{\overset{\displaystyle O}{\|}}{C}OH \qquad H\overset{\overset{\displaystyle O}{\|}}{C}OH$$

acetic acid   **formic acid**
pK_a = 4.76   pK_a = 3.75

Alcohols are much weaker organic acids, with pK_a values close to 16. Methanol and ethanol are examples of alcohols.

$$CH_3OH \qquad CH_3CH_2OH$$

methanol   **ethanol**
pK_a = 15.5   pK_a = 15.9

An alcohol can behave as an acid and donate a proton, or as a base and accept a proton.

$$CH_3OH + HO^- \rightleftharpoons CH_3O^- + H_2O$$
an acid

$$CH_3OH + H_3O^+ \rightleftharpoons CH_3\overset{+}{\underset{H}{O}}H + H_2O$$
a base

A carboxylic acid can behave as an acid and donate a proton, or as a base and accept a proton.

$$CH_3\overset{\overset{\displaystyle O}{\|}}{C}OH + HO^- \rightleftharpoons CH_3\overset{\overset{\displaystyle O}{\|}}{C}O^- + H_2O$$
an acid

$$CH_3\overset{\overset{\displaystyle O}{\|}}{C}OH + H_3O^+ \rightleftharpoons CH_3\overset{\overset{\displaystyle +OH}{\|}}{C}OH + H_2O$$
a base

A protonated alcohol or a protonated carboxylic acid is a very strong acid. For example, protonated methanol has a pK_a of -2.5, protonated ethanol has a pK_a of -2.4, and protonated acetic acid has a pK_a of -6.1.

$$CH_3\overset{+}{\underset{H}{O}}H \qquad CH_3CH_2\overset{+}{\underset{H}{O}}H \qquad CH_3\overset{\overset{\displaystyle +OH}{\|}}{C}OH$$

protonated methanol   **protonated ethanol**   **protonated acetic acid**
pK_a = -2.5 pK_a = -2.4 pK_a = -6.1

An amine can behave as an acid and donate a proton, or as a base and accept a proton.

$$CH_3NH_2 + HO^- \rightleftharpoons CH_3\overset{..}{N}H + H_2O$$
an acid

$$CH_3NH_2 + H_3O^+ \rightleftharpoons CH_3\overset{+}{N}H_3 + H_2O$$
a base

Amines, however, have such high pK_a values that they rarely behave as acids.

$$CH_3NH_2$$
methylamine
pK_a = 40

Amines are much more likely to act as bases. In fact, amines are the most common organic bases. Instead of talking about the strength of a base in terms of its pK_b (the basic equivalent of a pK_a), it is much easier to talk about the pK_a of its conjugate acid, remembering that the stronger the acid, the weaker its conjugate base. The pK_a values of protonated amines are about 10 to 11. Protonated methylamine is a stronger acid than protonated ethylamine, which means that methylamine is a weaker base than ethylamine.

$$CH_3\overset{+}{N}H_3 \qquad\qquad CH_3CH_2\overset{+}{N}H_3$$
protonated methylamine   **protonated ethylamine**
pK_a = 10.7   pK_a = 11.0

It is important to remember the approximate pK_a values of the various classes of organic compounds we have discussed. An easy way to remember them is in units of five, as shown in Table 1.8. Protonated alcohols, protonated carboxylic acids, and protonated water have pK_a values less than 0, carboxylic acids have pK_a values of about 5, protonated amines have pK_a values of about 10, and alcohols and water have pK_a values of about 15. These are also listed inside the back cover for easy reference.

Tutorial: Acid-base reaction

Strong reacts to give weak.

In determining the position of equilibrium for an acid-base reaction (i.e., whether reactants or products are favored at equilibrium), remember that the equilibrium favors *reaction* of the strong acid and strong base and *formation* of the weak acid and weak base. In other words, *strong reacts to give weak.* Thus, the equilibrium lies away from the stronger acid and toward the weaker acid.

$$CH_3-\overset{\overset{\displaystyle O}{\|}}{C}-OH \;+\; NH_3 \;\rightleftharpoons\; CH_3-\overset{\overset{\displaystyle O}{\|}}{C}-O^- \;+\; \overset{+}{N}H_4$$

stronger acid   stronger base   weaker base   weaker acid
pK_a = 4.8   pK_a = 9.4

$$CH_3CH_2OH \;+\; CH_3NH_2 \;\rightleftharpoons\; CH_3CH_2O^- \;+\; CH_3\overset{+}{N}H_3$$

weaker acid   weaker base   stronger base   stronger acid
pK_a = 15.9   pK_a = 10.7

TABLE 1.8 Approximate pK_a Values			
p$K_a < 0$	**p$K_a \sim 5$**	**p$K_a \sim 10$**	**p$K_a \sim 15$**
$CH_3\overset{+}{O}H_2$	$CH_3\overset{\overset{\displaystyle O}{\|}}{C}OH$	$CH_3\overset{+}{N}H_3$	CH_3OH
^+OH			H_2O
$CH_3\overset{\|}{C}OH$			
H_3O^+			

The precise value of the equilibrium constant (K_{eq}) can be calculated by dividing the K_a of the reactant acid by the K_a of the product acid. The equilibrium constant for the reaction of acetic acid with ammonia (top reaction) is 4.0×10^4. The equilibrium constant for the reaction of ethanol with methylamine (bottom reaction) is 6.3×10^{-6}.

$$K_{eq} = \frac{K_a \text{ reactant acid}}{K_a \text{ product acid}}$$

$$K_{eq} = \frac{10^{-4.8}}{10^{-9.4}} = 10^{4.6} = 4.0 \times 10^4$$

$$K_{eq} = \frac{10^{-15.9}}{10^{-10.7}} = 10^{-5.2} = 6.3 \times 10^{-6}$$

PROBLEM 28

Show that $K_{eq} = \dfrac{K_a \text{ reactant acid}}{K_a \text{ product acid}} = \dfrac{[\text{products}]}{[\text{reactants}]}$

PROBLEM 29 ◆

a. Which is a stronger base, CH_3COO^- or $HCOO^-$? (The pK_a of CH_3COOH is 4.8; the pK_a of $HCOOH$ is 3.8.)

b. Which is a stronger base, HO^- or $^-NH_2$? (The pK_a of H_2O is 15.7; the pK_a of NH_3 is 36.)

c. Which is a stronger base, H_2O or CH_3OH? (The pK_a of H_3O^+ is -1.7; the pK_a of $CH_3OH_2^+$ is -2.5.)

PROBLEM 30

a. For each of the acid-base reactions in Section 1.17, compare the pK_a values of the acids on either side of the equilibrium arrows and convince yourself that the position of equilibrium is in the direction indicated. (The pK_a values you need can be found in Section 1.17 or in Problem 29.

b. Do the same thing for the equilibria in Section 1.16 (the pK_a of $^+NH_4$ is 9.4).

PROBLEM 31 ◆

Using the pK_a values in Section 1.17, rank the following species in order of decreasing base strength:

$$CH_3NH_2 \quad CH_3NH^- \quad CH_3OH \quad CH_3O^- \quad CH_3\overset{\displaystyle O}{\overset{\|}{C}}O^-$$

PROBLEM 32 ◆

Calculate the equilibrium constant for acid-base reactions between the following pairs of reactants.

a. $HCl + H_2O$

b. $CH_3COOH + H_2O$

c. $CH_3NH_2 + H_2O$

d. $CH_3\overset{+}{N}H_3 + H_2O$

1.18 THE EFFECT OF STRUCTURE ON pK_a

Acid strength is determined by the stability of the conjugate base that is formed when the acid gives up its proton—the more stable the base, the stronger its conjugate acid. A stable base is one that readily bears the electrons it formerly shared with a proton, which means it is less apt to share those electrons with a proton. In other words, it is a weak base so it has a strong conjugate acid (Section 1.16).

The elements in the second row of the periodic table are all about the same size, but they have very different electronegativities, which increase across the row from left to right.

relative electronegativities: C < N < O < F

If we look at the bases when hydrogens are attached to these elements, we see that the stabilities of the bases also increase as we move to the right, because the more electronegative atom is better able to bear its negative charge.

relative stabilities: $^-CH_3$ < $^-NH_2$ < HO^- < F^-

The stronger acid is the one that forms the more stable conjugate base, so hydrogen fluoride is a stronger acid than water, water is a stronger acid than ammonia, and ammonia is a stronger acid than methane (Table 1.9).

relative acidities: CH_4 < NH_3 < H_2O < HF

Therefore, we can conclude that when the atoms are similar in size, the more acidic compound will have the hydrogen attached to the more electronegative atom.

The relative acidities of these four compounds are apparent from their potential maps. Methane has essentially no positive charge density (blue area), ammonia's hydrogens bear a little positive charge, water's hydrogens have significant positive charge, but not as much as the hydrogen in hydrogen fluoride.

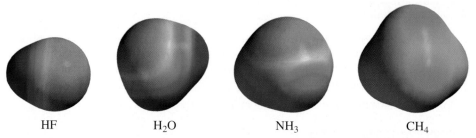

HF H_2O NH_3 CH_4

The effect that the electronegativity of the atom bonded to a hydrogen has on the acidity of that hydrogen can be appreciated when the pK_a values of alcohols and amines are compared. Because oxygen is more electronegative than nitrogen, an alcohol is

TABLE 1.9	The pK_a Values of Some Simple Acids		
CH_4	NH_3	H_2O	HF
pK_a = 50	pK_a = 36	pK_a = 15.7	pK_a = 3.2
		H_2S	HCl
		pK_a = 7.0	pK_a = −7
			HBr
			pK_a = −9
			HI
			pK_a = −10

about 24 pK_a units more acidic than an amine.

CH$_3$OH   CH$_3$NH$_2$

pK_a = 15.5   **pK_a = 40**

A protonated alcohol is about 13 pK_a units more acidic than a protonated amine. The difference in their pK_a values is reflected in their potential maps—compare the blue areas in protonated methanol and in protonated methylamine.

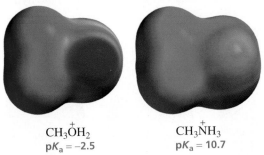

CH$_3\overset{+}{O}$H$_2$ CH$_3\overset{+}{N}$H$_3$

pK_a = −2.5   pK_a = 10.7

When comparing atoms that are very different in size, the *size* of the atom is more important than its *electronegativity* in determining how well it bears its negative charge. For example, as we proceed down a column in the periodic table, the electronegativity of the elements *decreases* but the stability of the base increases, so acid strength *increases*.

relative electronegativities:	I < Br < Cl < F
relative stabilities:	I$^-$ > Br$^-$ > Cl$^-$ > F$^-$
relative acidities:	HI > HBr > HCl > HF

Why does the size of an atom have such a significant effect on the stability of the base and, therefore, on the acidity of a hydrogen attached to it? As we move down a column, the orbitals that contain the valence electrons of a halide ion increase in volume. For example, the valence electrons of F$^-$ are in a $2p$ orbital, the valence electrons of Cl$^-$ are in a $3p$ orbital, those of Br$^-$ are in a $4p$ orbital, and those of I$^-$ are in a $5p$ orbital.

The volume of space occupied by a $3p$ orbital is significantly greater than the volume of space occupied by a $2p$ orbital because a $3p$ orbital extends out farther from the nucleus. The increased volume of a $3p$ orbital causes its average electron density to be less than that of a $2p$ orbital. Because its negative charge is spread over a larger volume of space, Cl$^-$ is more stable than F$^-$.

Thus, as the halide ion increases in size, its stability increases because its negative charge is spread over a larger volume of space. Therefore, HI is the strongest acid of the hydrogen halides because I$^-$ is the most stable halide ion even though iodine is the least electronegative of the halogens (Table 1.9). The potential maps illustrate the large difference in size of the halogens.

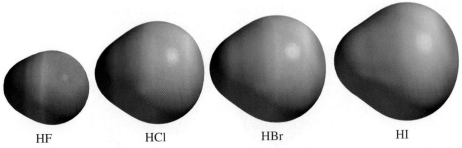

HF   HCl   HBr   HI

In summary, moving across a row of the periodic table, the orbitals have approximately the same volume so it is the electronegativity of the element that determines the stability of the base and, therefore, the ease with which an acid gives up a proton bonded to that base. Moving down a column of the periodic table, the volume of the orbitals increases, so electron density decreases, and the electron density of the orbital is more important than electronegativity in determining the stability of the base and, therefore, the ease with which the acid gives up a proton. That is, the less the electron density, the more stable the conjugate base, and the stronger the acid.

The acidic proton of each of the following five carboxylic acids is attached to an oxygen atom, but the five compounds have different acidities. Therefore, there must be a factor—other than the nature of the atom to which the hydrogen is bonded—that affects acidity.

3-D Molecule:
Acetic acid

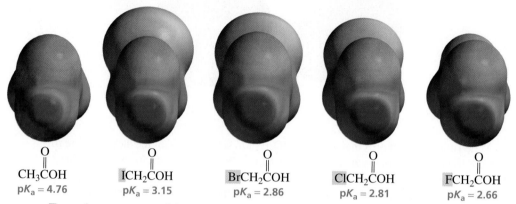

$$\underset{\substack{pK_a = 4.76}}{\overset{O}{\underset{\|}{CH_3COH}}} \qquad \underset{\substack{pK_a = 3.15}}{\overset{O}{\underset{\|}{ICH_2COH}}} \qquad \underset{\substack{pK_a = 2.86}}{\overset{O}{\underset{\|}{BrCH_2COH}}} \qquad \underset{\substack{pK_a = 2.81}}{\overset{O}{\underset{\|}{ClCH_2COH}}} \qquad \underset{\substack{pK_a = 2.66}}{\overset{O}{\underset{\|}{FCH_2COH}}}$$

From the structures of the five carboxylic acids, you can see that replacing one of the hydrogen atoms of the CH_3 group of the carboxylic acid with a halogen atom affects the acidity of the compound. (Chemists call this *substitution*, and the new atom is called a *substituent*.) All the halogens are more electronegative than hydrogen (Table 1.2), and an electronegative substituent increases the acidity of a carboxylic acid. This is because an electronegative substituent pulls the bonding electrons of the O—H bond away from the proton. This makes it easier for the compound to give up its proton because electron withdrawal stabilizes the conjugate base. Pulling electrons away through sigma (σ) bonds is called **inductive electron withdrawal.**

$$\overset{\text{H} \quad \text{O}}{\underset{\underset{\text{H}}{|}}{\text{Br}\leftarrow\text{C}\leftarrow\text{C}\leftarrow\text{O}\leftarrow\text{H}}}$$

inductive electron withdrawal

As the pK_a values of the five carboxylic acids show, the acidity increases as the electronegativity (the electron-withdrawing ability) of the halogen substituent increases.

Compare the potential maps for the conjugate base of acetic acid and the conjugate base of chloroacetic acid. Notice that the conjugate base of chloroacetic acid is better at dispersing the negative charge (it is more orange, less red). This is to be expected because if chloroacetic acid is a stronger acid than acetic acid, the conjugate base of chloroacetic acid must be a more stable (weaker) base than the conjugate base of acetic acid.

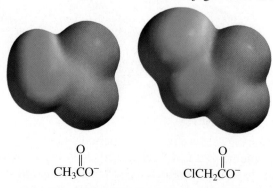

$$\underset{}{\overset{O}{\underset{\|}{CH_3CO^-}}} \qquad\qquad \underset{}{\overset{O}{\underset{\|}{ClCH_2CO^-}}}$$

The effect of a substituent on the acidity of a compound decreases as the substituent is moved farther away from the O—H bond.

$$\underset{\underset{\text{p}K_a = 2.97}{|}}{\underset{\text{Br}}{\overset{\overset{\text{O}}{\parallel}}{\text{CH}_3\text{CH}_2\text{CH}_2\text{CHCOH}}}} \qquad \underset{\underset{\text{p}K_a = 4.01}{|}}{\underset{\text{Br}}{\overset{\overset{\text{O}}{\parallel}}{\text{CH}_3\text{CH}_2\text{CHCH}_2\text{COH}}}} \qquad \underset{\underset{\text{p}K_a = 4.59}{|}}{\underset{\text{Br}}{\overset{\overset{\text{O}}{\parallel}}{\text{CH}_3\text{CHCH}_2\text{CH}_2\text{COH}}}} \qquad \underset{\underset{\text{p}K_a = 4.71}{|}}{\underset{\text{Br}}{\overset{\overset{\text{O}}{\parallel}}{\text{CH}_2\text{CH}_2\text{CH}_2\text{CH}_2\text{COH}}}}$$

PROBLEM-SOLVING STRATEGY

a. Which is a stronger acid?

$$\underset{\underset{\text{F}}{|}}{\text{CH}_3\text{CHCH}_2\text{OH}} \quad \text{or} \quad \underset{\underset{\text{Br}}{|}}{\text{CH}_3\text{CHCH}_2\text{OH}}$$

When you are asked to compare two items, such as two compounds, pay attention to how they differ; ignore where they are the same. These two compounds differ only in the halogen atom that is attached to the middle carbon of the molecule. Because fluorine is more electronegative than bromine, there is greater electron withdrawal from the O—H bond in the fluorinated compound, causing it to be the stronger acid.

b. Which is a stronger acid?

$$\underset{\underset{\text{Cl}}{|}}{\overset{\overset{\text{Cl}}{|}}{\text{CH}_3\text{CCH}_2\text{OH}}} \quad \text{or} \quad \underset{\underset{\text{Cl}}{|}}{\overset{\overset{\text{Cl}}{|}}{\text{CH}_2\text{CHCH}_2\text{OH}}}$$

These two compounds differ in the location of one of the chlorine atoms. Because the chlorine in the compound on the left is closer to the O—H bond than is the chlorine in the compound on the right, it is more effective at withdrawing electrons from the O—H bond, causing the compound on the left to be the stronger acid.

Now continue on to Problem 33.

PROBLEM 33 ◆

For each of the following compounds, indicate which is the stronger acid:

a. $\text{CH}_3\text{OCH}_2\text{CH}_2\text{OH}$ or $\text{CH}_3\text{CH}_2\text{CH}_2\text{CH}_2\text{OH}$?

b. $\text{CH}_3\text{CH}_2\text{CH}_2\overset{+}{\text{N}}\text{H}_3$ or $\text{CH}_3\text{CH}_2\text{CH}_2\overset{+}{\text{O}}\text{H}_2$?

c. $\text{CH}_3\text{OCH}_2\text{CH}_2\text{CH}_2\text{OH}$ or $\text{CH}_3\text{CH}_2\text{OCH}_2\text{CH}_2\text{OH}$?

d. $\underset{}{\overset{\overset{\text{O}}{\parallel}}{\text{CH}_3\text{CCH}_2\text{OH}}}$ or $\underset{}{\overset{\overset{\text{O}}{\parallel}}{\text{CH}_3\text{CH}_2\text{COH}}}$?

PROBLEM 34 ◆

List the following compounds in order of decreasing acidity:

$$\underset{\underset{\text{F}}{|}}{\text{CH}_2\text{CH}_2\text{OH}} \qquad \text{CH}_3\text{CH}_2\text{OH} \qquad \underset{\underset{\text{F}}{|}}{\text{CH}_3\text{CHOH}} \qquad \underset{\underset{\text{Cl}}{|}}{\text{CH}_2\text{CH}_2\text{OH}}$$

PROBLEM 35 / SOLVED

HCl is a weaker acid than HBr. Why then is ClCH_2COOH a stronger acid than BrCH_2COOH?

SOLUTION When comparing the acidities of HCl and HBr, we are comparing the ease of breaking two different kinds of bonds, an H—Cl bond and an H—Br bond to form Cl$^-$

and Br⁻, respectively. Because size is more important than electronegativity in determining stability, Br⁻ is more stable than Cl⁻. Therefore, HBr is a stronger acid than HCl.

In both carboxylic acids, the same kind of bond is broken (an O—H bond). Therefore the only factor to be considered is the electronegativities of the atoms that are pulling electrons away from the O—H bond. Because Cl is more electronegative than Br, Cl is better at pulling electrons away from the O—H bond. Thus it is better at stabilizing the base that is formed when the proton leaves.

PROBLEM 36 ◆

a. Which of the halide ions (F⁻, Cl⁻, Br⁻, I⁻) is the strongest base?

b. Which is the weakest base?

PROBLEM 37 ◆

a. Which is more electronegative, oxygen or sulfur?

b. Which is a stronger acid, H_2O or H_2S?

c. Which is a stronger acid, CH_3OH or CH_3SH?

PROBLEM 38 ◆

Using the table of pK_a values given in Appendix II, answer the following:

a. Which is the most acidic organic compound in the table?

b. Which is the least acidic organic compound in the table?

c. Which is the most acidic carboxylic acid in the table?

d. Which is more electronegative, an sp^3 hybridized oxygen or an sp^2 hybridized oxygen? (*Hint:* Pick a compound in Appendix II with an sp^2 hybridized oxygen and one with an sp^3 hybridized oxygen and compare their pK_a values.)

e. What are the relative electronegativities of sp^3, sp^2, and sp hybridized nitrogen atoms?

f. What are the relative electronegativities of sp^3, sp^2, and sp hybridized carbon atoms?

g. Which is more acidic, HNO_3 or HNO_2? Why?

1.19
AN INTRODUCTION TO DELOCALIZED ELECTRONS AND RESONANCE

We have seen that a carboxylic acid has a pK_a of about 5, whereas the pK_a of an alcohol is about 16. There are two factors that cause a carboxylic acid to be a much stronger acid than an alcohol. First, the carboxylic acid has a doubly-bonded oxygen in place of the two hydrogens of the alcohol. This electronegative oxygen pulls the bonding electrons of the O—H bond away from the proton, which makes it easier for the carboxylic acid to give up its proton.

$$\overset{\displaystyle O}{\overset{\displaystyle \|}{CH_3C}}O—H \qquad CH_3CH_2O—H$$
$$pK_a = 4.76 \qquad\qquad pK_a = 15.9$$

Second, the conjugate base of the carboxylic acid is considerably more stable than the conjugate base of the alcohol. When an alcohol loses a proton all the electron density resides on its single oxygen atom. In contrast, when a carboxylic acid loses a proton, the electron density is shared by both oxygen atoms—the electrons are delocalized. **Delocalized electrons** do not belong to a single atom, nor are they confined to a bond between two atoms. A compound with delocalized electrons is said to have **resonance.** The two structures shown next are called **resonance contributors.** Neither resonance contributor is the correct structure for a carboxylate ion—the actual structure is a composite of the two structures shown. The fact that neither of the two resonance contributors represents the actual structure is indicated by the double-headed arrow between them. Notice that the only thing different about the two resonance contributors is the location of their π electrons and lone-pair electrons; all the atoms stay in the same place. The actual structure is called a **resonance hybrid.** In the resonance hybrid, the negative charge is shared equally by the two oxygen atoms, and both carbon–oxygen bonds are the same length—they are not as long as a single bond but they are longer than a double bond. A resonance hybrid can be drawn by using dotted lines to show that the electrons are delocalized.

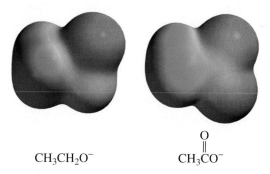

localized electrons   delocalized electrons

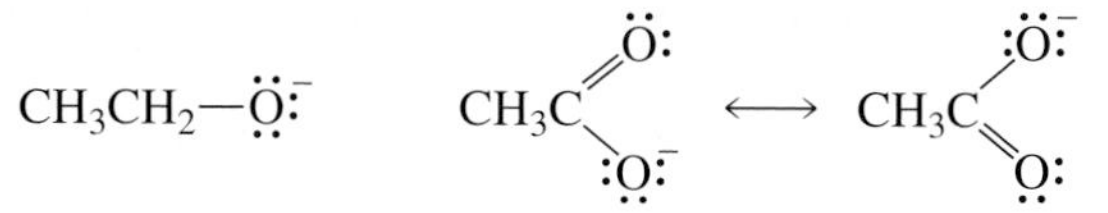

resonance contributors

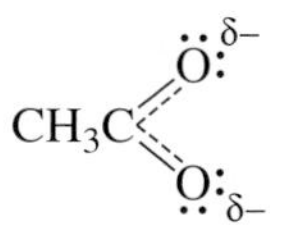

resonance hybrid

3-D Molecules:
Acetate;
Ethoxide

The following potential maps show that the electron density is on *one* oxygen in the conjugate base of the alcohol (electron-dense red region) and is shared by *both oxygens* in the conjugate base of the carboxylic acid (less electron-dense orange region). The ability of two atoms to share the negative charge is what makes the conjugate base of the carboxylic acid more stable.

$CH_3CH_2O^-$ CH_3CO^-

We will discuss delocalized electrons in greater detail in Chapter 6. By that time you will be thoroughly comfortable with compounds that have only localized electrons, and you can then further explore how delocalized electrons affect the stability and reactivity of organic compounds.

PROBLEM 39

Which compound would you expect to be a stronger acid? Why?

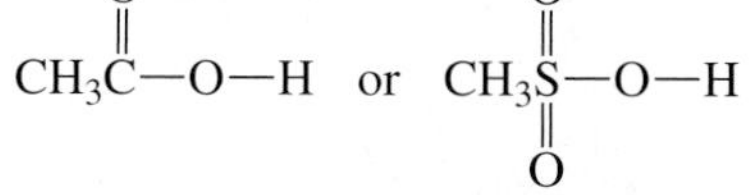

$$CH_3\overset{O}{\overset{\|}{C}}-O-H \quad \text{or} \quad CH_3\overset{O}{\underset{\overset{\|}{O}}{\overset{\|}{S}}}-O-H$$

1.20
THE EFFECT OF pH
ON THE STRUCTURE
OF AN ORGANIC
COMPOUND

Whether or not a given acid will lose a proton in an aqueous solution depends on the pK_a of the acid *and* on the pH of the solution. The relationship between the two is given by the **Henderson–Hasselbalch equation.** This is a very useful equation because it tells us whether a compound will exist in its acidic form (with its proton retained) or in its basic form (with its proton removed) at a particular pH.

the Henderson–Hasselbalch equation

$$pK_a = pH + \log \frac{[HA]}{[A^-]}$$

The Henderson–Hasselbalch equation tells us that when the pH of a solution equals the pK_a of the compound that undergoes dissociation, the concentration of the compound in its acidic form [HA] will equal the concentration of the compound in its basic form $[A^-]$ (because log 1 = 0). If the pH of the solution is less than the pK_a of the compound, the compound will exist primarily in its acidic form. If the pH of the solution is greater than the pK_a of the compound, the compound will exist primarily in its basic form. In other words, *compounds exist primarily in their acidic forms in solutions that are more acidic than their pK_a values, and they exist primarily in their basic forms in solutions that are more basic than their pK_a values.*

If the pH is less than the pK_a, the compound will exist primarily in its acidic form.

If we know the pH of the solution and the pK_a of the compound, the Henderson–Hasselbalch equation allows us to calculate precisely how much of the compound will be in its acidic form and how much will be in its basic form. For example, when a compound with a pK_a of 5.2 is in a solution of pH 5.2, half the compound will be in the acidic form and the other half will be in the basic form (Figure 1.15). If the pH is one unit less than the pK_a of the compound (pH = 4.2), there will be 10 times more compound present in the acidic form than in the basic form (because log 10 = 1). If the pH is two units less than the pK_a of the compound (pH = 3.2), there will be 100 times more compound present in the acidic form than in the basic form (because log 100 = 2). If the pH is changed to 6.2, there will be 10 times more compound in the basic form than in the acidic form, and at pH = 7.2 there will be 100 times more compound present in the basic form than in the acidic form.

If the pH is greater than the pK_a, the compound will exist primarily in its basic form.

The Henderson–Hasselbalch equation can be very useful in the laboratory when you must separate compounds from each other. Water and diethyl ether are immiscible liquids and, therefore, will form two layers when combined. The ether layer will lie over the more dense water layer. Charged compounds are more soluble in water, whereas neutral compounds are more soluble in diethyl ether. Two compounds, such as a carboxylic acid (RCOOH) with a pK_a of 4.0 and a protonated amine (RNH_3^+) with a pK_a of 10.0, dissolved in a mixture of water and diethyl ether, can be separated by

Tutorial: Effect of pH on structure

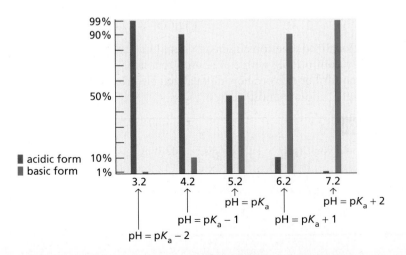

■ acidic form
■ basic form

Figure 1.15 ▶
The relative amounts of a compound with a pK_a of 5.2 in the acidic and basic forms at different pH values.

adjusting the pH of the water layer. (R means that the particular carboxylic acid or amine is not specified.) For example, if the pH of the water layer is 2, the carboxylic acid and the amine will both be in their acidic forms because the pH of the water is less than the pK_a's of both compounds. The acidic form of a carboxylic acid is neutral, whereas the acidic form of an amine is charged. Therefore, the carboxylic acid will be more soluble in the ether layer, whereas the protonated amine will dissolve in the water layer.

— ether
— water

$$\underset{\text{acidic form}}{\text{RCOOH}} \;\rightleftharpoons\; \underset{\text{basic form}}{\text{RCOO}^-} + \text{H}^+$$

$$\overset{+}{\text{RNH}_3} \;\rightleftharpoons\; \text{RNH}_2 + \text{H}^+$$

For the most effective separation, it is best if the pH of the water layer is at least two units away from the pK_a values of the compounds being separated so the relative amounts of the compounds in their acidic and basic forms will be at least 100:1 (Figure 1.15).

The Henderson–Hasselbalch equation is useful when working with buffer solutions. A **buffer solution** is a solution that maintains a nearly constant pH when small amounts of acid or base are added to it. The Henderson–Hasselbalch equation and buffer solutions are discussed in detail in the Solutions Manual/Study Guide. (Special Topic I).

DERIVATION OF THE HENDERSON–HASSELBALCH EQUATION

The Henderson–Hasselbalch equation can be derived from the expression that defines the acid dissociation constant:

$$K_a = \frac{[\text{H}_3\text{O}^+][\text{A}^-]}{[\text{HA}]}$$

Taking the logarithms of both sides of the equation and then multiplying both sides of the equation by -1, we obtain

$$\log K_a = \log [\text{H}_3\text{O}^+] + \log \frac{[\text{A}^-]}{[\text{HA}]}$$

$$-\log K_a = -\log [\text{H}_3\text{O}^+] - \log \frac{[\text{A}^-]}{[\text{HA}]}$$

Substituting and remembering that when a fraction is inverted, the sign of its log changes, we have

$$pK_a = \text{pH} + \log \frac{[\text{HA}]}{[\text{A}^-]}$$

BLOOD: A BUFFERED SOLUTION

Blood is the fluid that transports oxygen to all the cells of the human body. The normal pH of human blood is 7.35 to 7.45. Death will result if this pH decreases to less than ~6.8 or increases to a value greater than ~8.0 for even a few seconds. Oxygen is carried to cells by a protein in the blood called hemoglobin. When hemoglobin binds O_2, hemoglobin loses a proton, which would make the blood more acidic if it did not contain a buffer to maintain pH.

$$\text{HbH}^+ + \text{O}_2 \;\rightleftharpoons\; \text{HbO}_2 + \text{H}^+$$

The buffer that controls the pH of blood is a carbonic acid/ bicarbonate (H_2CO_3/HCO_3^-) buffer. An important feature of this buffer is that carbonic acid decomposes to CO_2 and H_2O.

Cells need a constant supply of O_2, especially during periods of strenuous exercise when very high levels of O_2 are needed. When O_2 is consumed, the hemoglobin equilibrium shifts to the left, so the concentration of H^+ decreases. However, increased metabolism during exercise produces large amounts of CO_2. This shifts the carbonate/bicarbonate equilibrium to the right, which increases the concentration of H^+. Significant amounts of lactic acid are also produced and this adds to the increased concentration of H^+. Receptors in the brain respond to the increased concentration of H^+ and trigger a reflex that increases the rate of breathing. This increases the concentration of O_2 available to cells. It also causes CO_2 to be eliminated by exhalation, which decreases the concentration of H^+.

$$\underset{\text{carbonic acid}}{\text{CO}_2 + \text{H}_2\text{O} \;\rightleftharpoons\; \text{H}_2\text{CO}_3} \;\rightleftharpoons\; \underset{\text{bicarbonate}}{\text{HCO}_3^-} + \text{H}^+$$

PROBLEM 40 ◆

As long as the pH is greater than _____, more than 50% of an amine with a pK_a of 10.4 will be in its neutral, nonprotonated form.

PROBLEM 41◆ / SOLVED

a. At what pH will 99% of a compound with a pK_a of 8.4 be in its basic form?

b. At what pH will 91% of a compound with a pK_a of 3.7 be in its acidic form?

c. At what pH will 9% of a compound with a pK_a of 5.9 be in its basic form?

d. At what pH will 50% of a compound with a pK_a of 7.3 be in its basic form?

e. At what pH will 1% of a compound with a pK_a of 7.3 be in its acidic form?

Solution to 41a If 99% is in the basic form and 1% is in the acidic form, the Henderson–Hasselbalch equation becomes:

$$pK_a = pH + \log \frac{1}{99}$$

$$8.4 = pH + \log .01$$

$$8.4 = pH - 2.0$$

$$pH = 10.4$$

A faster way to get the answer: If there is about 100 times more compound present in the basic form than in the acidic form, the pH will be two units more basic than the pK_a (pH = 8.4 + 2.0 = 10.4).

SOLUTION TO 41b If 91% is in the acidic form and 9% is in the basic form, there is about 10 times more compound present in the acidic form. Therefore, the pH is one unit more acidic than the pK_a (pH = 3.7 - 1.0 = 2.7).

PROBLEM 42 ◆

a. Indicate whether a carboxylic acid with a pK_a of 5 will be mostly charged or mostly neutral at each of the following pH values:

1. pH = 1   **3.** pH = 5   **5.** pH = 9   **7.** pH = 13

2. pH = 3   **4.** pH = 7   **6.** pH = 11

b. Answer the same question for a protonated amine with a pK_a of 9.

c. Answer the same question for an alcohol with a pK_a of 15.

PROBLEM 43 ◆

For each compound, indicate the pH at which:

a. 50% of the compound will be in a form that possesses a charge.

b. > 99% of the compound will be in a form that possesses a charge.

1. CH_3CH_2COOH (pK_a = 4.9)

2. $CH_3\overset{+}{N}H_3$ (pK_a = 10.7)

PROBLEM 44

The following compounds are all shown in their acidic forms. Draw the form in which each will predominately exist at pH = 7.

a. CH_3COOH ($pK_a = 4.76$)

b. $CH_3CH_2\overset{+}{N}H_3$ ($pK_a = 11.0$)

c. H_3O^+ ($pK_a = -1.7$)

d. CH_3CH_2OH ($pK_a = 15.9$)

e. $CH_3CH_2\overset{+}{O}H_2$ ($pK_a = -2.5$)

f. $\overset{+}{N}H_4$ ($pK_a = 9.4$)

g. $HC{\equiv}N$ ($pK_a = 9.1$)

h. HNO_2 ($pK_a = 3.4$)

i. HNO_3 ($pK_a = -1.3$)

j. HBr ($pK_a = -9$)

1.21 LEWIS ACIDS AND BASES

In 1923, G. N. Lewis offered new definitions for the terms "acid" and "base." He defined an acid as a species that accepts an electron pair and a base as a species that donates an electron pair. All Brønsted acids fit the Lewis definition because all Brønsted acids lose a proton and the proton accepts an electron pair.

$$H^+ \quad + \quad :NH_3 \;\rightleftharpoons\; H{-}\overset{+}{N}H_3$$

acid — accepts a share in a pair of electrons

base — shares a pair of electrons

The Lewis definition of an acid is much broader than the Brønsted–Lowry definition because it is not limited to compounds that donate protons. According to the Lewis definition, compounds such as aluminum chloride ($AlCl_3$), boron trifluoride (BF_3), and borane (BH_3) are acids because they have unfilled valence orbitals and thus can accept a pair of electrons. They react with a compound that has a lone pair of electrons just like a proton reacts with ammonia. (The curved arrows indicate where the pair of electrons starts from and where it ends up.) But they are not Brønsted acids because they are not proton donors. So the Lewis definition of an acid includes all Brønsted acids and some additional acids that are not Brønsted acids. Throughout this text, the term "acid" is used to mean a proton-donating acid (Brønsted acid) and

$$\underset{\substack{\text{aluminum trichloride}\\ \text{a Lewis acid}}}{\overset{\displaystyle \text{Cl}{-}\!\!\underset{\text{Cl}}{\overset{\text{Cl}}{\text{Al}}}}{}} \; + \; \underset{\substack{\text{dimethyl ether}\\ \text{a Lewis base}}}{CH_3\overset{..}{\underset{..}{O}}CH_3} \;\rightleftharpoons\; \text{Cl}{-}\underset{\underset{\displaystyle CH_3}{\underset{\displaystyle |}{\text{Cl}}}}{\overset{\overset{\displaystyle \text{Cl}}{\displaystyle |}}{\text{Al}}}{}^{-}{-}\overset{+}{\underset{..}{O}}{-}CH_3$$

$$\underset{\substack{\text{borane}\\ \text{a Lewis acid}}}{\text{H}{-}\!\!\underset{\text{H}}{\overset{\text{H}}{\text{B}}}} \; + \; \underset{\substack{\text{ammonia}\\ \text{a Lewis base}}}{:\!\underset{\text{H}}{\overset{\text{H}}{\text{N}}}{-}\text{H}} \;\rightleftharpoons\; \text{H}{-}\!\underset{\text{H}}{\overset{\text{H}}{\text{B}}}{}^{-}{-}\overset{+}{\underset{\text{H}}{\overset{\text{H}}{\text{N}}}}{-}\text{H}$$

Lewis base:
Have pair, will share.

Lewis acid:
Need two, from you.

the term **"Lewis acid"** is used to refer to non-proton-donating acids such as $AlCl_3$ or BF_3. All **Lewis bases** are Brønsted bases and *vice versa;* they have a pair of electrons that they can share. They can share the electrons either with an atom such as aluminum or boron, or they can share the electrons with a proton. For example, diethyl ether and ammonia are both Lewis bases *and* Brønsted bases.

The potential maps show that when borane reacts with ammonia, the electron-rich nitrogen of ammonia is attracted to the electron-poor boron of borane. In the reaction product, the negative charge resides on the BH_3 part of the molecule and the positive charge resides on the NH_3 part. (Notice that the hydrogen atoms bear most of the electron density even though the formal charge is assigned to boron.)

PROBLEM 45

What is the product of each of the following reactions?

a. $ZnCl_2 + CH_3OH \rightleftharpoons$

b. $FeBr_3 + Br^- \rightleftharpoons$

c. $AlCl_3 + Cl^- \rightleftharpoons$

d. $BF_3 + \overset{\displaystyle O}{\underset{}{\overset{\parallel}{H C H}}} \rightleftharpoons$

PROBLEM 46

Show how each of the following compounds reacts with HO^-.

a. CH_3OH

b. $\overset{+}{N}H_4$

c. $CH_3\overset{+}{N}H_3$

d. BF_3

e. $\overset{+}{C}H_3$

f. $FeBr_3$

g. $AlCl_3$

h. CH_3COOH

KEY TERMS

acid (page 40)
acid-base reaction (page 40)
acid dissociation constant (K_a) (page 42)
acidity (page 41)
antibonding molecular orbital (page 21)
atomic number (page 3)
atomic orbital (page 5)
atomic weight (page 4)
aufbau principle (page 5)
base (page 40)
basicity (page 41)
bond dissociation energy (page 21)
bond length (page 21)
bonding molecular orbital (page 21)
Brønsted acid (page 40)
Brønsted base (page 40)
buffer solution (page 53)
carbanion (page 14)
carbocation (page 14)
condensed structure (page 16)
conjugate acid (page 41)
conjugate base (page 41)
core electrons (page 7)
covalent bond (page 8)
degenerate orbitals (page 5)

delocalized electrons (page 50)
dipole (page 12)
dipole moment (μ) (page 12)
double bond (page 29)
electron affinity (page 7)
electronegative (page 7)
electronegativity (page 9)
electropositive (page 7)
electrostatic attraction (page 8)
electrostatic potential map (page 11)
equilibrium constant (page 41)
excited-state electronic configuration (page 5)
formal charge (page 13)
free radical (page 14)
ground-state electronic configuration (page 5)
Heisenberg uncertainty principle (page 18)
Henderson–Hasselbalch equation (page 52)
Hund's rule (page 6)
hybrid orbital (page 26)
hydride ion (page 9)
hydrogen ion (page 9)
inductive electron withdrawal (page 48)

ionic bond (page 8)
ionic compound (page 8)
ionization energy (page 7)
isotopes (page 4)
Kekulé structure (page 16)
Lewis acid (page 55)
Lewis base (page 55)
Lewis structure (page 13)
lone-pair electrons (page 13)
mass number (page 4)
molecular orbital (page 20)
molecular orbital (MO) theory (page 20)
node (page 19)
nonbonding electrons (page 13)
nonpolar covalent bond (page 9)
nonpolar molecule (page 25)
octet rule (page 7)
orbital (page 5)
orbital hybridization (page 26)
organic compound (page 2)
Pauli exclusion principle (page 6)
pH (page 42)
pi (π) bond (page 23)
pK_a (page 42)
polar covalent bond (page 9)
potential map (page 11)

PROBLEMS

47. Draw a Lewis structure for each of the following species:
  - **a.** H_2CO_3
  - **b.** CO_3^{2-}
  - **c.** H_2CO
  - **d.** N_2H_4
  - **e.** CH_3NH_2
  - **f.** $CH_3N_2^+$
  - **g.** CO_2
  - **h.** NO^+
  - **i.** H_2NO^-

48. Give the hybridization of the central atom of each of the following species and tell whether the bond arrangement around it is linear, trigonal planar, or tetrahedral.
  - **a.** NH_3
  - **b.** BH_3
  - **c.** $^-CH_3$
  - **d.** $\cdot CH_3$
  - **e.** $^+NH_4$
  - **f.** $^+CH_3$
  - **g.** HCN
  - **h.** $C(CH_3)_4$
  - **i.** H_3O^+

49. Draw a compound that contains only carbon atoms and hydrogen atoms and that has:
  - **a.** three sp^3 hybridized carbons.
  - **b.** one sp^3 hybridized carbon and two sp^2 hybridized carbons.
  - **c.** two sp^3 hybridized carbons and two sp hybridized carbons.

50. Predict the indicated bond angles:
  - **a.** the $C-N-H$ bond angle in $(CH_3)_2NH$
  - **b.** the $C-N-C$ bond angle in $(CH_3)_2NH$
  - **c.** the $C-N-C$ bond angle in $(CH_3)_2NH_2^+$
  - **d.** the $C-O-C$ bond angle in CH_3OCH_3
  - **e.** the $C-O-H$ bond angle in CH_3OH
  - **f.** the $H-C-H$ bond angle in $H_2C=O$
  - **g.** the $F-B-F$ bond angle in $^-BF_4$
  - **h.** the $C-C-N$ bond angle in $CH_3C\equiv N$

51. Give each atom the appropriate formal charge:

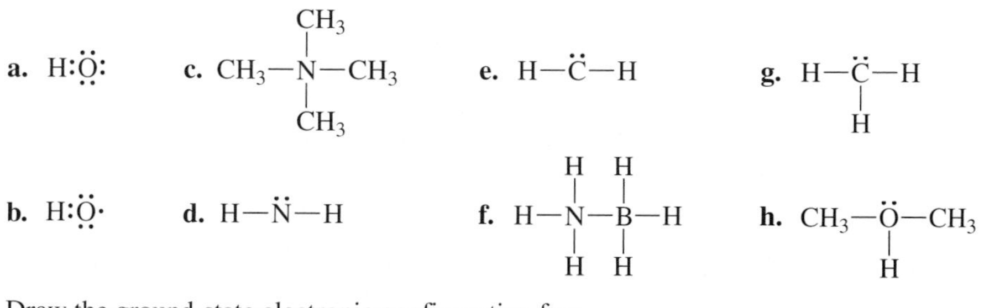

  a. $H\!:\!\ddot{O}\!:$ **c.** $CH_3-\overset{\displaystyle CH_3}{\underset{\displaystyle CH_3}{\overset{|}{\underset{|}{N}}}}-CH_3$ **e.** $H-\ddot{C}-H$ **g.** $H-\overset{\displaystyle}{\underset{\displaystyle H}{\overset{|}{\underset{|}{C}}}}-H$

  b. $H\!:\!\ddot{O}\cdot$ **d.** $H-\ddot{N}-H$ **f.** $H-\overset{\displaystyle H\ \ H}{\underset{\displaystyle H\ \ H}{\overset{|\ \ |}{\underset{|\ \ |}{N-B}}}}-H$ **h.** $CH_3-\overset{\displaystyle}{\underset{\displaystyle H}{\overset{|}{\underset{|}{\ddot{O}}}}}-CH_3$

52. Draw the ground-state electronic configuration for:
  - **a.** Ca
  - **b.** Ca^{2+}
  - **c.** Ar
  - **d.** Mg^{2+}

53. Write the Kekulé structure for each of the following compounds:
  - **a.** CH_3CHO
  - **b.** CH_3OCH_3
  - **c.** CH_3COOH
  - **d.** $(CH_3)_3COH$
  - **e.** $CH_3CH(OH)CH_2CN$
  - **f.** $(CH_3)_2CHCH(CH_3)CH_2C(CH_3)_3$

54. Show the direction of the dipole moment in each of the following bonds. (Use the electronegativities given in Table 1.3.)
  - **a.** CH_3-Br
  - **b.** CH_3-Li
  - **c.** $HO-NH_2$
  - **d.** $I-Br$
  - **e.** CH_3-OH
  - **f.** $(CH_3)_2N-H$

55. Indicate the hybridization of the indicated atom in each of the following compounds:

a. $CH_3\overset{\downarrow}{C}H=CH_2$ c. $CH_3CH_2\overset{\downarrow}{O}H$ e. $CH_3CH=\overset{\downarrow}{N}CH_3$

b. $CH_3\overset{O\leftarrow}{\overset{\|}{C}}CH_3$ d. $CH_3\overset{\downarrow}{C}\equiv N$ f. $CH_3\overset{\downarrow}{O}CH_2CH_3$

56. a. For each compound, which of the indicated bonds is shorter?
   b. Indicate the hybridization of the C, O, and N atoms in each of the molecules.

1. $CH_3\overset{\downarrow}{C}H=\overset{\downarrow}{C}HC\equiv CH$

4. $\overset{H}{\underset{H}{>}}C=CHC\equiv \overset{\downarrow}{C}-H$

2. $CH_3\overset{O}{\overset{\rightarrow\|}{C}}CH_2\overset{\downarrow}{-}OH$

5. $\overset{H}{\underset{H}{>}}C=CHC\equiv C-\overset{CH_3}{\underset{CH_3}{\overset{|}{\underset{|}{C}}}}-H$

3. $CH_3\overset{\downarrow}{N}H-CH_2CH_2\overset{\downarrow}{N}=CHCH_3$

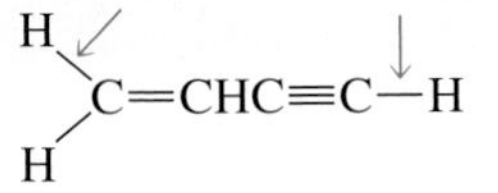

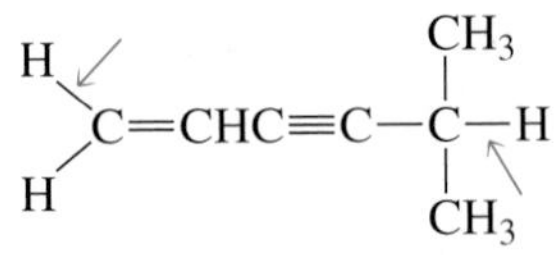

57. For each of the following compounds, draw the form in which it will predominantly exist at pH = 3, pH = 6, pH = 10, and pH = 14.

a. CH_3COOH
   pKa = 4.8

b. $CH_3CH_2\overset{+}{N}H_3$
   pKa = 11.0

c. CF_3CH_2OH
   pKa = 12.4

58. The following compound has two isomers. One isomer has a dipole moment of 0 D, and the other has a dipole moment of 2.95 D. Propose structures for the two isomers that are consistent with these data.

$$ClCH=CHCl$$

59. Do the sp^2 carbons and the indicated atoms lie in the same plane?

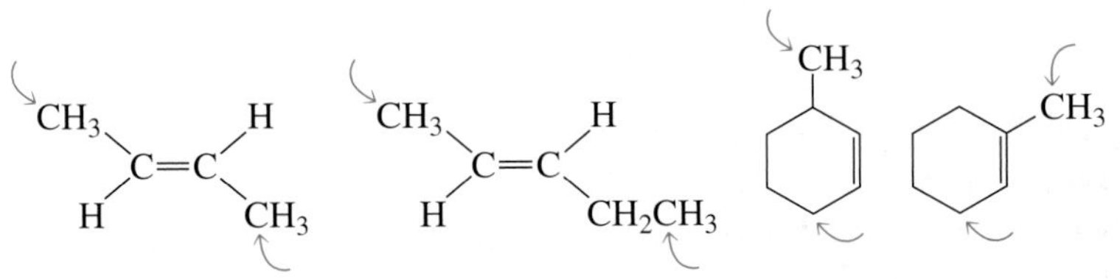

$\overset{CH_3}{\underset{H}{>}}C=C\overset{H}{\underset{CH_3}{<}}$ $\overset{CH_3}{\underset{H}{>}}C=C\overset{H}{\underset{CH_2CH_3}{<}}$

60. Give the products of the following acid-base reactions and indicate the direction of the equilibrium. (Use the pK_a values that are given in Section 1.17.)

a. $CH_3\overset{O}{\overset{\|}{C}}OH + CH_3O^- \rightleftharpoons$

c. $CH_3\overset{O}{\overset{\|}{C}}OH + CH_3NH_2 \rightleftharpoons$

b. $CH_3CH_2OH + {}^-NH_2 \rightleftharpoons$

d. $CH_3CH_2OH + HCl \rightleftharpoons$

61. For each of the following molecules, indicate the hybridization of each carbon atom and give approximate values to all the bond angles.
   a. $CH_3C\equiv CH$ b. $CH_3CH=CH_2$ c. $CH_3CH_2CH_3$

62. a. Estimate the pK_a value of each of the following acids without using a calculator (i.e., between 3 and 4, between 9 and 10, etc.).
   b. Determine the pK_a values using a calculator.
   c. Which is the strongest acid?
   1. nitrous acid (HNO_2), $K_a = 4.0 \times 10^{-4}$

2. nitric acid (HNO_3), $K_a = 22$
3. bicarbonate (HCO_3^-), $K_a = 6.3 \times 10^{-11}$
4. hydrogen cyanide (HCN), $K_a = 7.9 \times 10^{-10}$
5. formic acid (HCOOH), $K_a = 2.0 \times 10^{-4}$

63. a. List the following carboxylic acids in order of decreasing acidity.

   1. $CH_3CH_2CH_2COOH$
       $K_a = 1.52 \times 10^{-5}$

   3. $ClCH_2CH_2CH_2COOH$
       $K_a = 2.96 \times 10^{-5}$

   2. $CH_3CH_2CHCOOH$
            |
           Cl
     $K_a = 1.39 \times 10^{-3}$

   4. CH_3CHCH_2COOH
           |
         Cl
     $K_a = 8.9 \times 10^{-5}$

  b. How does the presence of an electronegative substituent such as Cl affect the acidity of a carboxylic acid?
  c. How does the location of the substituent affect the acidity of a carboxylic acid?

64. a. For each of the following pairs of reactions, indicate which one has the more favorable equilibrium constant (that is, which one most favors products).
  b. Which of the four reactions has the most favorable equilibrium constant?

   1. $CH_3CH_2OH + NH_3 \rightleftharpoons CH_3CH_2O^- + \overset{+}{N}H_4$

           or

       $CH_3OH + NH_3 \rightleftharpoons CH_3O^- + \overset{+}{N}H_4$

   2. $CH_3CH_2OH + NH_3 \rightleftharpoons CH_3CH_2O^- + \overset{+}{N}H_4$

           or

    $CH_3CH_2OH + CH_3NH_2 \rightleftharpoons CH_3CH_2O^- + CH_3\overset{+}{N}H_3$

65. Water and diethyl ether are immiscible liquids. Charged compounds dissolve in water and uncharged compounds dissolve in ether. $C_6H_{11}COOH$ has a pK_a of 4.8 and $C_6H_{11}NH_3^+$ has a pK_a of 10.7.

  a. What pH would you make the water layer to cause both compounds to dissolve in the water layer?
  b. What pH would you make the water layer to cause the acid to dissolve in the water layer and the amine to dissolve in the ether layer?
  c. What pH would you make the water layer to cause the acid to dissolve in the ether layer and the amine to dissolve in the water layer?

66. How could you separate a mixture of the following compounds? The reagents available to you are water, ether, 1.0 M HCl, and 1.0 M NaOH. (*Hint:* see problem 65.)

COOH

        OH

               $\overset{+}{N}H_3Cl^-$

          $\overset{+}{N}H_3Cl^-$

        Cl

$pK_a = 4.17$ $pK_a = 9.95$ $pK_a = 10.66$ $pK_a = 4.60$

67. Carbonic acid has a pK_a of 6.1 at physiological temperature. Is the carbonic acid/bicarbonate buffer system that maintains the pH of the blood (at pH $= 7.4$) better at neutralizing excess acid or excess base?

For help in answering Problems 67-69, see Special Topic I in the Solutions Manual/Study Guide.

68. a. If an acid with a pK_a of 5.3 is in an aqueous solution of pH 5.7, what percent of the acid is present in the acidic form?
  b. At what pH will 80% of the acid exist in the acidic form?

69. Calculate the pH values of the following solutions:
  a. a 1.0 M solution of acetic acid ($pK_a = 4.76$)
  b. a 0.1 M solution of protonated methylamine ($pK_a = 10.7$)
  c. a solution containing 0.3 M HCOOH and 0.1 M $HCOO^-$ (pK_a of HCOOH $= 3.76$)

CHAPTER

2

An Introduction to Organic Compounds:

Nomenclature, Physical Properties, and Representation of Structure

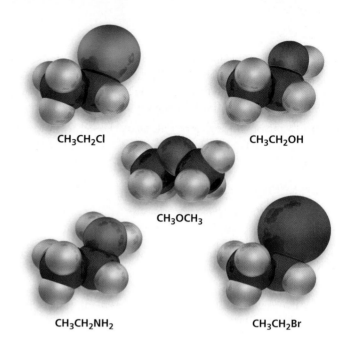

CH₃CH₂Cl

CH₃CH₂OH

CH₃OCH₃

CH₃CH₂NH₂

CH₃CH₂Br

This book organizes organic chemistry according to how organic compounds react. However, we must not forget that whenever an organic compound undergoes a reaction, a new organic compound is synthesized. In other words, while we are learning how organic compounds react, we will also be learning how to synthesize new compounds.

$$Y \longrightarrow Z$$

Y is reacting   **Z is being synthesized**

The main classes of compounds that are synthesized by the reactions we will study in Chapters 3–11 are alkanes, alkyl halides, ethers, alcohols, and amines. As we learn how to synthesize compounds, we will need to be able to refer to them by name. So we will begin our study of organic chemistry by learning how to name these five classes of compounds.

First we will learn how to name alkanes because they form the basis for the names of almost all organic compounds. **Alkanes** are composed of only carbon atoms and hydrogen atoms and contain only single bonds. (A **hydrocarbon** is a compound that is composed of only carbon and hydrogen, so an alkane is a hydrocarbon that has only single bonds.) Alkanes in which the carbons form a continuous chain with no branches are called **straight-chain alkanes.** The names of several straight-chain alkanes are given in Table 2.1. It is important that you learn the names of at least the first 10.

The family of alkanes shown in Table 2.1 is an example of a homologous series. A **homologous series** (*homos* is Greek for "the same as") is a family of compounds in which each member differs from the next by one **methylene (CH_2) group.** The members of a homologous series are called **homologs.** Propane ($CH_3CH_2CH_3$) and butane ($CH_3CH_2CH_2CH_3$) are homologs.

By looking at the relative numbers of carbon and hydrogen atoms in the alkanes listed in Table 2.1, you can conclude that the general molecular formula for an alkane is C_nH_{2n+2}, where n is any integer. So, if an alkane has one carbon atom, it must have four hydrogen atoms; if it has two carbon atoms, it must have six hydrogens. We have seen that carbon forms four covalent bonds and hydrogen forms only one covalent bond (Section 1.4). This means that there is only one possible structure for an alkane with molecular formula CH_4 (methane), and only one structure for an alkane with molecular formula C_2H_6 (ethane). We have examined the structures of these compounds in Section 1.7. There is also only one structure for an alkane with molecular formula C_3H_8 (propane).

name	Kekulé structure	condensed structure	ball-and-stick model
methane	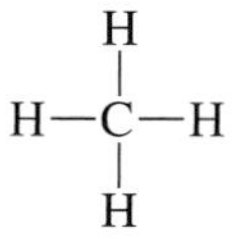 H—C—H (with H above and H below)	CH_4	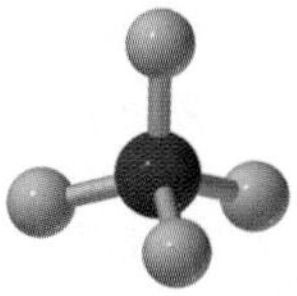
ethane	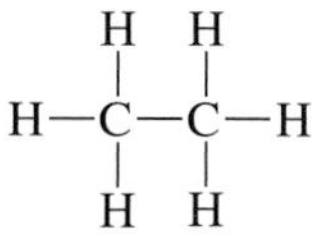 H—C—C—H (with H's)	CH_3CH_3	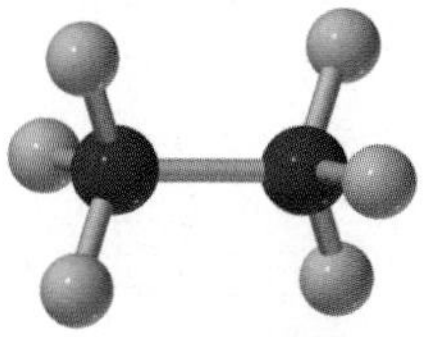
propane	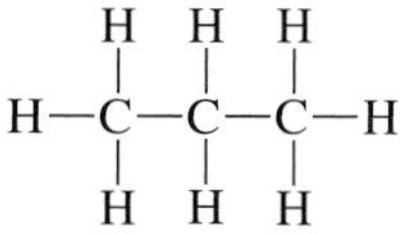 H—C—C—C—H (with H's)	$CH_3CH_2CH_3$	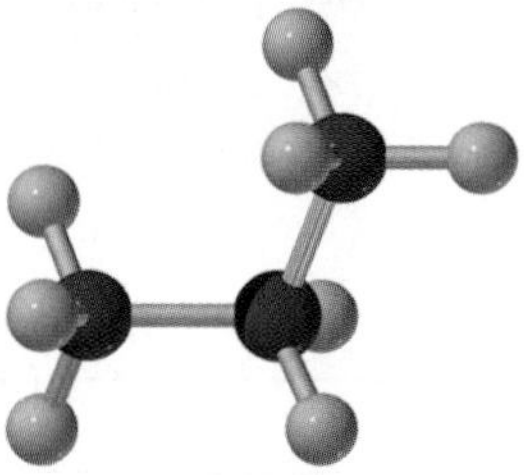
butane	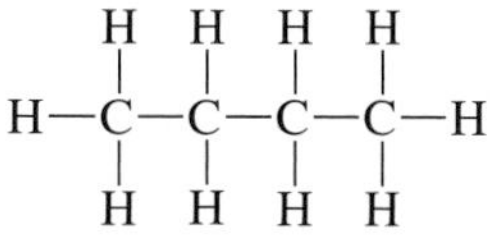 H—C—C—C—C—H (with H's)	$CH_3CH_2CH_2CH_3$	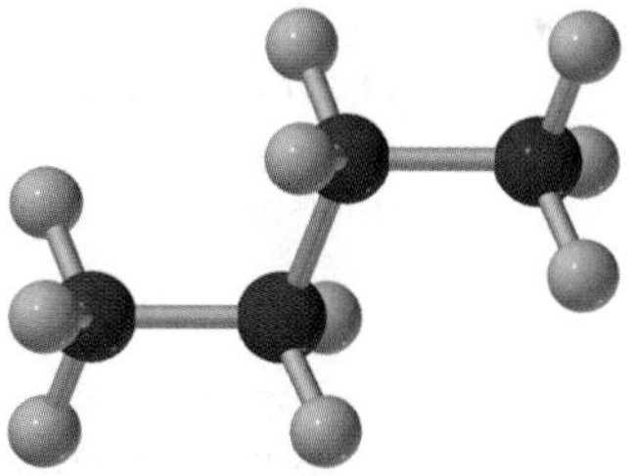

3-D Molecules: Propane; Butane

As the number of carbons in the alkane increases, the number of possible structures also increases. There are two possible structures for an alkane with molecular formula C_4H_{10}. In addition to butane—a straight-chain alkane—there is a branched butane called isobutane. Both of these structures fulfill the requirement that each carbon bonds to four atoms and each hydrogen forms only one bond.

Compounds such as butane and isobutane that have the same molecular formula but differ in the order in which the atoms are connected are called **constitutional isomers** (their molecules have different constitutions). In fact, isobutane got its name because it is an "iso"mer of butane. The structural unit with two methyl groups and a hydrogen bonded to a carbon that occurs in isobutane has come to be called "iso." Thus, the name isobutane tells you that it is a four-carbon alkane with an iso structural unit.

$CH_3CH_2CH_2CH_3$
butane

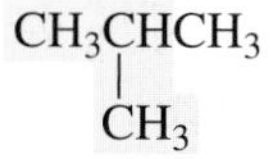

CH_3CHCH_3
|
CH_3
isobutane

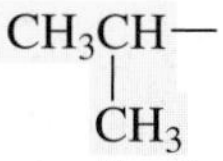

$CH_3CH—$
|
CH_3
an "iso"
structural unit

TABLE 2.1   Nomenclature and Physical Properties of Straight-Chain Alkanes

Number of carbons	Molecular formula	Name	Condensed structure	Boiling point (°C)	Melting point (°C)	Density[a] (g/mL)
1	CH_4	methane	CH_4	−167.7	−182.5	
2	C_2H_6	ethane	CH_3CH_3	−88.6	−183.3	
3	C_3H_8	propane	$CH_3CH_2CH_3$	−42.1	−187.7	
4	C_4H_{10}	butane	$CH_3CH_2CH_2CH_3$	−0.5	−138.3	
5	C_5H_{12}	pentane	$CH_3(CH_2)_3CH_3$	36.1	−129.8	0.5572
6	C_6H_{14}	hexane	$CH_3(CH_2)_4CH_3$	68.7	−95.3	0.6603
7	C_7H_{16}	heptane	$CH_3(CH_2)_5CH_3$	98.4	−90.6	0.6837
8	C_8H_{18}	octane	$CH_3(CH_2)_6CH_3$	127.7	−56.8	0.7026
9	C_9H_{20}	nonane	$CH_3(CH_2)_7CH_3$	150.8	−53.5	0.7177
10	$C_{10}H_{22}$	decane	$CH_3(CH_2)_8CH_3$	174.0	−29.7	0.7299
11	$C_{11}H_{24}$	undecane	$CH_3(CH_2)_9CH_3$	195.8	−25.6	0.7402
12	$C_{12}H_{26}$	dodecane	$CH_3(CH_2)_{10}CH_3$	216.3	−9.6	0.7487
13	$C_{13}H_{28}$	tridecane	$CH_3(CH_2)_{11}CH_3$	235.4	−5.5	0.7546
20	$C_{20}H_{42}$	eicosane	$CH_3(CH_2)_{18}CH_3$	343.0	36.8	0.7886
21	$C_{21}H_{44}$	heneicosane	$CH_3(CH_2)_{19}CH_3$	356.5	40.5	0.7917
30	$C_{30}H_{62}$	triacontane	$CH_3(CH_2)_{28}CH_3$	449.7	65.8	0.8097

[a]Density is temperature-dependent. The densities given are those determined at 20 °C ($d^{20°}$).

There are three alkanes with molecular formula C_5H_{12}. Pentane is the straight-chain alkane. Isopentane, as its name indicates, has an iso structural unit and five carbon atoms. The third isomer is called neopentane. The structural unit with a carbon surrounded by four other carbons is called "neo."

$$CH_3CH_2CH_2CH_2CH_3$$
pentane

$$CH_3CHCH_2CH_3 \atop CH_3$$
isopentane

$$CH_3 \atop CH_3CCH_3 \atop CH_3$$
neopentane

$$CH_3 \atop CH_3CCH_2— \atop CH_3$$
a "neo" structural unit

There are five isomers with molecular formula C_6H_{14}. We are now able to name three of them (hexane, isohexane, and neohexane), but we cannot name the other two without defining names for new structural units. (At this point, ignore the names written in blue.)

common name:
systematic name:

$$CH_3CH_2CH_2CH_2CH_2CH_3$$
hexane
hexane

$$CH_3CHCH_2CH_2CH_3 \atop CH_3$$
isohexane
2-methylpentane

$$CH_3 \atop CH_3CCH_2CH_3 \atop CH_3$$
neohexane
2,2-dimethylbutane

$$CH_3CH_2CHCH_2CH_3 \atop CH_3$$
3-methylpentane

$$CH_3CH—CHCH_3 \atop CH_3 \ \ CH_3$$
2,3-dimethylbutane

There are nine isomers with molecular formula C_7H_{16}. We can name only two of them (heptane and isoheptane) without defining new structural units. Notice that neo-heptane cannot be used as a name because three different heptanes have a carbon that is bonded to four other carbons and *a name must be specific to only one compound.*

A compound can have more than one name, but a name must specify only one compound.

$$CH_3CH_2CH_2CH_2CH_2CH_2CH_3$$

common name: heptane
systematic name: **heptane**

$$CH_3CHCH_2CH_2CH_2CH_3$$
$$|$$
$$CH_3$$

isoheptane
2-methylhexane

$$CH_3CH_2CHCH_2CH_2CH_3$$
$$|$$
$$CH_3$$
3-methylhexane

$$CH_3CH-CHCH_2CH_3$$
$$|\quad\quad|$$
$$CH_3\ \ CH_3$$
2,3-dimethylpentane

$$CH_3CHCH_2CHCH_3$$
$$|\quad\quad\quad|$$
$$CH_3\ \ \ \ CH_3$$
2,4-dimethylpentane

$$CH_3$$
$$|$$
$$CH_3CCH_2CH_2CH_3$$
$$|$$
$$CH_3$$
2,2-dimethylpentane

$$CH_3$$
$$|$$
$$CH_3CH_2CCH_2CH_3$$
$$|$$
$$CH_3$$
3,3-dimethylpentane

$$CH_3CH_2CHCH_2CH_3$$
$$|$$
$$CH_2CH_3$$
3-ethylpentane

$$CH_3\ \ CH_3$$
$$|\quad\quad|$$
$$CH_3C-CHCH_3$$
$$|$$
$$CH_3$$
2,2,3-trimethylbutane

The number of different structural units increases rapidly as the number of carbons in an alkane increases. For example, there are 75 alkanes with molecular formula $C_{10}H_{22}$ and 4347 alkanes with molecular formula $C_{15}H_{32}$. To avoid having to memorize the names of many different structural units, chemists have devised rules that name compounds on the basis of their structures. This way, only the rules have to be learned. Because the name is based on the structure, these rules make it possilbe to deduce the structure of a compound from its name.

This method of nomenclature is called **systematic nomenclature.** It is also called **IUPAC nomenclature** because it was designed by a commission of the International Union of Pure and Applied Chemistry (abbreviated IUPAC and pronounced "Eye-You-Pack") at a meeting in Geneva, Switzerland, in 1892. The IUPAC rules have been continually revised by the commission since then. Names such as isobutane and neopentane—nonsystematic names—are called **common names** and are shown in red in this text. The systematic or IUPAC names are shown in blue for all of the alkanes with six and seven carbon atoms. Before we can understand how an IUPAC name for an alkane is constructed, we must learn how to name alkyl substituents.

Removal of a hydrogen from an alkane results in an **alkyl substituent** (or an alkyl group). Alkyl substituents are named by replacing the "ane" ending of the alkane with "yl." The letter "R" is used to indicate any alkyl group.

2.1 NOMENCLATURE OF ALKYL SUBSTITUENTS

CH_3- CH_3CH_2- $CH_3CH_2CH_2-$ $CH_3CH_2CH_2CH_2-$
a methyl group   an ethyl group   a propyl group   a butyl group

$R-$
any alkyl group

If a hydrogen of an alkane is replaced by an OH, the compound becomes an **alcohol;** if it is replaced by an NH_2, the compound becomes an **amine;** and if it is replaced by a halogen, the compound becomes an **alkyl halide.**

$$R—OH \qquad R—NH_2 \qquad R—X \qquad X = F, Cl, Br, \text{ or } I$$
an alcohol   an amine   an alkyl halide

An alkyl group name followed by the name of the class of the compound (alcohol, amine, etc.) yields the common name of the compound. The following examples show how alkyl group names are used to build common names. Notice that there is a space between the name of the alkyl group and the name of the class of compound, except in the case of amines.

CH_3OH $CH_3CH_2NH_2$ $CH_3CH_2CH_2Br$ $CH_3CH_2CH_2CH_2Cl$
methyl alcohol   ethylamine   propyl bromide   butyl chloride

CH_3I CH_3CH_2OH $CH_3CH_2CH_2NH_2$ $CH_3CH_2CH_2CH_2OH$
methyl iodide   ethyl alcohol   propylamine   butyl alcohol

Two alkyl groups—a propyl and an isopropyl—contain three carbon atoms. A propyl group is obtained when a hydrogen is removed from a primary carbon of propane. A **primary carbon** is one that is bonded to only one other carbon. An isopropyl group is obtained when a hydrogen is removed from the secondary carbon of propane. A **secondary carbon** is one that is bonded to two other carbons. Notice that an isopropyl group, as its name indicates, has three carbon atoms arranged as an iso structural unit.

a propyl group   an isopropyl group

Build models of the two representations of isopropyl chloride and convince yourself that they represent the same compound.

Molecular structures can be drawn in different ways. Isopropyl chloride, for example, is drawn here in two ways. Both represent the same compound because the central carbon in each is bonded to a hydrogen, a chlorine, and two methyl groups.

3-D Molecules:
Propyl chloride;
Isopropyl choride

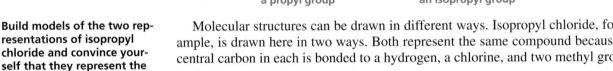

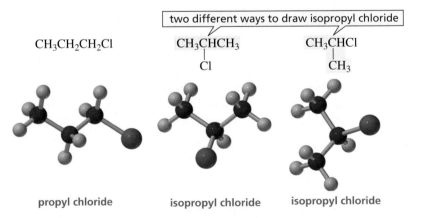

propyl chloride   isopropyl chloride   isopropyl chloride

There are four alkyl groups that contain four carbons. Butyl and isobutyl groups have both had a hydrogen removed from a primary (1°) carbon. A *sec*-butyl group has had a hydrogen removed from a secondary (2°) carbon (*sec-*, often abbreviated *s-*, stands for secondary); and a *tert*-butyl group has had a hydrogen removed from a tertiary (3°) carbon (*tert-*, sometimes abbreviated *t-*, stands for tertiary). A **tertiary carbon** is one that is bonded to three other carbons.

> A primary carbon is bonded to one carbon, a secondary carbon is bonded to two carbons, and a tertiary carbon is bonded to three carbons.

a primary carbon	a primary carbon	a secondary carbon	a tertiary carbon	CH_3

$CH_3CH_2CH_2CH_2-$ CH_3CHCH_2- CH_3CH_2CH- CH_3C-
                                   $|$ $|$ $|$
                                   CH_3 CH_3 CH_3

a butyl group   **an isobutyl group**   **a *sec*-butyl group**   **a *tert*-butyl group**

A straight-chain alkyl group is sometimes called *n*-butyl (*n* for normal) to emphasize that its four carbon atoms are in an unbranched chain. However, if the name does not have a prefix such as "*n*" or "iso," it is assumed that the carbons are in an unbranched chain.

$CH_3CH_2CH_2CH_2Br$ $CH_3CH_2CH_2CH_2CH_2F$

butyl bromide   pentyl fluoride

or   **or**

n-butyl bromide   *n*-pentyl fluoride

> 3-D Molecules:
> *n*-Butyl alcohol;
> *sec*-Butyl alcohol;
> *tert*-Butyl alcohol

The hydrogens in a molecule are also referred to as primary, secondary, and tertiary. **Primary hydrogens** are attached to primary carbons, **secondary hydrogens** are attached to secondary carbons, and **tertiary hydrogens** are attached to tertiary carbons.

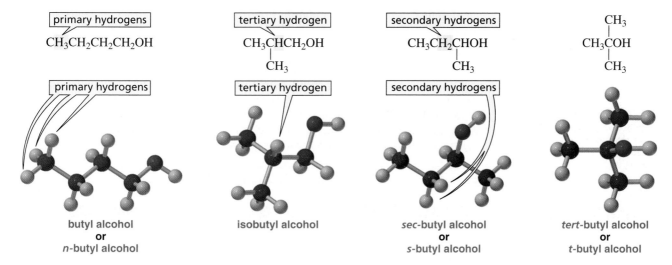

primary hydrogens	tertiary hydrogen	secondary hydrogens	CH_3

$CH_3CH_2CH_2CH_2OH$ CH_3CHCH_2OH CH_3CH_2CHOH CH_3COH
                                          $|$ $|$ $|$
                                          CH_3 CH_3 CH_3

butyl alcohol   isobutyl alcohol   *sec*-butyl alcohol   *tert*-butyl alcohol
or   **or**   **or**
n-butyl alcohol   *s*-butyl alcohol   *t*-butyl alcohol

Because a chemical name must apply to only one compound, the only time you will see the prefix "*sec*" is in *sec*-butyl. The name "*sec*-pentyl" cannot be used because pentane has two different secondary carbon atoms. Therefore, there are two different alkyl groups that result from the removal of a hydrogen from a secondary carbon of pentane.

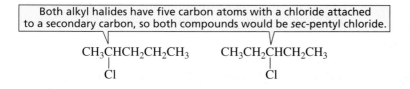

> Both alkyl halides have five carbon atoms with a chloride attached to a secondary carbon, so both compounds would be *sec*-pentyl chloride.

$CH_3CHCH_2CH_2CH_3$ $CH_3CH_2CHCH_2CH_3$
     $|$ $|$
     Cl Cl

The prefix "*tert*" is found in *tert*-butyl and *tert*-pentyl because each of these substituent names describes only one alkyl group. The name "*tert*-hexyl" cannot be used because it describes two different alkyl groups. (In older literature, you might find amyl used instead of pentyl to designate five carbons.)

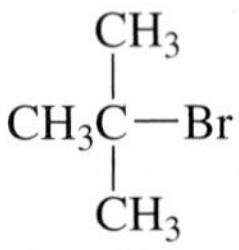

$$CH_3C-Br$$
with CH_3 above and CH_3 below
tert-butyl bromide

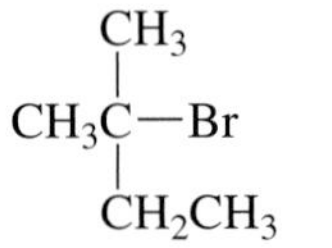

$$CH_3C-Br$$
with CH_3 above and CH_2CH_3 below
tert-pentyl bromide

$$CH_3CH_2C-Br$$
with CH_2CH_3 above and CH_3 below

$$CH_3CH_2CH_2C-Br$$
with CH_3 above and CH_3 below

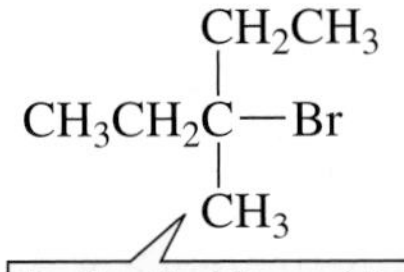

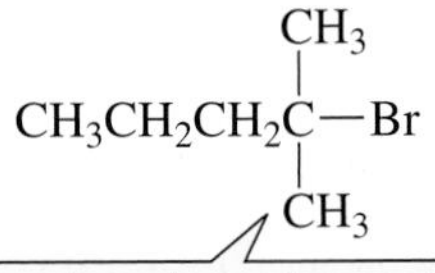

> Both alkyl bromides have six carbon atoms with a bromine attached to a tertiary carbon, so both compounds would be *tert*-hexyl bromide.

The prefix "iso" can be used regardless of the number of carbons in the alkyl group. This prefix means that there is an iso structural unit (a carbon bonded to two methyl groups and a hydrogen) at one end of the molecule, and any group replacing a hydrogen is at the other end. (Notice that an iso group has a methyl group on the next to the last carbon in the chain.)

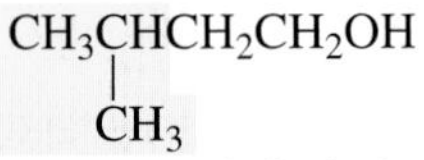

$$CH_3CHCH_2CH_2OH$$
with CH_3 below
isopentyl alcohol
or
isoamyl alcohol

$$CH_3CHCH_2CH_2CH_2Cl$$
with CH_3 below
isohexyl chloride

$$CH_3CHCH_2CH_2CH_2CH_2NH_2$$
with CH_3 below
isoheptylamine

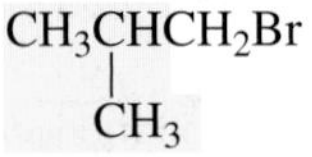

$$CH_3CHCH_2Br$$
with CH_3 below
isobutyl bromide

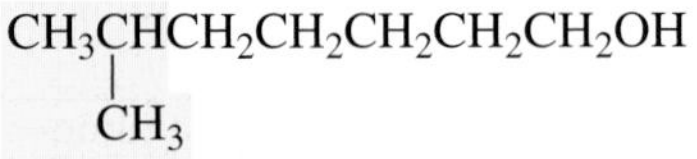

$$CH_3CHCH_2CH_2CH_2CH_2OH$$
with CH_3 below
isooctyl alcohol

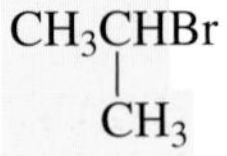

$$CH_3CHBr$$
with CH_3 below
isopropyl bromide

Notice that *all* isoalkyl compounds have the substituent (OH, Cl, NH_2, etc.) on a primary carbon except for isopropyl, which has the substituent on a secondary carbon. The isopropyl group could have been called a *sec*-propyl group. Either name would have been appropriate because the group has an iso structural unit and a hydrogen has been removed from a secondary carbon. Chemists decided to call it isopropyl, however, which means that "*sec*" is used only for *sec*-butyl.

Alkyl group names are used so frequently that you should learn them. Some of the most common alkyl group names are compiled in Table 2.2 for your convenience.

Tutorial: Common alkyl groups

TABLE 2.2   Names of Some Alkyl Groups

methyl	CH_3-	*sec*-butyl	CH_3CH_2CH- $\mid$ CH_3	neopentyl	CH_3CCH_2- with CH_3 above and CH_3 below
ethyl	CH_3CH_2-				
propyl	$CH_3CH_2CH_2-$			hexyl	$CH_3CH_2CH_2CH_2CH_2CH_2-$
isopropyl	CH_3CH- $\mid$ CH_3	*tert*-butyl	CH_3C- with CH_3 above and CH_3 below	isohexyl	$CH_3CHCH_2CH_2CH_2-$ $\mid$ CH_3
butyl	$CH_3CH_2CH_2CH_2-$	pentyl	$CH_3CH_2CH_2CH_2CH_2-$	heptyl	$CH_3CH_2CH_2CH_2CH_2CH_2CH_2-$
isobutyl	CH_3CHCH_2- $\mid$ CH_3	isopentyl	$CH_3CHCH_2CH_2-$ $\mid$ CH_3	isoheptyl	$CH_3CHCH_2CH_2CH_2CH_2-$ $\mid$ CH_3

PROBLEM 1◆

Write a structure for each of the following compounds:

a. isopropyl alcohol   **c.** *sec*-butyl iodide   **e.** *tert*-butylamine

b. isopentyl fluoride   **d.** neopentyl chloride   **f.** isooctyl bromide

---

The systematic name of an alkane is obtained using the following rules:

1. Determine the number of carbons in the longest continuous carbon chain. This chain is called the **parent hydrocarbon.** The alkane name indicating the number of carbons in the parent hydrocarbon becomes the alkane's "last name." Notice that the longest continuous chain is chosen regardless of how the molecule is written. Sometimes you have to "turn a corner" to obtain the longest continuous chain.

2.2 NOMENCLATURE OF ALKANES

Determine the number of carbons in the longest continuous chain.

$$\overset{8}{C}H_3\overset{7}{C}H_2\overset{6}{C}H_2\overset{5}{C}H_2\overset{4}{C}H\overset{3}{C}H_2\overset{2}{C}H_2\overset{1}{C}H_3 \qquad \overset{8}{C}H_3\overset{7}{C}H_2\overset{6}{C}H_2\overset{5}{C}H_2\overset{4}{C}HCH_3$$

$$CH_3 \qquad\qquad\qquad \underset{3}{C}H_2\underset{2}{C}H_2\underset{1}{C}H_3$$

three different ways to draw 4-methyloctane

$$\overset{4}{C}H_3\overset{3}{C}H\overset{2}{C}H_2\overset{1}{C}H_3$$

$$\underset{5}{C}H_2\underset{6}{C}H_2\underset{7}{C}H_2\underset{8}{C}H_3$$

2. The name of the alkyl substituent that hangs off the parent hydrocarbon is cited before the name of the parent hydrocarbon with a number to designate the carbon to which it is attached. The chain is numbered in the direction that gives the substituent as low a number as possible. The substituent name and the name of the parent hydrocarbon are joined in one word, and there is a hyphen between the number and the substituent name.

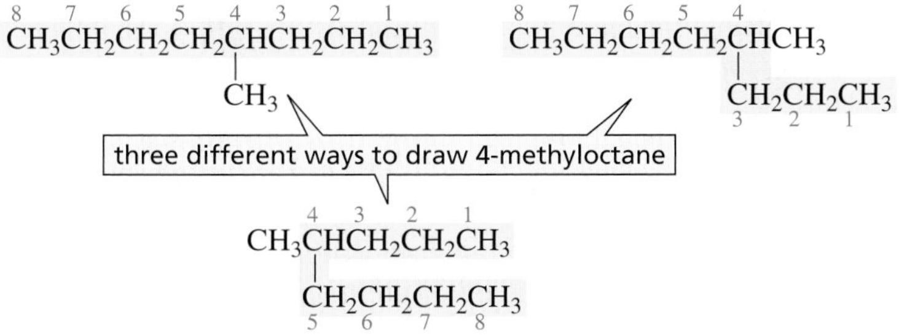

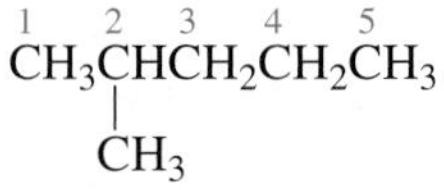

Number the chain so that the substituent gets the lowest possible number.

$$\overset{1}{C}H_3\overset{2}{C}H\overset{3}{C}H_2\overset{4}{C}H_2\overset{5}{C}H_3 \qquad \overset{6}{C}H_3\overset{5}{C}H_2\overset{4}{C}H_2\overset{3}{C}H\overset{2}{C}H_2\overset{1}{C}H_3$$

$$CH_3 \qquad\qquad\qquad\qquad CH_2CH_3$$

2-methylpentane 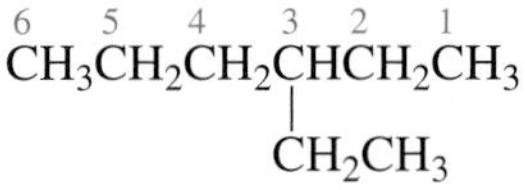   **3-ethylhexane**

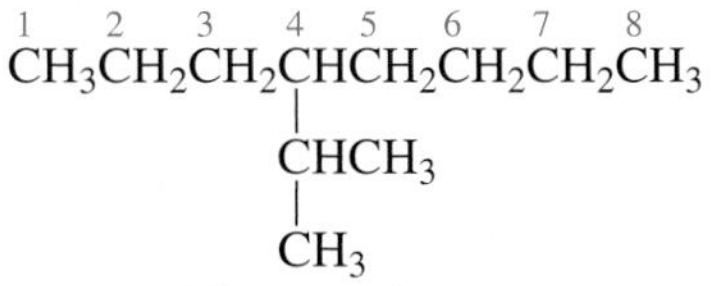

$$\overset{1}{C}H_3\overset{2}{C}H_2\overset{3}{C}H_2\overset{4}{C}H\overset{5}{C}H_2\overset{6}{C}H_2\overset{7}{C}H_2\overset{8}{C}H_3$$

$$CHCH_3$$

$$CH_3$$

4-isopropyloctane

Notice that only systematic names have numbers; common names never contain numbers.

Numbers are used only for systematic names and never for common names.

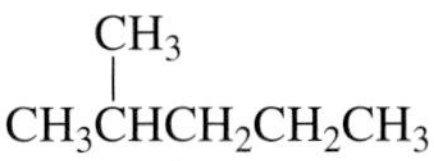

$$CH_3$$

$$CH_3CHCH_2CH_2CH_3$$

common name: isohexane
systematic name: 2-methylpentane

3. If more than one substituent is attached to the parent hydrocarbon, the chain is numbered in the direction that will result in the lowest possible number in the name of the compound. The substituents are listed in alphabetical order (not numerical order), with each substituent getting the appropriate number. In the following example, the correct name (5-ethyl-3-methyloctane) contains a 3 as its lowest number, while the incorrect name (4-ethyl-6-methyloctane) contains a 4 as its lowest number.

List substituents in alphabetical order.

$$CH_3CH_2CHCH_2CHCH_2CH_2CH_3$$
$$\quad\quad\quad | \quad\quad\quad |$$
$$\quad\quad\quad CH_3 \quad CH_2CH_3$$

5-ethyl-3-methyloctane
not
4-ethyl-6-methyloctane
because 3 < 4

If two or more substituents are the same, the prefixes di, tri, and tetra are used with the substituent names. The numbers indicating the locations of the identical substituents are cited together, separated by commas. Notice that there must be as many numbers in a name as there are substituents.

A number and a word are separated by a hyphen; numbers are separated by a comma.

$$CH_3CH_2CHCH_2CHCH_3$$
$$\quad\quad\quad | \quad\quad\quad |$$
$$\quad\quad\quad CH_3 \quad\quad CH_3$$

2,4-dimethylhexane

$$\quad\quad\quad\quad\quad CH_3 \quad CH_3$$
$$\quad\quad\quad\quad\quad | \quad\quad |$$
$$CH_3CH_2CH_2C—CCH_2CH_3$$
$$\quad\quad\quad\quad\quad | \quad\quad |$$
$$\quad\quad\quad\quad\quad CH_3 \quad CH_3$$

3,3,4,4-tetramethylheptane

di, tri, tetra, *sec,* and *tert* are ignored in alphabetizing.

iso, neo, and cyclo are not ignored in alphabetizing.

The prefixes di, tri, tetra, *sec,* and *tert* are ignored in alphabetizing substituent groups, but the prefixes iso, neo, and cyclo are not ignored.

$$\quad\quad\quad CH_2CH_3 \quad\quad CH_3$$
$$\quad\quad\quad | \quad\quad\quad\quad\quad |$$
$$CH_3CH_2CCH_2CH_2CHCHCH_2CH_2CH_3$$
$$\quad\quad\quad | \quad\quad\quad |$$
$$\quad\quad\quad CH_2CH_3 \quad CH_2CH_3$$

3,3,6-triethyl-7-methyldecane

$$\quad\quad\quad\quad\quad\quad\quad\quad\quad CH_3$$
$$\quad\quad\quad\quad\quad\quad\quad\quad\quad |$$
$$CH_3CH_2CH_2CHCH_2CH_2CHCH_3$$
$$\quad\quad\quad\quad\quad |$$
$$\quad\quad\quad\quad\quad CHCH_3$$
$$\quad\quad\quad\quad\quad |$$
$$\quad\quad\quad\quad\quad CH_3$$

5-isopropyl-2-methyloctane

4. When both directions lead to the same lowest number for one of the substituents, the direction is chosen that gives the lowest possible number to one of the remaining substituents.

$$\quad\quad CH_3$$
$$\quad\quad |$$
$$CH_3CCH_2CHCH_3$$
$$\quad\quad | \quad\quad |$$
$$\quad\quad CH_3 \quad CH_3$$

2,2,4-trimethylpentane
not
2,4,4-trimethylpentane
because 2 < 4

$$\quad\quad CH_2CH_3$$
$$\quad\quad |$$
$$CH_3CHCHCH_2CH_2CHCH_3$$
$$\quad\quad | \quad\quad\quad\quad\quad |$$
$$\quad\quad CH_3 \quad\quad\quad CH_3$$

3-ethyl-2,6-dimethylheptane
not
5-ethyl-2,6-dimethylheptane
because 3 < 5

Only if the same set of numbers is obtained in both directions does the first cited group get the lower number.

5. If the same substituent numbers are obtained in both directions, the first cited group receives the lower number.

$$\quad\quad Cl$$
$$\quad\quad |$$
$$CH_3CHCHCH_3$$
$$\quad\quad\quad |$$
$$\quad\quad\quad Br$$

2-bromo-3-chlorobutane
not
3-bromo-2-chlorobutane

$$\quad\quad\quad CH_2CH_3$$
$$\quad\quad\quad |$$
$$CH_3CH_2CHCH_2CHCH_2CH_3$$
$$\quad\quad\quad\quad\quad\quad\quad |$$
$$\quad\quad\quad\quad\quad\quad\quad CH_3$$

3-ethyl-5-methylheptane
not
5-ethyl-3-methylheptane

6. If a compound has two or more chains of the same length, the parent hydrocarbon is the chain with the greatest number of substituents.

In the case of two hydrocarbon chains with the same number of carbons, choose the one with the most substituents.

$$\overset{3}{CH_3}\overset{}{CH_2}\overset{4}{CH}\overset{5}{CH_2}\overset{6}{CH_2}CH_3$$

$$\underset{2}{|}\\CHCH_3$$

$$|\\\underset{1}{CH_3}$$

3-ethyl-2-methylhexane (two substituents)

$$\overset{1}{CH_3}\overset{2}{CH_2}\overset{3}{CH}\overset{4}{CH}\overset{5}{CH_2}\overset{6}{CH_2}CH_3$$

$$|\\CHCH_3$$

$$|\\CH_3$$

not
3-isopropylhexane (one substituent)

Rule 6 is why 3-ethyl-2,6-dimethylheptane (one of the examples shown for Rule 4) cannot be named 5-isopropyl-2-methylheptane.

7. Names such as isopropyl, *sec*-butyl, and *tert*-butyl are acceptable substituent names in the IUPAC system of nomenclature, but systematic substituent names are preferable. Systematic substituent names are obtained by numbering the substituent starting at the carbon that is attached to the parent hydrocarbon. This means that the carbon that is attached to the parent hydrocarbon is always the number 1 carbon of the substituent. In a compound such as 4-(1-methylethyl)octane, the substituent name is in parentheses; the number inside the parentheses indicates a position on the substituent, whereas the number outside the parentheses indicates a position on the parent hydrocarbon.

$$CH_3CH_2CH_2CH_2\overset{}{CH}CH_2CH_2CH_3$$

$$\underset{1}{|}\overset{2}{}\\CHCH_3$$

$$|\\CH_3$$

4-isopropyloctane
or
4-(1-methylethyl)octane

$$CH_3CH_2CH_2CH_2\overset{}{CH}CH_2CH_2CH_2CH_3$$

$$\underset{1}{|}\overset{2}{}\overset{3}{}\\CH_2CHCH_3$$

$$|\\CH_3$$

5-isobutyldecane
or
5-(2-methylpropyl)decane

Some substituents have only a systematic name.

$$CH_2CH_2CH_3$$

$$CH_3CH_2CH_2CH_2\overset{}{CH}CH_2\overset{}{CH}CH_2CH_2CH_3$$

$$\overset{2}{}\overset{3}{}\\CH_3CHCHCH_3$$

$$\underset{1}{|}\\CH_3$$

6-(1,2-dimethylpropyl)-4-propyldecane

$$\overset{}{CH_3}\qquad\overset{}{CH_3}$$

$$|\qquad\qquad\underset{1}{|}$$

$$CH_3CHCHCH_2\overset{}{CH}CH_2\overset{}{C}HCH_3$$

$$|\qquad\qquad\overset{2}{}\overset{3}{}$$

$$CH_3\qquad CH_2CH_2CH_2CH_3$$

2,3-dimethyl-5-(2-methylpropyl)nonane
or
5-isobutyl-2,3-dimethylnonane

These rules will allow you to name thousands of alkanes, and eventually you will learn the additional rules necessary to name many other kinds of compounds. These rules are also important if you want to look up a compound in scientific literature because it ususally will be listed by its systematic name. Nevertheless, you must still learn common names because they have been in existence for so long and are so entrenched in chemists' vocabulary that they are widely used in scientific conversation and are often found in the literature.

Look at the systematic names shown for the isomeric hexanes and isomeric heptanes at the beginning of this chapter to make sure you understand how they were constructed.

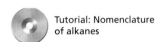
Tutorial: Nomenclature of alkanes

PROBLEM 2 ◆

Write the structure for each of the following compounds:

a. 2,3-dimethylhexane

b. 4-isopropyl-2,4,5-trimethylheptane

c. 4,4-diethyldecane

d. 2,2-dimethyl-4-propyloctane

e. 4-isobutyl-2,5-dimethyloctane

f. 4-(1,1-dimethylethyl)octane

PROBLEM 3 / SOLVED

a. Draw the 18 isomeric octanes.

b. Give each isomer its systematic name.

c. How many isomers have common names?

d. Which isomers contain an isopropyl group?

e. Which isomers contain a *sec*-butyl group?

f. Which isomers contain a *tert*-butyl group?

SOLUTION TO 3a   Start with the isomer with an 8-carbon continuous chain, then draw isomers with a 7-carbon continuous chain plus one methyl group, next isomers with a 6-carbon continuous chain plus two methyl groups or one ethyl group, then isomers with a 5-carbon continuous chain plus three methyl groups or one methyl group and one ethyl group, and finally a 4-carbon continuous chain with four methyl groups. (You will be able to tell whether or not you have drawn duplicate structures by your answers to **3b** because if two structures have the same systematic name, they are the same compound.)

PROBLEM 4◆

Give the systematic name for each of the following compounds:

a.
$$\begin{array}{cc} CH_3 & CH_3 \\ | & | \\ CH_3CH_2CHCH_2CCH_3 \\ & | \\ & CH_3 \end{array}$$

b. $CH_3CH_2C(CH_3)_3$

c.
$$\begin{array}{c} CH_3 \\ | \\ CH_3CHCH_2CH_2CHCH_3 \\ | \\ CH_2CH_3 \end{array}$$

d. $CH_3CH_2C(CH_2CH_3)_2CH_2CH_2CH_3$

e. $CH_3CH_2C(CH_2CH_3)_2CH(CH_3)CH(CH_2CH_3)_2$

f.
$$\begin{array}{c} CH_3CH_2CH_2CHCH_2CH_2CH_3 \\ | \\ CH_3CHCH_2CH_3 \end{array}$$

g.
$$\begin{array}{c} CH_3 \quad CH_2CH_2CH_3 \\ | \qquad | \\ CH_3C\!-\!CHCH_2CH_3 \\ | \\ CH_2CH_2CH_3 \end{array}$$

h.
$$\begin{array}{c} CH_3CH_2CH_2CH_2CHCH_2CH_2CH_3 \\ | \\ CH(CH_3)_2 \end{array}$$

2.3 NOMENCLATURE OF CYCLOALKANES

Cycloalkanes are alkanes with their carbons arranged in a ring. Because of the ring, a cycloalkane has two fewer hydrogens than an acyclic (noncyclic) alkane with the same number of carbons. This means that the general molecular formula for a cycloalkane is C_nH_{2n}. Cycloalkanes are named by adding the prefix "cyclo" to the alkane name that signifies the number of carbons in the ring.

$$\begin{array}{c} CH_2 \\ \diagup \ \diagdown \\ H_2C\!-\!CH_2 \end{array}$$
cyclopropane

$$\begin{array}{c} H_2C\!-\!CH_2 \\ | \qquad | \\ H_2C\!-\!CH_2 \end{array}$$
cyclobutane

$$\begin{array}{c} CH_2 \\ H_2C \diagup \ \diagdown CH_2 \\ | \qquad | \\ H_2C\!-\!CH_2 \end{array}$$
cyclopentane

$$\begin{array}{c} CH_2 \\ H_2C \diagup \ \diagdown CH_2 \\ | \qquad | \\ H_2C \diagdown \ \diagup CH_2 \\ CH_2 \end{array}$$
cyclohexane

Cycloalkanes are almost always written using **skeletal structures.** Skeletal structures show the carbon–carbon bonds as lines but do not show the carbons or the hydrogens bonded to carbons. Atoms other than carbon, and hydrogens bonded to atoms other than carbon are shown. Each vertex in a skeletal structure represents a carbon. It is understood that each carbon is bonded to the appropriate number of hydrogens to give it four bonds.

cyclopropane   cyclobutane   cyclopentane   cyclohexane

Acyclic molecules can also be represented by skeletal structures. In a skeletal structure, the carbon chains are represented by zigzag lines. Again, each vertex represents a carbon, and carbons are assumed to be present where a line begins or ends.

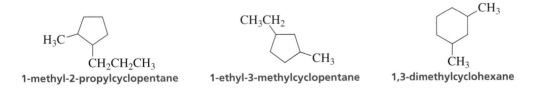

1-bromobutane   2-chlorohexane   3-methyl-4-propylheptane   6-ethyl-2,3-dimethylnonane

The rules for naming cycloalkanes resemble the rules for naming acyclic alkanes.

1. In the case of a cycloalkane with an attached alkyl substituent, the ring is the parent hydrocarbon unless the substituent has more carbons than the ring. In that case, the substituent is the parent hydrocarbon and the ring is named as a substituent. There is no need to number the position of a single substituent on a ring.

methylcyclopentane   ethylcyclohexane   1-cyclobutylpentane

2. If the ring has two different substituents, they are cited in alphabetical order and the number 1 position is given to the first cited substituent.

1-methyl-2-propylcyclopentane   1-ethyl-3-methylcyclopentane   1,3-dimethylcyclohexane

3. If there are more than two substituents on the ring, they are cited in alphabetical order. The substituent given the number 1 position is the one that results in a second substituent getting as low a number as possible. If two substituents have the same low numbers, the ring is numbered (either clockwise or counterclockwise) in the direction that gives the third substituent the lowest possible number. For example, the correct name of the following compound is 4-ethyl-2-methyl-1-propylcyclohexane, not 5-ethyl-1-methyl-2-propylcyclohexane.

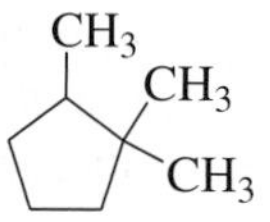

1,1,2-trimethylcyclopentane
not
1,2,2-trimethylcyclopentane
because 1 < 2
not
1,1,5-trimethylcyclopentane
because 2 < 5

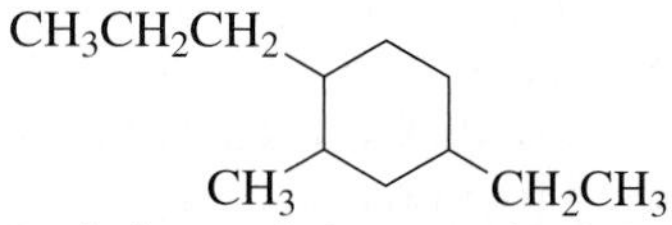

4-ethyl-2-methyl-1-propylcyclohexane
not
1-ethyl-3-methyl-4-propylcyclohexane
because 2 < 3
not
5-ethyl-1-methyl-2-propylcyclohexane
because 4 < 5

PROBLEM 5 ◆

Convert the following condensed structures into skeletal structures. (Remember that condensed structures show atoms but few, if any, bonds; skeletal structures show bonds but few, if any, atoms.

a. $CH_3CH_2CH_2CH_2CH_2CH_2OH$

b.
$\quad\quad\quad\; CH_3 \quad CH_3$
$\quad\quad\quad\;\; | \quad\quad\;\; |$
$CH_3CH_2CHCH_2CHCH_2CH_3$

c.
$\quad\;\; CH_3$
$\quad\;\; |$
$CH_3CHCH_2CH_2CHCH_3$
$\quad\quad\quad\quad\quad\quad\;\, |$
$\quad\quad\quad\quad\quad\quad\, Br$

d. $CH_3CH_2CH_2CH_2OCH_3$

PROBLEM 6 ◆

Give the systematic name for each of the following compounds:

a. 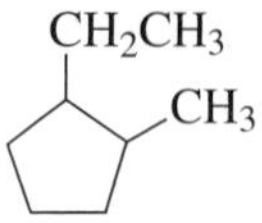

b. 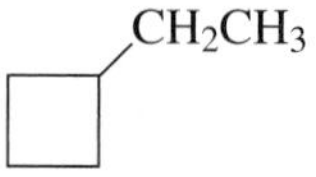

c. 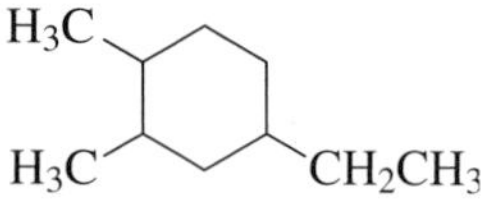

d. 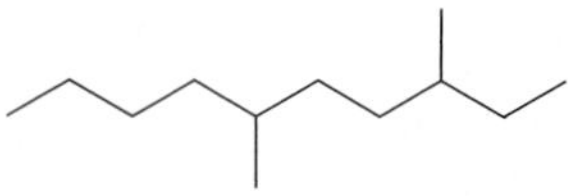

e. 

f. 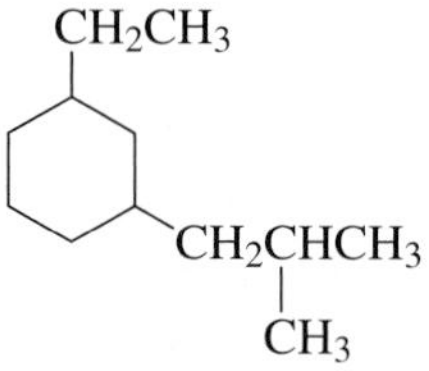

g. 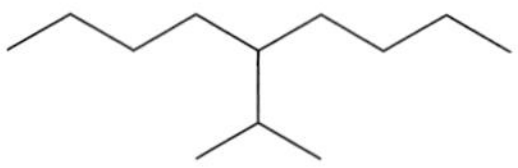

h. 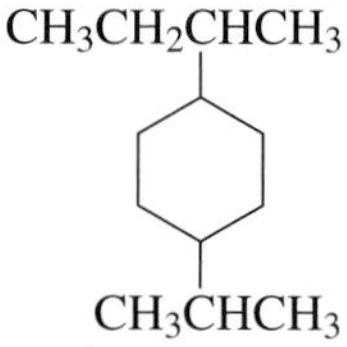

2.4 NOMENCLATURE OF ALKYL HALIDES

Alkyl halides are compounds in which a hydrogen of an alkane has been replaced by a halogen. Alkyl halides are classified as primary, secondary, or tertiary depending on the carbon to which the halogen is attached. The halogen is bonded to a primary carbon (a carbon bonded to one other carbon) in a **primary alkyl halide,** to a secondary carbon (a carbon bonded to two other carbons) in a **secondary alkyl**

halide, and to a tertiary carbon (a carbon bonded to three other carbons) in a **tertiary alkyl halide.** The nonbonding electrons on the halogens are generally not shown unless they are needed to draw your attention to some chemical property of the atom.

a primary carbon	a secondary carbon	a tertiary carbon

$$R—CH_2—Br$$

$$R—\underset{\underset{Br}{|}}{CH}—R$$

$$R—\underset{\underset{Br}{|}}{\overset{\overset{R}{|}}{C}}—R$$

a primary alkyl halide   a secondary alkyl halide   a tertiary alkyl halide

CH_3F
methyl fluoride

The common names of alkyl halides are obtained by citing the name of the alkyl group followed by the name of the halogen.

$$CH_3Cl$$
methyl chloride

$$CH_3CH_2F$$
ethyl fluoride

$$CH_3\underset{\underset{CH_3}{|}}{CHI}$$
isopropyl iodide

$$CH_3CH_2\underset{\underset{CH_3}{|}}{CHBr}$$
sec-butyl bromide

CH_3Cl
methyl chloride

In the IUPAC system, alkyl halides are named as substituted alkanes. The substituent prefix names for the halogens replace the "ine" ending in the name of the element with "o" (i.e., fluoro, chloro, bromo, iodo). Therefore, alkyl halides are often called haloalkanes.

$$CH_3CH_2\underset{\underset{Br}{|}}{\overset{\overset{CH_3}{|}}{CH}CH_2CH_2CHCH_3}$$
2-bromo-5-methylheptane

$$CH_3\underset{\underset{CH_3}{|}}{\overset{\overset{CH_3}{|}}{C}CH_2CH_2CH_2CH_2Cl}$$
1-chloro-5,5-dimethylhexane

CH_3Br
methyl bromide

4-bromo-2-chloro-1-methylcyclohexane

CH_3I
methyl iodide

PROBLEM 7◆

Give two names for each of the following compounds and tell whether each alkyl halide is primary, secondary, or tertiary.

a. $CH_3CH_2\underset{\underset{Cl}{|}}{CHCH_3}$

c.

b. $CH_3\underset{\underset{CH_3}{|}}{CHCH_2CH_2CH_2CH_2Cl}$

d. $CH_3\underset{\underset{F}{|}}{CHCH_3}$

PROBLEM 8◆

Draw **a–c** by substituting a chlorine for a hydrogen of methylcyclohexane. Name each of the alkyl halides.

a. primary alkyl halide

c. a tertiary alkyl halide

b. three different secondary alkyl halides

> **PROBLEM 9◆**
>
> Draw the structure and give the systematic name of a compound with a molecular formula C_5H_{12} that has:
>
> **a.** only primary and secondary hydrogens   **c.** one tertiary hydrogen
>
> **b.** only primary hydrogens

2.5 NOMENCLATURE OF ETHERS

Ethers are compounds in which an oxygen is bonded to two alkyl substituents. If the alkyl substituents are identical, the ether is a **symmetrical ether.** If the substituents are different, the ether is an **asymmetrical ether.**

R—O—R   R—O—R′

a symmetrical ether   an asymmetrical ether

Chemists sometimes neglect the prefix "di" when they name symmetrical ethers. Try not to make this oversight a habit.

The common name of an ether is obtained by citing the names of the two alkyl substituents (in alphabetical order) followed by the word "ether." The smallest ethers are almost always named by their common names.

$CH_3OCH_2CH_3$ $CH_3CH_2OCH_2CH_3$

ethyl methyl ether   diethyl ether
often called ethyl ether

$\overset{\displaystyle CH_3}{\underset{\displaystyle CH_3}{CH_3CHOCHCH_2CH_3}}$ $\overset{\displaystyle CH_3}{\underset{\displaystyle CH_3\quad CH_3}{CH_3CHCH_2OCCH_3}}$

sec-butyl isopropyl ether   tert-butyl isobutyl ether

$\underset{\displaystyle CH_3}{CH_3CHCH_2CH_2O-}\bigcirc$

cyclohexyl isopentyl ether

The IUPAC system names an ether as an alkane that has an RO substituent. The substituents are named by replacing the "yl" ending in the name of the alkyl substituent with "oxy."

dimethyl ether

$CH_3O—$ $CH_3CH_2O—$ $\underset{\displaystyle CH_3}{CH_3CHO—}$ $\underset{\displaystyle CH_3}{CH_3CH_2CHO—}$ $\overset{\displaystyle CH_3}{\underset{\displaystyle CH_3}{CH_3CO—}}$

methoxy   ethoxy   isopropoxy   sec-butoxy   tert-butoxy

$\underset{\displaystyle OCH_3}{CH_3CHCH_2CH_3}$ $\underset{\displaystyle CH_3}{CH_3CH_2CHCH_2CH_2OCH_2CH_3}$

2-methoxybutane   1-ethoxy-3-methylpentane

$\underset{\displaystyle CH_3}{CH_3CHOCH_2CH_2CH_2CH_2OCHCH_3}\;\underset{\displaystyle CH_3}{}$

1,4-diisopropoxybutane

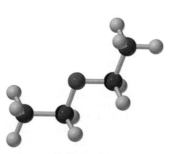

diethyl ether

PROBLEM 10◆

a. What is the systematic (IUPAC) name for each of the following ethers?

1. CH₃OCH₂CH₃

2. CH₃CH₂OCH₂CH₃

3. CH₃CH₂CH₂CH₂CHCH₂CH₂CH₃
              |
              OCH₃

4. CH₃CH₂CH₂OCH₂CH₂CH₂CH₃

5. CH₃CHOCHCH₂CH₂CH₃
              |
              CH₃   CH₃

(with CH₃ above)

6. CH₃CHOCH₂CH₂CHCH₃
          | |
          CH₃   CH₃

b. Do all of these ethers have common names?

c. What are their common names?

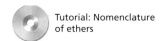

Tutorial: Nomenclature of ethers

Alcohols are compounds in which a hydrogen of an alkane has been replaced by an OH group. Alcohols are classified as **primary alcohols, secondary alcohols,** or **tertiary alcohols** depending on whether the OH group is bonded to a primary, secondary, or tertiary carbon. This is similar to the way alkyl halides are classified.

2.6 NOMENCLATURE OF ALCOHOLS

$$R-CH_2-OH \qquad R-CH-R \atop \qquad\qquad OH \qquad R-C-R \atop \qquad\qquad OH$$

a primary alcohol   a secondary alcohol   a tertiary alcohol

The common name of an alcohol is obtained by citing the name of the alkyl group to which the OH group is attached, followed by the word "alcohol."

CH₃CH₂OH   CH₃CH₂CH₂OH   CH₃CHOH   CH₃CCH₂OH
ethyl alcohol   propyl alcohol | |
                                            CH₃   CH₃
                                    isopropyl alcohol   neopentyl alcohol

(with CH₃ above neopentyl alcohol structure)

methyl alcohol

The **functional group** is the center of reactivity in a molecule. In an alcohol molecule, the OH is the functional group. The IUPAC system uses a suffix to denote certain functional groups. The systematic name of an alcohol, for example, is obtained by replacing the "e" at the end of the name of the parent hydrocarbon with the suffix "ol."

ethyl alcohol

CH₃OH   CH₃CH₂OH
methanol   ethanol

When necessary, the position of the functional group is indicated by a number immediately preceding the name of the alcohol, or immediately preceding the suffix. The most recently approved IUPAC names are those with the number immediately preceeding the suffix. However, names with the number preceding the name of the alcohol have been in use for a long time, so those are the ones most likely to

propyl alcohol

appear in the literature, on reagent bottles, and on standardized tests. Those will also appear most often in this book.

$$CH_3CH_2CHCH_2CH_3$$
$$|$$
$$OH$$

3-pentanol
or
pentan-3-ol

Use the following rules when naming a compound that has a functional group suffix:

1. The parent hydrocarbon is the longest continuous chain *containing the functional group.*

2. The parent hydrocarbon is numbered in the direction that gives the *functional group suffix the lowest possible number.*

$$\overset{1}{C}H_3\overset{2}{C}H\overset{3}{C}H_2\overset{4}{C}H_3$$
$$|$$
$$OH$$

2-butanol
or
butan-2-ol

$$\overset{5}{C}H_3\overset{4}{C}H_2\overset{3}{C}H_2\overset{2}{C}H\overset{1}{C}H_2OH$$
$$|$$
$$CH_2CH_3$$

2-ethyl-1-pentanol
or
2-ethylpentan-1-ol

$$CH_3CH_2CH_2CH_2O\overset{3}{C}H_2\overset{2}{C}H_2\overset{1}{C}H_2OH$$

3-butoxy-1-propanol
or
3-butoxypropan-1-ol

| The longest continuous chain has six carbons, but the longest continuous chain containing the OH functional group has five carbons so the compound is named as a pentanol. | The longest continuous chain has four carbons, but the longest continuous chain containing the OH functional group has three carbons, so the compound is named as a propanol. |

When there is only a substituent, the substituent gets the lowest number.

When there is only a functional group suffix, the functional group suffix gets the lowest number.

When there is a functional group suffix and a substituent, the functional group suffix gets the lowest number.

3. If there is a functional group suffix and a substituent, the functional group suffix gets the lowest possible number. Notice that a number is not needed to designate the position of a functional group suffix in a cyclic compound because it is assumed to be at the 1-position.

$$HO\overset{1}{C}H_2\overset{2}{C}H_2\overset{3}{C}H_2\overset{4}{C}H_2\overset{5}{C}H_2Br$$

5-bromo-1-pentanol

$$Cl\overset{4}{C}H_2\overset{3}{C}H_2\overset{2}{C}H\overset{1}{C}H_3$$
$$|$$
$$OH$$

4-chloro-2-butanol

3-methylcyclohexanol

4. If the same number for the functional group suffix is obtained in both directions, the chain is numbered in the direction that gives a substituent the lowest possible number.

$$CH_3CHCHCH_2CH_3$$
$$|\ \ |$$
$$Cl\ \ OH$$

2-chloro-3-pentanol
not
4-chloro-3-pentanol

$$CH_3CH_2CH_2CHCH_2CHCH_3$$
$$|\ \ \ \ \ \ |$$
$$OH\ \ \ \ CH_3$$

2-methyl-4-heptanol
not
6-methyl-4-heptanol

5. If there is more than one substituent, the substituents are cited in alphabetical order.

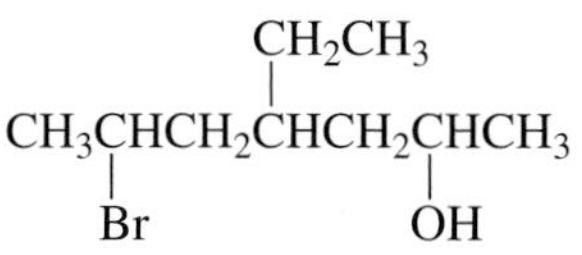

$$CH_3CHCH_2CHCH_2CHCH_3$$
with CH₂CH₃ on the third carbon, Br on first-shown carbon, OH on fifth-shown carbon
6-bromo-4-ethyl-2-heptanol

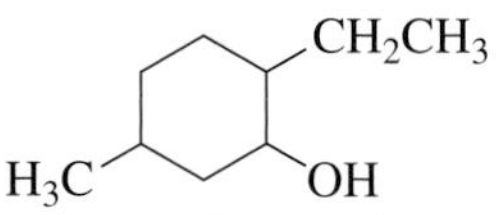

2-ethyl-5-methylcyclohexanol

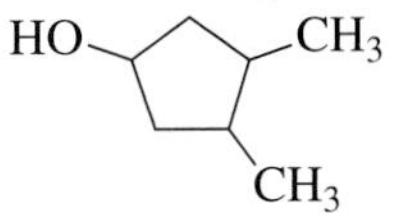

3,4-dimethylcyclopentanol

Remember that the name of a substituent is stated *before* the name of the parent hydrocarbon, and the functional group suffix is stated *after* the name of the parent hydrocarbon.

[Substituent] [Parent Hydrocarbon] [Functional Group Suffix]

PROBLEM 11

Write the structures of a homologous series of alcohols that have from one to six carbons, then give each of them a common name and a systematic name.

Tutorial: Nomenclature of alcohols

PROBLEM 12 ◆

Give each of the following compounds a systematic name, then indicate whether each is a primary, secondary, or tertiary alcohol:

a. $CH_3CH_2CH_2CH_2CH_2OH$

b. 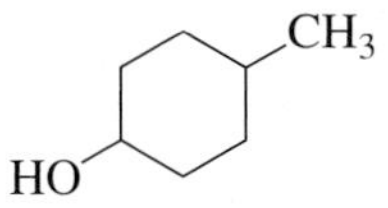 4-methylcyclohexanol structure (HO, CH₃)

c. 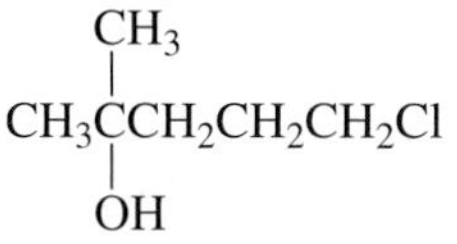 $CH_3\overset{CH_3}{\underset{OH}{C}}CH_2CH_2CH_2Cl$

d. 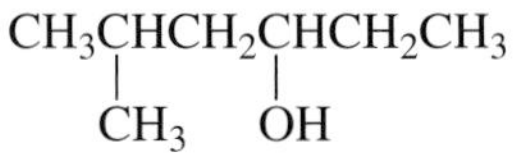 $CH_3\underset{CH_3}{C}HCH_2\underset{OH}{C}HCH_2CH_3$

e. 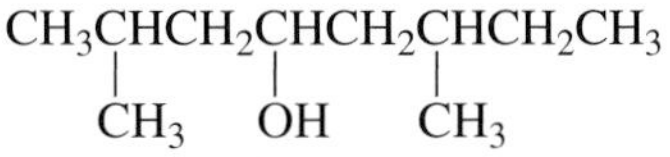 $CH_3\underset{CH_3}{C}HCH_2\underset{OH}{C}HCH_2\underset{CH_3}{C}HCH_2CH_3$

f. 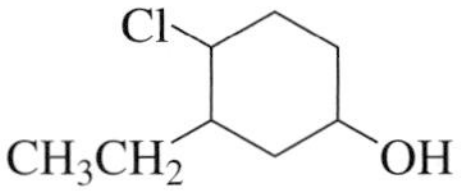 cyclohexane with Cl, CH₃CH₂, OH

PROBLEM 13 ◆

Write the structures of all the tertiary alcohols with molecular formula $C_6H_{14}O$. Give each one a systematic name.

Amines are compounds in which one or more of the hydrogens of ammonia have been replaced by alkyl groups. Smaller amines are characterized by their fishy odors. Fermented shark, for example, a traditional dish in Iceland, smells exactly like triethylamine. There are **primary amines, secondary amines,** and **tertiary amines.** The classification depends on how many alkyl groups are bonded to the nitrogen. Primary amines have one alkyl group bonded to the nitrogen, secondary amines have two, and tertiary amines have three.

2.7 NOMENCLATURE OF AMINES

$R-NH_2$
a primary amine

$R-NH-R$
a secondary amine

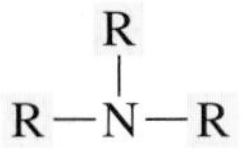

$R-\underset{R}{\overset{R}{N}}-R$
a tertiary amine

Notice that the number of alkyl groups attached to the nitrogen determines whether an amine is primary, secondary, or tertiary. For an alkyl halide or an alcohol, on the other hand, the number of alkyl groups attached to the carbon to which the halogen or the OH is bonded determines the classification.

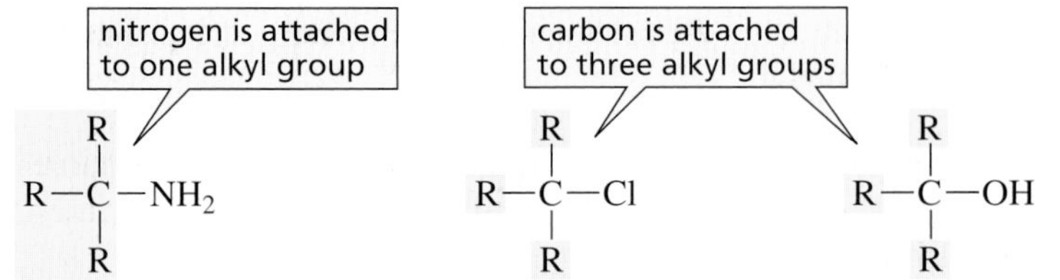

a primary amine   a tertiary alkyl chloride   a tertiary alcohol

The common name of an amine is obtained by citing the names of the alkyl groups bonded to the nitrogen in alphabetical order followed by "amine." The entire name is written as one word (unlike the names of alcohols, ethers, and alkyl halides, in which "alcohol," "ether," or "halide" are separate words).

CH_3NH_2
methylamine

$CH_3NHCH_2CH_2CH_3$
methylpropylamine

$CH_3CH_2NHCH_2CH_3$
diethylamine

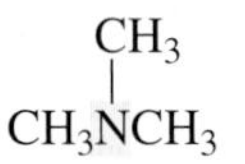

CH_3NCH_3
trimethylamine

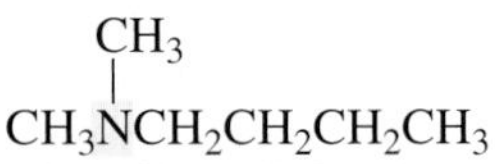

$CH_3NCH_2CH_2CH_2CH_3$
butyldimethylamine

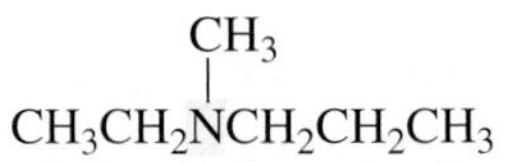

$CH_3CH_2NCH_2CH_2CH_3$
ethylmethylpropylamine

The IUPAC system uses a suffix to denote the amine functional group. This method is similar to the way in which alcohols are named. The "e" at the end of the name of the parent hydrocarbon is replaced by "amine." A number is used to identify the carbon to which the nitrogen is attached. The number can appear before the name of the parent hydrocarbon or before "amine." The name of any alkyl group bonded to nitrogen is preceded by an "*N*" (in italic) to indicate that the group is bonded to a nitrogen rather than to a carbon.

$CH_3CH_2CH_2CH_2NH_2$
1-butanamine
or
butan-1-amine

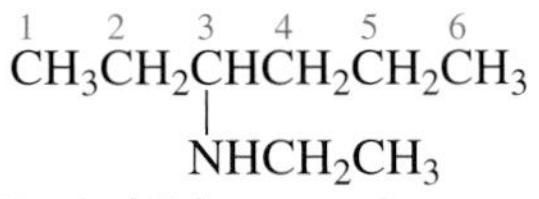

$CH_3CH_2CHCH_2CH_2CH_3$
$NHCH_2CH_3$
N-ethyl-3-hexanamine
or
N-ethylhexan-3-amine

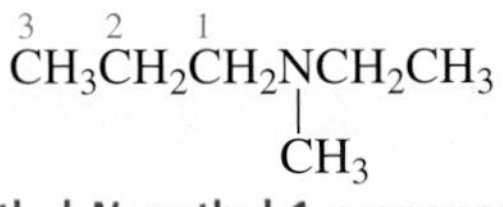

$CH_3CH_2CH_2NCH_2CH_3$
CH_3
N-ethyl-*N*-methyl-1-propanamine
or
N-ethyl-*N*-methylpropan-1-amine

All substituents (regardless of whether they are attached to the nitrogen or to the parent hydrocarbon) are listed in alphabetical order. The chain is numbered such that the functional group suffix gets the lowest possible number.

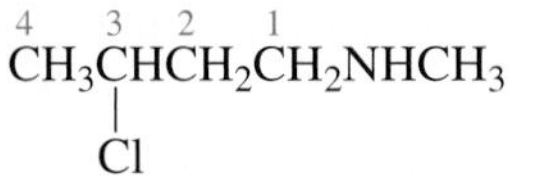

$CH_3CHCH_2CH_2NHCH_3$
Cl
3-chloro-*N*-methyl-1-butanamine

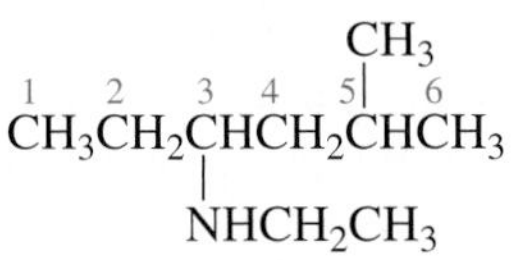

$CH_3CH_2CHCH_2CHCH_3$
CH_3
$NHCH_2CH_3$
N-ethyl-5-methyl-3-hexanamine

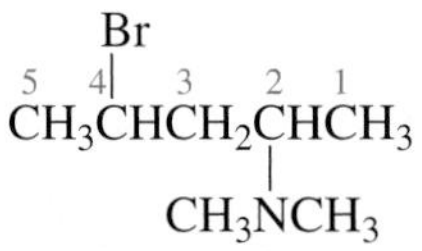

$CH_3CHCH_2CHCH_3$
Br
CH_3NCH_3
4-bromo-*N*,*N*-dimethyl-2-pentanamine

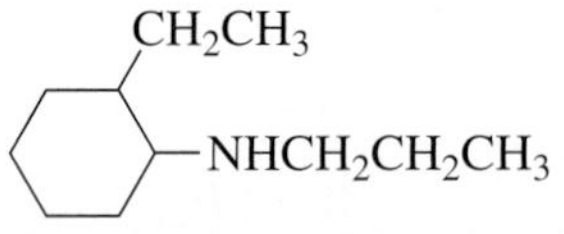

2-ethyl-*N*-propylcyclohexanamine

Nitrogen compounds with four alkyl groups bonded to the nitrogen—thereby placing a positive charge on the nitrogen—are called **quaternary ammonium salts.** Their names are obtained by citing the names of the alkyl groups in alphabetical order followed by "ammonium," then the name of the counter-ion as a separate word.

$$CH_3-\overset{\displaystyle CH_3}{\underset{\displaystyle CH_3}{\overset{|}{\underset{|}{N^{\pm}}}}}-CH_3 \quad HO^-$$

tetramethylammonium hydroxide

$$CH_3CH_2CH_2-\overset{\displaystyle CH_3}{\underset{\displaystyle CH_2CH_3}{\overset{|}{\underset{|}{N^{\pm}}}}}-CH_3 \quad Cl^-$$

ethyldimethylpropylammonium chloride

Table 2.3 summarizes the ways in which alkyl halides, ethers, alcohols, and amines are named.

TABLE 2.3	Nomenclature Summary	
	Systematic Name	**Common Name**
alkyl halide	substituted alkane	alkyl group to which halogen is attached plus *halide*
	CH_3Br bromomethane	CH_3Br methyl bromide
	CH_3CH_2Cl chloroethane	CH_3CH_2Cl ethyl chloride
ether	substituted alkane	alkyl groups attached to oxygen plus *ether*
	CH_3OCH_3 methoxymethane	CH_3OCH_3 dimethyl ether
	$CH_3CH_2OCH_3$ methoxyethane	$CH_3CH_2OCH_3$ ethyl methyl ether
alcohol	functional group suffix is *ol*	alkyl group to which OH is attached plus *alcohol*
	CH_3OH methanol	CH_3OH methyl alcohol
	CH_3CH_2OH ethanol	CH_3CH_2OH ethyl alcohol
amine	functional group suffix is *amine*	alkyl groups attached to N plus *amine*
	$CH_3CH_2NH_2$ ethanamine	$CH_3CH_2NH_2$ ethylamine
	$CH_3CH_2CH_2NHCH_3$ *N*-methyl-1-propanamine	$CH_3CH_2CH_2NHCH_3$ methylpropylamine

PROBLEM 14◆

Give common and systematic names for each of the following compounds.

a. $CH_3CH_2CH_2CH_2CH_2CH_2NH_2$

b. $CH_3CH_2CH_2NHCH_2CH_2CH_2CH_3$

c. $CH_3\underset{\displaystyle CH_3}{\overset{|}{CH}}CH_2NH\underset{\displaystyle CH_3}{\overset{|}{CH}}CH_2CH_3$

d. $CH_3CH_2CH_2\underset{\displaystyle CH_2CH_3}{\overset{|}{N}}CH_2CH_3$

e. cyclohexyl–NH_2

PROBLEM 15◆

Write the structure for each of the following compounds.

a. 2-methyl-*N*-propyl-1-propanamine

b. *N*-ethylethanamine

c. 5-methylhexan-1-amine

d. methyldipropylamine

e. *N,N*-dimethylpentan-3-amine

f. cyclohexylethylmethylamine

Tutorial: Summary of systematic nomenclature

PROBLEM 16◆

Give the systematic name and the common name (for those that have common names) for each of the following compounds. Indicate whether the amines are primary, secondary, or tertiary.

a. $CH_3CHCH_2CH_2CH_2CH_2CH_2NH_2$
   |
   CH_3

b. $CH_3CH_2CH_2NHCH_2CH_2CHCH_3$
   |
   CH_3

c. $(CH_3CH_2)_2NCH_3$

d.

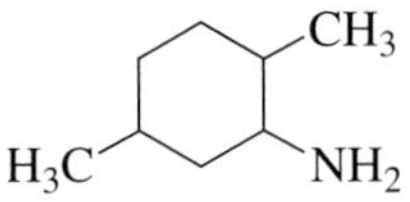

2.8
STRUCTURES OF ALKYL HALIDES, ALCOHOLS, ETHERS, AND AMINES

The carbon–halogen bond in an alkyl halide is formed from the overlap of an sp^3 orbital of carbon with a p orbital of the halogen (Section 1.13). Fluorine uses a $2p$ orbital, chlorine uses a $3p$ orbital, bromine uses a $4p$ orbital, and iodine uses a $5p$ orbital. Consequently, the carbon–halogen bond becomes longer and weaker as the atomic weight of the halogen increases (Table 2.4). Notice that this is the same trend shown by the hydrogen–halogen bond (Table 1.6).

An alcohol has the same geometry as water (Section 1.11). In fact, an alcohol molecule can be thought of as a water molecule with an alkyl group in place of one of the hydrogens. As in water, the oxygen atom in an alcohol is sp^3 hybridized. One of the sp^3 orbitals of oxygen overlaps an sp^3 orbital of a carbon, one sp^3 or-

TABLE 2.4 Carbon–Halogen Bond Lengths and Bond Strengths

	Orbital interactions		Bond lengths	Bond strength kcal/mol	$(DH°)$ kJ/mol
$H_3C—F$			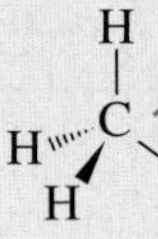 1.39 Å	108	451
$H_3C—Cl$			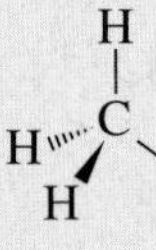 1.78 Å	84	350
$H_3C—Br$			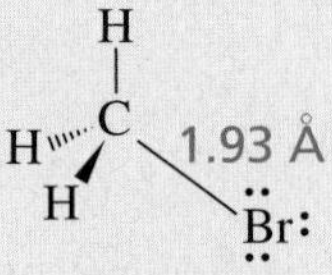 1.93 Å	70	294
$H_3C—I$			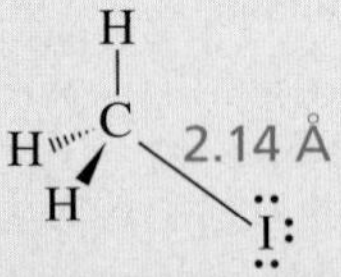 2.14 Å	57	239

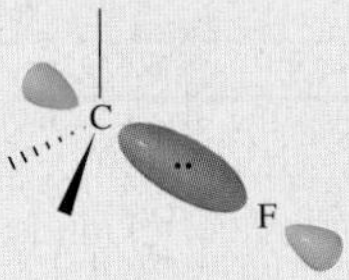

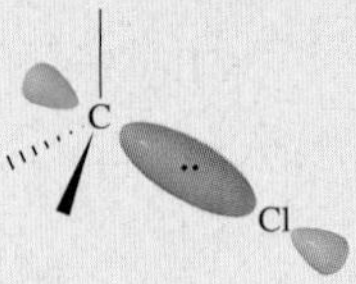

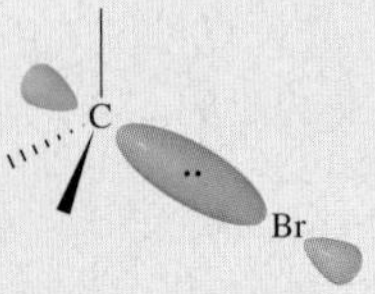

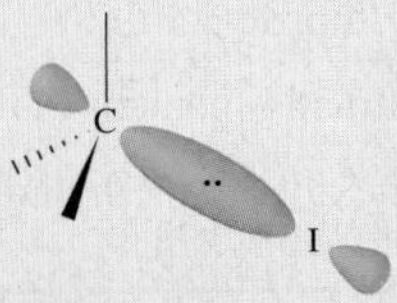

bital overlaps the *s* orbital of a hydrogen, and the other two *sp*³ orbitals each contain a pair of nonbonding electrons.

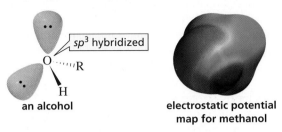

an alcohol

electrostatic potential
map for methanol

Ethers also have the same geometry as water. An ether molecule can be thought of as a water molecule with alkyl groups in place of both hydrogens.

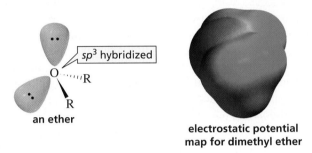

an ether

electrostatic potential
map for dimethyl ether

An amine has the same geometry as ammonia (Section 1.12). One, two, or three hydrogens may be replaced by alkyl groups. The number of hydrogens replaced by alkyl groups determines whether the amine is primary, secondary, or tertiary (Section 2.7).

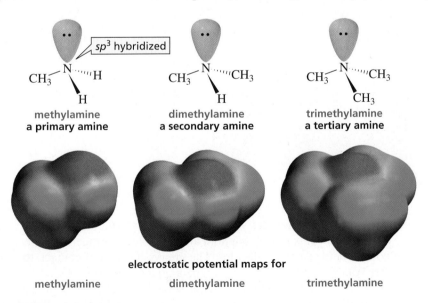

methylamine
a primary amine

dimethylamine
a secondary amine

trimethylamine
a tertiary amine

electrostatic potential maps for

methylamine   dimethylamine   trimethylamine

PROBLEM 17◆

Predict the approximate size of each of the following angles:

a. the C—O—C bond angle in an ether

c. the C—O—H bond angle in an alcohol

b. the C—N—C bond angle in a secondary amine

d. the C—N—C bond angle in a quaternary ammonium salt

2.9
PHYSICAL PROPERTIES OF ALKANES, ALKYL HALIDES, ALCOHOLS, ETHERS, AND AMINES

Johannes Diderik van der Waals (1837–1923) *was a Dutch physicist. He was born in Leiden, the son of a carpenter, and was largely self-taught when he entered the University of Leiden, from where he received a Ph.D. He was a professor of physics at the University of Amsterdam from 1877 to 1903. He won the 1910 Nobel Prize for his research on the gaseous and liquid states of matter.*

Fritz Wolfgang London (1900–1954) *was born in Germany. He left Germany when Hitler rose to power and, after spending some time in England and France, became a professor at Duke University in North Carolina, where he remained for the rest of his life.*

Boiling Points

The **boiling point (bp)** of a compound is the temperature at which the liquid form of the compound becomes a gas (vaporize). In other words, it is the temperature at which the compound's vapor pressure equals the surrounding pressure. In order for a compound to vaporize, the forces that hold the molecules close to each other in the liquid must be overcome. This means that the boiling point of a compound depends on the strength of the attractive forces between the individual molecules. If the molecules are held together by strong forces, it will take a lot of energy to pull the molecules away from each other and the compound will have a high boiling point. If, however, the molecules are held together by weak forces, only a small amount of energy will be needed to pull the molecules away from each other and the compound will have a low boiling point.

Relatively weak forces hold alkane molecules together. Alkanes contain only carbon and hydrogen atoms. Because the electronegativities of carbon and hydrogen are similar, the bonds in alkanes are nonpolar. Consequently, there are no significant partial charges on any of the atoms in an alkane. Alkanes are neutral molecules.

However, it is only the average charge distribution over the alkane molecule that is neutral. Electrons are continuously moving, so at any instant one side of the molecule can have slightly more electron density than the other side, giving the molecule a temporary dipole.

The temporary dipole in one molecule can induce a temporary dipole in a nearby molecule. As a result, the negative side of one molecule is adjacent to the positive side of another molecule, as shown in Figure 2.1. Because the dipoles in the molecules are induced, the interactions between the molecules are called **induced dipole–induced dipole interactions.** The molecules of an alkane are held together by these induced dipole–induced dipole interactions known as **van der Waals forces** or **London forces.** In order for an alkane to boil, these van der Waals forces must be overcome.

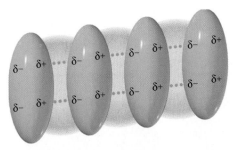

◀ **Figure 2.1**
Van der Waals forces are induced dipole–induced dipole interactions.

Van der Waals forces are the weakest of all the intermolecular attractions. The size of the van der Waals forces that hold alkane molecules together depends on the area of contact between the molecules. The greater the area of contact, the stronger the van der Waals forces and the greater the amount of energy needed to overcome these forces. If you look at the homologous series of alkanes in Table 2.1, you will see that the boiling points of alkanes increase as their size increases. This occurs because each additional methylene group increases the area of contact between the molecules. Because the four smallest alkanes have boiling points below room temperature (room temperature is about 25 °C), they exist as gases at room temperature. Pentane (bp = 36.1 °C) is the smallest alkane that is a liquid at room temperature.

Because the strength of van der Waals forces depends on the area of contact between molecules, branching in a compound lowers its boiling point because it reduces the area of contact. A branched compound has a more compact, nearly spherical shape. If you think of the unbranched alkane pentane as a cigar and branched neopentane as a tennis ball, you can see that branching decreases the area of contact between molecules. Two cigars make contact over a greater area than do two tennis balls. Thus, if two alkanes have the same molecular weight, the more highly branched alkane will have a lower boiling point.

$$CH_3CH_2CH_2CH_2CH_3$$
pentane
bp = 36.1 °C

$$CH_3CHCH_2CH_3$$
$$|$$
$$CH_3$$
isopentane
bp = 27.9 °C

$$CH_3$$
$$|$$
$$CH_3CCH_3$$
$$|$$
$$CH_3$$
neopentane
bp = 9.5 °C

The boiling points for any homologous series of compounds increase as their molecular weights increase because of the increase in van der Waals forces. So the boiling points of a homologous series of ethers, alkyl halides, alcohols, and amines increase with increasing molecular weight (see Appendix I). The boiling points of these compounds, however, are also affected by the polar character of C—N, C—O, and C—X(X = halogen) bonds because nitrogen, oxygen, and halogens are more electronegative than the carbon to which they are bonded.

$$R—\overset{|}{\underset{|}{C}}—\overset{\delta+\ \ \delta-}{Z} \qquad Z = N, O, F, Cl, or Br$$

The magnitude of the charge differential between the two bonded atoms is indicated by the bond dipole moment (Section 1.3).

> **The dipole moment of a bond is equal to the magnitude of the charge on one of the bonded atoms times the distance between the bonded atoms.**

$$H_3C—NH_2$$
0.2 D

$$H_3C—O—CH_3$$
0.7 D

$$H_3C—I$$
1.2 D

$$H_3C—OH$$
0.7 D

$$H_3C—F$$
1.6 D

$$H_3C—Br$$
1.4 D

$$H_3C—Cl$$
1.5 D

Molecules with dipole moments are attracted to one another because they can align themselves in such a way that the positive end of one dipole is adjacent to the negative end of another dipole. These electrostatic attractive forces, called **dipole–dipole interactions,** are stronger than van der Waals forces but not as strong as ionic or covalent bonds.

Ethers generally have higher boiling points than alkanes of comparable molecular weight because both van der Waals forces and dipole–dipole interactions must be overcome for an ether to boil (Table 2.5).

> **More extensive tables of physical properties can be found in Appendix I.**

cyclopentane
bp = 49.3 °C

tetrahydrofuran
bp = 65 °C

TABLE 2.5 Comparative Boiling Points of Alkanes, Ethers, Alcohols, and Amines (°C)			
$CH_3CH_2CH_3$ −42.1	CH_3OCH_3 −23.7	CH_3CH_2OH 78	$CH_3CH_2NH_2$ 16.6
$CH_3CH_2CH_2CH_3$ −0.5	$CH_3OCH_2CH_3$ 10.8	$CH_3CH_2CH_2OH$ 97.4	$CH_3CH_2CH_2NH_2$ 47.8
$CH_3CH_2CH_2CH_2CH_3$ 36.1	$CH_3CH_2OCH_2CH_3$ 34.5	$CH_3CH_2CH_2CH_2OH$ 117.3	$CH_3CH_2CH_2CH_2NH_2$ 77.8

Alcohols have much higher boiling points than alkanes or ethers of comparable molecular weight (Table 2.5) because, in addition to van der Waals forces and the dipole–dipole interactions of the carbon–oxygen bond, alcohols can form **hydrogen bonds.** A hydrogen bond is a special kind of dipole–dipole interaction that occurs between a hydrogen bonded to an oxygen, a nitrogen, or a fluorine, and the nonbonding electrons on an oxygen, nitrogen, or fluorine in another molecule.

The length of the covalent bond between oxygen and hydrogen is 0.96 Å. The hydrogen bond between an oxygen of one molecule and a hydrogen of another molecule is almost twice as long (1.69–1.79 Å). Although a hydrogen bond is not nearly as strong as an oxygen–hydrogen covalent bond, it is stronger than other dipole–dipole interactions. The strongest hydrogen bonds are linear—the two electronegative atoms and the hydrogen between them lie in a straight line.

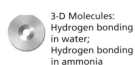
3-D Molecules: Hydrogen bonding in water; Hydrogen bonding in ammonia

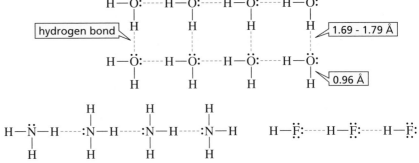

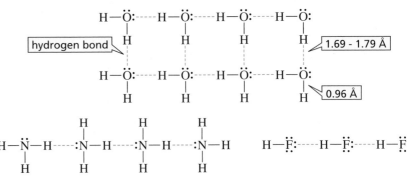

Although each individual hydrogen bond is weak (requiring about 5 kcal/mol to break), there are many such bonds holding alcohol molecules together. The extra energy required to break these hydrogen bonds is the reason alcohols have much higher boiling points than either alkanes or ethers with similar molecular weights.

The boiling point of water illustrates the dramatic effect hydrogen bonding has on boiling points. Water has a molecular weight of 18 and a boiling point of 100 °C. The alkane nearest in size is methane, with a molecular weight of 16. Methane boils at −167.7 °C.

Primary and secondary amines also form hydrogen bonds, so these amines have higher boiling points than alkanes with similar molecular weights. Nitrogen is not as electronegative as oxygen, however, which means that the hydrogen bonds between amine molecules are weaker than the hydrogen bonds between alcohol molecules. Amines, therefore, have lower boiling points than alcohols with similar molecular weights (Table 2.5).

Because primary amines have two N—H bonds, hydrogen bonding is more significant for primary amines than for secondary amines. Tertiary amines cannot form hydrogen bonds with each other because they do not have a hydrogen attached to the nitrogen. Consequently, if you compare amines with the same molecular weight and similar structures, primary amines have higher boiling points than secondary amines, and secondary amines have higher boiling points than tertiary amines.

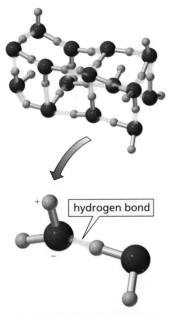
hydrogen bonding in water

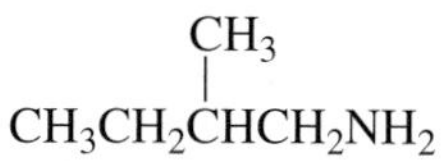

$$CH_3$$
$$|$$
$$CH_3CH_2CHCH_2NH_2$$

a primary amine
bp = 97 °C

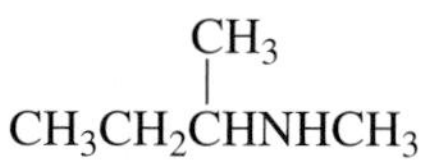

$$CH_3$$
$$|$$
$$CH_3CH_2CHNHCH_3$$

a secondary amine
bp = 84 °C

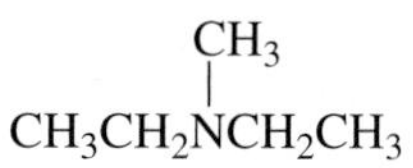

$$CH_3$$
$$|$$
$$CH_3CH_2NCH_2CH_3$$

a tertiary amine
bp = 65 °C

PROBLEM-SOLVING STRATEGY

a. Which of the following compounds (**1–3**) will form hydrogen bonds between its molecules?

b. Which will form hydrogen bonds with a solvent such as ethanol?

1. $CH_3CH_2CH_2OH$ **2.** $CH_3CH_2CH_2SH$ **3.** $CH_3OCH_2CH_3$

In solving this type of question, start by defining the kind of compound that will do what is being asked.

a. A hydrogen bond forms when a hydrogen that is attached to an O, N, or F of one molecule interacts with a lone pair of electrons on an O, N, or F of another molecule. Therefore, a compound that will form hydrogen bonds with itself must have a hydrogen bonded to an O, N, or F. Only compound **1** will be able to form hydrogen bonds with itself.

b. Ethanol has an H bonded to an O, so it will be able to form hydrogen bonds with a compound that has a lone pair of electrons on an O, N, or F. Compounds **1** and **3** will be able to form hydrogen bonds with ethanol.

Now continue on to Problem 18.

PROBLEM 18◆

a. Which of the following compounds (**1–6**) will form hydrogen bonds between its molecules?

b. Which will form hydrogen bonds with a solvent such as ethanol?

1. $CH_3CH_2CH_2COOH$ **4.** $CH_3CH_2CH_2NHCH_3$

2. $CH_3CH_2N(CH_3)_2$ **5.** $CH_3CH_2OCH_2CH_2OH$

3. $CH_3CH_2CH_2CH_2Br$ **6.** $CH_3CH_2CH_2CH_2F$

PROBLEM 19

Explain why:

a. H_2O has a higher boiling point than CH_3OH (65 °C).

b. H_2O has a higher boiling point than NH_3 (−33 °C).

c. H_2O has a higher boiling point than HF (20 °C).

PROBLEM 20

Explain why the dipole moments of methyl fluoride (1.56 D) and methyl chloride (1.51 D) are similar even though fluorine is considerably more electronegative than chlorine.

PROBLEM 21◆

List the following compounds in order of decreasing boiling point:

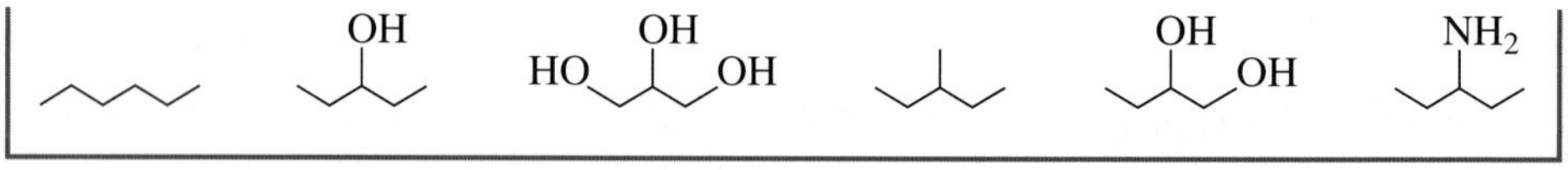

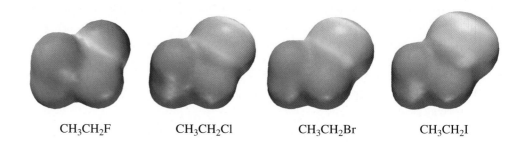

Both van der Waals forces and dipole–dipole interactions must be overcome for an alkyl halide to boil. As the halogen atom increases in size, the size of its electron cloud increases. This increases both the van der Waals contact area and the polarizability of the electron cloud.

CH₃CH₂F   CH₃CH₂Cl   CH₃CH₂Br   CH₃CH₂I

Polarizability indicates the ease with which an electron cloud can be distorted. The larger the atom, the more loosely it holds the electrons in its outermost shell, and the more they can be distorted. The more polarizable the atom, the stronger the van der Walls interactions. Therefore, an alkyl fluoride has a lower boiling point than an alkyl chloride with the same alkyl group. Similarly, alkyl chlorides have lower boiling points than alkyl bromides, which have lower boiling points than alkyl iodides (Table 2.6).

TABLE 2.6   Comparative Boiling Points of Alkanes and Alkyl Halides (°C)

			Y		
	H	**F**	**Cl**	**Br**	**I**
CH₃—Y	−161.7	−78.4	−24.2	3.6	42.4
CH₃CH₂—Y	−88.6	−37.7	12.3	38.4	72.3
CH₃CH₂CH₂—Y	−42.1	−2.5	46.6	71.0	102.5
CH₃CH₂CH₂CH₂—Y	−0.5	32.5	78.4	101.6	130.5
CH₃CH₂CH₂CH₂CH₂—Y	36.1	62.8	107.8	129.6	157.0

PROBLEM 22◆

List the following compounds in order of decreasing boiling point:

a. CH₃CH₂CH₂CH₂CH₂CH₂Br   CH₃CH₂CH₂CH₂Br   CH₃CH₂CH₂CH₂CH₂Br

b.
$$\begin{matrix} & CH_3 & CH_3 \\ & | & | \\ CH_3C & — & CCH_3 \\ & | & | \\ & CH_3 & CH_3 \end{matrix}$$ CH₃CH₂CH₂CH₂CH₂CH₂CH₂CH₃

CH₃CHCH₂CH₂CH₂CH₂CH₃
　　|
　　CH₃

c. CH₃CH₂CH₂CH₂CH₃   CH₃CH₂CH₂CH₂OH   CH₃CH₂CH₂CH₂Cl

CH₃CH₂CH₂CH₂CH₂OH

Melting Points

The **melting point (mp)** is the temperature at which a solid is converted into a liquid. If you examine the melting points of the alkanes in Table 2.1, you will see that the melting points increase (with a few exceptions) in a homologous series as the molecular weight increases. The increase in melting point is less regular than the increase in boiling point because **packing** influences the melting point of a compound. Packing is a property that determines how well the individual molecules in a solid fit together in the crystal lattice. The tighter the fit, the more energy required to break the lattice and melt the compound.

In Figure 2.2, you can see that the melting points of alkanes with even numbers of carbon atoms fall on a smooth curve (red line). The melting points of alkanes with odd numbers of carbon atoms also fall on a smooth curve (green line). The two curves do not overlap, however, because the methyl groups at the end of alkane chains with an odd number of carbons can avoid one another only by increasing the distance between their chains. In other words, alkanes with an odd number of carbon atoms pack less tightly than alkanes with an even number of carbon atoms. This decreases the intermolecular attractions between alkanes with odd numbers of carbon atoms, which decreases their melting points.

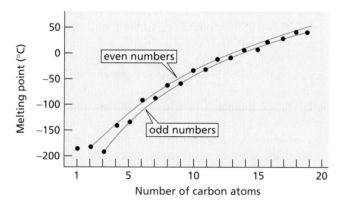

◀ **Figure 2.2**
Melting points of alkanes. Alkanes with even numbers of carbon atoms fall on a melting point curve that is higher than the melting point curve for alkanes with odd numbers of carbon atoms.

Solubility

Oil (nonpolar) and water (polar) don't mix. The general rule that explains **solubility** on the basis of the polarity of molecules is that "like dissolves like." In other words, polar compounds dissolve in polar solvents and nonpolar compounds dissolve in nonpolar solvents. This is generally true because a polar solvent such as water has partial charges that can interact with the partial charges on a polar compound. The negative poles of the solvent molecules surround the positive pole of the polar solute, and the positive poles of the solvent molecules surround the negative pole of the polar solute. Clustering of the solvent molecules around the solute molecules separates solute molecules from each other, which is what makes them soluble. The interaction between a solvent and a molecule or an ion that is dissolved in that solvent is called **solvation.**

Because nonpolar compounds have no net charge, polar solvents are not attracted to them. In order for a nonpolar molecule to dissolve in a polar solvent such as water, the nonpolar molecule would have to push the water molecules apart, disrupting their hydrogen bonding. Hydrogen bonding is so strong that it excludes the nonpolar compound. In contrast, nonpolar solutes dissolve in nonpolar solvents because of the van der Waals interactions between solvent and solute molecules.

Oil from a 70,000 ton oil spill in 1996 off the coast of Wales.

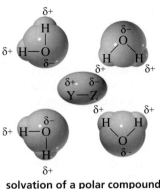

solvation of a polar compound
$$\overset{\delta+}{}\,\,\overset{\delta-}{}$$
(Y — Z) by water

Alkanes are nonpolar molecules, which causes them to be soluble in nonpolar solvents and insoluble in polar solvents such as water. The densities of alkanes (Table 2.1) increase with increasing molecular weight. But even a 30-carbon alkane such as triacontane (density at 20 °C = 0.8097 g/mL) is less dense than water (density at 20 °C = 0.9982 g/mL). This means that a mixture of an alkane and water will separate into two distinct layers with the less dense alkane floating on top. The Alaskan oil spill of 1989 and the Persian Gulf spill of 1991 are large-scale examples of this phenomenon. (Crude oil is primarily a mixture of alkanes.)

An alcohol has both a nonpolar alkyl group and a polar OH group. Therefore, is an alcohol molecule nonpolar or polar overall? Is it soluble in a nonpolar solvent or is it soluble in water? The answer depends on the size of the alkyl group. As the alkyl group increases in size, it becomes a more significant fraction of the alcohol molecule and the compound becomes less and less soluble in water. In other words, the molecule becomes more and more like an alkane. Four carbons tend to be the dividing line at room temperature. Alcohols with fewer than four carbons are soluble in water, but alcohols with more than four carbons are insoluble in water. In other words, an OH group can drag about three or four carbons into solution in water.

The four-carbon dividing line is only an approximate guide because the solubility of an alcohol also depends on the structure of the alkyl group. Branched alkyl groups are more water soluble than nonbranched alkyl groups with the same number of carbons because branching minimizes the contact surface of the nonpolar portion of the molecule. So, *tert*-butyl alcohol is more soluble than *n*-butyl alcohol in water.

The oxygen atom of an ether can drag only about three carbons into solution in water (Table 2.7).

Low-molecular-weight amines are soluble in water because amines can form hydrogen bonds with water. Comparing amines with the same number of carbons, primary amines are more soluble than secondary amines because primary amines have two hydrogens that can engage in hydrogen bonding. Tertiary amines, on the other

TABLE 2.7	Solubilities of Ethers in Water	
2 C's	CH_3OCH_3	soluble
3 C's	$CH_3OCH_2CH_3$	soluble
4 C's	$CH_3CH_2OCH_2CH_3$	slightly soluble (10 g/100 g H_2O)
5 C's	$CH_3CH_2OCH_2CH_2CH_3$	minimally soluble (1.0 g/100 g H_2O)
6 C's	$CH_3CH_2CH_2OCH_2CH_2CH_3$	insoluble (0.25 g/100 g H_2O)

hand, have lone-pair electrons that can accept hydrogen bonds but do not have hydrogens to donate for hydrogen bonds, so tertiary amines are less soluble in water than secondary amines with the same number of carbons.

Alkyl halides have some polar character, but only the alkyl fluorides have an atom that can form a hydrogen bond with water. This means that alkyl fluorides are the most water soluble of the alkyl halides. The other alkyl halides are less soluble in water than ethers or alcohols with the same number of carbons (Table 2.8).

TABLE 2.8 Solubilities of Alkyl Halides in Water			
CH_3F very soluble	CH_3Cl soluble	CH_3Br slightly soluble	CH_3I slightly soluble
CH_3CH_2F soluble	CH_3CH_2Cl slightly soluble	CH_3CH_2Br slightly soluble	CH_3CH_2I slightly soluble
$CH_3CH_2CH_2F$ slightly soluble	$CH_3CH_2CH_2Cl$ slightly soluble	$CH_3CH_2CH_2Br$ slightly soluble	$CH_3CH_2CH_2I$ slightly soluble
$CH_3CH_2CH_2CH_2F$ insoluble	$CH_3CH_2CH_2CH_2Cl$ insoluble	$CH_3CH_2CH_2CH_2Br$ insoluble	$CH_3CH_2CH_2CH_2I$ insoluble

PROBLEM 23◆

Rank the following groups of compounds in order of decreasing solubility in water.

a. $CH_3CH_2CH_2OH$ $CH_3CH_2CH_2CH_2Cl$ $CH_3CH_2CH_2CH_2OH$
$HOCH_2CH_2CH_2OH$

b. CH_3 NH_2 OH

PROBLEM 24◆

In which of the following solvents would cyclohexane have the lowest solubility: pentanol, diethyl ether, ethanol, or hexane?

A carbon–carbon single bond (a sigma bond) is formed when an sp^3 orbital of one carbon overlaps an sp^3 orbital of a second carbon. Because sigma bonds are cylindrically symmetrical (they are symmetrical about an imaginary line connecting the centers of the two atoms joined by the sigma bond), rotation about a carbon–carbon single bond is possible without any change in the amount of orbital overlap (Figure 2.3). The different spatial arrangements of the atoms that result from rotation about a single bond are called **conformations.** A specific conformation is called a **conformational isomer** or a **conformer.**

When rotation occurs about the carbon–carbon bond of ethane, two extreme conformations can result—a staggered conformation and an eclipsed conformation. An infinite number of conformations between these two extremes is also possible.

Compounds are three-dimensional, but we are limited to a two-dimensional sheet of paper when showing their structures. Perspective formulas, sawhorse projections, and Newman projections are methods chemists commonly use to represent on paper the three-dimensional spatial arrangements of the atoms that result from rotation about a sigma (σ) bond. In a **perspective formula,** solid lines are used for bonds that

2.10 CONFORMATIONS OF ALKANES: ROTATION ABOUT CARBON–CARBON BONDS

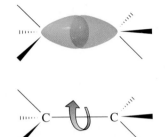

▲ Figure 2.3
A carbon–carbon bond is formed by the overlap of cylindrically symmetrical sp^3 orbitals. Therefore, rotation about the bond can occur without changing the amount of orbital overlap.

Melvin S. Newman (1908–1993) was born in New York. He received a Ph.D. from Yale University in 1932 and was a professor of chemistry at Ohio State University from 1936 to 1973.

3-D Molecule: Conformations of ethane

lie in the plane of the paper, solid wedges are used for bonds protruding out from the plane of the paper, and hatched wedges are used for bonds protruding into the paper. In a **sawhorse projection,** you are looking at the carbon–carbon bond from an oblique angle. In a **Newman projection,** you are looking down the length of a particular carbon–carbon bond. The carbon in front is represented by the point at which three bonds intersect, and the carbon in back is represented by a circle. The three lines emanating from each of the carbons represent the carbon's other three bonds. We will use Newman projections because they are easy to draw and they do a good job of representing the spatial relationships of the substituents on the two carbon atoms.

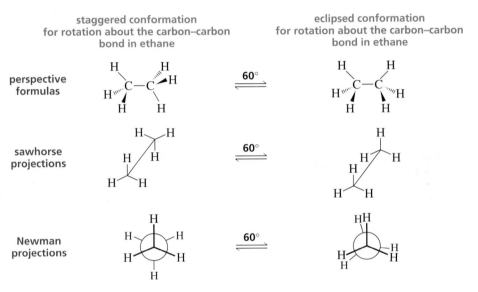

The electrons in a carbon–hydrogen bond will repel the electrons in another carbon–hydrogen bond if the bonds get too close to each other. The **staggered conformation,** therefore, is the most stable conformation because the carbon–hydrogen bonds are as far away from each other as possible. The **eclipsed conformation** is the least stable conformation because in no other conformation are the carbon–hydrogen bonds as close to one another. The extra energy of the eclipsed conformation is called torsional strain. **Torsional strain** is the name given to the repulsion felt by the bonding electrons of one substituent as they pass close to the bonding electrons of another substituent. The investigation of the various conformations of a compound and their relative stabilities is called **conformational analysis.**

Rotation about a carbon–carbon single bond is not completely free because of the energy difference between the staggered and eclipsed conformers. The eclipsed conformer is higher in energy, so an energy barrier must be overcome when rotation about the carbon–carbon bond occurs (Figure 2.4). But the barrier is small enough in ethane (2.9 kcal/mol or 12 kJ/mol) to allow the conformers to interconvert millions of times per second at room temperature. Figure 2.4 shows the potential energies of all the conformers of ethane obtained during one complete 360° rotation. Notice that the staggered conformers are at energy minima, whereas the eclipsed conformers are at energy maxima.

Butane has three carbon–carbon single bonds, and the molecule can rotate about each of them. For example, a staggered and an eclipsed conformer can be drawn for rotation about the C-1—C-2 bond. The carbon in the foreground in a Newman projection has the lower number.

the C-2—C-3 bond

$$\overset{1}{CH_3}\!-\!\overset{2}{CH_2}\!-\!\overset{3}{CH_2}\!-\!\overset{4}{CH_3}$$

butane

the C-1—C-2 bond

the C-3—C-4 bond

staggered conformation for rotation
about the C-1—C-2 bond in butane

eclipsed conformation for rotation
about the C-1—C-2 bond in butane

Although the staggered conformers resulting from rotation about the C-1—C-2 bond in butane all have the same energy, the staggered conformers resulting from rotation about the C-2—C-3 bond do not have the same energy. The staggered conformers for rotation about the C-2—C-3 bond in butane are shown on page 92. Conformer D, in which the two methyl groups are as far apart as possible, is more stable than the other two staggered conformers (B and F). The most stable of the staggered conformers (D) is called the **anti conformer,** and the other two staggered conformers (B and F) are called **gauche conformers** (*anti* is Greek for "opposite of"; *gauche* is French for "left"). In the anti conformer, the largest substituents are opposite each other; in the gauche conformer, they are adjacent. The two gauche conformers have the same energy, but each is less stable than the anti conformer.

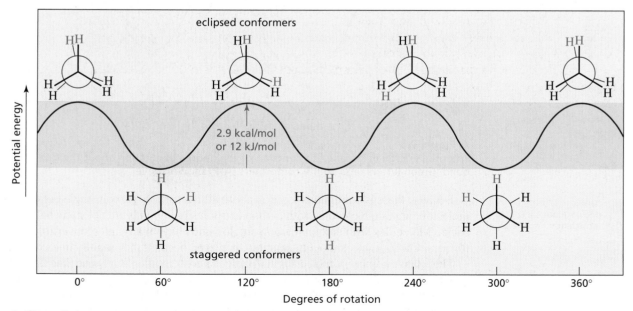

▲ **Figure 2.4**
Potential energy of ethane as a function of the angle of rotation about the carbon–carbon bond.

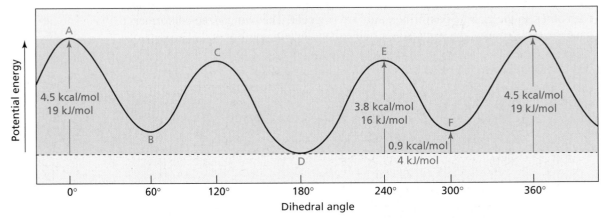

Anti and gauche conformers do not have the same energy because of steric strain. **Steric strain** (or **steric hindrance**) is the strain put on a molecule when atoms or groups are too close to each other, which results in repulsion between the electron clouds of the interacting atoms or groups. There is more steric strain in the gauche conformer than in the anti conformer because the two methyl groups are closer together in the gauche conformer. This is called a **gauche interaction.**

The eclipsed conformers resulting from rotation about the C-2—C-3 bond in butane also have different energies. The eclipsed conformer in which the two methyl groups are closest to each other (A) is less stable than the eclipsed conformers in which they are farther apart (C and E). The energies of the conformers obtained from rotation about the C-2—C-3 bond of butane are shown in Figure 2.5. [The dihedral angle is the angle between the CH_3—C—C and C—C—CH_3 planes. Therefore, the conformer in which one methyl group stands directly in front of the other (the least stable conformer) has a dihedral angle of 0°.] All the eclipsed conformers have both torsional and steric strain—torsional strain due to bond–bond repulsion, and steric strain due to the closeness of the eclipsing groups. In general, steric strain in molecules increases as the size of the eclipsing groups increases.

▲ Figure 2.5
Potential energy of butane as a function of the degree of rotation about the C-2—C-3 bond. Green letters refer to the conformers (A–F) shown above.

Because there is continuous rotation about all the carbon–carbon single bonds in a molecule, organic molecules with carbon–carbon single bonds are not static balls and sticks. Molecules with carbon–carbon single bonds have many interconvertible conformers. Conformers cannot be separated, however, because they rapidly interconvert.

The relative number of molecules in a particular conformation at any one time depends on the stability of the conformation—the more stable the conformation, the greater the fraction of molecules that will be in that conformation. Most molecules, therefore, are in staggered conformations, and more molecules are in an anti conformation than in a gauche conformation. The preference for a staggered conformation causes carbon chains to orient themselves in a zigzag fashion, as shown by the ball-and-stick model of decane.

ball-and-stick model of decane

PROBLEM 25

a. Draw all the staggered and eclipsed conformers that result from rotation about the C-2—C-3 bond of pentane.

b. Draw a potential energy diagram for rotation of the C-2—C-3 bond of pentane through 360°, starting with the least stable conformer.

PROBLEM 26◆

Using Newman projections, draw the most stable conformer for the following:

a. 3-methylpentane, considering rotation about the C-2—C-3 bond.

b. 3-methylhexane, considering rotation about the C-3—C-4 bond.

c. 3,3-dimethylhexane, considering rotation about the C-3—C-4 bond.

3-D Molecule:
Decane

Early chemists observed that cyclic compounds found in nature generally had five- or six-membered rings. Compounds with three- and four-membered rings were found much less frequently. This observation suggested that compounds with three- and four-membered rings were not as stable as compounds with five- or six-membered rings.

The German chemist Adolf von Baeyer first proposed in 1885 that the instability of three- and four-membered rings was due to angle strain. We know that an sp^3 hybridized carbon has ideal orbital angles of 109.5° (Section 1.7). Baeyer suggested that the stability of a cycloalkane could be predicted by determining how close the bond angle of a planar cycloalkane is to the optimal tetrahedral bond angle of 109.5°. The angles in a regular triangle are 60°. The bond angles in cyclopropane, therefore, are compressed from the desired tetrahedral angle of 109.5° to 60°. The deviation of the bond angle from the desired bond angle causes the strain known as **angle strain.**

The angle strain in a three-membered ring can be appreciated by looking at the orbitals that overlap to form the sigma (σ) bonds in cyclopropane (Figure 2.6). Normal σ bonds are formed by the overlap of two sp^3 orbitals that point directly toward

**2.11
CYCLOALKANES:
RING STRAIN**

a.

b.

good overlap
strong bond

poor overlap
weak bond

◀ **Figure 2.6**
(a) Overlap of sp^3 orbitals in a normal σ bond. (b) Overlap of sp^3 orbitals in cyclopropane.

banana bonds

each other. In cyclopropane, overlapping orbitals cannot point directly toward each other. Therefore, orbital overlap is less effective than in a normal carbon–carbon bond. The less effective orbital overlap, which causes the carbon–carbon bond to be weaker than a normal carbon–carbon bond, is the cause of angle strain. Because the carbon–carbon bonds in cyclopropane have shapes that resemble bananas, they are sometimes called **banana bonds.**

In addition to angle strain, three-membered rings also have torsional strain because all the adjacent carbon–hydrogen bonds are eclipsed.

The bond angles in planar cyclobutane would have to be compressed from 109.5° to 90°, the bond angle associated with a planar four-membered ring. Planar cyclobutane would be expected to have less angle strain than cyclopropane because the bond angles in cyclobutane are only 19.5° away from the normal tetrahedral bond angle.

HIGHLY STRAINED HYDROCARBONS

Organic chemists have been able to synthesize some highly strained cyclic hydrocarbons such as bicyclo[1.1.0]butane, cubane, and prismane[1]. Philip Eaton, the first to synthesize cubane, has recently synthesized octanitrocubane, which is expected to be the most powerful explosive known.[2]

bicyclo[1.1.0]butane

cubane

prismane

[1]Bicyclo[1.1.0]butane was synthesized by David Lemal, Fredric Menger, and George Clark at the University of Wisconsin (*Journal of the American Chemical Society* 1963, *85,* 2529). Cubane was synthesized by Philip Eaton and Thomas Cole, Jr. at the University of Chicago (*Journal of the American Chemical Society* 1964, *86,* 3157). Prismane was synthesized by Thomas Katz and Nancy Acton at Columbia University (*Journal of the American Chemical Society* 1973, *95,* 2738). [2] Mao-Xi Zhang, Philip Eaton, and Richard Gilardi, Angew. Chem. Int. Ed., 2000, *39 (2),* 401.

cyclopropane

Baeyer predicted that a five-membered ring compound would be the most stable of the cycloalkanes because its bond angles (108°) are closest to the tetrahedral bond angle. He predicted that six-membered ring compounds with bond angles of 120° would be less stable and, as the number of sides in the cycloalkanes increases, their stability would decrease.

"planar" cyclopentane
bond angles = 108°

"planar" cyclohexane
bond angles = 120°

"planar" cycloheptane
bond angles = 128.6°

cyclobutane

3-D Molecules:
Cyclopropane;
Cyclobutane

Contrary to what Baeyer predicted, however, cyclohexane is more stable than cyclopentane. Furthermore, cyclic compounds do not become less and less stable as the number of sides increases. The mistake that Baeyer made was to assume that all cyclic compounds are planar. Because three points define a plane, the carbons of cyclopropane must lie in a plane. The other cycloalkanes, however, are not planar. Cyclic compounds twist and bend in order to achieve a structure that minimizes the three different kinds of strain that can destabilize a cyclic compound.

1. *Angle strain* is the strain induced in a molecule when the bond angles are different from the desired tetrahedral bond angle of 109.5°.

2. *Torsional strain* is caused by repulsion of the bonding electrons of one substituent with the bonding electrons of a nearby substituent.

3. *Steric strain* is caused by atoms or groups of atoms approaching each other too closely.

Although planar cyclobutane has less angle strain than cyclopropane, it has more torsional strain because it has eight pairs of eclipsed hydrogens compared with the six pairs of cyclopropane. So cyclobutane is not planar—it is bent. One of its methylene groups is bent at an angle of about 25° from the plane of the other three methylene groups. This increases the angle strain, but the increase is more than compensated for by the decreased torsional strain that results when the adjacent hydrogens are no longer as eclipsed as they would be in a planar structure.

If cyclopentane were flat, as Baeyer had predicted, there would be essentially no angle strain but there would be 10 pairs of adjacent hydrogens that would experience considerable torsional strain. Instead, cyclopentane puckers so that the hydrogens become nearly staggered. In the process, cyclopentane acquires some angle strain. The puckered form of cyclopentane is called the envelope conformation because the shape resembles a squarish envelope with the flap up.

cyclopentane

PROBLEM 27◆

The bond angles of a regular polygon with n sides are equal to

$$180° - \frac{360°}{n}$$

a. What are the bond angles in a regular octagon?   **b.** In a regular nonagon?

3-D Molecule:
Cyclopentane

The cyclic compounds most commonly found in nature contain six-membered rings because six-membered rings can exist in a conformation that is almost completely free of strain. This conformation is called the **chair conformation** (Figure 2.7). In

2.12
CONFORMATIONS
OF CYCLOHEXANE

chair conformer of
cyclohexane

Newman projection of
the chair conformer

Figure 2.7 ▶
The chair conformer of cyclohexane and the Newman projection of the chair conformer, showing that all the bonds are staggered.

**ball-and-stick model of the
chair conformer of cyclohexane**

the chair conformation of cyclohexane all the bond angles are 111°, which is very close to the tetrahedral bond angle (109.5°), and all the adjacent carbon–hydrogen bonds are staggered. The chair conformation is such an important conformation that you should learn how to draw it.

1. Draw two parallel lines of the same length, slanted upward. Both lines should start at the same height.

2. Connect the tops of the lines with a V; the left-hand side of the V should be slightly longer than the right-hand side. Connect the bottoms of the lines with an inverted V; the lines of the V and the inverted V should be parallel. This completes the framework of the six-membered ring.

3. Each carbon has an axial and an equatorial bond. The **axial bonds** are vertical and alternate above and below the ring. The one on the uppermost carbon is up, the next is down, the next is up, and so on.

4. The **equatorial bonds** (blue circles) point outward from the ring. If the axial bond is up, the equatorial bond on the same carbon is below the plane perpendicular to the bottom of the axial bond. If the axial bond is down, the equatorial bond is above the plane perpendicular to the top of the axial bond.

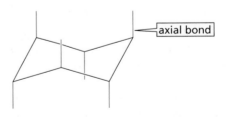

Notice that each equatorial bond is parallel to a ring bond two carbons over.

Remember that cyclohexane is viewed on edge. The lower bonds of the ring are in front and the upper bonds of the ring are in back.

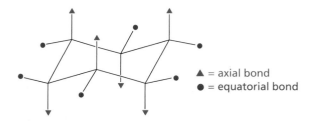

▲ = axial bond
● = equatorial bond

PROBLEM 28

Draw 1,2,3,4,5,6-hexamethylcyclohexane with:

a. all the methyl groups in axial positions.

b. all the methyl groups in equatorial positions.

If we assume that cyclohexane is completely free of strain, we can calculate the total strain energy (angle strain + torsional strain + steric strain) of the other cycloalkanes. Taking the heat of formation of cyclohexane (Table 2.9) and dividing by 6 for its six CH_2 groups gives us a value of -4.92 kcal/mol for a "strainless" CH_2 group ($-29.5/6 = -4.92$). (The heat of formation is the heat change when a compound is formed from its elements under standard conditions.) We can now calculate the heat of formation of a "strainless" cycloalkane by multiplying the number of CH_2 groups in its ring by -4.92 kcal/mol. The total strain in the compound is the difference between the "strainless" heat of formation and the actual heat of formation (Table 2.9). For example, cyclopentane has a "strainless" heat of formation of $(5)(-4.92) = -24.6$ kcal/mol.

TABLE 2.9 Heats of Formation and Total Strain Energies of Cycloalkanes

	Heat of formation		"Strainless" heat of formation		Total strain energy	
	(kcal/mol)	(kJ/mol)	(kcal/mol)	(kJ/mol)	(kcal/mol)	(kJ/mol)
cylopropane	+12.7	53.1	−14.6	−61.1	27.3	114.2
cyclobutane	+6.8	28.5	−19.7	−82.4	26.5	110.9
cyclopentane	−18.4	−77.0	−24.6	−102.9	6.2	25.9
cyclohexane	−29.5	−123.4	−29.5	−123.4	0	0
cycloheptane	−28.2	−118.0	−34.4	−143.9	6.2	25.9
cyclooctane	−29.7	−124.3	−39.4	−164.8	9.7	40.6
cyclononane	−31.7	−132.6	−44.3	−185.4	12.6	52.7
cyclodecane	−36.9	−154.4	−49.2	−205.9	12.3	51.5
cycloundecane	−42.9	−179.5	−54.1	−226.4	11.2	46.9

Because its actual heat of formation is -18.4 kcal/mol, cyclopentane has a total strain energy of 6.2 kcal/mol $[-18.4 - (-24.6) = 6.2]$.

PROBLEM 29

Calculate the total strain energy of cycloheptane.

As a result of the ease of rotation about its carbon–carbon single bonds, cyclohexane rapidly interconverts between two stable chair conformations. This interconversion is known as **ring-flip** (Figure 2.8). When the two chair conformers interconvert, bonds that are equatorial in one chair conformer become axial in the other chair conformer.

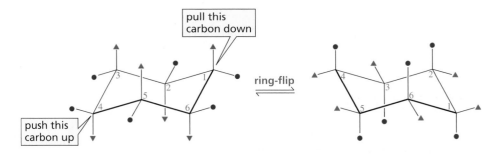

Figure 2.8 ▶
The bonds that are axial in one chair conformer are equatorial in the other chair conformer. The bonds that are equatorial in one chair conformer are axial in the other chair conformer.

Bonds that are equatorial in one chair conformer become axial in the other chair conformer.

Cyclohexane can also exist in a **boat conformation** shown in Figure 2.9. Like the chair conformer, the boat conformer is free of angle strain. However, the boat conformer is not as stable as the chair conformer because some of the bonds in the boat conformer are eclipsed, giving it torsional strain. The boat conformer is further destabilized by the close proximity of the **flagpole hydrogens** (the hydrogens at the "bow" and "stern" of the boat), which causes steric strain.

The conformations that cyclohexane can assume when interconverting from one chair conformer to the other are shown in Figure 2.10. To convert from the boat conformer to one of the chair conformers, one of the top-most carbons of the boat conformer must be pulled down so it becomes the bottom-most carbon. When the top-most carbon is pulled down just a little, the **twist-boat** (or **skew-boat**) **conformer** is obtained. The twist-boat conformer is more stable than the boat conformer because the flagpole hydrogens have moved away from each other, thus reducing the steric strain somewhat. When the top-most carbon is pulled down to the point where it is in the same plane as the sides of the boat, the very unstable **half-chair conformer** is obtained. Pulling the top-most carbon down farther produces the *chair conformer.* The graph in Figure 2.10 shows the energy of a cyclohexane molecule as it interconverts from one chair conformer to the other; the energy barrier for interconversion is 10.8 kcal/mol (45.2 kJ/mol). From this value, it can be calculated that cyclohexane

3-D Molecule:
Boat cyclohexane

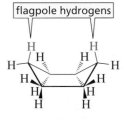
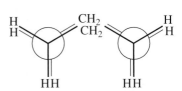

boat conformer of cyclohexane   **Newman projection of the boat conformer**   **ball-and-stick model of the boat conformer of cyclohexane**

Go to the Web site for better representations of the conformers of cyclohexane.

▲ **Figure 2.9**
The boat conformer of cyclohexane and the Newman projection of the boat conformer, showing that some of the bonds are eclipsed.

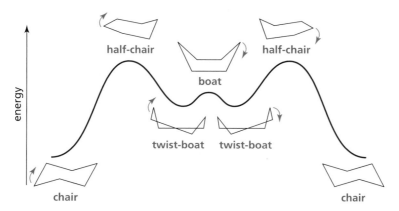

◀ **Figure 2.10**
Diagram showing the conformers of cyclohexane (and their relative energies) as one chair conformer interconverts to the other chair conformer.

undergoes 10^5 ring-flips per second at room temperature. In other words, the two chair conformers are in rapid equilibrium.

Because the chair conformers are the most stable of the conformers, at any instant more molecules of cyclohexane are in chair conformations than in any other conformation. It has been calculated that, for every thousand molecules of cyclohexane in the chair conformation, no more than two molecules are in the next most stable conformation—the twist boat.

Build a model of cyclohexane and convert it from one chair conformer to the other. To do this, pull the top-most carbon down and push the bottom-most carbon up.

Unlike cyclohexane, which has two equivalent chair conformers, the two chair conformers of a monosubstituted cyclohexane such as methylcyclohexane are not equivalent. The methyl substituent is in an equatorial position in one conformer and in an axial position in the other conformer (Figure 2.11) because substituents that are equatorial in one chair conformer are axial in the other chair conformer (Figure 2.8).

The chair conformer with the methyl substituent in an equatorial position is more stable because a substituent has more room and, therefore, fewer steric interactions when it is in an equatorial position. When the methyl group is in an equatorial position, it is anti to the C-3 and C-5 carbons (Figure 2.12). Therefore, the substituent extends into space away from the rest of the molecule.

In contrast, when the methyl group is in an axial position, it is gauche to the C-3 and C-5 carbons (Figure 2.13).

2.13 CONFORMATIONS OF MONOSUBSTITUED CYCLOHEXANES

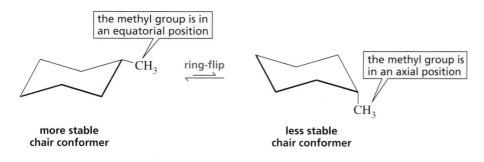

the methyl group is in an equatorial position

ring-flip

the methyl group is in an axial position

more stable chair conformer

less stable chair conformer

◀ **Figure 2.11**
A substituent is in the equatorial position in one chair conformer and in the axial position in the other chair conformer. The conformer with the substituent in the equatorial position is more stable.

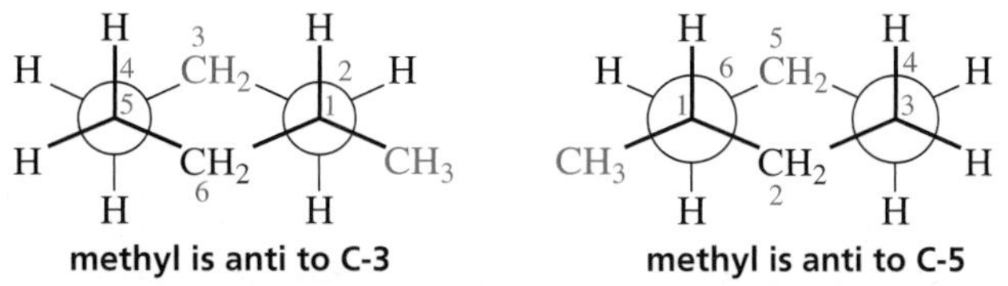

◀ **Figure 2.12**
An equatorial substituent on the C-1 carbon is anti to the C-3 and C-5 carbons.

methyl is anti to C-3   **methyl is anti to C-5**

Figure 2.13 ▶
An axial substituent on the C-1 carbon is gauche to the C-3 and C-5 carbons.

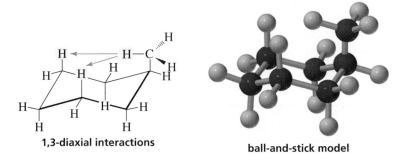

methyl is gauche to C-3   methyl is gauche to C-5

As a result, there are unfavorable steric interactions between the axial methyl group and both the axial substituent on C-3 and the axial substituent on C-5 (in this case hydrogens). In other words, the three axial bonds on the same side of the ring are parallel to each other, so any axial substituent will be too close to the axial substituents on the other two carbons. Because the interacting substituents are on 1,3-positions relative to each other, these unfavorable steric interactions are called **1,3-diaxial interactions.**

Build a model of methylcyclo-hexane and convert it from one chair conformer to the other.

3-D Molecule:
Chair conformers of methylcyclohexane

1,3-diaxial interactions   ball-and-stick model

The gauche conformer of butane and the axial conformer of methylcyclohexane are compared in Figure 2.14. Notice that the gauche interaction is the same in both—it is between a methyl group and a hydrogen bonded to a carbon gauche to the methyl group (Figure 2.14).

In Section 2.10, we saw that the gauche interaction between the methyl groups of butane caused the gauche conformer to be 0.9 kcal/mol (3.8 kJ/mole) less stable than the anti conformer. Because there are two such gauche interactions in the chair conformer of methylcyclohexane when the methyl group is in an axial position, this chair conformer is 1.8 kcal/mol (7.5 kJ/mole) less stable than the chair conformer with the methyl group in the equatorial position.

If you take a few minutes to build models, you will see that a substituent has more room if it is in an equatorial position than if it is in an axial position. Therefore, at any instant more monosubstituted cyclohexane molecules will be in the chair conformer with the substituent in the equatorial position than in the chair conformer with the substituent in the axial position. The relative amounts of the two chair conformers depend on the substituent (Table 2.10). The substituent with the greater bulk in the area of the 1,3-diaxial hydrogens will have a greater preference for the equa-

Figure 2.14 ▶
The steric strain of gauche butane is the same as the steric strain between an axial methyl group and an axial hydrogen.

gauche butane   axial
methylcyclohexane

TABLE 2.10   Equilibrium Constants for Several Monosubstituted Cyclohexanes at 25 °C

Substituent	Axial $\overset{K_{eq}}{\rightleftharpoons}$ Equatorial	Substituent	Axial $\overset{K_{eq}}{\rightleftharpoons}$ Equatorial
H	1	F	1.5
CH_3	18	Cl	2.4
CH_3CH_2	23		
$CH_3\overset{\displaystyle CH_3}{\underset{}{CH}}$	38	Br	2.2
		I	2.2
$CH_3\overset{\displaystyle CH_3}{\underset{\displaystyle CH_3}{C}}$	4000	HO	5.4

torial position because it will have stronger 1,3-diaxial interactions. For example, the equilibrium constant (K_{eq}) for the conformers of methylcyclohexane indicates that 95% of methylcyclohexane molecules have the methyl group in the equatorial position at 25 °C (see the following calculation). In the case of *tert*-butylcyclohexane, where the 1,3-diaxial interactions are even more destabilizing because a *tert*-butyl group is larger than a methyl group, more than 99.9% of the molecules have the *tert*-butyl group in the equatorial position.

$$K_{eq} = \frac{[\text{equatorial conformer}]}{[\text{axial conformer}]} = \frac{18}{1}$$

$$\% \text{ of equatorial conformer} = \frac{[\text{equatorial conformer}]}{[\text{equatorial conformer}] + [\text{axial conformer}]} \times 100$$

$$\% \text{ of equatorial conformer} = \frac{18}{18 + 1} \times 100 = 95\%$$

PROBLEM 30

Bromine is a larger atom than chlorine, but the equilibrium constants in Table 2.10 indicate that a chloro substituent has a greater preference for the equatorial position. Suggest an explanation.

If there are two substituents on a cyclohexane ring, both substituents have to be taken into account when determining which of the two chair conformers is the more stable. Let's start by looking at 1,4-dimethylcyclohexane. First of all, there are two different dimethylcyclohexanes. One has both methyl substituents on the *same side* of the cyclohexane ring; it is the **cis isomer** (*cis* is Latin for "on this side"). The other has the two methyl substituents on *opposite sides* of the ring; it is the **trans isomer** (*trans* is Latin for "across"). *cis*-1,4-Dimethylcyclohexane and *trans*-1,4-dimethylcyclohexane are called **geometric isomers** or **cis-trans stereoisomers**—they have the same atoms, and the atoms are linked in the same order, but they differ in the spatial (stereo) arrangement of the atoms.

**2.14
CONFORMATIONS
OF DISUBSTITUTED
CYCLOHEXANES**

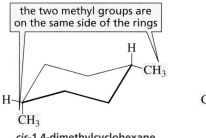

the two methyl groups are on the same side of the rings

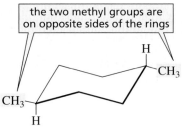

the two methyl groups are on opposite sides of the rings

***cis*-1,4-dimethylcyclohexane**   ***trans*-1,4-dimethylcyclohexane**

First we will determine which of the two chair conformers of *cis*-1,4-dimethyl-cyclohexane is more stable. One chair conformer has one methyl group in an equatorial position and one methyl group in an axial position. The other chair conformer also has one methyl group in an equatorial position and one methyl group in an axial position. Therefore, both chair conformers are equally stable.

***cis*-1,4-dimethylcyclohexane**

On the other hand, the two chair conformers of *trans*-1,4-dimethylcyclohexane have different stabilities because one has both methyl substituents in equatorial positions and the other has both methyl groups in axial positions.

more stable   less stable

***trans*-1,4-dimethylcyclohexane**

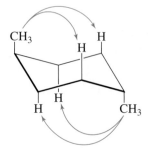

This chair conformer has four 1,3-diaxial interactions.

The chair conformer with both substituents in axial positions has four 1,3-diaxial interactions, causing it to be about 4 × 0.9 kcal/mol = 3.6 kcal mol (15.1 kJ/mol) less stable than the chair conformation with both methyl groups in equatorial positions. We can, therefore, predict that *trans*-1,4-dimethylcyclohexane will exist almost entirely in the more stable diequatorial conformation.

Both substituents of *cis*-1-*tert*-butyl-3-methylcyclohexane are in equatorial positions in one conformer and in axial positions in the other conformer. The conformer with both substituents in equatorial positions is more stable.

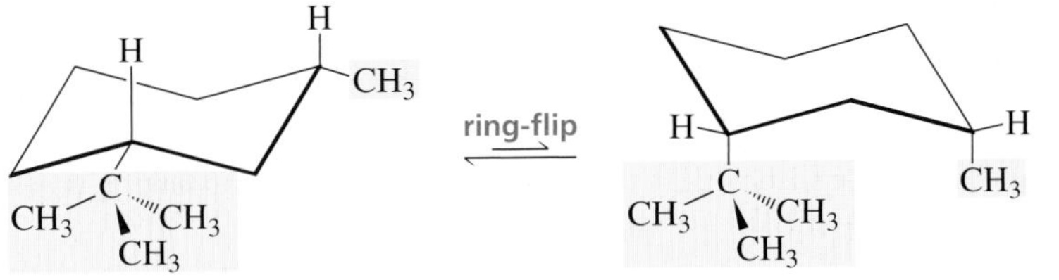

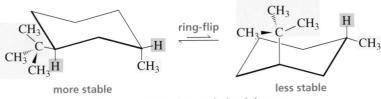

more stable   less stable

cis-1-*tert*-butyl-3-methylcyclohexane

cis-1-*tert*-Butyl-3-methylcyclohexane will have a higher percentage of the diequatorial conformer relative to the diaxial conformer than will *trans*-1,4-dimethylcyclohexane because of the greater steric bulk of the *tert*-butyl substituent.

Both conformers of *trans*-1-*tert*-butyl-3-methylcyclohexane have one substituent in an equatorial position and the other in an axial position. Because the *tert*-butyl group is larger than the methyl group, the 1,3-diaxial interactions will be stronger when the *tert*-butyl group is in the axial position. Therefore, the conformer with the *tert*-butyl group in the equatorial position is more stable.

more stable   less stable

trans-1-*tert*-butyl-3-methylcyclohexane

3-D Molecule:
trans-1-*tert*-Butyl-3-methylcyclohexane

PROBLEM-SOLVING STRATEGY

Is the conformer of 1,2-dimethylcyclohexane with one methyl group in an equatorial position and the other in an axial position the cis isomer or the trans isomer?

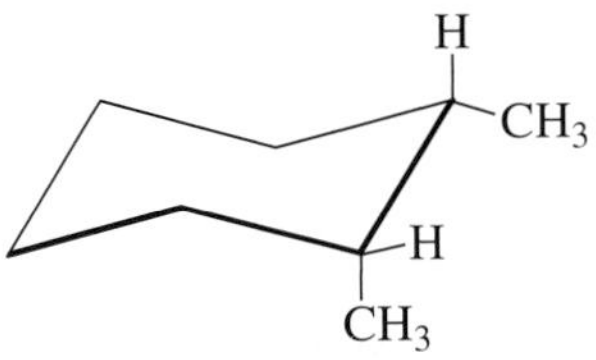

Is this the cis isomer or the trans isomer?

To solve this kind of problem you need to determine whether the two substituents are on the same side of the ring (cis) or on opposite sides of the ring (trans). If the bonds bearing the substituents are both pointing upward or both pointing downward, the compound is the cis isomer; if one bond is pointing upward and the other downward, the compound is the trans isomer. Because the conformer in question has both methyl groups attached to downward pointing bonds, it is the cis isomer.

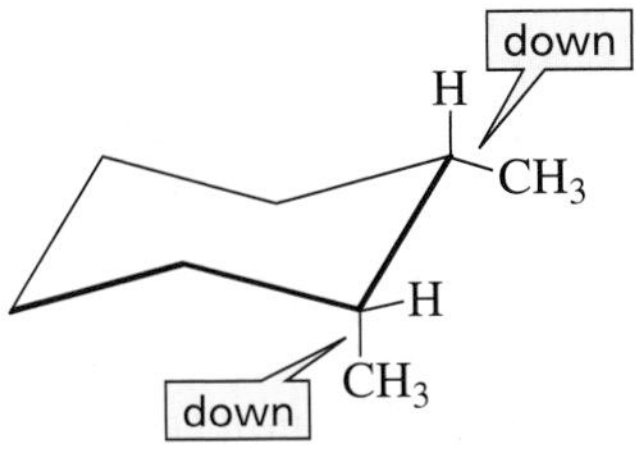

down   down

down

the cis isomer   up

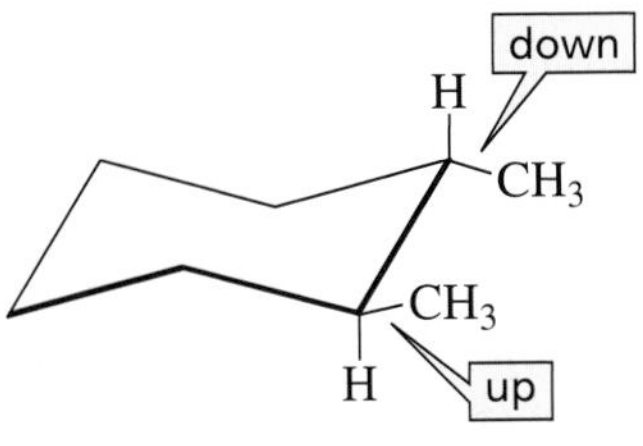

the trans isomer

The isomer that is the most misleading when it is drawn in two-dimensions is a *trans*-1,2-disubstituted isomer. At first glance, the methyl groups of *trans*-1,2-dimethylcyclohexane appear to be oriented in the same direction, so you might think it is the cis isomer. Closer inspection shows, however, that one bond is pointed upward and the other downward, so it is the trans isomer. (If you build a model of the compound, it is easier to see that it is the trans isomer.)

Now continue on to Problem 31.

PROBLEM 31 ◆

Determine whether each of the following compounds is a cis isomer or a trans isomer.

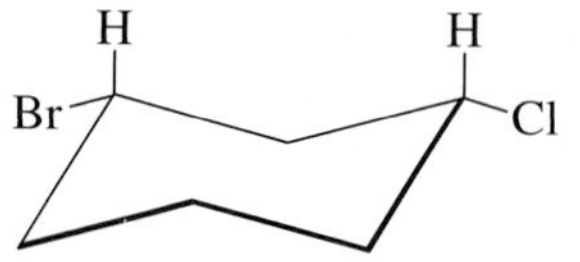

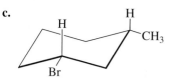

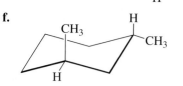

PROBLEM 32 / SOLVED

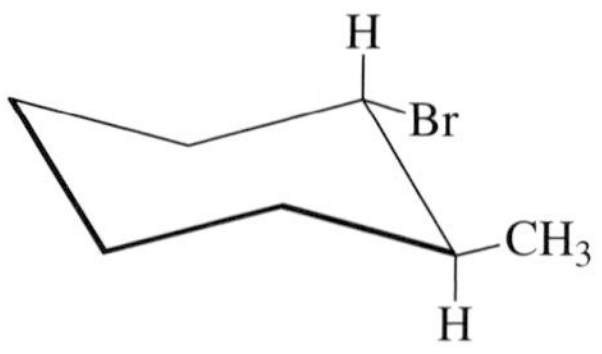

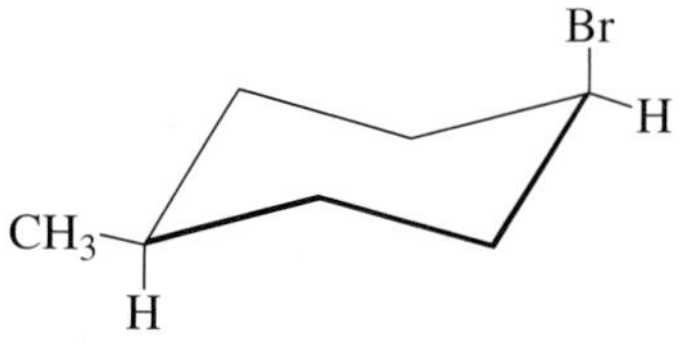

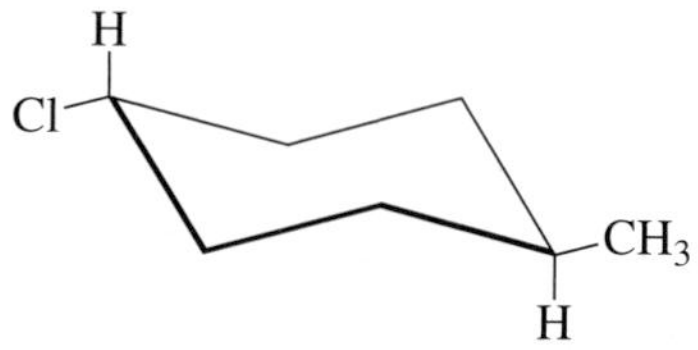

a. Draw the more stable chair conformer of *cis*-1-ethyl-2-methylcyclohexane.

b. Draw the more stable conformer of *trans*-1-ethyl-2-methylcyclohexane.

c. Which is more stable, *cis*-1-ethyl-2-methylcyclohexane or *trans*-1-ethyl-2-methylcyclohexane?

SOLUTION TO 32a If the two substituents of a 1,2-disubstituted cyclohexane are to be on the same side of the ring, one must be in an equatorial position and the other must be in an axial position. The more stable chair conformer is the one in which the larger of the two substituents (the ethyl group) is in the equatorial position.

PROBLEM 33 ◆

For each of the following disubstituted cyclohexanes, indicate whether the substituents in the two chair conformers would be both equatorial and both axial *or* one equatorial and one axial:

a. *cis*-1,2-

b. *trans*-1,2-

c. *cis*-1,3-

d. *trans*-1,3-

e. *cis*-1,4-

f. *trans*-1,4-

2.15
CONFORMATIONS
OF FUSED RINGS

When two cyclohexane rings are fused together, the second ring can be considered to be a pair of substituents bonded to the first ring. The two substituents can be either cis or trans. If the cyclohexane rings are drawn in their chair conformations, the trans isomer (one substituent bond is pointing upward and the other downward) will have both substituents in the equatorial position. The cis isomer will have one substituent in the equatorial position and one substituent in the axial position. **Trans-fused** rings, therefore, are more stable than **cis-fused** rings.

trans-decalin
trans-fused rings
more stable

cis-decalin
cis-fused rings
less stable

ring-flip

Trans-fused rings cannot undergo ring-flip (Section 2.12). Cis-fused rings, on the other hand, are much more flexible and undergo ring-flip almost as fast as cyclohexane.

KEY TERMS

alcohol (page 64)
alkane (page 60)
alkyl halide (page 64)
alkyl substituent (page 63)
amine (page 64)
angle strain (page 93)
anti conformer (page 91)
asymmetrical ether (page 74)
axial bond (page 96)
banana bond (page 94)
boat conformation (page 98)
boiling point (bp) (page 82)
chair conformation (page 95)
cis-fused (page 105)
cis isomer (page 101)
cis-trans stereoisomers (page 101)
common name (page 63)
conformation (page 89)
conformational analysis (page 90)
conformational isomer (page 89)
conformer (page 89)
constitutional isomers (page 61)
cycloalkane (page 70)
1,3-diaxial interaction (page 100)
dipole–dipole interaction (page 83)
eclipsed conformation (page 90)
equatorial bond (page 96)
ether (page 74)

flagpole hydrogen (page 98)
functional group (page 75)
gauche conformer (page 91)
gauche interaction (page 92)
geometric isomers (page 101)
half-chair conformer (page 98)
homolog (page 60)
homologous series (page 60)
hydrocarbon (page 60)
hydrogen bond (page 83)
induced dipole–induced dipole
    interaction (page 82)
IUPAC nomenclature (page 63)
London forces (page 82)
melting point (mp) (page 87)
methylene (CH_2) group (page 60)
Newman projection (page 90)
packing (page 87)
parent hydrocarbon (page 67)
perspective formula (page 89)
polarizability (page 86)
primary alcohol (page 75)
primary alkyl halide (page 72)
primary amine (page 77)
primary carbon (page 64)
primary hydrogen (page 65)
quaternary ammonium salt (page 79)
ring-flip (page 98)

sawhorse projection (page 90)
secondary alcohol (page 75)
secondary alkyl halide (page 72)
secondary amine (page 77)
secondary carbon (page 64)
secondary hydrogen (page 65)
skeletal structure (page 70)
skew-boat conformer (page 98)
solubility (page 87)
solvation (page 87)
staggered conformation (page 90)
steric hindrance (page 92)
steric strain (page 92)
straight-chain alkane (page 60)
symmetrical ether (page 74)
systematic nomenclature (page 63)
tertiary alcohol (page 75)
tertiary alkyl halide (page 72)
tertiary amine (page 77)
tertiary carbon (page 65)
tertiary hydrogen (page 65)
torsional strain (page 90)
trans-fused (page 105)
trans isomer (page 101)
twist-boat conformer (page 98)
van der Waals forces (page 82)

PROBLEMS

34. Write a structural formula for each of the following compounds.
  a. *sec*-butyl *tert*-butyl ether
  b. isoheptyl alcohol
  c. *sec*-butylamine
  d. neopentyl bromide

e. 1,1-dimethylcyclohexane
f. 4,5-diisopropylnonane
g. triethylamine
h. cyclopentylcyclohexane
i. 4-*tert*-butylheptane

j. 5,5-dibromo-2-methyloctane
k. 1-methylcyclopentanol
l. 3-ethoxy-2-methylhexane
m. 5-(1,2-dimethylpropyl)nonane
n. 3,4-dimethyloctane

35. Give the systematic name for each of the following compounds.

a. 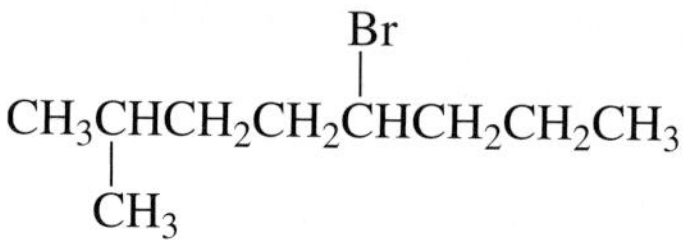
$$\underset{\underset{CH_3}{|}}{CH_3CHCH_2CH_2} \overset{\overset{Br}{|}}{CHCH_2CH_2CH_3}$$

f.
$$\underset{\underset{CH_2CH_2CH_2CH_3}{|}}{CH_3CH_2CHOCH_2CH_3}$$

b. $(CH_3)_3CCH_2CH_2CH_2CH(CH_3)_2$

g.

c. 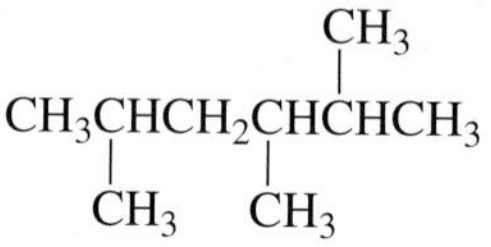
$$\underset{\underset{CH_3}{|} \ \underset{CH_3}{|}}{CH_3CHCH_2CHCHCH_3} \ \overset{\overset{CH_3}{|}}{}$$

h.

d. $(CH_3CH_2)_4C$

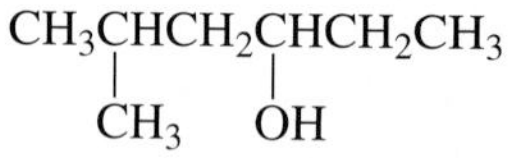

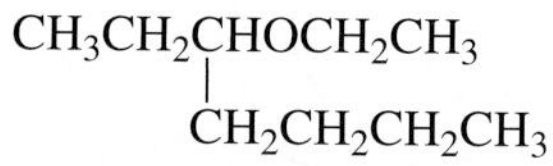

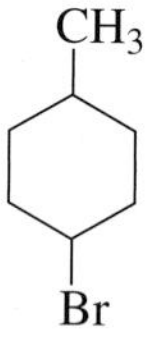

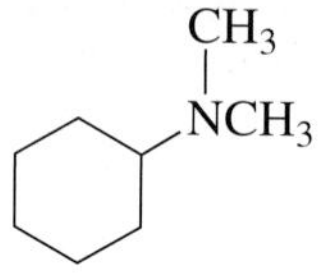

i.

e.
$$\underset{\underset{CH_3}{|} \ \underset{OH}{|}}{CH_3CHCH_2CHCH_2CH_3}$$

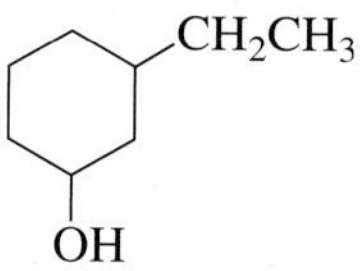

j. $CH_3OCH_2CH_2CH_2OCH_3$

36. Answer the questions for the following structure.
a. How many primary carbons does it have?
b. How many secondary carbons does it have?
c How many tertiary carbons does it have?

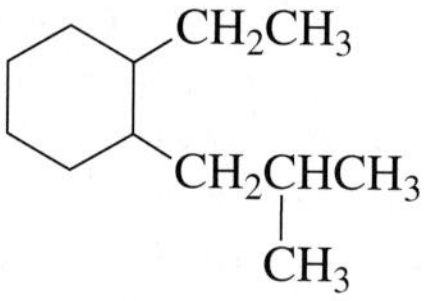

37. Draw the structural formula of an alkane that has
a. six carbons, all secondary carbons.
b. eight carbons and only primary hydrogens.
c. seven carbons with two isopropyl groups.

38. Give two names for each of the following compounds.
a. $CH_3CH_2CH_2OCH_2CH_3$

b. 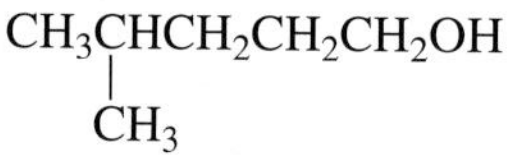
$$\underset{\underset{CH_3}{|}}{CH_3CHCH_2CH_2CH_2OH}$$

c. 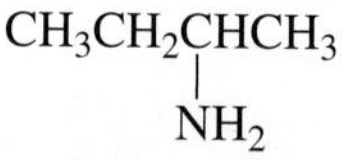
$$\underset{\underset{NH_2}{|}}{CH_3CH_2CHCH_3}$$

d. 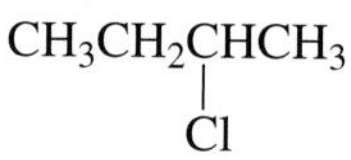
$$\underset{\underset{Cl}{|}}{CH_3CH_2CHCH_3}$$

e. 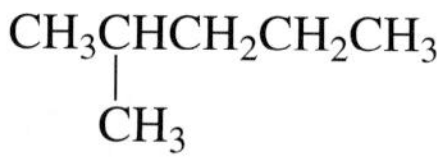
$$\underset{\underset{CH_3}{|}}{CH_3CHCH_2CH_2CH_3}$$

f. 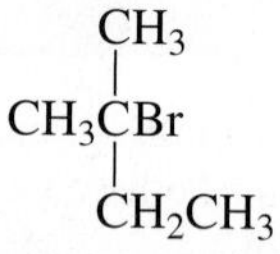
$$\underset{\underset{CH_2CH_3}{|}}{CH_3 \overset{\overset{CH_3}{|}}{C} Br}$$

g.

i. CH₃CHNH₂
        |
        CH₃

h.

39. Which of the following pairs of compounds has
   a. the higher boiling point: 1-bromopentane or 1-bromohexane?
   b. the higher boiling point: pentyl chloride or isopentyl chloride?
   c. the greater solubility in water: 1-butanol or 1-pentanol?
   d. the higher boiling point: 1-hexanol or 1-methoxypentane?
   e. the higher melting point: hexane or isohexane?
   f. the higher boiling point: 1-chloropentane or 1-pentanol?
   g. the higher boiling point: 1-bromopentane or 1-chloropentane?
   h. the higher boiling point: diethyl ether or butyl alcohol?
   i. the greater density: heptane or octane?
   j. the higher boiling point: isopentyl alcohol or isopentylamine?
   k. the higher boiling point: hexylamine or dipropylamine?

40. Al Kane was given the structural formulas of several compounds and was asked to give them systematic names. How many did Al name correctly? Correct those that are mis-named.
   a. 4-bromo-3-pentanol
   b. 2,2-dimethyl-4-ethylheptane
   c. 5-methylcyclohexanol
   d. 1,1-dimethyl-2-cyclohexanol
   e. 5-(2,2-dimethylethyl)nonane
   f. isopentyl bromide
   g. 3,3-dichlorooctane
   h. 5-ethyl-2-methylhexane
   i. 1-bromo-4-pentanol
   j. 3-isopropyloctane
   k. 2-methyl-2-isopropylheptane
   l. 2-methyl-*N*,*N*-dimethyl-4-hexanamine

41. Give systematic names for all the alkanes with molecular formula C₇H₁₆ that do not have any secondary hydrogens.

42. Draw skeletal structures of the following compounds.
   a. 5-ethyl-2-methyloctane
   b. 1,3-dimethylcyclohexane
   c. 2,3,3,4-tetramethylheptane
   d. propylcyclopentane
   e. 2-methyl-4-(1-methylethyl)octane
   f. 2,6-dimethyl-4-(2-methylpropyl)decane

43. Considering rotation about the C-3—C-4 bond of 2-methylhexane:
   a. draw the Newman projection of the most stable conformer.
   b. draw the Newman projection of the least stable conformer.
   c. about what other carbon–carbon bonds may rotation occur?
   d. how many of the carbon–carbon bonds in the compound have staggered conformers that are all equally stable?

44. Draw all the isomers that have molecular formula C₅H₁₁Br. (*Hint:* There are eight isomers.)
   a. Give the systematic name for each of the isomers.
   b. Give a common name for each isomer that has a common name.
   c. How many isomers do not have common names?
   d. How many of the isomers are primary alkyl halides?
   e. How many of the isomers are secondary alkyl halides?
   f. How many of the isomers are tertiary alkyl halides?

45. Give the systematic name for each of the following compounds.

   a.

   c.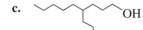

   b.

   d.

e.

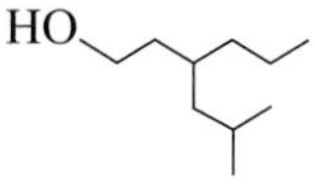

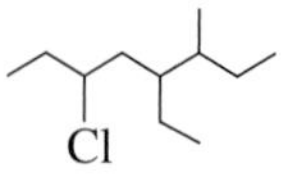

g.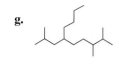

f. HO

46. Draw the two chair conformers of each compound and indicate which conformer is more stable.
   a. *cis*-1-ethyl-3-methylcyclohexane
   b. *trans*-1-ethyl-2-isopropylcyclohexane
   c. *trans*-1-ethyl-2-methylcyclohexane
   d. *trans*-1-ethyl-3-methylcyclohexane
   e. *cis*-1-ethyl-3-isopropylcyclohexane
   f. *cis*-1-ethyl-4-isopropylcyclohexane

47. Why are alcohols of lower molecular weight more water soluble than those of higher molecular weight?

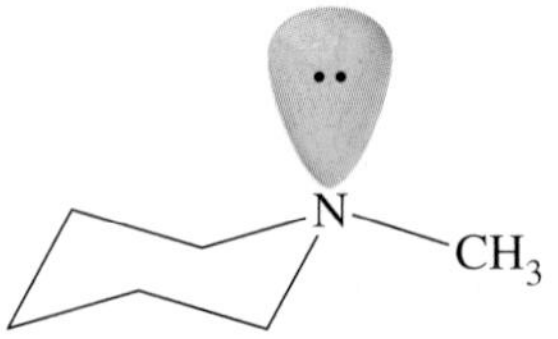

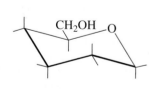

N-methylpiperidine

48. The most stable conformer of *N*-methylpiperidine is shown.
   a. Draw the other chair conformer.
   b. Which takes up more space, the nonbonded pair of electrons or the methyl group?

49. How many ethers have molecular formula $C_5H_{12}O$? Give the structural formula and systematic name for each. What are their common names?

50. Give the systematic name for each of the following compounds.

   a. $CH_3CH_2CHCH_2CH_2CHCH_3$
                 | |
               NHCH$_3$ CH$_3$

   b. $CH_3CH_2CHCH_2CHCH_2CH_3$
       with CH$_3$ above, CHCH$_3$ and CH$_3$ below

   c. $CH_3CHCHCH_2CH_2CH_2Cl$
       with CH$_2$CH$_3$ above and Cl below

   d. $CH_3CH_2CH_2CH_2CHCH_2CH_2CH_2CH_3$
       with CH$_3$CCH$_2$CH$_3$ and CH$_3$ below

   e. $CH_3CH_2CH_2CH_2CH_2CHCH_2CHCH_2CH_3$
       with CH$_2$CH$_3$ above, CH$_2$, CH$_3$CCH$_3$, CH$_2$CH$_3$ below

51. The most stable form of glucose (blood sugar) is a six-membered ring in a chair conformation with its five substituents all in equatorial positions. Starting with the structure on the right, draw the most stable form of glucose.

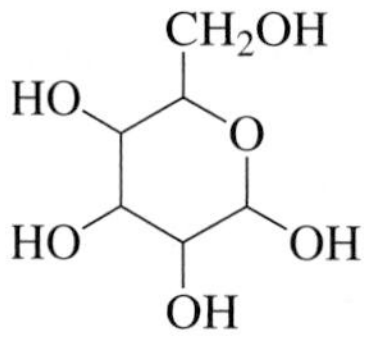

glucose

52. Explain the following:
   a. 1-Hexanol has a higher boiling point than 3-hexanol.
   b. Diethyl ether has only very limited solubility in water but tetrahydrofuran is essentially completely soluble.

tetrahydrofuran

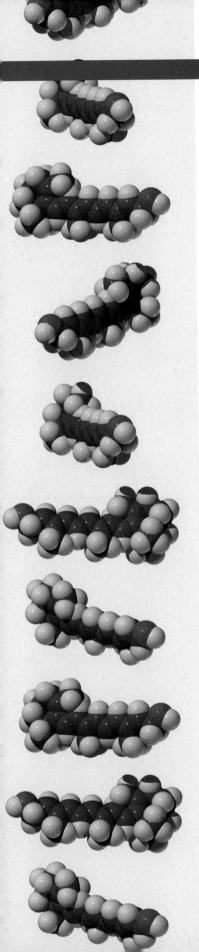

PART II

Hydrocarbons, Stereochemistry, And Resonance

Four of the next six chapters cover the reactions of hydrocarbons—compounds that contain only carbon and hydrogen. The other two chapters cover topics that are so important to the study of organic reactions that each deserves its own chapter. The first topic is stereochemistry and the second is electron delocalization and resonance.

In **Chapter 3** you will study the reactions of alkenes—*hydrocarbons that contain carbon–carbon double bonds.* You will learn how alkenes react and what kinds of products are formed from the reactions. You will also learn how to draw curved arrows to show how electrons move during the course of a reaction as new covalent bonds are formed and existing covalent bonds are broken. Although there are many different reactions in Chapter 3, you will see that they all take place by a similar pathway. Chapter 3 also discusses the principles of thermodynamics and kinetics—principles that are central to an understanding of how and why organic reactions take place.

Chapter 4 is all about stereochemistry. Here you will learn about the different kinds of isomers that are possible for organic compounds. Then we will revisit the reactions from Chapter 3 to determine whether the products of these reactions can exist as isomers and, if so, which isomers are formed.

Chapter 5 covers the reactions of alkynes—*hydrocarbons that contain carbon–carbon triple bonds.* Because alkenes and alkynes both have reactive carbon–carbon π bonds, you will discover that their reactions have many similarities. Chapter 5 will also introduce you to some of the techniques chemists use to design syntheses of organic compounds, and you will then have your first opportunity to design a multistep synthesis.

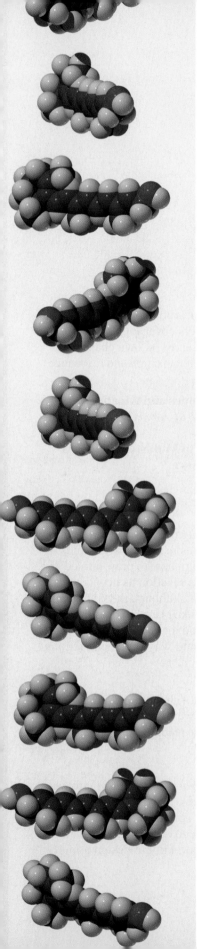

In **Chapter 6** you will learn more about delocalized electrons and the concept known as resonance—topics you were introduced to in Chapter 1. Then you will see how electron delocalization affects some of the things with which you are already familiar—acidity, the stability of carbocations and radicals, and the reactions of alkenes.

In **Chapter 7** you will learn about the reactions of dienes—*hydrocarbons that have two carbon–carbon double bonds.* You will see that if the two double bonds in a diene are sufficiently separated from one another, most reactions that occur are identical to the reactions of alkenes (Chapter 3). If, however, the double bonds are separated by only one single bond, resonance (Chapter 6) plays an important role in the reactions of that compound. The way in which Chapter 7 combines many of the concepts and theories you learned in previous chapters is a common theme in the study of organic chemistry.

Chapter 8 covers the reactions of alkanes—*hydrocarbons that contain only single bonds.* In the previous chapters you will have discovered that when an organic compound reacts, the weakest bond in the molecule is the one that usually breaks first. Alkanes, however, have only strong bonds. Therefore, you can correctly predict that alkanes undergo reactions only under extreme conditions.

3

Reactions of Alkenes. Thermodynamics and Kinetics

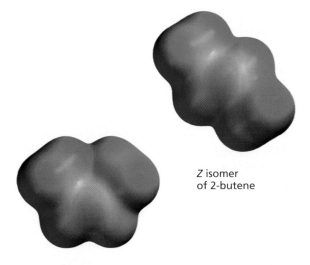

Z isomer
of 2-butene

E isomer
of 2-butene

Ethene is the hormone that causes tomatoes to ripen.

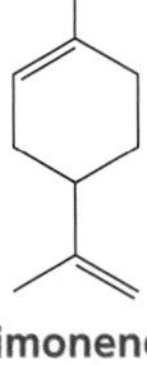

You learned in Chapter 2 that alkanes are hydrocarbons that contain only carbon–carbon single bonds. Hydrocarbons that contain a carbon–carbon double bond are called **alkenes.** Early chemists noted that an oily substance was formed when ethene ($H_2C=CH_2$), the smallest alkene, reacted with chlorine. Based on this observation, early organic chemists called alkenes **olefins** (oil forming).

Alkenes play many important roles in biology. Ethene, for example, is a plant hormone. A **hormone** is a compound that controls growth and other changes in tissues. Ethene affects seed germination, flower maturation, and fruit ripening.

Insects communicate by releasing **pheromones,** chemical substances that other insects of the same species detect with their antennae. There are sex, alarm, and trail pheromones, and many of these are alkenes. Interfering with an insect's ability to send or receive chemical signals is an environmentally safe way to control insect populations. For example, traps with synthetic sex attractants have been used to capture such crop-destroying insects as the gypsy moth and the bollweevil. A Mediterranean fruit fly that was caught in such a trap in a California citrus grove alerted the agricultural community to be prepared for an infestation.

Many of the flavors and fragrances produced by certain plants also belong to the alkene family.

limonene
from lemon and
orange oils

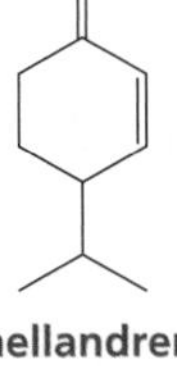

β-phellandrene
oil of eucalyptus

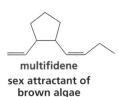

multifidene
sex attractant of
brown algae

3-D Molecules:
Limonene;
β-Phellandrene;
Multifidene

muscalure
sex attractant of the house fly

α-farnesene
found in the waxy coating
on apple skins

3.1
GENERAL
MOLECULAR
FORMULA FOR A
HYDROCARBON

**The general molecular
formula for a hydrocarbon is
C_nH_{2n+2} minus two hydrogens
for every π bond and/or ring
present in the molecule.**

In Chapter 2 you learned that the general molecular formula for a *noncyclic alkane* is C_nH_{2n+2}. You also learned that the general molecular formula for a *cyclic alkane* is C_nH_{2n} because the cyclic structure reduces the number of hydrogens by two. Noncyclic compounds are also called **acyclic** compounds (*"a"* is Greek for "non" or "not").

The general molecular formula for an *acyclic alkene* is also C_nH_{2n} and the general molecular formula for a *cyclic alkene* is C_nH_{2n-2} because, as a result of the carbon–carbon double bond, an alkene has two fewer hydrogens than an alkane with the same number of carbon atoms. Therefore, we can make the following general statement: *The general molecular formula for a hydrocarbon is C_nH_{2n+2} minus two hydrogens for every π bond and/or ring present in the molecule.*

$CH_3CH_2CH_2CH_2CH_3$	$CH_3CH_2CH_2CH=CH_2$		
an alkane	**an alkene**	**a cyclic alkane**	**a cyclic alkene**
C_5H_{12}	C_5H_{10}	C_5H_{10}	C_5H_8
C_nH_{2n+2}	C_nH_{2n}	C_nH_{2n}	C_nH_{2n-2}

Thus, for every *two* hydrogens that are missing from the general molecular formula C_nH_{2n+2}, a hydrocarbon has either a π bond or a ring. For example, a compound with a molecular formula of C_8H_{14} needs four more hydrogens to become C_8H_{18} ($C_8H_{2\times8+2}$). Therefore, the compound has either two double bonds, a ring and a double bond, two rings, or a triple bond. [Remember that a triple bond consists of two π bonds and a σ bond (Section 1.9).]

Several compounds with molecular formula C_8H_{14}

$CH_3CH=CH(CH_2)_3CH=CH_2$

$CH_3(CH_2)_5C\equiv CH$

Because alkanes contain the maximum number of carbon–hydrogen bonds possible—that is, they are saturated with hydrogen—they are called **saturated hydrocarbons.** Alkenes, because they have fewer than the maximum number of hydrogens, are called **unsaturated hydrocarbons.**

$CH_3CH_2CH_2CH_3$ $CH_3CH=CHCH_3$
a saturated hydrocarbon   **an unsaturated hydrocarbon**

PROBLEM 1 / SOLVED ◆

If you know the number of carbons in a hydrocarbon, you can determine its molecular formula if you know how many π bonds and rings it has ($\pi + r$).
Determine the molecular formula of each of the following:

a. a five-carbon compound with one π bond and one ring

b. a four-carbon compound with two π bonds and no rings

c. a ten-carbon compound with one π bond and two rings

d. an eight-carbon compound with three π bonds and one ring

SOLUTION TO 1a   For a five-carbon hydrocarbon with no π bonds and no rings, $C_nH_{2n+2} = C_5H_{12}$; a five-carbon hydrocarbon with $(\pi + r) = 2$, has four fewer hydrogens because two hydrogens are subtracted for every π bond or ring present in the compound. Therefore, its molecular formula is C_5H_8.

PROBLEM 2 / SOLVED ◆

Determine $(\pi + r)$ for the hydrocarbons with the following molecular formulas:

a. $C_{10}H_{16}$ **b.** $C_{20}H_{34}$ **c.** C_8H_{16} **d.** $C_{12}H_{20}$

SOLUTION TO 2a   For a ten-carbon hydrocarbon with no π bonds and no rings, $C_nH_{2n+2} = C_{10}H_{22}$. Thus, a ten-carbon compound with molecular formula $C_{10}H_{16}$ has six fewer hydrogens. Therefore, $(\pi + r) = 3$.

PROBLEM 3

Draw possible structures for compounds with the following molecular formulas:

a. C_3H_6 **b.** C_3H_4 **c.** C_4H_6

The systematic (IUPAC) name of an alkene is obtained by replacing the "ane" ending of the alkane with "ene." For example, a two-carbon alkene is called ethene and a three-carbon alkene is called propene. Ethene is frequently referred to by its common name (ethylene).

**3.2
NOMENCLATURE
OF ALKENES**

$H_2C = CH_2$ $CH_3CH = CH_2$ cyclopentene   cyclohexene

systematic name: ethene   propene
common name: ethylene   propylene

Most alkene names need a number to indicate the position of the double bond. (The previous names do not because there is no ambiguity.) The same IUPAC rules we learned in Chapter 2 are followed in naming alkenes.

1. The longest continuous chain containing the functional group (in this case, the carbon–carbon double bond) is numbered in a direction that gives the functional group suffix the lowest possible number. For example, 1-butene signifies that the double bond is between the first and second carbons of butene; 2-hexene signifies that the double bond is between the second and third carbons of hexene.

$\overset{4}{C}H_3\overset{3}{C}H_2\overset{2}{C}H=\overset{1}{C}H_2$ $\overset{1}{C}H_3\overset{2}{C}H=\overset{3}{C}H\overset{4}{C}H_3$ $\overset{1}{C}H_3\overset{2}{C}H=\overset{3}{C}H\overset{4}{C}H_2\overset{5}{C}H_2\overset{6}{C}H_3$

1-butene   **2-butene**   **2-hexene**

$\overset{6}{C}H_3\overset{5}{C}H_2\overset{4}{C}H_2\overset{3}{C}H_2\overset{2}{C}CH_2CH_2CH_3$
‖
$\underset{1}{C}H_2$

2-propyl-1-hexene

> the longest continuous chain has eight carbons but the longest continuous chain containing the functional group has six carbons, so the parent name of the compound is hexene

Notice that 1-butene does not have a common name. You might be tempted to call it "butylene," which is analogous to "propylene" for propene, but butylene is not an appropriate name. A name must be unambiguous and "buty-lene" could signify either 1-butene or 2-butene.

2. If the chain has substituents, it is still numbered in the direction that gives the functional group suffix the lowest possible number.

$$\overset{\text{CH}_3}{\underset{\text{4-methyl-2-pentene}}{\overset{1\quad2\quad3\ 4|\ 5}{\text{CH}_3\text{CH}=\text{CHCHCH}_3}}} \qquad \overset{\overset{2\ 1}{\text{CH}_2\text{CH}_3}}{\underset{\text{3-methyl-3-heptene}}{\overset{|3\ \ 4\ \ 5\ \ 6\ \ 7}{\text{CH}_3\text{C}=\text{CHCH}_2\text{CH}_2\text{CH}_3}}}$$

$$\underset{\text{4-pentoxy-1-butene}}{\overset{4\quad3\quad2\quad1}{\text{CH}_3\text{CH}_2\text{CH}_2\text{CH}_2\text{CH}_2\text{OCH}_2\text{CH}_2\text{CH}=\text{CH}_2}}$$

3. If a chain has more than one substituent, the substituents are cited in alpha-betical (not numerical) order, using the same rules for alphabetizing that you learned in Section 2.2 (the prefixes di, tri, *sec,* and *tert* are ignored in alpha-betizing, but iso, neo, and cyclo are not ignored).

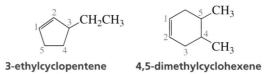

$$\overset{\text{Br Cl}}{\underset{7\quad6\quad5\quad4\quad3\ \ 2\quad1}{\text{CH}_3\text{CH}_2\text{CHCHCH}_2\text{CH}=\text{CH}_2}}$$

3,6-dimethyl-3-octene   5-bromo-4-chloro-1-heptene

4. If the same number for the alkene functional group suffix is obtained in both directions, the correct name is the one that contains the lowest substituent num-ber. For example, in 2,5-dimethyl-4-octene, the compound is a 4-octene whether the longest continuous chain is numbered from left to right or from right to left. If you number from left to right, the substituents are at positions 4 and 7, but if you number from right to left they are at positions 2 and 5. Of those four substituent numbers, 2 is the lowest number, so the compound is named 2,5-dimethyl-4-octene and not 4,7-dimethyl-4-octene.

$$\underset{\text{CH}_3\qquad\ \text{CH}_3}{\text{CH}_3\text{CH}_2\text{CH}_2\text{C}=\text{CHCH}_2\text{CHCH}_3} \qquad \underset{\text{Br}\qquad\ \text{CH}_3}{\text{CH}_3\text{CHCH}=\text{CCH}_2\text{CH}_3}$$

2,5-dimethyl-4-octene   **2-bromo-4-methyl-3-hexene**
not   **not**
4,7-dimethyl-4-octene   **5-bromo-3-methyl-3-hexene**
because 2 < 4   because 2 < 3

5. In cyclic compounds, a number is not needed to denote the position of the func-tional group because the ring is always numbered so that the double bond is be-tween carbons 1 and 2.

3-ethylcyclopentene   4,5-dimethylcyclohexene

Tutorial: Alkene nomenclature

In the following cyclohexenes, the double bond is between C-1 and C-2 re-gardless of whether you move around the ring clockwise or counterclock-wise. Therefore, you move around the ring in the direction that puts the lowest substituent number into the name, *not* in the direction that gives the lowest *sum* of the substituent numbers. For example, 1,6-dichlorocyclohexene is not called 2,3-dichlorocyclohexene even though the latter has the lowest sum of the substituent numbers ($1 + 6 = 7$ *versus* $2 + 3 = 5$); 1,6-dichlorocyclohexene is the correct name because it has the lowest substituent number (1).

1,6-dichlorocyclohexene
not
2,3-dichlorocyclohexene
because 1 < 2

5-ethyl-1-methylcyclohexene
not
4-ethyl-2-methylcyclohexene
because 1 < 2

6. If both directions lead to the same number for the alkene functional group suffix and the same low number(s) for one or more of the substituents, then those substituents are ignored and the direction is chosen that gives the lowest number to one of the remaining substituents.

$CH_3CHCH_2CH{=}CCH_2CHCH_3$
$\quad|\quad\quad\quad\quad|$
$\quad CH_3\quad\quad CH_2CH_3$

2-bromo-4-ethyl-7-methyl-4-octene
not
7-bromo-5-ethyl-2-methyl-4-octene
because 4 < 5

6-bromo-3-chloro-4-methylcyclohexene
not
3-bromo-6-chloro-5-methylcyclohexene
because 4 < 5

The sp^2 carbons of an alkene are called **vinylic carbons.** An sp^3 carbon that is adjacent to a vinylic carbon is called an **allylic carbon.**

| vinylic carbons |
$RCH_2{-}CH{=}CH{-}CH_2R$
| allylic carbons |

There are two groups that contain a carbon–carbon double bond that are used as substituent groups in common names—the **vinyl group** and the **allyl group.** The vinyl group is the smallest possible group that contains a vinylic carbon, and the allyl group is the smallest possible group that contains an allylic carbon. When allyl is used in nomenclature, the substituent must be attached to the allylic carbon.

$H_2C{=}CH{-}$
the vinyl group

$H_2C{=}CHCH_2{-}$
the allyl group

$H_2C{=}CHCl$
systematic name: chloroethene
common name: vinyl chloride

$H_2C{=}CHCH_2Br$
3-bromopropene
allyl bromide

Tutorial: Common names of alkyl groups

PROBLEM 4 ◆

Draw the structure for each of the following compounds.

a. 3,3-dimethylcyclopentene

b. 6-bromo-2,3-dimethyl-2-hexene

c. ethyl vinyl ether

d. allyl alcohol

PROBLEM 5 ◆

Give the systematic name for each of the following compounds.

a. 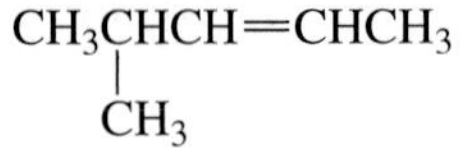 CH₃CHCH=CHCH₃
      |
      CH₃

d. 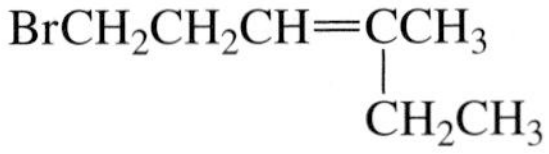 BrCH₂CH₂CH=CCH₃
               |
               CH₂CH₃

b. 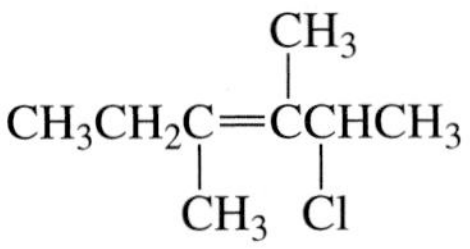
             CH₃
             |
CH₃CH₂C=CCHCH₃
         | |
         CH₃ Cl

e.

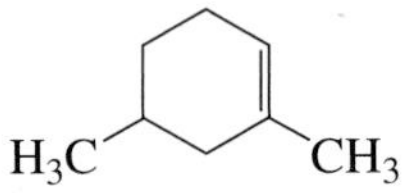

c. 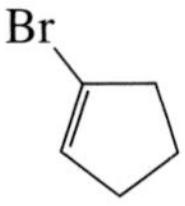 Br

f. CH₃CH=CHOCH₂CH₂CH₂CH₃

3.3
THE STRUCTURE OF ALKENES

The structure of the smallest alkene (ethene) was described in Section 1.8. Other alkenes have similar structures. Each double-bonded carbon of an alkene has three sp^2 orbitals that lie in a plane with angles of 120°. Each of these sp^2 orbitals overlaps an orbital of another atom to form a σ bond. Thus, one of the carbon–carbon bonds in a double bond is a σ bond, formed by the overlap of an sp^2 orbital of one carbon with an sp^2 orbital of the other carbon. The second carbon–carbon bond in the double bond (the π bond) is formed from side-to-side overlap of the remaining p orbitals on the sp^2 carbons. Because three points determine a plane, each sp^2 hybridized carbon and two of the three atoms bonded to it are in a plane. In order to achieve maximum orbital–orbital overlap, the two p orbitals must be parallel to each other. Therefore, all six atoms of the double bond system are in the same plane. A molecular orbital description of a carbon–carbon double bond is shown in Figure 1.6 on p. 24.

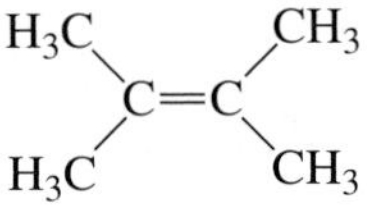

$$H_3C \qquad CH_3$$
$$\diagdown \; / $$
$$C=C$$
$$\diagup \; \diagdown$$
$$H_3C \qquad CH_3$$

the six carbon atoms in the molecule are in the same plane

It is important to remember that the π bond represents the cloud of electrons that is above and below the plane defined by the two sp^2 carbons and the four atoms bonded to them.

3-D Molecule:
2,3-Dimethyl-2-butene

***p* orbitals overlap to form a π bond**

electrostatic potential map for 2,3-dimethyl-2-butene

Because the two *p* orbitals that form the π bond must be parallel to one another to achieve maximum overlap, rotation about a double bond does not occur. If rotation about a double bond did occur, the two *p* orbitals would cease to overlap and the π bond would be destroyed (Figure 3.1).

3.4
CIS-TRANS ISOMERISM

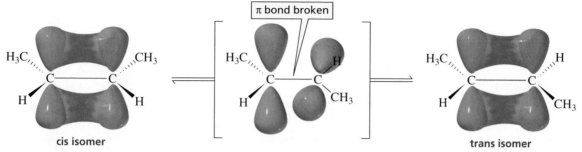

π bond broken

cis isomer   trans isomer

▲ **Figure 3.1**
Rotation about the carbon–carbon double bond would break the π bond.

Because there is no rotation about a carbon–carbon double bond, an alkene such as 2-butene can exist in two distinct forms. That is, the hydrogens bonded to the *sp*² carbons can be on the same side of the double bond or on opposite sides of the double bond. The isomer with the hydrogens on the same side of the double bond is called the **cis isomer,** and the isomer with the hydrogens on opposite sides of the double bond is called the **trans isomer.** A pair of isomers such as *cis*-2-butene and *trans*-2-butene are called **cis–trans isomers** or **geometric isomers.** Cis–trans isomers have the same molecular formula but differ in the way the atoms are arranged in space (Section 2.13).

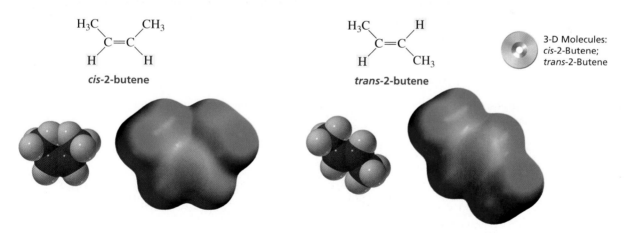

cis-**2-butene**   *trans*-**2-butene**

3-D Molecules:
cis-2-Butene;
trans-2-Butene

If one of the *sp*² carbons of the double bond is attached to two identical substituents, there is only one possible structure for the alkene. Cis and trans isomers are not possible for such a compound.

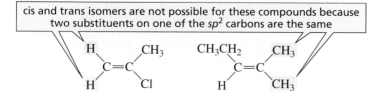

cis and trans isomers are not possible for these compounds because
two substituents on one of the *sp*² carbons are the same

3-D Molecule:
2-Methyl-2-pentene

Because there is no rotation about a double bond, cis and trans isomers cannot interconvert (except under extreme conditions). This means that they can be separated from each other. In other words, the two isomers are different compounds with different physical properties, such as different boiling points and different dipole moments. Notice that *trans*-2-butene and *trans*-1,2-dichloroethene have resultant dipole moments (μ) of zero because the bond dipole moments cancel. Notice the difference in the electrostatic potential maps for the cis and trans isomers. Recall that the red area in the electrostatic potential maps represents high electron density.

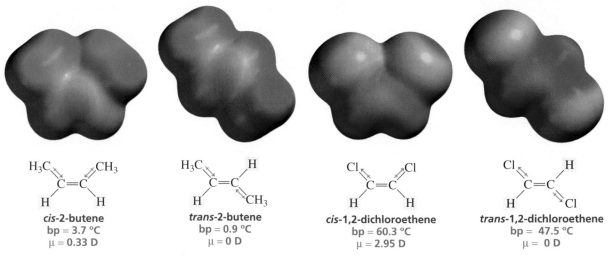

cis-2-butene
bp = 3.7 °C
μ = 0.33 D

trans-2-butene
bp = 0.9 °C
μ = 0 D

cis-1,2-dichloroethene
bp = 60.3 °C
μ = 2.95 D

trans-1,2-dichloroethene
bp = 47.5 °C
μ = 0 D

Cis and trans isomers can be interconverted (in the absence of any added reagents) only under extreme conditions. Therefore, cis–trans interconversion is not a practical laboratory process. Interconversion occurs when the molecule absorbs sufficient heat or light energy to cause the π bond to break. Once the π bond is broken, rotation can occur about the remaining σ bond (Section 2.10).

CIS–TRANS INTERCONVERSION IN VISION

When rhodopsin absorbs light, a double bond interconverts between the cis and trans forms. This plays an important role in vision (Section 24.7).

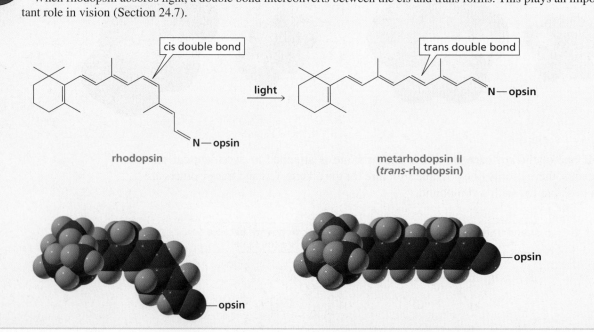

$$H_3C \quad CH_2CH_3$$
$$\underset{H}{\overset{}{\diagdown}} C=C \underset{H}{\overset{}{\diagup}}$$

cis-2-pentene

$$\xrightleftharpoons[\substack{or \\ h\nu}]{> 180\ °C}$$

$$H_3C \quad H$$
$$\underset{H}{\overset{}{\diagdown}} C=C \underset{CH_2CH_3}{\overset{}{\diagup}}$$

trans-2-pentene

3-D Molecules:
cis-Retinal;
trans-Retinal

PROBLEM 6 ◆

a. Which of the following compounds can exist as cis–trans isomers?

b. For those compounds, draw and label the cis and trans isomers.

1. $CH_3CH = CHCH_2CH_3$

2. $CH_3C = CHCH_3$
 $\qquad |$
 $\qquad CH_3$

3. $CH_3CH = CHCH_3$

4. $CH_3CH_2CH = CH_2$

As long as each of the sp^2 carbons of an alkene is bonded to only one substituent, we can use the terms cis and trans to designate the structure of the alkene. If the hydrogens are on the same side of the double bond, it is the cis isomer; if they are on opposite sides of the double bond, it is the trans isomer. But how would you determine cis and trans isomers for a compound such as 1-bromo-2-chloropropene?

$$Br \quad Cl \qquad\qquad Br \quad CH_3$$
$$\underset{H}{\overset{}{\diagdown}} C=C \underset{CH_3}{\overset{}{\diagup}} \qquad\qquad \underset{H}{\overset{}{\diagdown}} C=C \underset{Cl}{\overset{}{\diagup}}$$

Which isomer is cis and which is trans?

For a compound such as 1-bromo-2-chloropropene, the cis–trans system of nomenclature cannot be used because there are four different substituents on the two vinylic carbons. The *E, Z* system of nomenclature was devised for these kinds of situations.[1]

In order to name an isomer by the *E, Z* system, first determine the relative priorities of the two groups bonded to one of the sp^2 carbons and then the relative priorities of the two groups bonded to the other sp^2 carbon. (Rules for assigning relative priorities are explained next.) If the high-priority groups are on the same side of the double bond, the isomer is said to have the *Z* configuration (*Z* is for *zusammen*, German for "together"). If the high-priority groups are on opposite sides of the double bond, the isomer has the *E* configuration (*E* is for *entgegen*, German for "opposite").

The *Z* isomer has the high-priority groups on the same side.

high priority \ / high priority
           C=C
low priority / \ low priority

the *Z* isomer

high priority \ / low priority
           C=C
low priority / \ high priority

the *E* isomer

The following rules are used to determine the priorities of the groups bonded to the sp^2 carbons:

• **Rule 1.** The relative priorities of the groups depend on the atomic numbers of the atoms bonded directly to a particular sp^2 carbon. The greater the atomic number, the higher the priority. For example, in 1-bromo-2-chloropropene, one of the sp^2 carbons is bonded to a bromine and to a hydrogen. Bromine has a greater atomic number than hydrogen, so bromine has the higher priority. The other sp^2 carbon is bonded to a chlorine and to a carbon. Chlorine has a greater

---

[1]IUPAC prefers the *E* and *Z* designations because they can be used for all alkene isomers. However, many chemists continue to use the cis and trans designations for simple molecules.

3.5
THE *E, Z* SYSTEM OF NOMENCLATURE

atomic number than carbon, so chlorine has the higher priority. (Notice that you use the atomic number of carbon, not the formula weight of the methyl group, because the priorities are based on the atomic numbers of atoms and not on the formula weights of groups.) The isomer on the left has the high-priority groups (Br and Cl) on the same side of the double bond, so it is the **Z isomer** (Zee Groups are on Zee Zame Zide). The isomer on the right has the high-priority atoms on opposite sides of the double bond, so it is the **E isomer.**

$$
\begin{array}{cc}
\underset{\text{H}}{\overset{\text{Br}}{\diagdown}}\text{C}=\text{C}\underset{\text{CH}_3}{\overset{\text{Cl}}{\diagup}} & \underset{\text{H}}{\overset{\text{Br}}{\diagdown}}\text{C}=\text{C}\underset{\text{Cl}}{\overset{\text{CH}_3}{\diagup}} \\
\textbf{(Z)-1-bromo-2-chloropropene} & \textbf{(E)-1-bromo-2-chloropropene}
\end{array}
$$

- **Rule 2.** If the two substituents on an sp^2 carbon have the same atomic number (there is a tie), the atomic numbers of the atoms that are attached to the "tied" atoms must be considered. In the following pair of isomers, both atoms bonded to one of the sp^2 carbons are carbons (from an ethyl group and an isopropyl group), so there is a tie at this point. One carbon is bonded to C, C, and H (the isopropyl group); the other is bonded to C, H, and H (the ethyl group). One C cancels in each of the two groups, leaving C and H in the isopropyl group and H and H in the ethyl group. Carbon has a greater atomic number than hydrogen, so the isopropyl group has a higher priority than the ethyl group. The other sp^2 carbon is bonded to a chlorine and to a chloromethyl group. Chlorine has a greater atomic number than carbon, so the chloro group has the higher priority. The E and Z isomers are as shown.

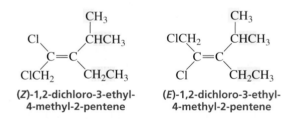

$$
\begin{array}{cc}
\textbf{(Z)-1,2-dichloro-3-ethyl-} & \textbf{(E)-1,2-dichloro-3-ethyl-} \\
\textbf{4-methyl-2-pentene} & \textbf{4-methyl-2-pentene}
\end{array}
$$

- **Rule 3.** If an atom is doubly bonded to another atom, the priority system treats it as if it were singly bonded to two of those atoms. If an atom is triply bonded to another atom, the priority system treats it as if it were singly bonded to three of those atoms. One of the sp^2 carbons in the following pair of isomers is bonded to an ethyl group and to a vinyl group. Because the atoms immediately bonded to the sp^2 carbon are both carbons, there is a tie. The carbon of the ethyl group is bonded to C, H, and H. The carbon of the vinyl group is bonded to an H and doubly bonded to a C. Therefore, the first carbon of the vinyl group is considered to be bonded to C, C, and H. Consequently, the vinyl group has a higher priority than the ethyl group. Both atoms that are bonded to the other sp^2 carbon are carbons, too, so there is a tie there as well. The carbon of the hydroxymethyl group is bonded to O, H, and H, and the carbon of the isopropyl group to C, C, and H. Of these six atoms, oxygen has the greatest atomic number, so the hydroxymethyl group has the higher priority. (Notice that you do not add the atomic numbers; you take the single atom with the greatest atomic number.)

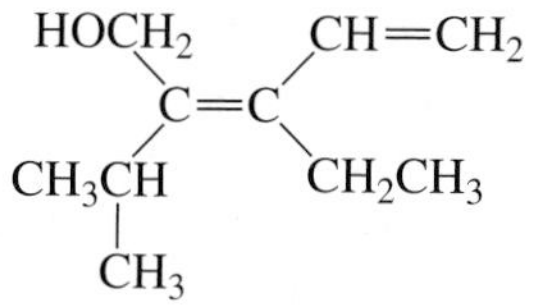

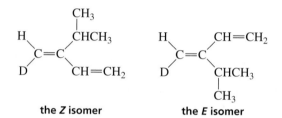

the Z isomer 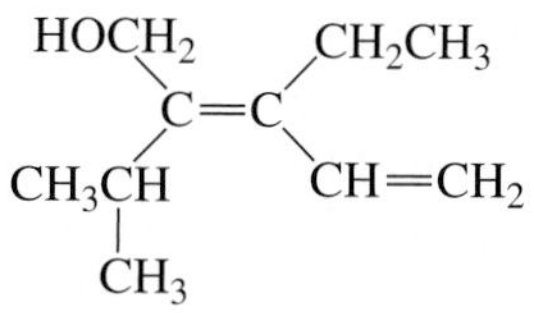   the E isomer

- **Rule 4.** In the case of isotopes (atoms with the same atomic number but different mass numbers), the mass number is used to determine their relative priorities. In the following structures, for example, one of the sp^2 carbons is bonded to deuterium (D) and hydrogen. Deuterium and hydrogen have the same atomic number, but deuterium has a greater mass number so it has the higher priority. The carbons that are bonded to the other sp^2 carbon are both bonded to C, C, and H, so you must go out to the next set of atoms to break the tie. The second carbon of the isopropyl group is bonded to H, H, and H, while the second carbon of the vinyl group is bonded to H, H, and C. Therefore, the vinyl group has the higher priority.

Tutorial: *E* and *Z* Nomenclature

the Z isomer   the E isomer

Notice that in all these examples you never count the atom bonded to the σ bond from which you originate. In differentiating between the isopropyl and vinyl groups in the last example, you did count the atom bonded to the π bond from which you originated. In the case of a triple bond, you can count the atoms bonded to both π bonds from which you originate.

PROBLEM 7

a. Draw and label the *E* and *Z* isomers of each of the following compounds.

$$CH_3CH_2CH_2CH_2$$

1. $CH_3CH_2CH{=}CHCH_3$

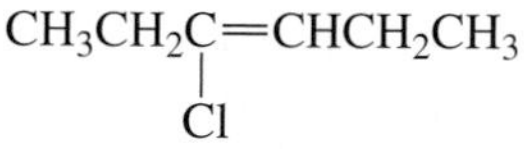

3. 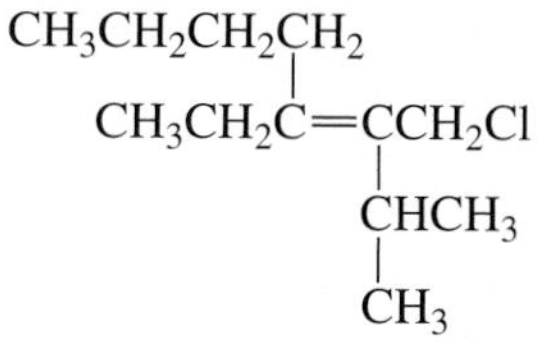 $CH_3CH_2\overset{|}{C}{=}CCH_2Cl$
$\overset{|}{C}HCH_3$
$\overset{|}{C}H_3$

2. $CH_3CH_2\overset{|}{C}{=}CHCH_2CH_3$
$\overset{|}{Cl}$

4. 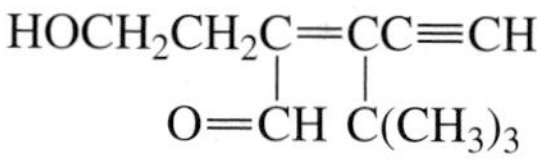 $HOCH_2CH_2C{=}CC{\equiv}CH$
$\overset{|}{O}{=}CH\ \overset{|}{C}(CH_3)_3$

b. Draw the structure of (*Z*)-3-isopropyl-2-heptene.

There are many millions of organic compounds. If you had to memorize how each of them reacts, studying organic chemistry would be a very unpleasant experience. Fortunately, organic compounds can be divided into families, and all the members

**3.6
REACTIVITY
CONSIDERATIONS**

of a family react in similar ways. What determines the family an organic compound belongs to is its functional group. The **functional group** is the center of reactivity of a molecule. You will find a Table of Common Functional Groups inside the front cover of this book. The functional group of an alkene is the carbon–carbon double bond. All compounds with a carbon–carbon double bond react in similar ways—whether it is a small molecule like ethene or a large molecule like cholesterol.

ethene

cholesterol

To further reduce the need to memorize, it is important to understand *why* a functional group reacts the way it does. In other words, it is not sufficient to know that a compound with a carbon–carbon double bond reacts with HBr to form a product in which the H and Br atoms have taken the place of the π bond. We need to understand *why* the compound reacts with HBr. In each of the chapters in this book where we discuss the reactivity of a particular functional group, we will see how the structure of the functional group allows us to predict the kind of reactions it will undergo. Thus, when you are confronted with a reaction you have never seen before, you will probably be able to predict the products of the reaction, if you understand how the structure of the molecule affects its reactivity.

Organic chemistry is all about the interaction between electron-rich atoms or molecules and electron-deficient atoms or molecules. It is these forces of attraction that make chemical reactions happen. Therefore, the following very important rule determines the reactivity of organic compounds: *Electron-rich atoms or molecules are attracted to electron-deficient atoms or molecules.* Each time you study the reactivity of a new functional group, remember that its reactivity can be explained by this very simple rule.

Before you study the reactivities of functional groups, however, you must understand more about electron-deficient and electron-rich atoms and molecules. An electron-deficient atom or molecule is called an **electrophile.** An electrophile can accept a pair of electrons or it can have an unpaired electron in need of a partner to complete its octet. An electrophile, therefore, looks for electrons. Literally, electrophile means electron loving (*phile* is the Greek suffix for "loving").

> **Electron-rich atoms or molecules are attracted to electron-deficient atoms or molecules.**

a few examples of electrophiles

$$H^+ \qquad CH_3\overset{+}{C}H_2 \qquad BH_3 \qquad\qquad :\ddot{B}r\cdot \qquad R\ddot{O}\cdot$$

these are electrophiles because they can accept a pair of electrons

these are electrophiles because they are seeking an electron

An electron-rich atom or molecule is called a **nucleophile.** A nucleophile has a pair of electrons it is willing to share. The following examples show that some nucleophiles are neutral and some are negatively charged. Because a nucleophile has electrons to share and an electrophile is looking for electrons, it should not be surprising that they attract each other. Thus, the following is another way of stating the preceding rule: *A nucleophile and an electrophile react with each other.*

A nucleophile reacts with an electrophile.

a few examples of nucleophiles

$$HO:^- \quad :Cl:^- \quad CH_3NH_2 \quad H_2O:$$

these are nucleophiles because they
have a pair of electrons to share

You know that a π bond is weaker than a σ bond (Section 1.14). Therefore, the π bond is the weakest bond in an alkene, which means that it is the bond most likely to break in a reaction. You also know that the π bond of an alkene consists of a cloud of electrons above and below the σ bond. As a result of this cloud of electrons, an alkene is an electron-rich molecule—it is a nucleophile. (Notice, for instance, the electron-rich red area in the electrostatic potential maps for *cis-* and *trans-*2-butene in Section 3.4.) You can, therefore, predict that an alkene will react with an electrophile. If a reagent such as hydrogen bromide is added to an alkene, the alkene will react with the partially positively charged hydrogen of hydrogen bromide and a carbocation will be formed. In the second step of the reaction, the positively charged carbocation (an electrophile) will react with the negatively charged bromide ion (a nucleophile) to form an alkyl halide.

$$CH_3CH=CHCH_3 \;+\; \overset{\delta+}{H}-\overset{\delta-}{Br} \longrightarrow CH_3CH\underset{+}{-}CHCH_3 \;+\; Br^- \longrightarrow CH_3CH-CHCH_3$$

a carbocation (with H below), 2-bromobutane / an alkyl halide (with Br and H below)

a carbocation

**2-bromobutane
an alkyl halide**

The description of the step-by-step process by which reactants (e.g., *alkene* $+$ HBr) are changed into products (e.g., alkyl halide) is called the **mechanism of the reaction.** The mechanism can be seen more clearly if the movement of the electrons in the reaction is depicted with curved arrows. The curved arrows are drawn to show how the electrons move as new covalent bonds are formed and existing covalent bonds are broken. Remember that the arrows show how the electrons move, so *they are drawn from an electron-rich center to an electron-deficient center.* An arrowhead with two barbs (⌢) represents the simultaneous movement of two electrons (an electron pair). We will see later that an arrowhead with one barb (⌢) represents the movement of one electron. (Notice that we have described these as "curved" arrows in order to distinguish them from the "straight" arrows used to link reactants with products in chemical reactions.)

$$CH_3CH=CHCH_3 \;+\; \overset{\delta+}{H}-\overset{\delta-}{Br}: \longrightarrow CH_3CH\underset{+}{-}CHCH_3 \;+\; :Br:^-$$

For the reaction of 2-butene with HBr, an arrow is drawn to show that the two electrons of the π bond are attracted to the partially positively charged hydrogen of HBr. The hydrogen, however, is not free to accept this pair of electrons because it is already bonded to a bromine, and you know that hydrogen can be bonded to only one atom at a time (Section 1.4). Therefore, as the π electrons of the alkene move toward the hydrogen, the H—Br bond breaks, with bromine keeping the bonding electrons. Notice that the π electrons are pulled away from one carbon but remain attached to the other. Thus, the two electrons that formerly formed the π bond of the alkene now form a σ bond between carbon and the hydrogen from HBr. The product of this first step in the reaction is a carbocation because the sp^2 carbon that did not form the new bond with hydrogen no longer shares the pair of π electrons. Therefore, it is positively charged.

In the second step of the reaction, a pair of nonbonding electrons on the negatively charged bromide ion forms a bond with the positively charged carbon of the carbocation. Notice that both steps of the reaction involve the reaction of an electrophile with a nucleophile.

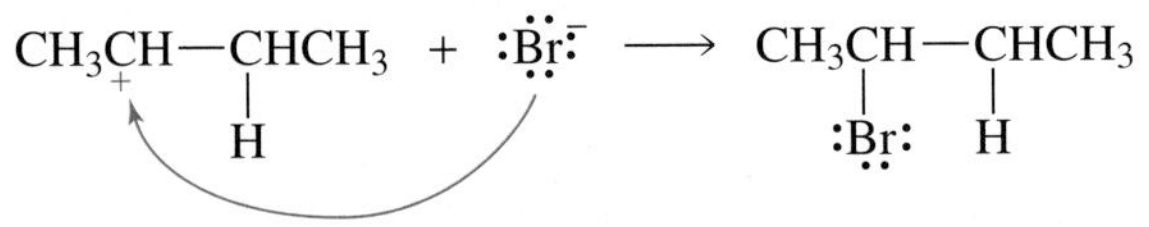

$$CH_3CH-CHCH_3 \ + \ :\ddot{B}r:^- \ \longrightarrow \ CH_3CH-CHCH_3$$

Solely from the knowledge that an electrophile reacts with a nucleophile, you have been able to predict that the product of the reaction of 2-butene and HBr is 2-bromobutane. Overall, the reaction involves the addition of 1 mole of HBr to 1 mole of the alkene. The reaction, therefore, is called an **addition reaction.** Because the first step of the reaction involves the addition of an electrophile (H^+) to the alkene, the reaction is more precisely called an **electrophilic addition reaction.** Electrophilic addition reactions are characteristic reactions of alkenes.

At this point you may think that it would be easier just to memorize that 2-bromobutane is the product of the reaction without trying to understand the mechanism of the reaction that explains why 2-bromobutane is the product. Keep in mind, however, that the number of reactions you encounter is going to increase substantially. Thus, you will find that having a thorough understanding of the mechanism of each of the reactions will teach you the unifying features behind organic chemistry, and mastering organic chemistry will become much easier and a lot more fun.

It will be helpful to do the exercise on drawing curved arrows in the Study Guide/Solution Manual (Special Topic II).

PROBLEM 8 ◆

Which of the following are electrophiles and which are nucleophiles?

$$H^- \qquad AlCl_3 \qquad CH_3O^- \qquad CH_3C{\equiv}CH \qquad CH_3\overset{+}{C}HCH_3 \qquad NH_3$$

PROBLEM 9

Use curved arrows to show the movement of electrons in each of the following reaction steps.

a. $CH_3\overset{O}{\overset{||}{C}}-O-H \ + \ H\ddot{O}:^- \ \longrightarrow \ CH_3\overset{O}{\overset{||}{C}}-O^- \ + \ H_2\ddot{O}:$

b. 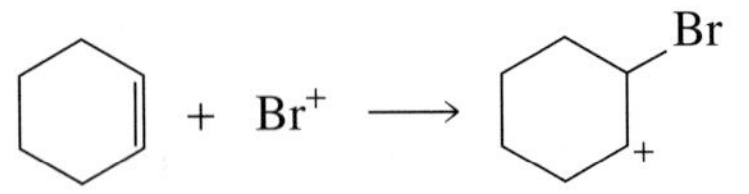 (cyclohexene) $+ \ Br^+ \ \longrightarrow$ (cyclohexane with Br and $+$)

c. $CH_3\overset{\ddot{O}:}{\overset{||}{C}}OH \ + \ H-\overset{+}{\underset{H}{\overset{|}{O}}}-H \ \longrightarrow \ CH_3\overset{+\ddot{O}H}{\overset{||}{C}}OH \ + \ H_2O$

d. $CH_3-\overset{CH_3}{\underset{CH_3}{\overset{|}{\underset{|}{C}}}}-Cl \ \longrightarrow \ CH_3-\overset{CH_3}{\underset{CH_3}{\overset{|}{\underset{|}{C}}}}{}^+ \ + \ Cl^-$

A FEW WORDS ABOUT CURVED ARROWS

1. Make certain that the arrows are drawn in the direction of the electron flow and never against the flow. This means that an arrow will always be drawn away from a negative charge and/or toward a positive charge.

correct   **incorrect**

$$CH_3-\underset{\underset{CH_3}{|}}{\overset{\overset{:\ddot{O}:^-}{|}}{C}}-\ddot{B}\ddot{r}: \longrightarrow CH_3-\overset{\overset{:\ddot{O}}{\|}}{\underset{CH_3}{C}} + :\ddot{B}\ddot{r}:^-$$

$$CH_3-\underset{\underset{CH_3}{|}}{\overset{\overset{:\ddot{O}:^-}{|}}{C}}-\ddot{B}\ddot{r}: \longrightarrow CH_3-\overset{\overset{:\ddot{O}}{\|}}{\underset{CH_3}{C}} + :\ddot{B}\ddot{r}:^-$$

$$CH_3-\underset{\underset{H}{|}}{\overset{+}{\ddot{O}}}-H \longrightarrow CH_3-\ddot{O}-H + H^+$$

$$CH_3-\underset{\underset{H}{|}}{\overset{+}{\ddot{O}}}-H \longrightarrow CH_3-\ddot{O}-H + H^+$$

2. Curved arrows are drawn to indicate the movement of electrons. Never use a curved arrow to indicate the movement of an atom (in this example the movement of a proton).

correct   **incorrect**

$$\underset{CH_3\overset{}{\underset{}{C}}CH_3}{\overset{+}{:\ddot{O}}-H} \longrightarrow \underset{CH_3\overset{:\ddot{O}}{\overset{\|}{C}}CH_3}{} + H^+$$

$$\underset{CH_3\overset{}{\underset{}{C}}CH_3}{\overset{+}{:\ddot{O}}-H} \longrightarrow \underset{CH_3\overset{:\ddot{O}}{\overset{\|}{C}}CH_3}{} + H^+$$

3. The arrow starts at the electron source. It does not start at an atom (in this example at a carbon atom).

correct

$$CH_3CH{=}CHCH_3 + H{-}\ddot{B}\ddot{r}: \longrightarrow CH_3\overset{+}{C}H-\underset{\underset{H}{|}}{C}HCH_3 + :\ddot{B}\ddot{r}:^-$$

incorrect

$$CH_3CH{=}CHCH_3 + H{-}\ddot{B}\ddot{r}: \longrightarrow CH_3\overset{+}{C}H-\underset{\underset{H}{|}}{C}HCH_3 + :\ddot{B}\ddot{r}:^-$$

Before you can understand the energy changes that take place in a reaction such as the addition of HBr to an alkene, you must have an understanding of *thermodynamics,* which describes a reaction at equilibrium, and an appreciation of *kinetics,* which deals with the rates of chemical reactions.

3.7 THERMODYNAMICS AND KINETICS

If we consider a reaction in which A is converted to B, the *thermodynamics* of the reaction tells us the relative amounts of A and B that are present when the reaction has reached equilibrium, whereas the *kinetics* of the reaction tells us how fast A is converted into B.

$$A \rightleftharpoons B$$

Reaction Coordinate Diagrams

The mechanism of a reaction shows the various steps that are believed to occur as reactants are converted into products. A **reaction coordinate diagram** describes the energy changes that take place in each of the steps of the mechanism. In a reaction coordinate diagram, the total energy of all species is plotted against the progress of the reaction. A reaction progresses from left to right as written in the chemical equation, so the energy of the reactants is plotted on the left-hand side of the *x*-axis, and the energy of the products is plotted on the right-hand side. *Remember that the more stable the species, the lower its energy.* A typical reaction coordinate diagram is shown in Figure 3.2.

> **The more stable the species, the lower its energy.**

$$A-B \ + \ C \ \rightleftharpoons \ A \ + \ B-C$$

reactants   products

Figure 3.2 ▶
A reaction coordinate diagram. The dashed lines in the transition state indicate bonds that are partially formed or partially broken.

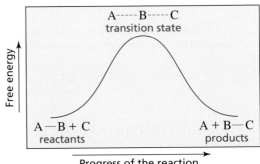

As the reactants are converted into products, the reaction passes through a *maximum* energy state called a **transition state.** The structure of the transition state lies somewhere between the structure of the reactants and the structure of the products. Bonds that break and bonds that form, as reactants are converted to products, are partially broken and partially formed in the transition state.

Thermodynamics

The field of chemistry that describes the properties of a system at equilibrium is called **thermodynamics.** The relative concentrations of reactants and products at equilibrium can be expressed numerically as an equilibrium constant, K_{eq}. For example, in a reaction in which *m* moles of A react with *n* moles of B to form *s* moles of C and *t* moles of D, K_{eq} is equal to the relative concentrations of products and reactants at equilibrium.

$$m\,A \ + \ n\,B \ \rightleftharpoons \ s\,C \ + \ t\,D$$

$$K_{eq} = \frac{[\text{products}]}{[\text{reactants}]} = \frac{[C]^s\,[D]^t}{[A]^m\,[B]^n}$$

The relative concentrations of products and reactants at equilibrium depend on their relative stabilities—*the more stable the compound, the greater its concentration at equilibrium.* Thus, if the products are more stable (have a lower free energy) than the reactants [Figure 3.3(a)], there will be a higher concentration of products than reactants at equilibrium and K_{eq} will be greater than 1. On the other hand, if the reactants are more stable than the products [Figure 3.3(b)], there will be a higher concentration of reactants than products at equilibrium and K_{eq} will be less than 1.

> **The more stable the compound, the greater its concentration at equilibrium.**

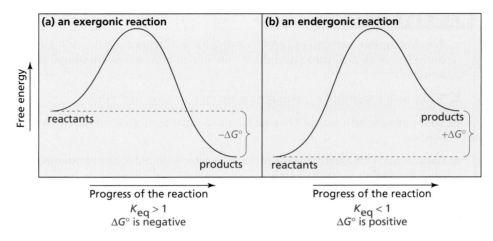

▲ Figure 3.3
Reaction coordinate diagrams for (a) a reaction in which the products are more stable than the reactants (an exergonic reaction) and (b) a reaction in which the products are less stable than the reactants (an endergonic reaction).

Josiah Willard Gibbs (1839–1903) was born in New Haven, Conn., the son of a Yale professor. In 1863 he obtained the first Ph.D. awarded by Yale in engineering. After studying in France and Germany, he returned to Yale to become a professor of mathematical physics. His work on free energy received little attention for more than 20 years because few chemists could understand his mathematical treatment and because Gibbs published it in Transactions of the Connecticut Academy of Sciences, a relatively obscure journal. In 1950 he was elected to the Hall of Fame for Great Americans.

Several thermodynamic parameters are used to describe a reaction at equilibrium. The difference between the free energy of the products and the free energy of the reactants at equilibrium under standard conditions is called the **Gibbs standard free energy change** ($\Delta G°$). The symbol $°$ indicates that the reaction takes place under standard conditions—all species at 1 M, temperature at 25 C$°$, and pressure at 1 atm.

$$\Delta G° = \text{(free energy of the products)} - \text{(free energy of the reactants)}$$

From this equation you can see that $\Delta G°$ will be negative if the products have a lower free energy (are more stable) than the reactants [(Figure 3.3(a)]. The reaction will release more energy into the system than it will consume from the system—it will be an **exergonic reaction.** If the products have a higher free energy (are less stable) than the reactants, $\Delta G°$ will be positive and the reaction will consume more energy from the system than it will release into the system—it will be an **endergonic reaction** [(Figure 3.3(b)]. (Notice that the terms *exergonic* and *endergonic* refer to whether the reaction has a negative $\Delta G°$ or a positive $\Delta G°$, respectively. Do not confuse these terms with *exothermic* and *endothermic,* which are defined later.)

Therefore, whether reactants or products are favored at equilibrium can be indicated either by the equilibrium constant (K_{eq}) or by the change in free energy ($\Delta G°$). These two quantities are related by the following equation:

$$\Delta G° = -RT \ln K_{eq}$$

where R is the gas constant (1.986×10^{-3} kcal mol^{-1} K^{-1} or 8.314×10^{-3} kJ mol^{-1} K^{-1} because 1 kcal = 4.184 kJ) and T is the temperature in Kelvin (K = $°$C + 273; therefore, 25 $°$C = 298 K). (By solving Problem 11, you will see that a small difference in $\Delta G°$ gives rise to a large difference in the relative concentrations of products and reactants.)

When products are favored at equilibrium, $\Delta G°$ is negative and K_{eq} is greater than 1.

When reactants are favored at equilibrium, $\Delta G°$ is positive and K_{eq} is less than 1.

The gas constant R *is thought to be named after **Henri Victor Regnault (1810–1878)**, who was commissioned by the French Minister of Public Works in 1842 to redetermine all the physical constants involved in the design and operation of the steam engine. Regnault was known for his work on the thermal properties of gases. Later, van't Hoff (p. 193) determined, while studying the thermodynamics of dilute solutions, that* R *could be used for all chemical equilibria.*

PROBLEM 10 ◆

a. Which of the monosubstituted cyclohexanes in Table 2.10 has a negative $\Delta G°$ for the conversion of an axial-substituted chair conformer to an equatorial-substituted chair conformer?

b. Which monosubstituted cyclohexane has the most negative $\Delta G°$ value?

c. Which monosubstituted cyclohexane has the greatest preference for the equatorial position?

d. Calculate $\Delta G°$ for interconversion of the axial-substituted and equatorial-substituted conformers of methylcyclohexane.

PROBLEM 11 / SOLVED

a. The $\Delta G°$ for conversion of "axial" fluorocyclohexane to "equatorial" fluorocyclohexane at 25 °C is −0.25 kcal/mol. Calculate the percentage of fluorocyclohexane molecules that have the fluoro substituent in the equatorial position.

b. Do the same calculation for isopropylcyclohexane (the $\Delta G°$ at 25 °C is −2.1 kcal/mol).

c. Why does isopropylcyclohexane have a greater percentage of conformer with the substituent in the equatorial position?

SOLUTION TO 11a

$$\text{fluorocyclohexane} \rightleftharpoons \text{fluorocyclohexane}$$
$$\text{axial} \qquad\qquad \text{equatorial}$$

$$\Delta G° = -0.25 \text{ kcal/mol at 25 °C}$$

$$\Delta G° = -RT \ln K_{eq}$$

$$-0.25 \frac{\text{kcal}}{\text{mol}} = -1.986 \times 10^{-3} \frac{\text{kcal}}{\text{mol K}} \times 298 \text{ K} \times \ln K_{eq}$$

$$\ln K_{eq} = 0.422$$

$$K_{eq} = 1.53 = \frac{[\text{fluorocyclohexane}]_{\text{equatorial}}}{[\text{fluorocyclohexane}]_{\text{axial}}} = \frac{1.53}{1}$$

Now we must determine the percentage of the total that is equatorial:

$$\frac{[\text{fluorocyclohexane}]_{\text{equatorial}}}{[\text{fluorocyclohexane}]_{\text{equatorial}} + [\text{fluorocyclohexane}]_{\text{axial}}} = \frac{1.53}{1.53 + 1} = \frac{1.53}{2.53} = .60 \text{ or } 60\%$$

The Gibbs standard free energy change ($\Delta G°$) has an enthalpy ($\Delta H°$) component and an entropy ($\Delta S°$) component.

$$\Delta G° = \Delta H° - T\Delta S°$$

The **enthalpy** term ($\Delta H°$) is the heat given off or the heat consumed during the course of a reaction. Heat is given off when bonds are formed, and heat is consumed when bonds are broken. Thus, $\Delta H°$ is a measure of the bond-making and bond-breaking processes that occur as reactants are converted into products.

If the bonds that are formed in a reaction are stronger than the bonds that are broken, more energy will be released as a result of bond formation than will be consumed in the bond-breaking process, and $\Delta H°$ will be negative. A reaction with a negative $\Delta H°$ is called an **exothermic reaction.** If the bonds that are formed are weaker than those that are broken, $\Delta H°$ will be positive. A reaction with a positive $\Delta H°$ is called an **endothermic reaction.**

$$\Delta H° = \text{(energy of the bonds being broken)} - \text{(energy of the bonds being formed)}$$

Entropy $(\Delta S°)$ is the degree of disorder. It is a measure of the freedom of motion in a system. Restricting the freedom of motion of a molecule decreases its entropy. For example, in a reaction in which two molecules come together to form a single molecule, the entropy in the product will be less than the entropy in the reactants because two individual molecules can move in ways that are not possible when the two are bound together in a single molecule. In such a reaction, $\Delta S°$ will be negative. In a reaction in which a single molecule is cleaved into two separate molecules, the products will have greater freedom of motion than the reactant, and $\Delta S°$ will be positive.

$$\Delta S° = \text{(freedom of motion of products)} - \text{(freedom of motion of reactants)}$$

A reaction with a negative $\Delta G°$ has a favorable equilibrium constant—the reaction is favored as written from left to right because the products are more stable than the reactants. If you examine the expression for the Gibbs standard free energy change, you will find that negative values of $\Delta H°$ and positive values of $\Delta S°$ contribute to make $\Delta G°$ negative. In other words, *the formation of products with stronger bonds and with greater freedom of motion causes $\Delta G°$ to be negative.*

The formation of products with stronger bonds and with greater freedom of motion causes $\Delta G°$ to be negative.

Values of $\Delta H°$ are relatively easy to calculate, so organic chemists frequently evaluate reactions only in terms of $\Delta H°$. If the reaction involves only a small change in entropy, the $T\Delta S°$ term will be small and the value of $\Delta H°$ will be very close to the value of $\Delta G°$. Ignoring the entropy term can be a dangerous practice, however, because many organic reactions occur with a significant change in entropy or occur at high temperatures and so have significant $T\Delta S°$ terms. It is permissible to use $\Delta H°$ values to approximate whether a reaction occurs with a favorable equilibrium constant, but if a precise answer is needed, $\Delta G°$ values must be used. When $\Delta G°$ values are used to construct reaction coordinate diagrams, the y-axis is free energy; when $\Delta H°$ values are used, the y-axis is potential energy.

PROBLEM 12

For which reaction will $\Delta S°$ be more significant?

a. $A \rightleftharpoons B$ or $A + B \rightleftharpoons C$

b. $A + B \rightleftharpoons C$ or $A + B \rightleftharpoons C + D$

PROBLEM 13 ◆

a. For a reaction with $\Delta H° = -12$ kcal mol^{-1} and $\Delta S = 0.01$ kcal K^{-1} mol^{-1}, calculate the $\Delta G°$ and the equilibrium constant: **1.** at 30 °C   **2.** at 150 °C

b. How does $\Delta G°$ change as T increases?

c. How does K_{eq} change as T increases?

Values of $\Delta H°$ can be calculated from bond dissociation energies (Table 3.1). For example, the $\Delta H°$ for the addition of HBr to ethene is calculated as shown here. The bond dissociation energy is indicated by the special term $DH°$. Recall from Section 1.14 that the value for the bond dissociation energy of the π bond is obtained by taking the bond dissociation energy of the double bond ($\sigma + \pi = 152$ kcal/mol) and subtracting the bond dissociation energy of the σ bond (91 kcal/mol).

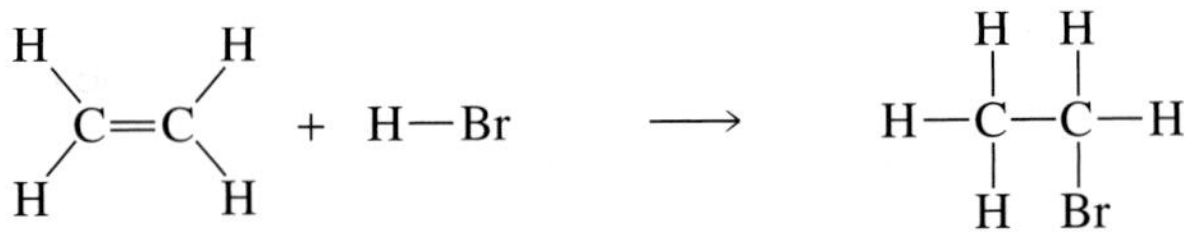

bonds being broken

π bond of ethene $\quad DH° = \quad 61$ kcal/mol

H—Br $\quad \underline{DH° = \quad 87 \text{ kcal/mol}}$

$DH°_{total} = 148$ kcal/mol

bonds being formed

C—H $\quad DH° = 101$ kcal/mol

C—Br $\quad \underline{DH° = \quad 69 \text{ kcal/mol}}$

$DH°_{total} = 170$ kcal/mol

$\Delta H°$ for the reaction $= DH°$ for bonds being broken $- DH°$ for bonds being formed

$= 148$ kcal/mol $- 170$ kcal/mol

$= -22$ kcal/mol

J. Berkowitz, G.B. Ellison, D. Gutman, J. Phys. Chem., **1994,** *98,* 2744.

TABLE 3.1 *Homolytic Bond Dissociation Energies ($DH°$) Y—Z $\rightarrow$ Y· + ·Z

Bond	$DH°$ kcal/mol	$DH°$ kJ/mol	Bond	$DH°$ kcal/mol	$DH°$ kJ/mol
CH_3—H	105	439	H—H	104	435
CH_3CH_2—H	101	423	F—F	38	159
$CH_3CH_2CH_2$—H	101	423	Cl—Cl	58	242
$(CH_3)_2CH$—H	99	414	Br—Br	46	192
$(CH_3)_3C$—H	97	406	I—I	36	150
			H—F	136	571
CH_3—CH_3	88	368	H—Cl	103	432
CH_3CH_2—CH_3	85	355	H—Br	87	366
$(CH_3)_2CH$—CH_3	84	351	H—I	71	298
$(CH_3)_3C$—CH_3	80	334			
			CH_3—F	108	451
$H_2C{=}CH_2$	152	636	CH_3—Cl	84	350
$HC{\equiv}CH$	200	837	CH_3CH_2—Cl	82	343
			$(CH_3)_2CH$—Cl	81	338
			$(CH_3)_3C$—Cl	79	330
HO—H	119	497	CH_3—Br	70	294
CH_3O—H	102	426	CH_3CH_2—Br	69	289
CH_3—OH	91	380	$(CH_3)_2CH$—Br	68	285
			$(CH_3)_3C$—Br	63	264
			CH_3—I	57	239
			CH_3CH_2—I	55	230

*See p. 163

The value of -22 kcal/mol for ΔH°—calculated by subtracting the ΔH° for the bonds being formed from the ΔH° for the bonds being broken—indicates that the addition of HBr to ethene is an exothermic reaction. But does this mean that the ΔG° for the reaction is also negative? Is the reaction exergonic as well as exothermic? Because ΔH° has a significant negative value (-22 kcal/mol), you can assume that ΔG° is also negative. If the value of ΔH° were close to 0, you could no longer assume that both ΔH° and ΔG° have the same sign.

Keep in mind that two assumptions are being made when you use ΔH° values to predict values of ΔG°. The first is that the entropy change in the reaction is small, with the result that the value of ΔH° is very close to the value of ΔG°; the second is that the reaction is taking place in the gas phase.

When reactions are carried out in solution, which is the case for the vast majority of organic reactions, the solvent molecules can interact with the reagents and with the products. Polar solvent molecules cluster around a charge (either a full charge or a partial charge) on the reactant or product so that the negative poles of the solvent molecules surround the positive charge and the positive poles of the solvent molecules surround the negative charge. The interaction between a solvent and a molecule (or ion) in solution is called **solvation.**

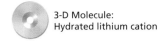

3-D Molecule:
Hydrated lithium cation

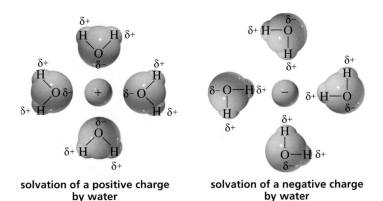

solvation of a positive charge   solvation of a negative charge
by water   by water

Solvation can have a large effect on the ΔH° of a reaction, and it can also affect the ΔS° of a reaction. In a reaction in which a polar reagent is solvated, for example, the ΔH° for breaking the dipole–dipole interactions between the solvent and the reagent has to be taken into account. In addition, solvation of a polar reagent or a polar product by a polar solvent can greatly reduce the freedom of motion of the solvent molecules, which will affect ΔS°.

PROBLEM 14 ◆

a. Using the bond dissociation energies in Table 3.1, calculate the ΔH° for the addition of HCl to ethene.

b. Calculate the ΔH° for the addition of H_2 to ethene.

c. Are the reactions exothermic or endothermic?

d. Do you expect the reactions to be exergonic or endergonic?

Kinetics

Knowing whether a given reaction is exergonic or endergonic will not tell you how fast the reaction occurs because the ΔG° of a reaction tells you only the difference between the stability of the reactants and the stability of the products. It does not tell you anything about the energy barrier of the reaction, which is the energy hill that must be climbed for the reactants to be converted into products. The higher the energy barrier, the slower the reaction. **Kinetics** is the field of chemistry that describes the rates of chemical reactions and the factors that affect those rates.

The energy barrier of a reaction, indicated in Figure 3.4 by $\Delta G^{\ddagger}$, is called the **free energy of activation.** It is the difference between the free energy of the transition state and the free energy of the reactants.

$$\Delta G^{\ddagger} = (\text{free energy of the transition state}) - (\text{free energy of the reactants})$$

The greater the $\Delta G^{\ddagger}$, the slower the reaction. Like ΔG°, $\Delta G^{\ddagger}$ has both an enthalpy component and an entropy component. Notice that any quantity that refers to the transition state is represented by the double dagger superscript (‡).

$$\Delta G^{\ddagger} = \Delta H^{\ddagger} - T\Delta S^{\ddagger}$$

$$\Delta H^{\ddagger} = (\text{enthalpy of the transition state}) - (\text{enthalpy of the reactants})$$

$$\Delta S^{\ddagger} = (\text{entropy of the transition state}) - (\text{entropy of the reactants})$$

Some exergonic reactions have small free energies of activation, so the reactions can take place at room temperature [Figure 3.4(a)]. Some exergonic reactions have free energies of activation that are so large that the reaction cannot take place without adding energy above that provided by the existing thermal conditions. [Figure 3.4(b)]. Endergonic reactions can also have either small free energies of activation as in Figure 3.4(c) or large free energies of activation as in Figure 3.4(d).

Notice that ΔG° relates to the equilibrium constant of the reaction, whereas $\Delta G^{\ddagger}$ relates to the rate of the reaction. The **thermodynamic stability** of a compound is indicated by ΔG°. If ΔG° is negative, for example, the product is *thermodynamically stable* compared with the reactant; and if ΔG° is positive, the product is *thermody-*

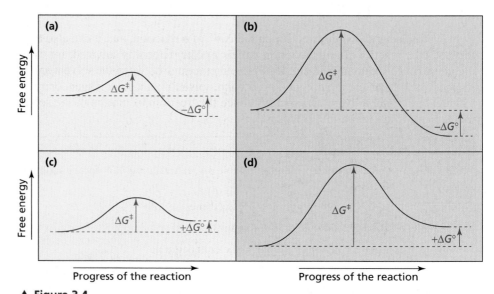

▲ **Figure 3.4**
Reaction coordinate diagrams for (a) a fast exergonic reaction; (b) a slow exergonic reaction; (c) a fast endergonic reaction; (d) a slow endergonic reaction. (The four reaction coordinates are drawn on the same scale.)

namically unstable compared with the reactant. The **kinetic stability** of a compound is indicated by $\Delta G^{\ddagger}$. If $\Delta G^{\ddagger}$ is large, the compound is *kinetically stable* because it is not very reactive. If $\Delta G^{\ddagger}$ is small, the compound is *kinetically unstable*—it is reactive. Generally when chemists use the term "stability," they are referring to "thermodynamic stability."

PROBLEM 15 ◆

a. Which of the reactions in Figure 3.4 has a thermodynamically stable product?

b. Which of the reactions in Figure 3.4 has the most kinetically stable product?

c. Which of the reactions in Figure 3.4 has the least kinetically stable product?

PROBLEM 16

Draw a reaction coordinate diagram for a reaction in which:

a. the product is thermodynamically unstable and kinetically unstable.

b. the product is thermodynamically unstable and kinetically stable.

The rate of a chemical reaction is the speed at which reacting substances are used up or the speed at which products are formed. The rate of a reaction depends on the following three factors.

1. *The number of collisions that take place between the reacting molecules in a given period of time.* The greater the number of collisions, the faster the reaction.

2. *The fraction of the collisions that occur with sufficient energy to get the reacting molecules over the energy barrier.* If the free energy of activation is small, more collisions will lead to reaction than if the free energy of activation is large.

3. *The fraction of the collisions that occur with the proper orientation.* For example, in the reaction of 2-butene with HBr, reaction will occur only if the molecules collide with the hydrogen of HBr approaching the π bond of 2-butene. If collision occurs with the hydrogen approaching a methyl group of 2-butene, no reaction will take place regardless of the energy of the collision.

$$\text{rate of a reaction} = \left(\begin{array}{c} \text{number of collisions} \\ \text{per unit of time} \end{array} \right) \times \left(\begin{array}{c} \text{fraction with} \\ \text{sufficient energy} \end{array} \right) \times \left(\begin{array}{c} \text{fraction with} \\ \text{proper orientation} \end{array} \right)$$

Increasing the concentration of the reactants increases the rate of a reaction because it increases the number of collisions that occur in a given period of time. Increasing the temperature at which the reaction is carried out also increases the rate of a reaction because it increases the frequency of collisions and the number of collisions that have sufficient energy to get the reacting molecules over the energy barrier.

For a reaction in which a single reactant molecule (A) is converted into a product molecule (B), the rate of the reaction is proportional to the concentration of A. If the concentration of A is doubled, the rate of the reaction will double; if the concentration of A is tripled, the rate of the reaction will triple. Because the rate of this reaction is proportional to the concentration of only one reactant, it is called a **first-order reaction.**

$$A \longrightarrow B$$

$$\text{rate} \propto [A]$$

We can replace the "proportional to" sign ($\propto$) with an "equal to" sign if we use a proportionality constant, k, which is called a **rate constant.** The rate constant of a first-order reaction is called a **first-order rate constant.**

$$\text{rate} = k[A]$$

A reaction whose rate depends on the concentrations of two reactants is called a **second-order reaction.** If the concentration of either A or B is doubled, the rate of the reaction will double. If the concentrations of both A and B are doubled, the rate of the reaction will quadruple. The rate constant, k, is a **second-order rate constant.**

$$A + B \longrightarrow C + D$$

$$\text{rate} = k[A][B]$$

A reaction in which two molecules of A combine to form a molecule of B is also a second-order reaction. If the concentration of A is doubled, the rate of the reaction will quadruple.

$$A + A \longrightarrow B$$

$$\text{rate} = k[A]^2$$

Do not confuse the *rate constant* of a reaction with the *rate* of a reaction. The *rate constant* tells us how easy it is to reach the transition state (how easy it is to get over the energy barrier). Low energy barriers are associated with large rate constants [Figures 3.4(a) and 3.4(c)], whereas high energy barriers have small rate constants [Figures 3.4(b) and 3.4(d)]. The reaction *rate* is a measure of the amount of product that is formed per unit of time. The preceding equations show that the "rate" is the product of the "rate constant" and the concentration(s) of the reactants. *Thus, reaction rates depend on concentration, whereas rate constants are independent of concentration.* Therefore, when we compare two reactions to see which one occurs more easily, we must compare their rate constants and not their concentration-dependent rates of reaction. (Appendix III explains how rate constants are determined.)

The smaller the rate constant, the slower the reaction.

RATE CONSTANTS

The rate of a reaction measures the change that has occurred in the concentration of some species in a period of time. Rate is generally expressed in moles per liter (molarity) per second (Ms^{-1}). Therefore, the rate constant of a first-order reaction has units of s^{-1}, so when it is multiplied by the molar concentration of the reactant (M), the rate will have the appropriate units (Ms^{-1}). Likewise, the rate constant of a second-order reaction has units of $M^{-1}s^{-1}$, so when it is multiplied by M^2, the rate will have units of Ms^{-1}.

First-order: $\quad \text{rate } (Ms^{-1}) = k(s^{-1})M$

Second-order: $\quad \text{rate } (Ms^{-1}) = k(M^{-1}s^{-1})MM$

Although rate constants are independent of concentration, they depend on temperature. The **Arrhenius equation** relates the rate constant of a reaction to the experimental energy of activation and to the temperature at which the reaction is carried out. A good rule of thumb is that an increase of 10 °C in temperature will double the rate constant for a reaction and, therefore, double the reaction rate.

The Arrhenius equation:

$$k = Ae^{-E_a/RT}$$

where k is the rate constant, E_a is the experimental energy of activation, R is the gas constant (1.986×10^{-3} kcal mol^{-1}K^{-1} or 8.314×10^{-3} kJ mol^{-1}K^{-1}), T is the absolute temperature (K), and A is the frequency factor. The frequency factor accounts for the fraction of collisions that occur with the proper orientation for reaction. The factor, $e^{-E_a/RT}$, corresponds to the fraction of the collisions that have the minimum energy (E_a) needed to react. Taking the logarithms of both sides, the equation becomes:

$$\ln k = \ln A - E_a/RT$$

THE DIFFERENCE BETWEEN $\Delta G^{\ddagger}$ AND E_a

Do not confuse the *free energy of activation,* $\Delta G^{\ddagger}$, with the **experimental energy of activation,** E_a, in the Arrhenius equation. The free energy of activation ($\Delta G^{\ddagger} = \Delta H^{\ddagger} - T\Delta S^{\ddagger}$) has both an enthalpy component and an entropy component, whereas the experimental energy of activation ($E_a = \Delta H^{\ddagger} + RT$) has only an enthalpy component. The experimental energy of activation is an approximate energy barrier to a reaction. The true energy barrier to a reaction is given by $\Delta G^{\ddagger}$ because some reactions are driven by a change in enthalpy, some by a change in entropy, but most by a change in both enthalpy and entropy.

PROBLEM 17

From the Arrhenius equation, predict how:

a. increasing the experimental activation energy will affect the rate constant of a reaction.

b. increasing the temperature will affect the rate constant of a reaction.

CALCULATING KINETIC PARAMETERS

In order to calculate E_a, $\Delta H^{\ddagger}$, and $\Delta S^{\ddagger}$ for a reaction, rate constants for the reaction must be obtained at several temperatures.

- E_a can be obtained from the Arrhenius equation (the slope of a plot of ln k versus $1/T$ because

$$\ln k_2 - \ln k_1 = -E_a/R\left(\frac{1}{T_2} - \frac{1}{T_1}\right)$$

- At a given temperature, $\Delta H^{\ddagger}$ can be determined from E_a because $\Delta H^{\ddagger} = E_a - RT$.

- $\Delta G^{\ddagger}$, in kJ/mol, can be determined from the following equation, which relates $\Delta G^{\ddagger}$ to the rate constant at a given temperature:

$$-\Delta G^{\ddagger} = RT \ln kh/Tk_B$$

where h is Planck's constant (6.62608×10^{-34} Js) and k_B is Boltzman's constant (1.38066×10^{-23} JK^{-1})

- The entropy of activation can be determined from the other two kinetic parameters, $\Delta S^{\ddagger} = (\Delta H^{\ddagger} - \Delta G^{\ddagger})/T$

Ludwig Edward Boltzmann (1844–1906), *the son of a civil servant, was born in Vienna and became a professor at the University of Vienna. He shares the credit for the kinetic theory of gases with Maxwell, although their work was done independently. He was the founder of statistical mechanics, a branch of physics that deals with the macroscopic behavior of a system based on the microscopic interactions of its constituents.*

PROBLEM 18 ◆

The rate constant for a reaction can be increased by _____ the stability of the reactant or by _____ the stability of the transition state.

PROBLEM 19 ◆

At 30 °C, the second-order rate constant for the reaction of methyl chloride and HO^- is $1.0 \times 10^{-5} \, M^{-1}s^{-1}$.

a. What is the rate of the reaction when $[CH_3Cl] = 0.10$ M and $[HO^-] = 0.10$ M?

b. If the concentration of methyl chloride is decreased to 0.01 M, what effect will this have on the *rate* of the reaction?

c. If the concentration of methyl chloride is decreased to 0.01 M, what effect will this have on the *rate constant* of the reaction?

How are the rate constants for a reaction related to the equilibrium constant? At equilibrium, the rate of the forward reaction must be equal to the rate of the reverse reaction because the amounts of reactants and products are not changing.

$$A \underset{k_{-1}}{\overset{k_1}{\rightleftharpoons}} B$$

forward rate = reverse rate

$$k_1 [A] = k_{-1} [B]$$

Therefore,

$$K_{eq} = \frac{k_1}{k_{-1}} = \frac{[B]}{[A]}$$

From this equation, you can see that the equilibrium constant for a reaction can be determined from the relative concentrations of the reactants and products at equilibrium or from the relative rate constants for the forward and reverse reactions. The reaction shown in Figure 3.3(a) has a large equilibrium constant because the products are much more stable than the reactants. You could also say that it has a large equilibrium constant because the rate constant of the forward reaction is much greater than the rate constant of the reverse reaction.

Reaction Coordinate Diagram for the Addition of HBr to 2-Butene

You have seen that the addition of HBr to 2-butene is a two-step reaction. The structure of the transition state for each of the steps is shown on page 137 in brackets. Notice that the bonds that break and the bonds that form during the course of the reaction are partially broken and partially formed in the transition state. Similarly, atoms that become charged or lose their charge during the course of the reaction are partially charged in the transition state. A reaction coordinate diagram can be drawn for each of the steps (Figure 3.5).

The first step of the reaction is the conversion of an alkene into a carbocation that is less stable than the reactants. The first step of the reaction, therefore, is endergonic ($\Delta G°$ is positive). In the second step of the reaction, the carbocation reacts with a nucleophile to form a product that is more stable than the carbocation reactant. This step, therefore, is exergonic ($\Delta G°$ is negative).

$$CH_3CH{=}CHCH_3 \;+\; HBr \;\longrightarrow\; \begin{bmatrix} \overset{\delta+}{CH_3CH}{=\!=}CHCH_3 \\ \vdots \\ H \\ \vdots \\ \underset{\delta-}{Br} \end{bmatrix}^{\ddagger} \;\longrightarrow\; CH_3\underset{+}{CH}CH_2CH_3 \;+\; Br^{-}$$

transition state

$$CH_3\underset{+}{CH}CH_2CH_3 \;+\; Br^{-} \;\longrightarrow\; \begin{bmatrix} \overset{\delta+}{CH_3CH}CH_2CH_3 \\ \vdots \\ \underset{\delta-}{Br} \end{bmatrix}^{\ddagger} \;\longrightarrow\; \underset{Br}{CH_3CHCH_2CH_3}$$

transition state

Mechanistic Tutorial:
Addition of HBr
to an alkene

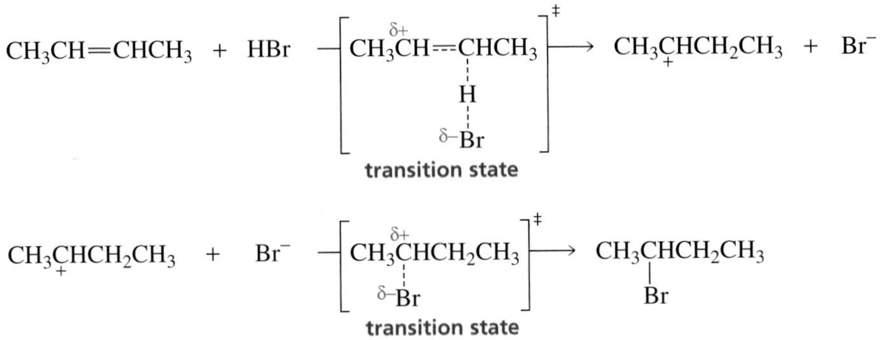

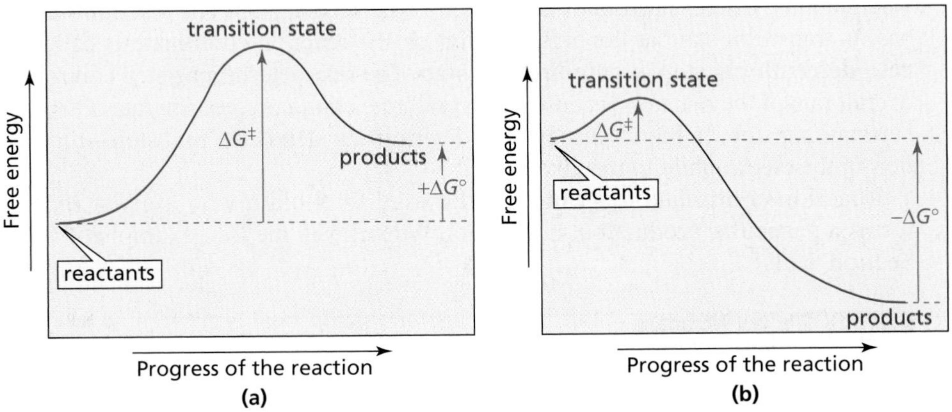

(a)
Free energy vs Progress of the reaction — transition state, $\Delta G^{\ddagger}$, reactants, products, $+\Delta G^{\circ}$

(b)
Free energy vs Progress of the reaction — transition state, $\Delta G^{\ddagger}$, reactants, $-\Delta G^{\circ}$, products

▲ **Figure 3.5**
Reaction coordinate diagrams for the two steps in the addition of HBr to 2-butene:
(a) the first step; (b) the second step.

Because the product of the first step is the reactant for the second step, we can hook the two reaction coordinate diagrams together to obtain the reaction coordinate diagram for the overall reaction (Figure 3.6). The ΔG° for the overall reaction is the difference between the free energy of the final products and the free energy of the initial reactants. Figure 3.6 shows that ΔG° for the overall reaction is negative. Therefore, the overall reaction is exergonic.

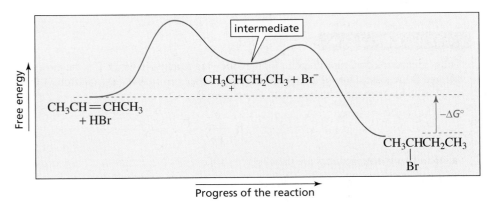

▲ **Figure 3.6**
Reaction coordinate diagram for the addition of HBr to 2-butene.

A chemical species that is the product of one step and the reactant for the next step is called an **intermediate.** The carbocation intermediate in this reaction is too unstable to be isolated, but some reactions have more stable intermediates that can be isolated. Transition states, in contrast, represent the highest-energy structures that are involved in the reaction. They exist only fleetingly and can never be isolated. Do not confuse transition states and intermediates: *Transition states have partially formed bonds, whereas intermediates have fully formed bonds.*

Transition states have partially formed bonds. Intermediates have fully formed bonds.

You can see from the reaction coordinate diagram that the free energy of activation for the first step of the reaction is greater than the free energy of activation for the second step. In other words, the rate constant for the first step is smaller than the rate constant for the second step. This is what you would expect because the molecules in the first step of this reaction must collide with sufficient energy to break covalent bonds, whereas no bonds are broken in the second step. The reaction step that has its transition state at the highest point on the reaction coordinate is called the **rate-determining step** or **rate-limiting step.** The rate-determining step controls the overall rate of the reaction because the overall rate cannot exceed the rate of the rate-determining step. In Figure 3.6 the rate-determining step is the first step—the addition of the electrophile to the alkene.

Reaction coordinate diagrams can be used to explain why a given reaction forms a particular product but not others. We will see the first example of this in Section 3.11.

PROBLEM 20

Draw a reaction coordinate diagram for a two-step reaction in which the first step is endergonic, the second step is exergonic, and the overall reaction is endergonic. Label reactants, products, intermediates, and transition states.

PROBLEM 21

Which step in the following reaction sequence is rate-determining?

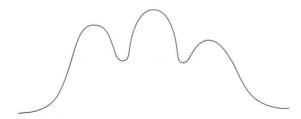

PROBLEM 22 ◆

Draw a reaction coordinate diagram for the following reaction in which C is the most stable and B the least stable of the three species, and the transition state going from A to B is more stable than the transition state going from B to C.

$$A \underset{k_{-1}}{\overset{k_1}{\rightleftharpoons}} B \underset{k_{-2}}{\overset{k_2}{\rightleftharpoons}} C$$

a. How many intermediates are there?

b. How many transition states are there?

c. Which is the faster step in the forward direction?

d. Which is the faster step in the reverse direction?

e. Of the four steps, which is the fastest?

f. Which is the rate-determining step in the forward direction?

g. Which is the rate-determining step in the reverse direction?

You have seen that an alkene such as 2-butene undergoes an electrophilic addition reaction with HBr. The first step in the reaction is the slow addition of the electrophilic proton to the nucleophilic alkene to form a carbocation intermediate. The positively charged intermediate reacts rapidly with the negatively charged ion (Br^-) in the second step of the reaction.

**3.8
GENERAL
MECHANISM FOR
ELECTROPHILIC
ADDITION
REACTIONS**

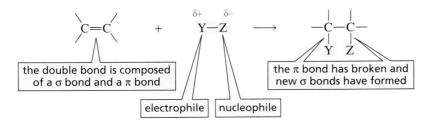

a carbocation
intermediate

In the following sections we will look at a wide variety of alkene reactions. You will see that many of the reactions form carbocation intermediates like the one formed when HBr reacts with an alkene. Some of the reactions form other kinds of intermediates and some form no intermediates at all. As you study each reaction, notice the feature that all alkene reactions have in common. *The relatively loosely held π electrons of the carbon–carbon double bond are attracted to an electrophile. Thus, each reaction starts with the addition of an electrophile to one of the sp^2 carbons of the alkene and concludes with the addition of a nucleophile to the other sp^2 carbon.* The end result is that the π bond breaks and the *sp^2* carbons form new σ bonds with the reagent.

$$\overset{\displaystyle\diagdown}{\underset{\displaystyle\diagup}{C}}=\overset{\displaystyle\diagup}{\underset{\displaystyle\diagdown}{C}} \quad + \quad \overset{\delta+ \quad \delta-}{Y-Z} \quad \longrightarrow \quad -\overset{|}{\underset{|}{C}}-\overset{|}{\underset{|}{C}}-$$

the double bond is composed
of a σ bond and a π bond

electrophile   nucleophile

the π bond has broken and
new σ bonds have formed

This reactivity makes alkenes an important class of organic compounds because they can be used to synthesize a wide variety of products (alkyl halides, alcohols, ethers, etc.). The particular product obtained from the reaction of an alkene depends only on the reagent that is used in the addition reaction (Y–Z).

If the electrophilic reagent that adds to an alkene is a hydrogen halide (HF, HCl, HBr, or HI), the product of the reaction will be an alkyl halide.

**3.9
ADDITION OF
HYDROGEN
HALIDES**

$$CH_2{=}CH_2 \ + \ HCl \ \longrightarrow \ CH_3CH_2Cl$$
ethene   ethyl chloride

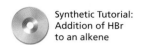

Synthetic Tutorial:
Addition of HBr
to an alkene

H₃C, CH₃ C=C, H₃C, CH₃ + HBr ⟶ CH₃CH—CCH₃ (CH₃ CH₃, Br)

2,3-dimethyl-2-butene   **2-bromo-2,3-dimethylbutane**

(cyclohexene) + HI ⟶ (iodocyclohexane, I)

cyclohexene   **iodocyclohexane**

Because the alkenes in the preceding reactions have the same substituents on the sp^2carbons, it is easy to determine the product of the reaction: The electrophile (H^+) adds to one of the sp^2carbons, and the nucleophile (X^-) adds to the other sp^2 carbon. But what happens if the alkene does not have the same substituents on the sp^2 carbons? Which sp^2 carbon gets the hydrogen? For example, does the addition of HCl to 2-methylpropene produce *tert*-butyl chloride or isobutyl chloride?

$$CH_3C=CH_2 + HCl \longrightarrow CH_3CCH_3 \quad or \quad CH_3CHCH_2Cl$$
(CH₃; Cl)

2-methylpropene   **tert-butyl chloride**   **isobutyl chloride**

To answer this question we need to look at the mechanism of the reaction. The first step of the reaction, the addition of H^+ to an sp^2 carbon to form either the *tert*-butyl cation or the isobutyl cation, is the rate-determining step. If there is any difference in the rate of formation of these two carbocations, the one that is formed faster will be the preferred product of the first step. Moreover, the particular carbocation that is formed in the first step determines the final product of the reaction. That is, if the *tert*-butyl cation is formed, it will quickly react with Cl^- to form *tert*-butyl chloride. On the other hand, if the isobutyl cation is formed, it will quickly react with Cl^- to form isobutyl chloride. It turns out that the only product of the reaction is *tert*-butyl chloride, so we know that the *tert*-butyl cation is formed faster than the isobutyl cation.

$CH_3C=CH_2 + HCl$

$CH_3\overset{+}{C}CH_3$ **tert-butyl cation** $\xrightarrow{Cl^-}$ CH_3CCH_3 (CH₃, Cl) **tert-butyl chloride only product formed**

$CH_3\overset{+}{C}HCH_2$ **isobutyl cation** $\xrightarrow{Cl^-}$ CH_3CHCH_2Cl (CH₃) **isobutyl chloride not formed**

Now we need to take a look at the factors that affect the stability of carbocations and, thus, the ease with which they are formed.

3.10
CARBOCATION
STABILITY

Carbocations are classified according to the number of alkyl substituents that are bonded to the positively charged carbon: A **primary carbocation** has one such substituent, a **secondary carbocation** has two, and a **tertiary carbocation** has three.

The stability of a carbocation increases as the number of alkyl substituents bonded to the positively charged carbon atom increases, so tertiary carbocations are more stable than secondary carbocations, and secondary carbocations are more stable than primary carbocations.

> **The greater the number of alkyl substituents bonded to the positively charged carbon, the more stable the carbocation will be.**

relative stabilities of carbocations

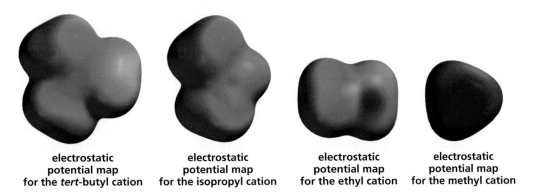

a tertiary carbocation > a secondary carbocation > a primary carbocation > methyl cation

increasing stability

Why does the stability of a carbocation increase as the number of alkyl substituents increases? Alkyl groups decrease the concentration of positive charge on the carbon—and decreasing the concentration of charge increases the stability of the carbocation. Notice that the blue—that represents positively-charged regions—is most intense for the least stable methyl cation and is least intense for the most stable *tert*-butyl cation.

electrostatic potential map for the *tert*-butyl cation   electrostatic potential map for the isopropyl cation   electrostatic potential map for the ethyl cation   electrostatic potential map for the methyl cation

Remember that the positive charge on a carbon signifies an empty p orbital (Section 1.10). The orbital of an adjacent carbon–hydrogen or carbon–carbon σ bond can overlap with the empty p orbital (Figure 3.7). The movement of the electrons from the σ bond toward the vacant p orbital causes a partial positive charge to develop on the carbon bonded by the σ bond. Therefore, the positive charge is no longer localized solely on one atom but is spread out over a greater volume of space. This stabilizes the carbocation because a charged species is more stable if its charge is spread out (delocalized) over more than one atom (Section 1.19). Delocalization of electrons by the overlap of a σ bond orbital with an empty p orbital is called **hyperconjugation.**

Hyperconjugation occurs only if the σ bond and the empty p orbital have the proper orientation. Achieving the proper orientation is not a problem, however, because there is free rotation about a carbon–carbon σ bond (Section 2.10). In the case of the *tert*-butyl cation, there are nine carbon–hydrogen σ bond orbitals that can potentially overlap with the empty p orbital of the positively charged carbon. The isopropyl cation has six such σ bond orbitals, and the ethyl cation has three σ bond orbitals that can overlap with the empty p orbital. Therefore, there is greater stabilization through hyperconjugation in the tertiary *tert*-butyl cation than in the secondary isopropyl cation, and greater stabilization in the secondary isopropyl cation than in the primary ethyl cation.

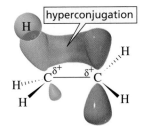

▲ **Figure 3.7**
Stabilization of a carbocation by hyperconjugation: The σ electrons in an adjacent carbon–hydrogen bond spread into the empty p orbital.

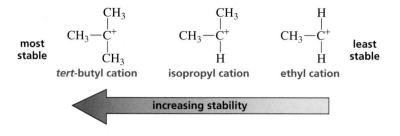

most stable — *tert*-butyl cation — isopropyl cation — ethyl cation — least stable

increasing stability

PROBLEM 23 ◆

List the carbocations in order of decreasing stability.

a.

$$CH_3CH_2\overset{+}{\underset{\underset{\textstyle}{}}{C}}CH_3 \text{ with } CH_3 \qquad CH_3CH_2\overset{+}{C}HCH_3 \qquad CH_3CH_2CH_2\overset{+}{C}H_2$$

b. $CH_3\underset{\underset{\textstyle Cl}{|}}{C}HCH_2\overset{+}{C}H_2 \qquad CH_3\underset{\underset{\textstyle CH_3}{|}}{C}HCH_2\overset{+}{C}H_2 \qquad CH_3CH_2\underset{\underset{\textstyle Cl}{|}}{C}H\overset{+}{C}H_2$

PROBLEM 24 ◆

a. How many carbon–hydrogen bond orbitals are available for overlap with the vacant *p* orbital in the methyl cation?

b. Which is more stable, a methyl cation or an ethyl cation?

3.11 THE STRUCTURE OF THE TRANSITION STATE

Knowing something about the structure of a transition state is important when you are trying to predict the products of a reaction. In Section 3.7 we said that the structure of the transition state lies between the structure of the reactants and the structure of the products. But what do we mean by "between"? Does the structure of the transition state lie *exactly* halfway between the structures of the reactants and products (as in II, below), or does it resemble the reactants more closely than it resembles the products (as in I), or is it more like the products than the reactants (as in III)?

$$A-B + C \quad \begin{bmatrix} \text{(I)} & A----B\cdots\cdots\cdots C \\ \text{(II)} & A------B------C \\ \text{(III)} & A\cdots\cdots\cdots B----C \end{bmatrix}^{\ddagger} \longrightarrow A + B-C$$

reactants   transition state   products

George Simms Hammond was born in Maine in 1921. He received a B.S. from Bates College in 1943 and a Ph.D. from Harvard University in 1947. He was a professor of chemistry at Iowa State University and at the California Institute of Technology and was a scientist at Allied Chemical Co.

According to the **Hammond postulate,** *the transition state will be more similar in structure to the species that it is more similar to in energy.* In the case of an exergonic reaction, the transition state is more similar in energy to the reactant than to the product (Figure 3.8, curve I). Therefore, the structure of the transition state will more closely resemble the structure of the reactant than that of the product. In an endergonic reaction (Figure 3.8, curve III), the transition state is more similar in energy to the product, so the structure of the transition state will more closely resemble the structure of the product. Only when the reactant and the product have identical energies (Figure 3.8, curve II) would we expect the structure of the transition state to be halfway between the structures of the reactant and the product.

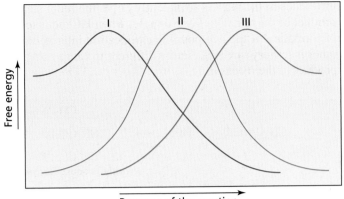

◀ **Figure 3.8**
Reaction coordinate diagrams for reactions with (I) an early transition state, (II) a mid-way transition state, and (III) a late transition state.

Now we can understand why the *tert*-butyl cation is formed faster than the isobutyl cation when 2-methylpropene reacts with HCl. Of the two possible products, the *tert*-butyl cation (a tertiary carbocation) is more stable than the isobutyl cation (a primary carbocation). Because formation of a carbocation is an endergonic reaction (Figure 3.9), the structure of the transition state will resemble the structure of the carbocation product. This means that the transition state will have a significant amount of positive charge on a carbon. The same factors that stabilize the positively charged product will stabilize the partially positively charged transition state. Therefore, the transition state leading to the *tert*-butyl cation is more stable than the transition state leading to the isobutyl cation, so the *tert*-butyl cation will be formed faster. It is important to keep in mind that the amount of positive charge in the transition state is not as great as the amount of positive charge in the carbocation product, so the difference in the stabilities of the two transition states is not as great as the difference in the stabilities of the two carbocation products.

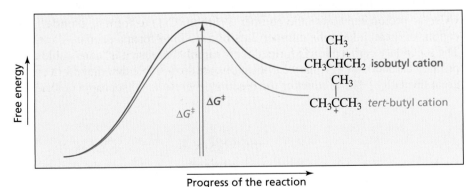

◀ **Figure 3.9**
Reaction coordinate diagram for the addition of H^+ to 2-methylpropene to form the primary isobutyl cation and the tertiary *tert*-butyl cation.

You have seen that the rate of a reaction is determined by the free energy of activation, which is the difference between the free energy of the transition state and the free energy of the reactant (Section 3.7). Because the free energy of activation for formation of the *tert*-butyl cation is less than that for formation of the isobutyl cation, the *tert*-butyl cation will be formed faster. In an electrophilic addition reaction, therefore, the more stable carbocation will be the one that is formed more rapidly.

The difference in the rate of formation of the two carbocations determines the relative amounts of the products that are formed. If the difference in the rates is small, both products will form, but the more stable product will be the major product. If the difference in the rates is sufficiently large, the more stable product will be the only product of the reaction. For example, when HCl adds to 2-methylpropene, the rates of formation of the two possible carbocation intermediates (one is primary and the other is tertiary) are sufficiently different to cause *tert*-butyl chloride to be the only product of the reaction.

$$CH_3\overset{\underset{\textstyle |}{CH_3}}{C}{=}CH_2 \; + \; HCl \; \longrightarrow \; CH_3\overset{\underset{\textstyle |}{CH_3}}{\underset{\textstyle \underset{|}{Cl}}{C}}CH_3 \qquad \left[CH_3\overset{\underset{\textstyle |}{CH_3}}{C}HCH_2Cl \right]$$

$$\text{only product formed} \qquad\qquad \text{not formed}$$

PROBLEM 25 ◆

Which compound will react faster with HCl? (*Hint:* the two alkenes have about the same stability.)

$$CH_3CH{=}CHCH_3 \quad \text{or} \quad CH_3\overset{\underset{\textstyle |}{CH_3}}{C}{=}CH_2$$

3.12 REGIOSELECTIVITY OF ELECTROPHILIC ADDITION REACTIONS: MARKOVNIKOV'S RULE

The sp^2 carbon that does *not* become attached to the proton is the one that is positively charged in the carbocation.

When an alkene that does not have the same substituents on the sp^2 carbons undergoes an electrophilic addition reaction, the electrophile can add to two different sp^2 carbon atoms. You have just seen in Section 3.11 that the major product of the reaction is the one obtained by adding the electrophile to the sp^2 carbon that results in formation of the more stable carbocation. For example, when propene reacts with HCl, the proton can add to the number 1 carbon (C-1) to form a secondary carbocation, or it can add to the number 2 carbon (C-2) to form a primary carbocation. The secondary carbocation is formed more rapidly because it is more stable than the primary carbocation. (Primary carbocations are so unstable that they form only with great difficulty.) The product of the reaction, therefore, is isopropyl chloride.

$$CH_3\overset{2}{C}H{=}\overset{1}{C}H_2 \; + \; H^+$$

$$CH_3\overset{+}{C}HCH_3 \xrightarrow{\;Cl^-\;} CH_3\overset{\underset{\textstyle |}{Cl}}{C}HCH_3$$

a secondary carbocation · isopropyl chloride

$$CH_3CH_2\overset{+}{C}H_2$$

a primary carbocation

The major product obtained from the addition of HI to 2-methyl-2-butene is 2-iodo-2-methylbutane; only a small amount of 2-iodo-3-methylbutane is obtained. The major product obtained from the addition of HBr to 1-methylcyclohexene is 1-bromo-1-methylcyclohexane. In both cases, the more stable tertiary carbocation is formed more rapidly than the less stable secondary carbocation, so the major product of each reaction is the one that results from forming the tertiary carbocation.

$$CH_3CH=CCH_3 \quad + \quad HI \quad \longrightarrow$$

2-methyl-2-butene

with CH_3 substituent

$$CH_3CH_2CCH_3 \quad + \quad CH_3CHCHCH_3$$

with CH_3 and I substituents

2-iodo-2-methylbutane
major product

2-iodo-3-methylbutane
minor product

1-methylcyclohexene $+$ HBr $\longrightarrow$

1-bromo-1-methyl-
cyclohexane
major product

$+$

1-bromo-2-methyl-
cyclohexane
minor product

The two different products in each of these reactions are called constitutional isomers. **Constitutional isomers** have the same molecular formula but differ in how their atoms are linked. A reaction (such as those just shown), in which two or more constitutional isomers could be obtained as products but one of them predominates, is called a **regioselective reaction.**

There are different degrees of regioselectivity. A reaction can be *moderately regioselective, highly regioselective,* or *completely regioselective.* In a completely regioselective reaction, one of the possible products is not formed at all. The addition of a hydrogen halide to 2-methylpropene (where the two possible carbocations are tertiary and primary) is more highly regioselective than the addition of a hydrogen halide to 2-methyl-2-butene (where the two possible carbocations are tertiary and secondary) because the two carbocations in the 2-methyl-2-butene reaction are closer in stability.

The addition of HBr to 2-pentene is not regioselective. Both isomers are obtained because the addition of the proton to either of the sp^2 carbons produces a secondary carbocation. Both carbocation intermediates have the same stability, so both will be formed equally easily, resulting in the formation of approximately equal amounts of the two alkyl halides.

$$CH_3CH=CHCH_2CH_3 \quad + \quad HBr \quad \longrightarrow \quad CH_3CHCH_2CH_2CH_3 \quad + \quad CH_3CH_2CHCH_2CH_3$$

with Br substituents

2-bromopentane

3-bromopentane

You have seen that if you want to predict the major product of an electrophilic addition reaction, you must first determine the relative stabilities of the two possible carbocation intermediates. In 1865 when carbocations and their relative stabilities were not yet known, Vladimir Markovnikov published a paper in which he described a way to predict the major product obtained from the addition of a hydrogen halide to an asymmetrical alkene. His shortcut is known as **Markovnikov's rule:** "When a hydrogen halide adds to an asymmetrical alkene, the addition occurs such that the halogen attaches itself to the carbon atom of the alkene bearing the least number of hydrogen atoms." Because H$^+$ is the first species that adds to the alkene, most chemists nowdays state the rule in the following way: "The H$^+$ adds to the sp^2 carbon that is bonded to the greater number of hydrogens." Because electrophiles other than H$^+$ can add to alkenes, a more general statement of Markovnikov's rule (and the one that you should learn) is: *The electrophile adds to the sp^2 carbon that is bonded to the greater number of hydrogens.*

The electrophile adds to the sp^2 carbon that is bonded to the greater number of hydrogens.

Whether you determine the major product of an electrophilic addition reaction by determining relative carbocation stabilities or by using Markovnikov's rule, you will get the same answer. In the reaction shown next, for example, the H^+ adds preferentially to C-1 not C-2. Using relative carbocation stabilities, we would say H^+ adds to C-1 because that forms a secondary carbocation, which is more stable than the primary carbocation that would be formed if H^+ added to C-2. Using Markovnikov's rule, we would say the electrophile (H^+) adds preferentially to C-1 because C-1 is bonded to two hydrogens, whereas C-2 is bonded to only one hydrogen.

$$CH_3CH_2\overset{2}{C}H=\overset{1}{C}H_2 \ + \ HCl \ \longrightarrow \ CH_3CH_2\overset{\overset{\displaystyle Cl}{|}}{C}HCH_3$$

Vladimir Vasilevich Markovnikov (1837–1904) *was born in Russia, the son of an army officer. He was a professor of chemistry at Kazan, Odessa, and Moscow Universities. By synthesizing rings containing four carbons and seven carbons, he disproved the notion that carbon could only form five- and six-membered rings.*

PROBLEM 26 ◆

What would be the major product obtained from the addition of HBr to each of the following compounds?

a. $CH_3CH_2CH=CH_2$

b. $CH_3CH=\overset{\overset{\displaystyle CH_3}{|}}{C}CH_3$

c. 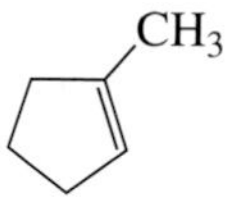 (cyclopentene with CH₃ substituent)

d. $CH_3CH=CHCH_3$

PROBLEM-SOLVING STRATEGY

a. What alkene should be used to synthesize 3-bromohexane?

$$? \ + \ HBr \ \longrightarrow \ CH_3CH_2\overset{\overset{\displaystyle }{}}{C}HCH_2CH_2CH_3$$
$$\underset{\textbf{3-bromohexane}}{\underset{\displaystyle Br}{|}}$$

To answer this kind of question, start by listing all the alkenes that could be used. Because you want to synthesize an alkyl halide that has a bromo substituent at the C-3 position, the alkene should have an sp^2 carbon at the C-3 position. Therefore, there are two alkenes that can be used to form 3-bromohexane: 2-hexene and 3-hexene.

$$CH_3CH=CHCH_2CH_2CH_3 \qquad CH_3CH_2CH=CHCH_2CH_3$$
$$\textbf{2-hexene} \qquad\qquad\qquad \textbf{3-hexene}$$

Because there are two possibilities, determine whether or not there is any advantage to using one over the other. The addition of H^+ to 2-hexene, for instance, can form two different carbocations. Because they are both secondary carbocations, they have the same stability and, therefore, approximately equal amounts of each will be formed. As a result, half of the product will be 3-bromohexane and half will be 2-bromohexane.

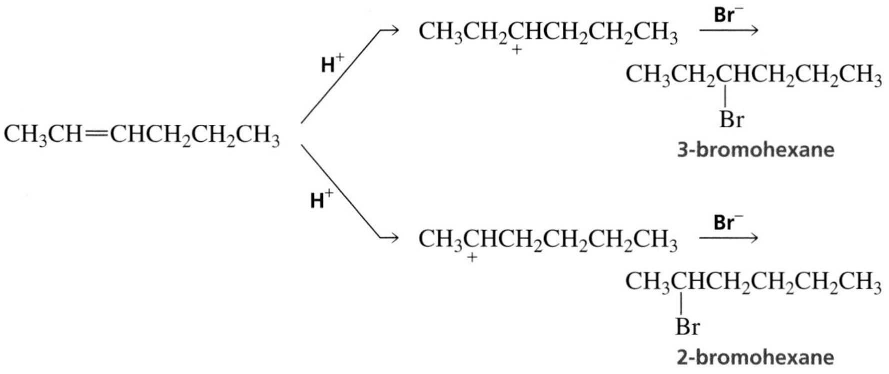

$$CH_3CH=CHCH_2CH_2CH_3$$

→ (H⁺) CH₃CH₂CHCH₂CH₂CH₃ (→ Br⁻)

CH₃CH₂CHCH₂CH₂CH₃
|
Br
3-bromohexane

→ (H⁺) CH₃CHCH₂CH₂CH₂CH₃ (→ Br⁻)

CH₃CHCH₂CH₂CH₂CH₃
|
Br
2-bromohexane

The addition of H⁺ to either of the *sp²* carbons of 3-hexene, on the other hand, forms the same carbocation because the alkene is symmetrical. Therefore, all of the product will be the desired 3-bromohexane.

CH₃CH₂CH=CHCH₂CH₃ —(H⁺)→ CH₃CH₂CHCH₂CH₂CH₃ —(Br⁻)→

CH₃CH₂CHCH₂CH₂CH₃
|
Br
3-bromohexane

Because all the alkyl halide formed from 3-hexene is 3-bromohexane, but only half the alkyl halide formed from 2-hexene is 3-bromohexane, 3-hexene is the best alkene to use to prepare 3-bromohexane.

b. What alkene should be used to synthesize 2-bromopentane?

Either 1-pentene or 2-pentene could be used because both have an *sp²* carbon at the C-2 position.

CH₂=CHCH₂CH₂CH₃   CH₃CH=CHCH₂CH₃
1-pentene   **2-pentene**

When H⁺ adds to 1-pentene, one of the carbocations that could be formed is secondary and the other is primary. A secondary carbocation is more stable than a primary carbocation, which is so unstable that little, if any, will be formed. Thus, 2-bromopentane will be the only product of the reaction.

CH₂=CHCH₂CH₂CH₃

→ (H⁺) CH₃CHCH₂CH₂CH₃ —(Br⁻)→ CH₃CHCH₂CH₂CH₃
|
Br
2-bromopentane

→ (H⁺) ✗ CH₂CH₂CH₂CH₂CH₃

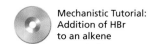
Mechanistic Tutorial:
Addition of HBr
to an alkene

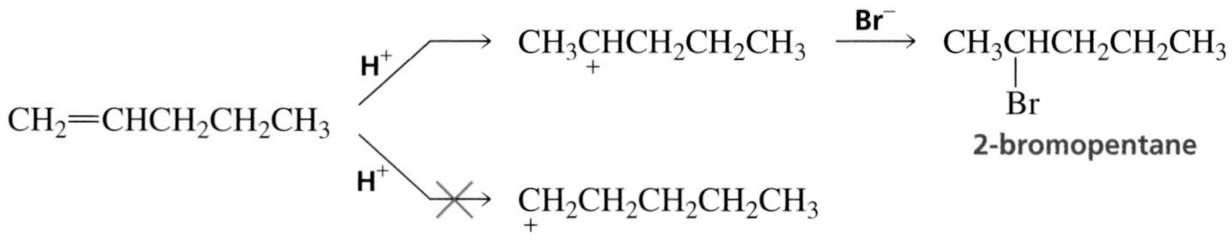

When H⁺ adds to 2-pentene, on the other hand, the two carbocations that can be formed are both secondary. Both are equally stable, so they will be formed in approximately equal amounts. Thus, only about half of the product of the reaction will be 2-bromopentane. The other half will be 3-bromopentane.

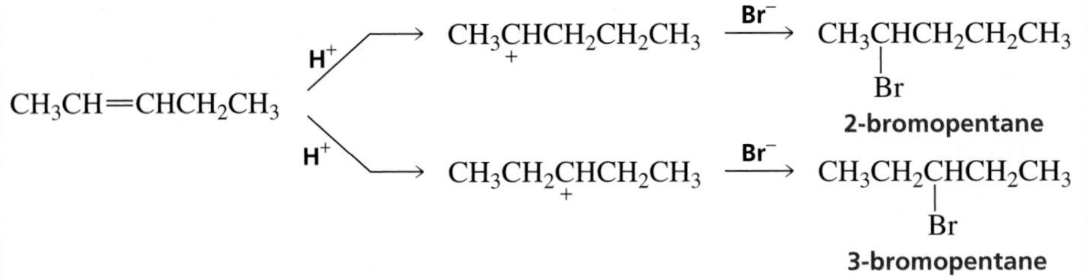

$$CH_3CH=CHCH_2CH_3$$

$$\xrightarrow{H^+} CH_3\overset{+}{C}HCH_2CH_2CH_3 \xrightarrow{Br^-} CH_3CHCH_2CH_2CH_3$$

with Br substituent

2-bromopentane

$$\xrightarrow{H^+} CH_3CH_2\overset{+}{C}HCH_2CH_3 \xrightarrow{Br^-} CH_3CH_2CHCH_2CH_3$$

with Br substituent

3-bromopentane

Because all the alkyl halide formed from 1-pentene is 2-bromopentane, but only half the alkyl halide formed from 2-pentene is 2-bromopentane, 1-pentene is the best alkene to use to prepare 2-bromopentane.

Now continue on to answer the questions in Problem 27.

PROBLEM 27 ◆

What alkene should be used to synthesize each of the following alkyl bromides?

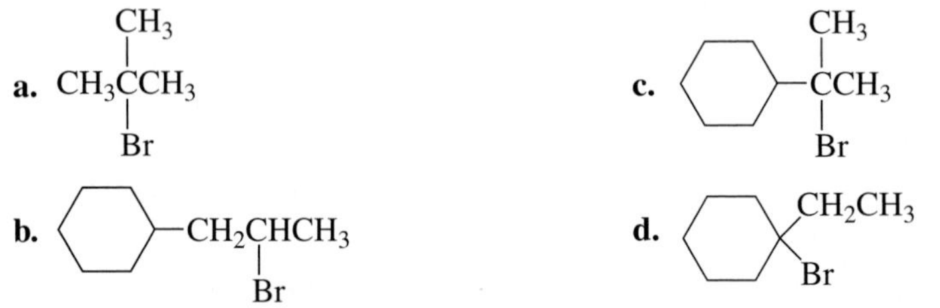

a.
$$\begin{array}{c} CH_3 \\ | \\ CH_3CCH_3 \\ | \\ Br \end{array}$$

b.
⬡—CH$_2$CHCH$_3$ with Br substituent

c.
⬡—$\overset{\displaystyle CH_3}{\underset{\displaystyle Br}{C}}CH_3$

d.
⬡ with CH$_2$CH$_3$ and Br substituents

PROBLEM 28 ◆

The addition of HBr to which of the following alkenes is more highly regioselective?

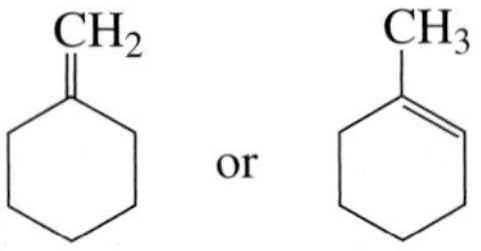

⬡=CH$_2$ or   ⬡ with CH$_3$ (cyclohexene)

3.13 ADDITION OF WATER AND ALCOHOLS

When you add water to an alkene, no reaction takes place because there is no electrophile to add to the nucleophilic alkene to start the reaction. The oxygen–hydrogen bonds of water are too strong (water is too weakly acidic) to allow the hydrogen to act as an electrophile for this reaction.

$$CH_3CH=CH_2 + H_2O \longrightarrow \text{no reaction}$$

If, however, you add acid (e.g., H_2SO_4) to the solution, water will add to the alkene, thereby forming an alcohol.

$$CH_3CH=CH_2 + H_2O \underset{}{\overset{H^+}{\rightleftharpoons}} CH_3CHCH_3 \text{ with OH substituent}$$

When acid is added to the aqueous solution, the acid provides a proton that can act as an electrophile. The proton will react with the alkene to form a carbocation, and the electrophilic carbocation will then react with the nucleophilic water. Addition of water to a molecule is called **hydration,** so we can say that an alkene will be *hydrated* in the presence of water and acid.

The first two steps of the mechanism for the addition of water are essentially the same as the mechanism for the addition of a hydrogen halide. The addition also follows Markovnikov's rule. That is, the electrophile (H^+) adds to the sp^2 carbon that is bonded to the most hydrogens, and the nucleophile (H_2O) adds to the other sp^2 carbon.

mechanism for the acid–catalyzed addition of water

$$CH_3CH{=}CH_2 \ + \ H{-}\overset{+}{\underset{H}{\overset{..}{O}H}} \ \underset{}{\overset{\text{slow}}{\rightleftharpoons}} \ CH_3\overset{+}{C}HCH_3 \ + \ H_2\overset{..}{O}{:} \ \overset{\text{fast}}{\rightleftharpoons} \ CH_3CHCH_3$$

$$CH_3CHCH_3$$
$$\overset{..}{\underset{|}{\overset{+}{O}H}}$$
$$\overset{|}{H}$$

$$H_2\overset{..}{O}{:}$$
$$\text{fast}$$

$$CH_3CHCH_3 \ + \ H_3\overset{+}{O}{:}$$
$$\underset{|}{\overset{|}{:}\overset{..}{O}H}$$

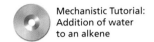
Mechanistic Tutorial:
Addition of water
to an alkene

As we saw in Section 3.7, the addition of the electrophile to the alkene is slow and the subsequent addition of the nucleophile to the carbocation occurs rapidly. The reaction of the carbocation with a nucleophile is so fast that the carbocation combines with whatever nucleophile it collides with first. In the previous hydration reaction there are two nucleophiles in solution: water and the counterion of the acid (HSO_4^-) that was used to start the reaction. (Notice that HO^- is not a nucleophile in this reaction because there is no appreciable HO^- in an acidic solution.)[2] Because the concentration of water is much greater than the concentration of the counterion, the carbocation is much more likely to collide with water. The product of the collision is a protonated alcohol. Because the pH of the solution is greater than the pK_a of the protonated alcohol (remember that protonated alcohols are very strong acids; see Sections 1.17 and 1.19), the protonated alcohol loses a proton, and the final product of the addition reaction is an alcohol.

A proton adds to the alkene in the first step but a proton is returned to the reaction mixture in the final step. Overall, a proton is not consumed. A species that increases the rate of a reaction and is not consumed during the course of the reaction is called a **catalyst.** Catalysts increase the reaction rate by decreasing the activation energy of the reaction (Section 22.0). Catalysts do *not* affect the equilibrium constant of the reaction. In other words, a catalyst increases the *rate* at which a product is formed but does not affect the *amount* of product formed. The catalyst in the hydration of an alkene is an acid (H^+), so the reaction is said to be an **acid-catalyzed reaction.**

Synthetic Tutorial:
Addition of water
to an alkene

PROBLEM 29 ◆

The pK_a of 2-propanol (the product of the previous reaction) is 16.0. As long as the pH is < _____, more than 50% of 2-propanol will be in its neutral, nonprotonated form.

---

[2]At pH = 4, for example, the concentration of HO^- is 1×10^{-10} M, whereas the concentration of water in an aqueous solution is 55.5 M.

PROBLEM 30

Give the structure of the alcohol that is formed from the acid-catalyzed hydration of each of the following alkenes.

a. $CH_3CH_2CH_2CH{=}CH_2$

c. $CH_3CH_2CH_2CH{=}CHCH_3$

b. ⬡

d. 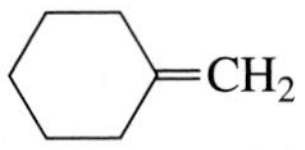 ⬡$={}CH_2$

Alcohols react with alkenes in the same way that water does. Like the addition of water, the addition of an alcohol requires an acid catalyst. The product of the reaction is an ether.

$$CH_3CH{=}CH_2 \;+\; CH_3OH \;\overset{H^+}{\rightleftharpoons}\; CH_3\underset{\underset{\textstyle OCH_3}{|}}{C}HCH_3$$

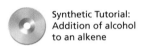

Synthetic Tutorial:
Addition of alcohol
to an alkene

The mechanism for the acid-catalyzed addition of an alcohol is essentially the same as the mechanism for the acid-catalyzed addition of water.

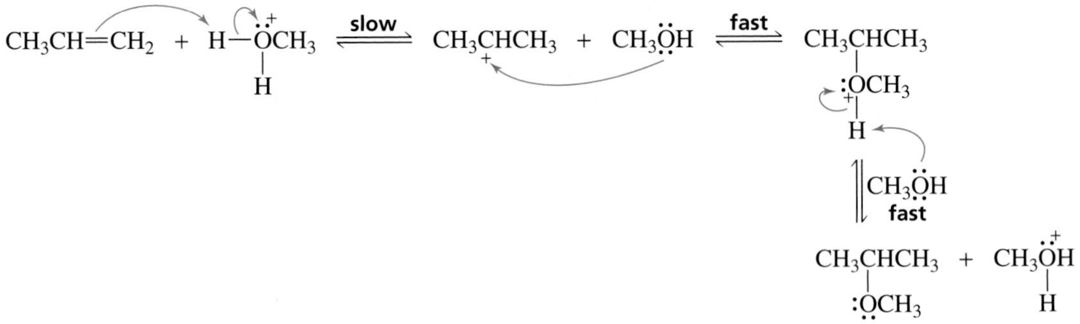

$$CH_3CH{=}CH_2 \;+\; H{-}\overset{+}{\underset{\underset{\textstyle H}{|}}{O}}CH_3 \;\overset{slow}{\rightleftharpoons}\; CH_3\overset{+}{C}HCH_3 \;+\; CH_3\overset{..}{O}H \;\overset{fast}{\rightleftharpoons}\; CH_3CHCH_3$$

$$CH_3CHCH_3 \;+\; CH_3\overset{+}{\underset{\underset{\textstyle H}{|}}{O}}H$$

PROBLEM 31

Propose a mechanism for the following reaction. (Remember to use curved arrows when showing a mechanism.)

$$CH_3\underset{\underset{\textstyle CH_3}{|}}{C}HCH_2CH_2OH \;+\; CH_3\underset{\underset{\textstyle CH_3}{|}}{C}{=}CH_2 \;\overset{H^+}{\longrightarrow}\; CH_3\underset{\underset{\textstyle CH_3}{|}}{C}HCH_2CH_2O\underset{\underset{\textstyle CH_3}{|}}{\overset{\overset{\textstyle CH_3}{|}}{C}}CH_3$$

PROBLEM 32

a. Give the products of the following reactions.

1. $CH_3\underset{\underset{\textstyle CH_3}{|}}{C}{=}CH_2 \;+\; HCl \longrightarrow$

3. $CH_3\underset{\underset{\textstyle CH_3}{|}}{C}{=}CH_2 \;+\; H_2O \;\overset{H^+}{\longrightarrow}$

2. $CH_3\underset{\underset{\textstyle CH_3}{|}}{C}{=}CH_2 \;+\; HBr \longrightarrow$

4. $CH_3\underset{\underset{\textstyle CH_3}{|}}{C}{=}CH_2 \;+\; CH_3OH \;\overset{H^+}{\longrightarrow}$

b. What do all the reactions have in common?

c. How do all the reactions differ?

PROBLEM 33

How could the following compounds be prepared using an alkene as one of the starting materials?

a.

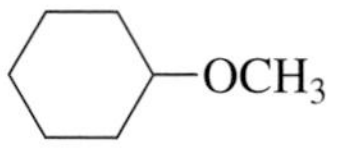

⬡—OCH₃

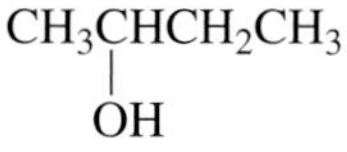

b. 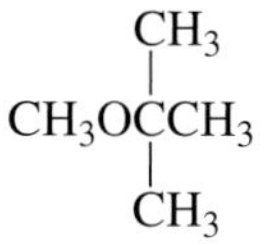 CH₃OCCH₃ with CH₃ above and CH₃ below

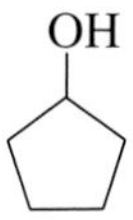

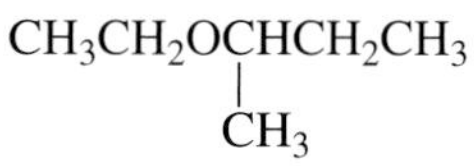

$$CH_3$$
b. CH_3OCCH_3
$$CH_3$$

c. $CH_3CH_2OCHCH_2CH_3$
         $\quad\quad\quad CH_3$

d. $CH_3CHCH_2CH_3$
         $\quad\quad OH$

e.

$\overset{OH}{\triangle}$ cyclopentane with OH

f. $CH_3CH_2CHCH_2CH_2CH_3$
         $\quad\quad\quad OH$

Some electrophilic addition reactions give products that are clearly not the result of the addition of an electrophile to the sp^2 carbon bonded to the most hydrogens and the addition of a nucleophile to the other sp^2 carbon. For example, the addition of HBr to 3-methyl-1-butene forms 2-bromo-3-methylbutane (minor product) and 2-bromo-2-methylbutane (major product). 2-Bromo-3-methylbutane is the product expected from the addition of H^+ to the sp^2 carbon bonded to the most hydrogens and Br^- to the other sp^2 carbon. 2-Bromo-2-methylbutane, on the other hand, is an "unexpected" product, even though it is the major product of the reaction.

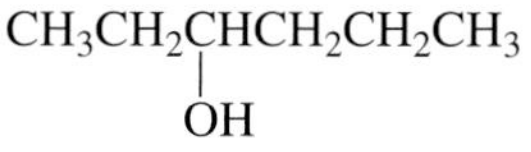

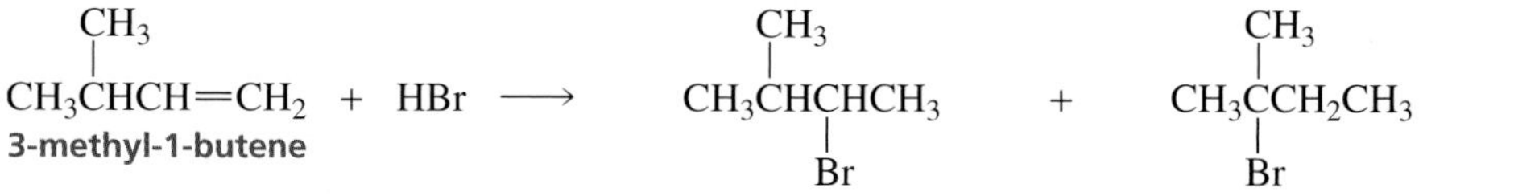

3.14 REARRANGEMENT OF CARBOCATIONS

$$CH_3CHCH=CH_2 + HBr \longrightarrow CH_3CHCHCH_3 + CH_3CCH_2CH_3$$
3-methyl-1-butene   **2-bromo-3-methylbutane** minor product   **2-bromo-2-methylbutane** major product

In another case, the addition of HCl to 3,3-dimethyl-1-butene forms both 3-chloro-2,2-dimethylbutane (an "expected" product) and 2-chloro-2,3-dimethylbutane (an "unexpected product"). Again, the unexpected product is obtained in greater yield.

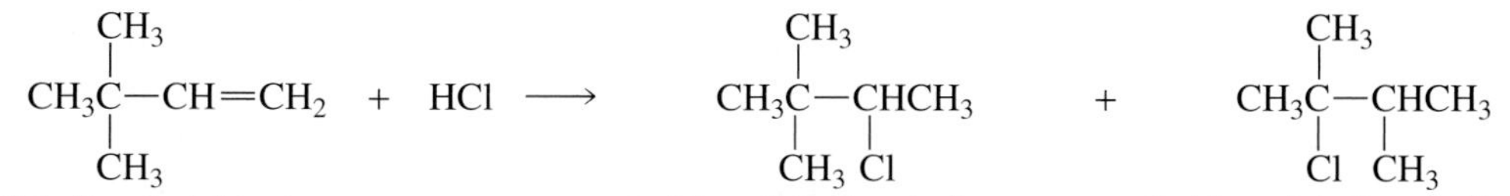

$$CH_3C-CH=CH_2 + HCl \longrightarrow CH_3C-CHCH_3 + CH_3C-CHCH_3$$
3,3-dimethyl-1-butene   **3-chloro-2,2-dimethylbutane** minor product   **2-chloro-2,3-dimethylbutane** major product

F. C. Whitmore was the first to suggest that the unexpected product results from a rearrangement of the carbocation intermediate. Not all carbocations rearrange. In fact, none of the carbocations that we have seen up to this point rearrange. Carbocations rearrange only if they become more stable as a result of the rearrangement. For example, when an electrophile adds to 3-methyl-1-butene, a *secondary* carbocation is formed initially. However, the secondary carbocation has a hydrogen that can shift (with its pair of electrons) to the adjacent positively charged carbon, creating a more stable *tertiary* carbocation.

Reaction scheme for 3-methyl-1-butene:

$$CH_3CH-CH=CH_2 \ + \ H-Br \longrightarrow$$

3-methyl-1-butene

forms a secondary carbocation, which via a 1,2-hydride shift forms a tertiary carbocation (major product), and via attack on unrearranged carbocation forms the minor product.

a secondary carbocation → attack on unrearranged carbocation → minor product

a 1,2-hydride shift → a tertiary carbocation → attack on rearranged carbocation → major product

Frank (Rocky) Clifford Whitmore (1887–1947) *was born in Massachusetts. He received a Ph.D. from Harvard University and was a professor of chemistry at Minnesota, Northwestern, and Pennsylvania State Universities. Whitmore never slept a full night; when he got tired, he took a one-hour nap. Consequently, he had the reputation of being an indefatigable worker; 20-hour work days were common. He generally had 30 graduate students working in his lab at a time, and he wrote an advanced textbook that was considered a milestone in the field of organic chemistry.*

As a result of this **carbocation rearrangement,** two alkyl halides are formed— one from attack of the nucleophile on the unrearranged carbocation and one from attack on the rearranged carbocation. The major product is the rearranged product. Because the rearrangement involves the shift of a hydrogen with its pair of electrons, it is called a hydride shift (H:$^-$ is a hydride ion). More specifically it is called a **1,2-hydride shift** because the hydride ion moves from one carbon to an adjacent carbon. (Notice that this does not mean that it moves from C-1 to C-2.)

3,3-Dimethyl-1-butene adds an electrophile to form a *secondary* carbocation. In this case there is a methyl group that can shift with its pair of electrons to the adjacent positively charged carbon to form a more stable *tertiary* carbocation. This is called a **1,2-methyl shift.**

Reaction scheme for 3,3-dimethyl-1-butene:

$$CH_3C-CH=CH_2 \ + \ H-Cl \longrightarrow$$

3,3-dimethyl-1-butene → a secondary carbocation → a 1,2-methyl shift → a tertiary carbocation

a secondary carbocation → attack on unrearranged carbocation → minor product

a tertiary carbocation → attack on rearranged carbocation → major product

A shift involves only the movement of a species from one carbon to an adjacent carbon—1,3-shifts normally do not occur. Furthermore, if the rearrangement does not lead to a more stable carbocation, then carbocation rearrangement does not occur. For example, when a proton adds to 4-methyl-1-pentene, a secondary carbocation is formed. A 1,2-hydride shift would form a different secondary carbocation. Because both carbocations are equally stable, there is no energetic advantage to the shift. Consequently, rearrangement does not occur and only one alkyl halide is formed.

$$CH_3CHCH_2CH=CH_2 \ + \ HBr \longrightarrow CH_3CHCH_2\overset{+}{C}HCH_3 \ \cancel{\longleftrightarrow} \ CH_3\overset{+}{C}HCHCH_2CH_3$$

4-methyl-1-pentene → the carbocation does not rearrange → CH_3CHCH_2CHCH_3 (with Br)

Carbocation rearrangements also can occur by ring expansion, which is another type of 1,2-shift. In the following example, a secondary carbocation is formed initially. Ring expansion leads to a tertiary carbocation, which is more stable and, because of the five-membered ring, has less angle strain.

In any reaction that forms a carbocation intermediate, check to see whether or not the carbocation will rearrange.

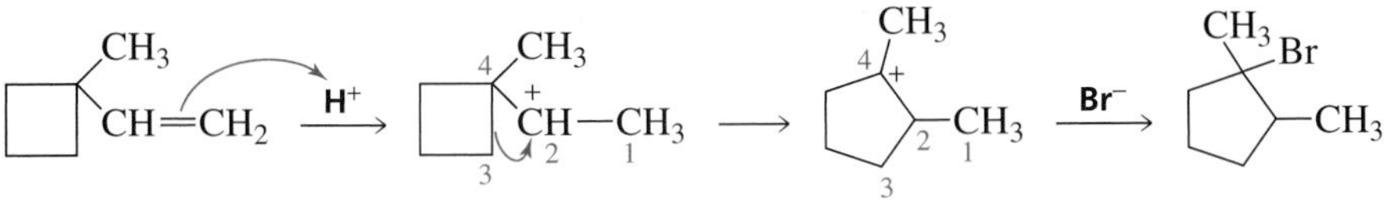

In subsequent chapters you will study other reactions that involve formation of carbocation intermediates. Keep in mind that *whenever a reaction leads to the formation of a carbocation, you must check its structure for the possibility of rearrangement.*

PROBLEM 34 / SOLVED

Which of the following carbocations would you expect to rearrange?

a.

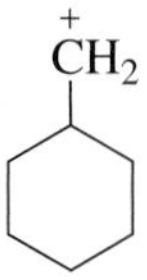

$\overset{+}{C}H_2$

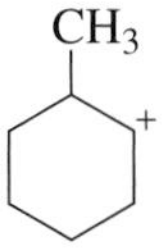

c.

CH_3

e. $CH_3CH_2\underset{+}{C}HCH_3$

b.

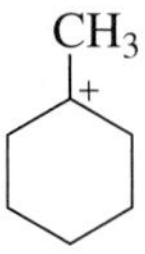

CH_3

d.

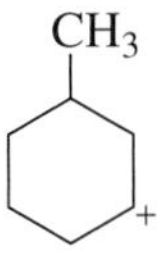

CH_3

f. 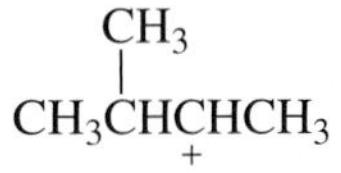 $CH_3\overset{\overset{\displaystyle CH_3}{|}}{C}H\underset{+}{C}HCH_3$

SOLUTION

a. This carbocation will rearrange because a 1,2-hydride shift will convert a primary carbocation into a tertiary carbocation.

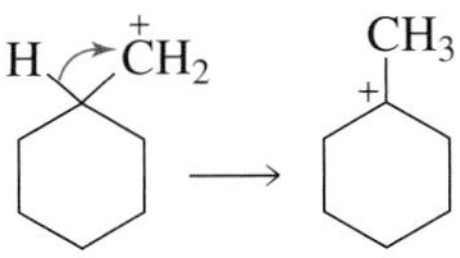

b. This carbocation will not rearrange because it is tertiary and its stability cannot be improved by a carbocation rearrangement.

c. This carbocation will rearrange because a 1,2-hydride shift will convert a secondary carbocation into a tertiary carbocation.

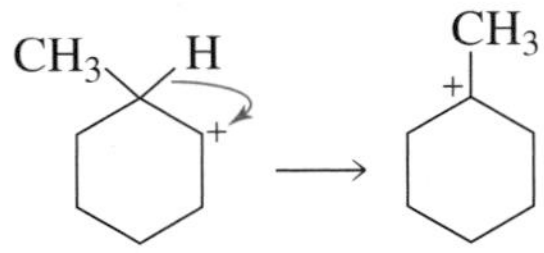

d. This carbocation will not rearrange because it is a secondary carbocation and a carbocation rearrangement would yield another secondary carbocation.

e. This carbocation will not rearrange because it is a secondary carbocation and a carbocation rearrangement would yield another secondary carbocation.

f. This carbocation will rearrange because a 1,2-hydride shift will convert a secondary carbocation into a tertiary carbocation.

$$CH_3\overset{\overset{\displaystyle CH_3}{|}}{\underset{\underset{\displaystyle H}{|}}{C}}\!-\!\underset{+}{C}HCH_3 \longrightarrow CH_3\overset{\overset{\displaystyle CH_3}{|}}{C}CH_2CH_3$$

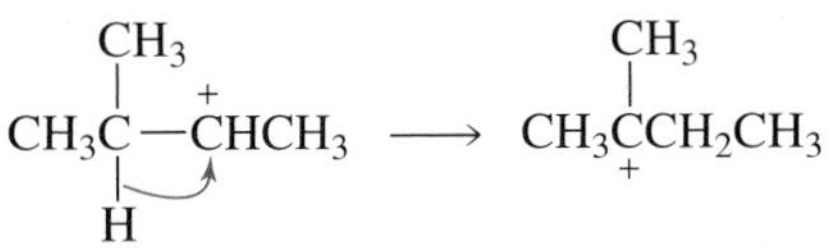

Give the major product(s) obtained from the reaction of each of the following with HBr.

a. $CH_3CHCH{=}CH_2$
$\quad\quad\quad\underset{CH_3}{|}$

b. (cyclohexane ring with CH_2 substituent)

c. (cyclohexene ring with CH_3 substituent)

d. (cyclohexene ring with CH_3 substituent)

e. (cyclohexene ring with CH_3 substituent)

f. $CH_2{=}CHCCH_3$
$\quad\quad\quad\quad\underset{CH_3}{\overset{CH_3}{|}}$

3.15 ADDITION OF HALOGENS

The halogens, Br_2 and Cl_2, add to alkenes. This may be surprising at first glance because it is not immediately apparent that an electrophile, necessary to start an electrophilic addition reaction, is present. However, the bond joining the two halogen atoms is relatively weak (see the bond dissociation energies in Table 3.1) and therefore easily broken. When the π electrons of the alkene approach a molecule of Br_2 or Cl_2, one of the halogen atoms accepts them and releases the shared electrons to the other halogen atom. Therefore, in an electrophilic addition reaction, Br_2 behaves as if it were Br^+ and Br^-, and Cl_2 behaves as if it were Cl^+ and Cl^-.

$$H_2C{=}CH_2 \;+\; :\!\overset{+}{Br}{-}\overset{..}{Br}: \;\longrightarrow\; \left[H_2\overset{+}{C}{-}CH_2\right] \;\longrightarrow\; H_2C{-}CH_2 \;+\; :\!\overset{-}{Br}: \;\longrightarrow\; :\!\overset{..}{Br}CH_2CH_2\overset{..}{Br}:$$

a bromonium ion   1,2-dibromoethane
a vicinal dibromide

bromonium ion of ethene

The carbocation product of the first step is shown in brackets because it is never actually formed. Instead, a nonbonding pair of electrons on bromine forms a bond with the positively charged carbon, thereby forming a cyclic bromonium ion. The bromonium ion is more stable than the carbocation because all the atoms in the bromonium ion have complete octets (except hydrogen), whereas the positively charged carbon of the carbocation does not have a complete octet. (To review the octet rule, see Section 1.3.) In the second step of the reaction, Br^- attacks a carbon atom of the bromonium ion. This releases the strain in the three-membered ring and forms a vicinal dibromide. **Vicinal** indicates that the two bromine atoms are on adjacent carbons (*vicinus* is the Latin word for "near"). The electrostatic potential maps for the cyclic bromonium ions show that the positively charged region (the blue area) encompasses the carbons even though the formal positive charge is on bromine.

When Cl_2 adds to an alkene, a cyclic chloronium ion intermediate is formed. The final product of the reaction is a vicinal dichloride.

bromonium ion of cis-2-butene

$$\underset{\underset{CH_3}{|}}{CH_3C{=}CH_2} \;+\; Cl_2 \;\xrightarrow{CH_2Cl_2}\; \underset{\underset{Cl}{|}}{\overset{\overset{CH_3}{|}}{CH_3CCH_2Cl}}$$

2-methylpropene

1,2-dichloro-2-methylpropane
a vicinal dichloride

Because a carbocation is not formed when Br_2 or Cl_2 adds to an alkene, carbocation rearrangements do not occur.

Mechanistic Tutorial: Addition of halogens to alkenes

$$\underset{\substack{\text{3-methyl-1-butene}}}{\overset{\overset{\displaystyle CH_3}{|}}{CH_3CHCH=CH_2}} + Br_2 \xrightarrow{CH_2Cl_2} \underset{\substack{\overset{\displaystyle Br}{|}}}{\overset{\overset{\displaystyle CH_3}{|}}{CH_3CHCHCH_2Br}}$$

1,2-dibromo-3-methylbutane
a vicinal dibromide

PROBLEM 36

a. How does the first step in the reaction of ethene with Br_2 differ from the first step in the reaction of ethene with HBr?

b. To understand why Br^- attacks a carbon atom of the bromonium ion and does not attack the positively charged bromine atom, draw the product that would be obtained if Br^- *did* attack the bromine atom.

Reactions of alkenes with Br_2 or Cl_2 are generally carried out by mixing the alkene and the halogen in an inert solvent, such as dichloromethane (CH_2Cl_2), which readily dissolves both reactants but does not participate in the reaction. The preceding reactions illustrate the way in which organic reactions are typically written. The reactants are placed to the left of the reaction arrow and the products are placed to the right of the arrow. The reaction conditions, such as solvent, temperature, or any required catalyst, are written above or below the arrow. Sometimes reactions are written by placing only the organic (carbon-containing) reagent on the left-hand side of the arrow and placing any inorganic reagent above the arrow.

$$CH_3CH=CHCH_3 \xrightarrow[CH_2Cl_2]{Cl_2} \underset{\substack{\overset{\displaystyle |\;\;|}{Cl\;\;Cl}}}{CH_3CHCHCH_3}$$

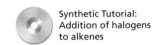
Synthetic Tutorial:
Addition of halogens
to alkenes

F_2 and I_2 are halogens, but are not used as reagents in electrophilic addition reactions. Fluorine reacts explosively with alkenes, so electrophilic addition of F_2 is not a synthetically useful reaction. Addition of I_2 to an alkene is a thermodynamically unfavorable reaction. Vicinal diiodides are unstable at room temperature, decomposing back to the alkene and I_2.

$$CH_3CH=CHCH_3 + I_2 \underset{CH_2Cl_2}{\rightleftharpoons} \underset{\substack{\overset{\displaystyle |\;\;|}{I\;\;\;I}}}{CH_3CHCHCH_3}$$

If H_2O is used as the solvent rather than CH_2Cl_2, the major product of the reaction will be a vicinal halohydrin. A **halohydrin** is an organic molecule that contains both an OH group and a halogen. In a vicinal halohydrin, the OH and halogen are bonded to adjacent carbons.

$$CH_3CH=CH_2 + Br_2 \xrightarrow{H_2O} \underset{\substack{\overset{\displaystyle |}{OH}}}{CH_3CHCH_2Br} + \underset{\substack{\overset{\displaystyle |}{Br}}}{CH_3CHCH_2Br} + HBr$$

propene

a bromohydrin   **minor product**
major product

$$\underset{\substack{\text{2-methyl-2-butene}}}{\overset{\overset{\displaystyle CH_3}{|}}{CH_3CH=CCH_3}} + Cl_2 \xrightarrow{H_2O} \underset{\substack{\overset{\displaystyle |\;\;|}{Cl\;\;OH}}}{\overset{\overset{\displaystyle CH_3}{|}}{CH_3CHCCH_3}} + \underset{\substack{\overset{\displaystyle |\;\;|}{Cl\;\;Cl}}}{\overset{\overset{\displaystyle CH_3}{|}}{CH_3CHCCH_3}} + HCl$$

a chlorohydrin   **minor product**
major product

The mechanism for halohydrin formation involves formation of a cyclic bromonium ion (or chloronium ion) in the first step of the reaction because Br^+ (or Cl^+) is the only electrophile in the reaction mixture. In the second step, the bromonium ion rapidly reacts with whatever nucleophile it bumps into. There are two nucleophiles present in solution, H_2O and Br^-. Because H_2O is the solvent, its concentration far exceeds the concentration of Br^-. Consequently, the cyclic bromonium ion is more likely to collide with a molecule of water than with Br^-. The protonated halohydrin that is formed is a strong acid (Section 1.19), so it loses a proton.

Synthetic Tutorial:
Halohydrin reaction

mechanism for halohydrin formation

$$CH_3CH{=}CH_2 \ + \ \ddot{\underset{..}{Br}}{-}\ddot{\underset{..}{Br}} \ \xrightarrow[]{slow} \ CH_3\overset{..}{\underset{+ \ :\ddot{Br}:^-}{CH{-}CH_2}} \ \xrightarrow[fast]{H_2\overset{..}{O}:} \ CH_3CHCH_2{-}\ddot{\underset{..}{Br}}: \ \xrightarrow[fast]{H_2\overset{..}{O}:} \ CH_3CHCH_2{-}\ddot{\underset{..}{Br}}: \ + \ H_3\overset{+}{\underset{..}{O}}:$$

In the preceding addition reaction, the electrophile (Br^+) ends up on the sp^2 carbon bonded to the most hydrogens. This occurs because, in the transition state for the second step of the reaction, breaking of the carbon–bromine bond has occurred to a greater extent than has formation of the carbon–oxygen bond. As a result, there is a partial positive charge on the carbon that is attacked by the nucleophile.

$$
\begin{array}{cc}
\overset{\delta+}{Br} & \overset{\delta+}{Br} \\
\overset{\delta+}{CH_3CH}{-}CH_2 & CH_3CH{-}\overset{\delta+}{CH_2} \\
\overset{\delta+}{:O}{-}H & \overset{\delta+}{:O}{-}H \\
H & H \\
\textbf{more stable transition state} & \textbf{less stable transition state}
\end{array}
$$

Therefore, the more stable transition state is achieved by adding the nucleophile to the sp^2 carbon that is bonded to the fewer hydrogens because the partial positive charge will be on a secondary carbon rather than on a primary carbon. So this reaction also follows Markovnikov's rule—the electrophile adds to the sp^2 carbon that is bonded to the most hydrogens. In this case, the electrophile is Br^+ rather than H^+.

When nucleophiles other than H_2O are added to the reaction mixture, they, too, change the product of the reaction, just as water changed the product of Br_2 addition from a vicinal dibromide to a vicinal bromohydrin. Because the concentration of the added nucleophile will be greater than the concentration of the halide ion generated from Br_2 or Cl_2, the added nucleophile will be most likely to participate in the second step of the reaction. Markovnikov's rule will be followed in order to achieve the most stable transition state for the second step of the reaction. (Ions such as Na^+ and K^+ cannot form covalent bonds, so they do not react with organic compounds. They serve only as counterions to negatively charged species, so their presence generally is ignored when writing chemical equations.)

Do not memorize the products of alkene addition reactions. Instead, for each reaction ask yourself: "What is the electrophile?" and "What nucleophile is present in the greatest concentration?"

$$
\begin{array}{c}
\qquad\qquad CH_3 \\
CH_3CH{=}CCH_3 \ + \ Cl_2 \ + \ CH_3OH \ \longrightarrow \ CH_3CHCCH_3 \ + \ HCl \\
\qquad\qquad\qquad\qquad\qquad\qquad\qquad\qquad\qquad Cl \ \ OCH_3
\end{array}
$$

$$
CH_3CH{=}CH_2 \ + \ Br_2 \ + \ NaCl \ \longrightarrow \ CH_3CHCH_2Br \ + \ NaBr \\
\qquad\qquad\qquad\qquad\qquad\qquad\qquad\qquad Cl
$$

PROBLEM 37 ◆

There are two nucleophiles in each of the following reactions. For each reaction, explain why there is a greater concentration of one nucleophile than the other. What will be the major product of each reaction?

a. $CH_2{=}\overset{\underset{\displaystyle |}{CH_3}}{C}{-}CH_3$ + Cl_2 $\xrightarrow{CH_3OH}$

b. $CH_2{=}CHCH_3$ + $2\ NaI$ + HBr $\longrightarrow$

c. $CH_3CH{=}CHCH_3$ + HCl $\xrightarrow{H_2O}$

d. $CH_3CH{=}CHCH_3$ + HBr $\xrightarrow{CH_3OH}$

PROBLEM 38

Why are Na^+ and K^+ unable to form covalent bonds?

PROBLEM 39 ◆

What will be the product of the addition of I-Cl to 1-butene? (*Hint:* chlorine is more electronegative than iodine.)

PROBLEM 40

What would be the major product obtained from the reaction of Br_2 with 1-butene if the reaction were carried out in:

a. dichloromethane?   c. ethyl alcohol?

b. water?   d. methyl alcohol?

In Section 3.13 you saw that water can add to an alkene if an acid catalyst is present. This is the way alkenes are converted into alcohols industrially. However, under normal laboratory conditions, water is generally added to an alkene by a procedure known as **oxymercuration–demercuration.** Addition of water by oxymercuration–demercuration has two advantages over the acid-catalyzed addition of water: It does not require acidic conditions that are harmful to many organic molecules, and carbocation rearrangements do not occur.

In oxymercuration, the alkene is treated with mercuric acetate in aqueous tetrahydrofuran (THF). When reaction with that reagent is complete, sodium borohydride and hydroxide ion are added to the reaction mixture. (The numbers 1 and 2 in front of the reagents above and below the arrow in the chemical equation indicate two sequential reactions; the second reagent is not added until reaction with the first reagent is completely over.)

3.16 OXYMERCURATION–DEMERCURATION

$$R{-}CH{=}CH_2 \xrightarrow[\text{2. NaBH}_4,\ \text{HO}^-]{\text{1. Hg(OAc)}_2,\ \text{H}_2\text{O, THF}} R\underset{\underset{\displaystyle OH}{|}}{-}CH{-}CH_3$$

In the first step of the oxymercuration mechanism, the electrophilic mercury of mercuric acetate adds to the double bond. (Two of mercury's 5*d* electrons are shown.) Because carbocation rearrangements do not occur, we can conclude that the product of the addition reaction is a cyclic mercurinium ion rather than a carbocation. The reaction is analogous to the addition of Br_2 to an alkene to form a cyclic bromonium ion.

mechanism for oxymercuration

In the second step of the reaction, water attacks the most substituted carbon (the one bonded to the *fewest* hydrogens) of the cyclic mercurinium ion for the same reason that it attacks the most substituted carbon of the bromonium ion (Section 3.15). That is, attack at the more substituted carbon leads to the more stable transition state.

Mechanistic Tutorial:
Oxymercuration-demercuration

Synthetic Tutorial:
Oxymercuration-demercuration

more stable transition state   less stable transition state

Sodium borohydride ($NaBH_4$) converts the carbon–mercury bond into a carbon–hydrogen bond. Because the reaction results in the loss of mercury, it is called *demercuration*.

$$CH_3CHCH_2\text{—}Hg\text{—}OAc \xrightarrow[\text{HO}^-]{\text{NaBH}_4} CH_3CHCH_3 + Hg + AcO^-$$

The overall reaction (oxymercuration–demercuration) results in the addition of water to the double bond in a manner that would be predicted by Markovnikov's rule: Hydrogen adds to the sp^2 carbon bonded to the most hydrogens, and OH adds to the other sp^2 carbon.

You have seen that alkenes react with alcohols in the presence of an acid catalyst to form ethers (Section 3.13). Just as addition of water works better in the presence of mercuric acetate than in the presence of a strong acid, addition of an alcohol works better in the presence of mercuric acetate. [Mercuric trifluoroacetate, $Hg(O_2CCF_3)_2$, works even better.] This reaction is called **alkoxymercuration-demercuration.**

1. $Hg(O_2CCF_3)_2$, CH_3OH
2. $NaBH_4$, HO^-

1-methylcyclohexene

**1-methoxy-1-methylcyclohexane
an ether**

The mechanisms for oxymercuration and alkoxymercuration are essentially identical. The only difference is that water is the nucleophile in oxymercuration and an alcohol is the nucleophile in alkoxymercuration. Therefore, the product of oxymercuration-demercuration is an alcohol, whereas the product of alkoxymercuration-demercuration is an ether.

PROBLEM 41

How could each of the following compounds be synthesized from an alkene?

a. (cyclopentane with OH)

b. (cyclopentane with OCH_2CH_3)

c. $CH_3\overset{\underset{\displaystyle |}{CH_3}}{\underset{\underset{\displaystyle OH}{|}}{C}}CH_2CH_3$

d. $CH_3\overset{\underset{\displaystyle |}{CH_3}}{\underset{\underset{\displaystyle OCH_3}{|}}{C}}CH_2CH_3$

An atom or molecule does not have to be positively charged to be an electrophile. Borane (BH_3), a neutral molecule, is an electrophile because boron has an empty $2p$ orbital and therefore readily accepts a pair of electrons in order to have a complete octet. So alkenes undergo electrophilic addition reactions with borane serving as the electrophile. When the addition reaction is over, an aqueous solution of sodium hydroxide and hydrogen peroxide is added to the reaction mixture and the resulting product is an alcohol. The addition of borane to an alkene followed by reaction with hydroxide ion and hydrogen peroxide is called **hydroboration–oxidation.** This reaction was first reported by H. C. Brown in 1959.

$$CH_2{=}CH_2 \quad \xrightarrow[\text{2. HO}^-,\, \text{H}_2\text{O}_2,\, \text{H}_2\text{O}]{\text{1. BH}_3} \quad CH_3CH_2OH$$

The alcohol that is formed from hydroboration–oxidation of an alkene has the H and OH groups on opposite carbons compared with the alcohol that is formed from the acid-catalyzed addition of water (Section 3.13). In other words, the product looks as if water has added to the alkene in a way that violates Markovnikov's rule. However, you will see that the more general statement of Markovnikov's rule is not violated: The electrophile adds to the sp^2 carbon that is bonded to the greater number of hydrogens. The product appears to violate Markovnikov's rule because when Markovnikov originally proposed the rule, he considered only reactions in which H^+ is the electrophile. In hydroboration–oxidation, H^+ is not the electrophile; $H{:}^-$ is the nucleophile.[3]

$$CH_3CH{=}CH_2 \quad \xrightarrow[\text{2. HO}^-,\, \text{H}_2\text{O}_2,\, \text{H}_2\text{O}]{\text{1. BH}_3} \quad CH_3CH_2CH_2OH$$

propene   1-propanol

$$CH_3CH{=}CH_2 \quad \xrightarrow[\text{H}_2\text{O}]{\text{H}^+} \quad CH_3\underset{\underset{\displaystyle OH}{|}}{C}HCH_3$$

propene   2-propanol

Because borane is a flammable, toxic, and explosive gas, a solution of borane in an ether such as tetrahydrofuran (THF) is a more convenient and less dangerous

3.17
ADDITION OF BORANE: HYDROBORATION–OXIDATION

Herbert Charles Brown *was born in London in 1921 and was brought to the United States by his parents at age two. He received a Ph.D. from the University of Chicago and has been a professor of chemistry at Purdue University since 1947. For his studies on boron-containing organic compounds, he shared the 1979 Nobel Prize in chemistry with G. Wittig.*

---

[3] In some chemistry literature you will see the term "anti-Markovnikov" used to describe the addition of borane since the addition appears to violate Markovnikov's rule.

BORANE AND DIBORANE

In the gaseous state, borane exists primarily as diborane because, in diborane, the two electrons in a hydrogen–boron bond can be shared by two boron atoms, giving each boron a share in eight electrons and satisfying boron's requirement for additional electrons. Diborane is a **dimer**—a molecule formed by the joining of two identical molecules. The hydrogen bridges in diborane are shown as dotted lines to indicate that the bond is made up of fewer than the normal two electrons.

3-D Molecule:
Diborane

Movie:
Borane-THF complex

reagent. A pair of nonbonding electrons on the ether oxygen atom satisfies boron's immediate requirement for electrons. So the reagent actually used for the first step of the hydroboration–oxidation reaction is a borane–THF complex that reacts in a manner similar to BH_3.

tetrahydrofuran
THF

To understand why hydroboration–oxidation of propene forms 1-propanol, the mechanism of the reaction must be examined. Borane is electron deficient, so it is the electrophile that reacts with the nucleophilic alkene. As boron accepts the π electrons from the alkene, it eliminates a hydride ion that also adds to the alkene. In all the addition reactions that we have seen up to this point, the electrophile adds to the alkene in the first step and the nucleophile adds to the positively charged intermediate in the second step. In the addition of borane to an alkene, the addition of the electrophilic boron and the nucleophilic hydride ion take place in one step. Therefore, no intermediate is formed.

$$CH_3CH{=}CH_2 \longrightarrow CH_3CH_2CH_2$$
$$H{-}BH_2 \qquad\qquad BH_2$$
an alkylborane

The addition of borane to an alkene is an example of a concerted reaction. A **concerted reaction** is a reaction in which all the bond-making and bond-breaking processes occur in a single step. Addition of borane to an alkene is also an example of a pericyclic reaction. (Pericyclic means "around the circle.") A **pericyclic reaction** is a concerted reaction that takes place as the result of a cyclic rearrangement of electrons.

The electrophilic boron adds to the sp^2 carbon that is bonded to the most hydrogens. The electrophiles that we have looked at previously (e.g., H^+) also added to the sp^2 carbon bonded to the most hydrogens in order to form the most stable carbocation intermediate. However, hydroboration is a concerted reaction, so no intermediate is formed. Why, then, does boron show the same preference in its addition?

To answer this question, we need to examine the transition state for the addition of borane. In the transition state, the sp^2 carbon that does not become attached to boron has a partial positive charge. The addition of borane to the sp^2 carbon of the alkene that is bonded to the most hydrogens results in the formation of a more stable transition state because the partial positive charge is on the more substituted carbon—in this case a secondary carbon. If boron had added to the other sp^2 carbon, the partial positive charge would have been on a primary carbon. So, even though a carbocation intermediate is not formed, a carbocation-like transition state is formed. Thus, both the addition of borane and the addition of an electrophile such as H^+ occur at the same sp^2 carbon and for the same reason—to form the more stable carbocation-like transition state.

addition of BH₃		addition of HBr	
more stable transition state	less stable transition state	more stable transition state	less stable transition state

The alkylborane formed in the first step of the reaction reacts with another molecule of alkene to form a dialkylborane, which then reacts with yet another molecule of alkene to form a trialkylborane. In each of these reactions, boron adds to the sp^2 carbon bonded to the most hydrogens and the nucleophilic hydride ion adds to the other sp^2 carbon.

Mechanistic Tutorial: Hydroboration-oxidation

$$CH_3CH{=}CH_2 \ + \ \underset{\text{an alkylborane}}{RBH_2} \ \longrightarrow \ \underset{\text{a dialkylborane}}{CH_3CH_2CH_2BHR}$$

$$CH_3CH{=}CH_2 \ + \ \underset{\text{a dialkylborane}}{R_2BH} \ \longrightarrow \ \underset{\text{a trialkylborane}}{CH_3CH_2CH_2BR_2}$$

The alkylborane, BH_2R, is a bulkier molecule than BH_3 because R is a larger substituent than H. The dialkylborane with two R groups, BHR_2, is even bulkier than the alkylborane. Thus, there are now two reasons for the alkylborane and the dialkylborane to add to the sp^2 carbon that is bonded to the most hydrogens: first, to achieve the *most stable carbocation-like transition state* and, second, because there is *more room* at this carbon for the bulky group to attach itself. **Steric effects** are space-filling effects. **Steric hindrance** refers to bulky groups at the site of the reaction that make it difficult for the reactants to approach each other. Steric hindrance associated with the alkylborane and particularly with the dialkylborane causes the addition to occur at the sp^2 carbon that is bonded to the most hydrogens because that is the least sterically hindered of the two sp^2 carbons. Therefore, in each of the three successive additions to the alkene, boron adds to the sp^2 carbon that is bonded to the most hydrogens and H^- adds to the other sp^2 carbon.

When the hydroboration addition reaction is over, aqueous sodium hydroxide and hydrogen peroxide are added to the reaction mixture. Notice that both the hydroxide ion and the hydroperoxide ion are reagents in the reaction.

$$HOOH + HO^- \ \rightleftharpoons \ HOO^- + H_2O$$

The end result is replacement of boron by an OH group. Because the replacement of boron by an OH group is an oxidation reaction (oxygen has been added to the compound), the overall reaction is called hydroboration–oxidation.

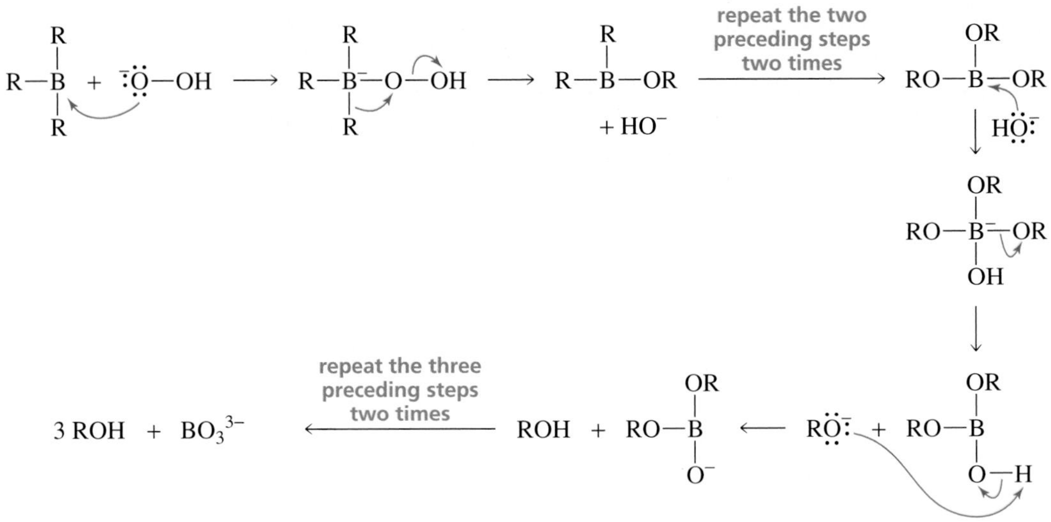

In the overall hydroboration–oxidation reaction, three moles of alkene react with one mole of BH_3 to form three moles of alcohol. The OH ends up on the sp^2 carbon that was bonded to the greater number of hydrogens because it replaces boron, which was the original electrophile in the reaction.

$$3\ CH_3CH{=}CH_2\ +\ BH_3\ \xrightarrow{THF}\ (CH_3CH_2CH_2)_3B\ \xrightarrow[H_2O]{HO^-,\ H_2O_2}\ 3\ CH_3CH_2CH_2OH\ +\ BO_3{}^{3-}$$

Synthetic Tutorial:
Hydroboration-oxidation

Because carbocation intermediates are not formed in the hydroboration reaction, carbocation rearrangements do not occur.

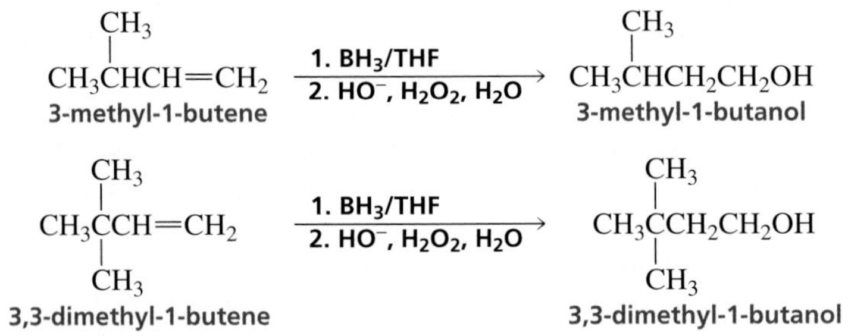

$$\underset{\text{3-methyl-1-butene}}{\overset{\overset{\displaystyle CH_3}{|}}{CH_3CHCH{=}CH_2}} \xrightarrow[\text{2. HO}^-,\ H_2O_2,\ H_2O]{\text{1. BH}_3/\text{THF}} \underset{\text{3-methyl-1-butanol}}{\overset{\overset{\displaystyle CH_3}{|}}{CH_3CHCH_2CH_2OH}}$$

$$\underset{\substack{|\\ CH_3 \\ \text{3,3-dimethyl-1-butene}}}{\overset{\overset{\displaystyle CH_3}{|}}{CH_3CCH{=}CH_2}} \xrightarrow[\text{2. HO}^-,\ H_2O_2,\ H_2O]{\text{1. BH}_3/\text{THF}} \underset{\substack{|\\ CH_3 \\ \text{3,3-dimethyl-1-butanol}}}{\overset{\overset{\displaystyle CH_3}{|}}{CH_3CCH_2CH_2OH}}$$

PROBLEM 42 ◆

How many moles of BH_3 are needed to react with two moles of 1-pentene?

PROBLEM 43 ◆

What product would be obtained from hydroboration–oxidation of the following alkenes?

a. 2-methyl-2-butene

b. 1-methylcyclohexene

A **carbene** is an unusual carbon-containing chemical species. It has a carbon with a lone pair of electrons and an empty orbital. The empty orbital makes the carbene very reactive. Methylene ($:CH_2$) is the simplest carbene. It can be generated by heating diazomethane. Propose a mechanism for the following reaction.

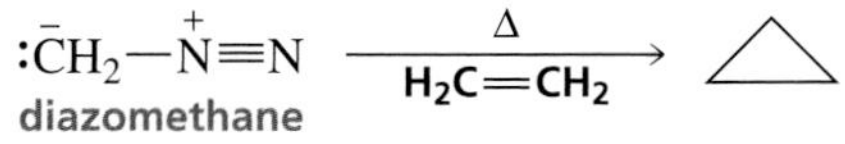

$$:\bar{C}H_2 - \overset{+}{N} \equiv N \xrightarrow[\substack{H_2C=CH_2}]{\Delta} \triangle$$
diazomethane

The information provided in the question is all you need to write a mechanism. First, because you know the structure of methylene, you can see that it can be generated by breaking the carbon–nitrogen bond of diazomethane. Second, because methylene has an empty orbital, it is an electrophile and, therefore, will react with ethene (a nucleophile). Now the question is, what is the nucleophile that reacts with the other sp^2 carbon of the alkene? Because you know that cyclopropane is the product of the reaction, the nucleophile must be the lone pair electrons of methylene.

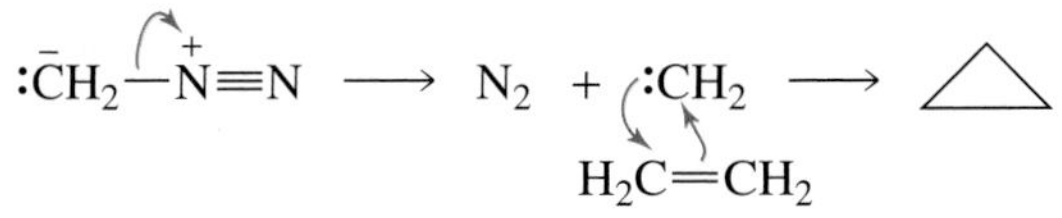

$$:\bar{C}H_2 - \overset{+}{N} \equiv N \longrightarrow N_2 + :CH_2 \longrightarrow \triangle$$
$$H_2C=CH_2$$

Note: Diazomethane must be handled with great care because it is both explosive and toxic.

The addition of HBr to 1-butene forms 2-bromobutane. But what if you wanted to synthesize 1-bromobutane? Formation of 1-bromobutane requires the addition of HBr in a manner that would violate Markovnikov's rule. If peroxide is added to the reaction mixture, however, the product of the addition reaction will be the desired 1-bromobutane. Either hydrogen peroxide (HOOH) or an alkyl peroxide (ROOR) can be used. The presence of peroxide in the reaction mixture causes bromine to add to the sp^2 carbon that is bonded to the most hydrogens and hydrogen to add to the other sp^2 carbon. That is, peroxide reverses the order of addition of the H and Br because it changes the mechanism of the reaction in a way that causes Br· to be the electrophile rather than H^+.

**3.18
ADDITION OF
RADICALS. THE
RELATIVE
STABILITIES OF
RADICALS**

$$\underset{\textbf{1-butene}}{CH_3CH_2CH=CH_2} + HBr \longrightarrow \underset{\textbf{2-bromobutane}}{CH_3CH_2\overset{\overset{\displaystyle Br}{|}}{C}HCH_3}$$

$$\underset{\textbf{1-butene}}{CH_3CH_2CH=CH_2} + HBr \xrightarrow{\textbf{peroxide}} \underset{\textbf{1-bromobutane}}{CH_3CH_2CH_2CH_2Br}$$

When a bond breaks such that both electrons in the bond stay with one of the atoms, the process is called **heterolytic bond cleavage** or **heterolysis.** When a bond breaks such that each of the atoms retains one of the bonding electrons, the process is called **homolytic bond cleavage** or **homolysis.** Remember that an arrowhead with two barbs signifies the movement of two electrons, whereas an arrowhead with one barb, sometimes called a fishhook, signifies the movement of a single electron.

An arrowhead with two barbs signifies the movement of two electrons.

An arrowhead with one barb signifies the movement of one electron.

heterolytic bond cleavage

$$H-\ddot{B}r: \longrightarrow H^+ + :\ddot{B}r:^-$$

homolytic bond cleavage

$$H-\ddot{B}r: \longrightarrow H\cdot + \cdot\ddot{B}r:$$

Either hydrogen peroxide or an alkyl peroxide can be used to reverse the order of addition of the H and Br to an alkene. Both contain a weak oxygen–oxygen single bond that is readily broken homolytically in the presence of light or heat to form radicals. A **radical** (also known as a **free radical**) is an atom or a group of atoms with an unpaired electron.

$$H\ddot{O}-\ddot{O}H \xrightarrow{\text{light}} 2 \, H\ddot{O}\cdot$$
hydrogen peroxide   **hydroxyl radicals**

$$R\ddot{O}-\ddot{O}R \xrightarrow{\text{light}} 2 \, R\ddot{O}\cdot$$
an alkyl peroxide   **alkoxyl radicals**

A radical is very reactive because it seeks an electron so it can complete its octet. To complete its octet, the hydroxyl (or alkoxyl) radical removes a hydrogen atom from a molecule of HBr. This forms a bromine radical.

$$R-\ddot{O}\cdot + H-\ddot{B}r: \longrightarrow R-\ddot{O}-H + \cdot\ddot{B}r:$$
                                                a bromine radical

The bromine radical now seeks an electron to complete its octet. Because the double bond of an alkene is electron-rich, the bromine radical completes its octet by combining with one of the electrons of the π bond of the alkene to form a bromine–carbon bond. The second electron of the π bond is the unpaired electron in the resulting alkyl radical. If the bromine radical adds to the sp^2 carbon in 1-butene that is bonded to the most hydrogens, a secondary alkyl radical is formed. If the bromine radical adds to the other sp^2 carbon, a primary alkyl radical is formed. Like carbocations, radicals are stabilized by electron-donating alkyl groups, so a **tertiary alkyl radical** is more stable than a **secondary alkyl radical,** and a secondary alkyl radical is more stable than a **primary alkyl radical.** The bromine radical, therefore, adds to the sp^2 carbon that is bonded to the most hydrogens in order to form the more stable secondary radical. The alkyl radical that is formed removes a hydrogen atom from another molecule of HBr to produce a molecule of the alkyl halide product and another bromine radical. Because the first species that adds to the alkene is a radical $(Br\cdot)$, the addition of HBr in the presence of peroxides is called a **radical addition reaction.**

$$:\ddot{B}r\cdot + CH_2{=}CHCH_2CH_3 \longrightarrow :\ddot{B}rCH_2\dot{C}HCH_2CH_3$$
                                                an alkyl radical

$$BrCH_2\dot{C}HCH_2CH_3 + H-\ddot{B}r: \longrightarrow :\ddot{B}rCH_2CH_2CH_2CH_3 + \cdot\ddot{B}r:$$

When HBr reacts with an alkene in the absence of peroxide, the electrophile (the first species to add to the alkene) is H^+. In the presence of peroxide, the electrophile is $Br\cdot$. In both cases, the electrophile adds to the sp^2 carbon that is bonded to the most hydrogens, so both reactions follow the more general statement of Markovnikov's rule (*the electrophile adds to the sp² carbon that is bonded to the greater number of hydrogens*). The addition of peroxide to the reaction mixture changes the order of the addition of the reagents to the double bond, making the reaction *appear* to violate Markovnikov's rule. This is why the presence of peroxide in the reaction mixture is said to cause the *apparent* anti-Markovnikov addition of HBr.

Because the addition of HBr in the presence of peroxide forms a radical intermediate rather than a carbocation intermediate, rearrangement of the intermediate does not occur. Radicals do not rearrange as readily as do carbocations.

$$\underset{\textbf{3-methyl-1-butene}}{CH_3\overset{\overset{\displaystyle CH_3}{|}}{C}HCH{=}CH_2} + HBr \xrightarrow{\textbf{peroxide}} \underset{\textbf{1-bromo-3-methylbutane}}{CH_3\overset{\overset{\displaystyle CH_3}{|}}{C}HCH_2CH_2Br}$$

As just mentioned, the relative stabilities of primary, secondary, and tertiary alkyl radicals are in the same order as the relative stabilities of primary, secondary, and tertiary carbocations. However, the difference in energy between the radicals is quite a bit smaller than the difference in energy between the carbocations.

$$\underset{\textbf{tertiary radical}}{R-\overset{\overset{\displaystyle R}{|}}{\underset{\underset{\displaystyle R}{|}}{C}}\cdot} \quad > \quad \underset{\textbf{secondary radical}}{R-\overset{\overset{\displaystyle R}{|}}{\underset{\underset{\displaystyle H}{|}}{C}}\cdot} \quad > \quad \underset{\textbf{primary radical}}{R-\overset{\overset{\displaystyle H}{|}}{\underset{\underset{\displaystyle H}{|}}{C}}\cdot} \quad > \quad \underset{\textbf{methyl radical}}{H-\overset{\overset{\displaystyle H}{|}}{\underset{\underset{\displaystyle H}{|}}{C}}\cdot}$$

increasing stability ⟵

The relative stabilities of primary, secondary, and tertiary alkyl radicals are reflected in the transition states leading to their formation. Consequently, the more stable the radical, the less energy required to make it. This explains why the bromine radical adds to the *sp²* carbon of 1-butene that is bonded to the most hydrogens to form a secondary alkyl radical rather than adding to the other *sp²* carbon to form a primary alkyl radical. The secondary radical is more stable than the primary radical and is therefore easier to form—the energy barrier is lower for its formation.

The following mechanism for the addition of HBr to an alkene in the presence of peroxide involves seven steps. The steps in a radical chain reaction can be divided into initiation steps, propagation steps, and termination steps.

1. $R\ddot{O}{-}\ddot{O}R \longrightarrow 2\ R\ddot{O}\cdot$

    $\left. \begin{array}{l} \\ \\ \end{array} \right\}$ **initiation steps**

2. $R\ddot{O}\cdot + H{-}\ddot{B}r{:} \longrightarrow R\ddot{O}H + \cdot\ddot{B}r{:}$

3. $CH_3\overset{\overset{\displaystyle CH_3}{|}}{C}{=}CH_2 + \cdot\ddot{B}r{:} \longrightarrow CH_3\overset{\overset{\displaystyle CH_3}{|}}{\overset{\displaystyle \cdot}{C}}CH_2\ddot{B}r{:}$

    $\left. \begin{array}{l} \\ \\ \\ \\ \end{array} \right\}$ **propagation steps**

4. $CH_3\overset{\overset{\displaystyle CH_3}{|}}{\overset{\displaystyle \cdot}{C}}CH_2\ddot{B}r{:} + H{-}\ddot{B}r{:} \longrightarrow CH_3\overset{\overset{\displaystyle CH_3}{|}}{C}HCH_2\ddot{B}r{:} + \cdot\ddot{B}r{:}$

Mechanistic Tutorial:
Addition of HBr in
presence of peroxide

5. $2 \, :\ddot{Br}\cdot \longrightarrow Br_2$

6.
$$\underset{\underset{CH_3}{|}}{CH_3\dot{C}CH_2\ddot{Br}:} \; + \; :\ddot{Br}\cdot \; \longrightarrow \; \underset{\underset{:\ddot{Br}:}{|}}{\underset{\underset{CH_3}{|}}{CH_3\dot{C}CH_2\ddot{Br}:}}$$

7.
$$2 \, \underset{\underset{CH_3}{|}}{CH_3\dot{C}CH_2\ddot{Br}:} \; \longrightarrow \; \underset{\underset{CH_3 \; CH_3}{|\quad|}}{:\ddot{Br}CH_2\overset{\overset{CH_3 \; CH_3}{|\quad|}}{C-C}CH_2\ddot{Br}:}$$

} termination steps

- **Initiation Steps.** The first step is an **initiation step** because it creates radicals. The second step is also an initiation step because it forms the chain-propagating radical (Br·).
- **Propagation Steps.** Steps 3 and 4 are **propagation steps.** In step 3, a radical (Br·) reacts to produce another radical. In step 4, the radical produced in the first propagation step reacts to form the radical (Br·) that was the reactant in the first propagation step. The two propagation steps are repeated over and over. Hence, the reaction is called a **radical chain reaction.** These two steps propagate the chain.
- **Termination Steps.** Steps 5, 6, and 7 are **termination steps.** In a termination step, two radicals combine to produce a molecule in which all the electrons are paired, thus ending the radical chain reaction. Any two radicals present in the reaction mixture can combine in a termination step, so radical reactions produce a mixture of products.

PROBLEM 44

Write out the propagation steps that occur when HBr adds to 1-methylcyclohexene in the presence of peroxide.

Peroxide is a **radical initiator** because it creates radicals. If peroxide were excluded from the reaction mixture, the preceding radical reaction would not occur. Any reaction that occurs in the presence of a radical initiator and does not occur in its absence must take place by a mechanism that involves radicals as intermediates. Any compound that can readily undergo homolysis (dissociate to form radicals) can act as a radical initiator.

While radical initiators cause radical reactions to occur, **radical inhibitors** have the opposite effect. They trap radicals as they are formed, preventing reactions that take place by mechanisms involving radicals. The reason radical inhibitors are able to trap radicals is discussed in Section 8.8.

Peroxide has no effect on the addition of HCl or HI to an alkene. In the presence of peroxide, addition occurs just as it does in the absence of peroxide.

$$CH_3CH{=}CH_2 \; + \; HCl \; \xrightarrow{\text{peroxide}} \; \underset{\underset{Cl}{|}}{CH_3CHCH_3}$$

$$\underset{\underset{CH_3}{|}}{CH_3C{=}CH_2} \; + \; HI \; \xrightarrow{\text{peroxide}} \; \underset{\underset{I}{|}}{\overset{\overset{CH_3}{|}}{CH_3CCH_3}}$$

Why is the **peroxide effect** observed for the addition of HBr but not for the addition of HCl or HI? This question can be answered by calculating the $\Delta H°$ for the two propagation steps in the radical chain reaction (using the bond dissociation energies in Table 3.1).

$Cl\cdot \; + \; CH_2{=}CH_2 \; \longrightarrow \; ClCH_2\overset{\cdot}{C}H_2$ $\Delta H° = 61 - 82 = -21$ kcal/mol (−88 kJ/mol)

$ClCH_2\overset{\cdot}{C}H_2 \; + \; HCl \; \longrightarrow \; ClCH_2CH_3 \; + \; Cl\cdot$ $\Delta H° = 103 - 101 = +2$ kcal/mol (+8 kJ/mol)

$Br\cdot \; + \; CH_2{=}CH_2 \; \longrightarrow \; BrCH_2\overset{\cdot}{C}H_2$ $\Delta H° = 61 - 69 = -8$ kcal/mol (−33 kJ/mol)

$BrCH_2\overset{\cdot}{C}H_2 \; + \; HBr \; \longrightarrow \; BrCH_2CH_3 \; + \; Br\cdot$ $\Delta H° = 87 - 101 = -14$ kcal/mol (−59 kJ/mol)

$I\cdot \; + \; CH_2{=}CH_2 \; \longrightarrow \; ICH_2\overset{\cdot}{C}H_2$ $\Delta H° = 61 - 55 = +6$ kcal/mol (+25 kJ/mol)

$ICH_2\overset{\cdot}{C}H_2 \; + \; HI \; \longrightarrow \; ICH_2CH_3 \; + \; I\cdot$ $\Delta H° = 71 - 101 = -30$ kcal/mol (−126 kJ/mol)

For the radical addition of HCl, the first propagation step is exothermic and the second is endothermic. For the radical addition of HI, the first propagation step is endothermic and the second is exothermic. Only for the radical addition of HBr are both propagation steps exothermic. In a radical reaction, the steps that propagate the chain reaction compete with the steps that terminate it. Termination steps are always exothermic because only bond making (and no bond breaking) occurs. Therefore, only when both propagation steps are exothermic can propagation compete with termination. When HCl or HI adds to an alkene in the presence of peroxides, any chain reaction that is initiated is terminated rather than propagated because propagation cannot compete with termination. Consequently, the radical chain reaction does not occur. Only ionic addition (H^+ followed by Cl^- or I^-) occurs.

In the presence of a metal catalyst such as platinum, palladium, or nickel, hydrogen (H_2) adds to the double bond of an alkene to form an alkane. Without the catalyst, the energy barrier to the reaction would be enormous because the hydrogen–hydrogen bond is so strong (Table 3.1). The catalyst decreases the energy of activation by breaking the hydrogen–hydrogen bond. Platinum and palladium are used in a finely divided state adsorbed on charcoal (Pt/C, Pd/C). The platinum catalyst is frequently used in the form of PtO_2, which is known as Adams's catalyst.

3.19
ADDITION OF HYDROGEN. THE RELATIVE STABILITIES OF ALKENES

$$CH_3CH{=}CHCH_3 \; + \; H_2 \; \xrightarrow{\text{Pt/C}} \; CH_3CH_2CH_2CH_3$$

    2-butene   **butane**

$$\underset{\text{2-methylpropene}}{\overset{\overset{\displaystyle CH_3}{|}}{CH_3C}{=}CH_2} \; + \; H_2 \; \xrightarrow{\text{Pd/C}} \; \underset{\text{2-methylpropane}}{\overset{\overset{\displaystyle CH_3}{|}}{CH_3CHCH_3}}$$

$+ \; H_2 \; \xrightarrow{\text{Ni}}$

    cyclohexene   **cyclohexane**

The addition of hydrogen is called **hydrogenation.** Because the preceding reactions require a catalyst, they are examples of **catalytic hydrogenations.** The metal

catalysts are insoluble in the reaction mixture and therefore are classified as **heterogeneous catalysts.** Catalysts that are soluble in a reaction mixture, such as the acid catalyst used in the hydration of alkenes in Section 3.13, are called **homogeneous catalysts.** A heterogeneous catalyst can easily be separated from the reaction mixture by filtration. It can then be recycled, which is an important feature because metal catalysts tend to be expensive.

The details of the mechanism of catalytic hydrogenation are not completely understood. We know that hydrogen is adsorbed on the surface of the metal and that the alkene complexes with the metal by overlapping its *p* orbitals with vacant orbitals of the metal. Breaking the π bond of the alkene and the σ bond of H_2 and forming the carbon–hydrogen σ bonds all occur on the surface of the metal. The alkane product diffuses away from the metal surface as it is formed (Figure 3.10).

Hydrogenation reactions are exothermic (they have negative $\Delta H°$ values). The heat released in a hydrogenation reaction $(-\Delta H°)$ is called the **heat of hydrogenation.** Thus, the heat of hydrogenation is the absolute value of $\Delta H°$.

$$\underset{\text{2-methyl-2-butene}}{CH_3\overset{\overset{\displaystyle CH_3}{|}}{C}=CHCH_3} + H_2 \xrightarrow{Pt/C} CH_3\overset{\overset{\displaystyle CH_3}{|}}{C}HCH_2CH_3 \qquad \Delta H° = -26.9 \text{ kcal/mol } (-113 \text{ kJ/mol})$$

$$\underset{\text{2-methyl-1-butene}}{CH_2=\overset{\overset{\displaystyle CH_3}{|}}{C}CH_2CH_3} + H_2 \xrightarrow{Pt/C} CH_3\overset{\overset{\displaystyle CH_3}{|}}{C}HCH_2CH_3 \qquad \Delta H° = -28.5 \text{ kcal/mol } (-119 \text{ kJ/mol})$$

$$\underset{\text{3-methyl-1-butene}}{CH_3\overset{\overset{\displaystyle CH_3}{|}}{C}HCH=CH_2} + H_2 \xrightarrow{Pt/C} CH_3\overset{\overset{\displaystyle CH_3}{|}}{C}HCH_2CH_3 \qquad \Delta H° = -30.3 \text{ kcal/mol } (-127 \text{ kJ/mol})$$

The preceding three catalytic hydrogenation reactions all form the same alkane product, so the energy of the *product* is the same for each of the three reactions. The three reactions have different heats of hydrogenation, however, so the three *reactants* must have different energies (Figure 3.11). The reaction associated with the *least* negative $\Delta H°$ (-26.9 kcal/mol) has the *most* stable reactant, and the reaction with the most negative $\Delta H°$ (-30.3 kcal/mol) has the *least* stable reactant. (If $\Delta H°$ is -26.9 kcal/mol, the heat of hydrogenation is 26.9 kcal/mole. Therefore we can say either that the most stable alkene has the least negative $\Delta H°$, or the most stable alkene has the smallest heat of hydrogenation.)

The most stable alkene has the smallest heat of hydrogenation.

Examination of the structures of the three alkene reactants shows that the most stable alkene has two alkyl substituents bonded to one of the sp^2 carbons and one alkyl

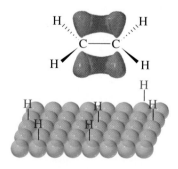
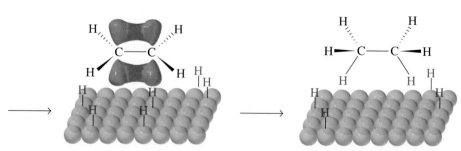

▲ **Figure 3.10**
Catalytic hydrogenation of an alkene.

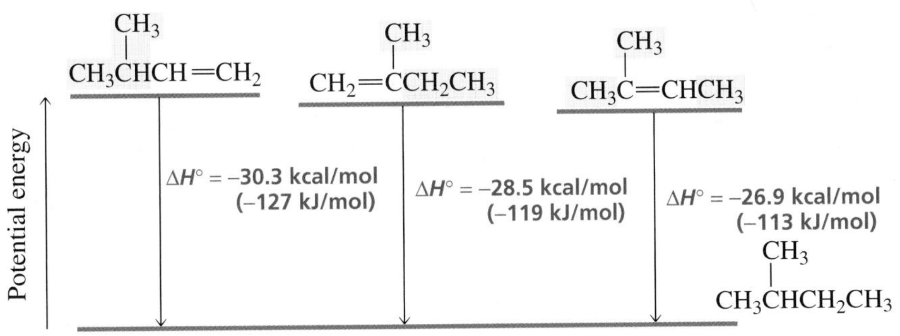

$$CH_3CHCH=CH_2 \qquad CH_2=CCH_2CH_3 \qquad CH_3C=CHCH_3$$

Potential energy

$\Delta H° = -30.3$ kcal/mol
$(-127$ kJ/mol$)$

$\Delta H° = -28.5$ kcal/mol
$(-119$ kJ/mol$)$

$\Delta H° = -26.9$ kcal/mol
$(-113$ kJ/mol$)$

$$CH_3CHCH_2CH_3$$

◀ **Figure 3.11**
The relative energy levels
of three alkenes that can be
catalytically hydrogenated to
2-methylbutane.

substituent bonded to the other sp^2 carbon, for a total of three alkyl substituents (three methyl groups) bonded to the two sp^2 carbons. The alkene of intermediate stability has a total of two alkyl substituents (a methyl group and an ethyl group) bonded to the sp^2 carbons, and the least stable of the three alkenes has only one alkyl substituent (an isopropyl group) bonded to the sp^2 carbons. It is apparent that alkyl substituents bonded to the sp^2 carbons of an alkene have a stabilizing effect on the alkene. We can therefore make the following statement: *The more alkyl substituents that are bonded to the* sp^2 *carbons of an alkene, the more stable it is.* (Some students find it easier to look at the number of hydrogens bonded to the sp^2 carbons. Stated in terms of hydrogens: *The fewer hydrogens that are bonded to the* sp^2 *carbons of an alkene, the more stable it is.*)

relative stabilities of alkyl-substituted alkenes

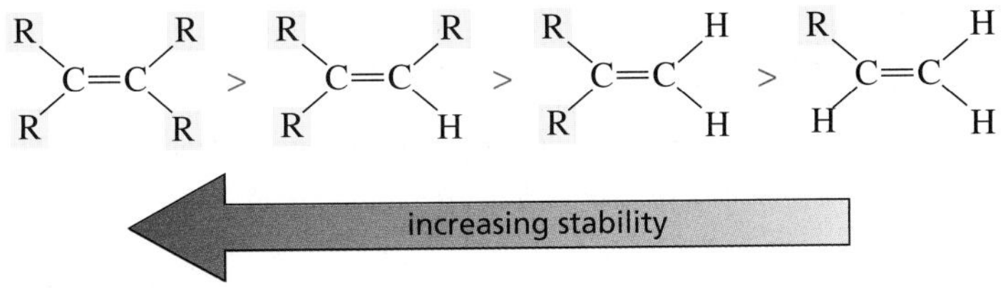

increasing stability

The fewer hydrogens bonded to the *sp²* carbons of an alkene, the more stable it is.

Alkyl substituents stabilize both alkenes *and* carbocations.

Both *trans*-2-butene and *cis*-2-butene have two alkyl groups bonded to the sp^2 carbons, but *trans*-2-butene has a smaller heat of hydrogenation (a less negative $\Delta H°$). This means that the trans isomer, in which the large substituents are farther apart, is more stable than the cis isomer, in which the large substituents are closer together.

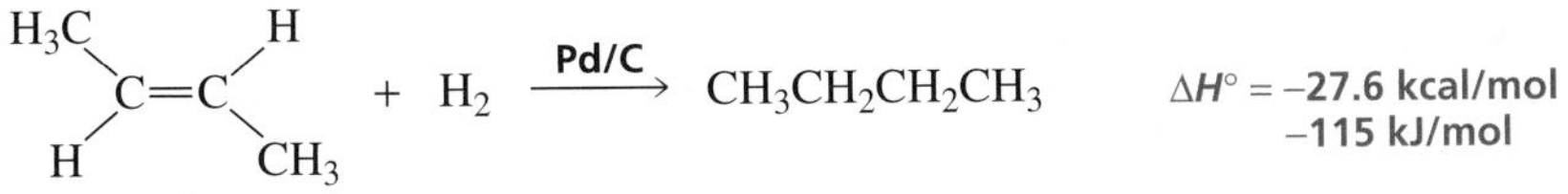

$$\text{trans-2-butene} + H_2 \xrightarrow{Pd/C} CH_3CH_2CH_2CH_3 \qquad \Delta H° = -27.6 \text{ kcal/mol}$$
$$-115 \text{ kJ/mol}$$

trans-2-butene

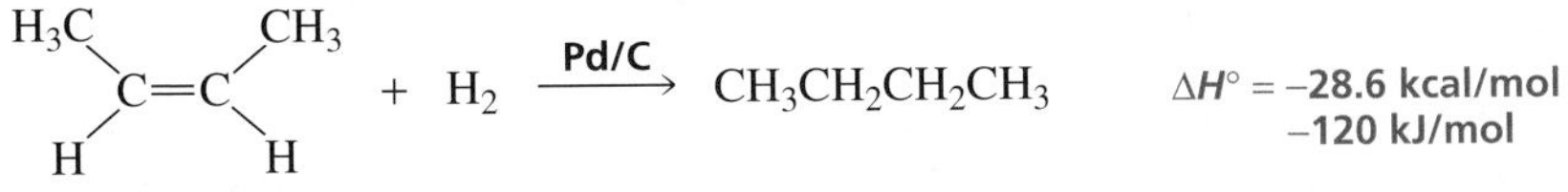

$$\text{cis-2-butene} + H_2 \xrightarrow{Pd/C} CH_3CH_2CH_2CH_3 \qquad \Delta H° = -28.6 \text{ kcal/mol}$$
$$-120 \text{ kJ/mol}$$

cis-2-butene

When the large substituents are on the same side of the molecule, their electron clouds can interfere with each other. This causes strain in the molecule and makes the molecule less stable. You saw in Section 2.11 that this kind of strain is called *steric strain*. When the large substituents are on opposite sides of the molecule, their electron clouds cannot interact, and the molecule has less steric strain.

Platinum and palladium are expensive metals, so the accidental finding by **Paul Sabatier (1854–1941)** *that nickel, a much cheaper metal, can catalyze hydrogenation reactions made hydrogenation a feasible large-scale industrial process. The conversion of plant oils to margarine is one such hydrogenation reaction. Sabatier was born in France and was a professor at the University of Toulouse. He shared the 1912 Nobel Prize in chemistry with Victor Grignard (page xxx).*

cis-2-butene
the cis isomer has steric strain

trans-2-butene
the trans isomer is free of steric strain

The heat of hydrogenation of *cis*-2-butene, in which the two alkyl substituents are on the *same side* of the double bond, is essentially identical to that of 2-methylpropene, in which the two alkyl substituents are on the *same carbon*. The three dialkyl-substituted alkenes are all *less* stable than a trialkyl-substituted alkene and are all *more* stable than a monoalkyl-substituted alkene.

relative stabilities of dialkyl-substituted alkenes

$$H_3C \quad H \qquad H_3C \quad CH_3 \qquad H_3C \quad H$$
$$\underset{H}{} C=C \underset{CH_3}{} \;>\; \underset{H}{} C=C \underset{H}{} \;=\; \underset{H_3C}{} C=C \underset{H}{}$$

alkyl substituents
are trans

alkyl substituents
are cis

alkyl substituents
are on the same sp^2 carbon

PROBLEM 45 ◆

The same alkane is obtained from the catalytic hydrogenation of both alkene A and alkene B. The heat of hydrogenation of alkene A is 29.8 kcal/mol (125 kJ/mol), and the heat of hydrogenation of alkene B is 31.4 kcal/mol (131 kJ/mol). Which alkene is more stable?

3.20 REACTIONS AND SYNTHESIS

This chapter has been concerned with the reactions of alkenes. We have looked at why alkenes react, the kinds of reagents with which they react, the mechanisms by which the reactions occur, and the products that are formed. It is important to remember that when you are studying reactions, you are simultaneously studying synthesis. When you learn that compound A reacts with a certain reagent to form compound B, you are learning not only about the reactivity of A, but also about one way in which compound B can be synthesized.

$$A \longrightarrow B$$

For example, you have seen that alkenes can add many different reagents and that, as a result of adding these reagents, compounds such as alkyl halides, vicinal dihalides, halohydrins, alcohols, ethers, and alkanes are synthesized.

Although you have seen how alkenes react and the kinds of compounds synthesized when alkenes undergo reactions, you have not yet seen how alkenes are synthesized. Reactions of alkenes involve the addition of atoms (or groups of atoms) to the two sp^2 carbons of the double bond. Reactions that lead to the synthesis of alkenes are exactly the opposite. They involve the elimination of atoms (or groups of atoms) from two adjacent sp^3 carbons.

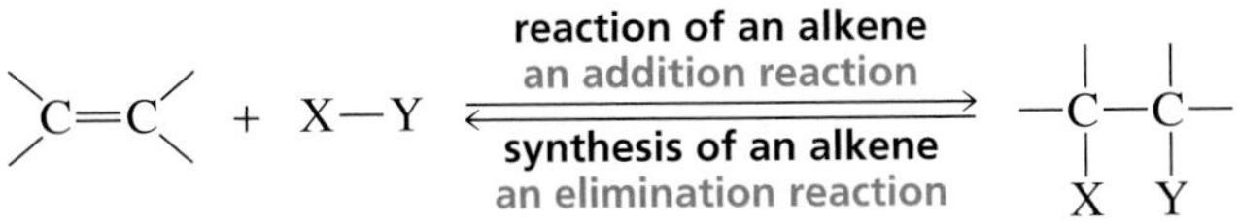

$$\overset{|}{\underset{|}{C}}=\overset{|}{\underset{|}{C} + X-Y} \quad \underset{\substack{\text{synthesis of an alkene} \\ \text{an elimination reaction}}}{\overset{\substack{\text{reaction of an alkene} \\ \text{an addition reaction}}}{\rightleftharpoons}} \quad \overset{|}{\underset{X}{-C}}-\overset{|}{\underset{Y}{C}}-$$

You will learn how alkenes are synthesized when you study compounds that undergo elimination reactions. The various reactions that result in the synthesis of alkenes are listed in Appendix IV.

PROBLEM 46 / SOLVED

Starting with an alkene, indicate how each of the following compounds can be synthesized.

a. 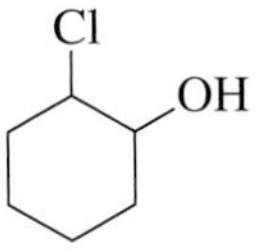 [structure: cyclohexane ring with Cl and OH]
b. 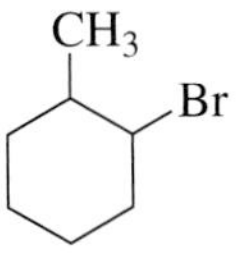 [structure: cyclohexane ring with CH₃ and Br]
c. 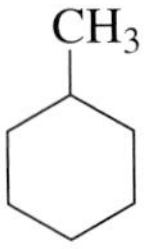 [structure: cyclohexane ring with CH₃]

SOLUTION

a. The only alkene that can be used is cyclohexene. To get the desired substituents on the ring, cyclohexene must react with Cl_2 in an aqueous solution so that water will be the nucleophile.

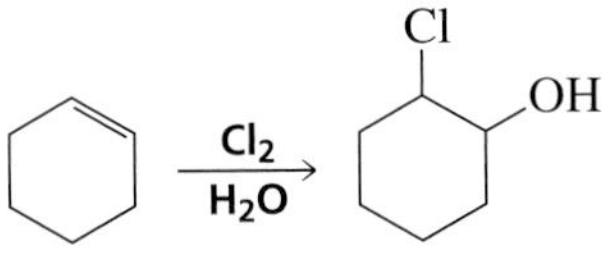

[reaction: cyclohexene + Cl₂/H₂O → chlorohydroxy cyclohexane with Cl and OH]

b. The alkene that should be used is 1-methylcyclohexene. To get the substituents in the desired locations, the electrophile in the reaction must be a bromine radical. Therefore, the reagents required to react with 1-methylcyclohexene are HBr and peroxide.

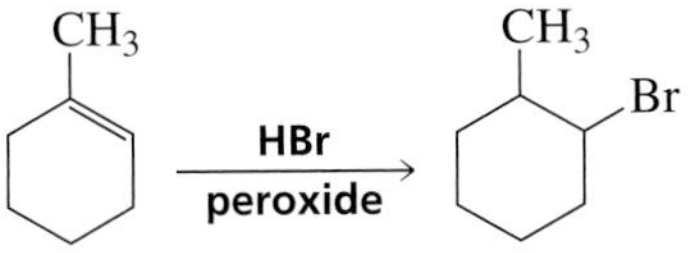

[reaction: 1-methylcyclohexene + HBr/peroxide → methylbromocyclohexane with CH₃ and Br]

c. In order to synthesize an alkane from an alkene, the alkene must undergo catalytic hydrogenation. Several alkenes could be used for the synthesis.

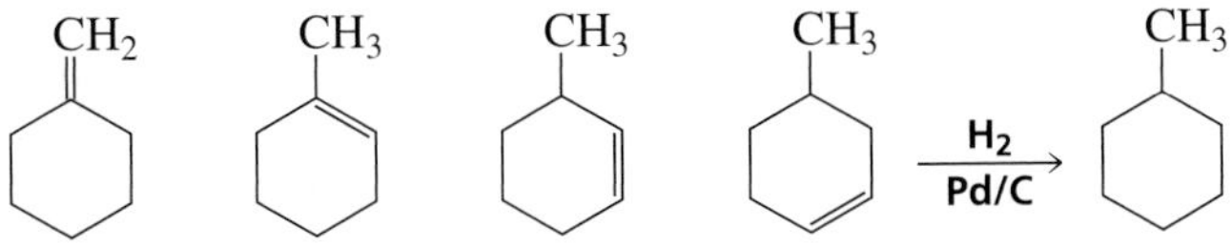

[reaction: methylenecyclohexane, and methylcyclohexene isomers + H₂/Pd/C → methylcyclohexane]

PROBLEM 47

Why should you avoid using 3-methylcyclohexene as the starting material in Problem 46b?

PROBLEM 48

Starting with an alkene, indicate how each of the following compounds can be synthesized.

a. CH_3CHOCH_3
        |
       CH_3

b. $CH_3CH_2CHCHCH_3$
          | |
          Br   Br

c. 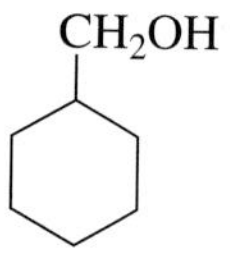 (cyclohexane with CH$_2$OH)

e. 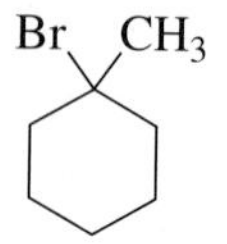 (cyclohexane with Br, CH$_3$)

d. 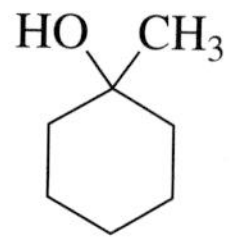 (cyclohexane with HO, CH$_3$)

f. 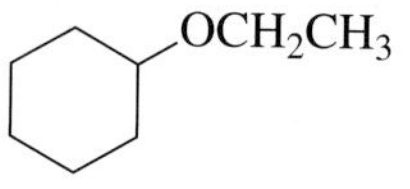 (cyclohexane with OCH$_2$CH$_3$)

SUMMARY OF REACTIONS

1. Electrophilic addition reactions

   a. Addition of hydrogen halides: ionic (Markovnikov orientation) (Section 3.9)

   $$RCH{=}CH_2 \ + \ HX \ \longrightarrow \ \underset{\underset{X}{|}}{R}CHCH_3$$

   HX = HF, HCl, HBr, HI

   b. Addition of hydrogen bromide: radical (apparent anti-Markovnikov orientation) (Section 3.18)

   $$RCH{=}CH_2 \ + \ HBr \ \xrightarrow{\text{peroxide}} \ RCH_2CH_2Br$$

   c. Addition of halogen (Section 3.15)

   $$RCH{=}CH_2 \ + \ Cl_2 \ \xrightarrow{\textbf{CH}_2\textbf{Cl}_2} \ \underset{\underset{Cl}{|}}{R}CHCH_2Cl$$

   $$RCH{=}CH_2 \ + \ Br_2 \ \xrightarrow{\textbf{CH}_2\textbf{Cl}_2} \ \underset{\underset{Br}{|}}{R}CHCH_2Br$$

   d. Addition of water and alcohols: acid-catalyzed (Markovnikov orientation) (Section 3.13)

   $$RCH{=}CH_2 \ + \ H_2O \ \underset{}{\overset{\textbf{H}^+}{\rightleftharpoons}} \ \underset{\underset{OH}{|}}{R}CHCH_3$$

   $$RCH{=}CH_2 \ + \ CH_3OH \ \underset{}{\overset{\textbf{H}^+}{\rightleftharpoons}} \ \underset{\underset{OCH_3}{|}}{R}CHCH_3$$

   e. Addition of water and alcohols: oxymercuration–demercuration and alkoxy-mercuration–demercuration (Markovnikov orientation) (Section 3.16)

   $$RCH{=}CH_2 \ \xrightarrow[\textbf{2. NaBH}_4\textbf{, HO}^-]{\textbf{1. Hg(OAc)}_2\textbf{, H}_2\textbf{O,THF}} \ \underset{\underset{OH}{|}}{R}CHCH_3$$

   $$RCH{=}CH_2 \ \xrightarrow[\textbf{2. NaBH}_4\textbf{, HO}^-]{\textbf{1. Hg(O}_2\textbf{CCF}_3)_2\textbf{, CH}_3\textbf{OH}} \ \underset{\underset{OCH_3}{|}}{R}CHCH_3$$

f. Addition of water: hydroboration–oxidation (apparent anti-Markovnikov orientation) (Section 3.17)

$$RCH=CH_2 \xrightarrow[\text{2. HO}^-, \text{H}_2\text{O}_2, \text{H}_2\text{O}]{\text{1. BH}_3/\text{THF}} RCH_2CH_2OH$$

2. Addition of hydrogen (Section 3.19)

$$RCH=CH_2 + H_2 \xrightarrow{\text{Pd/C, Pt/C, or Ni}} RCH_2CH_3$$

KEY TERMS

acid-catalyzed reaction (page 149)
acyclic (page 112)
addition reaction (page 124)
alkene (page 111)
alkoxymercuration-demercuration (page 158)
allyl group (page 115)
allylic carbon (page 115)
Arrhenius equation (page 135)
carbene (page 163)
carbocation rearrangement (page 152)
catalyst (page 149)
catalytic hydrogenation (page 167)
cis isomer (page 117)
cis–trans isomers (page 117)
concerted reaction (page 160)
constitutional isomers (page 145)
dimer (page 160)
E isomer (page 120)
electrophile (page 122)
electrophilic addition reaction (page 124)
endergonic reaction (page 127)
endothermic reaction (page 129)
enthalpy (page 128)
entropy (page 129)
exergonic reaction (page 127)
exothermic reaction (page 129)
experimental energy of activation (page 135)
first-order rate constant (page 134)
first-order reaction (page 134)
free energy of activation (page 132)
free radical (page 164)

functional group (page 122)
geometric isomers (page 117)
Gibbs standard free energy change (page 127)
halohydrin (page 155)
Hammond postulate (page 142)
heat of hydrogenation (page 168)
heterogeneous catalyst (page 168)
heterolysis (page 163)
heterolytic bond cleavage (page 163)
homogeneous catalyst (page 168)
homolysis (page 163)
homolytic bond cleavage (page 163)
hormone (page 111)
hydration (page 149)
1,2-hydride shift (page 152)
hydroboration–oxidation (page 159)
hydrogenation (page 167)
hyperconjugation (page 141)
initiation step (page 166)
intermediate (page 138)
kinetics (page 132)
kinetic stability (page 133)
Markovnikov's rule (page 145)
mechanism of the reaction (page 123)
1,2-methyl shift (page 152)
nucleophile (page 122)
olefin (page 111)
oxymercuration–demercuration (page 157)
pericyclic reaction (page 160)
peroxide effect (page 167)
pheromone (page 111)
primary alkyl radical (page 164)

primary carbocation (page 140)
propagation step (page 166)
radical (page 164)
radical addition reaction (page 164)
radical chain reaction (page 166)
radical inhibitor (page 166)
radical initiator (page 166)
rate constant (page 134)
rate-determining step (page 138)
rate-limiting step (page 138)
reaction coordinate diagram (page 126)
regioselective reaction (page 145)
saturated hydrocarbon (page 112)
secondary alkyl radical (page 164)
secondary carbocation (page 140)
second-order rate constant (page 134)
second-order reaction (page 134)
solvation (page 131)
steric effects (page 161)
steric hindrance (page 161)
termination step (page 166)
tertiary alkyl radical (page 164)
tertiary carbocation (page 140)
thermodynamics (page 126)
thermodynamic stability (page 132)
trans isomer (page 117)
transition state (page 126)
unsaturated hydrocarbon (page 112)
vicinal (page 154)
vinyl group (page 115)
vinylic carbon (page 115)
Z isomer (page 120)

PROBLEMS

49. Give the systematic name for each of the following compounds.

a. $CH_3CH_2CHCH=CHCH_2CH_2CHCH_3$
$||$
$BrBr$

b. 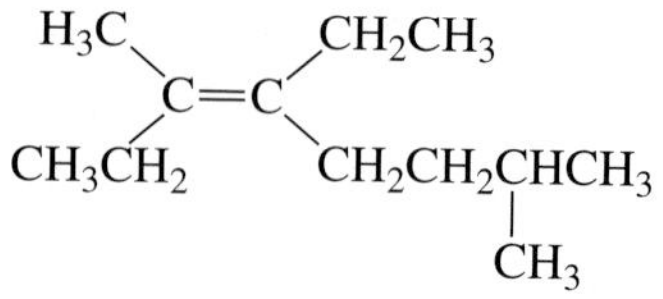 CH$_3$CH$_2$
H$_3$C—C=C—CH$_2$CH$_3$ / CH$_2$CH$_2$CHCH$_3$ with CH$_3$

c. (cyclopentene with CH$_3$ and CH$_3$)

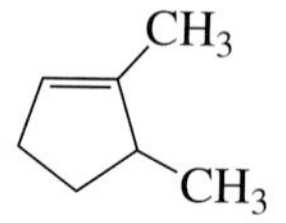

d. 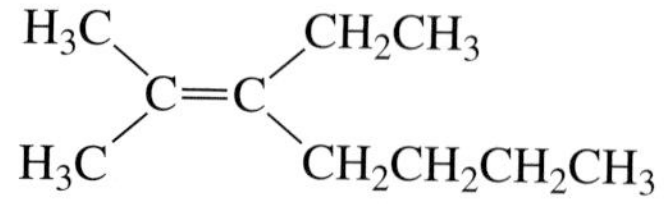 H$_3$C / H$_3$C—C=C—CH$_2$CH$_3$ / CH$_2$CH$_2$CH$_2$CH$_3$

50. Give the structure of a hydrocarbon that has six carbon atoms and:
   a. three vinylic hydrogens and two allylic hydrogens.
   b. three vinylic hydrogens and one allylic hydrogen.
   c. three vinylic hydrogens and no allylic hydrogens.

51. Draw the structure for each of the following.
   a. (Z)-1,3,5-tribromo-2-pentene
   b. (Z)-3-methyl-2-heptene
   c. (E)-1,2-dibromo-3-isopropyl-2-hexene
   d. vinyl bromide
   e. 1,2-dimethylcyclopentene
   f. diallylamine

52. Give the structures and the systematic names for all alkenes with molecular formula C_6H_{12}, ignoring cis–trans isomers. (*Hint:* There are 13.)

53. What will be the major product of the reaction of 2-methyl-2-butene with each of the following reagents?
   a. HBr
   b. HBr + peroxide
   c. HI
   d. HI + peroxide
   e. ICl
   f. H_2/Pd
   g. Br_2 + excess NaCl
   h. $Hg(OAc)_2$, H_2O followed by $NaBH_4/HO^-$
   i. H_2O + trace HCl
   j. Br_2/CH_2Cl_2
   k. Br_2/H_2O
   l. Br_2/CH_3OH
   m. BH_3/THF followed by H_2O_2/HO^-
   n. $Hg(O_2CCF_3)_2$, CH_3OH followed by $NaBH_4/HO^-$

54. Name the following compounds.

   a.

   b.

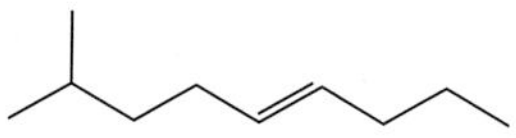

   c.

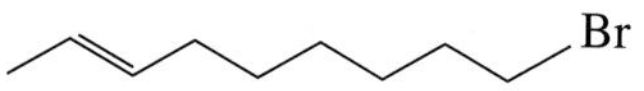

   d.

   e.

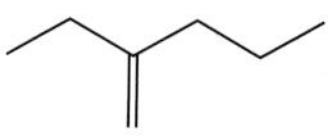

   f.

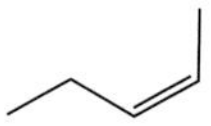

55. Draw curved arrows to show the flow of electrons responsible for the conversion of the reactants into the products.

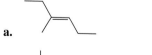
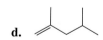

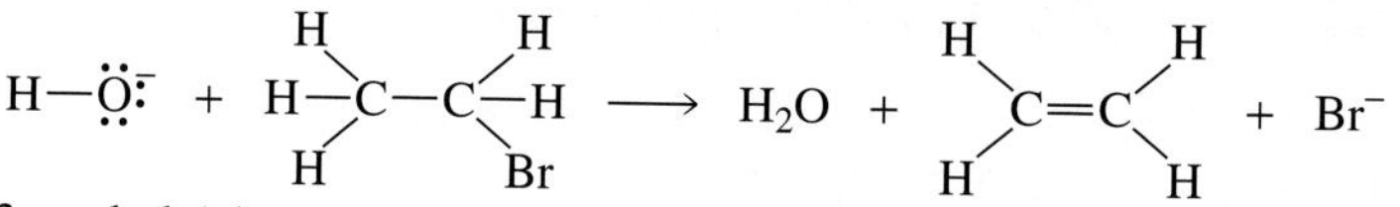

56. When 3-methyl-1-butene reacts with HBr, two alkyl halides are formed, 2-bromo-3-methylbutane and 2-bromo-2-methylbutane. Propose a mechanism that explains the formation of these products.

57. In a reaction in which reactant A is in equilibrium with product B at 25 °C, what are the relative amounts of A and B present at equilibrium if $\Delta G°$ at 25 °C is:
  a. 2.72 kcal/mol?   **c.** −2.72 kcal/mol?
  b. 0.65 kcal/mol?   **d.** −0.65 kcal/mol?

58. Give the reagents that would be required to carry out the following syntheses.

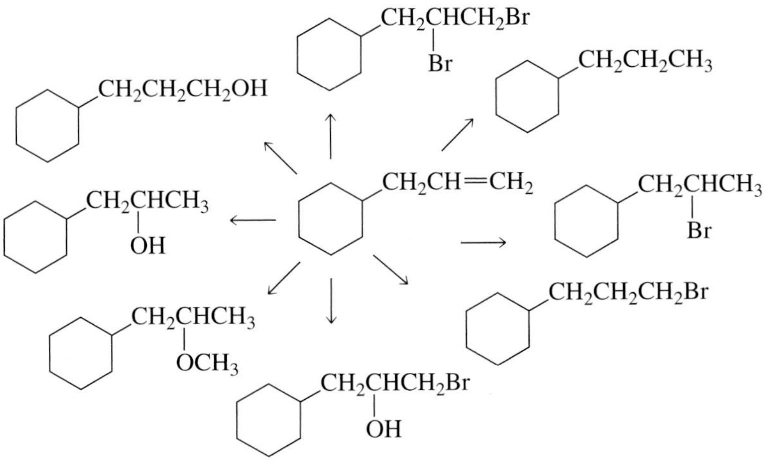

59. Several studies have shown that β-carotene, a precursor of vitamin A, may play a role in preventing cancer. β-Carotene has a molecular formula of $C_{40}H_{56}$, and it contains two rings and no triple bonds. How many double bonds does it have?

60. Which of the indicated bonds has the greater bond strength? Briefly explain why.

  a. $CH_3{-}Cl$ or $CH_3{-}Br$

  b. $CH_3CH_2CH_2$ or CH_3CHCH_3
       | |
       H   H

  c. $CH_3{-}CH_3$ or $CH_3{-}CH_2CH_3$

  d. I—Br or Br—Br

61. How could you prepare the following compounds? You can use any alkene and any other reagents.

  a.

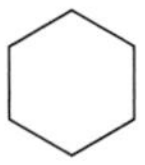

  b. $CH_3CH_2CH_2CHCH_3$
                    |
                    Cl

  c. CH_2Br

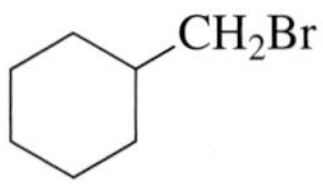

  d. 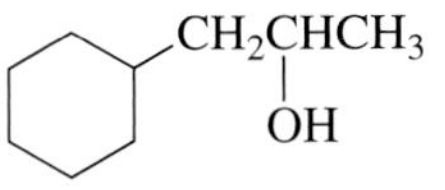 CH_2CHCH_3
                            |
                            OH

  e. $CH_3CH_2CHCHCH_2CH_3$
              | |
              Br OH

  f. $CH_3CH_2CHCHCH_2CH_3$
              | |
              Br Cl

62. Which of the following compounds is the most stable? Which would you expect to have the largest heat of hydrogenation? Which would you expect to have the smallest heat of hydrogenation?

  3,4-dimethyl-2-hexene; 2,3-dimethyl-2-hexene; 4,5-dimethyl-2-hexene

63. Tell whether each of the following compounds has the E or the Z configuration.

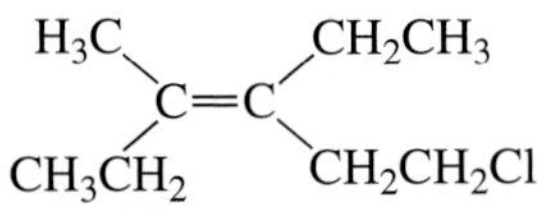

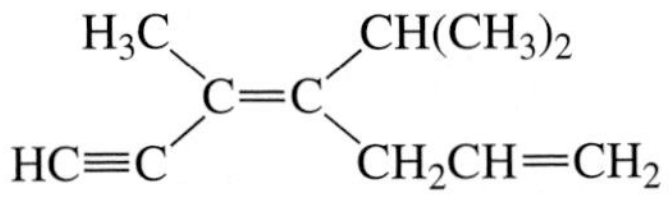

c.

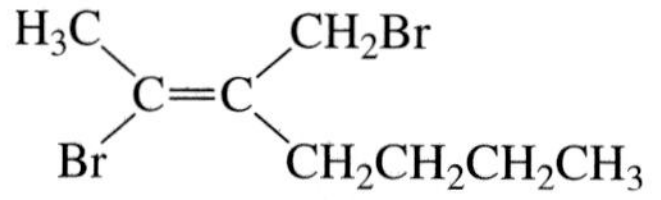

$$\underset{Br}{\overset{H_3C}{}}C{=}C\underset{CH_2CH_2CH_2CH_3}{\overset{CH_2Br}{}}$$

d.

$$\underset{HOCH_2}{\overset{CH_3\overset{\displaystyle O}{\overset{\|}{C}}}{}}C{=}C\underset{CH_2CH_2Cl}{\overset{CH_2Br}{}}$$

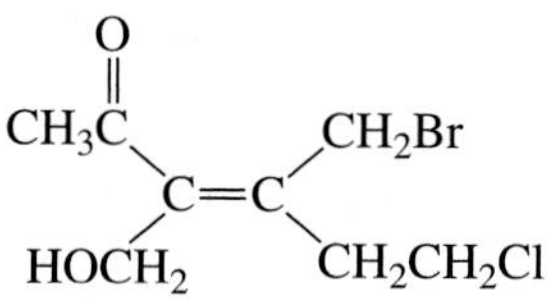

64. Squalene is a hydrocarbon with molecular formula $C_{30}H_{50}$ that is obtained from shark liver (*squalus* is Latin for "shark"). If squalene is not a cyclic compound, how many π bonds does it have?

65. Assign priorities to each set of substituents.
  a. $-Br, -I, -OH, -CH_3$
  b. $-CH_2CH_2OH, -OH, -CH_2Cl, -CH{=}CH_2$
  c. $-CH_2CH_2CH_3, -CH(CH_3)_2, -CH{=}CH_2, -CH_3$
  d. $-CH_2NH_2, -NH_2, -OH, -CH_2OH$
  e. $-COCH_3, -CH{=}CH_2, -Cl, -C{\equiv}N$

66. There are two alkenes that react with HBr to give 1-bromo-1-methylcyclohexane.
  a. Identify the alkenes.
  b. Will both alkenes give the same product when they react with HBr/peroxide?
  c. With HCl?
  d. With HCl/peroxide?

67. Indicate which member of each of the following pairs is the more stable.

  a. 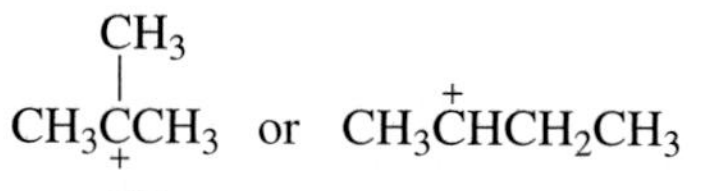 $CH_3\overset{CH_3}{\underset{|}{C}}CH_3$ or $CH_3\overset{+}{C}HCH_2CH_3$

  b. 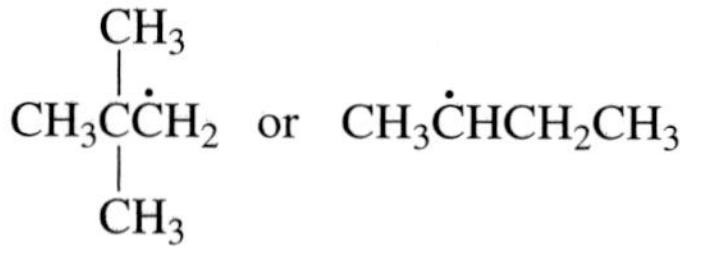 $CH_3\overset{CH_3}{\underset{\underset{CH_3}{|}}{\overset{|}{\dot{C}}}}CH_2$ or $CH_3\dot{C}HCH_2CH_3$

  c. $CH_3\overset{+}{C}HCH_3$ or $CH_3\overset{+}{C}HCH_2Cl$

  d. $CH_3CH_2CH_2\dot{C}HCH_3$ or $CH_3CH_2CH_2CH_2\dot{C}H_2$

  e. 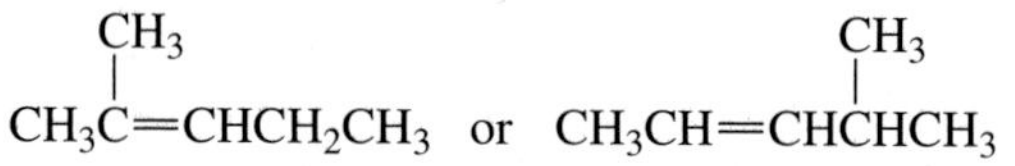 $CH_3\overset{CH_3}{\underset{|}{C}}{=}CHCH_2CH_3$ or $CH_3CH{=}CH\overset{CH_3}{\underset{|}{C}}HCH_3$

  f. 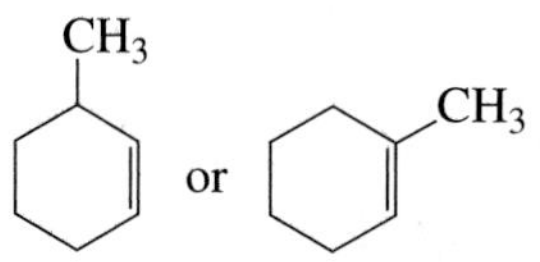 [cyclohexene with CH_3 substituent] or [cyclohexene with CH_3 on double-bond carbon]

  g. 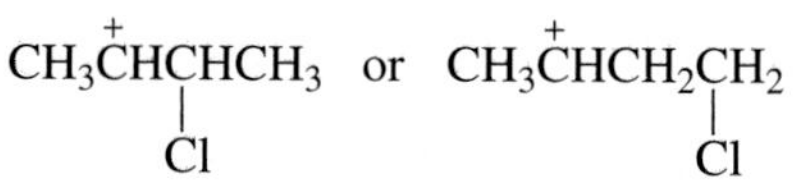 $CH_3\overset{+}{C}H\overset{}{\underset{|}{C}}HCH_3$ or $CH_3\overset{+}{C}HCH_2\overset{}{\underset{|}{C}}H_2$
  with Cl and Cl substituents respectively

68. Molly Kule was a lab technician who was asked by her supervisor to add names to the labels on a collection of alkenes that had only structures on the labels. How many did Molly get right? Correct the incorrect names.
  a. 3-pentene
  b. 2-octene
  c. 2-vinylpentane
  d. 1-ethyl-1-pentene
  e. 5-ethylcyclohexene
  f. 5-chloro-3-hexene
  g. 5-bromo-2-pentene
  h. (*E*)-2-methyl-1-hexene
  i. 2-methylcyclopentene
  j. 2-ethyl-2-butene

69. Given the following reaction coordinate diagram for the reaction of A to give D, answer the following questions.

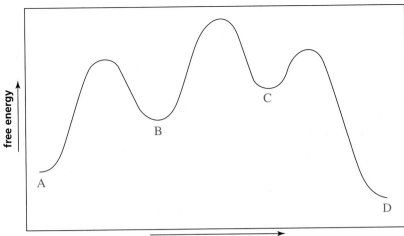

a. How many intermediates are there in the reaction?
b. How many transition states are there?
c. What is the fastest step in the reaction?
d. Which is more stable, A or D?
e. What is the reactant of the rate-determining step?
f. Is the first step of the reaction exergonic or endergonic?
g. Is the overall reaction exergonic or endergonic?

70. The second-order rate constant $(M^{-1}s^{-1})$ for acid-catalyzed hydration at 25 °C is given for each of the following alkenes.

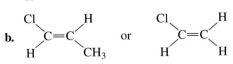

| 4.95×10^{-8} | 8.32×10^{-8} | 3.51×10^{-8} | 2.15×10^{-4} | 3.42×10^{-4} |

a. Calculate their relative rates of hydration.
b. Why does (Z)-2-butene react faster than (E)-2-butene?
c. Why does 2-methyl-2-butene react faster than (Z)-2-butene?
d. Why does 2,3-dimethyl-2-butene react faster than 2-methyl-2-butene?

71. Which compound has the greater dipole moment?

a. [structure] or [structure]   c. [structure] or [structure]

b. [structure] or [structure]

72. a. How many alkenes could you treat with H_2/Pt in order to prepare methylcyclopentane?
   b. Which alkene is the most stable?
   c. Which alkene has the smallest heat of hydrogenation?

73. a. What five-carbon alkene will form the same product whether it reacts with HBr in the *presence* of peroxides or with HBr in the *absence* of peroxides?
   b. Give three alkenes containing six carbon atoms that form the same product whether they react with HBr in the *presence* of peroxides or with HBr in the *absence* of peroxides.

74. a. What piece of information do you need to know before you can predict whether 2-methylpropene or 1-butene will react faster with HCl?

b. Knowing that 2-methylpropene reacts faster than 1-butene with HCl, predict which of the following will react faster with HCl.

1. ethene or propene? Explain.

2. 2-methyl-1-butene or 3-methyl-1-butene? Explain.

75. a. What is the equilibrium constant of a reaction that is carried out at 25 °C (298 K) with $\Delta H° = -20$ kcal/mol and $\Delta S° = -25$ cal deg^{-1} mol^{-1}?

b. What is the equilibrium constant of the same reaction carried out at 125 °C?

76. Mark Onikoff was about to turn in the products he had obtained from the reaction of HI with 3,3,3-trifluoropropene when he realized that the labels had fallen off his flasks. He didn't know which label belonged to which flask. Another student in the next lab told him that because the product obtained by following Markovnikov's rule was 1,1,1-trifluoro-2-iodopropane, he should put that label on the flask that contained the most product and label the flask with the least product 1,1,1-trifluoro-3-iodopropane, the anti-Markovnikov product. Should Mark follow the student's advice?

77. a. For a reaction that is carried out at 25 °C, how much must $\Delta G°$ change in order to increase the equilibrium constant by a factor of 10?

b. How much must $\Delta H°$ change if $\Delta S° = 0$ eu?

c. How much must $\Delta S°$ change if $\Delta H° = 0$ kcal?

78. a. Propose a mechanism for the following reaction. (Show all curved arrows.)

$$CH_3CH_2CH=CH_2 \; + \; CH_3OH \; \xrightarrow{\text{H}^+} \; CH_3CH_2\underset{\underset{\displaystyle OCH_3}{|}}{C}HCH_3$$

b. Which step is the rate-determining step?

c. What is the electrophile in the first step?

d. What is the nucleophile in the first step?

e. What is the electrophile in the second step?

f. What is the nucleophile in the second step?

79. a. What product is obtained from the reaction of HCl with 1-butene? With 2-butene?

b. Which of the two reactions has the greater free energy of activation?

c. Which of the two alkenes reacts more rapidly with HCl?

d. Which compound reacts more rapidly with HCl, (Z)-2-butene or (E)-2-butene?

80. a. Propose a mechanism for the following reaction.

b. Is the initially formed carbocation primary, secondary, or tertiary?

c. Is the rearranged carbocation primary, secondary, or tertiary?

d. Why does the rearrangement occur?

81. When the following compound is hydrated in the presence of acid, the unreacted alkene is found to have retained the deuterium atoms. What does this tell you about the mechanism of hydration?

82. Propose a mechanism for each of the following reactions.

a. 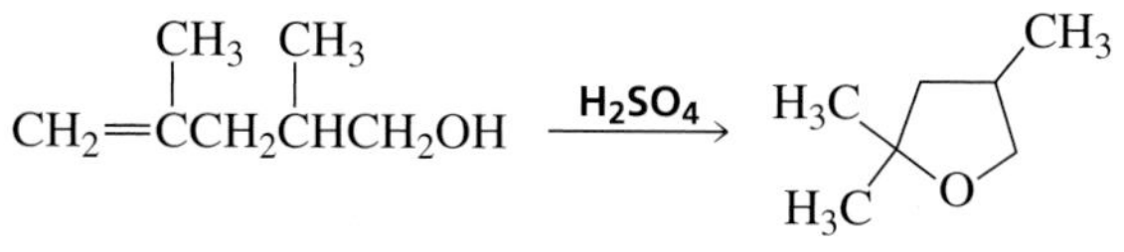 $CH_2=\overset{\underset{\displaystyle CH_3}{|}}{C}CH_2\overset{\underset{\displaystyle CH_3}{|}}{C}HCH_2OH$ $\xrightarrow{H_2SO_4}$

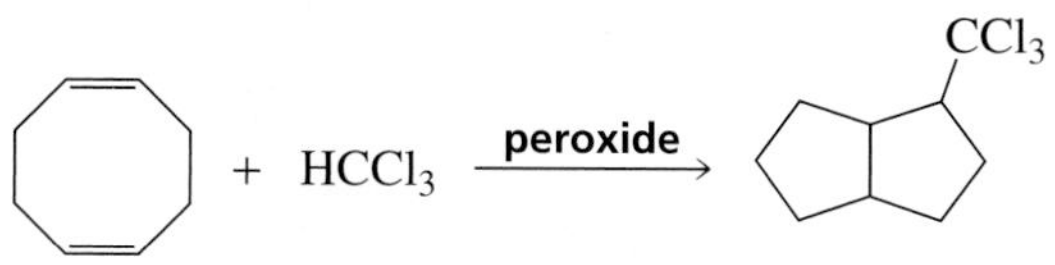

b. $+$ $HCCl_3$ $\xrightarrow{\textbf{peroxide}}$

c. $+$ H_2O $\xrightarrow{H_2SO_4}$

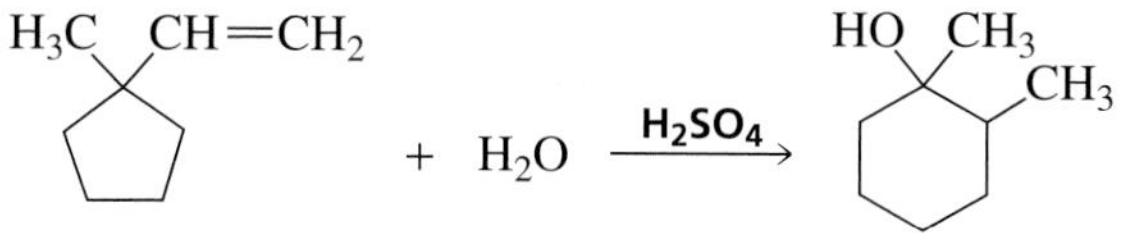

d. 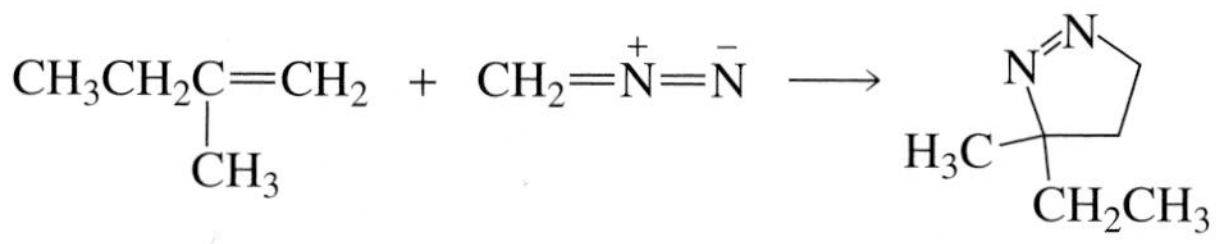 $CH_3CH_2\overset{\underset{\displaystyle CH_3}{|}}{C}=CH_2$ $+$ $CH_2=\overset{+}{N}=\overset{-}{N}$ $\longrightarrow$

83. Rate constants for a reaction were determined at five temperatures. From the following data, calculate the experimental energy of activation. Then calculate $\Delta G^{\ddagger}$, $\Delta H^{\ddagger}$, and $\Delta S^{\ddagger}$ for the reaction at 30 °C.

Temperature	Observed rate constant
31.0 °C	$2.11 \times 10^{-5}\ s^{-1}$
40.0 °C	$4.44 \times 10^{-5}\ s^{-1}$
51.5 °C	$1.16 \times 10^{-4}\ s^{-1}$
59.8 °C	$2.10 \times 10^{-4}\ s^{-1}$
69.2 °C	$4.34 \times 10^{-4}\ s^{-1}$

84. a. Dichlorocarbene can be generated by heating chloroform with HO^-. Propose a mechanism for the reaction.

$$CHCl_3 \quad + \quad HO^- \quad \xrightarrow{\Delta} \quad Cl_2C\text{:} \quad + \quad H_2O \quad + \quad Cl^-$$
chloroform   **dichlorocarbene**

b. Dichlorocarbene can also be generated by heating sodium trichloroacetate. Propose a mechanism for the reaction.

$$Cl_3C\overset{\overset{\displaystyle O}{\|}}{C}O^-\ Na^+ \quad \xrightarrow{\Delta} \quad Cl_2C\text{:} \ + \ CO_2 \ + \ Na^+\ Cl^-$$
sodium trichloroacetate

CHAPTER

4

Stereochemistry:

The Arrangement of Atoms in Space; The Stereochemistry of Addition Reactions

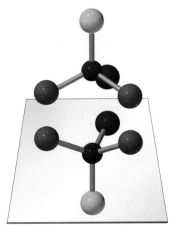

nonsuperimposable mirror images

Compounds that have the same molecular formula but are not identical are called **isomers.** The two main classes of isomers are constitutional isomers and stereoisomers. In Chapter 2 you learned that **constitutional isomers** differ in the way their atoms are connected. For example, ethanol and dimethyl ether are constitutional isomers because they have the same molecular formula, C_2H_6O, but the atoms in each compound are connected differently. The oxygen in ethanol is bonded to a carbon and to a hydrogen, whereas the oxygen in dimethyl ether is bonded to two carbons.

constitutional isomers

CH_3CH_2OH and CH_3OCH_3
ethanol **dimethyl ether**

$CH_3\overset{\overset{\displaystyle O}{\|}}{C}CH_3$ and $CH_3CH_2\overset{\overset{\displaystyle O}{\|}}{C}H$
acetone **propionaldehyde**

$CH_3CH_2CH_2CH_2CH_3$ and $CH_3\overset{\overset{\displaystyle CH_3}{|}}{C}HCH_2CH_3$
pentane **isopentane**

$CH_3CH_2CH_2CH_2Cl$ and $CH_3CH_2\overset{\overset{\displaystyle Cl}{|}}{C}HCH_3$
1-chlorobutane **2-chlorobutane**

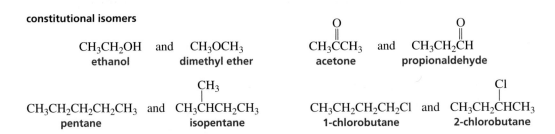

Unlike the atoms in constitutional isomers, the atoms in stereoisomers are connected in the same way. **Stereoisomers** differ in the way their atoms are arranged in space. There are two kinds of stereoisomers, conformational isomers and configurational isomers.

Conformational isomers can interconvert rapidly at room temperature. Because they interconvert, conformational isomers *cannot* be separated. Conformational isomers are also called **conformers.** There are two kinds of conformational isomers. The first result from rotation about single bonds, and the second result from amine inversion.

Configurational isomers do not readily interconvert. As a result, configurational isomers *can* be separated. There are two kinds of configurational isomers, too. The first are **cis–trans isomers** and the second are **isomers that contain chirality** (ky-RAL-i-tee) **centers.** In the following sections, we will meet the members of the stereoisomer family tree.

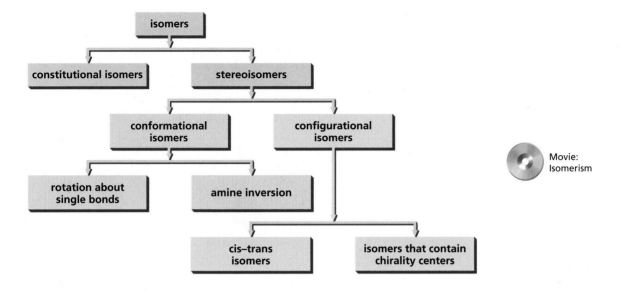

PROBLEM 1 ◆

a. Draw three constitutional isomers with molecular formula C_3H_8O.

b. How many constitutional isomers can you draw for $C_4H_{10}O$?

One kind of conformational isomer results from rotation about carbon–carbon single bonds. Because each of the carbon–carbon single bonds in a molecule rotates continually, compounds that contain carbon–carbon single bonds have many interconvertible conformational isomers (Section 2.10).

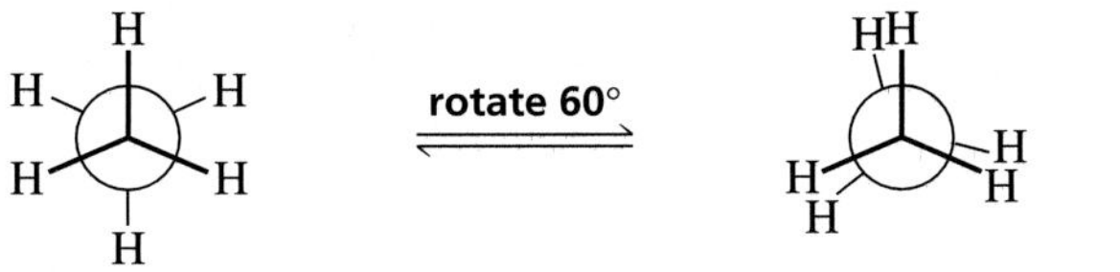

4.1 CONFORMATIONAL ISOMERS

Movie: Isomerism

staggered conformation of ethane   rotate 60°   **eclipsed conformation of ethane**

As a result of rotation about its carbon–carbon single bonds, methylcyclohexane rapidly interconverts between two chair conformers (Section 2.13).

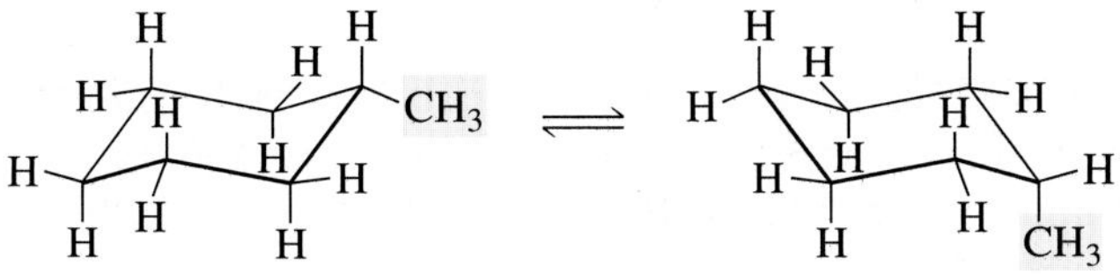

The second kind of conformational isomer occurs because nitrogen has a pair of nonbonding electrons that allow it to turn "inside out" rapidly at room temperature. This is called **amine inversion.** One way to picture amine inversion is to compare it to an umbrella that turns inside out in a windstorm.

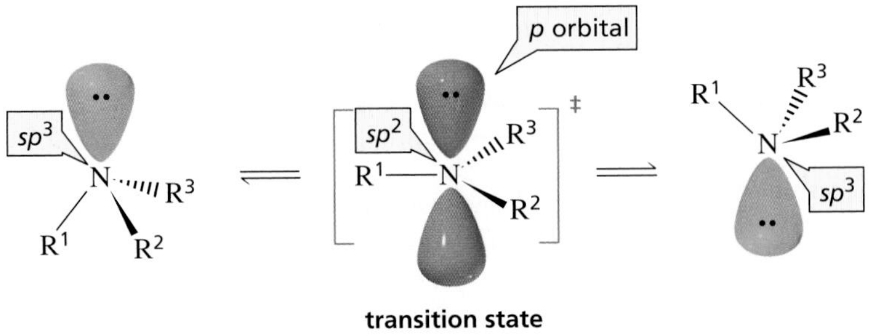

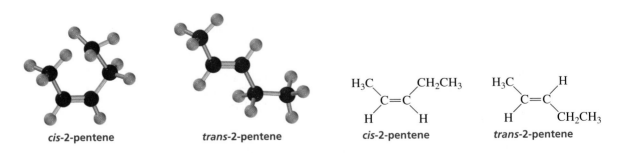

transition state

Because of rapid inversion, the individual stereoisomers cannot be isolated. The pair of nonbonding electrons is required for amine inversion. Quaternary ammonium ions—ions with four bonds to nitrogen and hence no nonbonding electrons—do not invert (Section 4.15).

Notice that amine inversion takes place through a transition state in which the sp^3 nitrogen becomes an sp^2 nitrogen. The three groups bonded to the sp^2 nitrogen are coplanar with bond angles of 120°, and the nonbonding electrons are in a p orbital. The energy required for amine inversion is approximately 6 kcal/mol (25 kJ/mol), about twice the amount of energy required for rotation about a carbon–carbon single bond, but still low enough to allow the conformers to interconvert rapidly at room temperature.

4.2 CONFIGURATIONAL ISOMERS: CIS-TRANS ISOMERS

Cis–trans isomers result from restricted rotation (Section 3.4). Restricted rotation can be caused either by a double bond or by a cyclic structure. Because there is no rotation about a carbon–carbon double bond, an alkene such as 2-pentene can exist as cis and trans isomers. The cis and trans isomers can be separated, so they are configurational isomers.

cis-2-pentene *trans*-2-pentene *cis*-2-pentene *trans*-2-pentene

3-D Molecules:
cis-2-Pentene;
trans-2-Pentene

Cyclic compounds can also have cis and trans isomers because the cyclic system prevents free rotation about the single bonds. The **cis isomer** has substituents on the same side of the ring, whereas the **trans isomer** has substituents on opposite sides of the ring (Section 2.14).

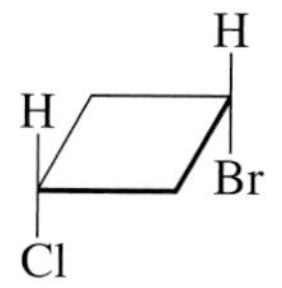

cis-1-bromo-3-chlorocyclobutane 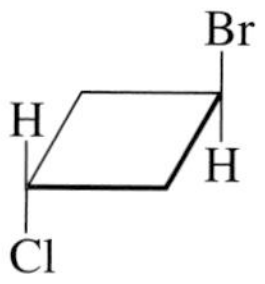   *trans*-1-bromo-3-chlorocyclobutane

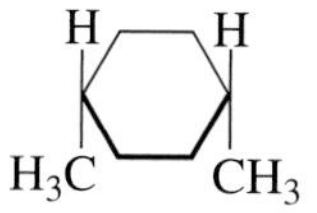

cis-1,4-dimethylcyclohexane 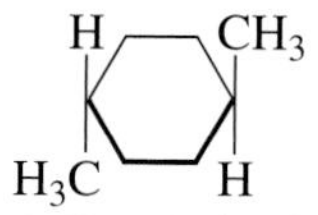   *trans*-1,4-dimethylcyclohexane

---

PROBLEM 2 ◆

Draw the cis and trans isomers for the following compounds:

a. 1-ethyl-3-methylcyclobutane

b. 3,4-dimethyl-3-heptene

c. 1-bromo-4-chlorocyclohexane

d. 1,3-dibromocyclobutane

A carbon bonded to four different groups is called a **chirality center**.[1] The chirality center in each of the following compounds is indicated by an asterisk. For example, the fourth carbon in 4-octanol is a chirality center because it is bonded to four different groups (H, OH, $CH_2CH_2CH_3$, and $CH_2CH_2CH_2CH_3$). Notice that the difference in the groups bonded to the chirality center is not necessarily right next to the chirality center. For example, the propyl and butyl groups are different even though the point at which they differ is somewhat removed from the chirality center. The starred carbon in 2,4-dimethylhexane is a chirality center because it is bonded to four different groups—methyl, ethyl, isobutyl, and hydrogen.

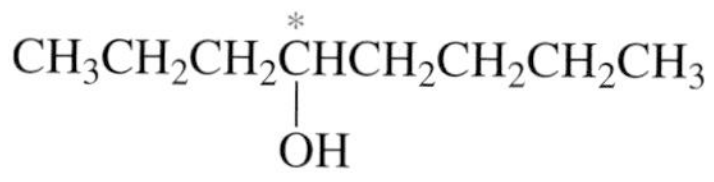

4.3
CONFIGURATIONAL ISOMERS: ISOMERS WITH ONE CHIRALITY CENTER

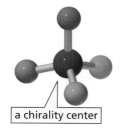

a chirality center

$$CH_3CH_2CH_2\overset{*}{C}HCH_2CH_2CH_2CH_3$$
$$\underset{OH}{|}$$
4-octanol

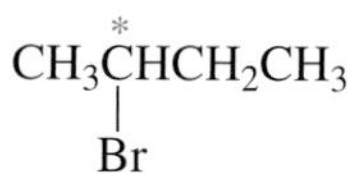

$$CH_3\overset{*}{C}HCH_2CH_3$$
$$\underset{Br}{|}$$
2-bromobutane

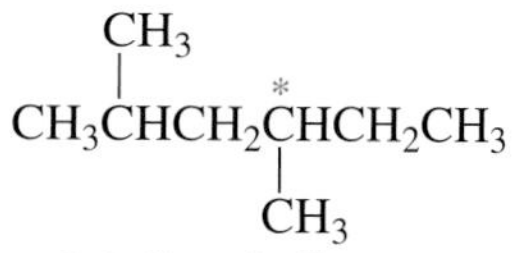

$$\overset{\overset{\displaystyle CH_3}{|}}{CH_3CHCH_2}\overset{*}{C}HCH_2CH_3$$
$$\underset{CH_3}{|}$$
2,4-dimethylhexane

---

[1]The term "chirality center" is the most recently IUPAC-approved name (Pure and Applied Chemistry, Vol. 68, No. 12, p. 2193, **1996**). At various times it has been called a stereogenic center, a stereocenter, a chiral center, and an asymmetric carbon.

Notice that the only carbons that can be chirality centers are sp^3 hybridized carbons; sp^2 and sp hybridized carbons cannot be chirality centers because they cannot have four groups attached to them.

PROBLEM 3 ◆

Which of the following compounds have chirality centers?

a. CH$_3$CH$_2$CHCH$_3$
             |
             Cl

d. CH$_3$CH$_2$OH

b. CH$_3$CH$_2$CHCH$_3$
             |
             CH$_3$

e. CH$_3$CH$_2$CHCH$_2$CH$_3$
                   |
                   Br

             CH$_3$
             |
c. CH$_3$CH$_2$CCH$_2$CH$_2$CH$_3$
             |
             Br

f. CH$_2$=CHCHCH$_3$
                  |
                  NH$_2$

Movie:
Nonsuperimposable
mirror image

A compound with a chirality center, such as 2-bromobutane, can exist as two different isomers. Because the two isomers are different, they cannot be superimposed. The two isomers are analogous to a left and a right hand. You cannot superimpose your left hand on your right hand. When you try to superimpose them, either the thumb of one hand lies on top of the little finger of the other hand or the palms and backs face opposite directions.

left hand

right hand

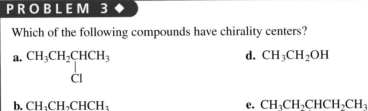

CH$_3\overset{*}{C}$HCH$_2$CH$_3$
     |
     Br
2-bromobutane

the two isomers of 2-bromobutane
a pair of enantiomers

The solid wedges represent bonds that point out of the plane of the paper toward the viewer.

Take a break and convince yourself that the two 2-bromobutane isomers are not identical by building ball-and-stick models using four different-colored balls to represent the four different groups bonded to the chirality center. Try to superimpose them.

If you imagine a mirror between the two isomers, you can see that they are mirror images of each other. A nonidentical (nonsuperimposable) mirror image is called an **enantiomer** of the original molecule. "Enantiomer" comes from the Greek *enantion,* which means "opposite." Each of the isomers of 2-bromobutane is the enantiomer of the other. A pair of nonidentical mirror images is called a **pair of enantiomers.** The two isomers of 2-bromobutane are a pair of enantiomers.

The hatched wedges represent bonds that point back from the plane of the paper away from the viewer.

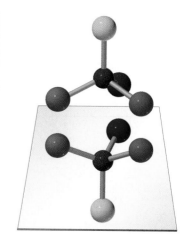

Which of the compounds in Problem 3 can exist as a pair of enantiomers?

A molecule that has a nonidentical mirror image, such as either enantiomer of 2-bromobutane, is said to be **chiral** (ky-ral). A chiral molecule does not contain a plane of symmetry. A **plane of symmetry** is a plane that cuts a molecule into two halves, each of which is the mirror image of the other.

Objects can be chiral, too. Chiral objects do not contain a plane of symmetry. Perhaps the most familiar chiral objects are your hands. If you imagine a mirror between your two hands, you can see that they are mirror images of each other. However, there is no place you can put a plane that cuts one of your hands into two halves, each of which is the mirror image of the other half. Instead, each hand is the nonidentical mirror image of the other, so each hand is chiral. A pair of enantiomers, therefore, like a pair of hands, are nonidentical mirror images of each other. Interestingly, the term "chiral" comes from the Greek word *cheir,* which means "hand." Other familiar chiral objects are feet, shoes, and ears.

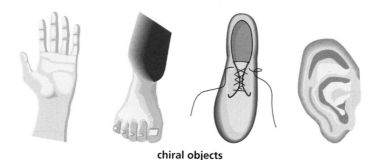

chiral objects

A chiral molecule has at least one chirality center and does not have a plane of symmetry.

A molecule or object that does contain a plane of symmetry is said to be **achiral** (ay-ky-ral). If you cut the object in two halves along the plane of symmetry, the left half is the mirror image of the right half. A fork and a table each has a plane of symmetry, so they are achiral. An achiral molecule and its mirror image are identical (they can be superimposed). Therefore, an achiral compound cannot have an enantiomer.

Movie:
Plane of symmetry

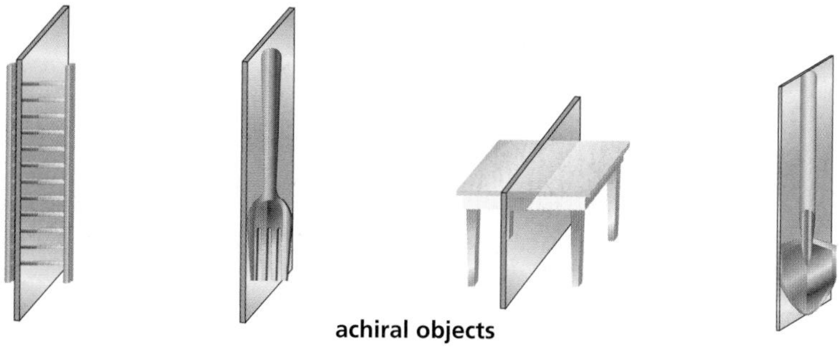

achiral objects

An achiral molecule has a plane of symmetry.

Notice that chirality is a property of an entire object or an entire molecule. A chirality center is what causes a molecule to be chiral.

Which of the following are chiral?

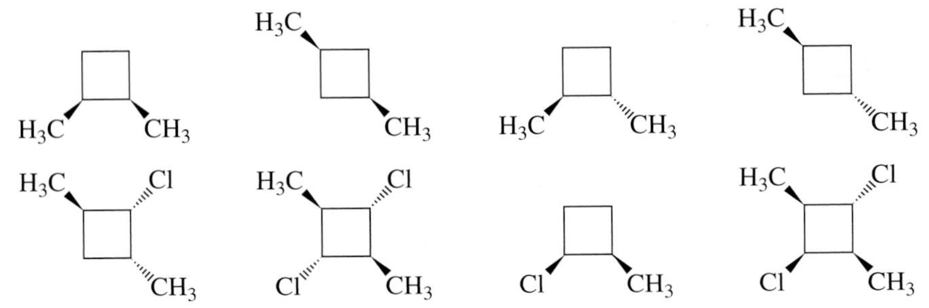

SOLUTION To be chiral, a molecule must not have plane of symmetry. Therefore, only the following compounds are chiral.

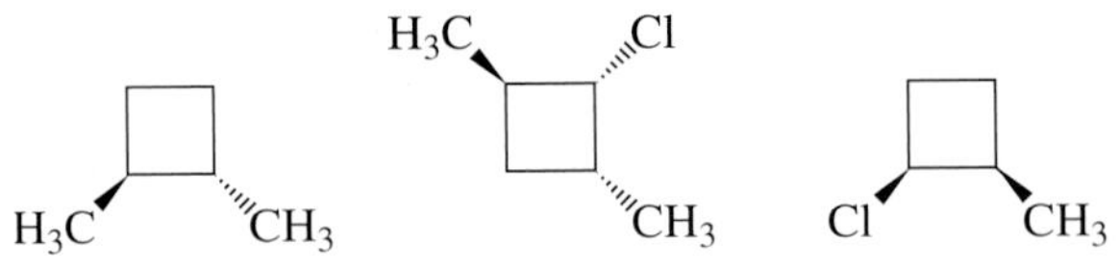

In the top row of compounds, only the third compound is chiral. The first, second, and fourth compounds each have a plane of symmetry.

In the bottom row of compounds, the first and third compounds are chiral. The second and fourth compounds each have a plane of symmetry.

PROBLEM 6 ◆

a. Name five capital letters that are chiral.

b. Name five capital letters that are achiral.

4.4 DRAWING ENANTIOMERS

Chemists use two different ways to draw enantiomers, *perspective formulas* and *Fischer projections*. **Perspective formulas** show two of the bonds to the chirality center in the plane of the paper, one bond as a solid wedge coming out of the paper, and the fourth bond as a hatched wedge projecting back from the paper. You can draw the first enantiomer by putting the four groups bonded to the chirality center in any order. Draw the second enantiomer by making the mirror image of the first enantiomer.

3-D Molecules:
(*R*)-2-Bromobutane;
(*S*)-2-Bromobutane

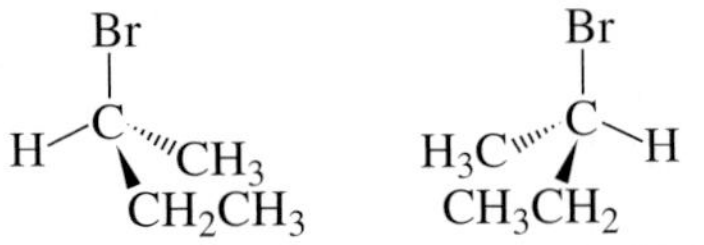

perspective formulas of the enantiomers of 2-bromobutane

Make certain when you draw a perspective formula that the two bonds in the plane of the paper are adjacent to one another; neither the solid wedge nor the hatched wedge should be drawn between them.

A shortcut for showing the three-dimensional arrangement of groups bonded to a chirality center was devised in the late 1800s by Emil Fischer. Fischer used a single dot to represent a bond that was slanted away from the viewer and a solid line for a bond that was directed toward the viewer. Several years later, Viktor Meyer modified Fischer's method. He used the point of intersection of two perpendicular lines to represent the chirality center; horizontal lines represent the bonds that project out of the plane of the paper toward the viewer, and vertical lines represent the bonds that project back from the plane of the paper away from the viewer. The carbon chain is drawn vertically with C-1 at the top of the chain. This method of representing the arrangement of groups bonded to a chirality center is known as a **Fischer projection.**

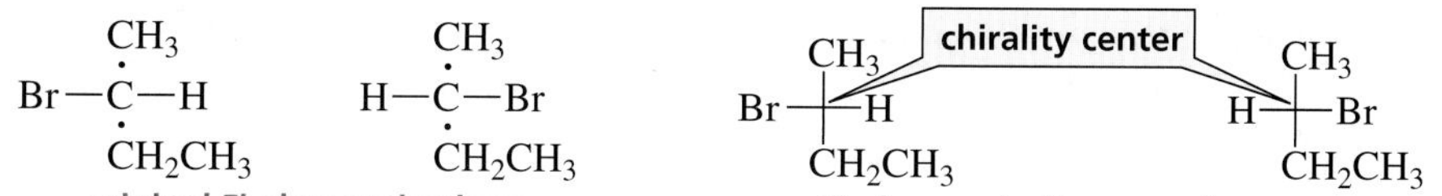

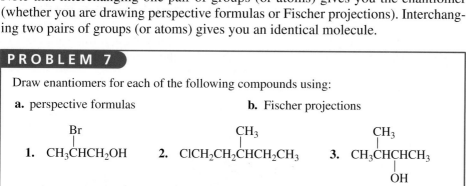

Emil Fischer (1852–1919) was born in a village near Cologne, Germany. He became a chemist against the wishes of his father, a successful merchant, who wanted him to enter the family business. He was a professor of chemistry at the Universities of Erlangen, Würzburg, and Berlin. In 1902 he received the Nobel Prize in chemistry for his work on sugars. During World War I, he organized German chemical production. Two of his three sons died in that war.

To draw enantiomers using a Fischer projection, draw the first enantiomer by arranging the four groups (or atoms) bonded to the chirality center in any order. Draw the second enantiomer by interchanging two of the groups (or atoms). It does not make any difference which two groups you interchange. (It is a good idea to make models to convince yourself that this is true.) Many organic chemists prefer to interchange the two horizontal groups because the enantiomers then look like mirror images on paper.

Note that interchanging one pair of groups (or atoms) gives you the enantiomer (whether you are drawing perspective formulas or Fischer projections). Interchanging two pairs of groups (or atoms) gives you an identical molecule.

PROBLEM 7

Draw enantiomers for each of the following compounds using:

a. perspective formulas   **b.** Fischer projections

$$\text{Br}$$
1. CH_3CHCH_2OH **2.** $ClCH_2CH_2CHCH_2CH_3$ **3.** $CH_3CHCHCH_3$
          (Br above)   (CH_3 above)   (CH_3 above, OH below)

Now that we know a compound such as 2-bromobutane has two stereoisomers, we need a way to name the individual stereoisomers so we can distinguish them. In other words, we need a system of nomenclature that will indicate the **configuration** (arrangement) of the groups (or atoms) about the chirality center. Chemists use the letters *R* and *S* to indicate the configuration of a chirality center. For any pair of enantiomers with a chirality center, one will have the **R configuration** and the other will have the **S configuration.** The *R,S* system of nomenclature was devised by Cahn, Ingold, and Prelog.

Let's first look at how you would determine the configuration of a compound for which you had a three-dimensional molecular model.

4.5
NAMING ENANTIOMERS: THE *R,S* SYSTEM OF NOMENCLATURE

Viktor Meyer (1848–1897) was born in Germany. To prevent him from becoming an actor, his parents persuaded him to enter the University of Heidelberg, where he earned a Ph.D. at the age of 19. He was a professor of chemistry at the Universities of Stuttgart and Heidelberg. He coined the term "stereochemistry" for the study of molecular shapes, and was the first to describe the effect of steric hindrance on a reaction.

1. Rank the groups (or atoms) that are bonded to the chirality center in order of priority. The priority depends on the atomic numbers of the atoms directly attached to the chirality center. The greater the atomic number, the higher the priority. This should remind you of the way that relative priorities were determined in Section 3.5 for the *E,Z* system of nomenclature because the system of priorities was originally devised for the *R,S* system of nomenclature and was later borrowed for the *E,Z* system. You may want to review how relative priorities are determined before you proceed with the *R,S* system.

2. Orient the molecule so that the group (or atom) with the lowest priority (4) is directed away from you. Then draw an imaginary arrow from the group (or atom) with the highest priority (1) to the group (or atom) with the next highest priority (2). If the arrow points in a clockwise direction, the chirality center has the *R* configuration (*R* is for *rectus*, which is Latin for "right"). If the arrow points in a counterclockwise direction, the chirality center has the *S* configuration (*S* is for *sinister*, which is Latin for "left").

left turn

If you forget which is which, you can imagine driving a car and turning the steering wheel to make a right or a left turn.

Because you may not have access to molecular models when you need to determine the configuration of a chirality center, we will now see how you can quickly determine the configuration of a compound whose structure is written on a two-dimensional piece of paper. First we will look at how to determine the configuration of a compound drawn as a perspective formula. As an example, we will look at the enantiomers of 2-bromobutane and determine which has the *R* configuration and which has the *S* configuration.

$$\text{CH}_3\text{CH}_2 \overset{\overset{\displaystyle \text{Br}}{|}}{\underset{\text{CH}_3}{\text{C}}} \text{''''H} \qquad \text{H} \text{''''} \overset{\overset{\displaystyle \text{Br}}{|}}{\underset{\text{CH}_3}{\text{C}}} \text{CH}_2\text{CH}_3$$

the enantiomers of 2-bromobutane

right turn

1. Rank the groups (or atoms) that are bonded to the chirality center in order of priority. In the following pair of enantiomers, bromine has the highest priority (1),

the ethyl group has the second highest priority (2), the methyl group is next (3), and hydrogen has the lowest priority (4).

2. If the group (or atom) with the lowest priority is bonded by a hatched wedge, draw a curved arrow from the group (or atom) with the highest priority (1) to the group (or atom) with the second highest priority (2). If the arrow points in a clockwise direction, the compound has the *R* configuration; if it points in a counterclockwise direction, the compound has the *S* configuration.

(*S*)-2-bromobutane   (*R*)-2-bromobutane

3. If the group (or atom) with the lowest priority (4) is not bonded by a hatched wedge, then first switch a pair of groups so group 4 is bonded by a hatched wedge. Then proceed as in step #2 (above): Draw a curved arrow from the group with the highest priority (1) to the group with the second highest priority (2). Because you have switched a pair of groups, you are now determining the configuration of *the enantiomer of the original molecule*. So if the arrow points in a clockwise direction, the enantiomer (with the switched pair) has the *R* configuration, which means the original molecule has the *S* configuration. Therefore, if the arrow points in a counterclockwise direction, the enantiomer (with the switched pair) has the *S* configuration, which means the original molecule has the *R* configuration.

switch
CH₃ and H

what is its configuration?

this compound has the *R* configuration, which means that the compound before the pair was switched had the *S* configuration

4. In drawing the arrow from group 1 to group 2, you can draw past the group with the lowest priority (4), but never draw past the group with the next lowest priority (3).

(*R*)-1-bromo-3-pentanol

Robert Sidney Cahn (1899–1981), was born in England and received an M.A. from Cambridge University and a doctorate in natural philosophy in France. He edited the Journal of the Chemical Society (London).

Sir Christopher Ingold (1893–1970) was born in Ilford, England and was knighted by Queen Elizabeth II. He was a professor of chemistry at Leeds University (1924–1930) and at University College, London (1930–1970).

Vladimir Prelog (1906–1998) was born in Sarajevo, Bosnia. He taught at the University of Zagreb from 1935 until 1941, when he fled to Switzerland just ahead of the invading German army. In 1929 he received a Dr.Ing. degree from the Institute of Technology in Prague, Czechoslovakia. He was a professor at the Swiss Federal Institute of Technology (ETH). For his work that contributed to an understanding of how living organisms carry out chemical reactions, he shared the 1975 Nobel Prize in chemistry with John Cornforth (page 232).

You can use the following rules to determine the configuration of a compound drawn as a Fischer projection.

1. Draw an arrow from the group (or atom) with the highest priority (1) to the group (or atom) with the next highest priority (2). If the arrow points in a clockwise direction, the enantiomer has the *R* configuration; if it points in a counterclockwise direction, the enantiomer has the *S* configuration, *provided that the group with the lowest priority (4) is on a vertical bond.*

(R)-3-chlorohexane   (S)-3-chlorohexane

2. If the group (or atom) with the lowest priority is on a *horizontal* bond, the answer you get by determining the direction of the arrow will be the opposite of the correct answer. For example, if the arrow points in a clockwise direction, suggesting that the chirality center has the *R* configuration, it actually has the *S* configuration; if the arrow points in a counterclockwise direction, suggesting that the chirality center has the *S* configuration, it actually has the *R* configuration. (Sometimes a mnemonic can help you remember these conventions. For example, if you assume that a clockwise arrow specifies an *R* configuration and a counterclockwise arrow specifies an *S* configuration, the answer you get is **V**ery true if the lowest priority substituent is on a **V**ertical bond and **H**orribly wrong if the lowest priority substituent is on a **H**orizontal bond.) In the following example the group with the lowest priority is on a horizontal bond, so clockwise signifies the *S* configuration, not the *R* configuration.

(S)-2-butanol   (R)-2-butanol

3. In drawing the arrow from group 1 to group 2, you can draw past the group with the lowest priority (4), but never draw past the group with the next lowest priority (3).

(S)-lactic acid   (R)-lactic acid

If you are working with three-dimensional molecular models, you can directly compare the models to tell whether two molecules are enantiomers (nonidentical mirror images) or identical molecules. If, however, you are working with structures on a two-dimensional piece of paper, the easiest way to determine whether two molecules with one chirality center and the same sets of substituents are enantiomers or identical molecules is by determining their configurations. If one has the *R* configuration and the other has the *S* configuration, they are enantiomers. If they both have the *R* configuration or both have the *S* configuration, they are identical.

When comparing two Fischer projections to see if they are the same or different, *never rotate one 90° or turn one over* because this is a quick way to get a wrong answer. A Fischer projection can be rotated 180° in the plane of the paper, but this is the only way to move it without risking an incorrect answer.

PROBLEM 8 ◆

Indicate whether each of the following structures has the *R* configuration or the *S* configuration:

a.

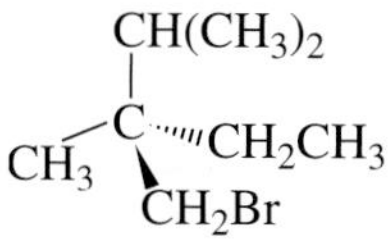

$$CH_3 \overset{\displaystyle CH(CH_3)_2}{\underset{\displaystyle CH_2Br}{\overset{|}{-}C\text{-}{\cdots}CH_2CH_3}}$$

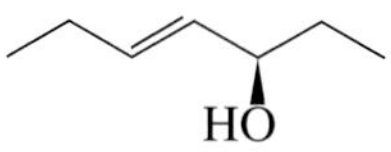

b.

$$CH_3CH_2 \overset{\displaystyle CH_2Br}{\underset{\displaystyle OH}{\overset{|}{-}C\text{-}{\cdots}CH_2CH_2Cl}}$$

c.

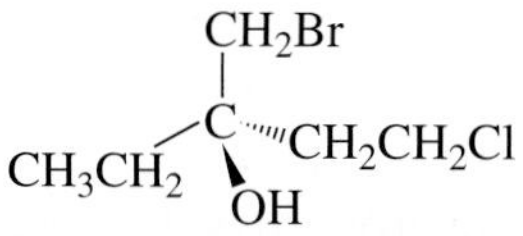

d.

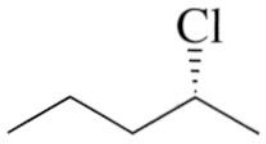

PROBLEM 9 / SOLVED ◆

Do the following structures represent identical molecules or a pair of enantiomers?

a.

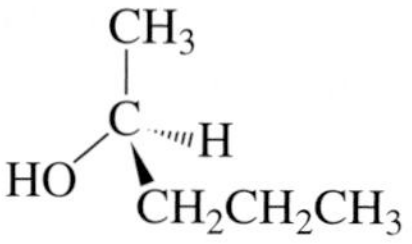

$$HO \overset{\displaystyle CH_3}{\underset{\displaystyle CH_2CH_2CH_3}{\overset{|}{-}C\text{-}{\cdots}H}} \quad \text{and} \quad CH_3CH_2CH_2 \overset{\displaystyle OH}{\underset{\displaystyle H}{\overset{|}{-}C\text{-}{\cdots}CH_3}}$$

b.

$$CH_3 \overset{\displaystyle CH_2Br}{\underset{\displaystyle CH_2CH_3}{\overset{|}{-}C\text{-}{\cdots}Cl}} \quad \text{and} \quad CH_3CH_2 \overset{\displaystyle Cl}{\underset{\displaystyle CH_2Br}{\overset{|}{-}C\text{-}{\cdots}CH_3}}$$

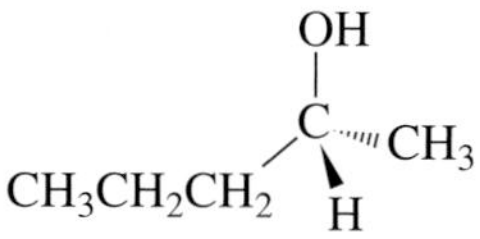

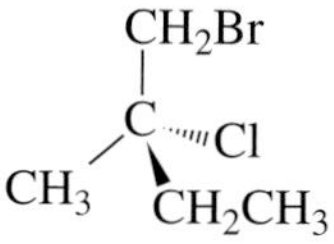

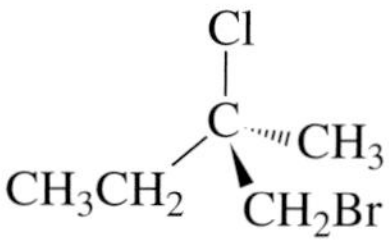

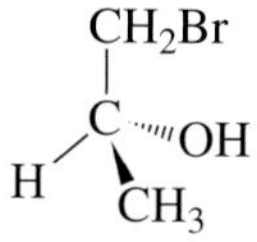

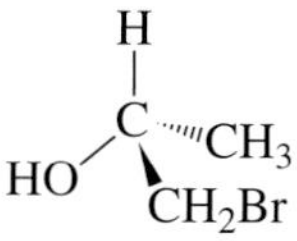

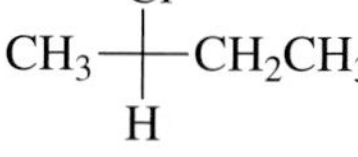

c.

$$H \overset{\displaystyle CH_2Br}{\underset{\displaystyle CH_3}{\overset{|}{-}C\text{-}{\cdots}OH}} \quad \text{and} \quad HO \overset{\displaystyle H}{\underset{\displaystyle CH_2Br}{\overset{|}{-}C\text{-}{\cdots}CH_3}}$$

d. $CH_3 \overset{\displaystyle Cl}{\underset{\displaystyle H}{-\!\!\!\!-\!\!\!\!-}} CH_2CH_3$ and $H \overset{\displaystyle CH_3}{\underset{\displaystyle CH_2CH_3}{-\!\!\!\!-\!\!\!\!-}} Cl$

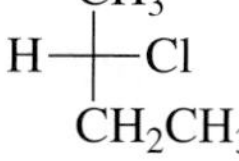

SOLUTION TO 9a   The first structure shown in part (a) has the *S* configuration and the second structure has the *R* configuration. Because they have opposite configurations, the structures represent a pair of enantiomers.

PROBLEM 10 ◆

Assign priority numbers to the following groups:

a. $-CH_2OH$ $-CH_3$ $-CH_2CH_2OH$ $-H$

b. $-CH{=}O$ $-OH$ $-CH_3$ $-CH_2OH$

c. $-CH(CH_3)_2$ $-CH_2CH_2Br$ $-Cl$ $-CH_2CH_2CH_2Br$

d. $-CH{=}CH_2$ $-CH_2CH_3$ 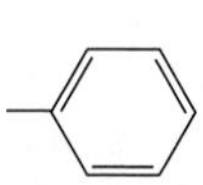 $-CH_3$

PROBLEM 11 ◆

Indicate whether each of the following structures has the *R* configuration or the *S* configuration.

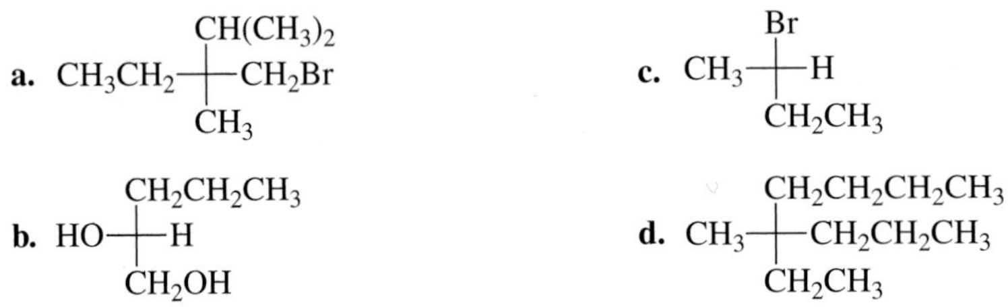

a.

$$CH_3CH_2 \overset{\displaystyle CH(CH_3)_2}{\underset{\displaystyle CH_3}{\vert}} CH_2Br$$

b.

$$HO \overset{\displaystyle CH_2CH_2CH_3}{\underset{\displaystyle CH_2OH}{\vert}} H$$

c.

$$CH_3 \overset{\displaystyle Br}{\underset{\displaystyle CH_2CH_3}{\vert}} H$$

d.

$$CH_3 \overset{\displaystyle CH_2CH_2CH_2CH_3}{\underset{\displaystyle CH_2CH_3}{\vert}} CH_2CH_2CH_3$$

4.6 OPTICAL ROTATION

Enantiomers share many of the same properties—they have the same boiling points, the same melting points, and the same solubilities. In fact, all the physical properties of enantiomers are the same except those properties that depend on how groups bonded to the chirality center are arranged in space. One of the properties that enantiomers do not share is the way they interact with polarized light.

What is polarized light? Normal light consists of electromagnetic waves that oscillate in all planes passing through the direction the light travels. **Plane-polarized light** oscillates only in a single plane passing through the direction the light travels. Plane-polarized light is produced by passing normal light through a polarizer such as a polarized lens or a Nicol prism.

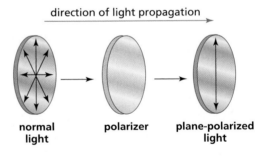

direction of light propagation

normal light   polarizer   plane-polarized light

When polarized lenses are at a 90° angle to each other, there is no transmission of light through them.

You can experience the effect of a polarized lens with polarized sunglasses. Polarized sunglasses allow only light oscillating in a single plane to pass through them, so they block reflections (glare) more effectively than nonpolarized sunglasses.

In 1815, the physicist Jean-Baptiste Biot discovered that certain natural organic substances such as camphor and oil of turpentine are able to rotate the plane of polarization. He noted that some compounds rotated the plane of polarization in a clockwise direction and some in a counterclockwise direction, while others did not rotate the plane of polarization at all. He predicted that the ability to rotate the plane of polarization was attributable to some asymmetry that existed in the molecule. Van't Hoff and Le Bel later determined that the molecular asymmetry was the result of having one or more chirality centers.

When plane-polarized light passes through a solution of achiral molecules, the light emerges from the solution with its plane of polarization unchanged because there is no asymmetry in the molecules. *An achiral compound does not rotate the plane of polarization. It is optically inactive.*

*Born in Scotland, **William Nicol (1768–1851)** was a professor at the University of Edinburgh. He developed the first prism that produced plane-polarized light. He also developed methods to produce thin slices of materials for use in microscopic studies.*

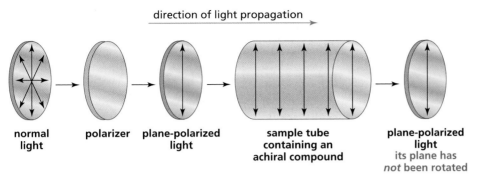

However, when plane-polarized light passes through a solution of a chiral compound, the light emerges with its plane of polarization changed because the molecules are asymmetric. Thus, *a chiral compound rotates the plane of polarization.* A chiral compound will rotate the plane of polarization in either a clockwise or a counterclockwise direction. If one enantiomer rotates the plane of polarization in a clockwise direction, its mirror image will rotate the plane of polarization exactly the same amount in a counterclockwise direction.

Jacobus Hendricus van't Hoff (1852–1911), a Dutch chemist, was a professor of chemistry at the University of Amsterdam and later at the University of Berlin. He received the first Nobel Prize in chemistry (1901) for his work on solutions.

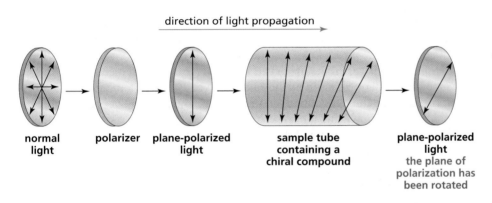

A compound that rotates the plane of polarization is said to be **optically active.** In other words, chiral compounds are optically active and achiral compounds are **optically inactive.**

If an optically active compound rotates the plane of polarization in a clockwise direction, it is called **dextrorotatory,** indicated by (+). If an optically active compound rotates the plane of polarization in a counterclockwise direction, it is called **levorotatory,** indicated by (−). *Dextro* and *levo* are Latin prefixes for "to the right" and "to the left," respectively.

Do not confuse (+) and (−) with *R* and *S.* The (+) and (−) symbols indicate the direction in which an optically active compound rotates plane-polarized light, whereas *R* and *S* indicate the arrangement of the groups about a chirality center. Some compounds with the *R* configuration are (+) and some are (−).

The amount that an optically active compound rotates the plane of polarization can be measured with an instrument called a **polarimeter** (Figure 4.1). Because the amount of rotation depends on the wavelength of the light used, the light source for a polarimeter must produce light with a single wavelength (monochromatic light). Most polarimeters use light from a sodium arc (called the sodium D-line; wavelength = 589 nm). In a polarimeter, monochromatic light passes through a polarizer and emerges as plane-polarized light. The plane-polarized light then passes through an empty sample tube (or one filled with an optically inactive solvent) and emerges

*Born in France, **Jean-Baptiste Biot (1774–1862)** was imprisoned for taking part in a street riot during the French Revolution. He became a professor of mathematics at the University of Beauvais and later a professor of physics at the Collège de France. He was awarded the Legion of Honor by Louis XVIII.*

Joseph Achille Le Bel (1847–1930), a French chemist, inherited his family's fortune and established his own laboratory. He and van't Hoff independently arrived at the reason for the optical activity of certain molecules. Although van't Hoff's explanation was more precise, they are both given credit for the work.

Movie:
Optical activity

Figure 4.1 ▶
Schematic of a polarimeter.

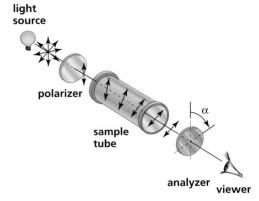

with its plane of polarization unchanged. The light finally passes through an analyzer. The analyzer is a second polarizer mounted on an eyepiece with a dial marked in degrees. The user rotates the analyzer until he or she sees total darkness. At this point the analyzer is at a right angle to the first polarizer, so no light passes through. This analyzer setting corresponds to zero rotation.

The sample to be measured is then placed in the sample tube. If the sample is optically active, it will rotate the plane of polarization. The analyzer will no longer block all the light, and light reaches the user's eye. The user then rotates the analyzer again until no light passes through. The amount the analyzer is rotated can be read from the dial and represents the difference between an inactive sample and the active sample. This is called the **observed rotation** (α), and it is measured in degrees. The amount of rotation caused by the sample depends on the number of optically active molecules the light encounters in the sample. This, in turn, depends on the concentration of the sample and the length of the sample tube. The observed rotation also depends on the temperature and the wavelength of the light source.

Each optically active compound has a characteristic specific rotation. The **specific rotation** is the number of degrees of rotation caused by a solution of 1.0 g of the compound per mL of solution in a sample tube 1.0 dm long at a specified temperature and wavelength. The specific rotation can be calculated from the observed rotation using the following formula:

$$[\alpha]_{\lambda}^{T} = \frac{\alpha}{l \times c}$$

where $[\alpha]$ is the specific rotation; T is temperature in °C; λ is the wavelength of the incident light (when the sodium D-line is used, λ is indicated as D); α is the observed rotation; l is the length of the sample tube in decimeters; and c is the concentration of the sample in grams per milliliter of solution.

For example, one enantiomer of 2-methyl-1-butanol has been found to have a specific rotation of +5.75°. Because its mirror image rotates the plane of polarization the same amount but in the opposite direction, the specific rotation of the other enantiomer must be −5.75°.

CH$_2$OH   CH$_2$OH

(R)-2-methyl-1-butanol   (S)-2-methyl-1-butanol

$[\alpha]_{D}^{20\,°C} = +5.75°$ $[\alpha]_{D}^{20\,°C} = -5.75°$

PROBLEM 12 ◆

The observed rotation of 2.0 g of a compound in 10 mL of solution in a polarimeter tube 25 cm long is +134°. What is the specific rotation of the compound?

Knowing whether a chiral molecule has the R configuration or the S configuration does not tell us the direction in which the compound rotates the plane of polarization because some compounds with the R configuration rotate the plane to the right (+), and some rotate the plane to the left (−). We can tell by looking at the structure of a compound whether it has the R configuration or the S configuration, but the only way we can tell whether a compound is dextrorotatory (+) or levorotatory (−) is to put the compound in a polarimeter. For example, (S)-lactic acid and (S)-sodium lactate have the same configuration, but (S)-lactic acid is dextrorotatory whereas (S)-sodium lactate is levorotatory. When we know the direction in which an optically active compound rotates the plane of polarization, we can incorporate (+) or (−) into its name.

CH3 | C ""H   HO   COOH
(S)-(+)-lactic acid

CH3 | C ""H   HO   COO⁻Na⁺
(S)-(−)-sodium lactate

PROBLEM 13 ◆

a. Is (R)-lactic acid dextrorotatory or levorotatory?

b. Is (R)-sodium lactate dextrorotatory or levorotatory?

A mixture of equal amounts of a pair of enantiomers is called a **racemic mixture,** a **racemic modification,** or a **racemate.** Racemic mixtures do not rotate plane-polarized light. They are optically inactive because for every molecule in a racemic mixture that rotates the plane of polarization in one direction, there is a mirror-image molecule that rotates the plane in the opposite direction. As a result, the light emerges from a racemic mixture with its plane of polarization unchanged. The symbol (±) is used to specify a racemic mixture. Thus, (±)-2-bromobutane indicates a mixture of (+)-2-bromobutane and an equal amount of (−)-2-bromobutane.

PROBLEM 14 ◆

(S)-(+)-Monosodium glutamate (MSG) is a flavor enhancer used in many foods. Some people have an allergic reaction to MSG (headache, chest pains, and an overall feeling of weakness). "Fast food" often contains substantial amounts of MSG, and it is widely used in Chinese food as well. MSG has a specific rotation of +24°.

COO⁻ Na⁺ | C ""H   HOOCCH₂CH₂   ⁺NH₃
(S)-(+)-monosodium glutamate

a. What is the specific rotation of (R)-(−)-monosodium glutamate?

b. What is the specific rotation of a racemic mixture of MSG?

4.7
OPTICAL PURITY

Whether a particular sample consists of a single enantiomer or a mixture of enantiomers can be determined by its observed specific rotation. For example, an **enantiomerically pure** sample—meaning only one enantiomer is present—of (S)-(+)-2-bromobutane will have an observed specific rotation of +23.1° because the specific rotation of (S)-(+)-2-bromobutane is +23.1°. If the sample of 2-bromobutane has an observed specific rotation of 0°, however, we will know that the compound is a racemic mixture. If the observed specific rotation is positive but less than +23.1°, we will know that we have a mixture of enantiomers and the mixture contains more of the enantiomer with the S configuration than the enantiomer with the R configuration.

From the observed specific rotation, we can calculate the amount of each enantiomer present in the mixture. First we must calculate the optical purity using the following formula.

$$\text{optical purity} = \frac{\text{observed specific rotation}}{\text{specific rotation of the pure enantiomer}}$$

For example, if a sample of 2-bromobutane has an observed specific rotation of +9.2°, its optical purity is 0.40. In other words, it is 40% optically pure.

$$\text{optical purity} = \frac{+9.2°}{+23.1°} = 0.40 \text{ or } 40\%$$

Because the observed specific rotation is positive, we know that the solution contains excess (S)-(+)-2-bromobutane. The **optical purity** tells us how much excess (S)-(+)-2-bromobutane is in the mixture. The mixture is 40% optically pure, which means that 40% of the mixture is excess S enantiomer and 60% is a racemic mixture. Half of the racemic mixture plus the amount of excess S enantiomer equals the amount of the S enantiomer present in the mixture. Thus, 70% of the mixture is the S enantiomer (1/2 × 60 + 40) and 30% is the R enantiomer. (The optical purity is also called the **enantiomeric excess.**)

PROBLEM 15 ◆

(+)-Mandelic acid has a specific rotation of +158°. What would be the observed specific rotation of each of the following mixtures?

a. 25% (−)-mandelic acid and 75% (+)-mandelic acid

b. 50% (−)-mandelic acid and 50% (+)-mandelic acid

c. 75% (−)-mandelic acid and 25% (+)-mandelic acid

PROBLEM 16 ◆

The specific rotation of (R)-(+)-glyceraldehyde is +8.7°. If the observed specific rotation of a mixture of (R)-glyceraldehyde and (S)-glyceraldehyde is +1.4°, what percent of glyceraldehyde is present as the R enantiomer?

PROBLEM 17 / SOLVED

A solution prepared by mixing 10 mL of a 0.10 M solution of the R enantiomer and 30 mL of a 0.10 M solution of the S enantiomer was found to have an observed specific rotation of +4.8°. What is the specific rotation of each of the enantiomers?

SOLUTION One mmol (millimole; 10 mL × 0.10 M) of the R enantiomer is mixed with 3 mmol of the S enantiomer; 1 mmol of the R enantiomer plus 1 mmol of the S enantiomer will form 2 mmol of a racemic mixture. There will be 2 mmol of S enantiomer left over. Therefore, 2 mmol out of 4 mmol is excess S enantiomer ($2/4 = 0.50$). The solution is 50% optically pure.

$$\text{optical purity} = 0.50 = \frac{\text{observed specific rotation}}{\text{specific rotation of the pure enantiomer}}$$

$$0.50 = \frac{+4.8°}{x}$$

$$x = +9.6°$$

The S enantiomer has a specific rotation of $+9.6°$; the R enantiomer has a specific rotation of $-9.6°$.

Many organic compounds have more than one chirality center. The more chirality centers a compound has, the more stereoisomers it has. If we know how many chirality centers a compound has, we can calculate the maximum number of stereoisomers for that compound. *A compound can have a maximum of 2^n stereoisomers, where* n *equals the number of chirality centers.* For example, 3-chloro-2-butanol has two chirality centers. Therefore, it can have as many as four ($2^2 = 4$) stereoisomers. In each of the following Fischer projections, all four horizontal bonds point out of the paper toward the viewer and the two vertical bonds point behind the paper away from the viewer. Groups can rotate freely about the carbon–carbon single bonds, but Fischer projections show the stereoisomers in their eclipsed conformations.

4.8
ISOMERS WITH MORE THAN ONE CHIRALITY CENTER

$$\overset{*}{C}H_3\overset{*}{C}HCHCH_3$$
$$| \quad |$$
$$Cl \ \ OH$$

3-chloro-2-butanol

CH₃	CH₃	CH₃	CH₃
H——OH	HO——H	H——OH	HO——H
H——Cl	Cl——H	Cl——H	H——Cl
CH₃	CH₃	CH₃	CH₃
1	2	3	4
erythro enantiomers		**threo enantiomers**	

Stereoisomers **1** and **2** are a pair of nonidentical mirror images. Therefore, stereoisomers **1** and **2** are enantiomers. Stereoisomers **3** and **4** are also enantiomers. The four stereoisomers of 3-chloro-2-butanol consist of two pairs of enantiomers. When Fischer projections are drawn (with the carbon chain aligned vertically) for stereoisomers with two adjacent chirality centers, the pair of enantiomers with similar groups on the same side of the carbon chain is called the **erythro enantiomers.** The pair of enantiomers with similar groups on opposite sides is called the **threo enantiomers.** Therefore, stereoisomers **1** and **2** are the erythro enantiomers of 3-chloro-2-butanol (the hydrogens are on the same side), whereas stereoisomers **3** and **4** are the threo enantiomers.

Stereoisomers **1** and **3** are not identical and they are not mirror images. Such stereoisomers are called diastereomers. **Diastereomers** are configurational isomers that are not enantiomers. Stereoisomers **1** and **4, 2** and **3,** and **2** and **4** are also

Diastereomers are configurational isomers that are not enantiomers.

diastereomers. (Notice that cis–trans isomers are also considered to be diastereomers because they are configurational isomers that are not enantiomers.)

Enantiomers have identical physical properties (except for the direction in which they rotate plane-polarized light) and identical chemical properties—they react at the same rate with a given achiral reagent. Diastereomers have different physical properties (different melting points, different boiling points, different solubilities, different specific rotations, and so on) and different chemical properties—they react with the same reagent at different rates.

Fischer projections are useful because they make it possible to readily represent stereoisomers in two dimensions. Unfortunately, they represent the molecule in its relatively unstable, eclipsed conformation. Perspective formulas are harder for some people to visualize, but they represent the molecule in its more stable, staggered conformation, so they provide a more accurate representation of its structure. We will use both kinds of structures to depict the three-dimensional arrangement of groups bonded to a chirality center.

erythro enantiomers   threo enantiomers

perspective formulas of the stereoisomers of 3-chloro-2-butanol

3-D Molecules:
(2S,3R)-3-Chloro-2-butanol;
(2S,3S)-3-Chloro-2-butanol;
(2R,3S)-3-Chloro-2-butanol;
(2S,3S)-3-Chloro-2-butanol;

stereoisomers of 3-chloro-2-butanol

PROBLEM 18 ◆

a. Configurational isomers with two chirality centers are called _____ if the configuration of both chirality centers in one isomer is the opposite of the configuration of the chirality centers in the other isomer.

b. Configurational isomers with two chirality centers are called _____ if the configuration of both chirality centers in one isomer is the same as the configuration of the chirality centers in the other isomer.

c. Configurational isomers with two chirality centers are called _____ if one of the chirality centers has the same configuration in both isomers and the other chirality center has the opposite configuration in the two isomers.

PROBLEM 19/ SOLVED

Tetracycline is called a broad-spectrum antibiotic because it is active against a wide variety of bacteria. How many chirality centers does tetracycline have?

SOLUTION First, locate all the sp^3 hybridized carbons in tetracycline. (They are numbered in red.) Because a chirality center has four different groups attached to it, only sp^3 hybridized carbons can be chirality centers (sp^2 and sp hybridized carbons cannot be bonded to four groups). Tetracycline has nine sp^3 hybridized carbons. Four of them (#1, #2, #5, and #8) are not chirality centers because they are not bonded to four different groups. Tetracycline, therefore, has five chirality centers.

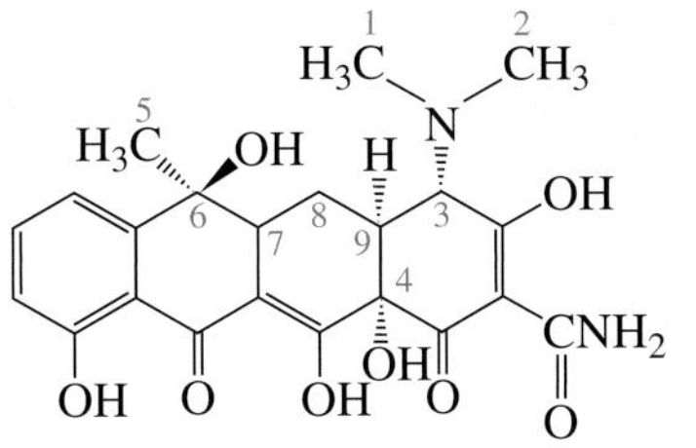

H₃C, ₁ ₂ CH₃

tetracycline

PROBLEM 20 ◆

a. How many chirality centers does cholesterol have?

b. What is the maximum number of stereoisomers that cholesterol can have?

c. How many of these stereoisomers are found in nature?

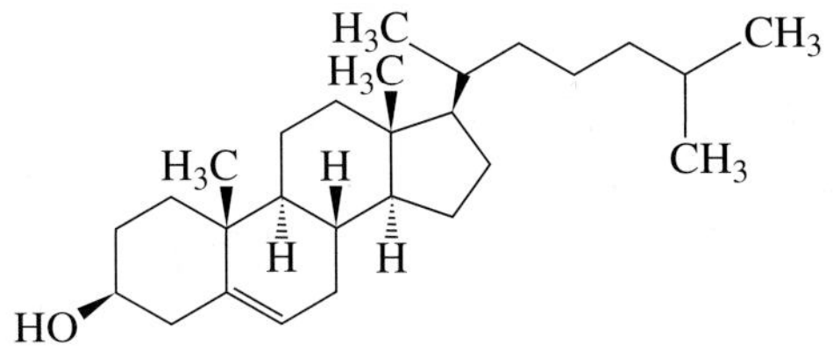

cholesterol

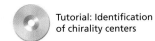

Tutorial: Identification
of chirality centers

PROBLEM 21

The following compound has only one chirality center. Why then does it have four stereoisomers?

$$CH_3CH_2\overset{*}{C}HCH_2CH=CHCH_3$$
$$|$$
$$Br$$

2,4-Dichlorohexane is another example of a compound with two chirality centers and four stereoisomers. Stereoisomers **1** and **2** are enantiomers; **3** and **4** are also enantiomers. Stereoisomers **1** and **3, 1** and **4, 2** and **3**, and **2** and **4** are diastereomers.

$$CH_3\overset{*}{C}HCH_2\overset{*}{C}HCH_2CH_3$$
$$|\quad\quad\ |$$
$$Cl\quad\quad Cl$$

2,4-dichlorohexane

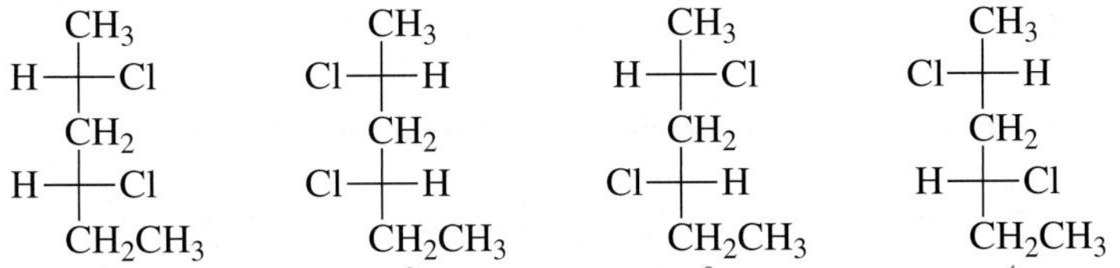

CH₃	CH₃	CH₃	CH₃
H——Cl	Cl——H	H——Cl	Cl——H
CH₂	CH₂	CH₂	CH₂
H——Cl	Cl——H	Cl——H	H——Cl
CH₂CH₃	CH₂CH₃	CH₂CH₃	CH₂CH₃
1	2	3	4

1-Bromo-2-methylcyclopentane also has two chirality centers and four stereoisomers. Because the compound is cyclic, the substituents can be in either the cis or the trans configuration. The cis isomer exists as a pair of enantiomers, and the trans isomer exists as a pair of enantiomers.

3-D Molecules:
(1*R*,2*S*)-*cis*-1-Bromo-2-
methylcyclopentane;
(1*S*,2*R*)-*cis*-1-Bromo-2-
methylcyclopentane;
(1*S*,2*R*)-*trans*-1-Bromo-2-
methylcyclopentane;
(1*S*,2*S*)-*trans*-1-Bromo-2-
methylcyclopentane

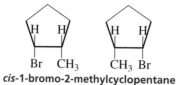

cis-1-bromo-2-methylcyclopentane

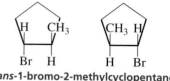

trans-1-bromo-2-methylcyclopentane

1-Bromo-3-methylcyclobutane does not have any chirality centers. C-1 has a bromine and a hydrogen attached to it, but its other two groups ($-CH_2CH(CH_3)CH_2-$)are identical; C-3 has a methyl group and a hydrogen attached to it, but its other two groups ($-CH_2CH(Br)CH_2-$) are identical. Because the compound does not have a carbon with four different groups attached to it, it has only two stereoisomers, the cis isomer and the trans isomer. The cis and trans isomers do not have enantiomers. Notice that each of them has a plane of symmetry. (In Section 4.3 we saw that a compound with a plane of symmetry cannot have an enantiomer.)

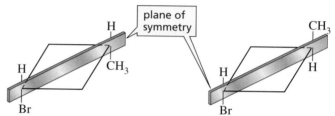

cis-1-bromo-3-methylcyclobutane   **trans-1-bromo-3-methylcyclobutane**

1-Bromo-3-methylcyclohexane has two chirality centers. The carbon that is bonded to a hydrogen and a bromine is also bonded to two different carbon-containing groups ($-CH_2CH(CH_3)CH_2CH_2CH_2-$ and $-CH_2CH_2CH_2CH(CH_3)CH_2-$), so it is a chirality center. The carbon that is bonded to a hydrogen and a methyl group is also bonded to two different carbon-containing groups, so it is also a chirality center.

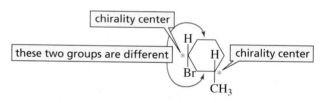

Because the compound has two chirality centers, it has four stereoisomers. The cis isomer exists as a pair of enantiomers, and the trans isomer exists as a pair of enantiomers.

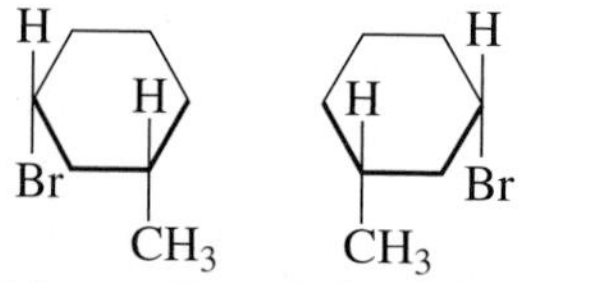

cis-1-bromo-3-methylcyclohexane 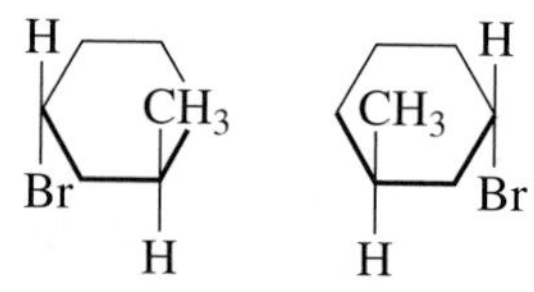   **trans-1-bromo-3-methylcyclohexane**

1-Bromo-4-methylcyclohexane has no chirality centers. Therefore, the compound has only one cis isomer and one trans isomer.

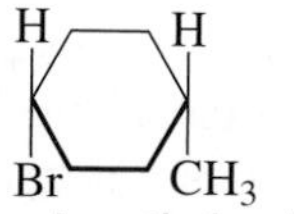

cis-1-bromo-4-methylcyclohexane 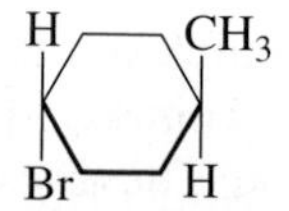   **trans-1-bromo-4-methylcyclohexane**

PROBLEM 22

Draw all possible stereoisomers for each of the following compounds:

a. 2-chloro-3-hexanol

b. 2-bromo-4-chlorohexane

c. 2,3-dichloropentane

d. 1,3-dibromopentane

PROBLEM 23

Draw the stereoisomers of 1-bromo-3-chlorocyclohexane.

PROBLEM 24 ◆

a. Draw the two chair conformations for each of the stereoisomers of *trans*-1-*tert*-butyl-3-methylcyclohexane.

b. For each pair, indicate which conformation is more stable.

PROBLEM 25 ◆

Of all the possible cyclooctanes that have one chloro substituent and one methyl substituent, which ones do not have any chirality centers?

PROBLEM 26

Draw a diastereomer for each of the following.

a.
$$CH_3$$
H——OH
H——OH
$$CH_3$$

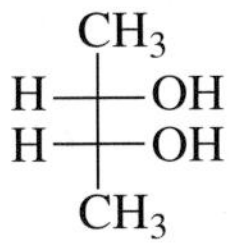

c.
$$\begin{array}{cc} H_3C & CH_3 \\ & C{=}C \\ H & H \end{array}$$

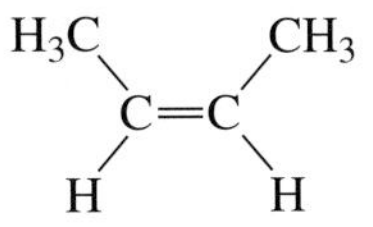

b.
Cl   Cl
H‴‴—C—C⟍H
CH₃CH₂   CH₃

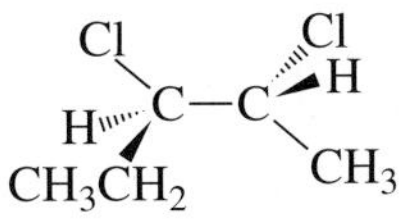

d.
$$\begin{array}{cc} H & H \\ HO & CH_3 \end{array}$$

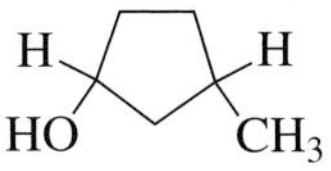

In the examples we have just seen, each compound with two chirality centers has four stereoisomers. However, some compounds with two chirality centers have only three stereoisomers. This is why we emphasized in Section 4.8 that the *maximum* number of stereoisomers a compound with n chirality centers can have is 2^n, instead of stating that a compound with n chirality centers has 2^n stereoisomers.

An example of a compound with two chirality centers that has only three stereoisomers is 2,3-dibromobutane.

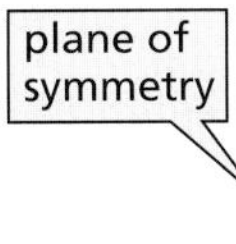

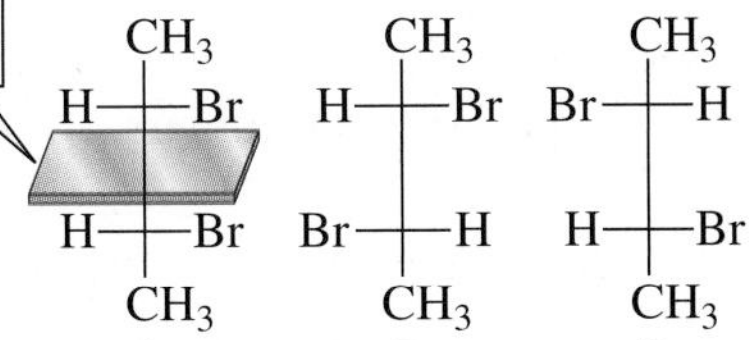

4.9
MESO COMPOUNDS

plane of symmetry

$$CH_3 \qquad CH_3 \qquad CH_3$$
H——Br   H——Br   Br——H
H——Br   Br——H   H——Br
$$CH_3 \qquad CH_3 \qquad CH_3$$
1   2   3

The "missing" stereoisomer is the mirror image of stereoisomer **1.** Stereoisomer **1** has a plane of symmetry, which means that it does *not* have a nonidentical mirror image. If we draw the mirror image of stereoisomer **1,** we find that it and stereoisomer **1** are identical. To convince yourself that the two structures are identical, rotate by 180° the mirror image of stereoisomer **1** that you just drew and you will see that it is identical to stereoisomer **1.** *(Remember, you can move Fischer projections only by rotating them 180° in the plane of the paper.)*

$$
\begin{array}{cc}
\text{CH}_3 & \text{CH}_3 \\
\text{H}-\!\!\!\!+\!\!\!\!-\text{Br} & \text{Br}-\!\!\!\!+\!\!\!\!-\text{H} \\
\text{H}-\!\!\!\!+\!\!\!\!-\text{Br} & \text{Br}-\!\!\!\!+\!\!\!\!-\text{H} \\
\text{CH}_3 & \text{CH}_3
\end{array}
$$

superimposable mirror images

A meso compound is an achiral compound that has two or more chirality centers.

Stereoisomer **1** is called a meso compound. Even though a **meso compound** has chirality centers, it is an achiral molecule because it has a plane of symmetry. *Mesos* is the Greek word for "middle." Because of the plane of symmetry, a meso compound does not rotate the plane of polarized light. It is optically inactive. A meso compound can be recognized by the fact that it has two or more chirality centers and a plane of symmetry. *If a compound has a plane of symmetry, it will not be optically active even though it has chirality centers.*

plane of symmetry

$$
\begin{array}{c}
\text{CH}_3 \\
\text{H}-\!\!\!\!+\!\!\!\!-\text{Br} \\
\text{H}-\!\!\!\!+\!\!\!\!-\text{Br} \\
\text{CH}_3
\end{array}
\qquad
\begin{array}{c}
\text{CH}_2\text{CH}_3 \\
\text{H}-\!\!\!\!+\!\!\!\!-\text{OH} \\
\text{CH}_2 \\
\text{H}-\!\!\!\!+\!\!\!\!-\text{OH} \\
\text{CH}_2\text{CH}_3
\end{array}
$$

meso compounds

Whenever a compound has two chirality centers and the four groups bonded to one chirality center are identical to the four groups bonded to the other chirality center, the compound will have three stereoisomers. One stereoisomer will be a meso compound, and the other two will be a pair of enantiomers.

If a compound with two chirality centers has the same four groups bonded to each of the chirality centers, one of its stereoisomers will be a meso compound.

$$
\begin{array}{c}
\text{CH}_3 \\
\text{H}-\!\!\!\!+\!\!\!\!-\text{OH} \\
\text{H}-\!\!\!\!+\!\!\!\!-\text{OH} \\
\text{CH}_3
\end{array}
\qquad
\begin{array}{c}
\text{CH}_3 \\
\text{H}-\!\!\!\!+\!\!\!\!-\text{OH} \\
\text{HO}-\!\!\!\!+\!\!\!\!-\text{H} \\
\text{CH}_3
\end{array}
\qquad
\begin{array}{c}
\text{CH}_3 \\
\text{HO}-\!\!\!\!+\!\!\!\!-\text{H} \\
\text{H}-\!\!\!\!+\!\!\!\!-\text{OH} \\
\text{CH}_3
\end{array}
$$

a meso compound   a pair of enantiomers

a meso compound   a pair of enantiomers

In the case of cyclic compounds, the cis isomer will be the meso compound and the trans isomer will exist as a pair of enantiomers.

3-D Molecules:
cis-1,3-Dimethyl-cyclopentane and its mirror image

a meso compound   a pair of enantiomers

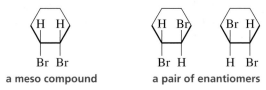

a meso compound   a pair of enantiomers

The preceding structure for *cis*-1,2-dibromocyclohexane suggests that the compound has a plane of symmetry. Cyclohexane, however, is not a flat hexagon—it exists preferentially in the chair conformation, and the chair conformer of *cis*-1,2-dibromocyclohexane does not have a plane of symmetry. Only the much less stable boat conformer of *cis*-1,2-dibromocyclohexane has a plane of symmetry. Then, is *cis*-1,2-dibromocyclohexane a meso compound? The answer is yes. As long as any one conformer of a compound has a plane of symmetry, the compound will be achiral, and an achiral compound with two chirality centers is a meso compound.

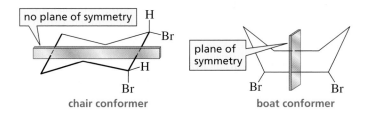

chair conformer   boat conformer

This holds for acyclic compounds as well. We have just seen that 2,3-dibromobutane is an achiral meso compound because it has a plane of symmetry. When we made that determination, however, we were looking at a relatively unstable eclipsed conformer. The more stable staggered conformer does not have a plane of symmetry. 2,3-Dibromobutane is still a meso compound, however, because it has a conformer that has a plane of symmetry.

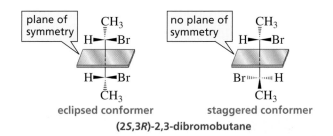

eclipsed conformer   staggered conformer

(2S,3R)-2,3-dibromobutane

PROBLEM-SOLVING STRATEGY

Which of the following compounds has a stereoisomer that is a meso compound?

a. 2,3-dimethylbutane   **e.** 1,4-dimethylcyclohexane

b. 3,4-dimethylhexane   **f.** 1,2-dimethylcyclohexane

c. 2-bromo-3-methylpentane   **g.** 3,4-diethylhexane

d. 1,3-dimethylcyclohexane   **h.** 1-bromo-2-methylcyclohexane

Check each compound to see if it has the necessary requirements to have a stereoisomer that is a meso compound. That is, does it have two chirality centers with the same four substituents attached to each of the chirality centers?

Compounds **a, e,** and **g** do *not* have a stereoisomer that is a meso compound because they don't have any chirality centers.

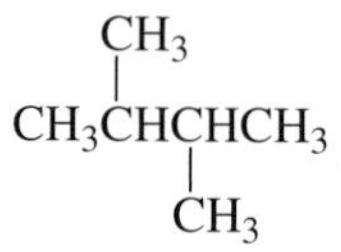

$$CH_3$$
$$CH_3CHCHCH_3$$
$$CH_3$$
(a)

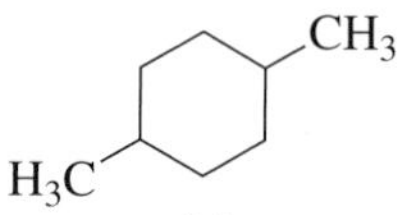

(e)

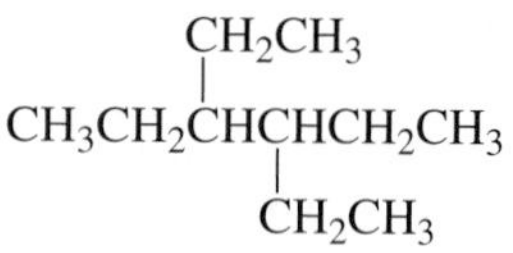

$$CH_2CH_3$$
$$CH_3CH_2CHCHCH_2CH_3$$
$$CH_2CH_3$$
(g)

Compounds **c** and **h** each have two chirality centers. They do *not* have a stereoisomer that is a meso compound because each of the chirality centers is *not* bonded to the same four substituents.

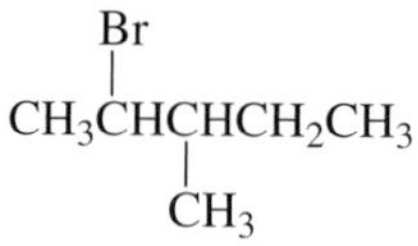

$$Br$$
$$CH_3CHCHCH_2CH_3$$
$$CH_3$$
(c)

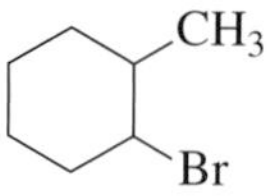

(h)

Compounds **b, d,** and **f** have a stereoisomer that is a meso compound.

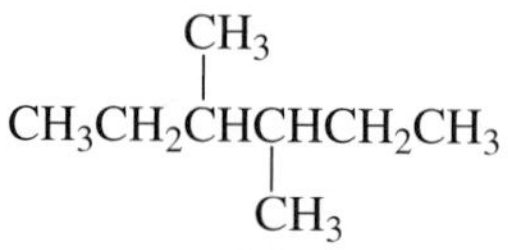

$$CH_3$$
$$CH_3CH_2CHCHCH_2CH_3$$
$$CH_3$$
(b)

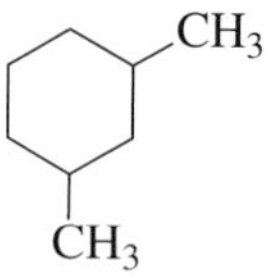

(d)

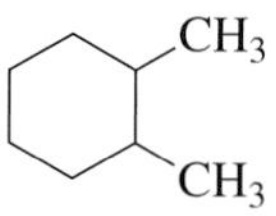

(f)

The isomer that is the meso compound is the one with a plane of symmetry when an acyclic compound is drawn as a Fischer projection **(b),** or when a cyclic compound is drawn with a flat ring (**d** and **f**).

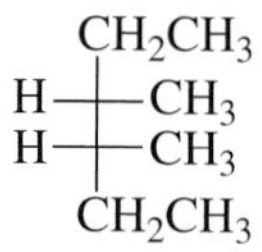

(b)

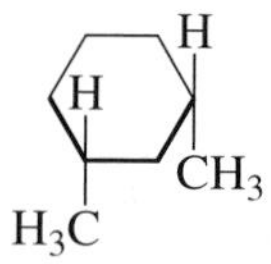

(d)

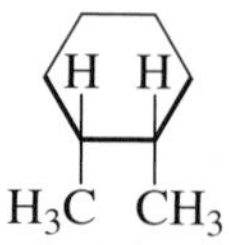

(f)

Now continue on to Problem 27.

PROBLEM 27 ◆

Which of the following compounds has a stereoisomer that is a meso compound?

a. 2,4-dibromohexane

b. 2,4-dibromopentane

c. 2,4-dimethylpentane

d. 1,3-dichlorocyclohexane

e. 1,4-dichlorocyclohexane

f. 1,2-dichlorocyclobutane

PROBLEM 28

Draw all the stereoisomers for each of the following compounds:

a. 1-bromo-2-methylbutane

b. 1-chloro-3-methylpentane

c. 2-methyl-1-propanol

d. 2-bromo-1-butanol

e. 3-chloro-3-methylpentane	**j.** 1,2-dichlorocyclobutane
f. 3-bromo-2-butanol	**k.** 1,3-dichlorocyclohexane
g. 3,4-dichlorohexane	**l.** 1,4-dichlorocyclohexane
h. 2,4-dichloropentane	**m.** 1-bromo-2-chlorocyclobutane
i. 2,4-dichloroheptane	**n.** 1-bromo-3-chlorocyclobutane

If a compound has more than one chirality center, the steps used to determine whether a chirality center has the *R* configuration or the *S* configuration must be applied to each of the chirality centers individually. As an example, let's name one of the stereoisomers of 3-bromo-2-butanol.

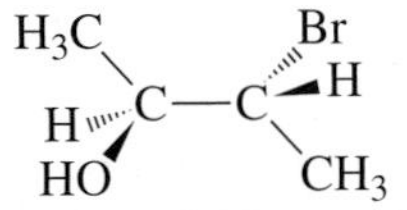

a stereoisomer of 3-bromo-2-butanol

First we will determine the configuration at C-2. The OH group has the number 1 priority, the C-3 carbon (the C attached to Br, C, H) is number 2, CH_3 is number 3, and H is number 4. Because the group with the lowest priority is bonded by a hatched wedge, we can immediately draw an arrow from the group with the highest priority to the group with the next highest priority. Because that arrow points in a counterclockwise direction, the configuration at C-2 is *S*.

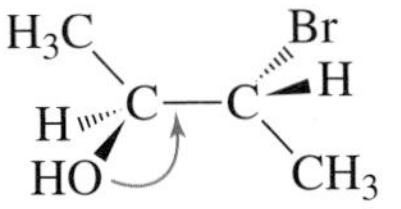

Now we can determine the configuration at C-3. Because the group with the lowest priority (H) is not bonded by a hatched wedge, we must put it there by temporarily switching a pair of groups.

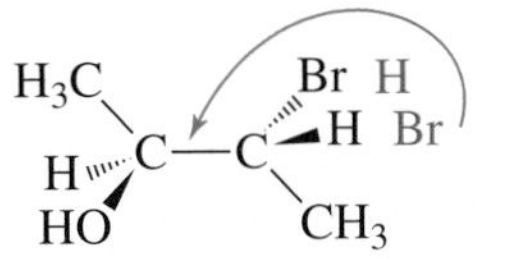

The arrow going from the highest priority group (Br) to the next highest priority group (the C attached to O, C, H) points in a counterclockwise direction, so our initial answer is "*S*"; however, because we switched a pair of groups before we drew the arrow, C-3 has the opposite configuration—the *R* configuration.

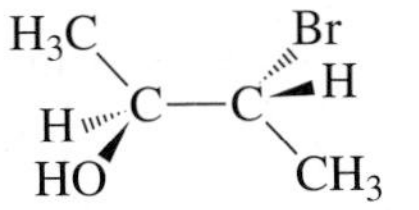

(2*S*,3*R*)-3-bromo-2-butanol

4.10
THE *R,S* SYSTEM OF NOMENCLATURE FOR ISOMERS WITH MORE THAN ONE CHIRALITY CENTER

Fischer projections with two chirality centers can be named in a similar manner. Just apply the steps to each chirality center that you learned for a Fischer projection with one chirality center. For C-2, the arrow from the group with the highest priority to the group with the next highest priority points in a clockwise direction, so our answer is "*R*." But because the group with the lowest priority is on a **h**orizontal bond, that answer is "**H**orribly wrong." You can conclude, then, that C-2 has the *S* configuration.

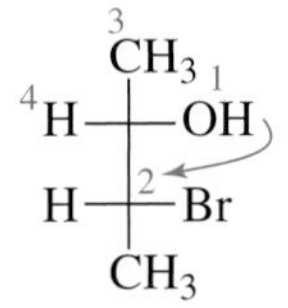

$$
\begin{array}{c}
\overset{3}{\text{C}}\text{H}_3 \;\overset{1}{}\\
{}^4\text{H} \!-\!\!\!-\!\!\!-\!\! \text{OH} \\
\text{H} \!-\!\!\overset{2}{}\!\!-\!\! \text{Br} \\
\text{CH}_3
\end{array}
$$

By repeating these steps for C-3, you will find that it has the *R* configuration. So the isomer is named (2*S*,3*R*)-3-bromo-2-butanol.

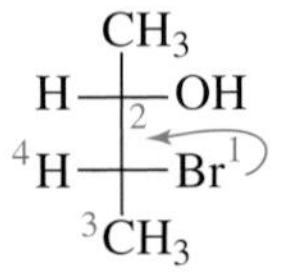

$$
\begin{array}{c}
\text{CH}_3 \\
\text{H} \!-\!\!\overset{2}{}\!\!-\!\! \text{OH} \\
{}^4\text{H} \!-\!\!-\!\! \text{Br}^1 \\
{}^3\text{CH}_3
\end{array}
$$

(2*S*,3*R*)-3-bromo-2-butanol

The four stereoisomers of 3-bromo-2-butanol are named as shown here. Take a few minutes to verify their names.

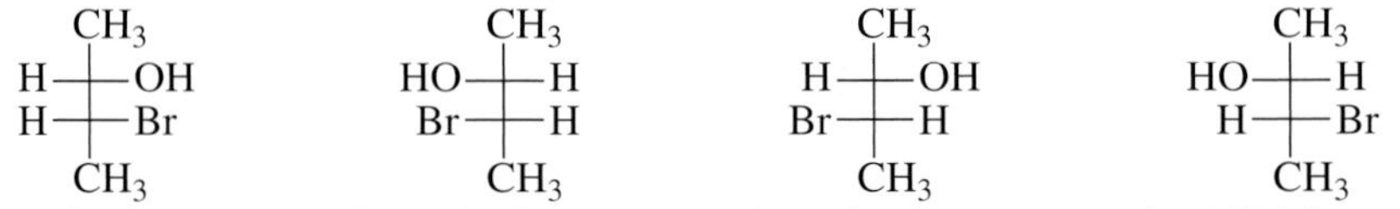

$$
\begin{array}{cccc}
\text{CH}_3 & \text{CH}_3 & \text{CH}_3 & \text{CH}_3 \\
\text{H}\!-\!\!-\!\text{OH} & \text{HO}\!-\!\!-\!\text{H} & \text{H}\!-\!\!-\!\text{OH} & \text{HO}\!-\!\!-\!\text{H} \\
\text{H}\!-\!\!-\!\text{Br} & \text{Br}\!-\!\!-\!\text{H} & \text{Br}\!-\!\!-\!\text{H} & \text{H}\!-\!\!-\!\text{Br} \\
\text{CH}_3 & \text{CH}_3 & \text{CH}_3 & \text{CH}_3 \\
\text{(2S,3R)-3-bromo-} & \text{(2R,3S)-3-bromo-} & \text{(2S,3S)-3-bromo-} & \text{(2R,3R)-3-bromo-} \\
\text{2-butanol} & \text{2-butanol} & \text{2-butanol} & \text{2-butanol}
\end{array}
$$

Fischer projections of the stereoisomers of 3-bromo-2-butanol

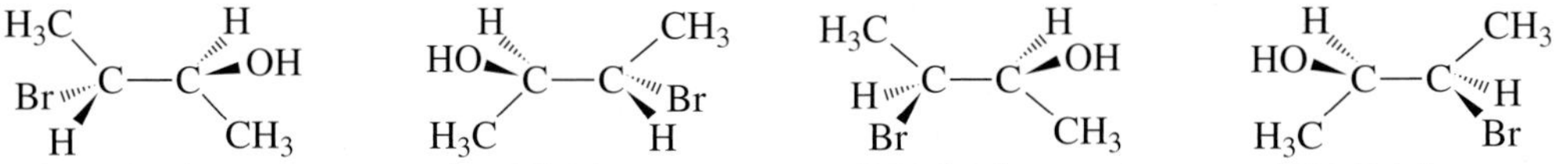

$$
\begin{array}{cccc}
\text{(2S,3R)-3-bromo-} & \text{(2R,3S)-3-bromo-} & \text{(2S,3S)-3-bromo-} & \text{(2R,3R)-3-bromo-} \\
\text{2-butanol} & \text{2-butanol} & \text{2-butanol} & \text{2-butanol}
\end{array}
$$

perspective formulas of the stereoisomers of 3-bromo-2-butanol

Notice that pairs of enantiomers have the opposite configuration at both chirality centers, whereas pairs of diastereomers have the opposite configuration at one chirality center and the same configuration at the other chirality center.

PROBLEM 29

Draw and name the four stereoisomers of 1,3-dichloro-2-butanol using:

a. perspective formulas.   **b.** Fischer projections.

Tartaric acid has three stereoisomers because its two chirality centers have the same set of four substituents. The meso compound and the pair of enantiomers are named as shown.

COOH
H——OH
H——OH
COOH
(2R,3S)-tartaric acid
a meso compound

COOH
H——OH
HO——H
COOH
(2R,3R)-tartaric acid

COOH
HO——H
H——OH
COOH
(2S,3S)-tartaric acid

a pair of enantiomers

Fischer projections of the stereoisomers of tartaric acid

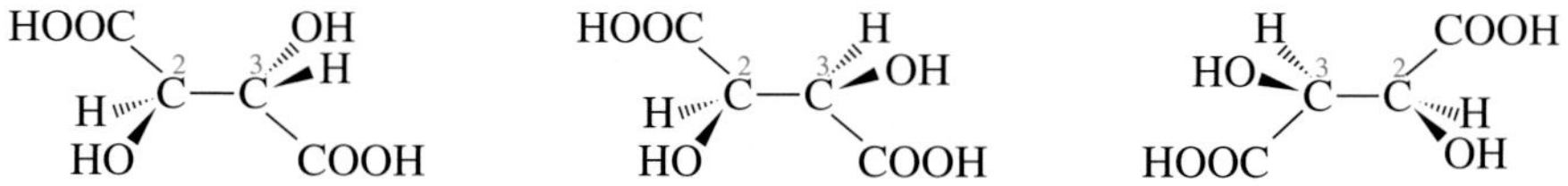

(2R,3S)-tartaric acid

(2R,3R)-tartaric acid

(2S,3S)-tartaric acid

perspective formulas of the stereoisomers of tartaric acid

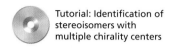
Tutorial: Identification of stereoisomers with multiple chirality centers

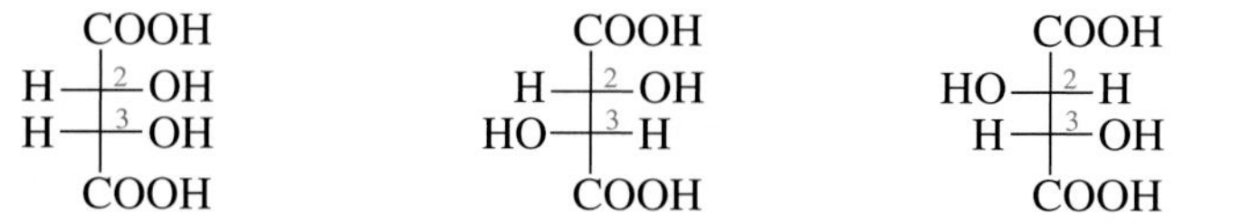

The physical properties of the three stereoisomers of tartaric acid are listed in Table 4.1. The meso compound and either one of the enantiomers constitute a pair of diastereomers. Notice that the physical properties of the enantiomers are identical, whereas the physical properties of the diastereomers are different. Notice also that the physical properties of the racemic mixture differ from the physical properties of the meso compound and either of the enantiomers.

TABLE 4.1 Physical Properties of the Stereoisomers of Tartaric Acid			
	Melting Point, °C	$[\alpha]_D^{25\,°C}$	**Solubility, g/100 g H_2O at 15 °C**
(2R,3R)-(+)-tartaric acid	170	+11.98°	139
(2S,3S)-(−)-tartaric acid	170	−11.98°	139
(2R,3S)-tartaric acid	140	0°	125
(±)-tartaric acid	206	0°	20.6

PROBLEM-SOLVING STRATEGY

Draw perspective formulas for the following compounds:

a. (*R*)-2-butanol

b. (2*S*,3*R*)-3-chloro-2-pentanol

a. First draw the compound so that you know what groups are bonded to the chirality center.

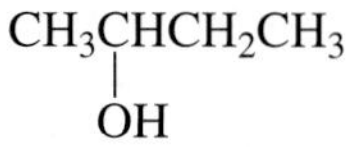

$$CH_3CHCH_2CH_3$$
$$|$$
$$OH$$
2-butanol

Draw the bonds about the chirality center.

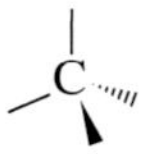

Put the group with the lowest priority on the hatched wedge. Put the group with the highest priority on any remaining bond.

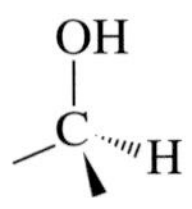

OH
|
C''''H

Because you have been asked to draw the *R* enantiomer, draw an arrow clockwise from the group with the highest priority to the next available bond and put the group with the next highest priority on that bond.

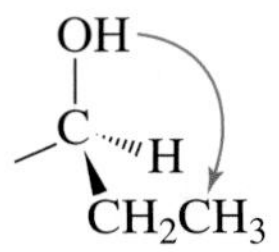

OH
|
C''''H
CH₂CH₃

Put the remaining substituent on the last available bond.

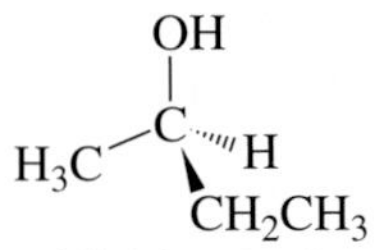

OH
|
H₃C—C''''H
CH₂CH₃
(R)-2-butanol

b. First draw the compound.

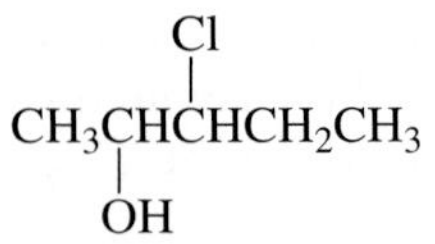

Cl
|
CH₃CHCHCH₂CH₃
|
OH
3-chloro-2-pentanol

Draw the bonds about the chirality centers.

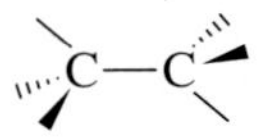

''''C—C''''

For each chirality center, put the group with the lowest priority on the hatched wedge.

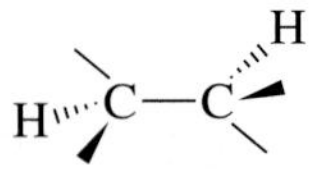

H
''''
H''''C—C''''

For each chirality center, put the group with the highest priority on a bond such that an arrow points clockwise (if you want the *R* configuration) or counterclockwise (if you want the *S* configuration) to the group with the next highest priority.

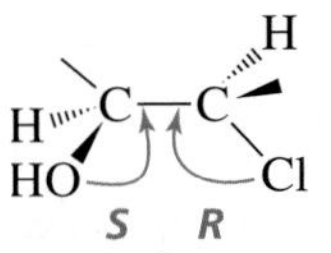

H
''''
H''''C—C''''
HO   Cl
  S   *R*

Put the remaining substituents on the last available bonds.

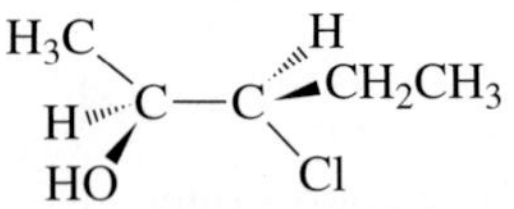

H₃C   H
     \\   ''''
H''''C—C''''CH₂CH₃
HO   Cl
(2S,3R)-3-chloro-2-pentanol

Now continue on to Problem 30.

PROBLEM 30

Draw perspective formulas for the following compounds:

a. (*S*)-3-chloro-1-pentanol

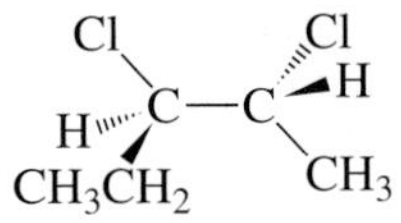

b. (2*R*,3*R*)-2,3-dibromopentane

c. (2*S*,3*R*)-3-methyl-2-pentanol

d. (*R*)-1,2-dibromobutane

PROBLEM 31 ◆

For many centuries, the Chinese have used extracts from a family of herbs known as Ephedra to treat asthma. Chemists have been able to isolate a compound from these herbs, which they named ephedrine, that is a potent dilator of air passages in the lungs.

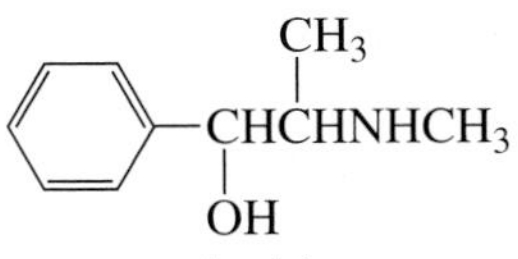

ephedrine

a. How many stereoisomers are possible for ephedrine?

b. The stereoisomer shown here is the one that is pharmacologically active. What is the configuration of each of the chirality centers?

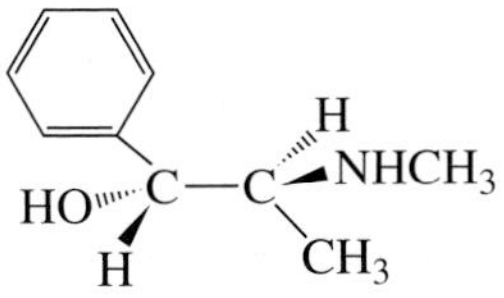

PROBLEM 32 ◆

Name the following compounds.

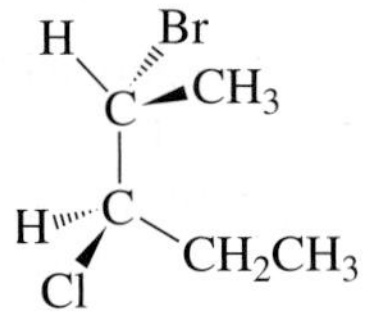

When a compound that contains a chirality center undergoes a reaction, what happens to the configuration of the chirality center depends on the specific reaction. If the reaction *does not break* any of the four bonds to the chirality center, then the relative positions of the groups bonded to the chirality center will not change. For example, when (*S*)-1-chloro-3-methylhexane reacts with hydroxide ion, the relative positions of the groups bonded to the chirality center remain the same because the reaction does not break any of the bonds to the chirality center.

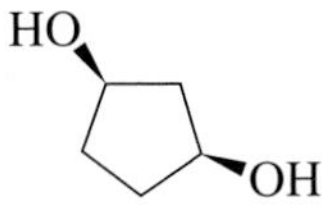

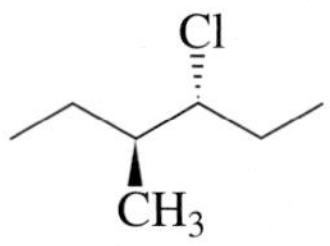

4.11
REACTIONS OF COMPOUNDS THAT CONTAIN A CHIRALITY CENTER

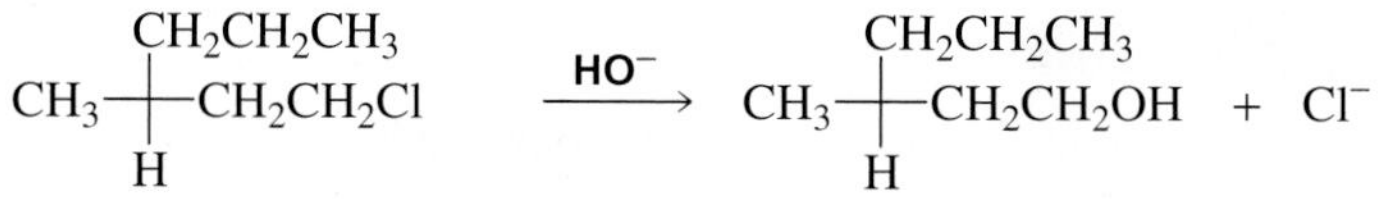

$$\underset{\substack{\text{(S)-1-chloro-3-methylhexane}}}{\text{CH}_3 \overset{\text{CH}_2\text{CH}_2\text{CH}_3}{\underset{\text{H}}{\rule{0pt}{0pt}}} \text{CH}_2\text{CH}_2\text{Cl}} \xrightarrow{\text{HO}^-} \underset{\substack{\text{(S)-3-methyl-1-hexanol}}}{\text{CH}_3 \overset{\text{CH}_2\text{CH}_2\text{CH}_3}{\underset{\text{H}}{\rule{0pt}{0pt}}} \text{CH}_2\text{CH}_2\text{OH}} + \text{Cl}^-$$

A word of warning: If the four groups bonded to the chirality center maintain their relative positions, it does not necessarily mean that an *S* reactant will always yield an *S* product. In the following reaction, for example, the relative positions of the groups are the same in the reactant and the product. In other words, the reactant and the product have the same **relative configurations.** However, the reactant has the *S* configuration, whereas the product has the *R* configuration (LiAlH$_4$ is a reagent that substitutes an H for a Cl). Although the groups maintained their relative positions during the reaction, their relative priorities as defined by the Cahn–Ingold–Prelog rules changed (Section 4.5). The change in priorities—not the change in positions of the groups—is what caused the *S* reactant to become an *R* product.

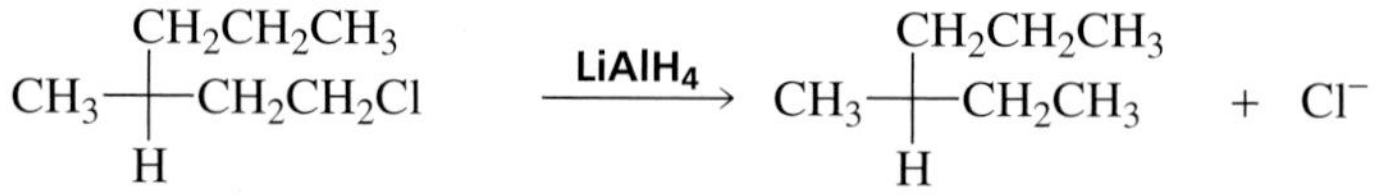

$$\underset{\substack{\text{(S)-1-chloro-3-methylhexane}}}{\text{CH}_3 \overset{\text{CH}_2\text{CH}_2\text{CH}_3}{\underset{\text{H}}{\rule{0pt}{0pt}}} \text{CH}_2\text{CH}_2\text{Cl}} \xrightarrow{\text{LiAlH}_4} \underset{\substack{\text{(R)-3-methylhexane}}}{\text{CH}_3 \overset{\text{CH}_2\text{CH}_2\text{CH}_3}{\underset{\text{H}}{\rule{0pt}{0pt}}} \text{CH}_2\text{CH}_3} + \text{Cl}^-$$

If the reaction *does break* a bond to the chirality center, the product can have the same relative configuration as the reactant or it can have the opposite relative configuration. Which of the products is actually obtained depends on the mechanism of the reaction. Therefore, we cannot predict what the configuration of the product will be unless we know the mechanism of the reaction.

$$\underset{\substack{}}{\text{CH}_3 \overset{\text{CH}_2\text{CH}_3}{\underset{\text{H}}{\rule{0pt}{0pt}}} \text{Y}} \xrightarrow{\text{Z}^-} \underset{\substack{}}{\text{CH}_3 \overset{\text{CH}_2\text{CH}_3}{\underset{\text{H}}{\rule{0pt}{0pt}}} \text{Z}} + \underset{\substack{}}{\text{Z} \overset{\text{CH}_2\text{CH}_3}{\underset{\text{H}}{\rule{0pt}{0pt}}} \text{CH}_3} + \text{Y}^-$$

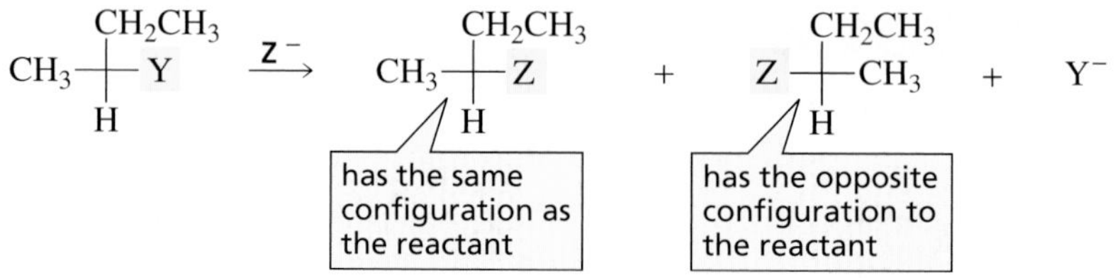

has the same configuration as the reactant

has the opposite configuration to the reactant

An achiral reagent reacts identically with both enantiomers. An achiral sock fits on either foot.

Enantiomers have the same chemical properties, so they react with *achiral* reagents at the same rate. Thus, hydroxide ion (an achiral reagent) reacts with (*R*)-2-bromobutane at the same rate that it reacts with (*S*)-2-bromobutane. If the reagent is *chiral*, however, the enantiomers react at different rates. One example of a chiral reagent is an enzyme. The enzyme D-amino acid oxidase, for example, reacts only with the *R* enantiomer, leaving the *S* enantiomer unchanged. If you imagine an enzyme to be a right-hand glove and the enantiomers to be a pair of hands, the enzyme typically reacts with only one enantiomer because only the right hand fits into the right-hand glove.

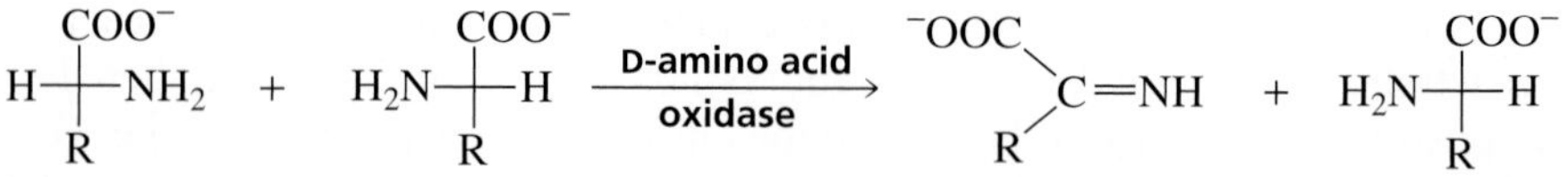

A chiral reagent reacts differently with each enantiomer. A chiral shoe fits on only one foot.

$$\underset{\substack{R\text{-enantiomer}}}{\text{H} \overset{\text{COO}^-}{\underset{\text{R}}{\rule{0pt}{0pt}}} \text{NH}_2} + \underset{\substack{S\text{-enantiomer}}}{\text{H}_2\text{N} \overset{\text{COO}^-}{\underset{\text{R}}{\rule{0pt}{0pt}}} \text{H}} \xrightarrow[\text{oxidase}]{\text{D-amino acid}} \underset{\substack{\text{oxidized} \\ R\text{-enantiomer}}}{\overset{^-\text{OOC}}{\underset{\text{R}}{\rule{0pt}{0pt}}} \text{C}=\text{NH}} + \underset{\substack{\text{unreacted} \\ S\text{-enantiomer}}}{\text{H}_2\text{N} \overset{\text{COO}^-}{\underset{\text{R}}{\rule{0pt}{0pt}}} \text{H}}$$

Receptors are proteins that bind particular molecules. Because a receptor is chiral, it will bind one enantiomer better than the other. In Figure 4.2, the receptor binds the *R*-enantiomer but it does not bind the *S*-enantiomer.

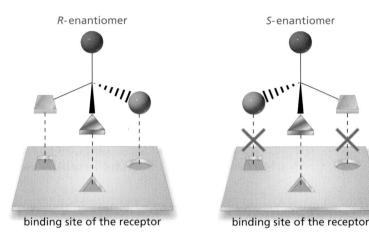

R-enantiomer   S-enantiomer

binding site of the receptor   binding site of the receptor

◀ **Figure 4.2**
Schematic diagram showing why only one enantiomer is bound by a receptor. One enantiomer fits into the binding site and one does not.

Because a receptor will recognize only one of a pair of enantiomers, enantiomers can have different physiological properties. Receptors located on the exterior of nerve cells in the nose, for example, are able to perceive and differentiate the estimated 10,000 smells to which they are exposed. (*R*)-(−)-carvone is found in spearmint oil, and (*S*)-(+)-carvone is the main constituent of caraway seed oil. The reason these two enantiomers have such different odors is that each fits into a different receptor.

(*R*)-(−)-carvone
spearmint oil

(*S*)-(+)-carvone
caraway seed oil

$[\alpha]_D^{20\,°C} = -62.5°$ $[\alpha]_D^{20\,°C} = +62.5°$

Suppose we had a sample of (+)-1-bromo-3-phenyl-2-propanamine and we wanted to know its configuration. In other words, is it the *R* enantiomer or the *S* enantiomer? We can draw the *R* and *S* enantiomers, but we cannot tell which one is (+) by looking at their structures.

4.12
DETERMINING THE CONFIGURATION

Which enantiomer is dextrorotatory?

(*R*)-1-bromo-
3-phenyl-2-propanamine

(*S*)-1-bromo-
3-phenyl-2-propanamine

The easiest way to determine the configuration of an optically active compound is to find a compound whose configuration is known that can be converted to the compound whose configuration you want to know by reactions that do not break any bonds to the chirality center. For example, the configuration of amphetamine is known: The dextrorotatory (+) isomer has the *S* configuration, and the levorotatory (−) isomer has the *R* configuration.

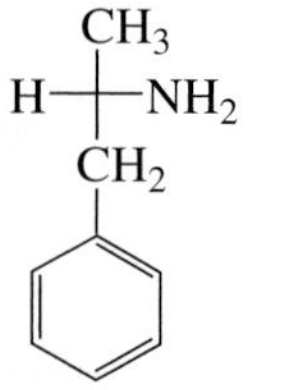

(S)-(+)-amphetamine 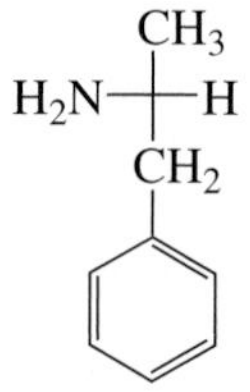   (R)-(−)-amphetamine

If we allow our sample of (+)-1-bromo-3-phenyl-2-propanamine to react with a reagent that will substitute H for Br, we will obtain one of the enantiomers of amphetamine.

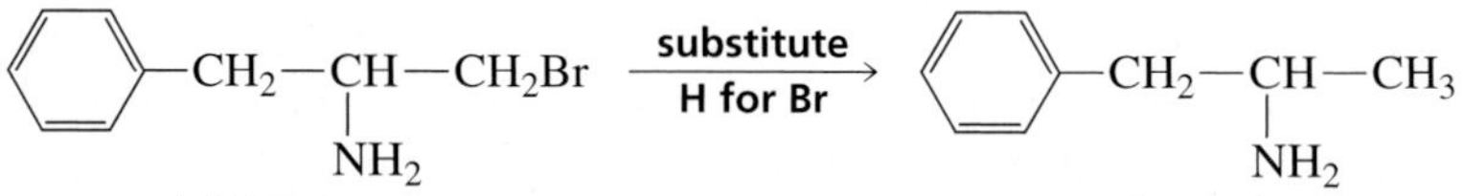

(+)-1-bromo-
3-phenyl-2-propanamine
the configuration is not known

amphetamine

If we now place the amphetamine formed in this reaction into a polarimeter to determine in which direction it rotates plane-polarized light, we will know whether it has the *R* or the *S* configuration. For example, if we find that it is the dextrorotatory isomer, we will know that it has the *S* configuration. Because our sample of amphetamine was synthesized in a reaction that did not break any bonds to the chirality center of the reactant, it has to have the same relative configuration as the reactant. In other words, the relative positions of the groups bonded to the chirality center must be the same in both compounds. Thus, we would know that the Fischer projection of our sample of (+)-1-bromo-3-phenyl-2-propanamine has the NH_2 group on the right, the benzyl group on the bottom, the CH_2Br group on top, and the H on the left—it has the *R* configuration.

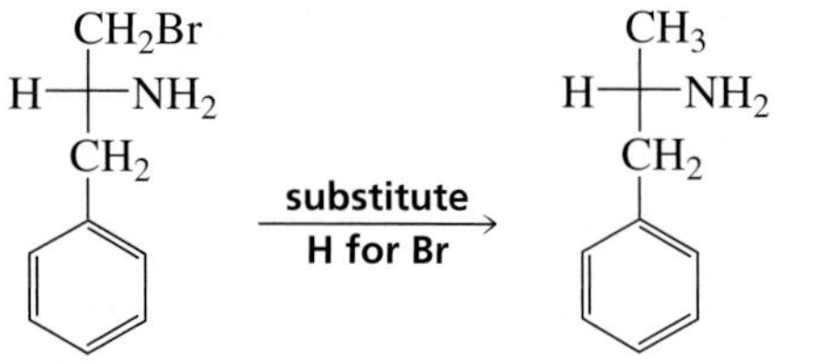

(R)-(+)-1-bromo-
3-phenyl-2-propanamine

(S)-(+)-amphetamine

On the other hand, if our sample of amphetamine were levorotatory, we would know that (+)-1-bromo-3-phenyl-2-propanamine has the *S* configuration.

We have determined the configuration of (+)-1-bromo-3-phenyl-2-propanamine by relating it to a compound whose configuration was already known. Once we knew that (+)-1-bromo-3-phenyl-2-propanamine and (+)-amphetamine had the same relative configuration, knowing the actual configuration of one of them allowed us to conclude what the actual configuration of the other one is. The actual configuration is frequently called the **absolute configuration** to indicate that the configuration is known in an absolute sense rather than in a relative sense.

Knowing the *absolute configuration* of a compound means that you know whether it has the *R* or the *S* configuration. Knowing that two compounds have the same *relative configuration* means that they have the same relative positions of their substituents; if you know the absolute configuration of either one of them, you can determine the absolute configuration of the other one.

PROBLEM 33 / SOLVED

(*S*)-(−)-2-Methyl-1-butanol can be oxidized to (+)-2-methylbutanoic acid without breaking any of the bonds to the chirality center. What is the configuration of (−)-2-methylbutanoic acid?

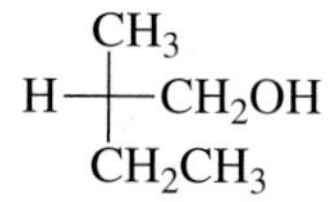

$$
\begin{array}{cc}
\underset{\text{(S)-(−)-2-methyl-1-butanol}}{\overset{\displaystyle CH_3}{\underset{\displaystyle CH_2CH_3}{H\!-\!\!|\!-\!CH_2OH}}}
&
\underset{\text{(+)-2-methylbutanoic acid}}{\overset{\displaystyle CH_3}{\underset{\displaystyle CH_2CH_3}{H\!-\!\!|\!-\!COOH}}}
\end{array}
$$

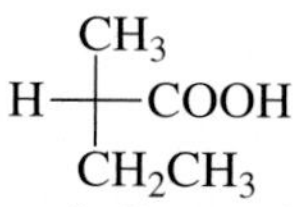

SOLUTION. We know that (+)-2-methylbutanoic acid has the configuration shown because it was formed from (*S*)-(−)-2-methyl-1-butanol without breaking any bonds to the chirality center. Therefore, we know that (+)-2-methylbutanoic acid has the *S* configuration. We can conclude then that (−)-2-methylbutanoic acid has the *R* configuration.

PROBLEM 34 ◆

The stereoisomer of 1-iodo-2-methylbutane with the *S* configuration rotates plane-polarized light in a counterclockwise direction. The following reaction results in an alcohol that rotates plane-polarized light in a clockwise direction. What is the configuration of (−)-2-methyl-1-butanol?

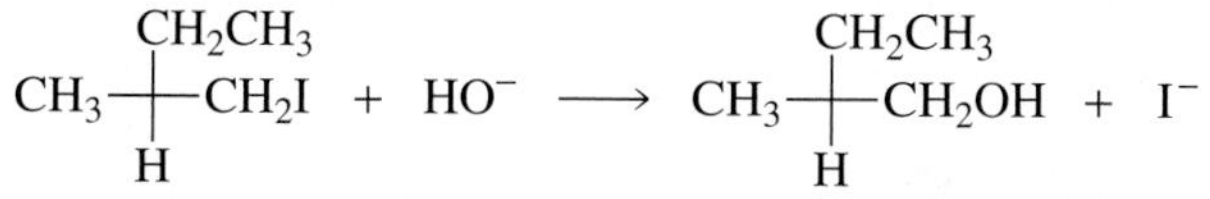

$$
\overset{\displaystyle CH_2CH_3}{\underset{\displaystyle H}{CH_3\!-\!\!|\!-\!CH_2I}} \; + \; HO^- \; \longrightarrow \; \overset{\displaystyle CH_2CH_3}{\underset{\displaystyle H}{CH_3\!-\!\!|\!-\!CH_2OH}} \; + \; I^-
$$

Glyceraldehyde has one chirality center and, therefore, has a pair of enantiomers. The absolute configuration of glyceraldehde was not known until 1951. Until then, chemists did not know whether (+)-glyceraldehyde had the *R* or the *S* configuration. They had arbitrarily decided that (+)-glyceraldehyde had the *R* configuration. They had a 50-50 chance of being correct.

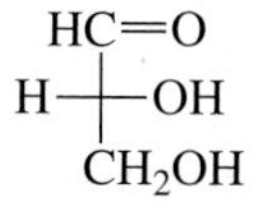

$$
\begin{array}{cc}
\underset{\text{(R)-(+)-glyceraldehyde}}{\overset{\displaystyle HC\!=\!O}{\underset{\displaystyle CH_2OH}{H\!-\!\!|\!-\!OH}}}
&
\underset{\text{(S)-(−)-glyceraldehyde}}{\overset{\displaystyle HC\!=\!O}{\underset{\displaystyle CH_2OH}{HO\!-\!\!|\!-\!H}}}
\end{array}
$$

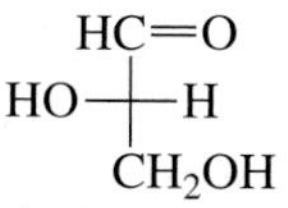

The configurations of many organic compounds were "determined" by synthesizing them from (+)- or (−)-glyceraldehyde or by converting them to (+)- or (−)-glyceraldehyde, always using reactions that did not break any of the bonds to the chirality center. For example, (−)-lactic acid was assumed to have the following configuration because it could be related to (+)-glyceraldehyde through the reactions indicated and it was assumed that (+)-glyceraldehyde was the *R* enantiomer. The configurations assigned to these molecules were relative configurations, not absolute configurations. They were relative to (+)- or (−)-glyceraldehyde and were based on the *assumption* that (+)-glyceraldehyde had the *R* configuration.

$$
\begin{array}{ccccccccc}
\text{HC} = \text{O} & & \text{COOH} & & \text{COOH} & & \text{COOH} & & \text{COOH} \\
\text{H}\!-\!\text{OH} & \xrightarrow{\text{HgO}} & \text{H}\!-\!\text{OH} & \xleftarrow[\text{H}_2\text{O}]{\text{HNO}_2} & \text{H}\!-\!\text{OH} & \xrightarrow[\text{HBr}]{\text{NaNO}_2} & \text{H}\!-\!\text{OH} & \xrightarrow{\text{Zn, H}^+} & \text{H}\!-\!\text{OH} \\
\text{CH}_2\text{OH} & & \text{CH}_2\text{OH} & & \text{CH}_2\text{NH}_2 & & \text{CH}_2\text{Br} & & \text{CH}_3
\end{array}
$$

(+)-glyceraldehyde   (−)-glyceric acid   (+)-isoserine   (−)-3-bromo-2-hydroxypropanoic acid   (−)-lactic acid

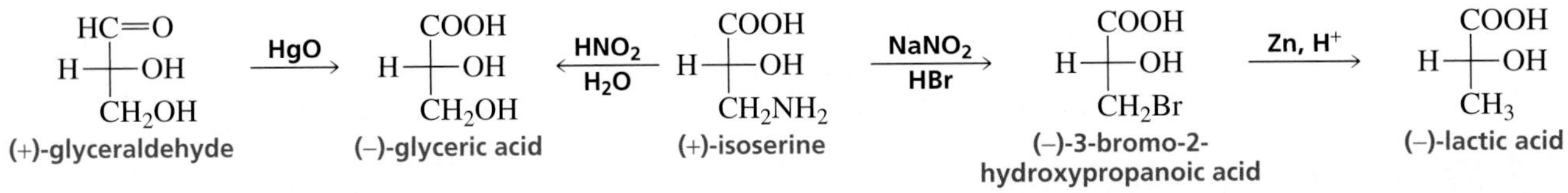

In 1951, the Dutch chemists J. M. Bijvoet, A. F. Peerdeman, and A. J. van Bommel, using X-ray crystallography and a new technique known as anomalous dispersion, determined that the sodium rubidium salt of (+)-tartaric acid had the *R,R* configuration. Because (+)-tartaric acid could be synthesized from (−)-glyceraldehyde, (−)-glyceraldehyde had to be the *S* enantiomer. The assumption, therefore, that (+)-glyceraldehyde had the *R* configuration was right!

$$
\begin{array}{ccc}
\text{HC}=\text{O} & & \text{COOH} \\
\text{HO}\!-\!\text{H} & \xrightarrow{\text{several steps}} & \text{H}\!-\!\text{OH} \\
\text{CH}_2\text{OH} & & \text{HO}\!-\!\text{H} \\
& & \text{COOH}
\end{array}
$$

(−)-glyceraldehyde   (+)-tartaric acid

Eilhardt Mitscherlich (1794–1863), a German chemist, first studied medicine so he could travel to Asia—a way to satisfy his interest in Oriental languages. He later became fascinated by chemistry. He was a professor of chemistry at the University of Berlin and wrote a successful chemistry textbook that was published in 1829.

The work of these chemists immediately provided absolute configurations for all those compounds whose relative configurations had been determined by relating them to (+)- or (−)-glyceraldehyde. Thus, (−)-lactic acid has the configuration shown above. If (+)-glyceraldehyde had been the *S* enantiomer, (−)-lactic acid would have had the opposite configuration.

PROBLEM 35 ◆

What is the absolute configuration of:

a. (−)-glyceric acid?   **b.** (+)-isoserine?   **c.** (−)-glyceraldehyde?

4.13 SEPARATION OF ENANTIOMERS

Enantiomers cannot be separated by the usual separation techniques such as fractional distillation or crystallization because their identical boiling points and solubilities cause them to distill or crystallize simultaneously. Louis Pasteur was the first to separate a pair of enantiomers successfully. While working with crystals of sodium ammonium tartrate, he noted that the crystals were not identical. Some of the crystals were "right-handed" and some were "left-handed," so he painstakingly separated the two kinds of crystals with a pair of tweezers. He found that a solution of the "right-handed" crystals rotated plane-polarized light in a clockwise direction, while a solution of the "left-handed" crystals rotated plane-polarized light in a counterclockwise direction.

Pasteur was only 26 years old at the time and was unknown in scientific circles. He was concerned about the accuracy of his observations because a few years earlier, the well-known German organic chemist Eilhardt Mitscherlich had reported that crystals of the same salt were all identical. Pasteur immediately reported his findings to Jean-Baptiste Biot and repeated the experiment with Biot present. Biot was convinced that Pasteur had successfully separated sodium ammonium tartrate into a pair of enantiomers. Pasteur's experiment also created a new chemical term. Tartaric acid is obtained from grapes and thus tartaric acid was also called racemic acid because the term *racemus* is Latin for "a bunch of grapes." When Pasteur found that tartaric acid was actually a mixture of enantiomers, he called it a "racemic mixture." Separation of enantiomers is called the **resolution of a racemic mixture.**

Later, chemists recognized how lucky Pasteur had been. Sodium ammonium tartrate forms asymmetric crystals only under certain conditions—precisely the conditions that Pasteur had employed. Under other conditions, the symmetrical crystals that had fooled Mitscherlich are formed. But to quote Pasteur, "Chance favors the prepared mind."

Separating enantiomers by hand, as Pasteur did, is not a universally useful method of resolving a racemic mixture because few compounds form asymmetric crystals. A more commonly used method is to convert the enantiomers into diastereomers. Diastereomers can be separated because they have different physical properties. After separation, the individual diastereomers are converted back into the original enantiomers.

For example, because an acid reacts with a base to form a salt, a racemic mixture of a carboxylic acid reacts with a naturally occurring optically pure (a single enantiomer) base to form two diastereomeric salts. Morphine, strychnine, and brucine are naturally occurring chiral bases commonly used for this purpose. The chiral base exists as a single enantiomer because when a chiral compound is synthesized in a living system, generally only one enantiomer is formed (Section 4.18). When an *R* acid reacts with an *S* base, an *R,S* salt will be formed; when an *S* acid reacts with an *S* base, an *S,S* salt will be formed.

*The French chemist and microbiologist **Louis Pasteur (1822–1895)** was the first to demonstrate that microbes cause specific diseases. He developed an antitoxin against rabies. Asked by the French wine industry to find out why wine often went sour while aging, he showed that microorganisms cause grape juice to ferment, producing wine, and cause wine to slowly become sour. Gently heating the wine after fermentation, a process called pasteurization, kills the organisms so they cannot sour the wine.*

COO⁻ Na⁺
H——OH
HO——H
COO⁻ NH₄⁺

sodium ammonium tartrate
right-handed crystals

COO⁻ Na⁺
HO——H
H——OH
COO⁻ NH₄⁺

sodium ammonium tartrate
left-handed crystals

Crystals of potassium hydrogen tartrate that have precipitated on the inner surface of the cork from a bottle of wine. Grapes are unusual in that they produce large quantities of tartaric acid, whereas most fruits produce citric acid.

One of the chirality centers in the *R,S* salt is identical to a chirality center in the *S,S* salt, and the other chirality center in the *R,S* salt is the mirror image of a chirality center in the *S,S* salt. Therefore, the salts are diastereomers and have different physical properties. They can be separated by fractional crystallization. After separation they can be converted back into the carboxylic acids by adding a strong acid such as HCl. The chiral base can be separated from the carboxylic acid and used again.

Enantiomers can also be separated by a technique called **chromatography.** In this method, the mixture to be separated is dissolved in a solvent and the solvent is then passed through a column packed with material that tends to adsorb organic compounds. If the chromatographic column is packed with *chiral* material, the two enantiomers can be expected to move through the column at different rates because they will have different affinities for the chiral material (a right hand prefers a right-hand glove), so one enantiomer will emerge from the column before the other.

THE ENANTIOMERS OF THALIDOMIDE

About half the commercially available drugs have one or more chirality centers. Most of the drugs obtained from natural sources are single enantiomers. When drugs are synthesized in the laboratory, however, they are usually obtained as racemic mixtures. With few exceptions, these drugs have been marketed as racemic mixtures because of the high cost of separating the enantiomers. Enantiomers can have the same physiological activities, different degrees of the same activity, or very different activities. Thalidomide was approved as a sedative for use in Europe and Canada in 1956. It was not used in the United States because some neurological side effects had been noted. The dextrorotatory isomer had stronger sedative properties, but the commercial drug was a racemic mixture. However, it wasn't recognized that the levorotatory isomer was highly teratogenic, until it was noticed that women who were given the drug during the first three months of pregnancy gave birth to babies with a wide variety of birth defects. It was eventually determined that the dextrorotatory isomer also had mild teratogenic activity and that the enantiomers racemized in vivo. Thus, it is not clear whether giving those women only the dextrorotatory isomer would have decreased the severity of the birth defects. Thalidomide recently has been approved—with restrictions—to treat leprosy.

chirality center

thalidomide

CHIRAL DRUGS

In 1992, the Federal Drug Administration (FDA) issued a policy statement that encouraged drug companies to use recent advances in synthetic and separation techniques to develop single enantiomer drugs. Such drugs are becoming more common. (*S*)-(−)-Ketamine is four times more potent an anesthetic than (*R*)-(−)-ketamine and, even more important, the disturbing side effects appear to be associated only with the (*R*)-(−)-enantiomer. The active ingredient of ibuprofen, the popular analgesic marketed as Advil, Nuprin, and Motrin, resides primarily in the (*S*)-(+)-enantiomer. The FDA has some concern about approving the drug as a single enantiomer because of potential drug overdoses. Can people who are used to taking two pills be convinced to take only one? It has recently been found that heroin addicts can be maintained with (−)-α-acetylmethadol for a 72-hour period compared to 24 hours with racemic methadone. This means less frequent visits to the clinic, and a single dose can get an addict through an entire weekend. The FDA may require drug companies that develop a drug as a single enantiomer rather than as the already approved racemic mixture to show that the missing enantiomer does not provide some type of beneficial effect.

If a carbon is bonded to two hydrogens and to two different groups, the two hydrogens are called **enantiotopic hydrogens.** For example, the two hydrogens (H_a and H_b) in the CH_2 group of ethanol are enantiotopic hydrogens because the other two groups bonded to the carbon (CH_3 and OH) are not identical. They are called enantiotopic hydrogens because replacing one of them by a deuterium (or any other atom or group other than CH_3 or OH) would make the compound an enantiomer. The two hydrogens (H_a and H_b) in the CH_2 group of propane are not enantiotopic hydrogens because the other two groups bonded to the carbon (CH_3 and CH_3) are identical. The H_a and H_b hydrogens of propane are called **homotopic hydrogens.**

**4.14
ENANTIOTOPIC
HYDROGENS,
DIASTEREOTOPIC
HYDROGENS, AND
PROCHIRALITY
CENTERS**

$$CH_3-\underset{\underset{H_b}{|}}{\overset{\overset{H_a}{|}}{C}}-OH \qquad\qquad CH_3-\underset{\underset{H_b}{|}}{\overset{\overset{H_a}{|}}{C}}-CH_3$$

H_a and H_b	H_a and H_b
are enantiotopic hydrogens	are homotopic hydrogens

If one of the enantiotopic hydrogens in ethanol were replaced by a deuterium, the carbon to which the enantiotopic hydrogens are attached would become a chirality center. If the H_a hydrogen were replaced by a deuterium, the chirality center would have the R configuration. Thus, the H_a hydrogen is called the **pro-R-hydrogen.** The H_b hydrogen is called the **pro-S-hydrogen** because if it were replaced by a deuterium, the chirality center would have the S configuration.

$$CH_3-\underset{\underset{H_b}{|}}{\overset{\overset{H_a}{|}}{}}-OH$$

$H_a \longrightarrow \boxed{\text{pro-}R\text{-hydrogen}}$

$H_b \longrightarrow \boxed{\text{pro-}S\text{-hydrogen}}$

The carbon to which the enantiotopic hydrogens are attached is called a **prochirality center** because it would become a chirality center if one of the hydrogens were replaced by a deuterium (or any group other than CH_3 or OH) because four different groups would then be bonded to the carbon. The molecule containing the prochirality center is called a prochiral molecule because it would become a chiral molecule if one of the hydrogens were replaced.

The pro-R- and pro-S-hydrogens are chemically equivalent, so they have the same chemical reactivity and cannot be distinguished by achiral chemical reagents. For example, when ethanol is oxidized (by PCC) to acetaldehyde, one of the enantiotopic hydrogens is removed. Because the two hydrogens are chemically equivalent, half the product results from removing the H_a hydrogen and the other half results from removing the H_b hydrogen.

$$CH_3-\underset{\underset{H_b}{|}}{\overset{\overset{H_a}{|}}{C}}-OH \xrightarrow[\text{pyridine}]{\textbf{PCC}} CH_3-\overset{\overset{O}{\|}}{C}-H_b \; + \; CH_3-\overset{\overset{O}{\|}}{C}-H_a$$

$$ \qquad\qquad\qquad 50\% \qquad\qquad 50\%$$

Enantiotopic hydrogens, however, are not chemically equivalent in enzyme-catalyzed reactions. An enzyme can distinguish between them because an enzyme is chiral (Section 4.11). For example, when the oxidation of ethanol to acetaldehyde is catalyzed by the enzyme alcohol dehydrogenase, only the H_a hydrogen is removed.

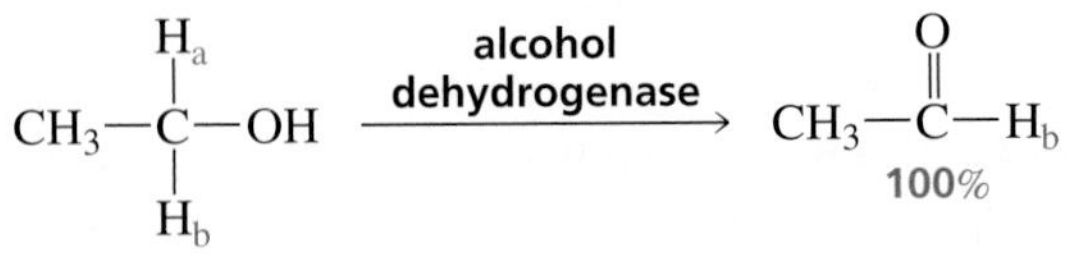

$$CH_3-\underset{\underset{H_b}{|}}{\overset{\overset{H_a}{|}}{C}}-OH \xrightarrow[\text{dehydrogenase}]{\text{alcohol}} CH_3-\overset{\overset{O}{||}}{C}-H_b$$
100%

If a carbon is bonded to two hydrogens and replacing each of them in turn with deuterium (or another group) creates a pair of diastereomers, the hydrogens are called **diastereotopic hydrogens.**

Tutorial: Common terms in stereochemistry

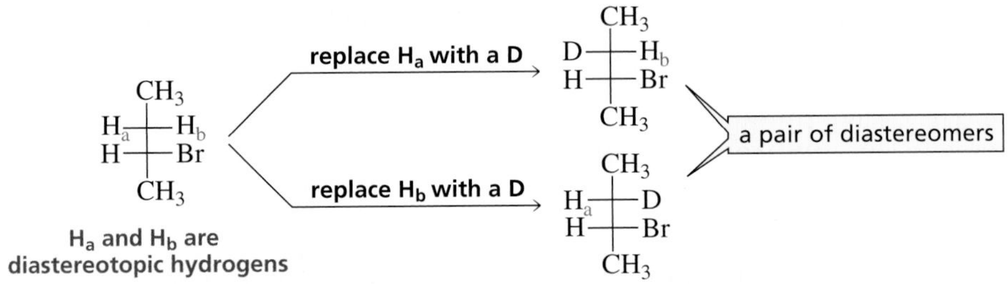

replace H$_a$ with a D

replace H$_b$ with a D

H$_a$ and H$_b$ are diastereotopic hydrogens

a pair of diastereomers

Unlike enantiotopic hydrogens, diastereotopic hydrogens do not have the same reactivity with achiral reagents. For example, in Chapter 10 we will see that because *trans*-2-butene is more stable than *cis*-2-butene (Section 3.19), removal of H$_b$ and Br to form *trans*-2-butene occurs faster than removal of H$_a$ and Br to form *cis*-2-butene.

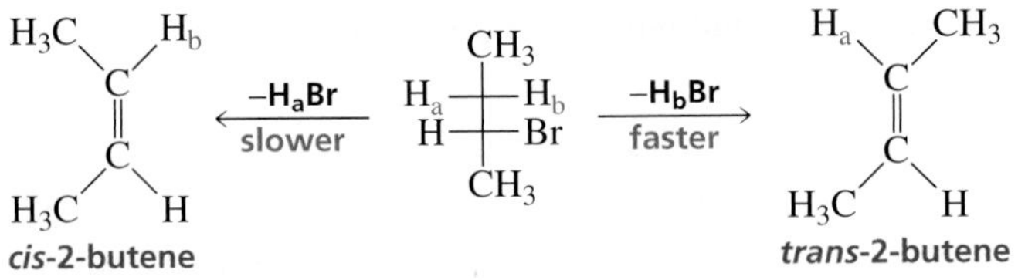

$-H_a Br$
slower

$-H_b Br$
faster

cis-2-butene

trans-2-butene

PROBLEM 36 ◆

Tell whether the H$_a$ and H$_b$ hydrogens in each of the following compounds are homotopic, enantiotopic, or diastereotopic.

a.

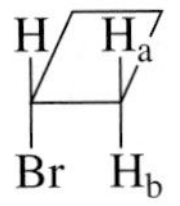

b.

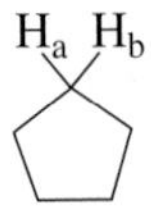

c. 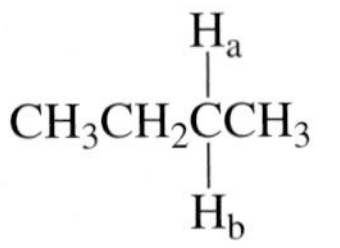 CH$_3$CH$_2$CCH$_3$

d. 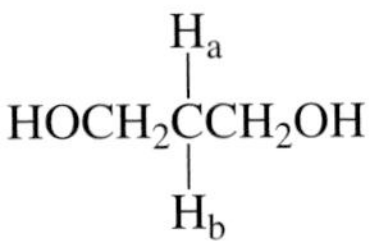 HOCH$_2$CCH$_2$OH

e.

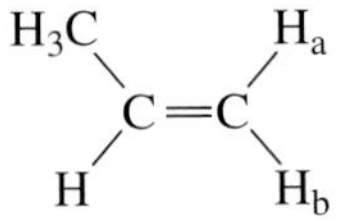

f.

(CH$_3$)$_3$C

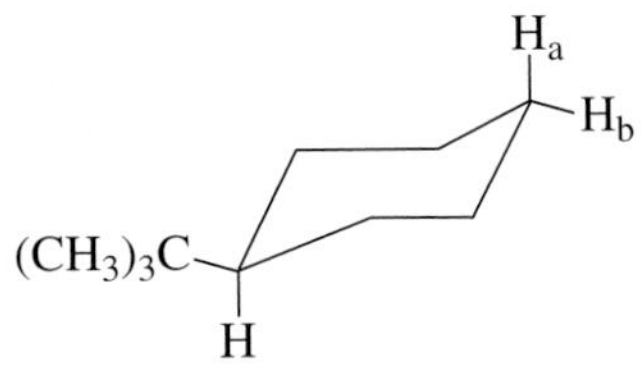

4.15
NITROGEN AND PHOSPHORUS CHIRALITY CENTERS

Atoms other than carbon can be chirality centers. When an atom such as nitrogen or phosphorus has four different groups or atoms attached to it (i.e., it has a tetrahedral geometry), it is a chirality center. These kinds of compounds can exist as a pair of enantiomers, and the enantiomers can be separated.

$$
\begin{array}{cc}
\underset{\underset{\text{CH}_3\text{CH}_2\text{CH}_2}{\big|}}{\overset{\text{CH}_3}{\big|}}\text{Br}^-\ \ \overset{+}{\text{N}}\cdots\text{H} & \underset{\text{CH}_3\text{CH}_2}{\overset{\text{CH}_3}{\big|}}\ \text{H}\cdots\overset{+}{\text{N}}\ \ \text{Br}^-
\end{array}
$$

CH$_3$CH$_2$CH$_2$ CH$_2$CH$_3$ CH$_3$CH$_2$ CH$_2$CH$_2$CH$_3$

a pair of enantiomers

$$
\underset{\text{CH}_3\text{CH}_2\text{O}}{\overset{\displaystyle O}{\underset{\displaystyle \|\,}{P}}}\cdots\text{H} \qquad \underset{\text{CH}_3\text{O}}{\overset{\displaystyle O}{\|}}\,\text{OCH}_3
$$

CH$_3$CH$_2$O   OCH$_3$ CH$_3$O   OCH$_2$CH$_3$

a pair of enantiomers

If one of the four "groups" attached to nitrogen is a pair of nonbonding electrons, the enantiomers cannot be separated because rapid inversion between the two enantiomers takes place at room temperature (Section 4.1).

PROBLEM 37

Compound I has two stereoisomers, but compounds II and III exist as single compounds. Explain.

$$
\begin{array}{ccc}
\underset{\underset{\text{CH}_2\text{CH}_3}{\big|}}{\overset{\text{CH}=\text{CH}_2}{\big|}}\text{CH}_3-\overset{+}{\text{N}}-\text{H}\ \ \text{Cl}^- &
\underset{\underset{\text{CH}_3}{\big|}}{\overset{\text{CH}=\text{CH}_2}{\big|}}\text{CH}_3-\overset{+}{\text{N}}-\text{H}\ \ \text{Cl}^- &
\underset{\underset{}{}}{\overset{\text{CH}=\text{CH}_2}{\big|}}\text{CH}_3-\overset{\cdot\cdot}{\text{N}}-\text{H}
\end{array}
$$

I   II   III

In Chapter 3, you learned that alkenes undergo addition reactions. We looked at the different kinds of reagents that add to alkenes, we examined the step-by-step process by which the reactions occur (the reaction mechanism), and we determined what products are formed. However, we did not consider the stereochemistry of the reactions. In other words, we did not determine which stereoisomers are formed.

Stereochemistry is the field of chemistry that deals with the structures of molecules in three dimensions. When we study the stereochemistry of a reaction, we are concerned with the following questions:

1. If stereoisomers are possible for a reaction product, does the reaction produce a single stereoisomer, a set of particular stereoisomers, or all possible stereoisomers?

2. If stereoisomers are possible for the reactant, do all stereoisomers react to form the same stereoisomeric product, or does each reactant form a different stereoisomer or a different set of stereoisomers?

Before we examine the stereochemistry of addition reactions, we need to become familiar with some terms used in describing the stereochemistry of a reaction.

We learned in Section 3.12 that a **regioselective** reaction is one in which two *constitutional isomers* can be obtained as products but more of one is obtained than of the other. A regioselective reaction selects for a particular constitutional isomer. Recall that a reaction can be *moderately regioselective, highly regioselective,* or *completely regioselective* depending on the relative amounts of the constitutional isomers formed in the reaction.

a regioselective reaction

$$
A \longrightarrow B + C
$$

more B is formed than C where B and C are constitutional isomers

**4.16
STEREOCHEMISTRY
OF REACTIONS:
REGIOSELECTIVE,
STEREOSELECTIVE,
AND
STEREOSPECIFIC
REACTIONS**

A regioselective reaction forms more of one constitutional isomer than of another.

Stereoselective is a similar term, but it refers to the preferential formation of a *stereoisomer* rather than a *constitutional isomer.* If a reaction that generates a carbon–carbon double bond or a chirality center in a product leads to the preferential formation of one stereoisomer over another, it is a stereoselective reaction. In other words, it selects for a particular stereoisomer. Depending on the degree of preference for a particular stereoisomer, a reaction can be described as being *moderately stereoselective, highly stereoselective,* or *completely stereoselective.*

A stereoselective reaction forms more of one stereoisomer than of another.

a stereoselective reaction

$$A \longrightarrow B + C$$

more B is formed than C where B and C are stereoisomers

In a stereospecific reaction, each stereoisomeric reactant forms a different stereoisomeric product or a different set of stereoisomeric products.

A reaction is **stereospecific** if the reactant can exist as stereoisomers and each stereoisomeric reactant leads to a different stereoisomeric product or a different set of stereoisomeric products.

a stereospecific reaction

$$A \longrightarrow B$$

$$C \longrightarrow D$$

A and C are stereoisomers
B and D are stereoisomers

A stereospecific reaction is also stereoselective. A stereoselective reaction is not necessarily stereospecific.

In the preceding reaction, stereoisomer A forms stereoisomer B but does not form D, so the reaction is stereoselective in addition to being stereospecific. *All stereospecific reactions, therefore, are also stereoselective. All stereoselective reactions are not stereospecific,* however, because there are stereoselective reactions in which the reactant does not have a carbon–carbon double bond or a chirality center, so it cannot exist as stereoisomers.

4.17 STEREOCHEMISTRY OF ALKENE ADDITION REACTIONS

Now that you are familiar with addition reactions and with stereoisomers, we will combine the two topics and take a look at the stereoisomers that are formed in an addition reaction.

In Chapter 3, we saw that when an alkene reacts with an electrophilic reagent such as HBr, the major product of the addition reaction can be predicted by Markovnikov's rule: The electrophile (H^+) adds to the sp^2 carbon bonded to the most hydrogens and the nucleophile (Br^-) adds to the other sp^2 carbon. For example, the major product obtained from the reaction of propene with HBr is 2-bromopropane. This particular product does not have stereoisomers. Therefore, we do not have to be concerned with the stereochemistry of this reaction.

$$CH_3CH=CH_2 \xrightarrow{\text{H}^+} CH_3\overset{+}{C}HCH_3 \xrightarrow{\text{Br}^-} CH_3CHCH_3$$

propene

$$\underset{\underset{\text{Br}}{|}}{CH_3CHCH_3}$$

2-bromopropane
major product

If, however, the reaction creates a product with a chirality center, we need to know which stereoisomers are formed. For example, the reaction of HBr with 1-butene forms 2-bromobutane, a compound with a chirality center. What is the configuration of the product? Do we get the R enantiomer, the S enantiomer, or both?

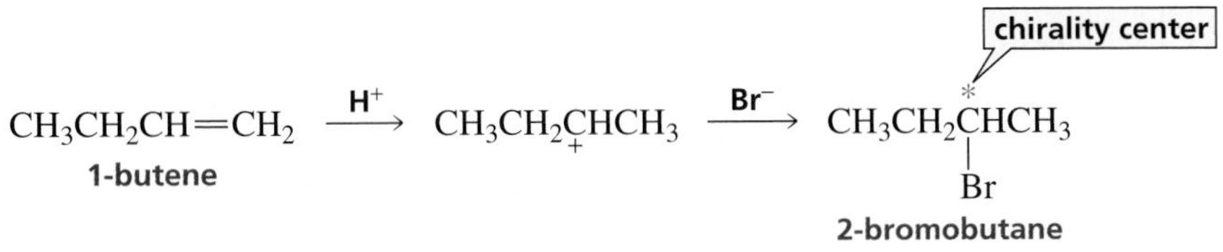

chirality center

$$CH_3CH_2CH = CH_2 \xrightarrow{H^+} CH_3CH_2\overset{+}{C}HCH_3 \xrightarrow{Br^-} CH_3CH_2\overset{*}{C}HCH_3$$

1-butene

$$\underset{\text{Br}}{|}$$

2-bromobutane

In discussing the stereochemistry of addition reactions, we will look first at reactions that form a product with one chirality center. Then we will look at reactions that form a product with two chirality centers.

Addition Reactions That Form One Chirality Center

When a reactant that does not have a chirality center undergoes an addition reaction that forms a product with *one* chirality center, the product will be a racemic mixture. For example, the reaction of 1-butene with HBr forms identical amounts of (R)-2-bromobutane and (S)-2-bromobutane. Thus, an addition reaction that forms a compound with one chirality center from a reactant without any chirality centers is not stereoselective because it does not select for a particular stereoisomer.

We can understand why a racemic mixture is obtained if we examine the structure of the carbocation formed in the first step of the reaction. The positively charged carbon is sp^2 hybridized, so the three atoms to which it is bonded lie in a plane (Section 1.10). When the bromide ion attacks the carbocation from above the plane, one enantiomer is formed, but when it attacks the carbocation from below the plane, the other enantiomer is formed. Because the bromide ion can attack the planar carbocation intermediate from above just as well as it can attack it from below, identical amounts of the R and S enantiomers are obtained from the addition reaction.

Tutorial: Review additon
of HBr (eChapter)

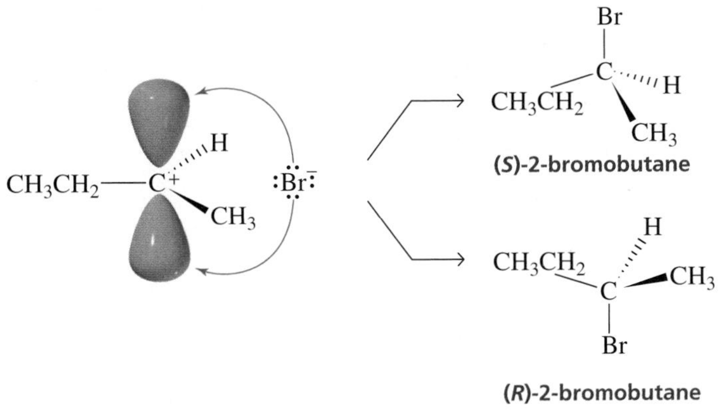

(S)-2-bromobutane

(R)-2-bromobutane

When HBr adds to 2-methyl-1-butene in the presence of peroxide, the product has one chirality center. Therefore, we can predict that identical amounts of the *R* and *S* enantiomers are obtained.

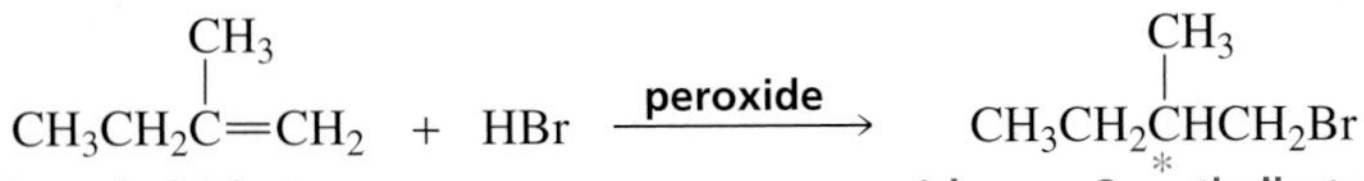

$$CH_3CH_2\overset{\underset{|}{CH_3}}{C}{=}CH_2 \;+\; HBr \;\xrightarrow{\text{peroxide}}\; CH_3CH_2\overset{\underset{|}{CH_3}}{\underset{*}{C}}HCH_2Br$$

2-methyl-1-butene   **1-bromo-2-methylbutane**

The product is a racemic mixture because the carbon in the radical intermediate that bears the unpaired electron is sp^2 hybridized. This means that the three atoms bonded to it are all in the same plane (Section 1.10). Consequently, the *R* enantiomer and the *S* enantiomer will be obtained in identical amounts because HBr has equal access to both sides of the radical. (Although the unpaired electron is shown to be in the top lobe of the *p* orbital, it really is spread out equally through both lobes.)

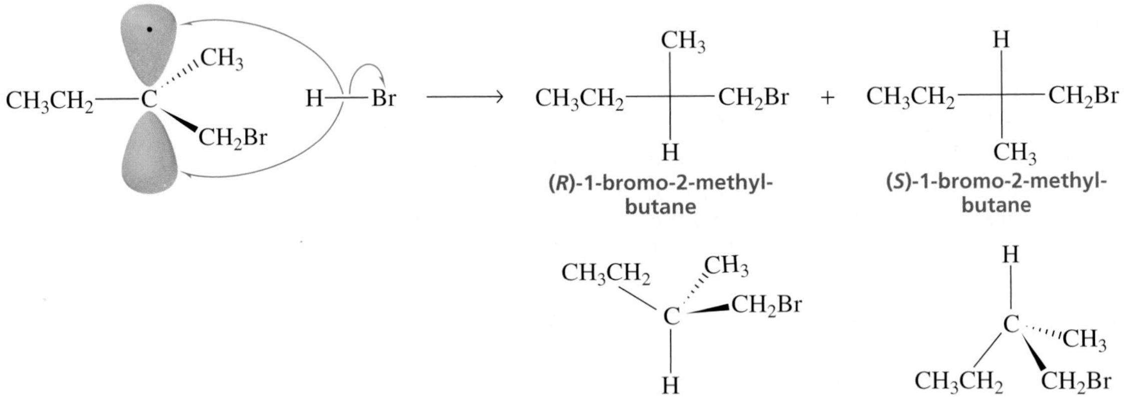

(R)-1-bromo-2-methyl-
butane

(S)-1-bromo-2-methyl-
butane

PROBLEM 39

What stereoisomers are obtained in each of the following reactions?

a. $CH_3CH_2CH_2CH{=}CH_2 + HCl \longrightarrow$

b. 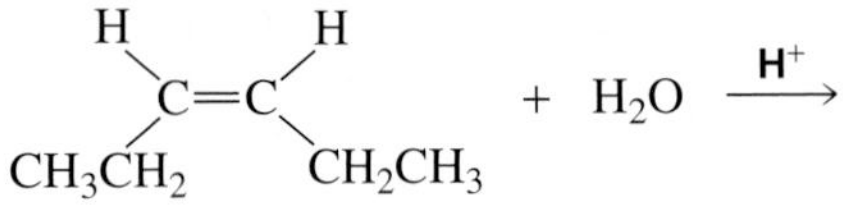 $+\; H_2O \xrightarrow{H^+}$

c. 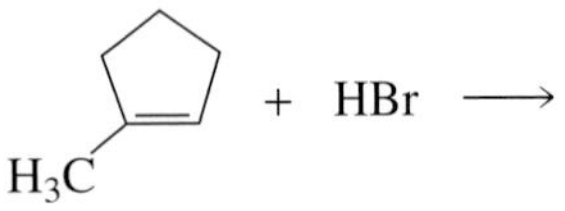 $+\; HBr \longrightarrow$

d. 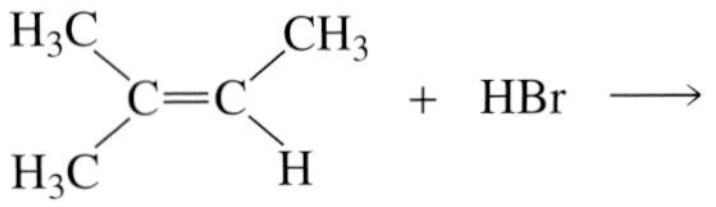 $+\; HBr \longrightarrow$

e. 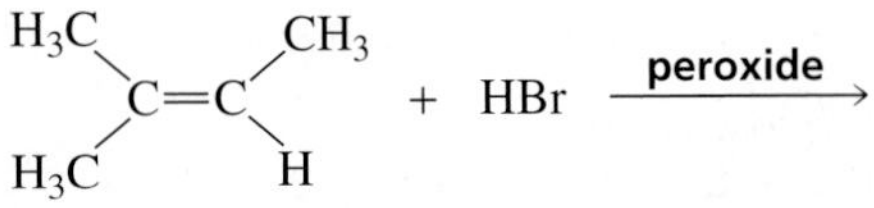 $+\; HBr \xrightarrow{\text{peroxide}}$

f.

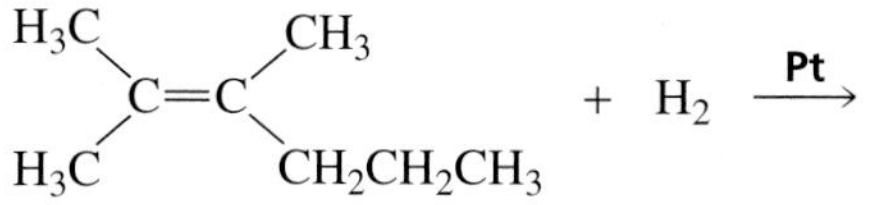

$$\underset{H_3C}{\overset{H_3C}{\diagdown}}C=C\underset{CH_2CH_2CH_3}{\overset{CH_3}{\diagup}} \quad + \quad H_2 \quad \xrightarrow{\textbf{Pt}}$$

If an addition reaction creates a chirality center in a compound that already has a chirality center, a pair of diastereomers will be formed. Because none of the bonds to the chirality center in (R)-3-chloro-1-butene is broken during the addition of HBr, the configuration of this chirality center does not change. Because the bromide ion can attack the planar carbocation intermediate from either the top or the bottom in the process of creating the new chirality center, two stereoisomers result. The stereoisomers are diastereomers because one of the chirality centers has the same configuration in both isomers and the other has opposite configurations in the two isomers.

stereochemistry of the product

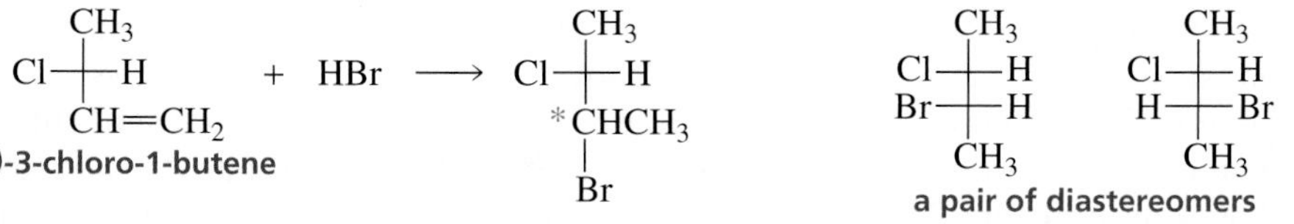

(R)-3-chloro-1-butene

a pair of diastereomers

The carbocation intermediate formed in the preceding reaction is chiral and therefore does not have a plane of symmetry. If the chirality center is near the positively charged carbon, one face of the carbocation will be more sterically hindered than the other. The incoming bromide ion will have greater access to the less sterically hindered face. Therefore, different amounts of the two diastereomers will be formed. The reaction is stereoselective—more of one stereoisomer is formed than of the other.

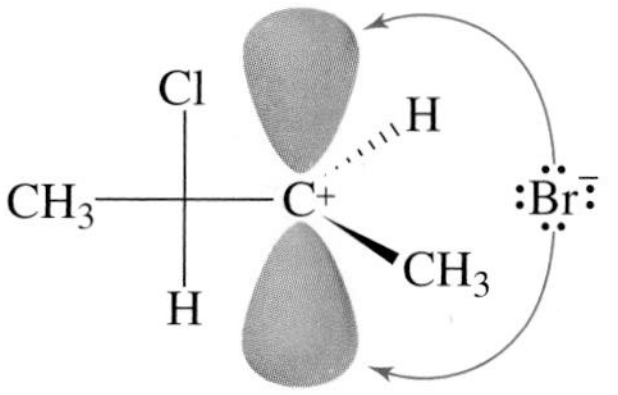

Addition Reactions That Form Products With Two Chirality Centers

When a reactant that does not have a chirality center undergoes an addition reaction that forms a product with *two* chirality centers, the stereoisomers that are formed depend on the mechanism of the addition reaction.

1. Addition Reactions That Form a Carbocation Intermediate or a Radical Intermediate

If two chirality centers are created as the result of an addition reaction that forms a carbocation intermediate, four stereoisomers can be obtained as products.

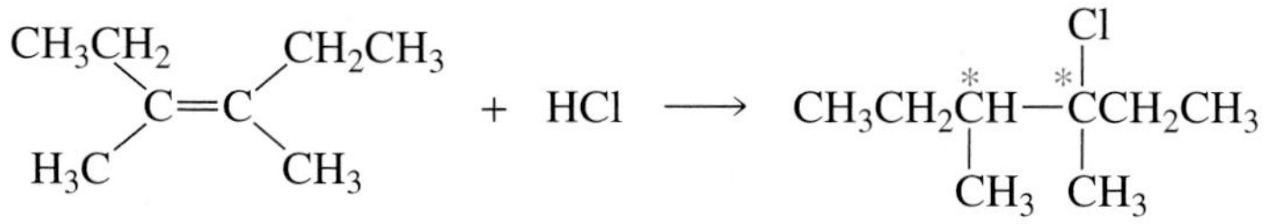

$$\underset{H_3C}{\overset{CH_3CH_2}{\diagdown}}C=C\underset{CH_3}{\overset{CH_2CH_3}{\diagup}} \quad + \quad HCl \quad \longrightarrow \quad \underset{\underset{CH_3\ CH_3}{|\quad\ |}}{CH_3CH_2\overset{*}{C}H-\overset{\overset{Cl}{|}}{\overset{*}{C}}CH_2CH_3}$$

cis-3,4-dimethyl-3-hexene

3-chloro-3,4-dimethylhexane

stereochemistry of the product

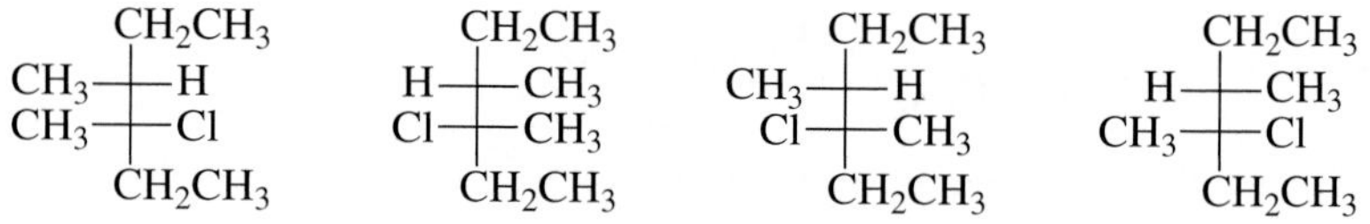

Fischer projections of the stereoisomers of the product

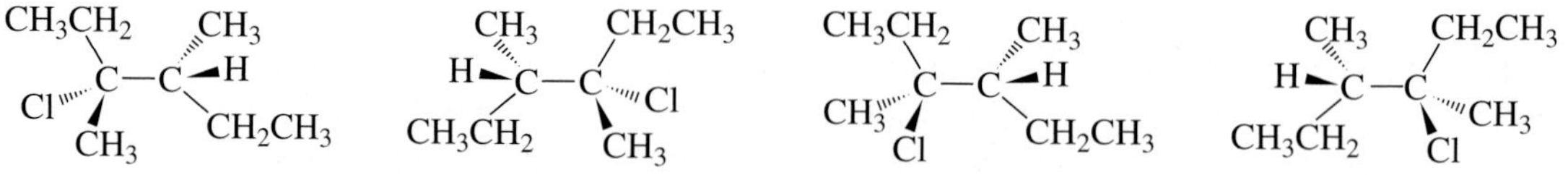

perspective formulas of the stereoisomers of the product

In the first step of the reaction, the proton can approach the plane containing the doubly bonded carbons of the alkene from above or below to form the carbocation. Once the carbocation is formed, the chloride ion can attack from above or below. As a result, four stereoisomers are obtained as products: The two substituents can add from above-above, above-below, below-above, or below-below. When the two substituents add to the same side of the double bond, the addition is called **syn addition.** When the two substituents add to opposite sides of the double bond, the addition is called **anti addition.** Both syn and anti addition occur in alkene addition reactions that take place by way of a carbocation intermediate. Because the four stereoisomers formed by the cis alkene are identical to the four stereoisomers formed by the trans alkene, the reaction is not stereospecific.

Similarly, if two chirality centers are created as the result of an addition reaction that forms a radical intermediate, four stereoisomers can be formed because both syn and anti addition are possible. And because the stereoisomers formed by the cis isomer are identical to those formed by the trans isomer, the reaction is not stereospecific.

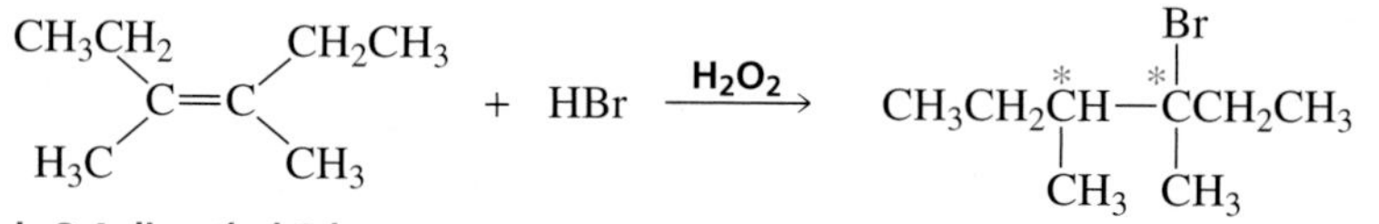

$$\text{cis-3,4-dimethyl-3-hexene} + HBr \xrightarrow{H_2O_2} \text{3-bromo-3,4-dimethylhexane}$$

cis-3,4-dimethyl-3-hexene

3-bromo-3,4-dimethylhexane

stereochemistry of the product

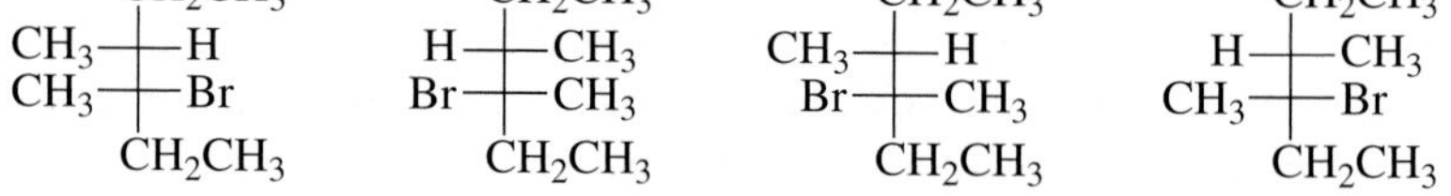

Fischer projections of the stereoisomers of the product

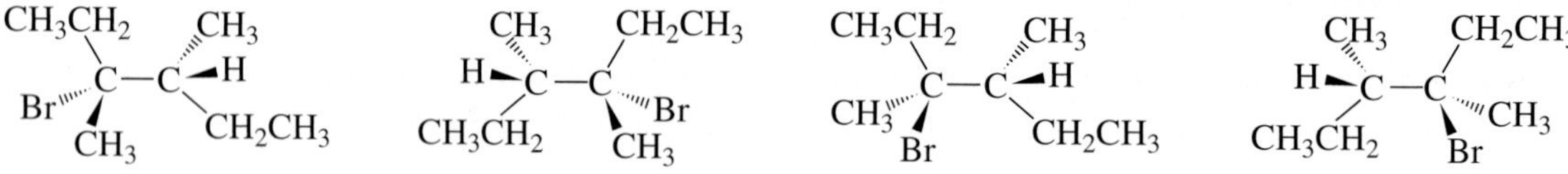

perspective formulas of the stereoisomers of the product

2. Stereochemistry of Hydrogen Addition

When hydrogen (H_2) adds to an alkene, both hydrogen atoms add to the same side of the double bond (Section 3.19), which makes the addition of H_2 a syn addition reaction.

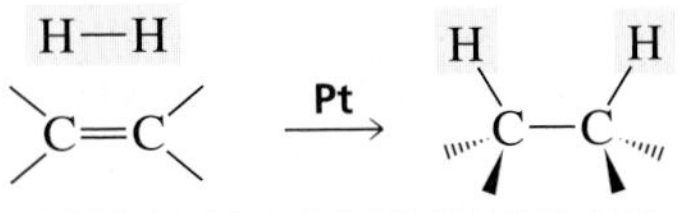

addition of H$_2$ is a syn addition

If addition of hydrogen to an alkene forms a product with two chirality centers, only two of the four possible stereoisomers are obtained because only syn addition can occur. (The other two stereoisomers would have to come from anti addition.) One stereoisomer results from addition of both hydrogens from above the plane of the double bond, and the other stereoisomer results from addition of both hydrogens from below the plane. The particular pair of stereoisomers that is formed depends on whether the reactant is a cis alkene or a trans alkene. Syn addition of H$_2$ to a cis alkene forms only the erythro pair of enantiomers. (In Section 4.8, we saw that the erythro pair of enantiomers is the pair with identical groups on the same side of the carbon chain in the Fischer projections.) Thus, addition of hydrogen is stereoselective and stereospecific.

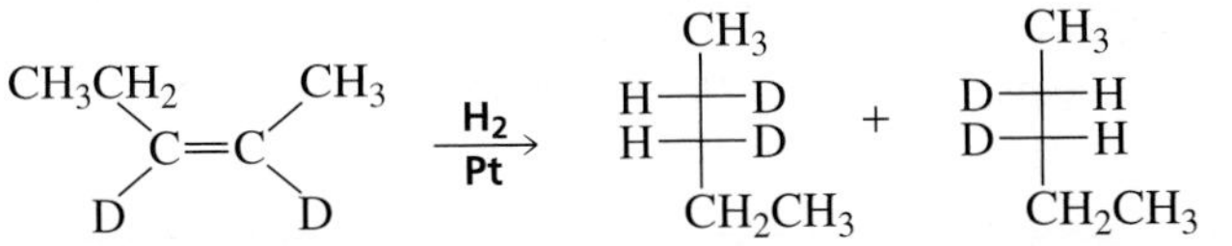

cis-2,3-dideuterio-
2-pentene

erythro enantiomers

Fischer projections of the products

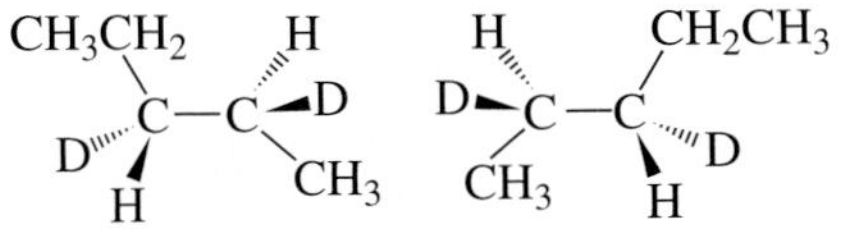

perspective formulas of the products

If the two chirality centers have the same set of substituents, a meso compound will be obtained instead of the pair of erythro enantiomers.

Syn addition of H$_2$ to a trans alkene forms only the threo pair of enantiomers. Thus the addition of hydrogen is a stereospecific reaction—the product obtained from addition to the cis isomer is different from the product obtained from addition to the trans isomer.

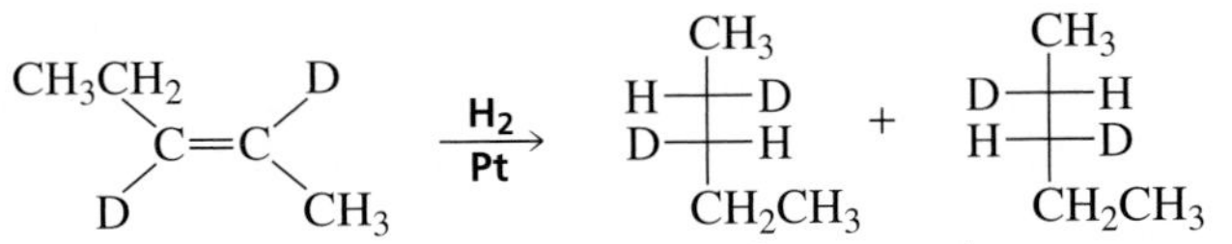

trans-2,3-dideuterio-
2-pentene

threo enantiomers

Fischer projections of the products

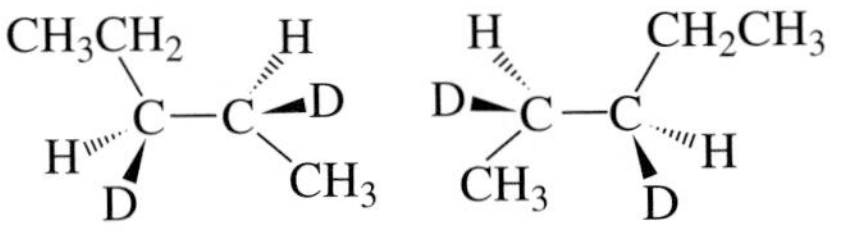

> If you have trouble determining the configuration of a product, make a model.

perspective formulas of the products

It is a little easier to see the outcome of syn addition with cyclic compounds. Because addition of H$_2$ is syn, only the cis enantiomers are formed.

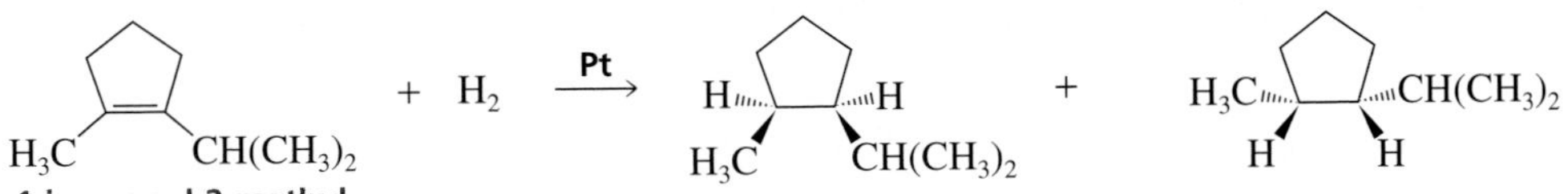

1-isopropyl-2-methyl-
cyclopentene

Syn addition of H_2 to 1,2-dideuteriocyclopentene forms only one product. Addition from above the plane of the double bond forms the same product as addition from below the plane. The product is a meso compound.

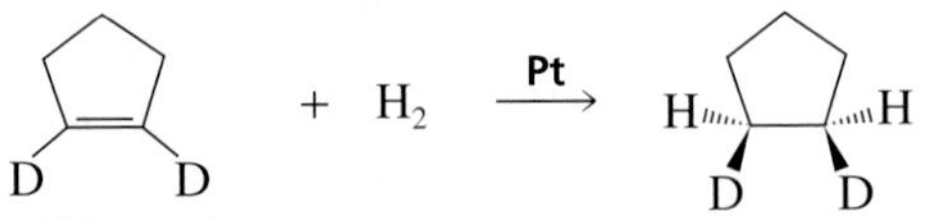

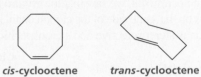

1,2-dideuterio-
cyclopentene

CONFIGURATION OF CYCLOALKENES

Cyclic alkenes with fewer than eight carbon atoms, such as cyclopentene and cyclohexene, can exist only in the cis configuration because they do not have enough carbons to incorporate a trans double bond. Therefore, it is not necessary to use the "cis" designation with their names. Both cis and trans isomers are possible for rings containing eight or more carbons, however, so the configuration of the compound must be specified in its name.

cis-cyclooctene   *trans*-cyclooctene

PROBLEM 40

a. What stereoisomers are formed in the following reaction?

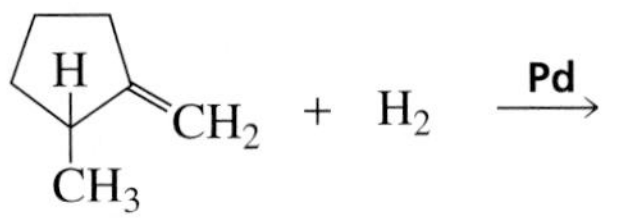

$$\text{(cyclopentane with } H, CH_3, CH_2 \text{)} \;+\; H_2 \xrightarrow{\text{Pd}}$$

b. What stereoisomer is formed in greater yield?

---

3. Stereochemistry of Hydroboration–Oxidation

You cannot predict the configuration of the product unless you know the mechanism of the reaction.

The addition of borane to an alkene is a concerted reaction. The boron and the hydride ion add to the two sp^2 carbons of the double bond at the same time (Section 3.17). Because the two species add simultaneously, they must add to the same side of the double bond. So the addition of borane to an alkene, like the addition of hydrogen, is a syn addition.

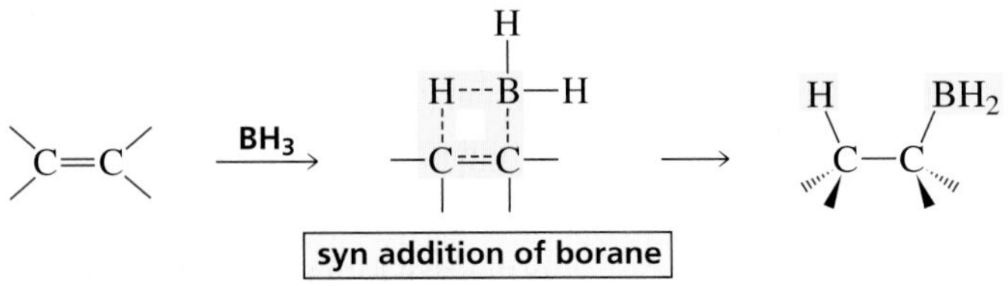

$$\text{C=C} \xrightarrow{\text{BH}_3} \text{C}\cdots\text{C} \text{ (H---B—H)} \longrightarrow \text{C—C (H, BH}_2)$$

syn addition of borane

When the alkylborane is oxidized by reaction with hydrogen peroxide and hydroxide ion, the OH group ends up in the same position as the boron group it replaces. Consequently, the overall hydroboration–oxidation reaction amounts to a syn addition of water to a carbon–carbon double bond.

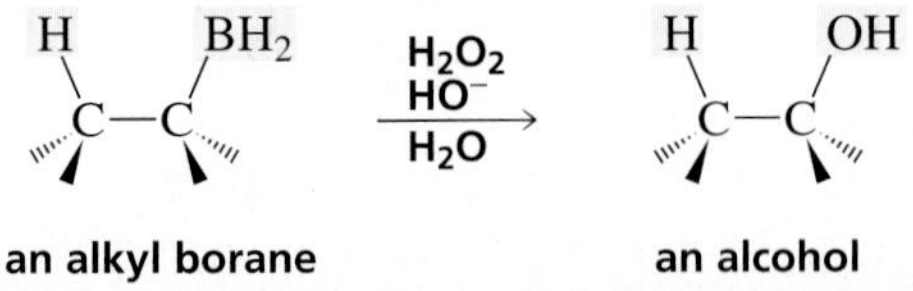

$$\text{(H, BH}_2) \xrightarrow[\text{H}_2\text{O}]{\substack{\text{H}_2\text{O}_2 \\ \text{HO}^-}} \text{(H, OH)}$$

an alkyl borane   an alcohol

Because only syn addition occurs, hydroboration–oxidation is stereoselective—only two of the four possible stereoisomers are formed. As we saw when we looked at the addition of H_2 to a cycloalkene, syn addition results in the formation of only the pair of enantiomers that has the added groups on the same side of the ring.

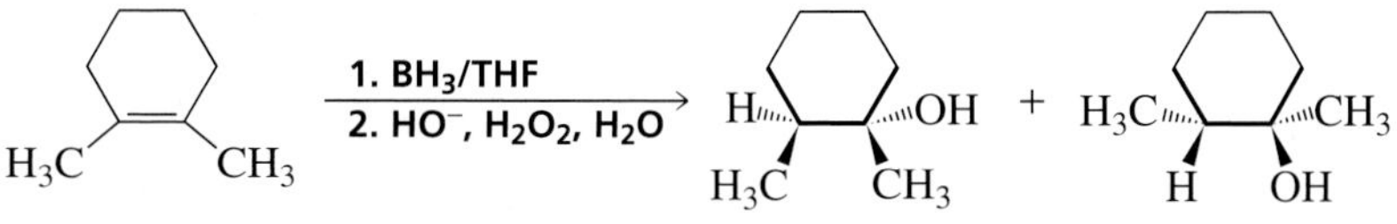

PROBLEM 41 ◆

Give the products that would be obtained from hydroboration–oxidation of the following compounds:

a. cyclohexene
b. 1-methylcyclohexene
c. 1,2-dimethylcyclopentene
d. *cis*-2-butene

PROBLEM 42

Reaction of 2-ethyl-1-pentene with Br_2, with H_2/Pt, or with BH_3 followed by $HO^- + H_2O_2$ leads to a racemic mixture. Explain why a racemic mixture is obtained in each case.

4. Addition Reactions That Form a Bromonium Ion Intermediate

If two chirality centers are created as the result of an addition reaction that forms a bromonium ion intermediate, only one pair of enantiomers will be formed. In other words, the addition of Br_2 is a stereoselective reaction. The addition of Br_2 to the cis alkene forms only the threo pair of enantiomers.

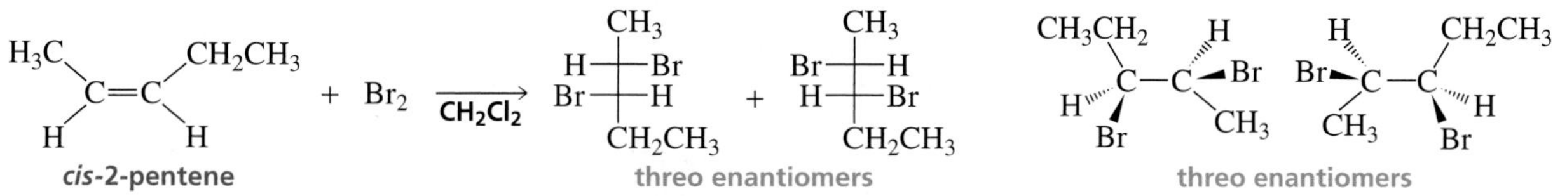

cis-2-pentene — threo enantiomers, Fischer projections — threo enantiomers, perspective formulas

Similarly, the addition of Br_2 to the trans alkene forms only the erythro pair of enantiomers. Because the cis and trans isomers form different products, the reaction is stereospecific as well as stereoselective.

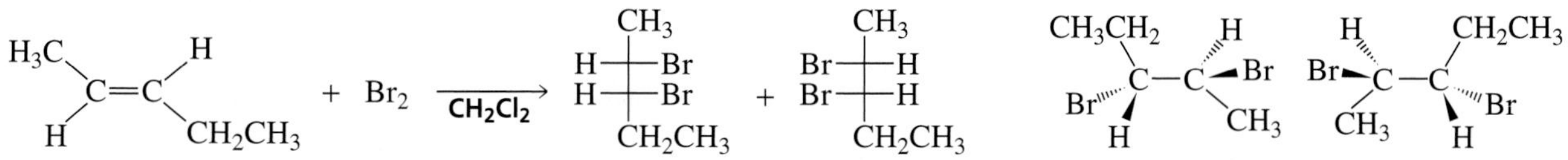

trans-2-pentene — erythro enantiomers, Fischer projections — erythro enantiomers, perspective formulas

Because the addition of Br_2 to the cis alkene forms the threo pair of enantiomers, we know that addition of Br_2 is an example of anti addition because syn addition to the cis alkene would have formed the erythro pair of enantiomers. The addition of Br_2 is anti because the reaction intermediate is a cyclic bromonium ion (Section 3.15). Once the bromonium ion is formed, the bridged bromine atom blocks that side of the ion from further attack. As a result, the negatively charged bromide ion must attack from the opposite side (following either the green arrows *or* the red arrows). Thus, the two bromine atoms add to opposite sides of the double bond. Because syn addition of Br_2 cannot occur, only two of the four possible stereoisomers are obtained.

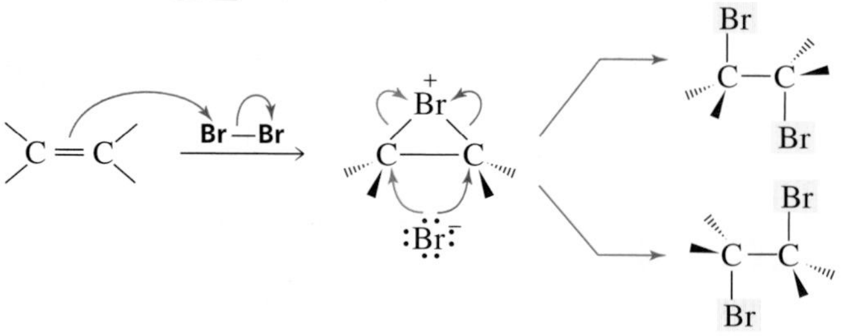

addition of Br_2 is an anti addition

The electrostatic potential map shows that the bromine atom is the most electron dense region in the bromonium ion even though it bears a formal positive charge.

Tutorial: Review halogenation (eChapter 3)

the bromonium ion formed
from the reaction of Br_2
with *cis*-2-butene

PROBLEM 43

How could you prove, using a sample of *trans*-2-butene, that addition of Br_2 forms a cyclic bromonium ion intermediate rather than a carbocation intermediate?

If the two chirality centers in the product each have the same four substituents, the erythro isomers are identical and constitute a meso compound. Therefore, addition of Br_2 to *trans*-2-butene forms a meso compound.

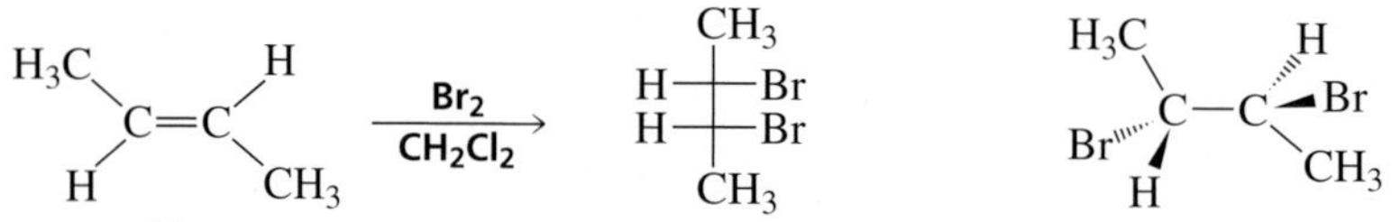

trans-2-butene   →   a meso compound
Fischer projection

a meso compound
perspective formula

Because only anti addition occurs, addition of Br_2 to a cyclic alkene forms only the pair of enantiomers that has the added bromine atoms on opposite sides of the ring.

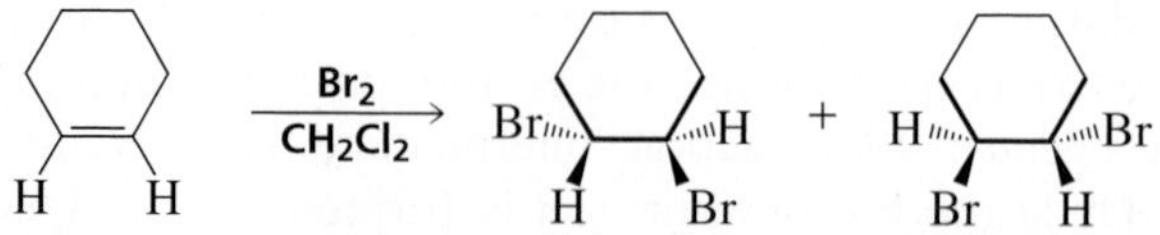

One way to determine what stereoisomers are obtained from a reaction that creates a product with two chirality centers is the mnemonic **"CIS-SYN-ERYTHRO,"** which is easy to remember because all three terms mean "on the same side." You can change any two of the terms, but you cannot change just one. (For example, **TRANS-ANTI-ERYTHRO, TRANS-SYN-THREO,** and **CIS-ANTI-THREO** are allowed, but TRANS-SYN-ERYTHRO is not allowed.)

A summary of the stereochemistry of the products obtained from addition reactions to alkenes is given in Table 4.2.

TABLE 4.2   Stereochemistry of Alkene Addition Reactions

Reaction	Type of addition	Products formed
Addition reactions that create one chirality center in the product		1. If the reactant does not have a chirality center, a pair of enantiomers will be obtained (equal amounts of the R and S isomers).
		2. If the reactant has a chirality center, unequal amounts of a pair of diastereomers will be obtained.
Addition reactions that create two chirality centers in the product		
Addition of reagents that form a carbocation or radical intermediate	syn and anti	Four stereoisomers can be obtained[a] (the *cis* and *trans* isomers form the same products)
Addition of H_2	syn	cis $\longrightarrow$ erythro enantiomers[a]
Addition of borane		trans $\longrightarrow$ threo enantiomers
Addition of Br_2	anti	cis $\longrightarrow$ threo enantiomers
		trans $\longrightarrow$ erythro enantiomers[a]

[a]If the two chirality centers have the same substituents, a meso compound will be obtained instead of the pair of erythro enantiomers.

PROBLEM-SOLVING STRATEGY

Give the configuration of the products obtained from the following reactions:

a. 1-butene + HCl

b. 1-butene + HBr + peroxide

c. 3-methyl-3-hexene + HBr + peroxide

d. *cis*-3-heptene + Br_2

e. *trans*-3-heptene + Br_2

f. *trans*-3-hexene + Br_2

Start by drawing the product without regard to stereochemistry to check whether or not the reaction has created any chirality centers. Then determine the configuration of the products, paying attention to the configuration (if any) of the reactant, how many chirality centers are formed, and the mechanism of the reaction. Let's start with **a.**

a. $CH_3CH_2CHCH_3$
$\quad\quad\quad\quad\;\; |$
$\quad\quad\quad\quad\;\; Cl$

The product has one chirality center, so equal amounts of the R and S enantiomers will be formed.

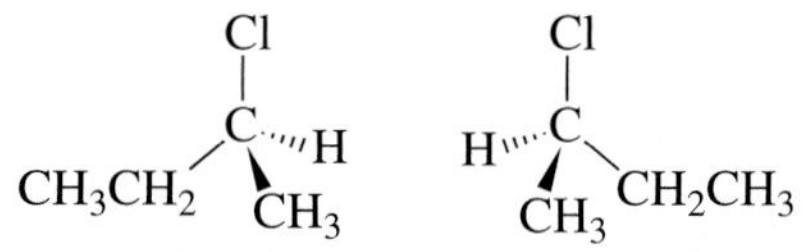

b. $CH_3CH_2CH_2CH_2Br$

The product does not have a chirality center, so there are no stereoisomers.

$$\underset{\displaystyle Br}{\overset{\displaystyle CH_3}{CH_3CH_2CHCHCH_2CH_3}}$$

c. 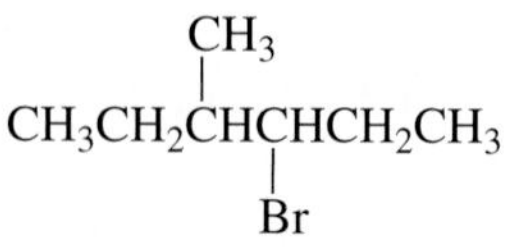 $CH_3CH_2\overset{CH_3}{\underset{Br}{CHCH}}CH_2CH_3$

Two chirality centers have been created in the product. Because the reaction forms a radical intermediate, two pairs of enantiomers are formed.

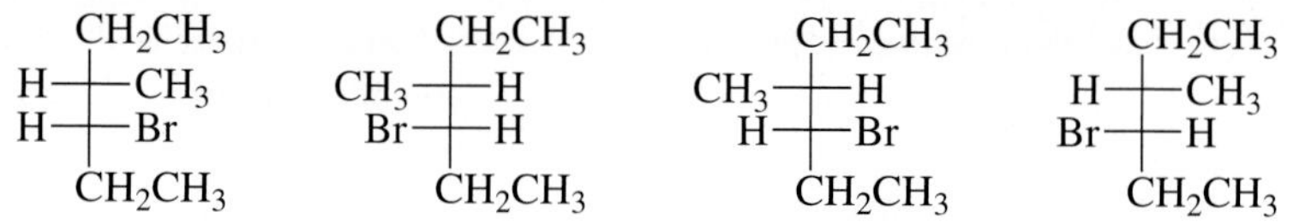

CH_2CH_3	CH_2CH_3	CH_2CH_3	CH_2CH_3
H——CH$_3$	CH$_3$——H	CH$_3$——H	H——CH$_3$
H——Br	Br——H	H——Br	Br——H
CH_2CH_3	CH_2CH_3	CH_2CH_3	CH_2CH_3

d. 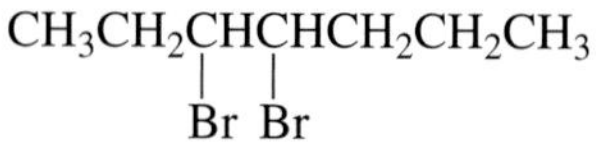 $CH_3CH_2\underset{\displaystyle Br\ Br}{CHCH}CH_2CH_2CH_3$

Two chirality centers have been created in the product. Because the reactant is cis and addition of Br_2 is anti, the threo pair of enantiomers is formed.

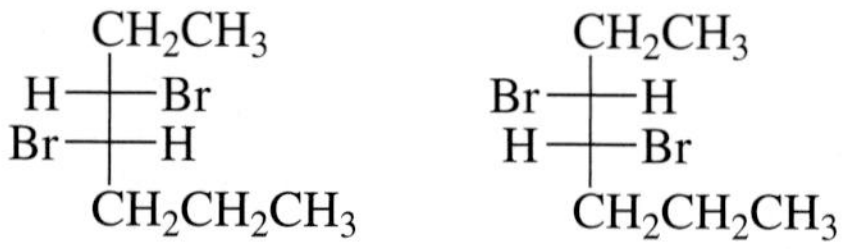

CH_2CH_3	CH_2CH_3
H——Br	Br——H
Br——H	H——Br
$CH_2CH_2CH_3$	$CH_2CH_2CH_3$

e. 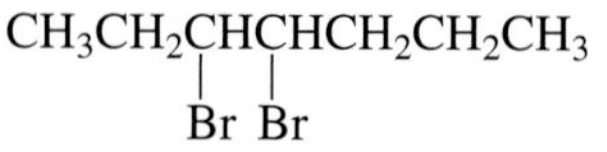 $CH_3CH_2\underset{\displaystyle Br\ Br}{CHCH}CH_2CH_2CH_3$

Two chirality centers have been created in the product. Because the reactant is trans and addition of Br_2 is anti, the erythro pair of enantiomers is formed.

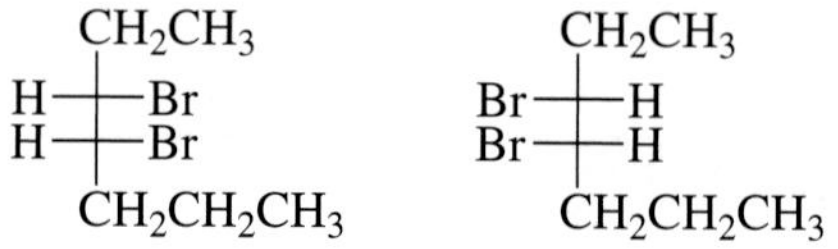

CH_2CH_3	CH_2CH_3
H——Br	Br——H
H——Br	Br——H
$CH_2CH_2CH_3$	$CH_2CH_2CH_3$

f. 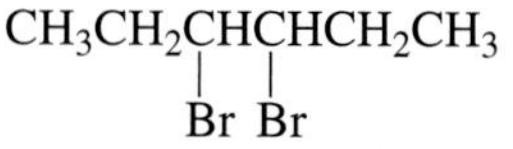 $CH_3CH_2\underset{\displaystyle Br\ Br}{CHCH}CH_2CH_3$

Two chirality centers have been created in the product. Because the reactant is trans and addition of Br_2 is anti, one would expect the erythro pair of enantiomers, but because the two chirality centers are bonded to the same four groups, the erythro product is a meso compound. Thus, only one stereoisomer is formed.

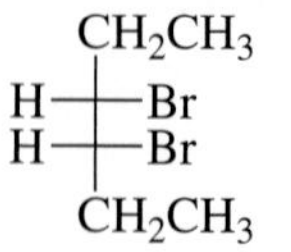

CH_2CH_3
H——Br
H——Br
CH_2CH_3

Now continue on to Problem 44.

PROBLEM 44 ◆

Give the configuration of the products obtained from the following reactions:

a. *trans*-2-butene + HBr + peroxide

b. *cis*-3-hexene + HBr

c. (Z)-3-methyl-2-pentene + HBr

d. (Z)-3-methyl-2-pentene + HBr + peroxide

e. *cis*-2-pentene + Br_2

f. 1-hexene + Br_2

PROBLEM 45

When Br_2 adds to an alkene with different substituents on the two sp^2 carbons, such as *cis*-2-heptene, identical amounts of the two threo enantiomers are obtained even though Br^- is more likely to attack the less sterically hindered carbon atom of the bromonium ion. Explain why identical amounts of the stereoisomers are obtained.

PROBLEM 46

a. What products would be obtained from the addition of Br_2 to cyclohexene if the solvent were H_2O instead of CH_2Cl_2?

b. Propose a mechanism for the reaction.

PROBLEM 47 ◆

What stereoisomers would you expect to obtain from each of the following reactions?

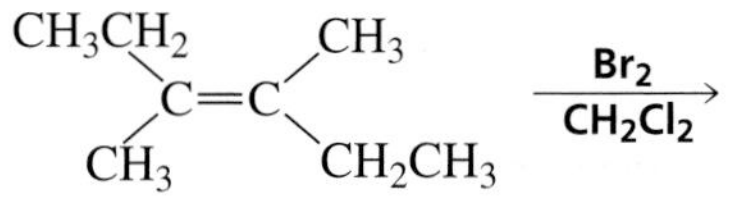

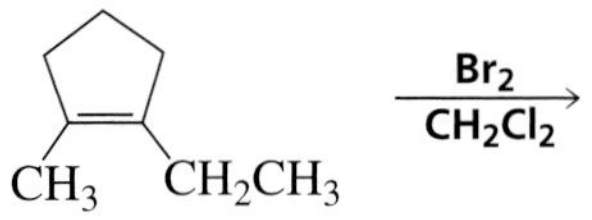

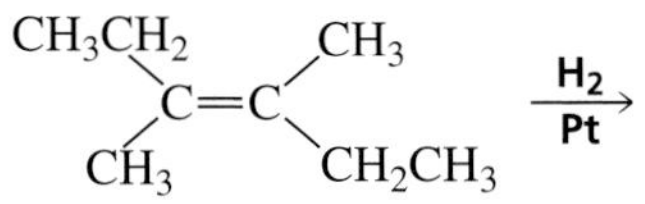

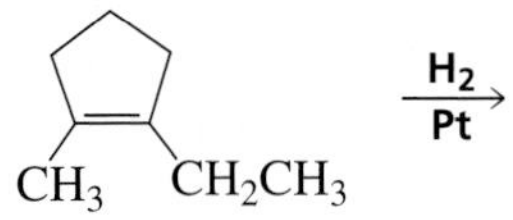

PROBLEM 48

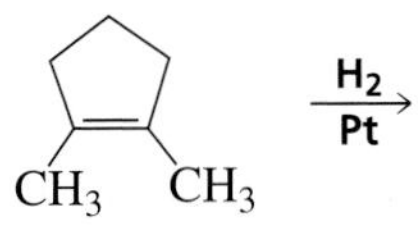

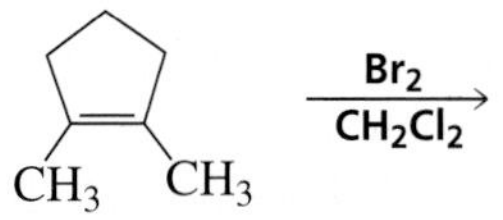

a. What is the major product obtained from the reaction of propene and Br_2 plus excess Cl^-?

b. Indicate the relative amounts of the stereoisomers obtained.

The chemistry associated with living organisms is called **biochemistry.** When you study biochemistry, you study the structures and functions of the molecules found in the biological world and the reactions involved in the synthesis and degradation of these molecules. Because most of the molecules in living organisms are organic molecules, you should expect that many of the reactions you encounter in organic chemistry will also be found in the chemistry of biological systems. Living cells

**4.18
STEREOCHEMISTRY
OF ENZYME-
CATALYZED
REACTIONS**

do not contain molecules such as Cl_2, HBr, or BH_3, so you would not expect to find the addition of such reagents to alkenes in biological systems. Living cells do contain water, however, so hydration reactions (Section 3.13) occur in biological systems.

Almost all the organic reactions that occur in biological systems are catalyzed by **enzymes.** Enzyme-catalyzed reactions are almost always completely stereoselective. In other words, enzymes catalyze reactions that form only a single stereoisomer. For example, the enzyme fumarase, which catalyzes the addition of water to fumarate, forms only (*S*)-malate. The *R* enantiomer is not formed.

fumarate   (*S*)-malate

An enzyme-catalyzed reaction forms only one stereoisomer because an enzyme possesses a chiral binding site. The chiral binding site allows reagents to be delivered to only one side of the functional group of the reactant. Consequently, only one stereoisomer is formed.

Enzyme-catalyzed reactions are also stereospecific. An enzyme typically catalyzes the reaction of only one stereoisomer. For example, fumarase catalyzes the addition of water to fumarate (the trans isomer) but not to maleate (the cis isomer).

maleate

An enzyme is able to differentiate between stereoisomeric reactants because of its chiral binding site (Figure 4.2). The enzyme will bind only the stereoisomer that has its substituents in the correct positions to interact with substituents in the chiral binding site. Other stereoisomers do not have substituents in the proper positions, so they cannot bind to the enzyme. An enzyme's stereospecificity can be thought of as a right-handed glove being able to differentiate between a right hand and a left hand.

*For studies on the stereo-chemistry of enzyme-catalyzed reactions, **Sir John Cornforth** received the Nobel Prize in chemistry in 1975 (sharing it with Vladimir Prelog, page 189). Born in Australia in 1917, he studied at the University of Sydney and received a Ph.D. from Oxford. His major research was carried out in laboratories at Britain's Medical Research Council and in laboratories at Shell Research Ltd. He was knighted in 1977.*

*Fundamental work on the stereo-chemistry of enzyme-catalyzed reactions was done by **Frank H. Westheimer.** Westheimer was born in Baltimore in 1912 and received his graduate training at Harvard University. He served on the faculty of the University of Chicago and subsequently returned to Harvard as Professor of Chemistry.*

KEY TERMS

absolute configuration (page 213)
achiral (page 185)
amine inversion (page 182)
anti addition (page 224)
biochemistry (page 231)
chiral (page 185)
chirality center (page 183)
cis isomer (page 182)
cis–trans isomers (page 180)
chromatography (page 216)
configuration (page 187)

configurational isomers (page 180)
conformational isomers (page 180)
conformer (page 180)
constitutional isomers (page 180)
dextrorotatory (page 193)
diastereomer (page 197)
diastereotopic hydrogens (page 218)
enantiomer (page 184)
enantiomerically pure (page 196)
enantiomeric excess (page 196)
enantiotopic hydrogens (page 217)

enzyme (page 232)
erythro enantiomers (page 197)
Fischer projection (page 187)
homotopic hydrogens (page 217)
isomers (page 180)
isomers that contain chirality centers
  (page 180)
levorotatory (page 193)
meso compound (page 202)
observed rotation (page 194)
optical purity (page 196)

optically active (page 193)
optically inactive (page 193)
pair of enantiomers (page 184)
perspective formula (page 186)
plane of symmetry (page 185)
plane-polarized light (page 192)
polarimeter (page 193)
pro-*R*-hydrogen (page 217)
pro-*S*-hydrogen (page 217)

prochirality center (page 217)
racemate (page 195)
racemic mixture (page 195)
racemic modification (page 195)
R configuration (page 187)
regioselective (page 219)
relative configuration (page 210)
resolution of a racemic mixture (page 215)

S configuration (page 187)
specific rotation (page 194)
stereochemistry (page 219)
stereoisomers (page 180)
stereoselective (page 220)
stereospecific (page 220)
syn addition (page 224)
threo enantiomers (page 197)
trans isomer (page 182)

PROBLEMS

49. Neglecting stereoisomers, give the structures of all compounds with molecular formula C_5H_{10}. Which ones can exist as stereoisomers?

50. Draw all possible stereoisomers for each of the following compounds. State if no stereoisomers are possible.
   a. 1-bromo-2-chlorocyclohexane
   b. 2-bromo-4-methylpentane
   c. 1,2-dichlorocyclohexane
   d. 2-bromo-4-chloropentane
   e. 3-heptene
   f. 1-bromo-4-chlorocyclohexane
   g. 1,2-dimethylcyclopropane
   h. 4-bromo-2-pentene
   i. 3,3-dimethylpentane
   j. 3-chloro-1-butene
   k. 1-bromo-2-chlorocyclobutane
   l. 1-bromo-3-chlorocyclobutane

51. a. Draw the staggered and eclipsed conformers for rotation about the C-2—C-3 bond of 1-chloropentane.
   b. Which conformer is the most stable?
   c. Which conformer is the least stable?

52. Name each of the following compounds using *R,S* and *E,Z* designations where necessary:

a. 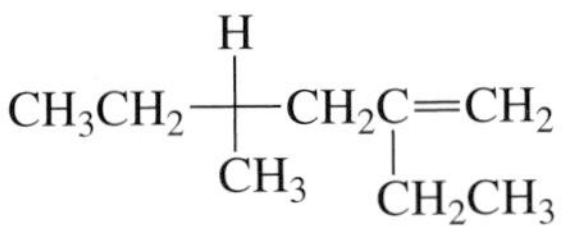 CH₃CH₂—| H / CH₃ |—CH₂C=CH₂ | CH₂CH₃

e. 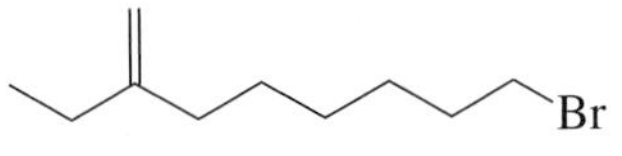 (structure) Br

b. F, Cl \ C=C / I, Br

f. BrCH₂CH₂, Br \ C=C / CH₂CH₂CHCH₃ (CH₃), CH₃

c. F, Cl \ C=C / H, Br

g. HO—|CH₂OH / CH₂CH₂CH₂OH|—CH₃

d. CH₃CH (CH₃), H₃C \ C=C / CH₂CH₂CH₂CH₃, CH₂CH₂Cl

h. HO—|CH₃|—H ; H—| |—Cl ; CH₂CH₃

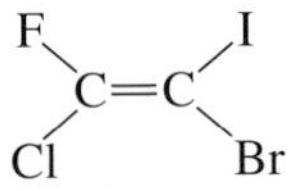

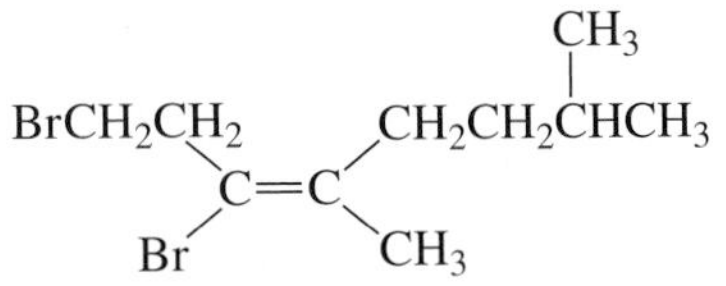

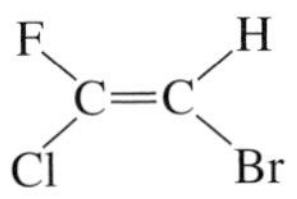

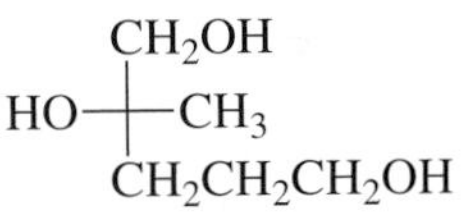

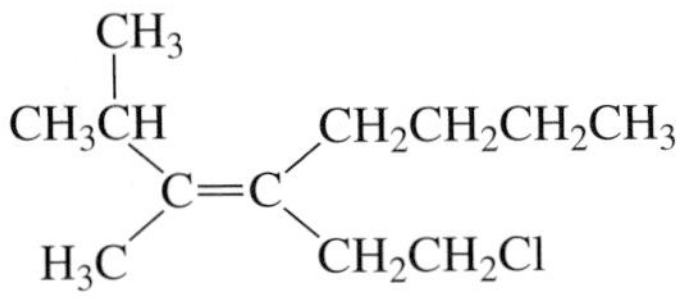

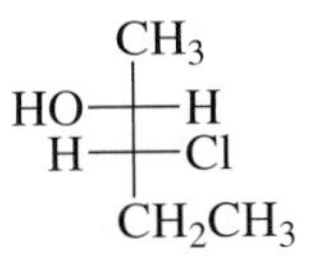

53. Mevacor is used clinically to lower serum cholesterol levels. How many chirality centers does Mevacor have?

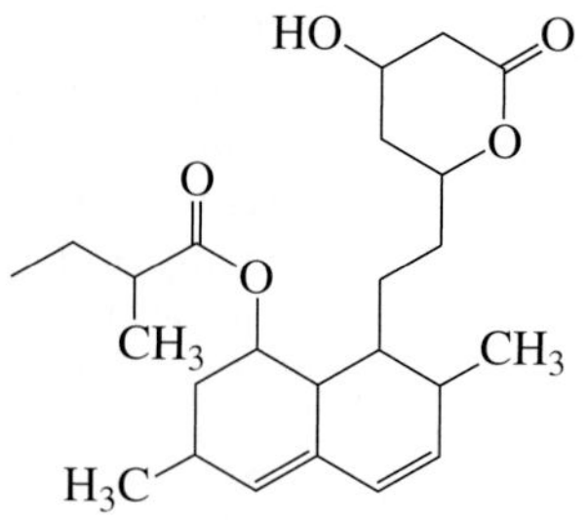

Mevacor

54. Indicate whether each of the following pairs of compounds are identical or are enantiomers, diastereomers, or constitutional isomers:

a.

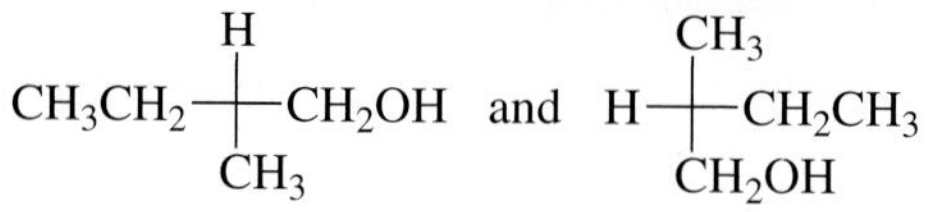

$$CH_3CH_2-\overset{\displaystyle H}{\underset{\displaystyle CH_3}{|}}-CH_2OH \quad \text{and} \quad H-\overset{\displaystyle CH_3}{\underset{\displaystyle CH_2OH}{|}}-CH_2CH_3$$

b.

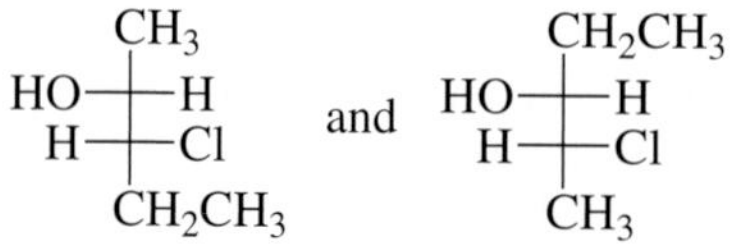

$$\begin{array}{c} CH_3 \\ HO-\!\!\!-H \\ H-\!\!\!-Cl \\ CH_2CH_3 \end{array} \quad \text{and} \quad \begin{array}{c} CH_2CH_3 \\ HO-\!\!\!-H \\ H-\!\!\!-Cl \\ CH_3 \end{array}$$

c.

$$\begin{array}{c} CH_3 \\ HO-\!\!\!-H \\ H-\!\!\!-Cl \\ CH_3 \end{array} \quad \text{and} \quad \begin{array}{c} CH_3 \\ H-\!\!\!-OH \\ Cl-\!\!\!-H \\ CH_3 \end{array}$$

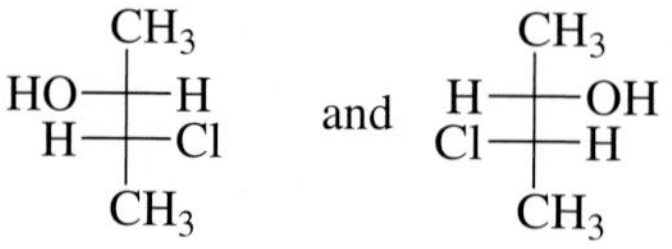

d.

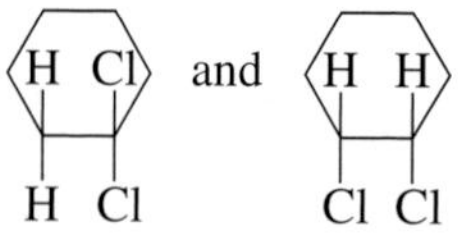

e.

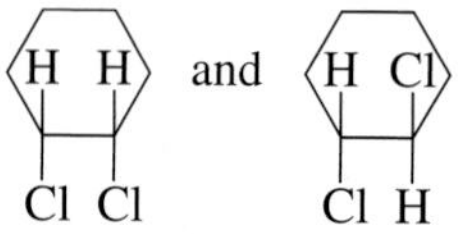

f.

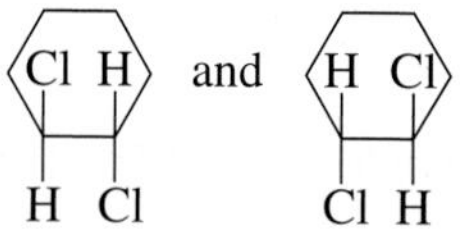

g.

$$CH_3CH_2\!\!\!-\!\!\overset{\displaystyle CH_2Cl}{\underset{\displaystyle H}{C}}\!\!\!-\!\!CH_3 \quad \text{and} \quad CH_3\!\!\!-\!\!\overset{\displaystyle CH_2CH_3}{\underset{\displaystyle H}{C}}\!\!\!-\!\!CH_2Cl$$

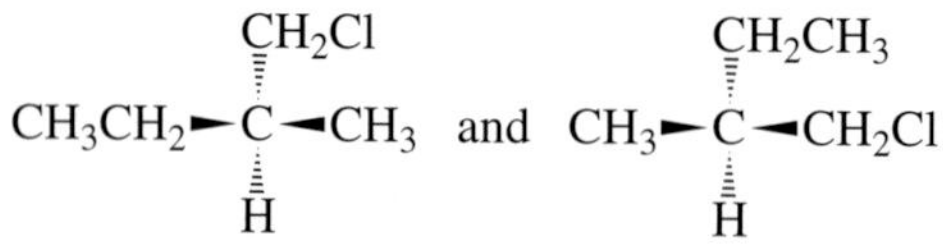

h.

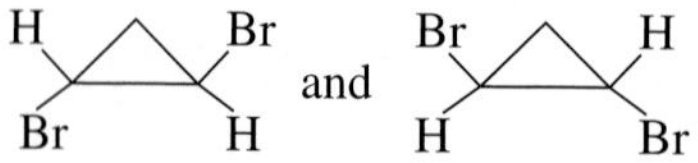

i.

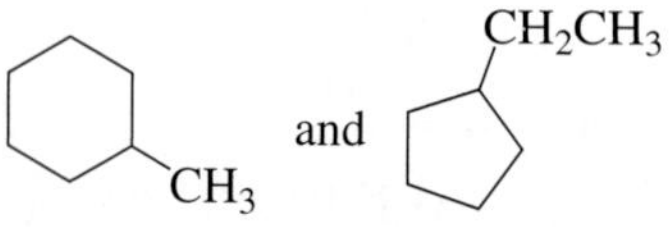

j.

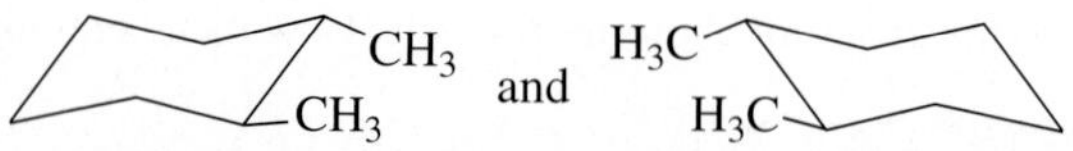

k. 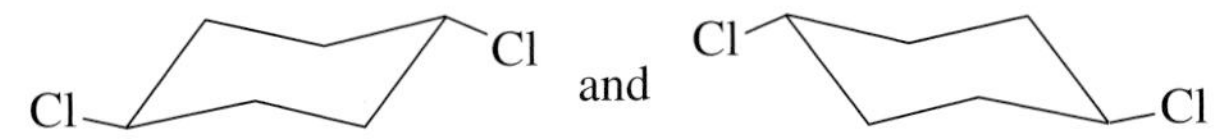 and

l. 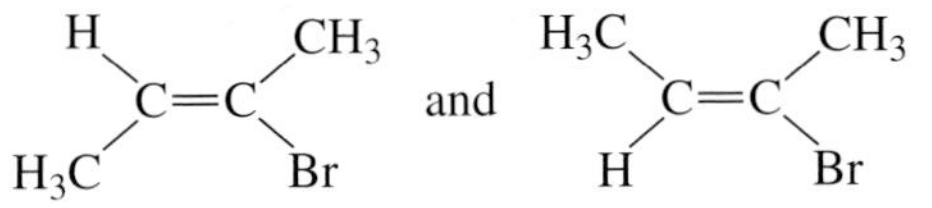 and

m. 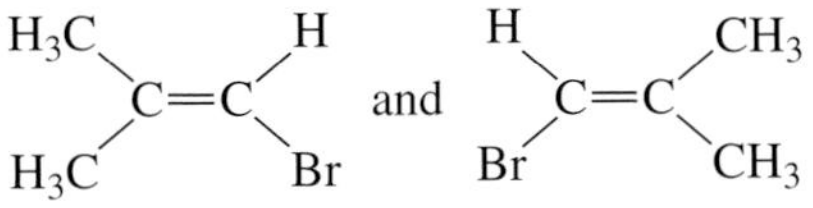 and

n. 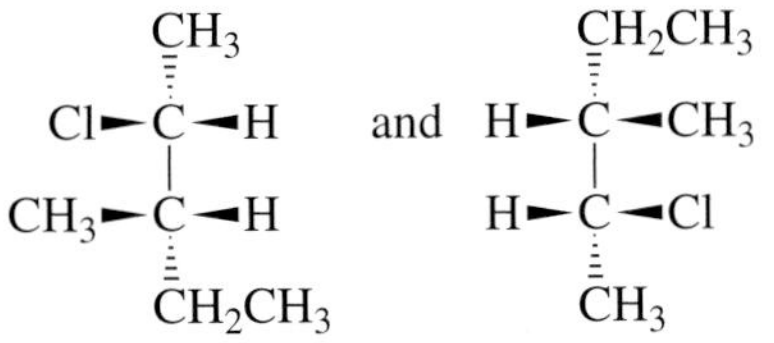 and

o. 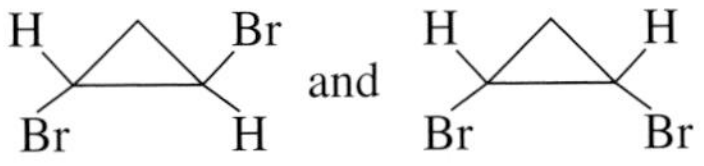 and

p. 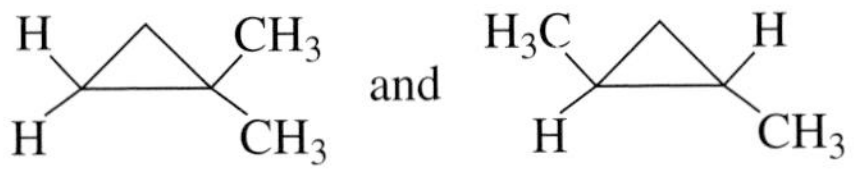 and

55. **a.** Give the product(s) that would be obtained from the reaction of *cis*-2-butene and *trans*-2-butene with each of the following reagents. If the products can exist as stereoisomers, show which stereoisomers are obtained.

    1. HCl
    2. BH_3 followed by HO^-, H_2O_2
    3. HBr + peroxide
    4. Br_2 in CH_2Cl_2

    5. Br_2 + H_2O
    6. H_2/Pt
    7. HCl + H_2O
    8. HCl + CH_3OH

    b. With which reagents do the two alkenes react to give different products?

56. Which of the following compounds have a stereoisomer that is achiral?

    a. 2,3-dichlorobutane
    b. 2,3-dichloropentane
    c. 2,3-dichloro-2,3-dimethylbutane
    d. 1,3-dichlorocyclopentane
    e. 1,3-dibromocyclobutane

    f. 2,4-dibromopentane
    g. 2,3-dibromopentane
    h. 1,4-dimethylcyclohexane
    i. 1,2-dimethylcyclopentane
    j. 1,2-dimethylcyclobutane

57. Citrate synthase, one of the enzymes in the series of enzyme-catalyzed reactions known as the Krebs cycle, catalyzes the synthesis of citric acid from oxaloacetic acid and acetyl-CoA. If the synthesis is carried out with acetyl-CoA that has radioactive carbon (^{14}C) in the indicated position, the isomer shown here is obtained.

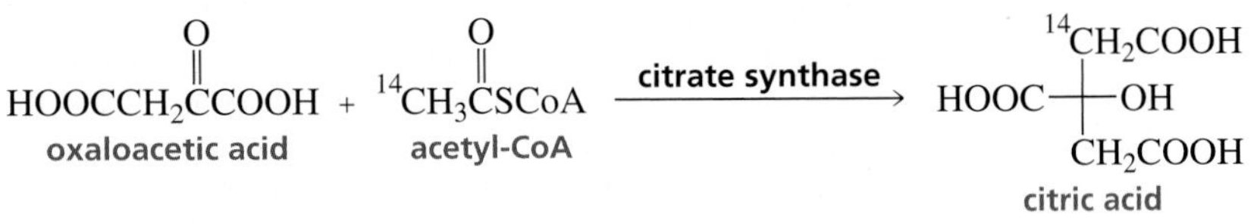

    a. Which stereoisomer of citric acid is synthesized, *R* or *S*?
    b. Why is the other stereoisomer not obtained?
    c. If the acetyl-CoA used in the synthesis does not contain ^{14}C, will the product of the reaction be chiral or achiral?

58. Give the products of the following reactions. If the products can exist as stereoisomers, show which stereoisomers are obtained.

a. *cis*-2-pentene + HCl
b. *trans*-2-pentene + HCl
c. 1-ethylcyclohexene + H_3O^+
d. 1,2-diethylcyclohexene + H_3O^+
e. 1,2-dimethylcyclohexene + HCl

f. 1,2-dideuteriocyclohexene + H_2/Pt
g. 3,3-dimethyl-1-pentene + Br_2/CH_2Cl_2
h. (*E*)-3,4-dimethyl-3-heptene + H_2/Pt
i. (*Z*)-3,4-dimethyl-3-heptene + H_2/Pt
j. 1-chloro-2-ethylcyclohexene + H_2/Pt

59. Indicate whether each of the following structures is (*R*)-2-chlorobutane or (*S*)-2-chlorobutane. (Use models, if necessary.)

a. [structure]

c. 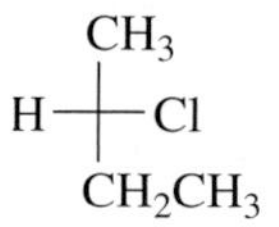 [structure]

e. 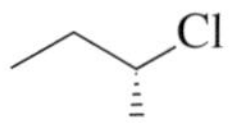 [structure]

b. 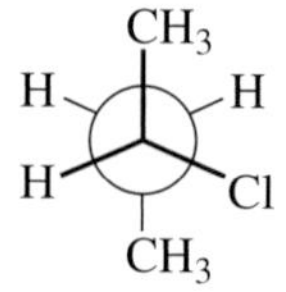 [structure]

d. 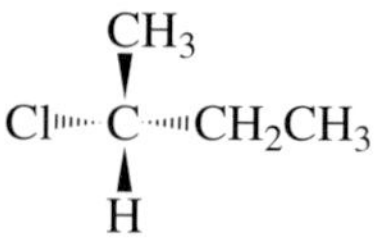 [structure]

f. 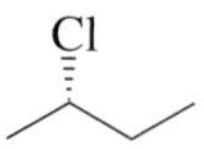 [structure]

60. A solution of an unknown compound (3.0 g of the compound in 20 mL of solution), when placed in a polarimeter tube 2.0 dm long, was found to rotate the plane of polarized light 1.8° in a counterclockwise direction. What is the specific rotation of the compound?

61. Butaclamol is a potent antipsychotic that has been used clinically in the treatment of schizophrenia. Although patients are given a racemic mixture of the drug, only the (+)-enantiomer has pharmacological activity. How many chirality centers does Butaclamol have?

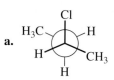

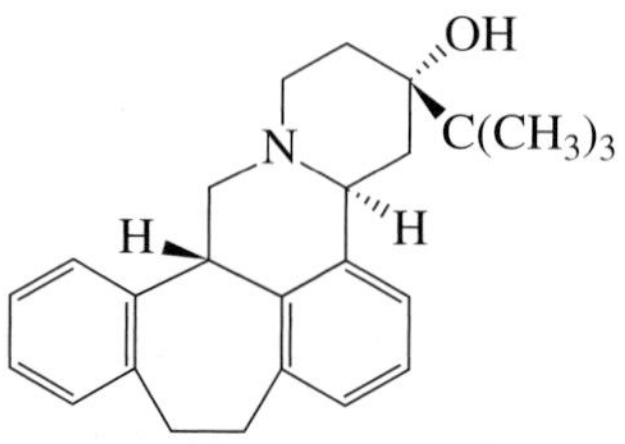

Butaclamol

62. Which of the following objects are chiral?

a. a mug with DAD written on one side
b. a mug with MOM written on one side
c. a mug with DAD written opposite the handle
d. a mug with MOM written opposite the handle

e. a nail
f. a wine glass
g. a screw

63. Explain how *R* and *S* are related to (+) and (−).

64. Give the products of the following reactions. If the products can exist as stereoisomers, show which stereoisomers are obtained.

a. *cis*-2-pentene + Br_2/CH_2Cl_2
b. *trans*-2-pentene + Br_2/CH_2Cl_2
c. 1-butene + HCl
d. 1-butene + HBr + peroxide
e. *trans*-3-hexene + Br_2/CH_2Cl_2
f. *cis*-3-hexene + Br_2/CH_2Cl_2

g. 3,3-dimethyl-1-pentene + HBr
h. *cis*-2-butene + HBr + peroxide
i. (*Z*)-2,3-dichloro-2-butene + H_2/Pt
j. (*E*)-2,3-dichloro-2-butene + H_2/Pt
k. (*Z*)-3,4-dimethyl-3-hexene + H_2/Pt
l. (*E*)-3,4-dimethyl-3-hexene + H_2/Pt

65. a. Draw all possible stereoisomers for the following compound.

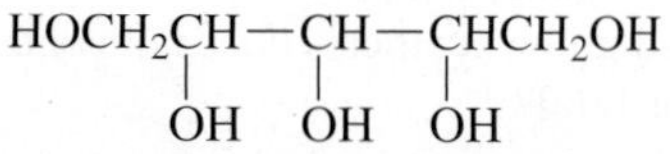

$$HOCH_2CH-CH-CHCH_2OH$$
$$\quad\quad\;\; | \quad\;\; | \quad\;\; |$$
$$\quad\quad OH \;\; OH \;\; OH$$

b. Which isomers are optically inactive (will not rotate plane-polarized light)?

66. Indicate the configuration of the chirality centers in the following molecules:

a. CH₃CH₂CH₂ —|— H with CH₂CH₂Br above and Br below

b. CH=O above; H —|— OH; HO —|— H; Br below

c. CH₃ above; H —|— Br; H —|— Br; CH₂CH₃ below

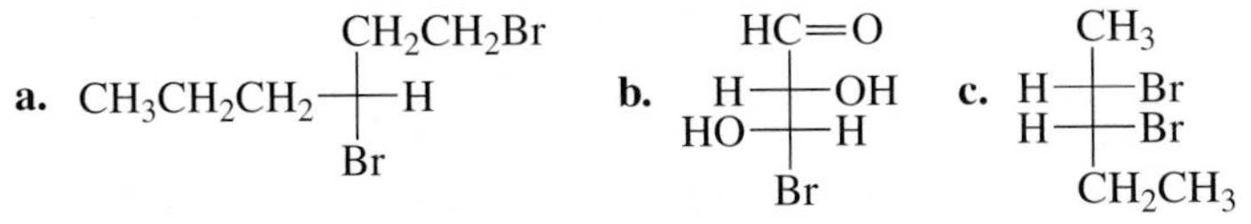

67. a. Draw all the isomers with molecular formula C_6H_{12} that contain a cyclobutane ring. (*Hint:* There are seven.)
   b. Name the compounds without specifying the configuration of any chirality centers.
   c. Identify:
   1. constitutional isomers
   2. stereoisomers
   3. cis–trans isomers
   4. chiral compounds
   5. achiral compounds
   6. meso compounds
   7. enantiomers
   8. diastereomers

68. Cholesterol has a specific rotation of $-39.0°$. A solution of a cholesterol sample (0.187 g/mL) has an observed rotation of $-6.52°$ when placed in a polarimeter tube 10 cm long. What is the percent optical purity of the cholesterol in the sample?

69. Draw structures for each of the following molecules:
   a. (*S*)-1-bromo-1-chlorobutane
   b. (2*R*,3*R*)-2,3-dichloropentane
   c. an achiral isomer of 1,2-dimethylcyclohexane
   d. a chiral isomer of 1,2-dibromocyclobutane
   e. two achiral isomers of 3,4,5-trimethylheptane

70. The enantiomers of 1,2-dimethylaziridine can be separated even though one of the "groups" attached to nitrogen is a pair of nonbonding electrons. Explain.

enantiomers of 1,2-dimethylaziridine

71. Of the possible monobromination products shown for the following reaction, circle any that would not be formed.

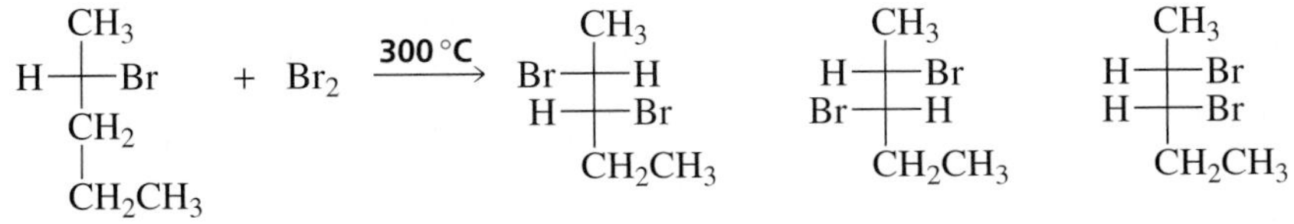

CH₃ H—|—Br, CH₂, CH₂CH₃ + Br₂ → (300 °C) → CH₃ Br—|—H, H—|—Br, CH₂CH₃ ; CH₃ H—|—Br, Br—|—H, CH₂CH₃ ; CH₃ H—|—Br, H—|—Br, CH₂CH₃

72. A sample of (*S*)-(+)-lactic acid was found to have an optical purity of 72%. How much *R* isomer is present in the sample?

73. Give the products of the following reactions and their configuration:

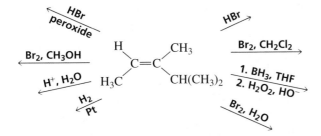

74. a. Using the wedge-and-dash notation, draw the nine configurational isomers of 1,2,3,4,5,6-hexachlorocyclohexane.
b. From the nine configurational isomers, identify one pair of enantiomers.
c. Draw the most stable conformation of the most stable configurational isomer.

75. Sherry O. Eismer decided that the configuration of the chirality centers in a sugar such as D-glucose could be determined rapidly by simply assigning the *R* configuration to a chirality center with an OH group on the right and the *S* configuration to a chirality center with an OH group on the left. Is she correct? (We will see in Chapter 19 that the "D" in D-glucose means the OH group on the bottom-most chirality center is on the right.)

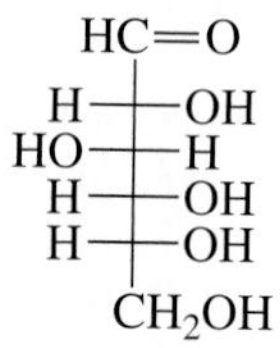

$$HC=O$$
$$H \long OH$$
$$HO \long H$$
$$H \long OH$$
$$H \long OH$$
$$CH_2OH$$
D-glucose

76. Cyclohexene exists only in the cis form, while cyclodecene exists in both the cis and trans forms. Explain. (*Hint:* Molecular models are helpful for this problem.)

77. When fumarate reacts with D_2O in the presence of the enzyme fumarase, only one isomer of the product is formed. Its structure is shown. Is the enzyme catalyzing a syn or an anti addition of D_2O?

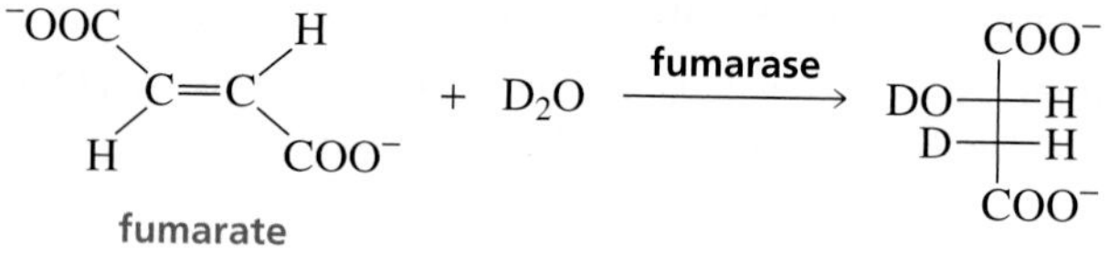

fumarate

78. Two stereoisomers are obtained from the reaction of HBr with (3*S*,4*S*)-4-bromo-3-methyl-1-pentene. One of the stereoisomers is optically active, and the other is not. Give the structures of the stereoisomers, indicating their absolute configurations, and explain the difference in optical properties.

79. When (*S*)-(+)-1-chloro-2-methylbutane reacts with chlorine, one of the products formed is (−)-1,4-dichloro-2-methylbutane. Does this product have the *R* or the *S* configuration?

80. Indicate the configuration of the chirality centers in the following molecules:

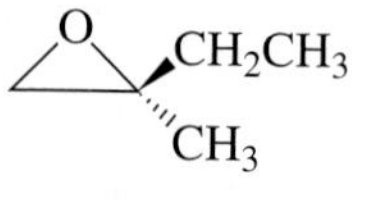

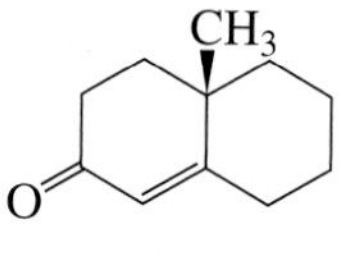

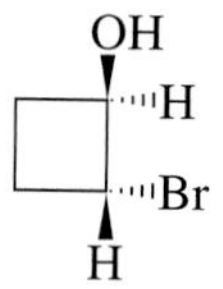

81. a. Do the following compounds have any chirality centers?
1. $CH_2=C=CH_2$ **2.** $CH_3CH=C=CHCH_3$
b. Are they chiral? (*Hint:* Make models.)

5

Reactions of Alkynes.
Introduction to
Multistep Synthesis

1-butyne + **2HCl** $\longrightarrow$ **2,2-dichlorobutane**

Alkynes are hydrocarbons that contain a carbon–carbon triple bond. Because of its triple bond, an alkyne has four fewer hydrogens than the corresponding alkane. Therefore, the general molecular formula for an alkyne is C_nH_{2n-2}. There are only a few naturally occurring alkynes. Examples include capillin, which has fungicidal activity, and ichthyothereol, a convulsant used by the Amazon Indians for poisoned arrowheads. A naturally occurring alkyne typically has at least one other functional group in addition to one or more triple bonds.

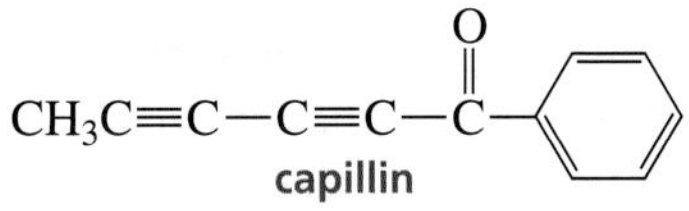

$$CH_3C{\equiv}C-C{\equiv}C-\overset{O}{\overset{\|}{C}}-\bigcirc$$
capillin

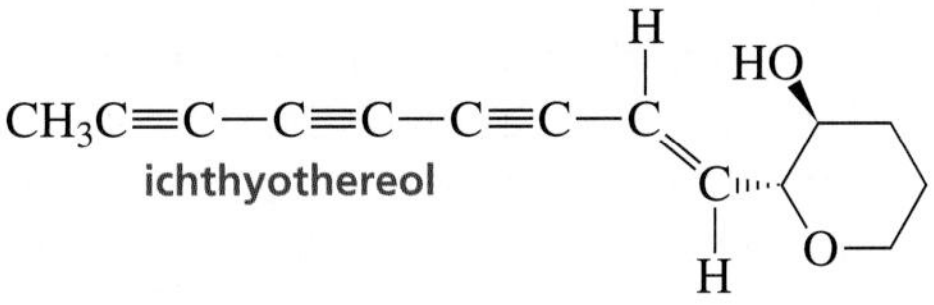

$$CH_3C{\equiv}C-C{\equiv}C-C{\equiv}C-\overset{H}{\underset{}{C}}$$
ichthyothereol

There are a few drugs that contain alkyne functional groups, but they are not naturally occurring compounds. They exist only because chemists have been able to synthesize them. Their trade names are shown in green. Trade names are always capitalized and can be used for commercial purposes only by the owner of the registered trademark (Section 29.1).

Parsal
Sinovial

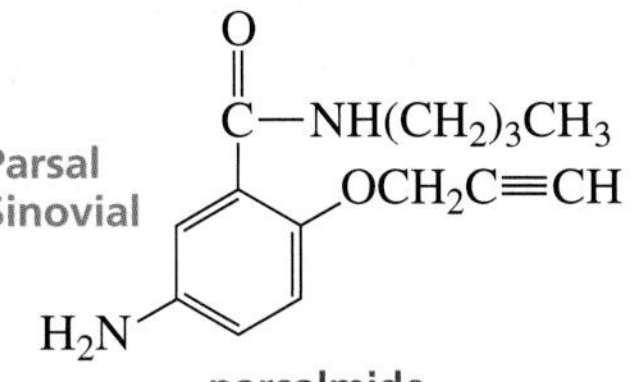

$$\overset{O}{\overset{\|}{C}}-NH(CH_2)_3CH_3$$
$$OCH_2C{\equiv}CH$$
$$H_2N$$

parsalmide
an analgesic

Eudatin
Supirdyl

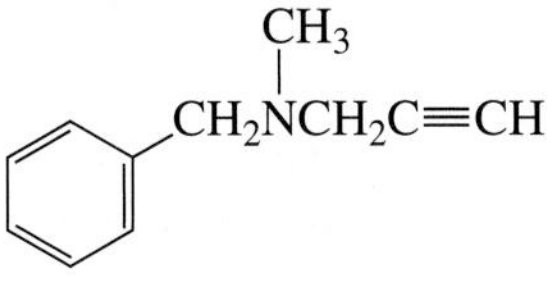

$$\underset{}{\bigcirc}-CH_2N\underset{CH_2C{\equiv}CH}{\overset{CH_3}{\underset{}{}}}$$

pargyline
an antihypertensive

Norquen
Ovastol

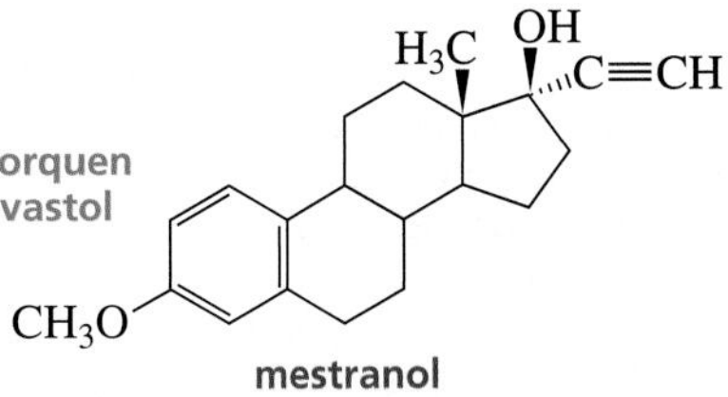

$$H_3C\quad \overset{OH}{\underset{}{C}}{\equiv}CH$$
$$CH_3O$$

mestranol
a component in oral contraceptives

Acetylene (HC≡CH), the common name for the smallest alkyne, may be a familiar word to you because of the oxyacetylene torch used in welding. Acetylene is supplied to the torch from one high-pressure gas tank and oxygen is supplied from another. Burning acetylene produces a high-temperature flame capable of melting and vaporizing iron and steel.

PROBLEM 1 ◆

What is the molecular formula for a cyclic alkyne with 14 carbons and 2 triple bonds?

5.1 NOMENCLATURE OF ALKYNES

The systematic name of an alkyne is obtained by replacing the "ane" ending of the alkane name with "yne." Analogous to naming compounds with other functional groups, the longest continuous chain containing the carbon–carbon triple bond is numbered in a direction that gives the alkyne functional group suffix as low a number as possible. If the triple bond is at the end of the chain, the alkyne is classified as a **terminal alkyne.** Alkynes with triple bonds located elsewhere along the chain are called **internal alkynes.** For example, 1-butyne is a terminal alkyne whereas 2-pentyne is an internal alkyne.

3-D Molecules:
1-Hexene
3-Hexene

Systematic: HC≡CH $\overset{4}{C}H_3\overset{3}{C}H_2\overset{2}{C}\equiv\overset{1}{C}H$ $\overset{1}{C}H_3\overset{2}{C}\equiv\overset{3}{C}\overset{4}{C}H_2\overset{5}{C}H_3$ $\overset{5}{C}H_2\overset{6}{C}H_3$
Common: **ethyne**   **1-butyne**   **2-pentyne** $\overset{4}{C}H_3\overset{3}{C}HC\equiv\overset{2}{C}\overset{1}{C}H_3$
      acetylene   ethylacetylene   ethylmethylacetylene   **4-methyl-2-hexyne**
                       a terminal alkyne   **an internal alkyne**   *sec*-butylmethylacetylene

1-hexyne

In common nomenclature, alkynes are named as substituted acetylenes. The common name is obtained by naming the alkyl groups, in alphabetical order, that have replaced the hydrogens of acetylene. Acetylene is an unfortunate common name for the smallest alkyne because its "ene" ending is characteristic of a double bond rather than a triple bond.

If the same number for the alkyne functional group suffix is obtained in both directions along the carbon chain, the correct systematic name is the one that contains the lowest substituent number. If the compound contains more than one substituent, the substituents are listed in alphabetical order.

3-hexyne

        Cl Br   CH₃
        | | |
$CH_3\overset{}{C}H\overset{}{C}HC\equiv CCH_2CH_2CH_3$ $CH_3\overset{}{C}HC\equiv CCH_2CH_2Br$
      1   2   3   4   5 6   7   8   6   5   4   3 2   1
    3-bromo-2-chloro-4-octyne   **1-bromo-5-methyl-3-hexyne**
    not **6-bromo-7-chloro-4-octyne**   *not* **6-bromo-2-methyl-3-hexyne**
        because 2 < 6   because 1 < 2

The triple-bond-containing propargyl group is used in common nomenclature. It is analogous to the double-bond-containing allyl group that you saw in Section 3.2.

> A substituent receives the lowest possible number only if there is no functional group suffix, or if the same number for the functional group suffix is obtained in both directions.

       HC≡CCH₂—   H₂C=CHCH₂—
       propargyl group   **allyl group**

       HC≡CCH₂Br   H₂C=CHCH₂OH
       propargyl bromide   allyl alcohol

PROBLEM 2 ◆

Draw the structure for each of the following compounds.

a. 1-chloro-3-hexyne

b. cyclooctyne

c. isopropylacetylene

d. propargyl chloride

e. 4,4-dimethyl-1-pentyne

f. dimethylacetylene

PROBLEM 3 ◆

Give the systematic name for each of the following compounds.

a. $BrCH_2CH_2C\equiv CCH_3$

b. $CH_3CH_2CHC\equiv CCH_2CHCH_3$
　　　 $\underset{Br}{|}$ 　　　 $\underset{Cl}{|}$

c. $CH_3OCH_2C\equiv CCH_2CH_3$

d. $CH_3CH_2CHC\equiv CH$
　　　　 $\underset{CH_2CH_2CH_3}{|}$

PROBLEM 4 ◆

Draw the structures and give the common and systematic names for the seven alkynes with molecular formula C_6H_{10}.

PROBLEM 5

Which would you expect to be more stable, an internal alkyne or a terminal alkyne? Why?

All hydrocarbons have similar physical properties. In other words, alkenes and alkynes have physical properties similar to those of alkanes (Section 2.9). All are insoluble in water and all are soluble in solvents with low polarity such as benzene and ether. They are less dense than water and, like other homologous series, have boiling points that increase with increasing molecular weight (Table 5.1). Alkynes are more linear than alkenes, and a triple bond is more polarizable than a double bond

5.2 PHYSICAL PROPERTIES OF UNSATURATED HYDROCARBONS

TABLE 5.1 Boiling Points of the Smallest Hydrocarbons					
	bp (°C)		**bp (°C)**		**bp (°C)**
CH_3CH_3 ethane	−88.6	$H_2C\!=\!CH_2$ ethene	−104	$HC\!\equiv\!CH$ ethyne	−84
$CH_3CH_2CH_3$ propane	−42.1	$CH_3CH\!=\!CH_2$ propene	−47	$CH_3C\!\equiv\!CH$ propyne	−23
$CH_3CH_2CH_2CH_3$ butane	−0.5	$CH_3CH_2CH\!=\!CH_2$ 1-butene	−6.5	$CH_3CH_2C\!\equiv\!CH$ 1-butyne	8
$CH_3(CH_2)_3CH_3$ pentane	36.1	$CH_3CH_2CH_2CH\!=\!CH_2$ 1-pentene	30	$CH_3CH_2CH_2C\!\equiv\!CH$ 1-pentyne	39
$CH_3(CH_2)_4CH_3$ hexane	68.7	$CH_3CH_2CH_2CH_2CH\!=\!CH_2$ 1-hexene	63.5	$CH_3CH_2CH_2CH_2C\!\equiv\!CH$ 1-hexyne	71
		$CH_3CH\!=\!CHCH_3$ cis-2-butene	3.7	$CH_3C\!\equiv\!CCH_3$ 2-butyne	27
		$CH_3CH\!=\!CHCH_3$ trans-2-butene	0.9	$CH_3CH_2C\!\equiv\!CCH_3$ 2-pentyne	55

(Section 2.9). These two features cause alkynes to have stronger van der Waals interactions. As a result, an alkyne has a higher boiling point than an alkene containing the same number of carbon atoms.

Internal alkenes have higher boiling points than terminal alkenes. Similarly, internal alkynes have higher boiling points than terminal alkynes. The boiling point of *cis*-2-butene is slightly higher than that of *trans*-2-butene, because the cis isomer has a small dipole moment, whereas the dipole moment of the trans isomer is zero (Section 3.4).

5.3
THE STRUCTURE OF ALKYNES

The structure of ethyne was discussed in Section 1.9. Each carbon is *sp* hybridized, so each has two *sp* orbitals and two *p* orbitals. One *sp* orbital overlaps the *s* orbital of a hydrogen, and the other overlaps an *sp* orbital of the other carbon. Because the *sp* orbitals are oriented as far from each other as possible to minimize electron repulsion, ethyne is a linear molecule with bond angles of 180° (Section 1.9).

3-D Molecule:
Ethyne

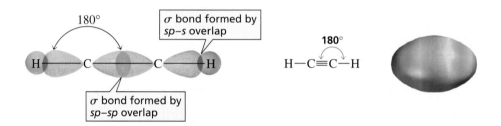

The two remaining *p* orbitals on each carbon are oriented at right angles to one another and to the *sp* orbitals (Figure 5.1). Each of the two *p* orbitals on one carbon overlaps the parallel *p* orbital on the other carbon to form two π bonds. One pair of overlapping *p* orbitals results in a cloud of electrons above and below the σ bond, and the other pair results in a cloud of electrons in front of and behind the σ bond. The electrostatic potential map of ethyne shows that the end result can be thought of as a cylinder of electrons wrapped around the σ bond.

Figure 5.1 ▶
(a) Each of the two π bonds of a triple bond is formed by side-to-side overlap of a *p* orbital of one carbon with a *p* orbital of the adjacent carbon.
(b) A triple bond consists of a σ bond formed by *sp–sp* overlap (yellow) and two π bonds formed by *p–p* overlap (blue and mauve).

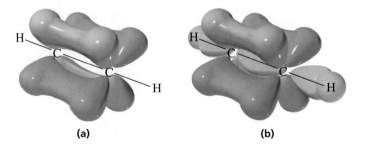

(a)   (b)

If we subtract the σ bond energy from the total bond energy of the triple bond (200 − 91 = 109 kcal/mol), we find that the average bond strength of each of the π bonds in the triple bond is less than 55 kcal/mol (230 kJ/mol). (The value of 91 kcal/mol is that for a σ bond formed by sp^2–sp^2 overlap. The σ bond formed by *sp–sp* overlap is expected to be somewhat stronger than that.) The π bonds, therefore, are much weaker than the σ bond. The weak π bonds allow alkynes to react easily.

With a cloud of electrons completely surrounding the σ bond, an alkyne is an electron-rich molecule. In other words, it is a nucleophile, so we expect it to react with electrophiles. Consequently, if a reagent such as HCl is added to an alkyne, a π bond will break because the π electrons are attracted to the electrophilic hydrogen. (We just saw that an alkyne's π bonds are weak compared with its σ bond.) In the second step of the reaction, the positively charged carbocation intermediate reacts rapidly with the negatively charged chloride ion.

5.4 REACTIVITY CONSIDERATIONS

3-D Molecule: Vinyl cation

$$CH_3C{\equiv}CCH_3 \;+\; H-\overset{..}{\underset{..}{Cl}}: \;\longrightarrow\; CH_3\overset{+}{C}{=}CHCH_3 \;+\; :\overset{..}{\underset{..}{Cl}}:^- \;\longrightarrow\; CH_3\overset{\overset{\displaystyle Cl}{|}}{C}{=}CHCH_3$$

Thus alkynes, like alkenes, undergo electrophilic addition reactions. We will see that the same electrophilic reagents that add to alkenes also add to alkynes and that—again like alkenes—electrophilic addition to a terminal alkyne is regioselective, following Markovnikov's rule. There is, however, an added complication in the reactions of alkynes. Because the product of the addition of an electrophilic reagent to an alkyne is an alkene, a second electrophilic addition reaction can occur.

$$CH_3C{\equiv}CCH_3 \;\xrightarrow{\text{HCl}}\; CH_3\overset{\overset{\displaystyle Cl}{|}}{C}{=}CHCH_3 \;\xrightarrow{\text{HCl}}\; CH_3\overset{\overset{\displaystyle Cl}{|}}{\underset{\underset{\displaystyle Cl}{|}}{C}}CH_2CH_3$$

An alkyne is less reactive than an alkene. This might at first seem surprising because the π bond of an alkyne is weaker ($<$ 55 kcal/mol or 230 kJ/mol) than the π bond of an alkene (61 kcal/mol or 255 kJ/mol), which makes the alkyne less stable (Figure 5.2). However, reactivity depends on $\Delta G^{\ddagger}$, which in turn depends on the stability of the reactant *and* the stability of the transition state (Section 3.7). For an alkyne to be both less stable and less reactive than an alkene two conditions must hold: the transition state for the first step (the rate-limiting step) of an electrophilic addition reaction for an alkyne must be less stable than the transition state for the first step of an electrophilic addition reaction for an alkene, *and* the difference in the stabili-

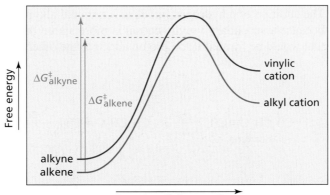

◀ **Figure 5.2**
Comparison of the free energies of activation for the addition of an electrophile to an alkyne and to an alkene. Because an alkyne is less reactive than an alkene toward electrophilic addition, we know that $\Delta G^{\ddagger}$ for the reaction of an alkyne is greater than the $\Delta G^{\ddagger}$ for the reaction of an alkene.

Alkynes are less reactive than alkenes in electrophilic addition reactions.

ties of the transition states must be greater than the difference in the stabilities of the reactants so that $\Delta G^{\ddagger}_{\text{alkyne}} > \Delta G^{\ddagger}_{\text{alkene}}$ (Figure 5.2).

> ### PROBLEM 6 ◆
>
> Under what circumstances can you assume that the less stable reactant will be the more reactive reactant?

Why is the transition state for the first step of an electrophilic addition reaction for an alkyne less stable than that for an alkene? The Hammond postulate predicts that the structure of the transition state will resemble the structure of the intermediate (Section 3.11). The intermediate formed when an electrophile (such as H$^+$) adds to an alkyne is a vinylic cation, whereas the intermediate formed when an electrophile adds to an alkene is an alkyl cation. A **vinylic cation** has a positive charge on a vinylic carbon. A vinylic cation is less stable than a similarly substituted alkyl cation. In other words, a primary vinylic cation is less stable than a primary alkyl cation.

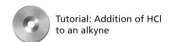

Tutorial: Addition of HCl to an alkyne

relative stabilities of carbocations

$$
\underset{\substack{\text{a tertiary}\\\text{carbocation}}}{R-\overset{\overset{R}{|}}{\underset{\underset{R}{|}}{C^+}}} \;>\; \underset{\substack{\text{a secondary}\\\text{carbocation}}}{R-\overset{\overset{R}{|}}{\underset{\underset{H}{|}}{C^+}}} \;>\; \underset{\substack{\text{a secondary}\\\text{vinylic cation}}}{RCH=\overset{+}{C}-R} \;\approx\; \underset{\substack{\text{a primary}\\\text{carbocation}}}{R-\overset{\overset{H}{|}}{\underset{\underset{H}{|}}{C^+}}} \;>\; \underset{\substack{\text{a primary}\\\text{vinylic cation}}}{RCH=\overset{+}{C}-H} \;\approx\; \underset{\substack{\text{the methyl}\\\text{cation}}}{H-\overset{\overset{H}{|}}{\underset{\underset{H}{|}}{C^+}}}
$$

← increasing stability

A vinylic cation is less stable because the positive charge is on an sp carbon, which we will see is more electronegative than the sp^2 carbon of an alkyl cation (Section 5.9). Therefore, a vinylic cation is less able to bear a positive charge. In addition, hyperconjugation is less effective in stabilizing the charge on a vinylic cation than on an alkyl cation (Section 3.10).

5.5 ADDITION OF HYDROGEN HALIDES AND ADDITION OF HALOGENS

The addition of a hydrogen halide to a terminal alkyne follows Markovnikov's rule because a secondary vinylic cation is more stable than the primary vinylic cation that would be formed if the proton added to the other sp carbon.

$$
\underset{\text{1-butyne}}{CH_3CH_2C\equiv CH} \xrightarrow{H^+} CH_3CH_2\overset{+}{C}=CH_2 \xrightarrow{Br^-} \underset{\substack{\text{2-bromo-1-butene}\\\text{a halo-substituted alkene}}}{CH_3CH_2\overset{\overset{\displaystyle Br}{|}}{C}=CH_2}
$$

$$
\underset{\substack{\text{a secondary vinylic cation}}}{CH_3CH_2\overset{+}{C}=CH_2} \;>\; \underset{\substack{\text{a primary vinylic cation}}}{CH_3CH_2CH=\overset{+}{C}H}
$$

Although the addition of a hydrogen halide to an alkyne can generally be stopped after the addition of one equivalent[1] of hydrogen halide, a second addition reaction will take place if excess hydrogen halide is present. The product of the second addition reaction is a **geminal dihalide,** a molecule with two halogens on the same carbon. "Geminal" comes from *geminus,* which is Latin for "twin."

$$CH_3CH_2C=CH_2 \xrightarrow{\text{HBr}} CH_3CH_2CCH_3$$

with Br above and Br below on product

2-bromo-1-butene   **2,2-dibromobutane**
                                a geminal dihalide

Addition of the second equivalent of hydrogen halide also follows Markovnikov's rule. The carbocation formed by following Markovnikov's rule is more stable than the carbocation that would be formed if the rule were not followed because bromine can share the positive charge with the carbon by donating a pair of nonbonding electrons.

$$CH_3CH_2\overset{+}{C}-CH_3 \longleftrightarrow CH_3CH_2\overset{+}{C}-CH_3$$

**carbocation formed by
following Markovnikov's rule**

The addition of a hydrogen halide to an alkyne can generally be stopped after the addition of one equivalent of HBr or HCl because, although an alkyne is less reactive than an alkene, an alkyne is more reactive than the halo-substituted alkene that is the product of the first addition reaction. This is because the electron-withdrawing halogen substituent decreases the nucleophilic character of the double bond.

In describing the mechanism for addition of a hydrogen halide, we have shown that the intermediate is a vinylic cation. It turns out that this mechanism may not be completely correct. A secondary vinylic cation is about as stable as a primary carbocation, and generally primary carbocations are too unstable to be formed. Therefore, can a secondary vinylic cation be formed? Some chemists think that a **pi-complex** rather than a vinylic cation is formed as an intermediate.

$$\overset{\delta-\ Cl}{\underset{HC\equiv CH}{\overset{|}{\underset{}{\overset{\delta+\ H}{\wedge}}}}}$$

a pi-complex

3-D Molecule:
pi-complex

Support for the intermediate being a pi-complex comes from the observation that many (but not all) alkyne addition reactions are stereoselective. For example, the addition of HCl to 2-butyne forms only (*Z*)-2-chloro-2-butene which means that only anti addition of H and Cl occurs. Clearly, the nature of the intermediate in alkyne addition reactions is not completely understood.

$$CH_3C\equiv CCH_3 \xrightarrow{\text{HCl}} \overset{H}{\underset{H_3C}{>}}C=C\overset{CH_3}{\underset{Cl}{<}}$$

2-butyne   **(*Z*)-2-chloro-2-butene**

[1]One equivalent means the same number of moles as the other reactant.

Addition of hydrogen halide to an internal alkyne forms two geminal dihalides because the initial addition of the proton can occur equally easily to either of the *sp* carbons.

$$CH_3CH_2C\equiv CCH_3 \ + \ HCl \ \longrightarrow \ \underset{\underset{Cl}{|}}{CH_3CH_2CH_2\overset{\overset{Cl}{|}}{C}CH_3} \ + \ \underset{\underset{Cl}{|}}{CH_3CH_2\overset{\overset{Cl}{|}}{C}CH_2CH_3}$$

| 2-pentyne | excess | 2,2-dichloropentane | 3,3-dichloropentane |

However, if the internal alkyne has identical groups attached to the *sp* carbons, only one geminal dihalide is obtained.

$$CH_3CH_2C\equiv CCH_2CH_3 \ + \ HBr \ \longrightarrow \ \underset{\underset{Br}{|}}{CH_3CH_2CH_2\overset{\overset{Br}{|}}{C}CH_2CH_3}$$

| 3-hexyne | excess | |
3,3-dibromohexane

Hydrogen peroxide (or an alkyl peroxide) has the same effect on the addition of HBr to an alkyne that it has on the addition of HBr to an alkene (Section 3.18). In the presence of peroxide, HBr adds to a terminal alkyne in an apparent anti-Markovnikov manner because Br · rather than H⁺ is the electrophile.

$$CH_3CH_2C\equiv CH \ + \ HBr \ \xrightarrow{\textbf{peroxide}} \ CH_3CH_2CH=CHBr$$

| 1-butyne | | 1-bromo-1-butene |

The mechanism of the reaction is the same as that for the addition of HBr to an alkene in the presence of peroxide. That is, peroxide is a radical initiator and creates a bromine radical that adds to the *sp* carbon bonded to the hydrogen. The resulting vinylic radical abstracts a hydrogen atom from HBr and regenerates the bromine radical.

mechanism for the addition of HBr in the presence of peroxide

$$R\ddot{O}-\ddot{O}R \ \longrightarrow \ 2\,R\ddot{O}\cdot$$

$$R\ddot{O}\cdot \ + \ H-\ddot{B}r: \ \longrightarrow \ R\ddot{O}H \ + \ \cdot\ddot{B}r:$$

$$CH_3CH_2C\equiv CH \ + \ \cdot\ddot{B}r: \ \longrightarrow \ \underset{:\ddot{B}r:}{CH_3CH_2\overset{}{C}=CH}$$

$$CH_3CH_2\underset{\underset{Br}{|}}{C}=CH \ + \ H-\ddot{B}r: \ \longrightarrow \ CH_3CH_2CH=CHBr \ + \ \cdot\ddot{B}r:$$

The halogens Cl_2 and Br_2 add to alkynes. In the presence of excess halogen, a second addition reaction occurs. Typically the solvent is CH_2Cl_2.

$$CH_3CH_2C\equiv CCH_3 \ \xrightarrow[\textbf{CH}_2\textbf{Cl}_2]{\textbf{Cl}_2} \ \underset{\underset{Cl}{|}}{CH_3CH_2\overset{\overset{Cl}{|}}{C}=CCH_3} \ \xrightarrow[\textbf{CH}_2\textbf{Cl}_2]{\textbf{Cl}_2} \ \underset{\underset{Cl}{|}\,\underset{Cl}{|}}{CH_3CH_2\overset{\overset{Cl}{|}\,\overset{Cl}{|}}{C}-CCH_3}$$

$$CH_3C\equiv CH \ \xrightarrow[\textbf{CH}_2\textbf{Cl}_2]{\textbf{Br}_2} \ \underset{\underset{Br}{|}}{CH_3\overset{\overset{Br}{|}}{C}=CH} \ \xrightarrow[\textbf{CH}_2\textbf{Cl}_2]{\textbf{Br}_2} \ \underset{\underset{Br}{|}\,\underset{Br}{|}}{CH_3\overset{\overset{Br}{|}\,\overset{Br}{|}}{C}-CH}$$

PROBLEM 7 ◆

Give the major product of each of the following reactions:

a. $HC\equiv CCH_2CH_3 + HBr \longrightarrow$

b. $HC\equiv CCH_2CH_3 + 2HBr \longrightarrow$

c. $HC\equiv CCH_2CH_3 + HBr \xrightarrow{\text{peroxide}}$

d. $HC\equiv CCH_2CH_3 + Br_2 \xrightarrow{CH_2Cl_2}$

e. $HC\equiv CCH_2CH_3 + 2\,Cl_2 \xrightarrow{CH_2Cl_2}$

PROBLEM 8 ◆

From what you know about the stereochemistry of alkene addition reactions, predict the configurations of the products that would be obtained from the reaction of 2-butyne with the following:

a. one equivalent of Br_2 in CH_2Cl_2 **b.** one equivalent of HBr + peroxide

---

In Section 3.13 we saw that alkenes undergo acid-catalyzed addition of water. The product of the reaction is an alcohol.

**5.6
ADDITION
OF WATER**

$$CH_3CH_2CH{=}CH_2 + H_2O \xrightarrow{H_2SO_4} CH_3CH_2CHCH_3$$
1-butene

(with OH below), 2-butanol

Alkynes also undergo acid-catalyzed addition of water. The product of the reaction is an enol. (The ending "ene" signifies the double bond, and "ol" the OH. When the two endings are joined, the final *e* of "ene" is dropped to avoid two consecutive vowels, but it is pronounced as if the *e* were there, "ene-ol.")

$$CH_3C\equiv CCH_3 + H_2O \xrightarrow{H_2SO_4} CH_3C{=}CHCH_3 \rightleftharpoons CH_3C{-}CH_2CH_3$$
an enol   a ketone

The enol immediately rearranges to a ketone. A carbon doubly bonded to an oxygen is called a carbonyl ("car-bo-nil") group. A ketone is a compound that has two alkyl groups bonded to a carbonyl group. An aldehyde is a compound that has at least one hydrogen bonded to a carbonyl group.

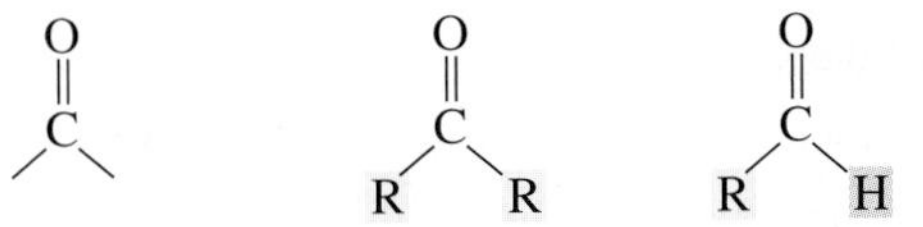

a carbonyl group   **a ketone**   **an aldehyde**

Terminal alkynes are less reactive than internal alkynes toward the addition of water. Terminal alkynes will add water if mercuric ion (Hg^{2+}) is added to the acidic mixture. The mercuric ion acts as a catalyst to increase the rate of the addition reaction.

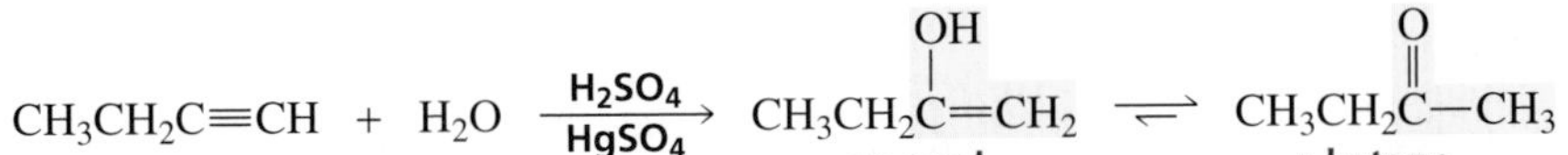

$$CH_3CH_2C\equiv CH + H_2O \xrightarrow[HgSO_4]{H_2SO_4} CH_3CH_2\underset{\text{an enol}}{C}=CH_2 \rightleftharpoons CH_3CH_2\overset{O}{\underset{}{C}}-CH_3$$

an enol   a ketone

The first step in the mercuric-ion-catalyzed hydration of an alkyne is formation of a cyclic mercurinium ion. (Two of the electrons in mercury's filled $5d$ atomic orbital are shown.) In the second step, water attacks the most substituted carbon of the cyclic intermediate (Section 3.16). Oxygen loses a proton to form a mercuric enol, which immediately rearranges to a mercuric ketone. Loss of the mercuric ion forms an enol, which rearranges to a ketone. Notice that the overall addition of water follows Markovnikov's rule.

mechanism for the mercuric-ion-catalyzed hydration of an alkyne

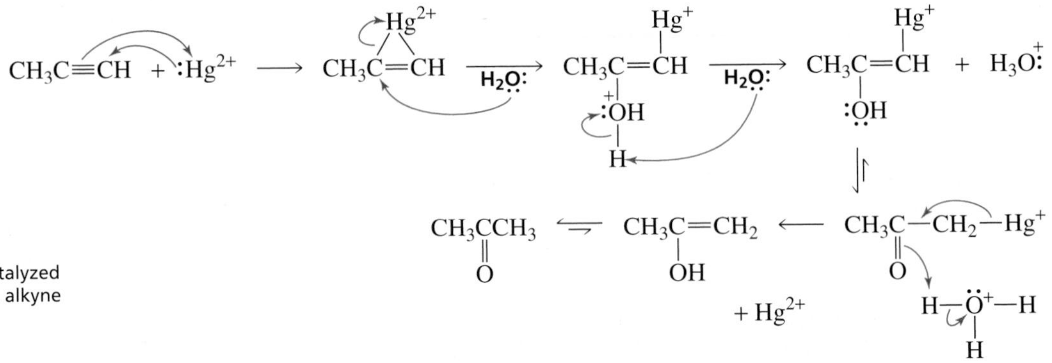

Tutorial:
Mercuric-ion catalyzed
hydration of an alkyne

A ketone and an enol differ from each other only in the location of a double bond and in the location of a hydrogen. Such isomers are called **tautomers.** The ketone and enol are called **keto-enol tautomers.** The conversion of one tautomer into the other tautomer is called **tautomerization.** We will examine the mechanism of this reaction in Chapter 19. For now, the important thing to remember is that the keto and enol tautomers come to equilibrium in solution, and the keto tautomer, because it is usually much more stable than the enol tautomer, predominates at equilibrium.

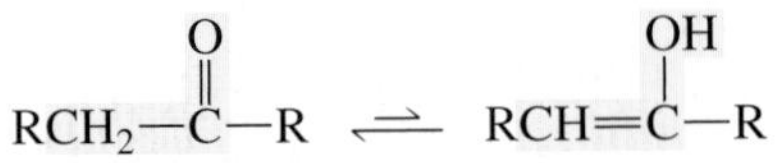

$$RCH_2-\overset{O}{\underset{}{C}}-R \rightleftharpoons RCH=\overset{OH}{\underset{}{C}}-R$$

keto tautomer   **enol tautomer**

Addition of water to an internal alkyne with identical groups attached to the sp carbons forms a single ketone as a product. But if the two groups are not identical, two ketones are formed because the initial addition of the proton can occur to either of the sp carbons.

$$CH_3CH_2C≡CCH_2CH_3 + H_2O \xrightarrow{H_2SO_4} CH_3CH_2\overset{\displaystyle O}{\overset{\|}{C}}CH_2CH_2CH_3$$

$$CH_3C≡CCH_2CH_3 + H_2O \xrightarrow{H_2SO_4} CH_3\overset{\displaystyle O}{\overset{\|}{C}}CH_2CH_2CH_3 + CH_3CH_2\overset{\displaystyle O}{\overset{\|}{C}}CH_2CH_3$$

PROBLEM 9 ◆

What ketones would be formed from the acid-catalyzed hydration of 3-heptyne?

PROBLEM 10 ◆

Which alkyne would be the best reagent to use for the synthesis of each of the following ketones?

a. $CH_3\overset{\displaystyle O}{\overset{\|}{C}}CH_3$ **b.** $CH_3CH_2\overset{\displaystyle O}{\overset{\|}{C}}CH_2CH_2CH_3$ **c.** $CH_3\overset{\displaystyle O}{\overset{\|}{C}}-$⬡

PROBLEM 11 ◆

Draw the enol tautomers for each of the ketones in Problem 10.

Borane adds to alkynes in the same way it adds to alkenes. That is, one mole of BH_3 reacts with three moles of alkyne to form one mole of boron-substituted alkene (Section 3.17). When the addition reaction is over, aqueous sodium hydroxide and hydrogen peroxide are added to the reaction mixture. The end result, as in the case of alkenes, is replacement of the boron by an OH group. The enol product immediately rearranges to a ketone.

5.7 ADDITION OF BORANE: HYDROBORATION-OXIDATION

$$3 \; CH_3C≡CCH_3 + BH_3 \xrightarrow{THF} \underset{\substack{H \quad\quad B-R \\ |\\ R}}{\overset{CH_3 \quad\quad CH_3}{C=C}} \xrightarrow[H_2O]{HO^-, \, H_2O_2} 3 \; \underset{\substack{H \quad\quad OH \\ \textbf{an enol}}}{\overset{CH_3 \quad\quad CH_3}{C=C}}$$

boron-substituted alkene

$$↕$$

$$3 \; CH_3CH_2\overset{\displaystyle O}{\overset{\|}{C}}CH_3$$

3-D Molecule:
Disiamylborane

BH_3

disiamylborane

   In order to obtain the enol as the product of the addition reaction, only one equivalent of BH_3 can be allowed to add to the alkyne. In other words, the reaction must stop at the alkene stage. In the case of internal alkynes, the bulky groups on the boron-substituted alkene prevent the second addition from occurring. However, there is less steric hindrance in a terminal alkyne, so it is harder to stop the addition reaction at the alkene stage. A special reagent called disiamylborane has been developed for use with terminal alkynes ("siamyl" stands for **s**econdary **iso** **amyl**; amyl is a common name for a five-carbon fragment). The bulky alkyl groups of disiamylborane prevent a second addition to the boron-substituted alkene. So borane can be used to hydrate internal alkynes, but disiamylborane is preferred for the hydration of terminal alkynes.

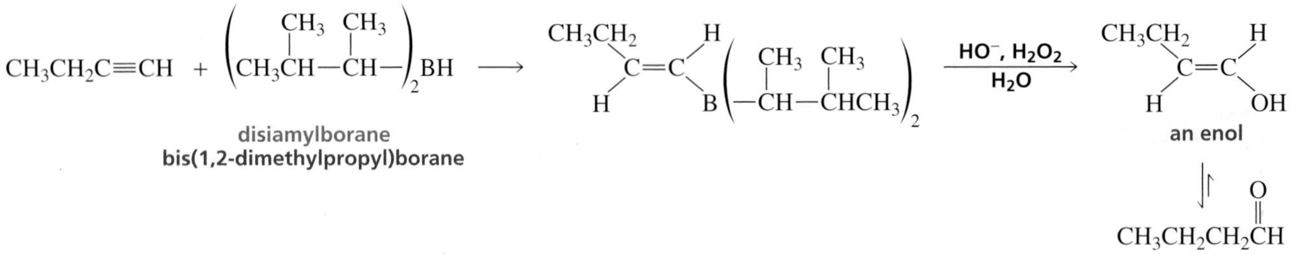

$$CH_3CH_2C{\equiv}CH \ + \ \left(CH_3\overset{\overset{\displaystyle CH_3}{|}}{CH}-\overset{\overset{\displaystyle CH_3}{|}}{CH}-\right)_2 BH \longrightarrow$$

disiamylborane
bis(1,2-dimethylpropyl)borane

an enol

$$CH_3CH_2CH_2\overset{\overset{\displaystyle O}{\|}}{CH}$$

The addition of borane (or disiamylborane) to a terminal alkyne exhibits the same regioselectivity seen in borane addition to an alkene. Boron, with its electron-seeking empty orbital, adds preferentially to the sp carbon bonded to the hydrogen. Because the second step of the hydroboration–oxidation reaction results in replacement of boron by an OH group, the overall reaction is an apparent anti-Markovnikov addition of water to the alkyne. The product appears to violate Markovnikov's original statement of the rule because the original statement was created for a reaction in which H^+ is the electrophile. In hydroboration–oxidation, H^+ is not the electrophile, $H:^-$ is the nucleophile. Consequently, mercuric-ion-catalyzed addition of water to a terminal alkyne produces a ketone (the carbonyl group is *not* on the terminal carbon), whereas hydroboration–oxidation of the alkyne produces an aldehyde (the carbonyl group *is* on the terminal carbon).

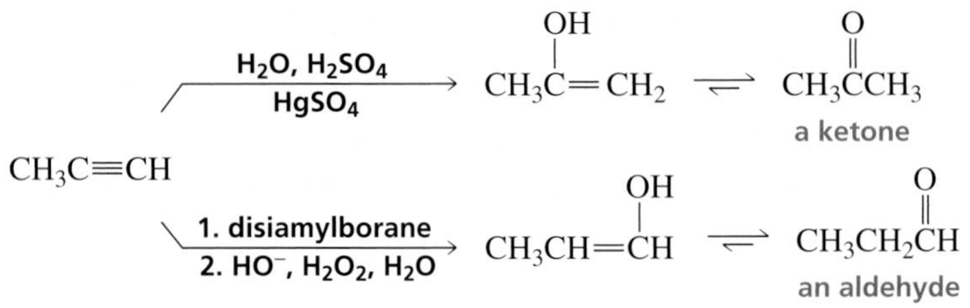

a ketone

an aldehyde

PROBLEM 12 ◆

Give the products of (1) mercuric-ion-catalyzed addition of water and (2) hydroboration–oxidation for the following:

a. 1-butyne **b.** 2-butyne **c.** 2-pentyne

PROBLEM 13 ◆

There is only one alkyne that forms an aldehyde when it undergoes either acid- or mercuric-ion-catalyzed addition of water. Identify the alkyne.

**5.8
ADDITION OF
HYDROGEN**

Hydrogen adds to an alkyne in the presence of a metal catalyst such as palladium, platinum, or nickel in the same manner in which it adds to an alkene. Because hydrogen readily adds to alkenes, the reaction cannot be stopped at the alkene stage. The product of the hydrogenation reaction, therefore, is an alkane.

$$CH_3CH_2C≡CH \xrightarrow[Pt/C]{H_2} CH_3CH_2CH=CH_2 \xrightarrow[Pt/C]{H_2} CH_3CH_2CH_2CH_3$$

The relative reactivities of double and triple bonds toward the addition of hydrogen depend on the metal catalyst. In the presence of a partially deactivated metal catalyst, addition of hydrogen to an alkene is much slower than addition of hydrogen to an alkyne. So if an alkene is the desired product of the reaction, such a catalyst must be used. The most common partially deactivated metal catalyst is Lindlar's catalyst. Because the alkyne sits on the surface of the metal catalyst and the hydrogens are delivered to the triple bond from the surface of the catalyst, only syn addition of hydrogen occurs (Section 3.19). Syn addition of hydrogen to an internal alkyne forms a cis alkene.

Herbert H. M. Lindlar was born in Switzerland in 1909 and received a Ph.D. from the University of Bern. He worked at Hoffmann–La Roche and Co. in Basel, Switzerland, and he authored many patents. His last patent was a procedure for isolating the carbohydrate xylose from the waste produced in paper mills.

$$CH_3CH_2C≡CCH_3 + H_2 \xrightarrow[\text{catalyst}]{\text{Lindlar's}}$$

2-pentyne

H   H
  \\   /
   C=C
  /   \\
CH₃CH₂   CH₃

cis-2-pentene

Tutorial:
Hydrogenation/
Lindlar's catalyst

LINDLAR'S CATALYST

Lindlar's catalyst can be prepared by precipitating palladium on calcium carbonate and treating it with lead(II) acetate and quinoline. This treatment partially deactivates the catalyst. In the deactivated state, it is much more effective at catalyzing the addition of hydrogen to a triple bond than to a double bond.

$(CH_3COO^-)_2Pb^{2+}$
lead(II) acetate

quinoline

Alkynes can also be converted into alkenes using an equivalent of sodium (or lithium) in liquid ammonia because sodium (or lithium) reacts more rapidly with triple bonds than with double bonds. This reaction converts an internal alkyne into a trans alkene. Ammonia $(bp = -33\,°C)$ is a gas at room temperature, so it is kept in the liquid state by using a dry ice/acetone mixture $(bp = -78\,°C)$.

$$CH_3C≡CCH_3 \xrightarrow[\substack{NH_3\ (liq) \\ -78\,°C}]{Na\ or\ Li}$$

2-butyne

CH₃   H
  \\   /
   C=C
  /   \\
H   CH₃

trans-2-butene

Sodium and lithium have a strong tendency to lose the single electron in their outer-shell *s* orbital (Section 1.3). Thus, the first step in the mechanism of this reaction is transfer of the *s* orbital electron from sodium (or lithium) to an *sp* carbon to form a **radical anion** (a species with a negative charge and an unpaired electron). The radical anion abstracts a proton from ammonia. This results in the formation of a **vinylic radical** (the unpaired electron is on a vinylic carbon). Another single electron transfer from sodium (or lithium) to the vinylic radical forms a vinylic anion, which abstracts a proton from another molecule of ammonia. The product is the trans alkene.

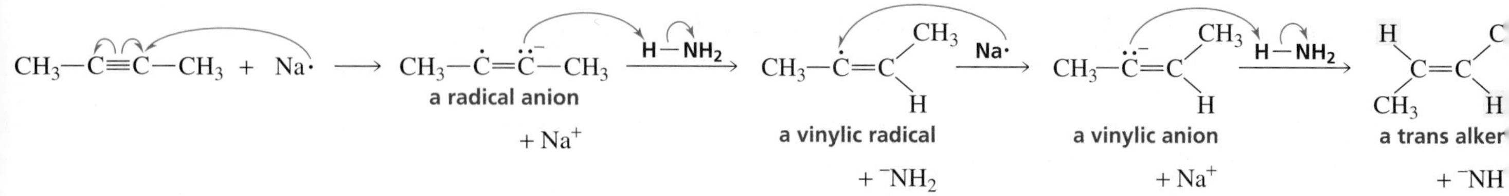

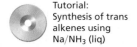

The vinylic anion intermediate can have either the cis or the trans configuration. The cis and trans configurations are in equilibrium, but the equilibrium favors the more stable trans configuration. The reaction forms the trans alkene because the rate-determining step occurs prior to the step in which the vinylic anion reacts. If reaction of the vinylic anion were the rate-determining step, the less stable cis vinylic anion could react faster than the trans vinylic anion, and the cis product would be formed.

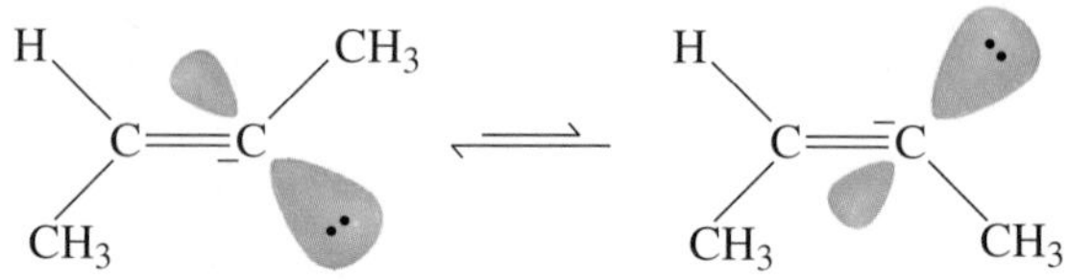

Tutorial:
Synthesis of trans alkenes using Na/NH₃ (liq)

trans vinylic anion
more stable

cis vinylic anion
less stable

PROBLEM 14 ◆

What alkyne would you start with and what reagents would you use if you wanted to synthesize:

a. pentane? **b.** *cis*-2-butene? **c.** *trans*-2-pentene? **d.** 1-hexene?

5.9
ACIDITY OF A HYDROGEN BONDED TO AN *sp* HYBRIDIZED CARBON

Carbon forms very strong covalent bonds with hydrogen because their similar electronegativities cause them to share the bonding electrons almost equally. However, not all carbon atoms have the same electronegativity. An *sp* hybridized carbon is more electronegative than an sp^2 hybridized carbon, which is more electronegative than an sp^3 hybridized carbon, which is just slightly more electronegative than a hydrogen. (Review Chapter 1, Problem 38, parts d, e, and f.)

relative electronegativities of carbon atoms

$$sp > sp^2 > sp^3$$

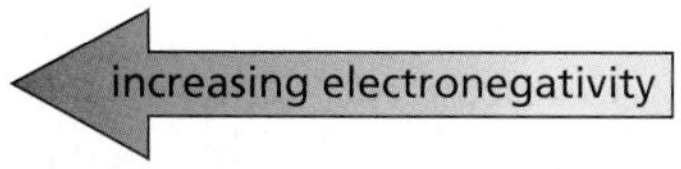

increasing electronegativity

Why does the type of hybridization affect the electronegativity of the carbon atom? Electronegativity is a measure of the ability of an atom to pull the bonding electrons toward itself. The closer the bonding electrons are to the nucleus, the more electronegative the atom. The average distance of a 2s electron from the nucleus is less than the average distance of a 2p electron from the nucleus. Therefore, the electrons in an *sp* hybrid orbital (50% *s* character) are closer, on average, to the nucleus than those in an sp^2 hybrid orbital (33.3% s character). In turn, sp^2 electrons are closer to the nucleus than sp^3 electrons (25% s character). The sp hybridized carbon, therefore, is the most electronegative.

In Section 1.18, we saw that the acidity of a hydrogen attached to some of the second-row elements depends on the electronegativity of the atom to which the hydrogen is attached. The greater the electronegativity, the greater the acidity of the hydrogen (i.e., the stronger the acid). (Don't forget, the stronger the acid, the lower its pK_a.)

relative electronegativities   N   <   O   <   F

relative acid strengths NH_3 < H_2O <   HF
$pK_a = 36$ $pK_a = 15.7$ $pK_a = 3.2$

increasing acid strength and electronegativity

Because the electronegativity of carbon atoms follows the order $sp > sp^2 > sp^3$, ethyne is a stronger acid than ethene, and ethene is a stronger acid than ethane. Notice that some positive charge (blue area) is evident on the hydrogens of ethyne, but essentially none is visible on the hydrogens of ethene or ethane, indicative of their much higher pK_a values.

$HC{\equiv}CH$ $H_2C{=}CH_2$ CH_3CH_3
$pK_a = 25$ $pK_a = 44$ $pK_a = 50$

We can compare the acidities of these compounds with the acidities of hydrogens attached to other second-row elements. Notice that the intensity of the blue area increases as the acidity increases (Section 1.3).

relative acid strengths

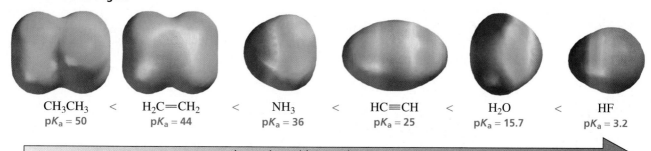

CH_3CH_3 < $H_2C{=}CH_2$ < NH_3 < $HC{\equiv}CH$ < H_2O <   HF
$pK_a = 50$ $pK_a = 44$ $pK_a = 36$ $pK_a = 25$ $pK_a = 15.7$ $pK_a = 3.2$

increasing acid strength

The conjugate bases have the following relative base strengths because *the stronger the acid, the weaker its conjugate base.*

relative base strengths

$CH_3CH_2^-$ > $H_2C{=}CH^-$ > H_2N^- > $HC{\equiv}C^-$ > HO^- > F^-

increasing base strength

The stonger the acid, the weakier its conjugate base.

In order to remove a proton from an acid, the base that is removing the proton must be stronger than the base generated as a result of proton removal (Section 1.17). In other words, you must start with a stronger base than the base you are making. The amide ion ($^-NH_2$) can remove a hydrogen bonded to an *sp* carbon because it is a stronger base than the acetylide ion that is formed. The reagent actually used to remove the hydrogen is sodium amide ($Na^+ \; ^-NH_2$) in liquid ammonia.

$$RC\equiv CH \quad + \quad ^-NH_2 \quad \rightleftharpoons \quad RC\equiv C^- \quad + \quad NH_3$$

	amide anion		acetylide anion	
stronger acid	stronger base		weaker base	weaker acid

Hydroxide ion is not a strong enough base to remove a hydrogen bonded to an *sp* carbon because it is a weaker base than the acetylide ion that would be formed.

$$RC\equiv CH \quad + \quad HO^- \quad \rightleftharpoons \quad RC\equiv C^- \quad + \quad H_2O$$

	hydroxide anion		acetylide anion	
weaker acid	weaker base		stronger base	stronger acid

The amide ion cannot remove a hydrogen bonded to an sp^2 or an sp^3 carbon. Only a hydrogen bonded to an *sp* carbon is sufficiently acidic to be removed by the amide ion. As a result, a hydrogen bonded to an *sp* carbon is referred to as an "acidic" hydrogen. The fact that terminal alkynes are "acidic" is one way in which their chemistry differs from that of alkenes. We must be careful not to misinterpret what we mean when we say that a hydrogen bonded to an *sp* carbon is acidic. It is more acidic than most other carbon-bound hydrogens but it is much less acidic than a hydrogen of a water molecule, and we normally do not consider water to be an acidic compound.

SODIUM AMIDE AND SODIUM

Take care not to confuse sodium amide ($Na^+ \; ^-NH_2$) in liquid ammonia with sodium (Na) in liquid ammonia. Sodium amide is the strong base used to remove a proton from a terminal alkyne. Sodium is used as a source of electrons in the reduction of an alkyne to a trans alkene (Section 5.8).

PROBLEM 15 ◆

Any base whose conjugate acid has a pK_a greater than _____ can remove a proton from a terminal alkyne to form an acetylide anion (in a reaction that favors products).

PROBLEM 16 ◆

Which carbocation in each of the following pairs is more stable?

a. $CH_3\overset{+}{C}H_2$ or $H_2C=\overset{+}{C}H$ **b.** $H_2C=\overset{+}{C}H$ or $HC\equiv\overset{+}{C}$

PROBLEM-SOLVING STRATEGY

a. List the following compounds in order of decreasing acidity:

$CH_3CH_2\overset{+}{N}H_3$ $CH_3CH=\overset{+}{N}H_2$ $CH_3C\equiv\overset{+}{N}H$

To compare the acidities of a group of compounds, first look at how they differ. These three compounds differ in the hybridization of the nitrogen to which the acidic hydrogen is attached. Now, recall what you know about hybridization and acidity. You know that hybridization of an atom affects its electronegativity (sp is more electronegative than sp^2, and sp^2 is more electronegative than sp^3), and you know that the more electronegative the atom to which a hydrogen is attached, the more acidic the hydrogen. Now you can answer the question.

relative acidities $CH_3C{\equiv}\overset{+}{N}H$ > $CH_3CH{=}\overset{+}{N}H_2$ > $CH_3CH_2\overset{+}{N}H_3$

b. Draw the conjugate bases and list them in order of decreasing basicity.

Simply remove a proton from each acid to get the structures of the conjugate bases, and recall that the stronger the acid, the weaker its conjugate base.

relative basicities $CH_3CH_2NH_2$ > $CH_3CH{=}NH$ > $CH_3C{\equiv}N$

Reactions that form new carbon–carbon bonds are important in the synthesis of organic compounds. Without such reactions, we could not convert molecules with small carbon skeletons into larger ones. Instead, the product of a reaction would always have the same number of carbons as the starting material.

One reaction that forms a new carbon–carbon bond is the reaction of an acetylide ion with an alkyl halide. Only primary alkyl halides or methyl halides should be used in this reaction.

**5.10
SYNTHESIS USING
ACETYLIDE IONS**

$$CH_3CH_2C{\equiv}C^- \; + \; CH_3CH_2CH_2Br \; \longrightarrow \; CH_3CH_2C{\equiv}CCH_2CH_2CH_3 \; + \; Br^-$$
3-heptyne

The mechanism of this reaction is well understood. Bromine is more electronegative than carbon, and as a result the electrons in the carbon–bromine bond are not shared equally by the two atoms. This means that the carbon–bromine bond is polar, with a partial negative charge on bromine and a partial positive charge on carbon. The negative charge on the acetylide ion (red area) is attracted to the partial positive charge on the carbon (bluish area) of the alkyl halide. As the electrons of the acetylide ion approach the carbon to form a new carbon–carbon bond, they push out the bromine and its bonding electrons because carbon can bond to no more than four atoms at a time. This is called an **alkylation reaction** because an alkyl group has been attached to the acetylide ion.

$$CH_3CH_2C{\equiv}\ddot{C}^- \; + \; CH_3CH_2CH_2\overset{\delta+}{-}\overset{\delta-}{Br} \; \longrightarrow \; CH_3CH_2C{\equiv}CCH_2CH_2CH_3 \; + \; Br^-$$

3-D Molecules:
Anion of
1-butyne;
1-bromobutane

The mechanism of this and similar reactions will be discussed in greater detail in Chapter 9. At that time, we will also see why the reaction works best with primary alkyl halides and methyl halides.

Simply by choosing an alkyl halide of the appropriate structure, terminal alkynes can be converted into internal alkynes of any desired chain length.

$$CH_3CH_2CH_2C\equiv CH \xrightarrow[\text{2. } CH_3CH_2CH_2CH_2CH_2Cl]{\text{1. NaNH}_2} CH_3CH_2CH_2C\equiv CCH_2CH_2CH_2CH_2CH_3$$

1-pentyne · 4-decyne

PROBLEM 17 / SOLVED

A chemist wants to synthesize 4-decyne but cannot find any 1-pentyne, the starting material used in the synthesis just described. How else can 4-decyne be synthesized?

SOLUTION   The *sp* carbons of 4-decyne are bonded to a propyl group and to a pentyl group. Therefore, to obtain 4-decyne the acetylide ion of 1-pentyne can react with a pentyl halide or the acetylide ion of 1-heptyne can react with a propyl halide.

$$CH_3CH_2CH_2CH_2CH_2C\equiv CH \xrightarrow[\text{2. } CH_3CH_2CH_2Cl]{\text{1. NaNH}_2} CH_3CH_2CH_2C\equiv CCH_2CH_2CH_2CH_2CH_3$$

1-heptyne · 4-decyne

5.11
DESIGNING A SYNTHESIS I: AN INTRODUCTION TO MULTISTEP SYNTHESIS

So far in our study of organic chemistry, we have paid attention to *why* a reaction occurs, *how* it occurs, and the products that are formed. A good way to review these reactions is to design syntheses because, in so doing, you must be able to recall many of the reactions you have learned.

Synthetic chemists consider time, cost, and yield in designing syntheses. In the interest of time, a well-designed synthesis should require as few steps (sequential reactions) as possible, and those steps should each involve a reaction that is easy to carry out. If two chemists in a pharmaceutical company were each asked to prepare a new drug, and one synthesized the drug in three simple steps while the other used 20 difficult steps, which chemist would be out looking for a new job? In addition, each step in the synthesis should result in the greatest possible yield of the desired product. The more reactant needed to synthesize one gram of product, the more expensive it is to make. Sometimes it is preferable to design a synthesis involving several steps if the starting materials are inexpensive, the reactions are easy to carry out, and the yield of each step is high. This would be better than designing a synthesis with fewer steps that require expensive starting materials and reactions that are more difficult or give lower yields.

The following examples will give you an idea of the type of thinking required for the design of a successful synthesis. This kind of problem will appear repeatedly throughout this book because working such problems is a good way to learn organic chemistry.

Example 1. Starting with 1-butyne, how could you make the following ketone? You can use any organic and inorganic reagents.

$$CH_3CH_2C\equiv CH \xrightarrow{?} CH_3CH_2\overset{\overset{\displaystyle O}{\|}}{C}CH_2CH_2CH_3$$

Many students find that the easiest way to design a synthesis is to work backward. Instead of looking at the starting material and deciding how to do the first step of the synthesis, look at the product and decide how to do the last step. The product is a ketone. At this point the only reaction you know that forms a ketone is the addition of water (in the presence of a catalyst) to an internal alkyne. If the alkyne used in the reaction has identical substituents on the *sp* carbons, only one ketone will be obtained. Thus, 3-hexyne is the best alkyne to use for the synthesis of the desired ketone.

$$CH_3CH_2C\equiv CCH_2CH_3 \xrightarrow[\text{H}_2\text{SO}_4]{\text{H}_2\text{O}} \underset{\text{OH}}{CH_3CH_2\overset{|}{C}=CHCH_2CH_3} \rightleftharpoons CH_3CH_2\overset{\overset{\text{O}}{\|}}{C}CH_2CH_2CH_3$$

3-Hexyne can be obtained from the starting material by removing the hydrogen bonded to the *sp* carbon followed by alkylation. To obtain the desired product, a two-carbon alkyl halide must be used in the alkylation reaction.

$$CH_3CH_2C\equiv CH \xrightarrow[\text{2. CH}_3\text{CH}_2\text{Br}]{\text{1. NaNH}_2} CH_3CH_2C\equiv CCH_2CH_3$$

Designing a synthesis by working backward from product to reactant is not simply a technique taught to organic chemistry students. It is done so frequently by experienced synthetic chemists that it has been given a name—**retrosynthetic analysis.** Chemists indicate they are working backward by using open arrows. Typically the reagents needed to carry out each step are not included until the reaction is written in the forward direction. For example, the route to the synthesis of the previous ketone can be arrived at by the following retrosynthetic analysis.

Elias James Corey coined the term "retrosynthetic analysis" (Section 28.2). He was born in Massachusetts in 1928 and is a professor of chemistry at Harvard University. He received the Nobel Prize in 1990 for his contribution to synthetic organic chemistry.

retrosynthetic analysis

$$CH_3CH_2\overset{\overset{\text{O}}{\|}}{C}CH_2CH_2CH_3 \Longrightarrow CH_3CH_2C\equiv CCH_2CH_3 \Longrightarrow CH_3CH_2C\equiv CH$$

Once the sequence of reactions has been worked out by retrosynthetic analysis, the synthetic scheme can be shown by reversing the steps and including the reagents required to carry out each step.

synthesis

$$CH_3CH_2C\equiv CH \xrightarrow[\text{2. CH}_3\text{CH}_2\text{Br}]{\text{1. NaNH}_2} CH_3CH_2C\equiv CCH_2CH_3 \xrightarrow[\text{H}_2\text{SO}_4]{\text{H}_2\text{O}} CH_3CH_2\overset{\overset{\text{O}}{\|}}{C}CH_2CH_2CH_3$$

Retrosynthetic analysis will be discussed again in other chapters and will be covered in much greater detail in Chapter 29.

Example 2. Starting with ethyne, how could you make 1-bromopentane?

$$HC\equiv CH \xrightarrow{?} CH_3CH_2CH_2CH_2CH_2Br$$

A primary alkyl halide can be prepared from a terminal alkene (using HBr in the presence of peroxide). A terminal alkene can be prepared from a terminal alkyne. A terminal alkyne can be prepared from ethyne and an alkyl halide of the appropriate length.

retrosynthetic analysis

$$CH_3CH_2CH_2CH_2CH_2Br \implies CH_3CH_2CH_2CH=CH_2 \implies CH_3CH_2CH_2C\equiv CH \implies HC\equiv CH$$

Now we can write the synthetic scheme.

synthesis

$$HC\equiv CH \xrightarrow[\text{2. } CH_3CH_2CH_2Br]{\text{1. } NaNH_2} CH_3CH_2CH_2C\equiv CH \xrightarrow[\text{Lindlar's catalyst}]{H_2} CH_3CH_2CH_2CH=CH_2$$

$$\xrightarrow[\text{peroxide}]{HBr} CH_3CH_2CH_2CH_2CH_2Br$$

Example 3. How could 2,6-dimethylheptane be prepared from an alkyne and an alkyl halide?

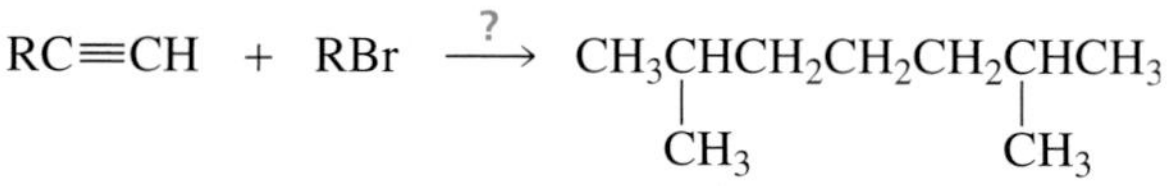

$$RC\equiv CH \ + \ RBr \ \xrightarrow{?} \ CH_3CHCH_2CH_2CH_2CHCH_3$$
$$\qquad\qquad\qquad\qquad\qquad\quad | \qquad\qquad\quad |$$
$$\qquad\qquad\qquad\qquad\qquad\quad CH_3 \qquad\qquad CH_3$$

2,6-Dimethyl-3-heptyne is the only alkyne that will form the desired alkane upon hydrogenation. This alkyne can be dissected in two different ways. In one case, the alkyne could be prepared from the reaction of an acetylide ion with a primary alkyl halide; in the other case, the acetylide ion would have to react with a secondary alkyl halide.

retrosynthetic analysis

$$CH_3CHCH_2CH_2CH_2CHCH_3 \implies CH_3CHCH_2C\equiv CCHCH_3$$
$$\qquad | \qquad\qquad\quad | \qquad\qquad\qquad\qquad | \qquad\qquad |$$
$$\qquad CH_3 \qquad\qquad CH_3 \qquad\qquad\qquad\quad CH_3 \qquad\quad CH_3$$

$$CH_3CHCH_2Br \ + \ CH_3CHC\equiv CH \quad \text{or} \quad CH_3CHBr \ + \ CH_3CHCH_2C\equiv CH$$
$$\quad | \qquad\qquad\qquad\quad | \qquad\qquad\qquad\qquad\quad | \qquad\qquad\qquad\quad |$$
$$\quad CH_3 \qquad\qquad\quad CH_3 \qquad\qquad\qquad\qquad CH_3 \qquad\qquad\quad CH_3$$

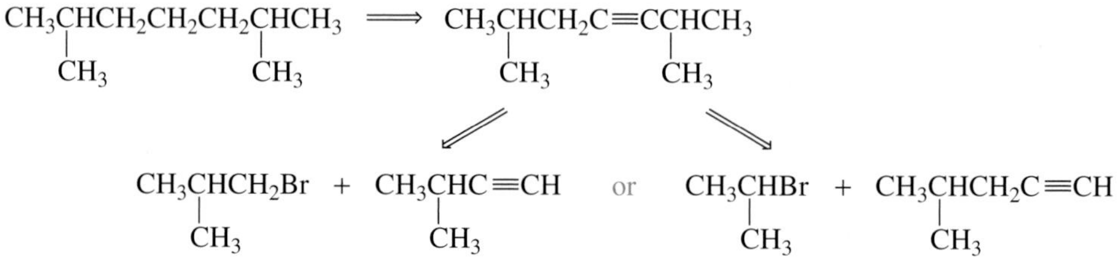

Because we know that the reaction of an acetylide ion with an alkyl halide works best with primary alkyl halides and methyl halides, we know how to proceed:

synthesis

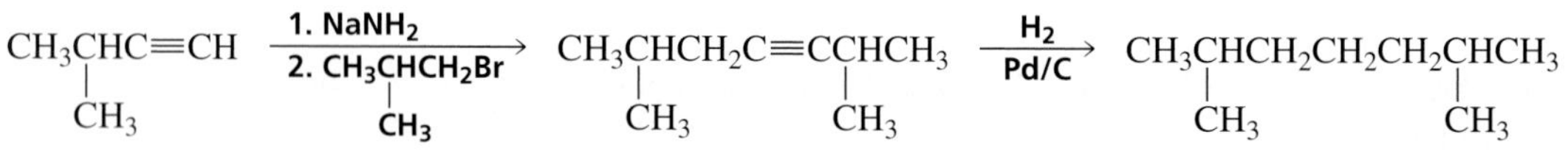

$$CH_3CHC\equiv CH \xrightarrow[\text{2. } CH_3CHCH_2Br]{\text{1. } NaNH_2} CH_3CHCH_2C\equiv CCHCH_3 \xrightarrow[\text{Pd/C}]{H_2} CH_3CHCH_2CH_2CH_2CHCH_3$$
$$\quad | \qquad\qquad\qquad\quad | \qquad\qquad\qquad\quad | \qquad\quad | \qquad\qquad\qquad\qquad\quad | \qquad\qquad\qquad | $$
$$\quad CH_3 \qquad\qquad\qquad CH_3 \qquad\qquad\quad CH_3 \quad CH_3 \qquad\qquad\qquad\qquad CH_3 \qquad\qquad CH_3$$

Example 4. How could you carry out the following synthesis using the given starting material?

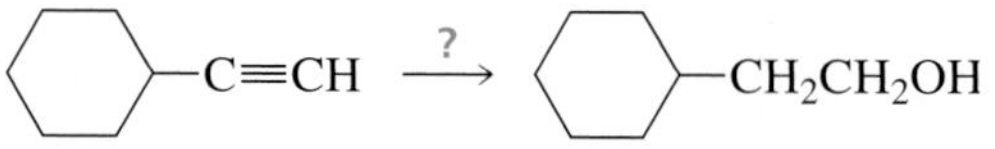

$$\bigcirc\!\!-C\equiv CH \ \xrightarrow{?} \ \bigcirc\!\!-CH_2CH_2OH$$

An alcohol can be prepared from an alkene, and an alkene can be prepared from an alkyne.

retrosynthetic analysis

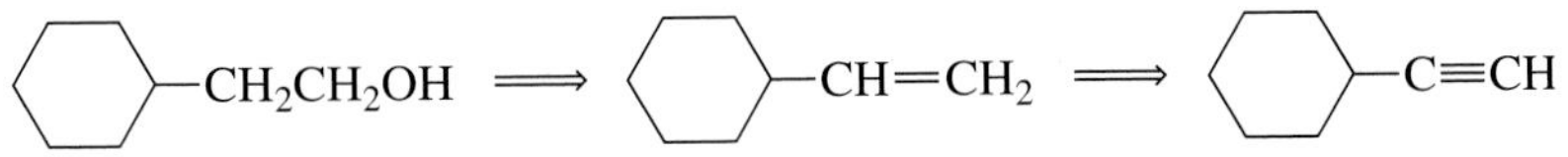

You can use either of the two methods you know to convert an alkyne into an alkene. Hydroboration–oxidation must be used to convert the alkene into the desired alcohol.

synthesis

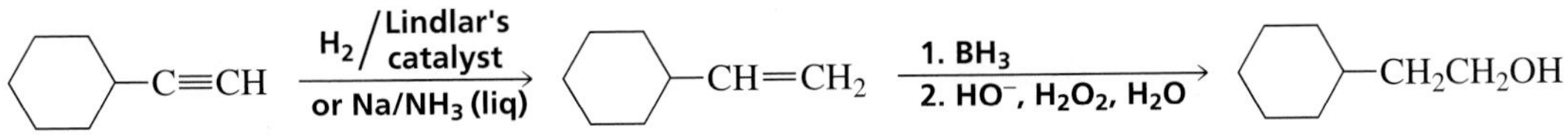

Example 5. How could you prepare (*E*)-2-pentene from ethyne?

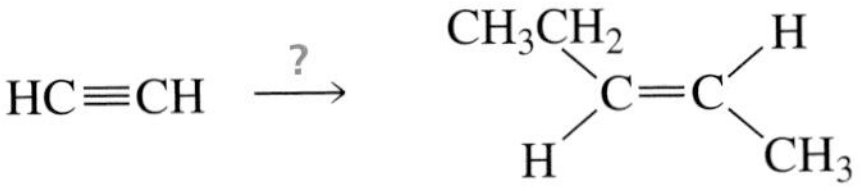

A trans alkene can be prepared from an internal alkyne. The alkyne needed to synthesize the desired alkene can be prepared from 1-butyne and a methyl halide. 1-Butyne can be prepared from ethyne and an ethyl halide.

retrosynthetic analysis

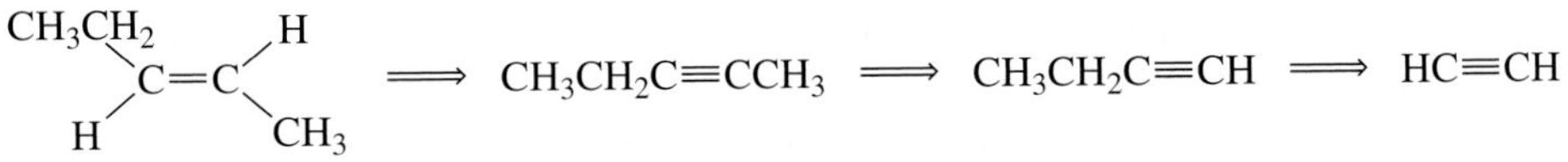

synthesis

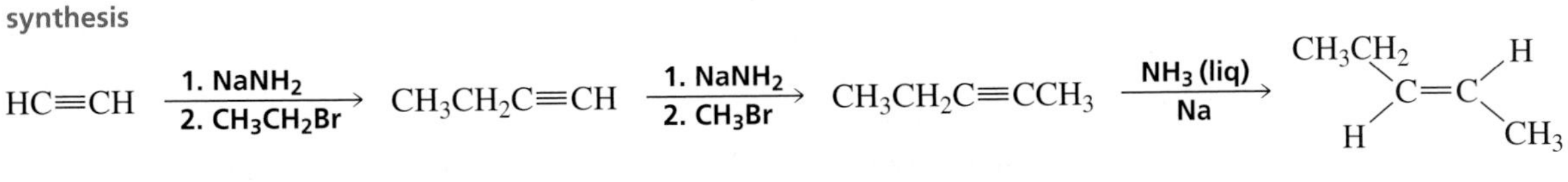

Example 6. How could you prepare 3,3-dibromohexane from reagents that contain no more than two carbon atoms?

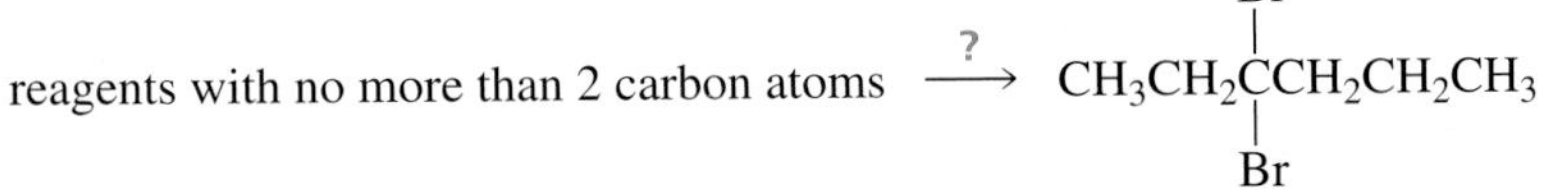

$$\text{reagents with no more than 2 carbon atoms} \xrightarrow{?} CH_3CH_2\overset{\displaystyle Br}{\underset{\displaystyle Br}{C}}CH_2CH_2CH_3$$

A geminal dibromide can be prepared from an alkyne. 3-Hexyne is the desired alkyne because it will form one geminal bromide, whereas 2-hexyne would form two different geminal dibromides. 3-Hexyne can be prepared from 1-butyne and ethyl bromide and 1-butyne can be prepared from ethyne and ethyl bromide.

retrosynthetic analysis

$$\underset{\substack{| \\ Br}}{\overset{\substack{Br \\ |}}{CH_3CH_2CCH_2CH_2CH_3}} \implies CH_3CH_2C\equiv CCH_2CH_3 \implies CH_3CH_2C\equiv CH \implies HC\equiv CH$$

synthesis

$$HC\equiv CH \xrightarrow[\substack{\textbf{2. CH}_3\textbf{CH}_2\textbf{Br}}]{\substack{\textbf{1. NaNH}_2}} CH_3CH_2C\equiv CH \xrightarrow[\substack{\textbf{2. CH}_3\textbf{CH}_2\textbf{Br}}]{\substack{\textbf{1. NaNH}_2}} CH_3CH_2C\equiv CCH_2CH_3 \xrightarrow{\textbf{excess HBr}} \underset{\substack{| \\ Br}}{\overset{\substack{Br \\ |}}{CH_3CH_2CCH_2CH_2CH_3}}$$

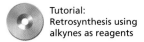
Tutorial:
Retrosynthesis using
alkynes as reagents

PROBLEM 18

Starting with acetylene, how could the following compounds be synthesized?

a. $CH_3CH_2CH_2C\equiv CH$

b. $CH_3CH=CH_2$

c. $CH_3CH_2CH_2CH_2\overset{\overset{\displaystyle O}{\|}}{C}H$

d. $\underset{\substack{| \\ Cl}}{\overset{\substack{Cl \\ |}}{CH_3CCH_3}}$

e. $\underset{\substack{| \\ Br}}{CH_3CHCH_3}$

f. $\underset{\substack{H \quad\quad H}}{\overset{\substack{CH_3 \quad\quad CH_3}}{C=C}}$

5.12 COMMERCIAL USE OF ETHYNE

Most of the ethyne produced commercially is used as a starting material for polymers that you encounter daily such as vinyl flooring, plastic piping, Teflon, and acrylics. Polymers are large molecules that are made by linking together many small molecules known as monomers. The mechanisms by which monomers are converted into polymers will be discussed in Chapter 26.

Poly(vinyl chloride), the polymer produced from the polymerization of vinyl chloride, is known as PVC. Poly(vinyl chloride) is hard and rather brittle, but it can be softened by adding certain compounds known as plasticizers. It then becomes the soft, elastic vinyl commonly used both as a substitute for leather and in the manufacture of such things as garbage bags and shower curtains. Poly(acrylonitrile) looks like wool when it is made into fibers. It is marketed under the trade names Orlon (DuPont), Creslan (Sterling Fibers), and Acrilan (Monsanto).

$$HC\equiv CH + HCl \longrightarrow \underset{\textbf{vinyl chloride}}{H_2C=CHCl} \xrightarrow{\textbf{polymerization}} \underset{\substack{\textbf{poly(vinyl chloride)} \\ \textbf{PVC}}}{\left[\begin{matrix} CH_2-CH \\ | \\ Cl \end{matrix}\right]_n}$$

$$HC\equiv CH + HCN \longrightarrow \underset{\textbf{acrylonitrile}}{H_2C=CHCN} \xrightarrow{\textbf{polymerization}} \underset{\substack{\textbf{poly(acrylonitrile)} \\ \textbf{Orlon}}}{\left[\begin{matrix} CH_2-CH \\ | \\ CN \end{matrix}\right]_n}$$

ETHYNE CHEMISTRY OR THE FORWARD PASS?

Father Julius Arthur Nieuwland (1878–1936) did much of the early work on the polymerization of ethyne. He was born in Belgium and settled with his parents in South Bend, Ind., two years later. He became a priest and professor of botany and chemistry at the University of Notre Dame, where Knute Rockne (the inventor of the forward pass) worked for him as a research assistant. Rockne also taught chemistry at Notre Dame, but when he received an offer to coach the football team he switched fields, in spite of Father Nieuwland's attempts to convince him to continue his work as a scientist.

Knute Rockne in his uniform during the year he was captain of the Notre Dame football team.

SUMMARY OF REACTIONS

1. Electrophilic addition reactions

   a. Addition of hydrogen halides (Markovnikov orientation) (Section 5.5)

$$RC\equiv CH \xrightarrow{\textbf{HX}} \underset{\underset{X}{|}}{RC}=CH_2 \xrightarrow{\textbf{excess HX}} \underset{\underset{X}{|}}{\overset{\overset{X}{|}}{RC}}-CH_3$$

   HX = HF, HCl, HBr, HI

   b. Addition of hydrogen bromide in the presence of peroxide (apparent anti-Markovnikov orientation) (Section 5.5)

$$RC\equiv CH \; + \; HBr \xrightarrow{\textbf{peroxide}} RCH=CHBr$$

   c. Addition of halogens (Section 5.5)

$$RC\equiv CH \xrightarrow[\textbf{CH}_2\textbf{Cl}_2]{\textbf{Cl}_2} \underset{\underset{Cl}{|}}{\overset{\overset{Cl}{|}}{RC}}=CH \xrightarrow[\textbf{CH}_2\textbf{Cl}_2]{\textbf{Cl}_2} \underset{\underset{Cl}{|}\;\underset{Cl}{|}}{\overset{\overset{Cl}{|}\;\overset{Cl}{|}}{RC}}-CH$$

$$RC\equiv CCH_3 \xrightarrow[\textbf{CH}_2\textbf{Cl}_2]{\textbf{Br}_2} \underset{\underset{Br}{|}}{\overset{\overset{Br}{|}}{RC}}=CCH_3 \xrightarrow[\textbf{CH}_2\textbf{Cl}_2]{\textbf{Br}_2} \underset{\underset{Br}{|}\;\underset{Br}{|}}{\overset{\overset{Br}{|}\;\overset{Br}{|}}{RC}}-CCH_3$$

   d. Addition of water/hydroboration–oxidation (Sections 5.6 and 5.7)

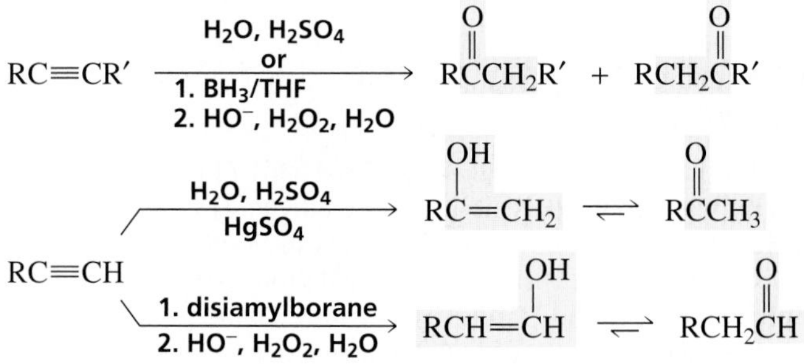

$$RC\equiv CR' \xrightarrow[\substack{\text{1. BH}_3\text{/THF} \\ \text{2. HO}^-,\ \text{H}_2\text{O}_2,\ \text{H}_2\text{O}}]{\substack{\text{H}_2\text{O, H}_2\text{SO}_4 \\ \text{or}}} RCCH_2R' + RCH_2CR'$$

$$RC\equiv CH$$

$$\xrightarrow[\text{HgSO}_4]{\text{H}_2\text{O, H}_2\text{SO}_4} RC{=}CH_2 \rightleftharpoons RCCH_3$$

$$\xrightarrow[\substack{\text{2. HO}^-,\ \text{H}_2\text{O}_2,\ \text{H}_2\text{O}}]{\text{1. disiamylborane}} RCH{=}CH \rightleftharpoons RCH_2CH$$

2. Addition of hydrogen (Section 5.8)

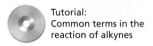

Tutorial:
Common terms in the
reaction of alkynes

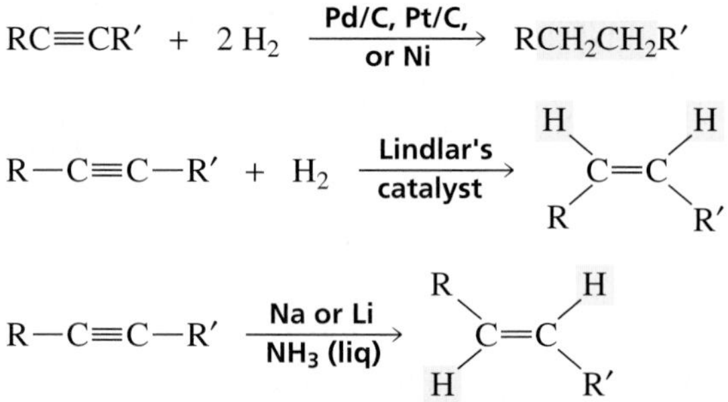

$$RC\equiv CR' + 2\ H_2 \xrightarrow[\text{or Ni}]{\text{Pd/C, Pt/C,}} RCH_2CH_2R'$$

$$R-C\equiv C-R' + H_2 \xrightarrow[\text{catalyst}]{\text{Lindlar's}} \underset{R}{\overset{H}{\diagdown}}C{=}C\underset{R'}{\overset{H}{\diagup}}$$

$$R-C\equiv C-R' \xrightarrow[\text{NH}_3\ \text{(liq)}]{\text{Na or Li}} \underset{H}{\overset{R}{\diagdown}}C{=}C\underset{R'}{\overset{H}{\diagup}}$$

3. Removal of a proton, followed by alkylation (Sections 5.9 and 5.10)

$$RC\equiv CH \xrightarrow{\text{NaNH}_2} RC\equiv C^- \xrightarrow{\text{R'CH}_2\text{Br}} RC\equiv CCH_2R'$$

KEY TERMS

alkylation reaction (page 255)
alkyne (page 239)
geminal dihalide (page 245)
internal alkyne (page 240)
keto-enol tautomers (page 248)

pi-complex (page 245)
radical anion (page 251)
retrosynthetic analysis (page 257)
tautomerization (page 248)
tautomers (page 248)

terminal alkyne (page 240)
vinylic cation (page 244)
vinylic radical (page 251)

PROBLEMS

19. Draw a structure for each of the following compounds:
   a. 2-hexyne
   b. 5-ethyl-3-octyne
   c. methylacetylene
   d. vinylacetylene
   e. methoxyethyne
   f. *sec*-butyl-*tert*-butylacetylene
   g. 1-bromo-1-pentyne
   h. propargyl bromide
   i. diethylacetylene
   j. di-*tert*-butylacetylene
   k. cyclopentylacetylene
   l. 5,6-dimethyl-2-heptyne

20. Give the systematic name for each of the following structures:

a. 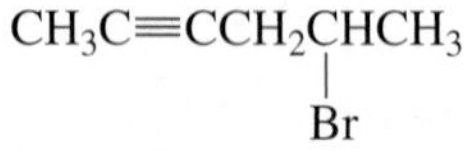 $CH_3C\equiv CCH_2CHCH_3$
       |
       Br

b. 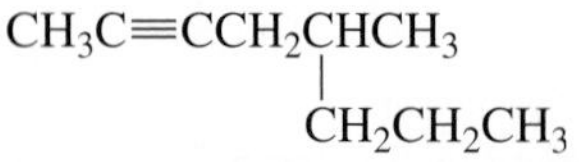 $CH_3C\equiv CCH_2CHCH_3$
       |
       $CH_2CH_2CH_3$

$$
\text{c. } CH_3C\equiv CCH_2\overset{\overset{\displaystyle CH_3}{|}}{\underset{\underset{\displaystyle CH_3}{|}}{C}}CH_3
$$

$$
\text{d. } CH_3\overset{\overset{\displaystyle }{}}{\underset{\underset{\displaystyle Cl}{|}}{C}}HCH_2C\equiv C\overset{\overset{\displaystyle }{}}{\underset{\underset{\displaystyle CH_3}{|}}{C}}HCH_3
$$

21. What reagents would be required to carry out the following syntheses?

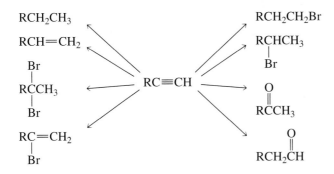

22. How could the following compounds be synthesized, starting with a hydrocarbon that has the same number of carbon atoms as the desired product?

$$
\text{a. } CH_3CH_2CH_2CH_2\overset{\overset{\displaystyle O}{\|}}{C}H
$$

$$
\text{c. } CH_3CH_2CH_2\overset{\overset{\displaystyle O}{\|}}{C}CH_2CH_2CH_2CH_3
$$

b. $CH_3CH_2CH_2CH_2OH$

d. $CH_3CH_2CH_2CH_2CH_2Br$

23. What reagents would you use for the following syntheses?
   a. (*Z*)-3-hexene from 3-hexyne
   b. (*E*)-3-hexene from 3-hexyne
   c. hexane from 3-hexyne

24. What is the molecular formula of a hydrocarbon that has 1 triple bond, 2 double bonds, 1 ring, and 32 carbons?

25. What will be the major product of the reaction of 1 mol of propyne with each of the following reagents?
   a. HBr (1 mol)
   b. HBr (2 mol)
   c. Br_2 (1 mol)/CH_2Cl_2
   d. Br_2 (2 mol)/CH_2Cl_2
   e. aqueous H_2SO_4, $HgSO_4$
   f. disiamylborane followed by H_2O_2/HO^-
   g. $HBr + H_2O_2$
   h. excess H_2/Pt
   i. $H_2/$Lindlar's catalyst
   j. sodium in liquid ammonia
   k. sodium amide in liquid ammonia
   l. product of **k** followed by 1-chloropentane

26. Answer Problem 25 using 2-butyne as the starting material instead of propyne.

27. a. Starting with isopropylacetylene, how could you prepare the following alcohols?
   1. 2-methyl-2-pentanol   **2.** 4-methyl-2-pentanol
   b. In each case, a second alcohol would also be obtained. What alcohol would it be?

28. How many of the following names are correct? Correct the incorrect names.
   a. 4-heptyne
   b. 2-ethyl-3-hexyne
   c. 4-chloro-2-pentyne
   d. 2,3-dimethyl-5-octyne
   e. 4,4-dimethyl-2-pentyne
   f. 2,5-dimethyl-3-hexyne

29. Which of the following pairs are keto-enol tautomers?

$$
\text{a. } CH_3CH_2CH=CHCH_2OH \quad \text{and} \quad CH_3CH_2CH_2CH_2\overset{\overset{\displaystyle O}{\|}}{C}H
$$

b. CH_3CHCH_3 and CH_3CCH_3
      | ‖
     OH   O

c. CH_3CH_2CH=$CHOH$ and $CH_3CH_2CH_2CH$ (with C=O)

d. $CH_3CH_2CH_2CH$=$CHOH$ and $CH_3CH_2CH_2CCH_3$ (with C=O)

e. $CH_3CH_2CH_2C$=CH_2 and $CH_3CH_2CH_2CCH_3$ (with C=O)
               |
              OH

30. Using ethyne as the starting material, how can the following compounds be prepared?

a. $CH_3CH_2CHCH_2Br$
             |
           Br

c. CH_3CCH_3 (with C=O)

e.

b. CH_3CH (with C=O)

d.

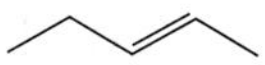

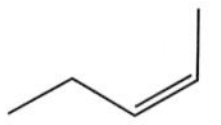

f.

31. Give the stereoisomers obtained from the reaction of 2-butyne with the following reagents:
a. 1. H_2/Lindlar's catalyst **2.** Br_2/CH_2Cl_2
b. 1. Na/NH_3 (liq) **2.** Br_2/CH_2Cl_2
c. 1. Cl_2/CH_2Cl_2 **2.** Br_2/CH_2Cl_2

32. Draw the keto tautomer for each of the following:

        OH
        |
a. CH_3CH=CCH_3

c. (cyclohexene ring)—OH

         OH
         |
b. $CH_3CH_2CH_2C$=CH_2

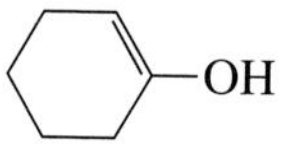

d. 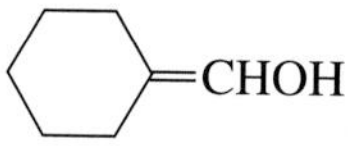 (cyclohexane ring)=CHOH

33. Show how each of the following compounds could be prepared using the given starting material, any necessary inorganic reagents, and any necessary organic compound that has no more than four carbon atoms:

a. HC≡CH $\longrightarrow$ $CH_3CH_2CH_2CH_2CCH_3$ (with C=O)

b. HC≡CH $\longrightarrow$ $CH_3CH_2CHCH_3$
                           |
                         Br

c. HC≡CH $\longrightarrow$ $CH_3CH_2CH_2CHCH_3$
                             |
                           OH

d. 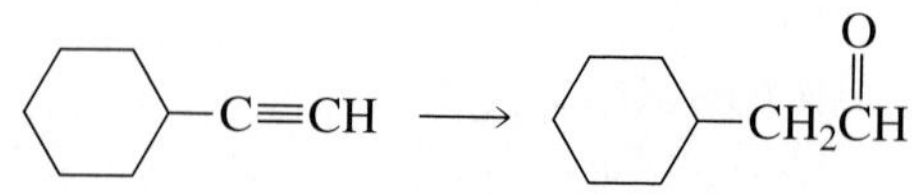 (cyclohexane ring)—C≡CH $\longrightarrow$ (cyclohexane ring)—CH_2CH (with C=O)

e.

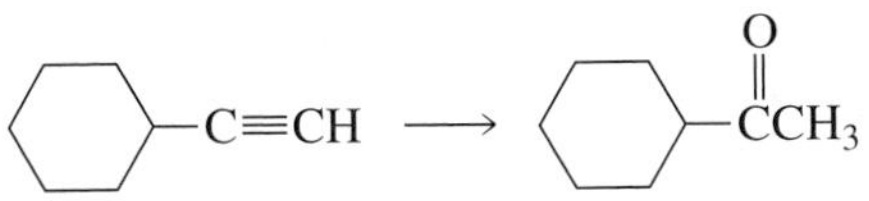

f.

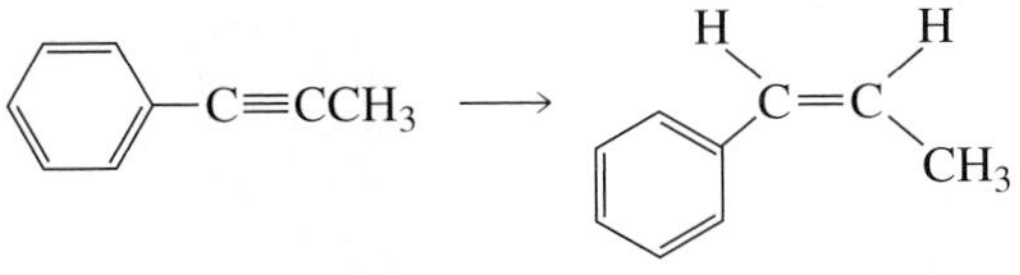

34. Dr. Iona Ford was planning to synthesize 3-octyne by adding 1-bromobutane to the product obtained from the reaction of 1-butyne with sodium amide. Unfortunately, however, she had forgotten to order 1-butyne. How else can she prepare 3-octyne?

35. The female of a certain species of worms produces a sex attractant called spodoptol, in low concentrations, which attracts the male worm. The population of these worms can be controlled by putting synthetic spodoptol in traps. Spodoptol can be synthesized by the following series of reactions.

$$HOCH_2CH_2CH_2CH_2CH_2CH_2CH_2CH_2C{\equiv}CH \xrightarrow[\text{NaNH}_2]{\text{excess}} A \xrightarrow{CH_3CH_2CH_2CH_2Br} B \xrightarrow{\text{EtOH}} C \xrightarrow[\substack{\text{Lindlar's} \\ \text{catalyst}}]{H_2} \text{spodoptol}$$

  a. Draw the structure of spodoptol.
  b. Why is excess $NaNH_2$ used in the first step?

36. a. Explain why a single pure product is obtained from hydroboration–oxidation of 2-butyne, whereas two products are obtained from hydroboration–oxidation of 2-pentyne.
  b. Name two other internal alkynes that will yield only one product on hydroboration–oxidation.

37. In Section 5.4 it was stated that hyperconjugation is less effective in stabilizing the charge on a vinylic cation than the charge on an alkyl cation. Why do you think this is so?

6

Electron Delocalization and Resonance

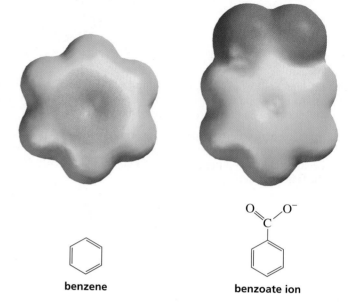

benzene

benzoate ion

E lectrons that are restricted to a particular region are called **localized electrons.** Localized electrons either belong to a single atom or are confined to a bond between two atoms.

$$CH_3\text{---}NH_2 \qquad CH_3\text{---}CH\text{=\!=}CH_2$$

localized electrons localized electrons

Many organic compounds contain delocalized electrons. **Delocalized electrons** neither belong to a single atom nor are confined to a bond between two atoms—they are electrons that are shared by more than two atoms. You were first introduced to delocalized electrons in Section 1.19. In this chapter you will learn to recognize compounds that contain delocalized electrons, and to draw structures that represent the electron distribution in molecules with delocalized electrons. You will also be introduced to some of the special characteristics of compounds that have delocalized electrons. You will then be able to understand the wide-ranging effects that delocalized electrons have on the reactivity of organic compounds.

We will begin by taking a look at benzene, a compound whose properties chemists could not explain until they recognized that electrons in organic molecules could be delocalized.

6.1
DELOCALIZED ELECTRONS: THE STRUCTURE OF BENZENE

The structure of benzene puzzled early organic chemists. They knew that benzene had a molecular formula of C_6H_6 and that it was an unusually stable compound that did not undergo the addition reactions characteristic of alkenes (Section 3.8). They also knew the following facts:

1. When a different atom is substituted for one of the hydrogen atoms of benzene, only one product is obtained.

2. When the substituted product undergoes a second substitution, three products
are obtained: **A, B,** and **C.**

$$C_6H_6 \xrightarrow[\text{with an X}]{\text{replace a hydrogen}} C_6H_5X \xrightarrow[\text{with an X}]{\text{replace a hydrogen}} \underset{A}{C_6H_4X_2} + \underset{B}{C_6H_4X_2} + \underset{C}{C_6H_4X_2}$$

What kind of structure would we arrive at for benzene if we knew only what the
early chemists knew? We can tell from its molecular formula (C_6H_6) that benzene
has eight fewer hydrogens than an acyclic (noncyclic) alkane with six carbons
($C_nH_{2n+2} = C_6H_{14}$). Therefore, benzene is either an acyclic compound with four π
bonds, a cyclic compound with three π bonds, a bicyclic compound with two π
bonds, a tricyclic compound with one π bond, or a tetracyclic compound (Section 3.1).
All the hydrogens in benzene must be identical because only one product is obtained
if any one hydrogen is replaced with another atom. Two structures that fit these requirements are shown here.

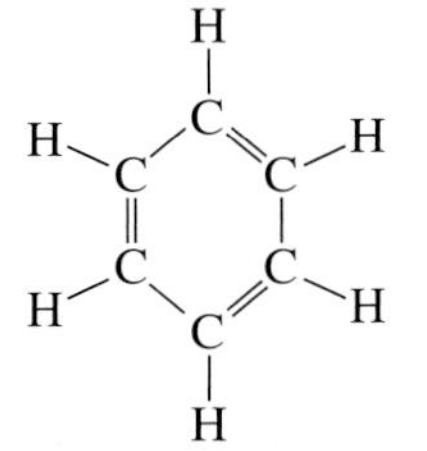

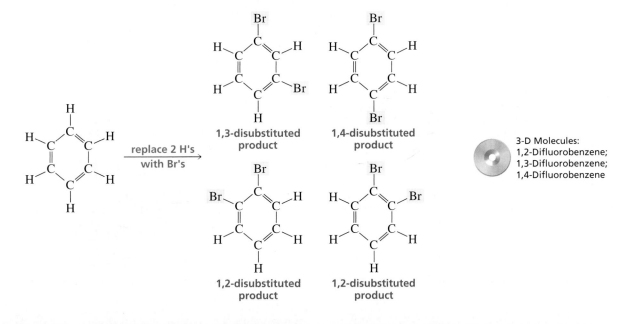

Neither of these structures is consistent with the observation that three compounds are obtained if a second hydrogen is replaced. The acyclic structure yields
two disubstituted products.

$$CH_3C\equiv C-C\equiv CCH_3 \xrightarrow[\text{with Br's}]{\text{replace 2 H's}} CH_3C\equiv C-C\equiv CCHBr \quad \text{and} \quad BrCH_2C\equiv C-C\equiv CCH_2Br$$
$$|$$
$$Br$$

The cyclic structure yields the following four disubstituted products: a 1,3-disub-
stituted product, a 1,4-disubstituted product, and two 1,2-disubstituted products—
because the two substituents can be placed either on two adjacent carbons joined by
a single bond or on two adjacent carbons joined by a double bond.

1,3-disubstituted product

1,4-disubstituted product

1,2-disubstituted product

1,2-disubstituted product

replace 2 H's
with Br's

3-D Molecules:
1,2-Difluorobenzene;
1,3-Difluorobenzene;
1,4-Difluorobenzene

In 1865, the German chemist Friedrich Kekulé suggested an answer to this dilemma. He proposed that benzene was not a single compound but a mixture of two compounds in rapid equilibrium.

Kekulé structures of benzene

This would explain why only three disubstituted products are obtained when a substituted benzene undergoes a second substitution. According to Kekulé, there actually *are* four disubstituted products, but the two 1,2-disubstituted products interconvert too rapidly to be distinguished and separated from each other.

The Kekulé structures of benzene account for the molecular formula of benzene and for the number of isomers obtained as a result of substitution. However, they fail to account for the unusual stability of benzene and for the observation that the double bonds of benzene do not undergo the addition reactions characteristic of alkenes. Confirmation that benzene had a six-membered ring came in 1901, when Paul Sabatier (Section 3.19) found that hydrogenation of benzene produced cyclohexane. This, however, did not solve the puzzle of benzene's structure.

$$\text{benzene} \xrightarrow[\text{150–250 °C, 25 atm}]{\text{H}_2,\ \text{Ni}} \text{cyclohexane}$$

benzene   **cyclohexane**

KEKULÉ'S DREAM

Friedrich August Kekulé von Stradonitz (1829–1896) was born in Germany. He entered the University of Giessen to study architecture but switched to chemistry after taking a chemistry course. He was a professor of chemistry at the University of Heidelberg, at the University of Ghent in Belgium, and then at the University of Bonn. In 1890, he gave an extemporaneous speech at the twenty-fifth anniversary celebration of his first paper on the cyclic structure of benzene. In this speech he claimed that he had arrived at the Kekulé structures as a result of dozing off in front of a fire while working on a textbook. He dreamed of chains of carbon atoms twisting and turning in a snake-like motion, when suddenly the head of one snake seized hold of its own tail and formed a spinning ring. Recently, the veracity of his snake story has been questioned by those who point out that there is no written record of the dream from the time he experienced it in 1861 until the time he related it in 1890. Others counter that dreams are not the kind of evidence one publishes in scientific papers. But it is not uncommon for scientists to report moments of creativity through the unconscious, when they were not thinking about science. After relating his dream, Kekulé said, "Let us learn to dream, and perhaps then we shall learn the truth. But let us also beware not to publish our dreams until they have been examined by the wakened mind." In 1895, he was made a nobleman by Emperor William II. This allowed him to add "von Stradonitz" to his name. Kekulé's students received three of the first five Nobel Prizes in chemistry: van't Hoff in 1901 (page 193), Fischer in 1902 (page 187), and Baeyer in 1905 (page 93).

Friedrich August Kekulé von Stradonitz

The controversy over the structure of benzene continued until the 1930s, when the new techniques of X-ray and electron diffraction produced a surprising result. *These methods showed that benzene is a planar molecule in which all six carbon–carbon bonds have the same length.* The length of each of the carbon–carbon bonds in benzene is 1.39 Å, which is shorter than a carbon–carbon single bond (1.54 Å) but a little longer than a carbon–carbon double bond (1.33 Å) (Section 1.14). If all the carbon–carbon bonds have the same length, then they must also have the same electron density. This can be true only if the π electrons of benzene are delocalized around the ring rather than having each of the π electrons localized between two carbon atoms. To gain a clearer understanding of delocalized electrons, let's take a close look at the bonding in benzene.

PROBLEM 1◆

a. How many monosubstituted products would each of the following compounds have? (Notice that each compound has the same molecular formula as benzene.)

   1. $HC \equiv CC \equiv CCH_2CH_3$ **2.** $CH_2 = CHC \equiv CCH = CH_2$

b. How many disubstituted products would each of these compounds have? (Do not include stereoisomers.)

PROBLEM 2

Between 1865 and 1890, other possible structures were proposed for benzene. Two of these are shown here. Considering what nineteenth-century chemists knew about benzene, which is a better proposal for the structure of benzene, Dewar benzene or Ladenburg benzene? Why?

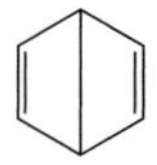

Dewar benzene 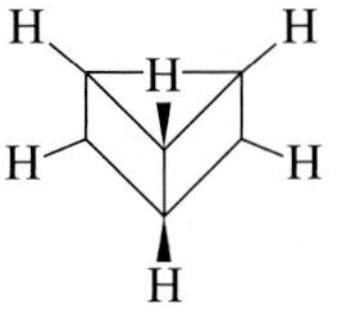   Ladenburg benzene

Sir James Dewar (1842–1923) was born in Scotland, the son of an inn keeper. After studying under Kekulé, he became a professor at Cambridge University and then at the Royal Institution in London. Dewar's most important work was in the field of low-temperature chemistry. He used double-wall flasks with evacuated space between the walls in order to reduce heat transmission. These flasks are now called Dewar flasks—better known to nonchemists as thermos bottles.

Albert Ladenburg (1842–1911) was born in Germany. He was a professor of chemistry at the University of Kiel.

Benzene is a planar molecule. Each of its six carbons is sp^2 hybridized. An sp^2 hybridized carbon has bond angles of 120°, identical to the angles of a planar hexagon. Each of the carbons in benzene uses two sp^2 orbitals to bond to two other carbons. Its third sp^2 orbital overlaps the s orbital of a hydrogen [Figure 6.1(a)]. Each carbon also has a p orbital at right angles to the sp^2 orbitals. Because benzene is planar, the six p orbitals are parallel [Figure 6.1(b)]. The p orbitals are close enough for side-to-side overlap, so each p orbital overlaps the p orbitals on both adjacent carbons. As a result, the overlapping p orbitals form a continuous doughnut-shaped cloud of π electrons above, and another doughnut-shaped cloud of π electrons below, the plane of the benzene ring [Figure 6.1(c)]. The electrostatic potential map [Figure 6.1(d)] shows that all the carbon–carbon bonds have the same electron density.

Each π electron is, therefore, neither localized on a single carbon, nor in a bond between two carbons (as in an alkene). Instead, each π electron is shared by all six carbons. The six π electrons are delocalized—they roam freely within the doughnut-shaped clouds that lie over and under the ring of carbon atoms. Consequently, benzene can be represented by a hexagon containing a circle that symbolizes the six delocalized π electrons.

6.2
BONDING IN BENZENE

3-D Molecule: Benzene

From this type of representation, it is clear that benzene does not contain any double bonds. We see now that Kekulé's structure for benzene was pretty close to the correct structure. The actual structure of benzene is a Kekulé structure with delocalized electrons.

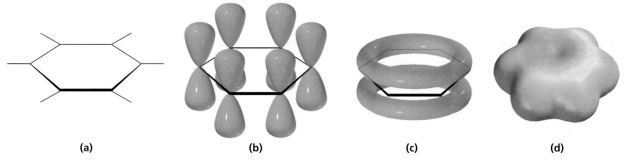

(a)   (b)   (c)   (d)

▲ **Figure 6.1**
(a) The carbon–carbon and carbon–hydrogen σ bonds in benzene.
(b) The *p* orbital on each carbon of benzene can overlap with two adjacent *p* orbitals.
(c) The clouds of π electrons above and below the plane of the benzene ring.
(d) The electrostatic potential map for benzene.

6.3 DELOCALIZED ELECTRONS AND RESONANCE

A significant disadvantage to using structures that represent electrons as delocalized is that these structures do not tell us how many π electrons are present in the molecule. For example, the circle inside the hexagon in the representation of the structure of benzene shows that the π electrons are shared equally by all six carbons and that all the carbon–carbon bonds have the same length, but it does not show how many π electrons are in the ring. Consequently, chemists prefer to use structures containing localized electrons to approximate the actual structure with delocalized electrons. A compound with delocalized electrons is said to have **resonance.** The approximate structure using localized electrons is called a **resonance contributor,** a **resonance structure,** or a **contributing resonance structure.** The actual structure, drawn using delocalized electrons, is called a **resonance hybrid.** Each resonance contributor of benzene clearly indicates that there are six π electrons in the ring.

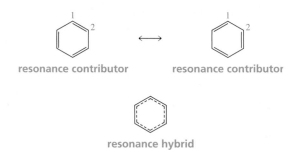

resonance contributor   resonance contributor

resonance hybrid

Resonance contributors are shown with a double-headed arrow separating them. The double-headed arrow does *not* mean that the structures are in equilibrium with one another. Rather, it indicates that the actual structure lies somewhere between the structures of the resonance contributors. The resonance contributors are only a convenient way to show the π electrons—they do not depict any real electron distribution. For example, the bond between C-1 and C-2 in benzene is not a double bond, as shown in the resonance contributor on the left, nor is it a single bond, as shown in the resonance contributor on the right. It is midway between what is represented by the two resonance contributors. Neither of the contributing resonance structures accurately represents the structure of benzene. The actual structure of benzene is given by the average of the two resonance contributors (the resonance hybrid).

The following analogy illustrates the difference between resonance contributors and the resonance hybrid. Imagine that you are trying to describe to a friend what a rhinoceros looks like. You might tell your friend that a rhinoceros looks like a cross between a unicorn and a dragon. The unicorn and the dragon don't really exist, so they are like the resonance contributors. They are not in equilibrium—a rhinoceros does not jump back and forth between the two resonance contributors, looking like a unicorn one minute and a dragon the next. The rhinoceros always looks like a rhinoceros, so it is like the resonance hybrid. The unicorn and the dragon are simply ways to represent what the actual structure (the rhinoceros) looks like. *Resonance contributors, like unicorns and dragons, are imaginary, not real. Only the resonance hybrid, like the rhinoceros, is real.*

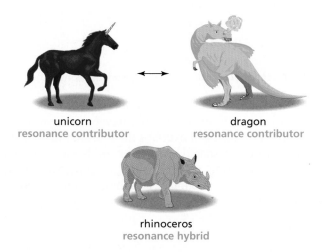

unicorn
resonance contributor

dragon
resonance contributor

rhinoceros
resonance hybrid

Electron delocalization occurs only if all the atoms sharing the delocalized electrons lie in or close to the same plane so their *p* orbitals can effectively overlap. For example, cyclooctatetraene is not planar—it is tub-shaped. Therefore, the *p* orbitals cannot overlap, so each pair of π electrons is *localized* between two carbons rather than being *delocalized* over the entire ring of eight carbon atoms. Compare the red areas in the electrostatic potential maps for cyclooctatetraene and benzene—one shows localized electrons, while the other shows delocalized electrons.

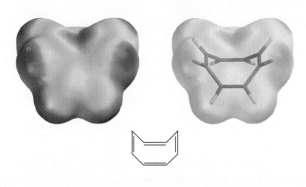

3-D Molecule:
Cyclooctatetraene

cyclooctatetraene

6.4
HOW TO DRAW RESONANCE CONTRIBUTORS

An organic compound with delocalized electrons is generally represented as a structure with localized electrons. For example, nitroethane is represented as having a nitrogen–oxygen double bond and a nitrogen–oxygen single bond.

$$CH_3CH_2 \overset{+}{-} N \overset{O}{\underset{O^-}{\diagup\!\!\diagdown}}$$

nitroethane

However, the two nitrogen–oxygen bonds in nitroethane are identical (they have the same bond length). A more accurate description of the molecule's structure is obtained by drawing two resonance contributors. Both resonance contributors show the compound with a nitrogen–oxygen double bond and a nitrogen–oxygen single bond, but the double bond in one contributor is the single bond in the other contributor, and vice versa. In other words, the electrons are delocalized.

$$CH_3CH_2 \overset{+}{-} N \overset{O}{\underset{O^-}{\diagup\!\!\diagdown}} \quad \longleftrightarrow \quad CH_3CH_2 \overset{+}{-} N \overset{O^-}{\underset{O}{\diagup\!\!\diagdown}}$$

resonance contributor   **resonance contributor**

Delocalized electrons result from a p orbital overlapping the p orbitals of more than one adjacent atom.

The resonance hybrid shows that the two nitrogen–oxygen bonds are identical and that the negative charge is shared by both oxygens. The resonance hybrid also shows that the p orbital of nitrogen overlaps the p orbital of each oxygen. In other words, the two π electrons are shared by three atoms. Notice that the resonance contributors tell us where the formal charges reside in a molecule, and they also tell us the approximate bond orders. But only by visualizing all the resonance contributors can we appreciate what the actual molecule—the resonance hybrid—looks like.

$$CH_3CH_2 \overset{+}{-} N \overset{\overset{\delta-}{O}}{\underset{\underset{\delta-}{O}}{\diagup\!\!\diagdown}}$$

resonance hybrid

Rules for Drawing Resonance Contributors

In order to draw resonance contributors, the electrons in one resonance contributor are moved to generate the next resonance contributor. As you draw contributing resonance structures, notice the following:

1. Only electrons move. The nuclei of the atoms never move.
2. The only electrons that can move are π electrons and nonbonding electrons.
3. The total number of electrons in the molecule does not change, and neither do the numbers of paired and unpaired electrons.

The electrons can be moved in one of the following ways:

1. Move π electrons toward a positive charge or toward a π bond (Figures 6.2 and 6.3).
2. Move a nonbonding pair of electrons toward a π bond (Figure 6.4).
3. Move a single nonbonding electron toward a π bond (Figure 6.5).

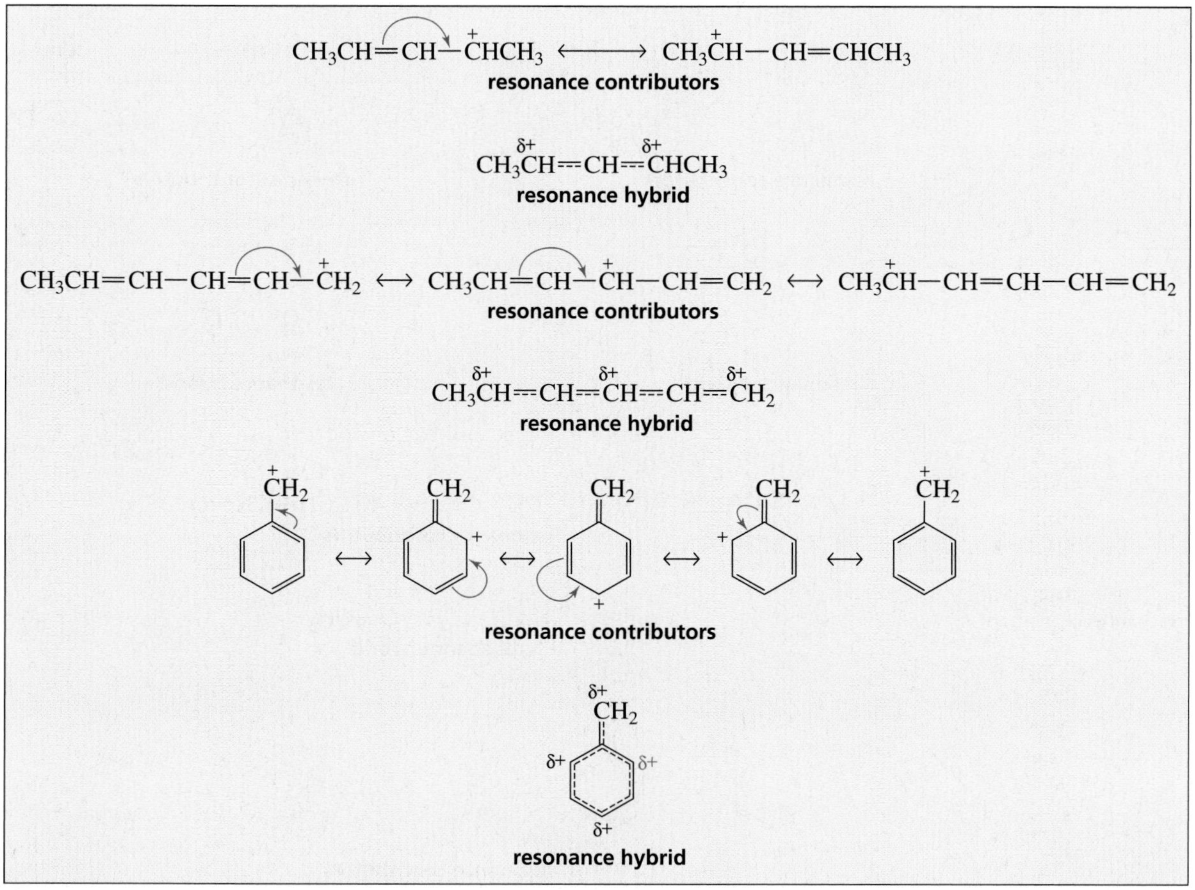

$$CH_3CH = CH - \overset{+}{C}HCH_3 \longleftrightarrow CH_3\overset{+}{C}H - CH = CHCH_3$$
resonance contributors

$$\overset{\delta+}{CH_3CH} = CH = \overset{\delta+}{CHCH_3}$$
resonance hybrid

$$CH_3CH = CH - CH = CH - \overset{+}{C}H_2 \longleftrightarrow CH_3CH = CH - \overset{+}{C}H - CH = CH_2 \longleftrightarrow CH_3\overset{+}{C}H - CH = CH - CH = CH_2$$
resonance contributors

$$\overset{\delta+}{CH_3CH} = CH = \overset{\delta+}{CH} = CH = \overset{\delta+}{CH_2}$$
resonance hybrid

resonance contributors

resonance hybrid

▲ **Figure 6.2**
Resonance contributors are obtained by moving π electrons toward a positive charge.

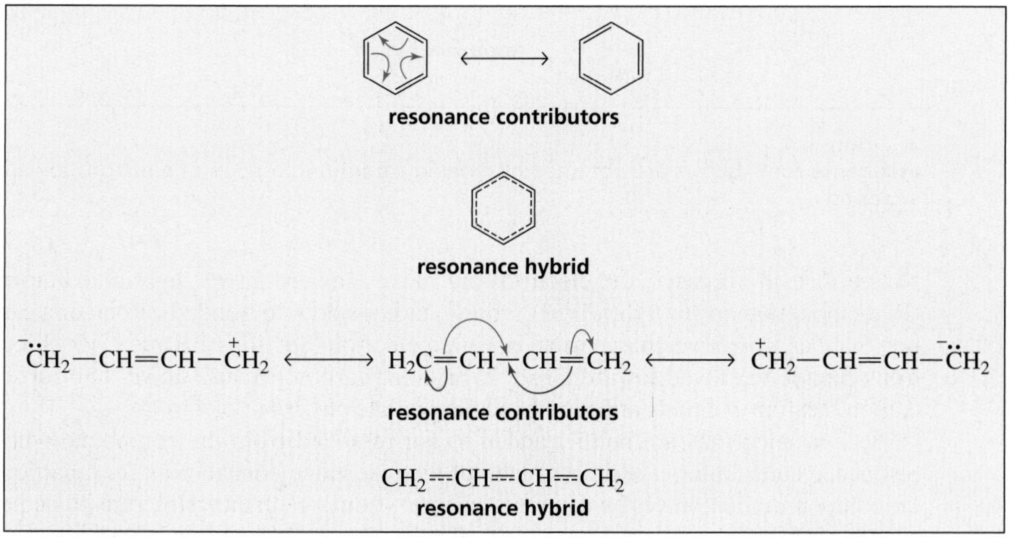

resonance contributors

resonance hybrid

$$\overset{-}{\ddot{C}}H_2 - CH = CH - \overset{+}{C}H_2 \longleftrightarrow H_2C = CH - CH = CH_2 \longleftrightarrow \overset{+}{C}H_2 - CH = CH - \overset{-}{\ddot{C}}H_2$$
resonance contributors

$$CH_2 = CH = CH = CH_2$$
resonance hybrid

▲ **Figure 6.3**
Resonance contributors are obtained by moving π electrons toward a π bond. (In the second example, the red arrows lead to the resonance contributor on the right, and the blue arrows lead to the resonance contributor on the left.)

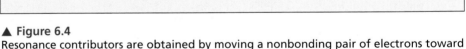

Tutorial: Drawing
resonance contributors

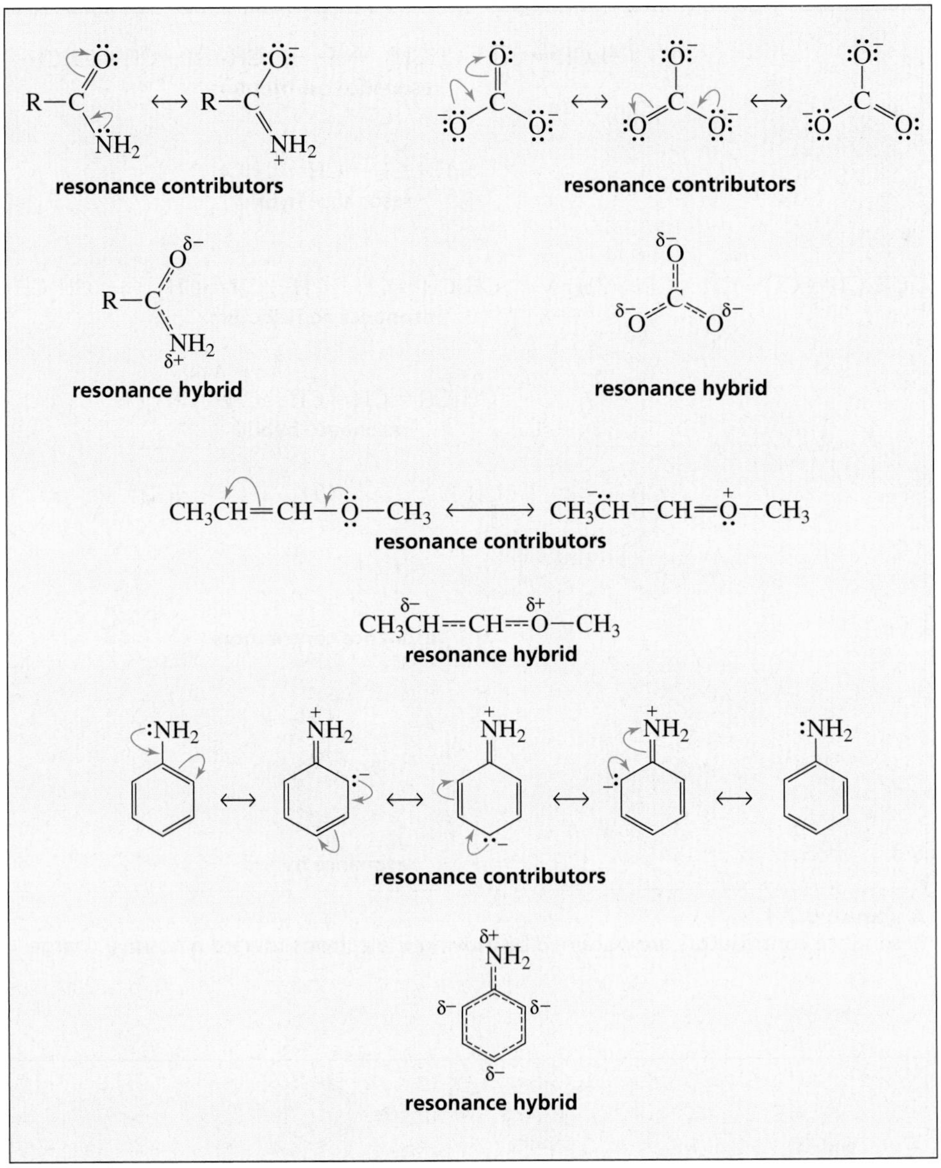

▲ **Figure 6.4**
Resonance contributors are obtained by moving a nonbonding pair of electrons toward
a π bond.

Notice that, in all cases, the electrons are moved toward an sp^2 hybridized atom.
(Remember that an sp^2 hybridized carbon is either a double-bonded carbon or a carbon that has a positive charge or an unpaired electron; Sections 1.8 and 1.10.) Electrons cannot be moved toward an sp^3 hybridized carbon because an sp^3 hybridized
carbon has four σ bonds and cannot accommodate any more electrons.

Because electrons are neither added to nor removed from the compound when
resonance contributors are drawn, each of the resonance contributors for a particular compound must have the same net charge. If one resonance structure has a net
charge of −1, all the others must also have net charges of −1; if one has a net charge
of 0, all the others must also have net charges of 0. (A net charge of 0 does not necessarily mean that there is no charge on any of the atoms—a molecule with a positive charge on one atom and a negative charge on another atom has a net charge of 0.)

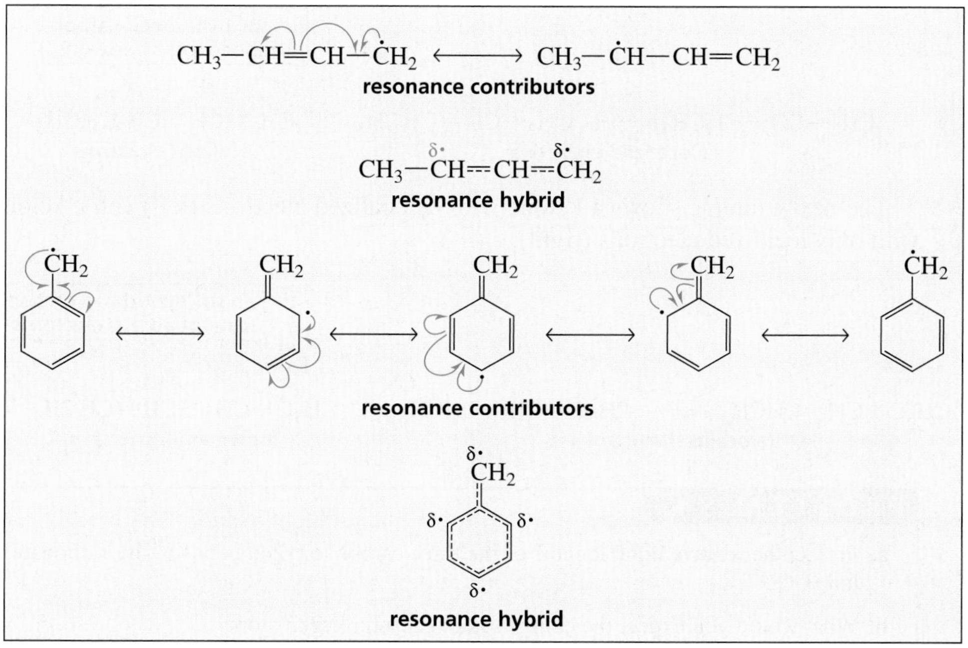

$$CH_3-CH=CH-\overset{\cdot}{C}H_2 \quad\longleftrightarrow\quad CH_3-\overset{\cdot}{C}H-CH=CH_2$$
resonance contributors

$$CH_3-\overset{\delta\cdot}{C}H\text{==}CH\text{==}\overset{\delta\cdot}{C}H_2$$
resonance hybrid

resonance contributors

resonance hybrid

▲ **Figure 6.5**
Resonance structures for an allylic radical and for the benzyl radical.

Radicals can also have delocalized electrons if the unpaired electron is on a carbon that is adjacent to an sp^2 hybridized atom (Figure 6.5). The arrows in Figure 6.5 are half-headed arrows because they denote the movement of only one electron (Section 3.6).

One way to recognize compounds with delocalized electrons is to compare them with similar compounds in which all the electrons are localized. The compound on the left in the first example has delocalized electrons because the nonbonding pair of electrons on nitrogen can be shared with the adjacent sp^2 carbon (because the carbon–carbon π bond can be broken). In contrast, all the electrons in the compound on the right are localized. The nonbonding pair of electrons on nitrogen cannot be shared with the adjacent sp^3 carbon because carbon cannot form five bonds. The octet rule requires that second-row elements be surrounded by no more than eight electrons, so sp^3 hybridized carbons cannot accept electrons. Because an sp^2 hybridized carbon has a π bond that can break, a positive charge, or an unpaired electron, it can accept electrons without violating the octet rule.

Tutorial: Localized and delocalized electrons

an sp^3 hybridized carbon cannot accept electrons

$$CH_3CH=CH-\overset{\cdot\cdot}{N}HCH_3 \quad\longleftrightarrow\quad CH_3\overset{-}{\overset{\cdot\cdot}{C}}H-CH=\overset{+}{N}HCH_3 \qquad CH_3CH=CH-CH_2-\overset{\cdot\cdot}{N}H_2$$
delocalized electrons   localized electrons

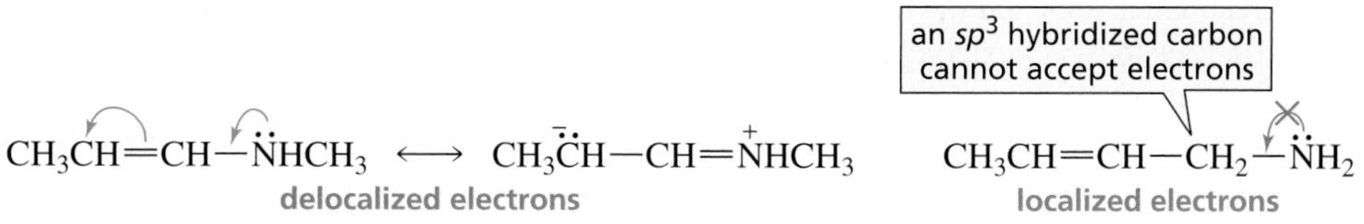

The carbocation shown on the left in the next example has delocalized electrons because the π electrons can move into the empty p orbital of the adjacent sp^2 carbon. We know that this carbon has an empty p orbital because it has a positive charge. The electrons in the carbocation on the right are localized because the π electrons cannot move. The carbon they would move to is sp^3 hybridized, and sp^3 hybridized carbons cannot accept electrons.

an sp^3 hybridized carbon cannot accept electrons

$$CH_2\!\!=\!\!CH\!-\!\overset{+}{C}HCH_3 \longleftrightarrow \overset{+}{C}H_2\!-\!CH\!\!=\!\!CHCH_3 \qquad CH_2\!\!=\!\!CH\!-\!CH_2\overset{+}{C}HCH_3$$

delocalized electrons   localized electrons

The next example shows a ketone with delocalized electrons (left) and a ketone with only localized electrons (right).

an sp^3 hybridized carbon cannot accept electrons

$$CH_3\overset{\ddot{O}:}{\overset{\|}{C}}\!-\!CH\!\!=\!\!CHCH_3 \longleftrightarrow CH_3\overset{:\ddot{O}:^-}{\overset{|}{C}}\!\!=\!\!CH\!-\!\overset{+}{C}HCH_3 \qquad CH_3\overset{\ddot{O}:}{\overset{\|}{C}}\!-\!CH_2\!-\!CH\!\!=\!\!CHCH_3$$

delocalized electrons   localized electrons

PROBLEM 3 ◆

a. Predict the relative bond lengths of the three carbon–oxygen bonds in the carbonate ion ($CO_3{}^{2-}$).

b. What would you expect the charge to be on each oxygen atom?

PROBLEM 4

a. Which of the following compounds have delocalized electrons?

   1. $CH_3CH_2NHCH_2CH\!\!=\!\!CH_2$ **4.** $CH_2\!\!=\!\!CHCH_2CH\!\!=\!\!CH_2$

   2. NH_2 (benzene ring)   **5.** (pyran ring with O and + charge)

   3. CH_2NH_2 (benzene ring)   **6.** $CH_3CH\!\!=\!\!CHCH\!\!=\!\!CH\overset{+}{C}H_2$

b. Draw the contributing resonance structures for these compounds.

6.5
THE RESONANCE HYBRID

All resonance contributors do not necessarily contribute equally to the resonance hybrid. The degree to which each resonance contributor contributes depends on its predicted stability. Because the resonance contributors are not real, their stabilities cannot be measured. Therefore, stabilities of resonance contributors have to be predicted on the basis of molecular features that are found in real molecules. *The greater the predicted stability of the resonance contributor, the more it contributes to the resonance hybrid.* And the more it contributes to the resonance hybrid, the more similar the contributor is to the real molecule. The following examples illustrate these points.

The two resonance contributors for a carboxylic acid are labeled **A** and **B**. Structure **B** has separated charges. To say that a structure has **separated charges** means that it has a positive charge and a negative charge that can be neutralized by the movement of electrons. We can predict that resonance contributors with separated charges are relatively unstable because it takes energy to keep opposite charges separated. Because structure **A** does not have separated charges, it has a considerably greater predicted stability.

Because structure **A** is predicted to be more stable than structure **B,** structure **A** makes a greater contribution to the resonance hybrid—the resonance hybrid looks more like **A** than **B.**

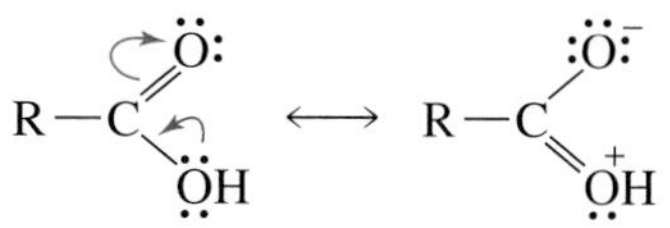

a carboxylic acid

The two resonance contributors for a carboxylate ion are shown next. Structures **C** and **D** are predicted to be equally stable and therefore are expected to contribute equally to the resonance hybrid.

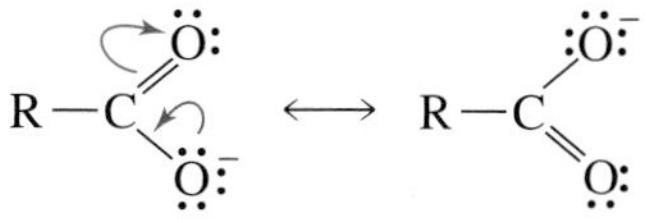

a carboxylate ion

When there are two possible directions in which electrons can be moved, they are always moved toward the more electronegative atom. For example, structure **G** in the next example results from moving π electrons toward oxygen, the more electronegative of the atoms. Structure **E** results from moving π electrons toward carbon, the less electronegative atom.

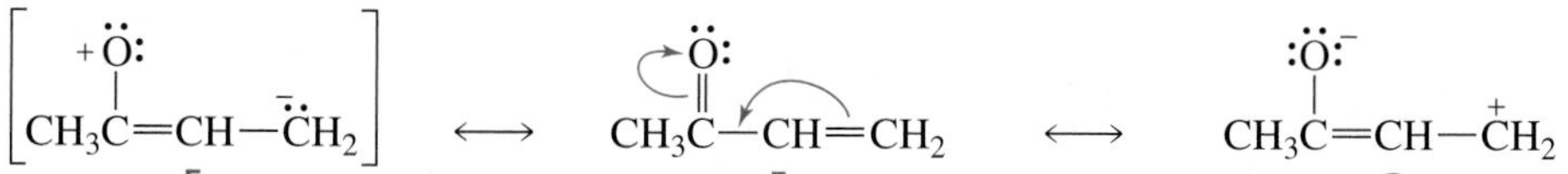

resonance contributor obtained by moving π electrons away from the more electronegative atom

resonance contributor obtained by moving π electrons toward the more electronegative atom

We can predict that structure **G** will make only a small contribution to the resonance hybrid because it has separated charges as well as an atom with an incomplete octet. Structure **E** also has separated charges and an atom with an incomplete octet, but its predicted stability is even less than that of structure **G** because it has a positive charge on the electronegative oxygen. Its contribution to the resonance hybrid is so insignificant that we do not need to include it as one of the resonance contributors. The resonance hybrid, therefore, looks very much like structure **F.**

The only time resonance contributors obtained by moving electrons away from the more electronegative atom should be shown is when this is the only way the electrons can be moved. In other words, movement of the electrons away from the more electronegative atom is better than no movement at all, because electron delocalization makes the molecule more stable (Section 6.6). For example, the only resonance contributor that can be drawn for the following molecule requires movement of the electrons away from oxygen. Structure **I** is predicted to be relatively unstable because it has separated charges and its most electronegative atom is the one with the positive charge. Therefore, the structure of the resonance hybrid is very similar to structure **H,** with only a small contribution from structure **I.**

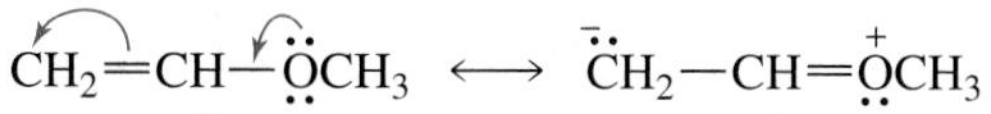

> The greater the predicted stability of the resonance contributor, the more it contributes to the structure of the resonance hybrid.

Of the two contributing resonance structures for the enolate anion, structure **J** has a negative charge on carbon and structure **K** has a negative charge on oxygen. Oxygen is more electronegative than carbon so oxygen can better accommodate the negative charge. Consequently, structure **K** is predicted to be more stable than structure **J**. The resonance hybrid, therefore, more closely resembles structure **K**; that is, the resonance hybrid has a greater concentration of negative charge on oxygen than on carbon.

3-D Molecule:
An enolate ion

$$\text{R}-\overset{\overset{\displaystyle ::\!\ddot{\text{O}}}{\|}}{\text{C}}-\overset{..}{\text{C}}\text{HCH}_3 \longleftrightarrow \text{R}-\overset{\overset{\displaystyle ::\!\ddot{\text{O}}:^-}{|}}{\text{C}}=\text{CHCH}_3$$

$$\qquad\quad \mathbf{J} \qquad\qquad\qquad\qquad \mathbf{K}$$

an enolate ion

We can summarize the features that decrease the predicted stability of a contributing resonance structure as follows:

1. an atom with an incomplete octet
2. a negative charge that is not on the most electronegative atom or a positive charge that is not on the least electronegative (most electropositive) atom
3. charge separation

When we compare the relative stabilities of structures, each of which has only one of these features, an atom with an incomplete octet (feature 1) generally makes a structure more unstable than does either feature 2 or feature 3.

PROBLEM 5 / SOLVED

Draw contributing resonance structures for each of the following species. Rank the structures in order of decreasing contribution to the hybrid.

a. $\text{CH}_3\overset{+}{\underset{\underset{\displaystyle \text{CH}_3}{|}}{\text{C}}}-\text{CH}=\text{CHCH}_3$

b. $\text{CH}_3\overset{\overset{\displaystyle \text{O}}{\|}}{\text{C}}\text{OCH}_3$

c. (cyclohexene ring with $^-$)$=\text{O}$

d. (cyclohexene ring)$=\text{O}$

e. $\text{CH}_3-\overset{\overset{\displaystyle \overset{+}{\text{O}}\text{H}}{\|}}{\text{C}}-\underset{\underset{\displaystyle \text{CH}_3}{|}}{\text{N}}\text{CH}_3$

f. $\text{CH}_3\overset{+}{\text{C}}\text{H}-\text{CH}=\text{CHCH}_3$

SOLUTION TO 5a Structure **A** is more stable than structure **B** because the positive charge is on a tertiary carbon in **A** and on a secondary carbon in **B**.

$$\text{CH}_3\overset{+}{\underset{\underset{\displaystyle \text{CH}_3}{|}}{\text{C}}}-\text{CH}=\text{CHCH}_3 \longleftrightarrow \text{CH}_3\underset{\underset{\displaystyle \text{CH}_3}{|}}{\text{C}}=\text{CH}-\overset{+}{\text{C}}\text{HCH}_3$$

$$\qquad\qquad \mathbf{A} \qquad\qquad\qquad\qquad\qquad \mathbf{B}$$

A compound with delocalized electrons is more stable than it would be if all its electrons were localized. The extra stability a compound gains from having delocalized electrons is called **delocalization energy** or **resonance energy.** Since it is awkward to think of "energy" as "stability," you might think of resonance energy as the energy the hybrid does *not* contain.

6.6 RESONANCE ENERGY

To better understand the concept of resonance energy, let's take a look at the resonance energy of benzene. In other words, let's determine how much more stable benzene (with three pairs of delocalized π electrons) is than the unknown, unreal, hypothetical compound "cyclohexatriene" (with three pairs of localized π electrons).

The heat of hydrogenation $(-\Delta H^{\circ})$ of cyclohexene, a compound with one localized double bond, has been experimentally determined to be 28.6 kcal/mol. We would expect "cyclohexatriene," the hypothetical compound with three localized double bonds, to have a heat of hydrogenation three times that of cyclohexene. Therefore, we can calculate that the heat of hydrogenation of "cyclohexatriene" is $3 \times 28.6 = 85.8$ kcal/mol (Section 3.19).

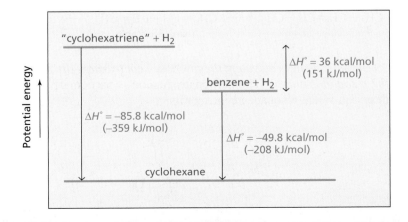

When the heat of hydrogenation was experimentally determined for benzene, it was found to be 49.8 kcal/mol, much smaller than that calculated for hypothetical "cyclohexatriene."

Because the hydrogenation of "cyclohexatriene" and the hydrogenation of benzene both form cyclohexane, the only way to account for the difference in the heats of hydrogenation is for "cyclohexatriene" and benzene to have different stabilities. Figure 6.6 shows that benzene must be 36 kcal/mol more stable than "cyclohexatriene" because the experimental heat of hydrogenation for benzene is 36 kcal/mol less than the heat of hydrogenation calculated for "cyclohexatriene" $(85.8 - 49.8 = 36)$.

◀ **Figure 6.6**
The difference in the energy levels of cyclohexane and "cyclohexatriene" + hydrogen, and the difference in the energy levels of cyclohexane and benzene + hydrogen.

Because benzene and "cyclohexatriene" have different stabilities, they must be different compounds. Benzene has six delocalized π electrons, whereas hypothetical "cyclohexatriene" has six localized π electrons. The difference in energy between them is the resonance energy of benzene. The resonance energy tells us *how much more stable a compound with delocalized electrons is than it would be if its electrons were localized.* Benzene, with six delocalized π electrons, is 36 kcal/mol more stable than hypothetical "cyclohexatriene" with six localized π electrons. Now we can understand why nineteenth-century chemists, who didn't know about delocalized electrons, were puzzled by benzene's unusual stability (Section 6.1).

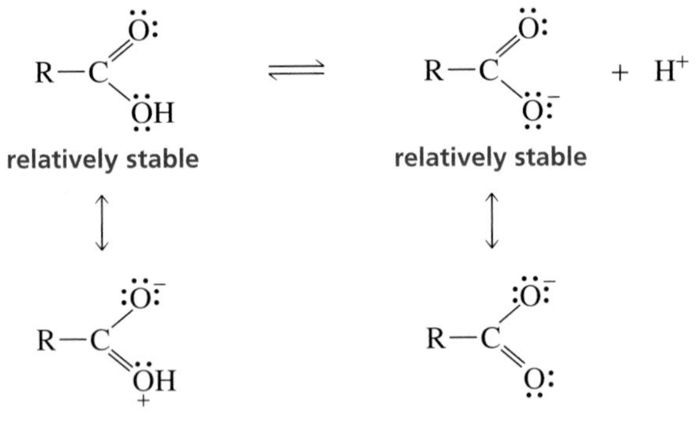

The greater the number of relatively stable resonance contributors, the greater the resonance energy.

Since the ability to delocalize electrons increases the stability of a molecule, we can conclude that *a resonance hybrid is more stable than the predicted stability of any of its resonance contributors.* The resonance energy associated with a compound that has delocalized electrons depends on the number *and* predicted stability of the resonance contributors. *The greater the number of relatively stable resonance contributors, the greater the resonance energy.* For example, the resonance energy of a carboxylate ion with two relatively stable resonance contributors is significantly greater than that of a carboxylic acid that has only one relatively stable resonance contributor.

resonance contributors for a carboxylic acid

resonance contributors for a carboxylate ion

Note that it is the number of *relatively stable* resonance contributors—not the total number of resonance contributors—that is important in determining the resonance energy. For example, the resonance energy of a carboxylate ion with two relatively stable resonance contributors is greater than the resonance energy of the compound in the next example because, even though this compound has three resonance contributors, only one of them is relatively stable.

$$\overset{-}{C}H_2{-}CH{=}CH{-}\overset{+}{C}H_2 \longleftrightarrow CH_2{=}CH{-}CH{=}CH_2 \longleftrightarrow \overset{+}{C}H_2{-}CH{=}CH{-}\overset{-}{C}H_2$$

relatively unstable — relatively stable — relatively unstable

The more nearly equivalent the resonance contributors are in structure, the greater the resonance energy. The carbonate dianion is particularly stable because it has three equivalent resonance contributors.

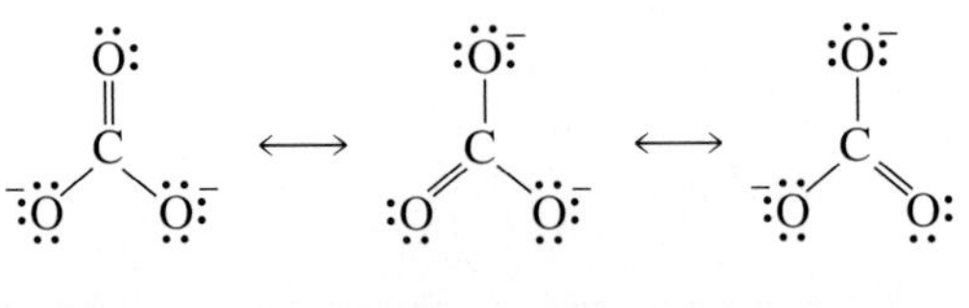

We can now summarize what we know about contributing resonance structures:

1. The greater the predicted stability of a resonance contributor, the more it contributes to the resonance hybrid.
2. The greater the number of relatively stable resonance contributors, the greater the resonance energy.
3. The more nearly equivalent the resonance contributors, the greater the resonance energy.

Allylic and benzylic cations have delocalized electrons, so they are more stable than similarly substituted carbocations with localized electrons. An **allylic carbon** is an sp^3 carbon that is adjacent to an sp^2 carbon of an alkene. An **allylic cation** is a carbocation with the positive charge on an allylic carbon. The allyl cation is an unsubstituted allylic cation. A **benzylic carbon** is an sp^3 carbon that is bonded to a benzene ring. A **benzylic cation** is a carbocation with the positive charge on a benzylic carbon. The benzyl cation is an unsubstituted benzylic cation.

6.7
STABILITY OF ALLYLIC AND BENZYLIC CATIONS

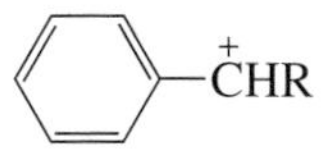

An allylic cation has two resonance contributors. The positive charge is not localized on a single carbon but is shared by two carbons.

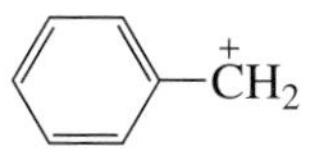

A benzylic cation has five resonance contributors—the positive charge is shared by four carbons.

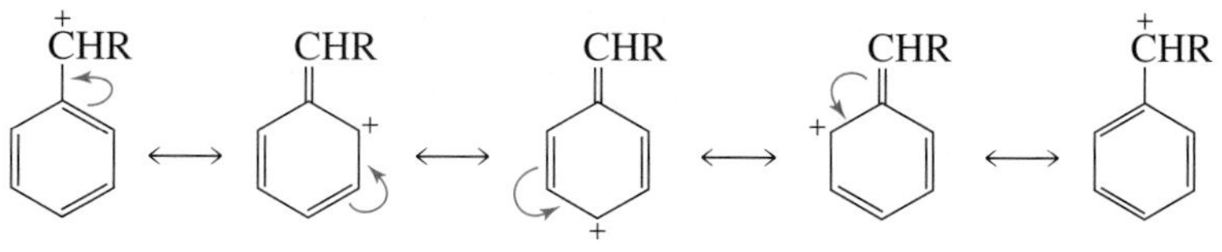

Because allyl and benzyl cations have delocalized electrons, they are more stable than other primary alkyl carbocations. [The relative stabilities are based on their relative heats of formation (Section 2.12)]. The more electron deficient the carbon, the greater the intensity of the blue area in the electrostatic potential map. We can add the benzyl and allyl cations to the group of carbocations whose relative stabilities were shown in Sections 3.10 and 5.4.

Not all benzylic and allylic cations have the same stability. Just as a tertiary alkyl carbocation is more stable than a secondary alkyl carbocation, a tertiary allylic cation is more stable than a secondary allylic cation, which in turn is more stable than the primary allyl cation. Similarly, a tertiary benzylic cation is more stable than a secondary benzylic cation, which is more stable than the primary benzyl cation.

relative stabilities of carbocations

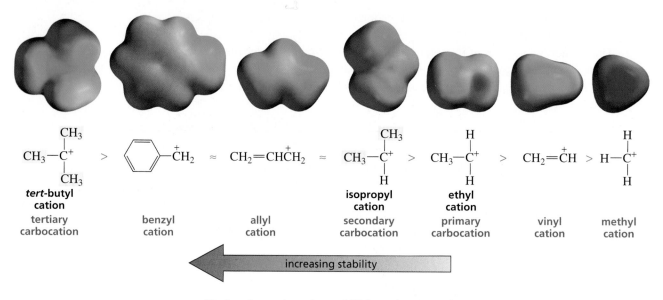

tert-butyl
cation

tertiary
carbocation

benzyl
cation

allyl
cation

isopropyl
cation

secondary
carbocation

ethyl
cation

primary
carbocation

vinyl
cation

methyl
cation

increasing stability

Notice than when the stabilities of carbocations are compared, it is the *primary* benzyl and allyl cations that have about the same stability as a secondary alkyl carbocation. Secondary benzylic and allylic cations as well as tertiary benzylic and allylic cations are even more stable than the primary benzyl and allyl cations.

3-D Molecules:
allyl cation;
benzyl cation

relative stabilities

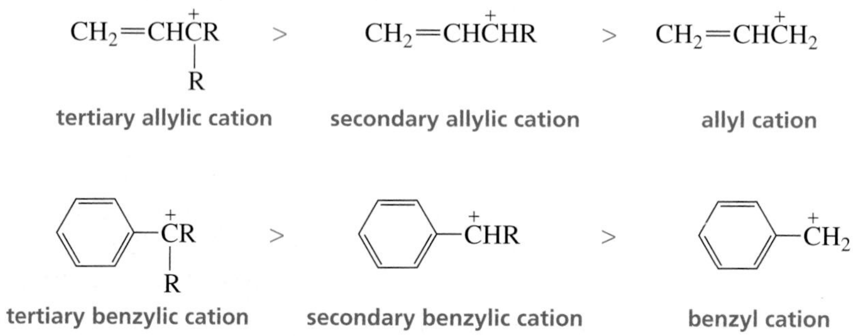

$$CH_2=CH\overset{+}{C}R \quad > \quad CH_2=CH\overset{+}{C}HR \quad > \quad CH_2=CH\overset{+}{C}H_2$$
$$\quad\quad\;\; |$$
$$\quad\quad\; R$$

tertiary allylic cation

secondary allylic cation

allyl cation

tertiary benzylic cation

secondary benzylic cation

benzyl cation

increasing stability

PROBLEM 6 ◆

Which carbocation in each of the following pairs is more stable?

a. $CH_3O\overset{+}{C}H_2$ or $CH_3N\overset{+}{H}CH_2$

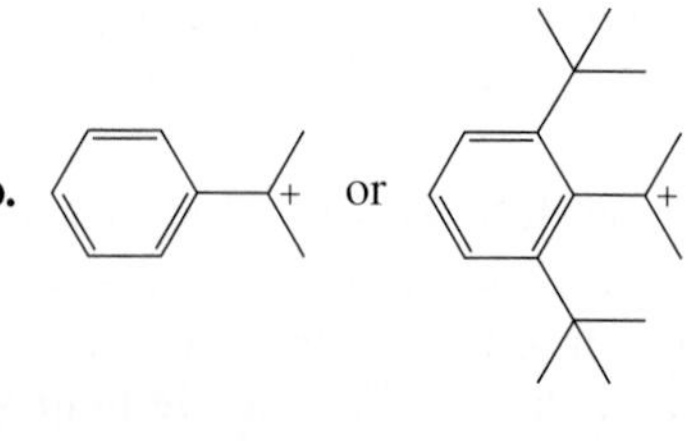

c. $CH_3OCH_2\overset{+}{C}H_2$ or $CH_3O\overset{+}{C}H_2$

b. or

d. or

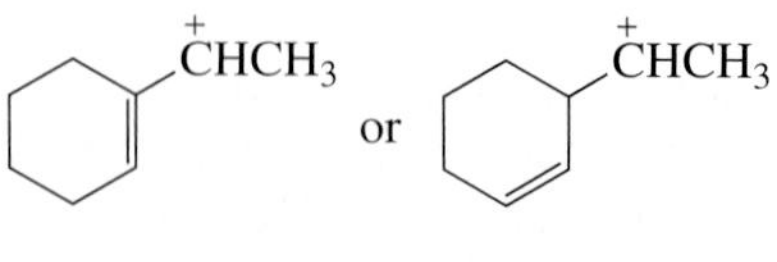

An allylic radical has an unpaired electron on an allylic carbon and, like an allylic cation, has two contributing resonance structures.

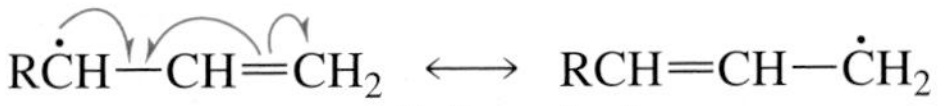

$$R\dot{C}H\text{—}CH\text{=}CH_2 \quad \longleftrightarrow \quad RCH\text{=}CH\text{—}\dot{C}H_2$$
an allylic radical

A benzylic radical has an unpaired electron on a benzylic carbon and, like a benzylic cation, has five contributing resonance structures.

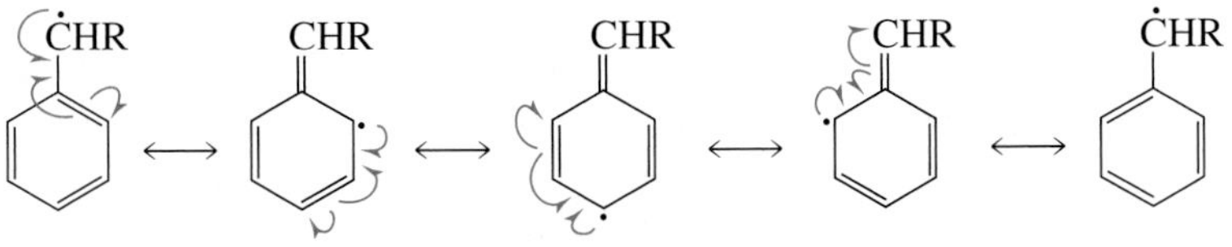

a benzylic radical

Because of their delocalized electrons, allyl and benzyl radicals are both more stable than other primary radicals.

relative stabilities of radicals

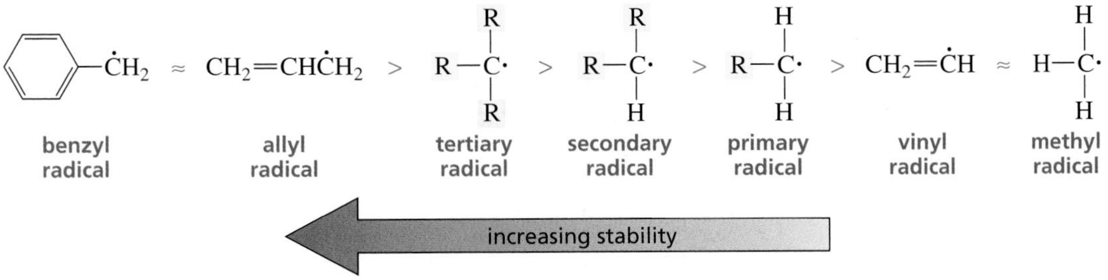

increasing stability

Our ability to predict the correct product of an organic reaction often depends on our ability to recognize when organic molecules have delocalized electrons. For example, in the following reaction both sp^2 carbons of the alkene are bonded to the same number of hydrogens. Therefore Markovnikov's rule predicts that approximately equal amounts of the two addition products will be formed. When the reaction is carried out, however, only one of the products is obtained.

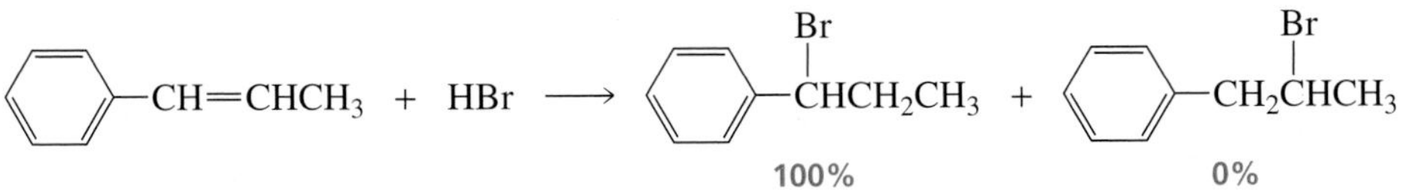

Markovnikov's rule leads us to an incorrect prediction of the product of this reaction because it does not take resonance (electron delocalization) into consideration. It assumes that both carbocation intermediates are equally stable since they are both secondary carbocations. The rule does not recognize that one intermediate is an ordinary secondary carbocation while the other is a secondary benzylic cation. Because the secondary benzylic cation is stabilized by resonance, it is formed more readily, and therefore only one addition product is obtained.

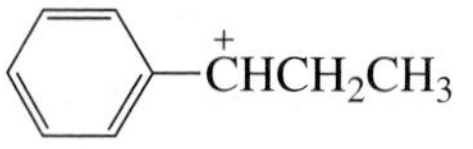

$$\text{C}_6\text{H}_5-\overset{+}{\text{C}}\text{HCH}_2\text{CH}_3 \qquad \text{C}_6\text{H}_5-\text{CH}_2\overset{+}{\text{C}}\text{HCH}_3$$

a secondary benzylic cation 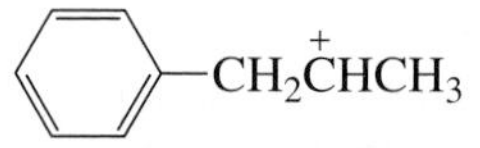   **a secondary carbocation**

This example serves as a warning. Do not use Markovnikov's rule in reactions in which the carbocations can be stabilized by resonance. In such cases, you must look at the stability of the individual carbocations in order to predict the product of the reaction.

The following reaction also demonstrates the importance of recognizing the presence of delocalized electrons in predicting the outcome of a reaction. The addition of a proton to the alkene yields a secondary carbocation. A carbocation rearrangement occurs because a more stable secondary benzylic cation is formed as a result of a 1,2-hydride shift. It is electron delocalization that causes the benzylic secondary cation to be more stable than the initially formed carbocation. If we had neglected electron delocalization, we would not have anticipated the carbocation rearrangement, and we would not have correctly predicted the product of the reaction.

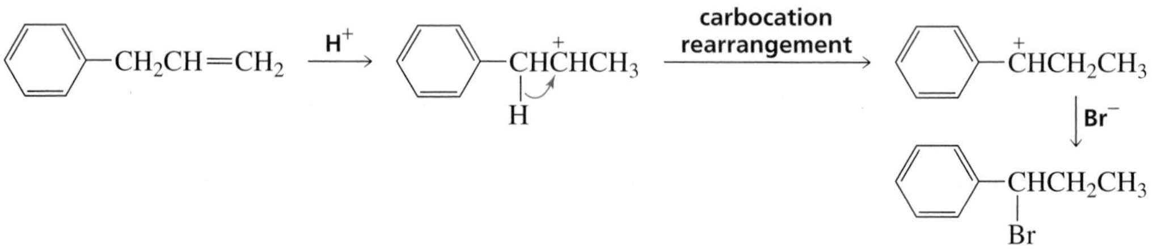

The relative rates at which alkenes **A, B,** and **C** undergo an electrophilic addition reaction with a reagent such as HBr demonstrate the effect delocalized electrons can have on the reactivity of a compound.

relative reactivities toward addition of HBr

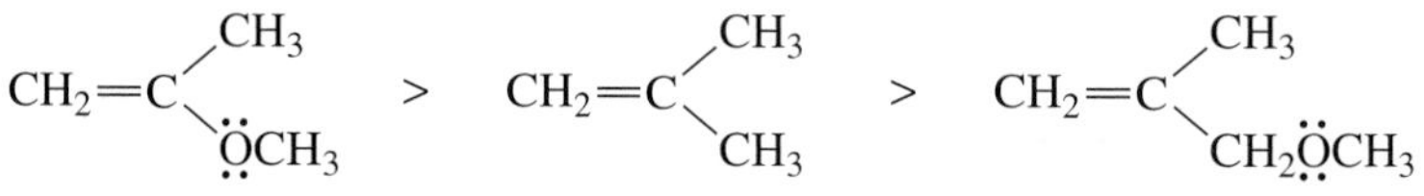

A is the most reactive of the three alkenes. The positive charge of the carbocation intermediate formed in the rate-limiting step is shared by carbon and oxygen. Therefore, it is a more stable carbocation—and, therefore, easier to form—than the carbocations formed by **B** and **C** that have their positive charge localized on a single atom.

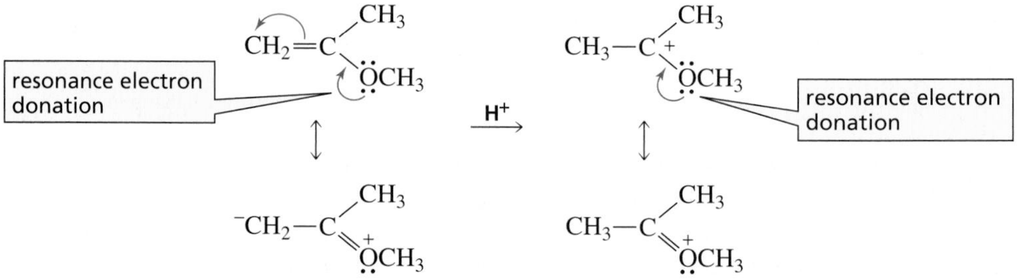

B reacts more rapidly than **C** with HBr because the carbocation formed by **C** is destabilized by the OCH₃ group that withdraws electrons inductively (through the σ bonds) from the positively charged carbon of the carbocation intermediate.

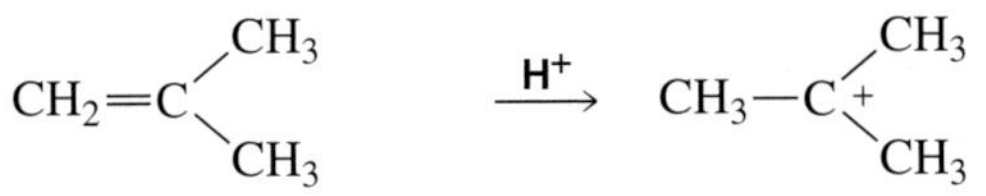

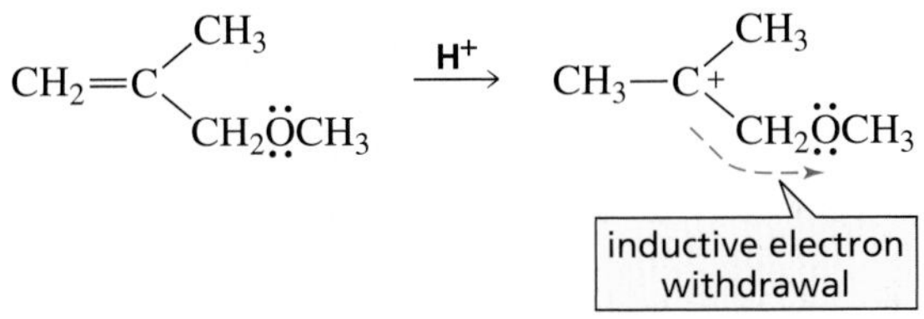

$$CH_2{=}C\overset{CH_3}{\underset{CH_3}{\diagdown}} \quad \xrightarrow{\ H^+\ } \quad CH_3{-}\overset{CH_3}{\underset{CH_3}{C{+}}}$$

$$CH_2{=}C\overset{CH_3}{\underset{CH_2\ddot{O}CH_3}{\diagdown}} \quad \xrightarrow{\ H^+\ } \quad CH_3{-}\overset{CH_3}{\underset{CH_2\ddot{O}CH_3}{C{+}}}$$

> inductive electron
> withdrawal

Notice that the OCH$_3$ group in **C** can only withdraw electrons inductively, whereas the OCH$_3$ group in **A** is positioned so that in addition to withdrawing electrons inductively, it can also donate the lone-pair electrons to stabilize the carbocation. This is called **resonance electron donation.** Because stabilization by resonance electron donation outweighs destabilization by inductive electron withdrawal, the overall effect of the OCH$_3$ group in **A** is stabilization of the carbocation intermediate.

PROBLEM 7 / SOLVED

Predict the sites on each of the following compounds where the reaction can occur.

a. $CH_3CH{=}CHOCH_3$ + H^+

c. 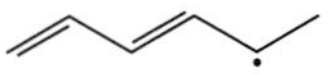 [structure] + Br$^\bullet$

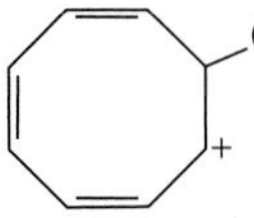

b. [structure with Cl] + HO$^-$

d. 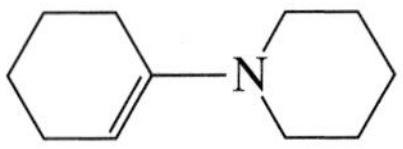 [structure with N] + H$^+$

SOLUTION TO 7a   The contributing resonance structures show that there are two sites that can be protonated, the nonbonded electrons on oxygen and the nonbonded electrons on carbon.

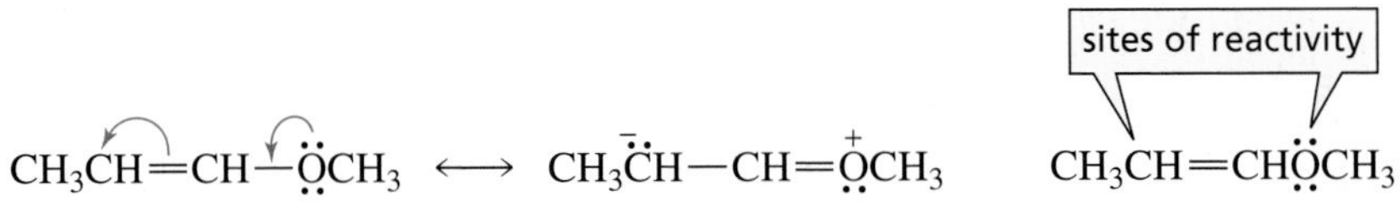

> sites of reactivity

$$CH_3\overset{\frown}{CH}{=}CH\overset{\frown}{-}\ddot{O}CH_3 \ \longleftrightarrow\ CH_3\overset{-}{\ddot{C}}H{-}CH{=}\overset{+}{O}CH_3 \qquad CH_3CH{=}CH\ddot{O}CH_3$$

resonance contributors

We have seen that a carboxylic acid is a much stronger acid than an alcohol. For example, the pK_a of acetic acid is 4.76, whereas the pK_a of ethyl alcohol is 15.9 (Section 1.17). Now we can explain more fully the reason for this difference in acidity.

**6.10
EFFECT OF
DELOCALIZED
ELECTRONS ON pK_a**

$$\overset{\displaystyle O}{\overset{\|}{CH_3COH}} \qquad CH_3CH_2OH$$

acetic acid   ethyl alcohol
pK_a = 4.76   pK_a = 15.9

The difference in acidity is attributable to two factors. First, a substituent that causes the electrons of the O—H bond to be drawn away from the proton increases the acidity of a compound because electron withdrawal stabilizes the conjugate base. (Section 1.18).

An alcohol such as ethyl alcohol has no electron-withdrawing substituents, but both resonance contributors of a carboxylic acid have an electron-withdrawing group close to the OH group (Section 6.6): One contributor has an electron-withdrawing oxygen two atoms away from the OH group, and the other has an electron-withdrawing positively charged oxygen directly attached to the acidic hydrogen.

$$CH_3C\underset{\ddot{O}H}{\overset{\ddot{O}:}{\diagdown}} \longleftrightarrow CH_3C\underset{\overset{+}{O}H}{\overset{:\ddot{O}:^-}{\diagdown}}$$

Another factor that makes a carboxylic acid more acidic than an alcohol is the resonance energy of the carboxylate ion. The carboxylate ion has greater resonance energy than the carboxylic acid because the anion has two equivalent resonance contributors that are predicted to be relatively stable, whereas the carboxylic acid has only one (Section 6.6). Therefore, loss of a proton from a carboxylic acid is accompanied by an increase in resonance energy (Figure 6.7). Remember that an increase in resonance energy means an increase in stability, and we saw in Section 1.18 that the greater the stability of the base that is formed when a proton is removed from an acid, the stronger the acid.

$$R{-}C\underset{\ddot{O}H}{\overset{\ddot{O}:}{\diagdown}} \rightleftharpoons R{-}C\underset{\ddot{O}:^-}{\overset{\ddot{O}:}{\diagdown}} + H^+$$

a carboxylic acid   a carboxylate ion

In contrast, all the electrons in an alcohol (such as ethanol) and its conjugate base are localized, so loss of a proton from an alcohol is not accompanied by an increase in resonance energy.

$$CH_3CH_2\ddot{O}H \rightleftharpoons CH_3CH_2\ddot{O}:^- + H^+$$

ethanol

Figure 6.7 ▶
One factor that makes a carboxylic acid more acidic than an alcohol is the greater resonance energy of the carboxylate ion compared to that of the carboxylic acid, which increases the K_a (and therefore decreases the pK_a).

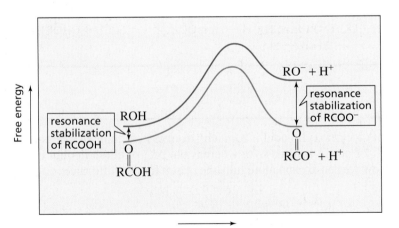

Progress of the reaction

Phenol, a compound in which an OH group is bonded to an sp^2 carbon of a benzene ring, is a stronger acid than an alcohol such as ethanol or cyclohexanol, compounds in which an OH group is bonded to an sp^3 carbon.

Tutorial: Acidity and electron delocalization

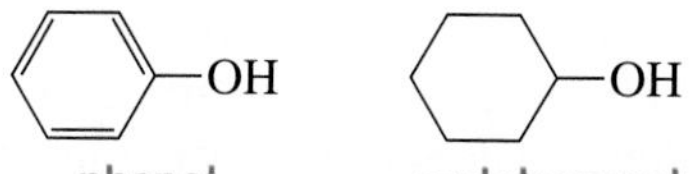

phenol
pK_a = 10

cyclohexanol
pK_a = 16

ethanol
pK_a = 16

CH_3CH_2OH

One reason for the increased acidity of phenol is that the electronegativity of an sp^2 carbon is greater than that of an sp^3 carbon. Also, the O—H bond of phenol is weakened by resonance contributors that have electron-withdrawing, positively charged oxygens. Finally, there is electron delocalization in both phenol and the phenoxide ion, but the resonance energy of the phenoxide ion is greater than the resonance energy of phenol because three of phenol's resonance contributors have separated charges. The greater resonance energy of the phenoxide ion contributes to the increased acidity of phenol compared with that of an alcohol such as cyclohexanol.

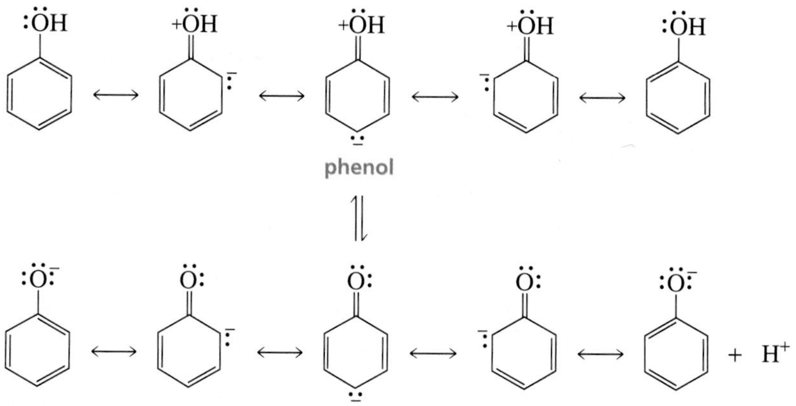

phenol

phenoxide ion

+ H$^+$

The electron withdrawal in phenol is not as great as the electron withdrawal in a carboxylic acid, and the resonance energy of a phenoxide ion is not as great as the resonance energy of a carboxylate ion, where the negative charge is shared equally by two oxygens. The reduced electron withdrawal in phenol compared with a carboxylic acid and the reduced resonance energy in the phenoxide ion compared with a carboxylate ion makes phenol a weaker acid than a carboxylic acid.

A protonated amine such as protonated aniline, in which the nitrogen atom is attached to an sp^2 carbon of a benzene ring, is a stronger acid than a protonated amine such as protonated cyclohexylamine, in which the nitrogen atom is attached to an sp^3 carbon.

protonated aniline
pK_a = 4.60

protonated cyclohexylamine
pK_a = 11.2

Part of the increased acidity of protonated aniline can be attributed to the greater electronegativity of an sp^2 carbon compared with an sp^3 carbon. But electron delocalization also plays an important role. An amine such as cyclohexylamine does not have any delocalized electrons either in the protonated form or in the unprotonated form.

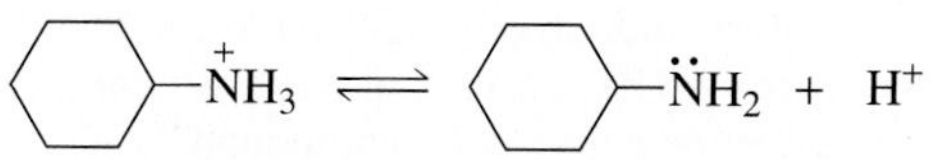

The nitrogen of protonated aniline also does not have any nonbonding electrons that can be delocalized. When it loses a proton, however, the nonbonding electron pair that held the proton can be delocalized. The increased stability of aniline achieved as a result of this electron delocalization contributes to the increased acidity of protonated aniline compared with a protonated amine in which loss of a proton does not result in any electron delocalization.

protonated aniline

aniline

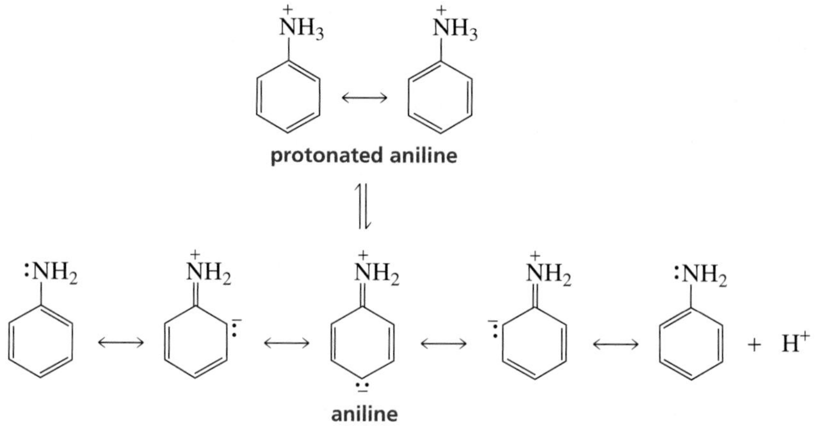

We will add the approximate pK_a values of phenol and protonated aniline to the classes of organic compounds whose approximate pK_a values you should know (Table 6.1). They are also listed inside the back cover for easy reference.

TABLE 6.1   Approximate pK_a Values

$pK_a < 0$	$pK_a \approx 5$	$pK_a \approx 10$	$pK_a \approx 15$
$\overset{+}{R}\overset{}{O}H$ H	$\overset{O}{\underset{}{\overset{\|\|}{RCOH}}}$	$R\overset{+}{N}H_3$	ROH
$+OH$ $\overset{\|\|}{RCOH}$	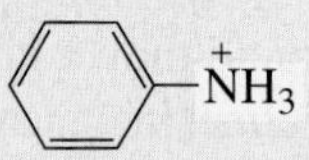 ⬡—$\overset{+}{N}H_3$	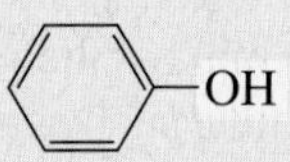 ⬡—OH	H_2O
H_3O^+			

PROBLEM 8 / SOLVED

Which of the following would you predict to be the stronger acid?

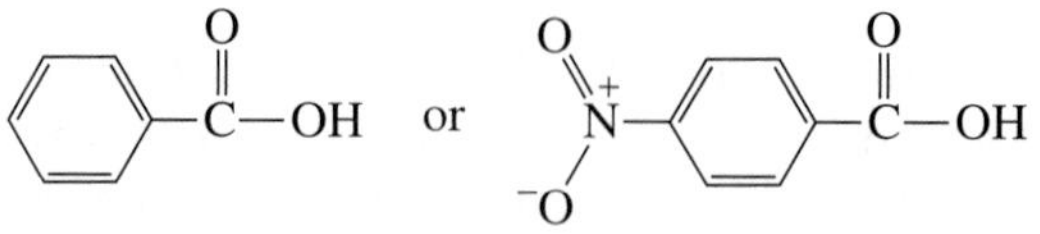

SOLUTION   The nitro substituted compound is the stronger acid because the nitro substituent can withdraw electrons inductively (through the sigma bonds), and it can withdraw electrons by resonance. We have seen that a substituent that draws electrons away from the OH group increases the acidity of a compound.

3-D Molecule:
4-Nitrobenzoic acid

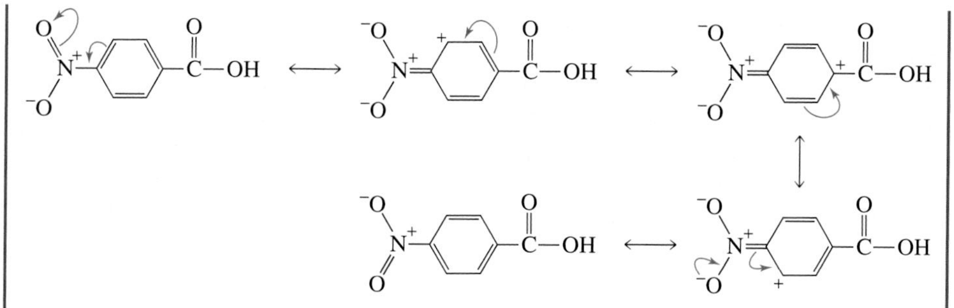

PROBLEM 9 ◆

Which is a stronger acid?

a. $CH_3CH_2CH_2OH$ or $CH_3CH=CHOH$

b. $CH_3CH=CHCH_2OH$ or $CH_3CH=CHOH$

PROBLEM 10 ◆

Which is a stronger base?

a. ethylamine or aniline

b. ethylamine or ethoxide ion $(CH_3CH_2O^-)$

c. phenoxide ion or ethoxide ion

PROBLEM 11 ◆

Rank the following compounds in order of decreasing acid strength.

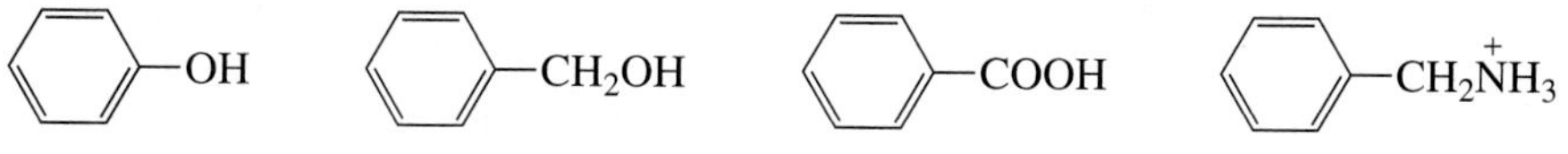

An Introduction to Molecular Orbital Theory

We have used contributing resonance structures to describe the stabilities of compounds with delocalized electrons. The stabilities of these compounds can also be described using molecular orbital (MO) theory.

In Section 1.6 you saw that the two lobes of a p orbital have opposite phases. You also learned that when two in-phase p orbitals overlap, a covalent bond is formed, and that when two out-of-phase p orbitals overlap, they cancel each other and produce a node between the two nuclei. A *node* is a region (in this case a plane) in which there is zero probability of finding an electron.

A molecular orbital description of ethene is shown in Figure 6.8. The two p orbitals in ethene can be either in-phase or out-of-phase. (The different phases are indicated by different colors.) Notice that the number of orbitals is conserved; that is, two atomic orbitals overlap to produce two molecular orbitals. Side-to-side overlapping of in-phase p orbitals (lobes of the same color) produces a **bonding π molecular orbital** designated ψ_1 (the Greek letter psi). The bonding π molecular orbital encompasses both carbons. In other words, each electron in the π molecular orbital spreads over both carbons. The bonding π molecular orbital is lower in energy than the p atomic orbitals.

6.11
A MOLECULAR ORBITAL DESCRIPTION OF STABILITY

Take a few minutes to review Section 1.6.

Side-to-side overlapping of out-of-phase p orbitals produces an **antibonding π molecular orbital, ψ_2**, which is higher in energy than the p atomic orbitals. An antibonding molecular orbital can be designated by * (i.e., a π* molecular orbital). The antibonding orbital has a node between the lobes of opposite phases. The bonding molecular orbital arises from additive interaction of the atomic orbitals, whereas the antibonding molecular orbital arises from subtractive interaction. In other words, the overlap of in-phase orbitals holds atoms together, while the overlap of out-of-phase orbitals pulls atoms apart.

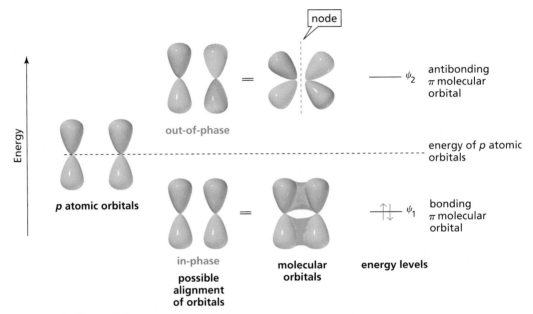

▲ **Figure 6.8**
The distribution of electrons in ethene. Overlapping of in-phase p orbitals produces a bonding π molecular orbital that is lower in energy than the p atomic orbitals. Overlapping of out-of-phase p orbitals produces an antibonding π molecular orbital that is higher in energy than the p atomic orbitals.

Molecular orbitals that result from side-to-side overlapping of p orbitals contain only π electrons. The π electrons are placed in molecular orbitals according to the aufbau principle (orbitals are filled in order of increasing energy), the Pauli exclusion principle (each orbital can hold two electrons of opposite spin), and Hund's rule (an electron will occupy an empty degenerate orbital before it will pair up with an electron already present in an orbital). These are the same rules that are followed when electrons are placed in atomic orbitals (Section 1.2). Each π electron is associated with all the atoms encompassed by the π molecular orbital.

The Allyl Cation, the Allyl Radical, and the Allyl Anion

Let's take a look at the molecular orbitals of the allyl cation, the allyl radical, and the allyl anion.

$$CH_2{=}CH{-}\overset{+}{C}H_2 \longleftrightarrow \overset{+}{C}H_2{-}CH{=}CH_2 \qquad CH_2{=}CH{-}\overset{..}{\overset{-}{C}}H_2 \longleftrightarrow \overset{..}{\overset{-}{C}}H_2{-}CH{=}CH_2$$
the allyl cation   the allyl anion

$$CH_2{=}CH{-}\overset{.}{C}H_2 \longleftrightarrow \overset{.}{C}H_2{-}CH{=}CH_2$$
the allyl radical

Each of the three carbon atoms contributes one p atomic orbital, and the three p orbitals combine to produce three molecular orbitals. Thus a molecular orbital can be described by the **linear combination of atomic orbitals (LCAO).** Each of the electrons that previously occupied a p orbital surrounding an individual carbon nucleus now surrounds the entire part of the molecule that is included in the overlapping p orbitals.

The allyl group has three π molecular orbitals, ψ_1, ψ_2, and ψ_3 (Figure 6.9). The bonding π molecular orbital (ψ_1) encompasses all the carbons in the π system. In a noncyclic system, the number of bonding molecular orbitals always equals the number of antibonding molecular orbitals. Therefore, when there is an odd number of molecular orbitals, one of them must be a nonbonding molecular orbital. In an allyl system, ψ_2 is a nonbonding molecular orbital. As the energy of the molecular orbital increases, the number of nodes increases. Therefore, the ψ_2 molecular orbital must have a node, in addition to the one that bisects the p orbitals, and the only symmetrical position for a node is for it to pass through the middle carbon. You can see from Figure 6.9 why it is called a **nonbonding molecular orbital**—there is no overlap between the p orbital on the middle carbon and the p orbital on either of the end carbons. Notice that a nonbonding molecular orbital has the same energy as the isolated atomic p orbitals. The third molecular orbital (ψ_3) is an antibonding π molecular orbital.

The two π electrons of the allyl cation are in the bonding π molecular orbital, which means they are spread over all three carbons. Consequently, the two carbon–carbon bonds in the allyl cation are identical, with each having some double-bond character. This is another way of showing that the stability of the allyl cation is due to electron delocalization.

The allyl radical has two π electrons in the bonding π molecular orbital and a third π electron in the nonbonding molecular orbital. The molecular orbital picture shows that the third electron is shared equally by the end carbons, with none of the electron density on the middle carbon. This is in agreement with what the resonance contributors show—only the end carbons have radical character.

Finally, the allyl anion has two π electrons in the nonbonding molecular orbital. These two electrons are shared equally by the end carbon atoms.

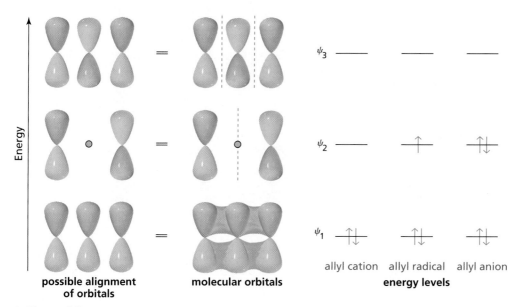

▲ Figure 6.9
Three p atomic orbitals overlap to produce three π molecular orbitals. The distribution of π electrons in the molecular orbitals of the allyl cation, the allyl radical, and the allyl anion.

1,3,5-Hexatriene and Benzene

1,3,5-Hexatriene, with six carbon atoms, has six p atomic orbitals.

$$CH_2=CH-CH=CH-CH=CH_2$$

1,3,5-hexatriene

The six p atomic orbitals combine to produce six π molecular orbitals, ψ_1, ψ_2, ψ_3, ψ_4, ψ_5, and ψ_6. (Figure 6.10) Half of the molecular orbitals are bonding π molecular orbitals (ψ_1, ψ_2, and ψ_3), and the other half are antibonding π molecular orbitals (ψ_4, ψ_5, and ψ_6). 1,3,5-Hexatriene's six π electrons occupy the three bonding π

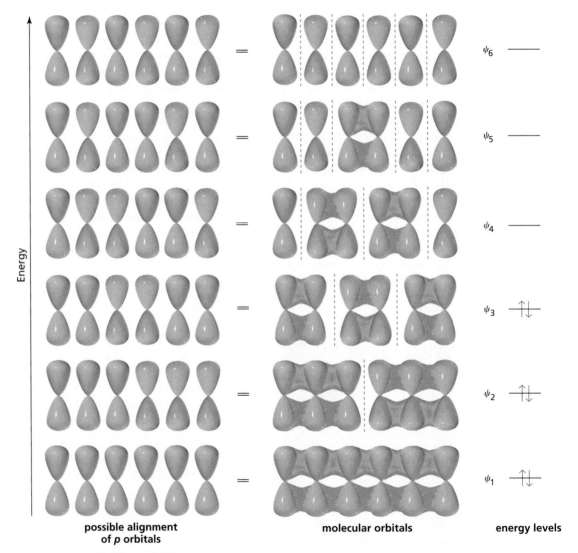

possible alignment of p orbitals   molecular orbitals   energy levels

▲ **Figure 6.10**
Six p atomic orbitals overlap to produce the six π molecular orbitals of 1,3,5-hexatriene. The six π electrons occupy the three bonding π molecular orbitals, ψ_1, ψ_2, and ψ_3.

molecular orbitals (ψ_1, ψ_2, and ψ_3). In other words, the six π electrons are delocalized over the six p orbitals. Thus, molecular orbital theory and resonance are two different ways of showing that the π electrons in 1,3,5-hexatriene are delocalized.

Notice that as the molecular orbitals increase in energy, the number of bonding interactions decreases and the number of nodes increases. For example, ψ_1 has *five* bonding interactions (one between each pair of *p* orbitals), ψ_2 has *four* bonding interactions and one antibonding interaction (because of the additional node[1]) for a net of *three* bonding interactions, and ψ_3 has three bonding interactions and two antibonding interactions (two additional nodes) for a net of *one* bonding interaction. We see then that ψ_2 and ψ_3 are bonding molecular orbitals but not as strongly bonding as ψ_1. A more complete description of molecular orbitals is presented in Section 28.2.

Like 1,3,5-hexatriene, benzene has six sp^2 carbon atoms and, therefore, six *p* orbitals. The six *p* orbitals combine to produce six π molecular orbitals (Figure 6.11). Three of the π molecular orbitals are bonding (ψ_1, ψ_2, and ψ_3) and three are antibonding (ψ_4, ψ_5, and ψ_6). Benzene's six π electrons occupy the three lowest energy π molecular orbitals (the bonding π molecular orbitals). Thus, the electrons are delocalized over the six carbon atoms. The method used to determine the relative energies of the molecular orbitals of a cyclic compound is described in Section 14.6.

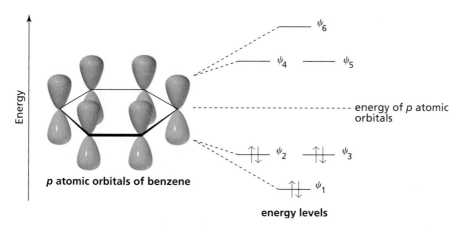

p atomic orbitals of benzene

energy levels

◀ **Figure 6.11**
Benzene has six π molecular orbitals, three bonding (ψ_1, ψ_2, ψ_3) and three antibonding (ψ_4, ψ_5, ψ_6). The six π electrons occupy the three bonding π molecular orbitals.

If you examine the lowest energy π molecular orbital (ψ_1) of benzene in Figure 6.12, you see that it has *six* bonding interactions—one more than the lowest energy π molecular orbital of 1,3,5-hexatriene (Figure 6.10). Thus, tying the three double bonds back into a ring is accompanied by increased stabilization. Each of the other two bonding π molecular orbitals of benzene (ψ_2 and ψ_3) has a node in addition to the node that bisects the *p* orbitals. These two orbitals are degenerate: ψ_2 has four bonding interactions and two antibonding interactions for a net of *two* bonding interactions, while ψ_3 has two bonding interactions and four nonbonding interactions for a net of *two* bonding interactions. Thus, ψ_2 and ψ_3 are bonding molecular orbitals but they are not as strongly bonding as ψ_1.

The energy levels of the molecular orbitals of ethene, 1,3,5-hexatriene, and benzene are compared in Figure 6.13. You can see that benzene is a particularly stable molecule. It is more stable than 1,3,5-hexatriene, and much more stable than three isolated ethene molecules. Compounds such as benzene that are unusually stable because of large delocalization energies are called **aromatic compounds.** The structural features that cause a compound to be aromatic are discussed in Section 14.1.

PROBLEM 12 ◆

How many bonding interactions are there in the ψ_1 and ψ_2 π molecular orbitals of:

a. 1,3-butadiene   **b.** 1,3,5,7-octatetraene

---

[1]The vertical node is in addition to the horizontal node that bisects the *p* orbitals.

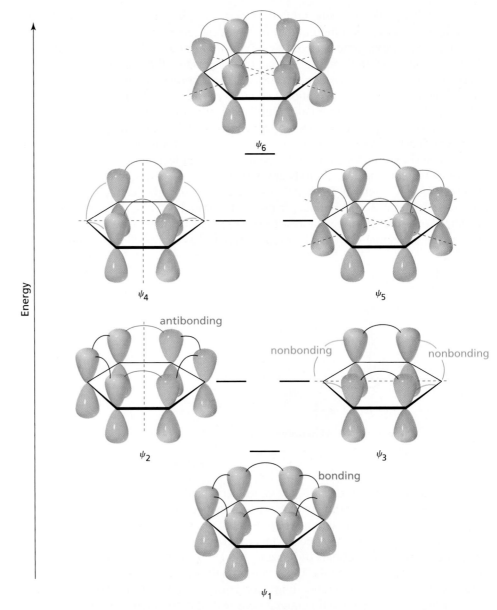

Tutorial:
Common terms

▲ **Figure 6.12**
As the energy of the π molecular orbitals increases, the number of nodes increases and the net number of bonding interactions decreases.

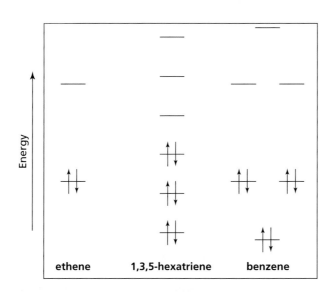

◀ **Figure 6.13**
A comparison of the energy
levels of the π molecular
orbitals of ethene, 1,3,5-
hexatriene, and benzene.

PROBLEMS

13. Which of the following compounds have delocalized electrons?

a. CH$_2$=CHCCH$_3$ (with O double bonded above the C)

g. (cyclopentene with radical)

b. (cyclohexadiene)

h. CH$_3$CH$_2$NHCH$_2$CH=CHCH$_3$

c. (cyclopentene with radical)

i. CH$_3$CH$_2$NHCH=CHCH$_3$

d. (tetrahydronaphthalene)

j. (dihydronaphthalene)

e. CH$_2$=CHCH$_2$CH=CH$_2$

k. CH$_3$C̈CH$_2$CH=CH$_2$ with CH$_3$ substituent and + charge

l. CH$_3$CH$_2$CHCH=CH$_2$ with + charge

f. (cyclohexene)

m. CH$_3$CH=CHOCH$_2$CH$_3$

14. a. Draw resonance contributors for the following species, showing all the nonbonding
pairs of electrons.
1. CH$_2$N$_2$ **2.** N$_2$O **3.** NO$_2^-$
b. For each species, indicate the most stable resonance contributor.

15. Draw resonance contributors for the following ions:

a. (structure with +)

b. (structure with +)

c.

d.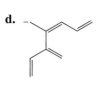

16. Are the following pairs of structures resonance contributors or different compounds?

a. 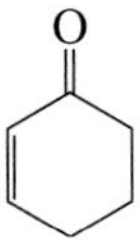 and

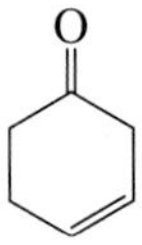

b. $CH_3CH=CH\overset{\cdot}{C}HCH=CH_2$ and $CH_3\overset{\cdot}{C}HCH=CHCH=CH_2$

c. $CH_3\overset{O}{\overset{\|}{C}}CH_2CH_3$ and $CH_3\overset{OH}{\overset{|}{C}}=CHCH_3$

d.  and

e. $CH_3\overset{+}{C}HCH=CHCH_3$ and $CH_3CH=CHCH_2\overset{+}{C}H_2$

17. Draw resonance contributors for the following species.
   a. Indicate which are major contributors and which are minor contributors to the resonance hybrid. Do not include structures that are so unstable that their contributions to the resonance hybrid would be negligible.

1. $CH_3CH=CHOCH_3$

2. $CH_3CH=CH\overset{+}{C}H_2$

3. $CH_3\overset{-}{C}HC\equiv N$

4.

5. 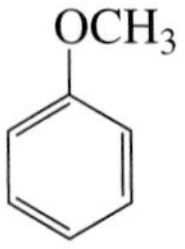 OCH$_3$

6. $CH_3-\overset{+}{N}\overset{O}{\diagdown_{O^-}}$

7. $CH_3CH_2\overset{O}{\overset{\|}{C}}OCH_2CH_3$

8. $CH_3CH=CHCH=CH\overset{\cdot}{C}H_2$

9. $H\overset{O}{\overset{\|}{C}}NHCH_3$

10. $CH_2=CH\overset{-}{C}H_2$

11. $CO_3{}^{2-}$

12. $H\overset{O}{\overset{\|}{C}}CH=CH\overset{-}{C}H_2$

13.

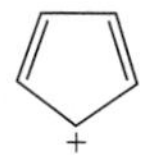

14. 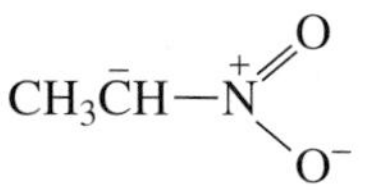 $CH_3\bar{C}H-\overset{+}{N}\overset{O}{\underset{O^-}{\diagdown}}$

15. 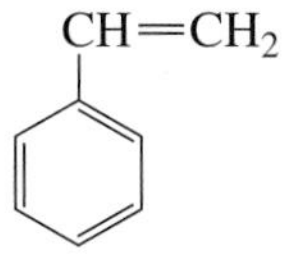 (with CH=CH$_2$)

16. 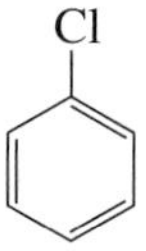 (with Cl)

17. $CH_3\overset{O}{\overset{\|}{C}}\bar{C}H\overset{O}{\overset{\|}{C}}CH_3$

18. $\bar{C}H_2\overset{O}{\overset{\|}{C}}OCH_2CH_3$

b. Do any of the species have resonance contributors that all contribute equally to the resonance hybrid?

18. Which resonance contributor makes the greater contribution to the resonance hybrid?

a. $CH_3\overset{+}{C}HCH=CH_2$ or $CH_3CH=CH\overset{+}{C}H_2$

b. 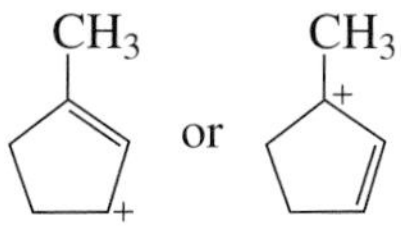   or

c. 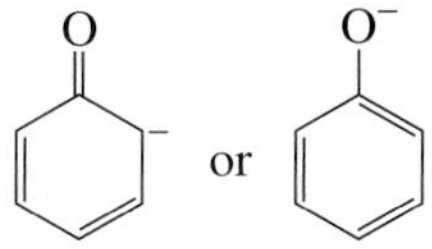   or

19. a. Which oxygen atom has the greater electron density?

$$CH_3\overset{O}{\overset{\|}{C}}OCH_3$$

b. Which compound has the greater electron density on its nitrogen atom?

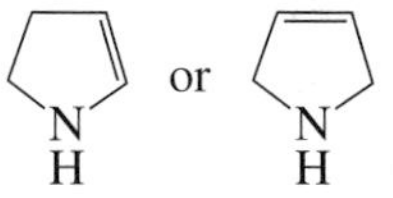

or

c. Which compound has the greater electron density on its oxygen atom?

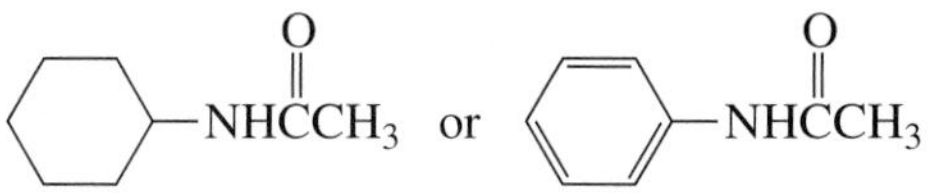

or

20. Which can lose a proton more readily, a methyl group bonded to cyclohexane or a methyl group bonded to benzene?

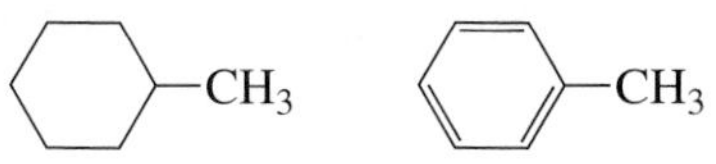

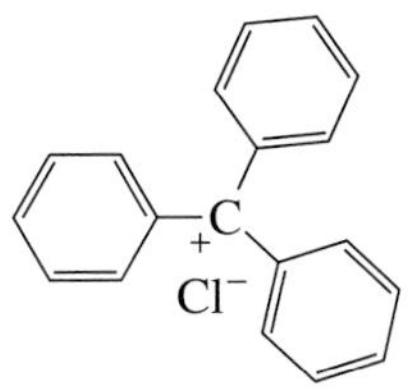

21. The triphenylmethyl cation is so stable that a salt such as triphenylmethyl chloride can be isolated and stored. Why is this carbocation so stable?

triphenylmethyl chloride

22. Draw the contributing resonance structures for the following anion and rank them in order of decreasing stability.

$$CH_3CH_2\overset{..}{\underset{..}{O}}{-}\overset{\overset{\displaystyle \ddot{O}:}{\|}}{C}{-}\overset{..}{C}H{-}C{\equiv}N:$$

23. Rank the following compounds in order of decreasing acidity.

24. In each of the following pairs, which species is more stable?

a. $CH_3CH_2O^-$ or $CH_3\overset{\overset{\displaystyle O}{\|}}{C}O^-$

b. $CH_3\overset{\overset{\displaystyle O}{\|}}{C}\bar{C}HCH_2\overset{\overset{\displaystyle O}{\|}}{C}H$ or $CH_3\overset{\overset{\displaystyle O}{\|}}{C}\overset{\overset{\displaystyle O}{\|}}{C}H\bar{C}CH_3$

c. $CH_3\bar{C}HCH_2\overset{\overset{\displaystyle O}{\|}}{C}CH_3$ or $CH_3CH_2\bar{C}H\overset{\overset{\displaystyle O}{\|}}{C}CH_3$

d. $CH_3\overset{\overset{\displaystyle NH_2}{|}}{C}HCH_3$ or $CH_3\overset{\overset{\displaystyle NH}{\|}}{C}NH_2$

e. $CH_3\overset{\overset{\displaystyle O}{\|}}{\underset{\underset{\displaystyle CH_3}{|}}{\bar{C}}}{-}CH$ or $CH_3\overset{\overset{\displaystyle CH_2}{\|}}{\underset{\underset{\displaystyle CH_3}{|}}{\bar{C}}}{-}CH$

f. or

25. In each of the pairs in Problem 24, which species is the stronger base?

26. We saw in Chapter 5 that ethyne reacts with an equivalent amount of HCl to form vinyl chloride. In the presence of excess HCl, the final reaction product is 1,1-dichloroethane. Why is 1,1-dichloroethane formed in preference to 1,2-dichloroethane?

$$HC{\equiv}CH \xrightarrow{\textbf{HCl}} H_2C{=}\underset{\underset{\displaystyle Cl}{|}}{C}H \xrightarrow{\textbf{HCl}} CH_3CHCl_2$$

27. Why is the resonance energy of pyrrole (21 kcal/mol) greater than the resonance energy of furan (16 kcal/mol)?

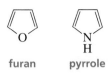

furan   pyrrole

28. Rank the following compounds in order of decreasing acidity of the indicated hydrogen:

$$CH_3\overset{O}{\overset{\|}{C}}CH_2CH_2\overset{O}{\overset{\|}{C}}CH_3 \qquad CH_3\overset{O}{\overset{\|}{C}}CH_2CH_2CH_2\overset{O}{\overset{\|}{C}}CH_3 \qquad CH_3\overset{O}{\overset{\|}{C}}CH_2\overset{O}{\overset{\|}{C}}CH_3$$

29. Explain why Markovnikov's rule is followed in reaction **a** but not in reaction **b**.

a. $CH_2{=}CHF \ + \ HF \ \longrightarrow \ CH_3CHF_2$

b. $CH_2{=}CHCF_3 \ + \ HF \ \longrightarrow \ FCH_2CH_2CF_3$

30. The acid dissociation constant (K_a) for loss of a proton from cyclohexanol is 1×10^{-16}.
   a. Draw an energy diagram for loss of a proton from cyclohexanol.

$$\text{⬡}{-}OH \ \underset{}{\overset{K_a = 1 \times 10^{-16}}{\rightleftharpoons}} \ \text{⬡}{-}O^- \ + \ H^+$$

   b. Draw the contributing resonance structures for phenol.
   c. Draw the contributing resonance structures for the phenoxide ion.
   d. Draw an energy diagram for loss of a proton from phenol on the same plot with the energy diagram for loss of a proton from cyclohexanol.

$$\text{⬡}{-}OH \ \rightleftharpoons \ \text{⬡}{-}O^- \ + \ H^+$$

   e. Which has a greater K_a, cyclohexanol or phenol?
   f. Which is a stronger acid, cyclohexanol or phenol?

31. Protonated cyclohexylamine has a K_a of 1×10^{-11}. Using the same sequence of steps as in Problem 30, determine which is a stronger base, cyclohexylamine or aniline.

$$\text{⬡}{-}\overset{+}{N}H_3 \ \rightleftharpoons \ \text{⬡}{-}NH_2 \ + \ H^+$$

$$\text{⬡}{-}\overset{+}{N}H_3 \ \rightleftharpoons \ \text{⬡}{-}NH_2 \ + \ H^+$$

7

Reactions of Dienes

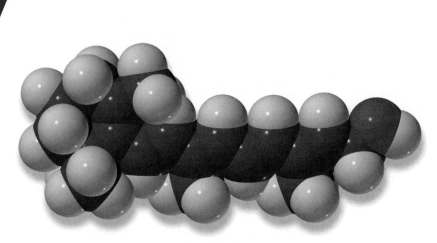

retinol
vitamin A

I n this chapter we will look at the reactions of compounds that have two double bonds. Hydrocarbons with two double bonds are called **dienes,** and those with three double bonds are called **trienes. Tetraenes** have four double bonds, and **polyenes** have many double bonds. Although, we will be concerned mainly with the reactions of dienes, the same considerations apply to hydrocarbons that contain more than two double bonds.

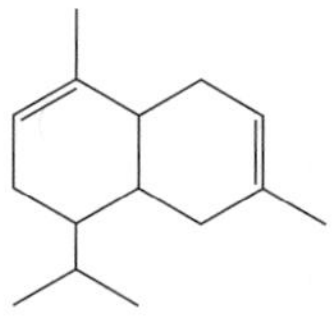

α-cadinene
oil of citronella
a diene

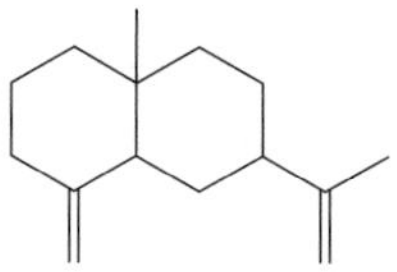

β-selinene
oil of celery
a diene

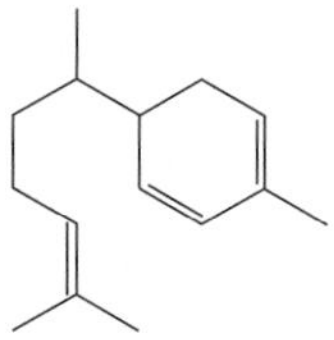

zingiberene
oil of ginger
a triene

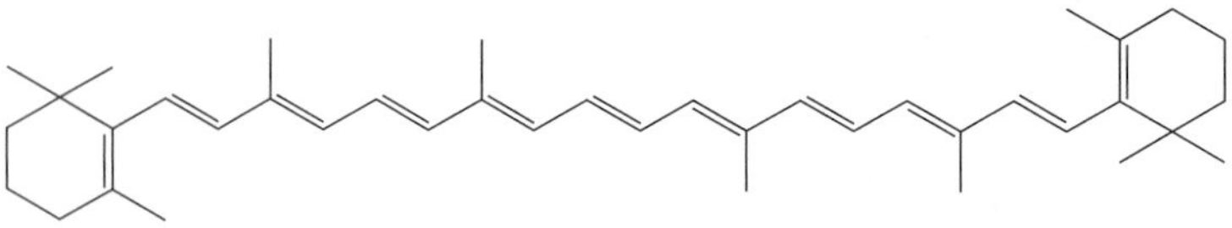

β-carotene
a polyene

Double bonds can be conjugated, isolated, or cumulated. **Conjugated double bonds** are separated by one single bond. **Isolated double bonds** are separated by more than one single bond. In other words, the double bonds are isolated from each other. **Cumulated double bonds** are adjacent to each other. Compounds with cumulated double bonds are called **allenes.**

$$CH_3CH=CH-CH=CHCH_3 \qquad CH_2=CH-CH_2-CH=CH_2$$
a conjugated diene   **an isolated diene**

$$CH_3-CH=C=CH-CH_3$$
a cumulated diene
an allene

The systematic name of a diene is obtained by designating the longest continuous chain that contains both double bonds by its alkane name, with the "ne" ending replaced by "diene." The chain is numbered in the direction that gives the double bonds the lowest possible numbers. The numbers indicating the locations of the double bonds are usually cited before the name of the parent compound. Substituents are cited in alphabetical order. Propadiene, the smallest member of the class of compounds known as allenes, is frequently called allene.

**7.1
NOMENCLATURE OF
ALKENES WITH
MORE THAN ONE
FUNCTIONAL GROUP**

	$CH_2=C=CH_2$	$\overset{1}{C}H_2=\overset{2}{C}-\overset{3}{C}H=\overset{4}{C}H_2$ with CH_3	5-bromo-1,3-cyclohexadiene
systematic:	propadiene	2-methyl-1,3-butadiene	
common:	allene	isoprene	

$$\overset{6}{C}H_3\overset{5}{C}H=\overset{4}{C}H\overset{3}{C}H_2\overset{2}{C}=\overset{1}{C}H_2 \quad (CH_3)$$
2-methyl-1,4-hexadiene

$$\overset{1}{C}H_3\overset{2}{C}=\overset{3}{C}H\overset{4}{C}H=\overset{5}{C}\overset{6}{C}H_2\overset{7}{C}H_3 \quad (CH_3, CH_2CH_3)$$
5-ethyl-2-methyl-2,4-heptadiene

To name an alkene in which the second functional group is not another double bond, the longest continuous chain containing both functional groups is chosen, and both functional group designations are cited at the end of the name. The "ene" ending is cited first, with the terminal "e" omitted in order to avoid two adjacent vowels. The location of the first cited functional group is usually cited before the name of the parent chain. The location of the second functional group is cited immediately before its suffix.

If the functional groups are a double bond and a triple bond, the chain is numbered in the direction that yields the lowest number in the name of the compound.

$$\overset{7}{C}H_3\overset{6}{C}H=\overset{5}{C}H\overset{4}{C}H_2\overset{3}{C}H_2\overset{2}{C}\equiv\overset{1}{C}H \qquad \overset{1}{C}H_2=\overset{2}{C}H\overset{3}{C}H_2\overset{4}{C}H_2\overset{5}{C}\equiv\overset{6}{C}\overset{7}{C}H_3$$
5-hepten-1-yne   **1-hepten-5-yne**
not **2-hepten-6-yne**   *not* **6-hepten-2-yne**
because 1 < 2   because 1 < 2

$$\overset{1}{C}H_2=\overset{2}{C}H\overset{3}{C}H\overset{4}{C}\equiv\overset{5}{C}\overset{6}{C}H_3 \quad (CH_2CH_2CH_2CH_3)$$
3-butyl-1-hexen-4-yne

> the longest continuous chain has 8 carbons but the 8-carbon chain does not contain both functional groups; the compound is named as a hexenyne because the longest continuous chain containing both functional groups has 6 carbons

If the same low number is obtained in both directions, the chain is numbered in the direction that gives the double bond the lower number.

$$\overset{1}{C}H_3\overset{2}{C}H=\overset{3}{C}H\overset{4}{C}\equiv\overset{5}{C}\overset{6}{C}H_3 \qquad \overset{6}{H}C\equiv\overset{5}{C}\overset{4}{C}H_2\overset{3}{C}H_2\overset{2}{C}H=\overset{1}{C}H_2$$
2-hexen-4-yne   **1-hexen-5-yne**
not **4-hexen-2-yne**   *not* **5-hexen-1-yne**

If the second functional group has a higher priority than the alkene (Table 7.1), the chain is numbered in the direction that assigns the lower number to the functional group with the higher priority.

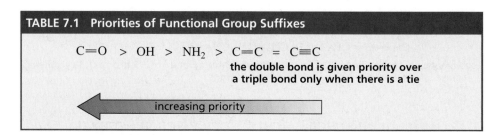

$CH_2\!=\!CHCH_2OH$ $CH_3\overset{\overset{\displaystyle CH_3}{|}}{C}\!=\!CHCH_2CH_2OH$ $CH_2\!=\!CHCH_2CH_2CH_2\overset{\overset{\displaystyle NH_2}{|}}{C}HCH_3$

2-propen-1-ol   4-methyl-3-penten-1-ol 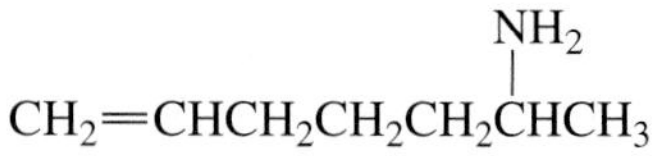   6-hepten-2-amine
not **1-propen-3-ol**

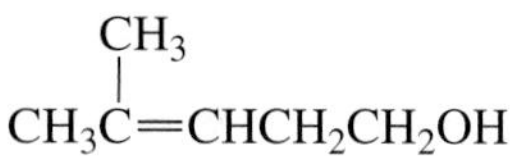

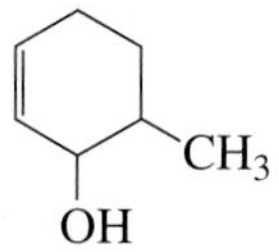

6-methyl-2-cyclohexenol 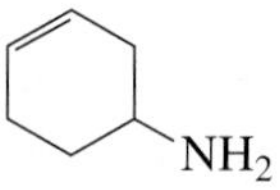   3-cyclohexenamine

TABLE 7.1   Priorities of Functional Group Suffixes

$$C\!=\!O \ > \ OH \ > \ NH_2 \ > \ C\!=\!C \ = \ C\!\equiv\!C$$

the double bond is given priority over a triple bond only when there is a tie

⟵ increasing priority

PROBLEM 1 ◆

Give the systematic name for each of the following compounds:

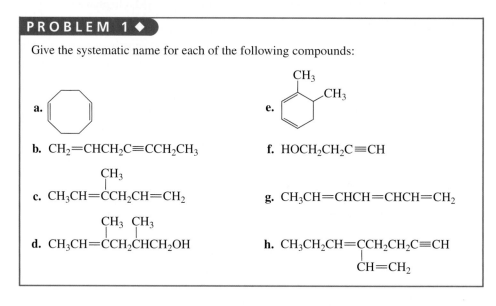

a.

b. $CH_2\!=\!CHCH_2C\!\equiv\!CCH_2CH_3$

c. $CH_3CH\!=\!\overset{\overset{\displaystyle CH_3}{|}}{C}CH_2CH\!=\!CH_2$

d. $CH_3CH\!=\!\overset{\overset{\displaystyle CH_3}{|}}{C}CH_2\overset{\overset{\displaystyle CH_3}{|}}{C}HCH_2OH$

e.

f. $HOCH_2CH_2C\!\equiv\!CH$

g. $CH_3CH\!=\!CHCH\!=\!CHCH\!=\!CH_2$

h. $CH_3CH_2CH\!=\!\overset{\overset{\displaystyle CH_2CH_2C\equiv CH}{|}}{C}\!\!\underset{\underset{\displaystyle CH=CH_2}{|}}{}$

7.2

CONFIGURATIONAL ISOMERS OF DIENES

A diene such as 1-chloro-2,4-heptadiene has four configurational isomers because each of the double bonds can have either the *E* configuration or the *Z* configuration. Thus there are *E-E*, *Z-Z*, *E-Z*, and *Z-E* isomers. The rules for determining the *E* and *Z* configurations are given in Section 3.5.

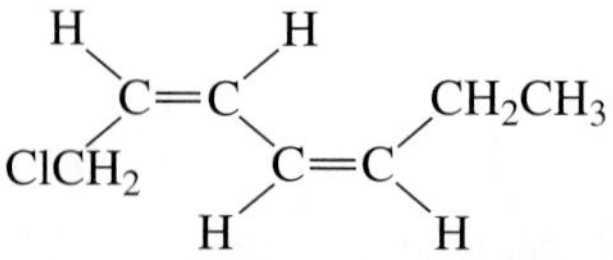

(2*Z*,4*Z*)-1-chloro-2,4-heptadiene 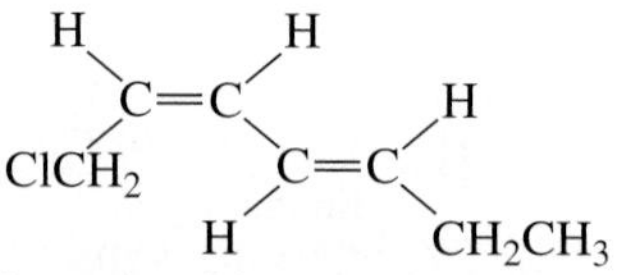   (2*Z*,4*E*)-1-chloro-2,4-heptadiene

$$
\begin{array}{c}
\underset{H}{\overset{ClCH_2}{>}}C=C\underset{\underset{H}{\overset{|}{C}}=C\underset{H}{\overset{H}{<}}}{\overset{CH_2CH_3}{<}}
\end{array}
$$

(2E,4Z)-1-chloro-2,4-heptadiene

$$
\begin{array}{c}
\underset{H}{\overset{ClCH_2}{>}}C=C\underset{\underset{H}{\overset{|}{C}}=C\underset{CH_2CH_3}{\overset{H}{<}}}{\overset{H}{<}}
\end{array}
$$

(2E,4E)-1-chloro-2,4-heptadiene

PROBLEM 2 ◆

Draw the configurational isomers for the following compounds. Name each one.

a. 2-methyl-2,4-hexadiene   **b.** 2,4-heptadiene   **c.** 1,3-pentadiene

---

In Section 3.19, we saw that the relative stabilities of substituted alkenes can be determined by their relative heats of hydrogenation ($-\Delta H°$). Remember that the least stable alkene has the greatest heat of hydrogenation—it gives off the most heat when hydrogenated because it contains more energy to begin with. The heat of hydrogenation of 2,3-pentadiene (a cumulated diene) is greater than the heat of hydrogenation of 1,4-pentadiene (an isolated diene) which, in turn, is greater than the heat of hydrogenation of 1,3-pentadiene (a conjugated diene). From the relative heats of hydrogenation of the three pentadienes, we can conclude that conjugated dienes are more stable than isolated dienes, which are more stable than cumulated dienes.

7.3 RELATIVE STABILITIES OF DIENES

$$
CH_3CH=C=CHCH_3 \;+\; 2\,H_2 \xrightarrow{\text{Pt}} CH_3CH_2CH_2CH_2CH_3 \qquad \Delta H° = -70.5 \text{ kcal/mol } (-295 \text{ kJ/mol})
$$
2,3-pentadiene

$$
CH_2=CHCH_2CH=CH_2 \;+\; 2\,H_2 \xrightarrow{\text{Pt}} CH_3CH_2CH_2CH_2CH_3 \qquad \Delta H° = -60.2 \text{ kcal/mol } (-252 \text{ kJ/mol})
$$
1,4-pentadiene

$$
CH_2=CHCH=CHCH_3 \;+\; 2\,H_2 \xrightarrow{\text{Pt}} CH_3CH_2CH_2CH_2CH_3 \qquad \Delta H° = -54.1 \text{ kcal/mol } (-226 \text{ kJ/mol})
$$
1,3-pentadiene

relative stabilities of dienes

conjugated diene   >   isolated diene   >   cumulated diene

◀ increasing stability

Why is a conjugated diene more stable than an isolated diene? Two factors contribute to the stability of conjugated dienes. One is the hybridization of the orbitals forming the carbon–carbon single bonds. The carbon–carbon single bond in 1,3-butadiene is formed from the overlap of an sp^2 orbital with another sp^2 orbital, whereas the carbon–carbon single bonds in 1,4-pentadiene are formed from the overlap of an sp^3 orbital with an sp^2 orbital.

single bond formed by sp^2–sp^2 overlap	single bonds formed by sp^3–sp^2 overlap
$CH_2=CH-CH=CH_2$	$CH_2=CH-CH_2-CH=CH_2$
1,3-butadiene	**1,4-pentadiene**

In Section 1.14, you saw that the length and strength of a bond depends on how close the electrons in the bonding orbital are to the nucleus—the closer to the nucleus, the shorter and stronger the bond. Because a $2s$ electron is closer, on average, to the nucleus than a $2p$ electron, a bond formed by sp^2–sp^2 overlap is shorter and stronger

TABLE 7.2 Dependence of the Length of a Carbon–Carbon Single Bond on the Hybridization of the Orbitals Used in Its Formation		
Compound	**Hybridization**	**Bond length (Å)**
H_3C-CH_3	sp^3-sp^3	1.54
$H_3C-\overset{\overset{\displaystyle H}{\vert}}{C}=CH_2$	sp^3-sp^2	1.50
$H_2C=\overset{\overset{\displaystyle H}{\vert}}{C}-\overset{\overset{\displaystyle H}{\vert}}{C}=CH_2$	sp^2-sp^2	1.47
$H_3C-C\equiv CH$	sp^3-sp	1.46
$H_2C=\overset{\overset{\displaystyle H}{\vert}}{C}-C\equiv CH$	sp^2-sp	1.43
$HC\equiv C-C\equiv CH$	$sp-sp$	1.37

Tutorial: Orbital overlap in C—C single bonds

than one formed by sp^3–sp^2 overlap (Table 7.2). (An sp^2 orbital has 33.3% *s* character, whereas an sp^3 orbital has 25% *s* character.) Because a conjugated diene has one stronger single bond than an isolated diene, the conjugated diene is more stable.

The second factor that causes a conjugated diene to be more stable than an isolated diene is resonance, which means that the compound has delocalized electrons. The π electrons in each of the double bonds of an isolated diene are *localized* between two carbons. In contrast, the four π electrons in a conjugated diene are *delocalized* over four carbons. As we discovered in Section 6.6, electron delocalization increases a molecule's stability.

$$\overset{-}{C}H_2-CH=CH-\overset{+}{C}H_2 \longleftrightarrow CH_2=CH-CH=CH_2 \longleftrightarrow \overset{+}{C}H_2-CH=CH-\overset{-}{C}H_2$$

resonance structures

$$CH_2\text{---}CH\text{---}CH\text{---}CH_2$$

resonance hybrid

The resonance hybrid shows that the single bond in 1,3-butadiene is not a pure single bond but has partial double-bond character. Thus, both electron delocalization and the orbitals involved in bond formation account for the fact that a bond formed by sp^2–sp^2 overlap is shorter and stronger than one formed by sp^3–sp^2 overlap.

PROBLEM 3 ◆

Name the following dienes and rank them in order of decreasing stability. (Alkyl groups stabilize dienes in the same way they stabilize alkenes; Section 3.19.)

$$CH_3CH=CHCH=CHCH_3 \qquad CH_2=CHCH_2CH=CH_2$$

$$\overset{\overset{\displaystyle CH_3}{\vert}}{CH_3C}=CHCH=\overset{\overset{\displaystyle CH_3}{\vert}}{CCH_3} \qquad CH_3CH=CHCH=CH_2$$

While a conjugated diene is more stable than an isolated diene, a cumulated diene is less stable than an isolated diene. Cumulated dienes are unlike other dienes in that the central carbon is *sp* hybridized because it has two π bonds. (All the double-bonded carbons of isolated dienes and conjugated dienes are sp^2 hybridized.) This sp hybridization gives the cumulated dienes unique properties. For example, the heat of hydrogenation of allene is very similar to the heat of hydrogenation of propyne, a compound with two sp hybridized carbons.

$$CH_2{=}C{=}CH_2 \ + \ 2\,H_2 \ \xrightarrow{\text{Pt}} \ CH_3CH_2CH_3 \qquad \Delta H° = -70.5 \text{ kcal/mol} \ (-295 \text{ kJ/mol})$$

allene

$$CH_3C{\equiv}CH \ + \ 2\,H_2 \ \xrightarrow{\text{Pt}} \ CH_3CH_2CH_3 \qquad \Delta H° = -69.9 \text{ kcal/mol} \ (-292 \text{ kJ/mol})$$

propyne

One of the p orbitals of the central carbon of allene overlaps a p orbital of an adjacent sp^2 carbon. The second p orbital of the central carbon overlaps a p orbital of the other sp^2 carbon (Figure 7.1). The two p orbitals of the central carbon are perpendicular to each other. Therefore, the plane containing one H-C-H group is perpendicular to the plane containing the other H-C-H group. We will not consider the reactions of cumulated dienes because they are rather specialized and are more appropriately covered in an advanced course in organic chemistry.

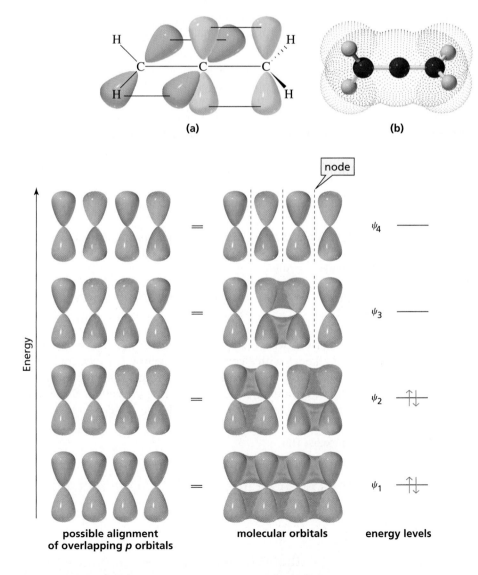

◀ **Figure 7.1**
(a) Double bonds are formed by p orbital–p orbital overlap. The two p orbitals on the central carbon are perpendicular. **(b)** The perpendicular p orbitals cause allene to be a nonplanar molecule.

(a) (b)

◀ **Figure 7.2**
Distribution of electrons in 1,3-butadiene. Four p atomic orbitals overlap to produce four π molecular orbitals.

node

ψ_4 _____

ψ_3 _____

ψ_2

ψ_1

Energy

possible alignment of overlapping p orbitals   **molecular orbitals**   **energy levels**

7.4
A MOLECULAR ORBITAL DESCRIPTION OF 1,3-BUTADIENE

A molecular orbital description of 1,3-butadiene is shown in Figure 7.2. The four p atomic orbitals of 1,3-butadiene combine to produce four π molecular orbitals, ψ_1, ψ_2, ψ_3, and ψ_4. Half the molecular orbitals are bonding π molecular orbitals (ψ_1 and ψ_2), and the other half are antibonding π molecular orbitals (ψ_3 and ψ_4). Notice that as the MOs (molecular orbitals) increase in energy, the number of nodes increases and the number of bonding interactions decreases.

The four π electrons of 1,3-butadiene are delocalized over the four p orbitals (Figure 7.2). In contrast, two of the π electrons of 1,4-pentadiene are completely separate from the other two π electrons (Figure 7.3). In other words, the electrons are localized. Molecular orbital theory and resonance are two different ways of showing that the π electrons in 1,3-butadiene are delocalized. A more complete description of molecular orbitals is presented in Section 28.2.

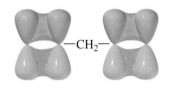

▲ **Figure 7.3**
Distribution of electrons in 1,4-pentadiene. Each pair of π electrons is isolated from the other.

PROBLEM 4 ◆

What is the total number of nodes in the ψ_3 and ψ_4 molecular orbitals of 1,3-butadiene?

7.5
REACTIVITY CONSIDERATIONS

Dienes, like alkenes and alkynes, are nucleophiles due to the electron density of their π bonds. This means they will react with electrophilic reagents. Therefore, like alkenes and alkynes, dienes undergo electrophilic addition reactions.

Until now, we have been concerned with the reactions of compounds that have only one functional group. Compounds with two or more functional groups exhibit reactions characteristic of the individual functional groups if the groups are sufficiently separated from each other. If they are close enough to allow electron delocalization, one functional group can affect the reactivity of the other. In other words, we will see that the reactions of isolated dienes are the same as the reactions of alkenes, whereas the reactions of conjugated dienes are a little different because of the proximity of the two double bonds.

7.6
ELECTROPHILIC ADDITION REACTIONS OF ISOLATED DIENES

The reactions of isolated dienes are the same as the reactions of alkenes. If an excess of the electrophilic reagent is present, two independent addition reactions will occur, each following Markovnikov's rule.

$$CH_2{=}CHCH_2CH_2CH{=}CH_2 \ + \ HBr \ \longrightarrow \ CH_3CHCH_2CH_2CHCH_3$$

1,5-hexadiene   **excess**   Br   Br

The reaction proceeds exactly as we would predict from our knowledge of the mechanism for the reaction of alkenes with electrophilic reagents. The electrophile (H^+) adds to the electron-rich double bond in a manner that produces the more stable carbocation (Section 3.12). The bromide ion then adds to the carbocation. Because there is an excess of the electrophilic reagent, both double bonds undergo addition.

mechanism for the reaction of 1,5-hexadiene with excess HBr

$$CH_2{=}CHCH_2CH_2CH{=}CH_2 \ + \ H{-}\ddot{B}r{:} \ \longrightarrow \ CH_3\overset{+}{C}HCH_2CH_2CH{=}CH_2 \ \xrightarrow{:\ddot{B}r\bar{:}} \ CH_3\overset{Br}{\underset{|}{C}}HCH_2CH_2CH{=}CH_2$$

$$\Big\downarrow H{-}\ddot{B}r{:}$$

$$\overset{Br}{\underset{|}{C}}H_3CHCH_2CH_2\overset{Br}{\underset{|}{C}}HCH_3 \ \xleftarrow{:\ddot{B}r\bar{:}} \ CH_3\overset{Br}{\underset{|}{C}}HCH_2CH_2\overset{+}{C}HCH_3$$

If there is only enough electrophilic reagent to add to one of the double bonds, it will add preferentially to the more reactive double bond. For example, in the reaction of 2-methyl-1,5-hexadiene with HCl, addition of HCl to the double bond on the left forms a secondary carbocation, while addition to the double bond on the right forms a tertiary carbocation. Because the transition state leading to formation of a tertiary carbocation is more stable than that leading to a secondary carbocation, the tertiary carbocation forms faster (Section 3.11). So, in the presence of a limited amount of HCl, the major product of the reaction will be 5-chloro-5-methyl-1-hexene.

$$CH_2=CHCH_2CH_2\overset{\overset{\displaystyle CH_3}{|}}{C}=CH_2 \quad + \quad HCl \quad \longrightarrow \quad CH_2=CHCH_2CH_2\overset{\overset{\displaystyle CH_3}{|}}{\underset{\underset{\displaystyle Cl}{|}}{C}}CH_3$$

2-methyl-1,5-hexadiene   **1 mol**
1 mol

5-chloro-5-methyl-1-hexene
major product

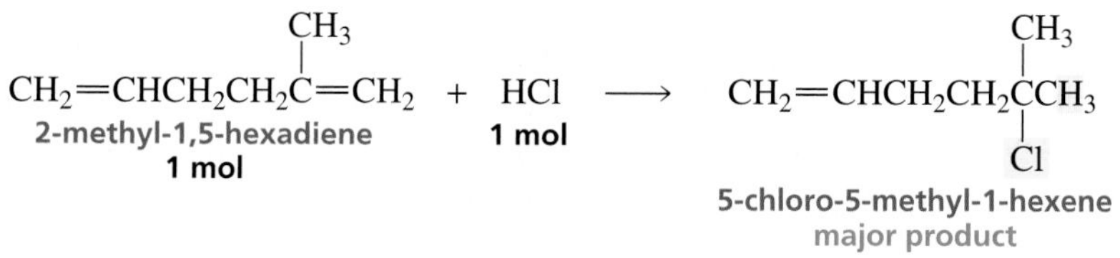

PROBLEM 5 ◆

Show what stereoisomers will be obtained from the two reactions in Section 7.6. (*Hint:* Review Section 4.17).

PROBLEM 6

Give the major product of each of the following reactions and show the stereoisomers that would be obtained. Equivalent amounts of reagents are used in each case.

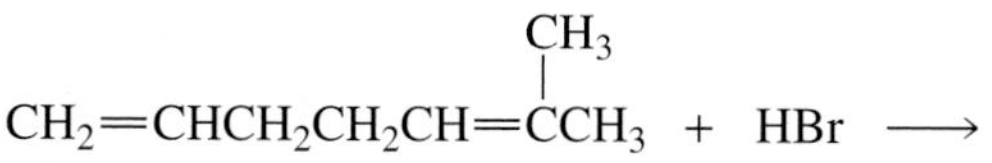

a. (cyclohexene ring with CH₃)   +   HCl ⟶

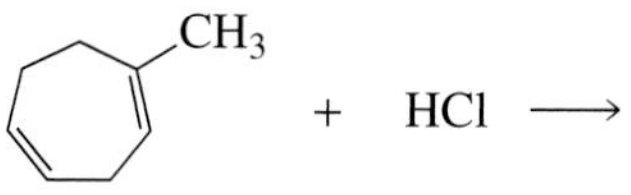

b. $CH_2=CHCH_2CH_2CH=\overset{\overset{\displaystyle CH_3}{|}}{C}CH_3$ +   HBr ⟶

c. $CH_2=CHCH_2CH_2\overset{\overset{\displaystyle CH_3}{|}}{C}=CH_2$ +   HBr $\xrightarrow{\text{peroxide}}$

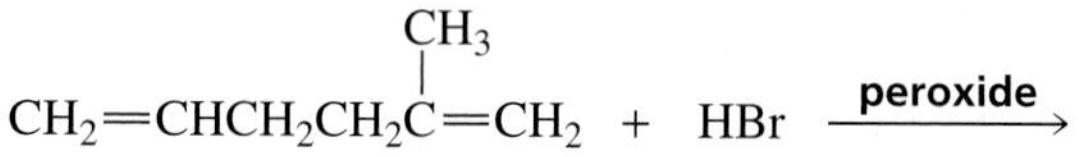

d. $HC\equiv CCH_2CH_2CH=CH_2$ + Cl_2 ⟶

PROBLEM 7 ◆

Which of the double bonds in zingiberene (its structure is given on the first page of this chapter) is the most reactive in an electrophilic addition reaction?

If a conjugated diene, such as 1,3-butadiene, reacts with a limited amount of electrophilic reagent so that addition can occur at only one of the double bonds, two addition products are formed. One is a 1,2-addition product, which is a result of addition at the 1- and 2-positions. The other is a 1,4-addition product—the result of addition at the 1- and 4-positions.

Addition at the 1- and 2-positions is called **1,2-addition** or **direct addition.** Addition at the 1- and 4-positions is called **1,4-addition** or **conjugate addition.** We expect the 1,2-addition product to form, based on our knowledge of how electrophilic

**7.7
ELECTROPHILIC
ADDITION
REACTIONS OF
CONJUGATED
DIENES**

$$CH_2{=}CH{-}CH{=}CH_2 + Cl_2 \longrightarrow$$

1,3-butadiene   1 mol
1 mol

Cl Cl
| |
$$CH_2CHCH{=}CH_2$$ + $$CH_2CH{=}CHCH_2$$
| |
Cl   Cl

3,4-dichloro-1-butene   **1,4-dichloro-2-butene**
1,2-addition product   1,4-addition product

$$CH_2{=}CH{-}CH{=}CH_2 + HBr \longrightarrow$$

1,3-butadiene   1 mol
1 mol

Br   Br
| |
$$CH_3CHCH{=}CH_2$$ + $$CH_3CH{=}CHCH_2$$

3-bromo-1-butene   **1-bromo-2-butene**
1,2-addition product   1,4-addition product

reagents add to double bonds. The 1,4-addition product, however, may be surprising because the reagent did not add to adjacent carbons and because a double bond has changed its position. The double bond in the product is between the 2- and 3-positions, whereas the reactant has a single bond in this position.

When we refer to addition at the 1- and 2-positions or at the 1- and 4-positions, we mean that addition occurs at the 1- and 2- or 1- and 4-positions of the four-carbon conjugated system. The 1-position is one of the sp^2 carbons at the end of the conjugated system—it is not necessarily the first carbon in the molecule.

$$R{-}\overset{1}{C}H{=}\overset{2}{C}H{-}\overset{3}{C}H{=}\overset{4}{C}H{-}R$$
the conjugated system

$$CH_3CH{=}CHCH{=}CHCH_3 \xrightarrow{\text{Br}_2}$$

2,4-hexadiene

Br Br
| |
$$CH_3CHCHCH{=}CHCH_3$$ + $$CH_3CHCH{=}CHCHCH_3$$
| |
Br   Br

4,5-dibromo-2-hexene   **2,5-dibromo-3-hexene**
1,2-addition product   1,4-addition product

To understand why both 1,2-addition and 1,4-addition products are obtained from the reaction of a conjugated diene with a limited amount of electrophilic reagent, we must look at the mechanism of the reaction. In the first step of the addition of HBr to 1,3-butadiene, the electrophilic proton adds to C-1, which forms an allylic cation. Recall that an allylic cation has a positive charge on a carbon that is next to a double-bonded carbon. Allylic means "next to a carbon–carbon double bond." Notice that, because 1,3-butadiene is symmetrical, adding to C-1 is the same as adding to C-4. The proton does not add to C-2 or C-3 because this would form a primary carbocation, which is less stable than an allylic cation.

The allylic cation has two contributing resonance structures. The positive charge on the carbocation is not localized on C-2, but is shared by C-2 and C-4 as a result of delocalization of the π electrons. Consequently, in the second step of the reaction, the bromide ion can attack either C-2 (direct addition) or C-4 (conjugate addition) to form the 1,2-addition product or the 1,4-addition product, respectively.

mechanism for the reaction of 1,3-butadiene with HBr

$$CH_2{=}CHCH{=}CH_2 + H{-}\ddot{B}r{:} \longrightarrow CH_3\overset{+}{C}HCH{=}CH_2 \longleftrightarrow CH_3CH{=}CH\overset{+}{C}H_2$$

1,3-butadiene   **an allylic cation**

$${:}\ddot{B}r{:}^-$$

Br
|
$$CH_3CHCH{=}CH_2$$ $$CH_3CH{=}CHCH_2$$
                              |
                              Br

3-bromo-1-butene   **1-bromo-2-butene**
1,2-addition product   1,4-addition product

The two contributing resonance structures and the resonance hybrid of the allylic cation are shown here (also see Figure 7.4).

$$CH_3\overset{+}{C}H-CH=CH_2 \longleftrightarrow CH_3CH=CH-\overset{+}{C}H_2$$
resonance contributors

$$CH_3\overset{\delta+}{CH}\!=\!\!=\!\!CH\!=\!\!=\!\overset{\delta+}{CH_2}$$
resonance hybrid

electrostatic potential map of
$CH_3-\overset{+}{C}H-CH=CH_2$

▲ **Figure 7.4**
The electrostatic potential map for the carbocation formed by adding a proton to 1,3-butadiene.

As we look at more examples, notice that the first step in all electrophilic additions to conjugated dienes is addition of the electrophile to one of the sp^2 carbons at the end of the conjugated system. This is the only way to obtain a carbocation that is stabilized by resonance. If the electrophile were to add to one of the internal sp^2 carbons, the resulting carbocation would not be stabilized by resonance.

Now we can compare addition to an isolated diene with addition to a conjugated diene. The carbocation formed by addition of an electrophile to an isolated diene is not stabilized by resonance. The positive charge is localized on a single carbon, so only direct (1,2-) addition occurs.

addition to an isolated diene

$$CH_2=CHCH_2CH_2CH=CH_2 \xrightarrow{\ H^+\ } CH_3\overset{+}{C}HCH_2CH_2CH=CH_2 \xrightarrow{\ Br^-\ } \underset{\textbf{5-bromo-1-hexene}}{CH_3\overset{\overset{\displaystyle Br}{|}}{C}HCH_2CH_2CH=CH_2}$$
1,5-hexadiene

The carbocation formed by addition of an electrophile to a conjugated diene, in contrast, is stabilized by resonance. The positive charge is spread over two carbons and, as a result, both direct (1,2-) and conjugate (1,4-) addition occur.

addition to a conjugated diene

$$CH_3CH=CHCH=CHCH_3 \xrightarrow{\ H^+\ } CH_3CH_2\overset{+}{C}HCH=CHCH_3 \longleftrightarrow CH_3CH_2CH=CH\overset{+}{C}HCH_3$$
2,4-hexadiene

$\downarrow$ **Br⁻**

$$\underset{\substack{\textbf{4-bromo-2-hexene}\\\textbf{1,2-addition product}}}{CH_3CH_2\underset{\underset{\textstyle Br}{|}}{C}HCH=CHCH_3} \qquad \underset{\substack{\textbf{2-bromo-3-hexene}\\\textbf{1,4-addition product}}}{CH_3CH_2CH=CH\underset{\underset{\textstyle Br}{|}}{C}HCH_3}$$

If the diene is asymmetrical, the major products of the reaction are those obtained by adding the electrophile to whichever terminal sp^2 carbon results in formation of the more stable carbocation. For example, in the reaction of 2-methyl-1,3-butadiene with HBr, the proton adds preferentially to C-1 because the positive charge on the resulting carbocation is shared by a tertiary allylic and a primary allylic carbon. Adding the proton to C-4 would form a carbocation with the positive charge shared by a secondary allylic and a primary allylic carbon. Therefore, addition to C-1 forms the more stable carbocation and 3-bromo-3-methyl-1-butene and 1-bromo-3-methyl-2-butene are the major products of the reaction.

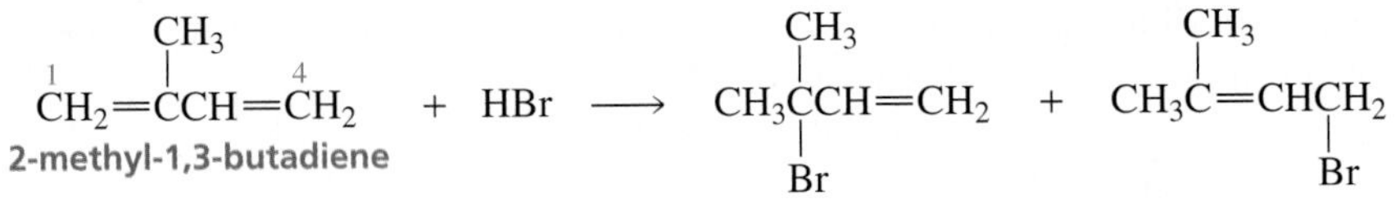

$$CH_2=CCH=CH_2 \quad + \quad HBr \quad \longrightarrow \quad CH_3CCH=CH_2 \quad + \quad CH_3C=CHCH_2$$

2-methyl-1,3-butadiene

3-bromo-3-methyl-1-butene

1-bromo-3-methyl-2-butene

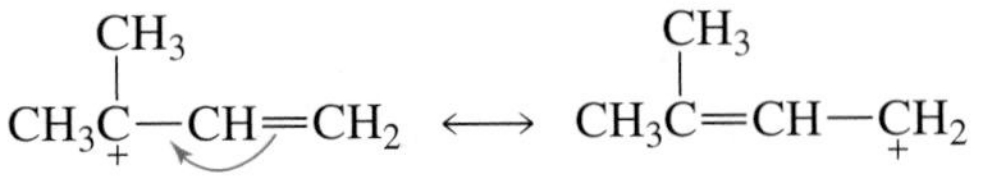

$$CH_3C—CH=CH_2 \quad \longleftrightarrow \quad CH_3C=CH—CH_2$$

carbocation formed by adding H⁺ to C-1

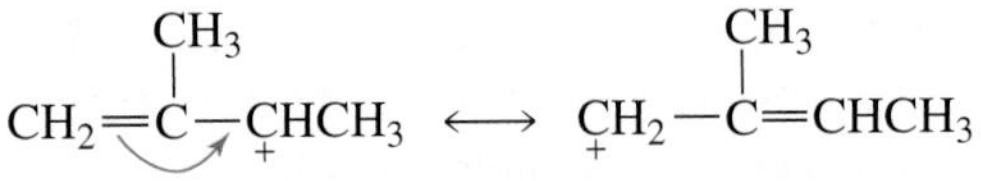

$$CH_2=C—CHCH_3 \quad \longleftrightarrow \quad CH_2—C=CHCH_3$$

carbocation formed by adding H⁺ to C-4

PROBLEM 8

What products would be obtained from the reaction of 1,3,5-hexatriene with one equivalent of HBr?

PROBLEM 9 ◆

Give the products of the following reactions. Equivalent amounts of reagents are used in each case.

a. $CH_3CH=CH—CH=CHCH_3 \quad + \quad Cl_2 \quad \longrightarrow$

b. 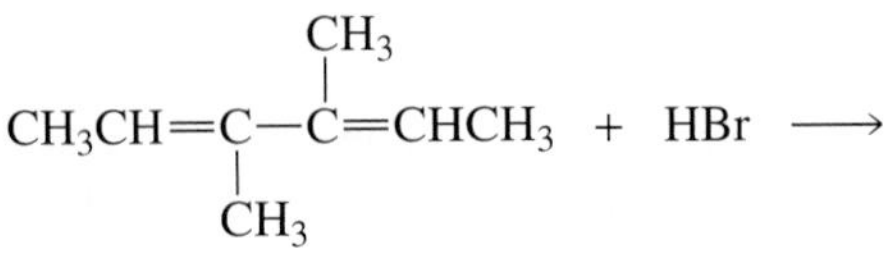 $CH_3CH=C—\overset{\displaystyle CH_3}{\underset{\displaystyle CH_3}{C}}=CHCH_3 \quad + \quad HBr \quad \longrightarrow$

c. $CH_3CH=CH—\underset{\displaystyle CH_3}{C}=CHCH_3 \quad + \quad HBr \quad \longrightarrow$

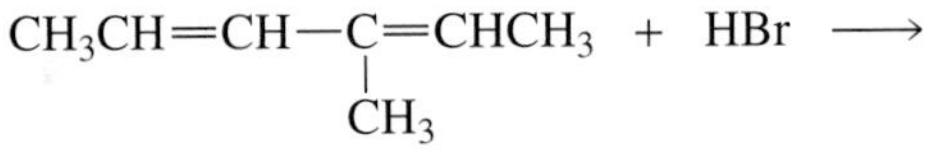

d. 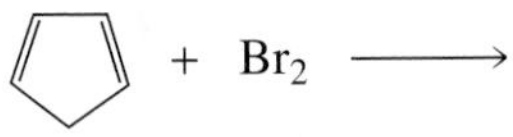 ⬠ $+ \quad Br_2 \quad \longrightarrow$

7.8
THERMODYNAMIC VERSUS KINETIC CONTROL OF REACTIONS

The thermodynamic product is the most stable product.

The kinetic product is the product formed fastest.

When a conjugated diene undergoes an electrophilic addition reaction, two factors—the structure of the reactant and the temperature at which the reaction is carried out—determine whether the 1,2-addition product or the 1,4-addition product will be the major product of the reaction.

When a reaction produces more than one product, the product that is formed most rapidly is called the **kinetic product**. The most stable product is called the **thermodynamic product**. Reactions that produce the kinetic product as the major product are said to be kinetically controlled. Reactions that produce the thermodynamic product as the major product are said to be thermodynamically controlled.

For many organic reactions, the kinetic product and the thermodynamic product are one and the same. Electrophilic addition to 1,3-butadiene is an example of a reaction in which the kinetic product and the thermodynamic product are *not* the same.

$$CH_2=CHCH=CH_2 \ + \ HBr \ \longrightarrow \ \underset{\text{1,2-addition product}}{CH_3\overset{\displaystyle Br}{\overset{|}{C}}HCH=CH_2} \ + \ \underset{\text{1,4-addition product}}{CH_3CH=CH\overset{\displaystyle Br}{\overset{|}{C}}H_2}$$

1,3-butadiene

Let's first determine which product in the preceding reaction is more stable (the thermodynamic product). We saw in Section 3.19 that the relative stability of an alkene can be determined by the number of alkyl groups that are bonded to the sp^2 carbons—the greater the number of alkyl groups bonded to the sp^2 carbons, the more stable the alkene. The two products formed from the reaction of 1,3-butadiene with one equivalent of HBr have different stabilities—the 1,2-product has one alkyl group bonded to the sp^2 carbons, while the 1,4-product has two alkyl groups bonded to the sp^2 carbons. So the 1,4-product is more stable than the 1,2-product. Therefore, the 1,4-product is the thermodynamic product.

Now let's determine which of the products is formed faster (the kinetic product). The first step of the addition reaction, addition of a proton to C-1, is the same whether the 1,2-product or the 1,4-product is being formed. The second step is different, how-ever, because the second step is where the bromide ion attacks either C-2 or C-4. Therefore, to determine which of the products is formed faster, we must look at the relative stabilities of the transition states for the second step of the reaction.

$$CH_2=CHCH=CH_2 \ \xrightarrow{\textbf{HBr}} \ CH_3\overset{+}{C}HCH=CH_2 \ \longleftrightarrow \ CH_3CH=CH\overset{+}{C}H_2$$

$$\Big\downarrow \textbf{Br}^-$$

$$\left[\begin{array}{c} CH_3\overset{\delta^+}{C}HCH=CH_2 \\ \overset{|}{\underset{\delta^-}{:}Br:} \end{array} \right]^{\ddagger} \qquad\qquad \left[\begin{array}{c} CH_3CH=CH\overset{\delta^+}{C}H_2 \\ \overset{|}{\underset{\delta^-}{:}Br:} \end{array} \right]^{\ddagger}$$

**transition state for
formation of the 1,2-product**   **transition state for
formation of the 1,4-product**

$\downarrow$ $\downarrow$

$$\underset{\text{Br}}{CH_3\overset{|}{C}HCH=CH_2} \qquad\qquad \underset{\text{Br}}{CH_3CH=CH\overset{|}{C}H_2}$$

1,2-product   **1,4-product**

kinetic product   **thermodynamic product**

Tutorial: Thermodynamic
product vs. kinetic product

The reactant for the second step is an allylic cation whose structure is represent-ed by two contributing resonance structures. As the bromide ion approaches the al-lylic cation, the positive charge becomes concentrated on the carbon that the bromide ion approaches. This means that the transition state for formation of the 1,2-product resembles the resonance structure in which the positive charge is on a secondary al-lylic carbon, whereas the transition state for formation of the 1,4-product resembles the resonance structure in which the positive charge is on a primary allylic carbon. Because a secondary allylic cation is more stable than a primary allylic cation, the transition state for formation of the 1,2-product is the more stable transition state (Figure 7.5). As a result, the 1,2-product is formed faster than the 1,4-product. The 1,2-product, therefore, is the kinetic product.

For a reaction in which the kinetic and thermodynamic products are not the same, the product that predominates depends on the conditions under which the reaction is carried out. If the reaction is carried out under sufficiently mild (low-temperature) conditions so that the reaction is *irreversible*, the major product will be the *kinetic product*. For example, when addition of HBr to 1,3-butadiene is carried out at −80 °C, the major product is the 1,2-product.

$$CH_2=CHCH=CH_2 + HBr \xrightarrow{\textbf{−80 °C}}$$

Br   Br
| |
$CH_3CHCH=CH_2$ + $CH_3CH=CHCH_2$
1,2-addition product **1,4-addition product**
80%   **20%**

At −80 °C, there is enough energy for the reactants to overcome the energy barrier for the first step of the reaction to form the carbocation, and enough energy for the carbocation to form the two addition products (Figure 7.5). However, there is not enough energy for the reverse reaction to occur. The products cannot overcome the large energy barriers separating them from the carbocation. Consequently, at −80 °C, the relative amounts of the two products obtained reflect the relative energy barriers to the second step of the reaction. The energy barrier to formation of the 1,2-product is lower than the energy barrier to formation of the 1,4-product, so the major product is the 1,2-product.

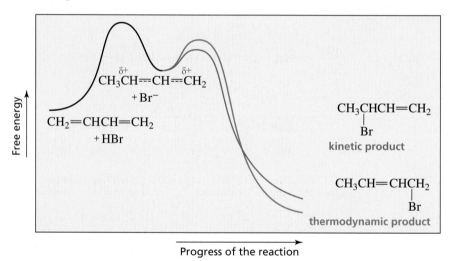

▲ **Figure 7.5**
Reaction coordinate diagram for the addition of HBr to 1,3-butadiene.

If, on the other hand, the reaction is carried out under sufficiently vigorous (high-temperature) conditions that the reaction is *reversible*, the major product will be the *thermodynamic product*. For example, when the reaction of 1,3-butadiene with HBr is carried out at 45 °C, the major product is the 1,4-product.

$$CH_2=CHCH=CH_2 + HBr \xrightarrow{\textbf{45 °C}}$$

Br   Br
| |
$CH_3CHCH=CH_2$ + $CH_3CH=CHCH_2$
1,2-addition product **1,4-addition product**
15%   **85%**

At 45 °C, there is enough energy for the reactants to get over the energy barriers to form the products and enough energy for one or more of the products to go back to the carbocation. The carbocation is called a **common intermediate** because it is an intermediate that both products have in common. The ability to return to a common intermediate allows the products to interconvert. Because the products can inter-convert, the relative amounts of the two products at equilibrium will depend on their relative stabilities. The thermodynamic product reverses less readily because it has a higher energy barrier to the common intermediate, so it gradually comes to predominate in the product mixture.

Thus, when the reaction is *irreversible* under the conditions used in the experiment, it is said to be under *kinetic control*. When a reaction is under **kinetic control,** the relative amounts of the products *depend on the rates* at which they are formed.

The kinetic product predomi-nates when the reaction is irreversible.

kinetic control:
the major product is the
one formed most rapidly

A reaction is said to be under *thermodynamic control* when there is enough energy available to allow it to be *reversible*. When a reaction is under **thermodynamic control,** the relative amounts of the products *depend on their stabilities*. Because a reaction must be reversible to be under thermodynamic control, thermodynamic control is also called **equilibrium control**.

The thermodynamic product predominates when the reaction is reversible.

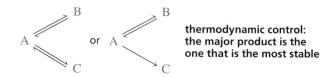

thermodynamic control:
the major product is the
one that is the most stable

The temperature at which a reaction changes from being kinetically controlled to being thermodynamically controlled depends on the reaction. Each reaction that is irreversible under mild conditions and reversible under more vigorous conditions has a temperature at which the changeover happens.

The particular temperature required to shift the reaction of a conjugated diene with an electrophilic reagent from kinetic control to thermodynamic control depends on both the diene and the electrophilic reagent. For example, the reaction of 1,3-butadiene with HCl remains under kinetic control at 45 °C even though addition of HBr to 1,3-butadiene is under thermodynamic control at that temperature. Because a carbon–chlorine bond is stronger than a carbon–bromine bond, a higher temperature is required for the products to undergo the reverse reaction. Thermodynamic control is achieved only when there is sufficient energy to allow the reaction to be reversible.

Do not assume, however, that the 1,2-product is the kinetic product and the 1,4-product is the thermodynamic product for all conjugated dienes. As the following examples show, the structure of the conjugated diene ultimately determines which product is the kinetic product and which is the thermodynamic product.

In the reaction of HBr with 4-methyl-1,3-pentadiene, the 1,2-product is the thermodynamic product because it has three alkyl groups bonded to the sp^2 carbons, whereas the 1,4-product has only two alkyl groups bonded to its sp^2 carbons. The 1,4-product is the kinetic product because the positive charge in the transition state leading to its formation is concentrated on a tertiary allylic carbon rather than on a secondary allylic carbon as it is in the transition state leading to the 1,2-product.

The two products obtained from the addition of HCl to 2,4-hexadiene have the same stability. Therefore, approximately equal amounts of the two products are obtained at temperatures high enough for the reaction to be reversible. The two transition states also

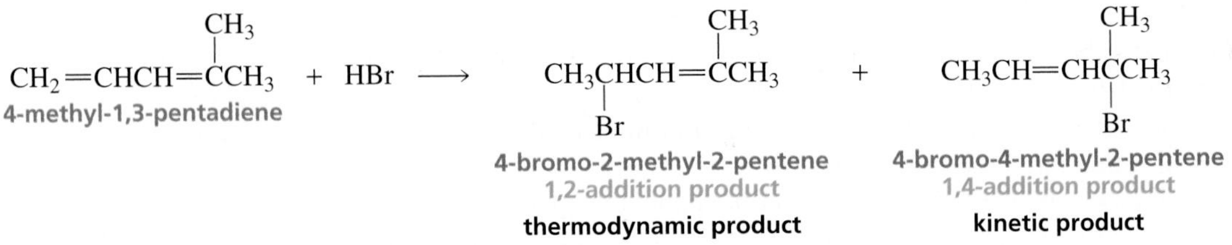

$$\underset{\text{4-methyl-1,3-pentadiene}}{CH_2{=}CHCH{=}\overset{\overset{\displaystyle CH_3}{|}}{C}CH_3} \;+\; HBr \;\longrightarrow\; \underset{\substack{\text{4-bromo-2-methyl-2-pentene}\\\text{1,2-addition product}\\\textbf{thermodynamic product}}}{CH_3\overset{\underset{\displaystyle Br}{|}}{C}HCH{=}\overset{\overset{\displaystyle CH_3}{|}}{C}CH_3} \;+\; \underset{\substack{\text{4-bromo-4-methyl-2-pentene}\\\text{1,4-addition product}\\\textbf{kinetic product}}}{CH_3CH{=}CH\overset{\overset{\displaystyle CH_3}{|}}{\underset{\underset{\displaystyle Br}{|}}{C}}CH_3}$$

have the same stability because, in both cases, the positive charge is shared by two secondary allylic carbons. Therefore, we also expect approximately equal amounts of the two products when the reaction is carried out at low temperatures.

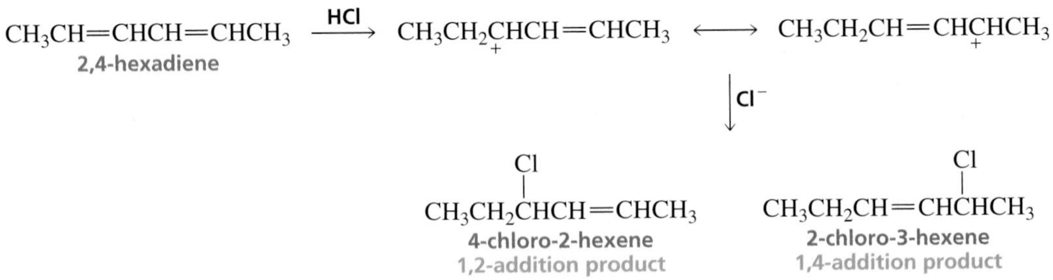

$$\underset{\text{2,4-hexadiene}}{CH_3CH{=}CHCH{=}CHCH_3} \xrightarrow{\;\textbf{HCl}\;} CH_3CH_2\overset{+}{C}HCH{=}CHCH_3 \;\longleftrightarrow\; CH_3CH_2CH{=}CH\overset{+}{C}HCH_3$$

$$\Big\downarrow Cl^-$$

$$\underset{\substack{\text{4-chloro-2-hexene}\\\text{1,2-addition product}}}{CH_3CH_2\overset{\overset{\displaystyle Cl}{|}}{C}HCH{=}CHCH_3} \qquad\qquad \underset{\substack{\text{2-chloro-3-hexene}\\\text{1,4-addition product}}}{CH_3CH_2CH{=}CH\overset{\overset{\displaystyle Cl}{|}}{C}HCH_3}$$

PROBLEM 10 ◆

a. How many products can be obtained from the reaction of one mole of 1,3-pentadiene with one mole of HBr?

b. What is the major product?

PROBLEM 11 / SOLVED

For each of the following reactions, give the major 1,2- and 1,4-addition products, and indicate which is the kinetic product and which is the thermodynamic product.

a. + HCl ⟶

c. + HCl ⟶

b. $CH_3CH{=}CH\overset{\overset{\displaystyle}{|}}{C}{=}CH_2$ with CH_3 substituent + HCl ⟶

d. + HCl ⟶

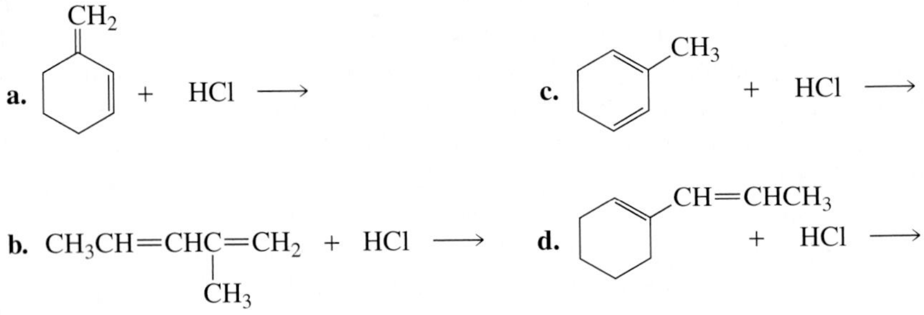

SOLUTION TO 11a The proton adds to the indicated sp^2 carbon because a carbocation is formed with the positive charge shared by a tertiary allylic and a secondary allylic carbon. If the proton were to add to the other end of the conjugated system, the carbocation would be less stable because the positive charge would be shared by a primary allylic and a secondary allylic carbon. Therefore, 3-chloro-3-methylcyclohexene is the 1,2-product and 3-chloro-1-methylcyclohexene is the 1,4-product. 3-Chloro-3-methylcyclohexene is the kinetic product because, in the transition state for its formation, the positive charge is concentrated on a tertiary carbon. 3-Chloro-1-methylcyclohexene is the thermodynamic product because it is more stable due to its more highly substituted double bond.

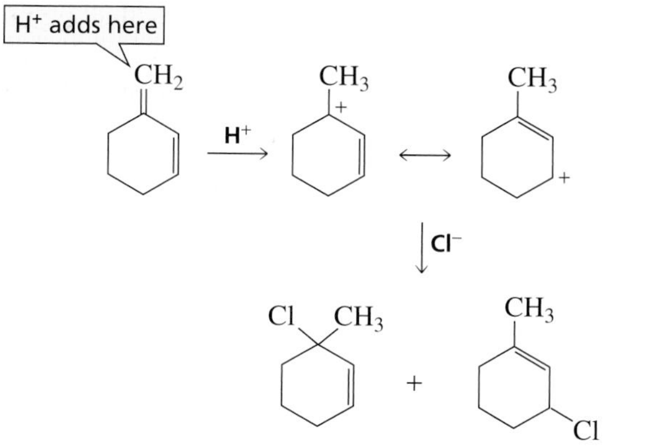

H⁺ adds here

CH_2

H^+

CH_3 +

CH_3 +

Cl^-

Cl CH_3

+

CH_3

Cl

3-chloro-3-
methylcyclohexene

3-chloro-1-
methylcyclohexene

kinetic
product

thermodynamic
product

Reactions that create new carbon–carbon bonds are very important to synthetic organic chemists because it is only through such reactions that small carbon skeletons can be converted into larger ones (Section 5.10). The **Diels–Alder reaction** is a particularly important reaction because it creates two new carbon–carbon bonds in a way that forms a cyclic molecule. In recognition of the importance of this reaction to synthetic organic chemistry, Otto Diels and Kurt Alder received the Nobel Prize in chemistry in 1950.

In the Diels–Alder reaction, a conjugated diene reacts with a compound containing a carbon–carbon double bond. This compound is called a **dienophile** because it "loves a diene." (Δ signifies heat.)

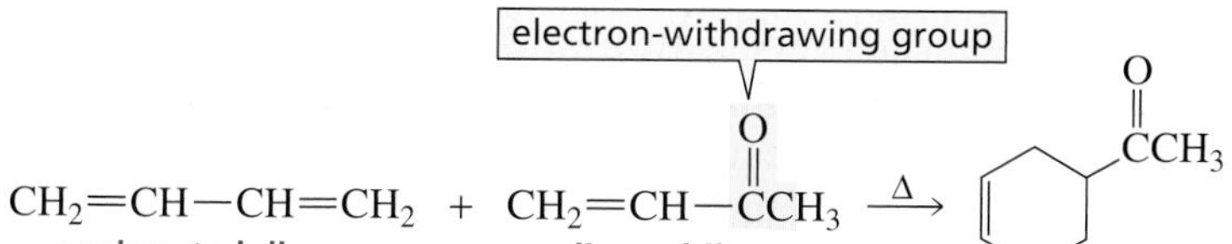

7.9
ADDITION OF A DIENOPHILE TO A CONJUGATED DIENE: THE DIELS–ALDER REACTION

electron-withdrawing group

O

CCH_3

CH_2=CH—CH=CH_2 + CH_2=CH—$\overset{\overset{\displaystyle O}{\|}}{C}CH_3$ $\xrightarrow{\Delta}$

conjugated diene   dienophile

An electron-withdrawing group bonded to one of the sp^2 carbons increases the reactivity of the dienophile. The electron-withdrawing group withdraws electrons from the double bond, making the Diels–Alder reaction easier to initiate by putting a partial positive charge on a carbon of the dienophile (Figure 7.6).

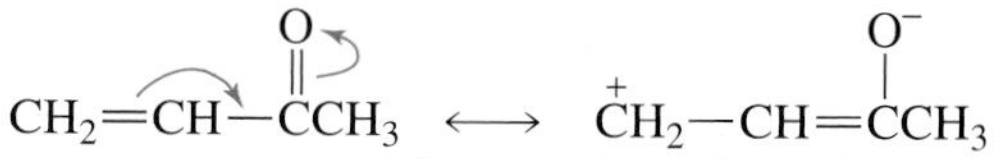

CH_2=CH—$\overset{\overset{\displaystyle O}{\|}}{C}CH_3$ $\longleftrightarrow$ $\overset{+}{C}H_2$—CH=$\overset{\overset{\displaystyle O^-}{|}}{C}CH_3$

resonance contributors of the dienophile

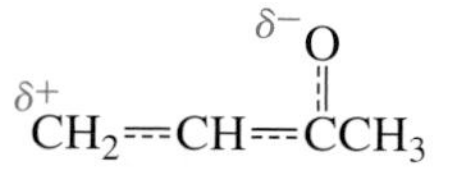

$\overset{\delta^+}{C}H_2$=$=$=CH=$=$=$\overset{\overset{\displaystyle \overset{\delta^-}{O}}{\|}}{C}CH_3$

resonance hybrid

The partially positively charged sp^2 carbon of the dienophile can be thought of as the electrophile that is attacked by electrons from C-1 of the conjugated diene. The other sp^2 carbon of the dienophile is the nucleophile that adds to C-4 of the diene.

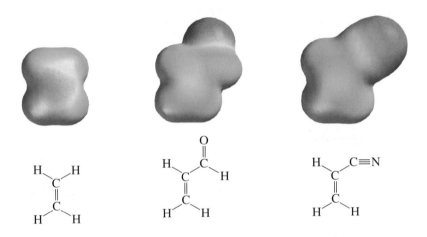

a 1,4-addition reaction to 1,3-butadiene

Figure 7.6 ▶
By comparing the electrostatic potential maps for ethene, propenal, and propenenitrile, you can see that an electron-withdrawing substituent decreases the electron density of the carbon-carbon double bond.

The Diels–Alder reaction may not look like any reaction that you have seen before, but it is simply a 1,4-addition to a conjugated diene. As with other 1,4-addition reactions, the double bond in the product is between C-2 and C-3 of what was the conjugated diene.

A significant difference between a Diels–Alder reaction and the 1,4-addition reactions that we have seen previously, however, is that the Diels–Alder reaction is a concerted reaction—the addition occurs in a single step. The other 1,4-addition reactions we have studied occur in two successive steps: the electrophile adds to the diene in the first step, and the nucleophile adds to the carbocation in the second step.

The Diels–Alder reaction is another example of a **pericyclic reaction,** a reaction that takes place in one step by a cyclic shift of electrons (Section 3.17). It is also a cycloaddition reaction. A **cycloaddition reaction** is one in which two reactants add together to form a cyclic product. More precisely, the Diels–Alder reaction is a **[4 + 2] cycloaddition reaction** because, of the six π electrons involved in the cyclic transition state, *four* come from the conjugated diene and *two* come from the dienophile.

diene
four π electrons

dienophile
two π electrons

transition state
six π electrons

new bond

new double bond

new bond

Stereochemistry of the Diels–Alder Reaction

If a Diels-Alder reaction creates a chirality center in the product, identical amounts of the R and S enantiomers will be formed. In other words, the product will be a racemic mixture (Section 4.17).

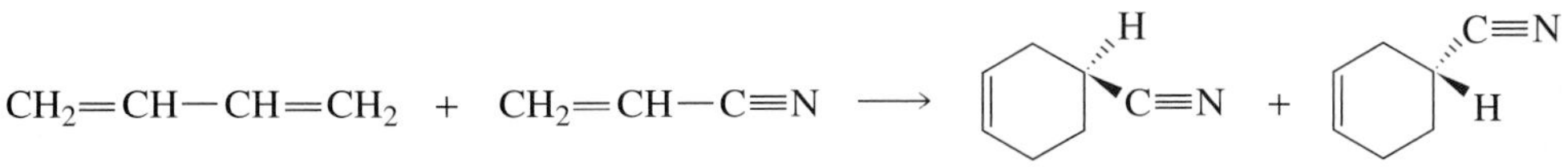

$$CH_2=CH-CH=CH_2 \ + \ CH_2=CH-C{\equiv}N \longrightarrow$$

The Diels–Alder reaction is stereospecific because the configuration of the reactants is maintained during the course of the reaction. If the substituents in the dienophile are cis, they will be cis in the product; if the substituents in the dienophile are trans, they will be trans in the product. The configuration of the reactants is maintained because the reaction is concerted. The stereochemistry of the reaction will be discussed in more detail in Section 28.4.

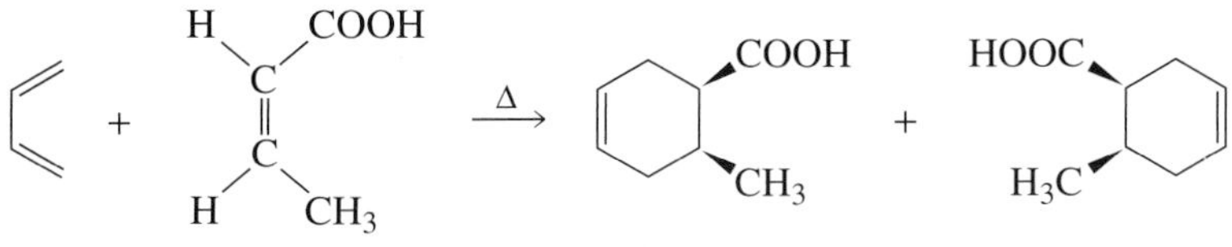

cis dienophile   cis products

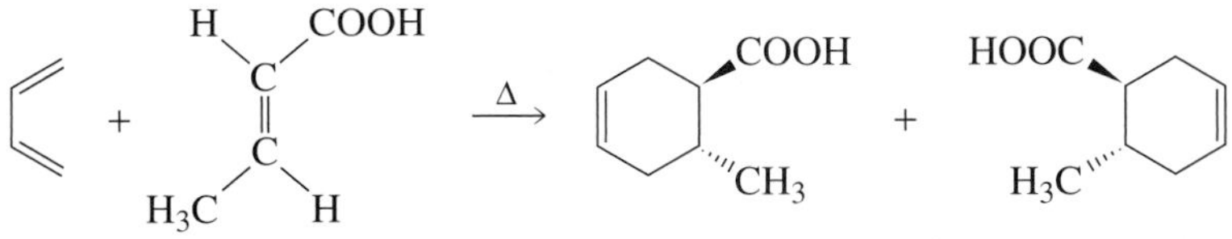

trans dienophile   trans products

Otto Paul Hermann Diels (1876–1954) *was born in Germany, the son of a professor of classical philology at the University of Berlin. He received a Ph.D. from that university, working with Emil Fischer. Diels was a professor of chemistry at the University of Berlin and later at the University of Kiel. Two of his sons died in World War II. He retired in 1945 after his home and laboratory were destroyed in bombing raids. He received the 1950 Nobel Prize in chemistry, sharing it with his former student, Kurt Alder.*

Each of the two previous reactions forms a product with two new chirality centers. Thus, each product has four stereoisomers. Because the configuration of the reactants is maintained, each reaction forms only two of the stereoisomers.

A wide variety of cyclic compounds can be obtained by varying the structures of the conjugated diene and the dienophile.

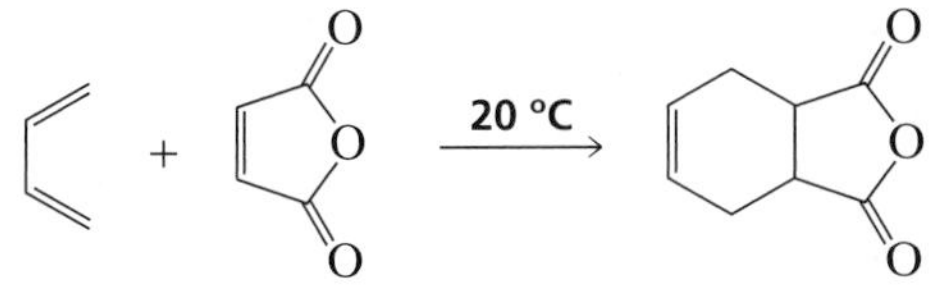

20 °C

Compounds containing carbon–carbon triple bonds can also be used as dienophiles in Diels–Alder reactions to prepare compounds with two double bonds.

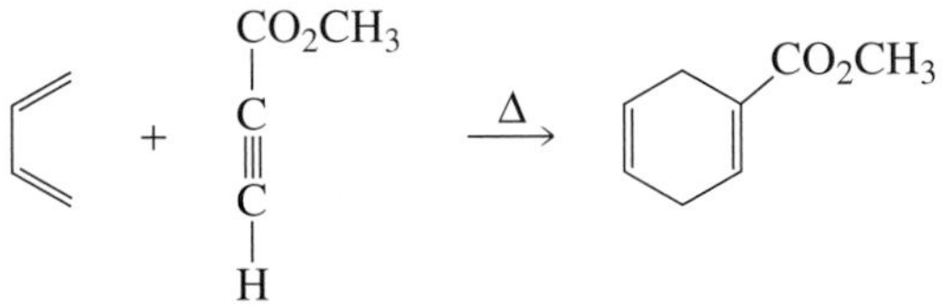

Δ

Kurt Alder (1902–1958) was born in a part of Germany that is now Poland. After World War I, he and his family moved to Germany when they were expelled from their home region when it became Polish. After receiving his Ph.D. under Diels in 1926, Alder continued working with him, and in 1928 they discovered the Diels–Alder reaction. Alder was a professor of chemistry at the University of Kiel and at the University of Cologne. He received the 1950 Nobel Prize in chemistry, sharing it with his mentor, Otto Diels.

If the dienophile has two carbon–carbon double bonds, two successive Diels–Alder reactions can occur if excess diene is available.

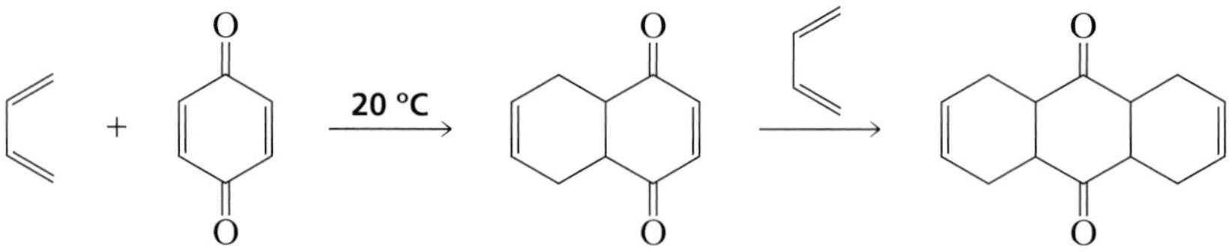

PROBLEM 12 ◆

Give the products of each of the following reactions:

a. $CH_2{=}CH{-}CH{=}CH_2$ + $CH_3\overset{O}{\overset{\|}{C}}{-}C{\equiv}C{-}\overset{O}{\overset{\|}{C}}CH_3$ $\xrightarrow{\Delta}$

b. $CH_2{=}CH{-}CH{=}CH_2$ + $HC{\equiv}C{-}C{\equiv}N$ $\xrightarrow{\Delta}$

c. $CH_2{=}CH{-}CH{=}CH{-}CH_3$ + $HC{\equiv}C{-}C{\equiv}N$ $\xrightarrow{\Delta}$

d. $CH_2{=}CH{-}\underset{\underset{CH_3}{|}}{C}{=}CH_2$ + $HC{\equiv}C{-}C{\equiv}N$ $\xrightarrow{\Delta}$

PROBLEM 13

Explain why the following are not optically active:

a. The product obtained from the reaction of 1,3-butadiene with *cis*-1,2-dichloroethene.

b. The product obtained from the reaction of 1,3-butadiene with *trans*-1,2-dichloroethene.

Predicting the Product When Both Reagents Are Unsymmetrically Substituted

In each of the preceding Diels–Alder reactions, only one product is formed (ignoring stereoisomers) because in each of these reactions at least one of the reacting molecules is symmetrically substituted. If both the diene and the dienophile are unsymmetrically substituted, however, two products are possible.

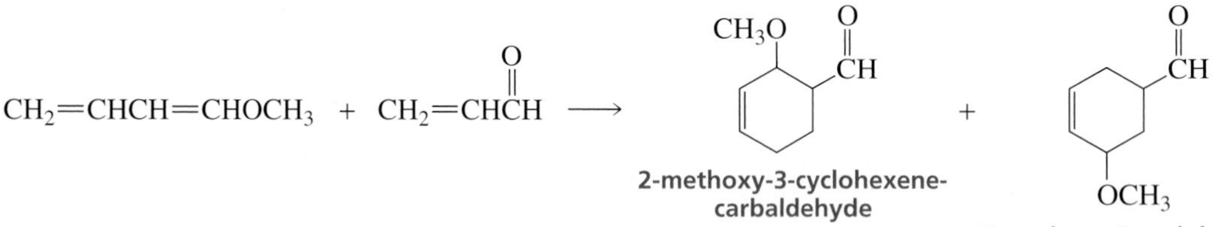

$$CH_2{=}CHCH{=}CHOCH_3 + CH_2{=}CHCH\overset{O}{\overset{\|}{}} \longrightarrow$$

2-methoxy-3-cyclohexene-carbaldehyde

+

5-methoxy-3-cyclohexene-carbaldehyde

The two products result from the reactants being aligned relative to each other in the following two different ways:

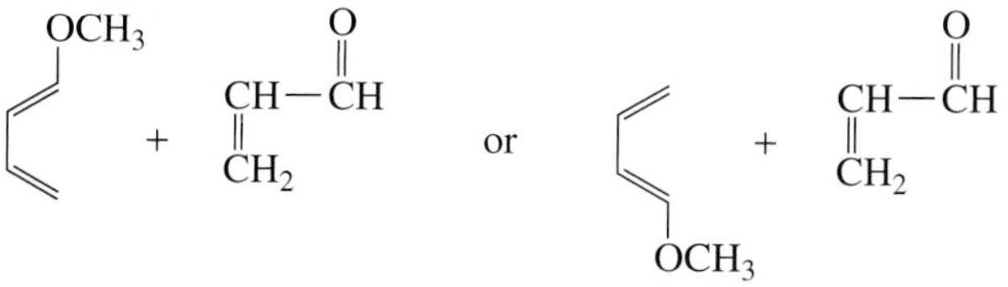

Which of the two products will be formed (or be formed in greater yield) depends on the charge distribution in each of the reactants. It is necessary to draw contributing resonance structures to determine the charge distribution. In the preceding example, the methoxy group of the diene is capable of donating electrons by resonance. As a result, the terminal carbon atom bears a partial negative charge. The aldehyde group of the dienophile, on the other hand, withdraws electrons by resonance, so its terminal carbon has a partial positive charge.

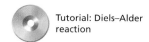

Tutorial: Diels–Alder reaction

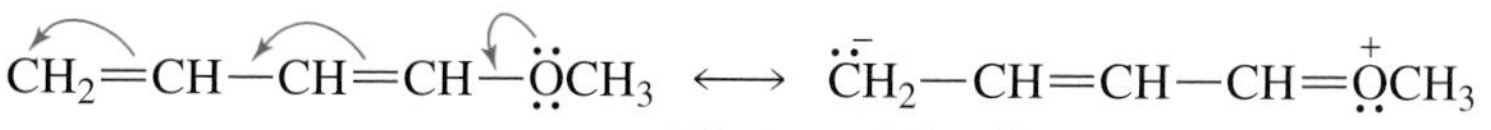

resonance contributors of the diene

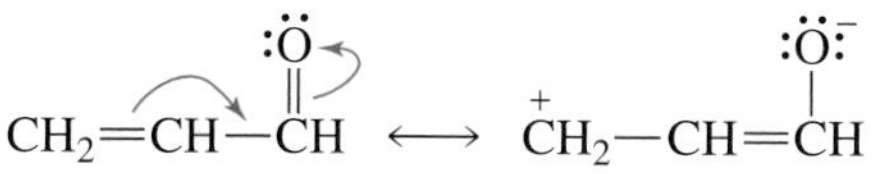

resonance contributors of the dienophile

The partially positively charged carbon atom of the dienophile will be attacked preferentially by the partially negatively charged carbon of the diene. Therefore, 2-methoxy-3-cyclohexenecarbaldehyde is the major product.

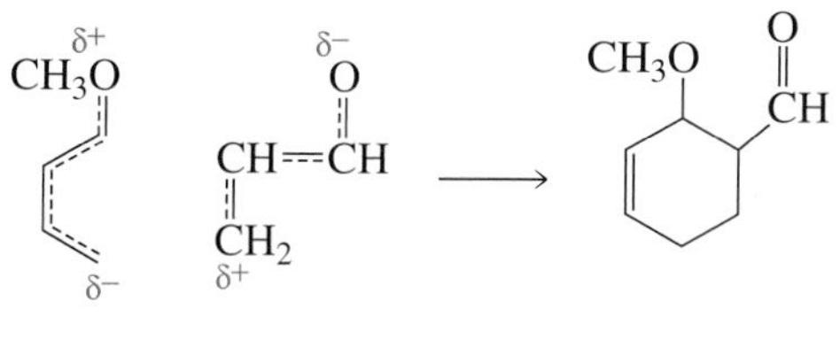

PROBLEM 14 ◆

What would be the major product if the methoxy substituent in the preceding reaction were bonded to C-2 of the diene rather than to C-1?

Conformations of the Diene

A conjugated diene can exist in two different conformations, an **s-cis conformation** and an **s-trans conformation.** By s-cis we mean that the double bonds are cis about the single bond (s = single). In order to participate in a Diels–Alder reaction, the conjugated diene must be in an s-cis conformation. In an s-trans conformation, the number 1 and number 4 carbons are too far apart to react with the dienophile. The rotational barrier between the s-cis and s-trans conformations is only slightly higher than that for a normal single bond, so the two can rapidly interconvert at room temperature.

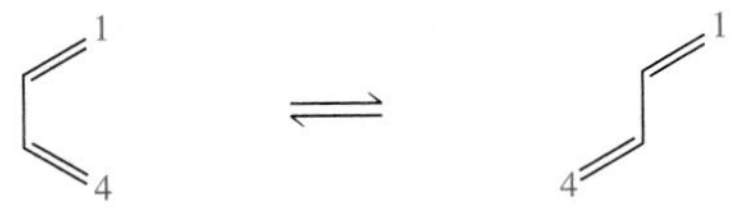

s-cis conformation   *s-trans conformation*

A conjugated diene that is locked in an *s*-trans conformation cannot undergo a Diels–Alder reaction.

$$+ \quad \begin{array}{c} CH_2 \\ \| \\ CHCO_2CH_3 \end{array} \quad \longrightarrow \quad \text{no reaction}$$

s-trans conformation

A conjugated diene that is locked in an *s*-cis conformation, such as 1,3-cyclopentadiene is very reactive in a Diels–Alder reaction. When the diene is a cyclic compound, the product of a Diels–Alder reaction is a bridged bicyclic compound. A **bridged bicyclic compound** contains two rings that share two nonadjacent carbons.

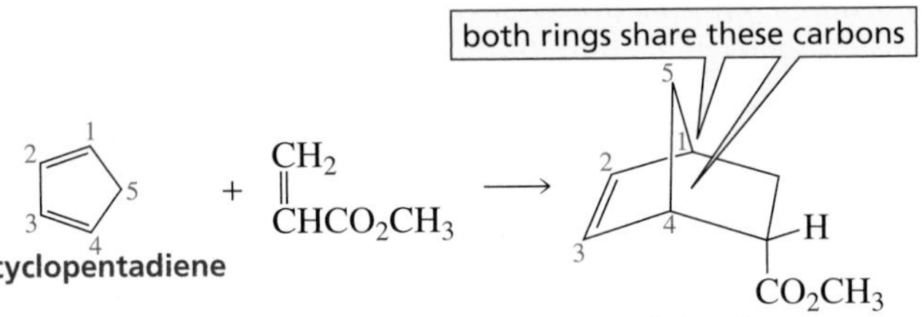

both rings share these carbons

$$\text{1,3-cyclopentadiene} \quad + \quad \begin{array}{c} CH_2 \\ \| \\ CHCO_2CH_3 \end{array} \quad \longrightarrow$$

1,3-cyclopentadiene

H
CO$_2$CH$_3$
bridged bicyclic compound

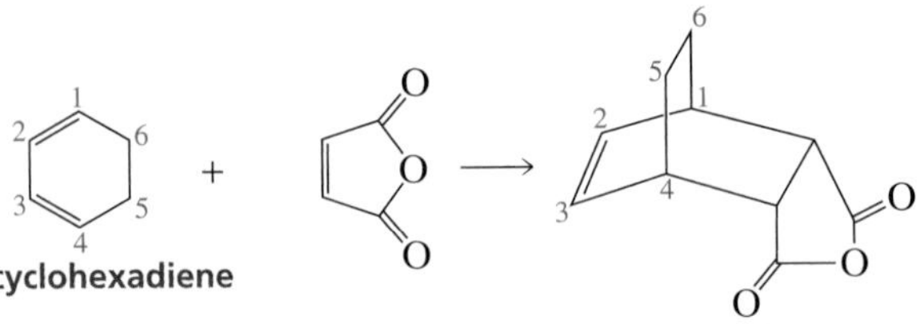

1,3-cyclohexadiene +

bridged bicyclic compound

There are two possible configurations for bridged bicyclic compounds. A substituent on a bridge is **endo** if it is closer to the longer (or more unsaturated) of the two other bridges; it is **exo** if it closer to the shorter (or more saturated) bridge.

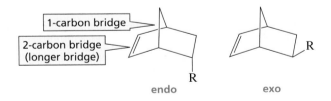

1-carbon bridge

2-carbon bridge
(longer bridge)

R

R

endo   *exo*

Notice in the two preceding reactions (i.e., those with 1,3-cyclopentadiene and 1,3-cyclohexadiene as dienes) that only the endo product is obtained. Notice, too, that each of the alkenes acting as a dienophile has a substituent with π electrons. When the dienophile has π electrons (other than the π electrons of its carbon–carbon double bond), only the endo product is obtained because the transition state for formation of the endo product is stabilized by the interaction of the p orbitals of the dienophile with the p orbitals of the new double bond formed in the diene. This interaction is called **secondary orbital overlap.** It can occur only if the substituent with the p orbitals lies underneath (endo) rather than away from (exo) the diene.

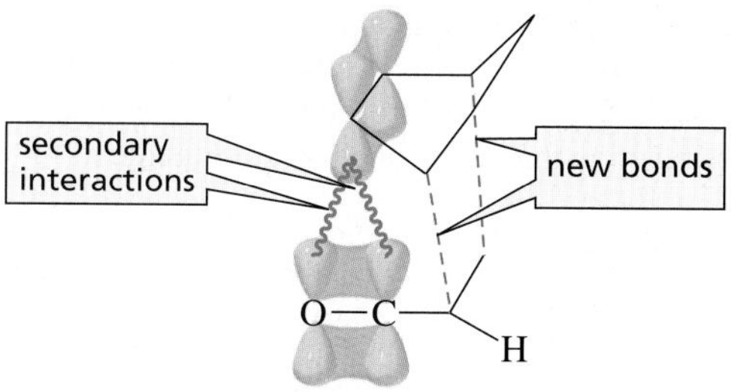

secondary interactions

new bonds

secondary orbital overlap
in the endo transition state

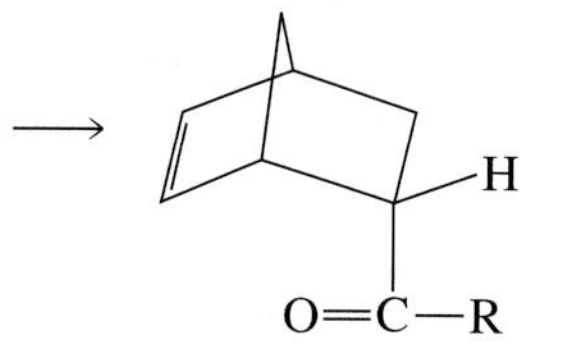

endo product

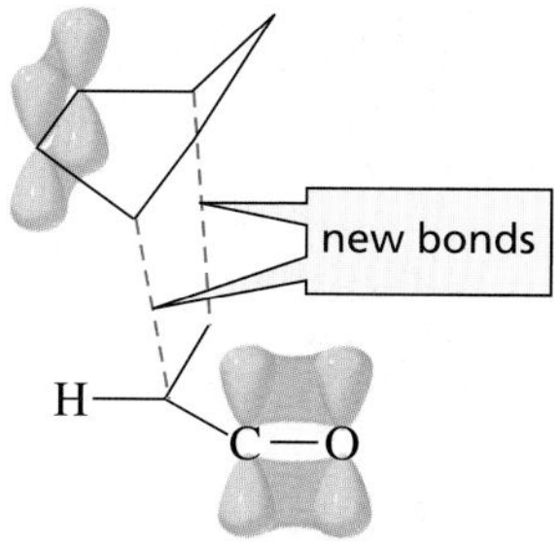

new bonds

exo product

no secondary orbital
overlap in the exo transition state

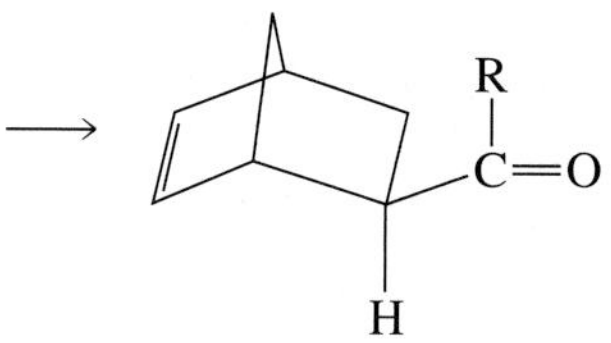

PROBLEM 15 ◆

Which of the following conjugated dienes would not react with a dienophile in a Diels–Alder reaction?

a.

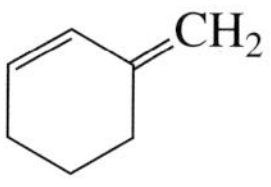

c.

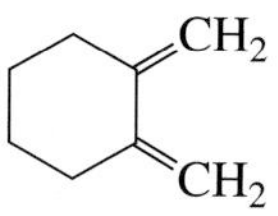

e.

b.

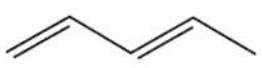

d.

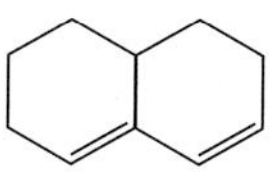

f.

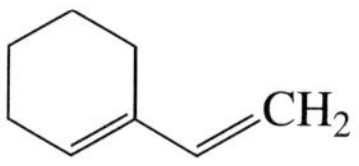

PROBLEM 16 / SOLVED

List the following dienes in order of decreasing reactivity in a Diels–Alder reaction.

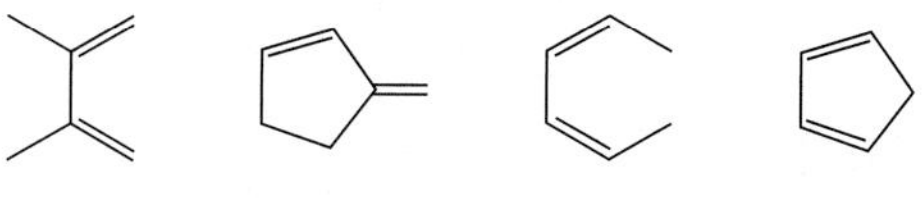

SOLUTION   The most reactive diene has the double bonds locked in the *s*-cis conformation, whereas the least reactive diene cannot achieve the required *s*-cis conformation because it is locked in the *s*-trans conformation.

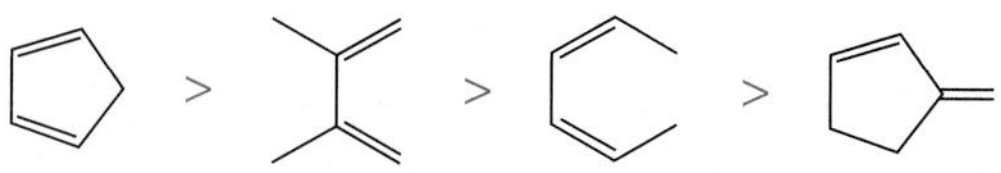

2,3-Dimethyl-1,3-butadiene and 2,4-hexadiene are of intermediate reactivity because they can exist both in the *s*-cis and *s*-trans conformations. 2,4-Hexadiene is less apt to be in the required *s*-cis conformation because of steric interference between the terminal methyl groups. Consequently, it is less reactive than 2,3-dimethyl-1,3-butadiene.

| *s*-cis | *s*-trans | *s*-cis | *s*-trans |

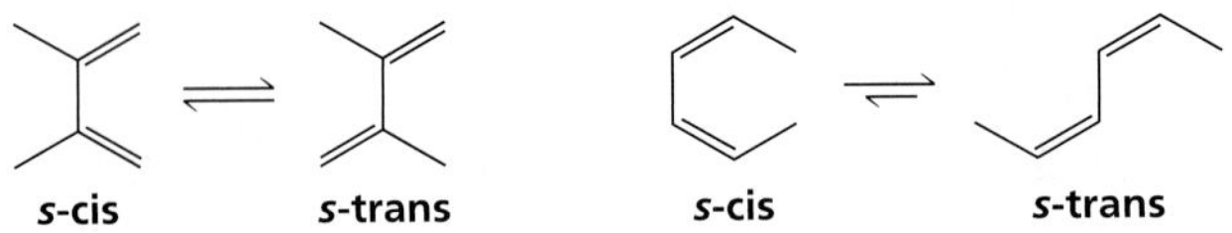

PROBLEM-SOLVING STRATEGY

What diene and what dienophile were used to synthesize the following compound?

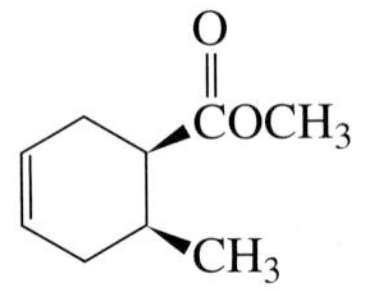

The diene that was used to form the cyclic product had double bonds on either side of the double bond in the product, so put in the π bonds and remove the π bond between them.

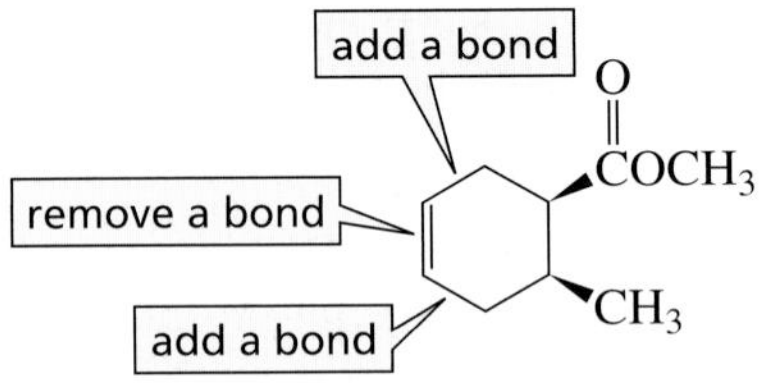

The new σ bonds are now on either side of the double bonds.

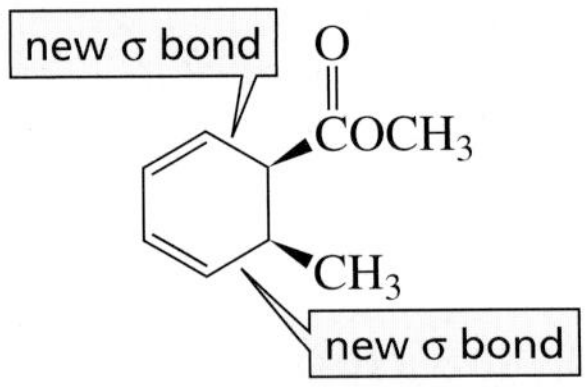

Erasing these σ bonds and putting a π bond between the two carbons that were attached by the σ bonds gives the diene and the dienophile.

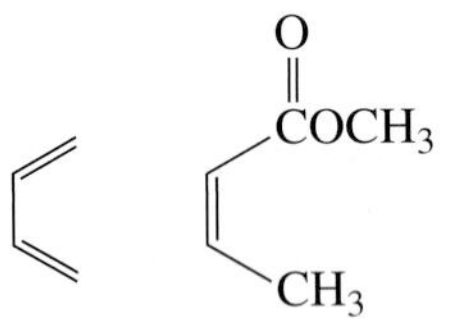

Now continue on to Problem 17.

PROBLEM 17 ◆

What diene and what dienophile should be used to synthesize the following compounds?

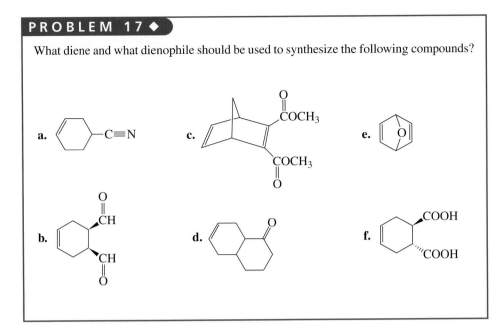

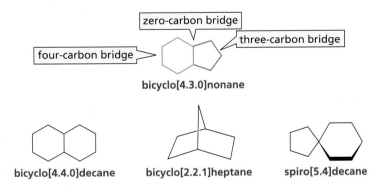

Bicyclic compounds are compounds that contain two rings. If the two rings share *one carbon*, the compound is a **spirocyclic compound.** If the two rings share *two adjacent carbons*, the compound is a **fused bicyclic compound.** Finally, if the rings share *two nonadjacent carbons*, the compound is a **bridged bicyclic compound.**

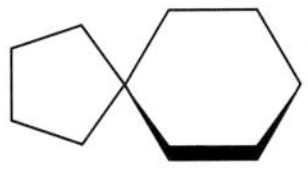

7.10
NOMENCLATURE OF BICYCLIC COMPOUNDS

Tutorial:
Bicyclic compounds

spirocyclic 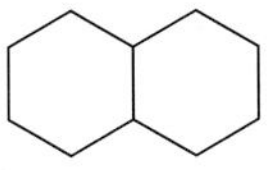   fused bicyclic 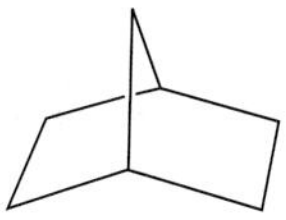   bridged bicyclic

   Bicyclic compounds are named by using the alkane name to designate the total number of carbons and the prefix "spiro" or "bicyclo" to indicate the number of shared carbons. "Spiro" indicates one shared carbon and "bicyclo" indicates two shared carbons. After the prefix come brackets that contain numbers indicating the number of carbons in each bridge. These are listed in order of decreasing bridge length. In bicyclo[4.3.0]nonane, the green bridge has four carbons, the blue bridge has three carbons, and the red bridge has zero carbons. Notice that the **bridgehead carbons** (the carbons shared by two rings) aren't counted.

zero-carbon bridge

three-carbon bridge

four-carbon bridge

bicyclo[4.3.0]nonane

bicyclo[4.4.0]decane   bicyclo[2.2.1]heptane   spiro[5.4]decane

To determine the position of a substituent, the carbons of the rings are numbered, starting at a shared carbon (a bridgehead carbon) and numbering around the longest bridge first.

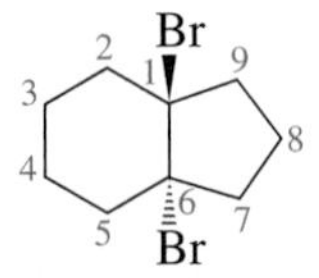

trans-1,6-dibromo-
bicyclo[4.3.0]nonane

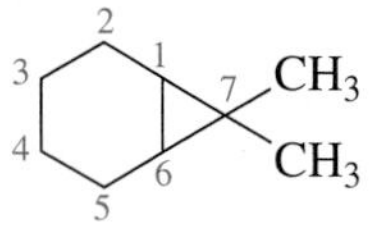

7,7-dimethyl-
bicyclo[4.1.0]heptane

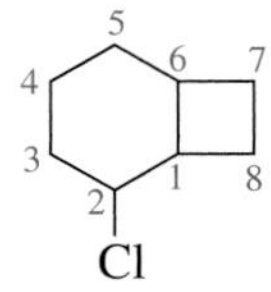

2-chloro-
bicyclo[4.2.0]octane

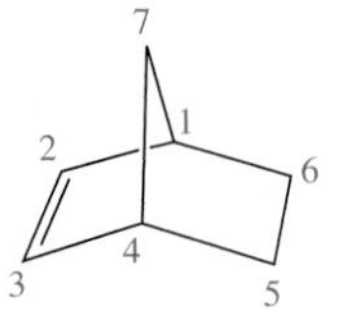

bicyclo[2.2.1]-
hept-2-ene

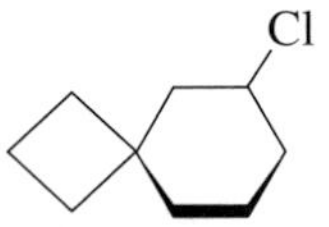

3-chlorospiro[5.3]nonane

PROBLEM 18 ◆

Name the following compounds:

a.

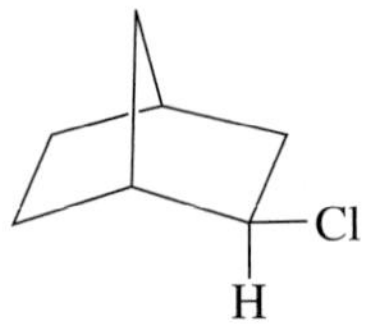

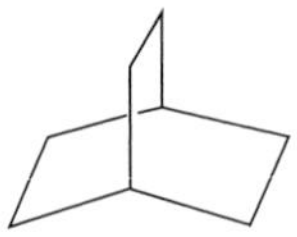

b.

c.

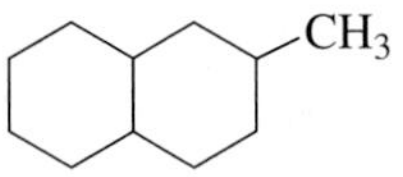

d.

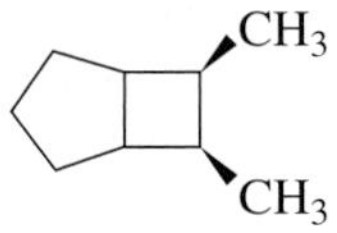

SUMMARY OF REACTIONS

1. If there is excess electrophilic reagent, both double bonds of an *isolated diene* will undergo electrophilic addition.

$$CH_2=CHCH_2CH_2\overset{\overset{\displaystyle CH_3}{|}}{C}=CH_2 \quad + \quad HBr \quad \longrightarrow \quad CH_3\overset{}{C}HCH_2CH_2\overset{\overset{\displaystyle CH_3}{|}}{C}CH_3$$

excess

$$\underset{Br}{|} \qquad \underset{Br}{|}$$

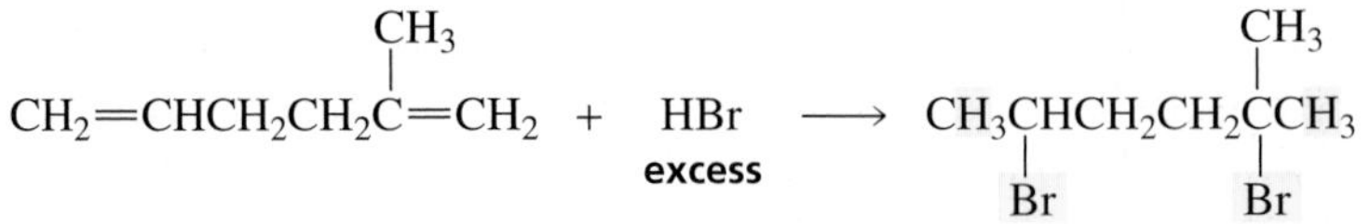

If there is only one equivalent of electrophilic reagent, only the most reactive of the double bonds of an *isolated diene* will undergo electrophilic addition (Section 7.6).

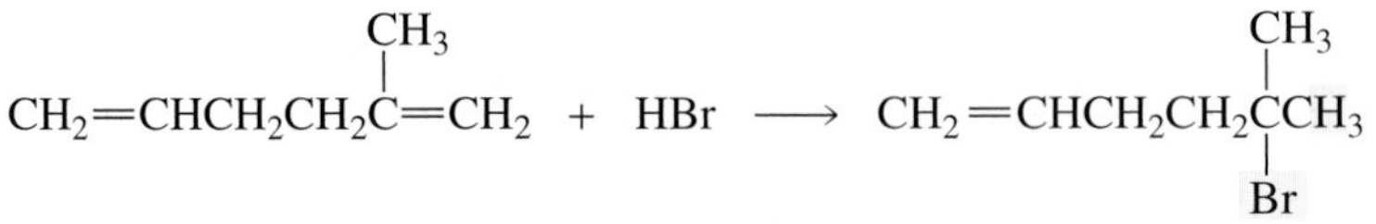

$$CH_2=CHCH_2CH_2\overset{\overset{\displaystyle CH_3}{|}}{C}=CH_2 \quad + \quad HBr \quad \longrightarrow \quad CH_2=CHCH_2CH_2\overset{\overset{\displaystyle CH_3}{|}}{C}CH_3$$

$$\underset{Br}{|}$$

2. *Conjugated dienes* undergo 1,2- and 1,4-addition with one equivalent of an electrophilic reagent (Section 7.7).

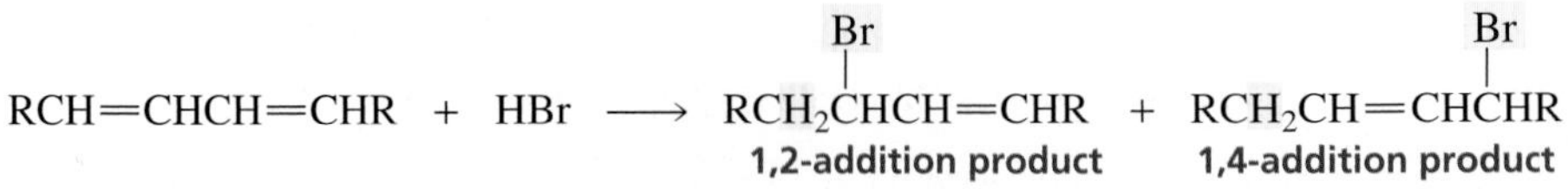

$$RCH=CHCH=CHR \quad + \quad HBr \quad \longrightarrow \quad RCH_2\overset{\overset{\displaystyle Br}{|}}{C}HCH=CHR \quad + \quad RCH_2CH=CH\overset{\overset{\displaystyle Br}{|}}{C}HR$$

1,2-addition product   1,4-addition product

3. *Conjugated dienes* undergo 1,4-cycloaddition with a dienophile (Section 7.9).

a Diels–Alder reaction

$$CH_2=CH-CH=CH_2 \;+\; CH_2=CH-\overset{\displaystyle O}{\overset{\|}{C}}-R \;\xrightarrow{\;\Delta\;}\; $$

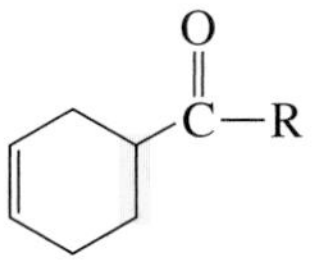

KEY TERMS

1,2-addition (page 307)
1,4-addition (page 307)
allene (page 300)
bicyclic compound (page 323)
bridged bicyclic compound (page 320)
bridgehead carbon (page 323)
common intermediate (page 312)
conjugate addition (page 307)
conjugated double bonds (page 300)
cumulated double bonds (page 300)
cycloaddition reaction (page 316)

[4 + 2] cycloaddition reaction
   (page 316)
Diels–Alder reaction (page 315)
diene (page 300)
dienophile (page 315)
direct addition (page 307)
endo (page 320)
equilibrium control (page 313)
exo (page 320)
fused bicyclic compound (page 323)
isolated double bonds (page 300)

kinetic control (page 313)
kinetic product (page 310)
pericyclic reaction (page 316)
polyene (page 300)
secondary orbital overlap (page 320)
s-cis conformation (page 319)
s-trans conformation (page 319)
spirocyclic compound (page 323)
tetraene (page 300)
thermodynamic control (page 313)
thermodynamic product (page 310)
triene (page 300)

PROBLEMS

19. Give the systematic name for each of the following compounds:

a. $CH_3C{\equiv}CCH_2CH_2CH_2CH{=}CH_2$

d.

b.
$$HOCH_2CH_2 \quad CH_2CH_3$$
$$C{=}C$$
$$H \qquad H$$

e.

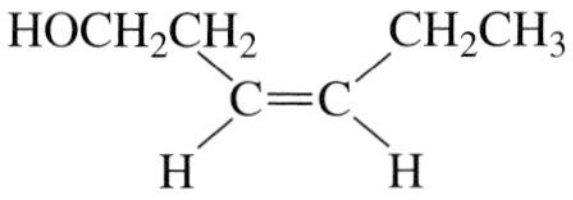

c. $CH_3CH_2C{\equiv}CCH_2CH_2C{\equiv}CH$

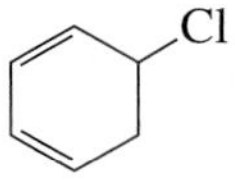

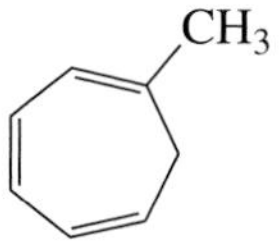

20. Draw structures for the following compounds:
  a. (2*E*,4*E*)-1-chloro-3-methyl-2,4-hexadiene
  b. (3*Z*,5*E*)-4-methyl-3,5-nonadiene
  c. (3*Z*,5*Z*)-4,5-dimethyl-3,5-nonadiene
  d. (3*E*,5*E*)-2,5-dibromo-3,5-octadiene

21. a. How many linear dienes are there with molecular formula C_6H_{10}? (Ignore cis–trans isomers.)
  b. How many of these are conjugated dienes?
  c. How many of these are isolated dienes?

22. Which compound would you expect to have the greater heat of hydrogenation: 1,2-pentadiene or 1,4-pentadiene? (*Hint:* Remember that the heat of hydrogenation $= -\Delta H°$.)

23. In Chapter 3 we learned that α-farnesene is found in the waxy coating of apple skins. What is its systematic name?

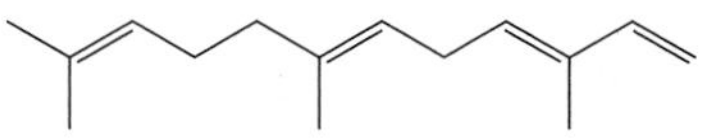

α-**farnesene**

24. Diane O'File treated 1,3-cyclohexadiene with Br_2 and obtained two products (ignoring stereoisomers). Her lab partner treated 1,3-cyclohexadiene with HBr and was surprised to find that only one product was formed (ignoring stereoisomers). Account for these results.

25. Give the major product of each of the following reactions. One equivalent of each reagent is used.

a. 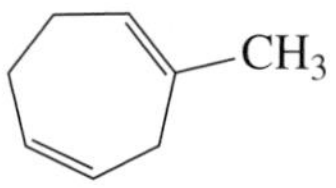 + HBr $\longrightarrow$

c. + HBr $\longrightarrow$

b. 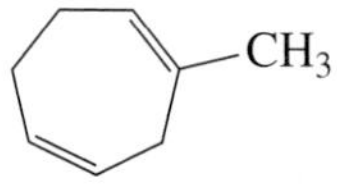 + HBr $\xrightarrow{\text{peroxide}}$

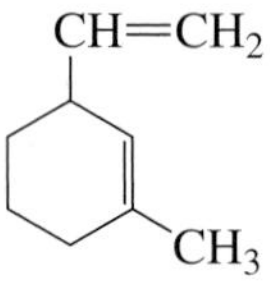

d. 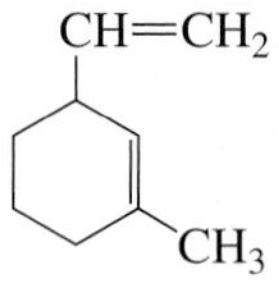 + HBr $\xrightarrow{\text{peroxide}}$

26. a. Give the products obtained from the reaction of one mole of HBr with one mole of 1,3,5-hexatriene.
  b. Which product(s) will predominate if the reaction is under kinetic control?
  c. Which product(s) will predominate if the reaction is under thermodynamic control?

27. How could the following compounds be synthesized using a Diels–Alder reaction?

a.

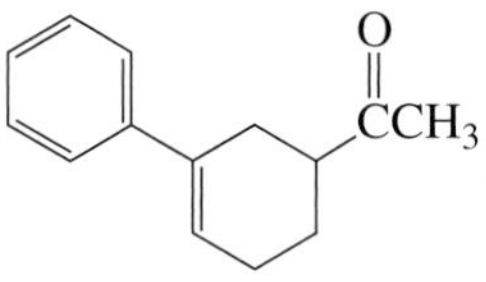

c.

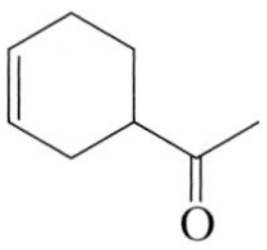

b.

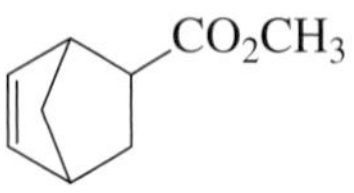

d.

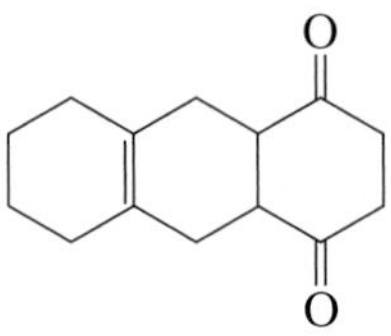

28. a. How could each of the following compounds be prepared from a hydrocarbon in a single step?

1. 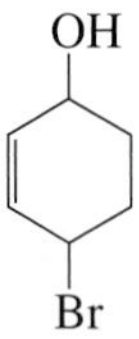
2. 
3.

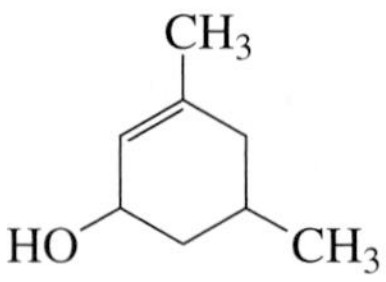

  b. What other organic compound would be obtained from each synthesis?

29. How would the following substituents affect the rate of a Diels–Alder reaction?
  a. an electron-donating substituent in the diene
  b. an electron-donating substituent in the dienophile
  c. an electron-withdrawing substituent in the diene

30. Give the major products obtained from the reaction of one equivalent of HCl with the following:
  a. 2,3-dimethyl-1,3-pentadiene
  b. 2,4-dimethyl-1,3-pentadiene

For each reaction, indicate the kinetic and thermodynamic products.

31. Give the product or products that would be obtained from each of the following reactions:

a. 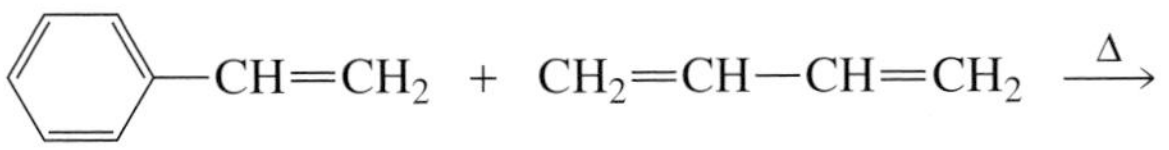 ⬡—CH=CH₂ + CH₂=CH—CH=CH₂ $\xrightarrow{\Delta}$

b. 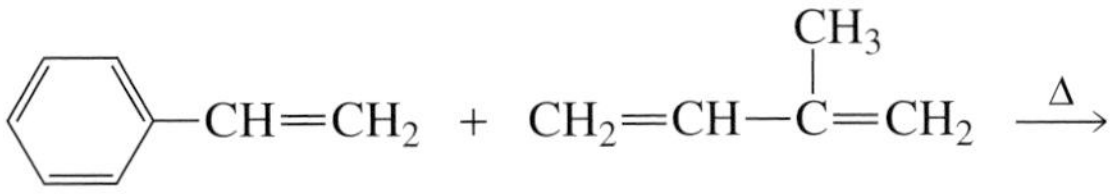 ⬡—CH=CH₂ + CH₂=CH—$\overset{\text{CH}_3}{\underset{|}{\text{C}}}$=CH₂ $\xrightarrow{\Delta}$

c. CH₂=CH—C=CH₂ + CH₂=CH$\overset{\text{O}}{\overset{\|}{\text{C}}}$CH₃ $\xrightarrow{\Delta}$
    (with phenyl group below)

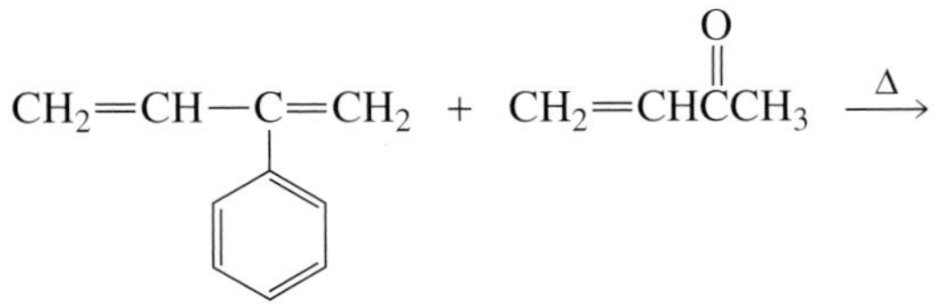

d. 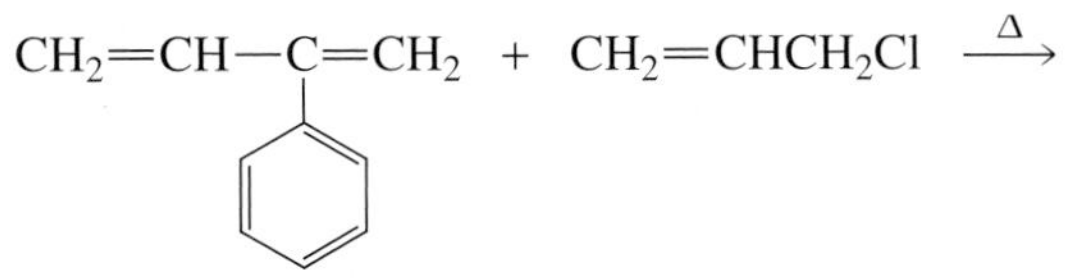 CH₂=CH—C=CH₂ + CH₂=CHCH₂Cl $\xrightarrow{\Delta}$
    (with phenyl group below)

32. Give two sets of a conjugated diene and a dienophile that could be used to prepare the following compound.

$\overset{\text{O}}{\overset{\|}{\text{C}}}$CH₃ (on cyclohexadiene ring)

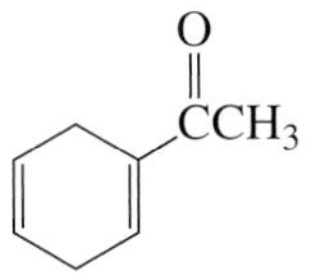

33. a. Which dienophile is more reactive in a Diels–Alder reaction?

1.   CH₂=CH$\overset{\text{O}}{\overset{\|}{\text{C}}}$H   or   CH₂=CHCH₂$\overset{\text{O}}{\overset{\|}{\text{C}}}$H

2.   CH₂=CH$\overset{\text{O}}{\overset{\|}{\text{C}}}$H   or   CH₂=CHCH₃

b. Which diene is more reactive in a Diels–Alder reaction?

CH₂=CHCH=CHOCH₃   or   CH₂=CHCH=CHCH₂OCH₃

34. Cyclopentadiene can react with itself in a Diels–Alder reaction. Draw the endo and exo products.

35. Which diene and which dienophile could be used to prepare each of the following compounds?

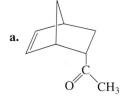

a.   b. 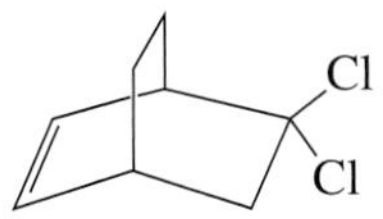   c. 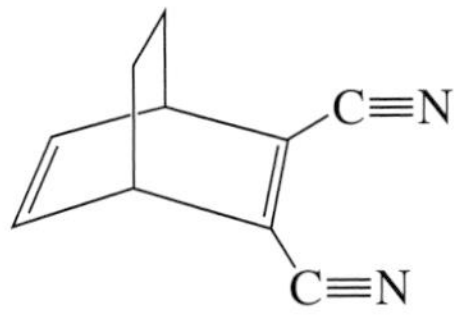   d.

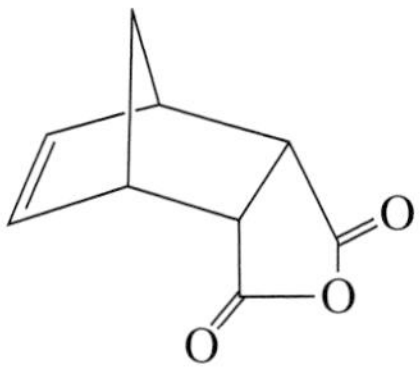

(a. structure with C(=O)CH₃ group; b. structure with Cl, Cl; c. structure with C≡N, C≡N; d. structure with O, O)

36. As many as nine different Diels–Alder products can be obtained by heating a mixture of 1,3-butadiene and 2-methyl-1,3-butadiene. Identify the products.

37. a. Give the products of the following reaction:

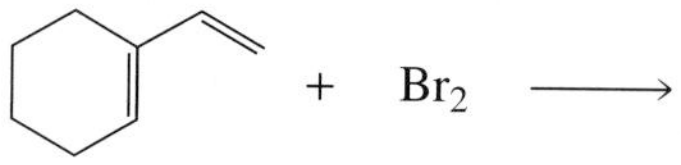

$$\text{(cyclohexenyl ethylene structure)} \quad + \quad Br_2 \quad \longrightarrow$$

b. How many stereoisomers of each product could be obtained?

38. On the same graph, draw the reaction coordinate for the addition of one equivalent of HBr to 2-methyl-1,3-pentadiene and one equivalent of HBr to 2-methyl-1,4-pentadiene. Which reaction is faster?

39. While attempting to recrystallize maleic anhydride, Professor Nots O. Kareful dissolved it in freshly distilled cyclopentadiene rather than in freshly distilled cyclopentane. Was his recrystallization successful?

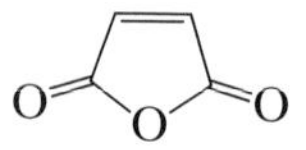

$$\text{(maleic anhydride structure)}$$

maleic anhydride

40. The following equilibrium is driven to the right if the reaction is carried out in the presence of maleic anhydride. What is the function of maleic anhydride?

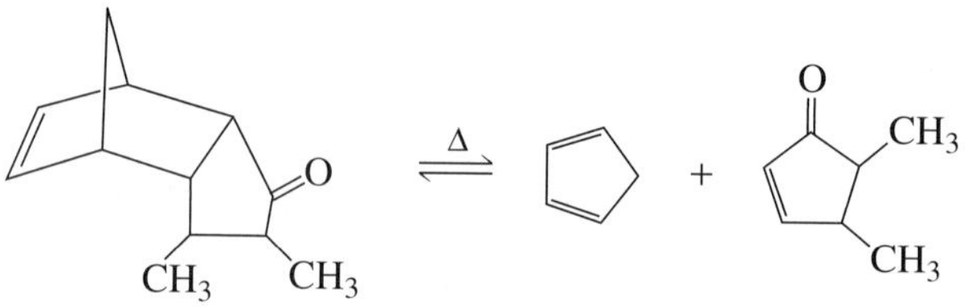

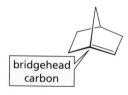

41. In 1935 J. Bredt, a German chemist, proposed that a bicycloalkene cannot have a double bond at a bridgehead carbon unless one of the rings contains at least 8 carbon atoms. This is known as Bredt's rule. Explain why there cannot be a double bond at this position.

bridgehead
carbon

8

Reactions of Alkanes:
Radicals

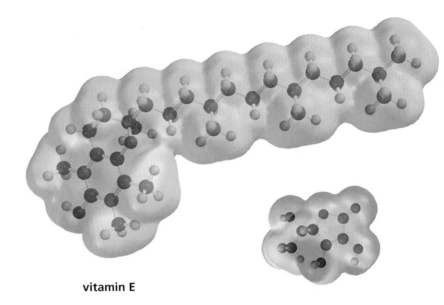

vitamin E

vitamin C

Y ou have learned that there are three classes of hydrocarbons—*alkanes*, which contain only single bonds; *alkenes*, which contain double bonds; and *alkynes*, which contain triple bonds. So far we have studied the chemistry of alkenes in Chapter 3 and alkynes in Chapter 5. Now we will take a look at the chemistry of alkanes.

Alkanes are called **saturated hydrocarbons** because they are saturated with hydrogen. In other words, they do not contain any double or triple bonds. A few examples of alkanes are shown here. Their nomenclature is discussed in Section 2.2.

$CH_3CH_2CH_2CH_3$
butane

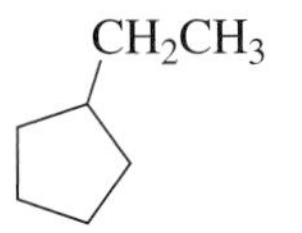

CH_2CH_3
ethylcyclopentane

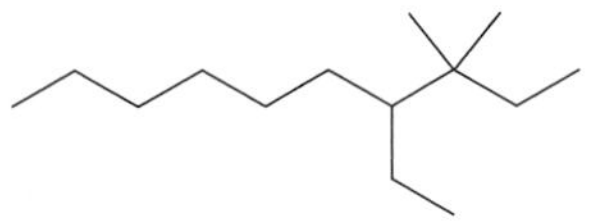

4-ethyl-3,3-dimethyldecane

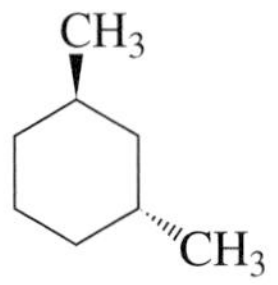

CH_3
CH_3
trans-1,3-dimethyl-cyclohexane

Alkanes are widely distributed both on Earth and on other planets. The atmospheres of Jupiter, Saturn, Uranus, and Neptune contain large quantities of methane (CH_4), the smallest alkane, which is an odorless and flammable gas. In fact, the blue colors of Uranus and Neptune are due to methane in their atmospheres. Alkanes on Earth are found in natural gas and petroleum. Natural gas and petroleum are formed by the decomposition of plant and animal material that has been buried for long periods of time in Earth's crust, an environment with little oxygen. Natural gas and petroleum, therefore, are known as *fossil fuels*.

Natural gas consists of about 75% methane. The remaining 25% is composed of small alkanes such as ethane, propane, and butane. In the 1950s, natural gas replaced coal as the main energy source for domestic and industrial heating in many parts of the United States.

Petroleum is a complex mixture of alkanes and cycloalkanes that can be separated by distillation into fractions. The fraction that boils at the lowest temperature (hydrocarbons containing three and four carbons) is a gas that can be liquefied under pressure. This gas is used as a fuel for cigarette lighters, camp stoves, and barbecues. The fraction that boils at the next higher temperature (hydrocarbons containing 5 to 11 carbons) is gasoline; the next fraction (9 to 16 carbons) includes kerosene and jet fuel. The fraction with 15 to 25 carbons is used for heating oil and diesel oil, and the highest boiling fraction is used for lubricants and greases. The nonpolar nature of these compounds is what gives them their oily feel. After distillation, a nonvolatile residue is left behind, which we call asphalt or tar.

The 5- to 11-carbon fraction that is used for gasoline is, in fact, a poor fuel for internal combustion engines and requires a process known as catalytic cracking to become a high performance gasoline. Catalytic cracking converts straight-chain hydrocarbons that are poor fuels into branched-chain compounds that are high performance fuels. Originally, cracking (also called pyrolysis) involved heating gasoline to very high temperatures in order to obtain hydrocarbons with three to five carbons. Modern cracking methods use catalysts to accomplish the same thing at much lower temperatures. The small hydrocarbons are then catalytically recombined to form highly branched hydrocarbons.

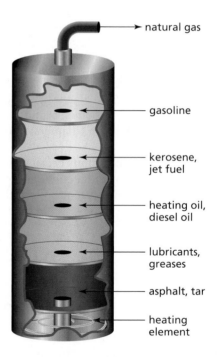

When poor fuels are used in an engine, combustion can be initiated before the spark plug fires. This produces a "pinging" or "knocking" in the running engine. As the quality of a fuel improves, the engine is less likely to knock. The quality of a fuel is indicated by its octane number. Thus, 92-octane fuel is higher quality than 87-octane fuel.

Straight-chain hydrocarbons have low octane numbers and make poor fuels. Heptane, for example, has an arbitrarily assigned octane number of 0, and causes engines to knock badly. Branched-chain fuels burn more slowly, thereby reducing knocking, because they have more primary hydrogens. Consequently, branched-chain alkanes have high octane numbers. 2,2,4-Trimethylpentane, for example, does not cause knocking and has arbitrarily been assigned an octane number of 100.

$$CH_3CH_2CH_2CH_2CH_2CH_2CH_3$$
heptane
octane number = 0

$$\begin{array}{cc} CH_3 & CH_3 \\ | & | \\ CH_3CCH_2CHCH_3 \\ | \\ CH_3 \end{array}$$
2,2,4-trimethylpentane
octane number = 100

The octane number of a gasoline is determined by comparing its knocking tendency to the knocking tendency of mixtures of heptane and 2,2,4-trimethylpentane. The octane number given to the gasoline corresponds to the percent of 2,2,4-trimethylpentane in the matching mixture. The fact that 2,2,4-trimethylpentane contains eight carbons explains the origin of the term "octane number."

FOSSIL FUELS: A PROBLEMATIC ENERGY SOURCE

We face three major problems as a consequence of our dependence on fossil fuels for energy. First, fossil fuels are a nonrenewable resource and the world's supply is continually decreasing. Second, a group of Middle Eastern and South American countries control a large portion of the world's supply of petroleum. These countries have formed a cartel known as the Organization of Petroleum Exporting Countries (OPEC), which controls both the supply and the price of crude oil. Any political instability in these countries, such as the Persian Gulf War in 1991, can seriously affect the world oil supply. Third, burning fossil fuels increases the con-

centrations of CO_2 and SO_2 in the atmosphere. Scientists have established experimentally that atmospheric SO_2 causes "acid rain," a threat to the Earth's plants and therefore to our food and oxygen supplies. Some scientists predict that the CO_2 currently being released into the atmosphere will cause an increase in the Earth's temperature as a result of the absorption of infrared radiation by the CO_2 (the so-called "greenhouse effect"). A steady increase in the Earth's temperature can have devastating consequences, including the creation of new deserts, the melting of polar ice caps (with a concomitant rise in sea level), and starvation as a result of crop failure. Clearly, what we need is a renewable, nonpolitical, nonpolluting, and economically affordable source of energy.

8.1 REACTIVITY CONSIDERATIONS

The double and triple bonds of alkenes and alkynes are composed of strong σ bonds and relatively weak π bonds. We have seen that the reactivity of alkenes and alkynes is the result of an electrophile being attracted to the cloud of electrons that constitutes the π bond.

Alkanes contain only strong σ bonds. Because the carbon and hydrogen atoms of an alkane have approximately the same electronegativity, the electrons in the carbon–hydrogen and carbon–carbon σ bonds are shared equally by the bonding atoms. Consequently, none of the atoms in an alkane have any significant charge. This means that neither nucleophiles nor electrophiles are attracted to them. As a result of having only strong σ bonds and having atoms without any partial charge, alkanes are very unreactive compounds. Because of their lack of reactivity, early organic chemists called them **paraffins** from the Latin *parum affinis,* which means "little affinity" (for other compounds).

8.2 CHLORINATION AND BROMINATION OF ALKANES

Alkanes react with molecular chlorine (Cl_2) or bromine (Br_2) to form alkyl chlorides or alkyl bromides. These reactions take place only at high temperatures or in the presence of light (symbolized by $h\nu$). These are the only reactions that alkanes undergo—with the exception of **combustion,** a reaction with oxygen that takes place at high temperatures and converts alkanes to carbon dioxide and water.

$$CH_4 + Cl_2 \xrightarrow[\substack{or \\ h\nu}]{\Delta} \underset{\text{methyl chloride}}{CH_3Cl} + HCl$$

$$CH_3CH_3 + Br_2 \xrightarrow[\substack{or \\ h\nu}]{\Delta} \underset{\text{ethyl bromide}}{CH_3CH_2Br} + HBr$$

The mechanism for the halogenation of an alkane is well understood. The high temperature (or light) supplies the energy required to break the chlorine–chlorine or bromine–bromine bond homolytically. **Homolytic bond cleavage** is the **initiation step** of the reaction because it creates the radical that is used in the first propagation step (Section 3.18). Remember that an arrowhead with one barb signifies the movement of one electron (Section 3.6).

homolytic cleavage

$$:\overset{..}{\underset{..}{Cl}}-\overset{..}{\underset{..}{Cl}}: \xrightarrow[\substack{or \\ h\nu}]{\Delta} 2\ :\overset{..}{\underset{..}{Cl}}\cdot$$

$$:\overset{..}{\underset{..}{Br}}-\overset{..}{\underset{..}{Br}}: \xrightarrow[\substack{or \\ h\nu}]{\Delta} 2\ :\overset{..}{\underset{..}{Br}}\cdot$$

A **radical** (often called a **free radical**) is a species containing an atom with an unpaired electron. A radical is highly reactive because it wants to acquire an electron and thereby complete its octet. In the mechanism for the monochlorination of methane, the chlorine radical formed in the initiation step abstracts a hydrogen atom from methane, forming HCl and a methyl radical. The methyl radical abstracts a chlorine atom from Cl_2, forming methyl chloride and another chlorine radical, which can abstract a hydrogen atom from another molecule of methane. These two steps are called **propagation steps** because the radical created in the first propagation step reacts in the second propagation step to produce a radical that can repeat the first propagation step. Thus, the two propagation steps are repeated over and over. The first propagation step is the rate-determining step of the overall reaction. Because the reaction has radical intermediates and repeating propagation steps, it is called a **radical chain reaction.**

3-D Molecule:
Methyl radical

mechanism for the monochlorination of methane

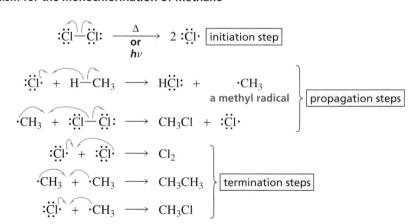

Any two radicals in the reaction mixture can combine to form a molecule in which all the electrons are paired. The combination of two radicals is called a **termination step** because it brings the reaction to an end by decreasing the number of radicals available to propagate the reaction. Radical chlorination of alkanes other than methane follows the same mechanism. A radical chain reaction, with its characteristic initiation, propagation, and termination steps, was first described in Section 3.18.

The reaction of an alkane with chlorine or bromine to form an alkyl halide is called a **radical substitution reaction** because radicals are involved as intermediates, and the end result is the substitution of a halogen atom for one of the hydrogen atoms of the alkane.

In order to maximize the amount of monohalogenated product obtained, a radical substitution reaction should be carried out in the presence of excess alkane. The presence of excess alkane in the reaction mixture ensures that there is a greater probability of the halogen radical colliding with a molecule of alkane than with a molecule of alkyl halide. This is true even toward the end of the reaction, by which time a considerable amount of alkyl halide will have been formed. If the halogen radical abstracts a hydrogen from a molecule of alkyl halide rather than from a molecule of alkane, a dihalogenated product will be obtained.

Tutorial: Radical chain reaction

$$Cl\cdot \ + \ CH_3Cl \ \longrightarrow \ \cdot CH_2Cl \ + \ HCl$$

$$\cdot CH_2Cl \ + \ Cl_2 \ \longrightarrow \ CH_2Cl_2 \ + \ Cl\cdot$$

Bromination of alkanes follows the same mechanism as chlorination. The only difference is that chlorination produces alkyl chlorides, whereas bromination forms alkyl bromides.

mechanism for the monobromination of ethane

$$Br_2 \ \xrightarrow{\Delta} \ 2 \ Br\cdot \qquad \boxed{\text{initiation step}}$$

$$Br\cdot \ + \ CH_3CH_3 \ \longrightarrow \ CH_3\dot{C}H_2 \ + \ HBr$$

$$CH_3\dot{C}H_2 \ + \ Br_2 \ \longrightarrow \ CH_3CH_2Br \ + \ Br\cdot \qquad \left.\right\} \ \boxed{\text{propagation steps}}$$

3-D Molecule: Ethyl radical

$$Br\cdot \ + \ Br\cdot \ \longrightarrow \ Br_2$$

$$CH_3\dot{C}H_2 \ + \ CH_3\dot{C}H_2 \ \longrightarrow \ CH_3CH_2CH_2CH_3 \ \left.\right\} \ \boxed{\text{termination steps}}$$

$$CH_3\dot{C}H_2 \ + \ Br\cdot \ \longrightarrow \ CH_3CH_2Br$$

PROBLEM 1

Show the initiation, propagation, and termination steps for the monochlorination of cyclohexane.

PROBLEM 2

Write the mechanism for the formation of carbon tetrachloride, CCl_4, from the reaction of methane with $Cl_2 + h\nu$.

PROBLEM 3 / SOLVED

If cyclopentane reacts with excess Cl_2 at a high temperature, how many dichlorocyclopentanes would you expect to obtain as products?

Solution   Seven dichlorocyclopentanes would be obtained as products. Only one isomer is possible for the 1,1-dichloro compound. The 1,2- and 1,3-dichloro compounds have two chirality centers. Each has three stereoisomers—the cis isomer is a meso compound and the trans isomer is a pair of enantiomers.

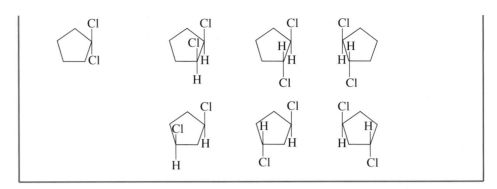

8.3 FACTORS THAT DETERMINE PRODUCT DISTRIBUTION

Two different alkyl halides are obtained from the monochlorination of butane. Substitution of a hydrogen bonded to one of the terminal carbons produces 1-chlorobutane, while substitution of a hydrogen bonded to one of the internal carbons forms 2-chlorobutane.

$$CH_3CH_2CH_2CH_3 + Cl_2 \xrightarrow{h\nu} CH_3CH_2CH_2CH_2Cl + CH_3CH_2\overset{Cl}{\underset{}{C}}HCH_3 + HCl$$

butane   1-chlorobutane   2-chlorobutane

expected = 60%   expected = 40%
experimental = 29%   experimental = 71%

The expected distribution of products is 60% 1-chlorobutane and 40% 2-chlorobutane because six of butane's ten hydrogens can be substituted to form 1-chloro-butane, whereas only four can be substituted to form 2-chlorobutane. This assumes, however, that all of the carbon–hydrogen bonds in butane are equally easy to break, so the relative amounts of the two products would depend only on probability. In other words, the probability of a chlorine radical colliding with a primary hydrogen compared with a secondary hydrogen is 6:4. When we carry out the reaction in the laboratory and analyze the product, however, we find that the product is 29% 1-chlorobutane and 71% 2-chlorobutane. Therefore, probability alone does not determine the relative amounts of the two products. Because more 2-chlorobutane is obtained than expected and the rate-determining step of the overall reaction is abstraction of a hydrogen atom, we can conclude that it must be easier to abstract a hydrogen atom from a secondary carbon than from a primary carbon.

Alkyl radicals have different stabilities (Section 3.18), and the more stable the radical, the more easily it is formed. Thus it is easier to remove a hydrogen atom from a secondary carbon to form a secondary radical, than it is to remove a hydrogen atom from a primary carbon to form a primary radical.

relative stabilities of alkyl radicals

$$\underset{\substack{\text{tertiary radical}}}{R-\overset{R}{\underset{R}{C}}\cdot} \quad > \quad \underset{\substack{\text{secondary radical}}}{R-\overset{R}{\underset{H}{C}}\cdot} \quad > \quad \underset{\substack{\text{primary radical}}}{R-\overset{H}{\underset{H}{C}}\cdot} \quad > \quad \underset{\substack{\text{methyl radical}}}{H-\overset{H}{\underset{H}{C}}\cdot}$$

← increasing stability

When a chlorine radical reacts with butane, it can abstract a hydrogen atom from an internal carbon, thereby forming a secondary alkyl radical, or it can abstract a hydrogen atom from a terminal carbon, thereby forming a primary alkyl radical. Because it is easier to form the more stable secondary alkyl radical, 2-chlorobutane is formed faster than 1-chlorobutane.

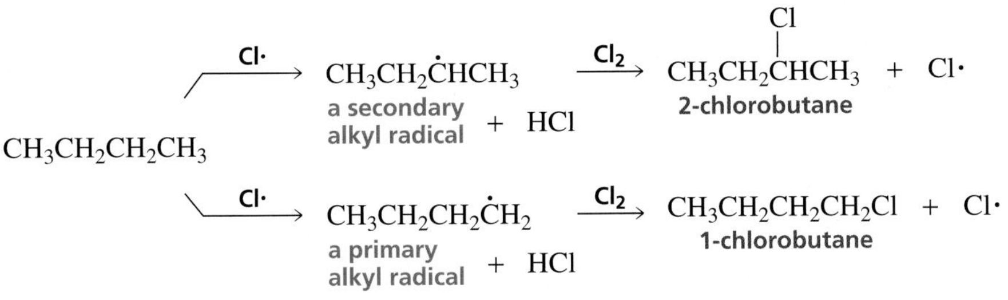

$$CH_3CH_2CH_2CH_3$$

$$\xrightarrow{Cl\cdot} \quad CH_3CH_2\dot{C}HCH_3 \quad \xrightarrow{Cl_2} \quad CH_3CH_2\overset{\overset{\displaystyle Cl}{|}}{C}HCH_3 \;+\; Cl\cdot$$
a secondary alkyl radical + HCl   2-chlorobutane

$$\xrightarrow{Cl\cdot} \quad CH_3CH_2CH_2\dot{C}H_2 \quad \xrightarrow{Cl_2} \quad CH_3CH_2CH_2CH_2Cl \;+\; Cl\cdot$$
a primary alkyl radical + HCl   1-chlorobutane

After experimentally determining the amount of each chlorination product obtained from various hydrocarbons, chemists were able to calculate that *at room temperature* it is 5.0 times easier for a chlorine radical to abstract a hydrogen atom from a tertiary carbon than from a primary carbon, and it is 3.8 times easier to abstract a hydrogen atom from a secondary carbon than from a primary carbon. (See Problem 16.) The precise ratios differ at different temperatures.

relative rates of alkyl radical formation by a chlorine radical at room temperature

tertiary > secondary > primary
5.0   3.8   1.0

⬅ increasing rate of formation

To determine the relative amounts of products obtained from radical chlorination of an alkane, both *probability* (the number of hydrogens that can be abstracted that will lead to the formation of the particular product) and *reactivity* (the relative rate at which a particular hydrogen is abstracted) must be taken into account. When both factors are considered, the calculated amounts of 1-chlorobutane and 2-chlorobutane agree with the amounts obtained experimentally.

relative amount of 1-chlorobutane

number of hydrogens × reactivity
$6 \times 1.0 = 6.0$

percent yield $= \dfrac{6.0}{21} = 29\%$

relative amount of 2-chlorobutane

number of hydrogens × reactivity
$4 \times 3.8 = 15$

percent yield $= \dfrac{15}{21} = 71\%$

The percent yield of each product is calculated by dividing the relative amount of the particular product by the sum of the relative amounts of all products ($6 + 15 = 21$).

Radical monochlorination of 2,2,5-trimethylhexane results in the formation of five monochlorination products. Because the relative amounts of the five products total 35 ($9.0 + 7.6 + 7.6 + 5.0 + 6.0 = 35$), the expected percent of each product can be calculated as shown here.

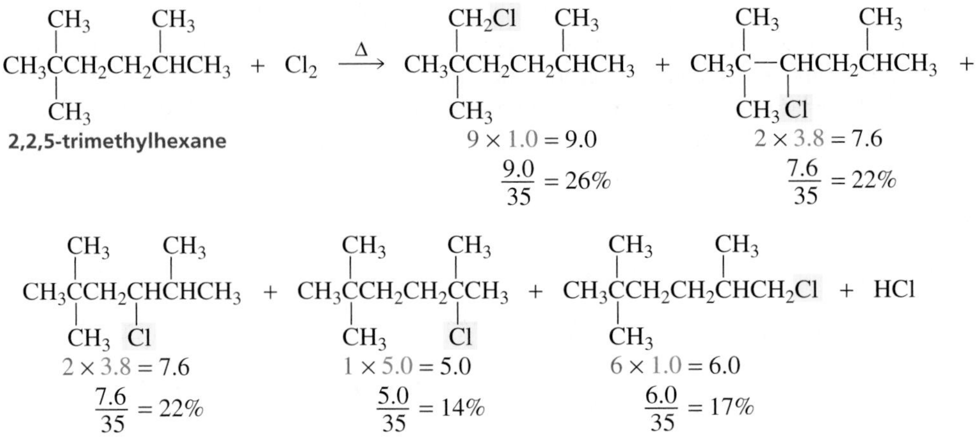

$$CH_3CCH_2CH_2CHCH_3 + Cl_2 \xrightarrow{\Delta} CH_3CCH_2CH_2CHCH_3 + CH_3C-CHCH_2CHCH_3 +$$

2,2,5-trimethylhexane

$9 \times 1.0 = 9.0$

$\dfrac{9.0}{35} = 26\%$

$2 \times 3.8 = 7.6$

$\dfrac{7.6}{35} = 22\%$

$$CH_3CCH_2CHCHCH_3 + CH_3CCH_2CH_2CCH_3 + CH_3CCH_2CH_2CHCH_2Cl + HCl$$

$2 \times 3.8 = 7.6$

$\dfrac{7.6}{35} = 22\%$

$1 \times 5.0 = 5.0$

$\dfrac{5.0}{35} = 14\%$

$6 \times 1.0 = 6.0$

$\dfrac{6.0}{35} = 17\%$

Because radical halogenation of an alkane can yield several different monosubstitution products as well as products that contain more than one halogen atom, it is not a good method to use when you want to synthesize an alkyl halide. Addition of HCl or HBr to an alkene (Section 3.9) or conversion of an alcohol to an alkyl halide (a reaction we will study in Chapter 11) is a much better way to make an alkyl halide. Nevertheless, radical halogenation of an alkane is still a useful reaction because it is the only way to convert an inert alkane into a reactive compound. In Chapter 9 we will see that once the halogen is introduced into the alkane, it can be replaced by a variety of other substituents.

PROBLEM 4

When 2-methylpropane is monochlorinated in the presence of light at room temperature, 36% of the product is 2-chloro-2-methylpropane and 64% is 1-chloro-2-methylpropane. From these data, calculate how much easier it is to abstract a hydrogen atom from a tertiary carbon than from a primary carbon under these conditions.

PROBLEM 5 ◆

How many alkyl halides can be obtained from monochlorination of the following alkanes? Neglect stereoisomers.

a. $CH_3CH_2CH_2CH_2CH_3$

b. 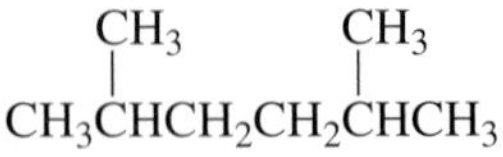 $CH_3CHCH_2CH_2CHCH_3$ (with CH_3 groups)

c. 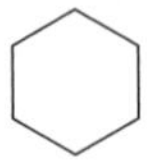 (cyclohexane)

d. 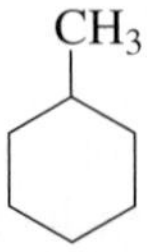 (methylcyclohexane, CH_3)

e. 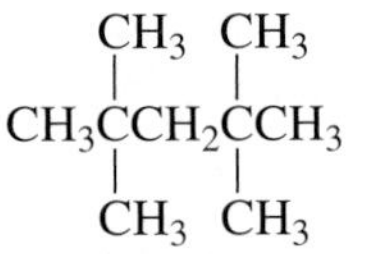 $CH_3CCH_2CCH_3$ (with CH_3, CH_3, CH_3, CH_3 groups)

f. 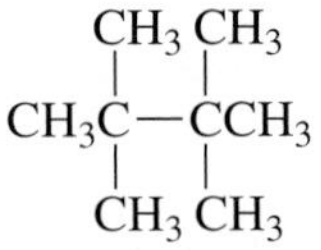 CH_3C-CCH_3 (with CH_3, CH_3, CH_3, CH_3 groups)

g. 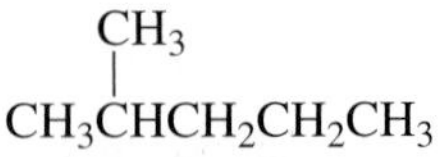 $CH_3CHCH_2CH_2CH_3$ (with CH_3)

h. 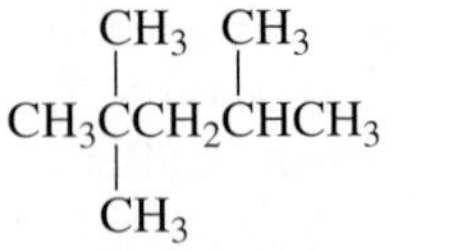 $CH_3CCH_2CHCH_3$ (with CH_3, CH_3, CH_3 groups)

PROBLEM 6 ◆

Calculate the percent of each product that you would expect to obtain in Problem 5a, b, and g if chlorination were carried out in the presence of light at room temperature.

The relative rates of radical formation when a bromine radical is used to abstract a hydrogen atom are very different from the relative rates of radical formation when a chlorine radical is used. At 125 °C, a bromine radical removes a hydrogen atom from a tertiary carbon 1600 times faster than from a primary carbon, and removes a hydrogen atom from a secondary carbon 82 times faster than from a primary carbon.

8.4
THE REACTIVITY–SELECTIVITY PRINCIPLE

relative rates of radical formation by a bromine radical at 125 °C

$$\text{tertiary} \quad > \quad \text{secondary} \quad > \quad \text{primary}$$
$$1600 \qquad\qquad 82 \qquad\qquad 1$$

increasing rate of formation

When a bromine radical is the hydrogen-abstracting agent, the differences in reactivity are so great that the reactivity factor is vastly more important than the probability factor. For example, radical bromination of butane gives a 98% yield of 2-bromobutane compared with the 71% yield of 2-chlorobutane obtained when butane is chlorinated (Section 8.3).

$$CH_3CH_2CH_2CH_3 \ + \ Br_2 \ \xrightarrow{h\nu} \ CH_3CH_2CH_2CH_2Br \ + \ CH_3CH_2\overset{\displaystyle Br}{\underset{\displaystyle |}{C}}HCH_3 \ + \ HBr$$

1-bromobutane   **2-bromobutane**
        2%   98%

Similarly, bromination of 2,2,5-trimethylhexane gives an 82% yield of the product in which bromine replaces the tertiary hydrogen. Chlorination of the same alkane results in a 14% yield of the tertiary alkyl chloride (Section 8.3).

$$CH_3\overset{\displaystyle CH_3}{\underset{\displaystyle CH_3}{C}}CH_2CH_2\overset{\displaystyle CH_3}{\underset{}{C}}HCH_3 \ + \ Br_2 \ \xrightarrow{h\nu} \ CH_3\overset{\displaystyle CH_3}{\underset{\displaystyle CH_3}{C}}CH_2CH_2\overset{\displaystyle CH_3}{\underset{\displaystyle Br}{C}}CH_3 \ + \ HBr$$

2,2,5-trimethylhexane   **2-bromo-2,5,5-trimethylhexane**
                                              82%

PROBLEM 7 ◆

Carry out the calculations that predict that:

a. 2-bromobutane will be obtained in 98% yield.

b. 2-bromo-2,5,5,-trimethylhexane will be obtained in 82% yield.

Tutorial: Radical bromination

Why are the relative rates of radical formation so different when a bromine radical is used as the hydrogen-abstracting reagent rather than a chlorine radical? To answer this question, we must compare the $\Delta H°$ values for the formation of primary, secondary, and tertiary radicals from reaction with a chlorine radical and from reaction with a bromine radical. These $\Delta H°$ values can be calculated using the bond dissociation energies in Table 3.1 on p. 130. (Remember that $\Delta H°$ is equal to the energy of the bond being broken minus the energy of the bond being formed.)

		$\Delta H°$ (kcal/mol)	$\Delta H°$ (kJ/mol)
Cl· + CH₃CH₂CH₃ ⟶ CH₃CH₂ĊH₂ + HCl		$101 - 103 = -2$	-8
Cl· + CH₃CH₂CH₃ ⟶ CH₃ĊHCH₃ + HCl		$99 - 103 = -4$	-17
Cl· + CH₃CHCH₃ (CH₃) ⟶ CH₃ĊCH₃ (CH₃) + HCl		$97 - 103 = -6$	-25

		$\Delta H°$ (kcal/mol)	$\Delta H°$ (kJ/mol)
Br· + CH₃CH₂CH₃ ⟶ CH₃CH₂ĊH₂ + HBr		$101 - 87 = 14$	59
Br· + CH₃CH₂CH₃ ⟶ CH₃ĊHCH₃ + HBr		$99 - 87 = 12$	50
Br· + CH₃CHCH₃ (CH₃) ⟶ CH₃ĊCH₃ (CH₃) + HBr		$97 - 87 = 10$	42

We must also be aware that bromination is much slower than chlorination. The activation energy for abstraction of a hydrogen atom by a bromine radical has been found experimentally to be about 4.5 times greater than that for abstraction of a hydrogen atom by a chlorine radical. Using the calculated $\Delta H°$ values and the experimental activation energies, we can draw reaction coordinate diagrams for the formation of primary, secondary, and tertiary radicals by means of abstraction of a hydrogen atom by a chlorine radical and by a bromine radical. These diagrams are shown in Figure 8.1.

Because the reaction of a chlorine radical with an alkane to form a primary, secondary, or tertiary radical is exothermic, the transition states resemble the reactants more

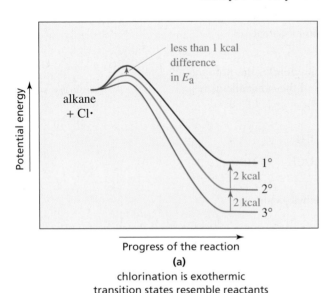

(a)
chlorination is exothermic
transition states resemble reactants

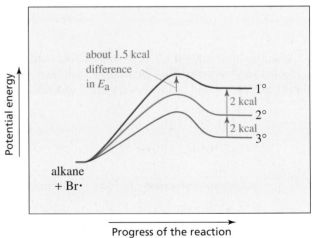

(b)
bromination is endothermic
transition states resemble products

▲ **Figure 8.1**
(a) Reaction coordinate diagrams for the formation of primary, secondary, and tertiary alkyl radicals as a result of abstraction of a hydrogen atom by a chlorine radical. The transition states have relatively little radical character because they resemble the reactants.
(b) Reaction coordinate diagrams for the formation of primary, secondary, and tertiary alkyl radicals as a result of abstraction of a hydrogen atom by a bromine radical. The transition states have a relatively high degree of radical character because they resemble the products.

than the products (the Hammond postulate, Section 3.11). The reactants all have approximately the same energy, so there will be only a small difference in the activation energies for removal of a hydrogen atom from a primary, secondary, or tertiary carbon. In contrast, the reaction of a bromine radical with an alkane is endothermic, so the transition states resemble the products more than the reactants. Because there is a significant difference in the energies of the product radicals, depending on whether they are primary, secondary, or tertiary, there is a significant difference in the activation energies.

Figure 8.1 shows that a chlorine radical makes primary, secondary, and tertiary radicals with almost equal ease, whereas a bromine radical has a clear preference for formation of the easiest-to-form tertiary radical. In other words, because a bromine radical is relatively unreactive, it is very selective about which hydrogen atom it abstracts. In contrast, the much more reactive chlorine radical is much less selective. These observations illustrate the **reactivity–selectivity principle,** which states that *the greater the reactivity of a species, the less selective it will be.*

The more reactive a species is, the less selective it will be.

Because chlorination is relatively nonselective, it is a useful reaction only when there is just one kind of hydrogen in the molecule.

$$
\bigcirc + \text{Cl}_2 \xrightarrow{\ h\nu\ } \overset{\text{Cl}}{\bigcirc} + \text{HCl}
$$

PROBLEM 8 ◆

If 2-methylpropane is brominated at 125 °C in the presence of light, what percent of the product will be 2-bromo-2-methylpropane? Compare your answer with the percent given in Problem 4 for chlorination.

PROBLEM 9 ◆

Take the same alkanes whose percents of monochlorination products you calculated in Problem 6, and calculate what the percents of monobromination products would be if bromination were carried out at 125 °C.

Tutorial: Reactivity-selectivity

By comparing the $\Delta H°$ values for the sum of the two propagating steps for the monohalogenation of methane, we can understand why alkanes undergo chlorination and bromination but not iodination, and why fluorination is too violent a reaction to be useful.

$$
\begin{aligned}
&\text{F·} + \text{CH}_4 \longrightarrow \text{·CH}_3 + \text{HF} && 105 - 136 = -31 \\
&\text{·CH}_3 + \text{F}_2 \longrightarrow \text{CH}_3\text{F} + \text{F·} && \underline{38 - 108 = -70} \\
& && \Delta H° = -101 \text{ kcal/mol} \quad (-423 \text{ kJ/mol})
\end{aligned}
$$

$$
\begin{aligned}
&\text{Cl·} + \text{CH}_4 \longrightarrow \text{·CH}_3 + \text{HCl} && 105 - 103 = 2 \\
&\text{·CH}_3 + \text{Cl}_2 \longrightarrow \text{CH}_3\text{Cl} + \text{Cl·} && \underline{58 - 84 = -26} \\
& && \Delta H° = -24 \text{ kcal/mol} \quad (-100 \text{ kJ/mol})
\end{aligned}
$$

$$
\begin{aligned}
&\text{Br·} + \text{CH}_4 \longrightarrow \text{·CH}_3 + \text{HBr} && 105 - 87 = 18 \\
&\text{·CH}_3 + \text{Br}_2 \longrightarrow \text{CH}_3\text{Br} + \text{Br·} && \underline{46 - 70 = -24} \\
& && \Delta H° = -6 \text{ kcal/mol} \quad (-25 \text{ kJ/mol})
\end{aligned}
$$

$$
\begin{aligned}
&\text{I·} + \text{CH}_4 \longrightarrow \text{·CH}_3 + \text{HI} && 105 - 71 = 34 \\
&\text{·CH}_3 + \text{I}_2 \longrightarrow \text{CH}_3\text{I} + \text{I·} && \underline{36 - 57 = -21} \\
& && \Delta H° = 13 \text{ kcal/mol} \quad (54 \text{ kJ/mol})
\end{aligned}
$$

F₂

Cl₂

Br₂

I₂
Halogens

A fluorine radical is the most reactive of the halogen radicals, and it reacts violently with alkanes ($\Delta H° = -31$ kcal/mol). Because fluorination is so difficult to control, it is not a synthetically useful reaction. In contrast, the iodine radical is the least reactive of the halogen radicals. It is so unreactive ($\Delta H° = 34$ kcal/mol) that it is unable to abstract a hydrogen atom from an alkane. Consequently, it reacts with another iodine radical and re-forms I₂.

PROBLEM-SOLVING STRATEGY

Will chlorination or bromination produce a better yield of 1-halo-1-methylcyclohexane?

In solving this kind of problem, first draw the structures of the compounds being discussed.

$$\underset{\text{hν}}{\overset{\textbf{X}_2}{\longrightarrow}}$$

The compound is a tertiary alkyl halide, so the question becomes: "Will bromination or chlorination produce a better yield of a tertiary alkyl halide?" Because bromination is more selective, it will produce a greater yield of the desired compound. Chlorination will form some of the tertiary alkyl halide, but it will also form significant amounts of primary and secondary alkyl halides.

Now continue on to Problem 10.

PROBLEM 10 ◆

a. Would chlorination or bromination produce a better yield of 1-halo-2,3-dimethylbutane?

b. Would chlorination or bromination produce a better yield of 2-halo-2,3-dimethylbutane?

c. Would chlorination or bromination be a better way to make 1-halo-2,2-dimethylpropane?

8.5 RADICAL SUBSTITUTION OF BENZYLIC AND ALLYLIC HYDROGENS

Benzyl and allyl radicals are more stable than alkyl radicals because their unpaired electrons are delocalized. We have seen that electron delocalization increases the stability of a molecule (Sections 6.6 and 6.8).

We know that the more stable a radical is, the faster it can be formed. This means that a hydrogen bonded to either a benzylic carbon or an allylic carbon will be substituted for preferentially in a halogenation reaction. The percent of substitution at the benzylic or allylic carbon is greater in the case of bromination than in the case of chlorination because a bromine radical is more selective.

relative stabilities of radicals

benzyl radical ≈ allyl radical > tertiary radical > secondary radical > primary radical > vinyl radical ≈ methyl radical

increasing stability

benzylic substituted product

$$CH_2CH_3 \text{ (phenyl)} + X_2 \xrightarrow{\Delta} \overset{X}{\underset{}{CHCH_3}} \text{ (phenyl)} + HX \quad (X = Cl \text{ or } Br)$$

$$CH_3CH{=}CH_2 + X_2 \xrightarrow{\Delta} \overset{X}{\underset{}{CH_2CH{=}CH_2}} + HX$$

allylic substituted product

3-D Molecule:
Benzyl radical

N-Bromosuccinimide (NBS) is frequently used to brominate allylic positions because NBS allows a radical substitution reaction to be carried out without having to use a relatively high concentration of Br_2 that could add to the double bond. Homolytic cleavage of the N—Br bond generates the bromine radical needed to initiate the radical reaction. Light or heat can be used to promote the reaction. The solvent used is generally CH_2Cl_2.

$$\text{(cyclohexene)} + \overset{O}{\underset{O}{\text{N—Br}}} \xrightarrow[CH_2Cl_2]{h\nu \text{ or } \Delta} \overset{Br}{\text{(bromocyclohexene)}} + \overset{O}{\underset{O}{\text{N—H}}}$$

N-bromosuccinimide
NBS

succinimide

When a radical abstracts a hydrogen atom from an allylic carbon, the allylic radical that is formed is stabilized by resonance (it has two resonance contributors). The radical reacts with Br_2 generated in the reaction mixture from the reaction of NBS with HBr. Because the groups attached to the sp^2 carbons are the same in both resonance contributors, only one substitution product is formed.

Tutorial: Common terms in radicals

$$Br\cdot + CH_3CH{=}CH_2 \longrightarrow \dot{C}H_2CH{=}CH_2 \longleftrightarrow CH_2{=}CH\dot{C}H_2$$
$$\downarrow Br_2$$
$$BrCH_2CH{=}CH_2 + Br\cdot$$
3-bromopropene

If, however, the groups attached to the two sp^2 carbons are not the same in both resonance contributors, two substitution products are formed.

3-D Molecule:
Allyl radical

$$Br\cdot + CH_3CH_2CH{=}CH_2 \longrightarrow CH_3\dot{C}HCH{=}CH_2 \longleftrightarrow CH_3CH{=}CH\dot{C}H_2$$
$$\downarrow Br_2$$
$$\overset{Br}{\underset{}{CH_3CHCH{=}CH_2}} + CH_3CH{=}CHCH_2Br + Br\cdot$$
3-bromo-1-butene **1-bromo-2-butene**

PROBLEM 11

Two products are formed when methylenecyclohexane reacts with NBS. Explain how each is formed.

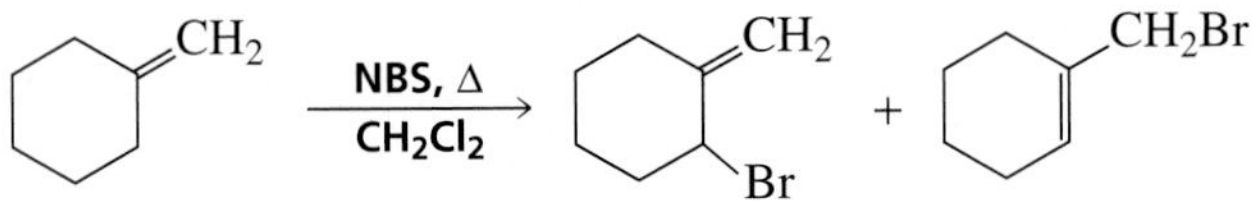

$$\text{(methylenecyclohexane)} \xrightarrow[\text{CH}_2\text{Cl}_2]{\textbf{NBS, } \Delta} \text{(bromo product)} + \text{(CH}_2\text{Br product)}$$

PROBLEM 12 / SOLVED

How many allylic substituted bromoalkenes are formed from the reaction of 2-pentene with NBS? Ignore stereoisomers.

SOLUTION The bromine radical will abstract a secondary allylic hydrogen from C-4 of 2-pentene in preference to a primary allylic hydrogen from C-1. The resonance contributors of the resulting radical intermediate have the same groups attached to the sp^2 carbons, so only one bromoalkene is formed. Because of the high selectivity of the bromine radical, an insignificant amount of radical will be formed by abstraction of a hydrogen from the less reactive primary allylic position.

$$CH_3CH=CHCH_2CH_3 \xrightarrow[\text{CH}_2\text{Cl}_2]{\textbf{NBS, } \Delta} CH_3CH=CH\dot{C}HCH_3 \longleftrightarrow CH_3\dot{C}HCH=CHCH_3$$

$$\downarrow$$

$$CH_3CH=CHCHCH_3$$
$$\overset{|}{\underset{Br}{}}$$

8.6 STEREOCHEMISTRY OF RADICAL SUBSTITUTION REACTIONS

If a radical substitution reaction forms a product with a chirality center, both the R and S enantiomers will be formed.

$$CH_3CH_2CH_2CH_3 \;+\; Br_2 \xrightarrow{\;h\nu\;} CH_3CH_2\underset{\overset{|}{Br}}{\overset{\text{chirality center}}{C}}HCH_3 \;+\; HBr$$

configuration of the product

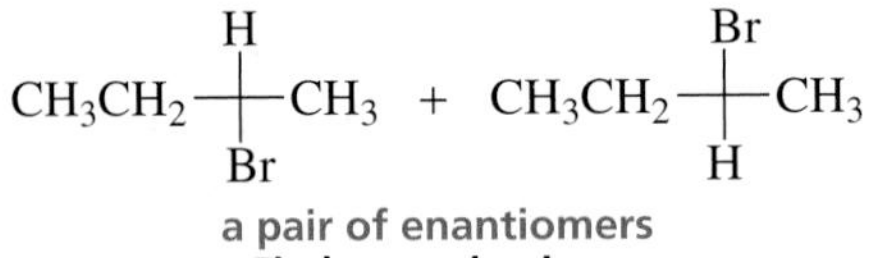

$$CH_3CH_2\overset{H}{\underset{Br}{\rule{0pt}{0pt}}}\!\!-\!\!CH_3 \;+\; CH_3CH_2\overset{Br}{\underset{H}{\rule{0pt}{0pt}}}\!\!-\!\!CH_3$$

a pair of enantiomers
Fischer projections

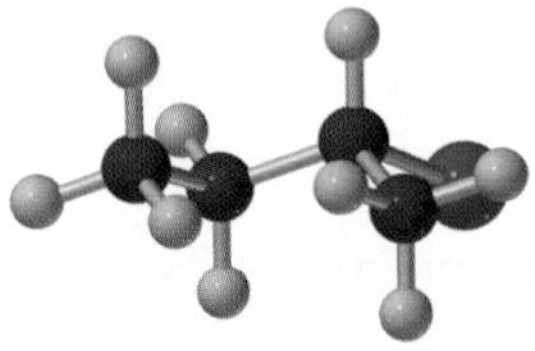

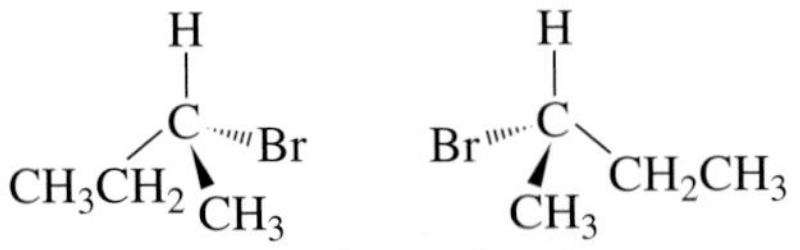

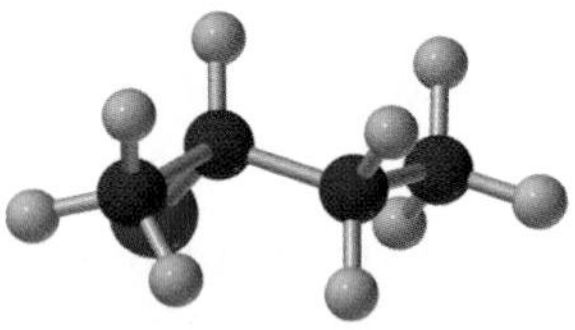

$$CH_3CH_2\overset{H}{\underset{CH_3}{\overset{|}{C}}}{}^{\cdots Br} \qquad Br^{\cdots}\overset{H}{\underset{CH_3}{\overset{|}{C}}}CH_2CH_3$$

a pair of enantiomers
perspective formulas

To understand why both enantiomers are formed, we must look at the propagation steps of the radical substitution reaction. In the first propagation step, the bromine radical removes a hydrogen atom from the alkane, creating a radical intermediate. The carbon bearing the unpaired electron is sp^2 hybridized and, therefore, the three groups to which it is bonded lie in a plane. In the second propagation step, the incoming halogen has equal access to both sides of the plane. As a result, both the *R* and *S* enantiomers are formed, and they are formed in equal amounts.

Notice that the stereochemical outcome of a radical substitution reaction is identical to the stereochemical outcome of a radical addition reaction (Section 4.17). This is because both reactions form a radical intermediate, and it is the reaction of the intermediate that determines the configuration of the products.

What happens if the reactant already has a chirality center and the radical substitution reaction creates a second chirality center? In this case, a pair of diastereomers will be formed, and they will be formed in unequal amounts. Diastereomers are formed because the new chirality center created in the product can have either the *R* or the *S* configuration, but the configuration of the chirality center in the reactant will be unchanged in the product because none of the bonds to that chirality center is broken during the course of the reaction.

a radical intermediate

3-D Molecule:
sec-Butyl radical

stereochemistry of the product

a pair of diastereomers
Fischer projections

a pair of diastereomers
perspective formulas

If the original chirality center is near the carbon bearing the unpaired electron, more of one diastereomer will be formed than the other because the incoming Br_2 will have greater access to one side of the radical intermediate than to the other due to the presence of the original chirality center.

Br_2 has greater access to "the back" of the radical

8.7
REACTIONS OF
CYCLIC COMPOUNDS

Cyclic compounds undergo the same reactions acyclic compounds undergo. For example, cyclic alkanes, like acyclic alkanes, undergo radical substitution reactions with chlorine or bromine.

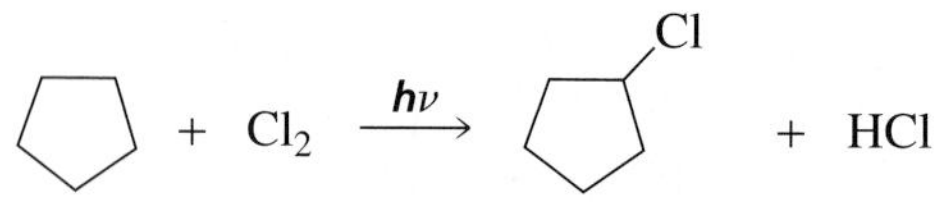

chlorocyclopentane

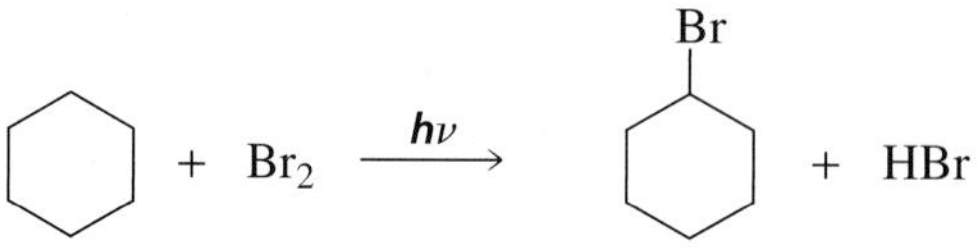

bromocyclohexane

Cyclic alkenes undergo the same reactions as acyclic alkenes.

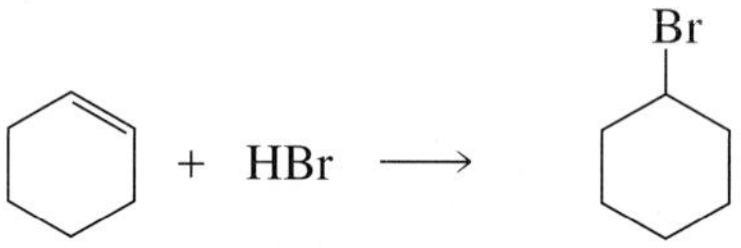

bromocyclohexane

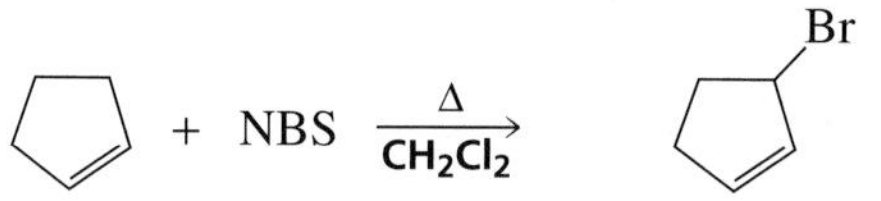

3-bromocyclopentene

In other words, the chemistry of a compound usually depends solely on its functional group, not on whether it is cyclic or acyclic.

PROBLEM 13 ◆

a. Give the major product(s) of the reaction of 1-methylcyclohexene with the following reagents, ignoring stereoisomers:

   1. $NBS/\Delta/CH_2Cl_2$ **2.** Br_2/CH_2Cl_2 **3.** HBr   **4.** HBr/peroxide

b. Give the configuration of the products.

Cyclopropane is one notable exception to the generalization that cyclic and acyclic compounds undergo the same reactions. Although it is an alkane, cyclopropane undergoes electrophilic addition reactions as if it were an alkene.

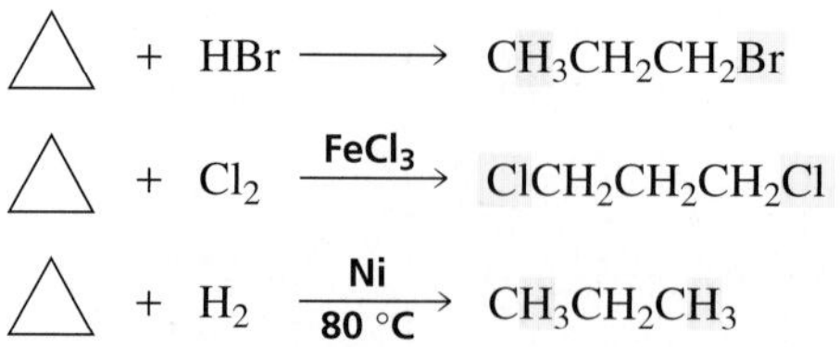

Cyclopropane is more reactive than propene toward addition of acids such as HBr and HCl but is less reactive toward addition of Cl_2 and Br_2, so a Lewis acid ($FeCl_3$) is needed to catalyze halogen addition (Section 1.21).

It is the strain in the small ring that makes it possible for cyclopropane to undergo electrophilic addition reactions (Section 2.11). Because of the 60° bond angles in the three-membered ring, the sp^3 orbitals cannot overlap head-on, which decreases the effectiveness of the orbital overlap. Thus, the bonds are considerably weaker than normal carbon–carbon σ bonds (Figure 2.6). Consequently, three-membered rings undergo ring-opening reactions with electrophilic reagents.

$$\triangle \;+\; \text{XY} \;\longrightarrow\; \underset{\overset{|}{X}\quad\overset{|}{Y}}{H_2C \overset{\displaystyle CH_2}{\diagdown}\; CH_2}$$

3-D Molecule:
Cyclopropyl radical

There is less angle strain in cyclobutane than in cyclopropane (Section 2.11). Therefore, the sp^3 orbitals in cyclobutane overlap more effectively, so the σ bonds are stronger. Cyclobutane, therefore, does *not* undergo electrophilic addition reactions. It behaves as a normal cycloalkane except that it will undergo a ring-opening reaction with hydrogen and a metal catalyst. However, a higher temperature is needed for the hydrogenation of cyclobutane than for the hydrogenation of cyclopropane (200 °C vs. 80 °C).

$$\square \;+\; H_2 \;\xrightarrow[\textbf{200 °C}]{\textbf{Ni}}\; CH_3CH_2CH_2CH_3$$
$$\text{cyclobutane} \qquad\qquad\qquad \text{butane}$$

Cycloalkanes with more than four carbons in the ring do not undergo electrophilic addition reactions because they have considerably less angle strain than four-membered rings (Table 2.9 in Section 2.12).

8.8 RADICAL REACTIONS IN BIOLOGICAL SYSTEMS

For a long time scientists assumed that radical reactions were not important in biological systems because a large amount of energy is required to initiate a radical reaction and it is difficult to control the repeating propagation steps of the chain reaction once initiation occurs. However, it is now widely recognized that there are biological reactions that involve radicals. Instead of being generated by heat or light, the radicals are formed by the interaction of organic molecules with metal ions. These radical reactions take place in the active site of an enzyme. Containing the reaction in a specific site allows the reaction to be controlled.

One radical reaction that takes place in a biological system is the conversion of toxic hydrocarbons to nontoxic alcohols. The hydroxylation of the hydrocarbon is carried out in the liver, catalyzed by an iron-porphyrin-containing enzyme called cytochrome P_{450} (Section 11.7). An alkyl radical intermediate is created when $Fe^{V}O$ abstracts a hydrogen atom from an alkane. In the next step, $Fe^{IV}OH$ dissociates homolytically into Fe^{III} and $HO^{\cdot}$, and the $HO^{\cdot}$ immediately combines with the radical intermediate to form the alcohol.

$$Fe^{V}{=}O \;+\; H{-}\overset{|}{\underset{|}{C}}{-} \;\longrightarrow\; Fe^{IV}{-}OH \;+\; {\cdot}\overset{|}{\underset{|}{C}}{-} \;\longrightarrow\; Fe^{III} \;+\; HO{-}\overset{|}{\underset{|}{C}}{-}$$
$$\quad\;\;\text{an alkane} \qquad\qquad\qquad \text{a radical} \qquad\qquad\qquad \text{an alcohol}$$
$$\qquad\qquad\qquad\qquad\qquad\qquad \text{intermediate}$$

This reaction can also have the opposite toxicological effect. That is, instead of converting a toxic hydrocarbon into a nontoxic alcohol, substituting an OH for an H in some compounds causes a nontoxic compound to become toxic. Therefore, compounds that are nontoxic *in vitro* are not necessarily nontoxic *in vivo*. For example, studies done on animals showed that substituting an OH for an H caused methylene chloride (CH_2Cl_2) to become a carcinogen when it is inhaled.

DECAFFEINATED COFFEE AND THE CANCER SCARE

Animal studies showing that methylene chloride becomes a carcinogen when inhaled caused some concern because methylene chloride was the solvent used to extract caffeine from coffee beans in the manufacture of decaffeinated coffee. However, when methylene chloride was added to drinking water fed to laboratory rats and mice, researchers found no toxic effects. They observed no toxicological responses of any kind either in rats that had consumed an amount of methylene chloride equivalent to the amount that would be ingested by drinking 120,000 cups of decaffeinated coffee per day or in mice that had consumed an amount equivalent to drinking 4.4 million cups of decaffeinated coffee per day. In addition, no increased risk of cancer was found in a study of thousands of workers exposed daily to inhaled methylene chloride—studies done on humans do not always agree with those done on animals. Because of the initial concern, however, researchers sought alternative methods for extracting caffeine from coffee beans. It turned out that extraction by CO_2 at supercritical temperature and pressure is not only safer but better. This method is better because it extracts caffeine without simultaneously extracting some of the flavor compounds that are removed when methylene chloride is used. There is essentially no difference in flavor between regular coffee and coffee decaffeinated with CO_2.

Another important biological reaction shown to involve a radical intermediate is the conversion of a ribonucleotide into a deoxyribonucleotide. The biosynthesis of ribonucleic acid (RNA) requires ribonucleotides, whereas the biosynthesis of deoxyribonucleic acid (DNA) requires deoxyribonucleotides (Section 25.1). The first step in the conversion of a ribonucleotide to the deoxyribonucleotide needed for DNA biosynthesis involves abstraction of a hydrogen atom from the ribonucleotide to form a radical intermediate.

RO \quad\quad\quad several steps \quad\quad\quad RO

HO OH \quad\quad\quad HO OH \quad\quad\quad HO H
a ribonucleotide \quad a radical intermediate \quad a deoxyribonucleotide

Unwanted radicals in biological systems must be destroyed before they have an opportunity to cause damage to cells. Cell membranes, for example, are susceptible to the same kind of radical reactions that cause butter to become rancid (Section 24.3). Imagine the state of your cell membranes if radical reactions could occur readily. Radical reactions in biological systems also have been implicated in the aging process. Unwanted radical reactions are prevented by **radical inhibitors,** compounds that destroy reactive radicals by creating unreactive radicals or compounds with only paired electrons. Hydroquinone is an example of a radical inhibitor. When hydroquinone traps a radical, it forms semiquinone. Semiquinone is stabilized by resonance and is therefore less reactive than other radicals. Furthermore, semiquinone can trap another radical and form quinone, a compound whose electrons are all paired.

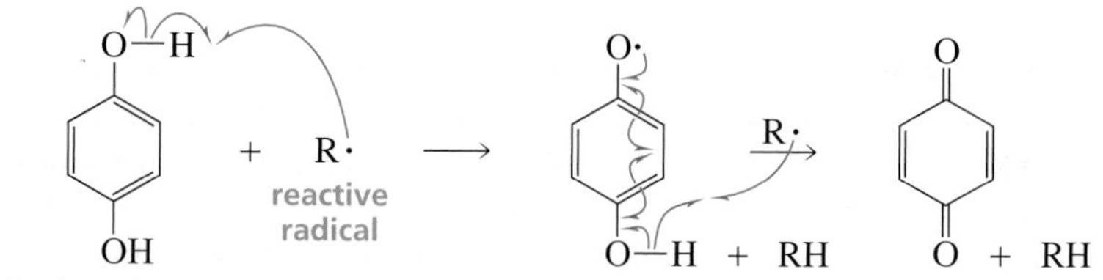

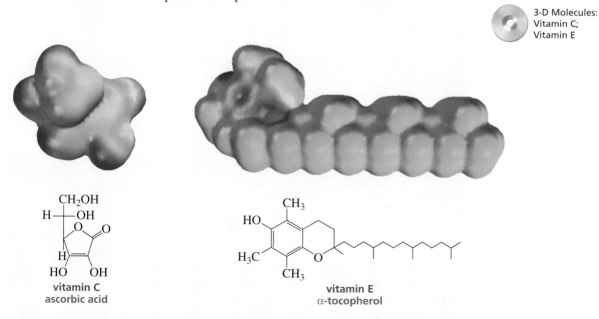

hydroquinone   reactive   semiquinone   quinone
                    radical

FOOD PRESERVATIVES

Radical inhibitors that are present in food are known as preservatives or antioxidants. They preserve food by preventing unwanted radical reactions. Vitamin E is a naturally occurring preservative found in vegetable oil. BHA and BHT are synthetic preservatives that are added to many packaged foods.

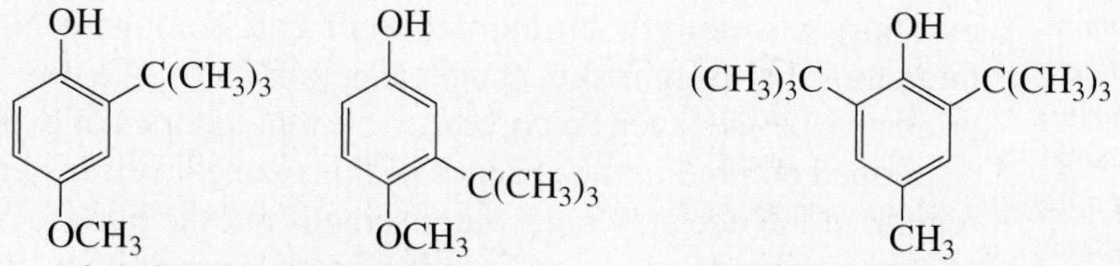

butylated hydroxyanisole
BHA

butylated hydroxytoluene
BHT

food preservatives

Two examples of radical inhibitors that are present in biological systems are vitamin C and vitamin E. Like hydroquinone, they form relatively stable radicals. Vitamin C (also known as ascorbic acid) is a water-soluble compound that traps radicals formed in the aqueous environment of the cell and in blood plasma. Vitamin E (also known as α-tocopherol) is a water-insoluble (fat-soluble) compound that traps radicals formed in nonpolar membranes. Why one vitamin functions in aqueous environments and the other in nonaqueous environments should be apparent from their structures and electrostatic potential maps.

3-D Molecules:
Vitamin C;
Vitamin E

vitamin C
ascorbic acid

vitamin E
α-tocopherol

8.9 RADICALS AND STRATOSPHERIC OZONE

Ozone, a major constituent of smog, is a health hazard at ground level. In the stratosphere, however, a layer of ozone shields the Earth from harmful solar radiation. The greatest concentration of ozone occurs between 12 and 15 miles above the Earth's surface. The layer of ozone is thinnest at the equator and densest towards the poles. Ozone is formed in the atmosphere from the interaction of molecular oxygen with very short-wavelength ultraviolet light.

$$O_2 \xrightarrow{h\nu} O + O$$

$$O + O_2 \longrightarrow O_3$$

*In 1995, the Nobel Prize in chemistry was awarded to **Sherwood Rowland, Mario Molina,** and **Paul Crutzen** for their pioneering work in explaining the chemical processes responsible for depletion of the ozone layer in the stratosphere. Their work demonstrated that human activities could interfere with global processes that support life. This was the first time that a Nobel Prize was presented for work in the environmental sciences.*

The stratospheric ozone layer acts as a filter for biologically harmful ultraviolet radiation that otherwise would reach Earth's surface. Among other effects, high-energy short-wavelength ultraviolet light can damage DNA in skin cells, causing mutations that trigger skin cancer (Section 28.6). We owe our very existence to this protective ozone layer. According to current theories of evolution, life could not have developed on land in the absence of this ozone layer. Instead, life would have had to remain in the ocean, where water screens out the harmful ultraviolet radiation.

Since about 1985, scientists have noted a precipitous drop in stratospheric ozone over Antarctica. This area of ozone depletion, known as the "ozone hole," is unprecedented in the history of ozone observations. Scientists subsequently noted a similar decrease in ozone over Arctic regions, and in 1988 they detected depletion of ozone over the United States for the first time. Three years later, scientists determined that the rate of ozone depletion was two to three times faster than originally anticipated. The recently observed increases in cataracts and genetic damage and reduced plant growth are being blamed on the ultraviolet radiation that is able to penetrate the reduced ozone layer. It has been predicted that erosion of the protective ozone layer will cause an additional 200,000 deaths from skin cancer over the next 50 years.

F. Sherwood Rowland was born in Ohio in 1927. He received a Ph.D. from the University of Chicago and is a professor of chemistry at the University of California, Irvine.

Growth of the Antarctic ozone hole, located mostly over the continent of Antarctica, since 1979. The images were made from data supplied by total ozone mapping spectrometers (TOMS). The color scale depicts the total ozone values in Dobson Units. Lowest ozone densities are represented by purple and blues, highest ozone densities by reds and greens.

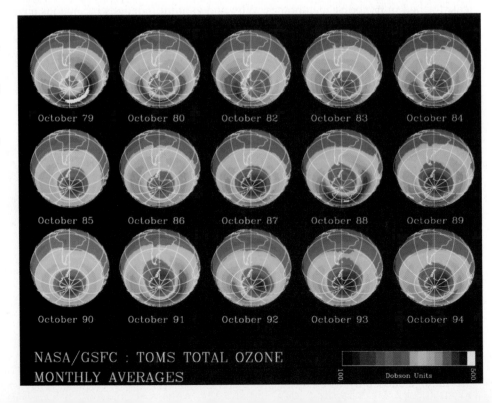

October 79 | October 80 | October 82 | October 83 | October 84

October 85 | October 86 | October 87 | October 88 | October 89

October 90 | October 91 | October 92 | October 93 | October 94

NASA/GSFC : TOMS TOTAL OZONE
MONTHLY AVERAGES

Dobson Units

Strong circumstantial evidence implicates synthetic chlorofluorocarbons (CFCs)—alkanes in which all the hydrogens have been replaced by fluorine and chlorine, such as $CFCl_3$ and CF_2Cl_2—as a major cause of ozone depletion. These gases, known commercially as Freons, have been extensively used as cooling fluids in refrigerators and in home and automobile air conditioners, in industrial cleaning solvents, and in the manufacture of some plastic foams (Section 26.6). They were once widely used as propellants in aerosol spray cans (deodorant, hair spray, etc.) because of their odorless, nontoxic, and nonflammable properties and because they are chemically inert and thus do not react with the contents of the can. Such use has now been banned.

Chlorofluorocarbons remain very stable in the atmosphere until they reach the stratosphere. There they encounter wavelengths of ultraviolet light that cause the homolytic cleavage that generates chlorine radicals.

Mario Molina *was born in Mexico in 1943 and subsequently became a U.S. citizen. He received a Ph.D. from the University of California, Berkeley, and then became a postdoctoral fellow in Rowland's laboratory. He is currently a professor of earth, atmospheric, and planetary sciences at the Massachusetts Institute of Technology.*

Paul Crutzen *was born in Amsterdam in 1933. He was trained as a meteorologist and taught himself chemistry. He is a professor at the Max Planck Institute for Chemistry in Mainz, Germany.*

Polar stratospheric clouds increase the rate of ozone destruction. These clouds form over Antarctica during the cold winter months. Ozone depletion in the Arctic is less severe because it generally does not get cold enough for the polar stratospheric clouds to form.

Movie: Chlorofluoro-carbons and ozone

The chlorine radicals are the ozone-removing agents. They react with ozone to form chlorine monoxide radicals and molecular oxygen. The chlorine monoxide radicals react with ozone to regenerate chlorine radicals. These two propagating steps are repeated over and over, destroying a molecule of ozone in each step. It has been calculated that each chlorine atom destroys 100,000 ozone molecules.

$$Cl\cdot \ + \ O_3 \ \longrightarrow \ ClO\cdot \ + \ O_2$$
$$ClO\cdot \ + \ O_3 \ \longrightarrow \ Cl\cdot \ + \ 2\,O_2$$

Because of their remarkable chemical stability, chlorofluorocarbons have half-lives of 70 to 120 years. A frightening consequence of such long half-lives is that once all CFC production is stopped, it will take more than 100 years to recover from the damage caused by CFCs already in the atmosphere. Recent studies have led to the development of environmentally sound CFC substitutes that will not destroy the ozone layer.

An international agreement known as the Montreal Protocol, which was signed in 1987 and which went into effect in 1989, mandated a reduction in the production of CFCs and total elimination of production by the year 2000. When scientists found that the ozone layer was eroding faster than anticipated in the Northern Hemisphere, the deadline for total elimination was moved up to the end of 1995. By 1993, scientists noted a marked decline in the rate of increase of CFCs in the atmosphere, but the protective ozone layer over the United States was still thinning at a substantial rate. The total ozone concentration measured in October 1994 above the Halley Bay Research Station in Antarctica was less than half the concentration measured 20 years earlier. By October 1998, the Antarctic ozone hole was larger than the North American continent. Fortunately, the ozone hole was a little smaller in October 1999 than it was in 1998. The ozone hole is a strong reminder that human activities can have a profound effect on our environment.

THE CONCORDE AND OZONE DEPLETION

Supersonic aircraft cruise in the lower stratosphere, and their jet engines convert molecular oxygen and nitrogen into nitrogen oxides such as NO and NO_2. Like CFCs, nitrogen oxides react with stratospheric ozone. Fortunately, the supersonic Concorde, built jointly by England and France, makes only a limited number of flights each week.

SUMMARY OF REACTIONS

1. *Alkanes* undergo radical substitution reactions with Cl_2 or Br_2 in the presence of heat or light (Sections 8.2, 8.3, and 8.4).

$$CH_3CH_3 \ + \ Cl_2 \ \xrightarrow{\Delta \text{ or } h\nu} \ CH_3CH_2Cl \ + \ HCl$$
excess

$$CH_3CH_3 \ + \ Br_2 \ \xrightarrow{\Delta \text{ or } h\nu} \ CH_3CH_2Br \ + \ HBr$$
excess
bromination is more selective than chlorination

2. *Alkenes* undergo radical halogenation at the allylic positions (Section 8.5).

$$RCH_2CH{=}CH_2 \ + \ Br_2 \ \xrightarrow{h\nu} \ \underset{\underset{Br}{|}}{RCHCH{=}CH_2} \ + \ \underset{\underset{Br}{|}}{RCH{=}CHCH_2} \ + \ HBr$$

Alkyl-substituted benzenes undergo radical halogenation at the benzylic position (Section 8.5).

$$\text{C}_6\text{H}_5{-}CH_2R \ + \ Br_2 \ \xrightarrow{h\nu} \ \text{C}_6\text{H}_5{-}\underset{\underset{Br}{|}}{CHR} \ + \ HBr$$

NBS is used for radical bromination at the allylic position (Section 8.5).

$$\text{cyclopentene} \ + \ NBS \ \xrightarrow[CH_2Cl_2]{h\nu} \ \text{bromocyclopentene}$$

3. *Cyclopropane* undergoes electrophilic addition reactions (Section 8.7).

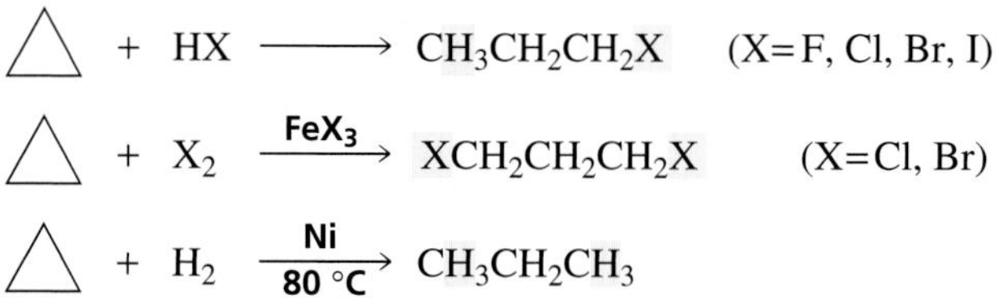

△ + HX ⟶ $CH_3CH_2CH_2X$ (X = F, Cl, Br, I)

△ + X_2 $\xrightarrow{FeX_3}$ $XCH_2CH_2CH_2X$ (X = Cl, Br)

△ + H_2 $\xrightarrow[80\ °C]{Ni}$ $CH_3CH_2CH_3$

KEY TERMS

alkane (page 329)
combustion (page 331)
free radical (page 332)
homolytic bond cleavage (page 332)
initiation step (page 332)

paraffin (page 331)
propagation step (page 332)
radical (page 332)
radical chain reaction (page 332)
radical inhibitor (page 346)

radical substitution reaction (page 333)
reactivity–selectivity principle
  (page 339)
saturated hydrocarbon (page 329)
termination step (page 332)

PROBLEMS

14. Give the product(s) of each of the following reactions:

a. $CH_2{=}CHCH_2CH_2CH_3$ + Br_2 $\xrightarrow{h\nu}$

e. △ + Br_2 $\xrightarrow{FeBr_3}$

b.
$$CH_3\overset{\overset{\displaystyle CH_3}{|}}{C}{=}CHCH_3$$
+ NBS $\xrightarrow[CH_2Cl_2]{\Delta}$

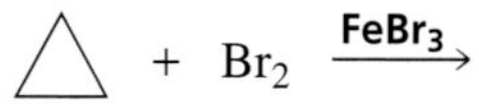

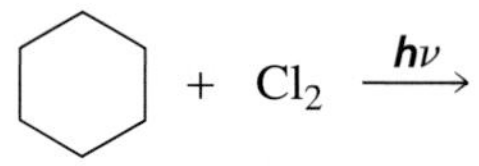

f. ⬡ + Cl_2 $\xrightarrow{h\nu}$

c.  △ + HCl ⟶

g. ⬠ + Cl_2 $\xrightarrow{FeCl_3}$

d.
$$CH_3CH_2\overset{\overset{\displaystyle CH_3}{|}}{C}HCH_2CH_2CH_3$$
+ Br_2 $\xrightarrow{h\nu}$

h. (cyclopentane with CH_3) + Cl_2 $\xrightarrow{h\nu}$

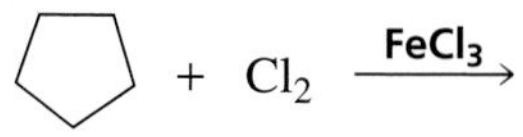

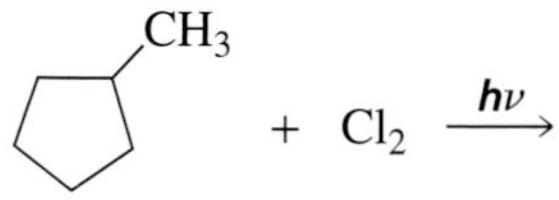

15. a. An alkane with molecular formula C_5H_{12} forms only one monochlorinated product when heated with Cl_2. Give the systematic name of this alkane.
  b. An alkane with molecular formula C_7H_{16} forms seven monochlorinated products (ignore stereoisomers) when heated with Cl_2. Give the systematic name of this alkane.

16. Dr. Al Cahall wanted to determine experimentally the relative ease of removal of a hydrogen atom from a tertiary, a secondary, and a primary carbon by a chlorine radical. He allowed 2-methylbutane to undergo chlorination at 300 °C and obtained as products 29% 1-chloro-2-methylbutane, 24% 2-chloro-2-methylbutane, 33% 2-chloro-3-methylbutane, and 14% 1-chloro-3-methylbutane. What values did he obtain for the relative ease of removal of tertiary, secondary, and primary hydrogen atoms by a chlorine radical under the conditions of his experiment?

17. At 600 °C, the ratio of the relative rates of formation of a tertiary, a secondary, and a primary radical is 2.6 : 2.1 : 1. Explain the change in the degree of regioselectivity compared with what Dr. Al Cahall found in Problem 16.

18. Iodine (I_2) does not react with ethane even though I_2 is more easily cleaved homolytically than the other halogens. Explain.

19. a. How many monobromination products would be obtained from the radical bromination of methylcyclohexane?
  b. Which one would be obtained in greatest yield? Explain.

20. Give the major product of each of the following reactions:

a.  (cyclopentene) + NBS $\xrightarrow[\text{CH}_2\text{Cl}_2]{\Delta}$

e. $\overset{\text{CH}_3}{\underset{\quad}{\text{CH}_3\text{CHCH}_3}}$ + Br$_2$ $\xrightarrow{h\nu}$

b. $\text{CH}_2{=}\text{CHCH}_2\text{CH}_2\text{CH}_3$ + NBS $\xrightarrow[\text{CH}_2\text{Cl}_2]{\Delta}$

f. 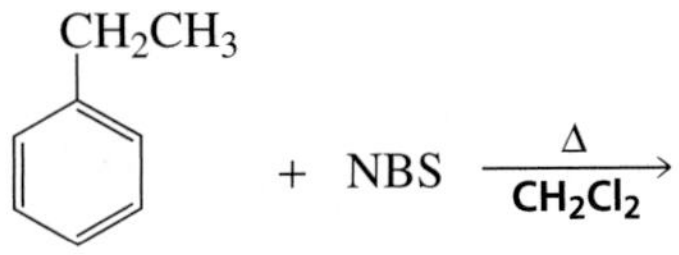 (benzene with CH$_2$CH$_3$) + NBS $\xrightarrow[\text{CH}_2\text{Cl}_2]{\Delta}$

c. (cyclohexene with CH$_3$) + NBS $\xrightarrow[\text{CH}_2\text{Cl}_2]{\Delta}$

g.  (cyclopentane) + NBS $\xrightarrow[\text{CH}_2\text{Cl}_2]{\Delta}$

d. $\overset{\text{CH}_3}{\underset{\quad}{\text{CH}_3\text{CHCH}_3}}$ + Cl$_2$ $\xrightarrow{h\nu}$

h. (cyclopentene with two CH$_3$ groups) + NBS $\xrightarrow[\text{CH}_2\text{Cl}_2]{\Delta}$

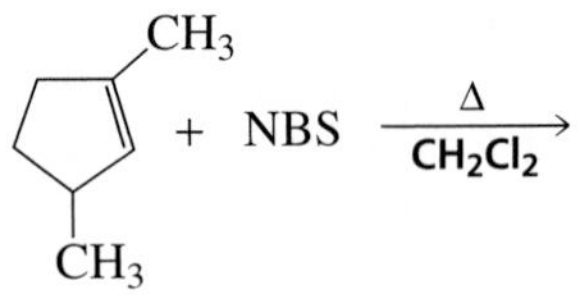

21. The deuterium kinetic isotope effect for chlorination of an alkane is defined here. Predict whether chlorination or bromination would have a greater deuterium kinetic isotope effect.

$$\frac{\text{deuterium kinetic}}{\text{isotope effect}} = \frac{\text{rate of homolytic cleavage of a C—H bond by Cl·}}{\text{rate of homolytic cleavage of a C—D bond by Cl·}}$$

22. The rate of bromination of methane is decreased if HBr is added to the reaction mixture. Explain.

23. Propose a mechanism for the following reaction:

$$\text{CH}_3\text{CH}_3 + \overset{\text{CH}_3}{\underset{\text{CH}_3}{\text{CH}_3{-}\text{C}{-}\text{OCl}}} \xrightarrow{\Delta} \text{CH}_3\text{CH}_2\text{Cl} + \overset{\text{CH}_3}{\underset{\text{CH}_3}{\text{CH}_3{-}\text{C}{-}\text{OH}}}$$

24. a. Calculate the $\Delta H°$ for the following reaction:

$$\text{CH}_4 + \text{Cl}_2 \xrightarrow{h\nu} \text{CH}_3\text{Cl} + \text{HCl}$$

b. Calculate the $\Delta H°$ for the two propagation steps:

$$\text{CH}_3{-}\text{H} + \text{·Cl} \longrightarrow \text{·CH}_3 + \text{H}{-}\text{Cl}$$
$$\text{·CH}_3 + \text{Cl}{-}\text{Cl} \longrightarrow \text{CH}_3{-}\text{Cl} + \text{·Cl}$$

c. Why do both calculations give you the same value of $\Delta H°$?

25. A possible alternative mechanism to that shown in Problem 24 for the monochlorination of methane would involve the following propagation steps.

$$\text{CH}_3{-}\text{H} + \text{·Cl} \longrightarrow \text{CH}_3{-}\text{Cl} + \text{·H}$$
$$\text{·H} + \text{Cl}{-}\text{Cl} \longrightarrow \text{H}{-}\text{Cl} + \text{·Cl}$$

How do you know that the reaction does not take place by this mechanism?

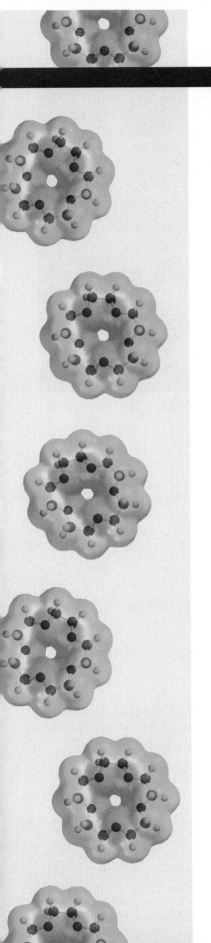

PART III

Substitution And Elimination Reactions

The three chapters in Part III discuss the reactions of compounds that have an electron-withdrawing atom or group (frequently called a leaving group) bonded to an sp^3 hybridized carbon. These compounds undergo two different kinds of reactions—substitution reactions and elimination reactions.

Chapter 9 discusses the substitution reactions of alkyl halides. Of the different compounds that undergo substitution and elimination reactions, alkyl halides are discussed first because they have relatively good leaving groups.

Chapter 10 covers the elimination reactions of alkyl halides. Because alkyl halides can undergo both substitution and elimination reactions, Chapter 10 also discusses the factors that determine whether a given alkyl halide will undergo a substitution reaction, an elimination reaction, or both substitution and elimination reactions.

Chapter 11 discusses compounds other than alkyl halides that undergo substitution and elimination reactions. You will see that because alcohols and ethers have relatively poor leaving groups compared with the leaving groups of alkyl halides, alcohols and ethers must be activated before the groups can be substituted or eliminated. Several methods commonly used to activate leaving groups will be examined. Other compounds that undergo substitution and/or elimination reactions, such as epoxides, thiols, and quaternary ammonium salts, are also discussed. Finally, this chapter will introduce you to a class of compounds that is very important to synthetic organic chemists—organometallic compounds.

9

Reactions at an sp^3 Hybridized Carbon I:

Substitution Reactions of Alkyl Halides

a sea hare

Organic compounds in which an sp^3 hybridized carbon is bonded to an electronegative atom or group can undergo two types of reactions. They can undergo **substitution reactions,** in which the electronegative atom or group is replaced by another atom or group. They can also undergo **elimination reactions,** in which the electronegative atom or group is eliminated along with a hydrogen from an adjacent carbon. The electronegative atom or group that is *substituted* or *eliminated* in these reactions is called the **leaving group.**

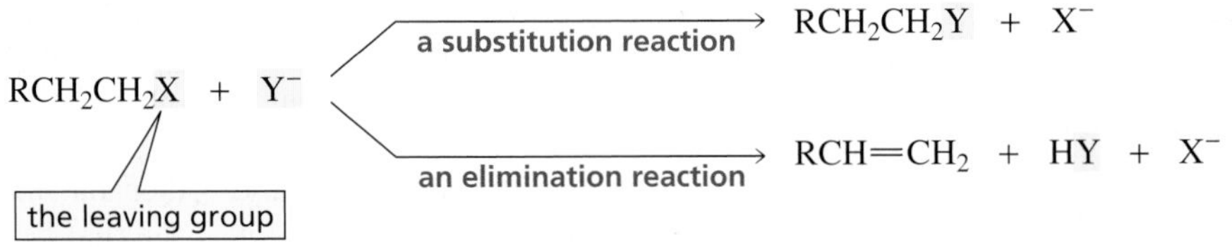

$$RCH_2CH_2X \ + \ Y^- \qquad\qquad \xrightarrow{\text{a substitution reaction}} \ RCH_2CH_2Y \ + \ X^-$$

$$\xrightarrow{\text{an elimination reaction}} \ RCH{=}CH_2 \ + \ HY \ + \ X^-$$

the leaving group

This chapter focuses on the substitution reactions of alkyl halides, compounds in which the leaving group is a halide ion (F^-, Cl^-, Br^-, or I^-). The nomenclature of alkyl halides was discussed in Section 2.4.

alkyl halides

R—F	R—Cl	R—Br	R—I
an alkyl fluoride	an alkyl chloride	an alkyl bromide	an alkyl iodide

In Chapter 10 we will discuss the elimination reactions of alkyl halides and the factors that determine whether substitution or elimination will prevail when an alkyl halide undergoes a reaction.

Because halide ions are relatively good leaving groups (are easily displaced), alkyl halides are a good family of compounds with which to start our study of substitution and elimination reactions. We will then be prepared to discuss the substitution and elimination reactions of compounds with leaving groups other than halide ions in Chapter 11.

Substitution reactions are important reactions in organic chemistry because they make it possible to convert readily available alkyl halides into a wide variety of other compounds. Substitution reactions are also important in the cells of plants and animals. Because cells exist in predominantly aqueous environments and alkyl halides are insoluble in water, biological systems use compounds in which the group that is substituted is more polar than a halogen and therefore more water-soluble. The reactions of some of these biological compounds are discussed in this chapter.

ALKYL HALIDES AS SURVIVAL COMPOUNDS

The few alkyl halides that exist in nature are synthesized by marine organisms that live in salt water (sponges, coral, algae) and are surrounded by a high concentration of chloride ion. Because alkyl halides tend to be toxic, they deter predators. For example, red algae synthesize a foul-tasting alkyl halide that keeps predators from eating them. The sea hare, however, is not deterred. After consuming red algae, it converts the original alkyl halide into a structurally similar alkyl halide it uses for its own defense. Unlike other mollusks, the sea hare does not have a shell. It surrounds itself with a slimy material that contains the alkyl halide and in this way protects itself from carnivorous fish.

synthesized by red algae   **synthesized by the sea hare**

a sea hare

A halogen is more electronegative than carbon. Consequently, the two atoms do not share their bonding electrons equally. Because the more electronegative halogen has a larger share of the electrons, it has a partial negative charge and the carbon to which it is bonded has a partial positive charge.

9.1 REACTIVITY CONSIDERATIONS

$$\overset{\delta+}{R}CH_2 - \overset{\delta-}{X}$$

X = F, Cl, Br, I

The polar carbon–halogen bond causes alkyl halides to undergo substitution and elimination reactions. There are two important mechanisms for the substitution reaction:

1. A nucleophile is attracted to the partially positively charged carbon. As the nucleophile approaches the carbon, it causes the carbon–halogen bond to break heterolytically (the halogen keeps both of the bonding electrons).

$$\overset{..}{Nu}^{-} + \overset{\delta+}{-}\overset{|}{\underset{|}{C}}\overset{\delta-}{-}X \longrightarrow -\overset{|}{\underset{|}{C}}-Nu + X^{-}$$

2. The carbon–halogen bond breaks heterolytically without any assistance from the nucleophile, forming a carbocation. The carbocation then reacts with the nucleophile to form the substitution product.

$$-\overset{|}{\underset{|}{C}}\overset{\delta+}{-}\overset{\delta-}{X} \longrightarrow -\overset{|}{C^+} + X^-$$

$$-\overset{|}{C^+} + \overset{..}{\underset{..}{Nu}}^- \longrightarrow -\overset{|}{\underset{|}{C}}-Nu$$

Regardless of the mechanism by which a substitution reaction occurs, it is called a **nucleophilic substitution reaction** because a nucleophile is substituted for the halogen. *The mechanism that predominates depends on the following:*

- the structure of the alkyl halide,
- the reactivity and structure of the nucleophile,
- the concentration of the nucleophile, and
- the solvent in which the reaction is carried out.

9.2
THE MECHANISM OF S_N2 REACTIONS

How is the mechanism of a reaction determined? We can learn a great deal about the mechanism of a reaction by determining the factors that affect the rate of the reaction. These factors are called the **kinetics** of the reaction.

The rate of a nucleophilic substitution reaction such as the reaction of methyl bromide with hydroxide ion depends on the concentrations of both reagents. If the concentration of methyl bromide in the reaction mixture is doubled, the rate of the nucleophilic substitution reaction doubles. If the concentration of hydroxide ion is doubled, the rate of the reaction also doubles. If the concentrations of both reactants are doubled, the rate of the reaction quadruples.

$$\underset{\text{methyl bromide}}{CH_3Br} + HO^- \longrightarrow \underset{\text{methyl alcohol}}{CH_3OH} + Br^-$$

When you know the relationship between the rate of a reaction and the concentration of the reactants, you can write a **rate law** for the reaction.

$$\text{rate} \propto \text{[alkyl halide][nucleophile]}$$

Recall from Section 3.7 that the "proportional to" sign ($\propto$) can be replaced by an "equal" sign and a proportionality constant (k).

$$\text{rate} = k\text{[alkyl halide][nucleophile]}$$

Because the rate of this reaction depends on the concentration of two reactants, it is a **second-order reaction** (Section 3.7).

The rate law tells us what molecules are involved in the transition state of the rate-determining step of the reaction. From the rate law for the reaction of methyl bromide with hydroxide ion, we know that both methyl bromide and hydroxide ion are involved in the rate-determining transition state. The transition state, therefore, is **bimolecular** because it involves two molecules. As we saw in Section 3.7, the proportionality constant (k) is called the rate constant. The rate constant describes how difficult it is to overcome the energy barrier of the reaction (how hard

Movie:
Biomolecular reaction

it is to reach the transition state). The larger the rate constant, the easier it is to reach the transition state.

The reaction of methyl bromide with hydroxide ion is an example of an **S$_N$2 reaction** where S stands for substitution, N stands for nucleophilic, and 2 stands for bimolecular. In 1937, Edward Hughes and Christopher Ingold proposed a mechanism for an S$_N$2 reaction. Recall that a mechanism describes the step-by-step process by which reactants are converted into products. It is a theory that fits the experimental evidence that has been accumulated concerning the reaction. Hughes and Ingold based their mechanism for an S$_N$2 reaction on the following three pieces of experimental evidence:

1. The rate of the reaction depends on the concentration of the alkyl halide *and* on the concentration of the nucleophile. This means that both reactants are involved in the transition state of the rate-determining step.

2. When the hydrogens of methyl bromide are successively replaced with methyl groups, the rate of the reaction with a given nucleophile becomes progressively slower (Table 9.1).

3. The reaction of an alkyl halide in which the halogen is bonded to a chirality center leads to the formation of only one stereoisomer, and its configuration is inverted relative to the configuration of the reacting alkyl halide.

Edward Davies Hughes (1906–1963) was born in North Wales. He earned two doctoral degrees, a Ph.D. from the University of Wales and a D.Sc. from the University of London, working with Sir Christopher Ingold. He was a professor of chemistry at University College, London.

Sir Christopher Ingold (1893–1970) was born in Ilford, England. In addition to determining the mechanism of the S$_N$2 reaction, he was a member of the group that developed nomenclature for enantiomers (see page 187), and he participated in developing the theory of resonance.

TABLE 9.1   Relative Rates of S$_N$2 Reactions for Several Alkyl Bromides

$$R\text{---}Br + Cl^- \xrightarrow{S_N2} R\text{---}Cl + Br^-$$

Alkyl bromide	Class	Relative rate
CH$_3$—Br	methyl	1200
CH$_3$CH$_2$—Br	primary	40
CH$_3$CH$_2$CH$_2$—Br	primary	16

CH$_3$CH—Br
   |
  CH$_3$
	secondary	1

CH$_3$
  |
CH$_3$C—Br
  |
  CH$_3$
	tertiary	too slow to measure

The mechanism proposed by Hughes and Ingold for an S$_N$2 reaction is a concerted reaction—the reaction takes place in a single step so no intermediates are formed. The nucleophile attacks the carbon bearing the leaving group and displaces the leaving group. Because the nucleophile hits the carbon on the side opposite to the side bonded to the leaving group, the carbon is said to undergo **back side attack.** Back side attack occurs because the orbital of the nucleophile that contains its nonbonding electrons interacts with the empty σ^* molecular orbital associated with the C—Br bond. This orbital has its larger lobe on the side of the carbon directed away from the C—Br bond. Consequently, the best overlap of the interacting orbitals is achieved through back side attack. An S$_N$2 reaction is also called a **direct displacement reaction** because the nucleophile displaces the leaving group in a single step.

mechanism of the S$_N$2 reaction

$$HO\!:^- + CH_3\text{---}\ddot{B}r\!: \longrightarrow CH_3\text{---}OH + :\!\ddot{B}\ddot{r}\!:^-$$

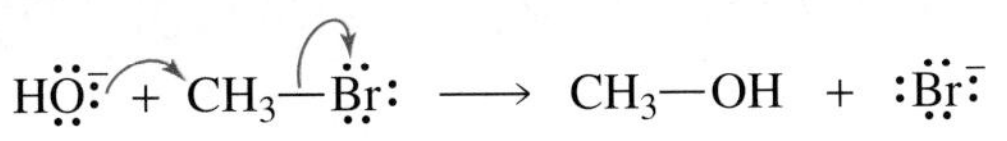

How does this mechanism account for the three observed pieces of experimental evidence? The mechanism shows the alkyl halide and the nucleophile coming together in the transition state of the one-step reaction. Therefore, increasing the concentration of either of them makes their coming together more probable. Thus, the reaction will follow second-order kinetics, exactly as observed.

Because the nucleophile attacks the back side of the carbon that is bonded to the halogen, bulky substituents attached to this carbon will make it harder for the nucleophile to get to the back side and, therefore, will decrease the rate of the reaction (Figure 9.1). This explains why substituting methyl groups for the hydrogens in methyl bromide progressively slows the rate of the substitution reaction (Table 9.1).

Figure 9.1 ▶
Increasing the bulk of the substituents bonded to the carbon undergoing nucleophilic attack decreases access to the back side of the carbon, thereby decreasing the rate of an S_N2 reaction.

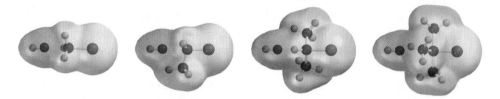

Effects due to groups occupying a certain volume of space are called **steric effects.** A steric effect that decreases reactivity is called **steric hindrance.** Steric hindrance results from groups getting in the way at the reaction site. Because of steric hindrance, alkyl halides have the following relative reactivities in an S_N2 reaction. The steric crowding in tertiary alkyl halides is so great that they are unable to undergo S_N2 reactions.

relative reactivities of alkyl halides in an S_N2 reaction

methyl halide > 1° alkyl halide > 2° alkyl halide > 3° alkyl halide

◀ increasing reactivity in an S_N2 reaction

It is not just the *number* of alkyl groups attached to the carbon undergoing nucleophilic attack that determines the rate of an S_N2 reaction—the *size* of the alkyl groups is also important. For example, ethyl bromide and propyl bromide are both primary alkyl halides, but ethyl bromide is more than twice as reactive in an S_N2 reaction because the methyl group of ethyl bromide provides less steric hindrance to back side attack than does the ethyl group of propyl bromide (Table 9.1). Reaction coordinate diagrams for an S_N2 reaction with methyl bromide and with a sterically hindered alkyl halide are shown in Figure 9.2.

As the nucleophile approaches the back side of the carbon of methyl bromide, the carbon–hydrogen bonds begin to move away from the nucleophile and its attacking electrons. By the time the transition state is reached, the carbon–hydrogen

3-D Molecules:
Methyl chloride;
t-Butyl chloride

transition state

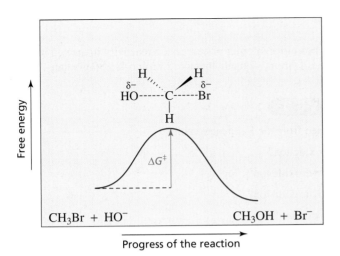

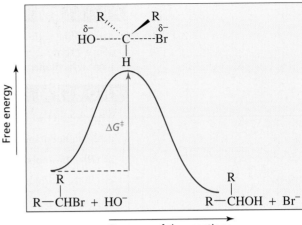

▲ **Figure 9.2**
Reaction coordinate diagrams for:
(a) the S$_N$2 reaction of methyl bromide with hydroxide ion.
(b) an S$_N$2 reaction of a sterically hindered alkyl bromide with hydroxide ion.

bonds are all in the same plane and the carbon is pentacoordinate (fully bonded to three atoms and partially bonded to two) rather than tetrahedral. As the nucleophile gets closer to the carbon and the bromine moves farther away from it, the carbon–hydrogen bonds continue to move in the same direction. Eventually, the bond between the carbon and the nucleophile is fully formed and the bond between the carbon and bromine is completely broken—the carbon is once again tetrahedral.

Tutorial: S$_N$2

One way to visualize movement of the groups bonded to the carbon at which substitution occurs is to picture an umbrella that turns inside out. This is called **inversion of configuration.** The carbon at which substitution occurs has inverted its configuration during the course of the reaction just as an umbrella has a tendency to invert in a windstorm. The inversion is known as a *Walden inversion,* since Paul Walden was the first to discover that compounds could invert their configurations as a result of substitution reactions.

Because an S$_N$2 reaction takes place with inversion of configuration, only one substitution product is formed when an alkyl halide—that has the halogen leaving group bonded to a chirality center—undergoes an S$_N$2 reaction. The configuration of that product is inverted relative to the configuration of the alkyl halide. In other words, the product of the reaction of hydroxide ion with (*R*)-2-bromopentane is (*S*)-2-pentanol. Therefore, the proposed mechanism accounts for the observed configuration of the product.

Paul Walden (1863–1957) was born in Riga, Latvia, the son of a farmer. His parents died when he was still a child, and he supported himself at Riga University and St. Petersburg University by working as a tutor. He received a Ph.D. from the University of Leipzig, and returned to Latvia to become a professor of chemistry at Riga University. Following the Russian Revolution, he returned to Germany and was a professor at the University of Rostock and later at the University of Tübingen.

the configuration of the product is inverted relative to the configuration of the reactant

---

PROBLEM 1 ◆

Does increasing the height of the energy barrier to an S$_N$2 reaction increase or decrease the magnitude of the rate constant for the reaction?

PROBLEM 2 ◆

Arrange the following alkyl bromides in order of decreasing reactivity in an S_N2 reaction: 1-bromo-2-methylbutane, 1-bromo-3-methylbutane, 2-bromo-2-methylbutane, and 1-bromopentane.

PROBLEM 3/SOLVED

What product would be formed from the S_N2 reaction of:

a. 2-bromobutane and hydroxide ion?

b. (*R*)-2-bromobutane and hydroxide ion?

c. (*S*)-3-chlorohexane and hydroxide ion?

d. 3-iodopentane and hydroxide ion?

SOLUTION TO 3a The product is 2-butanol. Because it is an S_N2 reaction, we know that the configuration of the product is inverted relative to the configuration of the reactant. The configuration of the reactant is not specified, however, so we cannot specify the configuration of the product.

9.3
THE S_N2 REACTION

The Leaving Group

If an alkyl iodide, an alkyl bromide, an alkyl chloride, and an alkyl fluoride (all with the same alkyl group) were allowed to react with the same nucleophile under the same conditions, we would find that the alkyl iodide is the most reactive and the alkyl fluoride is the least reactive.

relative rates of reaction

$HO^- + RCH_2I \longrightarrow RCH_2OH + I^-$	30,000
$HO^- + RCH_2Br \longrightarrow RCH_2OH + Br^-$	10,000
$HO^- + RCH_2Cl \longrightarrow RCH_2OH + Cl^-$	200
$HO^- + RCH_2F \longrightarrow RCH_2OH + F^-$	1

The only difference among these four reactions is the nature of the leaving group. Apparently, the iodide ion is the best leaving group and the fluoride ion is the worst. This brings us to an important rule in organic chemistry that will keep coming up in this text—*the weaker the basicity of a group, the better its leaving ability.* Leaving ability depends on basicity because a weak base does not share its electrons as well as a strong base does. Consequently, a weak base is not bonded as strongly to the carbon as a strong base would be, and a weaker bond is more easily broken.

> **The weaker the base, the better it is as a leaving group.**

We have seen that the hydrogen halides have the following relative acidities (see Section 1.18).

relative acidities of the hydrogen halides

$$HI > HBr > HCl > HF$$

⬅ increasing acidity

Because we know that the stronger the acid, the weaker its conjugate base, we know that the halide ions have the following relative basicities.

relative basicities of the halide ions

$$I^- \; < \; Br^- \; < \; Cl^- \; < \; F^-$$

increasing basicity

Furthermore, because weaker bases are better leaving groups, the halide ions have the following relative leaving abilities.

relative leaving abilities of the halide ions

$$I^- \; > \; Br^- \; > \; Cl^- \; > \; F^-$$

increasing leaving ability

As a consequence of the relative leaving abilities of the halide ions, alkyl halides have the following relative reactivities in an S$_N$2 reaction. (Recall from Section 1.13 that the bond formed by a halogen becomes longer and weaker as the atomic weight of the halogen increases.)

relative reactivities of alkyl halides in an S$_N$2 reaction

$$RI \; > \; RBr \; > \; RCl \; > \; RF$$

increasing reactivity

The Nucleophile

When we talk about atoms or molecules that have nonbonding electrons, sometimes we call them bases and sometimes we call them nucleophiles. What is the difference between a base and a nucleophile?

A **base** shares its nonbonding electrons with a proton. **Basicity** is a measure of how well the base shares those electrons with a proton. The stronger the base, the better it shares its electrons. Basicity is measured by an *equilibrium constant* (the acid dissociation constant, K_a) that indicates the tendency of the conjugate acid of the base to lose a proton (Section 1.17).

A **nucleophile** uses its nonbonding electrons to attack an electron-deficient atom other than a proton. **Nucleophilicity** is a measure of how readily the nucleophile is able to attack such an atom. It is measured by a *rate constant* (k). In the case of an S$_N$2 reaction, nucleophilicity is a measure of how readily the nucleophile attacks an sp^3 hybridized carbon bonded to a leaving group. Thus, *a base forms a new bond with a proton, whereas a nucleophile forms a new bond with an atom other than a proton.*

When comparing molecules *with the same attacking atom,* there is a direct relationship between basicity and nucleophilicity. Stronger bases are better nucleophiles. For example, a species with a negative charge is a stronger base *and* a better nucleophile than a species with the same attacking atom that is neutral. Thus HO$^-$ is a stronger base and a better nucleophile than H$_2$O.

A base forms a new bond with a proton. A nucleophile forms a new bond with an atom other than a proton.

stronger base, better nucleophile		weaker base, poorer nucleophile
HO^-	>	H_2O
CH_3O^-	>	CH_3OH
$^-NH_2$	>	NH_3
$CH_3CH_2NH^-$	>	$CH_3CH_2NH_2$

When comparing molecules *with attacking atoms of approximately the same size,* the stronger bases are again the better nucleophiles. The atoms across the second row of the periodic table have approximately the same size. If hydrogens are attached to the second-row elements, the resulting compounds have the following relative acidities (Section 1.18).

relative acid strengths

$$CH_4 \ < \ NH_3 \ < \ H_2O \ < \ HF$$

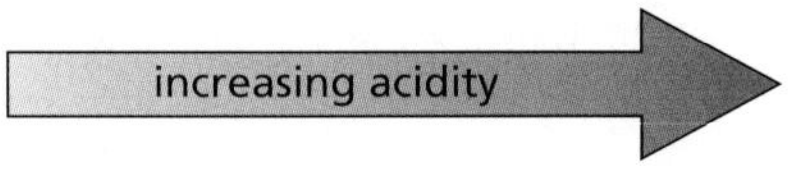

increasing acidity →

Consequently, the conjugate bases have the following relative base strengths and nucleophilicities. For example, the methyl anion is the strongest base as well as the best nucleophile.

relative base strengths and relative nucleophilicities

$$^-CH_3 \ > \ ^-NH_2 \ > \ HO^- \ > \ F^-$$

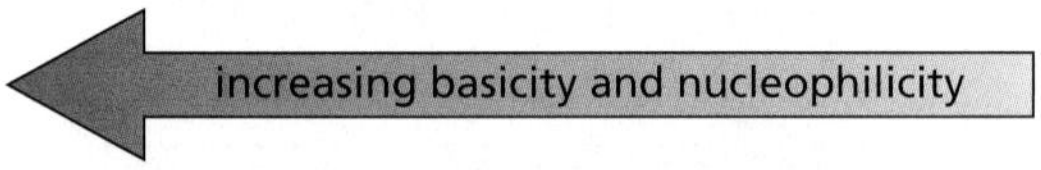

← increasing basicity and nucleophilicity

When comparing molecules *with attacking atoms that are very different in size,* the direct relationship between basicity and nucleophilicity is maintained if the reaction occurs in the gas phase. If, however, the reaction occurs in a solvent—as most organic reactions do—the relationship between basicity and nucleophilicity depends on the solvent.

If the solvent is protic (i.e., it has a hydrogen bonded to an oxygen or to a nitrogen, making it a hydrogen bond donor), the relationship between basicity and nucleophilicity becomes inverted. In other words, as basicity decreases, nucleophilicity increases. Iodide ion, therefore, which is the weakest base of the halogen family, is the best nucleophile in a protic solvent.

A protic solvent contains a hydrogen bonded to an oxygen or a nitrogen; it is a hydrogen bond donor.

An aprotic solvent does not contain a hydrogen bonded to an oxygen or a nitrogen; it is not a hydrogen bond donor.

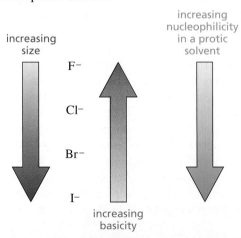

Why, in a protic solvent, is the largest atom the best nucleophile even though it is the weakest base? First of all, the larger the atom, the farther away the electrons are from the nucleus, and the more polarizable it is. (Recall from Section 2.9 that polarizability indicates how easily an atom's electron cloud can be distorted.) Because the electrons are farther away, they are not held as tightly and can, therefore, move more freely toward a positive charge. As a result, the electrons are able to overlap from farther away with the orbital of carbon, as shown in Figure 9.3. This results in a greater degree of bonding in the transition state, making it more stable. Secondly, the protic solvent makes the larger atoms better nucleophiles by making the smaller atoms less nucleophilic.

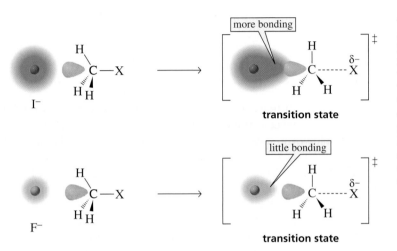

◀ **Figure 9.3**
An iodide ion is larger and more polarizable than a fluoride ion. Therefore, the relatively loosely held electrons of the iodide ion can overlap from farther away with the orbital of carbon undergoing nucleophilic attack. The tightly bound electrons of the fluoride ion cannot start to overlap until the atoms are closer together.

The Effect of the Solvent on Nucleophilicity

How does a protic solvent make the smaller atom less nucleophilic? When a negatively charged species is placed in a protic solvent, the ion becomes solvated (Section 2.9). The solvent molecules arrange themselves so that their partially positively charged hydrogens point toward the negatively charged species. The interaction between the ion and the dipole of the protic solvent is called an **ion–dipole interaction.** At least one of the ion–dipole interactions must be broken before the nucleophile can participate in an S$_N$2 reaction. Weak bases interact weakly with protic solvents, whereas strong bases interact more strongly because they are better at sharing their electrons. It is easier, therefore, to break the ion–dipole interactions between an iodide ion (a weak base) and the solvent than between a fluoride ion (a stronger base) and the solvent. As a result, iodide ion is a better nucleophile than fluoride ion in a protic solvent (Table 9.2).

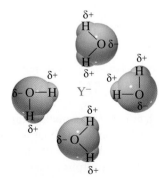

**ion–dipole interactions between
a nucleophile and water**

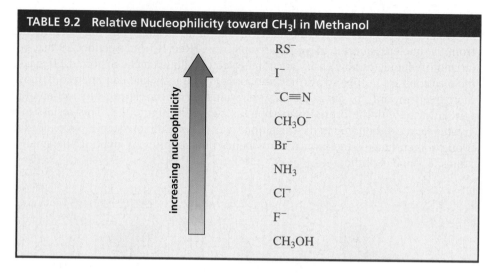

TABLE 9.2 Relative Nucleophilicity toward CH₃I in Methanol

increasing nucleophilicity

RS⁻

I⁻

⁻C≡N

CH₃O⁻

Br⁻

NH₃

Cl⁻

F⁻

CH₃OH

The nucleophilicity of a small species like a fluoride ion would increase in a nonpolar solvent because there are no ion–dipole interactions between the ion and the solvent. However, ions are insoluble in most nonpolar solvents but they can dissolve in aprotic polar solvents (see Table 9.7) such as dimethylformamide (DMF) or dimethylsulfoxide (DMSO). An aprotic polar solvent is not a hydrogen bond donor because it does not have a hydrogen attached to an oxygen or to a nitrogen, so there are no positively charged hydrogens to form ion–dipole interactions. Aprotic polar solvents have a partial negative charge on the surface of their molecules that can solvate cations, but the partial positive charge is on the inside of the molecule, which makes it less accessible. Therefore, the relatively "naked" anion can become a powerful nucleophile in an aprotic polar solvent. Thus fluoride ion is a better nucleophile in DMSO than it is in water.

3-D Molecules:
N,N-Dimethylformamide;
Dimethyl sulfoxide

$$\underset{\substack{\textbf{\textit{N,N}-dimethylformamide}\\ \textbf{DMF}}}{H-\overset{\overset{\displaystyle O}{\|}}{C}-\overset{\overset{\displaystyle CH_3}{|}}{N}-CH_3} \qquad \underset{\substack{\textbf{dimethyl sulfoxide}\\ \textbf{DMSO}}}{CH_3-\overset{\overset{\displaystyle O}{\|}}{S}-CH_3}$$

Nucleophilicity Is Affected by Steric Effects

Because a base removes a relatively unhindered proton, base strength is relatively unaffected by steric effects. The strength of a base depends only on how well the base shares its electrons with a proton. For example, *tert*-butoxide ion is a stronger base than ethoxide ion because *tert*-butanol (pK_a = 18) is a weaker acid than ethanol (pK_a = 15.9).

$$\underset{\substack{\textbf{ethoxide ion}\\ \textbf{better nucleophile}}}{CH_3CH_2O^-} \qquad \underset{\substack{\textbf{\textit{tert}-butoxide ion}\\ \textbf{stronger base}}}{CH_3\overset{\overset{\displaystyle CH_3}{|}}{\underset{\underset{\displaystyle CH_3}{|}}{C}}O^-}$$

Steric effects, on the other hand, do affect nucleophilicity. A bulky nucleophile cannot approach the back side of a carbon as easily as a less sterically hindered nucleophile can. Thus, the bulky *tert*-butoxide ion, with its three methyl groups, is a poorer nucleophile than ethoxide ion even though *tert*-butoxide ion is a stronger base. Therefore, if you want a species to act as a base, you should choose one that is bulky because it will be less apt to act as a nucleophile. On the other hand, if you want a species to act as a nucleophile, you should use a non-sterically hindered reagent.

PROBLEM 4 ◆

a. Which is a stronger base, RO^- or RS^-?

b. Which is a better nucleophile in an aqueous solution?

PROBLEM 5 / SOLVED

List the following species in order of *decreasing* nucleophilicity in an aqueous solution:

$$\text{C}_6\text{H}_5-\text{O}^- \quad CH_3OH \quad HO^- \quad CH_3\overset{\overset{\displaystyle O}{\|}}{C}O^- \quad CH_3S^-$$

SOLUTION Let's first divide the nucleophiles into groups. There is one nucleophile with a negatively charged sulfur, three with negatively charged oxygens, and one with a neutral oxygen. We know that in the polar aqueous solvent, the one with the negatively charged sulfur is the most nucleophilic because sulfur is larger than oxygen. We know that the poorest nucleophile is the one with the neutral oxygen atom. So now, to complete the problem, we need to rank the three nucleophiles with negatively charged oxygens in order of the pK_a's of their conjugate acids. A carboxylic acid is a stronger acid than phenol, which is a stronger acid than water. Because water is the weakest acid, its conjugate base is the strongest base and the best nucleophile. Thus, the relative nucleophilicities are:

$$CH_3S^- \;>\; HO^- \;>\; \text{C}_6\text{H}_5-\text{O}^- \;>\; CH_3\overset{\overset{\displaystyle O}{\|}}{C}O^- \;>\; CH_3OH$$

PROBLEM 6 ◆

Indicate whether each of the following solvents is protic or aprotic:

a. chloroform $(CHCl_3)$

b. diethyl ether $(CH_3CH_2OCH_2CH_3)$

c. acetic acid (CH_3COOH)

d. hexane $[CH_3(CH_2)_4CH_3]$

PROBLEM 7 ◆

For each of the following pairs of S$_N$2 reactions, indicate which occurs with the larger rate constant:

a. $CH_3CH_2Br \;+\; H_2O$ or $CH_3CH_2Br \;+\; HO^-$

b. $CH_3CHCH_2Br \;+\; HO^-$ or $CH_3CH_2CHBr \;+\; HO^-$
     $\quad\;\; |$ $\qquad\qquad\;\; |$
     $\quad\, CH_3$ $\qquad\qquad CH_3$

c. $CH_3CH_2Cl \;+\; CH_3O^-$ or $CH_3CH_2Cl \;+\; CH_3S^-$
        (in ethanol)

d. $CH_3CH_2Cl \;+\; I^-$ or $CH_3Br \;+\; I^-$

e. $BrCH_2CH_2CH_2NH_2$ or $\overset{+}{B}rCH_2CH_2CH_2CH_2NH_2$

9.4
THE REVERSIBILITY OF AN S_N2 REACTION

Because there are many different kinds of nucleophiles, a wide variety of organic compounds can be synthesized by means of S_N2 reactions. The following reactions show just a few of the many kinds of organic compounds that can be synthesized in this way:

$$CH_3CH_2Cl + HO^- \longrightarrow CH_3CH_2OH + Cl^-$$
$$\text{an alcohol}$$

$$CH_3CH_2Br + HS^- \longrightarrow CH_3CH_2SH + Br^-$$
$$\text{a thiol}$$

$$CH_3CH_2I + RO^- \longrightarrow CH_3CH_2OR + I^-$$
$$\text{an ether}$$

$$CH_3CH_2Br + RS^- \longrightarrow CH_3CH_2SR + Br^-$$
$$\text{a thioether}$$

$$CH_3CH_2Cl + {}^-NH_2 \longrightarrow CH_3CH_2NH_2 + Cl^-$$
$$\text{a primary amine}$$

$$CH_3CH_2Br + {}^-C\equiv CR \longrightarrow CH_3CH_2C\equiv CR + Br^-$$
$$\text{an alkyne}$$

$$CH_3CH_2I + {}^-C\equiv N \longrightarrow CH_3CH_2C\equiv N + I^-$$
$$\text{a nitrile}$$

$$CH_3CH_2CH_2Br + NH_3 \longrightarrow CH_3CH_2CH_2\overset{+}{N}H_3\ Br^- \rightleftharpoons CH_3CH_2CH_2NH_2 + HBr$$
$$\text{a primary amine}$$

$$CH_3I + CH_3CH_2\underset{\underset{CH_3}{|}}{N}H \longrightarrow CH_3CH_2\underset{\underset{CH_3}{|}}{\overset{+}{N}}HCH_3\ I^- \rightleftharpoons CH_3CH_2\underset{\underset{CH_3}{|}}{N}CH_3 + HI$$
$$\text{a tertiary amine}$$

The reverse of each of these reactions can also be visualized as a nucleophilic substitution reaction. In the first reaction, for example, ethyl chloride reacts with hydroxide ion to form ethyl alcohol and a chloride ion. The reverse reaction appears to satisfy the requirements for a nucleophilic substitution reaction because the chloride ion is a nucleophile and ethyl alcohol has an HO^- leaving group. But ethyl alcohol and chloride ion do not react.

Why does a nucleophilic substitution reaction take place in one direction but not in the other? We can answer this question by comparing the leaving tendency of Cl^- in the forward direction and the leaving tendency of HO^- in the reverse direction. In order to compare leaving tendencies, we must compare basicities. Most people find it easier to compare the acid strengths of the conjugate acids (Table 9.3), so that is what we will do. HCl is a much stronger acid than H_2O, which means that Cl^- is a much weaker base than HO^- (remember that the stronger the acid, the weaker its conjugate base). Because it is a weaker base, Cl^- is a better leaving group. Consequently, HO^- can displace Cl^- in the forward reaction, but Cl^- cannot displace HO^- in the reverse reaction. The reaction proceeds in the direction that allows the stronger base to displace the weaker base (the best leaving group).

If the difference between the basicities of the nucleophile and the leaving group is not very large, the reaction will be reversible. For example, in the reaction of ethyl bromide with iodide ion, Br^- is the leaving group in one direction and I^- is the leaving group in the other direction. Because the pK_a values of the conjugate acids of the two leaving groups are similar (pK_a of HBr $= -9$; pK_a of HI $= -10$), the reaction is reversible.

An S_N2 reaction proceeds in the direction that allows the stronger base to displace the weaker base.

TABLE 9.3   The Acidities of the Conjugate Acids of Some Leaving Groups

Acid	pK_a	Conjugate base (leaving group)
HI	−10.0	I⁻
HBr	−9.0	Br⁻
HCl	−7.0	Cl⁻
H_2SO_4	−5.0	⁻OSO_3H
$CH_3\overset{+}{O}H_2$	−2.5	CH_3OH
H_3O^+	−1.7	H_2O
⟨benzene⟩—SO_3H	−0.6	⟨benzene⟩—SO_3^-
HF	3.2	F⁻
$\overset{O}{\overset{\|}{CH_3COH}}$	4.8	$\overset{O}{\overset{\|}{CH_3CO^-}}$
H_2S	7.0	HS⁻
HC≡N	9.1	⁻C≡N
$\overset{+}{N}H_4$	9.4	NH_3
CH_3CH_2SH	10.5	$CH_3CH_2S^-$
$(CH_3)_3\overset{+}{N}H$	10.8	$(CH_3)_3N$
CH_3OH	15.5	CH_3O^-
H_2O	15.7	HO⁻
HC≡CH	25	HC≡C⁻
NH_3	36	⁻NH_2
H_2	very large	H⁻

$$CH_3CH_2Br + I^- \rightleftharpoons CH_3CH_2I + Br^-$$

You can drive a reversible reaction toward the desired products by removing one of the products as it is formed. **Le Châtelier's principle** states that *if an equilibrium is disturbed, the system will adjust to offset the disturbance.* In other words, if the concentration of product C is decreased, A and B will react to form more C and D so that the value of equilibrium constant does not change.

$$A + B \rightleftharpoons C + D$$

$$K_{eq} = \frac{[C][D]}{[A][B]}$$

For example, the reaction of ethyl chloride with methanol is reversible because the difference between the basicities of the nucleophile and the leaving group is not very large. If the reaction is carried out in a neutral solution, the protonated product

Henri Louis Le Châtelier (1850–1936) was born in France. He studied mining engineering and was particularly interested in learning how to prevent explosions. Because his father was France's inspector general of mines, Henri's interest in mine safety is understandable. Le Châtelier's research into preventing explosions led him to study heat and its measurement, which in turn led him to study thermodynamics.

will lose a proton (Section 1.20). This disturbs the equilibrium and drives the reaction toward the products.

If an equilibrium is disturbed, the system will adjust to offset the disturbance.

$$CH_3CH_2Cl + CH_3OH \rightleftharpoons CH_3CH_2\overset{+}{\underset{\underset{H + Cl^-}{|}}{O}}CH_3 \xrightarrow[\text{fast}]{-H^+} CH_3CH_2OCH_3$$

PROBLEM 8 / SOLVED

What product is obtained when ethylamine reacts with excess methyl iodide in a basic solution of potassium carbonate?

$$CH_3CH_2\overset{..}{N}H_2 + CH_3-I \xrightarrow[\text{excess}]{K_2CO_3} ?$$

SOLUTION   Methyl iodide and ethylamine undergo an S_N2 reaction. The initial product is a protonated secondary amine. The reaction is carried out in a basic solution of potassium carbonate so the amines will be present predominately in their nonprotonated forms. The secondary amine can undergo an S_N2 reaction with another equivalent of methyl iodide, forming a tertiary amine. The tertiary amine can react with methyl iodide in yet another S_N2 reaction. The final product of the reaction is a quaternary ammonium iodide.

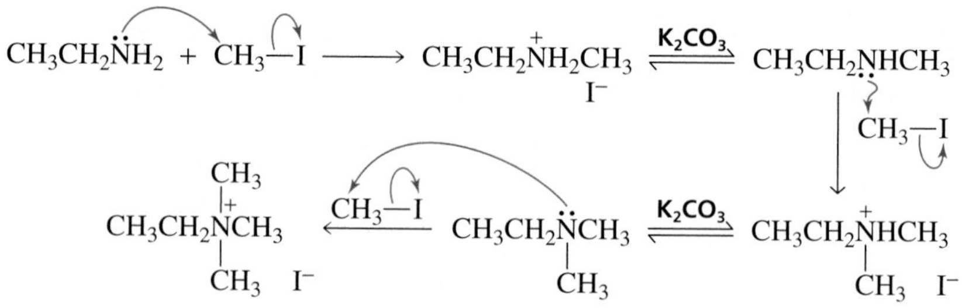

PROBLEM 9

Reaction of an alkyl halide with ammonia gives a low yield of primary amine. A much better yield of primary amine is obtained from the reaction of an alkyl halide with azide ion ($^-N_3$), followed by catalytic hydrogenation. Explain. (*Hint:* An alkyl azide is not nucleophilic.)

$$CH_3CH_2CH_2Br \xrightarrow{^-N_3} CH_3CH_2CH_2\overset{+}{N}=\overset{+}{N}=N^- \xrightarrow[\text{Pt}]{H_2} CH_3CH_2CH_2NH_2 + N_2$$
$$\text{an alkyl azide}$$

PROBLEM 10

Using the pK_a values listed in Table 9.3, convince yourself that each of the reactions on page 366 proceeds in the direction shown.

PROBLEM 11 ◆

What is the product of the reaction of ethyl bromide with each of the following nucleophiles?

a. CH_3OH b. $^-N_3$ c. $(CH_3)_3N$

PROBLEM 12

The reaction of an alkyl chloride with potassium iodide is generally carried out in acetone in order to maximize the amount of alkyl iodide that is formed. Why does the solvent increase the yield of alkyl iodide? (*Hint:* Potassium iodide is soluble in acetone, but potassium chloride is not.)

WHY CARBON INSTEAD OF SILICON?

There are two reasons why living organisms are composed primarily of carbon, oxygen, hydrogen, and nitrogen—*fitness* for specific roles in life processes and *availability* in the environment. Of the two reasons, fitness was probably more important than availability because carbon became the fundamental building block of living organisms instead of silicon—even though silicon is more than 140 times more abundant than carbon in the Earth's crust.

Abundance (atoms/100 atoms)

Element	Living organisms	Earth's crust
H	49	0.22
C	25	0.19
O	25	47
N	0.3	0.1
Si	0.03	28

Why are hydrogen, carbon, oxygen, and nitrogen so fit for the roles they play in living organisms? First and foremost, they are among the smallest atoms that form covalent bonds, and carbon, oxygen, and nitrogen can also form multiple bonds. Because the atoms are small and can form multiple bonds, they form the strongest bonds, thereby giving rise to the most stable molecules. The compounds that make up living organisms must be stable and, therefore, slow to react in order for organisms to survive.

Silicon has almost twice the diameter of carbon, so silicon forms longer and weaker bonds. Consequently, an S_N2 reaction at silicon would occur much more rapidly than an S_N2 reaction at carbon. Silicon also has empty d orbitals that are available to accept electrons, so silicon compounds would not exist very long in the presence of any compound with lone pair electrons, including water. This alone would eliminate silicon from consideration as the fundamental building block of living systems, but silicon has yet another problem. The end product of carbon metabolism is CO_2. The analogous product of silicon metabolism would be SiO_2. Because silicon is only single-bonded to oxygen in SiO_2, silicon dioxide molecules polymerize to form quartz (sea sand). Clearly, life based on animals exhaling CO_2 is preferable to life based on animals exhaling sea sand.

Given our understanding of the S_N2 reaction, we would expect the rate of reaction of *tert*-butyl bromide with water to be relatively slow because water is a poor nucleophile and *tert*-butyl bromide is sterically hindered to attack by a nucleophile. It turns out, however, that the reaction is surprisingly fast. In fact, it is over one million times faster than the reaction of methyl bromide—a compound with no steric hindrance—with water (Table 9.4). Clearly, the reaction must be taking place by a mechanism different from that of an S_N2 reaction.

9.5
THE MECHANISM OF S_N1 REACTIONS

$$CH_3-\underset{\underset{CH_3}{|}}{\overset{\overset{CH_3}{|}}{C}}-Br \;+\; H_2O \;\longrightarrow\; CH_3-\underset{\underset{CH_3}{|}}{\overset{\overset{CH_3}{|}}{C}}-OH \;+\; HBr$$

***tert*-butyl bromide**   ***tert*-butyl alcohol**

As we have seen, a study of the kinetics of a reaction is one of the first steps undertaken when investigating the mechanism of a reaction. If we were to investigate the kinetics of the reaction of *tert*-butyl bromide with water, we would find that doubling the concentration of the alkyl halide doubles the rate of the reaction. We would also find that changing the concentration of the nucleophile has no effect on the rate of the reaction. Knowing that the rate of this nucleophilic substitution reaction depends only on the concentration of the alkyl halide, we can write the following rate law for the reaction.

$$\text{rate} = k[\text{alkyl halide}]$$

Because the rate of the reaction depends on the concentration of only one reactant, it is a **first-order reaction.**

The rate law for the reaction of *tert*-butyl bromide with water differs from the rate law for the reaction of methyl bromide with hydroxide ion, so the two reactions must follow different mechanisms. We have seen that the reaction between methyl bromide and hydroxide ion is an S_N2 reaction (Section 9.2). The reaction between *tert*-butyl bromide and water is an **S_N1 reaction** where S stands for substitution, N stands for nucleophilic, and 1 stands for unimolecular. The mechanism of an S_N1 reaction is based on the following experimental evidence:

1. The rate law shows that the rate of the reaction depends only on the concentration of the alkyl halide. This means that we must be observing a reaction whose rate-determining transition state involves only the alkyl halide. The rate-determining transition state, therefore, is **unimolecular** because it involves only one molecule.

2. When the methyl groups of *tert*-butyl bromide are successively replaced by hydrogens, the rate of the S_N1 reaction decreases progressively (Table 9.4). This is opposite to the order of reactivity exhibited by alkyl halides in S_N2 reactions (Table 9.1).

3. The reaction of an alkyl halide in which the halogen is bonded to a chirality center forms two stereoisomers: one with the same relative configuration as the reacting alkyl halide, and the other with the inverted configuration.

TABLE 9.4 Relative Rates of S_N1 Reactions for Several Alkyl Bromides (Solvent Is H_2O, Nucleophile is H_2O)

Alkyl bromide	Class	Relative rate
CH_3 CH_3C-Br CH_3	tertiary	1,200,000
CH_3CH-Br CH_3	secondary	11.6
CH_3CH_2-Br	primary	1.00*
CH_3-Br	methyl	1.05*

*Although the rate of the S_N1 reaction of this compound with water is 0, a small rate is observed as a result of an S_N2 reaction.

An S_N1 reaction has two steps. In the first step, the carbon–halogen bond breaks heterolytically, with the halogen retaining the previously shared pair of electrons. In the second step, the nucleophile reacts rapidly with the carbocation formed in the first step.

mechanism of the S$_N$1 reaction

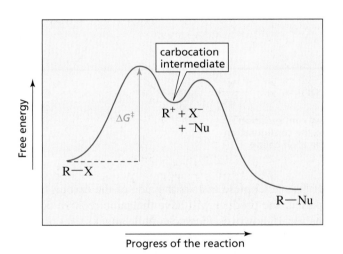

3-D Molecules:
t-Butyl bromide;
t-Butyl cation;
protonated
t-Butyl alcohol;
t-Butyl alcohol

Because the rate of an S$_N$1 reaction depends only on the concentration of the alkyl halide, the first step must be the slow and rate-determining step. The nucleophile, therefore, is not involved in the rate-determining step, so its concentration has no effect on the rate of the reaction. If you look at the reaction coordinate diagram in Figure 9.4, you will be able to see why increasing the rate of the second step will not make an S$_N$1 reaction go any faster.

◀ **Figure 9.4**
Reaction coordinate diagram for an S$_N$1 reaction.

How does the mechanism for an S$_N$1 reaction account for the three pieces of experimental evidence? First, because the alkyl halide is the only species that participates in the rate-limiting step, the mechanism agrees with the observation that the rate of the reaction depends on the concentration of the alkyl halide and does not depend on the concentration of the nucleophile.

Second, a carbocation is formed in the slow step of an S$_N$1 reaction. Because a tertiary carbocation is more stable and is therefore easier to form than a secondary carbocation, which in turn is more stable and easier to form than a primary carbocation (Section 3.10), tertiary alkyl halides are more reactive than secondary alkyl halides, which are more reactive than primary alkyl halides in an S$_N$1 reaction. Thus, the reactivity order agrees with the observation that the rate of an S$_N$1 reaction decreases as the methyl groups of *tert*-butyl bromide are successively replaced by hydrogens (Table 9.4).

relative reactivities of alkyl halides in an S_N1 reaction

$$3° \text{ alkyl halide} > 2° \text{ alkyl halide} > 1° \text{ alkyl halide}$$

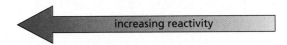

increasing reactivity

Actually, primary carbocations and methyl cations are so unstable that primary alkyl halides and methyl halides do not undergo S_N1 reactions. (The very slow reactions reported for ethyl bromide and methyl bromide in Table 9.4 are S_N2 reactions.)

The positively charged carbon of the carbocation intermediate is sp^2 hybridized, and the three bonds connected to an sp^2 hybridized carbon are in the same plane. In the second step of the S_N1 reaction, the nucleophile can approach the carbocation from either side of the plane.

Tutorial: S_N1

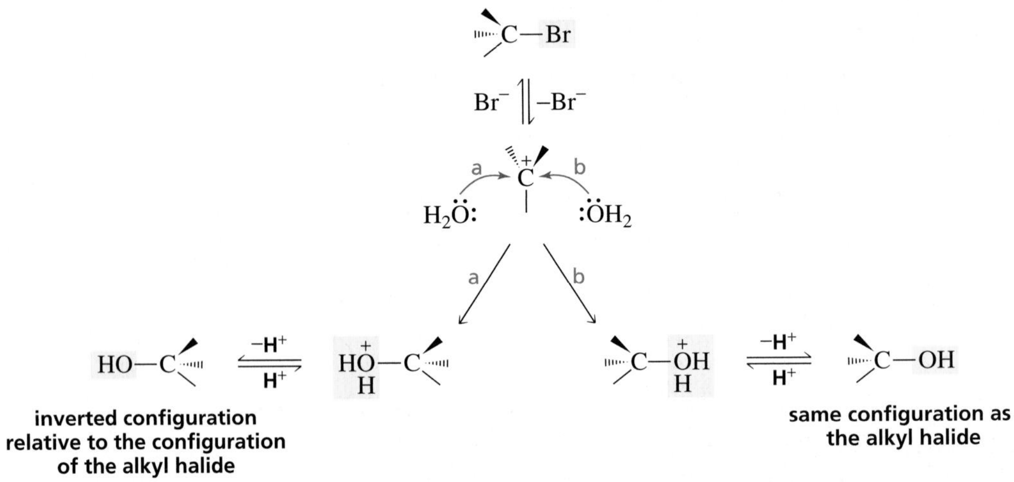

**inverted configuration
relative to the configuration
of the alkyl halide**

**same configuration as
the alkyl halide**

If the nucleophile attacks the side of the carbon from which the leaving group departed, the product will have the same relative configuration as the reacting alkyl halide. If, however, the nucleophile attacks the opposite side of the carbon, the product will have the inverted configuration relative to the configuration of the alkyl halide. We can now understand the third piece of experimental evidence for the mechanism of the S_N1 reaction. The S_N1 reaction of an alkyl halide, in which the leaving group is attached to a chirality center, forms two stereoisomers because attack of the nucleophile on one side of the planar carbocation forms one stereoisomer, and attack on the other side produces the other stereoisomer.

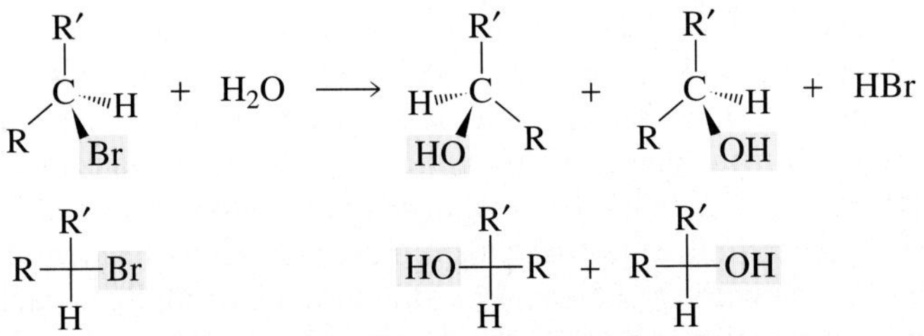

**if the leaving group in an S_N1 reaction is attached to a chirality center,
a pair of enantiomers will be formed as products**

PROBLEM 13 ◆

Arrange the following alkyl bromides in order of decreasing reactivity in an S$_N$1 reaction: isopropyl bromide, propyl bromide, *tert*-butyl bromide, methyl bromide.

The Leaving Group

Because the rate-determining step of an S$_N$1 reaction is the dissociation of the alkyl halide to form a carbocation, two factors affect the rate of an S$_N$1 reaction—the ease with which the leaving group dissociates from the carbon and the stability of the carbocation that is formed. In the preceding section, we saw that tertiary alkyl halides are more reactive than secondary alkyl halides, which are more reactive than primary alkyl halides. This is because the more substituted the carbocation is, the more stable it is and therefore the easier it is to form. But what about a series of alkyl halides with different leaving groups that dissociate to form the same carbocation? The answer is the same for the S$_N$1 reaction as for the S$_N$2 reaction. That is, the weaker the base, the less tightly it is bonded to the carbon and the easier it is to break the carbon–halogen bond. As a result, an alkyl iodide is the most reactive and an alkyl fluoride is the least reactive of the alkyl halides in both S$_N$1 and S$_N$2 reactions.

relative reactivities of alkyl halides in an S$_N$1 reaction

$$RI \; > \; RBr \; > \; RCl \; > \; RF$$

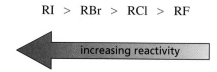

increasing reactivity

The Nucleophile

The nucleophile traps the carbocation that is formed in the rate-determining step of an S$_N$1 reaction. Because the nucleophile comes into play after the rate-determining step, the reactivity of the nucleophile has no effect on the rate of the S$_N$1 reaction.

In some S$_N$1 reactions, the solvent is the nucleophile. For example, the relative rates given in Table 9.4 are for the reactions of alkyl halides with water. Water serves as both the nucleophile and the solvent. When the solvent is the nucleophile, the reaction is called a **solvolysis reaction.** Thus, the relative rates in Table 9.4 are for the solvolysis reactions of the indicated alkyl bromides in water.

Carbocation Rearrangements

A carbocation intermediate is formed in an S$_N$1 reaction. In Section 3.14 we saw that a carbocation will rearrange if it becomes more stable in the process. If the carbocation formed in an S$_N$1 reaction can rearrange, S$_N$1 and S$_N$2 reactions of the same alkyl halide can produce different constitutional isomers as products because carbocations are not formed in an S$_N$2 reaction and, therefore, rearrangement of the carbon skeleton cannot occur. For example, the product obtained when 2-bromo-3-methylbutane undergoes an S$_N$1 reaction is different from the product obtained when it undergoes an S$_N$2 reaction. When the reaction is carried out under conditions that favor an S$_N$1 reaction, the initially formed secondary carbocation undergoes a 1,2-hydride shift, rearranging to a more stable tertiary carbocation.

$$
\begin{array}{c}
\text{CH}_3 \\
| \\
\text{CH}_3\text{CHCHCH}_3 \\
\underset{+}{} \\
\boxed{\text{secondary carbocation}}
\end{array}
\xrightarrow{\text{1,2-hydride shift}}
\begin{array}{c}
\text{CH}_3 \\
| \\
\text{CH}_3\overset{+}{\text{C}}\text{CH}_2\text{CH}_3 \\
\boxed{\text{tertiary carbocation}}
\end{array}
\xrightarrow{\text{H}_2\text{O}}
\begin{array}{c}
\text{CH}_3 \\
| \\
\text{CH}_3\text{CCH}_2\text{CH}_3 \\
\overset{+}{\text{O}}\text{H} \\
\text{H}
\end{array}
$$

$$
\begin{array}{c}
\text{CH}_3 \\
| \\
\text{CH}_3\text{CHCHCH}_3 \\
| \\
\text{Br} \\
\textbf{2-bromo-} \\
\textbf{3-methylbutane}
\end{array}
\xrightarrow[\text{HO}^-]{\text{S}_\text{N}2}
\begin{array}{c}
\text{CH}_3 \\
| \\
\text{CH}_3\text{CHCHCH}_3 \\
| \\
\text{OH} \\
\textbf{3-methyl-2-butanol}
\end{array}
$$

$$
\text{H}^+ \parallel -\text{H}^+
$$

$$
\begin{array}{c}
\text{CH}_3 \\
| \\
\text{CH}_3\text{CCH}_2\text{CH}_3 \\
| \\
\text{OH} \\
\textbf{2-methyl-2-butanol}
\end{array}
$$

When a reaction forms a carbocation intermediate, always check for the possibility of a carbocation rearrangement.

The product obtained from the reaction of 3-bromo-2,2-dimethylbutane with a nucleophile also depends on the conditions under which the reaction is carried out. The carbocation formed under conditions that favor an S$_\text{N}$1 reaction undergoes a 1,2-methyl shift. Because a carbocation is not formed under conditions that favor an S$_\text{N}$2 reaction, the carbon skeleton does not rearrange.

$$
\begin{array}{c}
\text{CH}_3 \\
| \\
\text{CH}_3\text{C}-\overset{+}{\text{C}}\text{HCH}_3 \\
| \\
\text{CH}_3 \\
\boxed{\text{secondary carbocation}}
\end{array}
\xrightarrow{\text{1,2-methyl shift}}
\begin{array}{c}
\text{CH}_3 \\
| \\
\text{CH}_3\overset{+}{\text{C}}-\text{CHCH}_3 \\
| \\
\text{CH}_3 \\
\boxed{\text{tertiary carbocation}}
\end{array}
\xrightarrow{\text{H}_2\text{O}}
\begin{array}{c}
\text{CH}_3 \\
| \\
\text{CH}_3\text{C}-\text{CHCH}_3 \\
\overset{+}{\text{O}}\text{H} \ \text{CH}_3 \\
\text{H}
\end{array}
$$

$$
\begin{array}{c}
\text{CH}_3 \\
| \\
\text{CH}_3\text{C}-\text{CHCH}_3 \\
| \\
\text{CH}_3 \ \text{Br} \\
\textbf{3-bromo-2,2-} \\
\textbf{dimethylbutane}
\end{array}
\xrightarrow[\text{HO}^-]{\text{S}_\text{N}2}
\begin{array}{c}
\text{CH}_3 \\
| \\
\text{CH}_3\text{C}-\text{CHCH}_3 \\
| \\
\text{CH}_3\text{OH} \\
\textbf{3,3-dimethyl-2-butanol}
\end{array}
$$

$$
\text{H}^+ \parallel -\text{H}^+
$$

$$
\begin{array}{c}
\text{CH}_3 \\
| \\
\text{CH}_3\text{C}-\text{CHCH}_3 \\
\text{OH} \ \text{CH}_3 \\
\textbf{2,3-dimethyl-2-butanol}
\end{array}
$$

We will see in Sections 9.9 and 9.10 that we can exercise some control over whether an S$_\text{N}$1 or an S$_\text{N}$2 reaction takes place by choosing appropriate reaction conditions.

PROBLEM 14 ◆

Arrange the following alkyl halides in order of decreasing reactivity in an S$_\text{N}$1 reaction: 2-bromopentane, 2-chloropentane, 1-chloropentane, 3-bromo-3-methylpentane.

PROBLEM 15 ◆

Which of the following alkyl halides form a substitution product from an S$_\text{N}$1 reaction that is different from the substitution product formed from an S$_\text{N}$2 reaction?

a.
$$
\begin{array}{c}
\text{CH}_3 \ \ \text{Br} \\
| \ \ \ \ | \\
\text{CH}_3\text{CHCHCHCH}_3 \\
| \\
\text{CH}_3
\end{array}
$$

b.

c.
$$
\begin{array}{c}
\text{CH}_3 \\
| \\
\text{CH}_3\text{CH}_2\text{C}-\text{CHCH}_3 \\
| \ \ \ \ | \\
\text{CH}_3 \ \text{Br}
\end{array}
$$

d.
$$
\begin{array}{c}
\text{CH}_3 \\
| \\
\text{CH}_3\text{CHCH}_2\text{CCH}_3 \\
| \ \ \ \ \ \ \ | \\
\text{Cl} \ \ \ \ \text{CH}_3
\end{array}
$$

PROBLEM 16 ◆

Two substitution products result from the reaction of 3-chloro-3-methyl-1-butene with sodium acetate $(CH_3COO^-Na^+)$ in acetic acid under S_N1 conditions. Identify the products. Which is the thermodynamic product? Which is the kinetic product?

The Stereochemistry of S_N2 Reactions

The substitution product obtained from the reaction of 2-bromopropane with hydroxide ion does not have a chirality center. Therefore, stereoisomers are not possible for 2-propanol.

$$CH_3CHCH_3 + HO^- \longrightarrow CH_3CHCH_3 + Br^-$$

	Br		OH	
	2-bromopropane		**2-propanol**	

The substitution product obtained from the reaction of 2-bromobutane with hydroxide ion has a chirality center. Therefore, 2-butanol can exist as a pair of enantiomers.

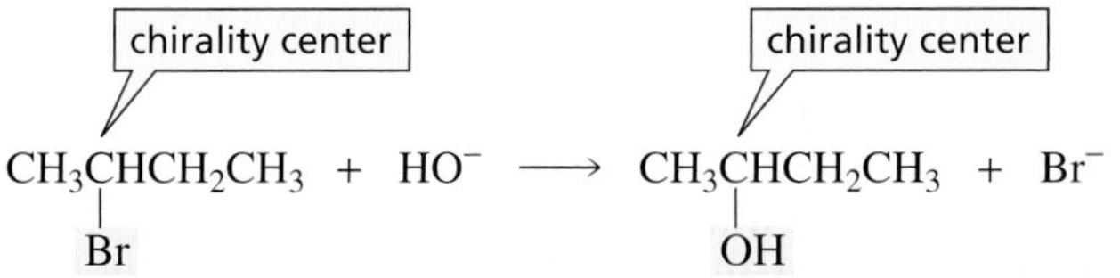

chirality center   chirality center

$$CH_3CHCH_2CH_3 + HO^- \longrightarrow CH_3CHCH_2CH_3 + Br^-$$

Br   OH

2-bromobutane   **2-butanol**

We cannot specify the configuration of the product formed from the reaction of 2-bromobutane with hydroxide ion unless we know the configuration of the alkyl halide and whether the reaction is an S_N2 or an S_N1 reaction. For example, if we know the reactant has the S configuration and the reaction is an S_N2 reaction, we know the product will be (R)-2-butanol because in an S_N2 reaction, the incoming hydroxide ion attacks the chirality center on the side opposite to where the bromine is bonded. This forms a product whose configuration is inverted relative to the configuration of the reactant—an S_N2 reaction takes place with *inversion of configuration*.

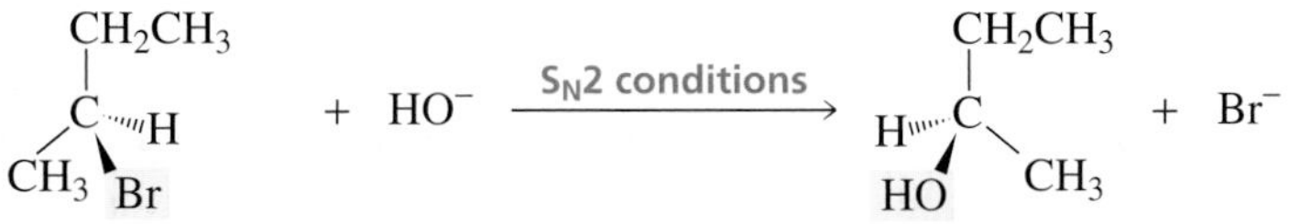

(S)-2-bromobutane (R)-2-butanol

The Stereochemistry of S_N1 Reactions

In the S_N1 reaction of (S)-2-bromobutane with water, two substitution products are formed—one has the same relative configuration as the reactant, and the other has the inverted configuration. In an S_N1 reaction, the leaving group leaves before the nucleophile attacks. This means the nucleophile is free to attack either side of the planar carbocation. If it attacks the side from which the bromide ion left, the product will have the same relative configuration as the reactant. If it attacks the opposite side, the product will have the inverted configuration.

Saul Winstein (1912–1969) was born in Montreal, Canada. He received a Ph.D. from the California Institute of Technology and was a professor of chemistry at the University of California, Los Angeles, from 1942 until his death.

Although you might expect that equal amounts of both products should be formed in an S_N1 reaction, a greater amount of the product with the inverted configuration is obtained in most cases. Typically, 50–70% of the product of an S_N1 reaction is the inverted product. If the reaction leads to equal amounts of the two stereoisomers, the reaction is said to take place with **complete racemization.** When more of the inverted product is formed, the reaction is said to take place with **partial racemization.**

Saul Winstein was the first to explain why extra inverted product generally is formed in an S_N1 reaction. He postulated that dissociation of the alkyl halide initially results in the formation of an **intimate ion pair.** In an intimate ion pair, the bond between the carbon and the leaving group has broken but the cation and anion remain next to each other. This species then forms an ion pair in which one or more solvent molecules have come between the cation and the anion. This is called a **solvent-separated ion pair.** Further separation between the two results in dissociated ions.

		solvent	
R—X $\longrightarrow$	R⁺ X⁻ $\longrightarrow$	R⁺ ▨ X⁻ $\longrightarrow$	R⁺ ▨ X⁻
undissociated molecule	intimate ion pair	solvent-separated ion pair	dissociated ions

The nucleophile can attack any of these four species. If the nucleophile attacks only the completely dissociated carbocation, the product will be completely racemized. If the nucleophile attacks the carbocation of either the intimate ion pair or the solvent-separated ion pair, the leaving group will be in position to partially block the approach of the nucleophile to that side of the carbocation. As a result, more of the product with the inverted configuration will be obtained. (If the nucleophile attacks the undissociated molecule, it will be an S_N2 reaction and all of the product will have the inverted configuration.)

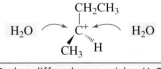

Br⁻ has diffused away, giving H₂O equal access to both sides of the carbocation	Br⁻ has not diffused away, so it blocks the approach of H₂O to one side of the carbocation

The difference between the products obtained from an S$_N$1 reaction and from an S$_N$2 reaction is a little easier to visualize in the case of cyclic compounds. When *cis*-1-bromo-4-methylcyclohexane undergoes an S$_N$2 reaction, only the trans product is obtained because the carbon bonded to the leaving group is attacked by the nucleophile only on its back side.

cis-**1-bromo-4-methylcyclohexane** → (HO⁻, S$_N$2 conditions) → *trans*-**4-methylcyclohexanol** + Br⁻

However, when *cis*-1-bromo-4-methylcyclohexane undergoes an S$_N$1 reaction, both the cis and the trans products are formed because the nucleophile can approach the carbocation intermediate from either side.

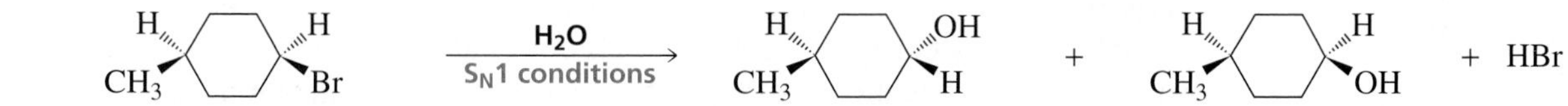

cis-**1-bromo-4-methylcyclohexane** → (H$_2$O, S$_N$1 conditions) → *trans*-**4-methylcyclohexanol** + *cis*-**4-methylcyclohexanol** + HBr

---

PROBLEM 17 ◆

If the products of the preceding reaction are not obtained in equal amounts, which stereoisomer would be present in excess?

PROBLEM 18

Give the products that will be obtained from the following reactions if:

a. the reaction is carried out under conditions that favor an S$_N$2 reaction.

b. the reaction is carried out under conditions that favor an S$_N$1 reaction.

    1. *trans*-1-bromo-4-methylcyclohexane + sodium methoxide/methanol

    2. *cis*-1-chloro-2-methylcyclobutane + sodium hydroxide/water

PROBLEM 19 ◆

Which of the following reactions will go faster if the concentration of the nucleophile is increased?

a. 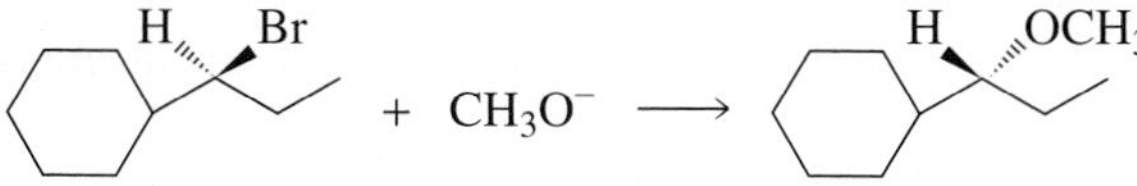 + CH$_3$O⁻ ⟶

b. 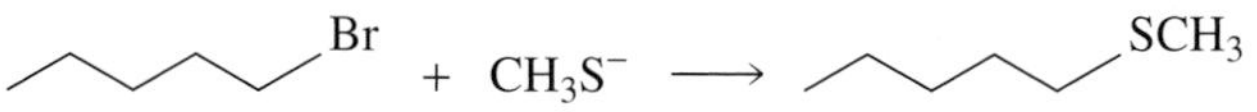 + CH$_3$S⁻ ⟶

c. 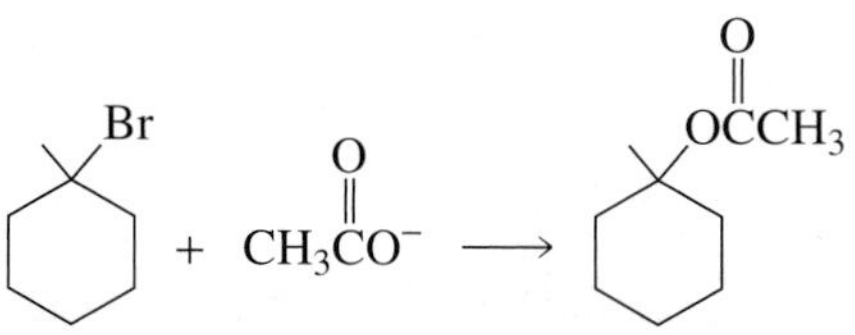 + CH$_3$CO⁻ ⟶

9.8
BENZYLIC HALIDES, ALLYLIC HALIDES, VINYLIC HALIDES, AND ARYL HALIDES

To this point, we have limited our discussion of substitution reactions to methyl halides and primary, secondary, and tertiary alkyl halides. But what about benzylic, allylic, vinylic, and aryl halides? Let's first consider benzylic and allylic halides. Benzylic and allylic halides readily undergo S_N2 reactions unless they are tertiary. Tertiary benzylic and tertiary allylic halides, like other tertiary halides, are unreactive in S_N2 reactions because of steric hindrance.

$$\text{benzyl chloride} \quad \longrightarrow\text{CH}_2\text{Cl} + \text{CH}_3\text{O}^- \xrightarrow{\text{S}_N\text{2 conditions}} \longrightarrow\text{CH}_2\text{OCH}_3 + \text{Cl}^-$$

benzyl chloride   benzyl methyl ether

$$\text{CH}_3\text{CH}=\text{CHCH}_2\text{Br} + \text{HO}^- \xrightarrow{\text{S}_N\text{2 conditions}} \text{CH}_3\text{CH}=\text{CHCH}_2\text{OH} + \text{Br}^-$$
1-bromo-2-butene   **2-buten-1-ol**
an allylic halide

Benzylic and allylic halides also undergo S_N1 reactions because they form relatively stable carbocations. While primary alkyl halides (such as CH_3CH_2Br and $CH_3CH_2CH_2Br$) cannot undergo S_N1 reactions because their carbocations are too unstable, primary benzylic and primary allylic halides readily undergo S_N1 reactions because their carbocations are stabilized by electron delocalization (Section 6.7).

3-D Molecule: Benzyl cation

$$\longrightarrow\text{CH}_2\text{Cl} \xrightarrow{\text{S}_N\text{1}} \longrightarrow\overset{+}{\text{C}}\text{H}_2 \xrightarrow{\text{CH}_3\text{OH}} \longrightarrow\text{CH}_2\text{OCH}_3 + \text{H}^+$$
$$+ \text{ Cl}^-$$

$$\text{CH}_2=\text{CHCH}_2\text{Br} \xrightarrow{\text{S}_N\text{1}} \text{CH}_2=\text{CH}\overset{+}{\text{C}}\text{H}_2 \xrightarrow{\text{H}_2\text{O}} \text{CH}_2=\text{CHCH}_2\text{OH} + \text{H}^+$$
$$+ \text{ Br}^-$$

If the resonance contributors of the allylic carbocation intermediate do not have the same groups bonded to their sp^2 carbons, two substitution products will be obtained.

$$\text{CH}_3\text{CH}=\text{CHCH}_2\text{Br} \xrightarrow{\text{S}_N\text{1}} \text{CH}_3\text{CH}=\text{CH}\overset{+}{\text{C}}\text{H}_2 \longleftrightarrow \text{CH}_3\overset{+}{\text{C}}\text{HCH}=\text{CH}_2$$

3-D Molecule: Allyl cation

$$\downarrow \text{H}_2\text{O} \qquad\qquad\qquad \downarrow \text{H}_2\text{O}$$

$$\begin{array}{cc} \text{CH}_3\text{CH}=\text{CHCH}_2\text{OH} & \text{CH}_3\text{CHCH}=\text{CH}_2 \\ + \text{ H}^+ & \text{OH} \quad + \text{ H}^+ \end{array}$$

Vinylic halides and aryl halides do not undergo S_N2 or S_N1 reactions. They do not undergo S_N2 reactions because as the nucleophile approaches the back side of the sp^2 carbon, it is repelled by the π electron cloud of the double bond or the aromatic ring.

a nucleophile is repelled by the π electron cloud

$$\begin{array}{c} \text{R} \quad\quad \text{Cl} \\ \text{C}=\text{C} \\ \text{H} \quad\quad \text{H} \\ \times \text{nucleophile} \end{array} \qquad \begin{array}{c} \text{—Br} \\ \times \text{nucleophile} \end{array}$$

a vinylic halide   **an aryl halide**

There are two reasons why vinylic halides and aryl halides do not undergo S_N1 reactions. First, vinylic and aryl cations are even more unstable than primary carbocations, and we saw that primary alkyl halides do not undergo S_N1 reactions because of the instability of their carbocations (Section 9.5). The instability of vinylic and aryl cations is the result of the positive charge being on an *sp* carbon. Because *sp* carbons are more electronegative than the sp^2 carbons that carry the positive charge of alkyl carbocations, *sp* carbons are more resistant to becoming positively charged. Second, we have seen that sp^2 carbons form stronger bonds than sp^3 carbons do (Section 1.14). As a result, it is harder to break the carbon–halogen bond when the halogen is bonded to an sp^2 carbon.

$$RCH{=}CHCl \not\longrightarrow RCH{=}\overset{+}{C}H \quad + \quad Cl^-$$

sp hybridized

vinyl cation
too unstable to be formed

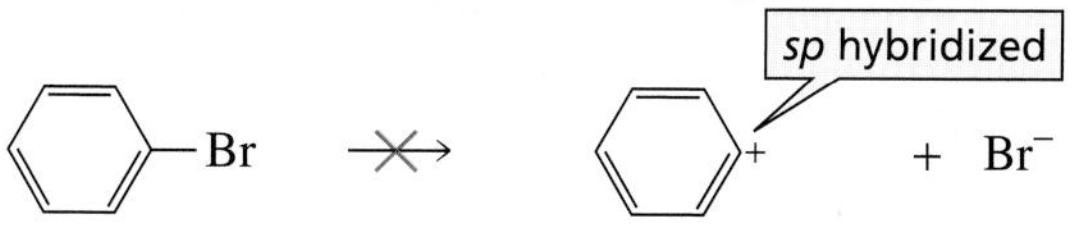

aryl cation
too unstable to be formed

PROBLEM 20 ◆

Which alkyl halide would you expect to be more reactive in an S_N2 reaction with a given nucleophile?

a. $CH_3CH_2CH_2Br$ or $CH_3CH_2CH_2I$

b. $CH_3CH_2CH_2Cl$ or CH_3OCH_2Cl

c. $CH_3CH_2\overset{CH_3}{\overset{|}{C}}HBr$ or $CH_3CH_2\overset{CH_2CH_3}{\overset{|}{C}}HBr$

d. $CH_3CH_2CH_2\overset{CH_3}{\overset{|}{C}}HBr$ or $CH_3CH_2\overset{CH_3}{\overset{|}{C}}HCH_2Br$

e. 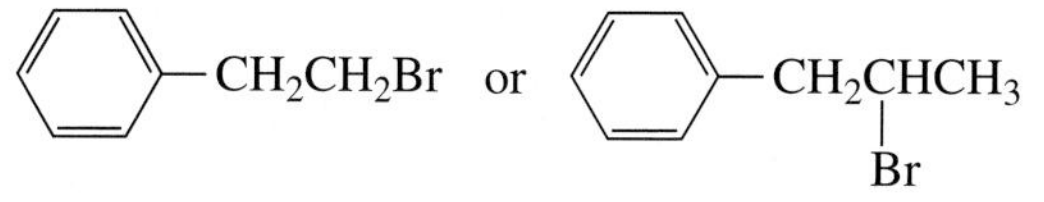

f. 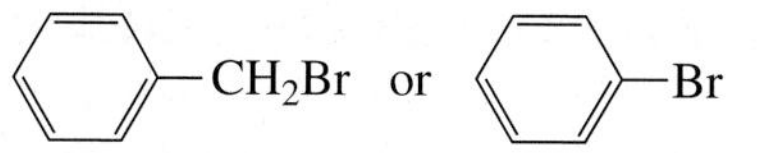

g. $CH_3CH{=}\underset{Br}{\underset{|}{C}}CH_3$ or $CH_3CH{=}CH\underset{Br}{\underset{|}{C}}HCH_3$

PROBLEM 21 ◆

For each of the pairs in Problem 20, which compound would be more reactive in an S_N1 reaction?

PROBLEM 22

Give the products of the following reaction:

a. under conditions that favor an S_N2 reaction.

b. under conditions that favor an S_N1 reaction.

$$CH_3CH{=}CHCH_2Br \quad + \quad CH_3O^- \xrightarrow{CH_3OH}$$

9.9
COMPETITION BETWEEN S$_N$2 AND S$_N$1 REACTIONS

The characteristics of the S$_N$2 and S$_N$1 reactions are summarized in Table 9.5. It is important to remember that the "2" in S$_N$2 and the "1" in S$_N$1 refer to the molecularity of the rate-determining step. Thus, the rate-determining step of an S$_N$2 reaction is bimolecular whereas the rate-determining step of an S$_N$1 reaction is unimolecular. These numbers do *not* refer to the number of steps in the mechanism. In fact, just the opposite is true—an S$_N$2 reaction proceeds by a one-step concerted mechanism whereas an S$_N$1 reaction proceeds by a two-step mechanism with a carbocation intermediate.

TABLE 9.5   Comparison of the S$_N$2 and S$_N$1 Reactions

S$_N$2	S$_N$1
A one-step mechanism	A two-step mechanism
A bimolecular rate-determining transition state	A unimolecular rate-determining transition state
No carbocation rearrangements	Carbocation rearrangements
Product has inverted configuration compared with reactant	Products have both identical and inverted configurations compared with reactant
Reactivity order: methyl $>$ 1° $>$ 2° $>$ 3°	Reactivity order: 3° $>$ 2° $>$ 1° $>$ methyl

We have seen that methyl halides and primary alkyl halides undergo only S$_N$2 reactions because methyl cations and primary carbocations, which would be formed in an S$_N$1 reaction, are too unstable to be formed. Tertiary alkyl halides undergo only S$_N$1 reactions because steric hindrance makes them very unreactive in an S$_N$2 reaction. Secondary alkyl halides as well as benzylic and allylic halides (unless they are tertiary) can undergo both S$_N$1 and S$_N$2 reactions because they form relatively stable carbocations and the amount of steric hindrance associated with these alkyl halides is generally not very great. Vinylic and aryl halides do not undergo either S$_N$1 or S$_N$2 reactions. These results are summarized in Table 9.6.

TABLE 9.6   Summary of the Reactivity of Alkyl Halides in Nucleophilic Substitution Reactions

methyl and 1° alkyl halides	S$_N$2 only	vinylic and aryl halides	neither S$_N$1 nor S$_N$2
2° alkyl halides	S$_N$1 and S$_N$2	1° and 2° benzylic and 1° and 2° allylic halides	S$_N$1 and S$_N$2
3° alkyl halides	S$_N$1 only	3° benzylic and 3° allylic halides	S$_N$1 only

When an alkyl halide can undergo both S$_N$1 and S$_N$2 reactions, both reactions take place simultaneously. The conditions under which the reaction is carried out determine which of the reactions predominates. Therefore, we have some experimental control over which reaction takes place.

What conditions favor an S$_N$1 reaction? What conditions favor an S$_N$2 reaction? These are important questions to synthetic chemists because an S$_N$2 reaction results in the formation of a single substitution product whereas an S$_N$1 reaction can form two substitution products if the leaving group is bonded to a chirality center. An S$_N$1 reaction is further complicated by carbocation rearrangements. In other words, an S$_N$2 reaction is a synthetic chemist's friend, but an S$_N$1 reaction can be a synthetic chemist's nightmare.

The conditions that determine whether the predominant reaction will be an S$_N$2 reaction or an S$_N$1 reaction are the concentration of the nucleophile, the reactivity of the nucleophile, and the solvent in which the reaction is carried out. To understand how the concentration of the nucleophile and the reactivity of the nucleophile affect whether an S$_N$2 or an S$_N$1 reaction predominates, we must first examine the rate laws for the two reactions. The rate constants have been given subscripts that indicate the reaction order.

Rate law for the S$_N$2 reaction: rate $= k_2$ [alkyl halide][nucleophile]

Rate law for the S$_N$1 reaction: rate $= k_1$ [alkyl halide]

The rate law for the reaction of an alkyl halide that can undergo both S$_N$2 and S$_N$1 reactions simultaneously is the sum of the individual rate laws.

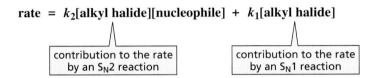

$$\text{rate} = k_2[\text{alkyl halide}][\text{nucleophile}] + k_1[\text{alkyl halide}]$$

| contribution to the rate by an S$_N$2 reaction | contribution to the rate by an S$_N$1 reaction |

From the rate law you can see that an increase in the concentration of the nucleophile increases the rate of an S$_N$2 reaction but has no effect on the rate of an S$_N$1 reaction. Therefore, when both reactions occur simultaneously, increasing the concentration of the nucleophile increases the fraction of the reaction that takes place by an S$_N$2 pathway. On the other hand, decreasing the concentration of the nucleophile decreases the fraction of the reaction that takes place by an S$_N$2 pathway.

The slow (and only) step of an S$_N$2 reaction is attack of the nucleophile on the alkyl halide. Increasing the *reactivity* of the nucleophile increases the rate of an S$_N$2 reaction by increasing the value of the rate constant (k_2), because more reactive nucleophiles are better able to displace the leaving group. The slow step of an S$_N$1 reaction is the dissociation of the alkyl halide. The carbocation formed in the slow step rapidly reacts in a second step with any nucleophile present in the reaction mixture. Increasing the rate of the fast step does not affect the rate of the prior slow, carbocation-forming step. This means that increasing the reactivity of the nucleophile has no effect on the rate of an S$_N$1 reaction. A good nucleophile, therefore, favors an S$_N$2 reaction over an S$_N$1 reaction. A poor nucleophile favors an S$_N$1 reaction, not by increasing the rate of the S$_N$1 reaction itself, but by decreasing the rate of the competing S$_N$2 reaction.

Tutorial: S$_N$2 promoting factors

- An S$_N$2 reaction is favored by a high concentration of a good nucleophile.
- An S$_N$1 reaction is favored by a low concentration of a nucleophile or by a poor nucleophile.

Notice in the previous sections that when S$_N$1 reactions are described, poor nucleophiles (H$_2$O, CH$_3$OH) are used, and when S$_N$2 reactions are described, good nucleophiles (HO$^-$, CH$_3$O$^-$) are used. In other words, a poor nucleophile is used to encourage an S$_N$1 reaction, and a good nucleophile is used to encourage an S$_N$2 reaction.

PROBLEM-SOLVING STRATEGY

This problem will give you practice determining whether a substitution reaction will take place by an S$_N$1 or an S$_N$2 pathway. (Keep in mind that good nucleophiles encourage S$_N$2 reactions, whereas poor nucleophiles encourage S$_N$1 reactions.)

Give the configuration(s) of the substitution product(s) that will be obtained from the reactions of the following secondary alkyl halides with the indicated nucleophile:

a.

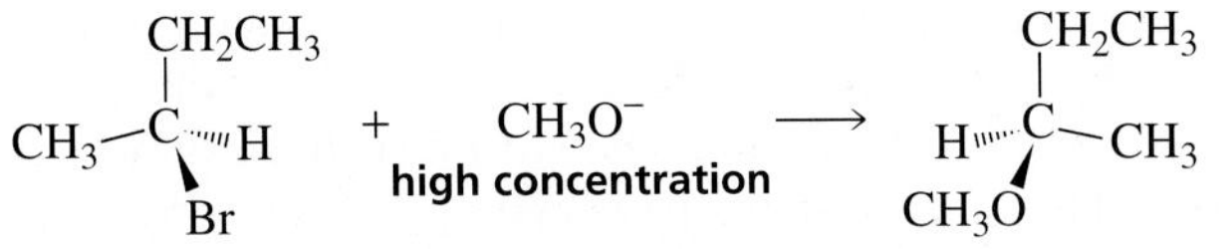

$$CH_3-\overset{\overset{\displaystyle CH_2CH_3}{|}}{\underset{\underset{\displaystyle Br}{|}}{C}}\text{''''}H \quad + \quad CH_3O^-\ \text{\textbf{high concentration}} \quad \longrightarrow \quad H\text{''''}\overset{\overset{\displaystyle CH_2CH_3}{|}}{\underset{\underset{\displaystyle CH_3O}{}}{C}}-CH_3$$

Because a high concentration of a good nucleophile is used, we can predict that the reaction is an S_N2 reaction. Therefore, the product will have the inverted configuration relative to the configuration of the reactant. (An easy way to draw the inverted product is to draw the mirror image of the reacting alkyl halide and then put the nucleophile in the same location as the leaving group.)

b.

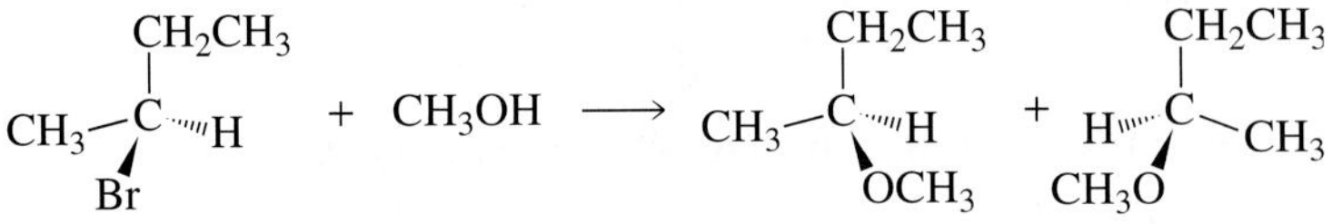

$$CH_3-\overset{\overset{\displaystyle CH_2CH_3}{|}}{\underset{\underset{\displaystyle Br}{|}}{C}}\text{''''}H \quad + \quad CH_3OH \quad \longrightarrow \quad CH_3-\overset{\overset{\displaystyle CH_2CH_3}{|}}{\underset{\underset{\displaystyle OCH_3}{}}{C}}\text{''''}H \quad + \quad H\text{''''}\overset{\overset{\displaystyle CH_2CH_3}{|}}{\underset{\underset{\displaystyle CH_3O}{}}{C}}CH_3$$

Because a poor nucleophile is used, we can predict that the reaction is an S_N1 reaction. Therefore, we will obtain two substitution products, one with the retained configuration and one with the inverted configuration relative to the configuration of the reactant.

c. 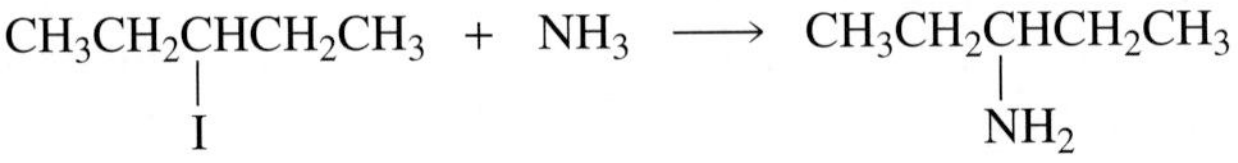 $CH_3CH_2\overset{\overset{\displaystyle }{|}}{\underset{\underset{\displaystyle I}{|}}{C}}HCH_2CH_3 \quad + \quad NH_3 \quad \longrightarrow \quad CH_3CH_2\overset{\overset{\displaystyle }{|}}{\underset{\underset{\displaystyle NH_2}{|}}{C}}HCH_2CH_3$

The relatively poor nucleophile suggests that the reaction is an S_N1 reaction. However, the product does not have a chirality center and so does not have stereoisomers. (The same substitution product would have been obtained if the reaction had been an S_N2 reaction.)

d.  $CH_3CH_2\overset{\overset{\displaystyle }{|}}{\underset{\underset{\displaystyle Cl}{|}}{C}}HCH_3 \quad + \quad HO^-\ \text{\textbf{high concentration}} \quad \longrightarrow \quad CH_3CH_2\overset{\overset{\displaystyle }{|}}{\underset{\underset{\displaystyle OH}{|}}{C}}HCH_3$

Because a high concentration of a good nucleophile is employed, we can predict that the reaction is an S_N2 reaction. Therefore, the product will have the inverted configuration relative to the configuration of the reactant. But since the configuration of the reactant is not indicated, we do not know the configuration of the product.

e. 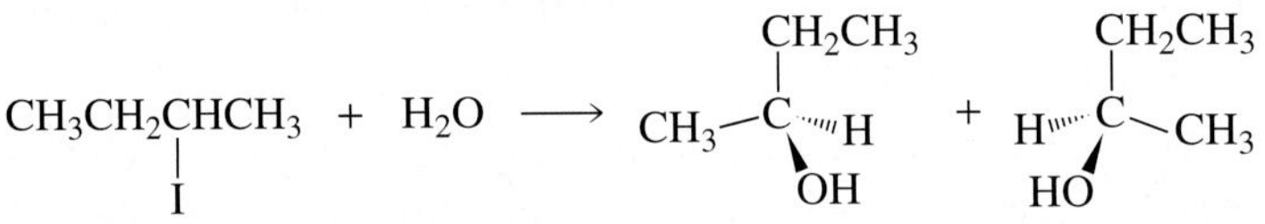 $CH_3CH_2\overset{\overset{\displaystyle }{|}}{\underset{\underset{\displaystyle I}{|}}{C}}HCH_3 \quad + \quad H_2O \quad \longrightarrow \quad CH_3-\overset{\overset{\displaystyle CH_2CH_3}{|}}{\underset{\underset{\displaystyle OH}{}}{C}}\text{''''}H \quad + \quad H\text{''''}\overset{\overset{\displaystyle CH_2CH_3}{|}}{\underset{\underset{\displaystyle HO}{}}{C}}CH_3$

Because a poor nucleophile is employed, we can predict that the reaction is an S_N1 reaction. Therefore, both stereoisomers will be formed regardless of the configuration of the reactant.

Now continue on to Problem 23.

PROBLEM 23

Give the configuration(s) of the substitution product(s) that will be obtained from the reactions of the following secondary alkyl halides with the indicated nucleophile:

a. 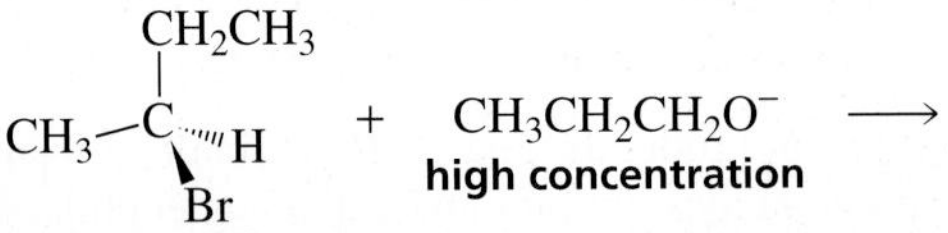 $CH_3-\overset{\overset{\displaystyle CH_2CH_3}{|}}{\underset{\underset{\displaystyle Br}{|}}{C}}\text{''''}H \quad + \quad CH_3CH_2CH_2O^-\ \text{\textbf{high concentration}} \quad \longrightarrow$

b. 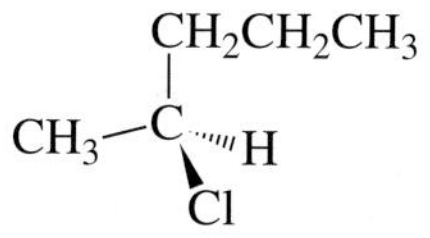 $CH_3-\overset{\overset{\displaystyle CH_2CH_2CH_3}{|}}{\underset{\underset{\displaystyle Cl}{|}}{C}}\text{\tiny{''''}}H$ + NH_3 $\longrightarrow$

c. 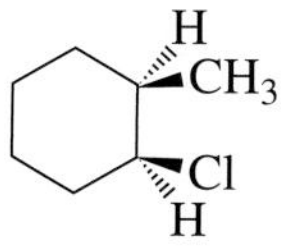 [cyclohexane ring with H, CH₃ up and Cl, H down]   + CH_3O^- **high concentration** $\longrightarrow$

d. 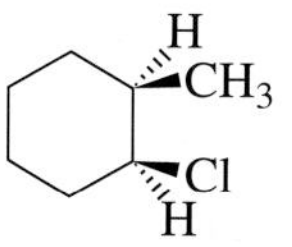 [cyclohexane ring with H, CH₃ up and Cl, H down]   + CH_3OH $\longrightarrow$

e. 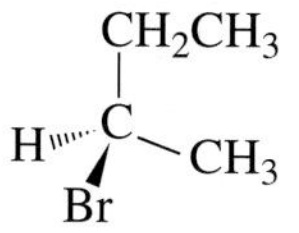 $H\text{\tiny{''''}}\overset{\overset{\displaystyle CH_2CH_3}{|}}{\underset{\underset{\displaystyle Br}{}}{C}}-CH_3$ + CH_3O^- **high concentration** $\longrightarrow$

f. 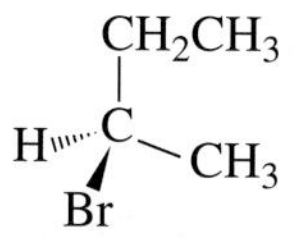 $H\text{\tiny{''''}}\overset{\overset{\displaystyle CH_2CH_3}{|}}{\underset{\underset{\displaystyle Br}{}}{C}}-CH_3$ + CH_3OH $\longrightarrow$

PROBLEM 24 / SOLVED

The rate law for the substitution reaction of 2-bromobutane and HO^- in 75% ethanol/25% water at 30 °C is

$$\text{rate} = 3.20 \times 10^{-5}\,[\text{2-bromobutane}][HO^-] + 1.5 \times 10^{-6}\,[\text{2-bromobutane}]$$

What percent of the reaction takes place by the S_N2 mechanism when:

a. $[HO^-] = 1.00$ M?

b. $[HO^-] = 0.001$ M?

SOLUTION to 24a

$$\text{percentage by } S_N2 = \frac{S_N2}{S_N2 + S_N1} \times 100$$

$$= \frac{3.20 \times 10^{-5}\,[\text{2-bromobutane}]\,(1\cdot00) \times 100}{3.20 \times 10^{-5}\,[\text{2-bromobutane}]\,(1\cdot00) + 1.5 \times 10^{-6}\,[\text{2-bromobutane}]}$$

$$= \frac{3.20 \times 10^{-5}}{3.20 \times 10^{-5} + 0.15 \times 10^{-5}} \times 100 = \frac{3.20 \times 10^{-5}}{3.35 \times 10^{-5}} \times 100$$

$$= 96\%$$

The solvent in which the nucleophilic substitution reaction is carried out also influences whether an S_N2 or an S_N1 reaction will predominate. Before we can understand how a particular solvent favors one reaction over another, however, we must understand how solvents stabilize organic molecules.

The **dielectric constant** of a solvent is a measure of how well the solvent can insulate opposite charges from one another. Solvent molecules insulate charges by

9.10
THE ROLE OF THE SOLVENT IN S_N2 AND S_N1 REACTIONS

clustering around a charge so that the positive poles of the solvent molecules surround negative charges while the negative poles of the solvent molecules surround positive charges. Recall that the interaction between a solvent and an ion or molecule dissolved in that solvent is called *solvation* (Section 2.9). When an ion interacts with a polar solvent, the charge is no longer localized solely on the ion but is spread out to the surrounding solvent molecules. Spreading out the charge stabilizes the charged species.

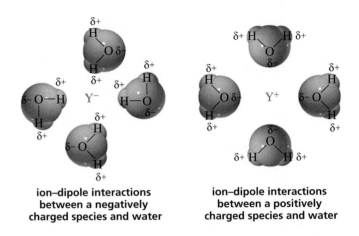

ion–dipole interactions between a negatively charged species and water

ion–dipole interactions between a positively charged species and water

Polar solvents have high dielectric constants and thus are very good at insulating (solvating) charges. Nonpolar solvents have low dielectric constants and are poor insulators. The dielectric constants of some common solvents are listed in Table 9.7. In this table, solvents are divided into two groups: protic solvents and aprotic solvents. Recall that **protic solvents** contain a hydrogen bonded to an oxygen or a nitrogen, so protic solvents are hydrogen bond donors. **Aprotic solvents,** on the other hand, do not have a hydrogen bonded to an oxygen or a nitrogen, so they are *not* hydrogen bond donors.

Stabilization of charges by solvent interaction plays an important role in organic reactions. For example, the first step in an S_N1 reaction is dissociation of the carbon–halogen bond of the alkyl halide to form a carbocation and a halide ion. Energy is required to break the bond but, with no bonds being formed, where does this energy come from? If the reaction is carried out in a polar solvent, the ions that are produced as products are solvated. The energy associated with a single ion–dipole interaction is small, but the additive effect of all the ion–dipole interactions involved in stabilizing a charged species by the solvent represents a great deal of energy. These ion–dipole interactions provide much of the energy necessary for dissociation of the carbon–halogen bond. So in an S_N1 reaction, the alkyl halide does not fall apart spontaneously, solvent molecules pull it apart. An S_N1 reaction, therefore, can take place in a polar solvent but not in a nonpolar solvent. An S_N1 reaction also cannot take place in the gas phase where there are no solvent molecules and consequently no solvation effects.

SOLVATION EFFECTS

The tremendous amount of energy that is provided by solvation can be appreciated by considering the energy required to break the crystal lattice of sodium chloride (Figure 1.1). In the absence of a solvent, sodium chloride must be heated to more than 800 °C to overcome the forces that hold the oppositely charged ions together. However, sodium chloride (table salt) readily dissolves in water at room temperature because solvation of the Na^+ and Cl^- ions by water provides the energy necessary to pull the ions apart.

TABLE 9.7   The Dielectric Constants of Some Common Solvents

Solvent	Structure	Dielectric constant (ε, at 25 °C)
Protic solvents		
Water	H_2O	79
Formic acid	HCOOH	59
Methanol	CH_3OH	33
Ethanol	CH_3CH_2OH	25
tert-Butyl alcohol	$(CH_3)_3COH$	11
Acetic acid	CH_3COOH	6
Aprotic solvents		
Dimethyl sulfoxide	$(CH_3)_2SO$	47
Acetonitrile	CH_3CN	38
Dimethylformamide	$(CH_3)_2NCHO$	37
Hexamethylphosphoramide	$[(CH_3)_2N]_3PO$	30
Acetone	$(CH_3)_2CO$	21
Dichloromethane	CH_2Cl_2	9.1
Ethyl acetate	$CH_3COOCH_2CH_3$	6
Diethyl ether	$CH_3CH_2OCH_2CH_3$	4.3
Benzene		2.3
Hexane	$CH_3(CH_2)_4CH_3$	1.9

The Effect of the Solvent on the Rate of a Reaction

One simple rule describes how a change in solvent will affect the rate of most chemical reactions—*increasing the polarity* of the solvent will *decrease* the rate of the reaction if one or more reactants in the rate-limiting step are charged and will *increase* the rate of the reaction if none of the reactants in the rate-limiting step is charged.

Now let's see why this rule is true. The rate of a reaction depends on the difference between the energy of the reactants and the energy of the transition state in the rate-limiting step of the reaction. We can therefore predict how changing the polarity of the solvent will affect the rate of a reaction. We do this simply by looking at the charge on the reactant(s) and the charge on the transition state of the rate-limiting step to determine which of these species will be more stabilized by a polar solvent. The greater the charge on the solvated molecule, the stronger it will interact with a polar solvent and the more the charge will be stabilized.

Therefore, if the charge on the reactants is greater than the charge on the rate-determining transition state, a polar solvent will interact more strongly with the reactants than with the transition state. As a result, a polar solvent will stabilize the reactants more than it will stabilize the transition state, increasing the difference in energy $(\Delta G^{\ddagger})$ between the reactants and the transition state. Therefore, increasing the polarity of the solvent will decrease the rate of the reaction, as shown in Figure 9.5.

Increasing the polarity of the solvent will decrease the rate of the reaction if one or more reactants in the rate-limiting step are charged.

Figure 9.5 ▶
Reaction coordinate diagram for a reaction in which the charge on the reactants is greater than the charge on the transition state.

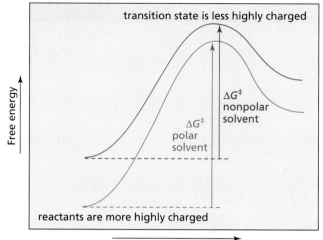

Increasing the polarity of the solvent will increase the rate of the reaction if none of the reactants in the rate-limiting step is charged.

On the other hand, if the charge on the rate-determining transition state is greater than the charge on the reactants, a polar solvent will interact more strongly with the transition state than with the reactants, stabilizing the transition state more than it stabilizes the reactants. Therefore, increasing the polarity of the solvent will decrease the difference in energy $(\Delta G^{\ddagger})$ between the reactants and the transition state, which will increase the rate of the reaction, as shown in Figure 9.6.

Figure 9.6 ▶
Reaction coordinate diagram for a reaction in which the charge on the transition state is greater than the charge on the reactants.

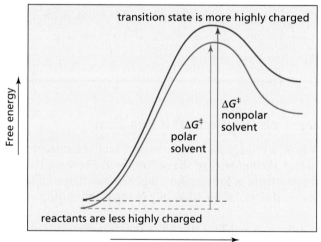

The Effect of the Solvent on the Rate of an S$_N$1 Reaction

Now let's see how increasing the polarity of the solvent affects the rate of an S$_N$1 reaction of an alkyl halide. The alkyl halide is the only reactant in the rate-determining step. It is a neutral molecule with a small dipole moment. The rate-determining transition state has a greater charge because as the carbon–halogen bond breaks, the carbon becomes more positive and the halogen becomes more negative. Since the charge is greater in the rate-determining transition state, increasing the polarity of the solvent will increase the rate of the S$_N$1 reaction (Figure 9.6 and Table 9.8).

rate-determining step of an S$_N$1 reaction

TABLE 9.8   The Effect of the Polarity of the Solvent on the Rate of Reaction of *tert*-Butyl Bromide in an S_N1 Reaction

Solvent	Relative rate
100% water	1200
80% water / 20% ethanol	400
50% water / 50% ethanol	60
20% water / 80% ethanol	10
100% ethanol	1

In Chapter 11 we will see that compounds other than alkyl halides undergo S_N1 reactions. As long as the compound undergoing an S_N1 reaction is neutral, increasing the polarity of the solvent will increase the rate of the S_N1 reaction because the polar solvent will stabilize the dispersed charges on the transition state more than it will stabilize the relatively neutral reactant (Figure 9.6). If, however, the compound undergoing an S_N1 reaction is charged, increasing the polarity of the solvent will decrease the rate of the reaction because the more polar solvent will stabilize the full charge on the reactant to a greater extent than it will stabilize the dispersed charge on the transition state (Figure 9.5).

The Effect of the Solvent on the Rate of an S_N2 Reaction

The way in which a change in the polarity of the solvent affects the rate of an S_N2 reaction depends on whether the reactants are charged or neutral, just as in an S_N1 reaction.

Most S_N2 reactions of alkyl halides involve a neutral alkyl halide and a charged nucleophile. Increasing the polarity of a solvent will have a strong stabilizing effect on the negatively charged nucleophile. The transition state also has a negative charge, but the charge is dispersed over two atoms. Consequently, the interactions between the solvent and the transition state are not as strong as the interactions between the solvent and the fully charged nucleophile. Therefore, a polar solvent stabilizes the nucleophile more than it stabilizes the transition state, so increasing the polarity of the solvent will decrease the rate of the reaction (Figure 9.5).

If, however, the S_N2 reaction involves an alkyl halide and a neutral nucleophile, the charge on the transition state will be larger than the charge on the neutral reactants, so increasing the polarity of the solvent will increase the rate of the substitution reaction (Figure 9.6).

In summary, the way in which a change in solvent affects the rate of a substitution reaction does not depend on the mechanism of the reaction. It depends *only* on whether or not the reactants are charged. *If a reactant in the rate-limiting step is charged, increasing the polarity of the solvent will decrease the rate of the reaction. If none of the reactants in the rate-limiting step is charged, increasing the polarity of the solvent will increase the rate of the reaction.*

In considering the solvation of charged species by a polar solvent, we have been discussing polar solvents, such as water or alcohols, that are hydrogen bond donors (protic polar solvents). There are also polar solvents (Table 9.7) such as *N,N*-dimethylformamide (DMF), dimethyl sulfoxide (DMSO), and hexamethylphosphoramide (HMPA) that are not hydrogen bond donors (aprotic polar solvents).

Ideally, one would like to carry out an S_N2 reaction with a negatively charged nucleophile in a nonpolar solvent because the charge on the reactants in such a reaction is greater than the charge on the transition state. However, a negatively charged nucleophile generally will not dissolve in a nonpolar solvent. Therefore, an aprotic polar solvent is used. Because aprotic polar solvents are not hydrogen bond donors, they are less effective than polar protic solvents in stabilizing negative charges. Thus, the rate

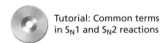

DMSO can solvate a cation better than it can solvate an anion.

Tutorial: Common terms in S$_N$1 and S$_N$2 reactions

of an S$_N$2 reaction involving a negatively charged nucleophile will be greater in an aprotic polar solvent than in a protic polar solvent. Consequently, an aprotic polar solvent is the solvent of choice for an S$_N$2 reaction in which the nucleophile is negatively charged, whereas a protic polar solvent is used if the nucleophile is a neutral molecule.

We have now seen that when an alkyl halide can undergo both S$_N$2 and S$_N$1 reactions, the S$_N$2 reaction will be favored by a high concentration of a good (negatively charged) nucleophile in an aprotic polar solvent, whereas the S$_N$1 reaction will be favored by a poor (neutral) nucleophile in a protic polar solvent.

PROBLEM 25 ◆

How will the rate of each of the following S$_N$2 reactions change if the polarity of a protic polar solvent is increased?

a. $CH_3CH_2CH_2CH_2Br + HO^- \longrightarrow CH_3CH_2CH_2CH_2OH + Br^-$

b. $CH_3\overset{+}{S}CH_3 + CH_3O^- \longrightarrow CH_3OCH_3 + CH_3SCH_3$
      $|$
     CH_3

c. $CH_3CH_2I + NH_3 \longrightarrow CH_3CH_2\overset{+}{N}H_3I^-$

PROBLEM 26 ◆

Which reaction in each of the following pairs will take place more rapidly?

a. $CH_3Br + HO^- \longrightarrow CH_3OH + Br^-$

    $CH_3Br + H_2O \longrightarrow CH_3OH + HBr$

b. $CH_3I + HO^- \longrightarrow CH_3OH + I^-$

    $CH_3Cl + HO^- \longrightarrow CH_3OH + Cl^-$

c. $CH_3Br + NH_3 \longrightarrow CH_3\overset{+}{N}H_3 + Br^-$

    $CH_3Br + H_2O \longrightarrow CH_3OH + HBr$

d. $CH_3Br + HO^- \xrightarrow{\text{DMSO}} CH_3OH + Br^-$

    $CH_3Br + HO^- \xrightarrow{\text{EtOH}} CH_3OH + Br^-$

e. $CH_3Br + NH_3 \xrightarrow{\text{Et}_2\text{O}} CH_3\overset{+}{N}H_3 + Br^-$

    $CH_3Br + NH_3 \xrightarrow{\text{EtOH}} CH_3\overset{+}{N}H_3 + Br^-$

PROBLEM 27 / SOLVED

Most of the pK_a values given throughout this text are values determined in water. How would the pK_a values of the following classes of compounds change if they were determined in a solvent less polar than water: carboxylic acids, alcohols, phenols, ammonium ions (RNH_3^+), and anilinium ions ($C_6H_5NH_3^+$)?

SOLUTION   A pK_a is equal to the negative log of an equilibrium constant, K_a. Because we are determining how changing the polarity of a solvent affects an equilibrium constant, we must look at how changing the polarity of the solvent affects the stability of the reactants and products.

$$K_a = \frac{[B^-][H^+]}{[HB]} \quad \text{or} \quad K_a = \frac{[B][H^+]}{[HB^+]}$$

Carboxylic acids, alcohols, and phenols are neutral in their acidic forms (HB) and charged in their basic forms (B^-). A polar protic solvent will stabilize B^- and H^+ more than it stabilizes HB, thereby increasing K_a. Therefore, K_a will be larger in water than in a less polar solvent, which means the pK_a value will be lower. As a result, the pK_a values of carboxylic acids, alcohols, and phenols determined in a less polar solvent will be higher than those determined in water.

Ammonium ions and anilinium ions are charged in their acidic forms (HB^+) and neutral in their basic forms (B). A polar solvent will stabilize HB^+ and H^+ more than it will stabilize B. Because the amount of stabilization is slightly greater for HB^+ than for H^+, the pK_a values of ammonium ions and anilinium ions determined in a less polar solvent will be slightly lower than those determined in water.

PROBLEM 28 ◆

Would you expect acetate ion $(CH_3CO_2^-)$ to be a more reactive nucleophile in an S_N2 reaction carried out in methanol or in dimethyl sulfoxide?

PROBLEM 29 ◆

Under which of the following reaction conditions would (R)-2-chlorobutane form the most (R)-2-butanol: HO^- in 50% water/50% ethanol or HO^- in 100% ethanol?

If an organic chemist wanted to put a methyl group on a nucleophile, methyl iodide would most likely be the methylating agent used. Of the methyl halides, methyl iodide has the most easily displaced leaving group because I^- is the weakest base of the halide ions. The reaction would be a simple S_N2 reaction.

9.11 BIOLOGICAL METHYLATING REAGENTS

$$\ddot{N}u \; + \; CH_3{-}I \longrightarrow CH_3{-}Nu \; + \; I^-$$

In a living cell, however, methyl iodide could not be used because it is only slightly soluble in water and therefore is not found in the predominantly aqueous environments of biological systems. Instead, biological systems use S-adenosylmethionine (SAM) and N^5-methyltetrahydrofolate as methylating agents, both of which are soluble in water. Although they look much more complicated than methyl iodide, they perform the same function—they transfer a methyl group to a nucleophile. Notice that the methyl group in each of these methylating agents is attached to a positively charged atom. This means the methyl groups are attached to very good leaving groups, so biological methylation can take place at a reasonable rate.

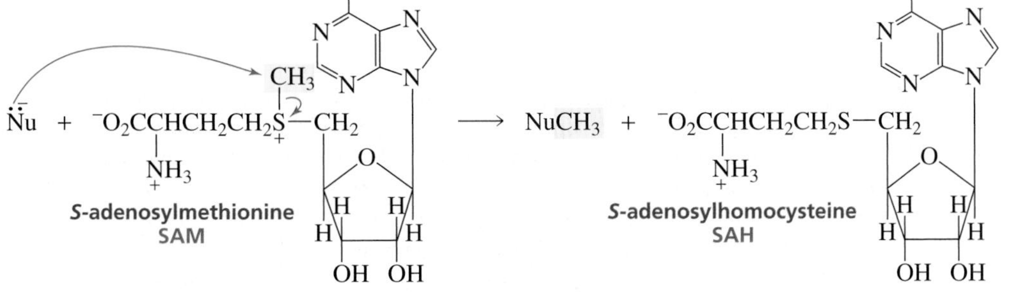

The reaction of N^5-methyltetrahydrofolate produces NuCH$_3$ + tetrahydrofolate.

N^5-methyltetrahydrofolate

tetrahydrofolate

SAM is used in biological systems to convert norepinephrine (noradrenaline) into epinephrine (adrenaline). This is a simple methylation reaction. Norepinephrine and epinephrine are hormones that are released into the bloodstream in response to stress. Epinephrine is the more potent hormone of the two.

norepinephrine
noradrenaline

epinephrine
adrenaline + H$^+$

The conversion of phosphatidylethanolamine, a component of cell membranes, into phosphatidylcholine requires three methylations by three equivalents of SAM. Biological cell membranes will be discussed in Section 24.4. The use of N^5-methyltetrahydrofolate as a biological methylating agent is discussed in more detail in Section 23.8.

phosphatidylethanolamine + 3 SAM ⟶ **phosphatidylcholine** + 3 SAH

S-ADENOSYLMETHIONINE: A NATURAL ANTIDEPRESSANT?

S-Adenosylmethionine is currently being sold in many health food and drug stores as a treatment for depression and arthritis. It is marketed under the name SAMe (pronounced Sammy). It works by methylating neurotransmitters, which in-

activates them. Although SAMe has been used clinically in Europe for more than two decades, it has not been rigorously evaluated in the United States and is not approved by the FDA. It can be sold, however, because the FDA does not prohibit the sale of most naturally occurring substances as long as the marketer does not make therapeutic claims.

SUMMARY OF REACTIONS

1. S_N2 reaction: a one-step mechanism

$$\overset{-}{\ddot{N}u} \; + \; -\overset{|}{\underset{|}{C}}-X \; \longrightarrow \; -\overset{|}{\underset{|}{C}}-Nu \; + \; X^-$$

Relative reactivities of alkyl halides: $CH_3X > 1° > 2° > 3°$.

Only the inverted product is formed.

2. S_N1 reaction: a two-step mechanism with a carbocation intermediate

$$-\overset{|}{\underset{|}{C}}-X \; \longrightarrow \; -\overset{|}{\underset{|}{C}}{}^{+} \xrightarrow{\overset{-}{\ddot{N}u}} \; -\overset{|}{\underset{|}{C}}-Nu$$
$$+ \; X^-$$

Relative reactivities of alkyl halides: $3° > 2° > 1° > CH_3X$.

Both the inverted and noninverted products are formed.

KEY TERMS

aprotic solvent (page 384)
back side attack (page 357)
base (page 361)
basicity (page 361)
bimolecular (page 356)
complete racemization (page 376)
dielectric constant (page 383)
direct displacement reaction
(page 357)
elimination reaction (page 354)
first-order reaction (page 370)
intimate ion pair (page 376)

inversion of configuration
(page 359)
ion–dipole interaction (page 363)
kinetics (page 356)
leaving group (page 354)
Le Châtelier's principle (page 367)
nucleophile (page 361)
nucleophilicity (page 361)
nucleophilic substitution reaction
(page 356)
partial racemization (page 376)
protic solvent (page 384)

rate law (page 356)
second-order reaction (page 356)
S_N1 reaction (page 370)
S_N2 reaction (page 357)
solvent-separated ion pair
(page 376)
solvolysis reaction (page 373)
steric effects (page 358)
steric hindrance (page 358)
substitution reaction (page 354)
unimolecular (page 370)

PROBLEMS

30. a. Indicate how each of the following factors affects an S_N1 reaction:
    b. Indicate how each of the following factors affects an S_N2 reaction:
        1. the structure of the alkyl halide   **3.** the concentration of the nucleophile
        2. the reactivity of the nucleophile   **4.** the solvent

31. Which is a better nucleophile in methanol?
    a. H_2O or HO^- **c.** H_2O or H_2S **e.** I^- or Br^-
    b. NH_3 or NH_2^- **d.** HO^- or HS^- **f.** Cl^- or Br^-

32. For each of the pairs in Problem 31, indicate which is a better leaving group.

33. What nucleophiles could be used to react with butyl bromide to prepare the following compounds?

    a. $CH_3CH_2CH_2CH_2OH$ **d.** $CH_3CH_2CH_2CH_2SCH_2CH_3$

    b. $CH_3CH_2CH_2CH_2OCH_3$ **e.** $CH_3CH_2CH_2CH_2NHCH_3$

    c. $CH_3CH_2CH_2CH_2SH$ **f.** $CH_3CH_2CH_2CH_2C{\equiv}N$

g. $CH_3CH_2CH_2CH_2\overset{\overset{\displaystyle O}{\|}}{O}CCH_3$ **h.** $CH_3CH_2CH_2CH_2C\equiv CCH_3$

34. Rank the following compounds in order of *decreasing* nucleophilicity.

a. $CH_3\overset{\overset{\displaystyle O}{\|}}{C}O^-$, $CH_3CH_2S^-$, $CH_3CH_2O^-$ in methanol

b. 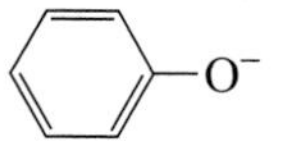 ⬡—O^- and ⬡—O^- in DMSO

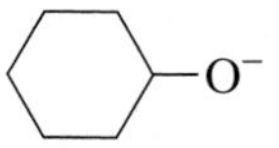

c. H_2O and NH_3 in methanol
d. Br^-, Cl^-, I^- in methanol

35. The pK_a of acetic acid in water is 4.76 (Section 1.17). What effect would a decrease in the polarity of the solvent have on the pK_a? Why?

36. Give the substitution products of the following reactions. If the products can exist as stereoisomers, show what stereoisomers are obtained:
a. (*R*)-2-bromopentane + high concentration of CH_3O^-
b. (*R*)-2-bromopentane + CH_3OH
c. *trans*-1-chloro-2-methylcyclohexane + high concentration of CH_3O^-
d. *trans*-1-chloro-2-methylcyclohexane + CH_3OH
e. 3-bromo-2-methylpentane + CH_3OH
f. 3-bromo-3-methylpentane + CH_3OH

37. Would you expect methoxide ion to be a better nucleophile if it were dissolved in CH_3OH or if it were dissolved in dimethyl sulfoxide (DMSO)? Why?

38. Which reaction in each of the following pairs will take place more rapidly?

a.

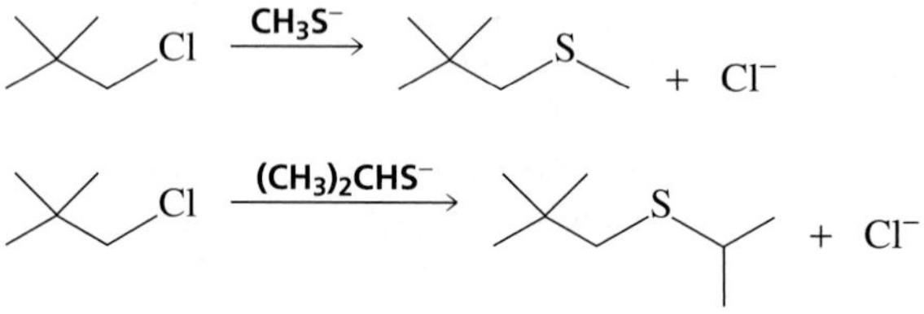

b.

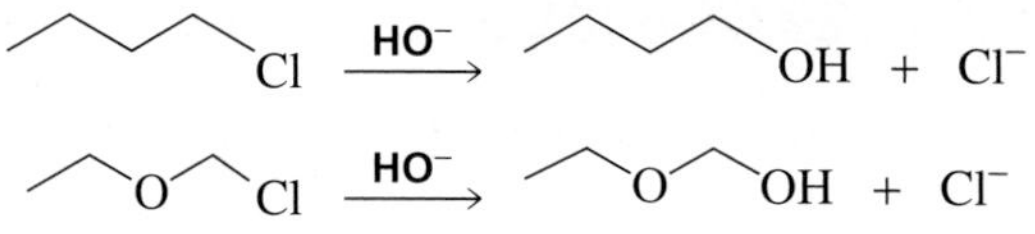

c.

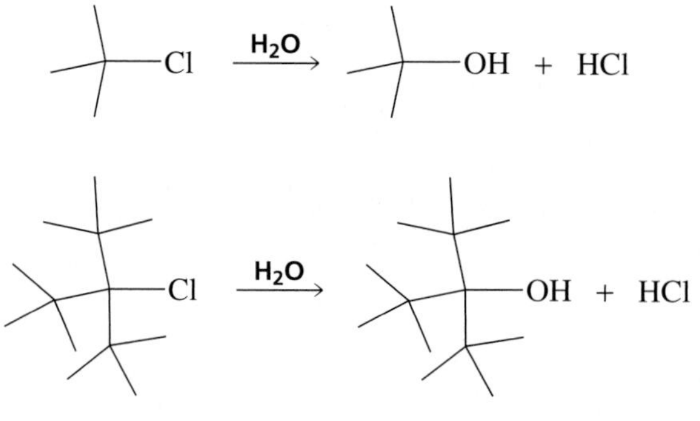

d. $(CH_3)_3CBr \xrightarrow{H_2O} (CH_3)_3COH + HBr$

$(CH_3)_3CBr \xrightarrow{CH_3CH_2OH} (CH_3)_3COCH_2CH_3 + HBr$

39. Which of the following compounds would you expect to be more reactive in an S_N2 reaction?

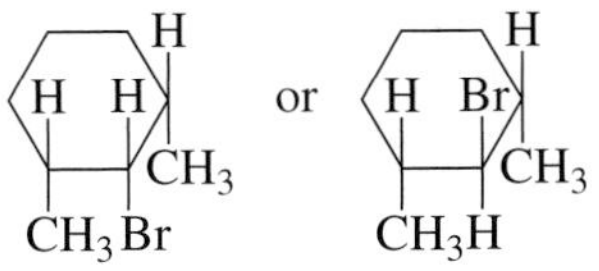

or

40. We saw that *S*-adenosylmethionine (SAM) methylates the nitrogen atom of noradrenaline to form adrenaline, a more potent hormone (Section 9.11). If SAM methylates an OH group on the benzene ring instead, it completely destroys noradrenaline's activity. Give the mechanism for the methylation of the OH group by SAM.

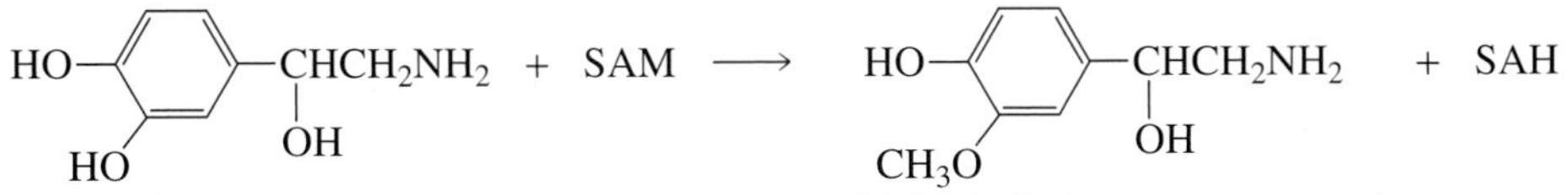

    norepinephrine   **a biologically inactive compound**

41. Give the substitution products of the following reactions. If the products can exist as stereoisomers, show what stereoisomers are obtained:
  a. (2S,3S)-2-chloro-3-methylpentane + high concentration of CH_3O^-
  b. (2S,3R)-2-chloro-3-methylpentane + high concentration of CH_3O^-
  c. (2R,3S)-2-chloro-3-methylpentane + high concentration of CH_3O^-
  d. (2R,3R)-2-chloro-3-methylpentane + high concentration of CH_3O^-
  e. 3-chloro-2,2-dimethylpentane + CH_3CH_2OH
  f. benzyl bromide + CH_3CH_2OH

42. 3-Bromo-3-methyl-1-butene forms two substitution products when it is added to a solution of sodium acetate in acetic acid.
  a. Give the structures of the substitution products.
  b. Which is the kinetically controlled product?
  c. Which is the thermodynamically controlled product?

43. The rate of reaction of methyl iodide with quinuclidine was measured in nitrobenzene, and then the rate of reaction of methyl iodide with triethylamine was measured in the same solvent.
  a. Which reaction had the larger rate constant?
  b. The same experiment was done using isopropyl iodide instead of methyl iodide. Which reaction had the larger rate constant?
  c. Which alkyl halide has the larger $k_{quinuclidine}/k_{triethylamine}$ ratio?

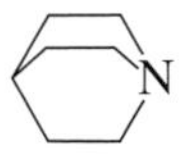

     quinuclidine 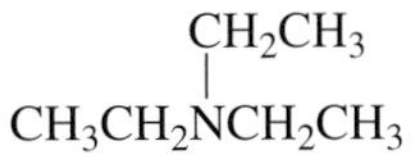   **triethylamine**

44. Only one bromoether (ignoring stereoisomers) is obtained from the reaction of the following alkyl dihalide with methanol. Give the structure of the ether.

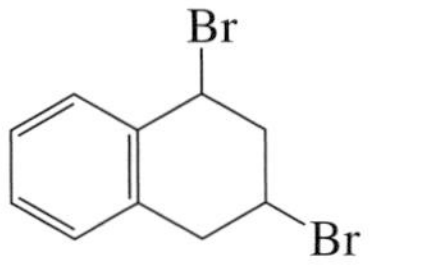

45. Starting with cyclohexane, how could the following compounds be prepared?
  a. cyclohexyl bromide   **b.** methoxycyclohexane   **c.** cyclohexanol

46. Give the substitution products of the following reactions, assuming that all the reactions are carried out under S_N2 conditions. If the products can exist as stereoisomers, show what stereoisomers are formed.
    a. (3S,4S)-3-bromo-4-methylhexane + CH_3O^-
    b. (3S,4R)-3-bromo-4-methylhexane + CH_3O^-
    c. (3R,4R)-3-bromo-4-methylhexane + CH_3O^-
    d. (3R,4S)-3-bromo-4-methylhexane + CH_3O^-

47. Tetrahydrofuran can solvate a positively charged species better than diethyl ether can. Explain.

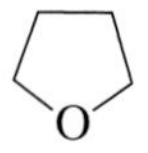

tetrahydrofuran $CH_3CH_2OCH_2CH_3$
                              diethyl ether

48. Propose a mechanism for each of the following reactions:

a.

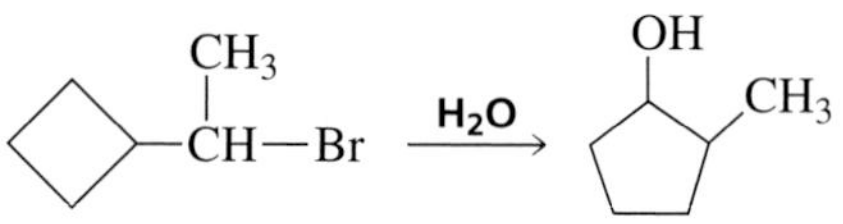

b.

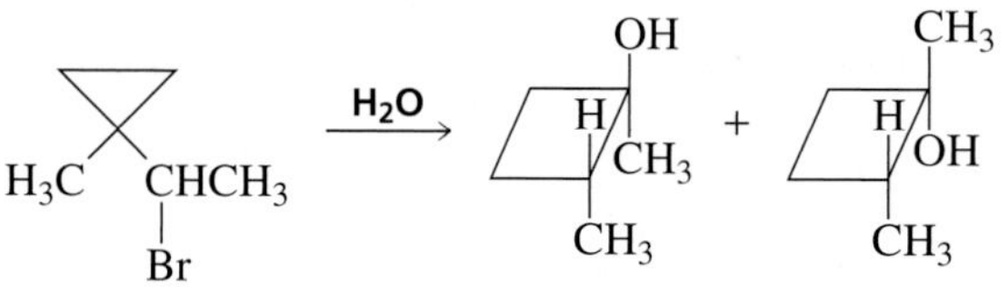

49. Which of the following will react faster in an S_N1 reaction?

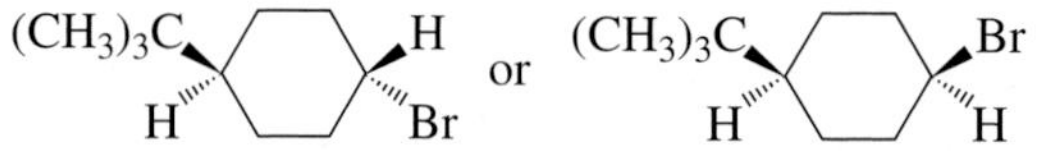

50. Alkyl halides have been used as insecticides since the discovery of DDT in 1939. DDT was the first compound to be found that had a high toxicity to insects and a relatively low toxicity to mammals. In 1972, DDT was banned in the United States because it is a long-lasting compound and its widespread use caused accumulation in substantial concentrations in wildlife. Chlordane is an alkyl halide insecticide that is used to protect wooden buildings from termites. Chlordane can be synthesized from two reactants in a one-step reaction. One of the reactants is hexachlorocyclopentadiene. What is the other reactant? (*Hint:* See Section 7.8.)

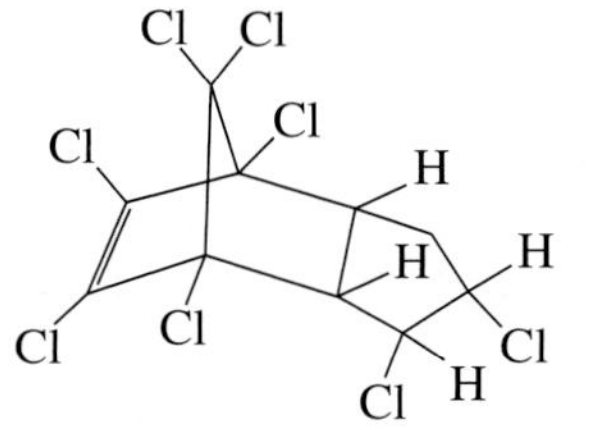

Chlordane

51. The following alkyl halide does not undergo a substitution reaction regardless of the conditions under which the reaction is run. Explain.

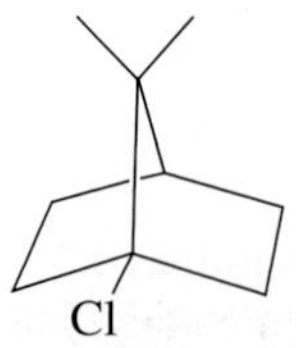

10

Reactions at an sp^3 Hybridized Carbon II:

Elimination Reactions of Alkyl Halides; Competition Between Substitution and Elimination

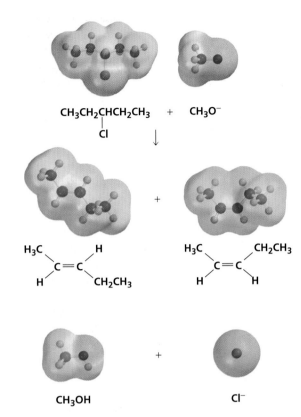

$$CH_3CH_2CHCH_2CH_3 \quad + \quad CH_3O^-$$
$$\overset{|}{Cl}$$

Alkyl halides, in addition to undergoing the nucleophilic substitution reactions you learned about in Chapter 9, also undergo elimination reactions. In an **elimination reaction,** groups are eliminated from the reactant. For example, when an alkyl halide undergoes an elimination reaction, the halogen (X) is removed from the carbon to which it is attached, and a proton is removed from an adjacent carbon. As a result, a double bond is formed between the two carbons from which the atoms are eliminated. Therefore, the product of an elimination reaction is an alkene.

$$CH_3CH_2CH_2X + Y^- \begin{cases} \xrightarrow{\textbf{substitution}} CH_3CH_2CH_2Y + X^- \\ \xrightarrow{\textbf{elimination}} CH_3CH=CH_2 + HY + X^- \end{cases}$$

In this chapter we will first discuss the elimination reactions of alkyl halides. Then we will examine the factors that determine whether a given alkyl halide will undergo a substitution reaction, an elimination reaction, or both substitution and elimination reactions.

There are two important mechanisms for an elimination reaction. The reaction of ethyl bromide with hydroxide ion is an example of an **E2 reaction,** where E stands for elimination and 2 stands for bimolecular. It is a second-order reaction because the rate of the reaction depends on the concentrations of both ethyl bromide and hydroxide ion (Section 9.2).

10.1 THE E2 REACTION

$$\underset{\textbf{ethyl bromide}}{CH_3CH_2Br} + HO^- \longrightarrow \underset{\textbf{ethene}}{CH_2=CH_2} + H_2O + Br^-$$

$$\textbf{rate} = k\textbf{[alkyl halide][base]}$$

The rate law tells us that both ethyl bromide and hydroxide ion are involved in the transition state of the rate-determining step of the reaction. The following mechanism agrees with the observed second-order kinetics. We see that the E2 reaction is a concerted, one-step reaction. The proton and the bromide ion are removed in the same step, so no intermediate is formed.

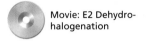
Movie: E2 Dehydro-halogenation

mechanism of the E2 reaction

$$HO^{-} \quad H \\ \quad | \\ CH_2{-}CH_2 \longrightarrow CH_2{=}CH_2 + H_2O + Br^{-} \\ \quad | \\ \quad Br$$

In an E2 reaction of an alkyl halide, a base removes a proton from a carbon adjacent to the carbon bonded to the halogen. As the proton is removed, the electrons that the hydrogen shared with carbon move toward the carbon bonded to the halogen. As these electrons move toward the carbon, the halogen leaves, taking its bonding electrons with it. The electrons that were bonded to the hydrogen in the reactant have formed the π bond of the double bond in the product. Removal of a proton and a halide ion is called **dehydrohalogenation.**

The carbon to which the halogen is attached is called the α-carbon. A carbon adjacent to an α-carbon is called a β-carbon. Because the elimination reaction is initiated by removing a proton from a β-carbon, an E2 reaction is sometimes called a **β-elimination reaction.** It is also called a **1,2-elimination reaction** because the atoms being removed are on adjacent carbons. (B:⁻ is any base.)

$$B^{-} \quad H \quad \boxed{\alpha\text{-carbon}} \\ \quad | \\ RCH{-}CHR \longrightarrow RCH{=}CHR + BH + Br^{-} \\ \quad | \\ \quad Br \\ \boxed{\beta\text{-carbon}}$$

2-Bromopropane has two β-carbons from which a proton can be removed in an E2 reaction. Because the two β-carbons are identical, the proton can be removed equally easily from either one. The product of this elimination reaction is propene.

$$CH_3CHCH_3 + CH_3O^{-} \longrightarrow CH_3CH{=}CH_2 + CH_3OH + Br^{-} \\ \quad | \qquad\qquad\qquad\qquad \text{propene} \\ \quad Br \\ \textbf{2-bromopropane}$$

2-Bromobutane has two structurally different β-carbons from which a proton can be removed. So when 2-bromobutane reacts with a base, two elimination products are formed, 2-butene and 1-butene.

$$\boxed{\beta\text{-carbons}} \\ CH_3CHCH_2CH_3 + CH_3O^{-} \xrightarrow[\textbf{CH}_3\textbf{OH}]{} CH_3CH{=}CHCH_3 + CH_2{=}CHCH_2CH_3 + CH_3OH + Br^{-} \\ \quad | \qquad\qquad\qquad\qquad\qquad \textbf{2-butene} \qquad\quad \textbf{1-butene} \\ \quad Br \qquad\qquad\qquad\qquad\qquad\qquad\quad 80\% \qquad\qquad\quad 20\% \\ \textbf{2-bromobutane} \qquad\qquad\qquad \text{(mixture of } E \text{ and } Z\text{)}$$

How do you know which of the two elimination products will be formed in greater yield? To answer this question, you must determine which of the alkenes is formed more easily—that is, which is formed faster. The reaction coordinate diagram for the reaction is shown in Figure 10.1.

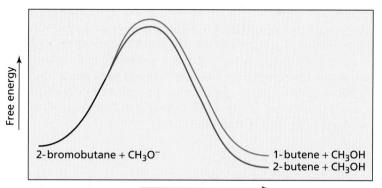

◄ **Figure 10.1**
Reaction coordinate diagram for the E2 reaction of 2-bromobutane and methoxide ion.

You know that alkene stability depends on the number of alkyl substituents bonded to the sp^2 carbons—the greater that number, the more stable the alkene (Section 3.19). For example, 2-butene, with a total of two methyl substituents bonded to its sp^2 carbons, is more stable than 1-butene, with one ethyl substituent. In the transition state leading to an alkene, the C—H and C—Br bonds are partially broken and the double bond is partially formed, giving the transition state an "alkene-like" structure. Because the transition state has an alkene-like structure, the one leading to 2-butene is more stable than the one leading to 1-butene for the same reason that 2-butene is more stable than 1-butene. The more stable transition state allows 2-butene to form faster than 1-butene.

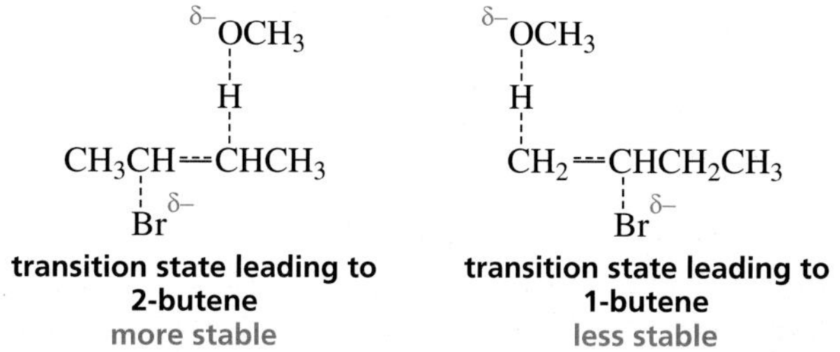

$$\overset{\delta-}{\text{OCH}_3}$$
$$\vdots$$
$$\text{H}$$
$$\vdots$$
$$\text{CH}_3\text{CH}\text{---}\text{CHCH}_3$$
$$\vdots \ \overset{\delta-}{}$$
$$\text{Br}$$

**transition state leading to
2-butene**
more stable

$$\overset{\delta-}{\text{OCH}_3}$$
$$\vdots$$
$$\text{H}$$
$$\vdots$$
$$\text{CH}_2\text{---}\text{CHCH}_2\text{CH}_3$$
$$\vdots \ \overset{\delta-}{}$$
$$\text{Br}$$

**transition state leading to
1-butene**
less stable

The difference in the rate of formation of the two alkenes is not very great (Figure 10.1). Consequently, both products are formed, but 2-butene (the more stable of the two alkenes) is the major product of the elimination reaction. Thus, an E2 reaction is *regioselective* because more of one constitutional isomer is formed than the other.

The reaction of 2-bromo-2-methylbutane with hydroxide ion produces both 2-methyl-2-butene and 2-methyl-1-butene. Because 2-methyl-2-butene is the more substituted alkene (it has a greater number of alkyl substituents bonded to its sp^2 carbons), it is the more stable of the two alkenes and, therefore, is the major product of the elimination reaction.

$$\underset{\underset{\text{Br}}{|}}{\overset{\overset{\text{CH}_3}{|}}{\text{CH}_3\text{CCH}_2\text{CH}_3}} \ + \ \text{HO}^- \ \xrightarrow[\text{H}_2\text{O}]{} \ \underset{\text{2-methyl-2-butene}}{\overset{\overset{\text{CH}_3}{|}}{\text{CH}_3\text{C}=\text{CHCH}_3}} \ + \ \underset{\text{2-methyl-1-butene}}{\overset{\overset{\text{CH}_3}{|}}{\text{CH}_2=\text{CCH}_2\text{CH}_3}} \ + \ \text{H}_2\text{O} \ + \ \text{Br}^-$$

2-bromo-2-methylbutane   70%   30%

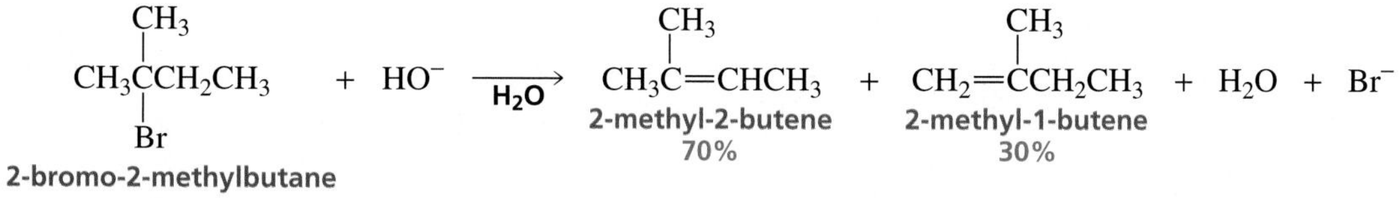

10.2
ZAITSEV'S RULE

Alexander M. Zaitsev (1841–1910) *was born in Kazan, Russia. The German transliteration, Saytzeff, is sometimes used for his family name. He received a Ph.D. from the University of Leipzig in 1866. He was a professor of chemistry first at the University of Kazan and later at the University of Kiev.*

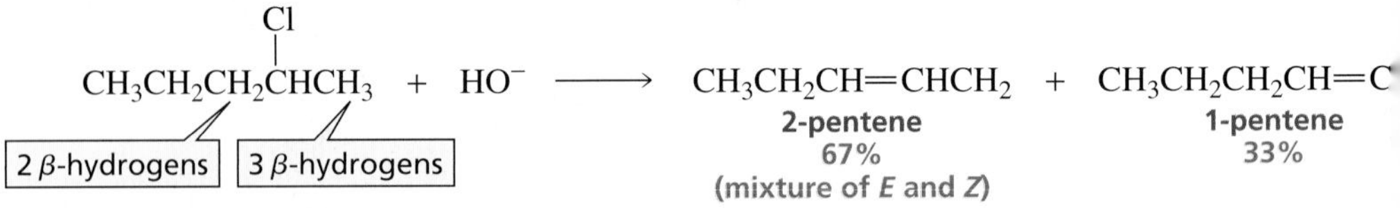

In the previous examples we saw that the major product of an elimination reaction is the more substituted alkene because it is the more easily formed product. Alexander M. Zaitsev, a nineteenth century Russian chemist, devised a shortcut to predict the more substituted alkene product. He pointed out that *the more substituted alkene product is obtained when a proton is removed from the* β-*carbon that is bonded to the fewest hydrogens.* This is called Zaitsev's rule. In 2-chloropentane, for example, there are three hydrogens bonded to one β-carbon and two hydrogens bonded to the other β-carbon. According to **Zaitsev's rule**, the more substituted alkene will be the one formed by removing a proton from the β-carbon that is bonded to two hydrogens rather than from the β-carbon that is bonded to three hydrogens. Keep in mind, though, that Zaitsev's rule is just a shortcut to determine which of the possible alkene products is the more substituted alkene.

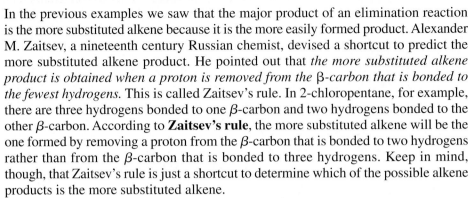

Alkyl halides have the following relative reactivities in an E2 reaction because elimination from a tertiary alkyl halide typically leads to a more substituted alkene than elimination from a secondary alkyl halide, and elimination from a secondary alkyl halide generally leads to a more substituted alkene than elimination from a primary alkyl halide.

relative reactivities of alkyl halides in an E2 reaction

    tertiary alkyl halide > secondary alkyl halide > primary alkyl halide

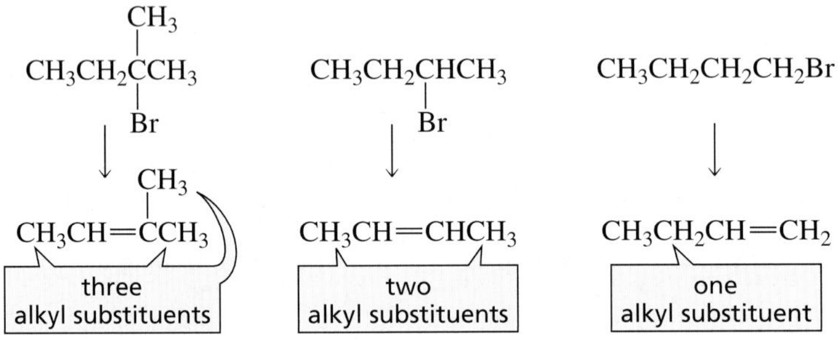

$$CH_3CH_2\overset{\overset{\displaystyle CH_3}{|}}{\underset{\underset{\displaystyle Br}{|}}{C}}CH_3 \qquad CH_3CH_2\overset{}{\underset{\underset{\displaystyle Br}{|}}{C}HCH_3} \qquad CH_3CH_2CH_2CH_2Br$$

$$CH_3CH{=}\overset{\overset{\displaystyle CH_3}{|}}{C}CH_3 \qquad CH_3CH{=}CHCH_3 \qquad CH_3CH_2CH{=}CH_2$$

| three alkyl substituents | two alkyl substituents | one alkyl substituent |

3-D Molecules:
2-Methyl-2-butene;
2-Methyl-1-butene

In most elimination reactions, the more substituted alkene is the major product of the reaction because it is the more easily formed product. There are some elimination reactions, however, in which the more substituted alkene is not the more easily formed product.

For example, if the base in an E2 reaction is sterically bulky, it will preferentially remove the most accessible hydrogen. In the following reaction, it is easier for the bulky *tert*-butoxide ion to remove one of the more exposed terminal hydrogens, which leads to the less substituted alkene. Because the less substituted alkene is the one that is more easily formed, it is the major product of the reaction.

$$CH_3\overset{\overset{\displaystyle CH_3}{|}}{\underset{\underset{\displaystyle Br}{|}}{C}}CH_2CH_3 \;+\; CH_3\overset{\overset{\displaystyle CH_3}{|}}{\underset{\underset{\displaystyle CH_3}{|}}{C}}O^- \;\xrightarrow{\;(CH_3)_3COH\;}\; CH_3\overset{\overset{\displaystyle CH_3}{|}}{C}{=}CHCH_3 \;+\; CH_2{=}\overset{\overset{\displaystyle CH_3}{|}}{C}CH_2CH_3 \;+\; CH_3\overset{\overset{\displaystyle CH_3}{|}}{\underset{\underset{\displaystyle CH_3}{|}}{C}}OH \;+\; Br^-$$

2-bromo-2-methylbutane   *tert*-butoxide ion   2-methyl-2-butene   2-methyl-1-butene
                                                                        28%   72%

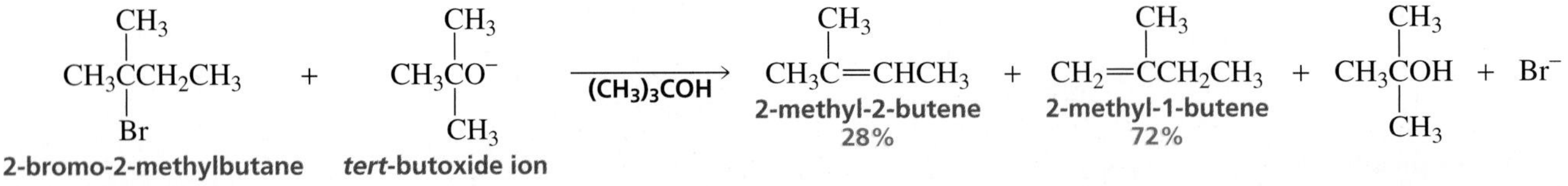

The data in Table 10.1 show that when an alkyl halide undergoes an E2 reaction with a variety of alkoxide ions, the percent of the less substituted alkene increases as the size of the base increases.

TABLE 10.1   Effect of the Steric Properties of the Base on the Distribution of Products in an E2 Reaction

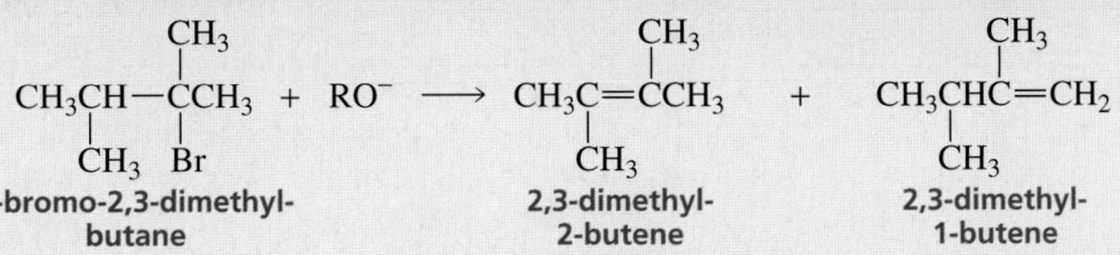

$$\underset{\substack{\text{2-bromo-2,3-dimethyl-}\\\text{butane}}}{\overset{\displaystyle CH_3}{\underset{\displaystyle CH_3 \;\; Br}{CH_3CH-CCH_3}}} + RO^- \longrightarrow \underset{\substack{\text{2,3-dimethyl-}\\\text{2-butene}}}{\overset{\displaystyle CH_3}{\underset{\displaystyle CH_3}{CH_3C=CCH_3}}} + \underset{\substack{\text{2,3-dimethyl-}\\\text{1-butene}}}{\overset{\displaystyle CH_3}{\underset{\displaystyle CH_3}{CH_3CHC=CH_2}}}$$

Base	More substituted product	Less substituted product
$CH_3CH_2O^-$	79%	21%
$\underset{\displaystyle CH_3}{\overset{\displaystyle CH_3}{CH_3CO^-}}$	27%	73%
$\underset{\displaystyle CH_2CH_3}{\overset{\displaystyle CH_3}{CH_3CO^-}}$	19%	81%
$\underset{\displaystyle CH_2CH_3}{\overset{\displaystyle CH_2CH_3}{CH_3CH_2CO^-}}$	8%	92%

PROBLEM 1

Draw a reaction coordinate diagram for the E2 reaction of 2-bromo-2,3-dimethylbutane with sodium *tert*-butoxide.

Although the major product of the E2 dehydrohalogenation of alkyl chlorides, alkyl bromides, and alkyl iodides is normally the more substituted alkene, the major product of the E2 dehydrohalogenation of alkyl fluorides is the less substituted alkene (Table 10.2).

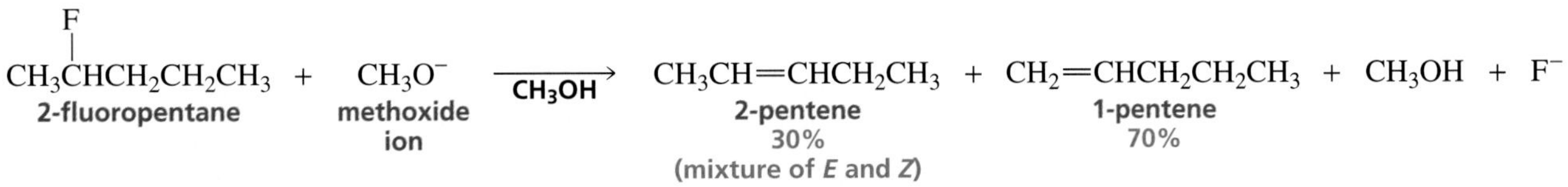

$$\underset{\text{2-fluoropentane}}{\overset{\displaystyle F}{CH_3CHCH_2CH_2CH_3}} + \underset{\substack{\text{methoxide}\\\text{ion}}}{CH_3O^-} \xrightarrow{\;CH_3OH\;} \underset{\substack{\text{2-pentene}\\30\%\\\text{(mixture of } E \text{ and } Z)}}{CH_3CH=CHCH_2CH_3} + \underset{\substack{\text{1-pentene}\\70\%}}{CH_2=CHCH_2CH_2CH_3} + CH_3OH + F^-$$

You have seen that when a hydrogen and a chlorine, bromine, or iodine are eliminated from an alkyl halide, the halogen starts to leave as soon as the base begins to remove the proton. Because the halogen begins to depart with its bonding electrons as soon as the proton begins to leave, there is no buildup of any negative charge on the carbon that is losing the proton. Thus, the transition state resembles an alkene

TABLE 10.2 Products Obtained from the E2 Reaction of CH_3O^- and 2-Halohexanes

			More substituted product	Less substituted product

$$\underset{\displaystyle CH_3CHCH_2CH_2CH_2CH_3}{\overset{\displaystyle \overset{X}{\mid}}{}} \; + \; CH_3O^- \; \longrightarrow \; \underset{\substack{\text{2-hexene}\\ \text{(mixture of } E \text{ and } Z)}}{CH_3CH\!=\!CHCH_2CH_2CH_3} \; + \; \underset{\text{1-hexene}}{CH_2\!=\!CHCH_2CH_2CH_3}$$

Leaving group	Conjugate acid	pK_a		
X = I	HI	−10	81%	19%
X = Br	HBr	−9	72%	28%
X = Cl	HCl	−7	67%	33%
X = F	HF	3.2	30%	70%

rather than a carbanion (Section 10.1). Of the halogen ions, the fluoride ion is the strongest base and, therefore, the poorest leaving group. So when a base begins to remove a proton from an alkyl fluoride, the fluoride ion has less tendency to leave than the other halide ions. As a result, negative charge develops on the carbon that is losing the proton, causing the transition state to resemble a carbanion rather than an alkene. To determine which of the carbanion-like transition states is more stable, we must determine which carbanion would be more stable.

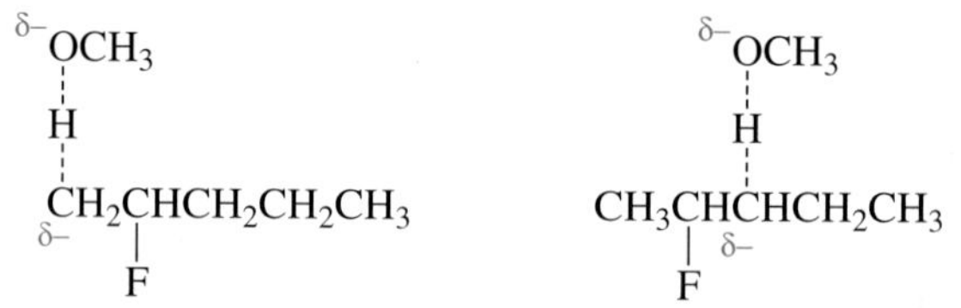

transition state leading to
1-pentene
more stable

transition state leading to
2-pentene
less stable

We have seen that carbocations, because they are positively charged, are stabilized by electron-donating groups. Because alkyl groups are more electron donating than a hydrogen, alkyl groups stabilize carbocations, so tertiary carbocations are the most stable and methyl cations are the least stable (Section 3.10).

relative stabilities of carbocations

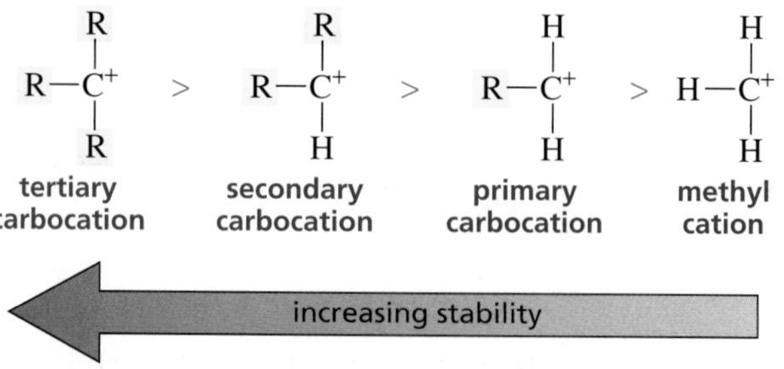

| tertiary carbocation | | secondary carbocation | | primary carbocation | | methyl cation |

increasing stability

Carbanions, on the other hand, are negatively charged, so they are destabilized by alkyl groups. Therefore, methyl anions are the most stable and tertiary carbanions are the least stable.

relative stabilities of carbanions

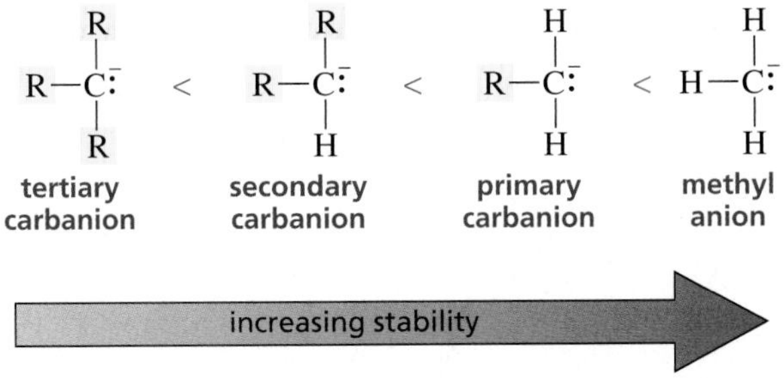

$$\underset{\substack{\text{tertiary}\\\text{carbanion}}}{\overset{\displaystyle R}{\underset{\displaystyle R}{R-\overset{..}{\underset{|}{C}}{:}}}} \;<\; \underset{\substack{\text{secondary}\\\text{carbanion}}}{\overset{\displaystyle R}{\underset{\displaystyle H}{R-\overset{..}{\underset{|}{C}}{:}}}} \;<\; \underset{\substack{\text{primary}\\\text{carbanion}}}{\overset{\displaystyle H}{\underset{\displaystyle H}{R-\overset{..}{\underset{|}{C}}{:}}}} \;<\; \underset{\substack{\text{methyl}\\\text{anion}}}{\overset{\displaystyle H}{\underset{\displaystyle H}{H-\overset{..}{\underset{|}{C}}{:}}}}$$

increasing stability ⟶

The transition state leading to 1-pentene has the developing negative charge on a primary carbon. This is more stable than the transition state leading to 2-pentene, which has the developing negative charge on a secondary carbon. Because the transition state leading to 1-pentene is more stable, 1-pentene is formed more rapidly and is the major product of the elimination reaction.

The data in Table 10.2 show that, as the halide ion increases in basicity (decreases in leaving ability), there is a decrease in the amount of the more substituted alkene product. However, the more substituted alkene remains the major elimination product in all cases except when the halogen is fluorine.

In the following reactions, Zaitsev's rule does not lead to the more stable alkene because the rule does not take into account the fact that conjugated double bonds are more stable than isolated double bonds (Section 7.3). Because the conjugated alkene is the more stable alkene, it is the one that is more easily formed and is, therefore, the major product of the reaction. So, if there is a double bond or a benzene ring in the alkyl halide, do not use Zaitsev's rule to predict the major product of the elimination reaction.

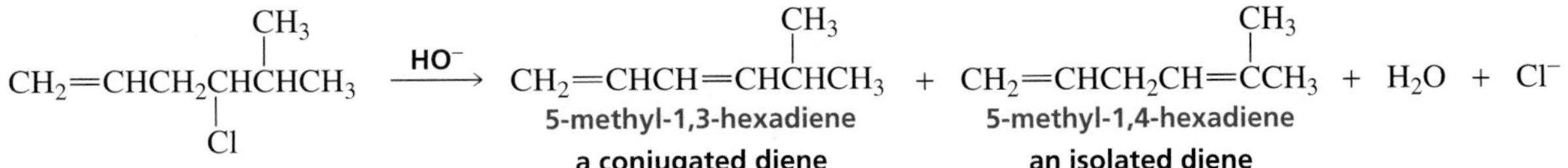

$$\underset{\substack{\text{4-chloro-5-methyl-1-hexene}}}{\overset{\displaystyle CH_3}{CH_2{=}CHCH_2\underset{\displaystyle Cl}{\overset{|}{CH}}\overset{|}{CH}CH_3}} \xrightarrow{HO^-} \underset{\substack{\textbf{5-methyl-1,3-hexadiene}\\\textbf{a conjugated diene}\\\text{major product}}}{\overset{\displaystyle CH_3}{CH_2{=}CHCH{=}CH\overset{|}{CH}CH_3}} + \underset{\substack{\textbf{5-methyl-1,4-hexadiene}\\\textbf{an isolated diene}\\\text{minor product}}}{\overset{\displaystyle CH_3}{CH_2{=}CHCH_2CH{=}\overset{|}{C}CH_3}} + H_2O + Cl^-$$

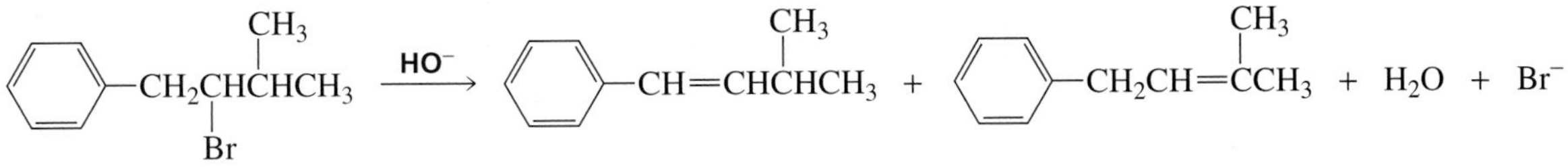

$$\underset{\substack{\text{2-bromo-3-methyl-}\\\text{1-phenylbutane}}}{C_6H_5{-}CH_2\underset{\displaystyle Br}{\overset{|}{CH}}\overset{\substack{\displaystyle CH_3\\|}}{CH}CH_3} \xrightarrow{HO^-} \underset{\substack{\textbf{3-methyl-1-phenyl-1-butene}\\\textbf{the double bond is}\\\textbf{conjugated with the}\\\textbf{benzene ring}\\\text{major product}}}{C_6H_5{-}CH{=}CH\overset{\substack{\displaystyle CH_3\\|}}{CH}CH_3} + \underset{\substack{\textbf{3-methyl-1-phenyl-2-butene}\\\textbf{the double bond is}\\\textbf{not conjugated with the}\\\textbf{benzene ring}\\\text{minor product}}}{C_6H_5{-}CH_2CH{=}\overset{\substack{\displaystyle CH_3\\|}}{C}CH_3} + H_2O + Br^-$$

We can summarize by saying that the major product of an E2 elimination reaction is the more substituted alkene *unless one of the following applies:*

- the base is large;
- the alkyl halide is an alkyl fluoride;
- the alkyl halide contains one or more double bonds.

Because Zaitsev's rule leads to the more substituted alkene, the rule cannot be used to predict the major product of an elimination reaction in these three situations. In Section 10.6 you will see that Zaitsev's rule also cannot be used with certain cyclic compounds.

PROBLEM 2 ◆

Give the major elimination product obtained from an E2 reaction of each of the following alkyl halides with hydroxide ion:

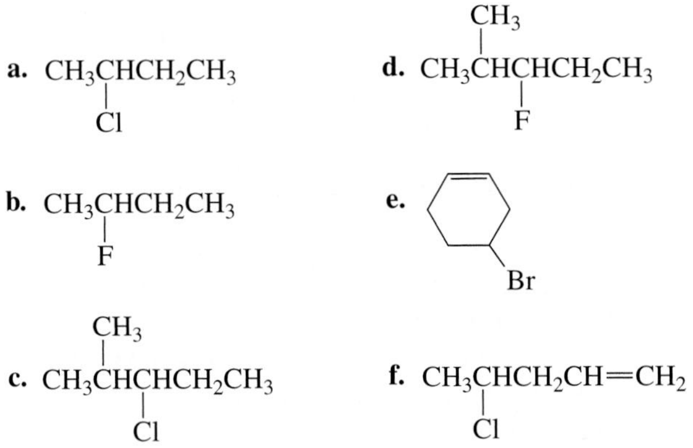

a. CH₃CHCH₂CH₃
    |
    Cl

b. CH₃CHCH₂CH₃
    |
    F

      CH₃
      |
c. CH₃CHCHCH₂CH₃
      |
      Cl

      CH₃
      |
d. CH₃CHCHCH₂CH₃
      |
      F

e. (cyclohexene ring with Br)

f. CH₃CHCH₂CH=CH₂
    |
    Cl

PROBLEM 3 ◆

Which alkyl halide would you expect to be more reactive in an E2 reaction?

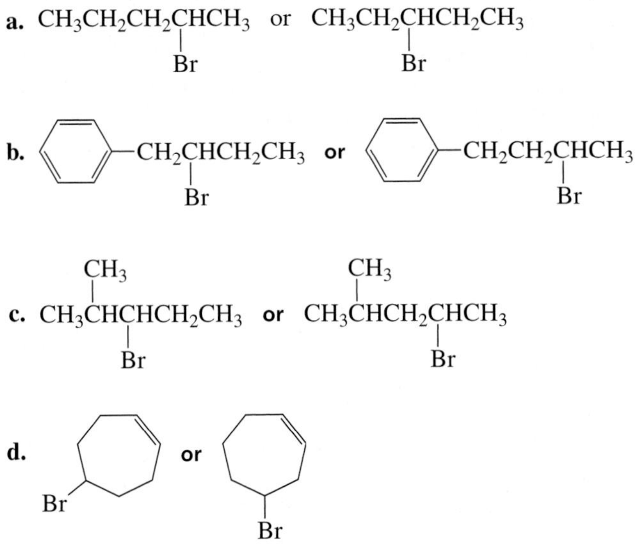

a. CH₃CH₂CH₂CHCH₃ or CH₃CH₂CHCH₂CH₃
           | |
           Br   Br

b. (phenyl)—CH₂CHCH₂CH₃ or (phenyl)—CH₂CH₂CHCH₃
               | |
               Br   Br

      CH₃   CH₃
      | |
c. CH₃CHCHCH₂CH₃ or CH₃CHCH₂CHCH₃
      | |
      Br   Br

d. (cycloheptene ring with Br) or (cycloheptene ring with Br)

10.3

THE E1 REACTION

The reaction of *tert*-butyl bromide with water to form 2-methylpropene is a first-order elimination reaction because the rate of the reaction depends only on the concentration of the alkyl halide. It is called an **E1 reaction,** where E stands for elimination and 1 stands for unimolecular.

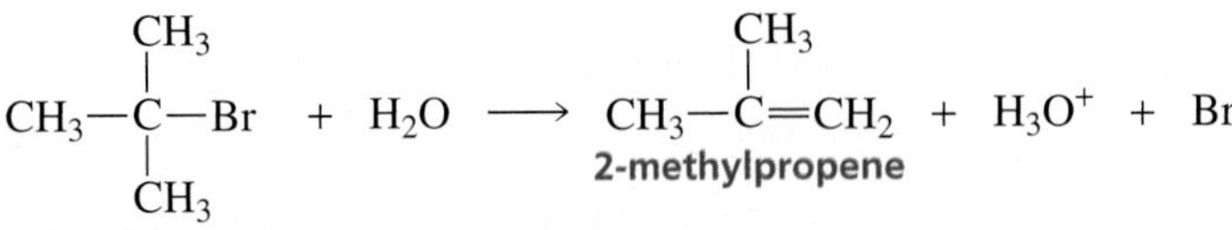

        CH₃   CH₃
        | |
CH₃—C—Br + H₂O ⟶ CH₃—C=CH₂ + H₃O⁺ + Br⁻
        | **2-methylpropene**
        CH₃
  tert-butyl bromide

rate = *k*[alkyl halide]

We know, then, that only the alkyl halide is involved in the transition state of the rate-determining step of the reaction. Therefore, there must be at least two steps in the reaction. The following mechanism agrees with the observed first-order kinetics.

Movie: E1 Elimination

mechanism of the E1 reaction

3-D Molecules:
t-Butylchloride;
t-Butylcation

An E1 reaction is a two-step reaction. In the first step, the alkyl halide dissociates heterolytically. This is the rate-determining step. In the second step of the reaction, the base forms an elimination product by removing a proton from a carbon adjacent to the positively charged carbon. Because the first step of the reaction is the rate-determining step, increasing the concentration of the base—which comes into play only in the second step of the reaction—has no effect on the rate of the reaction.

When two elimination products can be formed, the major product is generally the more substituted alkene.

$$CH_3CH_2\overset{\overset{\displaystyle CH_3}{|}}{\underset{\underset{\displaystyle Cl}{|}}{C}}CH_3 \ + \ H_2O \ \longrightarrow \ CH_3CH{=}\overset{\overset{\displaystyle CH_3}{|}}{C}CH_3 \ + \ CH_3CH_2\overset{\overset{\displaystyle CH_3}{|}}{C}{=}CH_2 \ + \ H_3O^+ \ + \ Cl^-$$

2-chloro-2-methylbutane

2-methyl-2-butene
major product

2-methyl-1-butene
minor product

You can see why by looking at the reaction coordinate diagram in Figure 10.2. The more substituted alkene is the more stable of the two alkenes and, therefore, has the more stable transition state. Because it has the more stable transition state, it is formed more rapidly. To obtain the more substituted alkene, the hydrogen is removed from the β-carbon bonded to the fewest hydrogens—following Zaitsev's rule.

3-D Molecule:
2-Chloro-2-methyl
butane

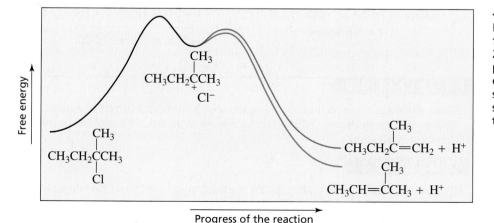

◀ **Figure 10.2**
Reaction coordinate diagram for an E1 reaction of 2-chloro-2-methylbutane. The major product is the more substituted alkene because its greater stability causes the transition state leading to its formation to be more stable.

Because the first step is the rate-determining step, the rate of an E1 reaction depends on both the ease with which the leaving group leaves and the stability of the carbocation that is formed. Thus, the relative reactivities of a series of alkyl halides with the same leaving group parallel the relative stabilities of the carbocations. A tertiary benzylic halide is the most reactive alkyl halide because a tertiary benzylic cation—the most stable carbocation—is the easiest to form (Sections 6.7 and 9.8).

relative reactivities of alkyl halides in an E1 reaction = **relative stabilities of carbocations**

3° benzylic > 3° allylic > 2° benzylic > 2° allylic > 3° > 1° benzylic ≈ 1° allylic ≈ 2° > 1° > vinyl

increasing reactivity and stability

Because the E1 reaction involves the formation of a carbocation intermediate, rearrangement of the carbon skeleton can occur before the proton is lost. For example, the secondary carbocation that is formed when a chloride ion dissociates from 3-chloro-2-methyl-2-phenylbutane undergoes a 1,2-methyl shift to form a more stable tertiary benzylic cation.

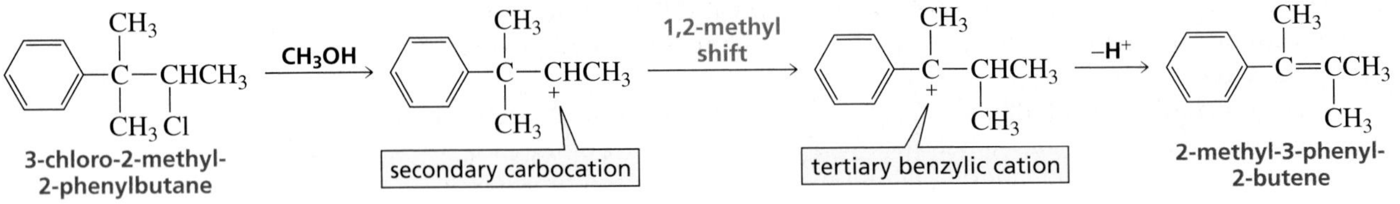

In the following example, the secondary carbocation undergoes a 1,2-hydride shift to form a more stable secondary allylic cation.

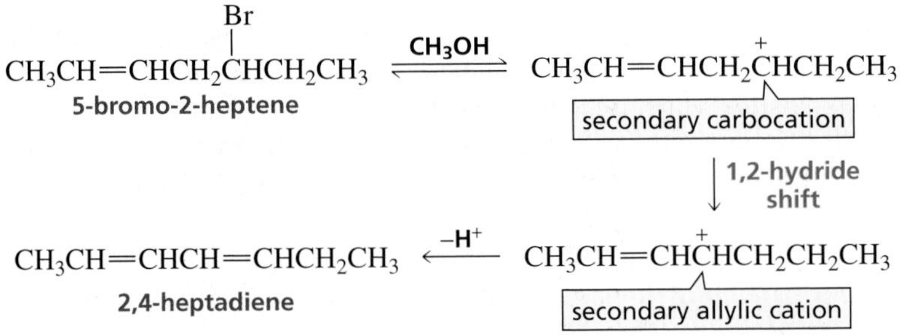

PROBLEM 4 ◆

Three alkenes are formed from the E1 reaction of 3-bromo-2,3-dimethylpentane. Give the structures of the alkenes and rank them according to the amount that would be formed. (Ignore stereoisomers.)

PROBLEM 5

If 2-fluoropentane were to undergo an E1 reaction, would you expect the major product to be the one predicted by Zaitsev's rule? Explain.

PROBLEM-SOLVING STRATEGY

Propose a mechanism for the following reaction:

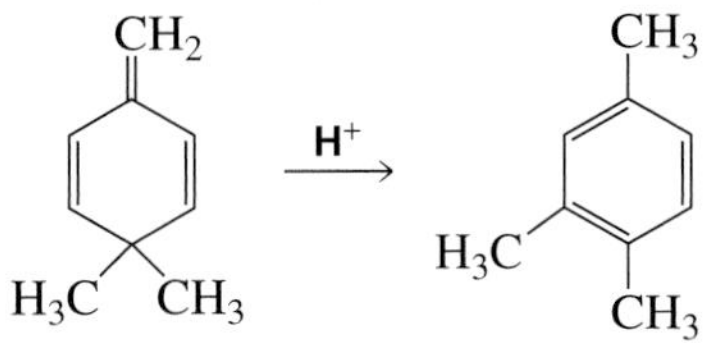

Because the given reagent is a proton, start by protonating the molecule at the position that forms the most stable carbocation. By protonating the CH_2 group, a tertiary allylic carbocation is formed in which the positive charge is delocalized over two other carbons. Then move the π electrons so that the 1,2-methyl shift required to obtain the product can take place. Loss of a proton gives the final product.

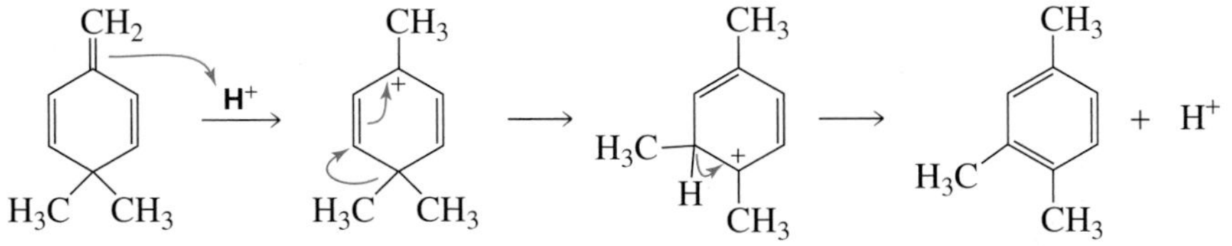

Now continue on to Problem 6.

PROBLEM 6

Propose a mechanism for the following reaction:

Primary alkyl halides undergo only E2 elimination reactions. They cannot undergo E1 reactions because of the difficulty encountered in forming primary carbocations. *Secondary* and *tertiary* alkyl halides undergo both E2 and E1 reactions (Table 10.3).

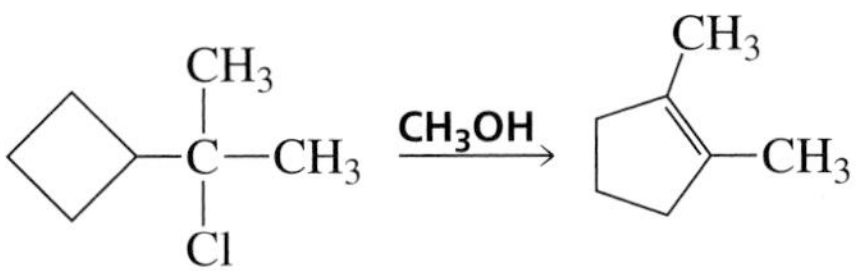

10.4 COMPETITION BETWEEN E2 AND E1 REACTIONS

TABLE 10.3   Summary of the Reactivity of Alkyl Halides in Elimination Reactions

primary alkyl halide	E2 only
secondary alkyl halide	E1 and E2
tertiary alkyl halide	E1 and E2

For those alkyl halides that can undergo both E2 and E1 reactions, the E2 reaction is favored by the same factors that favor an S_N2 reaction, and the E1 reaction is favored by the same factors that favor an S_N1 reaction. *An E2 reaction is favored by a high concentration of a strong base and an aprotic polar solvent (e.g., DMSO or DMF). An E1 reaction, on the other hand, is favored by a weak base and a protic polar solvent (e.g., H_2O or ROH).* How the solvent affects the mechanism of the reaction was discussed in Section 9.10.

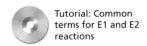

Tutorial: Common terms for E1 and E2 reactions

PROBLEM 7 ◆

For each of the following reactions, indicate whether elimination will occur via an E2 or an E1 reaction. Give the major elimination product of each reaction:

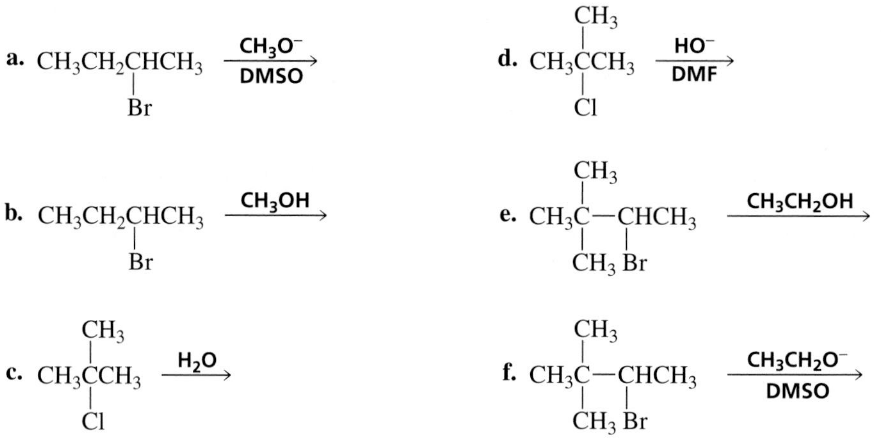

a. $CH_3CH_2\overset{\displaystyle |}{\underset{\displaystyle Br}{C}}HCH_3$ $\xrightarrow[\text{DMSO}]{CH_3O^-}$

d. $CH_3\overset{\displaystyle CH_3}{\underset{\displaystyle Cl}{\overset{\displaystyle |}{\underset{\displaystyle |}{C}}}}CH_3$ $\xrightarrow[\text{DMF}]{HO^-}$

b. $CH_3CH_2\overset{\displaystyle |}{\underset{\displaystyle Br}{C}}HCH_3$ $\xrightarrow{CH_3OH}$

e. $CH_3\overset{\displaystyle CH_3}{\underset{\displaystyle CH_3}{\overset{\displaystyle |}{\underset{\displaystyle |}{C}}}}\overset{}{\underset{\displaystyle Br}{C}}HCH_3$ $\xrightarrow{CH_3CH_2OH}$

c. $CH_3\overset{\displaystyle CH_3}{\underset{\displaystyle Cl}{\overset{\displaystyle |}{\underset{\displaystyle |}{C}}}}CH_3$ $\xrightarrow{H_2O}$

f. $CH_3\overset{\displaystyle CH_3}{\underset{\displaystyle CH_3}{\overset{\displaystyle |}{\underset{\displaystyle |}{C}}}}\overset{}{\underset{\displaystyle Br}{C}}HCH_3$ $\xrightarrow[\text{DMSO}]{CH_3CH_2O^-}$

PROBLEM 8 ◆

The rate law for the reaction of HO^- with *tert*-butyl bromide to form an elimination product in 75% ethanol/25% water at 30 °C is:

rate $= 7.1 \times 10^{-5}$[*tert*-butyl bromide][HO^-] $+ 1.5 \times 10^{-5}$[*tert*-butyl bromide]

What percent of the reaction takes place by the E2 pathway when:

a. $[HO^-] = 5.0$ M?

b. $[HO^-] = 0.0025$ M?

10.5 STEREOCHEMISTRY OF E2 AND E1 REACTIONS

Stereochemistry of the E2 Reaction

An E2 reaction involves the removal of two groups from adjacent carbons. It is a concerted reaction because the two groups are eliminated in the same step. The bonds to the eliminated groups (H and X) must be in the same plane because the sp^3 orbital of the carbon bonded to H and the sp^3 orbital of the carbon bonded to X become overlapping *p* orbitals in the alkene product. Therefore, the orbitals must overlap in the transition state. This overlap is optimal if the orbitals are parallel, and to be parallel they must be in the same plane.

There are two ways in which the C—H and C—X bonds can be in the same plane: They can be parallel to one another either on the same side of the molecule (**syn-periplanar**) or on opposite sides of the molecule (**anti-periplanar**).

substituents are syn-periplanar

an eclipsed conformer

substituents are anti-periplanar

a staggered conformer

If an elimination reaction removes two substituents from the same side of the C—C bond, the reaction is called a **syn elimination.** If the substituents are removed from opposite sides of the C—C bond, the reaction is called an **anti elimination.** Both can occur, but syn elimination is a much slower reaction, so anti elimination is favored in an E2 reaction. One reason anti elimination is favored is that syn elimination requires the molecule to be in an eclipsed conformation, while anti elimination requires it to be in a staggered conformation. Because the staggered conformer is more stable (Section 2.10), the transition state leading to elimination from the staggered conformer is more stable than that leading to elimination from the eclipsed conformer. Consequently, anti elimination occurs more rapidly.

You can understand the other reasons that anti elimination is preferred by drawing sawhorse projections for the two conformers. In syn elimination, the electrons of the departing hydrogen move to the *front* side of the carbon bonded to X. In anti elimination, on the other hand, the electrons move to the *back* side of the carbon bonded to X. You have seen that displacement reactions involve back side attack (Section 9.2). Finally, anti elimination avoids the repulsion experienced by the electron-rich base when it is on the same side of the molecule as the electron-rich departing halide ion.

syn elimination   **anti elimination**

In Section 10.1 you saw that an E2 reaction is *regioselective,* which means that more of one constitutional isomer is formed than the other. For example, the major product formed from E2 elimination of 2-bromopentane is 2-pentene.

$CH_3CH_2CH_2\overset{\underset{\textstyle |}{Br}}{C}HCH_3 \xrightarrow[\text{CH}_3\text{CH}_2\text{OH}]{\text{CH}_3\text{CH}_2\text{O}^-} CH_3CH_2CH=CHCH_3 + CH_3CH_2CH_2CH=CH_2$

2-bromopentane   **2-pentene**   **1-pentene**
                                              major product   **minor product**
                                         (mixture of *E* and *Z*)

The E2 reaction is also *stereoselective,* which means that more of one stereoisomer is formed than the other. For example, the 2-pentene obtained as the major product from the elimination reaction of 2-bromopentane can exist as a pair of stereoisomers, and more (*E*)-2-pentene is formed than (*Z*)-2-pentene.

(*E*)-2-pentene   **(*Z*)-2-pentene**
    41%   **14%**

We can make the following general statement about the stereochemistry of E2 reactions: If the reactant has two hydrogens bonded to the carbon from which a hydrogen is to be removed, both the *E* and *Z* products will be formed because there are two conformers in which the groups to be eliminated are anti. The alkene with the bulkiest groups on opposite sides of the double bond will be formed in greater yield. For example, more (*E*)-2-pentene is formed in the preceding reaction because the bulkiest groups (the methyl and ethyl groups) lie opposite each other.

The reason why the alkene with the *bulkiest groups* on *opposite sides of the double bond* is formed in greater yield is that it is more stable than the alkene with the bulkiest groups on the same side of the double bond. We saw in Section 3.19 that the decreased stability of the alkene with the bulkiest groups on the same side of the double bond is due to the steric strain that the isomer experiences as a result of the interacting electron clouds of the large substituents. Because the alkene with the largest groups on opposite sides of the double bond is more stable, it has the more stable transition state and therefore is formed more rapidly (Figure 10.3).

Figure 10.3 ▶
Reaction coordinate diagram for the E2 reaction of 2-bromopentane and ethoxide ion.

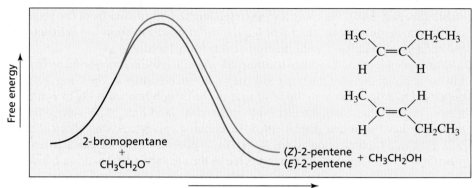

Elimination of HBr from 3-bromo-2,2,3-trimethylpentane leads predominantly to the *E* isomer because this stereoisomer has the methyl group, the bulkiest group on one sp^2 carbon, opposite the *tert*-butyl group, the bulkiest group on the other sp^2 carbon.

3-D Molecule:
3-Bromo-2,2,3-trimethylpentane

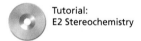
Tutorial:
E2 Stereochemistry

If the β-carbon from which a hydrogen is to be removed is bonded to only one hydrogen, only one alkene product can be formed because there is only one conformer in which the groups to be eliminated are anti. The particular isomer that is formed depends on the configuration of the reactant. For example, anti elimination of HBr from (2S,3S)-2-bromo-3-phenylbutane forms the *E* isomer, whereas anti elimination of HBr from (2S,3R)-2-bromo-3-phenylbutane forms the *Z* isomer.

Molecular models can be helpful whenever complex stereochemistry is involved.

PROBLEM 9 ◆

a. Determine the major product that would be obtained from an E2 reaction of each of the following alkyl halides. In each case, indicate the configuration of the product:

1.  CH$_3$CH$_2$CHCHCH$_3$ (Br, CH$_3$ substituents)

3. CH$_3$CH$_2$CHCH$_2$CH=CH$_2$ (Cl substituent)

2. CH$_3$CH$_2$CHCH$_2$— (phenyl) (Cl substituent)

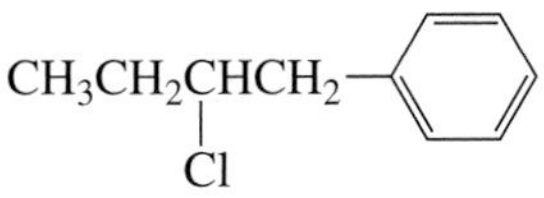

b. Does the product obtained depend on whether you started with the *R* or *S* enantiomer of the reactant?

PROBLEM 10 ◆

What alkene will be formed in an E2 reaction of each of the following compounds? Indicate its configuration.

a. (1*S*,2*S*)-1-bromo-1,2-diphenylpropane

b. (1*S*,2*R*)-1-bromo-1,2-diphenylpropane

Stereochemistry of the E1 Reaction

We have seen that an E1 elimination reaction takes place in two steps. The leaving group leaves in the first step, and a proton is lost from an adjacent carbon in the second step, following Zaitsev's rule in order to form the more stable alkene. The carbocation formed in the first step of an E1 reaction is planar. This means that the electrons from a departing proton can move toward the positively charged carbon from either side. Therefore, both syn and anti elimination can occur.

Because both syn and anti elimination can occur, an E1 reaction forms both the *E* and *Z* products regardless of whether the β-carbon from which the proton is removed is bonded to one or two hydrogens. The major product is the one with the bulkiest groups on opposite sides of the double bond because it is the more stable alkene.

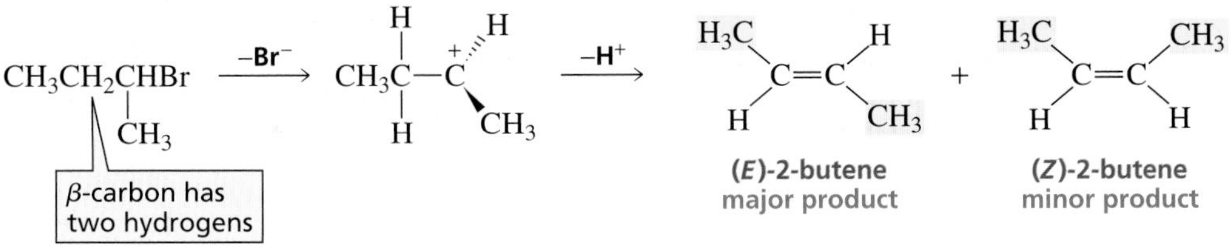

(E)-2-butene
major product

(Z)-2-butene
minor product

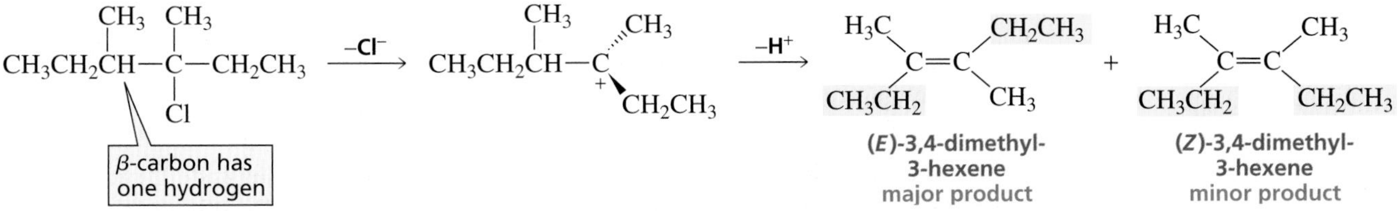

β-carbon has
two hydrogens

(E)-3,4-dimethyl-
3-hexene
major product

(Z)-3,4-dimethyl-
3-hexene
minor product

β-carbon has
one hydrogen

Recall that an E2 reaction forms both the *E* and *Z* products only if the β-carbon from which the proton is removed is bonded to two hydrogens. If it is bonded to only one hydrogen, only one product is obtained because only anti elimination can occur. The configuration of the product will depend on the configuration of the reactant.

PROBLEM 11/ SOLVED

Determine the major product that would be formed when each of the following alkyl halides undergoes an E1 reaction:

a. 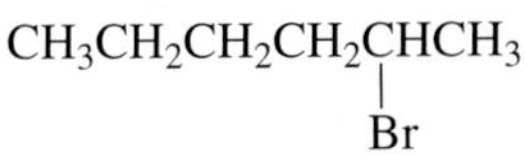 $CH_3CH_2CH_2CH_2CHCH_3$
          $|$
          Br

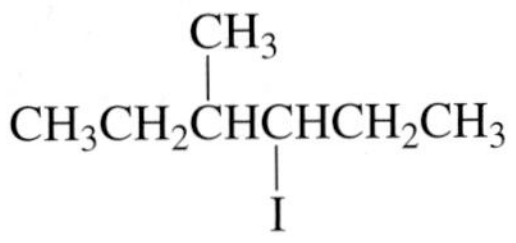

b. 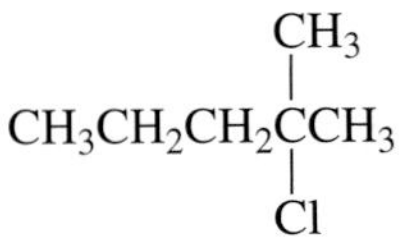 $CH_3CH_2CH_2CCH_3$
          $\overset{CH_3}{\underset{Cl}{|}}$

c. $CH_3CH_2CHCHCH_2CH_3$
          $\overset{CH_3}{|} \quad \underset{I}{|}$

d. 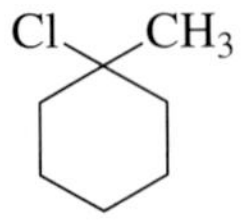
          (structure with Cl and CH₃ on cyclohexane)

SOLUTION TO 11a   First we must consider the regiochemistry of the reaction: More 2-hexene will be formed than 1-hexene because 2-hexene is more stable. Now the stereochemistry of the reaction must be considered: Of the 2-hexene that is formed, more (*E*)-2-hexene will be formed than (*Z*)-2-hexene because (*E*)-2-hexene is more stable. Thus, (*E*)-2-hexene is the major product of the reaction.

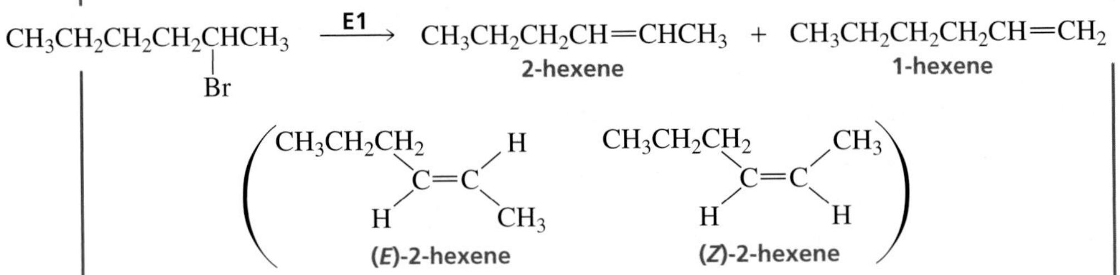

$CH_3CH_2CH_2CH_2CHCH_3 \xrightarrow{\textbf{E1}} CH_3CH_2CH_2CH=CHCH_3 + CH_3CH_2CH_2CH_2CH=CH_2$
$\qquad\qquad\quad |$
$\qquad\qquad\quad Br \qquad\qquad\qquad\qquad \textbf{2-hexene} \qquad\qquad\qquad\qquad \textbf{1-hexene}$

$$\left(\begin{array}{cc} \underset{H}{\overset{CH_3CH_2CH_2}{\diagdown}}C=C\overset{H}{\underset{CH_3}{\diagup}} & \underset{H}{\overset{CH_3CH_2CH_2}{\diagdown}}C=C\overset{CH_3}{\underset{H}{\diagup}} \\ \textbf{(\textit{E})-2-hexene} & \textbf{(\textit{Z})-2-hexene} \end{array} \right)$$

10.6 ELIMINATION FROM CYCLIC COMPOUNDS

E2 Elimination from Cyclic Compounds

Elimination from cyclic compounds follows the same stereochemical rules as elimination from open-chain compounds. To achieve the anti-periplanar geometry that is preferred for an E2 reaction, the two groups that are being eliminated from a six-membered ring must both be in axial positions. If either of the groups to be eliminated is in an equatorial position, an E2 reaction cannot occur.

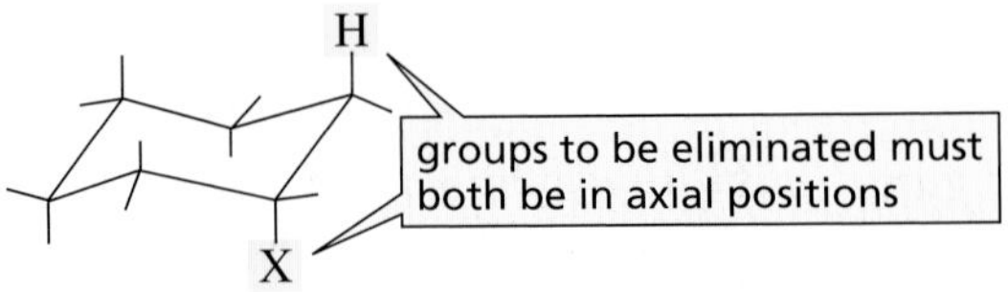

(chair cyclohexane structure with H and X labeled)

groups to be eliminated must both be in axial positions

The more stable conformer of chlorocyclohexane does not undergo an E2 reaction because the chloro substituent is in an equatorial position. (Recall from Section 2.13 that the more stable conformer of a monosubstituted cyclohexane is the one in which the substituent is in an equatorial position, because there is more room for a substituent in that position.) The less stable conformer, with the chloro substituent in the axial position, readily undergoes an E2 reaction.

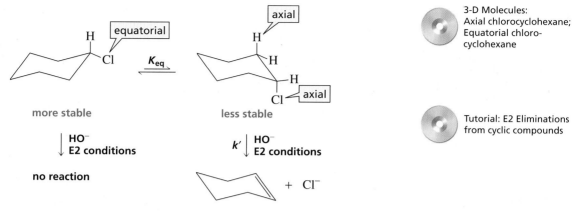

The rate of an elimination reaction is affected by the stability of the conformer that undergoes elimination. Because the rate constant of the reaction is given by $k'K_{eq}$, the reaction is faster if the equilibrium is favorable (K_{eq} is large). Because of the unfavorable equilibrium constant when a reaction has to take place by way of the less stable conformer (K_{eq} is small), the rate of elimination is slower than it would be if elimination could take place by way of the more stable conformer. For example, neomenthyl chloride undergoes an E2 reaction with ethoxide ion about 200 times faster than menthyl chloride. The more stable conformer of neomenthyl chloride undergoes elimination because when the Cl and H are in axial positions, the methyl and isopropyl groups are in equatorial positions.

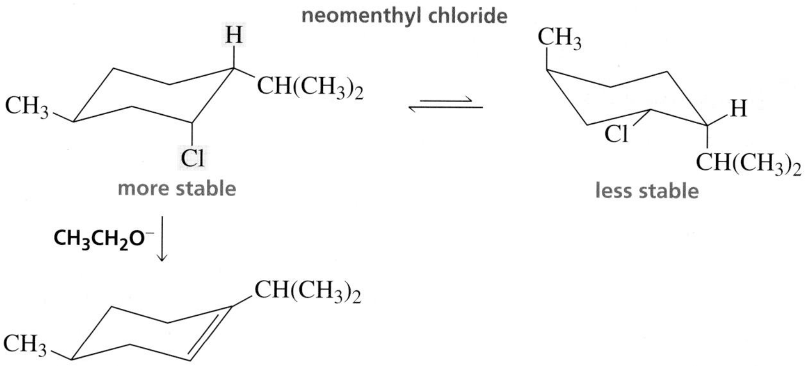

neomenthyl chloride

more stable   less stable

$CH_3CH_2O^-$

In contrast, the less stable conformer of menthyl chloride undergoes elimination because when the Cl and H are in axial positions, the methyl and isopropyl groups are also in axial positions.

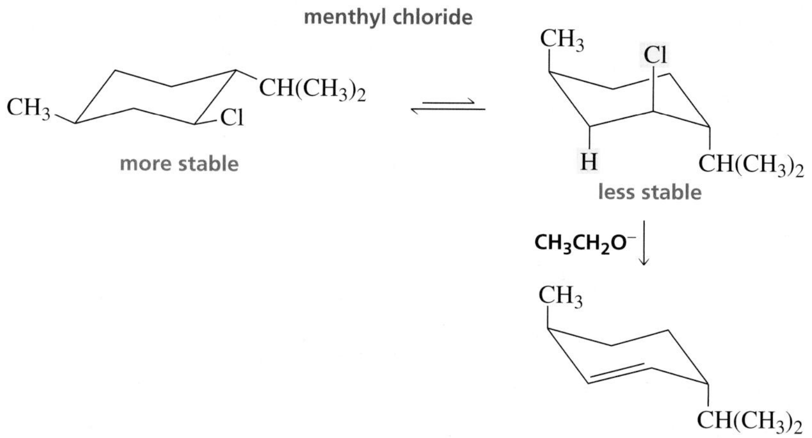

menthyl chloride

more stable   less stable

$CH_3CH_2O^-$

Sir Derek H. R. Barton (1918–1998) was the first to point out that the chemical reactivity of substituted cyclohexanes was controlled by their conformation. Barton was born in Gravesend, Kent, England. He received a Ph.D. and a D.Sc. from Imperial College, London in 1942 and became a faculty member there three years later. He was subsequently a professor at the University of London, the University of Glasgow, Institut de Chimie des Substances Naturelles, and Texas A & M University. For his work on the relationship of the three-dimensional structure of organic compounds to their chemical reactivity, he received the 1969 Nobel Prize in chemistry. Barton was knighted by Queen Elizabeth II in 1972.

(a)

(b)

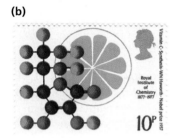

(c)

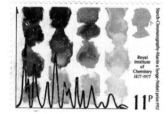

(d)

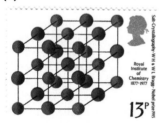

Stamps issued in honor of English Nobel Laureates:
(a) Sir Derek Barton for conformational analysis, 1969;
(b) Sir Walter Haworth for the synthesis of vitamin C, 1937;
(c) A. J. P. Martin and Richard L. M. Synge for chromatography, 1952;
(d) William H. Bragg and William L. Bragg for crystallography, 1915 (the only father and son to receive a Nobel prize).

Notice that when menthyl chloride or *trans*-1-chloro-2-methylcyclohexane undergoes an E2 reaction, the hydrogen that is eliminated is not removed from the β-carbon bonded to the fewest hydrogens. This may seem like a violation of Zaitsev's rule, but Zaitsev's rule states that *when there is a choice,* a hydrogen is removed from the β-carbon bonded to the fewest hydrogens. Because elimination can occur only if the two substituents that are eliminated are in axial positions, the axial hydrogen is removed even though it is not bonded to the β-carbon with the fewest hydrogens.

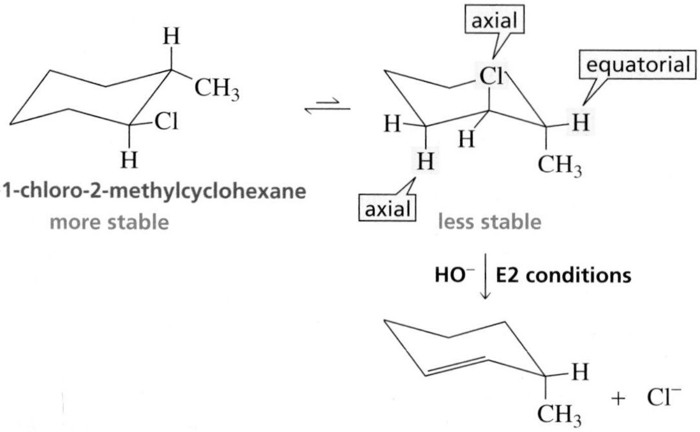

trans-1-chloro-2-methylcyclohexane
more stable

less stable

HO⁻ | E2 conditions

+ Cl⁻

PROBLEM 12

Why do *cis*-1-bromo-2-ethylcyclohexane and *trans*-1-bromo-2-ethylcyclohexane form different products when they undergo an E2 reaction?

PROBLEM 13

Which isomer reacts more rapidly in an E2 reaction, *cis*-1-bromo-4-*tert*-butylcyclohexane or *trans*-1-bromo-4-*tert*-butylcyclohexane? Explain your choice.

E1 Elimination from Cyclic Compounds

When a substituted cyclohexane undergoes an E1 reaction, the two groups that are eliminated do not have to both be in axial positions because the elimination reaction is not concerted. In the following reaction, a carbocation is formed in the first step. It then loses a proton, following Zaitsev's rule.

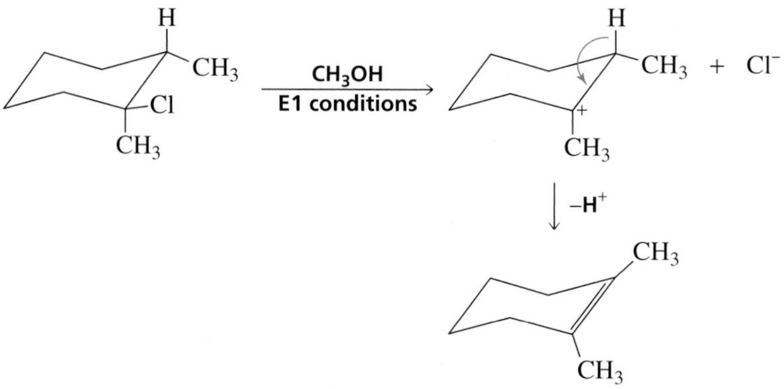

CH₃OH
E1 conditions

+ Cl⁻

–H⁺

Because a carbocation is formed in an E1 reaction, you must check for the possibility of a carbocation rearrangement before you use Zaitsev's rule to determine the elimination product. In the following reaction, the secondary carbocation undergoes a 1,2-hydride shift, forming a more stable tertiary carbocation.

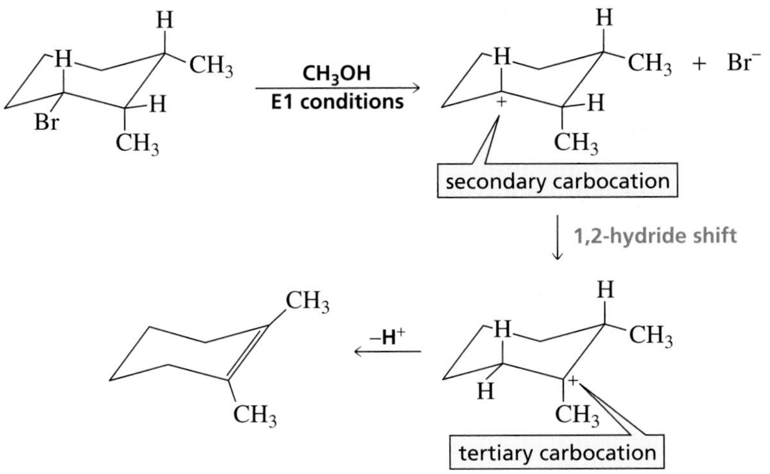

Table 10.4 summarizes the stereochemical outcome of substitution and elimination reactions.

TABLE 10.4   Stereochemistry of Substitution and Elimination Reactions

Mechanism	Products
S_N1	Both enantiomers are formed (more inverted than retained).
E1	Both *E* and *Z* stereoisomers are formed (more of the stereoisomer with the bulkiest groups on opposite sides of the double bond).
S_N2	Only the inverted product is formed.
E2	Both *E* and *Z* stereoisomers are formed (more of the stereoisomer with the bulkiest groups on opposite sides of the double bond), unless the β-carbon of the reactant has only one hydrogen, in which case only one stereoisomer is formed; its configuration depends on the configuration of the

PROBLEM 14

Give the substitution and elimination products for the following reactions, showing the configuration of each product.

a. (*S*)-2-chlorohexane $\xrightarrow[\textbf{S}_N\textbf{2/E2 conditions}]{\textbf{CH}_3\textbf{O}^-}$

b. (*S*)-2-chlorohexane $\xrightarrow[\textbf{S}_N\textbf{1/E1 conditions}]{\textbf{CH}_3\textbf{OH}}$

c. *trans*-1-chloro-2-methylcyclohexane $\xrightarrow[\textbf{S}_N\textbf{2/E2 conditions}]{\textbf{CH}_3\textbf{O}^-}$

d. *trans*-1-chloro-2-methylcyclohexane $\xrightarrow[\textbf{S}_N\textbf{1/E1 conditions}]{\textbf{CH}_3\textbf{OH}}$

e. 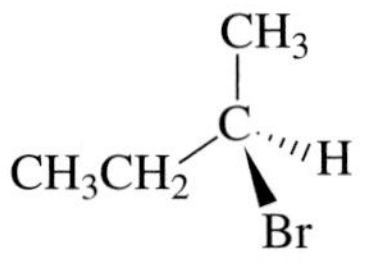 $\xrightarrow[\textbf{S}_N\textbf{2/E2 conditions}]{\textbf{CH}_3\textbf{O}^-}$ **f.** 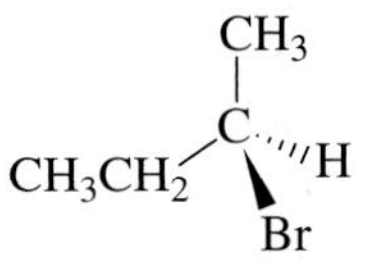 $\xrightarrow[\textbf{S}_N\textbf{1/E1 conditions}]{\textbf{CH}_3\textbf{OH}}$

10.7
A KINETIC ISOTOPE EFFECT

A mechanism is a model that accounts for all the experimental evidence known about a reaction. For example, the mechanisms of the S_N1 and S_N2 reactions are based on a knowledge of the order of the reaction, the relative reactivities of the reactants, and the structures of the products.

Another piece of experimental evidence that is helpful in determining the mechanism of a reaction is a kinetic isotope effect. A **kinetic isotope effect** is a comparison of the rate of reaction of a compound with the rate of reaction of the same compound in which one or more of the atoms has been replaced by an isotope.

A **deuterium kinetic isotope effect** is the ratio of the rate constant observed for a compound containing hydrogen to the rate constant observed for an identical compound in which one or more of the hydrogens has been replaced by deuterium. Deuterium is an isotope of hydrogen (Section 1.1). The nucleus of a deuterium contains a proton and a neutron, whereas the nucleus of a hydrogen contains only a proton.

$$\text{deuterium kinetic isotope effect} = \frac{k_H}{k_D} = \frac{\text{rate constant for H-containing reactant}}{\text{rate constant for D-containing reactant}}$$

The chemical properties of deuterium and hydrogen are similar. However, a carbon–deuterium bond is about 1.2 kcal/mol (5 kJ/mol) stronger than a carbon–hydrogen bond. Therefore, it is more difficult to break a C—D bond than a corresponding C—H bond.

When the rate constant (k_H) for elimination of HBr from 1-bromo-2-phenylethane at 30 °C is compared with the rate constant (k_D) for elimination of DBr from 2-bromo-1,1-dideuterio-1-phenylethane determined at the same temperature, k_H is found to be 7.1 times k_D. Therefore, the deuterium kinetic isotope effect is 7.1. The difference in the reaction rates is due to the difference in energy required to break a C—H bond compared with a C—D bond.

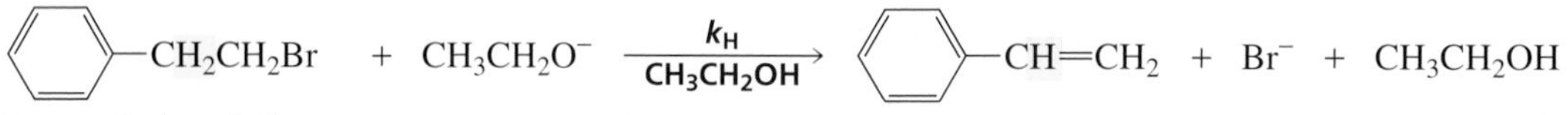

1-bromo-2-phenylethane

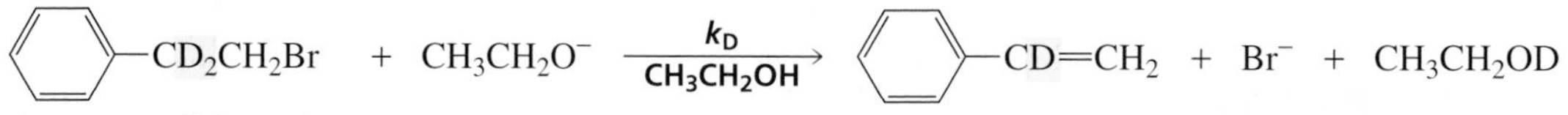

2-bromo-1,1-dideuterio-1-phenylethane

The fact that the deuterium kinetic isotope effect is greater than 1 for this reaction tells us that the carbon–hydrogen (or carbon–deuterium) bond must be broken in the rate-determining step. This is consistent with the mechanism proposed for an E2 reaction.

> **PROBLEM 15 ◆**
>
> If the two preceding reactions took place by an E1 mechanism, what value would you expect to obtain for the deuterium kinetic isotope effect?

10.8
COMPETITION BETWEEN SUBSTITUTION AND ELIMINATION

You have seen that alkyl halides can undergo four types of reactions: S_N2, S_N1, E2, and E1. At this point, it may seem a bit overwhelming to be given an alkyl halide and a nucleophile/base and asked to predict the products of the reaction. So now let's attempt to organize what you have learned about the reactions of alkyl halides to make it a little easier to predict the products of any given reaction.

The first thing you must decide is whether the reaction conditions favor S_N2/E2 or S_N1/E1 reactions. Fortunately, the conditions that favor an S_N2 reaction also favor an E2 reaction, and the conditions that favor an S_N1 reaction also favor an E1 reaction. This means you do not need to worry about $S_N2/E1$ or $S_N1/E2$ combinations— you must just decide whether the combination is going to be $S_N2/E2$ or $S_N1/E1$.

Your decision is easy if the reactant is a *primary* alkyl halide. Primary carbocations are too unstable to be formed, so primary alkyl halides cannot undergo $S_N1/E1$ reactions. They undergo only $S_N2/E2$ reactions.

If the reactant is a *secondary* or a *tertiary* alkyl halide, whether it undergoes $S_N2/E2$ or $S_N1/E1$ reactions depends on the reaction conditions. $S_N2/E2$ reactions are favored by a high concentration of a good nucleophile/strong base, whereas $S_N1/E1$ reactions are favored by a poor nucleophile/weak base (Sections 9.9 and 10.4). In addition, the solvent in which the reaction is carried out can affect the choice of mechanism (Section 9.10).

Once you have decided whether the conditions will lead to $S_N2/E2$ reactions or to $S_N1/E1$ reactions, you must decide how much of the product will be the substitution product and how much will be the elimination product. The relative amounts of substitution and elimination products depend on whether the alkyl halide is primary, secondary, or tertiary, and on the nature of the nucleophile/base. This is discussed in the following section and is later summarized in Table 10.6.

$S_N2/E2$ Conditions

Let's first consider conditions that lead to $S_N2/E2$ reactions (a high concentration of a good nucleophile/strong base). The negatively charged species can act as a nucleophile and hit the back side of the α-carbon to form the substitution product, or it can act as a base and remove a β-hydrogen, leading to the elimination product. Notice that both reactions occur for the same reason—the electron-withdrawing halogen causes the carbon to which it is bonded to have a partial positive charge.

$$CH_3-CH_2-Br \longrightarrow CH_3CH_2OH + Br^-$$
$$HO:^-$$

substitution
product

$$CH_2-CH_2-Br \longrightarrow CH_2=CH_2 + H_2O + Br^-$$
$$H \quad HO:^-$$

elimination
product

The relative reactivities of alkyl halides in S_N2 and E2 reactions are shown in Table 10.5. Because a *primary* alkyl halide is the most reactive in an S_N2 reaction and the least reactive in an E2 reaction, you can predict that a primary alkyl halide will form primarily the substitution product in a reaction carried out under conditions that favor $S_N2/E2$ reactions.

$$CH_3CH_2CH_2Br + CH_3O^- \xrightarrow{\text{CH}_3\text{OH}} CH_3CH_2CH_2OCH_3 + CH_3CH=CH_2 + CH_3OH + Br^-$$

propyl bromide   methyl propyl ether   propene
                              90%   10%

TABLE 10.5 Relative Reactivities of Alkyl Halides			
in an S_N2 reaction:	$1° > 2° > 3°$	in an S_N1 reaction:	$3° > 2° > 1°$
in an E2 reaction:	$3° > 2° > 1°$	in an E1 reaction:	$3° > 2° > 1°$

However, if either the primary alkyl halide or the nucleophile/base is sterically hindered, the nucleophile will have difficulty getting to the back side of the α-carbon and, as a result, the elimination product will predominate.

the primary alkyl halide is sterically hindered

$$CH_3CHCH_2Br \ + \ CH_3O^- \ \xrightarrow{\text{CH}_3\text{OH}} \ CH_3CHCH_2OCH_3 \ + \ CH_3C=CH_2 \ + \ CH_3OH \ + \ Br^-$$

1-bromo-2-methyl-
propane

isobutyl methyl ether
40%

2-methylpropene
60%

the nucleophile is sterically hindered

$$CH_3CH_2CH_2CH_2CH_2Br \ + \ CH_3CO^- \ \xrightarrow{\text{(CH}_3\text{)}_3\text{COH}} \ CH_3CH_2CH_2CH_2CH_2OCCH_3 \ + \ CH_3CH_2CH_2CH=CH_2$$

1-bromopentane

tert-butyl pentyl ether
15%

1-pentene
85%

$$+ \ CH_3COH \ + \ Br^-$$

A *secondary* alkyl halide can form both substitution and elimination products under $S_N2/E2$ conditions. The relative amounts of the two products depend on the base strength and the bulk of the nucleophile. *The stronger and bulkier the base, the greater the percent of the elimination product.* For example, acetic acid is a stronger acid ($pK_a = 4.76$) than ethanol ($pK_a = 15.9$), which means that acetate ion is a weaker base than ethoxide ion. The elimination product is the main product formed from the reaction of 2-chloropropane with the strongly basic ethoxide ion, whereas no elimination product is formed with the weakly basic acetate ion. The percent of elimination product produced would be further increased if the bulky *tert*-butoxide ion were used instead of ethoxide ion.

$$CH_3CHCH_3 \ + \ CH_3CH_2O^- \ \xrightarrow{\text{CH}_3\text{CH}_2\text{OH}} \ CH_3CHCH_3 \ + \ CH_3CH=CH_2 \ + \ CH_3CH_2OH \ + \ Br^-$$

2-chloropropane

ethoxide ion

2-ethoxypropane
25%

propene
75%

$$CH_3CHCH_3 \ + \ CH_3CO^- \ \xrightarrow{\text{acetic acid}} \ CH_3CHCH_3 \ + \ Cl^- \ + \ CH_3COH$$

2-chloropropane

acetate ion

isopropyl acetate
100%

Increasing the temperature at which the reaction is carried out increases the rates of both the substitution and elimination reactions, but the rate of the elimination reaction increases more. Therefore, if the substitution product is the desired product, the reaction should be carried out at a low temperature. If the elimination product is the desired product, high temperatures should be used.

A *tertiary* alkyl halide is the least reactive of the alkyl halides in an S_N2 reaction and the most reactive in an E2 reaction (Table 10.5). Consequently, only the elimination product is formed when a tertiary alkyl halide reacts with a nucleophile under $S_N2/E2$ conditions.

3-D Molecule:
2-Methyl-1-propene

$$CH_3CBr \ + \ CH_3CH_2O^- \ \xrightarrow{\text{CH}_3\text{CH}_2\text{OH}} \ CH_3C=CH_2 \ + \ CH_3CH_2OH \ + \ Br^-$$

2-bromo-2-methyl-
propane

2-methylpropene

PROBLEM 16 ◆

How would you expect the ratio of substitution product to elimination product formed from the reaction of propyl bromide with CH_3O^- in methanol to change when the nucleophile is changed to CH_3S^-?

PROBLEM 17

Only a substitution product is obtained when the following compound is treated with sodium methoxide. Explain why an elimination product is not obtained.

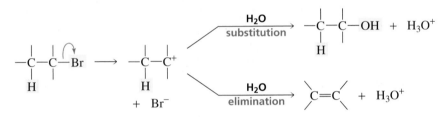

ethoxide ion

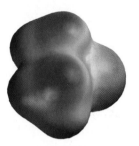

***tert*-butoxide ion**

S$_N$1/E1 Conditions

Let's now look at what happens when conditions lead to $S_N1/E1$ reactions (a poor nucleophile/weak base). In $S_N1/E1$ reactions, the alkyl halide dissociates to form a carbocation. The carbocation can either combine with the nucleophile to form the substitution product or lose a proton to form the elimination product.

Alkyl halides have the same order of reactivity in S_N1 reactions as they do in E1 reactions because both reactions have the same rate-determining step—dissociation of the alkyl halide (Table 10.5). This means that all alkyl halides that react under $S_N1/E1$ conditions will give both substitution and elimination products. Substitution is favored over elimination at lower temperatures, so increasing the temperature increases the percentage of the elimination product. Remember that primary alkyl halides do not undergo $S_N1/E1$ reactions because primary carbocations are too unstable to be formed.

Table 10.6 summarizes the products obtained when alkyl halides react with nucleophiles under $S_N2/E2$ and $S_N1/E1$ conditions.

TABLE 10.6 Summary of the Products Expected in Substitution/Elimination Reactions		
Class of alkyl halide	**S$_N$2 vs. E2**	**S$_N$1 vs. E1**
Primary alkyl halide	Primarily substitution unless there is steric hindrance in the alkyl halide or nucleophile, in which case elimination is favored	Cannot undergo S$_N$1/E1 reactions
Secondary alkyl halide	Both substitution and elimination; the stronger and bulkier the base and the higher the temperature, the greater the percentage of elimination	Both substitution and elimination; the higher the temperature, the greater the percentage of elimination
Tertiary alkyl halide	Only elimination	Both substitution and elimination; the higher the temperature, the greater the percentage of elimination

PROBLEM 18 ◆

Indicate whether the following alkyl halides will give mostly substitution products, mostly elimination products, or about equivalent amounts of substitution and elimination products when they react with:

a. methanol under $S_N1/E1$ conditions.

b. sodium methoxide under $S_N2/E2$ conditions.

    1. 1-bromobutane   **3.** 2-bromobutane

    2. 1-bromo-2-methylpropane   **4.** 2-bromo-2-methylpropane

PROBLEM 19

1-Bromo-2,2-dimethylpropane has difficulty undergoing either S_N2 or S_N1 reactions.

a. Explain why.   **b.** Can it undergo E2 and E1 reactions?

10.9 SUBSTITUTION AND ELIMINATION REACTIONS IN SYNTHESIS

When substitution or elimination reactions are used in synthesis, care must be taken to choose reactants and reaction conditions that will give the maximum possible yield of the desired product. In Section 9.4 you saw that a wide variety of organic compounds can be prepared by substitution reactions of alkyl halides. For example, ethers are synthesized by the reaction of an alkyl halide with an alkoxide ion. This reaction was discovered by Alexander Williamson in 1850 and is still considered to be one of the best ways to synthesize an ether.

Williamson ether synthesis

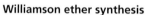

$$\underset{\text{alkyl halide}}{\text{R—Br}} \quad + \quad \underset{\text{alkoxide ion}}{\text{R—O}^-} \quad \longrightarrow \quad \underset{\text{ether}}{\text{R—O—R}} + \text{Br}^-$$

The alkoxide ion for the **Williamson ether synthesis** is prepared by using sodium metal or sodium hydride (NaH) to remove a proton from an alcohol.

$$\text{ROH} \quad + \quad \text{Na} \quad \longrightarrow \quad \text{RO}^- \quad + \quad \text{Na}^+ \quad + \quad \tfrac{1}{2}\text{H}_2$$

$$\text{ROH} \quad + \quad \text{NaH} \quad \longrightarrow \quad \text{RO}^- \quad + \quad \text{Na}^+ \quad + \quad \text{H}_2$$

Alexander William Williamson (1824–1904) was born in London to Scottish parents. As a child he lost an arm and the use of an eye. He was midway through his medical education when he changed his mind and decided to study chemistry. He received a Ph.D. from the University of Geissen in 1846. In 1849 he became a professor of chemistry at University College, London.

    The Williamson ether synthesis is a substitution reaction and, because it requires a high concentration of a good nucleophile (the alkoxide ion), it is an S_N2 reaction. If you want to synthesize an ether such as butyl propyl ether, you have a choice of starting materials. You can use either a propyl halide and butoxide ion or a butyl halide and propoxide ion.

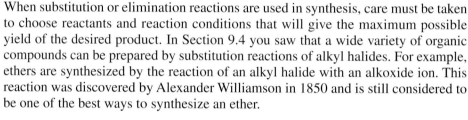

$$\underset{\text{propyl bromide}}{\text{CH}_3\text{CH}_2\text{CH}_2\text{Br}} + \underset{\text{butoxide ion}}{\text{CH}_3\text{CH}_2\text{CH}_2\text{CH}_2\text{O}^-} \longrightarrow \underset{\text{butyl propyl ether}}{\text{CH}_3\text{CH}_2\text{CH}_2\text{OCH}_2\text{CH}_2\text{CH}_3}} + \text{Br}^-$$

$$\underset{\text{butyl bromide}}{\text{CH}_3\text{CH}_2\text{CH}_2\text{CH}_2\text{Br}} + \underset{\text{propoxide ion}}{\text{CH}_3\text{CH}_2\text{CH}_2\text{O}^-} \longrightarrow \underset{\text{butyl propyl ether}}{\text{CH}_3\text{CH}_2\text{CH}_2\text{OCH}_2\text{CH}_2\text{CH}_3}} + \text{Br}^-$$

If, however, you want to synthesize *tert*-butyl ethyl ether, the starting materials must be an ethyl halide and *tert*-butoxide ion. If you tried to use a *tert*-butyl halide and ethoxide ion as reactants, you would obtain the elimination product and little or no ether because reaction of a tertiary alkyl halide under $S_N2/E2$ conditions forms primarily the elimination product. So the less hindered alkyl group should be obtained from the alkyl halide and the more hindered alkyl group should serve as the alkoxide ion.

Tutorial:
E2 Promoting factors

$$CH_3CH_2Br + CH_3\overset{\underset{\displaystyle |}{CH_3}}{\underset{\underset{\displaystyle CH_3}{|}}{C}}O^- \longrightarrow CH_3CH_2O\overset{\underset{\displaystyle |}{CH_3}}{\underset{\underset{\displaystyle CH_3}{|}}{C}}CH_3 + CH_2=CH_2 + CH_3\overset{\underset{\displaystyle |}{CH_3}}{\underset{\underset{\displaystyle CH_3}{|}}{C}}OH + Br^-$$

ethyl bromide   tert-butoxide ion   tert-butyl ethyl ether   ethene

$$CH_3CH_2O^- + CH_3\overset{\underset{\displaystyle |}{CH_3}}{\underset{\underset{\displaystyle CH_3}{|}}{C}}Br \longrightarrow CH_2=\overset{\underset{\displaystyle |}{CH_3}}{C}CH_3 + CH_3CH_2OH + Br^-$$

ethoxide ion   tert-butyl bromide   2-methylpropene

PROBLEM 20 ◆

What other organic product will be formed when the alkyl halide used in the synthesis of butyl propyl ether is:

a. propyl bromide?   **b.** butyl bromide?

PROBLEM 21

How could the following ethers be prepared using an alkyl halide and an alcohol?

a. $CH_3CH_2\overset{\underset{\displaystyle |}{CH_3}}{C}HOCH_2CH_2CH_3$

c. ⬡—CH_2O—⬡

b. ⬡—$O\overset{\underset{\displaystyle |}{CH_3}}{\underset{\underset{\displaystyle CH_3}{|}}{C}}CH_3$

d. $CH_3CH_2OCH_2CH_2CH_2CH_2CH_3$

In Section 5.10 you saw that alkynes can be synthesized by the reaction of an acetylide anion with an alkyl halide.

$$CH_3CH_2C\equiv C^- + CH_3CH_2CH_2Br \longrightarrow CH_3CH_2C\equiv CCH_2CH_2CH_3 + Br^-$$

Now that you know that this is an S_N2 reaction (the alkyl halide reacts with a high concentration of a good nucleophile), you can understand why it is best to use primary alkyl halides and methyl halides in the reaction. These alkyl halides are the only ones that form primarily the desired substitution product. A tertiary alkyl halide would form only the elimination product, and a secondary alkyl halide would form mainly the elimination product because the acetylide ion is a very strong base.

If the desired product is an alkene, you should choose the most hindered alkyl halide in order to maximize the amount of elimination product. For example, if you wanted to synthesize propene, 2-bromopropane would be a better starting material than 1-bromopropane because the secondary alkyl halide would give a higher yield of the desired elimination product and a lower yield of the competing substitution product.

$$CH_3\overset{\underset{\displaystyle |}{Br}}{C}HCH_3 + HO^- \longrightarrow CH_3CH=CH_2 + CH_3\overset{\underset{\displaystyle |}{OH}}{C}HCH_3 + H_2O + Br^-$$

2-bromopropane   major product   minor product

$$CH_3CH_2CH_2Br + HO^- \longrightarrow CH_3CH=CH_2 + CH_3CH_2CH_2OH + H_2O + Br^-$$

1-bromopropane   minor product   major product

To synthesize 2-methyl-2-butene from 2-bromo-2-methylbutane, you would use $S_N2/E2$ conditions (a high concentration of HO^- in an aprotic polar solvent) because a tertiary alkyl halide under these conditions gives only the elimination product. If $S_N1/E1$ conditions were used (a low concentration of HO^- in a protic polar solvent), both the elimination and substitution products would be obtained.

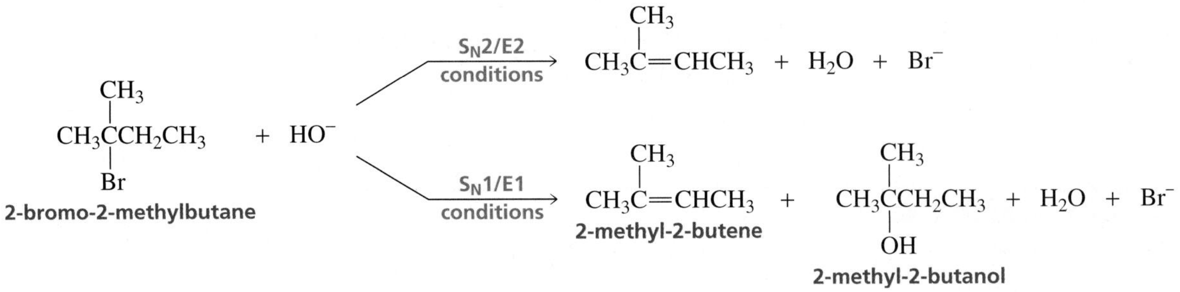

PROBLEM 22 ◆

Identify the three products that are formed when 2-bromo-2-methylpropane is dissolved in a mixture of 80% ethanol and 20% water.

PROBLEM 23

a. What products (including stereoisomers, if applicable) would be formed from the reaction of 3-bromo-2-methylpentane with HO^- under $S_N2/E2$ conditions and under $S_N1/E1$ conditions?

b. Answer the same question for 3-bromo-3-methylpentane.

10.10 CONSECUTIVE E2 ELIMINATION REACTIONS

Alkyl dihalides can undergo two consecutive dehydrohalogenations, giving products that contain two double bonds. In the following example, Zaitsev's rule predicts the most stable product of the first dehydrohalogenation but not the most stable product of the second. This is because the conjugated diene is more stable than the isolated diene.

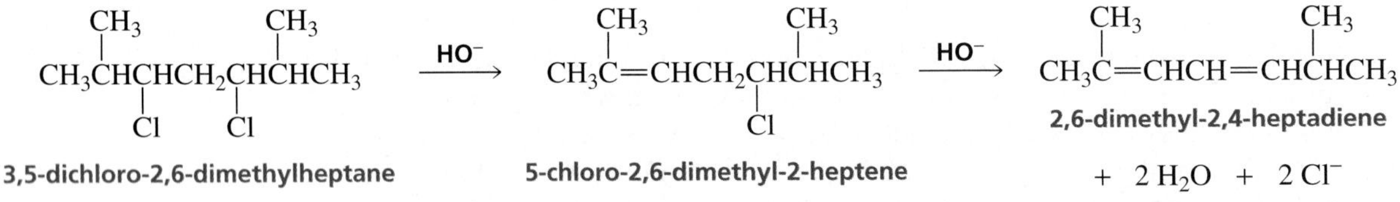

If the two halogens are on the same carbon (geminal dihalides) or on adjacent carbons (vicinal dihalides), the two consecutive E2 dehydrohalogenations can result in the formation of a triple bond. This is how alkynes are commonly synthesized.

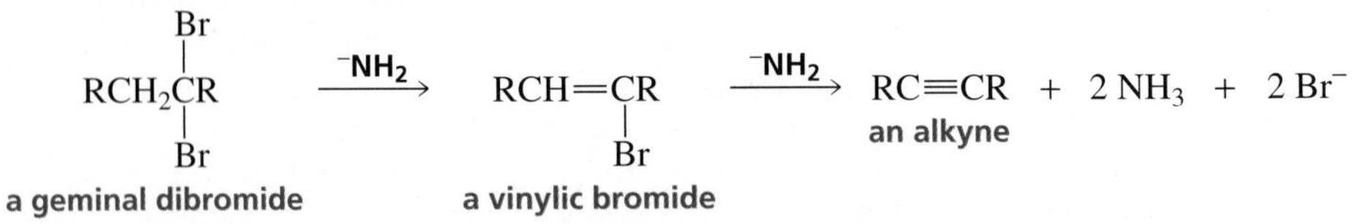

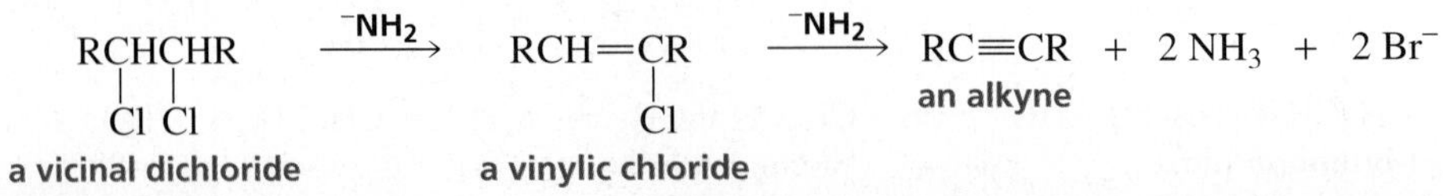

The vinylic halide intermediates in the preceding reactions are relatively unreactive. Consequently, a very strong base such as $^-NH_2$ is needed for the second elimination. If a weaker base such as HO^- is used at room temperature, the reaction will stop at the vinylic halide and no alkyne will be formed.

Because a vicinal dihalide is formed from the reaction of an alkene with Br_2 or Cl_2, you have just learned how to convert a double bond into a triple bond.

$$CH_3CH=CHCH_3 \xrightarrow[\text{CH}_2\text{Cl}_2]{\text{Br}_2} CH_3CHCHCH_3 \xrightarrow{^-\text{NH}_2} CH_3CH=CCH_3 \xrightarrow{^-\text{NH}_2} CH_3C\equiv CCH_3$$

2-butene   Br Br   Br   2-butyne

PROBLEM 24

Why isn't a cumulated diene formed in the preceding reaction?

A molecule with two functional groups is called a **bifunctional molecule.** If the two functional groups are able to react with each other, two kinds of reactions can occur. First, in the case of a molecule whose two functional groups are a nucleophile and a leaving group, the nucleophile of one molecule can displace the leaving group of a second molecule of the compound. Such a reaction is called an intermolecular reaction. *Inter* is Latin for "between," so an **intermolecular reaction** takes place between two molecules.

an intermolecular reaction

$$BrCH_2(CH_2)_nCH_2\ddot{\underset{..}{O}}\colon^- \quad Br-CH_2(CH_2)_nCH_2\ddot{\underset{..}{O}}\colon^- \longrightarrow BrCH_2(CH_2)_nCH_2\ddot{\underset{..}{O}}CH_2(CH_2)_nCH_2\ddot{\underset{..}{O}}\colon^- \ + \ Br^-$$

Alternatively, the nucleophile of a molecule can displace the leaving group of the same molecule, thereby forming a cyclic compound. Such a reaction is called an intramolecular reaction. *Intra* is Latin for "within," so an **intramolecular reaction** takes place within a single molecule.

an intramolecular reaction

$$Br-CH_2(CH_2)_nCH_2\ddot{\underset{..}{O}}\colon^- \longrightarrow \underset{\underset{\ddot{\underset{..}{O}}}{H_2C \diagdown \diagup CH_2}}{(CH_2)_n} \ + \ Br^-$$

Which reaction is more likely to occur when the nucleophile and the leaving group are parts of the same molecule—an intermolecular reaction or an intramolecular reaction? The answer depends on the *concentration* of the bifunctional molecule and the *size of the ring* that will be formed in the intramolecular reaction.

The intramolecular reaction has the advantage of reacting groups that are tethered close together and therefore do not have to wander through the solvent to find a group with which to react. As a result, a low concentration of reactant favors an intramolecular reaction because the two functional groups have a better chance of finding one another if they are in the same molecule. A high concentration of reactant, however, helps compensate for the advantage gained by tethering.

How much of an advantage an intramolecular reaction has over an intermolecular reaction depends on the length of the tether and therefore on the size of the ring that is formed. If the intramolecular reaction would form a five- or six-membered ring, it would be highly favored over the intermolecular reaction because five- and six-membered rings are stable and, therefore, easily formed.

10.11 INTERMOLECULAR VERSUS INTRAMOLECULAR REACTIONS

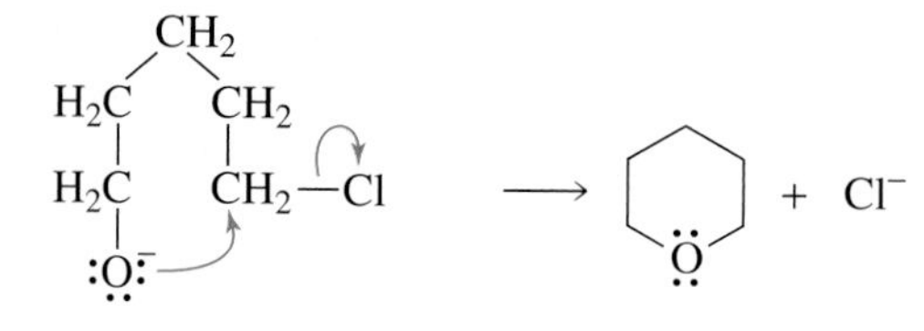

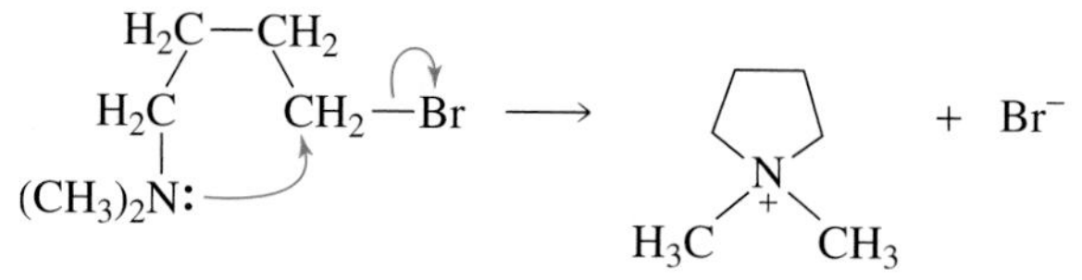

Three- and four-membered rings are strained so they are less stable than five- and six-membered rings. This means that the transition state leading to the formation of three- and four-membered rings is less stable than the transition state leading to the formation of five- and six-membered rings. The higher activation energy for the formation of three- and four-membered rings, therefore, cancels some of the advantage gained by tethering.

The likelihood that the reacting groups can find each other decreases sharply for the formation of seven-membered and larger rings, so the intramolecular reaction becomes less favored as the ring size increases beyond six members.

PROBLEM 25 ◆

Which compound, upon treatment with sodium hydride, would form a cyclic ether more rapidly?

a.  HO⌒⌒⌒⌒Br   or   HO⌒⌒⌒Br

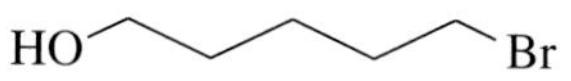

b. 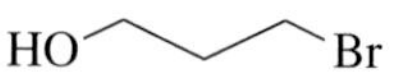 HO⌒⌒Br   or   HO⌒⌒⌒Br

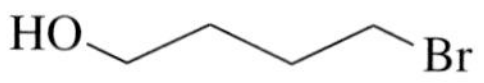

c.  HO⌒⌒⌒⌒Br   or   HO⌒⌒⌒⌒⌒Br

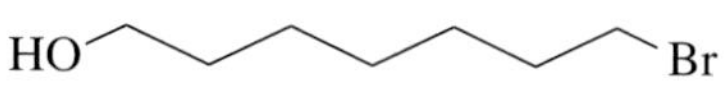

**10.12
DESIGNING A
SYNTHESIS II:
APPROACHING
THE PROBLEM**

When you are asked to design a synthesis, one way to approach the problem is to look at the given starting material to see if there is an obvious series of reactions that can get you started on the pathway to the **target molecule** (the desired product). This is often the best way to approach a simple synthesis. The following examples will give you practice in designing a successful synthesis.

Example 1. How could you prepare 1,3-cyclohexadiene from cyclohexane?

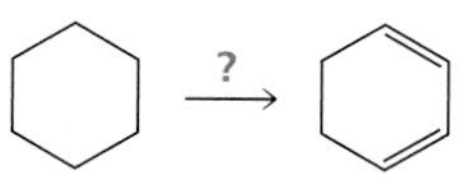

Since the only reaction an alkane can undergo is halogenation, deciding what the first reaction should be is easy. An E2 reaction using a high concentration of a strong and bulky base to encourage elimination over substitution will form cyclohexene. Therefore, *tert*-butoxide ion is used as the base and *tert*-butyl alcohol is used as the solvent. Bromination of cyclohexene will give an allylic bromide, which will form the desired target molecule by undergoing another E2 reaction.

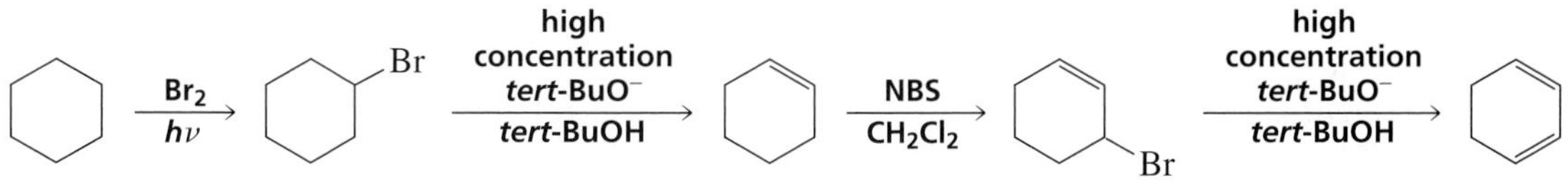

Example 2. Starting with methylcyclohexane, how could you prepare the following vicinal *trans*-dihalide?

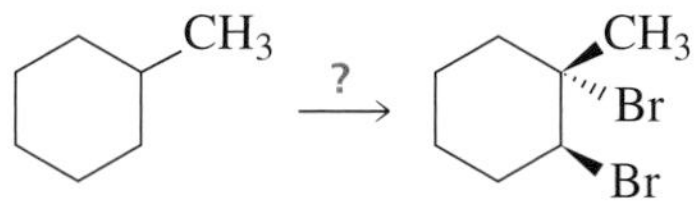

Again, since the starting material is an alkane, the first reaction must be a radical substitution. Bromination leads to selective substitution of the tertiary hydrogen. Under E2 conditions, tertiary alkyl halides undergo only elimination, so there will be no competing substitution product in the next reaction. Because addition of Br_2 involves only anti addition, the target molecule (and its enantiomer) is obtained.

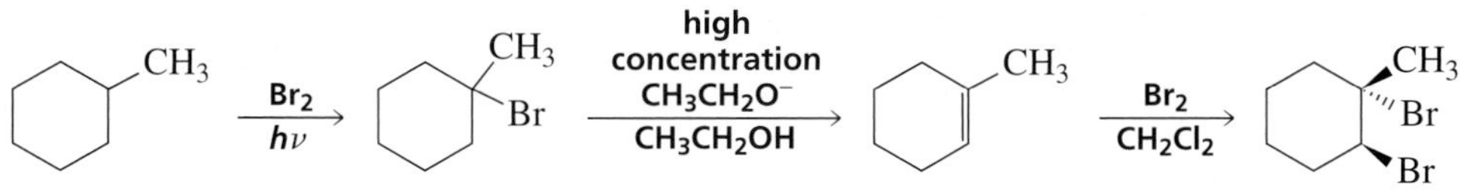

As you saw in Section 5.11, working backward can be a useful way to design a synthesis—particularly when it is not clear from the starting material how to proceed. Look at the target molecule and ask yourself how it could be prepared. Once you have an answer, work backward to the next compound, asking yourself how it could be prepared. Keep working backward one step at a time until you get to a readily available starting material. This technique is called *retrosynthetic analysis.*

Example 3. How could you prepare ethyl methyl ketone from 1-bromobutane?

$$CH_3CH_2CH_2CH_2Br \xrightarrow{?} CH_3CH_2\overset{\overset{\displaystyle O}{\|}}{C}CH_3$$

At this point you know only one method for synthesizing a ketone—adding water to an alkyne. The alkyne can be prepared from two successive E2 reactions of a vicinal dihalide. The dihalide, in turn, can be synthesized from an alkene. The desired alkene can be prepared from the given starting material by an elimination reaction.

retrosynthetic analysis

$$\underset{\text{target molecule}}{CH_3CH_2\overset{\overset{\displaystyle O}{\|}}{C}CH_3} \Longrightarrow CH_3CH_2C{\equiv}CH \Longrightarrow CH_3CH_2\underset{\underset{\displaystyle Br}{|}}{C}HCH_2Br \Longrightarrow CH_3CH_2CH{=}CH_2 \Longrightarrow CH_3CH_2CH_2CH_2Br$$

Now the reaction sequence can be written in the forward direction, indicating the reagents needed to carry out each reaction. A bulky base is used in the elimination reaction in order to maximize the amount of elimination product.

synthesis

$$CH_3CH_2CH_2CH_2Br \xrightarrow[\textit{tert-BuOH}]{\substack{\text{high} \\ \text{concentration} \\ \textit{tert}\text{-BuO}^-}} CH_3CH_2CH=CH_2 \xrightarrow[CH_2Cl_2]{Br_2} CH_3CH_2\underset{\underset{Br}{|}}{CH}CH_2Br \xrightarrow{^-NH_2} CH_3CH_2C\equiv CH$$

$$\xrightarrow[\text{HgSO}_4]{\substack{H_2SO_4 \\ }} \Big| H_2O$$

$$\underset{\text{target molecule}}{CH_3CH_2\overset{\overset{O}{\|}}{C}CH_3}$$

Example 4. How could the following cyclic ether be prepared from the given starting material?

$$BrCH_2CH_2CH_2CH=CH_2 \xrightarrow{?} \text{(cyclic ether)}-CH_3$$

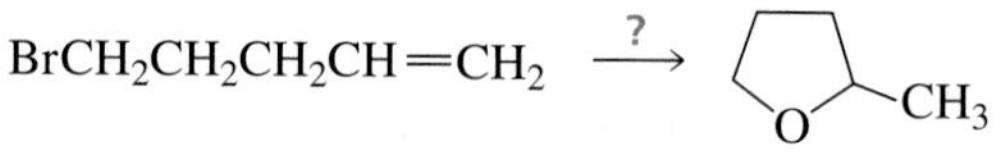

Tutorial: Retrosynthetic analysis

In order to obtain a cyclic ether, the two groups required for ether synthesis (the alkyl halide and the alcohol) must be in the same molecule. To obtain a five-membered ring, the carbons bearing the two groups must be separated by two additional carbons. Addition of water to the given starting material will give the required bifunctional compound.

retrosynthetic analysis

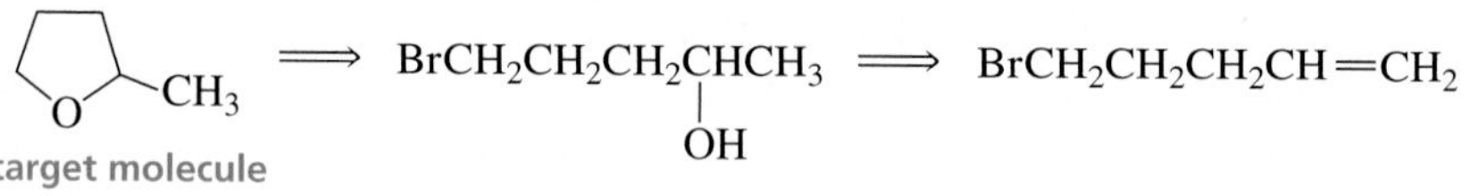

$$\underset{\text{target molecule}}{\text{(cyclic ether)}-CH_3} \Longrightarrow BrCH_2CH_2CH_2\underset{\underset{OH}{|}}{CH}CH_3 \Longrightarrow BrCH_2CH_2CH_2CH=CH_2$$

synthesis

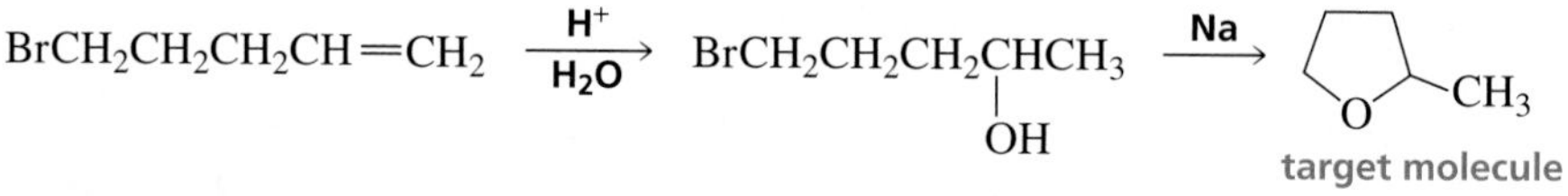

$$BrCH_2CH_2CH_2CH=CH_2 \xrightarrow[H_2O]{H^+} BrCH_2CH_2CH_2\underset{\underset{OH}{|}}{CH}CH_3 \xrightarrow{Na} \underset{\text{target molecule}}{\text{(cyclic ether)}-CH_3}$$

PROBLEM 26

Could you just as easily have used a precursor with the OH and Br reversed for the preceding synthesis? Explain.

PROBLEM 27

Describe how each of the following compounds could be prepared from the given starting material:

a. 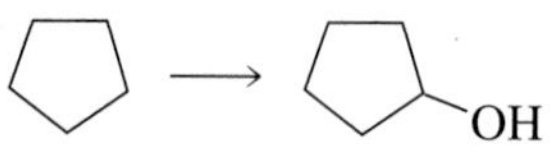 (cyclopentane) $\longrightarrow$ (cyclopentane)–OH

b. $CH_3CH_2CH_2CH_3 \longrightarrow CH_3\underset{\underset{OCH_2CH_3}{|}}{CH}CH_2CH_3$

c. 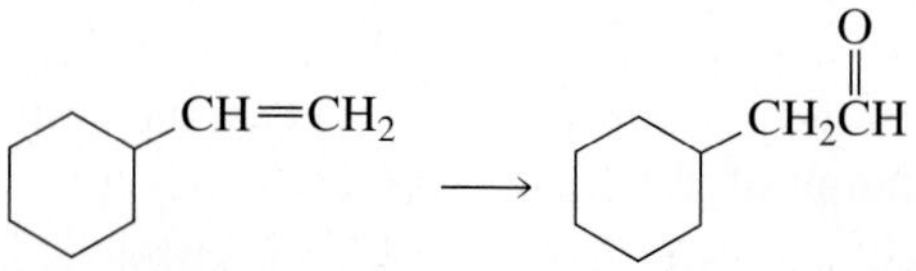 (cyclohexane)–CH=CH$_2$ $\longrightarrow$ (cyclohexane)–CH$_2\overset{\overset{O}{\|}}{C}$H

d. $CH_3CH_2CH_2CH_2Br \longrightarrow CH_3CH_2\overset{\overset{\displaystyle O}{\|}}{C}CH_2CH_2CH_3$

e. $BrCH_2CH_2CH_2CH_2Br \longrightarrow$ (cyclohexane with CH_2CH_3 substituent)

SUMMARY OF REACTIONS

1. E2 reaction: a one-step mechanism

$$B^- \;+\; \overset{\overset{\displaystyle H}{|}}{-\underset{|}{C}}-\overset{|}{\underset{|}{C}}-X \;\longrightarrow\; \overset{\diagdown}{\diagup}C{=}C\overset{\diagup}{\diagdown} \;+\; BH \;+\; X^-$$

Relative reactivities of alkyl halides: $3° > 2° > 1°$

Anti elimination; both E and Z stereoisomers are formed with a higher yield of the isomer with the bulkiest groups on opposite sides of the double bond. If the β-carbon from which the hydrogen is removed is bonded to only one hydrogen, only one elimination product is formed. Its configuration depends on the configuration of the reactant.

2. E1 reaction: a two-step mechanism with a carbocation intermediate

$$-\overset{|}{\underset{\underset{\displaystyle H}{|}}{C}}-\overset{|}{\underset{|}{C}}-X \;\longrightarrow\; -\overset{|}{\underset{\underset{\displaystyle H}{|}}{C}}-\overset{|}{C}{}^+ \;\xrightarrow{\;\textbf{B}\;}\; \overset{\diagdown}{\diagup}C{=}C\overset{\diagup}{\diagdown} \;+\; \overset{+}{B}H$$
$$+ \;\; X^-$$

Relative reactivities of alkyl halides: $3° > 2° > 1°$

Anti and syn elimination; both E and Z stereoisomers are formed, with a higher yield of the isomer with the bulkiest groups on opposite sides of the double bond.

Competing S_N2 and E2 Reactions
  Primary alkyl halides: primarily substitution
  Secondary alkyl halides: substitution and elimination
  Tertiary alkyl halides: only elimination

Competing S_N1 and E1 Reactions
  Primary alkyl halides: cannot undergo S_N1 or E1 reactions
  Secondary alkyl halides: substitution and elimination
  Tertiary alkyl halides: substitution and elimination

KEY TERMS

anti elimination (page 406)
anti-periplanar (page 406)
bifunctional molecule (page 421)
dehydrohalogenation (page 396)
deuterium kinetic isotope effect
  (page 414)
elimination reaction (page 395)

β-elimination reaction (page 396)
1,2-elimination reaction (page 396)
E1 reaction (page 402)
E2 reaction (page 395)
intermolecular reaction (page 421)
intramolecular reaction (page 421)
kinetic isotope effect (page 414)

syn elimination (page 406)
syn-periplanar (page 406)
target molecule (page 422)
Williamson ether synthesis
  (page 418)
Zaitsev's rule (page 398)

PROBLEMS

28. **a.** Indicate how each of the following factors affects an E1 reaction:
    b. Indicate how each of the following factors affects an E2 reaction:
    1. the structure of the alkyl halide   **3.** the concentration of the base
    2. the strength of the base   **4.** the solvent

29. Dr. Don T. Doit wanted to synthesize the anesthetic 2-ethoxy-2-methylpropane. He used ethoxide ion and 2-chloro-2-methylpropane for his synthesis and ended up with very little ether. What was the predominant product of his synthesis? What reagents should he have used?

30. Which reactant in each of the following pairs will undergo an elimination reaction more rapidly? Explain your choice.

    a. $(CH_3)_3CCl \xrightarrow[H_2O]{HO^-}$

    $(CH_3)_3CI \xrightarrow[H_2O]{HO^-}$

    b.

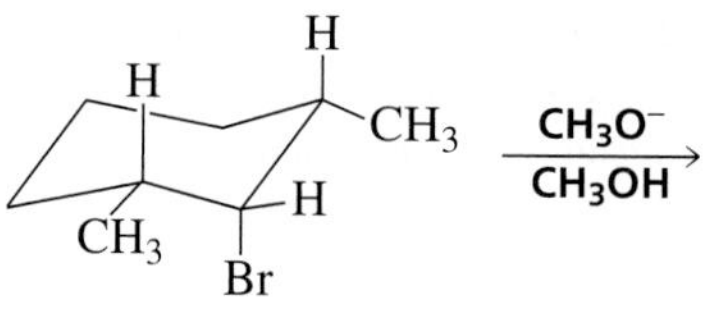

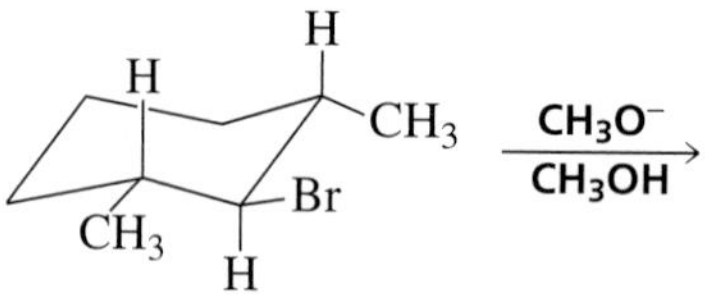

31. Give the elimination products of the following reactions. If the products can exist as stereoisomers, indicate which isomers are obtained.
    a. (R)-2-bromohexane + high concentration of HO^-
    b. (R)-2-bromohexane + H_2O
    c. *trans*-1-chloro-2-methylcyclohexane + high concentration of CH_3O^-
    d. *trans*-1-chloro-2-methylcyclohexane + CH_3OH
    e. 3-bromo-3-methylpentane + high concentration of HO^-
    f. 3-bromo-3-methylpentane + H_2O

32. Indicate which of the compounds in each pair would give a higher substitution/elimination ratio when it reacts with isopropyl bromide:
    a. ethoxide ion or *tert*-butoxide ion   **c.** Cl^- or Br^-
    b. ^-OCN or ^-SCN **d.** CH_3S^- or CH_3O^-

33. Rank the following compounds in order of decreasing rate in an E2 reaction.

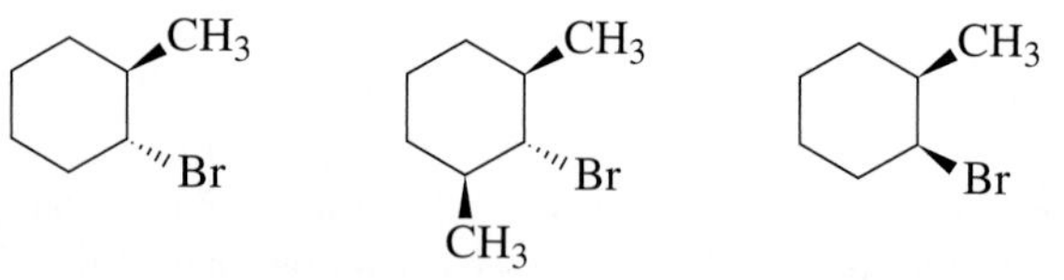

34. Using the given starting material and any necessary organic or inorganic reagents, indicate how the desired compounds could be synthesized:

a. 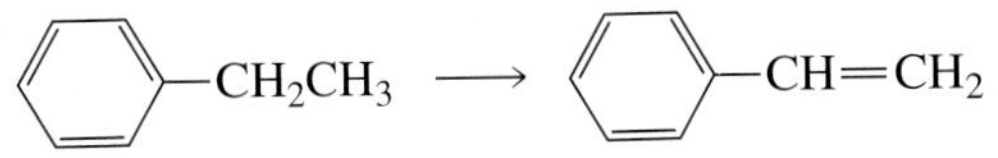 —CH₂CH₃ ⟶ —CH=CH₂

b. $CH_3CH_2CH=CH_2 \longrightarrow CH_3CH_2CH_2CH_2NH_2$

c. 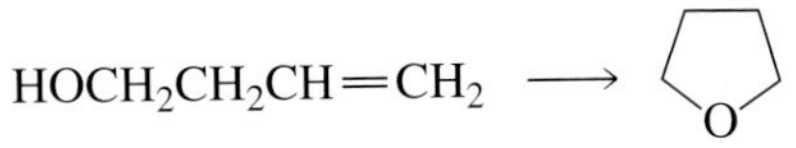 $HOCH_2CH_2CH=CH_2 \longrightarrow$

d. 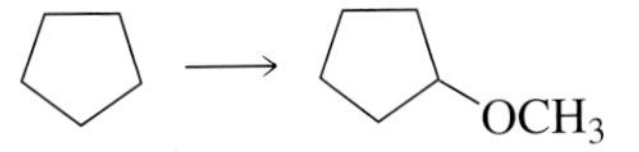 ⟶

e. $CH_3CH_2CH=CH_2 \longrightarrow CH_2=CHCH=CH_2$

35. When 2-bromo-2,3-dimethylbutane reacts with a base under E2 conditions, two alkenes (2,3-dimethyl-1-butene and 2,3-dimethyl-2-butene) are formed.
 a. Which of the following bases would give the highest percentage of the 1-alkene?
 b. Which would give the highest percentage of the 2-alkene?

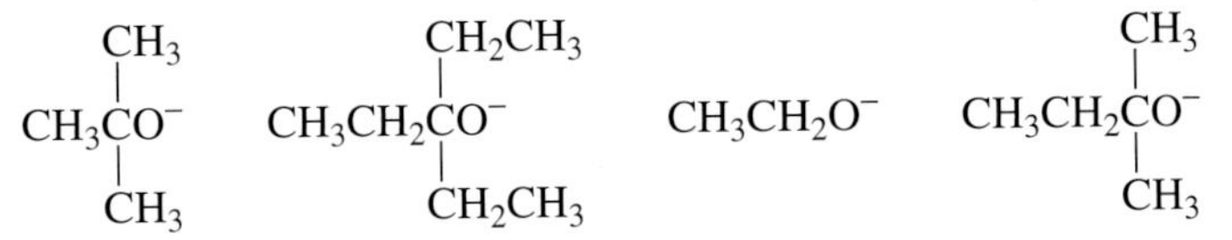

36. a. Give the structure of the products obtained from the reaction of each enantiomer of *cis*-1-chloro-2-isopropylcyclopentane with a high concentration of sodium methoxide in methanol.
 b. Are all the products optically active?
 c. How would the products differ if the starting material were the trans isomer? Are all these products optically active?
 d. Will the cis enantiomers or the trans enantiomers form substitution products more rapidly?
 e. Will the cis enantiomers or the trans enantiomers form elimination products more rapidly?

37. Starting with cyclohexane, how could the following compounds be prepared?
 a. *trans*-1,2-dichlorocyclohexane   **b.** 2-cyclohexenol

38. *cis*-1-Bromo-2-*tert*-butylcyclohexane and *trans*-1-bromo-2-*tert*-butylcyclohexane both react with sodium ethoxide in ethanol to give 1-*tert*-butylcyclohexene. The cis isomer reacts much more rapidly than the trans isomer. Explain.

39. Give the elimination products of the following reactions. If the products can exist as stereoisomers, indicate which isomers are obtained.
 a. (2S,3S)-2-chloro-3-methylpentane + high concentration of CH_3O^-
 b. (2S,3R)-2-chloro-3-methylpentane + high concentration of CH_3O^-
 c. (2R,3S)-2-chloro-3-methylpentane + high concentration of CH_3O^-
 d. (2R,3R)-2-chloro-3-methylpentane + high concentration of CH_3O^-
 e. 3-chloro-3-ethyl-2,2-dimethylpentane + high concentration of $CH_3CH_2O^-$

40. Which of the following hexachlorocyclohexanes is the least reactive in an E2 reaction?

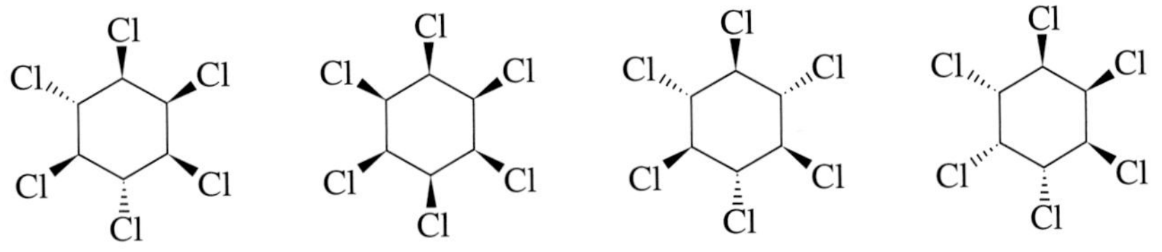

41. The rate of the reaction of 1-bromo-2-butene with ethanol is increased if silver nitrate is added to the reaction mixture. Explain.

42. Give the products of the following reactions, assuming that all the reactions are carried out under $S_N2/E2$ conditions. If the products can exist as stereoisomers, show which stereoisomers are formed.
   a. (3*S*,4*S*)-3-bromo-4-methylhexane + CH_3O^-
   b. (3*S*,4*R*)-3-bromo-4-methylhexane + CH_3O^-
   c. (3*R*,4*R*)-3-bromo-4-methylhexane + CH_3O^-
   d. (3*R*,4*S*)-3-bromo-4-methylhexane + CH_3O^-

43. Two elimination products are obtained from the following E2 reaction.

$$CH_3CH_2CHDCH_2Br \xrightarrow{\text{HO}^-}$$

   a. What are the elimination products?
   b. Which is formed in greater yield? Explain.

44. Three substitution products and three elimination products are obtained from the following reaction. Account for the formation of these products.

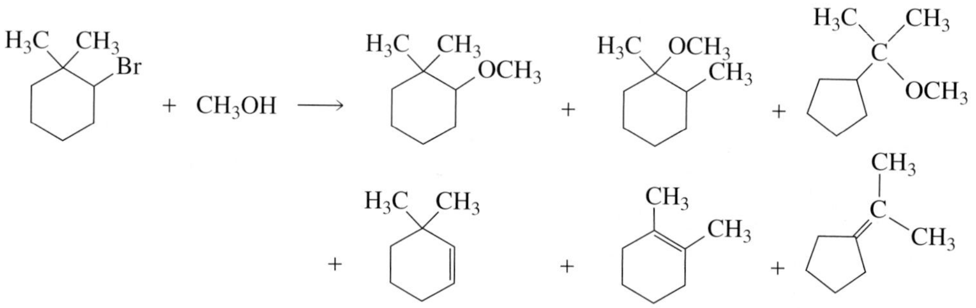

45. When the following stereoisomer of 2-chloro-1,3-dimethylcyclohexane reacts with methoxide ion in a solvent that encourages $S_N2/E2$ reactions, only one product is formed. When the same compound reacts with methoxide ion in a solvent that favors $S_N1/E1$ reactions, eight products are formed. Identify the products that are formed under the two sets of conditions.

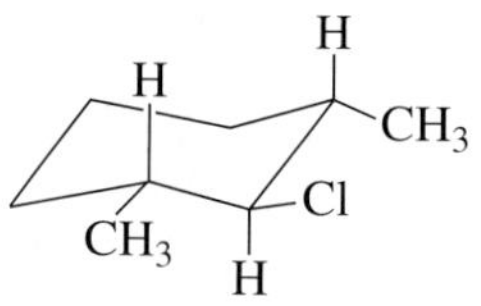

46. The rate constant of an intramolecular reaction depends on the size of the ring (*n*) that is formed. Explain the relative rates of formation of the cyclic secondary ammonium ions.

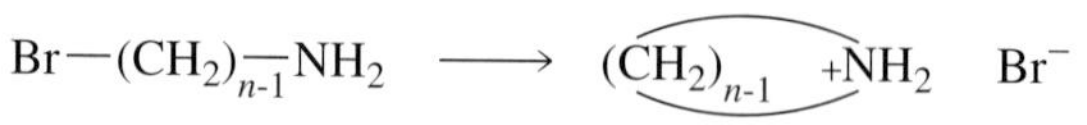

$$Br-(CH_2)_{n-1}-NH_2 \longrightarrow (CH_2)_{n-1}\overset{+}{N}H_2 \quad Br^-$$

$n =$	3	4	5	6	7	10
relative rate:	1×10^{-1}	2×10^{-3}	100	1.7	3×10^{-3}	1×10^{-8}

11

Reactions at an sp^3 Hybridized Carbon III:

Substitution and Elimination Reactions of Compounds With Leaving Groups Other Than Halogen. Organometallic Compounds

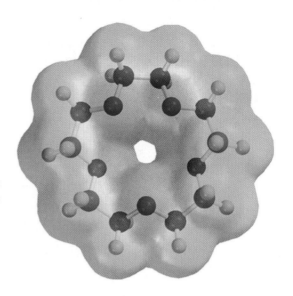

a crown ether

In Chapters 9 and 10 we saw that alkyl halides undergo substitution and elimination reactions because they have an electron-withdrawing halogen bonded to an sp^3 hybridized carbon. Compounds with an electron-withdrawing group other than a halogen bonded to an sp^3 hybridized carbon also undergo substitution and elimination reactions. The reactivity of these compounds compared to alkyl halides depends on the nature of the electron-withdrawing group. The weaker the basicity of the group, the better a leaving group it will be. In other words, the more easily it will be displaced. Remember that the stronger the acid, the weaker its conjugate base (Table 9.3).

Alcohols, ethers, and quaternary ammonium ions have leaving groups that are stronger bases ($^-$OH, $^-$OR, NR$_3$) than halide ions ($^-$X). Therefore, they are less reactive than alkyl halides in substitution and elimination reactions. You will see that quaternary ammonium ions can undergo an elimination reaction if they are heated in the presence of a strong base. However, alcohols and ethers, because of their strongly basic groups, have to be "activated" before they can undergo a substitution or elimination reaction.

The weaker the base, the more easily it can be displaced.

The stronger the acid, the weaker its conjugate base.

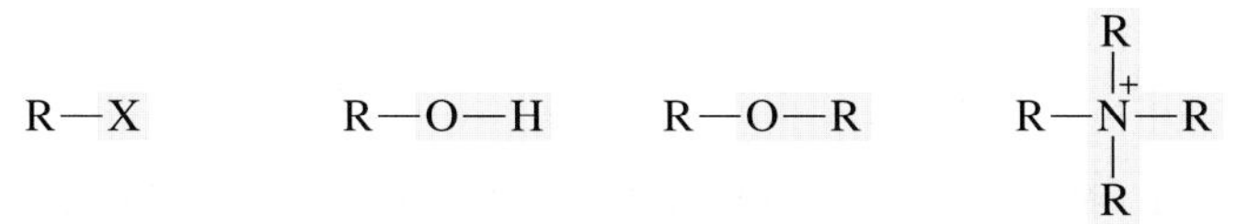

R—X   R—O—H   R—O—R $R-\overset{\displaystyle R}{\underset{\displaystyle R}{\overset{|}{\underset{|}{N^+}}}}-R$

an alkyl halide   an alcohol   an ether   a quaternary
X = F, Cl, Br, I   ammonium ion

11.1
SUBSTITUTION REACTIONS OF ALCOHOLS

An **alcohol** cannot undergo a nucleophilic substitution reaction because it has a strongly basic group ($^-$OH) that cannot be displaced by a nucleophile.

$$CH_3OH + Br^- \quad \xrightarrow{\quad\times\quad} \quad CH_3Br + HO^-$$
$$\text{strong base}$$

In order for an alcohol to undergo a nucleophilic substitution reaction, the OH group has to be converted into a weaker base. One way to convert it into a weaker base is to protonate it. Protonation changes the leaving group from $^-$OH (a strong base) to H_2O (a weak base). The water can then be displaced by a nucleophile. The substitution reaction is slow and requires heat for it to take place in a reasonable period of time.

$$CH_3OH + H^+ \rightleftharpoons CH_3\overset{H}{\underset{+}{O}}H \xrightarrow[\Delta]{Br^-} CH_3Br + H_2O$$
$$\text{weak base}$$

Because the OH group of the alcohol has to be protonated before it can be displaced by a nucleophile, only weakly basic nucleophiles (I^-, Br^-, Cl^-) can be used in the substitution reaction. Moderately and strongly basic nucleophiles (NH_3, RNH_2, CH_3O^-) would be protonated in the acidic solution and, once protonated, would no longer be nucleophiles or would be poor nucleophiles.

Primary, secondary, and tertiary alcohols can all undergo nucleophilic substitution reactions with HI, HBr, and HCl to form alkyl halides.

$$CH_3CH_2CH_2OH + HI \xrightarrow{\Delta} CH_3CH_2CH_2I + H_2O$$
1-propanol
a primary alcohol
1-iodopropane

cyclohexanol + HBr $\xrightarrow{\Delta}$ bromocyclohexane + H$_2$O

cyclohexanol
a secondary alcohol
bromocyclohexane

$$\overset{\displaystyle CH_3}{\underset{\displaystyle CH_3}{CH_3CH_2\overset{|}{\underset{|}{C}}OH}} + HCl \longrightarrow \overset{\displaystyle CH_3}{\underset{\displaystyle CH_3}{CH_3CH_2\overset{|}{\underset{|}{C}}Cl}} + H_2O$$

2-methyl-2-butanol
a tertiary alcohol
2-chloro-2-methylbutane

The mechanism of the substitution reaction depends on the structure of the alcohol. Secondary and tertiary alcohols undergo S_N1 reactions.

The carbocation intermediate formed in the S_N1 reaction has two possible fates: It can combine with a nucleophile and form a substitution product, or it can lose a proton and form an elimination product. However, the alkene formed in the elimination reaction can undergo an addition reaction with HX, so only the substitution product is actually obtained.

Because tertiary carbocations are easier to form than secondary carbocations, tertiary alcohols undergo substitution reactions with hydrogen halides faster than secondary alcohols. Thus, the reaction of a hydrogen halide with a tertiary alcohol

mechanism of the S$_N$1 reaction

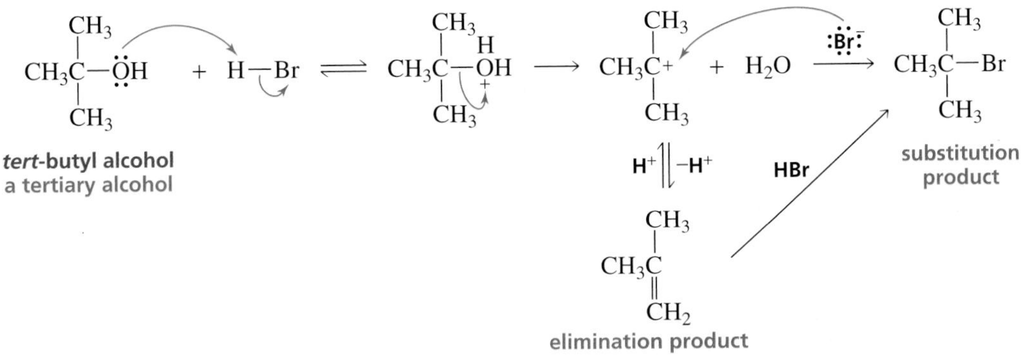

tert-butyl alcohol
a tertiary alcohol

elimination product

proceeds readily at room temperature, whereas the reaction of a hydrogen halide with a secondary alcohol requires heat.

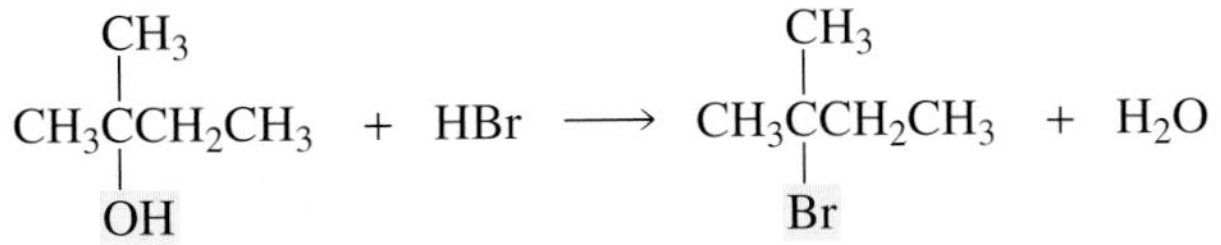

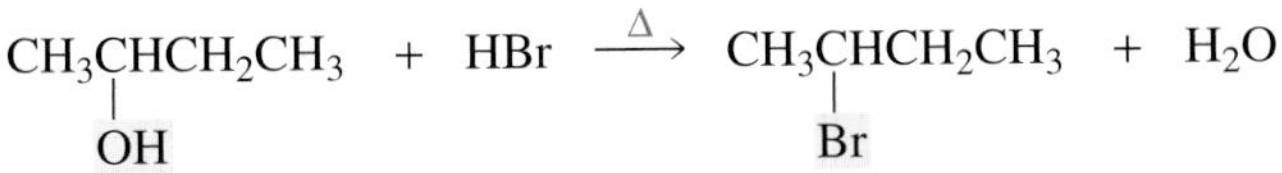

Primary alcohols cannot undergo S$_N$1 reactions because primary carbocations are too unstable to be formed. Primary alcohols, therefore, undergo S$_N$2 reactions.

Primary alcohols undergo S$_N$2 reactions with hydrogen halides. Secondary and tertiary alcohols undergo S$_N$1 reactions with hydrogen halides.

mechanism of the S$_N$2 reaction

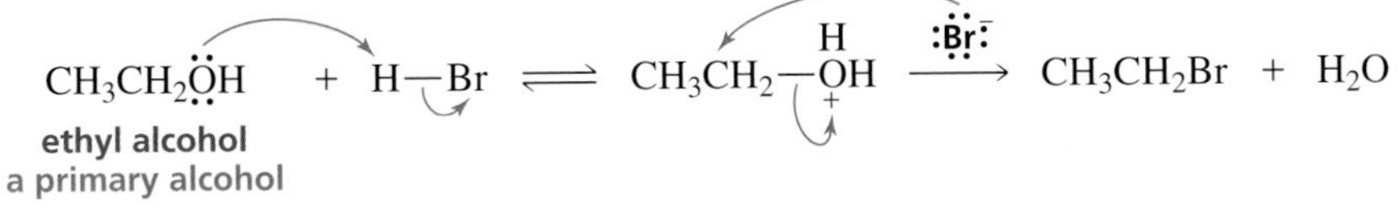

ethyl alcohol
a primary alcohol

Only the substitution product is obtained. No elimination product is formed because the halide ion, although a good nucleophile, is a weak base (Section 9.3). A strong base is needed in an E2 reaction to remove a proton from a β-carbon. (Remember that a β-carbon is adjacent to the carbon bonded to the leaving group.)

When HCl is used instead of HBr or HI, the S$_N$2 reaction is slower because Cl$^-$ is a poorer nucleophile than Br$^-$ or I$^-$ (Section 9.3). The rate of the reaction can be increased by using ZnCl$_2$ as a catalyst because the Zn^{2+} cation is a Lewis acid that complexes strongly with the nonbonding electrons on oxygen. This weakens the carbon–oxygen bond and creates a better leaving group.

$$CH_3CH_2CH_2OH + HCl \xrightarrow[\Delta]{ZnCl_2} CH_3CH_2CH_2Cl + H_2O$$

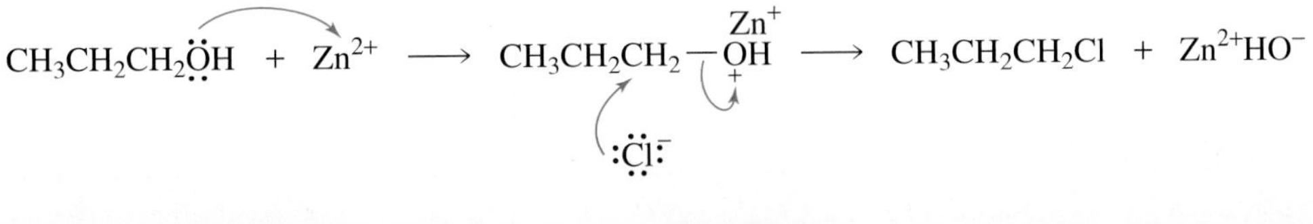

THE LUCAS TEST

Whether an alcohol is primary, secondary, or tertiary can be determined by taking advantage of the relative rates at which the three classes of alcohols react with HCl/ZnCl$_2$. This is known as the *Lucas test*. The alcohol is added to a mixture of concentrated HCl and ZnCl$_2$ (the Lucas reagent). Low-molecular-weight alcohols are soluble in the Lucas reagent, but the alkyl halide products are not, so the so-lution turns cloudy as the alkyl halide is formed. When the test is carried out at room temperature, the solution turns cloudy immediately if the alcohol is tertiary, in about five minutes if the alcohol is secondary, and remains clear if the alcohol is primary. Because the test relies on the complete solubility of the alcohol in the Lucas reagent, it is limited to alcohols with fewer than six carbons.

PROBLEM 1 ◆

The observed relative reactivities of primary, secondary, and tertiary alcohols toward reaction with a hydrogen halide are: Tertiary alcohols are more reactive than secondary alcohols, and secondary alcohols are more reactive than primary alcohols. If secondary alcohols underwent an S$_N$2 reaction with a hydrogen halide rather than an S$_N$1 reaction, what would be the relative reactivities of the three classes of alcohols? Explain.

Howard J. Lucas (1885–1963) was born in Ohio and earned BS and MS degrees from Ohio State University. He published a description of the Lucas test in 1930. He was a professor of chemistry at the California Institute of Technology.

Because the reaction of a secondary or a tertiary alcohol with a hydrogen halide is an S$_N$1 reaction, a carbocation is formed as an intermediate. Therefore, we must check for the possibility of a carbocation rearrangement when determining the product of the substitution reaction. Remember that a carbocation rearrangement will occur if it leads to formation of a more stable carbocation (Section 3.14). For example, the major product of the reaction of 3-methyl-2-butanol with HBr is 2-bromo-2-methylbutane because a 1,2-hydride shift converts the initially formed secondary carbocation into a more stable tertiary carbocation.

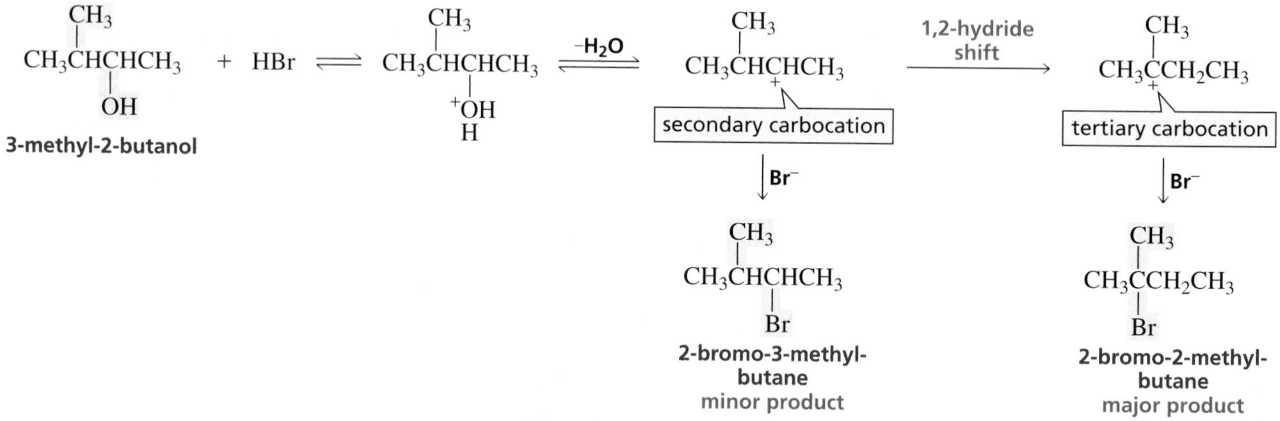

PROBLEM 2 ◆

Give the major product of each of the following reactions:

a. CH$_3$CH$_2$CHCH$_3$ + HBr $\xrightarrow{\Delta}$
   |
   OH

b. 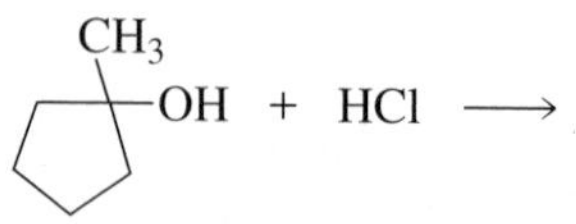 [cyclopentane structure]—CH$_3$—OH + HCl $\longrightarrow$

c. CH$_3$C—CHCH$_3$ + HBr $\xrightarrow{\Delta}$ (with CH$_3$ above and CH$_3$ OH below)

d. 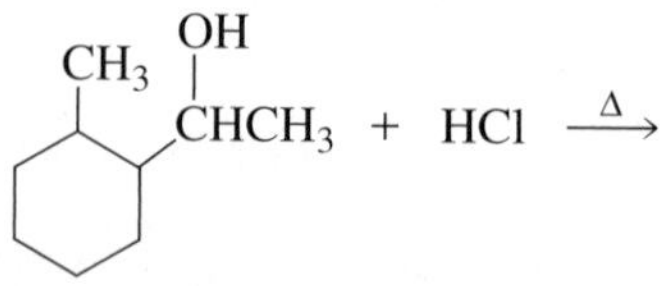 [cyclohexane structure with CH$_3$ and CHCH$_3$ with OH] + HCl $\xrightarrow{\Delta}$

GRAIN ALCOHOL AND WOOD ALCOHOL

When ethanol is ingested it acts on the central nervous system. Ingestion of moderate amounts affects judgment and lowers inhibitions. Higher concentrations interfere with motor coordination and cause slurred speech and amnesia. Even higher concentrations cause nausea and loss of consciousness. Ingestion of very large amounts interferes with spontaneous respiration and can be fatal.

The ethanol in alcoholic beverages is produced by the fermentation of glucose from grapes and from grains such as corn, rye, and wheat, which is why ethanol is also known as grain alcohol. Cooking the grain in the presence of malt (sprouted barley) converts much of its starch into glucose. Yeast is added to convert glucose into ethanol and carbon dioxide.

$$C_6H_{12}O_6 \xrightarrow{\text{yeast enzymes}} 2\ CH_3CH_2OH + 2\ CO_2$$
$$\text{glucose} \qquad\qquad \text{ethanol}$$

The kind of beverage produced (white or red wine, beer, scotch, bourbon, champagne) depends on the plant species being fermented, whether or not the CO_2 that is formed is allowed to escape, whether or not other substances are added, and how the beverage is purified (by sedimentation for wines; by distillation for scotch and brandy).

The tax on liquor would make ethanol a prohibitively expensive laboratory reagent. Because ethanol is needed in a wide variety of commercial processes, laboratory alcohol is not taxed, but it is carefully regulated by the federal government to make certain it is not used for the illegal preparation of alcoholic beverages. Denatured alcohol is ethanol that has been made undrinkable by adding a denaturant such as benzene or methanol. Denatured alcohol is not taxed, but the added impurities make it unfit for many laboratory uses.

Methanol, also known as wood alcohol because it can be obtained from the distillation of wood in the absence of oxygen, is toxic. Ingestion of even very small amounts can cause blindness. Ingestion of as little as an ounce has been fatal. The antidote to methanol poisoning is discussed in Section 18.11.

Alcohols are inexpensive and readily available compounds. However, they are not reactive toward substitution because the ⁻OH group is too basic to be displaced by a nucleophile. Alkyl halides, on the other hand, are very useful to synthetic chemists because the weakly basic halide ion can be displaced by a wide variety of nucleophiles, leading to a wide variety of products (Section 9.4). By converting alcohols into reactive alkyl halides, chemists are able to use the readily available alcohols as starting materials for the preparation of many organic compounds.

11.2 ADDITIONAL METHODS FOR CONVERTING ALCOHOLS INTO ALKYL HALIDES

$$R-OH \longrightarrow R-X \xrightarrow{\ ^-Nu\ } R-Nu$$
$$\text{alcohol} \qquad \text{alkyl halide}$$
$$X = Cl, Br, I$$

We have just seen that an alcohol can be converted into an alkyl halide by treating it with a hydrogen halide. However, better yields are obtained and carbocation rearrangements can be avoided if a phosphorus trihalide (PCl_3, PBr_3, or PI_3,)[1] or thionyl chloride ($SOCl_2$) is used to convert alcohols into alkyl halides. These reagents all act in the same way—they convert an alcohol into an intermediate with a leaving group that is readily displaced by a halide ion. For example, phosphorus tribromide converts the strongly basic leaving group of an alcohol into a weakly basic bromophosphite group that can be displaced by a bromide ion.

[1] Because of its instability, PI_3 is generated *in situ* (in the reaction mixture) from the reaction of phosphorus with iodine.

Thionyl chloride converts an OH group into a chlorosulfite leaving group that can be displaced by Cl⁻.

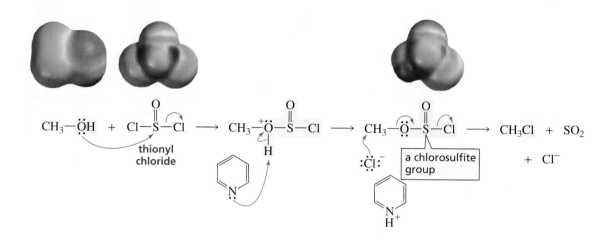

3-D Molecules:
Thionylchloride;
Pyridine;
Pyridinium ion

Pyridine is generally used as the solvent in these reactions because it prevents the formation of HCl and it is a poor nucleophile.

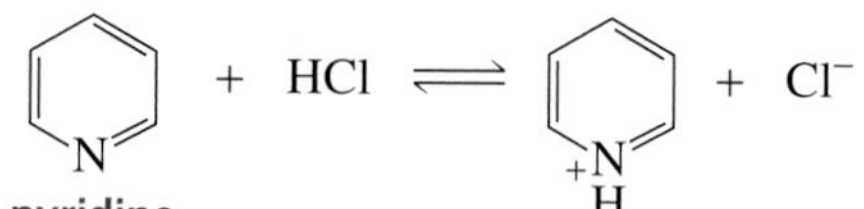

These reactions work well for primary and secondary alcohols, but tertiary alcohols give poor yields because the intermediate formed by a tertiary alcohol is sterically hindered to attack by the halide ion.

Table 11.1 shows some of the methods that can be used to convert alcohols into alkyl halides.

TABLE 11.1 Commonly Used Methods for Converting Alcohols into Alkyl Halides

ROH	+ HBr	$\xrightarrow{\Delta}$	RBr
ROH	+ HI	$\xrightarrow{\Delta}$	RI
ROH	+ HCl	$\xrightarrow{\Delta}$	RCl
ROH	+ PBr₃	$\xrightarrow{\text{pyridine}}$	RBr
ROH	+ PCl₃	$\xrightarrow{\text{pyridine}}$	RCl
ROH	+ SOCl₂	$\xrightarrow{\text{pyridine}}$	RCl

Another way to activate an alcohol for subsequent reaction with a nucleophile, other than converting it into an alkyl halide, is to convert it into a sulfonate ester. A **sulfonate ester** is formed when an alcohol reacts with a sulfonyl chloride.

11.3
CONVERSION OF ALCOHOLS INTO SULFONATE ESTERS

$$R'-\overset{\overset{O}{\|}}{\underset{\underset{O}{\|}}{S}}-Cl \quad + \quad ROH \quad \xrightarrow[\text{pyridine}]{} \quad R'-\overset{\overset{O}{\|}}{\underset{\underset{O}{\|}}{S}}-OR \quad + \quad HCl$$

a sulfonyl chloride   a sulfonate ester

A sulfonic acid is a strong acid $(pK_a \sim -1)$ because its conjugate base is particularly stable since the electrons holding the proton become delocalized over three oxygen atoms when they are no longer bonded to the proton. Notice that sulfur has an expanded octet (it is surrounded by 12 electrons). Because a sulfonic acid is a strong acid, its conjugate base is weak. The weak conjugate base gives the sulfonate ester an excellent leaving group.

$$R-\overset{\overset{O}{\|}}{\underset{\underset{O}{\|}}{S}}-OH \longrightarrow \left[R-\overset{\overset{O}{\|}}{\underset{\underset{O}{\|}}{S}}-O^- \longleftrightarrow R-\overset{\overset{O}{\|}}{\underset{\underset{O^-}{\|}}{S}}=O \longleftrightarrow R-\overset{\overset{O^-}{\|}}{\underset{\underset{O}{\|}}{S}}=O \right] + H^+$$

contributing resonance structures

Many sulfonyl chlorides are available to activate OH groups. The most common one is *para*-toluenesulfonyl chloride.

para-toluenesulfonyl chloride   methanesulfonyl chloride   trifluoromethanesulfonyl chloride

3-D Molecules:
Methanesulfonyl chloride;
Methanesulfonate
methyl ester

Once the alcohol has been activated by being converted into a sulfonate ester, the appropriate nucleophile is added, generally under conditions that favor S_N2 reactions. The reactions take place readily at room temperature because the leaving group is so good. A *para*-toluenesulfonate ion is about 100 times better than a chloride ion as a leaving group. The products are as varied as the nucleophiles chosen to react with the sulfonate ester.

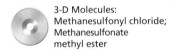

para-Toluenesulfonyl chloride is abbreviated to tosyl chloride (TsCl); esters of *para*-toluenesulfonic acid are called **alkyl tosylates** and are abbreviated ROTs. The leaving group, therefore, is ⁻OTs.

$$CH_3CH_2CH_2OTs + {}^-C{\equiv}N \longrightarrow CH_3CH_2CH_2C{\equiv}N + {}^-OTs$$
an alkyl tosylate

Tutorial: Leaving groups

If the nucleophile that reacts with a sulfonate ester is a halide ion, we have yet another method for converting an alcohol into an alkyl halide (Table 11.1).

$$CH_3CH_2CHCH_2CH_3 \underset{\textbf{pyridine}}{\overset{\textbf{TsCl}}{\longrightarrow}} CH_3CH_2CHCH_2CH_3 \overset{Cl^-}{\longrightarrow} CH_3CH_2CHCH_2CH_3 + {}^-OTs$$

OH	OTs	Cl

3-pentanol   **3-chloropentane**

PROBLEM 3 ◆

Give the structures of W, X, Y, and Z.

a.

CH₃
|
C—OH
/   \
CH₃CH₂   H

$\xrightarrow[\textbf{pyridine}]{\textbf{TsCl}}$ W $\xrightarrow{{}^-C{\equiv}N}$ X

b.

CH₃
|
C—OH
/   \
CH₃CH₂   H

$\xrightarrow[\textbf{pyridine}]{\textbf{PBr}_3}$ Y $\xrightarrow{{}^-C{\equiv}N}$ Z

PROBLEM 4

Show how 1-butanol can be converted into the following compounds:

a. $CH_3CH_2CH_2CH_2Br$

d. $CH_3CH_2CH_2CH_2NHCH_2CH_3$

b. $CH_3CH_2CH_2CH_2O\overset{\displaystyle O}{\overset{\|}{C}}CH_2CH_3$

e. $CH_3CH_2CH_2CH_2SH$

c. $CH_3CH_2CH_2CH_2OCH_3$

f. $CH_3CH_2CH_2CH_2C{\equiv}N$

11.4
DEHYDRATION OF ALCOHOLS

An alcohol can undergo an elimination reaction, losing an OH from one carbon and an H from an adjacent carbon. Overall, this amounts to the elimination of a molecule of water. Loss of water from a molecule is called **dehydration.** The dehydration of alcohols requires an acid catalyst and heat. Sulfuric acid (H_2SO_4) and phosphoric acid (H_3PO_4) are the most commonly used acid catalysts for this purpose.

$$CH_3CH_2CHCH_3 \underset{\Delta}{\overset{H_2SO_4}{\rightleftharpoons}} CH_3CH{=}CHCH_3 + H_2O$$
|
OH

In the first step of the dehydration reaction, the acid protonates the oxygen of the alcohol. As we saw earlier, protonation converts the very poor leaving group ($^-$OH) into a good leaving group (H_2O). In the next step, water departs and a carbocation is formed. A base in the reaction mixture (either hydrogen sulfate or water) removes a proton from a carbon adjacent to the positively charged carbon, forming the alkene product and regenerating the acid catalyst. Dehydration is an E1 reaction of a protonated alcohol.

mechanism of dehydration (E1)

$$CH_3CHCH_3 \ + \ H-OSO_3H \ \rightleftharpoons \ CH_3CHCH_3 \ \rightleftharpoons \ H-CH_2-\overset{+}{C}HCH_3 \ \rightleftharpoons \ CH_2=CHCH_3$$

:ÖH

$$\overset{+}{:}\underset{H}{O}H$$

$$+ \ H_2\ddot{O}:$$

$$+ \ H_3O^+$$

$$+ \ ^-OSO_3H$$

When more than one elimination product can be formed, the major product is the more substituted alkene—the one obtained by removing a proton from the β-carbon that is bonded to the fewest hydrogens (recall Zaitsev's rule, Section 10.2). The more substituted alkene is more stable, which means the transition state leading to its formation is more stable, allowing it to be formed more rapidly (Figure 11.1).

 Movie: Dehydration

$$
\begin{array}{c}
CH_3 \\
| \\
CH_3CCH_2CH_3 \\
| \\
OH
\end{array}
\xrightarrow[\Delta]{H_3PO_4}
\begin{array}{c}
CH_3 \\
| \\
CH_3C=CHCH_3 \\
84\%
\end{array}
\ + \
\begin{array}{c}
CH_3 \\
| \\
CH_2=CCH_2CH_3 \\
16\%
\end{array}
\ + \ H_2O
$$

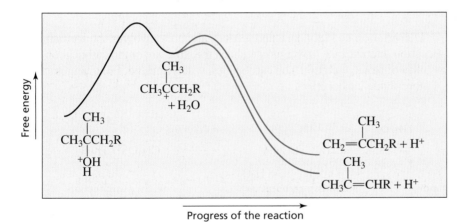

In Section 3.13 we saw that an alkene is hydrated (adds water) in the presence of an acid catalyst, thereby forming an alcohol. Hydration of an alkene is the exact reverse of the acid-catalyzed dehydration of an alcohol.

◀ **Figure 11.1**
The reaction coordinate diagram for the dehydration of a protonated alcohol. The major product is the more substituted alkene because its greater stability makes the transition state leading to its formation more stable.

$$
RCH_2CHR \ + \ H^+ \ \underset{\text{hydration}}{\overset{\text{dehydration}}{\rightleftharpoons}} \ RCH=CHR \ + \ H_2O \ + \ H^+
$$
OH

To prevent the alkene formed in the dehydration reaction from adding water and reverting back to the alcohol, the alkene can be removed by distillation as it is formed because it has a much lower boiling point than the alcohol. Removing a product displaces the reaction to the right (Le Châtelier's principle; Section 9.4).

> ### PROBLEM 5
>
> Why is the acid-catalyzed dehydration of an alcohol a reversible reaction whereas the base-promoted dehydrohalogenation of an alkyl halide is irreversible?

Because the rate-determining step in the dehydration of secondary or tertiary alcohols is the formation of a carbocation intermediate, the rate of dehydration parallels the ease with which the carbocation is formed. Tertiary alcohols are the easiest to dehydrate because tertiary carbocations are more stable and therefore are easier to form than secondary and primary carbocations (Section 3.10). In order to undergo dehydration, tertiary alcohols must be heated to about 50 °C in 5% H_2SO_4, secondary alcohols to about 100 °C in 75% H_2SO_4, and primary alcohols can be dehydrated only under extreme conditions (170 °C in 95% H_2SO_4) and by a different mechanism since primary carbocations are too unstable to be formed.

relative ease of dehydration

$$
\begin{array}{ccccc}
 & \text{R} & & \text{R} & \\
 & | & & | & \\
 & \text{RCOH} & > & \text{RCHOH} & > \quad \text{RCH}_2\text{OH} \\
 & | & & & \\
 & \text{R} & & & \\
\text{a tertiary alcohol} & & \text{a secondary alcohol} & & \text{a primary alcohol}
\end{array}
$$

← increasing ease of dehydration

Dehydration of secondary and tertiary alcohols involves the formation of a carbocation intermediate, so be sure to check the structure of the carbocation for the possibility of rearrangement. The carbocation will rearrange if rearrangement produces a more stable carbocation.

$$
\begin{array}{ccccc}
\text{CH}_3 & & \text{CH}_3 & & \text{CH}_3 \\
| & \xrightarrow[\Delta]{\text{H}_3\text{PO}_4} & | & \xrightarrow{\text{1,2-methyl shift}} & | \\
\text{CH}_3\text{C}-\text{CHCH}_3 & & \text{CH}_3\text{C}-\text{CHCH}_3 & & \text{CH}_3\overset{+}{\text{C}}-\text{CHCH}_3 \\
| \quad | & & \overset{+}{\;}| & & | \\
\text{CH}_3 \; \text{OH} & & \text{CH}_3 \; + \text{H}_2\text{O} & & \text{CH}_3
\end{array}
$$

3,3-dimethyl-2-butanol   secondary carbocation   tertiary carbocation

$\downarrow -H^+$ $\downarrow -H^+$

$$
\begin{array}{ccc}
\text{CH}_3 & \text{CH}_3 & \text{CH}_3 \\
| & | & | \\
\text{CH}_3\text{CCH}=\text{CH}_2 & \text{CH}_3\text{C}=\text{CCH}_3 \quad + & \text{CH}_2=\text{CCHCH}_3 \\
| & | & | \\
\text{CH}_3 & \text{CH}_3 & \text{CH}_3
\end{array}
$$

3,3-dimethyl-1-butene   2,3-dimethyl-2-butene   2,3-dimethyl-1-butene
3%   64%   33%

Tutorial: Carbocation rearrangements

The following is an example of a **ring-expansion rearrangement** (Section 3.14). Both the initially formed carbocation and the carbocation it rearranges to are secondary carbocations, but the initially formed carbocation is less stable because of the strain in the four-membered ring (Section 2.11). Rearrangement relieves this strain. The rearranged secondary carbocation can rearrange by a 1,2-hydride shift to an even more stable tertiary carbocation.

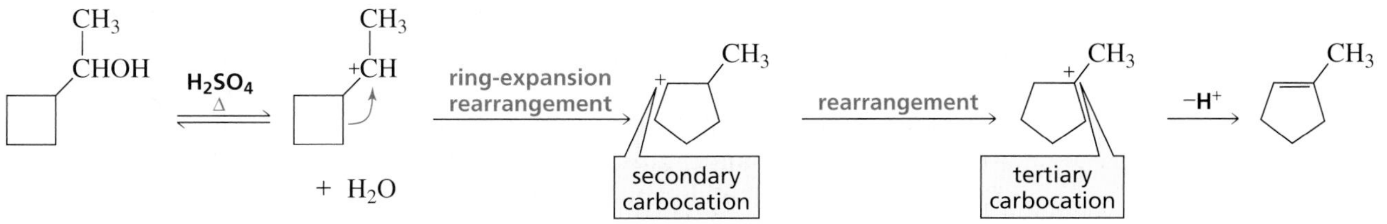

PROBLEM 6 ◆

What product would be formed if the preceding alcohol were heated with an equivalent amount of HBr rather than with a catalytic amount of H_2SO_4?

While dehydration of tertiary and secondary alcohols is an E1 reaction, dehydration of primary alcohols is an E2 reaction because of the difficulty encountered in forming primary carbocations. Any base in the reaction mixture (ROH, H_2O, HSO_4^-) can remove the proton in the elimination reaction. An ether is also obtained as the product of a competing S_N2 reaction.

Secondary and tertiary alcohols undergo dehydration by an E1 pathway.

Primary alcohols undergo dehydration by an E2 pathway.

mechanism of dehydration (E2)

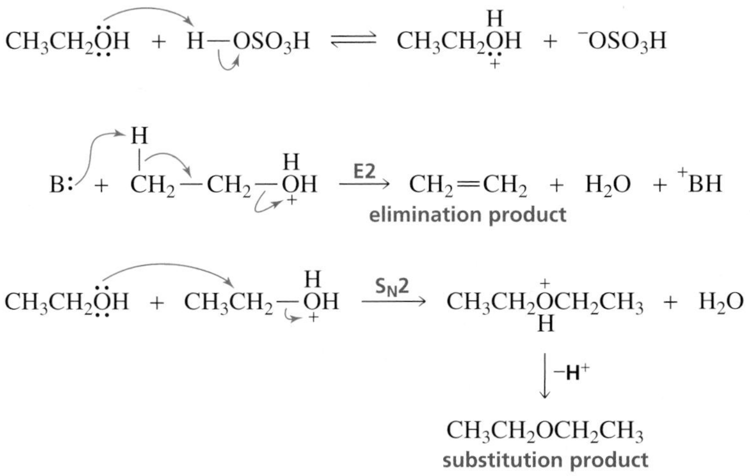

$$CH_3CH_2\overset{\cdot\cdot}{O}H \ + \ H{-}OSO_3H \ \rightleftharpoons \ CH_3CH_2\overset{H}{\underset{+}{O}}H \ + \ ^-OSO_3H$$

$$B{:} \ + \ CH_2{-}CH_2{-}\overset{H}{\underset{+}{O}}H \ \xrightarrow{E2} \ CH_2{=}CH_2 \ + \ H_2O \ + \ ^+BH$$

elimination product

$$CH_3CH_2\overset{\cdot\cdot}{O}H \ + \ CH_3CH_2{-}\overset{H}{\underset{+}{O}}H \ \xrightarrow{S_N2} \ CH_3CH_2\overset{+}{\underset{H}{O}}CH_2CH_3 \ + \ H_2O$$

$$\Big\downarrow {-}H^+$$

$$CH_3CH_2OCH_2CH_3$$

substitution product

PROBLEM 7

Heating an alcohol with sulfuric acid is a good method for preparing an ether with identical alkyl groups such as diethyl ether. It is not a good method for preparing an ether with different alkyl groups such as ethyl propyl ether.

a. Explain.

b. How would you synthesize ethyl propyl ether?

Although the dehydration of a primary alcohol is an E2 reaction and therefore does not form a carbocation intermediate, the product obtained in most cases is identical to the product that would be obtained if a carbocation had been formed in an E1 reaction and then had rearranged. For example, we would expect 1-butene to be the product of the E2 dehydration of 1-butanol. However, we find that the product is actually 2-butene, which would be the product of an E1 reaction in which the carbocation intermediate had rearranged. This occurs because after the E2 product (1-butene) is formed, a proton from the acidic solution adds to the double bond, following Markovnikov's rule, thereby forming a carbocation. Loss of a proton from the carbocation, following Zaitsev's rule, gives the final product of the reaction. This product is identical to the product that would have been formed if the primary alcohol had undergone an E1 dehydration.

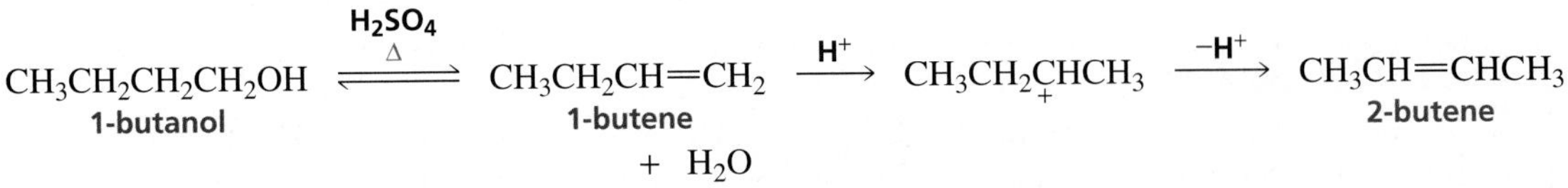

$$CH_3CH_2CH_2CH_2OH \underset{\Delta}{\overset{H_2SO_4}{\rightleftharpoons}} CH_3CH_2CH=CH_2 \overset{H^+}{\longrightarrow} CH_3CH_2\overset{+}{C}HCH_3 \overset{-H^+}{\longrightarrow} CH_3CH=CHCH_3$$

1-butanol   **1-butene**   **2-butene**

$+\ H_2O$

The stereochemical outcome of the E1 dehydration of an alcohol is identical to the stereochemical outcome of the E1 dehydrohalogenation of an alkyl halide. That is, both the E and Z isomers are obtained as products. More of the isomer with the bulky groups on opposite sides of the double bond is formed because it is the more stable isomer and, therefore, the transition state leading to its formation is more stable, causing it to be formed more rapidly (Section 10.5).

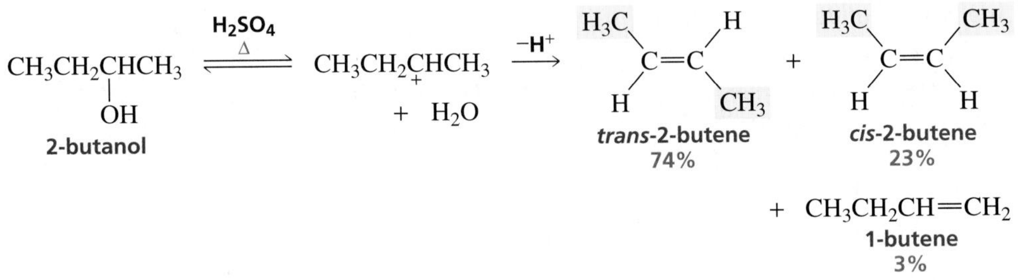

The relatively harsh conditions (acid and heat) required for alcohol dehydration and the structural changes resulting from carbocation rearrangements can cause low yields of a desired alkene to be obtained from the dehydration of an alcohol that has any structural complexity. The severe conditions required for dehydration of an alcohol can be avoided, however, if the reaction is carried out in the presence of phosphorus oxychloride ($POCl_3$) and pyridine.

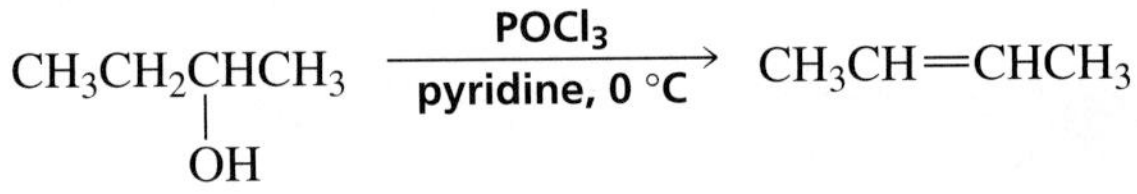

$$CH_3CH_2\underset{\underset{OH}{|}}{C}HCH_3 \xrightarrow[\text{pyridine, 0 °C}]{POCl_3} CH_3CH=CHCH_3$$

Reaction with $POCl_3$ converts the OH group of the alcohol into $OPOCl_2$, a good leaving group. Because the mildly basic reaction conditions favor an E2 reaction, a carbocation is not formed, so we do not have to be concerned with carbocation rearrangements. Pyridine serves as a base to prevent the formation of HCl and to remove the proton in the elimination reaction.

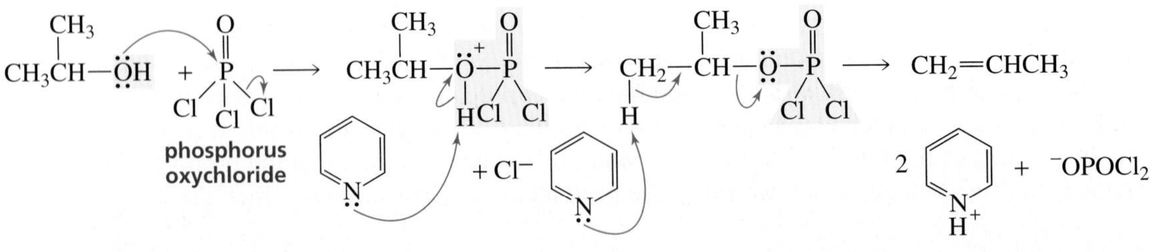

phosphorus oxychloride

PROBLEM-SOLVING STRATEGY

Propose a mechanism for the following reaction:

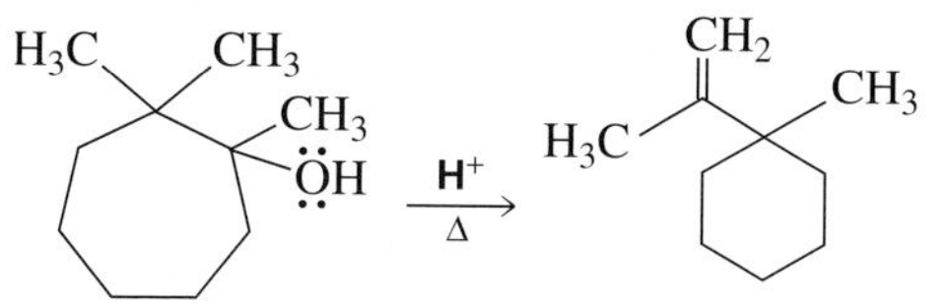

Even the most complicated-looking mechanism can be worked out if you proceed one step at a time, keeping in mind the structure of the final product. The oxygen is the only atom with lone pair electrons in the starting material, so it must react with the proton. Loss of water forms a tertiary carbocation.

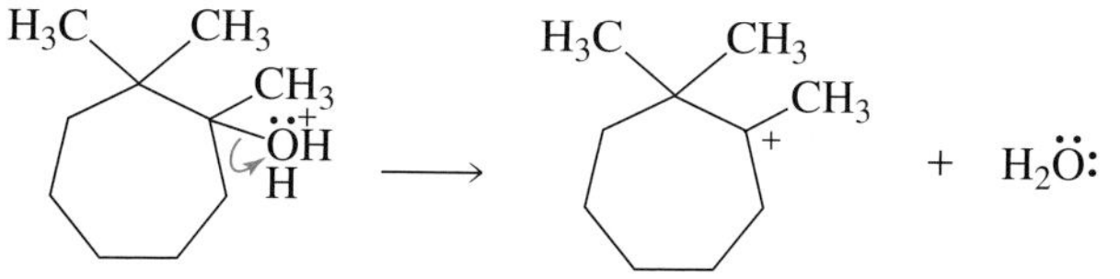

Because the starting material contains a seven-membered ring and the final product has a six-membered ring, a ring-contraction rearrangement must occur. Of the two possible pathways for ring contraction, one leads to a tertiary carbocation, and the other leads to a primary carbocation. It can be helpful to label the equivalent carbons in the reactant and product. Because a primary carbocation is unstable and this particular primary carbocation has a different arrangement of carbons than the product, the correct pathway must be the one that leads to the tertiary carbocation.

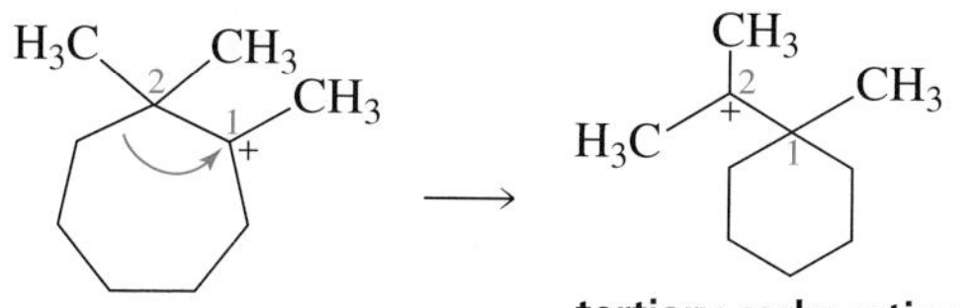

tertiary carbocation

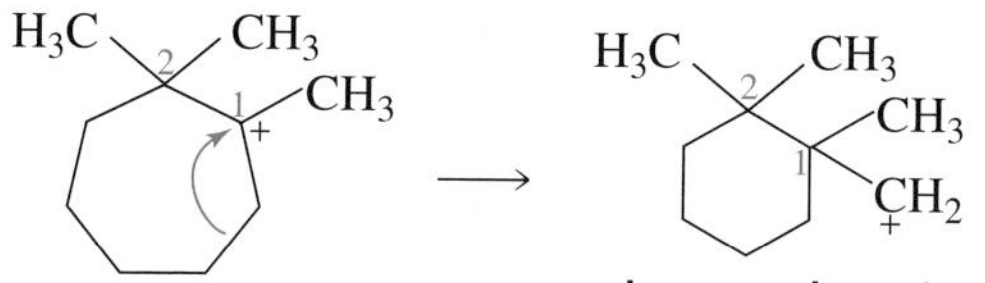

primary carbocation

The final product can now be obtained by removing a proton from the rearranged carbocation.

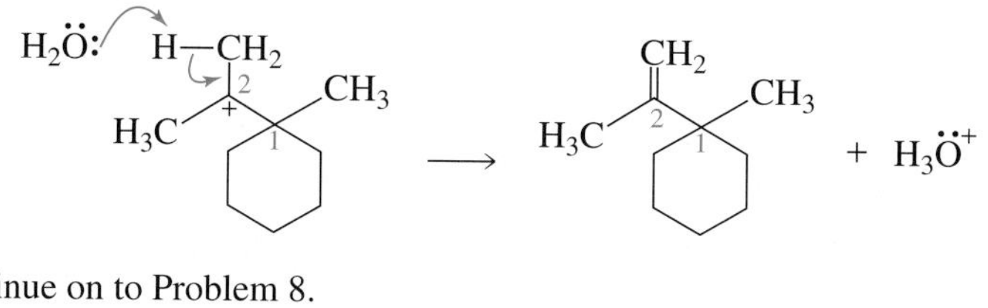

Now continue on to Problem 8.

PROBLEM 8

Propose a mechanism for each of the following reactions:

a.

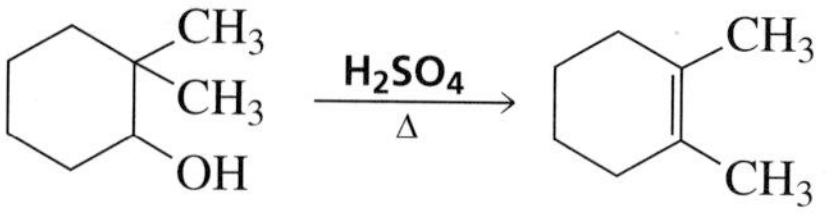

b. 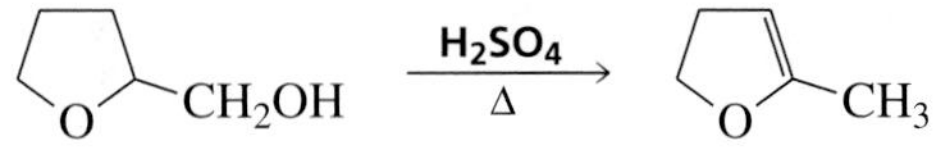 $\xrightarrow[\Delta]{H_2SO_4}$

c. 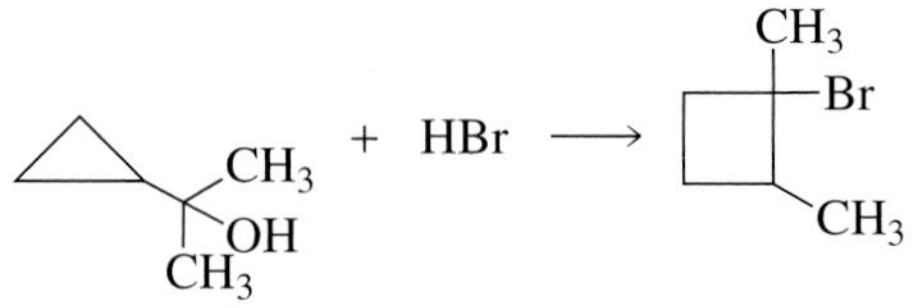 + HBr $\longrightarrow$

PROBLEM 9

Give the major product formed when each of the following alcohols is heated in the presence of H_2SO_4:

a. 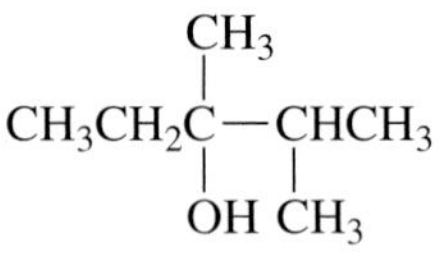 $CH_3CH_2\underset{\underset{OH}{|}}{\overset{\overset{CH_3}{|}}{C}}-\underset{\underset{CH_3}{|}}{C}HCH_3$

b. 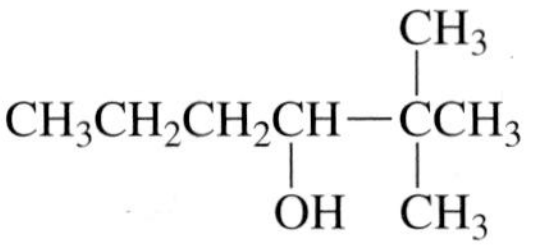 $CH_3CH_2CH_2\underset{\underset{OH}{|}}{C}H-\underset{\underset{CH_3}{|}}{\overset{\overset{CH_3}{|}}{C}}CH_3$

c.

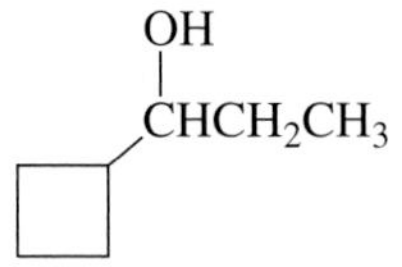

d.

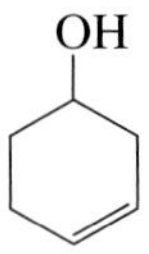

e.

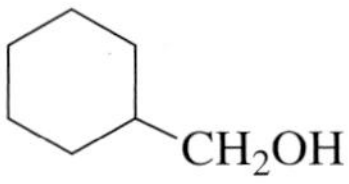

f. $CH_3CH_2CH_2CH_2CH_2OH$

BIOLOGICAL DEHYDRATIONS

Dehydration reactions are known to occur in many important biological processes. Fumarase, for example, is an enzyme that catalyzes the dehydration of malate in the citric acid cycle (page xxx). The citric acid cycle is a series of reactions in the terminal oxidation of carbohydrates, fatty acids, and amino acids.

Enolase, another enzyme, catalyzes the dehydration of α-phosphoglycerate in glycolysis (page xxx). Glycolysis is a series of reactions that prepares glucose for entry into the citric acid cycle.

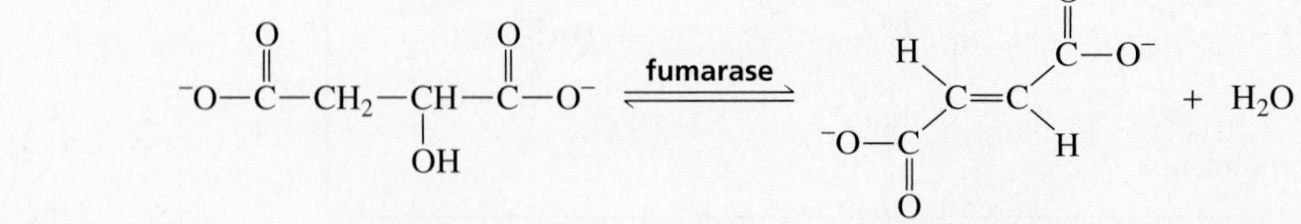

malate   fumarase   fumarate   + H_2O

α-phosphoglycerate   enolase   phosphoenolpyruvate   + H_2O

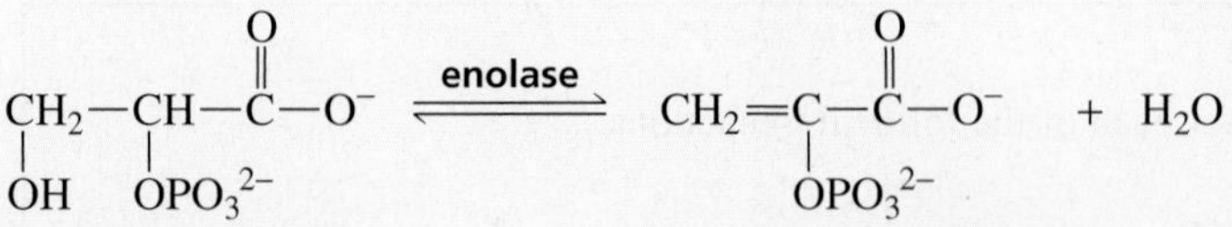

The leaving group of an **ether** (^-OR) and the leaving group of an alcohol (^-OH) have nearly the same basicity. For example, the pK_a of CH_3OH is 15.5 and the pK_a of H_2O is 15.7. Both leaving groups are strong bases, so both are very poor leaving groups. Consequently, alcohols and ethers are equally unreactive toward nucleophilic substitution.

11.5 SUBSTITUTION REACTIONS OF ETHERS

$$R\text{—}O\text{—}H \qquad R\text{—}O\text{—}R$$
$$\text{an alcohol} \qquad \text{an ether}$$

Many of the reagents that are used to activate alcohols toward nucleophilic substitution (e.g., $SOCl_2$, PCl_3) cannot be used to activate ethers. For example, when an alcohol reacts with an activating agent such as a sulfonyl chloride, a proton is lost in the second step of the reaction and a stable sulfonate ester results.

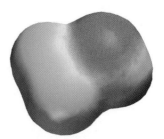

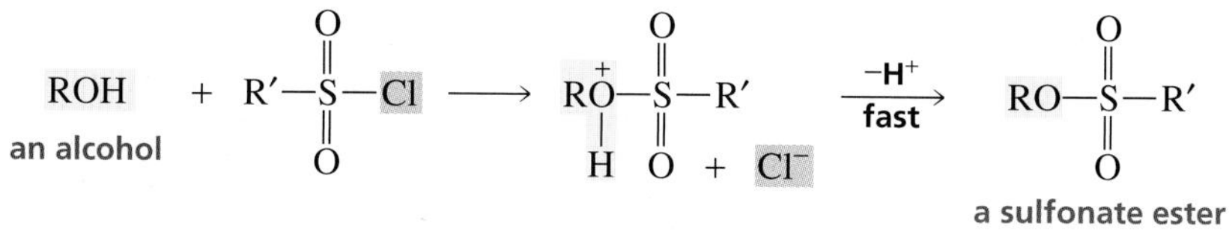

CH₃OH

When an ether reacts with a sulfonyl chloride, however, the R group cannot be lost, so a stable sulfonate ester cannot be formed. Instead, the intermediate returns to the starting materials.

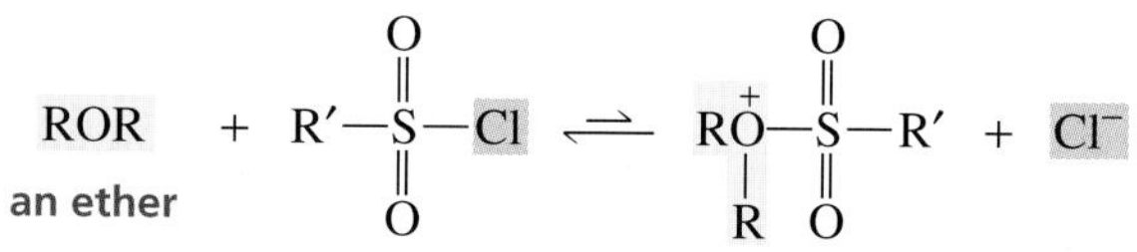

CH₃OCH₃

However, like alcohols, ethers can be activated by protonation. Therefore, in the presence of a high concentration of HI or HBr, ethers undergo a nucleophilic substitution reaction. Similar to what is seen for alcohols, the reaction of ethers with hydrogen halides is slow and the reaction mixture must be heated in order for the reaction to occur at a reasonable rate.

$$ROR' + HI \xrightarrow{\Delta} ROH + R'I$$

The first step in the cleavage of an ether by HI or HBr is protonation of the ether oxygen. This converts the very basic RO^- leaving group into the less basic ROH leaving group. What happens next in the mechanism depends on the structure of the ether.

If departure of the leaving group creates a relatively stable carbocation (e.g., a tertiary carbocation), an S_N1 reaction occurs—the leaving group departs, and the halide ion combines with the carbocation.

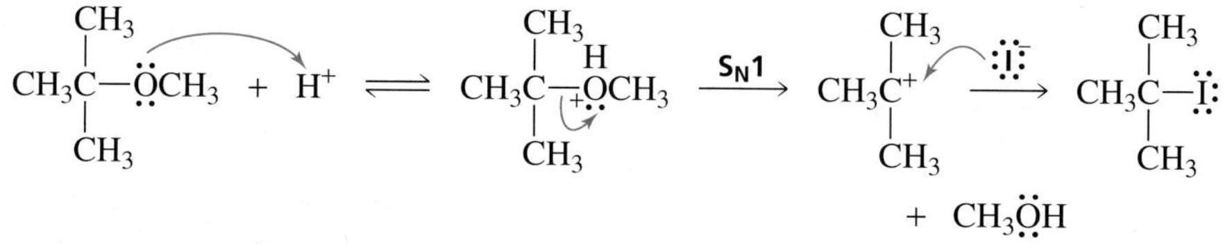

If departure of the leaving group would create an unstable carbocation (e.g., a methyl cation or a primary carbocation), the leaving group cannot depart. It has to be displaced by the halide ion. In other words, an S_N2 reaction occurs. The halide ion preferentially attacks the less sterically hindered of the two alkyl groups.

$$CH_3\overset{..}{\underset{..}{O}}CH_2CH_2CH_3 \ + \ H^+ \ \rightleftharpoons \ CH_3\overset{\overset{H}{|}}{\underset{+}{O}}CH_2CH_2CH_3 \ \xrightarrow{S_N2} \ CH_3\overset{..}{\underset{..}{I}}: \ + \ CH_3CH_2CH_2OH$$

$$:\overset{..}{\underset{..}{I}}:^-$$

If excess HI is used, the product alcohol will be converted into an alkyl halide.

$$\text{(phenyl)}-CH_2OCH_2CH_2CH_3 \ \xrightarrow[\Delta]{HI} \ \text{(phenyl)}-CH_2I \ + \ CH_3CH_2CH_2OH$$

$$\xrightarrow[\Delta]{HI}$$

$$CH_3CH_2CH_2I \ + \ H_2O$$

Cleavage of ethers by concentrated HI or HBr occurs more rapidly if the reaction can take place by an S_N1 pathway. If the instability of the carbocation causes the reaction to follow an S_N2 pathway, cleavage will be more rapid with HI than with HBr because I^- is a better nucleophile than Br^-. Only a substitution product is obtained because the bases present in the reaction mixture (halide ions and H_2O) are too weak

ANESTHETICS

Because diethyl ether ("ether") is a short-lived muscle relaxant, it has been widely used as an inhalation anesthetic. However, because it takes effect slowly and has a slow and unpleasant recovery period (ask anyone who has experienced it), other compounds such as enflurane, isoflurane, and halothane have replaced it as an anesthetic. Diethyl ether is still used where there is a lack of trained anesthesiologists since it is the safest anesthetic to administer by untrained hands. Anesthetics interact with the nonpolar molecules of cell membranes, causing the membranes to swell, which interferes with their permeability.

Amputation of a leg in 1528 without anesthetic.

$CH_3CH_2OCH_2CH_3$	$CF_3CHClOCHF_2$	$CHClFCF_2OCHF_2$	$CF_3CHClBr$
"ether"	isoflurane	enflurane	halothane

Sodium pentothal (also called thiopental sodium) is commonly used as an intravenous anesthetic. The onset of anesthesia and the loss of consciousness occur within seconds of its administration. Care must be taken when administering sodium pentothal because the dose for effective anesthesia is 75% of the lethal dose. Because of its toxicity, it cannot be used as the sole anesthetic. It is generally used to induce anesthesia before an inhalation anesthetic is administered. Propofol is an anesthetic that has all the properties of the "perfect anesthetic"—it can be used as the sole anesthetic by intravenous drip, it has a rapid and pleasant induction period, a wide margin of safety, and recovery is rapid and pleasant.

sodium pentothal

propofol

to abstract a proton in an E2 reaction, and any alkene formed in an E1 reaction would react with HBr or HI to form the substitution product. S$_N$2 cleavage reactions of ethers do not occur at all with concentrated HCl because Cl$^-$ is too poor a nucleophile.

PROBLEM 10 ◆

Can concentrated HF be used to cleave ethers? Explain.

3-D Molecules:
Diethyl ether;
Tetrahydrofuran

Ethers are relatively unreactive compounds. Because they do not react with most reagents, they are frequently used as solvents. Some common ether solvents are shown in Table 11.2.

TABLE 11.2   Some Ethers That Are Used as Solvents

$CH_3CH_2OCH_2CH_3$			$CH_3OCH_2CH_2OCH_3$	$CH_3OC(CH_3)_3$
diethyl ether "ether"	tetrahydrofuran THF	1,4-dioxane	1,2-dimethoxyethane DME	methyl *tert*-butyl ether MTBE

PROBLEM 11 / SOLVED

Give the major products that would be obtained from heating each of the following ethers with concentrated HI:

a. $CH_3CH{=}CHOCH_2CH_3$

c. $CH_3CH_2CH_2O\overset{\underset{\displaystyle CH_3}{\displaystyle |}}{\underset{\underset{\displaystyle CH_3}{\displaystyle |}}{C}}CH_2CH_3$

e. (cyclohexene with —OCH$_3$)

b. (phenyl—CH$_2$O—phenyl)

d. (tetrahydropyran)

f. (tetrahydrofuran ring with —CH$_3$ and CH$_3$ at 2-position)

SOLUTION TO 11a   The iodide ion attacks the carbon of the ethyl group because the other possibility is to attack an *sp^2* hybridized carbon and *sp^2* carbons are not attacked by nucleophiles (Section 9.8). Thus, the products are ethyl iodide and an enol that immediately rearranges to an aldehyde (Section 5.6).

$$CH_3CH{=}CH{-}O{-}CH_2CH_3 \xrightarrow[\Delta]{HI} CH_3CH{=}CH{-}\overset{+}{\underset{H}{O}}{-}CH_2CH_3 \longrightarrow CH_3CH{=}CH{-}OH + CH_3CH_2I$$

:Ï:

$$CH_3CH_2\overset{\displaystyle O}{\overset{\displaystyle \|}{C}}H$$

Ethers in which the oxygen atom is incorporated into a three-membered ring are called **epoxides** or **oxiranes**. The common name of an epoxide assumes that the oxygen atom is in place of the π bond of an alkene and uses the common name of the alkene followed by "oxide." The simplest epoxide is ethylene oxide.

**11.6
REACTIONS OF
EPOXIDES**

H₂C=CH₂
ethylene

H₂C—CH₂ (with O bridge)
ethylene oxide

H₂C=CHCH₃
propylene

H₂C—CHCH₃ (with O bridge)
propylene oxide

There are two systematic ways to name epoxides. One method calls the three-membered oxygen-containing ring "oxirane," with the oxygen occupying the 1-position. Thus, 2-ethyloxirane has an ethyl substituent at the 2-position of the oxirane ring. Alternatively, an epoxide can be named as an alkane, with "epoxy" and the numbers of the carbons to which the oxygen is attached immediately preceding the alkane name.

H₂C—CHCH₂CH₃ (with O bridge)
2-ethyloxirane
1,2-epoxybutane

CH₃CH—CHCH₃ (with O bridge)
2,3-dimethyloxirane
2,3-epoxybutane

H₂C—C (with O bridge), CH₃, CH₃
2,2-dimethyloxirane
1,2-epoxy-2-methylpropane

PROBLEM 12 ◆

Give a structure for each of the following:

a. 2-propyloxirane

b. cyclohexene oxide

c. 2,2,3,3-tetramethyloxirane

d. 2,3-epoxy-2-methylpentane

Although an epoxide and an ether have the same leaving group, epoxides are much more reactive than ethers because the strain in the three-membered ring is relieved when the ring opens (Figure 11.2). Epoxides, therefore, readily undergo ring-opening reactions with a wide variety of nucleophiles.

Epoxides, like other ethers, react with hydrogen halides. In the first step of the reaction, the oxygen is protonated. The protonated epoxide is then attacked by the halide ion. Because epoxides are so much more reactive than ethers, the reaction takes place readily at room temperature.

H₂C—CH₂ (O bridge) + H—Br: ⇌ H₂C—CH₂ (O⁺—H bridge) + :Br:⁻ ⟶ HOCH₂CH₂Br:

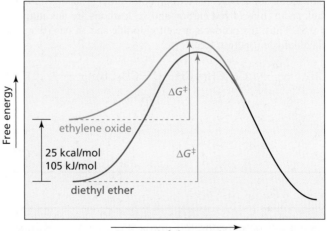

Figure 11.2 ▶
The reaction coordinate diagrams for nucleophilic attack of hydroxide ion on ethylene oxide and on diethyl ether. The greater reactivity of the epoxide is a result of the strain (ring strain and torsional strain) in the three-membered ring, which increases its free energy.

Free energy

ethylene oxide

25 kcal/mol
105 kJ/mol

diethyl ether

ΔG‡

ΔG‡

Progress of the reaction

Protonated epoxides can also be opened by poor nucleophiles such as H_2O and alcohols.

$$\underset{\text{H}_2\text{C}-\text{CH}_2}{\overset{\text{O}}{\triangle}} \overset{\text{H}^+}{\rightleftharpoons} \underset{\text{H}_2\text{C}-\text{CH}_2}{\overset{\overset{\text{H}}{\overset{+}{\text{O}}}}{\triangle}} \xrightarrow[\text{H}_2\ddot{\text{O}}:]{} \underset{\overset{|}{\underset{\text{H}}{\overset{+}{\text{OH}}}}}{\text{CH}_2\text{CH}_2\text{OH}} \overset{-\text{H}^+}{\rightleftharpoons} \text{HOCH}_2\text{CH}_2\text{OH}$$

1,2-ethanediol
ethylene glycol

$$\underset{\text{CH}_3\text{CH}-\text{CHCH}_3}{\overset{\text{O}}{\triangle}} \overset{\text{H}^+}{\rightleftharpoons} \underset{\text{CH}_3\text{CH}-\text{CHCH}_3}{\overset{\overset{\text{H}}{\overset{+}{\text{O}}}}{\triangle}} \xrightarrow[\text{CH}_3\ddot{\text{O}}\text{H}]{} \underset{\overset{|}{\underset{\text{H}}{\overset{+}{\text{OCH}_3}}}}{\overset{\text{OH}}{\overset{|}{\text{CH}_3\text{CHCHCH}_3}}} \overset{-\text{H}^+}{\rightleftharpoons} \underset{\overset{|}{\text{OCH}_3}}{\overset{\text{OH}}{\overset{|}{\text{CH}_3\text{CHCHCH}_3}}}$$

3-methoxy-2-butanol

If the protonated epoxide has different substituents on the two-ring carbon atoms (and the nucleophile is something other than H_2O), the product obtained from nucleophilic attack on the 2-position of the oxirane ring will be different from that obtained from nucleophilic attack on the 3-position. The major product is the one resulting from nucleophilic attack on the *more substituted* carbon.

$$\underset{\text{CH}_3\text{CH}-\text{CH}_2}{\overset{\text{O}}{\triangle}} \overset{\text{H}^+}{\rightleftharpoons} \underset{\text{CH}_3\text{CH}-\text{CH}_2}{\overset{\overset{\text{H}}{\overset{+}{\text{O}}}}{\triangle}} \xrightarrow{\text{CH}_3\text{OH}} \underset{}{\overset{\text{OCH}_3}{\overset{|}{\text{CH}_3\text{CHCH}_2\text{OH}}}} \;+\; \underset{}{\overset{\text{OH}}{\overset{|}{\text{CH}_3\text{CHCH}_2\text{OCH}_3}}}$$

2-methoxy-1-propanol   **1-methoxy-2-propanol**
major product   **minor product**

The more substituted carbon is preferentially attacked because after the epoxide is protonated, it is so reactive that one of the carbon–oxygen bonds begins to break before the nucleophile has a chance to attack. As the carbon–oxygen bond starts to break, a partial positive charge develops on the carbon that is losing its share of the oxygen's electrons. The protonated epoxide breaks preferentially in the direction that puts the partial positive charge on the more substituted carbon because the more substituted carbocation is more stable. (Recall that tertiary carbocations are more stable than secondary carbocations, which are more stable than primary carbocations.)

3-D Molecule:
Propylene oxide

$$\underset{\text{CH}_3\text{CH}-\text{CH}_2}{\overset{\overset{\text{H}}{\overset{+}{\text{O}}}}{\triangle}}$$

$$\underset{\text{CH}_3\overset{\delta+}{\text{CH}}-\text{CH}_2}{\overset{\overset{\delta+}{\text{O}}}{\cdots}} \xrightarrow[\text{CH}_3\ddot{\text{O}}\text{H}]{} \underset{}{\overset{\overset{\text{H}}{\overset{+}{\text{OCH}_3}}}{\text{CH}_3\text{CHCH}_2\text{OH}}} \overset{-\text{H}^+}{\longrightarrow} \underset{}{\overset{\text{OCH}_3}{\overset{|}{\text{CH}_3\text{CHCH}_2\text{OH}}}}$$

developing secondary carbocation

major product

$$\underset{\text{CH}_3\text{CH}-\overset{\delta+}{\text{CH}_2}}{\overset{\text{O}\delta+}{\cdots}} \xrightarrow[\text{CH}_3\ddot{\text{O}}\text{H}]{} \underset{}{\overset{\text{OH}}{\overset{|}{\text{CH}_3\text{CHCH}_2\overset{+}{\text{O}}\text{CH}_3}}} \overset{-\text{H}^+}{\longrightarrow} \underset{}{\overset{\text{OH}}{\overset{|}{\text{CH}_3\text{CHCH}_2\text{OCH}_3}}}$$

developing primary carbocation

minor product

The best way to describe the reaction is to say that it occurs by a pathway that is partially S_N1 and partially S_N2. It is not a pure S_N1 reaction because a carbocation intermediate is not fully formed. It is not a pure S_N2 reaction because the leaving group begins to depart before the compound is attacked by the nucleophile.

In Section 11.5 we saw that an ether does not undergo a nucleophilic substitution reaction unless the very basic ⁻OR leaving group is converted by protonation into a less basic ROH group. Because of the strain in the three-membered ring, epoxides are reactive enough to open without first being protonated. When a nucleophile attacks an unprotonated epoxide, the reaction is a pure S_N2 reaction. That is, the unprotonated epoxide does not begin to open until it is attacked by the nucleophile. In this case, the nucleophile is more likely to attack the *less substituted* carbon because the less substituted carbon is more accessible to attack (it is less sterically hindered). Thus, the site of nucleophilic attack on an asymmetrical epoxide under basic conditions (when the epoxide is not protonated) is different from the site of nucleophilic attack under acidic conditions (when the epoxide is protonated).

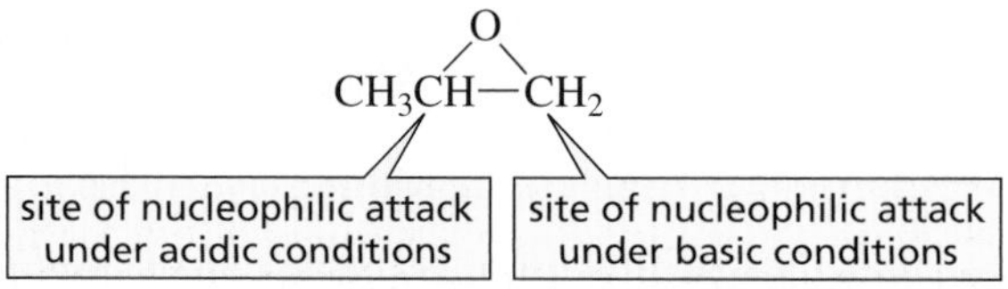

After the nucleophile has attacked the epoxide, the alkoxide ion can pick up a proton from the solvent or from an acid added after the reaction is over.

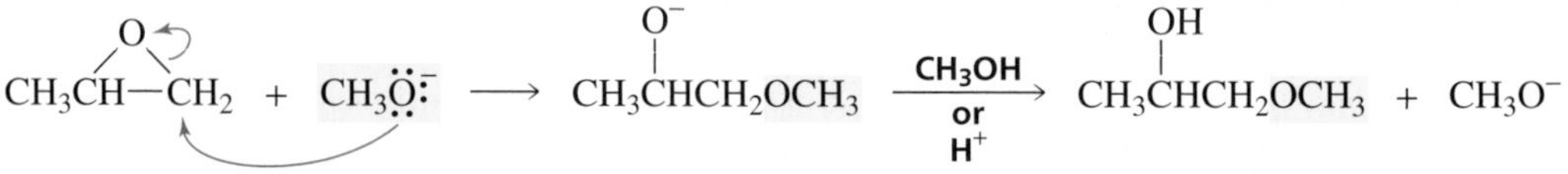

Epoxides are synthetically useful reagents because they react with a wide variety of nucleophiles, leading to the formation of a wide variety of products. They are important in biological processes because they are reactive enough to be attacked by nucleophiles under the conditions found in living systems.

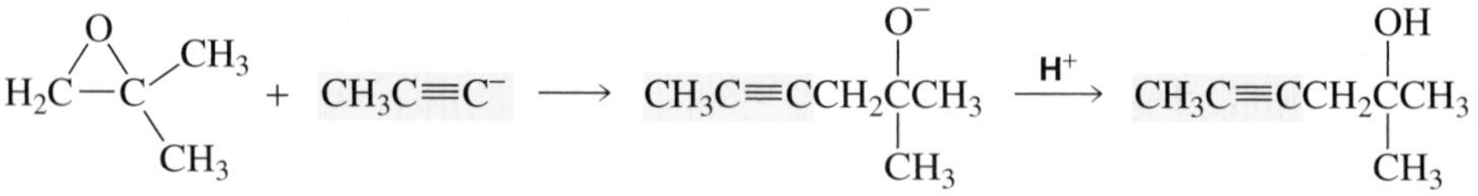

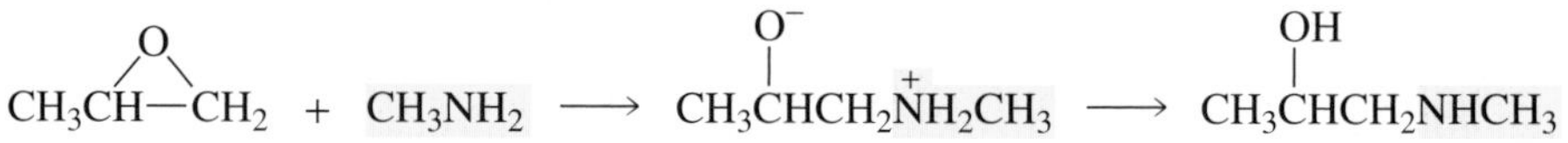

Notice that the reaction of cyclohexene oxide with a nucleophile leads to a trans product because an S_N2 reaction involves back side nucleophilic attack.

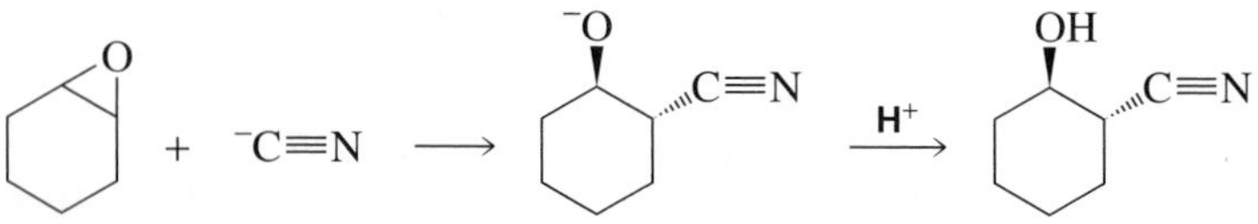

PROBLEM 13 ◆

How many stereoisomers would be obtained from the preceding reaction?

PROBLEM 14 ◆

Give the major product of each of the following reactions:

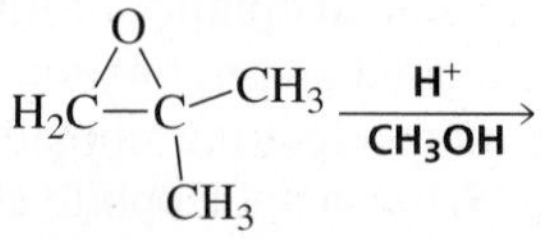

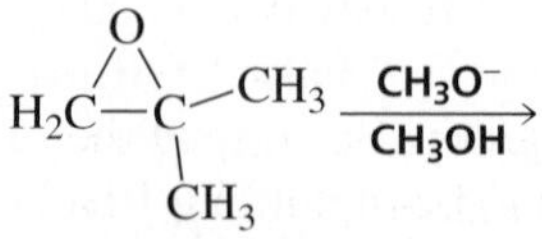

c. 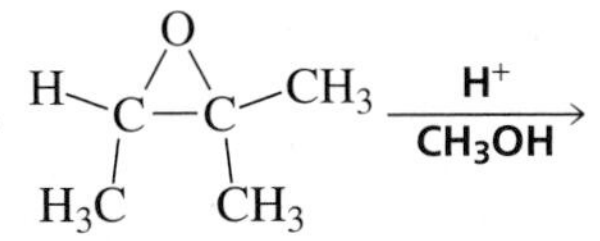 H—C—C—CH₃ $\xrightarrow[\text{CH}_3\text{OH}]{\text{H}^+}$ (with O epoxide bridge, H₃C and CH₃ substituents)

d. 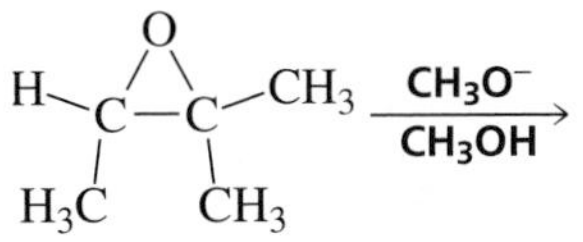 H—C—C—CH₃ $\xrightarrow[\text{CH}_3\text{OH}]{\text{CH}_3\text{O}^-}$ (with O epoxide bridge, H₃C and CH₃ substituents)

PROBLEM 15 ◆

Would you expect the reactivity of a five-membered ring ether such as tetrahydrofuran (Table 11.2) to be more similar to an epoxide or to a noncyclic ether?

When aromatic hydrocarbons are ingested or inhaled, they are enzymatically converted into arene oxides. An **arene oxide** is a compound in which one of the "double bonds" of the aromatic ring has been converted into an epoxide. Formation of an arene oxide is the first step in changing an aromatic compound that enters the body as a foreign substance (cigarette smoke, drugs, automobile exhaust) into a more water-soluble compound that can eventually be eliminated. The enzyme that detoxifies aromatic hydrocarbons by converting them into arene oxides is known as cytochrome P_{450}.

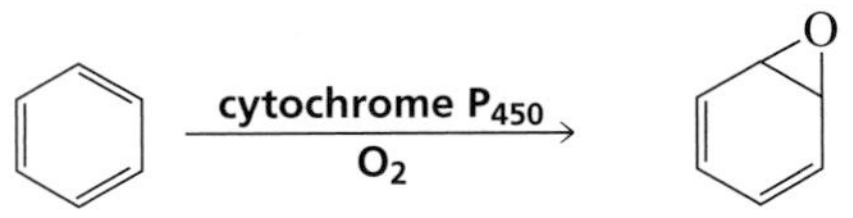

11.7
ARENE OXIDES

(benzene) $\xrightarrow[\text{O}_2]{\text{cytochrome } P_{450}}$ (benzene oxide)

benzene oxide
an arene oxide

An arene oxide can react in two different ways. It can react as a typical epoxide, undergoing attack by a nucleophile to form an addition product (Section 11.6). Alternatively, it can rearrange to form a phenol, something that epoxides such as ethylene oxide cannot do.

benzene

benzene oxide

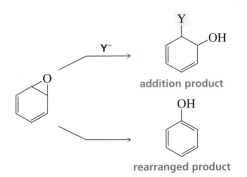

addition product

rearranged product

Arene oxides are also important intermediates in the biosynthesis of biochemically important phenols such as tyrosine and serotonin.

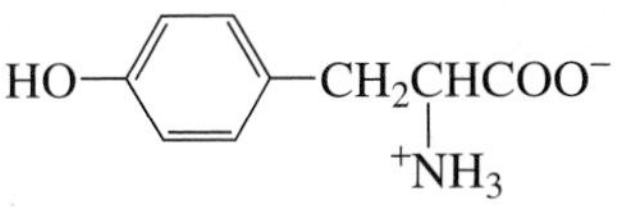

HO—⟨benzene ring⟩—CH₂CHCOO⁻
                              |
                            ⁺NH₃

tyrosine
an amino acid

HO—⟨indole ring⟩—CH₂CH₂NH₂ (with N—H)

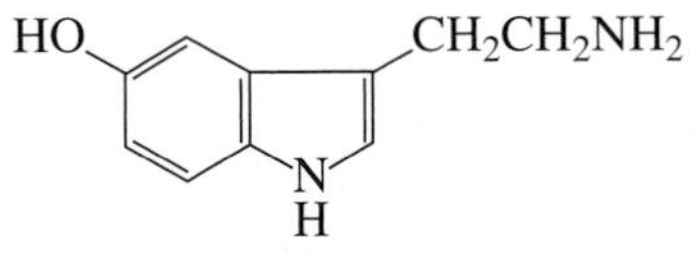

serotonin
a vasoconstrictor

When an arene oxide reacts with a nucleophile, the three-membered ring undergoes back side nucleophilic attack, forming an addition product.

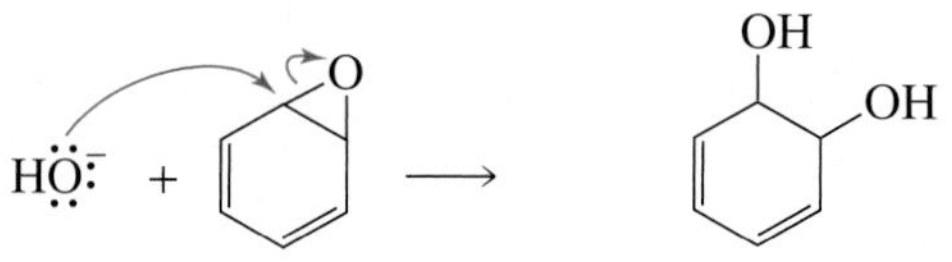

addition product

When an arene oxide undergoes rearrangement, the three-membered epoxide ring opens in the first step to form a carbocation. Instead of immediately losing a proton to form phenol, an enone is formed from a 1,2-hydride shift. This is called an *NIH shift* because it was first observed in a laboratory at the National Institutes of Health. Enolization of the enone gives phenol (Section 5.6).

benzene oxide → **a carbocation** →(**NIH shift**)→ **an enone** → **phenol**

X −H⁺

OH

3-D Molecule:
Benzene oxide

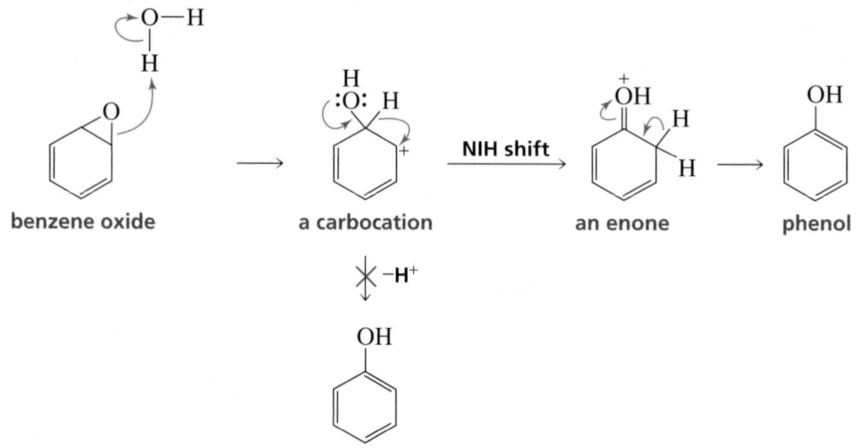

Because the first step is the rate-limiting step, the rate of formation of phenol depends on the stability of the carbocation. The more stable the carbocation, the easier it is to open the epoxide ring and form the phenol.

Only one arene oxide can be formed from naphthalene because the "double bond" shared by the two rings cannot be epoxidized. Recall from Section 6.11 that benzene rings are particularly stable, so it is much easier to epoxidize a position that will leave one of the benzene rings intact. Naphthalene oxide can rearrange to form either 1-naphthol or 2-naphthol. The carbocation leading to 1-naphthol is more stable because the carbocation leading to 2-naphthol can be stabilized by electron delocalization only if the aromaticity of the benzene ring on the left of the structure is destroyed. Consequently, rearrangement leads predominantly to 1-naphthol.

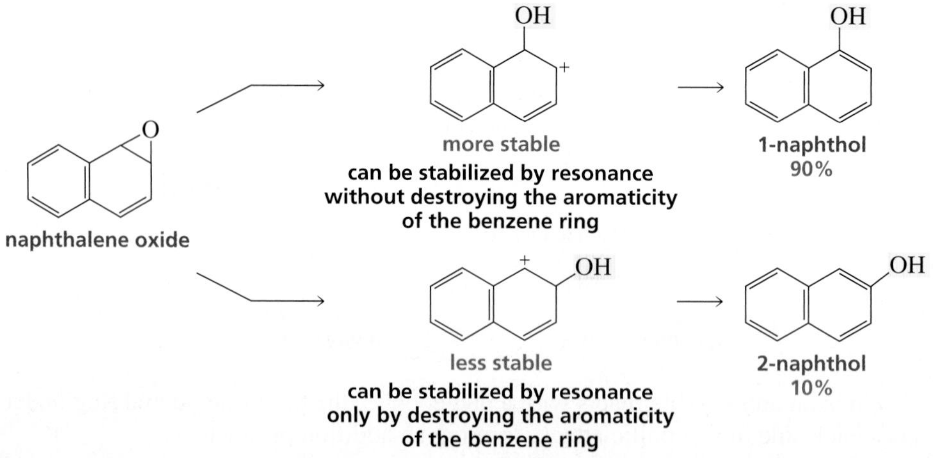

naphthalene oxide

more stable
can be stabilized by resonance
without destroying the aromaticity
of the benzene ring

1-naphthol
90%

less stable
can be stabilized by resonance
only by destroying the aromaticity
of the benzene ring

2-naphthol
10%

PROBLEM 16

Draw the resonance contributors for the two carbocations in the preceding reaction. Use the resonance contributors to explain why 1-napthol is the major product of the reaction.

PROBLEM 17 ◆

The existence of the NIH shift was established by the nature of the major product obtained from rearrangement of the following deuterated arene oxide. What would be the major product if the NIH shift did not occur? (*Hint:* Recall from Section 10.7 that a C—H bond is easier to break than a C—D bond.)

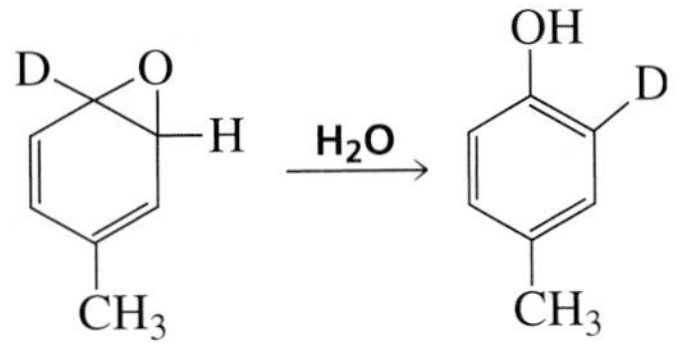

PROBLEM 18 ◆

How would the major products obtained from rearrangement of the following arene oxides differ?

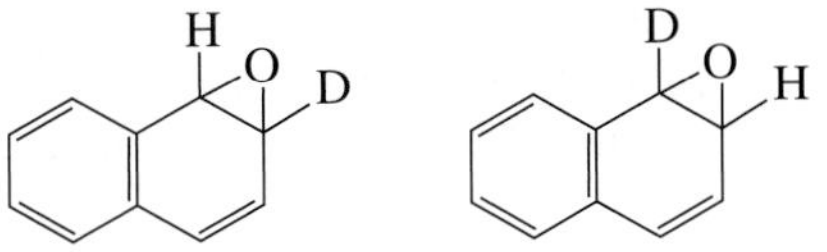

Some aromatic hydrocarbons are known to cause cancer. Investigation has revealed, however, that it is not the hydrocarbon itself that is carcinogenic, but rather the arene oxide into which the hydrocarbon is converted. How do arene oxides cause cancer? We have seen that nucleophiles react with epoxides to form addition products. 2′-Deoxyguanosine, a component of DNA (Section 25.1), is a nucleophile and reacts with certain arene oxides. Once 2′-deoxyguanosine becomes covalently attached to the arene oxide, 2′-deoxyguanosine can no longer fit into the DNA double helix. Thus the genetic code cannot be properly transcribed, which can lead to mutations that cause cancer. Cancer results when cells lose their ability to control their growth and reproduction.

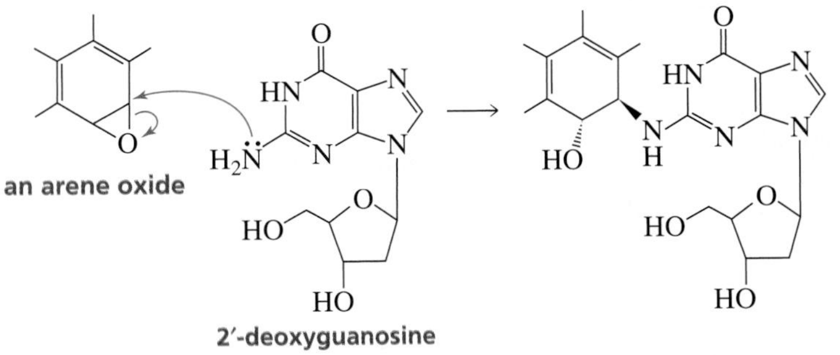

A segment of DNA.

Not all arene oxides are carcinogenic because not all of them react with nucleophiles. Because an arene oxide has two possible pathways for reaction, whether or not a particular arene oxide is carcinogenic depends on the relative rates of the two reaction pathways. Arene oxide rearrangement leads to phenolic products that are not carcinogenic, while formation of addition products from nucleophilic attack by DNA can lead to cancer-causing products. Thus, if the rate of arene oxide rearrangement is greater than the rate of nucleophilic attack by DNA, the arene oxide will be harmless. If, however, the rate of nucleophilic attack is greater than the rate of rearrangement, the arene oxide will be a carcinogen.

Because the rate of arene oxide rearrangement depends on the stability of the carbocation formed in the first step of the rearrangement, an arene oxide's cancer-causing potential depends on the stability of this carbocation. If the carbocation is relatively stable, rearrangement will be fast and the arene oxide will tend to be noncarcinogenic. If the carbocation is relatively unstable, rearrangement will be slow and the arene oxide will be more likely to undergo nucleophilic attack and be carcinogenic. This means that the more reactive the arene oxide (the more easily it opens to form a carbocation), the less likely it is to be carcinogenic.

BENZO[*a*]PYRENE AND CANCER

Benzo[*a*]pyrene is one of the most carcinogenic of the aromatic hydrocarbons. This hydrocarbon is formed whenever an organic compound undergoes incomplete combustion. For example, benzo[*a*]pyrene is found in cigarette smoke, in automobile exhaust, and in charcoal-broiled meat.

Several arene oxides can be formed from benzo[*a*]pyrene. The two most harmful are the 4,5-oxide and the 7,8-oxide. It has been suggested that people who develop lung cancer as a result of smoking may have a higher than normal concentration of cytochrome P_{450} in their lung tissue.

benzo[*a*]pyrene · 4,5-benzo[*a*]pyrene oxide · 7,8-benzo[*a*]pyrene oxide

a diol epoxide

The 4,5-oxide is harmful because it forms a carbocation that cannot be stabilized by electron delocalization without destroying the aromaticity of an adjacent benzene ring. Because the carbocation is relatively unstable, the epoxide will tend not to open until it is attacked by a nucleophile. The 7,8-oxide is harmful because it reacts with water to form a diol, which then forms a diol epoxide. The diol epoxide does not readily undergo rearrangement (the harmless pathway) because it opens to a carbocation that is destabilized by the electron-withdrawing OH groups. Therefore, the diol epoxide can exist long enough to be attacked by nucleophiles (the carcinogenic pathway).

CHIMNEY SWEEPS AND CANCER

In 1775, a British physician named Percival Potts was the first to recognize that environmental factors can cause cancer when he became aware that chimney sweeps had a higher incidence of scrotum cancer than the male population as a whole. He realized that something in the chimney soot was apparently causing cancer. We now know that it was benzo[*a*]pyrene.

Titch Cox, the chimney sweep responsible for cleaning the 800 chimneys in Buckingham Palace.

PROBLEM 19/SOLVED

For each of the following pairs of compounds, indicate the one that is more likely to be carcinogenic:

a. [structure with OCH₃] or [structure with NO₂]   **b.** [naphthalene oxide] or [dihydronaphthalene oxide]

SOLUTION TO 19a   The nitro-substituted compound is more likely to be carcinogenic. The electron-withdrawing nitro group destabilizes the carbocation formed when the ring opens, whereas the resonance electron-donating methoxy group stabilizes the carbocation. Carbocation formation leads to the harmless product, so the methoxy-substituted compound with a more stable (easier-to-form) carbocation will be more likely to form a harmless product. In addition, the electron-withdrawing nitro group increases the arene oxide's susceptibility to nucleophilic attack, the cancer-causing pathway.

[reaction schemes showing arene oxide + H₂O]

PROBLEM 20

Explain why the two arene oxides in Problem 19a open in opposite directions.

PROBLEM 21

Three arene oxides can be obtained from phenanthrene.

a. Give the structures of the three phenanthrene oxides.

b. What phenols can be obtained from each phenanthrene oxide?

phenanthrene

c. If a phenanthrene oxide can lead to the formation of more than one phenol, which phenol will be obtained in greater yield?

d. Which of the three phenanthrene oxides is most likely to be carcinogenic?

11.8 ORGANOMETALLIC COMPOUNDS

In compounds such as alcohols, ethers, and alkyl halides, carbon is bonded to a more electronegative atom. Carbon, therefore, is *electrophilic* and the alkyl group reacts with a nucleophile.

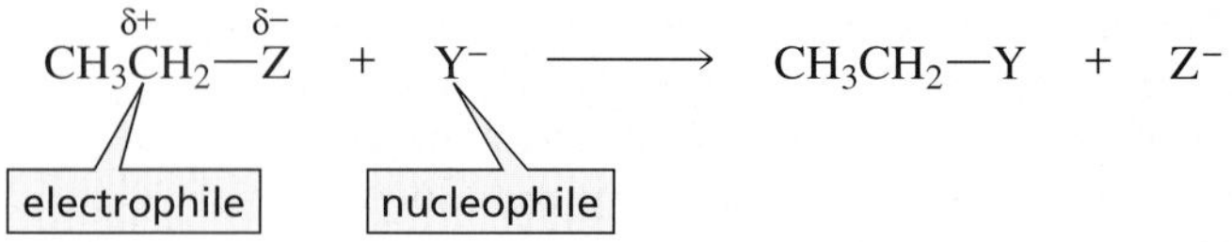

$$\overset{\delta+}{CH_3CH_2}\!\!-\!\!\overset{\delta-}{Z} \quad + \quad Y^- \quad \longrightarrow \quad CH_3CH_2\!\!-\!\!Y \quad + \quad Z^-$$

electrophile   nucleophile

But what if you wanted an alkyl group to react with an electrophile? To do so, you would need a compound in which carbon is nucleophilic. To be *nucleophilic,* carbon would have to be bonded to a less electronegative atom.

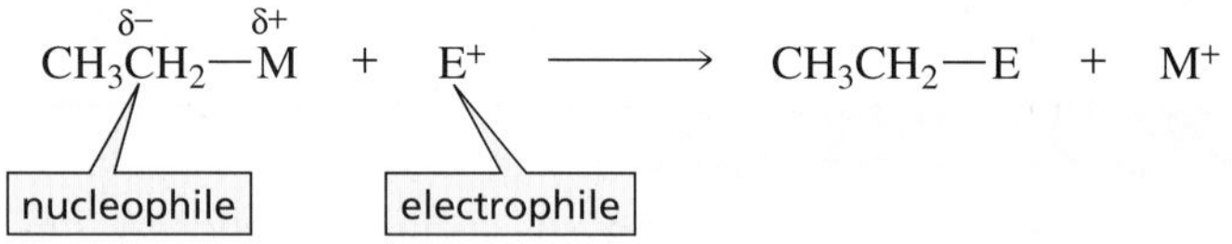

$$\overset{\delta-}{CH_3CH_2}\!\!-\!\!\overset{\delta+}{M} \quad + \quad E^+ \quad \longrightarrow \quad CH_3CH_2\!\!-\!\!E \quad + \quad M^+$$

nucleophile   electrophile

An **organometallic compound** is a compound that contains a carbon–metal bond. Organolithium compounds and organomagnesium compounds are two of the most common organometallic compounds. Because most metals are less electronegative than carbon (Table 11.3), the carbon bonded to the metal is nucleophilic. The electrostatic potential maps show that the carbon atom is electron-rich (red) in the organometallic compounds while it is electron-poor (blue-green) in the alkyl halide.

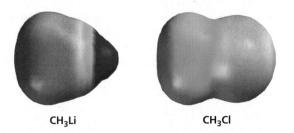

CH₃Li   CH₃Cl

TABLE 11.3 The Electronegativities of Some of the Elements[a]

IA	IIA	IB	IIB	IIIA	IVA	VA	VIA	VIIA
H 2.1								
Li 1.0	Be 1.5			B 2.0	C 2.5	N 3.0	O 3.5	F 4.0
Na 0.9	Mg 1.2			Al 1.5	Si 1.8	P 2.1	S 2.5	Cl 3.0
K 0.8	Ca 1.0	Cu 1.8	Zn 1.7	Ga 1.8	Ge 2.0			Br 2.8
		Ag 1.4	Cd 1.5		Sn 1.7			I 2.5
			Hg 1.5		Pb 1.6			

[a]From the scale devised by Linus Pauling

Organolithium compounds are prepared by adding lithium to an alkyl halide in a nonpolar solvent such as hexane.

3-D Molecules:
Chlorobenzene;
Phenyllithium

$$CH_3CH_2CH_2CH_2Br \ + \ 2\,Li \ \xrightarrow{\text{hexane}} \ CH_3CH_2CH_2CH_2Li \ + \ LiBr$$
1-bromobutane   butyllithium

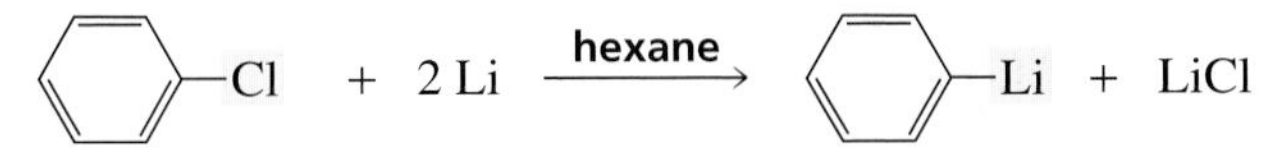

⬡—Cl + 2 Li $\xrightarrow{\text{hexane}}$ ⬡—Li + LiCl

chlorobenzene   phenyllithium

Organomagnesium compounds, frequently called **Grignard reagents** after their discoverer, are prepared by adding an alkyl halide to magnesium shavings being stirred in anhydrous diethyl ether or THF. The magnesium is inserted between the carbon and the halogen.

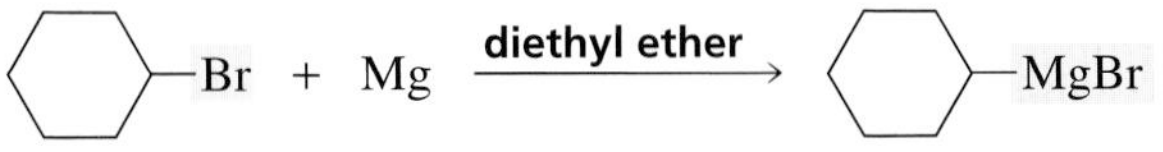

⬡—Br + Mg $\xrightarrow{\text{diethyl ether}}$ ⬡—MgBr

cyclohexyl   cyclohexylmagnesium
bromide   bromide

$$CH_2{=}CHBr \ + \ Mg \ \xrightarrow{\text{THF}} \ CH_2{=}CHMgBr$$
vinyl   vinylmagnesium
bromide   bromide

The solvent (usually diethyl ether or tetrahydrofuran) plays a crucial role in the formation of a Grignard reagent. Because the magnesium atom of a Grignard reagent is surrounded by only four electrons, it needs two more pairs of electrons to form an octet. Solvent molecules provide these electrons by coordinating (supplying electron pairs) to the metal. Coordination allows the Grignard reagent to dissolve in the solvent and prevents the Grignard reagent from coating the magnesium shavings, which would make them unreactive.

The following reactions show that the organometallic compounds form when the metal (Li or Mg) donates its valence electrons to the partially positively charged carbon of the alkyl halide.

Francis August Victor Grignard (1871–1935) was born in France, the son of a sailmaker. He received a Ph.D. from the University of Lyons in 1901. His synthesis of the first Grignard reagent was announced in 1900. During the next five years, some 200 papers were published about Grignard reagents. He was a professor of chemistry at the University of Nancy and later at the University of Lyons. He shared the Nobel Prize in chemistry in 1912 with Paul Sabatier (p. 169). During World War I, Grignard was drafted into the French army, where he developed a method to detect war gases.

$$\overset{\delta+}{CH_3CH_2}{-}\overset{\delta-}{Br} \ + \ 2\,Li \ \longrightarrow \ \overset{\delta-}{CH_3CH_2}{:}\overset{\delta+}{Li} \ + \ Li^+Br^-$$

$$\overset{\delta+}{CH_3CH_2}{-}\overset{\delta-}{Br} \ + \ Mg \ \longrightarrow \ \overset{\delta-}{CH_3CH_2}{:}\overset{\delta+}{MgBr}$$

Reaction with the metal has converted the alkyl halide into a compound that will react with electrophiles instead of with nucleophiles. Thus, organolithium and organomagnesium compounds react as if they were carbanions.

3-D Molecules:
Phenyllithium;
Ethylmagnesium bromide

CH_3CH_2MgBr reacts as if it were $CH_3\overset{..}{C}H_2 \ \overset{+}{MgBr}$
ethylmagnesium bromide

⬡—Li   reacts as if it were   ⬡:⁻ Li⁺

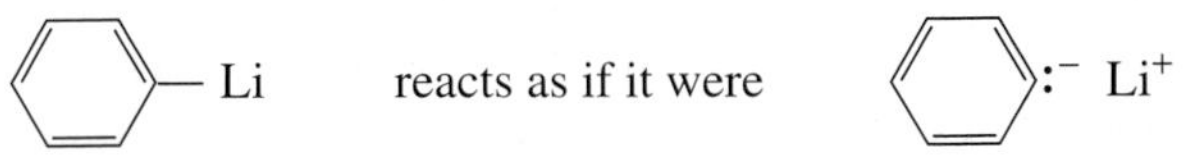

phenyllithium

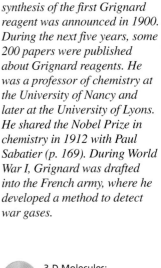

Alkyl halides, vinyl halides, and aryl halides can all be used to form organolithium and organomagnesium compounds. Alkyl bromides are the alkyl halides most often used to form organometallic compounds because they react more readily than alkyl chlorides and are less expensive than alkyl iodides.

Since organometallic compounds are nucleophiles, they can react with epoxides.

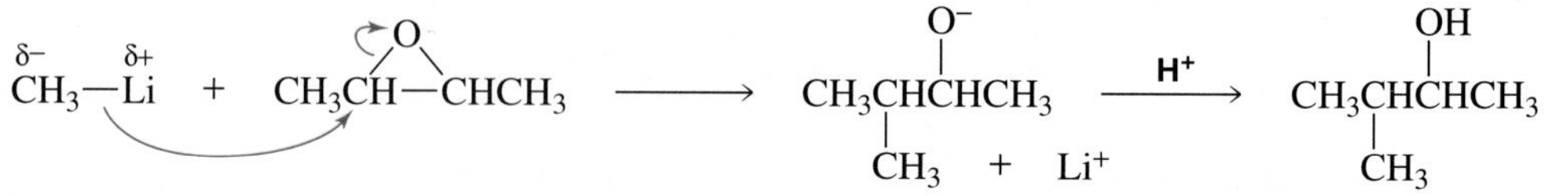

$$\overset{\delta-}{CH_3}-\overset{\delta+}{Li} \ + \ CH_3CH-CHCH_3 \ \longrightarrow \ CH_3CHCHCH_3 \ \xrightarrow{H^+} \ CH_3CHCHCH_3$$

Notice that when a Grignard reagent reacts with ethylene oxide, a primary alcohol containing two more carbons than the Grignard reagent is formed.

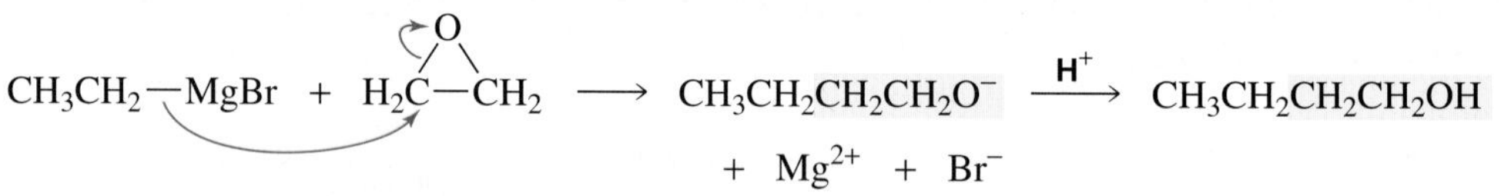

$$CH_3CH_2-MgBr \ + \ H_2C-CH_2 \ \longrightarrow \ CH_3CH_2CH_2CH_2O^- \ \xrightarrow{H^+} \ CH_3CH_2CH_2CH_2OH$$
$$+ \ Mg^{2+} \ + \ Br^-$$

PROBLEM 22 ◆

What alcohols would be formed from the reaction of ethylene oxide with the following Grignard reagents?

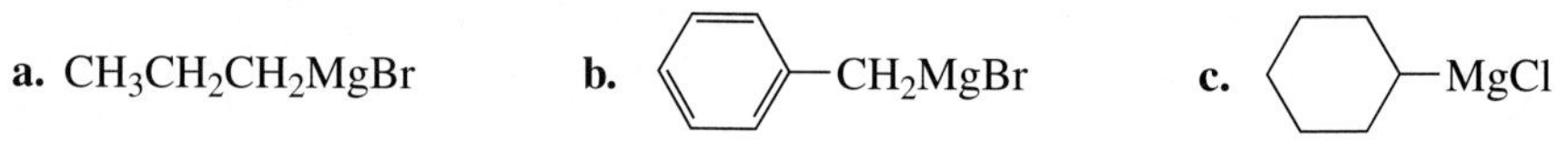

a. $CH_3CH_2CH_2MgBr$

b. ⬡—CH_2MgBr

c. ⬡—$MgCl$

PROBLEM 23

How could the following compounds be prepared using cyclohexene oxide as a starting material?

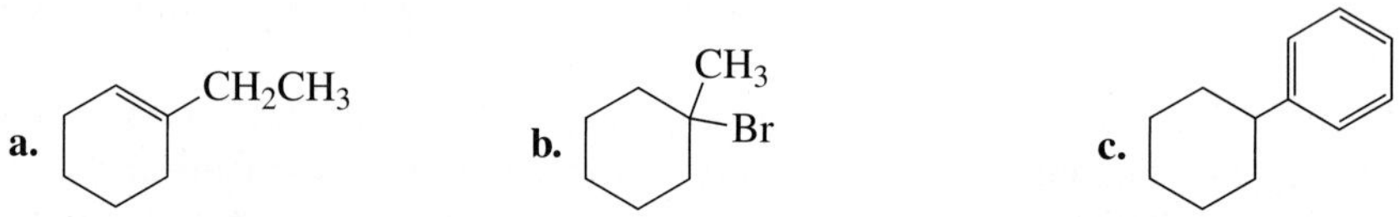

a. (structure with CH_2CH_3)

b. (structure with CH_3 and Br)

c. (cyclohexane with phenyl group)

Organomagnesium and organolithium compounds are such strong bases that they will react immediately with any acid present in the reaction mixture, even with very weak acids such as water and alcohols. When this happens, the organometallic compound is converted into an alkane. If D_2O is used instead of H_2O, a deuterated compound will be obtained.

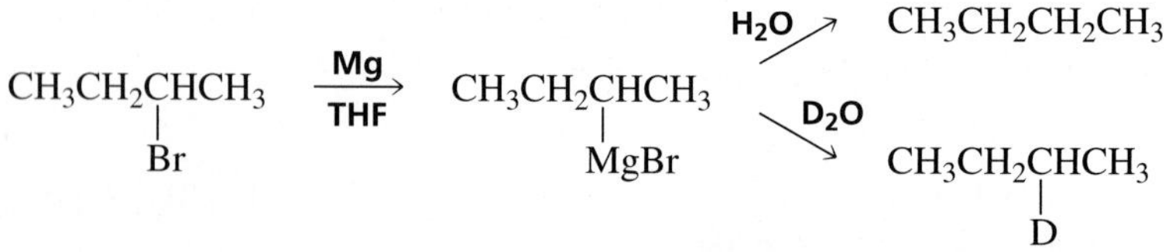

$$CH_3CH_2CHCH_3 \ \xrightarrow[THF]{Mg} \ CH_3CH_2CHCH_3 \ \begin{array}{c} \xrightarrow{H_2O} CH_3CH_2CH_2CH_3 \\ \xrightarrow{D_2O} CH_3CH_2CHCH_3 \end{array}$$

This means that Grignard reagents cannot be prepared from compounds that contain acidic groups ($-OH$, $-NH_2$, $-NHR$, $-SH$, or $-COOH$ groups). Because even trace amounts of moisture can destroy an organometallic compound, it is important that all the reagents be dry when organometallic compounds are being synthesized and when they react with other reagents.

PROBLEM 24 / SOLVED

Show, using any necessary reagents, how the following compounds could be prepared using ethylene oxide as one of the reactants:

a. $CH_3CH_2CH_2CH_2OH$

b. $CH_3CH_2CH_2CH_2Br$

c. $CH_3CH_2CH_2CH_2D$

d. $CH_3CH_2CH_2CH_2CH_2CH_2OH$

SOLUTION

a. $CH_3CH_2Br \xrightarrow[\text{Et}_2\text{O}]{\text{Mg}} CH_3CH_2MgBr \xrightarrow[\text{2. H}^+]{\text{1.} \overset{\text{O}}{\triangle}} CH_3CH_2CH_2CH_2OH$

b. product of **a** $\xrightarrow{\text{PBr}_3} CH_3CH_2CH_2CH_2Br$

c. product of **b** $\xrightarrow[\text{Et}_2\text{O}]{\text{Mg}} CH_3CH_2CH_2CH_2MgBr \xrightarrow{\text{D}_2\text{O}} CH_3CH_2CH_2CH_2D$

d. the same reaction sequence as in **a**, using butyl bromide in the first step.

PROBLEM 25 ◆

Which of the following reactions will occur? For the pK_a values necessary to do this problem, see Appendix II.

$CH_3MgBr \quad + \quad H_2O \quad \longrightarrow \quad CH_4 \quad + \quad HOMgBr$

$CH_3MgBr \quad + \quad CH_3OH \quad \longrightarrow \quad CH_4 \quad + \quad CH_3OMgBr$

$CH_3MgBr \quad + \quad NH_3 \quad \longrightarrow \quad CH_4 \quad + \quad H_2NMgBr$

$CH_3MgBr \quad + \quad CH_3NH_2 \quad \longrightarrow \quad CH_4 \quad + \quad CH_3NHMgBr$

$CH_3MgBr \quad + \quad HC\equiv CH \quad \longrightarrow \quad CH_4 \quad + \quad HC\equiv CMgBr$

There are many different organometallic compounds. As long as the metal is less electronegative than carbon, the carbon bonded to the metal will be nucleophilic.

$$\overset{\delta-\quad\delta+}{C-Mg} \qquad \overset{\delta-\quad\delta+}{C-Li} \qquad \overset{\delta-\quad\delta+}{C-Cu} \qquad \overset{\delta-\quad\delta+}{C-Cd} \qquad \overset{\delta-\quad\delta+}{C-Si}$$

$$\overset{\delta-\quad\delta+}{C-Zn} \qquad \overset{\delta-\quad\delta+}{C-Al} \qquad \overset{\delta-\quad\delta+}{C-Pb} \qquad \overset{\delta-\quad\delta+}{C-Hg} \qquad \overset{\delta-\quad\delta+}{C-Sn}$$

The reactivity of an organometallic compound toward an electrophile depends on the polarity of the carbon–metal bond—the greater the polarity of the bond, the more reactive the compound is as a nucleophile. The polarity of the bond depends on the difference in electronegativity between the metal and carbon (Table 11.3). For example, magnesium has an electronegativity of 1.2 compared with 2.5 for carbon. This large difference in electronegativity makes the carbon–magnesium bond very polar. (The carbon–magnesium bond is about 52% ionic.) Lithium (1.0) is less electronegative than magnesium (1.2). Thus the carbon–lithium bond is even more polar than the carbon–magnesium bond. Therefore, an organolithium reagent is a more reactive nucleophilic reagent than a Grignard reagent.

The names of organometallic compounds usually begin with the name of the alkyl group followed by the name of the metal.

CH_3CH_2MgBr

ethylmagnesium bromide

$CH_3CH_2CH_2CH_2Li$

butyllithium

$(CH_3CH_2CH_2)_2Cd$

dipropylcadmium

$(CH_3CH_2)_4Pb$

tetraethyllead

A Grignard reagent will undergo **transmetallation** (metal exchange) if it is added to a metal halide whose metal is more electronegative than magnesium. In other words, metal exchange will occur if it results in a less polar carbon–metal bond. For example, cadmium (1.5) is more electronegative than magnesium (1.2). Consequently, a carbon–cadmium bond is less polar than a carbon–magnesium bond, so metal exchange occurs.

$$2\ CH_3CH_2MgCl\ +\ CdCl_2\ \longrightarrow\ (CH_3CH_2)_2Cd\ +\ 2\ MgCl_2$$

ethylmagnesium chloride   **diethylcadmium**

PROBLEM 26 ◆

What organometallic compound will be formed from the reaction of methylmagnesium chloride and $SiCl_4$? (*Hint:* See Table 11.3.)

Henry Gilman (1893–1986) was born in Boston. He received his A.B. and Ph.D. degrees from Harvard University. He joined the faculty at Iowa State University in 1919, where he remained for his entire career. He published more than 1000 research papers. More than half of these papers were published after he lost almost all of his sight as the result of a detached retina and glaucoma in 1947. His wife, Ruth, acted as his eyes for 40 years.

Gilman reagents, also called **organocuprates,** are prepared from the reaction of an organolithium reagent with cuprous iodide in diethyl ether or THF.

$$2\ CH_3Li\ +\ CuI\ \xrightarrow{\text{ether}}\ (CH_3)_2CuLi\ +\ LiI$$

organolithium reagent   **Gilman reagent**

Gilman reagents are very useful to synthetic chemists. When a Gilman reagent reacts with an alkyl halide (with the exception of alkyl fluorides, which do not undergo this reaction), one of the alkyl groups of the Gilman reagent replaces the halogen. This is called a **coupling reaction** because two alkyl groups are joined (coupled together). The precise mechanism of the reaction is unknown but is thought to involve radicals.

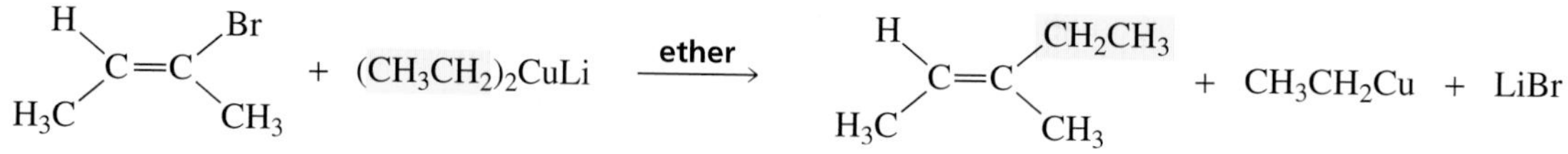

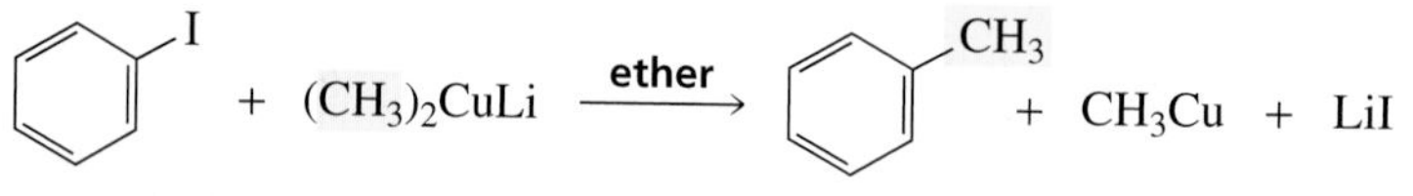

Tutorial: Organometallic compounds

Gilman reagents can even replace halogens in compounds that contain other functional groups.

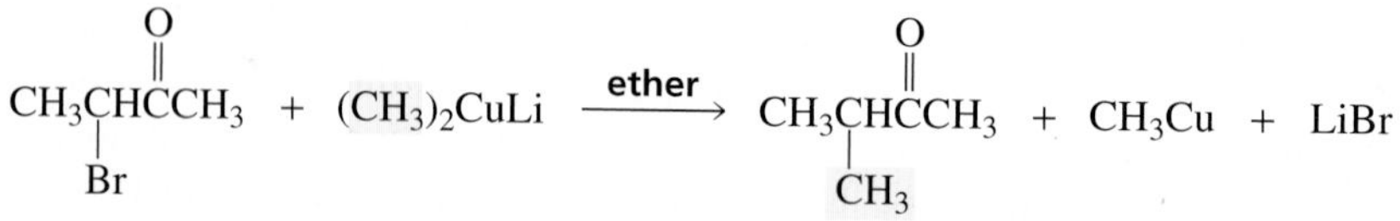

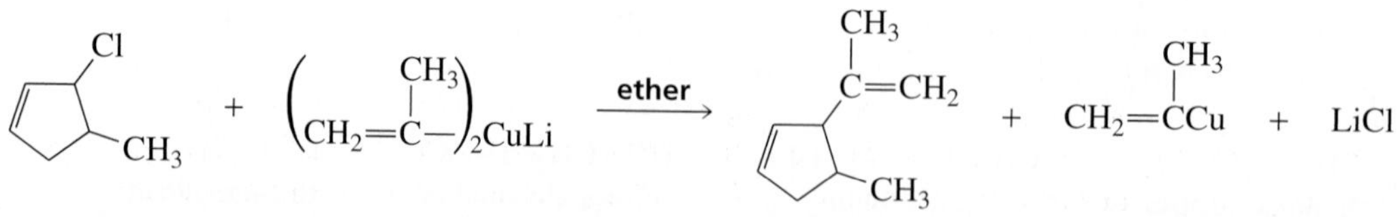

PROBLEM 27

Muscalure is the sex attractant of the common housefly. Flies are lured to traps by filling them with fly bait containing both muscalure and an insecticide. Eating the bait is fatal.

$$CH_3(CH_2)_7 \quad (CH_2)_{12}CH_3$$
$$C=C$$
$$H \qquad H$$

muscalure

How could you synthesize muscalure using the following reagents?

$$CH_3(CH_2)_7 \quad (CH_2)_8Br$$
$$C=C \qquad \text{and} \quad CH_3(CH_2)_4Br$$
$$H \qquad H$$

11.9
CROWN ETHERS

Crown ethers are cyclic compounds that have several ether linkages. A crown ether specifically binds certain metal ions or organic molecules depending on the size of its cavity. The crown ether is called the "host" and the species it binds is called the "guest." Because the ether linkages are chemically inert, the crown ether can bind the guest without reacting with it. The **crown–guest complex** is called an **inclusion compound.** Crown ethers are named [X]-crown-Y, where X is the total number of atoms in the ring and Y is the number of oxygen atoms in the ring. [15]-Crown-5 specifically binds Na^+ because the crown ether has a cavity diameter of 1.7 to 2.2 Å and Na^+ has an ionic diameter of 1.80 Å. Binding occurs as a result of interaction of the positively charged ion with the nonbonding electrons of the oxygens that point into the cavity of the crown ether.

For their work in the field of crown ethers, Charles J. Pedersen, Donald J. Cram, and Jean-Marie Lehn shared the 1987 Nobel Prize in chemistry.

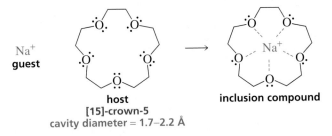

Na^+
guest

host
[15]-crown-5
cavity diameter = 1.7–2.2 Å

inclusion compound

Li^+ (ionic diameter, 1.20 Å) is too small to be bound by [15]-crown-5, but binds neatly in [12]-crown-4; K^+ (ionic diameter, 2.66 Å), on the other hand, is too large to fit into [15]-crown-5, but is bound specifically by [18]-crown-6.

3-D Molecules:
[15]-Crown-5;
[12]-Crown-4

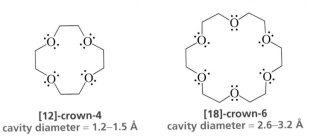

[12]-crown-4
cavity diameter = 1.2–1.5 Å

[18]-crown-6
cavity diameter = 2.6–3.2 Å

A remarkable property of crown ethers is that they allow inorganic salts to be dissolved in nonpolar organic solvents, thus permitting many reactions to be carried out in nonpolar solvents that otherwise would be able to take place only in polar protic solvents. For example, potassium permanganate ($K^+MnO_4^-$), a common oxidizing agent

Charles J. Pedersen (1904–1989) was born in Korea as a Norwegian citizen. He received a B.S. in chemical engineering from the University of Dayton and an M.S. in organic chemistry from MIT. He joined DuPont in 1927 and retired from there in 1969.

Donald J. Cram *was born in Vermont in 1919. He received a B.S. from Rollins College, an M.S. from the University of Nebraska, and a Ph.D. from Harvard University. He is a professor of chemistry at the University of California, Los Angeles.*

(Section 18.2), is not soluble in a nonpolar solvent because it is ionic. How, then, can potassium permanganate oxidize compounds that are soluble only in nonpolar solvents?

Potassium permanganate can be dissolved in a nonpolar solvent such as benzene if [18]-crown-6 is added to the solution. The crown ether binds potassium in its cavity, and the nonpolar crown ether–potassium complex dissolves in benzene. In order to maintain electrical neutrality, the permanganate ion accompanies the complexed potassium ion into the benzene layer, permitting permanganate to act as an oxidizing agent in the nonpolar solvent.

"purple benzene"
the [18]-crown-6–KMnO₄ complex is soluble in benzene

In this example, the crown ether is acting as a phase transfer catalyst. A **phase transfer catalyst** is a compound that catalyzes a reaction by transferring a reagent (usually an inorganic ion) into the phase in which it is needed (Section 11.12).

3-D Molecule:
[18]-Crown-6 with
potassium ion

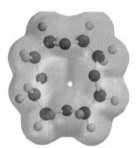

[12]-crown-4

[15]-crown-5

[18]-crown-6

AN IONOPHOROUS ANTIBIOTIC

An **antibiotic** is a compound that interferes with the growth of microorganisms. Nonactin is a naturally occurring antibiotic that owes its biological activity to its ability to disrupt the carefully maintained electrolyte balance between the inside and outside of a cell. It does this by acting like a crown ether. For proper cell function, the cell must maintain a higher concentration of K^+ inside the cell and a higher concentration of Na^+ outside the cell. Nonactin's diameter is such that it specifically binds potassium ions and then transports them out of the cell. The decreased concentration of K^+ within the bacterial cell causes the bacterium to die. Nonactin is an example of an ionophorous antibiotic. An **ionophore** is a compound that transports metal ions by binding them tightly.

nonactin

The ability of a host to bind only certain guests is an example of **molecular recognition**. The very precise reactions of enzymes and other biological molecules are made possible by molecular recognition. Only recently have chemists been able to design and synthesize organic molecules that exhibit molecular recognition, although the specificity of these synthetic compounds for their guests is generally less highly developed than the specificity exhibited by many biological molecules. The hexaammonium ion shown below was designed to bind adenosine triphosphate (ATP), a molecule with four negatively charged oxygens. ATP plays an important role in many cellular reactions (Section 25.2).

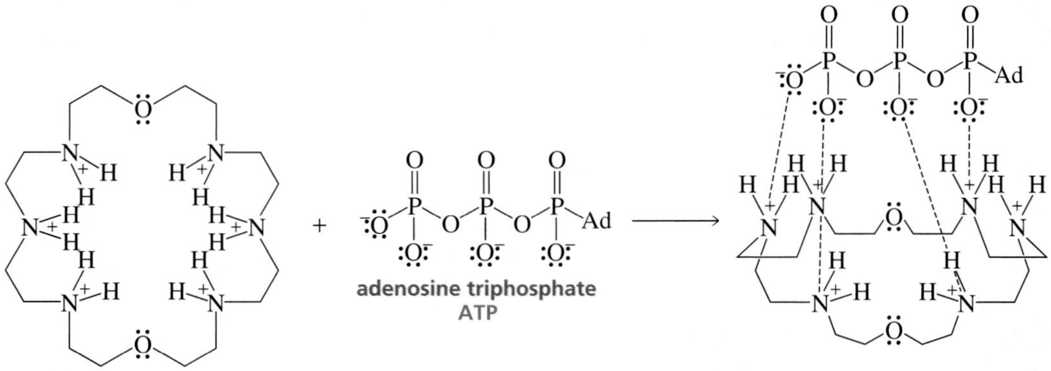

adenosine triphosphate
ATP

Polycyclic compounds are now being designed that are highly specific in their binding. These compounds use the nonbonding electrons of oxygen and nitrogen atoms to bind guests. The three-dimensional compounds are called **cryptands** (*kryptos* is the Greek word for "hidden"). The complex that results when the cryptand binds its substrate is called a **cryptate.** The diaminohexaether shown here is a cryptand that has been specifically designed to bind a potassium ion. It is better than [18]-crown-6 at binding K^+. [18]-Crown-6 is a floppy molecule that becomes rigid when it binds potassium, so some entropy is lost in the binding process. The three-dimensional cryptand is more rigid than the crown ether, so less entropy is lost on binding.

Jean-Marie Lehn was born in France in 1939. He initially studied philosophy and then switched to chemistry. He received a Ph.D. from the University of Strasbourg. As a postdoctoral fellow, he worked with R. B. Woodward at Harvard on the total synthesis of vitamin B_{12} (see pages 1156 and 1221). He is a professor of chemistry at the Université Louis Pasteur in Strasbourg, France, and at the Collège de France in Paris.

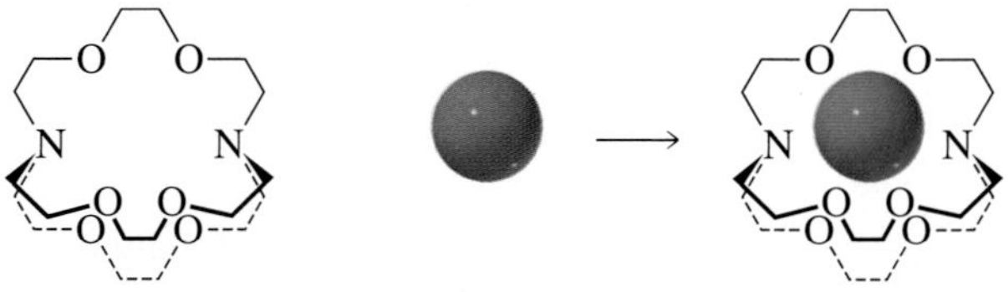

a cryptand   **K^+**   **a cryptate**

Thiols are sulfur analogs of alcohols. Thiols were formerly called mercaptans because they form strong complexes with heavy metal cations such as mercury and arsenic (they capture mercury).

11.10
THIOLS, SULFIDES, AND SULFONIUM SALTS

$$2 \ CH_3CH_2SH \ + \ Hg^{2+} \ \longrightarrow \ CH_3CH_2S-Hg-SCH_2CH_3 \ + \ 2 \ H^+$$
$$\text{mercuric ion}$$

Thiols are named by adding the suffix "thiol" to the parent hydrocarbon. The SH group is called a **mercapto group.**

3-D Molecule: Methanethiol

CH_3CH_2SH $CH_3CH_2CH_2SH$ $CH_3\overset{\overset{\displaystyle CH_3}{|}}{C}HCH_2CH_2SH$ $HSCH_2CH_2OH$
ethanethiol   **1-propanethiol**   **3-methyl-1-butanethiol**   **2-mercaptoethanol**

CH₃CH₂OH

CH₃CH₂SH

Low-molecular-weight thiols are noted for their strong and pungent odors, such as those associated with onions, garlic, and skunks. Natural gas is completely odorless and can cause deadly explosions if a leak goes undetected. Therefore, a small amount of a thiol is added to natural gas to make gas leaks easily detectable.

Sulfur is larger than oxygen, so a thiolate ion is more stable than an alkoxide ion (Section 1.18). Therefore, thiols are stronger acids ($pK_a = 10$) than alcohols ($pK_a = 16$). Consequently, thiolate ions are weaker bases than alkoxide ions. The larger thiolate ions are less well solvated than alkoxide ions, so in protic solvents thiolate ions are better nucleophiles than alkoxide ions (Section 9.3).

$$CH_3\ddot{S}: \; + \; CH_3CH_2{-}Br \; \longrightarrow \; CH_3\ddot{S}CH_2CH_3 \; + \; Br^-$$

Because sulfur is not as electronegative as oxygen, thiols are not good at hydrogen bonding. Therefore, they have considerably lower boiling points than alcohols. For example, the boiling point of CH_3CH_2SH is 37 °C, whereas the boiling point of CH_3CH_2OH is 78 °C.

The sulfur analogs of ethers are called **sulfides** or **thioethers.** Because sulfur is an excellent nucleophile, sulfides react readily with alkyl halides to form **sulfonium salts.**

$$CH_3\ddot{S}CH_3 \; + \; CH_3{-}I \; \longrightarrow \; \overset{\overset{\displaystyle CH_3}{|}}{CH_3\overset{+}{\ddot{S}}CH_3} \; I^-$$

dimethyl   trimethylsulfonium iodide
sulfide   a sulfonium salt

Since it has a weakly basic leaving group, a sulfonium ion readily undergoes nucleophilic substitution reactions. As with other S_N2 reactions, the reaction works best if the group undergoing nucleophilic attack is a methyl group or a primary alkyl group.

$$H\ddot{O}: \; + \; CH_3\overset{+}{\ddot{S}}CH_3 \; \longrightarrow \; CH_3\ddot{O}H \; + \; CH_3\ddot{S}CH_3$$

MUSTARD GAS

Chemical warfare was used for the first time in 1915 when Germany released chlorine gas against French and British forces in the battle of Ypres. For the remainder of World War I, both sides used a variety of chemical agents. One of the more common war gases was mustard gas, a reagent that produces blisters over the surface of the body. It is a very reactive compound because the highly nucleophilic sulfur easily displaces a chloride ion by an intramolecular S_N2

reaction, forming a cyclic sulfonium salt that rapidly reacts with a nucleophile. The sulfonium salt is particularly reactive because of the strained three-membered ring.

The toxicity of mustard gas is due to the high local concentrations of HCl that are produced when water, or any other nucleophile, reacts with the gas when it comes into contact with the skin and/or lungs. An international treaty in the 1980s banned its use and required all the mustard gas that had been stockpiled to be destroyed.

$$ClCH_2CH_2\ddot{S}CH_2CH_2{-}Cl \; \longrightarrow \; ClCH_2CH_2\overset{+}{\ddot{S}}\overset{CH_2}{\underset{CH_2}{\diagup}} \; \xrightarrow{H_2\ddot{O}:} \; Cl{-}CH_2CH_2\ddot{S}CH_2CH_2OH \; + \; H^+$$

mustard gas   sulfonium salt
+ Cl⁻

$$H^+ \; + \; HOCH_2CH_2\ddot{S}CH_2CH_2OH \; \xleftarrow{H_2\ddot{O}:} \; \overset{CH_2}{\underset{CH_2}{\diagup}}\overset{+}{\diagdown}SCH_2CH_2OH \; + \; Cl^-$$

ANTIDOTE TO A WAR GAS

Lewisite is a war gas developed in 1917 by W. Lee Lewis, an American scientist. It rapidly penetrates clothing and skin and is poisonous because it contains arsenic, which combines with thiol groups on enzymes, thereby inactivating the enzymes. During World War II, the Allies were concerned that the Germans would use lewisite. British scientists developed an antidote to lewisite that the Allies called "British anti-lewisite" (BAL). BAL contains two thiol groups and reacts with lewisite, preventing it from reacting with the thiol groups of enzymes.

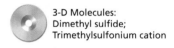

$$
\begin{array}{ccc}
\text{CH}_2\text{SH} & & \\
\mid & & \\
\text{CHSH} & + & \\
\mid & & \\
\text{CH}_2\text{OH} & & \\
\textbf{BAL} & & \textbf{lewisite}
\end{array}
$$

PROBLEM 28

How would you prepare the following compounds using an alkyl halide and a thiol as starting materials?

a. $CH_3CH_2SCH_2CH_3$

b. $CH_3\overset{\overset{\displaystyle CH_3}{\mid}}{\underset{\underset{\displaystyle CH_3}{\mid}}{S}}CCH_3$

c. $CH_2{=}CHCH\overset{\overset{\displaystyle CH_3}{\mid}}{}SCH_2CH_3$

d. ⬡—SCH_2—⬡

> 3-D Molecules:
> Dimethyl sulfide;
> Trimethylsulfonium cation

We have seen that alcohols are much less reactive than alkyl halides in substitution reactions. Amines are even less reactive than alcohols. The relative reactivities of an alkyl fluoride (the least reactive of the alkyl halides), an alcohol, and an amine can be appreciated by comparing the pK_a values of the conjugate acids of their leaving groups, recalling that the stronger the base, the weaker its conjugate acid. The leaving group of an amine ($^-NH_2$) is such a strong base that amines cannot undergo substitution or elimination reactions.

11.11
REACTIONS OF QUATERNARY AMMONIUM COMPOUNDS

relative reactivities

$$RCH_2F \quad > \quad RCH_2OH \quad > \quad RCH_2NH_2$$

$$
\begin{array}{ccc}
\text{HF} & \text{H}_2\text{O} & \text{NH}_3 \\
pK_a = 3.2 & pK_a = 15.7 & pK_a = 36
\end{array}
$$

Protonation of the amino group makes it a better leaving group, but not nearly as good a leaving group as a protonated alcohol. (Notice that protonated ethanol is more than 13 pK_a units more acidic than protonated ethylamine.) So unlike the leaving group of a protonated alcohol, the leaving group of a protonated amine cannot dissociate to form a carbocation or be replaced by a halide ion. Protonated amino groups also cannot be displaced by strongly basic nucleophiles such as HO$^-$ because the nucleophile would react immediately with the acidic hydrogen, which would convert it into a poor nucleophile (Section 11.1).

$$
\begin{array}{cc}
CH_3CH_2\overset{+}{O}H_2 & > \quad CH_3CH_2\overset{+}{N}H_3 \\
pK_a = -2.4 & pK_a = 11.2
\end{array}
$$

August Wilhelm von Hofmann (1818–1892) was born in Germany. He first studied law, then changed to chemistry. He founded the German Chemical Society. Hofmann taught at the Royal College of Chemistry in London for 20 years, then returned to Germany to teach at the University of Berlin. He was one of the founders of the German dye industry. He was married four times—three times left a widower—and had 11 children.

> **PROBLEM 29**
>
> Why is it that a halide ion such as Br^- can react with a protonated primary alcohol but cannot react with a protonated primary amine?

The Hofmann Elimination

Because the leaving group of a **quaternary ammonium ion** has about the same leaving tendency as a protonated amino group but does not have an acidic hydrogen, a quaternary ammonium ion can undergo reaction with a strong base. The reaction of a quaternary ammonium ion with hydroxide ion is known as a **Hofmann elimination reaction.** The leaving group in a Hofmann elimination reaction is a tertiary amine.

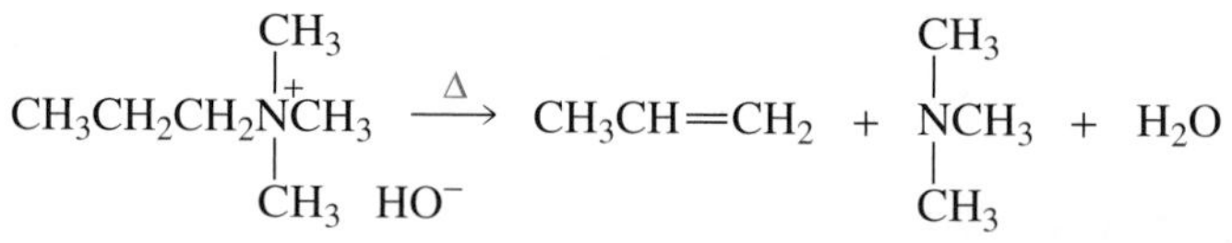

$$CH_3CH_2CH_2\overset{\overset{\displaystyle CH_3}{|}}{\underset{\underset{\displaystyle CH_3 \;\; HO^-}{|}}{\overset{+}{N}}}CH_3 \;\xrightarrow{\Delta}\; CH_3CH=CH_2 \;+\; \overset{\overset{\displaystyle CH_3}{|}}{\underset{\underset{\displaystyle CH_3}{|}}{N}}CH_3 \;+\; H_2O$$

Because a tertiary amine is only a moderately good leaving group, the reaction requires heat. It is an E2 reaction. Very little substitution product is formed.

mechanism of the Hofmann elimination

$$CH_3\overset{\frown}{CH}-CH_2-\overset{\overset{\displaystyle CH_3}{|}}{\underset{\underset{\displaystyle CH_3}{|}}{\overset{+}{N}}}CH_3 \;\longrightarrow\; CH_3CH=CH_2 \;+\; \overset{\overset{\displaystyle CH_3}{|}}{\underset{\underset{\displaystyle CH_3}{|}}{N}}CH_3 \;+\; H_2O$$

$$H\ddot{O}:^-$$

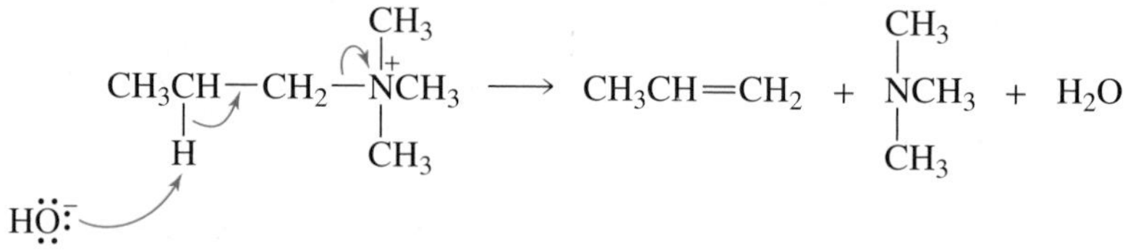

> **PROBLEM 30**
>
> What is the difference between the reaction that occurs when isopropyltrimethylammonium hydroxide is heated and the reaction that occurs when 2-bromopropane is treated with hydroxide ion?

The carbon to which the tertiary amine is attached is designated as the α-carbon, so the adjacent carbon, from which the proton is removed, is called the β-carbon. If the quaternary ammonium ion has more than one β-carbon, the major alkene product is the one obtained by removing a proton from the β-carbon bonded to the most hydrogens. In the following reaction, the major alkene product is obtained by removing a hydrogen from the β-carbon bonded to three hydrogens, and the minor alkene product results from removing a hydrogen from the β-carbon bonded to two hydrogens.

$$\boxed{\beta\text{-carbon}}\quad\boxed{\beta\text{-carbon}}$$
$$CH_3CHCH_2CH_2CH_3 \;\xrightarrow{\Delta}\; CH_2=CHCH_2CH_2CH_3 \;+\; CH_3CH=CHCH_2CH_3$$
$$\underset{\underset{\underset{\displaystyle CH_3 \;\; HO^-}{|}}{\overset{+}{N}}}{|}CH_3NCH_3$$

1-pentene major product   **2-pentene** minor product

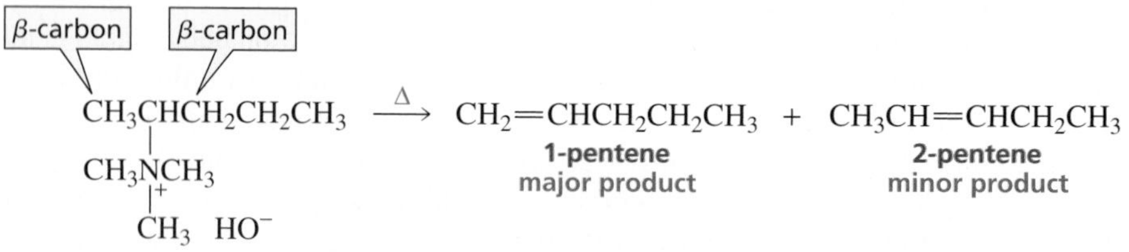

In the next reaction, the major alkene product comes from removing a hydrogen from the β-carbon bonded to two hydrogens because the other β-carbon is bonded to only one hydrogen.

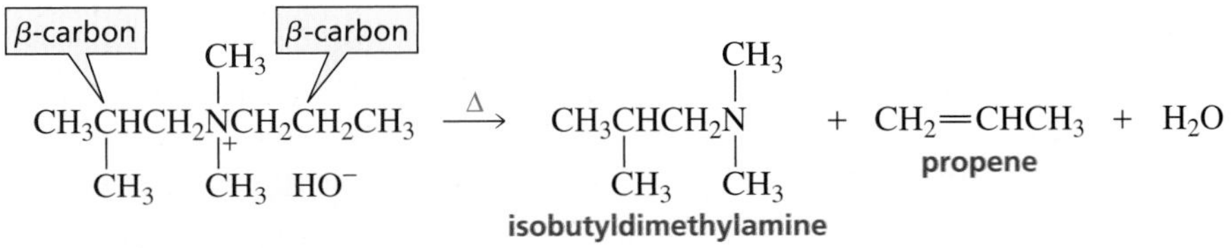

$$\underset{\substack{\text{CH}_3 \\ | \\ \underset{\underset{\text{CH}_3 \;\; \text{CH}_3 \; \text{HO}^-}{|}}{\text{CH}_3\text{CHCH}_2\overset{+}{\text{N}}\text{CH}_2\text{CH}_2\text{CH}_3}}{\boxed{\beta\text{-carbon}} \qquad \boxed{\beta\text{-carbon}}} \xrightarrow{\Delta} \underset{\substack{| \qquad | \\ \text{CH}_3 \;\; \text{CH}_3 \\ \textbf{isobutyldimethylamine}}}{\text{CH}_3\text{CHCH}_2\overset{\text{CH}_3}{\text{N}}} + \underset{\textbf{propene}}{\text{CH}_2{=}\text{CHCH}_3} + \text{H}_2\text{O}$$

PROBLEM 31 ◆

What are the minor products in the preceding Hofmann elimination reaction?

We saw that in an E2 reaction of an alkyl halide, a hydrogen is removed from the β-carbon bonded to the *fewest* hydrogens (Zaitsev's rule). Now we see that in an E2 reaction of a quaternary ammonium ion, the hydrogen is removed from the β-carbon bonded to the *most* hydrogens (anti-Zaitsev elimination).

Why do alkyl halides follow Zaitsev's rule, while quaternary amines violate the rule? When hydroxide ion starts to remove a proton from an alkyl halide, the leaving group immediately starts to depart and a transition state with an *alkene-like* structure results. The proton is removed from the β-carbon bonded to the fewest hydrogens in order to achieve the most stable alkene-like transition state. However, when hydroxide ion starts to remove a proton from a quaternary ammonium ion, the leaving group does not immediately start to leave because a tertiary amine is not as good a leaving group as Cl^-, Br^-, or I^-. As a result, a partial negative charge builds up on the carbon from which the proton is being removed. This gives the transition state a *carbanion-like* structure rather than an alkene-like structure. By removing a proton from the β-carbon with the most hydrogens, the most stable carbanion-like transition state is achieved. Recall from Section 10.2 that primary carbanions are more stable than secondary carbanions, which are more stable than tertiary carbanions. Steric factors in the Hofmann reaction also favor anti-Zaitsev elimination.

Zaitsev elimination "alkene-like" transition state		anti-Zaitsev elimination "carbanion-like" transition state	

$$\underset{\substack{| \\ \underset{\text{Br}}{\overset{\delta-}{}}}}{\overset{\overset{\delta-}{\text{OH}}}{\overset{|}{\underset{|}{\text{H}}}}} \qquad \text{CH}_3\text{CH}{=\!=}\text{CHCH}_3$$

CH₃CH===CHCH₃   CH₃CH₂C===CH₂   CH₂CHCH₂CH₂CH₃   CH₃CHCHCH₂CH₃

more stable   less stable   more stable   less stable

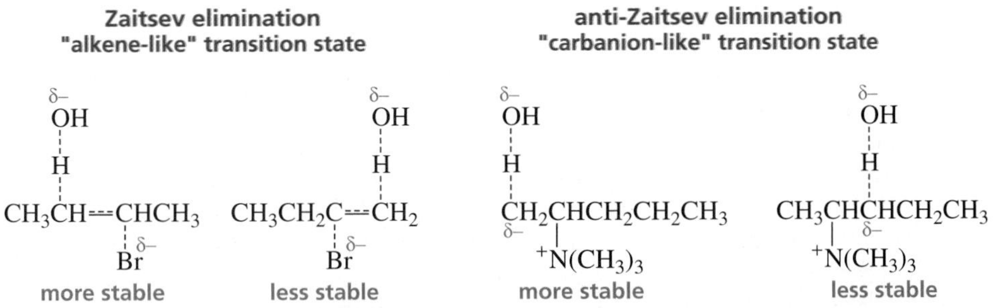

Because the Hofmann elimination reaction occurs in an anti-Zaitsev manner, anti-Zaitsev elimination is also referred to as **Hofmann elimination.** We have previously seen anti-Zaitsev elimination in the E2 reactions of alkyl fluorides as a result of fluoride ion being a poorer leaving group than chloride, bromide, or iodide ions. The poor leaving group causes the transition state to have a carbanion-like structure rather than an alkene-like structure (Section 10.2).

PROBLEM 32 ◆

Give the major products of each of the following reactions:

a. 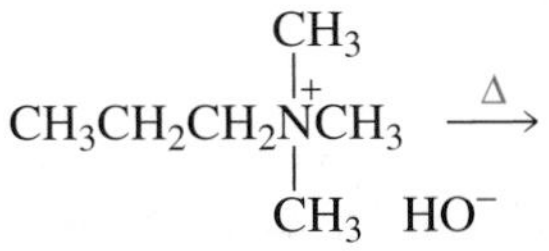 $\underset{\substack{| \\ \text{CH}_3 \;\; \text{HO}^-}}{\text{CH}_3\text{CH}_2\text{CH}_2\overset{\overset{\text{CH}_3}{|}}{\underset{}{\text{N}}}\text{CH}_3} \xrightarrow{\Delta}$

b. 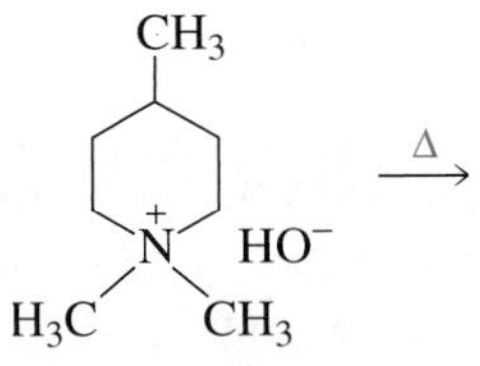 $\xrightarrow{\Delta}$

c. 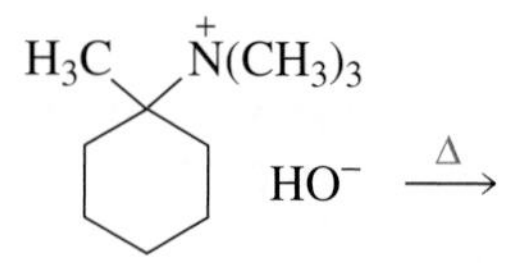 (structure: cyclohexane ring with H_3C and $\overset{+}{N}(CH_3)_3$ substituents, HO^-, $\xrightarrow{\Delta}$)

d. 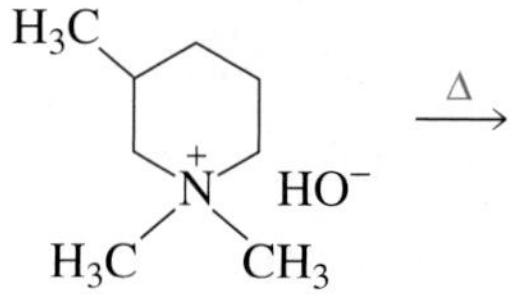 (structure: piperidine ring with H_3C, N^+ bearing H_3C and CH_3, HO^-, $\xrightarrow{\Delta}$)

In order for a quaternary ammonium ion to undergo an elimination reaction, the counterion must be hydroxide ion because a strong base is needed to start the reaction by removing a proton from a β-carbon. Because halide ions are weak bases, quaternary ammonium halides cannot undergo Hofmann eliminations. However, a quaternary ammonium halide can be converted into a quaternary ammonium hydroxide by treating it with silver oxide and water. The silver halide precipitates, and the halide ion is replaced by hydroxide ion. The compound can now undergo elimination.

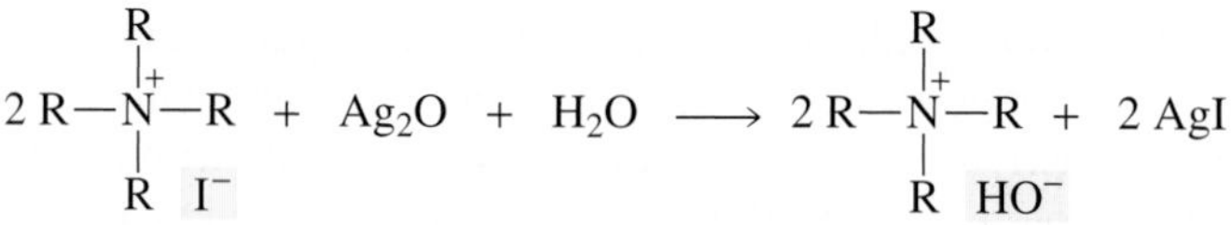

$$2\ R\overset{\overset{R}{|}}{\underset{\underset{R\ \ I^-}{|}}{N^+}}R\ +\ Ag_2O\ +\ H_2O\ \longrightarrow\ 2\ R\overset{\overset{R}{|}}{\underset{\underset{R\ \ HO^-}{|}}{N^+}}R\ +\ 2\ AgI\downarrow$$

The reaction of an amine with sufficient methyl iodide to convert the amine into a quaternary ammonium iodide is called **exhaustive methylation** (See Problem 8 in Section 9.4).

exhaustive methylation

$$CH_3CH_2CH_2NH_2\ +\ \underset{\text{excess}}{CH_3I}\ \xrightarrow{K_2CO_3}\ CH_3CH_2CH_2\overset{\overset{CH_3}{|}}{\underset{\underset{CH_3\ \ I^-}{|}}{N^+}}CH_3$$

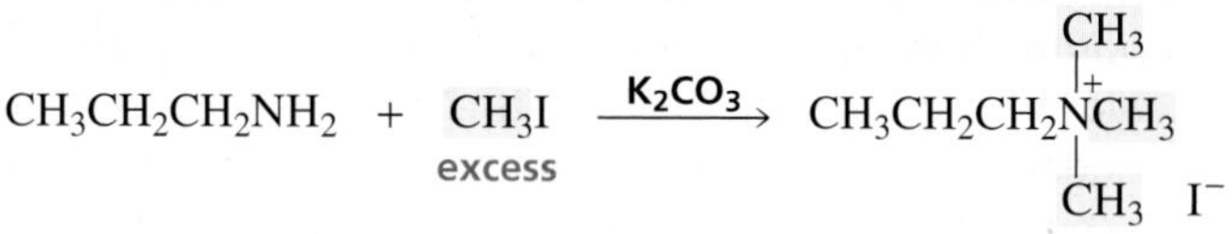

A USEFUL BAD-TASTING COMPOUND

Several practical uses have been found for Bitrex, a quaternary ammonium salt, because it is one of the most bitter-tasting substances known and it is nontoxic. It is put on bait to encourage deer to look elsewhere for their food, it is put on the backs of animals to keep them from biting one another, it is put on childrens' fingers to persuade them to stop sucking their thumbs or biting their fingernails, and it is added to toxic substances to keep them from being injested accidentally.

(structure of **Bitrex**)

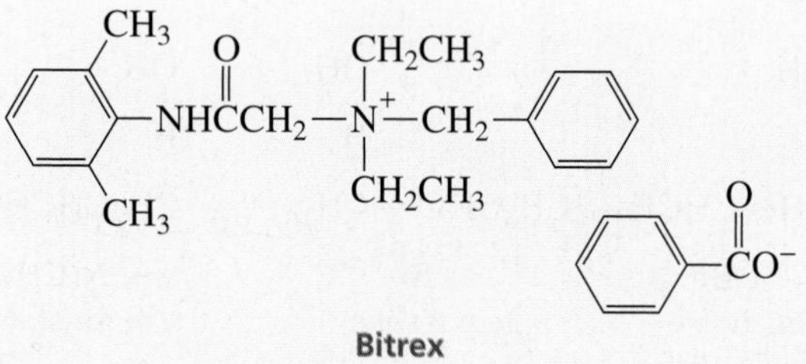

Bitrex

PROBLEM 33

Propose a mechanism for the following reaction:

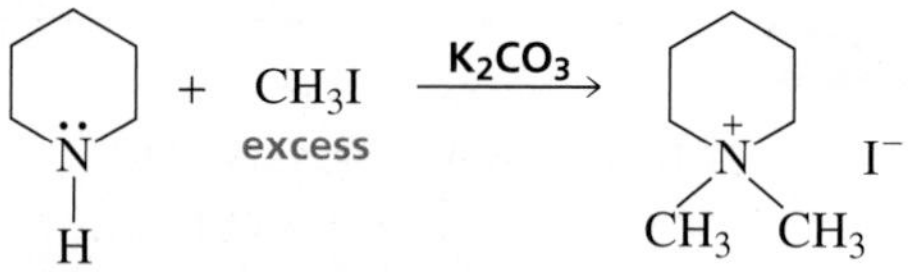

(piperidine, N–H) $+\ \underset{\text{excess}}{CH_3I}\ \xrightarrow{K_2CO_3}$ (N,N-dimethylpiperidinium ring with $CH_3\ \ CH_3$ and I^-)

PROBLEM 34 / SOLVED

Describe a synthesis for each of the following compounds, using the given starting material and any necessary reagents:

a. $CH_3CH_2CH_2CH_2NH_2 \longrightarrow CH_3CH_2CH=CH_2$

b. $CH_3CH_2CH_2\underset{\underset{Br}{|}}{C}HCH_3 \longrightarrow CH_3CH_2CH_2CH=CH_2$

c.

$\longrightarrow CH_2=CH-CH=CH_2$

SOLUTION TO 34a An amine cannot undergo an elimination reaction, but a quaternary ammonium hydroxide can undergo elimination. The amine, therefore, must be converted into a quaternary ammonium hydroxide. Reaction with excess methyl iodide converts the amine into a quaternary ammonium iodide and treatment with aqueous silver oxide forms the quaternary ammonium hydroxide. Heat is required for the elimination reaction.

$$CH_3CH_2CH_2CH_2NH_2 \xrightarrow[\text{K}_2\text{CO}_3]{\overset{\text{CH}_3\text{I}}{\underset{\text{excess}}{}}} CH_3CH_2CH_2CH_2\overset{+}{N}(CH_3)_3 \quad I^- \xrightarrow[\text{H}_2\text{O}]{\text{Ag}_2\text{O}} CH_3CH_2CH_2CH_2\overset{+}{N}(CH_3)_3 \quad HO^-$$

$$\downarrow \Delta$$

$$CH_3CH_2CH=CH_2 \quad + \quad H_2O$$

The Cope Elimination

The **Cope elimination reaction** is similar to the Hofmann elimination reaction. In the Cope elimination, a tertiary amine oxide rather than a quaternary ammonium ion undergoes elimination.

$$CH_3CH_2CH_2\underset{\underset{O^-}{\overset{|}{\overset{+}{|}}}}{\overset{\overset{\displaystyle CH_3}{|}}{N}}CH_3 \xrightarrow{\Delta} CH_3CH=CH_2 \quad + \quad \underset{\underset{OH}{|}}{\overset{\overset{\displaystyle CH_3}{|}}{N}}CH_3$$

a tertiary amine oxide   a hydroxylamine

Arthur C. Cope (1909–1966) was born in Indiana. He received a Ph.D. from the University of Wisconsin and was a professor of chemistry at Bryn Mawr College, Columbia University, and MIT.

The negatively charged oxygen of the amine oxide is the base that removes a proton from the β-carbon. Because the base and the leaving group are in the same molecule, the Cope elimination is an intramolecular E2 reaction. The reaction involves syn elimination.

mechanism of the Cope elimination

$$\underset{CH_3CH}{\overset{CH_2-\overset{\overset{\displaystyle CH_3}{|}}{\overset{+}{N}}-CH_3}{\underset{\underset{H}{}}{\overset{..}{:O:}}}} \xrightarrow{\Delta} CH_3CH=CH_2 \quad + \quad \underset{\underset{:\overset{..}{O}H}{|}}{\overset{\overset{\displaystyle CH_3}{|}}{N}-CH_3}$$

The major product of the Cope elimination, like that of the Hofmann elimination, is the one obtained by removing a hydrogen from the β-carbon bonded to the most hydrogens.

$$CH_3CH_2\underset{\underset{O^-}{\overset{|}{\overset{+}{|}}}}{\overset{\overset{\displaystyle CH_3}{|}}{N}}CH_2CH_2CH_3 \xrightarrow{\Delta} CH_2=CH_2 \quad + \quad \underset{\underset{OH}{|}}{\overset{\overset{\displaystyle CH_3}{|}}{N}}CH_2CH_2CH_3$$

An amine oxide can be synthesized by treating a tertiary amine with hydrogen peroxide.

$$\underset{\substack{\text{a tertiary}\\\text{amine}}}{\overset{\displaystyle R}{\underset{\displaystyle \overset{|}{\underset{\cdot\cdot}{R\!N\!R}}}{}}} + \underset{\substack{\text{hydrogen}\\\text{peroxide}}}{HO\!-\!OH} \longrightarrow \overset{\displaystyle R}{\underset{\displaystyle \overset{\overset{+}{|}}{\underset{\displaystyle \overset{|}{OH}}{R\!N\!R}}}{}} + HO^- \longrightarrow \underset{\substack{\text{an amine}\\\text{oxide}}}{\overset{\displaystyle R}{\underset{\displaystyle \overset{\overset{+}{|}}{\underset{\displaystyle \overset{|}{O^-}}{R\!N\!R}}}{}}} + H_2O$$

PROBLEM 35 ◆

Does the Cope elimination have an alkene-like transition state or a carbanion-like transition state?

PROBLEM 36 ◆

Give the products that would be obtained by treating the following tertiary amines with hydrogen peroxide followed by heat:

a. CH$_3$NCH$_2$CH$_2$CH$_3$ with CH$_3$ substituent

b. CH$_3$NCH$_2$CH$_2$CH$_3$ with phenyl group

c. CH$_3$CH$_2$NCH$_2$CHCH$_3$ with CH$_3$ substituents

d. piperidine ring with N-CH$_3$ and 2-CH$_3$

Tutorial: Common terms

11.12 PHASE TRANSFER CATALYSIS

A problem that faces organic chemists in the laboratory is finding a solvent that will dissolve all the reactants. For example, if we want cyanide ion to react with 1-bromohexane, we encounter a problem because sodium cyanide is an ionic compound that is soluble only in water, whereas the alkyl halide is insoluble in water.

$$\underset{\textbf{1-bromohexane}}{CH_3CH_2CH_2CH_2CH_2CH_2Br} + {}^-C\!\equiv\!N \overset{?}{\longrightarrow} CH_3CH_2CH_2CH_2CH_2CH_2C\!\equiv\!N + Br^-$$

If we mix an aqueous solution of sodium cyanide with a solution of 1-bromohexane in a nonpolar solvent, the two solutions will form two layers because they are immiscible. How, then, can sodium cyanide react with the alkyl halide? The two compounds will be able to react with each other if a catalytic amount of a phase transfer catalyst is added to the reaction mixture.

$$CH_3CH_2CH_2CH_2CH_2CH_2Br + {}^-C\!\equiv\!N \overset{\overset{\text{phase transfer}}{\text{catalyst}}}{\underset{R_4\overset{+}{N}\ HSO_4^-}{\longrightarrow}} CH_3CH_2CH_2CH_2CH_2CH_2C\!\equiv\!N + Br^-$$

Quaternary ammonium salts are the most common phase transfer catalysts. However, we saw in Section 11.9 that crown ethers can also be used as phase transfer catalysts.

phase transfer catalysts

$$
\underset{\substack{\text{tetrabutylammonium} \\ \text{hydrogen sulfate}}}{\overset{\displaystyle \overset{\text{CH}_2\text{CH}_2\text{CH}_2\text{CH}_3}{\underset{\underset{\text{CH}_2\text{CH}_2\text{CH}_2\text{CH}_3}{|}}{\overset{|+}{\text{CH}_3\text{CH}_2\text{CH}_2\text{CH}_2\overset{}{\text{N}}\text{CH}_2\text{CH}_2\text{CH}_2\text{CH}_3}}}}{\text{HSO}_4^-}}
\qquad
\underset{\substack{\text{hexadecyltrimethylammonium} \\ \text{hydrogen sulfate}}}{\overset{\displaystyle \overset{\text{CH}_3}{\underset{\underset{\text{CH}_3}{|}}{\overset{|+}{\text{CH}_3(\text{CH}_2)_{14}\text{CH}_2\overset{}{\text{N}}\text{CH}_3}}}}{\text{HSO}_4^-}}
\qquad
\underset{\substack{\text{benzyltriethylammonium} \\ \text{hydrogen sulfate}}}{\overset{\displaystyle \overset{\text{CH}_2\text{CH}_3}{\underset{\underset{\text{CH}_2\text{CH}_3}{|}}{\overset{|+}{\text{—CH}_2\overset{}{\text{N}}\text{CH}_2\text{CH}_3}}}}{\text{HSO}_4^-}}
$$

How does addition of a phase transfer catalyst allow the reaction of cyanide ion with 1-bromohexane to take place? The quaternary ammonium salt, because of its nonpolar alkyl groups, is soluble in nonpolar solvents, but it is also soluble in water because of its charge. This means that it can act as a mediator between two immiscible solvents. When a phase transfer catalyst such as tetrabutylammonium hydrogen sulfate passes into the nonpolar layer, it must carry a counterion with it in order to balance its positive charge. The counterion can be either its original counterion (hydrogen sulfate) or another ion present in the solution (in the reaction under discussion, it will be cyanide ion). Because there is more cyanide ion than hydrogen sulfate ion in the aqueous layer, cyanide ion will more often be the accompanying ion. Once it is in the nonpolar layer, cyanide ion can react with the alkyl halide. (When hydrogen sulfate is transported into the nonpolar layer, it is unreactive because it is both a weak base and a poor nucleophile.) The quaternary ammonium ion will pass back into the aqueous layer carrying with it, as a counterion, either hydrogen sulfate or bromide ion. The reaction continues with the phase transfer catalyst shuttling back and forth between the two phases. **Phase transfer catalysis** has been successfully used in a wide variety of organic reactions.

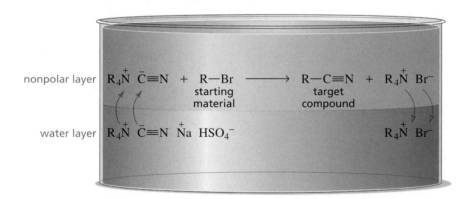

SUMMARY OF REACTIONS

1. Conversion of an *alcohol* to an *alkyl halide* (Sections 11.1 and 11.2)

$$\text{ROH} + \text{HBr} \xrightarrow{\Delta} \text{RBr}$$

$$\text{ROH} + \text{HI} \xrightarrow{\Delta} \text{RI}$$

$$\text{ROH} + \text{HCl} \xrightarrow[\Delta]{\text{ZnCl}_2} \text{RCl}$$

$$\text{ROH} + \text{PBr}_3 \xrightarrow[\text{pyridine}]{} \text{RBr}$$

$$\text{ROH} + \text{PCl}_3 \xrightarrow[\text{pyridine}]{} \text{RCl}$$

$$\text{ROH} + \text{SOCl}_2 \xrightarrow[\text{pyridine}]{} \text{RCl}$$

2. Conversion of an *alcohol* to a *sulfonate ester* (Section 11.3).

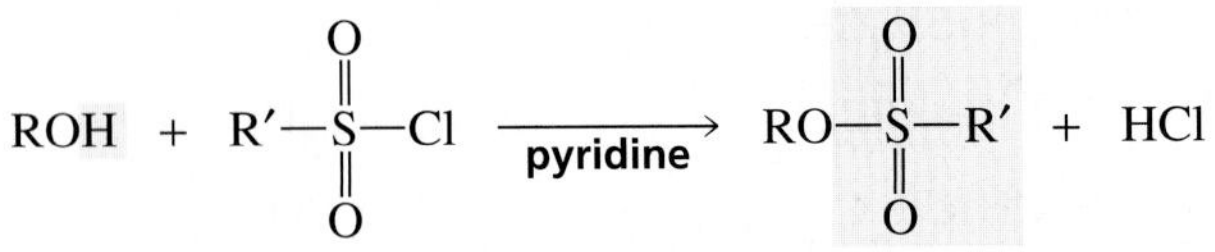

$$\text{ROH} + \text{R'} - \overset{\overset{\text{O}}{\|}}{\underset{\overset{\|}{\text{O}}}{\text{S}}} - \text{Cl} \xrightarrow{\text{pyridine}} \text{RO} - \overset{\overset{\text{O}}{\|}}{\underset{\overset{\|}{\text{O}}}{\text{S}}} - \text{R'} + \text{HCl}$$

3. Conversion of an *activated alcohol* (an alkyl halide or a sulfonate ester) to a *compound with a new group bonded to the* sp³ *carbon* (Section 11.3).

$$\text{RBr} + \text{Y}^- \longrightarrow \text{RY} + \text{Br}^-$$

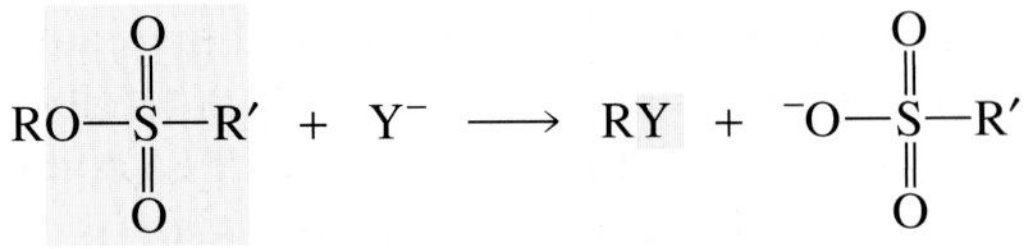

$$\text{RO} - \overset{\overset{\text{O}}{\|}}{\underset{\overset{\|}{\text{O}}}{\text{S}}} - \text{R'} + \text{Y}^- \longrightarrow \text{RY} + {}^-\text{O} - \overset{\overset{\text{O}}{\|}}{\underset{\overset{\|}{\text{O}}}{\text{S}}} - \text{R'}$$

4. Dehydration of *alcohols* (Section 11.4).

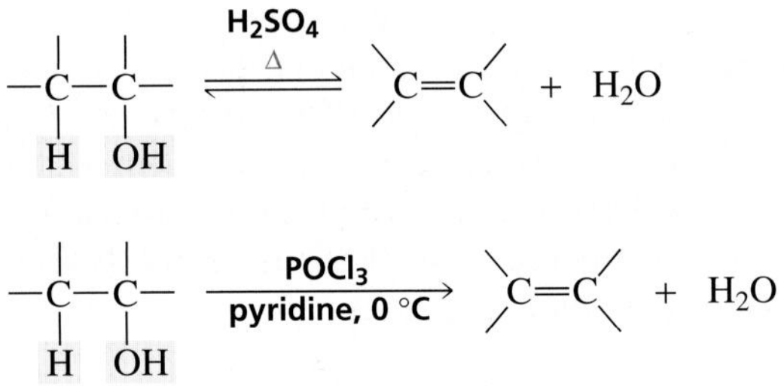

$$-\overset{|}{\underset{|}{\text{C}}}-\overset{|}{\underset{|}{\text{C}}}- \underset{\text{H}\quad\text{OH}}{} \overset{\text{H}_2\text{SO}_4}{\underset{\Delta}{\rightleftharpoons}} \text{C}=\text{C} + \text{H}_2\text{O}$$

$$-\overset{|}{\underset{|}{\text{C}}}-\overset{|}{\underset{|}{\text{C}}}- \underset{\text{H}\quad\text{OH}}{} \xrightarrow[\text{pyridine, 0 °C}]{\text{POCl}_3} \text{C}=\text{C} + \text{H}_2\text{O}$$

relative rate: tertiary > secondary > primary

5. Cleavage of *ethers* (Section 11.5).

$$\text{ROR'} + \text{HX} \xrightarrow{\Delta} \text{ROH} + \text{R'X}$$

HX = HBr or HI

6. Ring-opening reactions of *epoxides* (Section 11.6)

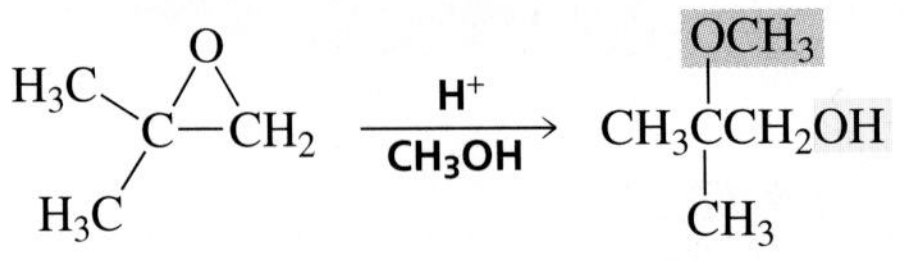

under acidic conditions, the nucleophile attacks the most substituted carbon

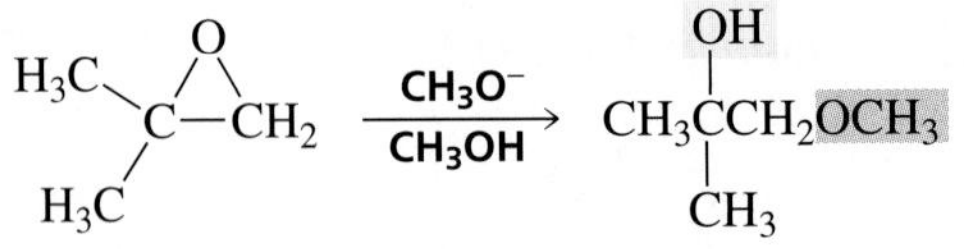

under basic conditions, the nucleophile attacks the least sterically hindered carbon

7. Reactions of *arene oxides:* ring opening and rearrangement (Section 11.7).

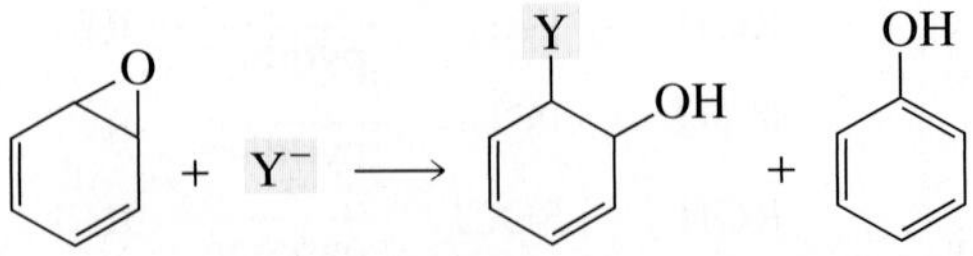

8. Reaction of a *Grignard reagent* with an *epoxide* (Section 11.8).

$$RBr \xrightarrow[\text{ether}]{\textbf{Mg}} RMgBr$$

a Grignard reagent

$$RMgBr + H_2C\overset{O}{-}CH_2 \longrightarrow RCH_2CH_2O^- \xrightarrow{H^+} RCH_2CH_2OH$$

product alcohol contains two more carbons than the Grignard reagent

9. Reaction of a *Gilman reagent* with an alkyl halide (Section 11.8).

$$2\ RLi + CuI \xrightarrow{\textbf{ether}} R_2CuLi + LiI$$

Gilman reagent

$$CH_3CH_2CH_2X + R_2CuLi \xrightarrow{\textbf{ether}} CH_3CH_2CH_2R + RCu + LiX$$

X = Cl, Br, or I

10. Reactions of *thiols, sulfides,* and *sulfonium salts* (Section 11.10).

$$2\ RSH + Hg^{2+} \longrightarrow RS-Hg-SR + 2\ H^+$$

$$RS^- + R'-Br \longrightarrow RSR' + Br^-$$

$$RSR + R'I \longrightarrow R\overset{R'}{\underset{+}{S}}R\ I^-$$

$$R\overset{R}{\underset{+}{S}}R + Y^- \longrightarrow RY + RSR$$

11. Elimination reactions of *quaternary ammonium hydroxides* or *tertiary amine oxides* (Section 11.11).

$$RCH_2CH_2\overset{CH_3}{\underset{\underset{CH_3}{HO^-}}{\overset{+}{N}}}CH_3 \xrightarrow[\textbf{Hofmann elimination}]{\Delta} RCH=CH_2 + \overset{CH_3}{\underset{CH_3}{N}}CH_3 + H_2O$$

$$RCH_2CH_2\overset{CH_3}{\underset{}{N}}CH_3 \xrightarrow{\textbf{H}_2\textbf{O}_2} RCH_2CH_2\overset{CH_3}{\underset{O^-}{\overset{+}{N}}}CH_3 \xrightarrow[\textbf{Cope elimination}]{\Delta} RCH=CH_2 + \overset{CH_3}{\underset{OH}{N}}CH_3$$

in both eliminations, the proton is removed from the β-carbon bonded to the most hydrogens

KEY TERMS

alcohol (page 430)
alkyl tosylate (page 435)
antibiotic (page 460)
arene oxide (page 449)
Cope elimination reaction (page 467)
coupling reaction (page 458)
crown ether (page 459)
crown–guest complex (page 459)
cryptand (page 461)
cryptate (page 461)
dehydration (page 436)
epoxide (page 445)
ether (page 443)
exhaustive methylation (page 466)

Gilman reagent (page 458)
Grignard reagent (page 455)
Hofmann elimination (page 465)
Hofmann elimination reaction (page 464)
inclusion compound (page 459)
ionophore (page 460)
mercapto group (page 461)
molecular recognition (page 460)
organocuprates (page 458)
organolithium compound (page 455)
organomagnesium compound (page 455)
organometallic compound (page 454)

oxirane (page 445)
phase transfer catalysis (page 469)
phase transfer catalyst (page 460)
pinacol rearrangement (page 477)
quaternary ammonium ion (page 464)
ring-expansion rearrangement (page 438)
sulfide (page 462)
sulfonate ester (page 435)
sulfonium salt (page 462)
thioether (page 462)
thiol (page 461)
transmetallation (page 458)
vicinal diol (page 477)

PROBLEMS

37. Give the product of each of the following reactions:

a. $CH_3CH_2CH_2OH$ $\xrightarrow[\substack{\text{2. } CH_3\overset{\displaystyle O}{\overset{\|}{C}}O^-}]{\text{1. methanesulfonyl chloride}}$

b. $CH_3CH_2CH_2CH_2OH$ + PBr_3 $\xrightarrow{\text{pyridine}}$

c. 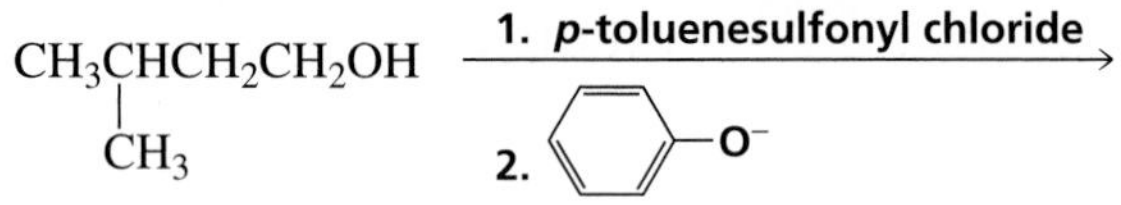 $\underset{\substack{|\\ CH_3}}{CH_3CHCH_2CH_2OH}$ $\xrightarrow[\substack{\text{2. } \langle\text{benzene}\rangle-O^-}]{\text{1. } p\text{-toluenesulfonyl chloride}}$

d. 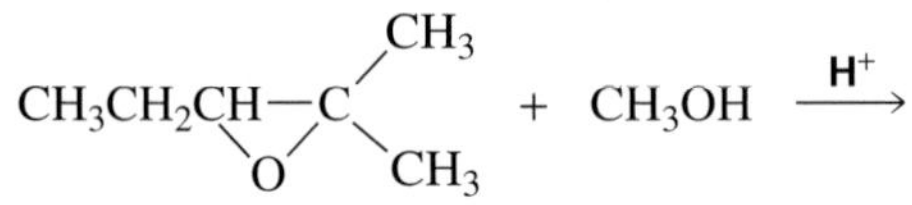 $CH_3CH_2CH-\underset{\substack{|\\O}}{\overset{\substack{CH_3\\|}}{C}}{\underset{CH_3}{}}$ + CH_3OH $\xrightarrow{H^+}$

e. 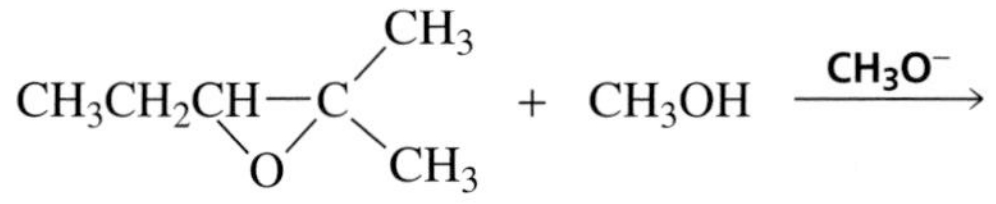 $CH_3CH_2CH-\underset{\substack{|\\O}}{\overset{\substack{CH_3\\|}}{C}}{\underset{CH_3}{}}$ + CH_3OH $\xrightarrow{CH_3O^-}$

f. $CH_3CH_2CH_2CH_2OH$ $\xrightarrow[\Delta]{H_2SO_4}$

g. 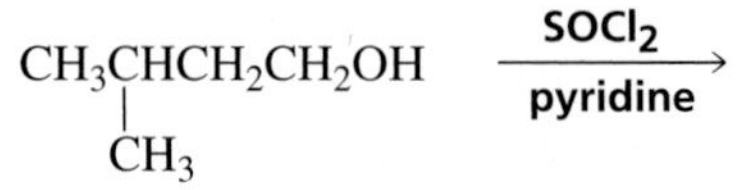 $\underset{\substack{|\\ CH_3}}{CH_3CHCH_2CH_2OH}$ $\xrightarrow[\text{pyridine}]{SOCl_2}$

h. 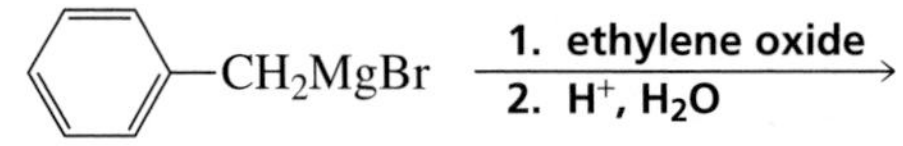 $\langle\text{benzene}\rangle-CH_2MgBr$ $\xrightarrow[\text{2. } H^+, H_2O]{\text{1. ethylene oxide}}$

i. 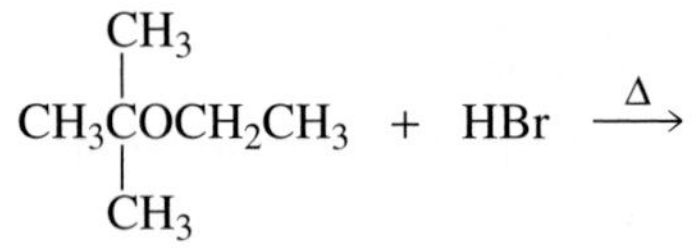 $\underset{\substack{|\\ CH_3}}{\overset{\substack{CH_3\\|}}{CH_3COCH_2CH_3}}$ + HBr $\xrightarrow{\Delta}$

j. 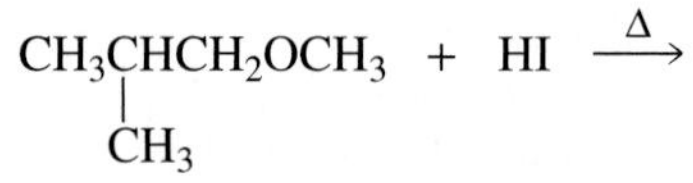 $\underset{\substack{|\\ CH_3}}{CH_3CHCH_2OCH_3}$ + HI $\xrightarrow{\Delta}$

k. 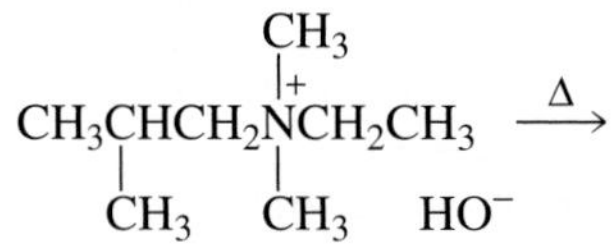 $\underset{\substack{|\\ CH_3}}{CH_3CHCH_2}\overset{\substack{CH_3\\|+}}{\underset{\substack{|\\ CH_3}}{N}}CH_2CH_3$ $\underset{HO^-}{\xrightarrow{\Delta}}$

l. 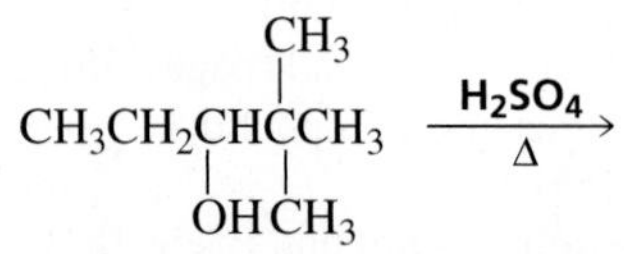 $\underset{\substack{|\\ OHCH_3}}{CH_3CH_2CH}\overset{\substack{CH_3\\|}}{C}CH_3$ $\xrightarrow[\Delta]{H_2SO_4}$

m. 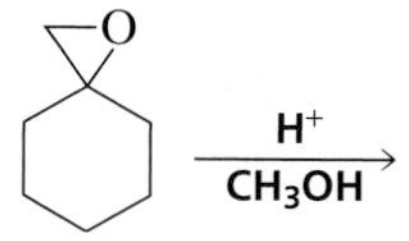 $\xrightarrow[\text{CH}_3\text{OH}]{\text{H}^+}$

n. 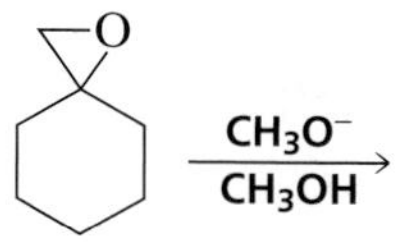 $\xrightarrow[\text{CH}_3\text{OH}]{\text{CH}_3\text{O}^-}$

38. Indicate which alcohol, when heated with H_2SO_4, will undergo dehydration more rapidly.

a. 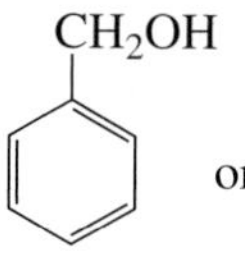 or

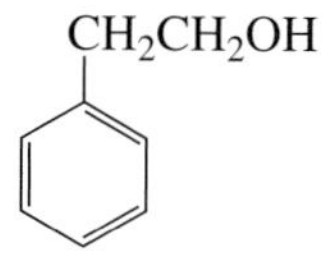

d. 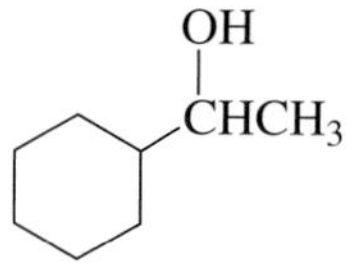 or

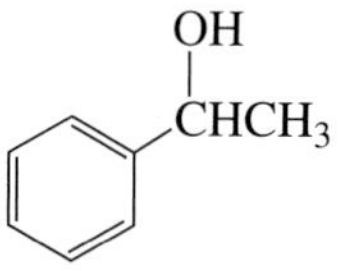

b. 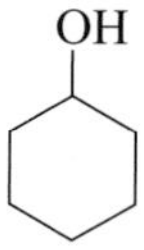 or

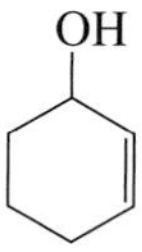

e. 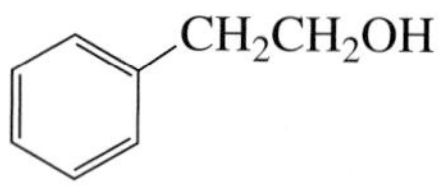 or

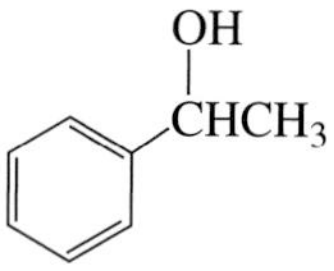

c. 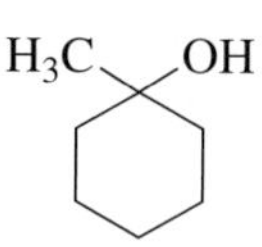 or

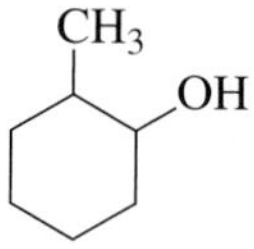

f. 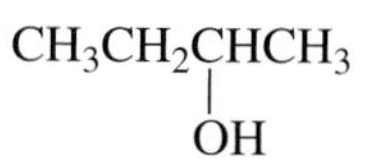 $CH_3CH_2\underset{\underset{OH}{|}}{C}HCH_3$ or $CH_3\underset{\underset{OH}{|}}{\overset{\overset{CH_3}{|}}{C}}CH_2CH_3$

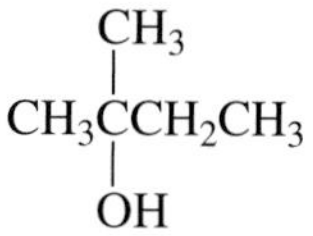

39. Which of the following alkyl halides could be successfully used to form a Grignard reagent?

a. $HOCH_2CH_2CH_2CH_2Br$

c. 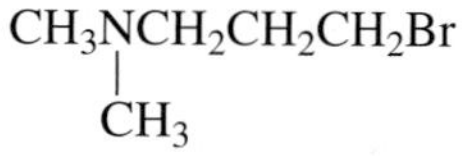 $CH_3\underset{\underset{CH_3}{|}}{N}CH_2CH_2CH_2Br$

b.  $BrCH_2CH_2CH_2\overset{\overset{O}{||}}{C}OH$

d. $H_2NCH_2CH_2CH_2Br$

40. Starting with (*R*)-1-deuterio-1-propanol, how could you prepare:
   a. (*S*)-1-deuterio-1-propanol?   **c.** (*R*)-1-deuterio-1-methoxypropane?
   b. (*S*)-1-deuterio-1-methoxypropane?

41. What alkenes would you expect to be obtained from the acid-catalyzed dehydration of 1-hexanol?

42. When heated with H_2SO_4, both 3,3-dimethyl-2-butanol and 2,3-dimethyl-2-butanol are dehydrated to form 2,3-dimethyl-2-butene. Which alcohol dehydrates more rapidly?

43. Give the product of each of the following reactions.

a. 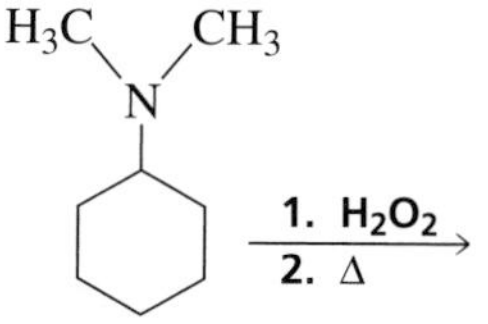 $\xrightarrow[\text{2. }\Delta]{\text{1. H}_2\text{O}_2}$

b. 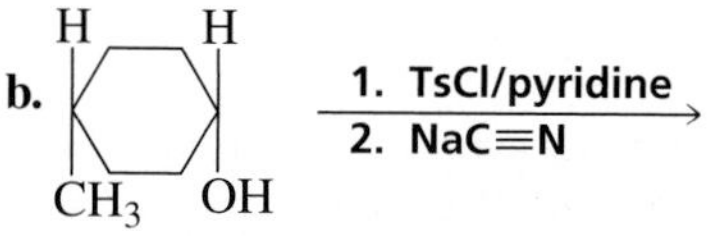 $\xrightarrow[\text{2. NaC}\equiv\text{N}]{\text{1. TsCl/pyridine}}$

c. 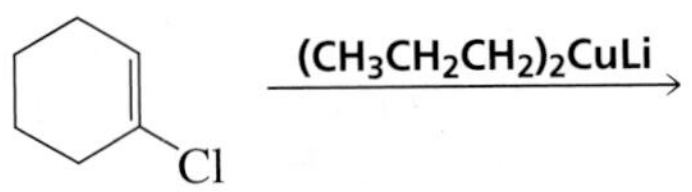 $\xrightarrow{(CH_3CH_2CH_2)_2CuLi}$

d. 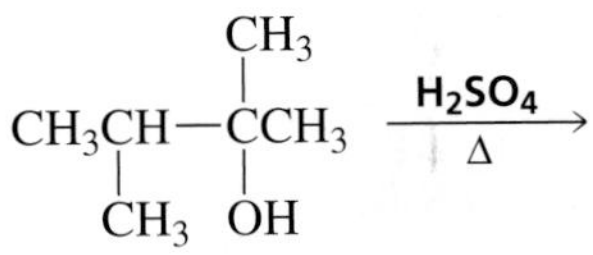 $CH_3\overset{\displaystyle CH_3}{\underset{\displaystyle CH_3}{\underset{\displaystyle OH}{CH-C}}}CH_3 \xrightarrow[\Delta]{H_2SO_4}$

44. Using the given starting material, any necessary inorganic reagents, and any carbon-containing compounds with no more than two carbon atoms, indicate how the following syntheses could be carried out:

a. 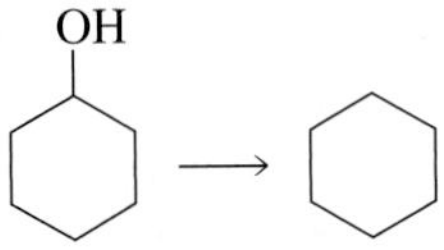 $\longrightarrow$

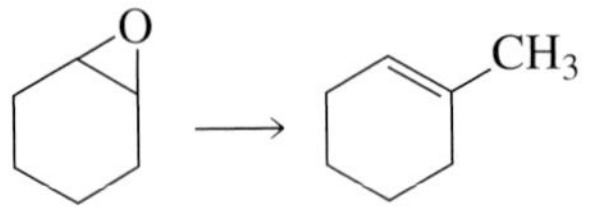

c. 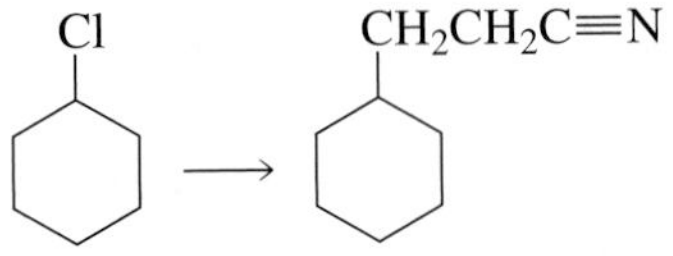 $\longrightarrow$

b. $\longrightarrow$

d. $CH_3CH_2C\equiv CH \longrightarrow CH_3CH_2C\equiv CCH_2CH_2OH$

e. 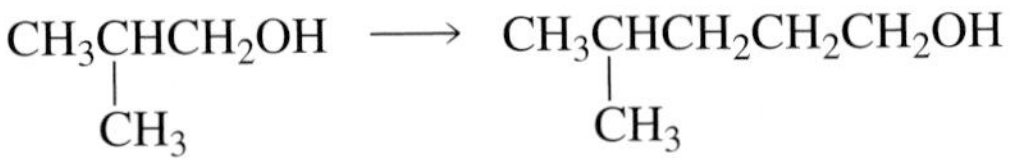 $CH_3\underset{\displaystyle CH_3}{CHCH_2OH} \longrightarrow CH_3\underset{\displaystyle CH_3}{CHCH_2CH_2CH_2OH}$

45. Propose a mechanism for the following reaction:

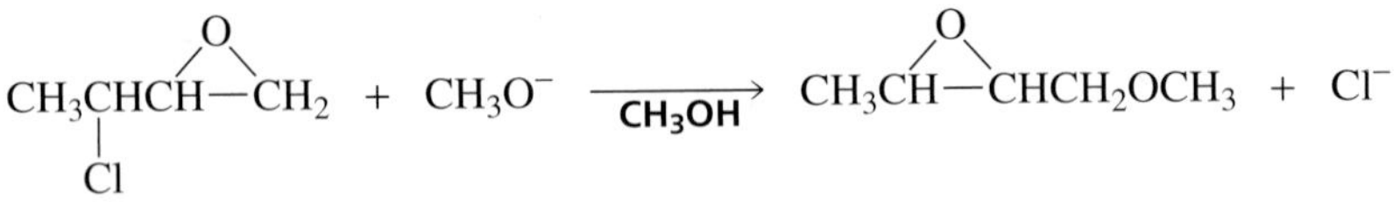

$$CH_3CH\overset{\displaystyle O}{\overset{\diagup\!\diagdown}{CH}}{-}CH_2 \; + \; CH_3O^- \xrightarrow{CH_3OH} CH_3\overset{\displaystyle O}{\overset{\diagup\!\diagdown}{CH}}{-}CHCH_2OCH_3 \; + \; Cl^-$$
(with Cl on the first carbon)

46. When deuterated phenanthrene oxide undergoes an epoxide rearrangement in water, 81% of the deuterium is retained in the product (*J. Am. Chem. Soc.,* 1976, *98,* 2965).

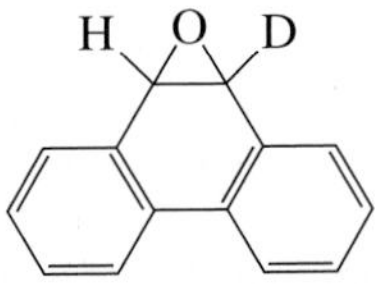

a. What percentage of the deuterium will be retained if an NIH shift occurs?
b. What percentage of the deuterium will be retained if an NIH shift does not occur?

47. When 3-methyl-2-butanol is heated with concentrated HBr, a rearranged product is obtained. When 2-methyl-1-propanol reacts under the same conditions, a rearranged product is not obtained. Explain.

48. When the following seven-membered ring alcohol is dehydrated, three alkenes are formed. Propose a mechanism for their formation.

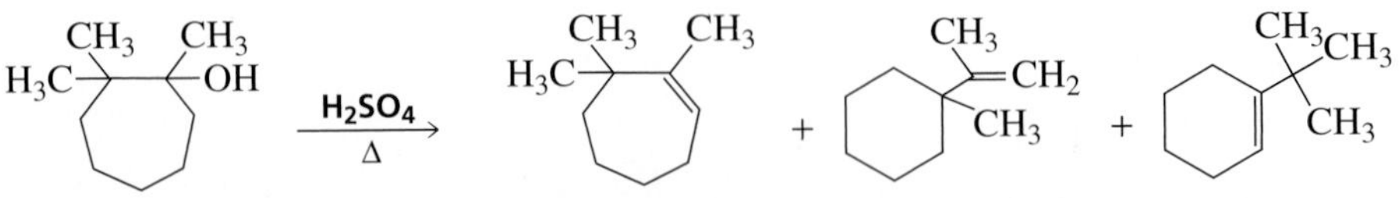

$\xrightarrow[\Delta]{H_2SO_4}$ + +

49. How could you synthesize isopropyl propyl ether using isopropyl alcohol as the only carbon-containing reagent?

50. When piperidine undergoes the indicated series of reactions, 1,4-pentadiene is obtained as the product. When the four different methyl-substituted piperidines undergo the same series of reactions, each forms a different diene. The dienes obtained are 1,5-hexadiene, 1,4-pentadiene, 2-methyl-1,4-pentadiene, and 3-methyl-1,4-pentadiene. Which methyl-substituted piperidine yields which diene?

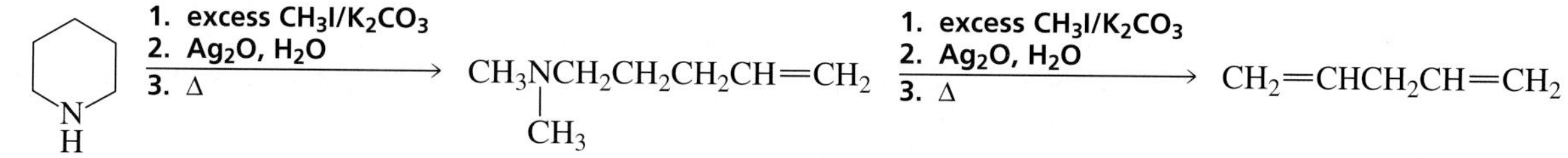

piperidine

$$\text{piperidine} \xrightarrow[\text{3. } \Delta]{\begin{array}{l}\textbf{1. excess CH}_3\textbf{I/K}_2\textbf{CO}_3\\ \textbf{2. Ag}_2\textbf{O, H}_2\textbf{O}\end{array}} CH_3NCH_2CH_2CH_2CH=CH_2 \xrightarrow[\text{3. } \Delta]{\begin{array}{l}\textbf{1. excess CH}_3\textbf{I/K}_2\textbf{CO}_3\\ \textbf{2. Ag}_2\textbf{O, H}_2\textbf{O}\end{array}} CH_2=CHCH_2CH=CH_2$$

$$\overset{|}{CH_3}$$

51. Ethylene oxide reacts with HO^- because of the strain in the three-membered ring. Cyclopropane has approximately the same amount of strain, but it does not react with HO^-. Explain.

52. Which of the following ethers would be obtained in greatest yield directly from alcohols?

$$CH_3OCH_2CH_2CH_3 \qquad CH_3CH_2OCH_2CH_2CH_3 \qquad CH_3CH_2OCH_2CH_3 \qquad \underset{\underset{CH_3}{|}}{\overset{\overset{CH_3}{|}}{CH_3OCCH_3}}$$

53. Propose a mechanism for each of the following reactions:

a. 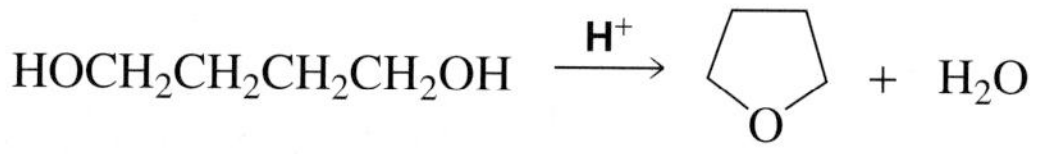 $HOCH_2CH_2CH_2CH_2OH \xrightarrow{H^+}$ [tetrahydrofuran ring] $+ H_2O$

b. 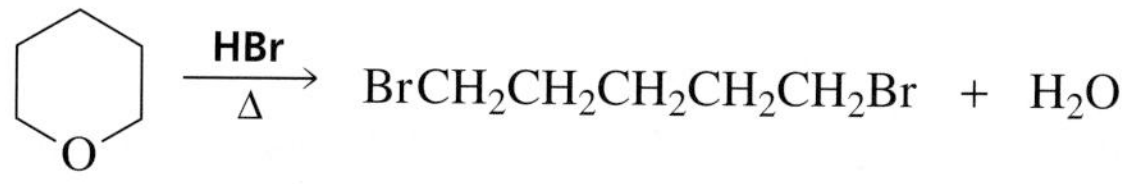 [tetrahydropyran ring] $\xrightarrow[\Delta]{HBr} BrCH_2CH_2CH_2CH_2CH_2Br + H_2O$

54. Indicate how each of the following compounds could be prepared using the given starting material:

a. 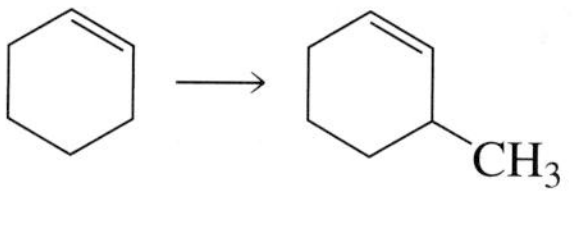 [cyclohexene] $\longrightarrow$ [methylcyclohexene with CH_3]

b. 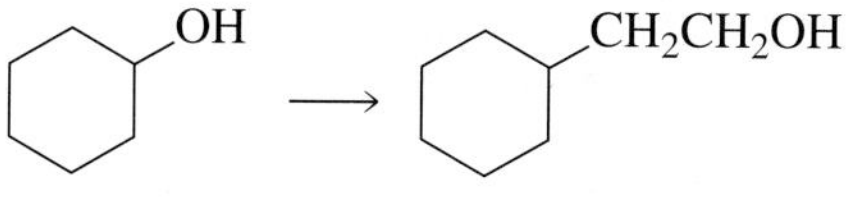 [cyclohexanol, OH] $\longrightarrow$ [cyclohexane with CH_2CH_2OH]

c. $\underset{\underset{OH}{|}}{\overset{\overset{CH_3}{|}}{CH_3CCH_2CH_2CH_3}} \longrightarrow \underset{\underset{Br}{|}}{\overset{\overset{CH_3}{|}}{CH_3CHCHCH_2CH_3}}$

d. [structure with OH] $\longrightarrow$ [structure with D]

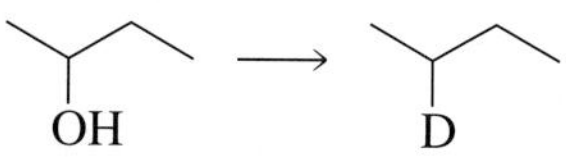

e. $CH_3CH_2CH=CH_2 \longrightarrow CH_3CH_2CH_2CH_2CH_2CH_2CH_2CH_3$

55. Triethylene glycol is one of the products obtained from the reaction of ethylene oxide and hydroxide ion. Propose a mechanism for its formation.

$$H_2C\overset{O}{\overset{\diagup\diagdown}{-}}CH_2 + HO^- \longrightarrow HOCH_2CH_2OCH_2CH_2OCH_2CH_2OH$$
$$\textbf{triethylene glycol}$$

56. Give the major product expected from the reaction of 2-ethyloxirane with each of the following reagents:
    a. 0.1 M HCl   **d.** 0.1 M NaOH
    b. CH_3OH/H^+ **e.** CH_3OH/CH_3O^-
    c. ethyl magnesium bromide in ether followed by 0.1 M HCl

57. When ethyl ether is heated with excess HI for several hours, the only organic product obtained is ethyl iodide. Explain why ethyl alcohol is not obtained as a product.

58. a. Propose a mechanism for the following reaction:

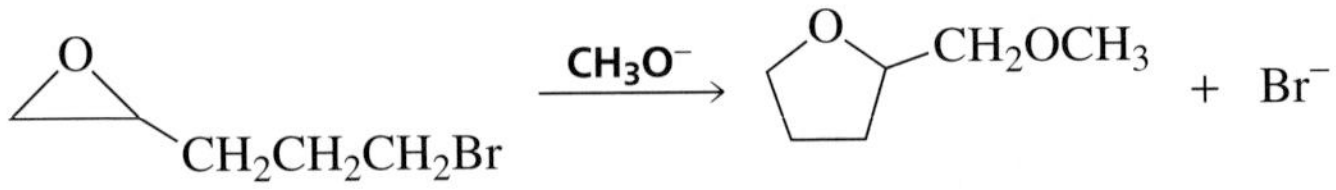

    b. A small amount of a product containing a six-membered ring is also formed. Give the structure of that product.
    c. Why is so little six-membered ring product formed?

59. Identify A through H.

$$CH_3Br \xrightarrow[\textbf{B}]{\textbf{A}} \textbf{C} \xrightarrow[\textbf{2. E}]{\textbf{1. D}} CH_3CH_2CH_2OH \xrightarrow[\substack{\textbf{2. G}\\\textbf{3. H}}]{\textbf{1. F}} CH_3CH_2CH_2OCH_2CH_2OH$$

60. Greg Nard added an equivalent of 3,4-epoxy-4-methylcyclohexanol to an ether solution of methyl magnesium bromide and then added dilute hydrochloric acid. He expected that the product would be a diol. He did not get any of the expected product. What product did he get?

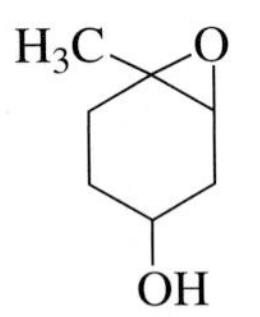

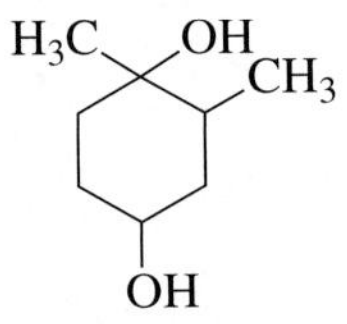

    3,4-epoxy-4-methyl-   **1,2-dimethyl-**
      cyclohexanol   **1,4-cyclohexanediol**

61. An ion with a positively charged nitrogen atom in a three-membered ring is called an aziridinium ion. The following aziridinium ion reacts with sodium methoxide to form A and B. If a small amount of aqueous Br_2 is added to A, the reddish color of Br_2 persists, but the color disappears when Br_2 is added to B. When the aziridinium ion reacts with methanol, only A is formed. Identify A and B.

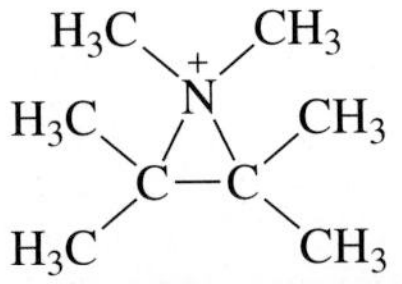

    an aziridinium ion

62. Dimerization is a side reaction that occurs during the preparation of a Grignard reagent. Propose a mechanism that accounts for the formation of the dimer.

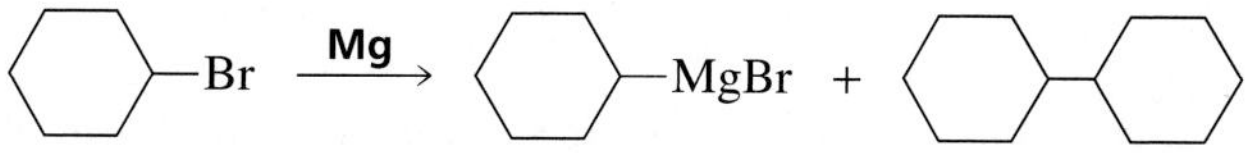

a dimer

63. Propose a mechanism for each of the following reactions:

a.

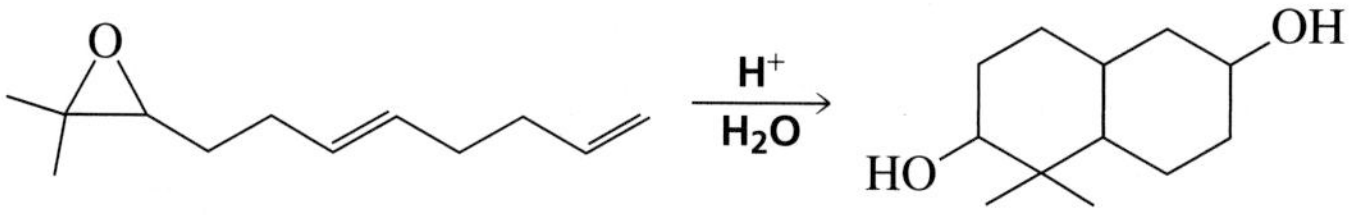

$\xrightarrow[\text{H}_2\text{O}]{\text{H}^+}$

b.

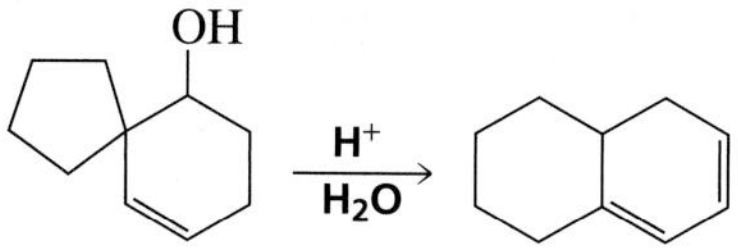

$\xrightarrow[\text{H}_2\text{O}]{\text{H}^+}$

c.

$\xrightarrow[\Delta]{\text{H}_2\text{SO}_4}$

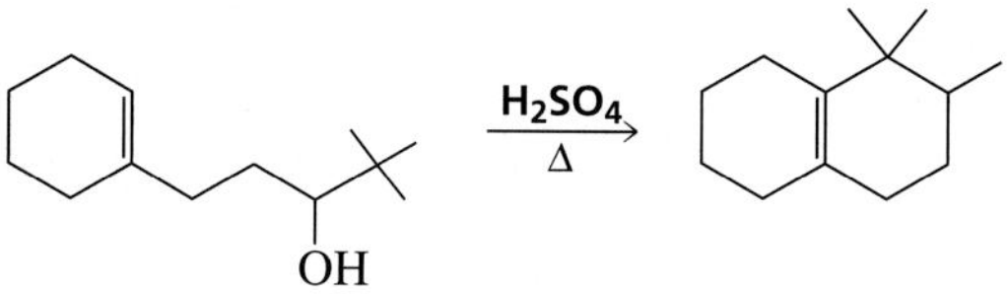

64. One method used for preparing an epoxide is to treat an alkene with an aqueous solution of Br_2, followed by an aqueous solution of sodium hydroxide.
   a. Propose a mechanism for the conversion of cyclohexene into cyclohexene oxide by this method.
   b. How many products are formed when cyclohexene oxide reacts with methoxide ion in methanol? Draw their structures.

65. Which of the following reactions occurs most rapidly? Why?

a.

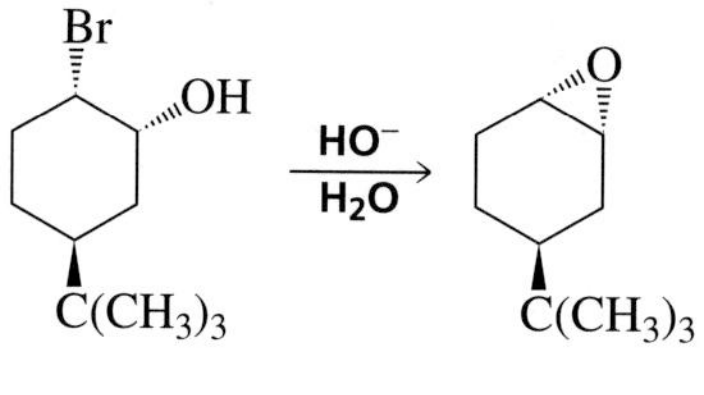

$\xrightarrow[\text{H}_2\text{O}]{\text{HO}^-}$

c.

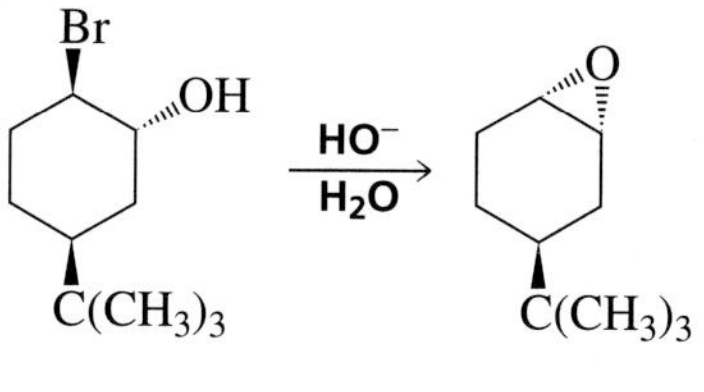

$\xrightarrow[\text{H}_2\text{O}]{\text{HO}^-}$

b.

$\xrightarrow[\text{H}_2\text{O}]{\text{HO}^-}$

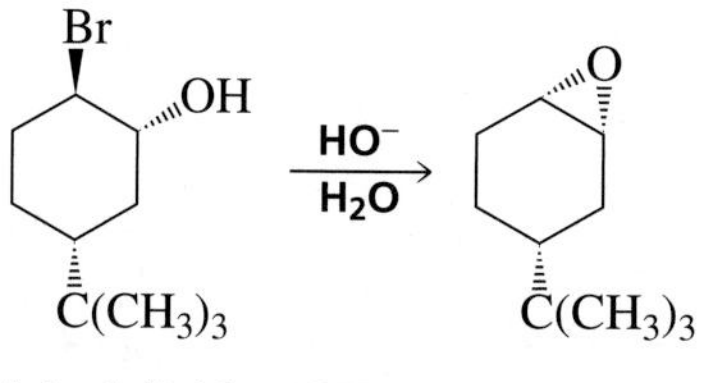

66. A vicinal diol has OH groups on adjacent carbons. The dehydration of a **vicinal diol** is accompanied by a rearrangement called the **pinacol rearrangement.** Propose a mechanism for this reaction.

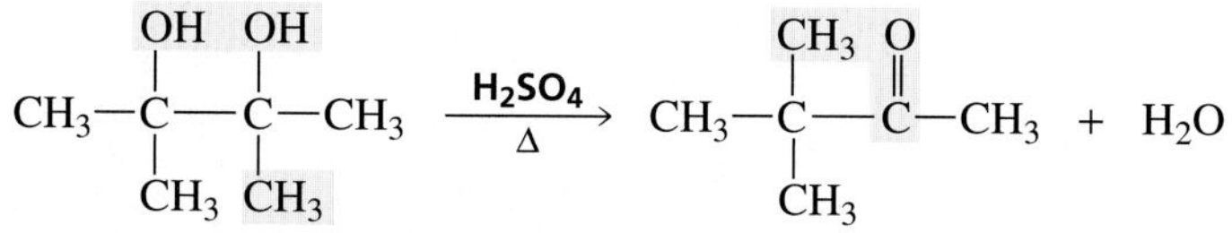

67. Although 2-methyl-1,2-propanediol is an asymmetrical vicinal diol, only one product is obtained when it is dehydrated in the presence of acid.
   a. What is this product?   **b.** Why is only one product formed?

68. What product is obtained when the following vicinal diol is heated in an acidic solution?

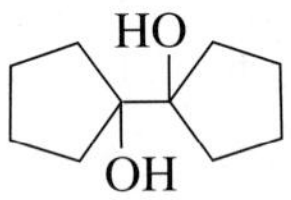

69. Propose a mechanism for each of the following reactions:

a.

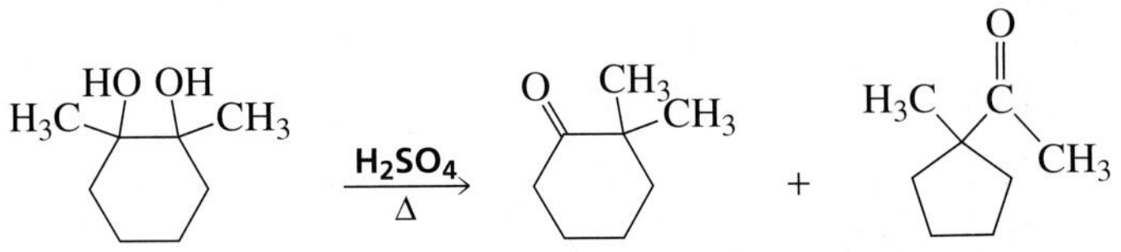

b.

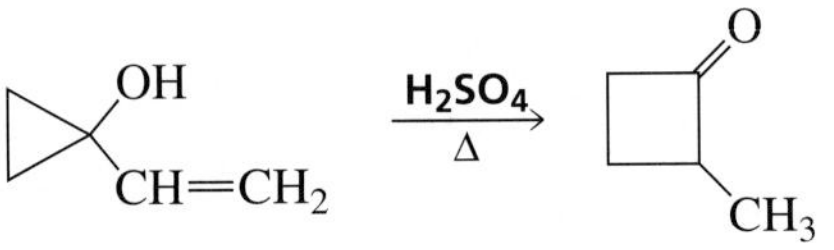

PART

IV

Identification of Organic Compounds

You have now worked through many problems that asked you to design the synthesis of an organic compound. But if you were actually to go into the laboratory to carry out a synthesis you designed, how would you know that the compound you obtained was the one you set out to prepare? Clearly, organic chemists must be able to identify organic compounds. In **Chapters 12 and 13** you will learn about four instrumental techniques that chemists use to identify compounds.

Mass spectrometry is used to determine the molecular weight and the molecular formula of an organic compound as well as certain structural features of the compound.

Infrared (IR) spectroscopy is used to determine the kinds of functional groups in an organic compound.

Ultraviolet/Visible (UV/Vis) spectroscopy gives information about organic compounds with conjugated double bonds.

Nuclear magnetic resonance (NMR) spectroscopy helps to identify the carbon–hydrogen framework of an organic compound.

12

Mass Spectrometry, Infrared Spectroscopy, and Ultraviolet/Visible Spectroscopy

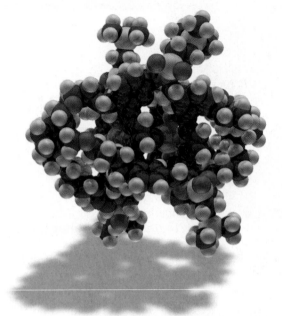

"football porphyrin"

In previous chapters you learned about various methods used to synthesize organic compounds. For example, you saw that an aldehyde is formed when a terminal alkyne undergoes hydroboration-oxidation (Section 5.7). But how do chemists know that the product of that reaction is actually an aldehyde?

Scientists search the world for new compounds with physiological activity. If a promising compound is found, its structure needs to be determined. Without knowing its structure, chemists cannot design ways to synthesize it, nor can they undertake studies to provide insights into its biological behavior.

Clearly, it is essential for chemists to be able to determine the structures of the compounds with which they work. Before a compound can be identified, it must be isolated. For example, if the product of a reaction is to be identified, it must first be isolated from the solvent and from any unreacted starting materials as well as from any side products that might have formed. Any compound found in nature must be isolated from the source from which it was obtained.

Isolating and analyzing products used to be daunting tasks. The only tools chemists had to isolate products were distillation (for liquids) and fractional recrystallization (for solids). Now a variety of chromatographic techniques allow compounds to be isolated relatively easily. You will learn about these when you take a laboratory course in organic chemistry.

At one time, the only methods organic chemists had to analyze products were simple color and solubility tests and physical properties, such as melting points. For example, when an aldehyde is added to a test tube containing a solution of silver oxide in ammonia, a silver mirror is formed on the inside of the test tube. Only aldehydes do this. If a mirror forms, you can conclude that the unknown compound is an aldehyde; if a mirror does not form, you know that the compound is not an aldehyde. An example of a solubility test is the Lucas test, which distinguishes primary, secondary, and tertiary alcohols by how rapidly they turn cloudy after the addition of Lucas' reagent (Section 11.1).

Now, however, compounds are commonly identified by means of instrumental techniques. These techniques can be performed quickly on small amounts of a compound and can provide much more information about the compound's structure than either color or solubility tests can provide. In this chapter we will look at three of these techniques: mass spectrometry, infrared (IR) spectroscopy, and ultraviolet/visible (UV/Vis) spectroscopy. **Mass spectrometry** allows us to determine the *molecular weight* and the *molecular formula* of a compound as well as certain *structural features*. **Infrared spectroscopy** allows us to determine the *kinds of functional groups* in a compound. **UV/Vis spectroscopy** gives us information about organic compounds with conjugated double bonds. Of these instrumental techniques, mass spectrometry is the only one that does not involve electromagnetic radiation. Thus, it is called *spectrometry* while the others are called *spectroscopy.*

We will be referring to different classes of organic compounds as we discuss the various instrumental techniques; these are listed in Table 12.1. They are also listed inside the front cover for easy reference.

TABLE 12.1   Classes of Organic Compounds

Class	Structure		Class	Structure
Alkane	$-\overset{\displaystyle |}{\underset{\displaystyle |}{C}}-$	contains only C—C and C—H bonds	Aldehyde	$\overset{\displaystyle O}{\overset{\displaystyle \|}{RCH}}$
Alkene	$\overset{\diagdown}{\diagup}C{=}C\overset{\diagup}{\diagdown}$		Ketone	$\overset{\displaystyle O}{\overset{\displaystyle \|}{RCR}}$
Alkyne	$-C{\equiv}C-$		Carboxylic acid	$\overset{\displaystyle O}{\overset{\displaystyle \|}{RCOH}}$
Nitrile	$-C{\equiv}N$		Ester	$\overset{\displaystyle O}{\overset{\displaystyle \|}{RCOR}}$
Alkyl halide	RX	where X = F, Cl, Br, or I	Amides	$\overset{\displaystyle O}{\overset{\displaystyle \|}{RCNH_2}}$
Ether	ROR			$\overset{\displaystyle O}{\overset{\displaystyle \|}{RCNHR}}$
Alcohol	ROH			$\overset{\displaystyle O}{\overset{\displaystyle \|}{RCNR_2}}$
Phenol	⬡—OH		Amine (primary)	RNH_2
			Amine (secondary)	R_2NH
Aniline	⬡—NH₂		Amine (tertiary)	R_3N

In mass spectrometry, a sample of a compound is introduced into an instrument called a mass spectrometer, where it is vaporized and then ionized by removing an electron. There are several methods to accomplish this—the most common is bombarding the vaporized molecules with a beam of high-energy electrons. The energy of the electron beam can be varied, but a beam of about 70 eV (electron volts) is commonly used. When the electron beam hits a molecule, it knocks out an electron. This produces a *molecular ion.* A molecular ion is a **radical cation,** a species with an unpaired electron

**12.1
MASS
SPECTROMETRY**

and a positive charge. The symbol $\dot{+}$ indicates that the molecule has lost an electron—it has an unpaired electron and is positively charged.

$$
\underset{\textbf{molecule}}{M} \xrightarrow{\textbf{70 eV}} \underset{\substack{\textbf{molecular ion}\\ \textit{a radical cation}}}{M\dot{+}} + \underset{\textbf{electron}}{e^-}
$$

The electron beam transfers so much energy to the molecules that many of the molecular ions break apart into cations, radicals, neutral molecules, and other radical cations. Not surprisingly, the bonds most likely to break are the weakest bonds and those that break to form the most stable products. All the *positively charged fragments* of the molecule pass between two negatively charged plates, which accelerate the fragments into an analyzer tube (Figure 12.1). Neutral fragments are not attracted to the negatively charged plates and therefore are not accelerated. They are eventually pumped out of the spectrometer.

The analyzer tube is surrounded by a magnet. Its magnetic field deflects the positively charged fragments in a curved path. At a given magnetic field strength, the radius of the path traveled by a fragment depends on its mass-to-charge ratio (m/z). The smaller the m/z of the fragment, the more the magnet deflects it. If a fragment's path matches the curvature of the analyzer tube, the fragment will pass through the tube and out the ion exit slit. A collector records the relative number of fragments with a particular m/z passing through the slit. The more stable the fragment, the more likely it will make it to the collector. By slowly increasing the strength of the magnetic field, fragments with progressively larger values of m/z can be guided through the tube and out the exit slit.

A **mass spectrum** consists of a plot of the relative abundance of each fragment versus its m/z value. Because the charge (z) on essentially all the fragments that

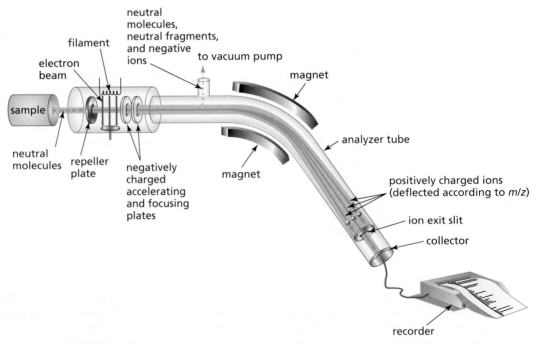

▲ **Figure 12.1**
Schematic of a mass spectrometer.

reach the collector plate is $+1$, m/z is the molecular weight (m) of the fragment. *Remember that only positively charged fragments reach the detector.*

The mass spectrum of pentane is shown in Figure 12.2. Each m/z value is the nominal molecular mass of the fragment. The **nominal molecular mass** is the mass to the nearest whole number. The peak in the spectrum at $m/z = 72$ is due to the fragment that results when an electron is ejected from pentane. It is the **molecular ion** (M) of pentane. The peak in the spectrum with the largest m/z value is the molecular ion, unless all the molecular ions break into fragments before they reach the collector plate. The m/z value of the molecular ion gives the molecular weight of the compound. Peaks with smaller m/z values represent positively charged fragments of the molecule. (For now, ignore the tiny peak at $m/z = 73$. Its significance will be discussed in Section 12.3.)

12.2
THE MASS SPECTRUM. FRAGMENTATION

$$CH_3CH_2CH_2CH_2CH_3 \xrightarrow{\textbf{70 eV}} [CH_3CH_2CH_2CH_2CH_3]^{+\cdot} + e^-$$
$$\textbf{molecular ion}$$
$$m/z = \textbf{72}$$

The **base peak** is the one with the greatest intensity. The base peak is assigned a relative intensity of 100%, and the relative intensity of each of the other peaks is reported as a percentage of the base peak. Mass spectra can be shown either as bar graphs or in tabular form.

The way a particular molecular ion fragments depends on the strength of its bonds and the stability of the fragments. Weak bonds break in preference to strong bonds, and bonds that break to form more stable fragments break in preference to those that form less stable fragments.

The carbon–carbon bonds in the molecular ion formed from pentane have about the same strength. However, the C-2—C-3 bond is more likely to break than the C-1—C-2 bond because C-2—C-3 fragmentation leads to a *primary* carbocation and a *primary* radical, which are more stable than the *primary* carbocation and *methyl* radical (or *primary* radical and *methyl* cation) obtained from C-1—C-2 fragmentation.

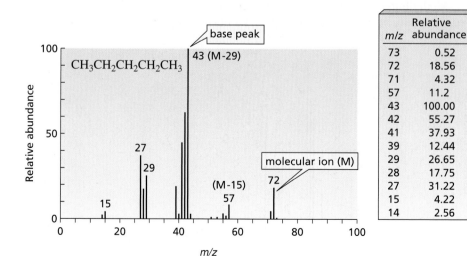

◀ **Figure 12.2**
The mass spectrum of pentane shown as a bar graph and in tabular form.

m/z	Relative abundance
73	0.52
72	18.56
71	4.32
57	11.2
43	100.00
42	55.27
41	37.93
39	12.44
29	26.65
28	17.75
27	31.22
15	4.22
14	2.56

C-2—C-3 fragmentation forms ions with $m/z = 43$ or 29, and C-1—C-2 fragmentation forms ions with $m/z = 57$ or 15. The base peak of 43 in the mass spectrum of pentane indicates the preference for C-2—C-3 fragmentation.

$$[\overset{1}{CH_3}-\overset{2}{CH_2}-\overset{3}{CH_2}-\overset{4}{CH_2}-\overset{5}{CH_3}]^{+\cdot}$$
molecular ion of pentane

$$[CH_3CH_2CH_2CH_2CH_3]^{+\cdot}$$
molecular ion
$m/z = 72$

$\longrightarrow CH_3CH_2\overset{+}{C}H_2 \; + \; CH_3\dot{C}H_2$
$\qquad\qquad m/z = 43$

$\longrightarrow CH_3CH_2\dot{C}H_2 \; + \; CH_3\overset{+}{C}H_2$
$\qquad\qquad\qquad m/z = 29$

$\longrightarrow CH_3CH_2CH_2\overset{+}{C}H_2 \; + \; \dot{C}H_3$
$\qquad\qquad m/z = 57$

$\longrightarrow CH_3CH_2CH_2\dot{C}H_2 \; + \; \overset{+}{C}H_3$
$\qquad\qquad\qquad\qquad m/z = 15$

A method commonly used to identify fragment ions is to determine the difference between the m/z value of a given ion and that of the molecular ion. For example, the ion with $m/z = 43$ in the mass spectrum of pentane is 29 units smaller than the molecular ion ($M - 29 = 43$). An ethyl group has a molecular weight of 29, so the peak at 43 can be attributed to the molecular ion minus an ethyl radical. Similarly, the peak at $m/z = 57$ ($M - 15$) can be attributed to the molecular ion minus a methyl radical. Peaks at $m/z = 15$ and $m/z = 29$ are readily recognizable as due to methyl and ethyl cations, respectively. Appendix V contains a table of common fragments lost and a table of common fragment ions.

Peaks are commonly observed at m/z values one and two units less than the m/z values of the carbocations because further fragmentation of the carbocation causes the loss of one or two hydrogen atoms.

3-D Molecules:
Propane;
Propane radical cation

$$CH_3CH_2\overset{+}{C}H_2 \xrightarrow{-H^{\cdot}} [CH_3CHCH_2]^{+\cdot} \xrightarrow{-H^{\cdot}} \overset{+}{C}H_2CH=CH_2$$
$\quad m/z = 43 \qquad\qquad\qquad m/z = 42 \qquad\qquad\qquad m/z = 41$

2-Methylbutane has the same molecular formula as pentane, so it also has a molecular ion with $m/z = 72$ (Figure 12.3). Its mass spectrum is very similar to that of

Figure 12.3 ▶
The mass spectrum of
2-methylbutane.

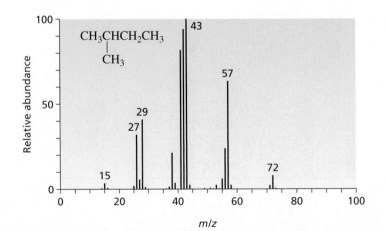

pentane, with the one notable exception that the peak at $m/z = 57\,(M - 15)$ is much more intense. 2-Methylbutane is more likely than pentane to lose a methyl radical because when it does, a *secondary* carbocation is formed. In contrast, when pentane loses a methyl radical, a less-stable *primary* carbocation is formed.

$$[CH_3\overset{\underset{\displaystyle CH_3}{|}}{C}HCH_2CH_3]^{+\cdot} \longrightarrow CH_3\overset{+}{C}HCH_2CH_3 + \cdot CH_3$$

molecular ion
$m/z = 72$
                                  $m/z = 57$

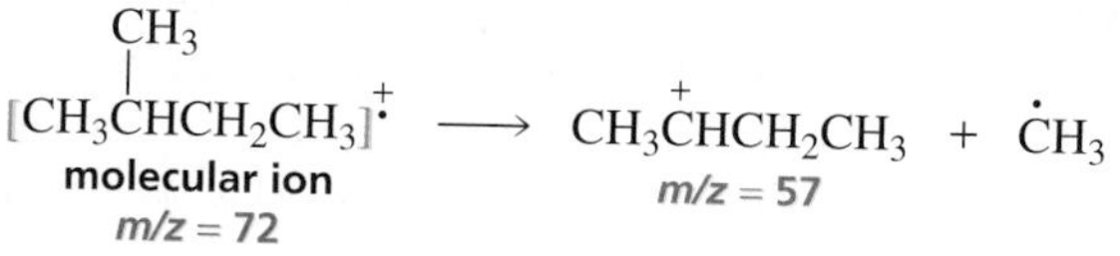

PROBLEM 1

How could you distinguish the mass spectrum of 2,2-dimethylpropane from those of pentane and 2-methylbutane?

PROBLEM 2◆

What m/z value would you predict for the base peak in the mass spectrum of 3-methylpentane?

PROBLEM 3/SOLVED

The mass spectra of two different cycloalkanes both show a molecular ion peak at $m/z = 98$. One spectrum shows a base peak at $m/z = 69$, and the other shows a base peak at $m/z = 83$. Identify the cycloalkanes.

SOLUTION   The molecular formula for a cycloalkane is C_nH_{2n}. Because the molecular weight of both cycloalkanes is 98, their molecular formulas must be C_7H_{14}. A base peak of 69 means loss of an ethyl substituent ($98 - 69 = 29$), whereas a base peak of 83 means loss of a methyl substituent ($98 - 83 = 15$). A seven-carbon cycloalkane with a base peak signifying loss of an ethyl substituent must be ethylcyclopentane. A seven-carbon cycloalkane with a base peak signifying loss of a methyl substituent must be methylcyclohexane.

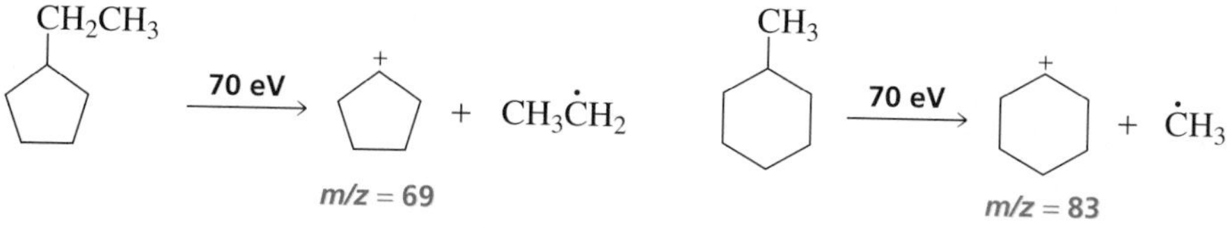

PROBLEM 4

The "nitrogen rule" states that if a compound has an odd-mass molecular ion, the compound contains an odd number of nitrogen atoms.

a. Explain why the rule holds.

b. State this rule in terms of an even-mass molecular ion.

Although the molecular ions of pentane and 2-methylbutane both have m/z values of 72, each spectrum shows a very small peak at $m/z = 73$ (Figures 12.2 and 12.3). This is called an M + 1 peak because the ion responsible for this peak is one unit

12.3
ISOTOPES IN MASS SPECTROMETRY

heavier than the molecular ion. The M + 1 peak occurs because there are two naturally occurring isotopes of carbon: 98.89% of natural carbon is ^{12}C and 1.11% is ^{13}C (Section 1.1). The M + 1 fragment results from molecular ions that contain one ^{13}C instead of a ^{12}C.

Peaks attributable to isotopes can help identify the compound responsible for a mass spectrum. For example, if a compound contains five carbon atoms, the relative abundance of the M + 1 ion should be 5(1.1%) = 5.5%, multiplied by the relative abundance of the molecular ion. This means that the number of carbon atoms in a compound can be calculated if the relative intensities of both the M and M + 1 peaks are known.

$$\text{number of carbon atoms} = \frac{\text{relative intensity of M + 1 peak}}{0.011 \times (\text{relative intensity of M peak})}$$

The isotopic distributions of several elements commonly found in organic compounds are shown in Table 12.2. Table 12.2 shows that the M + 1 peak can be used to determine the number of carbon atoms in a compound because the contributions to the M + 1 peak by isotopes of H, O, and the halogens are very small. This formula does not work as well in predicting the number of carbon atoms in a nitrogen-containing compound because the natural abundance of ^{15}N is relatively high.

Mass spectra can show M + 2 peaks as a result of a contribution from ^{18}O, or from having two heavy isotopes in the same molecule (say, ^{13}C and ^{2}H). Most of the time, though, the M + 2 peak is smaller than the M + 1 peak. The presence of a large M + 2 peak is evidence of a compound containing either chlorine or bromine, because each of these elements has a high percentage of a naturally occurring isotope two units heavier than the most abundant isotope. From the natural abundance of the isotopes of chlorine and bromine in Table 12.2, you can conclude that if the

TABLE 12.2   The Natural Abundance of Isotopes Commonly Found in Organic Compounds

Element	Natural abundance		
carbon	^{12}C 98.89%	^{13}C 1.11%	
oxygen	^{16}O 99.76%	^{17}O 0.036%	^{18}O 0.204%
nitrogen	^{14}N 99.63%	^{15}N 0.37%	
hydrogen	^{1}H 99.99%	^{2}H 0.01%	
fluorine	^{19}F 100%		
chlorine	^{35}Cl 75.77%		^{37}Cl 24.23%
bromine	^{79}Br 50.69%		^{81}Br 49.31%
iodine	^{127}I 100%		

M + 2 peak is one-third the height of the molecular ion peak, the compound contains a chlorine atom because the natural abundance of ^{37}Cl is one-third that of ^{35}Cl. If the M and M + 2 peaks are about the same height, the compound contains a bromine atom because the natural abundances of ^{79}Br and ^{81}Br are about the same.

In calculating the molecular weights of molecular ions and fragments, the monoisotopic atomic weights of the atoms must be used (Cl = 35 or 37, etc.); the atomic weights in the periodic table (Cl = 35.453) cannot be used because they are the *weighted averages* of all the naturally occurring isotopes for that element and mass spectrometry measures the m/z value of an *individual* fragment.

PROBLEM 5◆

The mass spectrum of an unknown compound has a molecular ion peak with a relative intensity of 43.27% and an M + 1 peak with a relative intensity of 3.81%. How many carbon atoms are in the compound?

All the mass spectra shown in this text were determined using a low-resolution mass spectrometer. Such spectrometers determine the *nominal molecular mass* of a fragment—the mass to the nearest whole number. High-resolution mass spectrometers can determine the *exact molecular mass* of a fragment with a precision of 0.0001 amu. If we know the exact molecular mass, we can determine a compound's molecular formula. For example, many compounds have a nominal molecular mass of 122 amu, but each of them has a different exact molecular mass. The exact molecular masses of some common isotopes are listed in Table 12.3. Computer programs exist that allow you to determine the molecular formula of a compound from its exact molecular mass.

12.4
DETERMINATION OF MOLECULAR FORMULAS: HIGH-RESOLUTION MASS SPECTROMETRY

Some Compounds With a Nominal Molecular Mass of 122 amu and Their Exact Molecular Masses

Molecular Formula	C_9H_{14}	$C_7H_{10}N_2$	$C_8H_{10}O$	$C_7H_6O_2$	$C_4H_{10}O_4$	$C_4H_{10}S_2$
Exact Molecular Mass (amu)	122.1096	122.0845	122.0732	122.0368	122.0579	122.0225

TABLE 12.3   The Exact Masses of Some Common Isotopes

^1H	1.007825 amu	^{32}S	31.9721 amu
^{12}C	12.00000 amu	^{35}Cl	34.9689 amu
^{14}N	14.0031 amu	^{79}Br	78.9183 amu
^{16}O	15.9949 amu		

PROBLEM 6◆

Which molecular formula has an exact mass of 86.1096 amu: C_6H_{14}, $C_4H_{10}N_2$, or $C_4H_6O_2$?

There are characteristic fragmentation patterns associated with specific functional groups that can help identify a substance based on its mass spectrum. These patterns were worked out after studying the mass spectra of many compounds that contain a particular functional group. We will look at the fragmentation patterns shown by alkyl halides, ethers, alcohols, and ketones as examples.

12.5
FRAGMENTATION AT FUNCTIONAL GROUPS

Alkyl Halides

Tutorial: Fragmentation
of alkyl halides

Let's look first at the mass spectrum of 1-bromopropane shown in Figure 12.4. The relative heights of the M and M + 2 peaks are about equal, so we can conclude that the compound contains a bromine atom. If a molecule contains nonbonding electrons, electron bombardment is most likely to dislodge one of the nonbonding electrons because a molecule does not hold onto its nonbonding electrons as tightly as it holds onto its bonding electrons.

$$CH_3CH_2CH_2Br \xrightarrow{-e^-} CH_3CH_2CH_2 \overset{+}{\underset{..}{^{79}Br}} + CH_3CH_2CH_2 \overset{+}{\underset{..}{^{81}Br}} \longrightarrow CH_3CH_2\overset{+}{CH_2} + \overset{.}{Br}$$

1-bromopropane	*m/z* = 122	*m/z* = 124	*m/z* = 43

The easiest bond to break in the resulting molecular ion is the carbon–bromine bond because it is the weakest bond. The bond breaks heterolytically, with both electrons going to the more electronegative of the atoms that were joined by the bond, thus forming a propyl cation and a bromine atom. As a result, the base peak in the mass spectrum of 1-bromopropane is at $m/z = 43\ [M - 79\ or\ (M + 2) - 81]$. The propyl cation shows the same fragmentation pattern that it showed when it was formed from the cleavage of pentane (Figure 12.2).

The mass spectrum of 2-chloropropane is shown in Figure 12.5. We know the compound contains a chlorine atom because the M + 2 peak is one-third the height of the molecular ion peak. The base peak at $m/z = 43$ results from *heterolytic cleavage* of the carbon–chlorine bond. The peaks at $m/z = 63$ and $m/z = 65$ have a 3:1 ratio, indicating that these fragments contain a chlorine atom. They result

Figure 12.4 ▶
The mass spectrum of
1-bromopropane.

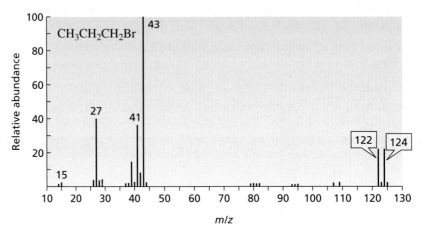

Figure 12.5 ▶
The mass spectrum of
2-chloropropane.

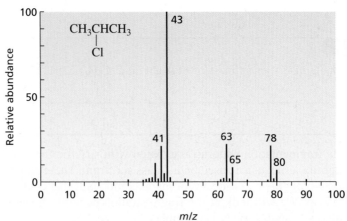

from *homolytic cleavage* of a carbon–carbon bond at the α-carbon (the carbon bonded to the chlorine). This is known as **α-cleavage.** α-Cleavage occurs because it leads to a resonance-stabilized cation (CH_3^+CH—$Cl \longleftrightarrow CH_3CH$=$Cl^+$).

$$CH_3CH_3\overset{\overset{\displaystyle CH_3}{|}}{\underset{+}{CH}} + \cdot Cl$$
$$m/z = 43$$

3-D Molecules:
sec-Propyl cation;
2-Chloropropane;
2-Chloropropane
radical cation

$$CH_3CH\overset{35}{-}\overset{+}{Cl} \quad + \quad CH_3CH\overset{37}{-}\overset{+}{Cl}$$
$$m/z = 78 \qquad\qquad m/z = 80$$

$$\overset{\overset{\displaystyle CH_3}{|}}{CH_3CH}-Cl$$
2-chloropropane

$$-e^-$$

Recall that an arrowhead with one barb ⌢ represents the movement of one electron.

$$CH_3CH-\overset{\cdot}{Cl}{}^{35} \quad + \quad CH_3CH-\overset{\cdot}{Cl}{}^{37}$$
$$m/z = 78 \qquad\qquad m/z = 80$$

$$CH_3CH\overset{35}{=\!=}\overset{+}{Cl} \quad + \quad CH_3CH\overset{37}{=\!=}\overset{+}{Cl} \quad + \quad \cdot CH_3$$
$$m/z = 63 \qquad\qquad m/z = 65$$

PROBLEM 7

Sketch the mass spectrum of 1-chloropropane.

Ethers

The mass spectrum of *sec*-butyl isopropyl ether is shown in Figure 12.6. We will see that the fragmentation pattern of an ether has many of the characteristics of the fragmentation pattern of an alkyl halide.

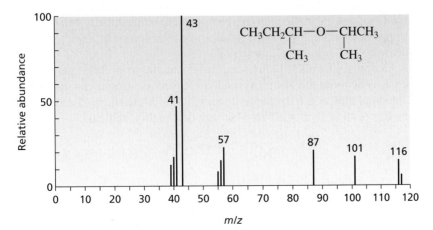

◀ **Figure 12.6**
The mass spectrum of
sec-butyl isopropyl ether.

Tutorial: Fragmentation of ethers

1. Electron bombardment dislodges a nonbonding electron from oxygen.
2. Fragmentation of the resulting molecular ion occurs in two principal ways:
    a. A carbon–oxygen bond is cleaved *heterolytically* with the electrons going to the more electronegative oxygen atom.

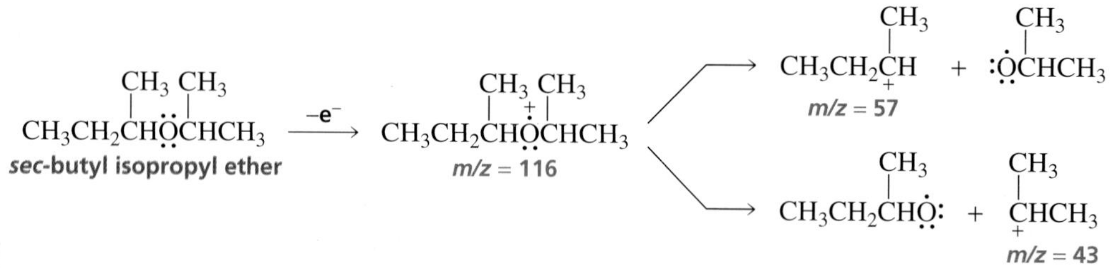

$$CH_3CH_2CHOCHCH_3 \xrightarrow{-e^-} CH_3CH_2CHOCHCH_3$$

sec-butyl isopropyl ether $m/z = 116$

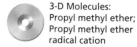

3-D Molecules:
Propyl methyl ether;
Propyl methyl ether
radical cation

    b. A carbon–carbon bond is cleaved *homolytically* at the α-position because it leads to a resonance-stabilized cation. The largest α-substituent is the one most readily cleaved.

$$CH_3CH_2-CH-\overset{+}{O}-CHCH_3 \xrightarrow{\alpha\text{-cleavage}} CH=\overset{+}{O}-CHCH_3 + CH_3\dot{C}H_2$$
$$m/z = 87$$

$$CH_3CH_2CH-\overset{+}{O}-CHCH_3 \xrightarrow{\alpha\text{-cleavage}} CH_3CH_2CH=\overset{+}{O}-CHCH_3 + \dot{C}H_3$$
$$m/z = 101$$

$$CH_3CH_2CH-\overset{+}{O}-CHCH_3 \xrightarrow{\alpha\text{-cleavage}} CH_3CH_2CH-\overset{+}{O}=CHCH_3 + \dot{C}H_3$$
$$m/z = 101$$

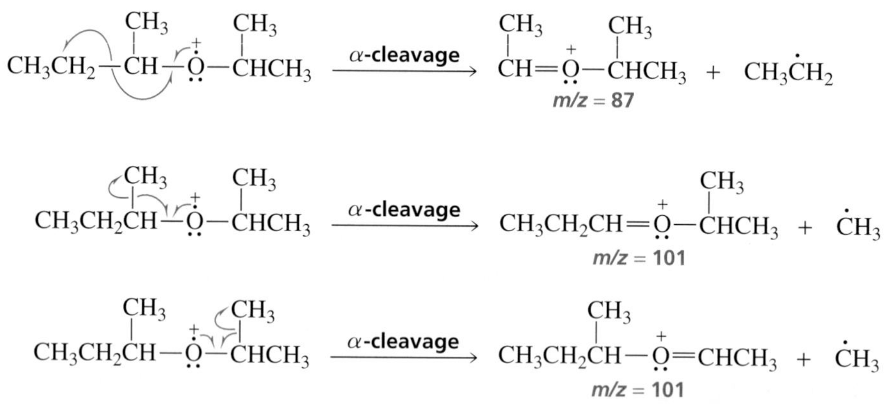

PROBLEM 8◆

The mass spectra of 1-methoxybutane, 2-methoxybutane, and 2-methoxy-2-methylpropane are shown in Figure 12.7. Match the compounds with the spectra.

Alcohols

The molecular ions obtained from alcohols fragment so readily that few of them survive to reach the collector. As a result, the mass spectra of primary and secondary alcohols show weak molecular ion peaks, and the molecular ions from tertiary alcohols are not detectable. Notice the small molecular ion peak at $m/z = 102$ in the mass spectrum of 2-hexanol (Figure 12.8).

Alcohols, like alkyl halides and ethers, undergo α-cleavage, with the largest α-substituent being the one most readily cleaved. Consequently, the mass spectrum of 2-hexanol shows a base peak at $m/z = 45$ (α-cleavage of a butyl radical) and a smaller peak at $m/z = 87$ (α-cleavage of a methyl radical).

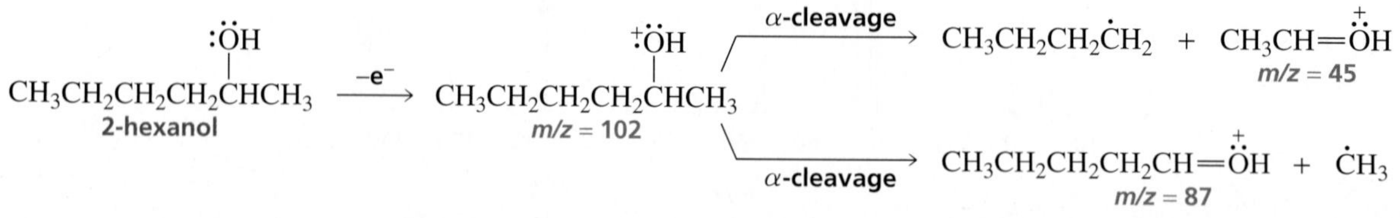

$$CH_3CH_2CH_2CH_2CHCH_3 \xrightarrow{-e^-} CH_3CH_2CH_2CH_2CHCH_3$$

2-hexanol $m/z = 102$

$$\xrightarrow{\alpha\text{-cleavage}} CH_3CH_2CH_2\dot{C}H_2 + CH_3CH=\overset{+}{O}H$$
$$m/z = 45$$

$$\xrightarrow{\alpha\text{-cleavage}} CH_3CH_2CH_2CH_2CH=\overset{+}{O}H + \dot{C}H_3$$
$$m/z = 87$$

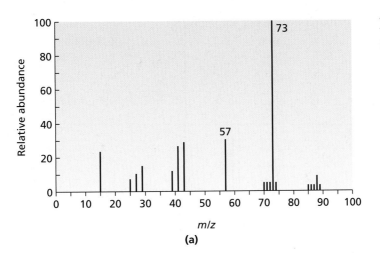

(a)

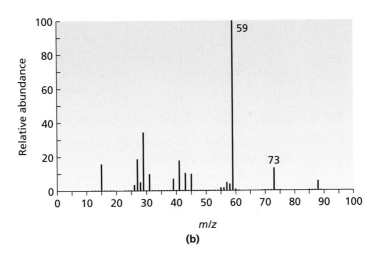

(b)

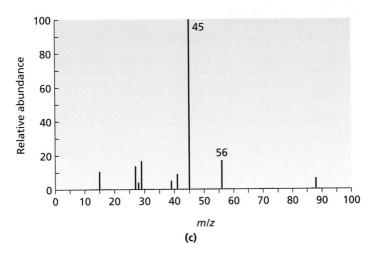

(c)

◀ **Figure 12.7**
The mass spectra for
Problem 8.

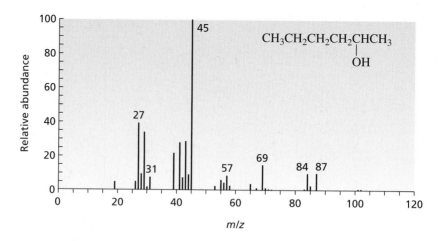

re 12.8 ▶
.e mass spectrum of
∠-hexanol.

Alcohols show a peak at $m/z = M - 18$ due to loss of water. The water that is eliminated comes from the OH group of the alcohol and a γ-hydrogen.

$$CH_3CH_2\overset{\gamma}{C}H\overset{\beta}{C}H_2\overset{\alpha}{C}HCH_3 \longrightarrow CH_3CH_2\dot{C}HCH_2\overset{+}{C}HCH_3 + H_2O$$
$$m/z = (102 - 18) = 84$$

Tutorial: Fragmentation of alcohols

Notice that alkyl halides, ethers, and alcohols show similar fragmentation behavior:

1. A bond between carbon and a *more electronegative* atom (a halogen or an oxygen) breaks heterolytically.

2. A bond between carbon and an atom of *similar electronegativity* (a carbon or a hydrogen) breaks homolytically.

3. The bonds most likely to break are those that lead to the most stable cation. (Look for fragmentation that results in the formation of resonance-stabilized cations.)

PROBLEM 9 ◆

Primary alcohols have a strong peak at $m/z = 31$. What fragment is responsible for this peak?

Ketones

3-D Molecules:
2-Hexanone;
2-Hexanone radical cation;
Acetone enol radical cation

The mass spectrum of a ketone generally has an intense molecular ion peak. Ketones fragment homolytically at the carbon–carbon bond adjacent to the carbon–oxygen double bond (α-cleavage), which results in the formation of a resonance-stabilized cation. The larger alkyl group is the one most easily cleaved.

$$CH_3CH_2CH_2\overset{\overset{\ddot{O}:}{\|}}{C}CH_3 \xrightarrow{-e^-} CH_3CH_2CH_2\overset{\overset{\ddot{O}:^+}{\|}}{C}CH_3$$
2-pentanone $m/z = 86$

$\xrightarrow{\alpha\text{-cleavage}} CH_3CH_2\dot{C}H_2 + CH_3C\overset{+}{\equiv}O:$ $m/z = 43$

$\xrightarrow{\alpha\text{-cleavage}} CH_3CH_2CH_2C\overset{+}{\equiv}O: + \dot{C}H_3$ $m/z = 71$

If one of the alkyl groups attached to the carbonyl carbon has a γ-hydrogen, a cleavage known as a **McLafferty rearrangement** may occur. In this rearrangement, the bond between the α-carbon and the β-carbon breaks homolytically and a hydrogen atom from the γ-carbon migrates to the oxygen atom. Again, fragmentation has occurred in a way the produces a resonance stabilized cation.

Fred Warren McLafferty was born in Evanston, Ill., in 1923. He received a B.S. and an M.S. from the University of Nebraska and a Ph.D. from Cornell University in 1950. He was a scientist at Dow Chemical Company until he joined the faculty at Purdue University in 1964. He has been a professor of chemistry at Cornell University since 1968.

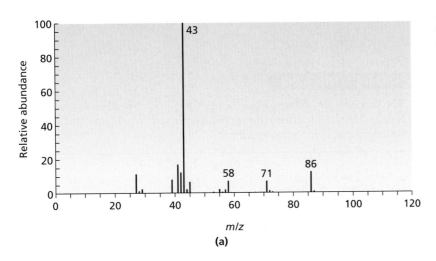

PROBLEM 10◆

Identify the ketones that are responsible for the mass spectra shown in Figure 12.9.

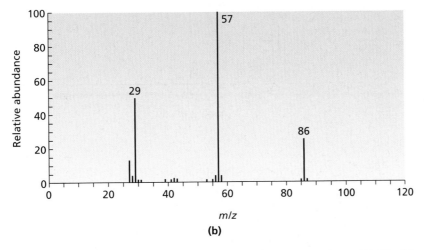

◀ **Figure 12.9**
The mass spectra for Problem 10.

Notice that the mass spectra of alkanes tend to have peaks at $m/z = 15, 29, 43$, and so on, due to methyl, ethyl, propyl, etc., cations. Ethers and alcohols, because of the presence of an oxygen atom, tend to form ions 16 units greater than alkanes: $m/z = (15 + 16) = 31, (29 + 16) = 45, (43 + 16) = 59$, etc. Ketones also have an oxygen atom but, because of the double bond, form ions two units less than those of alcohols and ethers. So, instead of having peaks at $m/z = 31, 45$, and 59, the mass spectra of ketones have peaks at $m/z = 29, 43$, and 57, which means they have peaks with m/z values similar to those of alkanes.

Tutorial: Fragmentation of ketones

PROBLEM 11

Using curved arrows, show the principal fragments that would be observed in the mass spectrum of each of the following compounds:

a. $CH_3CH_2CH_2CH_2CH_2OH$

b. $CH_3\overset{\displaystyle O}{\overset{\displaystyle \|}{C}}CH_2CH_2CH_2CH_3$

c. $CH_3-\overset{\displaystyle CH_3}{\underset{\displaystyle CH_3}{C}}-Br$

d. $CH_3CH_2O\overset{\displaystyle CH_2CH_3}{\underset{\displaystyle CH_3}{C}}CH_2CH_2CH_3$

e. $CH_3CH_2\overset{\displaystyle }{\underset{\displaystyle CH_3}{C}HCl}$

f. $CH_3CH_2\overset{\displaystyle }{\underset{\displaystyle OH}{C}HCH_2CH_2CH_2CH_3}$

PROBLEM 12◆

Two products are obtained from the reaction of *(Z)*-2-pentene with water and a trace of H_2SO_4. The mass spectra of these products are shown in Figure 12.10. Identify the compounds responsible for the spectra.

Figure 12.10 ▶
The mass spectra for Problem 12.

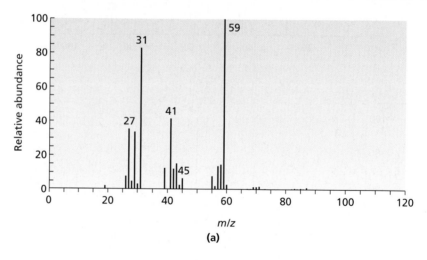

(a)

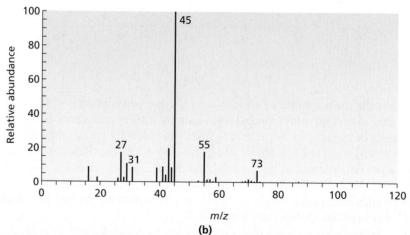

(b)

Spectroscopy is the study of the interaction of matter and electromagnetic radiation. **Electromagnetic radiation** is radiant energy that displays the properties of both particles and waves. The different types of electromagnetic radiation constitute the electromagnetic spectrum (Figure 12.11). Visible light is the electromagnetic radiation with which we are most familiar because we can see it. We will examine three different spectrophotometric techniques used to identify compounds. Each uses a different type of electromagnetic radiation. In this chapter we will look at infrared (IR) spectroscopy and ultraviolet/visible (UV/Vis) spectroscopy. In the next chapter we will see how compounds can be identified using nuclear magnetic resonance (NMR) spectroscopy.

A particle of electromagnetic radiation is called a photon. The relationship between the energy (E) of a photon and the frequency (ν) of the electromagnetic radiation is described by the following equation:

$$E = h\nu$$

where h is the proportionality constant called Planck's constant (Section 3.7). Frequency has units of hertz (Hz) or cycles per second (cps).

- *Cosmic rays*, a type of radiation that originates in outer space, have the highest energy.
- *γ-Rays* (gamma rays) are emitted from the nuclei of certain radioactive elements and, because of their high energy, can severely damage biological organisms.
- *X-rays,* somewhat lower in energy than γ-rays, are less harmful except in high doses. Low-dose X-rays are used to examine the internal structure of organisms. The denser the tissue, the more it blocks X-rays.
- *Ultraviolet (UV) light* is responsible for sunburns, and repeated exposure can cause skin cancer by damaging DNA molecules in skin cells (Section 28.6).

12.6 SPECTROSCOPY AND THE ELECTROMAGNETIC SPECTRUM

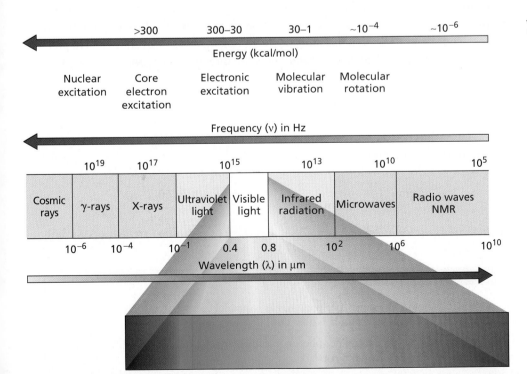

◀ **Figure 12.11**
The electromagnetic spectrum.

- *Visible light* is the electromagnetic radiation we see.
- We feel *infrared radiation* as heat.
- We cook with *microwaves* and use them in radar.
- *Radio waves* have the lowest energy (lowest frequency). We use them for radio and television communication. Radio waves are also used in NMR spectroscopy and in magnetic resonance imaging (MRI) (Chapter 13).

Because electromagnetic radiation has wave-like properties as well as particle-like properties, it can be described by its wavelength (λ). The **wavelength** is the distance from any point on one wave to the corresponding point on the next wave. Wavelength can be measured in *micrometers*. One micrometer (μm) is one millionth (10^{-6}) of a meter. In older literature, micrometers are called *microns* (μ).

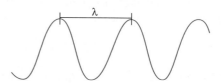

The **frequency** (ν) of the radiation is equal to the velocity (c) of the wave divided by its wavelength.

$$\nu = \frac{c}{\lambda} \qquad c = 3 \times 10^{10} \text{ cm/s}$$

Since the velocity (c) is constant, electromagnetic radiation can be described either by its frequency or by its wavelength. The energy of electromagnetic radiation is *directly* proportional to frequency and *inversely* proportional to wavelength.

$$E = h\nu = \frac{hc}{\lambda}$$

Wavenumber ($\tilde{\nu}$) is another way to describe the *frequency* of electromagnetic radiation, and the one most often used in infrared spectroscopy. It is the number of waves in one centimeter so it has units of cm^{-1}. Scientists use wavenumbers in preference to wavelengths because wavenumbers are directly proportional to energy.

$$\tilde{\nu}(\text{cm}^{-1}) = \frac{1}{\lambda(\text{cm})}$$

The relationship between wavenumber (in cm^{-1}) and wavelength (in μm) is given by the following equation:

$$\tilde{\nu}(\text{cm}^{-1}) = \frac{10^4}{\lambda(\mu\text{m})} \qquad \text{because } 1 \ \mu\text{m} = 10^{-4} \text{ cm}$$

High frequencies, large wavenumbers, and short wavelengths are associated with high energy.

So *high frequencies, large wavenumbers,* and *short wavelengths* are associated with *high energy.*

PROBLEM 13◆

a. Which is higher in energy per photon, electromagnetic radiation with wavenumber 100 cm^{-1} or with wavenumber 2000 cm^{-1}?

b. Which is higher in energy per photon, electromagnetic radiation with wavelength 9 μm or with wavelength 8 μm?

c. Which is higher in energy per photon, electromagnetic radiation with wavenumber 3000 cm^{-1} or with wavelength 2 μm?

PROBLEM 14 ◆

a. Light of what wavenumber has a wavelength of 4 μm?

b. Light of what wavelength has a wavenumber of 200 cm^{-1}?

Stretching and Bending Vibrations

Organic molecules are constantly vibrating. So when we say that a bond between two atoms has a certain length, we are citing an average because the bond behaves as if it were a vibrating spring connecting two atoms. The bond vibrates with both stretching and bending motions. A *stretch* is a vibration occurring along the line of the bond. A *bend* is a vibration that does *not* occur along the line of the bond. A diatomic molecule such as H—Cl can undergo only a **stretching vibration.**

The vibrations of a molecule containing three atoms are more complex (Figure 12.12). There are symmetric and asymmetric stretches and bends, and bending vibrations can be either in-plane or out-of-plane. **Bending vibrations** are often referred to by the descriptive terms *rock, scissor, wag,* and *twist.*

Each stretching and bending vibration of a molecule occurs with a characteristic frequency. When a compound is bombarded with radiation of a frequency that exactly matches the frequency of one of its vibrations, the molecule will absorb energy. This allows the bonds to stretch and bend a bit more. Thus, the absorption of

**12.7
INFRARED
SPECTROSCOPY**

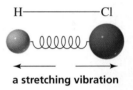

a stretching vibration

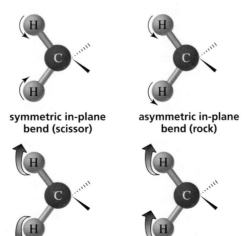

Stretching vibrations

symmetric stretch   asymmetric stretch

Bending vibrations

symmetric in-plane bend (scissor)   asymmetric in-plane bend (rock)

symmetric out-of-plane bend (twist) asymmetric out-of-plane bend (wag)

◀ **Figure 12.12**
Stretching and bending vibrations of bonds in organic molecules.

Tutorial:
IR stretching and bending

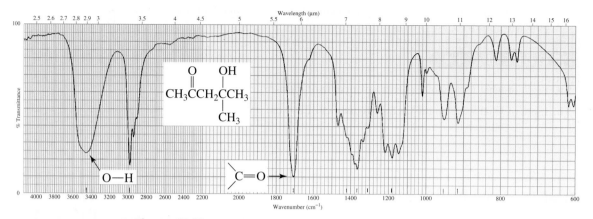

▲ **Figure 12.13**
The infrared spectrum of 4-hydroxy-4-methyl-2-pentanone.

energy increases the *amplitude* of the vibration but does not change its *frequency*. By experimentally determining the wavenumbers of light that are absorbed by a particular compound, we can determine what kinds of bonds it has. For example, the stretching vibration of a C=O bond absorbs light of wavenumber ~ 1700 cm^{-1}, whereas the stretching vibration of an O—H bond absorbs light of wavenumber ~ 3450 cm^{-1} (Figure 12.13).

$$\begin{array}{ccc} \diagup\!\!\!\!\diagdown\!\text{C}\!=\!\text{O} & & \text{O}\!-\!\text{H} \\ \text{~1700 cm}^{-1} & & \text{~3450 cm}^{-1} \end{array}$$

The Functional Group and Fingerprint Regions

Electromagnetic radiation with wavenumbers from 4000 to 600 cm^{-1} has just the right energy to correspond to stretching and bending vibrations in organic molecules. Electromagnetic radiation with this energy is known as **infrared radiation** because it is just below the "red region" of visible light. (*Infra* is Latin for "below.")

An IR spectrum can be divided into two areas. The left-hand two-thirds of an IR spectrum (4000–1400 cm^{-1}) is where most of the functional groups show absorption bands. This is called the **functional group region.** The right-hand third (1400–600 cm^{-1}) is called the **fingerprint region.** The overall pattern in the fingerprint region is characteristic of the compound as a whole because each compound shows a unique pattern in this region. For example, 2-pentanol and 3-pentanol have the same functional groups, so they show the same absorption bands in the functional group region. Their fingerprint regions are different, however, because the compounds are different (Figure 12.14). So a compound can be positively identified by comparing its fingerprint region with the fingerprint region of the actual spectrum of the compound.

Obtaining an Infrared Spectrum

An **infrared spectrum** is obtained by passing infrared radiation through a sample of the compound. A detector generates a plot of percent transmission of radiation versus the wavenumber or wavelength of the radiation transmitted. At 100% transmission, all the energy of the radiation passes through the molecule. At lower values of percent transmission, some of the energy is being absorbed by the compound. Each spike in the infrared (IR) spectrum represents absorption of energy. These spikes are called **absorption bands.** Most chemists report the location of absorption bands using wavenumbers.

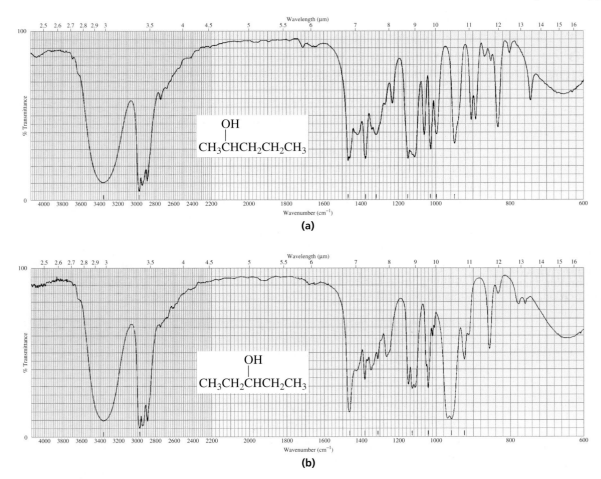

▲ **Figure 12.14**
The IR spectra of (a) 2-pentanol and (b) 3-pentanol.

An IR spectrum can be taken of a gas, a solid, or a liquid sample. Gases are expanded into an evacuated cell (a small container). Solids can be compressed with anhydrous KBr into a disc that is placed in the light beam. A solid can also be examined as a mull. A mull is prepared by mixing a few milligrams of the solid with a drop or two of mineral oil. In the case of liquid samples, a spectrum can be obtained of the neat (undiluted) liquid by placing a few drops of the liquid between two optically polished plates of sodium chloride that are placed in the light beam. A small container—a cell—with optically polished NaCl or KBr windows is used to hold samples dissolved in solvents. KBr discs and NaCl or KBr cells are used because, unlike glass or quartz, these materials have no covalent bonds so they do not absorb IR radiation.

When solvents are used, they must be solvents with few absorption bands in the region of interest. Commonly used solvents are CH_2Cl_2, $CHCl_3$, and CS_2. In a double-beam spectrophotometer, the IR radiation is split into two beams, one that passes through the sample cell, and the other that passes through a cell containing only the solvent. Any absorptions of the solvent are canceled out, so the absorption spectrum is that of the solute alone. In more sophisticated spectrophotometers, pulse techniques can enhance sensitivity.[1]

[1]The spectra in this text are FT-IR spectra. The information is digitized and Fourier transformed by a computer to produce the FT-IR spectrum. FT-IR techniques allow a spectrum to be taken in as little as 10 seconds.

12.8 INFRARED ABSORPTION BANDS

Each of the different kinds of stretching and bending vibrations shown in Figure 12.12 can give rise to an absorption band, so IR spectra can be quite complex. Because of this complexity, chemists generally do not try to identify all the absorption bands in an IR spectrum. We will look at some characteristic bands so you will be able to tell something about the structure of the compound that gives a particular IR spectrum. However, there is a lot more to infrared spectroscopy than we will be able to cover. You can find an extensive table of characteristic group frequencies in Appendix V. When identifying an unknown compound, IR spectroscopy is often used in conjunction with information obtained from other spectrophotometric techniques. Many of the problems in Chapters 12 and 13 give you the opportunity to identify compounds using information from two or more instrumental methods.

It takes more energy for a stretching vibration than for a bending vibration.

Because it takes more energy to stretch a bond than to bend it, absorption bands for stretching vibrations are found in the functional group region (4000–1400 cm^{-1}), whereas absorption bands for bending vibrations are typically found in the fingerprint region (1400–600 cm^{-1}). Stretching vibrations are therefore the most useful in determining what kinds of bonds a molecule has. The IR **stretching frequencies** associated with different types of bonds are shown in Table 12.4.

TABLE 12.4 Important IR Stretching Frequencies		
Type of bond	**$\tilde{\nu}$(cm^{-1})**	**Intensity**
C≡N	2260–2220	medium
C≡C	2260–2100	medium to weak
C=C	1680–1600	medium
C=N	1650–1550	medium
⬡	~1600 and ~1500–1430	strong to weak
C=O	1780–1650	strong
C—O	1250–1050	strong
C—N	1230–1020	medium
O—H (alcohol)	3650–3200	strong, broad
O—H (acid)	3300–2500	strong, very broad
N—H	3500–3300	medium, broad
C—H	3300–2700	medium

12.9 INTENSITY OF ABSORPTION BANDS

The intensity of an absorption band depends on the size of the dipole moment change associated with the vibration. In other words, it depends on the polarity of the vibrating bond. For example, the stretching vibration of an O—H bond is more intense than that of a N—H bond because the O—H bond is more polar. Likewise, the stretching vibration of a N—H bond is more intense than that of a C—H bond because the N—H bond is more polar.

The intensity of an absorption band also depends on the number of bonds responsible for the absorption. For example, the absorption band for the C—H stretch will be more intense for a compound such as octyl iodide, which has 17 C—H bonds, than for methyl iodide, which has only three C—H bonds. The concentration of the sample used to obtain an IR spectrum also affects the intensity of the absorption bands. Concentrated samples have greater numbers of absorbing molecules and therefore more intense absorption bands.

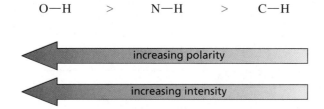

$$O—H \quad > \quad N—H \quad > \quad C—H$$

increasing polarity

increasing intensity

Hooke's Law

The amount of energy required to stretch a bond depends on the *strength* of the bond and the *masses* of the bonded atoms. The approximate wavenumber of an absorption can be calculated using the following equation derived from **Hooke's law,** which describes the motion of a vibrating spring:

$$\widetilde{\nu} = \frac{1}{2\pi c}\left[\frac{f(m_1 + m_2)}{m_1 m_2}\right]^{1/2}$$

The equation relates the wavenumber of the stretching vibration ($\widetilde{\nu}$) to the force constant of the bond (f) and the masses of the atoms (in grams) joined by the bond (m_1 and m_2). The force constant is a measure of the strength of the bond. The equation shows that stronger bonds and lighter atoms give rise to higher frequencies. For typographic reasons, we will use $[\ ]^{1/2}$ to signify the more familiar square root sign ($\sqrt{\ }$).

The stronger the bond, the greater the energy required to stretch it because a stronger bond corresponds to a tighter spring. The frequency of the vibration is inversely related to the mass of the atoms attached to the spring. Thus, vibrations of heavier atoms occur at lower frequencies.

**12.10
POSITION OF
ABSORPTION
BANDS**

Stronger bonds and lighter atoms give rise to higher wavenumbers.

THE ORIGINATOR OF HOOKE'S LAW

Robert Hooke (1635–1703) was born on the Isle of Wight off the southern coast of England. A brilliant scientist, he contributed to almost every field of science. He was the first to suggest that light had wave-like properties. He discovered that Gamma Arietis is a double star and he discovered Jupiter's Great Red Spot. In a lecture published posthumously, he suggested that earthquakes are caused by the cooling and contracting of the Earth. He examined cork under a microscope and coined the term "cell" to describe what he saw. He wrote about evolutionary development based on his studies of microscopic fossils, and his studies of insects were highly regarded as well. Hooke also invented the balance spring for watches and the universal joint currently used in cars.

Robert Hooke's drawing of a "blue fly," which appeared in *Micrographia*, the first book on microscopy, published by Hooke in 1665.

Effect of Bond Order

Bond order affects bond strength, so bond order affects the position of absorption bands. A $C\equiv C$ bond is stronger than a $C=C$ bond, so a $C\equiv C$ bond stretches at a higher frequency (~ 2100 cm^{-1}) than a $C=C$ bond (~ 1650 cm^{-1}). $C-C$ bonds show stretching vibrations in the region 1200 to 800 cm^{-1}, but these are weak and of little value in identifying compounds. Similarly, a $C=O$ bond stretches at a higher frequency (~ 1700 cm^{-1}) than a $C-O$ bond (~ 1100 cm^{-1}), and a $C\equiv N$ bond stretches at a higher frequency (~ 2200 cm^{-1}) than a $C=N$ bond (~ 1600 cm^{-1}), which stretches at a higher frequency than a $C-N$ bond (~ 1100 cm^{-1}) (Table 12.4).

PROBLEM 15◆

a. Which will occur at a higher frequency:

  1. a $C-O$ stretch or a $C=O$ stretch?

  2. a $C-H$ stretch or a $C-H$ bend?

  3. a $C\equiv C$ stretch or a $C=C$ stretch?

b. Assuming the force constants are the same, which will occur at a higher frequency:

  1. a $C-O$ stretch or a $C-Cl$ stretch?

  2. a $C-O$ stretch or a $C-C$ stretch?

Resonance and Inductive Electronic Effects

Table 12.4 shows a range of wavenumbers for each stretch because the actual position of the absorption band depends on other structural features of the molecule, such as electron delocalization, the electronic effect of neighboring substituents, and hydrogen bonding. Important details about the structure of a compound can be revealed by the exact position of the absorption band.

For example, the IR spectrum in Figure 12.15 shows that the carbonyl group ($C=O$) of 2-pentanone absorbs at 1720 cm^{-1}, whereas the IR spectrum in Figure 12.16 shows that the carbonyl group of 2-cyclohexenone absorbs at a lower frequency (1680 cm^{-1}). 2-Cyclohexenone absorbs at a lower frequency because the carbonyl group has greater single bond character due to electron delocalization. Because a double bond is stronger than a single bond, a carbonyl group with significant single bond character will stretch at a lower frequency than one with little or no single bond

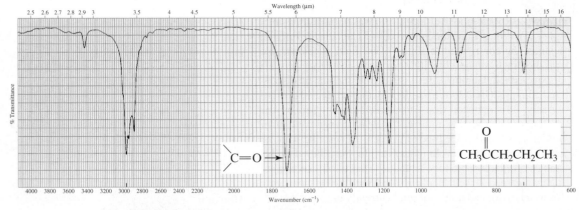

▲ **Figure 12.15**
The IR spectrum of 2-pentanone.

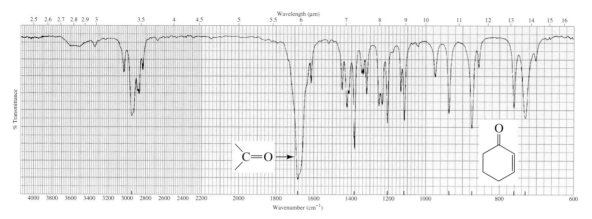

▲ **Figure 12.16**
The IR spectrum of 2-cyclohexenone.

character. Consequently, the carbonyl group of 2-cyclohexenone stretches at a lower frequency.

Putting an atom other than carbon next to the carbonyl group also causes the position of the carbonyl absorption band to shift. Whether it shifts to a lower or to a higher frequency depends on whether the predominant effect of the atom is to donate electrons by resonance or to withdraw electrons inductively.

resonance electron donation   **inductive electron withdrawal**

The predominant effect of the nitrogen of an amide (Figure 12.17) is electron donation by resonance. Therefore, the carbonyl group of an amide has less double bond character than the carbonyl group of a ketone, so it is weaker and stretches more easily (1650 cm^{-1}). In contrast, the predominant effect of the oxygen of an ester is inductive electron withdrawal (Figure 12.18). Therefore, the carbonyl group of an ester has more double bond character than the carbonyl group of a ketone, so it is stronger and harder to stretch (1740 cm^{-1}).

A C—O bond shows a stretch between 1250 and 1050 cm^{-1}. If the C—O bond is in an alcohol (Figure 12.19) or an ether, the stretch will occur toward the lower end of the range. If, however, the C—O bond is in a carboxylic acid (Figure 12.20), the stretch will occur at the higher end of the range. The position of the C—O absorption varies because the C—O bond in an alcohol is a pure single bond, whereas the C—O bond in a carboxylic acid has partial double bond character due to resonance electron donation. Esters show C—O stretches at both ends of the range (Figure 12.18) because esters have two C—O bonds—one that is a pure single bond and one that has partial double bond character.

$$CH_3CH_2—OH \qquad CH_3CH_2—O—CH_2CH_3 \qquad CH_3\overset{O}{\overset{\|}{C}}—OH \qquad CH_3\overset{O}{\overset{\|}{C}}—O—CH_3$$

~1050 cm⁻¹   ~1050 cm⁻¹   ~1250 cm⁻¹   ~1250 cm⁻¹ and ~1050 cm⁻¹

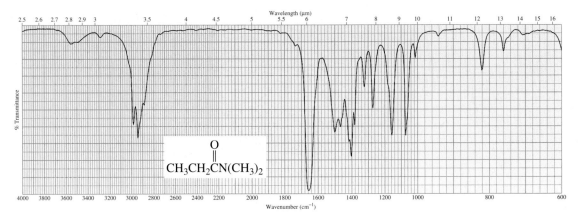

▲ **Figure 12.17**
The IR spectrum of *N*,*N*-dimethylpropanamide.

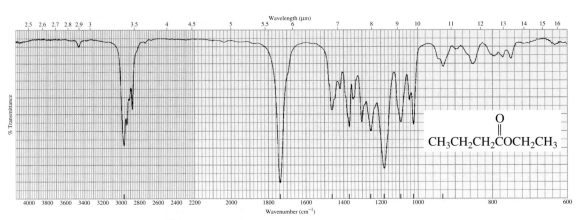

▲ **Figure 12.18**
The IR spectrum of ethyl butanoate.

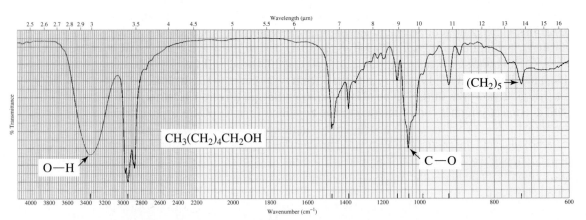

▲ **Figure 12.19**
The IR spectrum of 1-hexanol.

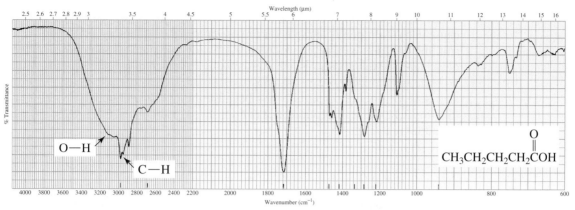

▲ **Figure 12.20**
The IR spectrum of pentanoic acid.

PROBLEM 16◆

Which will occur at a higher frequency:

a. the C—N stretch of an amine or the C—N stretch of an amide?

b. the C—O stretch of phenol or the C—O stretch of cyclohexanol?

c. the C=O stretch of a ketone or the C=O stretch of an amide?

d. the stretch or the bend of the C—O bond in ethanol?

PROBLEM 17◆

Would a carbonyl group bonded to an sp^3 hybridized carbon or a carbonyl group bonded to an sp^2 hybridized carbon show an absorption band at a higher frequency?

PROBLEM 18◆

List the following compounds in order of decreasing frequency of the C=O absorption band:

a.

b.

c. $CH_3-\overset{O}{\underset{||}{C}}-CH_3$ $H-\overset{O}{\underset{||}{C}}-H$ $CH_3-\overset{O}{\underset{||}{C}}-H$

O—H Absorption Bands

O—H absorptions are easy to detect because O—H bonds are polar, so they show intense absorption bands, and the bands generally are quite broad (Figures 12.19 and 12.20). The position, the breadth, and the intensity of an O—H absorption band depend on the concentration of the sample. The more concentrated the sample, the more likely it is for the OH-containing molecules to form intermolecular hydrogen bonds. It is easier to stretch the O—H bond of a hydrogen-bonded OH group because each hydrogen is attracted to the oxygen of a neighboring molecule. As a result, the O—H stretch of a concentrated (hydrogen-bonded) solution of an alcohol occurs at 3550 to 3200 cm^{-1}, whereas the O—H stretch of a dilute solution (little or no hydrogen bonding) occurs at 3650 to 3590 cm^{-1}. Hydrogen-bonded OH groups also have broader absorption bands because the hydrogen bonds vary in strength (Section 2.11). The absorption bands of non-hydrogen-bonded OH groups are sharper and less intense.

H
|
R—O—H------O—R

concentrated solution
a hydrogen-bonded O—H bond
3550–3200 cm^{-1}

R—O—H

dilute solution
a non-hydrogen-bonded O—H bond
3650–3590 cm^{-1}

PROBLEM 19◆

Will ethanol dissolved in carbon disulfide or an undiluted sample of ethanol show an O—H stretch at a higher frequency?

12.11
C—H ABSORPTION BANDS

Tutorial: IR spectra

Stretching Vibrations

The strength of a C—H bond depends on the hybridization of the carbon because the more *s* character the bond has the stronger it is (Section 1.14). Therefore, an *sp* C—H bond is stronger than an *sp*^{2} C—H bond, which is stronger than an *sp*^{3} C—H bond. More energy is needed to stretch a stronger bond, so the C—H stretch of an *sp* hybridized carbon occurs at ~3300 cm^{-1}, the C—H stretch of an *sp*^{2} hybridized carbon occurs at ~3100 cm^{-1}, and the C—H stretch of an *sp*^{3} hybridized carbon occurs at ~2900 cm^{-1} (Table 12.5).

A useful step in the analysis of a spectrum entails looking at the absorption bands in the vicinity of 3000 cm^{-1}. Figures 12.21, 12.22, and 12.23 show the IR spectra for methylcyclohexane, cyclohexene, and ethylbenzene, respectively. The only absorption band in the vicinity of 3000 cm^{-1} in Figure 12.21 is slightly below 3000 cm^{-1}. We know, therefore, that the compound has hydrogens bonded to *sp*^{3} carbons and no hydrogens bonded to *sp*^{2} or *sp* carbons. The spectra in Figures 12.22 and 12.23 show absorption bands slightly above and slightly below 3000 cm^{-1}, which means that the compounds responsible for the spectra contain hydrogens bonded to *sp*^{2} and *sp*^{3} carbons.

Once we know the compound has hydrogens bonded to *sp*^{2} carbons, we can then ask whether they are *sp*^{2} carbons of an alkene or *sp*^{2} carbons of a benzene ring. Because sharp absorption bands at both ~1600 cm^{-1} and between ~1500–1430 cm^{-1} indicate a benzene ring, whereas a band only at ~1600 cm^{-1} indicates an alkene (Table 12.4), we know that the compound with the spectrum shown in Figure 12.22 is an alkene and the compound with the spectrum shown in Figure 12.23 has a benzene ring. Be aware, however, that N—H bending vibrations also occur at ~1600 cm^{-1}, so absorption at that wavelength does not always indicate the presence of a C=C bond. However, absorption bands resulting from N—H bends tend to be broader and more intense (see Figure 12.25) than those resulting from C=C stretches.

The stretch of the C—H bond in an aldehyde group occurs at ~2700 cm^{-1} (Figure 12.24). This makes aldehydes relatively easy to identify because essentially no other absorption occurs at this frequency.

TABLE 12.5   IR Absorptions of Carbon–Hydrogen Bonds

Carbon–Hydrogen Stretching Vibrations	$\tilde{\nu}\,(cm^{-1})$
C≡C—H	~3300
C=C—H	3100–3020
C—C—H	2960–2850
$\overset{\overset{\displaystyle O}{\parallel}}{C}$—C—H	~2700

Carbon–Hydrogen Bending Vibrations		$\tilde{\nu}\,(cm^{-1})$
CH$_3$— —CH$_2$— —$\overset{\mid}{C}$H—		1450–1420
CH$_3$—		1385–1365
$\underset{R}{\overset{H}{>}}C=C\underset{H}{\overset{R}{<}}$	trans	980–960
$\underset{H}{\overset{R}{>}}C=C\underset{H}{\overset{R}{<}}$	cis	730–675
$\underset{R}{\overset{R}{>}}C=C\underset{H}{\overset{R}{<}}$	trisubstituted	840–800
$\underset{R}{\overset{R}{>}}C=C\underset{H}{\overset{H}{<}}$	terminal alkene	890
$\underset{H}{\overset{R}{>}}C=C\underset{H}{\overset{H}{<}}$	terminal alkene	990 and 910

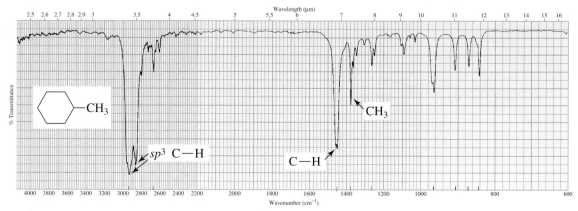

▲ Figure 12.21
The IR spectrum of methylcyclohexane.

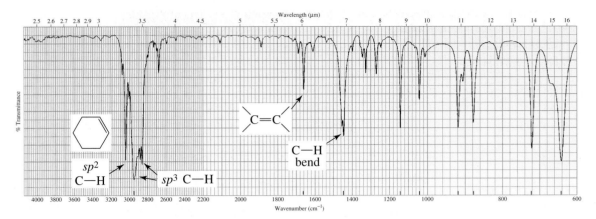

▲ **Figure 12.22**
The IR spectrum of cyclohexene.

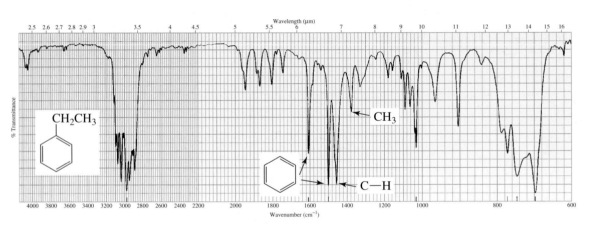

▲ **Figure 12.23**
The IR spectrum of ethylbenzene.

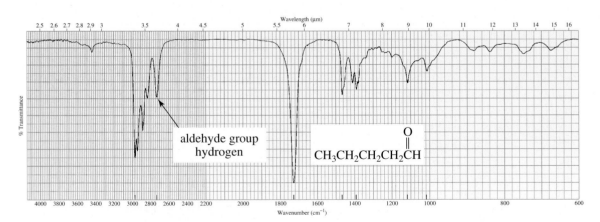

▲ **Figure 12.24**
The IR spectrum of pentanal.

Bending Vibrations

If a compound has sp^3 carbons, a look at ~ 1400 cm^{-1} will tell you whether it has a methyl group. All hydrogens bonded to sp^3 hybridized carbons show a C—H bending vibration at slightly *greater* than 1400 cm^{-1}. Only methyl groups show a C—H bending vibration at slightly *less* than 1400 cm^{-1}. So if the compound has a methyl group, absorption bands will appear at both ~1440 cm^{-1} and ~1380 cm^{-1}. If it does not have a methyl group, only the band at ~1440 cm^{-1} will be present. You can see evidence for a methyl group in Figure 12.21 (methylcyclohexane) and in Figure 12.23 (ethylbenzene), but no evidence for a methyl group in Figure 12.22 (cyclohexene). Two methyl groups attached to the same carbon can sometimes be detected by a split in the methyl peak at ~1380 cm^{-1} (Figure 12.25).

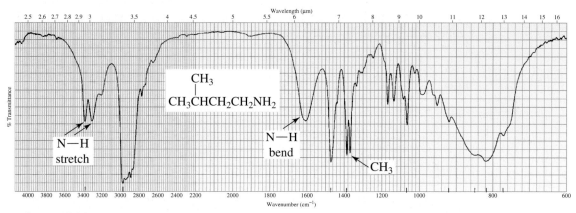

▲ **Figure 12.25**
The IR spectrum of isopentylamine.

C—H bending vibrations for hydrogens bonded to sp^2 carbons give rise to absorption bands in the 1000–600 cm^{-1} region (Table 12.5). As the table shows, the frequency of the C—H bending vibrations of an alkene depends on the number of alkyl groups attached to the double bond and the configuration of the alkene. It is important to realize that these absorption bands can be shifted out of the characteristic regions if strongly electron-withdrawing or electron-donating substituents are close to the double bond (Section 12.10). Open-chain compounds with more than four adjacent methylene (CH_2) groups show a characteristic absorption band at 720 cm^{-1} due to in-phase rocking of the methylene groups (Figure 12.19).

**12.12
SHAPE OF
ABSORPTION
BANDS**

The shape of an absorption band can be helpful in identifying the compound responsible for an IR spectrum. For example, both O—H and N—H bonds stretch at wavenumbers above 3100 cm^{-1}. It is apparent from the IR spectra of 1-hexanol (Figure 12.19), pentanoic acid (Figure 12.20), and isopentylamine (Figure 12.25) that a N—H stretching vibration (~3300 cm^{-1}) is narrower and not as intense as an O—H stretching vibration (~3300 cm^{-1}), and that the O—H stretching vibration of a carboxylic acid (~3300–2500 cm^{-1}) is broader than the O—H stretching vibration of an alcohol (Sections 12.9 and 12.10). Notice that two absorption bands are detectable in Figure 12.25 for the N—H stretch because there are two N—H bonds in the compound.

PROBLEM 20

a. Why is an O—H stretch more intense than an N—H stretch?

b. Why is an O—H stretch of a carboxylic acid broader than the O—H stretch of an alcohol?

The absence of an absorption band can be as useful as the presence of a band in identifying a compound by IR spectroscopy. For example, the spectrum in Figure 12.26 shows a strong absorption at ~1100 cm^{-1} indicating the presence of a C—O bond. It is not an alcohol because there is no strong absorption above 3100 cm^{-1}. It is not an ester or any other kind of carbonyl compound because there is no absorption at ~1700 cm^{-1}. It has no C≡C, C=C, C≡N, C=N, or C—N bonds. Thus, the compound must be an ether. Its C—H absorption bands show it has only sp^3 hybridized carbons and it has a methyl group. We also know that the compound has fewer than four adjacent methylene groups because there is no absorption at ~720 cm^{-1}. The compound is actually diethyl ether.

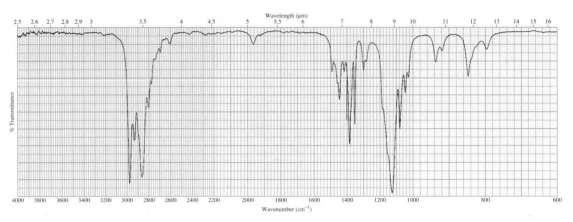

▲ **Figure 12.26**
The IR spectrum of diethyl ether.

PROBLEM 21◆

An oxygen-containing compound shows an absorption band at ~1700 cm^{-1} and no absorption bands at ~3300 cm^{-1}, ~2700 cm^{-1}, or 1100^{-1}. What class of compound is it?

PROBLEM 22

How could you use IR spectroscopy to distinguish between:

a. a primary amine and a tertiary amine?   **c.** *cis*-2-hexene and *trans*-2-hexene?

b. a cyclic ketone and an open-chain ketone?   **d.** cyclohexene and cyclohexane?

PROBLEM 23

For each of the following pairs of compounds, give one absorption band that could be used to distinguish between them:

a. $CH_3CH_2CH_2CH_2CH_3$ and $CH_3CH_2CH_2OCH_2CH_3$

b. $CH_3CH_2CH_2\overset{\displaystyle O}{\overset{\displaystyle \|}{C}}OH$ and $CH_3CH_2CH_2CH_2OH$ **d.**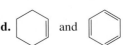

c. $CH_3CH_2CH_2\overset{\overset{\displaystyle O}{\|}}{C}OCH_3$ and $CH_3CH_2CH_2\overset{\overset{\displaystyle O}{\|}}{C}OH$

e. (hexagon) and (methylcyclohexane with CH_3)

12.14 INFRARED INACTIVE VIBRATIONS

Not all vibrations result in absorption bands. In order for a vibration to absorb IR radiation, the dipole moment of the molecule must change when the vibration occurs. In other words, the vibrating bond must be polar. For example, 1-butene has a dipole moment along its C=C bond. Recall from Section 1.3 that the dipole moment is equal to the magnitude of the charge on the atoms multiplied by the distance between them. When the C=C bond stretches, the distance between the atoms increases. This, in turn, increases the dipole moment. The change in dipole moment allows absorption of IR radiation. Consequently, an absorption band is observed for the C=C stretching vibration.

H₂C=CH—CH₂CH₃ **1-butene**

(H₃C)₂C=C(CH₃)₂ **2,3-dimethyl-2-butene**

(H₃C)₂C=C(CH₃)CH₂CH₂CH₂CH₃ **2,3-dimethyl-2-heptene**

On the other hand, 2,3-dimethyl-2-butene is a symmetrical molecule, so it does not have a dipole moment along its C=C bond. When the bond stretches, it still has no dipole moment. Because stretching is not accompanied by a change in dipole moment, an absorption band is not observed for the C=C stretching vibration. The vibration is *infrared inactive*. 2,3-Dimethyl-2-heptene experiences a very small change in dipole moment when its C=C bond stretches, so only an extremely weak absorption band (if any) will be detected for the stretching vibration of its C=C bond.

PROBLEM 24 ◆

Which of the following compounds has a vibration that is infrared inactive: acetone, CO_2, CO, 1-butyne, 2-butyne, H_2, H_2O, Cl_2, ethene?

PROBLEM 25 ◆

The mass spectrum and infrared spectrum of an unknown compound are shown in Figures 12.27 and 12.28, respectively. Identify the compound.

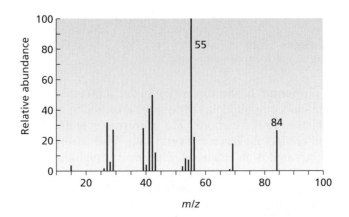

◀ **Figure 12.27**
The mass spectrum for Problem 5.

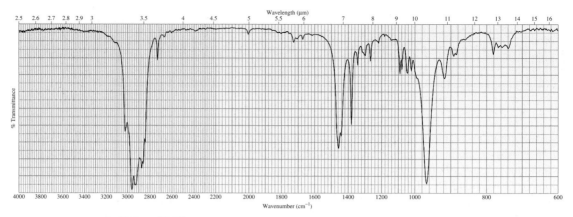

▲ **Figure 12.28**
The IR spectrum for Problem 25.

12.15 IDENTIFYING INFRARED SPECTRA

We will now look at some IR spectra and see what we can determine about the structure of the compounds that give rise to the spectra.

Compound 1. The absorptions in the 3000 cm^{-1} region in Figure 12.29 indicate that the compound has hydrogens attached to sp^2 carbons (3075 cm^{-1}) and to sp^3 carbons (2950 cm^{-1}). Let's now look to see if the sp^2 carbons are those of an alkene or those of a benzene ring. The absorption at 1650 cm^{-1} and the absorption at ~890 cm^{-1} suggest that the compound is a terminal alkene with two alkyl substituents at the 2-position. The lack of absorption at ~720 cm^{-1} indicates that the compound has fewer than four adjacent methylene groups. We cannot identify the compound precisely, but when we find out that the compound is 2-methyl-1-pentene, we are not surprised.

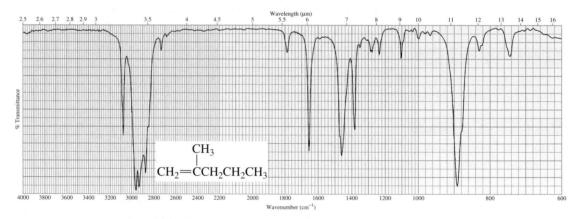

▲ **Figure 12.29**
The IR spectrum of Compound 1.

Compound 2. The absorption in the 3000 cm^{-1} region in Figure 12.30 indicates that the compound has hydrogens attached to sp^2 carbons (3050 cm^{-1}), but none attached to sp^3 carbons. The absorptions at 1600 cm^{-1} and 1460 cm^{-1} indicate the compound has a benzene ring. The absorption at 2740 cm^{-1} indicates the compound is an aldehyde. The absorption band for the carbonyl group C=O is lower (1700 cm^{-1}) than normal, so the carbonyl group has partial single bond character. Thus, it must be attached directly to the benzene ring. The compound is benzaldehyde.

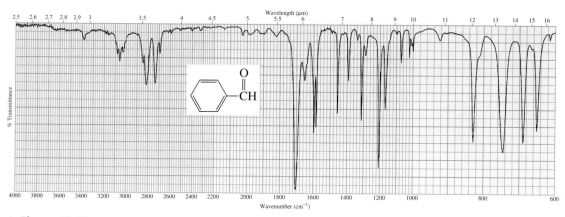

▲ **Figure 12.30**
The IR spectrum of Compound 2.

Compound 3. The absorptions in the 3000 cm^{-1} region in Figure 12.31 indicate that the compound has hydrogens attached to sp^3 carbons (2950 cm^{-1}), but none attached to sp^2 carbons. The shape of the strong absorption band at 3300 cm^{-1} is characteristic of an O—H group. The absorption at 2100 cm^{-1} indicates the compound has a triple bond. The sharp absorption band at 3300 cm^{-1} indicates the compound is a terminal alkyne. The compound is 2-propyn-1-ol.

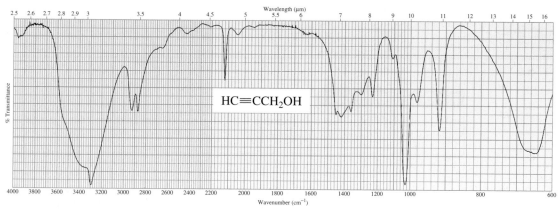

▲ **Figure 12.31**
The IR spectrum of Compound 3.

Compound 4. The absorption in the 3000 cm^{-1} region in Figure 12.32 indicates that the compound has hydrogens attached to sp^3 carbons (2950 cm^{-1}). The medium strength absorption band at 3300 cm^{-1} suggests that the compound has one N—H bond. The presence of the N—H bond is confirmed by the absorption at 1580 cm^{-1}. The C=O absorption at 1660 cm^{-1} indicates the compound is an amide. The compound is *N*-methylacetamide.

Compound 5. The absorptions in the 3000 cm^{-1} region in Figure 12.33 indicate that the compound has hydrogens attached to sp^2 carbons) (>3000 cm^{-1}) and to sp^3 carbons (<3000 cm^{-1}). The absorptions at 1605 cm^{-1} and 1500 cm^{-1} indicate the compound contains a benzene ring. The absorption at 1720 cm^{-1} for the carbonyl group indicates that the compound is a ketone and the carbonyl group is not directly attached to the benzene ring. The absorption at ~1380 cm^{-1} indicates the compound contains a methyl group. The compound is 1-phenyl-2-butanone.

Figure 12.32 ▶
The IR
spectrum of
Compound 4.

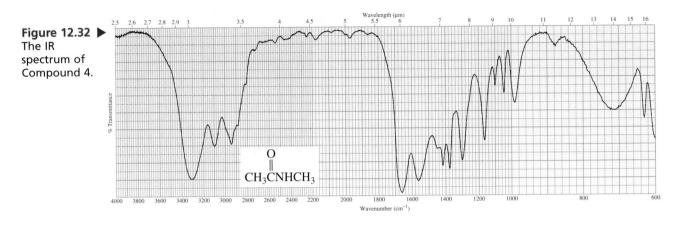

Figure 12.33 ▶
The IR
spectrum of
Compound 5.

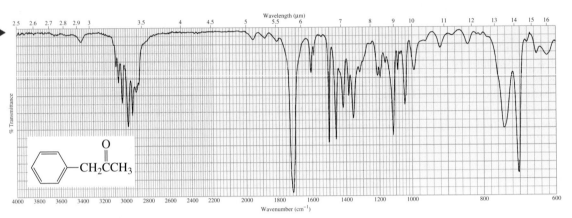

PROBLEM 26◆

A compound with molecular formula C_4H_6O gives the infrared spectrum shown in Figure 12.34. Identify the compound.

Figure 12.34 ▶
The IR
spectrum for
Problem 26.

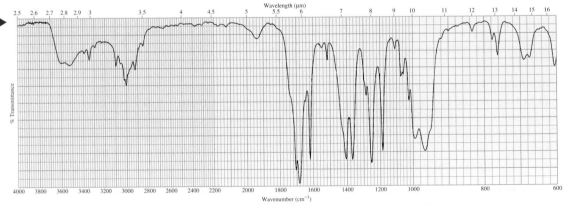

12.16
ULTRAVIOLET
AND VISIBLE
SPECTROSCOPY

Ultraviolet and visible (UV/Vis) spectroscopy gives us information about compounds with conjugated double bonds. Ultraviolet and visible light have just the right energy to cause an electronic transition—the promotion of an electron from one orbital to another of higher energy. Depending on the energy needed for the electronic transition, a molecule will absorb either ultraviolet or visible light. If it absorbs ultraviolet light, a UV spectrum is obtained; if it absorbs lower-energy visible light, a visible spectrum is obtained. **Ultraviolet light** is electromagnetic radiation with wavelengths ranging from 180 to 400 nm (nanometers). **Visible light** has wavelengths ranging from 400 to 780 nm. (One nanometer is 10^{-9} m or 10 Å. In older literature, millimicron (mμ) is used instead of nanometer, but interconversion is easy because 1 mμ = 1 nm.)

Recall from Section 12.6 that wavelength (λ) is related to the energy of the radiation by the following equation:

$$E = \frac{hc}{\lambda}$$

The shorter the wavelength, the greater the energy of the radiation.

The normal electronic configuration of a molecule is known as its **ground state**—all the electrons are in the available orbitals with the lowest energy.

When a molecule absorbs light of an appropriate wavelength, an electron can be promoted to a higher energy orbital. It is promoted from the **highest occupied molecular orbital (HOMO)** to the **lowest unoccupied molecular orbital (LUMO)**. Such a transition is called an **electronic transition**. The molecule is then in an **excited state**. The relative energies of the bonding, nonbonding, and antibonding orbitals are shown in Figure 12.35.

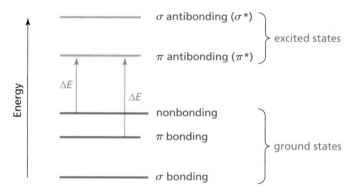

◀ **Figure 12.35**
Relative energies of the bonding, nonbonding, and antibonding orbitals.

Ultraviolet and visible light have sufficient energy to cause only the two electronic transitions shown in Figure 12.35. The electronic transition with the lowest energy is promotion of a nonbonding electron (n) into a π antibonding molecular orbital. This is called an $n \rightarrow \pi^*$ transition (stated as "n to pi star"). The higher energy electronic transition is promotion of an electron from a π bonding molecular orbital into a π antibonding molecular orbital, known as a $\pi \rightarrow \pi^*$ transition (stated as "pi to pi star"). This means that only compounds with π electrons or nonbonding electrons can produce UV/Vis spectra.

The UV spectrum of acetone is shown in Figure 12.36. Acetone has both π electrons and nonbonding electrons. Thus there are two absorption bands, one for the $\pi \rightarrow \pi^*$ transition and one for the $n \rightarrow \pi^*$ transition. The $\lambda_{\mathbf{max}}$ (stated as "lambda max") is the wavelength corresponding to the highest point (maximum) of the absorption band. The $\pi \rightarrow \pi^*$ transition has a $\lambda_{\max}$ at 187 nm, and the $n \rightarrow \pi^*$ transition has a $\lambda_{\max}$ at 270 nm. We know that the $\pi \rightarrow \pi^*$ transition corresponds to the $\lambda_{\max}$ at the shorter wavelength because this transition requires more energy than the $n \rightarrow \pi^*$ transition (Figure 12.35).

When a UV spectrum is recorded under normal conditions, absorption bands below about 210 nm cannot be detected because oxygen absorbs in that region and covers up all other absorption bands. To be able to see the absorption band at 187 nm,

Figure 12.36 ▶
The UV spectrum of acetone.

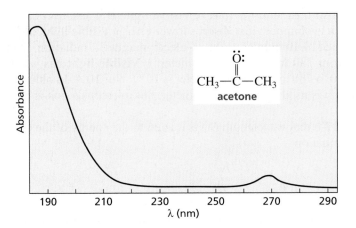

the spectrum of acetone must be obtained in a spectrophotometer purged with nitrogen or by using vacuum UV techniques.

PROBLEM 27◆

a. Which has greater energy, UV light or visible light?

b. Which has greater energy, visible light or IR radiation?

ULTRAVIOLET LIGHT AND SUNSCREENS

Exposure to ultraviolet light causes specialized cells in the skin to produce a black pigment known as melanin, which causes the skin to look tan. Melanin absorbs UV light, so it protects our bodies from the potentially harmful effects of the sun. If more UV light reaches the skin than the melanin can absorb, the light will burn the skin and can cause photochemical reactions that can result in skin cancer (Section 28.6). UV-A is the lowest energy UV light (315 to 400 nm) and does the least biological damage. Fortunately, most of the more dangerous, higher-energy UV light, UV-B (290 to 315 nm) and UV-C (180 to 290 nm), is filtered out by the ozone layer in the stratosphere. This is why there is such great concern about the diminishing ozone layer (Section 8.9).

Applying a sunscreen to skin can protect it against UV light. Some sunscreens contain an inorganic reagent such as zinc oxide that reflects the light as it reaches the skin. Others contain a compound that absorbs UV light. PABA was the first commercially available UV-absorbing sunscreen. PABA absorbs UV-B light, but it is not very soluble in oily skin lotions. More nonpolar compounds such as Padimate O are now commonly used. Giv Tan F absorbs both UV-B and UV-A light, so it gives better protection. Recent research has shown that sunscreens that absorb only UV-B light do not give adequate protection against skin cancer—both UV-A and UV-B protection are needed.

The amount of protection provided by a particular sunscreen is indicated by its SPF (sun protection factor). The higher the SPF, the greater the protection.

para-aminobenzoic acid
PABA

(2-ethylhexyl) 4-(dimethylamino)benzoate
Padimate O

(2-ethylhexyl) 3-(4-methoxyphenyl)-2-propenoate
Giv Tan F

Wilhelm Beer and Johann Lambert independently proposed that absorbance depends on the amount of absorbing species that the light encounters as it passes through a solution. In other words, absorbance depends on both the concentration of the sample and the length of the light path through the sample. The relationship among absorbance, concentration, and length of the light path is known as the **Beer–Lambert law.**

the Beer–Lambert law

$$A = cl\varepsilon$$

A = absorbance = $\log \dfrac{I_0}{I}$

$\quad I_0$ = intensity of the radiation entering the sample

$\quad I$ = intensity of the radiation emerging from the sample

$\quad c$ = concentration of the sample in moles/liter

$\quad l$ = length of the light path through the sample in centimeters

$\quad \varepsilon$ = molar absorptivity

The **molar absorptivity** (formerly called the extinction coefficient) is a constant that is characteristic of a compound at a particular wavelength. It is the absorbance that would be observed for a 1.00 M solution in a cell with a 1.00 cm path length. The molar absorptivity of acetone, for example, is 900 at 187 nm and 15 at 270 nm. The solvent in which the sample is dissolved when the spectrum is taken is reported because molar absorptivity is not exactly the same in all solvents. So the UV spectrum of acetone in hexane would be reported as λ_{max} 187 nm (ε = 900, hexane); λ_{max} 270 nm (ε = 15, hexane). Because absorbance is proportional to concentration, the concentration of a solution can be determined if the absorbance and molar absorptivity at a particular wavelength are known.

The two absorption bands of acetone in Figure 12.36 are very different in size because of the difference in molar absorptivity at the two wavelengths. Small molar absorptivities are characteristic of $n \rightarrow \pi^*$ transitions, so these transitions can be difficult to detect. Consequently, $\pi \rightarrow \pi^*$ transitions are much more useful in UV/Vis spectroscopy.

In order to obtain a UV or visible spectrum, the solution is placed in a cell. Most cells have path lengths of 1 cm. Either glass or quartz cells can be used for visible spectra, but quartz cells must be used for UV spectra because glass absorbs UV light.

PROBLEM 28◆

A solution of 4-methyl-3-penten-2-one in ethanol shows an absorbance of 0.52 at 236 nm in a cell with a 1 cm light path. Its molar absorptivity in ethanol at that wavelength is 12,600. What is the concentration of the compound?

12.17
THE BEER-LAMBERT LAW

Wilhelm Beer (1797–1850) was born in Germany. He was a banker whose hobby was astronomy. He was the first to make a map of the darker and lighter areas of Mars.

Johann Heinrich Lambert (1728–1777), a German-born mathematician, was the first to make accurate measurements of light intensities and to introduce hyperbolic functions into trigonometry.

*Although the equation that relates absorbance, concentration, and light path bears the names of Beer and Lambert, it is believed that **Pierre Bouguer (1698–1758)**, a French mathematician, first formulated this relationship in 1729.*

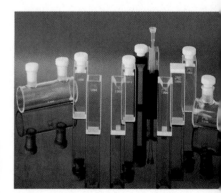

Cells used in UV/Vis spectroscopy.

The $n \rightarrow \pi^*$ transition for methyl vinyl ketone is at 324 nm, and the $\pi \rightarrow \pi^*$ transition is at 219 nm. Both λ_{max} values of methyl vinyl ketone are at longer wavelengths than the corresponding λ_{max}'s of acetone because methyl vinyl ketone has two conjugated double bonds. Because conjugation raises the energy of the HOMO and lowers the energy of the LUMO, less energy is required for an electronic transition

12.18
EFFECT OF CONJUGATION ON λ_{max}

Figure 12.37 ▶
Conjugation raises the energy of the HOMO and lowers the energy of the LUMO.

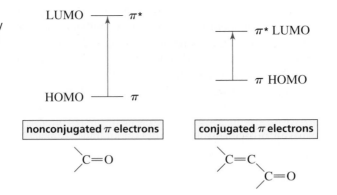

for a conjugated system than for a nonconjugated system (Figure 12.37). The more conjugated double bonds there are in a compound, the less energy required for the electronic transition and therefore the longer the wavelength at which the transition occurs.

$$
\begin{array}{cc}
\overset{\displaystyle O}{\underset{\parallel}{}} & \overset{\displaystyle O}{\underset{\parallel}{}} \\
CH_3\!-\!C\!-\!CH_3 & CH_3\!-\!C\!-\!CH=\!CH_2 \\
\textbf{acetone} & \textbf{methyl vinyl ketone}
\end{array}
$$

$n \longrightarrow \pi^*$ λ_{max} = 270 nm λ_{max} = 324 nm
$\pi \longrightarrow \pi^*$ λ_{max} = 187 nm λ_{max} = 219 nm

3-D Molecule:
Methyl vinyl ketone

The λ_{max} values of the $\pi \rightarrow \pi^*$ transition for several conjugated dienes are shown in Table 12.6. Notice that both the λ_{max} and the molar absorptivity increase as the number of conjugated double bonds increases. Thus, the λ_{max} of a compound can be used to predict the number of conjugated double bonds in the compound.

If a compound has enough double bonds, it will absorb visible light ($\lambda_{max} > 400$ nm) and the compound will be colored. β-Carotene, a precursor of vitamin A, is an orange substance found in carrots, apricots, and sweet potatoes. Lycopene is red and is found in tomatoes, watermelon, and pink grapefruit.

β-carotene
λ_{max} = 455 nm

lycopene
λ_{max} = 474 nm

A **chromophore** is that part of a molecule that is responsible for a UV or visible absorption spectrum. The carbonyl group is the chromophore of acetone. The four compounds on page 519 all have the same chromophore, so they all have approximately the same λ_{max}.

TABLE 12.6   Values of λ_{max} and ε for Ethylene and Conjugated Dienes

Compound	λ_{max} (nm)	ε
$H_2C=CH_2$	165	15,000
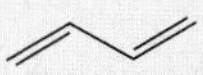	217	21,000
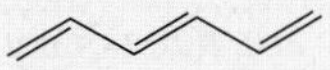	256	50,000
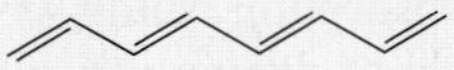	290	85,000
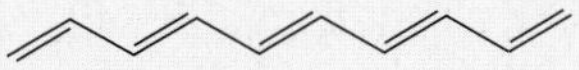	334	125,000
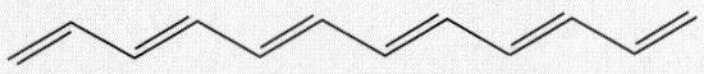	364	138,000

3-D Molecule:
1,3,5,7-Octatetraene

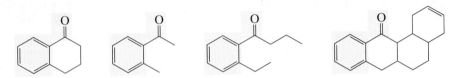

An **auxochrome** is a substituent that, when attached to a chromophore, alters the λ_{max} and the intensity of the absorption, usually increasing both of them; OH and NH_2 groups are auxochromes. The lone-pair (nonbonding) electrons on oxygen and nitrogen are available for interaction with the π electron cloud of the benzene ring and this increases the λ_{max}. Because the anilinium ion does not have an auxochrome, its λ_{max} is similar to that of benzene.

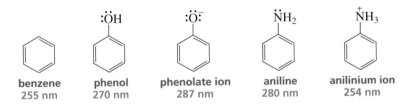

benzene	phenol	phenolate ion	aniline	anilinium ion
255 nm	270 nm	287 nm	280 nm	254 nm

3-D Molecules:
para-Aminobenzoic acid;
Phenol;
Phenolate ion

Removal of a proton from phenol (forming the phenolate ion) further increases the λ_{max} because of the additional pair of nonbonding electrons that becomes available for interaction with the ring. Protonating aniline (forming the anilinium ion) decreases its λ_{max} because the nonbonding electron pair is no longer available to interact with the benzene ring. Because wavelengths of red light are longer than those of blue light (Figure 12.11), a shift to a longer wavelength is called a **red shift,** and a shift to a shorter wavelength is called a **blue shift.** Deprotonation of phenol results in a red shift, while protonation of aniline produces a blue shift.

red shift →

← blue shift

200 nm   400 nm

PROBLEM 29◆

Rank the compounds in each group in order of decreasing λ_{max}.

a.

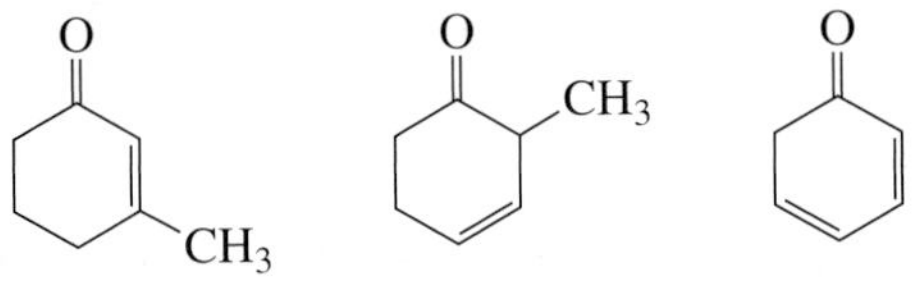

b.

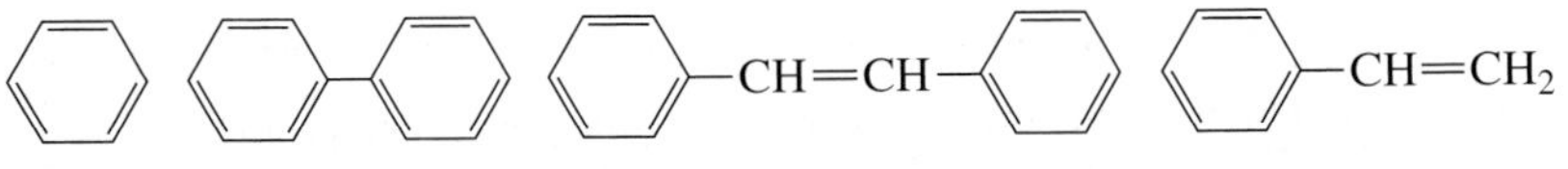

12.19 THE VISIBLE SPECTRUM AND COLOR

White light is a mixture of all wavelengths of visible light. If any color is removed from white light, the remaining light appears colored. So if a compound absorbs any visible light, the compound will appear colored. Its color depends on the color of the light transmitted to the eye. In other words, it depends on the color produced from the wavelengths of light that are *not* absorbed.

The relationship between the wavelengths of the light absorbed and the color observed is shown in Table 12.7. Notice that two absorption bands are necessary to produce green. Most colored compounds have fairly broad absorption bands; vivid colors have narrow absorption bands. The human eye is able to distinguish more than a million different shades of color.

3-D Molecule: Phenolphthalein

TABLE 12.7	Dependence of the Color Observed on the Wavelength of Light Absorbed
Wavelengths absorbed (nm)	**Observed color**
380–460	yellow
380–500	orange
440–560	red
480–610	purple
540–650	blue
380–420 and 610–700	green

Azobenzenes (benzene rings connected by a nitrogen–nitrogen double bond) have an extended conjugated system that causes them to absorb light from the visible region of the spectrum. Some substituted azobenzenes are used commercially as dyes. Varying the extent of conjugation and the substituents attached to the conjugated system provides a large number of different colors. Notice that the only difference between butter yellow and methyl orange is an $SO_3^- Na^+$ group. Methyl orange is a commonly used acid–base indicator. When margarine was first produced, it was colored with butter yellow to make it look more like butter. (White margarine would not have been very appetizing.) This dye was no longer used after it was found to be carcinogenic. β-Carotene is currently used to color margarine. There is some evidence that β-carotene is an anticancer agent.

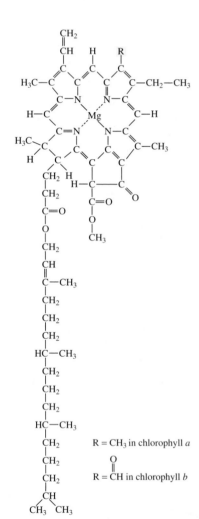

R = CH₃ in chlorophyll *a*

R = CH in chlorophyll *b* (with O double bonded)

Chlorophyll *a* and *b* are the pigments that make plants look green. These highly conjugated compounds absorb nongreen light. Therefore, plants reflect green light.

methyl orange
an azobenzene

butter yellow
an azobenzene

PROBLEM 30

a. At pH = 7, one of the following ions is purple and the other is blue. Which is which?

b. What would be the difference in the colors of the compounds at pH = 3?

ANTHOCYANINS: A COLORFUL CLASS OF COMPOUNDS

A class of highly conjugated compounds called anthocyanins are responsible for the red, purple, and blue colors of many flowers (poppies, peonies, cornflowers), fruits (cranberries, rhubarb, strawberries, blueberries), and vegetables (beets, radishes, red cabbage). In an acidic environment, an anthocyanin exists as a positively charged oxonium ion that absorbs light of wavelengths between 480 and 550 nm, due to the extended conjugation of the three rings. The exact wavelength of light absorbed depends on the substituents on the anthocyanin. Thus, the flower or fruit or vegetable appears red, purple, or blue.

In a neutral or basic solution, the conjugation between the second and third rings is disrupted. The anthocyanin is no longer colored because without the extended conjugation it absorbs ultraviolet light rather than visible light. (You can see this if you adjust the pH of cranberry juice from about 1 to about 6.)

red, blue, or purple   colorless   colorless

R = H, OH, or OCH$_3$
R' = H, OH, or OCH$_3$

UV/Vis spectroscopy is not nearly as useful as IR or NMR spectroscopy for structure determination. UV/Vis spectroscopy, however, has other uses that make it a very powerful analytical tool.

Reaction rates are commonly measured using UV/Vis spectroscopy. The rate of any reaction can be measured using UV/Vis spectroscopy as long as one of the reactants

12.20
USES OF UV/VIS SPECTROSCOPY

or one of the products absorbs UV or visible light at a wavelength at which the other reactants and products have little or no absorbance. For example, the anion of nitroethane has a λ_{max} at 240 nm, but neither H_2O nor the reactants show any significant absorbance at that wavelength. In order to measure the rate at which hydroxide ion removes a proton from nitroethane (that is, the rate at which the nitroethane anion is formed), the UV spectrophotometer is adjusted to measure absorbance at 240 nm as a function of time instead of absorbance as a function of wavelength. Nitroethane is added to a quartz cell containing a basic solution, and the rate of the reaction is determined by monitoring the increase in absorbance at 240 nm (Figure 12.38).

Figure 12.38 ▶
The rate at which a proton is removed from nitroethane is determined by monitoring the increase in absorbance at 240 nm.

$$CH_3CH_2NO_2 \ + \ HO^- \ \rightleftharpoons \ CH_3\bar{C}HNO_2 \ + \ H_2O$$

nitroethane   nitroethane anion
λ_{max} = 240 nm

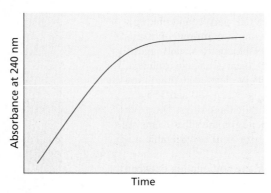

The enzyme lactate dehydrogenase catalyzes the reduction of pyruvate by NADH to form lactate. NADH is the only species in the reaction mixture that absorbs light at 340 nm, so the rate of the reaction can be determined by monitoring the decrease in absorbance at 340 nm (Figure 12.39).

Figure 12.39 ▶
The rate of reduction of pyruvate by NADH is measured by monitoring the decrease in absorbance at 340 nm.

$$\underset{\text{pyruvate}}{CH_3\overset{O}{\overset{\|}{C}}COO^-} \ + \ \underset{\lambda_{max} = 340 \text{ nm}}{NADH} \ + \ H^+ \ \xrightarrow{\underset{\text{dehydrogenase}}{\text{lactate}}} \ \underset{\text{lactate}}{CH_3\overset{OH}{\overset{|}{C}HCOO^-}} \ + \ NAD^+$$

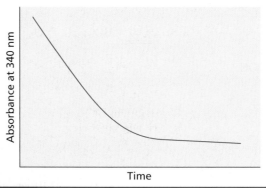

PROBLEM 31

Describe how one could determine the rate of the alcohol dehydrogenase catalyzed oxidation of ethanol by NAD^+ (Section 18.11).

The pK_a of a compound can be determined by UV/Vis spectroscopy if either the acidic form or the basic form of the compound absorbs UV or visible light. For example, the phenolate ion has a λ_{max} at 287 nm. If the absorbance at 287 nm is determined as a function of pH, the pK_a of phenol can be ascertained by determining the pH at which exactly one-half the increase in absorbance has occurred (Figure 12.40). At this pH, half of the phenol has been converted into phenolate. Recall from Section 1.20 that the Henderson–Hasselbalch equation states that the pK_a of the compound is the pH at which half the compound exists in its acidic form and half exists in its basic form ($[HA] = [A^-]$).

UV spectroscopy can also be used to estimate the nucleotide composition of DNA. The two strands of DNA are held together by both A–T base pairs and G–C base pairs (Section 25.4). When DNA is heated, the two strands break apart. Single-stranded DNA has a greater molar absorptivity at 260 nm than double-stranded DNA. The melting temperature (T_m) of DNA is the midpoint of an absorbance versus temperature curve (Figure 12.41). The T_m of double-stranded DNA increases with increasing content of G–C base pairs because they are held together by three hydrogen bonds whereas A–T base pairs are held together by only two hydrogen bonds. Therefore, the T_m can be used to estimate the G–C content. These are just a few examples of the many uses of UV/Vis spectroscopy.

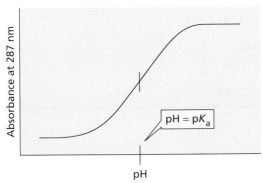

▲ **Figure 12.40**
The absorbance of an aqueous solution of phenol as a function of pH.

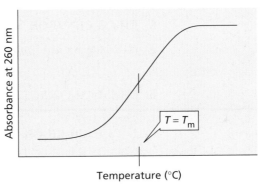

▲ **Figure 12.41**
The absorbance of a solution of DNA as a function of temperature.

KEY TERMS

absorption band (page 498)
auxochrome (page 519)
base peak (page 483)
Beer–Lambert law (page 517)
bending vibration (page 497)
blue shift (page 519)
chromophore (page 518)
α-cleavage (page 489)
electromagnetic radiation (page 495)
electronic transition (page 515)
excited state (page 515)
fingerprint region (page 498)
frequency (page 496)
functional group region (page 498)

ground state (page 515)
highest occupied molecular orbital
   (HOMO) (page 515)
Hooke's law (page 501)
infrared radiation (page 498)
infrared spectroscopy (page 481)
infrared spectrum (page 498)
λ_{max} (page 515)
lowest unoccupied molecular orbital
   (LUMO) (page 515)
mass spectrometry (page 481)
mass spectrum (page 482)
McLafferty rearrangement (page 492)
molar absorptivity (page 517)

molecular ion (page 483)
nominal molecular mass (page 483)
radical cation (page 481)
red shift (page 519)
spectroscopy (page 495)
stretching frequency (page 500)
stretching vibration (page 497)
ultraviolet light (page 515)
UV/Vis spectroscopy (page 481)
visible light (page 515)
wavelength (page 496)
wavenumber (page 496)

32. For each of the following pairs of compounds, give one absorption band that could be used to distinguish between them:

a. $CH_3CH_2CH_2\overset{\overset{\displaystyle O}{\|}}{C}H$ and $CH_3CH_2CH_2\overset{\overset{\displaystyle O}{\|}}{C}CH_3$

b. 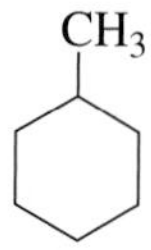 (cyclohexane with CH₃ substituent) and $CH_3CH_2CH_2CH_2CH_2CH_2CH_3$

c. $CH_3CH_2CH_2CH{=}CH_2$ and $CH_3CH_2CH_2CH{=}\overset{\overset{\displaystyle CH_3}{|}}{C}CH_3$

d. $CH_3CH_2\overset{\overset{\displaystyle O}{\|}}{C}NH_2$ and $CH_3CH_2\overset{\overset{\displaystyle O}{\|}}{C}OCH_3$

e. 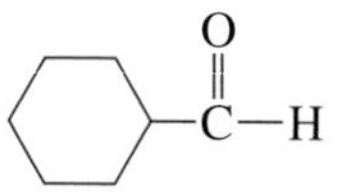 (cyclohexyl)$\overset{\overset{\displaystyle O}{\|}}{C}{-}H$ and (phenyl)$\overset{\overset{\displaystyle O}{\|}}{C}{-}H$

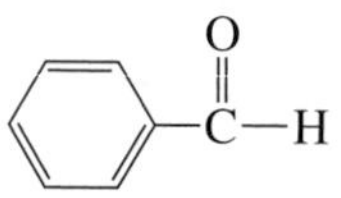

f. $CH_3CH_2CH_2CH_2OH$ and $CH_3CH_2CH_2OCH_3$

g. *cis*-2-butene and *trans*-2-butene

h. $CH_3CH_2CH_2\overset{\overset{\displaystyle O}{\|}}{C}OCH_3$ and $CH_3CH_2CH_2\overset{\overset{\displaystyle O}{\|}}{C}CH_3$

i. 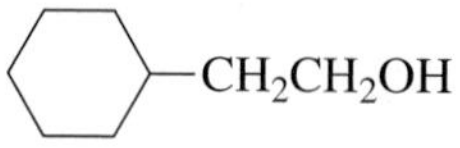 (phenyl)${-}CH_2CH_2OH$ and (phenyl)$\underset{\underset{\displaystyle OH}{|}}{C}HCH_3$

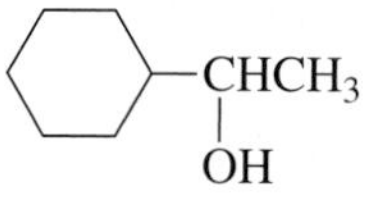

j. $CH_3CH_2CH{=}CHCH_3$ and $CH_3CH_2C{\equiv}CCH_3$

33. Which peak would be more intense in the mass spectrum of the following compounds, the peak at $m/z = 57$ or the peak at $m/z = 71$?
a. 3-methylpentane   b. 2-methylpentane

34. List three factors that influence the intensity of an IR absorption band.

35. 4-Methyl-3-penten-2-one has two absorption bands in its UV spectrum, one at 236 nm and one at 314 nm.
a. Why are there two absorption bands?
b. Which band shows the greater absorbance?

$$CH_3\overset{\overset{\displaystyle O}{\|}}{C}CH{=}\overset{\overset{\displaystyle CH_3}{|}}{C}CH_3$$
4-methyl-3-penten-2-one

36. How could you determine by IR spectroscopy that the following reaction had occurred?

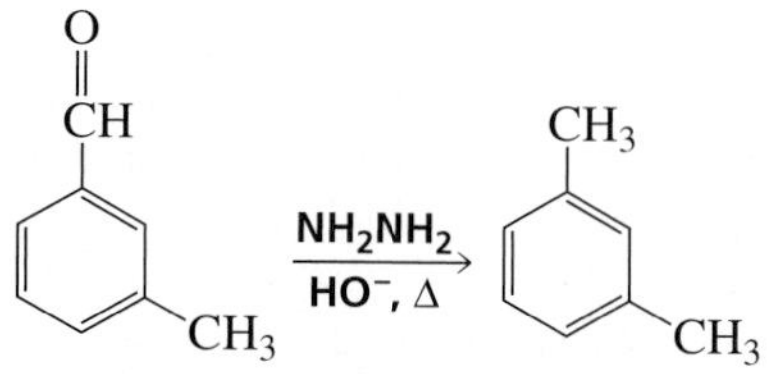

(structure: benzene ring with CHO at one position and CH₃ at another) $\xrightarrow[\text{HO}^-,\ \Delta]{\text{NH}_2\text{NH}_2}$ (structure: benzene ring with two CH₃ groups)

37. What identifying characteristic would be present in the mass spectrum of a compound containing two bromine atoms?

38. a. Methyl orange (its structure is given in Section 12.19) is an acid–base indicator. In solutions of pH < 4 it is red, and in solutions of pH > 4 it is yellow. Account for the change in color.

 b. Phenolphthalein is also an indicator, but it exhibits a much more dramatic color change. In solutions of pH < 8.5 it is colorless, and in solutions of pH > 8.5 it is deep red-purple. Account for the change in color.

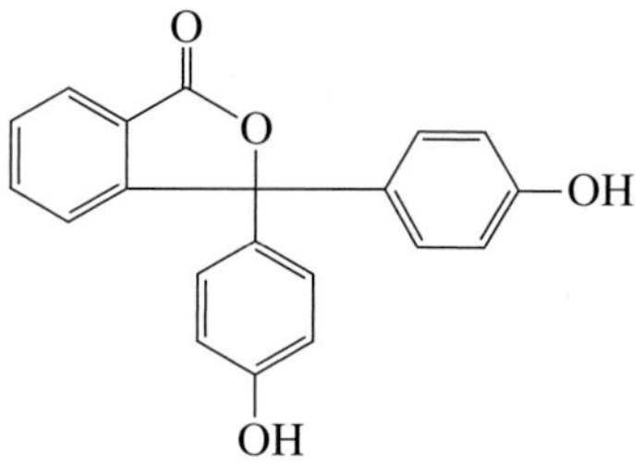

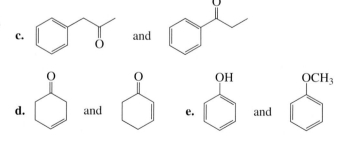

phenolphthalein

39. Assuming that the force constant is approximately the same for carbon–carbon, carbon–nitrogen, and carbon–oxygen single bonds, predict the relative positions of their stretching vibrations.

40. A mass spectrum shows significant peaks at m/z = 87, 115, 140, and 143. Which of the following compounds is responsible for that mass spectrum:

4,7-dimethyl-1-octanol, 2,6-dimethyl-4-octanol, or 2,2,4-trimethyl-4-heptanol?

41. How could you use IR spectroscopy to distinguish between 1,5-hexadiene and 2,4-hexadiene?

42. How could you distinguish between the compounds in each of the following pairs using UV spectroscopy?

 a. 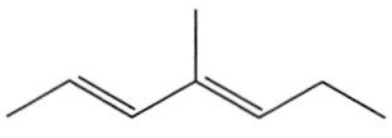   and

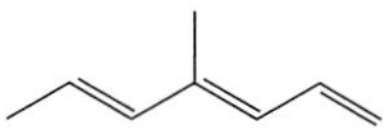

 b. CH_2=CHCH=CHCH=CH_2 and CH_2=CHCH=CHCCH$_3$

 c.   and

 d.   and   **e.**   and

43. For each of the IR spectra in Figures 12.42, 12.43, and 12.44, four compounds are shown. In each case, indicate which of the four compounds is responsible for the spectrum.

 a. $CH_3CH_2CH_2C$≡CCH_3 $CH_3CH_2CH_2CH_2OH$ $CH_3CH_2CH_2CH_2C$≡CH $CH_3CH_2CH_2COH$

 b. CH_3CH_2COH $CH_3CH_2COCH_2CH_3$ CH_3CH_2CH $CH_3CH_2CCH_3$

c. $C(CH_3)_3$ CH_2CH_2Br $CH=CH_2$ CH_2OH

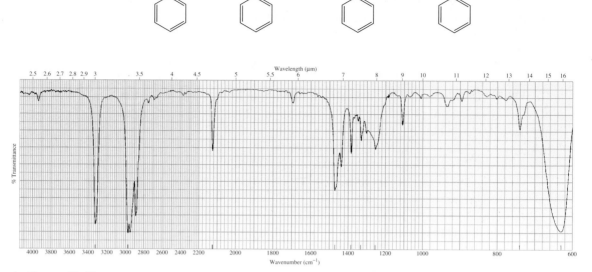

▲ **Figure 12.42**
The IR spectrum for Problem 43a.

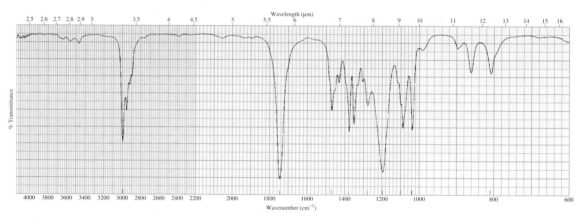

▲ **Figure 12.43**
The IR spectrum for Problem 43b.

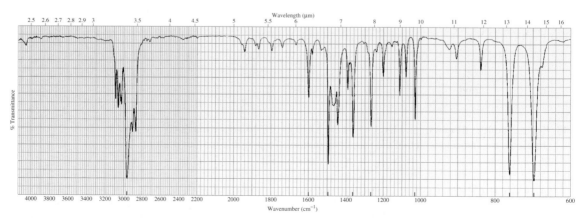

▲ **Figure 12.44**
The IR spectrum for Problem 43c.

44. What hydrocarbons will have a molecular ion peak at $m/z = 112$?

45. In the following boxes, list the types of bonds and the approximate wavenumber where each type of bond is expected to show an IR absorption.

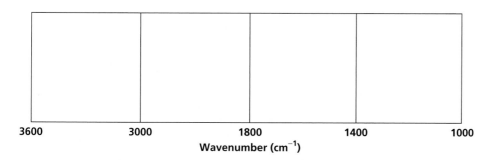

3600   3000   1800   1400   1000

Wavenumber (cm^{-1})

46. What peaks in their mass spectra would you use to distinguish between 4-methyl-2-pentanone and 2-methyl-3-pentanone?

47. A compound is known to be one of those shown here. Tell what absorption bands you would look for in the compound's IR spectrum that would allow you to identify the compound.

A   B   C

48. How could you use IR spectroscopy to distinguish among 1-hexyne, 2-hexyne, and 3-hexyne?

49. For each of the IR spectra in Figures 12.45, 12.46, and 12.47, indicate which of the five given compounds is responsible for the spectrum.

a.

$CH_3CH_2CH{=}CH_2$ $CH_3CH_2CH_2CH_2OH$ $CH_2{=}CHCH_2CH_2OH$ $CH_3CH_2CH_2OCH_3$ $CH_3CH_2CH_2\overset{\displaystyle O}{\overset{\displaystyle \|}{C}}OH$

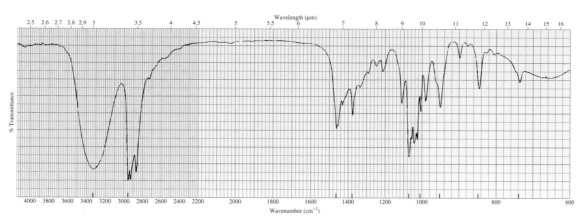

▲ **Figure 12.45**
The IR spectrum for Problem 49a.

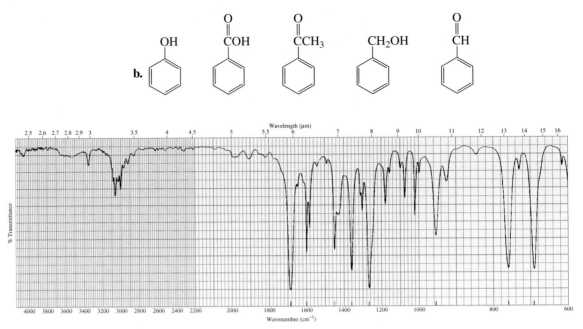

▲ **Figure 12.46**
The IR spectrum for Problem 49b.

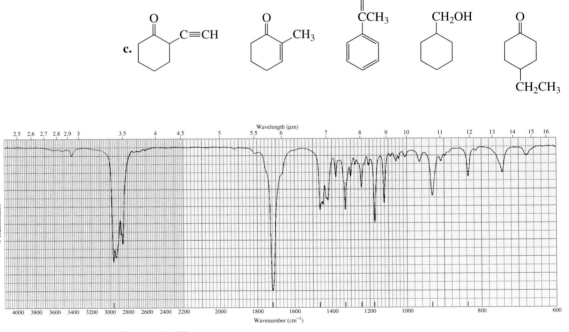

▲ **Figure 12.47**
The IR spectrum for Problem 49c.

50. Many credit card slips do not have carbon paper. Nevertheless, when you sign the slip, an imprint of your signature is made on the bottom copy. The carbonless paper contains tiny capsules that are filled with the colorless compound shown here. When you press on the paper, the capsules burst and the colorless compound comes into contact with the acid-treated paper, forming a highly colored compound. What is the structure of the colored compound?

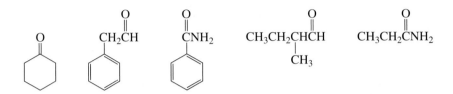

51. Each of the IR spectra shown in Figure 12.48 is the spectrum of one of the following compounds. Identify the spectra.

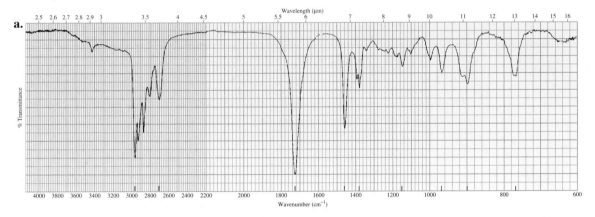

▼ **Figure 12.48**
The IR spectra for Problem 51.

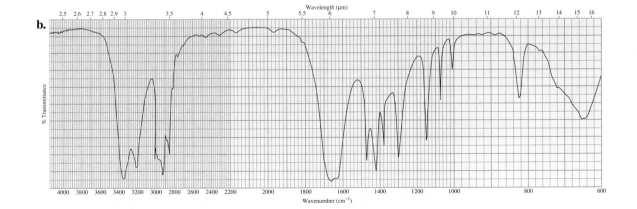

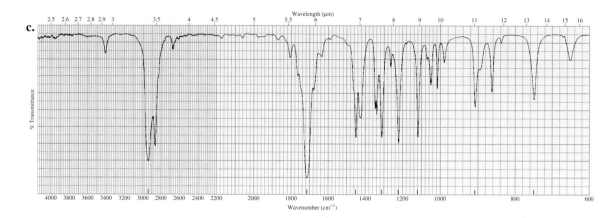

52. Predict the major characteristic IR absorption bands that would be given by each of the following compounds:

a. $CH_2{=}CHCH_2\overset{\overset{\displaystyle O}{\|}}{C}H$

b.   (structure) $\overset{\overset{\displaystyle O}{\|}}{C}OCH_2CH_3$

c. $CH_3CH_2\overset{\overset{\displaystyle O}{\|}}{C}CH_2CH_2NH_2$

d.   (cyclohexane ring) CH_2CH_2OH

e.   (cyclohexane ring) $CH_2C{\equiv}CH$

f.   (cyclohexane ring) $CH_2\overset{\overset{\displaystyle O}{\|}}{C}OH$

53. The IR spectrum of a compound with molecular formula C_5H_8O was obtained in CCl_4 and is shown in Figure 12.49. Identify the compound.

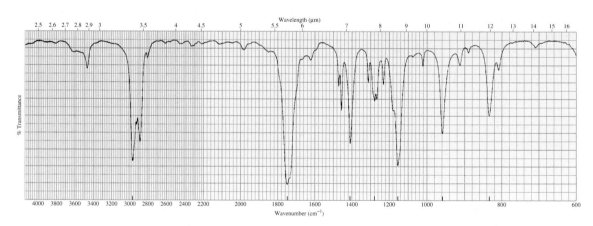

▲ **Figure 12.49**
The IR spectrum for Problem 53.

54. A solution of ethanol has been contaminated with benzene. Benzene has a molar absorptivity in ethanol of 230 at 260 nm, and ethanol shows no absorbance at 260 nm. How could you determine the concentration of benzene in the solution?

55. The IR spectrum shown in Figure 12.50 is the spectrum of one of the following compounds. Identify the compound.

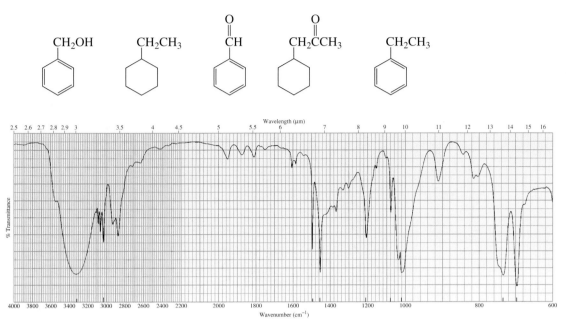

▲ **Figure 12.50**
The IR spectrum for Problem 55.

56. The IR spectrum shown in Figure 12.51 is the spectrum of one of the following compounds. Identify the compound.

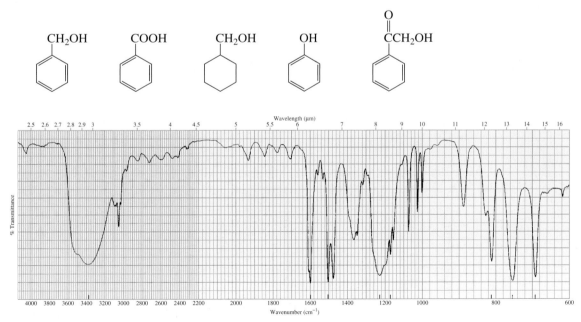

▲ **Figure 12.51**
The IR spectrum for Problem 56.

57. Determine the molecular formula of a saturated acyclic hydrocarbon with an M peak at $m/z = 100$ with a relative intensity of 27.32% and an M + 1 peak with a relative intensity of 2.10%.

58. Calculate the approximate wavenumber at which a C=C stretch will occur given that the force constant for the carbon–carbon double bond is 10×10^5 g s^{-2}.

59. Determine the structure of each of the compounds based on its mass spectrum and IR spectrum.

▼ **Figure 12.52**
The IR and mass spectra for Problem 59.

a.

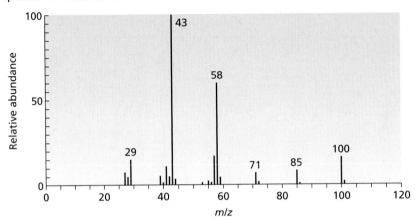

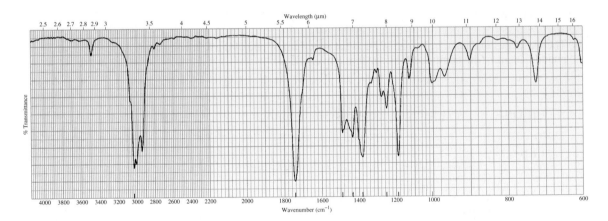

b.

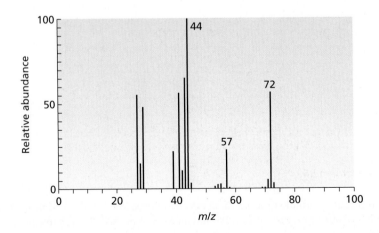

b. continued

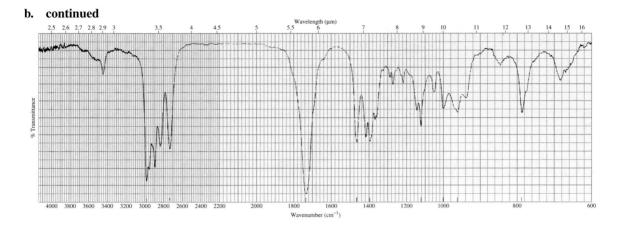

c.

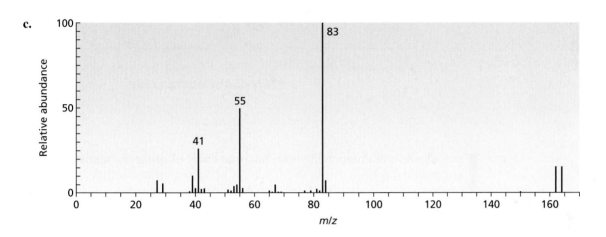

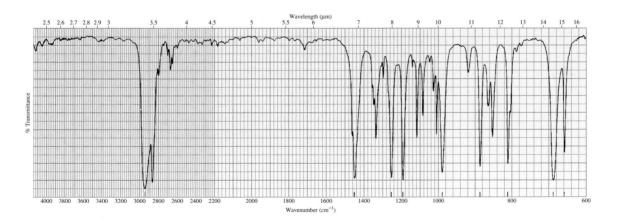

60. Given that the force constants are similar for carbon–hydrogen and carbon–carbon bonds, explain why the stretching vibration of a carbon–hydrogen bond occurs at a greater wavenumber.

NMR
Spectroscopy

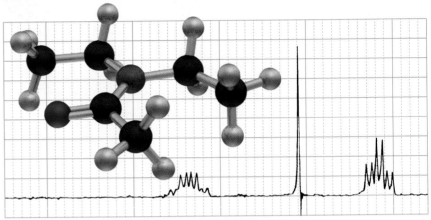

N,N-diethylethanamide

Identification of compounds is an important part of organic chemistry. After a compound has been synthesized, it must be identified to confirm its structure. Chemists who study natural products must determine the structure of a naturally occurring compound before they can either design a synthesis to produce the compound in greater quantities than nature can provide, or design and synthesize related compounds with modified properties.

In Chapter 12 you were introduced to three instrumental techniques used to identify organic compounds—mass spectrometry, infrared spectroscopy, and UV/Vis spectroscopy. Now we will look at nuclear magnetic resonance (NMR) spectroscopy, another instrumental technique that chemists use for structure determination. **NMR spectroscopy** helps to identify the carbon–hydrogen framework of an organic compound.

The power of NMR spectroscopy, compared to the other instrumental techniques we have studied, is that it not only makes it possible to identify the functionality at a specific carbon, but also lets us determine what the neighboring carbons look like. In many cases NMR spectroscopy can be used to determine the entire structure of a molecule.

13.1
INTRODUCTION TO
NMR SPECTROSCOPY

NMR spectroscopy was developed by physical chemists in the late 1940s to study the properties of atomic nuclei. In 1951, chemists realized that NMR spectroscopy could also be used to determine the structures of organic compounds. We have seen that electrons are charged, spinning particles with two allowed spin states, $+1/2$ and $-1/2$ (Section 1.2). Certain nuclei also have allowed spin states of $+1/2$ and $-1/2$, and this allows them to be studied by NMR. Examples of such nuclei are ^1H, ^{13}C, ^{19}F, and ^{31}P.

Because hydrogen nuclei (protons) were the first nuclei studied by *nuclear magnetic resonance,* the acronym "NMR" is generally assumed to mean **^1H NMR** (*proton magnetic resonance*). Spectrometers were later developed for **^{13}C NMR,** ^{15}N NMR, ^{19}F NMR, ^{31}P NMR, and other magnetic nuclei.

In the absence of a magnetic field, the nuclear spins are randomly oriented. However, when a sample is placed in a magnetic field (Figure 13.1), nuclei with spin $+1/2$ align with the applied field (in the lower energy α-**spin state**) and nuclei with spin $-1/2$ align against the applied field (in the higher energy β-spin state). There are more nuclei in the α-spin state than in the β-spin state. The difference in the populations is not great, but is sufficient to form the basis of NMR spectroscopy.

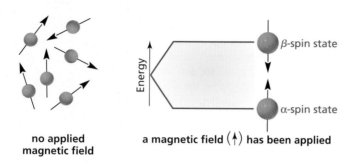

no applied
magnetic field

a magnetic field ($\uparrow$) has been applied

◀ **Figure 13.1**
In the absence of an applied magnetic field, the spins of the nuclei are randomly oriented. In the presence of an applied magnetic field, the spins of the nuclei line up with or against the field.

The energy difference (ΔE) between the α- and β-spin states depends on the strength of the **applied magnetic field** (B_0). The greater the strength of the applied magnetic field, the greater the difference in energy (Figure 13.2).

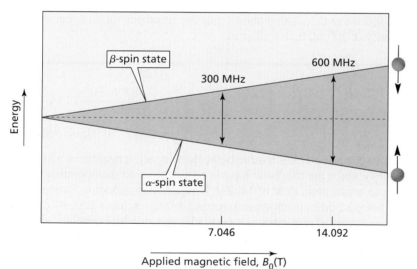

Applied magnetic field, B_0(T)

◀ **Figure 13.2**
The greater the strength of the applied magnetic field, the greater the difference in energy between the α- and β-spin states.

When the sample is subjected to a brief, intense pulse of radiation, nuclei in the α-spin state can be promoted to the β-spin state (called "flipping" the spin). The radiation required for currently available magnets is in the radiofrequency (rf) region of the electromagnetic spectrum, so it is called **rf radiation.** When the nuclei return to their original state, they emit signals whose frequency depends on the difference in energy (ΔE) between the α- and β-spin states. The NMR spectrometer detects these signals and displays them as a plot of signal frequency versus intensity—an NMR spectrum (see Figure 13.5). The term "nuclear magnetic resonance" comes from the fact that the nuclei are *in resonance* with the rf radiation. That is, they are flipping back and forth between the α- and β-spin states in response to the rf radiation.

Planck's constant, h, is the proportionality constant relating ΔE and frequency (ν). The following equation shows that the frequency of the signal depends on the strength

Edward Mills Purcell (1912–1997) and Felix Bloch did the work on the magnetic properties of nuclei that made the development of NMR spectroscopy possible. They shared the 1952 Nobel Prize in physics. Purcell was born in Illinois. He received a Ph.D. from Harvard University in 1938 and immediately was hired as a faculty member in the physics department.

NIKOLA TESLA
(1856–1943)

Nikola Tesla was born in Croatia, the son of a clergyman. He emigrated to the United States in 1884, becoming a citizen in 1891. He was a proponent of alternating current and bitterly fought Edison, who promoted direct current. Although he did not win his dispute with Marconi over which of them invented the radio, Tesla is given credit for developing neon and fluorescent lighting, the electron microscope, the motor for the refrigerator, and the Tesla coil, a type of transformer.

Nikola Tesla in his laboratory

of the magnetic field (B_0) measured in tesla (T)[1] and the **gyromagnetic ratio** (γ). The size of the gyromagnetic ratio depends on the particular kind of nucleus. In the case of the proton, it is $2.675 \times 10^8 \ T^{-1} \ s^{-1}$.

$$\Delta E = h\nu = h\frac{\gamma}{2\pi}B_0$$

If an NMR spectrometer is equipped with a magnet with a magnetic field of 7.046 T, the following calculation shows that the spectrometer will require an operating frequency of 300 MHz (megahertz).

$$\nu = \frac{\gamma}{2\pi}B_0$$

$$= \frac{2.675 \times 10^8}{2(3.1416)} T^{-1} s^{-1} \times 7.046 \ T$$

$$= 300 \times 10^6 \ Hz = 300 \ MHz$$

> Earth's magnetic field is 5×10^{-5} T, measured at the equator. Its maximum surface magnetic field is 7×10^{-5} T, measured at the south magnetic pole.

If the spectrometer has a more powerful magnet, it must have a higher operating frequency, since the magnetic field is proportional to the operating frequency. For example, a magnetic field of 14.092 T requires an operating frequency of 600 MHz.

Today's NMR spectrometers operate at frequencies of 200, 300, 500, and 600 MHz. The **operating frequency** of a particular spectrometer depends on the strength of the built-in magnet. The greater the operating frequency of the instrument—and the stronger the magnet—the better the resolution of the NMR spectrum (Section 13.17).

Because each kind of nucleus has its own gyromagnetic ratio, different energies are required to bring different kinds of nuclei into resonance. For example, an NMR spectrometer with a magnet requiring a frequency of 300 MHz to flip the spin of an ^1H nucleus requires a frequency of 75 MHz to flip the spin of a ^{13}C nucleus. NMR spectrometers are equipped with radiation sources that can be tuned to different frequencies so they can be used to obtain NMR spectra of different kinds of nuclei (^1H, ^{13}C, ^{15}N, ^{19}F, ^{31}P, etc.).

> *Felix Bloch (1905–1983) was born in Switzerland. His first academic appointment was at the University of Leipzig. He left Germany when Hitler came to power and worked at universities in Denmark, Holland, and Italy. He eventually came to the United States, becoming a citizen in 1939. He was a professor of physics at Stanford University and worked on the atomic bomb project at Los Alamos, New Mexico, during World War II.*

PROBLEM 1◆

What frequency (in MHz) is required to cause a proton to flip its spin when it is exposed to a magnetic field of 1 tesla?

---

[1]Until recently, the gauss (G) was the unit in which magnetic field strength was commonly measured ($1 \ T = 10^4 \ G$).

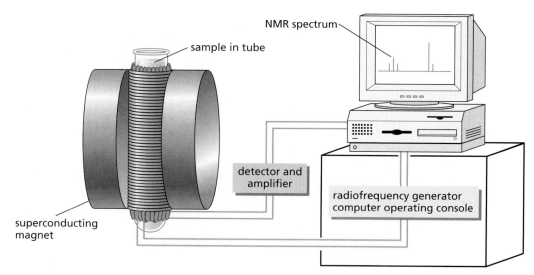

NMR spectrum

sample in tube

detector and amplifier

radiofrequency generator computer operating console

superconducting magnet

▲ **Figure 13.3**
Schematic of an NMR spectrometer.

PROBLEM 2◆

a. Calculate the magnetic field (in tesla) required to flip an ^{1}H nucleus in an NMR spectrometer that operates at 360 MHz.

b. What strength magnetic field is required when a 500-MHz instrument is used?

13.2 FOURIER TRANSFORM NMR

To obtain an NMR spectrum, a small amount of the compound is dissolved in about 0.5 mL of solvent and the solution is put into a long, thin glass tube, which is placed within a powerful magnetic field (Figure 13.3). The solvents used in NMR are discussed in Section 13.16. Spinning the sample tube about its long axis averages the position of the molecules in the magnetic field, which greatly increases the resolution of the spectrum.

In modern instruments called pulsed Fourier transform (FT) spectrometers, the magnetic field is kept constant and an rf pulse of short duration excites all the protons simultaneously. Because the short rf pulse covers a range of frequencies, the individual protons absorb the frequency required to come into resonance (flip their spin). As they return to equilibrium, they produce a sine wave emission at the frequency corresponding to ΔE. The intensity of the sine wave decays with time as nuclei resonating at that frequency return to equilibrium. A computer collects the intensity-versus-time data and converts it into intensity-versus-frequency information (a Fourier transform) to produce a spectrum called a **Fourier transform NMR (FT–NMR)** spectrum. An FT–NMR spectrum can be recorded in about 2 seconds using less than 5 mg of compound. The NMR spectra in this book are FT–NMR spectra that were taken on a spectrometer with an operating frequency of 300 MHz.

13.3 SHIELDING

We have seen that when a sample in a magnetic field is irradiated with rf radiation of the proper frequency, each proton[2] in an organic compound gives a signal at a frequency that depends on the energy difference (ΔE) between the α- and β-spin states—

---

[2]The terms "proton" and "hydrogen" are both used to describe covalently bound hydrogen in discussions of NMR spectroscopy.

*The 1991 Nobel Prize in chemistry was awarded to **Richard R. Ernst** for two important contributions: FT–NMR spectroscopy, and an NMR tomography method that forms the basis of magnetic resonance imaging (MRI). Ernst was born in 1933, received a Ph.D. from the Swiss Federal Institute of Technology [Eidgenössische Technische Hochschule (ETH)] in Zurich, and became a research scientist at Varian Associates in Palo Alto, Calif. In 1968 he returned to ETH, where he is a professor of chemistry.*

and ΔE is determined by the strength of the applied magnetic field (Figure 13.2). If all the protons in an organic compound had exactly the same environment, they would all give signals with the same frequency in response to a given applied magnetic field. If this were the case, all NMR spectra would consist of one signal, which would tell us nothing about the structure of the compound except that it contains protons.

A nucleus, however, is embedded in a cloud of electrons that partly shields it from the applied magnetic field. Fortunately for chemists, the **shielding** varies for different protons within a molecule. In other words, all the protons do not experience the same applied magnetic field.

What causes shielding? In an applied magnetic field, the electrons circulate about the nuclei and induce a local magnetic field that opposes (is subtracted from) the applied magnetic field. The **effective magnetic field,** therefore, is what the nuclei "sense" through the surrounding electronic environment.

$$B_{effective} = B_{applied} - B_{local}$$

This means that the greater the electron density of the environment in which the proton is located, the greater B_{local} will be, and the more the proton will be shielded from the applied magnetic field. This type of shielding is called **diamagnetic shielding.**

Thus, protons that are more shielded from the applied magnetic field sense a *smaller effective magnetic field*. Therefore, they will require a *lower frequency* to come into resonance (flip their spin) because ΔE is smaller (Figure 13.2). Protons that are less shielded from the applied magnetic field will sense a *larger effective magnetic field* and, therefore, will require a *higher frequency* to come into resonance because ΔE is larger.

We will see a signal in an NMR spectrum for each proton in a different environment. Protons in electron-rich environments are more shielded and will appear at lower frequencies—on the right-hand side of the spectrum (Figure 13.4). Protons in electron-poor environments are less shielded and will appear at higher frequencies—on the left-hand side of the spectrum.

Figure 13.4 ▶
Shielded nuclei come into resonance at lower frequencies than deshielded nuclei.

right-hand side of the spectrum:

**protons in electron-dense environments
low frequency
upfield
shielded**

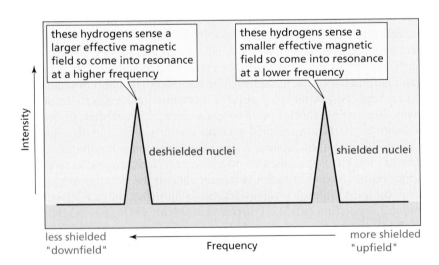

left-hand side of the spectrum:

**protons in electron-poor environments
high frequency
downfield
deshielded**

The terms "upfield" and "downfield," which came into use when continuous-wave spectrometers were used (before the advent of Fourier transform spectrometers), are so entrenched in the vocabulary of NMR that you should know what they mean—**upfield** means farther to the right-hand side of the spectrum and **downfield** means farther to the left-hand side of the spectrum. In contrast to FT–NMR techniques that hold magnetic field strength constant and vary frequency, continuous wave techniques hold frequency constant and vary magnetic field. Because higher magnetic fields are required for shielded protons to come into resonance at a given frequency, magnetic field increases going from left to right across a spectrum taken using continuous-wave techniques (Figure 13.4). Therefore, *upfield* is toward the right and *downfield* is toward the left.

13.4
THE NUMBER OF SIGNALS IN THE ^{1}H NMR SPECTRUM

Protons in the same environment are called **chemically equivalent protons.** For example, 1-bromopropane has three different sets of chemically equivalent protons. The three methyl protons are chemically equivalent because of rotation about the carbon–carbon bonds. The two methylene protons are chemically equivalent, and the two protons on the carbon bonded to the bromine atom make up the third set of chemically equivalent protons.

Each set of chemically equivalent protons in a compound should give rise to a signal in the ^{1}H NMR spectrum of that compound. (Sometimes the signals are not sufficiently separated and overlap each other. When this happens, one sees fewer signals than anticipated.) Because 1-bromopropane has three sets of chemically equivalent protons, it should have three signals in its ^{1}H NMR spectrum.

2-Bromopropane has two sets of chemically equivalent protons and, therefore, it has two signals in its ^{1}H NMR spectrum. The six methyl protons in 2-bromopropane are equivalent, so they give rise to only one signal. Ethyl methyl ether has three sets of chemically equivalent protons—the methyl protons on the carbon adjacent to the oxygen, the methylene protons on the carbon adjacent to the oxygen, and the methyl protons on the carbon one carbon removed from the oxygen. The chemically equivalent protons in the following compounds are designated by the same letter.

3-D Molecule:
Chlorocyclobutane

From the number of signals in an ^{1}H NMR spectrum, you can tell how many sets of chemically equivalent protons the compound has.

Sometimes two protons on the same carbon are not equivalent. For example, the ^{1}H NMR spectrum of chlorocyclobutane has five signals. Even though they are bonded to the same carbon, the H_a and H_b protons are not equivalent because they are not in the same environment: H_a is trans to Cl and H_b is cis to Cl. Similarly, the H_c and H_d protons are not equivalent.

chlorocyclobutane
the NMR spectrum has 5 signals
H_a and H_b are not equivalent
H_c and H_d are not equivalent

PROBLEM 3◆

How many signals would you expect to see in the ^{1}H NMR spectrum of each of the following compounds?

a. $CH_3CH_2CH_2\overset{\overset{\displaystyle O}{\|}}{C}CH_3$

b. $CH_2{=}CH\overset{\overset{\displaystyle O}{\|}}{C}H$

c. $CH_3CH_2\underset{\underset{\displaystyle Cl}{|}}{C}HCH_2CH_3$

d. $CH_2{=}CHCl$

e. $CH_3CH_2CH_2CH_3$

f. $CH_3\underset{\underset{\displaystyle CH_3}{|}}{C}HCH_2\underset{\underset{\displaystyle CH_3}{|}}{C}HCH_3$

g. $BrCH_2CH_2Br$

h. 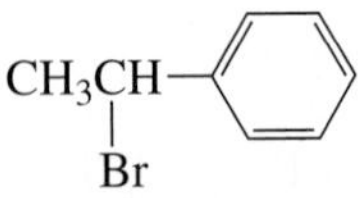 $CH_3\underset{\underset{\displaystyle Br}{|}}{C}H\!-\!\bigcirc$

i. 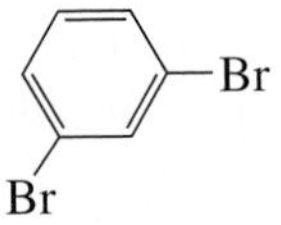 (benzene ring with two Br substituents, meta)

j. 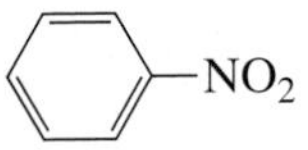 (benzene ring with NO_2)

k.  (cyclohexadiene ring)

l. 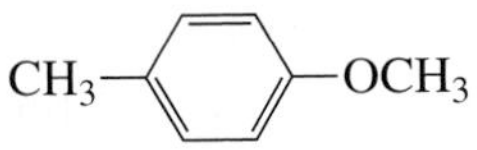 $CH_3\!-\!\bigcirc\!-\!OCH_3$

m.  $CH_3\!-\!\bigcirc\!-\!CH_3$

n. 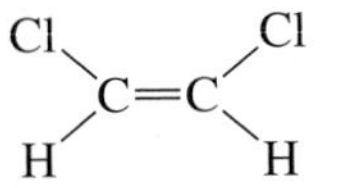 $\underset{\displaystyle H}{\overset{\displaystyle Cl}{}}C{=}C\underset{\displaystyle H}{\overset{\displaystyle Cl}{}}$

o. 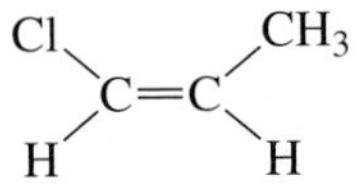 $\underset{\displaystyle H}{\overset{\displaystyle Cl}{}}C{=}C\underset{\displaystyle H}{\overset{\displaystyle CH_3}{}}$

PROBLEM 4

How could you distinguish among the ^1H NMR spectra of the following compounds?

a. $CH_3OCH_2OCH_3$ **b.** CH_3OCH_3 **c.** $CH_3OCH_2\underset{\underset{\displaystyle CH_3}{|}}{\overset{\overset{\displaystyle CH_3}{|}}{C}}CH_2OCH_3$

PROBLEM 5

There are three isomeric dichlorocyclopropanes. Their ^1H NMR spectra show one signal for isomer 1, two signals for isomer 2, and three signals for isomer 3. Draw the structures of isomers 1, 2, and 3.

**13.5
THE CHEMICAL
SHIFT**

A small amount of an inert **reference compound** is added to the sample tube containing the compound whose NMR spectrum is to be taken. The positions of the signals in an NMR spectrum are defined according to how far they are from the signal of the reference compound. The most commonly used reference compound is tetramethylsilane (TMS). TMS is a highly volatile compound, so—after the NMR spectrum is taken—it can be removed from the sample by evaporation.

The methyl protons of TMS are in a more electron-dense environment than most protons in organic molecules because silicon is less electronegative than carbon (electronegativity of silicon = 1.8; electronegativity of carbon = 2.5). Consequently, the signal for the methyl protons of TMS is at a lower frequency than most other signals.

The position in an NMR spectrum where a signal occurs is called the **chemical shift**. The chemical shift indicates how far the signal is from the TMS peak. The most common scale for chemical shifts is the δ (delta) scale. The TMS signal is used to define the zero position on the δ scale. The chemical shift is determined by measuring the distance from the TMS peak (in hertz) and dividing by the operating frequency of the instrument (in megahertz). Because the units are Hz/MHz, a chemical shift has units of parts per million (ppm) of the operating frequency. Most proton chemical shifts fall in the range 0 to 10 ppm.

tetramethylsilane
TMS

3-D Molecule:
Tetramethylsilane

$$\delta = \text{chemical shift (ppm)} = \frac{\text{distance downfield from TMS (Hz)}}{\text{operating frequency of the spectrometer (MHz)}}$$

The ^{1}H NMR spectrum for 1-bromo-2,2-dimethylpropane shows that the chemical shift of the methyl protons is at 1.05 ppm and the chemical shift of the methylene protons is at 3.28 ppm (Figure 13.5). *Notice that low frequency (upfield) signals have small ppm values, while high frequency (downfield) signals have large ppm values.*

The greater the chemical shift (δ), the higher the frequency.

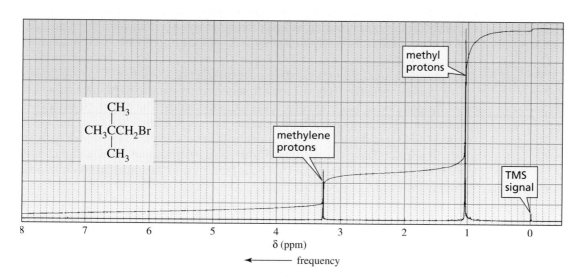

methyl protons

CH$_3$CCH$_2$Br with CH$_3$ above and CH$_3$ below

methylene protons

TMS signal

δ (ppm)

← frequency

The advantage of the δ scale is that the chemical shift of a given nucleus is *independent of the operating frequency of the NMR spectrometer.* The chemical shift of the methyl protons of 1-bromo-2,2-dimethylpropane is at 1.05 ppm in both a 60-MHz and a 360-MHz instrument. In contrast, if the chemical shift were reported in hertz, it would be at 63 Hz in a 60-MHz instrument and at 378 Hz in a 360-MHz instrument (63/60 = 1.05; 378/360 = 1.05).

▲ **Figure 13.5**
^{1}H NMR spectrum of 1-bromo-2,2-dimethylpropane.

Chemical shifts in ppm are independent of the operating frequency of the spectrometer.

PROBLEM 6◆

A signal has been reported to occur at 120 Hz downfield from TMS in an NMR spectrometer with a 300-MHz operating frequency.

a. What is its chemical shift?

b. What would its chemical shift be in an instrument operating at 100 MHz?

c. How many hertz downfield from TMS would the signal be in a 100-MHz spectrometer?

PROBLEM 7◆

a. If two signals differ by 1.5 ppm in a 300-MHz spectrometer, how do they differ in ppm in a 100-MHz spectrometer?

b. If two signals differ by 90 hertz in a 300-MHz spectrometer, how do they differ in hertz in a 100-MHz spectrometer?

PROBLEM 8◆

Where would you expect to find the ^{1}H NMR signal of $(CH_3)_2Mg$ relative to the TMS signal? (*Hint:* See Table 11.3.)

13.6
THE POSITION OF THE ^{1}H NMR SIGNALS

The ^{1}H NMR spectrum of 1-bromo-2,2-dimethylpropane has two signals because the compound has two different kinds of protons (Figure 13.5). The methylene protons are in a less electron-dense environment than the methyl protons because the methylene protons are closer to the electron-withdrawing bromine. Because the methylene protons are in a less electron-dense environment, they are less shielded from the applied magnetic field. The signal for these protons therefore occurs at a higher frequency than the signal for the more shielded methyl protons. *Remember that the right-hand side of an NMR spectrum is the lower frequency side, where protons with electron-dense environments (more shielded) show a signal. The left-hand side is the high frequency side, where less shielded protons show a signal (Figure 13.4).*

We would expect the ^{1}H NMR spectrum of nitropropane to have three signals because the compound has three different kinds of protons. The closer the protons are to the electron-withdrawing nitro group, the less shielded they are from the applied magnetic field and the higher the frequency (farther downfield) at which their signals will appear. Thus, the protons closest to the nitro group show a signal at the highest frequency (4.37 ppm) and the ones farthest from the nitro group show a signal at the lowest frequency (1.04 ppm).

Electron withdrawal causes NMR signals to appear at higher frequencies (at larger ppm values).

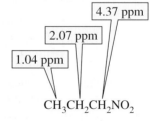

$$CH_3CH_2CH_2NO_2$$

with labels: 1.04 ppm, 2.07 ppm, 4.37 ppm

Compare the chemical shifts of the methylene protons immediately adjacent to the halogen in each of the following alkyl halides. The position of the signal depends on the electronegativity of the halogen. The signal for the methylene protons adjacent to fluorine (the most electronegative of the halogens) occurs at the highest frequency, while the signal for the methylene protons adjacent to iodine (the least electronegative of the halogens) occurs at the lowest frequency.

$CH_3CH_2CH_2CH_2CH_2F$ $CH_3CH_2CH_2CH_2CH_2Cl$ $CH_3CH_2CH_2CH_2CH_2Br$ $CH_3CH_2CH_2CH_2CH_2I$

4.50 ppm   3.50 ppm   3.40 ppm   3.20 ppm

Which set of protons in each of the following compounds is the least shielded?
Which set of protons in each compound is the most shielded?

a. $CH_3CH_2CH_2Cl$

b. $CH_3CH_2\overset{\overset{\displaystyle O}{\|}}{C}OCH_3$

c. $CH_3\underset{\underset{\displaystyle Br}{|}}{C}H\underset{\underset{\displaystyle Br}{|}}{C}HBr$

Approximate values of chemical shifts for different kinds of protons are shown in Table 13.1. (A more extensive compilation of chemical shifts is given in Appendix VI). An ^1H NMR spectrum can be divided into six regions. If you can remember the kinds of protons that are in each region, you will be able to tell what kinds of protons a molecule has from a quick look at its NMR spectrum.

13.7 CHARACTERISTIC VALUES OF CHEMICAL SHIFTS

$\overset{\overset{\displaystyle O}{\|}}{-C}-H$ $\overset{\overset{\displaystyle O}{\|}}{-C}-OH$			$\overset{\displaystyle H}{}$ $\overset{\displaystyle H}{}C=C\overset{\displaystyle H}{}$ vinylic	$\overset{\overset{\displaystyle Z}{\|}}{-C}-H$ $Z = O, N, halogen$	$\overset{\overset{\displaystyle O}{\|}}{C}-\overset{\overset{\displaystyle H}{\|}}{C}-$ $C=C-\overset{\overset{\displaystyle H}{\|}}{C}-$ allylic	$-\overset{\overset{\displaystyle |}{\|}}{C}-\overset{\overset{\displaystyle H}{\|}}{C}-H$ saturated

12   9.0   8.0   6.5   4.5   2.5   1.5   0

δ (ppm)

Table 13.1 shows that the chemical shift of methyl protons is at a lower frequency (0.9 ppm) than the chemical shift of methylene protons (1.3 ppm) in a similar environment, and that the chemical shift of methylene protons is at a lower frequency than the chemical shift of a methine proton (1.4 ppm) in a similar environment. (When an sp^3 carbon is bonded to only one hydrogen, the hydrogen is called a **methine hydrogen.**) For example, the ^1H NMR spectrum of butanone shows three signals. The *a* proton signal of butanone is the signal at the lowest frequency because the protons are farthest from the electron-withdrawing carbonyl group. (In illustrations of NMR spectra, the set of protons responsible for the signal at the lowest frequency will be labeled *a*, the next set will be labeled *b*, the next set *c*, etc.) The *b* and *c* protons are the same distance from the carbonyl group, but the signal for the *c* protons is at a higher frequency because methylene protons appear at a higher frequency than methyl protons in a similar environment.

In a similar environment, the signal for methyl protons occurs at a lower frequency than the signal for methylene protons, which occurs at a lower frequency than the signal for a methine proton.

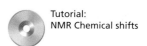

Tutorial:
NMR Chemical shifts

$$CH_3CH_2\overset{\overset{\displaystyle O}{\|}}{C}CH_3 \atop {}_{a\quad c\quad\;\; b}$$

butanone

$$\overset{b\quad\quad c\;\; a}{CH_3OCHCH_3} \atop {\underset{a}{\overset{|}{CH_3}}}$$

2-methoxypropane

The signal for the *a* protons of 2-methoxypropane is the signal at the lowest frequency in the ^1H NMR spectrum of this compound because these protons are farthest from the electron-withdrawing oxygen. The *b* and *c* protons are the same distance from the oxygen, but the signal for the *c* proton appears at a higher frequency because, in a similar environment, a methine proton signal appears at a higher frequency than a methyl proton signal.

TABLE 13.1 Approximate Values of Chemical Shifts for ^1H NMR[a]

Type of proton	Approximate chemical shift (ppm)	Type of proton	Approximate chemical shift (ppm)
$(CH_3)_4Si$	0	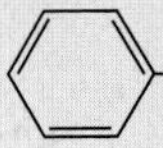 ⬡—H	6.5–8
—CH_3	0.9	$\overset{O}{\overset{\|\|}{-C-H}}$	9.0–10
—CH_2—	1.3	$I-\overset{\|}{\underset{\|}{C}}-H$	2.5–4
—$\overset{\|}{\underset{\|}{CH}}$—	1.4	$Br-\overset{\|}{\underset{\|}{C}}-H$	2.5–4
$-\overset{\|}{C}=\overset{\|}{C}-CH_3$	1.7		
$\overset{O}{\overset{\|\|}{-C-CH_3}}$	2.1	$Cl-\overset{\|}{\underset{\|}{C}}-H$	3–4
⬡—CH_3	2.3	$F-\overset{\|}{\underset{\|}{C}}-H$	4–4.5
—C≡C—H	2.4	RNH_2	variable, 1.5–4
R—O—CH_3	3.3	ROH	variable, 2–5
$R-\overset{\|}{\underset{R}{C}}=CH_2$	4.7	ArOH	variable, 4–7
$R-\overset{\|}{\underset{R}{C}}=\overset{\|}{\underset{R}{C}}-H$	5.3	$\overset{O}{\overset{\|\|}{-C-OH}}$	variable, 10–12

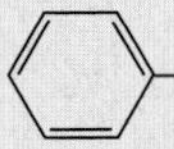

[a]The values are approximate because they are affected by neighboring substituents.

PROBLEM 10◆

In each of the following pairs of compounds, which of the underlined protons has the greater chemical shift (its signal is farther downfield; its signal will appear at a higher frequency)?

a. $CH_3CH_2C\underline{H}_2Cl$ or $CH_3CH_2C\underline{H}CH_3$
$\qquad\qquad\qquad\qquad\qquad\qquad\quad |$
$\qquad\qquad\qquad\qquad\qquad\qquad\; Cl$

b. $CH_3C\underline{H}_2CHCH_3$ or $CH_3CH_2C\underline{H}CH_3$
$\qquad\qquad\quad |$ $\qquad\qquad\qquad\qquad\quad |$
$\qquad\qquad\; Cl$ $\qquad\qquad\qquad\qquad\quad Cl$

c. $CH_3CH_2C\underline{H}_2Cl$ or $CH_3CH_2C\underline{H}_2Br$

d. $CH_3CH_2CH=C\underline{H}_2$ or $CH_3C\underline{H}_2CH=CH_2$

e. 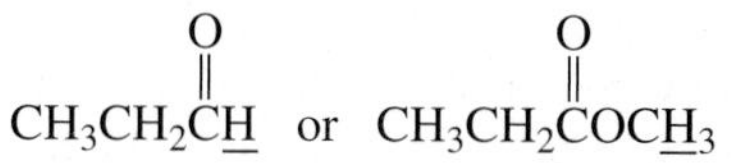 $CH_3CH_2C\underline{H}$ or $CH_3CH_2COC\underline{H}_3$ (each with O double-bonded to C)

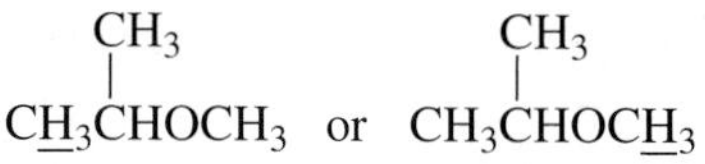

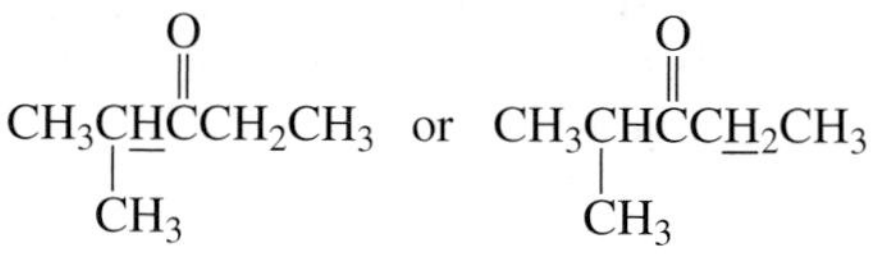

f. $C\underline{H}_3CHOCH_3$ or $CH_3CHOC\underline{H}_3$ (each with CH_3 substituent)

g. $CH_3C\underline{H}CCH_2CH_3$ or $CH_3CHCC\underline{H}_2CH_3$ (each with O double-bonded to C and CH_3 substituent)

h. $CH_3CHC\underline{H}Br$ or $CH_3CHC\underline{H}Br$ (each with Br Br substituents)

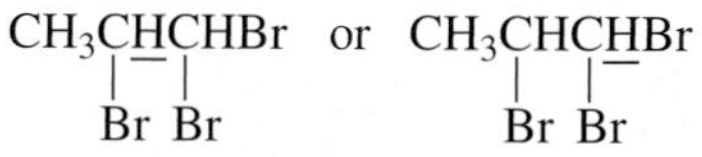

PROBLEM 11◆

Without referring to Table 13.1, label the protons in the following compounds. The proton that gives the signal at the lowest frequency should be labeled a, the next b, etc.

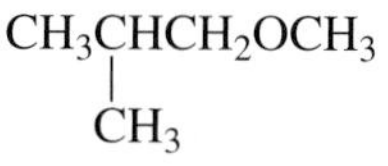

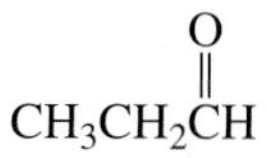

a. $CH_3CH_2CH_2Cl$

f. CH_3CH_2CH (with O double-bonded to C)

b. 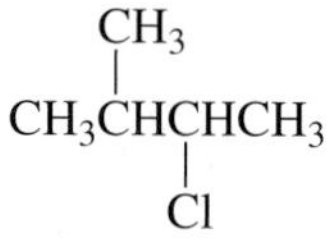 $CH_3CHCHCH_3$ (with CH_3 and Cl substituents)

g. $CH_3CH_2CH_2CCH_3$ (with O double-bonded to C)

c. $CH_3CH_2CH_2COCH_3$ (with O double-bonded to C)

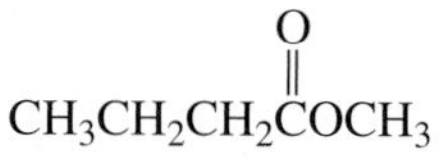

d. 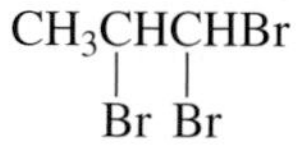 $CH_3CHCHBr$ (with Br Br substituents)

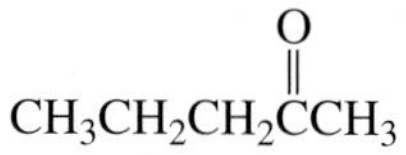

h. $ClCH_2CH_2CH_2Cl$

e. $CH_3CHCH_2OCH_3$ (with CH_3 substituent)

i. 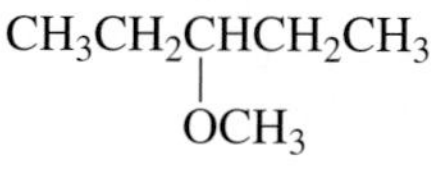 $CH_3CH_2CHCH_2CH_3$ (with OCH_3 substituent)

j. 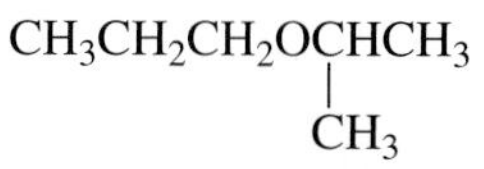 $CH_3CH_2CH_2OCHCH_3$ (with CH_3 substituent)

The two signals in the ^1H NMR spectrum of 1-bromo-2,2-dimethylpropane in Figure 13.5 are not the same size because the area under each signal is proportional to the number of protons that give rise to the signal. (The two signals are shown again in Figure 13.6.) The area under the signal occurring at the lower frequency is larger because the signal is caused by *nine* methyl protons, while the smaller higher frequency signal results from *two* methylene protons.

From your calculus course you probably remember that the area under a curve can be determined by integration. An ^1H NMR spectrometer is equipped with a computer that calculates the integrals electronically. Modern spectrometers print out

13.8

INTEGRATION OF THE NMR SIGNALS

the integrals as numbers on the spectrum. The integrals can also be displayed by a line of integration superimposed on the original spectrum (Figure 13.6). The height of each integration step is proportional to the area under that signal which, in turn, is proportional to the number of protons giving rise to the signal. By measuring the heights of the integration steps, you can determine that the ratio of the integrals is approximately $1.6 : 7.0 = 1 : 4.4$. (The measured integrals contain experimental error and thus the heights of the integration steps are approximate.) Because there can be only whole numbers of protons, the ratios are multiplied by a number that will cause all the numbers to be close to whole numbers (in this case we will multiply by 2). That means that the ratio of protons in the compound is $2 : 9$.

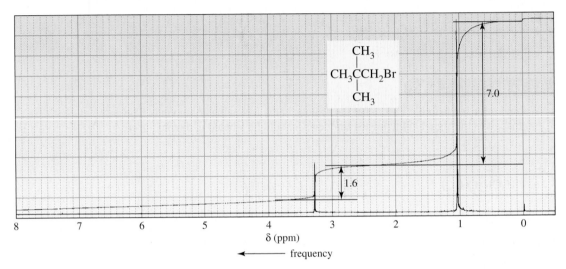

▲ **Figure 13.6**
Analysis of the integration line in the ^1H NMR spectrum of 1-bromo-2,2-dimethylpropane.

The integration tells us the *relative* number of protons that give rise to each signal, not the *absolute* number. For example, integration could not distinguish between 1,1-dichloroethane and 1,2-dichloro-2-methylpropane because both compounds would show integral ratios of $1 : 3$.

$$CH_3-CH-Cl$$
$$\quad\quad\; |$$
$$\quad\quad\; Cl$$

1,1-dichloroethane
ratio of protons = 1 : 3

$$\quad\quad\quad CH_3$$
$$\quad\quad\quad\; |$$
$$CH_3-C-CH_2Cl$$
$$\quad\quad\quad\; |$$
$$\quad\quad\quad\; Cl$$

1,2-dichloro-2-methylpropane
ratio of protons = 1 : 3

PROBLEM 12◆

How could you distinguish among the ^1H NMR spectra of the following compounds?

$$\quad\quad\quad CH_3 \quad\quad\quad\quad\quad CH_3 \quad\quad\quad\quad\quad CH_2Br$$
$$\quad\quad\quad\; | \quad\quad\quad\quad\quad\quad\; | \quad\quad\quad\quad\quad\quad\; |$$
$$CH_3-C-CH_2Br \quad CH_3-C-CH_2Br \quad CH_3-C-CH_2Br$$
$$\quad\quad\quad\; | \quad\quad\quad\quad\quad\quad\; | \quad\quad\quad\quad\quad\quad\; |$$
$$\quad\quad\quad CH_3 \quad\quad\quad\quad\quad Br \quad\quad\quad\quad\quad CH_2Br$$

PROBLEM 13/SOLVED

a. Calculate the ratios of the different kinds of protons in a compound with an integral ratio of 4 : 18.4 : 6 (going from left to right across the spectrum).

b. Give the structure of a compound that would give these relative integrals in the observed order.

SOLUTION

a. Divide each by the smallest number:

$$\frac{4}{4} = 1 \qquad\qquad \frac{18.4}{4} = 4.6 \qquad\qquad \frac{6}{4} = 1.5$$

Multiply by a number that will cause all the numbers to be close to whole numbers.

$$1 \times 2 = 2 \qquad\qquad 4.6 \times 2 = 9 \qquad\qquad 1.5 \times 2 = 3$$

The ratio of the different kinds of protons is then 2 : 9 : 3.

b. The methylene protons are the least shielded because they are bonded to a carbon adjacent to a carbon bonded to two oxygens. The *tert*-butyl methyl groups are in the middle because they are bonded to a carbon adjacent to a carbon bonded to one oxygen. The methyl group is the most shielded because it is farthest away from the oxygens.

$$\begin{array}{cc} & O \quad CH_3 \\ & \parallel \quad | \\ CH_3CH_2 & COCCH_3 \\ & | \\ & CH_3 \end{array}$$

PROBLEM 14◆

The ^1H NMR spectrum shown in Figure 13.7 corresponds to one of the following compounds. Which compound is responsible for this spectrum?

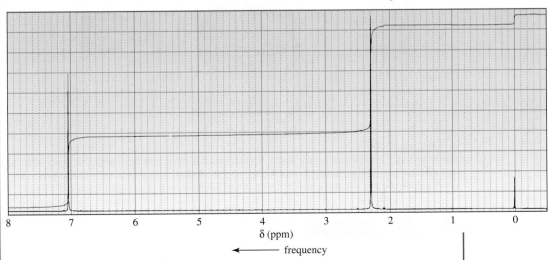

▲ **Figure 13.7**
^1H NMR spectrum for Problem 14.

13.9
DIAMAGNETIC
ANISOTROPY

The chemical shifts of hydrogens bonded to sp^2 hybridized carbons are at a higher frequency than one would predict, based on the electronegativity of the sp^2 carbons. For example, a hydrogen bonded to a terminal sp^2 carbon appears at 4.7 ppm, a hydrogen bonded to an internal sp^2 carbon appears at 5.3 ppm, and a hydrogen on a benzene ring appears at 6.5–8.0 ppm.

5.3 ppm

$$CH_3CH_2CH{=}CH_2$$

4.7 ppm

7.3 ppm

These unusual chemical shifts are due to **diamagnetic anisotropy.** *Anisotropic* is Greek for "different in different directions." Diamagnetic anisotropy means different magnetic fields at different points in space. Because π electrons are less tightly held by nuclei than σ electrons, they are more free to move in response to a magnetic field. When a magnetic field is applied to a compound with π electrons, the π electrons move in a circular path. This induced electron motion causes an induced magnetic field. How this induced magnetic field affects the chemical shift of a proton depends on the direction of the induced magnetic field—in the region where the proton is located—relative to the direction of the applied magnetic field.

The magnetic field induced by the π electrons of a benzene ring, in the region where benzene's protons are located, is oriented in the same direction as the applied field (Figure 13.8). The magnetic field induced by the π electrons of an alkene, in the region where the protons bonded to the sp^2 carbons of the alkene are located, is also oriented in the same direction as the applied field. Thus, in both cases, a larger effective magnetic field is sensed by the protons. Therefore, they will resonate at higher frequencies (because frequency is proportional to ΔE, which is proportional to magnetic field [Figure 13.2]) than they would have if the π electrons had not induced a magnetic field.

3-D Molecule:
Benzene

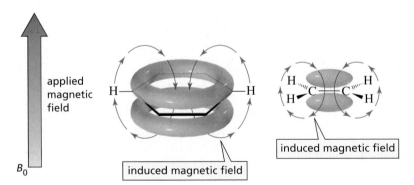

applied magnetic field

B_0

induced magnetic field

induced magnetic field

▲ **Figure 13.8**
The magnetic field induced by the π electrons of a benzene ring and the magnetic field induced by the π electrons of an alkene, in the area of space where the aromatic or vinylic protons are located, are in the same direction as the applied magnetic field.

PROBLEM 15◆

[18]-Annulene shows two signals in its ^{1}H NMR spectrum, one at 9.25 ppm and the other very far upfield at −2.88 ppm. What hydrogens are responsible for each of the signals? (*Hint:* Notice the direction of the induced magnetic field outside and inside the benzene ring in Figure 13.8.)

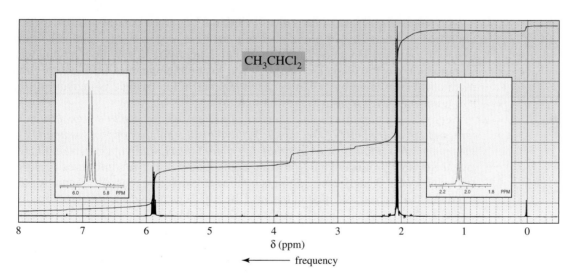

3-D Molecule:
[18]-Annulene

The signals in the ^{1}H NMR spectrum of 1,1-dichloroethane (Figure 13.9) look different from the signals in the ^{1}H NMR spectrum of 1-bromo-2,2-dimethylpropane (Figure 13.5). The signals in Figure 13.5 are both **singlets** (composed of a single peak), but the signal for the methyl protons of 1,1-dichloroethane (the lower frequency signal) is split into two peaks (a **doublet**) and the signal for the methine proton is split into four peaks (a **quartet**). (Magnifications of the doublet and quartet are shown as insets in Figure 13.9.)

13.10
SPLITTING OF THE SIGNALS

CH$_3$CHCl$_2$

6.0   5.8   PPM

2.2   2.0   1.8   PPM

8   7   6   5   4   3   2   1   0

δ (ppm)

←——— frequency

▲ **Figure 13.9**
^{1}H NMR spectrum of 1,1-dichloroethane.

The splitting of a signal is described by the **$N + 1$ rule,** where N is the number of equivalent protons bonded to an *adjacent* carbon. By equivalent protons we mean that the protons bonded to an adjacent carbon are equivalent to each other but not equivalent to the proton giving rise to the signal. For example, both signals in Figure 13.5 are singlets because the carbon adjacent to the methyl groups and adjacent to the methylene group in 1-bromo-2,2-dimethylpropane is not bonded to any protons $(N + 1 = 0 + 1 = 1)$. In contrast, the carbon adjacent to the methyl group in 1,1-dichloroethane is bonded to one proton, so the signal for the methyl protons is split into a doublet $(N + 1 = 1 + 1 = 2)$ (Figure 13.9). The carbon adjacent to the carbon bonded to the methine proton is bonded to three equivalent protons, so the signal for the methine proton is split into a quartet $(N + 1 = 3 + 1 = 4)$.

Because the methine proton and the methyl protons split each other's signal, the two sets of protons are called **coupled protons.** The number of peaks in a signal is called the **multiplicity** of the signal (Table 13.2).

Blaise Pascal (1623–1662) was born in France. At age 16 he published a book on geometry and at 19 invented a calculating machine. He propounded the modern theory of probability, developed the principle underlying the hydraulic press, and showed that atmospheric pressure decreases as altitude increases. In 1644, he narrowly escaped death when his carriage horses bolted. That scare caused him to devote the rest of his life to meditation and religious writings.

TABLE 13.2 Multiplicity of the Signal and Relative Intensities of the Peaks in the Signal

Number of equivalent protons causing splitting	Multiplicity of the signal	Relative peak intensities
0	singlet	1
1	doublet	1:1
2	triplet	1:2:1
3	quartet	1:3:3:1
4	quintet	1:4:6:4:1
5	sextet	1:5:10:10:5:1
6	septet	1:6:15:20:15:6:1

The relative peak intensities shown in Table 13.2 can be determined by figuring out how many ways the given number of neighboring protons can be aligned relative to the applied magnetic field. The relative intensities obey the mathematical mnemonic known as Pascal's triangle. (You may remember this from one of your math classes.) According to Pascal, each number in the triangle is the sum of the two numbers closest to it in the row above.

Keep in mind that it is not the number of protons giving rise to a signal that determines the multiplicity of the signal; rather, it is the number of protons bonded to the immediately adjacent carbons that determines the multiplicity. For example, the signal for the **a** protons in the following compound will be split into three peaks (a **triplet**) because the adjacent carbon is bonded to two hydrogens. The signal for the **b** protons will appear as a quartet because the adjacent carbon is bonded to three hydrogens, and the signal for the **c** protons will be a singlet.

$$\underset{a\quad\ b\qquad\ c}{CH_3CH_2\overset{\overset{\displaystyle O}{\|}}{C}OCH_3}$$

The splitting of signals is caused by **spin–spin coupling.** The frequency at which the methyl protons of 1,1-dichloroethane will show a signal is affected by the magnetic field of the methine proton. If the magnetic field of the methine proton is lined up with the applied magnetic field, it will add to the applied magnetic field, so the methyl protons will show a signal at a slightly higher frequency. On the other hand, if the magnetic field of the methine proton is lined up against the applied magnetic field, it will subtract from the applied magnetic field and the methyl protons will show a signal at a lower frequency (Figure 13.10). So the signal for the methyl protons is split into two peaks, one corresponding to the higher frequency and one corresponding to the lower frequency. Because about half the methine protons are lined up with the applied magnetic field and half are lined up against it, the two peaks of the *doublet* have approximately the same height and area (Table 13.2).

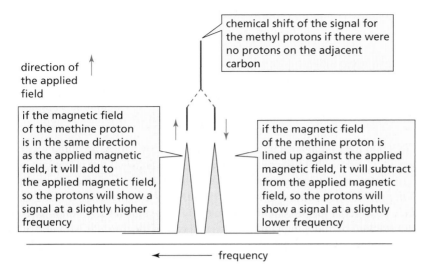

chemical shift of the signal for the methyl protons if there were no protons on the adjacent carbon

direction of the applied field

if the magnetic field of the methine proton is in the same direction as the applied magnetic field, it will add to the applied magnetic field, so the protons will show a signal at a slightly higher frequency

if the magnetic field of the methine proton is lined up against the applied magnetic field, it will subtract from the applied magnetic field, so the protons will show a signal at a slightly lower frequency

◄——— frequency

◄ **Figure 13.10**
The signal for the methyl protons of 1,1-dichloroethane is split into a doublet by the methine proton.

Similarly, the frequency at which the methine proton will show a signal is affected by the magnetic fields of the three protons bonded to the adjacent carbon. The magnetic fields of the three methyl protons can all line up with the applied magnetic field, two can line up with the field and one against it, one can line up with it and two against it, or all three can line up against it. Because the magnetic field the proton senses is affected in four different ways, the signal for the methine proton is a *quartet* (Figure 13.11).

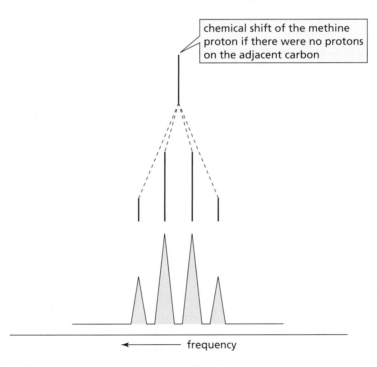

chemical shift of the methine proton if there were no protons on the adjacent carbon

◄——— frequency

◄ **Figure 13.11**
The signal for the methine proton of 1,1-dicloroethane is split into a quartet by the methyl protons.

A quartet has relative peak areas of 1 : 3 : 3 : 1 (Table 13.2). There is only one way to align the magnetic fields of three protons so they are all with the field, and only one way to align them so they are all against the field. Figure 13.12 shows, however, that there are three ways to align the magnetic fields of three protons so that two are lined up with the field and one is lined up against the field. That is, the first two can be with it and the third against it, the first and third with it and the second against it, or the second and third

Tutorial:
NMR Signal splitting

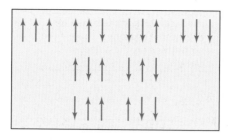

with it and the first against it. Figure 13.12 also shows there are three ways to align the magnetic fields so that two are lined up against the field and one is lined up with the field.

In summary, *equivalent* protons never split each other's signal. Normally, *nonequivalent* protons split each other's signal only if they are on *adjacent* carbons. Splitting is rarely observed if the protons are separated by more than three σ bonds. If, however, they are separated by more than three bonds and one of the bonds is a double or triple bond, small splitting is sometimes observable. This is called **long-range coupling.**

H_a and H_b will split each other because they are separated by 3 σ bonds

H_a and H_b will not split each other because they are separated by 4 σ bonds

H_a and H_b may split each other because they are separated by 4 bonds including one double bond

PROBLEM 16

Using a diagram like the one in Figure 13.12, predict:

a. the relative intensities of the peaks in a triplet.

b. the relative intensities of the peaks in a quintet.

PROBLEM 17 ◆

One of the spectra in Figure 13.13 is due to 1-chloropropane and the other to 1-iodopropane. Which is which?

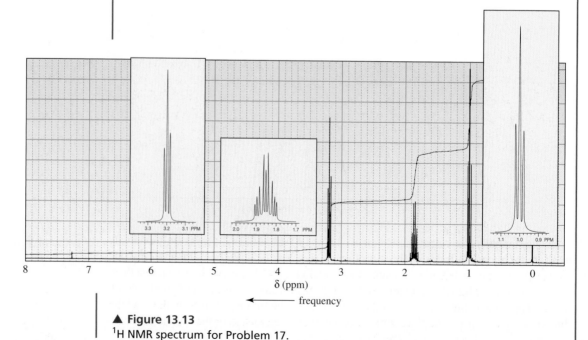

▲ **Figure 13.13**
¹H NMR spectrum for Problem 17.

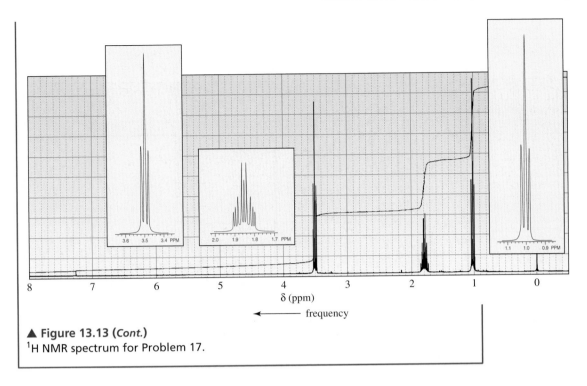

▲ **Figure 13.13 (Cont.)**
¹H NMR spectrum for Problem 17.

The ¹H NMR spectrum of bromomethane shows one singlet. The three methyl protons are chemically equivalent, and chemically equivalent protons do not split each other's signal. The four protons in 1,2-dichloroethane are chemically equivalent. Consequently, the ¹H NMR spectrum of 1,2-dichloroethane shows one singlet.

**13.11
MORE EXAMPLES
OF ¹H NMR SPECTRA**

$$CH_3Br \qquad\qquad ClCH_2CH_2Cl$$
bromomethane   1,2-dichloroethane
**both compounds have an NMR spectrum that shows one singlet because
equivalent protons do not split each other's signals**

There are two signals in the ¹H NMR spectrum of 1,3-dibromopropane (Figure 13.14). The signal for the H_b protons is split into a triplet by the two hydrogens on the adjacent carbon. The H_a protons have two adjacent carbons that are bonded to protons. The protons on one adjacent carbon are equivalent to the protons on the other adjacent carbon. Because the two sets of protons are equivalent, the $N + 1$ rule is applied to both sets at the same time (N is equal to the sum of the equivalent protons on both carbons). So the signal for the H_a protons is split into a quintet ($4 + 1 = 5$). The integration confirms that two methylene groups contribute to the H_b signal because twice as many protons give rise to the H_b signal as to the H_a signal.

The ¹H NMR spectrum of isopropyl butanoate shows five signals (Figure 13.15). The signal for the H_a protons is split into a triplet by the H_c protons. The signal for the H_b protons is split into a doublet by the H_e proton. The signal for the H_d protons is split into a triplet by the H_c protons, and the signal for the H_e proton is split into a septet by the H_b protons. The signal for the H_c protons is split by both the H_a and H_d protons. Because the H_a and H_d protons are not equivalent, the $N + 1$ rule has to be applied separately to each set. Thus, the signal for the H_c protons will be split into a quartet by the H_a protons, and each of these four peaks will be split into a triplet by the H_d protons:$(N_a + 1)(N_d + 1) = (4)(3) = 12$. As a result, the signal for the H_c protons is a **multiplet** (a signal with more than seven peaks). The reason we do not see 12 peaks is that some of the peaks overlap (Section 13.13).

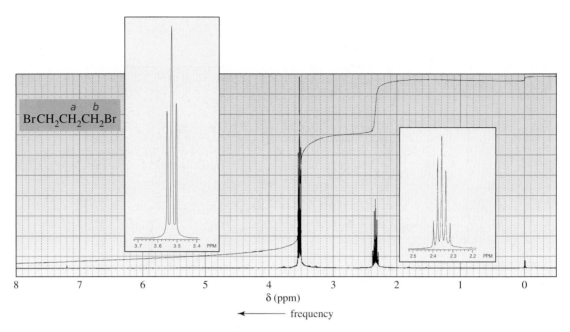

▲ **Figure 13.14**
^{1}H NMR spectrum of 1,3-dibromopropane.

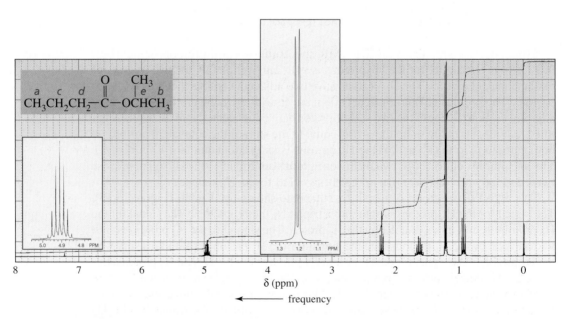

▲ **Figure 13.15**
^{1}H NMR spectrum of isopropyl butanoate.

PROBLEM 18

Indicate the number of signals and the multiplicity of each signal in the ^{1}H NMR spectrum of each of the following compounds:

    a. $ICH_2CH_2CH_2Br$ **b.** $ClCH_2CH_2CH_2Cl$ **c.** ICH_2CH_2CHBr
                                                                                                         Br

The ^{1}H NMR spectrum of 3-bromo-1-propene shows four signals (Figure 13.16). Although the H_b and H_c protons are bonded to the same carbon, they are not chemically equivalent (one is cis to the bromomethyl group, the other is trans to the bromomethyl group), so each produces a separate signal. The signal for the H_a protons is split into a doublet by the H_d proton. Notice that the signals for the three vinylic protons are at relatively high frequencies because of diamagnetic anisotropy (Section 13.9). The signal for the H_d proton is a multiplet because it is split separately by the H_a, H_b, and H_c protons.

Because the H_b and H_c protons are not equivalent, they can split one another. This means that the signal for the H_b proton can be split into a doublet by the H_d proton and that each of the peaks in the doublet can be split into a doublet by the H_c proton. Therefore, the signal for the H_b proton should be what is called a **doublet of doublets.** The signal for the H_c proton should also be a doublet of doublets. However, the mutual splitting of the signals of two nonidentical protons bonded to the same sp^2 carbon, called **geminal coupling,** is often too small to be observed (Section 13.12). Therefore, the signals for the H_b and H_c protons each appears as a doublet rather than as a doublet of doublets.

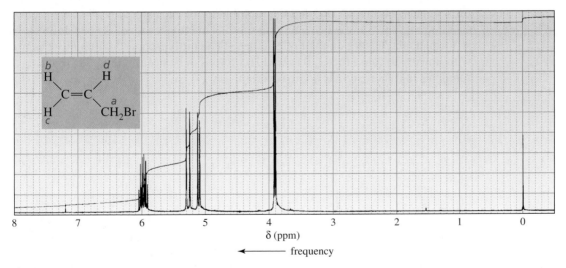

▲ **Figure 13.16**
^{1}H NMR spectrum of 3-bromo-1-propene.

There are five sets of chemically equivalent protons in ethylbenzene (Figure 13.17). We see the expected triplet for the H_a protons and the quartet for the H_b protons. (Notice that this is a characteristic pattern for an ethyl group.) We expect the signal for the H_c protons to be a doublet and the signal for the H_e proton to be a triplet. Because the H_c and H_e protons are not equivalent, they must be considered separately when determining the splitting of the signal for the H_d protons. Therefore, we expect the signal for the H_d protons to be split into a doublet by the H_c protons and each peak of the doublet to be split into another doublet by the H_e proton, forming a doublet of doublets. However, we do not see three distinct signals for the H_c, H_d, and H_e protons in Figure 13.17. Instead, we see overlapping signals. Apparently the electronic effect of an ethyl substituent is not sufficiently different from that of a hydrogen to cause a difference in the environments of the H_c, H_d, and H_e protons that is large enough to allow them to appear as separate signals.

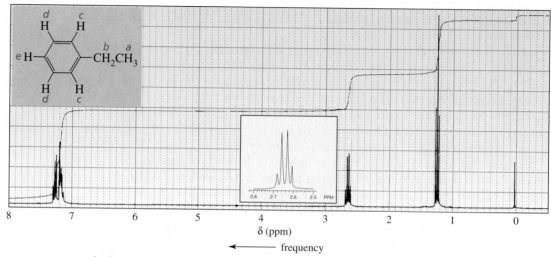

▲ **Figure 13.17**
^{1}H NMR spectrum of ethylbenzene.

Tutorial:
NMR Spectrum
assignment

The H_a, H_b, and H_c protons of nitrobenzene, on the other hand, show three distinct signals (Figure 13.18), and the multiplicity of each signal is what we predicted for the signals for the benzene ring protons in ethylbenzene (H_c is a doublet, H_b is a triplet, and H_a is a doublet of doublets). The nitro group is sufficiently electron withdrawing to cause the H_a, H_b, and H_c protons to be in sufficiently different environments that their signals do not overlap.

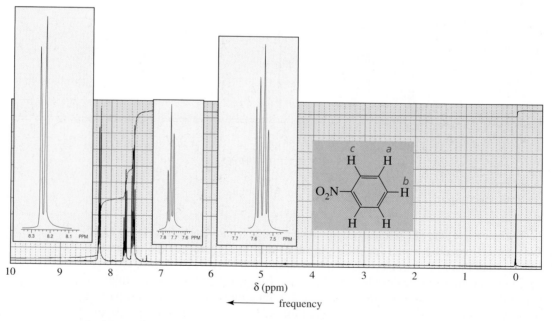

▲ **Figure 13.18**
^{1}H NMR spectrum of nitrobenzene.

Notice that the signals for the benzene ring protons in Figures 13.17 and 13.18 occur in the 6.5–8.0 ppm region. Other kinds of protons usually do not resonate in this region, so signals in this region of an ^{1}H NMR spectrum should suggest to you that the compound contains an aromatic ring.

PROBLEM 19

How would the ^{1}H NMR spectra for the four compounds with molecular formula $C_3H_6Br_2$ differ?

PROBLEM 20◆

Identify each compound from its molecular formula and its ^{1}H NMR spectrum.

a. C_9H_{12}

b. $C_5H_{10}O$

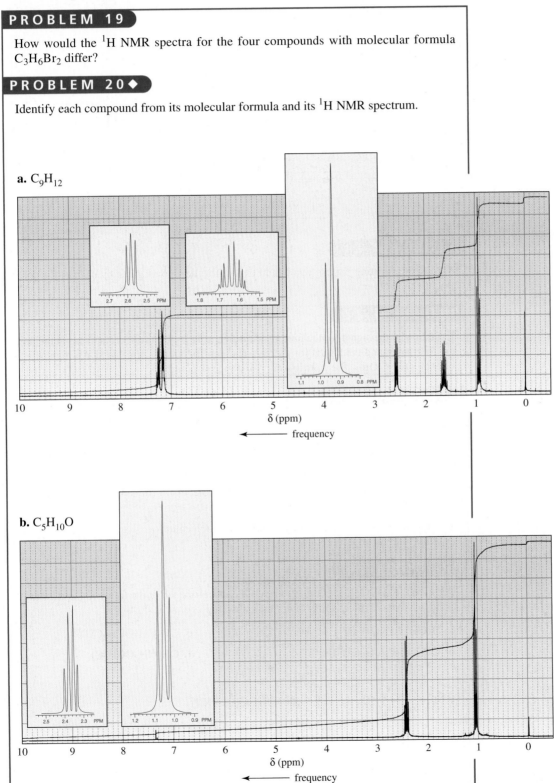

c. $C_9H_{10}O_2$

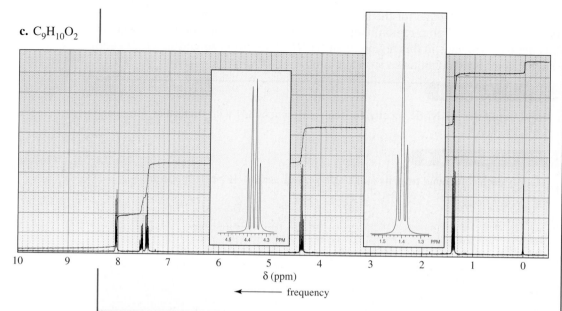

δ (ppm)

← frequency

PROBLEM 21

Predict the splitting pattern you would expect for the signals given by each of the compounds in Problem 3.

PROBLEM 22

Explain why the signal for the protons identified as H_a in Figure 13.18 appears at the lowest frequency and the signal for the protons identified as H_c appears at the highest frequency. (*Hint:* Draw the contributing resonance structures.)

PROBLEM 23◆

Identify the following compounds. (Relative integrals are stated going from left to right across the spectrum.)

a. The ^1H NMR spectrum of a compound with molecular formula $C_4H_{10}O_2$ has two singlets with an area ratio of 2 : 3.

b. The ^1H NMR spectrum of a compound with molecular formula $C_6H_{10}O_2$ has two singlets with an area ratio of 2 : 3.

c. The ^1H NMR spectrum of a compound with molecular formula $C_8H_6O_2$ has two singlets with an area ratio of 1 : 2.

PROBLEM 24

Describe the ^1H NMR spectrum you would expect for each of the following compounds, using relative chemical shifts rather than absolute chemical shifts.

a. $BrCH_2CH_2CH_2Br$

b. $CH_3OCH_2CH_2CH_2Br$

c. $CH_3CH_2OCH_2CH_3$

d. $CH_3CH_2OCH_2Cl$

e. 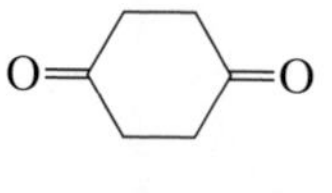 O=⟨⟩=O

$$\text{g. } CH_3\overset{O}{\overset{\|}{C}}CH_2\overset{O}{\overset{\|}{C}}OCH_3$$

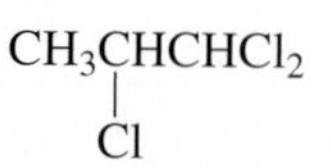

i. 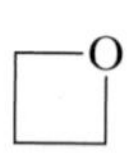 ▢—O

$$\text{f. } CH_3\overset{CH_3}{\underset{Br}{\overset{|}{\underset{|}{C}}}}CH_2CH_3$$

h. $CH_3\overset{}{\underset{Cl}{\overset{}{\underset{|}{C}H}}}CHCl_2$

$$\text{j. } CH_3\overset{CH_3}{\overset{|}{C}H}CH_2\overset{O}{\overset{\|}{C}H}$$

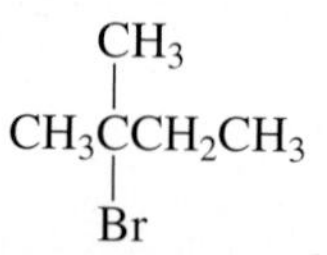

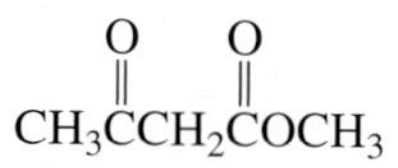

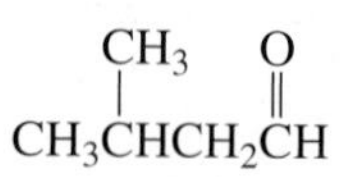

k.

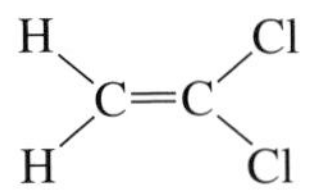

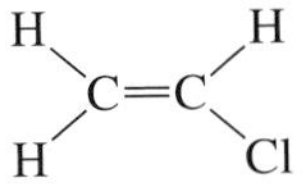

l.

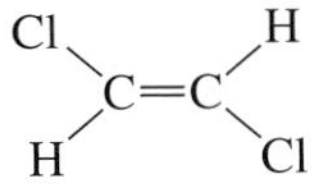

m.

n. $CH_3OCH_2CH_2CH_2OCH_3$

The distance (in hertz) between two adjacent peaks of a split NMR signal is called the **coupling constant** (denoted by J). The coupling constant for H_a being split by H_b is denoted by J_{ab}. The signals of coupled protons (protons that split each other's signal) have the same coupling constant—in other words, $J_{ab} = J_{ba}$ (Figure 13.19). Coupling constants are useful in analyzing complex NMR spectra because protons on adjacent carbons can be identified by the identical coupling constants of their signals.

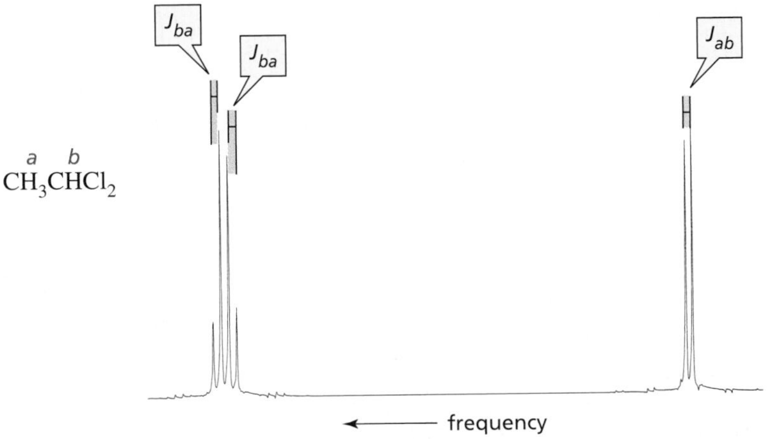

13.12 COUPLING CONSTANTS

◄ **Figure 13.19**
The H_a and H_b protons of 1,1-dichloroethane are coupled protons, so their signals have the same coupling constant, $J_{ab} = J_{ba}$.

The magnitude of a coupling constant is independent of the operating frequency of the spectrometer—the same coupling constant is obtained from a 300-MHz instrument as from a 600-MHz instrument. The magnitude of a coupling constant depends on the number and type of bonds that connect the coupled protons, as well as the geometric relationship of the protons. Characteristic coupling constants are shown in Table 13.3. They range from 0 to 15 Hz.

TABLE 13.3 Approximate Values of Coupling Constants

Approximate value of J_{ab} (Hz)		Approximate value of J_{ab} (Hz)	
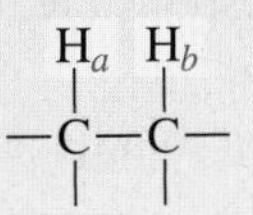	7	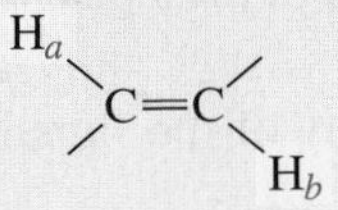	15 (trans)
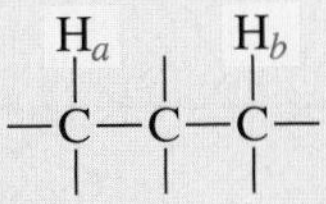	0	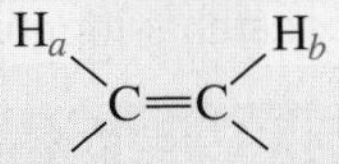	10 (cis)
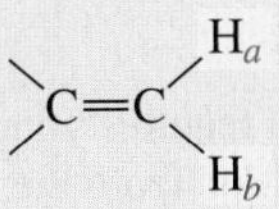	2 (geminal coupling)	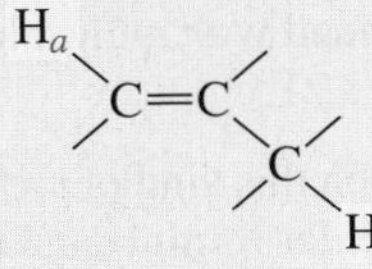	1 (long range coupling)

*The dependence of the coupling constant on the angle between the two C—H bonds is called the Karplus relationship after **Martin Karplus**, who first observed the relationship. Karplus was born in 1930. He received a B.A. from Harvard University and a Ph.D. from the California Institute of Technology. He is currently a professor of chemistry at Harvard University.*

The coupling constant for two nonequivalent hydrogens on the same sp^2 carbon is often too small to be observed (Figure 13.16), but it is large for nonequivalent hydrogens bonded to adjacent sp^2 carbons. Apparently the interaction between the hydrogens is strongly affected by the intervening π electrons. We have seen that π electrons also allow long-range coupling—that is, coupling through four or more bonds (Section 13.10).

Coupling constants can be used to distinguish between the ^{1}H NMR spectra of cis and trans alkenes. The coupling constant of *trans*-vinylic protons is significantly greater than the coupling constant of *cis*-vinylic protons (Figure 13.20) because the coupling constant depends on the dihedral angle between the two C—H bonds in the H—C=C—H unit. The coupling constant is greatest when the angle between the two C—H bonds is 180° (trans) and smaller when it is 0° (cis). Notice the difference between J_{bd} and J_{cd} in the spectrum of 3-bromo-1-propene (Figure 13.16).

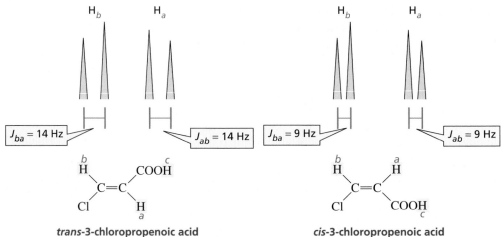

trans-3-chloropropenoic acid   cis-3-chloropropenoic acid

▲ **Figure 13.20**
The doublets observed for the H_a and H_b protons in the ^{1}H NMR spectra of *trans*-3-chloropropenoic acid and *cis*-3-chloropropenoic acid. The coupling constant for trans protons (14 Hz) is greater than the coupling constant for cis protons (9 Hz).

Let's now summarize the kind of information that can be obtained from an ^{1}H NMR spectrum.

1. The number of signals indicates the number of different kinds of protons that are in the compound.
2. The position of a signal indicates the kind of proton(s) responsible for the signal (1°, 2°, 3°, allylic, vinylic, aromatic, etc.) and the kinds of neighboring substituents.
3. The integration of the signal tells the relative number of protons responsible for the signal.
4. The multiplicity of the signal tells the number of protons bonded to adjacent carbons.
5. The coupling constants identify coupled protons.

Tutorial:
NMR Spectrum
interpretation

PROBLEM-SOLVING STRATEGY

Identify the compound with molecular formula $C_9H_{10}O$ that gives the IR and ^{1}H NMR spectra in Figure 13.21.

One way to approach this kind of problem is to identify whatever structural features you can from the molecular formula and IR spectrum and then use the information from the

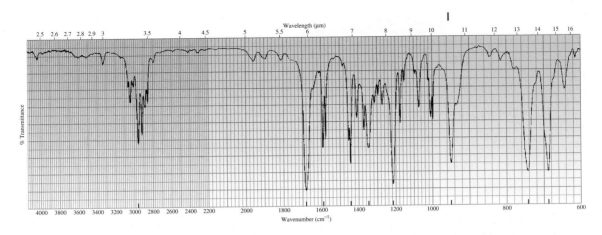

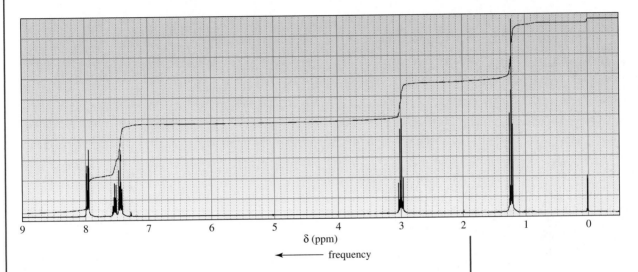

▲ **Figure 13.21**
IR and ^{1}H NMR spectra for the Problem-Solving Strategy.

NMR spectrum to expand on that knowledge. From the IR spectrum we learn that the compound is a ketone: It has a carbonyl group at ~1680 cm^{-1}, only one oxygen, and no absorption band at 2700 cm^{-1} that would indicate an aldehyde. That the carbonyl group absorption is at a lower frequency than normal suggests it has partial single bond character as a result of electron delocalization (it is attached to an sp^2 carbon). The compound contains a benzene ring (> 3000 cm^{-1}, ~1600 cm^{-1} and 1440 cm^{-1}), and it has hydrogens bonded to sp^3 carbons (< 3000 cm^{-1}). In the NMR spectrum, the triplet at ~1.2 ppm and the quartet at ~3.0 ppm indicate the presence of an ethyl group that is attached to an electron-withdrawing group. The signals in the 7.4–8.0 ppm region confirm the presence of a benzene ring. From this information we can conclude that the compound is the following ketone. The integration ratio (5 : 2 : 3) confirms this answer.

$$\text{C}_6\text{H}_5-\overset{\overset{\displaystyle O}{\|}}{\text{C}}\text{CH}_2\text{CH}_3$$

Now continue on to Problem 25.

PROBLEM 25◆

Identify the compound with molecular formula $C_8H_{10}O$ that gives the IR and ^1H NMR spectra shown in Figure 13.22.

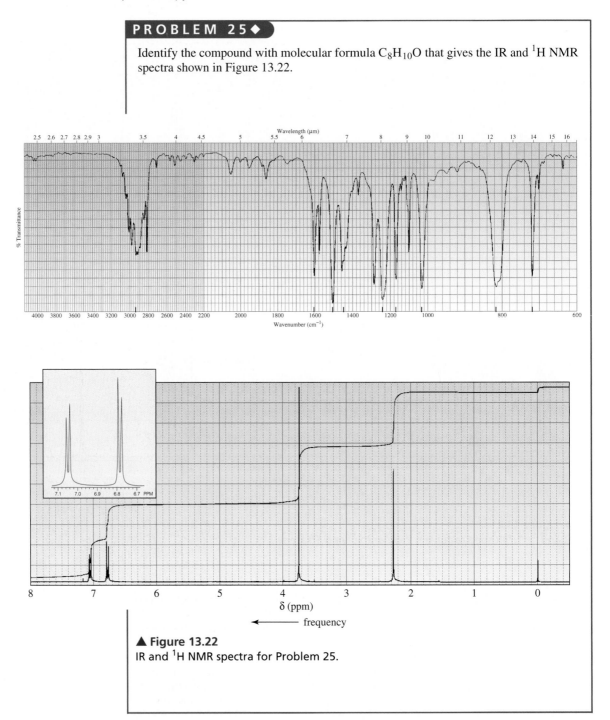

▲ **Figure 13.22**
IR and ^1H NMR spectra for Problem 25.

13.13 SPLITTING DIAGRAMS

The splitting pattern obtained when a signal is split by more than one set of protons can best be understood by using a splitting diagram. In a **splitting diagram** (also called a **splitting tree**), the NMR peaks are shown as vertical lines and the effect of each of the splittings is shown one at a time. For example, a splitting diagram is shown in Figure 13.23 for splitting of the signal for the H_c proton of 1,1,2-trichloro-3-methylbutane into a doublet of doublets by the H_b and H_d protons.

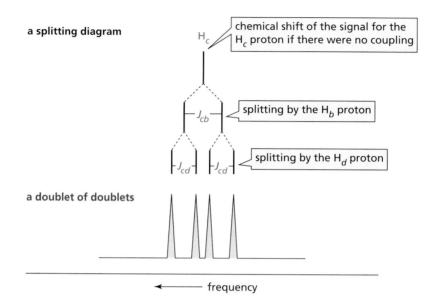

1,1,2-trichloro-3-methylbutane

◀ **Figure 13.23**
A splitting diagram for a
doublet of doublets.

Notice the difference between a quartet and a doublet of doublets. Both have four peaks. A quartet results from splitting by *three equivalent* adjacent protons, it has relative peak intensities of 1 : 3 : 3 : 1, and the individual peaks are equally spaced. A doublet of doublets, on the other hand, results from splitting by *two nonequivalent* adjacent protons, it has relative peak intensities of 1 : 1 : 1 : 1, and the individual peaks are not necessarily equally spaced (Figure 13.23).

a quartet
relative intensities: 1 : 3 : 3 : 1

a doublet of doublets
relative intensities: 1 : 1 : 1 : 1

The signal for the H_b protons of propyl bromide is split into a quartet by the H_a protons, and each of the resulting four peaks is split into a triplet by the H_c protons (Figure 13.24). How many of the 12 peaks are actually seen depends on the relative magnitudes of the two coupling constants, J_{ba} and J_{bc}. For example, Figure 13.24 shows that there are 12 peaks when J_{ba} is much greater than J_{bc}, 9 peaks when $J_{ba} = 2J_{bc}$, and only 6 peaks when $J_{ba} = J_{bc}$. As you can see, the number of peaks actually observed depends on how many overlap with each other. When peaks overlap, their intensities add together.

We would expect the signal for the H_a protons of 1-chloro-3-iodopropane to be a triplet of triplets (split into 9 peaks) because the signal would be split into a triplet

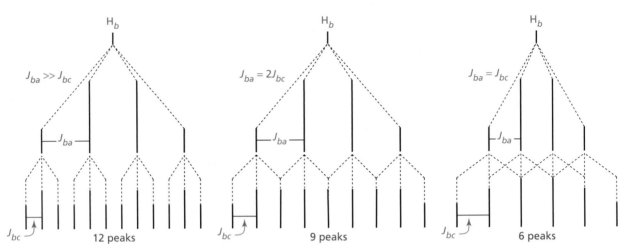

▲ **Figure 13.24**
A splitting diagram for a quartet of triplets. The number of peaks actually observed when a signal is split by two sets of protons depends on the relative magnitudes of the two coupling constants.

by the H_b protons and each of the resulting peaks into a triplet by the H_c protons. The signal, however, is a quintet (Figure 13.25).

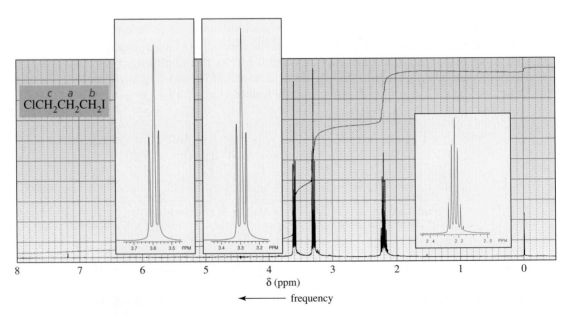

▲ **Figure 13.25**
^{1}H NMR spectrum of 1-chloro-3-iodopropane.

Finding that the signal for the H_a protons of 1-chloro-3-iodopropane is a quintet indicates that J_{ab} and J_{ac} have about the same value. The splitting diagram shows that a quintet results if $J_{ab} = J_{ac}$.

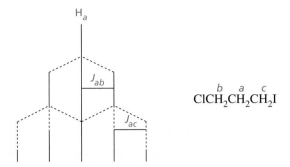

$$\overset{b}{Cl}CH_2\overset{a}{CH}_2\overset{c}{CH}_2I$$

From this we can conclude that when two different sets of protons split a signal, the multiplicity of the signal should be determined using the $N + 1$ rule separately for each set of hydrogens when the coupling constants for the two sets are different. When the coupling constants are similar, however, the multiplicity of a signal can be determined by treating both sets of hydrogens as if they were equivalent. In other words, the $N + 1$ rule can be applied to both sets simultaneously.

PROBLEM 26/SOLVED

The two hydrogens of a methylene group adjacent to a chirality center are not equivalent hydrogens because they are in different environments due to the chirality center. Therefore, the $N + 1$ rule must be applied to them separately when determining the multiplicity of the signal for the adjacent methyl hydrogens. We expect the signal to be a doublet of doublets. The signal, however, is a triplet. Explain, using a splitting diagram, why it is a triplet rather than a doublet of doublets.

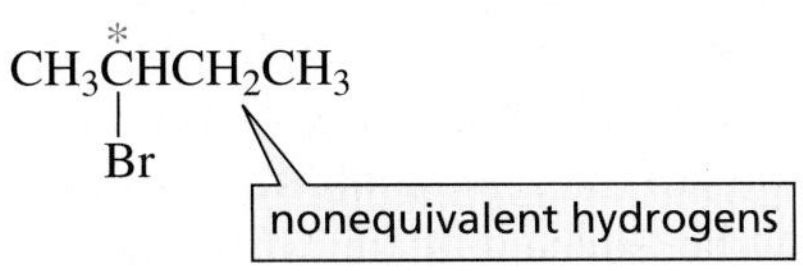

$$CH_3\overset{*}{C}HCH_2CH_3$$
$$|$$
$$Br$$

nonequivalent hydrogens

SOLUTION   The observation of a triplet means that the $N + 1$ rule did not have to be applied to the two protons separately but could have been applied to the two protons as a set ($N = 2$, so $N + 1 = 3$). This means that the coupling constant for splitting of the methyl signal by one of the methylene hydrogens is about the same as the coupling constant for splitting by the other methylene hydrogen.

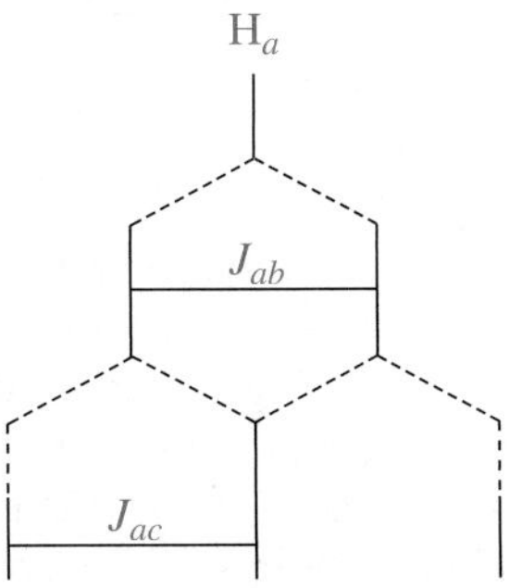

PROBLEM 27

Draw a splitting diagram for H_b where:

a. $J_{ba} = 12$ Hz and $J_{bc} = 6$ Hz.   **b.** $J_{ba} = 12$ Hz and $J_{bc} = 12$ Hz.

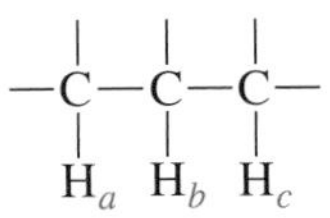

$$-\overset{|}{\underset{H_a}{C}}-\overset{|}{\underset{H_b}{C}}-\overset{|}{\underset{H_c}{C}}-$$

13.14
TIME DEPENDENCE
OF NMR
SPECTROSCOPY

We have seen that the three methyl hydrogens of ethyl bromide give rise to one signal in the ^{1}H NMR spectrum because they are chemically equivalent due to rotation about the carbon–carbon single bond. At any one time, however, the three hydrogens can be in quite different environments—one can be anti to the bromine, one can be gauche to the bromine, one can be eclipsed with the bromine, etc.

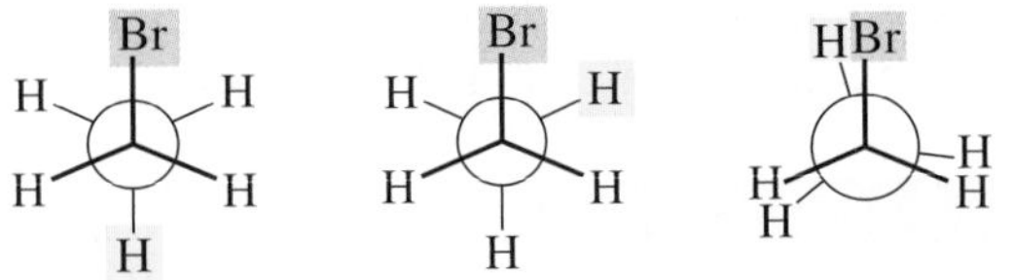

An NMR spectrometer is very much like a camera with a slow shutter speed—it is too slow to be able to detect these different environments, so what it sees is an average environment. Because each of the three methyl hydrogens has the same average environment, we see one signal for the methyl group in the ^{1}H NMR spectrum.

Similarly, the ^{1}H NMR spectrum of cyclohexane shows only one signal even though cyclohexane has both axial and equatorial protons. There is only one signal because the interconversion of the chair conformers of cyclohexane occurs too rapidly at room temperature to be detected by the NMR spectrometer. Because axial protons in one chair conformer are equatorial protons in the other chair conformer, all the protons in cyclohexane have the same average environment on the NMR time scale, so the NMR spectrum shows one signal.

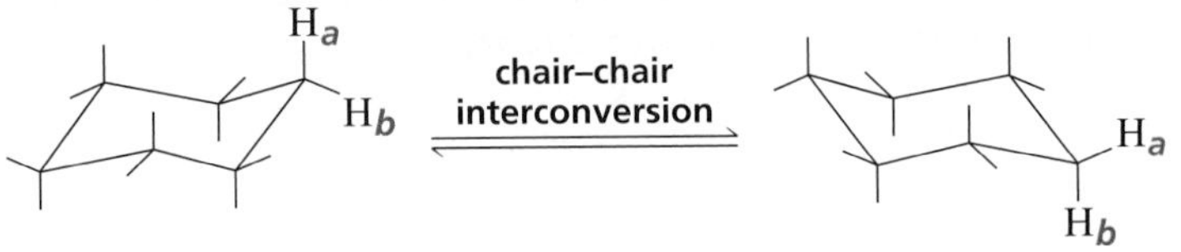

The rate of chair–chair interconversion is temperature dependent—the lower the temperature, the slower the rate of interconversion. Cyclohexane-d_{11} has 11 deuteriums, which means that it has only one hydrogen. ^{1}H NMR spectra of cyclohexane-d_{11} taken at various temperatures are shown in Figure 13.26. Cyclohexane with only one hydrogen was used for this experiment to prevent splitting of the signal, which would have complicated the spectrum. Deuterium signals are not detectable in ^{1}H NMR, and splitting by a deuterium on the same or on an adjacent carbon is not normally detectable at the operating frequency of an ^{1}H NMR spectrometer.

At room temperature, the ^{1}H NMR spectrum of cyclohexane-d_{11} shows one sharp signal, which is an average for the axial proton of one chair and the equatorial proton

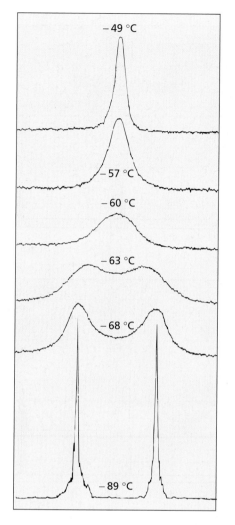

◀ **Figure 13.26**
^{1}H NMR spectra of cyclohexane-d_{11} at various temperatures. (Reproduced with permission from F. A. Bovey, *Nuclear Magnetic Resonance Spectroscopy.* New York: Academic Press, 1988.)

of the other chair. As the temperature decreases, the signal becomes broader and eventually separates into two signals, which are equidistant from the original signal. At −89 °C, two sharp singlets are observed because at this temperature, the rate of chair–chair interconversion has decreased sufficiently to allow the two kinds of protons (axial and equatorial) to be individually detected on the NMR time scale.

The chemical shift of a proton bonded to an oxygen or a nitrogen depends on the degree to which the proton is hydrogen bonded. For example, the chemical shift of the OH proton of an alcohol ranges from 2.0 to 5.5 ppm, the chemical shift of the OH proton of a carboxylic acid from 10 to 12 ppm, and the chemical shift of the NH proton of an amine from 1.5 to 4.0 ppm; the greater the extent of hydrogen bonding, the greater the chemical shift (the higher the frequency).

The ^{1}H NMR spectrum of pure, dry ethanol is shown in Figure 13.27(a), and the ^{1}H NMR spectrum of ethanol with a trace amount of acid is shown in Figure 13.27(b). The spectrum shown in Figure 13.27(a) is what we would predict from what we have learned so far. The signal for the proton bonded to oxygen is farthest downfield and is split into a triplet by the neighboring methylene protons; the signal for the

13.15
PROTONS BONDED TO OXYGEN AND NITROGEN

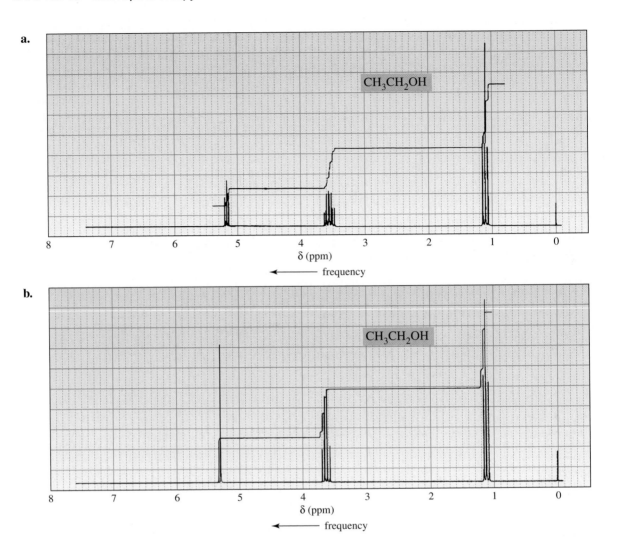

▲ **Figure 13.27**
(a) ^{1}H NMR spectrum of pure ethanol. (b) ^{1}H NMR spectrum of ethanol containing a trace amount of acid.

methylene protons is split into a multiplet by the combined effects of the methyl protons and the OH proton.

The spectrum shown in Figure 13.27(b) is the type of spectrum most often obtained for alcohols. The signal for the proton bonded to oxygen is not split, and this proton does not split the signal of the adjacent protons. So the signal for the OH proton is a singlet and the signal for the methylene protons is a quartet because it is split only by the methyl protons.

The difference in the two spectra is due to the fact that protons bonded to oxygen undergo **proton exchange,** which means that they are transferred from one molecule to another. Whether or not the OH proton and the methylene protons split each other's signals depends on how long a particular proton stays on the OH group.

In a sample of pure alcohol, the rate of proton exchange is very slow. This causes the spectrum to look identical to one that would be obtained if proton exchange did not occur. Acids and bases catalyze proton exchange, so if the alcohol is contaminated with acid or base, proton exchange becomes rapid. When proton exchange is rapid, the spectrum records only an average of all possible environments. A rapidly exchanging proton is recorded as a singlet and its effect on adjacent protons is averaged, which means that it does not cause splitting.

mechanism for acid-catalyzed proton exchange

$$RÖ—H + HÖH \rightleftharpoons RÖ—H + HÖH \rightleftharpoons RÖ: + HÖH$$

The signal for an OH proton is often easy to spot in an ^{1}H NMR spectrum because it is frequently somewhat broader than other signals (see the signal at δ 4.9 in Figure 13.29(b)). The broadening occurs because the rate of proton exchange is not slow enough to result in a cleanly split signal as in Figure 13.27(a), or fast enough for a cleanly averaged signal as in Figure 13.27(b).

PROBLEM 28◆

Which would show the signal for the OH proton at a greater chemical shift, the ^{1}H NMR spectrum of pure ethanol or the ^{1}H NMR spectrum of ethanol dissolved in CH_2Cl_2?

PROBLEM 29

Propose a mechanism for base-catalyzed proton exchange.

PROBLEM 30◆

Identify the compound with molecular formula C_3H_7NO responsible for the ^{1}H NMR spectrum in Figure 13.28.

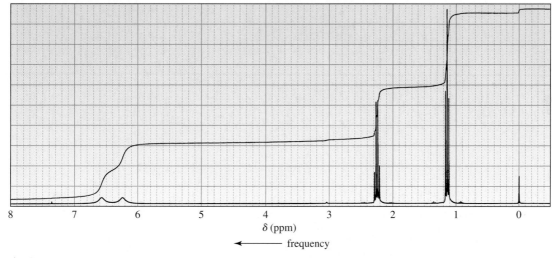

▲ **Figure 13.28**
^{1}H NMR spectrum for Problem 30.

Because deuterium signals are not seen in an ^{1}H NMR spectrum, substituting a deuterium for a hydrogen is a technique used to identify signals and to simplify ^{1}H NMR spectra (Section 13.14).

If, after obtaining an ^{1}H NMR spectrum of an alcohol, a few drops of D_2O are added to the sample and the spectrum is taken again, the OH signal in the first spectrum can be identified. It will be the signal that becomes less intense (or disappears) in the second spectrum because of the proton exchange process just discussed. This technique can be used with any proton that undergoes exchange.

$$R—O—H + D—O—D \longrightarrow R—O—D + D—O—H$$

13.16
USE OF DEUTERIUM IN ^{1}H NMR SPECTROSCOPY

If the ^{1}H NMR spectrum of $CH_3CH_2OCH_3$ were compared with the ^{1}H NMR spectrum of $CH_3CD_2OCH_3$, the signal at highest frequency in the first spectrum would be absent in the second spectrum, indicating that this signal corresponds to the methylene group. This can be a helpful technique for analysis of complicated ^{1}H NMR spectra.

The sample used to obtain an ^{1}H NMR spectrum is made by dissolving the compound in an appropriate solvent. Solvents with protons cannot be used because the signal for the protons of the solvent would be much more intense because there are more of them since there is much more solvent than compound in the solution. Therefore, deuterated solvents such as $CDCl_3$ and D_2O are commonly used in NMR spectroscopy.

13.17
HIGH RESOLUTION ^{1}H NMR SPECTROSCOPY

The ^{1}H NMR spectrum of 2-*sec*-butylphenol taken on a 60-MHz NMR spectrometer is shown in Figure 13.29(a), and the ^{1}H NMR spectrum of the same compound taken on a 300-MHz instrument is shown in Figure 13.29(b). Why is the resolution of the second spectrum so much better?

a.

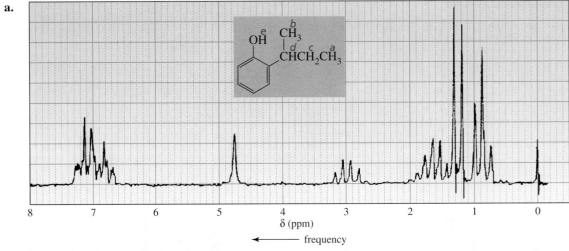

b.

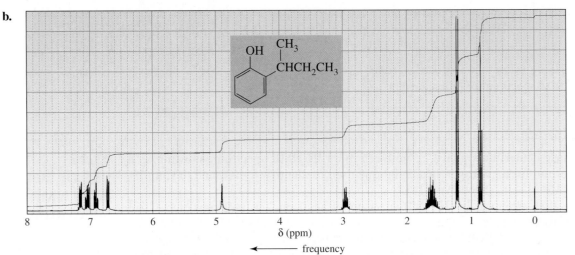

▲ Figure 13.29
(a) 60-MHz ^{1}H NMR spectrum of 2-*sec*-butylphenol.
(b) 300-MHz ^{1}H NMR spectrum of 2-*sec*-butylphenol.

To observe separate signals with "clean" splitting patterns, the difference in the chemical shifts of two adjacent protons ($\Delta\nu$ in Hz) must be at least 10 times the value of the coupling constant (J). As $\Delta\nu/J$ decreases, the two signals appear closer to each other, and the outer peaks of the signals become less intense while the inner peaks become more intense (Figure 13.30).

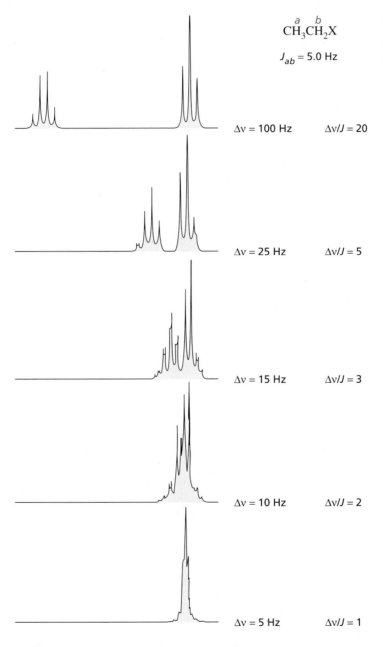

$$\overset{a\quad b}{CH_3CH_2X}$$

$J_{ab} = 5.0$ Hz

$\Delta\nu = 100$ Hz $\Delta\nu/J = 20$

$\Delta\nu = 25$ Hz $\Delta\nu/J = 5$

$\Delta\nu = 15$ Hz $\Delta\nu/J = 3$

$\Delta\nu = 10$ Hz $\Delta\nu/J = 2$

$\Delta\nu = 5$ Hz $\Delta\nu/J = 1$

◀ **Figure 13.30**
The splitting pattern of an ethyl group as a function of the $\Delta\nu/J$ ratio. (Reproduced with permission of J. Houser.)

The difference in the chemical shifts of the H_a and H_c protons of 2-sec-butylphenol is 0.8 ppm, which corresponds to 48 Hz in a 60-MHz spectrometer and 240 Hz in a 300-MHz spectrometer. Coupling constants are independent of the operating frequency of the NMR spectrometer, so J_{ac} is 7 Hz whether the spectrum is taken on a 60-MHz or a 300-MHz instrument. Only in the case of the 300-MHz spectrometer is the difference in chemical shift more than 10 times the value of the coupling constant, so only in the 300-MHz spectrometer do the signals show clean splitting patterns.

in a 300-MHz spectrometer

$$\frac{\Delta\nu}{J} = \frac{240}{7} = 34$$

in a 60-MHz spectrometer

$$\frac{\Delta\nu}{J} = \frac{48}{7} = 6.9$$

13.18
^{13}C NMR SPECTROSCOPY

The number of signals in a ^{13}C NMR spectrum tells you how many different kinds of carbons a compound has—just as the number of signals in an ^{1}H NMR spectrum tells you how many different kinds of hydrogens a compound has. The principles behind ^{1}H NMR and ^{13}C NMR spectroscopy are essentially the same. There are, however, some differences that make ^{13}C NMR easier to interpret.

The development of ^{13}C NMR spectroscopy as a routine analytical procedure was not possible until computers were developed that could carry out a Fourier transform (Section 13.2). ^{13}C NMR requires Fourier transform techniques because the signals obtained from a single scan are too weak to be distinguished from background electronic noise. However, FT–^{13}C NMR scans can be repeated rapidly, so a large number of scans can be recorded and added. Since electronic noise is random, its sum is close to zero, and the ^{13}C signals stand out when hundreds of scans are added. Without Fourier transform, it could take days to record the number of scans required for a ^{13}C NMR spectrum.

The weakness of the individual ^{13}C signals is due to the fact that the isotope of carbon (^{13}C) that gives rise to ^{13}C NMR signals constitutes only 1.11% of carbon atoms (Section 12.3). (The most abundant isotope of carbon, ^{12}C, has no nuclear spin and therefore cannot produce an NMR signal.) This means that the intensities of the signals in ^{13}C NMR compared with those in ^{1}H NMR are reduced by a factor of approximately 100. In addition, the gyromagnetic ratio (γ) of ^{13}C is about one-fourth that of ^{1}H, and intensity is proportional to γ^3. Therefore, the overall intensity of a ^{13}C signal is about 6400 times (100×64) less than the intensity of an ^{1}H signal.

One advantage of ^{13}C NMR spectroscopy is that the chemical shifts range over about 220 ppm, compared with about 12 ppm for ^{1}H NMR (Table 13.1). This means that signals are less likely to overlap. Another advantage is that carbons with no directly bonded hydrogens can be observed. The ^{13}C NMR chemical shifts of different kinds of carbons are shown in Table 13.4. TMS is used as the reference compound also in ^{13}C NMR.

A disadvantage of ^{13}C NMR spectroscopy is that, unless special techniques are used, the area under a ^{13}C NMR signal is not proportional to the number of atoms giving rise to the signal. This means that ^{13}C NMR spectra cannot be routinely analyzed by integration.

The ^{13}C NMR spectrum of 2-butanol is shown in Figure 13.31. 2-Butanol has carbons in four different environments, so there are four signals in the spectrum. The relative positions of the signals depend on the same factors that determine the relative positions of the proton signals in ^{1}H NMR. Carbons in electron-dense environments produce low frequency signals, and carbons close to electron-withdrawing groups produce high frequency signals. This means that the signals for the carbons of 2-butanol are in the same relative order that we would expect for the signals of the protons on those carbons in the ^{1}H NMR spectrum. Thus, the carbon of the methyl group farthest away from the electron-withdrawing OH group gives the lowest frequency signal. As the frequency increases, the other methyl carbon comes next, followed by the methylene carbon, and the carbon attached to the OH group gives the highest frequency signal.

The signals are not normally split by neighboring carbons because there is little likelihood of an adjacent carbon being a ^{13}C. The probability of two ^{13}C carbons being next to each other is $1.11\% \times 1.11\%$ (about 1 in 10,000). (Because ^{12}C does not have a magnetic moment, it cannot split the signal of an adjacent ^{13}C.)

The signals in a ^{13}C NMR spectrum can be split by nearby hydrogens. However, this splitting is not usually observed because the spectra are recorded using spin-decoupling. In other words, the carbon–proton interactions are decoupled, so all the signals are singlets in an ordinary ^{13}C NMR spectrum (Figure 13.31).

TABLE 13.4 Approximate Values of Chemical Shifts for ^{13}C NMR

Type of carbon	Approximate chemical shift (ppm)	Type of carbon	Approximate chemical shift (ppm)
$(CH_3)_4Si$	0	C—I	0–40
$R—CH_3$	8–35	C—Br	25–65
$R—CH_2—R$	15–50	C—Cl C—N C—O	35–80 40–60 50–80
R—CH—R (R)	20–60	$\underset{-N<}{\overset{R}{\diagdown}}C=O$	165–175
R—C—R (R, R)	30–40	$\underset{RO}{\overset{R}{\diagdown}}C=O$	165–175
≡C	65–85	$\underset{HO}{\overset{R}{\diagdown}}C=O$	175–185
=C	100–150	$\underset{H}{\overset{R}{\diagdown}}C=O$	190–200
(aromatic ring) C	110–170	$\underset{R}{\overset{R}{\diagdown}}C=O$	205–220

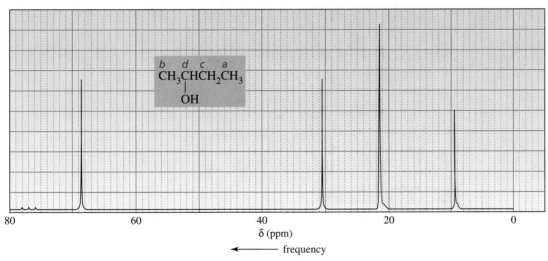

▲ **Figure 13.31**
Proton-decoupled ^{13}C NMR spectrum of 2-butanol.

   If the spectrometer is run in a *spin-coupled* mode, the signals show spin–spin coupling. The splitting is not caused by adjacent carbons, but by the hydrogens bonded to
the carbon that produces the signal. The multiplicity of the signal is determined by
the $N + 1$ rule. The **spin-coupled ^{13}C NMR spectrum** of 2-butanol is shown in
Figure 13.32. (The triplet at 78 ppm is produced by the solvent.) The signals for

the methyl carbons are each split into a quartet because each methyl carbon is bonded to three hydrogens $(3 + 1 = 4)$. The signal for the methylene carbon is split into a triplet $(2 + 1 = 3)$, and the signal for the carbon bonded to the OH group is split into a doublet $(1 + 1 = 2)$.

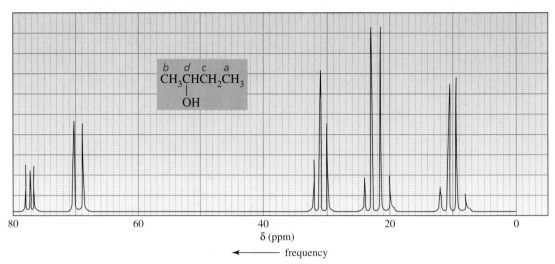

▲ **Figure 13.32**
Spin-coupled ^{13}C NMR spectrum of 2-butanol.

PROBLEM 31◆

Answer the following questions for each of the compounds:
a. How many signals would you expect to see in the ^{13}C NMR spectrum?
b. Which signal would you expect to be at the lowest frequency?

1. $CH_3CH_2CH_2Br$

2. $CH_3CH_2OCH_3$

3. CH_3CHCH_3
   $|$
   Br

4. CH_3COCH_3 (with CH_3 above and CH_3 below)

5. $CH_3CH_2COCH_3$ (O above C)

6. $(CH_3)_2C{=}CH_2$

7. $CH_2{=}CHBr$

8. CH_3CHCH (O above second C, CH_3 below first C)

9. benzene ring—Cl

10. $CH_3CCH_2CH_2CCH_3$ (O above each C)

PROBLEM 32

Describe the spin-coupled ^{13}C NMR spectrum for compounds 1 to 5 in Problem 31, showing relative values of chemical shifts (not absolute values).

PROBLEM 33

How can 1,2-, 1,3-, and 1,4-dinitrobenzene be distinguished by:
a. ^{1}H NMR spectroscopy?   **b.** ^{13}C NMR spectroscopy?

PROBLEM-SOLVING STRATEGY

A ^{13}C NMR spectrum is shown for a compound with a molecular formula of $C_9H_{10}O_2$.

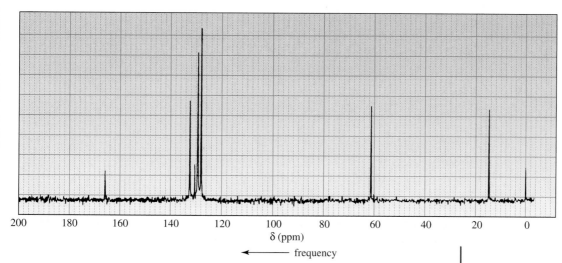

frequency

The two oxygen atoms in the molecular formula and the carbonyl carbon signal at 165 ppm indicate the compound is an ester. The four signals at about 130 ppm suggest the compound has a benzene ring with a single substituent. (One signal is for the carbon to which the substituent is attached, one is for the ortho carbons, one is for the meta carbons, and one is for the para carbon.) Subtracting C_6H_5 and CO_2 from the molecular formula leaves C_2H_5, the formula of an ethyl substituent. Therefore we know the compound is either ethyl benzoate or phenyl propanoate.

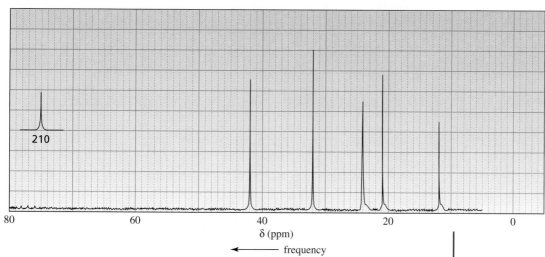

Seeing that the methylene group is at about 60 ppm indicates it is adjacent to an oxygen. Thus, the compound is ethyl benzoate.

Continue on to Problem 34.

PROBLEM 34 ◆

Identify each compound from its molecular formula and its ^{13}C NMR spectrum:

a. $C_{11}H_{22}O$

210

δ (ppm)

frequency

▲ **Figure 13.33**
The ^{13}C NMR spectrum for Problem 34.

b. C_8H_9Br

c. $C_3H_5Cl_3$

d. C_6H_{12}

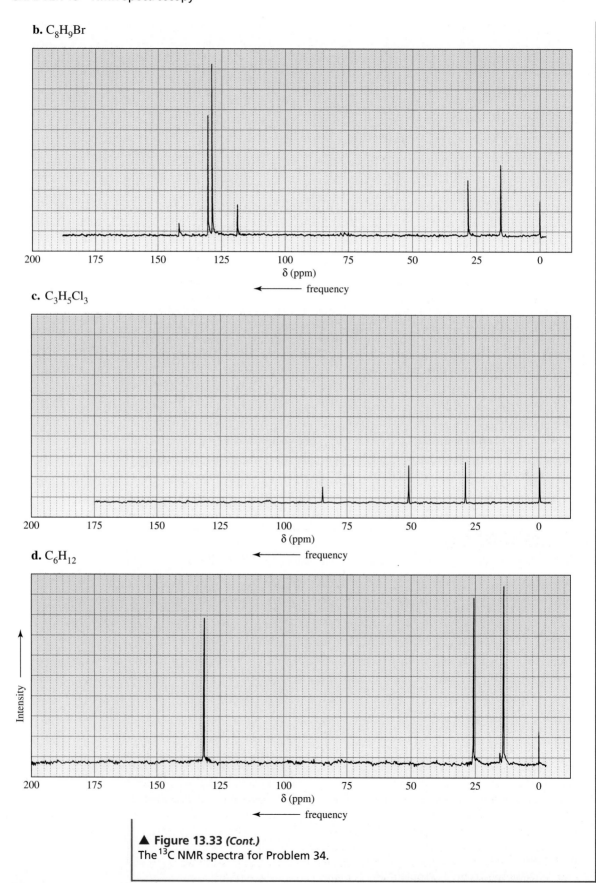

▲ **Figure 13.33** *(Cont.)*
The ^{13}C NMR spectra for Problem 34.

A technique called DEPT ^{13}C NMR (distortionless enhancement by polarization transfer) has been developed to distinguish among CH_3, CH_2, and CH groups. It is now much more widely used than **spin-coupling** to determine the number of hydrogens attached to a carbon.

The **DEPT ^{13}C NMR spectrum** of citronellal is shown in Figure 13.34. A DEPT ^{13}C NMR shows four spectra of the same compound. The bottom-most spectrum shows signals *for all carbons that are covalently bonded to hydrogens.* The next-to-the-bottom spectrum is run under conditions that allow only signals resulting from CH carbons to appear. The third spectrum is run under conditions that allow only signals produced by CH_2 carbons to appear, and the top spectrum shows signals produced by CH_3 carbons. Information from the upper three spectra lets you to determine whether a signal in a ^{13}C NMR spectrum is produced by a CH_3, CH_2, or CH group.

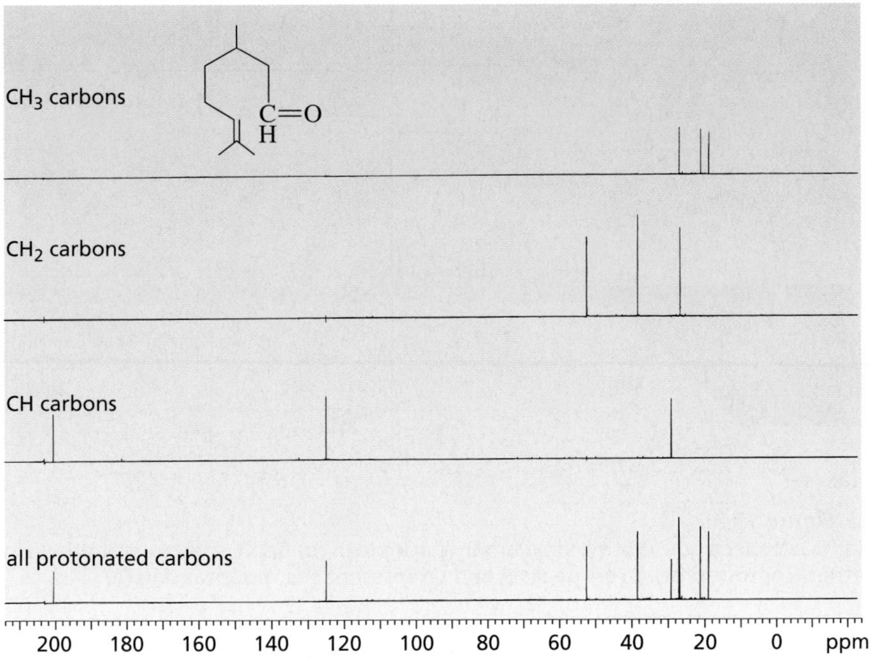

13.19 DEPT ^{13}C NMR SPECTRA

◀ **Figure 13.34**
DEPT ^{13}C NMR spectrum of citronellal.

Notice that a DEPT ^{13}C NMR spectrum does *not* show a signal for a carbon that is not attached to a hydrogen. For example, the ^{13}C NMR spectrum of 2-butanone will show four signals because it has four nonequivalent carbons, whereas the DEPT ^{13}C NMR spectrum of 2-butanone will show only three signals because the carbonyl carbon is not bonded to a hydrogen, so it will not produce a signal.

13.20 TWO-DIMENSIONAL NMR SPECTROSCOPY

Complex molecules such as proteins and nucleic acids are difficult to analyze by NMR because their signals overlap. Such compounds are now being analyzed by two-dimensional (2-D) NMR. **2-D NMR** techniques—unlike X-ray crystallography—allow scientists to determine the structures of complex molecules in solution. This is particularly important for biological molecules whose properties depend on how they fold in water. More recently, 3-D and 4-D NMR spectroscopy have been developed and can be used to determine the structures of very complex molecules. A thorough discussion of 2-D NMR is beyond the scope of this book, but the following discussion will give you a brief introduction to this increasingly important spectroscopic technique.

The ^{1}H NMR and ^{13}C NMR spectra discussed in the preceding sections have one frequency axis and one intensity axis; 2-D NMR spectra have two frequency axes and one intensity axis. The most common 2-D spectra involve ^{1}H–^{1}H shift correlations; they identify protons that are coupled to each other (i.e., that split each other's signal). This is called ^{1}H–^{1}H shift COrrelated SpectroscopY, which is known by the acronym COSY.

A portion of the **COSY spectrum** of ethyl vinyl ether is shown in Figure 13.35(a); it looks like a mountain range viewed from the air. These "mountain-like" spectra (known as *stack plots*) are not the spectra actually used to identify a compound. Instead, the compound is identified using a contour plot (Figure 13.35(b)), where each mountain in Figure 13.35(a) is represented by a large dot (as if its top had been cut off). The two mountains shown in Figure 13.35(a) correspond to the dots labeled B and C in Figure 13.35(b).

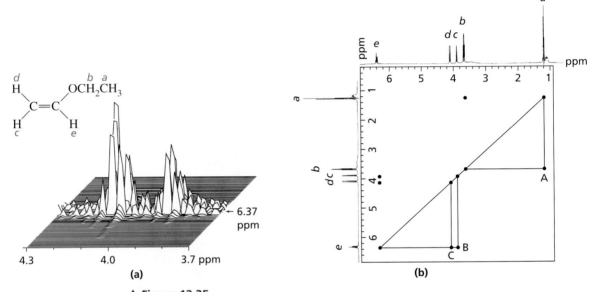

(a)   (b)

▲ **Figure 13.35**
(a) COSY spectrum of ethyl vinyl ether (stack plot). (b) COSY spectrum of ethyl vinyl ether (contour plot). Cross peaks B and C represent the mountains in (a).

Tutorial:
2-D NMR

In the contour plot, the usual one-dimensional ^{1}H NMR spectrum is plotted on both the x- and y-axes. To analyze the spectrum, a diagonal is drawn through the dots that bisect the spectrum. The dots that are not on the diagonal (A, B, C) are called *cross peaks*. Cross peaks indicate pairs of protons that are splitting each other (that are coupled). For example, if we start at the cross peak labeled A and draw a straight line parallel to the y-axis back to the diagonal, we hit the dot on the diagonal at ~1.1 ppm produced by the H_a protons. If we then go back to A and draw a straight line parallel to the x-axis back to the diagonal, we hit the dot on the diagonal at ~3.8 ppm produced by the H_b protons. This means that the H_a and H_b protons are coupled. If we then go to the cross peak labeled B and draw two perpendicular lines back to the diagonal, we see that the H_c and H_e protons are coupled; the cross peak labeled C shows that the H_d and H_e protons are coupled. Notice that we used only cross peaks below the diagonal. The cross peaks above the diagonal give the same information.

The COSY spectrum of 1-nitropropane is shown in Figure 13.36. Cross peak A shows that the H_a and H_b protons are coupled, and cross peak B shows that the H_b and H_c protons are coupled. Notice that the two triangles in Figure 13.36 have a common vertex since the H_b protons are coupled to both the H_a and H_c protons.

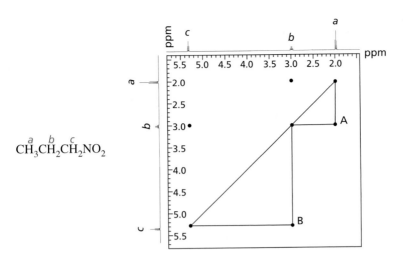

► **Figure 13.36**
COSY spectrum of
1-nitropropane.

PROBLEM 35

Identify pairs of coupled protons in 2-methyl-3-pentanone using the COSY spectrum in Figure 13.37.

Figure 13.37 ►
COSY spectrum for Problem 35.

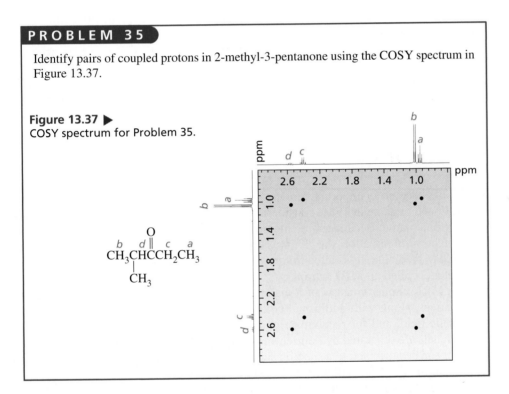

2-D NMR spectra that show $^{13}C–^{1}H$ shift correlations are called HETCOR (from HETeronuclear CORrelation) spectra. **HETCOR spectra** indicate coupling between protons and the carbon to which they are attached.

The HETCOR spectrum of 2-methyl-3-pentanone is shown in Figure 13.38. The ^{13}C NMR spectrum is shown on the x-axis and the ^{1}H NMR spectrum is shown on the y-axis. The cross peaks in a HETCOR spectrum identify which hydrogens are attached to which carbons (or vice versa). Cross peak A indicates that the hydrogens that show a signal at ~0.9 ppm in the ^{1}H NMR spectrum are bonded to the carbon that shows a signal at ~6 ppm in the ^{13}C NMR spectrum. Cross peak C shows that the hydrogens that show a signal at ~2.5 ppm are bonded to the carbon that shows a signal at ~34 ppm.

Clearly, 2-D NMR techniques are not necessary for assigning the signals in the NMR spectrum of a simple compound such as 2-methyl-3-pentanone. However, in

Figure 13.38 ▶
HETCOR spectrum of
2-methyl-3-pentanone.

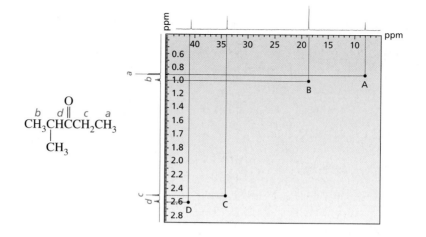

the case of many complicated molecules, signal assignment can be done only with the aid of 2-D NMR.

Spectra showing ^{13}C–^{13}C shift correlations (called 2-D ^{13}C INADEQUATE spectra) identify directly bonded carbons. There also are spectra in which chemical shifts are plotted on one of the frequency axes and coupling constants on the other. Other 2-D spectra involve the nuclear Overhauser effect (NOESY for very large molecules; ROESY for mid-sized molecules).[3] They are used to locate protons that are close together in space.

Albert Warner Overhauser was born in San Diego in 1925. He has been a professor of chemistry at Cornell University and Purdue University. From 1958 to 1973, he was at the Physical Science Laboratory of Ford Motor Company.

13.21 MAGNETIC RESONANCE IMAGING

NMR has become an important tool in medical diagnosis. When it was first introduced into clinical practice in 1981, considerable attention was given to the selection of an appropriate name. Because some members of the public associate nuclear process-es with harmful radiation, the "N" was dropped from NMR when this technique was applied to medicine. It is called **magnetic resonance imaging (MRI),** and the spec-trometer is called an **MRI scanner.**

An MRI scanner consists of a magnet sufficiently large to accommodate an en-tire patient, along with additional coils for exciting the nuclei, for modifying the magnetic field, and for receiving signals. Different tissues yield different signals. The signals are separated by Fourier transform analysis into components. Each com-ponent can be attributed to a specific site of origin in the patient. This allows a cross-sectional image of the patient's body to be constructed.

Most of the signals in an MRI scan originate from the hydrogens of water mole-cules because these hydrogens are far more abundant in tissues than are the hydro-gens of organic compounds. The difference in the way water is bound in different tissues is what produces much of the signal variation among organs, as well as the signal variation between healthy and diseased tissue (Figure 13.39).

MRI has become an important diagnostic tool for several reasons. First of all, an MRI scan can be obtained without exposing the patient to the harmful ioniz-ing radiation of X-rays. In addition, an MRI image may be obtained in any plane essentially independent of the patient's position. This allows optimum visualiza-tion of the anatomy of interest. By contrast, the plane of an X-ray image is defined by the physical position of the patient. Likewise, the plane of a CT (computed

---

[3]NOESY and ROESY are the acronyms for Nuclear Overhauser Effect Spectroscopy and Rotation-frame Overhauser Effect Spectroscopy.

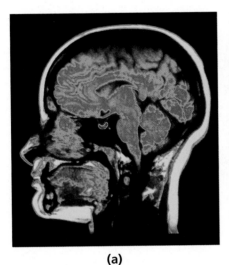 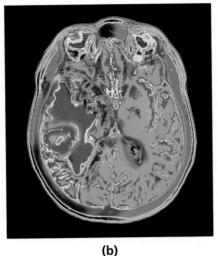

(a)   (b)

◀ **Figure 13.39**
(a) MRI of a normal brain. The pituitary is highlighted (pink). (b) MRI of an axial section through the brain showing a tumor (purple) surrounded by damaged, fluid-filled tissue (red).

tomography) scan is defined by the position of the patient within the machine and is usually perpendicular to the long axis of the body. CT scans in other planes may be obtained only if the patient is a skilled contortionist. Recent technology has allowed CT scans to be obtained independent of the patient's position, but so far the scans have low resolution.

Unlike X-rays or CT scans, MRI "sees through" hard tissues and provides images of soft tissues. This means that MRI scans can provide much more information than images obtained using other techniques. For example, MRI can provide detailed images of blood vessels. Flowing fluids such as blood respond differently to excitation in an MRI scanner than do stationary tissues and normally do not produce a signal. However, the data may be processed to eliminate signals from motionless structures instead, thereby showing signals only from flowing fluids. This technique, which is just beginning to be used clinically, will eventually replace more invasive methods of examining the vascular tree. NMR spectroscopy using ^{23}Na, ^{39}K, ^{31}P, and ^{15}N is not yet in routine clinical use, but is being used widely in clinical research and is of particular interest because of the importance of these elements in cellular metabolism and cell physiology.

KEY TERMS

applied magnetic field (page 535)
chemically equivalent protons (page 539)
chemical shift (page 541)
^{13}C NMR (page 534)
COSY spectrum (page 578)
coupled protons (page 550)
coupling constant (page 559)
DEPT ^{13}C NMR spectrum (page 577)
diamagnetic anisotropy (page 548)
diamagnetic shielding (page 538)
2-D NMR (page 577)
doublet (page 549)
doublet of doublets (page 555)
downfield (page 538)
effective magnetic field (page 538)

Fourier transform NMR (FT–NMR) (page 511)
geminal coupling (page 555)
gyromagnetic ratio (page 536)
HETCOR spectrum (page 579)
^{1}H NMR (page 534)
long-range coupling (page 552)
magnetic resonance imaging (MRI) (page 580)
methine hydrogen (page 543)
MRI scanner (page 580)
multiplet (page 553)
multiplicity (page 550)
N + 1 rule (page 549)
NMR spectroscopy (page 534)
operating frequency (page 536)

proton exchange (page 568)
quartet (page 549)
reference compound (page 540)
rf radiation (page 535)
shielding (page 538)
singlet (page 549)
spin-coupled ^{13}C NMR spectrum (page 573)
spin-coupling (page 577)
spin–spin coupling (page 550)
α-spin state (page 535)
β-spin state (page 535)
splitting diagram (page 562)
splitting tree (page 562)
triplet (page 550)
upfield (page 538)

36. How many signals would you expect for each of the following compounds in its:
   a. ^1H NMR spectrum?   **b.** ^{13}C NMR spectrum?

1. 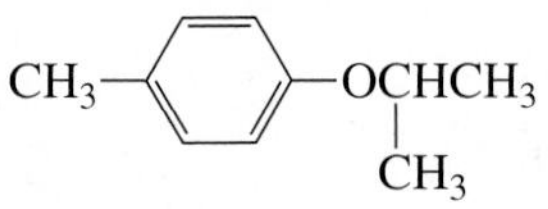 CH₃—⟨benzene ring⟩—OCHCH₃ with CH₃

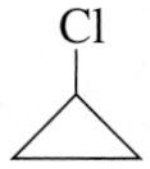

2. 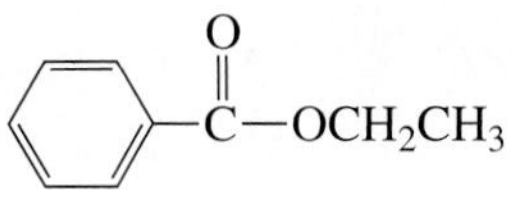 ⟨benzene ring⟩—C(=O)—OCH₂CH₃

3. ⟨cyclic structure⟩=O

4. 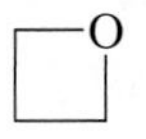 ⟨oxetane ring⟩—O

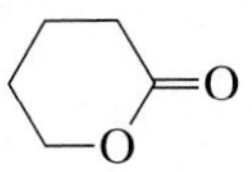

5. ⟨chlorocyclopropane⟩ Cl

6. 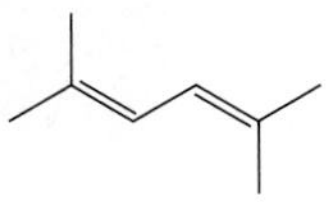 ⟨diene structure⟩

37. Draw a splitting diagram for the H_b proton and indicate its multiplicity if:
   a. $J_{ba} = J_{bc}$ **b.** $J_{ba} = 2 J_{bc}$

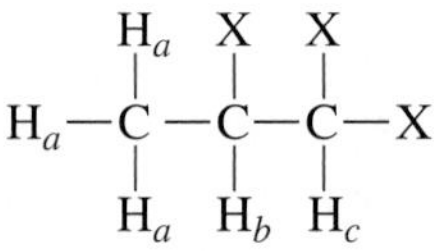

$$H_a-\underset{\underset{H_a}{|}}{\overset{\overset{H_a}{|}}{C}}-\underset{\underset{H_b}{|}}{\overset{\overset{X}{|}}{C}}-\underset{\underset{H_c}{|}}{\overset{\overset{X}{|}}{C}}-X$$

38. Label each set of chemically equivalent protons, using a for the set that will be at the lowest frequency (farthest upfield) in the ^1H NMR spectrum, b for the next, etc. Indicate the multiplicity of each signal.

a. 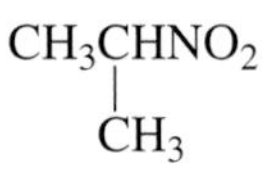 CH₃CHNO₂ with CH₃

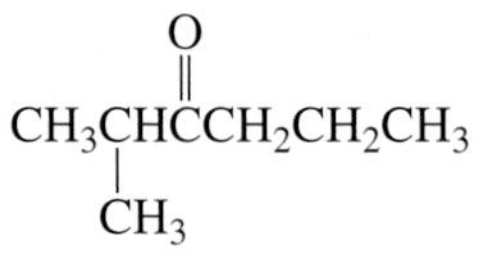

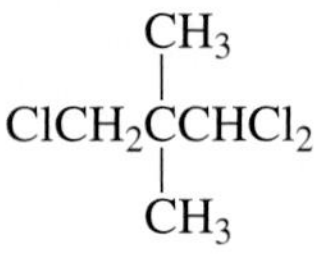

b. CH₃CH₂CH₂OCH₃

c. CH₃CHCCH₂CH₂CH₃ with O and CH₃

d. 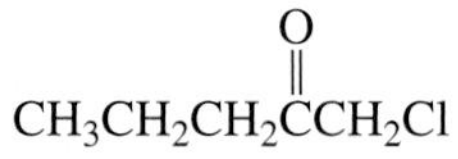 CH₃CH₂CH₂CCH₂Cl with O

e. ClCH₂CCHCl₂ with CH₃ and CH₃

f. ClCH₂CH₂CH₂CH₂CH₂Cl

39. Match each of the following ^1H NMR spectra with one of the following compounds:

CH₃CH₂CCH₃ with O

CH₃CNO₂ with CH₃ and CH₃

CH₃CH₂CCH₂CH₃ with O

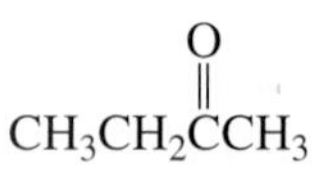

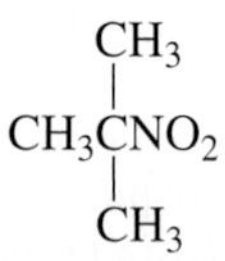

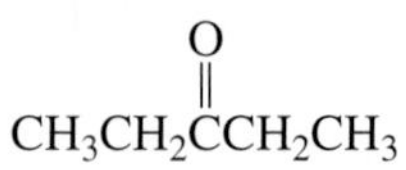

CH₃CH₂CH₂NO₂

CH₃CH₂NO₂

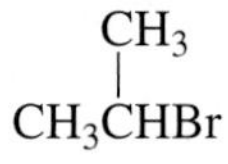

CH₃CHBr with CH₃

a.

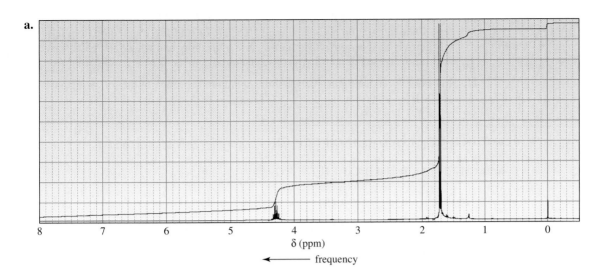

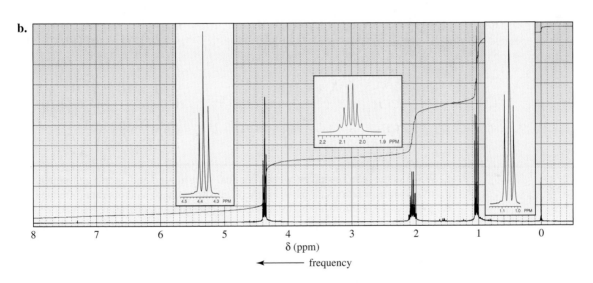

c.

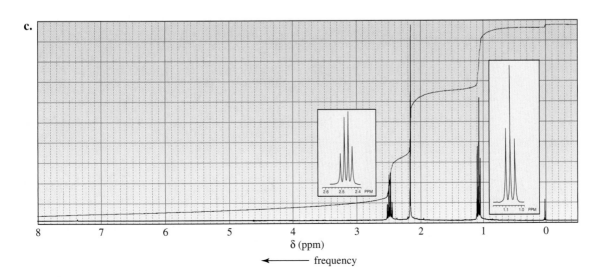

40. Determine the ratios of the magnetically nonequivalent hydrogens in a compound if the steps of the integration curves measure 40.5, 27, 13, and 118 mm, going from left to right across the spectrum. Give the structure of a compound whose ^{1}H NMR spectrum would show these integrals in the observed order.

41. How could you distinguish between the compounds in each of the following pairs by ^{1}H NMR?

a. $CH_3CH_2CH_2OCH_3$ and $CH_3CH_2OCH_2CH_3$

b. $BrCH_2CH_2CH_2Br$ and $BrCH_2CH_2CH_2NO_2$

c. 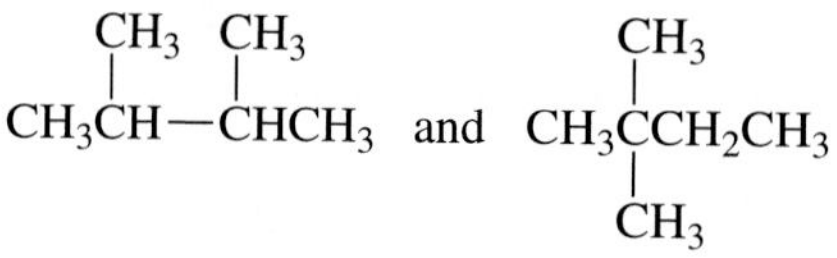 $CH_3CH{-}CHCH_3$ with two CH_3 groups above and $CH_3CCH_2CH_3$ with CH_3 above and CH_3 below

d. 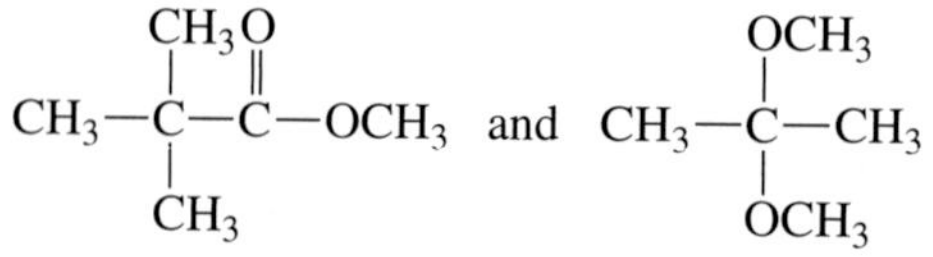 $CH_3{-}C{-}C{-}OCH_3$ (with CH_3, O, CH_3) and $CH_3{-}C{-}CH_3$ (with OCH_3, OCH_3)

e. 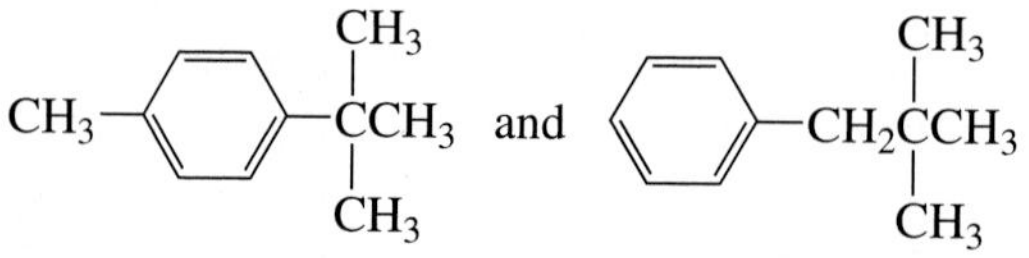 $CH_3{-}C_6H_4{-}CCH_3$ (with CH_3 and CH_3) and $C_6H_5{-}CH_2CCH_3$ (with CH_3 and CH_3)

f. 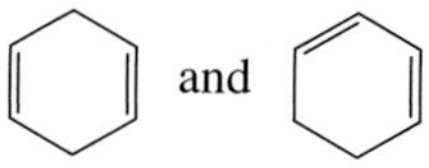 [cyclohexene ring] and [cyclohexadiene ring]

g. 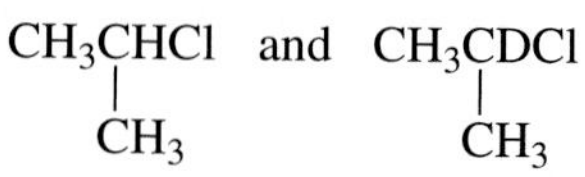 CH_3CHCl (with CH_3) and CH_3CDCl (with CH_3)

h. 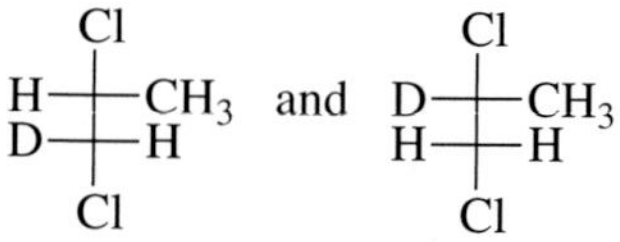 H${-}$Cl, CH_3, D, H, Cl (wedge structure) and D${-}$Cl, CH_3, H, H, Cl (wedge structure)

i. 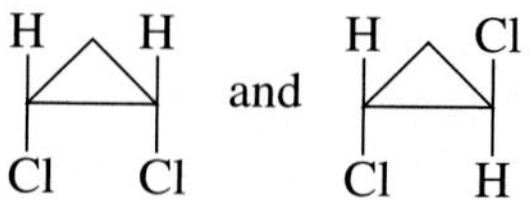 [cyclopropane with H, H, Cl, Cl] and [cyclopropane with H, Cl, Cl, H]

42. Answer the following:
   a. What is the relationship between chemical shift in ppm and operating frequency?
   b. What is the relationship between chemical shift in hertz and operating frequency?
   c. What is the relationship between coupling constant and operating frequency?
   d. How does the operating frequency in NMR spectroscopy compare with the operating frequency in IR and UV/Vis spectroscopy?

43. The ^{1}H NMR spectra of three isomers with molecular formula C_4H_9Br are shown here. Tell which isomer produces each spectrum.

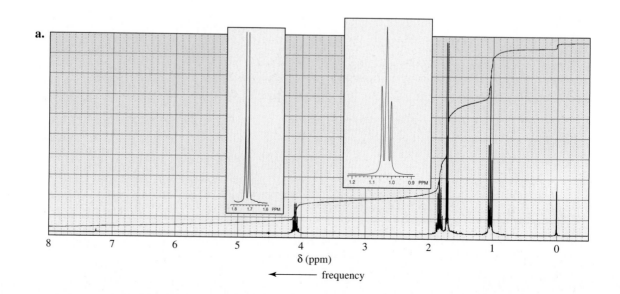

a.

δ (ppm)

$\longleftarrow$ frequency

b.

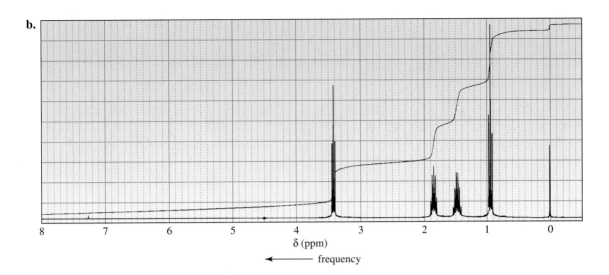

8
7
6
5
4
3
2
1
0
δ (ppm)
frequency

c.

3.4 3.3 3.2 PPM
1.2 1.1 1.0 PPM
8
7
6
5
4
3
2
1
0
δ (ppm)
frequency

44. Identify each of the following compounds from the ^1H NMR data and molecular formula. The number of hydrogens responsible for each signal is shown in parentheses.

a. $C_4H_8Br_2$ 1.97 ppm (6) singlet
3.89 ppm (2) singlet

b. C_8H_9Br 2.01 ppm (3) doublet
5.14 ppm (1) quartet
7.35 ppm (5) broad singlet

c. $C_5H_{10}O_2$ 1.15 ppm (3) triplet
1.25 ppm (3) triplet
2.33 ppm (2) quartet
4.13 ppm (2) quartet

45. Compound A, with molecular formula C_4H_9Cl, shows two signals in its ^{13}C NMR spectrum. Compound B, an isomer of compound A, shows four signals and, in the spin-coupled mode, the signal farthest downfield is a doublet. Identify compounds A and B.

46. The ^1H NMR spectra of three isomers with molecular formula $C_7H_{14}O$ are shown here. Tell which isomer produces each spectrum.

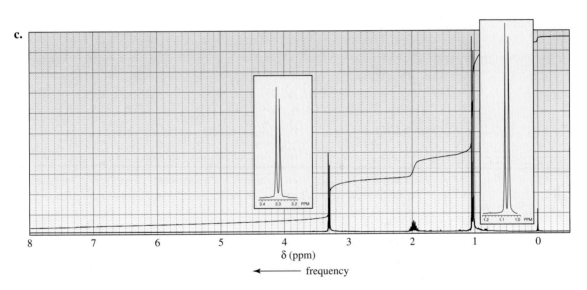

a.

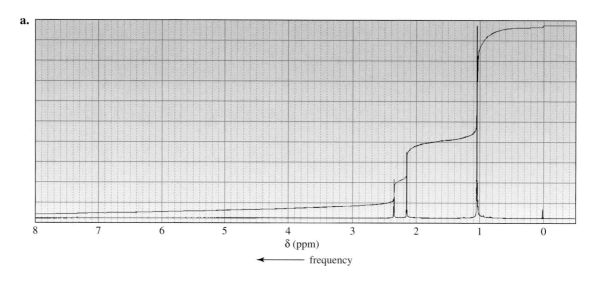

b.

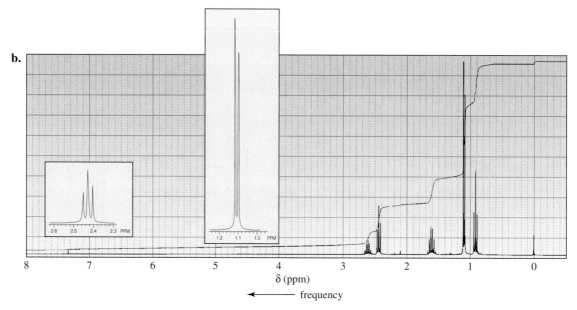

c.

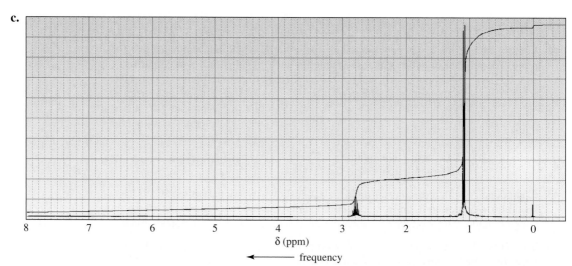

47. Would it be better to use ^{1}H NMR or ^{13}C NMR to distinguish among 1-butene, *cis*-2-butene, and 2-methylpropene? Explain your answer.

48. Determine the structure of each of the following unknown compounds based on its molecular formula and its IR and ^{1}H NMR spectra:

a. C_3H_6O

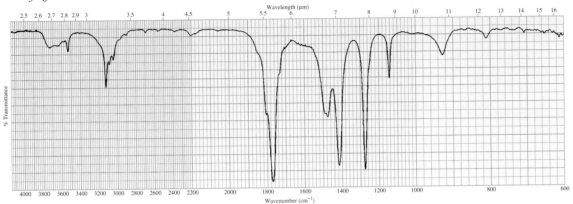

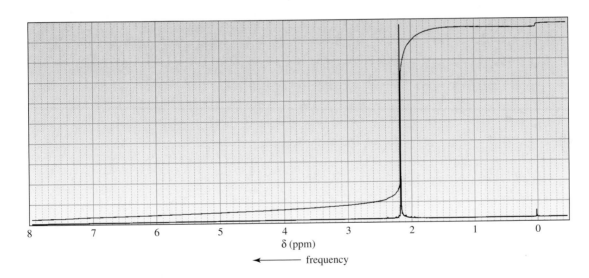

b. $C_5H_{12}O$

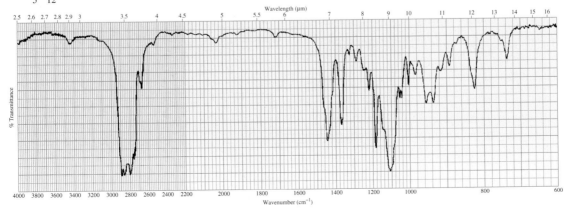

b. (*continued*)

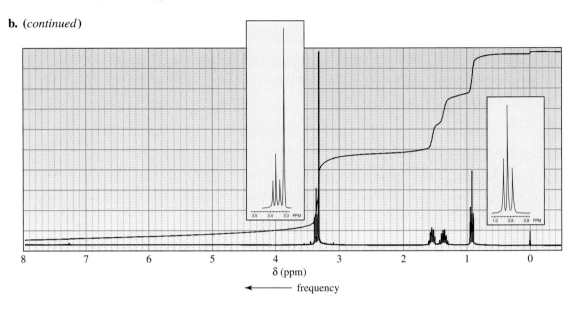

c. $C_6H_{12}O_2$

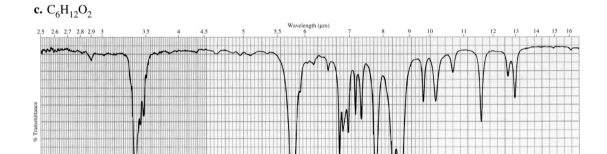

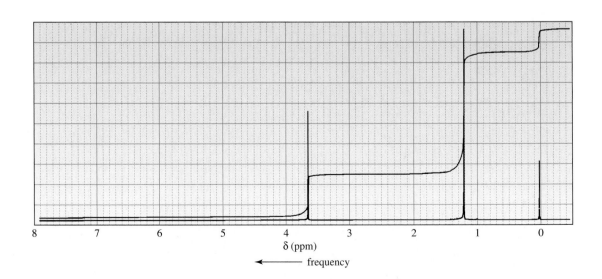

d. C$_4$H$_7$ClO$_2$

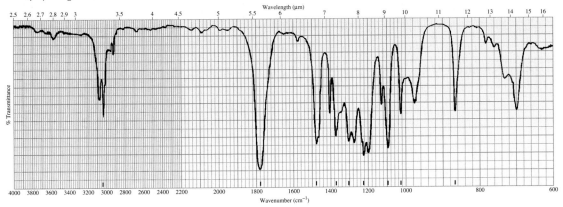

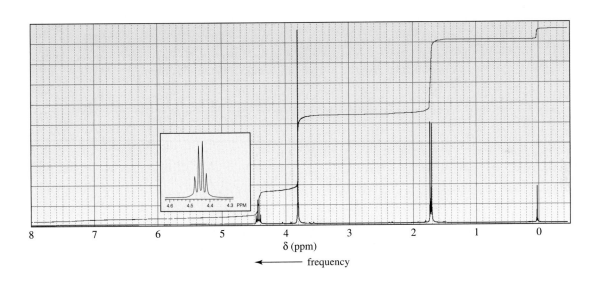

e. C$_4$H$_8$O$_2$

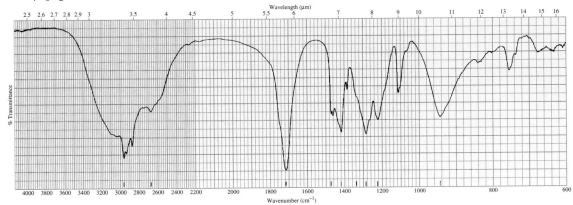

e. (*continued*)

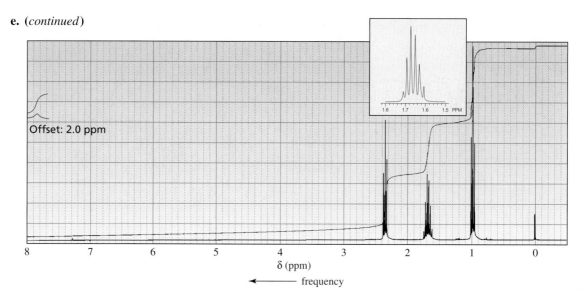

1.8 1.7 1.6 1.5 PPM
Offset: 2.0 ppm
8 7 6 5 4 3 2 1 0
δ (ppm)
frequency

49. There are four esters with molecular formula $C_4H_8O_2$. How could you distinguish among them by ^1H NMR?

50. The ^1H NMR spectrum of 2-propen-1-ol is shown here. Indicate the protons in the molecule that give rise to each of the signals in the spectrum.

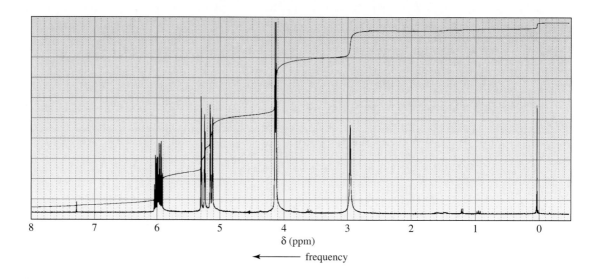

8 7 6 5 4 3 2 1 0
δ (ppm)
frequency

51. How could ^1H NMR be used to prove that the addition of HBr to propene follows Markovnikov's rule?

52. The ^1H NMR spectra for two compounds with molecular formula $C_{11}H_{16}$ are shown here. Identify the compounds.

a.

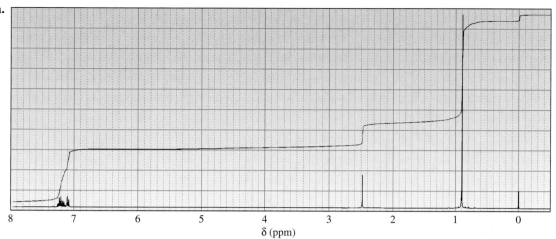

b.

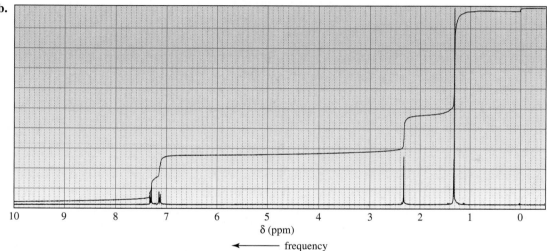

53. Draw a splitting diagram for the H_b proton if $J_{bc} = 10$ and $J_{ba} = 5$.

$$\underset{c}{\overset{Cl}{\underset{H}{\diagdown}}} C=C \overset{CH_2Cl}{\underset{\underset{b}{H}}{\diagup}} a$$

54. Sketch the following spectra that would be obtained for 2-chloroethanol:
 a. The ^1H NMR spectrum for a dry sample of the alcohol
 b. The ^1H NMR spectrum for a sample of the alcohol that contains a trace amount of acid
 c. The ^{13}C NMR spectrum
 d. The spin-coupled ^{13}C NMR spectrum
 e. The four parts of a DEPT ^{13}C NMR spectrum

55. Identify each of the following compounds from its molecular formula and its ^1H NMR spectrum:

a. C_8H_8

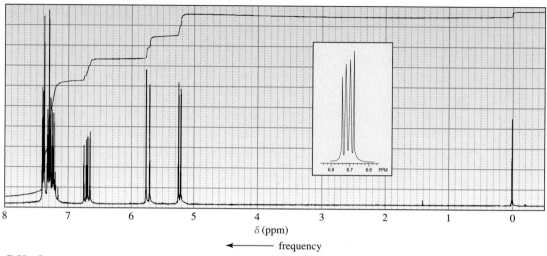

δ (ppm)

← frequency

b. $C_6H_{12}O$

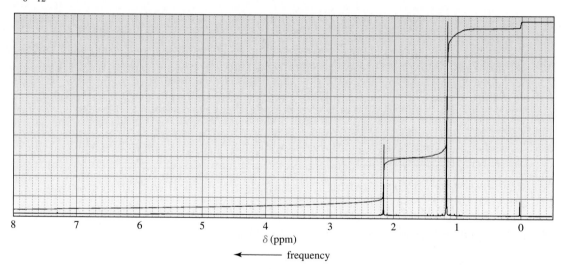

δ (ppm)

← frequency

c. $C_9H_{18}O$

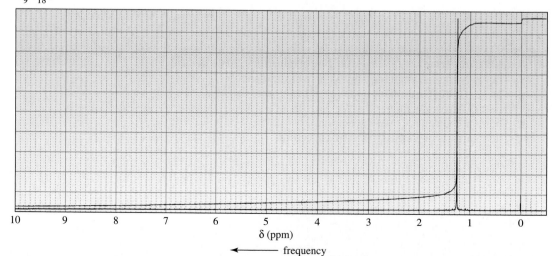

δ (ppm)

← frequency

d. C_4H_8O

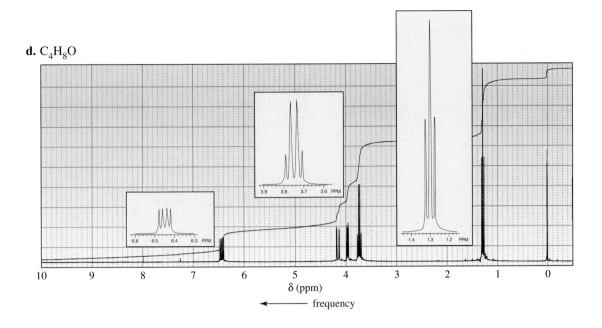

56. Dr. N. M. Arr was called in to help analyze the ^{1}H NMR spectrum of a mixture of compounds known to contain only C, H, and Br. The mixture showed two singlets, one at 1.8 ppm and the other at 2.7 ppm, with relative integrals of 1:6, respectively. Dr. Arr determined that the spectrum was that of a mixture of bromomethane and 2-bromo-2-methylpropane. What was the ratio of bromomethane to 2-bromo-2-methylpropane in the mixture?

57. Calculate the amount of energy (in calories) required to flip a ^{1}H nucleus in an NMR spectrometer that operates at 60 MHz.

58. The following ^{1}H NMR spectra are for four compounds with molecular formula $C_6H_{12}O_2$. Identify the compounds.

a.

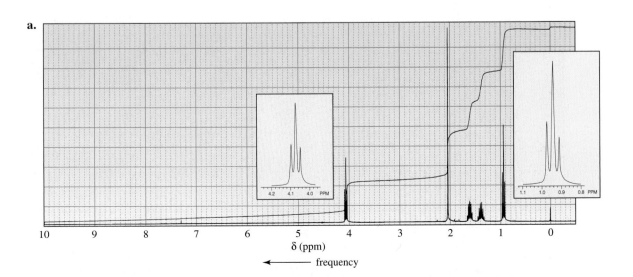

b.

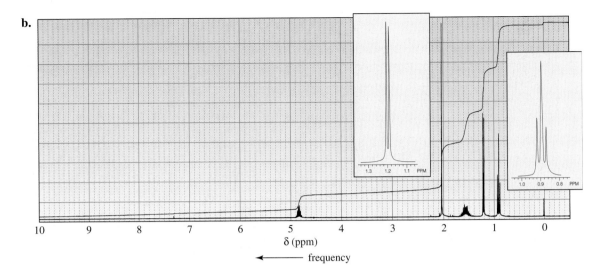

c.

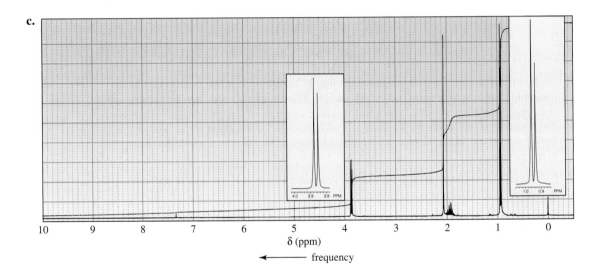

d.

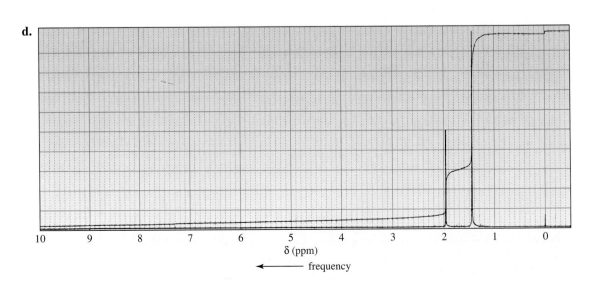

59. When compound A ($C_5H_{12}O$) is treated with HBr, it forms compound B ($C_5H_{11}Br$). The ^1H NMR spectrum of compound A has one singlet (1), two doublets (3, 6), and two multiplets (both 1). (The relative areas of the signals are indicated in parentheses.) The ^1H NMR spectrum of compound B has a singlet (6), a triplet (3), and a quartet (2). Identify compounds A and B.

60. Determine the structure of each of the following unknown compounds based on its molecular formula and its IR and ^1H NMR spectra:

a. $C_6H_{12}O$

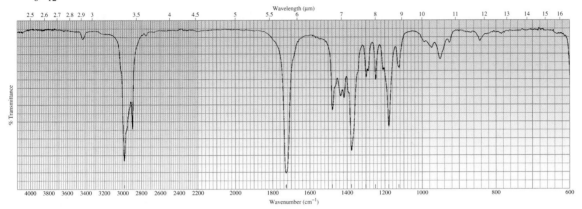

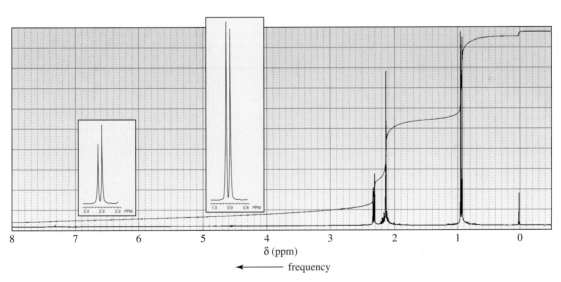

b. $C_6H_{14}O$

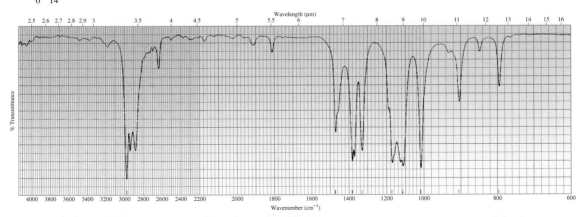

b. (*continued*)

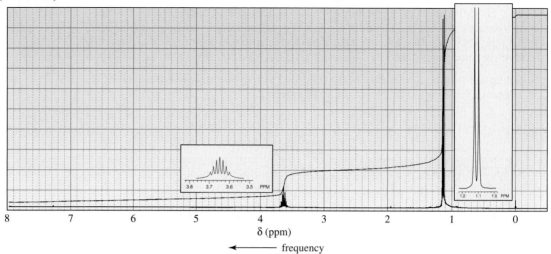

c. C$_{10}$H$_{13}$NO$_3$

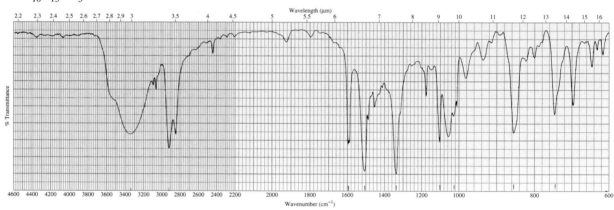

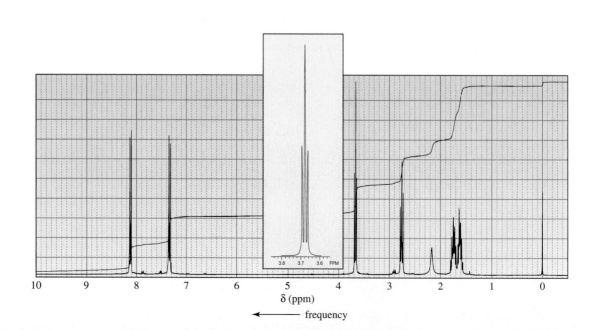

d. $C_{11}H_{14}O_2$

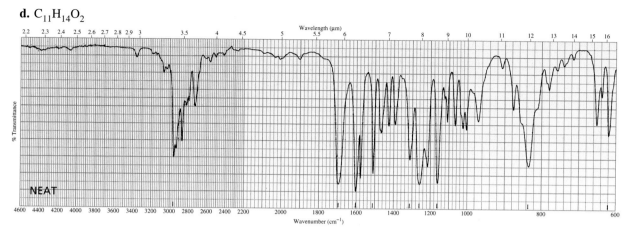

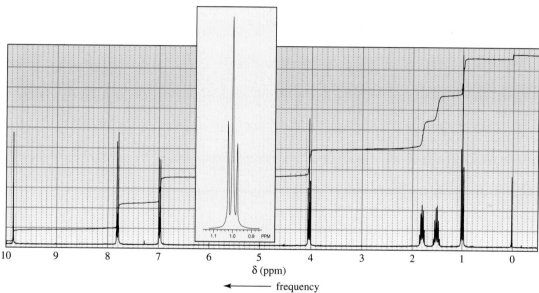

61. Identify the compound with molecular formula $C_7H_{14}O$ that gives the following spin-coupled ^{13}C NMR spectrum:

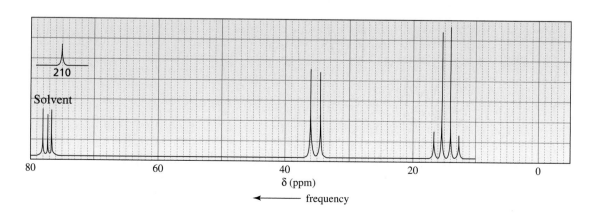

62. Determine the structure of each of the following unknown compounds based on its mass, IR, and ^{1}H NMR spectra:

a.

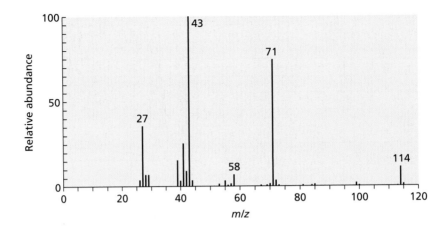

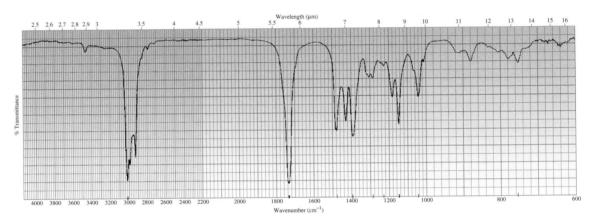

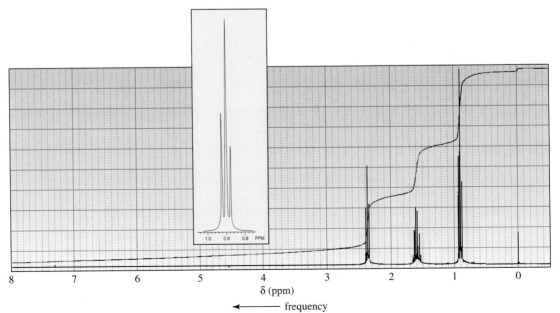

b.

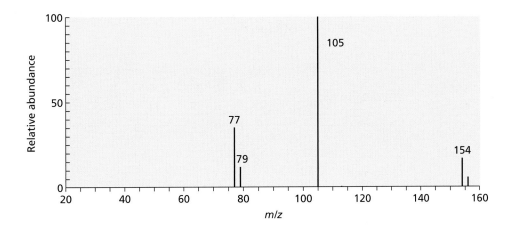

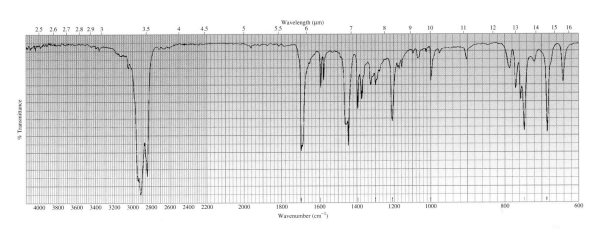

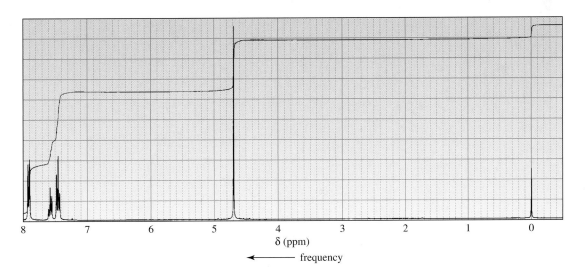

63. Identify the compound with molecular formula $C_6H_{10}O$ that is responsible for the following DEPT ^{13}C NMR spectrum.

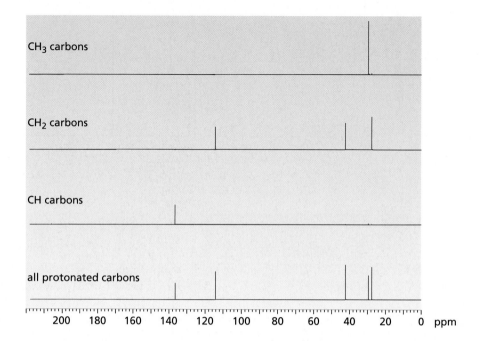

Aromatic Compounds

The two chapters in Part V discuss aromaticity and the reactions of aromatic compounds. You were first introduced to aromaticity in Chapter 6 where you saw that benzene, a compound with an unusually large resonance energy, is an aromatic compound. We will now look at the criteria that a compound must fulfill in order to be classified as an aromatic compound. Then we will look at the kinds of reactions that aromatic compounds undergo. In Chapter 27 we will return to aromatic compounds when we look at the reactions of aromatic compounds in which one of the ring atoms is an atom other than a carbon.

Chapter 14   **Aromaticity. Reactions of Benzene**
Chapter 15   **Reactions of Substituted Benzenes**

In **Chapter 14** we will examine the structural features that cause a compound to be aromatic. We will also look at the features that cause a compound to be antiaromatic. Then we will take a look at the reactions that benzene undergoes. You will see that although benzene, alkenes, and dienes are all nucleophiles because they all have carbon–carbon π bonds, benzene's aromaticity causes it to undergo reactions that are quite different from the reactions that alkenes and dienes undergo.

In **Chapter 15** we will look at the reactions of substituted benzenes. First we will study reactions that change the nature of the substituent on the benzene ring, and we will see how the nature of the substituent affects both the reactivity of the ring and the placement of any incoming substituent. You will then have the opportunity to design syntheses of compounds that contain benzene rings.

Aromaticity. Reactions of Benzene

pyridine

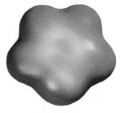

pyrrole

thiophene

Michael Faraday (1791–1867) was born in England, a son of a blacksmith. At the age of 14 he was apprenticed to a bookbinder and educated himself by reading the books that he bound. In 1833 he became a professor of chemistry at the Royal Institution. He is best known for his work on electricity.

The compound we know as benzene was first isolated in 1825 by Michael Faraday. He extracted the compound from a liquid residue left from heating whale oil under pressure to produce a gas used to illuminate buildings in London. Because of its origin, chemists suggested that it should be called "pheno" from the Greek word *phainein* ("to shine").

In 1834, Eilhardt Mitscherlich correctly determined the molecular formula of benzene (C_6H_6) and decided to call it benzin because of its relationship to benzoic acid, a known substituted form of the compound. Later its name was changed to benzene.

Compounds like benzene, with relatively few hydrogens compared to the number of carbons, are typically found in oils produced by trees and other plants. Early chemists called such compounds **aromatic compounds** because of their pleasing fragrances. In this way, they were distinguished from **aliphatic compounds,** which have higher hydrogen-to-carbon ratios, that were obtained from the chemical degradation of fats. The chemical meaning of the word "aromatic" now signifies certain kinds of chemical structures. We will now examine the criteria that classify a compound as aromatic.

14.1 CRITERIA FOR AROMATICITY

In Chapter 6 we saw that benzene is a planar, cyclic compound with a cyclic cloud of delocalized π electrons above and below the plane of the ring (Figure 14.1). Because its π electrons are delocalized, all the carbon–carbon bonds have the same length—partway between the length of a typical single and a typical double bond. We also saw that benzene is a particularly stable compound because it has an unusually large resonance energy (36 kcal/mol or 151 kJ/mol). Most compounds with delocalized electrons have much smaller resonance energies. Compounds such as benzene with unusually large resonance energies are called **aromatic** compounds. How can we tell by looking at the structure of a compound whether or not it is aromatic? In other words, what are the structural features that aromatic compounds have in common?

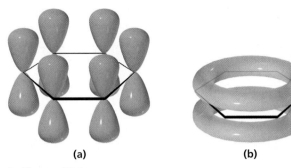

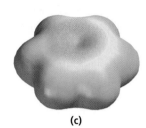

(a)   **(b)**   **(c)**

▲ **Figure 14.1**
(a) Each carbon of benzene has a *p* orbital. (b) The overlap of the *p* orbitals forms a cloud of π electrons above and below the plane of the benzene ring. (c) The electrostatic potential map for benzene shows that all the carbon–carbon bonds have the same electron density.

Eilhardt Mitscherlich (1794–1863) was born in Germany. He studied oriental languages at the University of Heidelberg and the Sorbonne, where he concentrated on Farsi, hoping that Napoleon would include him in a delegation he intended to send to Persia. That ambition ended with Napoleon's defeat. Mitscherlich returned to Germany to study science, simultaneously receiving a doctorate in Persian studies. He was a professor of chemistry at the University of Berlin.

For a compound to be classified as aromatic, it must fulfill both of the following criteria:

1. It must have an uninterrupted cyclic cloud of π electrons above and below the plane of the molecule (often called a π cloud). Let's look at what this means a little more closely.

    For the π cloud to be cyclic, the molecule must be *cyclic*.

    For the π cloud to be uninterrupted, *every atom in the ring must have a* p *orbital*.

    For the π cloud to form, each *p* orbital must be able to overlap with the *p* orbitals on either side of it. Therefore, the molecule must be *planar*.

2. The π cloud must contain an *odd* number of *pairs* of π electrons.

Benzene is an aromatic compound because it is cyclic and planar, every carbon in the ring has a *p* orbital, and the π cloud contains *three* pairs of π electrons.

The German chemist Erich Hückel was the first to recognize that an aromatic compound must have an odd number of pairs of π electrons. In 1931 he described this requirement by what has come to be known as **Hückel's rule,** or the **$4n + 2$ rule.** The rule states that for a planar, cyclic compound to be aromatic, its uninterrupted π cloud must contain $(4n + 2)$ π electrons, where *n* is any whole number. According to Hückel's rule, then, an aromatic compound must have 2 $(n = 0)$, 6 $(n = 1)$, 10 $(n = 2)$, 14 $(n = 3)$, 18 $(n = 4)$, etc., π electrons. Because there are two electrons in a pair, Hückel's rule requires that an aromatic compound must have 1, 3, 5, 7, 9, etc., pairs of π electrons. Thus, Hückel's rule is just a mathematical way of saying that an aromatic compound must have an *odd* number of pairs of π electrons.

For a compound to be aromatic, it must be cyclic and planar and have an uninterrupted cloud of π electrons. The cloud must contain an odd number of pairs of π electrons.

Erich Hückel (1896–1980) was born in Germany. He was a professor of chemistry at the University of Stuttgart and at the University of Marburg.

PROBLEM 1◆

a. What is the value of *n* in Hückel's rule when a compound has 9 pairs of π electrons?

b. Is such a compound aromatic?

Cyclobutadiene has two pairs of π electrons and cyclooctatetraene has four pairs of π electrons. Therefore, these compounds are not aromatic because they have an even number of pairs of π electrons. There is an additional reason why cyclooctatetraene is not aromatic—it is not planar, it is tub-shaped. We saw that for an eight-membered ring

**14.2
AROMATIC AND
NONAROMATIC
CYCLIC
HYDROCARBONS**

to be planar, it must have bond angles of 135° (Chapter 2, Problem 27), and we know that sp^2 carbons have 120° bond angles. Therefore, if cyclooctatetraene were planar, it would have considerable angle strain. Because cyclobutadiene and cyclooctatetrene are not aromatic, they do not have the unusual stability of aromatic compounds.

cyclobutadiene   benzene   cyclooctatetraene
[4]-annulene   [6]-annulene   [8]-annulene

Monocyclic hydrocarbons with alternating single and double bonds are called **annulenes.** A prefix in brackets denotes the number of carbons in the ring.

Which of the following three-membered ring structures is aromatic? Cyclopropene is not aromatic because it does not have an uninterrupted ring of p orbital-bearing atoms. One of its ring atoms is sp^3 hybridized, and only sp^2 and sp hybridized carbons have p orbitals. Therefore, cyclopropene does not fulfill the first criterion for aromaticity.

cyclopropene   cyclopropenyl   cyclopropenyl
                      cation   anion

The cyclopropenyl cation is aromatic because it does have an uninterrupted ring of p orbital-bearing atoms and the π cloud contains one pair of delocalized π electrons. The cyclopropenyl anion is not aromatic because its π cloud has two (an even number) pairs of π electrons.

resonance contributors for the cyclopropenyl cation

resonance hybrid

Cycloheptatriene is not aromatic. Although it has the correct number of pairs of π electrons to be aromatic (three pairs), it does not have an uninterrupted ring of p orbital-bearing atoms because one of the ring atoms is sp^3 hybridized. Cyclopentadiene is also not aromatic. It has an even number of pairs of π electrons (two pairs) *and* it does not have an uninterrupted ring of p orbital-bearing atoms.

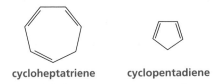

cycloheptatriene   cyclopentadiene

The criteria for determining whether a monocyclic hydrocarbon compound is aromatic can also be used to determine whether a polycyclic hydrocarbon compound is aromatic. Naphthalene (five pairs of π electrons), phenanthrene (seven pairs of π electrons), and chrysene (nine pairs of π electrons) are aromatic.

3-D Molecules:
Phenanthrene;
Naphthalene

naphthalene phenanthrene chrysene

Richard E. Smalley was born in 1943 in Akron, Ohio. He received a B.S. from the University of Michigan and a Ph.D. from Princeton University. He is a professor of chemistry at Rice University.

BUCKYBALLS AND AIDS

In addition to diamond and graphite, a third form of pure carbon was discovered while scientists were conducting experiments designed to understand how long-chain molecules are formed in outer space. R. E. Smalley, R. F. Curl, Jr., and H. W. Kroto, the discoverers of this new form of pure carbon, shared the 1996 Nobel Prize in chemistry for their discovery. They named it buckminsterfullerene (often shortened to fullerene) because it reminded them of the geodesic domes popularized by R. Buckminster Fuller, an American architect and philosopher. It is nicknamed "buckyball." Fullerene is the most symmetrical large molecule known, consisting of a hollow cluster of 60 carbons. Each molecule has 32 interlocking rings (20 hexagons and 12 pentagons) that fit together like the seams of a soccer ball. At first glance, buckyball would appear to be aromatic because of its benzene-like rings. However, it does not undergo electrophilic substitution reactions—it undergoes electrophilic addition reactions like an alkene. Its lack of aromaticity is apparently caused by the curvature of the ball, which prevents the molecule from fulfilling the first criterion for aromaticity (that it must be planar).

Buckyballs have extraordinary chemical and physical properties. They are exceedingly rugged and are capable of surviving the extreme temperatures of outer space. Because they are essentially hollow cages, they can be manipulated to make materials never before known. For example, when a buckyball is "doped" by inserting potassium or cesium into its cavity, it becomes the best organic superconductor known. These molecules are presently being studied for use in many other applications such as new polymers and catalysts and new drug delivery systems. The discovery of buckyballs is a strong reminder of the technological advances that can be achieved as a result of conducting basic research.

Scientists have even turned their attention to buckyballs in their quest for a cure for AIDS. An enzyme that is required for HIV to reproduce exhibits a nonpolar pocket in its three-dimensional structure. If this pocket is blocked, the production of virus ceases. Because buckyballs are nonpolar and have approximately the same diameter as the pocket of the enzyme, they are being considered as possible blockers. The first step in pursuing this possibility was to equip the buckyball with polar side chains to make it water soluble so that it could flow through the bloodstream. Scientists have now modified the arms so that they bind to the enzyme. It's still a long way from a cure for AIDS, but this represents one example of the many and varied approaches that scientists are taking to find a cure for this disease.

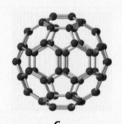
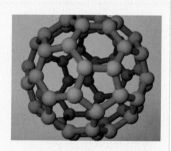

C_{60}
buckminsterfullerene
"buckyball"

PROBLEM 2 ◆

Which of the following are aromatic?

a.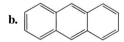

b.

c. cycloheptatrienyl cation

d.

e.

f.

g. cyclononatetraenyl anion

h. $CH_2\!=\!CHCH\!=\!CHCH\!=\!CH_2$

Robert F. Curl, Jr. was born in Texas in 1933. He received a B.A. from Rice University and a Ph.D. from the University of California, Berkeley. He is a professor of chemistry at Rice University.

Sir Harold W. Kroto was born in 1939 in England and is a professor of chemistry at the University of Sussex.

3-D Molecules:
1-Chloronaphthalene;
2-Chloronaphthalene

PROBLEM 3 / SOLVED

a. How many monobromonaphthalenes are there?

b. How many monobromophenanthrenes are there?

SOLUTION TO 3a   There are two monobromonaphthalenes.

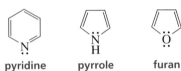

PROBLEM 4

The [10]- and [12]-annulenes have been synthesized, and neither has been found to be aromatic. Explain.

14.3 AROMATIC AND NONAROMATIC HETEROCYCLIC COMPOUNDS

A compound does not have to be a hydrocarbon to be aromatic. Many heterocyclic compounds are aromatic. A **heterocyclic compound** is a cyclic compound in which one or more of the ring atoms is an atom other than carbon. A ring atom that is not carbon is called a **heteroatom.** The name comes from the Greek word *heteros,* which means "different." The most common heteroatoms are N, O, and S.

heterocyclic compounds

| pyridine | pyrrole | furan | thiophene |

Pyridine is an aromatic heterocyclic compound. Each of the six ring atoms of pyridine is sp^2 hybridized so each has a p orbital, and the molecule contains three pairs of π electrons. Don't be confused by the nonbonding electrons on the nitrogen. They are not π electrons. Because nitrogen is sp^2 hybridized, it has three sp^2 orbitals and a p orbital. The p orbital is used to form the π bond. Two of the sp^2 orbitals overlap the sp^2 orbitals of adjacent carbon atoms and the third sp^2 orbital contains the nonbonding pair of electrons.

It is not immediately apparent that the electrons represented as lone-pair electrons on the nitrogen atom of pyrrole are π electrons. The resonance contributors, however, show that these electrons are in a p orbital that overlaps the p orbitals on adjacent carbons—thus they are π electrons. Pyrrole, therefore, has three pairs of π electrons and is aromatic.

pyridine

pyrrole

resonance contributors of pyrrole

Similarly, furan and thiophene are stable aromatic compounds. Both oxygen and sulfur have one pair of nonbonding electrons in an sp^2 orbital, and the second pair of electrons represented as lone-pair electrons are π electrons—they are in a p orbital that overlaps the p orbitals of adjacent carbons.

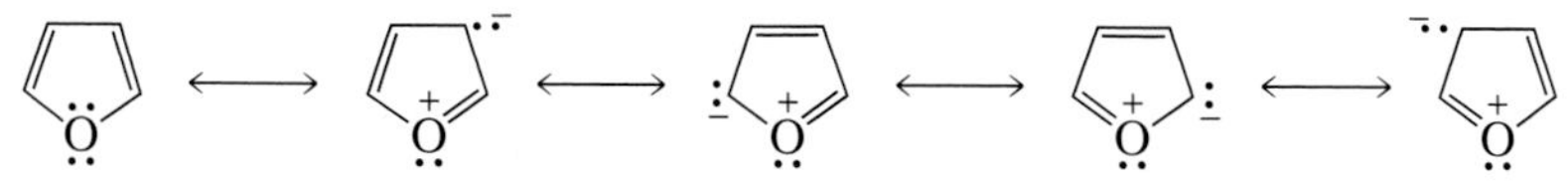

resonance contributors of furan

furan

Quinoline, indole, imidazole, purine, and pyrimidine are additional examples of heterocyclic aromatic compounds.

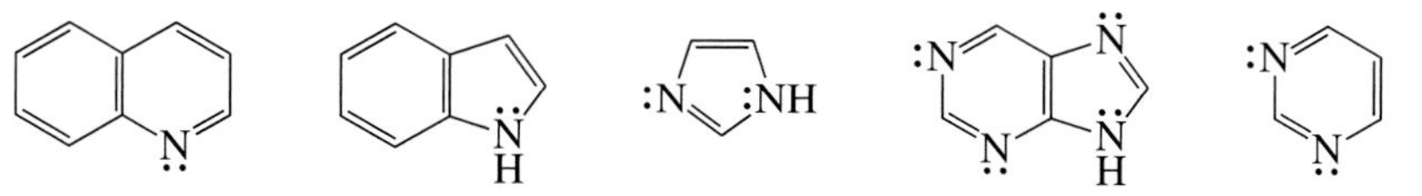

quinoline   indole   imidazole   purine   pyrimidine

thiophene

PROBLEM 5◆

In what orbitals are the electrons represented as lone-pair electrons when drawing the structures of quinoline, indole, imidazole, purine, and pyrimidine?

The pK_a of cyclopentadiene is 15, which is extraordinarily acidic for a hydrogen bonded to an sp^3 hybridized carbon. Ethane, for example, has a pK_a of 50.

14.4
SOME CHEMICAL CONSEQUENCES OF AROMATICITY

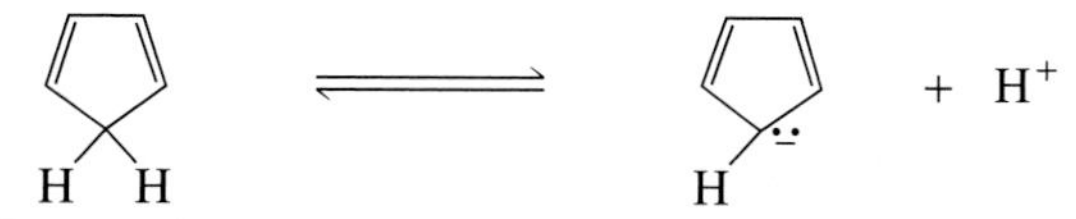

$$CH_3CH_3 \rightleftharpoons CH_3\overset{..}{C}H_2 + H^+$$
$$pK_a = 50$$

cyclopentadiene   cyclopentadienyl
$pK_a = 15$ anion

$+ H^+$

Why does cyclopentadiene have such a low pK_a? To answer this question, we must look at the anion that is formed when cyclopentadiene loses a proton. The cyclopentadienyl anion fulfills the requirements for aromaticity—it is cyclic and planar, each atom in the ring has a p orbital, and the π cloud has three pairs of π electrons. Notice that the negatively charged carbon in the cyclopentadienyl anion needs to be sp^2 hybridized. If it were sp^3 hybridized, the ion would not be aromatic. The resonance hybrid shows that all the carbons in the cyclopentadienyl anion are equivalent. Each carbon has exactly one-fifth of the negative charge associated with the anion.

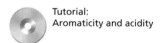

Tutorial:
Aromaticity and acidity

resonance contributors

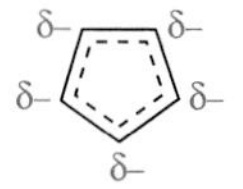

resonance hybrid

As a result of its aromaticity, the cyclopentadienyl anion is an unusually stable carbanion. This is why cyclopentadiene has an unusually low pK_a. In other words, the stability associated with the aromaticity of the cyclopentadienyl anion causes the hydrogen to be lost much more readily than hydrogens bonded to other sp^3 carbons. (Recall from Section 1.18 that the strength of an acid depends on the stability of its conjugate base).

PROBLEM 6

a. Draw arrows to show the movement of electrons in going from one resonance contributor to the next in:

    1. the cyclopentadienyl anion   **2.** pyrrole

b. How many ring atoms share the negative charge in the following?

    1. the cyclopentadienyl anion   **2.** pyrrole

PROBLEM 7◆

Predict the relative pK_a values of cyclopentadiene and cycloheptatriene.

Another example of the influence of aromaticity on chemical reactivity is the unusual chemical behavior exhibited by cycloheptatrienyl bromide. Recall from Section 2.9 that alkyl halides tend to be relatively nonpolar covalent compounds, so they are soluble in nonpolar solvents and insoluble in water. Cycloheptatrienyl bromide, on the other hand, is an alkyl halide that behaves like an ionic compound—it is insoluble in nonpolar solvents but is readily soluble in water.

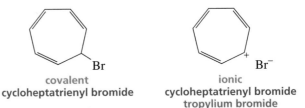

covalent
cycloheptatrienyl bromide

ionic
cycloheptatrienyl bromide
tropylium bromide

It turns out that cycloheptatrienyl bromide is an ionic compound because its cation is aromatic. It is *not* aromatic in the covalent form because it has an sp^3 hybridized carbon, so it does *not* have an uninterrupted ring of p orbital-bearing atoms. In the ionic form, the cycloheptatrienyl cation (also known as the tropylium cation) is aromatic because it is a planar cyclic ion, all the ring atoms are sp^2 hybridized (which means that each ring atom has a p orbital), and it has three pairs of delocalized π electrons. The stability associated with the aromatic cation causes the alkyl halide to exist in the ionic form.

resonance contributors

resonance hybrid

PROBLEM-SOLVING STRATEGY

Which of the following compounds has the greater dipole moment?

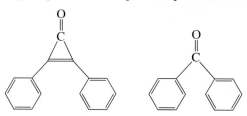

Before attempting to answer this kind of question, make sure you know exactly what the question is asking. You know that the dipole moment of these compounds results from the unequal sharing of electrons by carbon and oxygen. Therefore, the more unequal the sharing, the greater the dipole moment. Draw the resonance contributors with separated charges and determine their relative stabilities. In the case of the compound on the left, the three-membered ring becomes aromatic when the charges are separated. In the case of the compound on the right, the resonance contributor with separated charges does not have any additional aromaticity. Thus, the resonance contributor with separated charges is more stable and, therefore, makes a greater contribution to the hybrid for the compound on the left than for the compound on the right. The compound on the left, therefore, has the greater dipole moment.

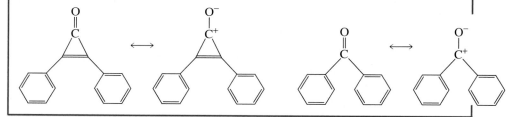

PROBLEM 8

Draw the resonance contributors of the cyclooctatrienyl dianion.

a. Which of the resonance contributors is the least stable?

b. Which of the resonance contributors makes the smallest contribution to the hybrid?

An aromatic compound is *more stable* than an analogous cyclic compound with localized electrons. An **antiaromatic** compound is *less stable* than an analogous cyclic compound with localized electrons. Aromaticity is characterized by stabilization, while antiaromaticity is characterized by destabilization.

14.5
ANTIAROMATICITY

relative stabilities

aromatic compound > cyclic compound with localized electrons > antiaromatic compound

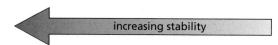

increasing stability

A compound is classified as being antiaromatic if it fulfills the first criterion for aromaticity but does not fulfill the second criterion. In other words, it must be a planar cyclic compound with an uninterrupted ring of p orbital-bearing atoms, but the π cloud must contain an *even* number of pairs of π electrons. Hückel would state that for a planar cyclic compound with an uninterrupted ring of *p* orbital-bearing atoms to be antiaromatic, it must have $4n$ π electrons, where *n* is any whole number.

Cyclobutadiene is a planar molecule with two pairs of π electrons. Hence it is antiaromatic and very unstable. The cyclopentadienyl cation also has two pairs of π electrons, so it, too, is antiaromatic.

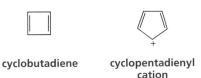

cyclobutadiene   cyclopentadienyl
                    cation

PROBLEM 9 ◆

a. Predict the relative pK_a values of cyclopropene and cyclopropane.

b. Which is more soluble in water, 3-bromocyclopropene or bromocyclopropane?

PROBLEM 10 ◆

Which of the compounds in Problem 2 are antiaromatic?

14.6
A MOLECULAR ORBITAL DESCRIPTION OF AROMATICITY AND ANTIAROMATICITY

Why are planar molecules with uninterrupted cyclic π electron clouds very stable (aromatic) if they have an odd number of pairs of π electrons and very unstable (antiaromatic) if they have an even number of pairs of π electrons? To answer this question, we must turn to molecular orbital theory.

The relative energies of the π molecular orbitals of planar molecules with uninterrupted cyclic π electron clouds can be determined, without having to use any math, by first drawing the cyclic compound with one of its vertices pointed down. The relative energies of the π molecular orbitals correspond to the relative levels of the vertices (Figure 14.2). Molecular orbitals below the midpoint of the cyclic

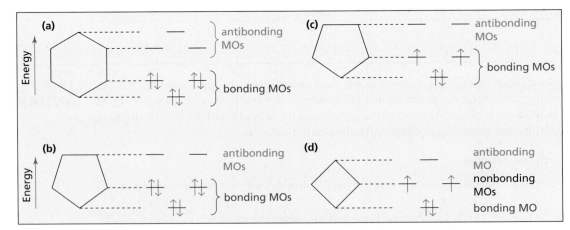

▲ **Figure 14.2**
The distribution of electrons in the π molecular orbitals of (a) benzene, (b) the cyclopentadienyl anion, (c) the cyclopentadienyl cation, and (d) cyclobutadiene. The relative energies of the π molecular orbitals in a cyclic compound correspond to the relative levels of the vertices. Molecular orbitals below the midpoint of the cyclic structure are bonding, those above the midpoint are antibonding, and those at the midpoint are nonbonding.

structure are bonding molecular orbitals, those above the midpoint are antibonding molecular orbitals, and any at the midpoint are nonbonding molecular orbitals. Notice that the number of π molecular orbitals is equal to the number of atoms in the ring since each ring atom contributes a p orbital.

The six π electrons of benzene occupy its three bonding π molecular orbitals, and the six π electrons of the cyclopentadienyl anion occupy its three bonding π molecular orbitals. Notice that there is always an odd number of bonding orbitals because one corresponds to the lowest vertex and the others come in degenerate pairs. This means that aromatic compounds (such as benzene and the cyclopentadienyl anion) with an odd number of pairs of π electrons have completely filled bonding orbitals and no electrons in either nonbonding or antibonding orbitals. This is what gives aromatic molecules their stability. (A more in-depth description of the molecular orbitals in benzene is given in Section 6.11.)

Antiaromatic compounds have an even number of pairs of π electrons. Therefore, either they are unable to fill their bonding orbitals (cyclopentadienyl cation) or they have a pair of π electrons left over after the bonding orbitals are filled (cyclobutadiene). Hund's rule requires that these two electrons go into two degenerate orbitals. The unpaired electrons are responsible for the destabilization of antiaromatic molecules.

Tutorial:
Molecular orbital
description of
aromaticity

PROBLEM 11

Following the instructions given for drawing the π molecular orbital energy levels for the compounds shown in Figure 14.2, draw the π molecular orbital energy levels for the cycloheptatrienyl cation, the cycloheptatrienyl anion, and the cyclopropenyl cation. For each compound, show the distribution of the π electrons. Which of the compounds are aromatic? Which are antiaromatic?

PROBLEM 12◆

How many bonding, nonbonding, and antibonding π molecular orbitals does cyclobutadiene have? In which molecular orbitals are the π electrons?

PROBLEM 13

Which is more stable, a cyclic compound with an odd number of pairs of delocalized π electrons or one with an even number of pairs of delocalized π electrons? Explain.

14.7 REACTIVITY CONSIDERATIONS

Benzene, as a consequence of the π electrons above and below the plane of its ring, is a nucleophile. An electrophile (Y^+) will be attracted to the π electrons. When an electrophile attaches itself to a benzene ring, a carbocation intermediate is formed.

carbocation
intermediate

This should remind you of the first step in an electrophilic addition reaction of an alkene—a nucleophilic alkene reacts with an electrophile, thereby forming a carbocation intermediate (Section 3.6). In the second step of an electrophilic addition reaction, the carbocation reacts with a nucleophile (Z^-) to form an addition product.

$$RCH=CHR \ + \ Y^+ \ \longrightarrow \ \underset{\underset{\underset{Y}{|}}{+}}{RCH-CHR} \ \xrightarrow{Z^-} \ \underset{\underset{Z \qquad Y}{|\ \ \ |}}{RCH-CHR}$$

carbocation intermediate

product of electrophilic addition

If the carbocation intermediate formed from the reaction of benzene with an electrophile were to react similarly with a nucleophile, the addition product would not be aromatic. Because there is a great deal of stabilization associated with an aromatic ring, the carbocation loses a proton from the site of electrophilic attack instead of adding a nucleophile. In this way, the aromaticity of the benzene ring is restored. The overall reaction, then, is a substitution reaction—an electrophile substitutes for one of the hydrogens attached to the benzene ring.

product of electrophilic addition

a nonaromatic compound

carbocation intermediate

product of electrophilic substitution

an aromatic compound

The reaction coordinate diagram in Figure 14.3 shows that the reaction of benzene to form a substituted benzene has a $\Delta G°$ close to zero. The reaction of benzene to form the much less stable nonaromatic addition product would have been a highly endergonic reaction. Consequently, benzene undergoes *electrophilic substitution reactions* that

Figure 14.3 ▶
Reaction coordinate diagrams for electrophilic substitution of benzene and for electrophilic addition to benzene.

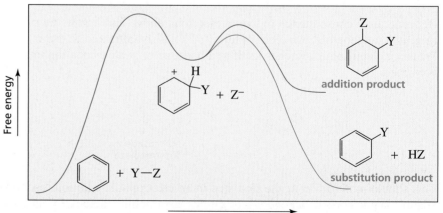

Free energy

Progress of the reaction

preserve aromatic stabilization rather than *electrophilic addition reactions,* the reactions characteristic of alkenes, which would destroy aromatic stabilization.

PROBLEM 14

If electrophilic addition to benzene is an endergonic reaction, how can electrophilic addition to an alkene be an exergonic reaction?

In an **electrophilic aromatic substitution reaction,** an electrophile substitutes for a hydrogen of an aromatic compound.

an electrophilic aromatic substitution reaction

$$\text{(benzene)}-\text{H} + \text{Y}^+ \longrightarrow \text{(benzene)}-\text{Y} + \text{H}^+$$

The five most common electrophilic aromatic substitution reactions are the following:

1. **Halogenation:** a bromine, chlorine, or iodine (Br, Cl, I) substitutes for a hydrogen.
2. **Nitration:** a nitro (NO_2) group substitutes for a hydrogen.
3. **Sulfonation:** a sulfonic acid (SO_3H) group substitutes for a hydrogen.
4. **Friedel–Crafts acylation:** an acyl ($RC=O$) group substitutes for a hydrogen.
5. **Friedel–Crafts alkylation:** an alkyl (R) group substitutes for a hydrogen.

All of these electrophilic aromatic substitution reactions take place by the same two-step mechanism. In the first step benzene reacts with an electrophile (Y^+), forming a carbocation intermediate. The structure of the carbocation intermediate can be approximated by three resonance contributors. In the second step of the reaction, a base in the reaction mixture pulls off a proton from the carbocation intermediate and the electrons that held the proton move into the ring to reestablish its aromaticity. Notice that the proton is always removed from the carbon that has formed the new bond with the electrophile.

14.8
GENERAL MECHANISM FOR ELECTROPHILIC AROMATIC SUBSTITUTION REACTIONS

In an electrophilic aromatic substitution reaction, an electrophile (Y^+) is put on the ring and an H^+ comes off the ring.

general mechanism for electrophilic aromatic substitution

$$\text{(benzene)} + \text{Y}^+ \xrightarrow{\text{slow}} \left[\text{(carbocation resonance contributors)} \right] \xrightarrow{\text{fast}} \text{(benzene)}-\text{Y} + \text{HB}^+$$

The first step is relatively slow and endergonic because an aromatic compound is being converted into a nonaromatic intermediate (Figure 14.3). The second step is fast and strongly exergonic because this step restores the stability-enhancing aromaticity.

We will look at each of these five electrophilic aromatic substitution reactions individually. As you study them, notice that they differ only in how the electrophile (Y^+) that is needed to start the reaction is generated. Once the electrophile is formed, all five reactions follow the same two-step mechanism for electrophilic aromatic substitution.

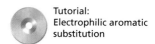

Tutorial:
Electrophilic aromatic substitution

14.9
HALOGENATION OF BENZENE

bromination

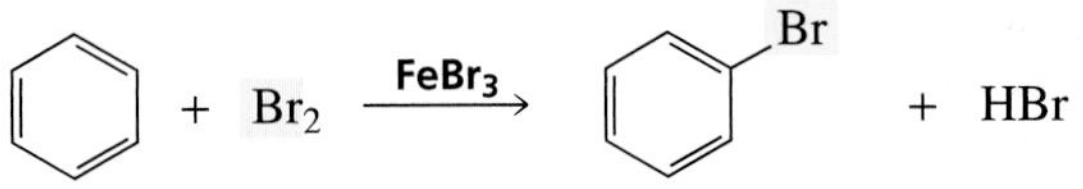

benzene + Br₂ →(FeBr₃) bromobenzene (Br) + HBr

bromobenzene

chlorination

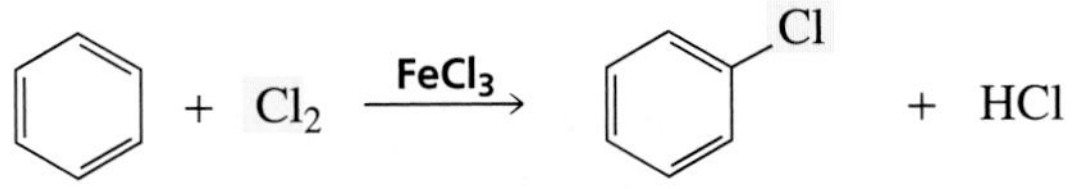

benzene + Cl₂ →(FeCl₃) chlorobenzene (Cl) + HCl

chlorobenzene

iodination

$$I_2 \xrightarrow{HNO_3} 2\,I^+$$

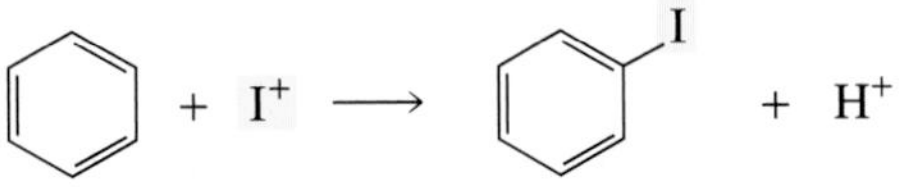

benzene + I⁺ → iodobenzene (I) + H⁺

iodobenzene

The bromination or chlorination of benzene requires a Lewis acid catalyst such as ferric bromide or ferric chloride. Recall that a *Lewis acid* is a compound that accepts a share in a pair of electrons (Section 1.21). In the first step of the bromination reaction, bromine donates a lone pair of electrons to the Lewis acid. This weakens the Br-Br bond, thereby providing the electrophile necessary for electrophilic aromatic substitution.

mechanism for bromination

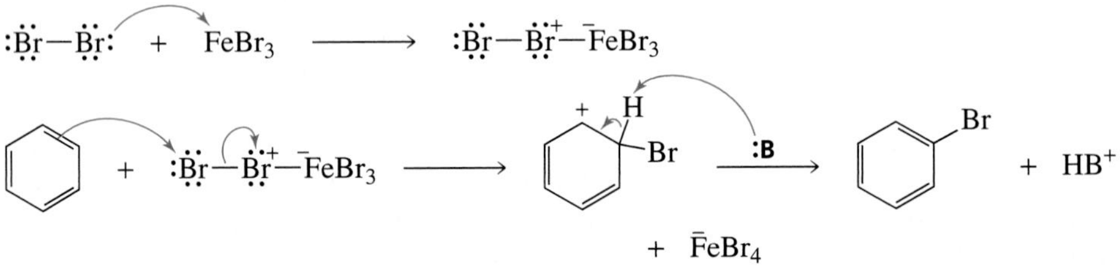

Movie: Bromination of benzene

$$:Br-Br: + FeBr_3 \longrightarrow :Br-Br^+-\bar{F}eBr_3$$

benzene + :Br—Br⁺—⁻FeBr₃ → carbocation intermediate (+H, Br) :B → bromobenzene (Br) + HB⁺

+ ⁻FeBr₄

In order to make the mechanisms as easy as possible to understand, only one of the three resonance contributors of the carbocation intermediate is shown in this and subsequent mechanisms. Bear in mind, however, that each carbocation intermediate actually has the three resonance contributors shown in Section 14.8. In the last step of the reaction, a base from the reaction mixture (⁻FeBr₄, Br⁻, or the solvent) removes a proton from the carbocation intermediate.
    Chlorination of benzene occurs by the same mechanism as bromination.

mechanism for chlorination

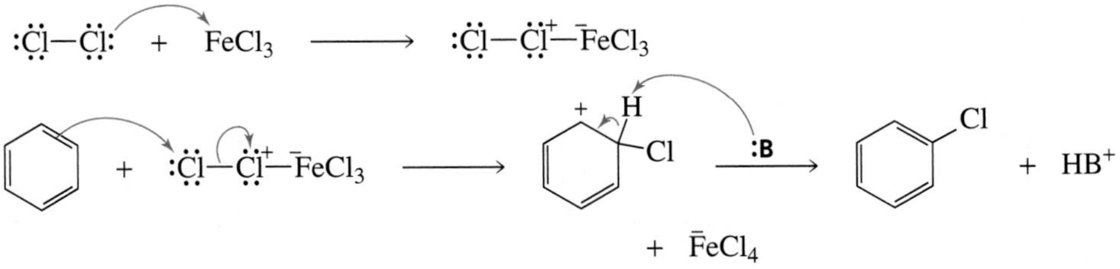

$$:Cl-Cl: + FeCl_3 \longrightarrow :Cl-Cl^+-\bar{F}eCl_3$$

benzene + :Cl—Cl⁺—⁻FeCl₃ → carbocation intermediate (+H, Cl) :B → chlorobenzene (Cl) + HB⁺

+ ⁻FeCl₄

Ferric bromide and ferric chloride readily react with moisture in the air during handling, which inactivates them as catalysts. Ferric bromide or ferric chloride, therefore, is generated *in situ* (in the reaction mixture) by adding iron filings and bromine or chlorine to the reaction mixture. Therefore, the reagent halogen matches the halogen in the catalyst (Cl_2 + $FeCl_3$, Br_2 + $FeBr_3$).

$$2\ Fe\ +\ 3\ Br_2\ \longrightarrow\ 2\ FeBr_3$$

PROBLEM 15◆

Why is hydrated $FeBr_3$ inactive as a Lewis acid catalyst?

Unlike the reaction of benzene with Br_2 or Cl_2, the reaction of an alkene with Br_2 or Cl_2 does not require a Lewis acid catalyst (Section 3.15). This is because an alkene is a better nucleophile than benzene and, as a result, the Br—Br or Cl—Cl bond does not have to be weakened to form a better electrophile.

Electrophilic iodine (I^+) is obtained by oxidizing I_2 with an oxidizing agent such as nitric acid.

mechanism for iodination

$$I_2 \xrightarrow{\text{oxidizing agent}} 2\ I^+\ +\ 2\ e^-$$

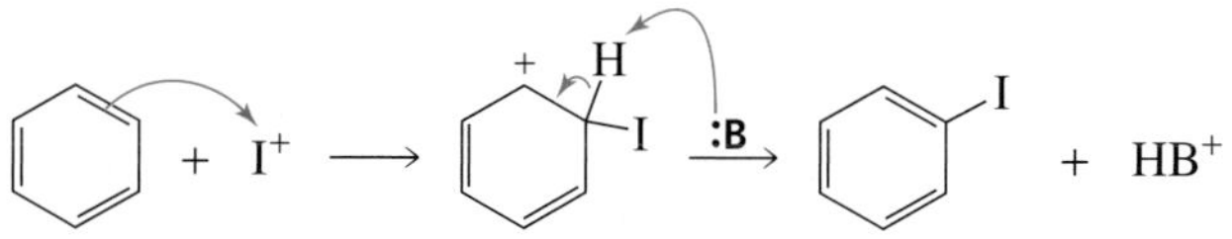

---

nitration

$$14.10$$
14.10
NITRATION OF
BENZENE

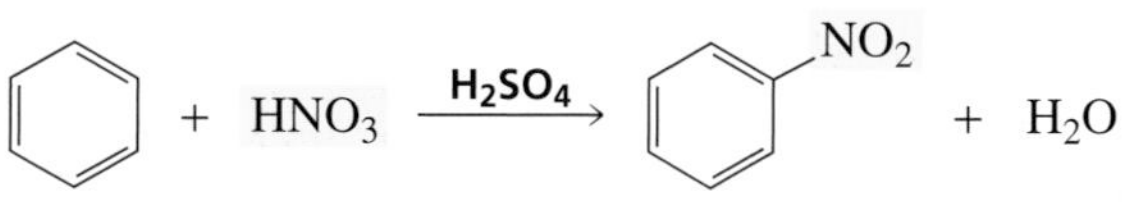

nitrobenzene

Nitration of benzene with nitric acid requires sulfuric acid as a catalyst. Sulfuric acid protonates nitric acid. Loss of water from protonated nitric acid forms a nitronium ion, the electrophile required for nitration. Remember that any base present in the reaction mixture (H_2O, HSO_4^-, solvent) can remove the proton in the second step of the aromatic substitution reaction.

mechanism for nitration

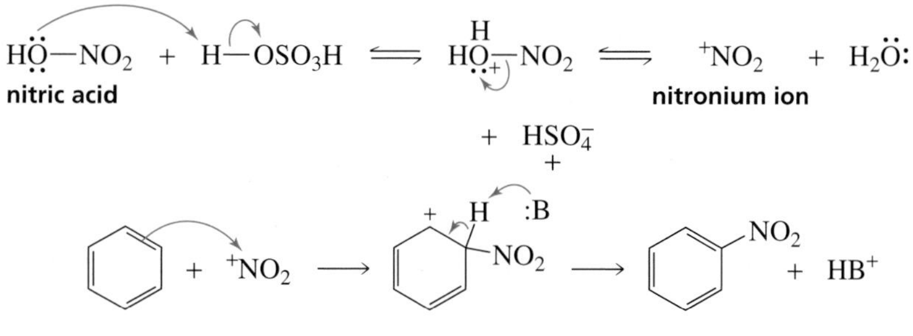

$$H\ddot{O}-NO_2\ +\ H-OSO_3H \rightleftharpoons H\underset{\overset{+}{\cdot}}{\ddot{O}}-NO_2 \rightleftharpoons {}^+NO_2\ +\ H_2\ddot{O}:$$

nitric acid   nitronium ion

$$+\ HSO_4^-$$

$$O=\overset{+}{N}=O$$

nitronium ion

14.11
SULFONATION OF BENZENE

sulfonation

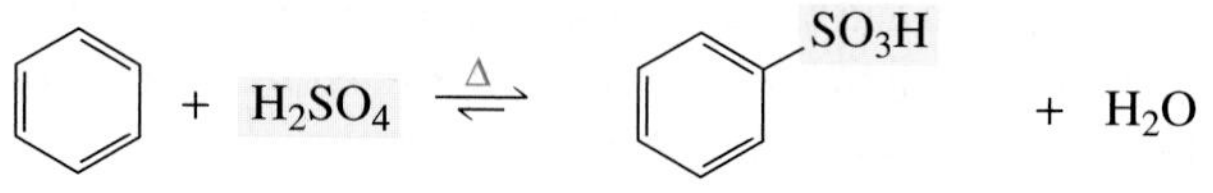

Benzene $+ H_2SO_4 \overset{\Delta}{\rightleftharpoons}$ benzenesulfonic acid (SO_3H) $+ H_2O$

benzenesulfonic acid

Fuming sulfuric acid (a solution of SO_3 in sulfuric acid) or concentrated sulfuric acid is used to sulfonate aromatic rings. As the following mechanism shows, a substantial amount of electrophilic sulfur trioxide (SO_3) is formed when concentrated sulfuric acid is heated, as a result of loss of water and a proton from protonated surfuric acid. Take a minute to note the similarities in the mechanisms for generating the $^+SO_3H$ electrophile for sulfonation and the $^+NO_2$ electrophile for nitration.

mechanism for sulfonation

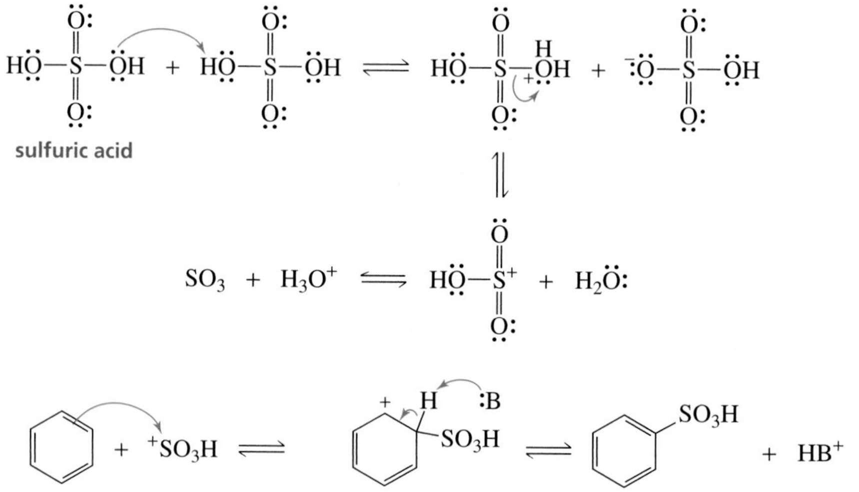

sulfuric acid

$$SO_3 + H_3O^+ \rightleftharpoons H\ddot{O}-\overset{O}{\underset{O}{\overset{\|}{S}}}^+ + H_2\ddot{O}:$$

Benzene $+ \ ^+SO_3H \rightleftharpoons$ [arenium ion with SO_3H and H, :B] $\rightleftharpoons$ benzenesulfonic acid (SO_3H) $+ HB^+$

Sulfonic acids are strong acids because of the three electron-withdrawing oxygen atoms and the electron delocalization in the sulfonate ion (Section 1.19).

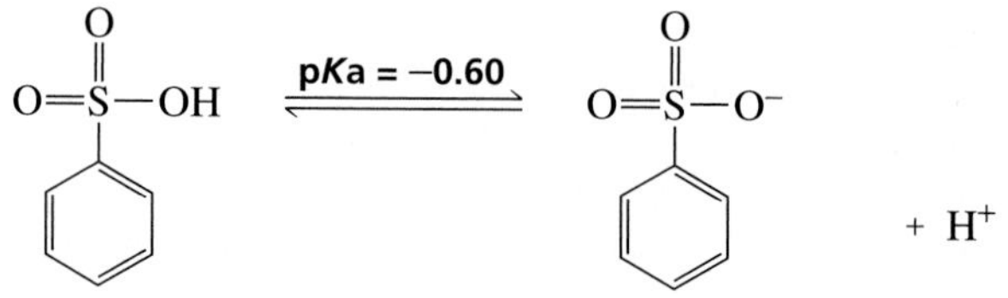

$$O=\overset{O}{\underset{}{\overset{\|}{S}}}-OH \quad \overset{pK_a = -0.60}{\rightleftharpoons} \quad O=\overset{O}{\underset{}{\overset{\|}{S}}}-O^- \quad + H^+$$

benzenesulfonic acid   benzenesulfonate ion

Sulfonation of benzene is a reversible reaction. If benzenesulfonic acid is heated in dilute acid, the reaction proceeds in the reverse direction.

benzenesulfonic acid (SO_3H) $\overset{H_3O^+ / 100\ °C}{\rightleftharpoons}$ benzene $+ SO_3 + H^+$

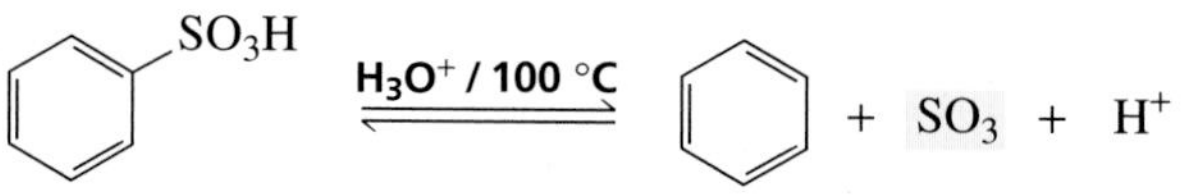

The **principle of microscopic reversibility** applies to all reactions. It states that the mechanism of a reaction in the reverse direction must retrace each step of the mechanism in the forward direction in microscopic detail. This means that the forward and reverse reactions must have the same intermediates and that the rate-determining "energy hill" is the same in both directions. For example, sulfonation is described by the reaction coordinate diagram in Figure 14.4 going from left to right. Therefore, desulfonation is described by the same reaction coordinate diagram going from right to left. In sulfonation, the rate-determining step is nucleophilic attack of benzene on the $^+SO_3H$ ion. In desulfonation, the rate-limiting step is loss of the $^+SO_3H$ ion from the benzene ring. An example of the usefulness of desulfonation to synthetic chemists is given in Chapter 15, Problem 23.

mechanism for desulfonation

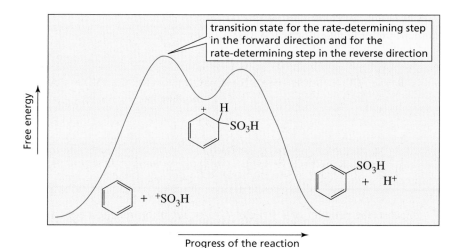

◀ **Figure 14.4**
Reaction coordinate diagram for the sulfonation of benzene (left to right) and for the desulfonation of benzenesulfonic acid (right to left).

PROBLEM 16

Sulfonation of benzene and desulfonation of benzenesulfonic acid are both two-step reactions. The rate-determining step for sulfonation is the step with the smaller rate constant. The rate-determining step for desulfonation is the step with the larger rate constant. Explain how the step with the larger rate constant can be the rate-determining step.

Two electrophilic substitution reactions bear the names of chemists Charles Friedel and James Crafts. *Friedel–Crafts acylation* places an acyl group on a benzene ring, and *Friedel–Crafts alkylation* places an alkyl group on a benzene ring.

**14.12
FRIEDEL–CRAFTS
ACYLATION OF
BENZENE**

$$\underset{\text{an acyl group}}{R-\overset{\overset{\displaystyle O}{\|}}{C}-} \qquad \underset{\text{an alkyl group}}{R-}$$

Either an acyl halide or an acid anhydride can be used for Friedel–Crafts acylation.

Friedel–Crafts acylation

Charles Friedel (1832–1899) was born in Strasbourg, France. He was a professor of chemistry and director of research at the Sorbonne. At one point his interest in mineralogy led him to attempt to make synthetic diamonds. He met Crafts when they both were doing research at L'Ecole de Médicine in Paris. They collaborated scientifically for most of their lives, discovering the Friedel–Crafts reactions in Friedel's laboratory in 1877.

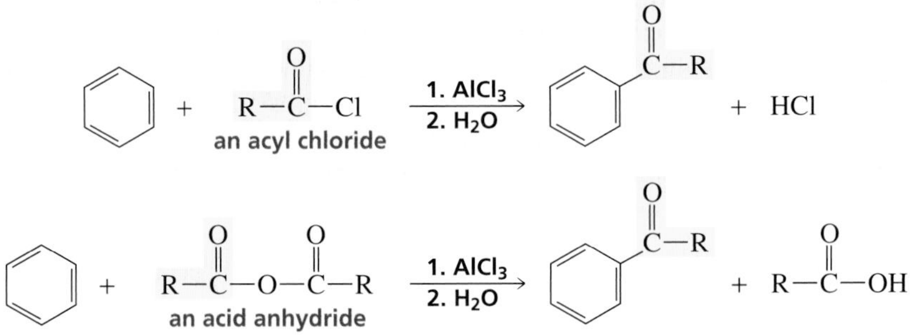

The acylium ion is the required electrophile in the Friedel–Crafts acylation reaction. This ion is formed by the reaction of an acyl chloride or an acid anhydride with $AlCl_3$, a Lewis acid.

mechanism for Friedel–Crafts acylation

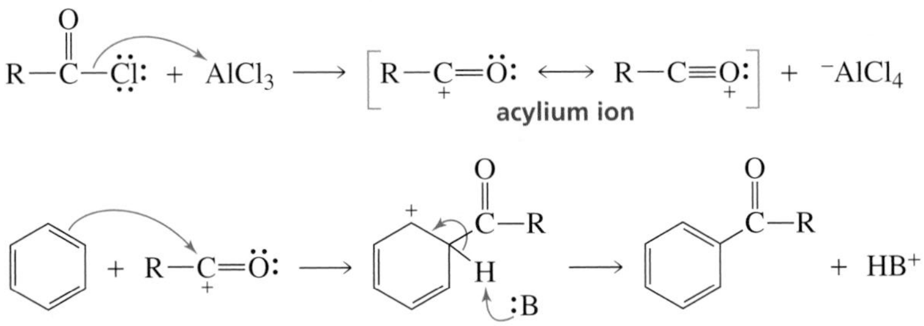

Because the product of a Friedel–Crafts acylation reaction contains a carbonyl group that can complex with $AlCl_3$, Friedel–Crafts acylation reactions must be carried out with more than one equivalent of $AlCl_3$. When the reaction is over, water is added to the reaction mixture to liberate the product from the complex by hydrolyzing the aluminum salts. (Hydrolyzing means reacting with water.)

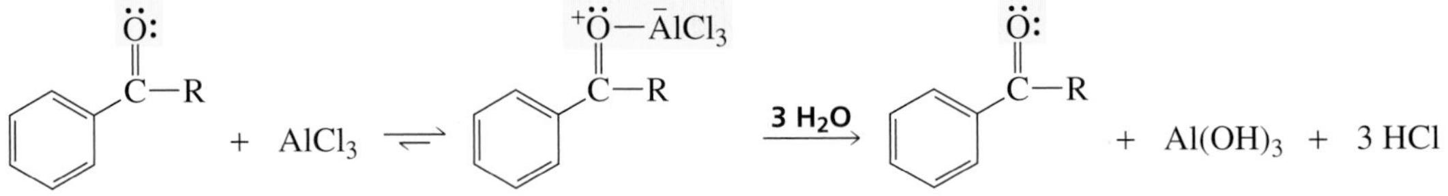

PROBLEM 17

Show the mechanism for the generation of the acylium ion if an anhydride is used instead of an acyl chloride in a Friedel–Crafts acylation reaction.

14.13 FRIEDEL–CRAFTS ALKYLATION OF BENZENE

Friedel–Crafts alkylation

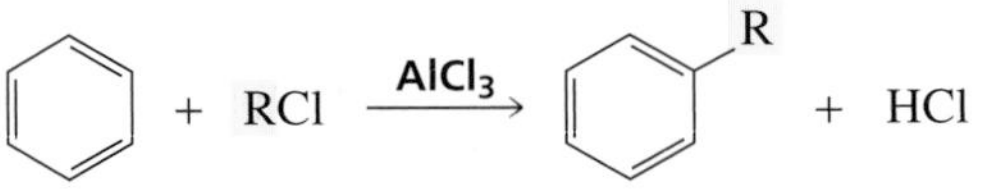

In the first step of a Friedel–Crafts alkylation reaction, a carbocation is formed from the reaction of an alkyl halide with $AlCl_3$. Alkyl fluorides, alkyl chlorides, alkyl bromides, and alkyl iodides can all be used. Vinyl halides and aryl halides cannot be used because their carbocations are too unstable to be formed (Section 9.8).

mechanism for the Friedel–Crafts alkylation

$$R-\ddot{\underset{\cdot\cdot}{C}l}: \ + \ AlCl_3 \longrightarrow R^+ \ + \ ^-AlCl_4$$

James Mason Crafts (1839–1917) was born in Boston, the son a of woolen goods manufacturer. He graduated from Harvard in 1858 and was a professor of chemistry at Cornell University and at the Massachusetts Institute of Technology. He was president of MIT from 1897 to 1900, when he was forced to retire because of chronic poor health.

Because an alkyl-substituted benzene is more reactive than benzene, to prevent further alkylation of the alkyl-substituted benzene, a large excess of benzene is used in Friedel–Crafts alkylation reactions to ensure that the electrophile is more likely to encounter a molecule of benzene than a molecule of alkyl-substituted benzene.

Recall that a carbocation will rearrange if rearrangement leads to a more stable carbocation (Section 3.14). If the carbocation rearranges in a Friedel–Crafts alkylation reaction, the major product will be the product with the rearranged alkyl group on the benzene ring. The relative amounts of rearranged and unrearranged product depend on the increase in carbocation stability achieved as a result of the rearrangement. For example, when benzene reacts with 1-chlorobutane, a primary carbocation rearranges to a secondary carbocation, and 65% of the product is the rearranged product.

1-chlorobutane $\xrightarrow[\text{0 °C}]{AlCl_3}$ **2-phenylbutane** rearranged alkyl substituent 65% + **1-phenylbutane** unrearranged alkyl substituent 35%

$CH_3CH_2CH_2CH_2Cl$

$CH_3CH_2\overset{+}{C}HCH_2$ with H below $\xrightarrow{\text{1,2-hydride shift}}$ $CH_3CH_2CH\overset{+}{C}H_3$

a primary carbocation   a secondary carbocation

On the other hand, when benzene reacts with 1-chloro-2,2-dimethylpropane, a primary carbocation rearranges to a tertiary carbocation. Thus, there is a greater increase in carbocation stability and, therefore, a greater amount of rearranged product—100% of the product has the rearranged alkyl substituent.

1-chloro-2,2-dimethylpropane $\xrightarrow{AlCl_3}$ **2-methyl-2-phenylbutane** rearranged alkyl substituent 100% + **2,2-dimethyl-1-phenylpropane** unrearranged alkyl substituent 0%

$CH_3\overset{CH_3}{\underset{CH_3}{C}}CH_2Cl$

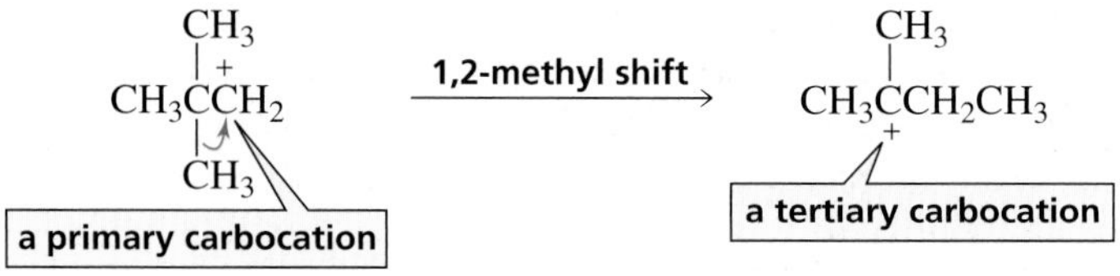

a primary carbocation —1,2-methyl shift→ a tertiary carbocation

In addition to reacting with carbocations generated from alkyl halides, benzene can also react with carbocations generated from alkenes (Section 3.8) and with carbocations generated from alcohols (Section 11.1).

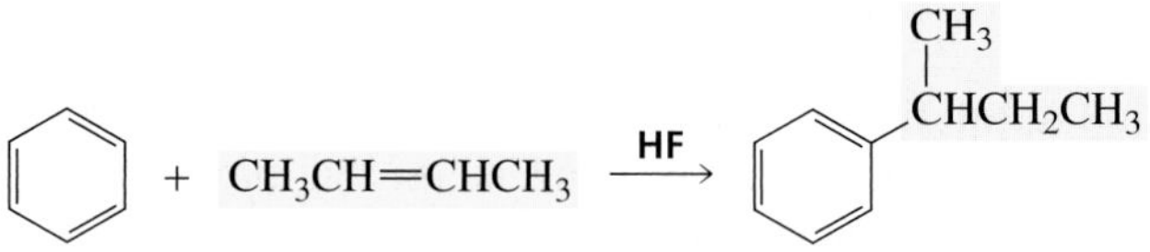

sec-butylbenzene

alkylation of benzene by an alkene

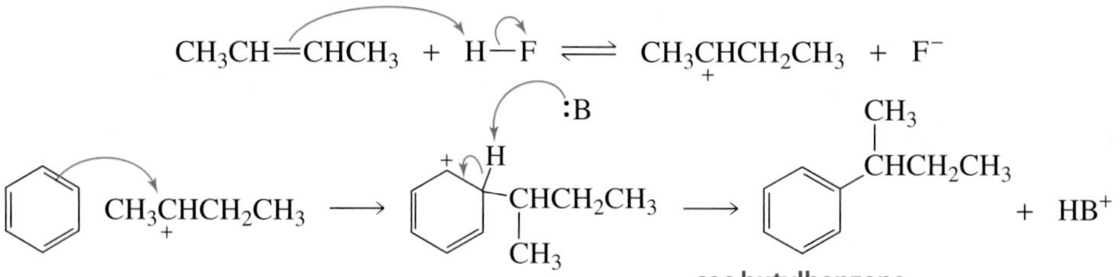

sec-butylbenzene

alkylation of benzene by an alcohol

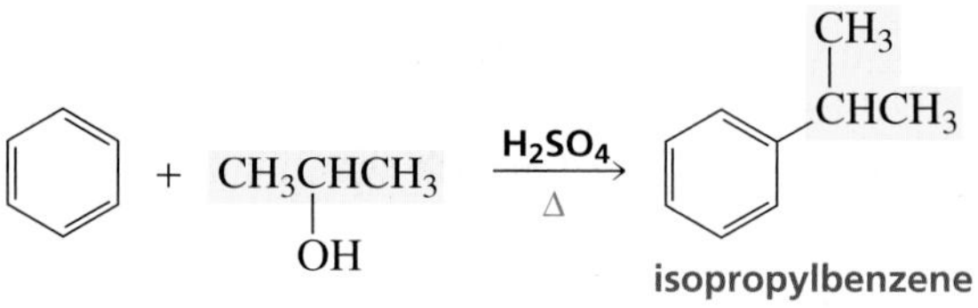

isopropylbenzene
cumene

INCIPIENT PRIMARY CARBOCATIONS

For simplicity, we have shown the formation of a primary carbocation in two of the preceding reactions. However, we learned in Section 9.5 that primary carbocations are too unstable to be formed in solution. We must conclude, then, that a true primary carbocation is never formed in a Friedel–Crafts alkylation reaction. The carbocation remains complexed with the catalyst—it is an "incipient" carbocation. The fact that a carbocation rearrangement still occurs means that the incipient carbocation has sufficient carbocation character to permit the rearrangement.

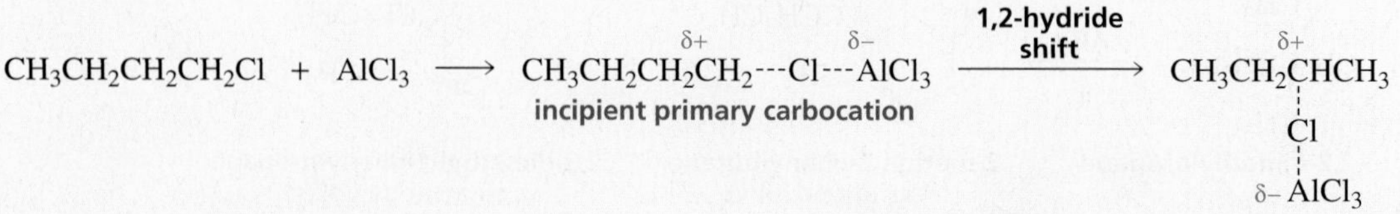

$CH_3CH_2CH_2CH_2Cl$ + $AlCl_3$ ⟶ $\overset{\delta+}{CH_3CH_2CH_2CH_2}$---$\overset{\delta-}{Cl}$---$AlCl_3$ —1,2-hydride shift→ $CH_3CH_2\overset{\delta+}{C}HCH_3$

incipient primary carbocation

PROBLEM 18◆

What would be the major product of a Friedel–Crafts alkylation reaction using the following alkyl halides?

a. CH_3CH_2Cl

b. $CH_3CH_2CH_2Cl$

c. $CH_3CH_2CH(Cl)CH_3$

d. $(CH_3)_3CCH_2Cl$

e. $(CH_3)_2CHCH_2Cl$

f. $CH_2{=}CHCH_2Cl$

It is not possible to obtain a good yield of an alkylbenzene containing a straight-chain alkyl group by using a Friedel–Crafts alkylation reaction because the incipient primary carbocation will rearrange to a more stable carbocation.

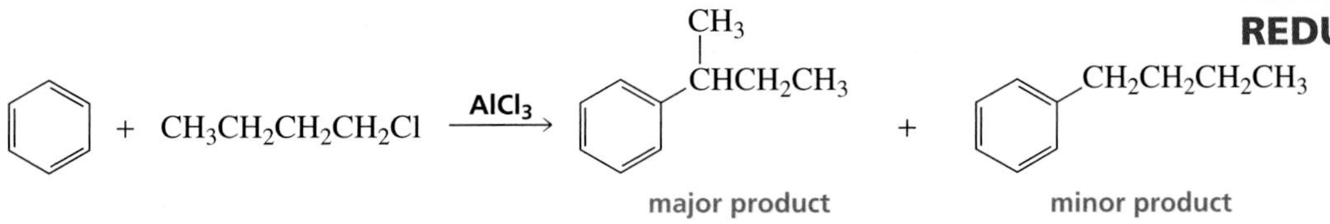

**14.14
ALKYLATION OF
BENZENE BY
ACYLATION–
REDUCTION**

major product   +   minor product

In contrast, acylium ions do not rearrange. Consequently, a straight-chain alkyl group can be placed on a benzene ring using a Friedel–Crafts acylation reaction, followed by reduction of the carbonyl group to a methylene group.

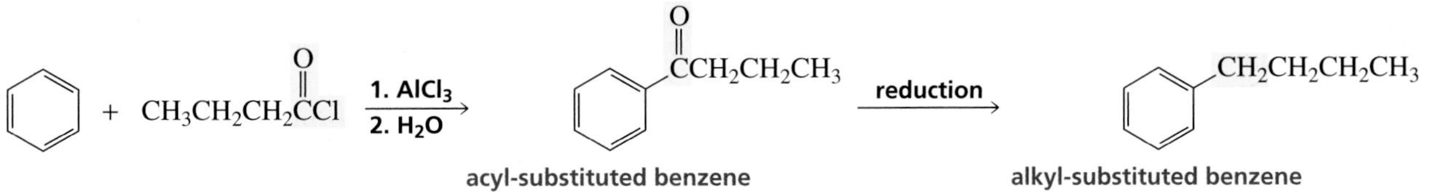

acyl-substituted benzene   alkyl-substituted benzene

Besides avoiding carbocation rearrangements, another advantage to preparing alkyl-substituted benzenes by acylation-reduction rather than by direct alkylation is that a large excess of benzene does not have to be used (Section 14.13) because, unlike alkyl-substituted benzenes, which are more reactive than benzene, acyl-substituted benzenes are less reactive than benzene so they will not undergo additional Friedel-Crafts reactions (Section 15.3).

Two methods are commonly used to reduce the carbonyl group of an aldehyde or a ketone to a methylene group. The **Clemmensen reduction** uses an acidic solution of zinc dissolved in mercury as the reducing reagent. The **Wolff–Kishner reduction** employs hydrazine (H_2NNH_2) under basic conditions. The mechanism of the Wolff–Kishner reaction is shown in Section 17.7.

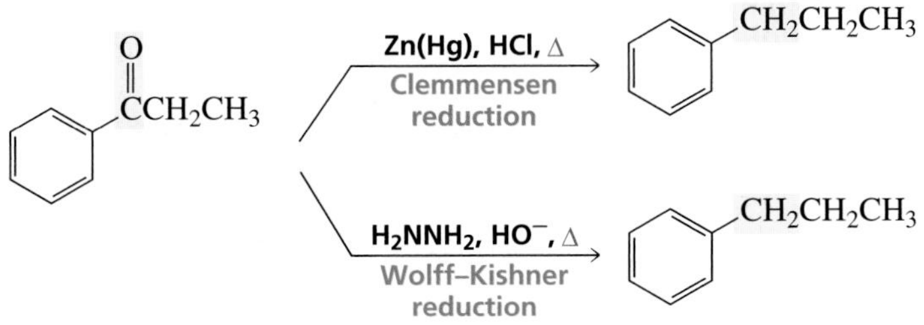

*E. C. Clemmensen
(1876–1941) was born in
Denmark and received a Ph.D.
from the University of
Copenhagen. He was a scientist at Clemmensen Corp. in
Newark, N. Y.*

Ludwig Wolff (1857–1919) *was born in Germany. He received a Ph.D. from the University of Strasbourg. He was a professor at the University of Jena in Germany.*

At this point, you may wonder why it is necessary to have two different ways to carry out the same reaction. It is because there may be another functional group in the molecule that you do not want to react with the reagents you are using to carry out the desired reaction. For example, heating the following compound with HCl (as required by the Clemmensen reduction) would cause the alcohol to undergo substitution (Section 11.1). Under the basic conditions of the Wolff–Kishner reduction, the alcohol group would remain unchanged.

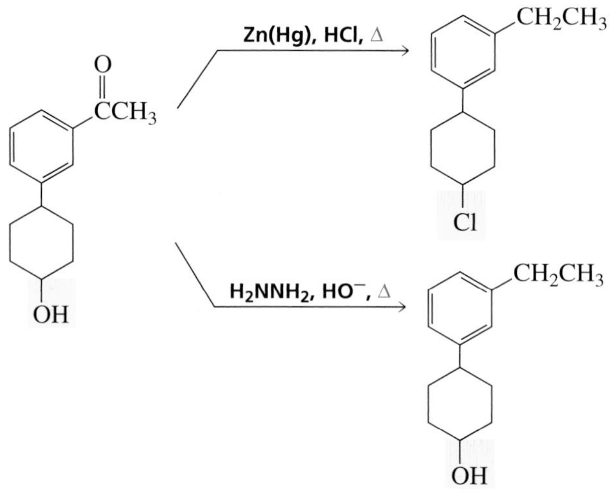
O
CCH3
OH
Zn(Hg), HCl, Δ
CH2CH3
Cl
H2NNH2, HO⁻, Δ
CH2CH3
OH

N. M. Kishner (1867–1935) *was born in Moscow. He received a Ph.D. from the University of Moscow under the direction of Markovnikov. He was a professor at the University of Tomsk and later at the University of Moscow.*

The quaternary ammonium ion shown below would undergo a Hofmann elimination reaction under the basic conditions required for the Wolff–Kishner reduction (Section 11.11), but it would be inert to the conditions of the Clemmensen reduction.

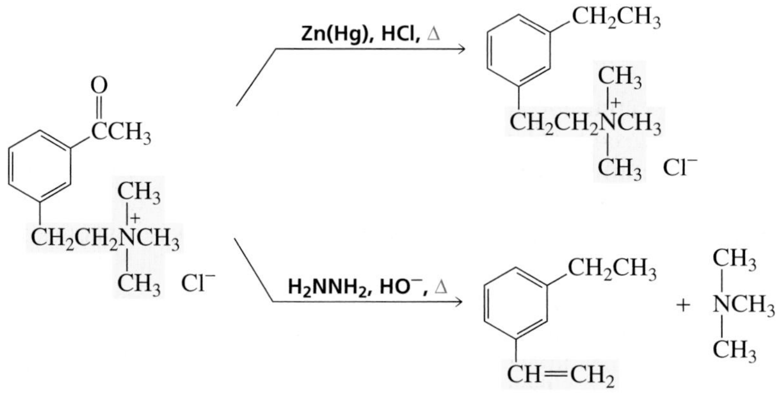
O
CCH3
CH3
CH2CH2NCH3
CH3 Cl⁻
Zn(Hg), HCl, Δ
CH2CH3
CH3
CH2CH2NCH3
CH3 Cl⁻
H2NNH2, HO⁻, Δ
CH2CH3
CH=CH2
+
CH3
NCH3
CH3

PROBLEM 19

Describe how the following compounds could be prepared from benzene:

a. 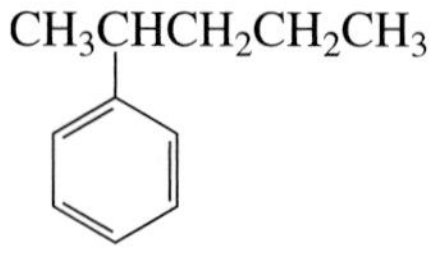
CH3CHCH2CH2CH3

b. 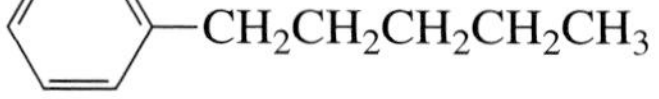
CH2CH2CH2CH2CH3

SUMMARY OF REACTIONS

1. Electrophilic aromatic substitution reactions.

   a. Halogenation (Section 14.9)

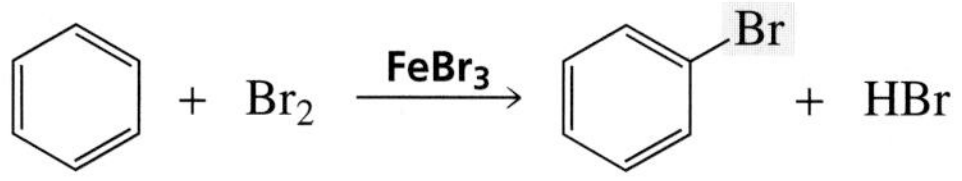

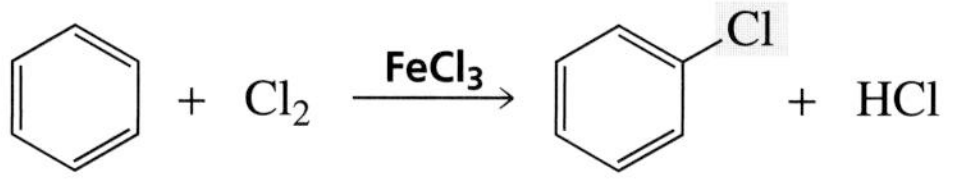

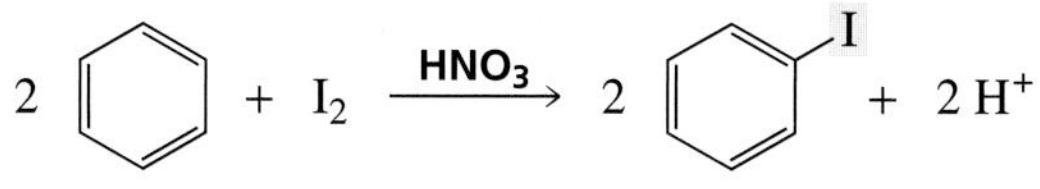

   $$\text{benzene} + Br_2 \xrightarrow{\textbf{FeBr}_3} \text{bromobenzene} + HBr$$

   $$\text{benzene} + Cl_2 \xrightarrow{\textbf{FeCl}_3} \text{chlorobenzene} + HCl$$

   $$2\,\text{benzene} + I_2 \xrightarrow{\textbf{HNO}_3} 2\,\text{iodobenzene} + 2\,H^+$$

   b. Nitration, sulfonation, and desulfonation (Sections 14.10 and 14.11)

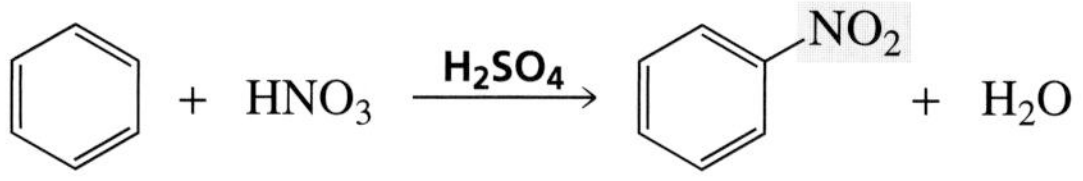

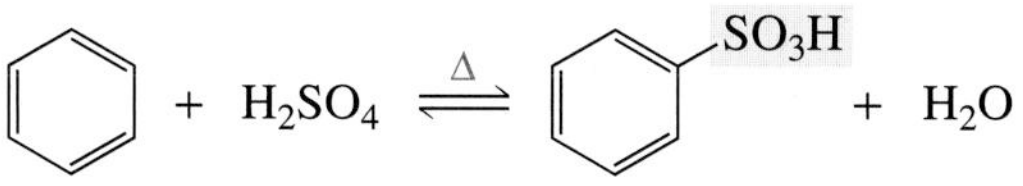

   $$\text{benzene} + HNO_3 \xrightarrow{\textbf{H}_2\textbf{SO}_4} \text{nitrobenzene} + H_2O$$

   $$\text{benzene} + H_2SO_4 \underset{}{\overset{\Delta}{\rightleftharpoons}} \text{benzenesulfonic acid} + H_2O$$

   c. Friedel–Crafts acylation and alkylation (Sections 14.12 and 14.13)

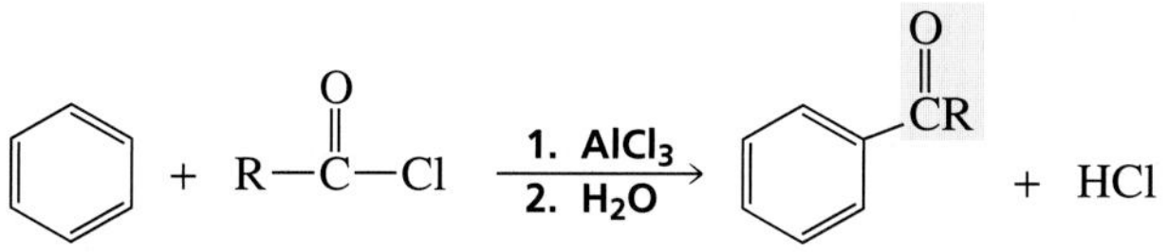

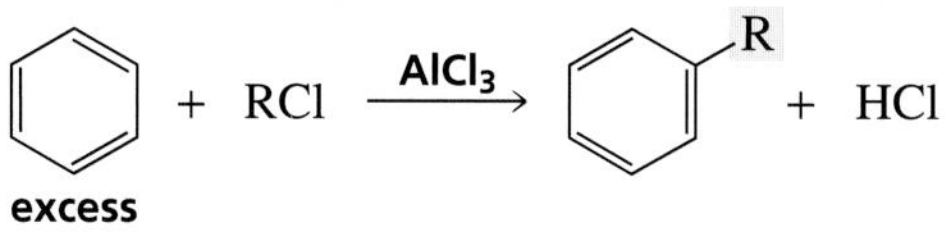

   $$\text{benzene} + R-\overset{\overset{\displaystyle O}{\|}}{C}-Cl \xrightarrow[\textbf{2. H}_2\textbf{O}]{\textbf{1. AlCl}_3} \text{(aryl ketone)} + HCl$$

   $$\underset{\textbf{excess}}{\text{benzene}} + RCl \xrightarrow{\textbf{AlCl}_3} \text{(alkylbenzene, R)} + HCl$$

2. Clemmensen reduction and Wolff–Kishner reduction (Section 14.14).

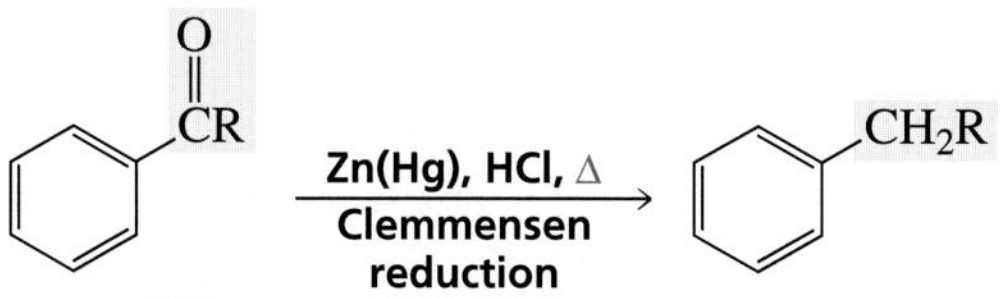

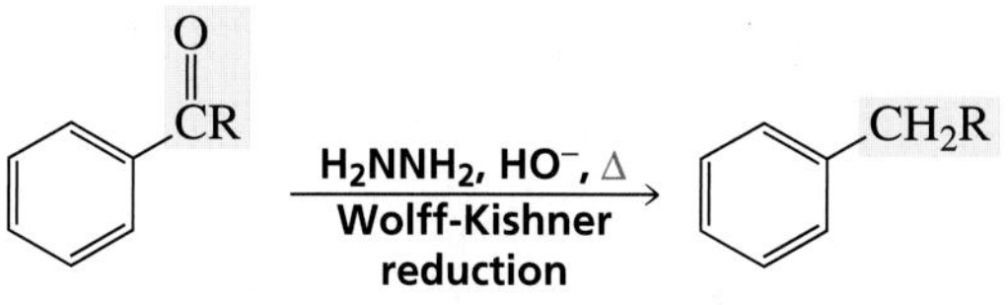

   $$\text{(aryl ketone, CR)} \xrightarrow[\substack{\textbf{Clemmensen}\\\textbf{reduction}}]{\textbf{Zn(Hg), HCl, }\Delta} \text{(CH}_2\text{R)}$$

   $$\text{(aryl ketone, CR)} \xrightarrow[\substack{\textbf{Wolff-Kishner}\\\textbf{reduction}}]{\textbf{H}_2\textbf{NNH}_2\textbf{, HO}^-\textbf{, }\Delta} \text{(CH}_2\text{R)}$$

aliphatic compound (page 602)
annulene (page 604)
antiaromatic (page 609)
aromatic (page 602)
aromatic compound (page 602)
Clemmensen reduction (page 621)
electrophilic aromatic substitution
  reaction (page 613)

Friedel–Crafts acylation (page 613)
Friedel–Crafts alkylation
  (page 613)
halogenation (page 613)
heteroatom (page 606)
heterocyclic compound (page 606)
Hückel's rule, or the $4n+2$ rule
  (page 603)

nitration (page 613)
principle of microscopic
  reversibility (page 616)
sulfonation (page 613)
Wolff–Kishner reduction (page 621)

PROBLEMS

20. Which of the following compounds are aromatic? Are any antiaromatic? (*Hint:* If possible, a ring will be nonplanar to avoid being antiaromatic.)

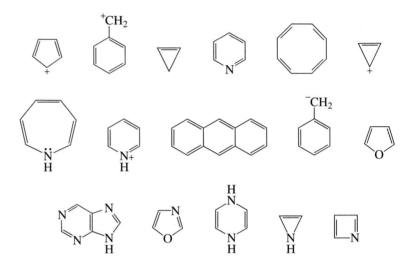

21. Give the product of the reaction of excess benzene with each of the following reagents:
  a. isobutyl chloride + $AlCl_3$
  b. propene + HF
  c. neopentyl chloride + $AlCl_3$
  d. dichloromethane + $AlCl_3$

22. Which ion in each of the following pairs is more stable?

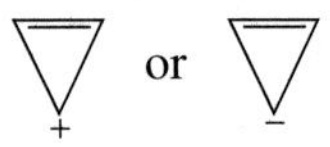

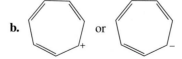

23. Which can lose a proton more readily, a methyl group bonded to cyclohexane or a methyl group bonded to benzene?

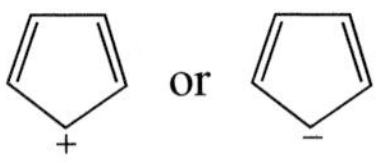

24. When this compound ionizes, is Cl^- or Cl^+ formed?

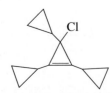

25. Benzene underwent a Friedel–Crafts acylation reaction followed by a Clemmensen reduction. The product gave the following ^{1}H NMR spectrum. What acyl chloride was used in the Friedel–Crafts acylation reaction?

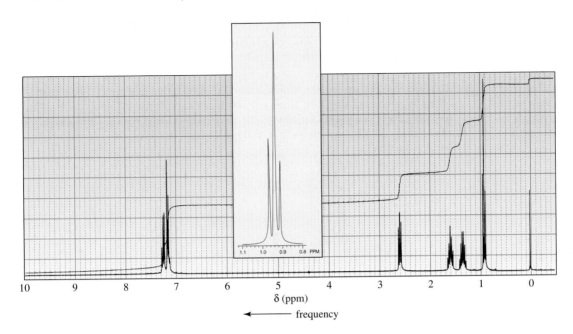

26. **a.** In what direction is the dipole moment in fulvene? Explain.
    b. In what direction is the dipole moment in calicene? Explain.

fulvene   **calicene**

27. Which compound in each of the following pairs is a stronger base? Why?

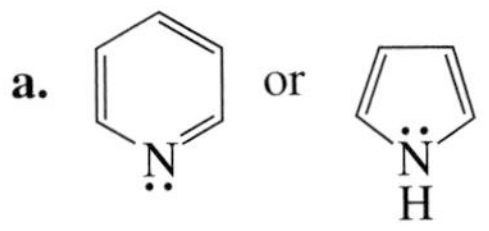

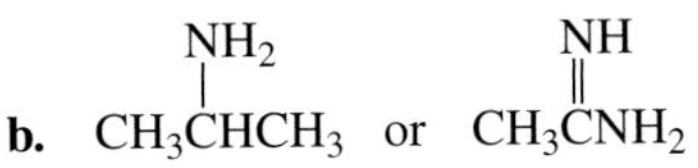

    a. [pyridine] or [pyrrole]

    b. $\overset{NH_2}{\underset{|}{CH_3CHCH_3}}$ or $\overset{NH}{\underset{\|}{CH_3CNH_2}}$

28. Propose a mechanism for each of the following reactions:

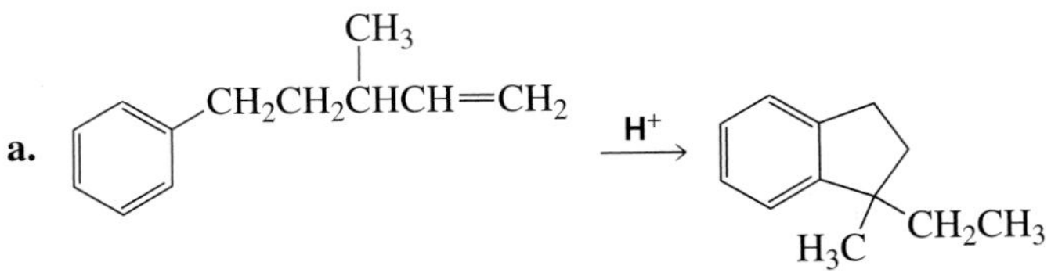

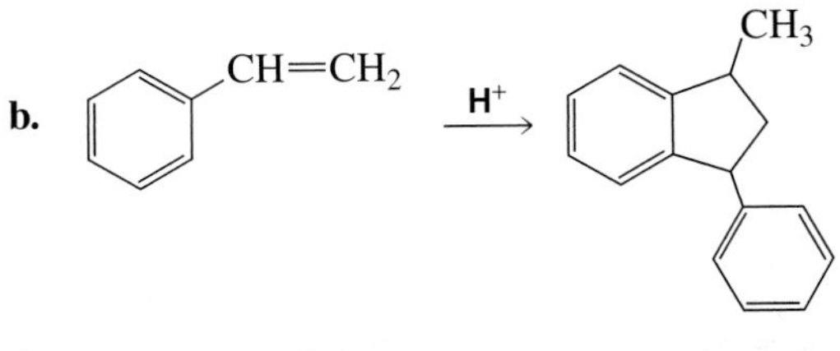

    a. [benzene ring]–CH$_2$CH$_2$CHCH=CH$_2$ (with CH$_3$ substituent) $\xrightarrow{H^+}$ [indane ring with H$_3$C and CH$_2$CH$_3$]

    b. [benzene ring]–CH=CH$_2$ $\xrightarrow{H^+}$ [indane ring with CH$_3$ and phenyl substituents]

29. Benzene can be partially reduced to 1,4-cyclohexadiene by an alkali metal (Na, Li, or K) in liquid ammonia and a low molecular weight alcohol. This reaction is called the Birch reduction. Propose a mechanism for this reaction. (*Hint:* See Section 5.8.)

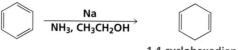

1,4-cyclohexadiene

30. The *principle of least motion*, which states that the reaction that involves the least change in atomic positions or electronic configuration (all else being equal) is favored, has been suggested to explain why the Birch reduction forms only 1,4-hexadiene. How does this account for the observation that no 1,3-cyclohexadiene is obtained from a Birch reduction?

15

Reactions of Substituted Benzenes

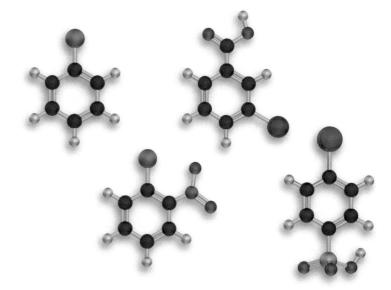

chlorobenzene, *meta*-bromobenzoic acid,
ortho-chloronitrobenzene, *para*-iodobenzenesulfonic acid

S ubstituted benzenes are quite prevalent in nature. A few examples that show physiological activity include adrenaline, mescaline, ephedrine, and chloramphenicol.

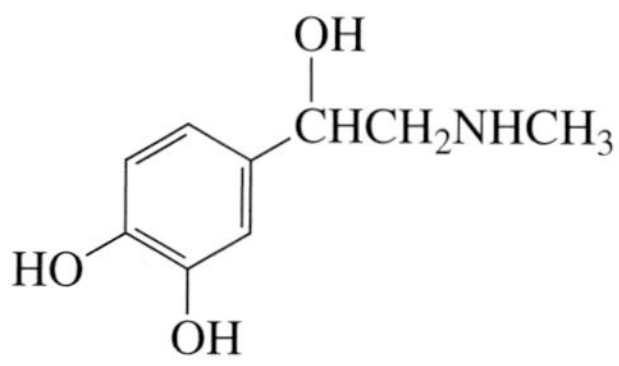

adrenaline
epinephrine

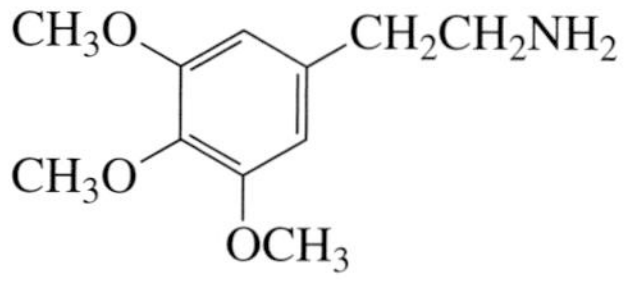

mescaline
active agent of the peyote cactus

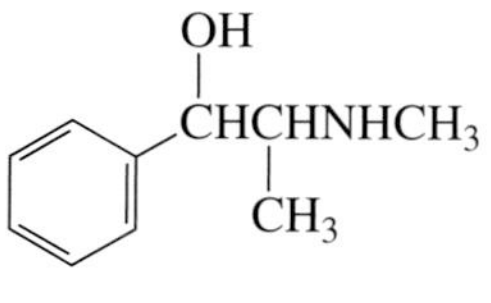

ephedrine
a bronchodilator

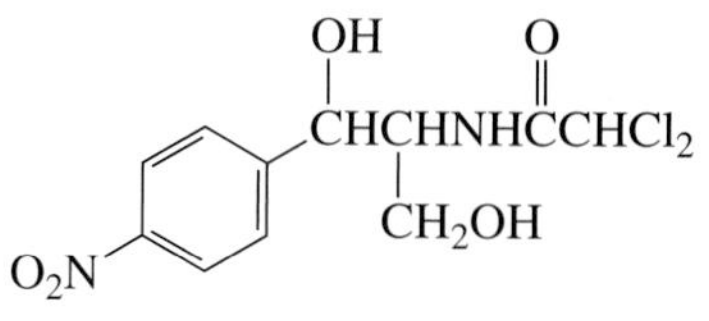

chloramphenicol
**an antibiotic used against
rickettsial infections such as
Rocky Mountain spotted fever;
particularly effective
against typhoid fever**

There are also many physiologically active substituted benzenes that are not found in nature but exist because chemists have synthesized them. The diet drug "fen-phen" is a mixture of two synthetic substituted benzenes, fenfluramine and phentermine. Agent Orange, a defoliant widely used in the 1960s during the Vietnam War, is also a mixture of two synthetic substituted benzenes, 2,4-D and 2,4,5-T. The compound TCDD (known as dioxin) is a contaminant formed during the manufacture of Agent Orange. It has been implicated as the causative agent behind various symptoms suffered by those exposed to Agent Orange during the war.

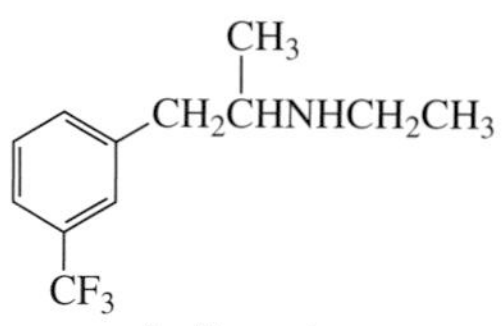

fenfluramine

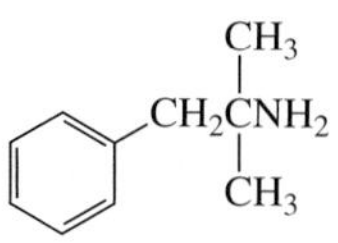

phentermine

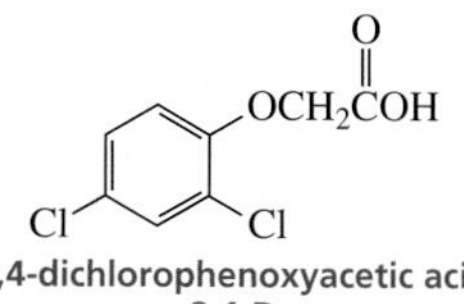

2,4-dichlorophenoxyacetic acid
2,4-D

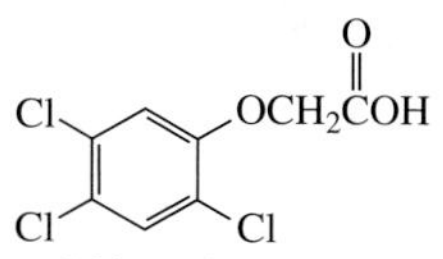

2,4,5-trichlorophenoxyacetic acid
2,4,5-T

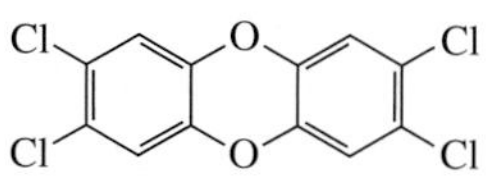

2,3,7,8-tetrachlorodibenzo[b,e][1,4]dioxin
TCDD

DIOXIN

The toxicity of a compound is indicated by its LD_{50} value. This is the quantity needed to kill 50% of the test animals. Dioxin, with an LD_{50} value of 0.0006 mg/kg for guinea pigs, is an extremely toxic compound. Strychnine (LD_{50} value of 0.96 mg/kg) and sodium cyanide (LD_{50} value of 15 mg/kg) are far less toxic. One of the most toxic agents known is the botulism toxin with an LD_{50} value of about 1×10^{-8} mg/kg.

Because of the known physiological activities of adrenaline (epinephrine) and mescaline, chemists have synthesized compounds with similar structures. One such compound is amphetamine, a central nervous system stimulant. Amphetamine and a close relative, methamphetamine, are used clinically as appetite suppressants. Methamphetamine is the street drug known as "speed" because of its rapid and intense psychological effects. Two other synthetic substituted benzenes, BHA and BHT, are preservatives (discussed in Section 8.8) found in a wide variety of packaged foods. These compounds represent just a few of the many substituted benzenes that have been synthesized for commercial use by the chemical and pharmaceutical industries.

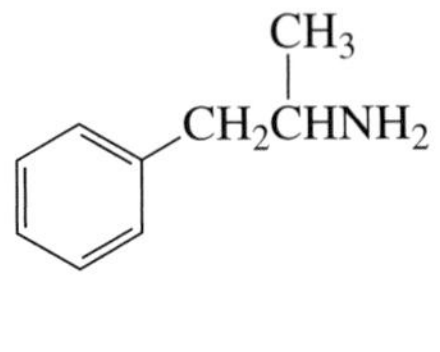

amphetamine

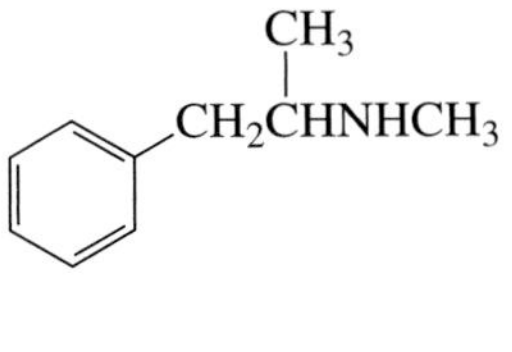

methamphetamine
"speed"

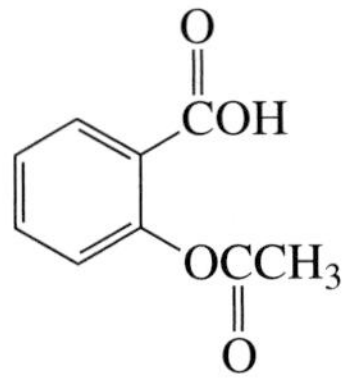

acetylsalicylic acid
aspirin

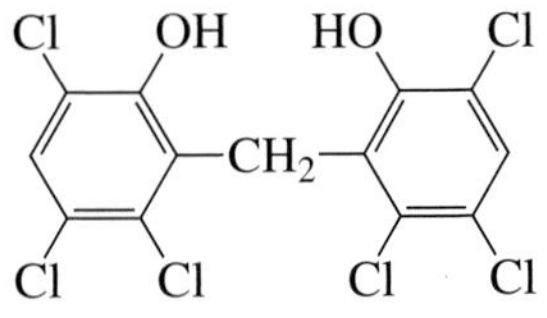

hexachlorophene
a disinfectant

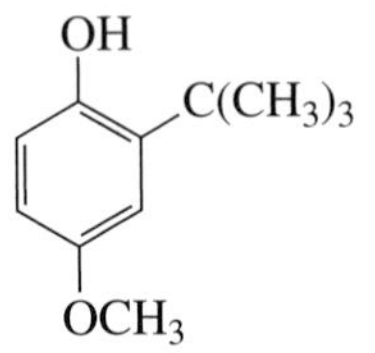

butylated
hydroxyanisole
BHA
a food antioxidant

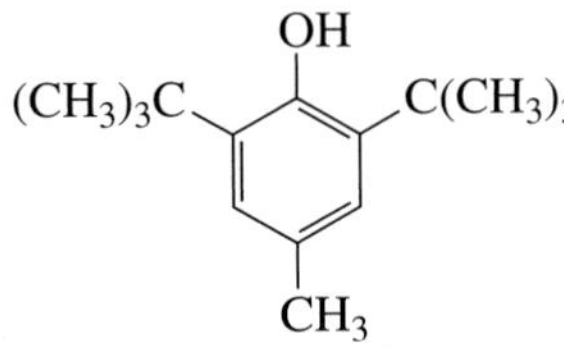

butylated
hydroxytoluene
BHT
a food antioxidant

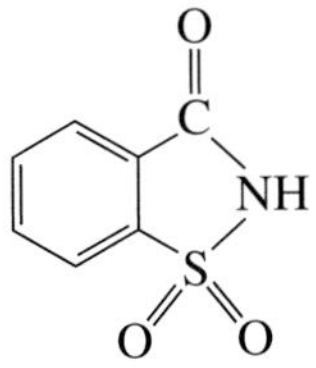

saccharin

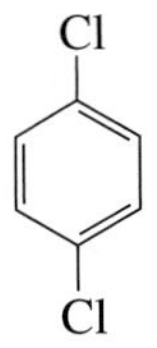

p-dichlorobenzene
**mothballs and
air fresheners**

In Chapter 14 we looked at the reactions benzene undergoes. Now we will look at the reactions of substituted benzenes. Before doing so, though, we must learn how to name substituted benzenes. The physical properties of several substituted benzenes are given in Appendix I.

PEYOTE CULTS

For several centuries, a peyote cult existed among the Aztecs in Mexico that later spread to many Native North American tribes. By 1880, a religion that combined Christian beliefs with the Native American use of the peyote cactus had developed in the southwestern United States, primarily among Native Americans. The followers of this religion believed that the cactus is divinely endowed to shape each person's life. Currently, the only people in the United States legally permitted to use peyote are members of the Native American Church, and they are allowed to use it only in their religious rites.

Monosubstituted Benzenes

Some monosubstituted benzenes are named simply by stating the name of the substituent followed by the word "benzene."

bromobenzene   chlorobenzene   nitrobenzene
**used as a solvent
in shoe polish**   ethylbenzene

Some monosubstituted benzenes have names that incorporate the name of the substituent. Unfortunately, these names have to be memorized.

toluene   phenol   aniline   benzenesulfonic acid

anisole   styrene   benzaldehyde   benzoic acid   benzonitrile

With the exception of toluene, benzene rings with an alkyl substituent are named as substituted benzenes.

isopropylbenzene   *sec*-butylbenzene   *tert*-butylbenzene
cumene

If the alkyl substituent cannot be unambiguously named, the compound must be named as a substituted alkane. The benzene ring, when it is a substituent, is called a **phenyl group,** and a benzene ring with a methylene group is called a **benzyl group.** The phenyl group gets its name from "pheno," the name that was rejected for benzene.

a phenyl group   a benzyl group

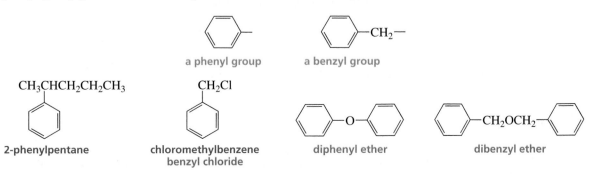

2-phenylpentane   chloromethylbenzene
benzyl chloride   diphenyl ether   dibenzyl ether

An aryl group (Ar) is the general term for either a phenyl group or a substituted phenyl group, just as an alkyl group (R) is the general term for a group derived from an alkane. In other words, ArOH could be used to designate any of the following phenols:

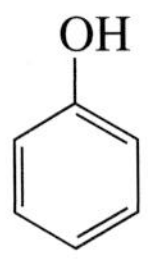

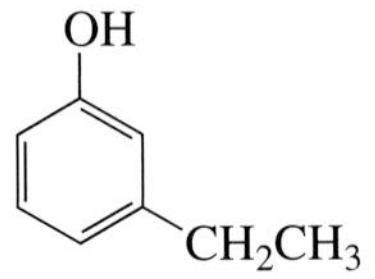

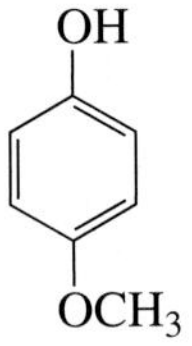

Disubstituted Benzenes

Either numbers or the prefixes *ortho, meta,* and *para* are used to indicate the relative positions of two substituents on a benzene ring. Adjacent substituents are called *ortho,* substituents separated by one carbon are called *meta,* and substituents located opposite one another are designated *para.* Often only their abbreviations (*o, m, p*) are used in naming compounds.

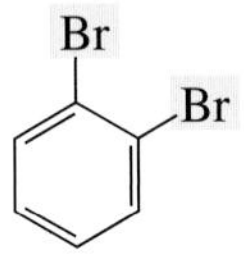

1,2-dibromobenzene
ortho-dibromobenzene
o-dibromobenzene

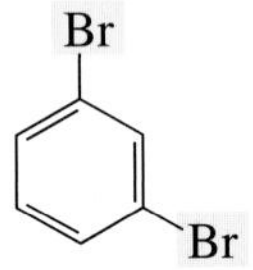

1,3-dibromobenzene
meta-dibromobenzene
m-dibromobenzene

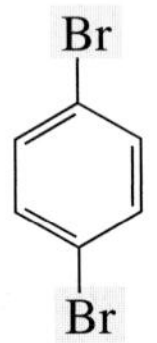

1,4-dibromobenzene
para-dibromobenzene
p-dibromobenzene

If the two substituents are different, they are listed in alphabetical order. The first stated substituent is given the 1-position and the ring is numbered in the direction that gives the second substituent the lowest possible number.

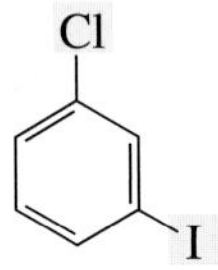

1-chloro-3-iodobenzene
meta-chloroiodobenzene
not
1-iodo-3-chlorobenzene
meta-iodochlorobenzene

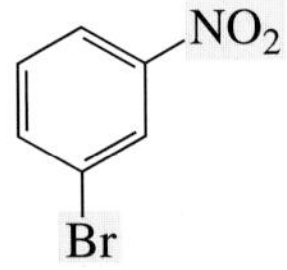

1-bromo-3-nitrobenzene
meta-bromonitrobenzene

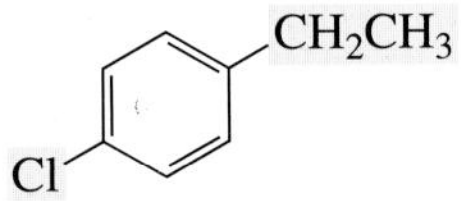

1-chloro-4-ethylbenzene
para-chloroethylbenzene

If one of the substituents can be incorporated into a name, that name is used and the incorporated substituent is given the 1-position.

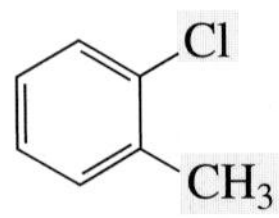

2-chlorotoluene
ortho-chlorotoluene
not
ortho-chloromethylbenzene

4-nitroaniline
para-nitroaniline
not
para-aminonitrobenzene

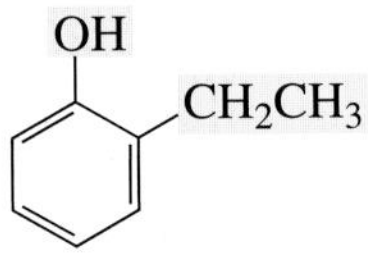

2-ethylphenol
ortho-ethylphenol
not
ortho-ethylhydroxybenzene

A few disubstituted benzenes have names that incorporate both substituents.

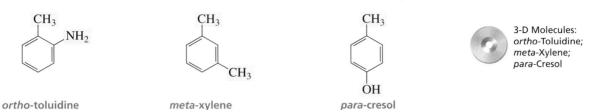

ortho-toluidine

meta-xylene

para-cresol
used as a wood preservative until prohibited for environmental reasons

3-D Molecules: *ortho*-Toluidine; *meta*-Xylene; *para*-Cresol

PROBLEM 1 ◆

Draw the structures of:

a. *para*-toluidine **b.** *meta*-cresol **c.** *para*-xylene

Polysubstituted Benzenes

If the benzene ring has more than two substituents, the substituents are numbered so the lowest possible numbers are used. The substituents are listed in alphabetical order with their appropriate numbers.

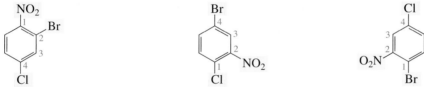

2-bromo-4-chloro-1-nitrobenzene

4-bromo-1-chloro-2-nitrobenzene

1-bromo-4-chloro-2-nitrobenzene

As with disubstituted benzenes, if one of the substituents can be incorporated into a name, that name is used and the incorporated substituent is given the 1-position. The ring is numbered in the direction that results in the lowest possible numbers in the name of the compound.

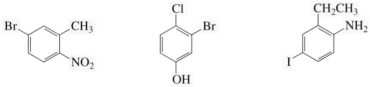

5-bromo-2-nitrotoluene

3-bromo-1-chlorophenol

2-ethyl-4-iodoaniline

PROBLEM 2 ◆

Draw the structure of each of the following compounds:

a. *m*-dichlorobenzene
b. *p*-bromophenol
c. *o*-nitroaniline
d. 2-bromo-4-iodo-1-nitrobenzene
e. 2-phenylhexane
f. 3-benzylpentane
g. *m*-chlorotoluene
h. 2,5-dinitrobenzaldehyde
i. *o*-xylene
j. *m*-chlorobenzonitrile

PROBLEM 3◆

Correct the following incorrect names:

a. 2,4,6-tribromobenzene

b. 3-hydroxynitrobenzene

c. *para*-methylbromobenzene

d. 1,6-dichlorobenzene

15.2 REACTIONS OF SUBSTITUENTS ON BENZENE

In Chapter 14 you learned how to prepare benzene rings with halo, nitro, sulfonic acid, alkyl, and acyl substituents.

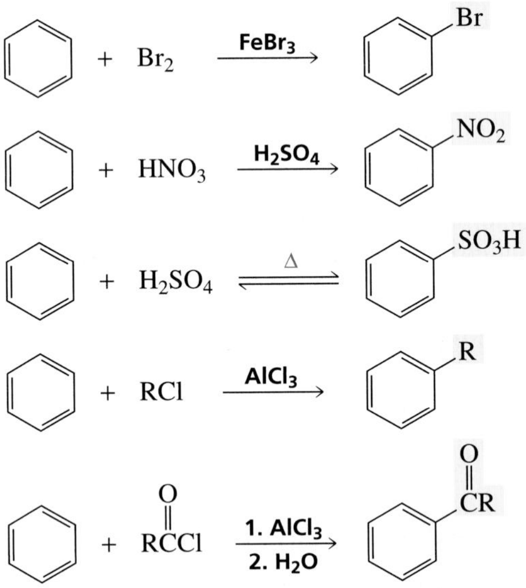

Benzene rings with other substituents can be prepared by first synthesizing one of these substituted benzenes and then chemically changing the substituent. Several of these reactions should be familiar.

Reactions of Alkyl Substituents

We have seen that a bromine will selectively substitute for a benzylic hydrogen in a radical substitution reaction. (NBS stands for *N*-bromosuccinimide; Section 8.5.)

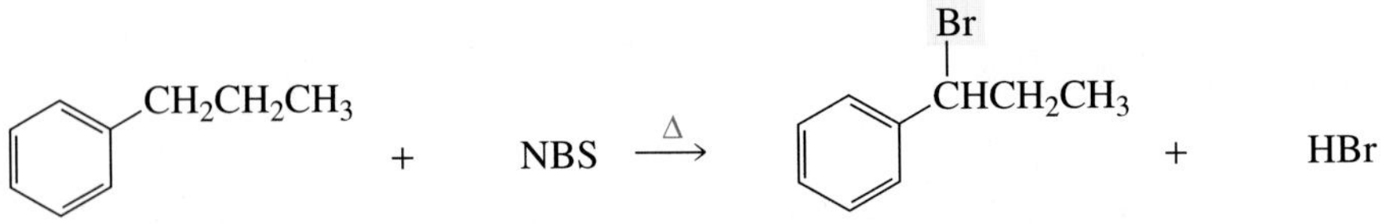

propylbenzene 1-bromo-1-phenylpropane

Once a halogen has been placed in the benzylic position, it can be replaced by a variety of nucleophiles by means of an S_N2 or an S_N1 reaction (Section 9.8). A wide variety of substituted benzenes can be prepared this way.

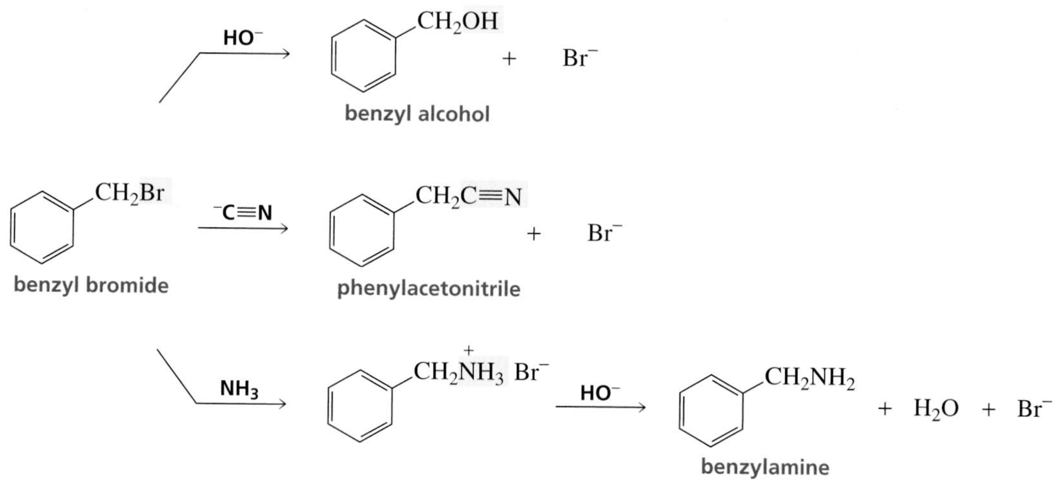

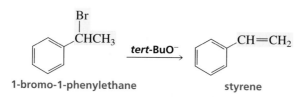

benzyl alcohol

benzyl bromide   phenylacetonitrile

benzylamine

Halo-substituted alkyl groups can also undergo E2 and E1 reactions (Section 10.8). Notice that a bulky base is used to encourage elimination over substitution.

3-D Molecules:
Benzyl bromide;
Styrene;
Benzonitrile;
Benzaldehyde

1-bromo-1-phenylethane   styrene

Substituents with double and triple bonds can be reduced by catalytic hydrogenation (Section 3.19). When an organic compound is *reduced*, the number of hydrogen atoms in the compound increases (see Section 18.0).

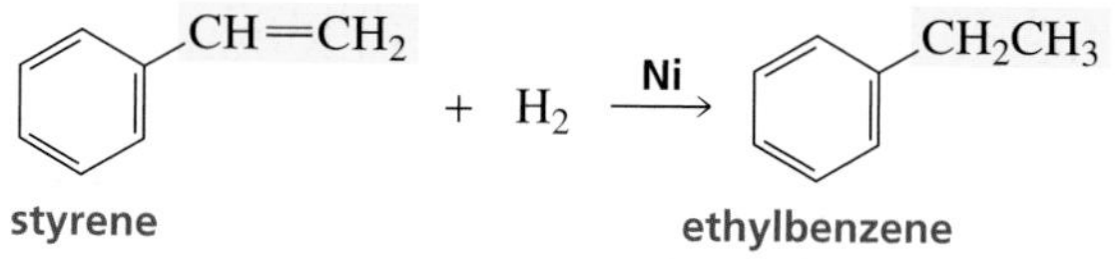

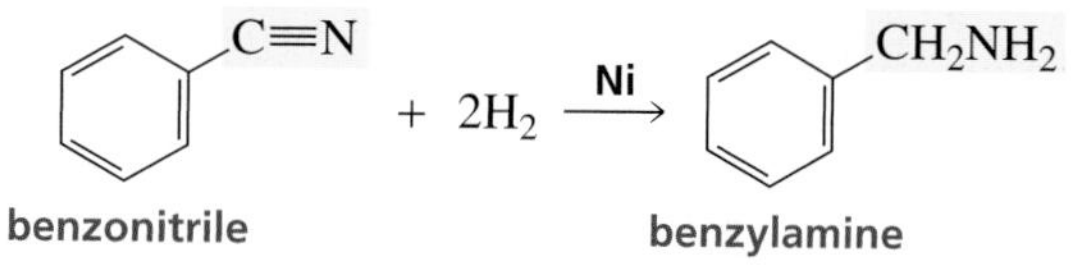

styrene   ethylbenzene

benzonitrile   benzylamine

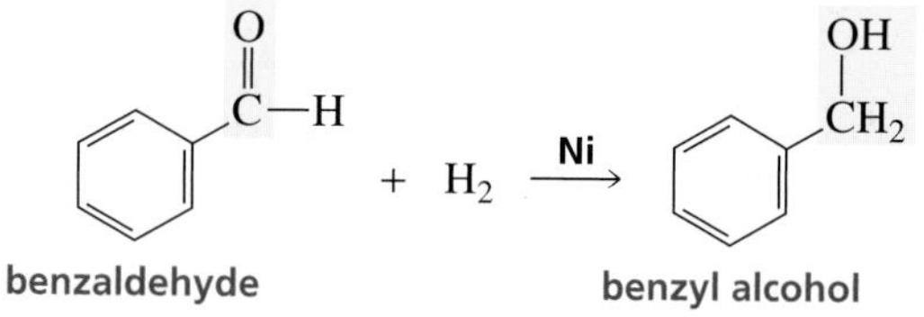

benzaldehyde   benzyl alcohol

Because of the stability of its aromatic ring, benzene can be reduced only under extreme conditions.

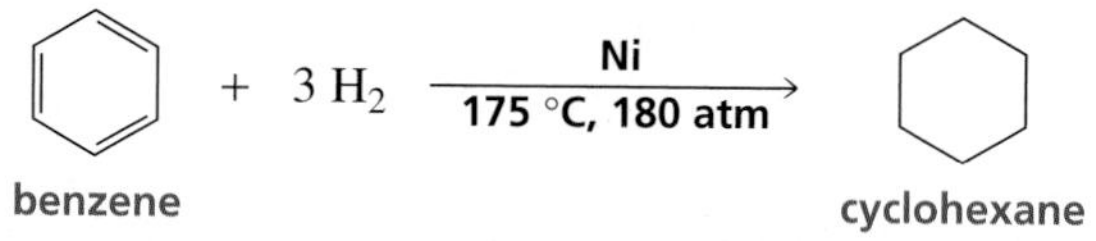

benzene   cyclohexane

An alkyl group bonded to a benzene ring can be oxidized to a carboxyl group. When an organic compound is *oxidized*, the number of oxygen atoms in the compound increases *or* the number of hydrogen atoms decreases (see Section 18.0). Commonly used oxidizing agents are potassium permanganate ($KMnO_4$) or acidic solutions of sodium dichromate ($Na_2Cr_2O_7$, ^+H). Because the benzene ring is so stable, it will not be oxidized—only the alkyl group will be oxidized.

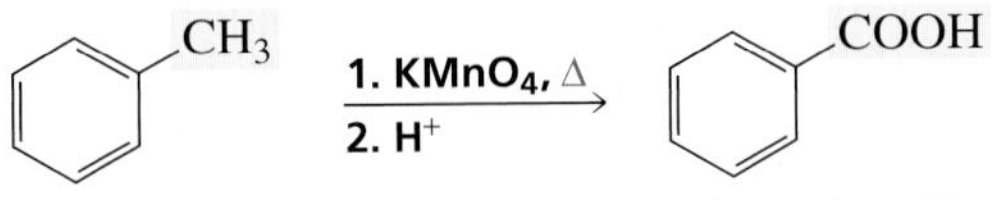

toluene   1. $KMnO_4$, Δ 2. H^+ benzoic acid

Regardless of the length of the alkyl substituent, it will be oxidized to a COOH group, provided that it has a hydrogen bonded to the benzylic carbon.

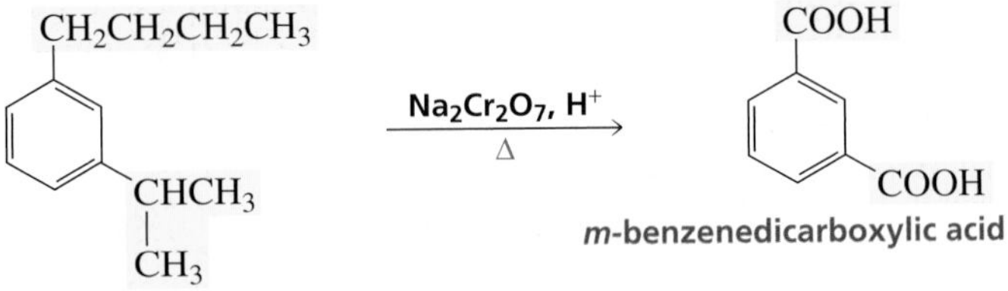

m-butylisopropylbenzene $Na_2Cr_2O_7$, H^+ / Δ *m*-benzenedicarboxylic acid

If the alkyl group does not have a benzylic hydrogen, the oxidation reaction will not occur because the first step in the oxidation reaction is removal of a hydrogen from the benzylic carbon.

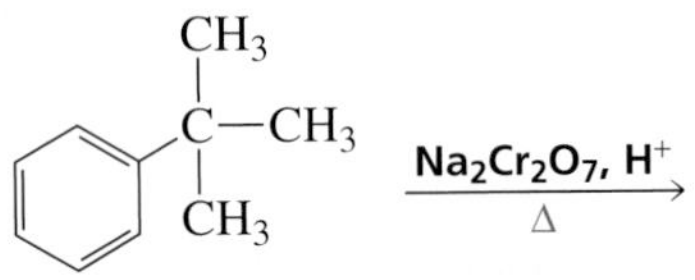

tert-butylbenzene $Na_2Cr_2O_7$, H^+ / Δ no reaction; *tert*-butylbenzene has no benzylic hydrogen

The same oxidizing reagents will oxidize benzylic alcohols to benzoic acid.

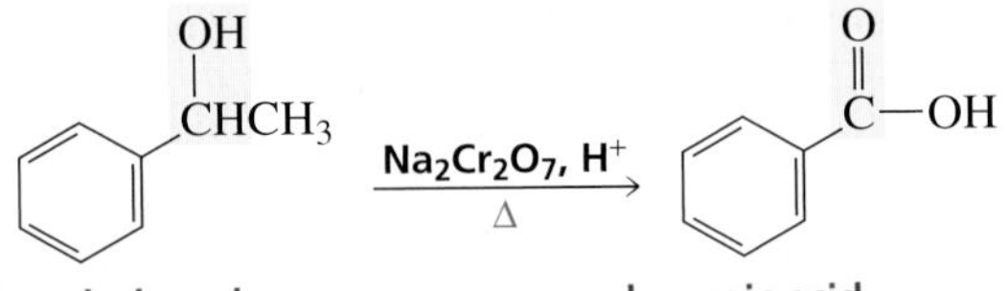

1-phenylethanol $Na_2Cr_2O_7$, H^+ / Δ benzoic acid

If, however, a mild oxidizing agent such as MnO_2 is used, benzylic alcohols are oxidized to aldehydes or ketones.

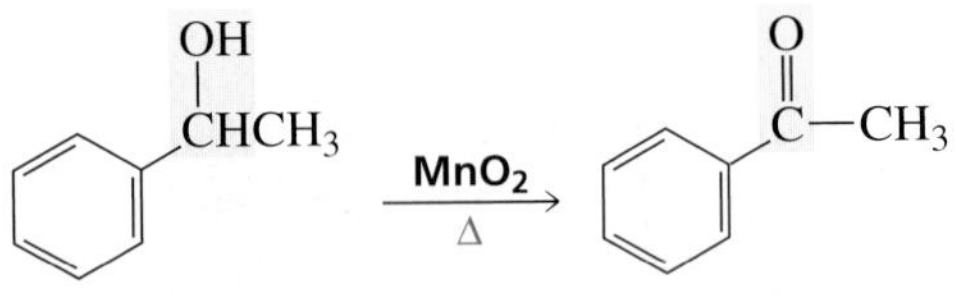

1-phenylethanol MnO_2 / Δ acetophenone

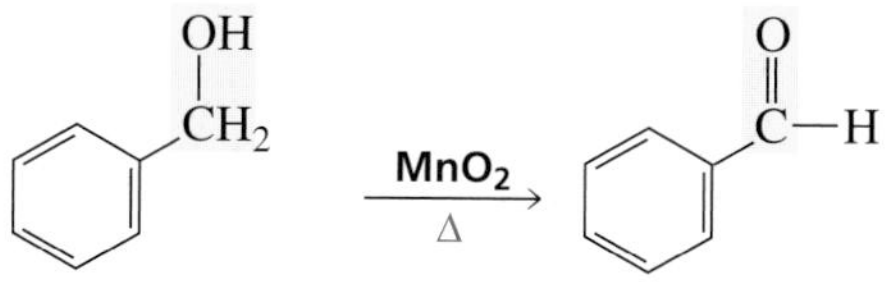

phenylmethanol
benzyl alcohol   benzaldehyde

Reducing a Nitro Substituent

A nitro substituent can be reduced to an amino substituent. Either catalytic hydrogenation or a metal (tin, iron, zinc) and an acid (HCl) can be used to carry out the reduction. Recall from Section 1.20 that if acidic conditions are employed in the reduction, the product will be in its acidic form (anilinium ion). When the reaction is over, base can be added to convert the product into its basic form (aniline).

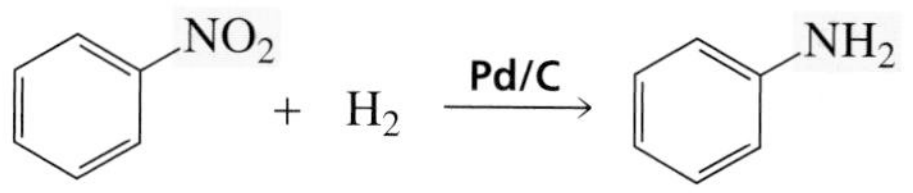

3-D Molecule:
Nitrobenzene

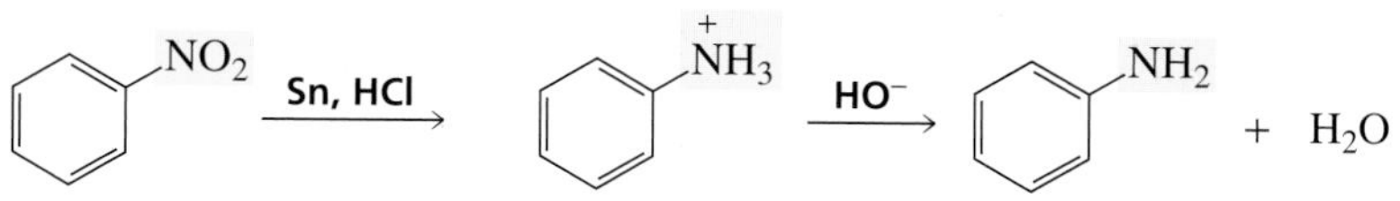

nitrobenzene   protonated   aniline
                        aniline
                        anilinium ion

It is possible to selectively reduce just one of two nitro groups.

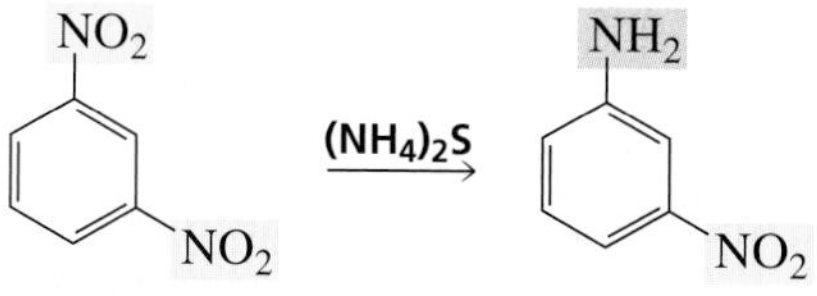

meta-dinitrobenzene   meta-nitroaniline

A FEW WORDS OF WARNING

Although benzene is widely used in chemical synthesis and is frequently used as a solvent, it is toxic. Its major toxic effect is on the central nervous system and on bone marrow. Chronic exposure to benzene causes leukemia and aplastic anemia. A higher than average incidence of leukemia has been found in industrial workers with long-term exposure to as little as 1 ppm benzene in the atmosphere. Toluene has replaced benzene as a solvent because, although it is a central nervous system depressant like benzene, it does not cause leukemia or aplastic anemia. "Glue sniffers" seek the narcotic central nervous system effects of solvents such as toluene. This can be a highly dangerous activity.

PROBLEM 4◆

Give the product of each of the following reactions:

a. 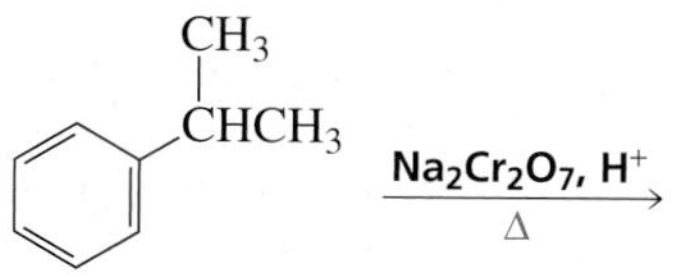 $Na_2Cr_2O_7$, H^+
          Δ

b. 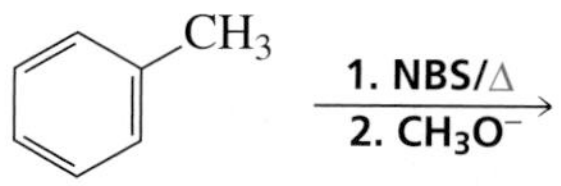   1. NBS/Δ
          2. CH_3O^-

c. (structure: benzene with CH₂CH₃ and CH₃ substituents) $\xrightarrow[\Delta]{\text{Na}_2\text{Cr}_2\text{O}_7,\ \text{H}^+}$

d. (structure: toluene, CH₃ on benzene) $\xrightarrow{\begin{array}{l}\text{1. NBS}/\Delta\\ \text{2. }^-\text{C}\equiv\text{N}\\ \text{3. H}_2/\text{Ni}\end{array}}$

PROBLEM 5 / SOLVED

Show how the following compounds could be prepared from benzene:

a. benzaldehyde   d. 2-phenyl-1-ethanol

b. styrene   e. aniline

c. 1-bromo-2-phenylethane   f. benzoic acid

SOLUTION TO 5a

(benzene) $\xrightarrow[\text{AlCl}_3]{\text{CH}_3\text{Cl}}$ (toluene, CH₃) $\xrightarrow[\Delta]{\text{NBS}}$ (CH₂Br) $\xrightarrow{\text{HO}^-}$ (CH₂OH) $\xrightarrow[\Delta]{\text{MnO}_2}$ (benzaldehyde, CH=O)

15.3
THE EFFECT OF SUBSTITUENTS ON REACTIVITY

Electron-donating substituents increase the reactivity of the benzene ring toward electrophilic aromatic substitution.

Electron-withdrawing substituents decrease the reactivity of the benzene ring toward electrophilic aromatic substitution.

Like benzene, substituted benzenes undergo the five electrophilic aromatic substitution reactions discussed in Chapter 14 and listed in Section 15.2—halogenation, nitration, sulfonation, alkylation, and acylation. Now we need to find out whether a substituted benzene is more reactive or less reactive than benzene itself. The answer depends on the substituent. Some substituents make the ring more reactive and some make it less reactive than benzene toward electrophilic aromatic substitution.

The slow step of an electrophilic aromatic substitution reaction is attack of a positively charged electrophile on the nucleophilic aromatic ring. Increasing the electron density of the benzene ring will increase its attractiveness to an electrophile. We can conclude, then, that *substituents capable of donating electrons into the benzene ring will increase the rate of electrophilic aromatic substitution, whereas substituents that withdraw electrons from the ring will decrease the rate of electrophilic aromatic substitution.*

relative rates of electrophilic substitution

(structure: benzene with Z) > (benzene) > (benzene with Y)

| Z donates electrons into the benzene ring | | Y withdraws electrons from the benzene ring |

There are two ways substituents can donate electrons into a benzene ring: *inductive* electron donation and electron donation by *resonance*. There are also two ways substituents can withdraw electrons from a benzene ring: *inductive* electron withdrawal and electron withdrawal by *resonance*.

Inductive Electron Donation and Withdrawal

If a substituent bonded to a benzene ring is *less electron-withdrawing than a hydrogen,* the electrons in the σ bond that attaches the substituent to the benzene ring will move toward the ring more readily than will those in the σ bond that attaches a hydrogen to the ring. Such a substituent, compared with a hydrogen, donates electrons inductively into the ring. Donation of electrons through a σ bond is called **inductive electron donation** (Section 1.18). Alkyl substituents (such as CH_3) donate electrons inductively compared with a hydrogen.

Notice that it is an alkyl *group* whose electron-donating ability is being compared with that of hydrogen. Carbon is actually slightly *more* electronegative than hydrogen (Table 1.3), but an alkyl group is *more* electron-donating than hydrogen (Section 3.10).

If a substituent is *more electron-withdrawing than a hydrogen,* it will withdraw the σ electrons away from the benzene ring more strongly than will a hydrogen. Withdrawal of electrons through a σ bond is called **inductive electron withdrawal.** The $^+NH_3$ group is a substituent that withdraws electrons inductively—the atom attached to the benzene ring is more electronegative than a hydrogen.

Tutorial:
Donation of electrons into a benzene ring

Resonance Electron Donation and Withdrawal

If a substituent has a pair of nonbonded electrons on the atom directly attached to the benzene ring, the nonbonded electrons can be delocalized into the ring through *p* orbital overlap (Section 6.10). Such substituents are said to donate electrons by resonance. Substituents such as NH_2, OH, OR, and Cl **donate electrons by resonance.** These substituents also withdraw electrons inductively because the atom attached to the benzene ring is more electronegative than a hydrogen.

donation of electrons into a benzene ring by resonance

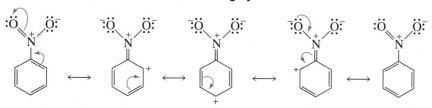

If a substituent is attached to the benzene ring by an atom that is doubly or triply bonded to a more electronegative atom, the π electrons of the ring can be delocalized onto the substituent. Such substituents are said to withdraw electrons by resonance. Substituents such as $C=O$, $C\equiv N$, and NO_2 **withdraw electrons by resonance.** These substituents also withdraw electrons inductively because the atom attached to the benzene ring is more electronegative than a hydrogen.

withdrawal of electrons from a benzene ring by resonance

anisole

nitrobenzene

Notice the difference in the electron densities of the benzene rings in the electrostatic potential maps for anisole and nitrobenzene.

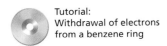

Tutorial:
Withdrawal of electrons
from a benzene ring

PROBLEM 6◆

For each of the following substituents, indicate whether it donates electrons inductively, withdraws electrons inductively, donates electrons by resonance, or withdraws electrons by resonance. Inductive effects should be compared with a hydrogen. (Remember that many substituents can be characterized in more than one way.)

a. Br   **b.** CH_2CH_3 **c.** $\overset{O}{\overset{\|}{C}}CH_3$ **d.** $NHCH_3$ **e.** OCH_3 **f.** $\overset{+}{N}(CH_3)_3$

Relative Reactivity of Substituted Benzenes

The substituents shown in Table 15.1 are listed according to how they affect the reactivity of the benzene ring toward electrophilic substitution compared with benzene itself. *The activating substituents make the benzene ring more reactive toward electrophilic substitution; the deactivating substituents make the benzene ring less reactive toward electrophilic substitution.* Remember that activating substituents donate electrons into the ring and deactivating substituents withdraw electrons from the ring.

All the *strongly activating substituents* donate electrons into the ring by resonance and withdraw electrons from the ring inductively. The fact that they have been found experimentally to be strong activators indicates that electron donation into the ring by resonance is more significant than electron withdrawal from the ring by the inductive effect.

strongly activating substituents

$:NH_2$ $:\overset{..}{O}H$ $:\overset{..}{O}R$

The *moderately activating substituents* also donate electrons into the ring by resonance and withdraw electrons from the ring inductively. However, they donate electrons into the ring by resonance less effectively than the strongly activating substituents.

moderately activating substituents

$\overset{..}{N}H\overset{O}{\overset{\|}{C}}R$ $:\overset{..}{O}\overset{O}{\overset{\|}{C}}R$

These substituents are less effective at donating electrons by resonance into the ring because unlike the strongly activating substituents that donate electrons by resonance only *into* the ring, these substituents can donate electrons by resonance in two competing directions—*into* the ring and *away* from the ring. The fact that these substituents are ac-

TABLE 15.1   The Effects of Substituents on the Reactivity of a Benzene Ring Toward Electrophilic Substitution

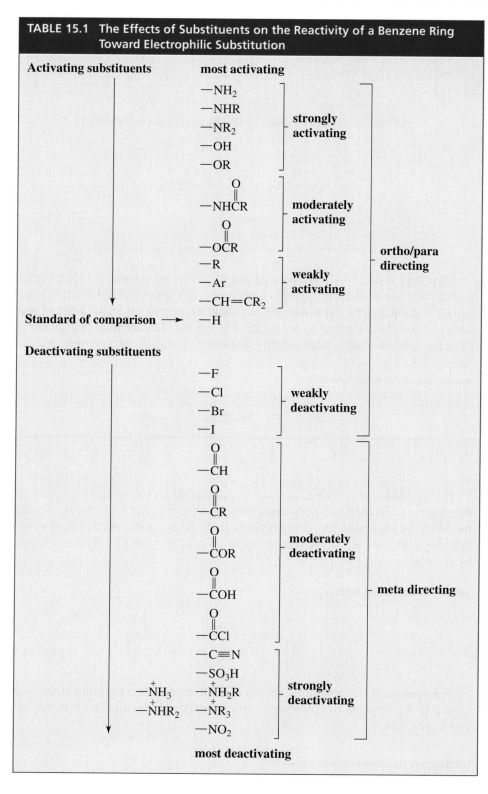

tivators indicates that, despite their diminished resonance electron donation into the ring, overall they donate electrons by resonance more strongly than they withdraw electrons inductively.

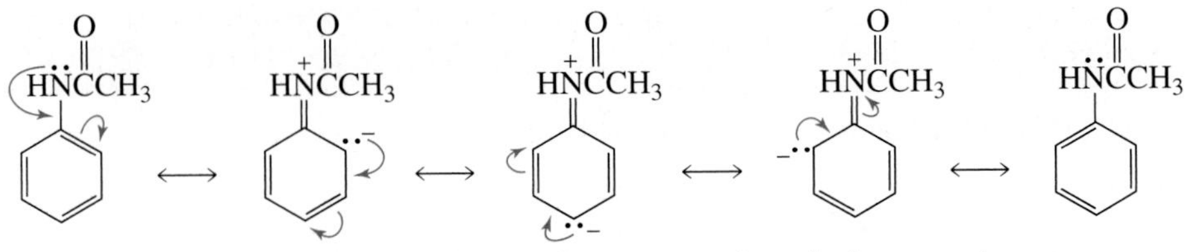

substituent donates electrons by resonance into the benzene ring

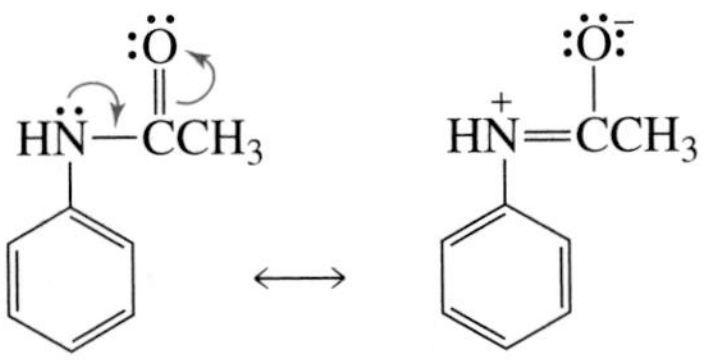

substituent donates electrons by resonance away from the benzene ring

Alkyl, aryl, and CH=CHR groups are *weakly activating substituents.* We have seen that an alkyl substituent, compared to a hydrogen, donates electrons inductively. Aryl and CH=CHR groups can donate electrons into the ring by resonance and can withdraw electrons from the ring by resonance. The fact that they are weak activators indicates that they are slightly more electron-donating than they are electron-withdrawing.

weakly activating substituents

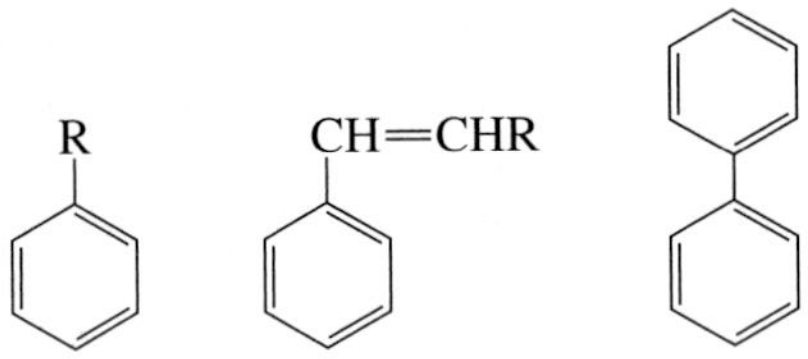

The halogens are *weakly deactivating substituents.* They donate electrons into the ring by resonance and withdraw electrons from the ring inductively. Because the halogens have been found experimentally to be deactivators, we can conclude that they withdraw electrons inductively more strongly than they donate electrons by resonance.

weakly deactivating substituents

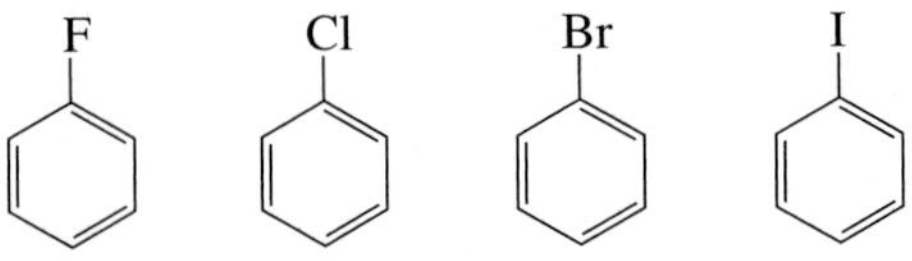

The *moderately deactivating substituents* all have carbonyl groups directly attached to the benzene ring. Carbonyl groups withdraw electrons both inductively and by resonance.

moderately deactivating substituents

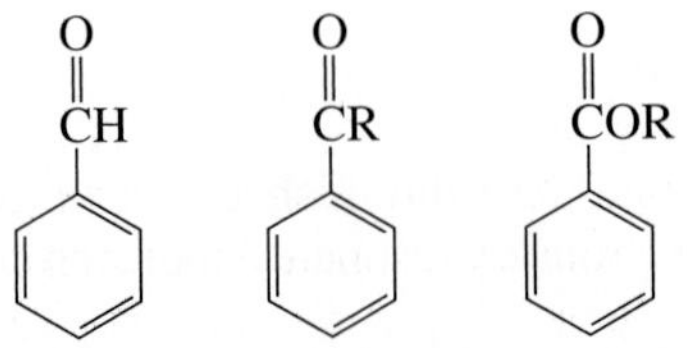

The *strongly deactivating substituents* are powerful electron withdrawers. With the exception of $^+NH_3$, $^+NH_2R$, $^+NHR_2$, and $^+NR_3$, they withdraw electrons both inductively and by resonance. The ammonium ions have no resonance effect, but the positive charge on the nitrogen atom causes them to strongly withdraw electrons inductively.

strongly deactivating substituents

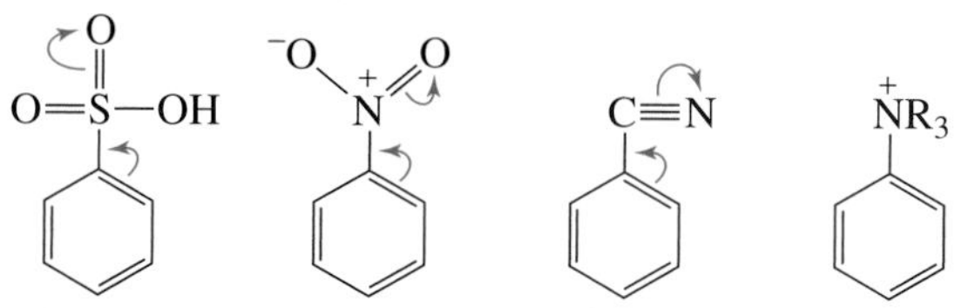

PROBLEM 7◆

List the members of the following sets of compounds in order of decreasing reactivity toward electrophilic substitution:

a. benzene, phenol, toluene, nitrobenzene, bromobenzene

b. dichloromethylbenzene, difluoromethylbenzene, toluene, chloromethylbenzene

---

PROBLEM 8/SOLVED

Explain why the halogens have the relative order of reactivity shown in Table 15.1.

SOLUTION   Table 15.1 shows that fluorine is the least deactivating of the halogen substituents and iodine is the most deactivating. We know that fluorine is the most electronegative of the halogens, which means that it is best at withdrawing electrons inductively. Fluorine is also best at donating electrons by resonance because its $2p$ orbital—compared with the $3p$ orbital of chlorine, the $4p$ orbital of bromine, or the $5p$ orbital of iodine—can better overlap with the $2p$ orbital of carbon. So the fluorine substituent is best at withdrawing electrons inductively and best at donating electrons by resonance. Because it is the weakest deactivator of the halogens, we can conclude that electron donation by resonance is the more important factor in determining the relative order of reactivity of halo-substituted benzenes.

When a substituted benzene undergoes an electrophilic substitution reaction, where does the new substituent attach itself? Is the product of the reaction the ortho isomer, the meta isomer, or the para isomer?

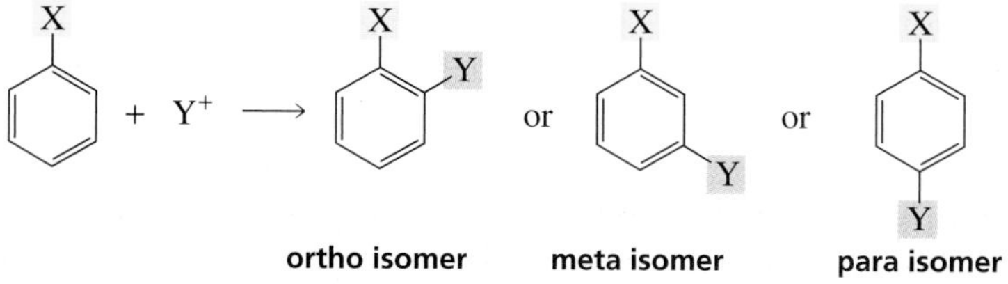

**15.4
THE EFFECT OF
SUBSTITUENTS ON
ORIENTATION**

The substituent already attached to benzene determines the location of the new substituent. There are two possibilities—a substituent will direct an incoming substituent either to the ortho *and* para positions or will direct an incoming substituent to the meta position. All activating substituents and the weakly deactivating halogens are

ortho/para directors, and all substituents that are more deactivating than the halogens are **meta directors.**

1. All activating substituents direct an incoming electrophile to the ortho and para positions.

All activating substituents are ortho/para directors.

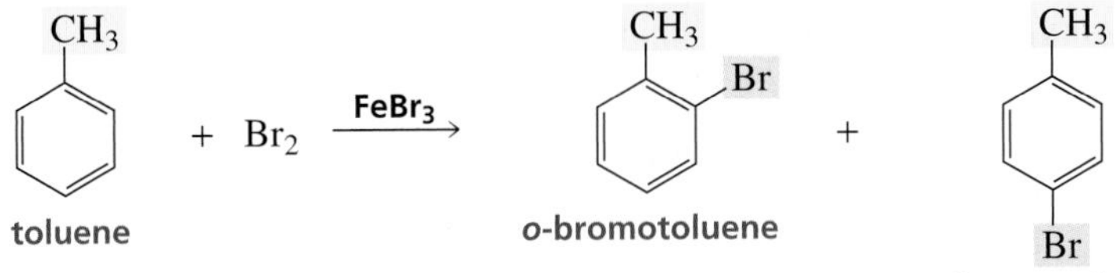

toluene + Br_2 → $\xrightarrow{FeBr_3}$ o-bromotoluene + p-bromotoluene

2. The weakly deactivating halogens also direct an incoming electrophile to the ortho and para positions.

The weakly deactivating halogens are ortho/para directors.

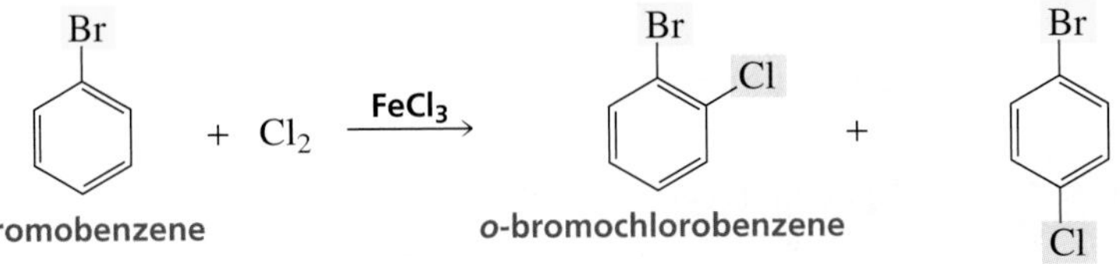

bromobenzene + Cl_2 $\xrightarrow{FeCl_3}$ o-bromochlorobenzene + p-bromochlorobenzene

3. All moderately deactivating and strongly deactivating substituents direct an incoming electrophile to the meta position.

All deactivating substituents (except the halogens) are meta directors.

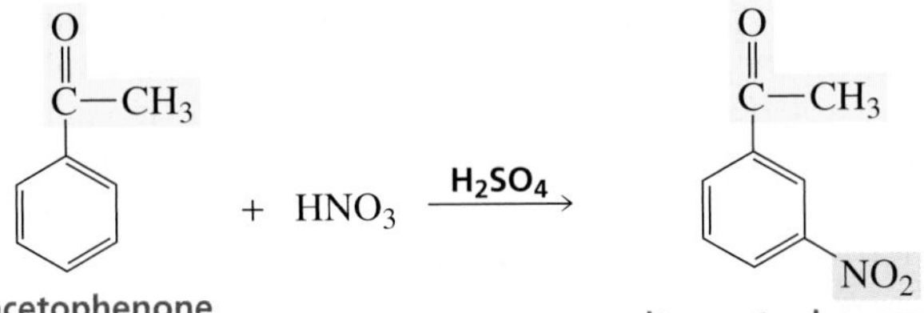

acetophenone + HNO_3 $\xrightarrow{H_2SO_4}$ m-nitroacetophenone

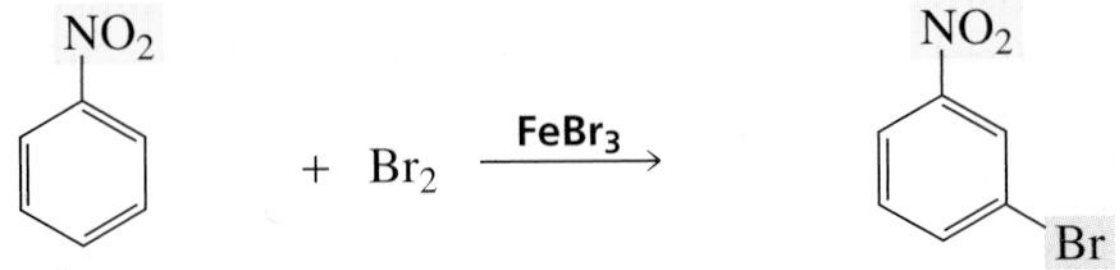

nitrobenzene + Br_2 $\xrightarrow{FeBr_3}$ m-bromonitrobenzene

To understand why a substituent directs an incoming electrophile to a particular position, we must look at the stability of the carbocation intermediate formed in the rate-determining step. When a substituted benzene undergoes an electrophilic substitution reaction, three different carbocation intermediates can be formed. Which one is more likely to be formed depends on their relative stabilities (Figure 15.1). Comparing the relative stabilities of the three carbocations allows us to determine the preferred pathway of the reaction because the more stable the carbocation, the less energy required to make it (Section 3.10).

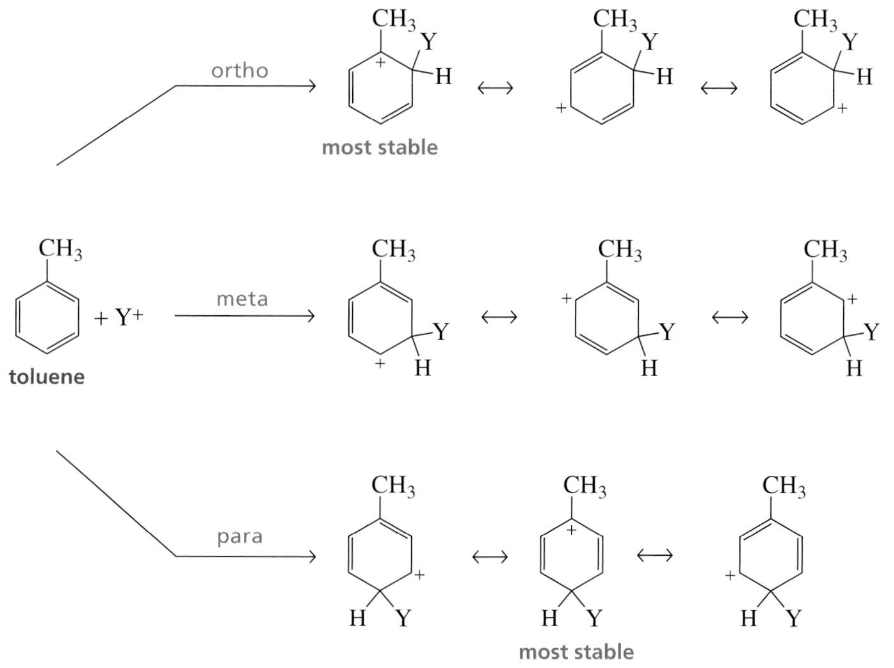

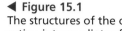

3-D Molecule: Anisole

If a substituent donates electrons inductively (a methyl group, for example), the indicated resonance contributors in Figure 15.1 are the most stable because the substituent is attached directly to the positively charged carbon, which it can stabilize by inductive electron donation. These relatively stable resonance contributors are obtained only when the substituent is directed to an ortho or para position. Therefore, the most stable carbocation is obtained by directing the incoming group to the ortho and para positions. This means that any substituent that donates electrons inductively is an ortho/para director.

If a substituent donates electrons by resonance, the carbocations formed by putting the incoming electrophile on the ortho and para positions have a fourth resonance contributor because the substituent can donate electrons by resonance (Figure 15.2). This is an especially stable resonance contributor because it is the only one whose atoms (except for hydrogen) all have complete octets. Therefore, *all groups that donate electrons by resonance are ortho/para directors*.

Substituents with a positive charge or a partial positive charge on the atom attached to the benzene ring withdraw electrons from the ring. All of these substituents withdraw electrons inductively and most withdraw electrons by resonance as well. For all such substituents, the indicated resonance contributors in Figure 15.3 are the least stable because they have positive charges on two adjacent atoms. So the most stable carbocation is formed when the incoming electrophile is directed to the meta position.

Notice that the three possible carbocation intermediates in Figure 15.1 and the three possible carbocation intermediates in Figure 15.3 are the same except for the substituent, and the fact that the resonance contributors with the substituent directly attached to the positively charged carbon are the most stable when the substituent donates electrons and the least stable when the substituent withdraws electrons.

The only deactivating substituents that are ortho/para directors are the halogens, which are the weakest of the deactivators. We have seen that they are deactivators because they inductively withdraw electrons from the ring more strongly than they donate electrons by resonance. Their ability to donate electrons by resonance,

Tutorial: Intermediates in electrophilic aromatic substitution

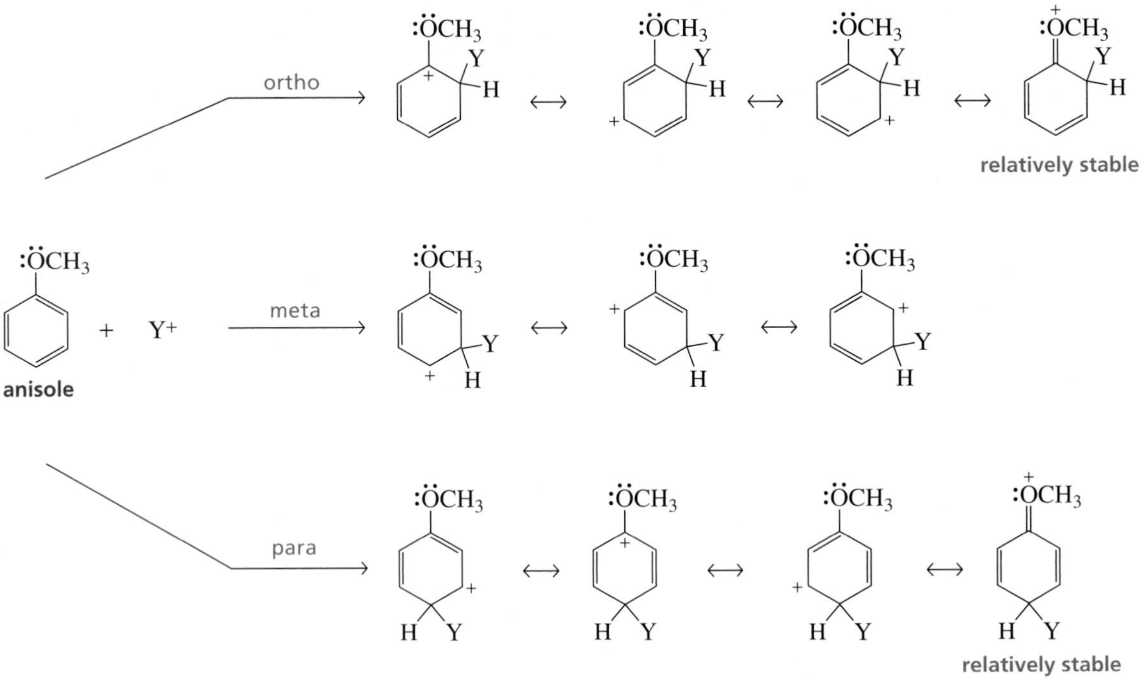

▲ Figure 15.2
The structures of the carbocation intermediates formed from the reaction of an electrophile with anisole at the ortho, meta, and para positions.

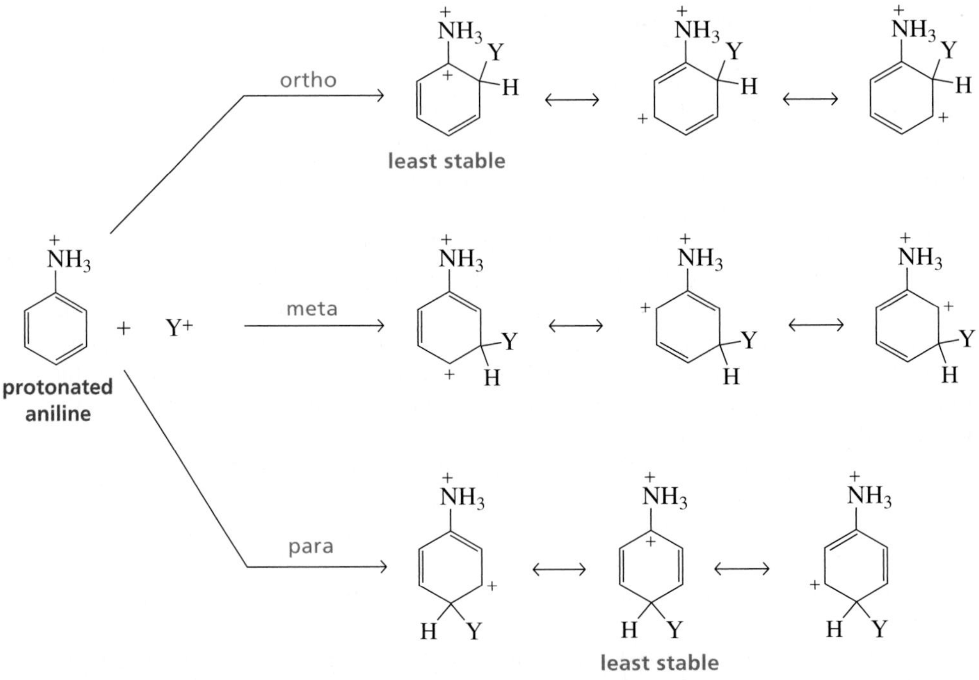

▲ Figure 15.3
The structures of the carbocation intermediates formed from the reaction of an electrophile with protonated aniline at the ortho, meta, and para positions.

however, causes them to be ortho/para directors—they can stabilize the transition states leading to reaction at the ortho and para positions by resonance electron donation (Figure 15.2).

In summary, the dividing line between ortho/para directors and meta directors is just below the halogens in Table 15.1. The halogens and all the activating substituents are ortho/para directors. All substituents more deactivating than the halogens are meta directors. Directing effects can be easily summarized—*all substituents that donate electrons into the ring inductively or by resonance are ortho/para directors, and all substituents that cannot donate electrons into the ring inductively or by resonance are meta directors.*

You don't need to resort to memorization to determine whether a substituent is an ortho/para director or a meta director. It is easy to tell them apart. All ortho/para directors have at least one lone pair of electrons on the atom directly attached to the ring (with the exception of alkyl, aryl, and CH=CHR groups). All meta directors have a positive charge or a partial positive charge on the atom attached to the ring. Take a few minutes to examine the substituents listed in Table 15.1 to convince yourself that this is true.

> **All substituents that donate electrons into the ring either inductively or by resonance are ortho/para directors.**

> **All substituents that cannot donate electrons into the ring either inductively or by resonance are meta directors.**

PROBLEM 9

a. Draw the resonance contributors for nitrobenzene.

b. Draw the resonance contributors for chlorobenzene.

PROBLEM 10 ◆

What product(s) would result from nitration of each of the following compounds?

a. propylbenzene

b. bromobenzene

c. benzaldehyde

d. benzenesulfonic acid

e. cyclohexylbenzene

f. benzonitrile

PROBLEM 11 ◆

Are the following substituents ortho/para directors or meta directors?

a. CF$_3$ **b.** N=O   **c.** NO$_2$

We have seen that electron-withdrawing substituents increase the acidity of a compound (Sections 1.18 and 6.10). The ability of a substituent to either withdraw electrons from a benzene ring or donate electrons into a benzene ring is reflected in the pK_a values of substituted phenols, benzoic acids, and protonated anilines. For example, the pK_a of phenol in H$_2$O at 25 °C is 9.95. The pK_a of *para*-nitrophenol is lower (7.14) because the nitro substituent withdraws electrons from the ring, whereas the pK_a of *para*-methyl phenol is higher (10.19) because the methyl substituent donates electrons into the ring.

15.5
THE EFFECT OF SUBSTITUENTS ON pK_a

OH / OCH$_3$	OH / CH$_3$	OH	OH / Cl	OH / HC=O	OH / NO$_2$
pK_a = 10.20	pK_a = 10.19	pK_a = 9.95	pK_a = 9.38	pK_a = 7.66	pK_a = 7.14

Take a minute to compare the effect of a substituent on the reactivity of a benzene ring toward electrophilic substitution with its effect on the pK_a of phenol. Notice that the more strongly deactivating the substituent, the lower the pK_a of the phenol and the more strongly activating the substituent, the higher the pK_a of the phenol. In other words, electron withdrawal decreases reactivity toward electrophilic substitution and increases acidity, whereas electron donation increases reactivity toward electrophilic substitution and decreases acidity.

A similar substituent effect on pK_a is observed for substituted benzoic acids and substituted protonated anilines.

The more deactivating (electron-withdrawing) the substituent, the more it increases the acidity of a COOH, OH, or $^+NH_3$ group attached to a benzene ring.

Tutorial:
Effect of substituents
on pKa

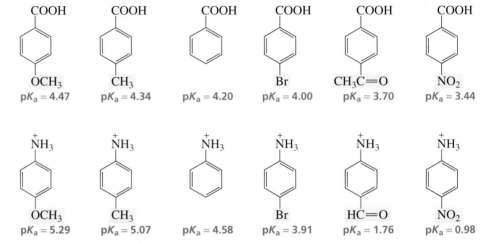

PROBLEM 12◆

Which of the compounds in each of the following pairs is more acidic?

a. $CH_3\overset{O}{\overset{\|}{C}}OH$ or $ClCH_2\overset{O}{\overset{\|}{C}}OH$

e. $H\overset{O}{\overset{\|}{C}}OH$ or $CH_3\overset{O}{\overset{\|}{C}}OH$

b. $O_2NCH_2\overset{O}{\overset{\|}{C}}OH$ or $O_2NCH_2CH_2\overset{O}{\overset{\|}{C}}OH$

f. (COOH with NH₂ ring) or (COOH ring)

c. $CH_3CH_2\overset{O}{\overset{\|}{C}}OH$ or $H_3\overset{+}{N}CH_2\overset{O}{\overset{\|}{C}}OH$

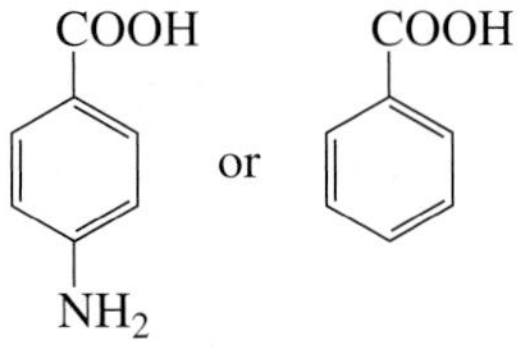

g. $FCH_2\overset{O}{\overset{\|}{C}}OH$ or $ClCH_2\overset{O}{\overset{\|}{C}}OH$

d. $HO\overset{O}{\overset{\|}{C}}CH_2\overset{O}{\overset{\|}{C}}OH$ or $^-O\overset{O}{\overset{\|}{C}}CH_2\overset{O}{\overset{\|}{C}}OH$

h. (COOH with F ring) or (COOH with Cl ring)

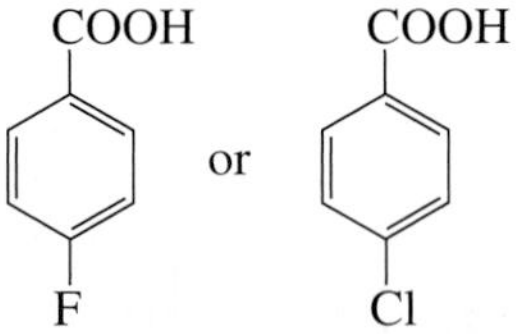

PROBLEM-SOLVING STRATEGY

The *para*-nitroanilinium ion is 3.60 pK_a units more acidic than the anilinium ion (pK_a = 4.58 versus 0.98), but *para*-nitrobenzoic acid is only 0.76 pK_a unit more acidic than benzoic acid (pK_a = 4.20 versus 3.44). Explain why the nitro substituent causes a large change in pK_a in one case and a small change in pK_a in the other.

Don't expect to be able to solve this kind of problem simply by reading it. First pay close attention to exactly what the question is asking. You are being asked why the addition of a nitro substituent has a greater effect on the acidity of one compound than it does on the acidity of another compound.

Now let's recall what causes compounds to have different acidities. We have seen that the acidity of a compound depends on the stability of its conjugate base (Sections 1.18 and 6.10). Next, draw structures of the compounds in question. Check to see how well the substituent that has changed the acidity of a given compound can stabilize the conjugate base.

The structures allow us to answer the question. When a proton is lost from the *para*-nitroanilinium ion, the electrons that are left behind can be delocalized onto the nitro substituent. When a proton is lost from *para*-nitrobenzoic acid, the electrons that are left behind cannot be delocalized onto the nitro substituent. In other words, loss of a proton leads to greater resonance stabilization in one case but not in the other. Therefore, the addition of a nitro substituent will have a greater affect on the acidity of an anilinium ion than on benzoic acid. Now continue on to Problem 13.

PROBLEM 13

p-Nitrophenol has a pK_a of 7.14, whereas the pK_a of *m*-nitrophenol is 8.39. Explain.

When a benzene ring with an ortho/para-directing substituent undergoes an electrophilic aromatic substitution reaction, what percentage of the product is the ortho isomer and what percentage is the para isomer? Based solely on probability, one would expect more of the ortho product because there are two ortho positions available to the incoming electrophile and only one para position. The ortho position, however, is sterically hindered and the para position is not. Consequently, the para isomer will be formed preferentially if either the substituent on the ring or the incoming electrophile is large. The following nitration reactions illustrate the decrease in the ortho/para ratio with an increase in size of the alkyl substituent.

15.6
THE ORTHO/PARA RATIO

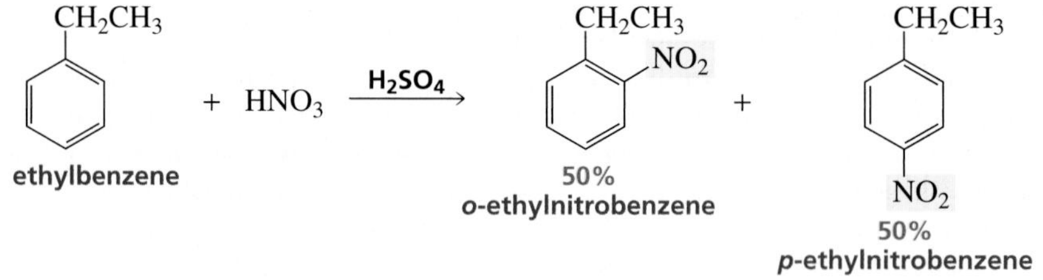

ethylbenzene + HNO$_3$ $\xrightarrow{\text{H}_2\text{SO}_4}$
50% o-ethylnitrobenzene + 50% p-ethylnitrobenzene

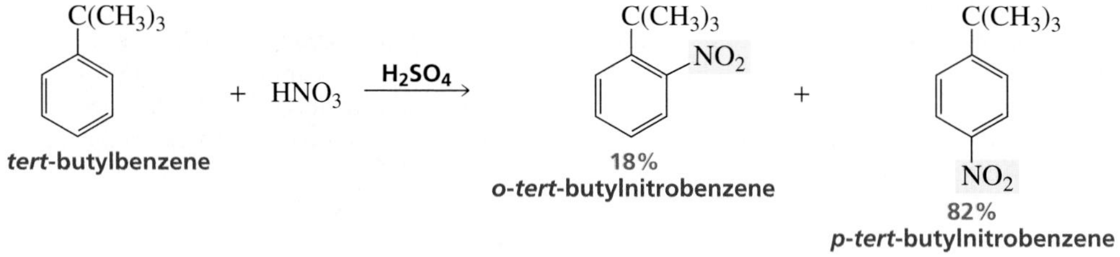

tert-butylbenzene + HNO$_3$ $\xrightarrow{\text{H}_2\text{SO}_4}$
18% o-tert-butylnitrobenzene + 82% p-tert-butylnitrobenzene

Fortunately, the difference in the physical properties of the ortho- and para-substituted isomers is sufficient to allow them to be easily separated. Consequently, electrophilic aromatic substitution reactions that lead to both ortho and para isomers are useful in synthetic schemes because the desired product can be readily separated from the reaction mixture.

15.7 ADDITIONAL CONSIDERATIONS REGARDING SUBSTITUENT EFFECTS

It is important to know whether a substituent is activating or deactivating in order to determine the conditions to use when carrying out a reaction. For example, sulfonation of benzene requires fuming sulfuric acid, but sulfonation of toluene can be done with concentrated sulfuric acid because the activating methyl group allows the reaction to proceed at a reasonable rate under less rigorous conditions.

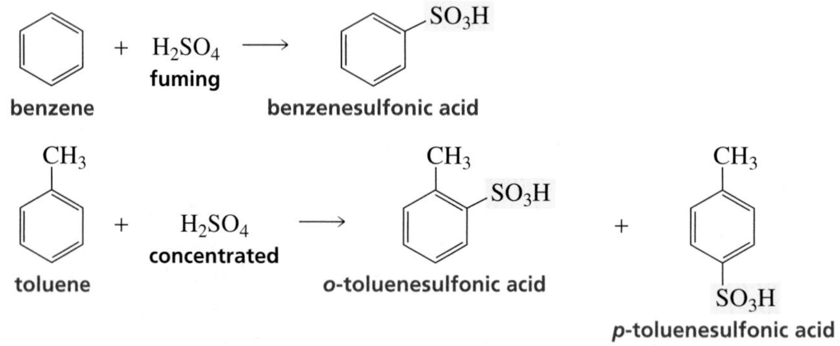

benzene + H$_2$SO$_4$ fuming $\longrightarrow$ benzenesulfonic acid

toluene + H$_2$SO$_4$ concentrated $\longrightarrow$ o-toluenesulfonic acid + p-toluenesulfonic acid

Of the five electrophilic aromatic substitution reactions listed in Section 14.8 and summarized in Section 15.2, halogenation occurs under the mildest conditions. The methoxy and hydroxy substituents are so strongly activating that halogenation is carried out without the Lewis acid (FeBr$_3$ or FeCl$_3$) catalyst.

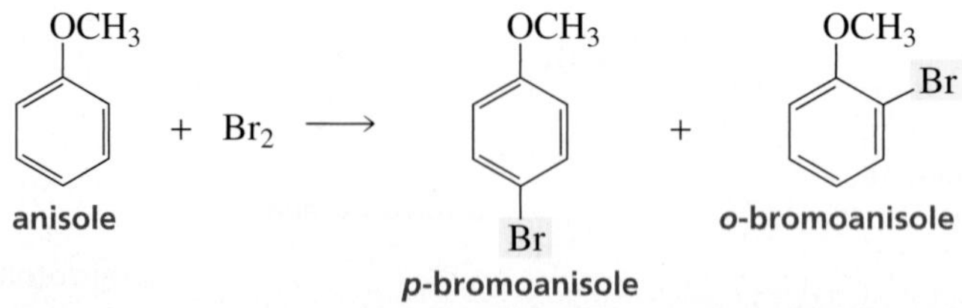

anisole + Br$_2$ $\longrightarrow$ p-bromoanisole + o-bromoanisole

If a catalyst and excess bromine are used, the tribromide is obtained.

$$\text{anisole} + 3\ Br_2 \xrightarrow{\ FeBr_3\ } \text{2,4,6-tribromoanisole}$$

Of the five electrophilic aromatic substitution reactions, the two Friedel–Crafts reactions are the most sluggish. Therefore, if there is a meta director on the ring (remember that meta directors are all moderate or strong deactivators), the ring will be too unreactive to undergo either Friedel–Crafts acylation or Friedel–Crafts alkylation.

$$\text{benzenesulfonic acid} + CH_3CH_2Cl \xrightarrow{\ AlCl_3\ } \text{no reaction}$$

$$\text{nitrobenzene} + CH_3\overset{O}{\overset{\|}{C}}Cl \xrightarrow{\ AlCl_3\ } \text{no reaction}$$

Aniline and *N*-substituted anilines also cannot undergo Friedel–Crafts reactions. The lone-pair electrons on the amino group will complex with the Lewis acid ($AlCl_3$) needed to catalyze the reaction. Complexation converts the NH_2 substituent into a deactivating meta director, and we have just seen that Friedel–Crafts reactions do not occur on benzene rings containing meta-directing substituents.

$$\text{aniline} \xrightarrow{\ AlCl_3\ } \text{substituent is a meta director}$$

Aniline also cannot be nitrated because nitric acid is an oxidizing agent and primary amines are easily oxidized. (Nitric acid and aniline can be an explosive combination.) Tertiary aromatic amines can be nitrated, however. Because the tertiary amino group is a strong activator, nitration is carried out using nitric acid in acetic acid, milder conditions than nitric acid in sulfuric acid. About twice as much para isomer is formed as ortho isomer.

3-D Molecules:
N,N-Dimethylaniline;
o-Nitro-*N,N*-dimethyl-aniline

$$\textit{N,N}\text{-dimethylaniline} \xrightarrow[\begin{array}{c}\textbf{CH}_3\textbf{COOH}\\\textbf{2. HO}^-\end{array}]{\textbf{1. HNO}_3} \textit{o}\text{-nitro-}\textit{N,N}\text{-dimethylaniline} + \textit{p}\text{-nitro-}\textit{N,N}\text{-dimethylaniline}$$

PROBLEM 14

When *N,N*-dimethylaniline is nitrated, some *meta*-nitro-*N,N*-dimethylaniline is formed. Why does the meta isomer form? (*Hint:* The pK_a of *N,N*-dimethylaniline is 5.07, and more meta isomer is formed if the reaction is carried out at pH = 3.5 than if it is carried out at pH = 4.5.)

PROBLEM 15 ◆

Give the products, if any, of each of the following reactions:

a. benzonitrile + methyl chloride + $AlCl_3$

b. aniline + 3 Br_2

c. benzoic acid + CH_3CH_2Cl + $AlCl_3$

d. benzene + 2 CH_3Cl + $AlCl_3$

15.8 DESIGNING A SYNTHESIS III: SYNTHESIS OF MONOSUBSTITUTED AND DISUBSTITUTED BENZENES

As the number of reactions we know increases, we have more reactions to choose from when we design a synthesis. For example, we can now design two very different routes for the synthesis of 2-phenylethanol from benzene.

2-phenylethanol

excess

2-phenylethanol

Which route we choose depends on how easy it is to do the required reactions and on the expected yield of the target molecule (the desired product). For example, the first route shown for the synthesis of 2-phenylethanol is the better procedure. The second route has more steps, it requires excess benzene (because the alkylated product is more reactive than benzene), it uses a radical reaction that can produce unwanted side products, the yield of the elimination reaction is not high (because some substitution product is formed as well), and hydroboration–oxidation is not an easy reaction to carry out.

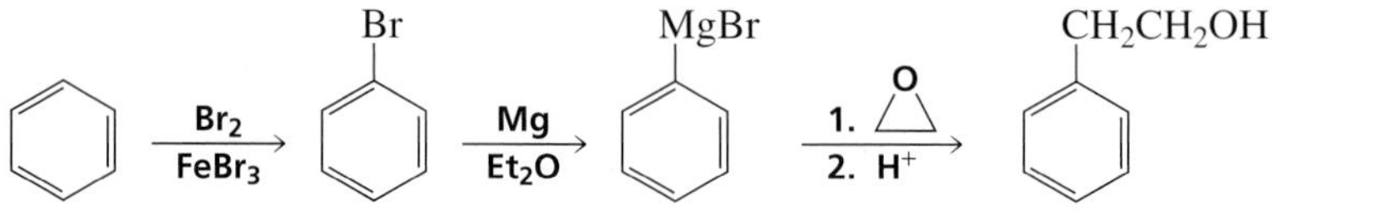

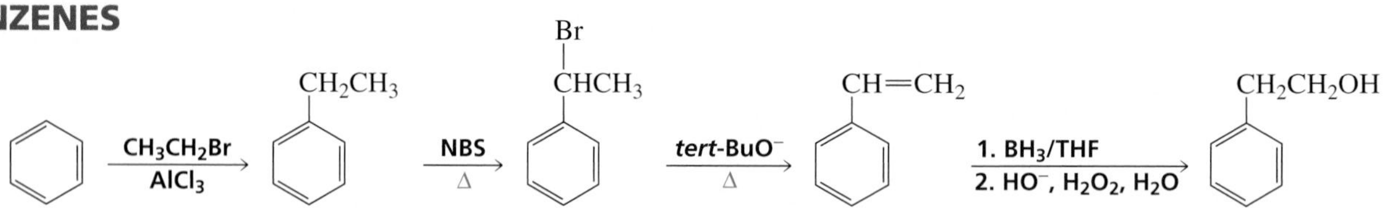

Tutorial: Multistep synthesis of disubstituted benzenes

In designing the synthesis of disubstituted benzenes, it is important to pay attention to the order in which the substituents are placed on the ring. For example, if you want to synthesize *meta*-bromobenzenesulfonic acid, the sulfonic acid group has to be placed on the ring first because that group will direct the bromo substituent to the desired meta position.

m-bromobenzenesulfonic acid

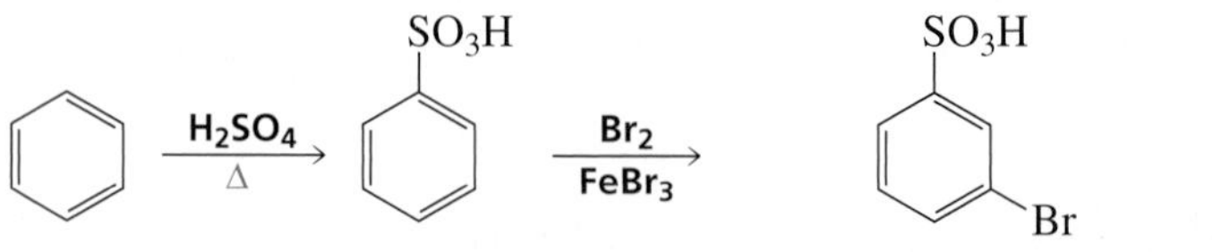

On the other hand, if the desired product is *para*-bromobenzenesulfonic acid, the order of the two reactions must be reversed because only the bromo substituent is an ortho/para director.

p-bromobenzenesulfonic acid

o-bromobenzenesulfonic acid

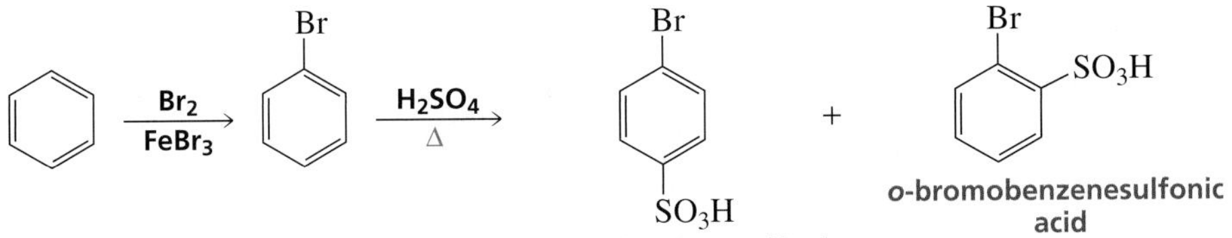

Both substituents in *meta*-nitroacetophenone are meta directors. However, the Friedel–Crafts acylation reaction must be carried out first because the ring of nitrobenzene is too deactivated to undergo a Friedel–Crafts reaction (Section 15.7).

m-nitroacetophenone

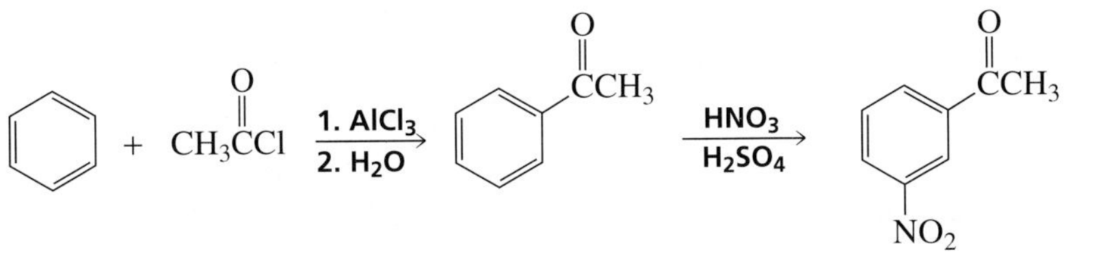

It is also important to determine the point in a reaction sequence at which a substituent should be chemically modified. In the synthesis of *para*-chlorobenzoic acid from toluene, the methyl group is oxidized after it directs the chloro substituent to the para position (*ortho*-chlorobenzoic acid is also formed in this reaction).

para-chlorobenzoic acid

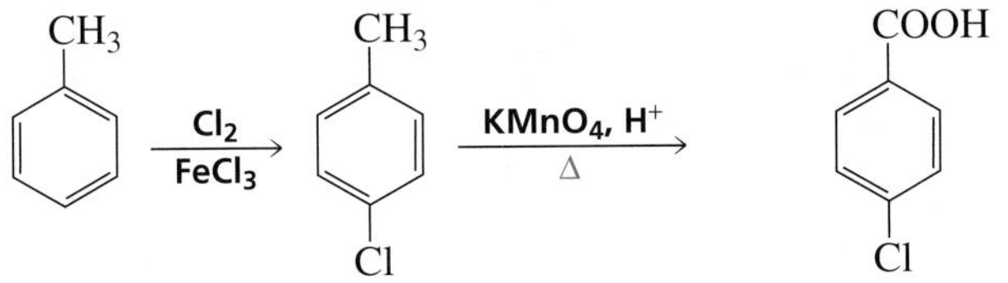

In the synthesis of *meta*-chlorobenzoic acid, the methyl group is oxidized before chlorination because a meta director is needed to obtain the desired product.

meta-chlorobenzoic acid

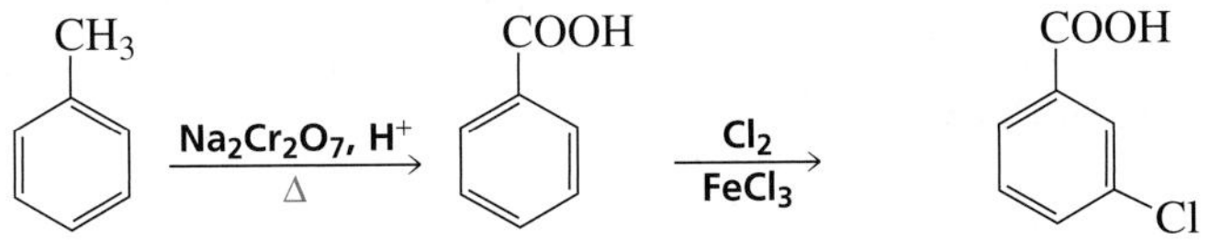

In the following synthesis of *para*-propylbenzenesulfonic acid, the type of reaction employed, the order in which the substituents are put on the benzene ring, and the point at which a substituent is chemically modified must all be considered. The straight-chain propyl substituent must be put on the ring by a Friedel–Crafts acylation reaction because of the carbocation rearrangement that would occur with a Friedel–Crafts alkylation reaction. The Friedel–Crafts acylation must be carried out before sulfonation because acylation could not be carried out on a ring with a strongly deactivating sulfonic acid substituent. Finally, the sulfonic acid group must be put on the ring after the carbonyl group is reduced to a methylene group so the sulfonic acid group will be directed primarily to the para position by the alkyl group.

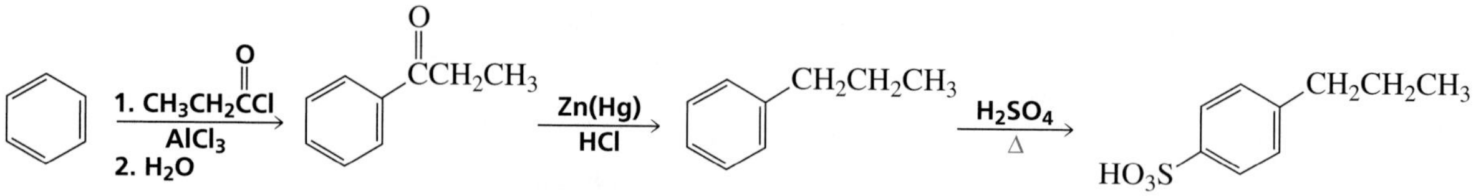

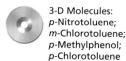

para-propylbenzenesulfonic acid

PROBLEM 16

Show how each of the following compounds can be synthesized from benzene:

a. *p*-chloroaniline

b. *m*-nitrobenzoic acid

c. *p*-nitrobenzoic acid

d. *m*-bromopropylbenzene

e. *o*-bromopropylbenzene

f. 1-phenyl-2-propanol

g. *o*-nitrophenol

h. *m*-nitrophenol

i. 2-phenylpropene

j. *m*-chloroaniline

15.9
SYNTHESIS OF TRISUBSTITUTED BENZENES

When disubstituted benzenes undergo electrophilic aromatic substitution reactions, the directing effect of both substituents has to be considered. If both substituents direct to the same position on the ring, the product of the reaction is easily predicted.

3-D Molecules:
p-Nitrotoluene;
m-Chlorotoluene;
p-Methylphenol;
p-Chlorotoluene

the methyl and nitro substituents direct the incoming substituent to the indicated positions

p-nitrotoluene + HNO$_3$ $\xrightarrow{\text{H}_2\text{SO}_4}$ 2,4-dinitrotoluene

Notice that three positions are activated in the following reaction, but the new substituent ends up on only two of the three positions. Steric hindrance makes the position between the substituents less accessible.

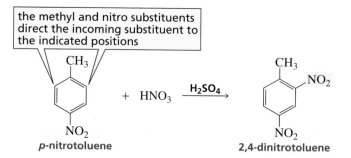

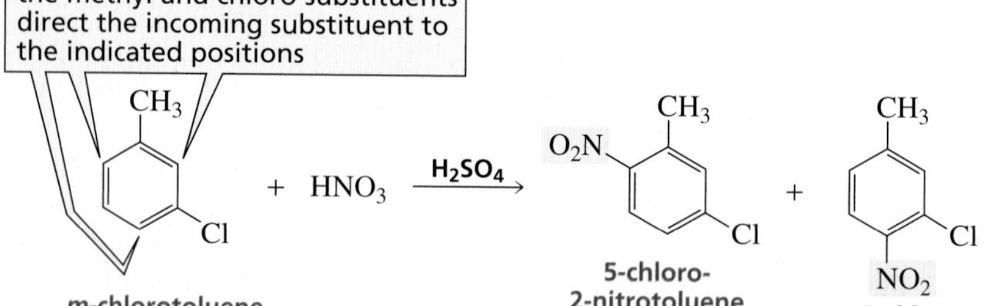

the methyl and chloro substituents direct the incoming substituent to the indicated positions

m-chlorotoluene + HNO$_3$ $\xrightarrow{\text{H}_2\text{SO}_4}$ 5-chloro-2-nitrotoluene + 3-chloro-4-nitrotoluene

If the two substituents direct to different positions, a strongly activating substituent will win out over a weakly activating substituent or a deactivating substituent.

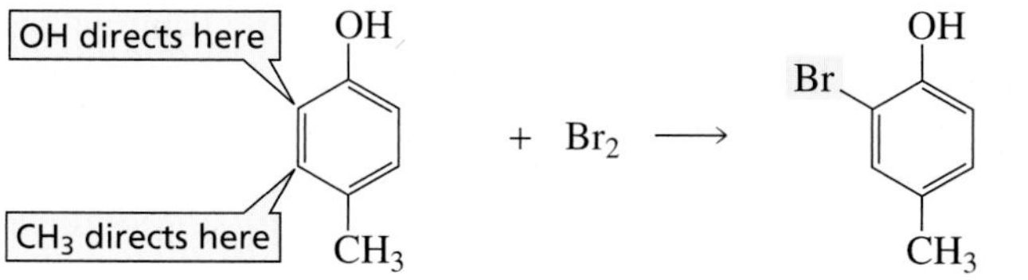

OH directs here — OH

CH₃ directs here — CH₃

p-methylphenol

+ Br₂ ⟶

OH, Br

CH₃

**2-bromo-
4-methylphenol**
major product

If the two substituents have similar activating properties, neither will dominate and a mixture of products will be obtained.

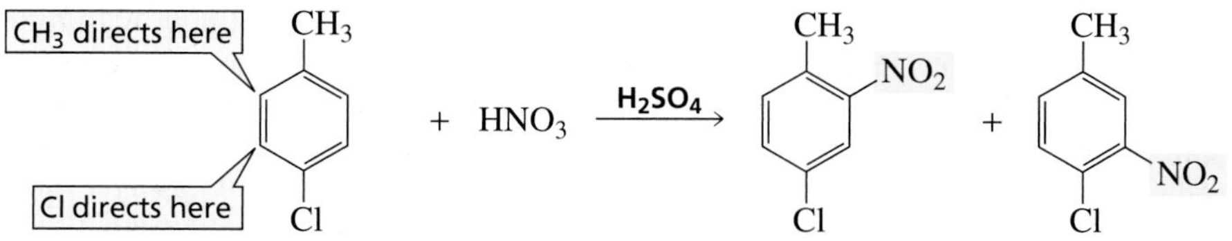

CH₃ directs here — CH₃

Cl directs here — Cl

p-chlorotoluene

+ HNO₃ $\xrightarrow{\text{H}_2\text{SO}_4}$

CH₃, NO₂

Cl

**4-chloro-
2-nitrotoluene**

+

CH₃

Cl, NO₂

**4-chloro-
3-nitrotoluene**

PROBLEM 17 ◆

Give the major product(s) of each of the following reactions:

a. nitration of *p*-fluoroanisole

b. chlorination of *o*-benzenedicarboxylic acid

c. bromination of *p*-chlorobenzoic acid

PROBLEM 18/SOLVED

When phenol is treated with Br₂, a mixture of monobromo-, dibromo-, and tribromophenols is obtained. Design a synthesis that would convert phenol primarily to *ortho*-bromophenol.

SOLUTION   The bulky sulfonic acid group will add preferentially to the para position. Both the OH and SO₃H groups will direct bromine to the position ortho to the OH group. Heating in dilute acid removes the sulfonic acid group (Section 14.11).

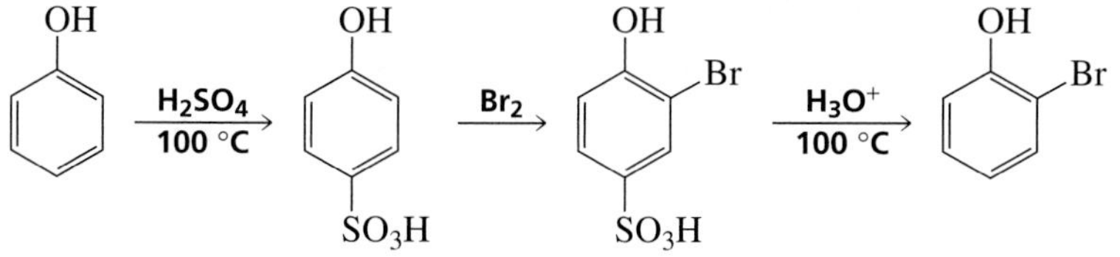

OH $\xrightarrow[\text{100 °C}]{\text{H}_2\text{SO}_4}$ OH, SO₃H $\xrightarrow{\text{Br}_2}$ OH, Br, SO₃H $\xrightarrow[\text{100 °C}]{\text{H}_3\text{O}^+}$ OH, Br

So far we have learned how to place a limited number of different substituents on a benzene ring—those substituents listed in Section 15.2, and those that can be obtained from these substituents by chemical conversion. However, the kinds of substituents that can be placed on benzene rings can be greatly expanded by the use of **arenediazonium salts.**

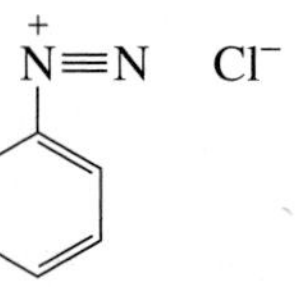

N≡N⁺ Cl⁻

an arenediazonium salt

**15.10
SYNTHESIS OF
SUBSTITUTED
BENZENES USING
ARENEDIAZONIUM
SALTS**

The drive to form a molecule of stable nitrogen gas (N_2) causes the leaving group of a diazonium ion to be easily displaced by a wide variety of nucleophiles. The mechanism by which a nucleophile displaces the diazonium group depends on the nucleophile—some displacements involve phenyl cations, others involve radicals.

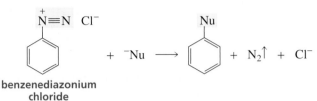

benzenediazonium
chloride

**benzenediazonium
ion**

A primary amine can be converted into a diazonium salt by treatment with nitrous acid (HNO_2). Nitrous acid is unstable, so it is formed *in situ* using an aqueous solution of sodium nitrite and HCl or HBr; N_2 is such a good leaving group that the diazonium salt must be synthesized at 0 °C and used immediately. (The mechanism for conversion of a primary amino group $[NH_2]$ to a diazonium group $[^+N\equiv N]$ is shown in Section 15.12.)

3-D Molecule:
Benzenediazonium
chloride

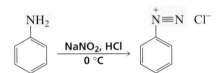

Nucleophiles such as $^-C\equiv N$, Cl^-, and Br^- will replace the diazonium group if the appropriate cuprous salt is added to the solution containing the arenediazonium salt. The reaction of an arenediazonium salt with a cuprous salt is known as a **Sandmeyer reaction.**

Sandmeyer reactions

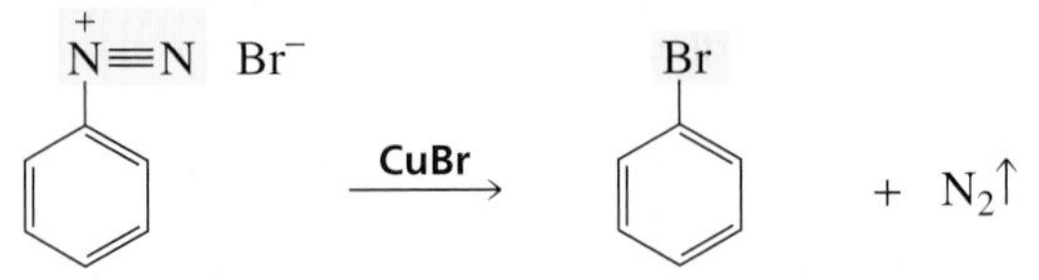

benzenediazonium
bromide

bromobenzene

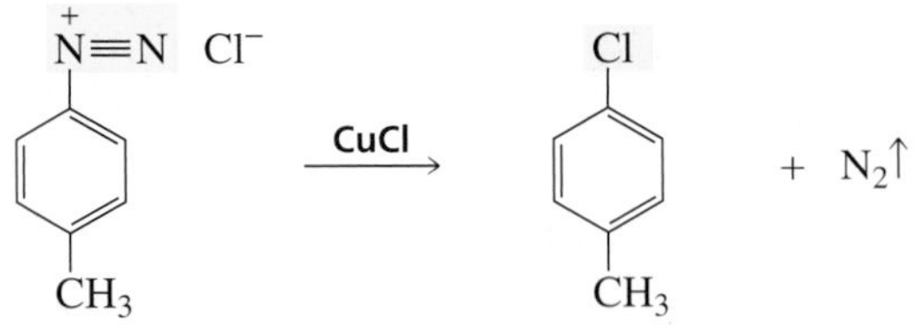

p-toluenediazonium
chloride

p-chlorotoluene

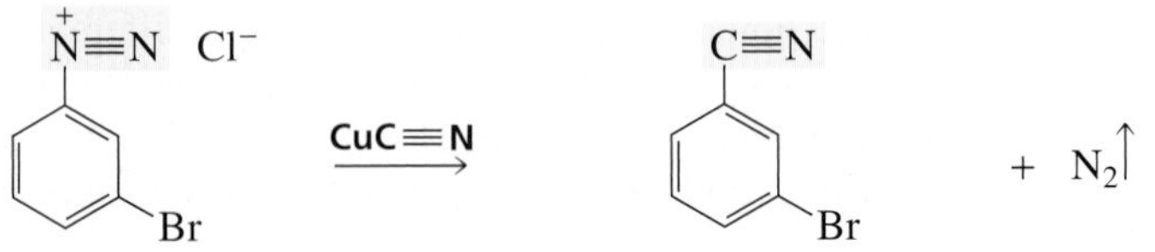

m-bromobenzenediazonium chloride

m-bromobenzonitrile

***Traugott Sandmeyer
(1854–1922)*** *was born in
Switzerland. He received a
Ph.D. from the University of
Heidelberg and discovered the
reaction that bears his name in
1884. He was a scientist at
Geigy Co. in Basel, Switzerland.*

KCl and KBr cannot be used in place of CuCl and CuBr in Sandmeyer reactions; the cuprous salts are required. This indicates that the cuprous ion is involved in the reaction, most likely by serving as a radical initiator.

Although chloro and bromo substituents can be placed directly on a benzene ring by halogenation, the Sandmeyer reaction can be a useful alternative. For example, if you wanted to make *para*-chloroethylbenzene, chlorination of ethylbenzene would lead to a mixture of the ortho and para isomers.

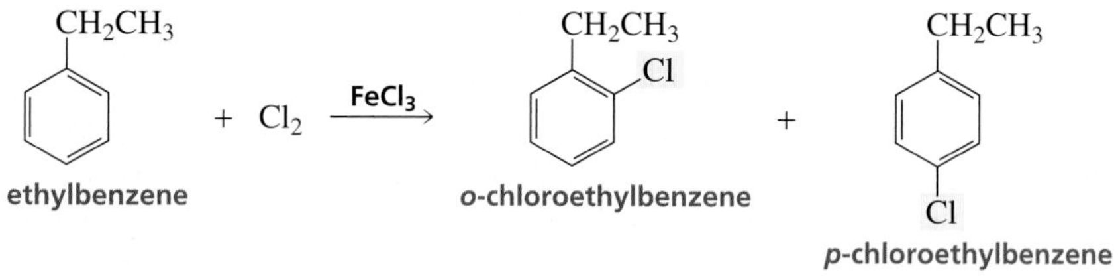

ethylbenzene   + Cl$_2$ $\xrightarrow{\text{FeCl}_3}$ o-chloroethylbenzene   +   p-chloroethylbenzene

However, if you started with *para*-ethylaniline and used a Sandmeyer reaction for chlorination, only the desired para product would be formed.

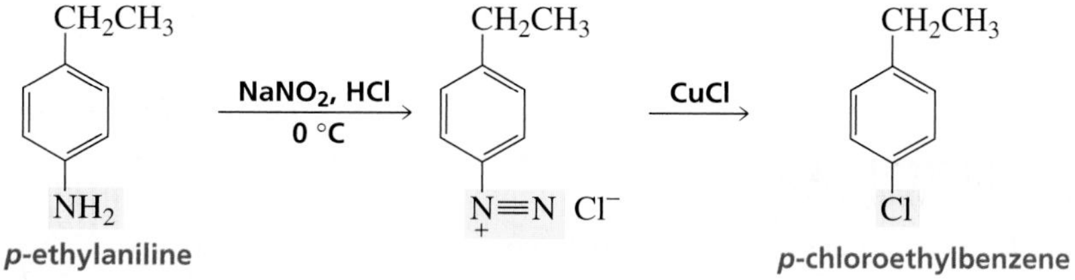

p-ethylaniline $\xrightarrow[\text{0 °C}]{\text{NaNO}_2, \text{HCl}}$ $\xrightarrow{\text{CuCl}}$ p-chloroethylbenzene

---

PROBLEM 19

Explain why HBr should be used to generate the benzenediazonium salt if bromobenzene is the desired product of the Sandmeyer reaction and HCl should be used if chlorobenzene is the desired product.

An iodo substituent will replace the diazonium group if potassium iodide is added to the solution containing the diazonium ion.

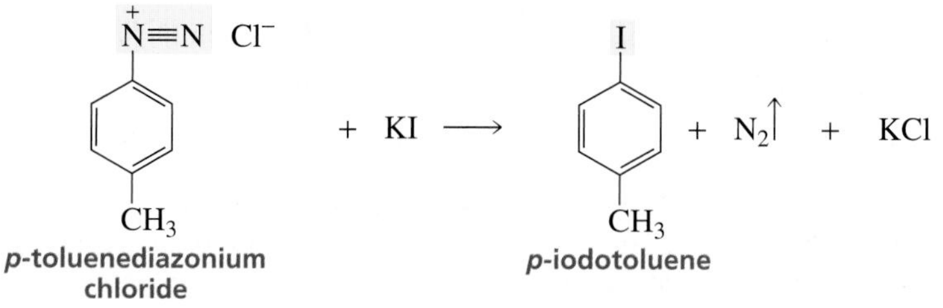

+ KI $\longrightarrow$ + N$_2\uparrow$ + KCl

p-toluenediazonium chloride   p-iodotoluene

Fluoro substitution occurs if the arenediazonium salt is heated with fluoroboric acid (HBF$_4$). This reaction is known as the **Schiemann reaction.**

Schiemann reaction

$\xrightarrow{\text{HBF}_4}$ $\xrightarrow{\Delta}$ fluorobenzene   + BF$_3$ + N$_2\uparrow$ + HCl

Günther Schiemann (1899–1969) was born in Germany. He was a professor of chemistry at the Technologische Hochschule in Hanover.

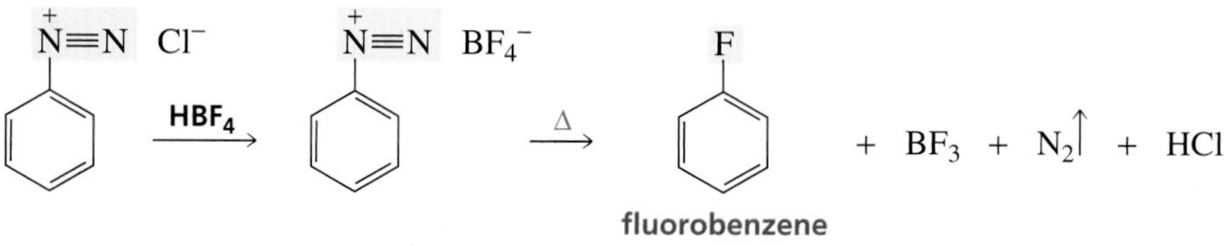

If the aqueous solution in which the diazonium salt has been synthesized is allowed to warm up to room temperature, an OH group will replace the diazonium group (H$_2$O is the nucleophile). This is a convenient way to synthesize phenols.

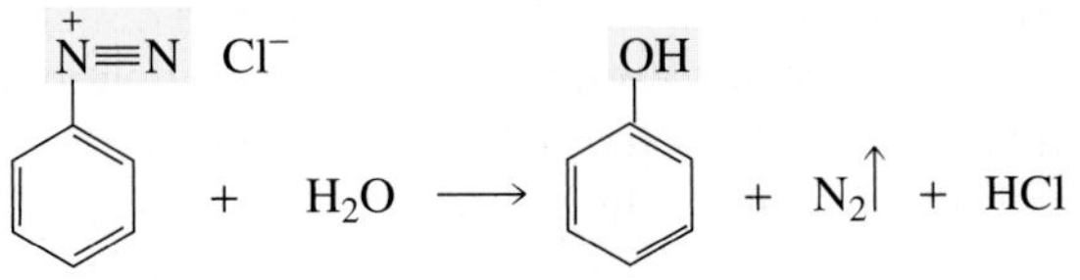

A hydrogen will replace a diazonium group if the diazonium salt is treated with hypophosphorous acid (H_3PO_2). This is a useful reaction if an amino group or a nitro group is needed for directing purposes and subsequently must be removed. It would be difficult to envision how 1,3,5-tribromobenzene could be synthesized without this reaction.

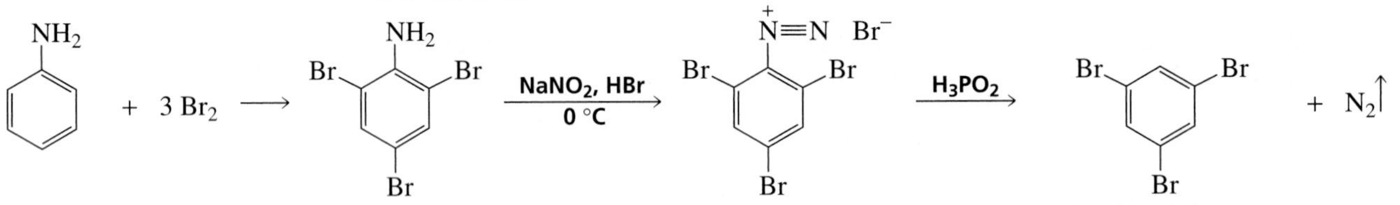

PROBLEM 20

Write the sequence of steps involved in the conversion of benzene into benzenediazonium chloride.

PROBLEM 21 ◆

In an attempt to synthesize benzonitrile, a student allowed a freshly prepared solution of benzenediazonium chloride to warm up to room temperature before adding CuCN.

a. Was the synthesis successful?

b. What was the major product of the reaction?

PROBLEM 22/SOLVED

Show how the following compounds could be synthesized from benzene:

a. *m*-dibromobenzene

b. *p*-methylbenzonitrile

c. *m*-bromophenol

d. *o*-chlorophenol

e. *m*-nitrotoluene

f. *meta*-chlorobenzaldehyde

SOLUTION TO 22a   A bromo substituent is an ortho/para director, so halogenation cannot be used to introduce both bromo substituents of *m*-dibromobenzene. If we recall that a bromo substituent can be placed on a benzene ring with a Sandmeyer reaction, and that the bromo substituent in a Sandmeyer reaction replaces what originally was a meta-directing nitro substituent, we have a route to the synthesis of the target compound.

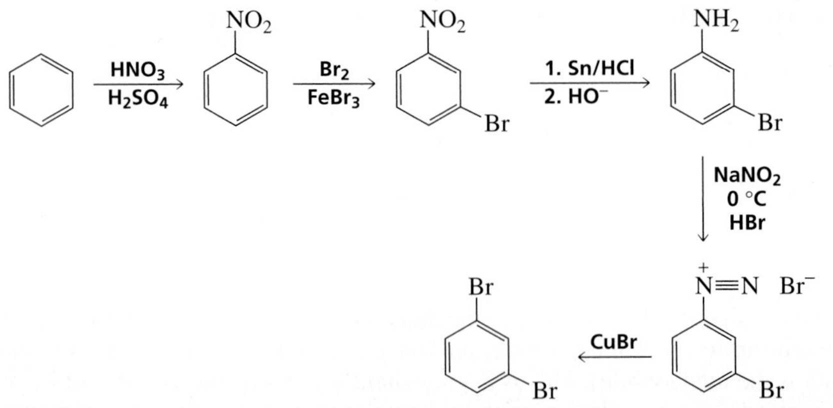

In addition to their use in the synthesis of substituted benzenes, arenediazonium ions can also be used as electrophiles in electrophilic aromatic substitution reactions. Because an arenediazonium ion is unstable at room temperature, it can be used as an electrophile only in electrophilic substitution reactions that can be carried out well below room temperature. In other words, only highly activated benzene rings (phenols, anilines, and *N*-alkylanilines) can undergo electrophilic substitution reactions with arenediazonium ion electrophiles. The product of the reaction is an *azo compound*. The —N=N— linkage is called an **azo linkage.** Because the electrophile is so large, substitution takes place preferentially at the less sterically hindered para position.

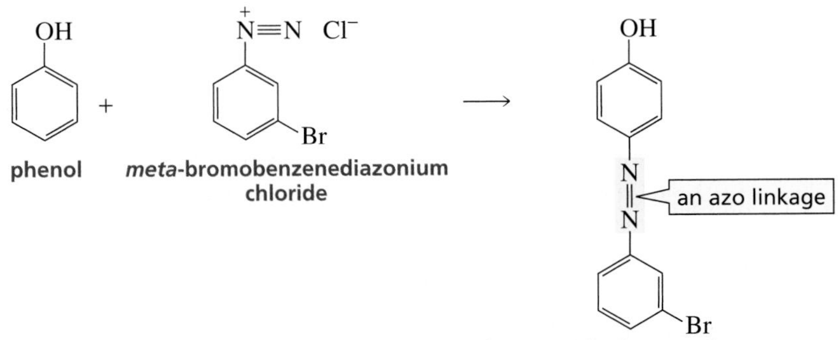

15.11 THE ARENEDIAZONIUM ION AS AN ELECTROPHILE

phenol   *meta*-bromobenzenediazonium chloride

an azo linkage

3-bromo-4'-hydroxyazobenzene
an azo compound

However, if the para position is blocked, substitution will occur at an ortho position.

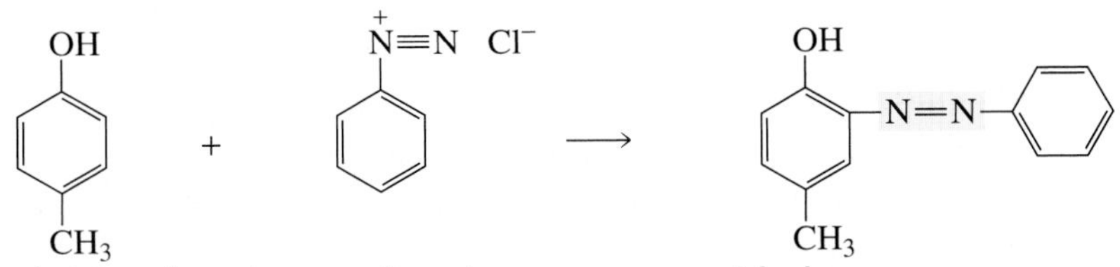

p-methylphenol   benzenediazonium chloride   2-hydroxy-5-methylazobenzene

The mechanism for electrophilic substitution using an arenediazonium ion electrophile is the same as the mechanism for electrophilic substitution using any other electrophile.

mechanism for electrophilic substitution using an arenediazonium ion electrophile

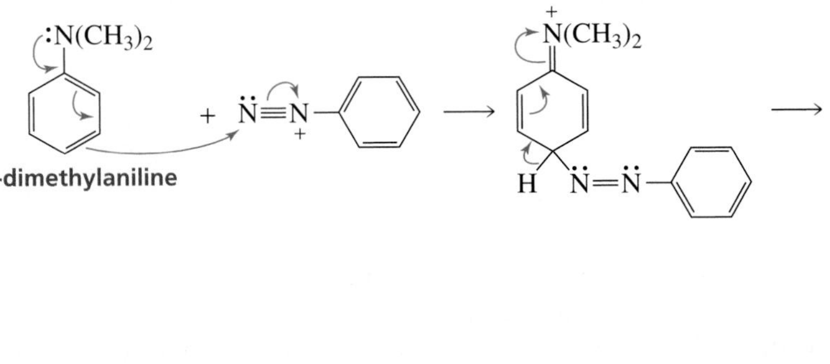

N,N-dimethylaniline

3-D Molecule:
para-N,N-Dimethylamino-azobenzene

p-N,N-dimethylaminoazobenzene

PROBLEM 23

In the mechanism for electrophilic substitution using a diazonium ion as the electrophile, why does nucleophilic attack occur on the terminal nitrogen atom of the diazonium ion rather than on the nitrogen atom bonded to the benzene ring?

PROBLEM 24

Explain why a diazonium group on a benzene ring cannot be used to direct an incoming substituent to the meta position.

Azo compounds, like alkenes, can exist in cis and trans forms. Because of steric strain, the trans isomer is considerably more stable than the cis isomer (Section 3.19).

trans-azobenzene   *cis*-azobenzene

We have seen that azobenzenes are colored compounds because of their extended conjugation and are used commercially as dyes (Section 12.19).

PROBLEM 25

Give the structure of the activated benzene ring and the diazonium ion used in the synthesis of the following compounds:

a. butter yellow   **b.** methyl orange

(The structures of these compounds can be found in Section 12.19.)

15.12 MECHANISM FOR THE REACTION OF AMINES WITH NITROUS ACID

We have seen that the reaction of a primary amine with nitrous acid produces a diazonium salt. Both aryl amines and alkyl amines undergo this reaction, and both follow the same mechanism.

$$\text{aryl—NH}_2 \xrightarrow[\text{0 °C}]{\text{NaNO}_2,\ \text{HCl}} \text{aryl—}\overset{+}{\text{N}}\equiv\text{N}\ \ \text{Cl}^-$$

$$\text{alkyl—NH}_2 \xrightarrow[\text{0 °C}]{\text{NaNO}_2,\ \text{HCl}} \text{alkyl—}\overset{+}{\text{N}}\equiv\text{N}\ \ \text{Cl}^-$$

The first step in the reaction is formation of nitrous acid from sodium nitrite and HCl. Loss of water from protonated nitrous acid generates the nitrosonium ion, which is the reactive species in the reaction of amines with nitrous acid.

$$\text{Na}^+\ \overset{\cdot\cdot}{\text{:O}}\text{—}\overset{\cdot\cdot}{\text{N}}\text{=}\overset{\cdot\cdot}{\text{O}}: \underset{-\text{H}^+}{\overset{\text{H}^+}{\rightleftharpoons}} \text{H}\overset{\cdot\cdot}{\text{O}}\text{—}\overset{\cdot\cdot}{\text{N}}\text{=}\overset{\cdot\cdot}{\text{O}}: \underset{-\text{H}^+}{\overset{\text{H}^+}{\rightleftharpoons}} \overset{+}{\text{H}}\overset{\cdot\cdot}{\text{O}}\text{—}\overset{\cdot\cdot}{\text{N}}\text{=}\overset{\cdot\cdot}{\text{O}}: + \text{Cl}^- \rightleftharpoons \overset{+}{\text{N}}\text{=}\overset{\cdot\cdot}{\text{O}}: + \text{H}_2\overset{\cdot\cdot}{\text{O}}:$$

sodium nitrite   nitrous acid   (H)   nitrosonium ion

$$+\ \ \text{Na}^+\text{Cl}^-$$

The nitrogen atom of the amine shares its lone-pair electrons with the nitrosonium ion. Loss of a proton from nitrogen forms a **nitrosamine** (also called an **N-nitroso compound** because a nitroso substituent is bonded to a nitrogen). Delocalization of nitrogen's lone-pair electrons and protonation of oxygen form a protonated *N*-hydroxyazo compound. This compound is in equilibrium with its non-protonated form, which can be reprotonated on nitrogen (reverse reaction) or proto-nated on oxygen (forward reaction). Elimination of water forms the diazonium ion.

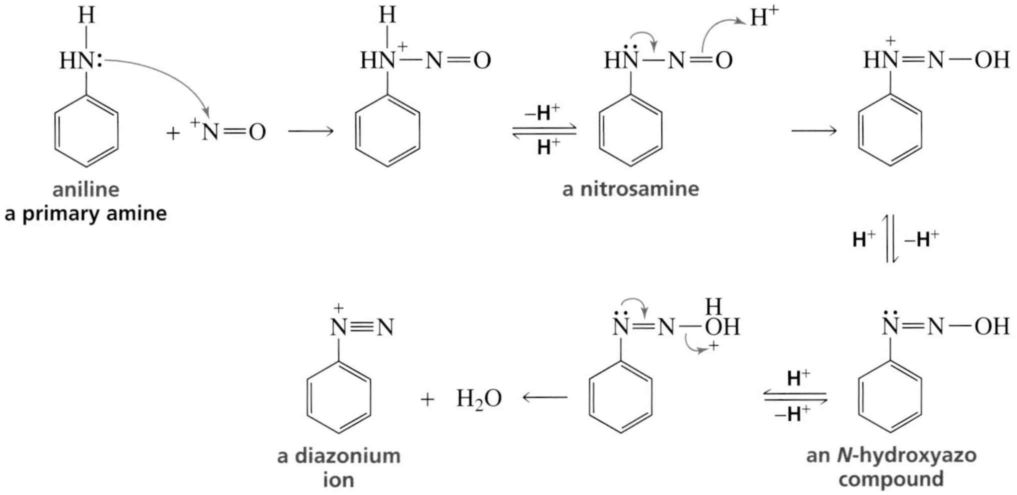

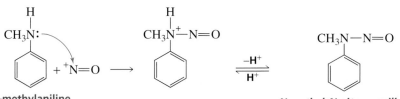

aniline
a primary amine

a nitrosamine

Notice how proton transfer steps are indicated:
$-H^+$ means loss of a proton;
H^+ means gain of a proton.

a diazonium
ion

an *N*-hydroxyazo
compound

Remember that reactions in which arenediazonium ions are involved must be carried out at 0 °C because they are unstable at higher temperatures. Alkanedia-zonium ions are even less stable. They lose molecular N_2 as they are formed, re-acting with whatever nucleophiles are present in the reaction mixture by both $S_N1/E1$ and $S_N2/E2$ mechanisms. Because of the mixture of products obtained, alkanediazonium ions are of limited synthetic use.

PROBLEM 26 ◆

What products would be formed from the reaction of isopropylamine with sodium nitrite and aqueous HCl?

PROBLEM 27

Diazomethane can be used to convert a carboxylic acid into a methyl ester. Propose a mechanism for this reaction.

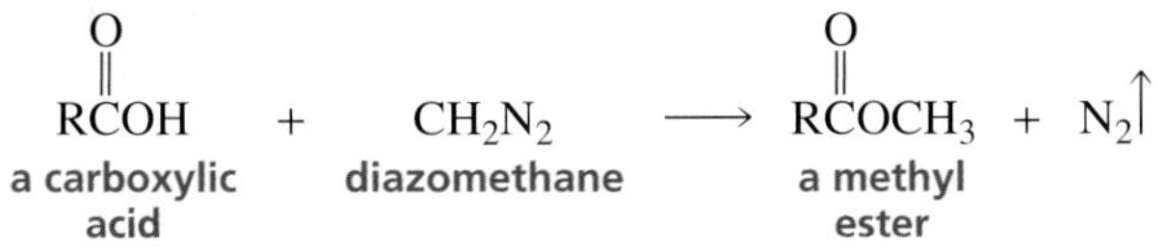

$$\underset{\substack{\text{a carboxylic}\\\text{acid}}}{\overset{O}{\underset{\|}{RCOH}}} + \underset{\substack{\text{diazomethane}}}{CH_2N_2} \longrightarrow \underset{\substack{\text{a methyl}\\\text{ester}}}{\overset{O}{\underset{\|}{RCOCH_3}}} + N_2\uparrow$$

Secondary aryl and alkyl amines react with the nitrosonium ion generated from nitrous acid to form nitrosamines. The mechanism of the reaction is similar to that for the reaction of a primary amine with a nitrosonium ion except that the reaction stops at the nitrosamine stage. The reaction stops because a secondary amine, un-like a primary amine, does not have the second proton that must be lost in order to generate the diazonium ion.

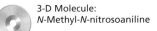

N-methylaniline
a secondary amine

N-methyl-*N*-nitrosoaniline
a nitrosamine

3-D Molecule:
N-Methyl-*N*-nitrosoaniline

The product obtained from the reaction of a tertiary amine with a nitrosonium ion cannot be stabilized by loss of a proton. A tertiary aryl amine, therefore, undergoes an electrophilic aromatic substitution reaction with a nitrosonium ion. The product of the reaction is primarily the para isomer because the bulky dialkylamino group blocks approach of the nitrosonium ion to the ortho position.

3-D Molecule:
para-Nitroso-*N,N*-
dimethylaniline

N,N-dimethylaniline
a tertiary amine

***para*-nitroso-*N,N*-dimethylaniline**
85%

NITROSAMINES AND CANCER

A 1962 outbreak of food poisoning in sheep in Norway was traced to their ingestion of nitrite-treated fish meal. This incident immediately raised concerns about human consumption of nitrite-treated foods. Sodium nitrite is used as a food preservative. It can react with naturally occurring secondary amines present in food to produce nitrosamines, which are known to be carcinogenic. Smoked fish, cured meats, and beer all contain nitrosamines. Nitrosamines have also been found in cheese, which is not surprising because some cheeses are preserved with nitrite, and cheese is rich in secondary amines. In the United States, consumer groups asked the Food and Drug Administration to ban the use of sodium nitrite as a preservative. This request was vigorously opposed by the meat-packing industry. Despite extensive investigations, it has not been determined whether or not the small amounts of nitrosamines present in our food pose a hazard to our health. Until this question can be answered, it will be hard to avoid sodium nitrite in our diet. It is worrisome to note, however, that Japan has one of the highest gastric cancer rates, and has the highest average ingestion of sodium nitrite. Some good news is that the concentration of nitrosamines present in bacon has been considerably reduced in recent years by adding ascorbic acid (a nitrosamine inhibitor) to the curing mixture. Additionally, improvements in the malting process have reduced the level of nitrosamines found in commercial beer. Dietary sodium nitrite does have a redeeming feature—there is some evidence that it protects against botulism (a type of severe food poisoning).

15.13 NUCLEOPHILIC AROMATIC SUBSTITUTION REACTIONS

We have seen (Section 9.9) that benzene does not react with nucleophiles under standard reaction conditions because the π electron clouds repel the approach of a nucleophile, and the leaving group would be a very basic hydride ion.

However, if the benzene ring has one or more substituents that strongly withdraw electrons from the ring by resonance and a good leaving group (such as a halogen), **nucleophilic aromatic substitution** reactions can occur without using extreme conditions. The electron-withdrawing groups must be positioned ortho or para to the leaving group. The greater the number of electron-withdrawing substituents, the easier it is to carry out the nucleophilic aromatic substitution reaction. Notice the different conditions under which the following reactions occur.

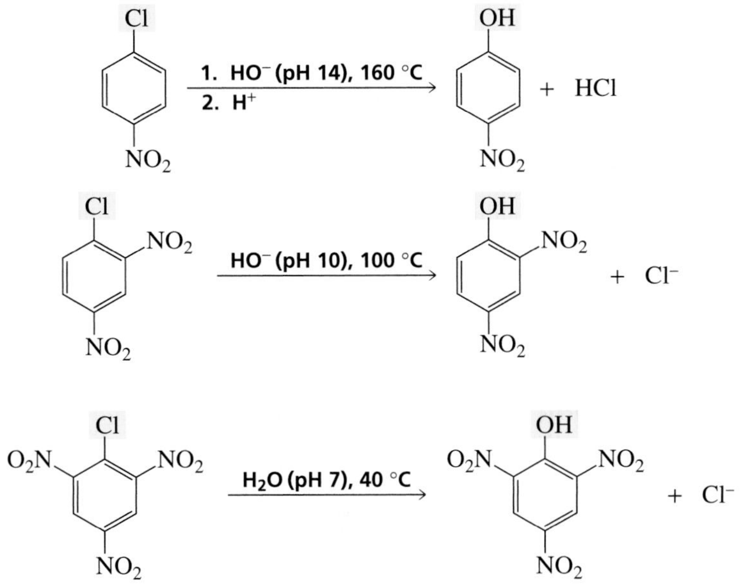

3-D Molecule:
1-Chloro-2,4,6-trinitro-benzene

Notice also that the strong electron-withdrawing substituents that activate the benzene ring toward nucleophilic aromatic substitution reactions are the same substituents that deactivate the ring toward electrophilic aromatic substitution. In other words, making the ring less electron rich makes it easier for a nucleophile to approach, but makes it more difficult for an electrophile to approach. Thus, any substituent that deactivates the ring toward electrophilic substitution activates the ring toward nucleophilic substitution, and vice versa.

Nucleophilic aromatic substitution takes place by a two-step reaction known as an **S$_N$Ar reaction** (substitution nucleophilic aromatic). In the first step, the nucleophile attacks the carbon bearing the leaving group from a trajectory nearly perpendicular to the aromatic ring. (Recall from Section 9.8 that leaving groups cannot be displaced from sp^2 carbon atoms by backside attack.) Nucleophilic attack forms a resonance-stabilized carbanion intermediate. In the second step of the reaction, the leaving group departs, reestablishing the aromaticity of the ring.

> **Electron-withdrawing substituents increase the reactivity of the benzene ring toward nucleophilic substitution and decrease the reactivity of the benzene ring toward electrophilic substitution.**

general mechanism for nucleophilic aromatic substitution

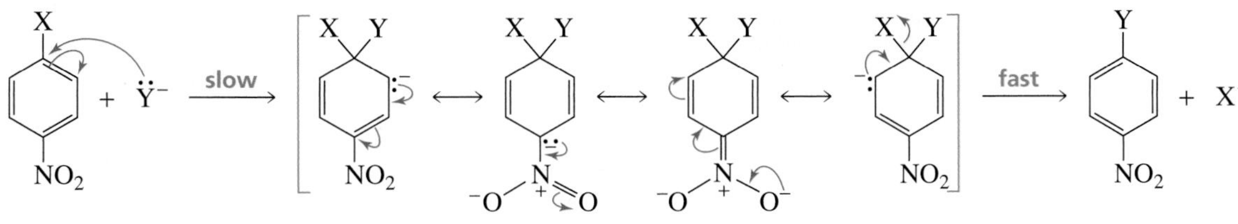

In a nucleophilic aromatic substitution reaction, the substituent that is replaced must be a weaker base than the incoming nucleophile so that it, rather than the incoming group, is eliminated from the intermediate.

The electron-withdrawing substituent must be ortho or para to the site of nucleophilic attack because the electrons of the attacking nucleophile can be delocalized onto the substituent only if the substituent is in one of those positions.

A variety of substituents can be placed on a benzene ring by means of nucleophilic aromatic substitution reactions. The only requirement is that the incoming group be a stronger base than the group that is being replaced.

3-D Molecule:
para-Fluoronitrobenzene

p-fluoronitrobenzene + CH₃O⁻ → (Δ) p-nitroanisole + F⁻

1-bromo-2,4-dinitrobenzene + CH₃CH₂NH₂ → (Δ) ... → (HO⁻) *N*-ethyl-2,4-dinitro-aniline + H₂O

PROBLEM 28

Draw resonance contributors for the carbanion that would be formed if *meta*-chloronitrobenzene could react with hydroxide ion. Why doesn't it react?

PROBLEM 29◆

a. List the following compounds in order of decreasing reactivity toward nucleophilic aromatic substitution:

chlorobenzene   1-chloro-2,4-dinitrobenzene   *p*-chloronitrobenzene

b. List the same compounds in order of decreasing reactivity toward electrophilic aromatic substitution.

PROBLEM 30

Show how each of the following compounds could be synthesized from benzene.

a. *o*-nitroaniline   **b.** *p*-nitro-*N*-methylaniline   **c.** *p*-bromoanisole

15.14
BENZYNE

An aryl halide such as chlorobenzene can undergo a nucleophilic substitution reaction in the presence of a very strong base such as ⁻NH₂. There are two surprising features about this reaction: The aryl halide does not have to contain an electron-withdrawing group, and the incoming substituent does not always end up on the carbon vacated by the leaving group. For example, when chlorobenzene—with the carbon to which the chlorine is attached isotopically labeled with ^{14}C—was treated with amide ion in liquid ammonia, aniline was obtained as the product. Half of the product had the amino group attached to the isotopically labeled carbon (denoted by the asterisk) as expected, but the other half had the amino group attached to the carbon adjacent to the labeled carbon.

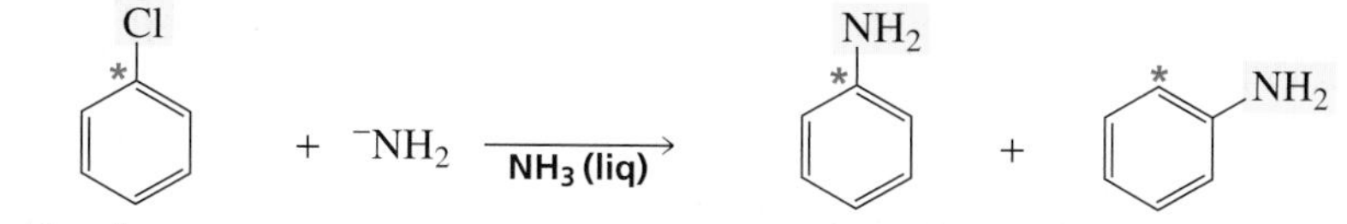

chlorobenzene

approximately equal amounts of the two products are obtained

These are the only products formed. Anilines with the amino group two or three carbons removed from the labeled carbon were not formed.

The fact that the two products are formed in approximately equal amounts indicates the reaction takes place by a mechanism that forms an intermediate in which the two carbons to which the amino group is attached in the product are equivalent. The mechanism that accounts for the experimental observations involves formation of a **benzyne intermediate.** Benzyne has an extra π bond between two adjacent carbon atoms of benzene. In the first step of the mechanism, the strong base ($^-$NH$_2$) removes a proton from the position ortho to the halogen. The resulting anion expels the halide ion, thereby forming benzyne.

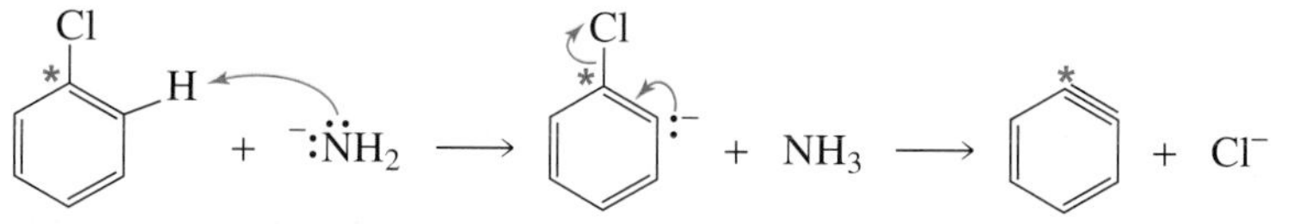

benzyne

The incoming nucleophile can attack either of the carbons of the "triple bond" of benzyne. Protonation of the resulting anion forms the substitution product. The overall reaction is an elimination-addition reaction: benzyne is formed in an elimination reaction and immediately undergoes an addition reaction.

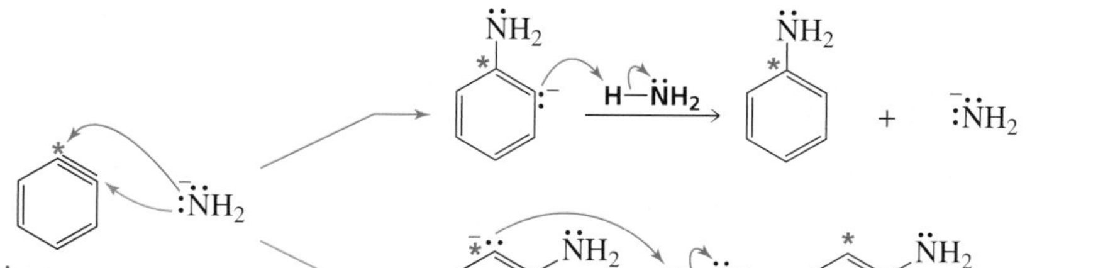

benzyne

Substitution at the carbon that was attached to the leaving group is called **direct substitution.** Substitution at the adjacent carbon is called **cine substitution** (cine comes from *kinesis,* which is Greek for "movement"). In the following reaction, *o*-toluidine is the direct substitution product; *m*-toluidine is the cine substitution product.

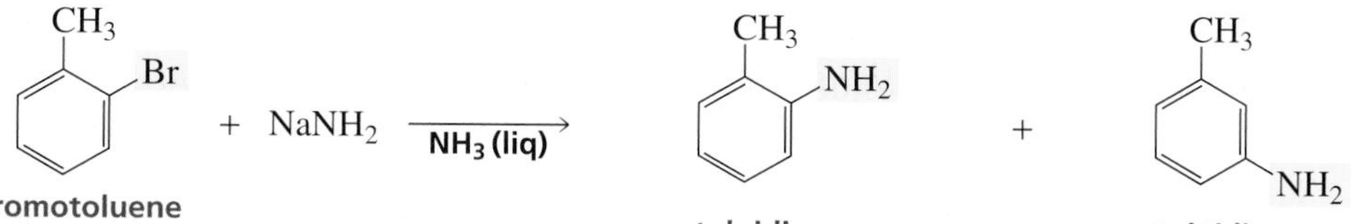

o-bromotoluene

o-toluidine
direct substitution product

m-toluidine
cine substitution product

Martin D. Kamen first isolated ^{14}*C. It immediately became the most useful of all the isotopes in chemical and biochemical research. Kamen was born in Toronto in 1913. He received a B.S. and a Ph.D. from the University of Chicago, becoming a U.S. citizen in 1938. He was a professor at the University of California, Berkeley, at Washington University, and at Brandeis University. He was one of the founding professors at the University of California, San Diego. In later years he became a member of the faculty at the University of Southern California. He received the Fermi medal in 1996.*

*The labeling experiment was done by **John D. Roberts** who was born in Los Angeles in 1918. He received both his B.A. and Ph.D. degrees from UCLA. He arrived at the California Institute of Technology in 1952 as a Guggenheim Fellow and has been a professor there since 1953.*

3-D Molecule:
Benzyne

Benzyne is an extremely reactive species. In Section 5.3 we saw that in a molecule with a triple bond, the two *sp* hybridized carbons and the atoms attached to these carbons (C—C≡C—C) are linear since the bond angles are 180°. Four linear atoms cannot be incorporated into a six-membered ring, so the C—C≡C—C system in benzyne is distorted—the original π bond is unchanged, but the orbitals that form the new π bond are not parallel to each other (Figure 15.4). Because of the distortion they cannot overlap as well as the orbitals that form a normal π bond, resulting in a weaker and more reactive π bond. Therefore, benzyne can be thought of as a highly distorted, very reactive alkyne.

Figure 15.4 ▶
Orbital pictures of the second π bond in (a) a normal triple bond and (b) the distorted "triple bond" in benzyne.

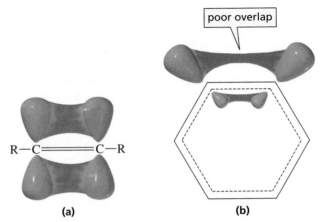

Although benzyne is too unstable to be isolated, evidence that it is formed can be obtained by a trapping experiment. When furan is added to a reaction that forms a benzyne intermediate, furan traps the benzyne intermediate by reacting with it in a Diels–Alder reaction (Section 7.9). The product of the Diels–Alder reaction can be isolated.

benzyne
a dienophile

furan
a diene

the product of the
Diels-Alder reaction

PROBLEM 31

Give the products that would be obtained from the reaction of the following compounds with sodium amide in liquid ammonia:

a. (CH₃, Cl on benzene ring)

b. (CH₂CH₃, Br on benzene ring)

c. (CH₃, Br, CH₃ on benzene ring)

**15.15
POLYCYCLIC
BENZENOID
HYDROCARBONS**

Polycyclic benzenoid hydrocarbons are compounds that contain two or more fused benzene rings. **Fused rings** share two adjacent carbons—naphthalene has two fused rings, anthracene and phenanthrene have three fused rings, and tetracene, triphenylene, pyrene, and chrysene have four fused rings. There are many polycyclic benzenoid hydrocarbons with more than four fused rings.

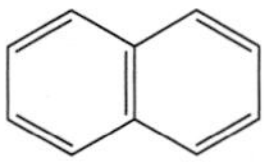

naphthalene 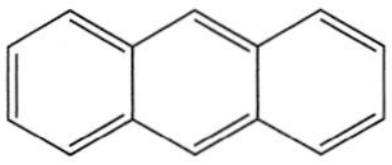   anthracene 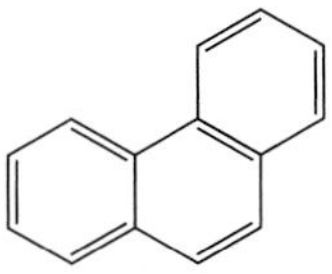   phenanthrene 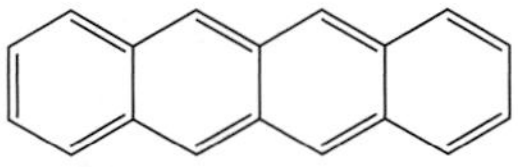   tetracene

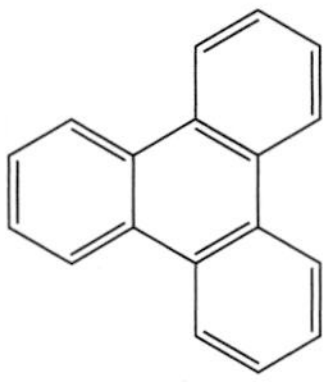

triphenylene 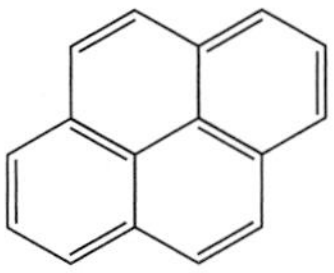   pyrene 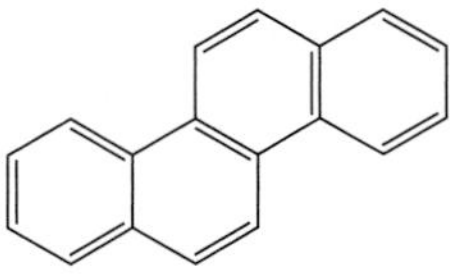   chrysene

   Like benzene, all the larger polycyclic benzenoid hydrocarbons undergo electrophilic substitution reactions. Some of these compounds are well-known carcinogens. The chemical reactions responsible for causing cancer and how you can predict which compounds are carcinogenic were discussed in Section 11.7.

Naphthalene, like benzene, is an aromatic hydrocarbon. The resonance hybrid of naphthalene has three resonance contributors.

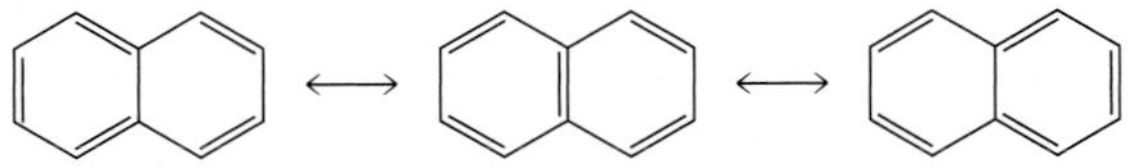

resonance contributors of naphthalene

Like benzene, naphthalene undergoes electrophilic aromatic substitution reactions. Substitution occurs preferentially at the 1-position. In common nomenclature, the 1-position is called the α-position, and the 2-position is called the β-position.

$$+ \ HNO_3 \ \xrightarrow{H_2SO_4} \ \text{...} \ + \ H_2O$$

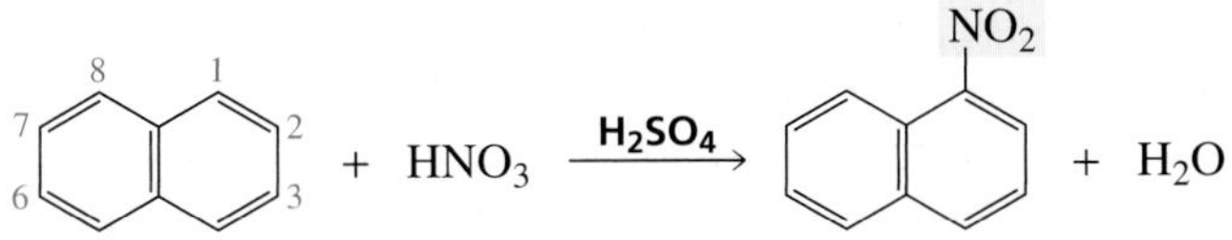

1-nitronaphthalene
α-nitronaphthalene

15.16 ELECTROPHILIC SUBSTITUTION REACTIONS OF NAPHTHALENE AND SUBSTITUTED NAPHTHALENES

   Naphthalene is more reactive than benzene toward electrophilic substitution because the carbocation intermediate is more stable and, therefore, easier to form than the analogous carbocation intermediate formed from benzene. Because of naphthalene's increased reactivity, a Lewis acid is not needed for bromination or chlorination.

3-D Molecule:
Naphthalene

$$+ \ Br_2 \ \longrightarrow \ \text{...} \ + \ HBr$$

1-bromonaphthalene

PROBLEM 32

Use the resonance contributors of naphthalene to predict whether all the carbon–carbon bonds have the same length, as they do in benzene.

PROBLEM 33

Draw the resonance contributors for the carbocation intermediates obtained from electrophilic aromatic substitution at the 1-position and at the 2-position of naphthalene. Use the resonance contribtuors to explain why substitution at the 1-position is preferred.

Sulfonation of naphthalene does not always lead to substitution at the 1-position. If the reaction is carried out under conditions where the reaction is not reversible (80 °C), substitution occurs at the 1-position.

$$\text{naphthalene} + H_2SO_4 \xrightarrow{80\ °C} \text{naphthalene-1-sulfonic acid} + H_2O$$

naphthalene-1-sulfonic acid

But if the reaction is carried out under conditions where sulfonation is readily reversible (160 °C), substitution occurs predominantly at the 2-position.

$$\text{naphthalene-1-sulfonic acid} \rightleftharpoons \text{naphthalene} + H_2SO_4 \xrightarrow{160\ °C} \text{naphthalene-2-sulfonic acid}$$

naphthalene-1-sulfonic acid
kinetic product

naphthalene-2-sulfonic acid
thermodynamic product

3-D Molecules:
Naphthalene-1-sulfonic acid;
Naphthalene-2-sulfonic acid

This is another example of a reaction whose product composition depends on whether the conditions used in the experiment cause the reaction to be irreversible—under kinetic control—or reversible—under thermodynamic control (Section 7.8). The 1-substituted product is the kinetic product because it is easier to form. It is the predominant product, therefore, when the reaction is carried out under conditions where the reaction is irreversible (mild reaction conditions). The 2-substituted product is the thermodynamic product because it is more stable. It is the predominant product under conditions where the reaction is reversible (higher temperatures).

The 1-substituted product is easier to form because the carbocation leading to its formation is more stable. The 2-substituted product is more stable because there is more room for the sulfonic acid group at the 2-position. In the 1-substituted product, the bulky sulfonic acid group is too close to the hydrogen at the 8-position.

an unfavorable
steric interaction

In the case of substituted naphthalenes, the nature of the substituent determines which ring will undergo electrophilic substitution. If the substituent is deactivating, the electrophile will attack the 1-position of the ring without the substituent because that ring is more reactive than the ring containing the deactivating substituent.

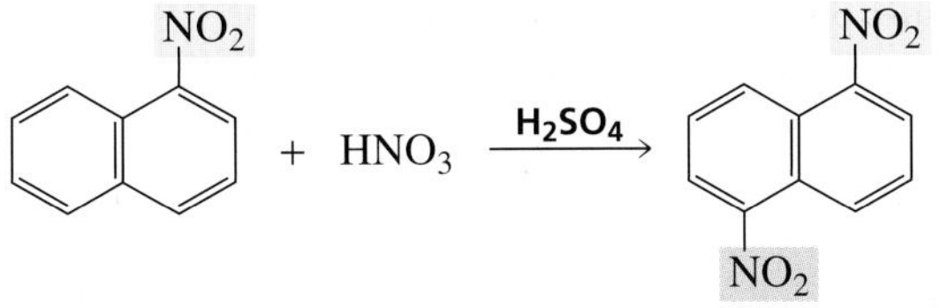

The reaction of 1-nitronaphthalene with HNO₃ and H₂SO₄ gives 1,5-dinitronaphthalene.

If the substituent is activating, the electrophile will attack the ring bearing the substituent. The incoming electrophile will be directed to a 1-position that is ortho or para to the substituent because all activating substituents are ortho/para directors (Section 15.4).

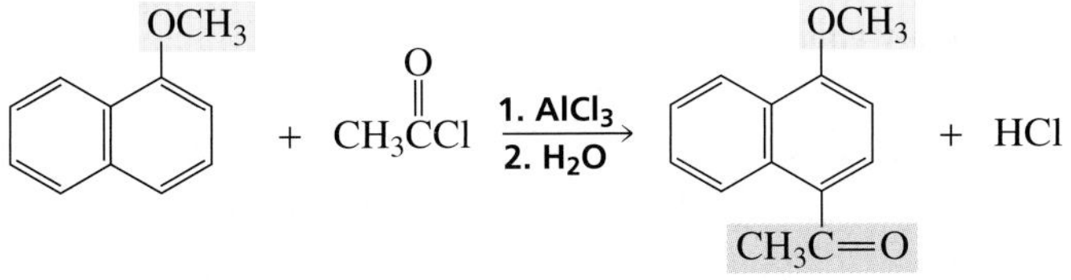

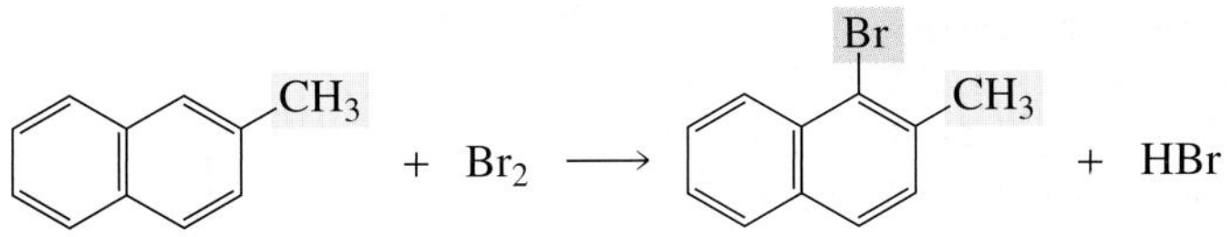

Reaction of 1-methoxynaphthalene with CH₃CCl, 1. AlCl₃, 2. H₂O gives the 4-acetyl product + HCl.

Reaction of 2-methylnaphthalene with Br₂ gives 1-bromo-2-methylnaphthalene + HBr.

PROBLEM 34

Give the products that would be obtained from the reaction of the following compounds with Cl₂:

a. 2-methoxynaphthalene (with OCH₃)

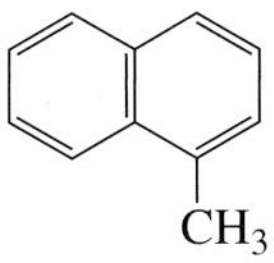

b. 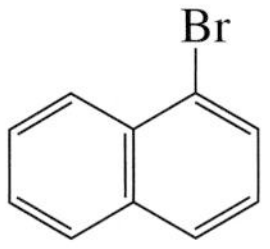 1-bromonaphthalene (with Br)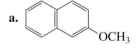

c. 1-methylnaphthalene (with CH₃)

d. 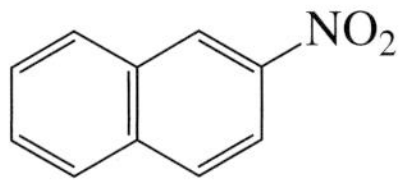 2-nitronaphthalene (with NO₂)

Tutorial:
Terms for the reactions of substituted benzenes

SUMMARY OF REACTIONS

1. Reactions of substituents on a benzene ring (Section 15.2)

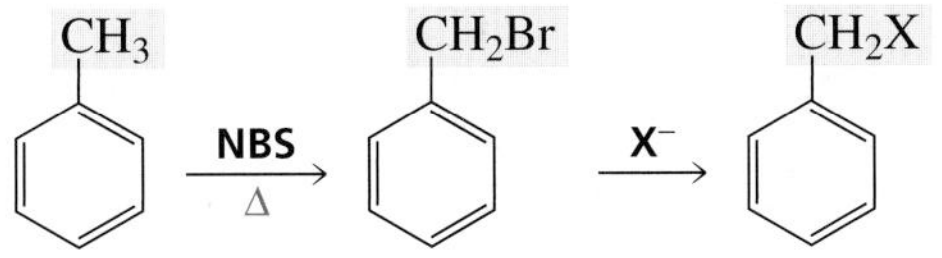

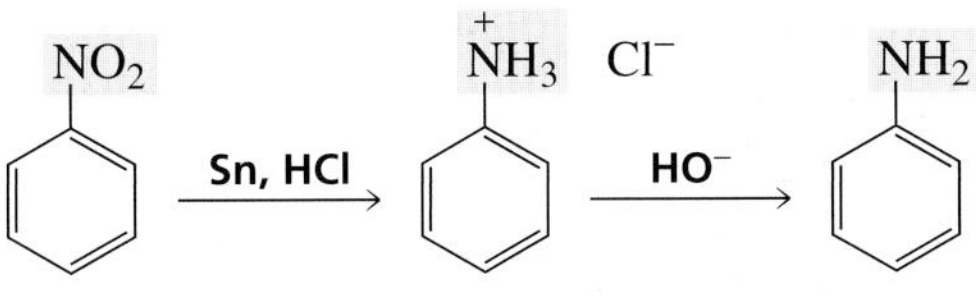

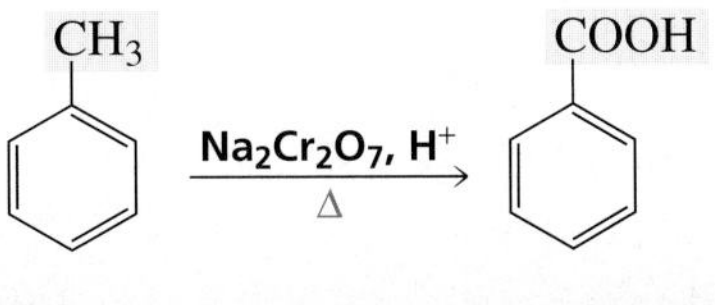

Toluene (CH₃) → NBS, Δ → benzyl bromide (CH₂Br) → X⁻ → benzyl halide (CH₂X)

Nitrobenzene (NO₂) → Sn, HCl → anilinium (⁺NH₃ Cl⁻) → HO⁻ → aniline (NH₂)

Toluene (CH₃) → Na₂Cr₂O₇, H⁺, Δ → benzoic acid (COOH)

2. Reaction of amines with nitrous acid (Section 15.12)

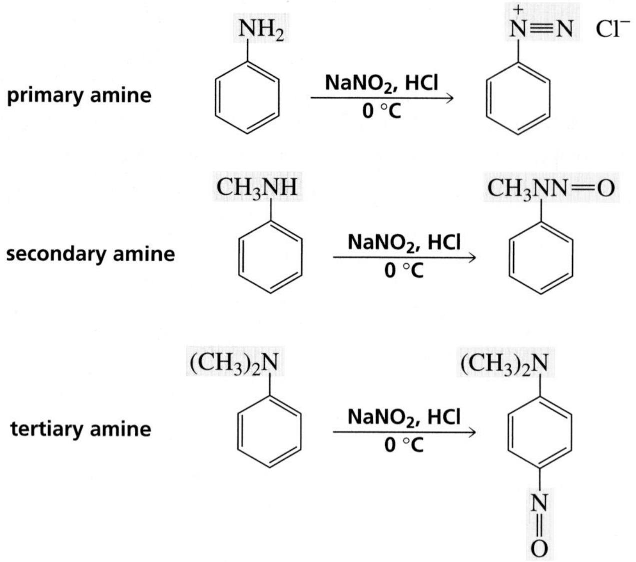

primary amine

$$\text{PhNH}_2 \xrightarrow[\text{0 °C}]{\textbf{NaNO}_2\textbf{, HCl}} \text{Ph}\overset{+}{N}\!\!\equiv\!\!N \ \ \text{Cl}^-$$

secondary amine

$$\text{PhNHCH}_3 \xrightarrow[\text{0 °C}]{\textbf{NaNO}_2\textbf{, HCl}} \text{PhN(CH}_3)\!\!=\!\!\text{O}$$

tertiary amine

$$\text{PhN(CH}_3)_2 \xrightarrow[\text{0 °C}]{\textbf{NaNO}_2\textbf{, HCl}} \text{(4-O=N)C}_6\text{H}_4\text{N(CH}_3)_2$$

3. Replacement of a diazonium group (Section 15.10)

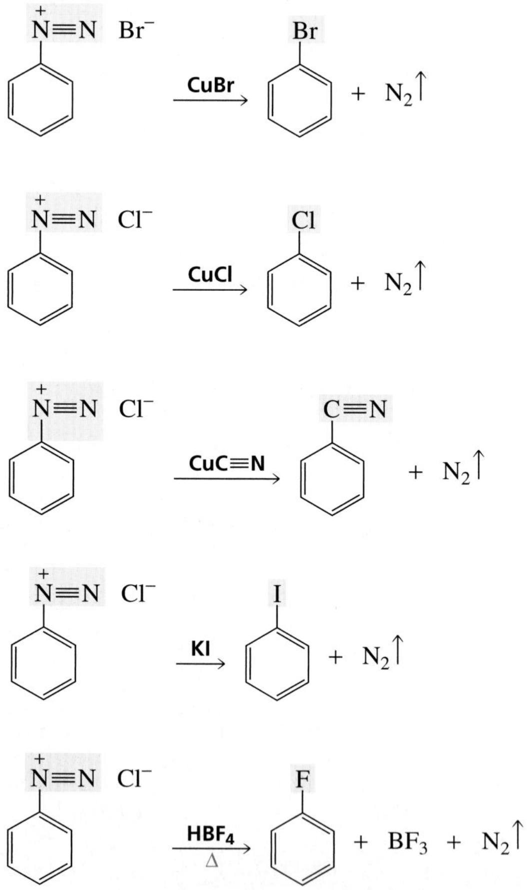

$$\text{Ph}\overset{+}{N}\!\!\equiv\!\!N \ \ \text{Br}^- \xrightarrow{\textbf{CuBr}} \text{PhBr} + \text{N}_2\uparrow$$

$$\text{Ph}\overset{+}{N}\!\!\equiv\!\!N \ \ \text{Cl}^- \xrightarrow{\textbf{CuCl}} \text{PhCl} + \text{N}_2\uparrow$$

$$\text{Ph}\overset{+}{N}\!\!\equiv\!\!N \ \ \text{Cl}^- \xrightarrow{\textbf{CuC}\equiv\textbf{N}} \text{PhC}\equiv\text{N} + \text{N}_2\uparrow$$

$$\text{Ph}\overset{+}{N}\!\!\equiv\!\!N \ \ \text{Cl}^- \xrightarrow{\textbf{KI}} \text{PhI} + \text{N}_2\uparrow$$

$$\text{Ph}\overset{+}{N}\!\!\equiv\!\!N \ \ \text{Cl}^- \xrightarrow[\Delta]{\textbf{HBF}_4} \text{PhF} + \text{BF}_3 + \text{N}_2\uparrow$$

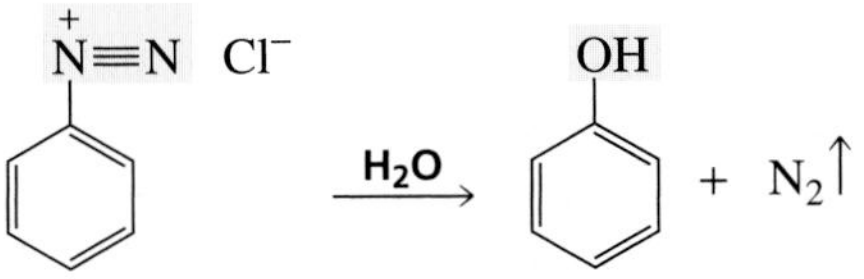

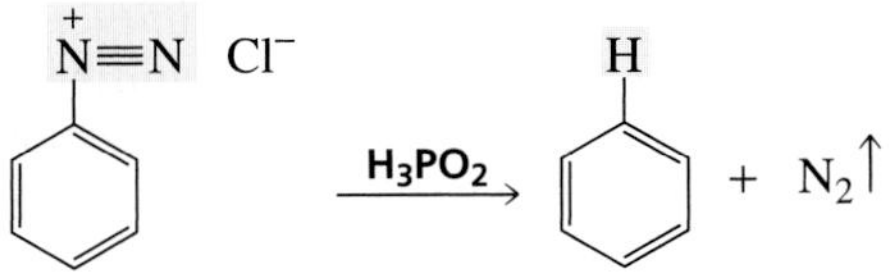

4. Formation of an azo compound (Section 15.11)

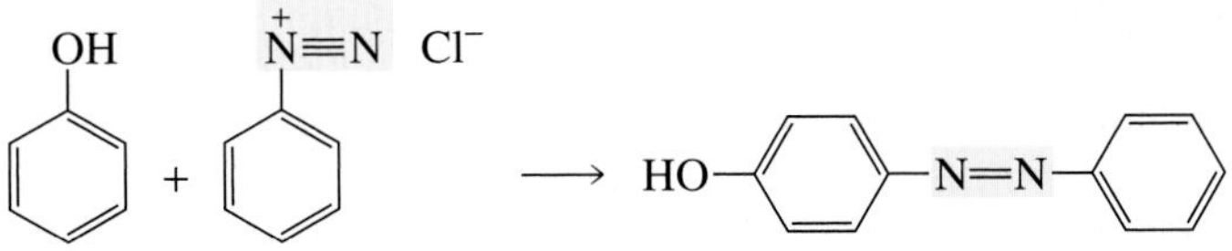

5. Nucleophilic aromatic substitution reactions (Sections 15.13)

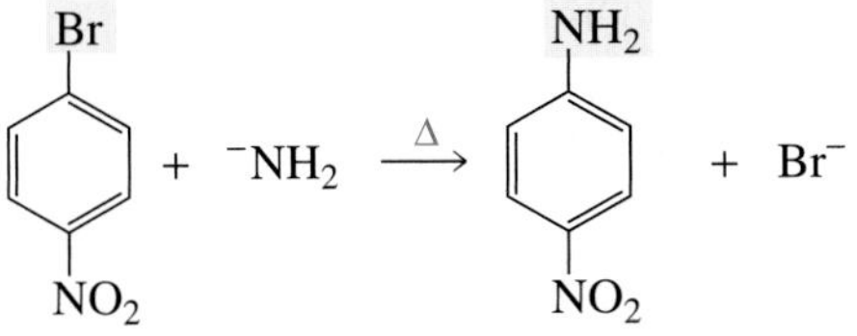

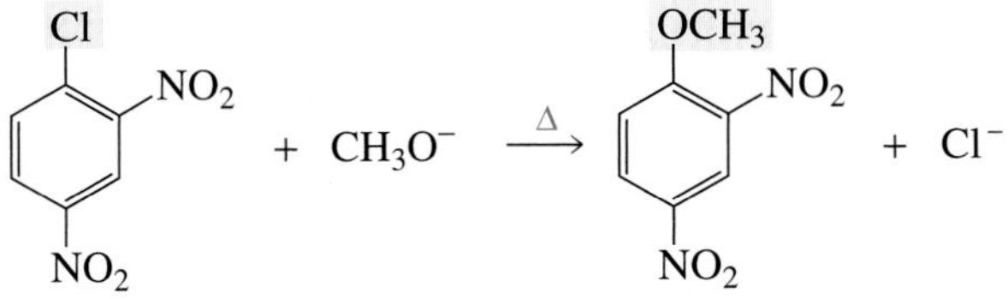

6. Nucleophilic substitution via a benzyne intermediate (Section 15.14)

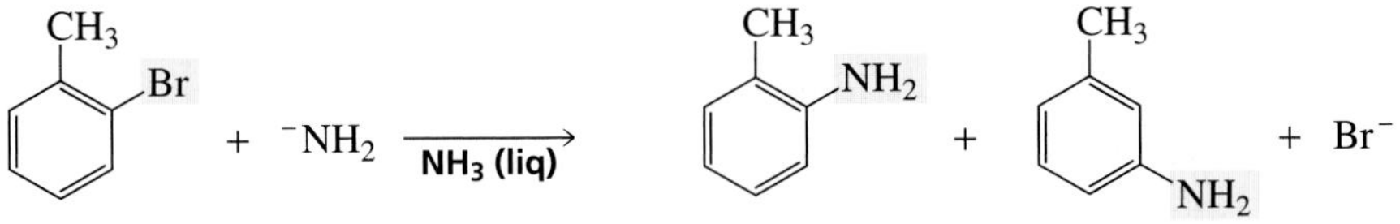

direct substitution product   cine substitution product

7. Electrophilic aromatic substitution reactions of naphthalene (Section 15.16)

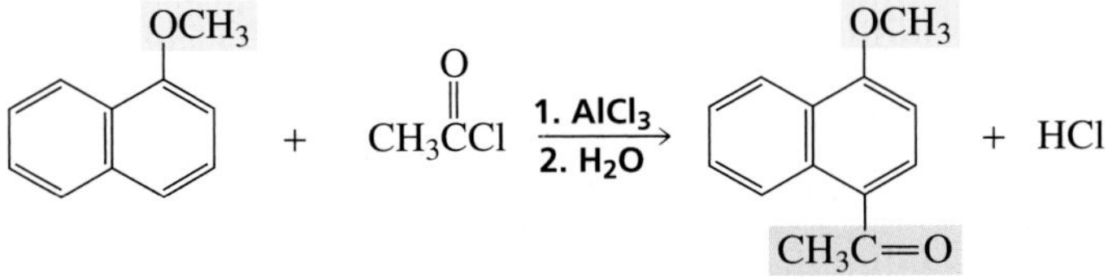

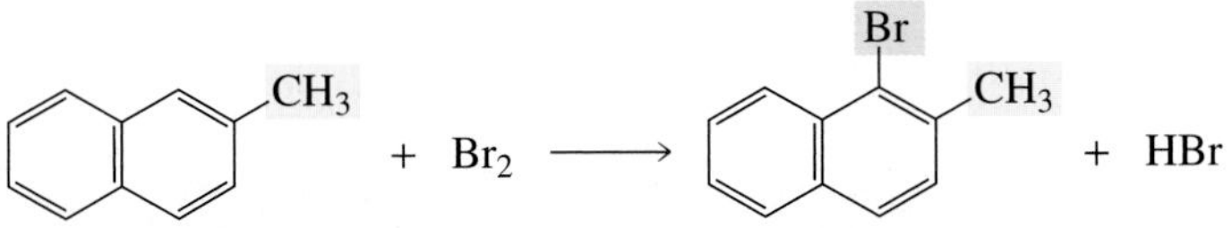

KEY TERMS

activating substituent (page 638)
arenediazonium salt (page 653)
azo linkage (page 657)
benzyl group (page 629)
benzyne intermediate (page 663)
cine substitution (page 663)
deactivating substituent (page 638)
direct substitution (page 663)
donate electrons by resonance (637)
fused rings (664)

inductive electron donation (page 637)
inductive electron withdrawal
   (page 637)
meta director (page 642)
nitrosamine (page 658)
N-nitroso compound (page 658)
nucleophilic aromatic substitution
   (page 660)
ortho/para director (page 642)
phenyl group (page 629)

resonance electron donation
   (page 637)
resonance electron withdrawal
   (page 637)
Sandmeyer reaction (page 654)
Schiemann reaction (page 655)
$S_N Ar$ reaction (page 661)
withdraw electrons by resonance (637)

PROBLEMS

35. Draw the structure of each of the following compounds:
   a. *m*-ethylphenol
   b. *p*-nitrobenzenesulfonic acid
   c. (*E*)-2-phenyl-2-pentene
   d. *o*-bromoaniline
   e. 2-chloroanthracene
   f. *m*-chlorostyrene
   g. *o*-nitroanisole
   h. 2,4-dichlorotoluene

36. Name the following compounds:

a. 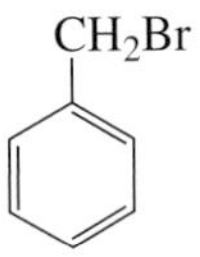 CH$_2$Br (benzene ring)

b.  CHBr$_2$ (benzene ring)

c. 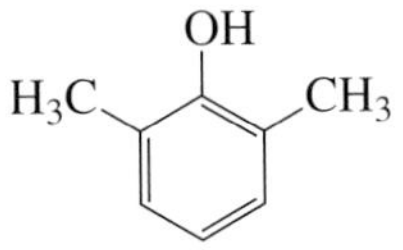 H$_3$C — (benzene ring with OH) — CH$_3$

d. COOH (benzene ring) Br

e. (benzene ring) CH$_3$ Br

f. OCH$_3$ (benzene ring) CH$_2$CH$_3$

g. SO$_3$H (benzene ring) Cl Cl

h. CH$_2$CH$_3$ (benzene ring) Cl — N=N — (benzene ring)

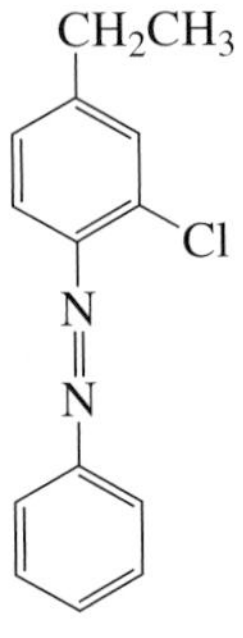

i. 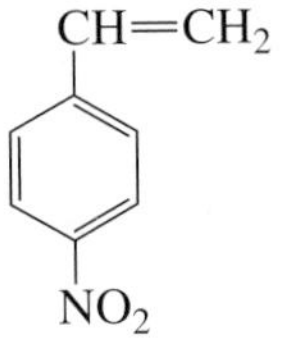 CH=CH$_2$ (benzene ring) NO$_2$

j. 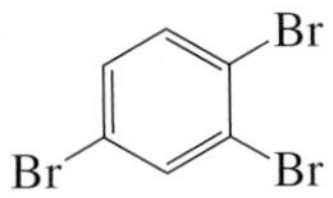 Br (benzene ring) Br Br

k. 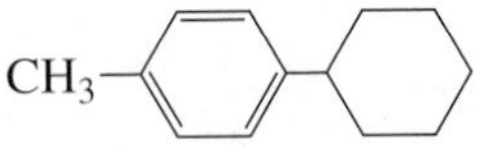 CH$_3$ — (benzene ring) — (cyclohexane ring)

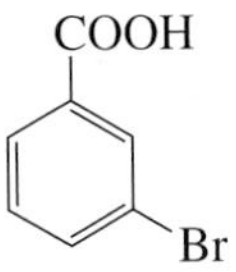

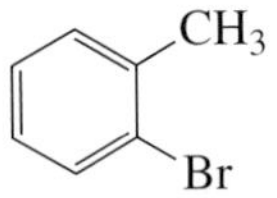

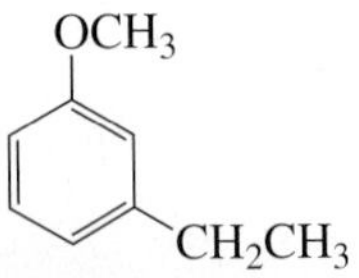

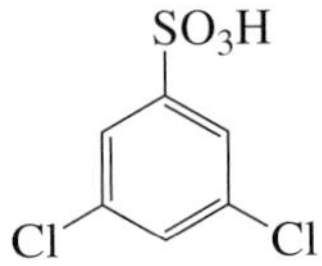

37. Provide the necessary reagents next to the arrows.

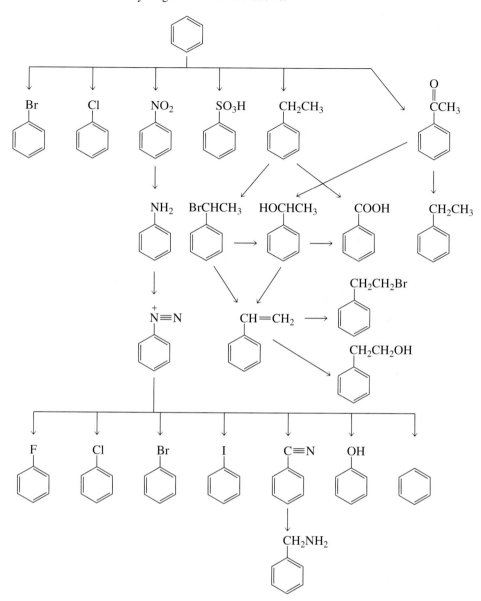

38. The pK_a values of a few ortho-, meta-, and para-substituted benzoic acids are shown here. The relative pK_a values depend on the substituent. For chloro-substituted benzoic acids, the ortho isomer is the most acidic and the para isomer is the least acidic; for nitro-substituted benzoic acids, the ortho isomer is the most acidic and the meta isomer is the least acidic; and for amino-substituted benzoic acids, the meta isomer is the most acidic and the ortho isomer is the least acidic. Explain these differences.

Cl: ortho > meta > para
NO$_2$: ortho > para > meta
NH$_2$: meta > para > ortho

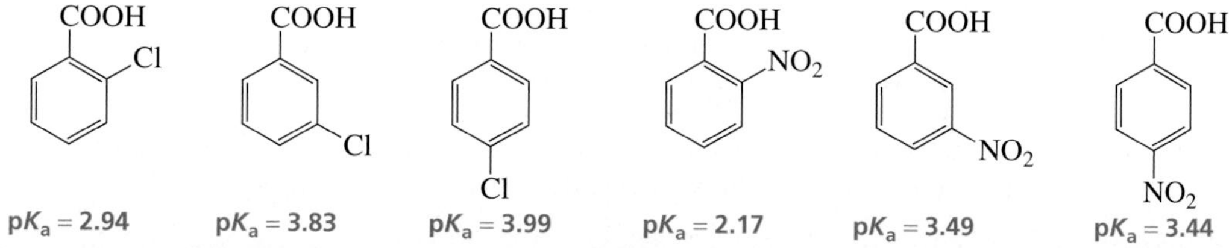

pK_a = 2.94   pK_a = 3.83   pK_a = 3.99   pK_a = 2.17   pK_a = 3.49   pK_a = 3.44

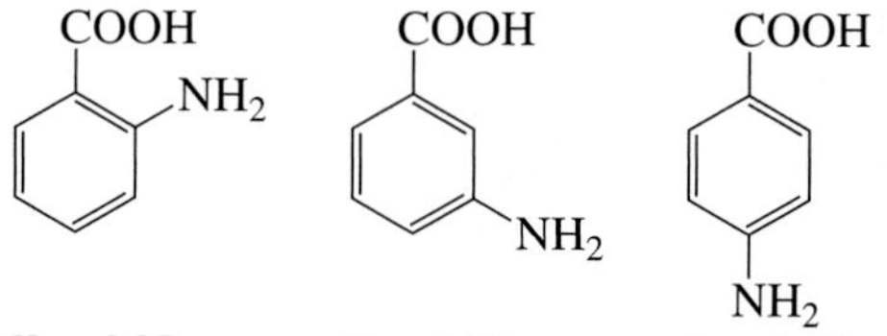

$pK_a = 4.95$ $pK_a = 4.73$ $pK_a = 4.89$

39. Give the product(s) of each of the following reactions:
  a. benzoic acid + HNO_3/H_2SO_4
  b. isopropylbenzene + cyclohexene + HF
  c. naphthalene + acetyl chloride + $AlCl_3$ followed by H_2O
  d. *o*-methylaniline + benzenediazonium chloride
  e. cyclohexyl phenyl ether + $Br_2/FeBr_3$
  f. phenol + H_2SO_4 + Δ
  g. ethylbenzene + $Br_2/FeBr_3$
  h. *m*-xylene + $Na_2Cr_2O_7$ + H^+ + Δ

40. Show how the following compounds could be synthesized from benzene:
  a. *m*-chlorobenzenesulfonic acid   **h.** 1-phenylpentane
  b. *m*-chloroethylbenzene   **i.** *m*-hydroxybenzoic acid
  c. benzyl alcohol   **j.** *m*-bromobenzoic acid
  d. *m*-bromobenzonitrile   **k.** *p*-cresol
  e. *N,N,N*-trimethylanilinium iodide   **l.** *p*-nitroaniline
  f. benzyl methyl ether   **m.** *m*-bromoiodobenzene
  g. *p*-benzylchlorobenzene   **n.** *p*-dideuteriobenzene

41. For each of the following groups of substituted benzenes, indicate:
  a. the one that would be the most reactive in an electrophilic substitution reaction.
  b. the one that would be the least reactive in an electrophilic substitution reaction.
  c. the one that would yield the highest percentage of meta product.

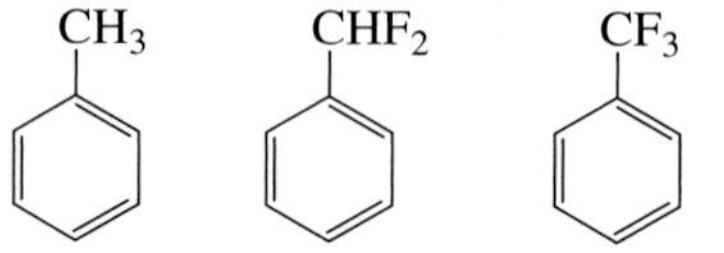

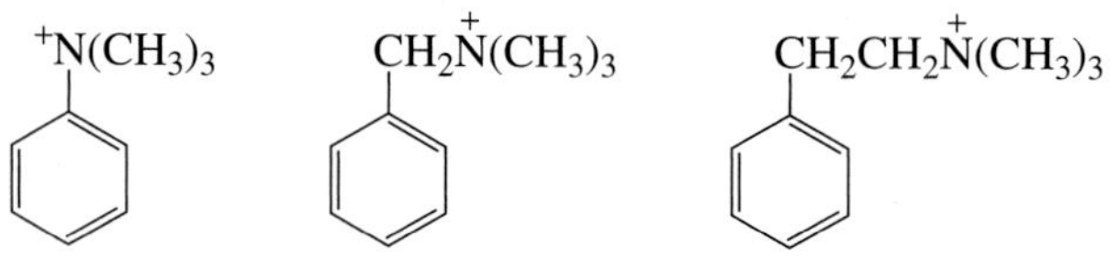

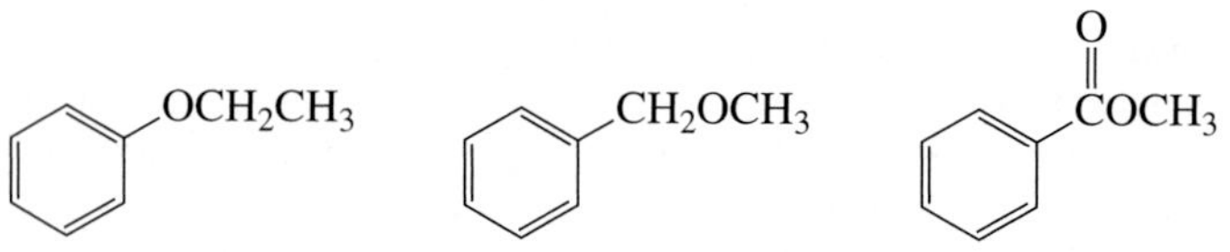

42. Arrange the following groups of compounds in order of decreasing reactivity toward electrophilic aromatic substitution:
  a. benzene, ethylbenzene, chlorobenzene, nitrobenzene, anisole
  b. 1-chloro-2,4-dinitrobenzene, 2,4-dinitrophenol, 2,4-dinitrotoluene
  c. toluene, *p*-cresol, benzene, *p*-xylene
  d. benzene, benzoic acid, phenol, propylbenzene
  e. *p*-nitrotoluene, 2-chloro-4-nitrotoluene, 2,4-dinitrotoluene, *p*-chlorotoluene
  f. bromobenzene, chlorobenzene, fluorobenzene, iodobenzene

43. Give the products of the following reactions:

a. 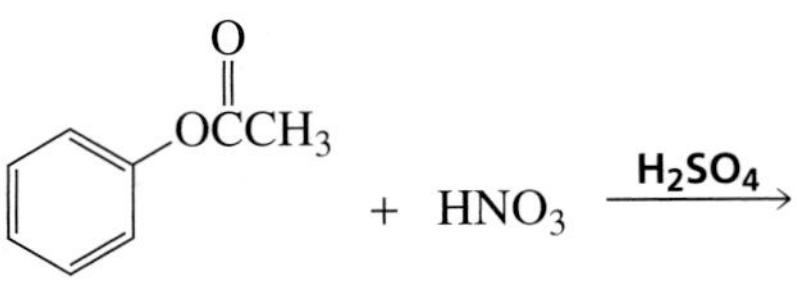 + HNO$_3$ $\xrightarrow{\text{H}_2\text{SO}_4}$

c. 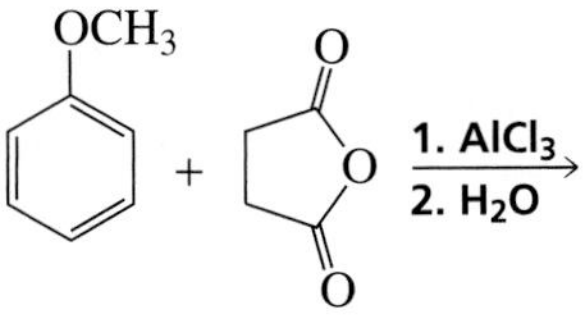 + $\xrightarrow[\text{2. H}_2\text{O}]{\text{1. AlCl}_3}$

e. 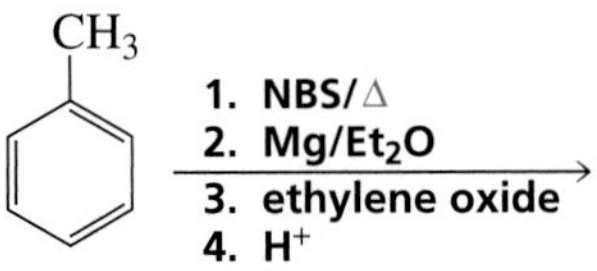 $\xrightarrow[\substack{\text{3. ethylene oxide}\\\text{4. H}^+}]{\substack{\text{1. NBS/}\triangle\\\text{2. Mg/Et}_2\text{O}}}$

b. 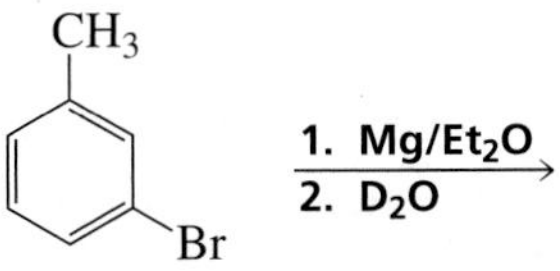 $\xrightarrow[\text{2. D}_2\text{O}]{\text{1. Mg/Et}_2\text{O}}$

d. + Br$_2$ $\longrightarrow$

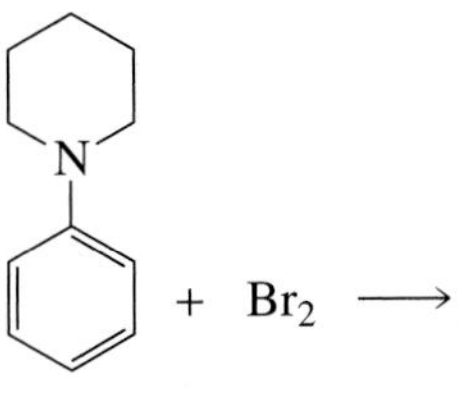

f. 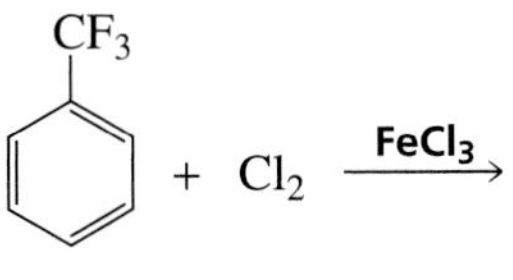 + Cl$_2$ $\xrightarrow{\text{FeCl}_3}$

44. For each of the statements in Column I, choose a substituent from Column II that fits the description for the compound at the right:

Column I

a. X donates electrons inductively but does not donate or withdraw electrons by resonance.

b. X withdraws electrons inductively and withdraws electrons by resonance.

c. X deactivates the ring and directs ortho/para.

d. X withdraws electrons inductively, donates electrons by resonance, and activates the ring.

e. X withdraws electrons inductively but does not donate or withdraw electrons by resonance.

Column II

OH

Br

$^+$NH$_3$

CH$_2$CH$_3$

NO$_2$

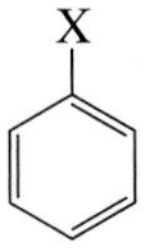

45. For each of the following compounds, indicate the ring carbon that would be nitrated if the compound were treated with HNO$_3$/H$_2$SO$_4$:

a.

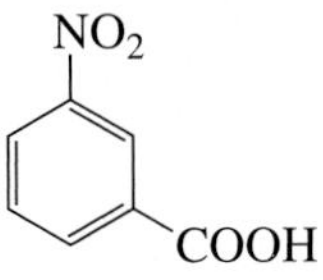

d.

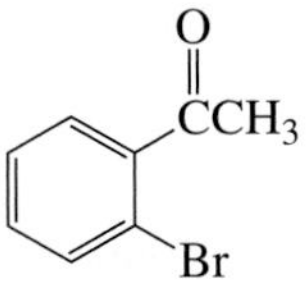

g.

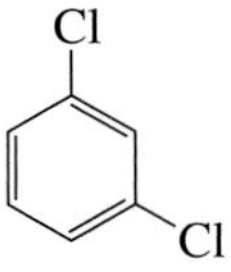

j.

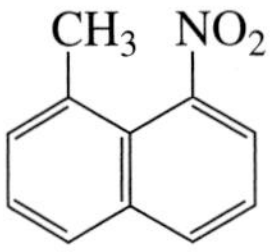

b.

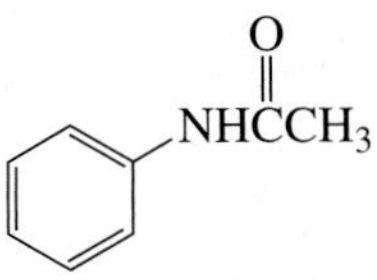

e.

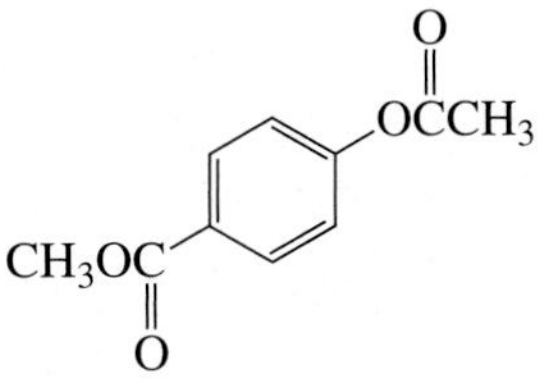

h.

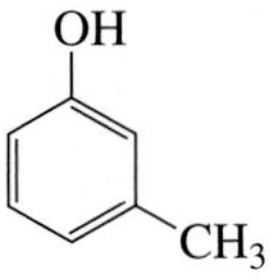

k.

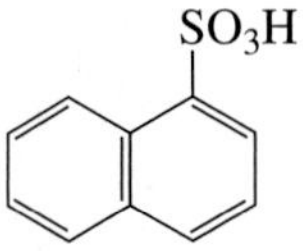

c.

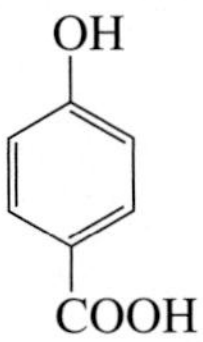

f.

i.

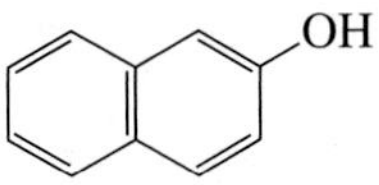

l.

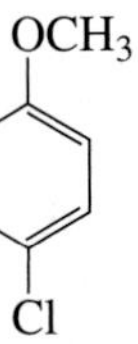

46. Give the product(s) obtained from the reaction of each of the following compounds with $Br_2/FeBr_3$:

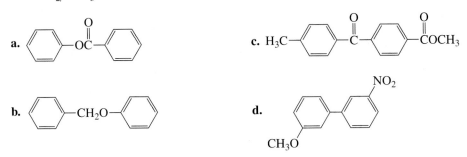

a.

b.

c. H_3C—

d.

47. Which would react more rapidly with Cl_2 + $FeCl_3$, *m*-xylene or *p*-xylene? Explain.

48. What products would be obtained from the reaction of the following compounds with $Na_2Cr_2O_7$ + H^+ + Δ?

a.

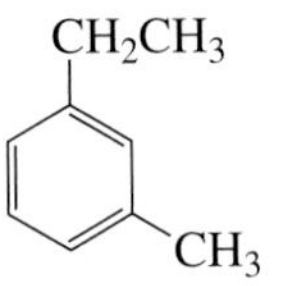

b.

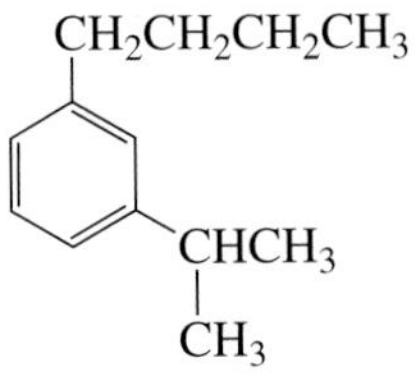

c.

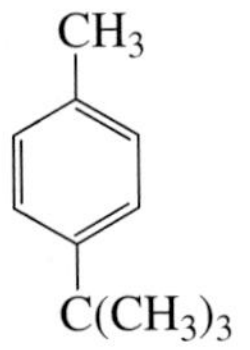

49. A student had prepared three ethyl-substituted benzaldehydes but had neglected to label them. The premed student at the next bench said they could be identified by brominating a sample of each and determining how many bromo-substituted products were formed. Is the premed student's advice sound?

50. Explain, using resonance contributors for the intermediate cation, why a phenyl group is an ortho/para director.

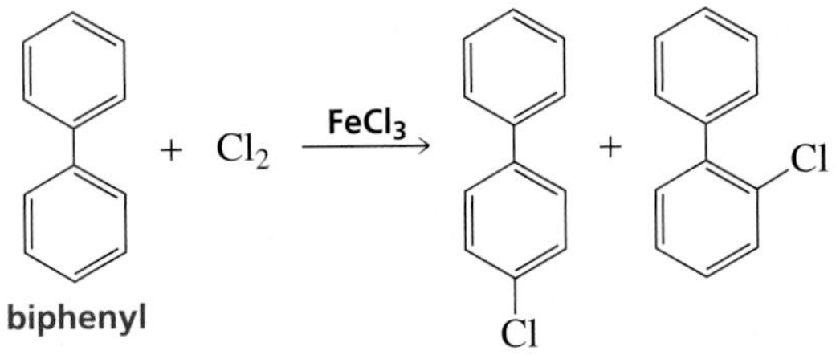

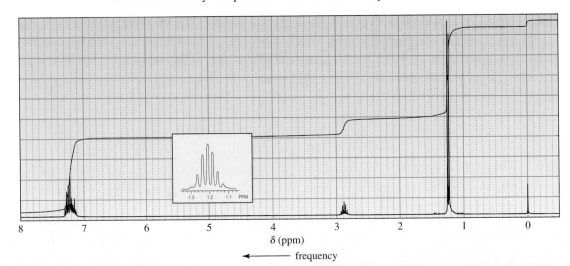

51. Compound A, when heated with an acidic solution of sodium dichromate, forms benzoic acid. Identify compound A from its ^1H NMR spectrum.

δ (ppm)

← frequency

52. Give two synthetic routes for the preparation of *p*-methoxyaniline using benzene as the starting material.

53. Using resonance contributors, answer the following:
    a. Which is a more stable carbocation intermediate?

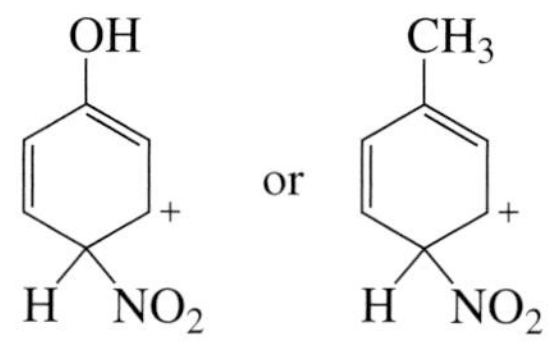

    b. Which is a more stable carbanion intermediate?

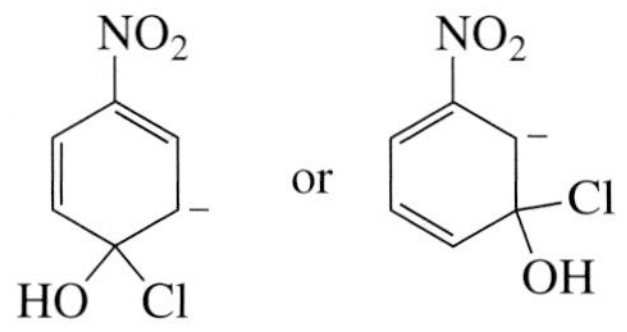

54. Show how the following compounds could be prepared from benzene:

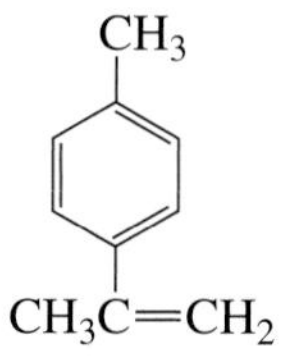

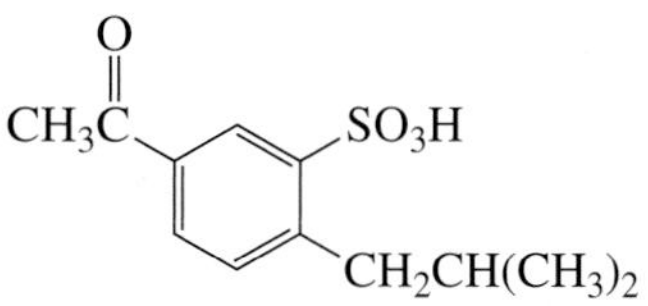

55. An OH substituent and an NH_2 substituent both strongly donate electrons into a benzene ring by resonance. However, while phenol is a good substrate for electrophilic substitution reactions, aniline is usually a poor substrate. Explain. How can the problem with aniline be circumvented?

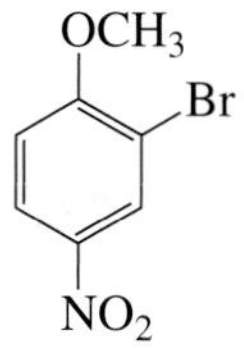

56. How could you distinguish among of the following compounds using:
    a. their infrared spectra?   **b.** their ^{1}H NMR spectra?

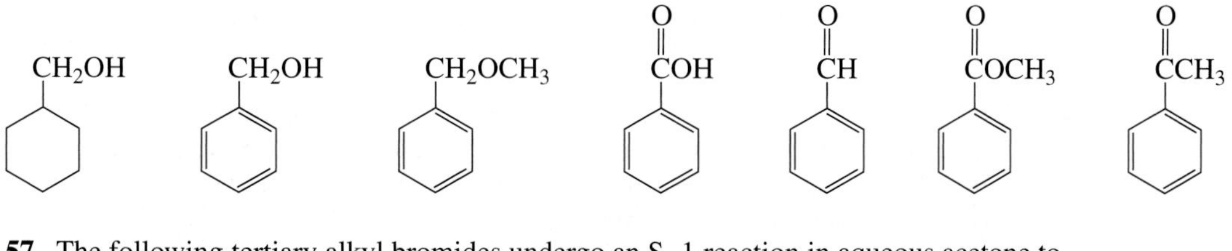

57. The following tertiary alkyl bromides undergo an S_N1 reaction in aqueous acetone to form the corresponding tertiary alcohols. List the alkyl bromides in order of decreasing reactivity.

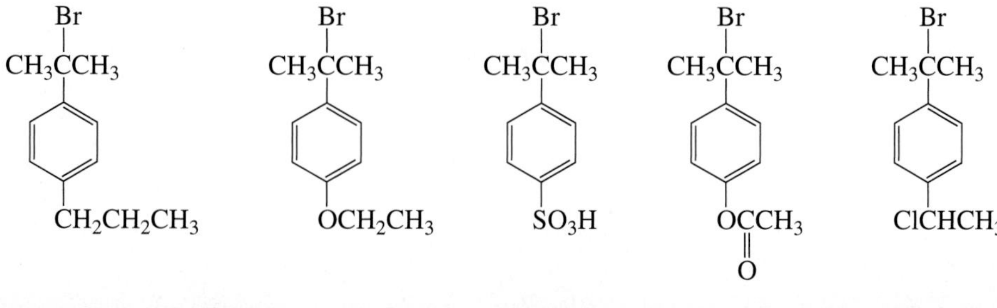

58. *p*-Fluoronitrobenzene is more reactive than *p*-chloronitrobenzene toward hydroxide ion. What does this tell you about the rate-determining step for nucleophilic aromatic substitution?

59. a. Explain why the following reaction leads to the products shown.

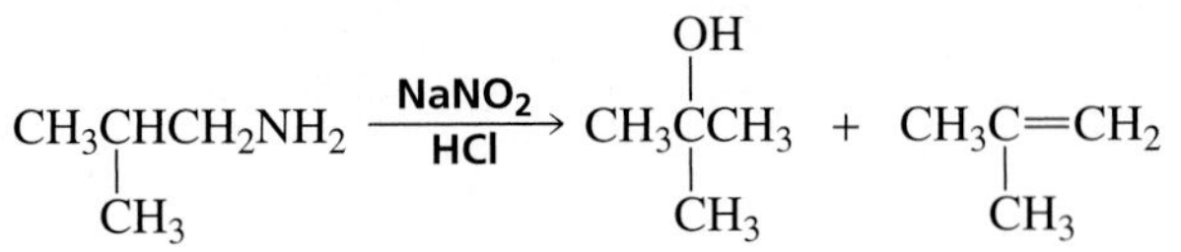

$$CH_3CHCH_2NH_2 \xrightarrow[HCl]{NaNO_2} CH_3\overset{OH}{\underset{CH_3}{C}}CH_3 \ + \ CH_3C{=}CH_2$$

with the reactant bearing CH₃ and the last product bearing CH₃.

b. What product would be obtained from the following reaction?

$$CH_3-\overset{OH}{\underset{CH_3}{C}}-\overset{NH_2}{\underset{CH_3}{C}}-CH_3 \xrightarrow[HCl]{NaNO_2}$$

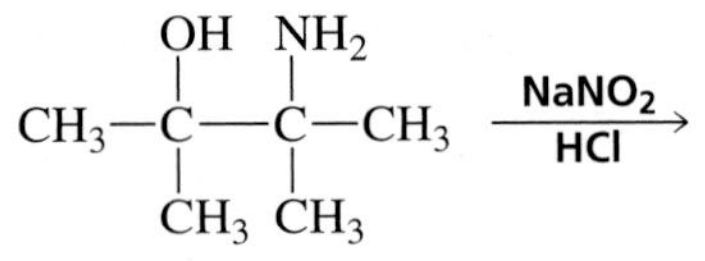

60. Describe how mescaline could be synthesized from benzene.

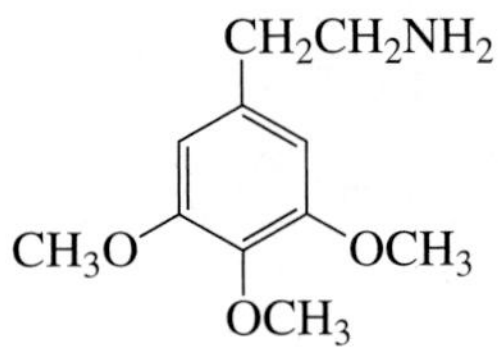

mescaline

61. Propose a mechanism for the following reaction that explains why the configuration of the chirality center in the reactant is retained in the product.

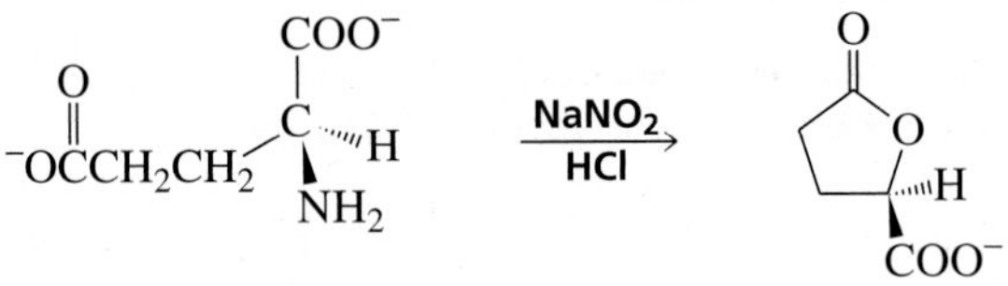

62. Hydroxide ion catalyzes the reaction of piperidine with 2,4-dinitroanisole but has no effect on the reaction of piperidine with 1-chloro-2,4-dinitrobenzene. Explain why hydroxide ion acts as a catalyst in one reaction but has no effect in the other reaction (*J. Am. Chem. Soc.,* 1965, *87*, 5209).

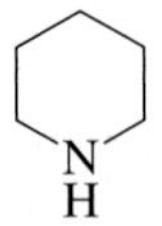

piperidine

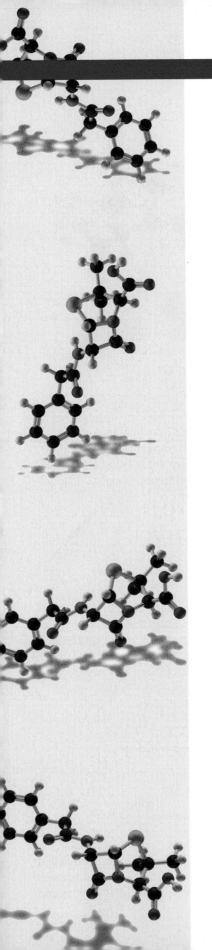

Carbonyl Compounds

The four chapters in Part VI focus on the reactions of carbonyl compounds. Carbonyl compounds can be placed in one of two groups—those that contain a group that can be replaced by another group (Class I), and those that do *not* contain a group that can be replaced by another group (Class II).

Chapter 16   **Carbonyl Compounds I:** *Reactions of Carboxylic Acids and Their Derivatives With Oxygen and Nitrogen Nucleophiles*

Chapter 17   **Carbonyl Compounds II:** *Reactions of Carbonyl Compounds With Carbon and Hydrogen Nucleophiles; Reactions of Aldehydes and Ketones with Oxygen and Nitrogen Nucleophiles; Reactions of α,β-Unsaturated Carbonyl Compounds*

Chapter 18   **More About Oxidation–Reduction Reactions**

Chapter 19   **Carbonyl Compounds III:** *Reactions at the α-Carbon*

Chapter 16 discusses the reactions of Class I carbonyl compounds with relatively poor nucleophiles. Carboxylic acids are the Class I carbonyl compounds most commonly available both in the laboratory and in biological systems. Because carboxylic acids are not very reactive, they must be activated to undergo reactions at reasonable rates. Chapter 16 describes the techniques chemists use to activate carboxylic acids and compares them to the techniques used by biological systems.

The first part of **Chapter 17** compares the reactivities of Class I and Class II carbonyl compounds by discussing the reactivity of each class with good nucleophiles. Chapter 17 goes on to discuss the reactions of Class II carbonyl compounds with the poorer nucleophiles whose reactions with Class I carbonyl compounds you studied in Chapter 16.

Chapter 18 covers oxidation-reduction reactions of organic compounds. You have seen many of these reactions in other chapters, but here they are discussed as a group, which gives you the opportunity to compare and contrast them. First you will study reduction reactions, where you will see that all reduction reactions follow one of three mechanisms. Then you will study oxidation reactions.

Many carbonyl compounds have two sites of reactivity, the carbonyl group *and* the α-carbon. Chapters 16 and 17 discuss the reactions carbonyl compounds undergo that take place at the carbonyl group. **Chapter 19** examines the reactions carbonyl compounds undergo that take place at the α-carbon.

16

Carbonyl Compounds I:

Reactions of Carboxylic Acids and Their Derivatives With Oxygen and Nitrogen Nucleophiles

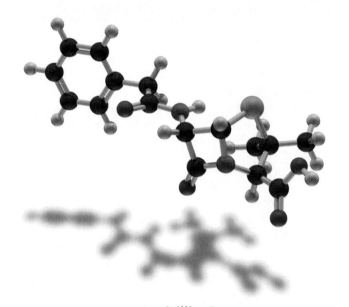

penicillin G

an acyl chloride

an ester

a carboxylic acid

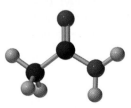

an amide

Compounds containing carbonyl groups abound in nature. Many play important roles in biological processes. Hormones, vitamins, amino acids, drugs, and flavorings are just a few of the carbonyl compounds that affect us daily. Recall that a **carbonyl group** is a carbon double-bonded to an oxygen. An **acyl group** consists of a carbonyl group attached to an alkyl group or to an aryl group.

a carbonyl group   **acyl groups**

Carbonyl compounds can be divided into two classes. One class includes compounds in which the acyl group is attached to an atom or group that *can* be replaced by another group. Carboxylic acids, acyl halides, acid anhydrides, esters, and amides belong to this class. All of these compounds contain a group ($-$OH, $-$Cl, $-$Br, $-$O(CO)R, $-$OR, $-$NH$_2$, $-$NHR, or $-$NR$_2$) that can be replaced by a nucleophile.

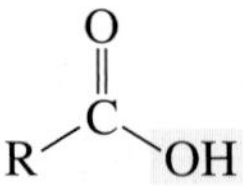

a carboxylic acid 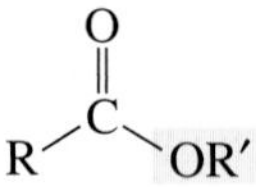   **an ester** 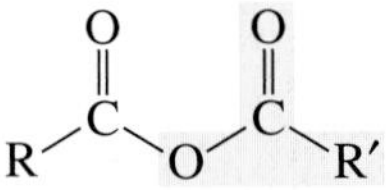   **an acid anhydride**

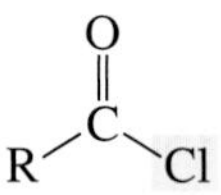

an acyl chloride 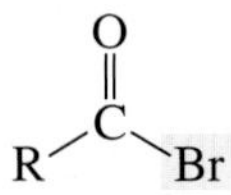   **an acyl bromide**   **amides**

acyl halides

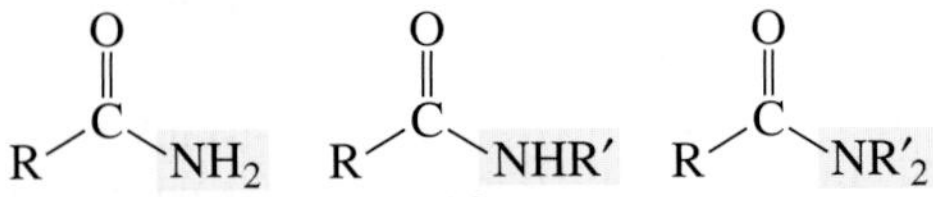

The second class consists of carbonyl compounds in which the acyl group is attached to a group that *cannot* be readily replaced by another group. Aldehydes and ketones belong to this class. The $-$H and alkyl or aryl ($-$R or $-$Ar) groups of aldehydes and ketones cannot be replaced by a nucleophile.

$$\underset{\text{an aldehyde}}{\overset{\overset{\displaystyle O}{\|}}{R-\underset{}{C}-H}} \qquad \underset{\text{a ketone}}{\overset{\overset{\displaystyle O}{\|}}{R-\underset{}{C}-R}}$$

In Chapter 9 we saw that the ability of a group to leave (that is, to be replaced by another group) depends on its basicity: *The weaker the basicity of a group, the better its leaving ability.* Recall from Section 9.3 that weak bases are good leaving groups because weak bases do not share their electrons as well as strong bases do.

The pK_a values of the conjugate acids of the leaving groups of various carbonyl compounds are listed in Table 16.1. Notice that the acyl groups of the carbonyl compounds in the first class are attached to weaker bases than are the acyl groups of the carbonyl compounds in the second class. (Remember that the lower the pK_a, the stronger the acid and the weaker its conjugate base.) The −H of an aldehyde and the alkyl or aryl (−R or −Ar) group of a ketone are too basic to be replaced by another group.

The weaker the base, the better it is as a leaving group.

3-D Molecules:
Acetyl chloride;
Methyl acetate;
Acetic acid;
Acetamide

TABLE 16.1	The pK_a Values of the Conjugate Acids of the Leaving Groups of Carbonyl Compounds		
Carbonyl compound	**Leaving group**	**Conjugate acid of the leaving group**	**pK_a**
Class I			
$R-\overset{O}{\overset{\|}{C}}-Br$	Br^-	HBr	-9
$R-\overset{O}{\overset{\|}{C}}-Cl$	Cl^-	HCl	-7
$R-\overset{O}{\overset{\|}{C}}-O-\overset{O}{\overset{\|}{C}}-R$	$^-O\overset{O}{\overset{\|}{C}}R$	$R\overset{O}{\overset{\|}{C}}OH$	$\sim3-5$
$R-\overset{O}{\overset{\|}{C}}-OR'$	$^-OR'$	$R'OH$	$\sim15-16$
$R-\overset{O}{\overset{\|}{C}}-OH$	^-OH	H_2O	15.7
$R-\overset{O}{\overset{\|}{C}}-NH_2$	$^-NH_2$	NH_3	36
Class II			
$R-\overset{O}{\overset{\|}{C}}-R$	R^-	RH	~50
$R-\overset{O}{\overset{\|}{C}}-H$	H^-	H_2	very large

In this chapter we will discuss the reactions of carbonyl compounds that belong to the first class. Because these compounds have an acyl group attached to a group that can be replaced by a nucleophile, we will see that they undergo substitution reactions. The reactions of aldehydes and ketones will be considered in Chapter 17. Because aldehydes and ketones have an acyl group that is attached to a group that cannot be replaced by a nucleophile, we can correctly predict that these compounds do not undergo substitution reactions.

16.1 NOMENCLATURE

Carboxylic Acids

In systematic nomenclature, a **carboxylic acid** is named by replacing the terminal "e" of the alkane name with "oic acid." For example, the one-carbon alkane is methan*e,* so the one-carbon carboxylic acid is methan*oic acid.* Long-chain carboxylic acids are called **fatty acids.**

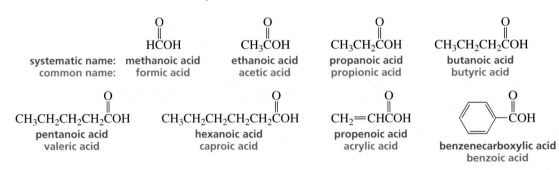

HCOH	CH₃COH	CH₃CH₂COH	CH₃CH₂CH₂COH
systematic name: methanoic acid	ethanoic acid	propanoic acid	butanoic acid
common name: formic acid	acetic acid	propionic acid	butyric acid

CH₃CH₂CH₂CH₂COH	CH₃CH₂CH₂CH₂CH₂COH	CH₂=CHCOH	⬡—COH
pentanoic acid	hexanoic acid	propenoic acid	benzenecarboxylic acid
valeric acid	caproic acid	acrylic acid	benzoic acid

Carboxylic acids containing six or fewer carbons are frequently called by their common names. These names were chosen by early chemists to describe some feature of the compound, usually its origin. For example, formic acid is found in ants, bees, and other stinging insects. Its name comes from *formica,* which is Latin for "ant." Acetic acid—contained in vinegar—got its name from *acetum,* the Latin word for "vinegar." Propionic acid is the smallest acid that shows some of the characteristics of the larger fatty acids. Its name comes from the Greek words *pro* ("the first") and *pion* ("fat"). Butyric acid is found in rancid butter—the Latin word for "butter" is *butyrum.* Caproic acid is found in goat's milk, and if you have the occasion to smell both a goat and caproic acid, you will find that they have similar odors. *Caper* is the Latin word for "goat."

In systematic nomenclature, the position of a substituent is designated by a number. The carbonyl carbon of a carboxylic acid is always the C-1 carbon. In common nomenclature, the position of a substituent is designated by a lowercase Greek letter, and the carbonyl carbon is not given a designation. The carbon adjacent to the carbonyl carbon is the **α-carbon**, the carbon adjacent to the α-carbon is the β-carbon, and so on.

α = alpha
β = beta
γ = gamma
δ = delta
ε = epsilon

CH₃CH₂CH₂CH₂CH₂COH
6   5   4   3   2   1
systematic nomenclature

CH₃CH₂CH₂CH₂CH₂COH
ε   δ   γ   β   α
common nomenclature

Take a careful look at the following examples to make sure you understand the difference between systematic (IUPAC) and common nomenclature.

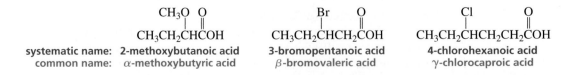

CH₃CH₂CHCOH	CH₃CH₂CHCH₂COH	CH₃CH₂CHCH₂CH₂COH
systematic name: 2-methoxybutanoic acid	3-bromopentanoic acid	4-chlorohexanoic acid
common name: α-methoxybutyric acid	β-bromovaleric acid	γ-chlorocaproic acid

The functional group of a carboxylic acid is called a **carboxyl group.**

$$\underset{\substack{\text{a carboxyl group}}}{-\overset{\displaystyle O}{\overset{\|}{C}}-OH} \qquad \underset{\substack{\textbf{carboxyl groups are frequently}\\ \textbf{shown in abbreviated forms}}}{-COOH \qquad -CO_2H}$$

Carboxylic acids in which a carboxyl group is attached to a ring are named by adding "carboxylic acid" to the name of the cyclic compound.

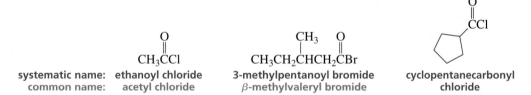

cyclohexanecarboxylic acid   *trans*-**3-methylcyclopentanecarboxylic acid**   1,2,4-benzenetricarboxylic acid

Acyl Halides

The most common **acyl halides** are acyl chlorides and acyl bromides. Acyl halides are named by using the acid name and replacing "ic acid" with "yl chloride (or bromide)." For acids ending with "carboxylic acid," "carboxylic acid" is replaced with "carbonyl chloride (or bromide)."

3-D Molecules:
trans-3-Methyl-
cyclopentane-
carboxylic acid;
Acetic anhydride

$\underset{\substack{O\\\|}}{CH_3\overset{\displaystyle O}{\overset{\|}{C}}Cl}$	$\underset{\substack{CH_3\\\|}}{CH_3CH_2CHCH_2\overset{\displaystyle O}{\overset{\|}{C}}Br}$	$\overset{\displaystyle O}{\overset{\|}{C}}Cl$

systematic name: ethanoyl chloride   3-methylpentanoyl bromide   cyclopentanecarbonyl
common name: acetyl chloride   *β*-methylvaleryl bromide   chloride

Acid Anhydrides

Loss of water from two molecules of a carboxylic acid results in an **acid anhydride.** Anhydride means "without water."

$$R-\overset{\displaystyle O}{\overset{\|}{C}}-O-H \quad H-O-\overset{\displaystyle O}{\overset{\|}{C}}-R \longrightarrow R-\overset{\displaystyle O}{\overset{\|}{C}}-O-\overset{\displaystyle O}{\overset{\|}{C}}-R \; + \; H_2O$$

an acid anhydride

If the two carboxylic acid molecules forming the acid anhydride are the same, the anhydride is a **symmetrical anhydride.** If the two carboxylic acid molecules are different, the anhydride is a **mixed anhydride.** Symmetrical anhydrides are named by using the acid name and replacing "acid" with "anhydride." Mixed anhydrides are named by stating the names of both acids in alphabetical order, followed by "anhydride."

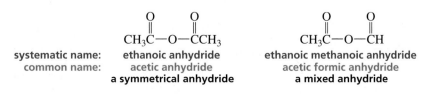

systematic name: ethanoic anhydride   ethanoic methanoic anhydride
common name: acetic anhydride   acetic formic anhydride
                    a symmetrical anhydride   **a mixed anhydride**

Esters

In naming an **ester**, the name of the group (R′) attached to the **carboxyl oxygen** is stated first, followed by the name of the acid with "ic acid" replaced by "ate."

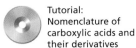

$CH_3COCH_2CH_3$	CH_3CH_2CO—⟨⟩	$CH_3CHCH_2COCH_3$ (with Br)	⟨⟩—$COCH_2CH_3$
systematic name: ethyl ethanoate			
common name: ethyl acetate	**phenyl propanoate** phenyl propionate	**methyl 3-bromobutanoate** methyl β-bromobutyrate	**ethyl cyclohexanecarboxylate**

Salts of carboxylic acids are named in the same way. The cation is named first, followed by the name of the acid, again with "ic acid" replaced by "ate."

Tutorial:
Nomenclature of
carboxylic acids and
their derivatives

$HCO^-\ Na^+$	$CH_3CO^-\ K^+$	⟨⟩—$CO^-\ Na^+$
systematic name: sodium methanoate	**potassium ethanoate**	
common name: sodium formate	potassium acetate	**sodium benzenecarboxylate** sodium benzoate

Cyclic esters are called **lactones.** The length of the carbon chain is designated by the common name of the carboxylic acid, and a Greek letter is used to indicate the carbon to which the carboxyl oxygen is attached. Thus, four-membered ring lactones are β-lactones (the carboxyl oxygen is on the β-carbon), five-membered ring lactones are γ-lactones, and six-membered ring lactones are δ-lactones.

3-D Molecules:
δ-Caprolactone;
cis-2-Ethylcyclohexane-
carboxamide

δ-caprolactone   **β-butyrolactone**   **γ-butyrolactone**   **γ-caprolactone**

PROBLEM 1

The word "lactone" comes from lactic acid, a three-carbon carboxylic acid with an OH group on the α-carbon. Ironically, lactic acid cannot form a lactone. Why not?

Amides

Amides are named by using the acid name, replacing "oic acid" or "ic acid" with "amide." For acids ending with "carboxylic acid," "ylic acid" is replaced with "amide."

CH_3CNH_2	$ClCH_2CH_2CH_2CNH_2$	⟨⟩—CNH_2	⟨⟩—CNH_2 (with CH_2CH_3)
systematic name: ethanamide	**4-chlorobutanamide**		
common name: acetamide	γ-chlorobutyramide	**benzenecarboxamide** benzamide	*cis*-2-ethylcyclohexanecarboxamide

If there is a substituent bonded to the nitrogen, the name of the substituent is stated first, followed by the name of the amide. The name of each substituent is preceded by a capital *N* (italic) to indicate that the substituent is bonded to a nitrogen.

$$CH_3CH_2\overset{\overset{\displaystyle O}{\|}}{C}NH-\bigcirc$$

N-cyclohexylpropanamide

$$CH_3CH_2CH_2CH_2\overset{\overset{\displaystyle O}{\|}}{C}-N\overset{\overset{\displaystyle CH_3}{|}}{}CH_2CH_3$$

N-ethyl-N-methylpentanamide

$$CH_3CH_2CH_2\overset{\overset{\displaystyle O}{\|}}{C}-N\overset{\overset{\displaystyle CH_2CH_3}{|}}{}CH_2CH_3$$

N,N-diethylbutanamide

Cyclic amides are called **lactams.** Their nomenclature is similar to that of lactones; that is, the length of the carbon chain is indicated by the common name of the carboxylic acid, and a Greek letter is used to indicate the carbon to which the nitrogen is attached.

δ-valerolactam   γ-valerolactam

Nitriles

3-D Molecule:
Acetonitrile

Nitriles are compounds that contain a C≡N functional group. They are considered to be carboxylic acid derivatives because, like all Class I carbonyl compounds, they react with water to form carboxylic acids (Section 16.16). In systematic nomenclature, nitriles are named by adding "nitrile" to the parent alkane name. Notice that the triple-bonded carbon of the nitrile group is counted in the number of carbons in the longest continuous chain. In common nomenclature, nitriles are named by replacing "ic acid" of the carboxylic acid name with "onitrile." They can also be named as alkyl cyanides.

$$CH_3C\equiv N$$ $$\bigcirc-C\equiv N$$ $$CH_3\overset{\overset{\displaystyle CH_3}{|}}{CH}CH_2CH_2CH_2C\equiv N$$ $$CH_2=CHC\equiv N$$

systematic name:	ethanenitrile	benzenecarbonitrile	5-methylhexanenitrile	propenenitrile
common name:	acetonitrile	benzonitrile		acrylonitrile
	methyl cyanide	phenyl cyanide	isohexyl cyanide	

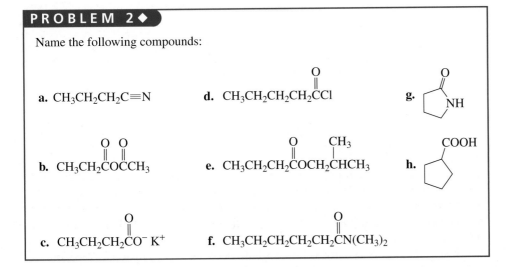

PROBLEM 2◆

Name the following compounds:

a. $CH_3CH_2CH_2C\equiv N$

b. $CH_3CH_2\overset{\overset{\displaystyle O}{\|}}{C}\overset{\overset{\displaystyle O}{\|}}{O}CCH_3$

c. $CH_3CH_2CH_2\overset{\overset{\displaystyle O}{\|}}{C}O^-\ K^+$

d. $CH_3CH_2CH_2CH_2\overset{\overset{\displaystyle O}{\|}}{C}Cl$

e. $CH_3CH_2CH_2\overset{\overset{\displaystyle O}{\|}}{C}OCH_2\overset{\overset{\displaystyle CH_3}{|}}{C}HCH_3$

f. $CH_3CH_2CH_2CH_2CH_2\overset{\overset{\displaystyle O}{\|}}{C}N(CH_3)_2$

g.

h.

16.2 STRUCTURES OF CARBOXYLIC ACIDS AND CARBOXYLIC ACID DERIVATIVES

Acyl halides, acid anhydrides, esters, and amides are all called **carboxylic acid derivatives** because they differ from a carboxylic acid only in the nature of the group that has replaced the OH group of the carboxylic acid.

The **carbonyl carbon** in carboxylic acids and carboxylic acid derivatives is sp^2 hybridized. It uses its three sp^2 orbitals to form σ bonds to the carbonyl oxygen, the α-carbon, and a substituent (Y). The three atoms attached to the carbonyl carbon are in the same plane, and their bond angles are each approximately 120°.

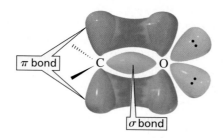

The **carbonyl oxygen** is also sp^2 hybridized. One of its sp^2 orbitals forms a σ bond with the carbonyl carbon, and each of the other two sp^2 orbitals contains a pair of nonbonding electrons. The remaining p orbital of the carbonyl oxygen overlaps with the remaining p orbital of the carbonyl carbon to form a π bond (Figure 16.1).

Figure 16.1 ▶
Bonding in a carbonyl group.

Esters, carboxylic acids, and amides each have two major resonance contributors.

acetic acid

acetamide

The resonance contributor on the right is more important for the amide than it is for the ester or the carboxylic acid because nitrogen is better than oxygen at sharing its electrons. Basicity is a measure of how well a group shares its electrons, and nitrogen is less electronegative than oxygen and, therefore, is better able to accommodate a positive charge.

The acid properties of carboxylic acids have been discussed previously (Sections 1.18 and 6.10). The acid properties of dicarboxylic acids are discussed in Section 16.19.

PROBLEM 3◆

Which is longer, the carbon–oxygen single bond in a carboxylic acid or the carbon–oxygen bond in an alcohol? Why?

PROBLEM 4◆

There are three carbon–oxygen bonds in methyl acetate.

a. What are their relative lengths?

b. What are the relative infrared (IR) stretching frequencies of these bonds?

PROBLEM 5◆

An ester has two oxygen atoms. Which is more basic, the carbonyl oxygen or the carboxyl oxygen?

PROBLEM 6

Match the compound with the appropriate carbonyl IR absorption band.

acyl chloride	$\sim$1800 and 1750 cm^{-1}
acid anhydride	$\sim$1640 cm^{-1}
ester	$\sim$1730 cm^{-1}
amide	$\sim$1800 cm^{-1}

The physical properties of carbonyl compounds are listed in Appendix I. For compounds with the same number of carbon atoms (in the case of acyl chlorides, the compound with one fewer carbons was chosen because of the mass of the Cl atom), the carbonyl compounds have the following relative boiling points:

16.3
PHYSICAL PROPERTIES OF CARBONYL COMPOUNDS

relative boiling points

amides > carboxylic acids > nitriles ≫ esters ~ acyl chlorides ~ aldehydes ~ ketones

The boiling points of esters, acyl chlorides, aldehydes, and ketones are higher than the boiling points of ethers because of the polar carbonyl group, but are lower than the boiling points of alcohols because their molecules can't form hydrogen bonds with each other.

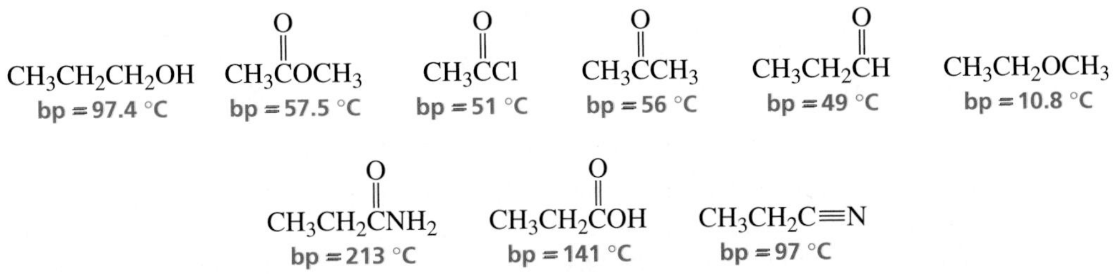

$$
\begin{array}{cccccc}
 & O & O & O & O & \\
 & \| & \| & \| & \| & \\
\text{CH}_3\text{CH}_2\text{CH}_2\text{OH} & \text{CH}_3\text{COCH}_3 & \text{CH}_3\text{CCl} & \text{CH}_3\text{CCH}_3 & \text{CH}_3\text{CH}_2\text{CH} & \text{CH}_3\text{CH}_2\text{OCH}_3 \\
\text{bp} = 97.4\ ^\circ\text{C} & \text{bp} = 57.5\ ^\circ\text{C} & \text{bp} = 51\ ^\circ\text{C} & \text{bp} = 56\ ^\circ\text{C} & \text{bp} = 49\ ^\circ\text{C} & \text{bp} = 10.8\ ^\circ\text{C}
\end{array}
$$

$$
\begin{array}{ccc}
O & O & \\
\| & \| & \\
\text{CH}_3\text{CH}_2\text{CNH}_2 & \text{CH}_3\text{CH}_2\text{COH} & \text{CH}_3\text{CH}_2\text{C} \equiv \text{N} \\
\text{bp} = 213\ ^\circ\text{C} & \text{bp} = 141\ ^\circ\text{C} & \text{bp} = 97\ ^\circ\text{C}
\end{array}
$$

The boiling point of a nitrile is similar to that of an alcohol because a nitrile has strong dipole–dipole interactions (Section 2.9). Carboxylic acids have relatively high boiling points because they can form hydrogen-bonded dimers, giving them larger effective molecular weights.

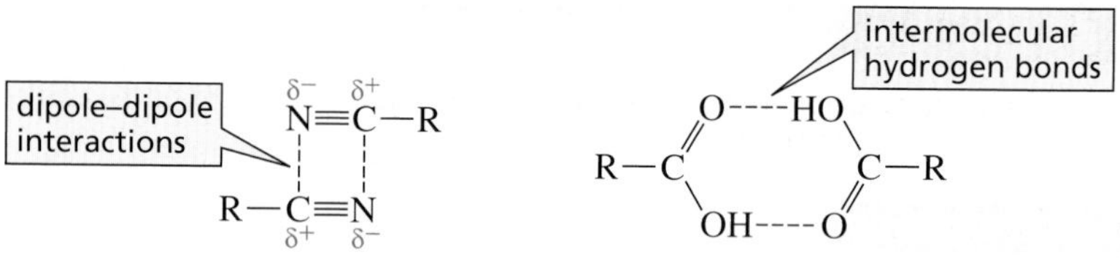

Amides have the highest boiling points. They have strong dipole–dipole interactions because the resonance contributor with separated charges contributes significantly to the overall structure of the compound (Section 16.2). If an amide has a hydrogen bonded to the nitrogen, the molecules also form intermolecular hydrogen bonds.

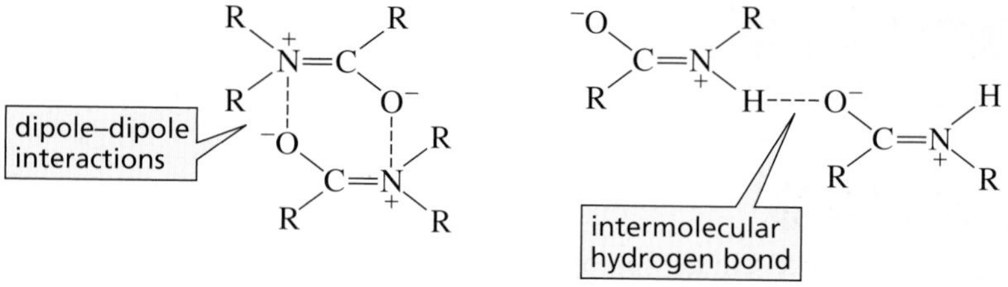

Carboxylic acid derivatives are soluble in solvents such as ethers, chlorinated alkanes, and aromatic hydrocarbons. Like alcohols and ethers, carbonyl compounds with fewer than four carbons are soluble in water.

Esters, *N,N*-disubstituted amides, and nitriles are often used as solvents because they are polar but do not have reactive hydroxyl or amino groups. We have seen that dimethylformamide (DMF) is a common aprotic polar solvent (Section 9.3).

16.4 REACTIVITY CONSIDERATIONS

Oxygen is more electronegative than carbon, so the carbonyl group (C=O) is polar. Because the carbonyl carbon has a partial positive charge, we can safely predict that it will be attacked by nucleophiles.

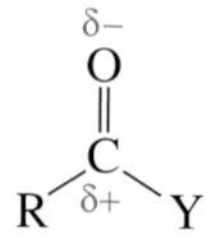

When a nucleophile attacks the carbonyl group of a carboxylic acid derivative, the carbon–oxygen π bond breaks. The resulting intermediate is called a **tetrahedral intermediate** because the trigonal (sp^2) carbon in the reactant has become a tetrahedral (sp^3) carbon in the intermediate.

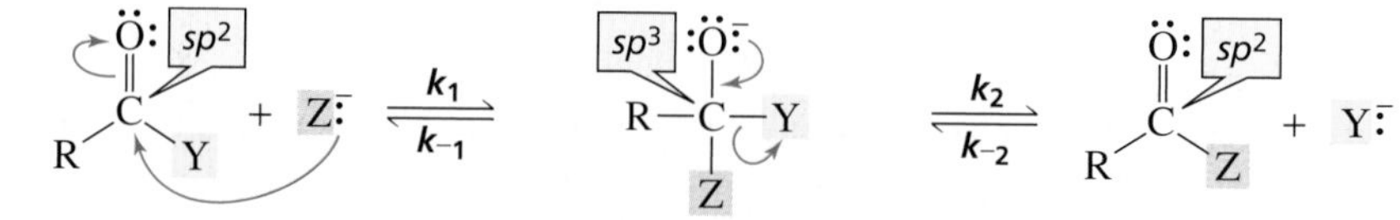

a tetrahedral intermediate

The tetrahedral intermediate formed when a nucleophile attacks the carbonyl carbon of a carboxylic acid derivative is not stable and cannot be isolated. It is not a final product—it is an intermediate formed on the way to the final product. A pair of nonbonding electrons on the oxygen re-forms the π bond, and either Y^- (k_2) or Z^- (k_{-1}) is expelled with its bonding electrons.

Whether Y^- or Z^- is expelled depends on their relative basicities. The weaker base is expelled preferentially, making this another example of the principle we first saw in Section 9.3—*the weaker the base, the better it is as a leaving group.* Because a weak base does not share its electrons as well as a strong base does, a weak base forms a weaker bond—one that is easier to break. If Z^- is a much weaker base than Y^-, Z^- will be expelled. In such a case, $k_{-1} \gg k_2$, and the reaction can be written as follows:

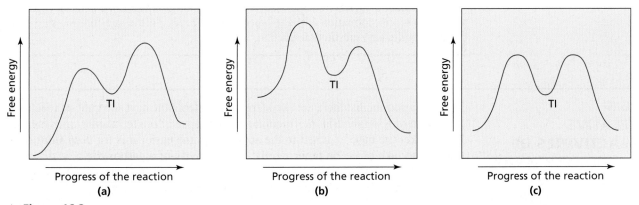

In this case, no new product is formed. The nucleophile attacks the carbonyl carbon, but the tetrahedral intermediate expels the attacking nucleophile and reforms the reactants.

On the other hand, if Y^- is a much weaker base than Z^-, Y^- will be expelled and a new product will be formed. In this case, $k_2 \gg k_{-1}$, and the reaction can be written as follows:

Tutorial:
Free energy diagrams for nucleophilic acyl substitution reactions

This is called a **nucleophilic acyl substitution reaction** because a nucleophile (Z^-) has replaced the substituent (Y^-) that was attached to the acyl group in the reactant.

If the basicities of Y^- and Z^- are similar, the values of k_{-1} and k_2 will be similar. Therefore, some molecules of the tetrahedral intermediate will expel Y^- and others will expel Z^-. When the reaction is over, reactant and product will both be present. The relative amounts of each depend on the relative basicities of Y^- and Z^- (that is, the relative values of k_2 and k_{-1}).

These three cases are illustrated by the reaction coordinate diagrams in Figure 16.2.

▲ **Figure 16.2**
Reaction coordinate diagrams for nucleophilic acyl substitution reactions in which:
(a) the nucleophile is a weaker base than the group attached to the acyl group in the starting material,
(b) the nucleophile is a stronger base than the group attached to the acyl group in the starting material, and
(c) the nucleophile and the group attached to the acyl group in the starting material have similar basicities.
TI is the tetrahedral intermediate.

1. If the incoming nucleophile is a weaker base than the group attached to the acyl group in the reactant [Figure 16.2(a)], the easier pathway is for the tetrahedral intermediate (TI) to expel the newly added group and re-form the reactants.

2. If the incoming nucleophile is a stronger base than the group attached to the acyl group in the reactant [Figure 16.2(b)], the easier pathway is for the tetrahedral intermediate to expel the group that was attached to the acyl group in the reactant and form a substitution product.

3. If the incoming nucleophile and the group attached to the acyl group in the reactant have similar basicities [Figure 16.2(c)], the tetrahedral intermediate can expel either group with similar ease. A mixture of reactants and substitution product will result.

> **For a carboxylic acid derivative to undergo a nucleophilic acyl substitution reaction, the incoming nucleophile must *not* be a much weaker base than the group that is to be replaced.**

We can make the following general statement about the reactions of carboxylic acid derivatives—*a carboxylic acid derivative will undergo a nucleophilic acyl substitution reaction provided that the incoming nucleophile is not a much weaker base than the substituent attached to the acyl group in the reactant.*

PROBLEM 7◆

Using the pK_a values in Table 16.1, predict the products of the following reactions:

a. $CH_3-\overset{\overset{\displaystyle O}{\|}}{C}-OCH_3$ + NaCl $\longrightarrow$

b. $CH_3-\overset{\overset{\displaystyle O}{\|}}{C}-Cl$ + $CH_3-\overset{\overset{\displaystyle O}{\|}}{C}-O^-\ Na^+$ $\longrightarrow$

c. $CH_3-\overset{\overset{\displaystyle O}{\|}}{C}-O-\overset{\overset{\displaystyle O}{\|}}{C}-CH_3$ + NaCl $\longrightarrow$

PROBLEM 8◆

Is the following statement true or false?

If the incoming nucleophile is a stronger base than the group attached to the acyl group in the reactant, formation of the tetrahedral intermediate is the rate-limiting step of a nucleophilic acyl substitution reaction.

16.5
RELATIVE REACTIVITIES OF CARBOXYLIC ACIDS, ACYL HALIDES, ACID ANHYDRIDES, ESTERS, AND AMIDES

We have just seen that there are two steps in a nucleophilic acyl substitution reaction—formation of a tetrahedral intermediate and collapse of the tetrahedral intermediate. The weaker the base attached to the acyl group, the easier it is for *both steps* of the reaction to take place. In other words, the reactivity of a carboxylic acid derivative depends on the basicity of the substituent attached to the acyl group—the less basic the substituent, the more reactive the carboxylic acid derivative.

relative basicities of the leaving groups

$$Cl^- \ < \ ^-O\overset{\overset{\displaystyle O}{\|}}{C}R \ < \ ^-OR \ \sim \ ^-OH \ < \ ^-NH_2$$

relative reactivities of carboxylic acid derivatives

$$\underset{\text{acyl chloride}}{R-\overset{\overset{\textstyle O}{\|}}{C}-Cl} \quad > \quad \underset{\text{acid anhydride}}{R-\overset{\overset{\textstyle O}{\|}}{C}-O-\overset{\overset{\textstyle O}{\|}}{C}-R} \quad > \quad \underset{\text{ester}}{R-\overset{\overset{\textstyle O}{\|}}{C}-OR'} \quad \sim \quad \underset{\text{carboxylic acid}}{R-\overset{\overset{\textstyle O}{\|}}{C}-OH} \quad > \quad \underset{\text{amide}}{R-\overset{\overset{\textstyle O}{\|}}{C}-NH_2}$$

← increasing reactivity

How does having a weak base attached to the acyl group make the first step of the nucleophilic substitution reaction easier? First of all, the weaker the base, the more electronegative the atom attached to the acyl group. Thus, weaker bases are better at withdrawing electrons inductively from the carbonyl carbon, which increases the carbonyl carbon's susceptibility to nucleophilic attack. The electrostatic potential maps show that the carbonyl carbon is more positive (more blue) and, therefore, more reactive in the acyl chloride than in the amide.

resonance contributors of a carboxylic acid or carboxylic acid derivative

Second, the weaker the basicity of Y, the smaller the contribution from the resonance contributor with a positive charge on Y (Section 16.2). The less the carboxylic acid derivative is stabilized by electron delocalization, the more reactive it will be.

inductive electron withdrawal increases the electrophilicity of the carbonyl carbon

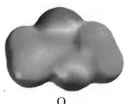

Having a weak base attached to the acyl group also makes the second step of the nucleophilic acyl substitution reaction easier because weak bases are easier to eliminate when the tetrahedral intermediate collapses.

$$R-\overset{\overset{\textstyle :\ddot{O}:^-}{|}}{\underset{\underset{\textstyle Z}{|}}{C}}-Y$$ the weaker the base, the easier it is to expel

Because the incoming nucleophile in a nucleophilic acyl substitution reaction must be a stronger base than the base that is already there (Section 16.4), a carboxylic acid derivative can be converted into a less reactive carboxylic acid derivative, but not into one that is more reactive. For example, an acyl chloride can be converted into an anhydride because a carboxylate ion is a stronger base than a chloride ion.

$$\underset{\text{an acyl chloride}}{R-\overset{\overset{\textstyle O}{\|}}{C}-Cl} \quad + \quad \underset{\text{carboxylate ion}}{R-\overset{\overset{\textstyle O}{\|}}{C}-O^-} \quad \longrightarrow \quad \underset{\text{an anhydride}}{R-\overset{\overset{\textstyle O}{\|}}{C}-O-\overset{\overset{\textstyle O}{\|}}{C}-R} \quad + \quad Cl^-$$

An anhydride, however, cannot be converted into an acyl chloride because a chloride ion is a weaker base than a carboxylate ion.

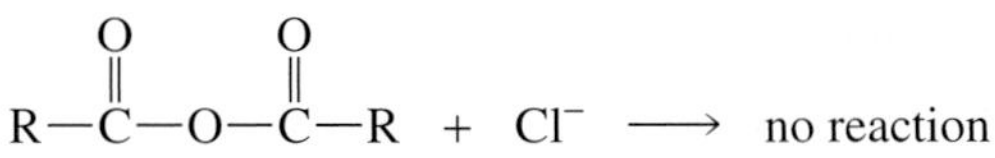

$$R-\overset{\overset{\text{O}}{\|}}{C}-O-\overset{\overset{\text{O}}{\|}}{C}-R \ + \ Cl^- \ \longrightarrow \ \text{no reaction}$$

an anhydride

Acyl halides and acid anhydrides are not found in nature because of their high reactivity. Carboxylic acids, on the other hand, are less reactive and are found widely in nature. For example, glucose is metabolized to pyruvic acid. (S)-(+)-Lactic acid is the compound responsible for the burning sensation felt in muscles during anaerobic exercise, and it is also found in sour milk. Spinach and other leafy green vegetables are rich in oxalic acid. Succinic acid and citric acid are important intermediates in the citric acid cycle (Section 23.1), a series of reactions that oxidize acetyl CoA to CO_2 in biological systems. Citrus fruits are rich in citric acid. The concentration is greatest in lemons, less in grapefruits, and still less in oranges. (S)-(−)-Malic acid is responsible for the sharp taste of unripe apples and pears. As the fruit ripens, the amount of malic acid in the fruit decreases and the amount of sugar increases. The inverse realtionship between the levels of malic acid and sugar is important for the propagation of the plant. Animals will not eat the fruit until it becomes ripe—at which time its seeds are mature enough to germinate when they are scattered about. Prostaglandins are locally acting hormones that have several different physiological functions (Sections 16.9 and 24.5), such as stimulating inflammation, causing hypertension, and producing pain and swelling.

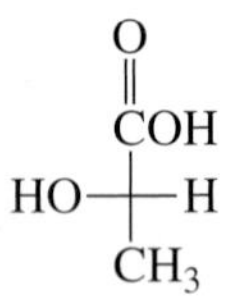

(S)-(+)-lactic acid

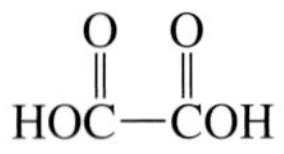

oxalic acid

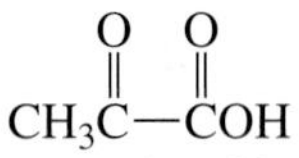

pyruvic acid

succinic acid

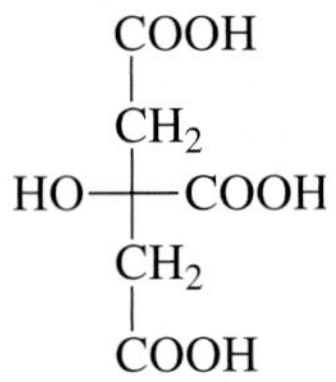

citric acid

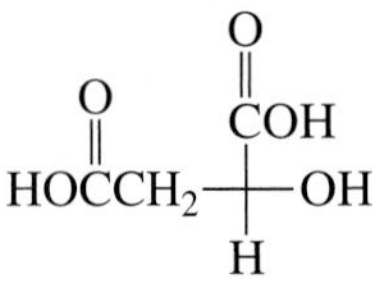

(S)-(–)-malic acid

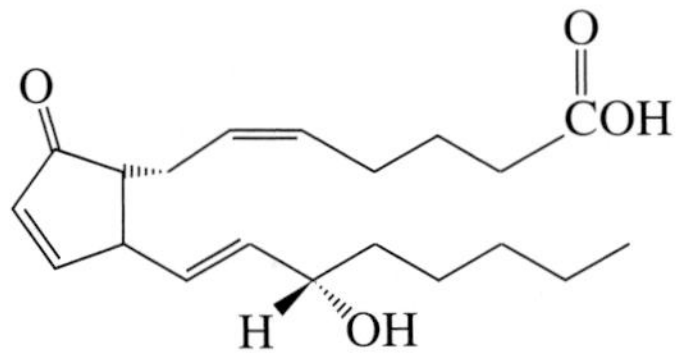

prostaglandin A₂

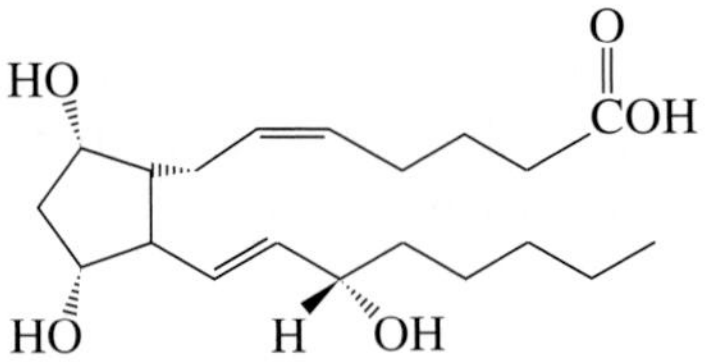

prostaglandin F₂α

Esters are also commonly found in nature. Many of the fragrances of flowers and fruits are due to esters.

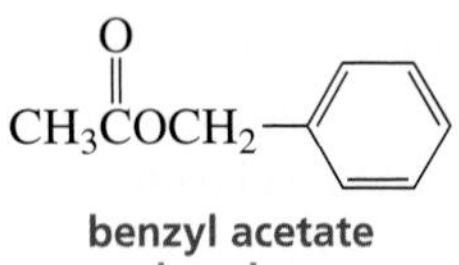

benzyl acetate
jasmine

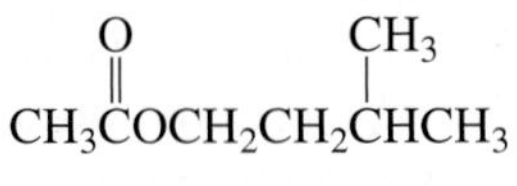

isopentyl acetate
banana

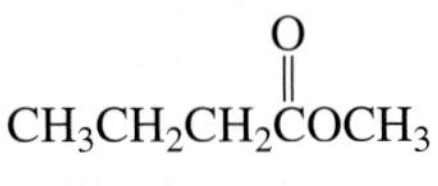

methyl butyrate
apple

Carboxylic acids with an amino group on the α-carbon are commonly called **amino acids.** Amino acids are linked together by amide bonds to form peptides and proteins (Section 21.7). Caffeine is another naturally occurring amide—it is found in cocoa and coffee beans. Penicillin G, a compound with two amide bonds (one of which is in a β-lactam ring), was first isolated from a mold in 1928 by Sir Alexander Fleming.

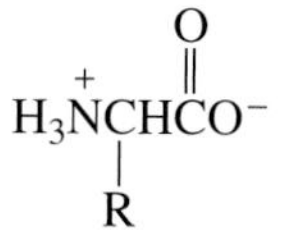

an amino acid

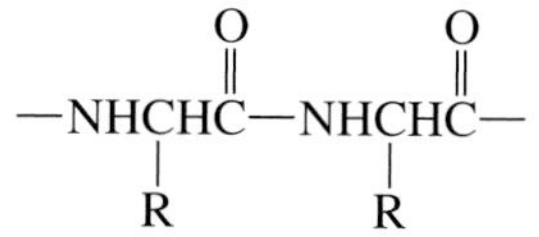

general structure for a peptide or a protein

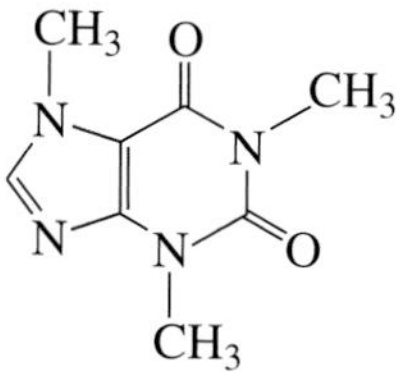

caffeine

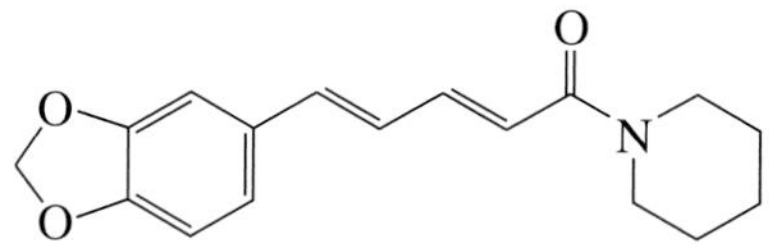

piperine
the major component of black pepper

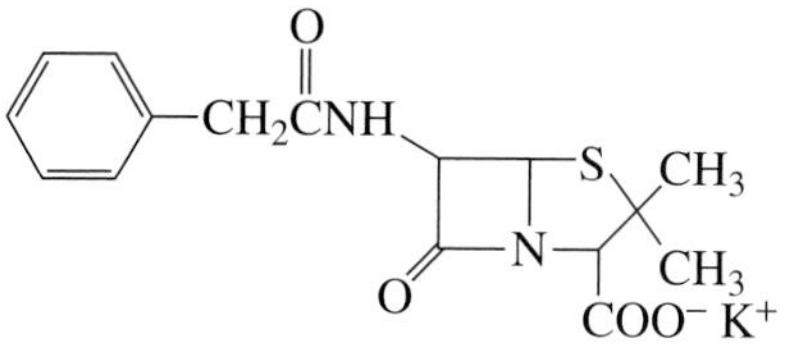

penicillin G

THE DISCOVERY OF PENICILLIN

Sir Alexander Fleming (1881–1955) was born in Scotland. He was a professor of bacteriology at University College, London. The story is told that one day Fleming was about to throw away a culture of staphylococcal bacteria that had been contaminated by a rare strain of the mold *Penicillium notatum*. He noticed that the bacteria had disappeared wherever there was a particle of mold. This suggested to him that the mold must have produced an antibacterial substance. He isolated an active extract of the mold, which was found 10 years later to be a mixture of seven to nine related compounds. The most reproducible of these was penicillin G, the first penicillin to be used by people (Section 16.14). Although Fleming is generally given credit for the discovery of penicillin, there is clear evidence that the germicidal activity of the mold was recognized in the nineteenth century by Lord Joseph Lister (1827–1912), the English physician renowned for the introduction of aseptic surgery.

DALMATIANS: DON'T TRY TO FOOL MOTHER NATURE

When amino acids are metabolized, the excess nitrogen is concentrated into uric acid—a compound with five amide bonds. A series of enzyme-catalyzed reactions degrades uric acid to ammonium ion. The extent to which this degradation process occurs depends on the species. Birds, reptiles, and insects excrete excess nitrogen as uric acid. Most mammals excrete excess nitrogen as allantoin. Excess nitrogen in aquatic animals is excreted as allantoic acid, as urea, or as ammonium salts.

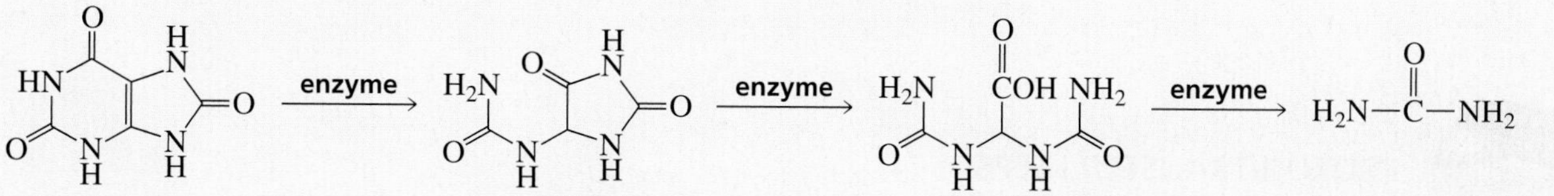

uric acid

excreted by: birds, reptiles, insects

allantoin

most mammals

allantoic acid

marine vertebrates

urea

cartilaginous fish, amphibia

$^+NH_4X^-$
ammonium salt
marine invertebrates

Dalmatians, unlike most mammals, excrete high levels of uric acid. Breeders of Dalmatians select dogs that have black spots with no white hairs, and the gene that determines the presence of white hairs is linked to the gene that converts uric acid to allantoin. Dalmatians, therefore, are susceptible to gout (painful deposits of uric acid in joints).

16.6
GENERAL MECHANISM FOR NUCLEOPHILIC ACYL SUBSTITUTION REACTIONS

All carboxylic acid derivatives undergo nucleophilic acyl substitution reactions by the same mechanism. If the nucleophile is negatively charged, the mechanism discussed in Section 16.4 is followed.

Movie:
Nucleophilic
acyl substitution

If the nucleophile is neutral, the mechanism has an additional step. A proton is lost from the tetrahedral intermediate formed in the first step, resulting in a tetrahedral intermediate equivalent to the one formed by negatively charged nucleophiles.

In Section 16.4, you learned that you will be able to determine the outcome of all the reactions of carboxylic acids and carboxylic acid derivatives discussed in this chapter simply by looking at the two leaving groups in the tetrahedral intermediate and remembering that the weaker base is the one that is preferentially eliminated.

1. If the new group in the tetrahedral intermediate is a weaker base than the group that was already there, the new group will be expelled and no reaction will take place.
2. If the new group in the tetrahedral intermediate is a stronger base than the group that was already there, the original group will be expelled and the overall result will be a nucleophilic acyl substitution reaction.
3. If the two groups in the tetrahedral intermediate have similar basicities, either of the groups can be expelled and an equilibrium mixture of reactant and product will be formed.

All carboxylic acid derivatives react by the same mechanism.

As you read the remaining sections of this chapter, remember that you are seeing specific examples of these general principles. It is important to remember that *all the reactions follow the same mechanism.*

PROTON-TRANSFER STEPS

Any source of protons present in solution (H_3O^+, HCl, H_2O) can serve as a proton donor. Therefore, we will not specify the proton donor in proton-transfer steps. Instead, we will show a protonation step by placing H^+ over the arrow and the reverse deprotonation step by placing $-H^+$ under the arrow.

Similarly, any base present in solution (HO^-, H_2O) can remove a proton. Therefore, we will not specify the base in proton-transfer steps. Instead, we will show a deprotonation step using $-H^+$ over the arrow and the reverse protonation step using H^+ under the arrow.

Acyl halides react with carboxylate ions to form anhydrides, with alcohols to form esters, with water to form carboxylic acids, and with amines to form amides because, in each case, the incoming nucleophile is a stronger base than the departing halide ion (Table 16.1). Notice that both alcohols and phenols can be used to prepare esters.

16.7 REACTIONS OF ACYL HALIDES

$$CH_3\overset{O}{\underset{}{C}}-Cl + CH_3\overset{O}{\underset{}{C}}O^- \longrightarrow CH_3\overset{O}{\underset{}{C}}-O-\overset{O}{\underset{}{C}}CH_3 + Cl^-$$

acetyl chloride   acetic anhydride

acetyl chloride

benzoyl chloride + CH_3OH ⟶ methyl benzoate + H^+ + Cl^-

propionyl chloride + phenol —OH ⟶ phenyl propionate + H^+ + Cl^-

3-D Molecule:
Benzoyl chloride

$$CH_3CH_2CH_2\overset{O}{\underset{}{C}}-Cl + H_2O \longrightarrow CH_3CH_2CH_2\overset{O}{\underset{}{C}}-OH + H^+ + Cl^-$$

butyryl chloride   butyric acid

phenylethanoyl chloride —$CH_2\overset{O}{\underset{}{C}}$—Cl + 2 CH_3NH_2 ⟶ N-methylphenylethanamide —$CH_2\overset{O}{\underset{}{C}}$—$NHCH_3$ + $CH_3\overset{+}{N}H_3$ Cl^-

All the reactions follow the general mechanism described in Section 16.6. In the conversion of an acyl chloride into an acid anhydride, the nucleophilic carboxylate ion attacks the carbonyl carbon of the acyl chloride. Because the resulting tetrahedral intermediate is unstable, the double bond is immediately reformed, expelling chloride ion because it is a weaker base than the carboxylate ion. The final product is an anhydride.

mechanism for the conversion of an acyl chloride into an acid anhydride

In the conversion of an acyl chloride into an ester, the nucleophilic alcohol attacks the carbonyl carbon of the acyl chloride. Because the protonated ether group is a strong acid (Section 1.17), the tetrahedral intermediate loses a proton. Chloride ion is expelled from the deprotonated tetrahedral intermediate because chloride ion is a weaker base than the alkoxide ion.

mechanism for the conversion of an acyl chloride into an ester

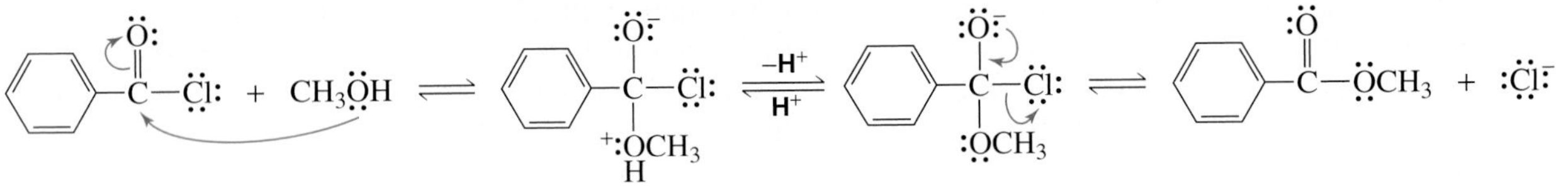

The reaction of an acyl chloride with ammonia or with a primary or secondary amine produces an amide and a proton (and Cl^-). The proton generated in the reaction will protonate unreacted ammonia or unreacted amine and, because the protonated amines are not nucleophiles, they cannot react with the acyl chloride. The reaction, therefore, must be carried out with twice as much ammonia or amine as acyl chloride so there will be enough amine to react with all the acyl halide.

$$CH_3\overset{O}{\overset{\|}{C}}{-}Cl + NH_3 \longrightarrow CH_3\overset{O}{\overset{\|}{C}}{-}NH_2 + H^+ + Cl^- \xrightarrow{NH_3} \overset{+}{N}H_4\,Cl^-$$

$$CH_3CH_2\overset{O}{\overset{\|}{C}}{-}Cl + CH_3NH_2 \longrightarrow CH_3CH_2\overset{O}{\overset{\|}{C}}{-}NHCH_3 + H^+ + Cl^- \xrightarrow{CH_3NH_2} CH_3\overset{+}{N}H_3\,Cl^-$$

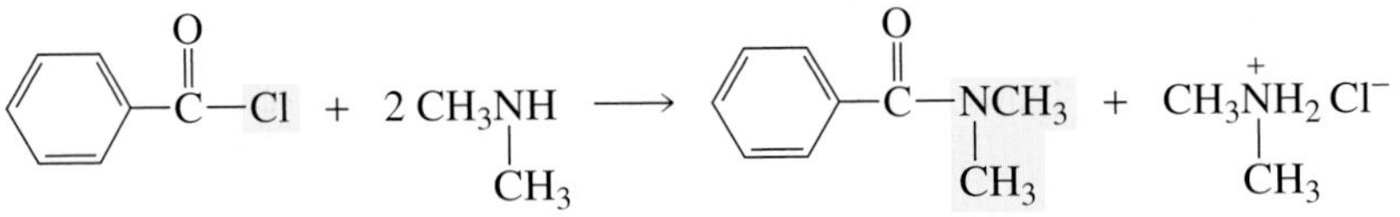

Because tertiary amines cannot form amides, an equivalent of a tertiary amine such as triethylamine or pyridine can be used instead of excess amine.

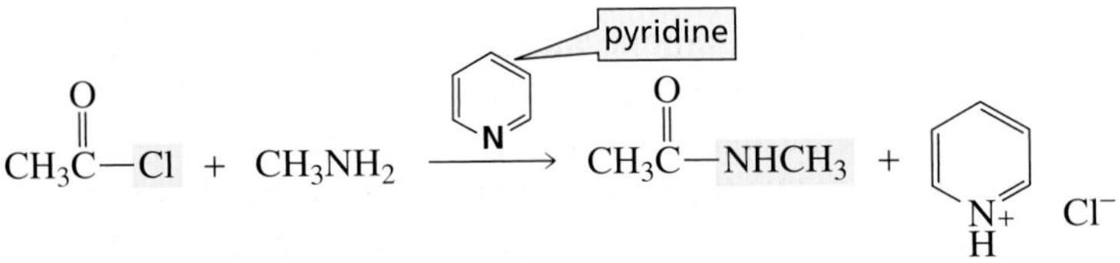

PROBLEM 9 / SOLVED

a. Two amides are obtained from the reaction of acetyl chloride with a mixture of ethylamine and propylamine. Identify the amides.

b. Only one amide is obtained from the reaction of acetyl chloride with a mixture of ethylamine and pyridine. Why is only one amide obtained?

SOLUTION TO 9a Either of the amines can react with acetyl chloride, so both *N*-ethylacetamide and *N*-propylacetamide are formed.

$$CH_3\overset{O}{\overset{\|}{C}}{-}Cl + CH_3CH_2NH_2 + CH_3CH_2CH_2NH_2 \longrightarrow \underset{\textit{N}\text{-ethylacetamide}}{CH_3\overset{O}{\overset{\|}{C}}{-}NHCH_2CH_3} + \underset{\textit{N}\text{-propylacetamide}}{CH_3\overset{O}{\overset{\|}{C}}{-}NHCH_2CH_2CH_3}$$

$$+ CH_3CH_2\overset{+}{N}H_3\,Cl^- + CH_3CH_2CH_2\overset{+}{N}H_3\,Cl^-$$

SOLUTION TO 9b   Initially, two amides are formed. However, the amide formed by the tertiary amine is very reactive because it has a positively charged nitrogen atom, which makes it an excellent leaving group. Therefore, it will react immediately with un-reacted ethylamine, making *N*-ethylacetamide the only amide obtained from the reaction.

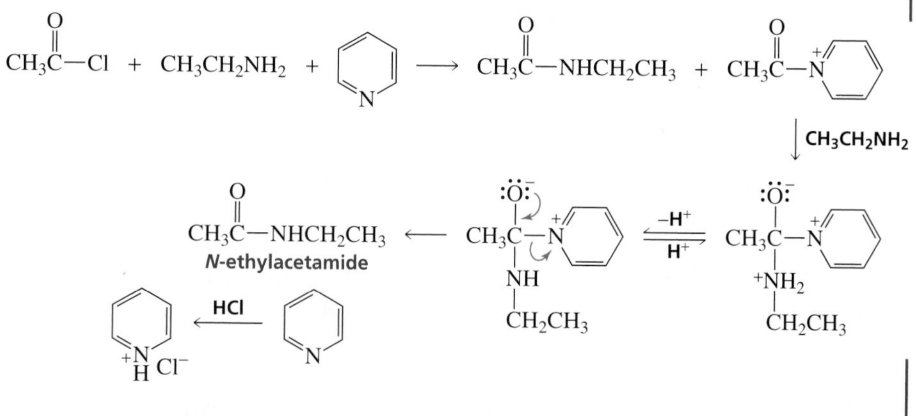

$$CH_3\overset{O}{\overset{\|}{C}}-Cl \; + \; CH_3CH_2NH_2 \; + \; \underset{N}{\bigcirc} \; \longrightarrow \; CH_3\overset{O}{\overset{\|}{C}}-NHCH_2CH_3 \; + \; CH_3\overset{O}{\overset{\|}{C}}-\overset{+}{N}\bigcirc$$

$$\downarrow CH_3CH_2NH_2$$

$$\underset{\substack{\textit{N}\text{-ethylacetamide}}}{CH_3\overset{O}{\overset{\|}{C}}-NHCH_2CH_3} \; \longleftarrow \; CH_3\overset{:\ddot{O}:^-}{\underset{\underset{CH_2CH_3}{NH}}{\overset{|}{C}}}-\overset{+}{N}\bigcirc \; \underset{\overset{H^+}{\rightleftharpoons}}{\overset{-H^+}{}} \; CH_3\overset{:\ddot{O}:^-}{\underset{\underset{CH_2CH_3}{^+NH_2}}{\overset{|}{C}}}-\overset{+}{N}\bigcirc$$

$$\underset{\overset{+}{\underset{H}{N}}\,Cl^-}{\bigcirc} \; \overset{HCl}{\longleftarrow} \; \underset{N}{\bigcirc}$$

PROBLEM 10

Although excess amine is necessary in the reaction of an acyl chloride with an amine, it is not necessary to use excess alcohol in the reaction of an acyl chloride with an alcohol. Explain.

PROBLEM 11

Write the mechanism for the following:

a. the reaction of acetyl chloride with water to form acetic acid.

b. the reaction of acetyl bromide with methyl amine to form *N*-methylacetamide.

PROBLEM 12◆

Starting with acetyl chloride, what nucleophile would you use to make each of the following?

a. $CH_3\overset{O}{\overset{\|}{C}}OCH_2CH_2CH_3$

d. $CH_3\overset{O}{\overset{\|}{C}}O\overset{O}{\overset{\|}{C}}CH_3$

b. $CH_3\overset{O}{\overset{\|}{C}}NHCH_2CH_3$

e. 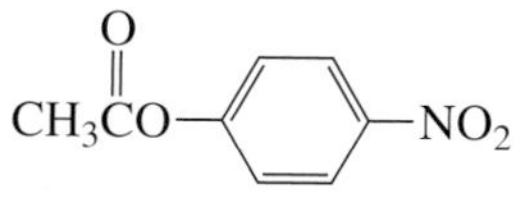 $CH_3\overset{O}{\overset{\|}{C}}O-\underset{}{\bigcirc}-NO_2$

c. $CH_3\overset{O}{\overset{\|}{C}}N(CH_3)_2$

f. $CH_3\overset{O}{\overset{\|}{C}}OH$

Acid anhydrides do not react with sodium chloride or with sodium bromide because the incoming halide ion is a weaker base than the departing carboxylate ion (Table 16.1).

16.8
REACTIONS OF ACID ANHYDRIDES

$$CH_3\overset{O}{\overset{\|}{C}}-O-\overset{O}{\overset{\|}{C}}CH_3 \; + \; Cl^- \; \longrightarrow \; \text{no reaction}$$

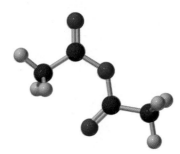

acetic anhydride

Because the incoming halide ion is the weaker base, it will be the substituent expelled from the tetrahedral intermediate.

$$CH_3\overset{\ddot{O}:}{\overset{\|}{C}}-O-\overset{O}{\overset{\|}{C}}CH_3 \; + \; :\ddot{C}l:^- \; \rightleftharpoons \; CH_3\overset{:\ddot{O}:^-}{\underset{Cl}{\overset{|}{C}}}-O-\overset{O}{\overset{\|}{C}}CH_3$$

An acid anhydride reacts with an alcohol to form an ester and a carboxylic acid, it reacts with water to form two equivalents of a carboxylic acid, and it reacts with an amine to form an amide and a carboxylic acid. In each case, the incoming nucleophile (after it loses a proton) is a stronger base than the departing carboxylate ion. In the reaction of an amine with an anhydride, two equivalents of the amine or one equivalent of the amine plus one equivalent of a tertiary amine such as pyridine must be used so that sufficient amine is present to react with the proton produced in the reaction.

$$\underset{\text{acetic anhydride}}{CH_3\overset{O}{\overset{\|}{C}}-O-\overset{O}{\overset{\|}{C}}CH_3} \; + \; CH_3CH_2OH \; \longrightarrow \; \underset{\text{ethyl acetate}}{CH_3\overset{O}{\overset{\|}{C}}-OCH_2CH_3} \; + \; \underset{\text{acetic acid}}{CH_3\overset{O}{\overset{\|}{C}}OH}$$

$$\underset{\text{benzoic anhydride}}{\bigcirc\!\!-\!\!\overset{O}{\overset{\|}{C}}-O-\overset{O}{\overset{\|}{C}}\!\!-\!\!\bigcirc} \; + \; H_2O \; \longrightarrow \; 2 \; \underset{\text{benzoic acid}}{\bigcirc\!\!-\!\!\overset{O}{\overset{\|}{C}}-OH}$$

$$\underset{\text{propionic anhydride}}{CH_3CH_2\overset{O}{\overset{\|}{C}}-O-\overset{O}{\overset{\|}{C}}CH_2CH_3} \; + \; 2\,CH_3NH_2 \; \longrightarrow \; \underset{\text{N-methylpropionamide}}{CH_3CH_2\overset{O}{\overset{\|}{C}}-NHCH_3} \; + \; CH_3CH_2\overset{O}{\overset{\|}{C}}-O^- \; \overset{+}{H_3NCH_3}$$

All the reactions follow the general mechanism described in Section 16.6. For example, compare the following mechanism for conversion of an acid anhydride into an ester with the mechanism for conversion of an acyl chloride into an ester in Section 16.6.

mechanism for the conversion of an acid anhydride into an ester (and a carboxylic acid)

$$CH_3\overset{\ddot{O}:}{\overset{\|}{C}}-\ddot{O}-\overset{\ddot{O}:}{\overset{\|}{C}}CH_3 \; + \; CH_3\ddot{O}H \; \rightleftharpoons \; CH_3\underset{\underset{H}{\overset{+}{\underset{|}{\overset{|}{:}OCH_3}}}}{\overset{:\ddot{O}:^-}{\overset{|}{C}}}-\ddot{O}-\overset{\ddot{O}:}{\overset{\|}{C}}CH_3 \; \underset{H^+}{\overset{-H^+}{\rightleftharpoons}} \; CH_3\underset{:\ddot{O}CH_3}{\overset{:\ddot{O}:^-}{\overset{|}{C}}}-\ddot{O}-\overset{\ddot{O}:}{\overset{\|}{C}}CH_3$$

$$\downarrow$$

$$CH_3\overset{:\ddot{O}}{\overset{\|}{C}}-\ddot{O}CH_3 \; + \; ^-:\ddot{O}-\overset{\ddot{O}:}{\overset{\|}{C}}CH_3$$

PROBLEM 13

a. Propose a mechanism for the reaction of acetic anhydride with water.

b. How does this mechanism differ from the mechanism for the reaction of acetic anhydride with an alcohol?

PROBLEM 14

We have seen that acid anhydrides react with alcohols, water, and amines. In which one of these three reactions does the tetrahedral intermediate not have to lose a proton before it eliminates the carboxylate ion? Explain.

Esters do not react with halide ions or with carboxylate ions because these nucle-ophiles are much weaker bases than the RO^- leaving group of the ester (Table 16.1). An ester reacts with water to form a carboxylic acid and an alcohol—a reaction with water that converts one compound into two compounds is called **hydrolysis.** *(Lysis is Greek for "breaking down.")* An ester reacts with an alcohol to form a new ester and a new alcohol—such a reaction with an alcohol is called **alcoholysis.** This par-ticular alcoholysis reaction is also called a **transesterification reaction** because one ester is converted to another ester.

$$CH_3\overset{O}{\overset{\|}{C}}-OCH_3 \;+\; H_2O \;\overset{H^+}{\rightleftharpoons}\; CH_3\overset{O}{\overset{\|}{C}}-OH \;+\; CH_3OH$$

methyl acetate   acetic acid

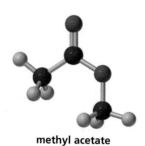

methyl acetate

methyl benzoate $\;\;$ $+$ CH_3CH_2OH $\overset{H^+}{\rightleftharpoons}$ ethyl benzoate $\;\;$ $+$ CH_3OH

methyl benzoate   ethyl benzoate

The hydrolysis or alcoholysis of an ester is a very slow reaction, so these reactions are always catalyzed when they are carried out in the laboratory. Both hydrolysis and alcoholysis of an ester can be catalyzed by acids (H^+) (see Section 16.10). The rate of hydrolysis can also be increased by hydroxide ion (HO^-), and the rate of alcoholysis can be increased by the conjugate base (RO^-) of the reactant alcohol (Section 16.11).

Esters also react with amines to form amides. A reaction with an amine that converts one compound into two compounds is called **aminolysis.** Notice that the aminolysis of an ester requires only one equivalent of amine, unlike the aminolysis of an acyl halide or an acid anhydride, which requires two equivalents (Sections 16.7 and 16.8). This is because the leaving group of an ester (RO^-) is more basic than the amine, so the alkox-ide ion—rather than unreacted amine—picks up the proton generated in the reaction.

$$CH_3CH_2\overset{O}{\overset{\|}{C}}-OCH_2CH_3 \;+\; CH_3NH_2 \;\longrightarrow\; CH_3CH_2\overset{O}{\overset{\|}{C}}-NHCH_3 \;+\; CH_3CH_2OH$$

ethyl propionate   *N*-methylpropionamide

The tetrahedral intermediate formed when an ester is hydrolyzed has both an RO^- leaving group and an HO^- leaving group. Because both groups have approximate-ly the same basicity (Table 16.1), they are equally likely to be expelled.

$$R-\overset{:\ddot{O}:^-}{\underset{OH}{\overset{|}{C}}}-OCH_3 \qquad\qquad R-\overset{:\ddot{O}:^-}{\underset{OH}{\overset{|}{C}}}-OCH_3$$

expels HO^- to
form the ester

expels CH_3O^- to
form the carboxylic acid

$$R-\overset{O}{\overset{\|}{C}}-OCH_3 \;+\; HO^- \qquad\qquad R-\overset{O}{\overset{\|}{C}}-OH \;+\; CH_3O^-$$

$$R-\overset{O}{\overset{\|}{C}}-O^- \;+\; CH_3OH$$

Notice that when CH_3O^- is expelled, the final products are not the carboxylic acid and methoxide ion because if only one species is protonated, it will be the more basic one. Because CH_3O^- is more basic than $RCOO^-$, the final products are the carboxylate ion and methanol. This irreversible acid-base reaction drives the equilibrium towards products.

Because the two leaving groups in the tetrahedral intermediate have approximately the same basicity, when the reaction comes to equilibrium about half the ester molecules will have been converted to carboxylic acid and the other half will remain as ester. The equilibrium can be driven to the right if the reaction is carried out in excess water (see Le Châtelier's principle in Section 9.4).

$$CH_3\overset{O}{\overset{\|}{C}}-OCH_3 \; + \; \underset{\text{excess}}{H_2O} \; \overset{H^+}{\rightleftharpoons} \; CH_3\overset{O}{\overset{\|}{C}}-OH \; + \; CH_3OH$$

An ester reacts with an alcohol to form a new ester and a new alcohol. This is called a transesterification reaction. Because both alcohols have approximately the same leaving tendency in the tetrahedral intermediate, an excess of the reactant alcohol is necessary to drive the equilibrium to the right. Or, if the boiling point of the product alcohol is significantly lower than the boiling points of the other reaction components, the reaction can be driven to the right by distilling off the product alcohol as it is formed. Either H^+ or the conjugate base (RO^-) of the reactant alcohol is used to catalyze the reaction (Sections 16.10 and 16.11).

$$\underset{\text{methyl acetate}}{CH_3\overset{O}{\overset{\|}{C}}-OCH_3} \; + \; \underset{\substack{\text{propyl alcohol}\\\text{excess}}}{CH_3CH_2CH_2OH} \; \overset{H^+}{\rightleftharpoons} \; \underset{\text{propyl acetate}}{CH_3\overset{O}{\overset{\|}{C}}-OCH_2CH_2CH_3} \; + \; \underset{\text{methyl alcohol}}{CH_3OH}$$

3-D Molecule:
Phenyl acetate

In Section 6.10 we saw that phenols are stronger acids than alcohols. Therefore, phenolate ions (ArO^-) are weaker bases than alkoxide ions (RO^-) which means that phenyl esters are more reactive than alkyl esters.

$$\underset{\substack{\text{phenyl acetate}}}{CH_3\overset{O}{\overset{\|}{C}}-O-\!\!\bigcirc} \quad \text{is more reactive than} \quad \underset{\text{methyl acetate}}{CH_3\overset{O}{\overset{\|}{C}}-OCH_3}$$

$$\underset{pK_a = 10.0}{\bigcirc\!\!-OH} \qquad\qquad \underset{pK_a = 15.5}{CH_3OH}$$

The reaction of an ester with an amine is also a slow reaction. However, unlike the reaction of an ester with water or an alcohol, the rate of the reaction of an ester with an amine cannot be increased by H^+ or by HO^- or RO^- (Problem 20). Aminolysis of an ester can be driven to completion by using excess amine or by distilling off the alcohol as it is formed.

$$CH_3\overset{O}{\overset{\|}{C}}-OCH_3 \; + \; \underset{\text{excess}}{CH_3(CH_2)_4NH_2} \; \longrightarrow \; CH_3\overset{O}{\overset{\|}{C}}-NH(CH_2)_4CH_3 \; + \; CH_3OH$$

ASPIRIN

A transesterification reaction that blocks prostaglandin synthesis is responsible for the activity of aspirin (acetylsalicylic acid) as an anti-inflammatory agent. Prostaglandins have several different biological functions, one of which is to stimulate inflammation. The enzyme prostaglandin synthase catalyzes the conversion of arachidonic acid into PGH_2, a precursor of prostaglandins and the related thromboxanes (Section 24.5).

$$\text{arachidonic acid} \xrightarrow{\textbf{prostaglandin synthase}} PGH_2 \begin{array}{c} \nearrow \text{prostaglandins} \\ \searrow \text{thromboxanes} \end{array}$$

Prostaglandin synthase is composed of two enzymes. One of the enzymes, cyclooxygenase, has a serine hydroxyl group that is necessary for enzyme activity. In the presence of aspirin, the serine hydroxyl group participates in a transesterification reaction and becomes acetylated. This inactivates the enzyme. Prostaglandin therefore cannot be synthesized, and inflammation is suppressed.

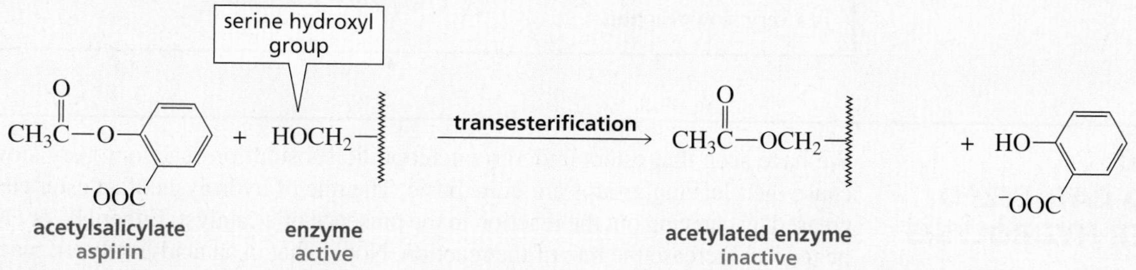

Thromboxanes stimulate platelet aggregation. Because aspirin inhibits the formation of PGH_2, it inhibits thromboxane production and therefore platelet aggregation. Presumably, this is why low levels of aspirin have been reported to reduce the incidence of strokes and heart attacks as a result of blood clot formation.

PROBLEM 15

Write a mechanism for the following reactions:

a. the noncatalyzed hydrolysis of methyl propionate.

b. the aminolysis of phenyl formate using methylamine.

PROBLEM 16 / SOLVED

a. List the following esters in order of decreasing reactivity toward hydrolysis:

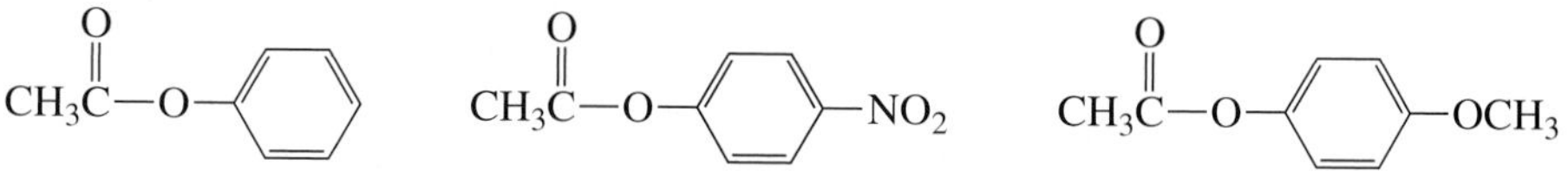

b. How would the rate of hydrolysis of the *para*-methylphenyl ester compare with the rate of hydrolysis of these three esters?

SOLUTION TO 16a Because the nitro group withdraws electrons from the benzene ring and the methoxy group donates electrons into the ring, the nitro-substituted ester will be the most susceptible and the methoxy-substituted ester the least susceptible to nucleophilic attack. Because electron withdrawal increases acidity and electron donation decreases acidity, *para*-nitrophenol is a stronger acid than phenol, which is a stronger acid than *para*-methoxyphenol. Therefore, the *para*-nitrophenolate ion is the weakest base and the best leaving group of the three, whereas the *para*-methoxyphenolate ion is the strongest base and the worst leaving group. Thus, both relatively slow steps of the hydrolysis reaction are fastest for the ester with the electron-withdrawing nitro substituent and slowest for the ester with the electron-donating methoxy substituent.

$$CH_3\overset{O}{\overset{\|}{C}}-O-\langle\rangle-NO_2 \quad > \quad CH_3\overset{O}{\overset{\|}{C}}-O-\langle\rangle \quad > \quad CH_3\overset{O}{\overset{\|}{C}}-O-\langle\rangle-OCH_3$$

SOLUTION TO 16b The methyl substituent donates electrons inductively into the benzene ring but donates electrons to a lesser extent than the resonance-donating methoxy substituent. Therefore the rate of hydrolysis of the methyl-substituted ester is slower than the rate of hydrolysis of the unsubstituted ester, but faster than the rate of hydrolysis of the methoxy-substituted ester.

$$CH_3\overset{O}{\overset{\|}{C}}-O-\langle\rangle \quad > \quad CH_3\overset{O}{\overset{\|}{C}}-O-\langle\rangle-CH_3 \quad > \quad CH_3\overset{O}{\overset{\|}{C}}-O-\langle\rangle-OCH_3$$

PROBLEM 17

State three factors that contribute to the fact that the noncatalyzed hydrolysis of an ester is a very slow reaction.

16.10 ACID-CATALYZED ESTER HYDROLYSIS

We have seen that esters undergo nucleophilic substitution reactions very slowly because their leaving groups are quite basic. The rate of hydrolysis of an ester can be increased by carrying out the reaction in the presence of a catalyst. Either H^+ or HO^- can be used to increase the rate of the reaction. Notice that in an acid-catalyzed reaction all organic reactants, intermediates, and products are positively charged or neutral; *there are no negatively charged organic reactants, intermediates, or products in acidic solutions.* Also notice that in a reaction in which HO^- is used to increase the rate of the reaction, all organic reactants, intermediates, and products are negatively charged or neutral; *there are no positively charged organic reactants, intermediates, or products in basic solutions.*

The first step in the mechanism for acid-catalyzed ester hydrolysis is protonation of the carbonyl oxygen by the acid.

$$CH_3\overset{O}{\overset{\|}{C}}-OCH_3 \underset{-H^+}{\overset{H^+}{\rightleftharpoons}} CH_3\overset{\overset{+}{O}H}{\overset{\|}{C}}-OCH_3$$

The carbonyl oxygen is protonated because it is the atom with the greatest electron density, as shown by the contributing resonance structures and the electrostatic potential map.

$$CH_3-\overset{\overset{\ddot{O}:}{\|}}{C}-\ddot{O}CH_3 \longleftrightarrow CH_3-\overset{\overset{:\ddot{O}:^-}{|}}{C}=\overset{+}{\ddot{O}}CH_3$$

contributing resonance structures of an ester

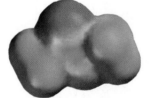
methyl acetate

In the second step of the mechanism, the nucleophile (H_2O) attacks the protonated carbonyl group. The resulting protonated tetrahedral intermediate (tetrahedral intermediate I) is in equilibrium with its nonprotonated form. Either the OH or the OR group of the nonprotonated intermediate can be protonated. Because the OH and OR groups have approximately the same basicity, both tetrahedral intermediate I (OH is protonated) and tetrahedral intermediate II (OR is protonated) are formed. (From Section 1.20 we know that the relative amounts of the protonated intermediates [I and II] compared with the nonprotonated intermediate depend on the pH of the solution and the pK_a values of the protonated intermediates.) When tetrahedral intermediate I collapses, it expels H_2O in preference to CH_3O^- because H_2O is a weaker base, thereby re-forming the ester. When tetrahedral intermediate II collapses, it expels CH_3OH rather than HO^- because CH_3OH is a weaker base, thereby forming the carboxylic acid.

mechanism for acid-catalyzed ester hydrolysis

$$CH_3-\overset{\overset{\displaystyle \ddot{O}:}{\|}}{C}-\ddot{O}CH_3 \underset{-H^+}{\overset{H^+}{\rightleftharpoons}} CH_3-\overset{\overset{\displaystyle \overset{+}{\ddot{O}}H}{\|}}{C}-\ddot{O}CH_3 + H_2\ddot{O}: \rightleftharpoons CH_3-\overset{\overset{\displaystyle :\ddot{O}H}{|}}{\underset{\underset{\displaystyle H}{\overset{+}{:}OH}}{C}}-\ddot{O}CH_3$$

tetrahedral intermediate I

$$H^+ \updownarrow -H^+$$

$$CH_3-\overset{\overset{\displaystyle :\ddot{O}H}{|}}{\underset{\underset{\displaystyle :\underset{\cdot\cdot}{O}H}{|}}{C}}-\ddot{O}CH_3$$

$$-H^+ \updownarrow H^+$$

$$CH_3-\overset{\overset{\displaystyle \ddot{O}:}{\|}}{C}-OH \underset{H^+}{\overset{-H^+}{\rightleftharpoons}} CH_3-\overset{\overset{\displaystyle \overset{+}{\ddot{O}}H}{\|}}{C}-OH + CH_3\ddot{O}H \rightleftharpoons CH_3-\overset{\overset{\displaystyle :\ddot{O}H}{|}}{\underset{\underset{\displaystyle :OH}{|}}{C}}-\underset{\underset{\displaystyle H}{}}{\overset{+}{\ddot{O}}}CH_3$$

tetrahedral intermediate II

Because H_2O and CH_3OH have approximately the same basicity, it will be as likely for tetrahedral intermediate I to collapse to re-form the ester as it will for tetrahedral intermediate II to collapse to form the carboxylic acid. Consequently, when the reaction has reached equilibrium, both ester and carboxylic acid will be present.

$$\overset{\overset{\displaystyle O}{\|}}{R C}-OCH_3 + H_2O \overset{H^+}{\rightleftharpoons} \overset{\overset{\displaystyle O}{\|}}{R C}-OH + CH_3OH$$

both ester and carboxylic acid will be present when the reaction has reached equilibrium

Tutorial:
Manipulating the equilibrium

Excess water will force the equilibrium to the right (Le Châtelier's principle).

$$\overset{\overset{\displaystyle O}{\|}}{R C}-OCH_3 + \underset{\text{excess}}{H_2O} \overset{H^+}{\rightleftharpoons} \overset{\overset{\displaystyle O}{\|}}{R C}-OH + CH_3OH$$

The mechanism for the acid-catalyzed reaction of a carboxylic acid and an alcohol to form an ester and water is the exact reverse of the mechanism for the acid-catalyzed hydrolysis of an ester to form a carboxylic acid and an alcohol. If the ester is the desired product, the reaction should be carried out under conditions that will drive the equilibrium to the left—using excess alcohol or removing water as it is formed (Section 16.12).

$$\overset{\overset{\displaystyle O}{\|}}{R C}-OCH_3 + H_2O \overset{H^+}{\rightleftharpoons} \overset{\overset{\displaystyle O}{\|}}{R C}-OH + \underset{\text{excess}}{CH_3OH}$$

PROBLEM 18

Referring to the mechanism for the acid-catalyzed hydrolysis of methyl acetate, propose a mechanism for the acid-catalyzed reaction of acetic acid and methanol to form methyl acetate.

Now let's see how H^+ increases the rate of ester hydrolysis. For a catalyst to increase the rate of a reaction, it must increase the rate of the slow step of the reaction. Changing the rate of a fast step will not affect the rate of the overall reaction. Four of the six steps in the mechanism for acid-catalyzed ester hydrolysis are proton transfer steps. Proton transfer to or from an electronegative atom such as oxygen or nitrogen is a fast step. So there are two relatively slow steps in the mechanism—formation of a tetrahedral intermediate and collapse of a tetrahedral intermediate. H^+ increases the rates of both of these steps.

H^+ increases the rate of formation of a tetrahedral intermediate by protonating the carbonyl oxygen. Protonated carbonyl groups are more susceptible than non-protonated carbonyl groups to nucleophilic attack because a positively charged oxygen is more electron withdrawing than a neutral oxygen. Increased electron withdrawal by the oxygen increases the electron deficiency of the carbonyl carbon, which increases its attractiveness to nucleophiles.

protonation of the carbonyl oxygen increases the susceptibility of the carbonyl carbon to nucleophilic attack

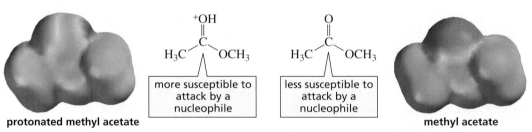

| protonated methyl acetate | methyl acetate |

H^+ increases the rate of collapse of a tetrahedral intermediate by decreasing the basicity of the leaving group, which makes it easier to eliminate. In the acid-catalyzed hydrolysis of an ester, the leaving group is ROH, a weaker base than the leaving group (RO^-) in the non-acid-catalyzed reaction.

The reaction of an ester with an alcohol (transesterification) is also catalyzed by H^+. The mechanism for transesterification is identical to the mechanism for hydrolysis except that the nucleophile is ROH rather than H_2O. As in hydrolysis, the leaving groups in the tetrahedral intermediate formed in transesterification have approximately the same basicity. Consequently, an excess of the reactant alcohol must be used to produce more of the desired product.

methyl benzoate + CH₃CH₂OH ⇌ (H⁺) ethyl benzoate + CH₃OH

methyl benzoate   ethyl alcohol   ethyl benzoate   methyl alcohol
              excess

PROBLEM 19

Write the mechanism for the acid-catalyzed transesterification reaction of methyl acetate with ethanol.

16.11 HYDROXIDE-ION-PROMOTED ESTER HYDROLYSIS

The rate of hydrolysis of an ester can also be increased by carrying out the reaction in a basic solution. Like H^+, hydroxide ion increases the rates of both slow steps of the reaction.

Hydroxide ion increases the rate of formation of the tetrahedral intermediate because HO^- is a better nucleophile than H_2O, so it more readily attacks the carbonyl carbon. Hydroxide ion also increases the rate of collapse of the tetrahedral intermediate because a smaller fraction of the negatively charged tetrahedral intermediate becomes protonated in a basic solution. A negatively charged oxygen can more readily expel the very basic leaving group (RO^-) because the oxygen does not develop a partial positive charge in the transition state.

mechanism for hydroxide-ion-promoted hydrolysis of an ester

$$CH_3-\overset{\overset{\displaystyle :\ddot{O}:}{\|}}{C}-\ddot{O}CH_3 + H\ddot{O}^{\overline{\cdot}} \; \rightleftharpoons \; CH_3-\overset{\overset{\displaystyle :\ddot{O}:^-}{|}}{\underset{\underset{\displaystyle :OH}{|}}{C}}-\ddot{O}CH_3 \; \rightleftharpoons \; CH_3-\overset{\overset{\displaystyle \ddot{O}:}{\|}}{C}-\ddot{O}H + CH_3\ddot{O}:^{\overline{\cdot}}$$

$$HO^-\;\|\;H_2O \qquad\qquad\qquad\qquad\qquad \downarrow$$

$$CH_3-\overset{\overset{\displaystyle :\ddot{O}H}{|}}{\underset{\underset{\displaystyle :OH}{|}}{C}}-\ddot{O}CH_3 \qquad\qquad CH_3-\overset{\overset{\displaystyle \ddot{O}:}{\|}}{C}-\ddot{O}:^{\overline{\cdot}} + CH_3\ddot{O}H$$

The final product of this reaction is a carboxylate ion rather than a carboxylic acid because the carboxylic acid is formed in a basic solution. Because carboxylate ions are negatively charged, they are not attacked by nucleophiles. Because carboxylate ions are not attacked by nucleophiles, the hydroxide-ion-promoted hydrolysis of an ester is not a reversible reaction.

This reaction is called a hydroxide-ion-promoted reaction rather than a base-catalyzed reaction because hydroxide ion increases the rate of the first step of the reaction by being a better nucleophile than water—not by being a stronger base than water—and because hydroxide ion is consumed in the overall reaction. A catalyst is regenerated, so hydroxide ion is actually a reagent rather than a catalyst. Therefore, it is more accurate to call the reaction a hydroxide-ion-*promoted* reaction than a hydroxide-ion-*catalyzed* reaction. We will use the former term from now on.

Hydroxide ion promotes only hydrolysis reactions, not transesterification reactions or aminolysis reactions. Hydroxide ion cannot promote reactions of carboxylic acid derivatives with alcohols or with amines because one function of hydroxide ion is to provide a strong nucleophile for the first step of the reaction. Thus, when the nucleophile is supposed to be an alcohol or an amine, nucleophilic attack by hydroxide ion would form a product different from the product that would be formed from nucleophilic attack by an alcohol or amine. In a hydrolysis reaction, however, the same product is formed whether the attacking nucleophile is H_2O or HO^-.

Reactions in which the nucleophile is an alcohol can be promoted by the conjugate base of the alcohol.

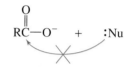

$$\text{(C}_6\text{H}_5)-\overset{\overset{\displaystyle O}{\|}}{C}-OCH_3 + CH_3CH_2OH \;\underset{\text{excess}}{\overset{CH_3CH_2O^-}{\rightleftharpoons}}\; \text{(C}_6\text{H}_5)-\overset{\overset{\displaystyle O}{\|}}{C}-OCH_2CH_3 + CH_3OH$$

PROBLEM 20◆

a. What species other than H^+ can be used to increase the rate of a transesterification reaction that converts methyl acetate to propyl acetate?

b. Explain why the rate of aminolysis of an ester cannot be increased by H^+, HO^-, or RO^-.

You have seen that nucleophilic acyl substitution reactions take place by a mechanism in which a tetrahedral intermediate is formed and subsequently collapses. The tetrahedral intermediate, however, is too unstable to be isolated. How then do we know it is formed? How do we know that the reaction doesn't take place by a one-step direct displacement mechanism—similar to the mechanism of an S_N2 reaction—without formation of a tetrahedral intermediate?

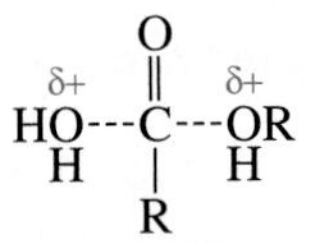

**transition state for a hypothetical one-step
direct displacement mechanism**

*Myron L. Bender (1924–1988)
was born in St. Louis. He was a
professor of chemistry at the
Illinois Institute of Technology
and at Northwestern University.*

To answer this question, Myron Bender investigated the hydroxide-ion-promoted hydrolysis of ethyl benzoate, with the carbonyl oxygen of ethyl benzoate labeled with ^{18}O. When he isolated ethyl benzoate from the reaction mixture, he found that some of the ester was no longer labeled. If the reaction had taken place by a direct displacement mechanism, all the isolated ester would have been labeled. If the mechanism involved a tetrahedral intermediate, on the other hand, some of the isolated ester would no longer be labeled. By this experiment, Bender provided evidence for the formation of a tetrahedral intermediate.

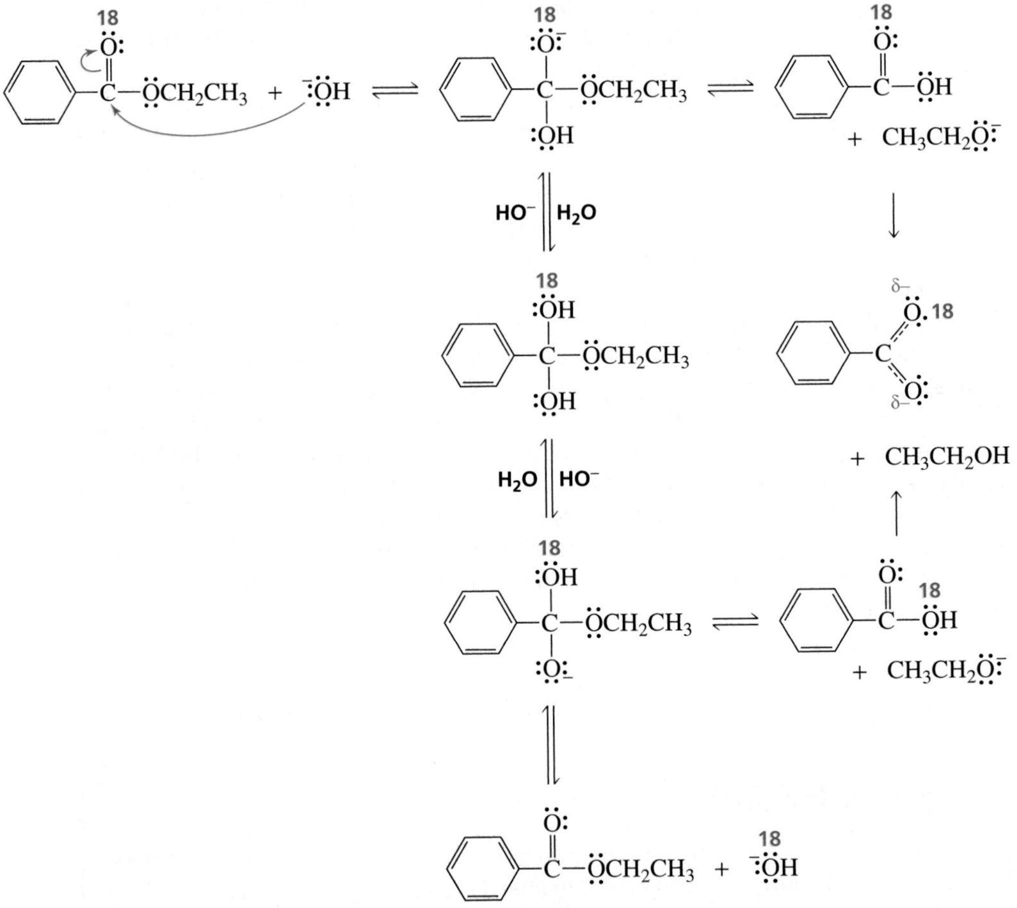

unlabeled ester

PROBLEM 21◆

D. N. Kursanov, a Russian chemist, proved that the bond broken in the hydroxide-ion-promoted hydrolysis of an ester is the acyl C—O bond rather than the alkyl C—O bond by studying the reaction of the following ester with HO^-/H_2O:

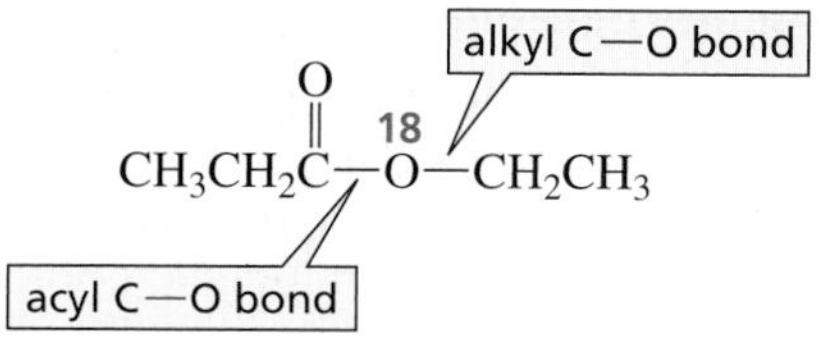
alkyl C—O bond
acyl C—O bond

a. Which of the products contained the ^{18}O label?

b. What product would have contained the ^{18}O label if the alkyl C—O bond had broken?

PROBLEM 22 / SOLVED

Early chemists could envision several possible mechanisms for hydroxide-ion-promoted ester hydrolysis:

1. a nucleophilic acyl substitution mechanism

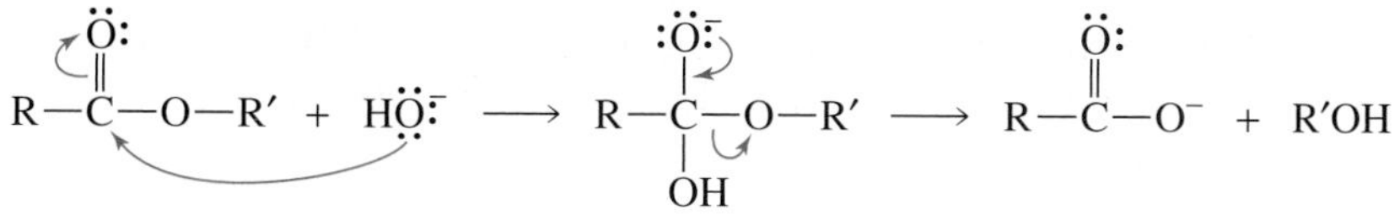

2. an S_N2 mechanism

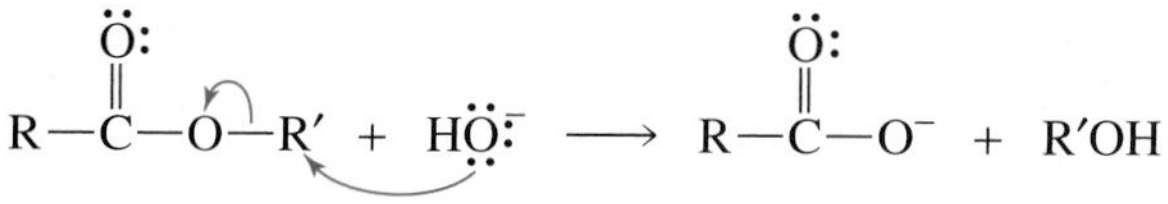

3. an S_N1 mechanism

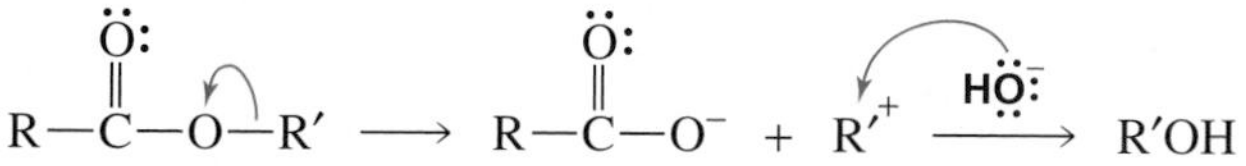

Devise an experiment that would distinguish among these three mechanisms.

SOLUTION Start with a single stereoisomer of an alcohol that has its OH group bonded to a chirality center, and determine the specific rotation of the alcohol. Convert the alcohol into an ester using a method that does not break any bonds to the chirality center. Hydrolyze the ester. Isolate the alcohol obtained from hydrolysis and determine its specific rotation.

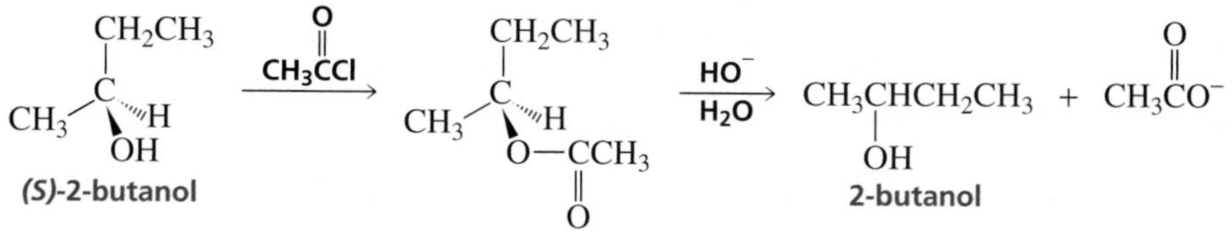
(S)-2-butanol
2-butanol

If the reaction is a nucleophilic acyl substitution reaction, the product alcohol will have the same specific rotation as the reactant alcohol because no bonds to the chirality center are broken.

If the reaction is an S_N2 reaction, the product alcohol and the reactant alcohol will have opposite specific rotations because the mechanism requires backside attack on the chirality center.

If the reaction is an S_N1 reaction, the product alcohol will have a small specific rotation because the mechanism requires carbocation formation, which leads to partial racemization of the alcohol.

16.12
REACTIONS OF
CARBOXYLIC ACIDS

The leaving group of a carboxylic acid (HO^-) has approximately the same basicity as the leaving group of an ester (RO^-). Therefore, carboxylic acids have approximately the same reactivity as esters. Like esters, carboxylic acids do not react with halide ions or with carboxylate ions.

A carboxylic acid reacts with an alcohol to form an ester. It is a very slow reaction, however, so it is always carried out in the presence of an acid catalyst (Section 16.9). Because the tetrahedral intermediate formed in this reaction has two potential leaving groups of approximately the same basicity (HO^- and RO^-), the reaction must be carried out with excess alcohol to drive it toward products. Emil Fischer (Section 4.4) was the first to discover that an ester could be prepared by treating a carboxylic acid with excess alcohol in the presence of an acid catalyst, so the reaction is called the **Fischer esterification reaction.**

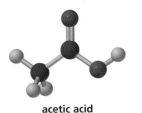

acetic acid

$$CH_3\overset{O}{\overset{\|}{C}}-OH + \underset{\substack{\text{methyl alcohol}\\\text{excess}}}{CH_3OH} \underset{}{\overset{H^+}{\rightleftharpoons}} CH_3\overset{O}{\overset{\|}{C}}-OCH_3 + H_2O$$

acetic acid   methyl alcohol   methyl acetate

A carboxylic acid is an acid—it donates a proton, and an amine is a base—it accepts a proton. Because a carboxylic acid has a lower pK_a than a protonated amine, the carboxylic acid immediately donates a proton to the amine when the two compounds are mixed. The protonated amine is not a nucleophile, so it cannot attack the carbonyl carbon. Therefore, the ammonium carboxylate salt is the final product of the reaction. When, however, the salt is heated, it loses water and forms the amide.

$$CH_3\overset{O}{\overset{\|}{C}}-OH + CH_3CH_2NH_2 \longrightarrow \underset{\substack{\text{an ammonium}\\\text{carboxylate salt}}}{CH_3\overset{O}{\overset{\|}{C}}-O^- \overset{+}{H_3N}CH_2CH_3} \overset{225\ °C}{\longrightarrow} CH_3\overset{O}{\overset{\|}{C}}-NHCH_2CH_3 + H_2O$$

an ammonium
carboxylate salt

$$CH_3CH_2\overset{O}{\overset{\|}{C}}-OH + NH_3 \longrightarrow CH_3CH_2\overset{O}{\overset{\|}{C}}-O^- \overset{+}{N}H_4 \overset{225\ °C}{\longrightarrow} CH_3CH_2\overset{O}{\overset{\|}{C}}-NH_2 + H_2O$$

The basic form of a carboxylic acid will not undergo nucleophilic acyl substitution reactions because the negatively charged carboxylate ion is resistant to nucleophilic attack. So carboxylic acids can undergo nucleophilic acyl substitution reactions only if they are in their acidic forms. When they are in their basic forms, they are even less reactive toward nucleophilic acyl substitution reactions than amides.

re ative reactivities toward nuc eophi ic acy substitution

$$R-\overset{O}{\overset{\|}{C}}-OH > R-\overset{O}{\overset{\|}{C}}-NH_2 > R-\overset{O}{\overset{\|}{C}}-O^-$$

PROBLEM-SOLVING STRATEGY

Propose a mechanism for the following reaction.

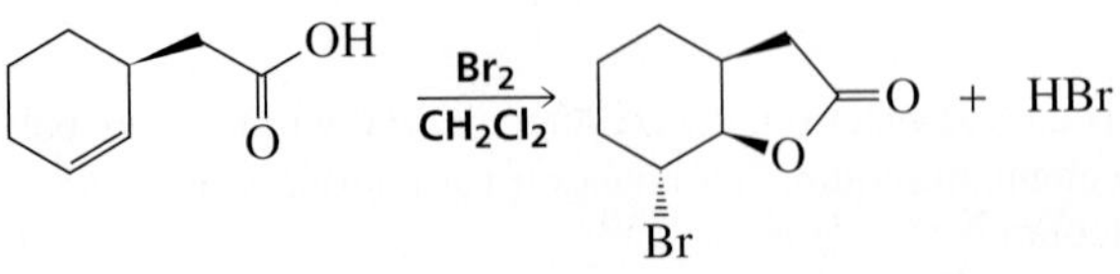

$$\underset{}{\text{cyclohexene-CH}_2\text{COOH}} \overset{Br_2}{\underset{CH_2Cl_2}{\longrightarrow}} \text{bicyclic lactone} =O + HBr$$

When you are asked to propose a mechanism, look carefully at the reagents to determine the first step of the mechanism. One of the given reagents has two functional groups, a carboxyl group and a carbon-carbon double bond. The other reagent, Br_2, does not react with carboxylic acids but does undergo electrophilic addition reactions with alkenes. One side of the alkene is sterically hindered by the carboxylic acid group. Br_2, therefore, will add to the other side of the double bond, forming a bromonium ion. We know that the second step of an alkene addition reaction is attack by a nucleophile. Of the two nucleophiles present, the carbonyl oxygen is positioned to attack the back side of the bromonium ion, resulting in a compound with the observed configuration. Loss of a proton gives the final product of the reaction.

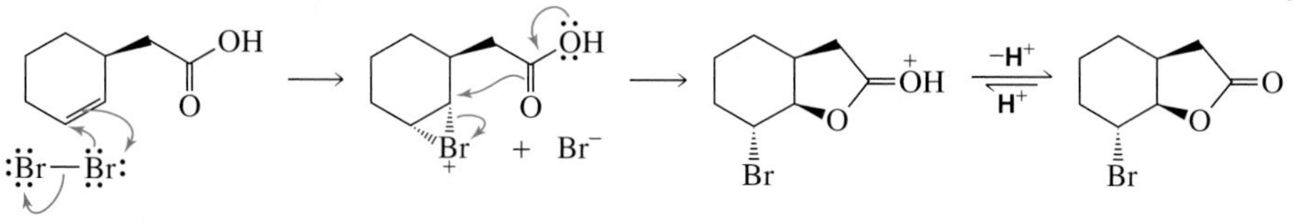

Now continue on to Problem 23.

PROBLEM 23

Propose a mechanism for the following reaction:

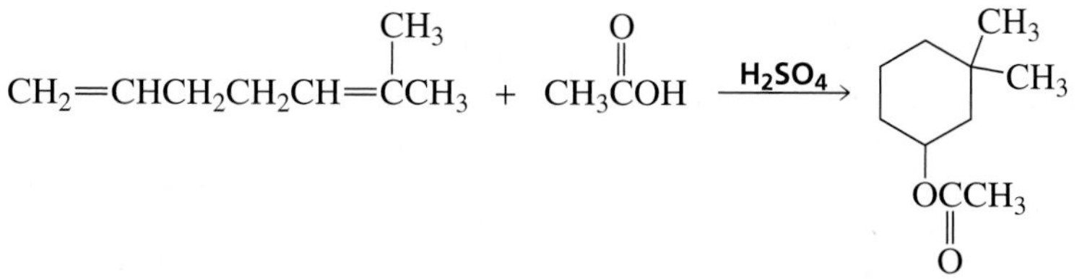

$$CH_2{=}CHCH_2CH_2CH{=}CCH_3 \ + \ CH_3\overset{O}{\overset{\|}{C}}OH \ \xrightarrow{H_2SO_4}$$

with CH_3 groups

Amides do not react with halide ions, carboxylate ions, alcohols, or water because in each case the incoming nucleophile is a weaker base than the leaving group of the amide (Table 16.1).

16.13
REACTIONS OF AMIDES

$$CH_3\overset{O}{\overset{\|}{C}}{-}NHCH_2CH_2CH_3 \ + \ Cl^- \ \longrightarrow \ \text{no reaction}$$
N-propylacetamide

$$CH_3CH_2\overset{O}{\overset{\|}{C}}{-}N(CH_3)_2 \ + \ CH_3\overset{O}{\overset{\|}{C}}O^- \ \longrightarrow \ \text{no reaction}$$
N,N-dimethylpropionamide

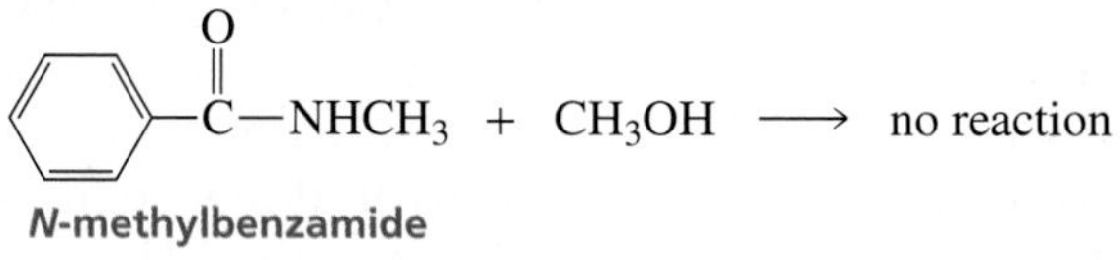

$$\text{C}_6\text{H}_5{-}\overset{O}{\overset{\|}{C}}{-}NHCH_3 \ + \ CH_3OH \ \longrightarrow \ \text{no reaction}$$
N-methylbenzamide

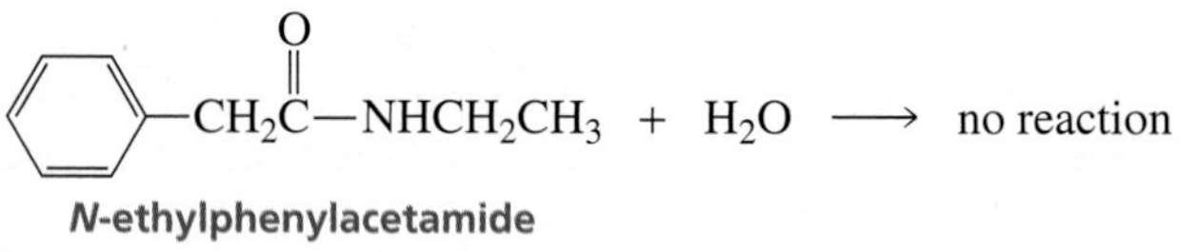

$$\text{C}_6\text{H}_5{-}CH_2\overset{O}{\overset{\|}{C}}{-}NHCH_2CH_3 \ + \ H_2O \ \longrightarrow \ \text{no reaction}$$
N-ethylphenylacetamide

acetamide

Amides, however, can react with water if the reaction mixture is heated in the presence of an acid or hydroxide ion. Amides can also react with an alcohol (ROH) if the reaction mixture is heated with H^+. The reason for this will be explained in Section 16.14.

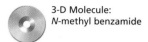

3-D Molecule:
N-methyl benzamide

$$CH_3\overset{O}{\overset{\|}{C}}-NHCH_2CH_3 + H_2O \xrightarrow[\Delta]{H^+} CH_3\overset{O}{\overset{\|}{C}}-OH + CH_3CH_2\overset{+}{N}H_3$$

N-ethylacetamide

$$\text{(phenyl)}-\overset{O}{\overset{\|}{C}}-NHCH_3 + H_2O \xrightarrow[\Delta]{HO^-} \text{(phenyl)}-\overset{O}{\overset{\|}{C}}-O^- + CH_3NH_2$$

N-methylbenzamide

An amide with an NH_2 group can be dehydrated to a nitrile. Dehydration reagents commonly employed for this purpose are P_2O_5, $POCl_3$, or $SOCl_2$.

$$CH_3CH_2\overset{O}{\overset{\|}{C}}-NH_2 \xrightarrow[85\ °C]{P_2O_5} CH_3CH_2C\equiv N$$

NATURE'S SLEEPING PILL

Melatonin, a naturally occuring amide, is a hormone synthesized by the pineal gland from the amino acid tryptophan (Section 21.2). Melatonin regulates the dark-light clock that governs such things as the sleep-wake cycle, body temperature, and hormone production.

Melatonin levels increase from evening to night and then decrease as morning approaches. People with high levels of melatonin sleep longer and more soundly than those with low levels. The concentration of the hormone in the blood varies with age—six-year-olds have more than five times the concentration that 80-year-olds have—the reason why young people have less trouble sleeping than older people. Melatonin supplements are used to treat insomnia, jet lag, and seasonal affective disorder.

melatonin

16.14
HYDROLYSIS OF AMIDES: ACID-CATALYZED HYDROLYSIS AND HYDROXIDE-ION-PROMOTED HYDROLYSIS

An amide cannot be hydrolyzed unless the reaction mixture is heated in the presence of an acid or hydroxide ion.

When the reaction is carried out under acidic conditions, the acid protonates the carbonyl oxygen, increasing the susceptibility of the carbonyl carbon to nucleophilic attack. Nucleophilic attack by water on the carbonyl carbon leads to tetrahedral intermediate I, which is in equilibrium with its nonprotonated form. Reprotonation can occur either on oxygen to reform tetrahedral intermediate I or on nitrogen to form tetrahedral intermediate II. Protonation on nitrogen is favored because the NH_2 group is a stronger base than the OH group. Tetrahedral intermediate II has two leaving groups, HO^- and NH_3. Because NH_3 is the weaker base, it is expelled, forming the carboxylic acid as the final product. (Because the reaction is carried out in an acidic solution, NH_3 will be protonated after it is expelled from the tetrahedral intermediate.)

mechanism for acid-catalyzed hydrolysis of an amide

tetrahedral intermediate I

tetrahedral intermediate II

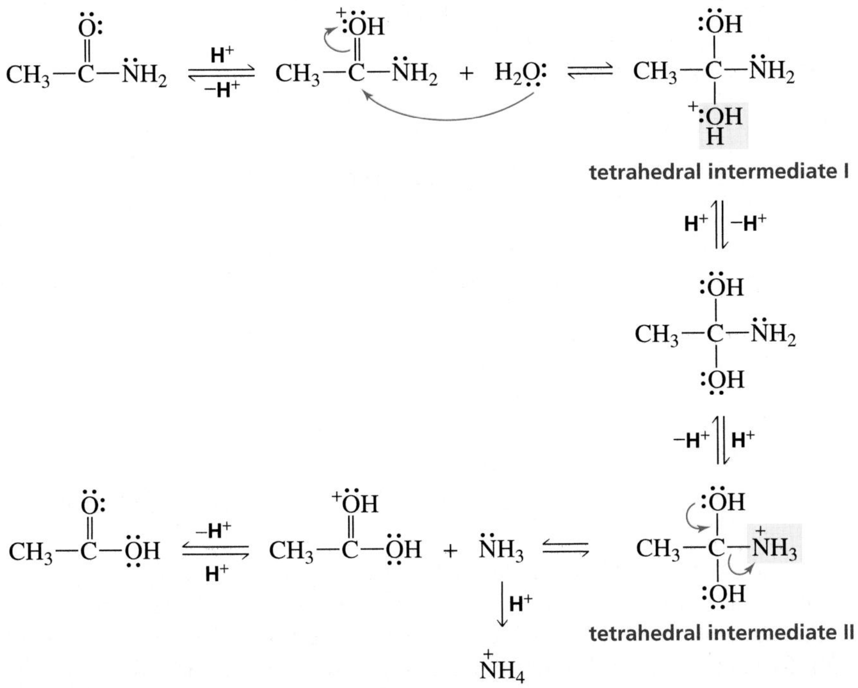

Let's take a minute to see why hydrolysis of an amide does not occur without a catalyst. In the non-acid-catalyzed reaction, the amide is not protonated. Therefore, water, a very poor nucleophile, must attack a neutral amide that is much less susceptible to nucleophilic attack than a protonated amide would be. Consequently, the tetrahedral intermediate is formed very slowly. In addition, the NH_2 group of the tetrahedral intermediate is not protonated in the non-acid-catalyzed reaction. As a result, the two leaving groups of the tetrahedral intermediate are HO^- and $^-NH_2$. Because HO^- is the weaker base, it is more easily expelled, which reforms the amide.

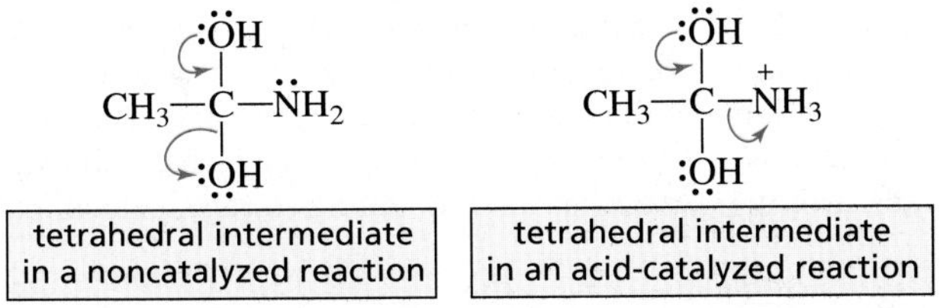

tetrahedral intermediate
in a noncatalyzed reaction

tetrahedral intermediate
in an acid-catalyzed reaction

The addition of acid increases the rate of formation of the tetrahedral intermediate by protonating the amide. The acid also changes the relative leaving abilities of the two groups. In an acidic solution, the two leaving groups are HO^- and NH_3. Because NH_3 is the weaker base, it is expelled and the hydrolysis product results.

In the hydroxide-ion-promoted hydrolysis of an amide, HO^- rather than water is the nucleophile. Because HO^- is a better nucleophile than water, it is better at forming the tetrahedral intermediate. The two leaving groups of the tetrahedral intermediate are HO^- and $^-NH_2$. Because HO^- is the weaker base, it is more easily expelled, but occasionally an $^-NH_2$ is ejected. When this happens, the carboxylic acid that is formed immediately loses a proton. Since this step is irreversible (the negatively charged carboxylate ion cannot be attacked by nucleophiles), it disturbs the equilibrium and drives the reaction toward products (Section 9.4). Because hydroxide ion is consumed in the reaction, it is a reagent, not a catalyst.

mechanism for hydroxide-ion-promoted hydrolysis of an amide

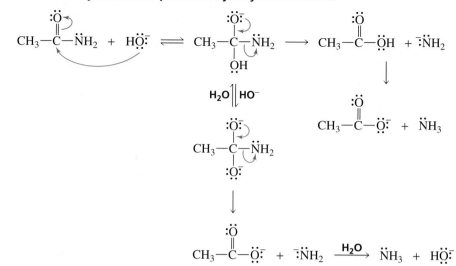

Tutorial:
Conversions between
carboxylic acid derivatives

In strongly basic solutions, the hydrolysis reaction is second-order in hydroxide ion. In other words, two equivalents of hydroxide ion participate in the reaction. The second equivalent of hydroxide ion removes a proton from the OH group of the initially formed tetrahedral intermediate. When the new tetrahedral intermediate collapses, the possible leaving groups are O^{2-} or $^-NH_2$. Because $^-NH_2$ is the weaker base, it is expelled and the hydrolysis reaction occurs. Expelling the strongly basic $^-NH_2$ group is helped by the driving force of the two negatively charged oxygens in the tetrahedral intermediate. Notice that the second equivalent of hydroxide ion is a catalyst because it is regenerated.

An amide reacts with an alcohol in the presence of H^+ for the same reason it reacts with water in the presence of H^+.

PENICILLIN AND DRUG RESISTANCE

Penicillin contains a strained β-lactam ring. The strain in the four-membered ring increases the amide's reactivity. It is thought that the antibiotic activity of penicillin results from its ability to acylate an OH group of an enzyme that is involved in the synthesis of bacterial cell walls. This in-activates the enzyme, and actively growing bacteria die because they are unable to produce functional cell walls. Penicillin has no effect on mammalian cells because mammalian cells are not enclosed by cell walls. To minimize hydrolysis of the β-lactam ring during storage, penicillins are refrigerated.

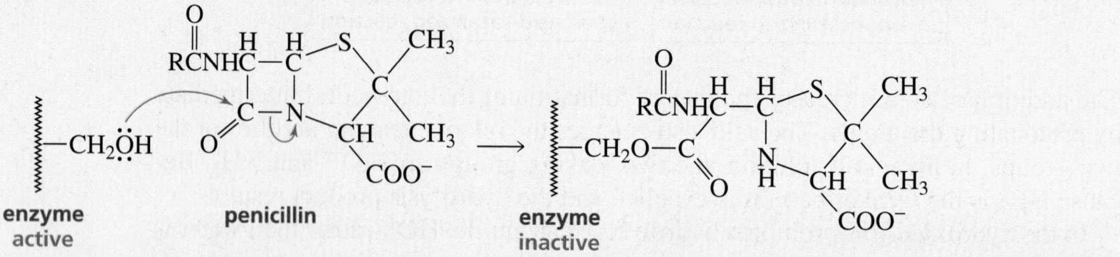

enzyme active   penicillin   enzyme inactive

Bacteria that are resistant to penicillin secrete a β-lactamase, an enzyme that catalyzes the hydrolysis of the β-lactam ring of penicillin. The ring-opened product has no antibacterial activity.

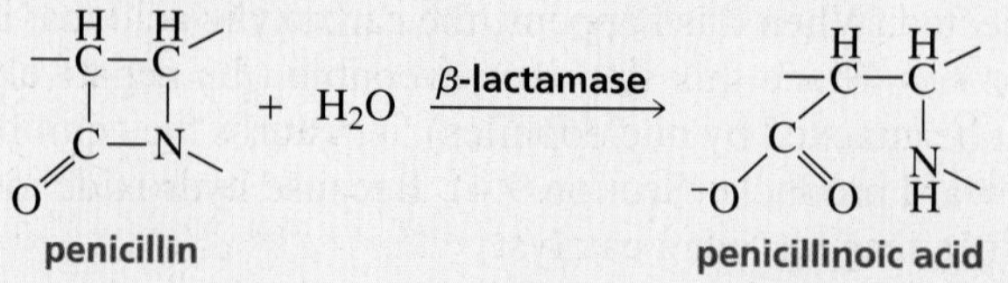

penicillin   penicillinoic acid

PENICILLINS IN CLINICAL USE

More than 10 different penicillins are currently in clinical use. They differ only in the group (R) attached to the carbonyl group. Some of these penicillins are shown here. In addition to their structural differences, the penicillins differ in the organisms against which they are most effective.

They also differ in their resistance to β-lactamase. For example, ampicillin, a synthetic penicillin, is clinically effective against bacteria that are resistant to penicillin G. Almost 19% of humans are allergic to penicillin G, a naturally occurring penicillin.

R
CH_2- **penicillin G**

R
$\text{CH}-$ **ampicillin**
$\quad\quad|$
$\quad\quad\text{NH}_2$

R
$\text{HO}-\text{CH}-$ **amoxicillin**
$\quad\quad\quad|$
$\quad\quad\quad\text{NH}_2$

oxacillin

cloxacillin

$\text{CH}_2{=}\text{CHCH}_2\text{SCH}_2-$ **penicillin O**

$-\text{OCH}_2-$ **penicillin V**

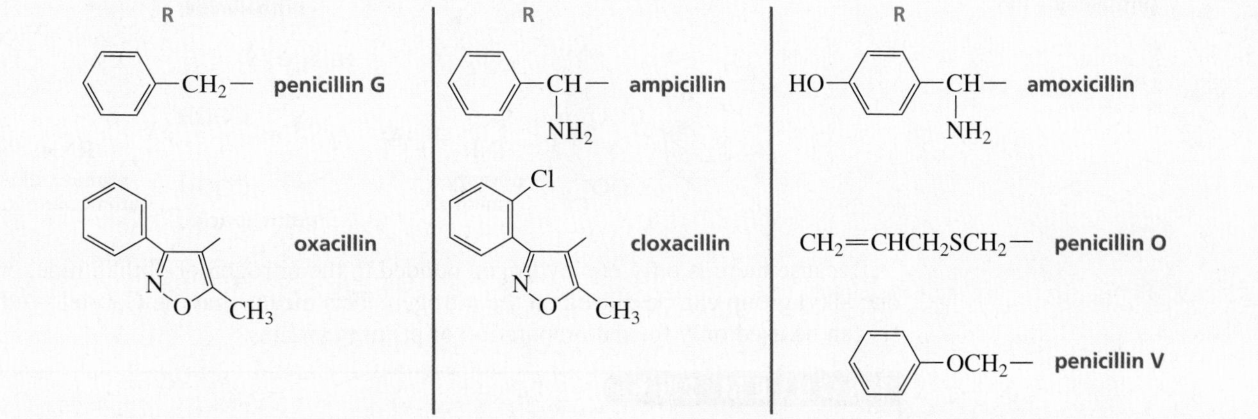

Penicillin V is a semisynthetic penicillin in clinical use. It is not a naturally occurring penicillin but it isn't a true synthetic penicillin either because chemists don't synthesize it. The

Penicillium mold synthesizes it after the mold is fed 2-phenoxyethanol, the compound it needs for the side chain.

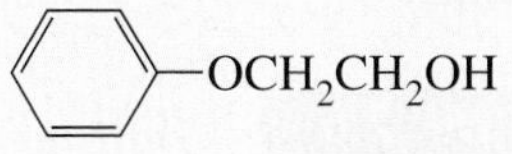

$-\text{OCH}_2\text{CH}_2\text{OH}$

2-phenoxyethanol

PROBLEM 24◆

List the following amides in order of decreasing reactivity toward hydroxide-ion-promoted hydrolysis:

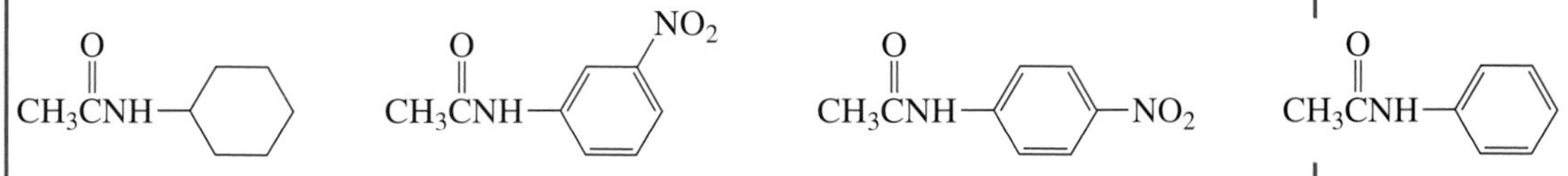

$$\overset{O}{\overset{\|}{\text{CH}_3\text{C}}}\text{NH}- \quad\quad \overset{O}{\overset{\|}{\text{CH}_3\text{C}}}\text{NH}-\overset{\text{NO}_2}{\bigcirc} \quad\quad \overset{O}{\overset{\|}{\text{CH}_3\text{C}}}\text{NH}-\text{NO}_2 \quad\quad \overset{O}{\overset{\|}{\text{CH}_3\text{C}}}\text{NH}-$$

16.15 THE GABRIEL SYNTHESIS OF PRIMARY AMINES

The **Gabriel synthesis,** which converts alkyl halides into primary amines, involves the hydrolysis of an **imide**—a compound with two acyl groups bonded to a nitrogen.

$$\text{RCH}_2\text{Br} \xrightarrow{\text{Gabriel synthesis}} \text{RCH}_2\text{NH}_2$$

In the first step of the reaction, a base removes a proton from the nitrogen of phthalimide. The resulting nucleophile reacts with an alkyl halide. Because this is an S_N2 reaction, it works best with primary alkyl halides (Section 9.2). Hydrolysis of the N-substituted phthalimide can be catalyzed by acid or promoted by hydroxide ion. Acid-catalyzed hydrolysis of the product forms the primary alkyl ammonium ion and phthalic acid. Neutralization of the primary alkyl ammonium

phthalimide ion

3-D Molecule:
Phthalimide

ion with base forms the primary amine. The alkyl group of the primary amine is identical to the alkyl group of the alkyl halide.

phthalimide   **an *N*-substituted phthalimide**

$$\text{H}^+, \text{H}_2\text{O} \Big| \Delta$$

+ RNH$_2$ ⟵$^{\text{HO}^-}$ + $\overset{+}{\text{RNH}}_3$

primary amine   **phthalic acid**   **primary alkyl ammonium ion**

Because there is only one hydrogen bonded to the nitrogen of phthalimide, only one alkyl group can be placed on the nitrogen. This means that the Gabriel synthesis can be used only for the preparation of primary amines.

PROBLEM 25◆

What alkyl bromide would you use in a Gabriel synthesis to prepare each of the following amines?

a. pentylamine

b. isohexylamine

c. benzylamine

d. cyclohexylamine

PROBLEM 26

Primary amines can also be prepared by the reaction of an alkyl halide with azide ion followed by catalytic hydrogenation. What advantage do this method and the Gabriel synthesis have over synthesis of a primary amine using an alkyl halide and ammonia?

$$\text{CH}_3\text{CH}_2\text{CH}_2\text{Br} \xrightarrow{^-\text{N}_3} \text{CH}_3\text{CH}_2\text{CH}_2\text{N}=\overset{+}{\text{N}}=\overset{-}{\text{N}} \xrightarrow[\text{Pt}]{\text{H}_2} \text{CH}_3\text{CH}_2\text{CH}_2\text{NH}_2 + \text{N}_2$$

16.16 HYDROLYSIS OF NITRILES

Nitriles are even harder to hydrolyze than amides. They are slowly hydrolyzed to carboxylic acids when heated with water and either H^+ or HO^-.

$$\text{CH}_3\text{CH}_2\text{C}\equiv\text{N} + \text{H}_2\text{O} \xrightarrow[\Delta]{\text{H}^+} \text{CH}_3\text{CH}_2\overset{\text{O}}{\overset{\|}{\text{C}}}\text{OH} + \overset{+}{\text{NH}}_4$$

$$\text{CH}_3\text{CH}_2\text{C}\equiv\text{N} + \text{H}_2\text{O} \xrightarrow[\Delta]{\text{HO}^-} \text{CH}_3\text{CH}_2\overset{\text{O}}{\overset{\|}{\text{C}}}\text{O}^- + \text{NH}_3$$

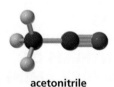

acetonitrile

In the first step of the acid-catalyzed hydrolysis of a nitrile, the acid protonates the nitrogen of the cyano group, making it easier for water to attack the carbon of the cyano group in the next step. Attack on the cyano group by water is analogous to attack on a carbonyl group by water. Because nitrogen is a stronger base than oxygen, oxygen loses a proton and nitrogen gains a proton, resulting in a product that is a protonated amide. Since an amide is easier to hydrolyze than a nitrile, the amide is immediately hydrolyzed to a carboxylic acid following the acid-catalyzed mechanism shown in Section 16.14.

mechanism for acid-catalyzed hydrolysis of a nitrile

$$R—C≡N: \underset{-H^+}{\overset{H^+}{\rightleftharpoons}} R—\overset{+}{C}=\overset{\cdot\cdot}{N}H + H_2\overset{\cdot\cdot}{O}: \rightleftharpoons R—C=\overset{\cdot\cdot}{N}H \underset{H^+}{\overset{-H^+}{\rightleftharpoons}} R—C=\overset{\cdot\cdot}{N}H$$

with :OH, $\overset{+}{H}$ and :OH beneath

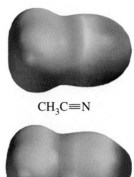

$CH_3C≡N$

$$R—\underset{\underset{\cdot\cdot}{\overset{\|}{O}:}}{C}—OH \underset{\text{(several steps)}}{\overset{H_2O}{\longleftarrow}} R—\underset{\underset{+}{\overset{\|}{OH}}}{C}—\overset{\cdot\cdot}{N}H_2 \longleftrightarrow R—\underset{:OH}{\overset{+}{C}}=\overset{+}{N}H_2$$

with $-H^+ \| H^+$ above the protonated amide

a carboxylic acid   **a protonated amide**

$CH_3C≡\overset{\cdot}{N}H$

The first step in the hydroxide-ion-promoted hydrolysis of a nitrile is attack of hydroxide ion on the carbon of the cyano group. Nitrogen gains a proton from water, oxygen loses a proton, and the product rearranges to an amide. The amide is further hydrolyzed to a carboxylate ion by means of the hydroxide-ion-promoted mechanism shown in Section 16.14.

mechanism for hydroxide-ion-promoted hydrolysis of a nitrile

$$R—C≡N: + H\overset{\cdot\cdot}{\overset{\cdot\cdot}{O}}: \rightleftharpoons R—\underset{:OH}{C}=\overset{\cdot\cdot}{\overset{\cdot\cdot}{N}}: \underset{HO^-}{\overset{H_2O}{\rightleftharpoons}} R—\underset{:OH}{C}=\overset{\cdot\cdot}{N}H \underset{H_2O}{\overset{HO^-}{\rightleftharpoons}} R—\underset{:\overset{\cdot\cdot}{O}:}{C}=\overset{\cdot\cdot}{N}H$$

with H—O: below

$$R—\underset{\overset{\|}{O}}{C}—O^- \underset{\text{(several steps)}}{\overset{H_2O}{\longleftarrow}} R—\underset{\overset{\|}{\underset{\cdot\cdot}{O}:}}{C}—\overset{\cdot\cdot}{N}H_2 + HO^-$$

a carboxylate ion   **an amide**

Because nitriles can be prepared from the reaction of an alkyl halide with cyanide ion (Section 9.4), we now know how to convert an alkyl halide into a carboxylic acid. Notice that the carboxylic acid has one more carbon than the alkyl halide.

$$CH_3CH_2Br \underset{DMF}{\overset{^-C≡N}{\longrightarrow}} CH_3CH_2C≡N \underset{\Delta}{\overset{H^+, H_2O}{\longrightarrow}} CH_3CH_2\overset{\overset{O}{\|}}{C}OH$$

PROBLEM 27◆

What alkyl halides form the following carboxylic acids after reaction with sodium cyanide and heating of the product in an acidic aqueous solution?

a. butyric acid   **c.** cyclohexanecarboxylic acid

b. isovaleric acid   **d.** succinic acid (Table 16.2, page 724)

Fats and **oils** are triesters of glycerol. Glycerol contains three alcohol groups and therefore can form three ester groups (Section 24.3). When the ester groups are hydrolyzed in a basic solution, glycerol and the salts of carboxylic acids are formed. The carboxylic acids that are bonded to glycerol in fats and oils have long, unbranched R groups (Section 24.3). Such carboxylic acids are called *fatty acids*.

**16.17
SOAPS,
DETERGENTS, AND
MICELLES**

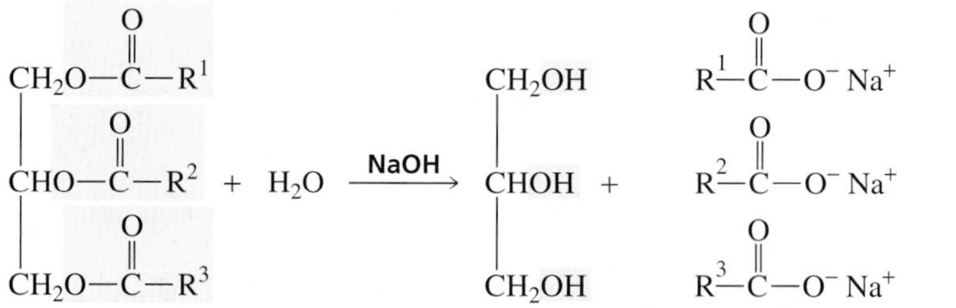

$$
\begin{array}{c}
\underset{\text{a fat or an oil}}{
\begin{array}{l}
CH_2O-\overset{\displaystyle O}{\overset{\|}{C}}-R^1 \\
CHO-\overset{\displaystyle O}{\overset{\|}{C}}-R^2 \\
CH_2O-\overset{\displaystyle O}{\overset{\|}{C}}-R^3
\end{array}}
\; + \; H_2O \;\xrightarrow{\textbf{NaOH}}\;
\underset{\text{glycerol}}{
\begin{array}{l}
CH_2OH \\
CHOH \\
CH_2OH
\end{array}}
\; + \;
\underset{\substack{\text{sodium salts of fatty acids}\\ \textbf{soap}}}{
\begin{array}{l}
R^1-\overset{\displaystyle O}{\overset{\|}{C}}-O^-\,Na^+ \\
R^2-\overset{\displaystyle O}{\overset{\|}{C}}-O^-\,Na^+ \\
R^3-\overset{\displaystyle O}{\overset{\|}{C}}-O^-\,Na^+
\end{array}}
$$

Soaps are sodium or potassium salts of fatty acids. Thus, soaps are obtained when fats or oils are hydrolyzed under basic conditions. Three of the most common soaps are shown below. Hydrolysis of an ester in a basic solution is called **saponification**—the Latin word for "soap" is *sapo*.

3-D Molecules:
Sodium stearate;
Sodium oleate;
Sodium linoleate

$$
\underset{\textbf{sodium stearate}}{CH_3(CH_2)_{16}\overset{\displaystyle O}{\overset{\|}{C}}O^-\,Na^+}
\qquad
\underset{\textbf{sodium oleate}}{CH_3(CH_2)_7CH=CH(CH_2)_7\overset{\displaystyle O}{\overset{\|}{C}}O^-\,Na^+}
$$

$$
\underset{\textbf{sodium linoleate}}{CH_3(CH_2)_4CH=CHCH_2CH=CH(CH_2)_7\overset{\displaystyle O}{\overset{\|}{C}}O^-\,Na^+}
$$

PROBLEM 28 / SOLVED

An oil obtained from coconuts is unusual in that all three fatty acid components are identical. The molecular formula of the oil is $C_{45}H_{86}O_6$. What is the molecular formula of the carboxylate ion obtained when the oil is saponified?

SOLUTION When the oil is saponified, it forms glycerol and three equivalents of carboxylate ion. In losing glycerol, the fat loses three carbons and five hydrogens. Thus, the three equivalents of carboxylate ion have a combined molecular formula of $C_{42}H_{81}O_6$. Dividing by three gives a molecular formula of $C_{14}H_{27}O_2$ for the carboxylate ion.

Long-chain carboxylate ions do not exist as individual ions in aqueous solution. Instead, they arrange themselves in spherical clusters called **micelles,** as shown in Figure 16.3. Each micelle contains 50 to 100 long-chain carboxylate ions. The polar carboxylate end of each ion is on the outside of the micelle because of its attraction for water. The nonpolar end is in the interior of the micelle to minimize its contact with water. The attractive forces of hydrocarbon chains for each other in water are called **hydrophobic interactions** (Section 21.14).

Because the surface of the micelle is negatively charged, the individual micelles repel each other instead of clustering to form larger aggregates. The cleansing ability of soap results from the fact that nonpolar oil molecules, which carry dirt, dissolve in the nonpolar interior of the micelle and are carried away with the soap during rinsing.

Soap lowers the surface tension of water, which is why soap solutions feel slippery. Lowering the surface tension enables soap to penetrate the fibers of a fabric, thus enhancing its cleaning ability. Compounds that lower the surface tension of water all have a polar head group and a long-chain nonpolar tail. They are called *surfactants.*

As water flows over and around rocks on its way to the water tap, it leaches out calcium and magnesium ions. The concentration of calcium and magnesium ions in

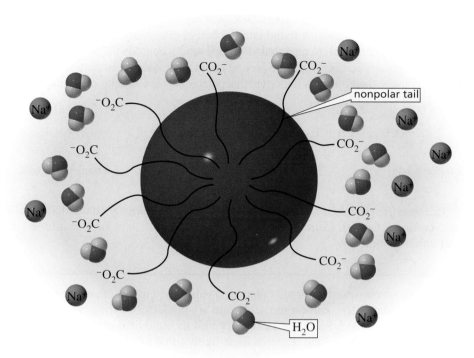

◀ **Figure 16.3**
In aqueous solution, soap forms micelles with the carboxylate groups on the surface and the nonpolar tails in the interior.

the water is described by its "hardness." Hard water contains high concentrations of these ions; soft water contains few if any calcium and magnesium ions. While micelles with sodium and potassium cations are dispersed in water, micelles with calcium and magnesium cations form aggregates. In hard water, therefore, soaps form a precipitate that we know as "bathtub ring" or "soap scum."

The formation of soap scum in hard water led to a search for synthetic materials that would have the cleansing properties of soap but would not form scum when they encountered calcium and magnesium ions. The synthetic "soaps" that were developed are known as detergents. **Detergents** are salts of sulfonic acids. Calcium and magnesium

MAKING SOAP

For thousands of years, soap was prepared by heating animal fat with wood ashes. Wood ashes contain potassium carbonate, which makes the solution basic. The modern commercial method of making soap involves boiling fats or oils in aqueous sodium hydroxide and adding sodium chloride to precipitate the soap. The soap is then dried and pressed into bars. Perfume can be added for scented soaps; dyes can be added for colored soaps; sand can be added for scouring soaps; and air can be blown into the soap to make it float in water.

sulfonate salts do not form aggregates. Detergent comes from the Latin *detergere,* which means "to wipe off." After the initial introduction of detergents into the marketplace, it was discovered that straight-chain alkyl groups are biodegradable, whereas branched-chain alkyl groups are not. In order to prevent nonbiodegradable detergents from polluting rivers and lakes, therefore, it is important that detergents have straight-chain alkyl groups.

a sulfonic acid   an anionic surfactant   a cationic surfactant
              a detergent   a germicide

Because the polar head of a soap or a detergent is negatively charged, it is called an *anionic* surfactant. *Cationic* surfactants are used widely as fabric softeners and hair conditioners.

16.18 SYNTHESIS OF CARBOXYLIC ACID DERIVATIVES

Of the various classes of carbonyl compounds discussed in this chapter (acyl halides, acid anhydrides, esters, carboxylic acids, and amides), carboxylic acids are the most commonly available both in the laboratory and in biological systems. This means that carboxylic acids are the reagents most likely to be available when a chemist or a cell needs to synthesize a carboxylic acid derivative. However, we have seen that carboxylic acids are relatively unreactive toward nucleophilic acyl substitution reactions because the OH group of a carboxylic acid is a strong base and therefore a poor leaving group. In neutral solutions (physiological pH = 7.3), a carboxylic acid is even more resistant to nucleophilic acyl substitution reactions because it exists predominantly in its negatively charged basic form (Sections 1.20 and 16.12). Therefore, both organic chemists and cells need a way to activate carboxylic acids so they can readily undergo nucleophilic acyl substitution reactions.

Activation of Carboxylic Acids for Nucleophilic Acyl Substitution Reactions in the Laboratory

Because acyl halides are the most reactive of the carboxylic acid derivatives, the easiest way to synthesize any other carboxylic acid derivative is to add the appropriate nucleophile to an acyl halide. Therefore, organic chemists activate carboxylic acids by converting them into acyl halides.

A carboxylic acid can be converted into an acyl chloride by heating it either with thionyl chloride ($SOCl_2$) or with phosphorus trichloride (PCl_3). Acyl bromides can be synthesized by using phosphorus tribromide (PBr_3).

$$CH_3\overset{O}{\overset{\|}{C}}\!-\!OH \ + \ SOCl_2 \ \xrightarrow{\Delta} \ CH_3\overset{O}{\overset{\|}{C}}\!-\!Cl \ + \ SO_2 \ + \ HCl$$

acetic acid   thionyl chloride   acetyl chloride

$$3 \ CH_3CH_2CH_2\overset{O}{\overset{\|}{C}}\!-\!OH \ + \ PCl_3 \ \xrightarrow{\Delta} \ 3 \ CH_3CH_2CH_2\overset{O}{\overset{\|}{C}}\!-\!Cl \ + \ H_3PO_3$$

butanoic acid   phosphorus trichloride   butanoyl chloride

benzoic acid   phosphorus tribromide   benzoyl bromide   phosphorous acid

All these reagents convert the OH group of the carboxylic acid into a better leaving group than the halide ion. Therefore, when the halide ion subsequently attacks the carbonyl carbon and forms a tetrahedral intermediate, the halide ion is *not* the group that is eliminated.

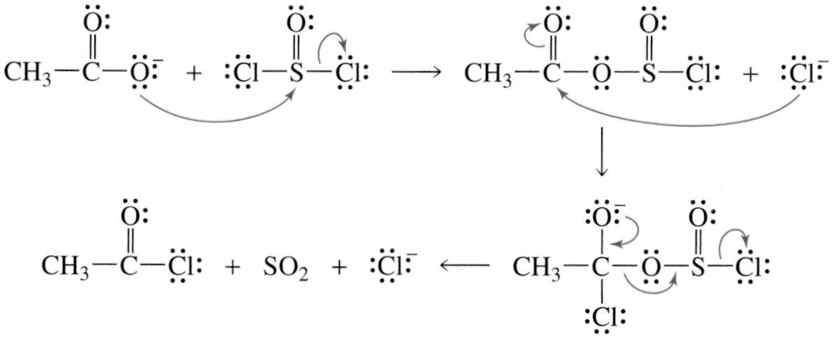

Notice that the reagents that cause the OH group of a carboxylic acid to be replaced by a halogen are the same reagents that cause the OH group of an alcohol to be replaced by a halogen (Section 11.2).

Once the acyl halide has been prepared, a wide variety of carboxylic acid derivatives can be synthesized by adding the appropriate nucleophile (Section 16.7). Because acyl halides are so reactive, they are generally prepared just prior to reaction with the nucleophile.

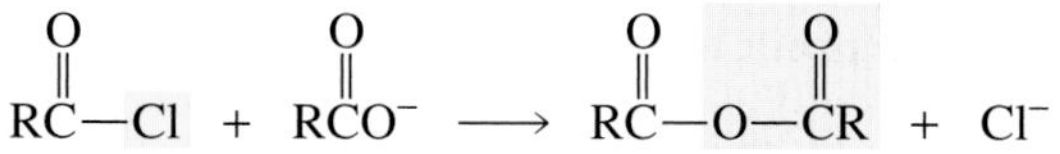

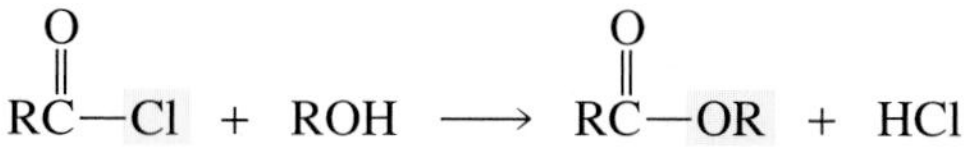

Carboxylic acids can also be activated for nucleophilic acyl substitution reactions by being converted into anhydrides. Treatment of a carboxylic acid with a strong dehydrating agent such as P_2O_5 yields an anhydride.

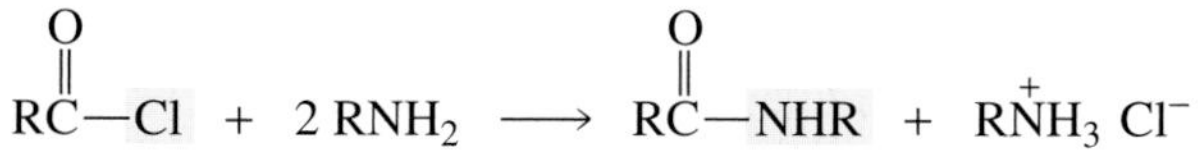

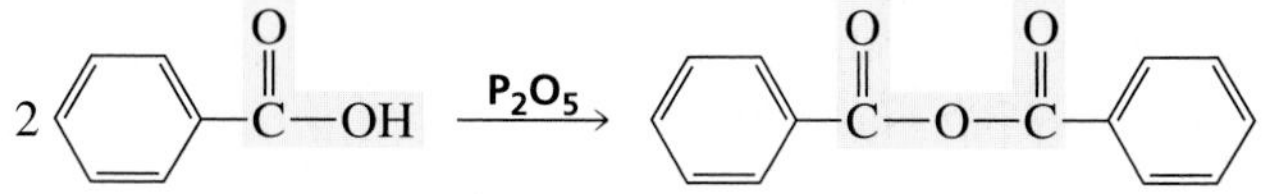

Carboxylic acids and carboxylic acid derivatives can also be prepared by methods other than nucleophilic acyl substitution reactions. A summary of the methods used to synthesize these compounds is given in Appendix IV.

Activation of Carboxylate Ions for Nucleophilic Acyl Substitution Reactions in Biological Systems

The synthesis of compounds by biological organisms is called **biosynthesis.** Acyl halides and acid anhydrides are too reactive to be used as reagents in biological systems. Cells live in a predominantly aqueous environment, and acyl halides and acid anhydrides are rapidly hydrolyzed in water. So living organisms must activate carboxylic acids in a different way.

Carboxylic acids are activated in cells by being converted to acyl phosphates, acyl pyrophosphates, acyl adenylates, or thioesters. An **acyl phosphate** is a mixed anhydride of a carboxylic acid and phosphoric acid; an **acyl pyrophosphate** is a mixed anhydride of a carboxylic acid and pyrophosphoric acid; an **acyl adenylate** is a mixed anhydride of a carboxylic acid and adenosine monophosphate (AMP). Because these mixed anhydrides are negatively charged, they are not readily approached by nucleophiles. One of the functions of enzymes that catalyze biological nucleophilic acyl substitution reactions is to neutralize the negative charges of the mixed anhydride (Section 25.5). The structure of adenosine triphosphate (ATP) is shown with "Ad" in place of the adenosyl group.

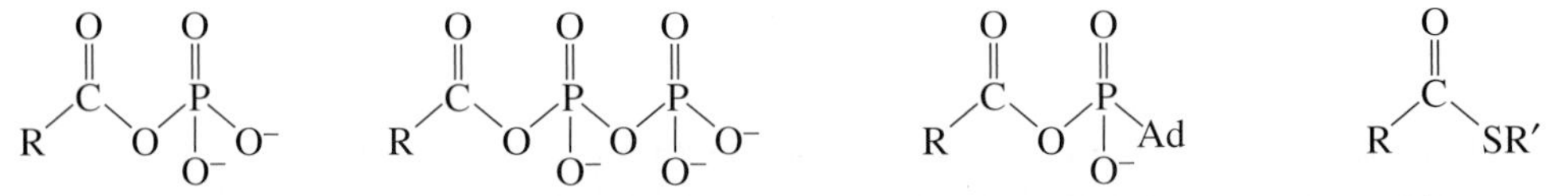

an acyl phosphate an acyl pyrophosphate an acyl adenylate a thioester

"activated" carboxylic acids

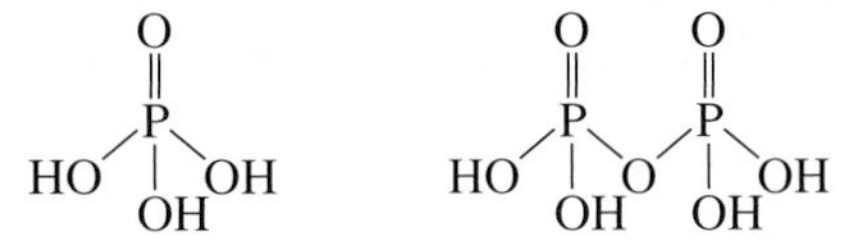

phosphoric acid   pyrophosphoric acid

3-D Molecule:
Adenosine triphosphate

adenosine triphosphate
ATP

Acyl phosphates are formed by nucleophilic attack of a carboxylate ion on the γ-phosphorus (the terminal phosphorus) of ATP. Notice that the reaction is a one-step reaction, very much like an S_N2 reaction. An intermediate is not formed because the π bond does not break. This reaction and the following reactions will be discussed in greater detail in Sections 25.3 and 25.4.

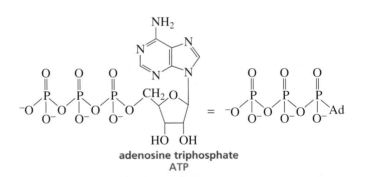

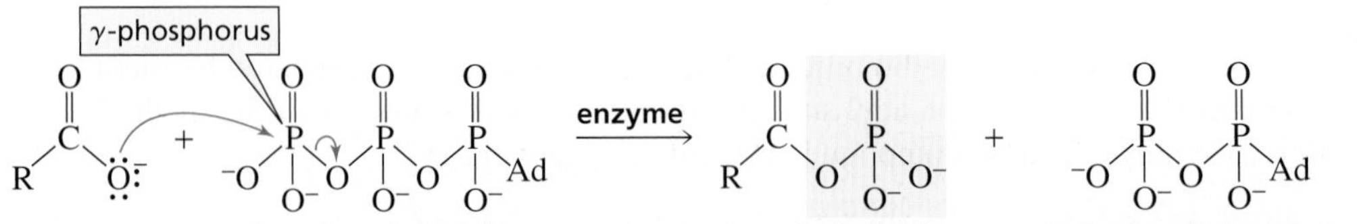

γ-phosphorus

adenosine triphosphate
ATP

enzyme

an acyl phosphate

adenosine diphosphate
ADP

Acyl pyrophosphates are formed by nucleophilic attack of a carboxylate ion on the β-phosphorus of ATP.

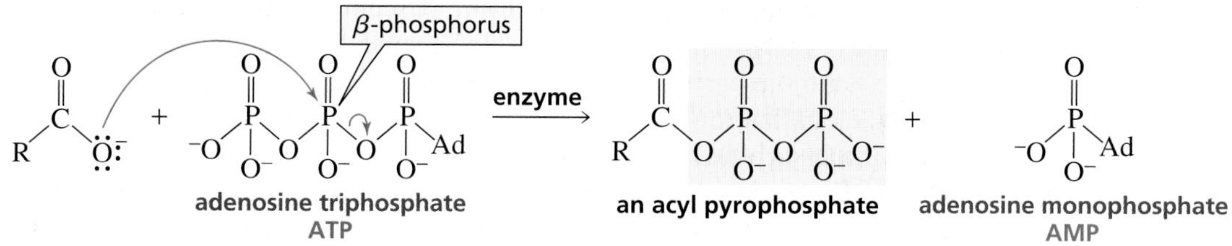

Acyl adenylates are formed by nucleophilic attack of a carboxylate ion on the α-phosphorus of ATP.

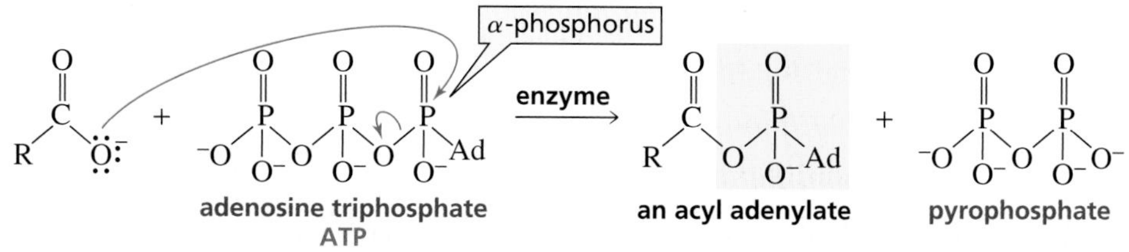

Acyl phosphates, acyl pyrophosphates, and acyl adenylates are used only in enzyme-catalyzed reactions where water can be excluded from the active site of the enzyme. Otherwise hydrolysis of the mixed anhydride would compete with the desired nucleophilic acyl substitution reaction.

USING ATP FOR BIOSYNTHESIS

An example of the biosynthesis of a carboxylic acid derivative is the biosynthesis of N-carbamoylaspartate from aspartate, bicarbonate, and ammonia. N-Carbamoylaspartate is a compound needed for the synthesis of pyrimidines. Substituted pyrimidines are found in DNA and RNA (Section 24.5). DNA is the genetic material that makes each living organism unique. The overall reaction requires the activation of both carboxyl oxygens of bicarbonate so that two amide bonds can be formed. Each of the carboxyl oxygens is activated by reacting with ATP to form an acyl phosphate. The biosynthesis is catalyzed by enzymes.

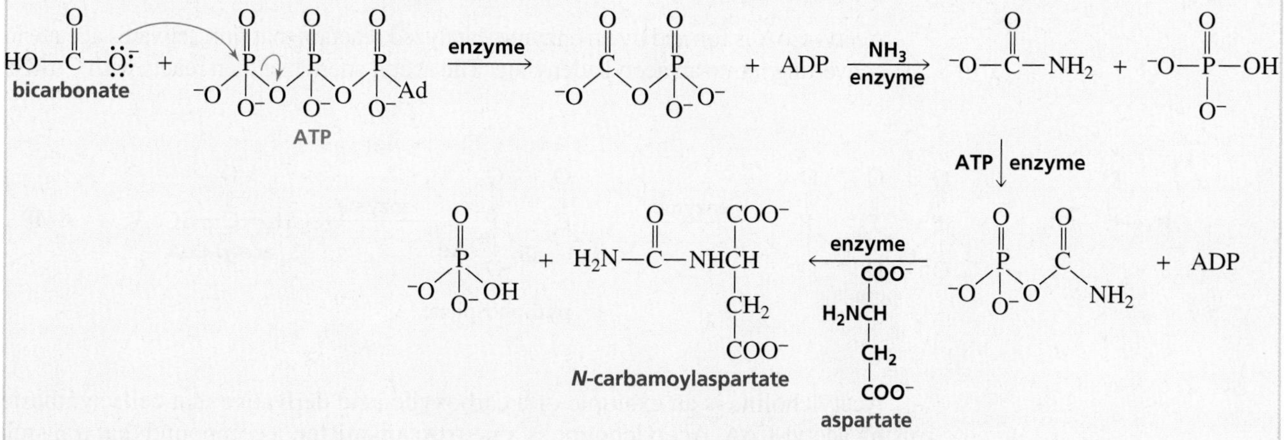

A **thioester** is an ester in which the leaving group is a thiol (RSH) instead of an alcohol (ROH). Thioesters are the most common activated carboxylic acids in a cell. What is remarkable about thioesters is that, although they hydrolyze at about the same rate as oxygen esters, they are much more reactive than oxygen esters toward attack by nitrogen and carbon nucleophiles. This allows a thioester to survive in the aqueous environment of the cell—without being hydrolyzed—waiting to be a substrate in a nucleophilic acyl substitution reaction.

Coenzyme A was discovered by ***Fritz A. Lipmann (1899–1986).*** *He also was the first to recognize its importance in intermediary metabolism. Lipmann was born in Germany. To escape the Nazis, he moved to Denmark in 1932 and to the United States in 1939, becoming a United States citizen in 1944. For his work on coenzyme A, he received the Nobel Prize in physiology and medicine in 1953, sharing it with Hans Krebs.*

The carbonyl carbon of a thioester is more susceptible to nucleophilic attack than the carbonyl carbon of an oxygen ester because there is much less electron delocalization onto the carbonyl oxygen when Y is S than when Y is O. There is less electron delocalization because there is less overlap between the $3p$ orbital of sulfur and the $2p$ orbital of carbon compared to the amount of overlap between the $2p$ orbital of oxygen and the $2p$ orbital of carbon (Section 16.2). In addition, a thiolate ion is a much weaker base and therefore a much better leaving group than an alkoxide ion.

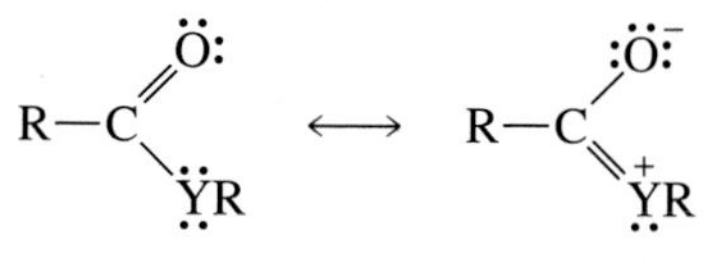

The thiol used in biological systems for the formation of thioesters is coenzyme A. The compound is written as CoASH to emphasize that the thiol group is the reactive part of the molecule.

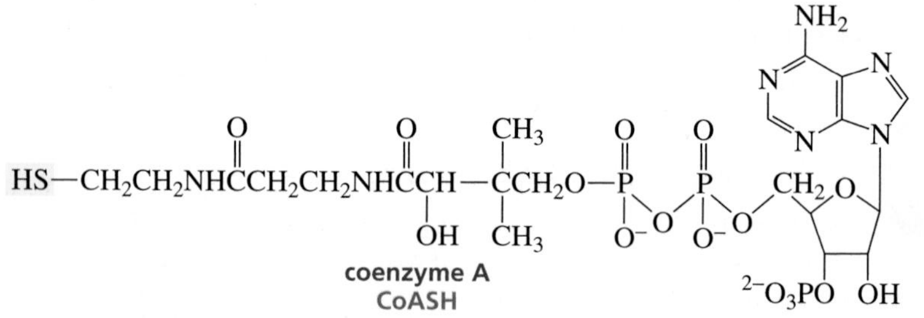

coenzyme A
CoASH

Acetyl-CoA is formed by an enzyme-catalyzed reaction that first activates acetate ion by converting it into an acetyl adenylate. The acetyl adenylate then reacts with CoASH.

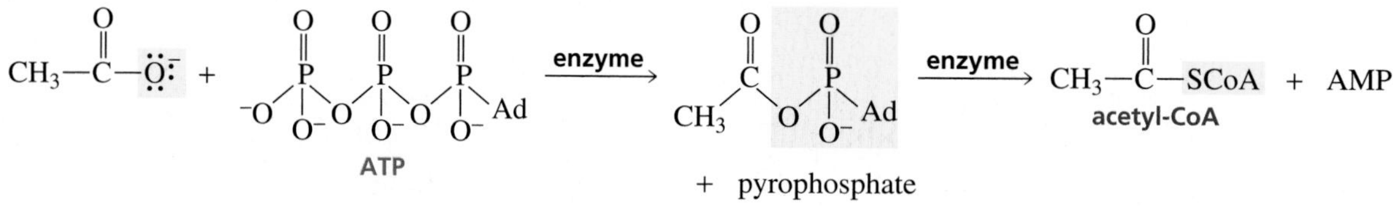

Acetylcholine is an example of a carboxylic acid derivative that cells synthesize using acetyl-CoA. Acetylcholine is a **neurotransmitter,** a compound that transmits nerve impulses across the synapses between nerve cells.

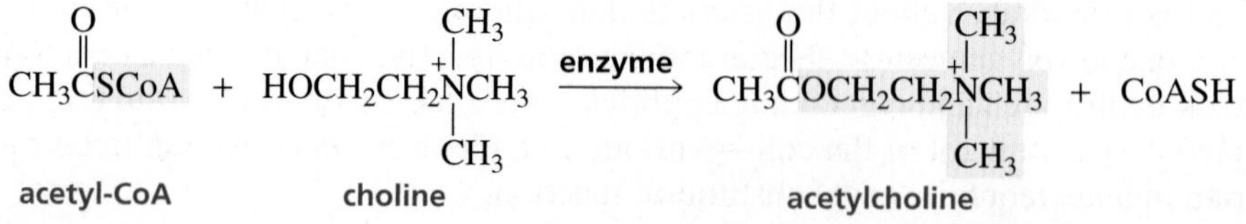

$$CH_3CSCoA + HOCH_2CH_2\overset{+}{N}CH_3 \xrightarrow{\text{enzyme}} CH_3COCH_2CH_2\overset{+}{N}CH_3 + CoASH$$

acetyl-CoA   **choline**   **acetylcholine**

NERVE IMPULSES, PARALYSIS, AND INSECTICIDES

After the nerve impulse is transmitted between cells, acetylcholine must be rapidly hydrolyzed so the receiving cell can get ready to receive another impulse.

$$
\underset{\substack{\\ \\ \\ \text{CH}_3}}{\overset{\overset{\overset{\text{O}}{\|}}{\text{CH}_3\text{C}}-\text{OCH}_2\text{CH}_2\overset{+}{\text{N}}\text{CH}_3}{}} + \text{H}_2\text{O} \xrightarrow{\textbf{acetylcholinesterase}} \underset{\substack{\\ \\ \\ \text{CH}_3}}{\overset{\overset{\overset{\text{O}}{\|}}{\text{CH}_3\text{C}}-\text{O}^- + \text{HOCH}_2\text{CH}_2\overset{+}{\text{N}}\text{CH}_3}{}}
$$

Acetylcholinesterase, the enzyme that catalyzes this hydrolysis, has a CH_2OH group that is necessary for its catalytic activity. Diisopropyl fluorophosphate (DFP), a military nerve gas, inhibits the enzyme by reacting with the CH_2OH group.

When the enzyme is inhibited, paralysis occurs because the nerve impulses cannot be transmitted properly. DFP is extremely toxic. Its LD_{50} (the lethal dose for 50% of the test animals) is only 0.5 mg/kg of body weight.

$$
\text{enzyme}-\text{CH}_2\text{OH} + \underset{\substack{\text{O} \\ | \\ \text{CH(CH}_3)_2}}{\overset{\substack{\text{CH(CH}_3)_2 \\ | \\ \text{O}}}{\text{F}-\text{P}=\text{O}}} \longrightarrow \text{enzyme}-\text{CH}_2\text{O}-\underset{\substack{\text{O} \\ | \\ \text{CH(CH}_3)_2}}{\overset{\substack{\text{CH(CH}_3)_2 \\ | \\ \text{O}}}{\text{P}=\text{O}}} + \text{HF}
$$

DFP

Malathion and parathion, compounds related to DFP, are used as insecticides. The LD_{50} of malathion is 2800 mg/kg.

Parathion is more toxic, with an LD_{50} of 2 mg/kg.

malathion

parathion

16.19 DICARBOXYLIC ACID DERIVATIVES

The structures of some common dicarboxylic acids and their pK_a values are shown in Table 16.2. Although the two carboxyl groups of a dicarboxylic acid are identical, the two pK_a values are different because the protons are lost one at a time and therefore leave from different species. The first proton is lost from a neutral molecule, whereas the second proton is lost from a negatively charged ion.

$$
\underset{}{\overset{\overset{\text{O} \quad \text{O}}{\| \quad \|}}{\text{HO}-\text{CCH}_2\text{C}-\text{OH}}} \underset{}{\overset{pK_{a1} = 2.86}{\rightleftharpoons}} \underset{+ \ \text{H}^+}{\overset{\overset{\text{O} \quad \text{O}}{\| \quad \|}}{\text{HO}-\text{CCH}_2\text{C}-\text{O}^-}} \underset{}{\overset{pK_{a2} = 5.70}{\rightleftharpoons}} \underset{+ \ \text{H}^+}{\overset{\overset{\text{O} \quad \text{O}}{\| \quad \|}}{{}^-\text{O}-\text{CCH}_2\text{C}-\text{O}^-}}
$$

A COOH group withdraws electrons (more strongly than an H) and therefore increases the acidity of the other COOH group. The pK_a values of the dicarboxylic acids show that the acid-strengthening effect of the COOH group decreases as the separation between the two carboxyl groups increases.

TABLE 16.2 Structures, Names, and pK_a Values of Some Simple Dicarboxylic Acids

Dicarboxylic Acid	Common Name	pK_{a1}	pK_{a2}
$HOCOH$ (with O double bond)	carbonic acid	3.58	6.35
$HOC-COH$ (with two O double bonds)	oxalic acid	1.27	4.27
$HOCCH_2COH$	malonic acid	2.86	5.70
$HOCCH_2CH_2COH$	succinic acid	4.21	5.64
$HOCCH_2CH_2CH_2COH$	glutaric acid	4.34	5.27
$HOCCH_2CH_2CH_2CH_2COH$	adipic acid	4.41	5.28
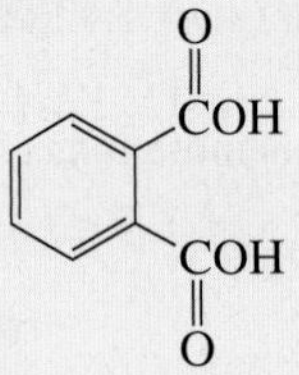 (benzene ring with two COH groups)	phthalic acid	2.95	5.41

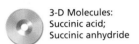

3-D Molecules:
Succinic acid;
Succinic anhydride

Dicarboxylic acids readily lose water when heated if they can form a cyclic anhydride with a five- or a six-membered ring.

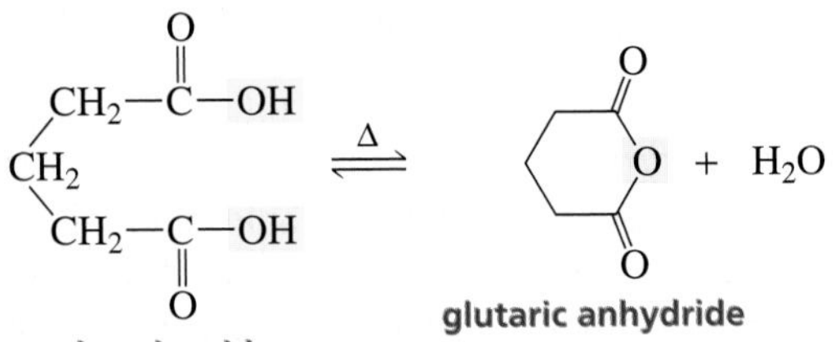

$$\text{glutaric acid} \;\overset{\Delta}{\rightleftharpoons}\; \text{glutaric anhydride} \;+\; H_2O$$

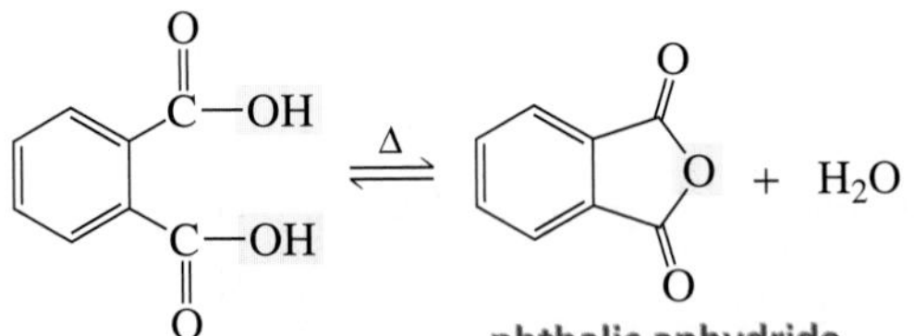

$$\text{phthalic acid} \;\overset{\Delta}{\rightleftharpoons}\; \text{phthalic anhydride} \;+\; H_2O$$

Cyclic anhydrides are more easily prepared if the dicarboxylic acid is heated in the presence of acetyl chloride or acetic anhydride.

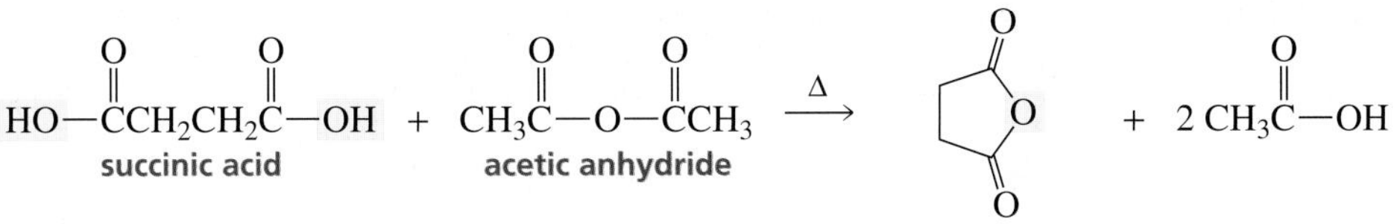

$$\underset{\text{succinic acid}}{\text{HO}-\overset{\overset{\displaystyle O}{\|}}{\text{C}}\text{CH}_2\text{CH}_2\overset{\overset{\displaystyle O}{\|}}{\text{C}}-\text{OH}} \; + \; \underset{\text{acetic anhydride}}{\text{CH}_3\overset{\overset{\displaystyle O}{\|}}{\text{C}}-\text{O}-\overset{\overset{\displaystyle O}{\|}}{\text{C}}\text{CH}_3} \; \xrightarrow{\Delta} \; \underset{\text{succinic anhydride}}{\text{[structure]}} \; + \; 2\,\text{CH}_3\overset{\overset{\displaystyle O}{\|}}{\text{C}}-\text{OH}$$

PROBLEM 29

a. Propose a mechanism for the formation of succinic anhydride in the presence of acetic anhydride.

b. How does acetic anhydride help in the formation of succinic anhydride?

Carbonic acid—a compound with two OH groups bonded to the carbonyl carbon—is unstable. It readily breaks down to CO_2 and H_2O. The reaction is reversible, so carbonic acid is formed when CO_2 is bubbled into water (Section 1.20).

Tutorial:
Common terms of carboxylic acids and their derivatives

$$\underset{\text{carbonic acid}}{\text{HO}-\overset{\overset{\displaystyle O}{\|}}{\text{C}}-\text{OH}} \; \rightleftharpoons \; CO_2 \; + \; H_2O$$

We have seen that the OH group of a carboxylic acid can be substituted to give a variety of carboxylic acid derivatives. Similarly, the OH groups of carbonic acid can be substituted by other groups.

$$\underset{\text{phosgene}}{\text{Cl}-\overset{\overset{\displaystyle O}{\|}}{\text{C}}-\text{Cl}} \qquad \underset{\text{dimethyl carbonate}}{\text{CH}_3\text{O}-\overset{\overset{\displaystyle O}{\|}}{\text{C}}-\text{OCH}_3} \qquad \underset{\text{urea}}{\text{H}_2\text{N}-\overset{\overset{\displaystyle O}{\|}}{\text{C}}-\text{NH}_2} \qquad \underset{\text{carbamic acid}}{\text{H}_2\text{N}-\overset{\overset{\displaystyle O}{\|}}{\text{C}}-\text{OH}} \qquad \underset{\text{methyl carbamate}}{\text{H}_2\text{N}-\overset{\overset{\displaystyle O}{\|}}{\text{C}}-\text{OCH}_3}$$

PROBLEM 30◆

What products would you expect to obtain from the following reactions?

a. phosgene + excess diethylamine

b. malonic acid + 2 acetyl chloride

c. methyl carbamate + methylamine

d. urea + water

e. urea + water + H^+

f. β-ethylglutaric acid + acetyl chloride + Δ

SUMMARY OF REACTIONS

1. Reactions of acyl halides (Section 16.7)

$$\underset{}{\text{R}\overset{\overset{\displaystyle O}{\|}}{\text{C}}\text{Cl}} \; + \; \text{CH}_3\overset{\overset{\displaystyle O}{\|}}{\text{C}}\text{O}^- \; \longrightarrow \; \text{R}\overset{\overset{\displaystyle O}{\|}}{\text{C}}\text{O}\overset{\overset{\displaystyle O}{\|}}{\text{C}}\text{CH}_3 \; + \; \text{Cl}^-$$

$$\text{R}\overset{\overset{\displaystyle O}{\|}}{\text{C}}\text{Cl} \; + \; \text{CH}_3\text{OH} \; \longrightarrow \; \text{R}\overset{\overset{\displaystyle O}{\|}}{\text{C}}\text{OCH}_3 \; + \; H^+ \; + \; \text{Cl}^-$$

$$\text{R}\overset{\overset{\displaystyle O}{\|}}{\text{C}}\text{Cl} \; + \; H_2O \; \longrightarrow \; \text{R}\overset{\overset{\displaystyle O}{\|}}{\text{C}}\text{OH} \; + \; H^+ \; + \; \text{Cl}^-$$

$$\text{R}\overset{\overset{\displaystyle O}{\|}}{\text{C}}\text{Cl} \; + \; 2\,\text{CH}_3\text{NH}_2 \; \longrightarrow \; \text{R}\overset{\overset{\displaystyle O}{\|}}{\text{C}}\text{NHCH}_3 \; + \; \text{CH}_3\overset{+}{\text{N}}\text{H}_3\,\text{Cl}^-$$

2. Reactions of acid anhydrides (Section 16.8)

$$\underset{\substack{\text{O} \quad \text{O} \\ \| \quad \| \\ \text{RCOCR}}}{} + \text{CH}_3\text{OH} \longrightarrow \underset{\substack{\text{O} \\ \| \\ \text{RCOCH}_3}}{} + \underset{\substack{\text{O} \\ \| \\ \text{RCOH}}}{}$$

$$\underset{\substack{\text{O} \quad \text{O} \\ \| \quad \| \\ \text{RCOCR}}}{} + \text{H}_2\text{O} \longrightarrow 2\,\underset{\substack{\text{O} \\ \| \\ \text{RCOH}}}{}$$

$$\underset{\substack{\text{O} \quad \text{O} \\ \| \quad \| \\ \text{RCOCR}}}{} + 2\,\text{CH}_3\text{NH}_2 \longrightarrow \underset{\substack{\text{O} \\ \| \\ \text{RCNHCH}_3}}{} + \underset{\substack{\text{O} \\ \| \\ \text{RCO}^-\;\text{H}_3\overset{+}{\text{N}}\text{CH}_3}}{}$$

3. Reactions of esters (Sections 16.9–16.11)

$$\underset{\substack{\text{O} \\ \| \\ \text{RCOR}}}{} + \text{CH}_3\text{OH} \overset{\text{H}^+}{\rightleftharpoons} \underset{\substack{\text{O} \\ \| \\ \text{RCOCH}_3}}{} + \text{ROH}$$

$$\underset{\substack{\text{O} \\ \| \\ \text{RCOR}}}{} + \text{H}_2\text{O} \overset{\text{H}^+}{\rightleftharpoons} \underset{\substack{\text{O} \\ \| \\ \text{RCOH}}}{} + \text{ROH}$$

$$\underset{\substack{\text{O} \\ \| \\ \text{RCOR}}}{} + \text{H}_2\text{O} \overset{\text{HO}^-}{\longrightarrow} \underset{\substack{\text{O} \\ \| \\ \text{RCO}^-}}{} + \text{ROH}$$

$$\underset{\substack{\text{O} \\ \| \\ \text{RCOR}}}{} + \text{CH}_3\text{NH}_2 \longrightarrow \underset{\substack{\text{O} \\ \| \\ \text{RCNHCH}_3}}{} + \text{ROH}$$

4. Reactions of carboxylic acids (Section 16.12)

$$\underset{\substack{\text{O} \\ \| \\ \text{RCOH}}}{} + \text{CH}_3\text{OH} \overset{\text{H}^+}{\rightleftharpoons} \underset{\substack{\text{O} \\ \| \\ \text{RCOCH}_3}}{} + \text{H}_2\text{O}$$

$$\underset{\substack{\text{O} \\ \| \\ \text{RCOH}}}{} + \text{CH}_3\text{NH}_2 \longrightarrow \underset{\substack{\text{O} \\ \| \\ \text{RCO}^-\;\text{H}_3\overset{+}{\text{N}}\text{CH}_3}}{} \overset{\textbf{225 °C}}{\longrightarrow} \underset{\substack{\text{O} \\ \| \\ \text{RCNHCH}_3}}{} + \text{H}_2\text{O}$$

5. Reactions of amides (Sections 16.13 and 16.14)

$$\underset{\substack{\text{O} \\ \| \\ \text{RCNH}_2}}{} + \text{H}_2\text{O} \overset{\text{H}^+}{\underset{\Delta}{\longrightarrow}} \underset{\substack{\text{O} \\ \| \\ \text{RCOH}}}{} + \overset{+}{\text{N}}\text{H}_4$$

$$\underset{\substack{\text{O} \\ \| \\ \text{RCNH}_2}}{} + \text{H}_2\text{O} \overset{\text{HO}^-}{\underset{\Delta}{\longrightarrow}} \underset{\substack{\text{O} \\ \| \\ \text{RCO}^-}}{} + \text{NH}_3$$

$$\underset{\substack{\text{O} \\ \| \\ \text{RCNH}_2}}{} \overset{\textbf{P}_2\textbf{O}_5}{\underset{\Delta}{\longrightarrow}} \text{RC} \equiv \text{N}$$

6. Gabriel synthesis of primary amines (Section 16.15)

$$\text{RCH}_2\text{Br} \xrightarrow[\substack{\textbf{2. H}^+\textbf{, H}_2\textbf{O, } \Delta \\ \textbf{3. HO}^-}]{\textbf{1. phthalimide, HO}^-} \text{RCH}_2\text{NH}_2$$

7. Reactions of nitriles (Section 16.16)

$$RC{\equiv}N \ + \ H_2O \ \xrightarrow[\Delta]{H^+} \ RCOH \ + \ \overset{+}{N}H_4$$

$$RC{\equiv}N \ + \ H_2O \ \xrightarrow[\Delta]{HO^-} \ RCO^- \ + \ NH_3$$

8. Activation of carboxylic acids (Section 16.18)

$$RCOH \ + \ SOCl_2 \ \xrightarrow{\Delta} \ RCCl \ + \ SO_2 \ + \ HCl$$

$$RCOH \ + \ PBr_3 \ \xrightarrow{\Delta} \ RCBr \ + \ H_3PO_3$$

$$RCOH \ + \ PCl_3 \ \xrightarrow{\Delta} \ RCCl \ + \ H_3PO_3$$

$$2\ RCOH \ \xrightarrow{P_2O_5} \ RCOCR$$

9. Dehydration of dicarboxylic acids (Section 16.19)

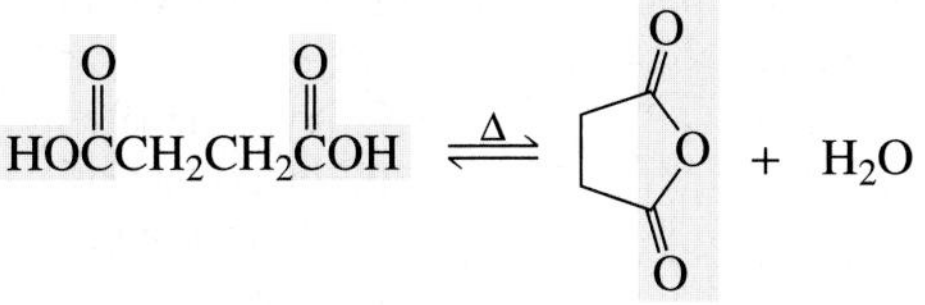

$$HOCCH_2CH_2COH \ \xrightleftharpoons{\Delta} \ \text{(cyclic anhydride)} \ + \ H_2O$$

KEY TERMS

acid anhydride (page 681)
acyl adenylate (page 718)
acyl group (page 678)
acyl halide (page 681)
acyl phosphate (page 718)
acyl pyrophosphate (page 718)
alcoholysis (page 697)
amide (page 682)
amino acid (page 690)
aminolysis (page 697)
biosynthesis (page 717)
α-carbon (page 680)
carbonyl carbon (page 684)
carbonyl compound (page 678)
carbonyl group (page 678)

carbonyl oxygen (page 684)
carboxyl group (page 681)
carboxylic acid (page 680)
carboxylic acid derivative (page 684)
carboxyl oxygen (page 682)
detergent (page 715)
ester (page 682)
fat (page 713)
fatty acid (page 680)
Fischer esterification reaction
  (page 706)
Gabriel synthesis (page 711)
hydrolysis (page 697)
hydrophobic interactions (page 714)
imide (page 711)

lactam (page 683)
lactone (page 682)
micelle (page 714)
mixed anhydride (page 681)
neurotransmitter (page 720)
nitrile (page 683)
nucleophilic acyl substitution reaction
  (page 687)
oil (page 713)
saponification (page 714)
soap (page 714)
symmetrical anhydride (page 681)
tetrahedral intermediate (page 686)
thioester (page 719)
transesterification reaction (page 697)

PROBLEMS

31. Write a structure for each of the following compounds:

   a. *N,N*-dimethylhexanamide

   b. 3,3-dimethylhexanamide

   c. cyclohexanecarbonyl chloride

   d. propanenitrile

   e. propionyl bromide

   f. sodium acetate

   g. benzoic anhydride

   h. β-valerolactone

   i. 3-methylbutanenitrile

   j. cycloheptanecarboxylic acid

32. Name the following compounds:

   a. CH₃CH₂CHCH₂CH₂CH₂COH (with CH₂CH₃ substituent and C=O)

   b. CH₃CH₂COCH₂CH₂CH₃

   c. CH₃CH₂CH₂CH₂C≡N

   d. CH₃CH₂COCCH₂CH₃

   e. CH₃CH₂CH₂CN(CH₃)₂

   f. CH₃CH₂CH₂CH₂CCl

   g. 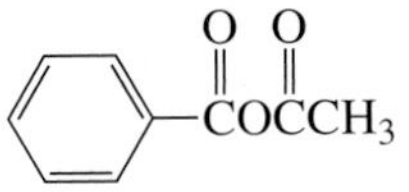 (benzene ring)—COCCH₃

   h. CH₂=CHCH₂CNHCH₃

   i. 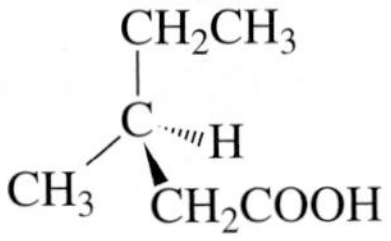 C with CH₂CH₃, H, CH₃, CH₂COOH

   j. 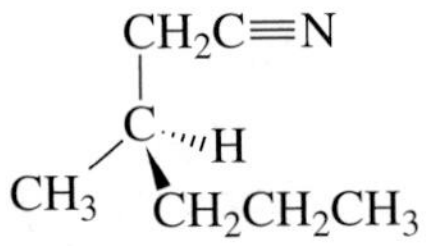 C with CH₂C≡N, H, CH₃, CH₂CH₂CH₃

33. What products would be formed from the reaction of benzoyl chloride with the following reagents?

   a. sodium acetate

   b. water

   c. dimethylamine

   d. aqueous HCl

   e. aqueous NaOH

   f. cyclohexanol

   g. benzylamine

   h. 4-chlorophenol

   i. isopropyl alcohol

   j. aniline

34. a. List the following esters in order of decreasing reactivity in the first step of a nucleophilic acyl substitution reaction (formation of the tetrahedral intermediate):

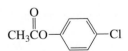

   b. List the same esters in order of decreasing reactivity in the second step of a nucleophilic acyl substitution reaction (collapse of the tetrahedral intermediate).

35. a. Which compound would you expect to have a higher dipole moment: methyl acetate or butanone?

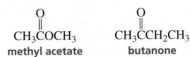

   methyl acetate   **butanone**

   b. Which would you expect to have a higher boiling point?

36. Using a carboxylic acid as a starting material, how could you synthesize the following esters?

a. ethyl phenylethanoate (odor of honey) **c.** isobutyl propionate (odor of rum)
b. ethyl butyrate (odor of pineapple) **d.** isopentyl acetate (odor of banana)

37. How could you use nuclear magnetic resonance (^{1}H NMR) spectroscopy to distinguish among the following esters?

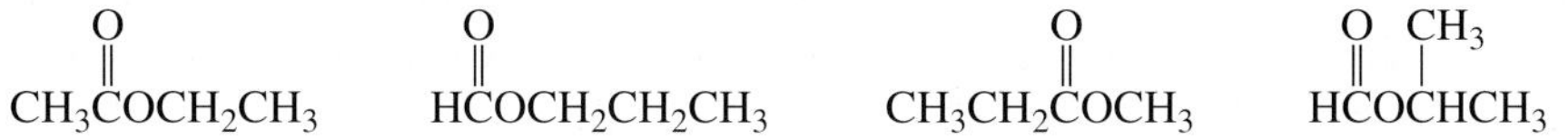

$$CH_3\overset{\displaystyle O}{\overset{\|}{C}}OCH_2CH_3 \qquad H\overset{\displaystyle O}{\overset{\|}{C}}OCH_2CH_2CH_3 \qquad CH_3CH_2\overset{\displaystyle O}{\overset{\|}{C}}OCH_3 \qquad H\overset{\displaystyle O}{\overset{\|}{C}}O\overset{\displaystyle CH_3}{\overset{|}{C}}HCH_3$$

38. If propionyl chloride is added to one equivalent of methylamine, only a 50% yield of *N*-methylpropanamide is obtained. If, however, the acyl chloride is added to two equivalents of methylamine, the yield of *N*-methylpropanamide is almost 100%. Explain these observations.

39. a. When a carboxylic acid is dissolved in isotopically labeled water, the label is incorporated into both oxygens of the acid. Propose a mechanism to account for this.

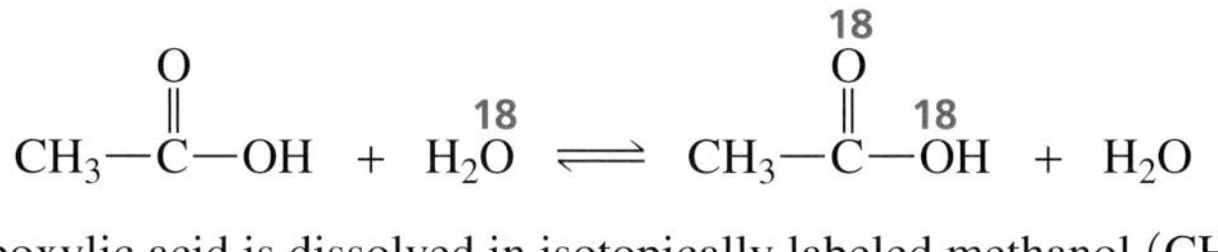

$$CH_3-\overset{\displaystyle O}{\overset{\|}{C}}-OH \ + \ H_2\overset{18}{O} \ \rightleftharpoons \ CH_3-\overset{\displaystyle \overset{18}{O}}{\overset{\|}{C}}-\overset{18}{O}H \ + \ H_2O$$

b. If a carboxylic acid is dissolved in isotopically labeled methanol ($CH_3\overset{18}{O}H$) and an acid catalyst is added, where will the label reside in the product?

40. What reagents would you use to convert methyl propanoate into the following compounds?
a. isopropyl propanoate **c.** *N*-ethylpropanamide
b. sodium propanoate **d.** propanoic acid

41. A compound with molecular formula $C_5H_{10}O_2$ gives the IR spectrum shown below. When it undergoes acid-catalyzed hydrolysis the compound with the ^{1}H NMR spectrum is formed. Identify the compounds.

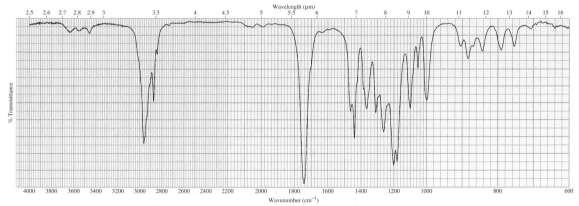

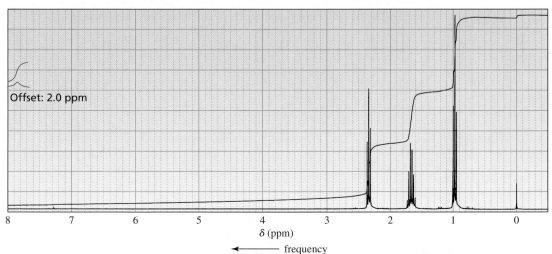

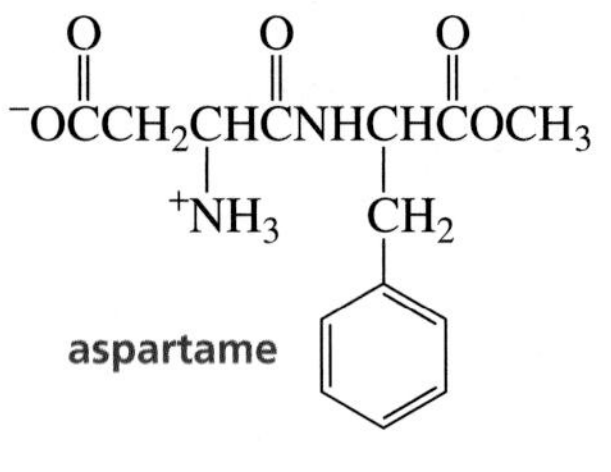

$$\overset{O}{\underset{}{\|}} \quad \overset{O}{\underset{}{\|}} \quad \overset{O}{\underset{}{\|}}$$

$$^-OCCH_2CHCNHCHCOCH_3$$

$$\overset{+}{N}H_3 \qquad CH_2$$

aspartame

42. Aspartame is 160 times sweeter than sucrose. It is the sweetener used in the commercial products known as NutraSweet and Equal. What products would be obtained if aspartame were hydrolyzed completely in an aqueous solution of HCl?

43. a. Which of the following reactions will not give the carbonyl product shown?
  b. Which of the reactions that do not occur can be made to occur if an acid catalyst is added to the reaction mixture?

$$\textbf{1.} \; CH_3\overset{O}{\overset{\|}{C}}OH \; + \; CH_3\overset{O}{\overset{\|}{C}}O^- \; \longrightarrow \; CH_3\overset{O}{\overset{\|}{C}}O\overset{O}{\overset{\|}{C}}CH_3$$

$$\textbf{2.} \; CH_3\overset{O}{\overset{\|}{C}}Cl \; + \; CH_3\overset{O}{\overset{\|}{C}}O^- \; \longrightarrow \; CH_3\overset{O}{\overset{\|}{C}}O\overset{O}{\overset{\|}{C}}CH_3$$

$$\textbf{3.} \; CH_3\overset{O}{\overset{\|}{C}}NH_2 \; + \; Cl^- \; \longrightarrow \; CH_3\overset{O}{\overset{\|}{C}}Cl$$

$$\textbf{4.} \; CH_3\overset{O}{\overset{\|}{C}}OH \; + \; CH_3NH_2 \; \longrightarrow \; CH_3\overset{O}{\overset{\|}{C}}NHCH_3$$

$$\textbf{5.} \; CH_3\overset{O}{\overset{\|}{C}}OCH_3 \; + \; CH_3NH_2 \; \longrightarrow \; CH_3\overset{O}{\overset{\|}{C}}NHCH_3$$

$$\textbf{6.} \; CH_3\overset{O}{\overset{\|}{C}}OCH_3 \; + \; Cl^- \; \longrightarrow \; CH_3\overset{O}{\overset{\|}{C}}Cl$$

$$\textbf{7.} \; CH_3\overset{O}{\overset{\|}{C}}NHCH_3 \; + \; CH_3\overset{O}{\overset{\|}{C}}O^- \; \longrightarrow \; CH_3\overset{O}{\overset{\|}{C}}O\overset{O}{\overset{\|}{C}}CH_3$$

$$\textbf{8.} \; CH_3\overset{O}{\overset{\|}{C}}Cl \; + \; H_2O \; \longrightarrow \; CH_3\overset{O}{\overset{\|}{C}}OH$$

$$\textbf{9.} \; CH_3\overset{O}{\overset{\|}{C}}NHCH_3 \; + \; H_2O \; \longrightarrow \; CH_3\overset{O}{\overset{\|}{C}}OH$$

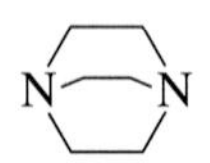

1,4-diazabicyclo[2.2.2]octane
DABCO

44. 1,4-Diazabicyclo[2.2.2]octane (abbreviated DABCO) is a tertiary amine that catalyzes transesterification reactions. Propose a mechanism to show how it does this.

45. Identify the major and minor products of the following reaction.

$$\overset{CH_2OH}{\underset{\underset{H}{N}}{\bigcirc}}\overset{\text{\tiny ''''}CH_2CH_3}{} \; + \; CH_3\overset{O}{\overset{\|}{C}}Cl \; \longrightarrow$$

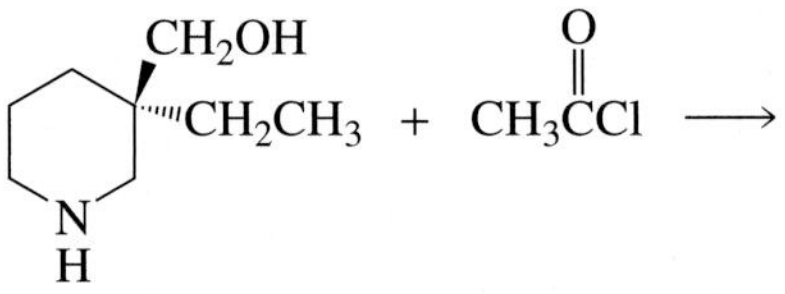

46. Two products (A and B) are obtained from the reaction of 1-bromobutane with NH_3. Compound A reacts with acetyl chloride to form C, and B reacts with acetyl chloride to form D. The IR spectra of C and D are shown. Identify A, B, C, and D.

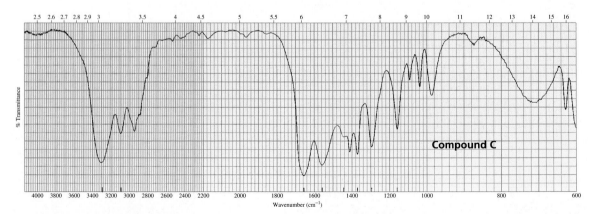

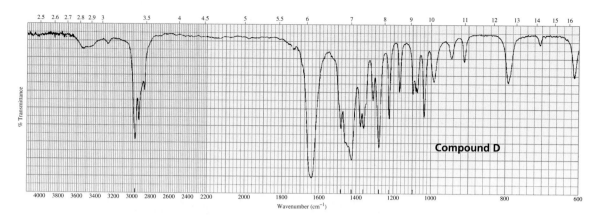

Compound D

47. a. Phosgene ($COCl_2$) was used as a poison gas in World War I. Give the product that would be formed from the reaction of phosgene with the following reagents:
  1. one equivalent of methanol
  2. excess methanol
  3. excess propylamine
  4. one equivalent of ethanol followed by one equivalent of methylamine
b. What is the mode of action of phosgene as a poison gas?

48. When Ethyl Ester treated butanedioic acid with thionyl chloride, she was surprised to find that the product she obtained was an anhydride rather than an acyl chloride. Propose a mechanism to explain why she obtained an anhydride.

49. Give the products of the following reactions:

a. CH_3CCl + KF $\longrightarrow$

b. 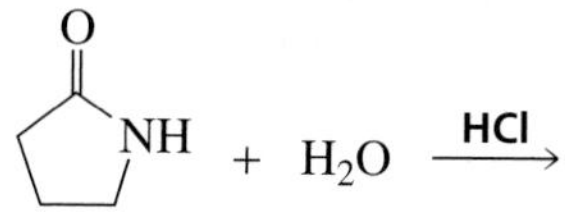 (pyrrolidinone) + H_2O $\xrightarrow{HCl}$

c. 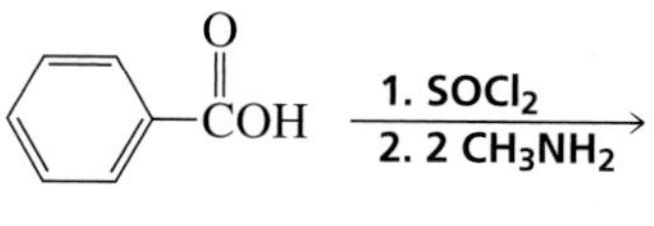 (benzoic acid) $\xrightarrow[\text{2. 2 CH}_3\text{NH}_2]{\text{1. SOCl}_2}$

d. 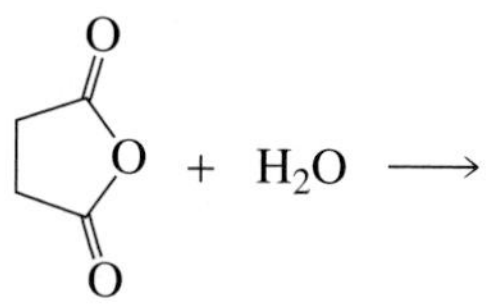 (succinic anhydride) + H_2O $\longrightarrow$

e. $ClCCl$ + (catechol) $\longrightarrow$

f. 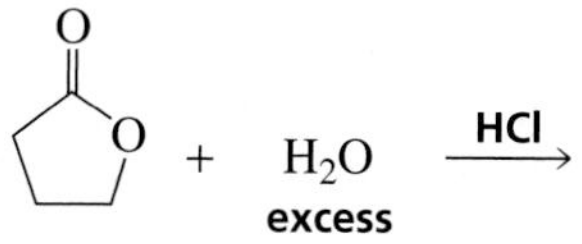 (γ-butyrolactone) + H_2O (excess) $\xrightarrow{HCl}$

g. $CH_3CCH_2OCCH_3$ + CH_3OH (excess) $\xrightarrow{CH_3O^-}$

h. 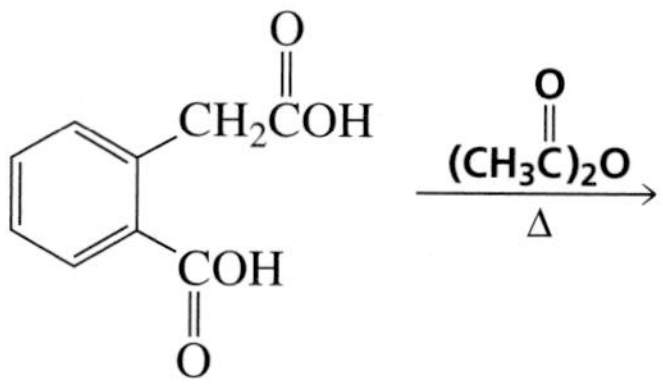 (2-(carboxymethyl)benzoic acid) $\xrightarrow[\Delta]{(CH_3C)_2O}$

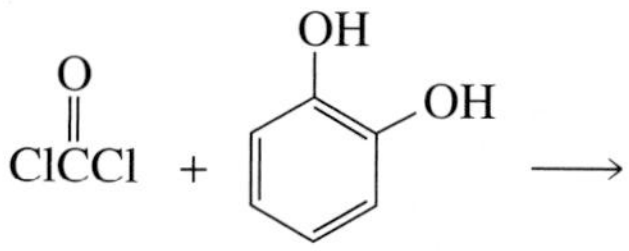

i. 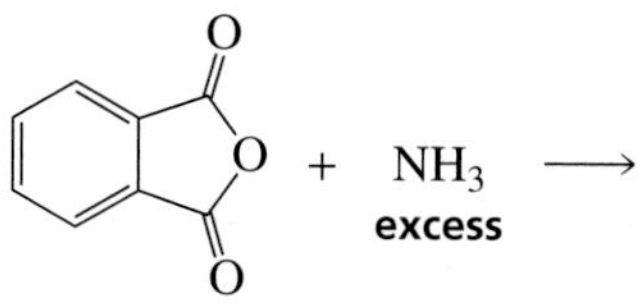 (phthalic anhydride) + NH_3 (excess) $\longrightarrow$

j. 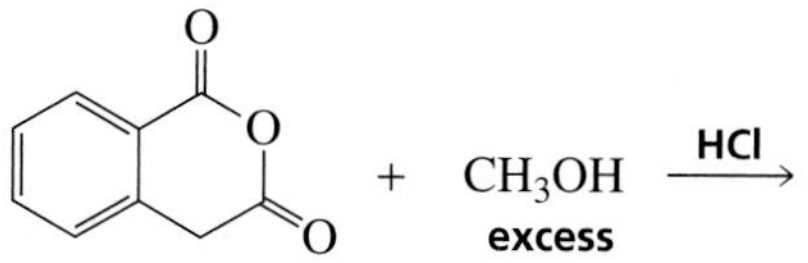 (isochroman-1,3-dione) + CH_3OH (excess) $\xrightarrow{HCl}$

50. Compound A (with molecular formula $C_4H_6Cl_2O$), when treated with an equivalent of methanol, forms the compound whose ^1H NMR spectrum is shown on the next page. Identify compound A.

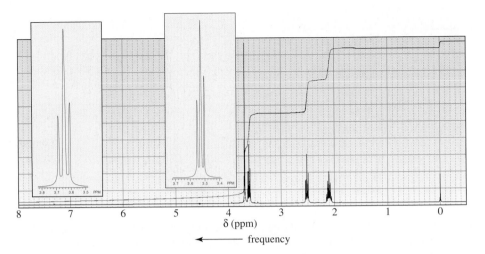

51. a. Identify the two products obtained from the following reaction:

$$\underset{\textbf{excess}}{CH_3\overset{O}{\overset{\|}{C}}O\overset{O}{\overset{\|}{C}}CH_3} \quad + \quad CH_3\overset{NH_2}{\overset{|}{C}H}CH_2CH_2OH \quad \longrightarrow$$

b. Eddie Amine carried out the reaction but stopped it before it was half over and isolated the major product. He was surprised to find that the product he isolated was neither of the products obtained when the reaction was allowed to go to completion. What product did he isolate?

52. An aqueous solution of a primary or secondary amine reacts with an acyl chloride to form an amide as the major product. However, if the amine is tertiary, an amide is not formed. What product is formed? Explain.

53. a. Ann Hydride did not obtain any ester when she added 2,4,6-trimethylbenzoic acid to an acidic solution of methanol. Why? (*Hint:* Build models.)
b. Would she have encountered the same problem if she had tried to synthesize the methyl ester of *p*-methylbenzoic acid in the same way?
c. How could she have prepared the methyl ester of 2,4,6-trimethylbenzoic acid? (*Hint:* See Section 15.12.)

54. List the following compounds in order of decreasing frequency of the carbon–oxygen double bond stretch:

$$\underset{CH_3\overset{O}{\overset{\|}{C}}OCH_3}{} \qquad \underset{CH_3\overset{O}{\overset{\|}{C}}Cl}{} \qquad \underset{CH_3\overset{O}{\overset{\|}{C}}H}{} \qquad \underset{CH_3\overset{O}{\overset{\|}{C}}NH_2}{}$$

55. a. If the equilibrium constant for the reaction of acetic acid and ethanol to form ethyl acetate is 4.02, what will be the concentration of ethyl acetate at equilibrium if the reaction is carried out with equal amounts of acetic acid and ethanol?
b. What will be the concentration of ethyl acetate at equilibrium if the reaction is carried out with 10 times more ethanol than acetic acid? (*Hint:* Recall the Quadratic Equation.)

$$\text{For:} \qquad ax^2 + bx + c = 0$$

$$x = \frac{-b \pm (b^2 - 4ac)^{1/2}}{2a}$$

c. What will be the concentration of ethyl acetate at equilibrium if the reaction is carried out with 100 times more ethanol than acetic acid?

56. The ^{1}H NMR spectra for two esters with molecular formula $C_8H_8O_2$ are shown on the next page. If each of the esters is added to an aqueous solution with a pH of 10, which of the esters will be hydrolyzed more completely when the hydrolysis reactions have reached equilibrium?

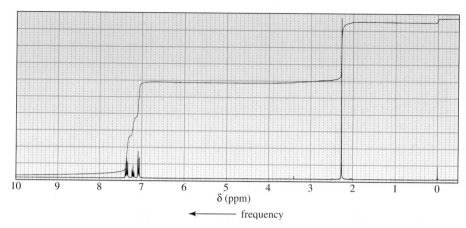

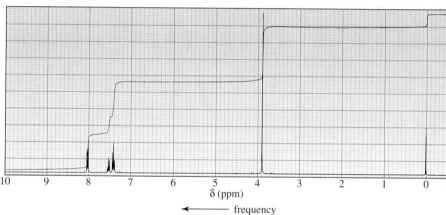

57. A study of the kinetics of the base-promoted hydrolysis of an amide has shown that the reaction is second-order in hydroxide ion. What does this tell you about the mechanism?

58. Show how the following compounds could be prepared from the given starting materials. You can use any necessary organic or inorganic reagents.

a. $CH_3CH_2\overset{O}{\overset{\|}{C}}NH_2 \longrightarrow CH_3CH_2\overset{O}{\overset{\|}{C}}Cl$

b. $CH_3CH_2CH_2CH_2OH \longrightarrow CH_3CH_2CH_2CH_2\overset{O}{\overset{\|}{C}}OH$

c. $CH_3(CH_2)_{10}\overset{O}{\overset{\|}{C}}OH \longrightarrow CH_3(CH_2)_{11}\!-\!\!\!\bigcirc\!\!\!-SO_3^- Na^+$

a detergent

d.

e.

f.

59. Is the acid-catalyzed hydrolysis of acetamide a reversible or an irreversible reaction? Explain.

60. The reaction of a nitrile with an alcohol in the presence of a strong acid forms a secondary amide. This reaction is known as the Ritter reaction. The Ritter reaction does not work with primary alcohols.

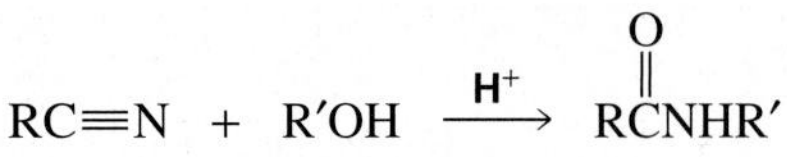

$$RC\equiv N + R'OH \xrightarrow{\text{H}^+} \overset{\overset{\displaystyle O}{\|}}{R\text{C}}NHR'$$

the Ritter reaction

    a. Propose a mechanism for the Ritter reaction.
    b. Why does the Ritter reaction not work with primary alcohols?
    c. How does the Ritter reaction differ from the acid-catalyzed hydrolysis of a nitrile to form a primary amide?

61. The intermediate shown here is formed during the hydroxide-ion-promoted hydrolysis of the ester group. Propose a mechanism for the reaction (*J. Am. Chem. Soc.*, 1988, 110, 8157).

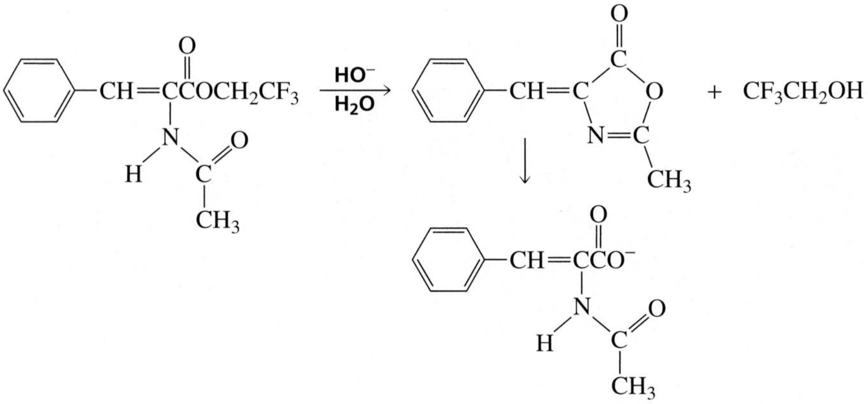

62. What product would you expect to obtain from each of the following reactions?

    a.  $CH_3CH_2\overset{\text{OH}}{\underset{|}{C}}HCH_2CH_2CH_2\overset{\overset{\displaystyle O}{\|}}{C}OH \xrightarrow{\text{H}^+}$

    b. (cyclopentane with $CH_2\overset{\overset{\displaystyle O}{\|}}{C}OCH_2CH_3$ and CH_2OH) $\xrightarrow{\text{H}^+}$

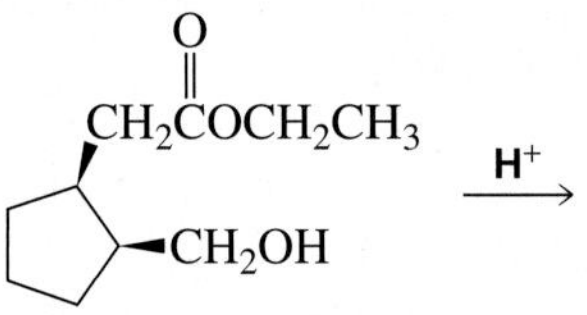

    c. (benzene with $CH_2CH_2\overset{\overset{\displaystyle O}{\|}}{C}OH$) $\xrightarrow[\substack{\text{2. AlCl}_3 \\ \text{3. H}_2\text{O}}]{\text{1. SOCl}_2}$

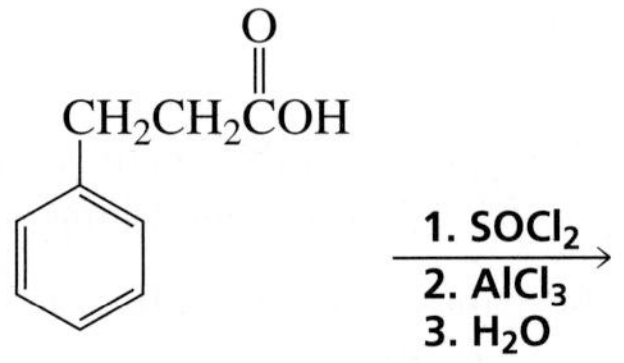

63. Sulfonamides, the first antibiotics, were introduced clinically in 1934 (Sections 23.8 and 30.4). Show how a sulfonamide can be prepared from benzene.

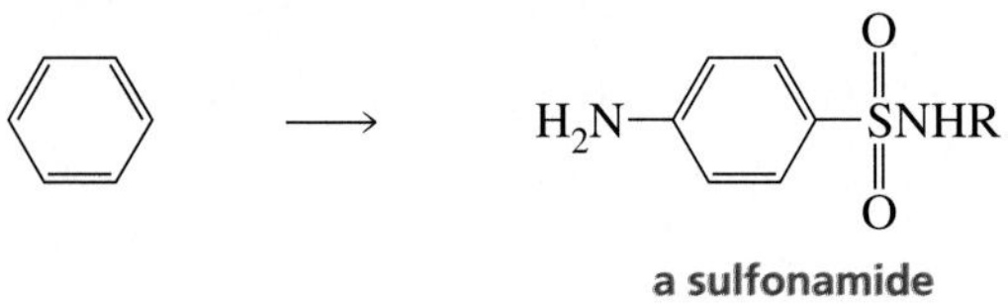

a sulfonamide

64. a. How could aspirin be synthesized starting with benzene?
    b. Ibuprofen is the active ingredient in pain relievers such as Advil, Motrin, and Nuprin. How could ibuprofen be synthesized starting with benzene?

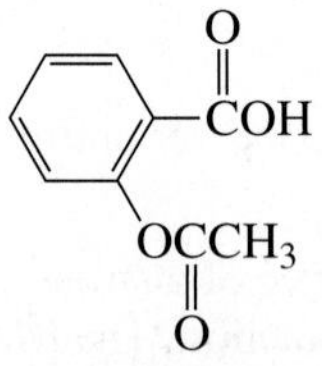

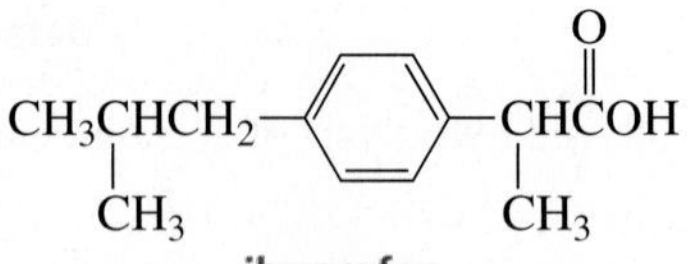

ibuprofen

aspirin

65. The following compound has been found to be an inhibitor of a β-lactamase. The enzyme can be reactivated by hydroxylamine (NH_2OH). Propose a mechanism to account for the inhibition and for the reactivation (*Science* 1989, 246, 917).

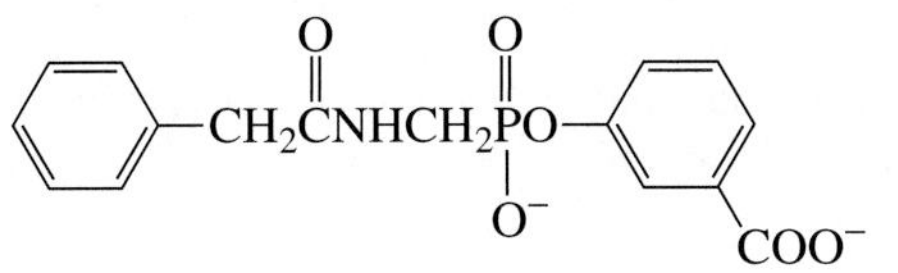

66. For each of the following reactions, propose a mechanism that will account for the formation of the product:

a.

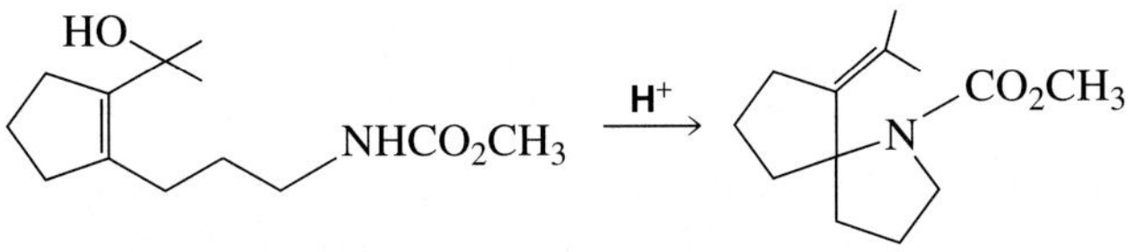

b.

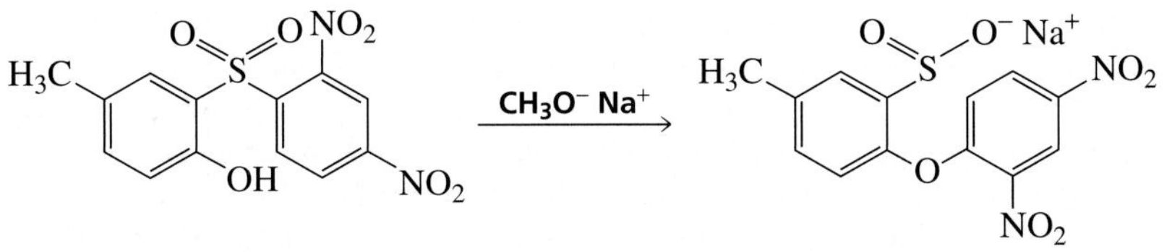

67. Show how Novocain, a pain killer used frequently by dentists (Section 30.3), can be prepared from benzene.

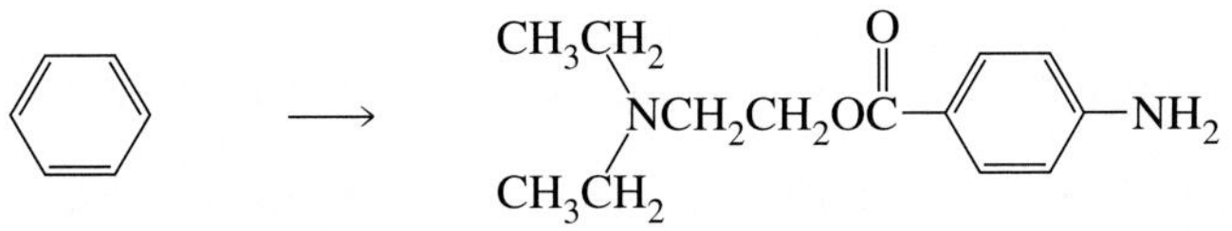

Novocain

68. Catalytic antibodies catalyze a reaction by binding to the transition state, thereby stabilizing it. This lowers the energy of activation and therefore the reaction goes faster. The synthesis of the antibody is carried out in the presence of a transition state analog. A transition state analog is a stable molecule that structurally resembles the transition state. This causes generation of an antibody that will recognize and bind to the transition state. For example, the following transition state analog has been used to generate a catalytic antibody that catalyzes the hydrolysis of the structurally similar ester.

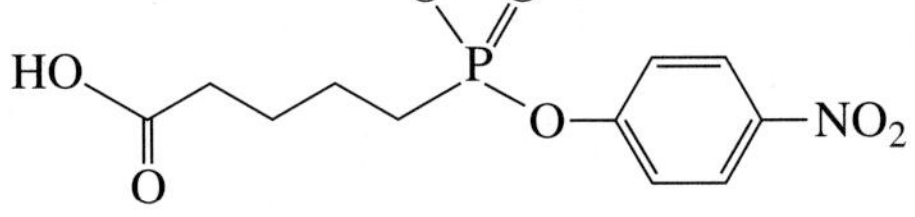

transition state analog

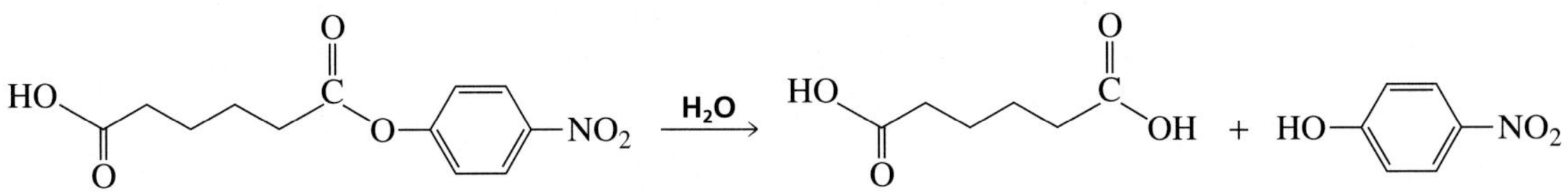

a. Draw the transition state for the hydrolysis reaction.
b. The following transition state analog is used to generate a catalytic antibody for the catalysis of ester hydrolysis. Give the structure of an ester whose rate of hydrolysis would be increased by this catalytic antibody.

c. Design a transition state analog that would catalyze amide hydrolysis at the amide group indicated by the arrow.

69. Information about the mechanism of reaction of a series of substituted benzenes can be obtained by plotting the logarithm of the observed rate constant obtained at a particular pH versus the Hammett substituent constant (σ) for the particular substituent. The Hammett substituent constant for hydrogen is 0. Electron-donating substituents have negative Hammett substituent constants, and electron-withdrawing substituents have positive Hammett substituent constants. The more strongly electron-donating the substituent, the greater the negative value of its substituent constant; the more strongly electron-withdrawing the substituent, the greater the positive value of its substituent constant. The slope of a plot of the log of the rate constant versus σ is called the ρ (rho) value. The ρ value for the hydroxide-ion-promoted hydrolysis of a series of *meta*- and *para*-substituted ethyl benzoates is $+2.46$; the ρ value for amide formation from the reaction of a series of *meta*- and *para*-substituted anilines with benzoyl chloride is -2.78.
    a. Why does one set of experiments give a positive ρ value while the other set of experiments gives a negative ρ value?
    b. Why do you think that ortho-substituted compounds were not included in the experiment?
    c. What would you predict the sign of the ρ value to be for the ionization of a series of *meta*- and *para*-substituted benzoic acids?

70. Saccharin, an artificial sweetener, is about 300 times sweeter than sucrose. Describe how saccharin could be prepared using benzene as the starting material.

saccharin

17

Carbonyl Compounds II:

Reactions of Carbonyl Compounds With Carbon and Hydrogen Nucleophiles; Reactions of Aldehydes and Ketones With Oxygen and Nitrogen Nucleophiles; Reactions of α,β-Unsaturated Carbonyl Compounds

vanillin

camphor

Aldehydes and ketones possess carbonyl groups ($C{=}O$), so they are carbonyl compounds. The carbonyl carbon of the simplest aldehyde, formaldehyde, is bonded to two hydrogens. In all other **aldehydes,** the carbonyl carbon is bonded to a *hydrogen* and to an *alkyl* (or an *aryl*) *group.* The carbonyl carbon of a **ketone** is bonded to *two alkyl* (or *aryl*) *groups.*

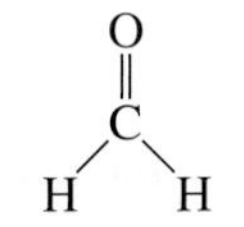

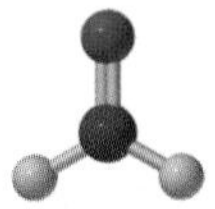

formaldehyde

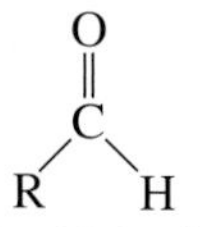

acetaldehyde

formaldehyde   an aldehyde 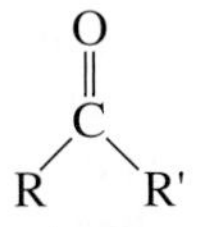   a ketone

Aldehydes and ketones differ from the carbonyl compounds discussed in Chapter 16 because they do not have a group that can be replaced by a nucleophile. That is, carbanions (R^-) and hydride ions (H^-) are too basic to be displaced by nucleophiles under normal conditions. The physical properties of aldehydes and ketones are discussed in Section 16.3 (also see Appendix I) and the methods used to prepare aldehydes and ketones are summarized in Appendix IV.

acetone

Many compounds found in nature have aldehyde or ketone functional groups. Aldehydes have pungent odors, while ketones tend to smell sweet. Vanilla and cinnamon flavorings are examples of naturally occurring aldehydes. The ketones carvone and camphor are responsible for the characteristic sweet odors of spearmint leaves, caraway seeds, and the camphor tree.

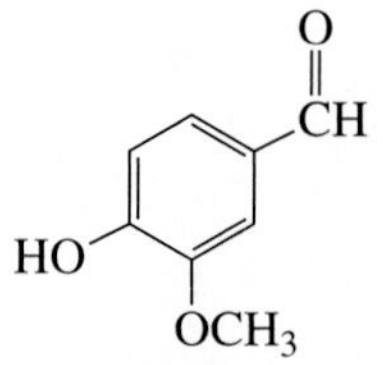

vanillin
vanilla flavoring

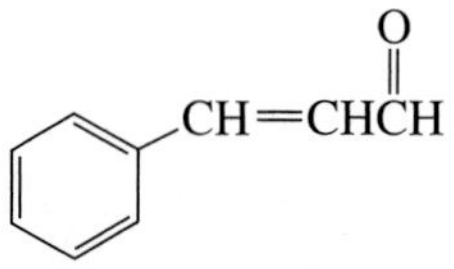

cinnamaldehyde
cinnamon flavoring

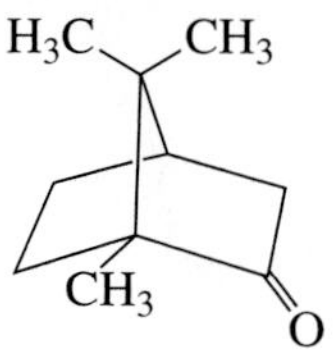

camphor

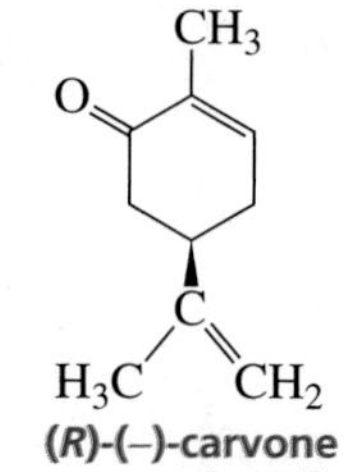

(R)-(−)-carvone
spearmint oil

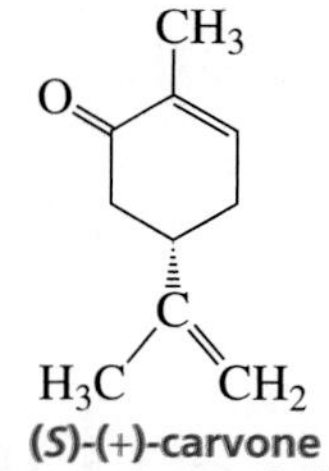

(S)-(+)-carvone
caraway seed oil

In ketosis, a pathological condition that can occur in people with diabetes, the body produces more acetoacetate than can be metabolized. The excess acetoacetate breaks down to acetone (a ketone) and CO_2. This condition can be recognized by the sweet smell of acetone on the breath.

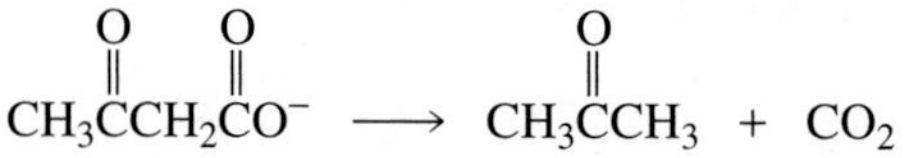

$$CH_3CCH_2CO^- \longrightarrow CH_3CCH_3 + CO_2$$

acetoacetate   acetone

Two ketones of biological importance illustrate how a small difference in structure can be responsible for a large difference in biological activity—progesterone is a female sex hormone synthesized primarily in the ovaries, while testosterone is a male sex hormone synthesized primarily in the testes.

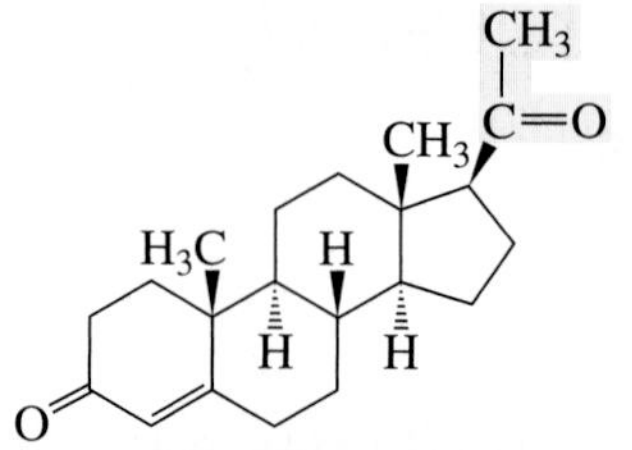

progesterone
a female sex hormone

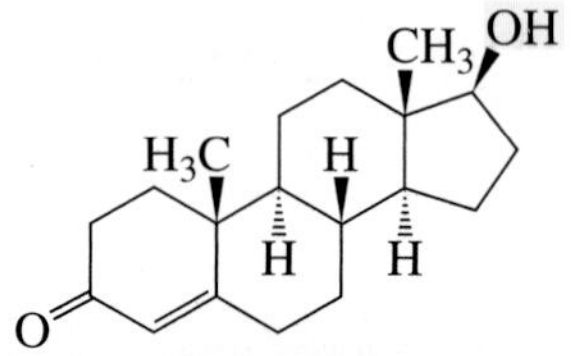

testosterone
a male sex hormone

17.1
NOMENCLATURE

Aldehydes

The systematic name of an aldehyde is obtained by removing the terminal "e" from the parent alkane name and adding "al." The position of the carbonyl carbon does not have to be designated because it is always at the end of the parent hydrocarbon chain and therefore always has the number 1 position.

The common name of an aldehyde is the same as the common name of the corresponding carboxylic acid (Section 16.1), except that "aldehyde" is substituted for "ic acid" (or "oic acid"). When common names are used, the position of a substituent is designated by a lowercase Greek letter. The carbonyl carbon is not given a designation—the carbon adjacent to the carbonyl carbon is the α-carbon.

HCH	CH_3CH	CH_3CHCH
systematic name: methanal	ethanal	2-bromopropanal
common name: formaldehyde	acetaldehyde	α-bromopropionaldehyde

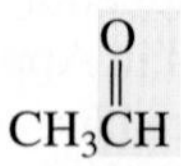

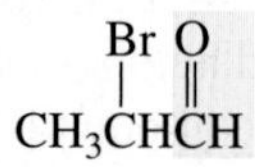

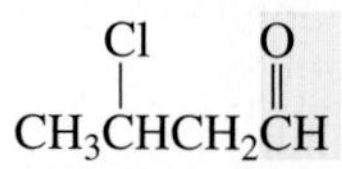

CH_3CHCH_2CH	CH_3CHCH_2CH	$HCCH_2CH_2CH_2CH_2CH$
systematic name: 3-chlorobutanal	3-methylbutanal	hexanedial
common name: β-chlorobutyraldehyde	isovaleraldehyde	

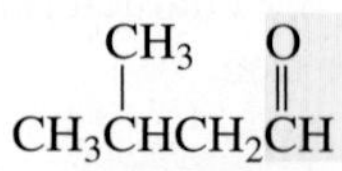

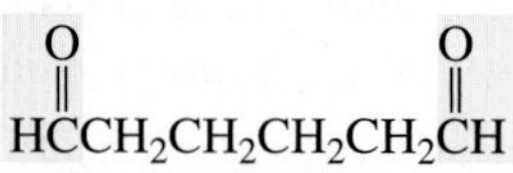

Notice that the terminal "e" is not removed in hexanedial. The "e" is removed only to avoid two successive vowels.

If the aldehyde group is attached to a ring, the aldehyde is named by adding "carbaldehyde" to the name of the cyclic compound.

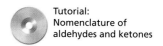

Tutorial:
Nomenclature of
aldehydes and ketones

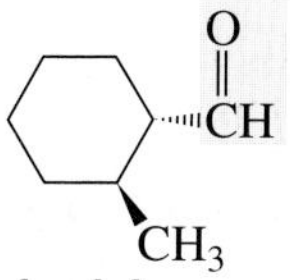

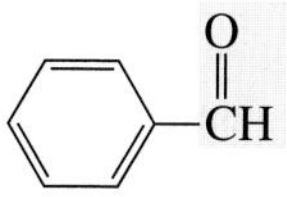

systematic name: *trans*-2-methylcyclohexanecarbaldehyde   benzenecarbaldehyde
common name: benzaldehyde

In Section 7.1 we saw that a carbonyl group has a higher nomenclature priority than an alcohol or an amine group. The list of functional group priorities in Table 17.1 shows the relative nomenclature priorities of the various carbonyl groups. If a compound has two functional groups, the one with the lower priority can be indicated by its prefix name. The prefix name of an aldehyde group is "formyl."

TABLE 17.1   Summary of Functional Group Nomenclature

Class	Suffix Name	Prefix Name
Carboxylic acid	-oic acid	Carboxy
Ester	-oate	Alkoxycarbonyl
Amide	-amide	Amido
Nitrile	-nitrile	Cyano
Aldehyde	-al	Formyl ($-CH=O$)
Ketone	-one	Oxo ($=O$)
Alcohol	-ol	Hydroxy
Amine	-amine	Amino
Alkene	-ene	Alkenyl
Alkyne	-yne	Alkynyl
Alkane	-ane	Alkyl
Ether	—	Alkoxy
Alkyl halide	—	Halo

increasing
priority

formaldehyde

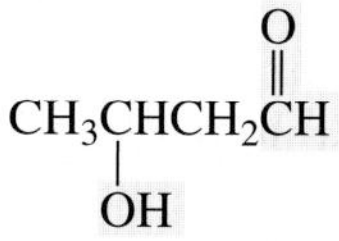

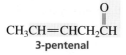

3-hydroxybutanal 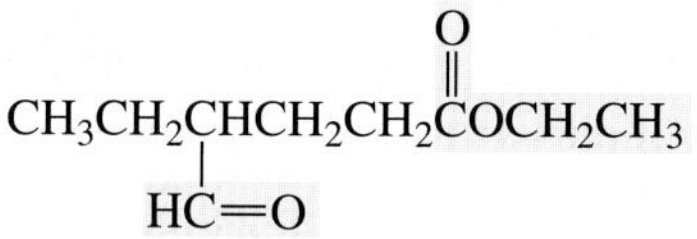   ethyl 4-formylhexanoate

If the compound has both an alkene and an aldehyde functional group, the alkene is cited first, with the "e" ending omitted to avoid two successive vowels (Section 7.1).

$$CH_3CH=CHCH_2CH$$
$$O$$
3-pentenal

acetaldehyde

Ketones

The systematic name of a ketone is obtained by removing the "e" from the parent alkane name and adding "one." The chain is numbered in the direction that gives the carbonyl carbon the smaller number. In the case of cyclic ketones, a number is not

acetone

necessary because the carbonyl carbon is assumed to be at the number 1 position. Frequently, derived names are used for ketones—the substituents attached to the carbonyl group are cited in alphabetical order followed by "ketone."

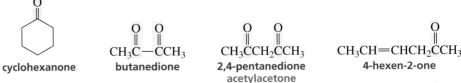

systematic name:	propanone	
common name:	acetone	
derived name:	dimethyl ketone	

3-hexanone

ethyl propyl ketone

6-methyl-2-heptanone

isohexyl methyl ketone

systematic name: cyclohexanone — butanedione — 2,4-pentanedione — 4-hexen-2-one

common name: — — acetylacetone —

PROBLEM 1

Why is a number not used to designate the position of the functional groups in propanone and butanedione?

Only a few ketones have common names. The smallest ketone, propanone, is usually referred to by its common name, acetone. Acetone is a common laboratory solvent. Common names are also used for some phenyl-substituted ketones; the number of carbons (other than those of the phenyl group) is indicated by the common name of the corresponding carboxylic acid, substituting "ophenone" for "ic acid."

common name:	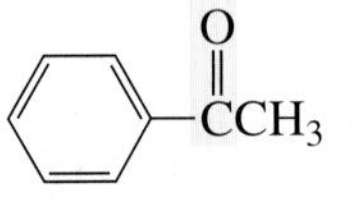 acetophenone	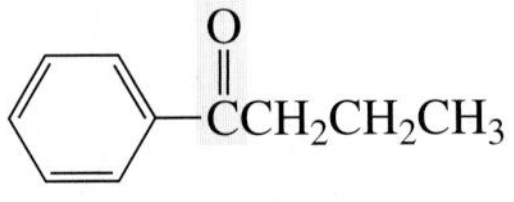 butyrophenone	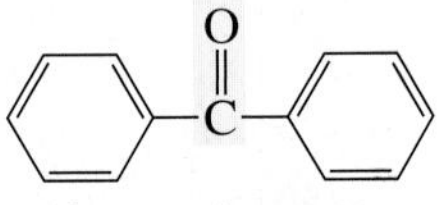 benzophenone
derived name:	methyl phenyl ketone	phenyl propyl ketone	diphenyl ketone

If the ketone has a second functional group of higher naming priority, the ketone oxygen is indicated by the prefix "oxo."

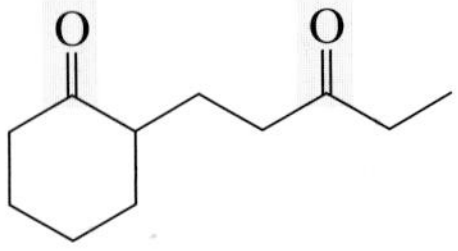

systematic name: 4-oxopentanal — methyl 3-oxobutanoate — 2-(3-oxopentyl)-cyclohexanone

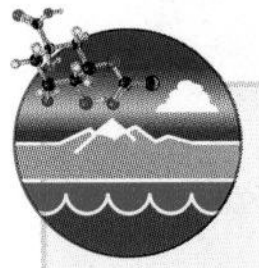

BUTANEDIONE: AN UNPLEASANT COMPOUND

Fresh perspiration is odorless. Bacteria that are always present on our skin produce lactic acid, thereby creating an acidic environment that allows other bacteria to break down the components of perspiration, producing compounds with the unappealing odors we associate with armpits and sweaty feet. One such compound is butanedione.

$$CH_3-\overset{O}{\underset{\|}{C}}-\overset{O}{\underset{\|}{C}}-CH_3$$

butanedione

PROBLEM 2◆

Give two names for each of the following compounds:

a. 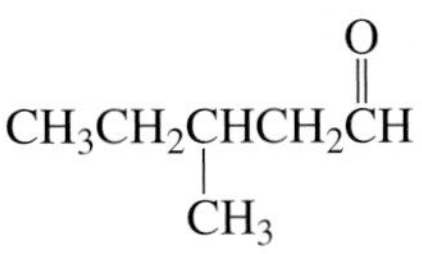 $\underset{\underset{CH_3}{|}}{CH_3CH_2CHCH_2}\overset{\overset{O}{\|}}{C}H$

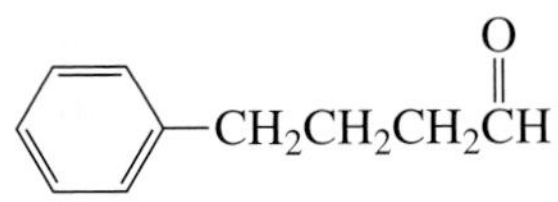

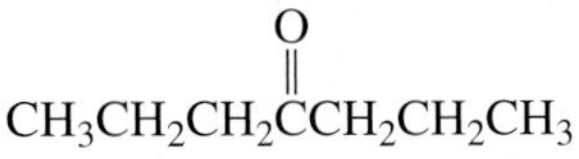

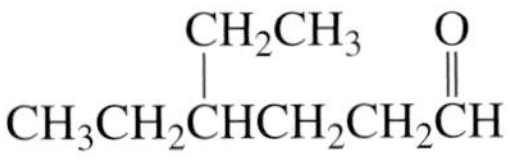

d. ⬡—$CH_2CH_2CH_2\overset{\overset{O}{\|}}{C}H$

b. $CH_3CH_2CH_2\overset{\overset{O}{\|}}{C}CH_2CH_2CH_3$

e. $\underset{\underset{CH_2CH_3}{|}}{CH_3CH_2CHCH_2CH_2}\overset{\overset{O}{\|}}{C}H$

c. $\underset{\underset{CH_3}{|}}{CH_3CHCH_2}\overset{\overset{O}{\|}}{C}CH_2CH_2CH_3$

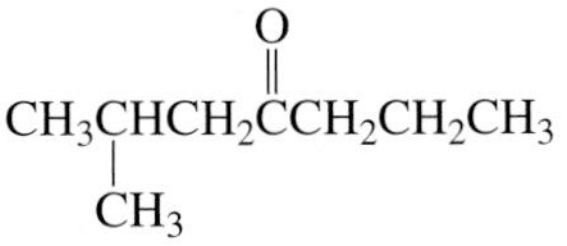

f.  $CH_2{=}CH\overset{\overset{O}{\|}}{C}CH_2CH_2CH_2CH_3$

PROBLEM 3◆

Name the following compounds:

a. 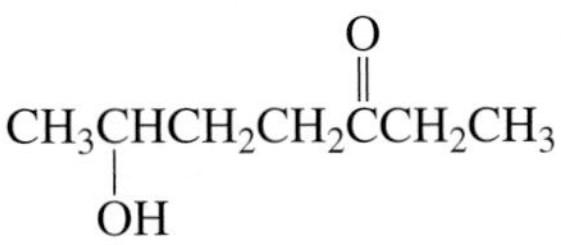 $\underset{\underset{OH}{|}}{CH_3CHCH_2CH_2}\overset{\overset{O}{\|}}{C}CH_2CH_3$

b.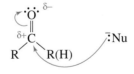

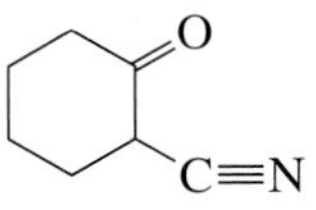

c. 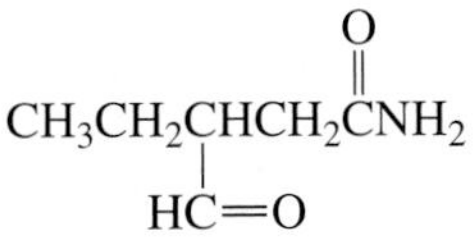 $\underset{\underset{HC=O}{|}}{CH_3CH_2CHCH_2}\overset{\overset{O}{\|}}{C}NH_2$

We have seen that the carbonyl group is polar because oxygen, being more electronegative than carbon, has a greater share of the electrons of the double bond (Section 16.4). The partial positive charge on the carbonyl carbon causes carbonyl compounds to be attacked by nucleophiles. The electron deficiency of the carbonyl carbon is indicated by the blue area in the electrostatic potential maps.

17.2
RELATIVE REACTIVITIES OF CARBONYL COMPOUNDS

Because a hydrogen is electron-withdrawing compared to an alkyl group, an aldehyde has a greater partial positive charge on its carbonyl carbon than a ketone. Therefore, an aldehyde is more reactive than a ketone toward nucleophilic attack.

relative reactivities

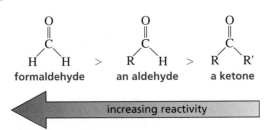

formaldehyde

3-D Molecules:
Formaldehyde;
Acetaldehyde;
Acetone

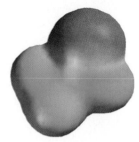

acetaldehyde

acetone

 3-D Molecules:
3-Methyl-2-butanone;
2,4-Dimethyl-3-pentanone

Steric factors also contribute to the greater reactivity of an aldehyde. Because the hydrogen attached to the carbonyl carbon of an aldehyde is smaller than the alkyl group attached to the carbonyl carbon of a ketone, the carbonyl carbon of an aldehyde is more accessible to the nucleophile than is the carbonyl carbon of a ketone. For the same reason, ketones with small alkyl groups bonded to the carbonyl carbon are more reactive than ketones with large alkyl groups.

relative reactivities

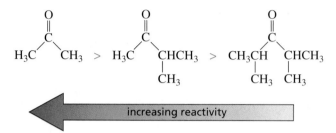

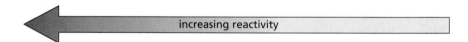

increasing reactivity

PROBLEM 4◆

Which ketone is more reactive?

a. 2-heptanone or 4-heptanone

b. *p*-nitroacetophenone or *p*-methoxyacetophenone

How does the reactivity of an aldehyde or a ketone toward nucleophiles compare with the reactivity of the carbonyl compounds whose reactions we studied in Chapter 16? Aldehydes and ketones are right in the middle. They are less reactive than acyl halides and acid anhydrides, but more reactive than esters, carboxylic acids, and amides.

relative reactivities of carbonyl compounds toward nucleophiles

acyl halide > acid anhydride > aldehyde > ketone > ester ~ carboxylic acid > amide

increasing reactivity

The carbonyl compounds discussed in Chapter 16 have a pair of nonbonding electrons on an atom attached to the carbonyl group. These electrons can be shared with the carbonyl carbon by resonance electron donation, which makes it less electron deficient. We have seen that the reactivity of these carbonyl compounds is related to the basicity of Y^-. The weaker the basicity of Y^-, the more reactive the carbonyl group because weak bases are better able to withdraw electrons inductively from the carbonyl carbon and are less able to donate electrons by resonance to the carbonyl carbon (Section 16.5).

Consequently, aldehydes and ketones are not as reactive as carbonyl compounds in which Y^- is a very weak base (acyl halides and acid anhydrides), but are more reactive than carbonyl compounds in which Y$^-$ is a relatively strong base (carboxylic acids, esters, and amides).

In Section 16.4 we saw that the carbonyl group of a carboxylic acid or a carboxylic acid derivative is attached to a group that can be replaced by another group. These compounds therefore react with nucleophiles to form substitution products.

17.3 REACTIVITY CONSIDERATIONS

product of nucleophilic substitution

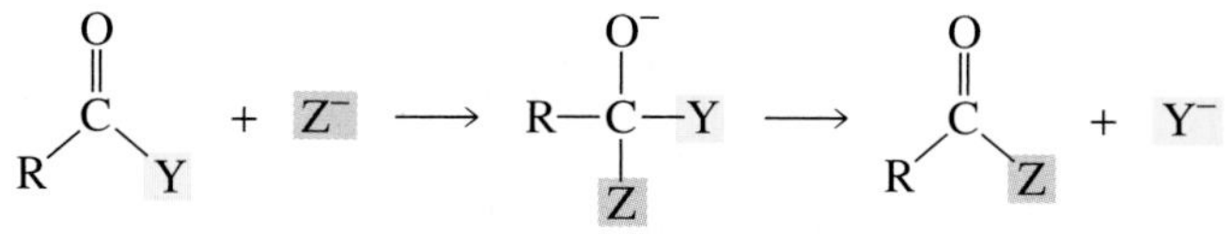

In contrast, the carbonyl group of an aldehyde or a ketone is attached to a group that is too strong a base (H$^-$ or R$^-$) to be eliminated under normal conditions, so it cannot be replaced by another group. Consequently, aldehydes and ketones react with nucleophiles to form addition products, not substitution products. Thus, aldehydes and ketones undergo **nucleophilic addition reactions,** whereas carboxylic acid derivatives undergo **nucleophilic acyl substitution reactions.**

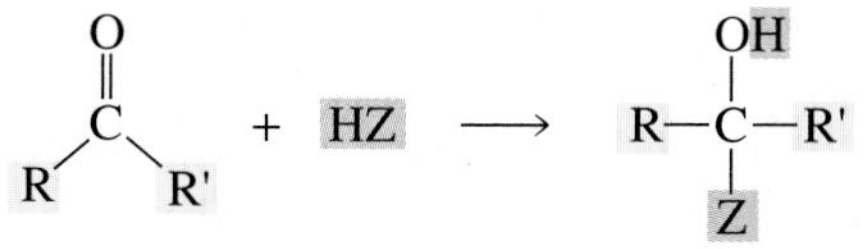

product of nucleophilic addition

If the nucleophile that adds to the aldehyde or ketone is a good nucleophile, it will readily attack the carbonyl carbon, forming an addition product—an alkoxide ion—that can be protonated either by the solvent or by added acid. The hybridization of the carbonyl carbon changes from sp^2 in the carbonyl compound to sp^3 in the addition product.

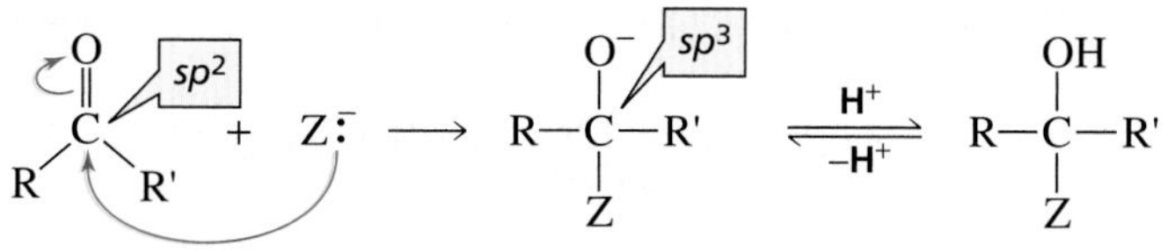

A poor nucleophile requires an acid catalyst to make the nucleophilic addition reaction occur at a reasonable rate. The acid protonates the carbonyl oxygen, which increases the susceptibility of the carbonyl carbon to nucleophilic attack (Figure 17.1).

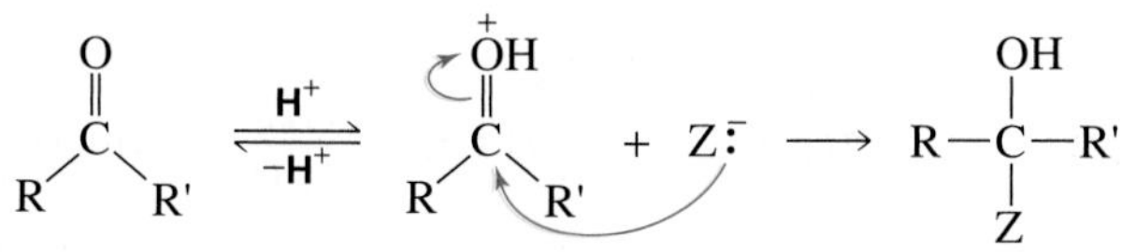

If the attacking atom of the nucleophile has a pair of nonbonding electrons in the addition product, water will be eliminated from the addition product. This is called a **nucleophilic addition–elimination reaction.** We will see that the fate of the dehydrated product depends on the identity of Z.

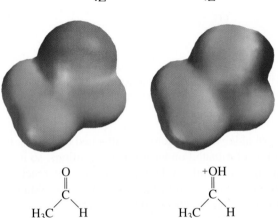

Figure 17.1 ▶
The electrostatic potential maps show that the carbonyl carbon of the protonated aldehyde is more susceptible to nucleophilic attack (the blue is more intense) than the carbonyl carbon of the unprotonated aldehyde.

17.4 ADDITION OF CARBON NUCLEOPHILES

Few reactions in organic chemistry result in formation of new carbon–carbon bonds. Consequently, those reactions that do are very important to synthetic organic chemists when they need to synthesize larger organic molecules from smaller molecules. The addition of a carbon nucleophile to a carbonyl compound is an example of a reaction that forms a new carbon–carbon bond, and therefore forms a product with more carbon atoms than the starting material.

Addition of Grignard Reagents

Addition of a Grignard reagent to a carbonyl compound is a versatile reaction that leads to the formation of a new carbon–carbon bond. Because both the structure of the carbonyl compound and the structure of the Grignard reagent can be varied, the reaction can produce compounds with a variety of structures. We saw in Section 11.8 that a Grignard reagent can be prepared by adding an alkyl halide to magnesium shavings in diethyl ether. We also saw that a Grignard reagent reacts as if it were a carbanion.

$$CH_3CH_2Br \xrightarrow[\text{Et}_2\text{O}]{\textbf{Mg}} CH_3CH_2MgBr$$

$$CH_3CH_2MgBr \quad \text{reacts as if it were} \quad CH_3\overset{..}{\overset{-}{C}}H_2 \quad \overset{+}{\text{Mg}}Br$$

Attack of a Grignard reagent on a carbonyl carbon forms an alkoxide ion that is complexed with magnesium ion. Addition of water or dilute acid breaks up the complex. When a Grignard reagent reacts with an aldehyde, the addition product is a secondary alcohol.

$$CH_3CH_2\overset{\overset{\displaystyle \overset{..}{\text{O}}:}{\|}}{C}H + CH_3CH_2CH_2-MgBr \longrightarrow CH_3CH_2CHCH_2CH_2CH_3 \xrightarrow[\text{H}_2\text{O}]{\text{H}^+} CH_3CH_2CHCH_2CH_2CH_3$$

propanal   propylmagnesium bromide   an oxyanion   3-hexanol a secondary alcohol

When a Grignard reagent reacts with a ketone, a tertiary alcohol is formed.

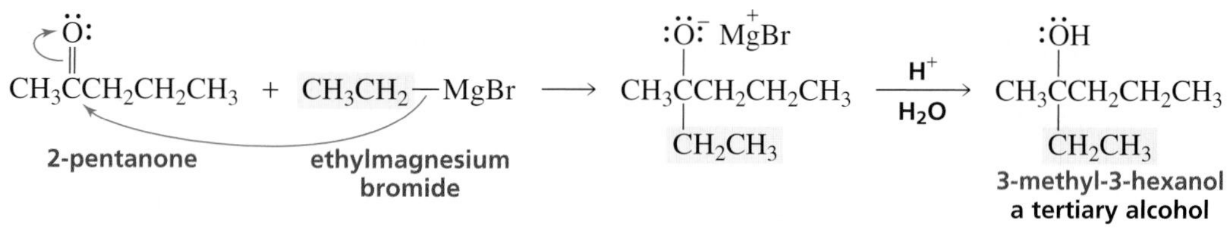

$$CH_3\overset{\overset{\displaystyle \ddot{O}:}{\|}}{C}CH_2CH_2CH_3 \;+\; CH_3CH_2\!-\!MgBr \longrightarrow CH_3\overset{\overset{\displaystyle :\ddot{O}:^{-}\ \overset{+}{MgBr}}{|}}{\underset{\underset{\displaystyle CH_2CH_3}{|}}{C}}CH_2CH_2CH_3 \;\xrightarrow[\text{H}_2\text{O}]{\text{H}^{+}}\; CH_3\overset{\overset{\displaystyle :\ddot{O}H}{|}}{\underset{\underset{\displaystyle CH_2CH_3}{|}}{C}}CH_2CH_2CH_3$$

2-pentanone   **ethylmagnesium bromide**   **3-methyl-3-hexanol a tertiary alcohol**

In the following reactions, numbers are used with the reagents to indicate that the acid is not added until reaction with the Grignard reagent is complete.

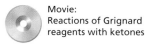

Movie: Reactions of Grignard reagents with ketones

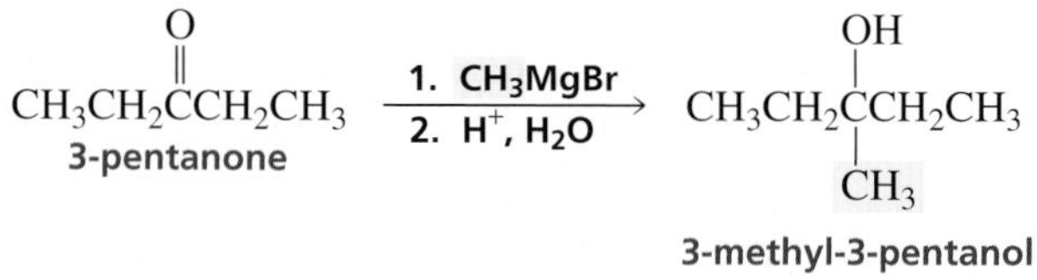

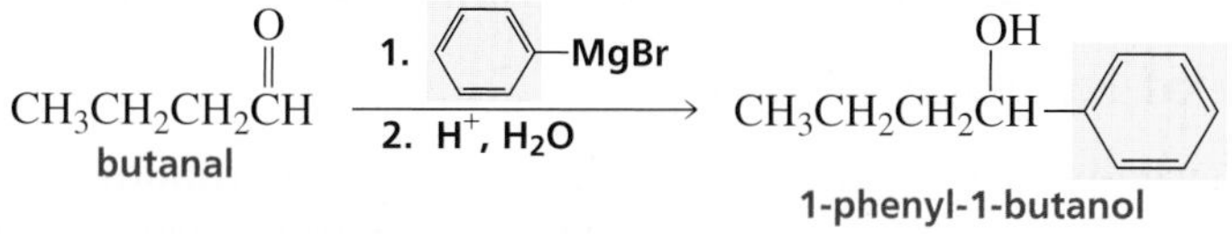

$$CH_3CH_2\overset{\overset{\displaystyle O}{\|}}{C}CH_2CH_3 \;\xrightarrow[\text{2. H}^{+},\ \text{H}_2\text{O}]{\text{1. CH}_3\text{MgBr}}\; CH_3CH_2\overset{\overset{\displaystyle OH}{|}}{\underset{\underset{\displaystyle CH_3}{|}}{C}}CH_2CH_3$$

3-pentanone   **3-methyl-3-pentanol**

$$CH_3CH_2CH_2\overset{\overset{\displaystyle O}{\|}}{C}H \;\xrightarrow[\text{2. H}^{+},\ \text{H}_2\text{O}]{\text{1.}\ \ \bigcirc\!-\!\text{MgBr}}\; CH_3CH_2CH_2\overset{\overset{\displaystyle OH}{|}}{C}H\!-\!\bigcirc$$

butanal   **1-phenyl-1-butanol**

Tutorial: Grignard reagents in synthesis

A Grignard reagent can react with carbon dioxide. The product of the reaction is a carboxylic acid with one more carbon atom than the Grignard reagent.

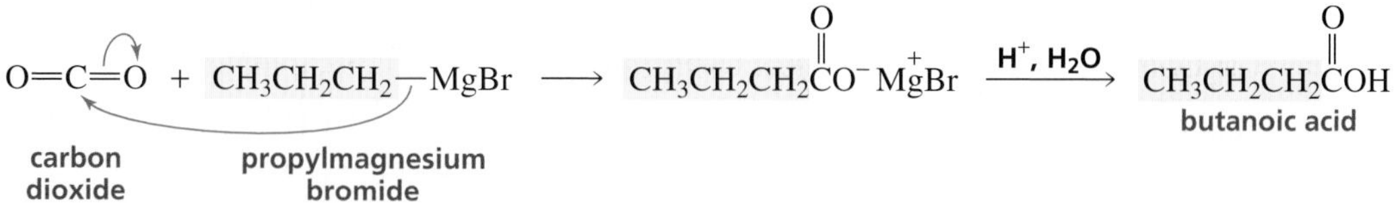

$$O\!=\!C\!=\!O \;+\; CH_3CH_2CH_2\!-\!MgBr \longrightarrow CH_3CH_2CH_2\overset{\overset{\displaystyle O}{\|}}{C}O^{-}\ \overset{+}{MgBr} \;\xrightarrow{\text{H}^{+},\ \text{H}_2\text{O}}\; CH_3CH_2CH_2\overset{\overset{\displaystyle O}{\|}}{C}OH$$

carbon dioxide   **propylmagnesium bromide**   **butanoic acid**

PROBLEM 5◆

a. How many isomers are obtained from the reaction of 2-pentanone with ethylmagnesium bromide followed by H^{+}, H_2O?

b. How many isomers are obtained from the reaction of 2-pentanone with methylmagnesium bromide followed by H^{+}, H_2O?

PROBLEM 6◆

We saw that 3-methyl-3-hexanol can be synthesized from the reaction of 2-pentanone with ethylmagnesium bromide. What two other combinations of ketone and Grignard reagent could be used to prepare the same tertiary alcohol?

PROBLEM 7◆

How can a primary alcohol be prepared from the reaction of a carbonyl compound with a Grignard reagent?

Addition of Acetylide Ions

We have seen that a terminal alkyne can be converted into an acetylide ion by a strong base (Section 5.9).

$$\text{CH}_3\text{C}{\equiv}\text{CH} \xrightarrow[\text{NH}_3]{\text{NaNH}_2} \text{CH}_3\text{C}{\equiv}\text{C:}^-$$

An acetylide ion is another example of a carbon nucleophile that adds to aldehydes and ketones. When the reaction is over, acid is added to the reaction mixture to protonate the alkoxide ion.

$$\text{CH}_3\text{CH}_2\overset{\overset{\textstyle :\ddot{\text{O}}:}{\textstyle \|}}{\text{C}}\text{H} + \text{CH}_3\text{C}{\equiv}\text{C:}^- \longrightarrow \text{CH}_3\text{CH}_2\overset{\overset{\textstyle :\ddot{\text{O}}:^-}{\textstyle |}}{\text{C}}\text{HC}{\equiv}\text{CCH}_3 \xrightarrow{\text{H}^+} \text{CH}_3\text{CH}_2\overset{\overset{\textstyle :\ddot{\text{O}}\text{H}}{\textstyle |}}{\text{C}}\text{HC}{\equiv}\text{CCH}_3$$

PROBLEM 8

Show how the following compounds could be prepared using ethyne as one of the starting materials. Explain why ethyne should be alkylated before rather than after nucleophilic addition.

a. 1-pentyn-3-ol   **b.** 1-phenyl-2-butyn-1-ol   **c.** 2-methyl-3-hexyn-2-ol

---

Addition of Hydrogen Cyanide

Hydrogen cyanide adds to aldehydes and ketones to form **cyanohydrins.** This reaction forms a product with one more carbon atom than the reactant. In the first step of the reaction, the cyanide ion attacks the carbonyl carbon. The alkoxide ion then accepts a proton from an undissociated molecule of hydrogen cyanide.

$$\underset{\textbf{acetone}}{\text{CH}_3\overset{\overset{\textstyle :\ddot{\text{O}}:}{\textstyle \|}}{\text{C}}\text{CH}_3} + {}^-\text{:C}{\equiv}\text{N} \rightleftharpoons \underset{\text{CH}_3}{\text{CH}_3\overset{\overset{\textstyle :\ddot{\text{O}}:^-}{\textstyle |}}{\underset{\textstyle |}{\text{C}}}\text{C}{\equiv}\text{N}} \underset{\text{H}-\text{C}{\equiv}\text{N}}{\rightleftharpoons} \underset{\underset{\textbf{acetone cyanohydrin}}{\text{CH}_3}}{\text{CH}_3\overset{\overset{\textstyle :\ddot{\text{O}}\text{H}}{\textstyle |}}{\underset{\textstyle |}{\text{C}}}\text{C}{\equiv}\text{N}} + {}^-\text{:C}{\equiv}\text{N}$$

Because hydrogen cyanide is a toxic gas, the best way to carry out this reaction is to generate hydrogen cyanide during the reaction by adding HCl to a mixture of the aldehyde or ketone and excess sodium cyanide. Excess sodium cyanide is used in order to make sure that some cyanide ion is available to act as a nucleophile.

Compared with other carbon nucleophiles, cyanide ion is a relatively weak base (pK_a of HC$\equiv$N is 9.14; pK_a of HC$\equiv$CH is 25; pK_a of CH$_3$CH$_3$ is ~50), which means that the cyano group is the most easily eliminated of the carbon nucleophiles from the addition product. Cyanohydrins, however, are stable because the OH group will not eliminate the cyano group since the transition state for the elimination reaction would be relatively unstable because the oxygen atom would bear a partial positive charge. If the OH group loses its proton, however, the cyano group will be eliminated because the oxygen atom would have a partial negative charge instead of a partial positive charge in the transition state of the elimination reaction. Therefore, in basic solutions, a cyanohydrin is converted back to the carbonyl compound.

$$\underset{\substack{\textbf{cyclohexanone} \\ \textbf{cyanohydrin}}}{\overset{\text{H}\ddot{\text{O}}:\ \text{C}{\equiv}\text{N}}{\bigcirc}} \underset{\text{H}_2\text{O}}{\overset{\text{HO}^-}{\rightleftharpoons}} \overset{\text{:}\ddot{\text{O}}:^-\ \text{C}{\equiv}\text{N}}{\bigcirc} \rightleftharpoons \overset{\ddot{\text{O}}:}{\bigcirc} + {}^-\text{:C}{\equiv}\text{N}$$

The addition of hydrogen cyanide to aldehydes and ketones is a synthetically use-ful reaction because of the subsequent reactions that can be carried out on the cyanohydrin. For example, the acid-catalyzed hydrolysis of a cyanohydrin forms an α-hydroxy carboxylic acid (Section 16.16).

$$\underset{\textbf{a cyanohydrin}}{\underset{\underset{\displaystyle CH_2CH_3}{|}}{\overset{\overset{\displaystyle OH}{|}}{CH_3CH_2C}}-C\equiv N} \quad \xrightarrow[\Delta]{\textbf{H}^+,\textbf{H}_2\textbf{O}} \quad \underset{\textbf{an }\alpha\textbf{-hydroxy carboxylic acid}}{\underset{\underset{\displaystyle CH_2CH_3}{|}}{\overset{\overset{\displaystyle OH\ \ O}{|\ \ \ ||}}{CH_3CH_2C}-COH}}$$

The catalytic addition of hydrogen to a cyanohydrin produces a primary amine with an OH group on the β-carbon.

$$\underset{}{\overset{\overset{\displaystyle OH}{|}}{CH_3CH_2CH_2CHC}\equiv N} \quad \xrightarrow[\textbf{Pt}]{\textbf{H}_2} \quad \overset{\overset{\displaystyle OH}{|}}{CH_3CH_2CH_2CHCH_2NH_2}$$

PROBLEM 9

Can a cyanohydrin be prepared by treating a ketone with sodium cyanide?

PROBLEM 10

Although aldehydes and ketones react with a weak acid such as hydrogen cyanide in the presence of $^-C\equiv N$, they do not react with strong acids such as HCl or H_2SO_4 in the presence of Cl^- or HSO_4^-. Explain.

PROBLEM 11 / SOLVED

How can the following compounds be prepared, starting with a carbonyl compound with one fewer carbon atom than the desired product?

a. $HOCH_2CH_2NH_2$

b. $\underset{\underset{\displaystyle OH}{|}}{\overset{\overset{\displaystyle O}{||}}{CH_3CHCOH}}$

SOLUTION TO 11a   The starting material for the synthesis of the two-carbon com-pound must be formaldehyde. Addition of hydrogen cyanide followed by addition of H_2 to the triple bond of the cyanohydrin forms the desired compound.

$$\overset{\overset{\displaystyle O}{||}}{HCH} \quad \xrightarrow[\textbf{HCl}]{\textbf{NaC}\equiv\textbf{N}} \quad HOCH_2C\equiv N \quad \xrightarrow[\textbf{Pt}]{\textbf{H}_2} \quad HOCH_2CH_2NH_2$$

SOLUTION TO 11b   The starting material for the synthesis of the three-carbon α-hydroxy carboxylic acid must be ethanal. Addition of hydrogen cyanide followed by hydrolysis of the cyanohydrin forms the target molecule.

$$\overset{\overset{\displaystyle O}{||}}{CH_3CH} \quad \xrightarrow[\textbf{HCl}]{\textbf{NaC}\equiv\textbf{N}} \quad \underset{\underset{\displaystyle OH}{|}}{CH_3CHC\equiv N} \quad \xrightarrow[\Delta]{\textbf{H}^+,\textbf{H}_2\textbf{O}} \quad \underset{\underset{\displaystyle OH}{|}}{\overset{\overset{\displaystyle O}{||}}{CH_3CHCOH}}$$

17.5
ADDITION OF HYDRIDE ION

Addition of hydride ion to an aldehyde or ketone forms an alkoxide ion. Subsequent protonation by water or a weak acid produces an alcohol. The overall reaction adds H_2 to the carbonyl group. Recall that the addition of hydrogen to an organic compound is a **reduction reaction** (Section 15.2).

$$\underset{\substack{R \quad R'}}{C}{=}O + :H^- \longrightarrow \underset{\substack{| \\ H}}{R-C-R'} \overset{H^+}{\rightleftharpoons} \underset{\substack{| \\ H}}{R-C-R'}$$

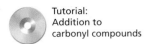

Tutorial:
Addition to
carbonyl compounds

Aldehydes and ketones can be reduced using sodium borohydride ($NaBH_4$) as the source of hydride ion. Aldehydes are reduced to primary alcohols, and ketones are reduced to secondary alcohols. Notice that the source of protons (water or a weak acid) is not added to the reaction mixture until reaction with the hydride donor is complete.

$$\underset{\substack{\text{butanal} \\ \text{an aldehyde}}}{CH_3CH_2CH_2\overset{O}{\overset{\|}{C}}H} \xrightarrow[\text{2. H}^+, \text{H}_2\text{O}]{\text{1. NaBH}_4} \underset{\substack{\text{1-butanol} \\ \text{a primary alcohol}}}{CH_3CH_2CH_2CH_2OH}$$

$$\underset{\substack{\text{2-pentanone} \\ \text{a ketone}}}{CH_3CH_2CH_2\overset{O}{\overset{\|}{C}}CH_3} \xrightarrow[\text{2. H}^+, \text{H}_2\text{O}]{\text{1. NaBH}_4} \underset{\substack{\text{2-pentanol} \\ \text{a secondary alcohol}}}{CH_3CH_2CH_2\overset{OH}{\overset{|}{C}}HCH_3}$$

> **PROBLEM 12◆**
>
> What alcohols are obtained from the reduction of the following compounds with sodium borohydride?
>
> **a.** 2-methylpropanal   **c.** benzaldehyde
>
> **b.** cyclohexanone   **d.** acetophenone

17.6
REACTIONS OF CARBONYL COMPOUNDS THAT HAVE LEAVING GROUPS WITH GRIGNARD REAGENTS AND HYDRIDE ION DONORS

Reaction With Grignard Reagents

In addition to reacting with aldehydes and ketones, Grignard reagents and hydride ion donors also react with carbonyl compounds that have leaving groups (the Class I carbonyl compounds discussed in Chapter 16).

The reaction of a carbonyl compound that has a leaving group with a Grignard reagent involves two successive reactions with the nucleophile. For example, when an ester reacts with a Grignard reagent, the first reaction is a nucleophilic acyl substitution reaction because an ester, unlike an aldehyde or a ketone, has a group that can be replaced by the Grignard reagent. The product of the reaction is a ketone. The reaction does not stop at the ketone stage, however, because ketones are more reactive than esters toward nucleophilic attack. Reaction of the ketone with a second molecule of the Grignard reagent forms a tertiary alcohol. Because the tertiary alcohol is formed as a result of two successive reactions with a Grignard reagent, the alcohol has two identical groups bonded to the tertiary carbon.

mechanism for the reaction of an ester with a Grignard reagent

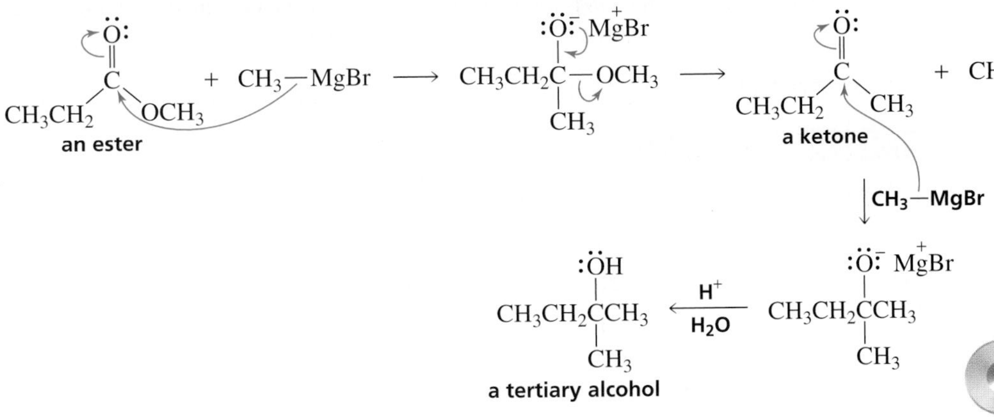

an ester

a ketone

a tertiary alcohol

Movie:
Reaction of a
Grignard reagent
with an ester

Tertiary alcohols are also formed from the reaction of two equivalents of a Grignard reagent with an acyl halide.

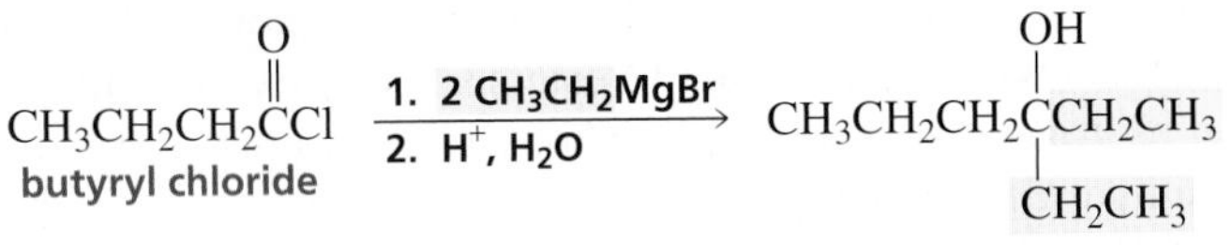

$$CH_3CH_2CH_2CCl \xrightarrow[\text{2. H}^+, \text{H}_2\text{O}]{\text{1. 2 CH}_3\text{CH}_2\text{MgBr}} CH_3CH_2CH_2CCH_2CH_3$$

butyryl chloride

3-ethyl-3-hexanol

In theory, we should be able to stop this reaction at the ketone stage because a ketone is less reactive than an acyl halide. However, the Grignard reagent is so reactive that it can be prevented from reacting with the ketone only under very carefully controlled conditions. There are better ways to synthesize ketones (Appendix IV).

PROBLEM 13◆

What product would be obtained from the reaction of one equivalent of a carboxylic acid with one equivalent of a Grignard reagent?

PROBLEM 14 / SOLVED

a. Which of the following tertiary alcohols cannot be prepared from the reaction of an ester with excess Grignard reagent?

1. $CH_3\underset{CH_3}{\overset{OH}{C}}CH_3$ **3.** $CH_3CH_2\underset{CH_3}{\overset{OH}{C}}CH_2CH_2CH_3$ **5.** $CH_3\underset{CH_2CH_3}{\overset{OH}{C}}CH_2CH_2CH_2CH_3$

2. $CH_3\underset{CH_3}{\overset{OH}{C}}CH_2CH_3$ **4.** $CH_3CH_2\underset{CH_3}{\overset{OH}{C}}CH_2CH_3$ **6.**

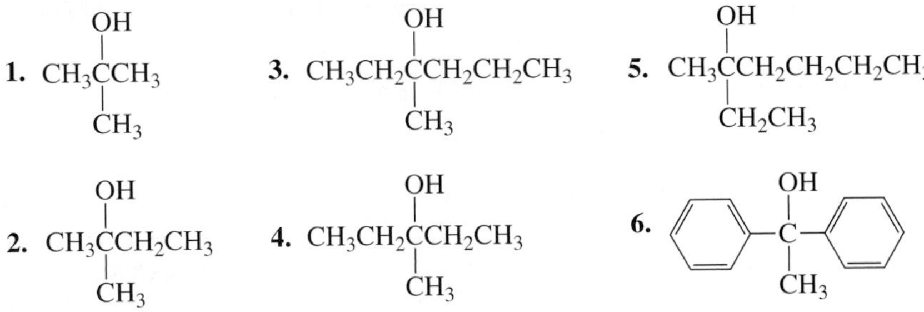

b. For those alcohols that can be prepared by the reaction of an ester with excess Grignard reagent, what ester and what Grignard reagent should be used?

SOLUTION TO 14a   A tertiary alcohol is obtained from the reaction of an ester with two equivalents of a Grignard reagent. Therefore, tertiary alcohols prepared in this way must have two identical substituents on the carbon to which the OH is bonded because two substituents come from the Grignard reagent. Alcohols (3) and (5) cannot be prepared in this way because they do not have two identical substituents.

SOLUTION TO 14b(2)   Methyl propanoate and excess methylmagnesium bromide.

PROBLEM 15◆

Which of the following secondary alcohols can be prepared from the reaction of methyl formate with excess Grignard reagent?

$$CH_3CH_2\underset{\underset{OH}{|}}{C}HCH_3 \qquad CH_3\underset{\underset{OH}{|}}{C}HCH_3 \qquad CH_3\underset{\underset{OH}{|}}{C}HCH_2CH_2CH_3 \qquad CH_3CH_2\underset{\underset{OH}{|}}{C}HCH_2CH_3$$

Reaction With a Hydride Ion Donor

The reaction of a carbonyl compound that has a leaving group with a hydride ion donor—like the reaction with a Grignard reagent—involves two successive reactions with the nucleophile. Sodium borohydride ($NaBH_4$) is not a sufficiently strong hydride donor to react with the less reactive esters and carboxylic acids (compared with aldehydes and ketones), so esters and carboxylic acids must be reduced using lithium aluminum hydride ($LiAlH_4$).

Although both sodium borohydride and lithium aluminum hydride are sources of hydride ion, sodium borohydride, because it is less reactive, is safer and easier to use. Lithium aluminum hydride must be used in a dry, aprotic solvent because it reacts violently with protic solvents. Depending on the reactant, $NaBH_4$ may be chosen for its ease of safe handling, or $LiAlH_4$ may be chosen for its greater reactivity.

The reaction of an ester with $LiAlH_4$ produces two alcohols, one corresponding to the acyl portion of the ester and one corresponding to the alkyl portion.

3-D Molecule:
Methyl propanoate

$$\underset{\substack{\text{methyl propanoate}\\ \text{an ester}}}{CH_3CH_2\overset{\overset{O}{\|}}{C}OCH_3} \xrightarrow[\text{2. } H^+,\ H_2O]{\text{1. } LiAlH_4} \underset{\text{1-propanol}}{CH_3CH_2CH_2OH} + \underset{\text{methanol}}{CH_3OH}$$

When an ester reacts with hydride ion, the first reaction is a nucleophilic acyl substitution reaction because an ester has a group that can be substituted by hydride ion. The product of this reaction is an aldehyde. The aldehyde then undergoes a nucleophilic addition reaction with a second equivalent of hydride ion, forming an alkoxide ion, which

mechanism for the reaction of an ester with hydride ion

$$\underset{\substack{\\ \text{an ester}}}{\underset{CH_3CH_2}{\overset{\overset{\ddot{O}:}{\|}}{C}}\diagdown_{OCH_3}} + H-\bar{A}lH_3 \longrightarrow CH_3CH_2\overset{\overset{:\ddot{O}:^-}{|}}{\underset{\underset{H}{|}}{C}}-OCH_3 \longrightarrow \underset{\substack{\\ \text{an aldehyde}}}{\underset{CH_3CH_2}{\overset{\overset{\ddot{O}:}{\|}}{C}}\diagdown_H} + CH_3O^-$$

$$\downarrow H-\bar{A}lH_3$$

$$\underset{\substack{\text{a primary alcohol}}}{CH_3CH_2CH_2OH} \underset{H_2O}{\overset{H^+}{\longleftarrow}} CH_3CH_2\overset{\overset{:\ddot{O}:^-}{|}}{\underset{\underset{H}{|}}{C}}H$$

when protonated gives a primary alcohol. The reaction cannot be stopped at the aldehyde stage because an aldehyde is more reactive than an ester toward nucleophilic attack.

If diisobutylaluminum hydride is used as the hydride donor at a low temperature, the reaction can be stopped after the addition of one equivalent of hydride ion. This reagent, therefore, makes it possible to convert esters into aldehydes. (A temperature of $-78\ °C$ is shown because that is the temperature of the dry ice/acetone bath that is typically used to cool the reaction mixture.)

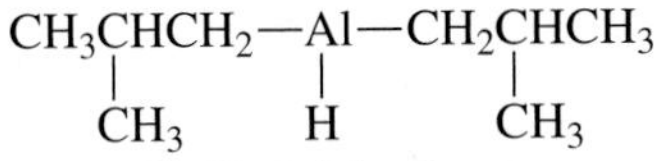

$$CH_3CHCH_2-Al-CH_2CHCH_3$$

with substituents CH_3, H, CH_3

**diisobutylaluminum
hydride**

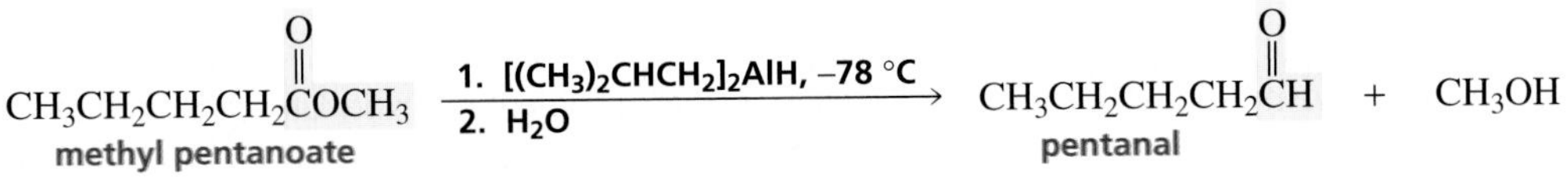

$$CH_3CH_2CH_2CH_2\overset{O}{\overset{\|}{C}}OCH_3 \xrightarrow[\text{2. } H_2O]{\text{1. } [(CH_3)_2CHCH_2]_2AlH,\ -78\ °C} CH_3CH_2CH_2CH_2\overset{O}{\overset{\|}{C}}H \ + \ CH_3OH$$

methyl pentanoate   **pentanal**

The reaction of a carboxylic acid with $LiAlH_4$ forms a single primary alcohol.

3-D Molecule:
Diisobutylaluminum hydride

$$CH_3\overset{O}{\overset{\|}{C}}OH \xrightarrow[\text{2. } H^+,\ H_2O]{\text{1. } LiAlH_4} CH_3CH_2OH$$

acetic acid   **ethanol**

$$\text{(benzene ring)}-\overset{O}{\overset{\|}{C}}OH \xrightarrow[\text{2. } H^+,\ H_2O]{\text{1. } LiAlH_4} \text{(benzene ring)}-CH_2OH$$

benzoic acid   **benzyl alcohol**

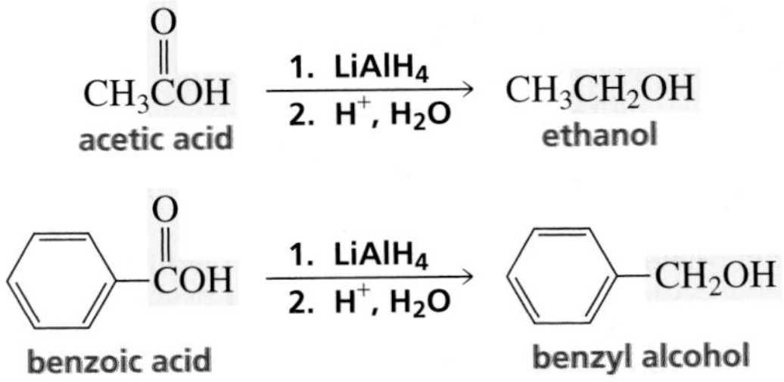

In the first step of the reaction, the hydride ion reacts with the acidic hydrogen of the carboxylic acid, forming H_2. Then, as in the reduction of an ester by $LiAlH_4$, two successive additions of hydride ion take place, with an aldehyde being formed as an intermediate. Notice that ^-OR is the leaving group in the reduction of an ester, while $^-OAlH_2$ is the leaving group in the reduction of a carboxylic acid.

mechanism for the reaction of a carboxylic acid with hydride ion

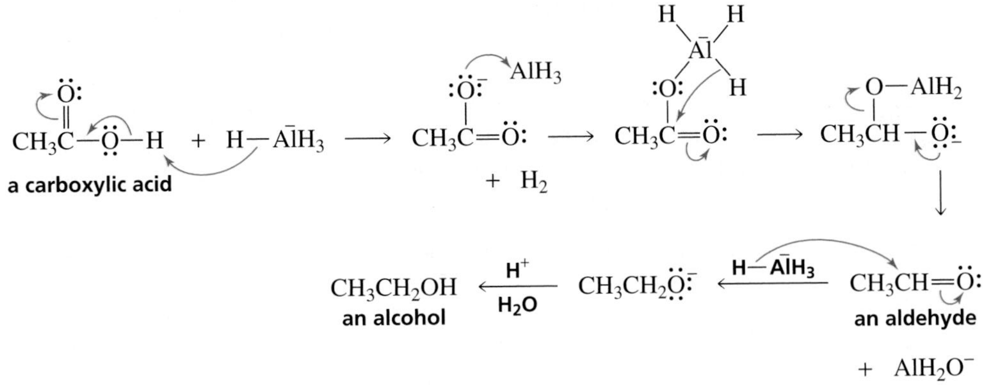

Acyl chlorides, like esters and carboxylic acids, undergo two successive additions of hydride ion when treated with $LiAlH_4$.

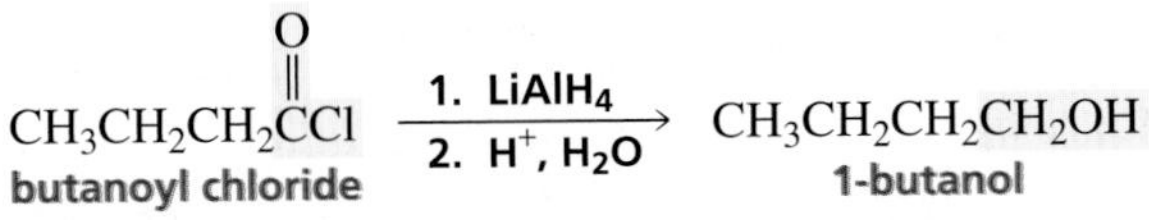

$$CH_3CH_2CH_2\overset{O}{\overset{\|}{C}}Cl \xrightarrow[\text{2. } H^+,\ H_2O]{\text{1. } LiAlH_4} CH_3CH_2CH_2CH_2OH$$

butanoyl chloride   **1-butanol**

$$CH_3-\overset{\overset{\displaystyle CH_3}{|}}{\underset{\underset{\displaystyle O}{|}}{C}}-CH_3$$

$$CH_3C\overset{\overset{\displaystyle CH_3}{|}}{\underset{\underset{\displaystyle CH_3}{|}}{O}}-\overset{}{\underset{\underset{\displaystyle H}{|}}{Al}}-\overset{\overset{\displaystyle CH_3}{|}}{\underset{\underset{\displaystyle CH_3}{|}}{O}}CCH_3$$

Li⁺

lithium tri-*tert*-butoxyaluminum hydride

3-D Molecule:
Lithium tri-*tert*-butoxy-aluminum hydride

If a less reactive, sterically bulky hydride donor, such as lithium tri-*tert*-butoxyaluminum hydride, is used at a low temperature, the reaction stops after one addition of hydride ion because the acyl halide is more reactive than the aldehyde. In this way, an acyl halide can be converted into an aldehyde.

$$CH_3CH_2CH_2\overset{\overset{\displaystyle O}{||}}{C}Cl \quad \xrightarrow[\text{2. } H_2O]{\text{1. LiAlH[OC(CH}_3)_3]_3, -78\ °C} \quad CH_3CH_2CH_2\overset{\overset{\displaystyle O}{||}}{C}H$$

butanoyl chloride   **butanal**

Amides also undergo two successive additions of hydride ion when they react with LiAlH₄. The product of the reaction is an amine. Primary, secondary, and tertiary amines can be formed, depending on the number of substituents bonded to the nitrogen of the amide. Overall, the reaction converts a carbonyl group into a methylene group. (Notice that H_2O rather than H^+, H_2O is used in the second step of the reaction. Therefore, the product is an amine rather than an ammonium ion.)

$$\text{benzamide} \quad \xrightarrow[\text{2. } H_2O]{\text{1. LiAlH}_4} \quad \text{benzylamine}$$

benzamide

benzylamine
a primary amine

$$CH_3\overset{\overset{\displaystyle O}{||}}{C}NHCH_3 \quad \xrightarrow[\text{2. } H_2O]{\text{1. LiAlH}_4} \quad CH_3CH_2NHCH_3$$

N-methylacetamide

ethylmethylamine
a secondary amine

$$\xrightarrow[\text{2. } H_2O]{\text{1. LiAlH}_4}$$

N-methy-γ-butyrolactam

N-methylpyrrolidine
a tertiary amine

The mechanism of the reaction shows why the product of the reaction is an amine. Take a minute to compare the mechanisms for the reaction of hydride ion with an N-substituted amide and with a carboxylic acid.

mechanism for the reaction of an *N*-substituted amide with hydride ion

$$CH_3\overset{\overset{\displaystyle \ddot{O}:}{||}}{C}-\overset{}{\underset{\underset{\displaystyle H}{|}}{\ddot{N}CH_3}} + H-\overset{}{\bar{A}}lH_3 \longrightarrow CH_3\overset{\overset{\displaystyle :\ddot{O}:^-}{|}}{C}=\overset{}{\ddot{N}CH_3} \longrightarrow CH_3\overset{\overset{\displaystyle :O:}{|}}{C}=\overset{}{\ddot{N}CH_3} \longrightarrow CH_3\overset{\overset{\displaystyle O-AlH_2}{|}}{C}H-\ddot{N}CH_3$$

an amide $+ H_2$

$$HO^- + CH_3CH_2NHCH_3 \xleftarrow{H_2O} CH_3CH_2\overset{}{\ddot{N}CH_3} \xleftarrow{H-\bar{A}lH_3} CH_3CH=\overset{}{\ddot{N}CH_3} + AlH_2O^-$$

an amine

mechanism for the reaction of an *N,N*-disubstituted amide with hydride ion

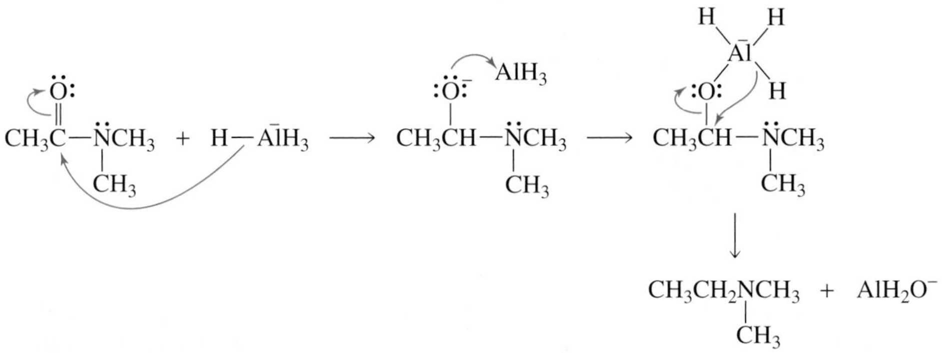

$$CH_3\overset{\overset{\displaystyle :\ddot{O}:}{\|}}{C}-\overset{\overset{\displaystyle |}{\ddot{N}CH_3}}{\underset{CH_3}{}} + H-\bar{Al}H_3 \longrightarrow CH_3\overset{\overset{\displaystyle :\ddot{O}:^-}{|}}{CH}-\overset{\overset{\displaystyle |}{\ddot{N}CH_3}}{\underset{CH_3}{}} \longrightarrow CH_3\overset{\overset{\displaystyle :O:}{|}}{CH}-\overset{\overset{\displaystyle |}{\ddot{N}CH_3}}{\underset{CH_3}{}}$$

$$\downarrow$$

$$CH_3CH_2\overset{\overset{\displaystyle |}{NCH_3}}{\underset{CH_3}{}} + AlH_2O^-$$

PROBLEM 16◆

What amides would you treat with LiAlH$_4$ in order to prepare the following amines?

a. benzylmethylamine   **c.** diethylamine

b. ethylamine   **d.** triethylamine

PROBLEM 17

Starting with *N*-benzylbenzamide, how would you make the following?

a. dibenzylamine   **c.** benzaldehyde

b. benzoic acid   **d.** benzyl alcohol

Addition of Ammonia, Primary Amines, and Other Ammonia Derivatives

17.7 ADDITION OF NITROGEN NUCLEOPHILES

Aldehydes and ketones react with primary amines $(R-NH_2)$ and with other ammonia derivatives $(Z-NH_2)$ to form imines. An **imine** is a compound with a carbon–nitrogen double bond.

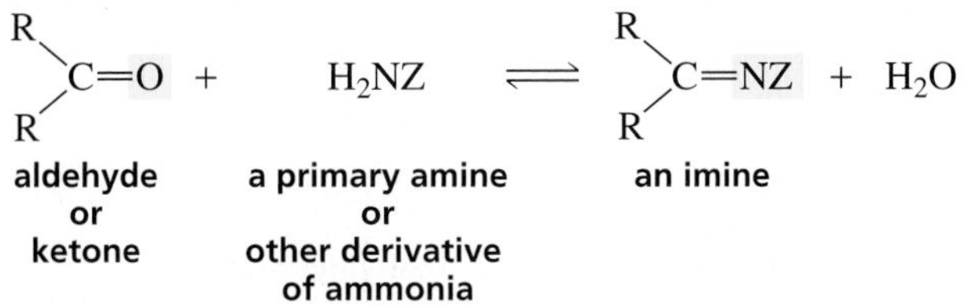

$$\underset{R}{\overset{R}{>}}C=O \;+\; H_2NZ \;\rightleftharpoons\; \underset{R}{\overset{R}{>}}C=NZ \;+\; H_2O$$

| aldehyde or ketone | a primary amine or other derivative of ammonia | an imine |

The orbital model of an imino group (Figure 17.2) is similar to the orbital model of a carbonyl group (Figure 16.1). The imine nitrogen is sp^2 hybridized. One of its sp^2 orbitals forms a σ bond with the imine carbon, one forms a σ bond with a substituent, and the third contains a pair of nonbonding electrons. The p orbital of nitrogen and the p orbital of carbon overlap to form a π bond.

Compounds such as hydroxylamine (NH_2OH), hydrazine (NH_2NH_2), semicarbazide $(NH_2NHCONH_2)$, and primary amines are all derivatives of ammonia because each has a substituent in place of one of the hydrogens of ammonia (NH_3).

The imine obtained from the reaction of a carbonyl compound and a primary amine is called a **Schiff base;** the imine obtained from the reaction with hydroxylamine is

Figure 17.2 ▶
Bonding in an imine.

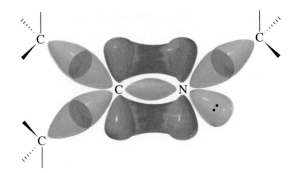

3-D Molecules:
N-methyl acetone imine;
Acetone oxime;
Acetone hydrazone;
Acetone semicarbazone

called an **oxime;** the imine obtained from the reaction with hydrazine is called a **hydrazone;** and the imine obtained from the reaction with semicarbazide is called a **semicarbazone.**

$$CH_3CH_2\underset{CH_3CH_2}{\overset{}{C}}=O \;+\; H_2NCH_2-\bigcirc \;\rightleftharpoons\; CH_3CH_2\underset{CH_3CH_2}{\overset{}{C}}=NCH_2-\bigcirc \;+\; H_2O$$

a primary amine   a Schiff base

$$\bigcirc-CH=O \;+\; H_2NOH \;\rightleftharpoons\; \bigcirc-CH=NOH \;+\; H_2O$$

hydroxylamine   an oxime

$$\bigcirc\underset{CH_3}{\overset{}{C}}=O \;+\; H_2NNH_2 \;\rightleftharpoons\; \bigcirc\underset{CH_3}{\overset{}{C}}=NNH_2 \;+\; H_2O$$

hydrazine   a hydrazone

$$\bigcirc=O \;+\; H_2NNH\overset{O}{\overset{\|}{C}}NH_2 \;\rightleftharpoons\; \bigcirc=NNH\overset{O}{\overset{\|}{C}}NH_2 \;+\; H_2O$$

semicarbazide   a semicarbazone

Tutorial:
Imine and oxime
reactions–synthesis

Phenyl-substituted hydrazines react with aldehydes and ketones to form **phenyl-hydrazones.**

$$\bigcirc=O \;+\; H_2NNH-\bigcirc \;\rightleftharpoons\; \bigcirc=NNH-\bigcirc \;+\; H_2O$$

phenylhydrazine   a phenylhydrazone

$$CH_3CH_2CH=O \;+\; H_2NNH-\bigcirc-NO_2 \;\rightleftharpoons\; CH_3CH_2CH=NNH-\bigcirc-NO_2 \;+\; H_2O$$
$$\qquad\qquad\qquad\qquad O_2N \qquad\qquad\qquad\qquad\qquad\qquad\qquad O_2N$$

2,4-dinitrophenylhydrazine   **a 2,4-dinitrophenylhydrazone**

In the first step of the mechanism for imine formation, the amine attacks the carbonyl carbon. Gain of a proton by the alkoxide ion and loss of a proton by the ammonium ion forms a neutral tetrahedral intermediate. The neutral tetrahedral intermediate is in equilibrium with two protonated forms. Protonation can take place on either the nitrogen or the oxygen atom. Loss of water from the oxygen-protonated intermediate forms a protonated imine that loses a proton to yield the imine.

mechanism for imine formation

[chemical mechanism scheme showing cyclohexanone reacting with RNH₂ through tetrahedral intermediates to form an imine]

neutral tetrahedral
intermediate

an imine   a protonated imine

Because nitrogen is more basic than oxygen, the equilibrium favors the nitrogen-protonated tetrahedral intermediate. The equilibrium can be forced toward the imine by removing water as it is formed or by precipitation of the imine. The formation of all imines (Schiff bases, semicarbazones, oximes, hydrazones, and phenylhydrazones) follows the same mechanism.

Overall, the addition of a nitrogen nucleophile to an aldehyde or a ketone is a nucleophilic addition-elimination reaction (that is, nucleophilic addition of an amine to form an unstable tetrahedral compound followed by elimination of water). The nonbonding electrons on nitrogen are what cause water to be eliminated. Loss of a proton from the resulting protonated imine forms a stable imine. On the other hand, when a carbon or hydrogen nucleophile adds to an aldehyde or ketone, the reaction is a simple addition because there are no nonbonding electrons on the carbon or hydrogen in the tetrahedral compound—so it is stable. Thus, aldehydes and ketones undergo *nucleophilic addition reactions* with carbon and hydrogen nucleophiles, and undergo *nucleophilic addition–elimination reactions* with nitrogen nucleophiles.

> Aldehydes and ketones undergo *nucleophilic addition reactions* with carbon and hydrogen nucleophiles, and undergo *nucleophilic addition–elimination reactions* with nitrogen nucleophiles.

The pH at which imine formation is carried out must be carefully controlled. There must be sufficient acid present to protonate the tetrahedral intermediate so that H_2O rather than the much more basic HO^- is the leaving group. However, if too much acid is present, it will protonate the reactant amine. Protonated amines are not nucleophiles, so they cannot react with carbonyl groups.

A plot of the observed rate constant for the reaction of acetone with hydroxylamine as a function of the pH of the reaction mixture is shown in Figure 17.3. This type of plot is called a **pH–rate profile.** The pH–rate profile in Figure 17.3 is a bell-shaped curve with the maximum rate occurring at about pH 4.5—1.5 pH units below the pK_a of hydroxylamine. As the acidity increases below pH 4.5, the rate of the reaction decreases because more and more of the amine becomes protonated. As a result, less and less of the amine is present in the nucleophilic nonprotonated form. As the acidity decreases above pH 4.5, the rate decreases because less and less of the tetrahedral intermediate is present in the reactive protonated form.

Imine formation is a reversible reaction. In aqueous acidic solutions, imines are hydrolyzed back to the carbonyl compound and amine. In an acidic solution, the amine is protonated and is unable to participate in the reverse reaction. Imine formation and hydrolysis are important reactions in biological systems (Sections 19.21, 22.9, and 23.6).

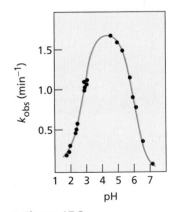

▲ **Figure 17.3**
Dependence of the rate of the reaction of acetone with hydroxylamine on the pH of the reaction mixture.

[chemical equation showing phenyl-CH=NCH₂CH₃ + H₂O with H⁺ giving phenyl-CH=O + CH₃CH₂N⁺H₃]

Imine hydrolysis is a necessary step in the conversion of a nitrile to a ketone. Reaction of a nitrile with a Grignard reagent forms an imine that can be hydrolyzed to a ketone.

$$CH_3CH_2C\equiv N: + R-MgBr \longrightarrow CH_3CH_2\overset{\overset{\ddot{N}:\,\overset{+}{M}gBr}{\|}}{C}-R \xrightarrow{H^+} CH_3CH_2\overset{\overset{+NH_2}{\|}}{C}-R \xrightarrow[H_2O]{H^+} CH_3CH_2\overset{\overset{O}{\|}}{C}-R + \overset{+}{N}H_4$$

PROBLEM 18

Semicarbazide has two NH_2 groups, but only one of them forms an imine. Explain.

NONSPECTROPHOTOMETRIC IDENTIFICATION OF ALDEHYDES AND KETONES

Before spectrophotometric techniques were available, unknown aldehydes and ketones were identified by preparing imine derivatives. For example, let's assume you have an unknown ketone whose boiling point you have determined to be 140 °C. This allows you to narrow the possibilities to five ketones (A to E, below) based on their boiling points. (Ketones boiling at 139 °C and 141 °C cannot be excluded unless your ther-

mometer is calibrated perfectly and your laboratory technique is sensational.) Adding 2,4-dinitrophenylhydrazine to a sample of the unknown ketone produces crystals of a 2,4-dinitrophenylhydrazone that melt at 102 °C. You can now narrow the choice to two ketones, B and D. Preparing the oxime of the unknown ketone will not distinguish between B and D because the oximes of B and D have similar melting points, but preparing the semicarbazone will allow you to identify the ketone. Finding that the semicarbazone of the unknown ketone has a melting point of 112 °C establishes that the unknown ketone is D.

Ketone	bp (°C)	2,4-Dinitrophenylhydrazone mp (°C)	Oxime mp (°C)	Semicarbazone mp (°C)
A	140	94	57	98
B	140	102	68	123
C	139	121	79	121
D	140	101	69	112
E	141	90	61	101

Addition of Secondary Amines

3-D Molecule:
N,N-Diethyl-1-cyclopentenamine

The reaction of an aldehyde or a ketone with a secondary amine produces an enamine (pronounced "ENE-amine"). An **enamine** is an α,β-unsaturated tertiary amine— it is a tertiary amine with a double bond in the α,β-position relative to the nitrogen atom. Notice that the double bond is in the part of the molecule that comes from the aldehyde or ketone. The name "enamine" comes from "ene" + "amine," with the "e" omitted in order to avoid two successive vowels.

cyclopentanone + diethylamine ⇌ *N,N*-diethyl-1-cyclopentenamine an enamine + H_2O

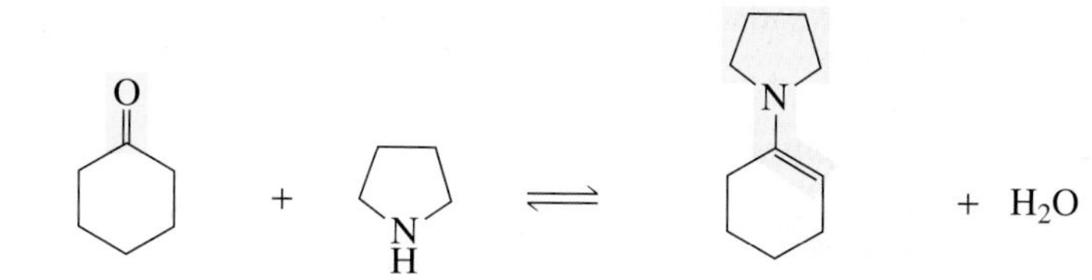

cyclohexanone   pyrrolidine   *N*-(1-cyclohexenyl)-
pyrrolidine
an enamine

The mechanism for enamine formation is exactly the same as the mechanism for imine formation except for the last step of the reaction. When a primary amine reacts with an aldehyde or a ketone, the protonated imine loses a proton from nitrogen in the last step of the reaction, forming a neutral imine. However, when the amine is secondary, the positively charged nitrogen is not bonded to a hydrogen. In order to obtain a stable neutral molecule, a proton must be removed from the α-carbon of the compound that was the carbonyl compound. An enamine results.

mechanism for enamine formation

neutral tetrahedral
intermediate

an enamine

this intermediate cannot lose
a proton from N, so it loses
a proton from an α-carbon

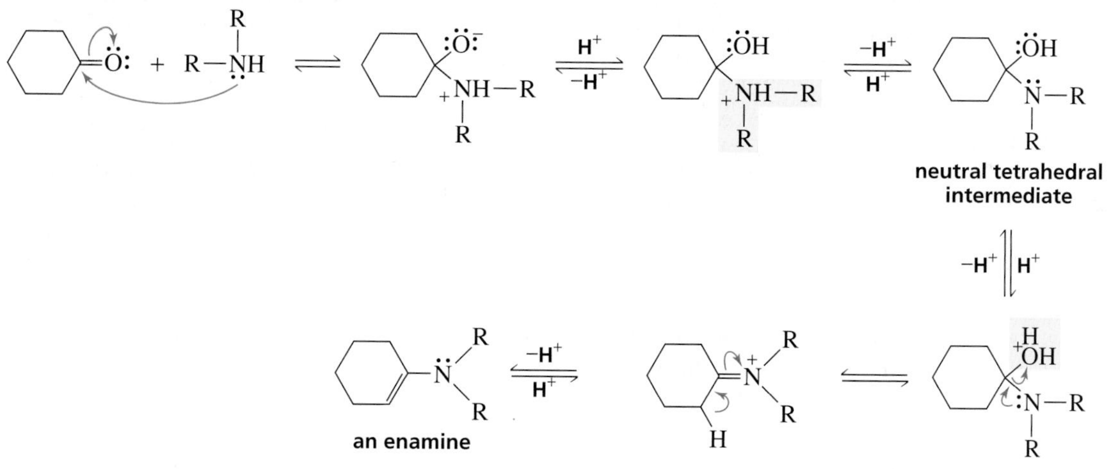

In aqueous acidic solutions, an enamine is hydrolyzed back to the carbonyl compound and secondary amine.

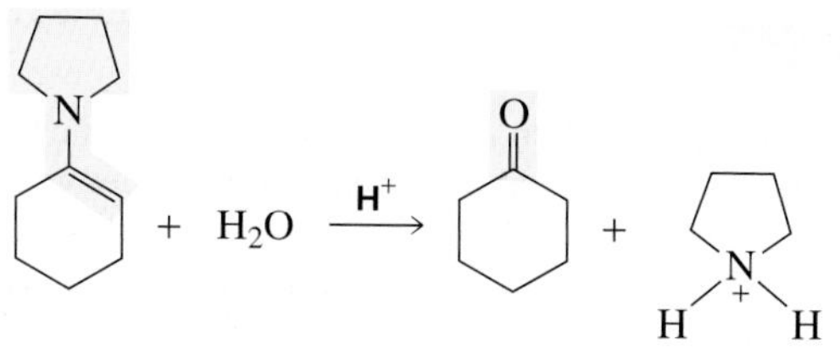

PROBLEM 19

a. Write the mechanism for the following reactions:

1. the acid-catalyzed hydrolysis of an imine to a carbonyl compound and a primary amine.

2. the acid-catalyzed hydrolysis of an enamine to a carbonyl compound and a secondary amine.

b. How do these mechanisms differ?

PROBLEM 20◆

Give the products of the following reactions:

a. cyclopentanone + ethylamine

b. cyclopentanone + diethylamine

c. acetophenone + hexylamine

d. acetophenone + cyclohexylamine

Addition of Ammonia and Hydrogen: Reductive Amination

If the imine formed from the reaction of an aldehyde or a ketone with ammonia does not have a substituent other than a hydrogen bonded to the nitrogen, it is relatively unstable. Nevertheless, such an imine is a useful intermediate. For example, if the reaction with ammonia is carried out in the presence of H_2 and an appropriate metal catalyst, H_2 will add to the $C=N$ bond as it is formed, forming a primary amine. The reaction of an aldehyde or a ketone with excess ammonia in the presence of a reducing agent is called **reductive amination.**

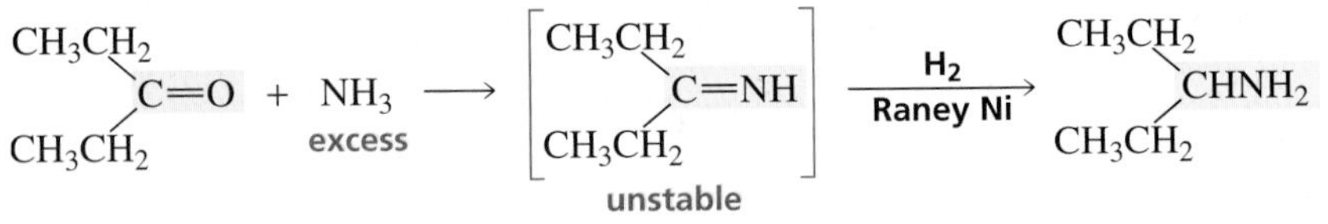

$$
\underset{CH_3CH_2}{\overset{CH_3CH_2}{>}}C=O \ + \ \underset{\textbf{excess}}{NH_3} \ \longrightarrow \ \left[\underset{CH_3CH_2}{\overset{CH_3CH_2}{>}}C=NH\right] \ \xrightarrow[\textbf{Raney Ni}]{H_2} \ \underset{CH_3CH_2}{\overset{CH_3CH_2}{>}}CHNH_2
$$

unstable

Secondary and tertiary amines can be prepared from imines and enamines by reduction of the imine or enamine. Sodium triacetoxyborohydride is a commonly used reducing agent for this reaction.

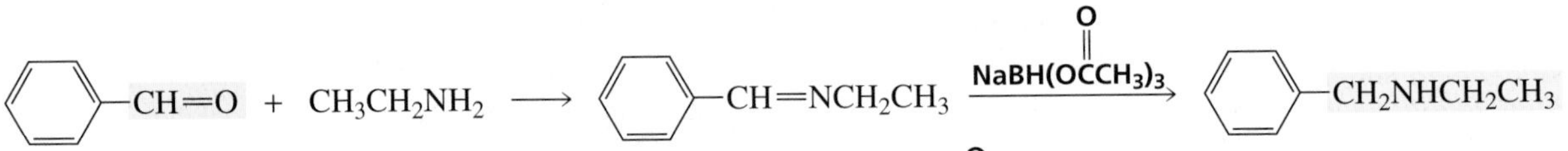

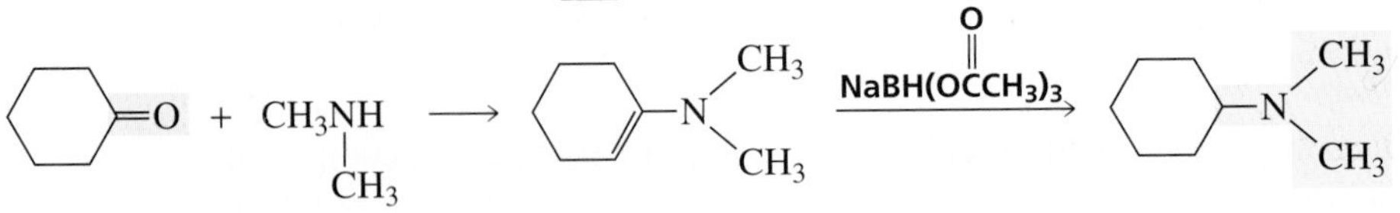

$$
\text{C}_6\text{H}_5-CH=O \ + \ CH_3CH_2NH_2 \ \longrightarrow \ \text{C}_6\text{H}_5-CH=NCH_2CH_3 \ \xrightarrow{NaBH(OCCH_3)_3} \ \text{C}_6\text{H}_5-CH_2NHCH_2CH_3
$$

$$
\text{cyclohexanone}=O \ + \ CH_3NH\text{—}CH_3 \ \longrightarrow \ \text{(enamine N(CH}_3)) \ \xrightarrow{NaBH(OCCH_3)_3} \ \text{(cyclohexyl—N(CH}_3)_2)
$$

PROBLEM 21

Excess ammonia must be used when a primary amine is synthesized by reductive amination. What product will be obtained if the reaction is carried out with an excess of the carbonyl compound instead?

The Wolff–Kishner Reduction

In Section 14.14 we saw that when a ketone or an aldehyde is heated in a basic solution of hydrazine, the carbonyl group is converted into a methylene group. This process is called **deoxygenation** because an oxygen is removed from the reactant. The reaction is known as the *Wolff–Kishner reduction.*

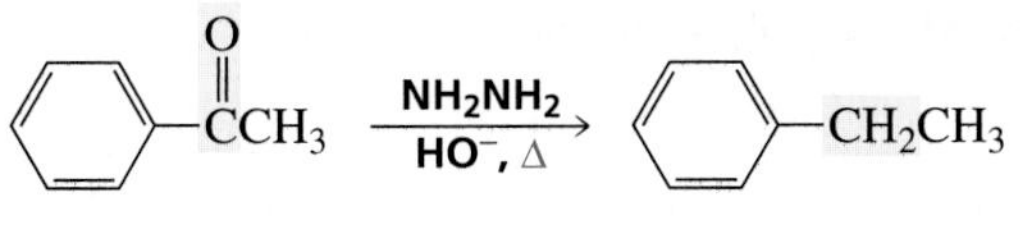

Hydroxide ion and heat differentiate the Wolff–Kishner reduction from ordinary hydrazone formation. Initially, the ketone reacts with hydrazine to form a hydrazone. After the hydrazone is formed, hydroxide ion removes a proton from the NH$_2$ group. Heat is required because these protons are not easily removed. The negative charge can be delocalized onto carbon, which abstracts a proton from water. The last two steps are repeated to form the deoxygenated product and nitrogen gas.

mechanism for the Wolff–Kishner reduction

a hydrazone

Addition of Water

Water adds to an aldehyde or to a ketone to form a hydrate. A **hydrate** is a molecule with two OH groups on the same carbon. Hydrates are also called **gem-diols** (*gem* comes from *geminus*, Latin for "twin").

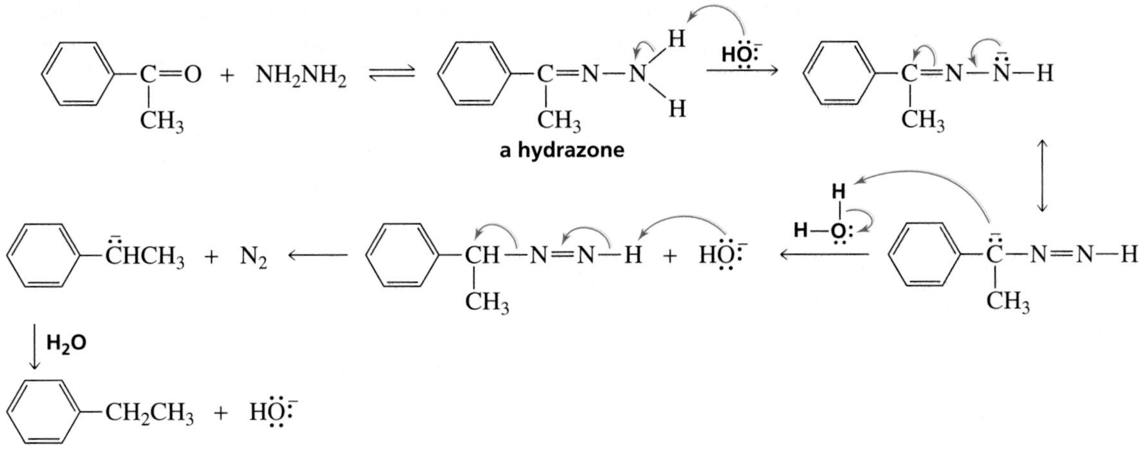

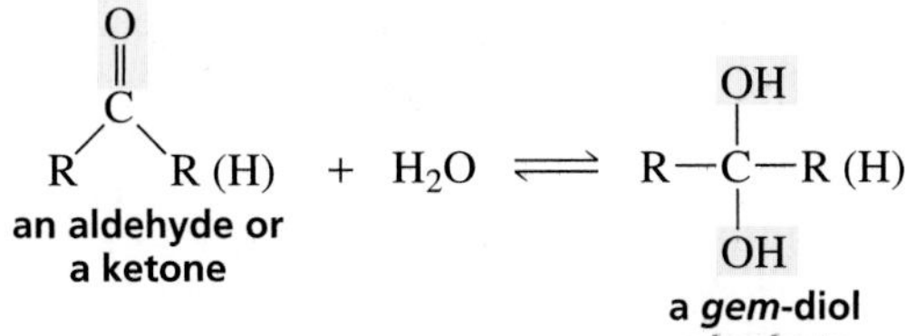

**17.8
ADDITION OF
OXYGEN
NUCLEOPHILES**

an aldehyde or
a ketone   +   H$_2$O   ⇌   a *gem*-diol
a hydrate

Water is a poor nucleophile and therefore adds relatively slowly to a carbonyl group. The rate of the reaction can be increased by an acid catalyst. Keep in mind that a catalyst has no effect on the position of the equilibrium. A catalyst affects the *rate* at which the equilibrium is achieved. In other words, the catalyst affects the rate at which an aldehyde or a ketone is converted to a hydrate—it has no effect on the *amount* of aldehyde or ketone converted to hydrate (Section 22.1).

mechanism for acid-catalyzed hydrate formation

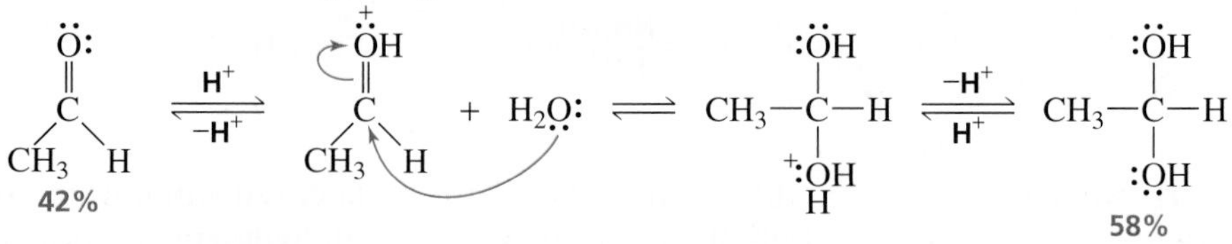

PROBLEM 22

Hydration of an aldehyde can also be catalyzed by hydroxide ion. Propose a mechanism for hydroxide-ion-catalyzed hydration.

The extent to which an aldehyde or a ketone is hydrated in an aqueous solution depends on the aldehyde or ketone. For example, only 0.2% of acetone is hydrated at equilibrium, but 99.9% of formaldehyde is hydrated. Why is there such a great difference?

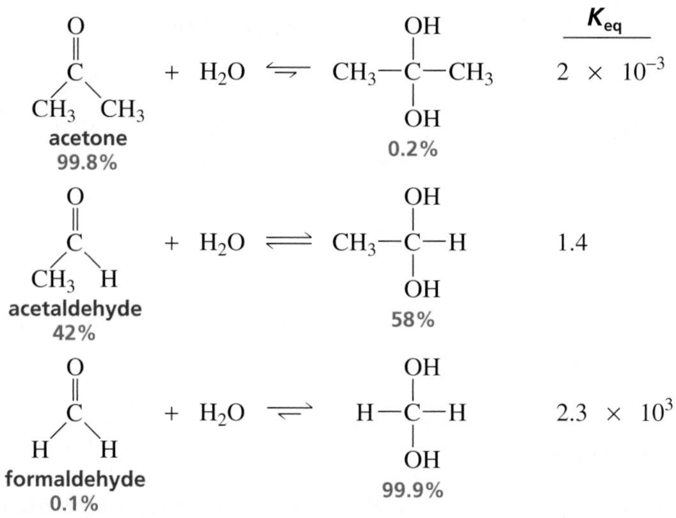

The equilibrium constant for hydrate formation depends on the *relative* stabilities of the carbonyl compound and the hydrate. We have seen that electron-donating alkyl groups make the carbonyl compound *more stable* (less reactive).

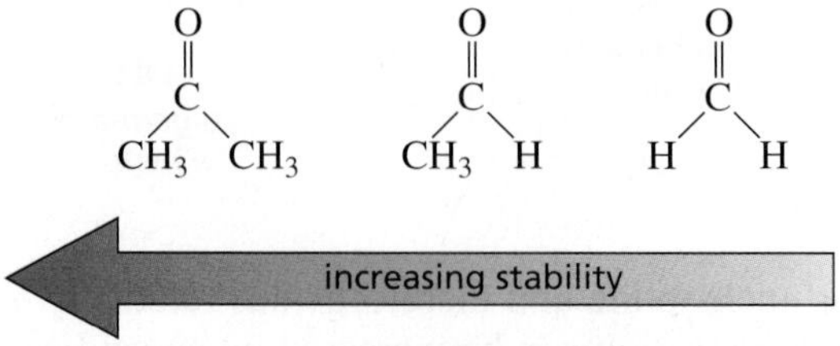

Alkyl groups make the hydrate *less stable*. Because they stabilize the carbonyl compound and destabilize the hydrate, alkyl groups shift the equilibrium to the left. As a result, less acetone than formaldehyde is hydrated at equilibrium.

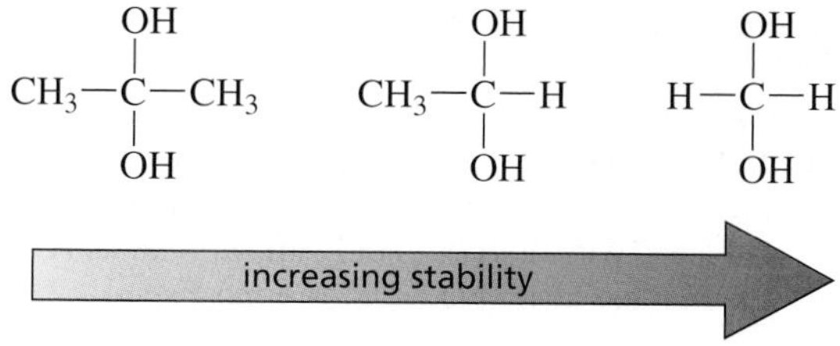

$$
\begin{array}{ccc}
\text{OH} & \text{OH} & \text{OH} \\
| & | & | \\
\text{CH}_3\text{—C—CH}_3 & \text{CH}_3\text{—C—H} & \text{H—C—H} \\
| & | & | \\
\text{OH} & \text{OH} & \text{OH}
\end{array}
$$

increasing stability

Steric interactions between the alkyl groups are responsible for the decreased stability of the hydrate. The electron clouds of the alkyl substituents do not interfere with each other in the carbonyl compound because the bond angles of the sp^2 hybridized carbon are 120°. However, the bond angles in the tetrahedral hydrate are 109.5° so the alkyl groups are closer to one other. Hydrates of aldehydes or ketones are generally too unstable to be isolated.

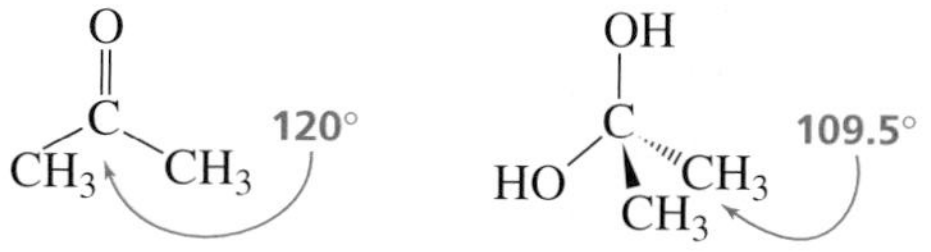

3-D Molecules:
Acetone;
Acetone hydrate

In conclusion, the percentage of hydrate present in solution at equilibrium depends on both electronic and steric effects. Electron donation and bulky substituents *decrease* the percentage of hydrate present at equilibrium, whereas electron withdrawal and small substituents *increase* it.

PRESERVING BIOLOGICAL SPECIMENS

A 37% solution of formaldehyde in water is known as formalin. It was commonly used to preserve biological specimens. Because formaldehyde is an eye and skin irritant, it has been replaced in most biology laboratories by other preservatives. One preservative frequently used is a solution of 2 to 5% phenol in ethanol with added antimicrobial agents.

If the amount of hydrate formed from the reaction of water with a ketone is too small to detect, how do we know that the reaction has even occurred? We can prove that it occurs by adding the ketone to ^{18}O-labeled water and isolating the ketone after equilibrium has been established. Finding that the label has been incorporated into the ketone proves that the reaction has occurred.

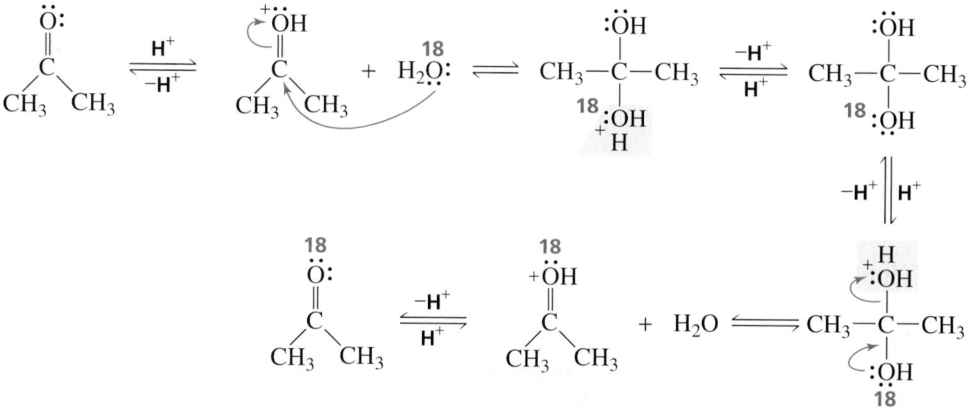

PROBLEM 23

Trichloroacetaldehyde has such a large equilibrium constant for its reaction with water that the reaction is essentially irreversible. Therefore, chloral hydrate, the product of the reaction, is one of the few hydrates that can be isolated. Chloral hydrate is a sedative that can be lethal. A cocktail laced with it is commonly known—in detective novels at least—as a "Mickey Finn." Explain the favorable equilibrium constant.

$$Cl_3C-\overset{\overset{\displaystyle O}{\|}}{C}-H \quad + \quad H_2O \quad \longrightarrow \quad Cl_3C-\overset{\overset{\displaystyle OH}{|}}{\underset{\underset{\displaystyle OH}{|}}{C}}-H$$

trichloroacetaldehyde   chloral hydrate

PROBLEM 24 ◆

Which of the following ketones has the largest equilibrium constant for the addition of water?

CH₃O⟶⟨C₆H₄⟩⟶C(=O)⟶⟨C₆H₄⟩⟶OCH₃   ⟨C₆H₅⟩⟶C(=O)⟶⟨C₆H₅⟩

O₂N⟶⟨C₆H₄⟩⟶C(=O)⟶⟨C₆H₄⟩⟶NO₂

Addition of Alcohol

The product formed when one equivalent of an alcohol adds to an aldehyde is called a **hemiacetal.** The product formed when a second equivalent of alcohol is added is called an **acetal.** An alcohol, like water, is a poor nucleophile, so the rate of the reaction is increased by an acid catalyst.

$$\underset{\underset{\displaystyle \text{an aldehyde}}{}}{\overset{\overset{\displaystyle O}{\|}}{\underset{CH_3\quad H}{C}}} \;+\; CH_3OH \;\overset{H^+}{\rightleftharpoons}\; \underset{\underset{\displaystyle \text{a hemiacetal}}{}}{CH_3-\overset{\overset{\displaystyle OH}{|}}{\underset{\underset{\displaystyle OCH_3}{|}}{C}}-H} \;\overset{CH_3OH,\ H^+}{\longrightarrow}\; \underset{\underset{\displaystyle \text{an acetal}}{}}{CH_3-\overset{\overset{\displaystyle OCH_3}{|}}{\underset{\underset{\displaystyle OCH_3}{|}}{C}}-H} \;+\; H_2O$$

When the carbonyl compound is a ketone instead of an aldehyde, the addition products are called a **hemiketal** and a **ketal,** respectively.

$$\underset{\underset{\displaystyle \text{a ketone}}{}}{\overset{\overset{\displaystyle O}{\|}}{\underset{CH_3\quad CH_3}{C}}} \;+\; CH_3OH \;\overset{H^+}{\rightleftharpoons}\; \underset{\underset{\displaystyle \text{a hemiketal}}{}}{CH_3-\overset{\overset{\displaystyle OH}{|}}{\underset{\underset{\displaystyle OCH_3}{|}}{C}}-CH_3} \;\overset{CH_3OH,\ H^+}{\longrightarrow}\; \underset{\underset{\displaystyle \text{a ketal}}{}}{CH_3-\overset{\overset{\displaystyle OCH_3}{|}}{\underset{\underset{\displaystyle OCH_3}{|}}{C}}-CH_3} \;+\; H_2O$$

Hemi is the Greek prefix for "half." When one equivalent of alcohol has added to an aldehyde or a ketone, the compound is halfway to the final acetal or ketal, which contains two groups from two equivalents of alcohol.

Acetal (or ketal) formation requires an acid catalyst. The acid protonates the carbonyl oxygen, making it easier for the alcohol to attack the carbonyl carbon. Loss of a proton from the protonated tetrahedral intermediate gives the hemiacetal (or hemiketal). Because the reaction is carried out in an acidic solution, the hemiacetal (or hemiketal) is in equilibrium with its protonated form. The two oxygen atoms of the hemiacetal (or hemiketal) are equally basic, so either one can be protonated. Loss of water from the tetrahedral intermediate with a protonated OH group forms a compound that is very reactive because of its electron deficient carbon. Nucleophilic attack on this compound by a second molecule of alcohol followed by loss of a proton forms the acetal (or ketal).

mechanism for acid-catalyzed acetal or ketal formation

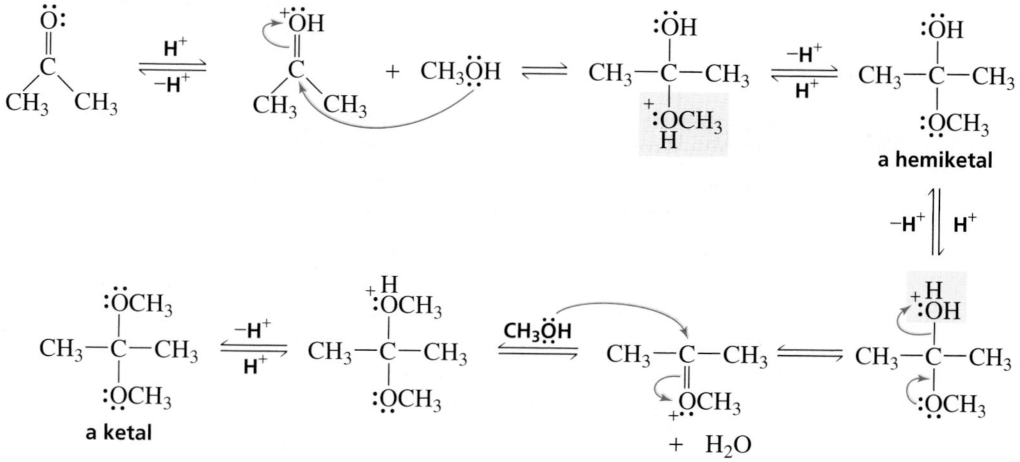

High yields of acetals and ketals are obtained if the water eliminated from the hemiacetal (or hemiketal) is removed from the reaction mixture as it is formed.

Because acetal (or ketal) formation is reversible, an acetal or ketal can be transformed back to the aldehyde or ketone in an acidic aqueous solution.

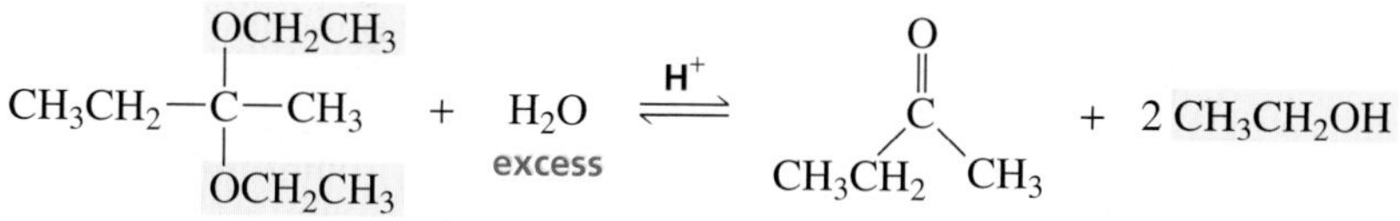

The mechanism for acetal (or ketal) formation is similar to the mechanism for imine formation. After the nucleophile (an alcohol in the case of acetal and ketal formation; an amine in the case of imine formation) has added to the carbonyl group, water is eliminated from the protonated tetrahedral intermediate. In imine formation, elimination of water is followed by loss of a proton from nitrogen to form a neutral imine. In acetal formation, the only way a neutral compound can be obtained after water is eliminated is by adding a second equivalent of alcohol.

imine formation

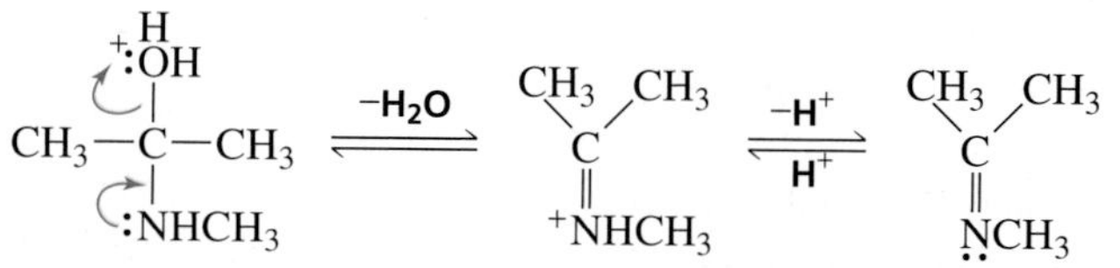

ketal formation

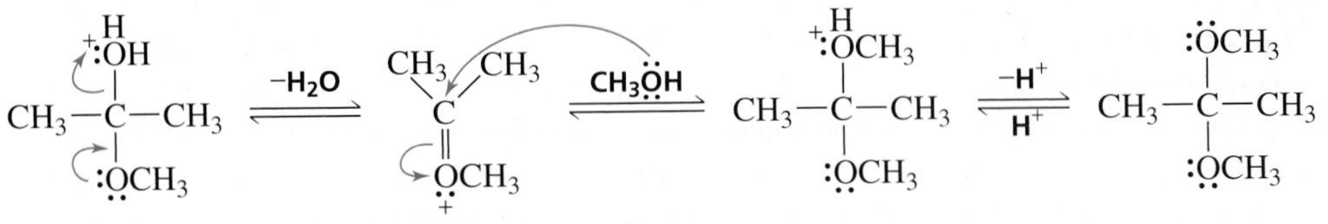

Hydrate formation also involves nucleophilic addition followed by elimination of water. Because the adding nucleophile is water, elimination of water gives back the original aldehyde or ketone. Notice that elimination of water occurs only with nucleophiles that still have a pair of nonbonding electrons on the attacking atom after forming the tetrahedral intermediate.

PROBLEM-SOLVING STRATEGY

Acetals and ketals are hydrolyzed back to the aldehyde or ketone in acidic aqueous solutions, but they are stable in basic aqueous solutions. Explain.

The easiest way to approach this kind of question is to write out the structures and the mechanism that describe what the question is asking. Once the mechanism is written, the answer should become apparent. In an acidic solution, the acid protonates an oxygen of the acetal. This creates a weak base (CH_3OH) that can be expelled by the other CH_3O group. Once the group is expelled, water can attack the reactive intermediate and you are then on your way back to the ketone (or aldehyde).

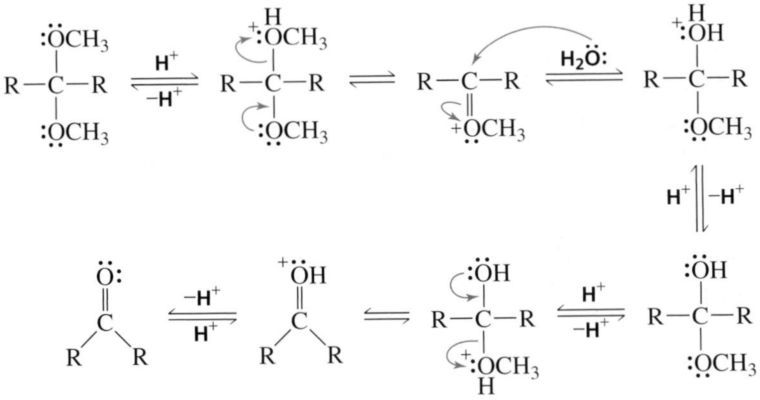

In a basic solution, however, the CH_3O group cannot be protonated. Therefore, the group that would have to be eliminated to re-form the ketone (or aldehyde) would be the very basic CH_3O^- group. A CH_3O^- group is too basic to be eliminated by the other CH_3O group, which has little driving force because of the positive charge that would be placed on its oxygen atom if elimination were to occur.

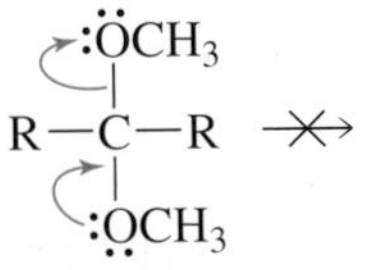

Now continue on to Problem 25.

> **PROBLEM 25**
>
> **a.** Would you expect hemiacetals to be stable in basic solutions? Explain.
>
> **b.** Acetal formation must be catalyzed by an acid. It cannot be catalyzed by CH_3O^-. Explain.
>
> **c.** Can hydrate formation be catalyzed by HO^- as well as by H^+? Explain.

17.9 PROTECTING GROUPS

Ketones (or aldehydes) react with 1,2-diols to form five-membered ring ketals (or acetals) and with 1,3-diols to form six-membered ring ketals (or acetals). Recall that five- and six-membered rings are formed relatively easily (Section 10.11). The mechanism is the same as that shown in Section 17.8 for acetal formation except that, instead of reacting with two separate molecules of alcohol, the carbonyl compound reacts with the two alcohol groups of a single molecule of the diol.

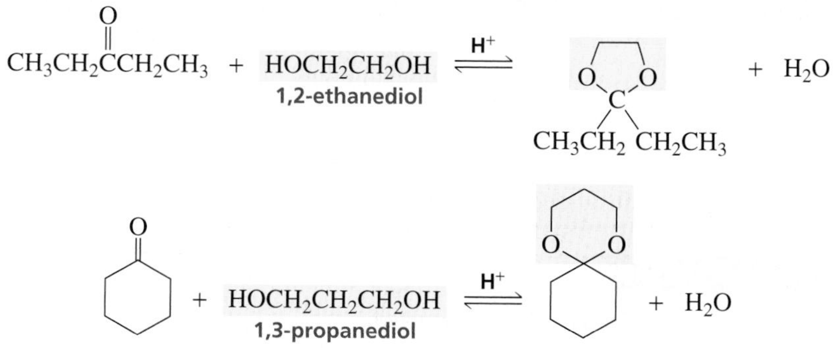

If a compound has more than one functional group that will react with a given reagent and you want only one of them to react, it is necessary to protect the second functional group from the reagent. A group that protects a functional group from a synthetic operation that it would not otherwise survive is called a **protecting group.**

If you have ever painted a room with a spray gun, you may have taped over the things you do not want to paint, such as baseboards and window frames. After you spray the room and remove the tape, the walls are painted but the areas that have been covered with tape are free of paint. In a similar way, 1,2-diols and 1,3-diols are used to protect the carbonyl group of aldehydes and ketones. The diol is like the tape. For example, in the synthesis of the following hydroxyketone from the keto ester, a problem arises because the keto group is more reactive than the ester group toward nucleophilic attack by hydride ion.

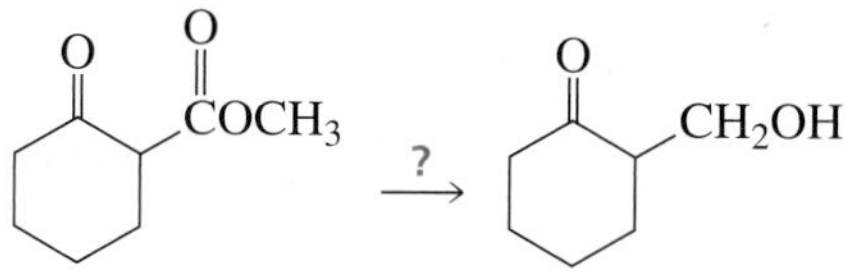

If the keto group is converted to a ketal, only the ester group will react with $LiAlH_4$. The protecting group can be removed by acid-catalyzed hydrolysis after the ester has been reduced. It is critical that the conditions used to remove a protecting group do not affect other groups in the molecule. Acetals and ketals are good protecting groups because, since they are ethers, they do not react with bases, reducing agents, or oxidizing agents.

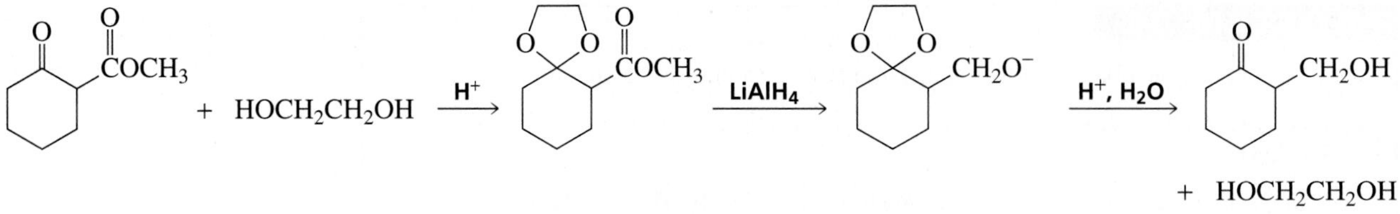

In the following reaction, the aldehyde carbonyl group reacts with the diol because aldehydes are more reactive than ketones. The Grignard reagent can now react only with the keto group. The protecting group can be removed by acid-catalyzed hydrolysis.

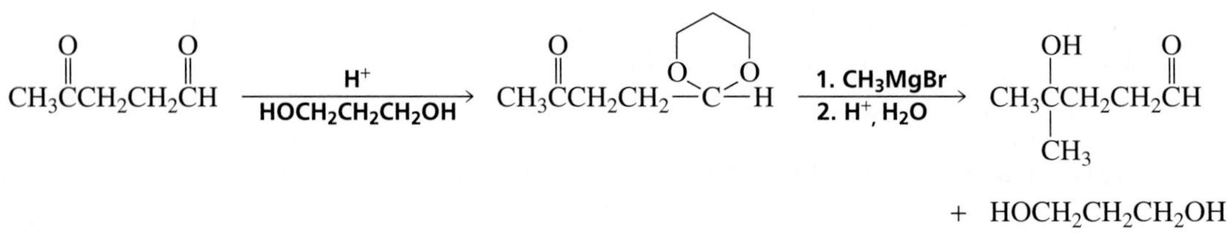

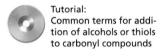

Protecting groups are discussed in greater detail in Section 29.4.

Tutorial:
Common terms for addition of alcohols or thiols to carbonyl compounds

---

PROBLEM 26

Propose a mechanism for the following reaction:

$$HOCH_2CH_2CH_2CH_2CH \xrightarrow[\text{CH}_3\text{OH}]{\text{H}^+}$$

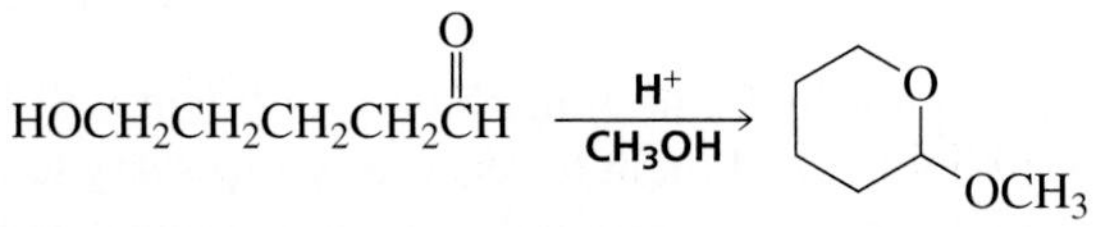

17.10 ADDITION OF SULFUR NUCLEOPHILES

Aldehydes and ketones react with thiols to form thioacetals and thioketals. The mechanism for addition of a thiol is exactly the same as the mechanism for addition of an alcohol. Recall that thiols are sulfur analogs of alcohols (Section 11.10).

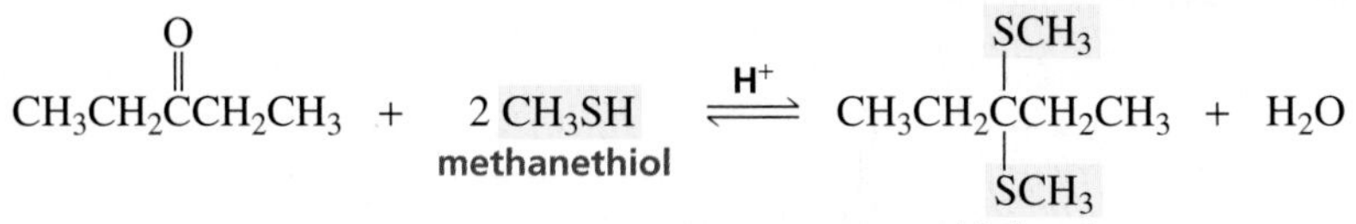

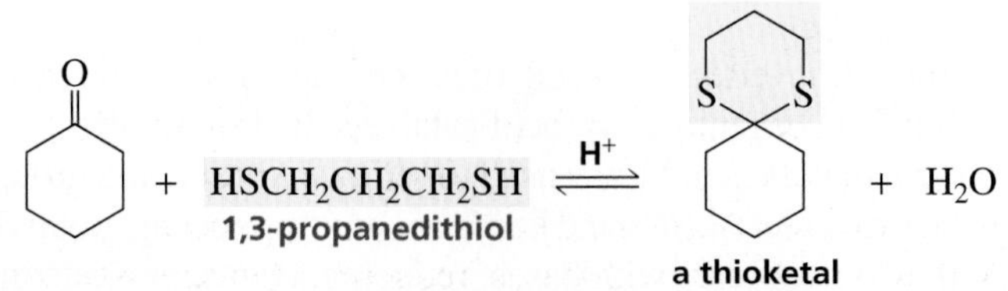

$$CH_3CH_2CCH_2CH_3 + \underset{\textbf{methanethiol}}{2\ CH_3SH} \xrightleftharpoons{\text{H}^+} \underset{\textbf{a thioketal}}{CH_3CH_2CCH_2CH_3} + H_2O$$

$$+ \underset{\textbf{1,3-propanedithiol}}{HSCH_2CH_2CH_2SH} \xrightleftharpoons{\text{H}^+} \underset{\textbf{a thioketal}}{} + H_2O$$

Thioacetal (or thioketal) formation is a synthetically useful reaction because a thioacetal (or thioketal) is desulfurized when it reacts with H_2 and Raney nickel. Desulfurization replaces the C—S bonds with C—H bonds.

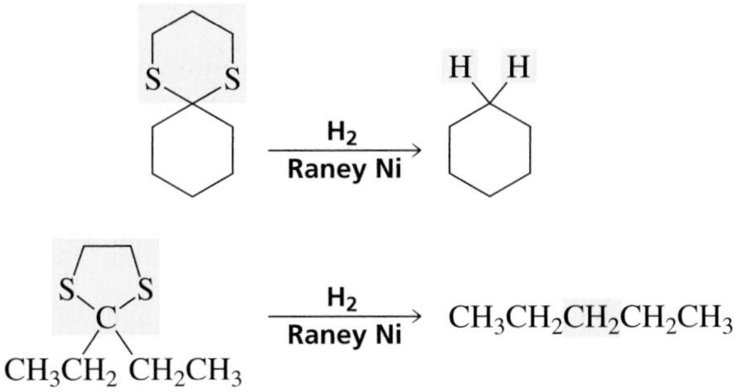

Thioketal formation followed by desulfurization provides us with a third method to convert the carbonyl group of a ketone into a methylene group. We have already seen the other two methods—the Clemmensen reduction and the Wolff–Kishner reduction (Sections 14.14 and 17.7).

An aldehyde or a ketone reacts with a phosphonium ylide (pronounced "ILL-id") to form an alkene. An **ylide** is a compound that has opposite charges on adjacent, covalently bonded atoms with complete octets of valence electrons (Section 1.3). The ylide can also be written in the double-bonded form because phosphorus can have more than eight valence electrons.

$$(C_6H_5)_3\overset{+}{P}-\overset{-}{C}H_2 \longleftrightarrow (C_6H_5)_3P=CH_2$$
a phosphonium ylide

**17.11
THE WITTIG
REACTION**

The reaction of an aldehyde or a ketone with a phosphonium ylide to form an alkene is called a **Wittig reaction.** The overall reaction amounts to interchanging the double-bonded oxygen of the carbonyl compound and the double-bonded carbon group of the phosphonium ylide.

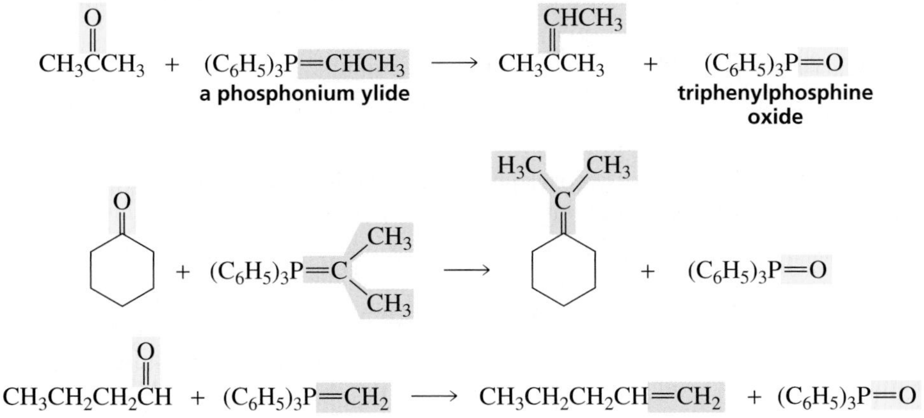

Georg Friedrich Karl Wittig (1897–1987) was born in Germany. He was a professor of chemistry at the University of Heidelberg, where he studied phosphorus-containing organic compounds. He received the Nobel Prize in chemistry in 1979, sharing it with H.C. Brown (Section 3.17).

In the first step of the Wittig reaction, the nucleophilic carbon of the ylide attacks the carbonyl carbon of the aldehyde or ketone. Nucleophilic attack results in formation of a dipolar intermediate known as a *betaine* (pronounced "BAY-tuh-ene"). Bond formation between oxygen and phosphorus forms a four-membered ring. Elimination of triphenylphosphine oxide forms the alkene product.

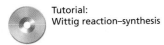
Tutorial:
Wittig reaction–synthesis

mechanism for the Wittig reaction

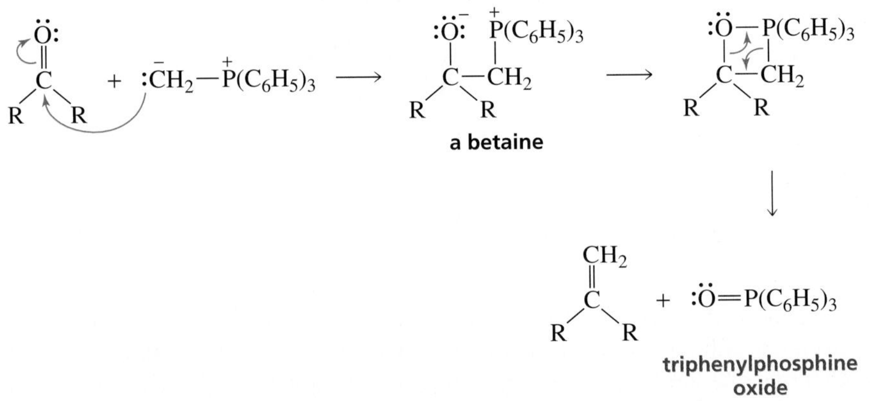

a betaine

triphenylphosphine
oxide

The phosphonium ylide needed for a particular synthesis is obtained by an S_N2 reaction between triphenylphosphine and an alkyl halide with the appropriate number of carbon atoms. A proton on the carbon adjacent to the positively charged phosphorus atom is sufficiently acidic $(pK_a = 35)$ to be removed by a strong base such as butyllithium (Section 11.8).

$(C_6H_5)_3P:$ + CH_3CH_2—Br →^{S_N2} $(C_6H_5)_3\overset{+}{P}$—CH_2CH_3 →$^{CH_3CH_2CH_2CH_2\ \overset{-}{Li}}$ $(C_6H_5)_3\overset{+}{P}$—$\overset{..}{\overset{-}{C}}HCH_3$

triphenylphosphine Br^- **a phosphonium ylide**

An advantage to using a Wittig reaction rather than an E2 elimination reaction to prepare an alkene is that a Wittig reaction is completely regioselective—that is, the product has the double bond only in one location (Section 4.16). An E2 reaction, on the other hand, often forms more than one constitutional isomer (Section 10.2).

$$CH_3CH_2CH_2\overset{\overset{\displaystyle O}{\|}}{C}H \ + \ (C_6H_5)_3P{=}CHCH_3 \ \longrightarrow \ CH_3CH_2CH_2CH{=}CHCH_3$$
2-hexene

$$CH_3CH_2CH_2\underset{\underset{\displaystyle Br}{|}}{C}HCH_2CH_3 \ \overset{HO^-}{\longrightarrow} \ CH_3CH_2CH_2CH{=}CHCH_3 \ + \ CH_3CH_2CH{=}CHCH_2CH_3$$
2-hexene   **3-hexene**

The Wittig reaction is also stereoselective. If the alkene product can exist as stereoisomers, more of one stereoisomer is obtained than of the other.

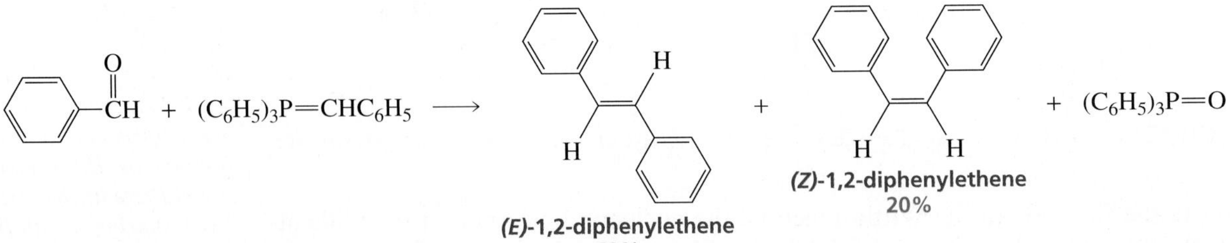

$$\text{PhCH(=O)} \ + \ (C_6H_5)_3P{=}CHC_6H_5 \ \longrightarrow$$

(E)-1,2-diphenylethene
62%

(Z)-1,2-diphenylethene
20%

$+ \ (C_6H_5)_3P{=}O$

β-CAROTENE

β-Carotene is found in yellow-orange fruits and vegetables such as apricots, mangos, carrots, and sweet potatoes. The synthesis of β-carotene from vitamin A is an important example of the use of the Wittig reaction in industry. β-Carotene is used in the food industry to color margarine. Many people take β-carotene as a dietary supplement because there is some evidence that high levels of β-carotene are associated with a low incidence of cancer. More recent evidence, however, suggests that β-carotene taken in pill form does not have the cancer-preventing effects of β-carotene obtained from vegetables.

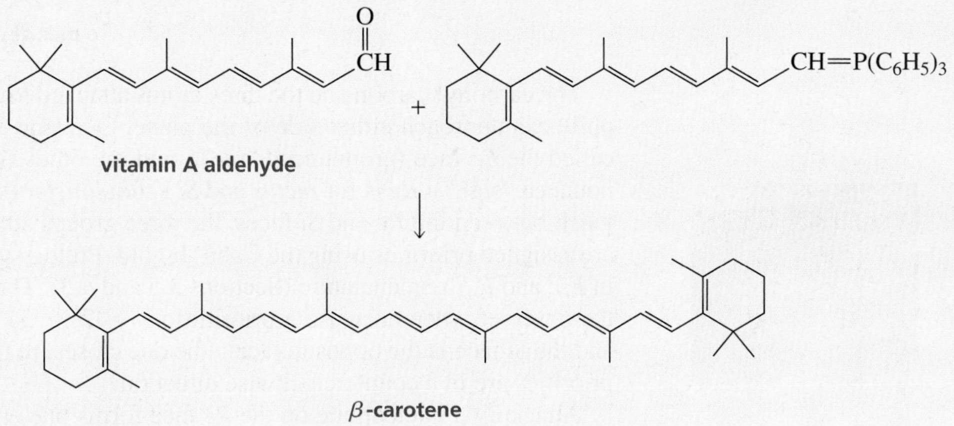

vitamin A aldehyde

+

CH=P(C_6H_5)_3

↓

β-carotene

PROBLEM 27 / SOLVED

a. What carbonyl compound and what phosphonium ylide are required for the synthesis of the following alkenes?

1. $CH_3CH_2CH_2CH{=}CCH_3$ (with CH_3 branch)

2. 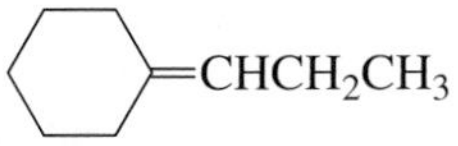 =CHCH_2CH_3

3. —CH=CH_2

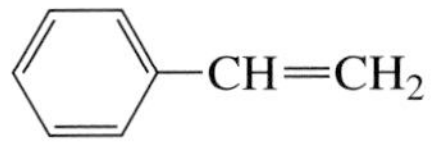

4. $(C_6H_5)_2C{=}CHCH_3$

b. What alkyl halide is required to prepare each of the phosphonium ylides?

SOLUTION TO 27a (1) The atoms on either side of the double bond can come from the carbonyl compound, so there are two possible pairs of compounds that can be used.

$$CH_3\overset{O}{\overset{\|}{C}}CH_3 + (C_6H_5)_3P{=}CHCH_2CH_2CH_3 \quad \text{or} \quad CH_3CH_2CH_2\overset{O}{\overset{\|}{C}}H + (C_6H_5)_3P{=}CCH_3 \text{ (CH}_3\text{)}$$

SOLUTION TO 27b (1) The alkyl halide required depends on which phosphonium ylide is used.

$$CH_3CH_2CH_2CH_2Br \quad \text{or} \quad CH_3CHCH_3 \text{ (Br)}$$

17.12 STEREOCHEMISTRY OF NUCLEOPHILIC ADDITION REACTIONS: *RE* AND *SI* FACES

A carbonyl carbon bonded to two different substituents is a **prochiral carbonyl carbon** because it will become a chirality center if it adds a group unlike either of the groups already bonded to it. The addition product will be a pair of enantiomers.

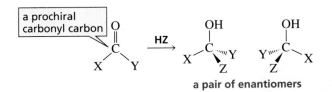

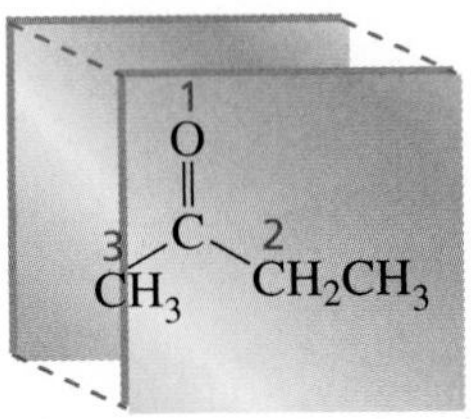

a pair of enantiomers

The carbonyl carbon and the three atoms attached to it define a plane. The nucleophile can approach either side of the plane. One side of the carbonyl compound is called the *Re* face (pronounced "ree"), and the other side is called the *Si* face (pronounced "sigh"); *Re* is for *rectus* and *Si* is for *sinister* (similar to *R* and *S*). To distinguish between the *Re* and *Si* faces, the three groups attached to the carbonyl carbon are assigned priorities using the Cahn–Ingold–Prelog system of priorities that is used in *E,Z* and *R,S* nomenclature (Sections 3.5 and 4.5). The *Re* face is the one closest to the observer when decreasing priorities (1 > 2 > 3) are in a clockwise direction, and the *Si* face is the opposite face—the one closest to the observer when decreasing priorities are in a counterclockwise direction.

Attack by a nucleophile on the *Re* face forms one enantiomer, whereas attack on the *Si* face forms the other enantiomer. For example, attack by hydride ion on the *Re* face of butanone forms *(S)*-2-butanol, and attack on the *Si* face forms *(R)*-2-butanol.

the *Si* face

the *Re* face

3-D Molecules:
2-Butanone;
(S)-2-Butanol;
(R)-2-Butanol

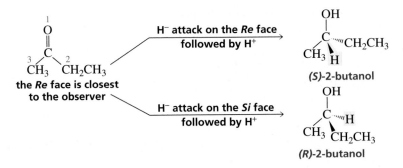

Whether attack on the *Re* face forms the *R* or *S* enantiomer depends on the priority of the attacking nucleophile compared with the priorities of the groups attached to the carbonyl carbon. For example, attack by hydride ion on the *Re* face of butanone forms *(S)*-2-butanol, but attack by a methyl Grignard reagent on the *Re* face of propanal forms *(R)*-2-butanol.

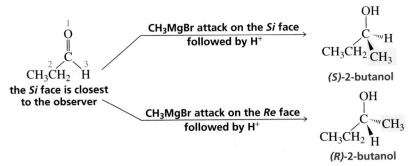

Because the carbonyl carbon and the three atoms attached to it define a plane, the *Re* and *Si* faces have an equal probability of being attacked. Consequently, an addition reaction forms equal amounts of the two enantiomers.

ENZYME-CATALYZED CARBONYL ADDITIONS

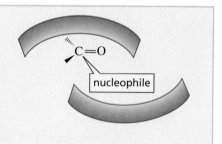

In an enzyme-catalyzed addition to a carbonyl compound, only one of the enantiomers is formed. The enzyme can block one face of the carbonyl compound so that it cannot be attacked, or it can position the nucleophile so that it is able to attack the carbonyl group from only one side of the molecule.

PROBLEM 28◆

Which enantiomer is formed when a methyl Grignard reagent attacks the *Re* face of each of the following carbonyl compounds?

a. propiophenone   **b.** benzaldehyde   **c.** 2-pentanone   **d.** 3-hexanone

---

Most of the reactions we have studied are *intermolecular reactions*—the two reacting groups are in *different* molecules. Cyclic compounds are formed from *intramolecular reactions*—the two reacting groups are in the *same* molecule. We have seen that intramolecular reactions are particularly favorable if the reaction forms a compound with a five- or a six-membered ring (Section 10.11).

In designing the synthesis of a cyclic compound, we must examine the target molecule to determine what kinds of reactive groups will be necessary for a successful synthesis. For example, we know that an ester is formed from the acid-catalyzed reaction of a carboxylic acid with an alcohol. Therefore, a cyclic ester (lactone) can be prepared from a reactant that has both a carboxylic acid group and an alcohol group in the same molecule. The size of the lactone ring will be determined by the number of carbon atoms between the carboxylic acid and alcohol groups.

17.13
DESIGNING A SYNTHESIS IV: THE SYNTHESIS OF CYCLIC COMPOUNDS

$$\text{HOCH}_2\text{CH}_2\text{CH}_2\text{CH}_2\overset{\displaystyle O}{\overset{\|}{\text{C}}}\text{OH} \xrightarrow{\text{H}^+}$$
4 intervening carbon atoms

$$\text{HOCH}_2\text{CH}_2\text{CH}_2\overset{\displaystyle O}{\overset{\|}{\text{C}}}\text{OH} \xrightarrow{\text{H}^+}$$
3 intervening carbon atoms

A compound with a ketone group attached to a benzene ring can be prepared using a Friedel–Crafts acylation reaction. Therefore, a cyclic ketone will result if a Lewis acid ($AlCl_3$) is added to a compound that contains both a benzene ring and an acyl chloride group separated by the appropriate number of carbon atoms.

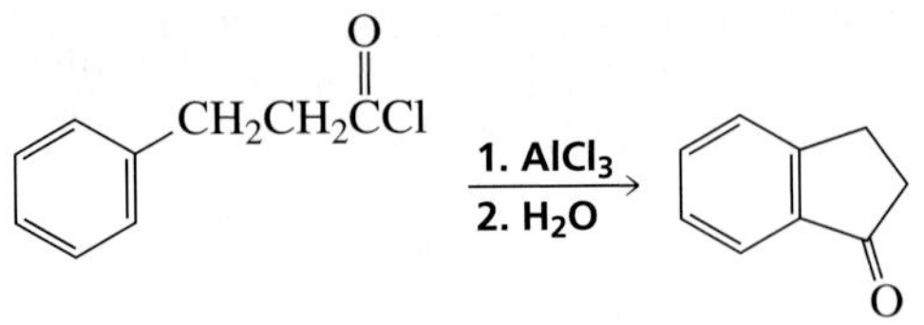

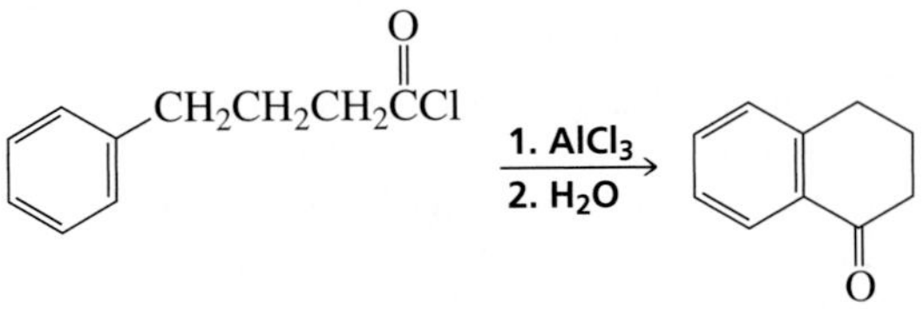

A cyclic secondary amine can be prepared by reductive amination of a compound that contains ketone and primary amine groups separated by the appropriate number of carbon atoms.

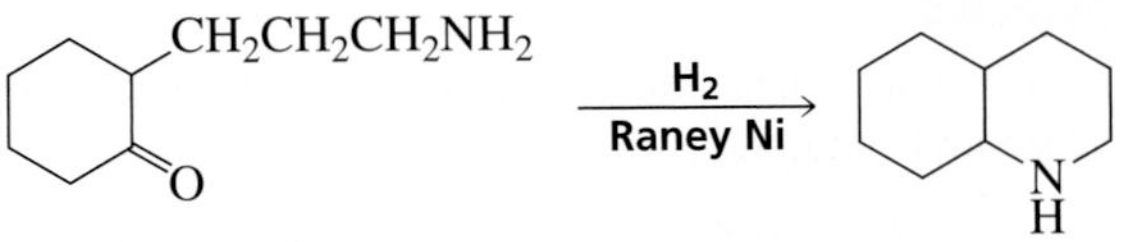

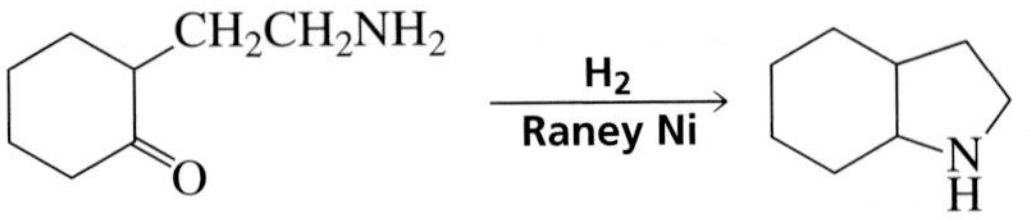

A cyclic ether can be prepared by an intramolecular Williamson ether synthesis (Section 10.9).

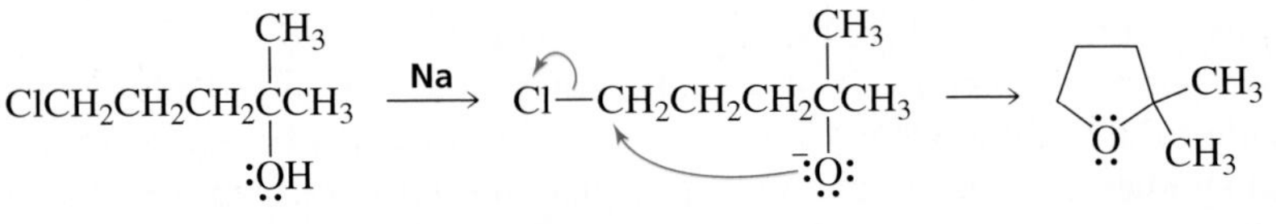

However, if the compound with both the alkyl halide and alcohol groups is not available, the ether could be prepared from the reaction of a Grignard reagent with a compound that has an alkyl halide group and a ketone group.

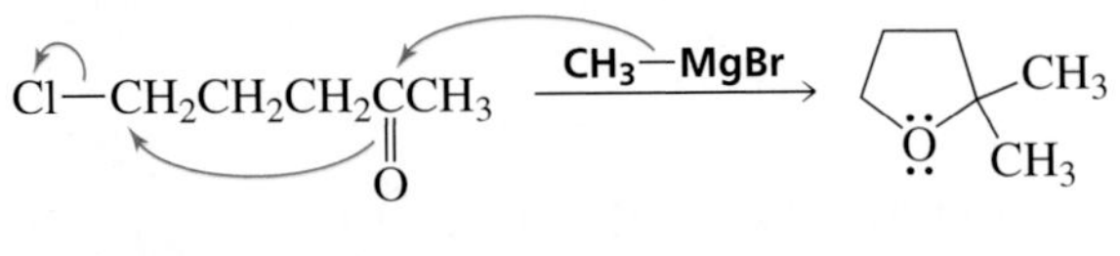

PROBLEM 29

Design a synthesis for each of the following compounds using an intramolecular reaction:

a.

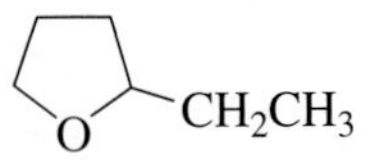

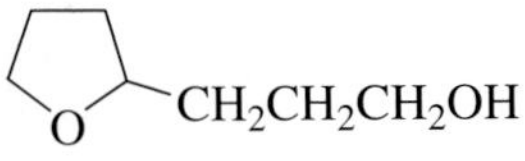

b.

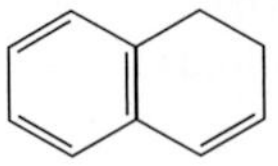

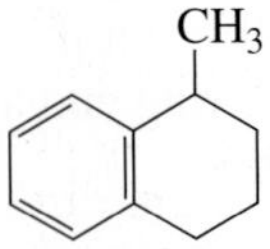

c.

d.

e.

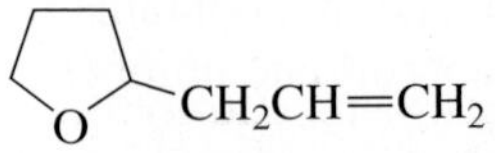

f.

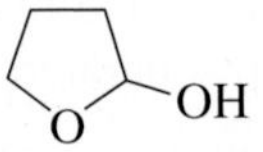

The contributing resonance structures for an α,β-unsaturated carbonyl compound show that the molecule has two electrophilic sites—the carbonyl carbon and the β-carbon.

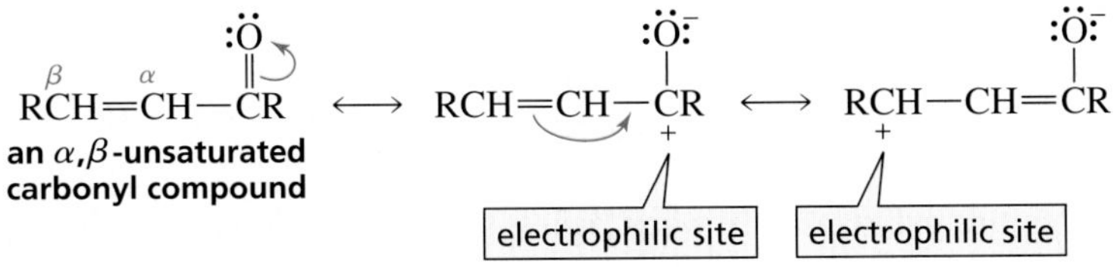

**17.14
NUCLEOPHILIC
ADDITION TO
α,β-UNSATURATED
ALDEHYDES AND
KETONES:
DIRECT ADDITION
VERSUS CONJUGATE
ADDITION**

This means that if an aldehyde or a ketone has a double bond in the α,β-position, a nucleophile can add either to the carbonyl carbon or to the β-carbon.

Nucleophilic addition to the carbonyl carbon is called **direct addition** or 1,2-addition.

direct addition

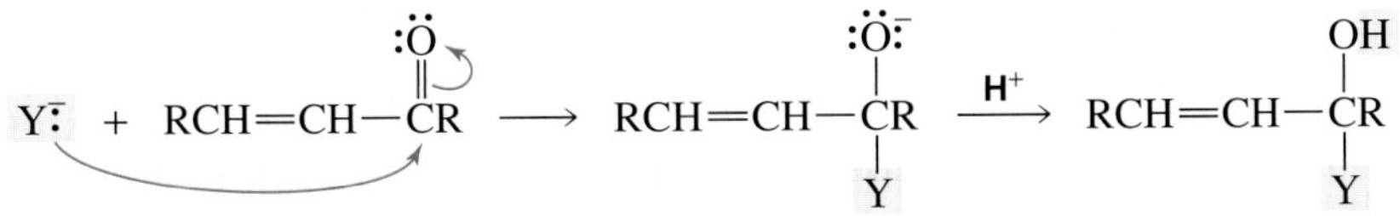

Nucleophilic addition to the β-carbon is called **conjugate addition,** because addition occurs across the conjugated system. Conjugate addition is also called 1,4-addition. The enol tautomerizes to a ketone (or to an aldehyde, Section 5.6), so the overall reaction amounts to addition to the carbon–carbon double bond, with the nucleophile adding to the β-carbon and a proton from the reaction mixture adding to the α-carbon. Compare these reactions to the 1,2- and 1,4-addition reactions you studied in Sections 7.7 and 7.8.

conjugate addition

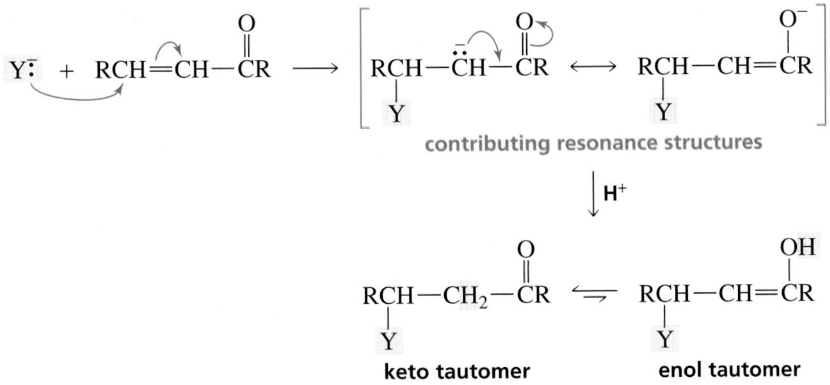

Whether the product obtained from nucleophilic addition to an α,β-unsaturated carbonyl compound is the direct addition product or the conjugate addition product depends on the nature of the nucleophile. Nucleophiles that are strong bases, such as organolithium reagents and hydride ion, tend to form direct addition products.

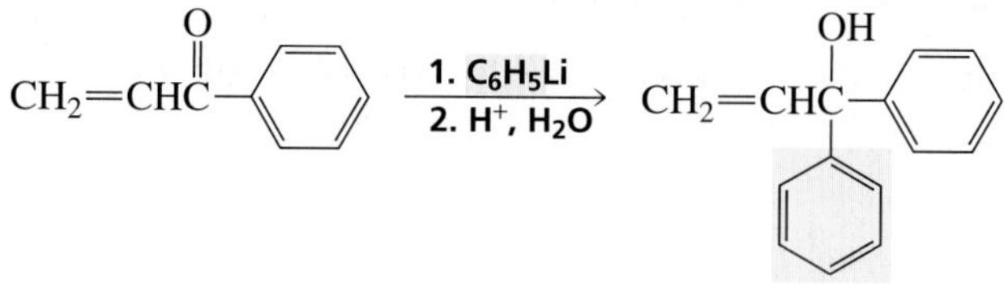

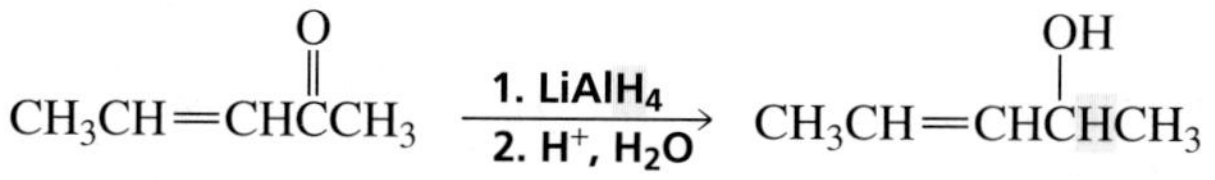

$$CH_2=CHC{-}\bigcirc \xrightarrow[\text{2. H}^+, \text{H}_2\text{O}]{\text{1. C}_6\text{H}_5\text{Li}} CH_2=CHC{-}\bigcirc$$

$$CH_3CH=CHCCH_3 \xrightarrow[\text{2. H}^+, \text{H}_2\text{O}]{\text{1. LiAlH}_4} CH_3CH=CHCHCH_3$$

Nucleophiles that are relatively weak bases, such as cyanide ion, amines, thiols, and halide ions, usually form conjugate addition products.

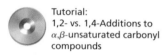

Tutorial:
1,2- vs. 1,4-Additions to
α,β-unsaturated carbonyl
compounds

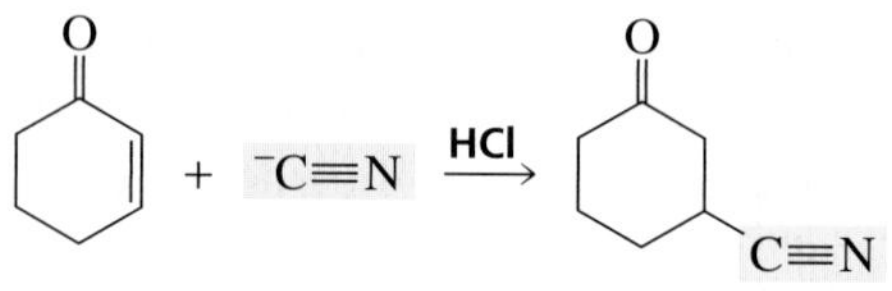

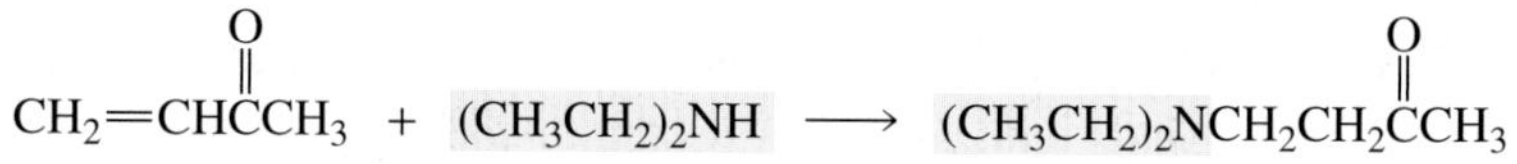

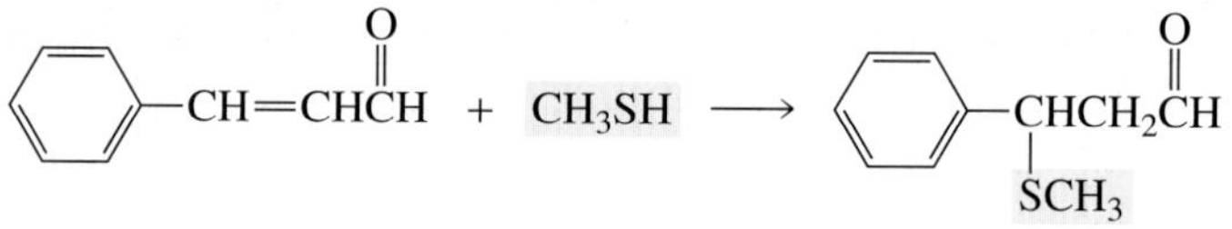

$$\bigcirc + \ {}^-C\equiv N \xrightarrow{\text{HCl}} \bigcirc{-}C\equiv N$$

$$CH_2=CHCCH_3 + (CH_3CH_2)_2NH \longrightarrow (CH_3CH_2)_2NCH_2CH_2CCH_3$$

$$\bigcirc{-}CH=CHCH + CH_3SH \longrightarrow \bigcirc{-}CHCH_2CH$$
$$\qquad\qquad\qquad\qquad\qquad\qquad\qquad\qquad SCH_3$$

   Why do strong bases form direct addition products and weak bases form conjugate addition products? Direct addition occurs more rapidly than conjugate addition. Therefore, the final product of the reaction depends on whether direct addition is irreversible or reversible. If direct addition is *irreversible,* the product of direct addition will be the final product of the reaction. If direct addition is *reversible,* the initially formed direct addition product can collapse back to the starting material, allowing conjugate addition to occur. Since conjugate addition is irreversible, the conjugate addition product will accumulate and eventually become the major product of the reaction.

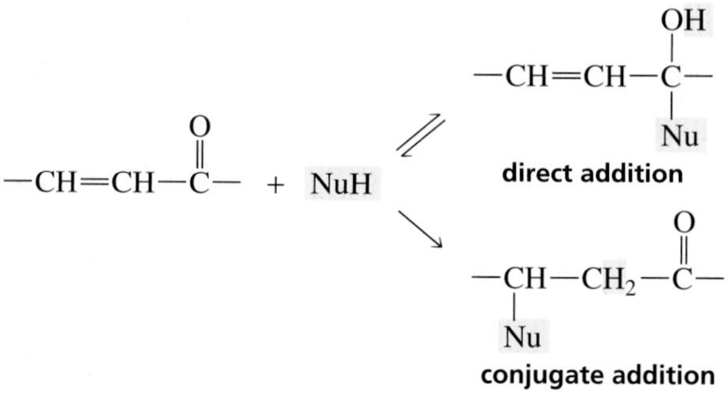

$$-CH=CH-C- + NuH$$

$$-CH=CH-C-$$
$$\qquad\qquad Nu$$
direct addition

$$-CH-CH_2-C-$$
$$Nu$$
conjugate addition

   Direct addition will be irreversible if the nucleophile is a strong base because strong bases are poor leaving groups and cannot be eliminated from the direct addition product. Direct addition will be reversible if the nucleophile is a weak base and, therefore, can be eliminated from the direct addition product. When direct addition is reversible, the conjugate addition product will be the major product of the reaction even though it is formed more slowly.

Because Grignard reagents are strong bases, they react with α,β-unsaturated aldehydes and most α,β-unsaturated ketones to form direct addition products.

$$CH_3CH=CHCH \xrightarrow[\text{2. H}^+\text{, H}_2\text{O}]{\text{1. CH}_3\text{MgBr}} CH_3CH=CHCHCH_3$$

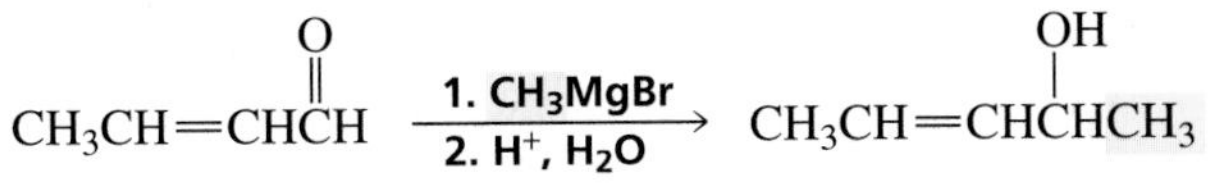

If, however, the carbonyl carbon is bonded to a bulky group, the conjugate addition product is formed preferentially. Apparently, the bulky group sufficiently slows the rate of direct addition to allow competition from conjugate addition.

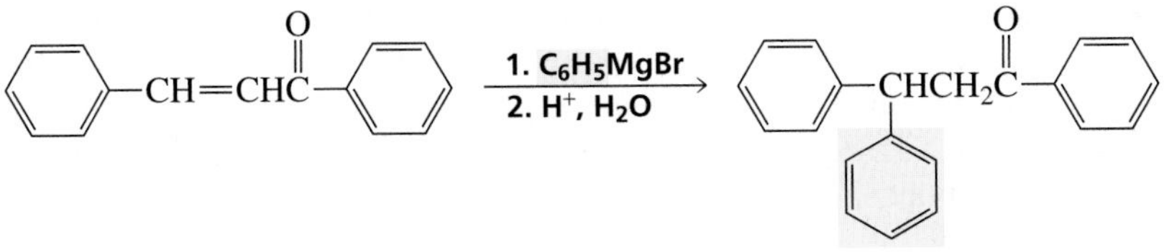

Only conjugate addition occurs when Gilman reagents (lithium dialkylcuprates, Section 11.8) react with unsaturated carbonyl compounds. Therefore, Grignard reagents should be used when you want to add an alkyl group to the carbonyl carbon, while Gilman reagents should be used when you want to add an alkyl group to the β-carbon.

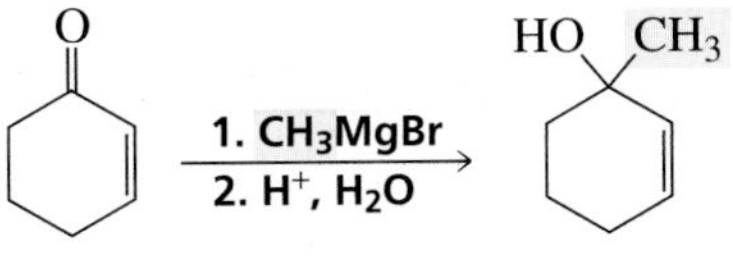

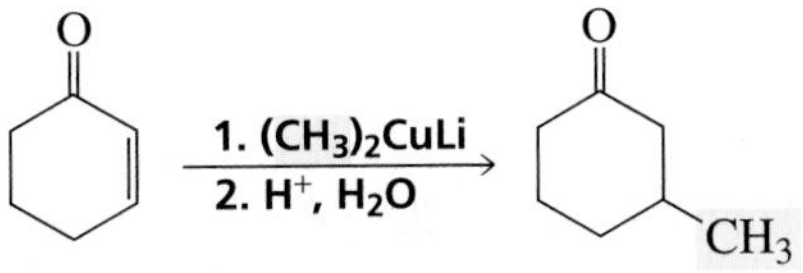

PROBLEM 30◆

Give the major product of each of the following reactions:

a.

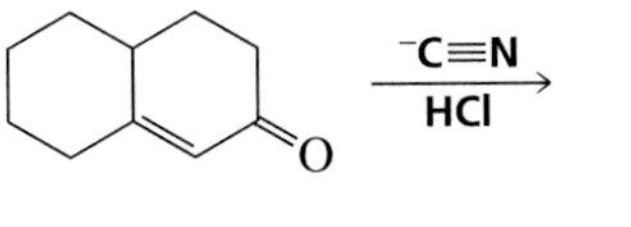

$$\xrightarrow[\text{HCl}]{^-\text{C}\equiv\text{N}}$$

b.

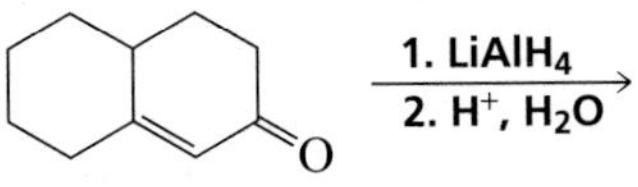

$$\xrightarrow[\text{2. H}^+\text{, H}_2\text{O}]{\text{1. LiAlH}_4}$$

c. $CH_3\overset{\underset{\displaystyle CH_3}{|}}{C}=CH\overset{\underset{\displaystyle O}{\|}}{C}CH_3 \xrightarrow[\text{2. H}^+\text{, H}_2\text{O}]{\text{1. CH}_3\text{MgBr}}$

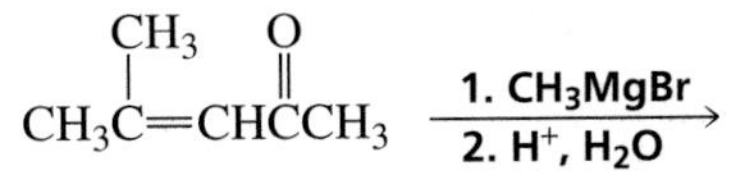

d. 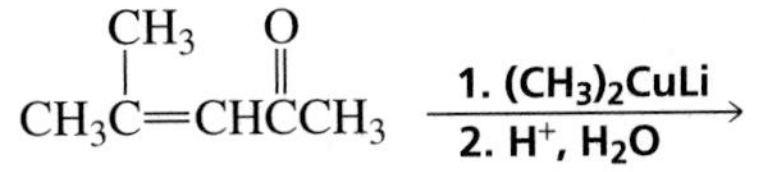 $CH_3\overset{\underset{\displaystyle CH_3}{|}}{C}=CH\overset{\underset{\displaystyle O}{\|}}{C}CH_3 \xrightarrow[\text{2. H}^+\text{, H}_2\text{O}]{\text{1. (CH}_3)_2\text{CuLi}}$

α,β-Unsaturated carboxylic acid derivatives, like α,β-unsaturated aldehydes and ketones, have two electrophilic sites for nucleophilic attack. As in the case of α,β-unsaturated aldehydes and ketones, nucleophiles that are weak bases form conjugate addition products. Nucleophiles that are strong bases form nucleophilic acyl substitution products. Notice that nucleophiles that are strong bases form *nucleophilic acyl substitution products* rather than *direct addition products* because there is a group in the reactant that can be replaced by a nucleophile.

**17.15
NUCLEOPHILIC
ADDITION TO
α,β-UNSATURATED
CARBOXYLIC ACID
DERIVATIVES**

$$CH_2\!\!=\!\!CHCOCH_3 \ + \ HBr \ \longrightarrow \ BrCH_2CH_2COCH_3$$

conjugate addition product

$$CH_3CH\!\!=\!\!CHCOCH_3 \ + \ HO^- \ \xrightarrow{\ H_2O\ } \ CH_3CH\!\!=\!\!CHCO^- \ + \ CH_3OH$$

nucleophilic acyl substitution product

17.16 ENZYME-CATALYZED ADDITIONS TO α,β-UNSATURATED CARBONYL COMPOUNDS

Several reactions in biological systems involve addition to α,β-unsaturated carbonyl compounds. Because the nucleophiles present in living organisms—such as H_2O and NH_3—tend to be weak bases, conjugate addition products are obtained from these enzyme-catalyzed addition reactions. The following are examples of conjugate addition reactions that occur in biological systems.

$$CH_2\!\!=\!\!CCO^- \ + \ H_2O \ \underset{}{\overset{enolase}{\rightleftharpoons}} \ CH_2CHCO^-$$

$$\underset{OPO_3{}^{2-}}{} \qquad \underset{OH \ \ OPO_3{}^{2-}}{}$$

$$\underset{}{\overset{}{\rightleftharpoons}} \ + \ H_2O \ \xrightarrow{fumarase} \ ^-OCCHCH_2CO^-$$

(H, CO^-, ^-OC, O — fumarate) + H_2O ⇌ $^-OCCHCH_2CO^-$ with OH

$$\underset{}{\overset{\beta\text{-methylaspartase}}{\rightleftharpoons}} \ ^-OCCH\!-\!CHCO^-$$

(H, CO^-, ^-OC, CH_3) + NH_3 ⇌ $^-OCCH\!-\!CHCO^-$ with $^+NH_3$ and CH_3

$$CH_3(CH_2)_nCH\!\!=\!\!CHCSCoA \ + \ H_2O \ \underset{}{\overset{crotonase}{\rightleftharpoons}} \ CH_3(CH_2)_nCHCH_2CSCoA$$

with OH

ENZYME-CATALYZED CIS–TRANS INTERCONVERSION

Enzymes that catalyze the interconversion of cis and trans isomers are called cis–trans isomerases. These isomerases are all known to contain thiol (SH) groups. Thiols are weak bases and therefore add to the β-carbon of an α,β-unsaturated carbonyl compound (conjugate addition). The resulting carbon–carbon single bond rotates before the enol is able to tautomerize to the ketone. When tautomerization occurs, the thiol is eliminated. Rotation results in cis–trans interconversion.

cis double bond

enzyme–SH

trans double bond

enzyme–SH

rotation

enzyme–S

enzyme–S   :ÖH

SUMMARY OF REACTIONS

1. Reaction of *carbonyl compounds* with Grignard reagents (Sections 17.4 and 17.6)

   a. Reaction of an *aldehyde* with a Grignard reagent forms a secondary alcohol:

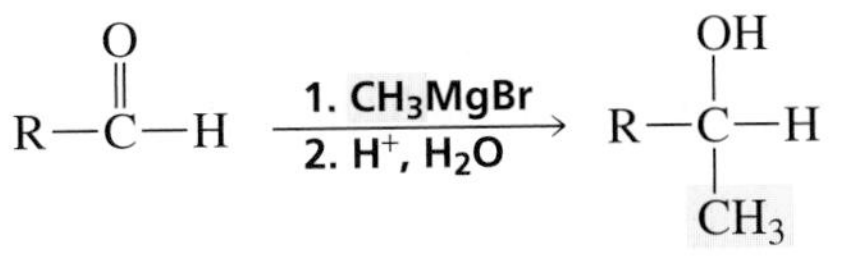

$$R-\overset{\overset{\displaystyle O}{\|}}{C}-H \quad\xrightarrow[\text{2. H}^+,\ \text{H}_2\text{O}]{\text{1. CH}_3\text{MgBr}}\quad R-\underset{\underset{\displaystyle CH_3}{|}}{\overset{\overset{\displaystyle OH}{|}}{C}}-H$$

   b. Reaction of a *ketone* with a Grignard reagent forms a tertiary alcohol:

$$R-\overset{\overset{\displaystyle O}{\|}}{C}-R' \quad\xrightarrow[\text{2. H}^+,\ \text{H}_2\text{O}]{\text{1. CH}_3\text{MgBr}}\quad R-\underset{\underset{\displaystyle CH_3}{|}}{\overset{\overset{\displaystyle OH}{|}}{C}}-R'$$

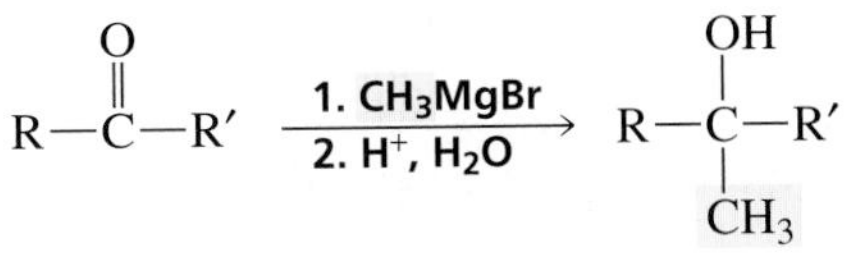

   c. Reaction of an *ester* with a Grignard reagent forms a tertiary alcohol with two identical substituents:

$$R-\overset{\overset{\displaystyle O}{\|}}{C}-OR' \quad\xrightarrow[\text{2. H}^+,\ \text{H}_2\text{O}]{\text{1. 2 CH}_3\text{MgBr}}\quad R-\underset{\underset{\displaystyle CH_3}{|}}{\overset{\overset{\displaystyle OH}{|}}{C}}-CH_3$$

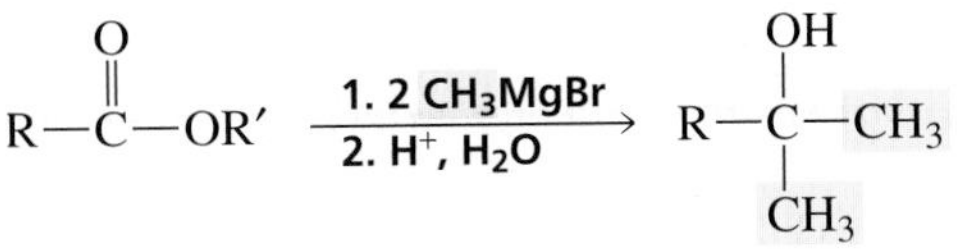

   d. Reaction of an *acyl chloride* with a Grignard reagent forms a tertiary alcohol with two identical substituents:

$$R-\overset{\overset{\displaystyle O}{\|}}{C}-Cl \quad\xrightarrow[\text{2. H}^+,\ \text{H}_2\text{O}]{\text{1. 2 CH}_3\text{MgBr}}\quad R-\underset{\underset{\displaystyle CH_3}{|}}{\overset{\overset{\displaystyle OH}{|}}{C}}-CH_3$$

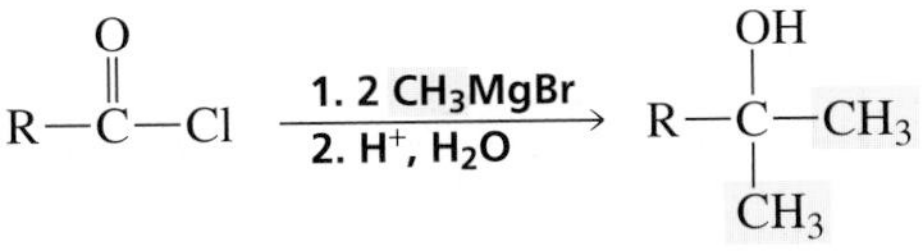

   e. Reaction of CO_2 with a Grignard reagent forms a carboxylic acid:

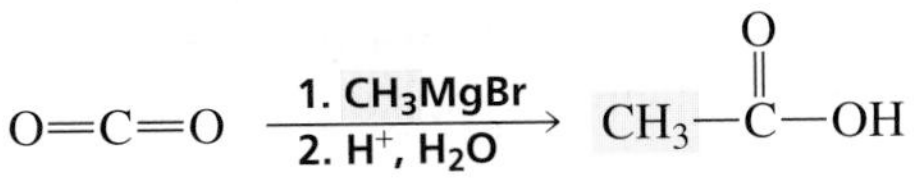

$$O=C=O \quad\xrightarrow[\text{2. H}^+,\ \text{H}_2\text{O}]{\text{1. CH}_3\text{MgBr}}\quad CH_3-\overset{\overset{\displaystyle O}{\|}}{C}-OH$$

2. Reactions of *aldehydes* and *ketones* with other carbon nucleophiles (Section 17.4)

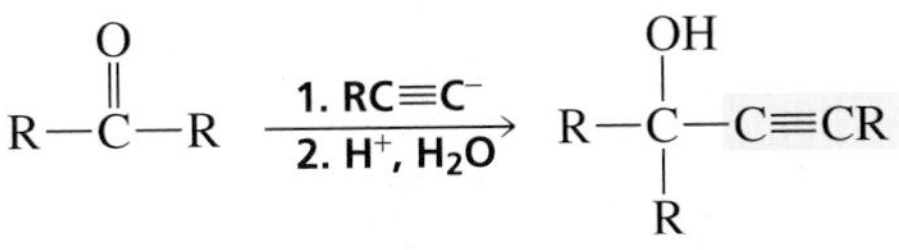

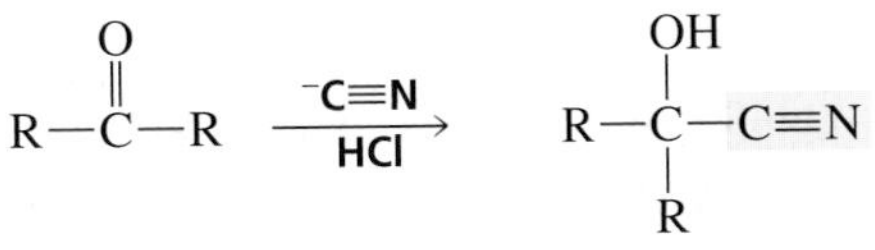

$$R-\overset{\overset{\displaystyle O}{\|}}{C}-R \quad\xrightarrow[\text{2. H}^+,\ \text{H}_2\text{O}]{\text{1. RC}\equiv\text{C}^-}\quad R-\underset{\underset{\displaystyle R}{|}}{\overset{\overset{\displaystyle OH}{|}}{C}}-C\equiv CR$$

$$R-\overset{\overset{\displaystyle O}{\|}}{C}-R \quad\xrightarrow[\text{HCl}]{^-\text{C}\equiv\text{N}}\quad R-\underset{\underset{\displaystyle R}{|}}{\overset{\overset{\displaystyle OH}{|}}{C}}-C\equiv N$$

3. Reactions of *carbonyl compounds* with hydride ion donors (Sections 17.5 and 17.6)

a. Reaction of an *aldehyde* with sodium borohydride forms a primary alcohol:

$$R-\overset{\overset{\displaystyle O}{\|}}{C}-H \quad \xrightarrow[\text{2. H}^+\text{, H}_2\text{O}]{\text{1. NaBH}_4} \quad RCH_2OH$$

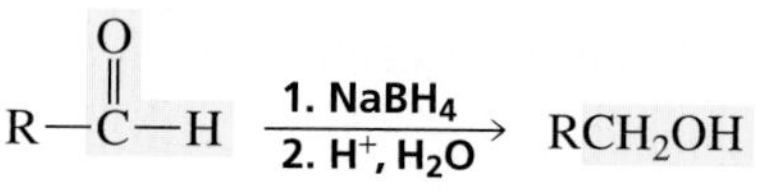

b. Reaction of a *ketone* with sodium borohydride forms a secondary alcohol:

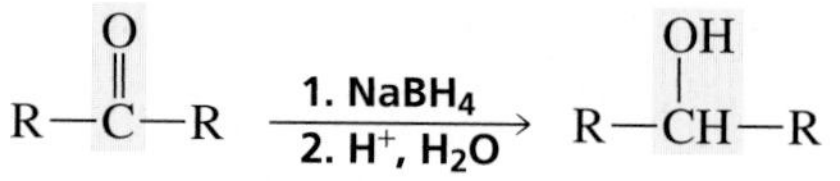

$$R-\overset{\overset{\displaystyle O}{\|}}{C}-R \quad \xrightarrow[\text{2. H}^+\text{, H}_2\text{O}]{\text{1. NaBH}_4} \quad R-\overset{\overset{\displaystyle OH}{|}}{CH}-R$$

c. Reaction of an *ester* with lithium aluminum hydride forms primary alcohols:

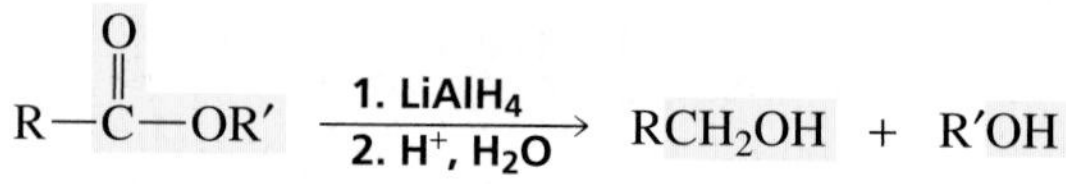

$$R-\overset{\overset{\displaystyle O}{\|}}{C}-OR' \quad \xrightarrow[\text{2. H}^+\text{, H}_2\text{O}]{\text{1. LiAlH}_4} \quad RCH_2OH \;+\; R'OH$$

d. Reaction of an *ester* with diisobutylaluminum hydride forms an aldehyde:

$$R-\overset{\overset{\displaystyle O}{\|}}{C}-OR' \quad \xrightarrow[\text{2. H}_2\text{O}]{\text{1. [(CH}_3)_2\text{CHCH}_2]_2\text{AlH, }-78\ °C} \quad R-\overset{\overset{\displaystyle O}{\|}}{C}-H$$

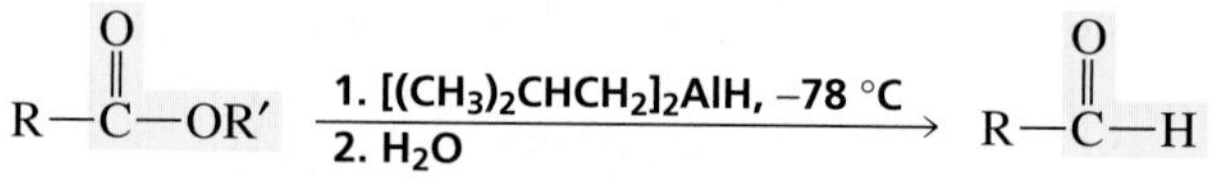

e. Reaction of a *carboxylic acid* with lithium aluminum hydride forms a primary alcohol:

$$R-\overset{\overset{\displaystyle O}{\|}}{C}-OH \quad \xrightarrow[\text{2. H}^+\text{, H}_2\text{O}]{\text{1. LiAlH}_4} \quad R-CH_2-OH$$

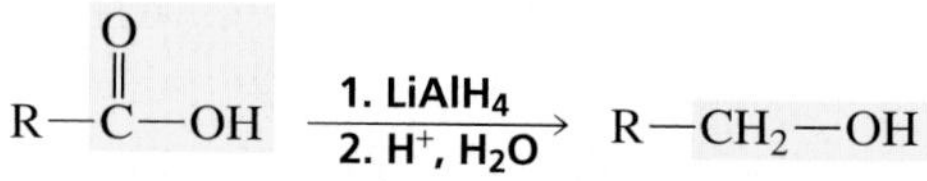

f. Reaction of an *acyl chloride* with lithium aluminum hydride forms a primary alcohol:

$$R-\overset{\overset{\displaystyle O}{\|}}{C}-Cl \quad \xrightarrow[\text{2. H}^+\text{, H}_2\text{O}]{\text{1. LiAlH}_4} \quad R-CH_2-OH$$

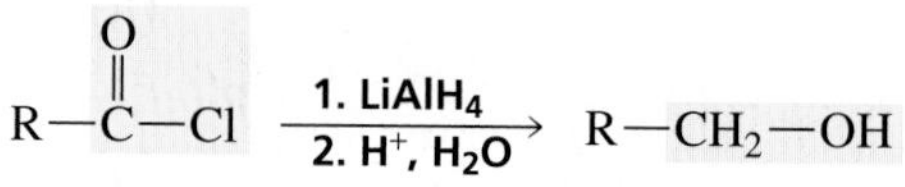

g. Reaction of an *acyl chloride* with lithium tri-*tert*-butoxyaluminum hydride forms an aldehyde:

$$R-\overset{\overset{\displaystyle O}{\|}}{C}-Cl \quad \xrightarrow[\text{2. H}_2\text{O}]{\text{1. LiAlH[OC(CH}_3)_3]_3\text{, }-78\ °C} \quad R-\overset{\overset{\displaystyle O}{\|}}{C}-H$$

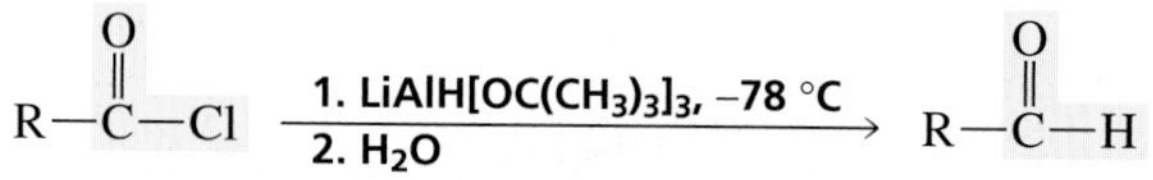

h. Reaction of an *amide* with lithium aluminum hydride forms an amine:

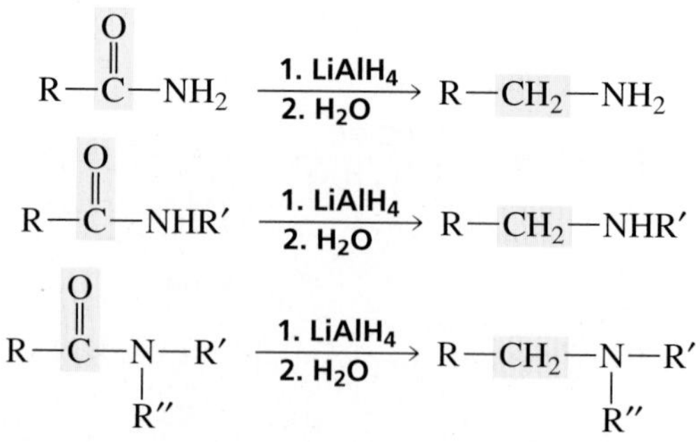

$$R-\overset{\overset{\displaystyle O}{\|}}{C}-NH_2 \quad \xrightarrow[\text{2. H}_2\text{O}]{\text{1. LiAlH}_4} \quad R-CH_2-NH_2$$

$$R-\overset{\overset{\displaystyle O}{\|}}{C}-NHR' \quad \xrightarrow[\text{2. H}_2\text{O}]{\text{1. LiAlH}_4} \quad R-CH_2-NHR'$$

$$R-\overset{\overset{\displaystyle O}{\|}}{C}-\underset{\underset{\displaystyle R''}{|}}{N}-R' \quad \xrightarrow[\text{2. H}_2\text{O}]{\text{1. LiAlH}_4} \quad R-CH_2-\underset{\underset{\displaystyle R''}{|}}{N}-R'$$

4. Reactions of *aldehydes* and *ketones* with amines (Section 17.7)

a. Reaction with a *primary amine* forms an imine:
When Z = R, the product is a Schiff base; Y can also be OH, NH_2, NHC_6H_5, $NHC_6H_3(NO_2)_2$, and $NHCONH_2$.

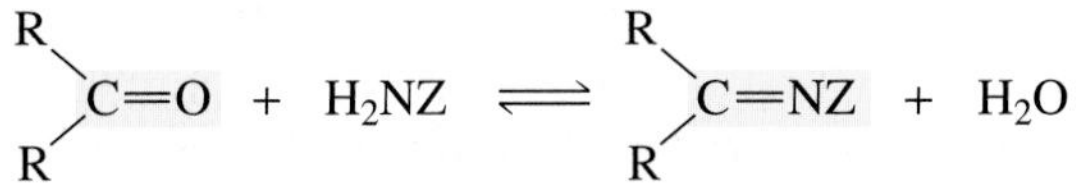

$$\underset{R}{\overset{R}{>}}C=O \ + \ H_2NZ \ \rightleftharpoons \ \underset{R}{\overset{R}{>}}C=NZ \ + \ H_2O$$

b. Reaction with a *secondary amine* forms an enamine:

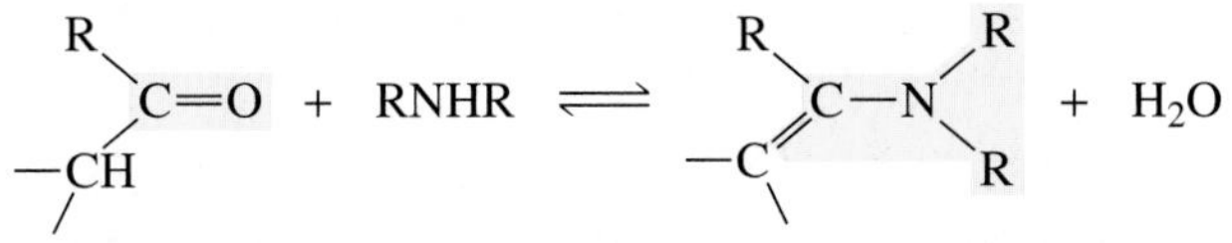

$$\underset{-CH}{\overset{R}{>}}C=O \ + \ RNHR \ \rightleftharpoons \ \underset{-C}{\overset{R}{>}}C-N\underset{R}{\overset{R}{<}} \ + \ H_2O$$

c. Reaction with excess *ammonia* plus H_2/metal catalyst or sodium triacetoxyborohydride forms a primary amine:

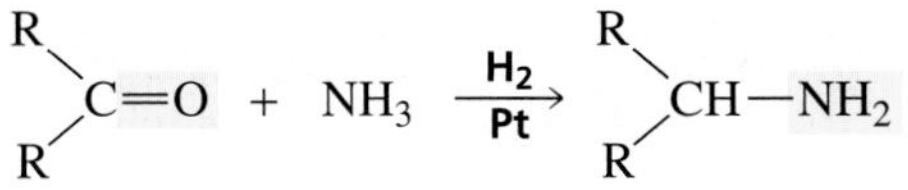

$$\underset{R}{\overset{R}{>}}C=O \ + \ NH_3 \ \xrightarrow[\text{Pt}]{H_2} \ \underset{R}{\overset{R}{>}}CH-NH_2$$

d. Reaction with excess *primary amine* followed by H_2/metal catalyst or sodium triacetoxyborohydride forms a secondary amine:

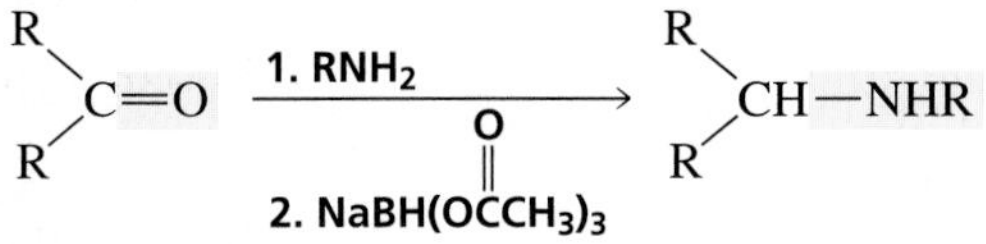

$$\underset{R}{\overset{R}{>}}C=O \ \xrightarrow[\text{2. NaBH(OCCH}_3)_3]{\text{1. RNH}_2} \ \underset{R}{\overset{R}{>}}CH-NHR$$

e. Reaction with a *secondary amine* followed by H_2/metal catalyst or sodium triacetoxyborohydride forms a tertiary amine:

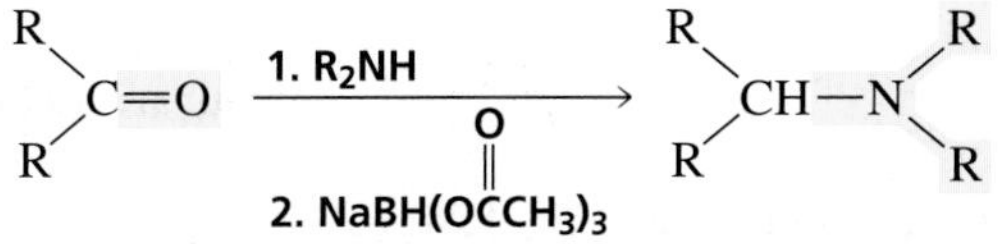

$$\underset{R}{\overset{R}{>}}C=O \ \xrightarrow[\text{2. NaBH(OCCH}_3)_3]{\text{1. R}_2\text{NH}} \ \underset{R}{\overset{R}{>}}CH-N\underset{R}{\overset{R}{<}}$$

f. The Wolff–Kishner reduction converts a carbonyl group to a methylene group:

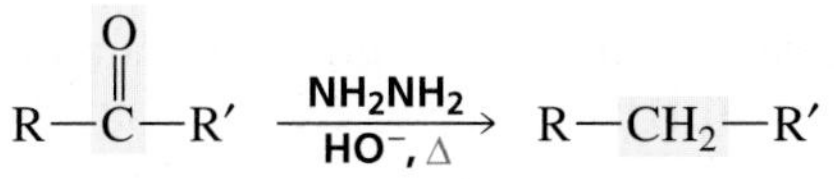

$$R-\overset{O}{\overset{\|}{C}}-R' \ \xrightarrow[\text{HO}^-, \Delta]{\text{NH}_2\text{NH}_2} \ R-CH_2-R'$$

5. Reaction of an *aldehyde* or a *ketone* with water forms a hydrate (Section 17.8)

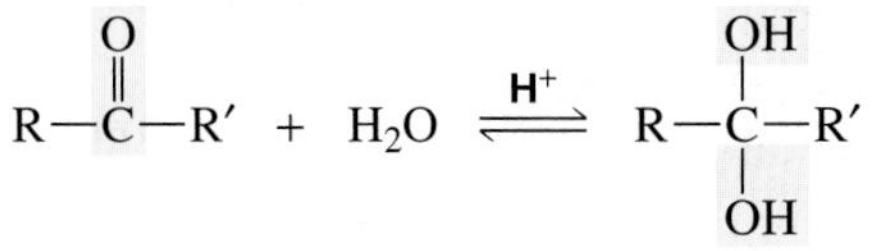

$$R-\overset{O}{\overset{\|}{C}}-R' \ + \ H_2O \ \xrightleftharpoons{H^+} \ R-\underset{OH}{\overset{OH}{\underset{|}{\overset{|}{C}}}}-R'$$

6. Reaction of an *aldehyde* or a *ketone* with excess alcohol forms an acetal or a ketal (Section 17.8)

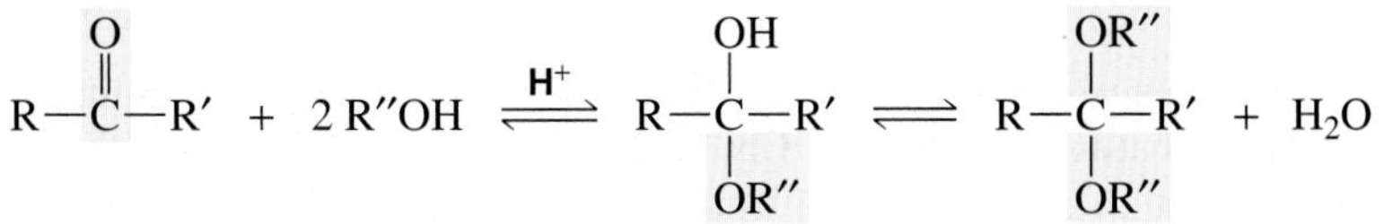

$$R-\overset{O}{\overset{\|}{C}}-R' \ + \ 2\,R''OH \ \xrightleftharpoons{H^+} \ R-\underset{OR''}{\overset{OH}{\underset{|}{\overset{|}{C}}}}-R' \ \rightleftharpoons \ R-\underset{OR''}{\overset{OR''}{\underset{|}{\overset{|}{C}}}}-R' \ + \ H_2O$$

7. Reaction of an *aldehyde* or a *ketone* with a thiol forms a thioacetal or a thioketal (Section 17.10)

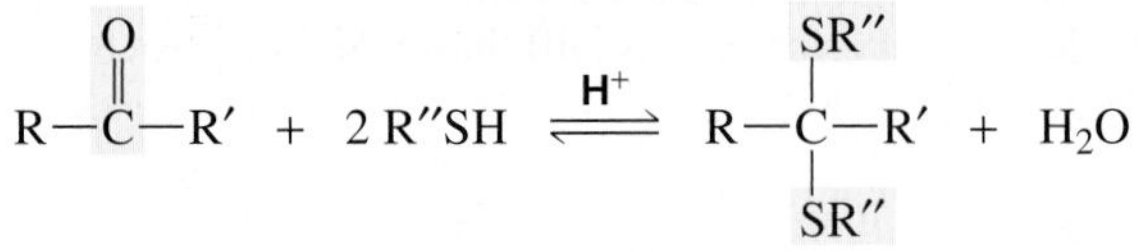

$$\underset{\substack{\| \\ R-C-R'}}{O} + 2\,R''SH \underset{}{\overset{H^+}{\rightleftharpoons}} \underset{\substack{| \\ R-C-R' \\ |}}{\overset{SR''}{\underset{SR''}{}}} + H_2O$$

8. Desulfurization of *thioacetals* and *thioketals* forms alkanes (Section 17.10)

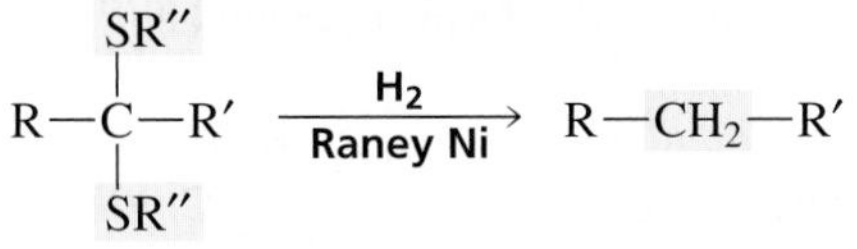

$$\underset{\substack{| \\ R-C-R' \\ |}}{\overset{SR''}{\underset{SR''}{}}} \xrightarrow[\text{Raney Ni}]{H_2} R-CH_2-R'$$

9. Reaction of an *aldehyde* or a *ketone* with a phosphonium ylide forms an alkene (Section 17.11)

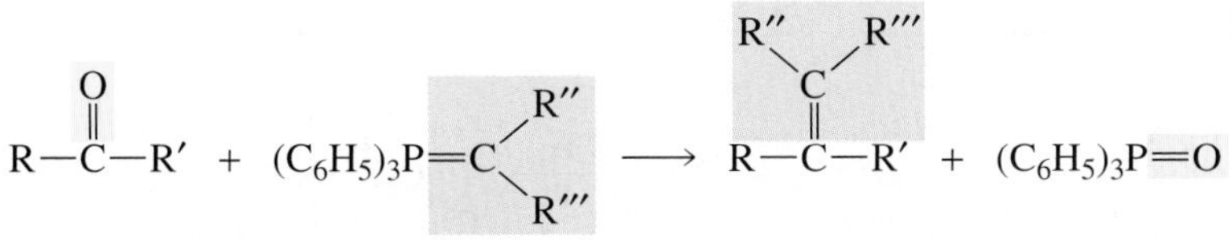

$$\underset{\substack{\| \\ R-C-R'}}{O} + (C_6H_5)_3P{=}C\underset{R'''}{\overset{R''}{<}} \longrightarrow \underset{\substack{\| \\ R-C-R'}}{\overset{R''\,\,\,R'''}{\underset{}{C}}} + (C_6H_5)_3P{=}O$$

10. Reactions of *α,β-unsaturated aldehydes* and *ketones* with nucleophiles (Section 17.14)

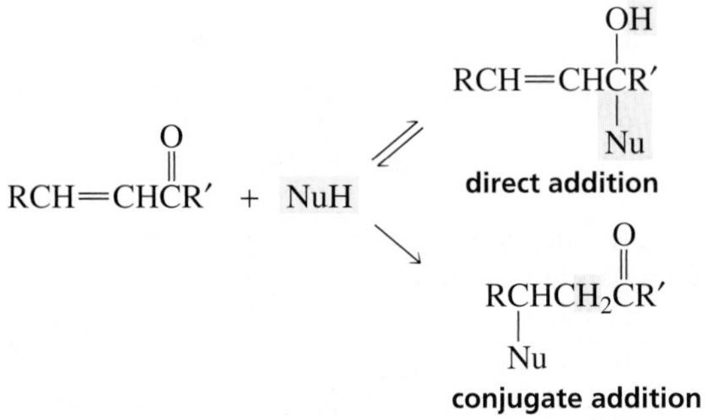

$$RCH{=}CHCR' + NuH$$

$$\overset{OH}{\underset{\substack{| \\ Nu}}{RCH{=}CHCR'}}$$
direct addition

$$\overset{O}{\underset{\substack{| \\ Nu}}{RCHCH_2CR'}}$$
conjugate addition

Nucleophiles that are strong bases (RLi, RMgBr, H⁻) form direct addition products. Nucleophiles that are weak bases (⁻CN, RNH₂, RSH, Br⁻, R₂CuLi) form conjugate addition products.

11. Reactions of *α,β-unsaturated carboxylic acid derivatives* with nucleophiles (Section 17.15)

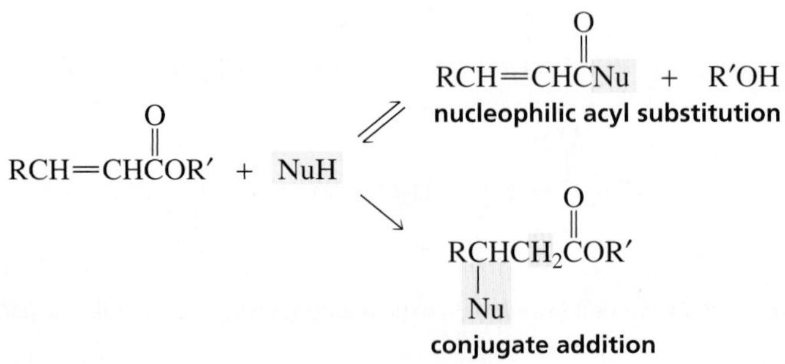

$$RCH{=}CHCOR' + NuH$$

$$\overset{O}{\underset{}{RCH{=}CHCNu}} + R'OH$$
nucleophilic acyl substitution

$$\overset{O}{\underset{\substack{| \\ Nu}}{RCHCH_2COR'}}$$
conjugate addition

Nucleophiles that are strong bases (RLi, RMgBr, H⁻) form nucleophilic substitution products. Nucleophiles that are weak bases (⁻CN, RNH₂, RSH, Br⁻, R₂CuLi) form conjugate addition products.

PROBLEMS

31. Give a structure for each of the following compounds:

   a. isobutyraldehyde
   b. 4-hexenal
   c. diisopentyl ketone
   d. 3-methylcyclohexanone
   e. 2,4-pentanedione

   f. 4-bromo-3-heptanone
   g. γ-bromocaproaldehyde
   h. 2-ethylcyclopentanecarbaldehyde
   i. 4-methyl-5-oxohexanal
   j. benzene-1,3-dicarbaldehyde

32. Give the products of each of the following reactions:

a. $CH_3CH_2\overset{\overset{\displaystyle O}{\|}}{C}H$ + CH_3CH_2OH (**excess**) $\xrightarrow{H^+}$

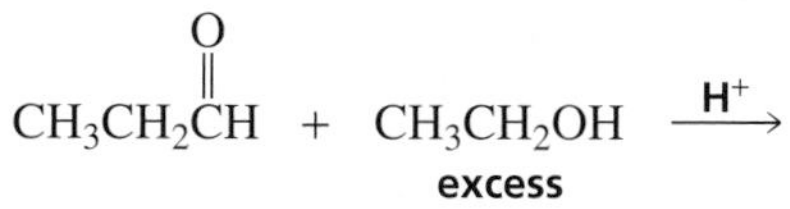

e. $CH_3CH_2\overset{\overset{\displaystyle O}{\|}}{C}CH_2CH_3$ + $NaC{\equiv}N$ (**excess**) $\xrightarrow{HCl}$

b. $C_6H_5\overset{\overset{\displaystyle O}{\|}}{C}CH_2CH_3$ + NH_2NH_2 $\longrightarrow$

f. $CH_3CH_2CH_2\overset{\overset{\displaystyle O}{\|}}{C}OCH_2CH_3$ $\xrightarrow[\text{2. } H^+,\, H_2O]{\text{1. LiAlH}_4}$

c. $C_6H_5\overset{\overset{\displaystyle O}{\|}}{C}CH_2CH_3$ + NH_2NH_2 $\xrightarrow[\Delta]{HO^-}$

g. $CH_3CH_2CH_2\overset{\overset{\displaystyle O}{\|}}{C}CH_3$ + $HOCH_2CH_2OH$ $\xrightarrow{H^+}$

d. $CH_3CH_2\overset{\overset{\displaystyle O}{\|}}{C}CH_3$ $\xrightarrow[\text{2. } H^+,\, H_2O]{\text{1. NaBH}_4}$

h. (cyclohexenone with CH₃) + $NaC{\equiv}N$ (**excess**) $\xrightarrow{HCl}$

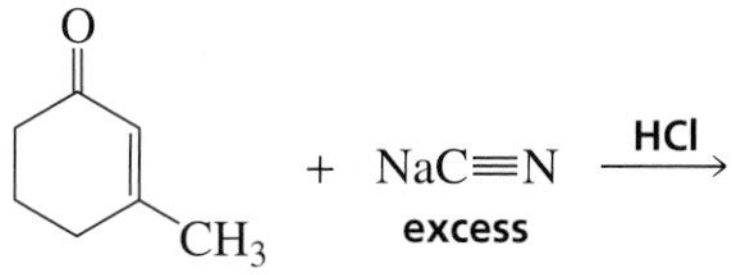

33. List the following compounds in order of decreasing reactivity toward nucleophilic attack:

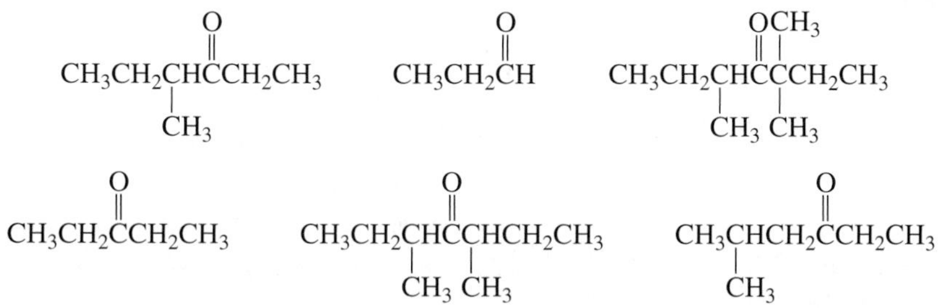

$$CH_3CH_2\underset{\underset{\displaystyle CH_3}{|}}{C}H\overset{\overset{\displaystyle O}{\|}}{C}CH_2CH_3 \qquad CH_3CH_2\overset{\overset{\displaystyle O}{\|}}{C}H \qquad CH_3CH_2\underset{\underset{\displaystyle CH_3}{|}}{C}H\overset{\overset{\displaystyle OCH_3}{\overset{\displaystyle |}{|}}}{C}\underset{\underset{\displaystyle CH_3}{|}}{C}CH_2CH_3$$

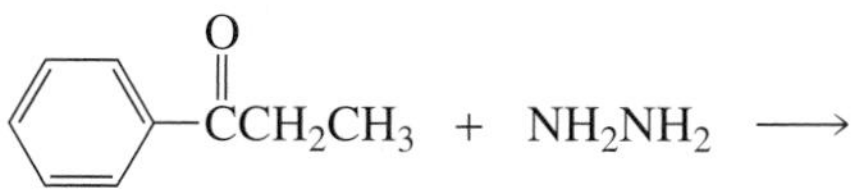

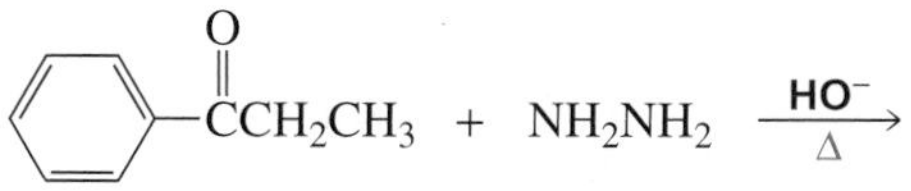

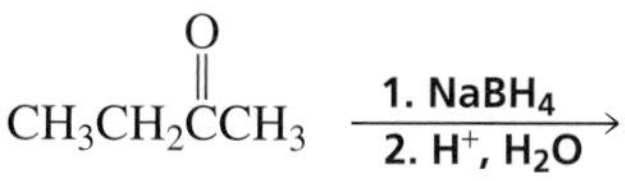

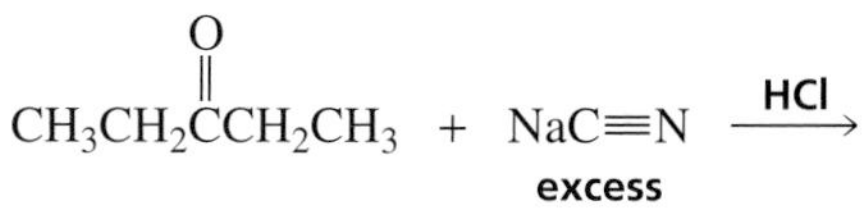

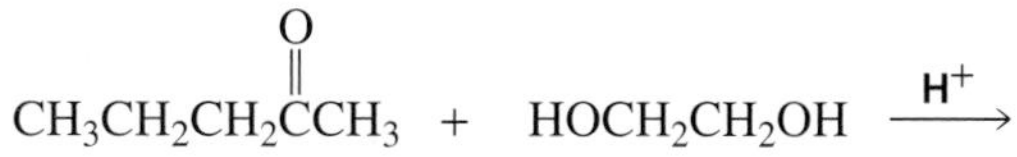

$$CH_3CH_2\overset{\overset{\displaystyle O}{\|}}{C}CH_2CH_3 \qquad CH_3CH_2\underset{\underset{\displaystyle CH_3}{|}}{C}H\overset{\overset{\displaystyle O}{\|}}{C}\underset{\underset{\displaystyle CH_3}{|}}{C}HCH_2CH_3 \qquad CH_3\underset{\underset{\displaystyle CH_3}{|}}{C}HCH_2\overset{\overset{\displaystyle O}{\|}}{C}CH_2CH_3$$

34. Using cyclohexanone as the starting material, describe how each of the following compounds could be synthesized:

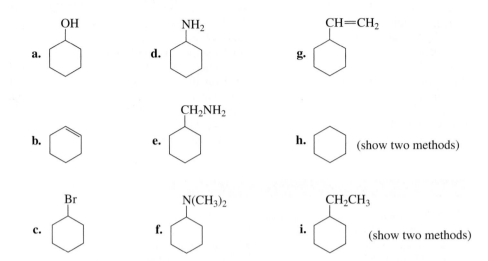

a. (OH on cyclohexane)

d. (NH₂ on cyclohexane)

g. (CH=CH₂ on cyclohexane)

b. (cyclohexene)

e. (CH₂NH₂ on cyclohexane)

h. (cyclohexane) (show two methods)

c. (Br on cyclohexane)

f. (N(CH₃)₂ on cyclohexane)

i. (CH₂CH₃ on cyclohexane) (show two methods)

35. List the following compounds in order of decreasing K_{eq} for hydrate formation:

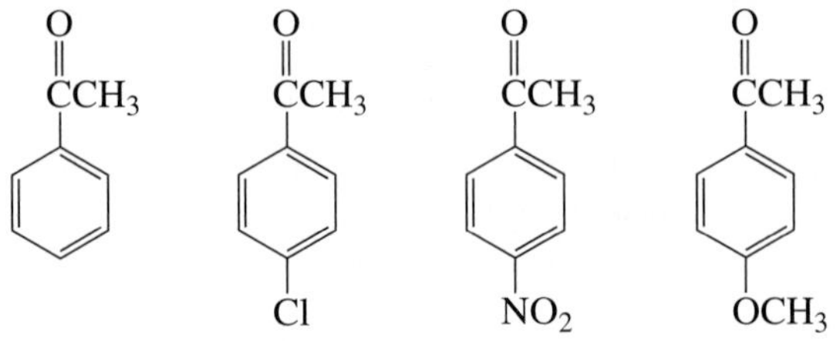

(four acetophenone-type structures: phenyl, 4-Cl-phenyl, 4-NO₂-phenyl, 4-OCH₃-phenyl with CCH₃ carbonyl groups)

36. Fill in the boxes.

a. CH_3OH ☐→ CH_3Br ☐→ ☐ 1.☐ / 2.☐ → CH_3CH_2OH

b. CH_4 ☐→ CH_3Br ☐→ ☐ 1.☐ / 2.☐ → $CH_3CH_2CH_2OH$

c. 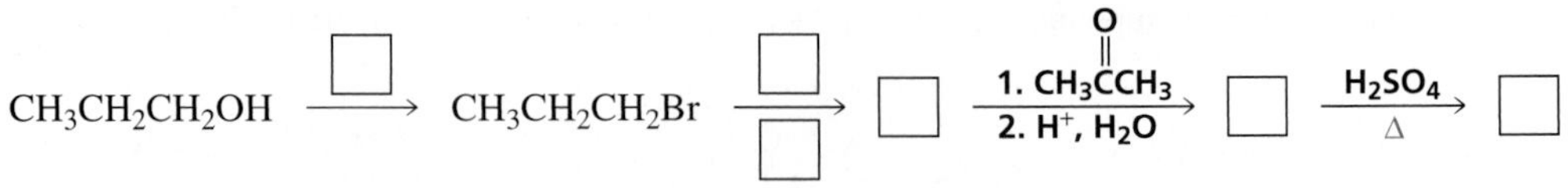 $CH_3CH_2CH_2OH$ ☐→ $CH_3CH_2CH_2Br$ ☐→ ☐ $\xrightarrow{\text{1. } CH_3\overset{O}{\overset{\|}{C}}CH_3,\ \text{2. } H^+, H_2O}$ ☐ $\xrightarrow[\Delta]{H_2SO_4}$ ☐

37. Give the products of each of the following reactions:

a. 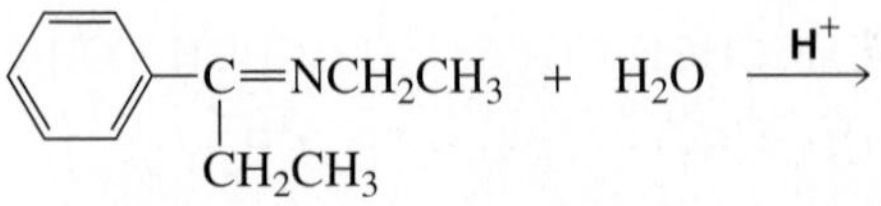 (phenyl)—C(=NCH₂CH₃)(CH₂CH₃) + H_2O $\xrightarrow{H^+}$

b. (cyclohexanone) + CH$_3$CH$_2$NH$_2$ ⟶

f. CH$_3$CH$_2$CCH$_3$ (ketone) $\xrightarrow{\text{1. CH}_3\text{CH}_2\text{MgBr}}{\text{2. H}^+, \text{H}_2\text{O}}$

c. (cyclohexanone) + (CH$_3$CH$_2$)$_2$NH ⟶

g. CH$_3$CH$_2$COCH$_3$ $\xrightarrow[\text{2. H}^+, \text{H}_2\text{O}]{\text{1. CH}_3\text{CH}_2\text{MgBr} \text{ excess}}$

d. (phenyl)–CCH$_2$CH$_3$ + NH$_3$ $\xrightarrow{\text{H}_2 \text{ Pt}}$

h. CH$_3$C=CHCCH$_3$ (with CH$_3$ substituent, O) + HBr ⟶

e. (cyclopentanone) + (C$_6$H$_5$)$_3$P=CHCH$_3$ ⟶

i. (pyrrolidinone) $\xrightarrow[\text{2. H}_2\text{O}]{\text{1. LiAlH}_4}$

38. Thiols can be prepared from the reaction of thiourea with an alkyl halide followed by hydroxide-ion-promoted hydrolysis.

H$_2$N–C–NH$_2$ (S) (thiourea) $\xrightarrow[\text{2. HO}^-, \text{H}_2\text{O}]{\text{1. CH}_3\text{CH}_2\text{Br}}$ H$_2$N–C–NH$_2$ (O) (urea) + CH$_3$CH$_2$SH (ethanethiol)

a. Propose a mechanism for the reaction.
b. What thiol would be formed if the alkyl halide employed were pentyl bromide?

39. Imines can exist as stereoisomers. The isomers are named by the *E,Z* system of nomenclature. (The unshared pair of electrons has the lowest priority.)

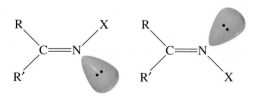

Give a structure for each of the following compounds:
a. *(E)*-benzaldehyde semicarbazone
b. *(Z)*-propiophenone oxime
c. cyclohexanone 2,4-dinitrophenylhydrazone

40. The only organic compound obtained when compound Z undergoes the following sequence of reactions gives the ^{1}H NMR spectrum shown. Identify compound Z.

Compound Z $\xrightarrow[\text{2. H}^+, \text{H}_2\text{O}]{\text{1. phenylmagnesium bromide}}$ $\xrightarrow[\Delta]{\text{MnO}_2}$

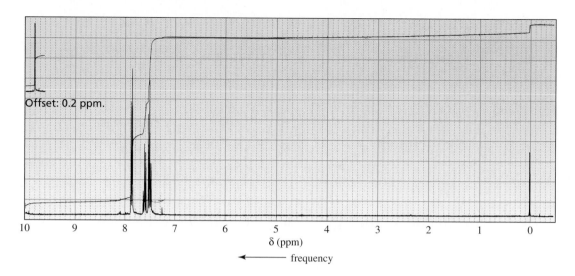

Offset: 0.2 ppm.

10 9 8 7 6 5 4 3 2 1 0

δ (ppm)

← frequency

41. Propose a mechanism for each of the following reactions:

a. 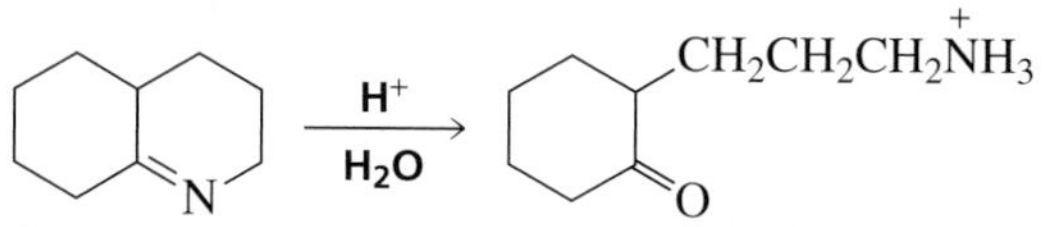 (structure) $\xrightarrow[H_2O]{H^+}$ (structure) $CH_2CH_2CH_2\overset{+}{N}H_3$

b. 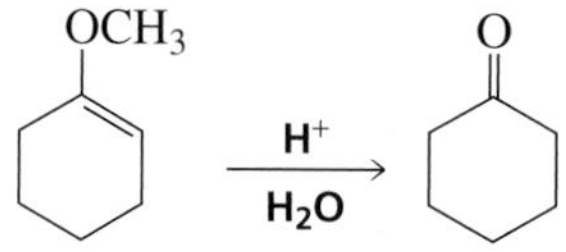 (structure with OCH₃) $\xrightarrow[H_2O]{H^+}$ (cyclohexanone)

c. 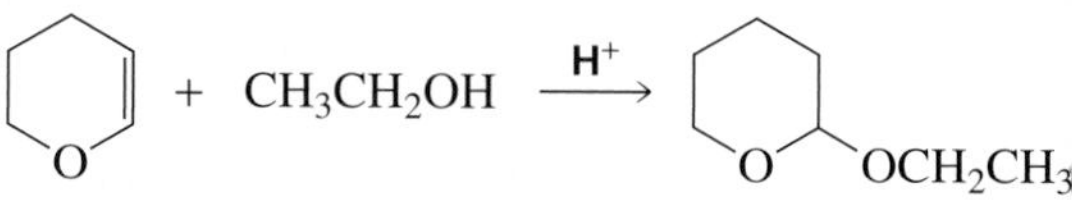 (structure) $+$ CH_3CH_2OH $\xrightarrow{H^+}$ (structure) OCH_2CH_3

42. How many signals would the product of the following reaction show in
a. its ^1H NMR spectrum? **b.** its ^{13}C NMR spectrum?

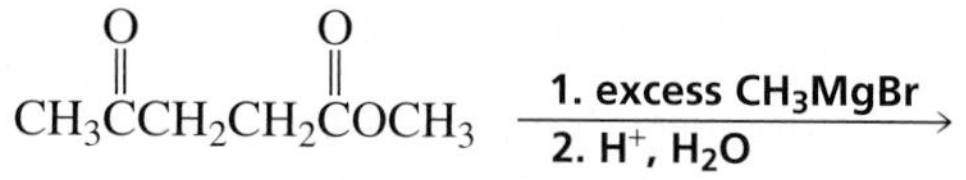

$$CH_3\overset{O}{\underset{\|}{C}}CH_2CH_2\overset{O}{\underset{\|}{C}}OCH_3 \xrightarrow[\text{2. H}^+, \text{H}_2\text{O}]{\text{1. excess CH}_3\text{MgBr}}$$

43. Each of the following reactions can form more than one stereoisomer. Show the stereoisomers.

a. 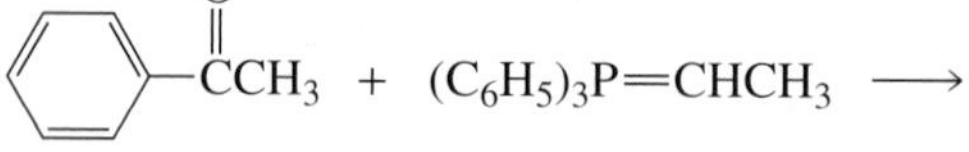 (structure) $\overset{O}{\underset{\|}{C}}CH_3$ $+$ $(C_6H_5)_3P{=}CHCH_3$ $\longrightarrow$

b. 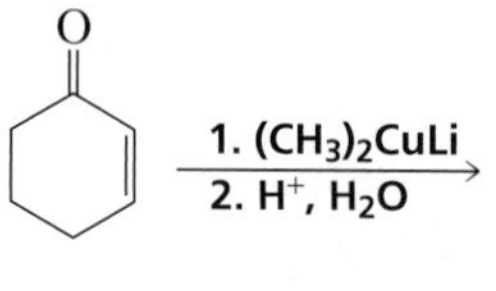 (structure) $\xrightarrow[\text{2. H}^+, \text{H}_2\text{O}]{\text{1. (CH}_3)_2\text{CuLi}}$

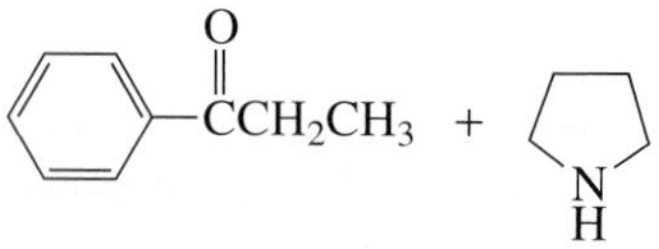

d. (structure) $\overset{O}{\underset{\|}{C}}CH_2CH_3$ $+$ (pyrrolidine structure)

c. (structure with CH₃) $\xrightarrow[\text{2. H}^+, \text{H}_2\text{O}]{\text{1. CH}_3\text{MgBr}}$

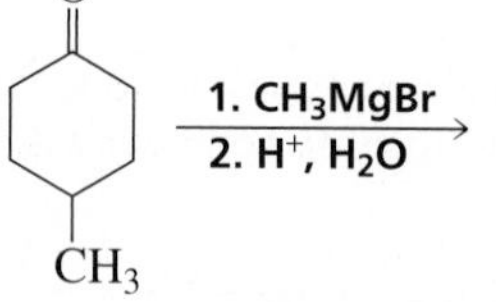

e. 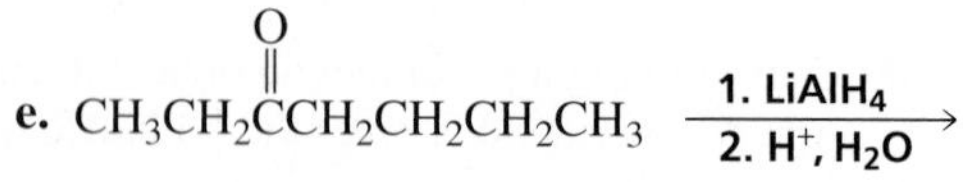 $CH_3CH_2\overset{O}{\underset{\|}{C}}CH_2CH_2CH_2CH_3$ $\xrightarrow[\text{2. H}^+, \text{H}_2\text{O}]{\text{1. LiAlH}_4}$

44. List three different sets of reagents (a carbonyl compound and a Grignard reagent) that could be used to prepare each of the following tertiary alcohols:

a.

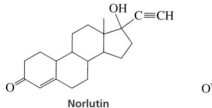

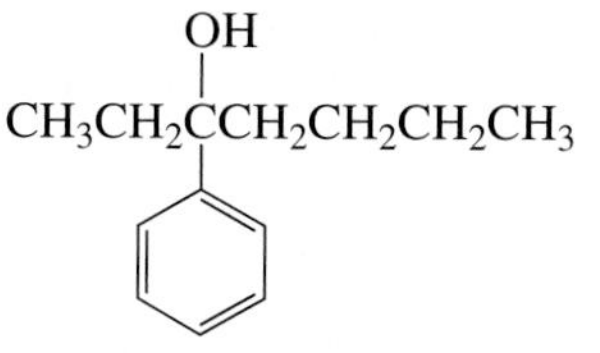

b. $CH_3CH_2C(OH)(CH_2CH_3)CH_2CH_2CH_3$

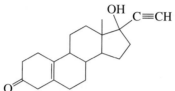

45. Give the product of the reaction of 3-methyl-2-cyclohexenone with each of the following reagents:

a. CH_3MgBr followed by H^+, H_2O
b. excess NaCN, HCl
c. H_2, Pd
d. $(CH_3)_2NH$
e. $(CH_3CH_2)_2CuLi$
f. CH_3CH_2SH

46. Norlutin and Enovid are ketones that suppress ovulation. Consequently, they have been used clinically as contraceptives. For which of these compounds would you expect the infrared carbonyl absorption (C═O stretch) to be at a higher frequency? Explain.

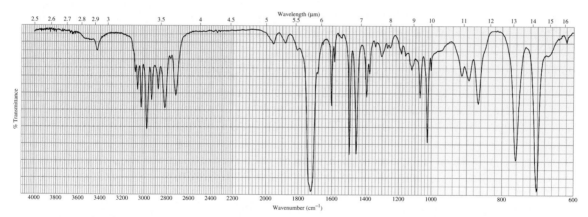

Norlutin

Enovid

47. A compound gives the following IR spectrum. Upon reaction with sodium borohydride followed by acidification, the product with the following ^1H NMR spectrum is formed. Identify the compounds.

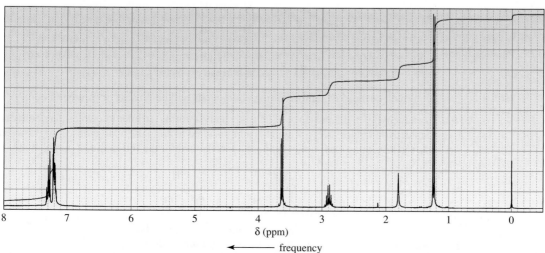

48. Unlike a phosphonium ylide that reacts with an aldehyde or ketone to form an alkene, a sulfonium ylide reacts with an aldehyde or ketone to form an epoxide. Explain why one ylide forms an alkene while the other forms an epoxide.

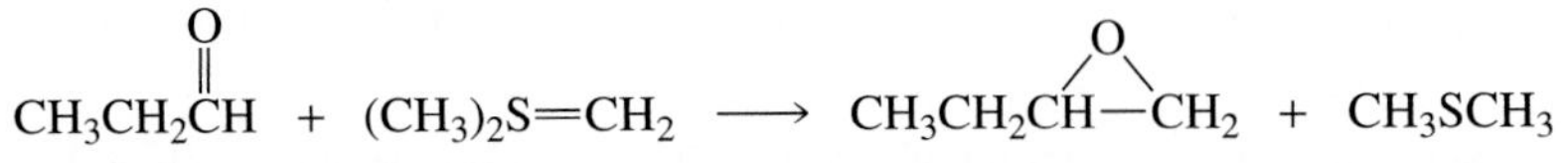

$$\underset{\text{O}}{\overset{\text{O}}{\underset{\|}{}}} \quad \text{CH}_3\text{CH}_2\overset{\text{O}}{\overset{\|}{\text{C}}}\text{H} \ + \ (\text{CH}_3)_2\text{S}=\text{CH}_2 \ \longrightarrow \ \text{CH}_3\text{CH}_2\overset{\text{O}}{\overset{\diagup\diagdown}{\text{CH}}}-\text{CH}_2 \ + \ \text{CH}_3\text{SCH}_3$$

49. Indicate how the following compounds could be prepared from the given starting materials:

a. 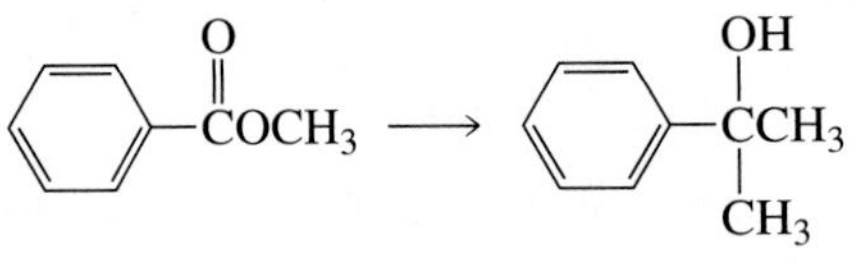 (benzene ring)–COCH₃ ⟶ (benzene ring)–C(OH)(CH₃)CH₃

d. 2-oxo-1-methylpiperidine ⟶ 2-methylpiperidine

b. (benzene ring)–COCH₃ ⟶ (benzene ring)–CH(=O)

e. cyclohexanol (OH) ⟶ N-methylcyclohexylamine (NHCH₃)

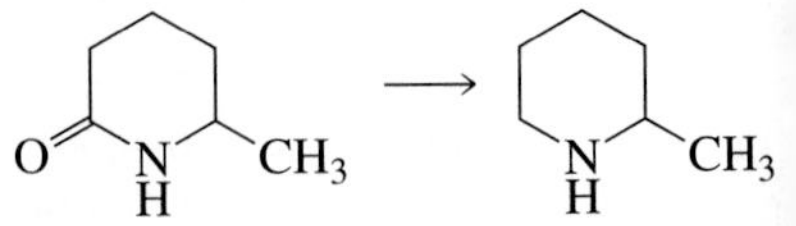

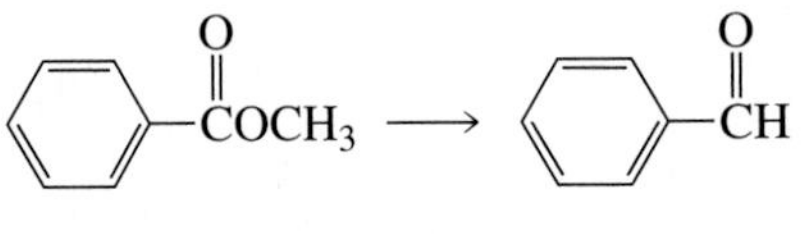

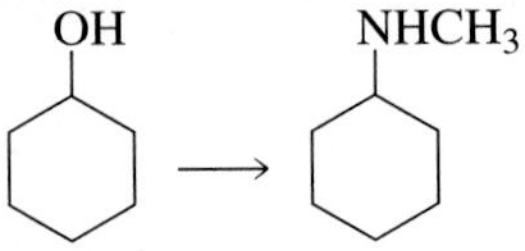

c. $\text{CH}_3\text{CH}_2\text{CH}_2\text{CH}_2\text{Br} \ \longrightarrow \ \text{CH}_3\text{CH}_2\text{CH}_2\text{CH}_2\overset{\text{O}}{\overset{\|}{\text{C}}}\text{OH}$

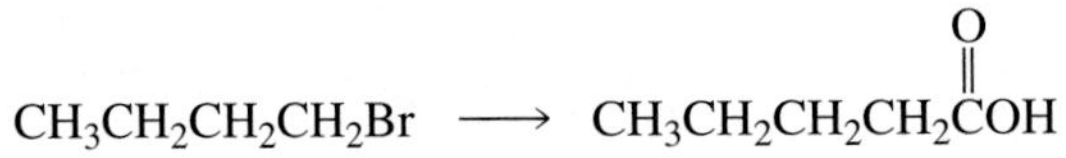

50. Propose a reasonable mechanism for each of the following reactions:

a. 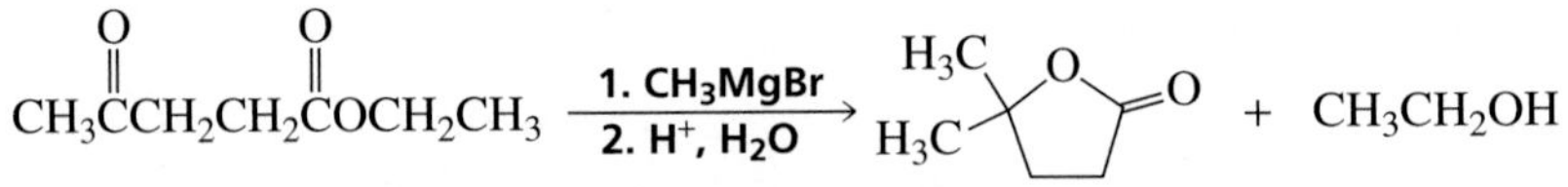 $\text{CH}_3\overset{\text{O}}{\overset{\|}{\text{C}}}\text{CH}_2\text{CH}_2\overset{\text{O}}{\overset{\|}{\text{C}}}\text{OCH}_2\text{CH}_3 \ \xrightarrow[\text{2. H}^+, \text{H}_2\text{O}]{\text{1. CH}_3\text{MgBr}} \ $ (lactone with (CH₃)₂ and H₃C) $+ \ \text{CH}_3\text{CH}_2\text{OH}$

b. 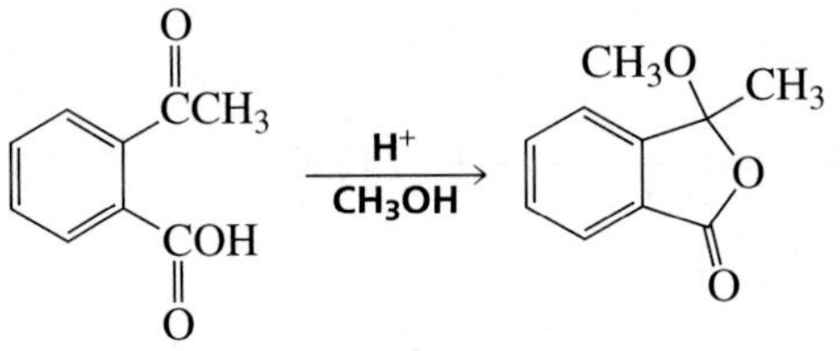 (benzene ring with COCH₃ and COH(=O) ortho substituents) $\xrightarrow[\text{CH}_3\text{OH}]{\text{H}^+}$ (bicyclic product with CH₃O, CH₃, O, =O)

51. a. In aqueous solution, D-glucose exists in equilibrium with two six-membered ring compounds. Give the structures of these compounds.

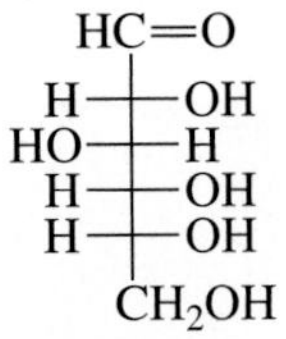

$$\begin{array}{c}
\text{HC}=\text{O} \\
\text{H}\!-\!\text{OH} \\
\text{HO}\!-\!\text{H} \\
\text{H}\!-\!\text{OH} \\
\text{H}\!-\!\text{OH} \\
\text{CH}_2\text{OH}
\end{array}$$

D-glucose

b. Which of the six-membered ring compounds will be present in greater amount?

52. Acetaldehyde, in the presence of an acid catalyst, forms a trimer known as paraldehyde. Paraldehyde induces sleep when it is administered to animals in large doses, so it is used as a sedative or as a hypnotic. Propose a mechanism for the formation of paraldehyde.

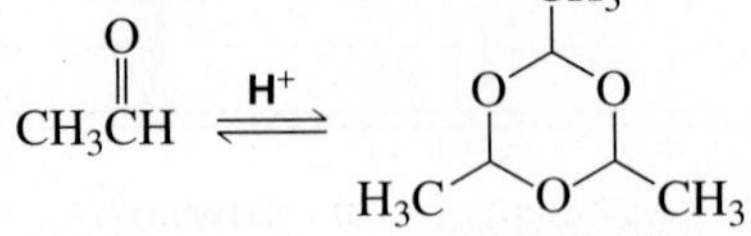

$$\text{CH}_3\overset{\text{O}}{\overset{\|}{\text{C}}}\text{H} \ \rightleftharpoons \ $$ (six-membered ring with three O and three CH₃ groups)

paraldehyde

53. The addition of hydrogen cyanide to benzaldehyde forms a compound called mandelonitrile. *(R)*-Mandelonitrile is formed from the hydrolysis of amygdalin, a compound found in the pits of peaches and apricots. Amygdalin is the principal constituent of Laetrile, a compound that was once highly touted as a treatment for cancer. It was subsequently found to be ineffective. Is *(R)*-mandelonitrile formed by attack of cyanide ion on the *Re* face or the *Si* face of benzaldehyde?

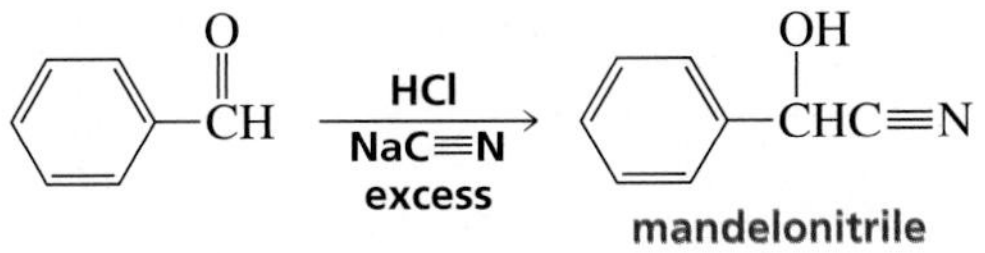

O
CH
HCl
NaC≡N
excess
OH
CHC≡N

mandelonitrile

54. What carbonyl compound and what phosphonium ylide are needed to synthesize the following compounds?

a. 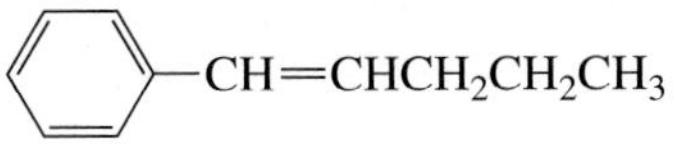

CH=CHCH2CH2CH3

c. 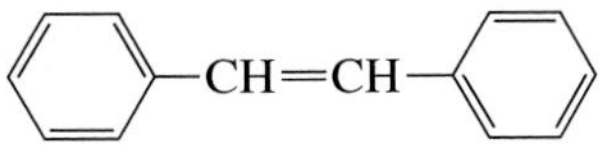

CH=CH

b. 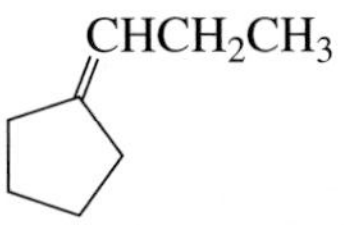

CHCH2CH3

d. 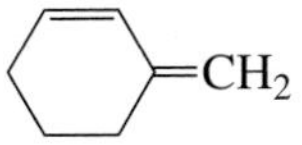

=CH2

55. Identify compounds A and B.

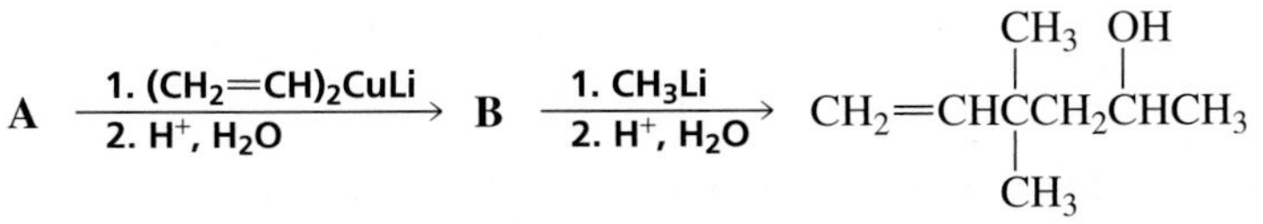

A
1. (CH2=CH)2CuLi
2. H+, H2O
B
1. CH3Li
2. H+, H2O
CH3 OH
CH2=CHCCH2CHCH3
CH3

56. Propose a reasonable mechanism for each of the following reactions:

a. 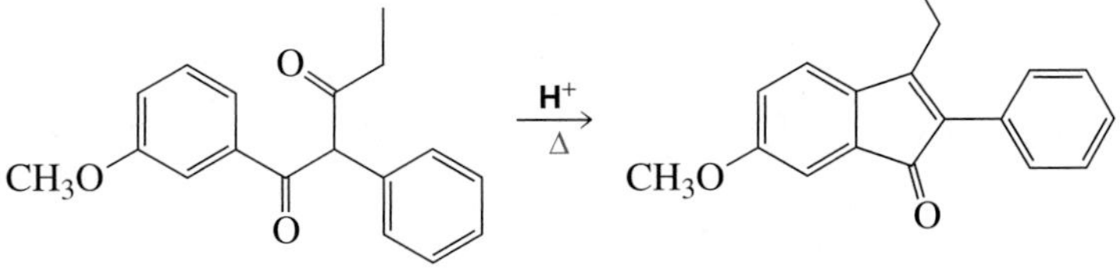

CH3O
O
O
H+
Δ
CH3O
O

b. 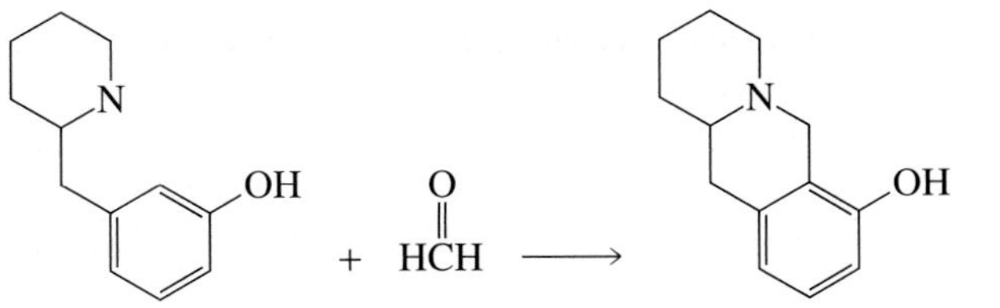

N
OH
O
+ HCH
N
OH

57. A compound reacts with methylmagnesium bromide followed by acidification to form the product with the following ^{1}H NMR spectrum. Identify the compound.

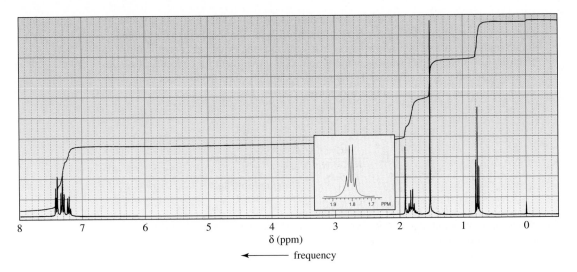

58. When a cyclic ketone reacts with diazomethane, the next larger cyclic ketone is formed. This is called a ring expansion reaction. Provide a mechanism for this reaction.

$$\text{cyclohexanone} + \overset{-}{CH_2}\overset{+}{N}\equiv N \longrightarrow \text{cycloheptanone} + N_2$$

      cyclohexanone   diazomethane   cycloheptanone

59. The pK_a values of oxaloacetic acid are 2.22 and 3.98.

$$\underset{\text{oxaloacetic acid}}{HO\overset{O}{\overset{\|}{C}}-\overset{O}{\overset{\|}{C}}CH_2\overset{O}{\overset{\|}{C}}OH}$$

The amount of hydrate present in an aqueous solution of oxaloacetic acid depends on the pH of the solution (*J. Am. Chem. Soc.*, 1983, *105*, 4982): 95% at pH = 0; 81% at pH = 1.3; 35% at pH = 3.1; 13% at pH = 4.7; 6% at pH = 6.7; and 6% at pH = 12.7. Explain this pH dependence.

60. The Horner–Emmons modification is a variation of a Wittig reaction in which a phosphonate-stabilized carbanion in used instead of a phosphonium ylide.

$$RCR + CH_3\overset{-}{C}-P(OEt)_2 \longrightarrow \underset{R}{\overset{R}{}}C=C\underset{CH_3}{\overset{CH_3}{}} + {}^-O-P(OEt)_2$$

$$Et = CH_3CH_2$$

The phosphonate-stabilized carbanion is prepared from an appropriate alkyl halide. This is called the Arbuzov reaction.

$$(EtO)_3P: + CH_3CHCH_3 \longrightarrow CH_3CH-P(OEt)_2 \xrightarrow{\text{strong base}} CH_3\overset{O}{\overset{\|}{C}}-P(OEt)_2$$

$$+ CH_3CH_2Br$$

Because the Arbuzov reaction can be carried out with an α-bromo ketone or an α-bromo ester (in which case it is called a Perkow reaction), it provides a way to synthesize α,β-unsaturated ketones and esters.

$$(EtO)_3P: \; + \; Br-CH_2CR \overset{O}{\underset{}{\parallel}} \longrightarrow (EtO)_2\overset{O}{\underset{}{\parallel}}P-CH_2CR \overset{O}{\underset{}{\parallel}} \; \overset{\text{strong base}}{\longrightarrow} \; (EtO)_2\overset{O}{\underset{}{\parallel}}P-\overset{-}{C}HCR \overset{O}{\underset{}{\parallel}}$$

$$+ \; CH_3CH_2Br$$

a. Propose a mechanism for the Arbuzov reaction.
b. Propose a mechanism for the Horner–Emmons modification.
c. Show how the following compounds can be prepared from the given starting material.

1. $CH_3CH_2\overset{O}{\underset{}{\parallel}}CH \longrightarrow CH_3CH_2CH=CH\overset{O}{\underset{}{\parallel}}CCH_3$

2.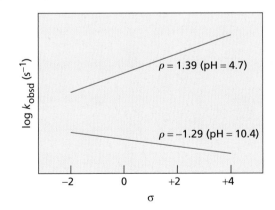

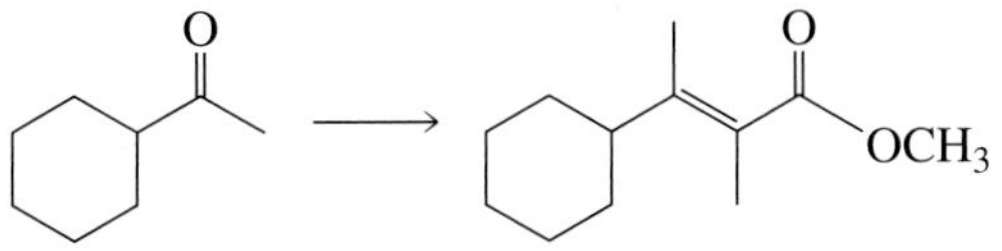

61. In order to solve this problem you must read the description of the Hammett σ, ρ treatment given in Chapter 16, Problem 69. When the rate constants for hydrolysis of several morpholine enamines of para-substituted propiophenones are determined at pH = 4.7, the ρ value is positive; however, when the rates of hydrolysis are determined at pH = 10.4, the ρ value is negative.

a. What is the rate-determining step of the hydrolysis reaction when it is carried out in basic solution?
b. What is the rate-determining step of the reaction when it is carried out in acidic solution? (*J. Am. Chem. Soc.*, 1970, *92*, 4261).

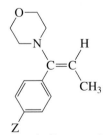

a morpholine enamine of
a para-substituted propiophenone

$\rho = 1.39$ (pH = 4.7)

$\rho = -1.29$ (pH = 10.4)

62. Propose a mechanism for each of the following reactions:

a. 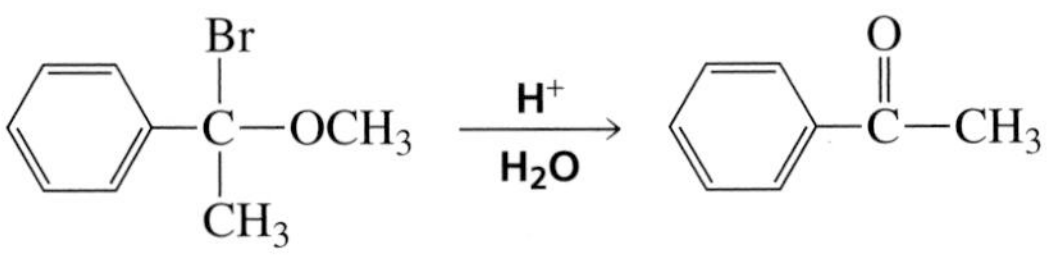 $\underset{CH_3}{\overset{Br}{\underset{|}{\overset{|}{C}}}}-OCH_3 \quad \overset{H^+}{\underset{H_2O}{\longrightarrow}} \quad \overset{O}{\underset{}{\parallel}}C-CH_3$

b. 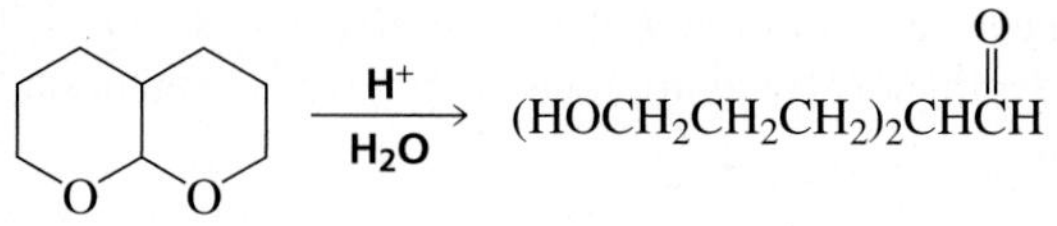 $\xrightarrow[\text{H}_2\text{O}]{\text{H}^+}$ $(\text{HOCH}_2\text{CH}_2\text{CH}_2)_2\text{CHCH}$
(with O above the final CH)

c. 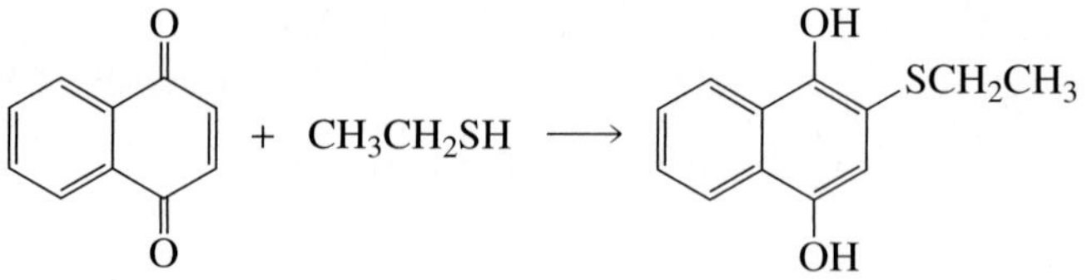 $+ \ \text{CH}_3\text{CH}_2\text{SH} \longrightarrow$

d. 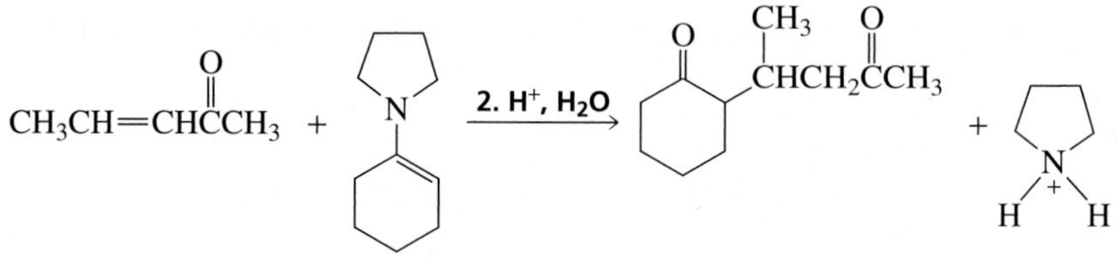 $\text{CH}_3\text{CH}=\text{CHCCH}_3$ (with O above second C) $+$ $\xrightarrow{\text{2. H}^+, \text{H}_2\text{O}}$ $+$

18

More About Oxidation–Reduction Reactions

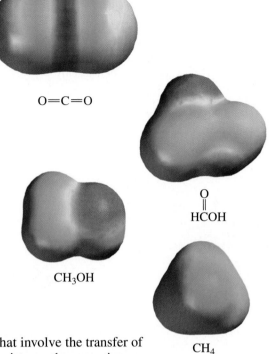

$O=C=O$

CH_3OH

$$\overset{\displaystyle O}{\underset{\displaystyle \|}{}}$$
HCOH

CH_4

A n important group of organic reactions are those that involve the transfer of electrons from one molecule to another. Organic chemists use these reactions—called **oxidation–reduction reactions** or **redox reactions**—to synthesize a large variety of compounds. Redox reactions are also important in biological systems because many of these reactions produce energy. You have seen a number of oxidation and reduction reactions in other chapters, but discussing them as a group will give you the opportunity to compare and contrast them.

In an oxidation–reduction reaction, one compound loses electrons and one compound gains electrons. The compound that loses electrons is oxidized, and the compound that gains electrons is reduced. One way to remember the difference between oxidation and reduction is with the phrase "LEO the lion says GER": *Loss of Electrons is Oxidation, Gain of Electrons is Reduction.*

The following is an example of an oxidation–reduction reaction involving inorganic reagents:

$$Cu^+ + Fe^{3+} \longrightarrow Cu^{2+} + Fe^{2+}$$

Cu^+ loses an electron in this reaction, so Cu^+ is oxidized. Fe^{3+}, on the other hand, gains an electron, so Fe^{3+} is reduced. This reaction demonstrates two important points about oxidation–reduction reactions. First, oxidation is always coupled with reduction. In other words, a compound cannot gain electrons (be reduced) unless another compound in the reaction simultaneously loses electrons (is oxidized). Second, the compound that is oxidized (Cu^+) is called the **reducing agent** because it loses the electrons that are used to reduce the other compound (Fe^{3+}). Similarly, the compound that is reduced (Fe^{3+}) is called the **oxidizing agent** because it gains the electrons given up by the other compound (Cu^+), thereby making it possible for the other compound to be oxidized.

In order to tell whether an organic compound has been oxidized or reduced, look at the change in the structure of the compound. If the number of carbon–hydrogen bonds has increased or if the number of carbon–oxygen, carbon–nitrogen, or carbon–halogen bonds has decreased, the compound has been reduced. If the number of carbon–hydrogen bonds has decreased, or if the number of carbon–oxygen, carbon–nitrogen, or carbon–halogen bonds has increased, the compound has been oxidized. Notice that the **oxidation state** of a carbon atom equals the number of its carbon–oxygen, carbon–nitrogen, and carbon–halogen bonds.

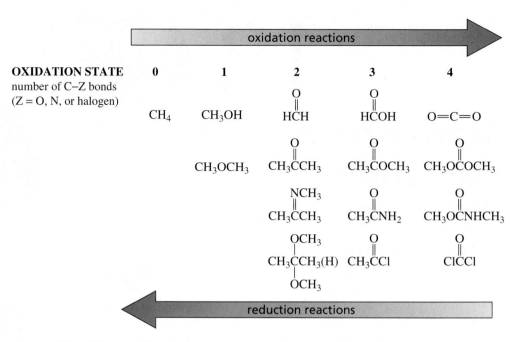

Tutorial:
Changes in
oxidation state

We will now take a look at some examples of organic oxidation–reduction reactions. In each of the following reactions, the product has more carbon–hydrogen bonds than the reactant. The alkene, aldehyde, and ketone, therefore, are being reduced. Hydrogen, sodium borohydride, and hydrazine are the reducing agents.

$$RCH{=}CHR \xrightarrow[\text{Pt}]{\text{H}_2} RCH_2CH_2R$$
an alkene

Reduction increases the number of carbon–hydrogen bonds or decreases the number of C—O, C—N, or C—X (X = halogen) bonds.

$$\underset{\text{an aldehyde}}{\overset{\overset{\displaystyle O}{\|}}{RCH}} \xrightarrow[\text{2. H}^+, \text{H}_2\text{O}]{\text{1. NaBH}_4} RCH_2OH$$

$$\underset{\text{a ketone}}{\overset{\overset{\displaystyle O}{\|}}{RCR}} \xrightarrow[\text{HO}^-, \Delta]{\text{H}_2\text{NNH}_2} RCH_2R$$

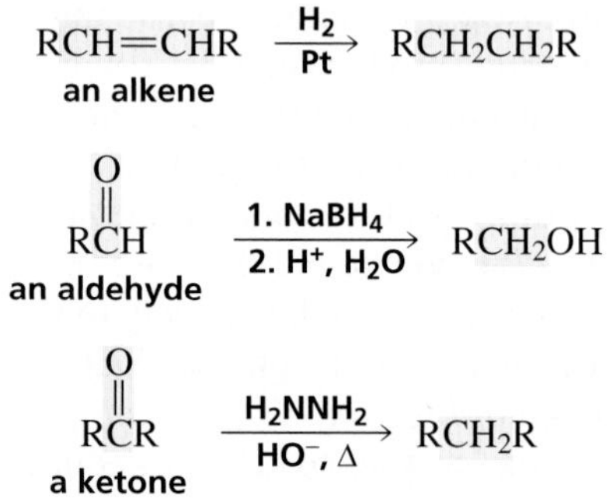

In the next group of reactions, the number of carbon–bromine bonds increases in the first reaction. In the second and third reactions, the number of carbon–hydrogen bonds decreases and the number of carbon–oxygen bonds increases. This means that the alkene, the aldehyde, and the alcohol are being oxidized. Bromine and chromic acid (H_2CrO_4) are the oxidizing agents. Notice that the increase in the number of carbon–oxygen bonds in the last reaction results from a carbon–oxygen single bond becoming a carbon–oxygen double bond.

$$RCH{=}CHR \xrightarrow{\text{Br}_2} \underset{\text{an alkene}}{\overset{\overset{\displaystyle \text{Br} \;\; \text{Br}}{| \;\; |}}{RCHCHR}}$$

Oxidation decreases the number of carbon–hydrogen bonds or increases the number of C—O, C—N, or C—X (X = halogen) bonds.

$$\underset{\text{an aldehyde}}{\overset{\overset{\displaystyle O}{\|}}{RCH}} \xrightarrow{\text{H}_2\text{CrO}_4} \underset{}{\overset{\overset{\displaystyle O}{\|}}{RCOH}}$$

$$\underset{\text{an alcohol}}{\overset{\overset{\displaystyle \text{OH}}{|}}{RCHR}} \xrightarrow{\text{H}_2\text{CrO}_4} \underset{}{\overset{\overset{\displaystyle O}{\|}}{RCR}}$$

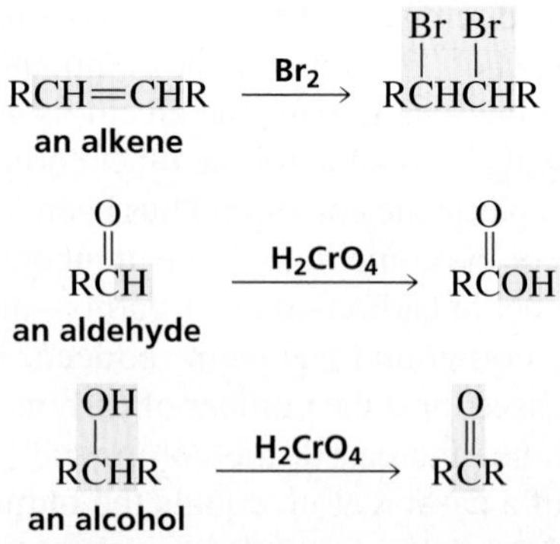

If water is added to an alkene, the product has one more carbon–hydrogen bond than the reactant, but it also has one more carbon–oxygen bond. In this reaction, one carbon is reduced and another is oxidized. The two processes cancel each other as far as the overall molecule is concerned, so the overall reaction is neither an oxidation nor a reduction.

$$RCH{=}CHR \xrightarrow[\text{H}_2\text{O}]{\text{H}^+} RCH_2\underset{\overset{|}{\text{OH}}}{CHR}$$

Many oxidizing reagents and many reducing reagents are available to organic chemists. This chapter highlights only a small fraction of the available reagents. The ones selected are some of the more common reagents that illustrate the types of transformations caused by oxidation and reduction.

PROBLEM 1◆

Indicate whether each of the following reactions is an oxidation reaction, a reduction reaction, or neither:

a. $CH_3\overset{\overset{\text{O}}{\|}}{C}Cl \xrightarrow[\substack{\text{partially} \\ \text{deactivated} \\ \text{Pd}}]{\text{H}_2} CH_3\overset{\overset{\text{O}}{\|}}{C}H$

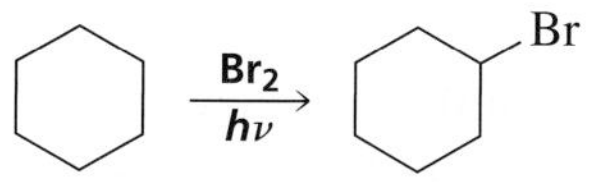

d. $CH_3CH_2OH \xrightarrow{\text{H}_2\text{CrO}_4} CH_3\overset{\overset{\text{O}}{\|}}{C}OH$

b. $RCH{=}CHR \xrightarrow{\text{HBr}} RCH_2\overset{\overset{\text{Br}}{|}}{C}HR$

e. $CH_3C{\equiv}N \xrightarrow{\overset{\text{H}_2}{\underset{\text{Pt}}{}}} CH_3CH_2NH_2$

c. ⬡ $\xrightarrow[h\nu]{\text{Br}_2}$ ⬡—Br

f. $CH_3CH_2CH_2Br \xrightarrow{\text{HO}^-} CH_3CH_2CH_2OH$

An organic compound is reduced when hydrogen (H_2) is added to it. We can think of H_2 as being composed of *two hydrogen atoms, two electrons and two protons,* or *a hydride ion and a proton.* In the following sections we will see that these three ways to describe H_2 correspond to the three mechanisms by which H_2 is added to an organic compound.

**18.1
REDUCTION
REACTIONS**

components of H:H

H· ·H	·⁻ H⁺ ·⁻ H⁺	H:⁻ H⁺
two hydrogen atoms	two electrons and two protons	a hydride ion and a proton

Reduction by Addition of Two Hydrogen Atoms

We have already seen that hydrogen can be added to carbon–carbon double bonds and to carbon–carbon triple bonds in the presence of a metal catalyst (Sections 3.19 and 5.8). These reactions are called **catalytic hydrogenations.** Catalytic hydrogenations are reduction reactions because there are more carbon–hydrogen bonds in the products than in the reactants. Alkenes and alkynes are both reduced to alkanes.

Movie: Catalytic hydrogenations

$$CH_3CH_2CH{=}CH_2 + H_2 \xrightarrow{\text{Pt, Pd, or Ni}} CH_3CH_2CH_2CH_3$$
$$\text{1-butene} \qquad\qquad\qquad\qquad\qquad \text{butane}$$

$$CH_3CH_2CH_2C{\equiv}CH + 2 H_2 \xrightarrow{\text{Pt, Pd, or Ni}} CH_3CH_2CH_2CH_2CH_3$$
$$\text{1-pentyne} \qquad\qquad\qquad\qquad\qquad \text{pentane}$$

In a catalytic hydrogenation, the hydrogen–hydrogen bond breaks homolytically (Section 3.18). This means that reduction occurs as a result of the addition of two hydrogen atoms to the organic molecule.

We have seen that the catalytic hydrogenation of an alkyne can be stopped at a *cis*-alkene if a partially deactivated catalyst is used (Section 5.8).

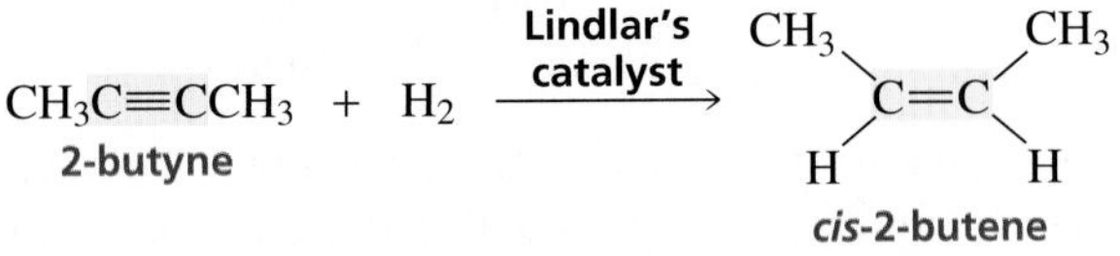

$$CH_3C{\equiv}CCH_3 \ + \ H_2 \ \xrightarrow{\text{Lindlar's catalyst}} \quad \underset{H \quad\quad H}{\overset{CH_3 \quad\quad CH_3}{C{=}C}}$$

2-butyne   *cis*-2-butene

Only the alkene substituent is reduced in the following reaction. The very stable benzene ring can be reduced only under special conditions.

3-D Molecules:
Stilbene;
Ethyl benzene

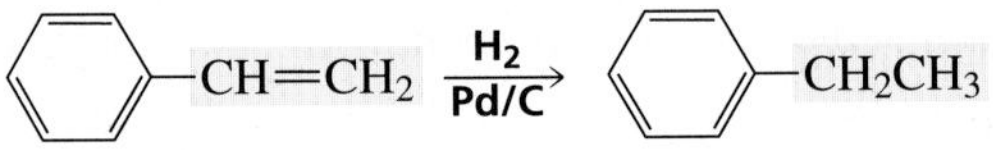

$$\bigcirc\!\!-CH{=}CH_2 \ \xrightarrow[\text{Pd/C}]{\textbf{H}_2} \ \bigcirc\!\!-CH_2CH_3$$

Catalytic hydrogenation can also be used to reduce carbon–nitrogen double bonds and carbon–nitrogen triple bonds. The reaction products are amines (Section 17.7).

$$CH_3CH_2CH{=}NCH_3 \ + \ H_2 \ \xrightarrow{\textbf{Pd/C}} \ CH_3CH_2CH_2NHCH_3$$
                                                         methylpropylamine

$$CH_3CH_2CH_2C{\equiv}N \ + \ 2\,H_2 \ \xrightarrow{\textbf{Pd/C}} \ CH_3CH_2CH_2CH_2NH_2$$
                                                         butylamine

The carbon–oxygen double bond (the carbonyl group) of ketones and aldehydes can be reduced by catalytic hydrogenation. The metal catalysts most commonly used for this reaction are Raney nickel (finely dispersed nickel with preadsorbed hydrogen), PtO_2, and Pd/C (Section 3.19). Ketones are reduced to secondary alcohols and aldehydes are reduced to primary alcohols.

Murray Raney (1885–1966) was born in Tennessee. He received a B.A. from the University of Tennessee. He worked at the Gilman Paint and Varnish Co. from 1925 to 1950, at which point he left to found the Raney Catalyst Co.

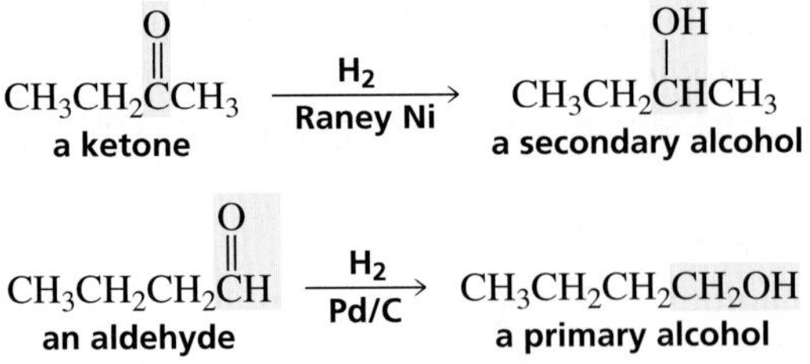

$$\underset{\text{a ketone}}{CH_3CH_2\overset{\overset{\displaystyle O}{\|}}{C}CH_3} \ \xrightarrow[\textbf{Raney Ni}]{\textbf{H}_2} \ \underset{\text{a secondary alcohol}}{CH_3CH_2\overset{\overset{\displaystyle OH}{|}}{C}HCH_3}$$

$$\underset{\text{an aldehyde}}{CH_3CH_2CH_2\overset{\overset{\displaystyle O}{\|}}{C}H} \ \xrightarrow[\textbf{Pd/C}]{\textbf{H}_2} \ \underset{\text{a primary alcohol}}{CH_3CH_2CH_2CH_2OH}$$

An acyl chloride is reduced by catalytic hydrogenation to an aldehyde, which is further reduced to a primary alcohol.

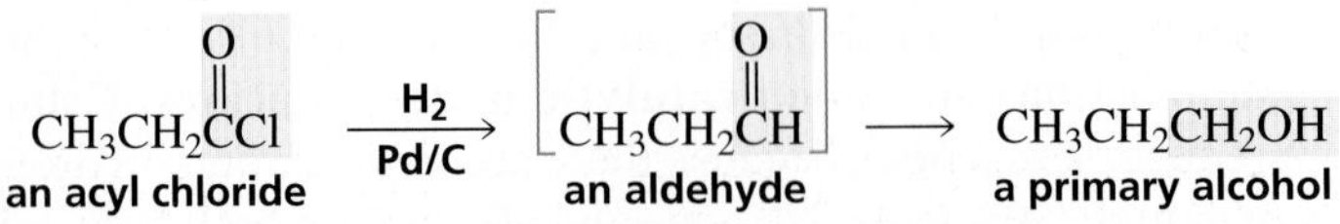

$$\underset{\text{an acyl chloride}}{CH_3CH_2\overset{\overset{\displaystyle O}{\|}}{C}Cl} \ \xrightarrow[\textbf{Pd/C}]{\textbf{H}_2} \ \left[\underset{\text{an aldehyde}}{CH_3CH_2\overset{\overset{\displaystyle O}{\|}}{C}H}\right] \ \longrightarrow \ \underset{\text{a primary alcohol}}{CH_3CH_2CH_2OH}$$

The reaction can be stopped at the aldehyde using a partially deactivated palladium catalyst. This reaction is known as the **Rosenmund reduction.** The catalyst for the Rosenmund reduction is similar to the partially deactivated palladium catalyst used to stop the reduction of an alkyne at a *cis*-alkene (Section 5.8).

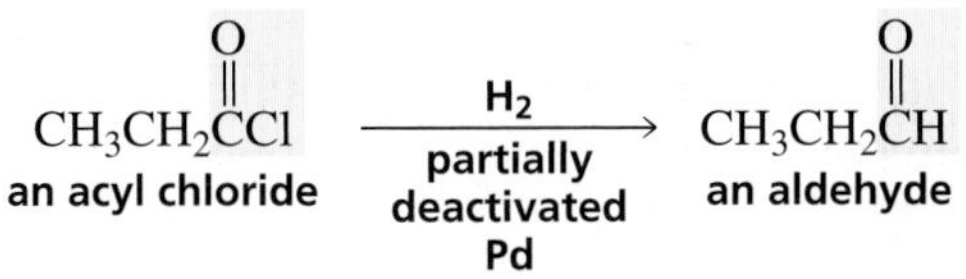

$$\underset{\textbf{an acyl chloride}}{CH_3CH_2\overset{\displaystyle O}{\overset{\displaystyle \|}{C}}Cl} \xrightarrow[\substack{\textbf{partially} \\ \textbf{deactivated} \\ \textbf{Pd}}]{\textbf{H}_2} \underset{\textbf{an aldehyde}}{CH_3CH_2\overset{\displaystyle O}{\overset{\displaystyle \|}{C}}H}$$

Karl W. Rosenmund (1884–1964) was born in Berlin. He was a professor of chemistry at the University of Kiel.

The carbon–oxygen double bonds of carboxylic acids, esters, and amides are less reactive, so they are harder to reduce than the carbon–oxygen double bonds of aldehydes and ketones (Section 17.2). They cannot be reduced by catalytic hydrogenation (except under extreme conditions). They can, however, be reduced by a method we will discuss later in this section. The relative ease of reduction of various functional groups by catalytic hydrogenation is shown in Table 18.1.

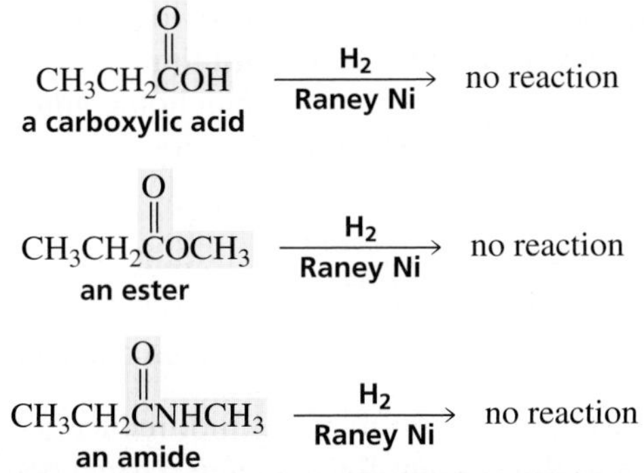

$$\underset{\textbf{a carboxylic acid}}{CH_3CH_2\overset{\displaystyle O}{\overset{\displaystyle \|}{C}}OH} \xrightarrow[\textbf{Raney Ni}]{\textbf{H}_2} \text{no reaction}$$

$$\underset{\textbf{an ester}}{CH_3CH_2\overset{\displaystyle O}{\overset{\displaystyle \|}{C}}OCH_3} \xrightarrow[\textbf{Raney Ni}]{\textbf{H}_2} \text{no reaction}$$

$$\underset{\textbf{an amide}}{CH_3CH_2\overset{\displaystyle O}{\overset{\displaystyle \|}{C}}NHCH_3} \xrightarrow[\textbf{Raney Ni}]{\textbf{H}_2} \text{no reaction}$$

TABLE 18.1 Relative Ease of Reduction of Functional Groups by Catalytic Hydrogenation

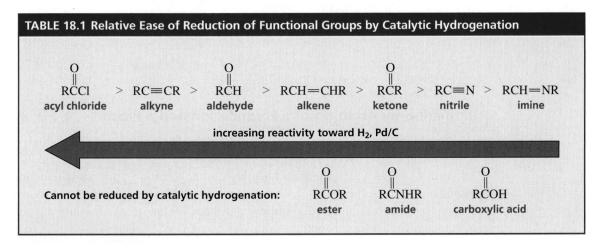

$R\overset{\displaystyle O}{\overset{\displaystyle \|}{C}}Cl$	>	$RC{\equiv}CR$	>	$R\overset{\displaystyle O}{\overset{\displaystyle \|}{C}}H$	>	$RCH{=}CHR$	>	$R\overset{\displaystyle O}{\overset{\displaystyle \|}{C}}R$	>	$RC{\equiv}N$	>	$RCH{=}NR$
acyl chloride		alkyne		aldehyde		alkene		ketone		nitrile		imine

increasing reactivity toward H₂, Pd/C

Cannot be reduced by catalytic hydrogenation: $R\overset{\displaystyle O}{\overset{\displaystyle \|}{C}}OR$ $R\overset{\displaystyle O}{\overset{\displaystyle \|}{C}}NHR$ $R\overset{\displaystyle O}{\overset{\displaystyle \|}{C}}OH$

ester   amide   carboxylic acid

PROBLEM 2◆

Give the products of the following reactions:

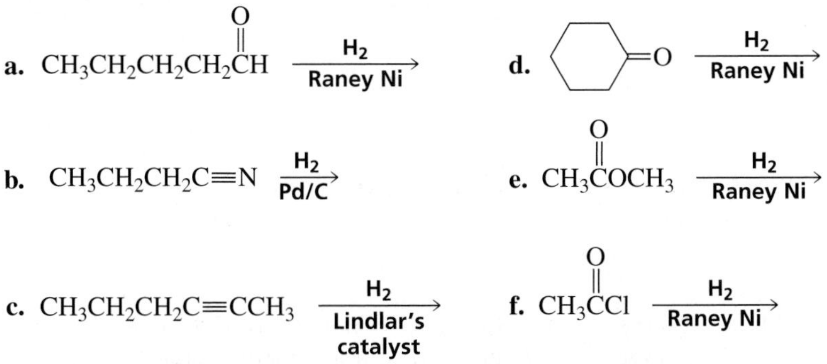

a. $CH_3CH_2CH_2CH_2\overset{\displaystyle O}{\overset{\displaystyle \|}{C}}H \xrightarrow[\textbf{Raney Ni}]{\textbf{H}_2}$

b. $CH_3CH_2CH_2C{\equiv}N \xrightarrow[\textbf{Pd/C}]{\textbf{H}_2}$

c. $CH_3CH_2CH_2C{\equiv}CCH_3 \xrightarrow[\substack{\textbf{Lindlar's} \\ \textbf{catalyst}}]{\textbf{H}_2}$

d. $\underset{}{\bigcirc}{=}O \xrightarrow[\textbf{Raney Ni}]{\textbf{H}_2}$

e. $CH_3\overset{\displaystyle O}{\overset{\displaystyle \|}{C}}OCH_3 \xrightarrow[\textbf{Raney Ni}]{\textbf{H}_2}$

f. $CH_3\overset{\displaystyle O}{\overset{\displaystyle \|}{C}}Cl \xrightarrow[\textbf{Raney Ni}]{\textbf{H}_2}$

$$\textbf{g.} \quad CH_3\overset{\overset{\displaystyle O}{\|}}{C}Cl \xrightarrow[\substack{\text{partially} \\ \text{deactivated} \\ \text{Pd}}]{H_2}$$

$$\textbf{h.} \quad \text{(cyclohexane)}=NCH_3 \xrightarrow{\substack{H_2 \\ \text{Pd/C}}}$$

Reduction by Addition of an Electron, a Proton, an Electron, a Proton

When a compound is reduced using sodium in liquid ammonia, sodium donates an electron to the compound and ammonia donates a proton. This sequence is repeated, so the overall reaction adds two electrons and two protons to the compound. Such a reaction is known as a **dissolving metal reduction.**

We saw that an alkyne is reduced by sodium in liquid ammonia to a *trans*-alkene. The mechanism for this reaction is shown in Section 5.8.

$$CH_3C\equiv CCH_3 \xrightarrow[\substack{\text{NH}_3 \text{ (liq)}}]{\text{Na or Li}} \begin{matrix} H_3C & & H \\ & C=C & \\ H & & CH_3 \end{matrix}$$

2-butyne

trans-**2-butene**

Sodium (or lithium) in liquid ammonia cannot reduce a carbon–carbon double bond. This makes it a useful reagent for reducing a triple bond in a compound that also contains a double bond.

$$\underset{\substack{| \\ CH_3}}{CH_3C}=CHCH_2C\equiv CCH_3 \xrightarrow[\substack{\text{NH}_3 \text{ (liq)}}]{\text{Na or Li}} \underset{\substack{| \\ CH_3}}{CH_3C}=CHCH_2 \begin{matrix} H & & CH_3 \\ & C=C & \\ & & H \end{matrix}$$

Reduction by Addition of a Hydride Ion and a Proton

Carbon–oxygen double bonds are easily reduced by sodium borohydride ($NaBH_4$) or lithium aluminum hydride ($LiAlH_4$). The actual reducing agent in these reductions is hydride ion (H^-). Hydride ion adds to the carbonyl carbon, and the alkoxide ion that is formed is subsequently protonated by water. In other words, the carbonyl group is reduced by adding an H^- followed by an H^+. The mechanism for reduction by these reagents is discussed in Sections 17.5 and 17.6.

$$H:^- \quad \overset{\overset{\displaystyle O}{\|}}{C} \longrightarrow \underset{|}{\overset{|}{\underset{\displaystyle O^-}{C}}} \xrightarrow{H_2O} \underset{|}{\overset{|}{\underset{\displaystyle OH}{C}}} + \quad HO^-$$

Aldehydes and ketones are reduced by sodium borohydride.

$$CH_3CH_2CH_2\overset{\overset{\displaystyle O}{\|}}{C}H \xrightarrow[\substack{\text{2. H}^+, \text{H}_2\text{O}}]{\text{1. NaBH}_4} CH_3CH_2CH_2CH_2OH$$

an aldehyde

a primary alcohol

$$CH_3CH_2CH_2\overset{\overset{\displaystyle O}{\|}}{C}CH_3 \xrightarrow[\substack{\text{2. H}^+, \text{H}_2\text{O}}]{\text{1. NaBH}_4} \underset{\substack{| \\ OH}}{CH_3CH_2CH_2CHCH_3}$$

a ketone

a secondary alcohol

Remember that the numbers in front of the reagents above or below a reaction arrow indicate that the second reagent is not added until reaction with the first reagent is completely over.

The metal–hydrogen bonds in lithium aluminum hydride are more polar than the metal–hydrogen bonds in sodium borohydride. As a result, $LiAlH_4$ is a stronger reducing agent than $NaBH_4$. Consequently, both $LiAlH_4$ and $NaBH_4$ reduce aldehydes and ketones, but $LiAlH_4$ is not often used for this purpose because $NaBH_4$ is safer and easier to use. $LiAlH_4$ is mainly used to reduce only those compounds that cannot be reduced by the milder reagent, such as carboxylic acids, esters, and amides.

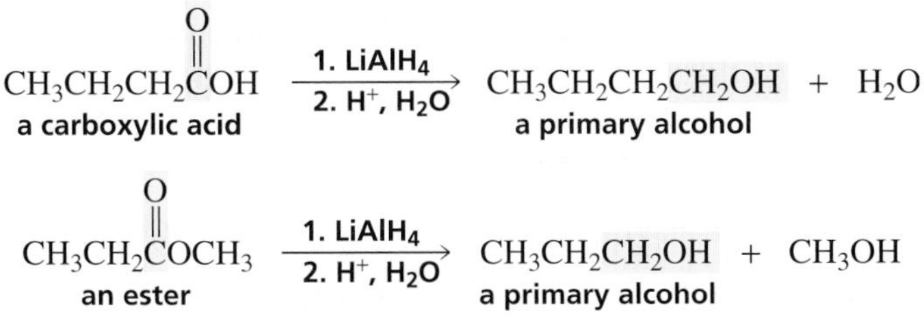

$$CH_3CH_2CH_2\overset{\overset{\displaystyle O}{\|}}{C}OH \xrightarrow[\text{2. H}^+, \text{H}_2\text{O}]{\text{1. LiAlH}_4} CH_3CH_2CH_2CH_2OH \ + \ H_2O$$
a carboxylic acid   a primary alcohol

$$CH_3CH_2\overset{\overset{\displaystyle O}{\|}}{C}OCH_3 \xrightarrow[\text{2. H}^+, \text{H}_2\text{O}]{\text{1. LiAlH}_4} CH_3CH_2CH_2OH \ + \ CH_3OH$$
an ester   a primary alcohol

The carbonyl group of an amide is reduced to a methylene group (CH_2) by lithium aluminum hydride. Primary, secondary, and tertiary amines are formed, depending on the number of substituents bonded to the nitrogen of the amide. In order to obtain a neutral (nonprotonated) amine, acid is not used in the second step of the reaction.

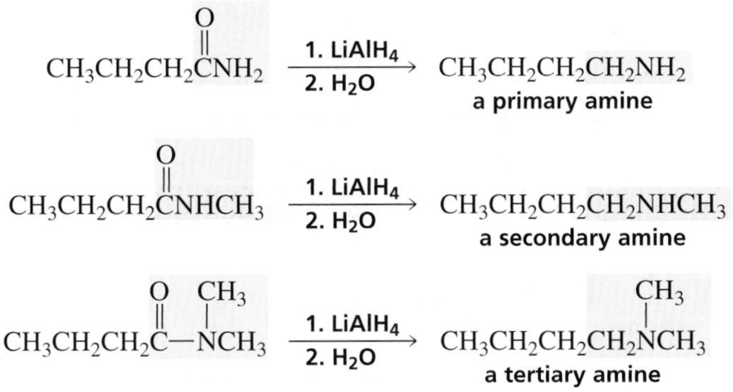

Tutorial:
Reductions

$$CH_3CH_2CH_2\overset{\overset{\displaystyle O}{\|}}{C}NH_2 \xrightarrow[\text{2. H}_2\text{O}]{\text{1. LiAlH}_4} CH_3CH_2CH_2CH_2NH_2$$
a primary amine

$$CH_3CH_2CH_2\overset{\overset{\displaystyle O}{\|}}{C}NHCH_3 \xrightarrow[\text{2. H}_2\text{O}]{\text{1. LiAlH}_4} CH_3CH_2CH_2CH_2NHCH_3$$
a secondary amine

$$CH_3CH_2CH_2\overset{\overset{\displaystyle O}{\|}}{C}\overset{\overset{\displaystyle CH_3}{|}}{-}NCH_3 \xrightarrow[\text{2. H}_2\text{O}]{\text{1. LiAlH}_4} CH_3CH_2CH_2CH_2\overset{\overset{\displaystyle CH_3}{|}}{N}CH_3$$
a tertiary amine

Because sodium borohydride cannot reduce an ester, an amide, or a carboxylic acid, it can be used to selectively reduce an aldehyde or a ketone group in a compound that also contains one of these less reactive groups. Acid is not used in the second step of the following reaction in order to avoid hydrolyzing the ester.

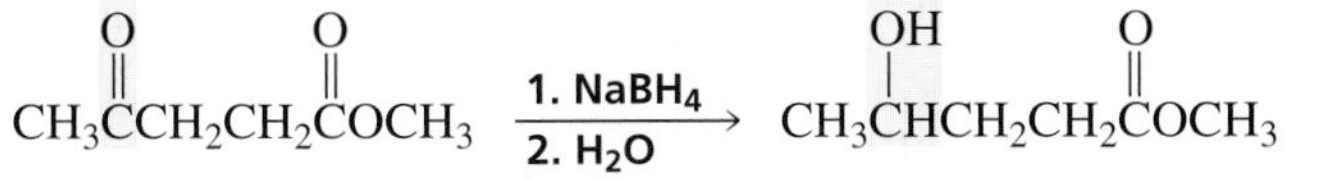

$$CH_3\overset{\overset{\displaystyle O}{\|}}{C}CH_2CH_2\overset{\overset{\displaystyle O}{\|}}{C}OCH_3 \xrightarrow[\text{2. H}_2\text{O}]{\text{1. NaBH}_4} CH_3\overset{\overset{\displaystyle OH}{|}}{C}HCH_2CH_2\overset{\overset{\displaystyle O}{\|}}{C}OCH_3$$

We have seen that sterically bulky hydride donors can be used to deliver only one equivalent of hydride ion. For example, diisobutylaluminum hydride reduces an ester to an aldehyde, and lithium tri-*tert*-butoxyaluminum hydride reduces an acyl chloride to an aldehyde (Section 17.6).

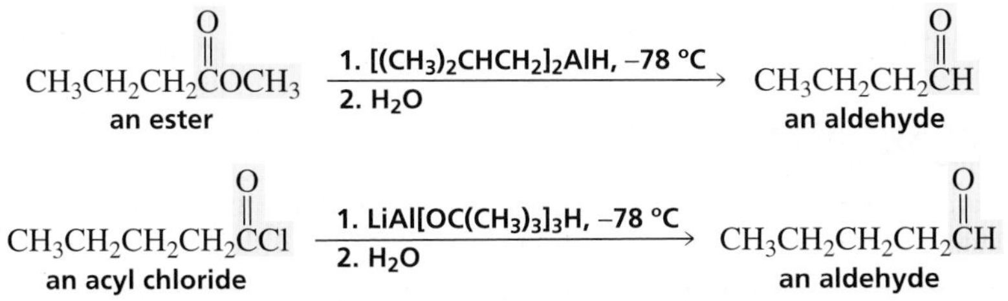

$$CH_3CH_2CH_2\overset{\overset{\displaystyle O}{\|}}{C}OCH_3 \xrightarrow[\text{2. H}_2\text{O}]{\text{1. [(CH}_3)_2\text{CHCH}_2]_2\text{AlH, }-78\ °\text{C}} CH_3CH_2CH_2\overset{\overset{\displaystyle O}{\|}}{C}H$$
an ester   an aldehyde

$$CH_3CH_2CH_2CH_2\overset{\overset{\displaystyle O}{\|}}{C}Cl \xrightarrow[\text{2. H}_2\text{O}]{\text{1. LiAl[OC(CH}_3)_3]_3\text{H, }-78\ °\text{C}} CH_3CH_2CH_2CH_2\overset{\overset{\displaystyle O}{\|}}{C}H$$
an acyl chloride   an aldehyde

The multiply bonded carbon atoms of alkenes and alkynes do not possess a partial positive charge and therefore will not react with reagents that reduce compounds by donating a hydride ion.

$$CH_3CH_2CH{=}CH_2 \xrightarrow{\textbf{NaBH}_4} \text{no reduction reaction}$$

$$CH_3CH_2C{\equiv}CH \xrightarrow{\textbf{NaBH}_4} \text{no reduction reaction}$$

Because sodium borohydride cannot reduce carbon–carbon double bonds, a carbonyl group in a compound that also has an alkene functional group can be selectively reduced. If the double bonds are conjugated, however, some product in which both functional groups are reduced is also formed (Section 17.14).

$$CH_3CH{=}CHCH_2\overset{\overset{\displaystyle O}{\|}}{C}CH_3 \xrightarrow[\textbf{2. H}_2\textbf{O}]{\textbf{1. NaBH}_4} CH_3CH{=}CHCH_2\overset{\overset{\displaystyle OH}{|}}{C}HCH_3$$

The relative ease of reduction of various functional groups by the addition of hydride ion is shown in Table 18.2.

TABLE 18.2 Relative Ease of Reduction of Functional Groups by Addition of Hydride Ion

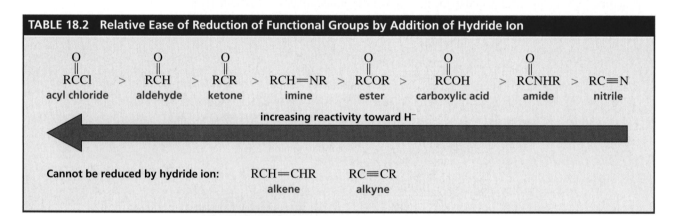

Cannot be reduced by hydride ion: RCH=CHR RC≡CR
alkene   alkyne

PROBLEM 3

Explain why terminal alkynes cannot be reduced by Na in liquid NH_3.

PROBLEM 4◆

Give the products of the following reactions:

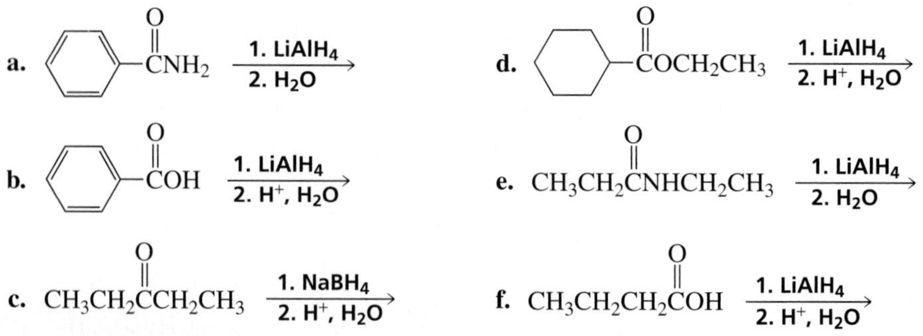

a. [benzene ring]—$\overset{\overset{\displaystyle O}{\|}}{C}NH_2 \xrightarrow[\textbf{2. H}_2\textbf{O}]{\textbf{1. LiAlH}_4}$

b. [benzene ring]—$\overset{\overset{\displaystyle O}{\|}}{C}OH \xrightarrow[\textbf{2. H}^+, \textbf{H}_2\textbf{O}]{\textbf{1. LiAlH}_4}$

c. $CH_3CH_2\overset{\overset{\displaystyle O}{\|}}{C}CH_2CH_3 \xrightarrow[\textbf{2. H}^+, \textbf{H}_2\textbf{O}]{\textbf{1. NaBH}_4}$

d. [cyclohexane ring]—$\overset{\overset{\displaystyle O}{\|}}{C}OCH_2CH_3 \xrightarrow[\textbf{2. H}^+, \textbf{H}_2\textbf{O}]{\textbf{1. LiAlH}_4}$

e. $CH_3CH_2\overset{\overset{\displaystyle O}{\|}}{C}NHCH_2CH_3 \xrightarrow[\textbf{2. H}_2\textbf{O}]{\textbf{1. LiAlH}_4}$

f. $CH_3CH_2CH_2\overset{\overset{\displaystyle O}{\|}}{C}OH \xrightarrow[\textbf{2. H}^+, \textbf{H}_2\textbf{O}]{\textbf{1. LiAlH}_4}$

PROBLEM 5

Can carbon–nitrogen double and triple bonds be reduced by lithium aluminum hydride? Explain your answer.

PROBLEM 6

Give the products of the following reactions. Assume excess reducing agent is used in **d**.

a.

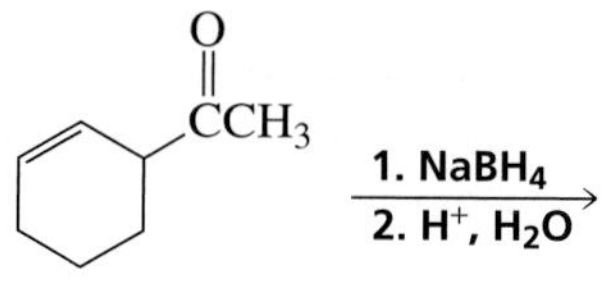

$$\underset{\text{CCH}_3}{\overset{O}{\parallel}}$$ → 1. NaBH₄ / 2. H⁺, H₂O

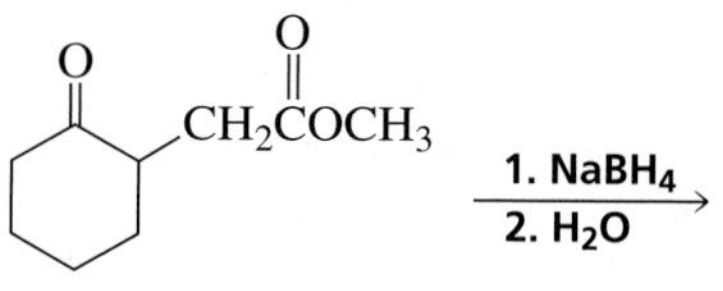

c.

$$\overset{O}{\parallel} \quad \text{CH}_2\text{COCH}_3 \overset{O}{\parallel}$$ → 1. NaBH₄ / 2. H₂O

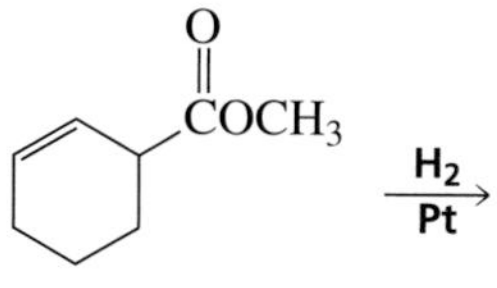

b.

$$\underset{\text{COCH}_3}{\overset{O}{\parallel}}$$ → H₂ / Pt

d.

$$\overset{O}{\parallel} \quad \text{CH}_2\text{COCH}_3 \overset{O}{\parallel}$$ → 1. LiAlH₄ / 2. H₂O

Oxidation is the reverse of **reduction**. For example, a ketone can be reduced to a secondary alcohol and the reverse reaction is the oxidation of a secondary alcohol to a ketone.

$$\text{ketone} \underset{\text{oxidation}}{\overset{\text{reduction}}{\rightleftarrows}} \text{secondary alcohol}$$

The reagent most commonly used to oxidize alcohols is chromic acid (H_2CrO_4), which is formed when chromium trioxide (CrO_3) or sodium dichromate $(Na_2Cr_2O_7)$ is dissolved in aqueous acid. These reactions are easily recognized as oxidations because the number of carbon–hydrogen bonds in the reactant decreases (and the number of carbon–oxygen bonds increases).

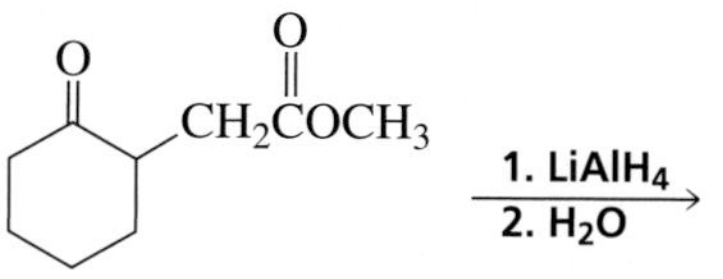

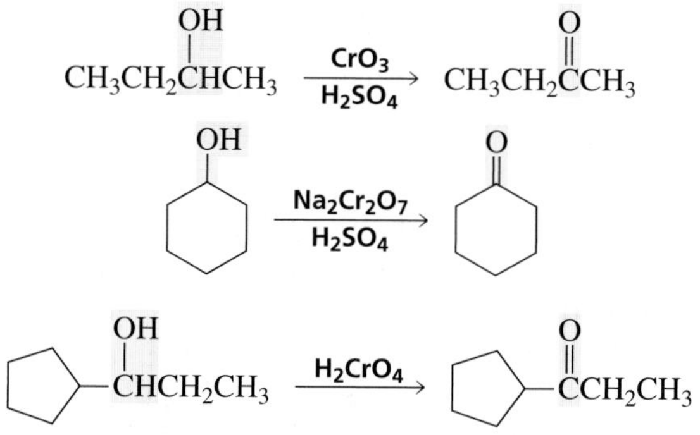

18.2

OXIDATION OF ALCOHOLS

3-D Molecules:
Chromic acid;
Chromium trioxide;
Sodium dichromate

$$\text{CH}_3\text{CH}_2\overset{\overset{\text{OH}}{|}}{\text{CHCH}_3} \xrightarrow[\text{H}_2\text{SO}_4]{\text{CrO}_3} \text{CH}_3\text{CH}_2\overset{\overset{O}{\parallel}}{\text{CCH}_3}$$

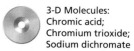

$$\xrightarrow[\text{H}_2\text{SO}_4]{\text{Na}_2\text{Cr}_2\text{O}_7}$$

$$\underset{\text{CHCH}_2\text{CH}_3}{\overset{\overset{\text{OH}}{|}}{}} \xrightarrow{\text{H}_2\text{CrO}_4} \underset{\text{CCH}_2\text{CH}_3}{\overset{\overset{O}{\parallel}}{}}$$

secondary alcohols   **ketones**

Primary alcohols are initially oxidized to aldehydes by these reagents. The reaction, however, does not stop at the aldehyde. The aldehyde is further oxidized to a carboxylic acid.

$$\underset{\textbf{a primary alcohol}}{\text{CH}_3\text{CH}_2\text{CH}_2\text{CH}_2\text{OH}} \xrightarrow{\text{H}_2\text{CrO}_4} \left[\underset{\textbf{an aldehyde}}{\text{CH}_3\text{CH}_2\text{CH}_2\overset{\overset{O}{\parallel}}{\text{CH}}}\right] \xrightarrow[\text{oxidation}]{\text{further}} \underset{\textbf{a carboxylic acid}}{\text{CH}_3\text{CH}_2\text{CH}_2\overset{\overset{O}{\parallel}}{\text{COH}}}$$

Notice that the oxidation of either a primary or a secondary alcohol involves removal of a hydrogen from the carbon to which the oxygen is attached. In a tertiary alcohol, the carbon bearing the OH group has no hydrogen, so the OH group of a tertiary alcohol cannot be oxidized to a carbonyl group.

BLOOD ALCOHOL CONTENT

As blood passes through the arteries in the lungs, an equilibrium is established between the alcohol in one's blood and the alcohol in one's breath. So if the concentration of one is known, the concentration of the other can be estimated. The test that law enforcement agencies use to approximate a person's blood alcohol level is based on the oxidation of breath ethanol by sodium dichromate. A sealed glass tube is used, which contains the oxidizing agent impregnated onto an inert material. The ends of the tube are broken off. One end of the tube is attached to a mouthpiece and the other to a balloon-type bag. The person undergoing the test blows into the mouthpiece until the bag is filled with air.

glass tube containing sodium dichromate-sulfuric acid coated on silica gel particles

person breathes into mouthpiece

as person blows into the tube, the plastic bag becomes inflated

Any ethanol in the breath is oxidized as it passes through the column. When ethanol is oxidized, the red-orange oxidizing agent $(Cr_2O_7^{2-})$ is reduced to green chromic ion. The greater the concentration of alcohol in the breath, the farther the green color spreads through the tube.

If the person fails this test—determined by the extent to which the green color spreads through the tube—a more accurate Breathalyzer test is administered. This test also depends on the oxidation of breath ethanol by sodium dichromate, but because it is quantitative, it provides more accurate results. Here, a known volume of breath is bubbled through an acidic solution of sodium dichromate. The concentration of chromic ion is measured precisely using a spectrophotometer.

$$CH_3CH_2OH \ + \ \underset{\textbf{red-orange}}{Cr_2O_7^{2-}} \ \xrightarrow{H^+} \ CH_3\overset{\overset{\textstyle O}{\|}}{C}OH \ + \ \underset{\textbf{green}}{Cr^{3+}}$$

Chromic acid and the other chromium-containing oxidizing reagents oxidize an alcohol by first forming a chromate ester. The carbonyl compound is formed when the chromate ester undergoes an E2 elimination.

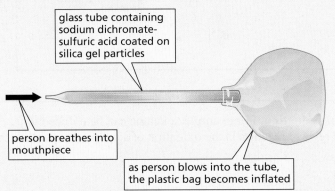

$$RCH{=}O \ + \ H_2CrO_3 \ + \ H_2O$$

The oxidation of a primary alcohol can be easily stopped at the aldehyde if *pyridinium chlorochromate* (PCC) is used as the oxidizing agent and the reaction is carried out in an anhydrous solvent such as dichloromethane.

$$\underset{\textbf{a primary alcohol}}{CH_3CH_2CH_2CH_2OH} \ \xrightarrow[\textbf{CH}_2\textbf{Cl}_2]{\textbf{PCC}} \ \underset{\textbf{an aldehyde}}{CH_3CH_2CH_2\overset{\overset{\textstyle O}{\|}}{C}H}$$

THE ROLE OF HYDRATES IN THE OXIDATION OF PRIMARY ALCOHOLS

When a primary alcohol is oxidized to a carboxylic acid, the alcohol is initially oxidized to an aldehyde. The aldehyde is in equilibrium with its hydrate because the reaction is carried out in an acidic aqueous solution. It is the hydrate that is subsequently oxidized to a carboxylic acid.

The oxidation reaction can be stopped at the aldehyde if the reaction is carried out with pyridinium chlorochromate (PCC) because PCC is used in an anhydrous solvent. If water is not present, the hydrate cannot be formed.

$$CH_3CH_2OH \xrightarrow{H_2CrO_4} CH_3\overset{O}{\overset{\|}{C}}H \underset{H_2O}{\overset{H^+}{\rightleftharpoons}} CH_3\underset{OH}{\overset{OH}{\underset{|}{\overset{|}{C}H}}} \xrightarrow{H_2CrO_4} CH_3\overset{O}{\overset{\|}{C}}OH$$

Because of the toxicity of reagents that contain chromium, other methods for the oxidation of alcohols have been developed. One of the most widely used methods—called the **Swern oxidation**—uses dimethyl sulfoxide $[(CH_3)_2SO]$, oxalyl chloride $[(COCl)_2]$, and triethylamine. Since the reaction is *not* carried out in an acidic aqueous solution, the oxidation of a primary alcohol (like PCC oxidation) stops at the aldehyde. Secondary alcohols are oxidized to ketones.

3-D Molecules:
Pyridinium chlorochromate (PCC);
Dimethyl sulfoxide;
Oxalyl chloride

$$CH_3CH_2CH_2OH \xrightarrow[\text{2. triethylamine}]{\substack{\text{1. } CH_3SCH_3, \quad Cl-\overset{O}{\overset{\|}{C}}-\overset{O}{\overset{\|}{C}}-Cl, -60\,°C}} CH_3CH_2\overset{O}{\overset{\|}{C}}H$$

a primary alchohol   an aldehyde

$$CH_3CH_2\underset{OH}{\overset{OH}{\underset{|}{\overset{|}{C}HCH_3}}} \xrightarrow[\text{2. triethylamine}]{\substack{\text{1. } CH_3SCH_3, \quad Cl-\overset{O}{\overset{\|}{C}}-\overset{O}{\overset{\|}{C}}-Cl, -60\,°C}} CH_3CH_2\overset{O}{\overset{\|}{C}}CH_3$$

a secondary alchohol   a ketone

The actual oxidizing agent in the Swern oxidation is dimethylchlorosulfonium ion, which is formed from the reaction of dimethyl sulfoxide and oxalyl chloride. Like chromic acid oxidation, the Swern oxidation uses an E2 reaction to form the aldehyde or ketone.

alcohol   dimethylchlorosulfonium ion $(CH_3CH_2)_3\overset{..}{N}$ triethylamine   aldehyde or ketone

To understand how dimethyl sulfoxide and oxalyl chloride react to form the dimethylchlorosulfonium ion, see Problem 56.

PROBLEM 7◆

Give the product formed from the reaction of each of the following alcohols with:

a. an acidic solution of sodium dichromate.   **b.** a Swern oxidation.

1. 3-pentanol   **3.** 2-methyl-2-pentanol   **5.** cyclohexanol

2. 1-pentanol   **4.** 2,4-hexanediol   **6.** 1,4-butanediol

PROBLEM 8 / SOLVED

How could butanone be prepared from butane?

$$CH_3CH_2CH_2CH_3 \longrightarrow CH_3\overset{\overset{\displaystyle O}{\|}}{C}CH_2CH_3$$

SOLUTION   Because the starting material is an alkane, we know that the first reaction has to be a radical halogenation because this is the only reaction that an alkane undergoes. Because of the greater selectivity of the bromine radical, bromination will lead to a greater yield of the desired 2-halo-substituted compound than would chlorination. Reaction of 2-bromobutane with HO^- forms a secondary alcohol, which forms the target compound when it is oxidized.

$$CH_3CH_2CH_2CH_3 \xrightarrow[h\nu]{Br_2} CH_3\overset{\overset{\displaystyle Br}{|}}{C}HCH_2CH_3 \xrightarrow{HO^-} CH_3\overset{\overset{\displaystyle OH}{|}}{C}HCH_2CH_3 \xrightarrow[H_2SO_4]{Na_2Cr_2O_7} CH_3\overset{\overset{\displaystyle O}{\|}}{C}C$$

PROBLEM 9

Propose a mechanism for the chromic acid oxidation of 1-propanol to propanoic acid.

18.3 OXIDATION OF ALDEHYDES AND KETONES

Aldehydes are oxidized to carboxylic acids. Because aldehydes are generally easier to oxidize than primary alcohols, any of the reagents described in the preceding section for oxidizing primary alcohols to carboxylic acids can be used to oxidize aldehydes to carboxylic acids.

$$CH_3CH_2\overset{\overset{\displaystyle O}{\|}}{C}H \xrightarrow[H_2SO_4]{Na_2Cr_2O_7} CH_3CH_2\overset{\overset{\displaystyle O}{\|}}{C}OH$$

aldehydes   **carboxylic acids**

Bernhard Tollens (1841–1918) was born in Germany. He was a professor of chemistry at the University of Göttingen, the same university from which he received a Ph.D.

Silver oxide is a mild oxidizing agent. A dilute solution of silver oxide in aqueous ammonia (*Tollens' reagent*) will oxidize an aldehyde, but it is too weak to oxidize an alcohol or any other functional group. An advantage to using Tollens' reagent to oxidize an aldehyde is that the reaction occurs under basic conditions. Therefore, you do not have to worry about harming other functional groups in the molecule that may react in an acidic solution.

$$CH_3CH_2\overset{\overset{\displaystyle O}{\|}}{C}H \xrightarrow[\text{2. } H^+, H_2O]{\text{1. } Ag_2O, NH_3} CH_3CH_2\overset{\overset{\displaystyle O}{\|}}{C}OH + \underset{\substack{\text{metallic} \\ \text{silver}}}{Ag}$$

The oxidizing agent in Tollens' reagent is Ag^+, which is reduced to metallic silver. This forms the basis of the **Tollens test.** If Tollens' reagent is added to a small amount of an aldehyde in a test tube, the inside of the test tube becomes coated with a shiny mirror of metallic silver. If a mirror is not formed when Tollens' reagent is added to a compound, we can conclude that the compound does not have an aldehyde functional group.

Ketones do not react with most of the reagents used to oxidize aldehydes. However, both aldehydes and ketones can be oxidized by a peroxyacid. Aldehydes are oxidized to carboxylic acids and ketones are oxidized to esters. A **peroxyacid** (also called a percarboxylic acid or an acyl hydroperoxide) contains one more oxygen than a carboxylic acid, and it is this oxygen that is inserted between the carbonyl carbon and the H of an aldehyde or the R of a ketone. This reaction is called a **Baeyer–Villiger oxidation.**

Johann Friedrich Wilhelm Adolf von Baeyer (1835–1917) started his study of chemistry under Bunsen and Kekulé at the University of Heidelberg and received a Ph.D. from the University of Berlin, studying under Hofmann (see also Section 2.11).

Victor Villiger (1868–1934) was Baeyer's student. They first published the Baeyer–Villiger oxidation in Chemische Berichte *in 1899.*

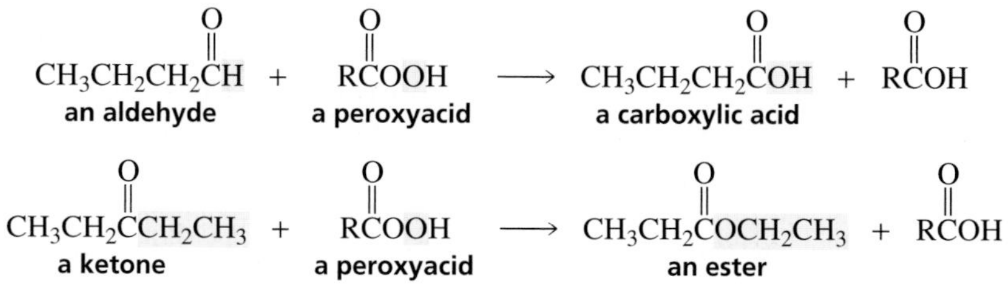

$$
\underset{\text{an aldehyde}}{CH_3CH_2CH_2\overset{O}{\overset{\|}{C}}H} + \underset{\text{a peroxyacid}}{R\overset{O}{\overset{\|}{C}}OOH} \longrightarrow \underset{\text{a carboxylic acid}}{CH_3CH_2CH_2\overset{O}{\overset{\|}{C}}OH} + R\overset{O}{\overset{\|}{C}}OH
$$

$$
\underset{\text{a ketone}}{CH_3CH_2\overset{O}{\overset{\|}{C}}CH_2CH_3} + \underset{\text{a peroxyacid}}{R\overset{O}{\overset{\|}{C}}OOH} \longrightarrow \underset{\text{an ester}}{CH_3CH_2\overset{O}{\overset{\|}{C}}OCH_2CH_3} + R\overset{O}{\overset{\|}{C}}OH
$$

If the two alkyl substituents attached to the carbonyl group of the ketone are not the same, on which side of the carbonyl carbon is the oxygen inserted? For example, does the oxidation of cyclohexyl methyl ketone form methyl cyclohexanecarboxylate or cyclohexyl acetate?

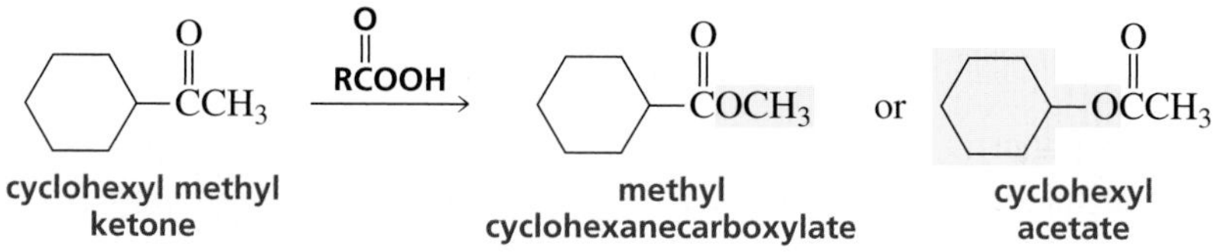

$$
\underset{\substack{\text{cyclohexyl methyl}\\\text{ketone}}}{\text{⬡}-\overset{O}{\overset{\|}{C}}CH_3} \xrightarrow{\overset{O}{\overset{\|}{R\text{C}}}OOH} \underset{\substack{\text{methyl}\\\text{cyclohexanecarboxylate}}}{\text{⬡}-\overset{O}{\overset{\|}{C}}OCH_3} \quad \text{or} \quad \underset{\substack{\text{cyclohexyl}\\\text{acetate}}}{\text{⬡}-O\overset{O}{\overset{\|}{C}}CH_3}
$$

Before we can answer this question, we must look at the mechanism of the reaction. The ketone and the peroxyacid react to form an unstable tetrahedral intermediate. The oxygen–oxygen bond in this intermediate is very weak. As it breaks heterolytically, one of the groups bonded to the carbonyl group of the ketone migrates to the oxygen. This is similar to the 1,2-shifts that occur when carbocations rearrange (Section 3.14).

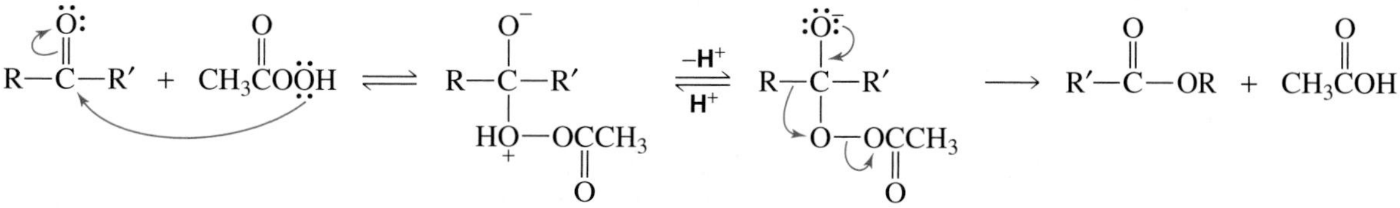

$$
\underset{}{R-\overset{\overset{\ddot{O}:}{\|}}{C}-R'} + CH_3\overset{O}{\overset{\|}{C}}\overset{..}{O}\overset{..}{H} \rightleftharpoons R-\overset{\overset{O^-}{|}}{\underset{\underset{\overset{\|}{O}}{HO-OCCH_3}}{C}}-R' \underset{H^+}{\overset{-H^+}{\rightleftharpoons}} R-\overset{\overset{:\ddot{O}:}{|}}{\underset{\underset{\overset{\|}{O}}{O-OCCH_3}}{C}}-R' \longrightarrow R'-\overset{O}{\overset{\|}{C}}-OR + CH_3\overset{O}{\overset{\|}{C}}OH
$$

Several studies have established the following order of group migration tendencies:

relative migration tendencies

$$H > \textit{tert}\text{-alkyl} > \textit{sec}\text{-alkyl} = \text{phenyl} > \text{primary alkyl} > \text{methyl}$$

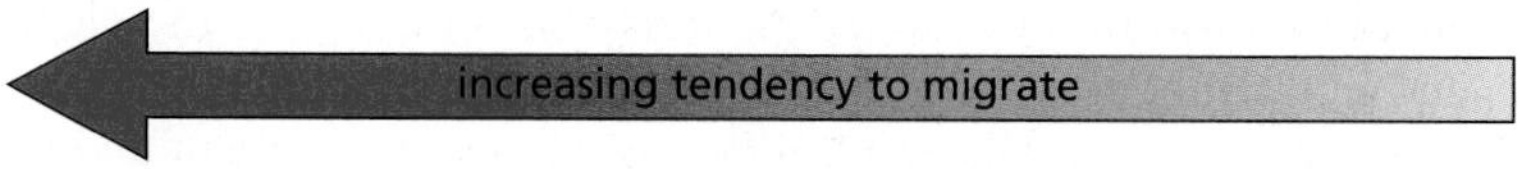

⬅ increasing tendency to migrate

Therefore, the product of the Baeyer–Villiger oxidation of cyclohexyl methyl ketone will be cyclohexyl acetate because a secondary alkyl group (the cyclohexyl group) is more likely to migrate than a methyl group. Because H has the greatest tendency to migrate, aldehydes are always oxidized to carboxylic acids.

PROBLEM 10

Give the products of the following reactions:

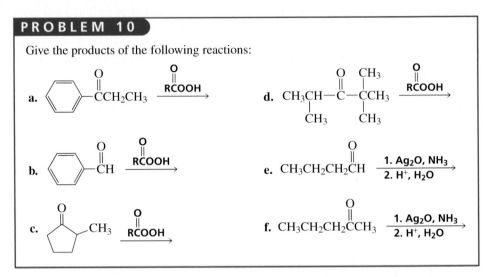

18.4
OXIDATION OF ALKENES WITH PEROXYACIDS

An alkene can be oxidized to an epoxide by a peroxyacid. The overall reaction amounts to the transfer of an oxygen atom from the peroxyacid to the alkene.

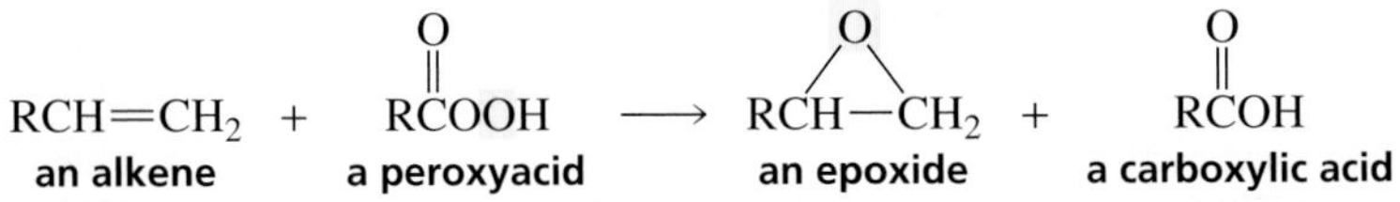

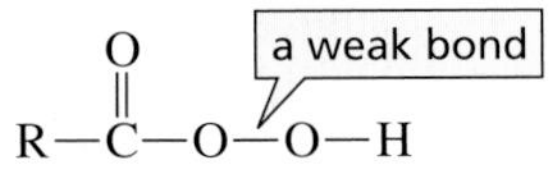

$$\underset{\text{an alkene}}{RCH=CH_2} + \underset{\text{a peroxyacid}}{RCOOH} \longrightarrow \underset{\text{an epoxide}}{RCH-CH_2} + \underset{\text{a carboxylic acid}}{RCOH}$$

Recall than an oxygen–oxygen single bond is weak and easily broken (Section 18.3). The oxygen atom of the OH group of the peroxyacid accepts a pair of electrons from the π bond of the alkene, causing the weak oxygen–oxygen single bond to break. The electrons from the oxygen–oxygen bond are delocalized onto the carbonyl group. The electrons left behind as the oxygen–hydrogen bond breaks add to the carbon of the alkene that becomes electron deficient when the π bond breaks. Notice that **epoxidation** of an alkene is a concerted reaction—all of the bond-forming and bond-breaking processes take place in a single step.

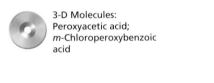

3-D Molecules:
Peroxyacetic acid;
m-Chloroperoxybenzoic acid

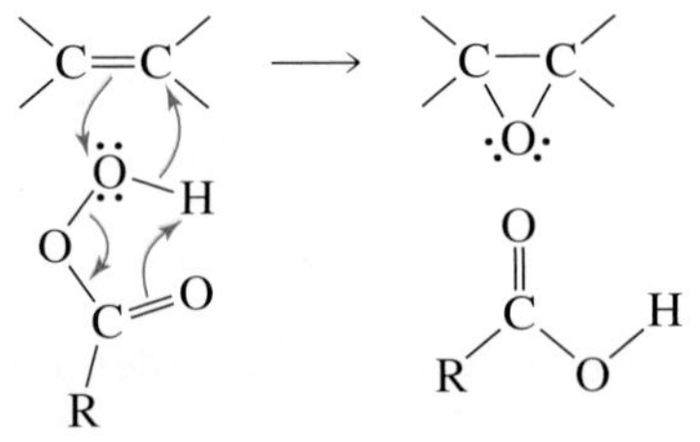

The mechanism for the addition of oxygen to a double bond to form an epoxide is analogous to the mechanism described in Section 3.15 for the addition of bromine to a double bond to form a cyclic bromonium ion. In one case the electrophile is oxygen, and in the other it is bromine. So the reaction of an alkene with a peroxyacid, like the reaction of an alkene with Br_2, is another example of an electrophilic addition reaction.

$$\text{>C=C<} \longrightarrow \text{>C—C<} + \quad :\ddot{Br}:^-$$

Because epoxidation results from the reaction of an electrophilic oxygen atom with a nucleophilic alkene, increasing the electron density of the double bond of the alkene increases the rate of epoxidation. Alkyl substituents increase the electron density of the double bond. Therefore, if a diene is treated with only enough peroxyacid to react with one of the double bonds, the most substituted double bond will be the bond that is epoxidized.

$$\text{limonene} \quad + \quad \underset{\text{one equivalent}}{\text{RCOOH}} \quad \longrightarrow \quad \text{(epoxide)} \quad + \quad \text{RCOH}$$

limonene

The addition of oxygen to an alkene is a stereospecific reaction. If the reactant has the cis configuration, the epoxide will also have the cis configuration. Similarly, a *trans*-alkene forms a *trans*-epoxide.

$$\underset{\text{cis-2-butene}}{\overset{H}{\underset{H_3C}{>}}C=C\overset{H}{\underset{CH_3}{<}}} \quad \xrightarrow{\text{RCOOH}} \quad \underset{\text{cis-2,3-dimethyloxirane}}{H\cdots\overset{O}{C}-\overset{}{C}\cdots H \atop H_3C \qquad CH_3}$$

$$\underset{\text{trans-2-butene}}{\overset{H}{\underset{H_3C}{>}}C=C\overset{CH_3}{\underset{H}{<}}} \quad \xrightarrow{\text{RCOOH}} \quad \underset{\text{trans-2,3-dimethyloxirane}}{H\cdots\overset{O}{C}-\overset{}{C}\cdots CH_3 \atop H_3C \qquad H}$$

3-D Molecules:
cis-2,3-Dimethyloxirane;
trans-2,3-Dimethyloxirane

PROBLEM 11◆

What alkene would you treat with a peroxyacid in order to obtain each of the following epoxides?

a. (cyclohexene oxide)

c. $H\cdots\overset{O}{C}-\overset{}{C}\cdots CH_2CH_3 \atop H_3C \qquad H$

b. $H_2C\overset{O}{-}CHCH_2CH_3$

d. $H\cdots\overset{O}{C}-\overset{}{C}\cdots H \atop H_3C \qquad CH_2CH_3$

PROBLEM 12

Give the major product of the reaction of each of the following compounds with one equivalent of a peroxyacid. Indicate the configuration of the product.

a. $\underset{H_3C}{\overset{H}{>}}C=C\overset{CH_3}{\underset{H}{<}}$

b. (cyclohexadiene with two CH₃ groups)

c. $CH_2{=}CH\underset{|\atop CH_3}{C}{=}CH_2$

d.

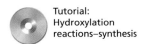

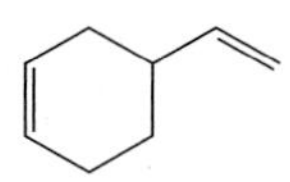

PROBLEM 13

Show how the following target molecules could be synthesized from propene:

a. 1-methoxy-2-propanol **b.** 2-butanol **c.** 2-butanone

PROBLEM 14

Explain why an epoxide is a relatively stable product while a bromonium ion is a reactive intermediate.

18.5 HYDROXYLATION OF ALKENES

An alkene can be oxidized to a 1,2-diol either by potassium permanganate ($KMnO_4$) in a cold basic solution or by osmium tetroxide (OsO_4). The solution of potassium permanganate must be basic and the oxidation must be carried out at room temperature or below. If the solution is heated or if it is acidic, the diol will be further oxidized (Section 18.7). A diol can also be called a **glycol.** The OH groups are on adjacent carbons in 1,2-diols, so 1,2-diols are also known as **vicinal diols** or **vicinal glycols.**

$$CH_3CH{=}CHCH_3 \xrightarrow[\text{cold}]{KMnO_4,\ HO^-,\ H_2O} CH_3\underset{}{\overset{OH}{\overset{|}{C}}}H{-}\overset{OH}{\overset{|}{C}}HCH_3$$

a vicinal diol

$$CH_3CH_2CH{=}CH_2 \xrightarrow[2.\ NaHSO_3,\ H_2O]{1.\ OsO_4} CH_3CH_2\overset{OH}{\overset{|}{C}}HCH_2OH$$

a vicinal diol

Tutorial: Hydroxylation reactions–synthesis

Both $KMnO_4$ and OsO_4 form a cyclic intermediate when they react with an alkene. The reactions occur because manganese and osmium are in a highly positive oxidation state (+7 and +8, respectively) and therefore attract electrons. Formation of the cyclic intermediate is a syn addition because both oxygens are delivered to the same side of the double bond. Therefore, the oxidation reaction is a stereoselective reaction—a cycloalkene forms only a *cis*-diol.

cyclopentene a cyclic manganate intermediate *cis*-1,2-cyclopentanediol

3-D Molecules: Potassium permanganate ($KMnO_4$); Osmium tetroxide (OsO_4)

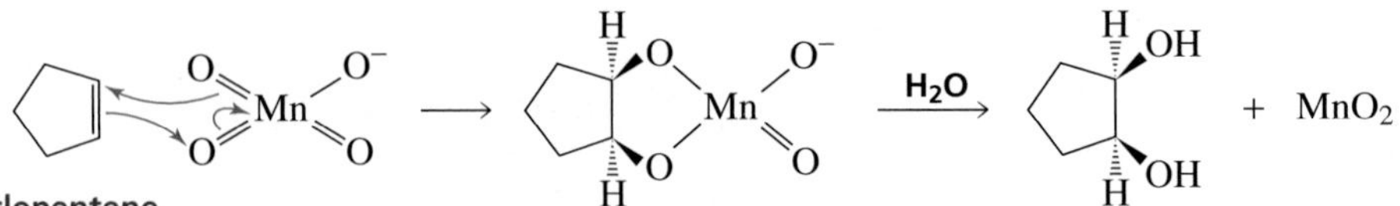

Notice that aqueous sodium bisulfite ($NaHSO_3$) is used to cleave the less reactive cyclic osmate intermediate.

cyclohexene a cyclic osmate intermediate *cis*-1,2-cyclohexanediol

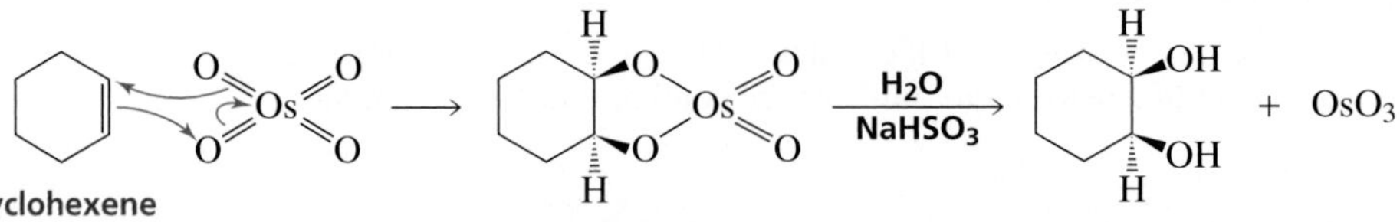

Because the cyclic osmate intermediate has less tendency to undergo side reactions, higher yields of the diol are obtained when osmium tetroxide is used. Osmium tetroxide, however, is considerably more expensive than potassium permanganate and is also toxic. Typically it is used only for small-scale laboratory preparations—potassium permanganate is preferred for large-scale preparations. To minimize the cost of an osmium tetroxide oxidation, a technique has been developed that uses a catalytic amount of OsO_4 together with an equivalent of a tertiary amine oxide (Section 11.11), which oxidizes the reduced oxidizing reagent back to OsO_4.

PROBLEM 15◆

Give the products that would be formed from the reaction of each of the following alkenes with OsO_4 followed by aqueous $NaHSO_3$:

a. $CH_3C=CHCH_2CH_3$
   |
   CH_3

b. 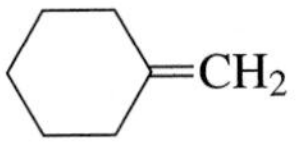 =CH_2

PROBLEM 16

What stereoisomers would be formed from the reaction of each of the following alkenes with OsO_4 followed by treatment with aqueous sodium bisulfite?

a. *trans*-2-butene

b. *cis*-2-butene

c. *cis*-2-pentene

d. *trans*-2-pentene

1,2-Diols are oxidized to ketones and/or aldehydes by periodic acid (HIO_4). Periodic acid reacts with the diol to form a cyclic intermediate. The reaction takes place because iodine is in a highly positive oxidation state ($+7$), so it readily accepts electrons. When the intermediate breaks down, the bond between the two carbons bonded to the OH groups breaks. If the carbon that is bonded to an OH group is also bonded to two R groups, the product will be a ketone. If it is bonded to an R and an H, the product will be an aldehyde. Because this oxidation reaction cuts the reactant into two pieces, it is called an **oxidative cleavage.**

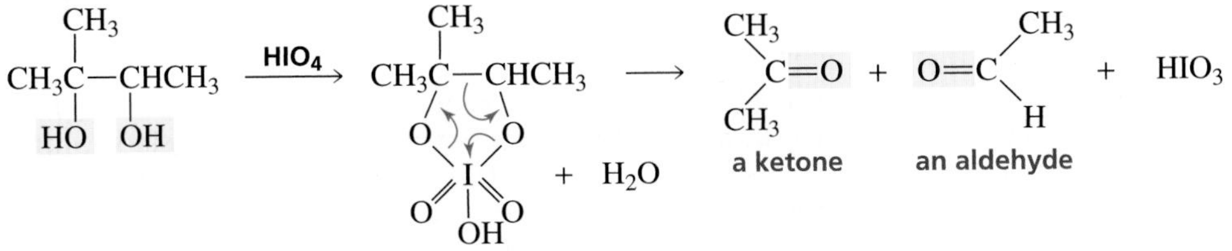

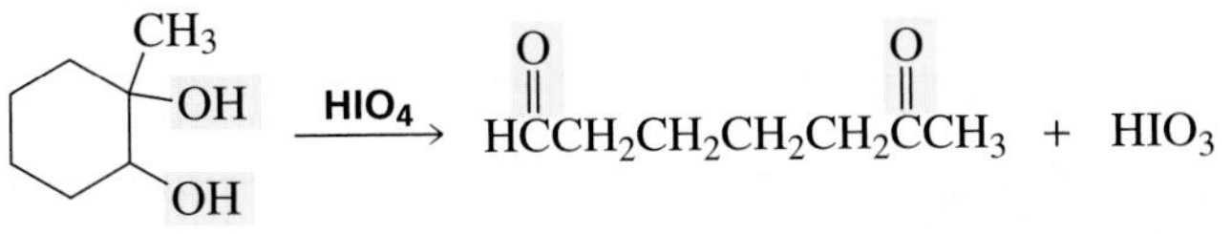

18.6
OXIDATIVE CLEAVAGE OF 1,2-DIOLS

PROBLEM 17◆

An alkene is treated with OsO_4 followed by aqueous $NaHSO_3$. When the resulting diol is treated with HIO_4, the only product obtained is an unsubstituted cyclic ketone with molecular formula $C_6H_{10}O$. What is the structure of the alkene?

PROBLEM-SOLVING STRATEGY

Of the five compounds shown, only **D** cannot be cleaved by periodic acid. Explain.

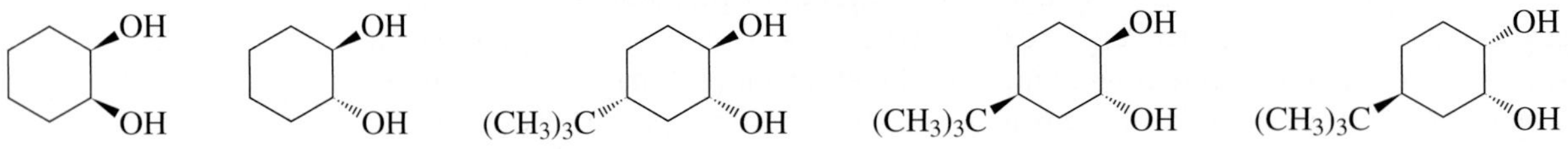

A   B   (CH₃)₃C **C**   (CH₃)₃C **D**   (CH₃)₃C **E**

In solving this kind of problem, first consider what kinds of compounds undergo the reaction and any stereochemical requirements of the reaction. We know that periodic acid
cleaves 1,2-diols. Because the reaction forms a cyclic intermediate, the two OH groups
of the diol must be positioned so they can form the intermediate. The two OH groups of
a cyclohexane can both be equatorial, can both be axial, or one can be equatorial and the
other axial.

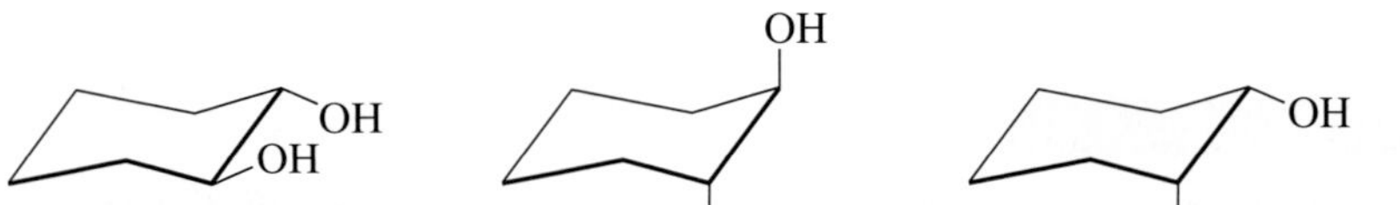

both equatorial
trans

both axial
trans

one equatorial, one axial
cis

In a *cis*-1,2-cyclohexanediol, one OH is equatorial and the other is axial. Because both
cis-1,2-diols (**A** and **E**) are cleaved, we know that the cyclic intermediate can be formed
when the OH groups are in these positions. In a *trans*-1,2-diol, both OH groups are equatorial *or* both are axial (Section 2.14). Two of the *trans*-1,2-diols can be cleaved (**B** and
C), and one cannot (**D**) be cleaved. We can conclude that the one that cannot be cleaved
must have both OH groups in axial positions because these would be too far from each
other to form a cyclic intermediate. Now we should draw the most stable conformation
for **B, C,** and **D** to see why only **D** has both OH groups in axial positions.

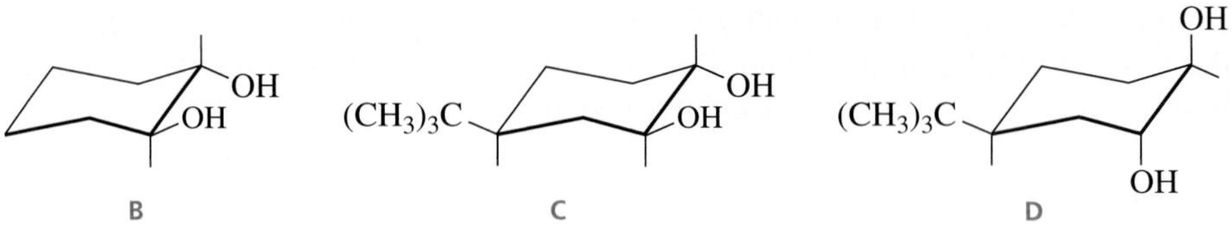

B   (CH₃)₃C **C**   (CH₃)₃C **D**

The most stable conformation for **B** is with both OH groups in equatorial positions. The
steric requirements of the bulky *tert*-butyl group force it into an equatorial position where
there is more room for such a large substituent. This causes both OH groups in compound **C** to be in equatorial positions and both OH groups in compound **D** to be in axial
positions. Therefore, **C** can be cleaved by periodic acid but **D** cannot.

Now continue on to Problem 18.

PROBLEM 18◆

Which of each pair of diols is cleaved more rapidly by periodic acid?

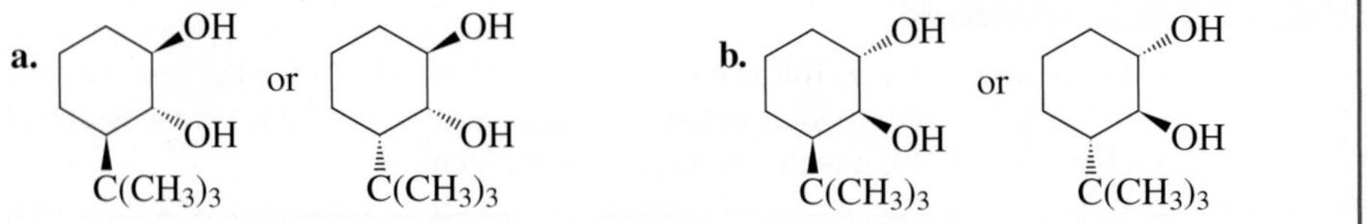

a.   or   C(CH₃)₃   C(CH₃)₃   **b.**   or   C(CH₃)₃   C(CH₃)₃

We have seen that alkenes can be oxidized to 1,2-diols, and that 1,2-diols can be further oxidized to aldehydes and ketones (Sections 18.5 and 18.6). Alternatively, alkenes can be directly oxidized to aldehydes and ketones by ozone (O_3). When an alkene is treated with ozone at low temperatures, the double bond breaks, and the carbons that were double-bonded to each other find themselves double-bonded to oxygens instead. This oxidation reaction is known as **ozonolysis.**

18.7 OXIDATIVE CLEAVAGE OF ALKENES: OZONOLYSIS

$$\text{>C=C<} \xrightarrow[\text{2. work-up}]{\text{1. O}_3, -78\ °C} \text{>C=O} + \text{O=C<}$$

Ozone is produced by passing oxygen gas through an electric discharge. Its structure can be represented by the following resonance contributors.

3-D Molecules: Ozone; Molozonide; Ozonide

resonance contributors of ozone

Ozone and the alkene undergo a concerted cycloaddition reaction—the oxygen atoms add to the two sp^2 carbons in a single step. Addition of ozone to the alkene should remind you of the electrophilic addition reactions of alkenes discussed in Chapter 3. An electrophile adds to one of the sp^2 carbons, and a nucleophile adds to the other. The electrophile is the oxygen at one end of the ozone molecule, and the nucleophile is the oxygen at the other end. The product of ozone addition to an alkene is an unstable **molozonide.** (The name "molozonide" indicates that one mole of ozone has added to the alkene.) The molozonide rearranges to a more stable **ozonide,** by a mechanism that we will not discuss here.

molozonide   ozonide   carbonyl compounds

Because ozonides are explosive, they are seldom isolated. The ozonide is easily cleaved in solution to form carbonyl compounds.

If the ozonide is cleaved in the presence of a reducing agent such as zinc or dimethyl sulfide, the products will be ketones and/or aldehydes. (The product will be a ketone if the sp^2 carbon of the alkene is bonded to two carbon-containing substituents. It will be an aldehyde if at least one of the substituents bonded to the sp^2 carbon is a hydrogen.) The reducing agent prevents aldehydes from being oxidized to carboxylic acids. Cleaving the ozonide in the presence of zinc or dimethyl sulfide is referred to as "working up the ozonide under reducing conditions."

ozonide

$$\xrightarrow[\text{(CH}_3)_2\text{S}]{\text{Zn, H}_2\text{O} \atop \text{or}} \quad \text{R}_2\text{C=O} + \text{O=C(R)(H)}$$

a ketone   an aldehyde

$$\xrightarrow{\text{H}_2\text{O}_2} \quad \text{R}_2\text{C=O} + \text{O=C(R)(OH)}$$

a ketone   a carboxylic acid

If the ozonide is cleaved in the presence of an oxidizing agent such as hydrogen peroxide (H_2O_2), the products will be ketones and/or carboxylic acids. Carboxylic acids are formed instead of aldehydes because any aldehyde initially formed will be immediately oxidized to a carboxylic acid by hydrogen peroxide. Cleavage in the presence of H_2O_2 is referred to as "working up the ozonide under oxidizing conditions."

The following reactions are examples of the oxidative cleavage of alkenes by ozonolysis:

$$CH_3CH_2CH{=}\underset{\underset{CH_3}{|}}{C}CH_2CH_3 \xrightarrow[\text{2. } H_2O_2]{\text{1. } O_3} CH_3CH_2\overset{\overset{O}{\|}}{C}OH \ + \ CH_3\overset{\overset{O}{\|}}{C}CH_2CH_3$$

$$CH_3CH_2CH{=}CHCH_2CH_3 \xrightarrow[\text{2. } (CH_3)_2S]{\text{1. } O_3} 2 \ CH_3CH_2\overset{\overset{O}{\|}}{C}H$$

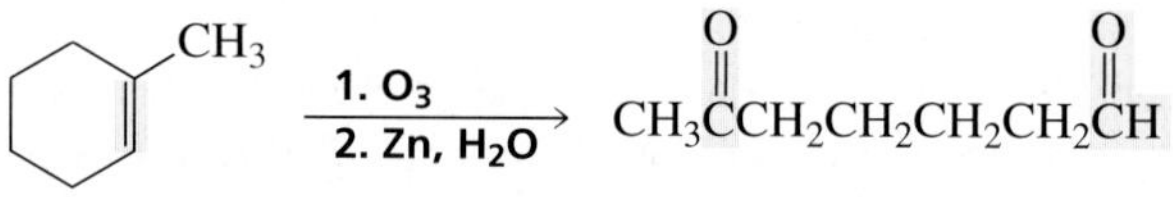

$$\text{(cyclohexene with } CH_3) \xrightarrow[\text{2. Zn, } H_2O]{\text{1. } O_3} CH_3\overset{\overset{O}{\|}}{C}CH_2CH_2CH_2CH_2\overset{\overset{O}{\|}}{C}H$$

The one-carbon fragment obtained from the reaction of a terminal alkene with ozone will be oxidized to formaldehyde if the ozonide is worked up under reducing conditions, and to CO_2 if it is worked up under oxidizing conditions.

> To determine the product of ozonolysis, replace C=C with C=O O=C. If work-up is done under oxidizing conditions, convert any aldehyde products to carboxylic acids.

$$CH_3CH_2CH_2CH{=}CH_2 \xrightarrow[\text{2. Zn, } H_2O]{\text{1. } O_3} CH_3CH_2CH_2\overset{\overset{O}{\|}}{C}H \ + \ H\overset{\overset{O}{\|}}{C}H$$

$$CH_3CH_2CH_2CH{=}CH_2 \xrightarrow[\text{2. } H_2O_2]{\text{1. } O_3} CH_3CH_2CH_2\overset{\overset{O}{\|}}{C}OH \ + \ CO_2$$

Only the side chain double bond will be oxidized in the following reaction because the stable benzene ring can be oxidized only under prolonged exposure to ozone.

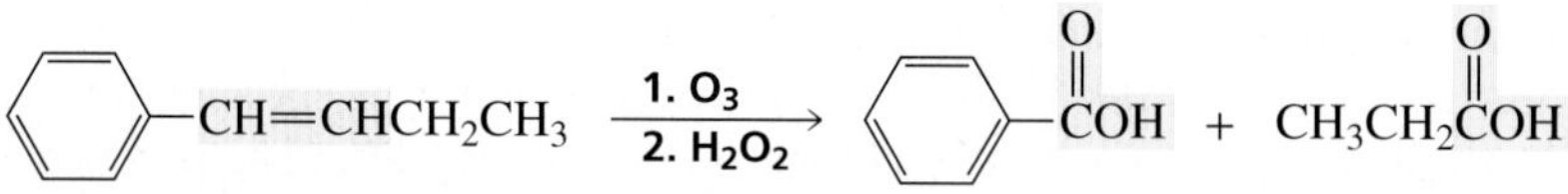

$$\text{(benzene)}{-}CH{=}CHCH_2CH_3 \xrightarrow[\text{2. } H_2O_2]{\text{1. } O_3} \text{(benzene)}{-}\overset{\overset{O}{\|}}{C}OH \ + \ CH_3CH_2\overset{\overset{O}{\|}}{C}OH$$

PROBLEM 19

Give an example of an alkene that will form the same ozonolysis products regardless of whether the ozonide is worked up under reducing conditions (Zn, H_2O) or oxidizing conditions (H_2O_2).

PROBLEM 20

Give the products that you would expect to obtain when the following compounds are treated with ozone followed by work-up with:

a. Zn, H_2O **b.** H_2O_2

1. $CH_3CH_2CH_2\underset{\underset{CH_3}{|}}{C}{=}CHCH_3$ **2.** $CH_2{=}CHCH_2CH_2CH_2CH_3$

3. ⬡—CH₃

5. $CH_3CH_2CH_2CH=CHCH_2CH_2CH_3$

4. ⬡=CH₂

6. (structure with CH_3)

Ozonolysis can be used to determine the structure of an unknown alkene. If you know what carbonyl compounds are formed by ozonolysis, you can mentally work backward to deduce the structure of the alkene. For example, if ozonolysis of an alkene followed by a work-up under reducing conditions forms acetone and butanal as products, you can conclude that the alkene was 2-methyl-2-hexene.

$$CH_3\overset{O}{\overset{\|}{C}}CH_3 \ + \ CH_3CH_2CH_2\overset{O}{\overset{\|}{C}}H \ \Longrightarrow \ CH_3\overset{CH_3}{\overset{|}{C}}=CHCH_2CH_2CH_3$$

| acetone | butanal | 2-methyl-2-hexene |

| ozonolysis products | alkene that underwent ozonolysis |

Tutorial:
Ozonolysis
reactions–synthesis

In Section 18.5 we saw that alkenes are oxidized to 1,2-diols by a basic solution of potassium permanganate at room temperature or below. However, if the basic reaction mixture is heated or if the potassium permanganate solution is acidic, the reaction will not stop at the diol. Instead, the alkene will be cleaved and the reaction products will be ketones and carboxylic acids. If the reaction is carried out under basic conditions, any carboxylic acid product will be in its basic form $(RCOO^-)$; if the reaction is carried out under acidic conditions, any carboxylic acid product will be in its acidic form $(RCOOH)$ (Section 1.20). Terminal alkenes form CO_2 as a product.

$$CH_3CH_2\overset{CH_3}{\overset{|}{C}}=CHCH_3 \ \xrightarrow[\Delta]{\textbf{KMnO}_4, \ \textbf{HO}^-} \ CH_3CH_2\overset{O}{\overset{\|}{C}}CH_3 \ + \ CH_3\overset{O}{\overset{\|}{C}}O^-$$

$$CH_3CH_2CH=CH_2 \ \xrightarrow[\textbf{H}^+]{\textbf{KMnO}_4} \ CH_3CH_2\overset{O}{\overset{\|}{C}}OH \ + \ CO_2$$

$$⬡=CH_2 \ \xrightarrow[\Delta]{\textbf{KMnO}_4, \ \textbf{HO}^-} \ ⬡=O \ + \ CO_2$$

The various methods used in Sections 18.4–18.7 to oxidize an alkene are summarized in Table 18.3. Notice that the products obtained from ozonolysis of an alkene followed by work-up under *reducing conditions* are the same as the products obtained from the oxidation of an alkene to a 1,2-diol with a cold basic solution of potassium permanganate (or with osmium tetroxide) followed by oxidation of the diol with periodic acid. Also notice that the products obtained from ozonolysis of an alkene followed by work-up under *oxidizing* conditions are the same as the products obtained from the oxidation of an alkene with an acidic (or a hot basic) solution of potassium permanganate. Ozonolysis, however, gives a better yield of product.

A peroxyacid, OsO₄, and (cold basic) KMnO₄ break only the π bond of the alkene. Ozone and acidic (or hot basic) KMnO₄ break both the π bond and the σ bond.

PROBLEM 21◆

a. What alkene would give only acetone as an ozonolysis product?

b. What alkenes would give only butanal as an ozonolysis product?

PROBLEM 22◆

What aspect of the structure of the alkene does ozonolysis not tell you?

PROBLEM 23 / SOLVED

The following products were obtained from ozonolysis of a diene followed by work-up under oxidizing conditions. Give the structure of the diene.

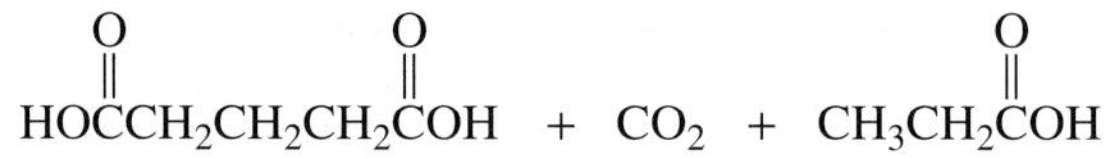

$$\underset{\text{HOCCH}_2\text{CH}_2\text{CH}_2\text{COH}}{\overset{\text{O}\qquad\qquad\text{O}}{\|\qquad\qquad\qquad\|}} \quad + \quad \text{CO}_2 \quad + \quad \underset{\text{CH}_3\text{CH}_2\text{COH}}{\overset{\text{O}}{\|}}$$

SOLUTION Because the first compound is a five-carbon dicarboxylic acid, the diene must contain five carbons flanked by two double bonds.

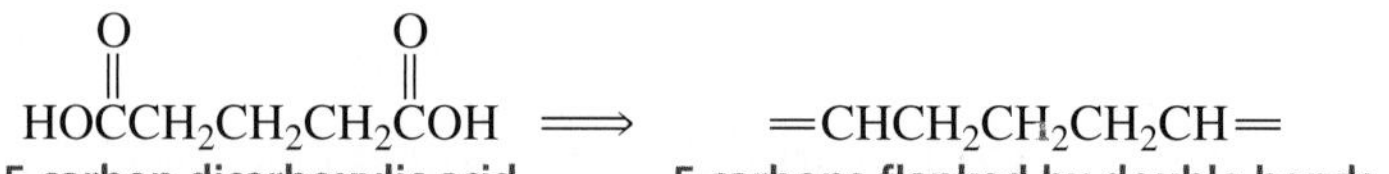

$$\underset{\substack{\text{HOCCH}_2\text{CH}_2\text{CH}_2\text{COH}\\ \textbf{5-carbon dicarboxylic acid}}}{\overset{\text{O}\qquad\quad\text{O}}{\|\qquad\qquad\|}} \implies \underset{\textbf{5 carbons flanked by double bonds}}{=\text{CHCH}_2\text{CH}_2\text{CH}_2\text{CH}=}$$

The other products obtained from ozonolysis are carbon dioxide (one carbon atom) and propanoic acid (three carbon atoms). Therefore, one carbon has to be added to one end of the diene and three carbons have to be added to the other end.

$$\text{CH}_2=\text{CHCH}_2\text{CH}_2\text{CH}_2\text{CH}=\text{CHCH}_2\text{CH}_3$$

TABLE 18.3 Summary of the Methods Used to Oxidize an Alkene

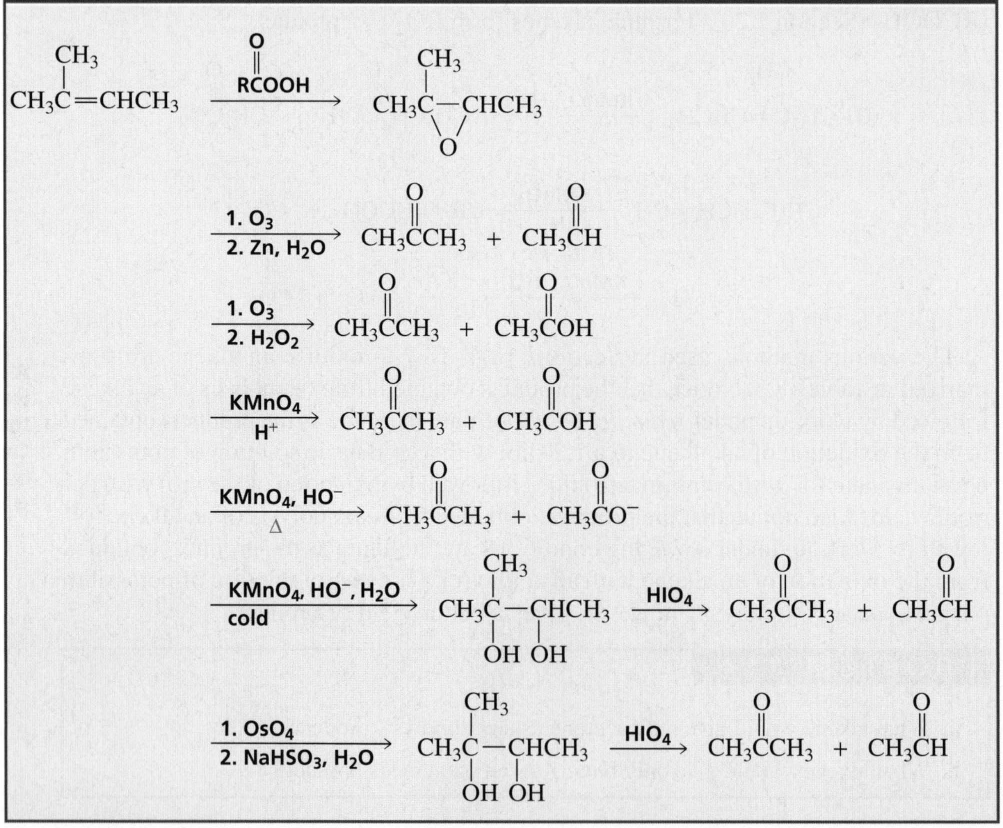

The same reagents that oxidize alkenes also oxidize alkynes. Alkynes are oxidized to diketones by a basic solution of $KMnO_4$ at room temperature, and are cleaved by ozonolysis to carboxylic acids. Ozonolysis requires neither oxidative nor reductive work-up—it is followed only by hydrolysis. Carbon dioxide is obtained from the CH group of a terminal alkyne.

18.8 OXIDATIVE CLEAVAGE OF ALKYNES

$$CH_3C{\equiv}CCH_2CH_3 \xrightarrow[\text{HO}^-]{\text{KMnO}_4} CH_3\overset{O}{\overset{\|}{C}}-\overset{O}{\overset{\|}{C}}CH_2CH_3$$
2-pentyne

$$CH_3C{\equiv}CCH_2CH_3 \xrightarrow[\text{2. H}_2\text{O}]{\text{1. O}_3} CH_3\overset{O}{\overset{\|}{C}}OH \ + \ CH_3CH_2\overset{O}{\overset{\|}{C}}OH$$
2-pentyne

$$CH_3CH_2CH_2C{\equiv}CH \xrightarrow[\text{2. H}_2\text{O}]{\text{1. O}_3} CH_3CH_2CH_2\overset{O}{\overset{\|}{C}}OH \ + \ CO_2$$
1-pentyne

PROBLEM 24◆

What is the structure of the alkyne that gives each of the following products upon ozonolysis followed by hydrolysis?

a. 
$$\text{(cyclohexane ring with COOH)} + CO_2$$

b. $HO\overset{O}{\overset{\|}{C}}CH_2CH_2CH_2\overset{O}{\overset{\|}{C}}OH \ + \ 2\ CH_3CH_2\overset{O}{\overset{\|}{C}}OH$

Converting one functional group into another is called **functional group interconversion.** Our knowledge of oxidation–reduction reactions has greatly expanded our ability to carry out functional group interconversions. For example, an aldehyde can be converted into a primary alcohol, an alkene, a secondary alcohol, a ketone, a carboxylic acid, an acyl chloride, an ester, an amide, or an amine.

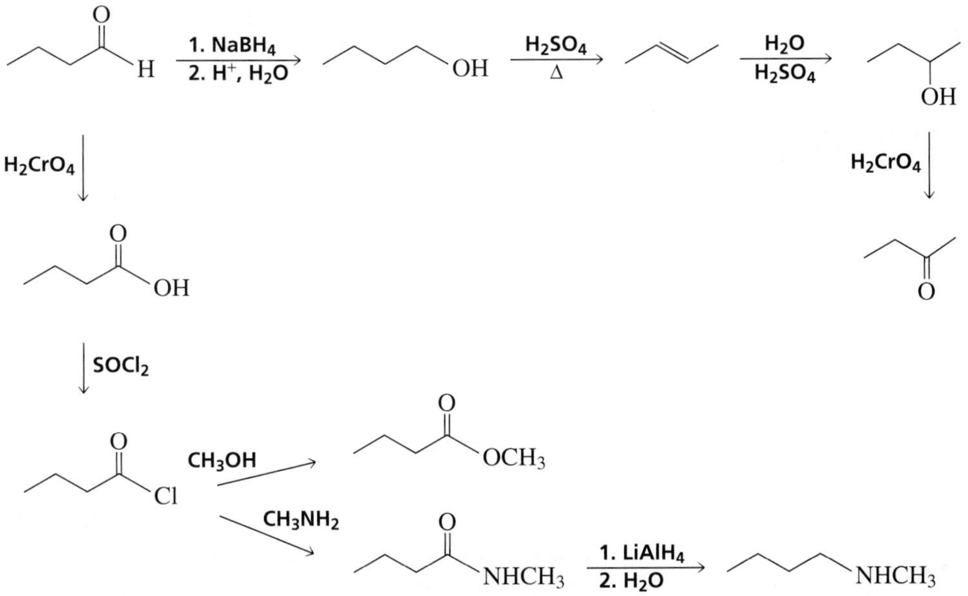

18.9 DESIGNING A SYNTHESIS V: FUNCTIONAL GROUP INTERCONVERSION

Tutorial: Multistep synthesis

A ketone can be converted into an ester or an alcohol.

Tutorial:
Common terms in oxidation–reduction reactions

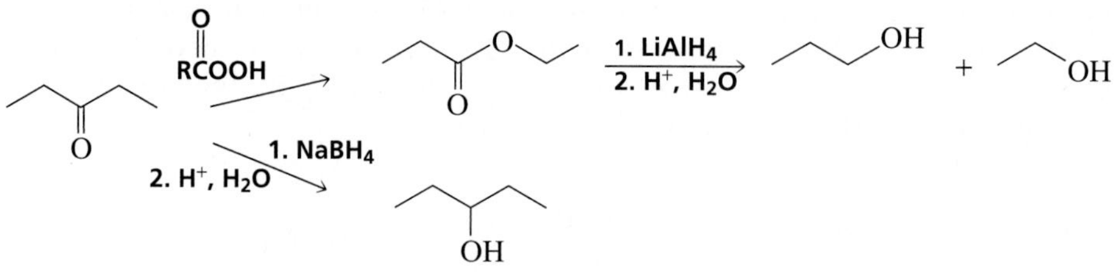

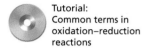

As the number of reactions you learn increases, so will the number of functional group interconversions you can do. You will also find that you have more than one route available when you design a synthesis. The route you actually decide to use will depend on the availability of the starting materials, the cost of the starting materials, and the ease with which the reactions in the synthetic pathway can be carried out.

PROBLEM 25

How many different functional groups can you use to synthesize a primary alcohol? (*Hint:* See Figure 29.1.)

PROBLEM 26 ◆

Add the necessary reagents over the reaction arrows:

a.

b.

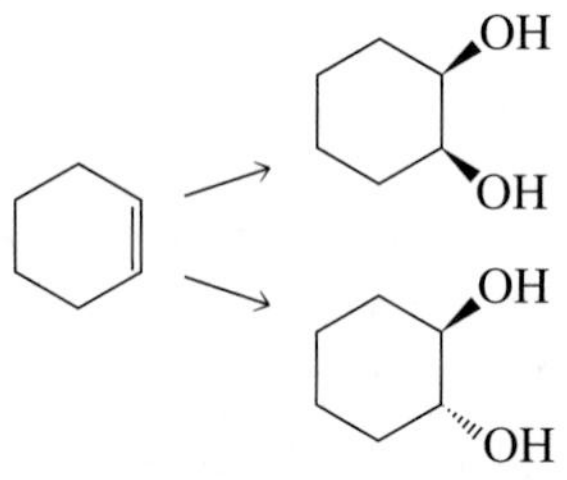

PROBLEM 27

a. Show two ways to convert an alkyl halide into an alcohol that contains one additional carbon atom.

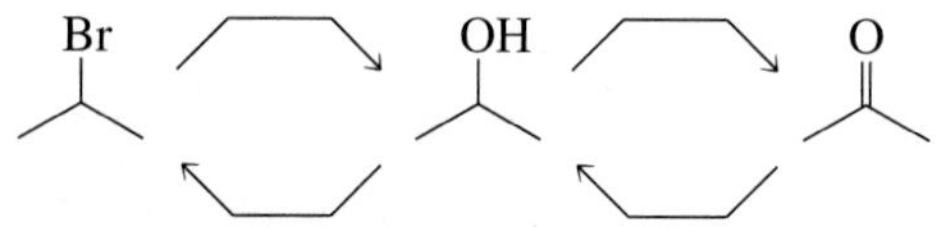

b. Show how a primary alkyl halide can be converted into an amine that contains one additional carbon atom.

c. Show how a primary alkyl halide can be converted into an amine that contains one fewer carbon atom.

PROBLEM 28

Show how each of the following compounds could be synthesized from the given starting material:

a.

c.

b.

d.

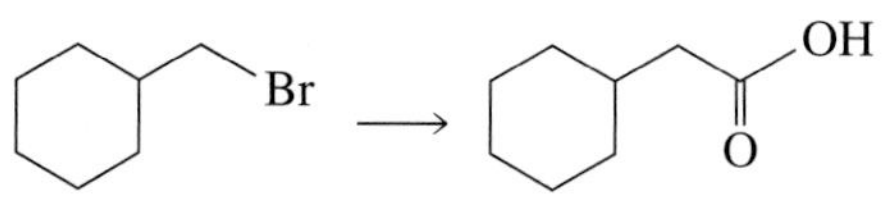

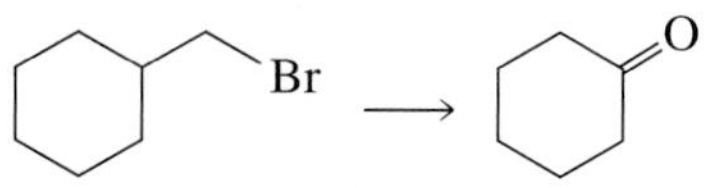

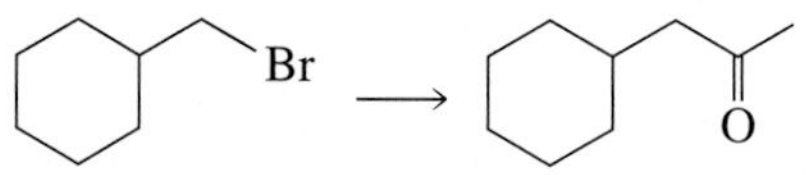

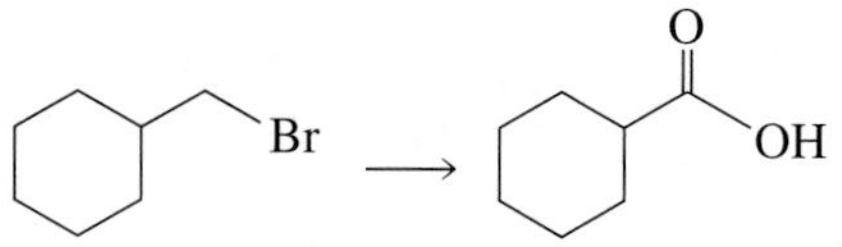

Oxidation and reduction reactions are important in living systems. An example of an oxidation reaction that takes place in animal cells is the oxidation of ethanol to acetaldehyde, a reaction catalyzed by the enzyme alcohol dehydrogenase. Ingestion of a moderate amount of ethanol lowers inhibitions and causes a lightheaded feeling, but the physiological effects of acetaldehyde are not as pleasant. Acetaldehyde is responsible for the feeling known as a hangover. (In Section 23.4, we will see how vitamin B_1 can help cure a hangover.)

18.10 BIOLOGICAL OXIDATION–REDUCTION REACTIONS

$$\underset{\text{ethanol}}{CH_3CH_2OH} + NAD^+ \xrightarrow[\text{dehydrogenase}]{\text{alcohol}} \underset{\text{acetaldehyde}}{CH_3\overset{O}{\overset{\|}{C}}H} + NADH + H^+$$

TREATING ALCOHOLICS WITH ANTABUSE

Disulfiram, most commonly known by one of its trade names, Antabuse, is used to treat alcoholics. It causes violently unpleasant effects when ethanol is consumed even when it is consumed a day or two after the drug is taken.

$$\underset{CH_3CH_2}{\overset{CH_3CH_2}{>}}N-\overset{\overset{S}{\|}}{C}-S-S-\overset{\overset{S}{\|}}{C}-N\underset{CH_2CH_3}{\overset{CH_2CH_3}{<}}$$

Antabuse

Antabuse inhibits aldehyde dehydrogenase, the enzyme responsible for oxidizing acetaldehyde to acetic acid, resulting in a buildup of acetaldehyde. Acetaldehyde causes the unpleasant physiological effects of intoxication—intense flushing, nausea, dizziness, sweating, throbbing headaches, decreased blood pressure, and ultimately shock. Consequently, Antabuse should be taken only under strict medical supervision.

$$\underset{\text{ethanol}}{CH_3CH_2OH} \xrightarrow[\text{dehydrogenase}]{\text{alcohol}} \underset{\text{acetaldehyde}}{CH_3\overset{O}{\overset{\|}{C}}H} \xrightarrow[\text{dehydrogenase}]{\text{aldehyde}} \underset{\text{acetic acid}}{CH_3\overset{O}{\overset{\|}{C}}OH}$$

In many people, aldehyde dehydrogenase is a nonfunctional enzyme. Their symptoms in response to ingestion of alcohol are nearly the same as those of individuals who are medicated with Antabuse.

Ethanol cannot be oxidized to acetaldehyde unless an oxidizing agent is present. Oxidizing agents used by organic chemists, such as chromate and permanganate salts, are not present in living systems. NAD^+ (nicotinamide adenine dinucleotide), the most common oxidizing agent available in living systems, is used by cells to oxidize alcohols to aldehydes (Section 23.2). Notice that NAD^+ is written with a positive charge to reflect the positive charge on the nitrogen atom of the pyridine ring.

NAD^+ is reduced to NADH when it oxidizes a compound. NADH is used by the cell as a reducing agent. When NADH reduces a compound, it is oxidized back to NAD^+, which can then be used for another oxidation. Although NAD^+ and NADH are complicated-looking molecules, the structural changes that occur when they act as oxidizing and reducing agents take place on a relatively small part of the molecule. The rest of the molecule is used to bind it to the proper site on the enzyme that catalyzes the reaction.

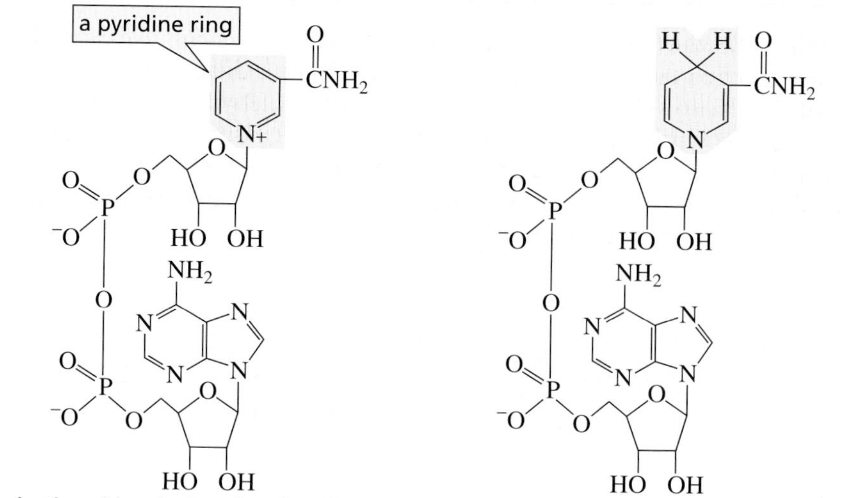

a pyridine ring

nicotinamide adenine dinucleotide
NAD⁺

reduced nicotinamide adenine dinucleotide
NADH

NAD^+ oxidizes a compound by accepting a hydride ion from it. In this way, the number of carbon–hydrogen bonds in the compound decreases (the compound is oxidized) and the number of carbon–hydrogen bonds in NAD^+ increases (NAD^+ is reduced). NAD^+ can accept a hydride ion at the 4-position of the pyridine ring because the electrons can be delocalized onto the positively charged nitrogen atom. Although NAD^+ could also accept a hydride ion at the 2-position, the hydride ion is always delivered to the 4-position in enzyme-catalyzed reactions.

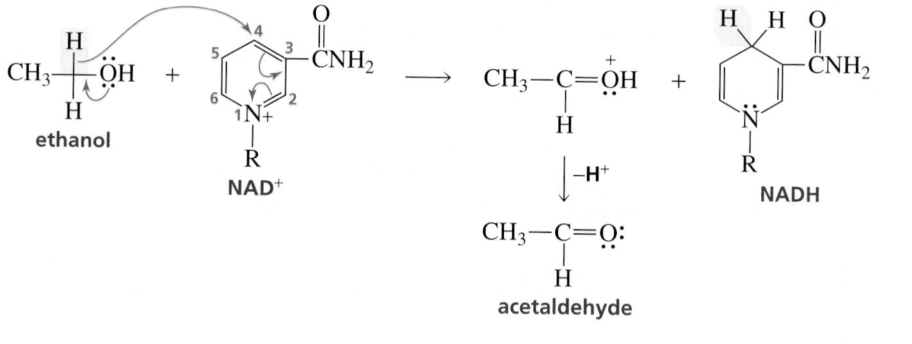

NADH reduces a compound by donating a hydride ion from the 4-position of the six-membered ring. Thus, NADH, $NaBH_4$, and $LiAlH_4$ all act as reducing agents in the same way—they donate a hydride ion.

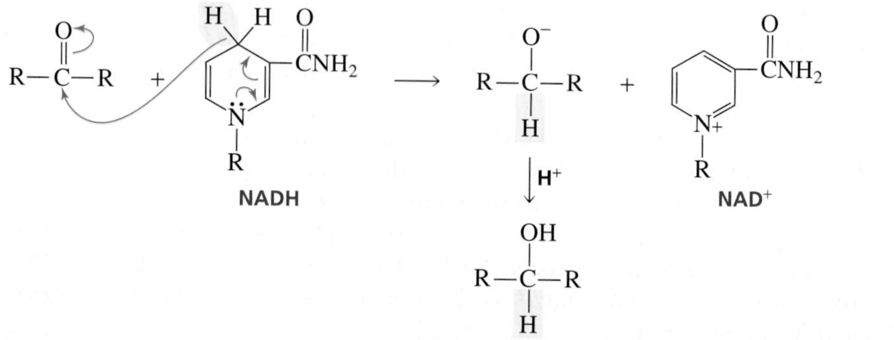

AN UNUSUAL ANTIDOTE

Alcohol dehydrogenase, the enzyme that catalyzes the oxidation of ethanol to acetaldehyde, catalyzes the oxidation of other alcohols as well. For example, it catalyzes the oxidation of methanol to formaldehyde.

Methanol itself is not harmful, but ingestion of methanol can be fatal because formaldehyde is extremely toxic. The treatment for methanol ingestion consists of giving the patient intravenous injections of ethanol. Alcohol dehydrogenase has 25 times the affinity for ethanol that it has for methanol. Thus, if the enzyme can be kept loaded with ethanol, methanol will be excreted before it has the opportunity to be oxidized.

$$\underset{\text{methanol}}{CH_3OH} + NAD^+ \xrightarrow{\text{alcohol dehydrogenase}} \underset{\text{formaldehyde}}{HCH(=O)} + NADH + H^+$$

FETAL ALCOHOL SYNDROME

The damage done to a human fetus when the mother drinks alcohol during her pregnancy is known as fetal alcohol syndrome. It has been shown that the harmful effects—growth retardation, decreased mental functioning, facial and limb abnormalities—are attributable to acetaldehyde formed from the oxidation of ethanol, which crosses the placenta and accumulates in the liver of the fetus.

18.11 OXIDATION OF HYDROQUINONES/ REDUCTION OF QUINONES

A *para*-benzenediol such as hydroquinone is easily oxidized to *para*-benzoquinone. Although a wide variety of oxidizing agents can be used, Fremy's salt (dipotassium nitrosodisulfonate) is the preferred oxidizing agent. The quinone can be easily reduced back to hydroquinone.

hydroquinone | *para*-benzoquinone, a *para*-quinone

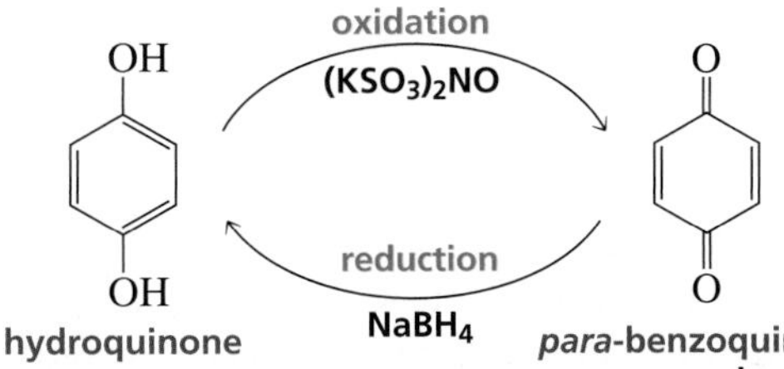

Similarly, *ortho*-benzenediols are oxidized to *ortho*-quinones.

ortho-benzoquinone, an *ortho*-quinone

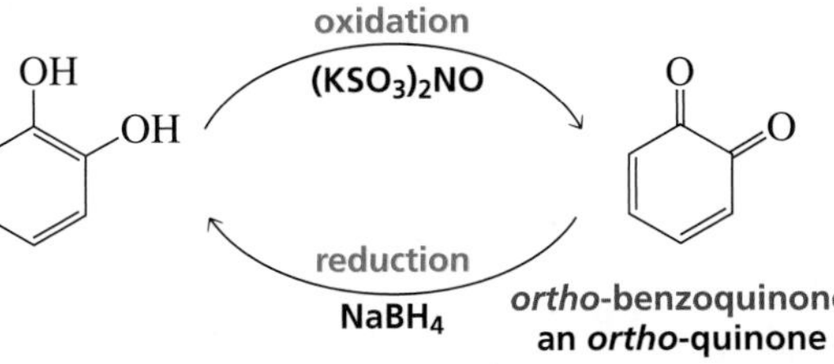

Overall, the oxidation reaction involves the loss of two hydrogen atoms and the reduction reaction involves the gain of two hydrogen atoms. In Section 8.8 we saw that the ability of a phenol to lose a hydrogen atom is why phenols are used as radical inhibitors.

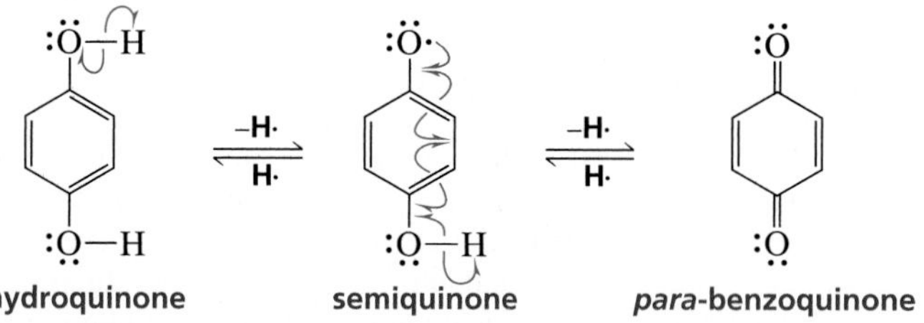

hydroquinone   semiquinone   *para*-benzoquinone

3-D Molecules:
Coenzyme Q
(oxidized form);
Coenzyme Q
(reduced form)

Coenzyme Q (CoQ) is a quinone found in the cells of all aerobic organisms. It is also called ubiquinone because it is ubiquitous (found everywhere) in nature. Its function is to carry electrons in the electron-transport chain. The oxidized form of CoQ accepts a pair of electrons from a biological reducing agent such as NADH and ultimately transfers them to O_2.

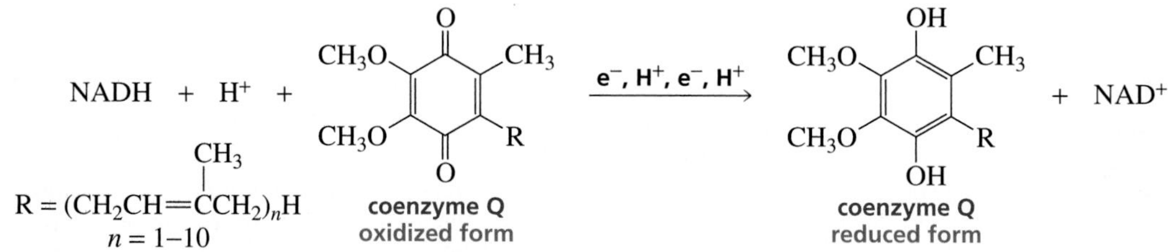

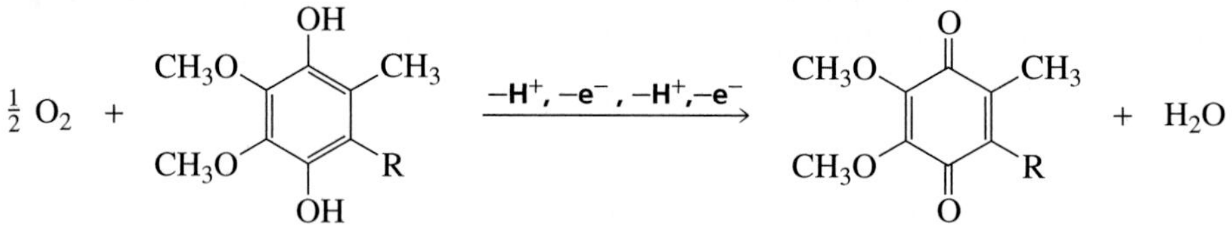

$R = (CH_2CH=\overset{\overset{\displaystyle CH_3}{|}}{C}CH_2)_nH$
$n = 1-10$

coenzyme Q
oxidized form

coenzyme Q
reduced form

In this way biological oxidizing agents are recycled—NAD^+ oxidizes a compound thereby forming NADH, which is oxidized back to NAD^+ by oxygen, via coenzyme Q, which is unchanged in the overall reaction.

$$NAD^+ + Substrate_{reduced} \longrightarrow Substrate_{oxidized} + NADH + H^+$$

$$NADH + H^+ + \tfrac{1}{2} O_2 \longrightarrow NAD^+ + H_2O$$

THE CHEMISTRY OF PHOTOGRAPHY

Black-and-white photography depends on the fact that hydroquinone is easily oxidized. Photographic film is covered by an emulsion of silver bromide. When light hits the film, the silver bromide is sensitized. Sensitized silver bromide is a better oxidizing agent than silver bromide that has not been exposed to light. When the exposed film is put into a solution of hydroquinone (a common pho-tographic developer), hydroquinone is oxidized to quinone by sensitized silver ion and the silver ion is reduced to silver metal, which remains in the emulsion. The exposed film is "fixed" by washing away unsensitized silver bromide with $Na_2S_2O_3/H_2O$. Black silver deposits are left in regions where light has struck the film. This is the black, opaque part of a photographic negative.

SUMMARY OF REACTIONS

1. Catalytic hydrogenation of double and triple bonds (Section 18.1)

$$\textbf{a.}\ RCH=CHR + H_2 \xrightarrow{\textbf{Pt, Pd, or Ni}} RCH_2CH_2R$$

$$RC\equiv CR + 2 H_2 \xrightarrow{\textbf{Pt, Pd, or Ni}} RCH_2CH_2R$$

$$RCH=NR + H_2 \xrightarrow{\textbf{Pt, Pd, or Ni}} RCH_2NHR$$

$$RC\equiv N + 2 H_2 \xrightarrow{\textbf{Pt, Pd, or Ni}} RCH_2NH_2$$

b. $\overset{\overset{\displaystyle O}{\|}}{RCH}$ + H$_2$ $\xrightarrow{\text{Pt, Pd, or Ni}}$ RCH$_2$OH

$\overset{\overset{\displaystyle O}{\|}}{RCR}$ + H$_2$ $\xrightarrow{\text{Pt, Pd, or Ni}}$ $\overset{\overset{\displaystyle OH}{|}}{RCHR}$

c. $\overset{\overset{\displaystyle O}{\|}}{RCCl}$ + H$_2$ $\xrightarrow{\text{Pt, Pd, or Ni}}$ RCH$_2$OH

$\overset{\overset{\displaystyle O}{\|}}{RCCl}$ + H$_2$ $\xrightarrow[\text{Pd}]{\substack{\text{partially} \\ \text{deactivated}}}$ $\overset{\overset{\displaystyle O}{\|}}{RCH}$

2. Reduction of alkynes to alkenes (Section 18.1)

$RC{\equiv}CR$ $\xrightarrow[\substack{\text{Lindlar's} \\ \text{catalyst}}]{H_2}$ $\underset{R}{\overset{H}{>}}C{=}C\underset{R}{\overset{H}{<}}$

$RC{\equiv}CR$ $\xrightarrow[\text{NH}_3\ (\text{liq})]{\text{Na or Li}}$ $\underset{R}{\overset{H}{>}}C{=}C\underset{H}{\overset{R}{<}}$

3. Reduction of carbonyl compounds with reagents that donate hydride ion (Section 18.1)

a. $\overset{\overset{\displaystyle O}{\|}}{RCH}$ $\xrightarrow[\text{2. H}^+,\text{H}_2\text{O}]{\text{1. NaBH}_4}$ RCH$_2$OH   **c.** $\overset{\overset{\displaystyle O}{\|}}{RCOH}$ $\xrightarrow[\text{2. H}^+,\text{H}_2\text{O}]{\text{1. LiAlH}_4}$ RCH$_2$OH

b. $\overset{\overset{\displaystyle O}{\|}}{RCR}$ $\xrightarrow[\text{2. H}^+,\text{H}_2\text{O}]{\text{1. NaBH}_4}$ $\overset{\overset{\displaystyle OH}{|}}{RCHR}$ **d.** $\overset{\overset{\displaystyle O}{\|}}{RCOR'}$ $\xrightarrow[\text{2. H}^+,\text{H}_2\text{O}]{\text{1. LiAlH}_4}$ RCH$_2$OH + R'OH

e. $\overset{\overset{\displaystyle O}{\|}}{RCNHR'}$ $\xrightarrow[\text{2. H}_2\text{O}]{\text{1. LiAlH}_4}$ RCH$_2$NHR'

f. $\overset{\overset{\displaystyle O}{\|}}{RCOR'}$ $\xrightarrow[\text{2. H}_2\text{O}]{\text{1. [(CH}_3)_2\text{CHCH}_2]_2\text{AlH, }-78°}$ $\overset{\overset{\displaystyle O}{\|}}{RCH}$ + R'OH

g. $\overset{\overset{\displaystyle O}{\|}}{RCCl}$ $\xrightarrow[\text{2. H}_2\text{O}]{\text{1. LiAl[OC(CH}_3)_3]_3\text{H, }-78°}$ $\overset{\overset{\displaystyle O}{\|}}{RCH}$

4. Oxidation of alcohols (Section 18.2)

primary alcohols   RCH$_2$OH $\xrightarrow{\text{H}_2\text{CrO}_4}$ $\left[\overset{\overset{\displaystyle O}{\|}}{RCH}\right]$ $\xrightarrow[\text{oxidation}]{\text{further}}$ $\overset{\overset{\displaystyle O}{\|}}{RCOH}$

RCH$_2$OH $\xrightarrow[\text{CH}_2\text{Cl}_2]{\text{PCC}}$ $\overset{\overset{\displaystyle O}{\|}}{RCH}$

RCH$_2$OH $\xrightarrow[\substack{-60\ °\text{C} \\ \text{2. triethylamine}}]{\text{1. CH}_3\text{SCH}_3,\ \overset{\overset{\displaystyle O}{\|}}{\text{ClC}}{-}\overset{\overset{\displaystyle O}{\|}}{\text{CCl}}}$ $\overset{\overset{\displaystyle O}{\|}}{RCH}$

$$\text{secondary alcohols } \underset{}{R\overset{OH}{\underset{|}{C}}HR} \xrightarrow[\text{H}_2\text{SO}_4]{\text{Na}_2\text{Cr}_2\text{O}_7} R\overset{O}{\overset{||}{C}}R$$

$$R\overset{OH}{\underset{|}{C}}HR \xrightarrow[\text{2. triethylamine}]{\overset{\text{1. CH}_3\overset{O}{\overset{||}{S}}\text{CH}_3,\ \text{ClC}\overset{O}{\overset{||}{-}}\overset{O}{\overset{||}{C}}\text{Cl}}{-60\ °C}} R\overset{O}{\overset{||}{C}}R$$

5. Oxidation of aldehydes and ketones (Section 18.3)

a. aldehydes $\quad R\overset{O}{\overset{||}{C}}H \xrightarrow[\text{H}_2\text{SO}_4]{\text{Na}_2\text{Cr}_2\text{O}_7} R\overset{O}{\overset{||}{C}}OH$

$$R\overset{O}{\overset{||}{C}}H \xrightarrow[\text{2. H}^+,\ \text{H}_2\text{O}]{\text{1. Ag}_2\text{O, NH}_3} R\overset{O}{\overset{||}{C}}OH \ + \ \underset{\substack{\text{metallic}\\\text{silver}}}{\text{Ag}}$$

$$R\overset{O}{\overset{||}{C}}H \xrightarrow{R'\overset{O}{\overset{||}{C}}OOH} R\overset{O}{\overset{||}{C}}OH \ + \ R'\overset{O}{\overset{||}{C}}OH$$

b. **ketones** $\quad R\overset{O}{\overset{||}{C}}R \xrightarrow{R'\overset{O}{\overset{||}{C}}OOH} R\overset{O}{\overset{||}{C}}OR \ + \ R'\overset{O}{\overset{||}{C}}OH$

6. Oxidation of alkenes (Sections 18.4, 18.5, 18.7)

a. $\quad R\overset{R}{\underset{|}{C}}{=}CHR' \xrightarrow[\text{2. Zn, H}_2\text{O}]{\text{1. O}_3,\ -78\ °C} R\overset{O}{\overset{||}{C}}R \ + \ R'\overset{O}{\overset{||}{C}}H$

$$\xrightarrow[\text{2. H}_2\text{O}_2]{\text{1. O}_3,\ -78\ °C} R\overset{O}{\overset{||}{C}}R \ + \ R'\overset{O}{\overset{||}{C}}OH$$

b. $\quad R\overset{R}{\underset{|}{C}}{=}CHR' \xrightarrow{\text{KMnO}_4,\ \text{H}^+} R\overset{O}{\overset{||}{C}}R \ + \ R'\overset{O}{\overset{||}{C}}OH$

c. $\quad R\overset{R}{\underset{|}{C}}{=}CHR' \xrightarrow[\text{cold}]{\text{KMnO}_4,\ \text{HO}^-,\ \text{H}_2\text{O}} R\overset{R}{\underset{\underset{\text{OH}}{|}}{C}}{-}\underset{\underset{\text{OH}}{|}}{C}HR' \xrightarrow{\text{HIO}_4} R\overset{O}{\overset{||}{C}}R \ + \ R'\overset{O}{\overset{||}{C}}H$

$$\xrightarrow[\text{2. NaHSO}_3,\ \text{H}_2\text{O}]{\text{1. OsO}_4} R\overset{R}{\underset{\underset{\text{OH}}{|}}{C}}{-}\underset{\underset{\text{OH}}{|}}{C}HR' \xrightarrow{\text{HIO}_4} R\overset{O}{\overset{||}{C}}R \ + \ R'\overset{O}{\overset{||}{C}}H$$

$$\xrightarrow{R\overset{O}{\overset{||}{C}}OOH} R\overset{\overset{O}{\triangle}}{\underset{R}{C}}{-}CHR'$$

7. Oxidation of 1,2-diols (Section 18.6)

$$R\overset{R}{\underset{\underset{\text{OH}}{|}}{C}}{-}\underset{\underset{\text{OH}}{|}}{C}HR' \xrightarrow{\text{HIO}_4} R\overset{O}{\overset{||}{C}}R \ + \ R'\overset{O}{\overset{||}{C}}H$$

8. Oxidation of alkynes (Section 18.8)

a. $RC{\equiv}CR' \xrightarrow[\text{HO}^-]{\text{KMnO}_4}$ RC$-$CR' (with two C=O groups)

b. $RC{\equiv}CR' \xrightarrow[\text{2. H}_2\text{O}]{\text{1. O}_3, -78\,°\text{C}}$ RCOH $+$ R'COH

c. $RC{\equiv}CH \xrightarrow[\text{2. H}_2\text{O}]{\text{1. O}_3, -78\,°\text{C}}$ RCOH $+$ CO$_2$

9. Oxidation of hydroquinones/reduction of quinones (Section 18.11)

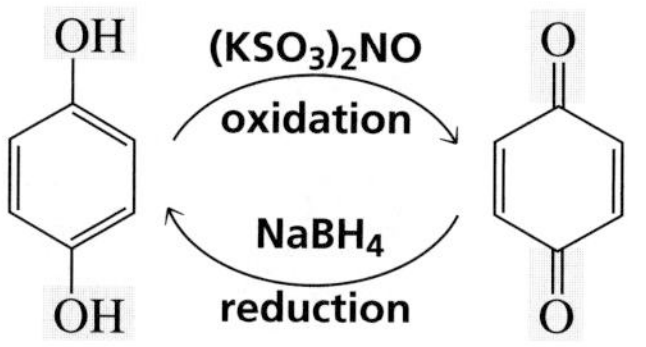

KEY TERMS

Baeyer–Villiger oxidation (page 801)
catalytic hydrogenation (page 791)
dissolving metal reduction (page 794)
epoxidation (page 802)
functional group interconversion
(page 811)
glycol (page 804)
molozonide (page 807)

oxidation (page 797)
oxidation–reduction reaction(page 789)
oxidation state (page 789)
oxidative cleavage (page 805)
oxidizing agent (page 789)
ozonide (page 807)
ozonolysis (page 807)
peroxyacid (page 801)

redox reaction (page 789)
reducing agent (page 789)
reduction (page 797)
Rosenmund reduction (page 792)
Swern oxidation (page 799)
Tollens test (page 801)
vicinal diol (page 804)
vicinal glycol (page 804)

PROBLEMS

29. Fill in the blank with "oxidized" or "reduced."
   a. Secondary alcohols are _____ to ketones.
   b. Acyl halides are _____ to aldehydes.
   c. Aldehydes are _____ to primary alcohols.
   d. Alkenes are _____ to aldehydes and/or ketones.
   e. Aldehydes are_____ to carboxylic acids.
   f. Alkenes are _____ to 1,2-diols.
   g. Alkenes are _____ to alkanes.

30. Give the products of the following reactions. Indicate whether each reaction is an oxidation or a reduction.

a. $CH_3CH_2CH_2CH_2CH_2OH \xrightarrow[\text{H}_2\text{SO}_4]{\text{Na}_2\text{Cr}_2\text{O}_7}$

b. 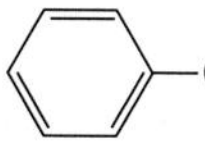 (benzene ring)$-CH{=}CH_2 \xrightarrow[\text{HO}^-, \Delta]{\text{KMnO}_4}$

c. $CH_3CH_2CH_2\overset{\text{O}}{\overset{\|}{C}}Cl \xrightarrow[\text{Pt}]{\text{H}_2}$

d. $CH_3CH_2C{\equiv}CH \xrightarrow{\begin{array}{l}\text{1. disiamylborane}\\ \text{2. H}_2\text{O}_2, \text{HO}^-, \text{H}_2\text{O}\\ \text{3. LiAlH}_4\\ \text{4. H}^+, \text{H}_2\text{O}\end{array}}$

e. $CH_3CH_2CH{=}CHCH_2CH_3 \xrightarrow[\text{2. Zn, H}_2\text{O}]{\text{1. O}_3}$

f. $CH_3CH_2CH_2\overset{\text{O}}{\overset{\|}{C}}NHCH_3 \xrightarrow[\text{2. H}_2\text{O}]{\text{1. LiAlH}_4}$

g. 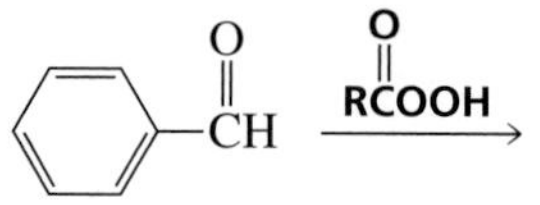 [benzaldehyde structure] $\overset{\overset{\displaystyle O}{\parallel}}{C}H$ $\xrightarrow{\overset{\overset{\displaystyle O}{\parallel}}{R}COOH}$

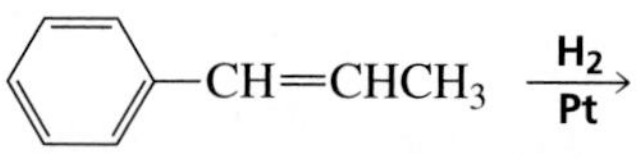

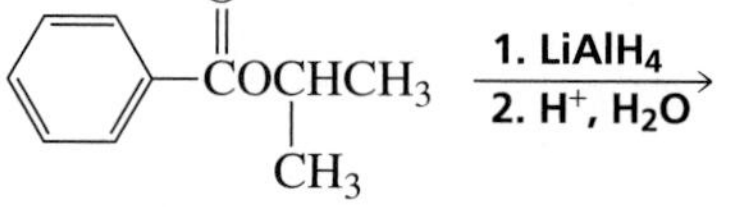

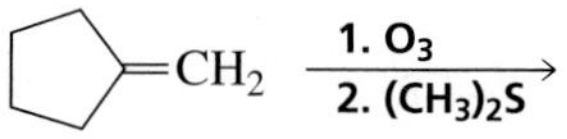

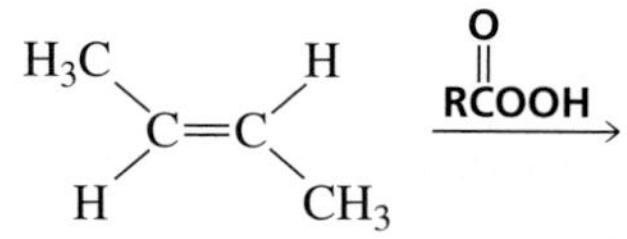

m. [benzene ring]—CH=CHCH$_3$ $\xrightarrow{\underset{Pt}{H_2}}$

h. [benzene ring]—$\overset{\overset{\displaystyle O}{\parallel}}{C}OCHCH_3$ with CH$_3$ below $\xrightarrow{\underset{2.\ H^+,\ H_2O}{1.\ LiAlH_4}}$

n. [cyclopentane ring]=CH$_2$ $\xrightarrow{\underset{2.\ (CH_3)_2S}{1.\ O_3}}$

i. $\overset{H_3C}{\underset{H}{}}C=C\overset{H}{\underset{CH_3}{}}$ $\xrightarrow{\overset{\overset{\displaystyle O}{\parallel}}{R}COOH}$

o. [cyclohexene ring] $\xrightarrow{\underset{\underset{3.\ H^+,\ H_2O}{2.\ CH_3MgBr}}{1.\ RCOOH}}$

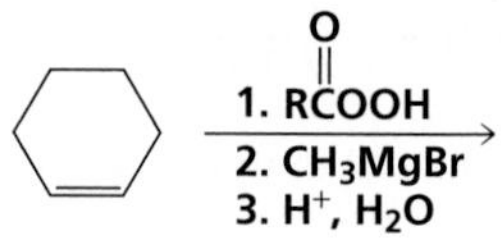

j. 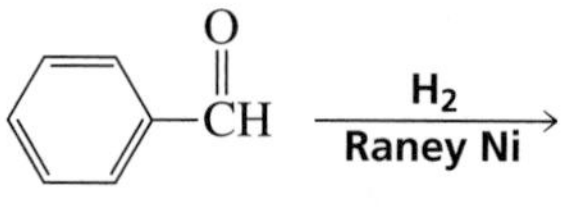 [benzaldehyde structure] $\overset{\overset{\displaystyle O}{\parallel}}{C}H$ $\xrightarrow{\underset{Raney\ Ni}{H_2}}$

p. 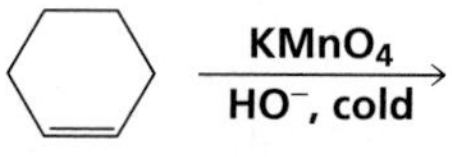 [cyclohexene ring] $\xrightarrow{\underset{HO^-,\ cold}{KMnO_4}}$

k. CH$_3$CH$_2$CH$_2$C≡CCH$_3$ $\xrightarrow{\underset{NH_3\ (liq)}{Na}}$

q.  [cyclohexene ring] $\xrightarrow{\underset{HO^-,\ \Delta}{KMnO_4}}$

l. CH$_3$CH$_2$CH$_2$C≡CCH$_3$ $\xrightarrow{\underset{2.\ H_2O}{1.\ O_3}}$

r. 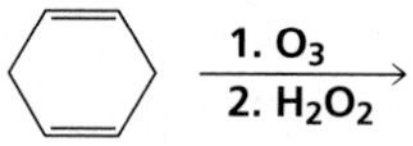 [cyclohexadiene ring] $\xrightarrow{\underset{2.\ H_2O_2}{1.\ O_3}}$

31. How could each of the following compounds be converted to CH$_3$CH$_2$CH$_2\overset{\overset{\displaystyle O}{\parallel}}{C}OH$?

a. CH$_3$CH$_2$CH$_2\overset{\overset{\displaystyle O}{\parallel}}{C}H$

c. CH$_3$CH$_2$CH$_2$CH$_2$Br

b. CH$_3$CH$_2$CH$_2$CH$_2$OH

d. CH$_3$CH$_2$CH=CH$_2$

32. Identify the alkene that would give each of the following products upon ozonolysis followed by treatment with hydrogen peroxide:

a. CH$_3$CH$_2$CH$_2\overset{\overset{\displaystyle O}{\parallel}}{C}OH$ + CH$_3\overset{\overset{\displaystyle O}{\parallel}}{C}CH_3$

d. [cyclohexanone ring] + CH$_3$CH$_2\overset{\overset{\displaystyle O}{\parallel}}{C}OH$

b. CH$_3\overset{\overset{\displaystyle O}{\parallel}}{C}CH_2CH_2CH_2CH_2\overset{\overset{\displaystyle O}{\parallel}}{C}CH_2CH_3$

e. [benzene ring]—$\overset{\overset{\displaystyle O}{\parallel}}{C}CH_3$ + CO$_2$

c. HO$\overset{\overset{\displaystyle O}{\parallel}}{\underset{\underset{\displaystyle O}{}}{}}$ + [structure]$\overset{\displaystyle O}{}$...$\overset{\displaystyle O}{}$OH

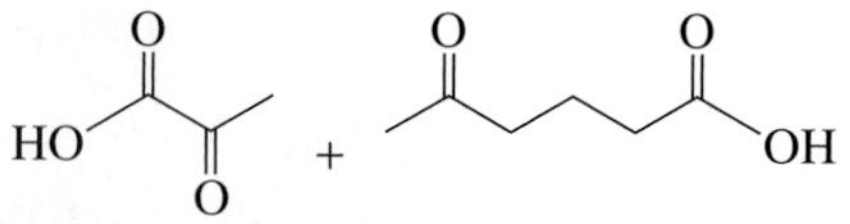

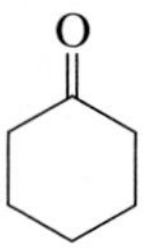

f. 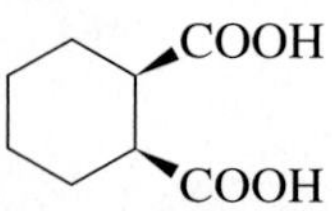 [cyclohexane ring with COOH and COOH] + HO—$\overset{\overset{\displaystyle O}{\parallel}}{C}$—$\overset{\overset{\displaystyle O}{\parallel}}{C}$—OH

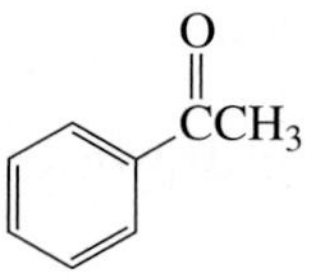

33. Fill in each box with the appropriate reagent.

a. 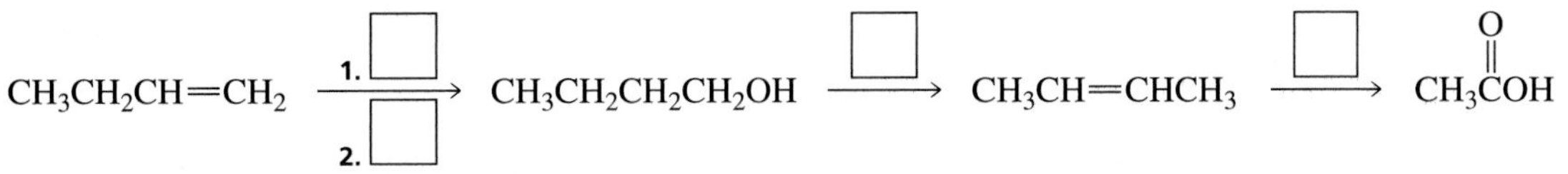 CH₃CH₂CH=CH₂ →(1. ☐ / 2. ☐)→ CH₃CH₂CH₂CH₂OH →(☐)→ CH₃CH=CHCH₃ →(☐)→ CH₃COH

b. 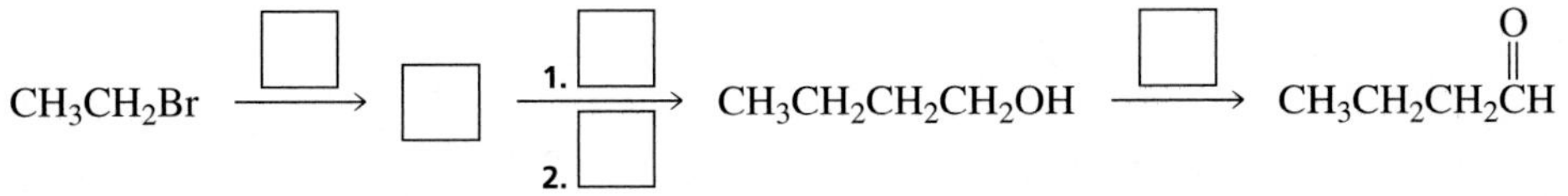 CH₃CH₂Br →(☐)→ ☐ →(1. ☐ / 2. ☐)→ CH₃CH₂CH₂CH₂OH →(☐)→ CH₃CH₂CH₂CH

c. 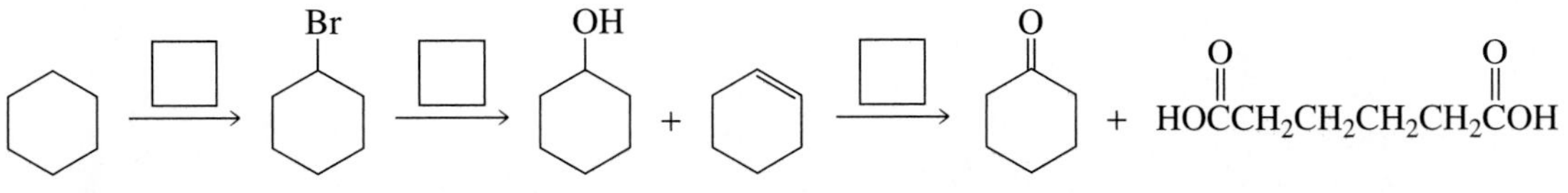 ⬡ →(☐)→ ⬡–Br →(☐)→ ⬡–OH + ⬡ →(☐)→ ⬡=O + HOCCH₂CH₂CH₂CH₂COH

34. Describe how 1-butyne can be converted into each of the following compounds:

a. 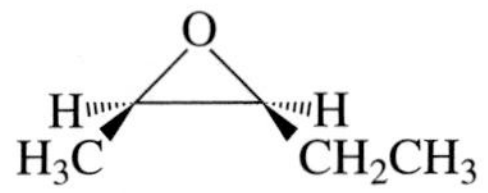 epoxide: H₃C, H (one side), CH₂CH₃, H (other side)

b. 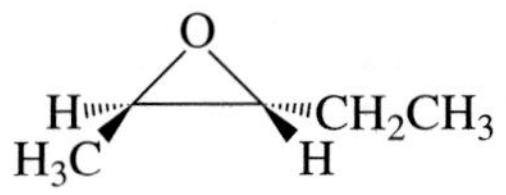 epoxide: H₃C, H (one side), CH₂CH₃, H (other side)

35. a. Give the products obtained from ozonolysis of each of the following compounds followed by work-up under oxidizing conditions:

1. ⬭(methyl cycloheptene) **2.** ⬭ **3.** ⬭ **4.** ⬭ **5.** ⬭

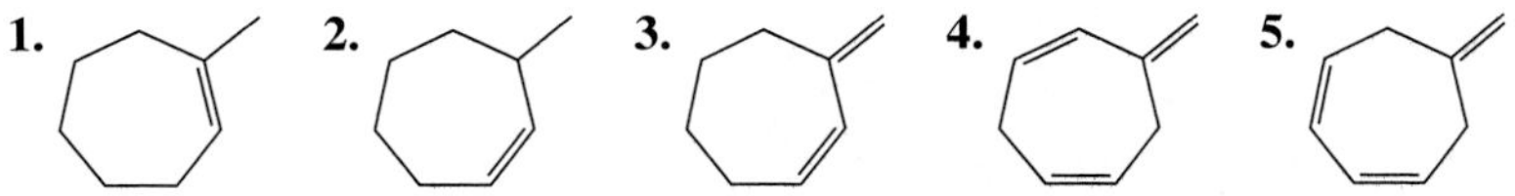

b. What compound would form the following products upon reaction with ozone followed by work-up under oxidizing conditions?

HOCCH₂CH₂C—COH + HO—C—C—OH + CO₂

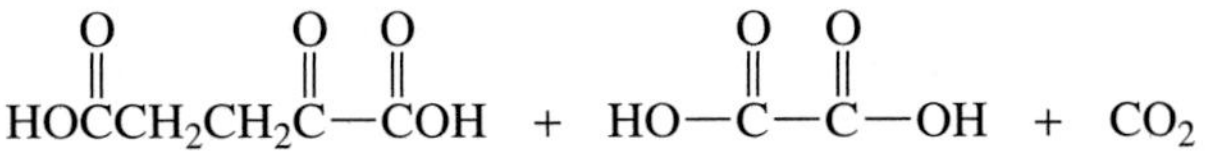

36. Show how each of the following compounds can be prepared from the given starting material:

a. 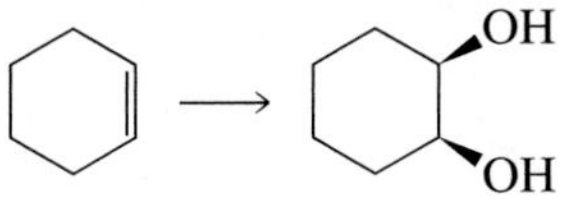 ⬡ (cyclohexene) → ⬡ with OH (wedge) and OH (wedge)

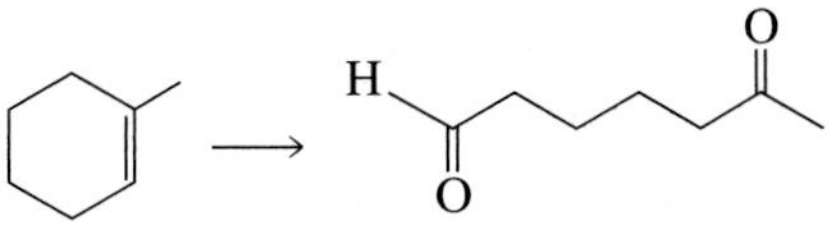

b. 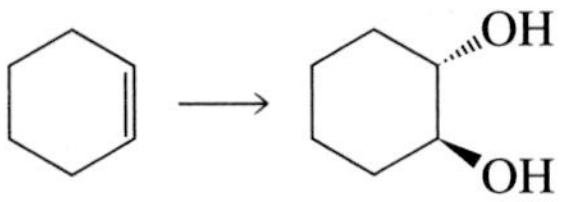 ⬡ → ⬡ with OH and OH (trans)

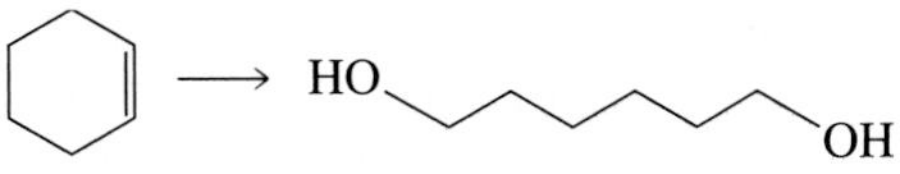

c. 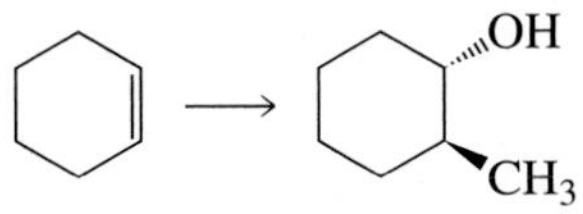 ⬡ → ⬡ with OH and CH₃

d. ⬡ (1-methylcyclohexene) → H—C(=O)...C(=O) open chain ketoaldehyde

e. ⬡ → HO~~~~OH (diol chain)

f. 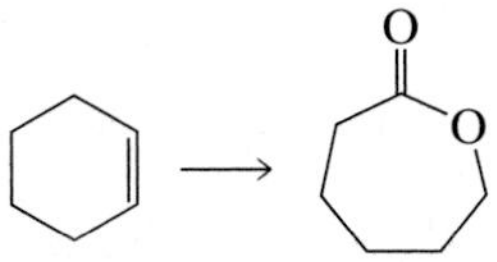 ⬡ → lactone (7-membered ring with O and C=O)

g. 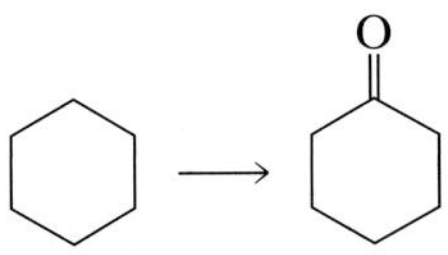 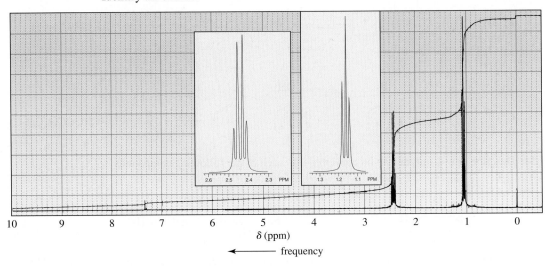 **h.**

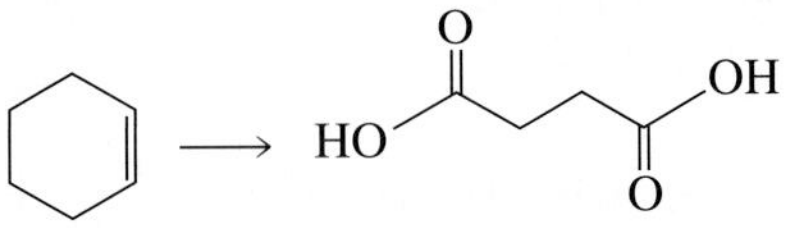

37. The ^{1}H NMR spectrum of the product obtained when an unknown alkene reacts with ozone and the ozonolysis product is worked up under oxidizing conditions is shown. Identify the alkene.

38. Identify **A–N**.

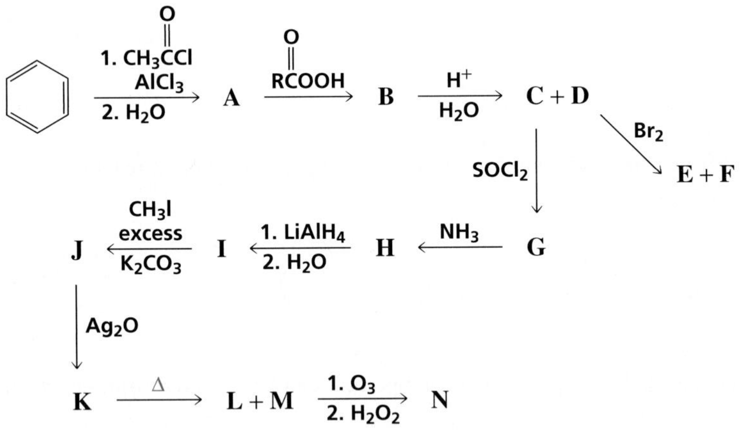

39. Chromic acid oxidizes 2-propanol six times faster than it oxidizes 2-deuterio-2-propanol. Explain. (*Hint:* See Section 10.7.)

40. Show how each of the following compounds could be prepared using the given starting material:

a. $CH_3CH_2\overset{O}{\overset{\|}{C}}H \longrightarrow CH_3CH_2\overset{O}{\overset{\|}{C}}OCH_2CH_2CH_3$

b. $CH_3CH_2CH_2CH_2OH \longrightarrow CH_3CH_2CH_2\overset{O}{\overset{\|}{C}}CH_2CH_3$

c.

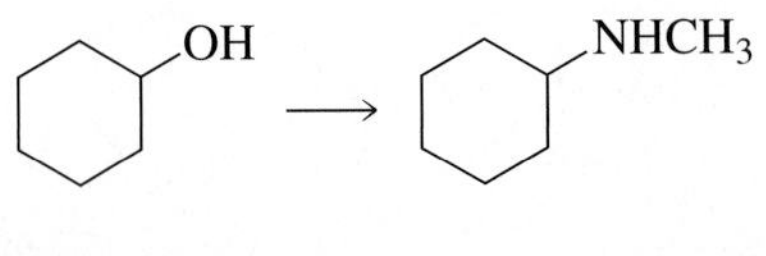

d. 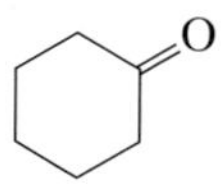 $\longrightarrow$ $HO\overset{O}{\overset{\|}{C}}CH_2CH_2CH_2CH_2\overset{O}{\overset{\|}{C}}OH$

e. 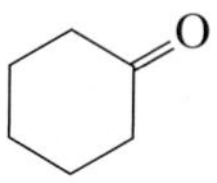 $\longrightarrow$ $HO\overset{O}{\overset{\|}{C}}CH_2CH_2CH_2CH_2\overset{O}{\overset{\|}{C}}CH_3$

41. An alkene, upon treatment with ozone followed by work-up with hydrogen peroxide, forms carbon dioxide and a compound that shows three signals in its ^1H NMR spectrum (a singlet, a triplet, and a quartet). Identify the alkene.

42. Which of the following compounds would be more rapidly cleaved by HIO_4?

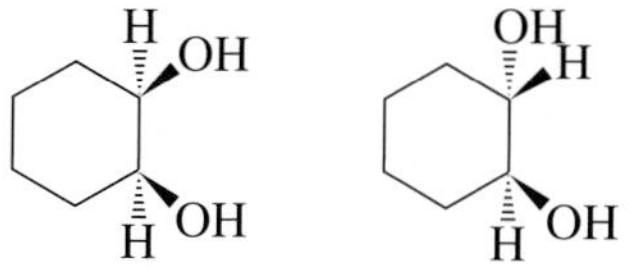

A   B

43. Show how cyclohexylacetylene can be converted into each of the following compounds:

a.

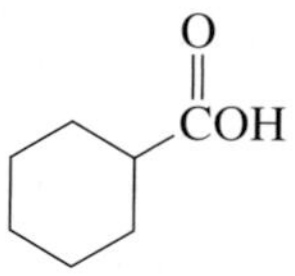

b.

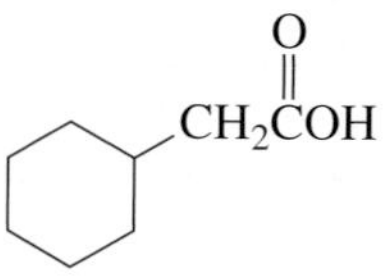

44. Catalytic hydrogenation of 0.5 g of a hydrocarbon at 25 °C consumed about 200 mL of H_2 under one atm of pressure. Reaction of the hydrocarbon with ozone followed by treatment with hydrogen peroxide gave one product, which was found to be a four-carbon carboxylic acid. Identify the hydrocarbon.

45. Tom Thumbs was asked to prepare the following compounds from the given starting materials. The reagents he chose to use for each synthesis are shown.
a. Which of his syntheses were successful?
b. What products did he obtain from the other syntheses?
c. In his unsuccessful syntheses, what reagents should he have used instead to obtain the desired product?

1. $CH_3CH_2\overset{CH_3}{\overset{|}{C}}=CHCH_3 \quad \xrightarrow[\text{H}_2\text{SO}_4]{\text{KMnO}_4} \quad CH_3CH_2\overset{CH_3}{\overset{|}{C}}-\overset{}{C}HCH_3$
$\qquad\qquad\qquad\qquad\qquad\qquad\qquad\quad \underset{\overset{|}{OH}\ \overset{|}{OH}}{}$

2. $CH_3CH_2\overset{O}{\overset{\|}{C}}OCH_3 \quad \xrightarrow[\text{2. H}^+,\ \text{H}_2\text{O}]{\text{1. NaBH}_4} \quad CH_3CH_2CH_2OH \ + \ CH_3OH$

3. 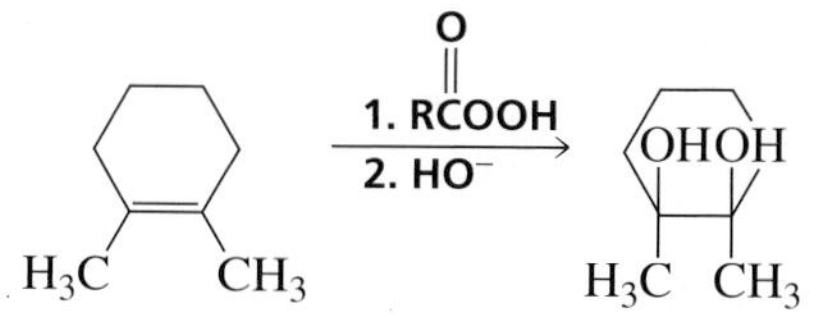 $\xrightarrow[\text{2. HO}^-]{\text{1. RCOOH}}$

4. $CH_3CH=CH\overset{O}{\overset{\|}{C}}Cl \quad \xrightarrow[\text{Pd/C}]{\text{excess H}_2} \quad CH_3CH_2CH_2CH_2OH$

46. Catalytic hydrogenation of compound A formed compound B. The IR spectrum of compound A and the ^{1}H NMR spectrum of compound B are shown. Identify the compounds.

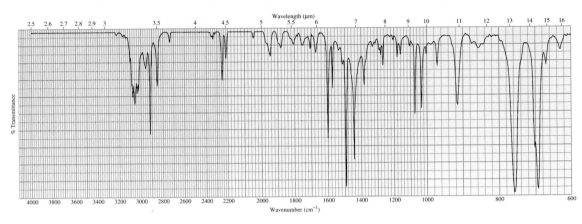

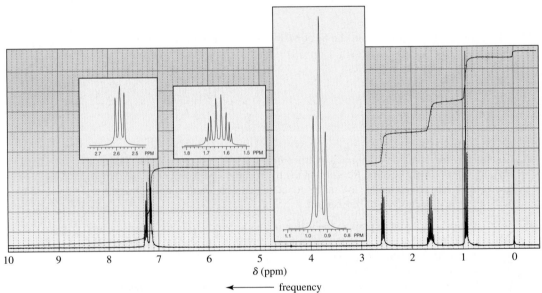

frequency

47. Diane Diol had worked for several days to prepare the compounds shown. She carefully labeled and went to lunch. To her horror, when she returned she found that the labels had fallen off the bottles and onto the floor. Gladys Glycol, the student at the next bench, told her that the diols could be easily distinguished by two experiments. All Diane had to do was determine which ones were optically active and how many products were obtained when each was treated with periodic acid. Diane did what Gladys suggested and found the following:

    1. Compounds A, E, and F are optically active, and B, C, and D are optically inactive.

    2. One product is obtained from the reaction of A, B, and D with periodic acid.

    3. Two products are obtained from the reaction of F with periodic acid.

    4. C and E do not react with periodic acid.

Will Diane be able to distinguish among the six diols and label them A to F with only this information? Label the structures.

48. Show how propyl propionate could be prepared using allyl alcohol as the only source of carbon.

49. Compound A has a molecular formula of $C_5H_{12}O$ and is oxidized by an acidic solution of sodium dichromate to give compound B, whose molecular formula is $C_5H_{10}O$. When compound A is heated with H_2SO_4, C and D are obtained. Considerably more D is obtained than C. Compound C reacts with O_3, followed by treatment with H_2O_2, to give two products—CO_2 and compound E whose molecular formula is C_4H_8O. Compound D reacts with O_3, followed by treatment with H_2O_2, to give compound F whose molecular formula is C_3H_6O, and compound G whose molecular formula is $C_2H_4O_2$. What are the structures of compounds A to G?

50. A compound forms *cis*-1,2-dimethycyclopropane when it is reduced with H_2 and Pd/C. The ^1H NMR spectrum of the compound shows only two singlets. What is the structure of the compound?

51. Show how you could convert:
  a. maleic acid to (2R,3S)-tartaric acid
  b. fumaric acid to (2R,3S)-tartaric acid
  c. maleic acid to (2R,3R)- and (2S,3S)-tartaric acid
  d. fumaric acid to (2R,3R)- and (2S,3S)-tartaric acid

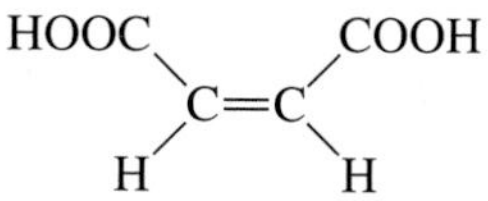

maleic acid 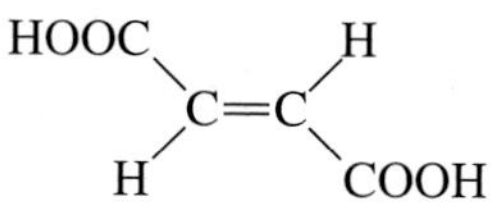 fumaric acid 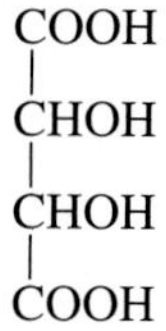 tartaric acid

52. Identify A–O.

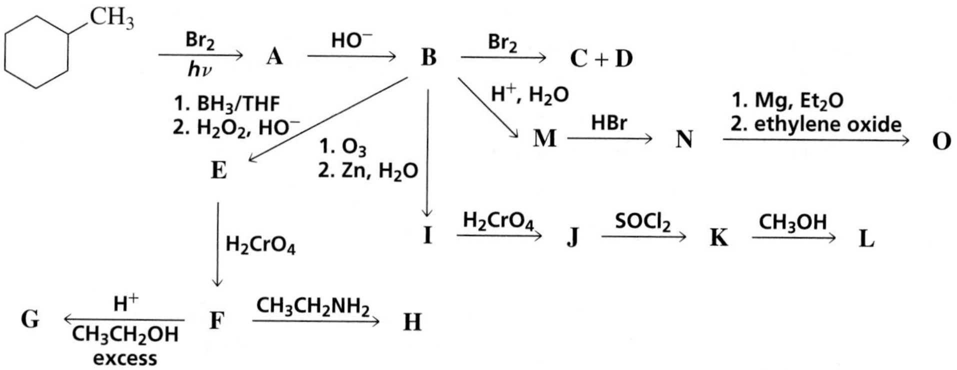

53. Show how the following compounds could be prepared using only the indicated starting material as the source of carbon:

a. $CH_3\overset{O}{\overset{\|}{C}}CH_3$ from $CH_3\overset{CH_3}{\overset{|}{C}H}CH_3$

b. $CH_3CH{=}\overset{CH_3}{\overset{|}{C}}CH_3$ using propane as the only source of carbon

c. $CH_3\overset{O}{\overset{\|}{C}}{-}\overset{CH_3}{\overset{|}{C}H}CH_3$ from propane and any molecule with two carbon atoms

d. CH₃CH₂CH from two molecules of ethane

(with a C=O double bond above the CH)

54. A primary alcohol can be oxidized only as far as the aldehyde stage if the alcohol is first treated with tosyl chloride (TsCl) and the resulting tosylate is allowed to react with dimethyl sulfoxide (DMSO). Propose a mechanism for this reaction. (*Hint:* See Section 18.2.)

$$CH_3CH_2CH_2CH_2OH \xrightarrow[\text{pyridine}]{\text{TsCl}} CH_3CH_2CH_2CH_2OTs \xrightarrow{\text{DMSO}} CH_3CH_2CH_2CH$$

55. Identify the alkene that gives each of the following products upon ozonolysis followed by treatment with dimethyl sulfide:

a.

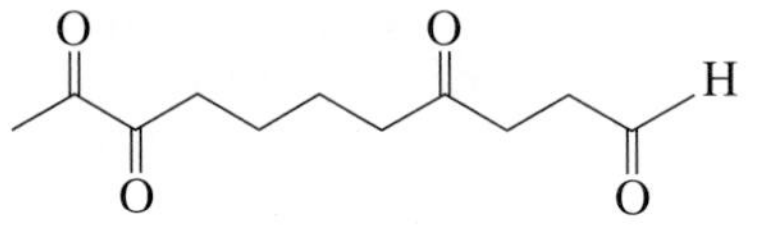

b.

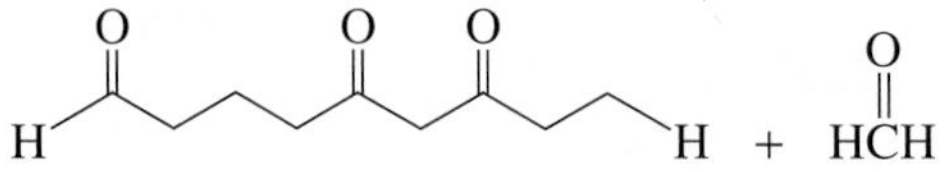

56. Propose a mechanism to explain how dimethyl sulfoxide and oxalyl chloride react to form the dimethylchlorosulfonium ion used as the oxidizing agent in the Swern oxidation.

$$CH_3-\overset{O}{\underset{\|}{S}}-CH_3 \; + \; Cl-\overset{O}{\underset{\|}{C}}-\overset{O}{\underset{\|}{C}}-Cl \; \longrightarrow \; CH_3-\overset{Cl}{\underset{+}{S}}-CH_3 + CO_2 + CO + Cl^-$$

dimethyl sulfoxide   oxalyl chloride   dimethylchlorosulfonium ion

57. Propose a mechanism for the following enzyme-catalyzed reaction. (*Hint:* Notice that Br is attached to the more substituted carbon.)

pregnenolone

58. Terpineol (C₁₀H₁₈O) is an optically active compound with one chirality center. It is used as an antiseptic. Reaction of terpineol with H₂/Pt forms an optically inactive compound (C₁₀H₂₀O). Heating the reduced compound in acid followed by ozonolysis and work-up under reducing conditions produces the following compounds. What is the structure of terpineol?

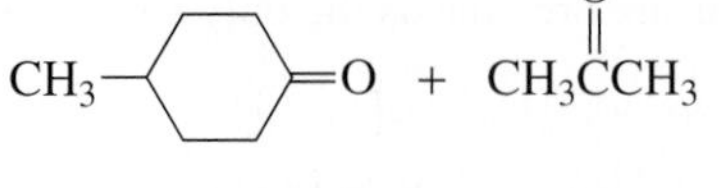

19

Carbonyl Compounds III: Reactions at the α-Carbon

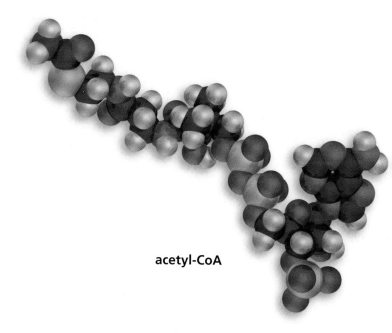

acetyl-CoA

When you studied the reactions of carbonyl compounds in Chapters 16 and 17, you saw that their site of reactivity is the partially positively charged carbonyl carbon, which is attacked by nucleophiles.

$$RCH_2-\overset{\overset{\displaystyle \overset{\delta -}{O}}{\|}}{\underset{\delta +}{C}}-R \quad \rightleftharpoons \quad RCH_2-\overset{\overset{\displaystyle O^-}{|}}{\underset{\underset{\displaystyle Nu}{|}}{C}}-R$$

$$Nu:^-$$

Aldehydes, ketones, esters, and amides have a second site of reactivity. A hydrogen bonded to *a carbon adjacent to a carbonyl carbon* is sufficiently acidic to be removed by a strong base. This carbon is called an *α*-**carbon.** A hydrogen bonded to an α-carbon is called an *α*-**hydrogen.**

an α-carbon

$$R-\overset{|}{\underset{\underset{\displaystyle H}{|}}{CH}}-\overset{\overset{\displaystyle O}{\|}}{C}-R \quad \rightleftharpoons \quad R-\overset{\cdot\cdot}{\underset{}{CH}}-\overset{\overset{\displaystyle O}{\|}}{C}-R$$

:base

baseH

an α-hydrogen

In Section 19.1, we will find out why a hydrogen bonded to a carbon adjacent to a carbonyl carbon is acidic. We will then look at reactions resulting from this acidity. At the end of this chapter, we will see that a proton is not the only substituent that can be removed from an α-carbon—a carboxyl group bonded to an α-carbon can be removed as CO_2. Finally, we will look at synthetic schemes that take advantage of the ability to remove protons and carboxyl groups from α-carbons.

19.1
ACIDITY OF
α-HYDROGENS

Hydrogen and carbon have similar electronegativities, which means that the electrons binding them together are shared almost equally by the two atoms. Consequently, a hydrogen bonded to a carbon is usually not acidic. This is particularly true for hydrogens bonded to sp^3 hybridized carbons because these carbons are the most similar to hydrogen in electronegativity (Section 5.9). The high pK_a of ethane is evidence of the low acidity of hydrogens bonded to sp^3 hybridized carbons.

$$CH_3CH_3 \quad \boxed{pK_a = 50}$$

A hydrogen bonded to an sp^3 hybridized carbon adjacent to a carbonyl carbon is much more acidic than hydrogens bonded to other sp^3 hybridized carbons. For example, the pK_a of a hydrogen bonded to the α-carbon of an aldehyde or a ketone ranges from 16 to 20, and the pK_a of a hydrogen bonded to the α-carbon of an ester is about 25 (Table 19.1). Notice that although an α-hydrogen is more acidic than most other carbon-bound hydrogens, it is less acidic than a hydrogen of water ($pK_a = 15.7$). A compound that contains a relatively acidic hydrogen bonded to a carbon is called a **carbon acid.**

$$\underset{\boxed{pK_a \sim 16-20}}{RCH_2\overset{\overset{O}{\|}}{C}H \qquad RCH_2\overset{\overset{O}{\|}}{C}R} \qquad \underset{\boxed{pK_a \sim 25}}{RCH_2\overset{\overset{O}{\|}}{C}OR}$$

TABLE 19.1 The pK_a Values of Some Carbon Acids

	pK_a		pK_a
$\underset{H}{\overset{\overset{O}{\|}}{CH_2CN(CH_3)_2}}$	30	$\underset{H}{N{\equiv}CCHC{\equiv}N}$	11.8
$\underset{H}{\overset{\overset{O}{\|}}{CH_2COCH_2CH_3}}$	25	$\underset{H}{CH_3\overset{\overset{O}{\|}}{C}CH\overset{\overset{O}{\|}}{C}OCH_2CH_3}$	10.7
$\underset{H}{CH_2C{\equiv}N}$	25	$\underset{H}{\bigcirc{-}\overset{\overset{O}{\|}}{C}CH\overset{\overset{O}{\|}}{C}CH_3}$	9.4
$\underset{H}{\overset{\overset{O}{\|}}{CH_2CCH_3}}$	20	$\underset{H}{CH_3\overset{\overset{O}{\|}}{C}CH\overset{\overset{O}{\|}}{C}CH_3}$	8.9
$\underset{H}{\overset{\overset{O}{\|}}{CH_2CH}}$	17	$\underset{H}{CH_3\overset{\overset{O}{\|}}{C}CH\overset{\overset{O}{\|}}{C}H}$	5.9
$\underset{H}{CH_3CHNO_2}$	8.6	$\underset{H}{O_2NCHNO_2}$	3.6

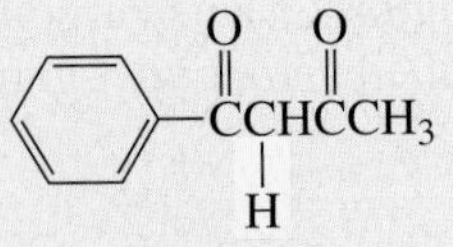

Why is a hydrogen bonded to a carbon adjacent to a carbonyl carbon so much more acidic than hydrogens bonded to other sp^3 hybridized carbons? An α-hydrogen is more acidic because the base formed when the proton is removed from the α-carbon is more stable than the base formed when a proton is removed from other sp^3 hybridized carbons, and acid strength is determined by the stability of the conjugate base that is formed when the acid gives up its proton (Section 1.17).

Why is the base more stable? When a proton is removed from ethane, the electrons left behind reside on a carbon atom. Because carbon is not very electronegative, a carbanion is unstable and therefore difficult to form. As a result, the pK_a of its conjugate acid is very high.

$$CH_3CH_3 \rightleftharpoons CH_3\overset{..}{C}H_2 + H^+$$

When a proton is removed from a carbon adjacent to a carbonyl carbon, two factors combine to increase the stability of the base that is formed. First, the electrons left behind when the proton is removed are delocalized, and electron delocalization increases the stability of a compound (Section 6.6). More importantly, the electrons are delocalized onto an oxygen, an atom that is better able to accommodate the electrons because it is more electronegative than carbon.

$$\underset{\underset{H}{|}}{RCH}-\overset{\overset{O}{||}}{C}-R \rightleftharpoons R\overset{..}{C}H-\overset{\overset{O}{||}}{C}-R \longleftrightarrow RCH=\overset{\overset{O^-}{|}}{C}-R + H^+$$

resonance contributors

PROBLEM 1

The pK_a of propene is 42, which is greater than the pK_a of the carbon acids in Table 19.1 but less than the pK_a of an alkane. Explain.

Now we can understand why aldehydes and ketones ($pK_a = 16$–20) are more acidic than esters ($pK_a = 25$). The electrons left behind when an α-hydrogen is removed from an ester are not as readily delocalized onto the carbonyl oxygen as the electrons left behind when an α-hydrogen is removed from an aldehyde or a ketone. That is because a pair of nonbonding electrons on the oxygen of the OR group of the ester can also be delocalized onto the carbonyl oxygen, so the two pairs of electrons compete for delocalization onto oxygen.

$$R\overset{..}{C}H-\overset{\overset{:\overset{..}{O}:^-}{|}}{C}=\overset{+}{\overset{..}{O}}R \xleftrightarrow[\text{on oxygen}]{\substack{\text{delocalization of} \\ \text{nonbonding electrons}}} R\overset{..}{C}H-\overset{\overset{\overset{..}{O}:}{||}}{C}-\overset{..}{O}R \xleftrightarrow[\text{on the α-carbon}]{\substack{\text{delocalization of} \\ \text{the negative charge}}} RCH=\overset{\overset{:\overset{..}{O}:^-}{|}}{C}-\overset{..}{O}R$$

contributing resonance structures

Nitroalkanes, nitriles, and *N,N*-disubstituted amides also have relatively acidic α-hydrogens (Table 19.1) because, in each case, the electrons left behind when the proton is removed can be delocalized onto an atom that is more electronegative than carbon.

$$CH_3CH_2NO_2 \qquad CH_3CH_2C{\equiv}N \qquad \overset{\overset{O}{||}}{CH_3CN(CH_3)_2}$$

nitroethane propanenitrile *N,N*-dimethylacetamide
$pK_a = 8.6$ $pK_a = 26$ $pK_a = 30$

If the α-carbon is between two carbonyl groups, the acidity of an α-hydrogen is even greater (Table 19.1). For example, an α-hydrogen of ethyl 3-oxobutyrate, a compound with an α-carbon between a ketone carbonyl group and an ester carbonyl group, has a pK_a of 10.7. An α-hydrogen of 2,4-pentanedione, a compound with an α-carbon between two ketone carbonyl groups, has a pK_a of 8.9. Ethyl 3-oxobutyrate is classified as a **β-keto ester** because the ester has a carbonyl group at the β-position. 2,4-Pentanedione is a **β-diketone.**

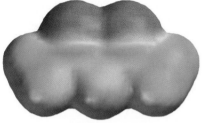

2,4-pentanedione

$$CH_3-\overset{O}{\underset{\|}{C}}-CH_2-\overset{O}{\underset{\|}{C}}-OCH_2CH_3$$

ethyl 3-oxobutyrate
ethyl acetoacetate
a β-keto ester
$pK_a = 10.7$

$$CH_3-\overset{O}{\underset{\|}{C}}-CH_2-\overset{O}{\underset{\|}{C}}-CH_3$$

2,4-pentanedione
acetylacetone
a β-diketone
$pK_a = 8.9$

The acidity of α-hydrogens bonded to carbons flanked by two carbonyl groups increases because the electrons left behind when the proton is removed can be delocalized onto *two* oxygen atoms. β-Diketones have lower pK_a values than β-keto esters because electrons are more readily delocalized onto ketone carbonyl groups than they are onto ester carbonyl groups.

$$CH_3-\overset{:\ddot{O}}{\underset{\|}{C}}-CH=\overset{:\ddot{O}:^-}{\underset{\|}{C}}-CH_3 \longleftrightarrow CH_3-\overset{:\ddot{O}}{\underset{\|}{C}}-\overset{..}{\underset{}{C}}H-\overset{:\ddot{O}}{\underset{\|}{C}}-CH_3 \longleftrightarrow CH_3-\overset{:\ddot{O}:^-}{\underset{}{C}}=CH-\overset{:\ddot{O}}{\underset{\|}{C}}-CH_3$$

resonance contributors for the 2,4-pentanedione anion

PROBLEM 2◆

Give an example of:

a. a β-keto nitrile

b. a β-diester

PROBLEM 3◆

List the compounds in each of the following groups in order of decreasing acidity:

a. $CH_2=CH_2$ CH_3CH_3 $CH_3\overset{O}{\underset{\|}{C}}H$ $HC\equiv CH$

b. $CH_3\overset{O}{\underset{\|}{C}}CH_2\overset{O}{\underset{\|}{C}}CH_3$ $CH_3O\overset{O}{\underset{\|}{C}}CH_2\overset{O}{\underset{\|}{C}}OCH_3$ $CH_3\overset{O}{\underset{\|}{C}}CH_2\overset{O}{\underset{\|}{C}}OCH_3$ $CH_3\overset{O}{\underset{\|}{C}}CH_3$

c.

19.2
KETO–ENOL
TAUTOMERISM

A ketone exists in equilibrium with its enol tautomer (Section 5.6). The two constitutional isomers are **tautomers** because they differ in the location of a double bond and a hydrogen.

$$RCH_2-\overset{O}{\underset{\|}{C}}-R \rightleftharpoons RCH=\overset{OH}{\underset{}{C}}-R$$

keto tautomer   **enol tautomer**

For most ketones, the enol tautomer is much less stable than the keto tautomer. For example, an aqueous solution of acetone exists as an equilibrium mixture of more than 99.9% keto tautomer and less than 0.1% enol tautomer.

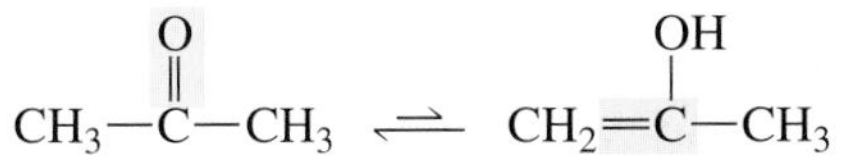

$$CH_3-\overset{\overset{\displaystyle O}{\|}}{C}-CH_3 \ \rightleftharpoons \ CH_2=\overset{\overset{\displaystyle OH}{|}}{C}-CH_3$$

\> 99.9%   < 0.1%

keto tautomer   enol tautomer

The fraction of enol tautomer is considerably greater for a β-diketone because the enol tautomer is stabilized by hydrogen bonding and by conjugation of the carbon–carbon double bond with the second carbonyl group.

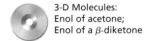

3-D Molecules:
Enol of acetone;
Enol of a β-diketone

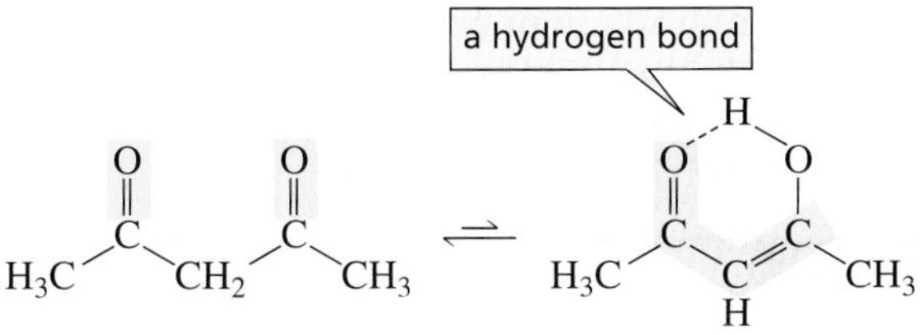

a hydrogen bond

85%   15%

keto tautomer   enol tautomer

Phenol is unusual in that its enol tautomer is *more* stable than its keto tautomer because the enol tautomer is aromatic while the keto tautomer is not.

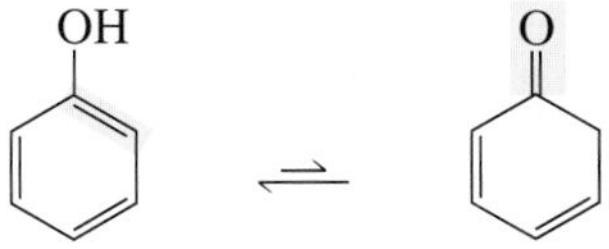

enol tautomer   keto tautomer

PROBLEM 4

Only 15% of 2,4-pentanedione exists as the enol tautomer in water, but 92% exists as the enol tautomer in hexane. Explain why this is so.

Now that we know that a hydrogen on a carbon adjacent to a carbonyl carbon is somewhat acidic, we can understand the mechanism for the interconversion of keto and enol tautomers we first encountered in Chapter 5. **Keto–enol interconversion** is also called **keto–enol tautomerization** or **enolization.** The interconversion of the tautomers can be catalyzed by either acids or bases.

In a basic solution, hydroxide ion removes a proton from the α-carbon of the keto tautomer, forming an enolate ion. Protonation on oxygen forms the enol tautomer, whereas protonation on the α-carbon re-forms the keto tautomer.

base-catalyzed keto–enol interconversion

keto tautomer   enolate ion   enol tautomer

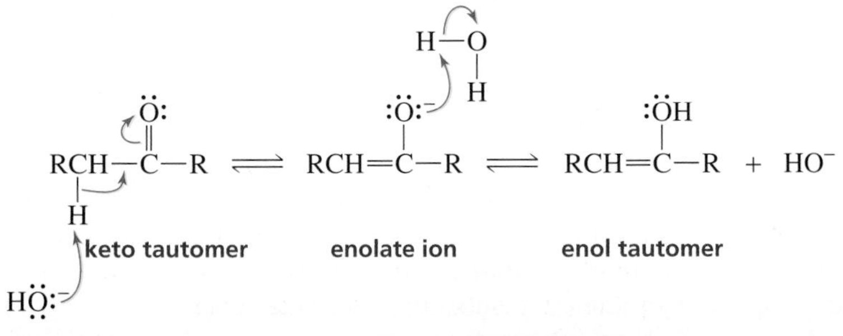

In an acidic solution, the carbonyl oxygen of the keto tautomer is protonated and water removes a proton from the α-carbon, forming the enol.

acid-catalyzed keto–enol interconversion

$$RCH_2{-}\overset{\overset{\ddot{O}:}{\|}}{C}{-}R \underset{-H^+}{\overset{H^+}{\rightleftharpoons}} RCH{-}\overset{\overset{+\,\ddot{O}H}{|}}{\underset{H}{C}}{-}R \rightleftharpoons RCH{=}\overset{\overset{:\ddot{O}H}{|}}{C}{-}R \;+\; H_3O^+$$

keto tautomer ⟶ H₂Ö: ⟶ enol tautomer

Notice that the steps are reversed in the base- and acid-catalyzed reactions. In the base-catalyzed reaction, the base removes the α-proton in the first step and the oxygen is protonated in the second step. In the acid-catalyzed reaction, the acid protonates the oxygen in the first step and the α-proton is removed in the second step.

PROBLEM 5◆

Draw the enol tautomers for each of the following compounds. For those compounds that have more than one enol tautomer, indicate which is more stable.

a. $CH_3CH_2\overset{\overset{O}{\|}}{C}CH_2CH_3$

b. [benzene ring]$\overset{\overset{O}{\|}}{C}CH_3$

c. [cyclohexanone with O]

d. [cyclohexane-1,3-dione: ring with O at top and O at lower right]

e. $CH_3CH_2\overset{\overset{O}{\|}}{C}CH_2\overset{\overset{O}{\|}}{C}CH_2CH_3$

f. [benzene ring]$-CH_2\overset{\overset{O}{\|}}{C}CH_3$

19.3 REACTIVITY CONSIDERATIONS

The carbon–carbon double bond of an enol suggests that it is a nucleophile (like an alkene). An enol is more electron-rich than an alkene because the oxygen atom donates electrons by resonance. As a result, an enol is a better nucleophile. The electrostatic potential map of the enol of acetone shows the electron-rich (red) α-carbon.

enol of acetone

$$RCH{=}\overset{\overset{:\ddot{O}H}{|}}{C}{-}R \longleftrightarrow \underset{\cdot\cdot}{R}CH{-}\overset{\overset{+\ddot{O}H}{\|}}{C}{-}R$$

electron-rich α-carbon

resonance contributors for an enol

Carbonyl compounds that form enols undergo substitution reactions at the α-carbon. When an α-substitution reaction takes place under acidic conditions, water removes a proton from the α-carbon of the protonated carbonyl compound. The nucleophilic enol then reacts with an electrophile. The overall reaction is an **α-substitution reaction**—one electrophile (E^+) substitutes for another (H^+).

enolate of acetone

Removing a proton from the α-carbon under basic conditions forms an enolate ion. Enolate ions are much better nucleophiles than enols because they are negatively charged. The resonance contributors of the enolate ion show that it has two electron-rich sites—the α-carbon and the oxygen. The enolate ion is an example of an ambident nucleophile (*ambi* is Latin for "both"; *dent* is Latin for "teeth"). An **ambident nucleophile** is a nucleophile with two nucleophilic sites ("two teeth").

resonance contributors for an enolate ion

Which nucleophilic site (C or O) will react with an electrophile depends on the electrophile and on the reaction conditions. Protonation occurs preferentially on oxygen because there is a greater concentration of negative charge on the more electronegative oxygen atom. However, when the electrophile is something other than a proton, carbon is more likely to be the nucleophile because carbon is a better nucleophile than oxygen.

When an α-substitution reaction takes place under basic conditions, a base removes a proton from the α-carbon and the nucleophilic enolate ion then reacts with an electrophile.

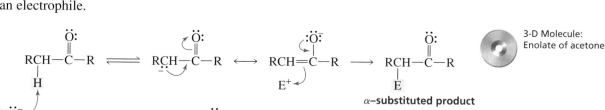

3-D Molecule: Enolate of acetone

Notice the similarity between keto–enol interconversion and α-substitution. Actually, keto–enol interconversion is an α-substitution reaction in which hydrogen serves as both the electrophile that is removed from the α-carbon and the electrophile that is added to the α-carbon (when the enol or enolate reverses back to the keto tautomer).

As various α-substitution reactions are discussed in this chapter, notice that they all follow basically the same mechanism—a proton is removed from an α-carbon and the α-carbon reacts with an electrophile. The reactions differ only in the nature of the base and the electrophile—and whether the reaction is carried out under acidic or basic conditions.

PROBLEM 6

Explain why the aldehyde hydrogen is not exchanged with deuterium.

$$CH_3CH \xrightarrow[\text{D}_2\text{O}]{^{-}\text{OD}} CD_3CH$$

19.4
HALOGENATION OF THE α-CARBON OF ALDEHYDES AND KETONES

Acid-Catalyzed Halogenation

When Br_2, Cl_2, or I_2 is added to an acidic solution of an aldehyde or a ketone, a halogen replaces one of the α-hydrogens of the carbonyl compound.

$$RCH_2CH{=}O + Br_2 \xrightarrow{\text{H}^+, \text{H}_2\text{O}} RCHCH{=}O\ (Br) + HBr$$

(cyclohexanone) $+ Cl_2 \xrightarrow{\text{H}^+, \text{H}_2\text{O}}$ (2-chlorocyclohexanone) $+ HCl$

(acetophenone $C_6H_5CCH_3$) $+ I_2 \xrightarrow{\text{H}^+, \text{H}_2\text{O}}$ ($C_6H_5CCH_2I$) $+ HI$

In the first step of this acid-catalyzed reaction, the carbonyl oxygen is protonated. Water is the base that removes a proton from the α-carbon, forming an enol that attacks an electrophilic bromine.

3-D Molecule: α-Bromoacetone

Base-Promoted Halogenation

When excess Br_2, Cl_2, or I_2 is added to a *basic* solution of an aldehyde or a ketone, the halogen replaces *all* the α-hydrogens.

$$R{-}CH_2{-}C({=}O){-}R + Br_2 \text{ (excess)} \xrightarrow{\text{HO}^-} R{-}C(Br)_2{-}C({=}O){-}R + 2\,Br^-$$

In the first step of this base-promoted reaction, hydroxide ion removes a proton from the α-carbon. The enolate ion then reacts with the electrophilic bromine. These two steps are repeated until all the α-hydrogens are replaced by bromine.

Each successive halogenation is more rapid than the previous one because the electron-withdrawing bromine increases the acidity of the remaining α-hydrogens. This is why *all* the α-hydrogens are replaced by bromines. Under acidic conditions, on the other hand, each successive halogenation is slower than the previous one because the electron-withdrawing bromine decreases the basicity of the carbonyl oxygen, thereby making protonatation of the carbonyl oxygen less favorable.

The Haloform Reaction

In the presence of excess base and excess halogen, a methyl ketone is converted first into a trihalo-substituted ketone and then into a carboxylic acid. After the trihalo-substituted ketone is formed, hydroxide ion attacks the carbonyl carbon. Because the trihalomethyl ion is a weaker base than hydroxide ion (the pK_a of CHI_3 is 14, the pK_a of H_2O is 15.7), the trihalomethyl ion is the group more easily expelled from the tetrahedral intermediate. The conversion of a methyl ketone to a carboxylic acid is called a **haloform reaction** because one of the products is haloform—$CHCl_3$ (chloroform), $CHBr_3$ (bromoform), or CHI_3 (iodoform).

the haloform reaction

PROBLEM 7◆

A ketone undergoes acid-catalyzed bromination, acid-catalyzed chlorination, and acid-catalyzed deuterium exchange at the α-carbon, all at about the same rate. What does this tell you about the mechanism of the reactions?

Carl Magnus von Hell (1849–1926) was born in Germany. He studied with H. Fehling at the University of Stuttgart and with Richard Erlenmeyer (1825–1909) at the University of Munich. Von Hell reported the HVZ reaction in 1881, and the reaction was independently confirmed by both Volhard and Zelinski in 1887.

Jacob Volhard (1834–1910) was also born in Germany. Brilliant but lacking direction, his parents sent him to England to be with August von Hofmann (Section 11.11), a family friend. After working with Hofmann, Volhard became a professor of chemistry—first at the University of Munich, then at the University of Erlangen, and later at the University of Halle. He was the first to synthesize sarcosine and creatine.

19.5 HALOGENATION OF THE α-CARBON OF CARBOXYLIC ACIDS: THE HELL–VOLHARD–ZELINSKI REACTION

Carboxylic acids do not undergo substitution reactions at the α-carbon because the OH group is much more acidic than the α-carbon. However, if a carboxylic acid is treated with PBr_3 and Br_2, then bromination at the α-carbon occurs. (Red phosphorus can be used in place of PBr_3 because P and excess Br_2 react to form PBr_3.) This halogenation reaction is known as the **Hell–Volhard–Zelinski reaction** or, more simply, as the **HVZ reaction.** We will see when we look at the mechanism of the HVZ reaction that α-substitution occurs because an acyl bromide, rather than a carboxylic acid, undergoes α-substitution.

In the first step of the HVZ reaction, PBr_3 converts the carboxylic acid into an acyl bromide by a mechanism similar to the one by which PBr_3 converts an alcohol into an alkyl bromide (Section 11.2). (In both reactions PBr_3 replaces an OH with a Br.) The acyl bromide is in equilibrium with its enol. Bromination of the enol forms the α-brominated acyl bromide, which is hydrolyzed to the α-brominated carboxylic acid.

Nikolai Dimitrievich Zelinski (1861–1953) was born in Moldavia. He was a professor of chemistry at the University of Moscow. In 1911 he left the university to protest the firing of the entire administration by the Ministry of Education. He went to St. Petersburg, where he directed the laboratory of the Ministry of Finances. In 1917, after the revolution, he returned to the University of Moscow.

mechanism for the Hell–Volhard–Zelinski reaction

$$CH_3CH_2-\overset{\overset{\displaystyle \ddot{O}:}{\|}}{C}-OH \xrightarrow{PBr_3} CH_3CH_2-\overset{\overset{\displaystyle \ddot{O}:}{\|}}{C}-Br \rightleftharpoons CH_3CH=\overset{\overset{\displaystyle :\ddot{O}H}{|}}{C}-Br$$

a carboxylic acid   an acyl bromide   enol

$$CH_3\overset{\underset{\displaystyle Br}{|}}{CH}-\overset{\overset{\displaystyle \ddot{O}:}{\|}}{C}-OH \xleftarrow{H_2O} CH_3\overset{\underset{\displaystyle Br}{|}}{CH}-\overset{\overset{\displaystyle \ddot{O}:}{\|}}{C}-Br \underset{H^+}{\overset{-H^+}{\rightleftharpoons}} CH_3\overset{\underset{\displaystyle Br}{|}}{CH}-\overset{\overset{\displaystyle ^+\ddot{O}H}{\|}}{C}-Br \;+\; Br^-$$

an α-brominated carboxylic acid   an α-brominated acyl bromide

19.6 α-HALOGENATED CARBONYL COMPOUNDS IN SYNTHESIS

We have seen that because a base can remove a hydrogen from an α-carbon of an aldehyde or a ketone (Section 19.2), the α-carbon becomes nucleophilic—it reacts with electrophiles.

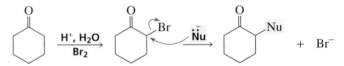

However, when the α-position is halogenated, the α-carbon becomes electrophilic—it reacts with nucleophiles. Therefore, α-halogenation greatly increases the kinds of substituents that can be placed on α-carbon atoms.

Tutorial: α-Halogenated carbonyl compounds in synthesis

α-Brominated carbonyl compounds are also useful to synthetic chemists because once a bromine has been introduced into the α-position of a carbonyl compound, an α,β-unsaturated carbonyl compound can be prepared by means of an (E2) elimination reaction, using a strong and bulky base to encourage elimination over substitution (Section 10.8).

an α,β-unsaturated carbonyl compound

PROBLEM 8

How would you prepare the following compounds from the given starting materials?

a. $CH_3CH_2\overset{\overset{\displaystyle O}{\|}}{C}H \longrightarrow CH_3\overset{\underset{\displaystyle N(CH_3)_2}{|}}{CH}\overset{\overset{\displaystyle O}{\|}}{C}H$

b. $CH_3CH_2\overset{\overset{\displaystyle O}{\|}}{C}H \longrightarrow CH_3\overset{\underset{\displaystyle OH}{|}}{CH}\overset{\overset{\displaystyle O}{\|}}{C}H$

c.

d.

PROBLEM 9

How could the following compounds be prepared from a carbonyl compound with no carbon–carbon double bonds?

a.  $CH_3CH=CHCCH_2CH_2CH_3$ (with C=O)

b. 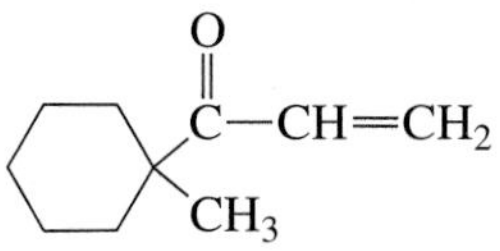 (cyclohexane ring with) $C-CH=CH_2$ and CH_3 (and C=O)

PROBLEM 10

How could the following compounds be prepared from cyclohexanone?

a. 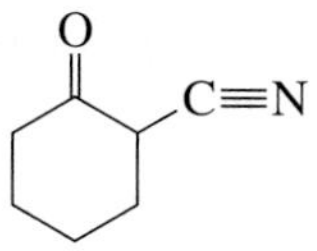 (cyclohexanone with) $C\equiv N$

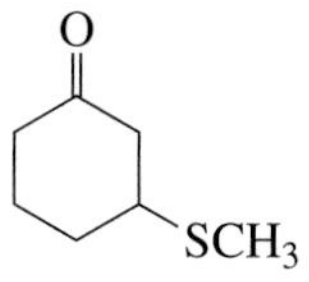

b. 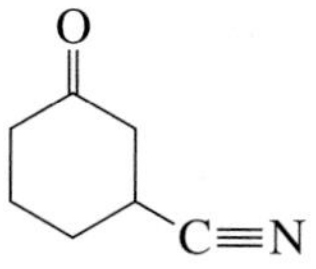 (cyclohexanone with) $C\equiv N$

c. (cyclohexanone with) SCH_3

d. 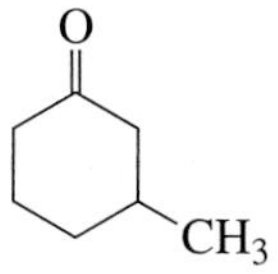 (cyclohexanone with) CH_3

The amount of carbonyl compound converted to enolate depends on the pK_a of the carbonyl compound and the base used to remove the α-proton. For example, when hydroxide ion (the pK_a of its conjugate acid is 15.7) is used to remove an α-proton from cyclohexanone ($pK_a = 17$), only a small amount of the carbonyl compound is converted into enolate because hydroxide ion is a weaker base than the base being formed. When lithium diisopropylamide (LDA) is used to remove the α-proton (the pK_a of its conjugate acid is about 35), essentially all the carbonyl compound is converted to enolate because LDA is a much stronger base than the base being formed (Section 1.17). Therefore, LDA is the base of choice for those reactions that require the carbonyl compound to be completely converted to enolate before it reacts with an electrophile.

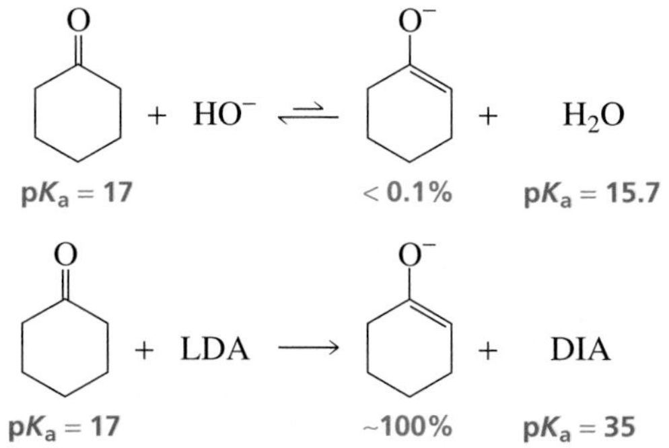

19.7

USING LITHIUM DIISOPROPYLAMIDE (LDA) TO FORM AN ENOLATE

$$\text{(cyclohexanone)} + HO^- \rightleftharpoons \text{(enolate)} + H_2O$$

$pK_a = 17$ $< 0.1\%$ $pK_a = 15.7$

$$\text{(cyclohexanone)} + LDA \longrightarrow \text{(enolate)} + DIA$$

$pK_a = 17$ $\sim 100\%$ $pK_a = 35$

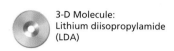

Using a nitrogen base to remove a proton can be a problem because a nitrogen base can also react as a nucleophile and attack the carbonyl carbon (Section 17.7). However, the two bulky alkyl substituents bonded to the nitrogen of LDA make it difficult for the nitrogen to get close enough to the carbonyl carbon to attack it. Consequently, LDA is a strong base but a poor nucleophile, so it removes an α-hydrogen much faster than it attacks a carbonyl carbon. LDA is easily prepared by adding butyllithium to diisopropylamine (DIA) in THF at −78 °C.

$$CH_3CHNHCHCH_3 \ + \ CH_3CH_2CH_2\overset{-}{C}H_2\overset{+}{Li} \ \xrightarrow[-78\ °C]{THF} \ CH_3CHN\overset{-}{C}HCH_3 \ + \ CH_3CH_2CH_2CH_3$$

diisopropylamine	butyllithium		lithium diisopropylamide	butane
$pK_a = 35$			LDA	$pK_a \sim 50$

19.8
ALKYLATION OF THE α-CARBON OF CARBONYL COMPOUNDS

Alkylation of the α-carbon of a carbonyl compound is an important reaction because it gives us another way to form a carbon–carbon bond. Alkylation is carried out by *first* removing a proton from the α-carbon with a strong base such as LDA and *then* adding the appropriate alkyl halide.

Ketones, esters, and nitriles can be alkylated at the α-carbon in this way. Aldehydes, however, give poor yields of α-alkylated products (Section 19.11).

Two different products can be formed when the ketone has two different R groups because both α-carbons can be alkylated. For example, methylation of 2-methylcyclohexanone with one equivalent of methyl iodide forms both 2,6-dimethylcyclohexanone and 2,2-dimethylcyclohexanone. The relative amounts of the two products depend on the reaction conditions.

2-methylcyclohexanone

2,6-dimethylcyclohexanone   **2,2-dimethylcyclohexanone**

The enolate leading to 2,6-dimethylcyclohexanone is the *kinetic* enolate because the α-proton that is removed to make this enolate is more accessible and slightly more acidic. So 2,6-dimethylcyclohexanone is formed faster and is the major product if the reaction is carried out at $-78\ ^\circ C$.

The enolate leading to 2,2-dimethylcyclohexanone is the *thermodynamic* enolate because it has the more substituted double bond, making it the more stable enolate. (Substitution increases enolate stability for the same reason that substitution increases alkene stability; Section 3.19.) Therefore, 2,2-dimethylcyclohexanone is the major product if the reaction is carried out under conditions that cause enolate formation to be reversible (high temperature).

The less substituted α-carbon can be alkylated—without having to control the conditions to make certain that the reaction does not become reversible—by first making the *N,N*-dimethylhydrazone of the ketone.

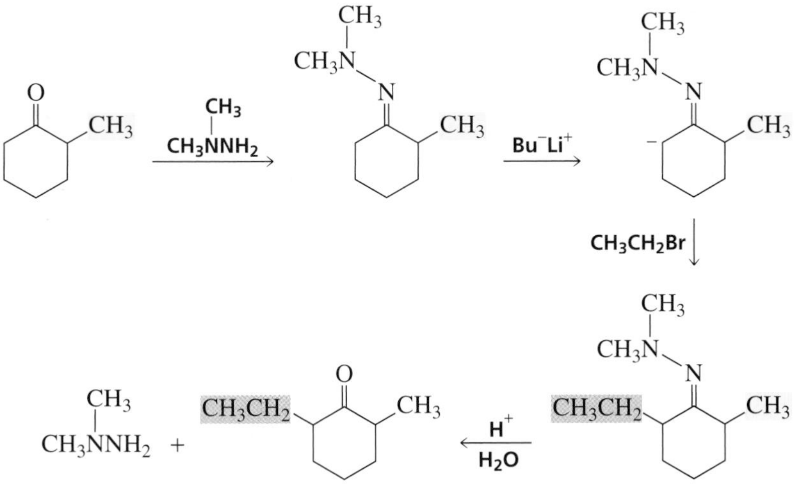

The *N,N*-dimethylhydrazone will form so that the dimethylamino group is pointing away from the more substituted α-carbon. The nitrogen of the dimethylamino group by coordinating with the lithium ion of butyllithium ($Bu^- Li^+$), the base generally employed in this reaction, directs the base to the less substituted carbon. Hydrolysis of the hydrazone re-forms the ketone (Section 17.7).

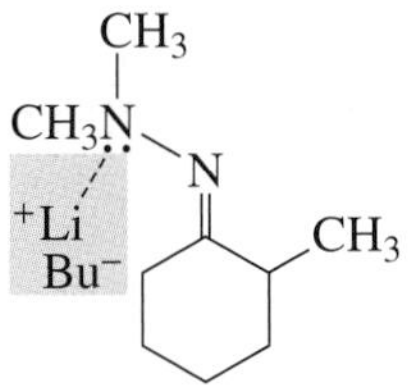

3-D Molecule:
N,N-Dimethylhydrazone
of 2-methylcyclohexanone

PROBLEM 11◆

What compound is formed when cyclohexanone is shaken with NaOD in D_2O for several hours?

PROBLEM 12

Explain why alkylation of an α-carbon works best if the alkyl halide used in the reaction is a primary alkyl halide and does not work at all if it is a tertiary alkyl halide.

> ### PROBLEM 13◆
>
> How could each of the following compounds be prepared from the given starting material?
>
> **a.** 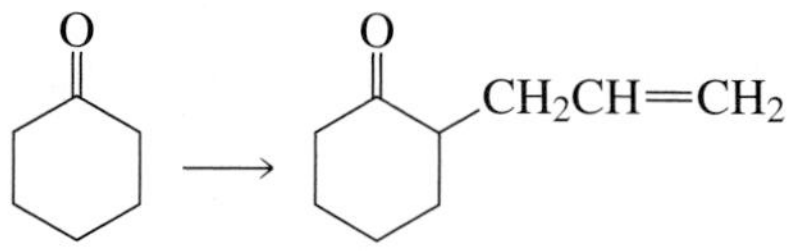
>
> **b.** 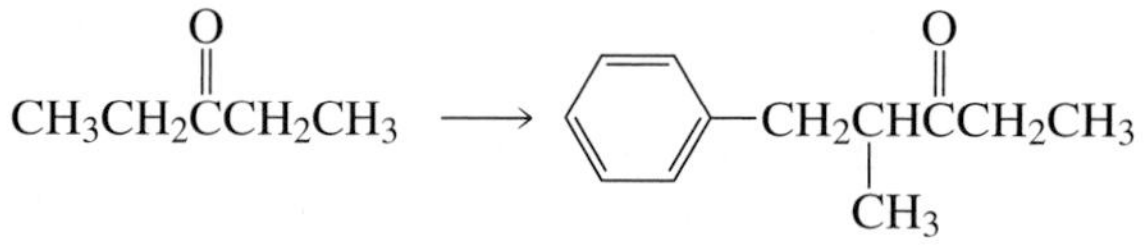 $CH_3CH_2CCH_2CH_3$ ⟶

19.9 ALKYLATION AND ACYLATION OF THE α-CARBON VIA AN ENAMINE INTERMEDIATE

We have seen that an enamine is formed when an aldehyde or a ketone reacts with a secondary amine (Section 17.7).

pyrrolidine
secondary amine

enamine

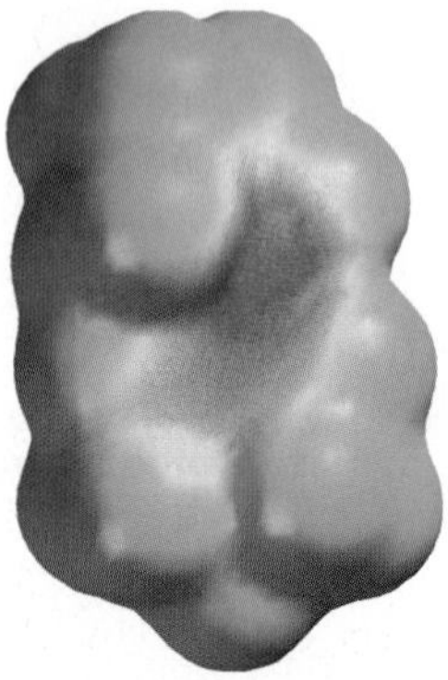

Enamines react with electrophiles in the same way that enolates do.

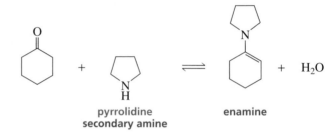

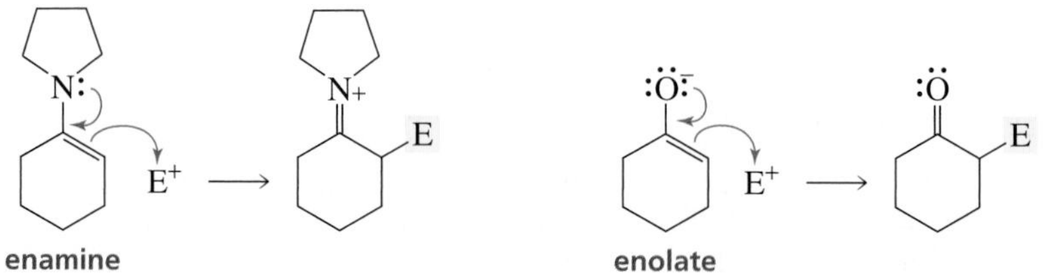

enamine

enolate

enamine formed by cyclohexanone and pyrrolidine

This means that electrophiles can be added to the α-carbon of an aldehyde or a ketone by first converting the carbonyl compound to an enamine (by treating the carbonyl compound with a secondary amine; Section 17.7), adding the electrophile, and then hydrolyzing the imine back to the ketone.

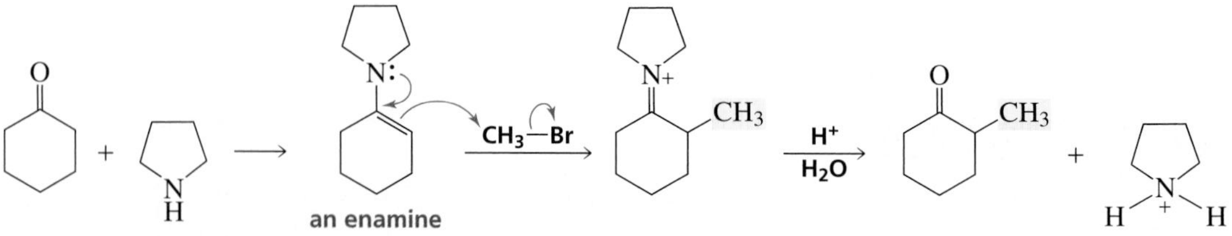

an enamine

Because the alkylation step is an S_N2 reaction, only primary alkyl halides or methyl halides should be used (Section 9.2).

One advantage to using an enamine intermediate to alkylate an aldehyde or a ketone is that only the monoalkylated product is formed.

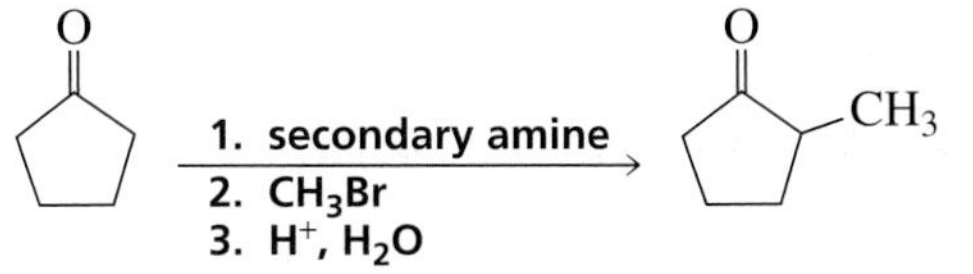

When a carbonyl compound is alkylated directly, dialkylated and O-alkylated products can also be formed.

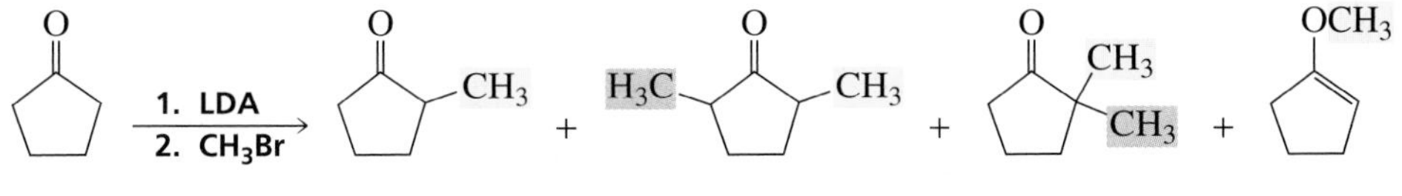

Aldehyde and ketones can also be acylated via an enamine intermediate.

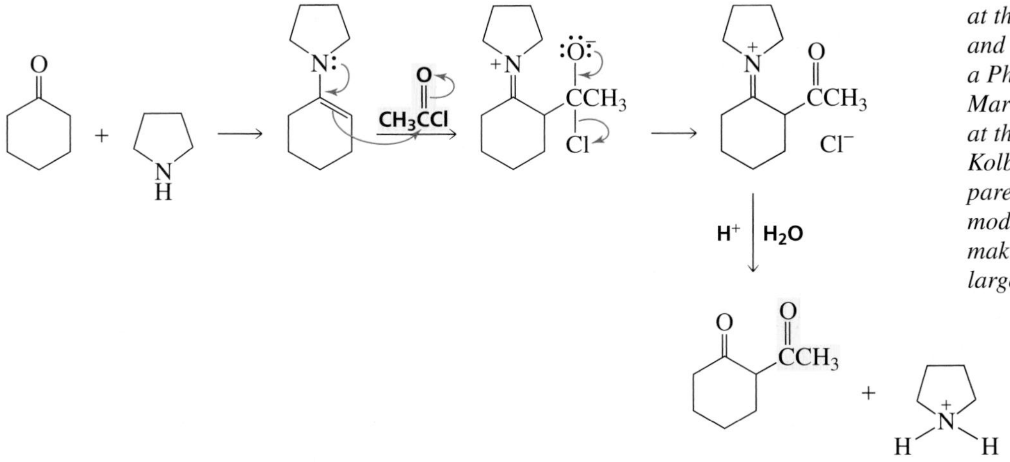

Hermann Kolbe (1818–1884) and *Rudolph Schmitt (1830–1898) were born in Germany. Kolbe was a professor at the Universities of Marburg and Leipzig. Schmitt received a Ph.D. from the University of Marburg and was a professor at the University of Dresden. Kolbe discovered how to prepare aspirin in 1859. Schmitt modified the synthesis in 1885, making aspirin available in large quantities at a low price.*

THE SYNTHESIS OF ASPIRIN

The first step in the industrial synthesis of aspirin is known as the **Kolbe–Schmitt carboxylation reaction.** The phenolate ion reacts with carbon dioxide under pressure to form *o*-hydroxybenzoic acid, also known as salicylic acid. Acetylation of salicylic acid forms acetylsalicylic acid (aspirin).

During World War I, the American subsidiary of the Bayer Company bought as much phenol as it could from the international market with the knowledge that eventually all the phenol could be converted into aspirin. This left little phenol available for other countries to purchase for the synthesis of 2,4,6-trinitrophenol, a common explosive.

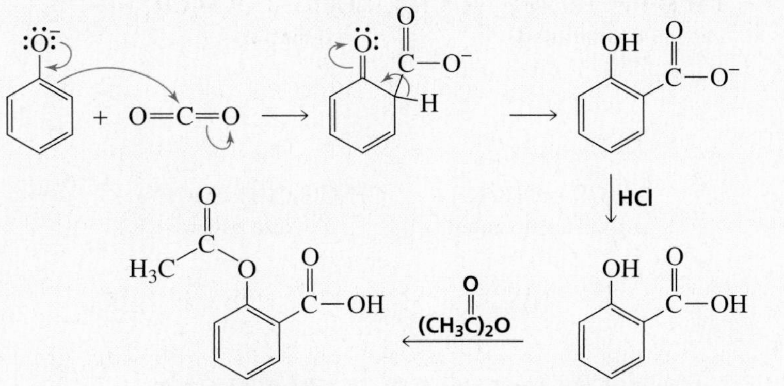

acetylsalicylic acid
aspirin

salicylic acid
o-hydroxybenzoic acid

PROBLEM 14

Describe how the following compounds could be prepared using an enamine intermediate:

a. 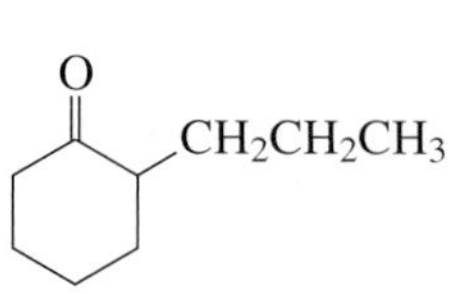 (cyclohexanone with CH₂CH₂CH₃ substituent)

b. 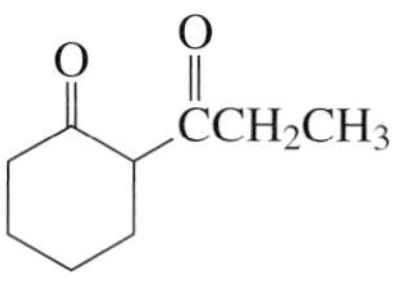 (cyclohexanone with CCH₂CH₃ ketone substituent, $\overset{O}{\overset{\|}{C}}$)

19.10 ALKYLATION OF THE β-CARBON: THE MICHAEL REACTION

In Section 17.14 we saw that nucleophiles react with α,β-unsaturated carbonyl compounds, forming either direct addition products or conjugate addition products. Nucleophiles that are strong bases (unless they are bulky) add preferentially to the carbonyl carbon, forming direct addition products. Weak bases and bulky strong bases add preferentially to the β-carbon, forming conjugate addition products.

$$RCH=CHCR \xrightarrow[\text{2. H}^+]{\text{1. Nu}} \underset{\underset{\text{Nu}}{|}}{RCH=CHCR} + \underset{\underset{\text{Nu}}{|}}{RCHCH_2CR}$$

direct addition   conjugate addition

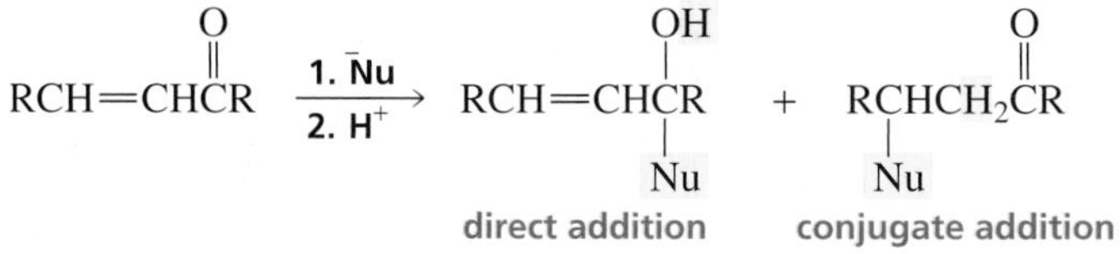

Arthur Michael (1853–1942) was born in Buffalo, New York. He studied at the University of Heidelberg, the University of Berlin, and L'École de Médicine in Paris. He was a professor of chemistry at Tufts and Harvard Universities, retiring from Harvard when he was 83.

Many different nucleophiles can add to α,β-unsaturated carbonyl compounds. When the nucleophile is an enolate, the addition reaction has a special name—it is called a **Michael reaction.** The enolates that work best in Michael reactions are those that are flanked by two electron-withdrawing groups—enolates of β-diketones, β-diesters, β-keto esters, and β-keto nitriles. Because these enolates are relatively weak bases and because they can be quite bulky, addition occurs at the β-carbon. Notice that a wide variety of α,β-unsaturated carbonyl compounds undergo Michael reactions, and that Michael reactions form 1,5-dicarbonyl compounds.

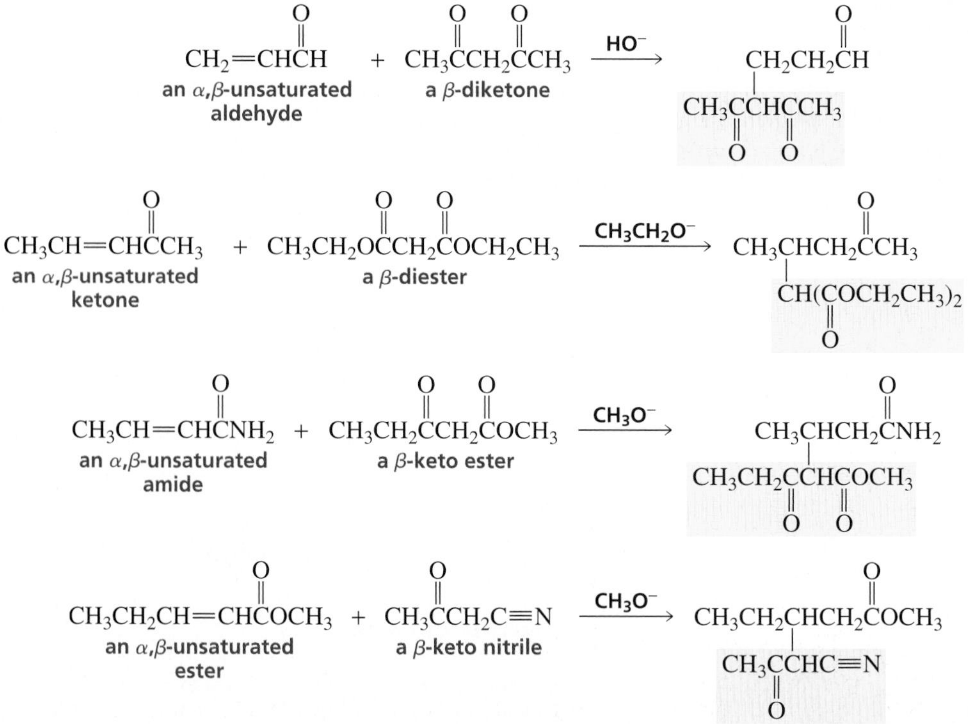

$$CH_2=CHCH + CH_3CCH_2CCH_3 \xrightarrow{HO^-} \underset{CH_3CCHCCH_3}{CH_2CH_2CH}$$

an α,β-unsaturated aldehyde   a β-diketone

$$CH_3CH=CHCCH_3 + CH_3CH_2OCCH_2COCH_2CH_3 \xrightarrow{CH_3CH_2O^-} CH_3CHCH_2CCH_3$$

an α,β-unsaturated ketone   a β-diester $CH(COCH_2CH_3)_2$

$$CH_3CH=CHCNH_2 + CH_3CH_2CCH_2COCH_3 \xrightarrow{CH_3O^-} CH_3CHCH_2CNH_2$$

an α,β-unsaturated amide   a β-keto ester $CH_3CH_2CCHCOCH_3$

$$CH_3CH_2CH=CHCOCH_3 + CH_3CCH_2C\equiv N \xrightarrow{CH_3O^-} CH_3CH_2CHCH_2COCH_3$$

an α,β-unsaturated ester   a β-keto nitrile $CH_3CCHC\equiv N$

All these reactions take place by the same mechanism. A base removes a proton from the α-carbon of the carbon acid. The enolate adds to the β-carbon of an α,β-unsaturated carbonyl compound, and the α-carbon obtains a proton from the solvent.

$$
\underset{\substack{\| \ \ \ \| \\ RCCH_2CR}}{\overset{O \ \ \ O}{}} \quad \overset{HO^-}{\rightleftharpoons} \quad \underset{\substack{\| \ \ \ \| \\ RCCHCR}}{\overset{O \ \ \ O}{}} \quad \underset{RCH=CH-CR}{\overset{\beta \quad \alpha \quad :\overset{..}{O}}{}} \longrightarrow \underset{\substack{| \\ RCCHCR \\ \| \ \ \| \\ O \ \ O}}{RCH-CH=CR} \longrightarrow \underset{\substack{| \\ RCCHCR \\ \| \ \ \| \\ O \ \ O}}{RCH-CH_2-CR} + HO^-
$$

Notice that if either of the reactants in a Michael reaction has an ester group, the base used to remove the α-proton is the same as the leaving group of the ester. This is done because the base, in addition to being able to remove an α-proton, can react as a nucleophile and attack the carbonyl group of the ester. If the nucleophile is identical to the OR group of the ester, nucleophilic attack on the carbonyl group will not change the reactant.

$$
\underset{}{CH_3-\overset{\overset{..}{O}:}{\underset{}{C}}-\overset{..}{O}CH_3} + CH_3\overset{..}{O}{:}^- \rightleftharpoons \underset{\underset{:OCH_3}{|}}{CH_3-\overset{:\overset{..}{O}:^-}{\underset{}{C}}-\overset{..}{O}CH_3}
$$

Enamines can be used in place of enolates in Michael reactions. When an enamine is used as a nucleophile in a Michael reaction, the reaction is called a **Stork enamine reaction.**

Gilbert Stork *was born in Belgium in 1921. He graduated from the University of Florida and received a Ph.D. from the University of Wisconsin. He was a professor of chemistry at Harvard University and has been a professor at Columbia University since 1953. He has been responsible for the development of many new synthetic procedures in addition to the one bearing his name.*

PROBLEM 15◆

What reagents would you use to prepare the following compounds?

a.

b. $CH_3\overset{O}{\overset{\|}{C}}CH_2CH_2CH(\overset{O}{\overset{\|}{C}}OCH_2CH_3)_2$

In Chapter 17 we saw that aldehydes and ketones are electrophiles and therefore can react with nucleophiles. In the preceding sections we have seen that when a proton is removed from the α-carbon of an aldehyde or a ketone, the resulting anion

19.11
THE ALDOL
ADDITION

is a nucleophile and therefore reacts with electrophiles. An **aldol addition** is a reaction in which *both* of these activities are observed: One molecule of a carbonyl compound—after a proton is removed from an α-carbon—reacts as a *nucleophile* and attacks the *electrophilic* carbonyl carbon of a second molecule of the carbonyl compound.

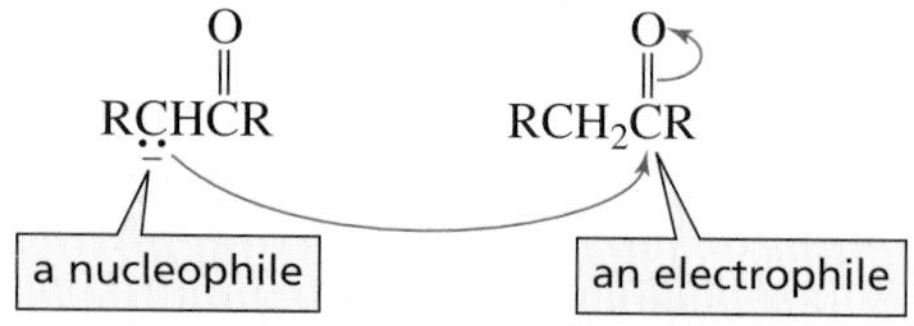

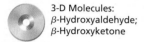

3-D Molecules: β-Hydroxyaldehyde; β-Hydroxyketone

An aldol addition is a reaction between two molecules of an *aldehyde* or two molecules of a *ketone*. When the reactant is an aldehyde, the addition product is a β-hydroxyaldehyde, which is why the reaction is called an aldol addition ("ald" for aldehyde, "ol" for alcohol). When the reactant is a ketone, the addition product is a β-hydroxyketone. Because the addition reaction is reversible, good yields of the addition product are obtained only if it is removed from the solution as it is formed.

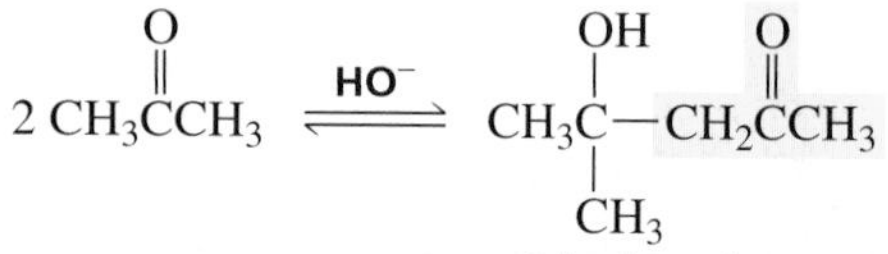

aldol additions

a β-hydroxyaldehyde

a β-hydroxyketone

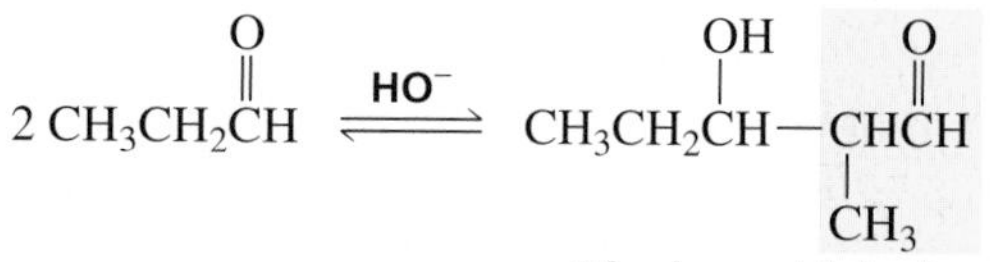

In the first step of an aldol addition, a base removes an α-proton from the carbonyl compound, creating an enolate. The enolate adds to the carbonyl carbon of a second molecule of carbonyl compound, and the resulting negatively charged oxygen is protonated by the solvent.

mechanism for the aldol addition

a β-hydroxyaldehyde

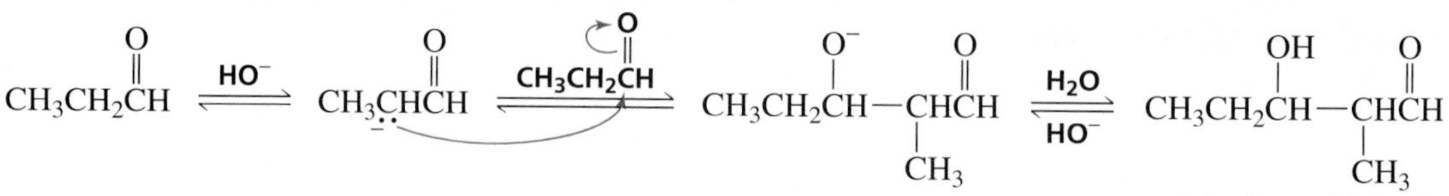

Ketones are less susceptible than aldehydes to attack by nucleophiles, so aldol additions occur more slowly with ketones. The relatively high reactivity of aldehydes in competing aldol addition reactions is what causes them to give low yields of α-alkylation products (Sections 19.8).

a β-hydroxyketone

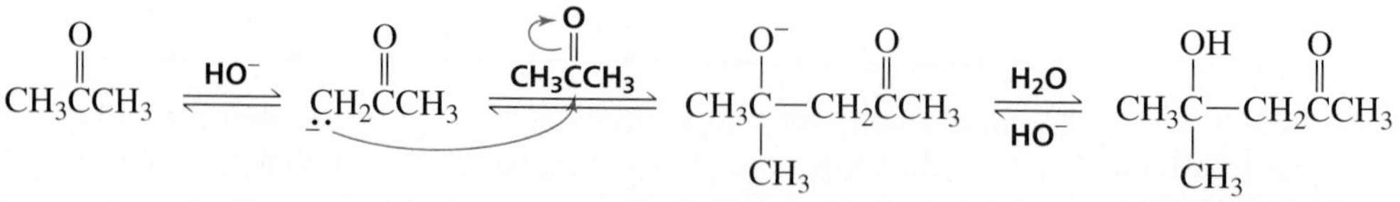

Because an aldol addition reaction occurs between two molecules of the same carbonyl compound, the product has twice as many carbons as the reacting aldehyde or ketone.

PROBLEM 16

Show the aldol addition product for each of the following compounds:

a. $CH_3CH_2CH_2CH_2CH$ with O double bonded at carbonyl

c. $CH_3CH_2CCH_2CH_3$ with O double bonded at carbonyl

b. $CH_3CHCH_2CH_2CH$ with CH_3 substituent and O double bonded at carbonyl

d. (cyclohexanone structure)

PROBLEM 17◆

For each of the following compounds, indicate the aldehyde or ketone from which it would be formed by an aldol addition:

a. 2-ethyl-3-hydroxyhexanal

c. 2,4-dicyclohexyl-3-hydroxybutanal

b. 4-hydroxy-4-methyl-2-pentanone

d. 5-ethyl-5-hydroxy-4-methyl-3-heptanone

PROBLEM 18

An aldol addition can be catalyzed by acids as well as by bases. Propose a mechanism for the acid-catalyzed aldol addition of propanal.

We have seen that alcohols are dehydrated when they are heated with acid (Section 11.4). The β-hydroxyaldehyde and β-hydroxyketone products of aldol addition reactions are easier to dehydrate than many other alcohols because the double bond formed as the result of dehydration is conjugated with a carbonyl group. Conjugation increases the stability of the product (Section 7.3) and therefore makes it easier to form. The product is called an enone—"ene" for the double bond and "one" for the carbonyl group.

$$2\ CH_3CH_2CH \overset{O}{\underset{}{\|}} \quad \underset{}{\overset{HO^-}{\rightleftharpoons}} \quad CH_3CH_2CH-CHCH \quad \underset{\Delta}{\overset{H^+}{\rightarrow}} \quad CH_3CH_2CH=CCH \quad +\ H_2O$$

a β-hydroxyaldehyde   an α,β-unsaturated aldehyde
an enone

If the product of an aldol addition is dehydrated, the overall reaction is called an **aldol condensation.** A **condensation reaction** is a reaction that combines two molecules while removing a small molecule (usually water or an alcohol).

β-Hydroxyaldehydes and β-hydroxyketones can also be dehydrated under basic conditions. So heating the aldol addition product in either acid or base leads to dehydration.

$$2\ CH_3CCH_3 \overset{O}{\underset{}{\|}} \quad \underset{}{\overset{HO^-}{\rightleftharpoons}} \quad CH_3C-CH_2CCH_3 \quad \underset{\Delta}{\overset{HO^-}{\rightarrow}} \quad CH_3C=CHCCH_3 \quad +\ H_2O$$

a β-hydroxyketone   an α,β-unsaturated ketone
an enone

19.12
DEHYDRATION OF ALDOL ADDITION PRODUCTS: FORMATION OF α,β-UNSATURATED ALDEHYDES AND KETONES

3-D Molecule:
2-Methyl-2-pentenal (an α,β-unsaturated aldehyde)

Dehydration sometimes occurs under the conditions in which the aldol addition is carried out, without additional heating. In such cases, the β-hydroxycarbonyl compound is an intermediate and the enone is the final product of the reaction. For example, the β-hydroxyketone formed from the aldol addition of acetophenone loses water as soon as it is formed because the double bond formed by loss of water is conjugated not only with the carbonyl group but also with the benzene ring. Conjugation stabilizes the dehydrated product and therefore makes it relatively easy to form.

PROBLEM 19 / SOLVED

How could you prepare the following compounds using a starting material containing no more than three carbon atoms?

$$\textbf{a. } CH_3CH_2CH_2\overset{\overset{\displaystyle O}{\|}}{C}OH$$

$$\textbf{b. } CH_3CH_2CH_2\overset{\overset{\displaystyle CH_3}{|}}{C}HCH_2OH$$

$$\textbf{c. } CH_3\overset{\overset{\displaystyle }{|}}{C}HCH_2\overset{\overset{\displaystyle O}{\|}}{C}CH_3 \quad (CH_3)$$

SOLUTION TO 19a   A compound with the correct four-carbon skeleton can be obtained if a two-carbon aldehyde undergoes an aldol addition. Dehydration of the addition product forms an α,β-unsaturated aldehyde. Catalytic hydrogenation forms an aldehyde. Some of the α,β-unsaturated aldehyde might be reduced to an alcohol, but that's all right because both the aldehyde and the alcohol can be oxidized to the target compound.

19.13
THE MIXED ALDOL
ADDITION

If two different carbonyl compounds are used in an aldol addition, four products can be formed because each carbonyl compound can react with itself as well as with the other carbonyl compound. In the following example, both carbonyl compound A and carbonyl compound B can lose a proton from an α-carbon to form nucleophiles A⁻ and B⁻; A⁻ can react with either A or B, and B⁻ can react with either A or B. This is called a **mixed aldol addition** or a **crossed aldol addition.** The four products have similar physical properties, making them difficult to separate. Consequently, a mixed aldol addition that forms four products is not a synthetically useful reaction.

Under certain conditions, a mixed aldol addition can lead primarily to one product. If one of the carbonyl compounds does not have any α-hydrogens, it cannot form an enolate. This reduces the number of possible products from four to two. A greater amount of one of the two products will be formed if the compound *without* α-hydrogens is always present in excess because the enolate will be more likely to

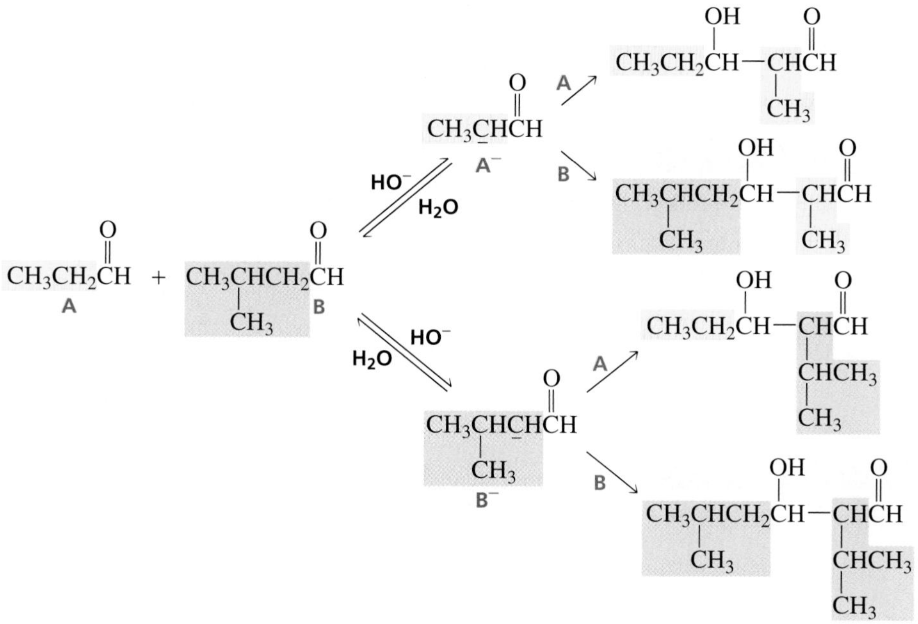

The reaction scheme showing mixed aldol additions:

$$CH_3CH_2CH(O) + CH_3CHCH_2CH(O)$$

with products:

$$\underset{OH}{CH_3CH_2CH}-\underset{O}{CHCH}$$ with CH_3

$$CH_3CHCH_2CH-CHCH$$ with CH_3, CH_3

$$CH_3CH_2CH-CHCH$$ with $CHCH_3$, CH_3

$$CH_3CHCH_2CH-CHCH$$ with CH_3, $CHCH_3$, CH_3

Reactants: CH_3CH_2CH (A) $\overset{O}{}$ + CH_3CHCH_2CH (B) with CH_3

Via CH_3CHCH (A) with HO⁻/H₂O

Via $CH_3CHCHCH$ (B) with CH_3, HO⁻/H₂O

react with it, rather than with another molecule of itself, if there is more of it in solution. Therefore, the compound with α-hydrogens should be added slowly to a basic solution of the compound without α-hydrogens.

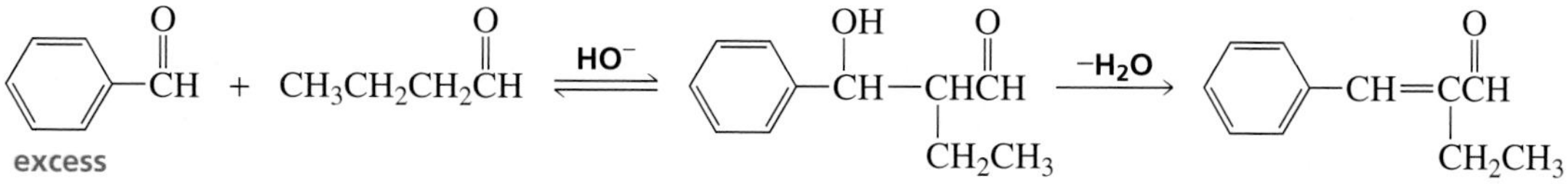

$$\text{(phenyl)}-CH(O) + CH_3CH_2CH_2CH(O) \underset{}{\overset{HO^-}{\rightleftharpoons}} \text{(phenyl)}-CH(OH)-CHCH(O) \text{ with } CH_2CH_3 \overset{-H_2O}{\rightarrow} \text{(phenyl)}-CH=CCH(O) \text{ with } CH_2CH_3$$

excess

If both carbonyl compounds have α-hydrogens, primarily one aldol product can be obtained if LDA is used to remove the α-proton that creates the enolate. Because LDA is a strong base (Section 19.7), all the carbonyl compound will be converted into an enolate so there will be no carbonyl compound left with which the enolate could react in an aldol addition—if the aldehyde is added slowly to a solution of LDA. The aldol addition will occur when the second carbonyl compound is added to the reaction mixture. If the second carbonyl compound is added slowly, the chance that it will form an enolate and then react with another molecule of itself in an aldol addition will be minimized.

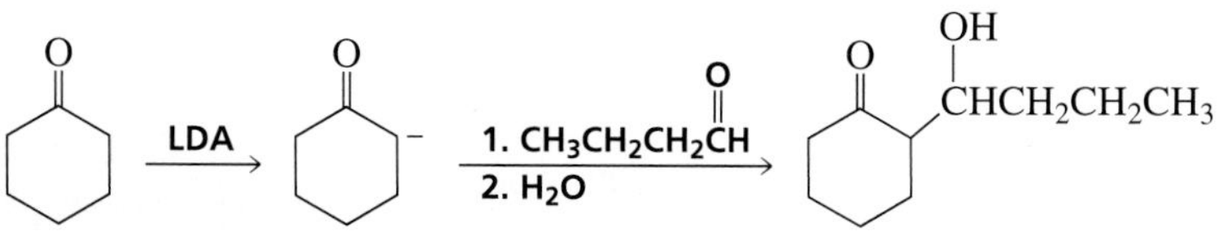

$$\text{(cyclohexanone)} \overset{LDA}{\rightarrow} \text{(cyclohexanone enolate)}^- \overset{\begin{array}{l}1.\ CH_3CH_2CH_2CH(O)\\2.\ H_2O\end{array}}{\longrightarrow} \text{(cyclohexanone)}-CHCH_2CH_2CH_3 \text{ with } OH$$

PROBLEM 20

Give the products obtained from mixed aldol additions of the following compounds:

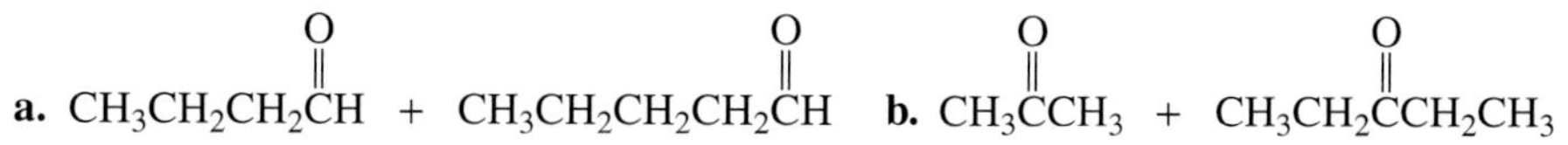

a. $CH_3CH_2CH_2CH(O)$ + $CH_3CH_2CH_2CH_2CH(O)$ **b.** $CH_3CCH_3(O)$ + $CH_3CH_2CCH_2CH_3(O)$

c. 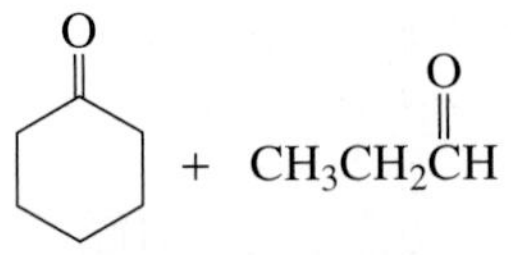 (cyclohexanone) + $CH_3CH_2\overset{O}{\overset{\|}{C}}H$ **d.** 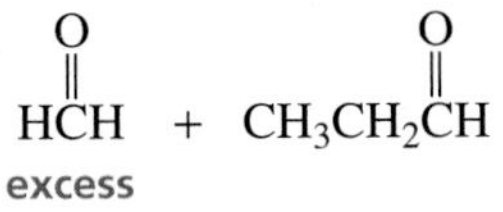 $H\overset{O}{\overset{\|}{C}}H$ + $CH_3CH_2\overset{O}{\overset{\|}{C}}H$
                                                                              excess

PROBLEM 21

Describe how the following compounds could be prepared using an aldol addition in the first step of the synthesis:

a. 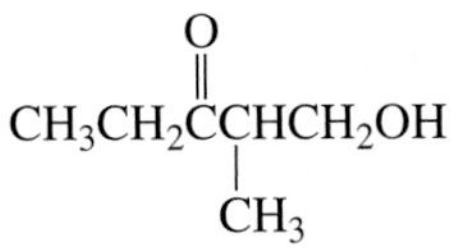 $CH_3CH_2\overset{O}{\overset{\|}{C}}\underset{\underset{CH_3}{|}}{C}HCH_2OH$

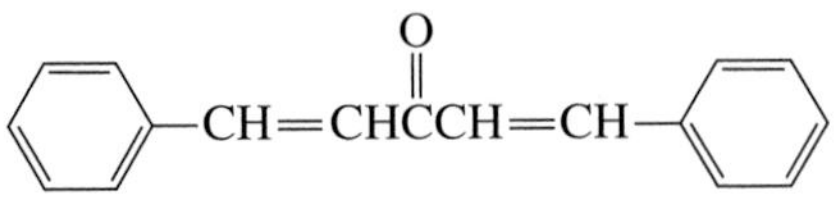

b. $\langle \rangle$—CH=CHС(=O)CH=CH—$\langle \rangle$

c. (a cyclohexanone with benzylidene substituent)

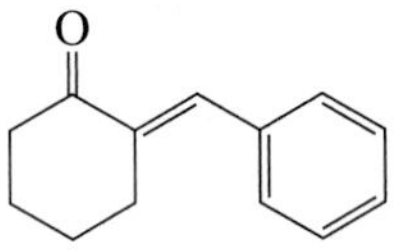

PROBLEM 22

Propose a mechanism for the following reaction:

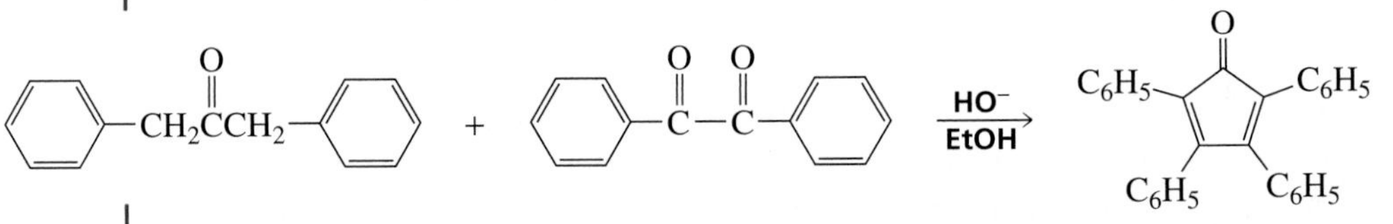

$\langle \rangle$—CH$_2$С(=O)CH$_2$—$\langle \rangle$ + $\langle \rangle$—С(=O)—С(=O)—$\langle \rangle$ $\xrightarrow[\text{EtOH}]{\text{HO}^-}$ (tetraphenylcyclopentadienone with C_6H_5 substituents)

19.14
THE CLAISEN
CONDENSATION

When two molecules of an *ester* undergo a condensation reaction, the reaction is called a **Claisen condensation.** The product of a Claisen condensation is a β-keto ester.

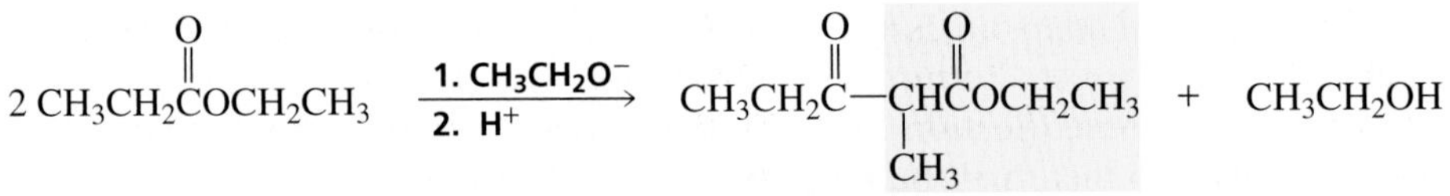

$$2\ CH_3CH_2\overset{O}{\overset{\|}{C}}OCH_2CH_3 \xrightarrow[\text{2. H}^+]{\text{1. CH}_3\text{CH}_2\text{O}^-} CH_3CH_2\overset{O}{\overset{\|}{C}}-\underset{\underset{CH_3}{|}}{C}H\overset{O}{\overset{\|}{C}}OCH_2CH_3 + CH_3CH_2OH$$

a β-keto ester

3-D Molecule:
β-Keto ester

As in the aldol addition, one molecule of carbonyl compound is converted into an enolate when an α-proton is removed by a strong base. The enolate attacks the carbonyl carbon of a second molecule of ester. The base employed corresponds to the leaving group of the ester so that the reactant is not changed (Section 19.10).

mechanism for the Claisen condensation

$$CH_3\underset{\underset{H}{|}}{\overset{\overset{\ddot{O}:}{\|}}{C}}HCOCH_3 \underset{CH_3\ddot{O}:^-}{\rightleftharpoons} CH_3\overset{\overset{\ddot{O}:}{\|}}{\ddot{C}}HCOCH_3 \underset{+ CH_3\ddot{O}H}{\overset{CH_3CH_2COCH_3}{\rightleftharpoons}} CH_3CH_2\overset{\overset{:\ddot{O}:^-}{|}}{\underset{\underset{CH_3\ddot{O}:\ \ CH_3}{|}}{C}}-\overset{\overset{\ddot{O}:}{\|}}{C}HCOCH_3$$

$$\Downarrow$$

$$CH_3CH_2\overset{\overset{\ddot{O}:}{\|}}{C}-\underset{\underset{CH_3}{|}}{C}H\overset{\overset{\ddot{O}:}{\|}}{C}OCH_3 + CH_3\ddot{O}:^-$$

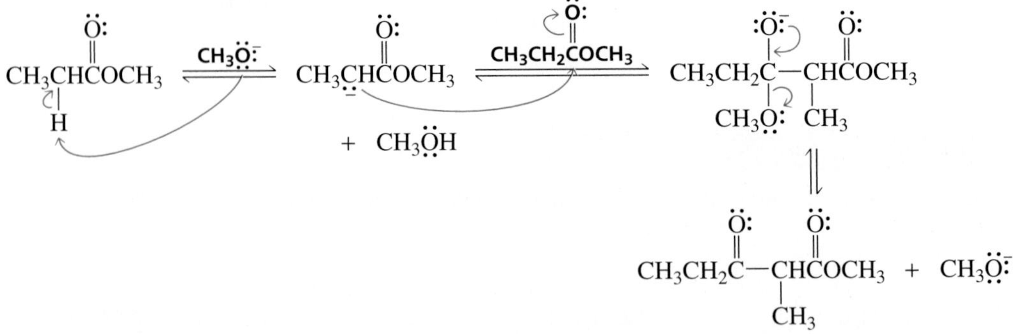

After nucleophilic attack, the Claisen condensation and the aldol addition differ. In the Claisen condensation, the negatively charged oxygen re-forms the carbon–oxygen π bond and expels the ⁻OR group. In the aldol addition, the negatively charged oxygen obtains a proton from the solvent.

The difference between the last step of the Claisen condensation and the last step of the aldol addition arises from the difference between esters and aldehydes or ketones. With esters, the carbon to which the negatively charged oxygen is bonded is also bonded to a group weakly basic enough to be expelled. With aldehydes or ketones, the carbon to which the negatively charged oxygen is bonded is not bonded to a group that can be expelled. Thus, the Claisen condensation is a substitution reaction, whereas the aldol addition is an addition reaction.

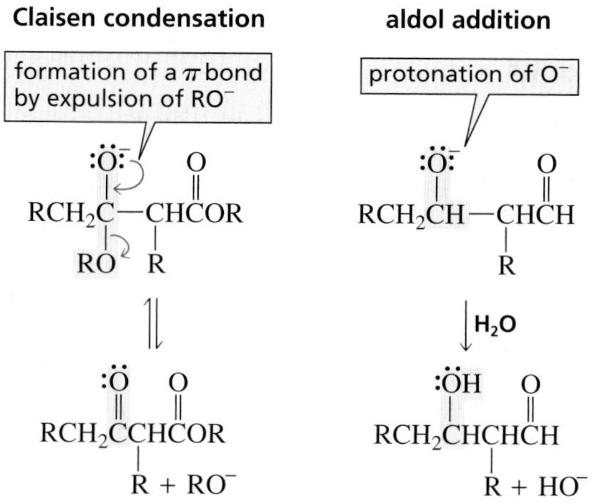

Ludwig Claisen (1851–1930) was born in Germany and received a Ph.D. from the University of Bonn, studying under Kekulé. He was a professor of chemistry at the University of Bonn, Owens College (Manchester, England), the University of Munich, the University of Aachen, the University of Kiel, and the University of Berlin.

Claisen condensation   **aldol addition**

Expulsion of the alkoxide ion is reversible because the alkoxide ion can readily attack the keto group of the β-keto ester to re-form the tetrahedral intermediate. The condensation reaction can be driven to completion, however, if a proton is removed from the β-keto ester. Removing a proton prevents the reverse reaction from occurring because the negatively charged alkoxide ion will not attack the negatively charged β-keto ester ion. It is easy to remove a proton from the β-keto ester because its central α-carbon is flanked by two carbonyl groups. Therefore, its α-hydrogen is much more acidic than the α-hydrogens of the ester starting material.

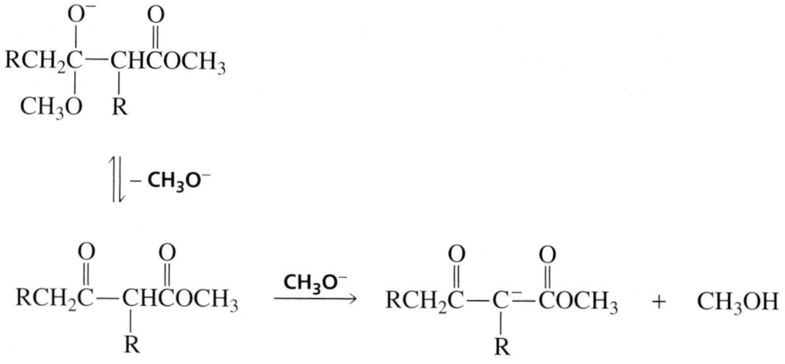

Consequently, a successful Claisen condensation requires an ester with two α-hydrogens and an equivalent amount of base rather than a catalytic amount of base. When the reaction is over, addition of acid to the reaction mixture reprotonates the β-keto ester anion. Any remaining alkoxide ion that could cause the reaction to reverse would also be protonated.

$$RCH_2\overset{O}{\overset{\|}{C}}-\overset{R}{\overset{|}{C}}{=}\overset{O}{\overset{\|}{C}}OCH_3 \xrightarrow{H^+} RCH_2\overset{O}{\overset{\|}{C}}-\overset{R}{\overset{|}{\underset{}{CH}}}-\overset{O}{\overset{\|}{C}}OCH_3$$

PROBLEM 23◆

Give the products of the following reactions:

a. $CH_3CH_2CH_2\overset{O}{\overset{\|}{C}}OCH_3 \xrightarrow[\text{2. H}^+]{\text{1. CH}_3\text{O}^-}$

b. $CH_3\underset{\overset{|}{CH_3}}{CH}CH_2\overset{O}{\overset{\|}{C}}OCH_2CH_3 \xrightarrow[\text{2. H}^+]{\text{1. CH}_3\text{CH}_2\text{O}^-}$

PROBLEM 24

Explain why a Claisen condensation product is not obtained from an ester that has only one α-hydrogen.

19.15
THE MIXED CLAISEN CONDENSATION

A **mixed Claisen condensation** is a condensation reaction between two different esters. Like a mixed aldol reaction, a mixed Claisen condensation is a useful reaction only if it is carried out under conditions that foster the formation of primarily one product. Otherwise, the result is a mixture of products that are difficult to separate. Primarily one product will be formed if one of the esters has no α-hydrogens (so it can't form an enolate) and the other ester is added slowly so that the ester without α-hydrogens is always in excess.

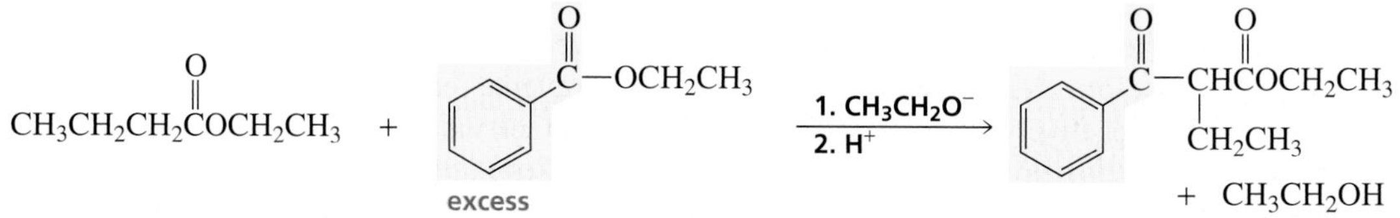

$$CH_3CH_2CH_2\overset{O}{\overset{\|}{C}}OCH_2CH_3 \ + \ \text{(phenyl)}\overset{O}{\overset{\|}{C}}{-}OCH_2CH_3 \xrightarrow[\text{2. H}^+]{\text{1. CH}_3\text{CH}_2\text{O}^-} \text{(phenyl)}\overset{O}{\overset{\|}{C}}{-}\underset{\overset{|}{CH_2CH_3}}{CH}\overset{O}{\overset{\|}{C}}OCH_2CH_3$$
$$+ \ CH_3CH_2OH$$

excess

Tutorial: Claisen reactions—synthesis

A reaction similar to a mixed Claisen condensation is the condensation of a ketone and an ester. Because the α-hydrogens of a ketone are more acidic than those of an ester, primarily one product is formed if the ketone and the base are each added slowly to the ester. The product is a β-diketone. Because of the difference in acidities of the α-hydrogens, primarily one condensation product is obtained even if both reagents have α-hydrogens.

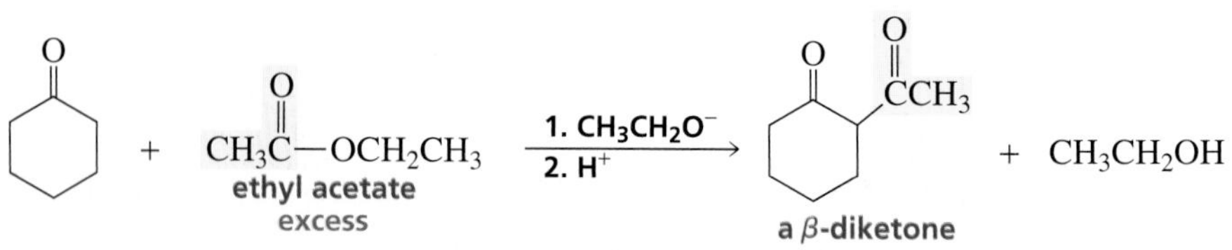

(cyclohexanone) $+ \ CH_3\overset{O}{\overset{\|}{C}}{-}OCH_2CH_3 \xrightarrow[\text{2. H}^+]{\text{1. CH}_3\text{CH}_2\text{O}^-}$ (β-diketone product) $+ \ CH_3CH_2OH$

ethyl acetate
excess

a β-diketone

A β-keto aldehyde is formed when a ketone condenses with a formate ester.

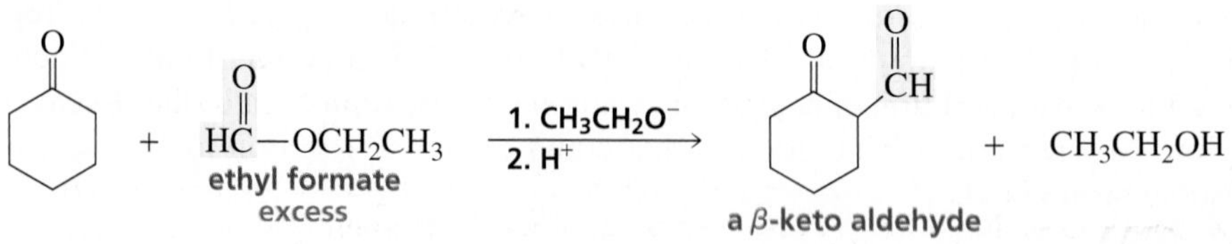

(cyclohexanone) $+ \ H\overset{O}{\overset{\|}{C}}{-}OCH_2CH_3 \xrightarrow[\text{2. H}^+]{\text{1. CH}_3\text{CH}_2\text{O}^-}$ (β-keto aldehyde product) $+ \ CH_3CH_2OH$

ethyl formate
excess

a β-keto aldehyde

A β-keto ester is formed when a ketone condenses with diethyl carbonate.

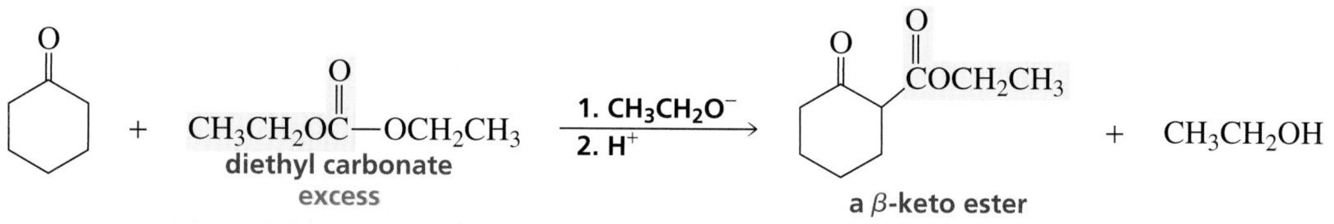

$$\text{cyclohexanone} + \underset{\substack{\text{diethyl carbonate}\\\text{excess}}}{CH_3CH_2O\overset{O}{\overset{\|}{C}}{-}OCH_2CH_3} \xrightarrow[\text{2. H}^+]{\text{1. CH}_3\text{CH}_2\text{O}^-} \underset{\text{a }\beta\text{-keto ester}}{} + CH_3CH_2OH$$

PROBLEM 25

Give the product of each of the following reactions:

a. $CH_3CH_2\overset{O}{\overset{\|}{C}}OCH_3 + CH_3\overset{O}{\overset{\|}{C}}OCH_3 \xrightarrow[\text{2. H}^+]{\text{1. CH}_3\text{O}^-}$

b. $+$ excess $\xrightarrow[\text{2. H}^+]{\text{1. CH}_3\text{CH}_2\text{O}^-}$

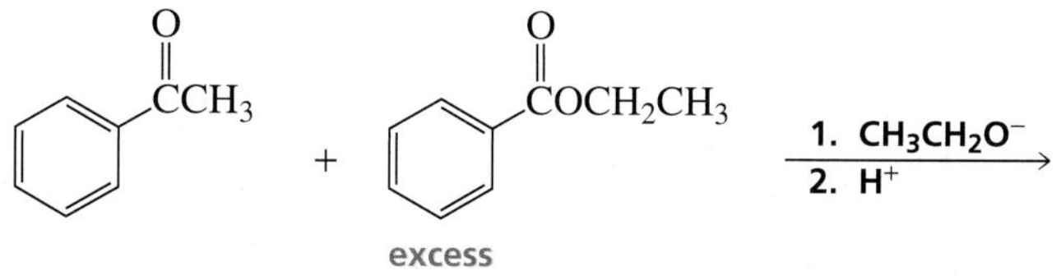

c. $\underset{\text{excess}}{H\overset{O}{\overset{\|}{C}}OCH_3} + CH_3CH_2CH_2\overset{O}{\overset{\|}{C}}OCH_3 \xrightarrow[\text{2. H}^+]{\text{1. CH}_3\text{O}^-}$

d. $\underset{\text{excess}}{CH_3CH_2O\overset{O}{\overset{\|}{C}}OCH_2CH_3} + CH_3CH_2\overset{O}{\overset{\|}{C}}OCH_2CH_3 \xrightarrow[\text{2. H}^+]{\text{1. CH}_3\text{CH}_2\text{O}^-}$

PROBLEM 26 / SOLVED

Show how the following compounds could be prepared:

a.   **b.**

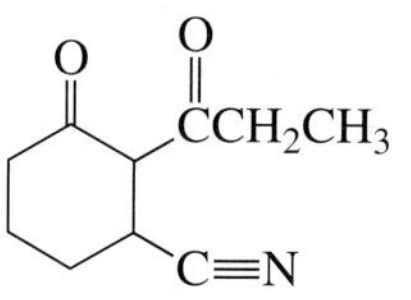

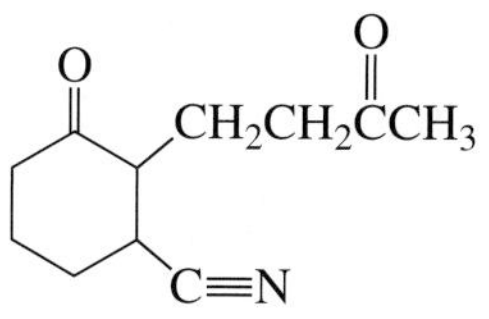

SOLUTION TO 26a Because it is a 1,3-dicarbonyl compound, it can be prepared by treating the enolate with an ester.

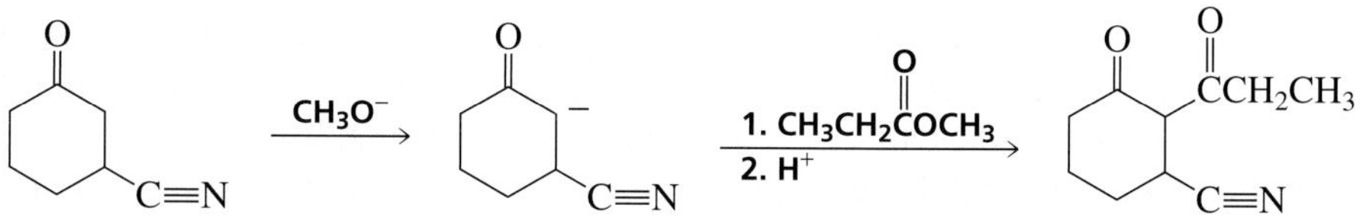

SOLUTION TO 26b Because it is a 1,5-dicarbonyl compound, it can be prepared by treating the enolate with an α,β-unsaturated carbonyl compound.

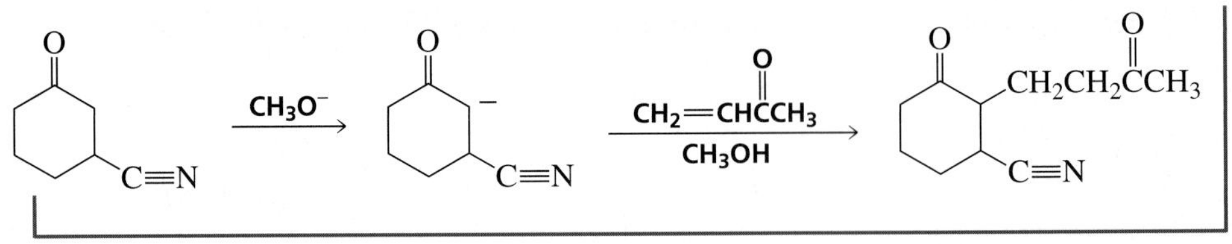

19.16 INTRAMOLECULAR CONDENSATION AND ADDITION REACTIONS

We have seen that if a compound has two functional groups that can react with each other, an intramolecular reaction readily occurs if the reaction leads to the formation of five- or six-membered rings (Section 10.11). Consequently, if base is added to a compound that contains two carbonyl groups, an intramolecular reaction occurs if a product with a five- or six-membered ring can be formed. Thus, a compound with two ester groups undergoes an intramolecular Claisen condensation and a compound with two aldehyde or ketone groups undergoes an intramolecular aldol addition.

Intramolecular Claisen Condensations

Walter Dieckmann (1869–1925) was born in Germany. He received a Ph.D. from the University of Munich and became a professor of chemistry there.

The addition of base to a 1,6-diester causes the diester to undergo an intramolecular Claisen condensation, thereby forming a five-membered ring β-keto ester. An intramolecular Claisen condensation is called a **Dieckmann condensation.**

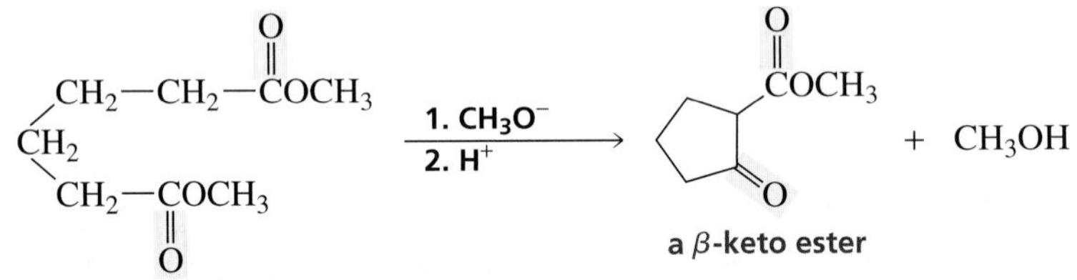

A six-membered ring β-keto ester is formed from a Dieckmann condensation of a 1,7-diester.

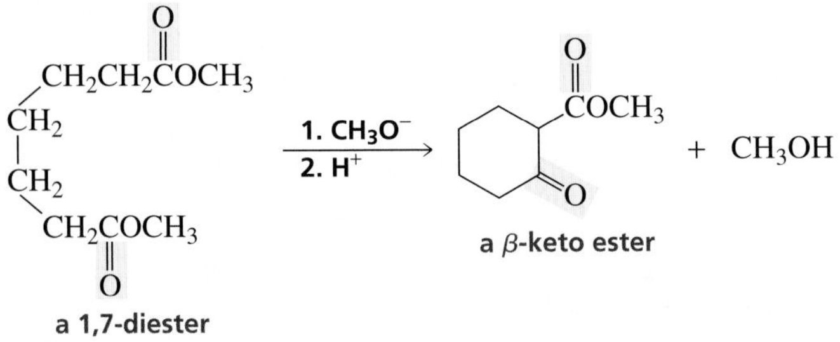

The mechanism of the Dieckmann condensation is the same as the mechanism of the Claisen condensation. The only difference in the two reactions is that the attacking enolate and the carbonyl group undergoing nucleophilic attack are in different molecules in the Claisen condensation but are in the same molecule in the Dieckmann condensation. The Dieckmann condensation, like the Claisen condensation, is driven to completion by carrying out the reaction with enough base to remove a proton from the α-carbon of the β-keto ester product. When the reaction is over, acid is added to reprotonate the condensation product.

mechanism for the Dieckmann condensation

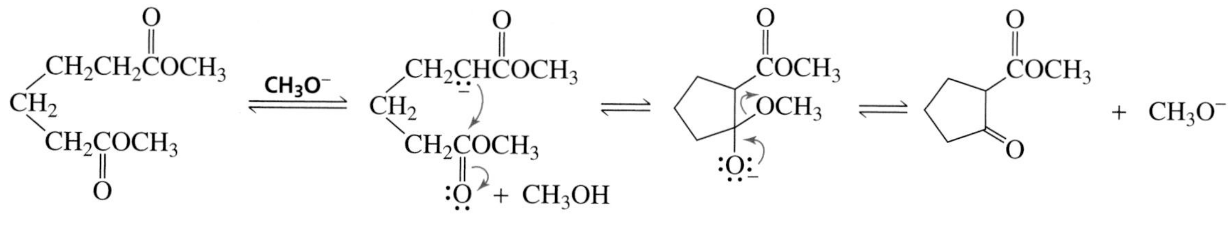

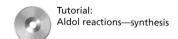

PROBLEM 27

Write the mechanism for the base-catalyzed formation of a cyclic β-keto ester from a 1,7-diester.

Intramolecular Aldol Additions

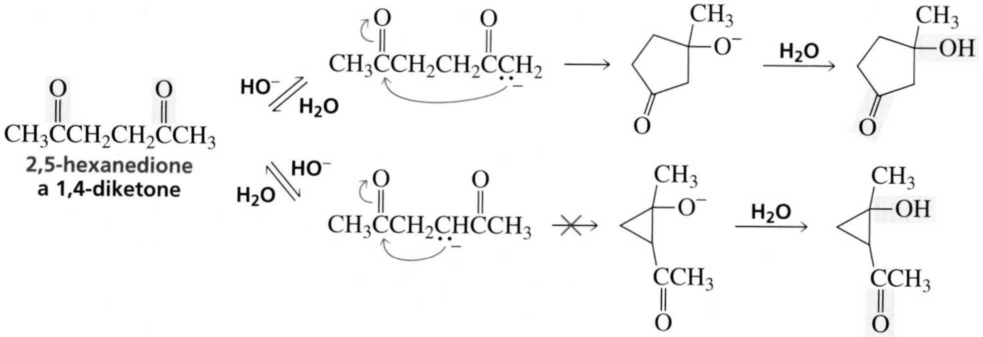

Tutorial: Aldol reactions—synthesis

Because a 1,4-diketone has two different sets of α-hydrogens, two different intramolecular addition products can potentially form—one product would contain a five-membered ring and the other a three-membered ring. The greater stability of five- and six-membered rings causes them to be formed preferentially (Section 2.11). The five-membered ring product is the only product formed from the intramolecular aldol addition of a 1,4-diketone.

2,5-hexanedione
a 1,4-diketone

not formed

An intramolecular aldol addition of a 1,6-diketone can lead to either a seven- or a five-membered ring product. Again, the more stable product—the five-membered ring product—is the only product of the reaction.

2,7-octanedione
a 1,6-diketone

not formed

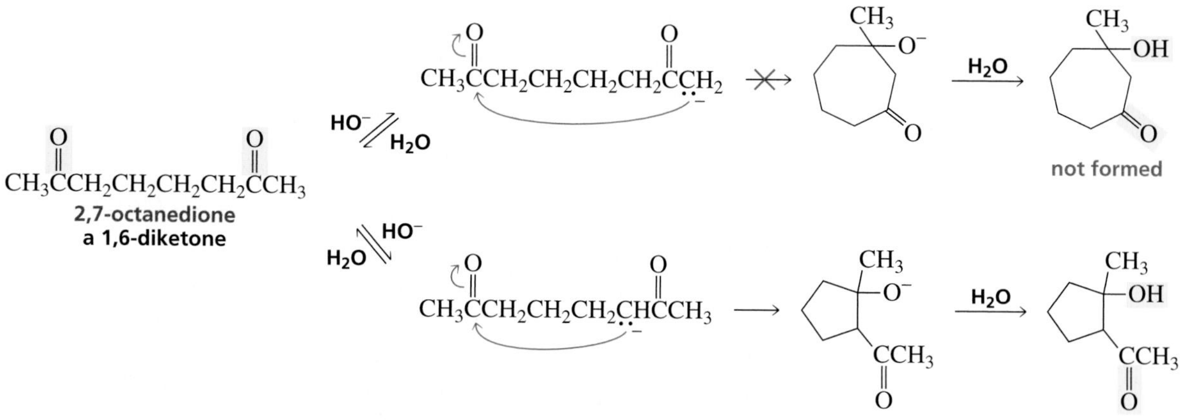

1,5-Diketones and 1,7-diketones undergo intramolecular aldol additions to form six-membered ring products.

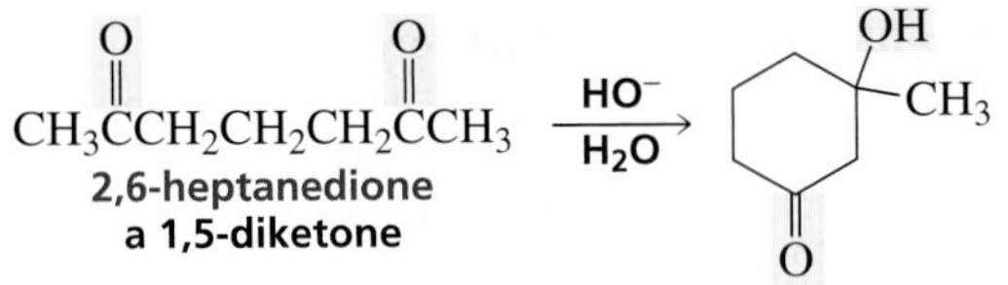

$$CH_3CCH_2CH_2CH_2CCH_3 \xrightarrow[\text{H}_2\text{O}]{\text{HO}^-}$$

2,6-heptanedione
a 1,5-diketone

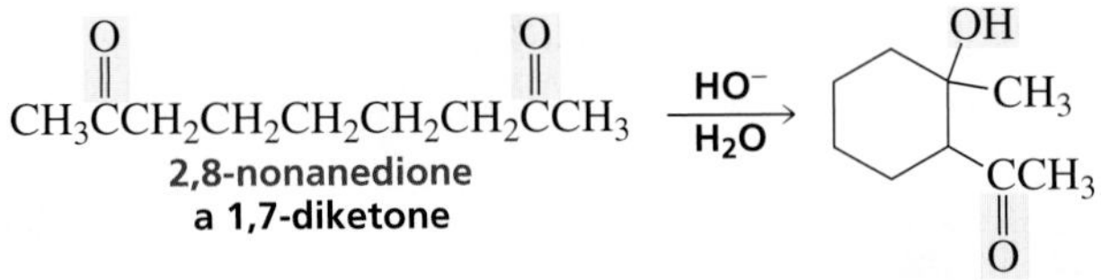

$$CH_3CCH_2CH_2CH_2CH_2CCH_3 \xrightarrow[\text{H}_2\text{O}]{\text{HO}^-}$$

2,8-nonanedione
a 1,7-diketone

PROBLEM 28◆

If the preference for formation of a six-membered ring were not so great, what other cyclic product would be formed from the intramolecular aldol addition of:

a. 2,6-heptanedione?　　　　　**b.** 2,8-nonanedione?

PROBLEM 29

Can 2,4-pentanedione undergo an intramolecular aldol addition? If so, why? If not, why not?

PROBLEM 30 / SOLVED

The following ketoaldehyde forms three cyclic products when treated with a base. Identify the products. Which is the major product?

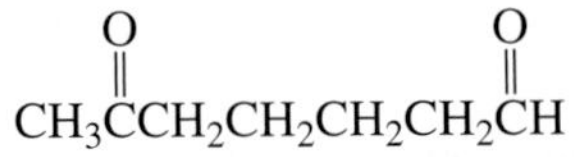

$$CH_3CCH_2CH_2CH_2CH_2CH$$

SOLUTION Three products can form because there are three different sets of α-hydrogens. More **B** and **C** are formed than **A** because a five-membered ring is formed in preference to a seven-membered ring. More **B** is formed than **C** because the double bond of the enolate leading to **B** is more substituted, which makes that enolate more stable and, therefore, more easily formed.

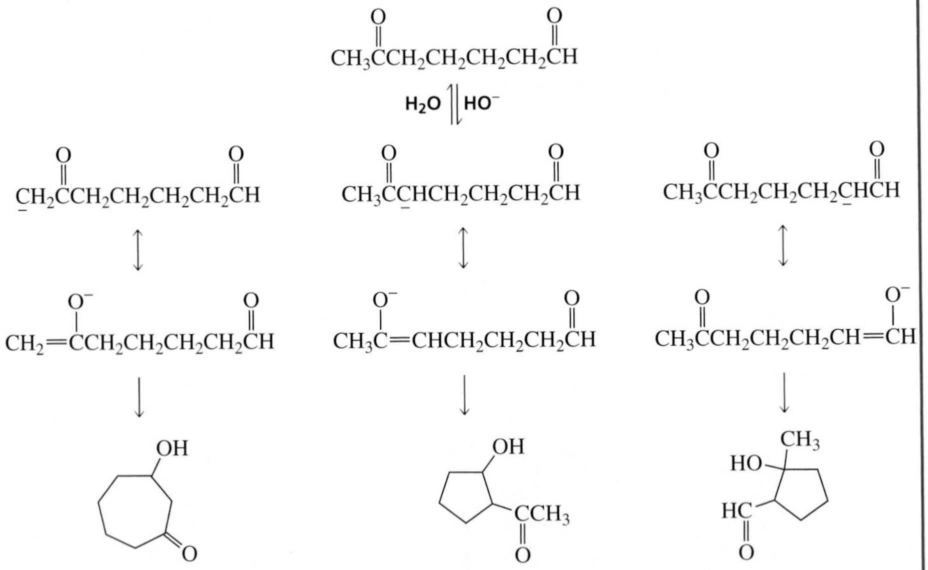

$$CH_3CCH_2CH_2CH_2CH_2CH$$

$$\xleftrightarrow[\text{HO}^-]{\text{H}_2\text{O}}$$

$$CH_2CCH_2CH_2CH_2CH \quad\quad CH_3CCHCH_2CH_2CH \quad\quad CH_3CCH_2CH_2CH_2CHCH$$

$$\updownarrow \quad\quad\quad\quad \updownarrow \quad\quad\quad\quad \updownarrow$$

$$CH_2=CCH_2CH_2CH_2CH \quad CH_3C=CHCH_2CH_2CH \quad CH_3CCH_2CH_2CH_2CH=CH$$

A　　　　　　　　　　　B　　　　　　　　　　　C

PROBLEM 31◆

Give the product of the reaction of each of the following compounds with base:

a.

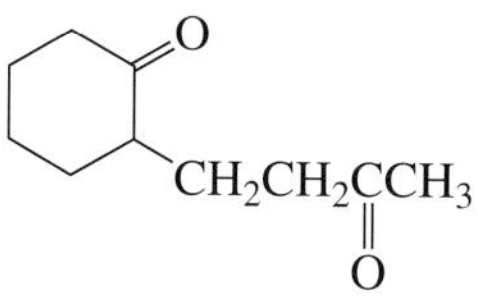

b.

c. $HCCH_2CH_2CH_2CH_2CH_2CH$

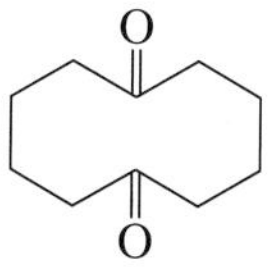

d.

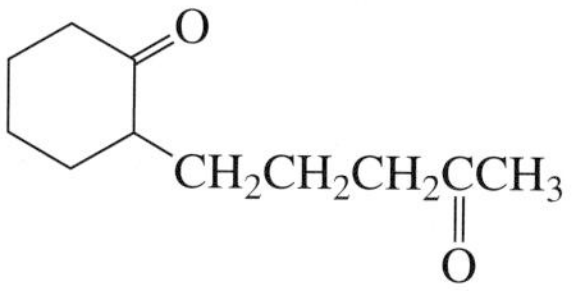

Sir Robert Robinson (1886–1975) was born in England, the son of a manufacturer. After receiving a Ph.D. from the University of Manchester, he joined the faculty at the University of Sydney in Australia. He returned to England three years later, becoming a professor of chemistry at Oxford in 1929. Robinson was an accomplished mountain climber. Knighted in 1939, he received the 1947 Nobel Prize in chemistry for his work on alkaloids.

The Robinson Annulation

Reactions that form carbon–carbon bonds are important to synthetic chemists. Without such reactions, large organic molecules could not be prepared from smaller ones. We have seen that Michael reactions and aldol additions are reactions that form carbon–carbon bonds. The **Robinson annulation** is a reaction that puts these two carbon–carbon bond forming reactions together, providing a route to the synthesis of many complicated organic molecules. Annulation comes from *annulus,* Latin for "ring." Thus, an **annulation reaction** is a ring-forming reaction.

The first stage of a Robinson annulation is a Michael reaction that forms a 1,5-diketone. We just saw that a 1,5-diketone undergoes an intramolecular aldol addition when treated with base—this is the second stage of the Robinson annulation. Notice that a Robinson annulation results in a product that has a 2-cyclohexenone ring.

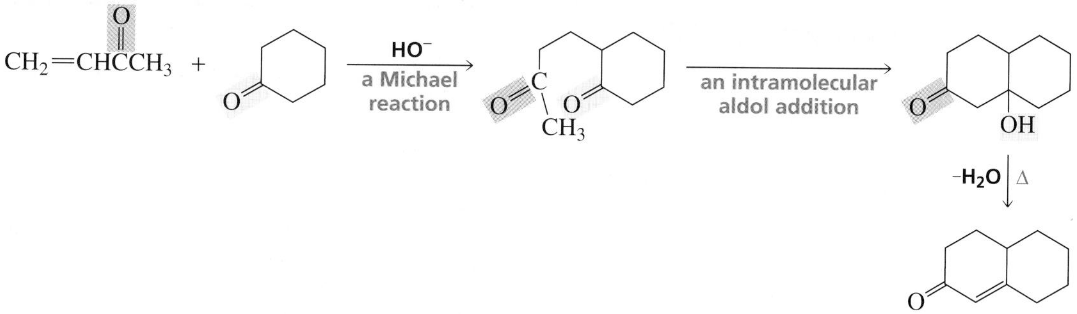

$CH_2=CHCCH_3$ +   → **HO⁻** / **a Michael reaction** →   → **an intramolecular aldol addition** →

−H₂O | Δ

PROBLEM-SOLVING STRATEGY

Propose a synthesis for each of the following compounds using a Robinson annulation:

a. 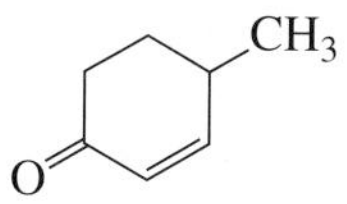   **b.**

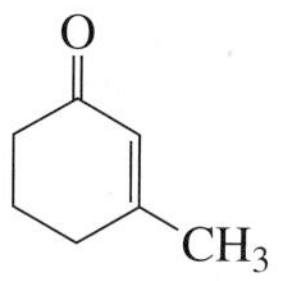

By analyzing a Robinson annulation, we will be able to determine which part of the molecule comes from which reactant. This, then, will allow us to choose appropriate reactants for any other Robinson annulation. Analysis shows that the keto group of the cyclohexenone comes from the α,β-unsaturated carbonyl compound, and the double bond results from

Tutorial: Robinson annulation—synthesis

attack of the enolate of the α,β-unsaturated carbonyl compound on the carbonyl group of the other reactant. Thus, we can arrive at the appropriate reactants by cutting through the double bond and between the β- and γ-carbons on the other side of the carbonyl group.

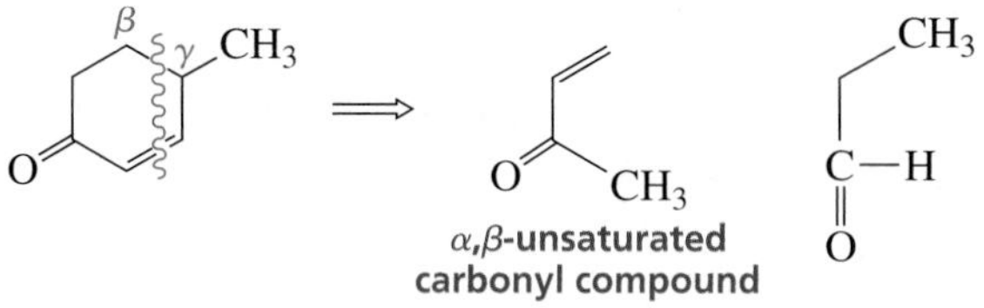

Therefore, the required reactants for **a** are:

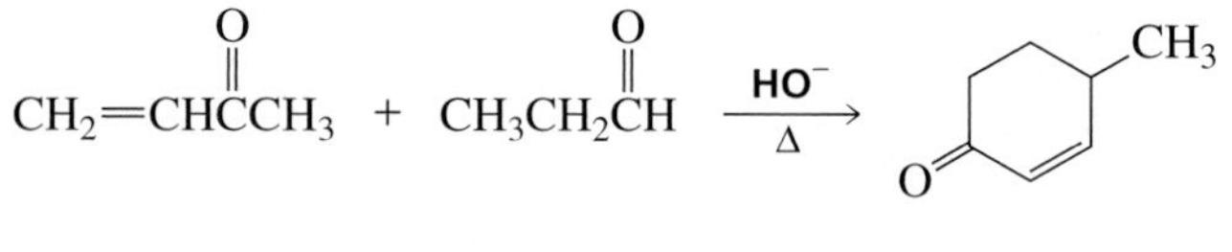

By cutting through **b,** we can determine the required reactants for its synthesis:

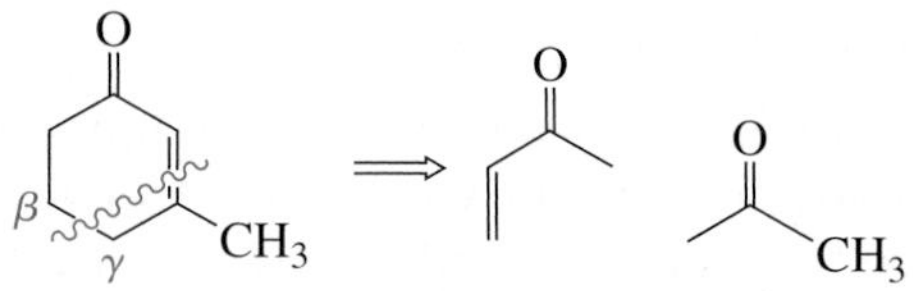

Therefore, the required reactants for **b** are:

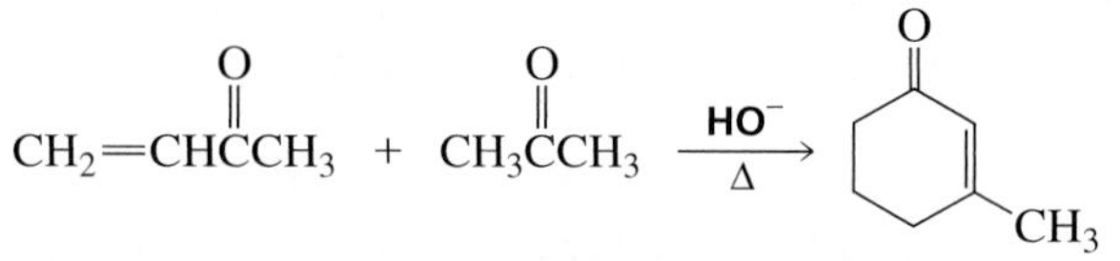

Now continue on to Problem 32.

PROBLEM 32

Propose a synthesis for each of the following compounds using a Robinson annulation:

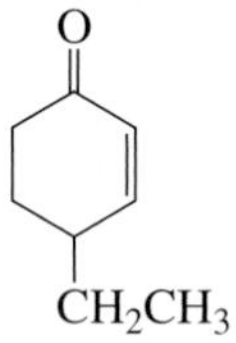

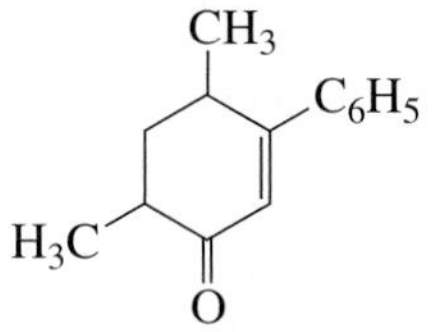

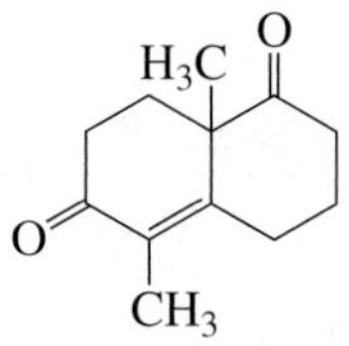

19.17
DECARBOXYLATION
OF 3-OXO-
CARBOXYLIC ACIDS

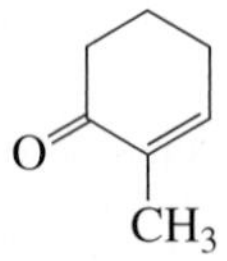

Carboxylate ions do not lose CO_2 for the same reason that alkanes such as ethane do not lose a proton. That is, carboxylate ions do not lose CO_2 and ethane does not lose a proton because the leaving group would be a carbanion. Carbanions are very strong bases and therefore very poor leaving groups.

$$CH_3CH_2{-}H \qquad CH_3CH_2{-}\overset{\displaystyle O}{\overset{\|}{C}}{-}\ddot{\underset{\cdot\cdot}{O}}{:}$$

If, however, the CO_2 group is bonded to a carbon adjacent to a carbonyl carbon, the CO_2 group can be removed because the electrons left behind can be delocalized onto the carbonyl oxygen. Consequently, 3-oxocarboxylate ions (carboxylate ions with a keto group at the 3-position) lose CO_2 when they are heated. Loss of CO_2 from a molecule is called **decarboxylation.**

removing CO$_2$ from an α-carbon

$$CH_3\overset{\displaystyle \ddot{O}:}{\overset{\|}{C}}{-}CH_2{-}\overset{\displaystyle O}{\overset{\|}{C}}{-}\ddot{\underset{\cdot\cdot}{O}}{:} \;\;\xrightarrow{\Delta}\;\; CH_3\overset{\displaystyle :\ddot{O}^{\bar{}}}{\overset{|}{C}}{=}CH_2 \;\longleftrightarrow\; CH_3\overset{\displaystyle O}{\overset{\|}{C}}\bar{C}H_2$$

3-oxobutanoate ion
acetoacetate ion
$$+\;\;CO_2$$

Notice the similarity between removal of CO_2 from a 3-oxocarboxylate ion and removal of a proton from an α-carbon. In both reactions a substituent—CO_2 in one case, H^+ in the other—is removed from an α-carbon and its bonding electrons are delocalized onto an oxygen.

removing a proton from an α-carbon

$$CH_3\overset{\displaystyle \ddot{O}:}{\overset{\|}{C}}{-}CH_2{-}H \;\;\rightleftharpoons\;\; CH_3\overset{\displaystyle :\ddot{O}^{\bar{}}}{\overset{|}{C}}{=}CH_2 \;\longleftrightarrow\; CH_3\overset{\displaystyle O}{\overset{\|}{C}}\bar{C}H_2$$

acetone
propanone
$$+\;\;H^+$$

Decarboxylation is even easier if the reaction is carried out under acidic conditions because the reaction is catalyzed by an intramolecular transfer of a proton from the carboxyl group to the carbonyl oxygen. The enol that is formed immediately tautomerizes to a ketone.

$$CH_3\overset{\displaystyle O}{\overset{\|}{C}}{-}CH_2{-}\overset{\displaystyle O}{\overset{|}{C}}{=}O \;\;\xrightarrow{\Delta}\;\; CH_3\overset{\displaystyle OH}{\overset{|}{C}}{=}CH_2 \;\;\xrightarrow[\text{tautomerization}]{} \;\; CH_3\overset{\displaystyle O}{\overset{\|}{C}}CH_3$$

3-oxobutanoic acid
acetoacetic acid
a β-keto acid
$$+\;CO_2$$

We saw in Section 19.1 that it is harder to remove a proton from an α-carbon if the electrons are delocalized onto the carbonyl group of an ester rather than onto the carbonyl group of a ketone. For the same reason, a higher temperature is required to decarboxylate a β-dicarboxylic acid such as malonic acid than to decarboxylate a β-keto acid.

$$HO\overset{\displaystyle O}{\overset{\|}{C}}{-}CH_2{-}\overset{\displaystyle O}{\overset{|}{C}}{=}O \;\;\xrightarrow{135\ ^\circ C}\;\; HO\overset{\displaystyle OH}{\overset{|}{C}}{=}CH_2 \;\;\xrightarrow[\text{tautomerization}]{}\;\; HO\overset{\displaystyle O}{\overset{\|}{C}}CH_3$$

malonic acid
$$+\;CO_2$$

Thus, carboxylic acids with a carbonyl group at the 3-position (both β-ketocarboxylic acids and β-dicarboxylic acids) lose CO_2 when they are heated.

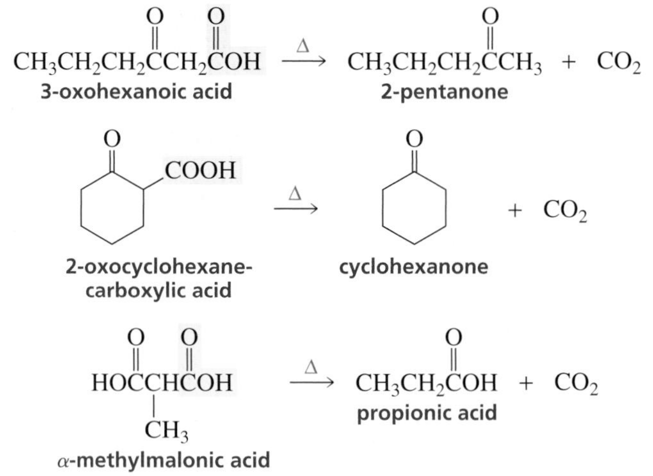

$$CH_3CH_2CH_2\overset{O}{\underset{\|}{C}}CH_2\overset{O}{\underset{\|}{C}}OH \xrightarrow{\Delta} CH_3CH_2CH_2\overset{O}{\underset{\|}{C}}CH_3 + CO_2$$

3-oxohexanoic acid   **2-pentanone**

2-oxocyclohexane-carboxylic acid   **cyclohexanone**

$$HO\overset{O}{\underset{\|}{C}}\underset{\underset{CH_3}{|}}{CH}\overset{O}{\underset{\|}{C}}OH \xrightarrow{\Delta} CH_3CH_2\overset{O}{\underset{\|}{C}}OH + CO_2$$

α-methylmalonic acid   **propionic acid**

Heinz (1904–1981) and Clare (1903–1995) Hunsdiecker were both born in Germany— Heinz in Cologne and Clare in Kiel. They both received Ph.D.s from the University of Cologne and spent their careers working in a private laboratory in Cologne.

THE HUNSDIECKER REACTION

Heinz and Clare Hunsdiecker found that a carboxylic acid can be decarboxylated if a heavy metal salt of the carboxylic acid is heated with bromine or iodine. The product is an alkyl halide with one less carbon than the starting carboxylic acid. The heavy metal can be silver ion, mercuric ion, or lead(IV). The reaction is now known as the **Hunsdiecker reaction.**

$$CH_3CH_2CH_2CH_2\overset{O}{\underset{\|}{C}}OH \xrightarrow[\text{2. Br}_2, \Delta]{\text{1. Ag}_2O} CH_3CH_2CH_2CH_2Br + CO_2 + AgBr$$

pentanoic acid   **1-bromobutane**

The mechanism of the reaction involves the initial formation of a hypobromite as a result of precipitation of AgBr. A radical reaction is initiated by homolytic cleavage of the oxygen–bromine bond of the hypobromite. The carboxyl radical loses CO_2, and the alkyl radical thus formed abstracts a bromine radical from the hypobromite to propagate the reaction.

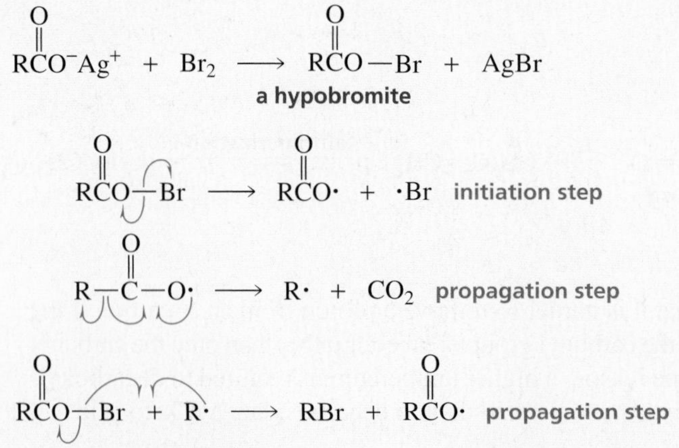

$$R\overset{O}{\underset{\|}{C}}O^-Ag^+ + Br_2 \longrightarrow R\overset{O}{\underset{\|}{C}}O-Br + AgBr$$

a hypobromite

$$R\overset{O}{\underset{\|}{C}}O-Br \longrightarrow R\overset{O}{\underset{\|}{C}}O\cdot + \cdot Br \quad \text{initiation step}$$

$$R-\overset{O}{\underset{\|}{C}}-O\cdot \longrightarrow R\cdot + CO_2 \quad \text{propagation step}$$

$$R\overset{O}{\underset{\|}{C}}O-Br + R\cdot \longrightarrow RBr + R\overset{O}{\underset{\|}{C}}O\cdot \quad \text{propagation step}$$

PROBLEM 33◆

Which of the following compounds would be expected to decarboxylate when heated?

a.

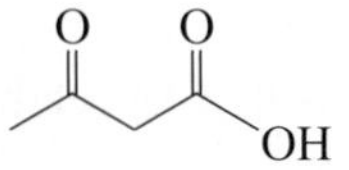

b.

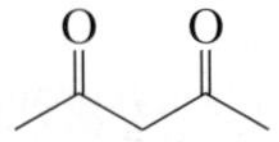

c.

HO—C(=O)—CH₂—CH₂—C(=O)—OH

d.

HO—C(=O)—CH₂—C(=O)—CH₂—CH₃

A combination of two of the reactions discussed in this chapter—alkylation of an α-carbon and decarboxylation of a β-dicarboxylic acid—can be used to prepare carboxylic acids of any desired chain length. The procedure is called the **malonic ester synthesis** because the starting material for the synthesis is the diethyl ester of malonic acid. The first two carbons of the carboxylic acid come from malonic ester, and the rest of the carboxylic acid comes from the alkyl halide used in the second step of the reaction.

**19.18
THE MALONIC
ESTER SYNTHESIS:
SYNTHESIS OF
CARBOXYLIC ACIDS**

$$C_2H_5O\overset{O}{\overset{\|}{C}}{-}CH_2{-}\overset{O}{\overset{\|}{C}}OC_2H_5 \xrightarrow[\substack{\text{2. RBr} \\ \text{3. H}^+, \text{H}_2\text{O}, \Delta}]{\text{1. CH}_3\text{CH}_2\text{O}^-} R{-}CH_2\overset{O}{\overset{\|}{C}}OH$$

**diethyl malonate
"malonic ester"**

from malonic ester

from the alkyl halide

In the first part of the malonic ester synthesis, the α-carbon of the diester is alkylated (Section 19.8). A proton is easily removed from the α-carbon because it is flanked by two ester groups ($pK_a = 13$). The resulting α-carbanion reacts with an alkyl halide, forming an α-substituted malonic ester. Heating the α-substituted malonic ester in an acidic aqueous solution hydrolyzes it to an α-substituted malonic acid, which upon further heating loses CO_2, forming a carboxylic acid with two more carbons than the alkyl halide.

Tutorial:
Malonic ester synthesis

$$C_2H_5O\overset{O}{\overset{\|}{C}}{-}CH_2{-}\overset{O}{\overset{\|}{C}}OC_2H_5 \xrightarrow{\text{CH}_3\text{CH}_2\text{O}^-} C_2H_5O\overset{O}{\overset{\|}{C}}{-}\overset{..}{C}H{-}\overset{O}{\overset{\|}{C}}OC_2H_5 \xrightarrow{R{-}Br} C_2H_5O\overset{O}{\overset{\|}{C}}{-}\underset{R}{CH}{-}\overset{O}{\overset{\|}{C}}OC_2H_5 + Br^-$$

an α-substituted malonic ester

$$H^+, H_2O \downarrow \Delta$$

$$R{-}CH_2{-}\overset{O}{\overset{\|}{C}}OH + CO_2 \xleftarrow{\Delta} HO\overset{O}{\overset{\|}{C}}{-}\underset{R}{CH}{-}\overset{O}{\overset{\|}{C}}OH + 2\,CH_3CH_2OH$$

an α-substituted malonic acid

PROBLEM 34◆

What alkyl bromide(s) should be used in the malonic ester synthesis of each of the following carboxylic acids?

a. propanoic acid

b. 2-methylpropanoic acid

c. 3-phenylpropanoic acid

d. pentanoic acid

e. 4-methylpentanoic acid

f. 2-methylpentanoic acid

Carboxylic acids with two substituents bonded to the α-carbon can be prepared by carrying out two successive α-carbon alkylations.

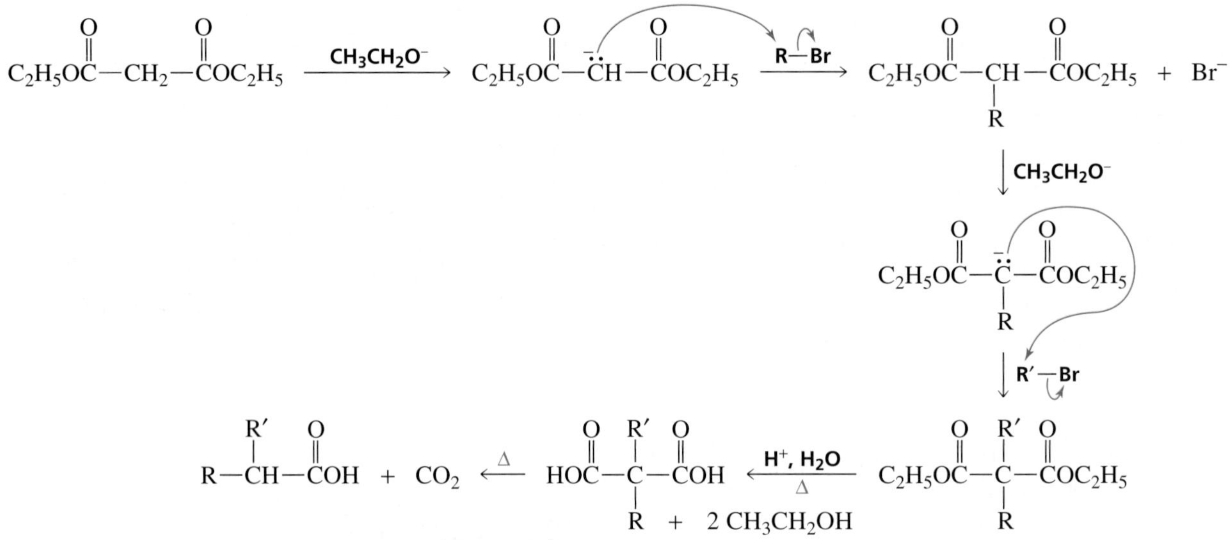

$$C_2H_5OC{-}CH_2{-}COC_2H_5 \xrightarrow{CH_3CH_2O^-} C_2H_5OC{-}\overset{..}{C}H{-}COC_2H_5 \xrightarrow{R{-}Br} C_2H_5OC{-}\underset{R}{CH}{-}COC_2H_5 + Br^-$$

$$\downarrow CH_3CH_2O^-$$

$$C_2H_5OC{-}\underset{R}{\overset{..}{C}}{-}COC_2H_5$$

$$\downarrow R'{-}Br$$

$$R{-}\underset{\overset{|}{R'}}{CH}{-}COH + CO_2 \xleftarrow{\Delta} HOC{-}\underset{R}{\overset{R'}{C}}{-}COH \xleftarrow[\Delta]{H^+, H_2O} C_2H_5OC{-}\underset{R}{\overset{R'}{C}}{-}COC_2H_5$$

$$R + 2\,CH_3CH_2OH$$

PROBLEM 35

Explain why the following carboxylic acids cannot be prepared by the malonic ester synthesis:

a. ⬡—CH_2COH (with O double bond)

b. $CH_2{=}CHCH_2COH$ (with O double bond)

c. $CH_3\overset{CH_3}{\underset{CH_3}{C}}CH_2COH$ (with O double bond)

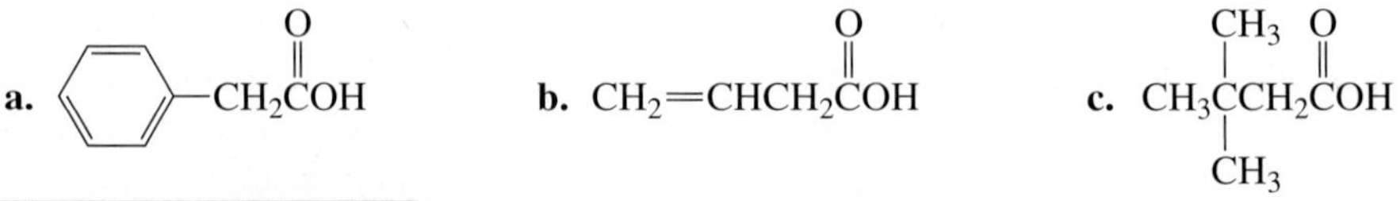

PROBLEM 36◆

a. What carboxylic acid would be formed if the malonic ester synthesis were carried out with one equivalent of malonic ester, one equivalent of 1,5-dibromopentane, and two equivalents of base?

b. What carboxylic acid would be formed if the malonic ester synthesis were carried out with two equivalents of malonic ester, one equivalent of 1,5-dibromopentane, and two equivalents of base?

19.19
THE ACETOACETIC ESTER SYNTHESIS: SYNTHESIS OF METHYL KETONES

The only difference between the acetoacetic ester synthesis and the malonic ester synthesis is the use of acetoacetic ester rather than malonic ester as the starting material. Because of the difference in starting material, the product of the **acetoacetic ester synthesis** is a methyl ketone rather than a carboxylic acid. The carbonyl group and the carbon atoms on either side of it come from acetoacetic ester, and the rest of the ketone comes from the alkyl halide used in the second step of the reaction.

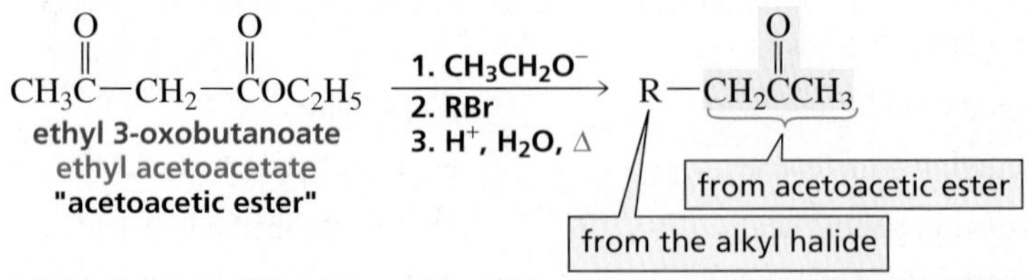

$$CH_3C{-}CH_2{-}COC_2H_5 \xrightarrow[\text{2. RBr}]{\text{1. }CH_3CH_2O^-\ \ \ \ \ \text{3. }H^+, H_2O, \Delta} R{-}CH_2CCH_3$$

ethyl 3-oxobutanoate
ethyl acetoacetate
"acetoacetic ester"

from acetoacetic ester

from the alkyl halide

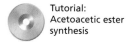

Tutorial:
Acetoacetic ester
synthesis

The mechanisms for the acetoacetic ester synthesis and the malonic ester synthesis are similar. The last step in the acetoacetic ester synthesis is decarboxylation of a substituted acetoacetic acid rather than a substituted malonic acid.

$$\underset{\substack{\text{O}\\\text{||}}}{CH_3C}-CH_2-\underset{\substack{\text{O}\\\text{||}}}{C}OC_2H_5 \xrightarrow{CH_3CH_2O^-} \underset{\substack{\text{O}\\\text{||}}}{CH_3C}-\overset{..}{C}H-\underset{\substack{\text{O}\\\text{||}}}{C}OC_2H_5 \xrightarrow{R-Br} \underset{\substack{\text{O}\\\text{||}}}{CH_3C}-\underset{\substack{R}}{CH}-\underset{\substack{\text{O}\\\text{||}}}{C}OC_2H_5 + Br^-$$

$$H^+, H_2O \downarrow \Delta$$

$$\underset{\substack{\text{O}\\\text{||}}}{CH_3C}-CH_2-R + CO_2 \xleftarrow{\Delta} \underset{\substack{\text{O}\\\text{||}}}{CH_3C}-\underset{\substack{R}}{CH}-\underset{\substack{\text{O}\\\text{||}}}{C}OH + CH_3CH_2OH$$

PROBLEM 37◆

What alkyl bromide should be used in the acetoacetic ester synthesis of each of the following methyl ketones?

a. 2-pentanone   **b.** 2-octanone   **c.** 4-phenyl-2-butanone

---

When you are planning the synthesis of a compound that requires the formation of a new carbon–carbon bond, first locate the new bond that must be made. For example, in the synthesis of the following β-diketone, the new bond is the one that makes the second five-membered ring.

19.20
DESIGNING A SYNTHESIS VI: MAKING NEW CARBON–CARBON BONDS

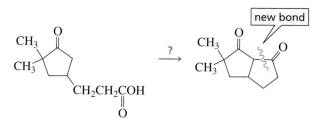

Next, determine which of the atoms that form the bond should be the nucleophile and which should be the electrophile. Because you know that a carbonyl carbon is an electrophile, it is easy to choose between the two possibilities in this case.

Now you need to determine what compound you could use that would give you the desired electrophilic and nucleophilic sites. If the starting material is indicated, use it as a clue to arrive at the desired compound. For example, an ester carbonyl group would be a good electrophile for this synthesis because it has a group that would be eliminated. Furthermore, the α-hydrogens of the ketone are more acidic than the α-hydrogens of the ester, so it would be easy to obtain the desired nucleophile. The ester can be easily prepared from the carboxylic acid starting material.

Tutorial:
Retronsynthesis

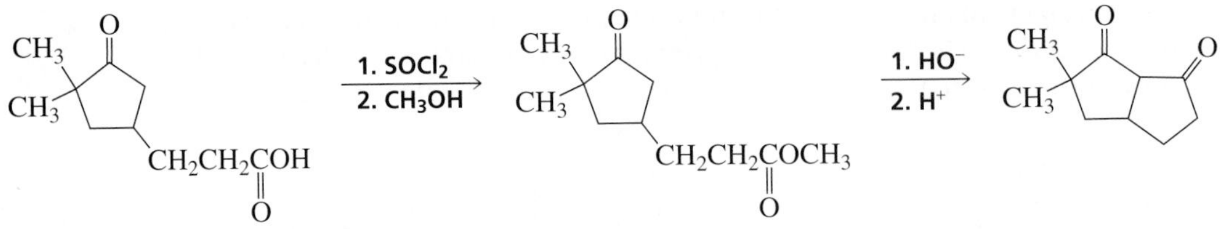

In the next synthesis, two new carbon–carbon bonds must be formed.

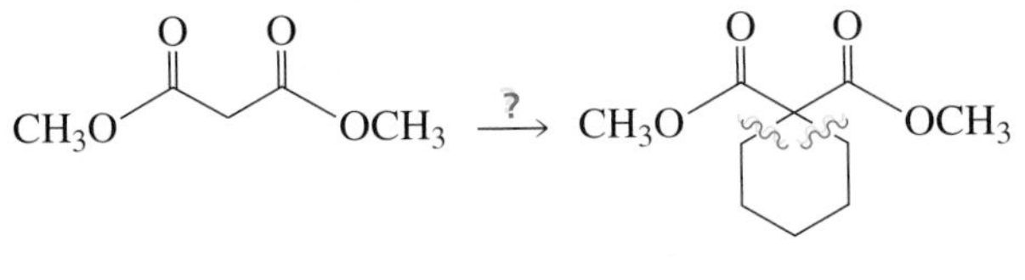

After determining the electrophilic and nucleophilic sites, you can see that two successive alkylations of a diester of malonic acid, using 1,5-dibromopentane as the alkyl halide, will produce the desired compound.

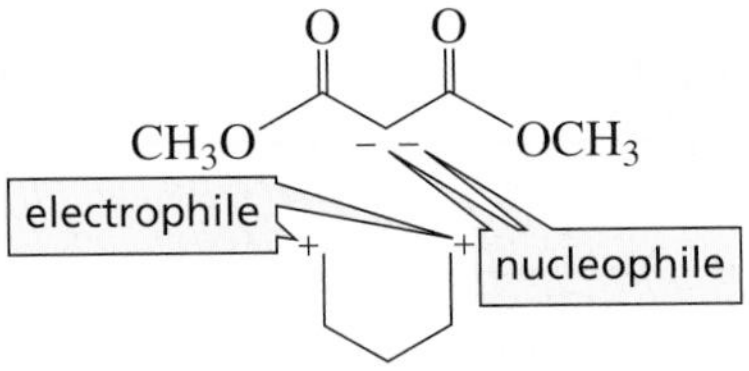

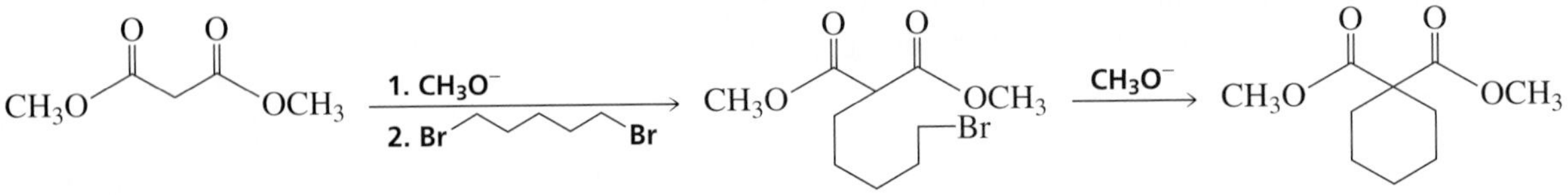

In planning the following synthesis, the diester that is given as the starting material suggests that you should use a Dieckmann condensation to obtain the cyclic compound.

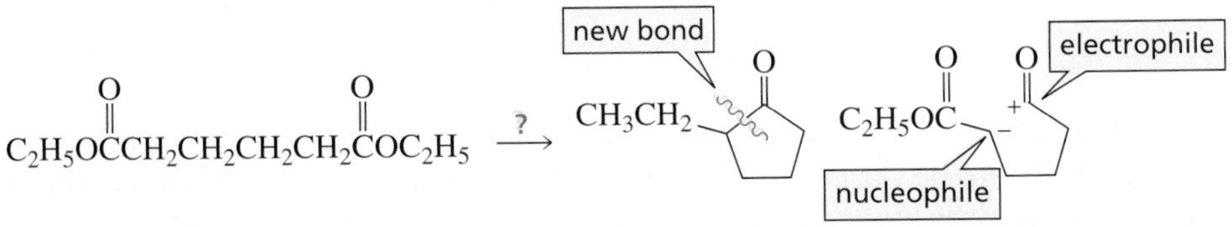

After the cyclopentanone ring is formed from a Dieckmann condensation of the starting material, alkylation of the α-carbon followed by hydrolysis of the β-keto ester and decarboxylation forms the desired product.

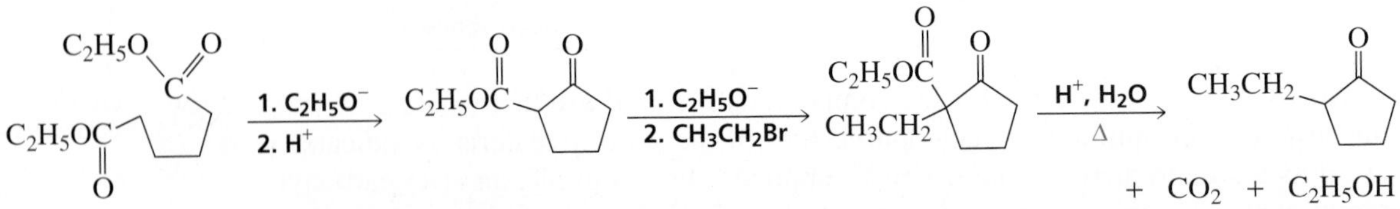

PROBLEM 38

Design a synthesis for each of the following compounds:

a. (diagram: acetophenone → propiophenone derivative)

b. (diagram: cyclohexene → 2-ethylcyclohexanol)

c. (diagram: benzaldehyde → cinnamic acid derivative)

d. $CH_3OCCH_2COCH_3 \longrightarrow$ (diagram: cyclopentane carboxylic acid)

Many reactions that occur in biological systems involve reactions at the α-carbon—the kinds of reactions you have studied in this chapter. We will now look at a few examples.

A Biological Aldol Addition

D-Glucose, the most abundant sugar found in nature, is synthesized by biological systems from two molecules of pyruvate. The series of reactions that convert two molecules of pyruvate into D-glucose is called **gluconeogenesis.** The reverse process, the breakdown of D-glucose into two molecules of pyruvate, is called **glycolysis.**

$$2\ CH_3\overset{O}{\underset{}{C}}{-}\overset{O}{\underset{}{C}}O^- \underset{\text{glycolysis}}{\overset{\text{gluconeogenesis}}{\rightleftharpoons}} \overset{\text{several steps}}{\rightleftharpoons\rightleftharpoons\rightleftharpoons\rightleftharpoons}$$

pyruvate

$$\begin{array}{c} HC{=}O \\ H{-\!\!-}OH \\ HO{-\!\!-}H \\ H{-\!\!-}OH \\ H{-\!\!-}OH \\ CH_2OH \end{array}$$

D-glucose

19.21 REACTIONS AT THE α-CARBON IN BIOLOGICAL SYSTEMS

Because D-glucose has twice as many carbons as pyruvate, it should not be surprising that one of the steps in the biosynthesis of D-glucose is an aldol addition. The aldol addition is catalyzed by an enzyme called aldolase. Aldolase catalyzes an aldol addition between dihydroxyacetone phosphate and D-glyceraldehyde-3-phosphate. The product is D-fructose-1,6-diphosphate, which is subsequently converted to D-glucose. The mechanism of this reaction is discussed in Section 22.9.

3-D Molecule:
Aldolase

$$\begin{array}{c} CH_2OPO_3{}^{2-} \\ | \\ C{=}O \\ | \\ CH_2OH \end{array}$$

dihydroxyacetone phosphate

$$\begin{array}{c} O \\ \| \\ C{-}H \\ H{-\!\!-}OH \\ CH_2OPO_3{}^{2-} \end{array}$$

D-glyceraldehyde-3-phosphate

$$\xrightarrow{\text{aldolase}}$$

$$\begin{array}{c} CH_2OPO_3{}^{2-} \\ | \\ C{=}O \\ HO{-\!\!-}H \\ H{-\!\!-}OH \\ H{-\!\!-}OH \\ CH_2OPO_3{}^{2-} \end{array}$$

D-fructose-1,6-diphosphate

PROBLEM 39

Propose a mechanism for the formation of D-fructose-1,6-diphosphate from dihydroxy-acetone phosphate and D-glyceraldehyde-3-phosphate using HO^- as the catalyst.

A Biological Aldol Condensation

Collagen is the most abundant protein in mammals, amounting to about one-fourth of the total protein. It is the major fibrous component of bone, teeth, skin, cartilage, and tendons. It also is responsible for holding groups of cells together in discrete units. Individual collagen molecules are called tropocollagen. Tropocollagen can be isolated only from tissues of young animals. As animals age, the individual molecules become cross-linked. This is why meat from older animals is tougher than meat from younger ones. Collagen cross-linking is an example of an aldol condensation.

Before collagen molecules can cross-link, the primary amino groups of the lysine residues of collagen must be converted to aldehyde groups. The enzyme that catalyzes this reaction is called lysyl oxidase. An aldol condensation between two aldehyde residues results in a cross-linked protein.

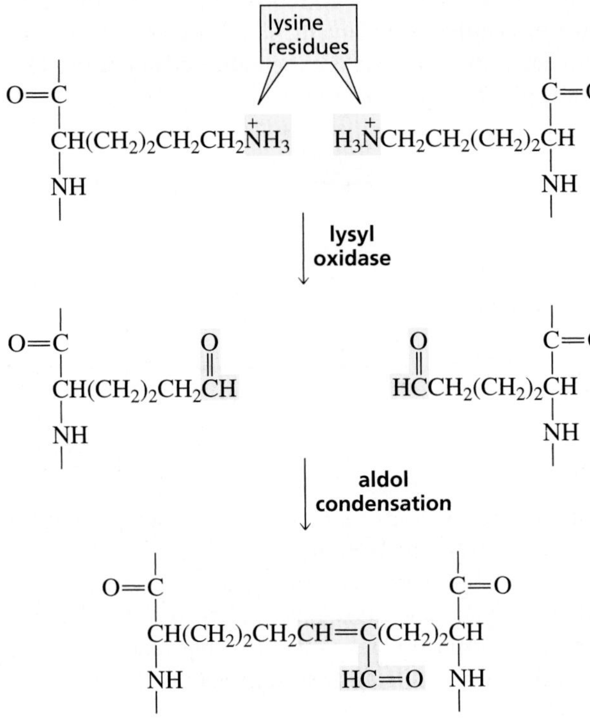

cross-linked collagen

A Biological Claisen Condensation

Fatty acids are long-chain, unbranched carboxylic acids (Section 16.17). Most naturally occurring fatty acids contain an even number of carbons because they are synthesized from acetate, an ion with two carbons.

In Section 16.18 we saw that carboxylic acids can be activated in biological systems by being converted to thioesters of coenzyme A.

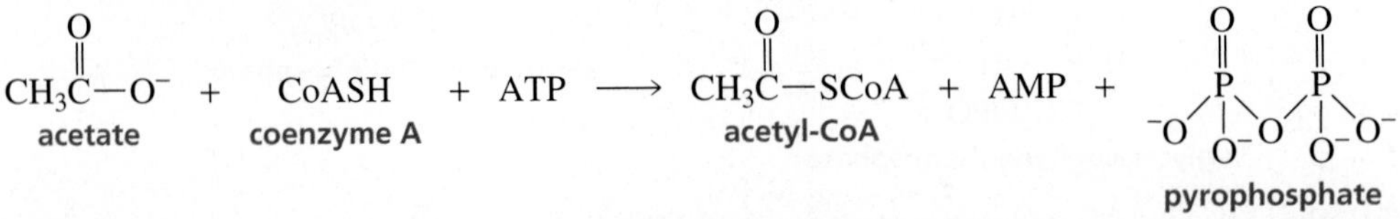

One of the necessary reactants for fatty acid synthesis is malonyl-CoA, which is obtained by carboxylation of acetyl-CoA. A possible mechanism for this reaction is discussed in Section 23.5.

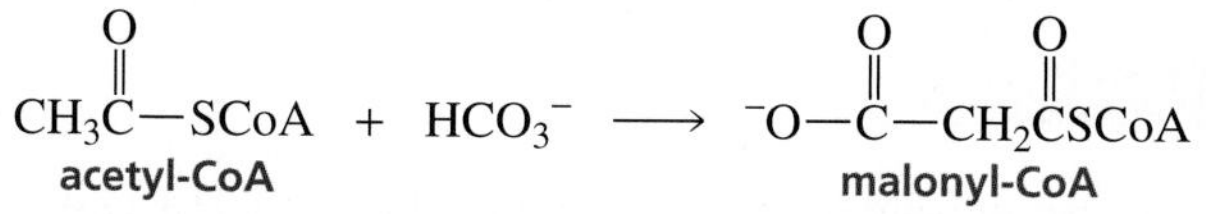

$$CH_3\overset{\overset{\displaystyle O}{\|}}{C}-SCoA \ + \ HCO_3^- \ \longrightarrow \ {}^-O-\overset{\overset{\displaystyle O}{\|}}{C}-CH_2\overset{\overset{\displaystyle O}{\|}}{C}SCoA$$

acetyl-CoA   **malonyl-CoA**

Before fatty acid synthesis can occur, however, the acyl groups of acetyl-CoA and malonyl-CoA are transferred to other thiols by means of a transesterification reaction.

$$CH_3\overset{\overset{\displaystyle O}{\|}}{C}-SCoA \ + \ RSH \ \longrightarrow \ CH_3\overset{\overset{\displaystyle O}{\|}}{C}-SR \ + \ CoASH$$

$${}^-O-\overset{\overset{\displaystyle O}{\|}}{C}-CH_2\overset{\overset{\displaystyle O}{\|}}{C}-SCoA \ + \ RSH \ \longrightarrow \ {}^-O-\overset{\overset{\displaystyle O}{\|}}{C}-CH_2\overset{\overset{\displaystyle O}{\|}}{C}-SR \ + \ CoASH$$

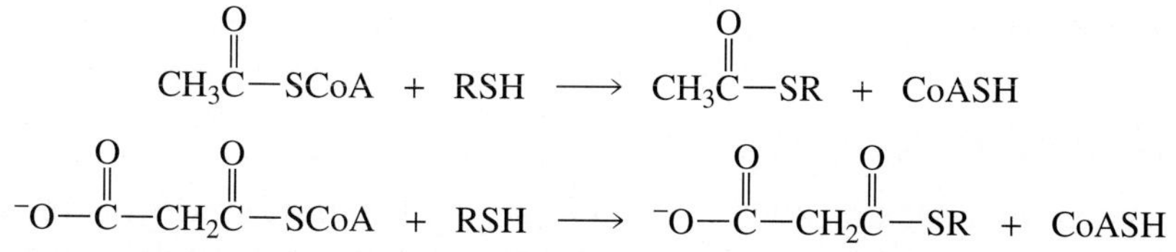

The first step in the biosynthesis of a fatty acid is a Claisen condensation between a molecule of acetyl thioester and a molecule of malonyl thioester. We have seen that in the laboratory the nucleophile needed for a Claisen condensation is obtained by using a strong base to remove an α-proton. A strong base is not available for biological reactions because they take place at neutral pH. So the required nucleophile is generated by removing CO_2 from the α-carbon of malonylthioester. Loss of CO_2 also serves to drive the condensation reaction to completion. The product of the condensation reaction undergoes a reduction, a dehydration, and a second reduction to give a four-carbon thioester.

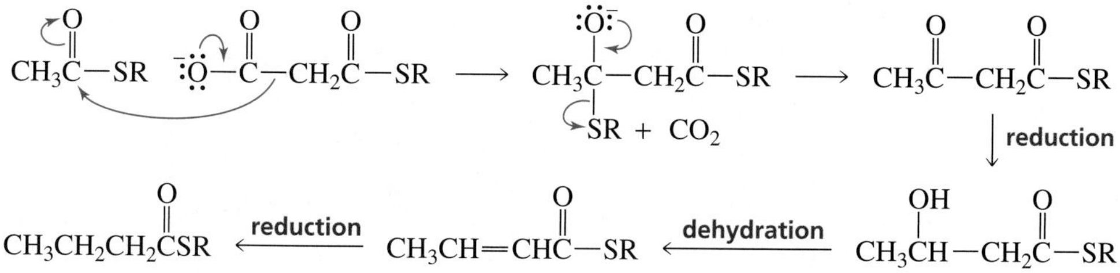

The four-carbon thioester undergoes a Claisen condensation with another molecule of malonyl thioester. Again, the product of the condensation reaction undergoes a reduction, a dehydration, and a second reduction to form a six-carbon thioester. This sequence of reactions is repeated, and each time two more carbons are added to the chain. This mechanism explains why most naturally occurring fatty acids contain an even number of carbons.

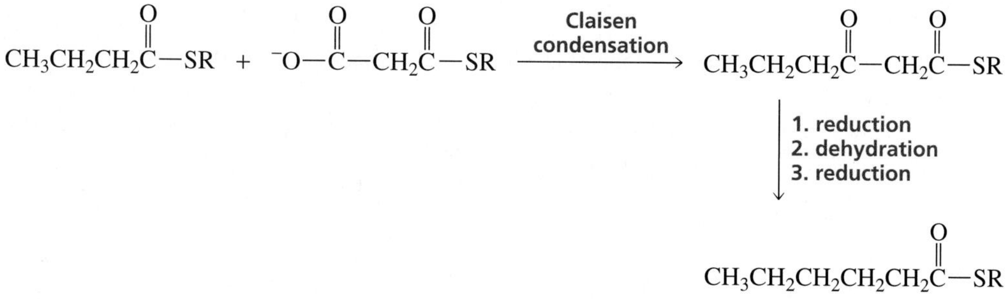

Once a thioester with the appropriate number of carbon atoms is obtained, it undergoes a transesterification reaction with glycerol in order to form fats, oils, and phospholipids (Sections 24.3 and 24.4).

PROBLEM 40◆

Palmitic acid (hexadecanoic acid) is a saturated 16-carbon fatty acid. How many moles of malonyl-CoA are required for the synthesis of one mole of palmitic acid?

PROBLEM 41◆

a. If the biosynthesis of palmitic acid were carried out with CD_3COSR and nondeuterated malonyl thioester, how many deuteriums would be incorporated into palmitic acid?

b. If the biosynthesis of palmitic acid were carried out with $^-OOCCD_2COSR$ and nondeuterated acetyl thioester, how many deuteriums would be incorporated into palmitic acid?

A Biological Decarboxylation

An example of a decarboxylation that occurs in biological systems is the decarboxylation of acetoacetate. Acetoacetate decarboxylase, the enzyme that catalyzes the reaction, first forms an imine with acetoacetate. The imine is protonated under physiological conditions and therefore readily accepts the pair of electrons left behind when the substrate loses CO_2. Decarboxylation leads to the formation of an enamine. Hydrolysis of the enamine produces the decarboxylated product (acetone) and regenerates the enzyme (Section 17.7).

Tutorial:
Common terms
for reactions at
the α-carbon

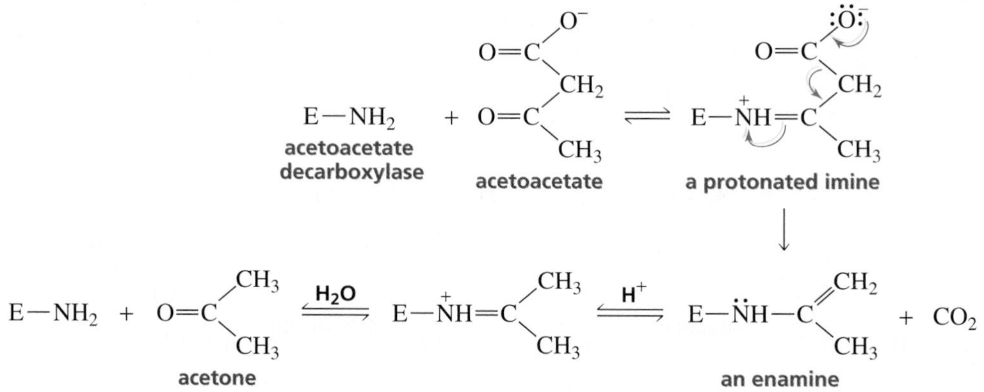

PROBLEM 42

When the enzymatic decarboxylation of acetoacetate is carried out in $H_2^{18}O$, the acetone that is formed contains ^{18}O. What does this tell you about the mechanism of the reaction?

SUMMARY OF REACTIONS

1. Halogenation of the α-carbon of aldehydes and ketones (Section 19.4)

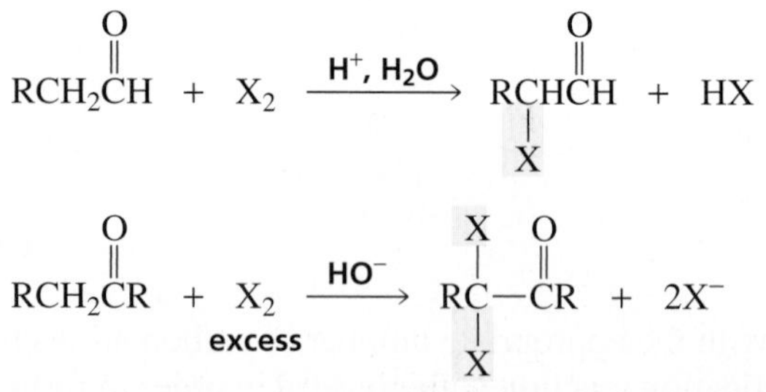

$$RCH_2CH + X_2 \xrightarrow{H^+, H_2O} RCHCH + HX$$

$$RCH_2CR + X_2 \xrightarrow[\text{excess}]{HO^-} RC-CR + 2X^-$$

$$X_2 = Cl_2, Br_2, \text{ or } I_2$$

2. Halogenation of the α-carbon of carboxylic acids: the Hell–Volhard–Zelinski reaction (Section 19.5)

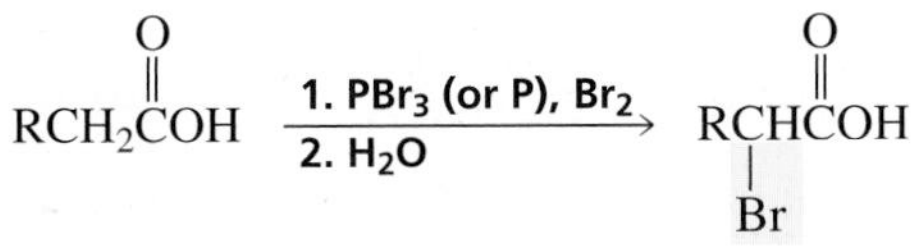

$$RCH_2\overset{O}{\overset{\|}{C}}OH \xrightarrow[\text{2. H}_2\text{O}]{\text{1. PBr}_3 \text{ (or P), Br}_2} RCH\overset{O}{\overset{\|}{C}}OH$$
$$\underset{Br}{|}$$

3. The α-carbon as a nucleophile and as an electrophile (Section 19.6)

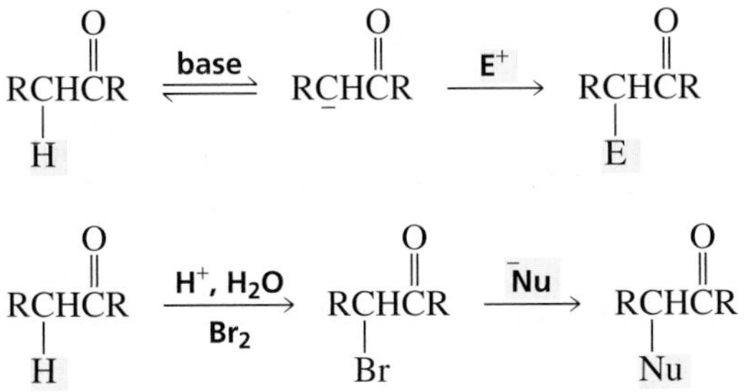

$$RCH\overset{O}{\overset{\|}{C}}R \underset{}{\overset{\text{base}}{\rightleftharpoons}} R\overset{O}{\underset{}{\overset{\|}{C}}}R \xrightarrow{E^+} RCH\overset{O}{\overset{\|}{C}}R$$

$$RCH\overset{O}{\overset{\|}{C}}R \xrightarrow[\text{Br}_2]{\text{H}^+, \text{H}_2\text{O}} RCH\overset{O}{\overset{\|}{C}}R \xrightarrow{\bar{\text{Nu}}} RCH\overset{O}{\overset{\|}{C}}R$$

4. Synthesis of α,β-unsaturated carbonyl compounds (Section 19.6)

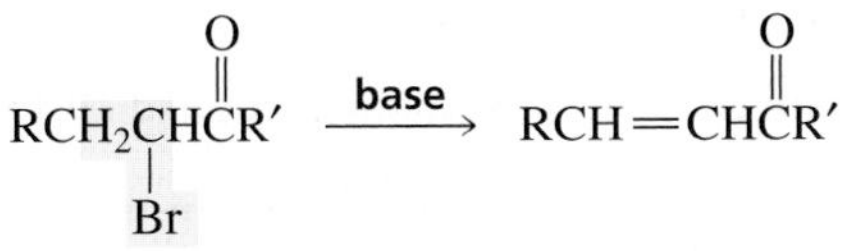

$$RCH_2CH\overset{O}{\overset{\|}{C}}R' \xrightarrow{\text{base}} RCH=CH\overset{O}{\overset{\|}{C}}R'$$
$$\underset{Br}{|}$$

5. Alkylation of the α-carbon of carbonyl compounds (Section 19.8)

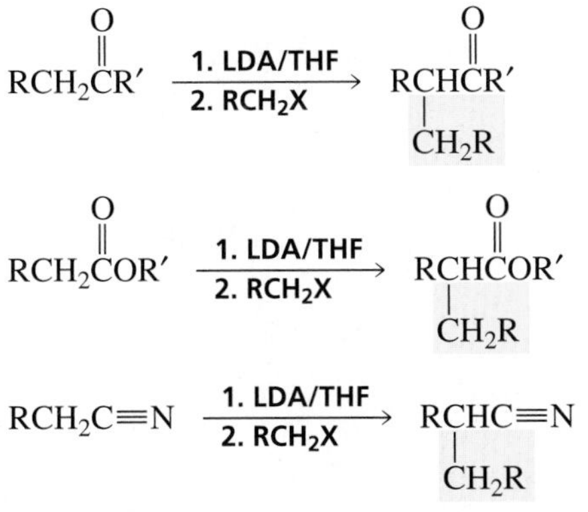

$$RCH_2\overset{O}{\overset{\|}{C}}R' \xrightarrow[\text{2. RCH}_2\text{X}]{\text{1. LDA/THF}} RCH\overset{O}{\overset{\|}{C}}R'$$
$$\underset{CH_2R}{|}$$

$$RCH_2\overset{O}{\overset{\|}{C}}OR' \xrightarrow[\text{2. RCH}_2\text{X}]{\text{1. LDA/THF}} RCH\overset{O}{\overset{\|}{C}}OR'$$
$$\underset{CH_2R}{|}$$

$$RCH_2C{\equiv}N \xrightarrow[\text{2. RCH}_2\text{X}]{\text{1. LDA/THF}} RCHC{\equiv}N$$
$$\underset{CH_2R}{|}$$

6. Alkylation and acylation of the α-carbon of aldehydes and ketones by means of an enamine intermediate (Sections 19.9 and 19.10).

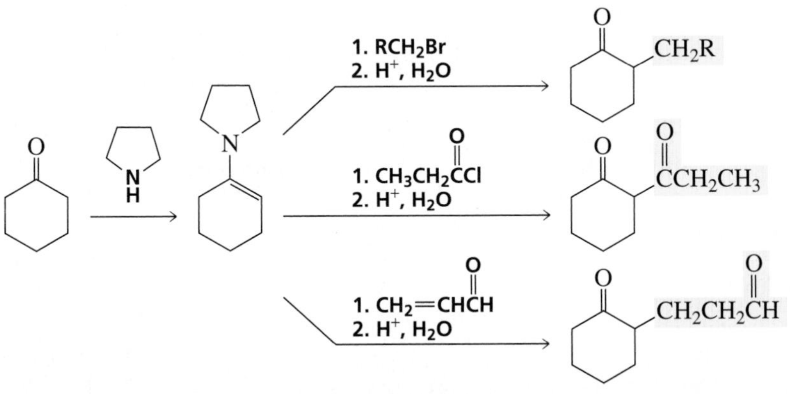

7. Michael reaction: attack of an enolate on an α,β-unsaturated carbonyl compound (Section 19.10)

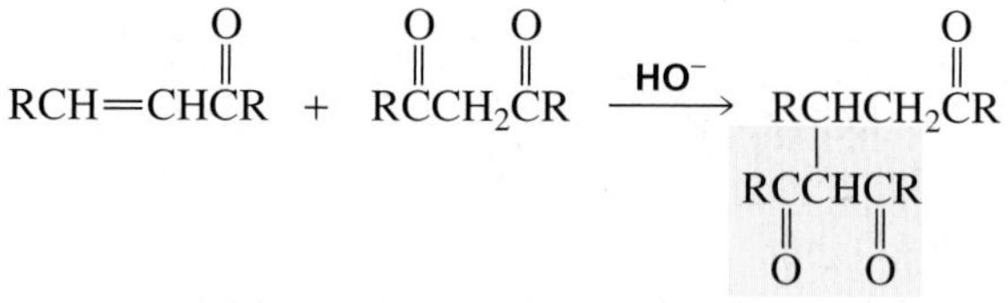

$$RCH{=}CHCR + RCCH_2CR \xrightarrow{HO^-} RCHCH_2CR$$

RCCHCR

8. Aldol addition of two aldehydes, two ketones, or an aldehyde and a ketone (Sections 19.11 and 19.13)

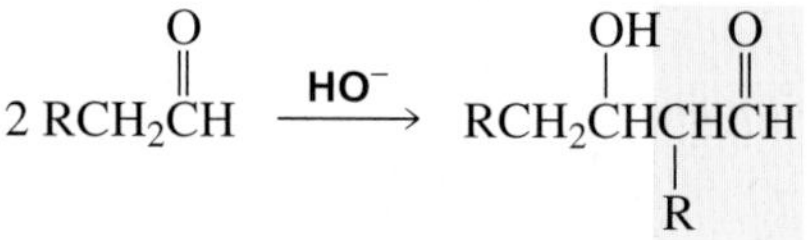

$$2\ RCH_2CH \xrightarrow{HO^-} RCH_2CHCHCH$$

9. Aldol condensation: dehydration of the product of an aldol addition (Section 19.12)

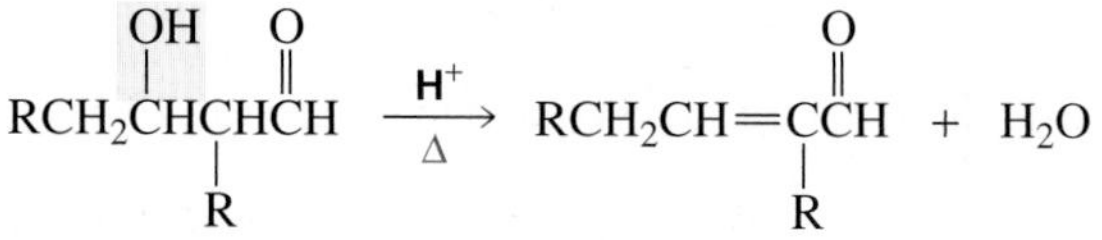

$$RCH_2CHCHCH \xrightarrow[\Delta]{H^+} RCH_2CH{=}CCH + H_2O$$

10. Claisen condensation of two esters (Sections 19.14 and 19.15)

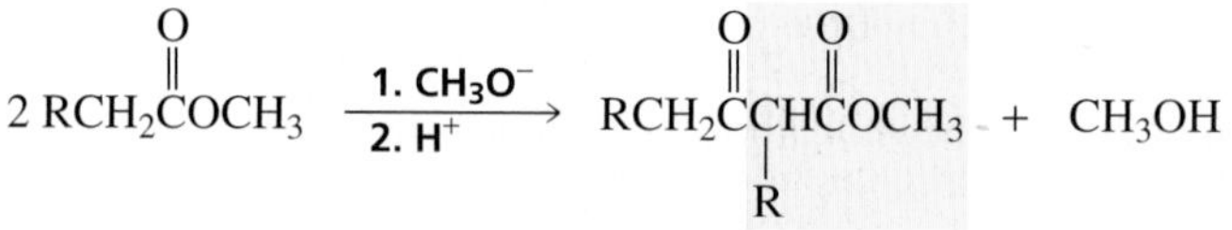

$$2\ RCH_2COCH_3 \xrightarrow[\text{2. } H^+]{\text{1. } CH_3O^-} RCH_2CCHCOCH_3 + CH_3OH$$

11. Condensation of a ketone and an ester (Section 19.15)

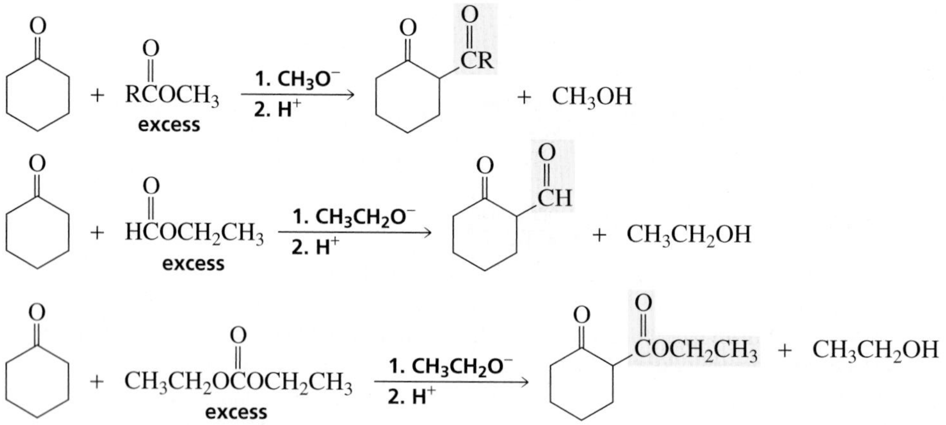

+ $RCOCH_3$ (excess) $\xrightarrow[\text{2. } H^+]{\text{1. } CH_3O^-}$ + CH_3OH

+ $HCOCH_2CH_3$ (excess) $\xrightarrow[\text{2. } H^+]{\text{1. } CH_3CH_2O^-}$ + CH_3CH_2OH

+ $CH_3CH_2OCOCH_2CH_3$ (excess) $\xrightarrow[\text{2. } H^+]{\text{1. } CH_3CH_2O^-}$ + $COCH_2CH_3$ + CH_3CH_2OH

12. Robinson annulation (Section 19.16)

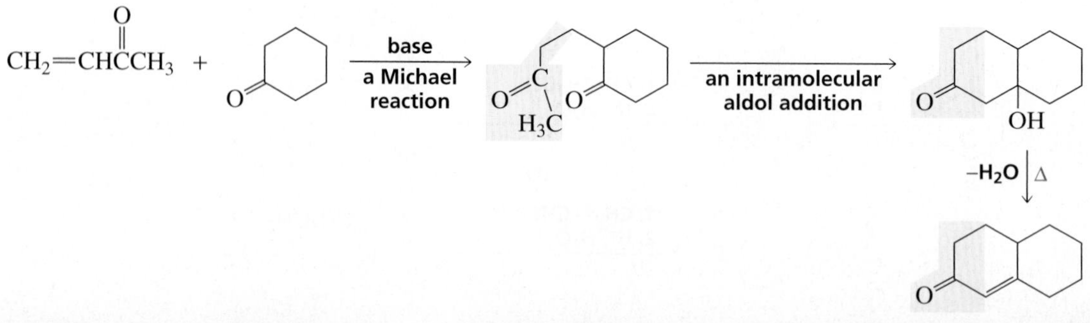

$CH_2{=}CHCCH_3$ + $\xrightarrow[\text{a Michael reaction}]{\text{base}}$ $\xrightarrow{\text{an intramolecular aldol addition}}$

$-H_2O \bigg\downarrow \Delta$

13. Decarboxylation of 3-oxocarboxylic acids (Section 19.17)

$$\underset{\text{RCCH}_2\text{COH}}{\overset{O\quad O}{\|\quad\|}} \xrightarrow{\ \Delta\ } \underset{\text{RCCH}_3}{\overset{O}{\|}} + CO_2$$

14. Malonic ester synthesis: preparation of carboxylic acids (Section 19.18)

$$\underset{\text{C}_2\text{H}_5\text{OCCH}_2\text{COC}_2\text{H}_5}{\overset{O\quad O}{\|\quad\|}} \xrightarrow[\substack{\textbf{2. RBr} \\ \textbf{3. H}^+, \textbf{H}_2\textbf{O},\,\Delta}]{\textbf{1. CH}_3\textbf{CH}_2\textbf{O}^-} \underset{\text{RCH}_2\text{COH}}{\overset{O}{\|}} + CO_2$$

15. Acetoacetic ester synthesis: preparation of methyl ketones (Section 19.19)

$$\underset{\text{CH}_3\text{CCH}_2\text{COC}_2\text{H}_5}{\overset{O\quad O}{\|\quad\|}} \xrightarrow[\substack{\textbf{2. RBr} \\ \textbf{3. H}^+, \textbf{H}_2\textbf{O},\,\Delta}]{\textbf{1. CH}_3\textbf{CH}_2\textbf{O}^-} \underset{\text{RCH}_2\text{CCH}_3}{\overset{O}{\|}} + CO_2$$

KEY TERMS

acetoacetic ester synthesis (page 860)
aldol addition (page 844)
aldol condensation (page 845)
ambident nucleophile (page 833)
α-carbon (page 827)
carbon acid (page 828)
Claisen condensation (page 848)
condensation reaction (page 845)
crossed aldol addition (page 846)
decarboxylation (page 857)
Dieckmann condensation (page 852)

β-diketone (page 830)
enolization (page 831)
gluconeogenesis (page 863)
glycolysis (page 863)
haloform reaction (page 835)
Hell–Volhard–Zelinski (HVZ)
   reaction (page 835)
Hunsdiecker reaction (page 858)
α-hydrogen (page 827)
keto–enol interconversion (page 831)
keto–enol tautomerization (page 831)
β-keto ester (page 830)

Kolbe–Schmitt carboxylation reaction
   (page 841)
malonic ester synthesis (page 859)
Michael reaction (page 842)
mixed aldol addition (page 846)
mixed Claisen condensation
   (page 850)
Robinson annulation (page 855)
Stork enamine reaction (page 843)
α-substitution reaction (page 832)
tautomers (page 830)

PROBLEMS

43. Write a structure for each of the following:
  a. ethyl acetoacetate
  b. α-methylmalonic acid
  c. a β-keto ester
  d. the enol tautomer of cyclopentanone
  e. the carboxylic acid obtained from the malonic ester
   synthesis when the alkyl halide is propyl bromide

44. Give the products of the following reactions:
  a. diethyl heptanedioate: (1) sodium ethoxide; (2) H$^+$, H$_2$O
  b. pentanoic acid + PBr$_3$ + Br$_2$ followed by hydrolysis
  c. acetone + ethyl acetate: (1) sodium ethoxide; (2) H$^+$, H$_2$O
  d. diethyl 2-ethylhexanedioate: (1) sodium ethoxide; (2) H$^+$, H$_2$O
  e. diethyl malonate: (1) sodium ethoxide; (2) isobutyl bromide; (3) H$^+$, H$_2$O + Δ
  f. acetophenone + diethyl carbonate: (1) sodium ethoxide; (2) H$^+$
  g. 1,3-cyclohexanedione + allyl bromide + sodium hydroxide
  h. dibenzyl ketone + methyl vinyl ketone + excess sodium hydroxide
  i. cyclopentanone: (1) pyrrolidine + Δ; (2) ethyl bromide; (3) H$^+$, H$_2$O
  j. γ-butyrolactone + LDA in THF followed by methyl iodide
  k. 2,7-octanedione + sodium hydroxide
  l. cyclohexanone + NaOD in D$_2$O
  m. diethyl 1,2-benzenedicarboxylate + ethyl acetate: (1) excess sodium ethoxide; (2) H$^+$

45. a. When (R)-2-methyl-1-phenyl-1-butanone is dissolved in an acidic or basic aqueous
   solution, a racemic mixture of 2-methyl-1-phenyl-1-butanone is formed. Explain.
  b. Give an example of another ketone that would undergo acid- or base-catalyzed
   racemization.

46. Identify A–L. (*Hint:* A gives a positive iodoform test and shows three singlets in its ^{1}H NMR spectrum with integral ratios 3 : 2 : 3.)

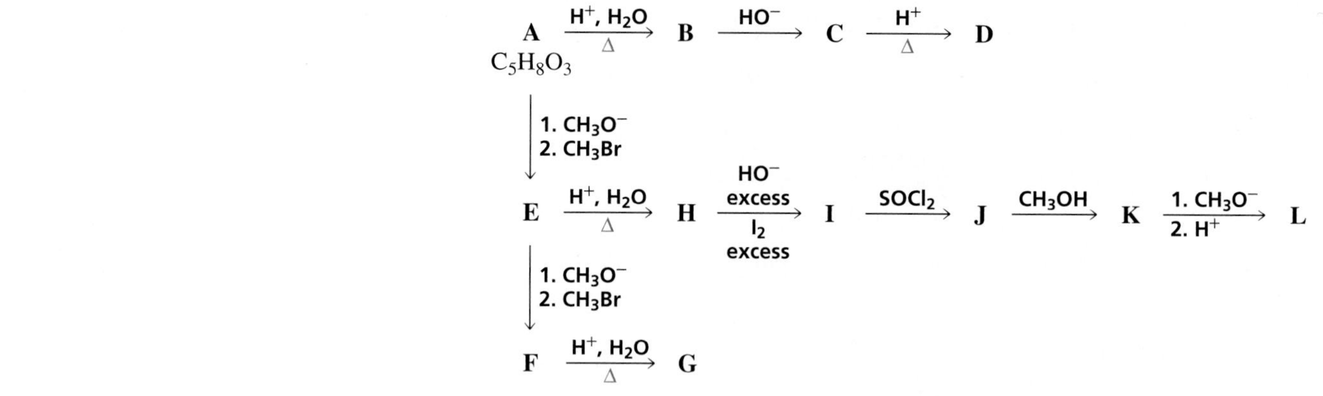

$$
\begin{array}{c}
\text{A} \xrightarrow[\Delta]{\text{H}^+,\ \text{H}_2\text{O}} \text{B} \xrightarrow{\text{HO}^-} \text{C} \xrightarrow[\Delta]{\text{H}^+} \text{D} \\
\text{C}_5\text{H}_8\text{O}_3
\end{array}
$$

$$\downarrow \begin{array}{l} \text{1. CH}_3\text{O}^- \\ \text{2. CH}_3\text{Br} \end{array}$$

$$
\text{E} \xrightarrow[\Delta]{\text{H}^+,\ \text{H}_2\text{O}} \text{H} \xrightarrow[\substack{\text{excess} \\ \text{I}_2 \\ \text{excess}}]{\text{HO}^-} \text{I} \xrightarrow{\text{SOCl}_2} \text{J} \xrightarrow{\text{CH}_3\text{OH}} \text{K} \xrightarrow[\text{2. H}^+]{\text{1. CH}_3\text{O}^-} \text{L}
$$

$$\downarrow \begin{array}{l} \text{1. CH}_3\text{O}^- \\ \text{2. CH}_3\text{Br} \end{array}$$

$$
\text{F} \xrightarrow[\Delta]{\text{H}^+,\ \text{H}_2\text{O}} \text{G}
$$

47. A β,γ-unsaturated carbonyl compound rearranges to a more stable conjugated α,β-unsaturated compound in the presence of either acid or base.
  a. Propose a mechanism for the base-catalyzed rearrangement.
  b. Propose a mechanism for the acid-catalyzed rearrangement.

a β,γ-unsaturated
carbonyl compound

an α,β-unsaturated
carbonyl compound

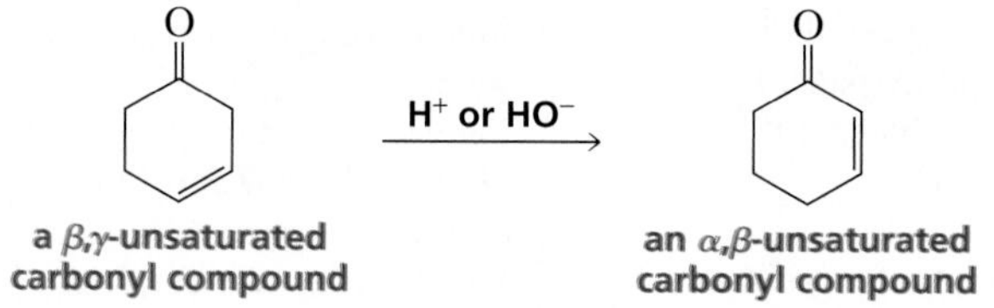

48. There are other condensation reactions similar to the aldol and Claisen condensations.
  a. The Perkin condensation is the condensation of an aromatic aldehyde and acetic anhydride. Give the product obtained from the following Perkin condensation:

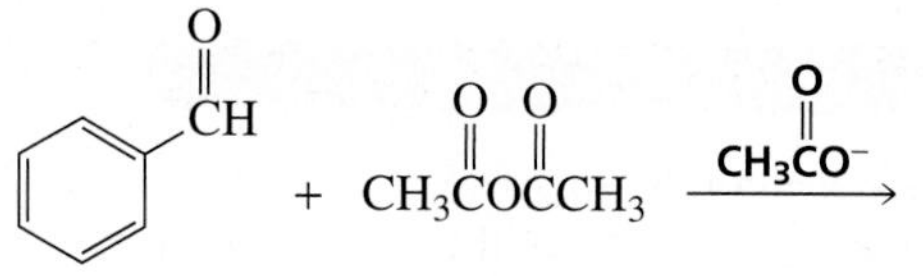

  b. What compound would result if water were added to the product of the Perkin condensation?

  c. The Knoevenagel condensation is the condensation of an aldehyde or a ketone that has no α-hydrogens and a compound such as diethyl malonate that has an α-carbon flanked by two electron-withdrawing groups. Give the product obtained from the following Knoevenagel condensation:

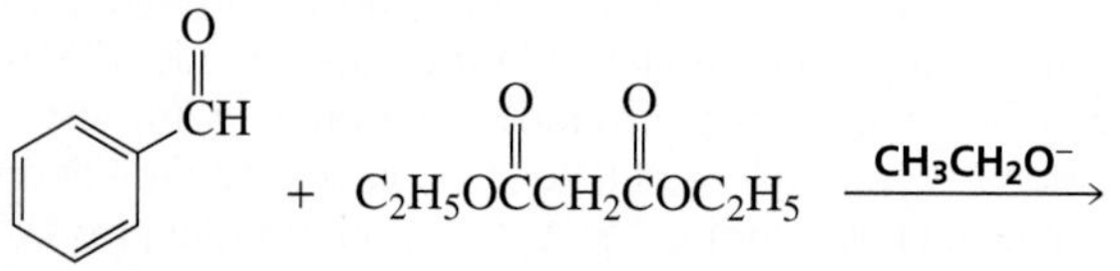

  d. What product would be obtained if the product of the Knoevenagel condensation were heated in an aqueous acidic solution?

49. The Reformatsky reaction is an addition reaction in which an organozinc reagent is used, instead of a Grignard reagent, to attack the carbonyl group of an aldehyde or a ketone. Because the organozinc reagent is less reactive than a Grignard reagent, a nucleophilic addition to the ester group does not occur. The organozinc reagent is prepared by treating an α-bromo ester with zinc.

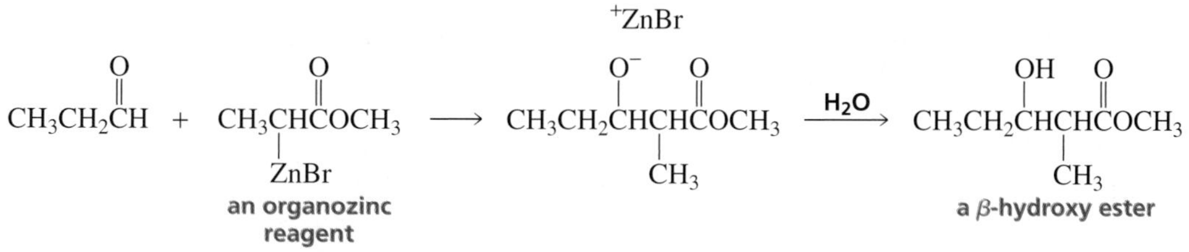

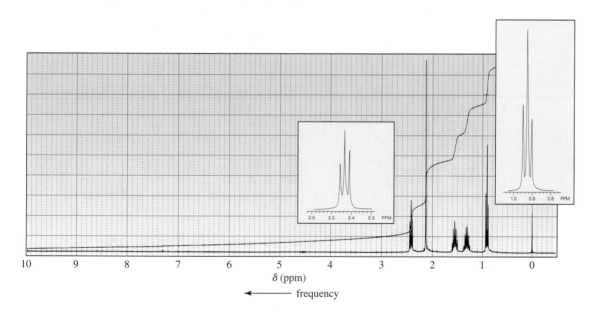

Describe how each of the following compounds could be prepared using a Reformatsky reaction:

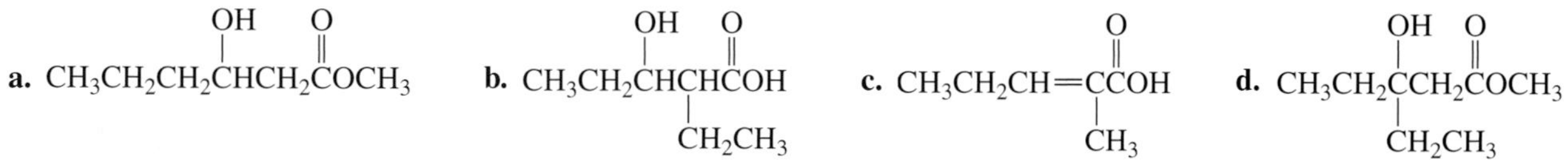

a. CH₃CH₂CH₂CHCH₂COCH₃ (with OH and O groups)

$$\text{a. } CH_3CH_2CH_2\underset{\underset{}{OH}}{CH}CH_2\overset{O}{C}OCH_3$$

$$\text{b. } CH_3CH_2\underset{\underset{CH_2CH_3}{OH}}{CH}CH\overset{O}{C}OH$$

$$\text{c. } CH_3CH_2CH=\underset{\underset{CH_3}{}}{C}\overset{O}{C}OH$$

$$\text{d. } CH_3CH_2\underset{\underset{CH_2CH_3}{OH}}{C}CH_2\overset{O}{C}OCH_3$$

50. The ketone whose ¹H NMR spectrum is shown here was obtained as the product of an acetoacetic ester synthesis. What alkyl halide was used in the synthesis?

10   9   8   7   6   5   4   3   2   1   0
δ (ppm)
← frequency

51. Indicate how the following compounds could be synthesized from cyclohexanone and any other necessary reagents:

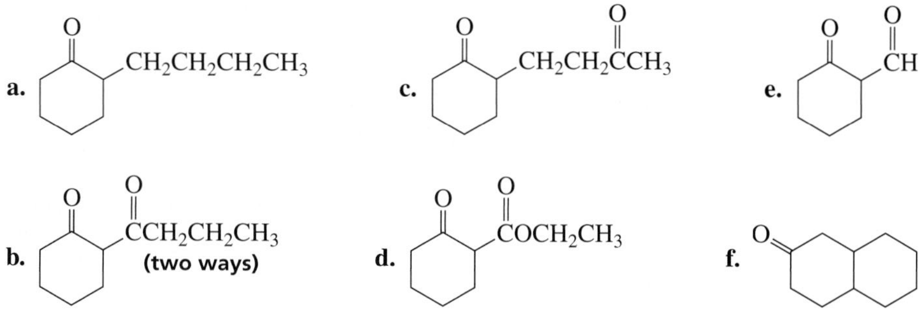

a. (cyclohexanone with) CH₂CH₂CH₂CH₃

c. (cyclohexanone with) CH₂CH₂CCH₃ (O)

e. (cyclohexanone with) CH (O)

b. (cyclohexanone with) CCH₂CH₂CH₃ (O) **(two ways)**

d. (cyclohexanone with) COCH₂CH₃ (O)

f. (fused bicyclic ketone)

52. A compound with molecular formula C₆H₁₀ has two peaks in its NMR spectrum, both of which are singlets (ratio 9 : 1). The compound reacts with an acidic aqueous solution containing mercuric sulfate to form a compound that also has an NMR spectrum that shows two singlets (ratio 3 : 1) and that gives a positive iodoform test. Identify this compound.

53. Indicate how each of the following compounds could be synthesized from the given starting material and any other necessary reagents:

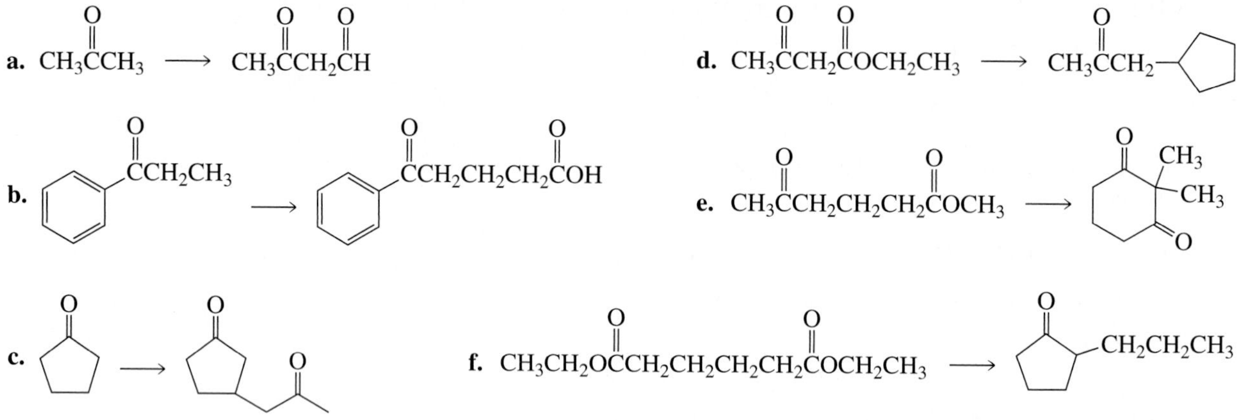

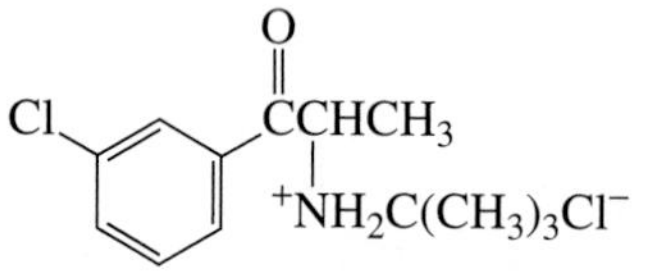

a. CH_3CCH_3 ⟶ CH_3CCH_2CH

d. $CH_3CCH_2COCH_2CH_3$ ⟶ CH_3CCH_2-⬠ (cyclopentyl)

b. Ph-CCH_2CH_3 ⟶ Ph-$CCH_2CH_2CH_2COH$

e. $CH_3CCH_2CH_2CH_2COCH_3$ ⟶ (1,2-dimethyl/diketo cyclohexanone with two CH₃)

c. cyclopentanone ⟶ cyclopentanone with $-CH_2-CCH_3$ side chain

f. $CH_3CH_2OCCH_2CH_2CH_2CH_2COCH_2CH_3$ ⟶ (cyclopentanone with $-CH_2CH_2CH_3$)

54. Bupropion hydrochloride is an antidepressant marketed under the trade name Wellbutrin. Propose a synthesis of bupropion hydrochloride starting with benzene.

Cl-(phenyl)-$CCHCH_3$ with $^+NH_2C(CH_3)_3\ Cl^-$

bupropion hydrochloride

55. What reagents would be required to carry out the following transformations?

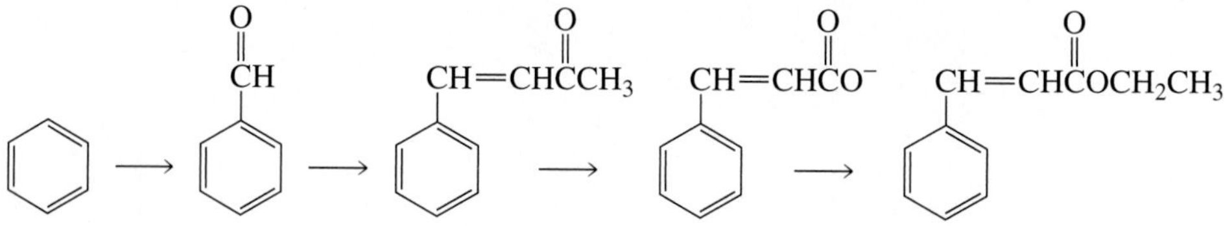

benzene ⟶ Ph-CH (O) ⟶ Ph-$CH=CHCCH_3$ (O) ⟶ Ph-$CH=CHCO^-$ (O) ⟶ Ph-$CH=CHCOCH_2CH_3$ (O)

56. Identify A–G.

benzene $\xrightarrow[\text{2. H}_2\text{O}]{\begin{array}{c}\text{1. CH}_3\text{CCl}\\ \text{AlCl}_3\end{array}}$ **A** $\xrightarrow{\text{HO}^-}$ **B** $\xrightarrow[\text{2. H}^+, \text{H}_2\text{O}]{\text{1. CH}_3\text{MgBr}}$ **C** $\xrightarrow{\Delta}$ **D** $\xrightarrow[\text{2. H}_2\text{O}_2]{\text{1. O}_3}$ **E** + **F** + **G**

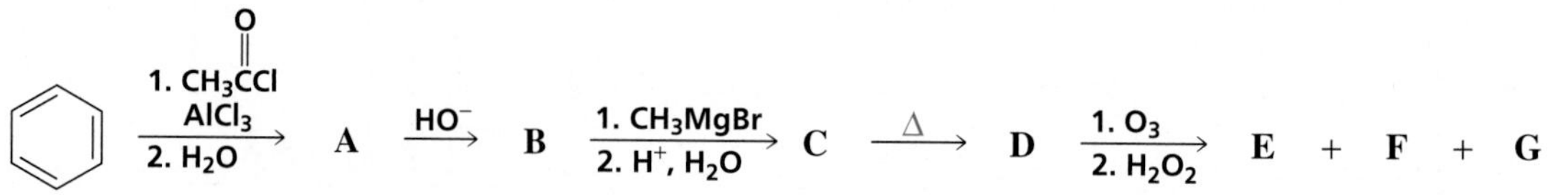

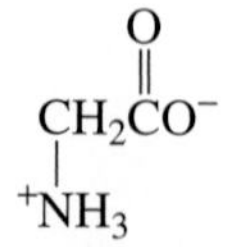

57. Give the products of the following reactions:

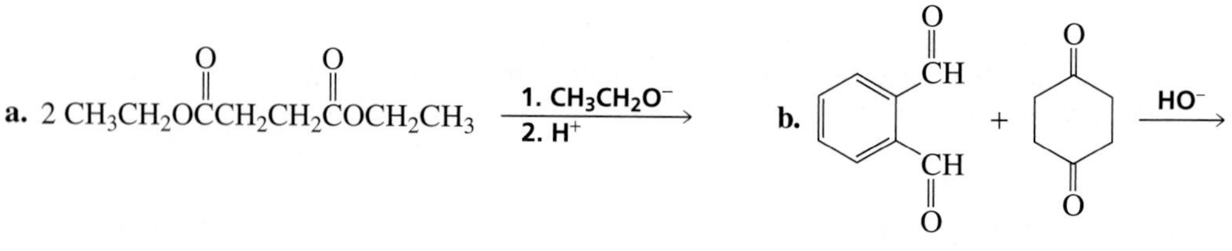

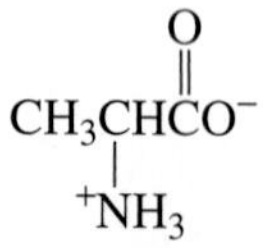

a. $2\ CH_3CH_2OCCH_2CH_2COCH_2CH_3$ $\xrightarrow[\text{2. H}^+]{\text{1. CH}_3\text{CH}_2\text{O}^-}$

b. (benzene-1,2-dicarbaldehyde) + (1,4-cyclohexanedione) $\xrightarrow{\text{HO}^-}$

58. a. Show how the amino acid alanine can be synthesized from propanoic acid.
b. Show how the amino acid glycine can be synthesized from phthalimide and diethyl 2-bromomalonate.

CH_3CHCO^- (O) with $^+NH_3$ — **alanine**

CH_2CO^- (O) with $^+NH_3$ — **glycine**

59. Cindy Synthon tried to prepare the following compounds using aldol condensations. Which of the compounds was she successful in synthesizing? Explain why the other syntheses were not successful.

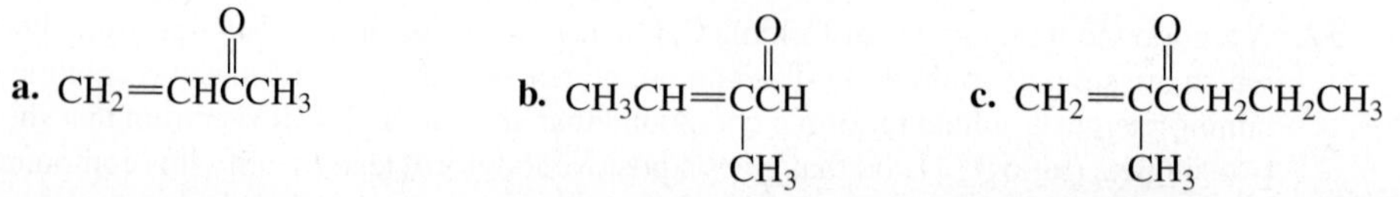

a. $CH_2=CHCCH_3$ (O)

b. $CH_3CH=CCH$ (O) with CH_3

c. $CH_2=CCCH_2CH_2CH_3$ (O) with CH_3

d. 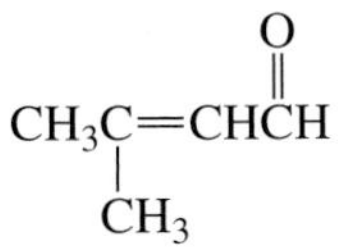 CH₃C=CHCH
          | ‖
          CH₃   O

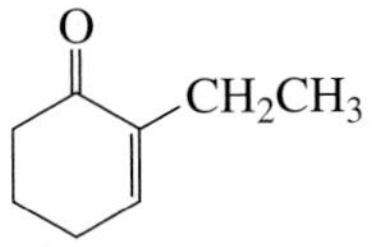

f. 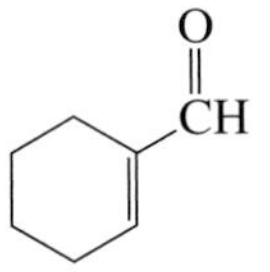 (structure: cyclohexene with CH=O group)

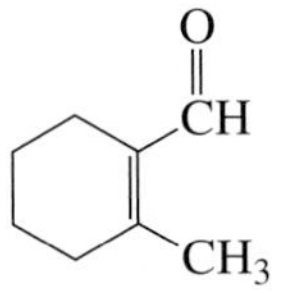

h. 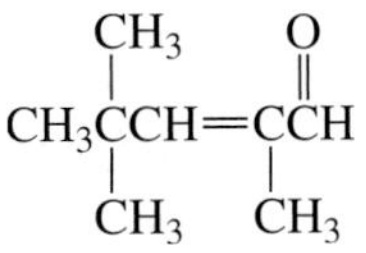   CH₃   O
          | ‖
   CH₃CCH=CCH
          |
          CH₃   CH₃

e. (structure: cyclohexenone with CH₂CH₃ group)

g. (structure: cyclohexene with CH=O and CH₃ groups)

i. 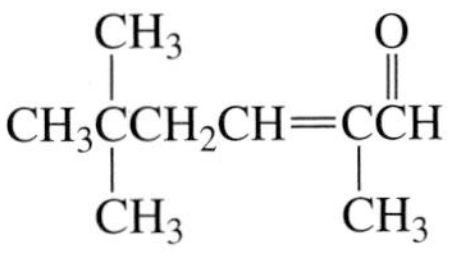   CH₃   O
          | ‖
   CH₃CCH₂CH=CCH
          |
          CH₃   CH₃

60. Explain why the following bromoketone forms different bicyclic compounds under different reaction conditions:

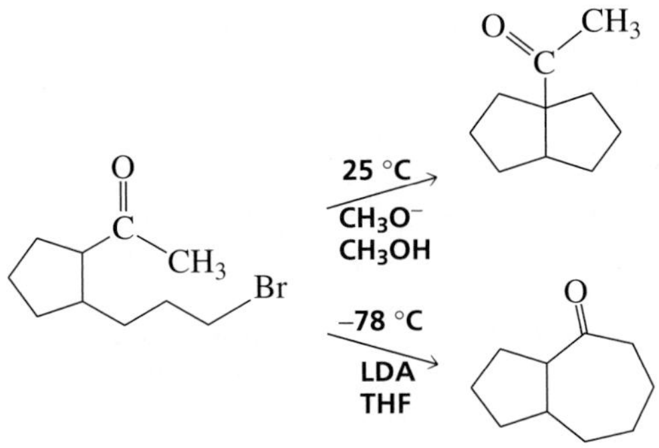

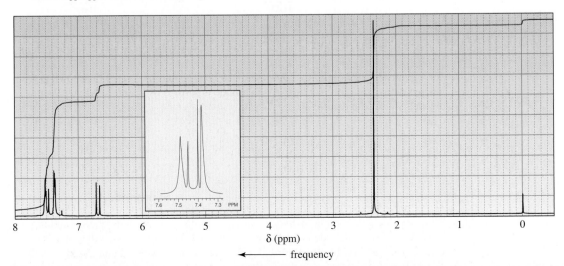

$$\text{(bromoketone)} \xrightarrow[\substack{\text{CH}_3\text{O}^- \\ \text{CH}_3\text{OH}}]{25\,°\text{C}} \text{(bicyclic product)}$$

$$\text{(bromoketone)} \xrightarrow[\substack{\text{LDA} \\ \text{THF}}]{-78\,°\text{C}} \text{(bicyclic product)}$$

61. Explain why the product obtained in the following reactions depends on the number of equivalents of base used in the reaction:

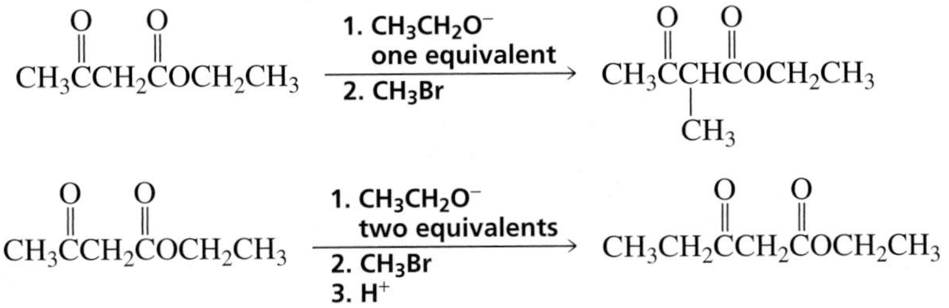

        O   O
        ‖   ‖   1. CH₃CH₂O⁻
 CH₃CCH₂COCH₂CH₃   ─────────────────→
                           one equivalent
                           2. CH₃Br

        O   O
        ‖   ‖
 CH₃CCHCOCH₂CH₃
        |
        CH₃

        O   O
        ‖   ‖   1. CH₃CH₂O⁻
 CH₃CCH₂COCH₂CH₃   ─────────────────→
                           two equivalents
                           2. CH₃Br
                           3. H⁺

        O   O
        ‖   ‖
 CH₃CH₂CCH₂COCH₂CH₃

62. What carbonyl compounds are required to prepare a compound with molecular formula $C_{10}H_{10}O$ whose ^1H NMR spectrum is shown?

(¹H NMR spectrum plotted against δ (ppm), with axis labeled from 8 to 0; inset shows expanded region 7.3–7.6 PPM; frequency arrow pointing left)

63. Fill in the boxes:

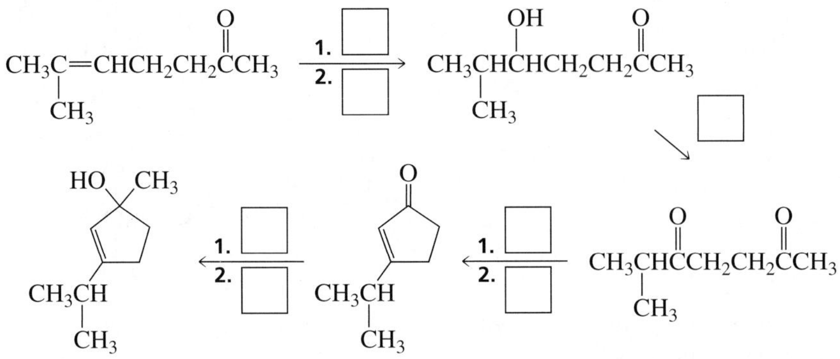

64. Ninhydrin reacts with an amino acid in basic solution to form a purple-colored compound. Propose a mechanism to account for the formation of the colored compound.

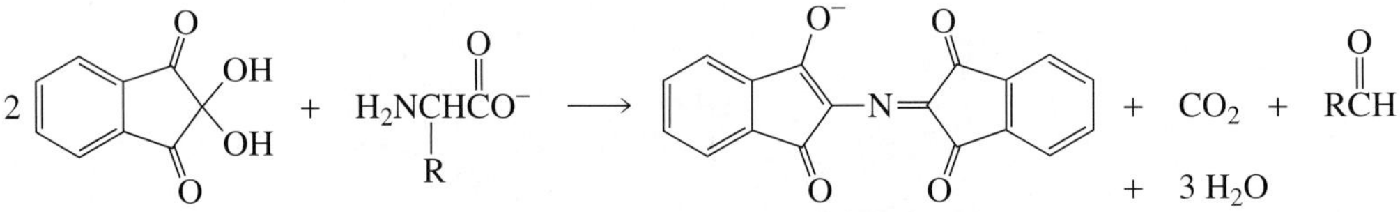

ninhydrin   an amino acid   purple-colored compound

65. A carboxylic acid is formed when an α-haloketone reacts with hydroxide ion. This reaction is called a Favorskii reaction. Propose a mechanism for the following Favorskii reaction. (*Hint:* In the first step, HO^- removes a proton from the α-carbon not bonded to Br; a three-membered ring is formed in the second step; and HO^- is a nucleophile in the third step.)

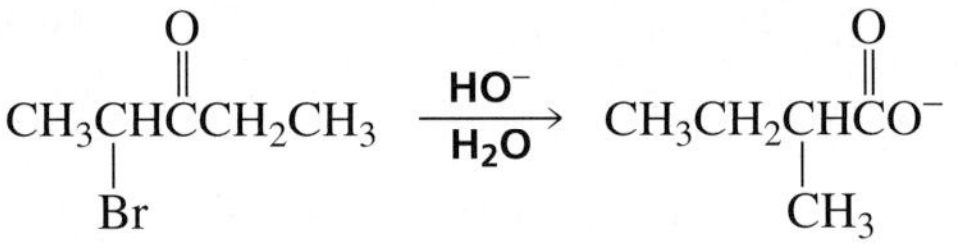

$$CH_3CHCCH_2CH_3 \quad \xrightarrow[\text{H}_2\text{O}]{\text{HO}^-} \quad CH_3CH_2CHCO^-$$

66. Give the products of the following reactions. (*Hint:* See Problem 65.)

a.

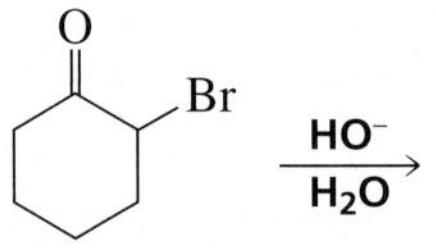

b.

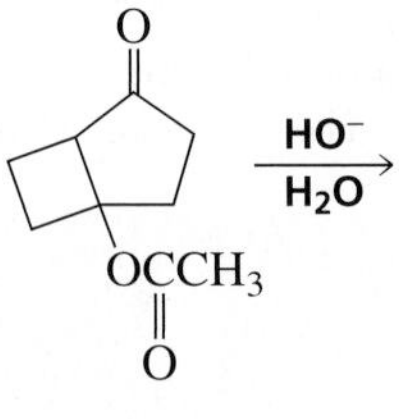

67. An α,β-unsaturated carbonyl compound can be prepared by a reaction known as a selenenylation-oxidation reaction. A selenoxide is formed as an intermediate. Propose a mechanism for the reaction.

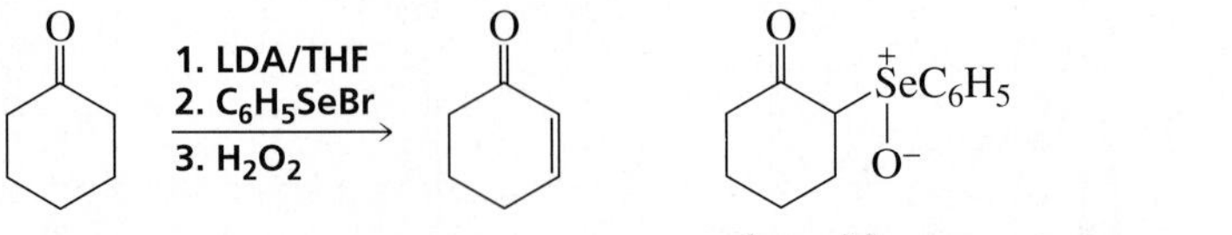

1. LDA/THF
2. C_6H_5SeBr
3. H_2O_2

a selenoxide

68. When an aldehyde with no α-hydrogens reacts with concentrated aqueous sodium hydroxide, half the aldehyde is converted to a carboxylic acid and the other half is converted to an alcohol. In other words, half the reactant is oxidized and the other half is reduced. This reaction is known as the Cannizzaro reaction. Propose a reasonable mechanism for the following Cannizzaro reaction:

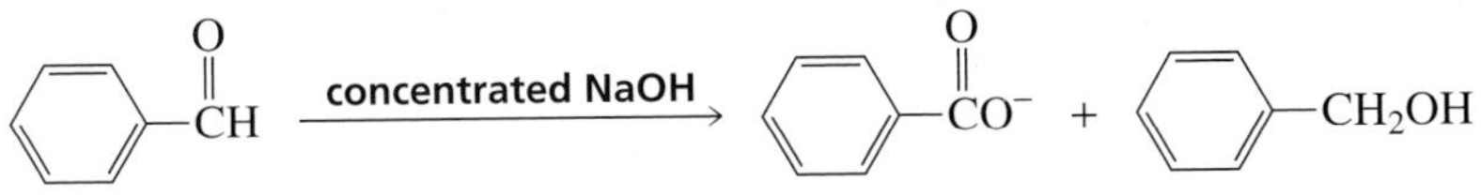

69. Propose a reasonable mechanism for each of the following reactions:

a.

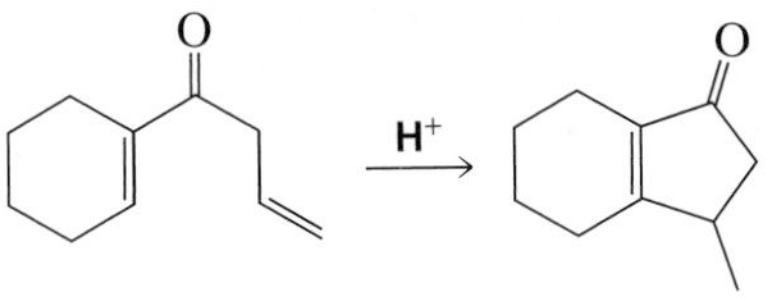

b.

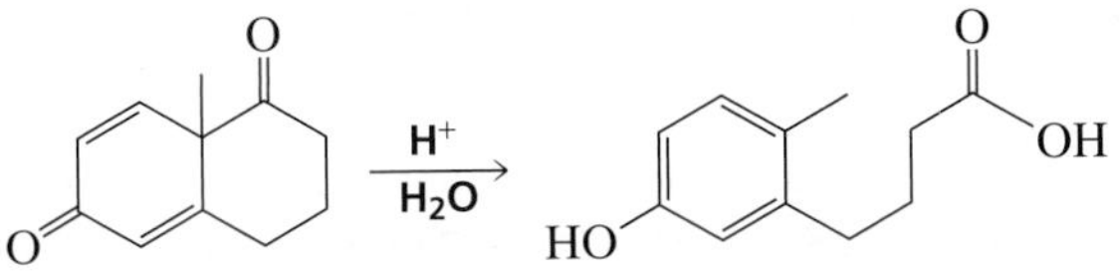

70. Show how the following compounds could be synthesized. The only carbon-containing reagents that are available to you for each synthesis are shown.

    a. $CH_3CH_2CH_2OH \longrightarrow CH_3CH_2CH_2CHCH_2OH$
                                                  CH_3

    b. $CH_3CHOH \longrightarrow CH_3CHCH_2CH_2CH_3$
             CH_3 CH_3

    c. $CH_3CH_2OH + CH_3CHOH \longrightarrow$
                               CH_3

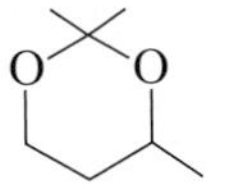

71. The following reaction is known as the benzoin condensation. The reaction will not take place if sodium hydroxide is used instead of sodium cyanide. Propose a mechanism for the reaction.

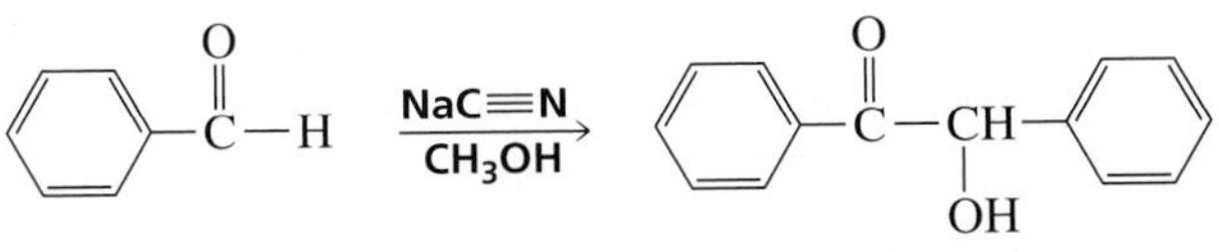

benzoin

72. Orsellinic acid, a common constituent of lichens, is synthesized from the condensation of acetyl thioester and malonyl thioester. If the lichen were grown on a medium containing acetate that was radioactively labeled with ^{14}C at the carbonyl carbon, which carbons would be labeled in orsellinic acid?

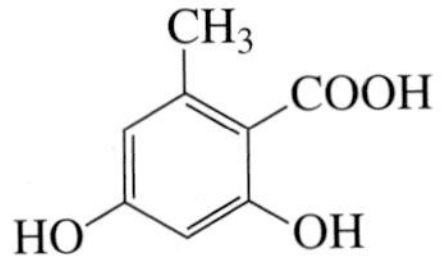

orsellinic acid

73. A compound known as Hagemann's ester can be prepared by treating a mixture of formaldehyde and ethyl acetoacetate first with base and then with acid and heat. Write the structure for the product of each of the steps.
    a. The first step is an aldol-like condensation.
    b. The second step is a Michael addition.
    c. The third step is an intramolecular aldol condensation.
    d. The fourth step is a hydrolysis followed by a decarboxylation.

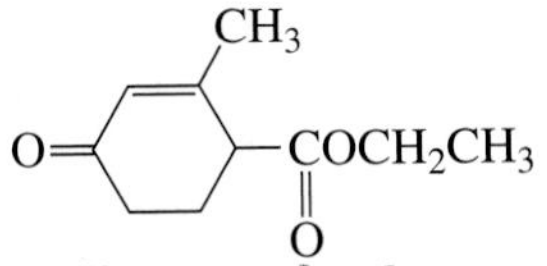

Hagemann's ester

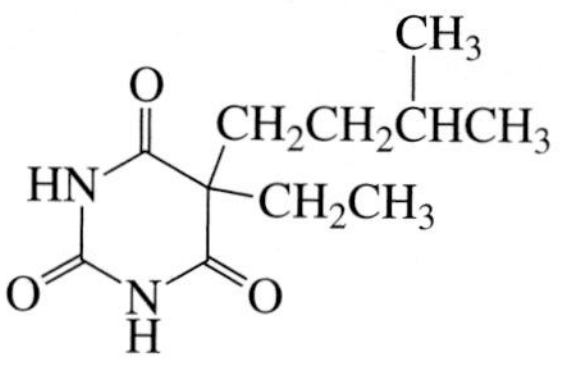

Amytal

74. Amobarbital is a sedative marketed under the trade name Amytal. Propose a synthesis of amobarbital using diethyl malonate and urea as two of the starting materials.

75. Propose a reasonable mechanism for each of the following reactions:

a. $CH_3\overset{O}{\overset{||}{C}}CH_2\overset{O}{\overset{||}{C}}OCH_2CH_3$ $\xrightarrow[\text{2.}]{\text{1. } CH_3CH_2O^-}$

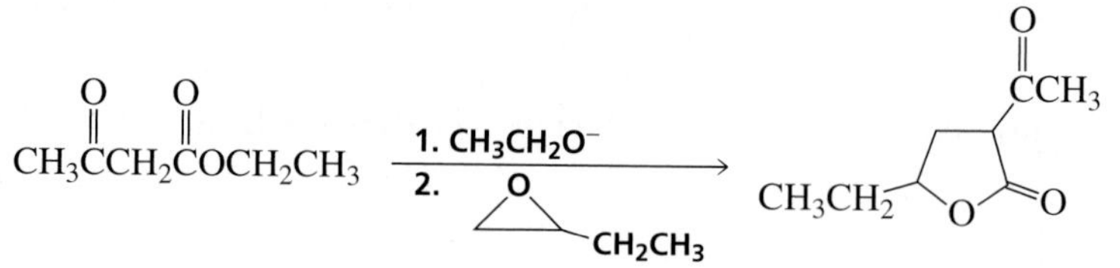

b.

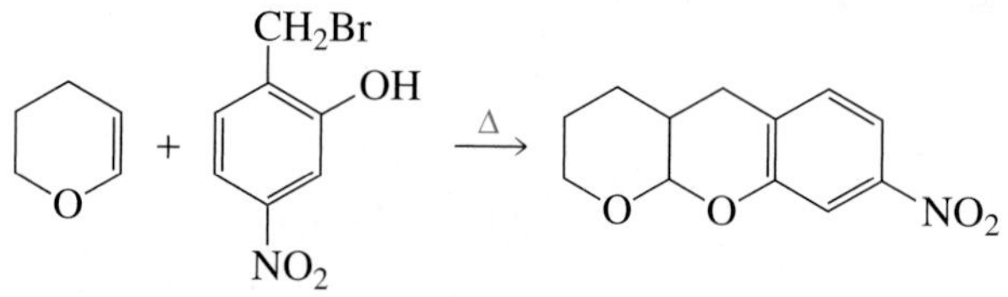

76. Tyramine is an alkaloid found in mistletoe, ripe cheese, and putrefied animal tissue. Dopamine is a neurotransmitter involved in the regulation of the central nervous system.

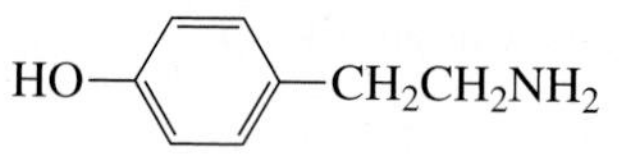

tyramine

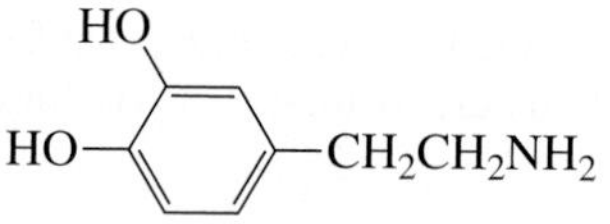

dopamine

a. Give two ways to prepare β-phenylethylamine from β-phenylethyl chloride.
b. How can β-phenylethylamine be prepared from benzyl chloride?
c. How can β-phenylethylamine be prepared from benzaldehyde?
d. How can tyramine be prepared from β-phenylethylamine?
e. How can dopamine be prepared from tyramine?

77. Show how estrone, a steroid hormone, can be prepared from the given starting materials. (*Hint:* Start with a Robinson annulation.)

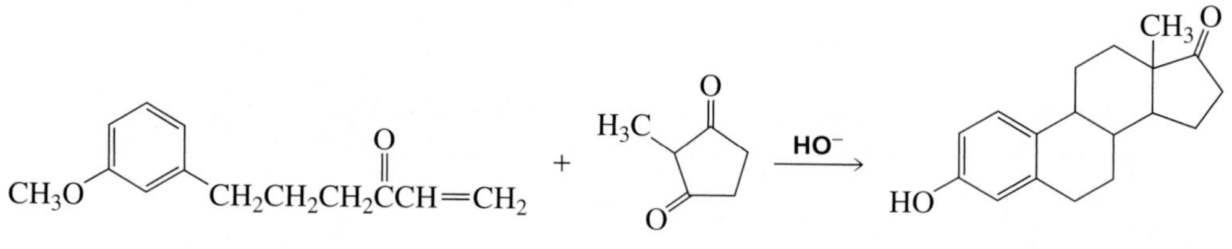

estrone

78. a. Ketoprofen, like ibuprofen, is an anti-inflammatory analgesic. How could ketoprofen be synthesized from the given starting material?

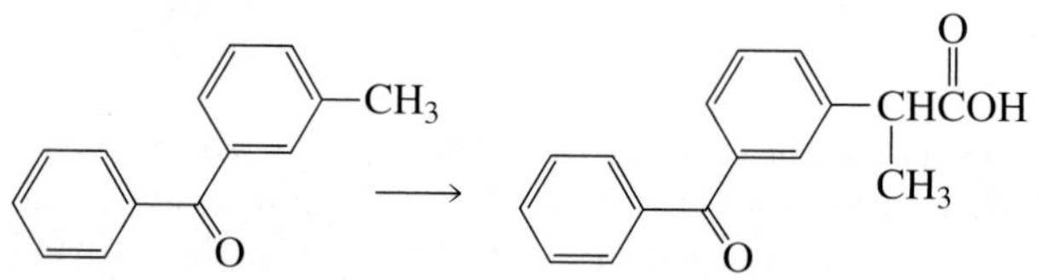

ketoprofen

b. Ketoprofen and ibuprofen both have a propanoic acid substituent (see Problem 64 in Chapter 16). Explain why the identical subunits are synthesized in different ways.

Bioorganic Compounds

Chapters 20 through 25 discuss the chemistry of organic compounds found in biological systems. Many of these compounds may be larger than the organic compounds you have seen up to this point and they may have more than one functional group, but the principles that govern their structure and reactivity are essentially the same as those that govern the structure and reactivity of the compounds you have been studying. These chapters, therefore, will give you the opportunity to review much of the organic chemistry you have learned as you apply your knowledge of organic reactions to compounds found in the biological world.

Chapter 20   **Carbohydrates**
Chapter 21   **Amino Acids, Peptides, and Proteins**
Chapter 22   **Catalysis**
Chapter 23   **The Organic Mechanisms of the Coenzymes. Metabolism**
Chapter 24   **Lipids**
Chapter 25   **Nucleosides, Nucleotides, and Nucleic Acids**

Chapter 20 introduces you to the chemistry of carbohydrates.

Chapter 21 discusses the chemistry of amino acids, peptides, and proteins.

Chapter 22 first describes the various ways that organic reactions are catalyzed, and then shows you how enzymes use the same principles in their catalysis of reactions that occur in biological systems.

Chapter 23 describes the chemistry of the coenzymes—organic compounds that some enzymes need to catalyze biological reactions. Coenzymes play a variety of chemical roles—some function as oxidizing and reducing agents, some allow electrons to be delocalized, some activate groups for further reaction, and some provide good nucleophiles or strong bases needed for reactions. Because coenzymes are derived from vitamins, you will see why vitamins are necessary for many of the organic reactions that occur in biological systems.

Chapter 24 discusses the chemistry of lipids. Lipids are water-insoluble compounds found in animals and plants.

Chapter 25 discusses the chemistry of nucleosides, nucleotides, and nucleic acids (RNA and DNA).

20

Carbohydrates

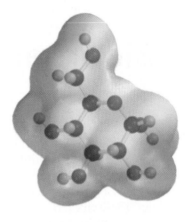

β-D-glucose

α-D-glucose

D-glucose

D-fructose

Carbohydrates are the first group of bioorganic compounds we will study. **Bioorganic compounds** are organic compounds found in biological systems. The structures of most bioorganic compounds are more complicated than the structures of many of the organic compounds you are used to seeing, but do not let the complicated-looking structures fool you into thinking their chemistry is equally complicated. Bioorganic compounds follow the same principles of structure and reactivity as the organic molecules we have discussed so far. One reason bioorganic compounds have complicated structures is because the compounds must be able to recognize each other, and much of their structure is for this purpose—a concept known as **molecular recognition.**

Carbohydrates are the most abundant class of compounds in the biological world, making up more than 50% of the dry weight of the Earth's biomass. Carbohydrates are important constituents of all living organisms, and have a variety of different functions. Some are important structural components of cells and some act as recognition sites on cell surfaces. Others serve as a major source of metabolic energy. For example, the leaves, fruits, seeds, stems, and roots of plants contain carbohydrates that plants use for their own metabolic needs and that serve the metabolic needs of the animals that eat them.

Early chemists noted that carbohydrates had molecular formulas that made them appear to be hydrates of carbon, $C_n(H_2O)_n$—hence the name. Later structural studies revealed that these compounds were not hydrates because they did not contain intact water molecules. However, the term "carbohydrate" persisted. It now refers either to polyhydroxy aldehydes such as D-glucose and polyhydroxy ketones such as D-fructose, or to compounds such as sucrose that can be hydrolyzed to polyhydroxy aldehydes or polyhydroxy ketones. The chemical structures of carbohydrates are commonly represented by wedge-and-dash structures or by Fischer projections. Notice that both D-glucose and D-fructose have molecular formulas $C_6H_{12}O_6$, which is consistent with $[C_6(H_2O)_6]$ and makes them appear to be hydrates of carbon.

$$
\begin{array}{cccc}
\text{HC}=\text{O} & \text{HC}=\text{O} & \text{CH}_2\text{OH} & \text{CH}_2\text{OH} \\
\text{H}\!-\!\text{C}\!-\!\text{OH} & \text{H}\!-\!\text{OH} & \text{C}=\text{O} & \text{C}=\text{O} \\
\text{HO}\!-\!\text{C}\!-\!\text{H} & \text{HO}\!-\!\text{H} & \text{HO}\!-\!\text{C}\!-\!\text{H} & \text{HO}\!-\!\text{H} \\
\text{H}\!-\!\text{C}\!-\!\text{OH} & \text{H}\!-\!\text{OH} & \text{H}\!-\!\text{C}\!-\!\text{OH} & \text{H}\!-\!\text{OH} \\
\text{H}\!-\!\text{C}\!-\!\text{OH} & \text{H}\!-\!\text{OH} & \text{H}\!-\!\text{C}\!-\!\text{OH} & \text{H}\!-\!\text{OH} \\
\text{CH}_2\text{OH} & \text{CH}_2\text{OH} & \text{CH}_2\text{OH} & \text{CH}_2\text{OH}
\end{array}
$$

wedge-and-dash   Fischer projection   wedge-and-dash   Fischer projection
structure   structure

D-glucose   **D-fructose**
a polyhydroxy aldehyde   **a polyhydroxy ketone**

3-D Molecules:
D-Glucose;
D-Fructose

The most abundant carbohydrate in nature is D-glucose. Cells of organisms oxidize D-glucose in the first of a series of processes that provide them energy. When animals have more D-glucose than they need for energy, they convert excess D-glucose into a polymer called glycogen. When an animal needs energy, glycogen is broken down into individual D-glucose molecules. Plants convert excess D-glucose into a polymer known as starch. Cellulose is another polymer of D-glucose. It is the major structural component of plants. Chitin, a carbohydrate similar to cellulose, makes up the exoskeletons of crustaceans, insects, and other arthropods, and the structural material of fungi.

Animals obtain glucose by eating plants or other food that contains glucose. Plants form glucose by a process known as **photosynthesis.** During photosynthesis, plants take up water through their roots and use carbon dioxide from the air to synthesize glucose and oxygen. Because photosynthesis is the opposite of the process used by organisms to obtain energy—the oxidation of glucose to carbon dioxide and water—plants require energy to carry out photosynthesis. Plants obtain the energy for photosynthesis from sunlight. Chlorophyll molecules in green plants capture the light energy used in photosynthesis. Notice that photosynthesis uses the CO_2 that animals exhale as waste and generates the O_2 that animals inhale to sustain life. In fact, nearly all the oxygen in the Earth's atmosphere has been released by photosynthetic processes.

$$
\underset{\text{glucose}}{C_6H_{12}O_6} \; + \; 6\,O_2 \; \underset{\text{photosynthesis}}{\overset{\text{oxidation}}{\rightleftharpoons}} \; 6\,CO_2 \; + \; 6\,H_2O \; + \; \text{energy}
$$

20.1 CLASSIFICATION OF CARBOHYDRATES

The terms "carbohydrate," "saccharide," and "sugar" are often used interchangeably. Saccharide comes from the word for "sugar" in several early languages (*sarkara* in Sanskrit, *sakcharon* in Greek, and *saccharum* in Latin).

There are two classes of carbohydrates—simple carbohydrates and complex carbohydrates. **Simple carbohydrates** are **monosaccharides** (single sugars), whereas **complex carbohydrates** contain two or more sugar subunits linked together. **Disaccharides** are complex carbohydrates with two sugar subunits linked together, **oligosaccharides** have three to ten sugar subunits (*oligos* is Greek for "few") linked together, and **polysaccharides** have more than ten sugar subunits linked together. Disaccharides, oligosaccharides, and polysaccharides can be broken down to monosaccharide subunits by hydrolysis.

$$
\underset{\textbf{polysaccharide}}{-\text{M}-\text{M}-\text{M}-\text{M}-\text{M}-\text{M}-\text{M}-\text{M}-\text{M}-} \; \overset{\text{hydrolysis}}{\longrightarrow} \; \underset{\textbf{monosaccharide}}{\text{M}}
$$

A *monosaccharide* can be a polyhydroxy aldehyde or a polyhydroxy ketone. Poly-hydroxy aldehydes are called **aldoses** ("ald" is for aldehyde; "ose" is the suffix for a sugar), while polyhydroxy ketones are called **ketoses.** Monosaccharides are also classified according to the number of carbons they contain: Monosaccharides with three carbons are **trioses;** those with four carbons are **tetroses;** those with five carbons are **pentoses;** and those with six and seven carbons are **hexoses** and **heptoses,** respectively. A six-carbon polyhydroxy aldehyde such as D-glucose is an aldohexose, whereas a six-carbon polyhydroxy ketone such as D-fructose is a ketohexose.

PROBLEM 1◆

Classify the following monosaccharides:

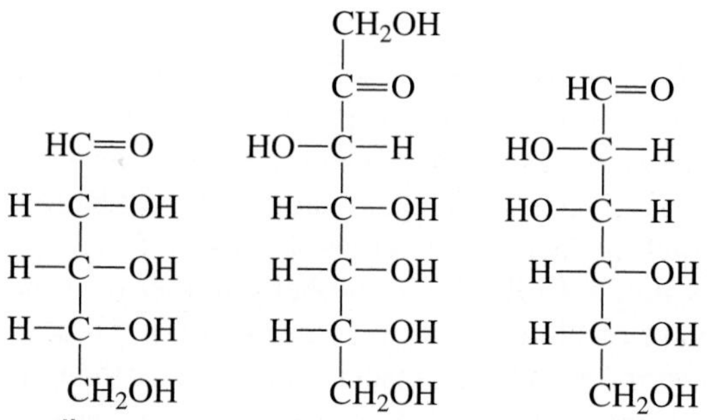

D-ribose   D-sedoheptulose   D-mannose

20.2 THE D AND L NOTATION

The smallest aldose, and the only one whose name does not end in "ose," is glycer-aldehyde, an aldotriose.

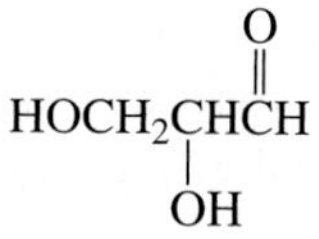

$$HOCH_2\overset{O}{\underset{OH}{CHCH}}$$
glyceraldehyde

Recall from Section 4.4 that horizontal bonds point toward the viewer and vertical bonds point away from the viewer in Fischer projections.

Because glyceraldehyde has a chirality center, it can exist as a pair of enantiomers.

(*R*)-(+)-glyceraldehyde   (*S*)-(−)-glyceraldehyde

Emil Fischer and his colleagues studied carbohydrates in the late nineteenth century, when techniques for determining the configurations of compounds were not available. Fischer arbitrarily assigned the configuration shown above on the left to the dextrorotatory isomer of glyceraldehyde that we call D-glyceraldehyde. He turned out to be correct—D-glyceraldehyde is (*R*)-(+)-glyceraldehyde, and L-glyceralde-hyde is (*S*)-(−)-glyceraldehyde (Section 4.12).

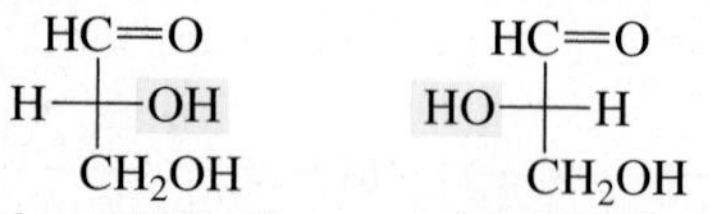

D-glyceraldehyde   L-glyceraldehyde

D and L notations can be used to describe the configuration of carbohydrates and amino acids (Section 21.2), so it is important to learn what D and L signify. In Fischer projections of monosaccharides, the carbonyl group is always placed on top (in the case of aldoses) or as close to the top as possible (in the case of ketoses). From its structure, you can see that galactose has four chirality centers (C-2, C-3, C-4, and C-5). If the OH group attached to the bottom-most chirality center (the second from the bottom carbon) is on the right, the compound is a D-sugar. If the OH group is on the left, the compound is an L-sugar. Almost all sugars found in nature are D-sugars. Notice that the mirror image of a D-sugar is an L-sugar.

> **Recall that a carbon with four different groups attached to it is a chirality center.**

```
        HC=O                          HC=O
     H──┼──OH                     HO──┼──H
    HO──┼──H                       H──┼──OH
    HO──┼──H                       H──┼──OH
     H──┼──OH                     HO──┼──H
        CH₂OH   the OH group          CH₂OH
    D-galactose  is on the right    L-galactose
                              mirror image of D-galactose
```

D and L, like *R* and *S*, indicate the configuration of a chirality center, but they do not indicate whether the compound rotates plane-polarized light to the right (+) or to the left (−). For example, D-glyceraldehyde is dextrorotatory but D-lactic acid is levorotatory. In other words, optical rotation, like melting or boiling points, is a physical property of a compound, whereas "*R*, *S*, D, and L" are conventions humans use to depict the structure of the molecule.

Tutorial:
D and L notation

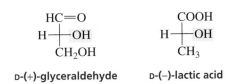

```
        HC=O                  COOH
    H──┼──OH              H──┼──OH
       CH₂OH                 CH₃

D-(+)-glyceraldehyde    D-(−)-lactic acid
```

The common name of the monosaccharide, together with the D or L designation, completely defines its structure because the relative configurations of all the chirality centers are implicit in the common name.

PROBLEM 2

Draw Fischer projections of L-glucose and L-fructose.

PROBLEM 3 ◆

Indicate whether each of the following is D-glyceraldehyde or L-glyceraldehyde (Section 4.5):

```
        HC=O                    H                      CH₂OH
a. HOCH₂──┼──OH        b. HO──┼──CH₂OH        c. HO──┼──H
           H                   HC=O                    HC=O
```

Aldotetroses have two chirality centers and therefore four stereoisomers. Two of the stereoisomers are D-sugars and two are L-sugars. The names of the erythro and threo pairs of enantiomers described in Section 4.8 originated from the names of the aldotetroses: erythrose and threose.

20.3
CONFIGURATIONS OF THE ALDOSES

```
   HC=O           HC=O           HC=O           HC=O
 H—OH          HO—H          HO—H          H—OH
 H—OH          HO—H          H—OH          HO—H
   CH₂OH          CH₂OH          CH₂OH          CH₂OH
 D-erythrose      L-erythrose      D-threose        L-threose
```

Aldopentoses have three chirality centers and therefore eight stereoisomers (four pairs of enantiomers), while aldohexoses have four chirality centers and 16 stereoisomers (eight pairs of enantiomers). The four D-aldopentoses and the eight D-aldohexoses are shown in Table 20.1.

Diastereomers that differ in configuration at only one chirality center are called **epimers.** For example, D-ribose and D-arabinose are C-2 epimers (they differ in configuration only at C-2), and D-idose and D-talose are C-3 epimers.

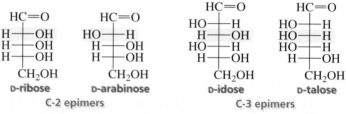

```
   HC=O           HC=O           HC=O           HC=O
 H—OH          HO—H          HO—H          HO—H
 H—OH          H—OH          H—OH          HO—H
 H—OH          H—OH          HO—H          HO—H
   CH₂OH          CH₂OH          H—OH          H—OH
 D-ribose        D-arabinose        CH₂OH          CH₂OH
      C-2 epimers              D-idose         D-talose
                                    C-3 epimers
```

TABLE 20.1 Configurations of the D-Aldoses

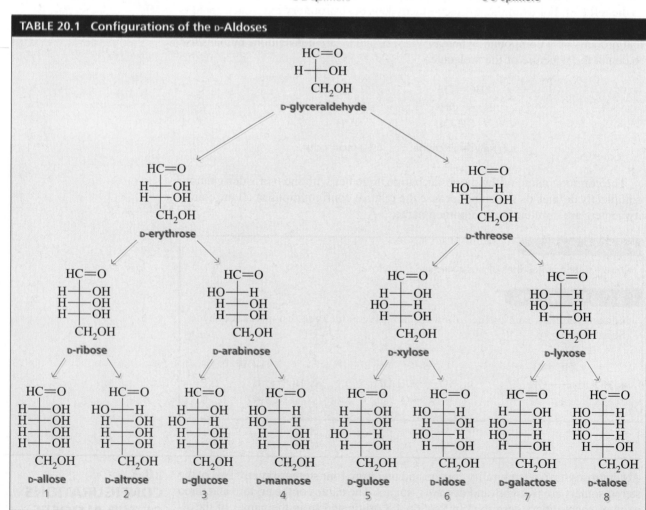

D-Glucose, D-mannose, and D-galactose are the most common aldohexoses in biological systems. An easy way to learn their structures is to memorize the structure of D-glucose and then remember that D-mannose is a C-2 epimer of D-glucose and D-galactose is a C-4 epimer of D-glucose. Sugars such as D-glucose and D-galactose are also diastereomers because they are configurational isomers that are not enantiomers (Section 4.8). An epimer is a particular kind of diastereomer.

D-Mannose is a C-2 epimer of D-glucose.

D-Galactose is a C-4 epimer of D-glucose.

Diastereomers are configurational isomers that are not enantiomers.

PROBLEM 4◆

a. Are D-erythrose and L-erythrose enantiomers or diastereomers?

b. Are L-erythrose and L-threose enantiomers or diastereomers?

PROBLEM 5◆

a. Which sugar is the C-3 epimer of D-xylose?

c. Which sugar is the C-4 epimer of L-gulose?

b. Which sugar is the C-5 epimer of D-allose?

PROBLEM 6◆

Give systematic names to the following compounds. Indicate the configuration (*R* or *S*) of each chirality center.

a. D-glucose

b. L-glucose

Naturally occurring ketoses have the ketone group in the 2-position. The configurations of the D-2-ketoses are shown in Table 20.2. A ketose has one fewer chirality center than does an aldose with the same number of carbon atoms. Therefore, a ketose has only half as many stereoisomers as an aldose with the same number of carbon atoms.

20.4 CONFIGURATIONS OF THE KETOSES

PROBLEM 7◆

Which sugar is the C-3 epimer of D-fructose?

PROBLEM 8◆

How many stereoisomers are possible for:

a. a 2-ketoheptose?   **b.** an aldoheptose?   **c.** a ketotriose?

Because monosaccharides contain *alcohol* functional groups and *aldehyde* (or *ketone*) functional groups, the reactions of monosaccharides are an extension of what you have already learned about the reactions of alcohols, aldehydes, and ketones. For example, an aldehyde group in a monosaccharide can be oxidized or reduced and can react with nucleophiles to form imines, hemiacetals, and acetals. When you read the sections that deal with the reactions of monosaccharides, you will find cross-references to the sections where the same reactivity for simple organic compounds is discussed. As you study, refer back to these sections—it will make learning about carbohydrates a lot easier and it will give you a good review of chemistry you have already learned.

20.5 REDOX REACTIONS OF MONOSACCHARIDES

TABLE 20.2 Configurations of the D-Ketoses

$$
\begin{array}{c}
\text{CH}_2\text{OH} \\
| \\
\text{C}=\text{O} \\
| \\
\text{CH}_2\text{OH}
\end{array}
$$

dihydroxyacetone

$$
\begin{array}{c}
\text{CH}_2\text{OH} \\
| \\
\text{C}=\text{O} \\
\text{H}\!-\!\!-\!\text{OH} \\
\text{CH}_2\text{OH}
\end{array}
$$

D-erythrulose

D-ribulose		D-xylulose

$$
\begin{array}{c}
\text{CH}_2\text{OH} \\
\text{C}=\text{O} \\
\text{H}\!-\!\!\text{OH} \\
\text{H}\!-\!\!\text{OH} \\
\text{CH}_2\text{OH}
\end{array}
\qquad
\begin{array}{c}
\text{CH}_2\text{OH} \\
\text{C}=\text{O} \\
\text{HO}\!-\!\!\text{H} \\
\text{H}\!-\!\!\text{OH} \\
\text{CH}_2\text{OH}
\end{array}
$$

D-ribulose   **D-xylulose**

$$
\begin{array}{c}
\text{CH}_2\text{OH} \\
\text{C}=\text{O} \\
\text{H}\!-\!\!\text{OH} \\
\text{H}\!-\!\!\text{OH} \\
\text{H}\!-\!\!\text{OH} \\
\text{CH}_2\text{OH}
\end{array}
\quad
\begin{array}{c}
\text{CH}_2\text{OH} \\
\text{C}=\text{O} \\
\text{HO}\!-\!\!\text{H} \\
\text{H}\!-\!\!\text{OH} \\
\text{H}\!-\!\!\text{OH} \\
\text{CH}_2\text{OH}
\end{array}
\quad
\begin{array}{c}
\text{CH}_2\text{OH} \\
\text{C}=\text{O} \\
\text{H}\!-\!\!\text{OH} \\
\text{HO}\!-\!\!\text{H} \\
\text{H}\!-\!\!\text{OH} \\
\text{CH}_2\text{OH}
\end{array}
\quad
\begin{array}{c}
\text{CH}_2\text{OH} \\
\text{C}=\text{O} \\
\text{HO}\!-\!\!\text{H} \\
\text{HO}\!-\!\!\text{H} \\
\text{H}\!-\!\!\text{OH} \\
\text{CH}_2\text{OH}
\end{array}
$$

D-psicose   **D-fructose**   **D-sorbose**   **D-tagatose**

Reduction

The carbonyl group of aldoses and ketoses can be reduced by the usual carbonyl-group reducing agents (H_2 + Pd/C, NaBH_4; Section 18.1). The product of the reduction is a polyalcohol, known as an **alditol.** Reduction of an aldose forms one alditol. Reduction of a ketose forms two alditols because the reaction creates a new chirality center in the product. D-Mannitol, the alditol formed from the reduction of D-mannose, is found in mushrooms, olives, and onions. The reduction of D-fructose forms D-mannitol and D-glucitol, the C-2 epimer of D-mannitol.

$$
\begin{array}{c}
\text{HC}=\text{O} \\
\text{HO}\!-\!\!\text{H} \\
\text{HO}\!-\!\!\text{H} \\
\text{H}\!-\!\!\text{OH} \\
\text{H}\!-\!\!\text{OH} \\
\text{CH}_2\text{OH}
\end{array}
\xrightarrow[\text{2. H}^+,\,\text{H}_2\text{O}]{\text{1. NaBH}_4}
\begin{array}{c}
\text{CH}_2\text{OH} \\
\text{HO}\!-\!\!\text{H} \\
\text{HO}\!-\!\!\text{H} \\
\text{H}\!-\!\!\text{OH} \\
\text{H}\!-\!\!\text{OH} \\
\text{CH}_2\text{OH}
\end{array}
\xleftarrow[\text{2. H}^+,\,\text{H}_2\text{O}]{\text{1. NaBH}_4}
\begin{array}{c}
\text{CH}_2\text{OH} \\
\text{C}=\text{O} \\
\text{HO}\!-\!\!\text{H} \\
\text{H}\!-\!\!\text{OH} \\
\text{H}\!-\!\!\text{OH} \\
\text{CH}_2\text{OH}
\end{array}
\xrightarrow[\text{2. H}^+,\,\text{H}_2\text{O}]{\text{1. NaBH}_4}
\begin{array}{c}
\text{CH}_2\text{OH} \\
\text{H}\!-\!\!\text{OH} \\
\text{HO}\!-\!\!\text{H} \\
\text{H}\!-\!\!\text{OH} \\
\text{H}\!-\!\!\text{OH} \\
\text{CH}_2\text{OH}
\end{array}
$$

D-mannose   **D-mannitol**   **D-fructose**   **D-glucitol**
 an alditol   **an alditol**

D-Glucitol is also obtained from the reduction of either D-glucose or L-gulose.

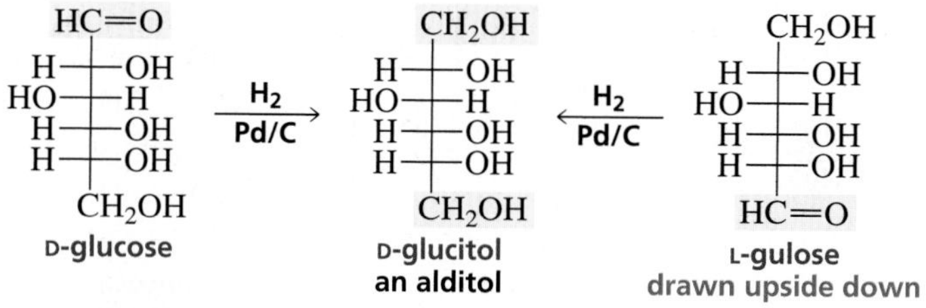

D-glucose   —H₂/Pd/C→   D-glucitol an alditol   ←H₂/Pd/C—   L-gulose drawn upside down

---

PROBLEM 9◆

What product is obtained from the reduction of:

a. D-idose?

b. D-sorbose?

PROBLEM 10◆

a. What other monosaccharide is reduced only to the alditol obtained from the reduction of:

1. D-talose?

2. D-galactose?

b. What monosaccharide is reduced to two alditols, one of which is the alditol obtained from the reduction of:

1. D-talose?

2. D-mannose?

Oxidation

Aldoses can be distinguished from ketoses by observing what happens to the color of an aqueous solution of bromine when it is added to the sugar. Br_2 is a mild oxidizing agent and easily oxidizes the aldehyde group but it cannot oxidize ketones or alcohols. Consequently, if a small amount of an aqueous solution of Br_2 is added to an unknown monosaccharide, the reddish-brown color of Br_2 will disappear if the monosaccharide is an aldose but will persist if the monosaccharide is a ketose. The product of the oxidation reaction is an **aldonic acid.**

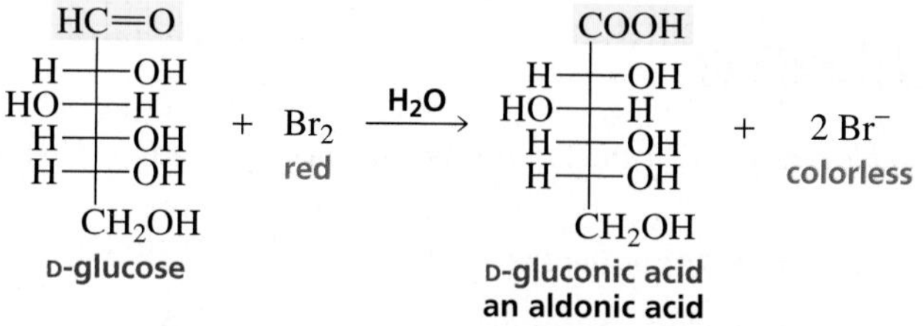

D-glucose + Br_2 (red) —H_2O→ D-gluconic acid an aldonic acid + $2\ Br^-$ colorless

Both aldoses and ketoses are oxidized to aldonic acids by Tollens' reagent (Ag^+, NH_3, HO^-), so Tollens' reagent cannot be used to distinguish between aldoses and ketoses. Recall from Section 18.3, however, that Tollens' reagent oxidizes aldehydes but not ketones. Why, then, are ketoses oxidized by Tollens' reagent when ketones aren't? Ketoses are oxidized because the reaction is carried out under basic conditions, and in a basic solution ketoses are converted into aldoses by enolization (Section 19.2). For example, the ketose D-fructose is in equilibrium with its enol. However, the enol of D-fructose is also the enol of D-glucose as well as the enol of D-mannose. Therefore, when the enol reketonizes, all three carbonyl compounds can be formed.

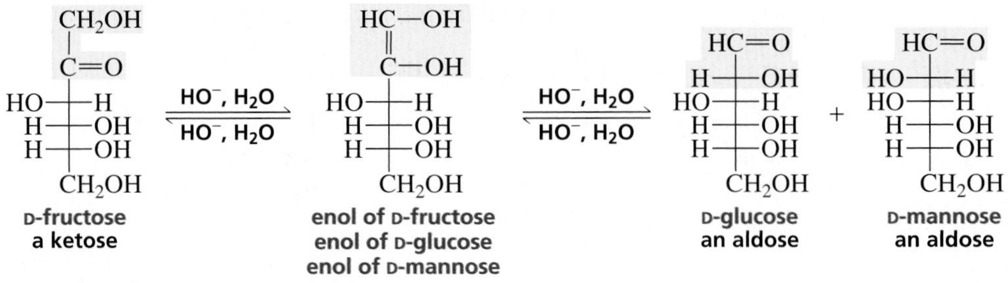

D-fructose
a ketose

$\xrightarrow[\text{HO}^-,\text{H}_2\text{O}]{\text{HO}^-,\text{H}_2\text{O}}$

enol of D-fructose
enol of D-glucose
enol of D-mannose

$\xrightleftharpoons[\text{HO}^-,\text{H}_2\text{O}]{\text{HO}^-,\text{H}_2\text{O}}$

D-glucose
an aldose

+

D-mannose
an aldose

PROBLEM 11

Write the mechanism for the base-catalyzed conversion of D-fructose into D-glucose and D-mannose.

PROBLEM 12◆

When D-tagatose is added to a basic aqueous solution, an equilibrium mixture of three monosaccharides is obtained. What are these monosaccharides?

If an oxidizing agent stronger than those discussed previously is used (such as HNO_3), one or more of the alcohol groups can be oxidized in addition to the aldehyde group. The primary alcohol is the one most easily oxidized. The product obtained when both the aldehyde and the primary alcohol groups are oxidized is called an aldaric acid. (In an aldonic acid, one end is oxidized. In an **aldaric acid,** both ends are oxidized.)

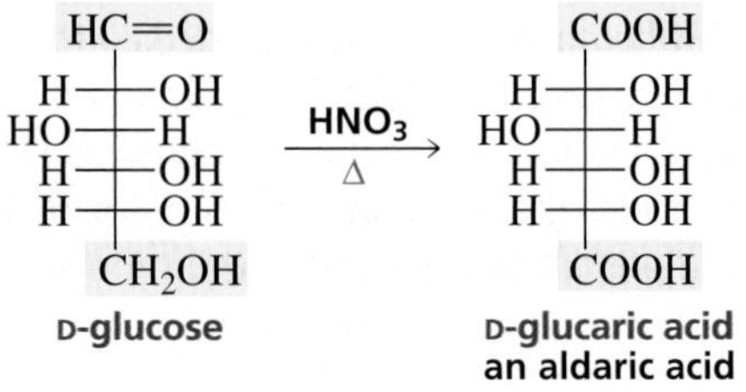

D-glucose

$\xrightarrow[\Delta]{\text{HNO}_3}$

D-glucaric acid
an aldaric acid

PROBLEM 13◆

a. Name an aldohexose other than D-glucose that is oxidized to D-glucaric acid by nitric acid.

b. What is another name for D-glucaric acid?

c. Name another pair of aldohexoses that are oxidized to identical aldaric acids.

20.6 OSAZONE FORMATION

The tendency of monosaccharides to form syrups that do not crystallize made the purification and isolation of monosaccharides difficult. Emil Fischer found that when phenylhydrazine is added to an aldose or a ketose, a yellow crystalline solid that is insoluble in water is formed. He called this derivative an **osazone** ("ose" for sugar; "azone" for hydrazone). Osazones are easily isolated and purified and were once used extensively to identify monosaccharides.

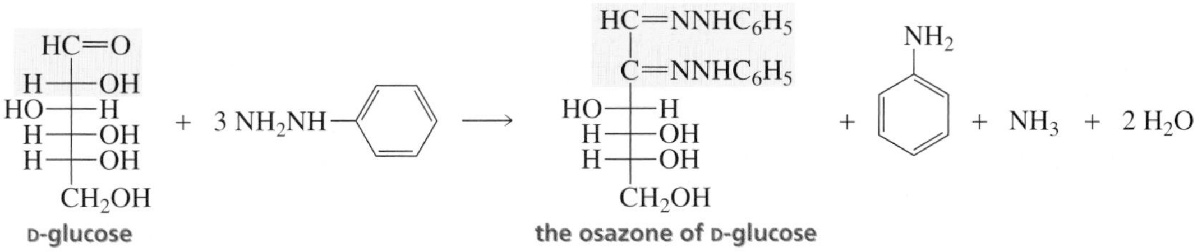

HC=O   HC=NNHC₆H₅ ... (the osazone of D-glucose)

the reaction of D-glucose + 3 NH₂NH—C₆H₅ → the osazone of D-glucose + aniline + NH₃ + 2 H₂O

Aldehydes and ketones react with one equivalent of phenylhydrazine, forming phenylhydrazones (Section 17.7). Aldoses and ketoses, in contrast, react with three equivalents of phenylhydrazine. One equivalent functions as an oxidizing agent, and is reduced to aniline and ammonia. Two equivalents form imines with carbonyl groups. The reaction stops at this point, regardless of how much phenylhydrazine is present.

Because the configuration of the number 2 carbon is lost during osazone formation, C-2 epimers form identical osazones. For example, D-idose and D-gulose, which are C-2 epimers, both form the same osazone.

C-2 epimers form identical osazones.

D-idose   3 NH₂NH—C₆H₅ → the osazone of D-idose and/or D-gulose ← 3 NH₂NH—C₆H₅   D-gulose

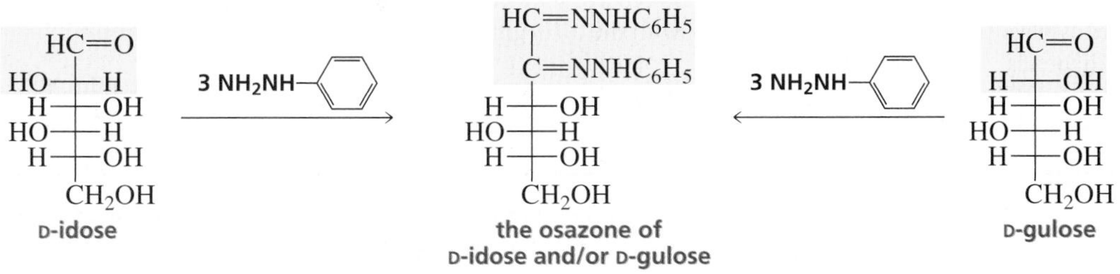

The number 1 and number 2 carbons of ketoses react with phenylhydrazine, too. Consequently, D-fructose, D-glucose, and D-mannose all form the same osazone.

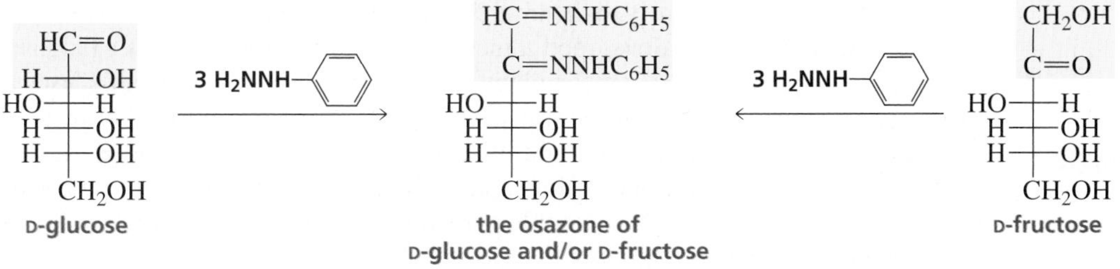

D-glucose   3 H₂NNH—C₆H₅ → the osazone of D-glucose and/or D-fructose ← 3 H₂NNH—C₆H₅   D-fructose

---

PROBLEM 14◆

Name a ketose and another aldose that form the same osazone as:

a. D-ribose   **c.** L-idose

b. D-altrose   **d.** D-galactose

PROBLEM 15◆

What monosaccharides form the same osazone as D-sorbose?

MEASURING THE BLOOD GLUCOSE LEVELS OF DIABETICS

Glucose reacts with an NH_2 group of hemoglobin to form an imine (Section 17.7). The imine undergoes an irreversible rearrangement to form a more stable α-aminoketone known as hemoglobin-A_{Ic}.

$$
\begin{array}{ccc}
\text{HC=O} & & \text{HC=N–hemoglobin} \\
\text{H—OH} & & \text{H—OH} \\
\text{HO—H} & \xrightarrow{\text{NH}_2\text{–hemoglobin}} & \text{HO—H} \\
\text{H—OH} & & \text{H—OH} \\
\text{H—OH} & & \text{H—OH} \\
\text{CH}_2\text{OH} & & \text{CH}_2\text{OH} \\
\text{D-glucose} & &
\end{array}
$$

$$
\xrightarrow{\text{rearrangement}}
\begin{array}{c}
\text{CH}_2\text{NH–hemoglobin} \\
\text{C=O} \\
\text{HO—H} \\
\text{H—OH} \\
\text{H—OH} \\
\text{CH}_2\text{OH} \\
\text{hemoglobin-A}_{Ic}
\end{array}
$$

Diabetes results when the body does not produce sufficient insulin, or when the insulin it produces does not properly stimulate its target cells. Because insulin is the hormone that maintains the proper level of glucose in the blood, diabetics have increased blood glucose levels. The amount of hemoglobin-A_{Ic} formed is proportional to the concentration of glucose in the blood, so diabetics have a higher concentration of hemoglobin-A_{Ic} than nondiabetics. Thus, measuring the hemoglobin-A_{Ic} level is a way to determine whether the blood glucose level of a diabetic is being adequately controlled. Cataracts, a common complication in diabetics, are caused by the reaction of glucose with the NH_2 group of proteins in the lens of the eye. It is thought that the arterial rigidity common in old age may be attributable to a similar reaction of glucose with the NH_2 group of proteins.

20.7 CHAIN ELONGATION: THE KILIANI–FISCHER SYNTHESIS

The Kiliani–Fischer synthesis leads to a pair of C-2 epimers.

Heinrich Kiliani (1855–1945) was born in Germany. He received a Ph.D. from the University of Munich, studying under Professor Emil Erlenmeyer. Kiliani became a professor of chemistry at the University of Freiburg.

The carbon chain of an aldose can be increased by one carbon by the **Kiliani–Fischer synthesis.** In other words, tetroses can be converted into pentoses, and pentoses can be converted into hexoses.

In the first step of the synthesis (the Kiliani portion), hydrogen cyanide is added to the aldose (Section 17.4). Addition of cyanide ion to the carbonyl group creates a new chirality center. Consequently, two cyanohydrins that differ only in configuration at C-2 are formed. The configurations of the other chirality centers do not change because no bond to any of the chirality centers is broken during the course of the reaction. Kiliani went on to hydrolyze the cyanohydrins to aldonic acids (Section 16.16), and Fischer had previously developed a method to convert aldonic acids to aldoses. This reaction sequence was used for many years, but the method currently used to convert the cyanohydrins to aldoses was developed by Serianni and Barker in 1979; it is referred to as the modified Kiliani–Fischer synthesis. Serianni and Barker reduced the cyanohydrins to imines using a partially deactivated palladium (on barium sulfate) catalyst so that the imines would not be further reduced to amines. The imines could then be hydrolyzed to aldoses (Section 17.7).

Notice that the synthesis leads to a pair of C-2 epimers because the first step of the reaction converts the carbonyl carbon in the starting material to a chirality center. Therefore the OH on C-2 in the product can be on the right or on the left in the Fischer projection. The two epimers are not obtained in equal amounts, however, because the first step of the reaction produces a pair of diastereomers and diastereomers are generally formed in unequal amounts (Section 4.16).

PROBLEM 16◆

What monosaccharides would be formed in a Kiliani–Fischer synthesis starting with:

a. D-xylose?   **b.** L-threose?

the modified Kiliani–Fischer synthesis

$$
\begin{array}{l}
\text{HC}=\text{O} \\
\text{H}-\text{OH} \quad + \ ^+\text{NH}_4 \\
\text{H}-\text{OH} \\
\text{H}-\text{OH} \\
\text{CH}_2\text{OH} \\
\textbf{D-ribose}
\end{array}
$$

H^+, H_2O

$$
\begin{array}{l}
\text{C}\equiv\text{N} \\
\text{H}-\text{OH} \\
\text{H}-\text{OH} \\
\text{H}-\text{OH} \\
\text{CH}_2\text{OH}
\end{array}
\quad \xrightarrow[\text{Pd/BaSO}_4]{\text{H}_2} \quad
\begin{array}{l}
\text{HC}=\text{NH} \\
\text{H}-\text{OH} \\
\text{H}-\text{OH} \\
\text{H}-\text{OH} \\
\text{CH}_2\text{OH}
\end{array}
$$

HCl

$$
\begin{array}{l}
\text{HC}=\text{O} \\
\text{H}-\text{OH} \\
\text{H}-\text{OH} \\
\text{CH}_2\text{OH} \\
\textbf{D-erythrose}
\end{array}
\quad + \ ^-\text{C}\equiv\text{N}
$$

HCl

$$
\begin{array}{l}
\text{C}\equiv\text{N} \\
\text{HO}-\text{H} \\
\text{H}-\text{OH} \\
\text{H}-\text{OH} \\
\text{CH}_2\text{OH}
\end{array}
\quad \xrightarrow[\text{Pd/BaSO}_4]{\text{H}_2} \quad
\begin{array}{l}
\text{HC}=\text{NH} \\
\text{HO}-\text{H} \\
\text{H}-\text{OH} \\
\text{H}-\text{OH} \\
\text{CH}_2\text{OH}
\end{array}
$$

H^+, H_2O

$$
\begin{array}{l}
\text{HC}=\text{O} \\
\text{HO}-\text{H} \quad + \ ^+\text{NH}_4 \\
\text{H}-\text{OH} \\
\text{H}-\text{OH} \\
\text{CH}_2\text{OH} \\
\textbf{D-arabinose}
\end{array}
$$

The **Ruff degradation** is the opposite of the Kiliani–Fischer synthesis. Thus the Ruff degradation shortens an aldose chain by one carbon—hexoses are converted into pentoses, and pentoses are converted into tetroses. In the Ruff degradation, the calcium salt of an aldonic acid is oxidized with hydrogen peroxide. Ferric ion catalyzes the reaction. The oxidation reaction cleaves the bond between C-1 and C-2, forming CO_2 and an aldehyde. It is known that the reaction involves the formation of radicals, but the precise mechanism is not well understood.

20.8 CHAIN SHORTENING: THE RUFF DEGRADATION

Otto Ruff (1871–1939) was born in Germany. He received a Ph.D. from the University of Berlin. He was a professor of chemistry at the University of Danzig and later at the University of Breslau.

the Ruff degradation

$$
\begin{array}{l}
\text{COO}^- \ (\text{Ca}^{2+})_{1/2} \\
\text{H}-\text{OH} \\
\text{HO}-\text{H} \\
\text{H}-\text{OH} \\
\text{H}-\text{OH} \\
\text{CH}_2\text{OH} \\
\textbf{calcium D-gluconate}
\end{array}
\quad + \ H_2O_2 \quad \xrightarrow{\text{Fe}^{3+}} \quad
\begin{array}{l}
\text{HC}=\text{O} \\
\text{HO}-\text{H} \\
\text{H}-\text{OH} \\
\text{H}-\text{OH} \\
\text{CH}_2\text{OH} \\
\textbf{D-arabinose}
\end{array}
\quad + \ CO_2
$$

The calcium salt of the aldonic acid necessary for the Ruff degradation is easily obtained by oxidizing an aldose with an aqueous solution of bromine and then adding calcium hydroxide to the reaction mixture.

$$
\begin{array}{l}
\text{HC}=\text{O} \\
\text{H}-\text{OH} \\
\text{HO}-\text{H} \\
\text{H}-\text{OH} \\
\text{H}-\text{OH} \\
\text{CH}_2\text{OH} \\
\textbf{D-glucose}
\end{array}
\quad \xrightarrow[\text{2. Ca(OH)}_2]{\text{1. Br}_2,\ \text{H}_2\text{O}} \quad
\begin{array}{l}
\text{COO}^- \ (\text{Ca}^{2+})_{1/2} \\
\text{H}-\text{OH} \\
\text{HO}-\text{H} \\
\text{H}-\text{OH} \\
\text{H}-\text{OH} \\
\text{CH}_2\text{OH} \\
\textbf{calcium D-gluconate}
\end{array}
$$

PROBLEM 17◆

Which two monosaccharides can be degraded to:

a. D-arabinose? **b.** D-glyceraldehyde? **c.** L-ribose?

20.9 STEREOCHEMISTRY OF GLUCOSE: THE FISCHER PROOF

In 1891, Emil Fischer determined the stereochemistry of glucose using one of the most brilliant examples of reasoning in the history of chemistry. He set out to determine the stereochemistry of (+)-glucose because it was the most common monosaccharide found in nature.

Fischer knew that (+)-glucose was an aldohexose, but 16 different structures can be written for an aldohexose. Which of them represents the structure of (+)-glucose? The 16 stereoisomers of the aldohexoses are actually eight pairs of enantiomers, so if you know the structures of one set, you automatically know the structures of the other set. Therefore, Fischer needed to consider only one set of eight. He considered the eight stereoisomers that had the C-5 OH group on the right in the Fischer projection (the stereoisomers that we now call the D-sugars). One of these would be one enantiomer of glucose, and its mirror image would be the other enantiomer. It was not possible to determine whether (+)-glucose was D-glucose or L-glucose until 1951 (Section 4.12). The D-sugars are shown in Table 20.1, numbered 1 to 8. Fischer used the following to determine the stereochemistry of glucose:

1. When the Kiliani–Fischer synthesis is done on the sugar that was known as (−)-arabinose, the two sugars that were known as (+)-glucose and (+)-mannose are obtained. This means that (+)-glucose and (+)-mannose are C-2 epimers. In other words, they have the same configuration at C-3, C-4, and C-5. Consequently, (+)-glucose and (+)-mannose have to be one of the following pairs from Table 20.1: sugars 1 and 2, 3 and 4, 5 and 6, or 7 and 8.
2. (+)-Glucose and (+)-mannose are both oxidized by nitric acid to optically active aldaric acids. The aldaric acids of sugars 1 and 7 would not be optically active because they have a plane of symmetry. Excluding sugars 1 and 7 means that (+)-glucose and (+)-mannose must be sugars 3 and 4 or 5 and 6.
3. Because (+)-glucose and (+)-mannose are the products obtained when the Kiliani–Fischer synthesis is carried out on (−)-arabinose, there are only two possibilities for the structure of (−)-arabinose. That is, if (+)-glucose and (+)-mannose are sugars 3 and 4, then (−)-arabinose has the structure shown below on the left; on the other hand, if (+)-glucose and (+)-mannose are sugars 5 and 6, then (−)-arabinose has the structure shown on the right.

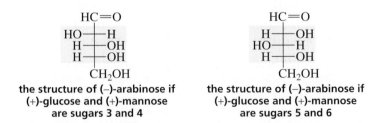

the structure of (−)-arabinose if (+)-glucose and (+)-mannose are sugars 3 and 4

the structure of (−)-arabinose if (+)-glucose and (+)-mannose are sugars 5 and 6

When (−)-arabinose is oxidized with nitric acid, the aldaric acid obtained is optically active. This means that the aldaric acid does *not* have a plane of symmetry. Therefore, (−)-arabinose must have the structure shown on the left because the aldaric acid of the sugar on the right has a plane of symmetry. Thus, (+)-glucose and (+)-mannose are represented by sugars 3 and 4.

4. The last step in the Fischer proof was to determine whether (+)-glucose is sugar 3 or 4. In order to answer this question, Fischer had to develop a chemical method to interchange the aldehyde and primary alcohol groups of an aldohexose. When he chemically interchanged the aldehyde and primary alcohol groups of the sugar known as (+)-glucose, he obtained an aldohexose that was different from (+)-glucose. When he chemically interchanged the aldehyde and primary alcohol groups of (+)-mannose, he still had (+)-mannose. Therefore, he concluded that (+)-glucose is sugar 3 because reversing the aldehyde and alcohol groups of sugar 3 leads to a different sugar (L-gulose).

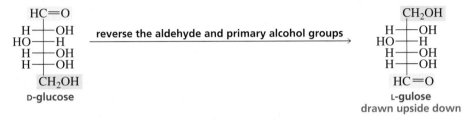

If (+)-glucose is sugar 3, (+)-mannose must be sugar 4. As predicted, when the aldehyde and primary alcohol groups of sugar 4 are reversed, the same sugar is obtained.

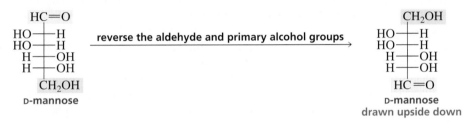

Using similar reasoning, Fischer went on to determine the stereochemistry of 14 of the 16 aldohexoses. He received the Nobel Prize in chemistry in 1902 for this achievement. His original guess that (+)-glucose is a D-sugar was later shown to be correct, so all of his structures are correct. If he had been wrong and (+)-glucose had been an L-sugar, his contribution to the stereochemistry of aldoses would still have had the same significance, but all his stereochemical assignments would have been reversed.

Jean-Baptiste-André Dumas (1800–1884) was born in France. Apprenticed to an apothecary, he left to study chemistry in Switzerland. He became a professor of chemistry at the University of Paris and at the Collège de France. He was the first French chemist to teach laboratory courses. In 1848, he left science for a political career. He became a senator, master of the French mint, and mayor of Paris.

GLUCOSE/DEXTROSE

André Dumas first used the term glucose (in 1838) to refer to the sweet compound that comes from honey and grapes. Later, Kekulé (Section 6.1) decided that it should be called dextrose because it was dextrorotatory. When Fischer studied the sugar, he called it glucose, and chemists have called it glucose ever since—although dextrose is often found on food labels.

PROBLEM 18 / SOLVED

Aldohexoses **A** and **B** form the same osazone. **A** is oxidized by nitric acid to an optically active aldaric acid and **B** is oxidized to an optically inactive aldaric acid. Ruff degradation of either **A** or **B** forms aldopentose **C**, which is oxidized by nitric acid to an optically active aldaric acid. Ruff degradation of **C** forms **D**, which is oxidized by nitric acid to an optically active aldaric acid. Ruff degradation of **D** forms (+)-glyceraldehyde. Identify **A**, **B**, **C**, and **D**.

SOLUTION This is the kind of problem that should be solved working backward. The bottommost chirality center in **D** must have the OH group on the right because **D** is degraded to (+)-glyceraldehyde. **D** must be D-threose because **D** is oxidized to an optically active aldaric acid. The two bottommost chirality centers in **C** and **D** have the same configuration because **C** is degraded to **D**. **C** must be D-lyxose because it is oxidized to an optically active aldaric acid. **A** and **B**, therefore, must be D-galactose and D-talose. Because **A** is oxidized to an optically active aldaric acid, it must be D-talose and **B** must be D-galactose.

PROBLEM 19◆

Identify **A, B, C,** and **D** in the preceding problem if **D** is oxidized to an optically *inactive* aldaric acid, **A, B,** and **C** are oxidized to optically active aldaric acids, and interchanging the aldehyde and alcohol groups of **A** leads to a different sugar.

20.10 CYCLIC STRUCTURE OF MONOSACCHARIDES: HEMIACETAL FORMATION

Movie:
Cyclization of a
monosaccharide

D-Glucose exists in three different forms. There is the open-chain form of D-glucose that we have been discussing and there are two cyclic forms—α-D-glucose and β-D-glucose. We know the two cyclic forms are different because they have different physical properties: α-D-glucose melts at 146 °C, whereas β-D-glucose melts at 150 °C; α-D-glucose has a specific rotation of +112.2°, and β-D-glucose has a specific rotation of +18.7°.

How can D-glucose exist in a cyclic form? In Section 17.8, we saw that an aldehyde reacts with an equivalent of an alcohol to form a hemiacetal. A monosaccharide such as D-glucose has an aldehyde group and several alcohol groups. The alcohol group bonded to C-5 of D-glucose reacts intramolecularly with the aldehyde group, forming a six-membered ring hemiacetal. Why are there two different cyclic forms? Two different hemiacetals are formed because the carbonyl carbon of the open-chain sugar becomes a new chirality center in the hemiacetal. If the OH group bonded to the new chirality center is on the right, it is α-D-glucose; if the OH group is on the left, it is β-D-glucose. The mechanism for cyclic hemiacetal formation is the same as the mechanism for hemiacetal formation between individual aldehyde and alcohol molecules (Section 17.8).

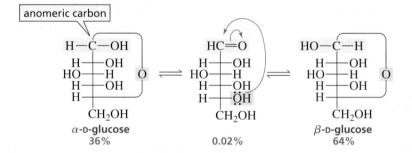

α-D-glucose
36%

0.02%

β-D-glucose
64%

α-D-Glucose and β-D-glucose are called anomers. **Anomers** are two sugars that differ in configuration only at the carbon that was the carbonyl carbon in the open-chain form. This is called the **anomeric carbon.** *Ano* is Greek for "upper." Thus, anomers differ in configuration at the uppermost chirality center. The anomeric carbon is the only carbon in the molecule bonded to two oxygens. The prefixes α and β denote the configuration of the anomeric carbon. Anomers, like epimers, are a particular kind of diastereomers.

In an aqueous solution, the open-chain compound is in equilibrium with the two cyclic hemiacetals. Formation of five- and six-membered ring cyclic hemiacetals proceeds nearly to completion (unlike formation of acyclic hemiacetals), so very little glucose exists in the open-chain form (about 0.02%). At equilibrium, there is almost twice as much β-D-glucose (64%) as α-D-glucose (36%). The sugar still undergoes the reactions discussed in previous sections (oxidation, reduction, and osazone formation) because the reagents can react with the small amount of open-chain aldehyde that is present. As the aldehyde reacts, the equilibrium shifts to form more aldehyde, which can then undergo reaction. Eventually, all the glucose molecules react by way of the open-chain aldehyde.

When crystals of pure α-D-glucose are dissolved in water, the specific rotation gradually changes from $+112.2°$ to $+52.7°$. When crystals of pure β-D-glucose are dissolved in water, the specific rotation gradually changes from $+18.7°$ to $+52.7°$. This change in rotation occurs because, in water, the hemiacetal opens to form the aldehyde and, when the aldehyde recyclizes, both α-D-glucose and β-D-glucose can be formed. Eventually, the three forms of glucose reach equilibrium concentrations. The specific rotation of the equilibrium mixture is $+52.7°$, which is why the same specific rotation results whether the crystals originally dissolved in water are α-D-glucose or β-D-glucose. A slow change in optical rotation to an equilibrium value is known as **mutarotation.**

If an aldose can form a five- or a six-membered ring, it will exist predominantly as a cyclic hemiacetal in solution. Whether a five- or a six-membered ring is formed depends on their relative stabilities. Six-membered ring sugars are called **pyranoses,** and five-membered ring sugars are called **furanoses.** These names come from pyran and furan, the names of the five- and six-membered ring cyclic ethers shown in the margin. Consequently, α-D-glucose is also called α-D-glucopyranose. The prefix α or β indicates the configuration of the anomeric carbon and "pyranose" indicates that the sugar exists as a six-membered ring cyclic hemiacetal.

Using Fischer projections to show the structure of a cyclic sugar is unsatisfactory because of the way the C—O—C bond is represented. A somewhat more satisfactory representation is given by a **Haworth projection.** In a Haworth projection of a D-pyranose, the six-membered ring is represented as being flat and is viewed edge on. The ring oxygen is always placed in the back right-hand corner of the ring, with the anomeric carbon (C-1) on the right-hand side and the primary alcohol group drawn *up* from the back left-hand corner (C-5). Groups on the *right* in a Fischer projection are *down* in a Haworth projection, while groups on the *left* in a Fischer projection are *up* in a Haworth projection.

The Haworth projection of a D-furanose is viewed edge on with the ring oxygen away from the viewer. The anomeric carbon is on the right-hand side of the molecule and the primary alcohol group is drawn *up* from the back left-hand corner.

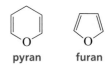

pyran   furan

Sir Walter Norman Haworth (1883–1950) was born in England. He received a Ph.D. in Germany from the University of Göttingen and later was a professor of chemistry at the Universities of Durham and Birmingham in Britain. He was the first to synthesize vitamin C and was the one who named it ascorbic acid. During World War II, he worked on the atomic bomb project. He received the Nobel Prize in Chemistry in 1937 and was knighted in 1947.

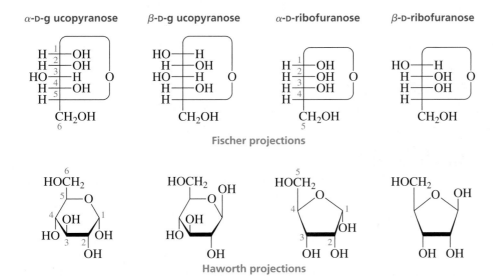

Fischer projections

Haworth projections

Groups on the *right* in a Fischer projection are *down* in a Haworth projection.

Groups on the *left* in a Fischer projection are *up* in a Haworth projection.

Ketoses also exist predominantly in cyclic forms. D-Fructose forms a five-membered ring hemiketal as a consequence of the C-5 OH group reacting with the ketone

carbonyl group (Section 17.8). If the new chirality center has the OH group on the right in a Fischer projection, the compound is α-D-fructose; if the OH group is on the left, the compound is β-D-fructose. These sugars can also be called α-D-fructo-furanose and β-D-fructofuranose. Notice that in fructose the anomeric carbon is C-2, not C-1 as in aldoses. D-Fructose can also form a six-membered ring by using the C-6 OH group. The pyranose form predominates in the monosaccharide, while the furanose form predominates when the sugar is in a disaccharide. (See the structure of sucrose in Section 20.17.)

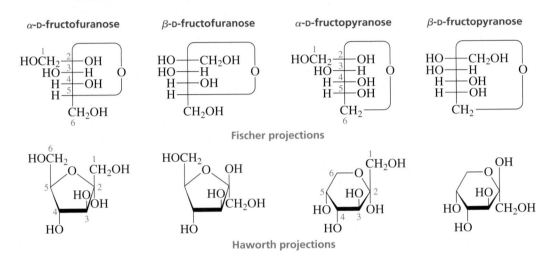

Fischer projections

Haworth projections

Haworth projections are useful because they allow us to see easily whether the OH groups on the ring are cis or trans to each other. Because five-membered rings are close to planar, furanoses are well represented by Haworth projections. However, Haworth projections are structurally misleading for pyranoses because six-membered rings are not flat but exist preferentially in a chair conformation (Section 2.12).

PROBLEM 20

4-Hydroxy- and 5-hydroxyaldehydes exist primarily in the cyclic hemiacetal form. Give the structure of the cyclic hemiacetal formed by each of the following:

a. 4-hydroxybutanal   **c.** 4-hydroxypentanal

b. 5-hydroxypentanal   **d.** 4-hydroxyheptanal

PROBLEM 21

Draw the following sugars as pyranoses using Haworth projections:

a. β-D-galactose   **b.** α-D-tagatose   **c.** α-L-glucose

PROBLEM 22

D-Glucose most often exists as a pyranose, but it can also exist as a furanose. Draw the Haworth projection of α-D-glucofuranose.

20.11
STABILITY OF
GLUCOSE

Drawing glucose in its chair conformation shows why it is the most common aldo-hexose in nature. To convert a Haworth projection into a chair conformation, start by placing the ring oxygen at the back right-hand corner and the primary alcohol group in the equatorial position. The primary alcohol group is the largest of all the substituents, and large substituents are more stable in the equatorial position because there is less steric strain in that position (Section 2.13). Because the OH group bonded to C-4 is trans to the primary alcohol group (this is easily seen in the Haworth projection), the C-4 OH group is also in the equatorial position. (Recall from Section 2.14 that 1,2-diequatorial substituents are trans to one another.) The C-3 OH group is trans to the C-4 OH group, so the C-3 OH group is also in the equatorial position. As you move around the ring, you will find that all the OH substituents in β-D-glucose are in equatorial positions. The axial positions are all occupied by hydrogens, which require little space and therefore experience little steric strain. No other aldohexose exists in such a strain-free conformation. This means that β-D-glucose is the most stable of all the aldohexoses, so it is not surprising that it is the most prevalent aldohexose in nature.

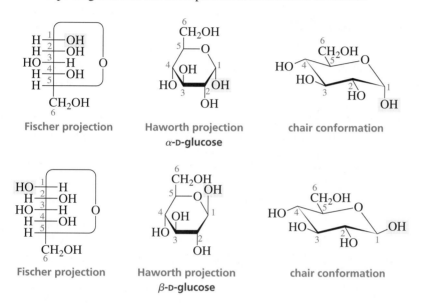

Fischer projection · Haworth projection · chair conformation

α-D-**glucose**

Fischer projection · Haworth projection · chair conformation

β-D-**glucose**

The α-**position is to the right in a Fischer projection; it is down in a Haworth projection; and it is axial in a chair conformation.**

The β-**position is to the left in a Fischer projection; it is up in a Haworth projection; and it is equatorial in a chair conformation.**

Why is there more β-D-glucose than α-D-glucose in an aqueous solution at equilibrium? The OH group bonded to the anomeric carbon is in the equatorial position in β-D-glucose, whereas it is in the axial position in α-D-glucose. Therefore, β-D-glucose is more stable than α-D-glucose, so β-D-glucose predominates at equilibrium in an aqueous solution.

If you remember that all the OH groups in β-D-glucose are in equatorial positions, it is easy to draw the chair conformation of any other pyranose. For example, if you want to draw α-D-galactose, you would put all the OH groups in equatorial positions except the OH groups at C-4 (because galactose is a C-4 epimer of glucose) and at C-1 (because it is the α-anomer). You would put these two OH groups in axial positions instead.

3-D Molecules:
α-D-Galactose;
β-D-Gulose;
β-L-Gulose

the OH at C-4 is axial

the OH at C-1 is axial (α)

α-D-**galactose**

To draw an L-pyranose, draw the D-pyranose first, and then draw its mirror image. For example, to draw β-L-gulose, first draw β-D-gulose (gulose differs from glucose at C-3 and C-4, so the OH groups at these positions are in axial positions). Then draw the mirror image of β-D-gulose to get β-L-gulose.

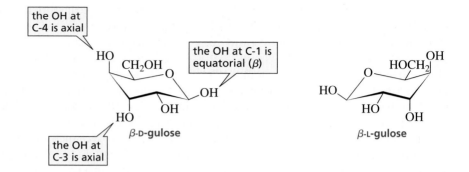

β-D-**gulose** β-L-**gulose**

the OH at C-4 is axial

the OH at C-1 is equatorial (β)

the OH at C-3 is axial

PROBLEM 23◆

Which OH groups are in the axial position in:

a. β-D-mannose? **b.** β-D-idose? **c.** α-D-allose?

20.12 ACYLATION AND ALKYLATION OF MONOSACCHARIDES

The OH groups of monosaccharides show the chemistry typical of alcohols. For example, they react with acetyl chloride or acetic anhydride to form esters (Sections 16.7 and 16.8).

β-D-**glucose** → penta-*O*-acetyl-β-D-**glucose**

The OH groups also react with methyl iodide/silver oxide to form ethers (Section 9.4). The OH group is a relatively poor nucleophile, so silver oxide is used to increase the leaving tendency of the iodide ion in the S$_N$2 reaction.

β-D-**glucose** → methyl tetra-*O*-methyl-β-D-**glucoside**

20.13 FORMATION OF GLYCOSIDES

In Section 17.8 we saw that, after an aldehyde reacts with an equivalent of an alcohol to form a hemiacetal, the hemiacetal reacts with a second equivalent of alcohol to form an acetal. Similarly, the cyclic hemiacetal (or hemiketal) formed by a monosaccharide can react with an alcohol to form an acetal (or ketal). The acetal (or ketal) of a sugar is called a **glycoside,** and the bond between the anomeric carbon and the alkoxy oxygen is called a **glycosidic bond.** Glycosides are named by replacing the "e" ending of the sugar's name with "ide." Thus, the glycoside of glucose is a glucoside, the glycoside of galactose is a galactoside, etc. If the pyranose or furanose name is used, the acetal is called a **pyranoside** or a **furanoside.**

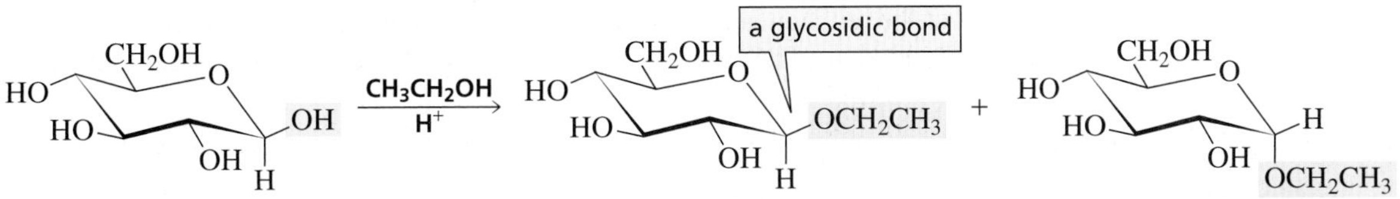

β-D-glucose
β-D-glucopyranose

ethyl β-D-glucoside
ethyl β-D-glucopyranoside

ethyl α-D-glucoside
ethyl α-D-glucopyranoside

Notice that the reaction of a single anomer with an alcohol leads to the formation of both the α- and β-glycosides. The mechanism of the reaction shows why both glycosides are formed. The OH group bonded to the anomeric carbon becomes protonated in the acidic solution, and a nonbonding pair of electrons on the ring oxygen helps expel a molecule of water. The anomeric carbon in the resulting oxocarbenium ion is sp^2 hybridized, so that part of the molecule is planar. (An **oxocarbenium ion** has a positive charge that is shared by a carbon and an oxygen.) When the alcohol comes in from the top of the plane, the β-glycoside is formed; when the alcohol comes in from the bottom of the plane, the α-glycoside is formed. Notice that the mechanism is the same as the mechanism shown for acetal formation shown in Section 17.8.

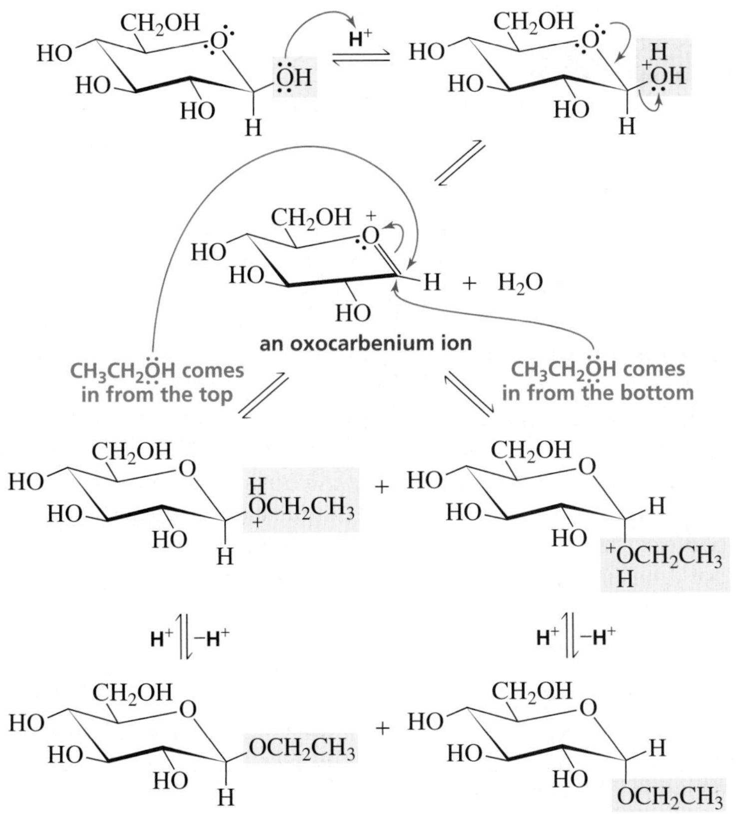

an oxocarbenium ion

CH_3CH_2OH comes in from the top

CH_3CH_2OH comes in from the bottom

a β-glycoside

an α-glycoside

Surprisingly, D-glucose forms more of the α-glycoside than the β-glycoside. The reason for this is explained in the next section.

A reaction similar to the reaction of a monosaccharide with an alcohol is the reaction of a monosaccharide with an amine in the presence of a trace amount of acid. The product of the reaction is an **N-glycoside**—it has a nitrogen in place of the oxygen at the glycosidic linkage. The subunits of DNA and RNA are β-*N*-glycosides (Section 25.1).

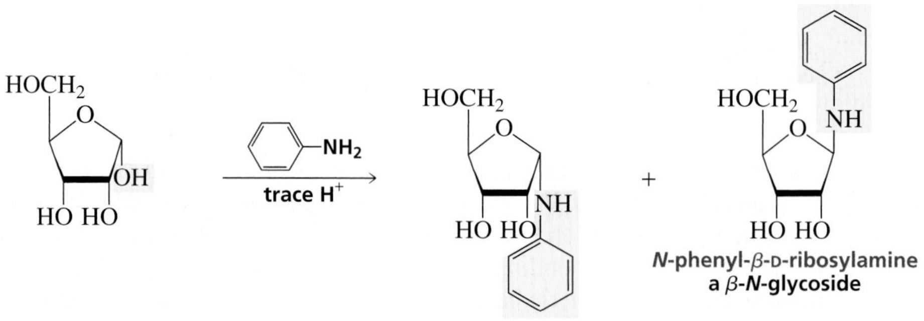

N-phenyl-α-D-ribosylamine
an α-*N*-glycoside

N-phenyl-β-D-ribosylamine
a β-*N*-glycoside

PROBLEM 24

Draw the resonance contributors for the oxocarbenium ion formed as an intermediate in glycoside formation. Which one is more stable?

PROBLEM 25 ◆

Why is only a trace amount of acid used in the formation of an *N*-glycoside?

20.14 THE ANOMERIC EFFECT

We have seen that β-D-glucose is more stable than α-D-glucose because there is more room for a substituent in the equatorial position. However, we just saw in Section 20.13 that when glucose reacts with an alcohol to form a glucoside, the major product is the α-glucoside. This means that the α-glucoside must be more stable than the β-glucoside. The preference for the axial position by certain substituents bonded to the anomeric carbon is called the **anomeric effect.**

What is responsible for the anomeric effect? One clue is that all the substituents that prefer the axial position have lone-pair electrons on the atom bonded to the ring. The lone-pair electrons of the anomeric substituent have repulsive interactions with the lone-pair electrons of the ring oxygen if the anomeric substituent is in the β-position, but not if it is in the α-position.

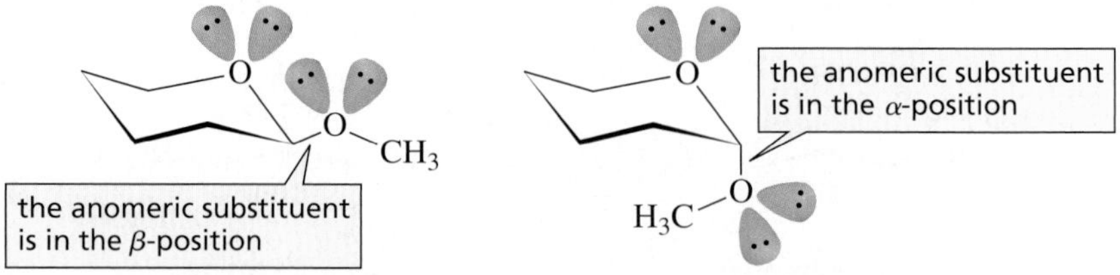

the anomeric substituent is in the β-position

the anomeric substituent is in the α-position

Apparently the attractive interaction of the hydrogen of the anomeric OH group of D-glucose with the lone-pair electrons of the ring oxygen decreases the importance of the anomeric effect, making β-D-glucose more stable than α-D-glucose. However, when the hydrogen is replaced by an alkyl group, the anomeric effect decreases the stability of the β-position, so α-glycosides are more stable than β-glycosides.

D-Mannose is a C-2 epimer of D-glucose, so the OH group at the 2-position in D-mannose is axial. The interaction of the hydrogen of the anomeric OH group with the lone-pair electrons of the ring oxygen apparently cannot make up for the additional repulsive interaction of the equatorial anomeric substituent with the oxygen atom at the 2-position, so α-D-mannose is more stable than β-D-mannose.

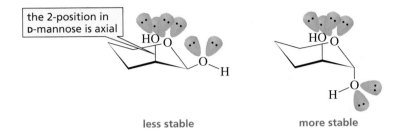

less stable   more stable

20.15 REDUCING AND NONREDUCING SUGARS

Because glycosides are acetals (or ketals), they are not in equilibrium with the open-chain aldehyde (or ketone) in aqueous solution. Because they are not in equilibrium with a compound with a carbonyl group, they cannot be oxidized by reagents such as Ag^+ or Br_2. Glycosides, therefore, are **nonreducing sugars**—they cannot reduce Ag^+ or Br_2.

Hemiacetals (or hemiketals) are in equilibrium with the open-chain sugars in aqueous solution. So as long as a sugar has an aldehyde, a ketone, a hemiacetal, or a hemiketal group, it is able to reduce an oxidizing agent and therefore is classified as a **reducing sugar.** Without one of these groups, it is a nonreducing sugar.

> A sugar with an aldehyde, ketone, hemiacetal, or hemiketal group is a reducing sugar. A sugar without one of these groups is a nonreducing sugar.

PROBLEM 26 / SOLVED

Name the following compounds and indicate whether each is a reducing sugar or a nonreducing sugar:

a. [structure: CH₂OH, HO, HO, OH, OCH₂CH₂CH₃]

b. [structure: HO, CH₂OH, HO, OH, OCH₃]

c. [structure: HO, CH₂OH, OH, HO, OH]

d. [structure: HOCH₂, O, OCH₂CH₃, CH₂OH, OH OH]

SOLUTION TO 26a The only OH group in an axial position in **a** is the one at C-3. Therefore, this sugar is the C-3 epimer of D-glucose, which is D-allose. The substituent at the anomeric carbon is in the β-position. Thus, the sugar's name is propyl β-D-alloside or propyl β-D-allopyranoside. Because the sugar is an acetal, it is a nonreducing sugar.

20.16 DETERMINATION OF RING SIZE

Two different procedures can be used to determine what size ring a monosaccharide forms. In the first procedure, treatment of the monosaccharide with excess methyl iodide and silver oxide converts all the OH groups to OCH_3 groups (Section 20.12). Acid-catalyzed hydrolysis of the acetal then forms a hemiacetal, which is in equilibrium with its open-chain form. The size of the ring can be determined from the structure of the open-chain form because the sole OH group is the one that had formed the cyclic hemiacetal.

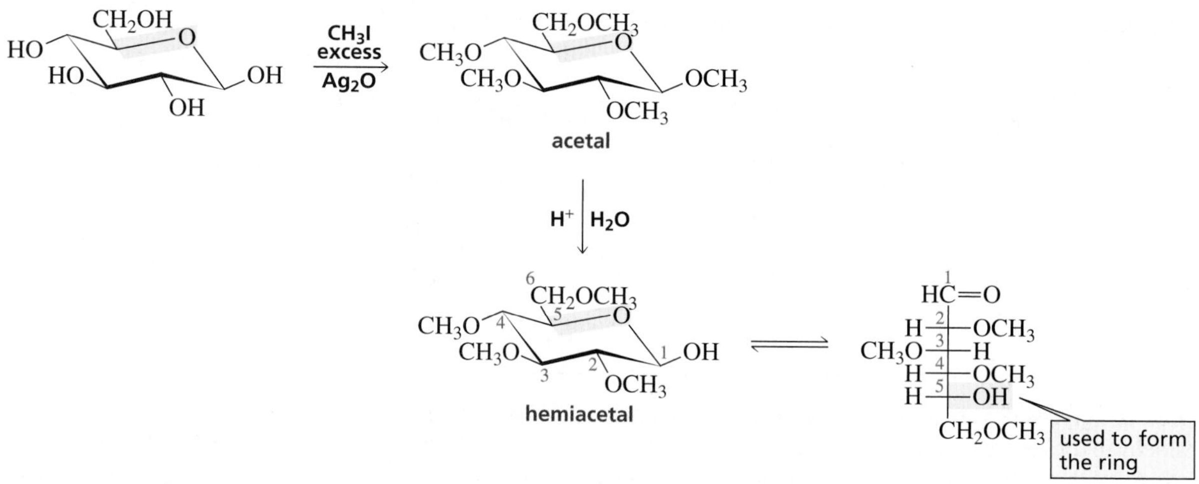

acetal

$H^+ \mid H_2O$

hemiacetal

used to form the ring

In the second procedure used to determine ring size, an acetal of the monosaccharide is oxidized with excess periodic acid. Recall from Section 18.6 that periodate cleaves 1,2-diols.

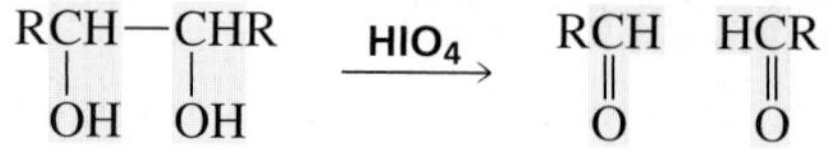

$$RCH-CHR \xrightarrow{HIO_4} RCH \quad HCR$$
$$\quad | \quad | \qquad\qquad \| \qquad \|$$
$$\quad OH \quad OH \qquad\qquad O \qquad O$$

The products obtained from oxidation of an acetal of six-membered-ring D-glucose are different than the products that would be obtained from oxidation of a five-membered ring acetal.

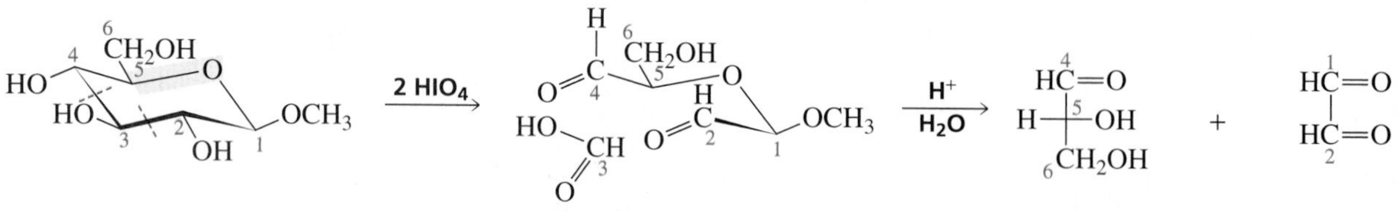

$2\ HIO_4$

$\dfrac{H^+}{H_2O}$

PROBLEM 27

What products would be obtained from oxidation with periodic acid if D-glucose had formed a furanose rather than a pyranose?

20.17
DISACCHARIDES

If the hemiacetal group of a monosaccharide forms an acetal by reacting with an alcohol group of another monosaccharide, the glycoside that is formed is a disaccharide. *Disaccharides* are compounds with two monosaccharide subunits hooked together by an acetal linkage. For example, maltose is a disaccharide obtained from the hydrolysis of starch. It contains two D-glucose subunits hooked together by an **α-1,4′-glycosidic linkage.** The linkage is between C-1 of one sugar subunit and C-4 of the other. The "prime" superscript indicates that C-4 is not in the same ring as C-1. It is an α-1,4′-glycosidic linkage because the oxygen atom involved in the glycosidic linkage is in the α-position. *Remember that the α-position is axial when a sugar is shown in a chair conformation and is down when the sugar is shown in a Haworth projection; the β-position is equatorial when a sugar is shown in a chair confor-mation and is up when the sugar is shown in a Haworth projection.*

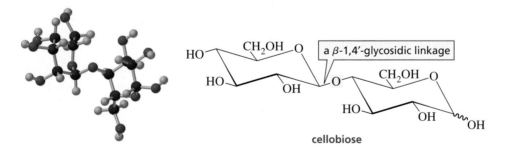

maltose

The structure of maltose is shown without specifying the configuration of the anomeric carbon that is not an acetal (the anomeric carbon of the subunit on the right), because maltose can exist in both the α and β forms. In α-maltose, the OH group bonded to this anomeric carbon is in the axial position. In β-maltose, this OH group is in the equatorial position. Because maltose can exist in both α and β forms, mutarotation occurs when crystals of one form are dissolved in a solvent. Maltose is a reducing sugar because the right-hand subunit is a hemiacetal and therefore is in equilibrium with the open-chain aldehyde that is easily oxidized.

3-D Molecules:
Maltose;
Cellobiose;
Lactose

Cellobiose, a disaccharide obtained from the hydrolysis of cellulose, also con-tains two D-glucose subunits. It differs from maltose in that the two glucose subunits are hooked together by a **β-1,4′-glycosidic linkage.** Thus, the only difference in the structures of maltose and cellobiose is the configuration of the glycosidic linkage. Like maltose, cellobiose exists in both α and β forms because the OH group bond-ed to the anomeric carbon not involved in acetal formation can be in either the axial position (α-cellobiose) or the equatorial position (β-cellobiose). Cellobiose is a re-ducing sugar because the subunit on the right is a hemiacetal.

cellobiose

Lactose is a disaccharide found in milk. It constitutes 4.5% of cow's milk by weight and 6.5% of human milk. One of the subunits of lactose is D-galactose, and the other is D-glucose. The D-galactose subunit is an acetal, and the D-glucose subunit is a hemi-acetal. The subunits are joined through a β-1,4′-glycosidic linkage. Because one of the subunits is a hemiacetal, lactose is a reducing sugar and undergoes mutarotation.

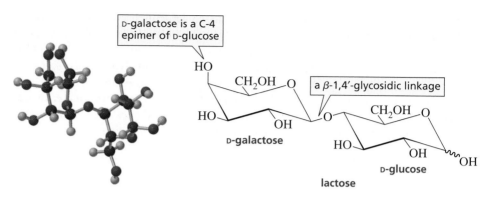

D-galactose is a C-4 epimer of D-glucose

a β-1,4'-glycosidic linkage

D-galactose

D-glucose

lactose

LACTOSE INTOLERANCE

Lactase is an enzyme that specifically breaks the β-1,4'-glycosidic linkage of lactose. Cats and dogs lose their intestinal lactase when they become adults and therefore are no longer able to digest lactose. Consequently, when they are fed milk or milk products, the undegraded lactose causes digestive problems such as bloating, abdominal pain, and diarrhea. These problems occur because only monosaccharides can pass into the bloodstream, so lactose has to pass undigested into the large intestine. When humans have stomach flu or other intestinal disturbances, they can temporarily lose their lactase, thereby becoming lactose intolerant. Some humans lose their lactase permanently as they mature. This occurs in approximately 10% of the caucasian population of the United States. It is much more common in people whose ancestors came from non-dairy-producing countries. For example, only 3% of Danes but 97% of Thais are lactose-intolerant.

GALACTOSEMIA

After lactose is degraded into glucose and galactose, the galactose must be converted into glucose before it can be used by cells. Individuals who do not have the enzyme that converts galactose into glucose have the genetic disease known as galactosemia. Without this enzyme, galactose accumulates in the bloodstream, which can cause mental retardation in infants and even death. Galactosemia is treated by excluding galactose from the diet.

The most common disaccharide is sucrose (table sugar). Sucrose is obtained from sugar beets and sugar cane. About 90 million tons of sucrose are produced in the world each year. Sucrose consists of a D-glucose subunit and a D-fructose subunit linked by a glycosidic bond between C-1 of glucose (in the α-position) and C-2 of fructose (in the β-position).

Sucrose, unlike the other disaccharides that have been discussed, is not a reducing sugar and does not exhibit mutarotation because the glycosidic bond is between the anomeric carbon of glucose and the anomeric carbon of fructose. Because sucrose does not have a hemiacetal or hemiketal group, it is not in equilibrium with the readily oxidized open-chain aldehyde or ketone form in aqueous solution.

3-D Molecule: Sucrose

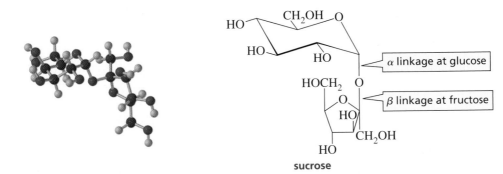

sucrose

Sucrose has a specific rotation of $+66.5°$. When it is hydrolyzed, the resulting equimolar mixture of glucose and fructose has a specific rotation of $-22.0°$. Because the sign of the rotation changes when sucrose is hydrolyzed, a 1:1 mixture of glucose and fructose is called invert sugar. The enzyme that catalyzes the hydrolysis of sucrose is called invertase. Honeybees have invertase, so the honey they produce is a mixture of sucrose, glucose, and fructose. Because fructose is sweeter than sucrose, invert sugar is sweeter than sucrose. Some foods are advertised as containing fructose instead of sucrose, which means they achieve the same level of sweetness with a lower sugar content.

20.18 POLYSACCHARIDES

Polysaccharides contain as few as 10 or as many as several thousand monosaccharide units joined together by glycosidic linkages. The molecular weight of the individual polysaccharide chains is variable. The most common polysaccharides are starch and cellulose.

Starch is the major component of flour, potatoes, rice, beans, corn, and peas. It is a mixture of two different polysaccharides—amylose (about 20%) and amylopectin (about 80%). Amylose is composed of unbranched chains of D-glucose units joined by α-1,4′-glycosidic linkages.

an α-1,4′-glycosidic linkage

3 subunits of amylose

Amylopectin is a branched polysaccharide. Like amylose, it is composed of chains of D-glucose units joined by α-1,4′-glycosidic linkages. Unlike amylose, however, amylopectin also contains α-1,6′-glycosidic linkages. These linkages create the branches in the polysaccharide. Amylopectin can contain up to 10^6 glucose units, making it one of the largest molecules found in nature.

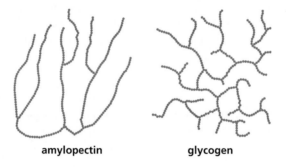

an α-1,6'-glycosidic linkage

5 subunits of amylopectin

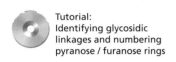

Animals store their excess glucose in a polysaccharide known as glycogen. Glycogen has a structure similar to that of amylopectin, but glycogen has more branches (Figure 20.1). The branch points in glycogen occur about every 10 residues, whereas those in amylopectin occur about every 25 residues. The high degree of branching in glycogen has important physiological effects. When the body needs energy, many individual glucose units can be simultaneously removed from the ends of many branches.

Figure 20.1 ▶
A comparison of the branching in amylopectin and glycogen.

amylopectin   glycogen

WHY THE DENTIST IS RIGHT

Bacteria found in the mouth have an enzyme that converts sucrose into a polysaccharide called dextran. Dextran is made up of glucose units joined mainly through α-1,3'- and α-1,6'-glycosidic linkages. About 10% of dental plaque is composed of dextran. This is the chemical reason why your dentist cautions you not to eat candy.

Cellulose is the structural material of higher plants. Cotton, for example, is composed of about 90% cellulose, and wood is about 50% cellulose. Like amylose, cellulose is composed of unbranched chains of D-glucose units. Unlike amylose, however, the glucose units in cellulose are joined by β-1,4'-glycosidic linkages, not α-1,4'-glycosidic linkages.

a β-1,4'-glycosidic linkage

3 subunits of cellulose

All mammals have the enzyme (α-glucosidase) that hydrolyzes the α-1,4'-glycosidic linkages that join glucose units, but they do not have the enzyme (β-glucosidase) that hydrolyzes β-1,4'-glycosidic linkages. As a result, mammals *cannot* obtain the glucose they need by eating cellulose. However, bacteria that possess β-glucosidase inhabit the digestive tracts of grazing animals, so cows can eat grass and horses can eat hay to meet their nutritional requirements for glucose. Termites also harbor bacteria that break down the cellulose in the wood they eat.

The different glycosidic linkages in starch and cellulose give these compounds very different physical properties. The α-linkages in starch cause amylose to form a helix that allows many of its OH groups to be hydrogen-bonded to water molecules (Figure 20.2). As a result, starch is soluble in water.

The β-linkages in cellulose allow the molecules to form intramolecular hydrogen bonds. Consequently, they line up in linear arrays (Figure 20.3) and intermolecular hydrogen bonds form between adjacent chains. These large aggregates cause cellulose to be insoluble in water. The strength of these bundles of polymer chains makes cellulose an effective structural material.

▲ **Figure 20.2**
The α-1,4'-glycosidic linkages in amylose cause it to form a left-handed helix.

◀ **Figure 20.3**
The β-1,4'-glycosidic linkages in cellulose cause it to form intramolecular hydrogen bonds and line up in linear arrays.

SYNTHETIC FIBERS

Cellulose is the raw material for some commercially important products. For example, after being specially treated, cellulose is spun into a fiber known as rayon. The OH groups are esterified with acetic anhydride, resulting in a fabric known as acetate. These cellulose products were the first synthetic fibers. The production of rayon and acetate is now decreasing because large amounts of polluting waste are generated during production. Processed cellulose is also used for the production of paper and cellophane.

Chitin is a polysaccharide structurally similar to cellulose. It is the major structural component of the shells of crustaceans (lobsters, crabs, shrimps) and the exoskeletons of insects. Like cellulose, chitin has β-1,4'-glycosidic linkages. It differs from cellulose, though, in that it has an *N*-acetylamino group instead of an OH group at the C-2 position. The β-1,4'-glycosidic linkages give chitin its structural rigidity.

A bright orange crab from Australia.

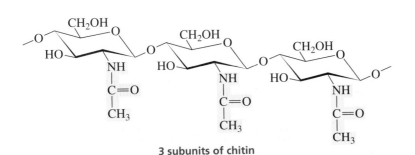

3 subunits of chitin

CONTROLLING FLEAS

Several different drugs have been developed to help pet owners control fleas. One of these is lufenuron, the active ingredient in Program. Lufenuron interferes with the production of chitin. Since the exoskeleton of a flea is composed primarily of chitin, a flea cannot live if it cannot make chitin.

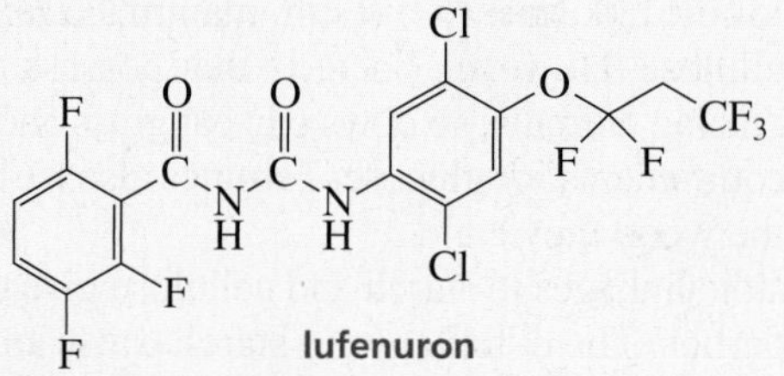

lufenuron

PROBLEM 28

What is the main structural difference between:

a. amylose and cellulose?   **c.** amylopectin and glycogen?

b. amylose and amylopectin?   **d.** cellulose and chitin?

20.19
SOME NATURALLY OCCURRING PRODUCTS DERIVED FROM CARBOHYDRATES

Deoxy sugars are sugars in which one of the OH groups is replaced by a hydrogen (deoxy means "without oxygen"). 2-Deoxyribose (it is missing the oxygen at the C-2 position) is an important example of a deoxy sugar. Ribose is the sugar component of RNA (ribonucleic acid), while 2-deoxyribose is the sugar component of DNA (deoxyribonucleic acid). RNA and DNA are *N*-glycosides—their subunits consist of an amine bonded to the β-position of the anomeric carbon of ribose or 2-deoxyribose. The subunits are linked by a phosphate group between C-3 of one sugar and C-5 of the next sugar (Section 25.1).[1]

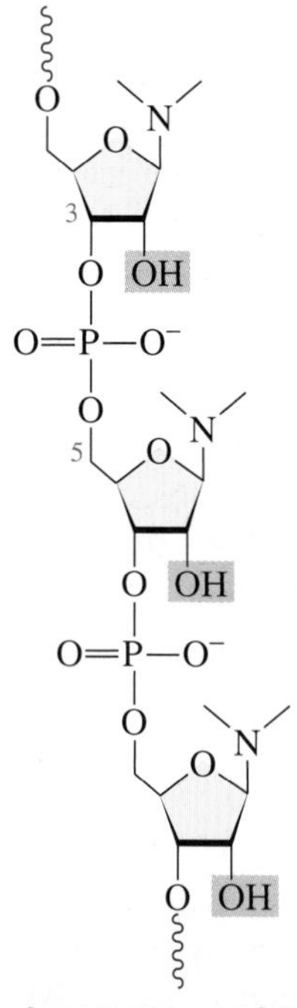

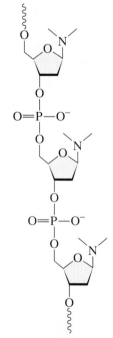

a short segment of RNA
the sugar component is D-ribose

a short segment of DNA
the sugar component is D-2′-deoxyribose

[1] When referring to the sugar found in DNA, it is called 2-deoxyribose. When numbering the sugar in a DNA molecule, it is 2′-deoxyribose because the non-primed numbers refer to positions on the heterocyclic amine components.

3-D Molecule:
Gentamicin

In **amino sugars** one of the OH groups is replaced by an amino group. *N*-Acetyl-glucosamine (the subunit of chitin and one of the subunits of certain bacterial cell walls; Sections 20.18 and 22.9) is an example of an amino sugar. Some important antibiotics contain amino sugars. For example, the three subunits of the antibiotic gentamicin are deoxyamino sugars. Notice that the middle subunit is missing the ring oxygen so it really isn't a sugar.

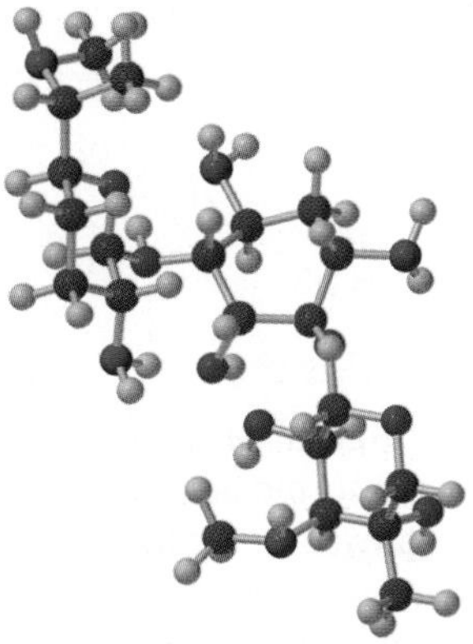

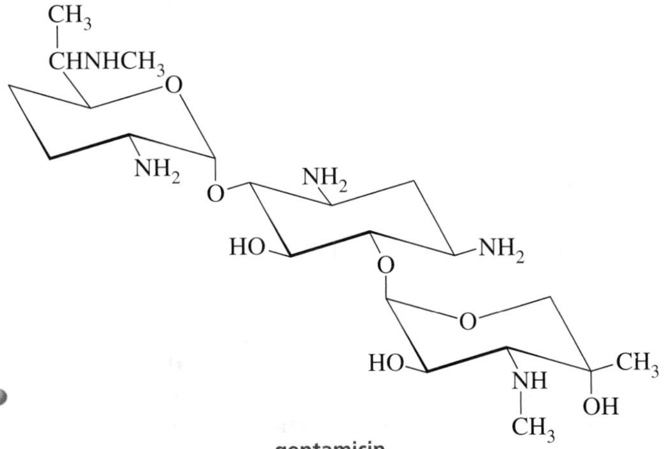

**gentamicin
an antibiotic**

HEPARIN

Heparin is an anticoagulant that is released to prevent excessive blood clot formation when an injury occurs. It is a polysaccharide made up of glucosamine, glucuronic acid, and iduronic acid subunits. The C-6 OH groups of the glucosamine subunits and the C-2 OH groups of the iduronic acid subunits are sulfonated. Some of the amino groups are sulfonated and some are acetylated. Thus heparin is a highly negatively charged molecule. Heparin is found principally in cells that line arterial walls. The longer the polysaccharide chain, the greater the anticoagulant activity. Heparin is widely used clinically as an anticoagulant.

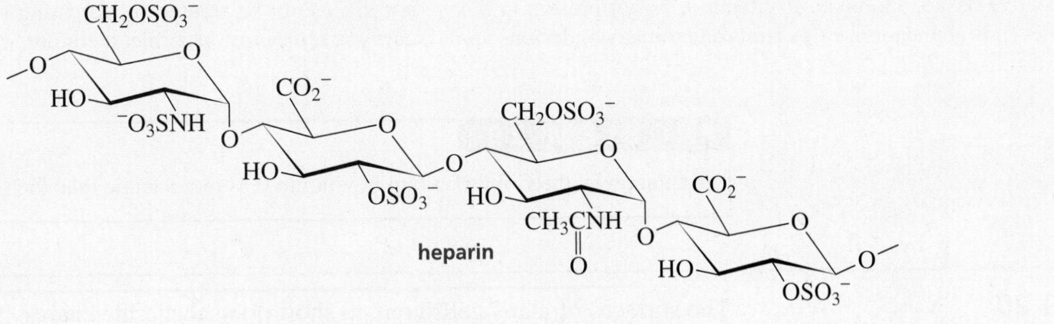

heparin

L-Ascorbic acid (vitamin C) is synthesized in plants and in the livers of most vertebrates. Humans, monkeys, and guinea pigs do not have the enzymes necessary for the biosynthesis of vitamin C, so they must obtain the vitamin in their diets. The biosynthesis of vitamin C involves the enzymatic conversion of D-glucose into L-gulonic acid—reminiscent of the last step in the Fischer proof. L-Gulonic acid is converted into a γ-lactone by the enzyme lactonase, and then an enzyme called oxidase oxidizes the lactone to L-ascorbic acid. The L-configuration of ascorbic acid refers to the configuration at C-5, which was C-2 in D-glucose.

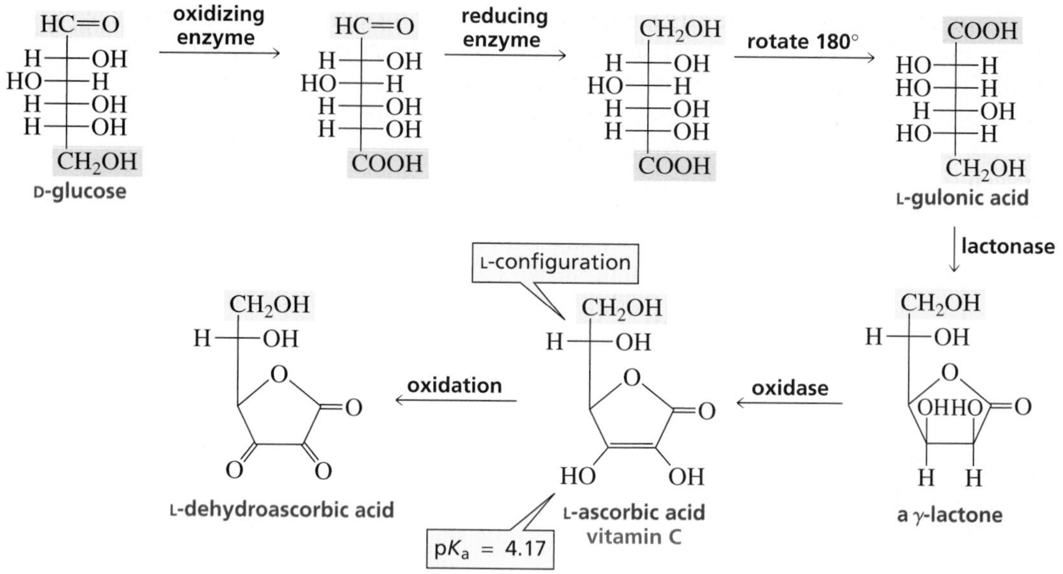

D-glucose → L-gulonic acid → a γ-lactone → L-ascorbic acid (vitamin C) → L-dehydroascorbic acid

Although L-ascorbic acid does not have a carboxylic acid group, it is an acidic compound because the pK_a of the C-3 OH group is 4.17. L-Ascorbic acid is readily oxidized to L-dehydroascorbic acid, which is also physiologically active. If the lactone ring is opened by hydrolysis, all vitamin C activity is lost. Therefore, not much intact vitamin C survives in food that has been thoroughly cooked. And if the food is cooked in water and then drained, the water-soluble vitamin is thrown out with the water.

VITAMIN C

Vitamin C traps radicals formed in aqueous environments (Section 8.8). It is an antioxidant because it prevents oxidation reactions by radicals. Not all the physiological functions of vitamin C are known. What is known, though, is that vitamin C is required for the synthesis of collagen, which is the structural protein of skin, tendons, connective tissue, and bone. If vitamin C is not present in the diet (it is abundant in citrus fruits and tomatoes), lesions appear on the skin, severe bleeding occurs about the gums, in the joints, and under the skin, and wounds heal slowly. The disease caused by a deficiency of vitamin C is known as scurvy. British sailors who shipped out to sea after the late 1700s were required to eat limes to prevent scurvy. This is how they came to be called "limeys." Thus, scurvy was the first disease to be treated by adjusting the diet. Latin for scurvy is *scorbutus;* ascorbic, therefore, means "no scurvy."

PROBLEM 29

Explain why the C-3 OH group of vitamin C is more acidic than the C-2 OH group.

20.20 CARBOHYDRATES ON CELL SURFACES

The surfaces of many cells contain short polysaccharide chains. These polysaccharides are linked to the cell surface by the reaction of an OH or an NH_2 group of a protein with the anomeric carbon of a cyclic sugar. Proteins bonded to polysaccharides are called **glycoproteins.** The percentage of carbohydrate is variable—some glycoproteins contain as little as 1% carbohydrate by weight, whereas others contain as much as 80% carbohydrate.

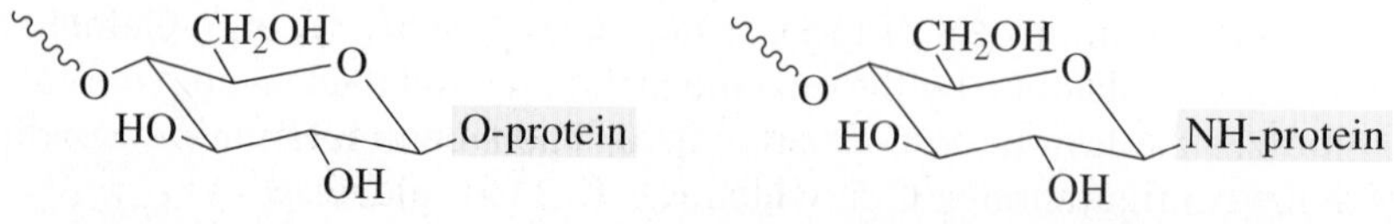

Many different types of proteins are glycoproteins. For example, structural proteins such as collagen, proteins found in mucous secretions, immunoglobulins, follicle-stimulating hormone and thyroid-stimulating hormone, interferon (an antiviral protein), and blood plasma proteins are all glycoproteins. One of the functions of the polysaccharide chain is to act as a receptor site on the cell surface in order to transmit signals from hormones and other molecules across the cell membrane into the cell. The carbohydrates on the surfaces of cells also serve as points of attachment for other cells, viruses, and toxins.

Carbohydrates on the surfaces of cells provide a way for cells to recognize one another. The interaction between surface carbohydrates has been found to play a role in inflammatory diseases such as rheumatoid arthritis and septic shock. The fact that several known antibiotics contain amino sugars suggests that they function by recognizing target cells. Carbohydrate interactions also are involved in the regulation of cell growth, so changes in membrane glycoproteins are thought to be correlated with malignant transformations.

Blood type (A, B, or O) is determined by the nature of the sugar bound to the protein on the outer surface of red blood cells. Each type of blood is associated with a different carbohydrate structure (Figure 20.4). Type AB blood has the carbohydrate structure of both type A and type B.

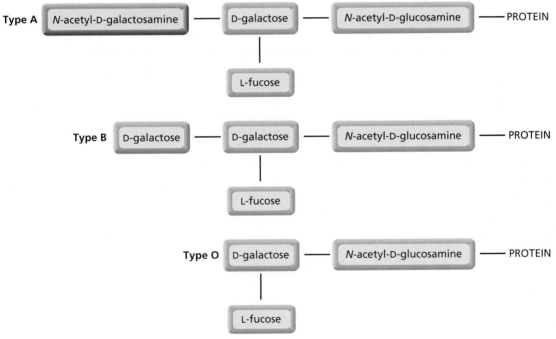

▲ **Figure 20.4**
Blood type is determined by the nature of the sugar on the surfaces of red blood cells.
Fucose is 6-deoxygalactose.

Antibodies are proteins that are synthesized by the body in response to a foreign substance, called an **antigen.** Interaction with the antibody causes the antigen to precipitate or flags it for destruction by immune-system cells. Blood cannot be transferred from one person to another unless the carbohydrate portions of the donor and acceptor are compatible—otherwise the blood will be considered a foreign substance.

From the nature of the carbohydrates bound to red blood cells, it is apparent why the immune system of type A people recognizes the blood from type B people as being foreign and vice versa. The immune system of people with type A, B, or AB blood does not recognize type O blood as being foreign, because the carbohydrate in type O blood is also a component of types A, B, and AB blood. Thus, anyone can accept type O blood—people with type O blood are called universal donors. Type AB people can accept types A, B, and O blood—people with type AB blood are referred to as universal acceptors.

PROBLEM 30◆

From the nature of the carbohydrate bound to red blood cells, answer the following questions:

a. People with type O blood can donate blood to anyone, but they cannot receive blood from everyone. From whom can they *not* receive blood?

b. People with type AB blood can receive blood from anyone, but they cannot give blood to everyone. To whom can they *not* give blood?

20.21 SYNTHETIC SWEETENERS

For a molecule to taste sweet, it must bind to a receptor on a taste bud cell of the tongue. The binding of this molecule causes a nerve impulse to pass from the taste bud to the brain, where the molecule is interpreted as being sweet. Sugars differ in their degree of "sweetness." The relative sweetness of glucose is 1.00, that of sucrose is 1.45, and that of fructose, the sweetest of all sugars, is 1.65. Developers of synthetic sweeteners must consider several factors in addition to taste, however, such as toxicity, stability, and cost.

Saccharin, the first synthetic sweetener, was discovered by Ira Remsen and his student Constantine Fahlberg at Johns Hopkins University in 1878. Fahlberg was studying the oxidation of ortho-substituted toluenes in Remsen's laboratory when he found that one of his newly synthesized compounds had an extremely sweet taste. As strange as it may seem today, at one time it was common for chemists to taste a compound in order to characterize it. He called this compound saccharin, and it was eventually found to be about 300 times sweeter than glucose. Notice that, in spite of its name, saccharin is not a saccharide.

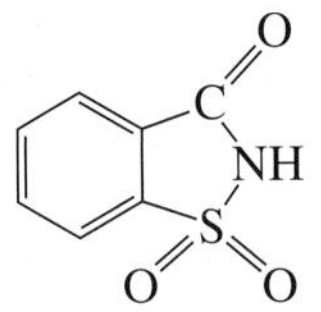

saccharin   dulcin   sodium cyclamate

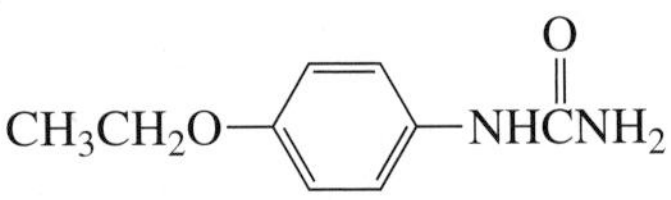

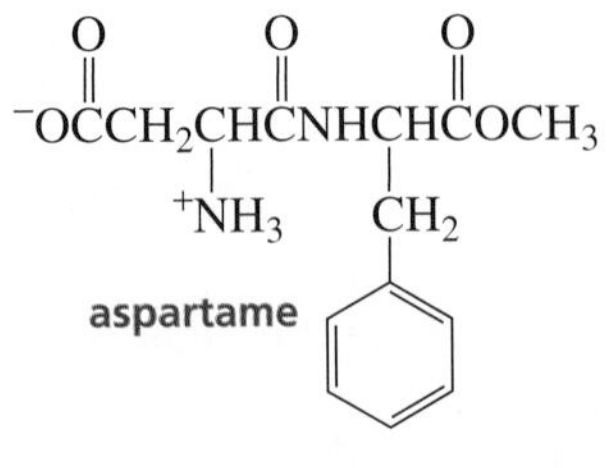

aspartame

Ira Remsen (1846–1927) was born in New York. After receiving an M.D. from Columbia University, he decided to become a chemist. He earned a Ph.D. in Germany, and then returned to the United States in 1872 to accept a faculty position at Williams College. In 1876, he became a professor of chemistry at the newly established Johns Hopkins University, where he initiated the first center for chemical research in the United States. He later became the second president of Johns Hopkins.

Because it has no caloric value, saccharin became an important substitute for sucrose when it became commercially available in 1885. Its use was spurred by the fact that the chief nutritional problem in the West was (and still is) overconsumption of sugar and its consequences—obesity, heart disease, and dental decay. Saccharin is also important to diabetics, who must limit their consumption of sucrose and glucose. Although careful toxicity studies had not been done when saccharin first became available to the public (our current concern with toxicity is a fairly recent development), the extensive studies done since then have shown saccharin to be a safe sugar substitute. In 1912, saccharin was temporarily banned in the United States, not because of any concern about its toxicity, but because of a concern that people would miss out on the nutritional benefits of sugar.

THE WONDER OF DISCOVERY

Ira Remsen gave the following account of why he became a scientist.[2] He was working as a physician and came across the statement "nitric acid acts upon copper" while reading a chemistry book. He decided to see what "acts on" meant. He poured nitric acid on a penny sitting on a table. "But what was this wonderful thing which I beheld? The cent had already changed and it was not small change either. A greenish blue liquid foamed and fumed over the cent and over the table. The air in the neighborhood of the performance became dark red. A great colored cloud arose. This was disagreeable and suffocating—how should I stop this? I tried to get rid of the objectionable mess by picking it up and throwing it out of the window, which I had meanwhile opened. I learned another fact—nitric acid not only acts upon copper but it acts upon my fingers. The pain led to another unpremeditated experiment. I drew my fingers across my trousers and another fact was discovered. Nitric acid also acts upon trousers. Taking everything else into consideration, that was the most impressive experiment, and, relatively, probably the most expensive experiment I ever performed. I tell of it even now with interest. It was a revelation to me. It resulted in a desire on my part to learn even more about that remarkable kind of action. Plainly the only way to learn about it was to see its results, to experiment, to work in a laboratory."

[2]*Chemical Demonstrations,* L. R. Summerlin, C. L. Bordford, and J. B. Ealy, 2nd ed., (Washington, D.C.: American Chemical Society, 1988).

Dulcin was the second synthetic sweetener to be discovered (in 1884). Even though it did not have the bitter, metallic aftertaste associated with saccharin, it never achieved much popularity. It was taken off the market in 1951 in response to questions about its toxicity.

Sodium cyclamate became a widely used non-nutritive sweetener in the 1950s but was banned in the United States some 20 years later in response to two studies that appeared to show that large amounts of sodium cyclamate cause liver cancer in mice.

Aspartame was approved by the U.S. Food and Drug Administration (FDA) in 1981. It is about 200 times sweeter than sucrose and is sold under the trade name NutraSweet (Section 21.8). Because NutraSweet contains a phenylalanine subunit, it should not be used by people with the genetic disease known as PKU (Section 23.6).

The fact that these four synthetic sweeteners have such different structures and that their structures are very different from those of monosaccharides indicates that the sensation of sweetness is not induced by a single molecular shape.

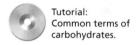

Tutorial:
Common terms of carbohydrates.

SUMMARY OF REACTIONS

1. Reduction (Section 20.5)

$$\begin{array}{c} \text{HC}=\text{O} \\ | \\ (\text{CHOH})_n \\ | \\ \text{CH}_2\text{OH} \end{array} \quad \xrightarrow[\text{Pd/C}]{\text{H}_2} \quad \begin{array}{c} \text{CH}_2\text{OH} \\ | \\ (\text{CHOH})_n \\ | \\ \text{CH}_2\text{OH} \end{array}$$

$$\begin{array}{c} \text{CH}_2\text{OH} \\ | \\ \text{C}=\text{O} \\ | \\ (\text{CHOH})_n \\ | \\ \text{CH}_2\text{OH} \end{array} \quad \xrightarrow[\text{2. H}^+, \text{ H}_2\text{O}]{\text{1. NaBH}_4} \quad \begin{array}{c} \text{CH}_2\text{OH} \\ | \\ \text{CHOH} \\ | \\ (\text{CHOH})_n \\ | \\ \text{CH}_2\text{OH} \end{array}$$

2. Oxidation (Section 20.5)

a.
$$\underset{\substack{\displaystyle | \\ (CHOH)_n \\ | \\ CH_2OH}}{HC=O} \xrightarrow[\text{HO}^-]{\text{Ag}^+,\ \text{NH}_3} \underset{\substack{\displaystyle | \\ (CHOH)_n \\ | \\ CH_2OH}}{COO^-} +\ Ag$$

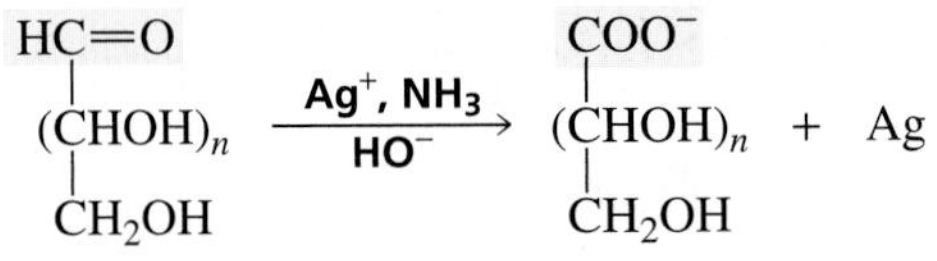

b.
$$\underset{\substack{\displaystyle | \\ C=O \\ | \\ (CHOH)_n \\ | \\ CH_2OH}}{CH_2OH} \xrightarrow[\text{HO}^-]{\text{Ag}^+,\ \text{NH}_3} \underset{\substack{\displaystyle | \\ CHOH \\ | \\ (CHOH)_n \\ | \\ CH_2OH}}{COO^-} +\ Ag$$

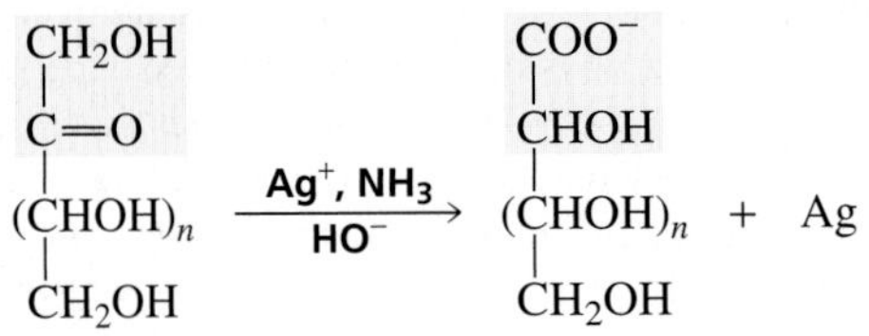

c.
$$\underset{\substack{\displaystyle | \\ (CHOH)_n \\ | \\ CH_2OH}}{HC=O} \xrightarrow[\text{H}_2\text{O}]{\text{Br}_2} \underset{\substack{\displaystyle | \\ (CHOH)_n \\ | \\ CH_2OH}}{COOH} +\ 2\ Br^-$$

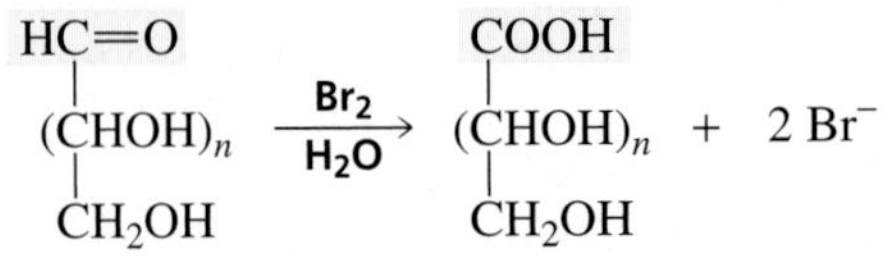

d.
$$\underset{\substack{\displaystyle | \\ (CHOH)_n \\ | \\ CH_2OH}}{HC=O} \xrightarrow[\Delta]{\text{HNO}_3} \underset{\substack{\displaystyle | \\ (CHOH)_n \\ | \\ COOH}}{COOH}$$

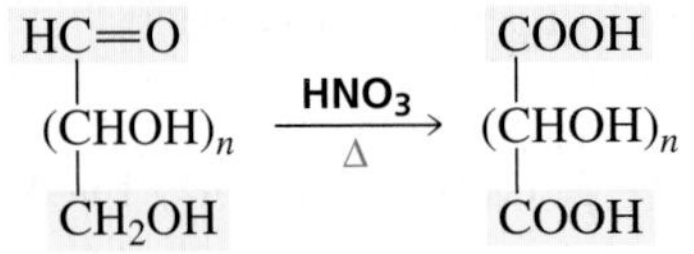

3. Enolization (Section 20.5)

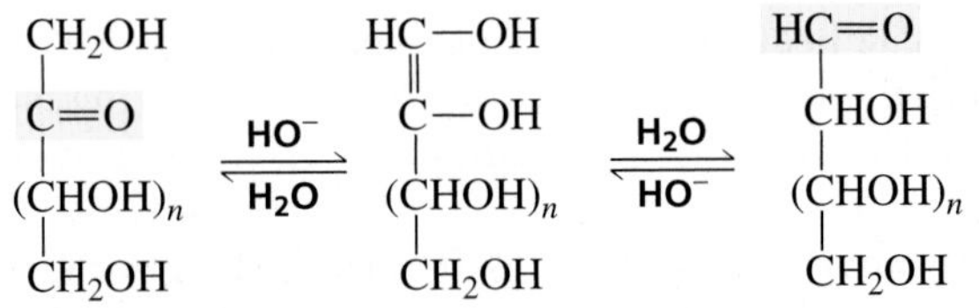

$$\underset{\substack{\displaystyle | \\ C=O \\ | \\ (CHOH)_n \\ | \\ CH_2OH}}{CH_2OH} \underset{\text{H}_2\text{O}}{\overset{\text{HO}^-}{\rightleftharpoons}} \underset{\substack{\displaystyle \| \\ C-OH \\ | \\ (CHOH)_n \\ | \\ CH_2OH}}{HC-OH} \underset{\text{HO}^-}{\overset{\text{H}_2\text{O}}{\rightleftharpoons}} \underset{\substack{\displaystyle | \\ CHOH \\ | \\ (CHOH)_n \\ | \\ CH_2OH}}{HC=O}$$

4. Osazone formation (Section 20.6)

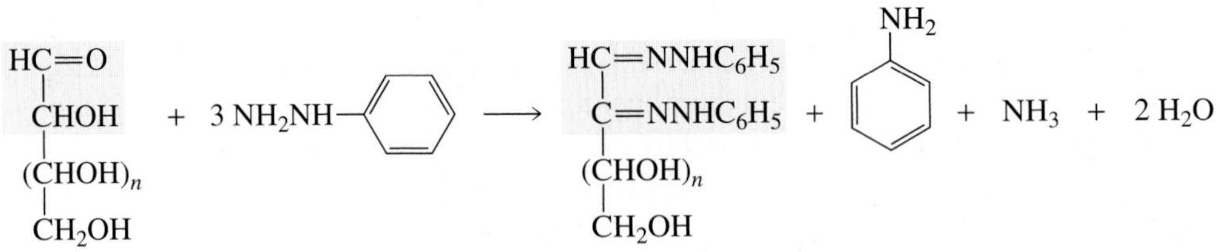

$$\underset{\substack{\displaystyle | \\ CHOH \\ | \\ (CHOH)_n \\ | \\ CH_2OH}}{HC=O} +\ 3\ NH_2NH-\!\!\bigcirc \longrightarrow \underset{\substack{\displaystyle | \\ C=NNHC_6H_5 \\ | \\ (CHOH)_n \\ | \\ CH_2OH}}{HC=NNHC_6H_5} +\ \bigcirc\!\!-NH_2 +\ NH_3 +\ 2\ H_2O$$

5. Chain elongation: the Kiliani–Fischer synthesis (Section 20.7)

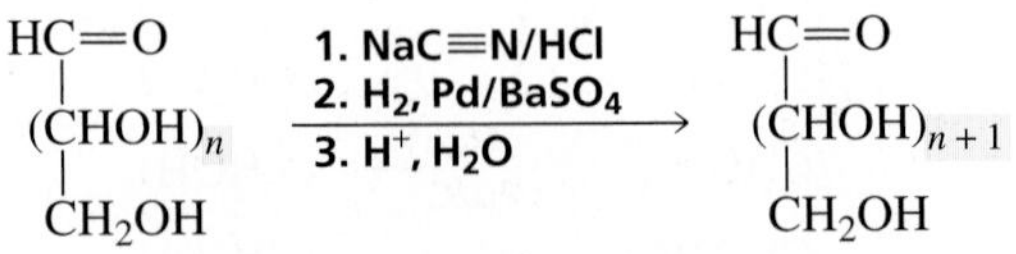

$$\underset{\substack{\displaystyle | \\ (CHOH)_n \\ | \\ CH_2OH}}{HC=O} \xrightarrow[\substack{\textbf{1. NaC} \equiv \textbf{N/HCl} \\ \textbf{2. H}_2,\ \textbf{Pd/BaSO}_4 \\ \textbf{3. H}^+,\ \textbf{H}_2\textbf{O}}]{} \underset{\substack{\displaystyle | \\ (CHOH)_{n+1} \\ | \\ CH_2OH}}{HC=O}$$

6. Chain shortening: the Ruff degradation (Section 20.8)

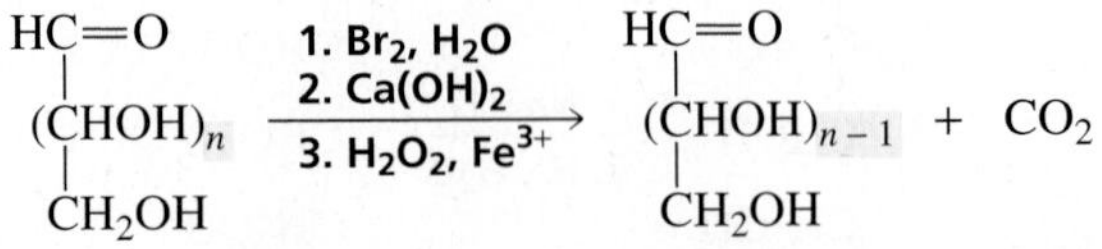

$$\underset{\substack{\displaystyle | \\ (CHOH)_n \\ | \\ CH_2OH}}{HC=O} \xrightarrow[\substack{\textbf{1. Br}_2,\ \textbf{H}_2\textbf{O} \\ \textbf{2. Ca(OH)}_2 \\ \textbf{3. H}_2\textbf{O}_2,\ \textbf{Fe}^{3+}}]{} \underset{\substack{\displaystyle | \\ (CHOH)_{n-1} \\ | \\ CH_2OH}}{HC=O} +\ CO_2$$

7. Acylation (Section 20.12)

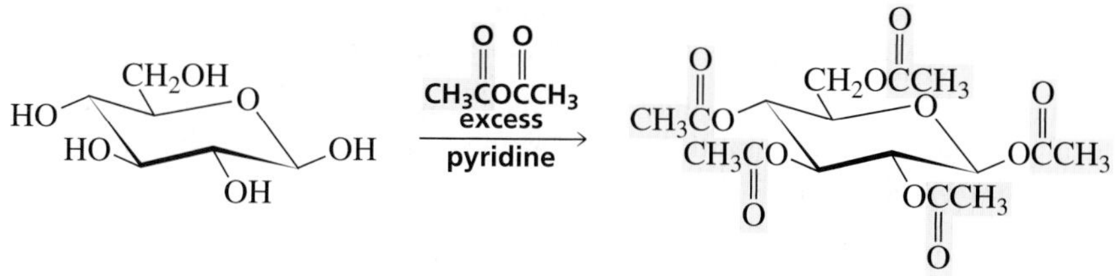

8. Alkylation (Section 20.12)

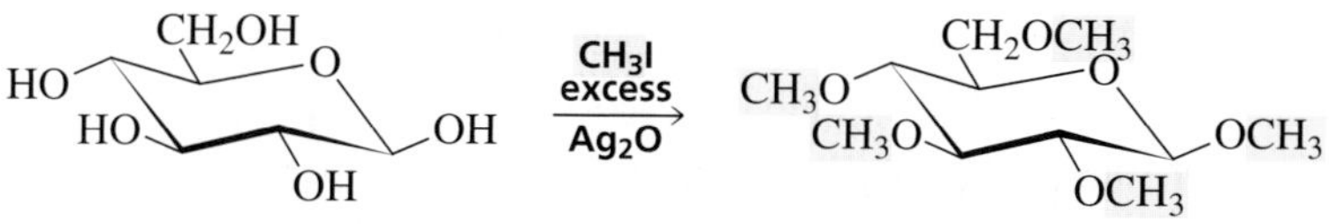

9. Acetal (and ketal) formation (Section 20.13)

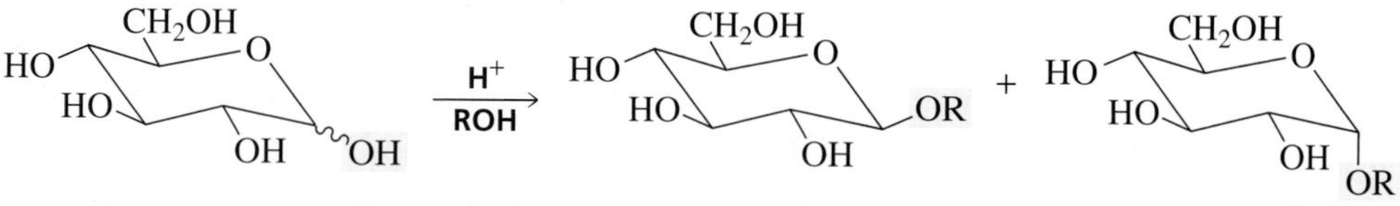

KEY TERMS

aldaric acid (page 886)
alditol (page 884)
aldonic acid (page 885)
aldose (page 880)
amino sugar (page 907)
anomeric carbon (page 892)
anomeric effect (page 898)
anomers (page 982)
antibody (page 909)
antigen (page 909)
bioorganic compound (page 878)
carbohydrate (page 878)
complex carbohydrate (page 879)
deoxy sugar (page 906)
disaccharide (page 879)
epimers (page 882)

furanose (page 893)
furanoside (page 896)
glycoprotein (page 908)
glycoside (page 896)
N-glycoside (page 898)
glycosidic bond (page 896)
α-1,4′-glycosidic linkage (page 901)
β-1,4′-glycosidic linkage (page 901)
Haworth projection (page 893)
heptose (page 880)
hexose (page 880)
ketose (page 880)
Kiliani–Fischer synthesis (page 888)
molecular recognition (page 878)
monosaccharide (page 879)
mutarotation (page 893)

nonreducing sugar (page 899)
oligosaccharide (page 879)
osazone (page 886)
oxocarbenium ion (page 897)
pentose (page 880)
photosynthesis (page 879)
polysaccharide (page 879)
pyranose (page 893)
pyranoside (page 896)
reducing sugar (page 899)
Ruff degradation (page 889)
simple carbohydrate (page 879)
tetrose (page 880)
triose (page 880)

PROBLEMS

31. Give the product(s) that are obtained when D-galactose reacts with the following:
  a. nitric acid
  b. Tollens' reagent
  c. $H_2/Pd/C$
  d. three equivalents of phenylhydrazine
  e. Br_2 in water
  f. ethanol + HCl
  g. product of reaction **e** (above) + $Ca(OH)_2$, Fe^{3+}, H_2O_2

32. Identify the following sugars:
  a. An aldopentose that is not D-arabinose forms D-arabinitol when it is reduced with $NaBH_4$.

b. A sugar forms the same osazone as D-galactose with phenylhydrazine but it is not oxidized by an aqueous solution of Br_2.

c. A sugar that is not D-altrose forms D-altraric acid when it reacts with nitric acid.

d. A ketose, when reduced with $H_2/Pd/C$, forms D-altritol and D-allitol.

33. Pectin is a polysaccharide obtained from fruits. It is used as a gelling agent in making jams and jellies. It can be synthesized by treating amylose with nitric acid. Draw a short segment of pectin.

34. Answer the following questions for the eight aldopentoses:

a. Which are enantiomers?

b. Which give identical osazones?

c. Which form an optically active compound when oxidized with nitric acid?

35. The reaction of D-ribose with an equivalent of methanol plus HCl forms four products. Give the structures of the products.

36. Determine the structure of D-galactose, using arguments similar to those used by Fischer to prove the structure of D-glucose.

37. Dr. Isent T. Sweet isolated a monosaccharide and determined that it had a molecular weight of 150. Much to his surprise, he found that it was not optically active. What is the structure of the monosaccharide?

38. The ^1H NMR spectrum of D-glucose in D_2O exhibits two high-frequency (low-field) doublets. What is responsible for these doublets?

39. Treatment with sodium borohydride converts aldose **A** into an optically inactive alditol. Ruff degradation of **A** forms **B**, whose alditol is optically inactive. Ruff degradation of **B** forms D-glyceraldehyde. Identify **A** and **B**.

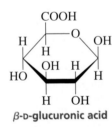

COOH
H
H
OH H
HO
H
H OH
OH

β-D-glucuronic acid

40. D-Glucuronic acid is found widely in plants and animals. One of its functions is to detoxify poisonous HO-containing compounds by reacting with them in the liver to form glucuronides. Glucuronides are water soluble and therefore readily excreted. After ingestion of a poison such as turpentine, morphine, or phenol, the glucuronides of these compounds are found in the urine. Give the structure of the glucuronide formed by the reaction of β-D-glucuronic acid and phenol.

41. Hyaluronic acid is a component of connective tissue and is the fluid that lubricates the joints. It is an alternating polymer of N-acetyl-D-glucosamine and D-glucuronic acid joined by β-1,3'-glycosidic linkages. Draw a short segment of hyaluronic acid.

42. In order to synthesize D-galactose, Professor Amy Losse went to the stockroom to get some D-lyxose to use as a starting material. She found that the labels had fallen off the bottles containing D-lyxose and D-xylose. How could she determine which bottle contains D-lyxose?

43. When D-fructose is dissolved in D_2O and the solution is made basic, the D-fructose recovered from the solution has an average of 1.7 deuterium atoms attached to carbon per molecule. Show the mechanism that accounts for the incorporation of these deuterium atoms into D-fructose.

44. A D-aldopentose is oxidized by nitric acid to an optically active aldaric acid. A Ruff degradation of the aldopentose leads to a monosaccharide that is oxidized by nitric acid to an optically inactive aldaric acid. Identify the D-aldopentose.

45. How many aldaric acids are obtained from the 16 aldohexoses?

46. Calculate the percentages of α-D-glucose and β-D-glucose present at equilibrium from the specific rotations of α-D-glucose, β-D-glucose, and the equilibrium mixture. Compare your values with those given in Section 20.10. (*Hint:* The specific rotation of the mixture = the specific rotation of α-D-glucose times the fraction of glucose present in the α-form + the specific rotation of β-D-glucose times the fraction of glucose present in the β-form.)

47. Predict whether D-altrose exists preferentially as a pyranose or a furanose. (*Hint:* The most stable arrangement for a five-membered ring is for all the adjacent substituents to be trans.)

48. Propose a mechanism for the rearrangement that converts an α-hydroxyimine into an α-aminoketone (Section 20.6).

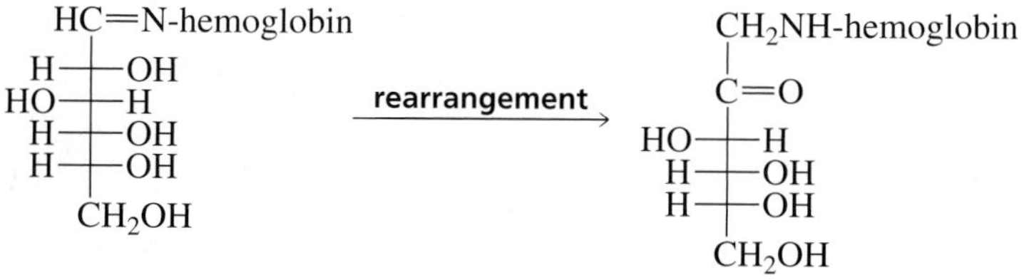

49. A disaccharide forms a silver mirror with Tollens' reagent and is hydrolyzed by a β-glycosidase. When the disaccharide is treated with excess methyl iodide in the presence of Ag_2O (a reaction that converts all the OH groups to OCH_3 groups) and then hydrolyzed with water under acidic conditions, 2,3,4-tri-*O*-methylmannose and 2,3,4,6-tetra-*O*-methylgalactose are formed.
  a. Draw the structure of the disaccharide.
  b. What is the function of Ag_2O?

50. Devise an experiment to prove that the hemiacetal group in lactose is in the glucose residue rather than in the galactose residue.

51. All the glucose units in dextran have six-membered rings. When a sample of dextran is treated with methyl iodide and silver oxide, and the product is hydrolyzed under acidic conditions, the products obtained are 2,3,4,6-tetra-*O*-methyl-D-glucose, 2,4,6-tri-*O*-methyl-D-glucose, 2,3,4-tri-*O*-methyl-D-glucose, and 2,4-di-*O*-methyl-D-glucose. Draw a short segment of dextran.

52. When a pyranose is in the chair conformation in which the CH_2OH group and the C-1 OH group are both in the axial position, the two groups can react to form an acetal. This is called the anhydro form of the sugar (it has lost water). The anhydro form of D-idose is shown here. In an aqueous solution at 100 °C, a large percentage of D-idose exists in the anhydro form (about 80%). Under the same conditions, only about 0.1% of D-glucose exists in the anhydro form. Explain.

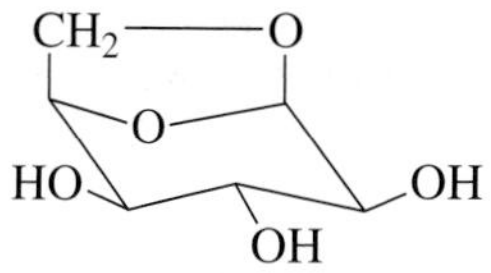

anhydro form of D-idose

53. Oxidation with periodic acid can be used to determine whether a glycoside has the pyranose or the furanose structure. Explain how the products obtained from the oxidation of methyl β-D-glucoside can be used to determine whether the glucoside has a five- or a six-membered ring.

54. Devise a method to convert D-glucose into D-allose.

21

Amino Acids, Peptides, and Proteins

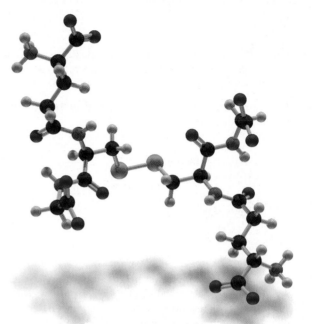

oxidized glutathione

alanine
an amino acid

The three kinds of polymers prevalent in nature are polysaccharides, proteins, and nucleic acids. You have already learned about polysaccharides, which are naturally occurring polymers of sugar subunits (Section 20.18), and nucleic acids are covered in Chapter 25. We will now look at proteins and the structurally similar but shorter peptides. **Peptides** and **proteins** are polymers of **amino acids** linked together by amide bonds. The repeating units (monomers) are called **amino acid residues.**

Amino acid polymers can be composed of any number of monomers. A **dipeptide** contains two amino acid residues, a **tripeptide** contains three, an **oligopeptide** contains three to ten, and a **polypeptide** contains many amino acid residues. Proteins are naturally occurring polypeptides that are made up of 40 to 4000 amino acid residues.

From the structure of an amino acid, we can see that the name is not very precise. The compounds commonly called amino acids are more precisely called α-aminocarboxylic acids.

an α-aminocarboxylic acid
an "amino acid"

amide bonds

amino acids are linked together by amide bonds

Proteins and peptides serve many functions in biological systems. Some protect organisms from their environment and/or impart strength to certain biological structures. Hair, horns, hoofs, feathers, fur, and the tough outer layer of skin are all composed largely of a protein called keratin. Keratin is a **structural protein.** Collagen, another structural protein, is a major component of bones, muscles, and tendons. Some proteins have other protective functions. Snake venoms and plant toxins, for example, protect their owners from other species, blood-clotting proteins protect the vascular system when it is injured, and antibodies and protein antibiotics protect us from disease. A group of proteins called **enzymes** catalyze the chemical reactions that occur in living organisms, and some of the hormones that regulate these reactions are peptides. Proteins are also responsible for many physiological functions such as the transport and storage of oxygen in the body and the contraction of muscles.

The structures of the 20 most common naturally occurring amino acids and the frequency with which each occurs in proteins are shown in Table 21.1. Other amino acids occur in nature, but only infrequently. All the amino acids except proline contain a primary amino group. Proline contains a secondary amino group incorporated into a five-membered ring. The amino acids differ only in the substituent (R) attached to the α-carbon. The wide variation in these substituents (side chains) is what gives proteins their great structural diversity and, as a consequence, their great functional diversity.

21.1 CLASSIFICATION AND NOMENCLATURE OF AMINO ACIDS

TABLE 21.1   The Most Common Naturally Occurring Amino Acids
The amino acids are shown in the form that predominates at physiological pH (pH = 7.3).

	Formula	Name	Abbreviations		Average relative abundance in proteins
Aliphatic side chain amino acids	H—CHCO⁻ with O double bond and ⁺NH₃	glycine	Gly	G	7.5%
	CH₃—CHCO⁻ with O double bond and ⁺NH₃	alanine	Ala	A	9.0%
	CH₃CH—CHCO⁻ with CH₃, ⁺NH₃ and O double bond	valine*	Val	V	6.9%
	CH₃CHCH₂—CHCO⁻ with CH₃, ⁺NH₃ and O double bond	leucine*	Leu	L	7.5%
	CH₃CH₂CH—CHCO⁻ with CH₃, ⁺NH₃ and O double bond	isoleucine*	Ile	I	4.6%
Hydroxy-containing amino acids	HOCH₂—CHCO⁻ with ⁺NH₃ and O double bond	serine	Ser	S	7.1%
	CH₃CH—CHCO⁻ with OH, ⁺NH₃ and O double bond	threonine*	Thr	T	6.0%
Sulfur-containing amino acids	HSCH₂—CHCO⁻ with ⁺NH₃ and O double bond	cysteine	Cys	C	2.8%

*essential amino acids

TABLE 21.1
(continued)

	Formula	Name	Abbreviations		Average relative abundance in proteins
	$CH_3SCH_2CH_2-CHCO^-$ $^+NH_3$ (with C=O)	methionine*	Met	M	1.7%
Acidic amino acids	$^-OCCH_2-CHCO^-$ $^+NH_3$	aspartate (aspartic acid)	Asp	D	5.5%
	$^-OCCH_2CH_2-CHCO^-$ $^+NH_3$	glutamate (glutamic acid)	Glu	E	6.2%
Amides of acidic amino acids	$H_2NCCH_2-CHCO^-$ $^+NH_3$	asparagine	Asn	N	4.4%
	$H_2NCCH_2CH_2-CHCO^-$ $^+NH_3$	glutamine	Gln	Q	3.9%
Basic amino acids	$H_3\overset{+}{N}CH_2CH_2CH_2CH_2-CHCO^-$ $^+NH_3$	lysine*	Lys	K	7.0%
	$\overset{+}{N}H_2$ $H_2NCNHCH_2CH_2CH_2-CHCO^-$ $^+NH_3$	arginine*	Arg	R	4.7%
Benzene-containing amino acids	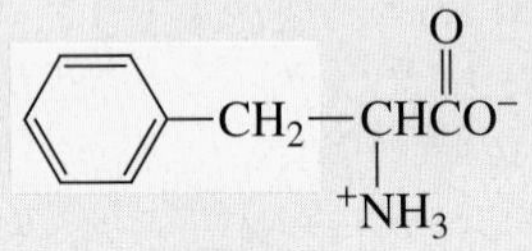 $C_6H_5-CH_2-CHCO^-$ $^+NH_3$	phenylalanine*	Phe	F	3.5%
	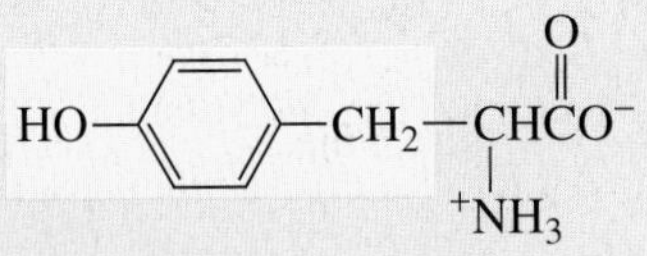 $HO-C_6H_4-CH_2-CHCO^-$ $^+NH_3$	tyrosine	Tyr	Y	3.5%
Heterocylic amino acids	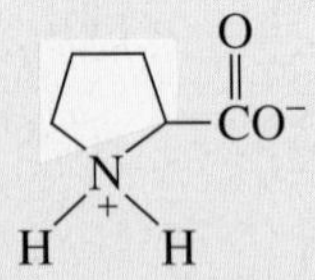 proline ring CO^-, $\overset{+}{N}H$	proline	Pro	P	4.6%

TABLE 21.1
(*continued*)

Formula	Name	Abbreviations		Average relative abundance in proteins
histidine* structure	histidine*	His	H	2.1%
tryptophan* structure	tryptophan*	Trp	W	1.1%

The amino acids are almost always called by their common names. Often the name tells you something about the amino acid. For example, glycine got its name because of its sweet taste (*glykos* is Greek for "sweet"), and valine, like valeric acid, has five carbon atoms. Asparagine was first found in asparagus, and tyrosine was isolated from cheese (*tyros* is Greek for "cheese").

Dividing the amino acids into classes makes them easier to learn. The aliphatic side chain amino acids include glycine, the amino acid in which $R = H$, and four amino acids with alkyl side chains. Alanine is the amino acid with a methyl side chain, and valine has an isopropyl side chain. Can you guess which amino acid—leucine or isoleucine—has an isobutyl side chain? If you gave the obvious answer, you guessed incorrectly. Isoleucine does *not* have an "iso" group. It is leucine that has an isobutyl substituent—isoleucine has a *sec*-butyl substituent. Each of the amino acids has a three-letter abbreviation (the first three letters of the name in most cases) and a single-letter abbreviation.

Two amino acid side chains—serine and threonine—contain alcohol groups. Serine is an HO-substituted alanine and threonine has a branched ethanol substituent. There are also two sulfur-containing amino acids. Cysteine is an HS-substituted alanine and methionine has a 2-methylthioethyl substituent.

There are two acidic amino acids (amino acids with two carboxylic acid groups)—aspartate and glutamate. Aspartate is a carboxy-substituted alanine and glutamate has one more methylene group than aspartate. (If their carboxyl groups are protonated they are called aspartic acid and glutamic acid.) Two amino acids—asparagine and glutamine—are amides of the acidic amino acids. Asparagine is the amide of aspartate and glutamine is the amide of glutamate. Notice that the obvious one-letter abbreviations cannot be used for these four amino acids because A and G are used for alanine and glycine. Aspartic acid and glutamic acid are abbreviated D and E, while asparagine and glutamine are abbreviated N and Q.

There are two basic amino acids (amino acids with two basic nitrogen-containing groups)—lysine and arginine. Lysine has an ϵ-amino group and arginine has a δ-guanidino group. At physiological pH these groups are protonated. The ϵ and δ can remind you how many methylene groups each amino acid has.

glycine

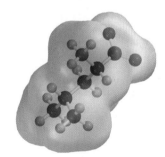

leucine

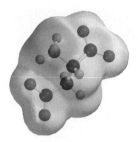

aspartate

lysine

$$\overset{+}{H_3N}-\overset{\epsilon}{CH_2}\overset{\delta}{CH_2}\overset{\gamma}{CH_2}\overset{\beta}{CH_2}\overset{\alpha}{CH}CO^-$$

with $^+NH_3$ group and an ϵ-amino group labeled.

lysine

$$H_2N-\overset{\overset{+NH_2}{\|}}{C}-NH-\overset{\delta}{CH_2}\overset{\gamma}{CH_2}\overset{\beta}{CH_2}\overset{\alpha}{CH}CO^-$$

with $^+NH_3$ group and a δ-guanidino group labeled.

arginine

Two amino acids—phenylalanine and tyrosine—contain benzene rings. As its name indicates, phenylalanine is a phenyl-substituted alanine. Tyrosine is phenylalanine with a *para*-hydroxy substituent.

Proline, histidine, and tryptophan are heterocyclic amino acids. Proline has its nitrogen incorporated into a five-membered ring—it is the only amino acid that contains a secondary amino group. Histidine is an imidazole-substituted alanine. Imidazole is an aromatic compound because it is cyclic, planar, and has three pairs of delocalized π electrons. The pK_a of a protonated imidazole ring is 6.0, so the ring will be protonated in acidic solutions and nonprotonated in basic solutions (Section 21.3).

$$HN\overset{+}{\underset{}{\diagup}}:NH \rightleftharpoons :N\overset{}{\underset{}{\diagup}}:NH + H^+$$

protonated imidazole   **imidazole**

indole

Tryptophan is an indole-substituted alanine (Section 27.2). Like imidazole, indole is an aromatic compound. Because indole needs the lone-pair electrons on nitrogen for its aromaticity, indole is a very weak base (the pK_a of protonated indole is -2.4). Therefore, the ring nitrogen in tryptophan is never protonated under physiological conditions.

Ten of the amino acids are **essential amino acids.** We humans must obtain these 10 amino acids from our diets because we either cannot synthesize them at all or cannot synthesize them in adequate amounts. For example, we must have a dietary source of phenylalanine, but we do not need tyrosine in our diets because we can synthesize the necessary amounts from phenylalanine. The essential amino acids are denoted by red asterisks (*) in Table 21.1. Although humans can synthesize arginine, it is needed for growth in greater amounts than can be synthesized. So arginine is an essential amino acid for children but a nonessential amino acid for adults. Not all proteins contain the same amino acids. Bean protein is deficient in methionine, for example, and wheat protein is deficient in lysine. They are *incomplete* proteins—they contain too little of one or more essential amino acids to support growth. Therefore, a balanced diet must contain proteins from different sources.

Dietary protein is hydrolyzed in the body to individual amino acids. Some of these amino acids are used to synthesize proteins needed by the body, some are broken down further to supply energy to the body, and some are used as starting materials for the synthesis of nonprotein compounds the body needs, such as adrenaline, thyroxine, and melanin (Section 23.6).

3-D Molecules:
Common naturally
occurring amino acids

PROBLEM 1

a. Explain why when the imidazole ring of histidine is protonated, the double-bonded nitrogen is the nitrogen that accepts the proton.

$$:N\overset{}{\underset{}{\diagup}}:NH \overset{CH_2CHCOO^-}{\underset{NH_2}{|}} + 2H^+ \rightleftharpoons HN\overset{+}{\underset{}{\diagup}}:NH \overset{CH_2CHCOO^-}{\underset{^+NH_3}{|}}$$

b. Explain why when the guanidino group of arginine is protonated, the double-bonded nitrogen is the nitrogen that accepts the proton.

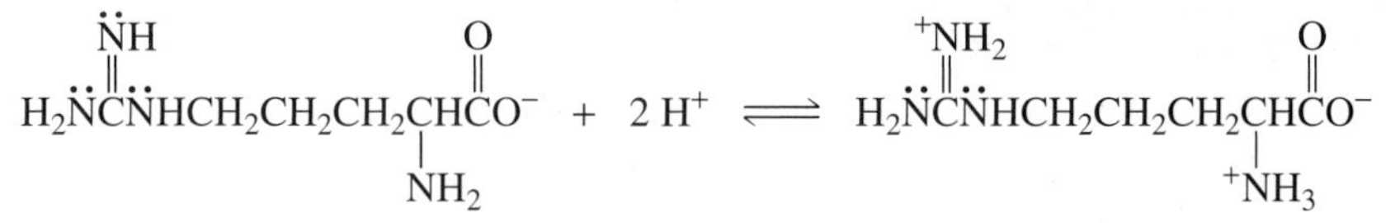

$$H_2\overset{\cdot\cdot}{N}\overset{\overset{\displaystyle\overset{\cdot\cdot}{N}H}{\|}}{C}\overset{\cdot\cdot}{N}HCH_2CH_2CH_2\overset{\overset{\displaystyle O}{\|}}{C}CO^- \;+\; 2\,H^+ \;\rightleftharpoons\; H_2\overset{\cdot\cdot}{N}\overset{\overset{\displaystyle\overset{+}{N}H_2}{\|}}{C}\overset{\cdot\cdot}{N}HCH_2CH_2CH_2\overset{\overset{\displaystyle O}{\|}}{C}CO^-$$

with NH_2 below the left structure and $^+NH_3$ below the right structure.

The α-carbon of all the naturally occurring amino acids except glycine is a chirality center. Therefore, 19 of the 20 amino acids listed in Table 21.1 can exist as enantiomers. The D and L notation used for monosaccharides (Section 20.2) is also used for amino acids. The D and L isomers of monosaccharides and amino acids are defined the same way. An amino acid drawn in a Fischer projection with the carboxyl group on the top and the R group on the bottom of the vertical axis is a **D-amino acid** if the amino group is on the right and an **L-amino acid** if it is on the left. Unlike monosaccharides, where the D isomer is the one found in nature, most amino acids found in nature have the L configuration. To date, D-amino acid residues have been found only in a few peptide antibiotics and in some small peptides attached to the cell walls of bacteria.

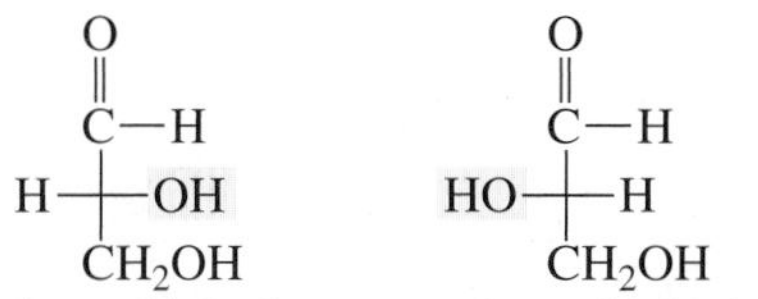

21.2 CONFIGURATION OF AMINO ACIDS

D-glyceraldehyde   L-glyceraldehyde

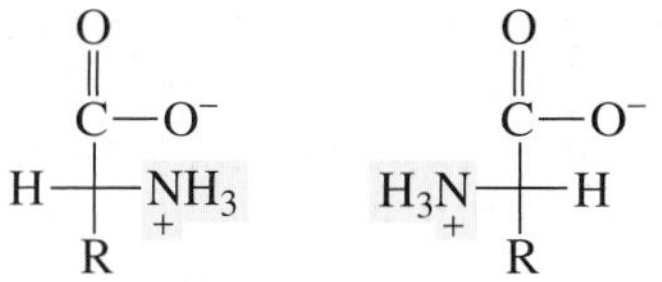

D-amino acid   L-amino acid

Why D-sugars and L-amino acids? It did not make any difference which isomer was selected to be the one synthesized in nature, but it was important that the same isomer was synthesized by all organisms. For example, if mammals ended up having L-amino acids, it was important for L-amino acids to be the isomers synthesized by the organisms that mammals depend on for food.

AMINO ACIDS AND DISEASE

The Chamorro people of Guam have a high incidence of a syndrome that resembles amyotropic lateral sclerosis (ALS) with elements of Parkinson's disease and dementia. This syndrome developed during World War II when, as a result of food shortages, the tribe ate large quantities of the seeds of *Cycas circinalis*. These seeds contain β-methylamino-L-alanine, an amino acid that binds to glutamate receptors. When monkeys are given β-methylamino-L-alanine, they develop some of the features of this syndrome. There is hope that, by studying the mechanism of action of β-methylamino-L-alanine, we may gain an understanding of how ALS and Parkinson's disease arise.

PROBLEM 2◆

a. Which isomer is D-alanine: (R)-alanine or (S)-alanine?

b. Which isomer is D-aspartate: (R)-aspartate or (S)-aspartate?

c. Can a general statement be made?

> **PROBLEM 3◆**
>
> Which amino acids in Table 21.1 have more than one chirality center?

21.3
ACID–BASE PROPERTIES OF AMINO ACIDS

Every amino acid has a carboxyl group and an amino group and each group can exist in an acidic form or a basic form depending on the pH of the solution in which the amino acid is dissolved. The carboxyl groups of the amino acids have pK_a values of approximately 2 and the protonated amino groups have pK_a values near 9 (Table 21.2). Therefore, both groups will be in their acidic forms in a very acidic solution (pH near 0). At pH $=$ 7, the pH of the solution is greater than the pK_a of the carboxyl group but less than the pK_a of the protonated amino group. The carboxyl group, therefore, will be in its basic form and the amino group will be in its acidic form. In a strongly basic solution (pH$\approx$11), both groups will be in their basic forms.

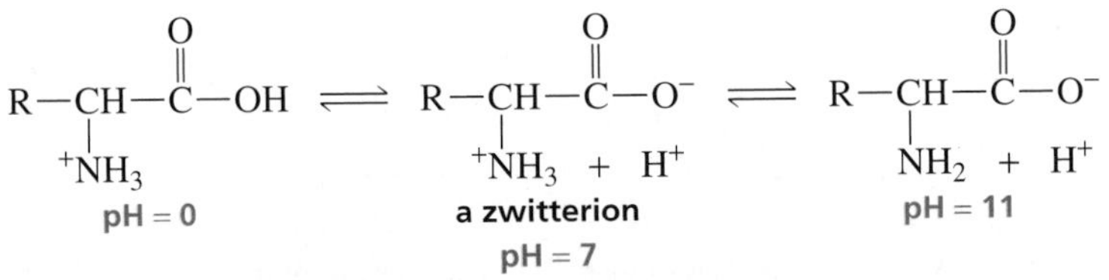

$$R-\underset{\underset{\text{pH} = 0}{\overset{|}{^+NH_3}}}{CH}-\overset{\overset{O}{\|}}{C}-OH \; \rightleftharpoons \; R-\underset{\underset{\underset{\text{pH} = 7}{\text{a zwitterion}}}{\overset{|}{^+NH_3} \; + \; H^+}}{CH}-\overset{\overset{O}{\|}}{C}-O^- \; \rightleftharpoons \; R-\underset{\underset{\text{pH} = 11}{\overset{|}{NH_2} \; + \; H^+}}{CH}-\overset{\overset{O}{\|}}{C}-O^-$$

TABLE 21.2   The pK_a Values of Amino Acids

Amino acid	pK_a α-COOH	pK_a α-NH$_3^+$	pK_a side chain
Alanine	2.34	9.69	—
Arginine	2.17	9.04	12.48
Asparagine	2.02	8.84	—
Aspartic acid	2.09	9.82	3.86
Cysteine	1.71	10.78	8.33
Glutamic acid	2.19	9.67	4.25
Glutamine	2.17	9.13	—
Glycine	2.34	9.60	—
Histidine	1.82	9.17	6.04
Isoleucine	2.36	9.68	—
Leucine	2.36	9.60	—
Lysine	2.18	8.95	10.79
Methionine	2.28	9.21	—
Phenylalanine	1.83	9.13	—
Proline	1.99	10.60	—
Serine	2.21	9.15	—
Threonine	2.63	9.10	—
Tryptophan	2.38	9.39	—
Tyrosine	2.20	9.11	10.07
Valine	2.32	9.62	—

Recall from the Henderson–Hasselbalch equation (Section 1.20) that the acidic form predominates if the pH of the solution is less than the pK_a of the compound, and the basic form predominates if the pH of the solution is greater than the pK_a of the compound.

Notice that an amino acid can never exist as an uncharged compound regardless of the pH of the solution. To be uncharged, an amino acid would have to lose a proton from an $^+NH_3$ group with a pK_a of about 9 more readily than it would lose a proton from a COOH group with a pK_a of about 2. This clearly is impossible—a weak acid cannot be more acidic than a strong acid. Therefore, at physiological pH (pH = 7.3) an amino acid exists as a dipolar ion, called a **zwitterion.** A zwitterion is a compound that has a negative charge on one atom and a positive charge on a nonadjacent atom. (The name comes from *zwitter,* German for "hermaphrodite" or "hybrid.")

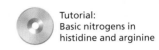

Tutorial:
Basic nitrogens in
histidine and arginine

A few amino acids have side chains with ionizable hydrogens (Table 21.2). The protonated imidazole side chain of histidine, for example, has a pK_a of 6.04. Histidine therefore can exist in four different forms, and the form that predominates depends on the pH of the solution.

histidine

PROBLEM 4 ◆

Why are the carboxylic acid groups of the amino acids so much more acidic ($pK_a \sim 2$) than a carboxylic acid such as acetic acid ($pK_a = 4.76$)?

PROBLEM 5 / SOLVED

Draw the form in which each of the following amino acids predominantly exists at physiological pH (pH = 7.3):

a. aspartic acid   **c.** glutamine   **e.** arginine

b. histidine   **d.** lysine   **f.** tyrosine

SOLUTION to 5a   Because the pH is greater than the pK_a of both carboxyl groups, these groups are in their basic forms. Because the pH is less than the pK_a of the protonated amino group, this group is in its acidic form.

PROBLEM 6 ◆

Draw the form in which glutamic acid predominantly exists in a solution with the following pH:

a. pH = 0   **b.** pH = 3   **c.** pH = 6   **d.** pH = 11

PROBLEM 7

a. Why is the pK_a of the glutamic acid side chain greater than the pK_a of the aspartic acid side chain?

b. Why is the pK_a of the arginine side chain greater than the pK_a of the lysine side chain?

21.4 THE ISOELECTRIC POINT

The **isoelectric point** (pI) of an amino acid is the pH at which it has no net charge. In other words, it is the pH at which the amount of negative charge on an amino acid exactly balances the amount of positive charge.

$$\text{pI (isoelectric point)} = \text{pH at which there is no net charge}$$

The pI of an amino acid that does not have an ionizable side chain—such as alanine—is midway between its two pK_a values. This is because at pH=2.34, half the molecules have a negatively charged carboxyl group and half have an uncharged carboxyl group, and at pH=9.69, half the molecules have a positively charged amino group and half have an uncharged amino group. As the pH increases from 2.34, the carboxyl group of more molecules becomes negatively charged; as the pH decreases from 9.69, the amino group of more molecules becomes positively charged. Therefore at the average of the two pK_a values, the number of negatively charged groups equals the number of positively charged groups.

> **Recall from the Henderson–Hasselbalch equation that when pH = pK_a, half the group is in its acidic form and half is in its basic form (Section 1.20).**

> **An amino acid will be positively charged if the pH of the solution is less than its pI and will be negatively charged if the pH of the solution is greater than its pI.**

$$\text{CH}_3\text{CHCOH} \quad pK_a = 2.34$$
$$\overset{|}{{}^+\text{NH}_3} \quad pK_a = 9.69$$
alanine

$$\text{pI} = \frac{2.34 + 9.69}{2} = \frac{12.03}{2} = 6.02$$

If an amino acid has an ionizable side chain, its pI is the average of the pK_a values of the similarly ionizing groups (positive ionizing to uncharged, or uncharged ionizing to negative). For example, the pI of lysine is the average of the pK_a values of the two groups that are positively charged in their acidic form and uncharged in their basic form. The pI of glutamate, on the other hand, is the average of the pK_a values of the two groups that are uncharged in their acidic form and negatively charged in their basic form.

$$\overset{+}{\text{H}_3}\text{NCH}_2\text{CH}_2\text{CH}_2\text{CH}_2\text{CHCOH} \quad pK_a = 2.18$$
$$pK_a = 10.79 \qquad \overset{|}{{}^+\text{NH}_3} \quad pK_a = 8.95$$
lysine

$$\text{pI} = \frac{8.95 + 10.79}{2} = \frac{19.74}{2} = 9.87$$

$$\text{HOCCH}_2\text{CH}_2\text{CHCOH} \quad pK_a = 2.19$$
$$pK_a = 4.25 \qquad \overset{|}{{}^+\text{NH}_3} \quad pK_a = 9.67$$
glutamic acid

$$\text{pI} = \frac{2.19 + 4.25}{2} = \frac{6.44}{2} = 3.22$$

PROBLEM 8

Explain why the pI of lysine is the average of the pK_a values of its two protonated amino groups.

PROBLEM 9◆

Calculate the pI of each of the following amino acids:

    a. asparagine   **b.** arginine   **c.** serine

PROBLEM 10◆

a. Which amino acid has the lowest pI value?

b. Which amino acid has the highest pI value?

c. Which amino acid has the greatest amount of negative charge at pH = 6.20?

d. Which amino acid—glycine or methionine—has a greater negative charge at pH = 6.20?

PROBLEM 11

Explain why the pI values of tyrosine and cysteine cannot be determined by the method just described.

Electrophoresis

A mixture of amino acids can be separated by several different techniques. **Electrophoresis** separates amino acids on the basis of their pI values. A few drops of a solution of an amino acid mixture are applied to the middle of a piece of filter paper or to a gel. When the paper (or the gel) is placed in a buffered solution between two electrodes and an electric field is applied, an amino acid with a pI greater than the pH of the solution will have an overall positive charge and will migrate toward the cathode (the negative electrode). The farther its pI is from the pH of the buffer, the more positive it will be and the farther it will migrate toward the cathode in a given amount of time. An amino acid with a pI less than the pH of the buffer will have an overall negative charge and will migrate toward the anode (the positive electrode). If two molecules have the same charge, the larger one will move more slowly during electrophoresis because the same charge has to move a greater mass.

21.5 SEPARATION OF AMINO ACIDS

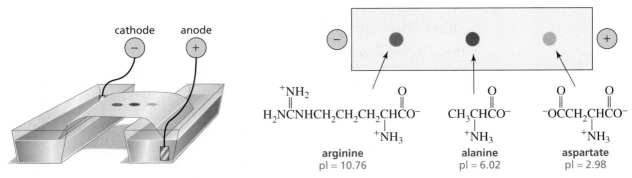

▲ **Figure 21.1**
Arginine, alanine, and aspartic acid separated by electrophoresis at pH = 5.

Amino acids are colorless, so how can we detect the separation of the amino acids in the mixture? When amino acids are heated with ninhydrin, they form a colored product. After electrophoretic separation of the amino acids, the filter paper is sprayed with ninhydrin and dried in a warm oven. Most amino acids form a purple product. The number of different kinds of amino acids in the mixture is determined by the number of colored spots on the filter paper (Figure 21.1[b]). The individual amino acids are identified by their location on the paper compared with a standard.

Tutorial:
Electrophoresis and pI

The mechanism for formation of the colored product is as shown, omitting the mechanisms for the steps involving dehydration, imine formation, and imine hydrolysis. (These mechanisms are shown in Sections 17.7 and 17.8.)

mechanism for the reaction of an amino acid with ninhydrin to form a colored product

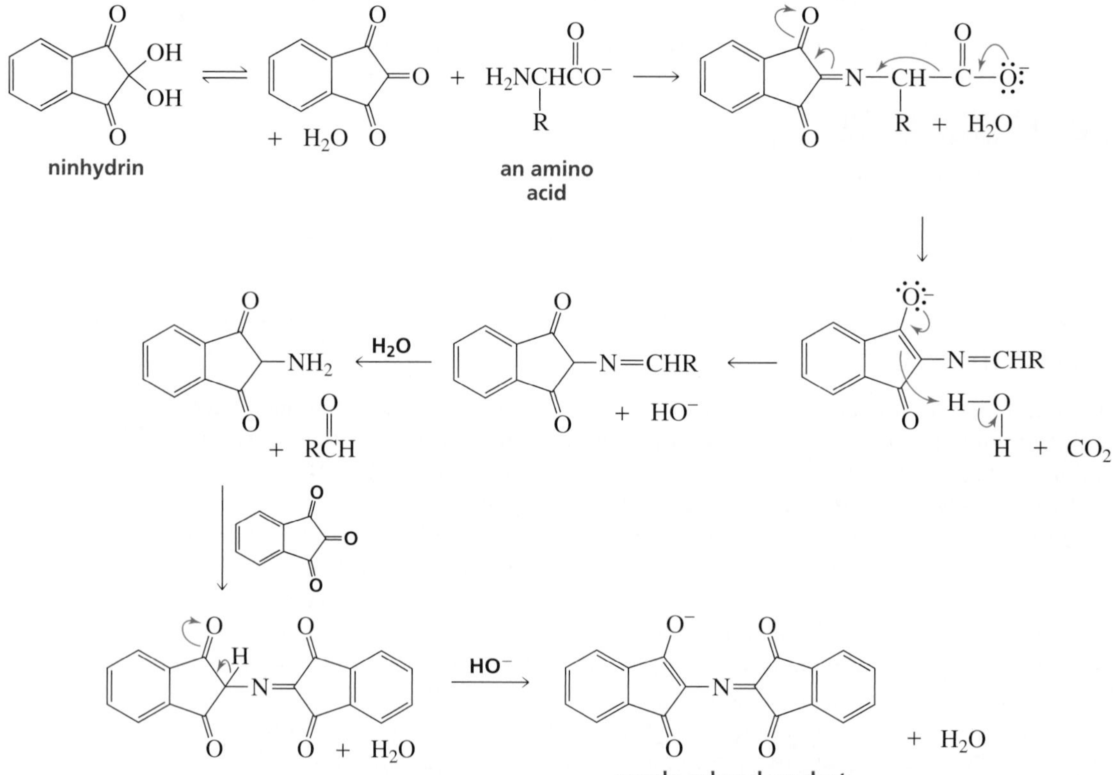

Paper Chromatography and Thin-Layer Chromatography

Paper chromatography once played an important role in biochemical analysis because it provided a method to separate amino acids using very simple equipment. Although more modern techniques are now more commonly used, we will describe the principles behind paper chromatography because many of the same principles are employed in modern separation techniques.

The technique of paper chromatography separates amino acids on the basis of polarity. A few drops of a solution of an amino acid mixture are applied to the bottom of a strip of filter paper. The edge of the paper is placed in a solvent (typically a mixture of water, acetic acid, and butanol). The solvent moves up the paper by capillary action, carrying the amino acids with it. Depending on their polarities, the amino acids have different affinities for the mobile (solvent) and stationary (paper) phases, and therefore travel up the paper at different rates. The more polar the amino acid, the more strongly it is adsorbed onto the relatively polar paper. The less polar amino acids travel up the paper more rapidly since they have a greater affinity for the mobile phase. Therefore, when the paper is developed with ninhydrin, the colored spot closest to the origin is the most polar amino acid and the spot farthest away from the origin is the least polar amino acid (Figure 21.2).

The most polar amino acids are those with charged side chains; the next most polar are those with side chains that can form hydrogen bonds; and the least polar are those with hydrocarbon side chains. For amino acids with hydrocarbon side

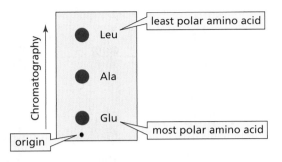

chains, the larger the alkyl group the less polar the amino acid. In other words, leucine is less polar than valine.

Paper chromatography has largely been replaced by **thin-layer chromatography** (TLC). TLC is similar to paper chromatography, but thin-layer chromatography uses a plate with a coating of solid material instead of filter paper. The physical property on which the separation is based depends on the solid material and the solvent chosen for the mobile phase.

PROBLEM 12◆

A mixture of seven amino acids (glycine, glutamate, leucine, lysine, alanine, isoleucine, and aspartate) is separated by thin-layer chromatography. Explain why only six spots show up when the chromatographic plate is sprayed with ninhydrin and heated.

Ion-Exchange Chromatography

Electrophoresis and thin-layer chromatography are analytical separations—small amounts of amino acids are separated for analysis. Preparative separation—in which larger amounts of amino acids are separated for use in subsequent processes—can be achieved using **ion-exchange chromatography.** This technique uses a column packed with an insoluble resin. A solution of a mixture of amino acids is loaded onto the top of the column and eluted with a buffer. The amino acids separate because they flow through the column at different rates

The resin is a chemically inert material with charged side chains. One commonly used resin is a copolymer of styrene and divinylbenzene with negatively charged sulfonic acid groups on some of the benzene rings (Figure 21.3). If a mixture of lysine and glutamate in a solution with a pH of 6 were loaded onto the column, glutamate would travel down the column rapidly because its negatively charged side chain would be repelled by the negatively charged sulfonic acid groups of the resin. The positively charged side chain of lysine would cause it to be retained on the column.

◀ **Figure 21.3**
A section of a cation-exchange resin. This particular resin is called Dowex 50.

$SO_3^-Na^+$ $SO_3^-Na^+$ $SO_3^-Na^+$

$-CH_2-CH-CH_2-CH-CH-CH_2-CH-CH_2-CH-CH_2-CH-CH_2-CH-$

$-CH_2-CH-CH_2-CH-CH_2-CH-CH_2-CH-CH_2-CH-CH_2-CH-$

$SO_3^-Na^+$ $SO_3^-Na^+$

Figure 21.4 ▶
Separation of amino acids by
ion-exchange chromatography.

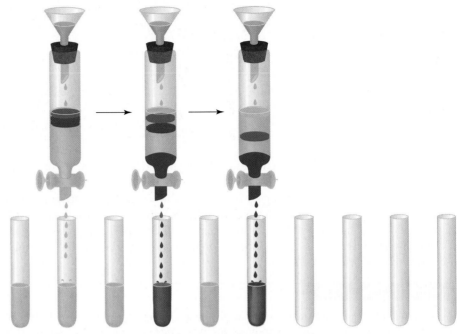

Movie:
Column chromatography

Fractions sequentially collected

**Cations bind most strongly
to cation-exchange resins.**

**Anions bind most strongly
to anion-exchange resins.**

This kind of resin is called a **cation-exchange resin** because it exchanges the Na^+ counterions of the SO_3^- groups for the positively charged species that are added to the column. In addition, the relatively nonpolar nature of the column causes it to retain nonpolar amino acids longer than polar amino acids. Resins with positively charged groups are called **anion-exchange resins** because they impede the flow of anions. A common anion-exchange resin (Dowex 1) has $CH_2N^+(CH_3)_3$ Cl^- groups in place of the SO_3^- Na^+ groups in Figure 21.3.

An **amino acid analyzer** is an instrument that automates ion-exchange chromatography. When a solution of an amino acid mixture passes through the column of an amino acid analyzer containing a cation-exchange resin, the amino acids move through the column at different rates depending on their overall charge. The solution leaving the column is collected in fractions. Fractions are collected often enough that a different amino acid ends up in each fraction (Figure 21.4). If ninhydrin is added to each of the fractions, the concentration of the amino acid in each fraction can be determined by the amount of absorption at 570 nm—because the colored compound formed by the reaction of an amino acid with ninhydrin has a λ_{max} of 570 (Section 12.17). In this way, the identity and the relative amount of each amino acid can be determined. (Figure 21.5).

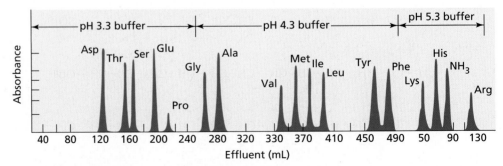

▲ **Figure 21.5**
A typical chromatogram obtained from separation of a mixture of amino acids by using an automated amino acid analyzer.

WATER SOFTENERS: EXAMPLES OF CATION-EXCHANGE CHROMATOGRAPHY

Water softeners contain a cation-exchange resin that has been flushed with concentrated sodium chloride. In Section 16.17 we saw that the presence of calcium and magnesium ions in water is what causes it to be "hard." When water passes through the column, the resin binds magnesium and calcium ions more tightly than it binds sodium ions. In this way, the water softener removes magnesium and calcium ions from water, replacing them with sodium ions. The resin must be recharged from time to time. This is done by flushing it with concentrated sodium chloride to replace the bound magnesium and calcium ions with sodium ions.

PROBLEM 13

Why are buffer solutions of increasingly higher pH used to elute the column that generates the chromatogram shown in Figure 21.5?

PROBLEM 14

Explain the order of elution (with a buffer of pH 4) of each of the following pairs of amino acids on a column packed with Dowex 50 (Figure 21.3):

a. aspartate before serine

c. valine before leucine

b. glycine before alanine

d. tyrosine before phenylalanine

PROBLEM 15◆

In what order would the following amino acids be eluted with a buffer of pH 4 from a column containing an anion-exchange resin?

histidine, serine, aspartate, valine

Chemists do not have to rely on nature to produce amino acids—they can synthesize them in the laboratory using a variety of methods. One of the oldest methods replaces an α-hydrogen of a carboxylic acid with a bromine using the Hell–Volhard–Zelinski reaction (Section 19.5). The resulting α-bromocarboxylic acid then undergoes an S_N2 reaction with ammonia to form the amino acid.

21.6 RESOLUTION OF RACEMIC MIXTURES OF AMINO ACIDS

$$RCH_2\overset{O}{\overset{\|}{C}}OH \xrightarrow[\text{2. H}^+, \text{H}_2\text{O}]{\text{1. Br}_2, \text{P}} R\overset{O}{\underset{\underset{Br}{|}}{\overset{\|}{C}H}}COH \xrightarrow[]{\text{excess} \atop \text{NH}_3} R\overset{O}{\underset{\underset{^+NH_3}{|}}{\overset{\|}{C}H}}CO^- + \overset{+}{N}H_4Br^-$$

a carboxylic acid   an amino acid

PROBLEM 16

Why is excess ammonia used in the preceding reaction?

When amino acids are synthesized in nature, only the L-enantiomer is formed (Section 4.18). However, when amino acids are synthesized in the laboratory, the product is usually a racemic mixture—a mixture of D and L enantiomers. If only one isomer is desired, the enantiomers must be separated. We have seen that a method commonly used to separate enantiomers is to convert them into diastereomers, separate the diastereomers, and convert the separated diastereomers back to enantiomers (Section 4.13). This is a time-consuming technique, and the yields are often low. Fortunately, there is a much easier way to separate a racemic mixture of amino acids.

Because enzymes are chiral, they react at a different rate with each of the enantiomers (Section 4.18). For example, pig kidney aminoacylase is an enzyme that catalyzes the hydrolysis of N-acetyl-L-amino acids but does not catalyze the hydrolysis of N-acetyl-D-amino acids. Therefore, if the racemic amino acid is converted into a pair of N-acetylamino acids and the N-acetylated mixture is hydrolyzed with pig kidney aminoacylase, the products will be the L-amino acid and N-acetyl-D-amino acid, which are easily separated. Because the resolution (separation) of the enantiomers depends on the difference in the rates of reaction of the enzyme with the two N-acetylated compounds, this technique is known as a **kinetic resolution.**

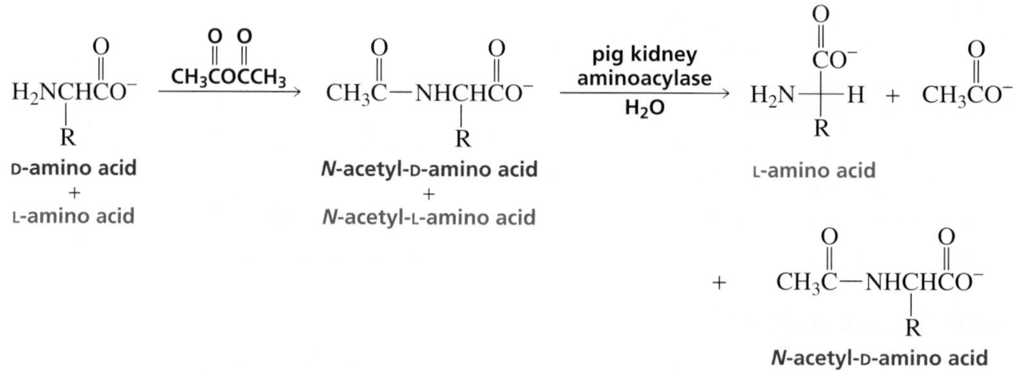

PROBLEM 17

Pig liver esterase is an enzyme that catalyzes the hydrolysis of esters. It hydrolyzes esters of L-amino acids more rapidly than esters of D-amino acids. How can this enzyme be used to separate a racemic mixture of amino acids?

PROBLEM 18 ◆

Amino acids can be synthesized by reductive amination of α-keto acids (Section 17.7).

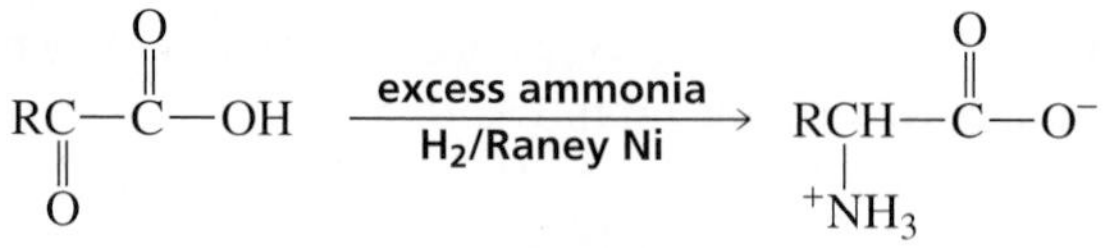

Biological organisms can also convert α-keto acids into amino acids but, because H_2 and metal catalysts are not available to the cell, they do so by a different mechanism (Section 23.6.)

a. What amino acid is obtained from the reductive amination of each of the following metabolic intermediates in the cell?

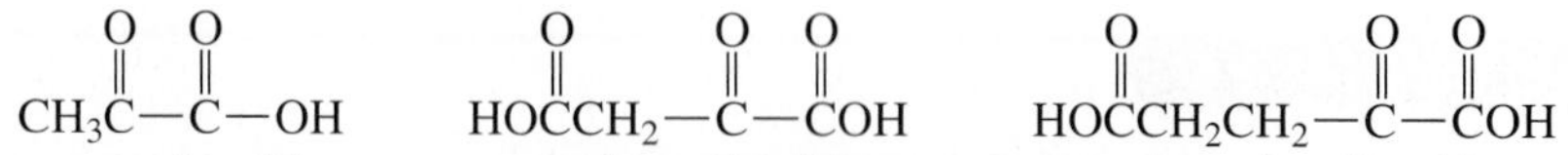

b. What amino acids are obtained from the same metabolic intermediates when they are synthesized in the laboratory?

Peptide Bonds

Peptide bonds and disulfide bonds are the only covalent bonds that hold amino acid residues together in a peptide or a protein. The amide bonds that link amino acid residues are called **peptide bonds.** By convention, peptides and proteins are written with the free amino group (the **N-terminal amino acid**) on the left and the free carboxyl group (the **C-terminal amino acid**) on the right.

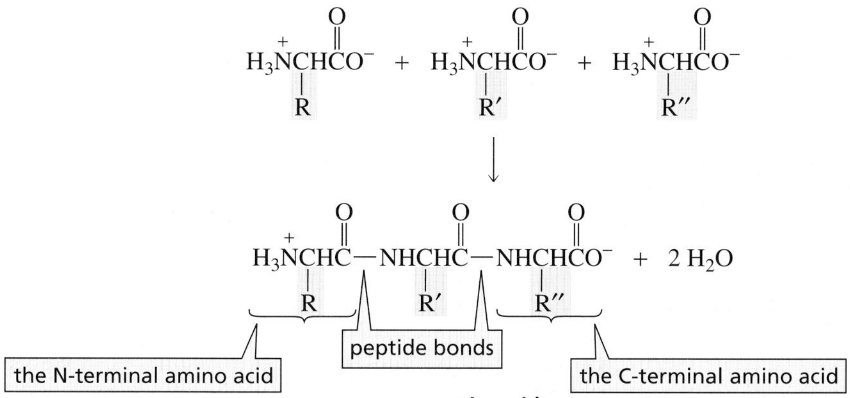

a tripeptide

When the identities of the amino acids in a peptide are known but their sequence is not known, the amino acids are written separated by commas. When the sequence of amino acids is known, the amino acids are written separated by hyphens. In the following pentapeptide, valine is the N-terminal amino acid and histidine is the C-terminal amino acid. The amino acids are numbered starting with the N-terminal end. The glutamate residue is referred to as Glu 4 because it is the fourth amino acid from the N-terminal end. In naming the peptide, adjective names (ending in "yl") are used for all the amino acids except the C-terminal amino acid. Thus, the pentapeptide shown here is named valylcysteylalanylglutamylhistidine.

Glu, Cys, His, Val, Ala	Val-Cys-Ala-Glu-His
the pentapeptide contains the indicated amino acids, but their sequence is not known	the amino acids in the pentapeptide have the indicated sequence

A peptide bond has about 40% double-bond character because of electron delocalization. Steric hindrance causes the trans configuration to be more stable than the cis configuration, so the α-carbons of adjacent amino acids are trans to each other (Section 3.19).

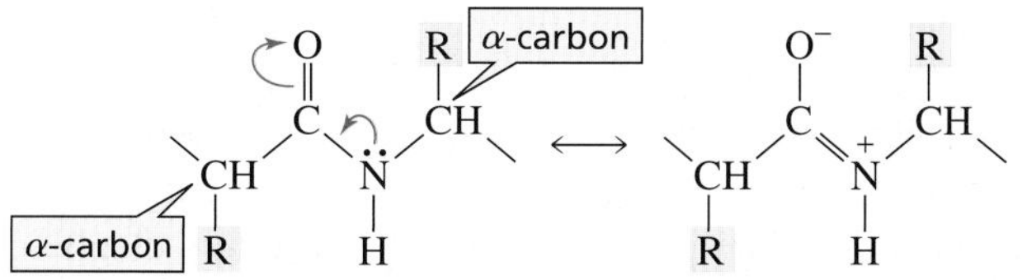

trans configuration

Free rotation is not possible about the peptide bond because of its partial double-bond character. The carbon and nitrogen atoms of the peptide bond and the two atoms to which each is attached are held rigidly in a plane (Figure 21.6). This

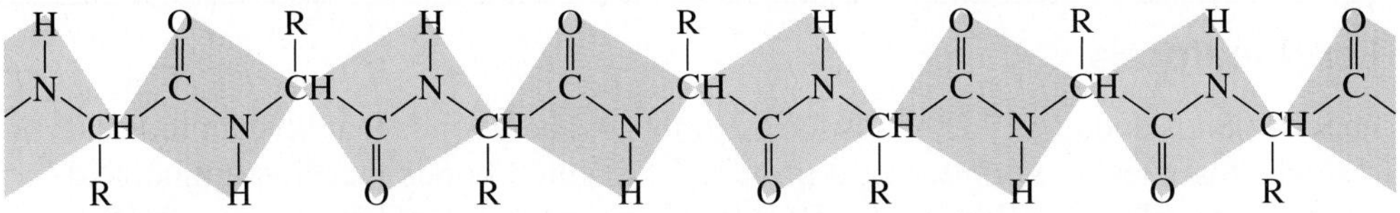

▲ **Figure 21.6**
A segment of a polypeptide chain. The plane defined by each peptide bond is indicated. Notice that the R groups bonded to the α-carbons are on alternate sides of the peptide backbone.

regional planarity affects the way a chain of amino acids can fold, so it has important implications for the three-dimensional shapes of peptides and proteins (Section 21.13).

PROBLEM 19

Draw a peptide bond in a cis configuration.

Disulfide Bonds

When thiols are oxidized under mild conditions, they form disulfides. A **disulfide** is a compound with an S—S bond.

$$2\,R\text{—SH} \xrightarrow{\text{mild oxidation}} RS\text{—}SR$$
$$\text{a thiol} \qquad\qquad\qquad \text{a disulfide}$$

An oxidizing agent commonly used for this reaction is Br_2 (or I_2) in a basic solution.

mechanism for oxidation of a thiol to a disulfide

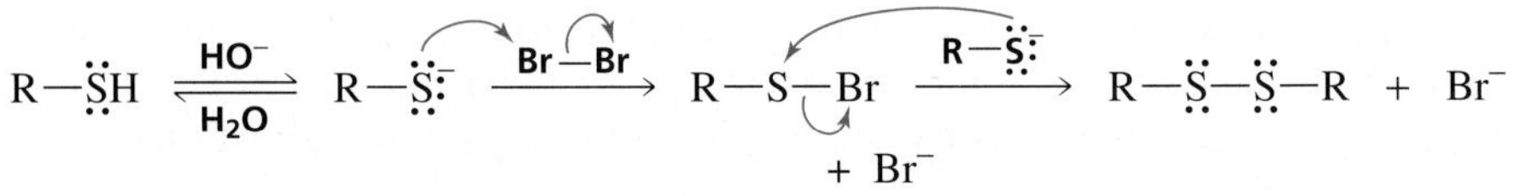

Because thiols can be oxidized to disulfides, disulfides can be reduced to thiols.

$$RS\text{—}SR \xrightarrow{\text{reduction}} 2\,R\text{—SH}$$
$$\text{a disulfide} \qquad\qquad \text{a thiol}$$

Cysteine is an amino acid that contains a thiol group. Two cysteine molecules therefore can be oxidized to a disulfide. This disulfide is called cystine.

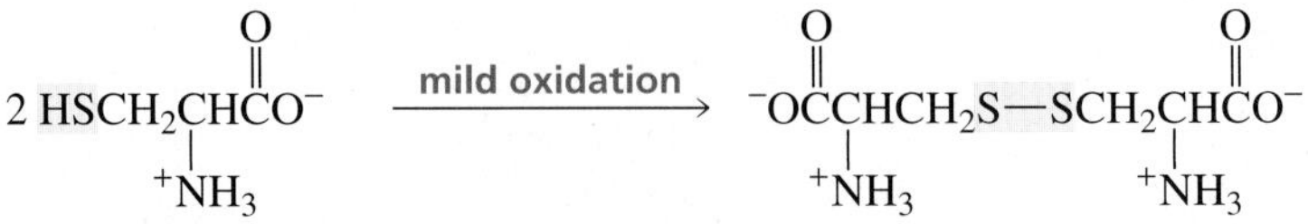

Two cysteine residues in a protein can be oxidized to a disulfide. This is known as a **disulfide bridge.** Disulfide bridges are the only covalent bonds that can form between nonadjacent amino acids. They contribute to the overall shape of a protein by holding the cysteine residues in close proximity, as shown in Figure 21.7.

Insulin is a hormone secreted by the pancreas that controls the level of glucose in the blood by regulating glucose metabolism. Insulin is a polypeptide with two

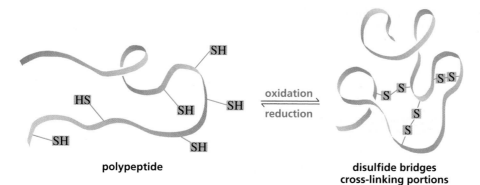

◀ **Figure 21.7**
Disulfide bridges cross-linking portions of a peptide.

polypeptide

$$\underset{\text{reduction}}{\overset{\text{oxidation}}{\rightleftharpoons}}$$

disulfide bridges
cross-linking portions
of a polypeptide

peptide chains. The short chain (the A-chain) contains 21 amino acids and the long chain (the B-chain) contains 30 amino acids. The two chains are held together by two disulfide bridges. These are **interchain disulfide bridges** (between the A- and B-chains). Insulin also has an **intrachain disulfide bridge** (within the A-chain).

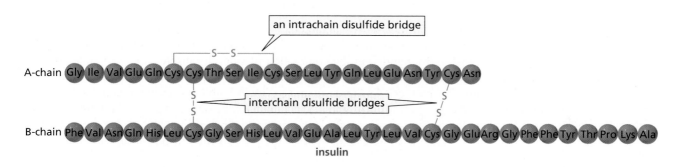

insulin

HAIR: STRAIGHT OR CURLY?

Hair is made up of a protein known as keratin. Keratin contains an unusually large number of cysteine residues (about 8%), which give it many disulfide bridges to maintain its three-dimensional structure. People can alter the structure of their hair (if they feel it is either too straight or too curly) by changing the location of these disulfide bridges. This is accomplished by first applying a reducing agent to the hair to reduce all the disulfide bridges in the protein strands. Then the hair is given the desired shape (using curlers to curl it or combing it straight to uncurl it), and an oxidizing agent is applied that forms new disulfide bridges. The new disulfide bridges maintain the hair's new shape. When this treatment is applied to straight hair, it is called a "permanent." It is called "hair straightening" when it is applied to curly hair.

curly hair

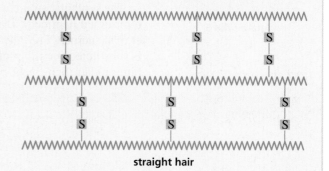

straight hair

PROBLEM 20◆

a. How many different octapeptides can be made from the 20 naturally occurring amino acids?

b. How many different proteins containing 100 amino acids can be made from the 20 naturally occurring amino acids?

PROBLEM 21◆

Which bonds in the backbone of a peptide can rotate freely?

21.8
SOME INTERESTING PEPTIDES

*Oxytocin was the first small peptide to be synthesized. This was achieved in 1953 by **Vincent du Vigneaud (1901–1978)**, who later synthesized vasopressin. Du Vigneaud was born in Chicago and was a professor at George Washington University Medical School and later at Cornell University Medical College. For synthesizing these nonapeptides, he received the Nobel Prize in chemistry in 1955.*

Enkephalins are pentapeptides synthesized by the body to control pain. They decrease the sensitivity to pain by binding to receptors in certain brain cells. Part of their three-dimensional structures must be similar to those of morphine and painkillers such as Darvon and Demerol because they bind to the same receptors.

Tyr-Gly-Gly-Phe-Leu   Tyr-Gly-Gly-Phe-Met
leucine enkephalin   **methionine enkephalin**

Bradykinin, vasopressin, and oxytocin are peptide hormones. They are all nonapeptides. Bradykinin inhibits the inflammation of tissues. Vasopressin controls blood pressure by regulating the contraction of smooth muscle. It is also an antidiuretic. Oxytocin induces labor in pregnant women and stimulates milk production in nursing mothers. Vasopressin and oxytocin both have an intrachain disulfide bond, and their C-terminal amino acids contain amide rather than carboxyl groups. Notice that the C-terminal amide group is indicated by writing "NH_2" after the name of the C-terminal amino acid. In spite of their very different physiological effects, vasopressin and oxytocin differ only by two amino acids.

bradykinin   Arg-Pro-Pro-Gly-Phe-Ser-Pro-Phe-Arg

vasopressin   Cys-Tyr-Phe-Gln-Asn-Cys-Pro-Arg-Gly-NH_2
                     S————————S

oxytocin   Cys-Tyr-Ile-Gln-Asn-Cys-Pro-Leu-Gly-NH_2
                     S————————S

Gramicidin S is an antibiotic produced by a strain of bacteria. It is a cyclic decapeptide. Notice that it contains the amino acids L-ornithine (L-Orn), D-ornithine (D-Orn), and also D-phenylalanine. Ornithine is not listed in Table 21.1 because it occurs rarely in nature. Ornithine resembles lysine but has one fewer methylene group in its side chain.

The synthetic sweetener aspartame, or NutraSweet (Section 20.21), is the methyl ester of a dipeptide of L-aspartate and L-phenylalanine. It is about 200 times sweeter than sucrose. The ethyl ester of the same dipeptide is not sweet. If a D-amino acid is substituted for either of the L-amino acids of aspartame, the resulting dipeptide is bitter rather than sweet.

```
        ┌─ L-Val ─┐
   L-Pro            L-Orn
   /                    \
 L-Phe                  L-Leu
   |                      |
 L-Leu                  D-Phe
   \                    /
   D-Orn            L-Pro
        └─ L-Val ─┘
```
gramicidin S

$$\overset{+}{H_3}NCH_2CH_2CH_2\overset{O}{\overset{\|}{CHCO^-}}$$
$$\underset{\overset{|}{\overset{+}{N}H_3}}{}$$

ornithine

$$\underset{\overset{|}{CH_2}}{\underset{\overset{|}{COO^-}}{\overset{+}{H_3}NCHC}}\overset{O}{\overset{\|}{}}-NHCHCOCH_3 \overset{O}{\overset{\|}{}}$$

**aspartame
NutraSweet**

Glutathione is a tripeptide of glutamate, cysteine, and glycine. Its function is to destroy harmful oxidizing agents in the body. Oxidizing agents are thought to be responsible for some of the effects of aging and are believed to play a role in cancer (Section 8.8). Glutathione removes oxidizing agents by reducing them. Consequently, glutathione is oxidized, forming a disulfide bond between two glutathione molecules. An enzyme subsequently reduces the disulfide bond, allowing glutathione to react with more oxidizing agents.

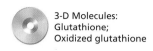

3-D Molecules:
Glutathione;
Oxidized glutathione

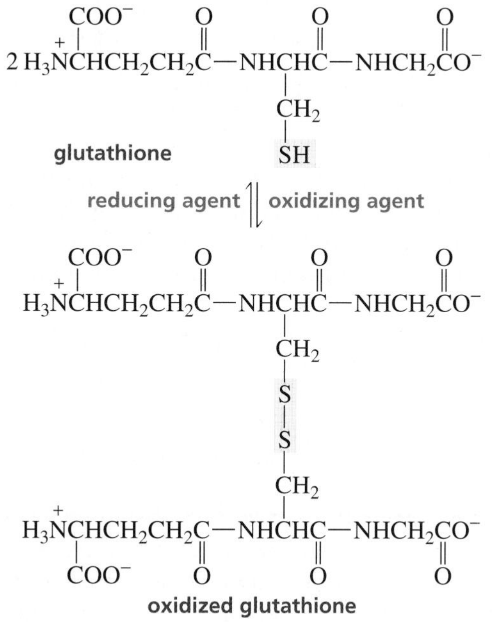

$$2 \; \overset{+}{H_3N}CHCH_2CH_2\overset{O}{\overset{\|}{C}}-NHCHC-NHCH_2CO^-$$

glutathione

reducing agent ‖ oxidizing agent

oxidized glutathione

PROBLEM 22

What is unusual about glutathione's structure? (If you can't answer this question, draw the structure you would expect for a tripeptide of glutamate, cysteine, and glycine and compare your structure with the structure of glutathione.)

Because amino acids have two functional groups, a problem arises when one attempts to make a particular peptide bond. Say, for example, you wanted to make the dipeptide Gly-Ala. Heating a mixture of glycine and alanine results in four dipeptides—the one you want and three others (Section 16.12).

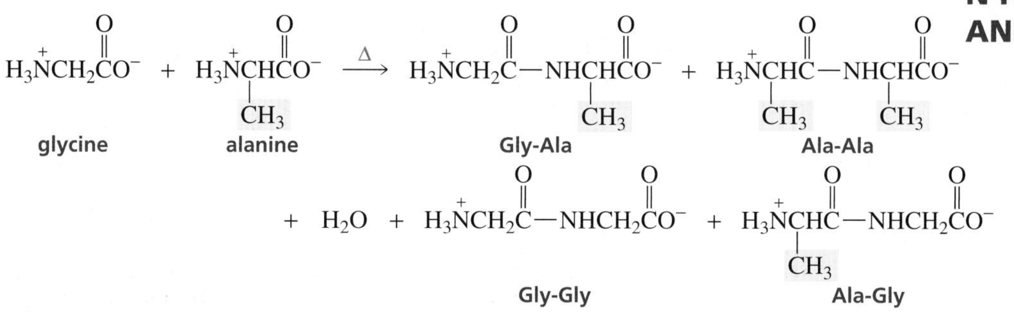

**21.9
STRATEGY OF
PEPTIDE BOND
SYNTHESIS:
N-PROTECTION
AND C-ACTIVATION**

glycine   alanine   Gly-Ala   Ala-Ala

$+ \; H_2O \; + \; $ Gly-Gly   Ala-Gly

If the amino group of the amino acid that is to be on the N-terminal end (in this case Gly) is protected, it will not be available to form a peptide bond. If the carboxyl group of this same amino acid is activated before the second amino acid is added, the amino group of the added amino acid (in this case Ala) will react with the activated carboxyl group of glycine in preference to reacting with a nonactivated carboxyl group of another alanine molecule.

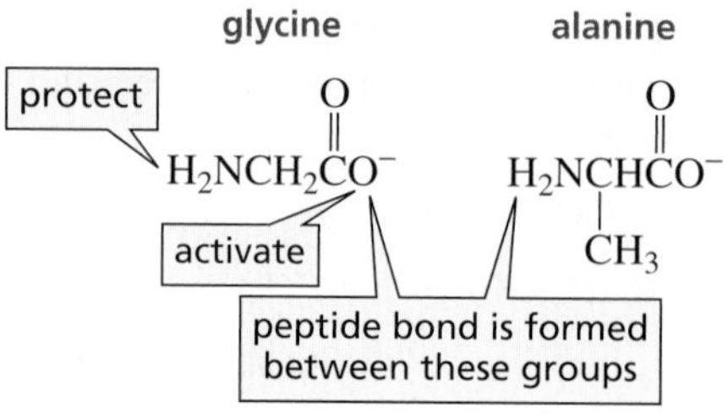

The reagent most often used to protect the amino group of an amino acid is di-*tert*-butyl dicarbonate. Its popularity is due to the ease with which the protecting group can be removed when the need for protection is over (see following). The protecting group is known by the acronym *t*-BOC (pronounced tee-boc).

di-*tert*-butyl dicarbonate **glycine** **N-protected glycine**

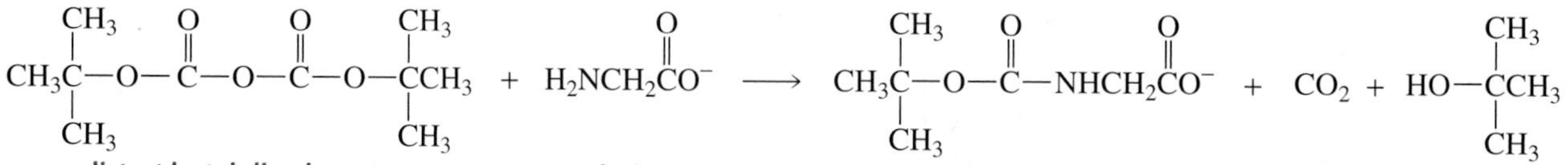

Carboxylic acids are generally activated by being converted into acyl chlorides (Section 16.18). Acyl chlorides, however, are so reactive that they can readily react with the substituents of some of the amino acids during peptide synthesis, creating unwanted products. The preferred method for activating the carboxyl group of an N-protected amino acid is to convert it into an imidate using dicyclohexylcarbodiimide (DCC). (You have probably noticed that biochemists are even more fond of acronyms than organic chemists are.) DCC activates a carboxyl group by putting a good leaving group on the carbonyl carbon.

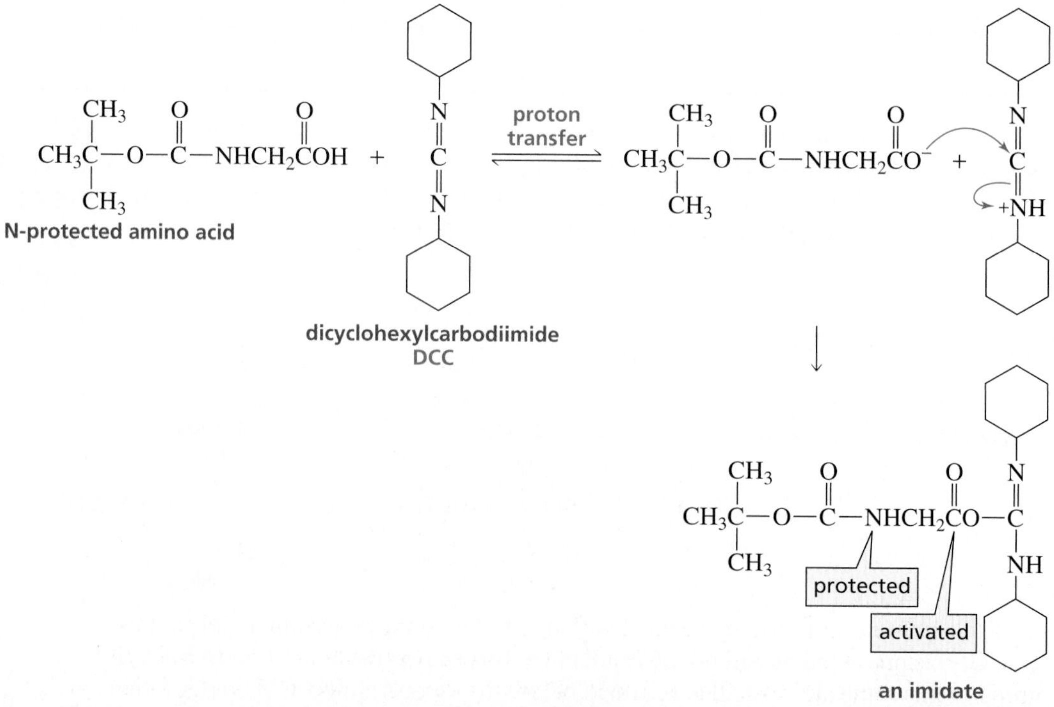

After the amino acid has its N-terminal group protected and its C-terminal group activated, the second amino acid is added to form the new peptide bond. The C—O bond of the tetrahedral intermediate is easily broken (the activated group is a good leaving group) because the bonding electrons are delocalized, forming dicyclohexylurea, a stable diamide.

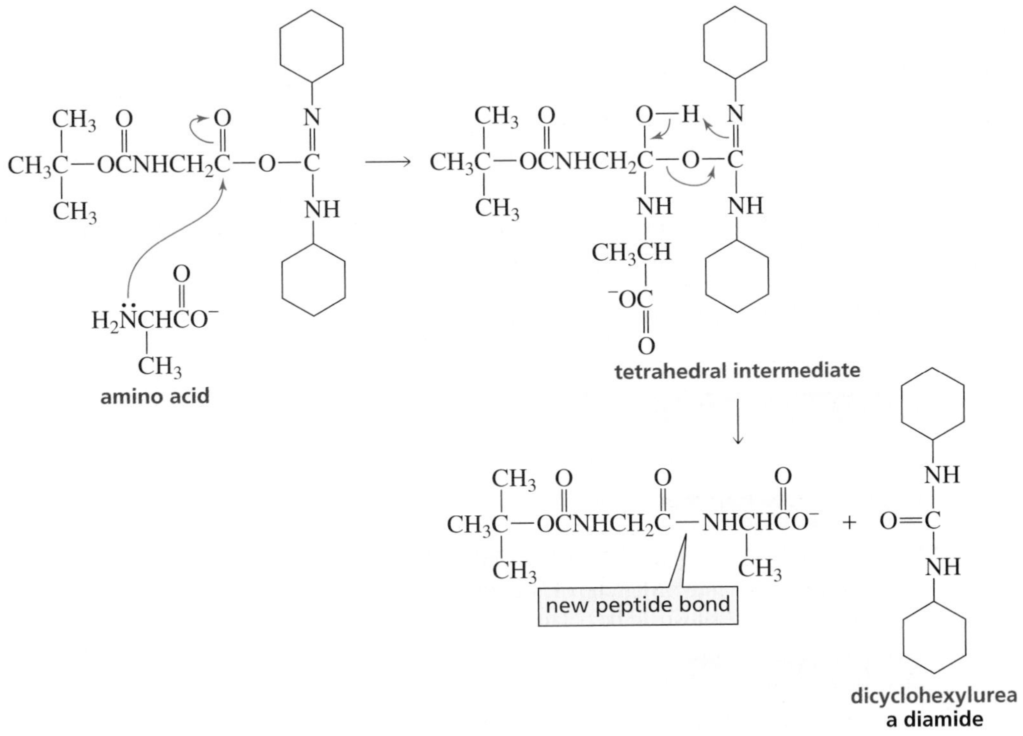

tetrahedral intermediate

amino acid

new peptide bond

**dicyclohexylurea
a diamide**

Amino acids can be added to the growing C-terminal end by repeating these two steps—activating the carboxyl group of the C-terminal amino acid of the peptide by treating it with DCC and then adding a new amino acid.

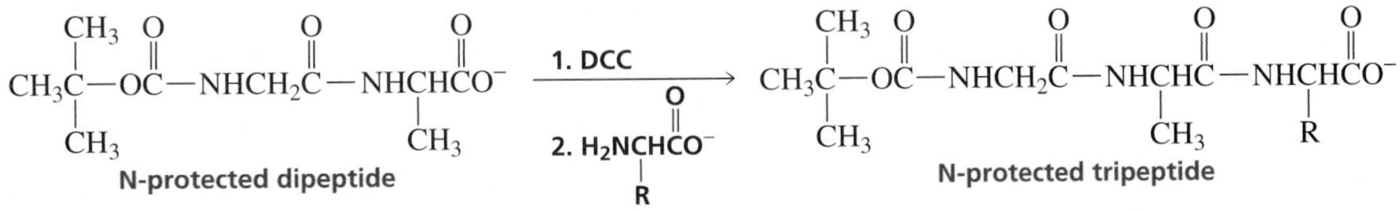

N-protected dipeptide

N-protected tripeptide

When the desired number of amino acids has been added to the chain, the protecting group on the N-terminal amino acid is removed. *t*-BOC is an ideal protecting group because it can be removed by washing with trifluoroacetic acid and methylene chloride, reagents that will not break any other covalent bonds. The protecting group is removed by an elimination reaction, forming isobutylene and carbon dioxide. Because these products are gases, they escape, driving the reaction to completion.

Theoretically, one should be able to make as long a peptide as desired using this technique. Reactions do not produce 100% yield, however, and the yield is further decreased during the purification process. After each step of the synthesis, the peptide must be purified to prevent subsequent unwanted reactions with leftover reagents. Assuming that each amino acid can be added to the growing end of the peptide chain in an 80% yield (a relatively high yield, as you can probably appreciate from your own experience in the laboratory), the overall yield of a nonapeptide such as bradykinin would be only 17%. It is clear that large polypeptides could never be synthesized in this way.

Number of amino acids	2	3	4	5	6	7	8	9
Overall yield	80%	64%	51%	41%	33%	26%	21%	17%

PROBLEM 23

What dipeptides would be formed by heating a mixture of valine and N-protected leucine?

PROBLEM 24

Assume you are trying to synthesize the dipeptide Val-Ser. Compare the product that would be obtained if the carboxyl group of N-protected valine were activated with thionyl chloride with the product that would be obtained if the carboxyl group were activated with DCC.

PROBLEM 25

Show the steps in the synthesis of the tetrapeptide Leu-Phe-Lys-Val.

PROBLEM 26◆

a. Calculate the overall yield of bradykinin if the yield for the addition of each amino acid to the chain were 70%.

b. What would be the overall yield of a peptide containing 15 amino acid residues if the yield for incorporation of each were 80%?

21.10 AUTOMATED PEPTIDE SYNTHESIS

In addition to the problem of low overall yields, the method of peptide synthesis described in Section 21.9 is extremely time consuming because the product must be purified at each step of the synthesis. In 1969, Bruce Merrifield described a method that revolutionized the synthesis of peptides because it provided a much faster way to produce peptides in much higher yields. Furthermore, because it is automated, the synthesis requires fewer hours of direct attention. Using this technique, bradykinin was synthesized with an 85% yield in 27 hours. Subsequent refinements in the technique now allow a peptide containing 100 amino acids to be synthesized in 4 days in a reasonable yield.

In the Merrifield method, the C-terminal amino acid is covalently attached to a solid support contained in a column. Each N-terminal blocked amino acid is added one at a time, along with other needed reagents, so the protein is synthesized from the C-terminal end to the N-terminal end. Notice that this is opposite to the way proteins are synthesized in nature (from the N-terminal end to the C-terminal end; Section 25.12). Because it uses a solid support and is automated, this method of protein synthesis is called **automated solid-phase peptide synthesis.**

The solid support to which the C-terminal amino acid is attached is a polystyrene resin similar to the one used in ion-exchange chromatography (Section 21.5) except that the benzene rings have chloromethyl substituents instead of sulfonic acid substituents. Before the C-terminal amino acid is attached to the resin, its amino group is protected with *t*-BOC to prevent the amino group from reacting with the resin. The C-terminal amino acid is attached to the resin by means of an S_N2 reaction—its carboxyl group attacks a benzyl carbon of the resin, displacing a chloride ion.

After the C-terminal amino acid is attached to the resin, the *t*-BOC protecting group is removed (Section 21.9). The next amino acid, with its amino group protected with *t*-BOC and its carboxyl group activated with DCC, is added to the column.

A huge advantage of the Merrifield method of peptide synthesis is that the growing peptide can be purified by washing the column with an appropriate solvent after each step of the procedure. The impurities are washed out of the column because they are not attached to the solid support. Because the peptide is covalently attached to the resin, none of it is lost in the purification step, leading to high yields of purified product.

R. Bruce Merrifield was born in 1921 and received a B.S. and a Ph.D. from the University of California, Los Angeles. He is a professor of chemistry at Rockefeller University. Merrifield received the 1984 Nobel Prize in chemistry for developing automated solid-phase peptide synthesis.

Merrifield automated solid-phase synthesis of a tripeptide

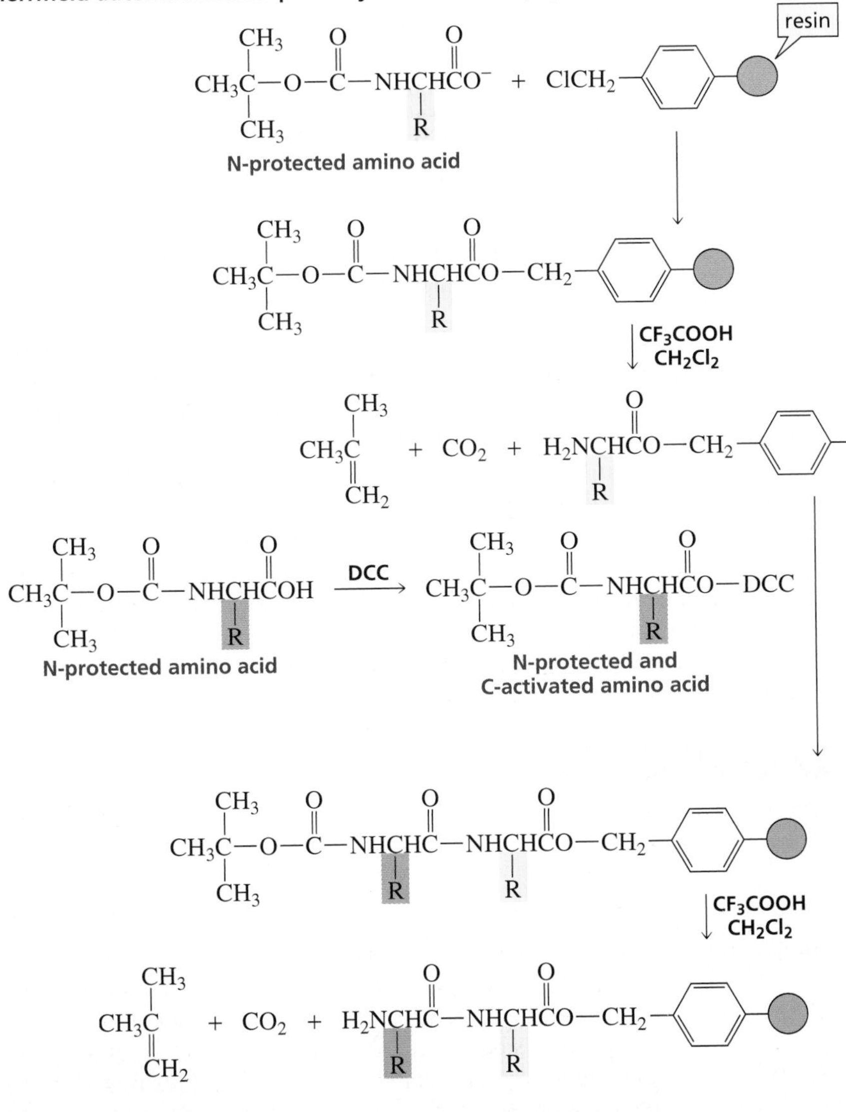

Movie:
Merrifield automated
solid-phase synthesis

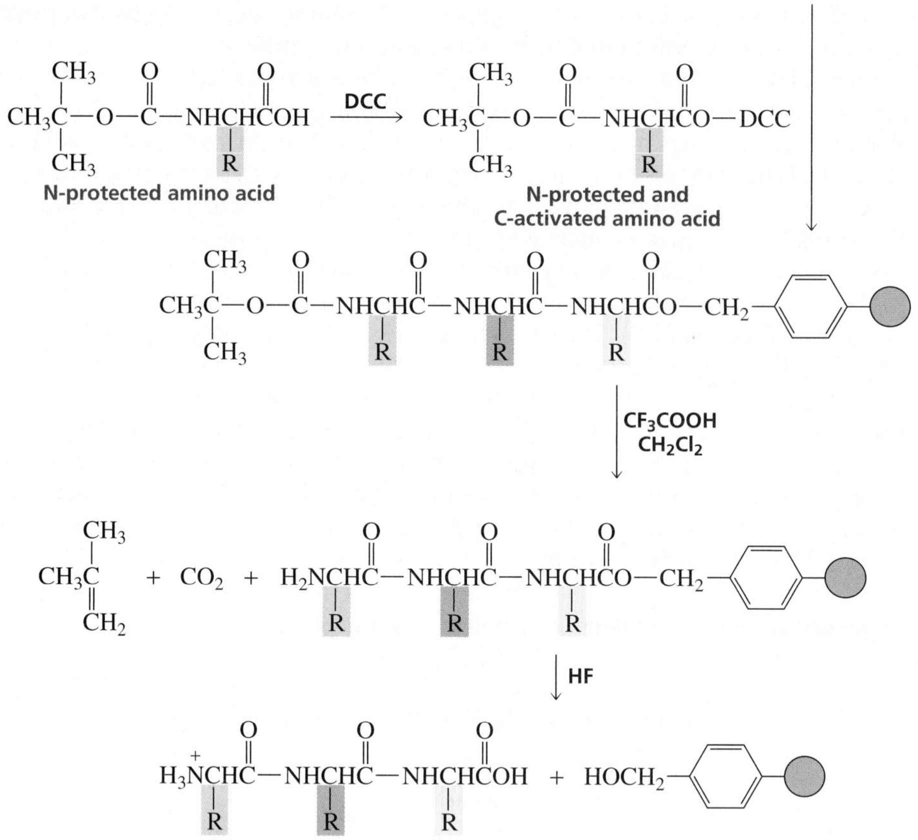

After the required amino acids have been added one by one, the peptide can be removed from the resin by treatment with HF under mild conditions that do not break the peptide bonds.

This technique is constantly being improved so that peptides can be made more rapidly and more efficiently. However, it cannot begin to compare with nature. A bacterial cell is able to synthesize a protein thousands of amino acids long in seconds, and it can simultaneously synthesize thousands of different proteins with no mistakes.

Since the early 1980s, it has been possible to synthesize proteins by genetic engineering techniques. Strands of DNA can be introduced into bacterial cells, which cause the cells to produce large amounts of a desired protein (Section 25.12). For example, mass quantities of human insulin are produced from genetically modified *E. coli*. Genetic engineering techniques also have been useful in synthesizing proteins that differ in one or a few amino acids from the natural protein. Such synthetic proteins have been used, for example, to determine how a single amino acid change affects the properties of a protein (Section 22.9).

PROBLEM 27

Show the steps in the synthesis of the peptide in Problem 25 using Merrifield's method.

21.11 PROTEIN STRUCTURE

Protein molecules are described by several levels of structure. The **primary structure** of a protein is the sequence of amino acids in the chain and the location of all the disulfide bridges. The **secondary structure** describes the regular conformation assumed by segments of the protein's backbone. In other words, the secondary structure describes how local regions of the backbone fold. The **tertiary structure** describes the three-dimensional structure of the entire polypeptide. If a protein has more than one polypeptide chain, it has quaternary structure. The **quaternary structure** of a protein is the way the individual protein chains are arranged with respect to each other.

Proteins can be divided roughly into two classes. **Fibrous proteins** contain long chains of polypeptides that occur in bundles. These proteins are insoluble in water. All the structural proteins described at the beginning of this chapter, such as keratin and collagen, are fibrous proteins. **Globular proteins** are soluble in water and tend to have roughly spherical shapes. Essentially all enzymes are globular proteins.

PRIMARY STRUCTURE AND EVOLUTION

When we examine the primary structures of proteins that carry out the same function in different organisms, we can relate the number of amino acid differences between the proteins to the taxonomic differences between the species. For example, cytochrome *c*, a protein that transfers electrons in biological oxidations, has about 100 amino acid residues. Yeast cytochrome *c* differs by 48 amino acids from horse cytochrome *c*, while duck cytochrome *c* differs by only two amino acids from chicken cytochrome *c*. Chickens and turkeys have cytochrome *c*'s with identical primary structures. Humans and chimpanzees also have identical cytochrome *c*'s, differing by one amino acid from the cytochrome *c* of the rhesus monkey.

The first step in determining the sequence of amino acids in a peptide or a protein is to reduce any disulfide bridges in the protein. A commonly used reducing agent is 2-mercaptoethanol, which is oxidized to a disulfide. Reaction of the protein thiol groups with iodoacetic acid prevents the disulfide bridges from re-forming as a result of oxidation by O_2.

21.12 DETERMINING THE PRIMARY STRUCTURE OF A PROTEIN

cleaving disulfide bridges

PROBLEM 28

Write the mechanism for the reaction of a cysteine residue with iodoacetic acid.

*Insulin was the first protein for which the primary sequence was determined. This was done in 1953 by **Frederick Sanger,** who received the 1958 Nobel Prize in chemistry for this work. Sanger was born in England in 1918 and received a Ph.D. from Cambridge University, where he has worked for his entire career. He also received a share of the 1980 Nobel Prize in chemistry (Section 25.15) for being the first to sequence a DNA molecule (with 5375 nucleotide pairs).*

The next step is to determine the number and kinds of amino acids in the peptide or protein. To do this, a sample of the peptide or protein is dissolved in 6 N HCl and heated at 100 °C for 24 hours. This treatment hydrolyzes all the amide bonds in the protein, including the amide bonds of asparagine and glutamine.

$$\text{protein} \xrightarrow[\substack{100\,°C \\ 24\,h}]{6\,N\,HCl} \text{amino acids}$$

The mixture of amino acids is then passed through an amino acid analyzer to determine the number and kind of each amino acid in the peptide or protein (Section 21.5).

Because all the asparagine and glutamine residues have been hydrolyzed to aspartate and glutamate residues, the number of aspartate or glutamate residues in the amino acid mixture tells us the number of aspartate plus asparagine—or glutamate plus glutamine—residues in the original protein. A separate technique must be used to distinguish between aspartate and asparagine or between glutamate and glutamine in the original protein.

The strongly acidic conditions used for hydrolysis destroy all the tryptophan residues because the indole ring is unstable in acid (Section 27.2). The tryptophan content of the protein can be determined by hydroxide ion-promoted hydrolysis of the protein. This is not a general method for peptide bond hydrolysis because the strongly basic conditions destroy several other amino acid residues.

There are several ways to identify the N-terminal amino acid of a peptide or protein. One of the most widely used methods is to treat the protein with phenyl isothiocyanate (PITC), more commonly known as **Edman's reagent.** Edman's reagent reacts with the N-terminal amino group and the resulting thiazolinone derivative is cleaved from the protein under mildly acidic conditions. The thiazolinone derivative is extracted into an organic solvent and, in the presence of acid, rearranges to a more stable phenylthiohydantoin (PTH).

Because each amino acid has a different substituent (R), each amino acid forms a different PTH. The particular PTH can be identified by chromatography using known standards. Several successive Edman degradations can be carried out on a protein. The entire primary sequence cannot be determined in this way, however, because side products accumulate that interfere with the results. An automated instru-

phenyl isothiocyanate
PITC
Edman's reagent

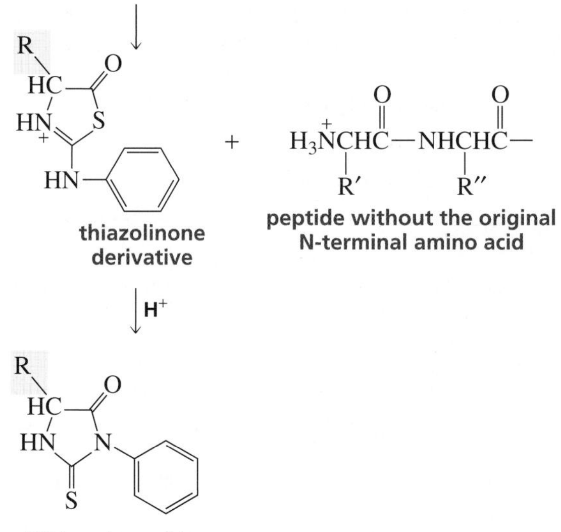

thiazolinone
derivative

peptide without the original
N-terminal amino acid

PTH–amino acid

ment known as a sequenator allows about 50 successive Edman degradations to be carried out on a protein.

The C-terminal amino acid of the peptide or protein can be identified by treating the protein with carboxypeptidase A. Carboxypeptidase A cleaves off the C-terminal amino acid as long as it is *not* arginine or lysine (Section 22.8). Carboxypeptidase B, on the other hand, cleaves it off *only* if it is arginine or lysine. Carboxypeptidases are exopeptidases. An **exopeptidase** is an enzyme that catalyzes the hydrolysis of a peptide bond at the end of a peptide chain.

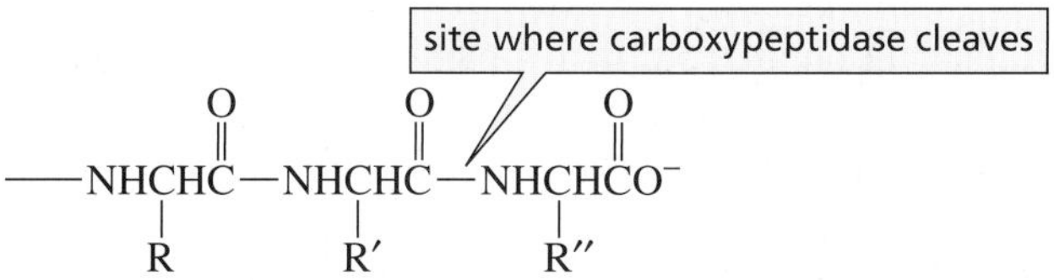

site where carboxypeptidase cleaves

Once the N-terminal and C-terminal amino acids have been identified, a sample of the protein is hydrolyzed with dilute acid. This treatment, called **partial hydrolysis,** hydrolyzes only some of the peptide bonds. The resulting fragments are separated, and the amino acid composition of each is determined. The N-terminal and C-terminal amino acids of each fragment can also be determined. The sequence of the original protein can then be determined by lining up the peptides, looking for points of overlap.

PROBLEM-SOLVING STRATEGY

A nonapeptide undergoes partial hydrolysis to give peptides whose amino acid compositions are shown. Reaction of the intact nonapeptide with Edman's reagent releases PTH-Leu. What is the sequence of the nonapeptide?

a. Pro, Ser   **c.** Met, Ala, Leu   **e.** Glu, Ser, Val, Pro   **g.** Met, Leu

b. Gly, Glu   **d.** Gly, Ala   **f.** Glu, Pro, Gly   **h.** His, Val

Let's start with the N-terminal amino acid. We know it is Leu. Now we need to look for a fragment that contains Leu. Fragment **g** tells us that Met is next to Leu and fragment **c** tells us that Ala is next to Met. Now look for a fragment that contains Ala. Fragment **d** contains Ala and tells us that Gly is next to Ala. From fragment **b** we know that Glu comes next. Glu is in both fragments **e** and **f**. Fragment **e** has two amino acids we have yet to place in the growing peptide, but fragment **f** has only one, so from fragment **f** we know that Pro is the next amino acid. Now we can use fragment **e**. Fragment **e** tells us that the next amino acid is Val, and fragment **h** tells us that His is the last (C-terminal) amino acid.

Leu-Met-Ala-Gly-Glu-Pro-Ser-Val-His

Now continue on to Problem 29.

PROBLEM 29◆

A decapeptide undergoes partial hydrolysis to give peptides whose amino acid compositions are shown. Reaction of the intact decapeptide with Edman's reagent releases PTH-Gly. What is the sequence of the decapeptide?

a. Ala, Trp **c.** Pro, Val **e.** Trp, Ala, Arg **g.** Glu, Ala, Leu

b. Val, Pro, Asp **d.** Ala, Glu **f.** Arg, Gly **h.** Met, Pro, Leu, Glu

---

The peptide or protein can also be partially hydrolyzed using endopeptidases. An **endopeptidase** is an enzyme that catalyzes the hydrolysis of a peptide bond that is not at the end of a peptide chain. Trypsin, chymotrypsin, and elastase are endopeptidases that catalyze the hydrolysis of only the specific peptide bonds listed in Table 21.3. Trypsin, for example, catalyzes the hydrolysis of the peptide bond on the C-side of only arginine or lysine residues.

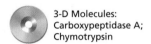

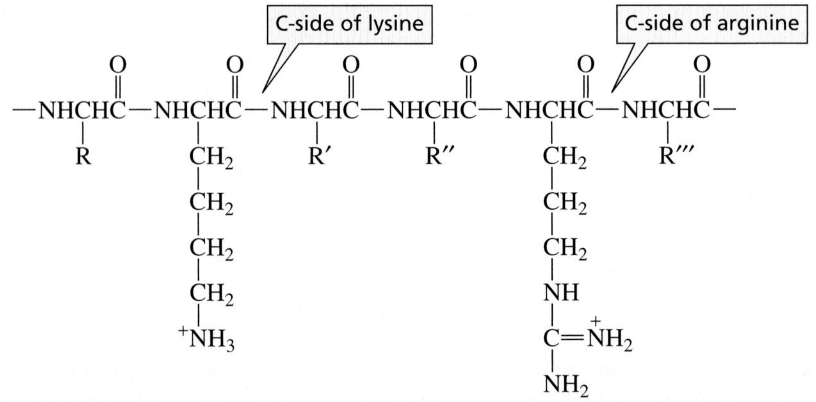

3-D Molecules: Carboxypeptidase A; Chymotrypsin

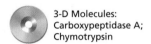

Table 21.3 Specificity of Peptide/Protein Cleavage	
Reagent	**Specificity**
Chemical reagents	
Edman's reagent	removes the N-terminal amino acid
Cyanogen bromide	hydrolyzes on the C-side of Met
Exopeptidases*	
Carboxypeptidase A	removes the C-terminal amino acid (not Arg or Lys)
Carboxypeptidase B	removes the C-terminal amino acid (only Arg or Lys)
Endopeptidases*	
Trypsin	hydrolyzes on the C-side of Arg and Lys
Chymotrypsin	hydrolyzes on the C-side of amino acids that contain aromatic six-membered rings (Phe, Tyr, Trp)
Elastase	hydrolyzes on the C-side of small amino acids (Gly and Ala)

*Cleavage will not occur if Pro is on either side of the bond to be hydrolyzed.

Thus, trypsin will catalyze the hydrolysis of three peptide bonds in the following peptide, creating a hexapeptide, a dipeptide, and two tripeptides.

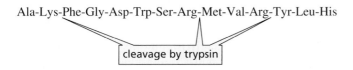

Chymotrypsin catalyzes the hydrolysis of the peptide bond on the C-side of amino acids that contain aromatic six-membered rings (Phe, Tyr, Trp).

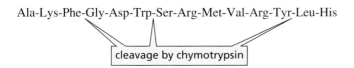

Elastase catalyzes the hydrolysis of peptide bonds on the C-side of small amino acids (Gly, Ala). Chymotrypsin and elastase are much less specific than trypsin. An explanation for the specificity of these enzymes is given in Section 22.8.

Ala-Lys-Phe-Gly-Asp-Trp-Ser-Arg-Met-Val-Arg-Tyr-Leu-His

cleavage by elastase

None of the exopeptidases or endopeptidases that we have mentioned will catalyze the hydrolysis of an amide bond if proline is at the hydrolysis site. These enzymes recognize the appropriate hydrolysis site by its shape and charge, and proline causes the hydrolysis site to have an unrecognizable three-dimensional shape.

Ala-Lys-Pro	Leu-Phe-Pro	Pro-Phe-Val
trypsin will not cleave	chymotrypsin will not cleave	chymotrypsin will cleave

Cyanogen bromide (BrC≡N) causes the hydrolysis of the amide bond on the C-side of a methionine residue. Cyanogen bromide is more specific about what peptide bonds it cleaves than the endopeptidases, so it provides more reliable information about the primary sequence. Because cyanogen bromide is not a protein and therefore does not recognize the substrate by its shape, cyanogen bromide will still cleave the peptide bond if proline is at the cleavage site.

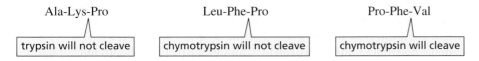

The first step in the mechanism for cyanogen bromide cleavage of a peptide bond is attack by the highly nucleophilic sulfur of methionine on cyanogen bromide. Formation of a five-membered ring with departure of the weakly basic leaving group is followed by acid-catalyzed hydrolysis, which cleaves the protein (Section 17.7). Further hydrolysis can cause the lactone (a cyclic ester) to open to a carboxyl group and an alcohol group (Section 16.9).

mechanism for the cleavage of a peptide bond by cyanogen bromide

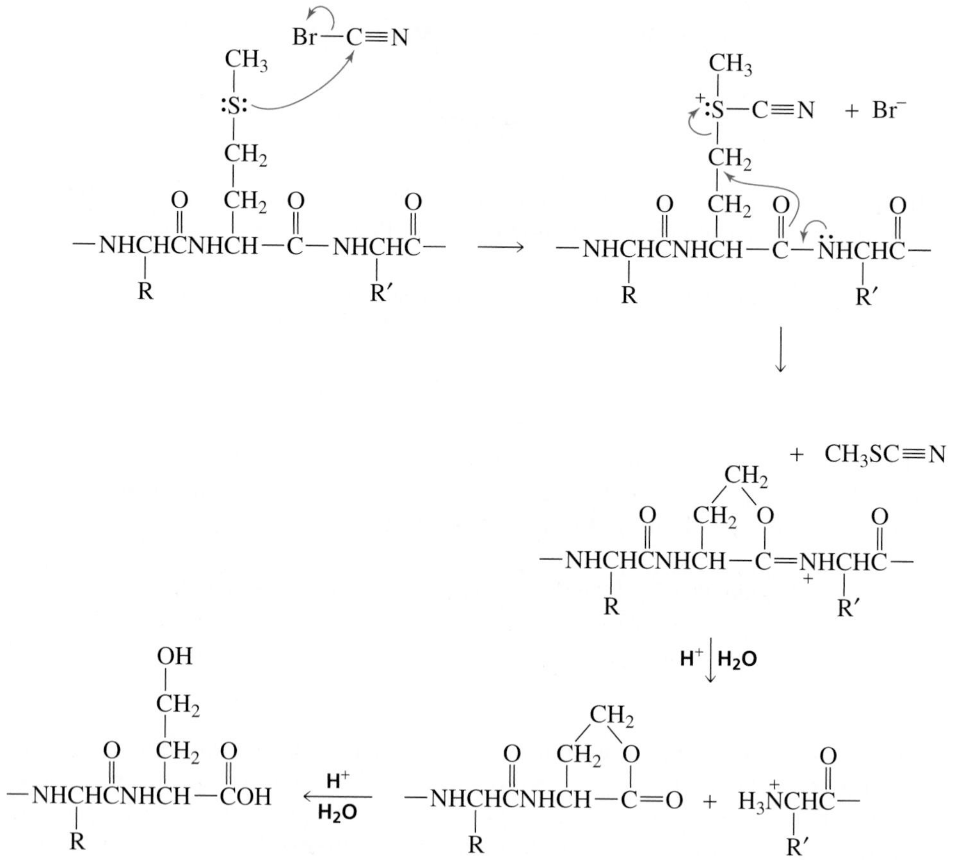

The last step in determining the primary structure of a protein is to figure out the location of any disulfide bonds. This is done by hydrolyzing a sample of the protein that has intact disulfide bonds. From a determination of the amino acids in the cysteine-containing fragments, the locations of the disulfide bonds in the protein can be established (Problem 47).

PROBLEM 30

Why won't cyanogen bromide cleave at cysteine residues?

PROBLEM 31◆

In determining the primary structure of insulin, what would lead you to conclude that it had more than one polypeptide chain?

PROBLEM 32 / SOLVED

Determine the amino acid sequence of a polypeptide from the following results:

Acid hydrolysis gives Ala, Arg, His, 2 Lys, Leu, 2 Met, Pro, 2 Ser, Thr, Val.

Carboxypeptidase A releases Val.

Edman's reagent releases PTH-Leu.

Cleavage with cyanogen bromide gives three peptides with the following amino acid compositions:

   1. His, Lys, Met, Pro, Ser   **2.** Thr, Val   **3.** Ala, Arg, Leu, Lys, Met, Ser

Trypsin-catalyzed hydrolysis gives three peptides and a single amino acid:

1. Arg, Leu, Ser

2. Met, Pro, Ser, Thr, Val

3. Lys

4. Ala, His, Lys, Met

SOLUTION   Acid hydrolysis shows that the polypeptide has 13 amino acids. The N-terminal amino acid is Leu (Edman's reagent), and the C-terminal amino acid is Val (carboxypeptidase A).

Leu __ __ __ __ __ __ __ __ __ __ __ Val

Because cyanogen bromide cleaves on the C-side of Met, any peptide containing Met must have Met as its C-terminal amino acid. The peptide that does not contain Met must be the C-terminal peptide. We know #3 is the N-terminal peptide because it contains Leu. Since it is a hexapeptide, we know the sixth amino acid in the 13-amino acid peptide is Met. We also know the eleventh amino acid is Met because cyanogen bromide cleavage gave the dipeptide Thr, Val.

        Ala, Arg, Lys, Ser   His, Lys, Pro, Ser

Leu __ __ __ __ Met __ __ __ __ Met Thr Val

Because trypsin cleaves on the C-side of Arg and Lys, any peptide containing Arg or Lys must have that amino acid as its C-terminal amino acid. Therefore, Arg is the C-terminal amino acid of #1, so we know the first three amino acids are Leu-Ser-Arg. We know the next two are Lys-Ala because if they were Ala-Lys, trypsin cleavage would give an Ala, Lys dipeptide. The trypsin data also identifies the positions of His and Lys.

                                 Pro, Ser

Leu Ser Arg Lys Ala Met His Lys __ __ Met Thr Val

Because trypsin successfully cleaves on the C-side of Lys, Pro cannot be adjacent to Lys.

Leu Ser Arg Lys Ala Met His Lys Ser Pro Met Thr Val

PROBLEM 33◆

Determine the primary structure of an octapeptide from the following data:

Acid hydrolysis gives 2 Arg, Leu, Lys, Met, Phe, Ser, Tyr.

Carboxypeptidase A releases Ser.

Edman's reagent releases Leu.

Cyanogen bromide forms two peptides with the following amino acid compositions:

1. Arg, Phe, Ser

2. Arg, Leu, Lys, Met, Tyr

Trypsin forms the following two peptides and two amino acids:

1. Arg

2. Ser

3. Arg, Met, Phe

4. Leu, Lys, Tyr

Secondary structure describes the conformation of segments of the backbone chain of a peptide or protein. In order to minimize energy, a polypeptide chain tends to fold in a repeating geometric structure. Three factors determine the choice of secondary structure:

**21.13
SECONDARY
STRUCTURE OF
PROTEINS**

$$\overset{\diagdown}{\underset{\diagup}{C}}=O\text{-----}H-\overset{\diagup}{\underset{\diagdown}{N}}$$

- the regional planarity about each peptide bond, which limits the possible conformations of the peptide chain (Section 21.7).
- maximizing the number of peptide groups that engage in hydrogen bonding (i.e., hydrogen bonding between the carbonyl oxygen of one amino acid residue and the amide hydrogen of another).
- adequate separation between nearby R groups to avoid steric hindrance and repulsion of like charges.

α-Helix

One type of secondary structure is the α-**helix.** In an α-helix, the backbone of the polypeptide coils around the long axis of the protein molecule (Figure 21.8). The helix is stabilized by hydrogen bonds—each hydrogen attached to an amide nitrogen is hydrogen bonded to a carbonyl oxygen of an amino acid four residues away. The substituents on the α-carbons of the amino acids protrude outward from the helix, thereby minimizing steric hindrance. Because the amino acids have the L-configuration, the α-helix is a right-handed helix. A right-handed helix rotates in a clockwise direction as it spirals down. Each turn of the helix contains 3.6 amino acid residues, and the repeat distance of the helix is 5.4 Å.

Figure 21.8 ▶
(a) A segment of a protein in an α-helix.
(b) Looking up the longitudinal axis of an α-helix.

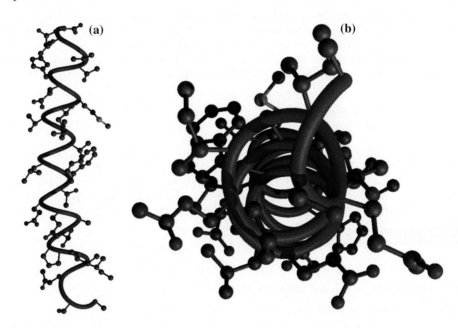

3-D Molecule:
An α-helix

Not all amino acids are able to fit into an α-helix. A proline residue, for example, forces a bend in a helix because the bond between the proline nitrogen and the α-carbon cannot rotate to enable it to fit readily into a helix. Two adjacent amino acids that have more than one substituent on a β-carbon (valine, isoleucine, or threonine) cannot fit into a helix because of steric crowding between the R groups. Two adjacent amino acids with like-charged substituents cannot fit into a helix because of electrostatic repulsion between the R groups. The percentage of amino acid residues coiled into an α-helix varies from protein to protein, but on average about 25% of the residues in globular proteins are in α-helices.

β-Pleated Sheet

The second type of secondary structure is the β-**pleated sheet.** In a β-pleated sheet, the polypeptide backbone is extended in a zigzag structure resembling a series of pleats. A β-pleated sheet is almost fully extended—the average two-residue repeat distance is 7.0 Å. The hydrogen bonding in a β-pleated sheet occurs between neighboring peptide chains. The adjacent hydrogen-bonded peptide chains can run in the same direction or in opposite directions. In a **parallel β-pleated sheet,** the adjacent chains run in the same direction. In an **antiparallel β-pleated sheet,** the adjacent chains run in opposite directions (Figure 21.9).

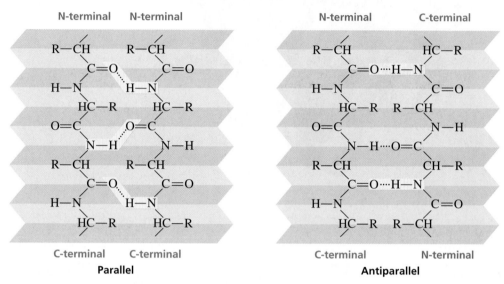

Parallel   **Antiparallel**

▲ **Figure 21.9**
Segment of a β-pleated sheet drawn to illustrate its pleated character.

Because the substituents (R) on the α-carbons of the amino acids on adjacent chains are close to each other, the chains can nestle closely together to maximize hydrogen-bonding interactions only if the substituents are small. Silk, for example, a protein with a large number of relatively small amino acids (glycine and alanine), has large segments of β-pleated sheets. The number of side-by-side strands in a β-pleated sheet ranges from 2 to 15 in a globular protein. The average strand in a β-pleated sheet section of a globular protein contains six amino acid residues.

3-D Molecule:
Antiparallel β-pleated
sheet

Wool and the fibrous protein of muscle are examples of proteins with secondary structures that are almost all α-helices. Consequently, these proteins can be stretched. In contrast, the secondary structures of silk and spider webs are predominantly β-pleated sheets. Because the β-pleated sheet is a fully extended structure, these proteins cannot be stretched.

Coil Conformation

Generally less than half of a globular protein is in an α-helix or β-pleated sheet (Figure 21.10). Most of the rest of the protein is still highly ordered but is difficult to describe. These polypeptide fragments are said to be in a **coil conformation** or a **loop conformation.**

PROBLEM 34◆

How long is an α-helix that contains 74 amino acids? Compare this with the length of a fully extended peptide chain containing the same number of amino acids. (The distance between consecutive amino acids in a fully extended chain is 3.5 Å.)

Figure 21.10 ▶
The backbone structure of carboxypeptidase A: α-helical segments are purple; β-pleated sheets are indicated by flat green arrows pointing in the N——▶C direction.

β-PEPTIDES: AN ATTEMPT TO IMPROVE ON NATURE

Chemists are currently studying β-peptides, which are polymers of β-amino acids. These peptides have backbones one carbon longer than the peptides nature synthesizes using α-amino acids. Therefore, each β-amino acid residue has two carbons to which side chains can be attached.

$$\overset{+}{H_3N}-\underset{\underset{R}{|}}{CH}-\overset{\overset{O}{\|}}{C}-O^-$$
α-amino acid

$$\overset{+}{H_3N}-\underset{\underset{R}{|}}{CH}-\underset{\underset{R'}{|}}{CH}-\overset{\overset{O}{\|}}{C}-O^-$$
β-amino acid

β-Polypeptides, like α-polypeptides, fold into relatively stable helical and pleated sheet conformations causing scientists to wonder if biological activity might be possible. Recently, a β-peptide with biological activity has been synthesized—one that mimics the activity of the hormone somatostatin. There is hope that β-polypeptides will provide a source of new drugs and catalysts. Surprisingly, the peptide bonds in β-polypeptides are resistant to the enzymes that catalyze the hydrolysis of peptide bonds in α-polypeptides. This resistance to hydrolysis means a β-polypeptide drug will have a longer duration of action in the blood stream.

21.14 TERTIARY STRUCTURE OF PROTEINS

Max Ferdinand Perutz and John Cowdery Kendrew were the first to determine the tertiary structure of a protein. Using X-ray diffraction, they determined the tertiary structure of myoglobin (1957) and hemoglobin (1959). For this work they shared the 1962 Nobel Prize in chemistry.

The *tertiary structure* of a protein is the three-dimensional arrangement of all the atoms in the protein. Proteins fold spontaneously in solution in order to maximize their stability. Every time there is a stabilizing interaction between two atoms, free energy is released. The more free energy released (the more negative the $\Delta G°$), the more stable the protein. So a protein tends to fold in a way that maximizes the number of stabilizing interactions (Figure 21.11).

The stabilizing interactions that occur in folding are covalent bonds, hydrogen bonds, electrostatic attractions, and hydrophobic (van der Waals) interactions. The interactions can occur between peptide groups (atoms in the backbone of the protein), between side-chain groups (α-substituents), and between peptide and side-chain groups. Because the side-chain groups help determine how a protein folds, the tertiary structure of a protein is determined by its primary structure.

Disulfide bonds are the only covalent bonds that can form when a protein folds. The other bonding interactions that occur in folding are much weaker but, because there are so many of them (Figure 21.12), they are the most important interactions in determining how a protein folds.

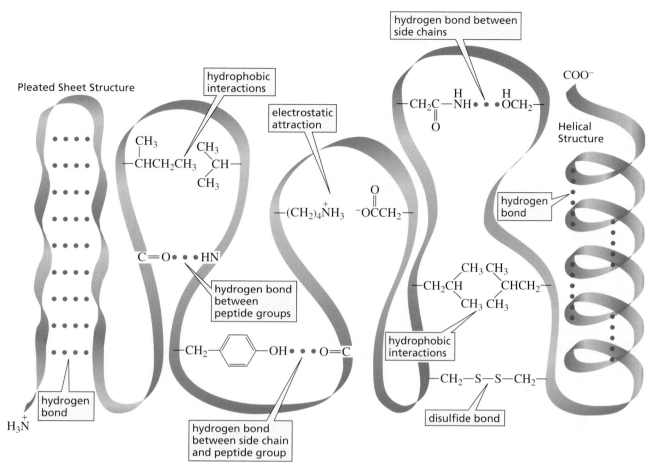

▲ Figure 21.11
Stabilizing interactions responsible for the tertiary structure of a protein.

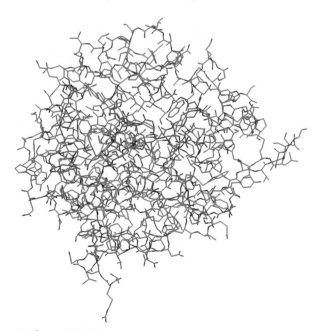

▲ Figure 21.12
The three-dimensional structure of carboxypeptidase A.

Max Perutz was born in Austria in 1914. In 1936, because of the rise of Nazism, he moved to England. He received a Ph.D. from and became a professor at Cambridge University. He worked on the three-dimensional structure of hemoglobin and assigned to John Kendrew (1917–1997) the work on myoglobin (a smaller protein). Kendrew was born in England and was educated at Cambridge University.

Most proteins exist in aqueous environments. Therefore, they tend to fold in a way that exposes the maximum number of polar groups to the aqueous environment and that buries the nonpolar groups in the interior of the protein, away from water.

The interactions between nonpolar groups are known as **hydrophobic interactions.** Hydrophobic interactions increase the stability of a protein by increasing the entropy of water molecules. Water molecules that surround nonpolar groups are highly structured. When two nonpolar groups come together, the surface area in contact with water decreases, decreasing the amount of structured water. Decreasing structure increases entropy. Increasing the entropy decreases the free energy, which increases the stability. (Recall that $\Delta G° = \Delta H° - T\Delta S°$.)

> **PROBLEM 35◆**
>
> How would a protein that resides in the interior of a membrane fold compared with the water-soluble protein just discussed?

21.15 QUATERNARY STRUCTURE OF PROTEINS

Proteins that have more than one peptide chain are called **oligomers.** The individual chains are called **subunits.** A protein with a single subunit is called a *monomer;* one with two subunits is called a *dimer;* one with three subunits is called a *trimer;* and one with four subunits is called a *tetramer.* Hemoglobin is an example of a tetramer. It has two different kinds of subunits and two of each kind. The quaternary structure of hemoglobin is shown in Figure 21.13.

The subunits are held together by the same kinds of interactions that hold the individual protein chains in a particular three-dimensional conformation—hydrophobic interactions, hydrogen bonding, and electrostatic attractions. The quaternary structure of a protein describes the way the subunits are arranged in space. Some of the possible arrangements of the six subunits of a hexamer are shown here.

possible quaternary structures for a hexamer

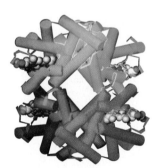

▲ **Figure 21.13**
Computer graphic representation of the quaternary structure of hemoglobin. The orange and pink subunits are identical; the green and purple subunits are identical. The cylindrical tubes represent the polypeptide chains. The beads represent the iron-containing porphyrin rings (Section 27.4).

> **PROBLEM 36◆**
>
> **a.** Which of the following water-soluble proteins would have the greatest percentage of polar amino acids: a spherical protein, a cigar-shaped protein, or a subunit of a hexamer?
>
> **b.** Which of these would have the smallest percentage of polar amino acids?

21.16 PROTEIN DENATURATION

Destroying the highly organized tertiary structure of a protein is called **denaturation.** Anything that breaks the bonds responsible for maintaining the three-dimensional shape of the protein will cause the protein to denature (unfold). Because these bonds are weak, proteins are easily denatured. The totally random conformation of a denatured protein is called a **random coil.**

- Changing the pH denatures proteins because it changes the charges on many of the side chains. This disrupts electrostatic attractions and hydrogen bonds.
- Certain reagents such as urea and guanidine hydrochloride denature proteins by forming hydrogen bonds to the protein groups that are stronger than the hydrogen bonds formed between the groups.
- Detergents such as sodium dodecyl sulfate denature proteins by associating with the nonpolar groups of the protein, thus interfering with the normal hydrophobic interactions.
- Organic solvents denature proteins by disrupting hydrophobic interactions.
- Proteins can also be denatured by heat or by agitation. Both increase molecular motion, which can disrupt the attractive forces. A well-known example is the change that occurs to the white of an egg when it is heated or whipped.

KEY TERMS

amino acid (page 916)
D-amino acid (page 921)
L-amino acid (page 921)
amino acid analyzer (page 928)
amino acid residue (page 916)
anion-exchange resin (page 928)
antiparallel β-pleated sheet (page 949)
automated solid-phase peptide
   synthesis (page 939)
cation-exchange resin (page 928)
coil conformation (page 949)
denaturation (page 952)
dipeptide (page 916)
disulfide (page 932)
disulfide bridge (page 932)
Edman's reagent (page 942)
electrophoresis (page 925)
endopeptidase (page 944)
enkephalin (page 934)

enzyme (page 916)
essential amino acid (page 920)
exopeptidase (page 943)
fibrous protein (page 941)
globular protein (page 941)
α-helix (page 948)
hydrophobic interactions (page 952)
interchain disulfide bridge (page 933)
intrachain disulfide bridge (page 933)
ion-exchange chromatography
   (page 927)
isoelectric point (page 924)
kinetic resolution (page 930)
loop conformation (page 949)
oligomer (page 952)
oligopeptide (page 916)
paper chromatography (page 926)
parallel β-pleated sheet (page 949)
partial hydrolysis (page 943)

peptide (page 916)
peptide bond (page 931)
β-pleated sheet (page 948)
polypeptide (page 916)
primary structure (page 940)
protein (page 916)
quaternary structure (page 941)
random coil (page 952)
secondary structure (page 940)
structural protein (page 916)
subunit (page 952)
C-terminal amino acid (page 931)
N-terminal amino acid (page 931)
tertiary structure (page 940)
thin-layer chromatography (page 927)
tripeptide (page 916)
zwitterion (page 923)

PROBLEMS

37. Unlike most amines and carboxylic acids, amino acids are insoluble in diethyl ether. Explain.

38. Indicate the peptides that would result from cleavage by the indicated reagent:
   a. His-Lys-Leu-Val-Glu-Pro-Arg-Ala-Gly-Ala by trypsin
   b. Leu-Gly-Ser-Met-Phe-Pro-Tyr-Gly-Val by chymotrypsin
   c. Val-Arg-Gly-Met-Arg-Ala-Ser by carboxypeptidase A
   d. Ser-Phe-Lys-Met-Pro-Ser-Ala-Asp by cyanogen bromide
   e. Arg-Ser-Pro-Lys-Lys-Ser-Glu-Gly by trypsin

39. Aspartame has a pI of 5.9. Draw its most prevalent form at physiological pH.

40. Draw the form of aspartic acid that predominates at
   a. pH = 1.0   **b.** pH = 2.6   **c.** pH = 6.0   **d.** pH = 11.0

41. Dr. Kim S. Tree was preparing a manuscript for publication in which she had reported that the pI of the tripeptide Lys-Lys-Lys was 10.6. One of her students pointed out that there must be an error in her calculations because the pK_a of the ε-amino group of lysine is 10.8 and the pI of the tripeptide has to be greater than any of its individual pK_a values. Was the student correct?

42. A mixture of amino acids that do not separate sufficiently when a single technique is used can often be separated by two-dimensional chromatography. In this technique the mixture of amino acids is applied to a piece of filter paper and separated by chromatographic techniques. The paper is rotated 90°, and the amino acids are further separated by electrophoresis. This type of chromatogram is called a fingerprint. Identify the spots in the chromatogram obtained from a mixture of Ser, Glu, Leu, His, Met, Thr.

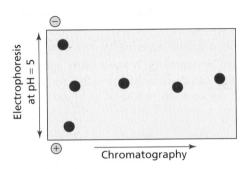

43. Explain the difference in the pK_a values of the carboxyl groups of alanine, serine, and cysteine.

44. Which would be a more effective buffer at physiological pH, a solution of 0.1 M glycylglycylglycylglycine or a solution of 0.2 M glycine?

45. Identify the location and type of charge on the hexapeptide Lys-Ser-Asp-Cys-His-Tyr at
   a. pH = 7   **b.** pH = 5   **c.** pH = 9

46. The following polypeptide was treated with 2-mercaptoethanol and then with iodoacetic acid. After reaction with maleic anhydride, the peptide was hydrolyzed by trypsin. Treatment with maleic anhydride causes trypsin to cleave a peptide only at arginine residues.

Gly-Ser-Asp-Ala-Leu-Pro-Gly-Ile-Thr-Ser-Arg-Asp-Val-Ser-Lys-Val-Glu-Tyr-Phe-Glu-Ala-Gly-Arg-Ser-Glu-Phe-Lys-Glu-Pro-Arg-Leu-Tyr-Met-Lys-Val-Glu-Gly-Arg-Pro-Val-Ser-Ala-Gly-Leu-Trp

   a. Why, after a peptide is treated with maleic anhydride, does trypsin no longer cleave it at lysine residues?
   b. How many fragments are obtained from the peptide?
   c. In what order would the fragments be eluted from an anion-exchange column using a buffer of pH = 5?

47. Treatment of a polypeptide with 2-mercaptoethanol yields two polypeptides with the following primary sequences.

Val-Met-Tyr-Ala-Cys-Ser-Phe-Ala-Glu-Ser

Ser-Cys-Phe-Lys-Cys-Trp-Lys-Tyr-Cys-Phe-Arg-Cys-Ser

Treatment of the original intact polypeptide with chymotrypsin yields the following peptides:
   a. Ala, Glu, Ser   **d.** Arg, Ser, Cys
   b. 2 Phe, 2 Cys, Ser   **e.** Ser, Phe, 2 Cys, Lys, Ala, Trp
   c. Tyr, Val, Met   **f.** Tyr, Lys

Determine the positions of the disulfide bridges in the original polypeptide.

48. Show how aspartame can be synthesized using DCC.

49. Reaction of a polypeptide with carboxypeptidase A releases Met. It undergoes partial hydrolysis to give the following peptides. What is the sequence of the polypeptide?
   a. Ser, Lys, Trp   **e.** Met, Ala, Gly   **i.** Lys, Ser
   b. Gly, His, Ala   **f.** Ser, Lys, Val   **j.** Glu, His, Val
   c. Glu, Val, Ser   **g.** Glu, His   **k.** Trp, Leu, Glu
   d. Leu, Glu, Ser   **h.** Leu, Lys, Trp   **l.** Ala, Met

50. Glycine has pK_a values of 2.3 and 9.6. Would you expect the pK_a values of glycyl-glycine to be higher or lower than these values?

51. A mixture of 15 amino acids gave the fingerprint shown in the margin. Identify the spots. (*Hint #1:* Pro reacts with ninhydrin to form a yellow color; Phe and Tyr form a yellow-green color. *Hint #2:* Count the number of spots before you start.)

52. Dithiothreitol reacts with disulfide bridges in the same way 2-mercaptoethanol does. With dithiothreitol, however, the equilibrium lies much more to the right. Explain.

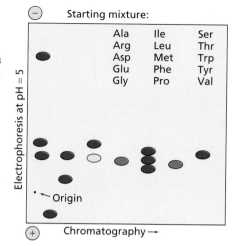

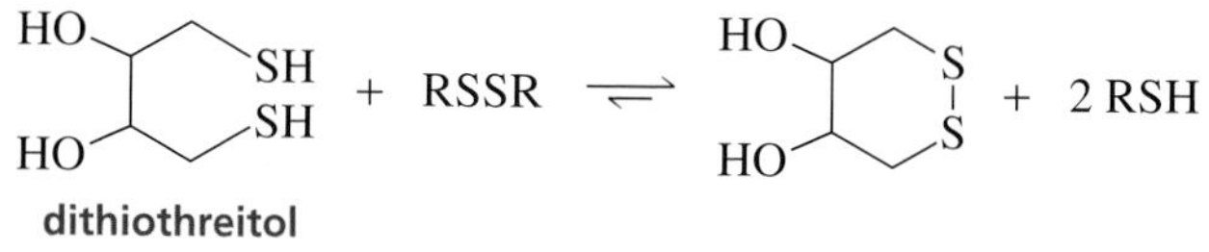

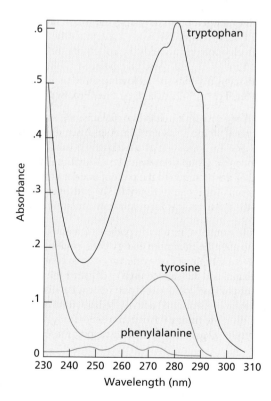

53. α-Amino acids can be prepared by treating an aldehyde with ammonia and hydrogen cyanide, followed by acid-catalyzed hydrolysis.
   a. Give the structures of the two intermediates formed in this reaction.
   b. What amino acid is formed when the aldehyde used is 3-methylbutanal?
   c. What aldehyde would be needed to prepare valine?

54. The UV spectra of tryptophan, tyrosine, and phenylalanine are shown here. Each spectrum is that of a 1×10^{-3} M solution of the amino acid buffered at pH = 6.0. Calculate the approximate molar absorptivity of each of the three amino acids at 280 nm.

55. A normal polypeptide and a mutant of the polypeptide were hydrolyzed by an endopeptidase under the same conditions. The normal and mutant differ by one amino acid residue. The fingerprints of the peptides obtained from the normal and mutant polypeptides are as shown. What kind of amino acid substitution occurred as a result of the mutation (i.e., is the substituted amino acid more or less polar than the original amino acid? Is its p*I* lower or higher?)

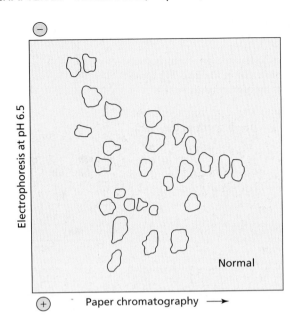

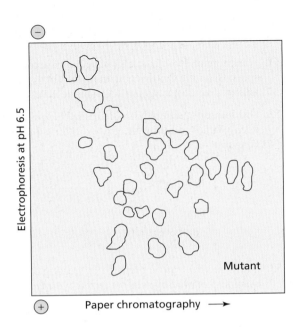

56. Determine the amino acid sequence of a polypeptide from the following results:
    a. Complete hydrolysis of the peptide yields the following amino acids: Ala, Arg, Gly, 2 Lys, Met, Phe, Pro, 2 Ser, Tyr, Val.
    b. Treatment with Edman's reagent gives PTH-Val.
    c. Carboxypeptidase A releases Ala.
    d. Treatment with cyanogen bromide yields the following two peptides:
       1. Ala, 2 Lys, Phe, Pro, Ser, Tyr   **2.** Arg, Gly, Met, Ser, Val
    e. Treatment with chymotrypsin yields the following three peptides:
       1. 2 Lys, Phe, Pro   **2.** Arg, Gly, Met, Ser, Tyr, Val   **3.** Ala, Ser
    f. Treatment with trypsin yields the following three peptides:
       1. Gly, Lys, Met, Tyr   **2.** Ala, Lys, Phe, Pro, Ser   **3.** Arg, Ser, Val

57. The C-terminal end of a protein extends into the aqueous environment surrounding the protein. The C-terminal amino acids are Gln, Asp, 2 Ser, and three nonpolar amino acids. Assuming that the $\Delta G°$ for formation of a hydrogen bond is -3 kcal/mol and the $\Delta G°$ for removal of a hydrophobic group from water is -4 kcal/mol, calculate the $\Delta G°$ for folding the C-terminal end of the protein into the protein interior under the following conditions:
    a. all the polar groups form one intramolecular hydrogen bond.
    b. all but two of the polar groups form intramolecular hydrogen bonds.

58. Professor Mary Gold wanted to test her hypothesis that the disulfide bridges that form in many proteins do so after the minimum energy conformation of the protein has been achieved. She treated a sample of lysozyme, an enzyme containing four disulfide bridges, with 2-mercaptoethanol and then added urea to denature the enzyme. She slowly removed these reagents so the enzyme could refold and reform the disulfide bridges. The lysozyme she recovered had 80% of its original activity. What would be the percent activity in the recovered enzyme if disulfide bridge formation were entirely random rather than determined by the tertiary structure? Does this experiment support Professor Gold's hypothesis?

22

Catalysis

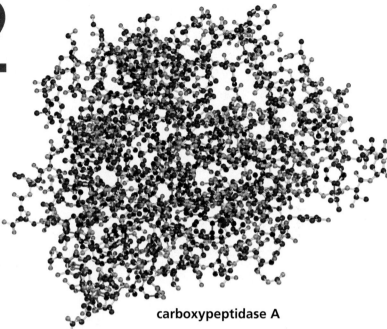

carboxypeptidase A

A **catalyst** is a substance that increases the rate of a chemical reaction without itself being consumed or changed in the overall reaction. We have seen that the rate of a chemical reaction depends on the height of the energy hill that must be overcome in the process of converting reactants into products (Section 3.7). The height of the hill is indicated by the free energy of activation ($\Delta G^{\ddagger}$). A catalyst increases the rate of a chemical reaction by providing a pathway with a lower $\Delta G^{\ddagger}$.

A catalyst can decrease $\Delta G^{\ddagger}$ in one of three ways: (1) the catalyzed and uncatalyzed reactions can follow a similar mechanism, with the catalyst providing a way to convert the reactant into a *less stable reactant* (Figure 22.1[a]); (2) the catalyzed and uncatalyzed reactions can follow a similar mechanism, with the catalyst providing a way to make *the transition state more stable* (Figure 22.1[b]); or (3) the catalyst can completely *change the mechanism* of the reaction, providing an alternative pathway with a smaller $\Delta G^{\ddagger}$ than that of the uncatalyzed reaction (Figure 22.2).

Just because a catalyst is neither consumed nor changed by the reaction does not mean that it does not participate in the reaction. How could a catalyst make the reaction go faster if it did not participate? The catalyst simply has the same form when the reaction is over that it had before the reaction started. Because the catalyst is not used up during the reaction—if it is used up in one step of the reaction, it must be regenerated in a subsequent step—only a small amount of the catalyst is needed. Therefore, a catalyst is added to a reaction mixture in small, *catalytic* amounts—much less than the number of moles of reactant.

Notice that the stability of the original reactants and final products is the same in the catalyzed and uncatalyzed reactions. In other words, the catalyst does not change the equilibrium constant of the reaction (ΔG° is the same for the catalyzed and uncatalyzed reactions in Figures 22.1[a], 22.1[b], and 22.2). Because the catalyst does not change the equilibrium constant, it does not change the *amount* of product formed during the reaction. It only changes the *rate* at which the product is formed.

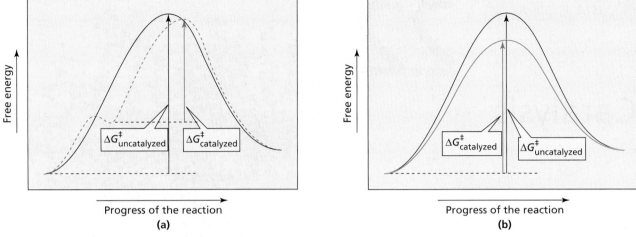

▲ Figure 22.1
Reaction coordinate diagrams for an uncatalyzed reaction and for a catalyzed reaction.
(a) The catalyst destabilizes the reactant. (b) The catalyst stabilizes the transition state.

Figure 22.2 ▶
Reaction coordinate diagrams
for an uncatalyzed reaction
and for a catalyzed reaction.
The catalyzed reaction takes
place by an alternative and
energetically more favorable
pathway.

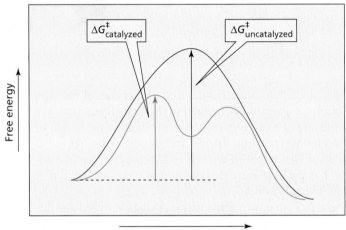

PROBLEM 1◆

Which of the following parameters would be different for a reaction carried out in the presence of a catalyst compared to the same reaction carried out in the absence of a catalyst? (*Hint:* See Section 3.7.)

$$\Delta G°, \Delta H^{\ddagger}, E_a, \Delta S^{\ddagger}, \Delta H°, K_{eq}, \Delta G^{\ddagger}, \Delta S°, k_{rate}$$

22.1
CATALYSIS IN ORGANIC REACTIONS

There are several ways for a catalyst to provide a more favorable pathway for an organic reaction:

- It can increase the susceptibility of an electrophile to nucleophilic attack.
- It can increase the reactivity of a nucleophile.
- It can increase the leaving ability of a group by converting it into a weaker base.

In this chapter we will look at some of the most common catalysts (nucleophilic catalysts, acid catalysts, base catalysts, and metal-ion catalysts) and the ways in which they provide an energetically more favorable pathway for an organic reaction. We will then see how the same modes of catalysis are used in enzyme-catalyzed reactions.

A **nucleophilic catalyst** increases the rate of a reaction by acting as a nucleophile— it forms an intermediate by forming a covalent bond with one of the reactants. Because a nucleophilic catalyst forms a covalent bond with a reactant, **nucleophilic catalysis** is also known as **covalent catalysis.** A nucleophilic catalyst increases the reaction rate by changing the mechanism of the reaction.

22.2 NUCLEOPHILIC CATALYSIS

In the following reaction, iodide ion increases the rate of conversion of ethyl chloride into ethyl alcohol by acting as a nucleophilic catalyst.

$$CH_3CH_2Cl \ + \ HO^- \ \xrightarrow[\text{H}_2\text{O}]{\text{I}^-} \ CH_3CH_2OH \ + \ Cl^-$$

To understand exactly how iodide ion catalyzes this reaction, we must compare the mechanism of the reaction in the absence of the catalyst with the mechanism of the reaction in the presence of the catalyst. In the absence of iodide ion, ethyl chloride is converted into ethyl alcohol in a one-step S_N2 reaction.

mechanism for the uncatalyzed reaction

$$HO^{\cdot \cdot -} \ + \ CH_3CH_2{-}Cl \ \longrightarrow \ CH_3CH_2OH \ + \ Cl^-$$

If iodide ion is present in the reaction mixture, the reaction takes place by two successive S_N2 reactions.

mechanism for the iodide-ion-catalyzed reaction

$$\vphantom{x}:\!I^{\cdot -} \ + \ CH_3CH_2{-}Cl \ \longrightarrow \ CH_3CH_2I \ + \ Cl^-$$
$$HO^{\cdot \cdot -} \ + \ CH_3CH_2{-}I \ \longrightarrow \ CH_3CH_2OH \ + \ I^-$$

The first S_N2 reaction in the catalyzed reaction is faster than the uncatalyzed reaction because in a protic solvent, iodide ion is a better nucleophile than hydroxide ion, which is the nucleophile in the uncatalyzed reaction (Section 9.3). The second S_N2 reaction in the catalyzed reaction is also faster than the uncatalyzed reaction because iodide ion is a weaker base and therefore a better leaving group than chloride ion, the leaving group in the uncatalyzed reaction. Thus, iodide ion increases the rate of formation of ethanol by changing a relatively slow one-step reaction into a reaction with two relatively fast steps (Figure 22.2).

Iodide ion is a nucleophilic catalyst because it reacts as a nucleophile, forming a covalent bond with the reactant. The iodide ion consumed in the first reaction is regenerated in the second, so iodide ion comes out of the reaction unchanged.

Another reaction in which a nucleophilic catalyst provides a more favorable pathway by changing the mechanism is the imidazole-catalyzed hydrolysis of phenyl acetate.

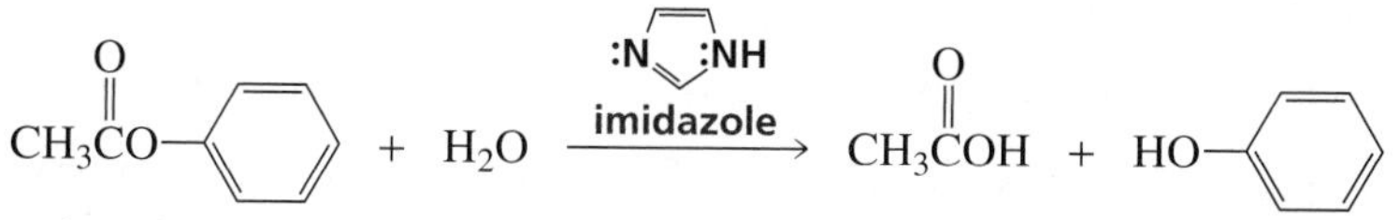

Imidazole is a better nucleophile than water, so imidazole reacts faster than water with the ester. The acyl imidazole that is formed is particularly reactive because the positively charged nitrogen makes imidazole a very good leaving group. Therefore, it is hydrolyzed much more rapidly than the ester would have been. Because formation of the acyl imidazole and its subsequent hydrolysis are both faster than ester hydrolysis, imidazole increases the rate of the overall reaction.

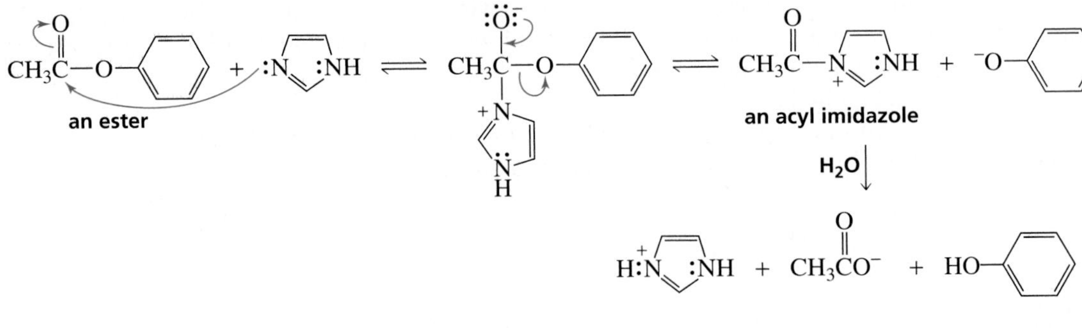

22.3 ACID CATALYSIS

An **acid catalyst** increases the rate of a reaction by donating a proton to a reactant. For example, the rate of hydrolysis of an ester is markedly increased by an acid catalyst (H^+).

$$CH_3\overset{O}{\overset{\|}{C}}OCH_3 + H_2O \overset{H^+}{\rightleftharpoons} CH_3\overset{O}{\overset{\|}{C}}OH + CH_3OH$$

A catalyst must increase the rate of a slow step. Increasing the rate of a fast step will not increase the rate of the overall reaction.

How an acid catalyzes the hydrolysis of an ester can be understood by examining the mechanism for the acid-catalyzed reaction. *Donation of a proton to and removal of a proton from an electronegative atom such as oxygen are fast steps.* Therefore, the reaction has two slow steps—formation of the tetrahedral intermediate and collapse of the tetrahedral intermediate. A catalyst must increase the rate of a slow step because increasing the rate of a fast step will not increase the rate of the overall reaction.

mechanism for the acid-catalyzed hydrolysis of an ester

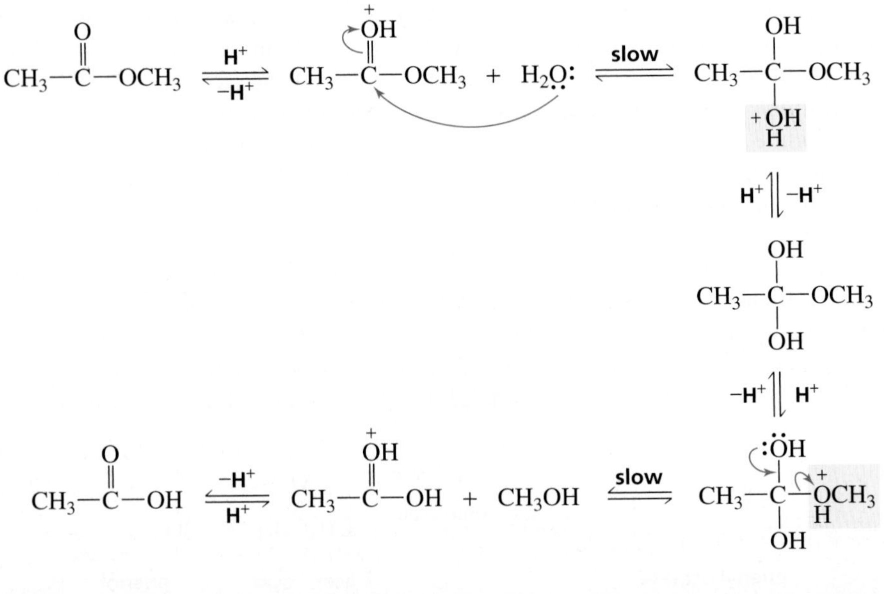

methyl acetate

protonated methyl acetate

The acid increases the rates of both slow steps of this reaction. It increases the rate of formation of the tetrahedral intermediate by protonating the carbonyl oxygen, thereby increasing the reactivity of the carbonyl group. We have seen that a protonated carbonyl group is more electrophilic than an unprotonated carbonyl group. In other words, the protonated carbonyl group is more susceptible to nucleophilic attack (Section 16.10). Increasing the reactivity of the carbonyl group by protonating it is an example of providing a way to make the reactant less stable—i.e., more reactive (Figure 22.1[a]).

acid-catalyzed first slow step   **uncatalyzed first slow step**

$$CH_3-\overset{\overset{+}{OH}}{\underset{\|}{C}}-OCH_3 \ + \ H_2\ddot{O}: \qquad CH_3-\overset{O}{\underset{\|}{C}}-OCH_3 \ + \ H_2\ddot{O}:$$

Alfred Bernhard Nobel

The acid increases the rate of the second slow step by changing the basicity of the group eliminated when the tetrahedral intermediate collapses. In the presence of an acid, methanol is eliminated; in the absence of an acid, methoxide ion is eliminated. Methanol is a weaker base than methoxide ion, so it is more easily eliminated.

acid-catalyzed second slow step   **uncatalyzed second slow step**

$$CH_3-\overset{:\ddot{O}H}{\underset{\underset{OH}{|}}{\overset{|}{C}}}-\overset{+}{\underset{H}{O}}CH_3 \qquad CH_3-\overset{:\ddot{O}H}{\underset{\underset{OH}{|}}{\overset{|}{C}}}-OCH_3$$

The mechanism for acid-catalyzed hydrolysis of an ester shows that the reaction can be divided into two distinct phases—formation of a tetrahedral intermediate and collapse of a tetrahedral intermediate. There are three steps in each phase. Notice that in each phase the first step is a fast protonation step, the second step is a slow catalyzed step that involves either breaking a π bond or forming a π bond, and the last step is a fast deprotonation step (to regenerate the catalyst).

PROBLEM 2

Compare each of the following mechanisms with the mechanism for each phase of the acid-catalyzed hydrolysis of an ester, indicating:

a. similarities.   **b.** differences.

(1) acid-catalyzed formation of a hydrate (Section 17.8)

(2) acid-catalyzed conversion of an aldehyde into a hemiacetal (Section 17.8)

(3) acid-catalyzed conversion of a hemiacetal into an acetal (Section 17.8)

(4) acid-catalyzed hydrolysis of an amide (Section 16.14)

The example of acid catalysis shown on page 960 is known as **specific-acid catalysis.** Both slow steps of the reaction are specific-acid-catalyzed. In specific-acid catalysis, the proton is fully transferred to the reactant *before* the slow step of the reaction begins (Figure 22.3[a]).

mixture could be molded into sticks that could not be set off without a detonating cap—thus he invented dynamite. He also invented blasting gelatin and smokeless powder. Although he was the inventor of explosives used by the military, he was a strong supporter of peace movements.

The 355 patents Nobel held made him a wealthy man. He never married and when he died his will stipulated that the bulk of his estate ($9,200,000) be used to establish prizes to be awarded to those who "have conferred the greatest benefit on mankind." He instructed that the money be invested and the interest earned each year be divided into five equal portions "to be awarded to the persons having made the most important contributions in the fields of chemistry, physics, physiology or medicine, literature, and to the one who had done the most or the best work toward fostering fraternity among nations, the abolition of standing armies, and the holding and

(continued)

A second kind of acid catalysis is called general-acid catalysis. In **general-acid catalysis,** the proton is transferred to the reactant *during* the slow step of the reaction (Figure 22.3[b]). Specific- and general-acid catalysts speed up a reaction in the same way—by donating a proton in order to make bond making and bond breaking easier. The two types of acid catalysts differ in the extent to which the proton is transferred in the transition state of the slow step of the reaction. In a specific-acid-catalyzed reaction, the transition state has a fully transferred proton, while in a general-acid-catalyzed reaction, the transition state has a partially transferred proton (Figure 22.3).

In the following pairs of examples, notice the difference in the amount of proton transfer that has occurred when the nucleophile attacks the reactant. In specific-acid-catalyzed attack by water on a carbonyl group, the nucleophile attacks a fully protonated carbonyl group. In general-acid-catalyzed attack by water on a carbonyl group, the carbonyl group becomes protonated as the nucleophile attacks it.

specific-acid-catalyzed attack by water

$$CH_3-\overset{\overset{\textstyle O}{\|}}{C}-OCH_3 + H^+ \rightleftharpoons CH_3-\overset{\overset{\textstyle +OH}{\|}}{C}-OCH_3 + H_2\ddot{O}: \longrightarrow CH_3-\overset{\overset{\textstyle OH}{|}}{\underset{\underset{\textstyle H}{+OH}}{C}}-OCH_3$$

general-acid-catalyzed attack by water

$$CH_3-\overset{\overset{\textstyle O}{\|}}{C}-OCH_3 + H_2\ddot{O}: \longrightarrow CH_3-\overset{\overset{\textstyle OH}{|}}{\underset{\underset{\textstyle H}{+OH}}{C}}-OCH_3$$

H—A   ⁻A

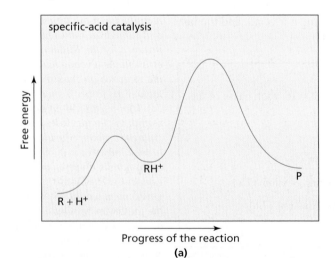

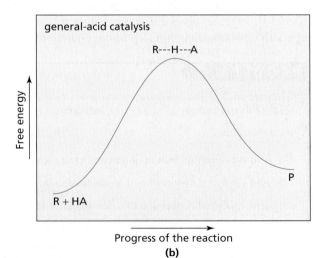

▲ **Figure 22.3**
(a) Reaction coordinate diagram for a specific-acid-catalyzed reaction (the proton is completely transferred to the reactant before the slow step of the reaction begins).
(b) Reaction coordinate diagram for a general-acid-catalyzed reaction (the proton is partially transferred to the reactant in the transition state of the slow step of the reaction).

In specific-acid-catalyzed collapse of a tetrahedral intermediate, a fully protonated leaving group is eliminated, while in general-acid-catalyzed collapse of a tetrahedral intermediate, the leaving group picks up a proton as it is eliminated.

specific-acid-catalyzed elimination of the leaving group

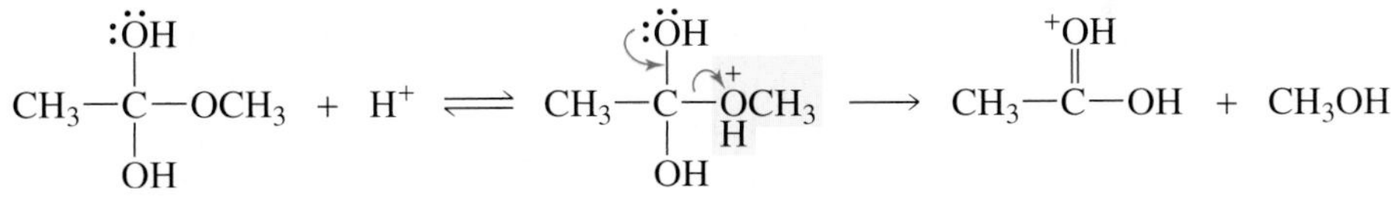

$$CH_3-\underset{\underset{OH}{|}}{\overset{\overset{:\ddot{O}H}{|}}{C}}-OCH_3 \ + \ H^+ \ \rightleftharpoons \ CH_3-\underset{\underset{OH}{|}}{\overset{\overset{:\ddot{O}H}{|}}{C}}-\overset{+}{\underset{H}{O}}CH_3 \ \longrightarrow \ CH_3-\overset{\overset{+OH}{\|}}{C}-OH \ + \ CH_3OH$$

general-acid-catalyzed elimination of the leaving group

$$CH_3-\underset{\underset{OH}{|}\;H-A}{\overset{\overset{:\ddot{O}H}{|}}{C}}-OCH_3 \ \longrightarrow \ CH_3-\overset{\overset{+OH}{\|}}{C}-OH \ + \ CH_3OH$$
$${}^-A$$

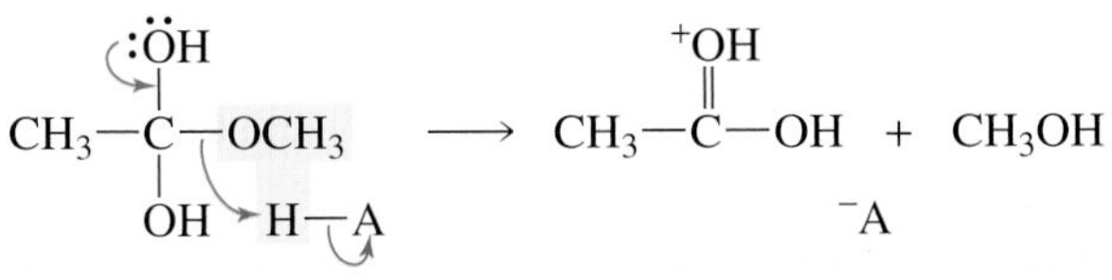

A specific-acid catalyst must be an acid strong enough to protonate the reactant fully before the slow step begins. A general-acid catalyst can be a weaker acid because it only partially transfers a proton in the transition state of the slow step.

PROBLEM 3

The following reaction occurs by a general-acid-catalyzed mechanism. Propose a mechanism for this reaction.

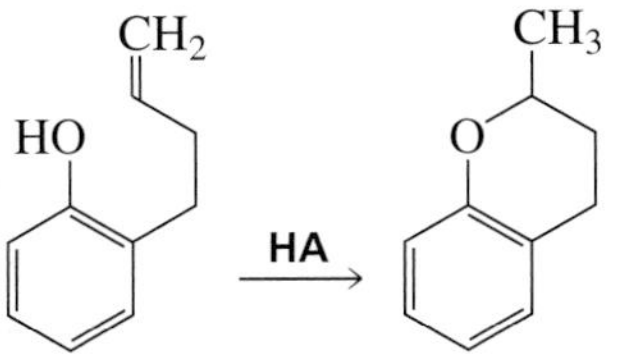

PROBLEM 4 / SOLVED

An alcohol will not react with aziridine unless the reaction is catalyzed by an acid. Why is a catalyst necessary?

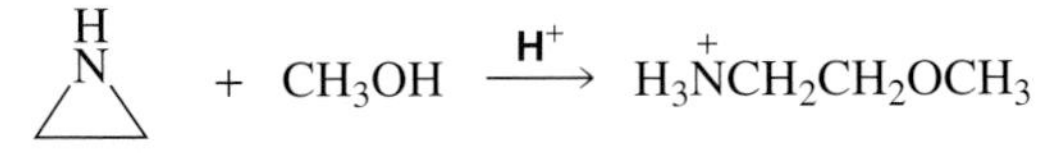

aziridine

$$+ \ CH_3OH \ \xrightarrow{\ H^+\ } \ H_3\overset{+}{N}CH_2CH_2OCH_3$$

SOLUTION Although relief of ring strain is sufficient by itself to cause an epoxide to undergo a ring-opening reaction (Section 11.6), it is not sufficient to cause an aziridine to undergo a ring-opening reaction. A negatively charged nitrogen is a stronger base and therefore a poorer leaving group than a negatively charged oxygen. Therefore, an acid must protonate the ring nitrogen (making it a better leaving group) before the reaction can take place.

A **base catalyst** increases the rate of a reaction by removing a proton from a reactant. For example, dehydration of a hydrate in the presence of hydroxide ion is a base-catalyzed reaction. Hydroxide ion (the base) increases the rate of the reaction by removing a proton from the neutral hydrate.

22.4
BASE CATALYSIS

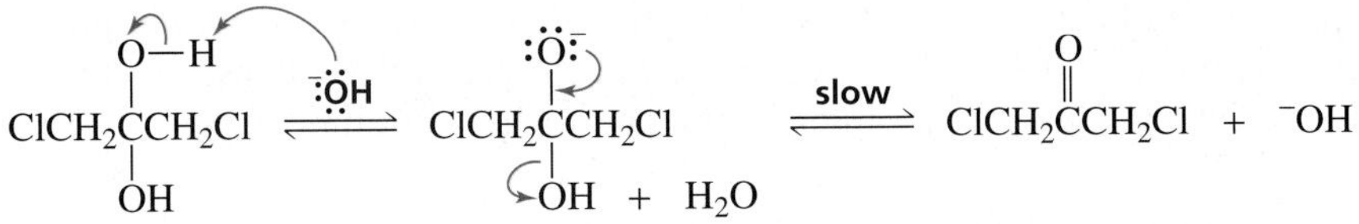

a hydrate

How does the base increase the rate of dehydration? It increases the rate of the re-action by providing a pathway with a more stable transition state. The transition state for elimination of ¯OH from a negatively charged tetrahedral intermediate is more stable than the transition state for elimination of ¯OH from a neutral tetrahedral in-termediate. The former transition state is more stable because a positive charge does not develop on the electronegative oxygen atom.

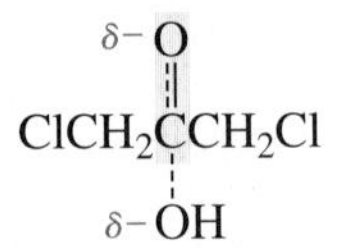

transition state for elimination of ¯OH from a negatively charged tetrahedral intermediate

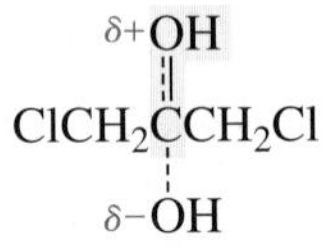

transition state for elimination of ¯OH from a neutral tetrahedral intermediate

A proton is donated to the reactant in an acid-catalyzed reaction.

A proton is removed from the reactant in a base-catalyzed reaction.

The preceding base-catalyzed dehydration of a hydrate is an example of specific-base catalysis. In **specific-base catalysis,** the proton is completely removed from the reactant *before* the slow step of the reaction begins. In **general-base catalysis,** on the other hand, the proton is removed from the reactant *during* the slow step of the reaction. Compare the amount of proton transfer in the slow step of the preceding specific-base-catalyzed dehydration with the amount of proton transfer in the slow step of the following general-base-catalyzed dehydration.

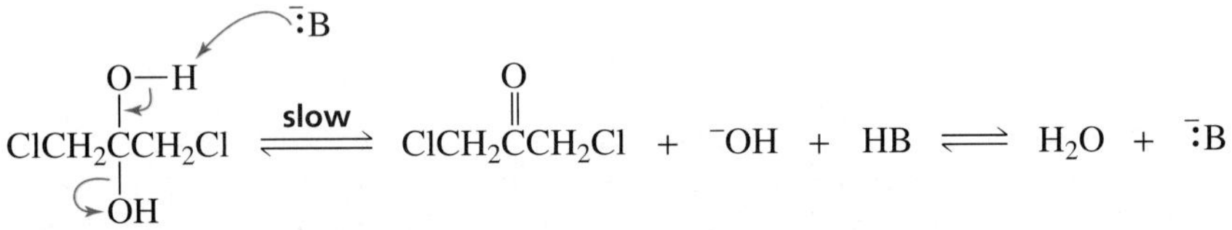

a hydrate

In specific-base catalysis, the base has to be strong enough to remove a proton from the reactant completely before the slow step begins. In general-base catalysis, the base can be weaker because the proton is only partially transferred to the base in the tran-sition state of the slow step. We will see that enzymes catalyze reactions using general-acid and general-base catalytic groups because at physiological pH (pH = 7.3) only a very small concentration ($\sim 1 \times 10^{-7}$ M) of H^+ for specific-acid catalysis or HO^- for specific-base catalysis is available.

PROBLEM 5◆

The mechanism for hydroxide-ion-promoted ester hydrolysis is shown in Section 16.11. What catalytic role does hydroxide ion play in this mechanism?

PROBLEM 6

The following reaction occurs by a mechanism involving general-base catalysis. Pro-pose a mechanism for this reaction.

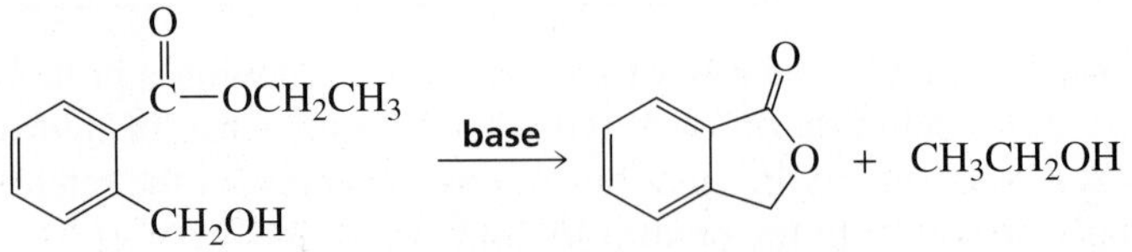

> **PROBLEM 7**
>
> Propose a mechanism for the general-base-catalyzed hydrolysis of an ester.

Metal ions exert their catalytic effect by coordinating (complexing) with atoms that have lone-pair electrons. In other words, metal ions are Lewis acids (Section 1.21). **Metal-ion catalysis** can increase the rate of a reaction in the following ways:

22.5 METAL-ION CATALYSIS

- A metal ion can make a reaction center more susceptible to receiving electrons, as in A in the following diagram.
- A metal ion can make a leaving group a weaker base and therefore a better leaving group, as in B.
- A metal ion can increase the rate of a hydrolysis reaction by forming a complex with water, thereby increasing water's acidity, as in C.

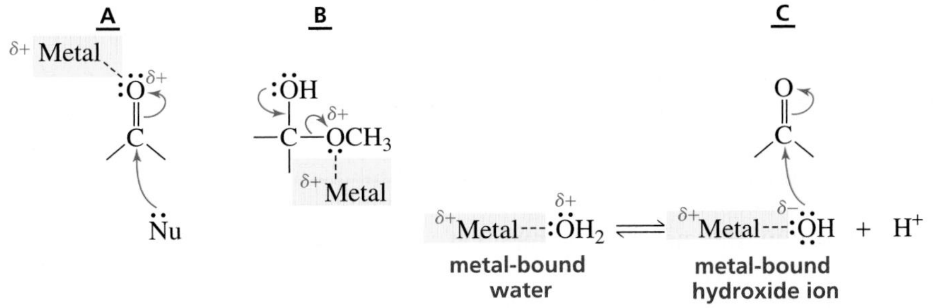

The pK_a of nonmetal-bound water is 15.7. The pK_a of metal-bound water depends on the metal atom (Table 22.1). When metal-bound water loses a proton, metal-bound hydroxide ion is formed. Metal-bound hydroxide ion, while not as good a nucleophile as hydroxide ion, is a better nucleophile than water. Thus, metal-ion complexation of water increases the reactivity of the attacking nucleophile in a hydrolysis reaction.

Table 22.1   The pK_a of Metal-Bound Water

M^{2+}	pK_a	M^{2+}	pK_a
Ca^{2+}	12.7	Co^{2+}	8.9
Mg^{2+}	11.8	Zn^{2+}	8.7
Cd^{2+}	11.6	Fe^{2+}	7.2
Mn^{2+}	10.6	Cu^{2+}	6.8
Ni^{2+}	9.4	Be^{2+}	5.7

In cases A and B above, the metal ion exerts the same kind of catalytic effect as a proton. In a reaction where the metal ion has the same catalytic effect as a proton (increasing the electrophilicity of a reaction center or decreasing the basicity of a leaving group), the metal ion is often called an **electrophilic catalyst.** Now we will look at some examples of metal-ion-catalyzed reactions.

Co^{2+} catalyzes the condensation of two molecules of the ethyl ester of glycine to form the ethyl ester of glycylglycine. The actual catalyst is a cobalt complex, $[Co(ethylenediamine)_2]^{2+}$.

$$2\ H_2NCH_2COCH_2CH_3 \xrightarrow{\ Co^{2+}\ } H_2NCH_2CNHCH_2COCH_2CH_3\ +\ CH_3CH_2OH$$

Coordination of the metal ion with the carbonyl oxygen makes the carbonyl group more susceptible to nucleophilic attack because it stabilizes the developing negative charge on the oxygen in the transition state.

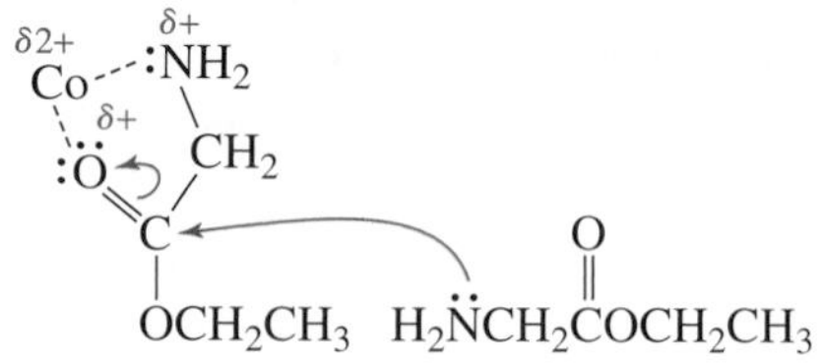

The decarboxylation of dimethyloxaloacetate can be catalyzed by either Cu^{2+} or Al^{3+}.

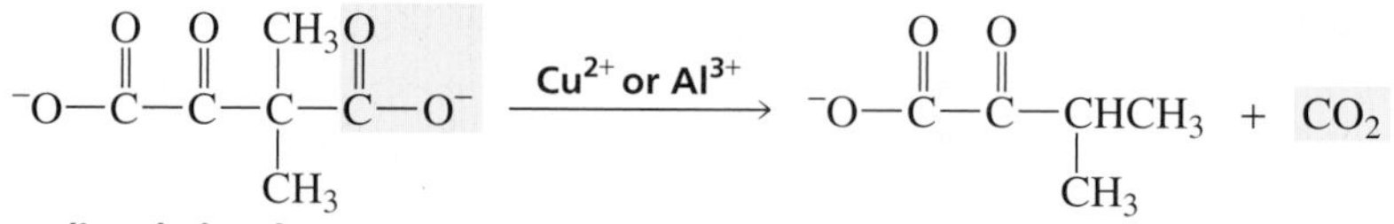

dimethyloxaloacetate

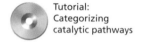

Tutorial:
Categorizing
catalytic pathways

In this reaction, the metal ion complexes with two oxygen atoms of the reactant. Complexation increases the rate of decarboxylation by making the carbonyl group more susceptible to receiving the electrons left behind when CO_2 is eliminated.

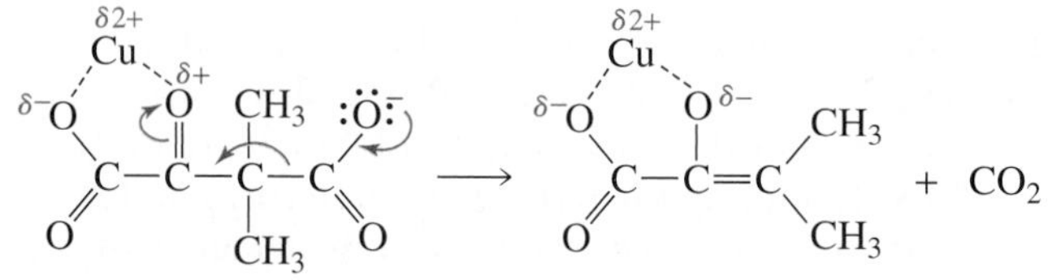

The hydrolysis of methyl trifluoroacetate has two slow steps. Zn^{2+} increases the rate of the first slow step by increasing the acidity of water, thereby providing metal-bound hydroxide ion, a better nucleophile than water. Zn^{2+} increases the rate of the second slow step by decreasing the basicity of the group that is eliminated from the tetrahedral intermediate.

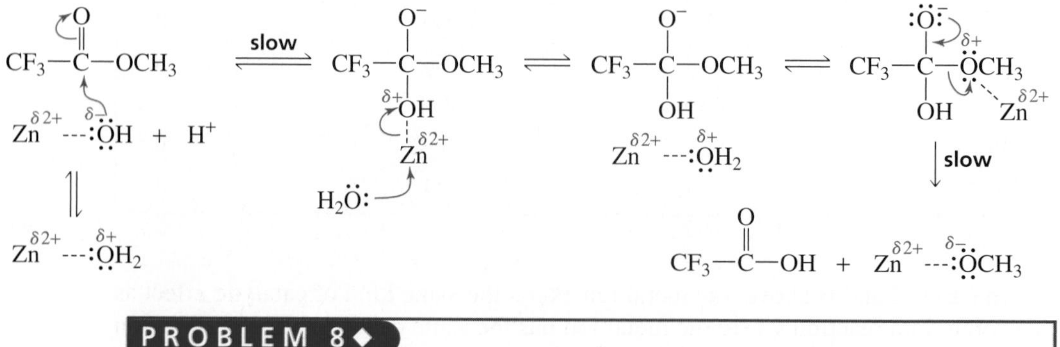

PROBLEM 8◆

The rate constant for the uncatalyzed reaction of two molecules of glycine ethyl ester to form glycylglycine ethyl ester is $0.6\ s^{-1}\ M^{-1}$. In the presence of $[Co(ethylenediamine)_2]^{2+}$, the rate constant is $1.5 \times 10^6 \, s^{-1} \, M^{-1}$. What rate enhancement does the catalyst provide?

PROBLEM 9

Although metal ions increase the rate of decarboxylation of dimethyloxaloacetate, they have no effect on the rate of decarboxylation of either the monoethyl ester of dimethyl-oxaloacetate or acetoacetate. Explain.

dimethyloxaloacetate

monoethyl ester of
dimethyloxaloacetate

acetoacetate

PROBLEM 10

The hydrolysis of glycinamide is catalyzed by $[Co(ethylenediamine)_2]^{2+}$. Propose a mechanism for this reaction.

$$H_2NCH_2CNH_2 \ + \ H_2O \ \xrightarrow{Co^{2+}} \ H_2NCH_2CO^- \ + \ ^+NH_4$$

The rate of a chemical reaction is determined by the number of collisions between two molecules with sufficient energy *and* with the proper orientation in a given period of time (Section 3.7).

**22.6
INTRAMOLECULAR
REACTIONS**

$$\textbf{rate of a reaction} = \frac{\textbf{number of collisions}}{\textbf{unit of time}} \times \frac{\textbf{fraction with}}{\textbf{sufficient energy}} \times \frac{\textbf{fraction with}}{\textbf{proper orientation}}$$

Because a catalyst decreases the energy barrier of a reaction, it increases the reaction rate by increasing the fraction of collisions that occur with sufficient energy to overcome the barrier. The rate of a reaction can also be increased by increasing the number of collisions *and* the fraction of those collisions that take place with the proper orientation. We have seen that an intramolecular reaction that results in the formation of a five- or a six-membered ring occurs more readily than the corresponding intermolecular reaction. This is because an intramolecular reaction has the advantage that the reacting groups are tied together in the same molecule. This gives them a better chance of finding each other than if they were in two different molecules in a solution of the same concentration (Section 10.11). As a result, the number of collisions increases.

If, in addition to being in the same molecule, the reacting groups are juxtaposed in a way that increases the probability they will collide with each other in the proper orientation, the rate of the reaction is further increased. The relative rates shown in Table 22.2 demonstrate the enormous increase that occurs in the rate of a reaction when the reacting groups are properly juxtaposed.

Rate constants for a series of reactions are generally compared using relative rates because relative rates allow one to see immediately how much faster one reaction is than another. **Relative rates** are obtained by dividing the rate constant for each of the reactions by the rate constant of the slowest reaction in the series. Because an intramolecular reaction is a first-order reaction (units of time^{-1}) and an intermolecular reaction is a second-order reaction (units of time^{-1} M^{-1}), the relative rates in Table 22.2 have units of molarity (M) (Section 3.7).

$$\text{relative rate} = \frac{\text{first-order rate constant}}{\text{second-order rate constant}} = \frac{\text{time}^{-1}}{\text{time}^{-1}\,\text{M}^{-1}} = \text{M}$$

The relative rates shown in Table 22.2 are also called effective molarities. The **effective molarity** is the advantage given to the reaction by having the reacting groups

TABLE 22.2   Relative Rates of an Intermolecular Reaction and Five Intramolecular Reactions

Reaction	Relative rate
A (intermolecular ester reaction)	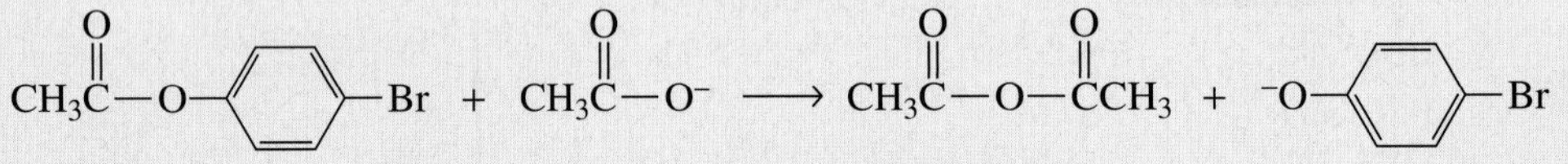 1.0
B (intramolecular reaction forming glutaric anhydride)	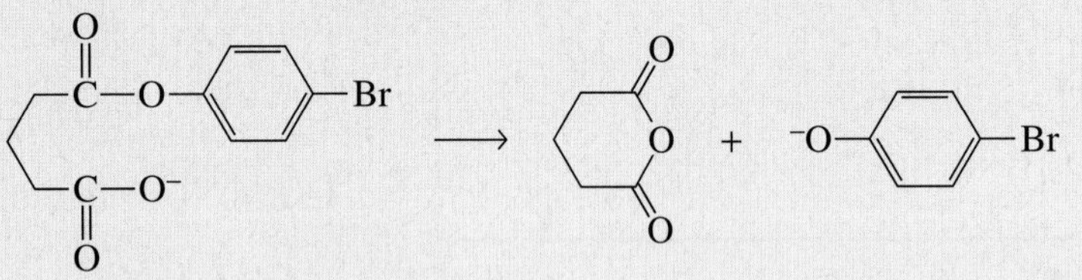 1×10^3 M
C (intramolecular reaction, R-substituted)	2.3×10^4 M, R = CH$_3$; 1.3×10^6 M, R = *iso*-C$_3$H$_7$
D (intramolecular reaction forming succinic anhydride)	2.2×10^5 M
E (intramolecular reaction forming maleic anhydride)	1×10^7 M
F (intramolecular reaction, bicyclic)	5×10^7 M

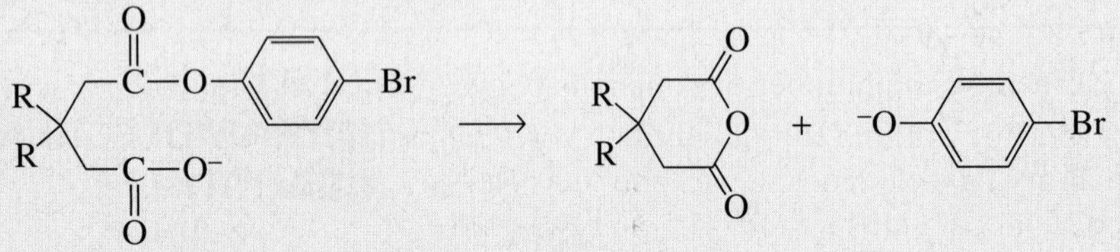

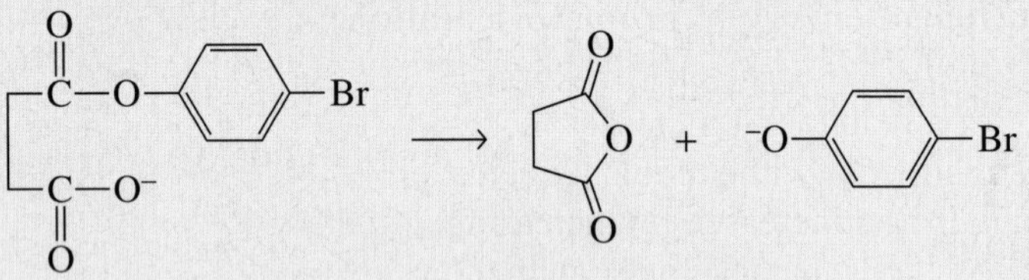

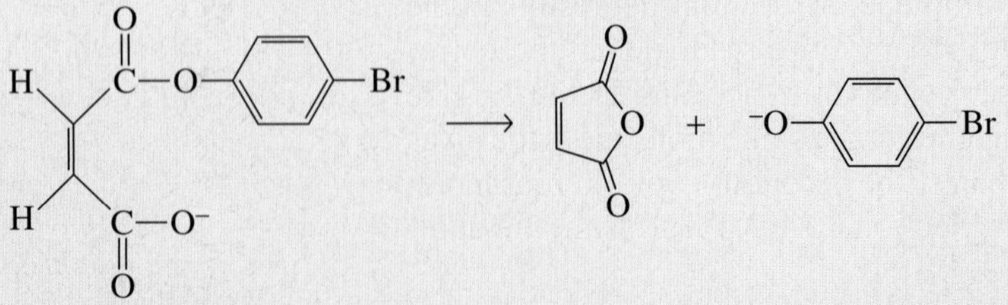

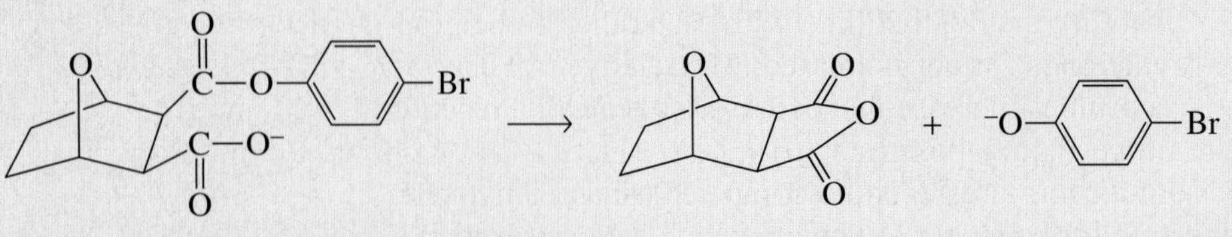

in the same molecule. Effective molarity is the concentration of the reactant that would be required in an *intermolecular* reaction for it to have the same rate as the *intramolecular* reaction. In some cases, juxtaposing the reacting groups provides such an enormous increase in rate that the effective molarity is greater than the concentration of the reactant in its solid state.

The first reaction shown in Table 22.2 (reaction **A**) is an intermolecular reaction between an ester and a carboxylate ion. The second reaction (reaction **B**) has the same reacting groups in a single molecule. The rate of the intramolecular reaction is 1000 times faster than the rate of the intermolecular reaction.

The reactant in reaction **B** has four carbon–carbon bonds that are free to rotate, whereas the reactant in reaction **D** has only three such bonds. Conformers in which the large groups are rotated away from each other are more stable. However, when these groups are pointed away from each other, they are in an unfavorable conformation for reaction. Because the reactant in reaction **D** has fewer bonds that are free to rotate, the groups are more apt to be in a conformation that is favorable for a reaction. Therefore, reaction **D** is faster than reaction **B**. The relative rate constants for the reactions shown in Table 22.2 are quantitatively related to the calculated probability of forming the conformation where the carboxylate ion is in position to attack the carbonyl carbon.

Reaction **C** is faster than reaction **B** because the alkyl substituents of the reactant in **C** decrease the available volume for rotation of the reactive groups away from each other. Thus, there is a greater probability that the molecule will be in a conformation with the reacting groups positioned for ring closure. This is called the **gem-dialkyl effect** because the two alkyl substituents are bonded to the same (geminal) carbon. If we compare the rate when the substituents are methyl groups to the rate when the substituents are isopropyl groups, we see that the rate is further increased when the size of the alkyl groups is increased.

The increased rate of reaction **E** is due to the double bond that prevents the reacting groups from rotating away from each other. The bicyclic compound in reaction **F** reacts at an even greater rate because the reacting groups are locked in the proper orientation for reaction.

PROBLEM 11◆

The relative rate of reaction of the cis alkene (**E**) is given in Table 22.2. What would you expect the relative rate of the trans isomer to be?

22.7 INTRAMOLECULAR CATALYSIS

Just as putting two reacting groups in the same molecule increases the rate of a reaction compared to having the groups in separate molecules, putting a *reacting group* and a *catalyst* in the same molecule increases the rate of a reaction compared to having them in separate molecules. When a catalyst is part of the reacting molecule, the catalysis is called **intramolecular catalysis.** Intramolecular nucleophilic catalysis, intramolecular general-acid or general-base catalysis, and intramolecular metal-ion catalysis all occur. Intramolecular catalysis is also known as **anchimeric assistance** (*anchimeric* means "adjacent parts" in Greek). Now we will look at some examples of intramolecular catalysis.

The following intramolecular general-base-catalyzed enolization reaction is much faster than the analogous intermolecular general-base-catalyzed reaction.

intramolecular general-base catalysis

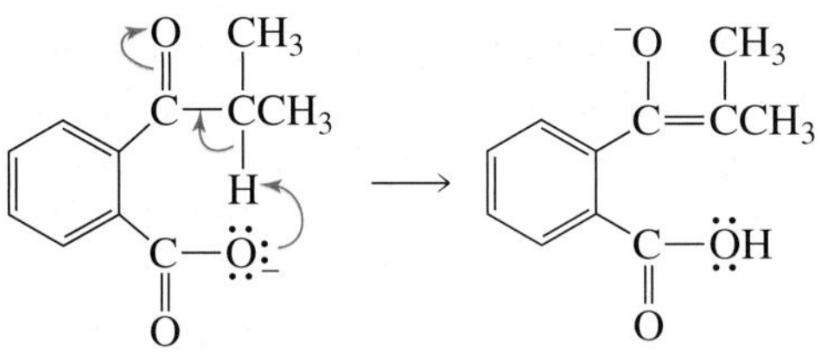

intermolecular general-base catalysis

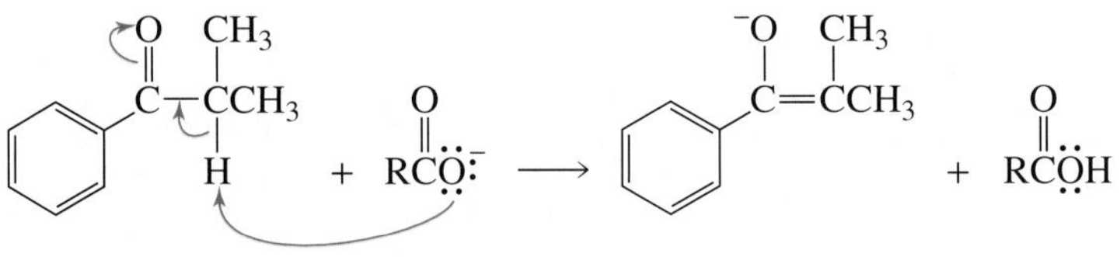

When chlorocyclohexane is added to an aqueous solution of ethanol, an alcohol and an ether are formed. Two products are formed because there are two nucleophiles (H_2O and CH_3CH_2OH) in the solution.

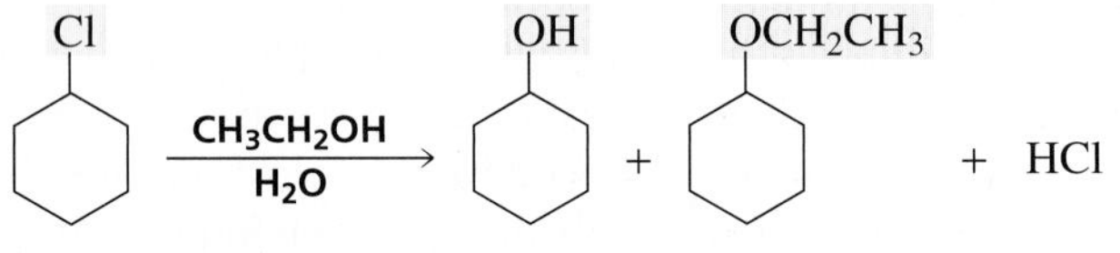

A 2-thio-substituted chlorocyclohexane undergoes the same reaction. However, the rate of the reaction depends on whether the substituent is cis or trans to the chloro substituent. If it is trans, the 2-thio-substituted compound reacts about 70,000 times faster than the unsubstituted compound. But if it is cis, the 2-thio-substituted compound reacts a little more slowly than the unsubstituted compound.

3-D Molecules: *cis*-2-Thiophenylchlorocyclohexane; *trans*-2-Thiophenylchlorocyclohexane

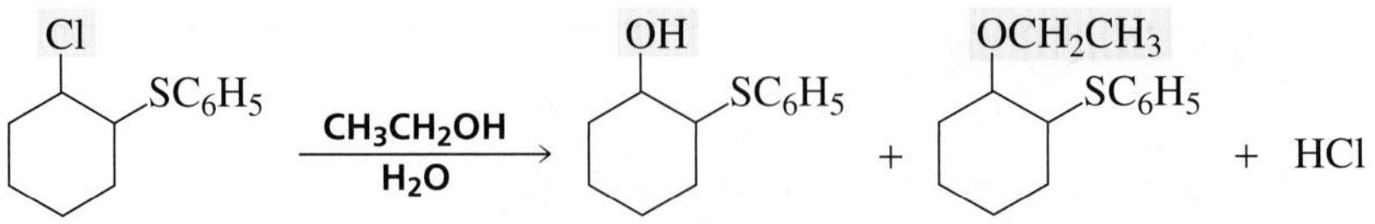

What accounts for the much faster reaction of the trans-substituted compound? In this reaction, the thio substituent is an intramolecular nucleophilic catalyst. It displaces the chloro substituent by attacking the back side of the carbon to which the chloro substituent is attached. Back side attack requires both substituents to be in axial positions, which means they must be trans to each other (Section 2.14). Subsequent attack by water or ethanol on the sulfonium ion is rapid because the positively charged sulfur is an excellent leaving group and breaking the ring releases the strain in the three-membered ring.

PROBLEM 12◆

a. How many disubstituted cyclohexane products are obtained from solvolysis of the trans-substituted compound shown above?

b. Give the products and their configurations.

The rate of hydrolysis of phenyl acetate is increased about 150-fold at neutral pH by the presence of a carboxylate ion in the ortho position. The *ortho*-carboxyl-substituted ester is commonly known as aspirin (Section 19.9).

The *ortho*-carboxylate group is an intramolecular general-base catalyst. It increases the nucleophilicity of water, thereby increasing the rate of formation of the tetrahedral intermediate.

3-D Molecule: Aspirin

If nitro groups are put on the benzene ring, the *ortho*-carboxyl substituent acts as an intramolecular nucleophilic catalyst instead of an intramolecular general-base catalyst. It increases the rate of the hydrolysis reaction by converting the ester into an anhydride, and an anhydride is more rapidly hydrolyzed than an ester.

an ester

an anhydride

PROBLEM 13 / SOLVED

Explain why adding nitro groups to the benzene ring changes the mode of catalysis for the hydrolysis of an *ortho*-carboxyl-substituted phenyl acetate.

SOLUTION The *ortho*-carboxyl substituent is in position to form a tetrahedral intermediate. If the carboxyl group in the tetrahedral intermediate is a better leaving group than the phenoxy group, the carboxyl group will be eliminated from the intermediate. This will re-form the starting material, which will be hydrolyzed by a general-base-catalyzed mechanism (path A). However, if the phenoxy group is a better leaving group than the carboxyl group, the phenoxy group will be eliminated. This will form the anhydride, and the reaction will have occurred by a mechanism involving a nucleophilic catalysis (path B).

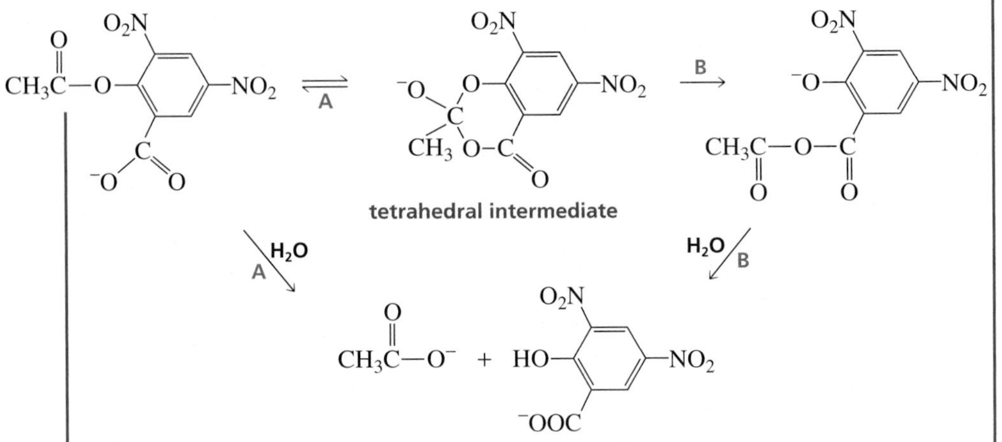

tetrahedral intermediate

PROBLEM 14 ◆

Why do the nitro groups change the relative leaving tendencies of the carboxyl and phenyl groups in the tetrahedral intermediate in Problem 13?

PROBLEM 15

Whether the *ortho*-carboxyl substituent acts as an intramolecular general-base catalyst or as an intramolecular nucleophilic catalyst can be determined by carrying out the hydrolysis of aspirin with ^{18}O-labeled water and determining whether ^{18}O is incorporated into salicylic acid. Explain the results that would be obtained with the two types of catalysis.

The following reaction is an example of intramolecular metal-ion catalysis. Hydrolysis of the ester is catalyzed by Ni^{2+}.

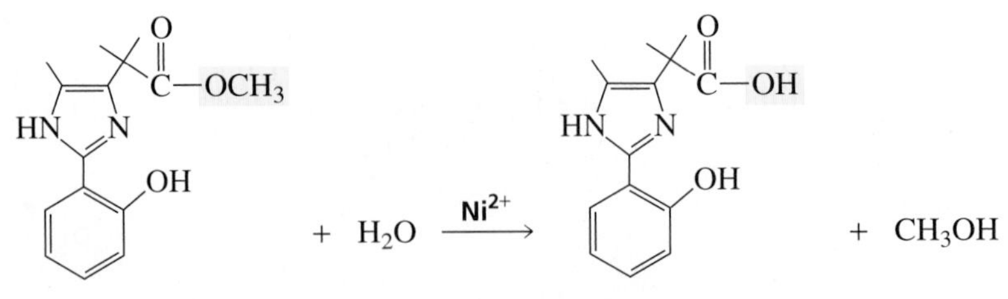

The metal ion complexes with an oxygen and a nitrogen of the reactant as well as with a molecule of water. The metal ion increases the rate of the reaction by positioning the water molecule and making it a stronger acid, which increases its nucleophilicity.

22.8 CATALYSIS IN BIOLOGICAL REACTIONS

Essentially all organic reactions that occur in biological systems require a catalyst. Most biological catalysts are **enzymes.** Enzymes are globular proteins. Each biological reaction is catalyzed by a different enzyme. Enzymes are extraordinarily good catalysts—they can increase the rate of an intermolecular reaction by as much as 10^{16}-fold. This can be compared with rate enhancements achieved by nonbiological catalysts in intermolecular reactions, which are seldom greater than 10^4-fold.

The reactant of an enzyme-catalyzed reaction is called a **substrate.** The enzyme has a pocket or cleft known as an active site. The substrate specifically fits and binds to the **active site,** and all the bond-making and bond-breaking steps of the reaction occur while the substrate is in this site. Enzymes differ from nonbiological catalysts in that they are specific for the reactant whose reaction they catalyze (Section 4.18). Not all enzymes have the same degree of specificity. Some are specific for a single compound and will not tolerate even the slightest variation in structure, while some catalyze the reaction of an entire family of compounds with related structures. The specificity of an enzyme for its substrate is an example of the phenomenon known as **molecular recognition**—one molecule "recognizes" another.

The specificity of an enzyme results from the conformation of the protein and the particular amino acid side chains that make up the active site. For example, a negatively charged side chain of an amino acid residue at the active site of an enzyme can associate with a positively charged group on the substrate; a hydrogen-bond donor on the enzyme can associate with a hydrogen-bond acceptor on the substrate; and hydrophobic groups on the enzyme associate with hydrophobic groups on the substrate. The specificity of an enzyme for its substrate has been described by the lock-and-key model. In the **lock-and-key model,** the substrate is said to fit the enzyme much as a key fits a lock.

The energy released as a result of binding the substrate to the enzyme can be used to induce a change in the shape of the enzyme, resulting in more precise binding between the substrate and the active site. This change in conformation of the enzyme is known as induced fit. In the **induced-fit model,** the shape of the active site does not become completely complementary to the shape of the substrate until the enzyme has bound the substrate.

substrate

+

active site

enzyme

enzyme-substrate complex

lock-and-key model
(a)

substrate

+

active site

enzyme

enzyme-substrate complex

induced-fit model
(b)

An example of induced fit is shown in Figure 22.4. The three-dimensional structure of the enzyme hexokinase is shown before and after binding glucose, its substrate. Notice the change in conformation that occurs upon binding the substrate.

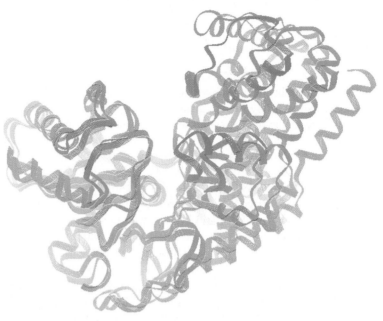

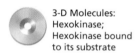

3-D Molecules:
Hexokinase;
Hexokinase bound
to its substrate

▲ **Figure 22.4**
The structure of hexokinase before binding its substrate is shown in red. The structure of hexokinase after binding its substrate is shown in green.

Some of the most important factors that account for the remarkable catalytic ability of enzymes are the following:

- All reacting groups are brought together at the active site in the proper orientation for reaction. This is analogous to the way that proper positioning of reacting groups increases the rate of intramolecular reactions (Section 22.6).
- Some of the amino acid side chains serve as catalytic groups. These side chains are in the proper position relative to the substrate to act as catalysts. This is analogous to the rate enhancements observed for intramolecular catalysis (Section 22.7).
- The conformational change an enzyme undergoes after binding a substrate can introduce strain into the substrate, making it more reactive (Figure 22.1[a]).
- Groups on the enzyme can stabilize an intermediate and, therefore, the transition state leading to the intermediate, by van der Waals interactions, electrostatic interactions, and hydrogen bonding (Figure 22.1[b]).

As we look at some examples of enzyme-catalyzed reactions, it is important to note that the functional groups on the enzyme side chains are the same functional groups you are used to seeing in simple organic compounds, and the modes of catalysis used by enzymes are exactly the same as the modes of catalysis used in organic reactions. One factor contributing to the remarkable catalytic ability of enzymes is their ability to use several modes of catalysis in the same reaction. Factors other than those just listed can contribute to the increased rate of enzyme-catalyzed reactions, but not all factors are employed by every enzyme. We will consider some of these factors when we discuss individual enzymes. We will now look at the mechanisms for four enzyme-catalyzed reactions.

Mechanism for Carboxypeptidase A

Carboxypeptidase A is an *exopeptidase*—it catalyzes the hydrolysis of the C-terminal peptide bond in peptides and proteins, releasing the C-terminal amino acid (Section 21.12).

$$\underset{R}{-\!\!\text{NHCHC}}\overset{O}{\underset{\|}{}}-\underset{R'}{\text{NHCHC}}\overset{O}{\underset{\|}{}}-\underset{R''}{\text{NHCHCO}^-}\overset{O}{\underset{\|}{}}\;+\;\text{H}_2\text{O}$$

$$\downarrow \textbf{carboxypeptidase A}$$

$$\underset{R}{-\!\!\text{NHCHC}}\overset{O}{\underset{\|}{}}-\underset{R'}{\text{NHCHCO}^-}\overset{O}{\underset{\|}{}}\;+\;\underset{R''}{\overset{+}{\text{H}_3\text{NCHCO}^-}}\overset{O}{\underset{\|}{}}$$

Carboxypeptidase A is a metalloenzyme. A *metalloenzyme* is an enzyme that contains a tightly bound metal ion. The metal ion in carboxypeptidase A is Zn^{2+}. In bovine pancreatic carboxypeptidase A, Zn^{2+} is bound to the enzyme at its active site by forming a complex with Glu 72, His 196, and His 69 as well as with a water molecule (Figure 22.5). (The source of the enzyme is specified because although carboxypeptidase A's from different sources follow the same mechanism, they have slightly different primary structures.)

Several groups at the active site of carboxypeptidase A participate in binding the substrate in the optimum position for reaction (Figure 22.5). Arg 145 forms two hydrogen bonds with the C-terminal carboxyl group of the substrate and Tyr 248 also hydrogen bonds with the carboxyl group. The side chain of the C-terminal amino acid is positioned in a hydrophobic pocket, which is why carboxypeptidase A is not active if the C-terminal amino acid is arginine or lysine. Apparently, the long, positively charged side chains of these amino acid residues cannot fit into the nonpolar pocket (Table 21.1).

3-D Molecule:
Carboxypeptidase A

- When the substrate binds to the active site, Zn^{2+} partially complexes with the oxygen of the carbonyl group of the amide that will be hydrolyzed (Figure 22.5). Zn^{2+} polarizes the carbon–oxygen double bond, making the carbonyl carbon more susceptible to nucleophilic attack and stabilizing the negative charge that develops on the oxygen atom in the transition state that leads to the tetrahedral intermediate. Arg 127 also increases the carbonyl group's electrophilicity and stabilizes the developing negative charge on the transition state. Zn^{2+} also complexes with water, which increases its acidity, thereby making the nucleophile more like hydroxide ion. Glu 270 functions as a general-base catalyst, further increasing water's nucleophilicity.

- In the second step of the catalytic reaction, Glu 270 functions as a general-acid catalyst, increasing the leaving tendency of the amino group. When the reaction is over, the amino acid (phenylalanine in this example) and the peptide with one less amino acid residue dissociate from the enzyme and another molecule of substrate binds to the active site. It has been suggested that the unfavorable electrostatic interaction between the negatively charged carboxyl group of the peptide product and the negatively charged Glu 270 residue facilitates release of the product from the enzyme.

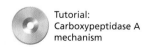
Tutorial:
Carboxypeptidase A
mechanism

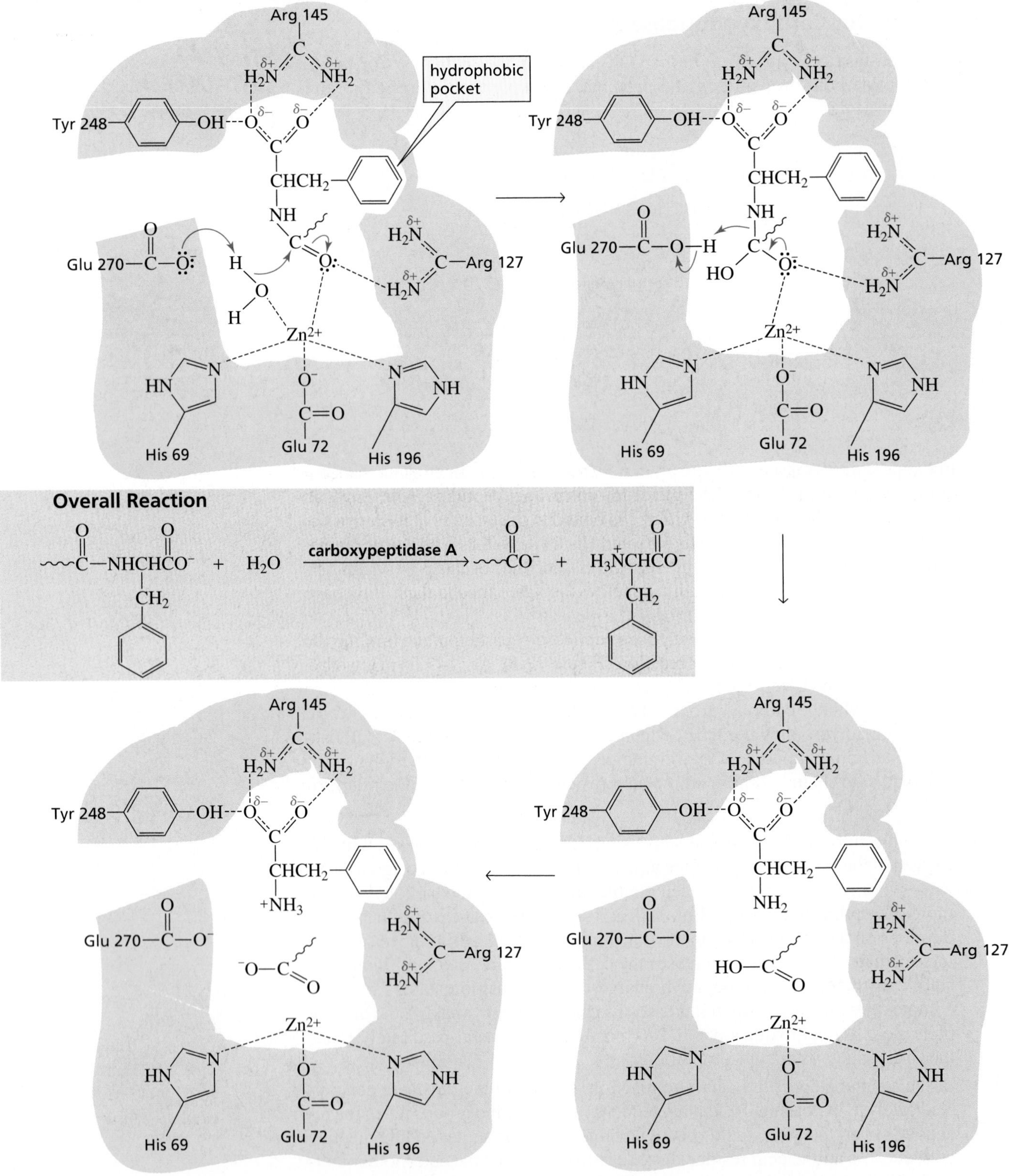

▲ Figure 22.5
Proposed mechanism for the carboxypeptidase A-catalyzed hydrolysis of a peptide bond.

PROBLEM 16◆

Which of the following C-terminal peptide bonds would be more readily cleaved by carboxypeptidase A? Explain.

Ser-Ala-Phe   or   Ser-Ala-Asp

PROBLEM 17

Carboxypeptidase A has esterase activity as well as peptidase activity. In other words, it can hydrolyze ester bonds as well as peptide bonds. When it hydrolyzes ester bonds, Glu 270 acts as a nucleophilic catalyst instead of a general-base catalyst. Propose a mechanism for the carboxypeptidase A-catalyzed hydrolysis of an ester bond.

Mechanism for the Serine Proteases

Trypsin, chymotrypsin, and elastase are members of a large group of *endopeptidases* known collectively as serine proteases. (Recall that an endopeptidase cleaves a peptide bond that is not at the end of a peptide chain.) They are called *proteases* because they catalyze the hydrolysis of protein peptide bonds. They are called *serine proteases* because they all have a serine residue at the active site that participates in the catalysis.

The various serine proteases have similar primary structures, suggesting that they are evolutionarily related. They all have the same three catalytic residues at the active site: an aspartate, a histidine, and a serine. But they have one important difference—the nature of the pocket at the active site that binds the side chain of the amino acid residue undergoing hydrolysis (Figure 22.6). This is what gives the serine proteases their different specificities (Section 21.12). The pocket in trypsin is narrow, with a serine and a negatively charged aspartate carboxyl group at the bottom

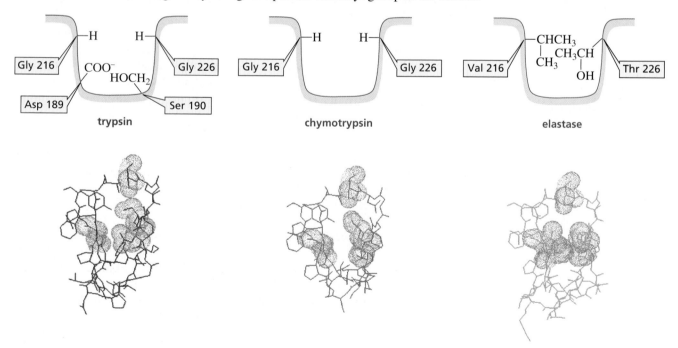

▲ **Figure 22.6**
The binding pockets in trypsin, chymotrypsin, and elastase. The negatively charged aspartate is shown in red, and the relatively nonpolar amino acids are shown in green. The structures of the binding pockets explain why trypsin binds long positively charged amino acids, chymotrypsin binds flat nonpolar amino acids, and elastase binds only small amino acids.

of the pocket. The shape and charge of the binding pocket cause it to bind long, positively charged amino acid side chains (Lys and Arg). This is why trypsin hydrolyzes peptide bonds on the C-side of arginine and lysine residues. The pocket in chymotrypsin is narrow and is lined with nonpolar amino acids, so chymotrypsin cleaves on the C-side of amino acids with flat, nonpolar side chains (Phe, Tyr, Trp). In elastase, two glycines on the sides of the pocket in trypsin and in chymotrypsin are replaced by relatively bulky valine and threonine residues. Consequently, only small amino acids can fit into the pocket. Elastase, therefore, hydrolyzes peptide bonds on the C-side of small amino acids (Gly and Ala).

The mechanism for bovine chymotrypsin-catalyzed hydrolysis of a peptide bond is shown in Figure 22.7. The other serine proteases follow the same mechanism.

- As a result of binding the flat, nonpolar side chain in the pocket, the amide linkage that is to be hydrolyzed is positioned very close to Ser 195. His 57 functions as a general-base catalyst, increasing the nucleophilicity of serine, which attacks the carbonyl group. This process is helped by Asp 102, which uses its negative charge to stabilize the resulting positive charge on His 57. The stabilization of a charge by an opposite charge is called **electrostatic catalysis.** Formation of the tetrahedral intermediate causes a slight change in the shape of the protein that allows the negatively charged oxygen to slip into a previously unoccupied area of the active site known as the *oxyanion hole.* Once in the oxyanion hole, the negatively charged oxygen can hydrogen bond with two peptide groups (Gly 193 and Ser 195), and in this way the tetrahedral intermediate is stabilized.

- In the next step, the tetrahedral intermediate collapses, expelling the amino group. This is a strongly basic group that cannot be expelled without the participation of His 57, which acts as a general-acid catalyst. The product of the second step is an **acyl-enzyme intermediate** because the serine group of the enzyme has been acylated.

- The third step is just like the first step except that water instead of serine is the nucleophile. Water attacks the acylated group of the acyl-enzyme intermediate, with His 57 functioning as a general-base catalyst to increase the nucleophilicity of water, and Asp 102 stabilizing the positively charged histidine residue.

- In the final step of the reaction, the tetrahedral intermediate collapses, expelling serine. His 57 functions as a general-acid catalyst in this step, increasing serine's leaving tendency.

Tutorial:
Serine protease
mechanism

The mechanism for chymotrypsin-catalyzed hydrolysis shows the importance of histidine as a catalytic group. Because the pK_a of the imidazole ring of histidine ($pK_a = 6.0$) is close to neutrality, histidine can act both as a general-acid catalyst and as a general-base catalyst (see also Figure 22.11).

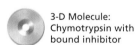

3-D Molecule:
Chymotrypsin with
bound inhibitor

Much information about the relationship between the structure of a protein and its function has been determined by **site-specific mutagenesis**—a technique that replaces one amino acid of a protein with another. For example, when Asp 102 of chymotrypsin is replaced with Asn 102, the enzyme's ability to bind the substrate is unchanged, but its ability to catalyze the reaction decreases to less than 0.05% of its value with the native enzyme. Clearly, Asp 102 must be involved in the catalytic process—in fact, its negative charge stabilizes histidine's positive charge.

side chain of an aspartate (Asp) residue   side chain of an asparagine (Asn) residue

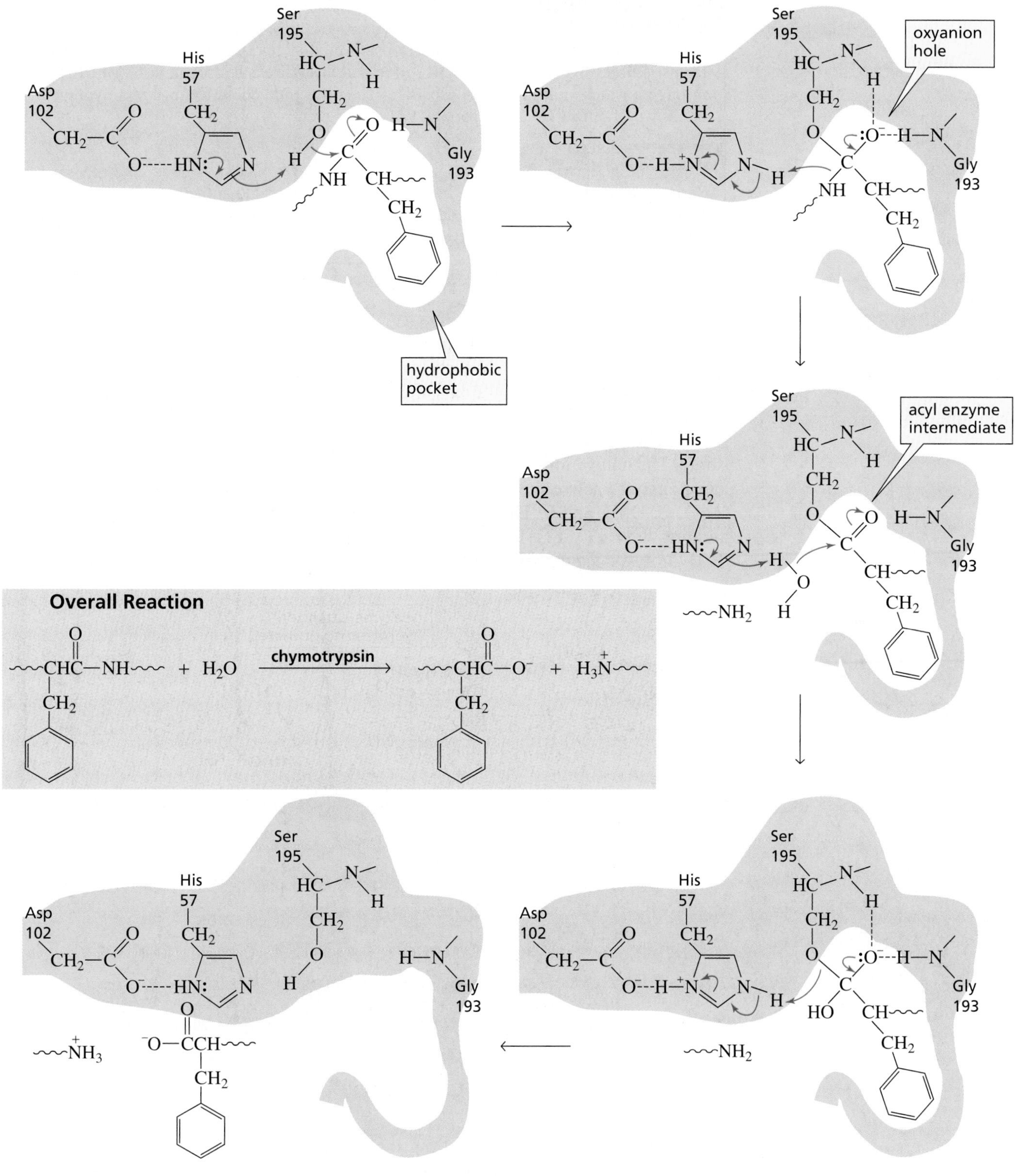

Overall Reaction

▲ Figure 22.7
Proposed mechanism for the chymotrypsin-catalyzed hydrolysis of a peptide bond.

PROBLEM 18◆

Arginine and lysine side chains fit into trypsin's binding pocket. One of these side chains forms a direct hydrogen bond with serine and an indirect hydrogen bond (mediated through a water molecule) with aspartate. The other side chain forms direct hydrogen bonds with both serine and aspartate. Which is which?

PROBLEM 19

Serine proteases do not catalyze hydrolysis if the amino acid at the hydrolysis site is a D-amino acid. Trypsin, for example, cleaves on the C-side of L-Arg and L-Lys but not on the C-side of D-Arg and D-Lys. Explain.

Mechanism for Lysozyme

Lysozyme is an enzyme that destroys bacterial cell walls. These cell walls are composed of alternating NAM (*N*-acetylmuramic acid) and NAG (*N*-acetylglucosamine) units. Lysozyme destroys the cell wall by catalyzing the hydrolysis of the NAM-NAG bond.

The active site of hen egg-white lysozyme binds six sugar residues of the substrate. The many amino acid residues involved in binding the substrate in the correct position in the active site are shown in Figure 22.8. The six sugar residues are labeled A, B, C, D, E, and F. The carboxylic acid substituent of NAM cannot fit into the binding site for C or E. This means that NAM units must be in the sites for B, D, and F. Hydrolysis occurs between D and E.

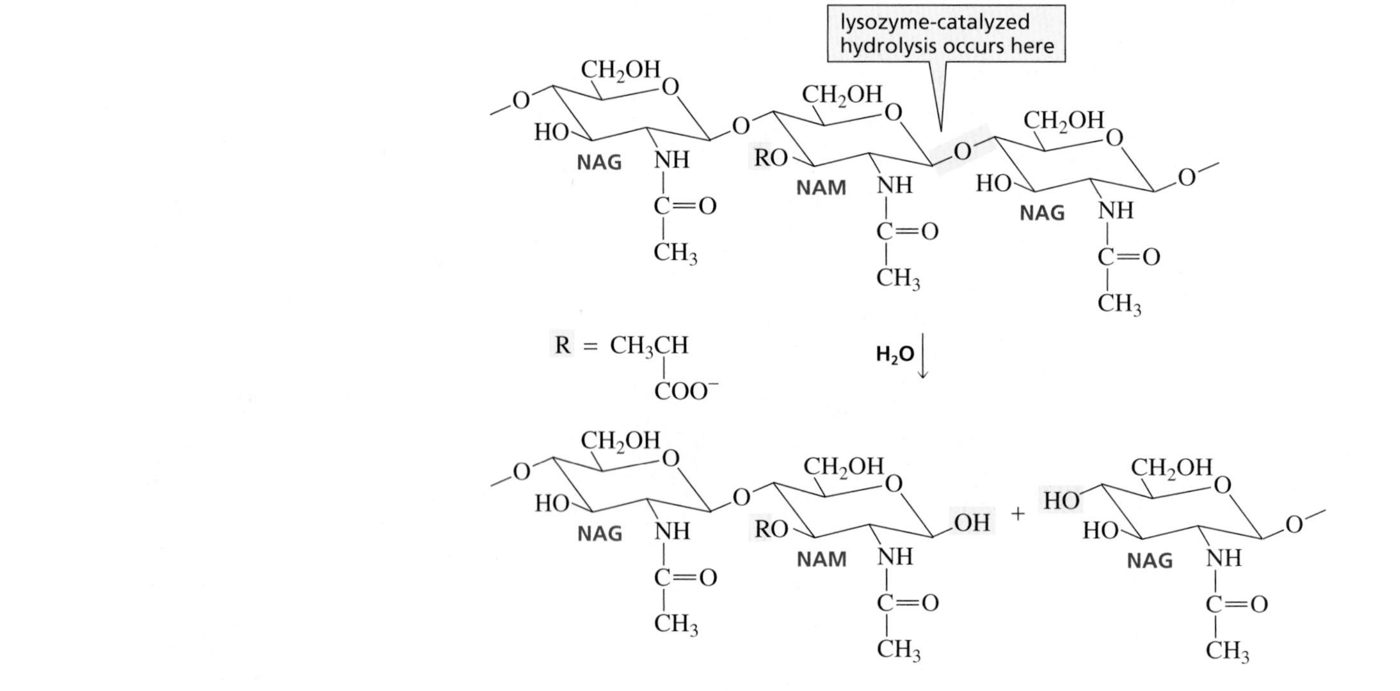

- Lysozyme has two catalytic groups at the active site, Glu 35 and Asp 52 (Figure 22.9). Glu 35 acts as a general-acid catalyst, protonating the leaving group and thereby making it a weaker base and a better leaving group. The developing positive charge on the oxygen atom is stabilized by Asp 52. This is another example of electrostatic catalysis. Site-specific mutagenesis studies show that when Glu 35 is replaced by Asp, the enzyme has only weak activity. Apparently Asp does not have the optimal distance and angle

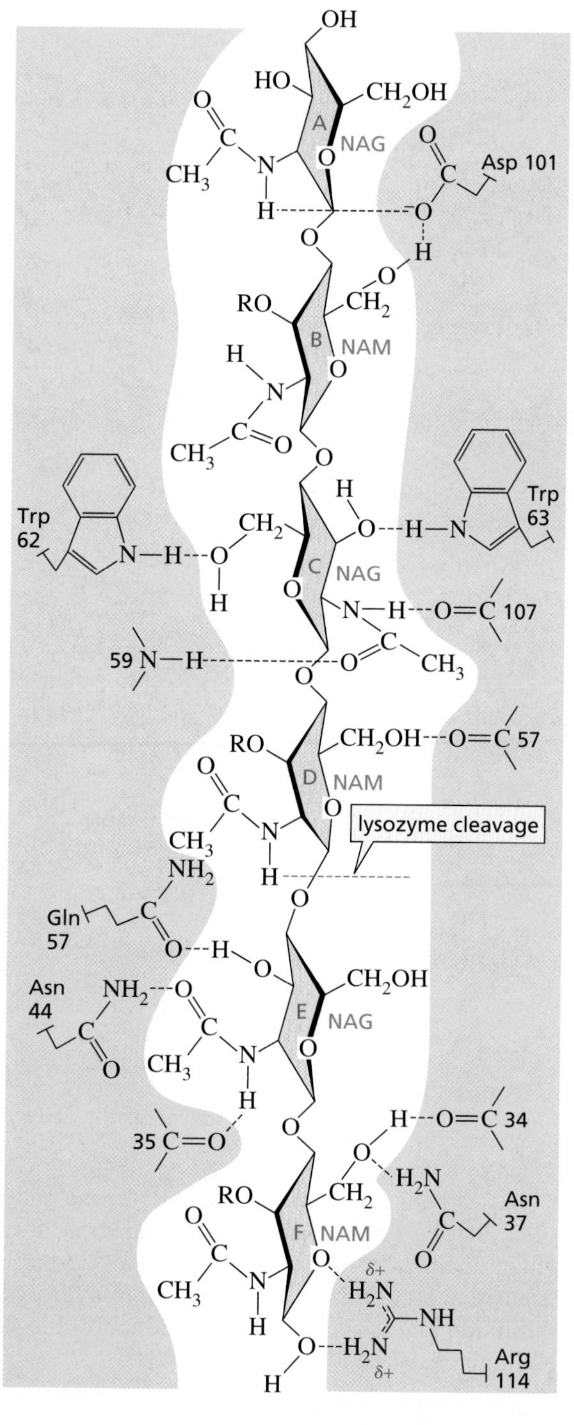

◀ **Figure 22.8**
The amino acids at the active site of lysozyme that are involved in binding the substrate. (Figure copyright by Irving Geis.)

to the oxygen atom that needs to be protonated. When Glu 35 is replaced by Ala, an amino acid that cannot act as an acid catalyst, the activity of the enzyme is completely abolished.

- In the second step of the reaction, water attacks the oxocarbenium ion intermediate with Glu 35 acting as a general-base catalyst. (The positive charge in an **oxocarbenium ion** is shared by a carbon and an oxygen.) The mechanism results in retention of configuration at C-1 of the NAM residue, because the enzyme shields the other side of the oxocarbenium ion intermediate from attack by water.

3-D Molecule:
Lysozyme with bound NAG

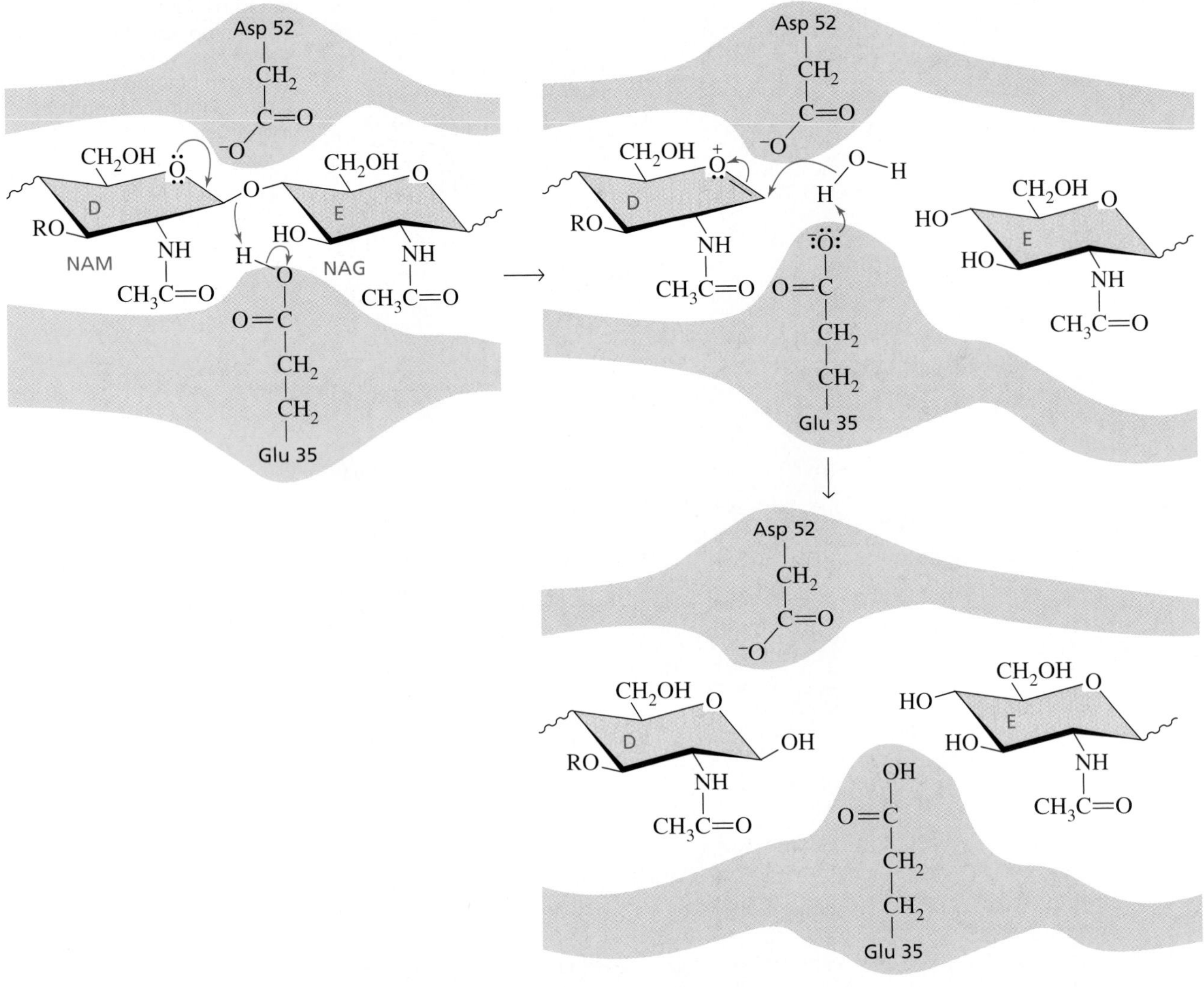

▲ **Figure 22.9**
Proposed mechanism for the lysozyme-catalyzed hydrolysis of a cell wall.

After lysozyme binds its substrate, the enzyme undergoes a change in shape that distorts the D residue of the substrate from a chair conformation to a half-chair conformation (Section 2.12). This destabilizes the reactant *and* stabilizes the transition state for the first step of the hydrolysis reaction because the oxocarbenium ion that is formed as the product of this step is in a half-chair conformation (Figure 22.1[a] and [b]).

PROBLEM 20◆

If H_2O^{18} were used to hydrolyze lysozyme, which ring would contain the label, NAM or NAG?

A plot of the activity of an enzyme as a function of the pH of the reaction mixture is called a **pH-activity profile** or a **pH-rate profile** (Section 17.7). The pH-activity profile for lysozyme is shown in Figure 22.10. It is a bell-shaped curve with the maximum rate occurring at about pH 5.3. The pH at which the enzyme is 50%

active is 3.8 on the ascending leg of the profile and 6.7 on the descending leg. These pH values correspond to the pK_a values of the enzyme's catalytic groups. (This is true for all bell-shaped pH-rate profiles provided the pK_a values are at least 2 pK_a units apart. If the difference between them is less than 2 pK_a units, the precise pK_a values of the catalytic groups must be determined in other ways.)

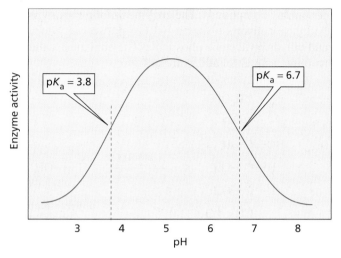

◀ **Figure 22.10**
Dependence of lysozyme activity on the pH of the reaction mixture.

The pK_a given by the ascending leg is the pK_a of a group that is catalytically active in its basic form. When that group is fully protonated, the enzyme is not active. As the pH of the reaction mixture increases, a larger fraction of the group is present in its basic form and, as a result, the enzyme shows increasing activity. Similarly, the pK_a given by the descending leg is the pK_a of a group that is catalytically active in its acidic form. There is maximum catalytic activity when the group is fully protonated, and the activity decreases with increasing pH because a greater fraction of the group lacks a proton.

From the lysozyme mechanism shown in Figure 22.9, we can conclude that Asp 52 is the group with a pK_a of 3.8 and Glu 35 is the group with a pK_a of 6.7. The pH-activity profile indicates that lysozyme is maximally active when Asp 52 is in its basic form and Glu 35 is in its acidic form.

In Table 21.2 we see that the pK_a of aspartic acid is 3.86 and the pK_a of glutamic acid is 4.25. The pK_a of the side chain of Asp 52 agrees with the pK_a of aspartic acid, but the pK_a of the side chain of Glu 35 is much greater than the pK_a of glutamic acid. Why is the pK_a of the glutamic acid side chain at the active site of the enzyme so much greater than the pK_a given for glutamic acid in Table 21.2? The pK_a values in Table 21.2 were determined in water. In the enzyme, Asp 52 is surrounded by polar groups, which means that its pK_a should be close to the pK_a determined in water, a polar solvent. Glu 35, however, is in a predominantly nonpolar microenvironment. The pK_a of a carboxylic acid is greater in a nonpolar solvent because there is less of a tendency to form charged species in nonpolar solvents (Section 9.10).

Part of the catalytic efficiency of lysozyme results from its ability to provide different solvent environments at the active site. This allows one catalytic group to exist in its acidic form at the same surrounding pH at which a second catalytic group exists in its basic form. This property is unique to enzymes. Chemists cannot provide different solvent environments for different parts of nonenzymatic systems.

PROBLEM 21◆

When cut apples are exposed to oxygen, an enzyme-catalyzed reaction causes them to turn brown. They can be prevented from turning brown by coating them with lemon juice (pH = ~3.5). Explain.

Mechanism for Aldolase

Glycolysis is the name given to the series of reactions responsible for converting D-glucose into two molecules of pyruvate (Sections 19.21 and 23.1). Because D-glucose contains six carbons and pyruvate contains three carbons, at some point in the reaction pathway a six-carbon compound must be cleaved into two three-carbon-containing compounds. The enzyme *aldolase* catalyzes this cleavage (Figure 22.11). Aldolase converts D-fructose-1,6-diphosphate into D-glyceraldehyde-3-phosphate and dihydroxyacetone phosphate. The enzyme is called aldolase because the reverse reaction is an aldol addition reaction.

$$
\begin{array}{l}
CH_2OPO_3{}^{2-} \\
| \\
C=O \\
| \\
HO-C-H \\
| \\
H-C-OH \\
| \\
H-C-OH \\
| \\
CH_2OPO_3{}^{2-}
\end{array}
\quad \underset{\text{aldolase}}{\rightleftharpoons} \quad
\begin{array}{l}
HC=O \\
| \\
H\!-\!\!-OH \\
CH_2OPO_3{}^{2-}
\end{array}
\;+\;
\begin{array}{l}
CH_2OPO_3{}^{2-} \\
| \\
C=O \\
| \\
CH_2OH
\end{array}
$$

D-fructose-1,6-diphosphate   D-glyceraldehyde-3-phosphate   dihydroxyacetone phosphate

3-D Molecule: Aldolase

- In the first step of the aldolase-catalyzed reaction, D-fructose-1,6-diphosphate forms an imine with a lysine residue at the active site of the enzyme (Section 17.7).
- A tyrosine residue functions as a general-base catalyst in the step that cleaves the bond between C-3 and C-4. The molecule of D-glyceraldehyde-3-phosphate formed in this step dissociates from the enzyme.
- The enamine intermediate rearranges to an imine with the tyrosine residue now functioning as a general-acid catalyst.
- Hydrolysis of the imine releases dihydroxyacetone phosphate, the other three-carbon product.

PROBLEM 22

Propose a mechanism for the aldolase-catalyzed cleavage of D-fructose-1,6-diphosphate if it did not form an imine with the substrate. What is the advantage gained by imine formation?

PROBLEM 23

In glycolysis, why must D-glucose-6-phosphate isomerize to D-fructose-6-phosphate before the cleavage reaction with aldolase occurs? (See page 995.)

PROBLEM 24 ◆

Aldolase shows no activity if it is incubated with iodoacetic acid before D-fructose-1,6-diphosphate is added to the reaction mixture. Suggest what could cause the loss of activity.

22.10 CATALYTIC ANTIBODIES AND ARTIFICIAL ENZYMES

The induced-fit model led chemists to propose that the most precise binding between an enzyme and its substrate occurs when the substrate is in the transition state. Stabilization of the transition state of an enzyme-catalyzed reaction contributes to the catalytic power of the enzyme. Because transition states are only transitory, the binding

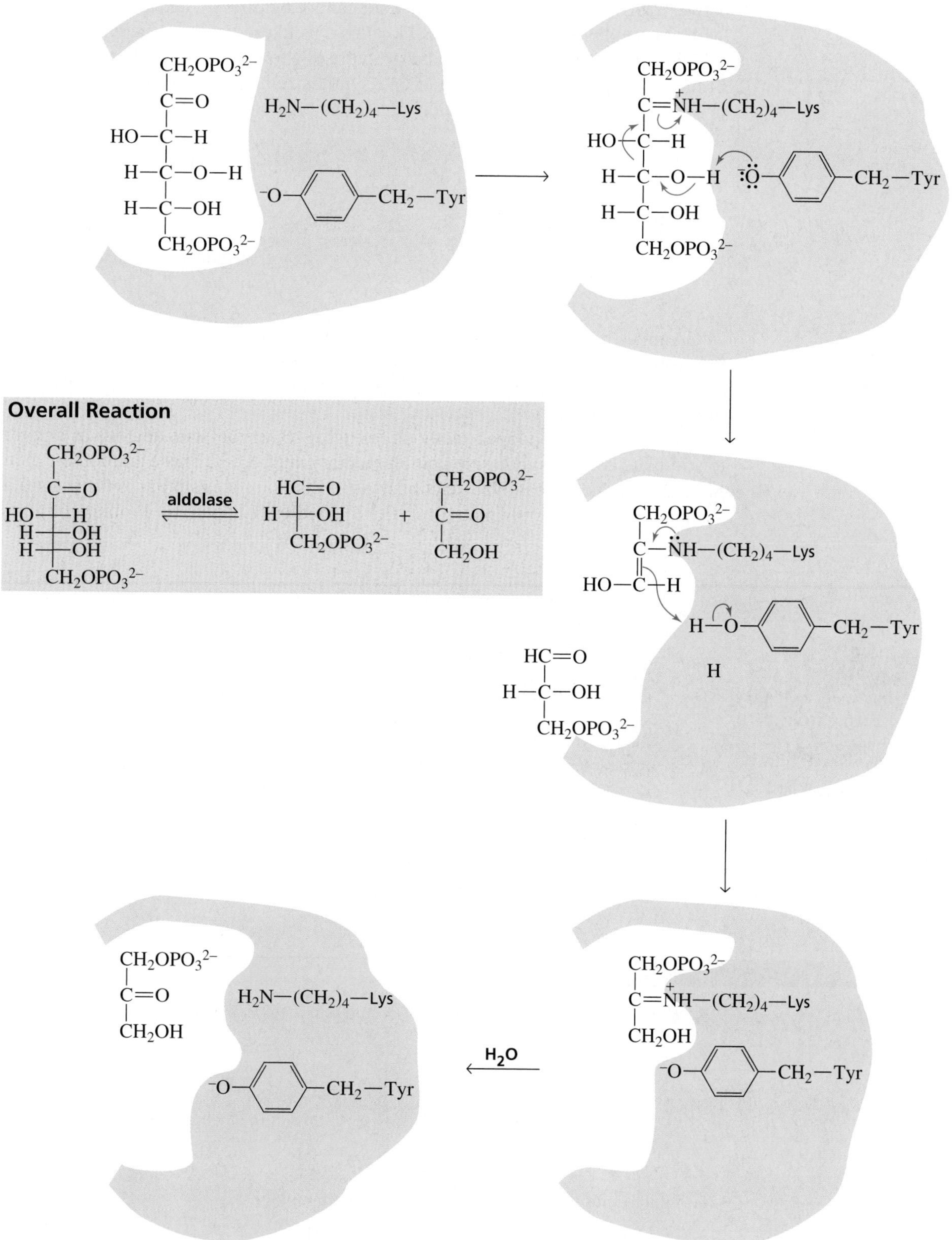

▲ Figure 22.11
Proposed mechanism for the aldolase-catalyzed cleavage of D-fructose-1,6-diphosphate.

between an enzyme and a transition state cannot be directly determined. However, the Hammond postulate (Section 3.11) allows us to predict the structure of a transition state. Knowing the approximate structure of a transition state allows stable transition-state analogs to be designed and synthesized. A **transition-state analog** is structurally similar to the transition state of an enzyme-catalyzed reaction.

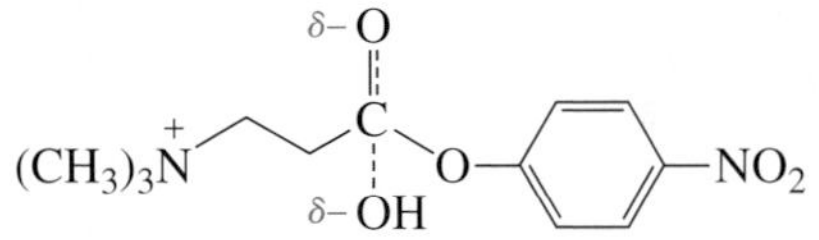

**transition state for attack
by HO⁻ on an ester** 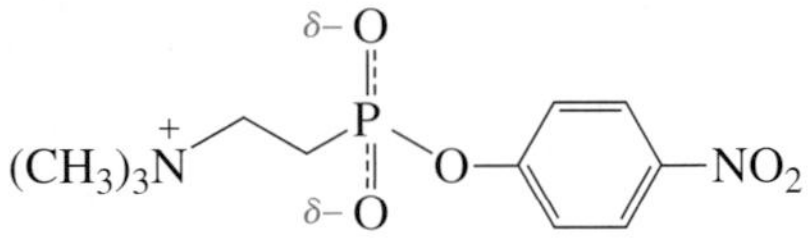   **transition-state analog**

Chemists have found that enzymes bind a transition-state analog 100 to 1 million times more tightly than they bind the substrate. This is good evidence for strong complementarity between an enzyme and the transition state of the reaction it catalyzes.

Chemists have developed methods for using transition-state analogs as antigens to stimulate cells to synthesize complementary antibodies. These antibodies, therefore, are designed to catalyze reactions with transition states similar to the transition-state analog and are known as **catalytic antibodies** (Chapter 16, Problem 68).

KEY TERMS

acid catalyst (page 960)
active site (page 973)
acyl-enzyme intermediate (page 978)
anchimeric assistance (page 970)
base catalyst (page 963)
catalyst (page 957)
catalytic antibody (page 986)
covalent catalysis (page 959)
effective molarity (page 968)
electrophilic catalyst (page 965)
electrostatic catalysis (page 978)

enzyme (page 973)
gem-dialkyl effect (page 969)
general-acid catalysis (page 962)
general-base catalysis (page 964)
induced-fit model (page 973)
intramolecular catalysis (page 969)
lock-and-key model (page 973)
metal-ion catalysis (page 965
molecular recognition (page 973)
nucleophilic catalysis (page 959)

nucleophilic catalyst (page 959)
oxocarbenium ion (page 981)
pH-activity profile (page 982)
pH-rate profile (page 982)
relative rate (page 967)
site-specific mutagenesis (page 978)
specific-acid catalysis (page 961)
specific-base catalysis (page 964)
substrate (page 973)
transition-state analog (page 986)

PROBLEMS

25. Which of the following compounds would eliminate HBr more rapidly?

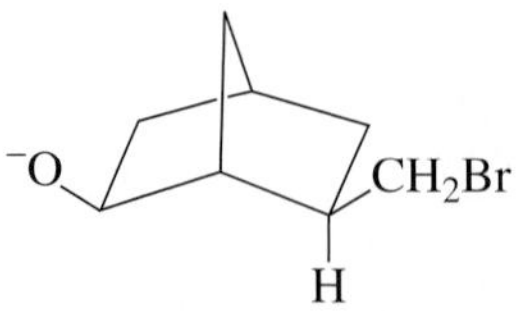

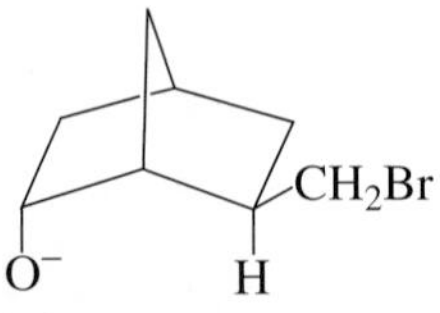

26. Which compound would form a lactone more rapidly?

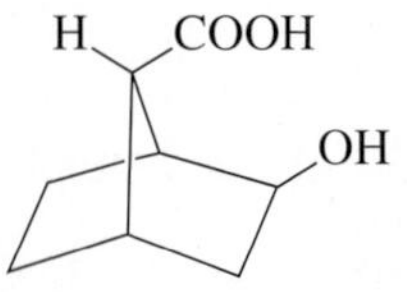

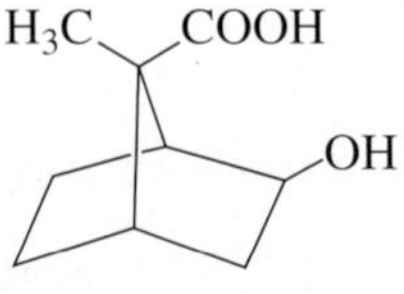

b.

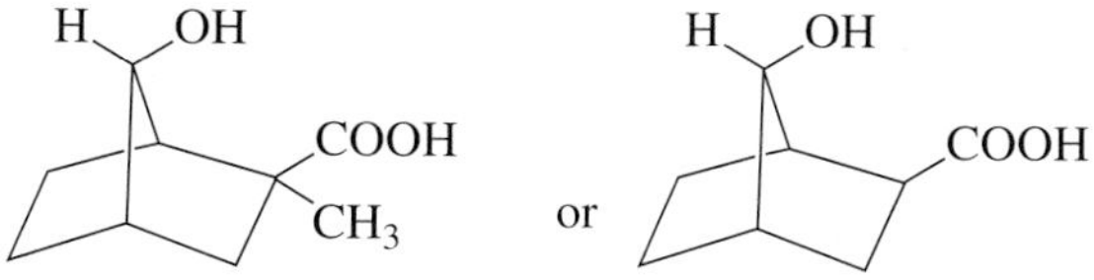

27. Which compound would form an anhydride more rapidly?

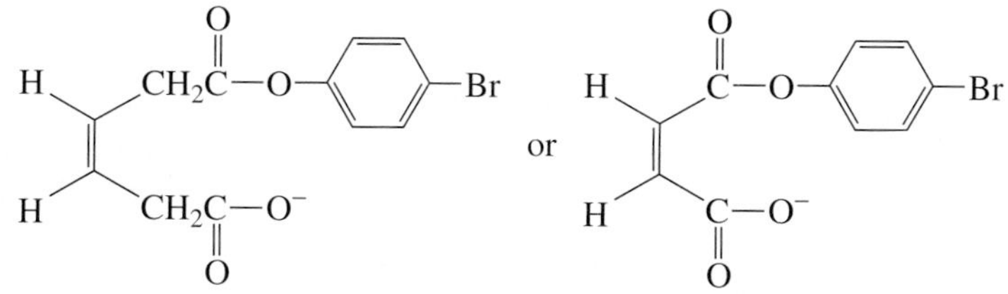

28. Which compound has the greatest rate of hydrolysis:
benzamide, *o*-carboxybenzamide, *o*-formylbenzamide, or *o*-hydroxybenzamide?

29. Indicate the type of catalysis occurring in the slow step in each of the following reaction sequences:

a. 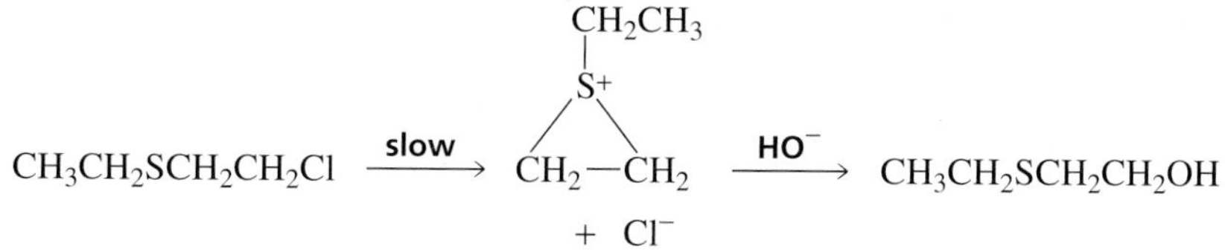 $CH_3CH_2SCH_2CH_2Cl$ $\xrightarrow{\text{slow}}$ $\xrightarrow{\text{HO}^-}$ $CH_3CH_2SCH_2CH_2OH$

b.

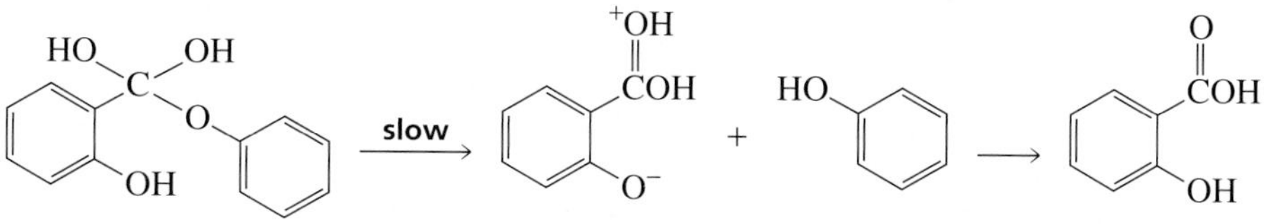

30. The deuterium kinetic isotope effect (k_{H_2O}/k_{D_2O}) for the hydrolysis of aspirin is 2.2. What does this tell you about the kind of catalysis exerted by the *ortho*-carboxyl substituent? (*Hint:* It is easier to break an O—H bond than an O—D bond.)

31. Draw the pH-activity profile for an enzyme with one catalytic group at the active site. The catalytic group is a general-acid catalyst with a pK_a of 5.6.

32. A Co^{2+} complex catalyzes the hydrolysis of the following lactam. Propose a mechanism for the metal-ion catalyzed reaction.

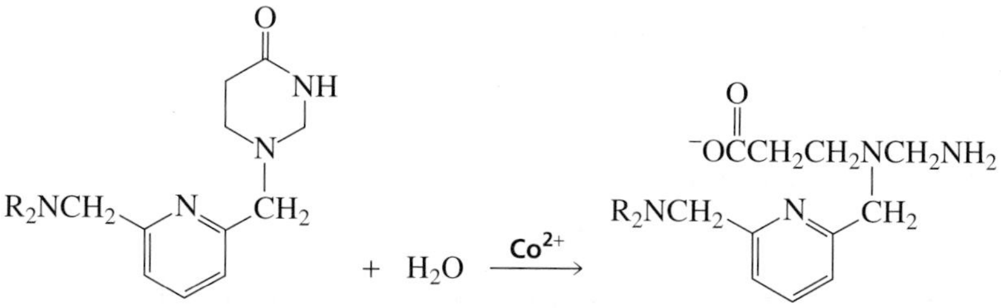

33. There are two kinds of aldolases. Class I aldolases are found in animals and plants, while class II aldolases are found in fungi, algae, and some bacteria. Only class I aldolases form an imine. Class II aldolases have a metal ion (Zn^{2+}) at the active site. The mechanism for the class I aldolases was given in Section 22.9. Propose a mechanism for the class II aldolases.

34. The hydrolysis of the ester shown here is catalyzed by morpholine, a secondary amine. Propose a mechanism for this reaction. (*Hint:* The pK_a of the conjugate acid of morpholine is 9.3, so it is too weak a base to function as a general-base catalyst in this reaction.)

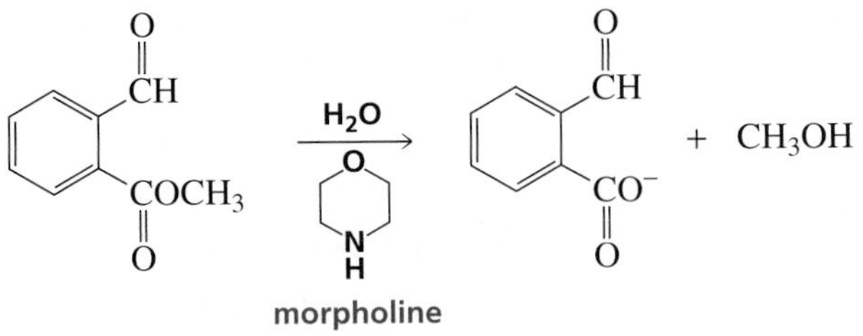

morpholine

35. The enzyme carbonic anhydrase catalyzes the conversion of carbon dioxide into bicarbonate ion (Section 1.20). It is a metalloenzyme with Zn^{2+} coordinated at the active site by three histidine residues. Propose a mechanism for this reaction.

$$CO_2 + H_2O \xrightarrow{\text{carbonic anhydrase}} HCO_3^- + H^+$$

36. At pH $= 12$, the rate of hydrolysis of ester A is faster than the rate of hydrolysis of ester B. At pH $= 8$, the relative rates reverse (ester B hydrolyzes faster than ester A). Explain these observations.

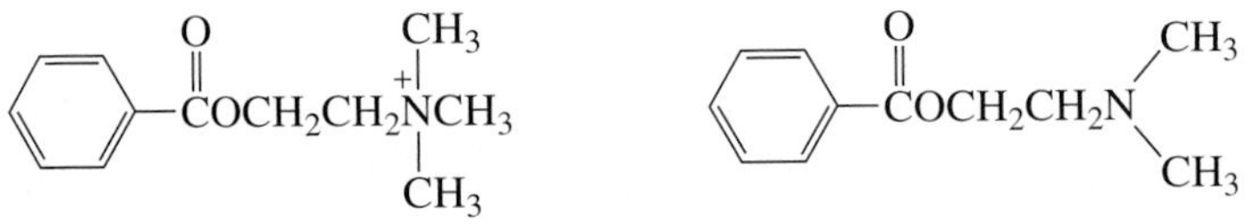

A

B

37. Reaction of 2-acetoxycyclohexyl tosylate with acetate ion forms 1,2-cyclohexanediol diacetate. The reaction is stereospecific; the stereoisomers obtained as products depend on the stereoisomer used as a reactant. Explain the following observations:
  a. Both cis reactants form an optically active trans product, but each cis reactant forms a different trans product.
  b. Both trans reactants form the same racemic mixture.
  c. A trans reactant is more reactive than a cis reactant.

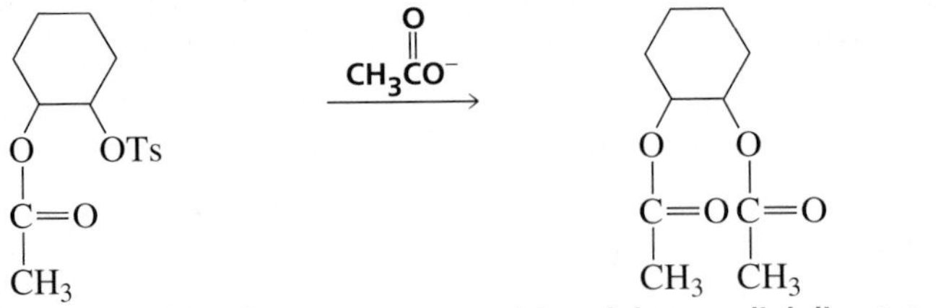

2-acetoxycyclohexyl tosylate   1,2-cyclohexanediol diacetate

38. *Staphylococcus* nuclease is an enzyme that catalyzes the hydrolysis of DNA. The overall hydrolysis reaction is shown here. The reaction is catalyzed by Ca^{2+}, Glu 43, and Arg 87. Propose a mechanism for this reaction.

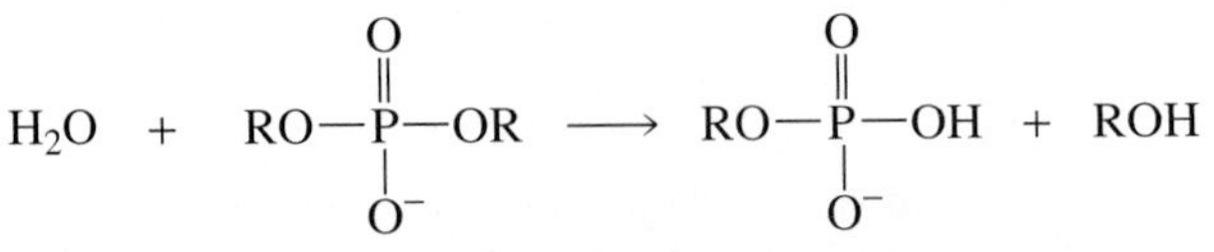

39. Proof for formation of an imine between aldolase and its substrate was obtained using D-fructose-1,6-diphosphate labeled at the C-2 position with ^{14}C as the substrate. $NaBH_4$ was added to the reaction mixture. The radioactive product was isolated from the reaction mixture and hydrolyzed in an acidic solution. A radioactive product was isolated. Draw its structure. (*Hint:* $NaBH_4$ reduces an imine linkage.)

40. 3-Amino-2-oxindole catalyzes the decarboxylation of α-keto acids.
a. Propose a mechanism for the catalyzed reaction.
b. Would 3-aminoindole be equally effective as a catalyst?

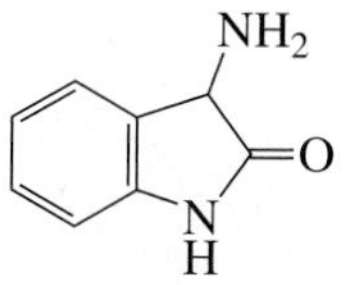

3-amino-2-oxindole

41. a. Explain why the alkyl halide shown here reacts much more rapidly than primary alkyl halides, such as butyl chloride and pentyl chloride, with guanine residues.
b. The alkyl halide can react with two guanine residues in two different chains, thereby cross-linking the chains. Propose a mechanism for the cross-linking reaction.

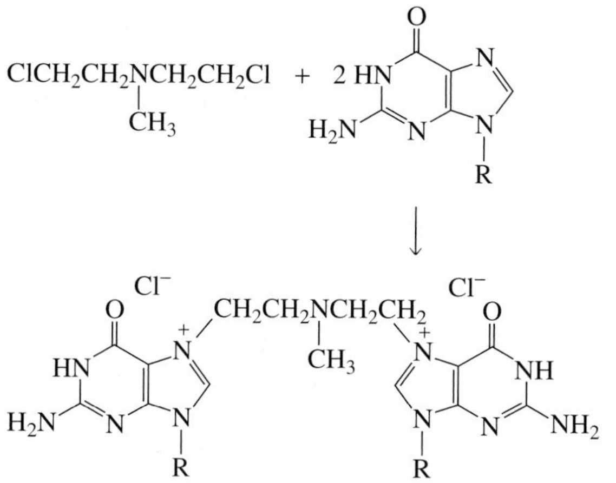

42. Triosephosphate isomerase catalyzes the conversion of dihydroxyacetone phosphate to glyceraldehyde-3-phosphate. The enzyme's catalytic groups are Glu 165 and His 95. In the first step of the reaction, these catalytic groups function as a general-base and a general-acid catalyst, respectively. Propose a mechanism for the reaction.

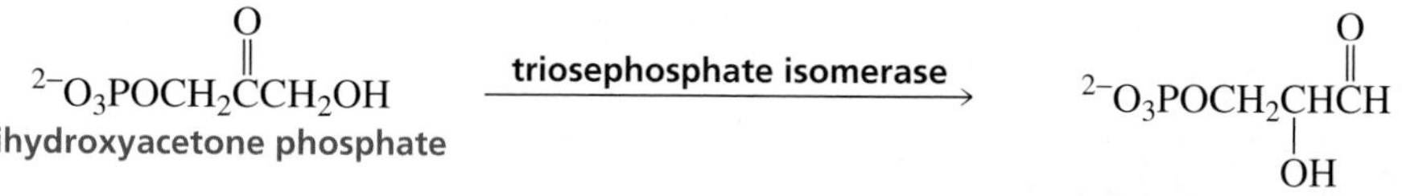

dihydroxyacetone phosphate → **glyceraldehyde-3-phosphate**

23

The Organic Mechanisms of the Coenzymes. Metabolism

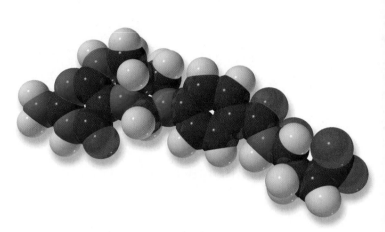

N^5,N^{10}-methylenetetrahydrofolate

Many enzymes cannot catalyze a reaction without the help of a cofactor. **Cofactors** assist enzymes in catalyzing a variety of reactions that cannot be catalyzed solely by the amino acid side chains of the protein. Some cofactors are metal ions, others are organic molecules.

A metal-ion cofactor acts as a Lewis acid in a variety of ways to maximize the catalytic activity of an enzyme. It can coordinate with groups on the enzyme, causing them to align in a geometry that is advantageous for reactivity; it can help bind the substrate to the active site of the enzyme; it can form a coordination complex with the substrate to increase its reactivity; or it can increase the nucleophilicity of water at the active site (Section 22.5). An enzyme that has a tightly bound metal ion (Co^{2+}, Cu^{2+}, Fe^{2+}, Mo^{2+}, Zn^{2+}) is known as a **metalloenzyme.** Carboxypeptidase A is an example of a metalloenzyme (Section 22.9).

PROBLEM 1 ◆

How does the metal ion in carboxypeptidase A increase its catalytic activity?

Casimir Funk (1884–1967) was born in Poland, received his medical degree from the University of Bern, and became a U.S. citizen in 1920. In 1923 he returned to Poland to direct the State Institute of Hygiene but returned to the United States permanently when World War II broke out.

Cofactors that are organic molecules are called **coenzymes.** Coenzymes are derived from organic compounds that are commonly known as **vitamins.** Table 23.1 lists the vitamins and their biochemically active coenzyme forms.

A vitamin is a substance that is needed in small amounts for normal body function that the body cannot synthesize. Sir Frederick Hopkins was the first to suggest that diseases such as rickets and scurvy might result from the absence of substances in the diet that are needed only in very small quantities. Because the first such compound recognized to be essential in the diet was an amine, Casimir Funk incorrectly concluded that all such compounds were amines and called them vitamines ("life-amines"). The *e* was later dropped from the name.

We have seen that enzymes catalyze reactions following the principles of organic chemistry (Section 22.9). Coenzymes use these same principles. We will see that

Table 23.1 The Vitamins, Their Coenzymes, and Their Chemical Functions

Vitamin	Coenzyme	Reaction catalyzed	Human deficiency disease
Water-Soluble Vitamins			
Niacin (niacinate)	NAD^+, $NADP^+$	Oxidation	Pellagra
	NADH, NADPH	Reduction	
Riboflavin (vitamin B_2	FAD, FMN	Oxidation	Skin inflammation
	$FADH_2$, $FMNH_2$	Reduction	
Thiamine (vitamin B_1)	Thiamine pyrophosphate (TPP)	Two-carbon transfer	Beriberi
Lipoic acid (lipoate)	Lipoate	Oxidation	—
	Dihydrolipoate	Reduction	
Pantothenic acid (pantothenate)	Coenzyme A (CoASH)	Acyl transfer	—
Biotin (vitamin H)	Biotin	Carboxylation	—
Pyridoxine (vitamin B_6)	Pyridoxal phosphate (PLP)	Decarboxylation	Anemia
		Transamination	
		Racemization	
		C_α—C_β Bond cleavage	
		α,β-Elimination	
		β-Substitution	
Vitamin B_{12}	Coenzyme B_{12}	Isomerization	Pernicious anemia
Folic acid (folate)	Tetrahydrofolate (THF)	One-carbon transfer	Megaloblastic anemia
Ascorbic acid (vitamin C)	—	—	Scurvy
Water-Insoluble (Lipid-Soluble) Vitamins			
Vitamin A	—	—	
Vitamin D	—	—	Rickets
Vitamin E	—	—	—
Vitamin K	Vitamin KH_2	Carboxylation	—

coenzymes play a variety of chemical roles that the amino acid side chains of enzymes cannot play. Some coenzymes function as oxidizing and reducing agents, some allow electrons to be delocalized, some activate groups for further reaction, and some provide good nucleophiles or strong bases needed for a reaction. Because it would be very inefficient for the body to use a compound only once and then discard it, coenzymes are recycled. Therefore, we will see that after a coenzyme has carried out a reaction, a second reaction converts the coenzyme back to its original form.

An enzyme plus the cofactor it requires to catalyze a reaction is called a **holoenzyme.** An enzyme that has had its cofactor removed is called an **apoenzyme.** Holoenzymes are catalytically active, whereas apoenzymes are catalytically inactive because they have lost their cofactors.

Early nutritional studies divided vitamins into two classes—water-soluble vitamins and water-insoluble vitamins (Table 23.1). Vitamins A, D, E, and K are water-insoluble. Vitamin K is the only water-insoluble vitamin currently recognized to function as a coenzyme. Vitamin A is required for proper vision, vitamin D regulates calcium and phosphate metabolism, and vitamin E is an antioxidant. Because they do not function as coenzymes, vitamins A, D, and E are not discussed in this chapter. Vitamins A and E are discussed in Sections 8.8 and 24.7, and vitamin D is discussed in Section 28.6.

Sir Frederick G. Hopkins (1861–1947) was born in England. His finding that one sample of a protein supported life while another did not led him to conclude that one sample contained a trace amount of a substance essential to life. His postulation later became known as the "vitamin concept," for which he received a share of the 1929 Nobel Prize in medicine or physiology. He also originated the concept of "essential amino acids."

All the water-soluble vitamins function as coenzymes except vitamin C. In spite of its name, vitamin C is actually not a vitamin because it is required in fairly high amounts and most mammals are able to synthesize it (Section 20.19). Humans and guinea pigs cannot synthesize it, however, so it must be included in their diets. We have seen that vitamin C and vitamin E are radical inhibitors and therefore act as antioxidants. Vitamin C traps radicals formed in aqueous environments, while vitamin E traps radicals formed in nonpolar environments (Section 8.8).

It is impossible to overdose on water-soluble vitamins because the body can readily eliminate any excess. However, you can overdose on water-insoluble vitamins because they are not easily eliminated by the body. As a result they can accumulate in cell membranes and other nonpolar components of the body. For example, excess vitamin D causes calcification of soft tissues. The kidneys are particularly susceptible to calcification, and this eventually leads to kidney failure. Vitamin D is formed in the skin as a result of a photochemical reaction caused by the ultraviolet rays from the sun (Section 28.6).

VITAMIN B₁

Christiaan Eijkman (1858–1930) was a member of a medical team that was sent to the East Indies to study beriberi in 1886. At that time, all diseases were thought to be caused by microorganisms. When the microorganism that caused beriberi could not be found, the team left the East Indies. Eijkman stayed behind to become the director of a new bacteriological laboratory. In 1896 he accidentally discovered the cause of beriberi when he noticed that chickens being used in the laboratory had developed symptoms characteristic of the disease. He found that the symptoms had developed when a cook had started feeding the chickens rice meant for hospital patients. The symptoms disappeared when a new cook resumed feeding them chicken feed. Later it was recognized that thiamine (vitamin B₁) is present in rice hulls but not in polished rice. For this work, Eijkman shared the 1929 Nobel Prize in medicine or physiology with Frederick Hopkins.

Christiaan Eijkman

Sir Frederick Hopkins

23.1 OVERALL VIEW OF METABOLISM

The reactions that living organisms carry out to obtain the energy they need and to synthesize the compounds they require are collectively known as **metabolism.** The process of metabolism can be divided into two parts, **catabolism** and **anabolism.** *Catabolic reactions* break down complex nutrient molecules to provide energy and simple precursor molecules for synthesis. *Anabolic reactions* require energy and result in the synthesis of complex biomolecules from simpler precursor molecules.

catabolism complex molecules ⟶ simple molecules + energy

anabolism simple molecules + energy ⟶ complex molecules

We will look at many different enzyme-catalyzed reactions in this chapter. It is important to remember that almost every reaction that occurs in a living system is catalyzed by an enzyme. The enzyme holds the reactants and any necessary cofactors in place so that the reacting functional groups and catalytic groups are oriented in a way that will cause a very specific and well-defined chemical reaction (Section 22.8). In some cases it may be helpful to see where a particular reaction fits into the overall metabolic scheme. So we will start by taking an over-all view of metabolism.

Catabolism can be divided into four stages. The *first stage of catabolism* is called digestion. The reactants required for all life processes ultimately come from our diet. We really are what we eat. In this first stage, fats, carbohydrates, and proteins are hydrolyzed into fatty acids, monosaccharides, and amino acids, respectively (Figure 23.1).

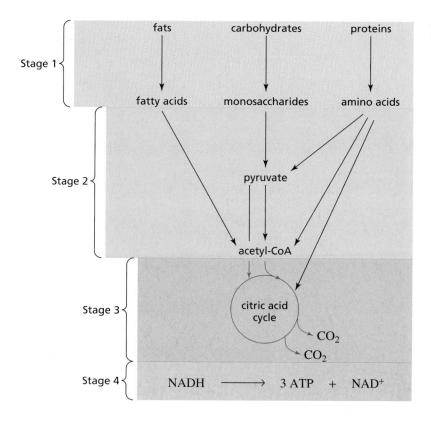

◀ **Figure 23.1**
The four stages of catabolism.

In *the second stage of catabolism,* these hydrolysis products (fatty acids, monosac-charides, and amino acids) are converted to compounds that can enter the citric acid cycle (also known as the Krebs cycle; Figure 23.2). To enter the citric acid cycle, a com-pound must be either one of the compounds in the citric acid cycle (called a citric acid cycle intermediate), or acetyl-CoA (the only non-citric acid cycle intermediate that can enter the cycle, Section 16.18), or pyruvate (because pyruvate can be converted to acetyl-CoA or to oxaloacetate, a citric acid cycle intermediate; Sections 23.4 and 23.5). Fatty acids are converted to acetyl-CoA; monosaccharides are converted to pyruvate via glycolysis (Figure 23.3); and amino acids are converted to acetyl-CoA, pyruvate, and/or citric acid cycle intermediates, depending on the amino acid.

The *third stage of catabolism* is the citric acid cycle. For every acetyl-CoA that enters the cycle, two molecules of CO_2 are formed.

Metabolic energy is measured in terms of adenosine triphosphate (ATP). How the body uses ATP is described in Sections 25.2 and 25.3. Very little ATP is formed in the first three stages of catabolism. However, in the *fourth stage of catabolism,* every NADH (see Section 23.2) that is formed in the process of carrying out oxidation re-actions in the earlier stages is converted into three ATPs in a process known as ox-idative phosphorylation. Thus, the bulk of the energy provided by fats, carbohydrates, and proteins is obtained in this fourth stage.

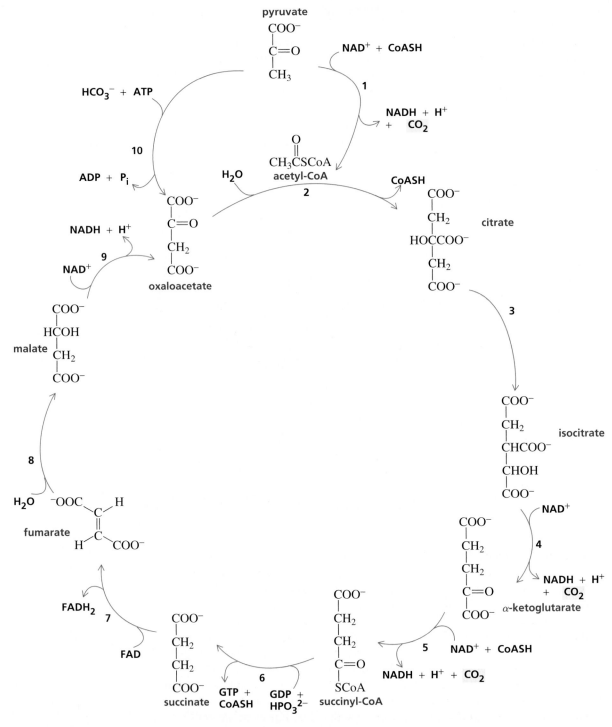

1   pyruvate dehydrogenase
2   citrate synthase
3   aconitase
4   isocitrate dehydrogenase

5   α-ketogutarate dehydrogenase
6   succinyl-CoA synthetase
7   succinate dehydrogenase

8   fumarase
9   malate dehydrogenase
10   pyruvate carboxylase

Figure 23.2 ▲
The citric acid cycle (also known as the Krebs cycle)—the series of enzyme-catalyzed reactions responsible for the oxidation of pyruvate and acetyl-CoA to CO_2, and the entry of pyruvate and acetyl-CoA into the cycle.

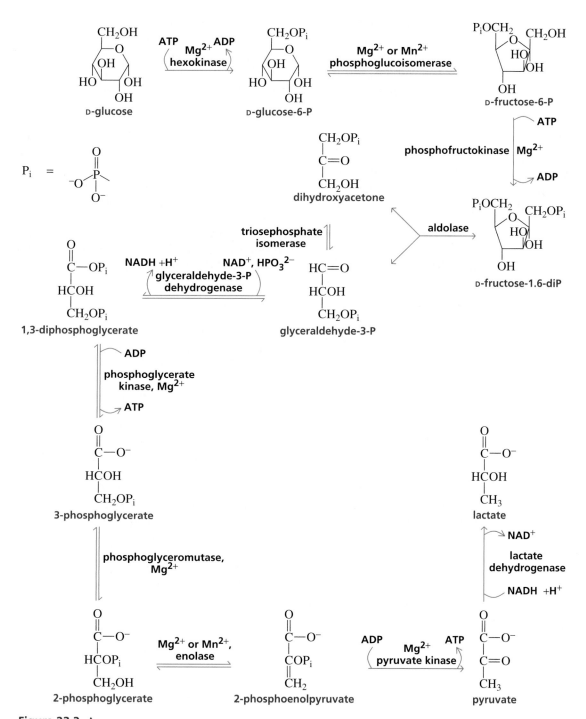

Figure 23.3 ▲
Glycolysis—the series of enzyme-catalyzed reactions responsible for the conversion of
1 mole of D-glucose into 2 moles of pyruvate.

Anabolism can be thought of as the reverse of catabolism. Acetyl-CoA, pyruvate, and citric acid cycle intermediates are the starting materials for the synthesis of fatty acids, monosaccharides, and amino acids. These compounds are then used to form fats, carbohydrates, and proteins. The mechanisms used by biological systems to synthesize fats and proteins are discussed in Sections 19.21 and 25.13.

23.2 NIACIN: THE VITAMIN NEEDED FOR MANY REDOX REACTIONS

An enzyme that catalyzes an oxidation or a reduction reaction requires a coenzyme because none of the amino acid side chains can act as oxidizing or reducing agents. Instead, the coenzyme must be the oxidizing or reducing agent. The enzyme's role is to hold the substrate and coenzyme together so the oxidation or reduction reaction can take place.

The coenzymes most commonly used by enzymes that catalyze oxidation reactions are **nicotinamide adenine dinucleotide (NAD$^+$)** and **nicotinamide adenine dinucleotide phosphate (NADP$^+$)**. (You were introduced to NAD$^+$ in Section 18.10.)

$$\text{substrate}_{\text{reduced}} + \text{NAD}^+ \underset{\text{enzyme}}{\overset{\text{enzyme}}{\rightleftharpoons}} \text{substrate}_{\text{oxidized}} + \text{NADH} + \text{H}^+$$

$$\text{substrate}_{\text{reduced}} + \text{NADP}^+ \underset{\text{enzyme}}{\overset{\text{enzyme}}{\rightleftharpoons}} \text{substrate}_{\text{oxidized}} + \text{NADPH} + \text{H}^+$$

When NAD$^+$ (or NADP$^+$) oxidizes a substrate, the coenzyme is reduced to NADH (or NADPH). NADH and NADPH are reducing agents. They are used as coenzymes by enzymes that catalyze reduction reactions. Enzymes that catalyze oxidation reactions bind NAD$^+$ (or NADP$^+$) more tightly than they bind NADH (or NADPH). When the oxidation reaction is over, the relatively loosely bound NADH (or NADPH) dissociates from the enzyme. Likewise, enzymes that catalyze reduction reactions bind NADH (or NADPH) more tightly than they bind NAD$^+$ (or NADP$^+$). When the reduction reaction is over, the relatively loosely bound NAD$^+$ (or NADP$^+$) dissociates from the enzyme.

NAD$^+$ and NADP$^+$ are oxidizing agents.

NADH and NADPH are reducing agents.

NAD$^+$ is composed of two nucleotides linked together through their phosphate groups. A **nucleotide** consists of a heterocycle attached, in a β-configuration, to C-1 of a phosphorylated ribose. A **heterocycle** is a cyclic compound in which one or more of the ring atoms is an atom other than carbon (see introduction to Chapter 27).

a nucleotide   adenine   niacinamide / nicotinamide   niacin / nicotinic acid

3-D Molecules:
Niacin;
Nicotinamide;
Reduced form
of nicotinamide

The heterocyclic component of one of the nucleotides of NAD$^+$ is nicotinamide, and the heterocyclic component of the other is adenine. This accounts for the coenzyme's name (**nicotinamide adenine dinucleotide**). The positive charge in the NAD$^+$ abbreviation indicates the positively charged nitrogen of the substituted pyridine ring.

The only way in which NADP$^+$ differs structurally from NAD$^+$ is the phosphate group bonded to the 2'-OH group of the ribose of the adenine nucleotide; this explains the addition of "P" to its name. NAD$^+$ and NADH are generally used as coenzymes in catabolic reactions, and the phosphorylated derivatives, NADP$^+$ and NADPH, are generally used as coenzymes in anabolic reactions.

ATP provides the adenine nucleotide for the coenzymes. The nicotinamide nucleotide is not synthesized by biological systems—it is derived from the vitamin known as niacin.

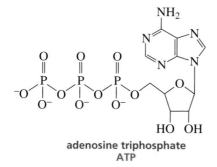

adenosine triphosphate
ATP

NIACIN DEFICIENCY

A deficiency in niacin causes pellagra, a disease that begins with dermatitis and ultimately causes insanity and death. More than 120,000 cases of pellagra were reported in the United States in 1927, mainly among poor people with unvaried diets. A factor known to be present in preparations of vitamin B prevented pellagra, but it was not until 1937 that the factor was identified as nicotinic acid.

When bread companies started adding nicotinic acid to their bread, they insisted that its name be changed to niacin because they thought that nicotinic acid sounded too much like nicotine and they did not want their vitamin-enriched bread to be associated with a harmful substance. Niacinamide is a nutritionally equivalent form of the vitamin.

The oxidation of the *secondary alcohol group* of malate to a *ketone group* is one of the reactions in the citric acid cycle (Figure 23.2). NAD^+ is the oxidizing reagent in this reaction. Many enzymes that catalyze oxidation reactions are called **dehydrogenases** (they remove hydrogen).

$$^-OCCH_2CHCO^- + NAD^+ \xrightleftharpoons{\text{malate dehydrogenase}} {}^-OCCH_2CCO^- + NADH + H^+$$

malate ⟶ oxaloacetate

Pyruvate is reduced to lactate in glycolysis (a catabolic pathway; Figure 23.3)—NADH is the reducing agent. β-Aspartate-semialdehyde is reduced to homoserine in an anabolic pathway—NADPH is the reducing agent.

$$CH_3C-CO^- + NADH + H^+ \xrightarrow[\substack{\text{a reaction that occurs} \\ \text{in a catabolic pathway}}]{\text{lactate dehydrogenase}} CH_3CH-CO^- + NAD^+$$

pyruvate ⟶ lactate

$$HCCH_2CHCO^- + NADPH + H^+ \xrightarrow[\substack{\text{a reaction that occurs} \\ \text{in an anabolic pathway}}]{\text{homoserine dehydrogenase}} HOCH_2CH_2CHCO^- + NADP^+$$

β-aspartate-semialdehyde ⟶ homoserine

The differentiation between the coenzymes used in catabolism and anabolism is maintained because the enzymes that catalyze these oxidation–reduction reactions exhibit strong specificity for a particular coenzyme. For example, an enzyme that catalyzes an oxidation reaction can readily tell the difference between NAD^+ and $NADP^+$ and, if it is an enzyme in a catabolic pathway, will bind NAD^+ but not $NADP^+$. In addition, the relative concentrations of the coenzymes in the cell encourage binding of the appropriate coenzyme. For example, because NAD^+ and NADH are catabolic coenzymes and catabolic reactions are most often oxidation reactions, the NAD^+ concentration in the cell is much greater than the NADH concentration (the cell maintains its $[NAD^+]/[NADH]$ ratio near 1000). Because $NADP^+$ and NADPH are anabolic coenzymes and anabolic pathways are predominantly reduction reactions, the concentration of NADPH in the cell is greater than the concentration of $NADP^+$ (the ratio of $[NADP^+]/[NADPH]$ is maintained at about 0.01).

Mechanisms for Pyridine Nucleotide Coenzymes

How do these oxidation–reduction reactions take place? All the chemistry of the pyridine nucleotide coenzymes (NAD^+, $NADP^+$, NADH, NADPH) takes place at the 4-position of the pyridine ring. The rest of the molecule is important for binding the coenzyme to the proper site on the enzyme. If a substrate is being *oxidized,* it donates a hydride ion (H^-) to the 4-position of the pyridine ring. A basic group of the enzyme can help the reaction by removing a proton from the oxygen in the substrate.

oxidation of substrate
reduction of coenzyme

Glyceraldehyde-3-phosphate dehydrogenase is an example of an enzyme that uses NAD^+ as an oxidizing coenzyme. The enzyme catalyzes the oxidation of the aldehyde group of glyceraldehyde-3-phosphate (GAP) to an anhydride of a carboxylic acid and phosphoric acid—one of the reactions in glycolysis (Figure 23.3).

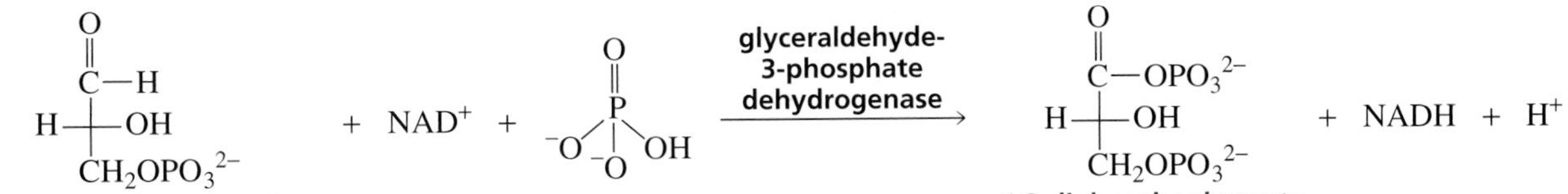

D-glyceraldehyde-3-phosphate

D-1,3-diphosphoglycerate

In the first step of the mechanism for this reaction, an SH group of a cysteine side chain at the active site of the enzyme reacts with glyceraldehyde-3-phosphate to form a tetrahedral intermediate. A group on the enzyme increases cysteine's nucleophilicity by acting as a general-base catalyst. The tetrahedral intermediate expels a hydride ion, transferring it to the 4-position of the pyridine ring of an NAD^+ that is bonded to the enzyme at an adjacent site. NADH dissociates from the enzyme, and the enzyme binds a new NAD^+. Phosphate reacts with the thioester, forming the anhydride product and releasing cysteine. Notice that at the end of the reaction the holoenzyme is exactly as it was at the beginning of the reaction, so the catalytic cycle can be repeated. The NADH produced in the reaction is reoxidized to NAD^+ in the fourth stage of catabolism (Figure 23.1).

The mechanism for reduction by NADH (or by NADPH) is the reverse of the mechanism for oxidation by NAD^+ (or by $NADP^+$). If a substrate is being *reduced,* the dihydropyridine ring donates a hydride ion from its 4-position to the substrate. An acidic group of the enzyme aids the reaction by donating a proton to the substrate.

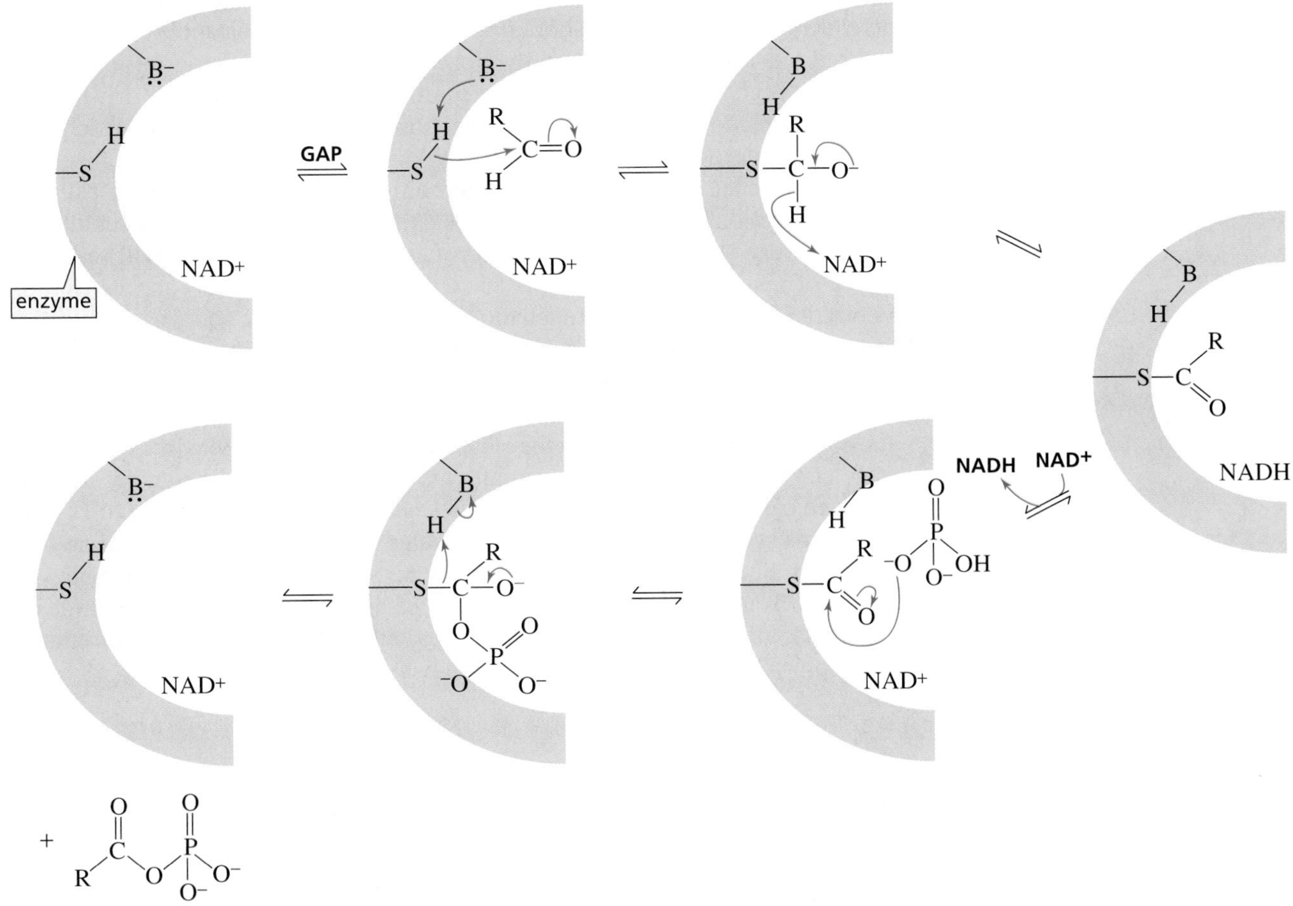

Tutorial: Mechanism of NAD⁺ and NADH dependent reactions

reduction of substrate
oxidation of coenzyme

Because NADH and NADPH reduce compounds by donating a hydride ion, they can be thought of as biological equivalents of $NaBH_4$ or $LiAlH_4$, which are hydride donors used as reducing reagents in nonbiological reactions (Section 18.1).

Why are biological redox (reducing and oxidizing) reagents so much more complicated than the redox reagents used to carry out the same reactions in the laboratory? NADH is certainly more complicated than $LiAlH_4$, although both reagents reduce compounds by donating a hydride ion. Much of the structural complexity of a coenzyme is for molecular recognition—to allow it to be recognized by the enzyme. *Molecular recognition* allows the enzyme to bind the substrate and the coenzyme in the proper orientation for reaction.

In addition, a redox reagent found in a biological system must be less reactive than a laboratory redox reagent because it must be selective. It cannot reduce just any reducible compound with which it comes into contact. Biological reactions are much more carefully controlled than that. Also, because a biological reducing agent must be recycled rather than having its oxidized form thrown away (as would be the case for a reducing agent used in a laboratory), the equilibrium constant between the oxidized and reduced forms is generally close to 1. Therefore, biological redox reactions are not highly exergonic—they are equilibrium reactions that are driven in the appropriate direction by the removal of reaction products as a result of participation in subsequent reactions.

Because the coenzymes are relatively unreactive compared to nonbiological redox agents, the reaction between the substrate and the coenzyme does not occur at all, or takes place very slowly, without the enzyme. For example, NADH will reduce an aldehyde or a ketone only in the presence of an enzyme. $NaBH_4$ and $LiAlH_4$ are more reactive hydride donors—much too reactive to exist in the aqueous environment of the cell. Similarly, NAD^+ is a much more selective oxidizing agent than the typical oxidizing agents used in the laboratory—NAD^+ will oxidize an alcohol only in the presence of an enzyme.

As you study the coenzymes in this chapter, don't let the complexity of their structures deter you. Notice that only a small part of the coenzyme is actually involved in the chemical reaction. Notice also that the coenzymes follow the same rules of organic chemistry as the simple organic molecules with which you are familiar.

We saw in Section 4.14 that an oxidizing enzyme can tell the difference between the two hydrogens on the carbon from which it catalyzes the removal of a hydride ion. For example, alcohol dehydrogenase removes only the pro-*R*-hydrogen of ethanol. H_a is the pro-*R*-hydrogen, and H_b is the pro-*S*-hydrogen.

Similarly, a reducing enzyme can tell the difference between the two hydrogens at the 4-position of the nicotinamide ring of NADH. An enzyme has a specific binding site for the coenzyme, and when it binds the coenzyme, the enzyme blocks one of the coenzyme's sides. If the enzyme blocks the B-side of NADH, the substrate will bind to the A-side and the H_a hydride ion will be transferred to the substrate. If the enzyme blocks the A-side of the coenzyme, the substrate will have to bind to the B-side and the H_b hydride ion will be transferred. Currently, 156 dehydrogenases are known to transfer H_a, and 121 are known to transfer H_b.

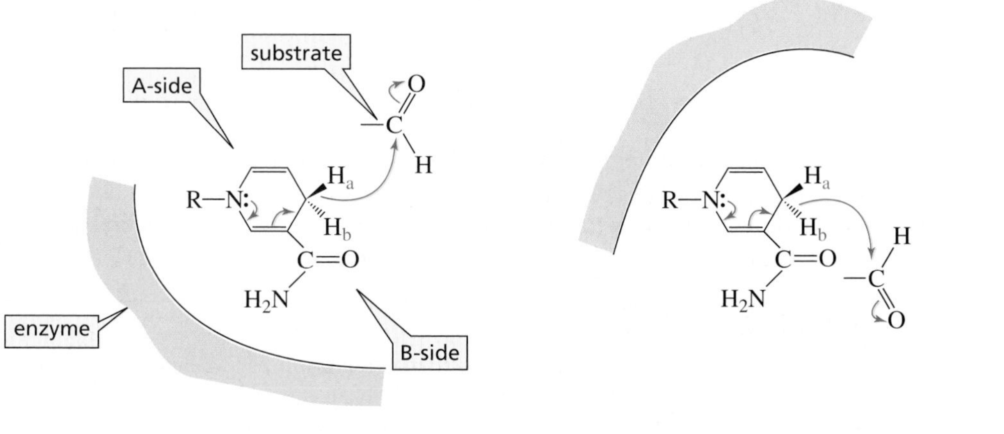

The enzyme blocks the B-side of the coenzyme, so the substrate binds to the A-side.

The enzyme blocks the A-side of the coenzyme, so the substrate binds to the B-side.

A *flavoprotein* is an enzyme that contains either **flavin adenine dinucleotide (FAD)** or **flavin mononucleotide (FMN)** as a coenzyme. FAD and FMN, like NAD$^+$ and NADP$^+$, are coenzymes used in oxidation reactions. As its name indicates, FAD is a dinucleotide in which one of the heterocyclic components is flavin and the other is adenine. FMN contains flavin but not adenine—it is a mononucleotide. (Flavin is a bright yellow compound; *flavus* is Latin for "yellow.") Notice that instead of ribose, the flavin nucleotide has a ribitol group (a reduced ribose). Flavin plus ribitol is called *riboflavin*. Riboflavin is also known as vitamin B$_2$. A vitamin B$_2$ deficiency causes inflammation of the skin.

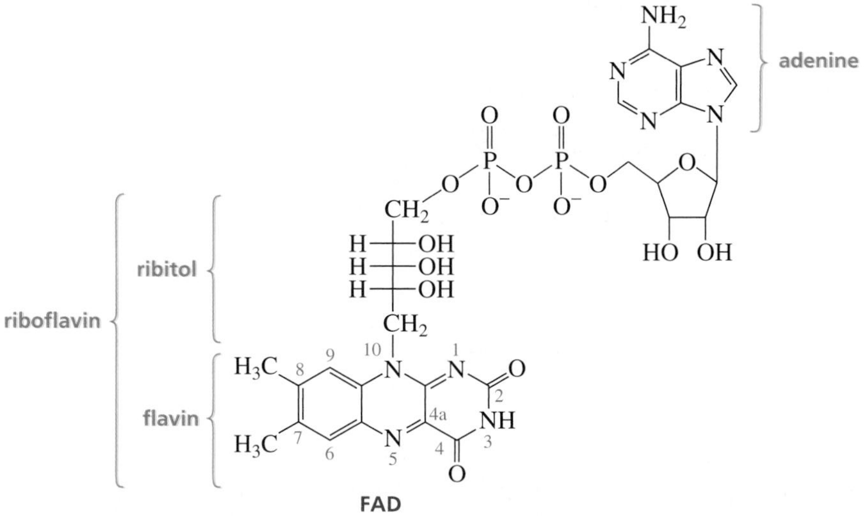

23.3
FLAVIN ADENINE DINUCLEOTIDE AND FLAVIN MONONUCLEOTIDE: VITAMIN B$_2$

FAD

3-D Molecules:
Flavin adenine
dinucleotide (FAD);
Flavin mononucleotide
(FMN)

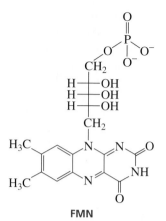

FMN

In many flavoproteins, FAD (or FMN) is bound quite tightly. Tight binding allows the enzyme to control the oxidation potential of the coenzyme. (The more positive the oxidation potential, the stronger the oxidizing agent.) Consequently, some flavoproteins are stronger oxidizing agents than others.

> ### PROBLEM 2 ◆
>
> FAD is obtained by an enzyme-catalyzed reaction that uses FMN and ATP as substrates. What is the other product of this reaction?

FAD and FMN are oxidizing agents.

How can we tell which enzymes use FAD (or FMN) rather than NAD^+ (or $NADP^+$) as the oxidizing coenzyme? A rough guideline is that NAD^+ and $NADP^+$ are the coenzymes used in enzyme-catalyzed oxidation reactions involving carbonyl compounds (alcohols being oxidized to ketones, aldehydes, or carboxylic acids), while FAD and FMN are the coenzymes used in other types of oxidations. For example, in the following reactions, FAD oxidizes a dithiol to a disulfide, an amine to an imine, and a saturated alkyl group to an unsaturated alkene, and FMN oxidizes NADH to NAD^+. This is only an approximate guideline, however, because FAD is involved in some oxidations that involve carbonyl compounds, while NAD^+ and $NADP^+$ are involved in some oxidations that do not involve carbonyl compounds.

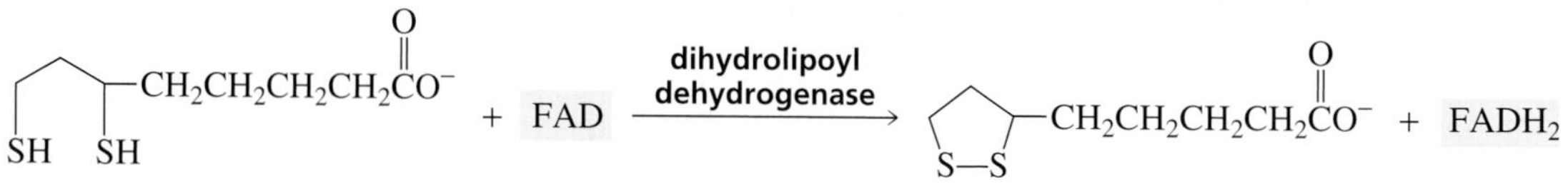

FADH₂ and FMNH₂ are reducing agents.

$$\underset{\substack{\text{SH} \quad \text{SH} \\ \textbf{dihydrolipoate}}}{\text{—CH}_2\text{CH}_2\text{CH}_2\text{CH}_2\overset{\text{O}}{\overset{\|}{\text{C}}}\text{O}^-} + \text{FAD} \xrightarrow[\text{dehydrogenase}]{\text{dihydrolipoyl}} \underset{\substack{\text{S—S} \\ \textbf{lipoate}}}{\text{—CH}_2\text{CH}_2\text{CH}_2\text{CH}_2\overset{\text{O}}{\overset{\|}{\text{C}}}\text{O}^-} + \text{FADH}_2$$

$$\underset{\substack{\text{NH}_2 \\ \textbf{D-amino acid or} \\ \textbf{L-amino acid}}}{\text{RCH}\overset{\text{O}}{\overset{\|}{\text{C}}}\text{O}^-} + \text{FAD} \xrightarrow[\text{L-amino acid oxidase}]{\text{D-amino acid oxidase or}} \underset{\text{NH}}{\text{RC}\overset{\text{O}}{\overset{\|}{\text{C}}}\text{O}^-} + \text{FADH}_2$$

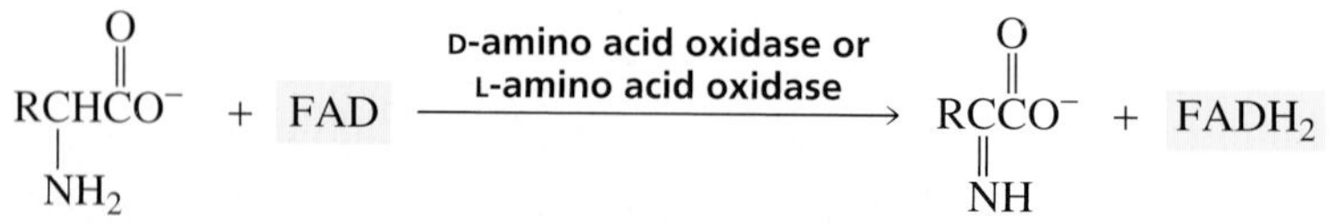

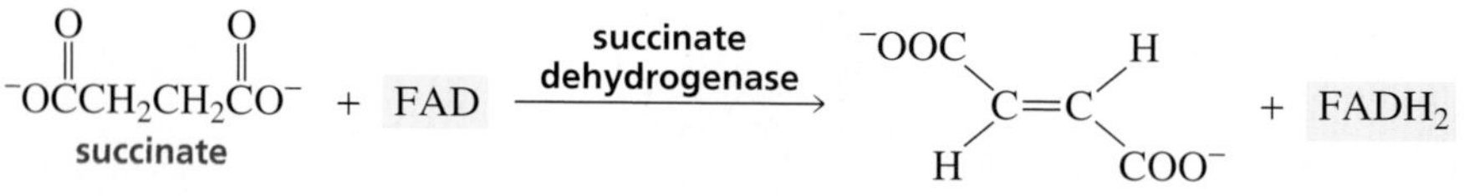

$$\underset{\textbf{succinate}}{{}^-\text{O}\overset{\text{O}}{\overset{\|}{\text{C}}}\text{CH}_2\text{CH}_2\overset{\text{O}}{\overset{\|}{\text{C}}}\text{O}^-} + \text{FAD} \xrightarrow[\text{dehydrogenase}]{\text{succinate}} \underset{\textbf{fumarate}}{\overset{{}^-\text{OOC}}{\underset{\text{H}}{}}\text{C}{=}\text{C}\overset{\text{H}}{\underset{\text{COO}^-}{}}} + \text{FADH}_2$$

$$\text{NADH} + \text{H}^+ + \text{FMN} \xrightarrow{\text{NADH dehydrogenase}} \text{NAD}^+ + \text{FMNH}_2$$

Mechanisms for Flavin Nucleotide Coenzymes

When FAD (or FMN) oxidizes a substrate (S), the coenzyme is reduced to FADH$_2$ (or FMNH$_2$). All the oxidation/reduction chemistry takes place on the flavin ring. Reduction of the flavin ring disrupts the conjugated system, so the reduced coenzymes are less colored than their oxidized forms.

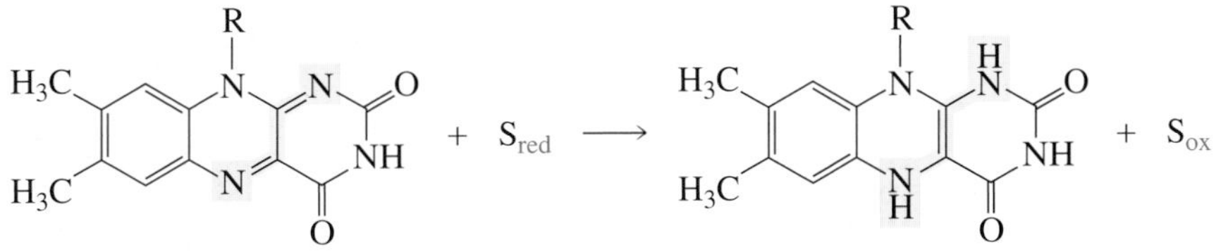

FAD
FMN

$+ \ S_{red} \longrightarrow$

FADH$_2$
FMNH$_2$

$+ \ S_{ox}$

PROBLEM 3 ◆

How many conjugated double bonds are there in:

a. FAD?   **b.** FADH$_2$?

In the first step of the mechanism for the FAD-catalyzed oxidation of dihydrolipoate to lipoate, the thiolate ion attacks the C-4a-position of the flavin ring. This reaction is general-acid-catalyzed—as the thiolate ion attacks the ring, a proton is donated to the N-5 nitrogen. A second nucleophilic attack by a thiolate ion, this time on the sulfur covalently attached to the coenzyme, generates the oxidized product and FADH$_2$. Where this FAD-catalyzed reaction fits into metabolism is discussed in Section 23.4.

mechanism for dihydrolipoyl dehydrogenase

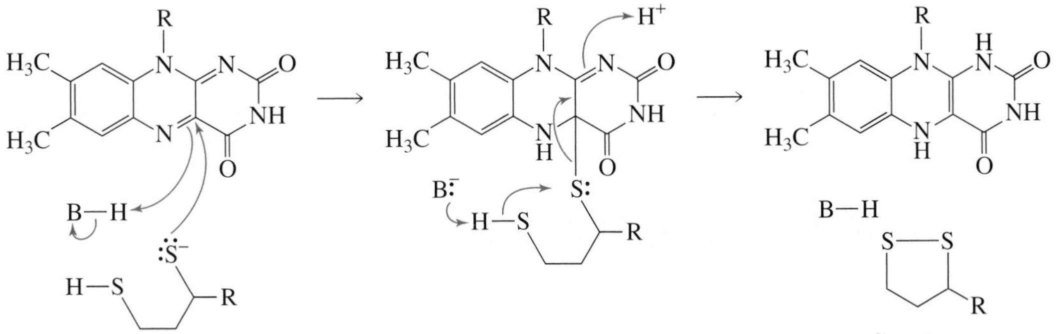

dihydrolipoate

lipoate

In the first step of the flavin-catalyzed oxidation of an amino acid to an imino acid, a basic group at the active site of the enzyme removes a proton from the α-carbon of the amino acid. The carbanion that is formed attacks the N-5 position of the flavin ring. Collapse of the resulting tetrahedral intermediate gives the oxidized amino acid and the reduced coenzyme (FADH$_2$).

Notice that FAD is a more versatile coenzyme than NAD$^+$. Unlike NAD$^+$, which always uses the same mechanism, flavin coenzymes can use several different mechanisms to carry out an oxidation. For example, we have just seen that when FAD oxidizes dihydrolipoate, nucleophilic attack occurs on the C-4a position of the flavin ring, but when it oxidizes an amino acid, nucleophilic attack occurs on the N-5 position.

mechanism for D- or L-amino acid oxidase

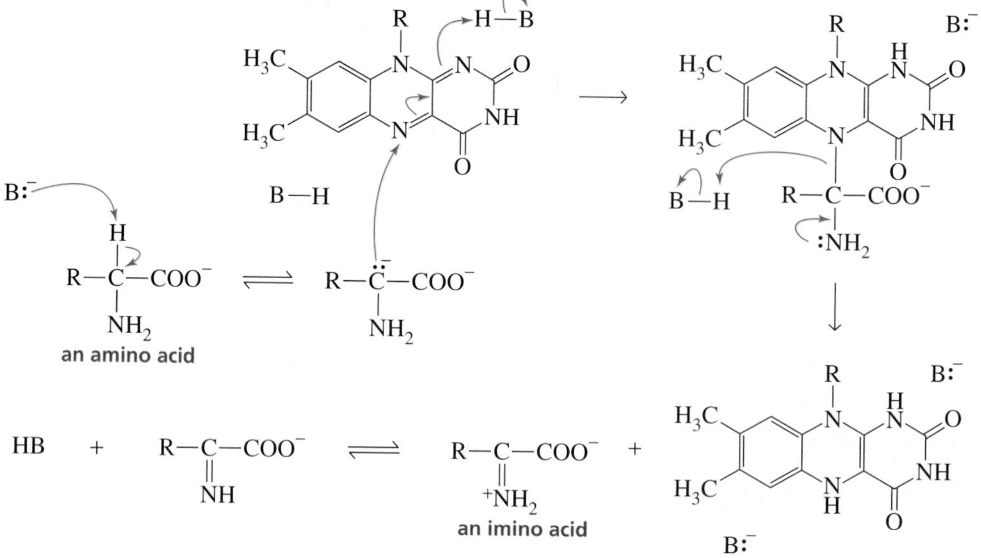

Cells contain very low concentrations of FAD and much higher concentrations of NAD^+. This difference in concentration is responsible for a significant difference in the behavior of enzymes (**E**) that use NAD^+ as an oxidizing agent and those that use FAD. Generally FAD is covalently bound to its enzyme and remains bound after being reduced to $FADH_2$. Because $FADH_2$ (unlike NADH) does not dissociate from the enzyme, $FADH_2$ must be reoxidized to FAD before the enzyme can enter another round of catalysis. The oxidizing agent used for this reaction is NAD^+ or O_2. Therefore, an enzyme that uses an oxidizing coenzyme other than NAD^+ may still require NAD^+ to reoxidize the reduced coenzyme. For this reason NAD^+ has been called the "common coinage" of biological oxidation–reduction reactions.

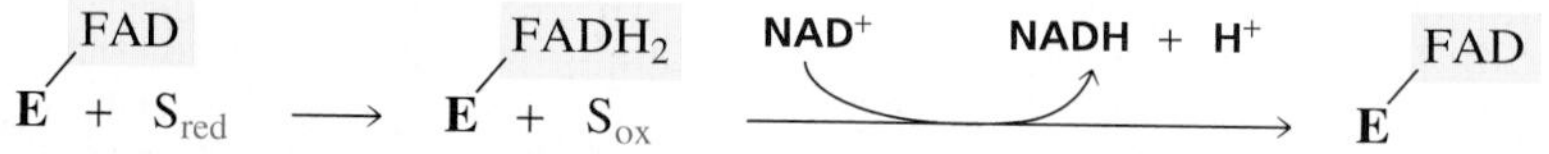

PROBLEM 4

Propose a mechanism for the reduction of lipoate by $FADH_2$.

PROBLEM 5 / SOLVED

A common way for FAD to become covalently bound to its enzyme is by having a proton removed from the C-8 methyl group and a proton donated to N-1. Then a cysteine or other nucleophilic side chain of the enzyme attacks the methylene carbon at C-8 as a proton is donated to N-5. Describe these events mechanistically.

SOLUTION

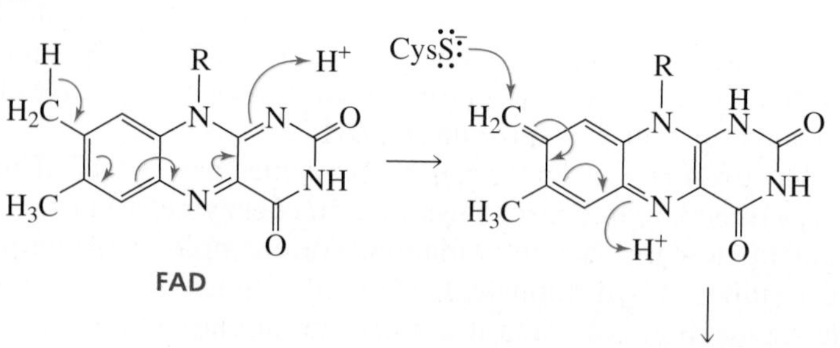

FAD

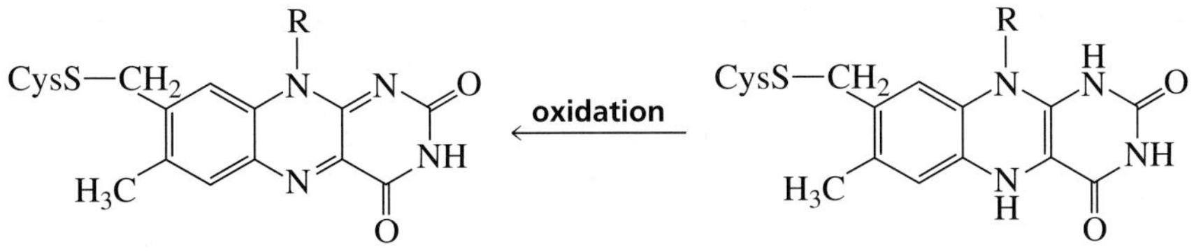

enzyme-FAD   enzyme-FADH$_2$

Notice that during the process of attaching FAD to the enzyme, FAD is reduced to FADH$_2$. It is subsequently oxidized back to FAD. Once the coenzyme is attached to the enzyme, it does not come off.

PROBLEM 6

Explain why the hydrogens of the methyl group bonded to flavin at C-8 are more acidic than those of the methyl group bonded at C-7.

Thiamine was the first of the B vitamins to be identified, so it became known as vitamin B$_1$. The absence of thiamine in the diet causes a disease called beriberi, which damages the heart and impairs nerve reflexes. In extreme cases, it causes paralysis. One major dietary source of vitamin B$_1$ is the hulls of rice kernels (page 992). A deficiency is therefore most likely to occur when highly polished rice is a major component of the diet. A deficiency is also seen in alcoholics who are severely malnourished.

The coenzyme form of vitamin B$_1$ is **thiamine pyrophosphate (TPP).** TPP is the coenzyme required by enzymes that catalyze the transfer of a two-carbon fragment from one species to another.

23.4 THIAMINE PYROPHOSPHATE: VITAMIN B$_1$

Thiamine pyrophosphate (TPP) is required by enzymes that catalyze the transfer of a two-carbon fragment from one species to another.

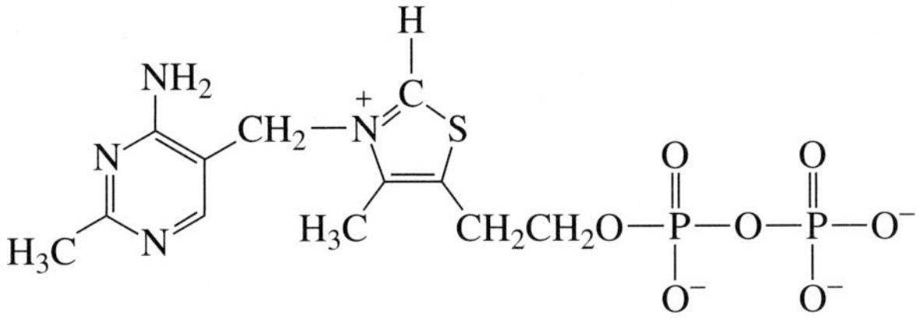

thiamine pyrophosphate
TPP

Pyruvate decarboxylase is an enzyme that requires thiamine pyrophosphate. Pyruvate decarboxylase catalyzes the decarboxylation of pyruvate and transfers the resulting two-carbon fragment to a proton, resulting in the formation of acetaldehyde.

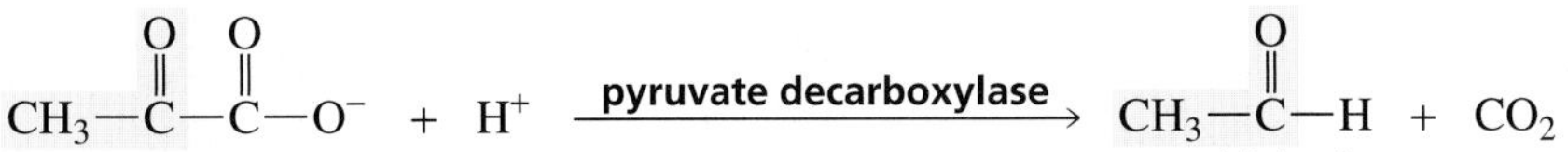

Finding that an α-keto acid such as pyruvate can be decarboxylated should be surprising because the electrons left behind when CO_2 is removed cannot be delocalized onto the carbonyl oxygen. We will see that the thiazolium ring of the coenzyme provides a site for the delocalization of the electrons. A site to which electrons can be delocalized is called an **electron sink.**

Mechanism for Pyruvate Decarboxylase

The hydrogen bonded to the imine carbon of TPP is relatively acidic ($pK_a = 12.7$) because the carbanion formed when the proton is removed is stabilized by the adjacent positively charged nitrogen. This carbanion formed when the proton is removed is a good nucleophile.

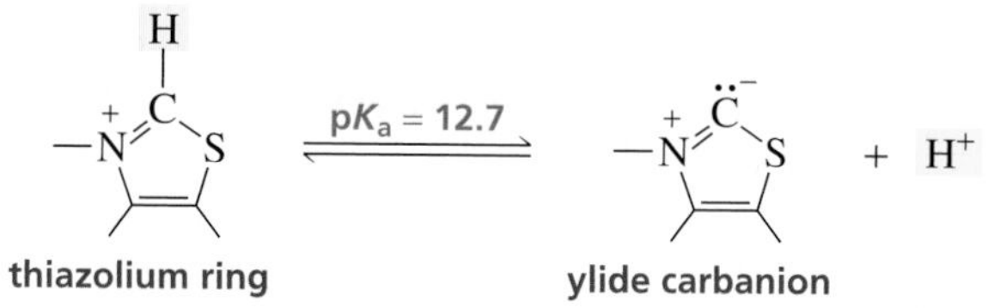

thiazolium ring   ylide carbanion

In the first step of the mechanism for pyruvate decarboxylase, the nucleophilic carbanion attacks the ketone group of the α-keto acid. The resulting intermediate can easily undergo decarboxylation because the electrons left behind when CO_2 is removed can be delocalized onto the positively charged nitrogen. The positively charged nitrogen of TPP is a more effective *electron sink* than the β-keto group of a β-keto acid, a class of compounds that are fairly easily decarboxylated (Section 19.17). The product of decarboxylation is stabilized by electron delocalization—one of the resonance contributors is neutral and the other has separated charges, which we will refer to as a "resonance-stabilized carbanion." Protonation of the resonance-stabilized carbanion and a subsequent elimination reaction form acetaldehyde and regenerate the coenzyme.

Tutorial: Mechanism of pyruvate decarboxylase

mechanism for pyruvate decarboxylase

resonance contributor

resonance contributor
resonance-stabilized carbanion

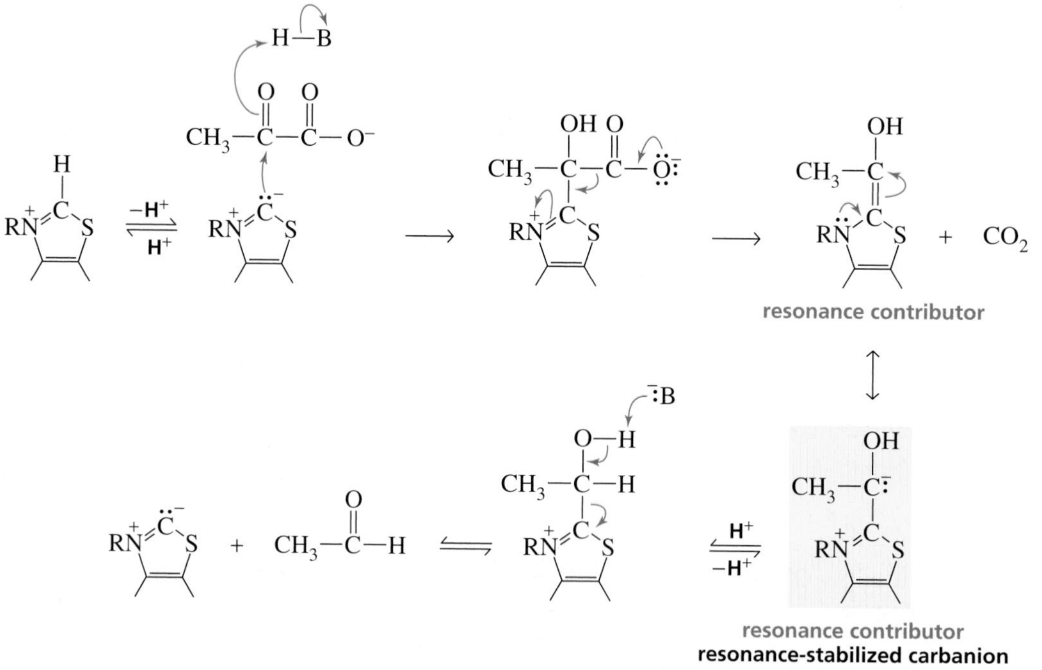

PROBLEM 7

Draw structures that show the similarity between decarboxylation of the pyruvate–TPP intermediate and decarboxylation of a β-keto acid.

PROBLEM 8

Acetolactate synthase is another TPP-requiring enzyme. It also catalyzes the decarboxylation of pyruvate but transfers the resulting two-carbon fragment to another molecule of pyruvate, forming acetolactate. This is the first step in the biosynthesis of the amino acids valine and leucine. Propose a mechanism for acetolactate synthase.

$$2\ CH_3-\overset{\overset{O}{\|}}{C}-\overset{\overset{O}{\|}}{C}-O^- \xrightarrow[\text{TPP}]{\substack{\text{acetolactate} \\ \text{synthase}}} CH_3-\overset{\overset{O}{\|}}{C}-\underset{\underset{CH_3}{|}}{\overset{\overset{OH}{|}}{C}}-\overset{\overset{O}{\|}}{C}-O^- +\ CO_2$$

pyruvate   acetolactate

PROBLEM 9

Acetolactate synthase can also transfer the two-carbon fragment from pyruvate to α-ketobutyrate, forming α-aceto-α-hydroxybutyrate. This is the first step in the formation of isoleucine. Propose a mechanism for this reaction.

$$CH_3\overset{\overset{O}{\|}}{C}-\overset{\overset{O}{\|}}{C}O^- +\ CH_3CH_2\overset{\overset{O}{\|}}{C}-\overset{\overset{O}{\|}}{C}O^- \xrightarrow[\text{TPP}]{\substack{\text{acetolactate} \\ \text{synthase}}} CH_3\overset{\overset{O}{\|}}{C}-\underset{\underset{CH_2CH_3}{|}}{\overset{\overset{OH}{|}}{C}}COO^- +\ CO_2$$

α-ketobutyrate α-aceto-α-hydroxybutyrate

Mechanism for the Pyruvate Dehydrogenase System

A compound must enter the citric acid cycle to be completely metabolized (Figure 23.1). Thus fats, carbohydrates, and proteins must be converted into compounds that are part of the citric acid cycle or can enter the cycle. Acetyl-CoA is the only compound capable of entering the citric acid cycle (Figure 23.2). The final product of carbohydrate metabolism is pyruvate (Figure 23.3). Pyruvate, therefore, must be converted into acetyl-CoA before it can enter the citric acid cycle (or pyruvate can be converted to oxaloacetate—a citric acid cycle intermediate; Section 23.5).

The pyruvate dehydrogenase system is a group of three enzymes responsible for the conversion of pyruvate to acetyl-CoA. The pyruvate dehydrogenase system requires TPP and four other coenzymes (lipoate, coenzyme A, FAD, and NAD$^+$).

$$CH_3-\overset{\overset{O}{\|}}{C}-\overset{\overset{O}{\|}}{C}-O^- +\ CoASH \xrightarrow[\text{system}]{\text{pyruvate dehydrogenase}} CH_3-\overset{\overset{O}{\|}}{C}-SCoA +\ CO_2$$

pyruvate   acetyl-CoA

The first enzyme in the system catalyzes the reaction of TPP with pyruvate to form the same resonance-stabilized carbanion formed by pyruvate decarboxylase and by the enzyme in Problems 8 and 9. The second enzyme of the system (E_2) requires **lipoate,** a coenzyme that is attached to its enzyme by an amide linkage to a lysine side chain. The disulfide linkage of lipoate is cleaved when it undergoes nucleophilic attack by the carbanion. In the next step, the TPP carbanion is eliminated from the tetrahedral intermediate. **Coenzyme A (CoASH)** reacts with the thioester in a transthioesterification reaction (one thioester is converted into another), substituting coenzyme A for dihydrolipoate. At this point the final reaction product (acetyl-CoA) has been formed. However, before another cycle of catalysis can occur, dihydrolipoate must be oxidized back to lipoate. This is done by the third enzyme

(E_3), an FAD-requiring enzyme (Section 23.3). Oxidation of dihydrolipoate by FAD forms enzyme-bound $FADH_2$. NAD^+ then oxidizes $FADH_2$ back to FAD.

mechanism for the pyruvate dehydrogenase system

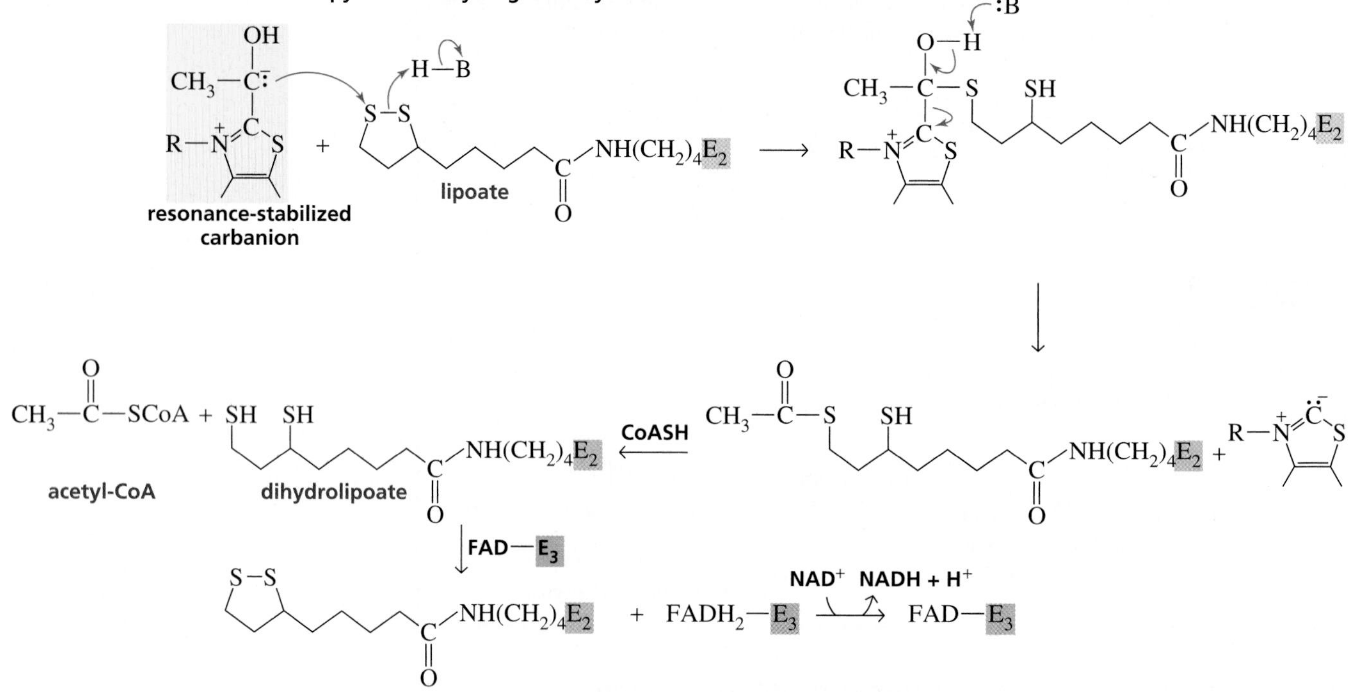

The vitamin precursor of coenzyme A is pantothenate. We have seen that CoASH is used in biological systems to activate carboxylic acids by converting them into thioesters, which are much more reactive toward nucleophilic acyl substitution reactions than carboxylic acids (Section 16.18). At physiological pH (pH = 7.3), a carboxylic acid would be present in its negatively charged basic form, which could not be approached by a nucleophile.

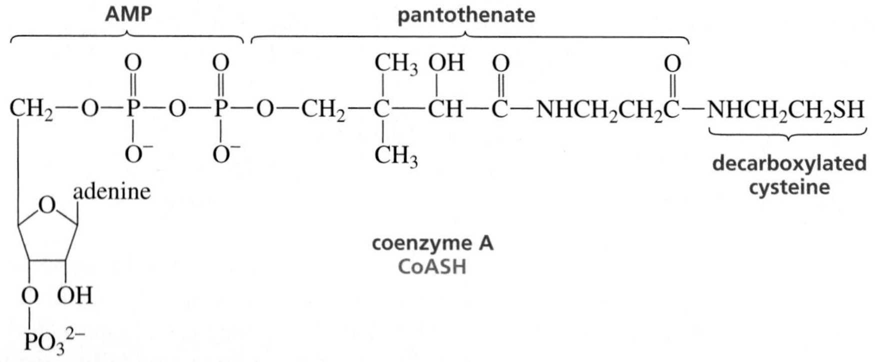

3-D Molecule:
Coenzyme A

**coenzyme A
CoASH**

PROBLEM 10 / SOLVED

TPP is a coenzyme for transketolase, the enzyme that catalyzes the conversion of a ketopentose (xylulose-5-phosphate) and an aldopentose (ribose-5-phosphate) into an aldotriose (glyceraldehyde-3-phosphate) and a ketoheptose (sedoheptulose-7-phosphate). Notice that the total number of carbon atoms in the sugars is conserved (5 + 5 = 3 + 7). Propose a mechanism for this reaction.

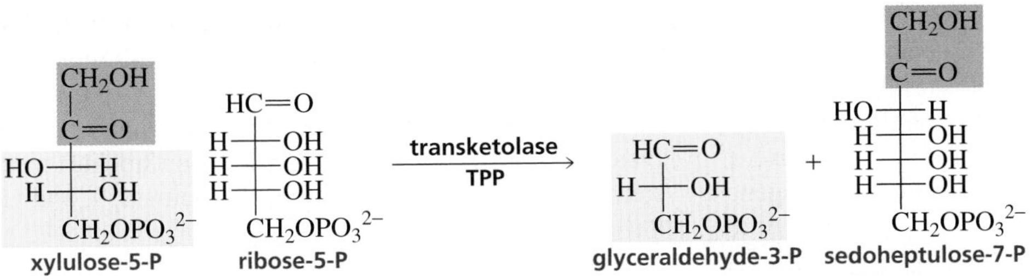

xylulose-5-P   **ribose-5-P** $\xrightarrow[\text{TPP}]{\text{transketolase}}$ **glyceraldehyde-3-P** + **sedoheptulose-7-P**

SOLUTION   The reaction shows that a two-carbon fragment is transferred from xylulose-5-phosphate to ribose-5-phosphate. Because TPP transfers two-carbon fragments, we know that TPP must remove the two-carbon fragment that is to be transferred from xylulose-5-phosphate. Thus, the reaction must start by TPP attacking the carbonyl group of xylulose-5-phosphate. We can add an acid group to accept the electrons from the carbonyl group and a basic group to aid in the removal of the two-carbon fragment. The two-carbon fragment, when attached to TPP, is a resonance-stabilized carbanion that adds to the carbonyl group of ribose-5-phosphate. Again, an acid group accepts the electrons from the carbonyl group and a basic group aids in the elimination of TPP.

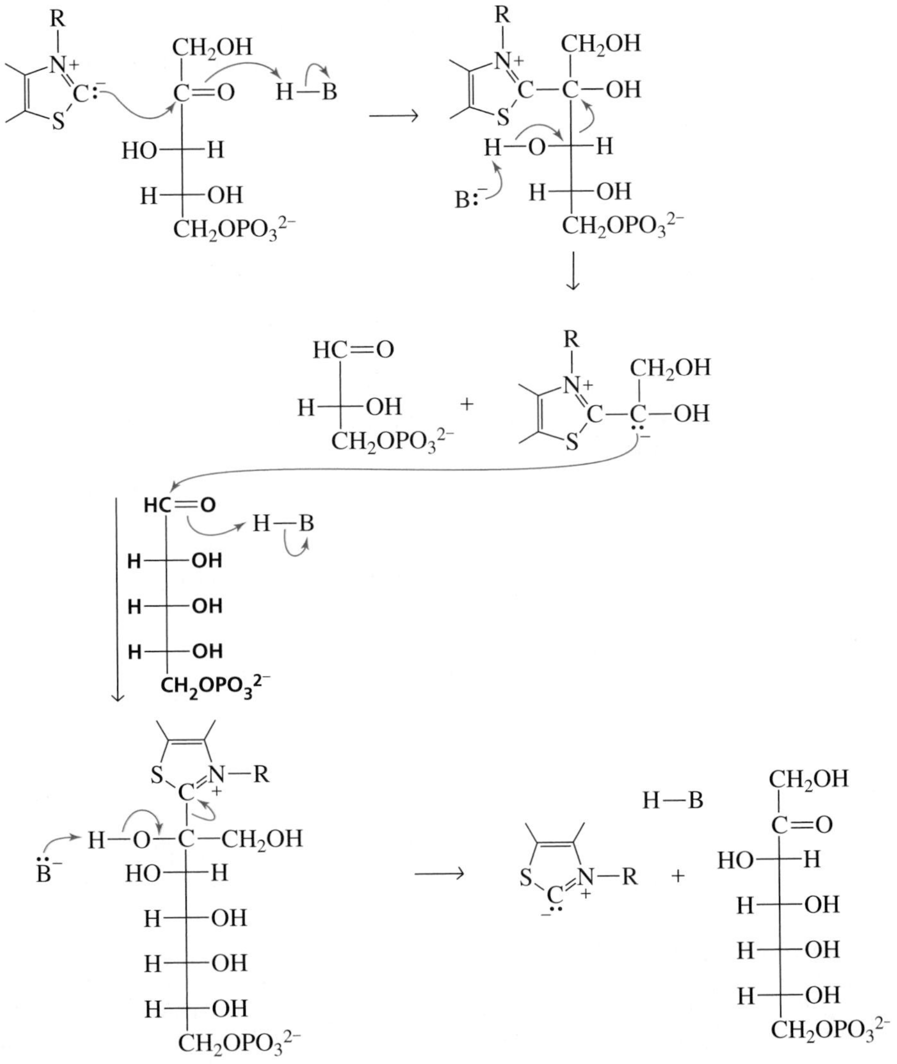

Notice the similar function of TPP in all TPP-requiring enzymes. In each reaction, the coenzyme attacks a carbonyl group of the substrate and allows a bond to that carbonyl group to be broken because the electrons left behind can be delocalized into the thiazolium ring. The resulting two-carbon fragment is then transferred—to a proton in the case of pyruvate decarboxylase, to coenzyme A (via lipoate) in the pyruvate dehydrogenase system, and to a carbonyl group in Problems 8, 9, and 10.

PROBLEM 11

An unfortunate effect of drinking too much alcohol—a hangover—is attributable to the acetaldehyde formed when ethanol is oxidized. There is some evidence that vitamin B_1 can cure a hangover. How can the vitamin do this?

23.5
BIOTIN:
VITAMIN H

Biotin (vitamin H) is an unusual vitamin because it can be synthesized by bacteria that live in the intestine. Consequently, biotin does not have to be included in our diet and deficiencies are rare. However, biotin deficiencies can be found in people who maintain a diet high in raw eggs. Egg whites contain a protein (avidin) that tightly binds biotin and thereby prevents it from acting as a coenzyme. When eggs are cooked, avidin is denatured, and the denatured protein does not bind biotin. Biotin is attached to its enzyme through an amide linkage with the amino group of a lysine side chain.

biotin

enzyme-bound biotin

3-D Molecule:
Biotin

Biotin is the coenzyme required by enzymes that catalyze carboxylation of a carbon adjacent to a carbonyl group. For example, pyruvate carboxylase converts pyruvate—the end product of carbohydrate metabolism—to oxaloacetate, a citric acid cycle intermediate (Figure 23.2). Acetyl-CoA carboxylase converts acetyl-CoA into malonyl-CoA, one of the reactions in the anabolic pathway that converts acetyl-CoA into fatty acids (Section 19.21). Biotin-requiring enzymes use bicarbonate (HCO_3^-) for the source of the carboxyl group that becomes attached to the substrate.

Biotin is required by enzymes that catalyze the carboxylation of a carbon adjacent to a carbonyl group.

pyruvate + HCO_3^- + ATP $\xrightarrow[\substack{Mg^{2+} \\ biotin}]{\text{pyruvate carboxylase}}$ oxaloacetate + ADP + phosphate

acetyl-CoA + HCO_3^- + ATP $\xrightarrow[\substack{Mg^{2+} \\ biotin}]{\text{acetyl-CoA carboxylase}}$ malonyl-CoA + ADP + phosphate

Mechanism for Biotin

In addition to bicarbonate, biotin-requiring enzymes also require Mg^{2+} and ATP. The function of Mg^{2+} is to decrease the overall negative charge on ATP by complexing with two of its negatively charged oxygens (Section 25.5). Unless its negative charge is reduced, ATP cannot be approached by a nucleophile. The function of ATP is to increase the reactivity of bicarbonate by converting it to "activated bicarbonate"—a compound with a good leaving group (Section 25.2). Notice that "activated bicarbonate" is a mixed anhydride of carbonic acid and phosphoric acid (Section 16.1).

Nucleophilic attack by biotin on activated bicarbonate forms carboxybiotin. Because the nitrogen of an amide is not nucleophilic, it is likely that the active form of biotin has an "enolate-like" structure. Nucleophilic attack by the substrate (in this case, the enol form of acetyl-CoA) on carboxybiotin transfers the carboxyl group from biotin to the substrate.

All biotin-requiring enzymes follow the same three steps: activation of bicarbonate by ATP, reaction of activated bicarbonate with biotin to form carboxybiotin, and transfer of the carboxyl group from carboxybiotin to the substrate.

23.6 PYRIDOXAL PHOSPHATE: VITAMIN B₆

Pyridoxal phosphate (PLP) is the coenzyme required by enzymes that catalyze certain transformations of amino acids. The coenzyme is derived from pyridoxine, the vitamin known as vitamin B₆. The name "pyridox**al**" indicates that the coenzyme is a pyridine aldehyde. A deficiency in vitamin B₆ causes anemia. Severe deficiencies can cause seizures and death.

Pyridoxal phosphate (PLP) is required by enzymes that catalyze certain transformations of amino acids.

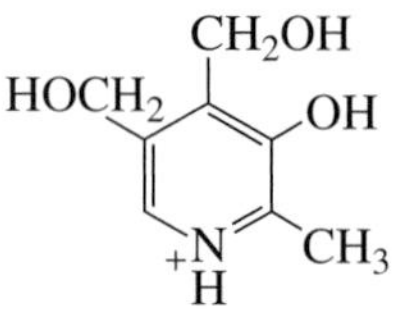

pyridoxine
vitamin B₆

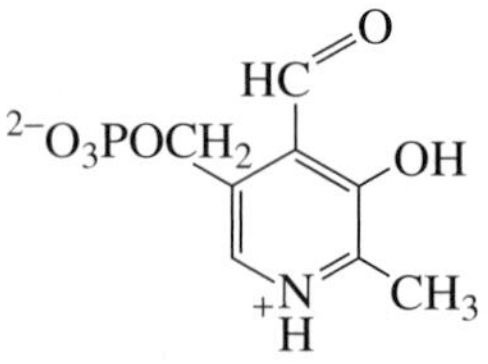

pyridoxal phosphate
PLP

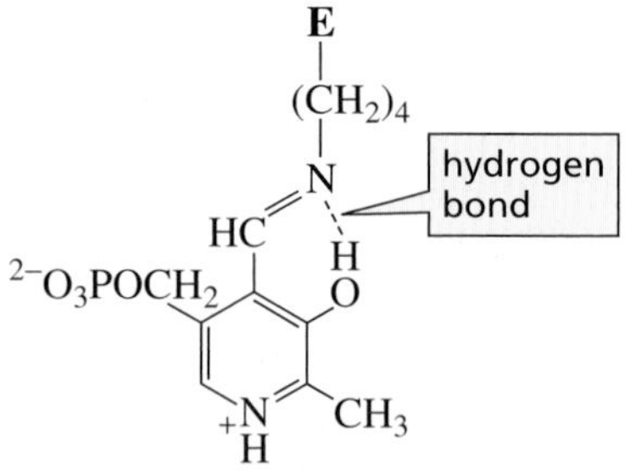

the coenzyme is bound to the enzyme by means of an imine linkage with a lysine residue

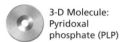
3-D Molecule:
Pyridoxal
phosphate (PLP)

There are several different kinds of amino acid transformations that PLP-requiring enzymes catalyze. The most common are decarboxylation, transamination, racemization (interconversion of L- and D-amino acids), C_α—C_β bond cleavage, and α,β-elimination.

decarboxylation

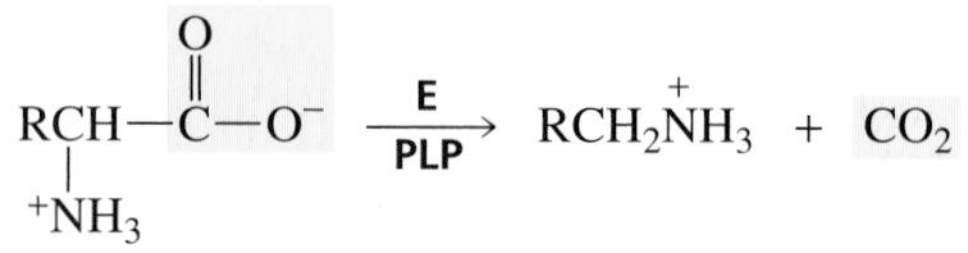

$$\underset{\overset{|}{{}^+NH_3}}{RCH}\overset{\overset{O}{\|}}{-C}-O^- \xrightarrow[\text{PLP}]{\text{E}} RCH_2\overset{+}{N}H_3 \ + \ CO_2$$

transamination

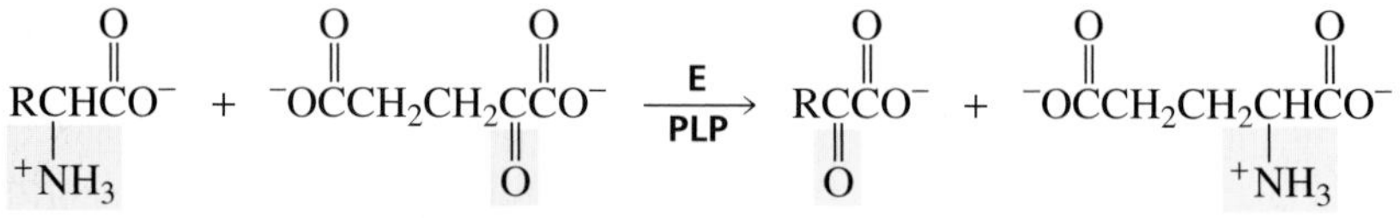

$$\underset{\overset{|}{{}^+NH_3}}{R\overset{\overset{O}{\|}}{C}HCO^-} + {}^-O\overset{\overset{O}{\|}}{C}CH_2CH_2\overset{\overset{O}{\|}}{C}CO^- \xrightarrow[\text{PLP}]{\text{E}} R\overset{\overset{O}{\|}}{C}\underset{\overset{\|}{O}}{C}O^- + {}^-O\overset{\overset{O}{\|}}{C}CH_2CH_2\underset{\overset{|}{{}^+NH_3}}{\overset{\overset{O}{\|}}{C}HCO^-}$$

α-ketoglutarate   glutamate

racemization

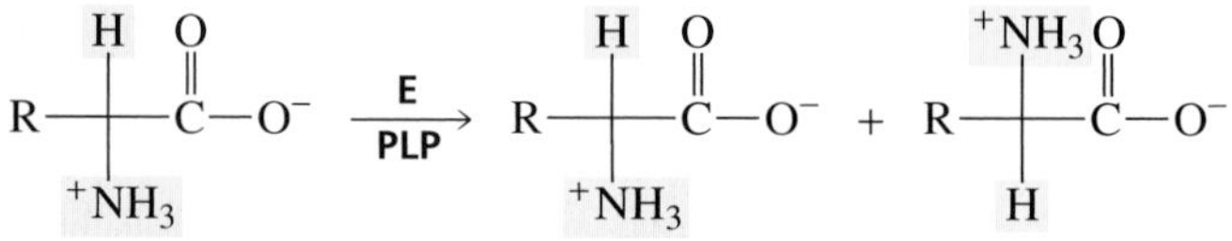

L-amino acid   L-amino acid   D-amino acid

C_α—C_β bond cleavage

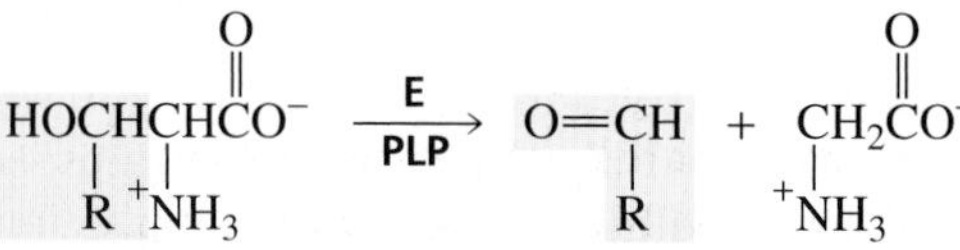

$$\underset{\overset{|}{R}\ \overset{|}{{}^+NH_3}}{HOCHCHCO^-}\overset{\overset{O}{\|}}{} \xrightarrow[\text{PLP}]{\text{E}} \underset{\overset{|}{R}}{O=CH} + \underset{\overset{|}{{}^+NH_3}}{CH_2\overset{\overset{O}{\|}}{C}O^-}$$

α,β-elimination

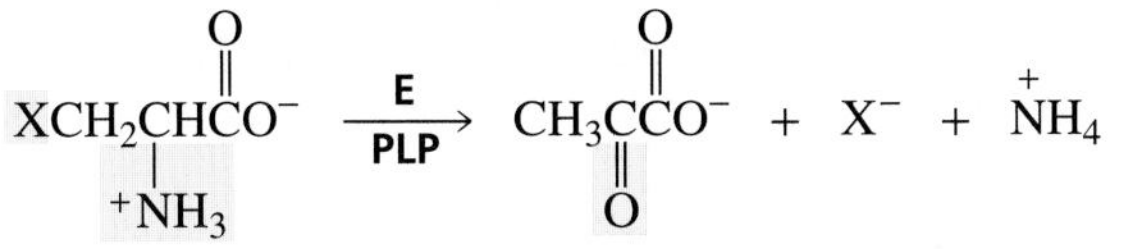

$$\text{XCH}_2\underset{\underset{\text{+NH}_3}{|}}{\text{CH}}\overset{\overset{\text{O}}{\|}}{\text{C}}\text{O}^- \xrightarrow[\text{PLP}]{\text{E}} \text{CH}_3\underset{\underset{\text{O}}{\|}}{\overset{\overset{\text{O}}{\|}}{\text{C}}}\text{CO}^- + \text{X}^- + \overset{+}{\text{N}}\text{H}_4$$

In each of these transformations, one of the bonds to the α-carbon of the amino acid substrate is broken in the first step of the reaction. Decarboxylation breaks the bond joining the carboxyl group to the α-carbon; transamination, racemization, and α,β-elimination break the bond joining the hydrogen to the α-carbon; and C_α—C_β bond cleavage breaks the bond joining the R group to the α-carbon.

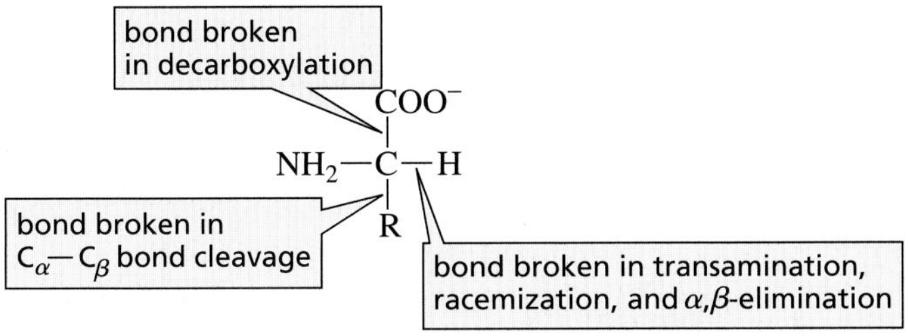

PLP attaches to its enzyme by forming an imine with the amino group of a lysine side chain. The first thing that happens in all PLP-requiring enzymes is a **transimination** reaction—one imine is converted into another imine. In the transimination reaction, the amino acid substrate reacts with the imine formed by *PLP and the enzyme,* forming a tetrahedral intermediate. The lysine group of the enzyme is expelled, forming a new imine between *PLP and the amino acid.*

transimination

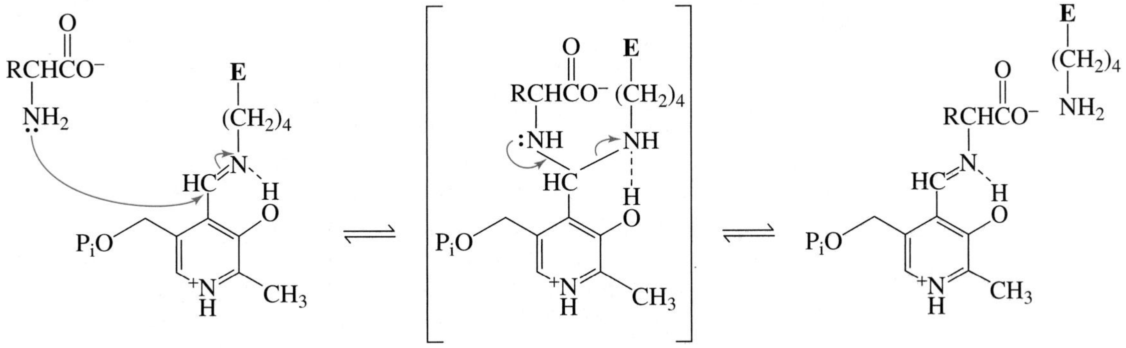

enzyme-bound PLP   amino acid-bound PLP

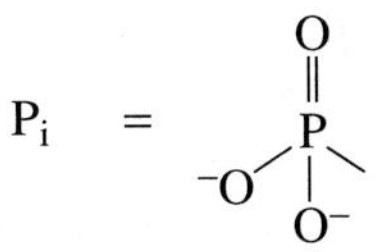

$$\text{P}_\text{i} \;=\; \overset{\overset{\text{O}}{\|}}{\underset{\underset{\text{O}^-}{|}}{\underset{}{\text{P}}}}\underset{}{{}^{-\text{O}}} \diagdown$$

Once the amino acid has formed an imine with PLP, a bond to the α-carbon can be broken because the electrons left behind when the bond breaks can be delocalized onto the positively charged protonated nitrogen of the pyridine ring. In other words, the protonated nitrogen of the pyridine ring is an electron sink. If the OH group is removed from the pyridine ring, the cofactor loses much of its activity. Apparently, the hydrogen bond formed by the OH group helps weaken the bond to the α-carbon.

Mechanism for Decarboxylation

If the PLP-catalyzed reaction is a decarboxylation, the carboxyl group is removed from the α-carbon of the amino acid. Electron rearrangement and protonation of the α-carbon of the decarboxylated intermediate by a protonated lysine or some other acid group reestablishes the aromaticity of the pyridine ring. Transimination with a lysine side chain releases the decarboxylated substrate and regenerates enzyme-bound PLP.

mechanism for PLP-catalyzed decarboxylation of an amino acid

Mechanism for Transamination

The first reaction in the catabolism of most amino acids is replacement of the amino group of the amino acid by a ketone group. This reaction is known as **transamination** because the amino group removed from the amino acid is not lost but is *transferred* to the ketone group of α-ketoglutarate, thereby forming glutamate. The enzymes that catalyze transamination are called *aminotransferases*. Transamination allows the amino groups of the various amino acids to be collected into a single amino acid (glutamate) so that excess nitrogen can be easily excreted. (Do not confuse *transamination* with *transimination* discussed previously.)

In the first step of transamination, a proton is removed from the α-carbon of the amino acid bound to PLP. Rearrangement of the electrons and protonation of the carbon attached to the pyridine ring, followed by hydrolysis of the imine, forms the α-keto acid and pyridoxamine. At this point the amino group has been removed from the amino acid but pyridoxamine has to be converted back to enzyme-bound PLP before another round of catalysis can occur. Pyridoxamine forms an imine with α-ketoglutarate, the second substrate of the reaction. Removal of a proton from the carbon attached to the pyridine ring, followed by rearrangement of the electrons and donation of a proton to the α-carbon of the substrate forms an imine that, when transiminated with a lysine side chain, releases glutamate and reforms enzyme-bound PLP.

Notice that the proton transfer steps are reversed in the two phases of the reaction. Transfer of the amino group of the amino acid to pyridoxal requires removal of the proton from the α-carbon and donation of a proton to the carbon bonded to the pyridine ring. Transfer of the amino group of pyridoxamine to α-ketoglutarate requires removal of the proton from the carbon bonded to the pyridine ring and donation of a proton to the α-carbon.

mechanism for PLP-dependent transamination of an amino acid

The mechanism for PLP-dependent transamination is shown through a series of reaction structures, including formation of **pyridoxamine**, **transaminated amino acid an α-keto acid**, reaction with **α-ketoglutarate**, and **transimination with E—(CH$_2$)$_4$NH$_2$** yielding **glutamate**.

$$^-OCCH_2CH_2CCO^-$$
$$\alpha\text{-ketoglutarate}$$

HEART ATTACKS:
ASSESSING THE DAMAGE

Damage to heart muscle after myocardial infarction allows aminotransferases and other enzymes to leak from the damaged cells of the heart into the bloodstream. After a heart attack, the severity of damage done to the heart can be determined from the concentrations of alanine aminotransferase and aspartate aminotransferase in the bloodstream.

PHENYLKETONURIA: AN INBORN ERROR OF METABOLISM

Tyrosine is a nonessential amino acid because the body can make it by hydroxylating phenylalanine. About 1 in every 20,000 babies, however, is born without phenylalanine hydroxylase, the enzyme that converts phenylalanine into tyrosine. This genetic disease is called phenylketonuria (PKU).

Without phenylalanine hydroxylase, the level of phenylalanine builds up and, when it reaches a high concentration, it is transaminated to phenylpyruvate. The high level of phenylpyruvate found in urine gave the disease its name.

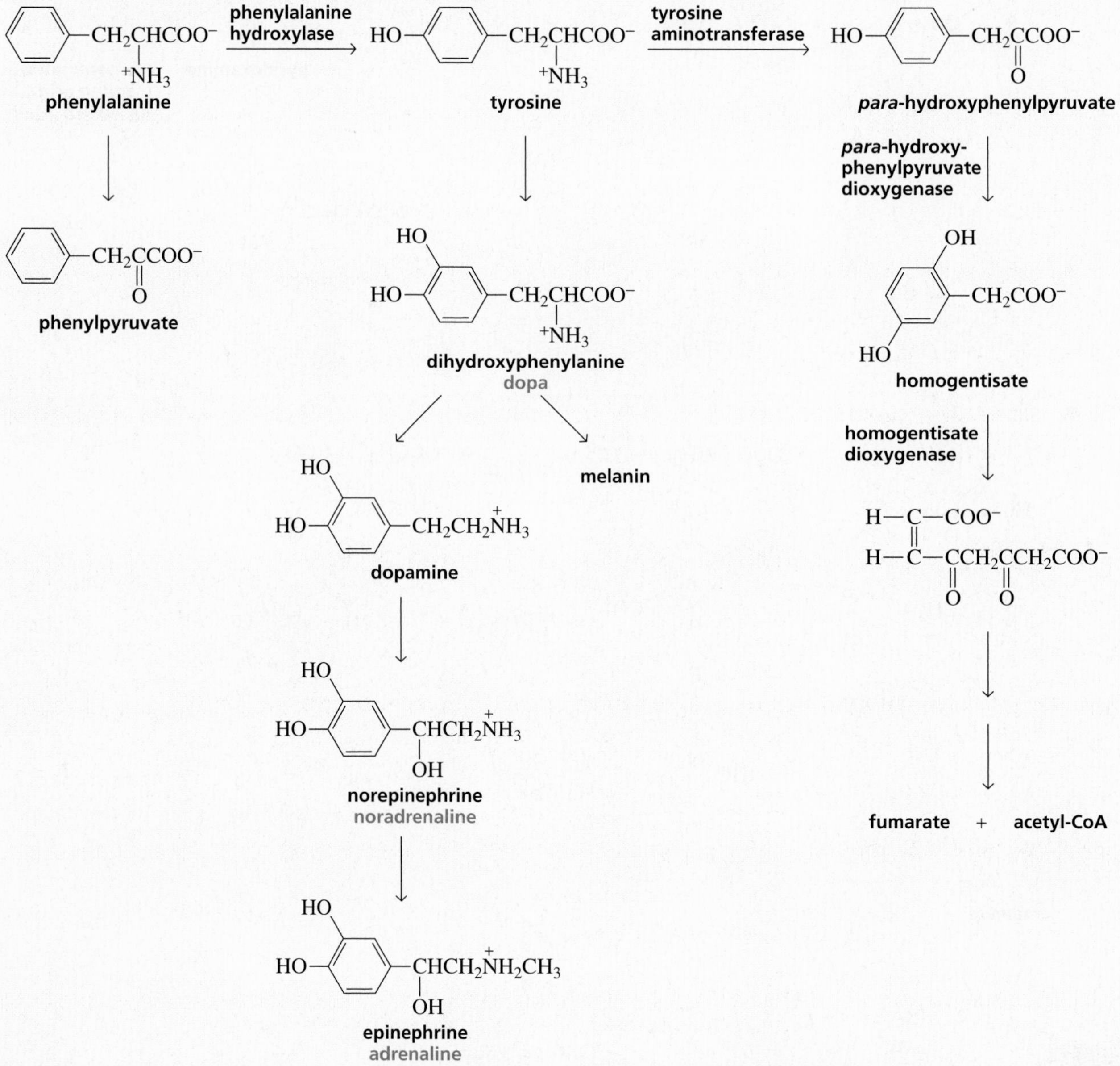

Metabolic pathway diagram:

phenylalanine → (phenylalanine hydroxylase) → tyrosine → (tyrosine aminotransferase) → *para*-hydroxyphenylpyruvate

phenylalanine → phenylpyruvate

tyrosine → dihydroxyphenylanine (dopa) → melanin

dopa → dopamine → norepinephrine (noradrenaline) → epinephrine (adrenaline)

para-hydroxyphenylpyruvate → (*para*-hydroxyphenylpyruvate dioxygenase) → homogentisate → (homogentisate dioxygenase) → fumarate + acetyl-CoA

All babies born in the United States are tested about 3 days after they begin drinking milk for high serum phenylalanine levels, which would indicate a build up of phenylalanine. If the baby is found to lack phenylalanine hydroxylase, he or she is immediately put on a diet low in phenylalanine and high in tyrosine. As long as the phenylalanine level is kept under careful control for the first 5 to 10 years of life, the baby will experience no adverse effects.

If the diet is not controlled, however, the baby will be severely mentally retarded by the time he or she is a few months old. Untreated children have paler skin and fairer hair than other members of their family because they don't synthesize tyrosine, so their melanin levels are low. Melanin is the compound responsible for skin pigmentation. Melanin is formed from dopa, which is formed from the transamination of tyrosine. Half of untreated phenylketonurics are dead by age 20. When a woman with PKU becomes pregnant, she must return to the low phenylalanine diet she had as a child because a high level of phenylalanine can cause abnormal development of the fetus.

Another genetic disease that results from a deficiency of an enzyme in the pathway for phenylalanine degradation is alcaptonuria, which is caused by lack of homogentisate dioxygenase. The only ill effect of this enzyme deficiency is black urine for those who are afflicted. Their urine turns black because the homogentisate they excrete immediately oxidizes in the air.

PROBLEM 12 ◆

Using the pathway for phenylalanine degradation, answer the following questions:

a. What coenzyme and what other organic compound are needed by tyrosine aminotransferase?

b. What bond in homogentisate is oxidized by homogentisate dioxygenase? (*Hint:* Keto-enol tautomerism occurs after oxidation.)

c. What compound is used to supply the methyl group needed to convert noradrenaline into adrenaline? (*Hint:* See Section 9.11.)

PROBLEM 13 ◆

α-Keto acids other than α-ketoglutarate can be used to accept the amino group from pyridoxamine in enzyme-catalyzed transaminations. What amino acids are formed from the following α-keto acids?

pyruvate   oxaloacetate

Mechanism for Racemization

The first step in the PLP-catalyzed racemization of an L-amino acid is the same as the first step in the PLP-catalyzed transamination of an amino acid—removal of a proton from the α-carbon of the amino acid bound to PLP. In the second step of racemization, reprotonation occurs on the α-carbon. The proton can be donated to the sp^2 hybridized α-carbon from either side of the plane defined by the double bond. Consequently, both D- and L-amino acids are formed. In other words, the L-amino acid is racemized.

mechanism for PLP-catalyzed racemization of an L-amino acid

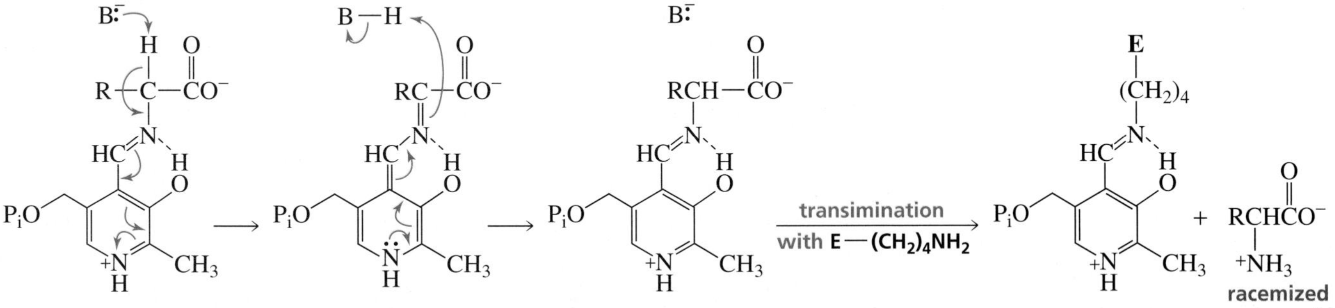

Compare the second step in a PLP-catalyzed transamination with the second step in a PLP-catalyzed racemization. In an enzyme that catalyzes transamination, an acidic group at the active site of the enzyme is in position to donate a proton to the carbon attached to the pyridine ring. The enzyme that catalyzes racemization does not have this acidic group, so the substrate is reprotonated at the α-carbon. In other words, the *coenzyme* is carrying out the chemical reaction but the *enzyme* is determining the course of the reaction.

Mechanism for C_α—C_β Bond Cleavage

In the first step of the mechanism for PLP-catalyzed C_α—C_β bond cleavage, a basic group at the active site of the enzyme removes a proton from an OH group bonded to the β-carbon of the amino acid. This causes the C_α—C_β bond to be cleaved. Serine and threonine are the only two amino acids that can serve as substrates for this reaction because they are the only amino acids that have an OH group bonded to their β-carbon. When serine is the substrate, the product of the cleavage reaction is formaldehyde (R = H); when threonine is the substrate, the product of the cleavage reaction is acetaldehyde (R = CH_3). Electron rearrangement and protonation of the α-carbon of the amino acid followed by transimination with a lysine side chain releases glycine.

mechanism for PLP-catalyzed C_α—C_β bond cleavage

The formaldehyde formed when serine undergoes C_α—C_β bond cleavage never leaves the active site of the enzyme; it is immediately transferred to tetrahydrofolate (Section 23.8).

PROBLEM 14

Propose a mechanism for a PLP-catalyzed α,β-elimination.

Choosing the Bond to Be Cleaved

If all PLP-requiring enzymes start with the same substrate—an amino acid bound to pyridoxal phosphate by an imine linkage—how can three different bonds be cleaved in the first step of the reaction? The bond cleaved in the first step depends on the conformation of the amino acid that the enzyme binds. There is free rotation about the C_α—N bond of the amino acid, and an enzyme can bind any of the possible confor-

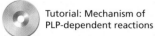

Tutorial: Mechanism of
PLP-dependent reactions

mations about this bond. The enzyme will bind the conformation in which the overlapping orbitals of the bond to be broken in the first step of the reaction lie parallel
to the *p* orbitals of the conjugated system. In this way the orbital containing the electrons left behind when the bond is broken can overlap with the orbitals of the conjugated system. If this overlap cannot occur, the electrons cannot be delocalized into
the conjugated system and the carbanion intermediate cannot be stabilized.

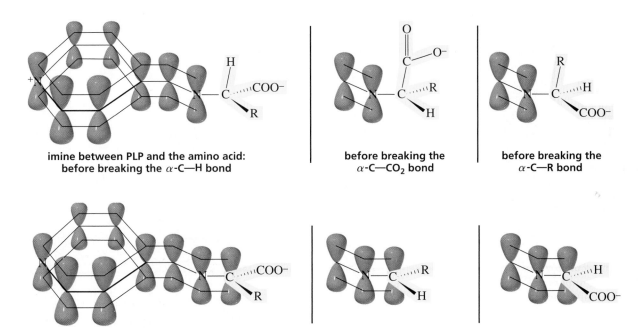

imine between PLP and the amino acid:
before breaking the α-C—H bond

before breaking the
α-C—CO$_2$ bond

before breaking the
α-C—R bond

electrons left behind have been delocalized into
the conjugated system

PROBLEM 15

Which of the following compounds is more easily decarboxylated?

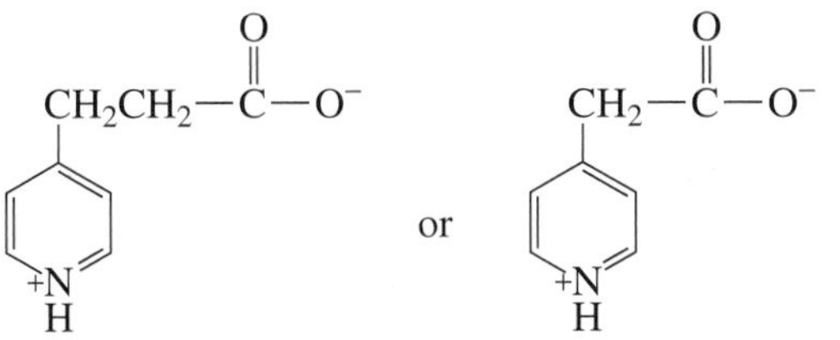

$$CH_2CH_2-\overset{\displaystyle O}{\overset{\|}{C}}-O^-$$

or

$$CH_2-\overset{\displaystyle O}{\overset{\|}{C}}-O^-$$

PROBLEM 16

If a PLP-requiring enzymatic reaction is carried out at a pH where the pyridine nitrogen
is not protonated, the ability of PLP to catalyze an amino acid transformation is greatly
reduced. Account for this observation.

PROBLEM 17

If the OH substituent of pyridoxal phosphate is replaced by an OCH$_3$ substituent, PLP's
ability to catalyze an amino acid transformation is greatly reduced. Account for this
observation.

23.7 COENZYME B₁₂: VITAMIN B₁₂

Enzymes that catalyze certain rearrangement reactions require **coenzyme B₁₂,** a coenzyme derived from vitamin B₁₂. The structure of vitamin B₁₂ was determined by Dorothy Crowfoot Hodgkin using X-ray crystallography. The vitamin has a cyano group (or HO⁻ or H₂O) coordinated to cobalt (Section 27.4). In coenzyme B₁₂, the cyano group is replaced by a 5′-deoxyadenosyl group.

Dorothy Crowfoot Hodgkin

Dorothy Crowfoot Hodgkin (1910–1994) *was born in Egypt to English parents. She received an undergraduate degree from Somerville College, Oxford University and earned a Ph.D. from Cambridge University. She determined the structures of penicillin, insulin, and vitamin B₁₂. For her work on vitamin B₁₂, she received the 1964 Nobel Prize in chemistry. She was a professor of chemistry at Somerville, where one of her graduate students was former Prime Minister Margaret Thatcher. Hodgkin was a founding member of Pugwash, an organization whose purpose was to further communication between scientists on both sides of the Iron Curtain.*

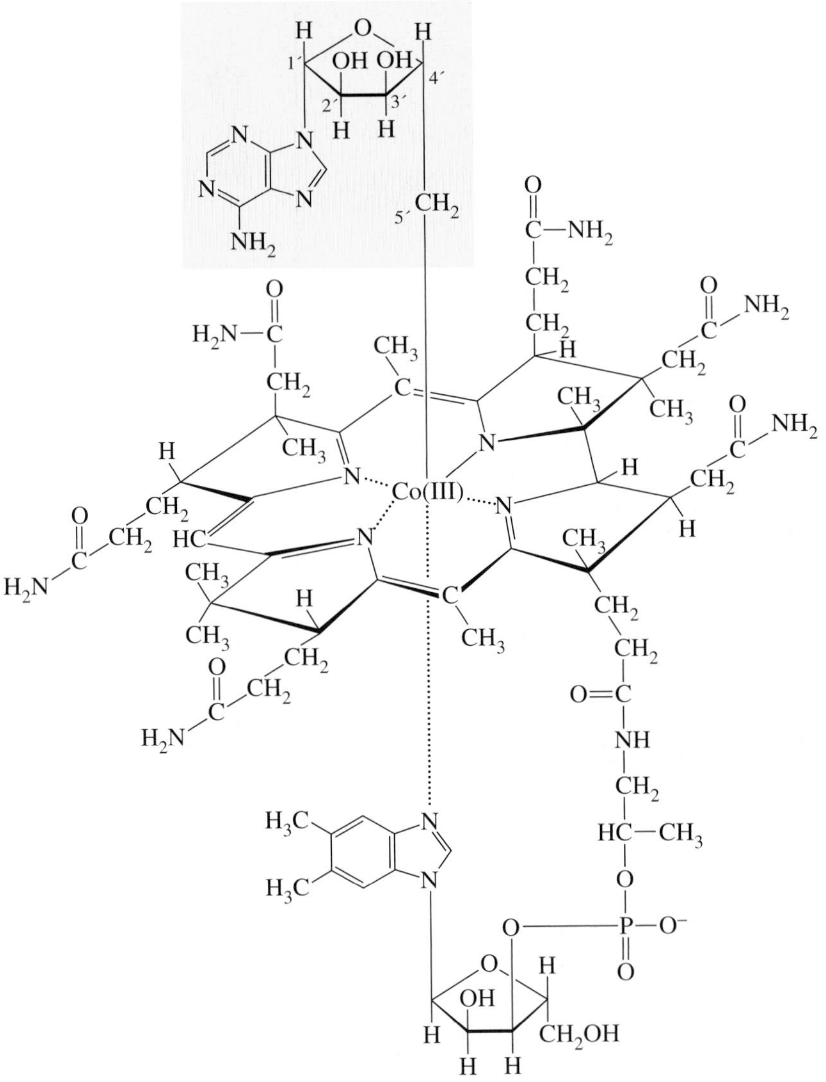

coenzyme B₁₂

Animals and plants cannot synthesize vitamin B₁₂. In fact, only a few microorganisms can synthesize it. Humans must obtain all their vitamin B₁₂ from their diet, particularly from meat. Because vitamin B₁₂ is needed in only very small amounts, deficiencies caused by consumption of insufficient amounts of the vitamin are rare but have been found in vegetarians who eat no animal products. A vitamin B₁₂ deficiency is most commonly caused by an inability to absorb the vitamin in the intestine. The deficiency causes pernicious anemia. The following are examples of enzyme-catalyzed reactions that require coenzyme B₁₂.

$$\underset{\substack{\text{β-methylaspartate}}}{\underset{\substack{|\\^+NH_3}}{CH_3\underset{|}{C}HCHCOO^-}} \overset{\substack{\textbf{glutamate}\\\textbf{mutase}}}{\underset{\textbf{coenzyme B}_{12}}{\rightleftharpoons}} \underset{\substack{\text{glutamate}}}{\underset{\substack{|\\^+NH_3}}{CH_2CH_2CHCOO^-}}$$

3-D Molecule:
Coenzyme B$_{12}$

$$\underset{\substack{\textbf{methylmalonyl-CoA}}}{\underset{\substack{|\\COO^-}}{CH_3\underset{}{C}H\overset{O}{\overset{||}{C}}SCoA}} \overset{\substack{\textbf{methylmalonyl-CoA}\\\textbf{mutase}}}{\underset{\textbf{coenzyme B}_{12}}{\rightleftharpoons}} \underset{\substack{\textbf{succinyl-CoA}}}{\underset{\substack{|\\COO^-}}{CH_2CH_2\overset{O}{\overset{||}{C}}SCoA}}$$

$$\underset{\substack{\textbf{1,2-propanediol}}}{\underset{\substack{|\\OH}}{CH_3CHCH_2OH}} \xrightarrow[\textbf{coenzyme B}_{12}]{\textbf{dioldehydrase}} \left[\underset{\substack{\textbf{a hydrate}}}{\underset{\substack{|\\OH}}{CH_3CH_2CHOH}}\right] \xrightarrow{-\textbf{H}_2\textbf{O}} \underset{\substack{\textbf{propanal}}}{CH_3CH_2\overset{O}{\overset{||}{C}}H}$$

In each of these coenzyme B$_{12}$-requiring reactions, a group (Y) bonded to one carbon changes places with a hydrogen bonded to an adjacent carbon.

$$\underset{\substack{|\quad|\\Y\;\;H}}{-\overset{|}{\underset{|}{C}}-\overset{|}{\underset{|}{C}}-} \overset{\substack{\textbf{a coenzyme B}_{12}\text{-}\\\textbf{requiring enzyme}}}{\rightleftharpoons} \underset{\substack{|\quad|\\H\;\;Y}}{-\overset{|}{\underset{|}{C}}-\overset{|}{\underset{|}{C}}-}$$

For example, glutamate mutase and methylmalonyl-CoA mutase both catalyze a reaction in which the COO$^-$ group bonded to one carbon changes places with a hydrogen of an adjacent methyl group. In the reaction catalyzed by dioldehydrase, an OH group changes places with a methylene hydrogen. The resulting product is a hydrate that loses water to form propanal.

Mechanism for Coenzyme B$_{12}$

The chemistry of coenzyme B$_{12}$ takes place at the bond joining the cobalt and the 5'-deoxyadenosyl group. The currently accepted mechanism for dioldehydrase involves an initial homolytic cleavage of this unusually weak bond (26 kcal/mol [109 kJ/mol] compared to 99 kcal/mol [414 kJ/mol] for a C—H bond). Breaking the bond forms a 5'-deoxyadenosyl radical and reduces Co(III) to Co(II). The 5'-deoxyadenosyl radical abstracts a hydrogen atom from the C-1 carbon of the substrate, thereby becoming 5'-deoxyadenosine. A hydroxyl radical ($^\cdot$OH) migrates from C-2 to C-1, creating a radical at C-2. This radical abstracts a hydrogen atom from 5'-deoxyadenosine, forming the rearranged product and regenerating the 5'-deoxyadenosyl radical. The 5'-deoxyadenosyl radical recombines with Co(II), which regenerates the coenzyme. The enzyme–coenzyme complex is then ready for another catalytic cycle. The initial product is a hydrate that loses water to form propanal, the final product of the reaction.

It is likely that all coenzyme B$_{12}$-requiring enzymes catalyze reactions by means of the same general mechanism. The role of the coenzyme is to provide a way to remove a hydrogen atom from the substrate. Once the hydrogen atom has been removed, an adjacent group can migrate to take its place. The coenzyme then gives back the hydrogen atom, delivering it to the carbon that lost the migrating group.

Coenzyme B$_{12}$ is required by enzymes that catalyze the exchange of a hydrogen bonded to one carbon with a group bonded to an adjacent carbon.

the role of 5′-deoxyadenosylcobalamin in a coenzyme B$_{12}$-requiring enzyme-catalyzed reaction

---

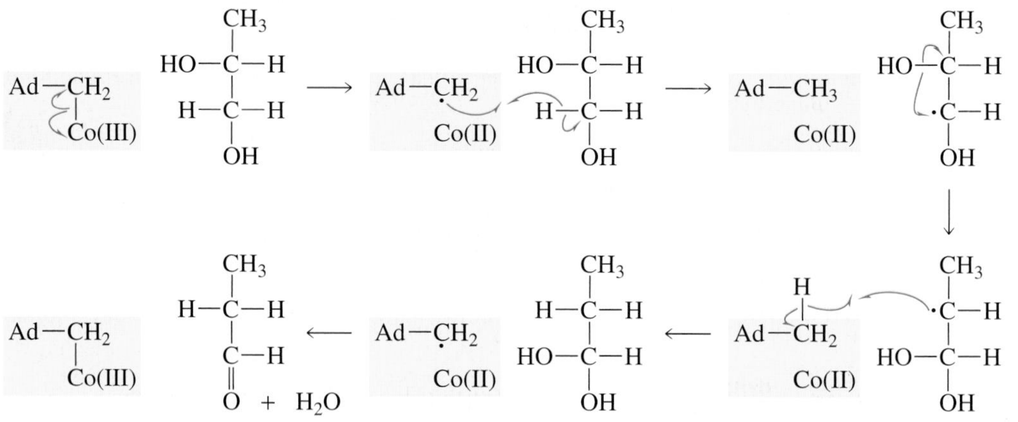

PROBLEM 18

Ethanolamine ammonia lyase, a coenzyme B$_{12}$-requiring enzyme, catalyzes the following reaction. Propose a mechanism for this reaction.

$$HOCH_2CH_2NH_2 \longrightarrow CH_3\overset{\displaystyle O}{\overset{\displaystyle \|}{C}}H + NH_3$$

PROBLEM 19 ◆

A fatty acid with an even number of carbon atoms is metabolized to acetyl-CoA, which can enter the citric acid cycle. A fatty acid with an odd number of carbon atoms is metabolized to acetyl-CoA and one equivalent of propionyl-CoA. Two coenzyme-requiring enzymes are needed to convert propionyl-CoA into succinyl-CoA, a citric acid cycle intermediate. Write the two enzyme-catalyzed reactions and indicate the required coenzymes.

23.8 TETRAHYDROFOLATE: FOLIC ACID

Tetrahydrofolate (THF) is the coenzyme used by enzymes catalyzing reactions that transfer a group containing a single carbon to their substrates. The one-carbon group can be a methyl group (CH_3), a methylene group (CH_2), or a formyl group ($HC=O$). Tetrahydrofolate results from the reduction of two double bonds of folic acid (folate), its precursor vitamin. Bacteria synthesize folate, but mammals cannot.

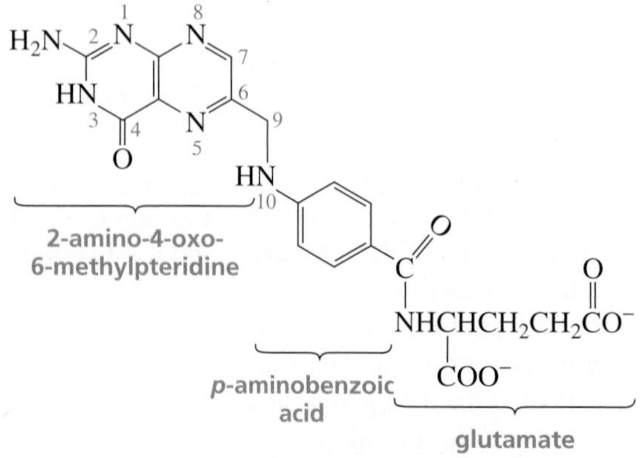

2-amino-4-oxo-6-methylpteridine

p-aminobenzoic acid

glutamate

folic acid (folate)

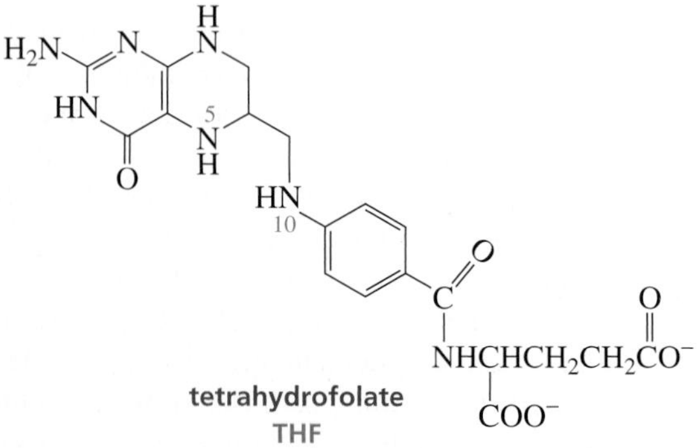

tetrahydrofolate THF

There are six different THF-coenzymes. N^5-Methyl-THF transfers a methyl group (CH_3), N^5,N^{10}-methylene-THF transfers a methylene group (CH_2), and the others transfer a formyl group ($HC{=}O$).

Tetrahydrofolate (THF) is the coenzyme required by enzymes that catalyze the transfer of a group containing one carbon to their substrates.

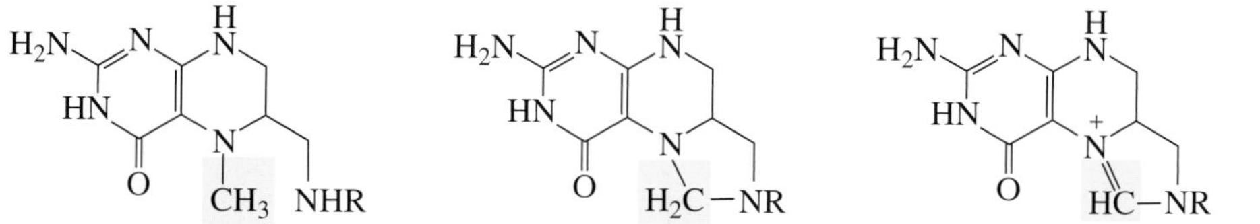

N^5-methyl-THF   *N*^{5},*N*^{10}-methylene-THF   *N*^{5},*N*^{10}-methenyl-THF

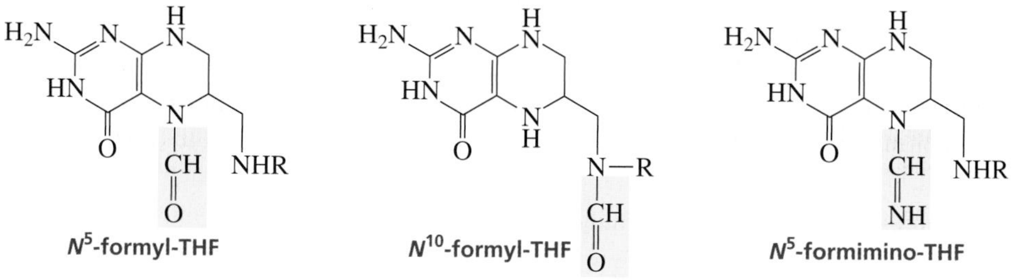

3-D Molecule:
Tetrahydrofolate (THF)

N^5-formyl-THF   *N*^{10}-formyl-THF   *N*^{5}-formimino-THF

Homocysteine methyl transferase and glycinamide ribonucleotide (GAR) transformylase are enzymes that require THF-coenzymes.

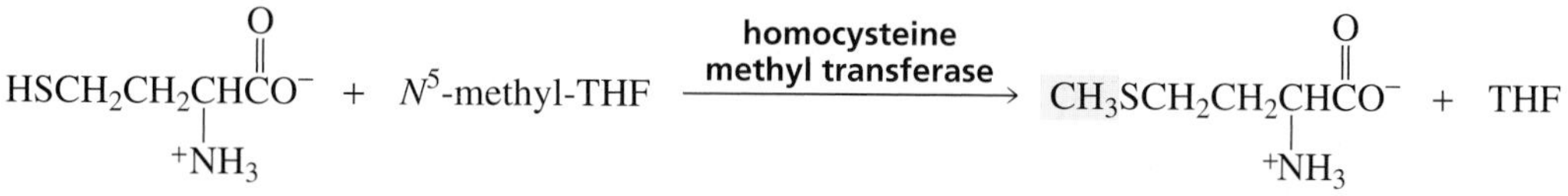

homocysteine   methionine

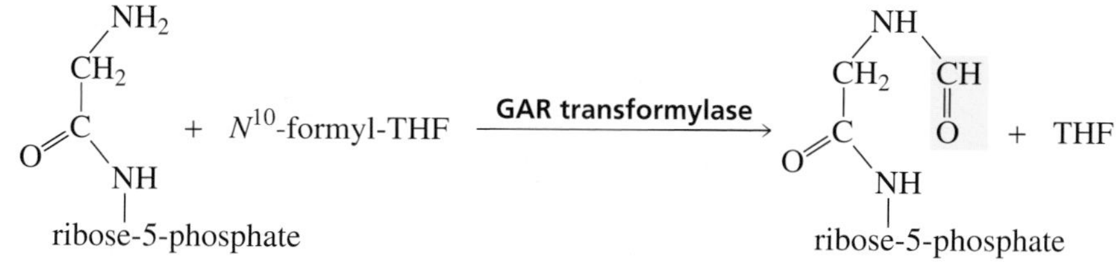

ribose-5-phosphate   ribose-5-phosphate

Thymidylate Synthase: The Enzyme That Converts U's into T's

RNA is a polymer of nucleotides in which the heterocyclic components are adenine, guanine, cytosine, and uracil (A, G, C, and U). DNA is a polymer of nucleotides in which the heterocyclic components are adenine, guanine, cytosine, and thymine (A, G, C, and T). In other words, the heterocyclic bases in RNA and DNA are the same except that RNA contains U's, while DNA contains T's (Sections 25.1, and 25.14). The T's used for the biosynthesis of DNA are synthesized from U's by thymidylate synthase, an enzyme that requires N^5,N^{10}-methylene-THF as a coenzyme. Even though the only structural difference between a U and a T is a *methyl* group, a T is synthesized by first transferring a *methylene* group to a U.

In the first step of the reaction catalyzed by thymidylate synthase, a nucleophilic cysteine group at the active site of the enzyme attacks the β-carbon of uridine (an

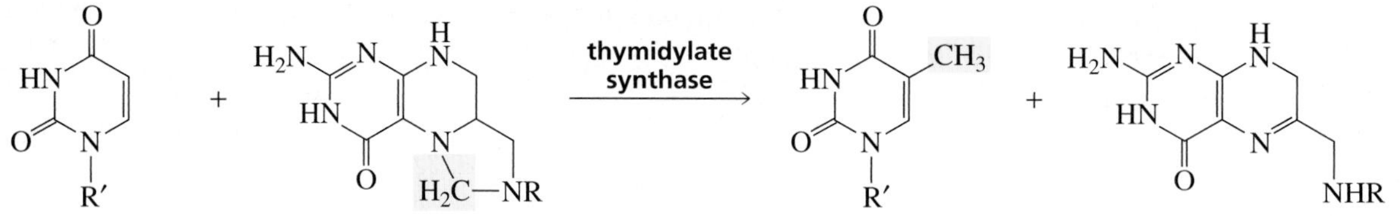

2'-deoxyuridine 5'-monophosphate dUMP

N^5,N^{10}-methylene-THF

thymidylate synthase

2'-deoxythymidine 5'-monophosphate dTMP

dihydrofolate DHF

R' = 2'-deoxyribose-5-phosphate

example of conjugate addition; Section 17.14). A subsequent nucleophilic attack by the α-carbon of uridine on the methylene group of N^5,N^{10}-methylene-THF forms a covalent bond between uridine and the coenzyme. A proton on the α-carbon of uridine is removed with the help of a basic group at the active site of the enzyme, eliminating the coenzyme. Transfer of a hydride ion from the coenzyme to uridine and elimination of the enzyme forms thymidine and dihydrofolate (DHF).

mechanism for catalysis by thymidylate synthase

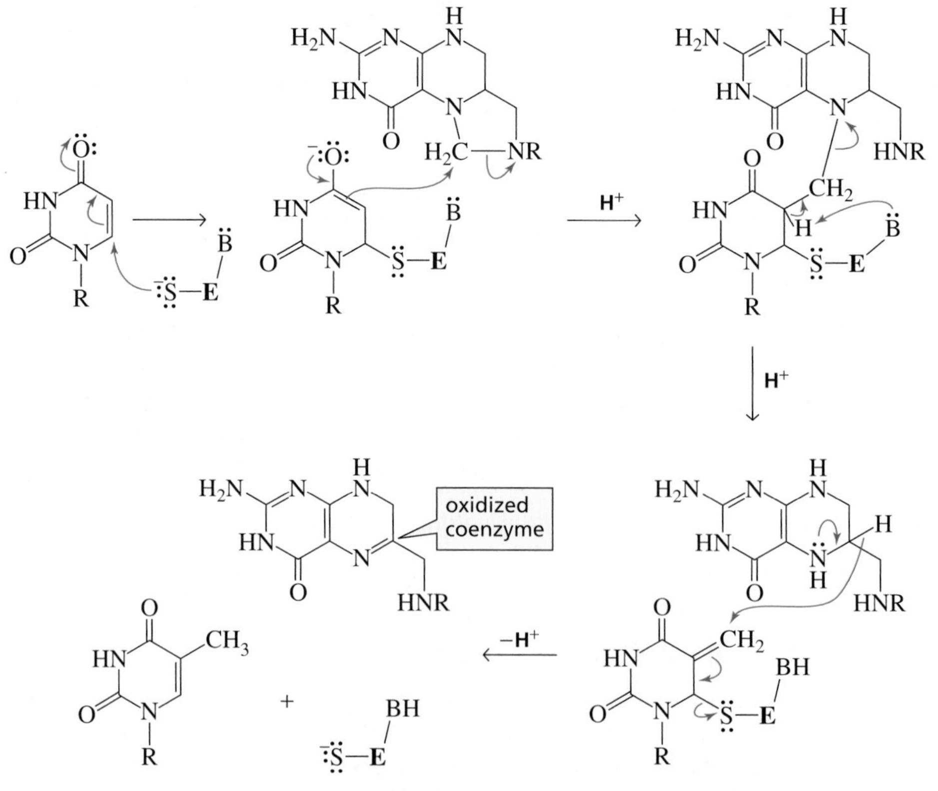

Notice that the coenzyme initially transfers a methylene group to the substrate. The methylene group is subsequently reduced to a methyl group. Because the coenzyme is the reducing agent, it is simultaneously oxidized. The oxidized coenzyme is dihydrofolate.

When the reaction is over, dihydrofolate must be converted back to N^5,N^{10}-methylene-THF so the coenzyme can undergo another catalytic cycle. This reaction occurs in two steps. In the first step, dihydrofolate is reduced to tetrahydrofolate. In the second step, serine hydroxymethyl transferase catalyzes the transfer of the hydroxymethyl group from serine to the coenzyme. Serine hydroxymethyl trans-

Tutorial: Mechanism for catalysis by thymidylate synthase

ferase is the PLP-requiring enzyme that cleaves the C_α—C_β bond of serine to form glycine and formaldehyde (Section 23.6). In other words, the formaldehyde cleaved off serine is immediately transferred to THF to form N^5,N^{10}-methylene-THF—this is fortunate because formaldehyde is cytotoxic (it kills cells).

$$\text{dihydrofolate + NADPH + H}^+ \xrightarrow{\text{dihydrofolate reductase}} \text{tetrahydrofolate + NADP}^+$$

$$\text{tetrahydrofolate + HOCH}_2\text{CHCOO}^- \xrightarrow[\text{PLP}]{\text{serine hydroxymethyl transferase}} N^5,N^{10}\text{-methylene-THF + CH}_2\text{COO}^-$$

$$\underset{\substack{\overset{|}{^+\text{NH}_3} \\ \text{serine}}}{} \qquad\qquad\qquad\qquad\qquad \underset{\substack{\overset{|}{^+\text{NH}_3} \\ \text{glycine}}}{}$$

Cancer Chemotherapy

Cancer is associated with rapidly growing and proliferating cells. Because cells cannot multiply if they cannot synthesize DNA, several cancer chemotherapeutic agents have been developed to inhibit thymidylate synthase and dihydrofolate reductase. If a cell cannot make thymidine, it cannot synthesize DNA. Inhibiting dihydrofolate reductase also prevents the synthesis of thymidine because cells have a limited amount of tetrahydrofolate. If they cannot convert dihydrofolate back to tetrahydrofolate, they cannot continue to synthesize thymidine.

A common anticancer drug that inhibits thymidylate synthase is 5-fluorouracil. 5-Fluorouracil and uracil react with thymidylate synthase in the same way. However, the fluorine at the 5-position cannot be removed by the base in the third step of the reaction because fluorine is too electronegative to come off as F^+. Because fluorine cannot be removed, the reaction stops at this point, leaving the enzyme permanently attached to the substrate. The active site of the enzyme is now blocked with 5-fluorouracil, so the enzyme cannot bind uracil, which means thymidine can no longer be synthesized. This means that the synthesis of DNA is also stopped. Unfortunately, most anticancer drugs cannot discriminate between diseased and normal cells. As a result, cancer chemotherapy is accompanied by terrible side effects. However, cancer cells undergo uncontrolled cell division, so they are dividing more rapidly than normal cells and thus are the hardest hit by cancer chemotherapeutic agents.

5-Fluorouracil is a **mechanism-based inhibitor**—it inactivates the enzyme by taking part in the normal catalytic mechanism. It is also called a **suicide inhibitor** because when the enzyme reacts with it, the enzyme "commits suicide." The use of 5-fluorouracil illustrates the importance of knowing the mechanism for an enzyme-catalyzed reaction. If you know the mechanism, you may be able to design an inhibitor to turn the reaction off at a certain step.

Aminopterin and methotrexate are anticancer drugs that are inhibitors of dihydrofolate reductase. Because their structures are similar to that of dihydrofolate, they compete with dihydrofolate for binding at the active site of the enzyme. Since they bind 1000 times more tightly to the enzyme than does dihydrofolate, they inhibit the enzyme. These two compounds are examples of **competitive inhibitors.**

Donald D. Woods (1912–1964) *was born in Ipswich, England, and received a B.A. and a Ph.D. from Cambridge University. He worked in Paul Flores's laboratory at the London Hospital Medical College.*

Paul B. Flores (1882–1971) *was born in London. He moved his laboratory to Middlesex Hospital Medical School when a bacterial chemistry unit was established there. He was knighted in 1946.*

aminopterin R = H
methotrexate R = CH$_3$

trimethoprim

Because these compounds inhibit the synthesis of THF, they interfere with the synthesis of any compound that requires a THF-coenzyme in one of the steps of its synthesis. Thus, not only do they prevent the synthesis of thymidine, they also inhibit the synthesis of adenine and guanine—other heterocyclic compounds needed for the synthesis of DNA—because their synthesis also requires a THF-coenzyme. One clinical technique used in cancer chemotherapy calls for the patient to be given a lethal dose of methotrexate and then to "save" him or her by administering N^5-formyl-THF.

Trimethoprim is used as an antibiotic because it binds to bacterial dihydrofolate reductase much more tightly than to mammalian dihydrofolate reductase.

THE FIRST ANTIBACTERIAL DRUGS

Sulfonamides—commonly known as sulfa drugs—were introduced clinically in 1934 as the first effective antibacterial drugs (Section 30.4). Donald Woods, a British bacteriologist, noticed that sulfanilamide, the initially most widely used sulfonamide, was structurally similar to *p*-aminobenzoic acid, a compound necessary for bacterial growth. He proposed that sulfanilamide had antibacterial properties because it blocked the normal utilization of *p*-aminobenzoic acid.

a sulfonamide   sulfanilamide   *p*-aminobenzoic acid

Woods and Paul Flores suggested that sulfanilamide acts by inhibiting the enzyme that incorporates *p*-aminobenzoic acid into folic acid. Because the enzyme cannot tell the difference between sulfanilamide and *p*-aminobenzoic acid, both compounds compete for the active site of the enzyme. Humans are not affected by the drug because they do not synthesize folate—they get all their folate from their diets.

PROBLEM 20 ◆

What is the source of the methyl group in thymidine?

Vitamin K is required for proper clotting of blood. The letter K comes from *koagulation,* which is German for "clotting." A series of reactions utilizing six proteins is involved in blood clotting. In order for blood to clot, the blood-clotting proteins must bind Ca^{2+}. Vitamin K is required for proper Ca^{2+} binding. Vitamin K deficiencies are rare because the vitamin is synthesized by intestinal bacteria. Vitamin K is also found in the leaves of green plants. **Vitamin KH₂** (the hydroquinone of vitamin K) is the coenzyme form of the vitamin (Section 18.11).

23.9 VITAMIN KH₂: VITAMIN K

**vitamin K
a quinone**

**vitamin KH₂
a hydroquinone**

Vitamin KH₂ is the coenzyme for the enzyme that catalyzes the carboxylation of the γ-carbon of glutamate side chains in proteins, forming γ-carboxyglutamates. γ-Carboxyglutamates complex Ca^{2+} much more effectively than glutamates do. The proteins involved in blood clotting all have several glutamates near their N-terminal ends. For example, prothrombin has glutamates at positions 7, 8, 15, 17, 20, 21, 26, 27, 30, and 33.

3-D Molecule:
Vitamin K

glutamate side chain   **γ-carboxyglutamate side chain**   **calcium complex**

The mechanism for the vitamin KH₂-catalyzed carboxylation of glutamate has puzzled chemists for quite some time because the γ-proton that must be removed from glutamate before it can attack CO_2 is not very acidic. Therefore, the mechanism must involve the creation of a strong base. The following mechanism has been proposed by Paul Dowd. The vitamin loses a proton from a phenolic OH group and the base that is formed attacks molecular oxygen. A dioxetane is formed that collapses to give a vitamin K base that is strong enough to remove a proton from the γ-carbon of glutamate. The glutamate carbanion attacks CO_2 to form γ-carboxyglutamate, and the protonated vitamin K base (a hydrate) loses water, forming vitamin K epoxide.

Vitamin KH₂ is required by the enzyme that catalyzes the carboxylation of the γ-carbon of a glutamate side chain in a protein.

mechanism for the vitamin KH₂-dependent carboxylation of glutamate

a dioxetane

Paul Dowd (1936–1996) was born in Brockton, Mass. He did his undergraduate work at Harvard University and received a Ph.D. from Columbia University. He was a professor of chemistry at Harvard University and then was a professor of chemistry from 1970 to 1996 at the University of Pittsburgh.

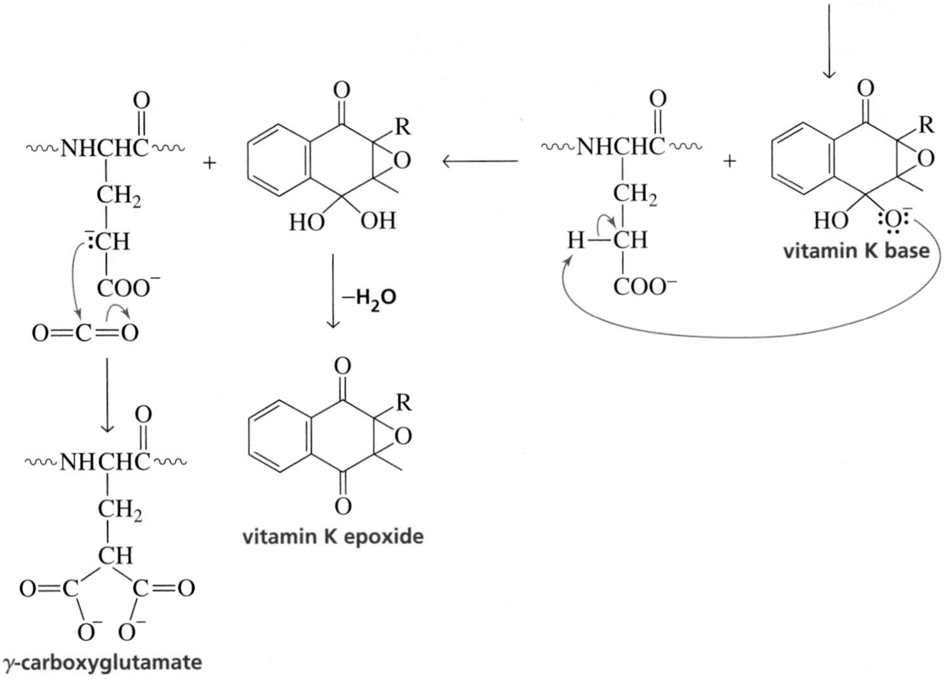

vitamin K epoxide

γ-carboxyglutamate

Vitamin K epoxide is reduced back to vitamin KH_2 by an enzyme that uses the coenzyme dihydrolipoate as the reducing agent. The epoxide is first reduced to vitamin K, which is further reduced to vitamin KH_2.

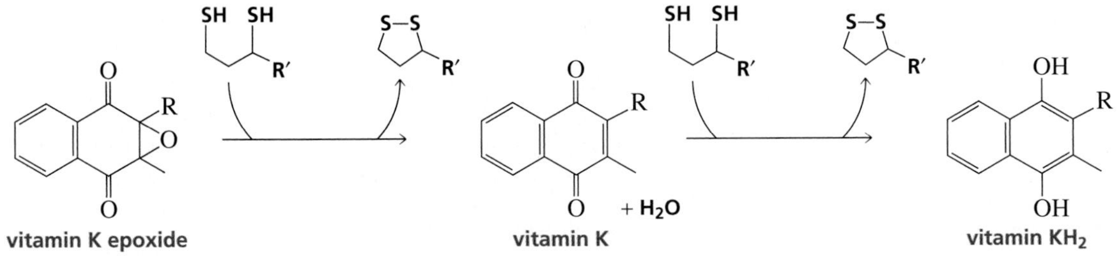

vitamin K epoxide   vitamin K   vitamin KH_2

Warfarin and dicoumarol are used clinically as anticoagulants. Warfarin is also a common rat poison, causing death by internal bleeding. These compounds prevent the clotting of blood by inhibiting the enzyme that reduces vitamin K epoxide to vitamin KH_2, thereby preventing carboxylation of glutamate. The enzyme cannot tell the difference between these two compounds and vitamin K epoxide, so the compounds act as *competitive inhibitors*.

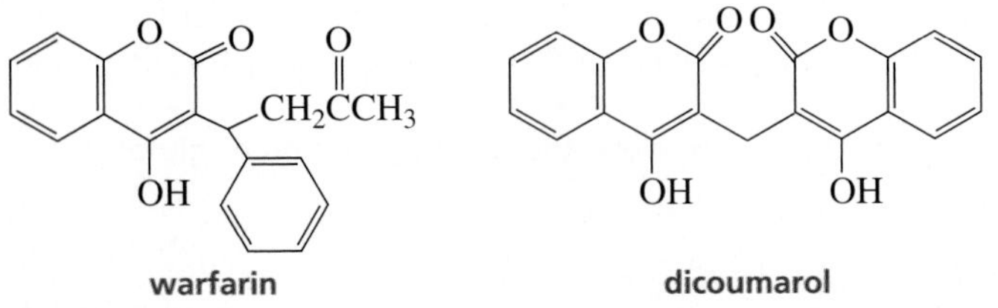

warfarin   dicoumarol

Vitamin E has recently been found to be an anticoagulant. Unlike warfarin and dicumarol, which inhibit the enzyme that returns vitamin K epoxide back to vitamin KH_2, vitamin E directly inhibits the enzyme that carboxylates glutamate residues.

TOO MUCH BROCCOLI

An article describing two women with diseases characterized by abnormal blood clotting reported that they did not improve when they were given warfarin. When questioned about their diets, one woman said that she ate at least a pound (0.45 kg) of broccoli every day, and the other ate broccoli soup and a broccoli salad every day. When broccoli was removed from their diets, warfarin became effective in preventing the abnormal clotting of their blood. Because broccoli is high in vitamin K, these patients had been getting enough dietary vitamin K to compete with the drug, thereby making the drug ineffective.

PROBLEM 21

Thiols such as ethanethiol and propanethiol can be used to reduce vitamin K epoxide back to vitamin KH$_2$, but they react much more slowly than dihydrolipoate. Explain.

KEY TERMS

anabolism (page 992)
apoenzyme (page 991)
biotin (page 1010)
catabolism (page 992)
coenzyme (page 990)
coenzyme A (CoASH) (page 1007)
coenzyme B$_{12}$ (page 1020)
cofactor (page 990)
competitive inhibitor (page 1026)
dehydrogenase (page 998)
electron sink (page 1005)
flavin adenine dinucleotide (FAD)
   (page 1001)

flavin mononucleotide (FMN)
   (page 1001)
heterocycle (page 996)
holoenzyme (page 991)
lipoate (page 1007)
mechanism-based inhibitor
   (page 1025)
metabolism (page 992)
metalloenzyme (page 990)
nicotinamide adenine dinucleotide
   (NAD$^+$) (page 996)
nicotinamide adenine dinucleotide
   phosphate (NADP$^+$) (page 996)

nucleotide (page 996)
pyridoxal phosphate (PLP)
   (page 1012)
suicide inhibitor (page 1025)
tetrahydrofolate (THF) (page 1022)
thiamine pyrophosphate (TPP)
   (page 1005)
transamination (page 1014)
transimination (page 1013)
vitamin (page 990)
vitamin KH$_2$ (page 1027)

PROBLEMS

22. Answer the following:
   a. Name six cofactors that act as oxidizing agents.
   b. What are the cofactors that donate one-carbon groups?
   c. What three one-carbon groups are various tetrahydrofolates capable of donating to substrates?
   d. What is the function of FAD in the pyruvate dehydrogenase complex?
   e. What is the function of NAD$^+$ in the pyruvate dehydrogenase complex?
   f. What is the reaction necessary for proper blood clotting catalyzed by vitamin KH$_2$?
   g. What coenzymes are used for decarboxylation reactions?
   h. What kinds of substrates do the decarboxylating coenzymes work on?
   i. What coenzymes are used for carboxylation reactions?
   j. What kinds of substrates do the carboxylating coenzymes work on?

23. Name the coenzymes that:
   a. allow electrons to be delocalized.   **c.** provide a good nucleophile.
   b. activate groups for further reaction.   **d.** provide a strong base.

24. For each of the following reactions, name the enzyme that catalyzes the reaction and name the required coenzyme:

a. 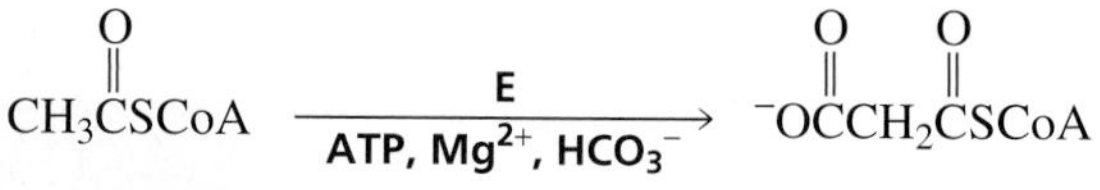 $CH_3\overset{O}{\underset{||}{C}}SCoA$ $\xrightarrow[\text{ATP, Mg}^{2+}\text{, HCO}_3^-]{E}$ $^-O\overset{O}{\underset{||}{C}}CH_2\overset{O}{\underset{||}{C}}SCoA$

b.

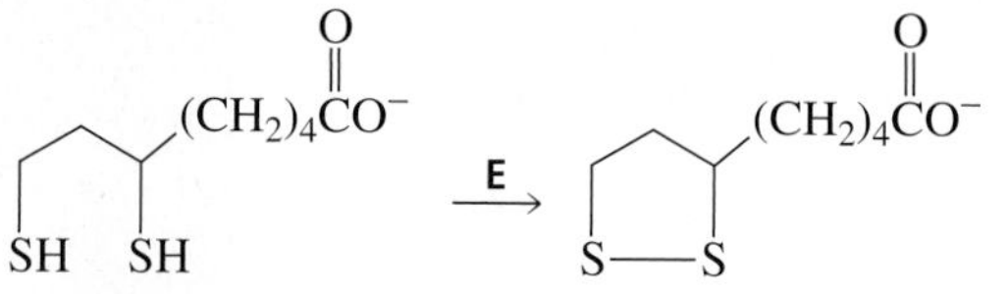

c. 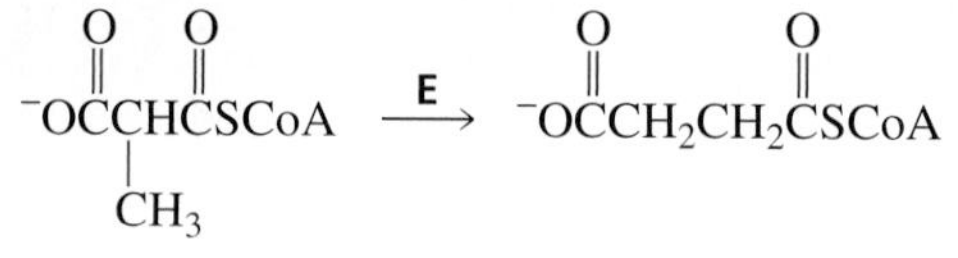 $^-O\overset{O}{\underset{||}{C}}\underset{\underset{CH_3}{|}}{C}H\overset{O}{\underset{||}{C}}SCoA$ $\xrightarrow{E}$ $^-O\overset{O}{\underset{||}{C}}CH_2CH_2\overset{O}{\underset{||}{C}}SCoA$

d. $CH_3\overset{O}{\underset{||}{C}}-\overset{O}{\underset{||}{C}}O^-$ $\xrightarrow[\text{a catabolic reaction}]{E}$ $CH_3\underset{\underset{}{\overset{OH}{|}}}{C}H-\overset{O}{\underset{||}{C}}O^-$

e. $^-O\overset{O}{\underset{||}{C}}CH_2\underset{\underset{^+NH_3}{|}}{C}H\overset{O}{\underset{||}{C}}O^-$ $\xrightarrow{E}$ $^-O\overset{O}{\underset{||}{C}}CH_2\overset{O}{\underset{||}{C}}\underset{\underset{O}{||}}{C}O^-$

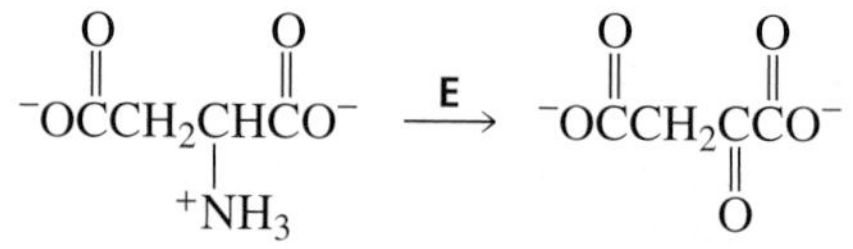

f. 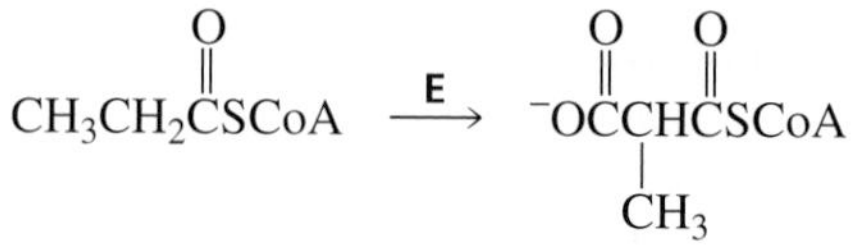 $CH_3CH_2\overset{O}{\underset{||}{C}}SCoA$ $\xrightarrow{E}$ $^-O\overset{O}{\underset{||}{C}}\underset{\underset{CH_3}{|}}{C}H\overset{O}{\underset{||}{C}}SCoA$

25. *S*-Adenosylmethionine (SAM) is formed from the reaction between ATP (Section 9.11) and methionine. The other product of the reaction is triphosphate. Propose a mechanism for this reaction.

26. Five coenzymes are required by α-ketoglutarate dehydrogenase, the enzyme in the citric acid cycle that converts α-ketoglutarate to succinyl-CoA.
    a. Identify the coenzymes.
    b. Propose a mechanism for this reaction.

$^-O\overset{O}{\underset{||}{C}}CH_2CH_2\overset{O}{\underset{||}{C}}-\overset{O}{\underset{||}{C}}O^-$ $\xrightarrow{\alpha\text{-ketoglutarate dehydrogenase}}$ $^-O\overset{O}{\underset{||}{C}}CH_2CH_2\overset{O}{\underset{||}{C}}SCoA$ + CO_2
α-ketoglutarate $\qquad\qquad\qquad\qquad\qquad\qquad\qquad\qquad\qquad\qquad$ succinyl-CoA

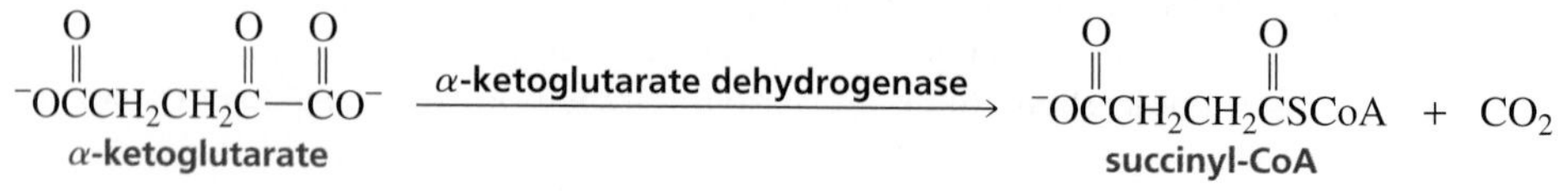

27. Give the products of the following reaction, where T is tritium. (*Hint:* Tritium is a hydrogen atom with two neutrons. Although a carbon–tritium bond breaks four times slower than a carbon–hydrogen bond, it is still the first bond in the substrate that breaks.)

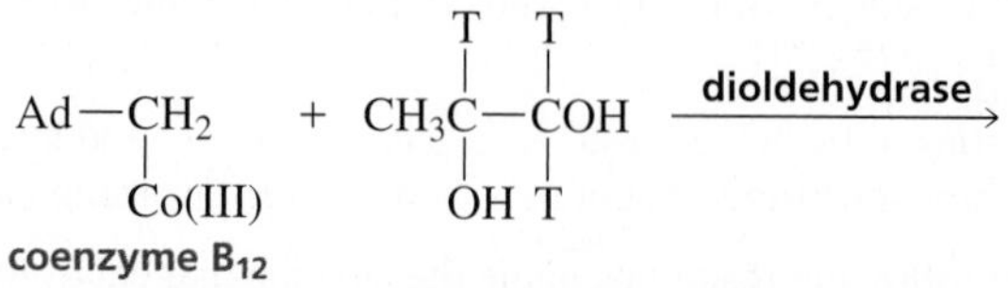

$Ad-CH_2$ + $CH_3\underset{\underset{OH}{|}}{\overset{\overset{T}{|}}{C}}-\underset{\underset{T}{|}}{\overset{\overset{T}{|}}{C}}OH$ $\xrightarrow{\text{dioldehydrase}}$
$\underset{\text{Co(III)}}{|}$
coenzyme B$_{12}$

28. Propose a mechanism for methylmalonyl-CoA mutase, the enzyme that converts methylmalonyl-CoA into succinyl-CoA.

29. When transaminated, the three branched-chain amino acids (valine, leucine, and isoleucine) form compounds that have the characteristic odor of maple syrup. An enzyme known as branched-chain α-keto acid dehydrogenase converts these compounds into CoA esters. People who do not have this enzyme have the genetic disease known as maple syrup urine disease—so-called because their urine smells like maple syrup.
   a. Give the structures of the compounds that smell like maple syrup.
   b. Give the structures of the CoA esters.
   c. Branched-chain α-keto acid dehydrogenase has five coenzymes. Identify the coenzymes.
   d. How can this disease be treated?

30. When UMP is dissolved in T_2O (T = tritium; see Problem 27), exchange of T for H occurs at the 5-position. Propose a mechanism for this exchange.

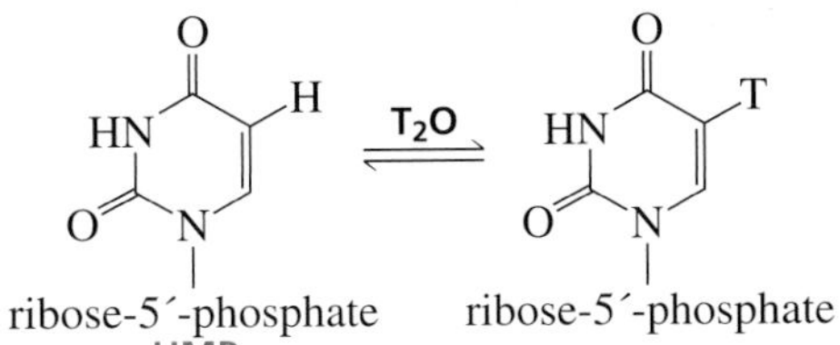

ribose-5′-phosphate   ribose-5′-phosphate
        UMP

31. Dehydratase is a pyridoxal-requiring enzyme that catalyzes an α,β-elimination reaction. Propose a mechanism for this reaction.

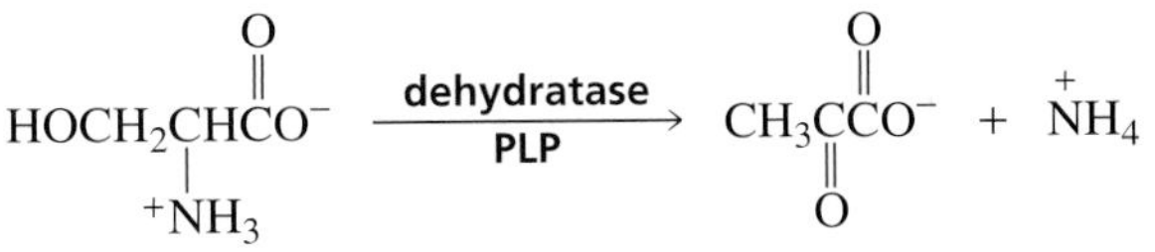

$$HOCH_2\underset{\underset{^+NH_3}{|}}{CH}\overset{\overset{O}{\|}}{C}O^- \xrightarrow[\text{PLP}]{\textbf{dehydratase}} CH_3\overset{\overset{O}{\|}}{\underset{\underset{O}{\|}}{C}}CO^- + \ ^+NH_4$$

32. In addition to the reactions mentioned in Section 23.6, PLP can also catalyze β-substitution reactions. Propose a mechanism for the following PLP-catalyzed β-substitution reaction:

$$XCH_2\underset{\underset{^+NH_3}{|}}{CH}\overset{\overset{O}{\|}}{C}O^- + Y^- \xrightarrow[\text{PLP}]{\textbf{E}} YCH_2\underset{\underset{^+NH_3}{|}}{CH}\overset{\overset{O}{\|}}{C}O^- + X^-$$

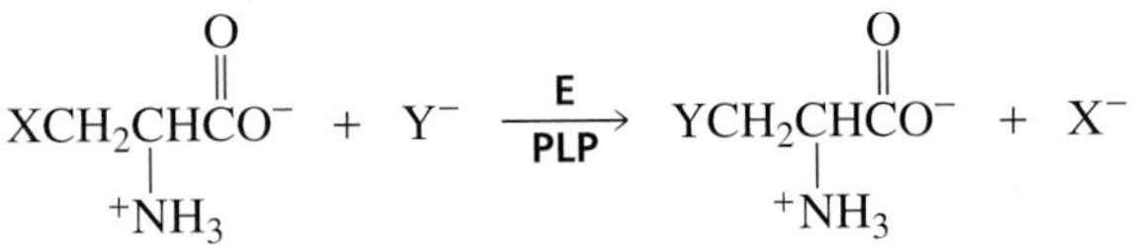

33. PLP can catalyze both α,β-elimination reactions (Problem 31) and β,γ-elimination reactions. Propose a mechanism for the following PLP-catalyzed β,γ-elimination:

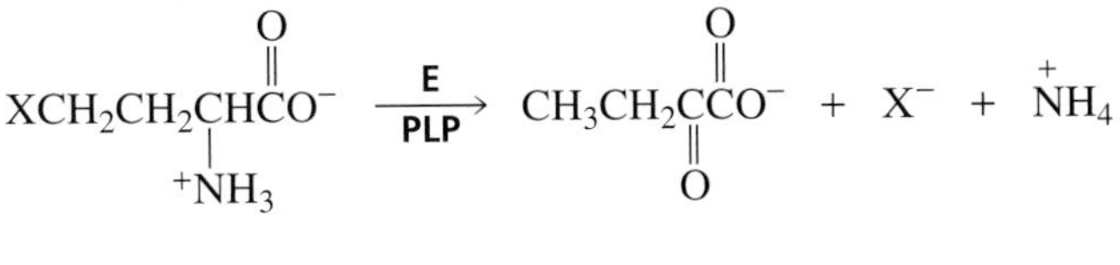

$$XCH_2CH_2\underset{\underset{^+NH_3}{|}}{CH}\overset{\overset{O}{\|}}{C}O^- \xrightarrow[\text{PLP}]{\textbf{E}} CH_3CH_2\overset{\overset{O}{\|}}{\underset{\underset{O}{\|}}{C}}CO^- + X^- + \ ^+NH_4$$

34. The glycine cleavage system is a group of four enzymes that together catalyze the following reaction:

glycine + THF $\xrightarrow{\textbf{glycine cleavage system}}$ N^5,N^{10}-methylene-THF + CO_2

Use the following information to determine the sequence of reactions involved in the glycine cleavage system:

a. The first enzyme involved in the reaction is a PLP-requiring decarboxylase.

b. The second enzyme is aminomethyltransferase. This enzyme has a lipoate coenzyme.

c. N^5,N^{10}-Methylene-THF synthesizing enzyme is the third enzyme. It catalyzes a reaction that forms $^+NH_4$ as one of the products.

d. The fourth enzyme is an FAD-requiring enzyme.

e. The cleavage system also requires NAD^+.

35. Nonenzyme-bound FAD is a stronger oxidizing agent than NAD^+. How, then, can NAD^+ oxidize the reduced flavoenzyme in the pyruvate dehydrogenase system?

36. $FADH_2$ reduces α,β-unsaturated thioesters to saturated thioesters. The reaction is thought to take place by a mechanism that involves radicals. Propose a mechanism for this reaction.

$$\underset{\substack{\\ }}{RCH}=\overset{\displaystyle O}{\overset{\displaystyle \|}{CHCSR}} \;+\; FADH_2 \;\longrightarrow\; RCH_2CH_2\overset{\displaystyle O}{\overset{\displaystyle \|}{C}}SR \;+\; FAD$$

Lipids

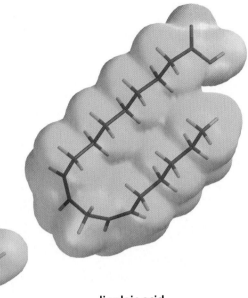

linoleic acid

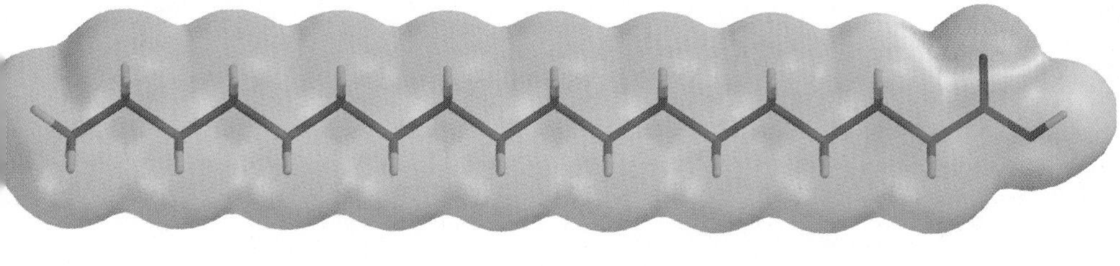

stearic acid

Lipids are organic compounds, found in living organisms, that are soluble in nonpolar organic solvents. Because compounds are classified as lipids as a result of their physical properties rather than their structures, lipids have a variety of structures and functions, as the examples here illustrate.

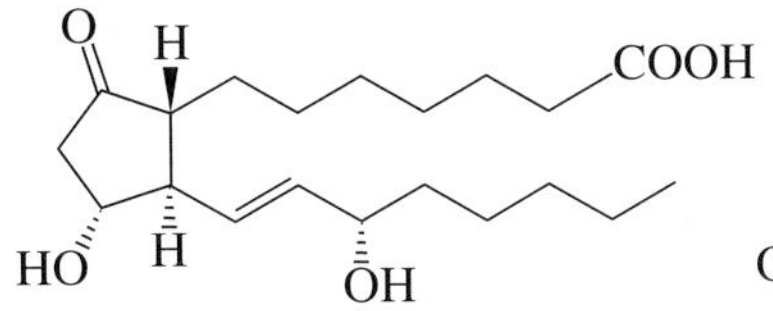

PGE₁
a vasodilator

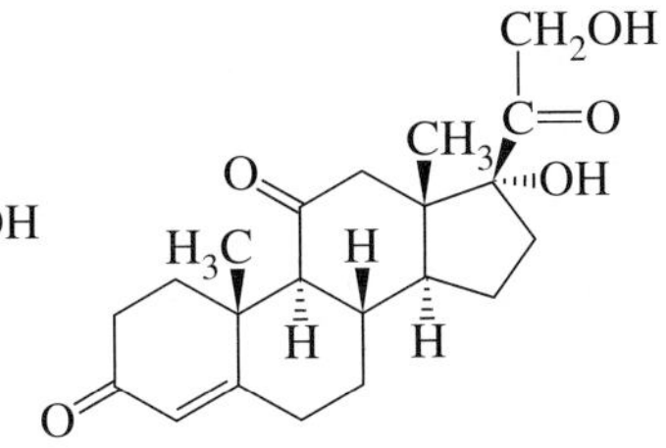

cortisone
a hormone

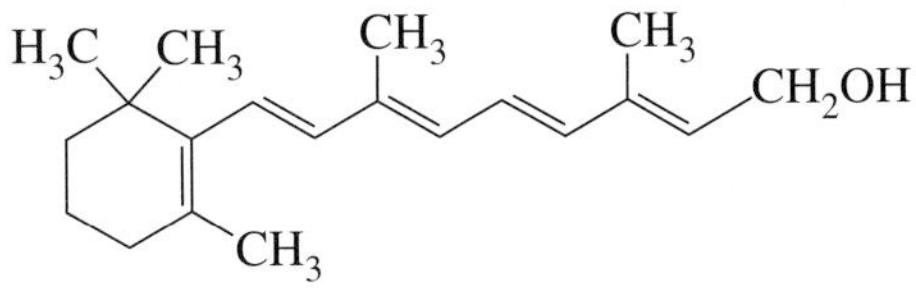

vitamin A
a vitamin

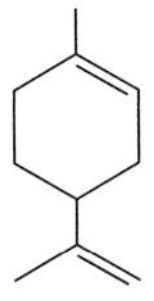

limonene
in orange and
lemon oils

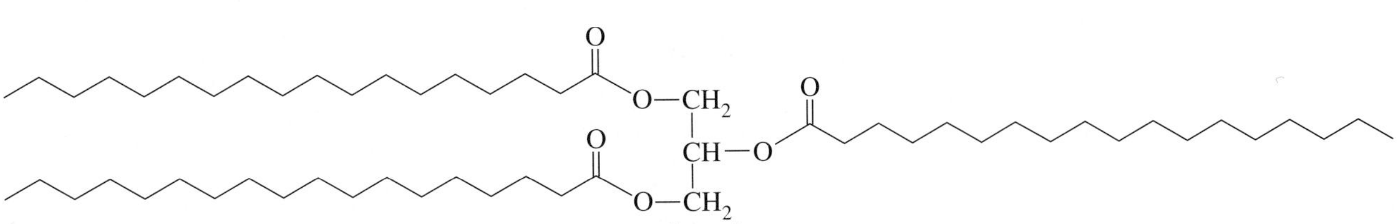

tristearin
a fat

The solubility of lipids in nonpolar organic solvents results from their significant hydrocarbon component. The hydrocarbon portion of the compound is responsible for the "oiliness" or "fattiness" associated with lipids. The word *lipid* comes from the Greek *lipos,* which means "fat."

24.1
FATTY ACIDS

Fatty acids are carboxylic acids with long hydrocarbon chains. The fatty acids most frequently found in nature are shown in Table 24.1. Because they are synthesized from acetate, a compound with two carbon atoms, most naturally occurring fatty acids contain an even number of carbon atoms and are unbranched. The mechanism for the biosynthesis of fatty acids is discussed in Section 19.21. Fatty acids can be saturated ("saturated" with hydrogen, therefore containing no carbon–carbon double bonds) or unsaturated (containing carbon–carbon double bonds). Fatty acids with more than one double bond are called **polyunsaturated fatty acids.** Double bonds in naturally occurring unsaturated fatty acids are never conjugated—they are always separated by one methylene group.

TABLE 24.1 Common Naturally Occurring Fatty Acids

Number of carbons	Common name	Systematic name	Structure	Melting point °C
Saturated				
12	lauric acid	dodecanoic acid	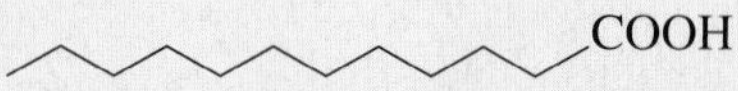 COOH	44
14	myristic acid	tetradecanoic acid	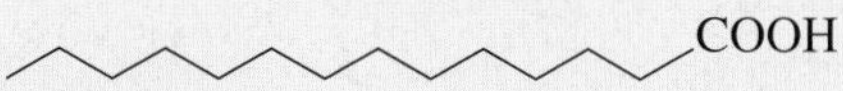 COOH	58
16	palmitic acid	hexadecanoic acid	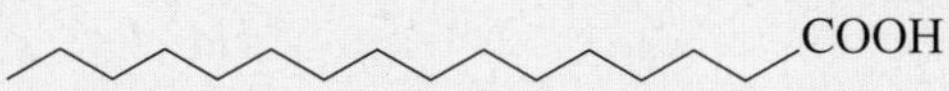 COOH	63
18	stearic acid	octadecanoic acid	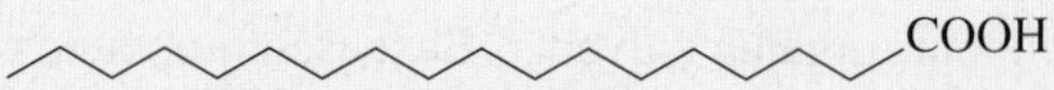 COOH	69
20	arachidic acid	eicosanoic acid	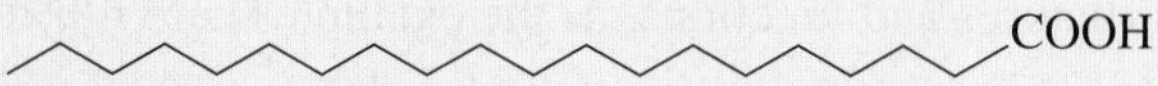 COOH	77
Unsaturated				
16	palmitoleic acid	(9Z)-hexadecenoic acid	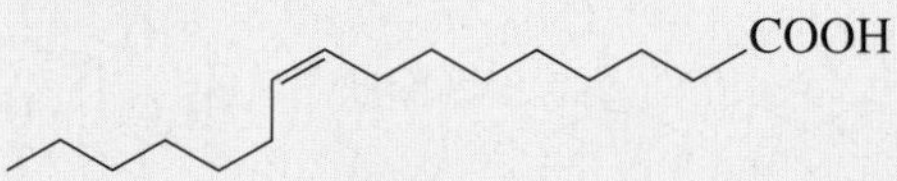 COOH	0
18	oleic acid	(9Z)-octadecenoic acid	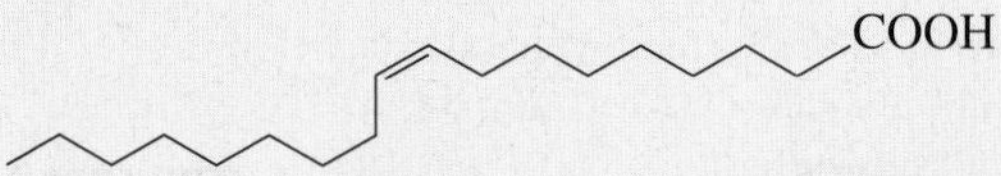 COOH	13
18	linoleic acid	(9Z,12Z)-octadecadienoic acid	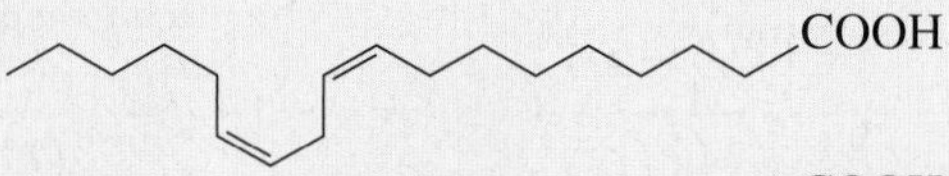 COOH	−5
18	linolenic acid	(9Z,12Z,15Z)-octadecatrienoic acid	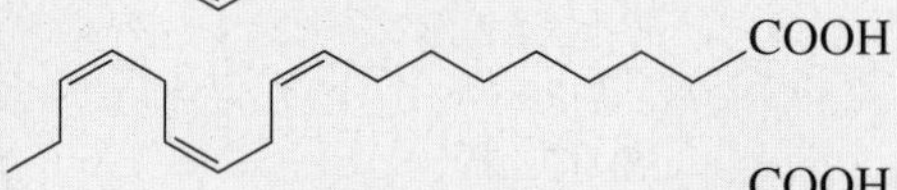 COOH	−11
20	arachidonic acid	(5Z,8Z,11Z,14Z)-eicosatetraenoic acid	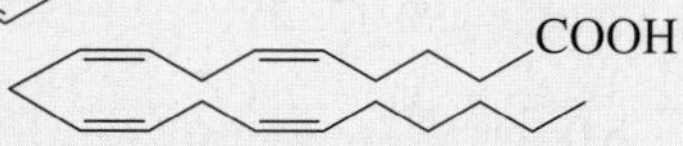 COOH	−50
20	EPA	(5Z,8Z,11Z,14Z,17Z)-eicosapentaeneoic acid	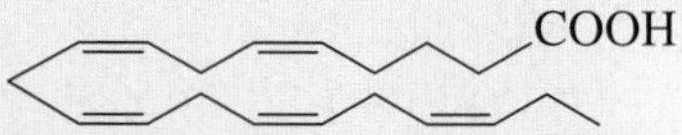 COOH	−50

The physical properties of a fatty acid depend on the length of the hydrocarbon chain and the degree of unsaturation. As expected, the melting points of saturated fatty acids increase with increasing molecular weight because of increased intermolecular van der Waals interactions (Section 2.9).

The double bonds in unsaturated fatty acids generally have the cis configuration. This produces a bend in the molecules, which prevents them from packing together as tightly as fully saturated fatty acids. As a result, unsaturated fatty acids have fewer intermolecular interactions and, therefore, lower melting points than saturated fatty acids with comparable molecular weights (Table 24.1). The melting points of the unsaturated fatty acids decrease as the number of double bonds increases.

3-D Molecules:
Stearic acid;
Oleic acid;
Linoleic acid;
Linolenic acid

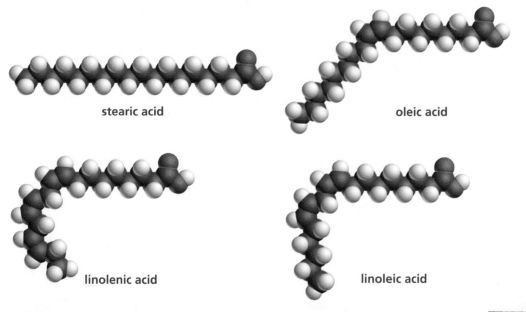

stearic acid

oleic acid

linolenic acid

linoleic acid

PROBLEM 1

Explain the difference in the melting points of the following:

a. palmitic acid and stearic acid   **c.** oleic acid and linoleic acid

b. palmitic acid and palmitoleic acid

PROBLEM 2◆

What products are formed when arachidonic acid reacts with excess ozone followed by treatment with H_2O_2? (*Hint:* See Section 18.7.)

Layers of honeycomb in a beehive.

Waxes are esters formed from long-chain carboxylic acids and long-chain alcohols. For example, beeswax, the structural material of beehives, has a 16-carbon carboxylic acid component and a 30-carbon alcohol component. The word *wax* comes from the Old English *weax,* meaning "material of the honeycomb." Carnauba wax is a particularly hard wax because of its relatively high molecular weight (a 32-carbon carboxylic acid component and a 34-carbon alcohol component). It is widely used in car waxes and floor polishes.

Waxes are common among living organisms. The feathers of birds are coated with wax to make them water-repellent. Some vertebrates secrete wax in order to keep their fur lubricated and water-repellent. Insects secrete a waterproof, waxy layer on the outside of their exoskeletons. Wax is also found on the surfaces of certain leaves and fruits, where it serves as a protectant against parasites and minimizes the evaporation of water.

24.2 WAXES

Raindrops on a feather.

$$CH_3(CH_2)_{24}CO(CH_2)_{29}CH_3$$
a major component of beeswax
structural material
of beehives

$$CH_3(CH_2)_{30}CO(CH_2)_{33}CH_3$$
a major component of carnauba wax
coating on the leaves
of a Brazilian palm

$$CH_3(CH_2)_{14}CO(CH_2)_{15}CH_3$$
a major component of spermaceti wax
from the heads of
sperm whales

24.3 FATS AND OILS

Triacylglycerols, also called triglycerides, are compounds in which the three OH groups of glycerol are esterified with fatty acids. If the three fatty acid components of a triacylglycerol are the same, the compound is called a **simple triacylglycerol. Mixed triacylglycerols,** on the other hand, contain two or three different fatty acid components and are more common than simple triacylglycerols. Not all triacylglycerol molecules from a single source are necessarily identical. Substances such as lard and olive oil, for example, are mixtures of several different triacylglycerols (Table 24.2).

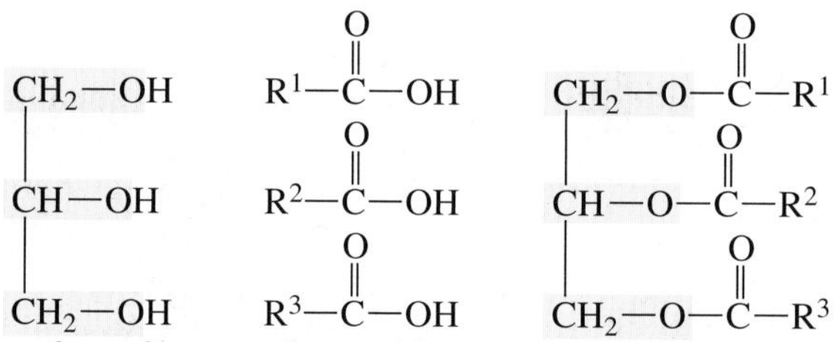

TABLE 24.2 Approximate Percentage of Fatty Acids in Some Common Fats and Oils

| | mp | Saturated fatty acids | | | | Unsaturated fatty acids | | |
| | | lauric | myristic | palmitic | stearic | oleic | linoleic | linolenic |
	°C	C_{12}	C_{14}	C_{16}	C_{18}	C_{18}	C_{18}	C_{18}
Animal fats								
butter	32	2	11	29	9	27	4	—
lard	30	—	1	28	12	48	6	—
human fat	15	1	3	25	8	46	10	—
whale blubber	24	—	8	12	3	35	10	—
Plant oils								
corn	20	—	1	10	3	50	34	—
cottonseed	−1	—	1	23	1	23	48	—
linseed	−24	—	—	6	3	19	24	47
olive	−6	—	—	7	2	84	5	—
peanut	3	—	—	8	3	56	26	—
safflower	−15	—	—	3	3	19	70	3
sesame	−6	—	—	10	4	45	40	—
soybean	−16	—	—	10	2	29	51	7

Triacylglycerols that are solids or semisolids at room temperature are called **fats.** Fats are usually obtained from animals and are composed largely of triacylglycerols with either saturated fatty acids or fatty acids with only one double bond. The saturated fatty acid tails pack closely together, giving the triacylglycerols relatively high melting points at room temperature.

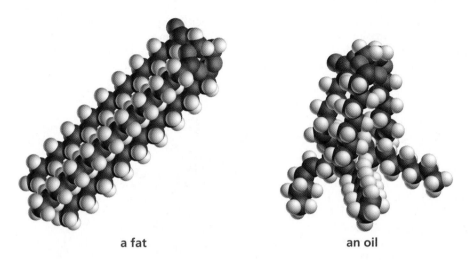

a fat   an oil

Liquid triacylglycerols are called **oils.** Oils typically come from plant products such as corn, soybeans, olives, and peanuts. They are composed primarily of triacylglycerols with unsaturated fatty acids that cannot pack tightly together. Consequently they have relatively low melting points, causing them to be liquids at room temperature. The approximate fatty acid compositions of some common fats and oils are listed in Table 24.2.

Some or all of the double bonds of polyunsaturated oils can be reduced by catalytic hydrogenation (Section 3.19). Margarine and shortening are prepared by hydrogenating vegetable oils such as soybean oil and safflower oil until they have the desired consistency. This process is called "hardening of oils." The hydrogenation reaction must be carefully controlled, however, because reducing all the carbon–carbon double bonds would produce a hard fat with the consistency of beef tallow.

$$RCH=CHCH_2CH=CHCH_2CH=CH- \quad \xrightarrow[Pt]{H_2} \quad RCH_2CH_2CH_2CH=CHCH_2CH_2CH_2-$$

Vegetable oils have become popular for food preparation because some studies have linked the consumption of saturated fats with heart disease. Recent studies, however, have shown that unsaturated fats may also be implicated in heart disease. A 20-carbon fatty acid with five double bonds (known as EPA), found in high concentrations in fish oils, is thought to lower the chance of developing certain forms of heart disease. Once consumed, dietary fat is hydrolyzed in the intestine, regenerating glycerol and fatty acids. We have seen that hydrolysis of fats under basic conditions forms glycerol and salts of fatty acids that are commonly known as *soap* (Section 16.17).

This puffin's diet is high in fish oil.

PROBLEM 3◆

Do all triacylglycerols have the same number of chirality centers?

PROBLEM 4◆

Which has a higher melting point, glyceryl tripalmitoleate or glyceryl tripalmitate?

Organisms store energy in the form of triacylglycerols. A fat provides about six times as much metabolic energy as an equal weight of hydrated glycogen because

OLESTRA—NONFAT WITH FLAVOR

Chemists have been searching for ways to reduce the caloric content of foods without decreasing flavor. Many people who believe that "no fat" is synonymous with "no flavor" can understand this problem. The Federal Food and Drug Administration (Section 30.13) approved the use of Olestra for limited use as a substitute for dietary fat in snack foods. Procter and Gamble spent 30 years and more than $2 billion to develop this compound. Its approval was based on the results of more than 150 studies.

Olestra is a semisynthetic compound. That is, Olestra does not exist in nature but its component parts do. Developing a compound that can be made from units that are a normal part of our diet decreases the potential toxic effects of the new compound. Olestra is made by esterifying all the OH groups of sucrose with fatty acids obtained from cottonseed oil and soybean oil. Therefore its component parts are table sugar and vegetable oil. Olestra works as a fat sub-stitute because its ester linkages are too hindered to be hydrolyzed by digestive enzymes. As a result, Olestra tastes like fat, but it has no caloric value since it cannot be digested.

olestra

fats are less oxidized than carbohydrates and, since fats are nonpolar, they do not bind water. In contrast, two-thirds of the weight of stored glycogen is water (Section 20.18).

Animals have a subcutaneous layer of fat cells that serves as both an energy source and an insulator. The fat content of the average man is about 21%, while the fat content of the average woman is about 25%. Humans can store sufficient fat to provide for the body's metabolic needs for two to three months but can store only enough carbohydrate to provide for less than 24 hours of the body's metabolic needs. Carbohydrates, therefore, are used primarily as a quick, short-term energy source.

Polyunsaturated fats and oils are easily oxidized by O_2 by means of a radical chain reaction. In the initiation step, a radical removes a hydrogen from a methylene group that is flanked by two double bonds. This is the most easily removed hydrogen because the resulting radical is resonance-stabilized by both double bonds. The resulting radical reacts with O_2, forming a peroxy radical with conjugated double bonds. The peroxy radical removes a hydrogen from a methylene group of another molecule of fatty acid, forming an alkyl hydroperoxide. The two propagating steps are repeated over and over.

3-D Molecule:
Olestra

$$RCH{=}CH{-}CH{-}CH{=}CH{-} \; + \; X\cdot \xrightarrow{\text{initiation}} \; RCH{=}CH{-}CH{-}CH{=}CH{-} \; + \; HX$$

$$\overset{|}{H}$$

resonance contributor
with isolated double bonds

$$\updownarrow$$

$$RCH{-}CH{=}CH{-}CH{=}CH{-}$$

resonance contributor
with conjugated double bonds

$$\cdot\overset{..}{\underset{..}{O}}{-}\overset{..}{\underset{..}{O}}\cdot \quad \Big| \; \text{propagation}$$

$$RCH{-}CH{=}CH{-}CH{=}CH{-}$$
$$\overset{|}{\underset{..}{O}}{-}\overset{..}{\underset{..}{O}}\cdot$$

a peroxy radical

$$RCH{=}CH{-}CH_2{-}CH{=}CH{-} \quad \Big| \; \text{propagation}$$

$$RCH{=}CH{-}\overset{\cdot}{C}H{-}CH{=}CH{-} \; + \; RCH{-}CH{=}CH{-}CH{=}CH{-}$$
$$\overset{|}{\underset{..}{O}}{-}\overset{..}{O}H$$

an alkyl hydroperoxide

The reaction of fatty acids with O_2 causes them to become rancid. The unpleasant taste and smell associated with rancidity are the results of further oxidation of the alkyl hydroperoxide to shorter-chain carboxylic acids that have strong odors. The same process contributes to the odor associated with sour milk.

> ### PROBLEM 5
>
> Draw the resonance contributors for the radical formed when a hydrogen atom is removed from the CH_2 group of a polyunsaturated fatty acid.

WHALES AND ECHOLOCATION

Whales have enormous heads, accounting for 33% of their total weight. They have large deposits of fat in their heads and lower jaws. This fat is very different from both the whale's normal body fat and its dietary fat. Because major anatomical modifications were necessary to accommodate this fat, it must have some important function for the animal. It is now believed that the fat is used for echolocation. Echolocation is the process of emitting sounds in pulses and gaining information by analyzing the returning echoes. The fat in the whale's head focuses the emitted sound waves in a directional beam, and the echoes are received by the fat organ in the lower jaw. This fat organ transmits the sound to the brain for processing and interpretation, providing the animal with information about the depth of the water, changes in the sea floor, and the position of the coastline. These fat deposits, therefore, give whales a unique acoustic sensory system and allow them to compete successfully for survival with sharks—who also have a well-developed sense of sound direction.

Humpback whale in Alaska

24.4
MEMBRANES

For biological systems to operate, some parts of organisms must be separated from other parts. On a cellular level, the outside of the cell must be separated from the inside of the cell. "Greasy" lipid **membranes** serve as the barrier. In addition to isolating the contents of the cell, these membranes allow the selective transport of ions and organic molecules into and out of the cell.

Phospholipids

Phosphoacylglycerols (also called **phosphoglycerides**) are the major components of cell membranes. They are similar to triacylglycerols except that a terminal OH group of glycerol is esterified with phosphoric acid rather than with a fatty acid, forming a **phosphatidic acid.** Because these are lipids that contain a phosphate group, they are classified as **phospholipids.** The C-2 carbon of glycerol in phosphoacylglycerols has the *R* configuration.

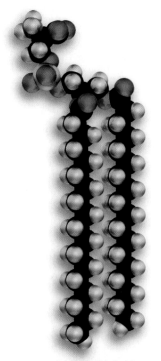

phosphatidylserine

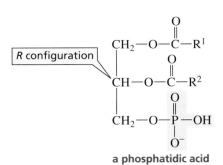

a phosphatidic acid

Phosphatidic acids are the simplest phosphoacylglycerols and are present only in small amounts in membranes. The most common phosphoacylglycerols in membranes have a second phosphate ester linkage. The alcohols most commonly used to form this second ester group are ethanolamine, choline, and serine. Phosphatidylethanolamines are also called **cephalins,** and phosphatidylcholines are called **lecithins.** Lecithins are used in food as emulsifying agents. They are added to foods such as mayonnaise to prevent the aqueous and fat components from separating.

a phosphatidylethanolamine
a cephalin

a phosphatidylcholine
a lecithin

a phosphatidylserine

Phosphoacylglycerols form membranes by arranging themselves in a **lipid bilayer.** The polar heads of the phosphoacylglycerols are on the outside of the bilayer, while the fatty acid chains and cholesterol—a membrane lipid discussed in Section 24.9—form the interior of the bilayer (Figure 24.1). A typical bilayer is about 50 Å thick. Compare the bilayer with the micelles formed by soap in aqueous solution (Section 16.17).

3-D Molecule:
Phosphatidic acid

polar
head

cholesterol
molecule

nonpolar
fatty acid
chains

$$CH_2O-\overset{\overset{\displaystyle O}{\|}}{C}-R^1$$

$$CHO-\overset{\overset{\displaystyle O}{\|}}{C}-R^2$$

$$CH_2O-\overset{\overset{\displaystyle O}{\|}}{\underset{\underset{\displaystyle O^-}{|}}{P}}-O(CH_2)_2\overset{+}{N}H_3$$

▲ **Figure 24.1**
A lipid bilayer.

The fluidity of a membrane is controlled by the fatty acid components of the phosphoacylglycerols. Long-chain saturated fatty acids decrease membrane fluidity because their hydrocarbon chains can pack closely together. Unsaturated fatty acids increase fluidity because they pack less closely together. Cholesterol also decreases fluidity (Section 24.9). Only animal membranes contain cholesterol, so animal membranes are more rigid than plant membranes.

The unsaturated fatty acid chains of phosphoacylglycerols are susceptible to reaction with O_2, similar to the reaction described on page 1039 for fats and oils. Oxidation of phosphoacylglycerols can lead to the degradation of membranes. Vitamin E is an important antioxidant that protects fatty acid chains from degradation via oxidation. Vitamin E, also called α-tocopherol, is classified as a lipid because it is soluble in nonpolar organic solvents. Because vitamin E reacts more rapidly with oxygen than triacylglycerols do, the vitamin prevents biological membranes from reacting with oxygen (Section 8.8). There are some who believe that vitamin E slows the aging process. Because vitamin E also reacts with oxygen more rapidly than fats do, it is added to many foods to prevent spoilage.

3-D Molecule:
Vitamin E

**α-tocopherol
vitamin E**

IS CHOCOLATE A HEALTH FOOD?

We have long been told that our diets should include lots of fruits and vegetables because they are good sources of antioxidants. Antioxidants protect against cardiovascular disease, cancer, and cataracts and they slow the effects of aging. Recent studies show that chocolate has high levels of antioxidants—complex mixtures of phenolic compounds (Section 8.8). On a weight basis, the concentration of antioxidants in chocolate is higher than the concentration in red wine or green tea and 20 times higher than the concentration in tomatoes. Dark chocolate contains more than twice the level of antioxidants as milk chocolate. Unfortunately, white chocolate contains no antioxidants. Another piece of good news is that stearic acid, the main fatty acid in chocolate, does not appear to raise blood cholesterol levels the way other saturated fatty acids do.

PROBLEM 6◆

Membranes contain proteins. Integral membrane proteins extend partly or completely through the membrane, whereas peripheral membrane proteins are found on the inner or outer surfaces of the membrane. What is the likely difference in the overall amino acid composition of integral and peripheral membrane proteins?

PROBLEM 7◆

A colony of bacteria accustomed to an environment at 25 °C was moved to an identical environment at 35 °C. The increased temperature increased the fluidity of the bacterial membranes. What could the bacteria do to regain their original membrane fluidity?

Sphingolipids

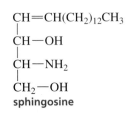

CH=CH(CH₂)₁₂CH₃
|
CH—OH
|
CH—NH₂
|
CH₂—OH
sphingosine

3-D Molecule:
Sphingosine

Sphingolipids are also found in membranes. They are the major lipid components in the myelin sheaths of nerve fibers. Sphingolipids contain sphingosine instead of glycerol. In sphingolipids, the amino group of sphingosine is bonded to the acyl group of a fatty acid.

Two of the most common kinds of sphingolipids are **sphingomyelins** and **cerebrosides.** In sphingomyelins, the primary OH group of sphingosine is bonded to phosphocholine or phosphoethanolamine, similar to lecithins and cephalins. In cerebrosides, the primary OH group of sphingosine is bonded to a sugar residue through a β-glycosidic linkage. Sphingomyelins are phospholipids because they contain a phosphate group. Cerebrosides, on the other hand, are not phospholipids.

CH=CH(CH₂)₁₂CH₃
|
CH—OH   O
| ||
CH—NH—C—R
| O   CH₃
| || |+
CH₂—O—P—OCH₂CH₂NCH₃
| | |
O⁻   CH₃

a sphingomyelin

CH=CH(CH₂)₁₂CH₃
|
CH—OH   O
| ||
CH—NH—C—R
|
CH₂
|
O

a glucocerebroside

MULTIPLE SCLEROSIS AND THE MYELIN SHEATH

The myelin sheath is a lipid-rich material that is wrapped around the axons of nerve cells. This sheath is composed largely of sphingomyelins and cerebrosides. The sheath's function is to increase the velocity of nerve impulses. Multiple sclerosis is a disease characterized by loss of the myelin sheath and a consequent slowing of nerve impulses and eventual paralysis.

PROBLEM 8

a. Draw the structures of three different sphingomyelins.

b. Draw the structure of a galactocerebroside.

PROBLEM 9

The membrane phospholipids in animals such as deer and elk have a higher degree of unsaturation in cells closer to the hoof than in cells closer to the body. Explain how this can be important for survival.

Prostaglandins are responsible for regulating a variety of physiological responses such as inflammation, blood pressure, blood clotting, fever, pain, induction of labor, and the sleep/wake cycle. All prostaglandins have a five-membered ring with a seven-carbon carboxylic acid substituent and an eight-carbon hydrocarbon substituent. The two substituents are trans to each other.

prostaglandin skeleton

Prostaglandins are named using the format PGX, where X designates the functional groups of the five-membered ring. PGAs, PGBs, and PGCs all contain a carbonyl group and a double bond in the five-membered ring. The location of the double bond determines whether a prostaglandin is a PGA, PGB, or PGC. PGDs and PGEs are β-hydroxy ketones, and PGFs are 1,3-diols. A subscript indicates the total number of double bonds in the side chains, and "α" and "β" indicate the configuration of the two OH groups in a PGF: "α" indicates a *cis*-diol and "β" indicates a *trans*-diol.

PGAs   PGBs   PGCs   PGDs

PGE$_1$ PGE$_2$

PGF$_{2\alpha}$

Prostaglandins are synthesized from arachidonic acid, a 20-carbon fatty acid with four cis double bonds. In the cell, arachidonic acid is found esterified to the 2-position of glycerol in many phospholipids. Arachidonic acid is synthesized from linoleic acid. Because linoleic acid cannot be synthesized by mammals, it must be included in the diet.

An enzyme called prostaglandin endoperoxide synthase catalyzes the conversion of arachidonic acid to PGH$_2$, the precursor of all prostaglandins. This enzyme has two activities—a cyclooxygenase activity and a hydroperoxidase activity. It uses its cyclooxygenase activity to form the five-membered ring. In the first step of this transformation, a hydrogen atom is removed from a carbon flanked by two double bonds. This hydrogen is removed relatively easily because the resulting radical is stabilized by electron delocalization. The radical reacts with oxygen to form a peroxy radical. Notice that these two steps are the same as the first two steps in the reaction that causes fats to become rancid (Section 24.3). The peroxy radical rearranges and reacts with a second molecule of oxygen. The enzyme then uses its hydroperoxidase activity to convert the OOH group into an OH group, forming PGH$_2$, which rearranges to form PGE$_2$, a prostaglandin.

24.5 PROSTAGLANDINS

Ulf von Euler

Ulf Svante von Euler (1905–1983) *first identified prostaglandins—from semen— in the early 1930s. He named them for their source, the prostate gland. By the time it was realized that all cells except red blood cells synthesize prostaglandins, their name had become entrenched. Von Euler was born in Stockholm and received an M.D. from the Karolinska Institute, where he remained as a member of the faculty after receiving his degree. He discovered noradrenaline and identified its function as a chemical intermediate in nerve transmission. For this work, he shared the 1970 Nobel Prize in medicine or physiology with Julius Axelrod and Sir Bernard Katz.*

For their work on prostaglandins, ***Sune Bergström, Bengt Ingemar Samuelsson,*** *and* ***John Robert Vane*** *shared the 1982 Nobel Prize in medicine or physiology. Bergström and Samuelsson were born in Sweden— Bergström in 1916 and Samuelsson in 1934. They are both at the Karolinska Institute. Vane was born in England in 1927 and is at the Wellcome Foundation in Beckenham, England.*

biosynthesis of prostaglandins, thromboxanes, and prostacyclins

arachidonic acid

cyclooxygenase

a peroxy radical

hydroperoxidase

a thromboxane

several steps

several steps

a prostacyclin

PGH$_2$

PGE$_2$
a prostaglandin

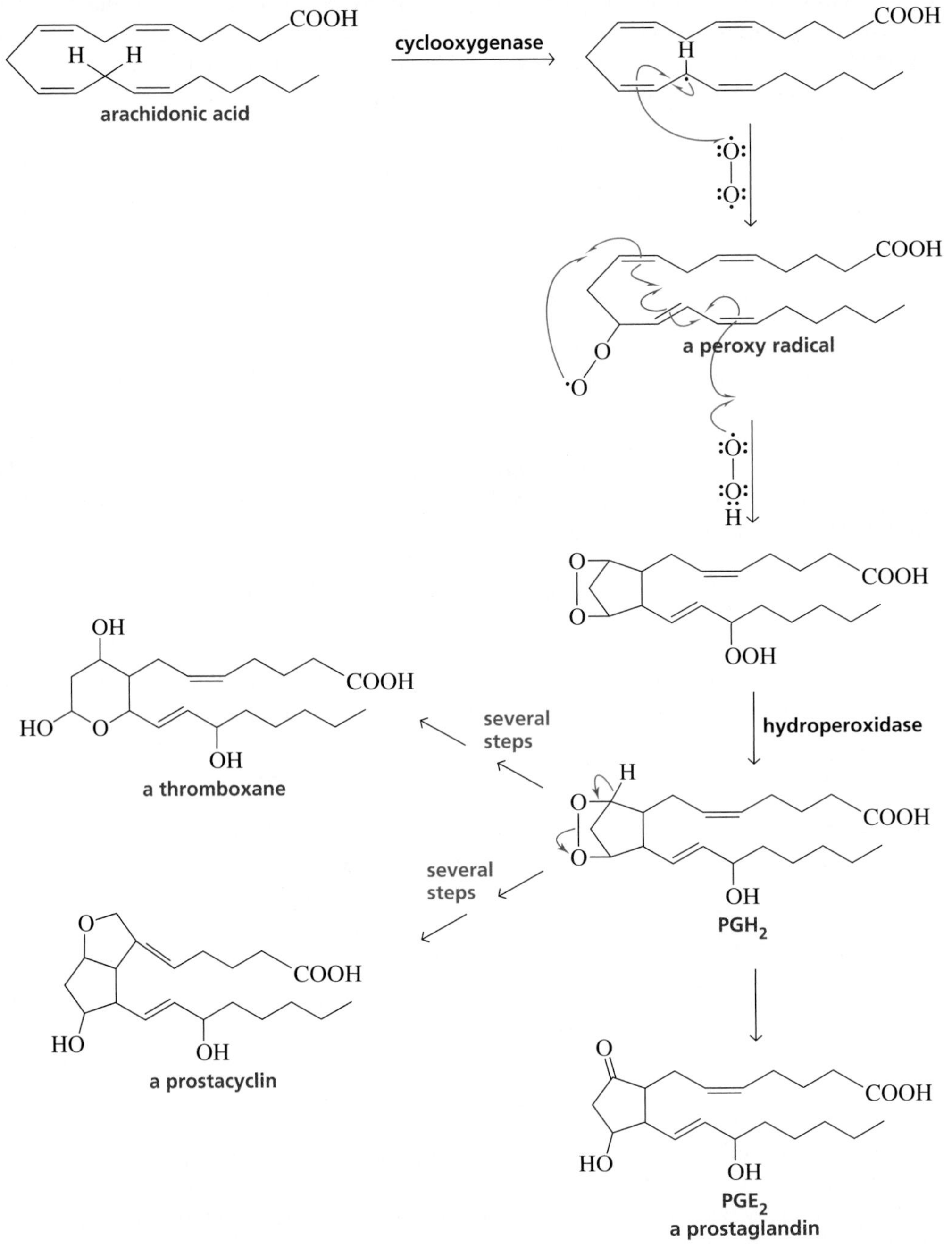

In addition to serving as a precursor for the synthesis of prostaglandins, PGH$_2$ is a precursor for the synthesis of **thromboxanes** and **prostacyclins.** Thromboxanes constrict blood vessels and stimulate platelet aggregation, the first step in blood clotting. Prostacyclins have the opposite effect. They dilate blood vessels and inhibit platelet aggregation. The levels of these two compounds must be carefully controlled to maintain the proper balance in the blood.

Aspirin (acetylsalicylic acid) inhibits the cyclooxygenase activity of prostaglandin endoperoxide synthase. It does this by transferring an acetyl group to a serine hydroxyl group of the enzyme (Section 16.9). Acetylating the serine residue inhibits the enzyme. Aspirin, therefore, inhibits the synthesis of prostaglandins and in this way decreases the inflammation produced by these compounds. Aspirin also inhibits the synthesis of thromboxanes and prostacyclins. Overall, this causes a slight decrease in the rate of blood clotting, which is why some doctors recommend one aspirin tablet every other day to reduce the chance of heart attacks and strokes caused by clotting in blood vessels.

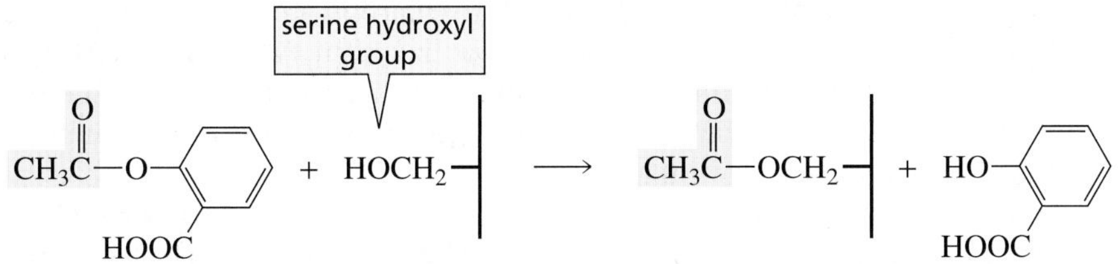

acetylsalicylic acid **cyclooxygenase** **cyclooxygenase**
aspirin active enzyme inactive enzyme

Other anti-inflammatory drugs such as ibuprofen (the active ingredient in Advil, Motrin, and Nuprin) and naproxen (the active ingredient in Aleve) also inhibit the synthesis of prostaglandins. They compete with either arachidonic acid or the peroxy radical for the enzyme's binding site.

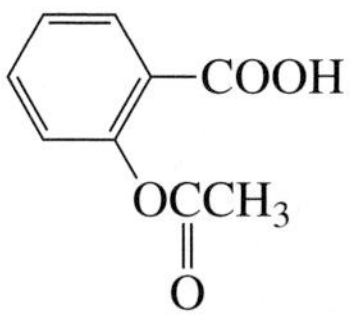

aspirin 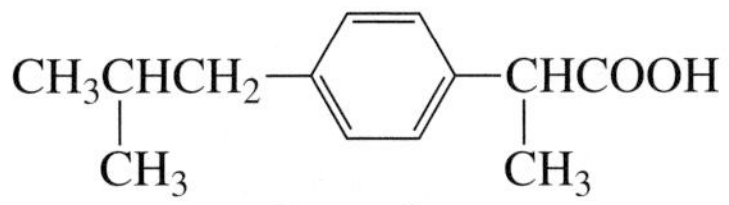 **ibuprofen** 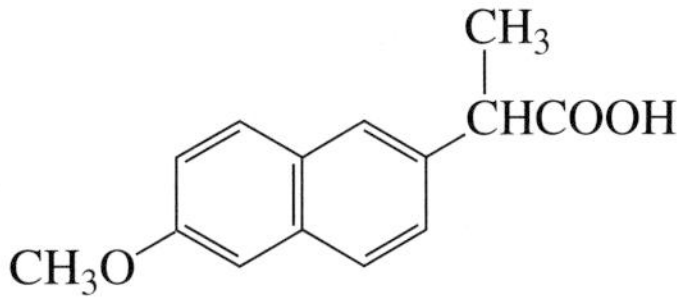 **naproxen**

Arachidonic acid can also be converted into a leukotriene. **Leukotrienes** contain three conjugated double bonds. Because they induce contraction of the muscle that lines the airways to the lungs, they are implicated in allergic reactions, inflammatory reactions, and heart attacks. Leukotrienes are the compounds that bring on the symptoms of asthma. They are also implicated in anaphylactic shock, a potentially fatal allergic reaction. There are several antileukotriene agents available for the treatment of asthma.

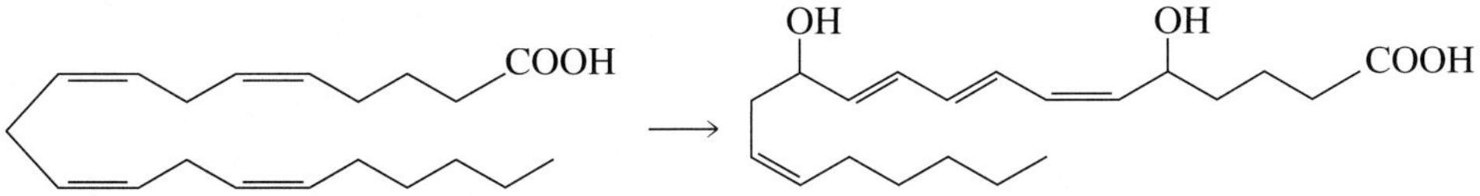

arachidonic acid **a leukotriene**

PROBLEM 10

Treating PGA_2 with a strong base such as sodium *tert*-butoxide followed by addition of acid converts it to PGC_2. Propose a mechanism for this reaction.

**24.6
TERPENES**

Terpenes are a very diverse class of lipids. More than 20,000 terpenes are known. They can be hydrocarbons or they can contain oxygen and be alcohols, ketones, or aldehydes. Oxygen-containing terpenes are sometimes called **terpenoids.** Certain terpenes and terpenoids have been used as spices, perfumes, and medicines for many thousands of years.

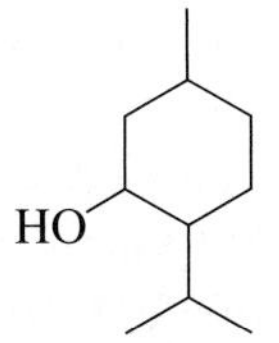

menthol **geraniol** **zingiberene** **β-selinene**
peppermint oil geranium oil oil of ginger oil of celery

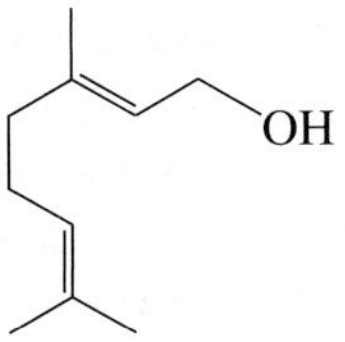

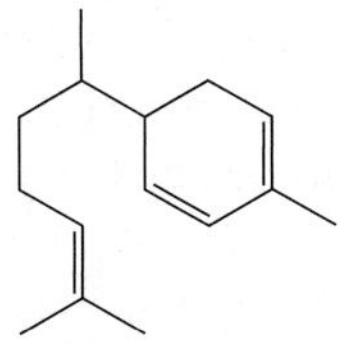

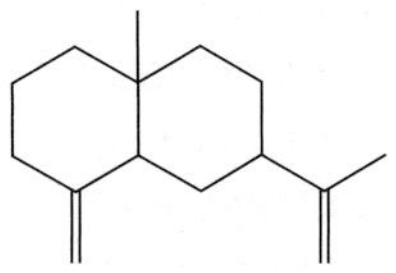

After analyzing a large number of terpenes, organic chemists realized that many of them share a common feature—they contain carbon atoms in multiples of 5. These naturally occurring compounds contain 10, 15, 20, 25, 30, and 40 carbon atoms, which suggests that there is a compound with five carbon atoms that serves as their building block. Further investigation showed that their structures are consistent with the assumption that they were made by joining together isoprene units usually in a "head-to-tail" fashion. Isoprene is the common name for 2-methyl-1,3-butadiene, a compound containing five carbon atoms.

$$CH_2{=}\overset{\overset{\displaystyle CH_3}{|}}{C}{-}CH{=}CH_2$$

2-methyl-1,3-butadiene
isoprene

The branched end of isoprene is called the "head," and the unbranched end is called the "tail." That isoprene units are linked in a head-to-tail fashion to form terpenes is known as the **isoprene rule.**

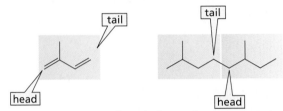

carbon skeleton of two isoprene units with a bond
between the tail of one and the head of another

Leopold Ružička

Leopold Stephen Ružička (1887–1976) was the first to recognize that many organic compounds contain multiples of five carbons. Ružička, a Croatian, attended college in Switzerland and became a Swiss citizen in 1917. He was a professor of chemistry at the University of Utrecht in The Netherlands and later at the Federal Institute of Technology in Zurich. For his work on terpenes, he shared the 1939 Nobel Prize in chemistry with Adolph Butenandt (page 1056).

In the case of cyclic compounds, the linkage of the head of one isoprene unit to the tail of another is followed by an additional linkage to form the ring. The second linkage is not necessarily head to tail but is whatever is necessary to form a stable (five- or six-membered) ring.

α-farnesene
a sesquiterpene found in the
waxy coating on apple skins

carvone
spearmint oil
a monoterpene

In Section 24.8, we will see that the compound actually used in the biosynthesis of terpenes is not isoprene but isopentenyl pyrophosphate, a compound that has the same carbon skeleton as isoprene. We will also look at the mechanism by which isopentenyl phosphate units are joined together in a head-to-tail fashion.

Terpenes are classified according to the number of carbons they contain (Table 24.3). **Monoterpenes** have 10 carbons and are therefore composed of two isoprene units. **Sesquiterpenes** have 15 carbons, so they are composed of three isoprene units. Many fragrances and flavorings found in plants are monoterpenes and sesquiterpenes. These compounds are known as **essential oils.**

Table 24.3   Classification of Terpenes

Carbon atoms	Classification	Carbon atoms	Classification
10	monoterpenes	25	sesterterpenes
15	sesquiterpenes	30	triterpenes
20	diterpenes	40	tetraterpenes

Triterpenes (six isoprene units) and **tetraterpenes** (eight isoprene units) have important biological roles. For example, **squalene,** a triterpene, is a precursor of steroid molecules (Section 24.9).

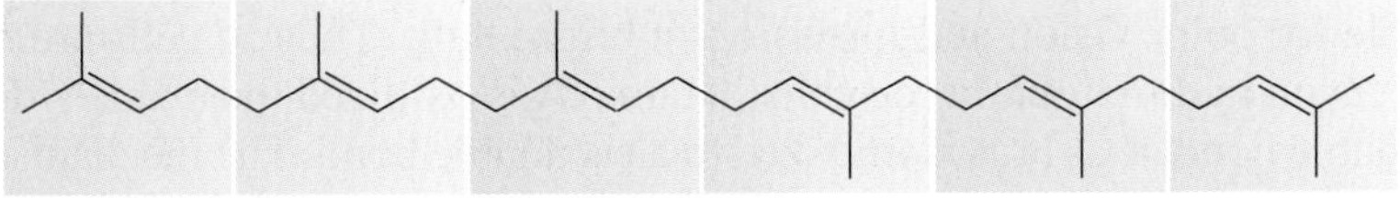

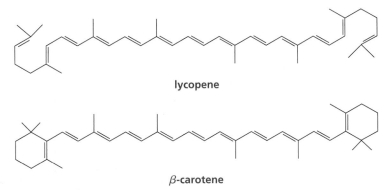

squalene

Carotenoids are tetraterpenes. Lycopene, the compound responsible for the red coloring of tomatoes and watermelon, and β-carotene, the compound that causes carrots and apricots to be orange, are examples of carotenoids. β-Carotene is also the coloring agent used in margarine. β-Carotene and other colored compounds are found in the leaves of trees, but their characteristic colors are usually obscured by the green color of chlorophyll. In the fall when chlorophyll degrades, the colors become apparent. It is the many conjugated double bonds in lycopene and β-carotene that cause the compounds to be colored (Section 12.18).

lycopene

β-carotene

Carotenoids become visible in autumn when chlorophyll degrades.

Tutorial: Isoprene units in terpenes

PROBLEM 11 / SOLVED

Mark off the isoprene units in menthol, zingiberene, β-selinene, and squalene.

SOLUTION   For zingiberene:

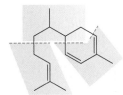

PROBLEM 12 ◆

One of the linkages in squalene is tail to tail, not head-to-head. What does this suggest about how squalene is synthesized in nature? (*Hint*: Locate the position of the tail-to-tail linkage.)

> ## PROBLEM 13
>
> Mark off the isoprene units in lycopene and β-carotene. Can you detect a similarity in the way in which squalene, lycopene, and β-carotene are biosynthesized?

24.7
VITAMIN A

Vitamins A, D, E, and K are lipids (Sections 23.9 and 28.6). Vitamin A is the only water-insoluble vitamin we have not already discussed. β-Carotene, which is cleaved to form two molecules of vitamin A, is the major dietary source of the vitamin. Vitamin A, also called retinol, plays an important role in vision.

The retina of the eye contains cone cells and rod cells. The cone cells are responsible for color vision and for vision in bright light. The rod cells are responsible for vision in dim light. In rod cells, vitamin A is oxidized to an aldehyde and the trans double bond at C-11 is isomerized to a cis double bond. The mechanism for the enzyme-catalyzed interconversion of cis and trans double bonds is discussed in Section 17.16. The protein *opsin* uses a lysine side chain (Lys 216) to form an imine with (11Z)-retinal, resulting in a complex known as *rhodopsin*. When rhodopsin absorbs visible light, it isomerizes to the trans isomer. This change in molecular geometry causes an electrical signal to be sent to the brain, where it is perceived as a visual image. The trans isomer is not stable and is hydrolyzed to (11E)-retinal and opsin. This reaction is referred to as *bleaching* of the visual pigment. (11E)-Retinal is then converted back to (11Z)-retinal to complete the visual cycle.

the chemistry of vision

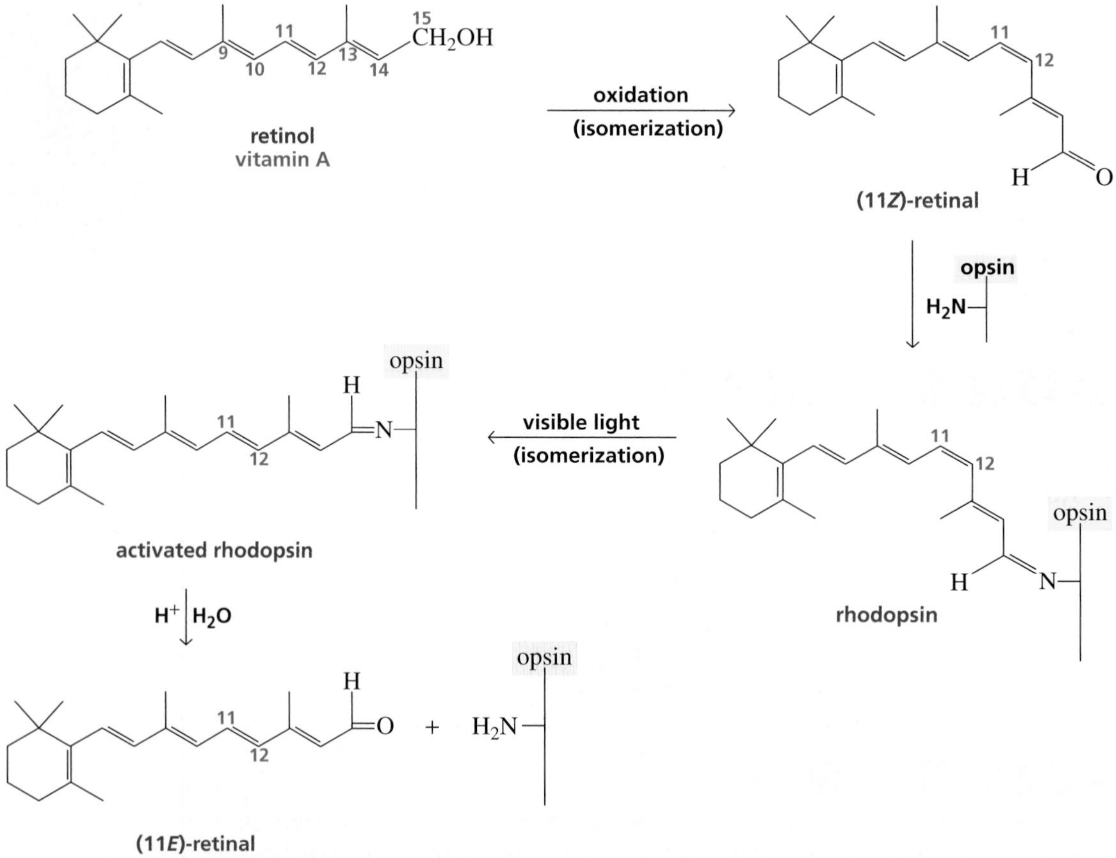

retinol
vitamin A

$\xrightarrow[\text{(isomerization)}]{\text{oxidation}}$

(11Z)-retinal

opsin

activated rhodopsin

$\xleftarrow[\text{(isomerization)}]{\text{visible light}}$

rhodopsin

$H^+ \downarrow H_2O$

(11E)-retinal

The details of how the previous sequence of reactions creates a visual image are not clearly understood. The fact that a simple change in configuration can be responsible for initiating a process as complicated as vision, though, is remarkable.

Biosynthesis of Isopentenyl Pyrophosphate

The five-carbon compound used for the biosynthesis of terpenes is 3-methyl-3-butenylpyrophosphate, loosely called isopentenyl pyrophosphate by biochemists. Each step in the biosynthesis is catalyzed by a different enzyme. The first step is the same Claisen condensation that occurs in the first step of the biosynthesis of fatty acids, except that the acetyl and malonyl groups remain attached to coenzyme A rather than being transferred to the acyl carrier protein (Section 19.21). The Claisen condensation is followed by an aldol addition with a second molecule of malonyl-CoA. The resulting thioester is reduced with two equivalents of NADPH to form mevalonic acid (Section 23.2). A pyrophosphate group is added by means of two successive phosphorylations with ATP. Decarboxylation and loss of the OH group result in isopentenyl pyrophosphate.

biosynthesis of isopentenyl pyrophosphate

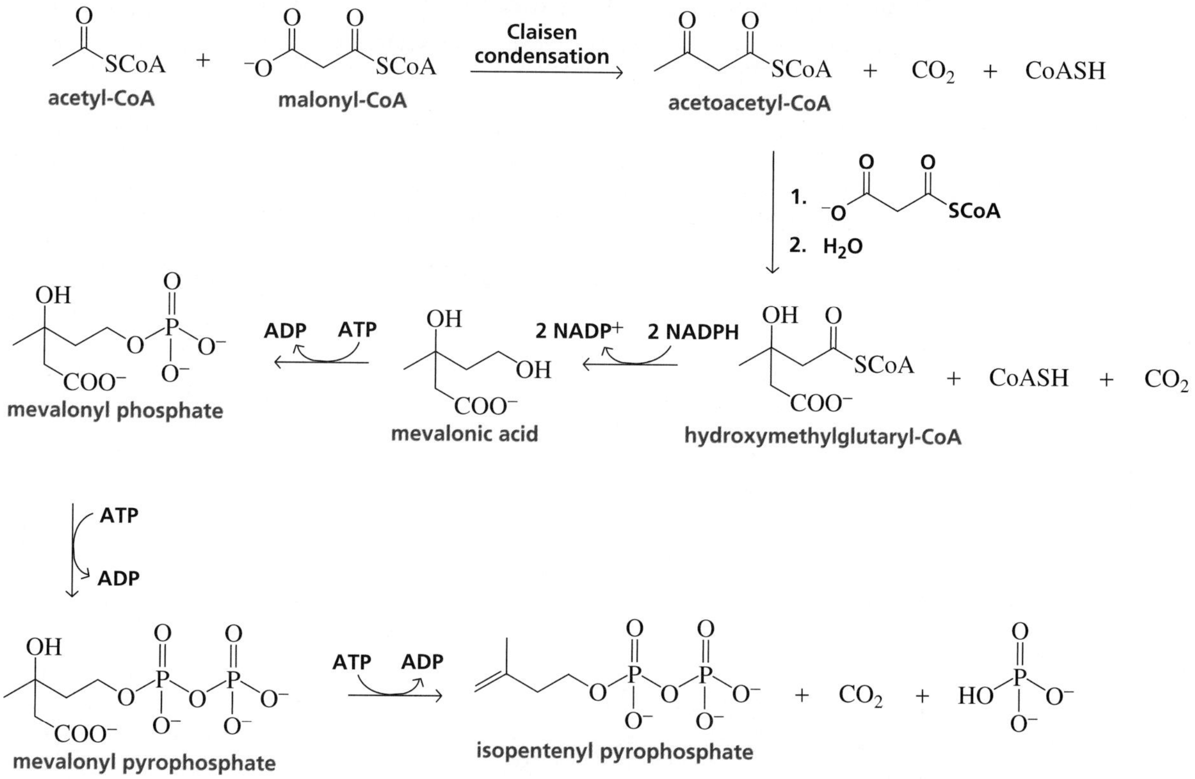

The mechanism for converting mevalonic acid into mevalonyl phosphate is essentially an S_N2 reaction with an adenosyl pyrophosphate leaving group. A second S_N2 reaction converts mevalonyl phosphate to mevalonyl pyrophosphate. ATP is an excellent phosphorylating reagent for nucleophiles because its phosphoanhydride bonds are easily broken. The reason that phosphoanhydride bonds are so easily broken is discussed in Section 25.4.

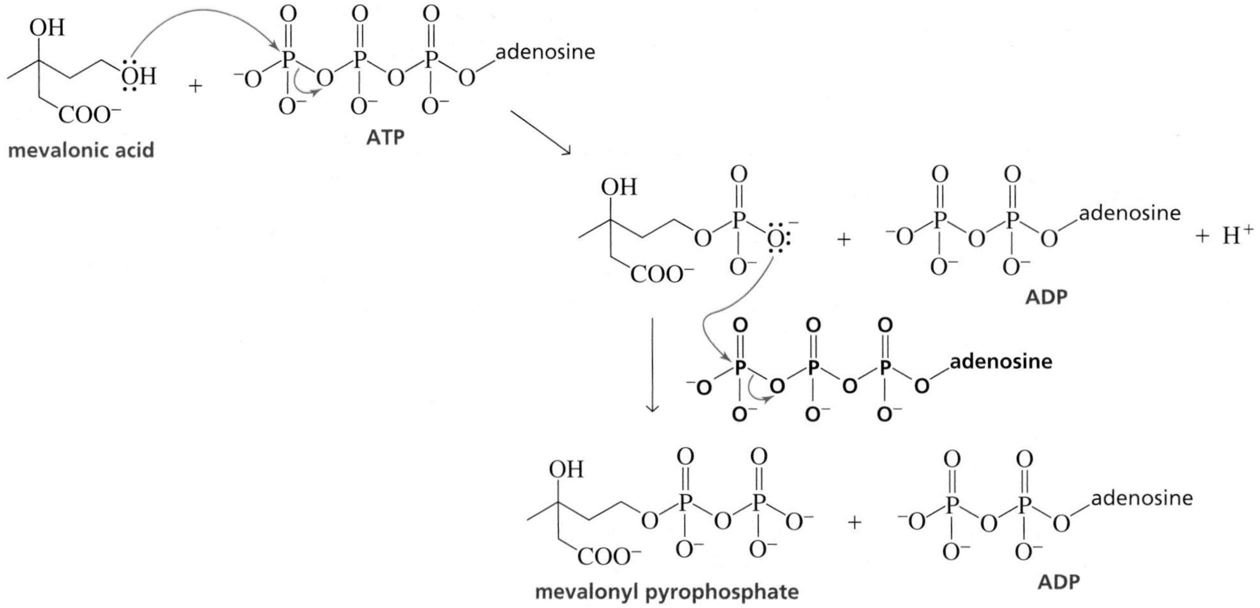

mevalonic acid + ATP

ADP + H$^+$

mevalonyl pyrophosphate + ADP

PROBLEM 14 / SOLVED

Give the mechanism for the last step in the biosynthesis of isopentenyl pyrophosphate, showing why ATP is required.

SOLUTION   In the last step of the biosynthesis of isopentenyl pyrophosphate, elimination of CO_2 is accompanied by elimination of an $^-$OH group. The $^-$OH is a strong base and therefore a poor leaving group. ATP is used to convert the OH group into a phosphate group, which is easily eliminated because it is a good leaving group.

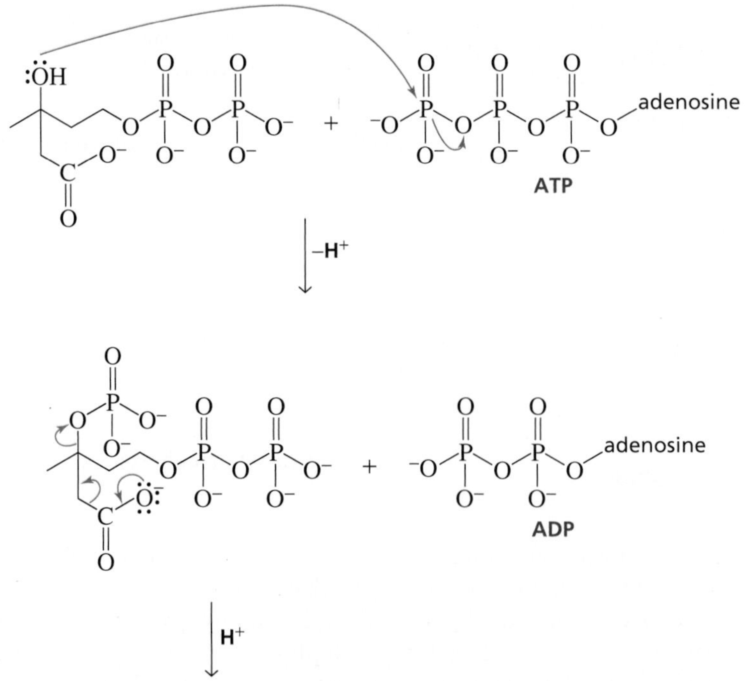

ATP

$-H^+$

ADP

H^+

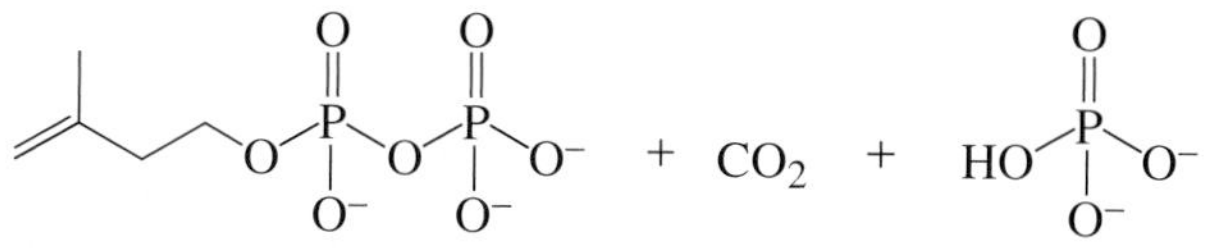

Give the mechanisms for the Claisen condensation and aldol addition that occur in the first two steps of the biosynthesis of isopentenyl pyrophosphate.

Biosynthesis of Dimethylallyl Pyrophosphate

Both isopentenyl pyrophosphate and dimethylallyl pyrophosphate are needed for the biosynthesis of terpenes. Therefore, some isopentenyl pyrophosphate is converted to dimethylallyl pyrophosphate by an enzyme-catalyzed isomerization reaction. The isomerization involves addition of a proton to isopentenyl pyrophosphate following Markovnikov's rule (Section 3.12) and elimination of a proton from the carbocation intermediate following Zaitsev's rule (Section 10.2).

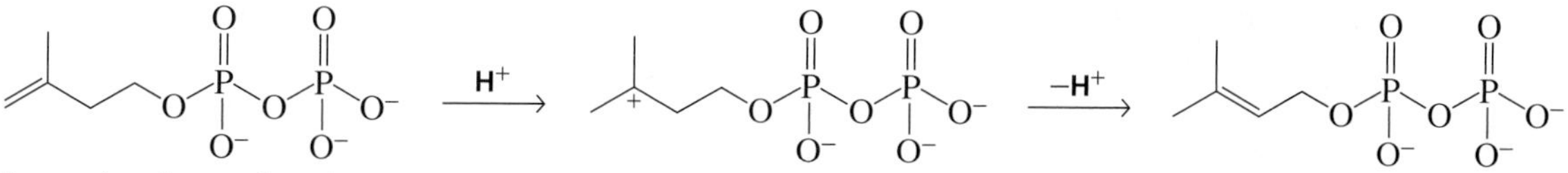

isopentenyl pyrophosphate

dimethylallyl pyrophosphate

Terpene Biosynthesis

The reaction of dimethylallyl pyrophosphate with isopentenyl pyrophosphate forms geranyl pyrophosphate, a 10-carbon compound. In the first step of the reaction, isopentenyl pyrophosphate acts as a nucleophile and displaces a pyrophosphate group from dimethylallyl pyrophosphate. Pyrophosphate is an excellent leaving group—its four OH groups have pK_a values of 0.9, 2.0, 6.6, and 9.4. Therefore, three of the four groups will be primarily in their basic forms at physiological pH (pH = 7.3). A proton is removed in the next step, resulting in the formation of geranyl pyrophosphate.

dimethylallyl pyrophosphate   isopentenyl pyrophosphate

geranyl pyrophosphate

pyrophosphate

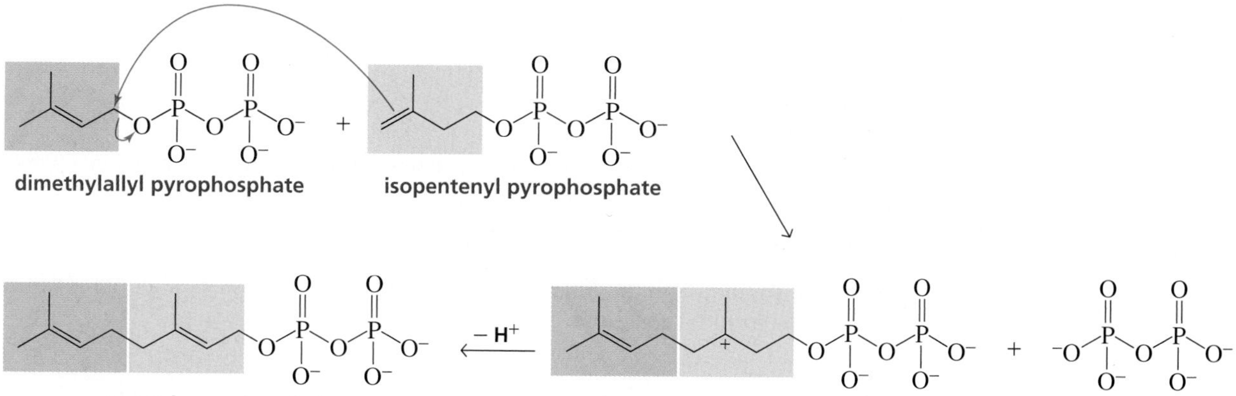

The following scheme shows how some of the many monoterpenes could be synthesized from geranyl pyrophosphate.

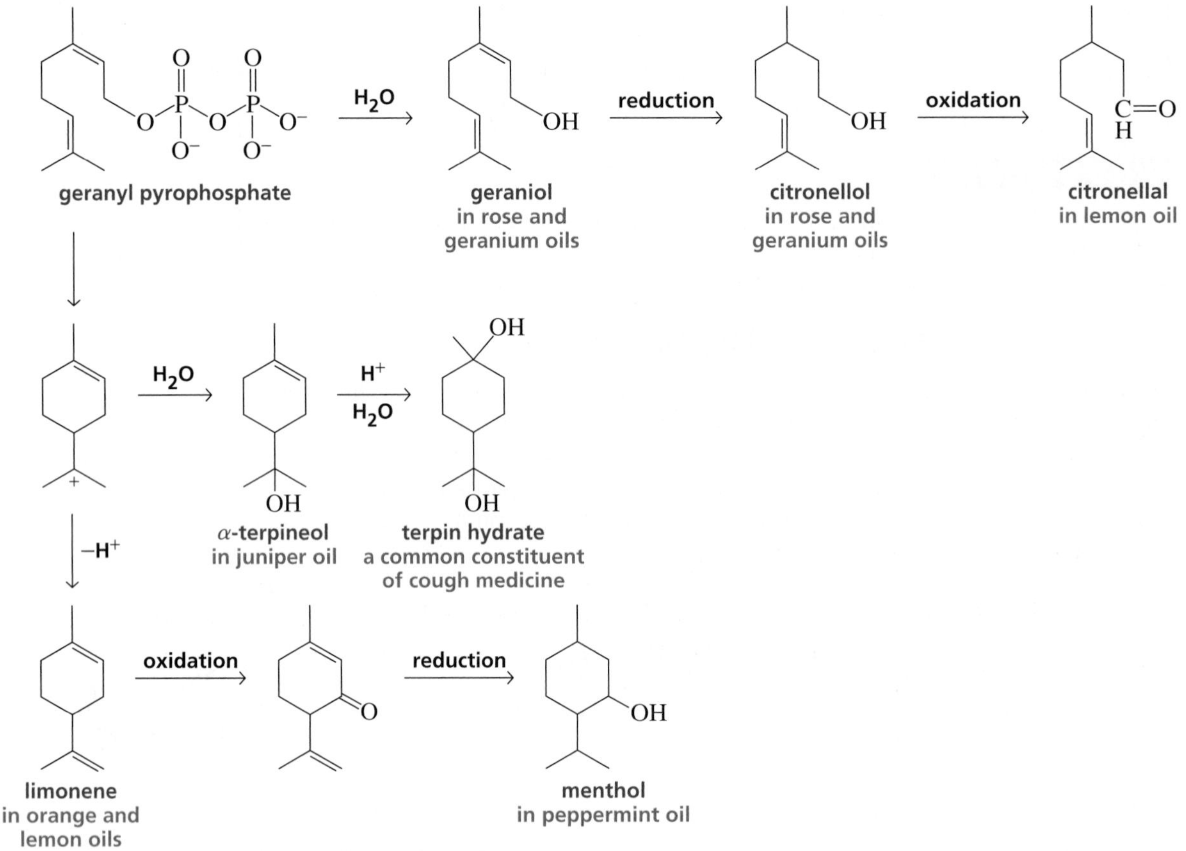

geranyl pyrophosphate
H_2O
geraniol
in rose and
geranium oils
reduction
citronellol
in rose and
geranium oils
oxidation
citronellal
in lemon oil
H_2O
H^+
H_2O
α-terpineol
in juniper oil
terpin hydrate
a common constituent
of cough medicine
$-H^+$
oxidation
reduction
limonene
in orange and
lemon oils
menthol
in peppermint oil

PROBLEM 16

Propose a mechanism for the conversion of the *E* isomer of geranyl pyrophosphate to the *Z* isomer.

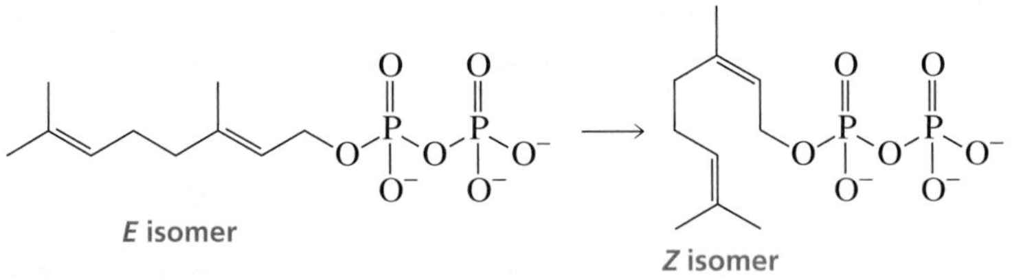

E isomer
Z isomer

PROBLEM 17

Propose mechanisms for the formation of α-terpineol and limonene from geranyl pyrophosphate.

Geranyl pyrophosphate can react with another molecule of isopentenyl pyrophosphate to form farnesyl pyrophosphate, a 15-carbon compound.

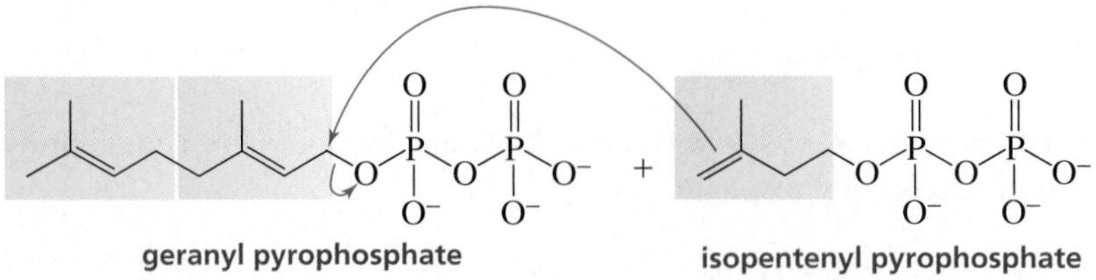

geranyl pyrophosphate
isopentenyl pyrophosphate

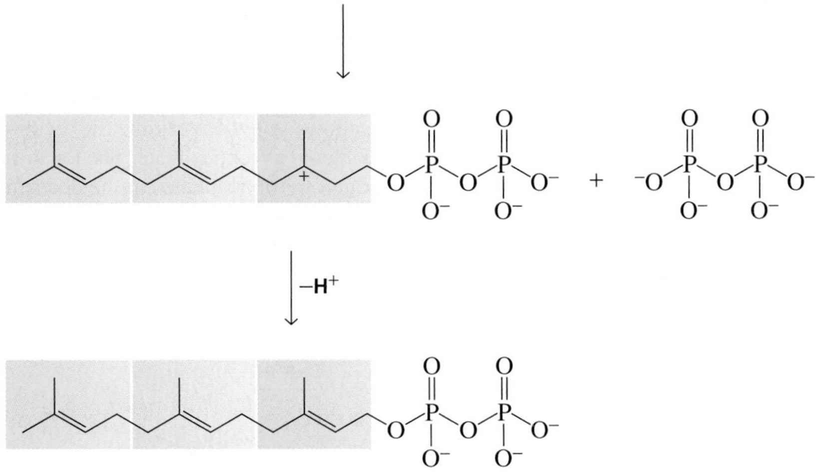

farnesyl pyrophosphate

Two molecules of farnesyl pyrophosphate form squalene, a 30-carbon compound. This reaction is catalyzed by the enzyme squalene synthase. Squalene synthase joins the two molecules in a tail-to-tail linkage. Squalene is the precursor of cholesterol, and cholesterol is the precursor of all other steroids.

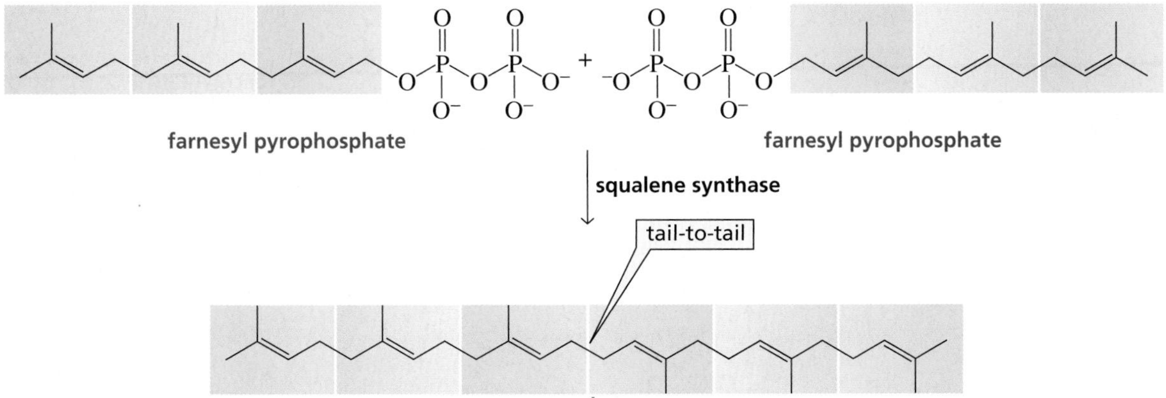

farnesyl pyrophosphate   **farnesyl pyrophosphate**

squalene synthase

tail-to-tail

squalene

Farnesyl pyrophosphate can react with another molecule of isopentenyl pyrophosphate to form geranylgeranyl pyrophosphate, a 20-carbon compound. Two geranylgeranyl pyrophosphates can join to form phytoene, a 40-carbon compound. Phytoene is the precursor of the carotenoid (tetraterpene) pigments of plants.

PROBLEM 18

In aqueous acidic solution, farnesyl pyrophosphate forms the following sesquiterpene. Propose a mechanism for this reaction.

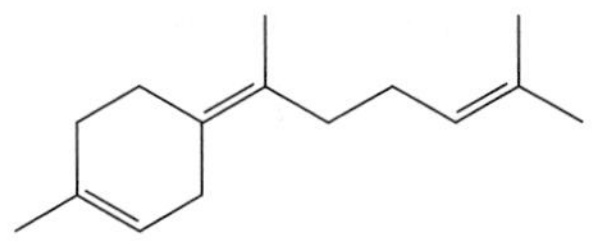

PROBLEM 19 / SOLVED

If squalene were synthesized in a medium containing acetate in which the carbonyl carbon of acetate were labeled with radioactive ^{14}C, which carbons in squalene would be labeled?

SOLUTION Acetate reacts with ATP to form acetyl adenylate, which then reacts with CoASH to form acetyl-CoA (Section 16.18). Because malonyl-CoA is prepared from acetyl-CoA, the thioester carbonyl carbon of malonyl-CoA will also be labeled. Examining each step of the mechanism for the biosynthesis of isopentenyl pyrophosphate from acetyl-CoA and malonyl-CoA allows you to determine the location of each of the radioactively labeled carbons in isopentenyl pyrophosphate. The locations of each of the radioactively labeled carbons in geranyl pyrophosphate can be determined from the mechanism for its biosynthesis from isopentenyl pyrophosphate. The locations of the radioactively labeled carbons in farnesyl pyrophosphate can be determined from the mechanism for its biosynthesis from geranyl pyrophosphate. Knowing that squalene is obtained from a tail-to-tail linkage of two farnesyl pyrophosphates gives you the answer.

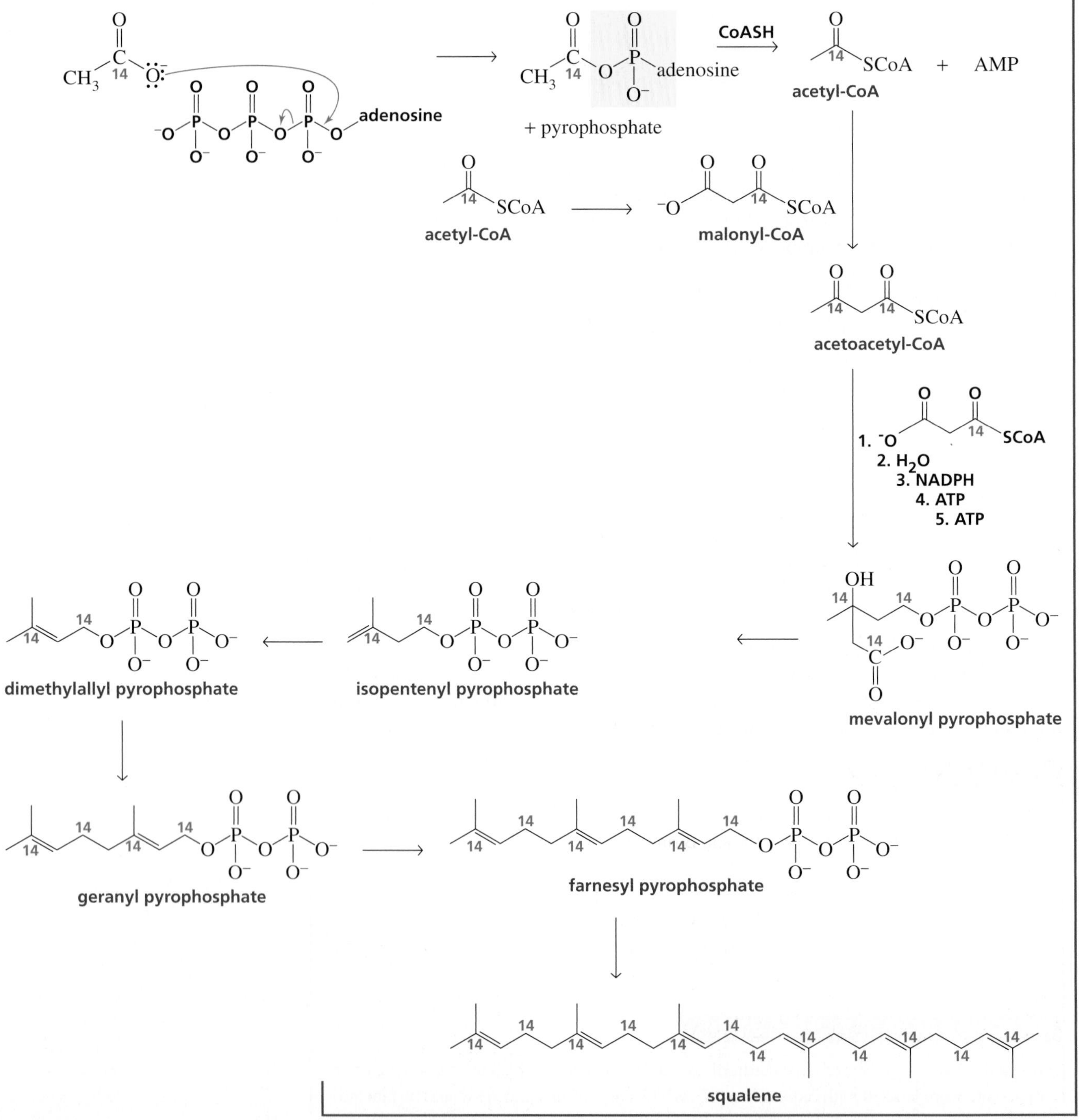

dimethylallyl pyrophosphate

isopentenyl pyrophosphate

mevalonyl pyrophosphate

geranyl pyrophosphate

farnesyl pyrophosphate

squalene

Hormones are chemical messengers. They are organic compounds that are synthesized in glands and delivered by the bloodstream to target tissues to stimulate or inhibit some process. Many hormones are **steroids.** Because steroids are nonpolar compounds, they are lipids. Their nonpolar character allows them to cross cell membranes, so they can leave the cells in which they are synthesized and enter their target cells.

All steroids contain a tetracyclic ring system. The four rings are designated A, B, C, and D. A, B, and C are six-membered rings and D is a five-membered ring. The carbons in the steroid ring system are numbered as shown.

We have seen that rings can be **trans fused** or **cis fused** and that trans fused rings are more stable (Section 2.15). In steroids, the B, C, and D rings are all trans fused. In most naturally occurring steroids, the A and B rings are also trans fused.

24.9
STEROIDS

the steroid ring system

CH₃ and H are trans

CH₃ and H are cis

angular methyl groups

A and B rings are trans fused

CH₃ and H are trans

A and B rings are cis fused

CH₃ and H are cis

Many steroids have methyl groups at the 10- and 13-positions. These are called **angular methyl groups.** When steroids are drawn, both angular methyl groups are shown to be above the plane of the steroid ring system. Substituents on the same side of the steroid ring system as the angular methyl groups are designated **β-substituents** (written with a solid wedge). Those on the opposite side of the plane of the ring system are **α-substituents** (indicated with a hatched wedge).

PROBLEM 20◆

A β-hydrogen at C-5 means that the A and B rings are _____ fused, while an α-hydrogen at C-5 means that the A and B rings are _____ fused.

The most abundant member of the steroid family in animals is **cholesterol.** There are many different steroid hormones, and cholesterol is the precursor for all of them. Cholesterol is biosynthesized from squalene, a triterpene (Section 24.6). Cholesterol is an important component of cell membranes (Figure 24.1). Its ring structure makes it more rigid than other membrane lipids. Because cholesterol has eight centers of chirality, 256 stereoisomers are possible but only one exists in nature (Chapter 4, Problem 20).

Two German chemists, **Heinrich Otto Wieland** *(1877–1957) and* **Adolf Windaus (1876–1959),** *each received a Nobel Prize in chemistry (Wieland in 1927 and Windaus in 1928) for work that led to the determination of the structure of cholesterol.*

Heinrich Wieland, *the son of a chemist, was a professor at the University of Munich, where he showed that bile acids were steroids and determined their individual structures. During World War II, he remained in Germany but was openly anti-Nazi.*

cholesterol

The steroid hormones can be divided into the following five classes: glucocorticoids, mineralocorticoids, androgens, estrogens, and progestins. Glucocorticoids and mineralocorticoids are synthesized in the adrenal cortex and are collectively known as **adrenal cortical steroids.** All adrenal cortical steroids have an oxygen at C-11.

Glucocorticoids, as their name suggests, are involved in glucose metabolism as well as in the metabolism of proteins and fatty acids. Cortisone is an example of a glucocorticoid. Because of its anti-inflammatory effect, it is used clinically to treat arthritis.

cortisone

aldosterone

Mineralocorticoids cause increased reabsorption of Na^+, Cl^-, and HCO_3^- by the kidneys, leading to an increase in blood pressure. Aldosterone is an example of a mineralocorticoid.

PROBLEM 21◆

Is the OH substituent of the A ring of cholesterol an α-substituent or a β-substituent?

PROBLEM 22

Aldosterone is in equilibrium with its cyclic hemiacetal. Draw the hemiacetal form of aldosterone.

The male sex hormones, known as **androgens,** are secreted by the testes. They are responsible for the development of male secondary sex characteristics during puberty. They also promote muscle growth. Testosterone and 5α-dihydrotestosterone are androgens.

Estradiol and estrone are female sex hormones, known as **estrogens.** They are secreted by the ovaries and are responsible for the development of female secondary sex characteristics. They also regulate the menstrual cycle. Progesterone is the hormone that prepares the lining of the uterus for implantation of an ovum and is essential for the maintenance of pregnancy. It also prevents ovulation during pregnancy. This hormone belongs to the class of hormones known as **progestins.** Progesterone, which

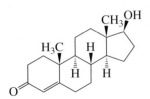

testosterone

5α-dihydrotestosterone

is biosynthesized from cholesterol, is a precursor for the glucocorticoids, the min-eralocorticoids, and the sex hormones.

estradiol   **estrone**   **progesterone**

Although the various steroid hormones have remarkably different physiological effects, their structures are quite similar. For example, the only difference between testosterone and progesterone is the substituent at C-17, and the only difference between 5α-dihydrotestosterone and estradiol is one carbon and six hydrogens, but these two compounds make the difference between being male or female. These examples illustrate the extreme specificity of biochemical reactions.

Michael Brown and Joseph Goldstein

> ## PROBLEM 23
>
> The acid component of a cholesterol ester is a fatty acid such as linoleic acid. Draw the structure of a cholesterol ester.

In addition to being the precursor of all the steroid hormones in animals, cholesterol is also the precursor of the **bile acids.** In fact, the word *cholesterol* is derived from the Greek words *chole* meaning "bile" and *stereos* meaning "solid." The bile acids—cholic acid and chenodeoxycholic acid—are synthesized in the liver, stored in the gallbladder, and secreted into the small intestine where they act as emulsifying agents so that fats and oils can be digested by water-soluble digestive enzymes. Cholesterol is also the precursor of vitamin D (Section 28.6).

Michael S. Brown and *Joseph Leonard Goldstein* shared the 1985 Nobel Prize in medicine or physiology for their work on the regulation of cholesterol metabolism and the treatment of disease caused by elevated cholesterol levels in the blood. Brown was born in New York in 1941, and Goldstein in South Carolina in 1940. They are both professors of medicine at the University of Texas Southwestern Medical Center.

CHOLESTEROL AND HEART DISEASE

Cholesterol is probably the best known lipid because of the correlation between cholesterol levels in the blood and heart disease. Cholesterol is synthesized in the liver and is found in almost all body tissues. Cholesterol is found in many foods, but we do not require it in our diet because the body can synthesize all we need. A diet high in cholesterol can lead to high levels of cholesterol in the bloodstream. This excess cholesterol can accumulate on the walls of arteries, restricting the flow of blood. This disease of the circulatory system is known as *atherosclerosis* and is a primary cause of heart disease. Cholesterol travels through the bloodstream packaged in particles containing cholesterol, cholesterol esters, phospholipids, and proteins. The particles are classified according to their density. LDL (low-density lipoprotein) particles transport cholesterol from the liver to other tissues. Receptors on the surfaces of cells bind LDL particles, allowing them to be brought into the cell so that the cell can use the cholesterol. HDL (high-density lipoprotein) is a cholesterol scavenger, removing cholesterol from the surfaces of membranes and delivering it back to the liver, where it is converted into bile acids. LDL is the so-called "bad" cholesterol, while HDL is the "good" cholesterol. The more cholesterol we eat, the less the body synthesizes. But this does not mean that the presence of dietary cholesterol has no effect on the total amount of cholesterol in the bloodstream. Unfortunately, dietary cholesterol also inhibits the synthesis of the LDL receptors. So the more cholesterol we eat, the less the body synthesizes but also the less the body can get rid of by bringing it into target cells.

cholic acid

chenodeoxycholic acid

PROBLEM 24◆

Are the three OH groups of cholic acid axial or equatorial?

CLINICAL TREATMENT OF HIGH CHOLESTEROL

Statins are the newest class of cholesterol-reducing drugs. They reduce serum cholesterol levels by inhibiting the enzyme that catalyzes the reduction of hydroxymethylglutaryl-CoA to mevalonic acid (Section 24.8). Decreasing the mevalonic acid concentration decreases the isopentenyl pyrophosphate concentration, so the biosynthesis of all terpenes, including cholesterol, is decreased. As a consequence of decreasing cholesterol synthesis in the liver, the liver expresses more LDL receptors—the receptors that help clear LDL from the bloodstream. Studies show that for every 10% that cholesterol is reduced, deaths from coronary heart disease are reduced by 15% and total death risk is reduced by 11%.

Compactin and lovastatin are natural statins used clinically under the trade names Zocor and Mevacor. Atorvastatin (Lipitor), a synthetic statin, is now the most popular statin. It has greater potency and a longer half-life because its metabolites are as active as the parent drug in reducing cholesterol levels. Therefore, smaller doses of the drug are necessary. The required dose is further reduced because Lipitor is marketed as a single enantiomer. In addition, it is more lipophilic than compactin and lovastatin, so it has a greater tendency to remain in the endoplasmic reticulum of the liver cells where it is needed.

lovastatin
Mevacor

simvastatin
Zocor

atorvastatin
Lipitor

24.10 BIOSYNTHESIS OF CHOLESTEROL

How is cholesterol, the precursor of all the steroid hormones, biosynthesized? The starting material for the biosynthesis is the triterpene squalene, which must first be converted to lanosterol. Lanosterol is converted to cholesterol in a series of 19 steps.

The first step in the conversion of squalene to lanosterol is epoxidation of the 2,3-double bond of squalene. Acid-catalyzed opening of the epoxide initiates a series of cyclizations resulting in the protosterol cation. Elimination of a C-9 proton from the cation initiates a series of 1,2-hydride and 1,2-methyl shifts, resulting in lanosterol.

biosynthesis of lanosterol and cholesterol

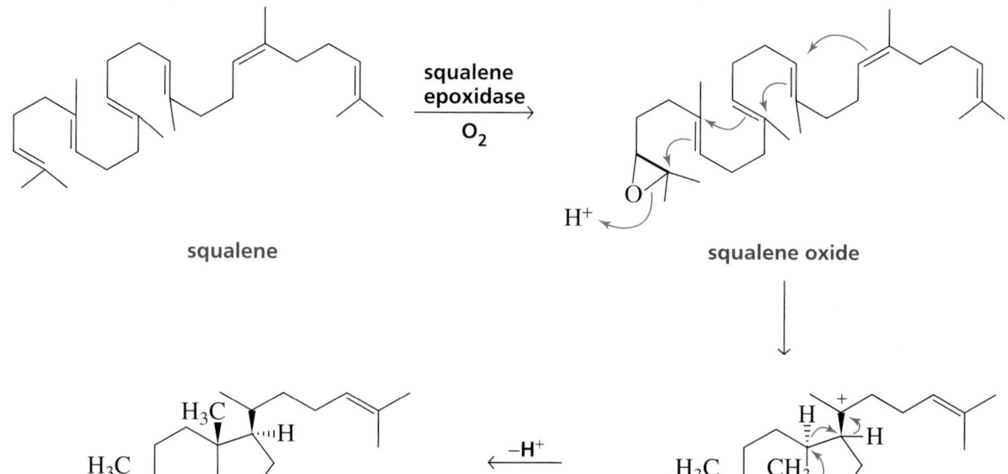

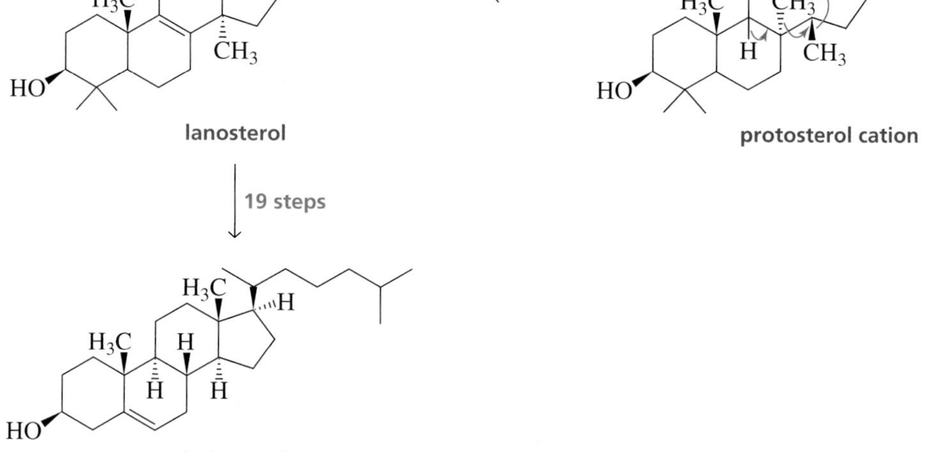

squalene

squalene oxide

$$\xrightarrow{\text{squalene epoxidase}} \quad O_2$$

lanosterol

protosterol cation

$$\xleftarrow{-H^+}$$

19 steps

cholesterol

Konrad Bloch *and* **Feodor Lynen** *shared the 1964 Nobel Prize in medicine or physiology. Bloch showed how fatty acids and cholesterol are biosynthesized from acetate. Lynen showed that the two-carbon acetate unit is actually acetyl-CoA, and he determined the structure of coenzyme A.*

Konrad Emil Bloch *was born in Upper Silesia (then Germany) in 1912, left Nazi Germany for Switzerland in 1934, and came to the United States in 1936, becoming a U.S. citizen in 1944. He received a Ph.D. from Columbia in 1938, taught at the University of Chicago, and became a professor of biochemistry at Harvard in 1954.*

Feodor Lynen (1911–1979) *was born in Germany, received a Ph.D. under Heinrich Wieland, and married Wieland's daughter. He was head of the Institute of Cell Chemistry at the University of Munich.*

Converting lanosterol to cholesterol requires removing three methyl groups from lanosterol (in addition to reducing two double bonds and creating a new double bond). Removing methyl groups from carbon atoms is not an easy process—many different enzymes are required to carry out the 19 steps. Why does nature bother? Why not just use lanosterol instead of cholesterol? Konrad Bloch answered that question when he found that membranes that contain lanosterol instead of cholesterol are much more permeable. Small molecules are able to pass easily through lanosterol-containing membranes. As each methyl group is removed from lanosterol, the membrane becomes less and less permeable.

PROBLEM 25

Draw the individual 1,2-hydride and 1,2-methyl shifts responsible for conversion of the pro-tosterol cation to lanosterol. How many hydride shifts are involved? How many methyl shifts?

The potent physiological effects of steroids led scientists, in their search for new drugs, to synthesize steroids not available in nature and to investigate their physiological effects. Stanozolol and Dianabol are drugs developed in this way. They have the same muscle-building effect as testosterone. Steroids that aid in the development of muscle are called **anabolic steroids.** These drugs are available by prescription

24.11 SYNTHETIC STEROIDS

and are used to treat people suffering from traumas accompanied by muscle deterioration. The same drugs have been administered to athletes and racehorses to increase muscle mass. Stanozolol was the drug detected in several athletes in the 1988 Olympics. Anabolic steroids, when taken in relatively high dosages, have been found to cause liver tumors, personality disorders, and testicular atrophy.

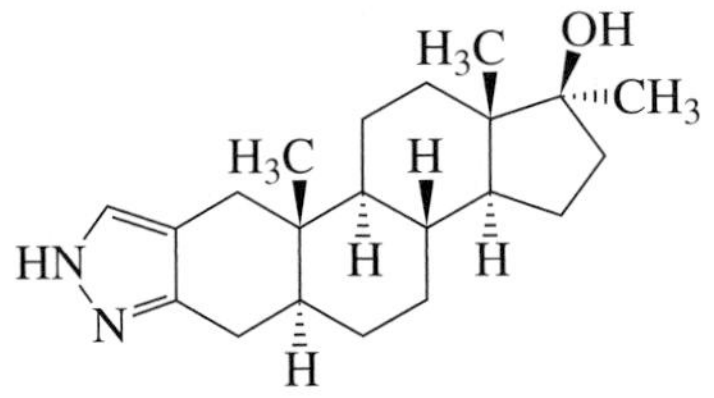

stanozolol

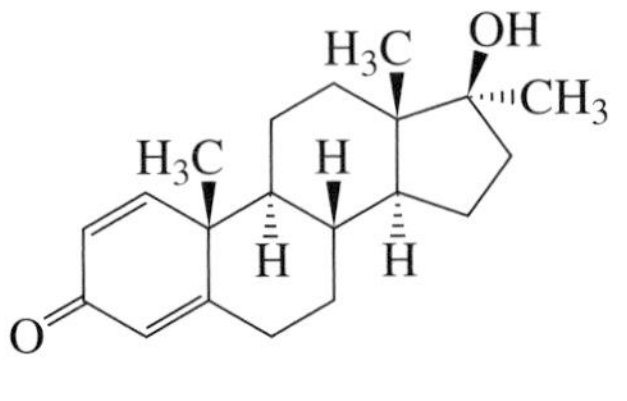

dianabol

Many synthetic steroids have been found to be much more potent than natural steroids. Norethindrone, for example, is better than progesterone in arresting ovulation. RU 486, when taken along with prostaglandins, terminates pregnancy within the first nine weeks of gestation. Notice that these oral contraceptives have structures similar to that of progesterone.

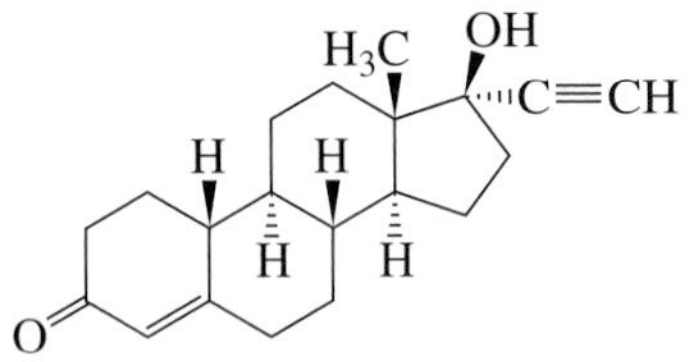

norethindrone

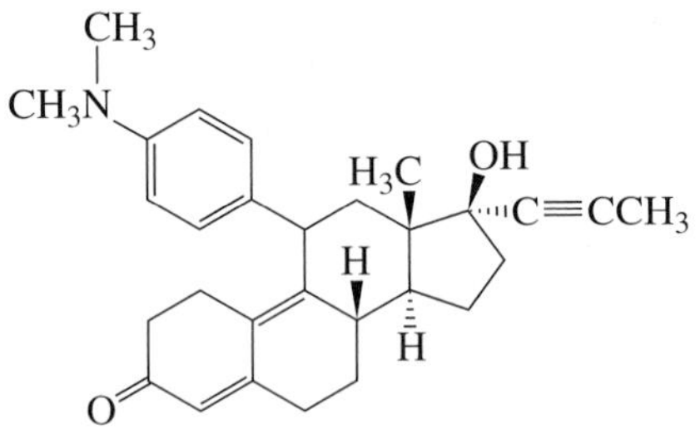

RU 486

KEY TERMS

adrenal cortical steroids (page 1056)
anabolic steroids (page 1059)
androgens (page 1056)
angular methyl group (page 1055)
bile acids (page 1057)
carotenoid (page 1047)
cephalin (page 1040)
cerebroside (page 1042)
cholesterol (page 1055)
cis fused (page 1055)
essential oil (page 1046)
estrogens (page 1056)
fat (page 1036)
fatty acid (page 1034)
hormone (page 1055)
isoprene rule (page 1046)

lecithin (page 1040)
leukotriene (page 1045)
lipid (page 1033)
lipid bilayer (page 1040)
membrane (page 1040)
mixed triacylglycerol (page 1036)
monoterpene (page 1046)
oil (page 1037)
phosphatidic acid (page 1040)
phosphoacylglycerol (page 1040)
phosphoglyceride (page 1040)
phospholipid (page 1040)
polyunsaturated fatty acid (page 1034)
progestins (page 1056)
prostacyclin (page 1044)
prostaglandin (page 1043)

sesquiterpene (page 1043)
simple triacylglycerol (page 1036)
sphingolipid (page 1042)
sphingomyelin (page 1042)
squalene (page 1047)
steroid (page 1055)
α-substituent (page 1055)
β-substituent (page 1055)
terpene (page 1045)
terpenoid (page 1045)
tetraterpene (page 1047)
thromboxane (page 1044)
trans fused (page 1055)
triacylglycerol (page 1036)
triterpene (page 1047)
wax (page 1035)

PROBLEMS

26. An optically active fat, when completely hydrolyzed, yields twice as much stearic acid as palmitic acid. Draw the structure of the fat.

27. a. How many different triacylglycerols are there in which one of the fatty acid components is lauric acid and two are myristic acid? (Include stereoisomers.)

b. How many different triacylglycerols are there in which one of the fatty acid components is lauric acid, one is myristic acid, and one is palmitic acid? (Include stereoisomers.)

28. Cardiolipins are found in some membranes. Give the products formed when a cardiolipin undergoes complete hydrolysis.

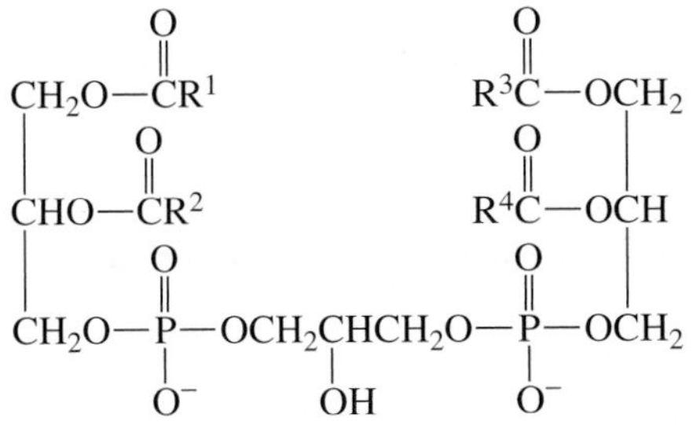

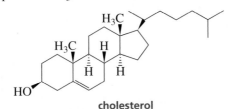

a cardiolipin

29. Nutmeg contains a simple, fully saturated triacylglycerol with a molecular weight of 723. Draw its structure.

30. Give the product that would be obtained from the reaction of cholesterol with each of the following reagents. (*Hint:* Because of steric hindrance from the angular methyl groups, the α-face is more susceptible to attack by reagents than the β-face.)

a. H_2O, H^+
b. BH_3 in THF followed by H_2O_2 + HO^-
c. H_2, Pd/C
d. Br_2 + H_2O
e. peroxyacetic acid
f. the product of part **e** + CH_3O^-

cholesterol

31. Dr. Chole S. Terol synthesized the following samples of mevalonic acid and fed them to a group of lemon trees. Which carbons in citronellal, which is isolated from lemon oil, will be labeled in trees that were fed

a. sample A?   **b.** sample B?   **c.** sample C?

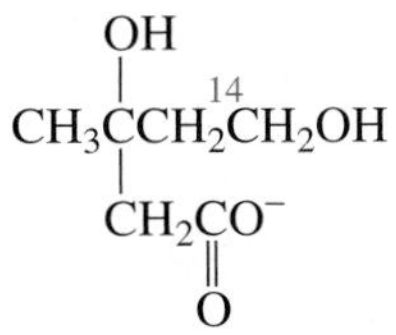

sample A 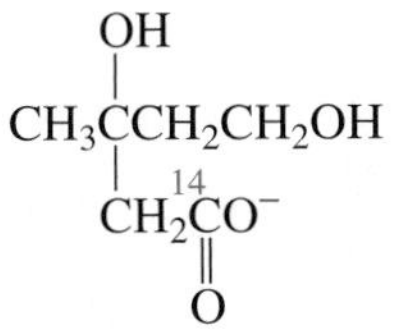   sample B 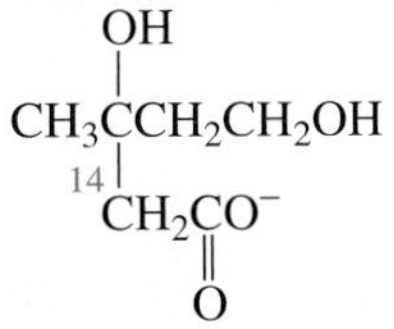   sample C

32. An optically active monoterpene (compound A) with molecular formula $C_{10}H_{18}O$ undergoes catalytic hydrogenation to form an optically inactive compound with molecular formula $C_{10}H_{20}O$ (compound B). When compound B is heated with acid, followed by reaction with O_3 and work-up under reducing conditions (Zn, H_2O), one of the products obtained is 4-methylcyclohexanone. Give possible structures for compound A.

33. If junipers were allowed to grow in a medium containing acetate in which the methyl carbon was labeled with ^{14}C, which carbons in α-terpineol would be labeled?

34. a. Propose a mechanism for the following reaction.
b. To what class of terpene does the starting material belong? Mark off the isoprene units in the starting material.

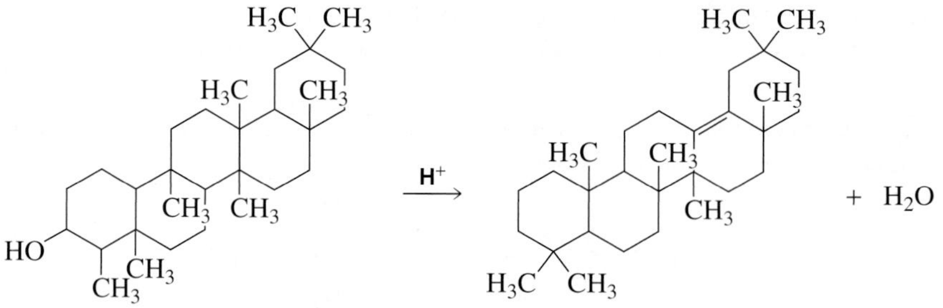

35. 5-Androstene-3,17-dione is isomerized to 4-androstene-3,17-dione by hydroxide ion. Propose a mechanism for this reaction (*J. Am. Chem. Soc.* 1989, *111,* 6419).

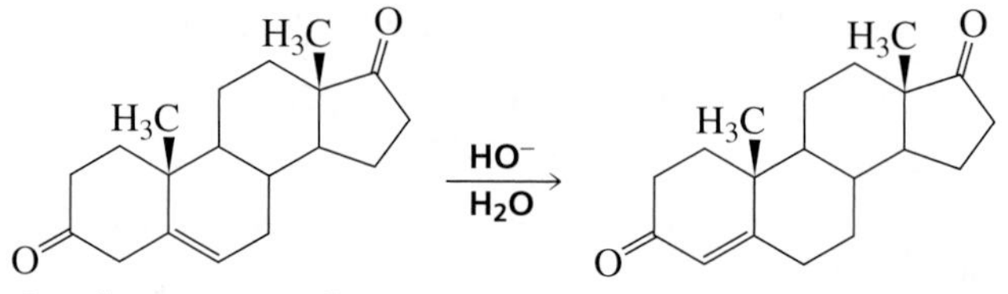

5-androstene-3,17-dione   4-androstene-3,17-dione

36. Both OH groups of one of the following steroid diols react with excess ethyl chloroformate, but only one OH group of the other steroid diol reacts under the same conditions. Explain this difference in reactivity.

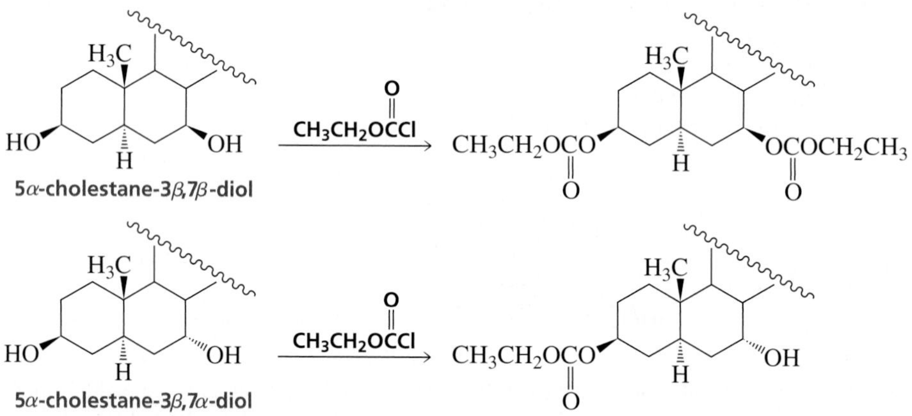

5α-cholestane-3β,7β-diol

5α-cholestane-3β,7α-diol

37. The acid-catalyzed dehydration of an alcohol to a rearranged alkene is known as a Wagner–Meerwein rearrangement. Propose a mechanism for the following Wagner–Meerwein rearrangement:

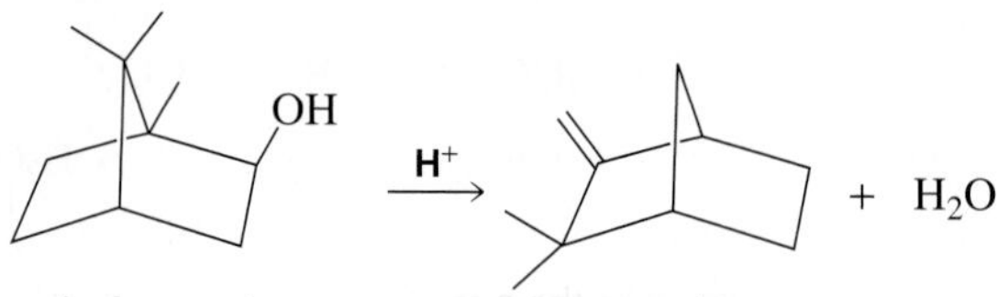

isoborneol   camphene

38. Diethylstilbestrol (DES) was given to pregnant women to prevent miscarriage until it was found that the drug caused cancer in both the mothers and their female children. DES has estradiol activity even though it is not a steroid. Draw DES so that it is structurally similar to estradiol.

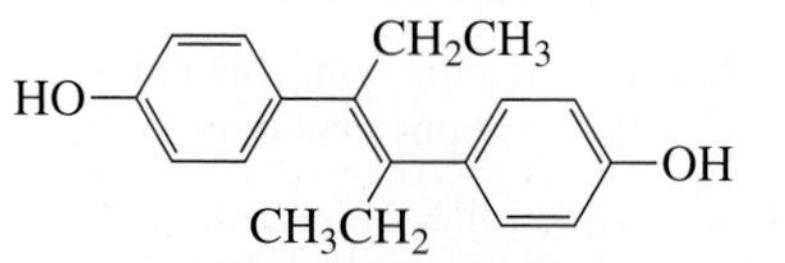

diethylstilbestrol
DES

25

Nucleosides, Nucleotides, and Nucleic Acids

an RNA catalyst

In previous chapters, we studied two of the three major kinds of biopolymers—polysaccharides and proteins. Now we will look at the third—nucleic acids. There are two types of nucleic acids—**deoxyribonucleic acid (DNA)** and **ribonucleic acid (RNA).** DNA encodes an organism's entire hereditary information and controls the growth and division of cells. In most organisms, the genetic information stored in DNA is transcribed into RNA. This information can then be translated for the synthesis of all the proteins needed for cellular structure and function.

DNA was first isolated in 1869 from the nuclei of white blood cells. Because this material was found in the nucleus and was acidic, it was called *nucleic acid.* Eventually, scientists found that the nuclei of all cells contain DNA, but it wasn't until 1944 that they realized that nucleic acids are the carriers of genetic information. In 1953, James Watson and Francis Crick described the three-dimensional structure of DNA—the famed double helix.

THE STRUCTURE OF DNA: WATSON, CRICK, FRANKLIN, AND WILKINS

James D. Watson was born in Chicago in 1928. He graduated from the University of Chicago at the age of 19 and received a Ph.D. three years later from Indiana University. In 1951, as a postdoctoral fellow at Cambridge University, Watson worked on determining the three-dimensional structure of DNA.

Francis H. C. Crick was born in Northampton, England, in 1916. Originally trained as a physicist, Crick was involved in radar research during World War II. After the war, he entered Cambridge University to study for a Ph.D. in chemistry, which he received in 1953. He was a graduate student when he carried out his portion of the work that led to the proposal of the double helical structure of DNA.

Francis Crick and James Watson

Rosalind Franklin was born in London in 1920. She graduated from Cambridge University and in 1942 quit her graduate studies to accept a position as a research officer in the British Coal Utilisation Research Association. After the war, she studied X-ray diffraction techniques in Paris. In 1951 she returned to England, accepting a position to develop an X-ray diffraction unit in the biophysics department at King's College. Her X-ray studies showed that DNA was a helix with phosphate groups on the outside of the molecule. Franklin died in 1958 without knowing the significance her work had played in determining the structure of DNA.

Watson and Crick shared the 1962 Nobel Prize in medicine or physiology with Maurice Wilkins for determining the double helical structure of DNA. Wilkins contributed X-ray studies that confirmed the double helical structure. Wilkins was born in New Zealand in 1916 and moved to England six years later with his parents. During World War II he joined other British scientists who were working with American scientists on the development of the atomic bomb.

Rosalind Franklin

25.1 NUCLEOSIDES AND NUCLEOTIDES

Nucleic acids are chains of five-membered-ring sugars linked by phosphate groups (Figure 25.1). The anomeric carbon of each sugar is bonded to a nitrogen of a heterocyclic compound in a β-glycosidic linkage. (Recall from Section 20.10 that a β-linkage is one in which the substituents at C-1 and C-4 are on the same side of the furanose ring.) Because the heterocyclic compounds are amines, they are commonly referred to as **bases** (Section 23.2). In RNA, the five-membered-ring sugar is D-ribose. In DNA, it is 2′-deoxy-D-ribose (D-ribose without an OH group in the 2-position).

Figure 25.1 ▶
Nucleic acids consist of a chain of five-membered-ring sugars linked by phosphate groups. Each sugar (D-ribose in RNA, 2′-deoxy-D-ribose in DNA) is bonded to a heterocyclic amine in a β-glycosidic linkage.

Studies that determined the structures of the nucleic acids and paved the way for the discovery of the DNA double helix were carried out by Phoebus Levene and elaborated by Sir Alexander Todd.

a β-glycosidic linkage

2′-OH group

RNA

DNA

Phosphoric acid links the sugars in both RNA and DNA. The acid has three dissociable OH groups with pK_a values of 2.1, 7.2, and 12.3. Each of the OH groups can react with an alcohol to form a phosphomonoester, a phosphodiester, or a phosphotriester. In nucleic acids, the phosphate group is a phosphodiester.

phosphoric acid   a phosphomonoester   a phosphodiester   a phosphotriester

The vast differences in heredity among species and among members of the same species are determined by the sequence of the bases in DNA. Surprisingly, there are only four bases in DNA—two are substituted purines (adenine and guanine), and two are substituted pyrimidines (cytosine and thymine).

purine   pyrimidine

adenine   guanine   cytosine   uracil   thymine

RNA also contains only four bases. Three (adenine, guanine, and cytosine) are the same as those in DNA, but the fourth base in RNA is uracil instead of thymine. Notice that thymine and uracil differ only by a methyl group—thymine is 5-methyluracil. The reason DNA contains thymine instead of uracil is explained in Section 25.14.

The purines and pyrimidines are bonded to the anomeric carbon of the furanose ring—purines at N-9 and pyrimidines at N-1—in a β-glycosidic linkage. A compound containing a base bonded to D-ribose or to 2′-deoxy-D-ribose is called a **nucleoside.** In a nucleoside, the ring positions of the sugar are indicated by primed numbers to distinguish them from the ring positions of the base. This is why the sugar component of DNA is referred to as 2′-deoxy-D-ribose. Notice on page 1066 the difference in the base names and their corresponding nucleoside names. For example, adenine is the base, whereas adenosine is the nucleoside. Similarly, cytosine is the base, whereas cytidine is the nucleoside, and so forth. Because uracil is found only in RNA, it is shown attached to D-ribose but not to 2′-deoxy-D-ribose; because thymine is found only in DNA, it is shown attached to 2′-deoxy-D-ribose but not to D-ribose.

Phoebus Aaron Theodor Levene (1869–1940) was born in Russia. When he emigrated to the United States with his family in 1891, his Russian name Fishel was changed to Phoebus. Because his medical school education had been interrupted, he returned to Russia to complete his studies. When he returned to the United States he took chemistry courses at Columbia University. Deciding to forgo medicine for a career in chemistry, he went to Germany to study under Emil Fischer. He was a professor of chemistry at the Rockefeller Institute (now Rockefeller University).

Alexander R. Todd (1907–1997) was born in Scotland. He received two Ph.D. degrees, one from Johann Wolfgang Goethe University in Frankfurt (1931) and one from Oxford University (1933). He was a professor of chemistry at the University of Edinburgh, at the University of Manchester, and from 1944 to 1971 at Cambridge University. He was knighted in 1954 and was made a baron in 1962 (Baron Todd of Trumpington). For his work on nucleotides, he was awarded the 1957 Nobel Prize in chemistry.

> **PROBLEM 1**
>
> In acidic solutions, nucleosides are hydrolyzed to a sugar and a heterocyclic base. Propose a mechanism for this reaction.

A **nucleotide** is a nucleoside with either the 5′- or the 3′-OH group bonded in an ester linkage to phosphoric acid. The nucleotides of RNA (where the sugar is D-ribose) are more precisely called **ribonucleotides,** while the nucleotides of DNA (where the sugar is 2′-deoxy-D-ribose) are called **deoxyribonucleotides.**

nucleosides

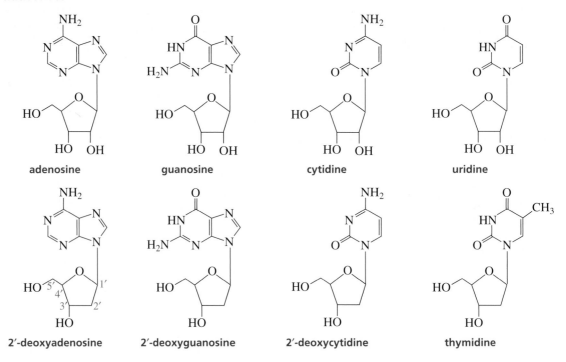

adenosine · guanosine · cytidine · uridine

2′-deoxyadenosine · 2′-deoxyguanosine · 2′-deoxycytidine · thymidine

nucleoside = base + sugar

nucleotide = base + sugar + phosphate

nucleotides

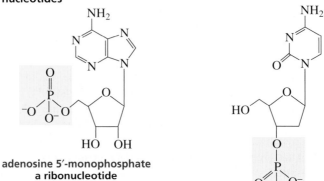

adenosine 5′-monophosphate
a ribonucleotide

2′-deoxycytidine 3′-monophosphate
a deoxyribonucleotide

3-D Molecules:
Bases;
Nucleosides;
Nucleotides

When phosphoric acid is heated with P_2O_5 it loses water, forming an anhydride called pyrophosphoric acid. Its name comes from *pyr,* the Greek word for "fire." Thus, pyrophosphoric acid is prepared by "fire"—that is, by heating. Triphosphoric acid and higher polyphosphoric acids are also formed.

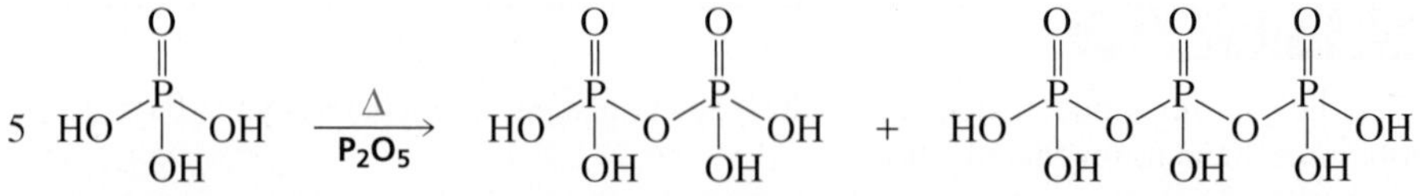

phosphoric acid → pyrophosphoric acid + triphosphoric acid

Because phosphoric acid can form an anhydride, nucleotides can exist as monophosphates, diphosphates, and triphosphates. They are named by adding *monophosphate* (or *diphosphate* or *triphosphate*) to the name of the nucleoside (Table 25.1).

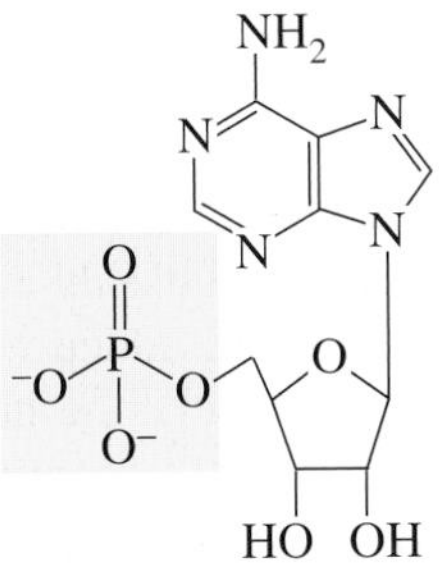

adenosine
5′-monophosphate
AMP

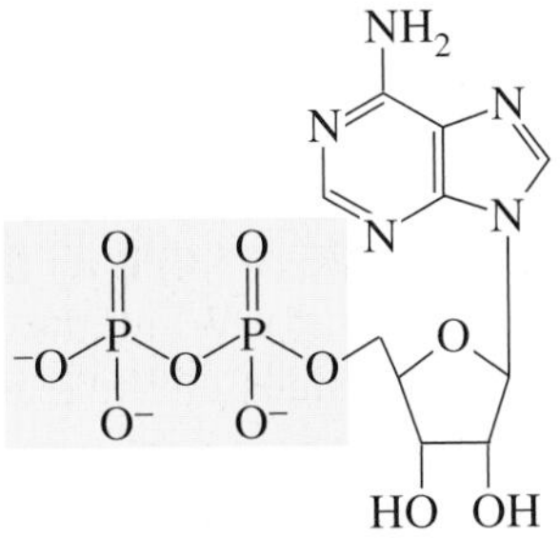

adenosine
5′-diphosphate
ADP

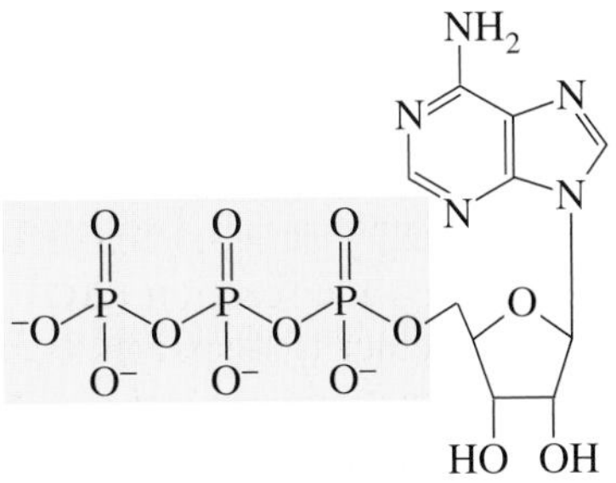

adenosine
5′-triphosphate
ATP

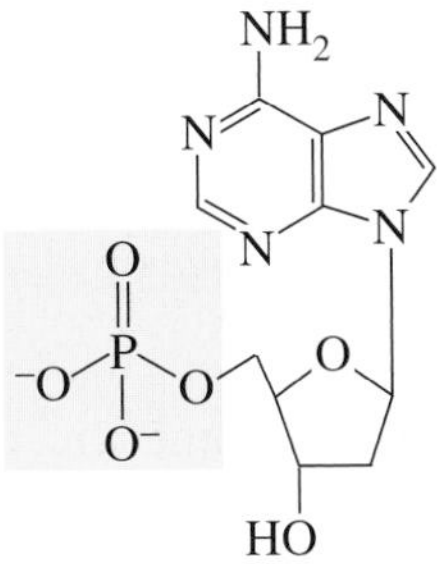

2′-deoxyadenosine
5′-monophosphate
dAMP

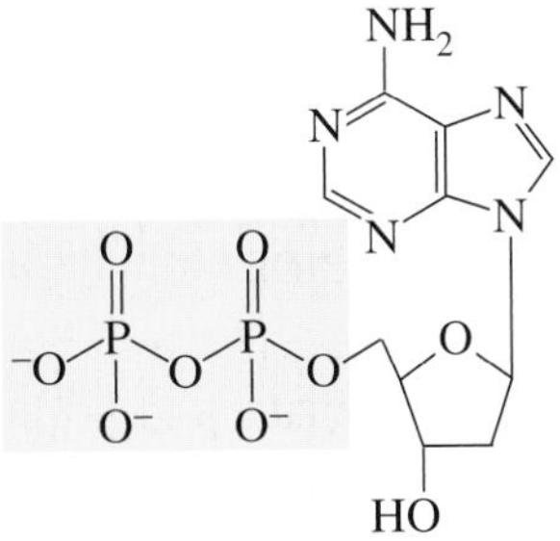

2′-deoxyadenosine
5′-diphosphate
dADP

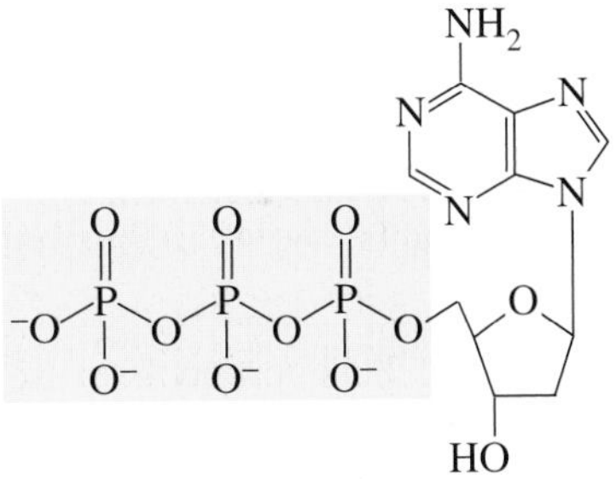

2′-deoxyadenosine
5′-triphosphate
dATP

Table 25.1   The Names of the Bases, the Nucleosides, and the Nucleotides

Base	Nucleoside	Deoxynucleoside	Ribonucleotide	Deoxyribonucleotide
adenine	adenosine	2′-deoxyadenosine	adenosine 5′-phosphate	2′-deoxyadenosine 5′-phosphate
guanine	guanosine	2′-deoxyguanosine	guanosine 5′-phosphate	2′-deoxyguanosine 5′-phosphate
cytosine	cytidine	2′-deoxycytidine	cytidine 5′-phosphate	2′-deoxycytidine 5′-phosphate
thymine	—	thymidine	—	thymidine 5′-phosphate
uracil	uridine	—	uridine 5′-phosphate	

PROBLEM 2

Draw the structure for each of the following:

a. dCDP

b. dTTP

c. dUMP

d. UDP

e. guanosine 5′-triphosphate

f. adenosine 3′-monophosphate

25.2
ATP: THE CARRIER OF CHEMICAL ENERGY

Adenosine 5′-triphosphate (ATP) is known as the universal carrier of chemical energy because, as it is commonly stated, "the energy of hydrolysis of ATP converts endergonic reactions into exergonic reactions." The importance of ATP to biological reactions is shown by its turnover rate in humans—each day, a person uses an amount of ATP equivalent to his or her body weight. Clearly, an enormous amount of ATP is continually being broken down and synthesized in biological organisms.

The ability of ATP to enable otherwise unfavorable reactions to occur is attributed to the large amount of energy released when ATP is hydrolyzed, which can be used to drive an endergonic reaction. For example, the reaction of D-glucose with hydrogen phosphate to form D-glucose-6-phosphate is an endergonic reaction ($\Delta G^{\circ\prime} = +3.3$ kcal/mol).[1] The hydrolysis of ATP, on the other hand, is a highly exergonic reaction ($\Delta G^{\circ\prime} = -7.3$ kcal/mol). When the two reactions are added together (the species occurring on both sides of the reaction arrow cancel), the net reaction is exergonic ($\Delta G^{\circ\prime} = -4.0$ kcal/mol or -16.7 kJ/mol). Thus, the energy released from the hydrolysis of ATP is more than enough to drive the phosphorylation of D-glucose. Two reactions in which the energy of one is used to drive the other are known as *coupled reactions*.

		$\Delta G^{\circ\prime}$	
D-glucose + hydrogen phosphate ⟶ D-glucose-6-phosphate + H_2O		+ 3.3 kcal/mol	(+13.8 kJ/mol)
ATP + H_2O ⟶ ADP + hydrogen phosphate		− 7.3 kcal/mol	(− 30.5 kJ/mol)
D-glucose + ATP ⟶ D-glucose-6-phosphate + ADP		− 4.0 kcal/mol	(− 16.7 kJ/mol)

3-D Molecules:
Adenosine
5'-triphosphate
(ATP)

This nonmechanistic description of ATP's power makes it look like some kind of magical source of energy. Let's look at the mechanism of the reaction to see what really happens. The reaction is a simple one-step nucleophilic substitution reaction. The 6-OH group of glucose attacks the terminal phosphate of ATP, breaking a **phosphoanhydride bond** without forming an intermediate. This is called an **in-line displacement mechanism.** Essentially it is an S_N2 reaction with an adenosine pyrophosphate leaving group. Notice that nucleophilic attack breaks the phosphoanhydride bond rather than the π bond—unlike nucleophilic attack on a carbonyl group where the π bond is the first bond to break. The reason the phosphoanhydride bond is so easily broken is discussed in Section 25.4.

a phosphoanhydride bond

D-glucose + ATP

↓

D-glucose-6-phosphate + ADP

Now we have a chemical understanding of why the phosphorylation of glucose requires ATP. Without ATP, the 6-OH group of D-glucose would have to displace a

[1]The prime in $\Delta G^{\circ\prime}$ indicates that two additional parameters have been added to the ΔG° defined in Section 3.7—the reaction occurs in aqueous solution at pH = 7 and the concentration of water is assumed to be constant.

very basic $^-$OH group from hydrogen phosphate. With ATP, the 6-OH group of D-glucose displaces the weakly basic ADP.

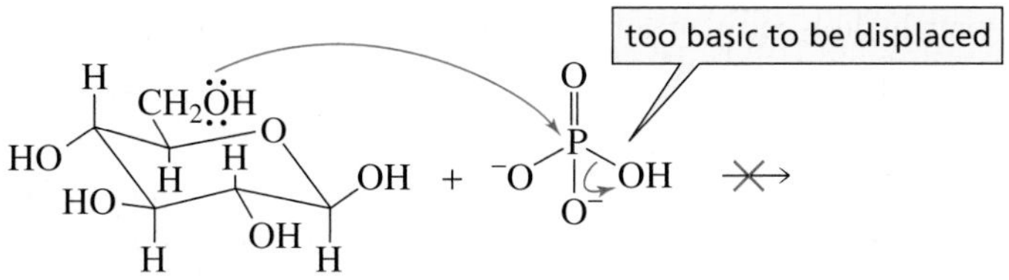

Although the phosphorylation of glucose is described as being driven by the hydrolysis of ATP, you can see from the mechanism that glucose does not react with phosphate and that ATP is not hydrolyzed because it does not react with water. In other words, neither of the coupled reactions actually occurs. Instead, the phosphate group of ATP is transferred directly to D-glucose.

The transfer of a phosphate group from ATP to D-glucose is an example of a **phosphoryl transfer reaction.** There are many phosphoryl transfer reactions in biological systems. In all of these reactions, the electrophilic phosphate group is transferred to a nucleophile as a result of breaking a phosphoanhydride bond. This example of a phosphoryl transfer reaction demonstrates the actual chemical function of ATP—it provides a reaction pathway involving a good leaving group for a reaction that cannot occur because of a poor leaving group.

PROBLEM 3 / SOLVED

The hydrolysis of phosphoenolpyruvate is so highly exergonic ($\Delta G^{\circ\prime} = -14.8 \text{ kcal/mol} = -61.9 \text{ kJ/mol}$) that it can be used to "drive the formation" of ATP from ADP ($\Delta G^{\circ\prime} = +7.3 \text{ kcal/mol} = +30.5 \text{ kJ/mol}$). Propose a mechanism for this reaction.

SOLUTION As we saw in the example with ATP, the overall reaction is hydrolysis of phosphoenolpyruvate, but phosphoenolpyruvate does not actually react with water. Just as ATP "drives the formation" of glucose-6-phosphate by supplying glucose with a phosphate that has a good leaving group (ADP), phosphoenolpyruvate "drives the formation" of ATP by supplying ADP with a phosphate that has a good leaving group (pyruvate).

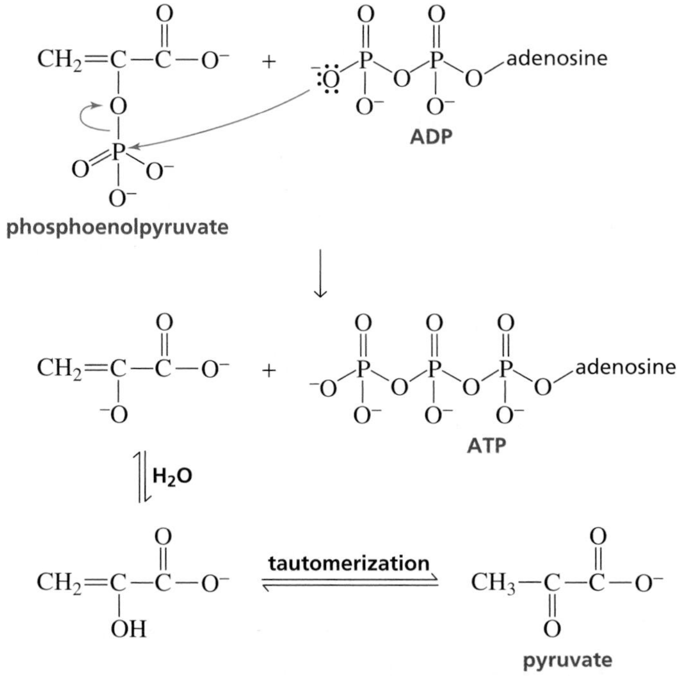

PROBLEM 4◆

Several important biomolecules and the $\Delta G°'$ values for their hydrolysis are listed here. Which of them hydrolyzes with sufficient energy to drive the formation of ATP?

glycerol-1-phosphate:
−2.2 kcal/mol (−9.2 kJ/mol)

phosphocreatine:
−11.8 kcal/mol (−49.4 kJ/mol)

fructose-6-phosphate:
−3.8 kcal/mol (−15.9 kJ/mol)

glucose-6-phosphate:
−3.3 kcal/mol (−13.8 kJ/mol).

25.3 THREE MECHANISMS FOR PHOSPHORYL TRANSFER REACTIONS

There are three possible mechanisms for phosphoryl transfer reactions. We will illustrate them using the following nucleophilic acyl substitution reaction.

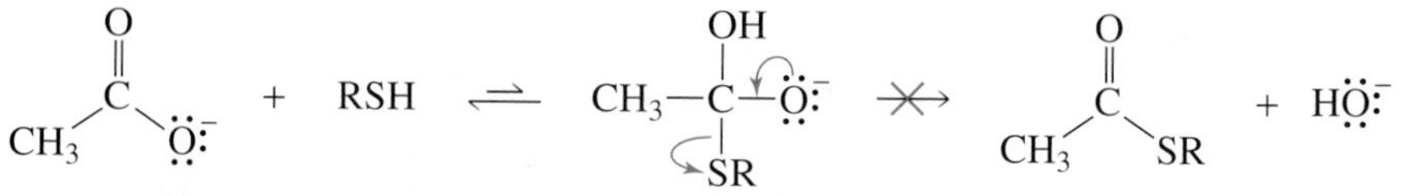

This reaction does not occur without ATP because the negatively charged carboxylate ion resists nucleophilic attack, and the incoming nucleophile is a weaker base than the base that would have to be expelled from the tetrahedral intermediate to form the thioester. In other words, the thiol would be expelled from the tetrahedral intermediate, re-forming the carboxylate ion (Section 16.4).

If ATP is added to the reaction mixture, the reaction occurs. The carboxylate ion attacks one of the phosphate groups of ATP, breaking a phosphoanhydride bond. This puts a leaving group on the carboxyl group that can be displaced by the thiol.

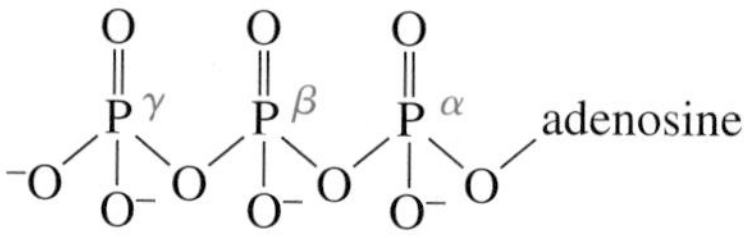

ATP

Tutorial:
Phosphoryl transfer reactions

There are three possible mechanisms for the reaction of a nucleophile with ATP, because each of the three phosphorus atoms of ATP can undergo nucleophilic attack. Each mechanism puts a different phosphate leaving group on the nucleophile.

If the carboxylate ion attacks the γ-phosphorus of ATP, an **acyl phosphate** is formed. The acyl phosphate reacts with the thiol in a nucleophilic acyl substitution reaction, forming the thioester.

nucleophilic attack on the γ-phosphorus

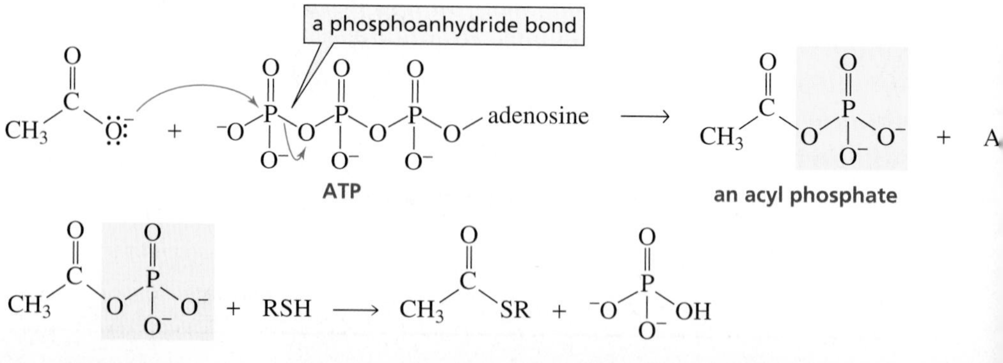

a phosphoanhydride bond

ATP   **an acyl phosphate**

overall reaction

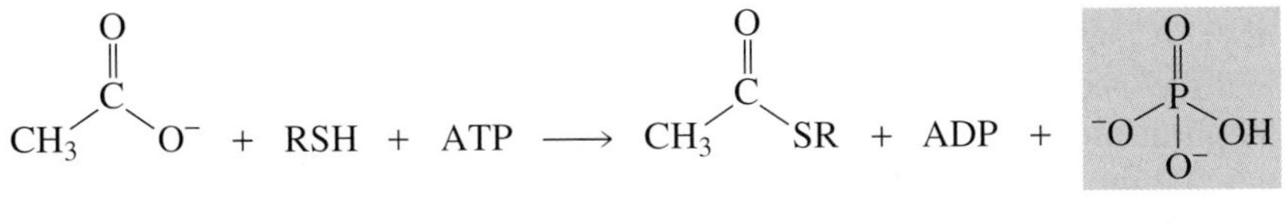

If the carboxylate ion attacks the β-phosphorus of ATP, an **acyl pyrophosphate** is formed. Attack by the thiol on the intermediate forms the thioester because pyrophosphate is a good leaving group.

nucleophilic attack on the β-phosphorus

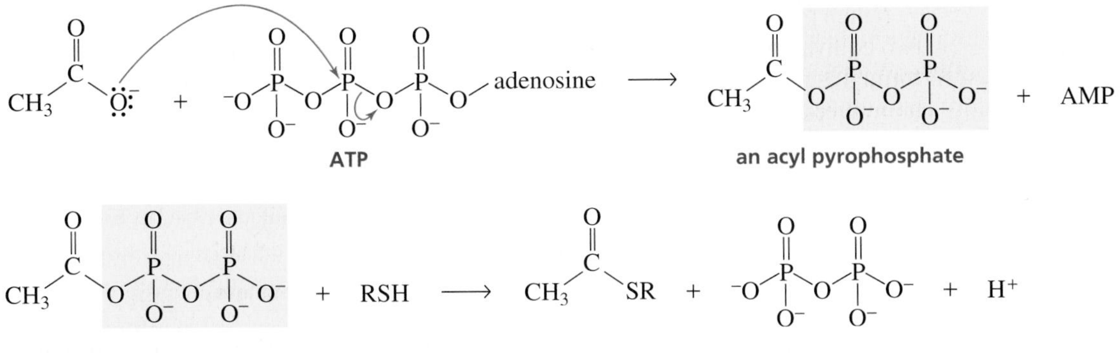

overall reaction

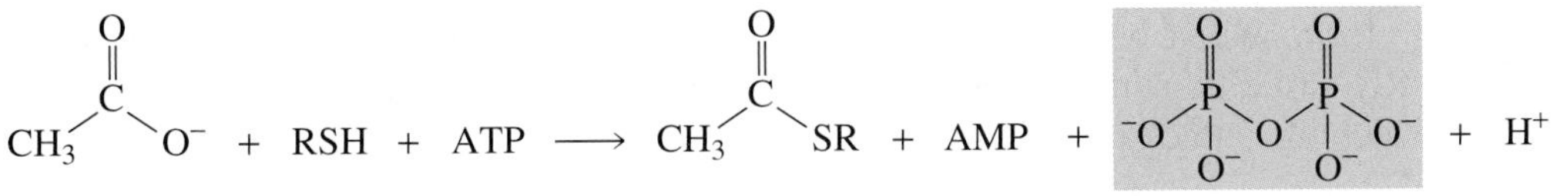

In the third possible mechanism, the carboxylate ion attacks the α-phosphorus of ATP, forming an **acyl adenylate.** Nucleophilic attack by the thiol on the acyl adenylate forms the thioester because AMP is a good leaving group.

nucleophilic attack on the α-phosphorus

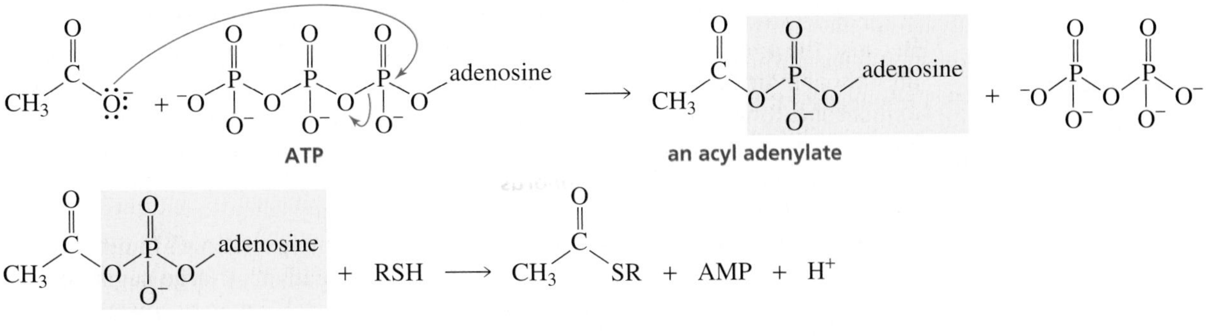

overall reaction

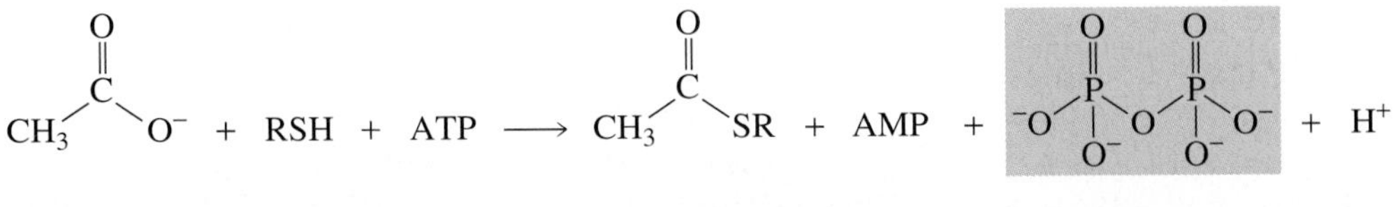

In Section 16.18, we saw that carboxylic acids in biological systems can be activated by being converted into acyl phosphates, acyl pyrophosphates, and acyl adenylates. Formation of these three classes of activated carboxylic acids corresponds to the three mechanisms just discussed for the reaction of a nucleophile with ATP. Nucleophilic attack of a carboxylate ion on the γ-phosphorus forms an acyl phosphate, nucleophilic attack on the β-phosphorus forms an acyl pyrophosphate, and nucleophilic attack on the α-phosphorus forms an acyl adenylate.

ATP provides a reaction pathway with a good leaving group for a reaction that cannot occur because of a poor leaving group.

Many different nucleophiles react with ATP in biological systems. Whether nucleophilic attack occurs on the α-, β-, or γ-phosphorus in any particular reaction depends on the enzyme catalyzing the reaction (Section 25.5). Mechanisms involving nucleophilic attack on the γ-phosphorus form ADP and phosphate as side products, while mechanisms involving nucleophilic attack on the α- or β-phosphorus form AMP and pyrophosphate as side products.

When pyrophosphate is one of the side products, it is subsequently hydrolyzed to two equivalents of phosphate. Consequently, in reactions in which pyrophosphate is formed as a product, its subsequent hydrolysis drives the reaction to the right, ensuring its irreversibility.

pyrophosphate   phosphate

Therefore, enzyme-catalyzed reactions in which irreversibility is important take place by one of the mechanisms that form pyrophosphate as a product (attack on the α- or β-phosphorus of ATP). For example, both the reaction that links nucleotide subunits to form nucleic acids (Section 25.7) and the reaction that binds an amino acid to a tRNA (the first step in translating RNA into a protein; Section 25.12) involve nucleophilic attack on the α-phosphorus of ATP.

PROBLEM 5

The β-phosphorus of ATP has two phosphoanhydride linkages, but only the one linking the β-phosphorus to the α-phosphorus is broken in phosphoryl transfer reactions. Explain why the one linking the β-phosphorus to the γ-phosphorus is never broken.

**25.4
THE "HIGH-ENERGY" CHARACTER OF PHOSPHOANHYDRIDE BONDS**

Because the hydrolysis of a phosphoanhydride bond is a highly exergonic reaction, phosphoanhydride bonds are called "**high-energy bonds.**" The term "high-energy" in this context means that a lot of energy is released when the bond is broken. Do not confuse it with "bond energy," the term chemists use to describe how difficult it is to break a bond. A bond with a *high bond energy* is hard to break, whereas a *high-energy bond* breaks readily.

Why is the hydrolysis of a phosphoanhydride bond so exergonic? In other words, why is the $\Delta G^{\circ\prime}$ value for its hydrolysis large and negative? A large negative $\Delta G^{\circ\prime}$ means that the products of the reaction are much more stable than the reactants. Let's look at ATP and its hydrolysis products to see why this is so.

ATP   phosphate   ADP

Three factors contribute to the greater stability of ADP and phosphate compared to ATP:

1. At physiological pH (pH $=$ 7.3), ATP has 3.3 negative charges, ADP has 2.8 negative charges, and phosphate has 1.6 negative charges. Because of ATP's greater negative charge, more electrostatic repulsions are present in ATP than in ADP or phosphate. Electrostatic repulsions destabilize a molecule.

2. Negatively charged ions are stabilized in an aqueous solution by solvation. Because the reactant has 3.3 negative charges, while the sum of the negative charges on the products is 4.4 (2.8 $+$ 1.6), there is more solvation in the products than in the reactant.

3. There is greater resonance stabilization in the products than in the reactant. Delocalization of a pair of nonbonding electrons on the oxygen joining the two phosphorus atoms is not very effective because it puts a positive charge on an oxygen that is next to a partially positively charged phosphorus atom. When the phosphoanhydride bond breaks, one additional lone pair can be effectively delocalized.

Similar factors explain the large, negative $\Delta G^{\circ\prime}$ when ATP is hydrolyzed to AMP and pyrophosphate, and when pyrophosphate is hydrolyzed to two equivalents of phosphate.

PROBLEM 6 / SOLVED

Do the calculation showing that at pH 7.3:

a. the charge on ATP is -3.3.

b. the charge on ADP is -2.8.

c. the charge on phosphate is -1.6.

ATP has pK_a values of 0.9, 1.5, 2.3, and 7.7; ADP has pK_a values of 0.9, 2.8, and 6.8; and phosphoric acid has pK_a values of 2.1, 7.2, and 12.3.

SOLUTION TO 6a Because pH 7.3 is much more basic than the pK_a values of the first three ionizations of ATP, we know that these three groups will be entirely in their basic forms at that pH, giving ATP three negative charges. We need to determine what fraction of the group with pK_a 7.7 will be in its basic form at pH 7.3.

$$\frac{\text{concentration in the basic form}}{\text{total concentration}} = \frac{[A^-]}{[A^-] + [HA]}$$

$$[A^-] = \text{concentration of the basic form}$$

$$[HA] = \text{concentration of the acidic form}$$

Because this equation has two unknowns, one of the unknowns must be expressed in terms of the other unknown. Using the definition of the acid dissociation constant (K_a), we can define [HA] in terms of $[A^-]$, K_a, and $[H^+]$.

$$K_a = \frac{[A^-][H^+]}{[HA]}$$

$$[HA] = \frac{[A^-][H^+]}{K_a}$$

$$\frac{[A^-]}{[A^-] + [HA]} = \frac{[A^-]}{[A^-] + \dfrac{[A^-][H^+]}{K_a}} = \frac{K_a}{K_a + [H^+]}$$

Now we can calculate the fraction of the group with pK_a 7.7 that will be in the basic form.

$$\frac{K_a}{K_a + [H^+]} = \frac{2.0 \times 10^{-8}}{2.0 \times 10^{-8} + 5.0 \times 10^{-8}} = 0.3$$

total negative charge on ATP $= 3.0 + 0.3 = 3.3$

25.5 KINETIC STABILITY OF ATP IN THE CELL

Although ATP reacts readily in enzyme-catalyzed reactions, it reacts quite slowly in the absence of an enzyme. For example, carboxylic acid anhydrides hydrolyze in a matter of minutes, but ATP takes several weeks to hydrolyze. The low rate of ATP hydrolysis is important because it allows ATP to exist in the cell until it is needed for an enzyme-catalyzed reaction.

The negative charges on ATP are what make it relatively unreactive. These negative charges repel the approach of nucleophiles. When ATP is bound at an active site of an enzyme, it complexes with magnesium (Mg^{2+}), which decreases the overall negative charge on ATP. (This is why ATP-requiring enzymes also require metal ions; Section 23.5.) The other two negative charges can be stabilized by positively charged groups such as arginine or lysine residues at the active site, as shown in Figure 25.2. In this form, ATP is readily approached by nucleophiles, so ATP reacts rapidly in enzyme-catalyzed nucleophilic substitution reactions, but only very slowly in the absence of the enzyme.

Figure 25.2 ▶
The interactions among ATP, Mg^{2+}, and arginine and lysine residues at the active site of an enzyme.

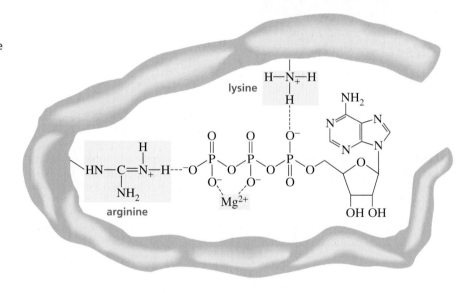

25.6 OTHER IMPORTANT NUCLEOTIDES

ATP is not the only biologically important nucleotide. Guanosine 5′-triphosphate (GTP) is used in place of ATP in some phosphoryl transfer reactions. We have also seen in Sections 23.2 and 23.3 that dinucleotides are used as oxidizing agents (NAD^+, $NADP^+$, FAD, FMN) and reducing agents (NADH, NADPH, $FADH_2$, $FMNH_2$).

Another important nucleotide is adenosine 3′,5′-phosphate, commonly known as cyclic AMP. Cyclic AMP is called a "second messenger" because it serves as a link between several hormones (the first messengers) and certain enzymes that regulate cellular function. Secretion of certain hormones, such as adrenaline, causes the activation of adenylate cyclase, the enzyme responsible for the synthesis of cyclic AMP from ATP. Cyclic AMP then activates an enzyme, generally by phosphorylating it. Cyclic nucleotides are so important in regulating cellular reactions that an entire scientific journal is devoted to these processes.

ATP → adenylate cyclase → cyclic AMP

PROBLEM 7

Propose a mechanism for the formation of cyclic AMP from ATP.

3-D Molecule:
Cyclic AMP

25.7 THE NUCLEIC ACIDS

Nucleic acids are composed of long strands of nucleotide subunits linked by phosphodiester bonds. These linkages join the 3′-OH group of one nucleotide to the 5′-OH group of the next nucleotide (Figure 25.1). A **dinucleotide** contains two nucleotide subunits, an **oligonucleotide** contains three to ten subunits, and a **polynucleotide** contains many subunits. DNA and RNA are polynucleotides. Notice that the nucleotide at one end of the strand has an unlinked 5′-triphosphate group, and the nucleotide at the other end of the strand has an unlinked 3′-hydroxyl group.

Nucleotide triphosphates are the starting materials for the biosynthesis of nucleic acids. DNA is synthesized by enzymes called DNA polymerases, and RNA is synthesized by enzymes called RNA polymerases. The nucleotide strand is formed as a result of nucleophilic attack by a 3′-OH group of one nucleotide triphosphate on the α-phosphorus of another nucleotide triphosphate, breaking a phosphoanhydride bond and eliminating pyrophosphate (Figure 25.3). This means that the growing polymer is synthesized in the 5′⟶3′ direction; in other words, new nucleotides are added to the 3′-end, so the strand grows in the 5′⟶3′ direction by attacking from 3′ to 5′. Pyrophosphate is subsequently hydrolyzed, which makes the reaction irreversibile (Section 25.3). RNA strands are biosynthesized in the same way, using ribonucleotides instead of 2′-deoxyribonucleotides. Like proteins, nucleic acids have both primary and secondary structures. The **primary structure** of a nucleic acid is the sequence of bases in the strand.

Watson and Crick concluded that DNA consists of two strands of nucleic acids with the sugar–phosphate backbone on the outside and the bases on the inside. The chains are held together by hydrogen bonds between the bases on one strand and the bases on the other strand (Figure 25.4). The distance between the two strands is relatively constant, so a purine must pair with a pyrimidine. If the larger purines paired, the strand would bulge; if the smaller pyrimidines paired, the strands would have to cave in to bring the two pyrimidines close enough to form hydrogen bonds.

Critical to Watson and Crick's proposal for the secondary structure of DNA were experiments carried out by Erwin Chargaff. His research showed that the number of

a trinucleotide

Figure 25.3 ▶
Addition of nucleotides to a growing strand of DNA. Biosynthesis occurs in the 5′⟶3′ direction.

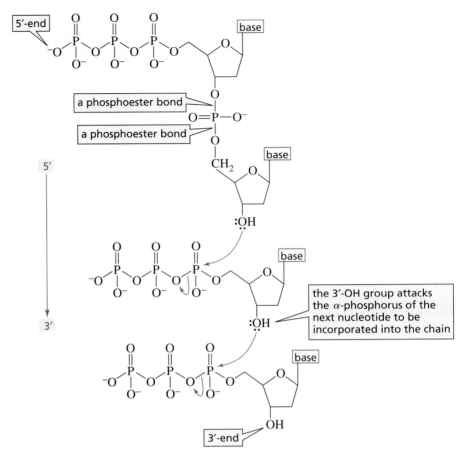

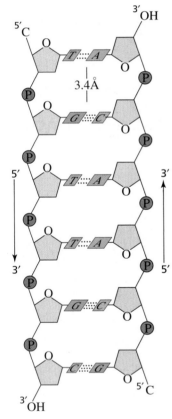

▲ **Figure 25.4**
Complementary base pairing in DNA.

[A] = [T]
[G] = [C]

adenines in DNA equals the number of thymines, and the number of guanines equals the number of cytosines. Chargaff also noted that the number of adenines and thymines relative to the number of guanines and cytosines is characteristic of a given species but varies from species to species. In human DNA, for example, 60.4% of the bases are adenines and thymines, while 74.2% of them are adenines and thymines in the DNA of the bacterium *Sarcina lutea*.

Chargaff's data showing that [adenine] = [thymine] and [guanine] = [cytosine] could be explained if adenine (A) always paired with thymine (T) and guanine (G) always paired with cytosine (C). This means the two strands are *complementary*—where there is an A in one strand, there is a T in the opposing strand, and where there is a G in one strand there is a C in the other strand (Figure 25.4). Thus, if you know the sequence of bases in one strand, you can figure out the sequence of bases in the other strand.

What causes adenine to pair with thymine rather than with cytosine (the other pyrimidine)? The base pairing is dictated by hydrogen bonding. Learning that the bases exist in the keto form allowed Watson to explain the pairing.[2] Adenine forms two hydrogen bonds with thymine but would form only one hydrogen bond with cytosine. Guanine forms three hydrogen bonds with cytosine but would form only one hydrogen bond with thymine (Figure 25.5). Notice that the hydrogen bonds holding the bases together are all about the same length (2.9 ± 0.1 Å).

The two DNA strands are antiparallel—they run in opposite directions, with the sugar–phosphate backbone on the outside and the bases on the inside (Figures 25.4

[2]Watson was having difficulty understanding the base pairing in DNA because he thought the bases existed in the enol form (see Problem 9). When Jerry Donohue, an American crystallographer, informed him that the bases more likely existed in the keto form, Chargaff's data could easily be explained by hydrogen bonding between adenine and thymine and between guanine and cytosine.

Erwin Chargaff *was born in Austria in 1905 and received a Ph.D. from the University of Vienna. To escape Hitler, he came to the United States in 1935, becoming a professor at Columbia University College of Physicians and Surgeons. He modified paper chromatography, a technique developed to identify amino acids (Section 21.5), so that it could be used to quantify the different bases in a sample of DNA.*

▲ **Figure 25.5**
Base pairing in DNA: adenine and thymine form two hydrogen bonds; cytosine and guanine form three hydrogen bonds.

and 25.6). By convention, base sequences of a portion of a nucleic acid are written in the 5′——→3′ direction (the 5′-end is on the left).

The DNA strands are not linear but are twisted into a helix around a common axis (see Figure 25.7[a]). The base pairs are planar and parallel to each other on the inside of the helix (Figure 25.7[b] and [c]). The secondary structure is therefore known as a **double helix.** The double helix resembles a ladder (the base pairs are the rungs) twisted around an axis running down through its rungs (Figure 25.4 and 25.7[c]). The sugar–phosphate backbone is wrapped around the bases. The phosphate OH group has a pK_a of about 2, so it is in its basic form (negatively charged) at physiological

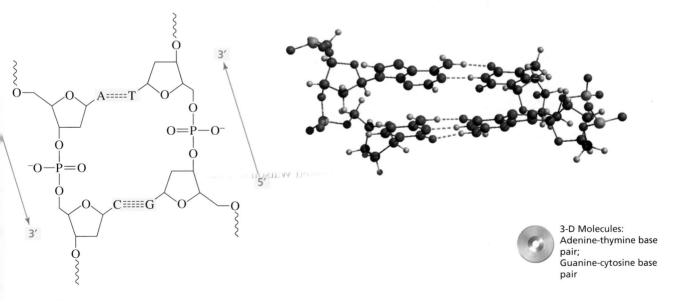

3-D Molecules:
Adenine-thymine base pair;
Guanine-cytosine base pair

▲ **Figure 25.6**
The sugar–phosphate backbone of DNA is on the outside, and the bases are on the inside, with A's pairing with T's and G's pairing with C's. The two strands are antiparallel—they run in opposite directions.

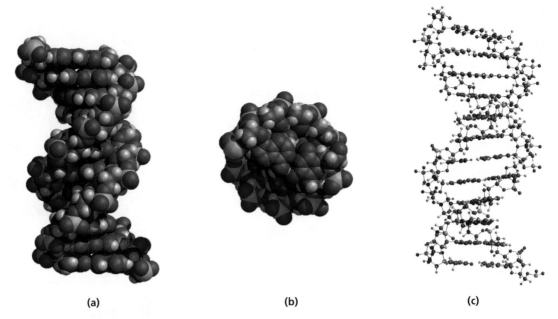

(a)   (b)   (c)

▲ **Figure 25.7**
(a) The DNA double helix. (b) View looking down the long axis of the helix.
(c) The bases are planar and parallel on the inside of the helix.

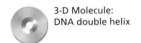

3-D Molecule:
DNA double helix

pH. The negatively charged backbone repels nucleophiles, thereby preventing cleavage of the phosphodiester bonds.

Unlike DNA, RNA is easily cleaved because the 2′-OH group of ribose can act as the nucleophile that cleaves the strand (Figure 25.8). This explains why the 2′-OH group is absent in DNA. To preserve the genetic information, DNA must remain intact throughout the life span of a cell. Cleavage of DNA would have disastrous consequences for the cell and for life itself. RNA, in contrast, is synthesized as it is needed and is degraded once it has served its purpose.

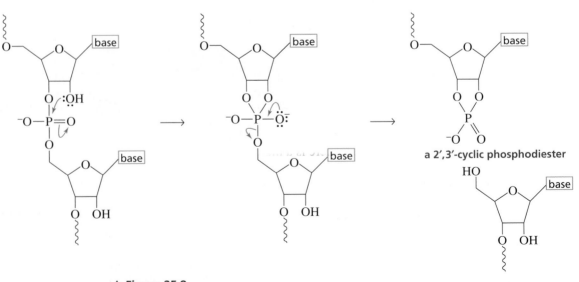

▲ **Figure 25.8**
Hydrolysis of RNA. The 2′-OH group acts as an intramolecular nucleophilic catalyst.
It has been estimated that RNA is hydrolyzed 3 billion times faster than DNA.

Hydrogen bonding between base pairs is just one of the forces holding the two strands of the DNA double helix together. The bases are planar, aromatic molecules that stack on top of one another. Each pair is slightly rotated with respect to the next pair, like a partially spread out hand of cards. There are favorable van der Waals interactions between the mutually induced dipoles of adjacent pairs of bases. These interactions, known as **stacking interactions,** are weak attractive forces, but when added together they contribute significantly to the stability of the double helix. Stacking interactions are strongest between two purines and weakest between two pyrimidines.

Confining the bases to the inside of the helix also stabilizes the helix because it reduces the surface area of the relatively nonpolar residues exposed to water. This increases the entropy of the surrounding water molecules (Section 21.14).

PROBLEM 8

Indicate whether each functional group of the five heterocyclic bases in nucleic acids can function as a hydrogen bond acceptor (A), a hydrogen bond donor (D), or both (D/A).

PROBLEM 9

Using the D, A, and D/A designations in Problem 8, explain how base pairing would be affected if the bases existed in the enol form.

PROBLEM 10

The 2′,3′-cyclic phosphodiester, which is formed when RNA is hydrolyzed (Figure 25.8), reacts with water, forming a mixture of nucleotide 2′- and 3′-phosphates. Propose a mechanism for this reaction.

PROBLEM 11◆

If one of the strands of DNA has the following sequence of bases running in the 5′⟶3′ direction,

$$5'-G-G-A-C-A-A-T-C-T-G-C-3'$$

a. What is the sequence of bases in the complementary strand?

b. What base is closest to the 5′ end in the complementary strand?

Naturally occurring DNA can exist in the three different helical forms shown in Figure 25.9. The B- and A-helices are both right-handed. The B-helix is the predominant form in aqueous solution, while the A-helix is the predominant form in nonpolar solvents. Nearly all the DNA in living organisms is in a B-helix. The Z-helix is a left-handed helix. It occurs in regions where there is a high content of G—C base pairs. The A-helix is shorter (for a given number of base pairs) and about 3% broader than the B-helix, which is shorter and broader than the Z-helix.

Helices are characterized by the number of bases per 360° turn and the distance (the rise) between adjacent base pairs. A-DNA has 11 base pairs per turn and a 2.3 Å rise; B-DNA has 10 base pairs per turn and a 3.4 Å rise; and Z-DNA has 12 base pairs per turn and a 3.8 Å rise.

If you examine Figure 25.9, you will see that there are two kinds of alternating grooves in DNA. In B-DNA, the **major groove** is wider than the **minor groove.** Cross sections of the double helix show that one side of each base pair faces into the major groove and the other side faces into the minor groove (Figure 25.10).

25.8 HELICAL FORMS OF DNA

Figure 25.9 ▶
The three helical forms of
DNA.

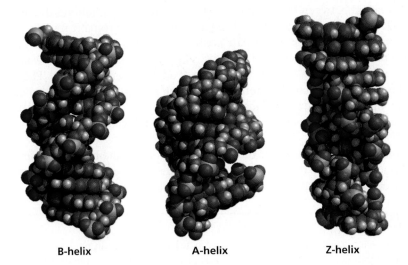

B-helix A-helix Z-helix

3-D Molecules:
B-helix;
A-helix;
Z-helix

Proteins and other molecules can bind to the grooves. The hydrogen-bonding properties of the functional groups facing into each groove determine what kind of molecules will bind to the groove. Distamycin is a naturally occurring compound that has been found to have both antibacterial activity and anticancer activity. It works by binding to the minor groove of DNA. It binds at regions rich in A's and T's (Section 30.10).

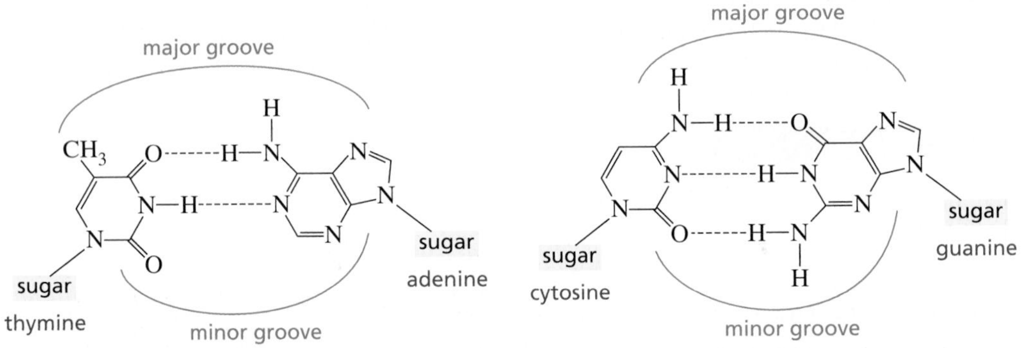

▲ **Figure 25.10**
One side of each base pair faces into the major groove, and the other side faces into the minor groove.

PROBLEM 12◆

Calculate the length of a turn in:

a. A-DNA   **b.** B-DNA   **c.** Z-DNA

25.9
BIOSYNTHESIS OF
DNA: REPLICATION

Watson and Crick's proposal for the structure of DNA was an exciting development because the structure immediately suggested how DNA is able to pass on genetic information to succeeding generations. Because the two strands are complementary, both strands carry the same genetic information. Both strands serve as templates for the synthesis of complementary new strands (Figure 25.11). The new (daughter) DNA molecules are identical to the original (parent) molecule—they contain all the original genetic information. The synthesis of identical copies of DNA is called **replication.**

All the reactions involved in nucleic acid synthesis are catalyzed by enzymes. The synthesis of DNA takes place in a region of the molecule where the strands

◀ **Figure 25.11**
Replication of DNA.

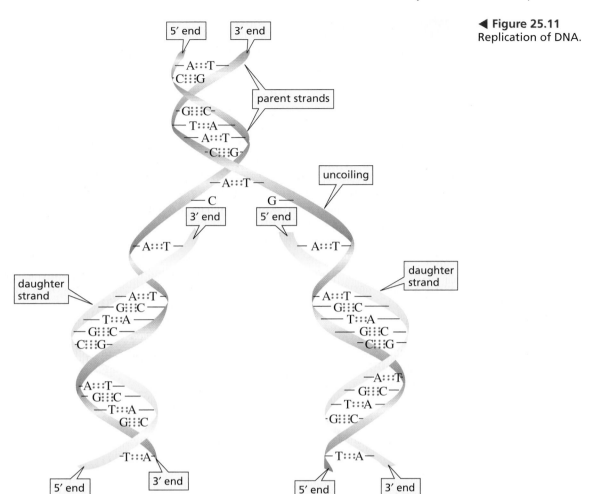

have started to separate, called a **replication fork.** Because a nucleic acid can be synthesized only in the 5′⟶3′ direction, only the daughter strand on the left in Figure 25.11 is synthesized continuously in a single piece (because it is synthesized in the 5′⟶3′ direction). Becuse the other daughter strand needs to grow in the 3′⟶5′ direction, it is synthesized discontinuously in small pieces. Each piece is synthesized in the 5′⟶3′ direction and the fragments are later joined together by an enzyme called DNA ligase. Each of the two daughter molecules of DNA that result contains one of the original strands plus a newly synthesized strand. This process is called **semiconservative replication.**

The genetic information of a human cell is contained in 23 pairs of chromosomes. Each chromosome is composed of several thousand **genes** (segments of DNA). The total DNA of a human cell (the **human genome**) contains 3.1 billion base pairs.

PROBLEM 13

Using a dark line for parental DNA and wavy lines for DNA synthesized from parental DNA, show what the population of DNA molecules would look like in the fourth generation.

PROBLEM 14◆

Assuming that the human genome, with its 3.1 billion base pairs, is entirely in a B-helix, how long is the DNA in a human cell?

> **PROBLEM 15**
>
> Why doesn't DNA unravel completely before replication begins?

25.10 BIOSYNTHESIS OF MESSENGER RNA: TRANSCRIPTION

*Severo Ochoa was the first to prepare synthetic strands of RNA by incubating nucleotides in the presence of enzymes that are involved in the biosynthesis of RNA. **Arthur Kornberg** prepared synthetic strands of DNA in a similar manner. For this work they shared the 1959 Nobel Prize in medicine or physiology.*

The sequence of DNA bases provides the blueprint for the synthesis of messenger RNA (mRNA). The synthesis of mRNA from a DNA blueprint, called **transcription,** takes place in the nucleus of the cell. The newly synthesized mRNA can then leave the nucleus, carrying the genetic information into the cytoplasm (the cell material outside the nucleus) where translation of this information into proteins takes place.

DNA contains sequences of bases known as promoter sites. The **promoter sites** are at the beginning of genes. An enzyme recognizes a promoter site and binds to it, initiating RNA synthesis. The DNA at a promoter site unwinds to give two single strands, exposing the bases. One of the strands is called the **sense strand** or **informational strand.** The complementary strand is called the **template strand** or **antisense strand.** The template strand is read in the $3' \longrightarrow 5'$ direction so that mRNA can be synthesized in the $5' \longrightarrow 3'$ direction (Figure 25.12). The bases in the template strand specify the bases that need to be incorporated into mRNA, following the same base pairing found in DNA. For example, each guanine in the template strand specifies the incorporation of a cytosine into mRNA, and each adenine in the template strand specifies the incorporation of a uracil into mRNA. (Recall that in RNA, uracil is used instead of thymine.)

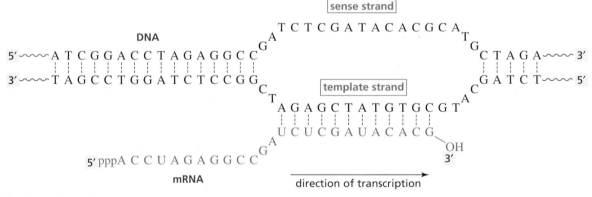

▲ **Figure 25.12**
Transcription: using DNA as a blueprint for mRNA.

Severo Ochoa (1905–1993) was born in Spain. He graduated from the University of Malaga in 1921 and received an M.D. from the University of Madrid. He spent the next four years studying in Germany and England and then joined the faculty at New York University College of Medicine. He became a U.S. citizen in 1956.

Because both the sense strand and mRNA are complementary to the template strand, the sense strand and mRNA have the same base sequence, except that mRNA has a uracil wherever the sense strand has a thymine. Just as there are promoter sites that signal the places to start mRNA synthesis, there are sites in DNA that signal that no more bases should be added to the growing strand of mRNA, at which point synthesis stops.

Surprisingly, a gene is not necessarily a continuous sequence of bases. Often the bases of a gene are interrupted by bases that appear to have no informational content. A stretch of bases representing a portion of a gene is called an **exon,** while a stretch of bases that contains no genetic information is called an **intron.** The mRNA that is synthesized is complementary to the entire sequence of DNA bases—exons and introns. So after the mRNA is synthesized, but before it leaves the nucleus, the so-called nonsense bases (encoded by the introns) are cut out and the informational fragments are spliced together, resulting in a much shorter mRNA molecule. This RNA processing step is known as **RNA splicing.** Scientists have found that only about 2% of DNA contains genetic information, while 98% consists of introns.

It has been suggested that the purpose of introns is to make mRNA more versatile. The originally synthesized long strand of mRNA can be spliced in different ways to create a variety of shorter mRNAs. Each short mRNA specifies the synthesis of a different protein (Section 25.13), so different proteins can be prepared from the same gene.

Arthur Kornberg was born in New York in 1918. He graduated from the College of the City of New York and received an M.D. from the University of Rochester. He is a member of the faculty of the biochemistry department at Stanford University.

PROBLEM 16

Why do both thymine and uracil specify the incorporation of adenine?

25.11 RIBOSOMAL RNA

RNA is much shorter than DNA and is generally single-stranded. Although DNA molecules can have billions of base pairs, RNA molecules rarely have more than 10,000 nucleotides. Messenger RNA (mRNA) is one of three kinds of RNA—the other two are ribosomal RNA (rRNA), a structural component of ribosomes, and transfer RNA (tRNA), the carriers of amino acids for protein synthesis.

Biosynthesis of proteins takes place on particles known as **ribosomes.** A ribosome is composed of about 40% protein and about 60% rRNA. There is increasing evidence that protein synthesis is catalyzed by rRNA molecules rather than by enzymes. RNA molecules—found in ribosomes—that act as catalysts are known as **ribozymes.** The protein molecules in the ribosome enhance the functioning of the rRNA molecules.

Ribosomes are made up of two subunits. The size of the subunits depends on whether they are found in prokaryotic organisms or eukaryotic organisms. **Prokaryotic organisms** (*pro,* Greek for "before"; *karyon,* Greek for "kernel" or "nut") are the earliest organisms. They are unicellular and do not have nuclei. A **eukaryotic organism** (*eu,* Greek for "well") is much more complicated. Eukaryotic organisms can be unicellular or multicellular, and their cells have nuclei. A prokaryotic ribosome is composed of a 50S subunit and a smaller 30S subunit; together they form a 70S ribosome. A eukaryotic ribosome has a 60S subunit and a 40S subunit; together they form an 80S ribosome. The S stands for the **sedimentation constant,** which designates where a given component sediments during centrifugation.[3]

Sidney Altman and Thomas R. Cech received the 1989 Nobel Prize in chemistry for their discovery of the catalytic properties of RNA.

Sidney Altman was born in Montreal in 1939. He received a B.S. from MIT and a Ph.D. from the University of Colorado, Boulder. He was a postdoctoral fellow in Francis Crick's laboratory at Cambridge University. He is a professor of biology at Yale University.

Thomas Cech was born in Chicago in 1947. He received a B.A. from Grinnell College and a Ph.D. from the University of California, Berkeley. He was a postdoctoral fellow at MIT. He is a professor of chemistry at the University of Colorado, Boulder.

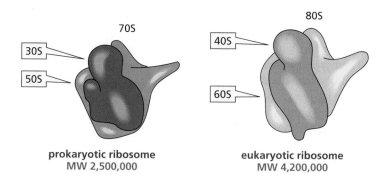

prokaryotic ribosome
MW 2,500,000

eukaryotic ribosome
MW 4,200,000

---

[3]Sedimentation constants are not additive, which is why a 50S and a 30S can combine to form a 70S.

25.12 TRANSFER RNA

Transfer RNA (tRNA) is much smaller than mRNA or rRNA. It contains only 70 to 90 nucleotides. The single strand of tRNA is folded into a characteristic cloverleaf structure strung out with three loops and a little bulge next to the right-hand loop (Figure 25.13). There are at least four regions with complementary base pairing. All tRNAs have a CCA sequence at the 3′-end. The three bases at the bottom of the loop directly opposite the 5′- and 3′-ends are called an **anticodon.**

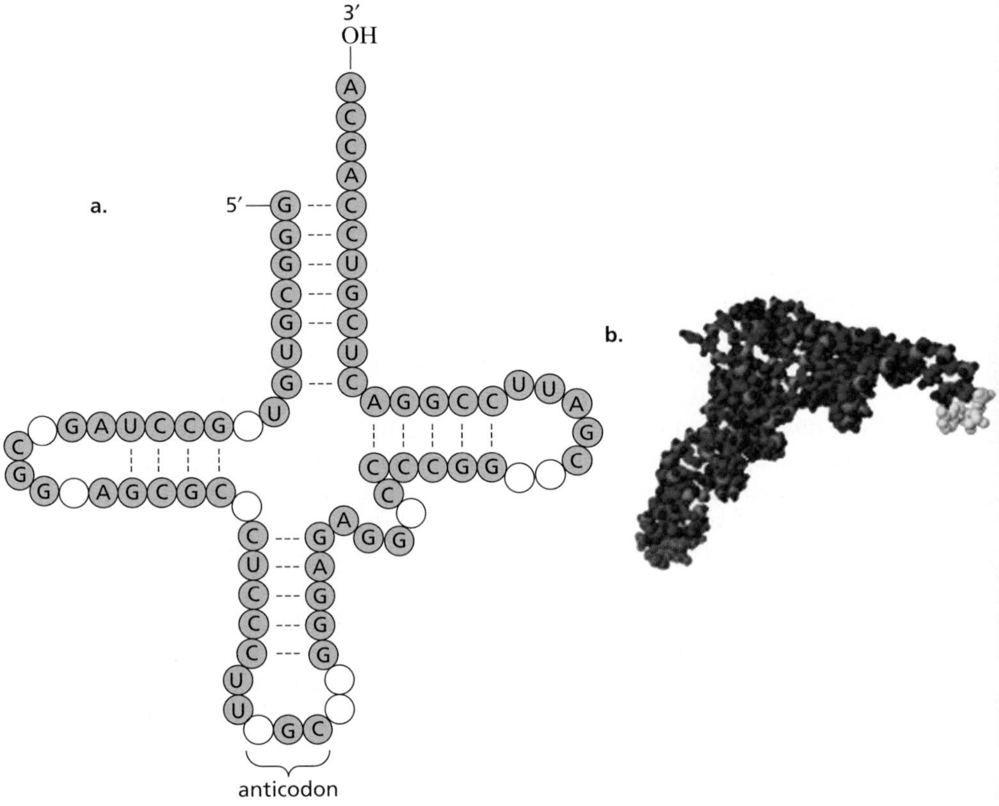

Figure 25.13 ▶
a. tRNA^Ala, a transfer RNA that carries alanine. Compared with other RNAs, tRNA contains a high percentage of unusual bases (shown as empty circles). These bases result from enzymatic modification of the four normal bases. **b.** tRNA^Ser. The anticodon is shown in red; the serine binding site is shown in yellow.

Elizabeth Keller (1918–1997) was the first to recognize that tRNA had a clover-leaf structure. She received a B.S. from the University of Chicago in 1940 and a Ph.D. from Cornell University Medical College in 1948. She worked at the Huntington Memorial Laboratory of Massachusetts General Hospital and at the United States Public Health Service. Later she became a professor at MIT and then at Cornell University.

3-D Molecule:
tRNA

anticodon

Each tRNA can carry an amino acid bound as an ester to its terminal 3′-OH group. The amino acid will be inserted into a protein during protein biosynthesis. Each tRNA can carry only one particular amino acid. A tRNA that carries alanine is designated as tRNA^Ala.

How does a tRNA molecule become attached to the amino acid? Attachment of the amino acid is catalyzed by an enzyme called aminoacyl-tRNA synthetase. In the first step of the enzyme-catalyzed reaction, the carboxyl group of the amino acid attacks the α-phosphorus of ATP, activating the carboxyl group by forming an acyl adenylate. The pyrophosphate that is expelled is subsequently hydrolyzed, ensuring the irreversibility of the phosphoryl transfer reaction (Section 25.3). Then a nucleophilic acyl substitution reaction occurs—the 3′-OH group of tRNA attacks the carbonyl carbon of the amino acid, forming a tetrahedral intermediate. The amino acyl tRNA is formed when AMP is expelled from the tetrahedral intermediate. All the steps take place at the active site of the enzyme.

Each amino acid has its own aminoacyl-tRNA synthetase. Each synthetase has two specific binding sites, one for the amino acid and one for the tRNA that carries that amino acid (Figure 25.14). It is critical that the correct amino acid is attached to the tRNA. Otherwise, the correct protein will not be synthesized. Fortunately, the synthetases correct their own mistakes. For example, valine and threonine are approximately the same size—valine has a CH_3 in place of the OH of threonine. Therefore, both can bind at the amino acid binding site of the aminoacyl-tRNA synthetase for valine, and both can then be activated by reacting with ATP to form an acyl adenylate. The aminoacyl-tRNA synthetase for valine has two adjacent catalytic sites, one for attaching the acyl adenylate to tRNA and one for hydrolyzing the acyl adenylate. The acylation site is hydrophobic, so valine is preferred over threonine for the tRNA acylation reaction. The hydrolytic site is polar, so threonine is preferred over valine for the hydrolysis reaction. Thus, if threonine is activated by the aminoacyl-tRNA synthetase for valine, it will be hydrolyzed rather than transferred to the tRNA.

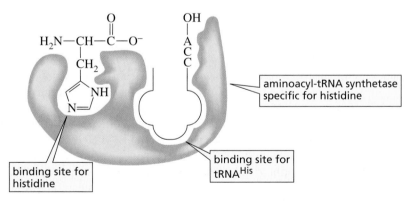

binding site for histidine

aminoacyl-tRNA synthetase specific for histidine

binding site for tRNA^{His}

◀ **Figure 25.14**
An aminoacyl-tRNA syn-thetase has a binding site for tRNA and a binding site for the particular amino acid that is to be attached to that tRNA. Histidine is the amino acid and tRNA^{His} is the tRNA molecule in this example.

A protein is synthesized from its N-terminal end to its C-terminal end by reading the bases along the mRNA strand in the $5' \longrightarrow 3'$ direction. A sequence of three bases, called a **codon,** specifies a particular amino acid that is to be incorporated into a pro-tein. The bases are read consecutively and are never skipped. A codon is written with the 5'-nucleotide on the left. Each amino acid is specified by a three-base sequence known as the **genetic code** (Table 25.2). For example, UCA on mRNA codes for the amino acid serine, while CAG codes for glutamine.

25.13 BIOSYNTHESIS OF PROTEINS: TRANSLATION

*The genetic code was worked out independently by **Marshall Nirenberg** and **Har Gobind Khorana**, for which they shared the 1968 Nobel Prize in medicine or physiology. **Robert Holley**, who worked on the structure of tRNA molecules, also shared that year's prize.*

__Marshall Nirenberg__ was born in New York in 1927. He received a bachelor's degree from the University of Florida and a Ph.D. from the University of Michigan. He is a scientist at the National Institutes of Health.

Table 25.2 The Genetic Code

5'-Position	Middle position				3'-Position
	U	C	A	G	
U	Phe	Ser	Tyr	Cys	U
	Phe	Ser	Tyr	Cys	C
	Leu	Ser	STOP	STOP	A
	Leu	Ser	STOP	Trp	G
C	Leu	Pro	His	Arg	U
	Leu	Pro	His	Arg	C
	Leu	Pro	Gln	Arg	A
	Leu	Pro	Gln	Arg	G
A	Ile	Thr	Asn	Ser	U
	Ile	Thr	Asn	Ser	C
	Ile	Thr	Lys	Arg	A
	Met	Thr	Lys	Arg	G
G	Val	Ala	Asp	Gly	U
	Val	Ala	Asp	Gly	C
	Val	Ala	Glu	Gly	A
	Val	Ala	Glu	Gly	G

__Robert W. Holley (1922–1993)__ was born in Illinois and received a bachelor's degree from the University of Illinois and a Ph.D. from Cornell University. During World War II he worked on the synthesis of penicillin at Cornell Medical School. He was a professor at Cornell and later at the University of California, San Diego. He was also a noted sculptor.

A sculpture done by Robert Holley.

Because there are four bases and the codons are triplets, $4^3 = 64$ different codons are possible. This is many more than are needed to specify the 20 different amino acids, so all the amino acids—except methionine and tryptophan—have more than one codon. It is not surprising, therefore, that methionine and tryptophan are the least abundant amino acids in proteins. Actually, 61 of the bases specify amino acids, and three bases are stop codons. **Stop codons** tell the cell to "stop protein synthesis here."

Translation is the process by which the genetic message in DNA that has been passed to mRNA is decoded and used to build proteins. Each of the approximately 100,000 proteins in the human body is synthesized from a different mRNA. Don't confuse transcription and translation—these words are used just as they are used in English. Transcription (DNA to RNA) is copying *within the same language* of nucleotides. Translation (RNA to protein) is *changing to another language*—the language of amino acids.

How the information in mRNA is translated into a polypeptide is shown in Figure 25.15. Figure 25.15 shows that serine was the last amino acid incorporated into the growing polypeptide chain. Serine was specified by the AGC codon because the anticodon of the tRNA that carries serine is GCU (3'-UCG-5'). (Remember that a base sequence is read in the 5' to 3' direction, so the sequence of bases in an anticodon must be read from right to left.) The next codon is CUU, signaling for a tRNA with an anticodon of AAG (3'-GAA-5'). That particular tRNA carries leucine. The amino group of leucine reacts, in an enzyme-catalyzed nucleophilic acyl substitution reaction, with the ester on the adjacent tRNA, displacing the tRNA. The next codon (GCC) brings in a tRNA carrying alanine. The amino group of alanine displaces the tRNA that brought in leucine. Subsequent amino acids are brought in one at a time in the same way, with the codon in mRNA specifying the amino acid to be incorporated by complementary base pairing with the anticodon of the tRNA that carries that amino acid.

Protein synthesis takes place on the ribosomes (Figure 25.16). The smaller subunit of the ribosome (30S in prokaryotic cells) has three binding sites for RNA molecules. It binds the mRNA whose base sequence is to be read, the tRNA carrying the

Tutorial: Translation

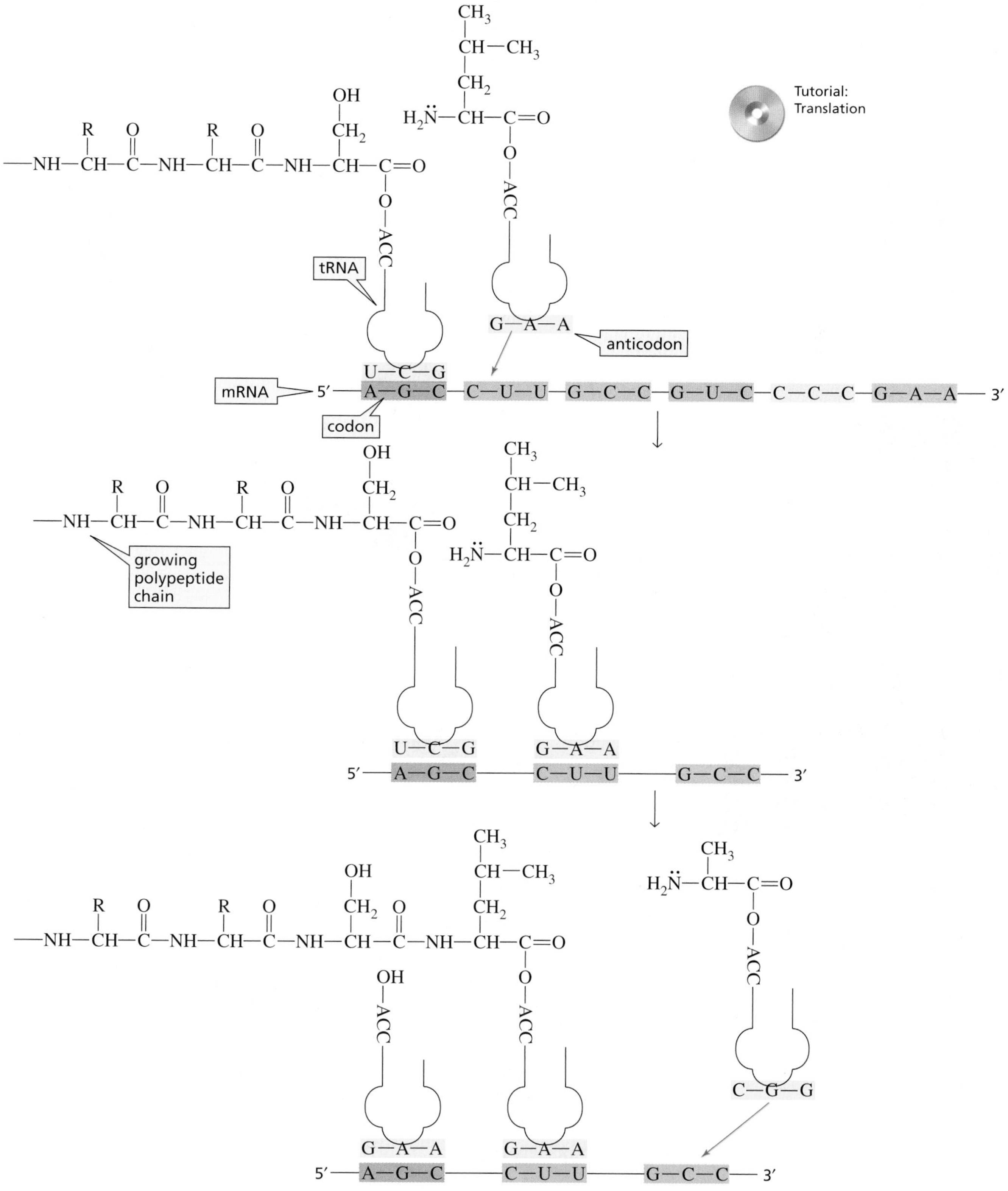

▲ **Figure 25.15**
Translation. The sequence of bases in mRNA determines the sequence of amino acids in a protein.

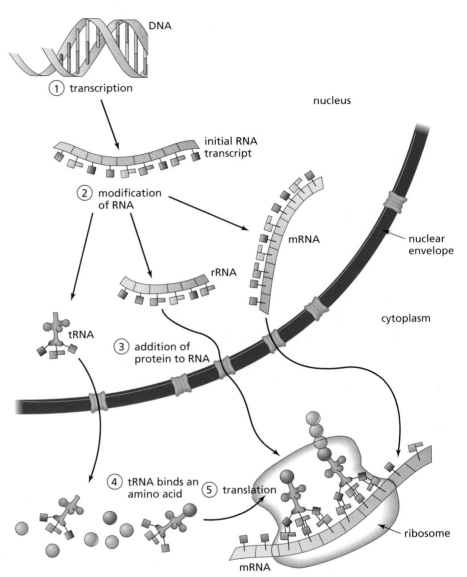

DNA⟶RNA

Translation:
mRNA⟶protein

Har Gobind Khorana was born in India in 1922. He received a bachelor's and a master's degree from Punjab University and a Ph.D. from the University of Liverpool. In 1960 he joined the faculty at the University of Wisconsin and later became a professor at MIT.

▲ **Figure 25.16**

1. Transcription of DNA occurs in the nucleus. The initial RNA transcript is the precursor of all RNA: tRNA, rRNA, and mRNA. 2. The initially formed RNA often must be chemically modified before it acquires biological activity. Modification can entail removing nucleotide segments, adding nucleotides to the 5′ or 3′ ends, or chemically altering certain nucleosides. 3. Proteins are added to rRNA to form ribosomal subunits. tRNA, mRNA, and ribosomal subunits leave the nucleus. 4. Each tRNA binds the appropriate amino acid. 5. tRNA, mRNA, and a ribosome work together to translate the mRNA information into a protein.

growing peptide chain (at its P site), and the tRNA carrying the next amino acid to be incorporated into the protein (at its A site). The larger subunit of the ribosome (50S in prokaryotic cells) catalyzes peptide bond formation.

PROBLEM 17◆

If methionine is the first amino acid incorporated into a heptapeptide, what is the sequence of the amino acids encoded for by the following stretch of mRNA?

5′—G—C—A—U—G—G—A—C—C—C—C—G—U—U—A—U—U—A—A—A—C—A—C—3′

SICKLE CELL ANEMIA

Sickle cell anemia is an example of the damage that can be caused by a change in a single base of DNA (Problem 55 in Chapter 21). It is a hereditary disease caused when a GAG triplet becomes a GTG triplet in the section of DNA that codes for the β-subunit of hemoglobin. As a consequence, the mRNA codon becomes GUG—which signals for incorporation of valine—rather than GAG—which would have signaled for incorporation of glutamic acid. The change from a polar glutamic acid to a nonpolar valine is sufficient to change the shape of the hemoglobin molecule and reduce its solubility, causing it to precipitate in red blood cells. This stiffens the cells, making it difficult for them to squeeze through a capillary. Blocking capillaries causes severe pain and can be fatal.

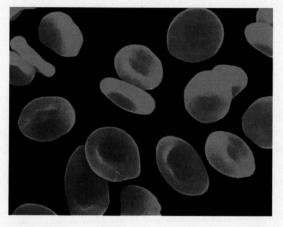

Normal red blood cells

Sickle red blood cells

ANTIBIOTICS THAT ACT BY INHIBITING TRANSLATION

Puromycin is a naturally occurring antibiotic. It is one of several antibiotics that act by inhibiting translation. Puromycin mimics the 3′-CCA-aminoacyl portion of a tRNA. If, during translation, the enzyme is fooled into transferring the growing peptide chain to the amino group of puromycin rather than to the amino group of the incoming 3′-CCA-aminoacyl tRNA, protein synthesis stops. Because puromycin blocks protein synthesis in eukaryotes as well as in prokaryotes, it is poisonous to humans and therefore is not a clinically useful antibiotic. To be clinically useful, an antibiotic must affect protein synthesis only in prokaryotic cells.

puromycin

Clinically useful antibiotics	Mode of action
Tetracycline	Prevents the aminoacyl-tRNA from binding to the ribosome
Erythromycin	Prevents the incorporation of new amino acids into the protein
Streptomycin	Inhibits the initiation of protein synthesis
Chloramphenicol	Prevents the new peptide bond from being formed

PROBLEM 18◆

Four C's occur in a row in the segment of mRNA in Problem 17. What polypeptide would be formed from the mRNA if one of the four C's were cut out of the strand?

3-D Molecules: Chloramphenicol complexed to acetyltransferase; Tetracycline

PROBLEM 19

UAA is a stop codon. Why does the UAA sequence in mRNA in Problem 17 not cause protein synthesis to stop?

PROBLEM 20◆

Write the sequences of bases in the sense strand of DNA that resulted in the mRNA in Problem 17.

PROBLEM 21

List the possible codons on mRNA that specify each amino acid in Problem 17 and the anticodon on the tRNA that carries that amino acid.

**25.14
WHY DNA
CONTAINS THYMINE
INSTEAD OF URACIL**

In Section 23.8, we saw that dTMP is formed when dUMP is methylated, with coenzyme N^5,N^{10}-methylenetetrahydrofolate supplying the methyl group. Because the incorporation of the methyl group into uracil oxidizes tetrahydrofolate to dihydrofolate, dihydrofolate must be reduced back to tetrahydrofolate to prepare the cofactor for another catalytic cycle. The reducing agent is NADPH. Every NADPH formed in a biological organism can drive the formation of three ATPs, so using an NADPH to reduce dihydrofolate comes at the expense of ATP. This means that the synthesis of thymine is energetically expensive, so there must be a good reason for DNA to contain thymine instead of uracil.

$$dUMP + N^5,N^{10}\text{-methylene-THF} \xrightarrow{\text{thymidylate synthase}} dTMP + \text{dihydrofolate}$$

R' = 2'-deoxyribose-5-P

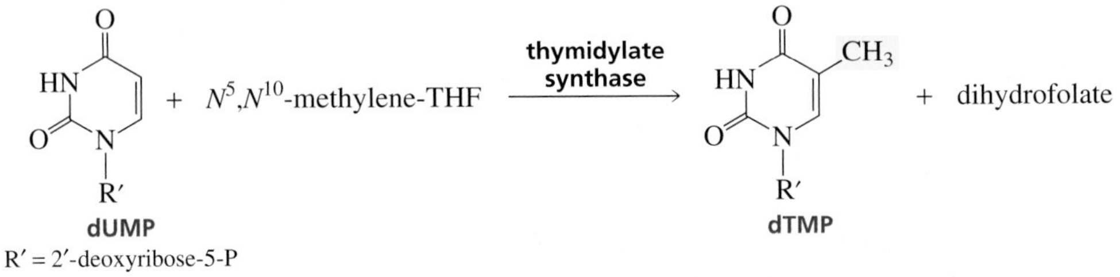

$$\text{dihydrofolate} + \text{NADPH} + \text{H}^+ \xrightarrow{\text{dihydrofolate reductase}} \text{tetrahydrofolate} + \text{NADP}^+$$

The presence of thymine instead of uracil in DNA prevents potentially lethal mutations. Cytosine can tautomerize to form an imine, which can be hydrolyzed to uracil (Section 17.7). The overall reaction is called a **deamination** since it removes an amino group.

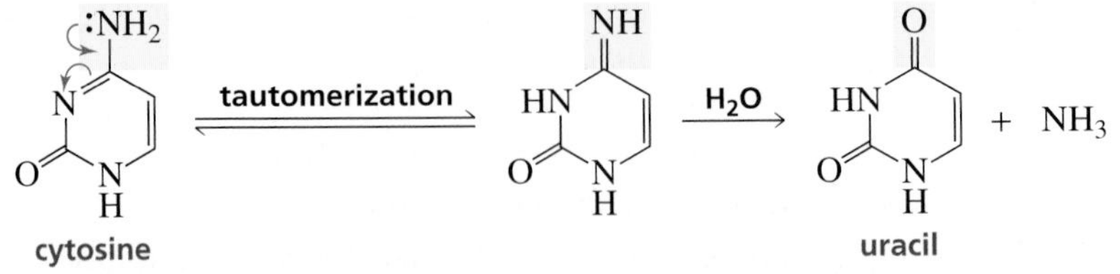

$$\text{cytosine} \xrightleftharpoons{\text{tautomerization}} \xrightarrow{\text{H}_2\text{O}} \text{uracil} + \text{NH}_3$$

If a cytosine in DNA is deaminated to a uracil, uracil will specify incorporation of an adenine into the daughter strand during replication instead of the guanine that would have been specified by cytosine. Fortunately, a U in DNA is recognized as a "mistake" by cell enzymes before an incorrect base can be inserted into the daughter strand. These enzymes cut out the U and replace it with a C. If U's were normally found in DNA, the enzymes could not distinguish between a normal U and a U formed by deamination of cytosine. Having T's in place of U's in DNA allows the U's that are found in DNA to be recognized as mistakes.

Unlike DNA, which replicates itself, any mistake in RNA does not survive for long because RNA is constantly being degraded and then resynthesized from the DNA template. Therefore, it is not worth spending the extra energy to incorporate T's into RNA.

PROBLEM 22◆

Adenine can be deaminated to hypoxanthine and guanine can be deaminated to xanthine. Draw structures for hypoxanthine and xanthine.

PROBLEM 23

Explain why thymine cannot be deaminated.

In June 2000, two teams of scientists (one from a private biotechnology company and one from the publicly funded Human Genome Project) announced that they had completed the first draft of the sequence of the 3.1 billion base pairs in human DNA. This is an enormous accomplishment. For example, if the sequence of 1 million base pairs were determined each day, it would take more than 10 years to complete the sequence of the human genome.

DNA molecules are too large to sequence as a unit, so DNA is first cleaved at specific base sequences and the resulting DNA fragments are sequenced. The enzymes that cleave DNA at specific base sequences are called **restriction endonucleases** and the DNA fragments that are formed are called **restriction fragments.** Several hundred restriction enzymes are now known. A few examples of restriction enzymes, the base sequence each recognizes, and the point of cleavage in that base sequence are shown here.

restriction enzyme	recognition sequence
*Alu*I	AGCT TCGA
*Fnu*DI	GGCC CCGG
*Pst*I	CTGCAG GACGTC

The base sequences that most restriction enzymes recognize are *palindromes*. A palindrome is a word or a group of words that reads the same forward and backward. "Toot" and "poor Dan in a droop" are examples of palindromes.[4] A restriction enzyme recognizes a piece of DNA in which *the template strand is a palindrome of the sense strand.* In other words, the sequence of bases in the template strand (reading from right to left) is identical to the sequence of bases in the sense strand (reading from left to right).

PROBLEM 24◆

Which of the following base sequences would most likely be recognized by a restriction endonuclease?

a. ACGCGT   **c.** ACGGCA   **e.** ACATCGT

b. ACGGGT   **d.** ACACGT   **f.** CCAACC

---

[4]Some other palindromes are "Mom," "Dad," "Bob," "Lil," "radar," "race car," "a man, a plan, a canal, Panama," "Sex at noon taxes," and "He lived as a devil, eh?"

The **dideoxy method** developed by Frederick Sanger is currently the preferred method to determine the sequence of bases in DNA. In the dideoxy method, a small piece of DNA, called a primer, labeled at the 5′-end with ^{32}P, is added to the restriction fragment whose sequence is to be determined. Next, the four 2′-deoxyribonucleoside triphosphates are added as well as DNA polymerase, the enzyme that adds nucleotides to a strand of DNA. In addition, a small amount of the 2′,3′-dideoxynucleoside triphosphate of one of the bases is also added to the reaction mixture. A 2′,3′-dideoxynucleoside triphosphate has no OH groups at the 2′- and 3′-positions.

Frederick Sanger (Section 21.12) and Walter Gilbert shared half of the 1980 Nobel Prize in chemistry for their work on DNA sequencing. The other half went to Paul Berg, who had developed a method of cutting nucleic acids at specific sites and recombining the fragments in new ways, a technique known as recombinant DNA technology.

Walter Gilbert was born in Boston in 1932. He received a master's degree in physics from Harvard and a Ph.D. in mathematics from Cambridge University. In 1958 he joined the faculty at Harvard, where he became interested in molecular biology.

Paul Berg was born in New York in 1926. He received a Ph.D. from Western Reserve University (now Case Western Reserve University). He joined the faculty at Washington University in St. Louis in 1955 and moved to Stanford in 1959.

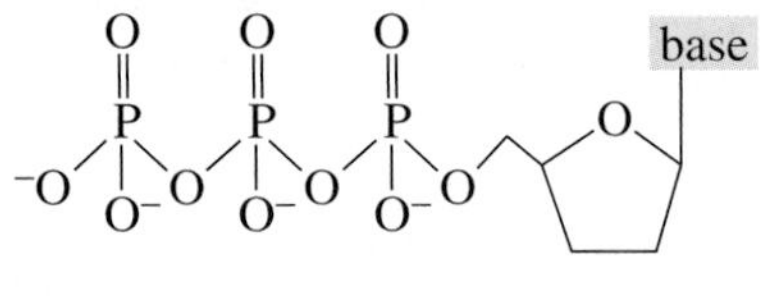

a 2′,3′-dideoxynucleoside triphosphate

Nucleotides will be added to the primer by base pairing with the restriction fragment. Synthesis will stop if the 2′,3′-dideoxy analog of dATP is added instead of dATP because the 2′,3′-dideoxy analog does not have a 3′-OH to which additional nucleotides can be added. Therefore, three different chain-terminated fragments will be obtained from the DNA restriction fragment shown here.

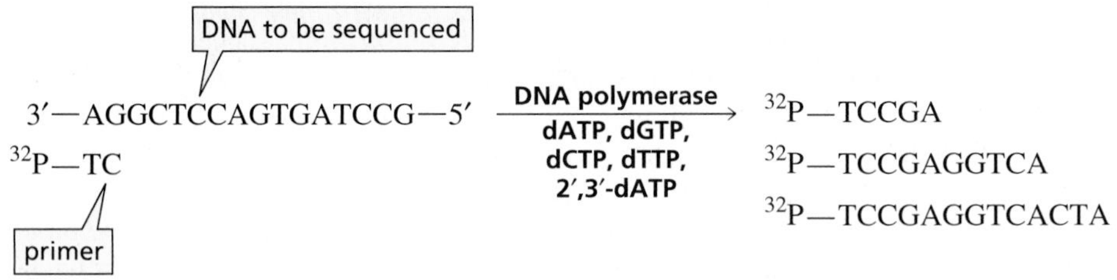

3′—AGGCTCCAGTGATCCG—5′ $\xrightarrow[\substack{\text{dCTP, dTTP,}\\ \text{2′,3′-dATP}}]{\substack{\text{DNA polymerase}\\ \text{dATP, dGTP,}}}$ ^{32}P—TCCGA

^{32}P—TC ^{32}P—TCCGAGGTCA

^{32}P—TCCGAGGTCACTA

The procedure is repeated three more times using a 2′,3′-dideoxy analog of dGTP, then a 2′,3′-dideoxy analog of dCTP, and then a 2′,3′-dideoxy analog of dTTP.

PROBLEM 25

What labeled fragments would be obtained from the segment of DNA shown above if a 2′,3′-dideoxy analog of dGTP had been added to the reaction mixture instead of a 2′,3′-dideoxy analog of dATP?

The chain-terminated fragments obtained from each of the four experiments are loaded onto separate lanes of a buffered polyacrylamide gel—the fragments obtained from using a 2′,3′-dideoxy analog of dATP are loaded onto one lane, the fragments obtained from using a 2′,3′-dideoxy analog of dGTP onto another lane, and so on. An electric field is applied across the ends of the gel, causing the negatively charged fragments to travel toward the positively charged electrode (the anode). The smaller fragments fit through the spaces in the gel relatively easily and therefore travel through the gel faster, while the larger fragments pass through the gel more slowly.

After separation, the gel is placed in contact with a photographic plate. Radiation from ^{32}P causes a dark spot to appear on the plate opposite the location of each labeled fragment in the gel. This technique is called **autoradiography,** and the exposed photographic plate is known as an **autoradiograph** (Figure 25.17).

The sequence of bases in the original restriction fragment can be read directly from the autoradiograph. The identity of each base is determined by noting the column where each successive dark spot (larger piece of labeled fragment) appears, starting at the bottom of the gel. The sequence of the fragment of DNA responsible for the autoradiograph in Figure 25.17 is shown on the left-hand side of the figure.

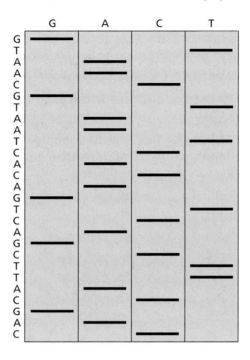

◀ **Figure 25.17**
An autoradiograph.

Once the sequence of bases in a restriction fragment is determined, the results can be checked by determining the base sequence of the complementary strand. The base sequence in the original piece of DNA can be determined by repeating the entire procedure with a different restriction endonuclease and noting overlapping fragments.

DNA FINGERPRINTING

The base sequence of the human genome varies from individual to individual, generally by a single base change every few hundred base pairs. Because some of these changes occur in base sequences recognized by restriction endonucleases, the fragments formed when human DNA reacts with a particular restriction endonuclease vary in size depending on the individual. It is this variation that forms the basis of DNA fingerprinting (also called DNA profiling or DNA typing). This technique is used by forensic chemists to compare DNA samples collected at the scene of a crime with the DNA of the suspected perpetrator. The most powerful technique for DNA identification analyzes restriction fragment length polymorphisms (RFLPs) obtained from regions of DNA in which individual variations are most common. This technique takes four to six weeks and requires a blood stain about the size of a dime. The chance of identical results from two different persons is thought to be one in a million. The second type of DNA profiling uses a polymerase chain reaction (PCR), which amplifies a specific region of DNA and compares differences at that site among individuals. This technique can be done in less than a week and requires only 1% of the amount required for RFLP, but does not discriminate as well among individuals. The chance of identical results from two different people is 1 in 500 to 1 in 2000. DNA fingerprinting is also being used to establish paternity, accounting for about 100,000 DNA profiles a year.

25.16 LABORATORY SYNTHESIS OF DNA STRANDS

There is a great deal of interest in the synthesis of oligonucleotides with specific base sequences. This would allow scientists to synthesize genes that could be inserted into the DNA of microorganisms, causing the organisms to synthesize a particular protein. Alternatively, a synthetic gene could be inserted into the DNA of an organism defective in that gene—a process known as **gene therapy.**

Synthesizing an oligonucleotide with a particular base sequence is an even more challenging task than synthesizing a polypeptide with a specific amino acid sequence. The nucleotides must be strung together in a particular order, and each nucleotide has several groups that must be protected and then deprotected at the proper times. The approach taken was to develop an automated method similar to automated peptide synthesis (Section 21.10). The growing nucleotide is attached to a solid support so that it can be purified by flushing the reaction container with an appropriate solvent. Therefore, none of the synthesized product will be lost during purification.

Phosphoramidite Monomers

One current method synthesizes oligonucleotides using phosphoramidite monomers. The 5′-OH group of each phosphoramidite monomer is attached to a *para*-dimethoxytrityl (DMTr) protecting group. The particular group (Pr) used to protect the base depends on the base.

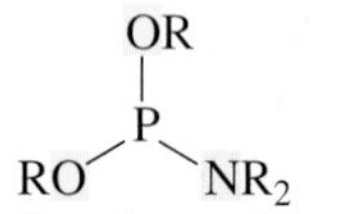

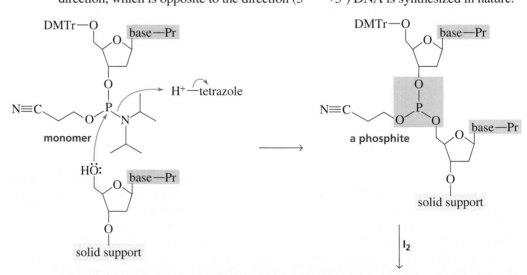

a phosphoramidite

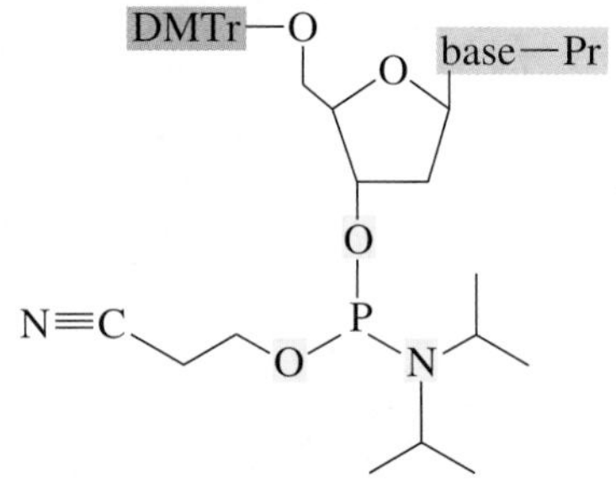

a phosphoramidite monomer used for oligonucleotide synthesis

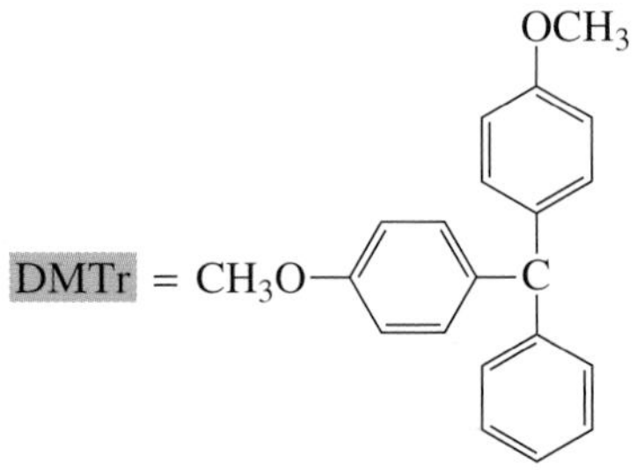

$DMTr = CH_3O$—

The 3′-nucleoside of the oligonucleotide being synthesized is attached to a controlled-pore glass solid support by an ester bond, and the oligonucleotide is synthesized from the 3′-end. When a monomer is added to the nucleoside attached to the solid support, the only nucleophile in the reaction mixture is the 5′-OH group of the sugar bonded to the solid support. This nucleophile attacks the phosphorus of the phosphoramidite, displacing the amine and forming a phosphite. The amine is too strong a base to be expelled without being protonated. Protonated tetrazole is the acid used for protonation because it is strong enough to protonate the diisopropylamine leaving group, but not strong enough to remove the DMTr protecting group. The phosphite is oxidized to a phosphate using I_2 or *tert*-butylhydroperoxide. The DMTr protecting group on the 5′-end of the dinucleotide is removed with mild acid. The cycle of (1) monomer addition, (2) oxidation, and (3) deprotection with acid is repeated over and over until a polymer of the desired length and sequence is obtained. Notice that the DNA polymer is synthesized in the 3′⟶5′ direction, which is opposite to the direction (5′⟶3′) DNA is synthesized in nature.

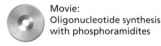

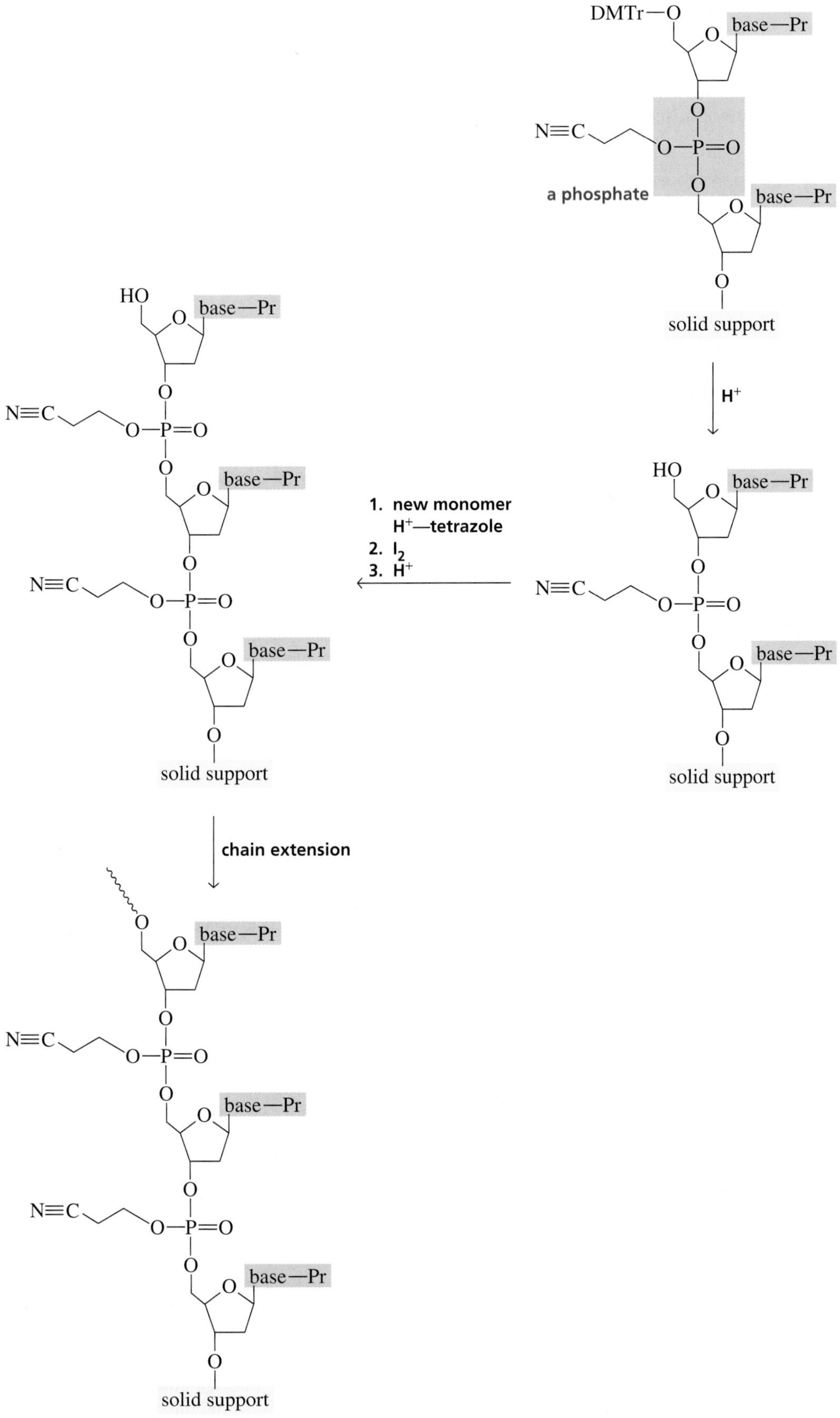
DMTr—O
base—Pr
N≡C
O—P=O
a phosphate
base—Pr
solid support
H+
HO
base—Pr
N≡C
O—P=O
base—Pr
solid support
1. new monomer
H+—tetrazole
2. I2
3. H+
HO
base—Pr
N≡C
O—P=O
base—Pr
N≡C
O—P=O
base—Pr
solid support
chain extension
base—Pr
N≡C
O—P=O
base—Pr
N≡C
O—P=O
base—Pr
solid support

The NH_2 groups of cytosine, adenine, and guanine are nucleophiles and therefore must be protected to prevent them from reacting with a newly added monomer. The NH_2 groups of cytosine and adenine are protected as benzamides and the NH_2 group of guanine is protected as a butanamide. Thymine does not contain any nucleophilic groups so it does not have to be protected.

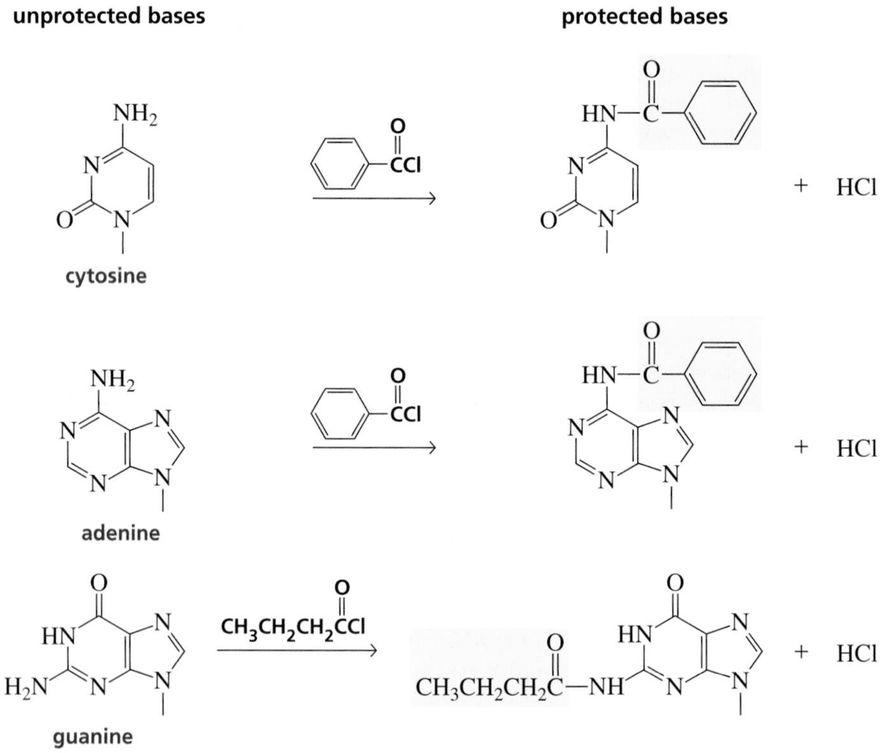

unprotected bases   **protected bases**

cytosine

adenine

guanine

After the oligonucleotide is synthesized, the protecting groups on the phosphates and the protecting groups on the bases must be removed, and the oligonucleotide has to be detached from the solid support. This can all be done in a single step using aqueous ammonia. Because a hydrogen bonded to a carbon adjacent to a cyano group is relatively acidic (Section 19.1), ammonia can be used as the base in the elimination reaction that removes the phosphate protecting group.

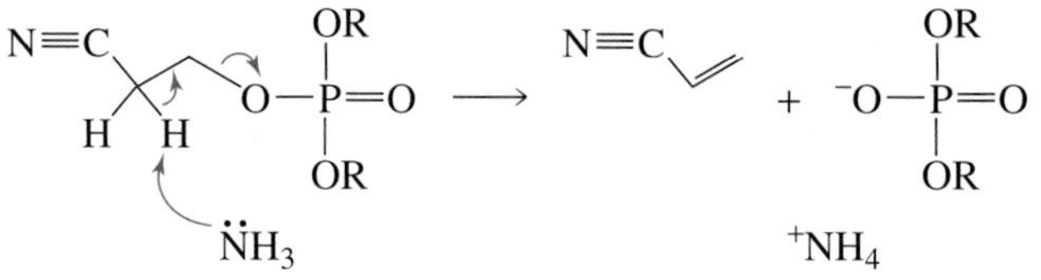

Currently, automated DNA synthesizers can synthesize nucleic acids containing as many as 130 nucleotides in acceptable yields, adding one nucleotide every 2 to 3 minutes. Longer nucleic acids can be prepared by splicing together two or more individually prepared strands. To ensure a good overall yield of oligonucleotide, the addition of each monomer must occur in high yield (>98%). This can be accomplished by using a large excess of monomer. This, however, makes oligonucleotide synthesis very expensive because the unreacted monomers are wasted.

H-Phosphonate Monomers

A second method, using H-phosphonate monomers, has the advantage over the phosphoramidite method in that the monomers are easier to handle and phosphate protecting groups are not needed. However, the yields are not as good. The H-phosphonate monomers are activated by being converted into anhydrides as a result of reacting with an acyl chloride. The 5'-OH group of the nucleoside attached to the solid support reacts with the anhydride, forming a dimer. The DMTr protecting group is removed with acid under mild conditions, and a second activated monomer is added. Monomers are added one at a time in this way until the strand is complete. Oxidation with I_2 (or *tert*-butylhydroperoxide) converts the H-phosphonate groups to phosphate groups. The base protecting groups are removed by aqueous ammonia as in the phosphoramidite method.

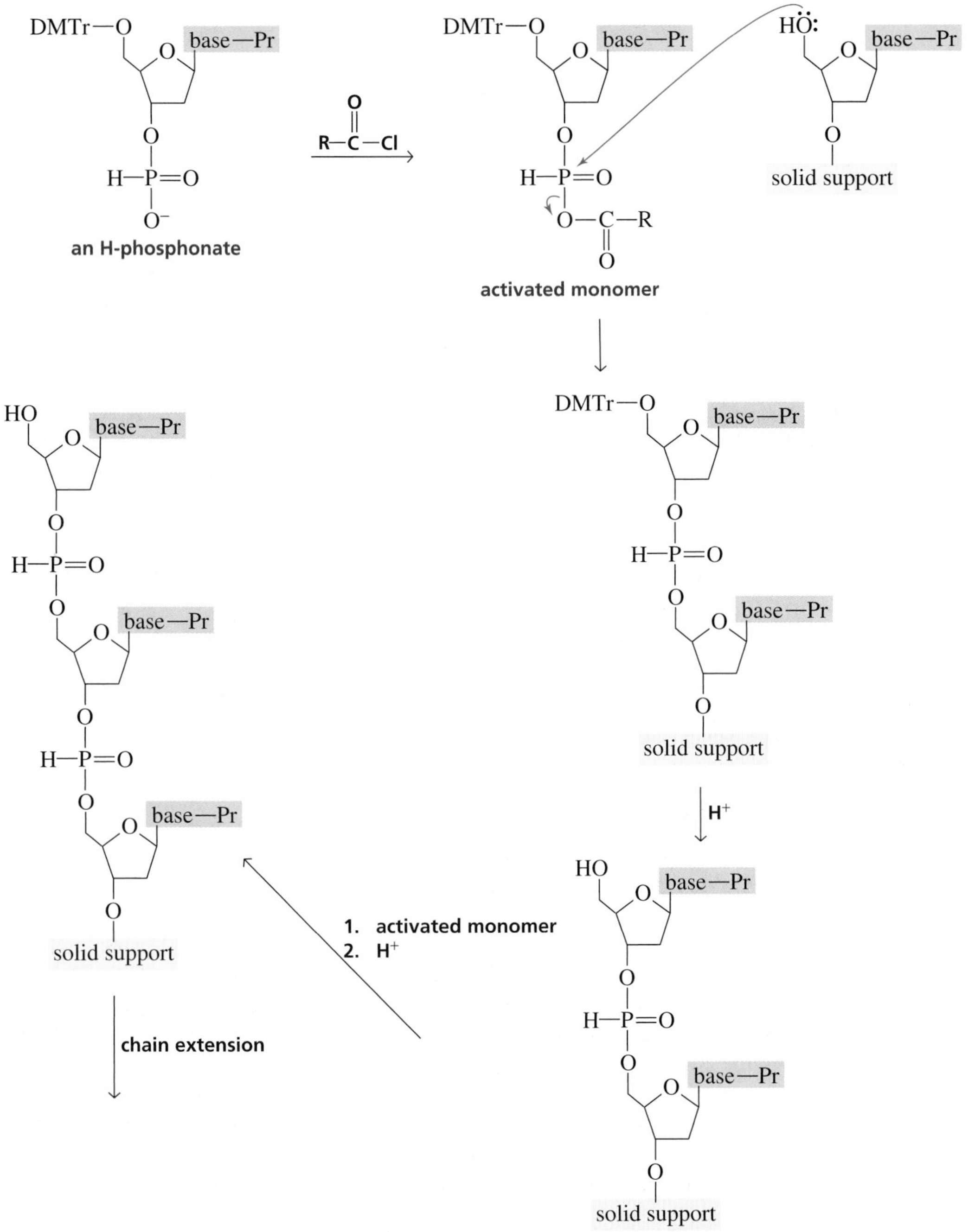

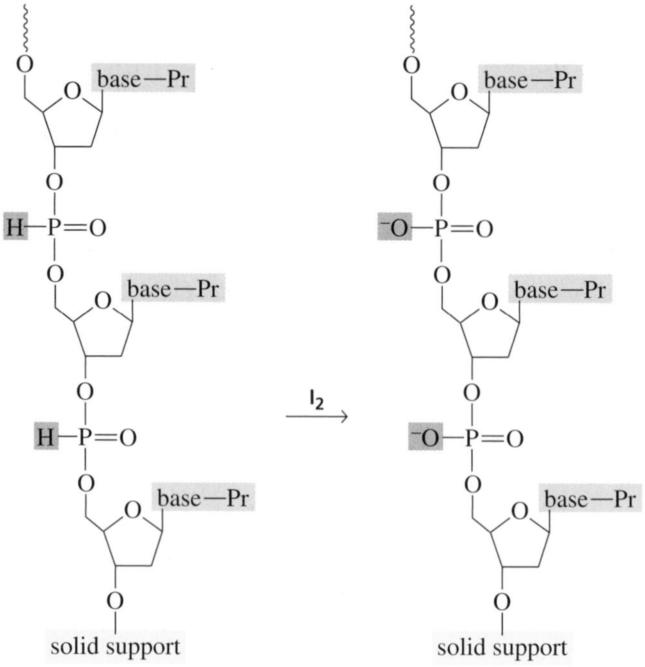

Notice that in the phosphoramidite method an oxidation is done each time a monomer is added, whereas in the H-phosphonate method a single oxidation is carried out after the entire strand has been synthesized.

PROBLEM 26

Propose a mechanism for removal of the DMTr protecting group by treatment with acid.

**25.17
RATIONAL DRUG
DESIGN**

Certain diseases such as acquired immunodeficiency syndrome (AIDS) and herpes are caused by **retroviruses.** The genetic information of a retrovirus is contained in its RNA. The retrovirus uses the sequence of bases in RNA as a template to synthesize DNA. It is called a retrovirus because its genetic information flows from RNA to DNA instead of the more typical flow from DNA to RNA.

Drugs that interfere with the synthesis of DNA by retroviruses have been designed and developed. If the retrovirus cannot synthesize DNA, it cannot take over the genetic machinery of the cell to produce more retroviral RNA and retroviral proteins. Designing drugs with particular structures to achieve specific purposes is called **rational drug design.** AZT is perhaps the best known of the drugs designed to interfere with retroviral DNA synthesis. AZT is taken up by the T lymphocytes, cells that are particularly susceptible to human immunodeficiency virus (HIV), the retrovirus that causes AIDS. Once in the cell, AZT is converted to AZT–triphosphate. The retroviral enzyme (reverse transcriptase) that catalyzes DNA formation from RNA binds AZT–triphosphate more tightly than it binds dTTP. Therefore, AZT rather than T is added to the growing DNA chain. Because AZT does not have a 3'-OH group, no additional nucleotides can be added to the chain and DNA synthesis comes to a sudden halt. Fortunately, the concentration of AZT required to affect reverse transcription is too low to affect most cellular DNA replication. A newer drug, 2',3'-dideoxyinosine (ddI), has a similar mechanism of action.

3-D Molecule:
AZT

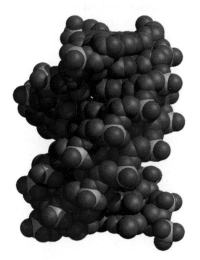

3′-azido-2′-deoxythymidine
AZT

2′,3′-dideoxyinosine
ddI

Because viruses, bacteria, and cancer cells all require DNA to grow and reproduce, chemists are trying to design compounds that will bind to the DNA of invading organisms and interfere with their reproduction. Chemists are also attempting to design compounds that will bind to specific sequences of human DNA. Such compounds could disrupt the expression of a gene (interfere with its transcription into RNA). For example, there is hope that compounds can be designed that will interfere with the expression of genes that contribute to the development of cancer. Polymers that bind to DNA are called **antigene agents;** those that bind to mRNA are called **antisense agents.**

For a compound to target a particular gene, the compound must be able to recognize a specific sequence of 15 to 20 bases. A sequence that long might occur only once in the human genome, so the compound would be specific for a particular site on DNA. In contrast, if the compound recognizes a sequence of only six bases, it could affect the human genome at more than a million locations because that sequence could occur once in every 2000 bases. However, since only 10% of the genes are expressed in most cells, a compound that recognizes a specific sequence of 10 to 12 bases may confer a gene-specific effect.

One approach to **site-specific recognition** uses a synthetic strand of oligonucleotides. When a strand of oligonucleotides is added to natural double-stranded DNA, the strand wraps around the DNA, forming a triple helix (Figure 25.18). The hope is that if the DNA sequence of a particular gene is known, a deoxyribonucleotide can be synthesized that will bind to that gene.

3-D Molecule:
Triple helix

◀ **Figure 25.18**
A triple helix. A synthetic strand of oligonucleotides (purple atoms) is wrapped around double-stranded DNA.

A triple helix is formed through Hoogsteen base pairing between the existing base pairs in DNA and bases in the third synthetic strand. In **Hoogsteen base pairing,** a T in the synthetic strand binds to an A of an A — T base pair, while a protonated cytosine in the synthetic strand binds to a G of a G—C base pair (Figure 25.19). Thus, oligonucleotides can be prepared with sequences that will base-pair to the sense strand of DNA at the desired location. Several methods are being investigated that will cut out the piece of targeted DNA after the synthetic strand binds to the double helix.

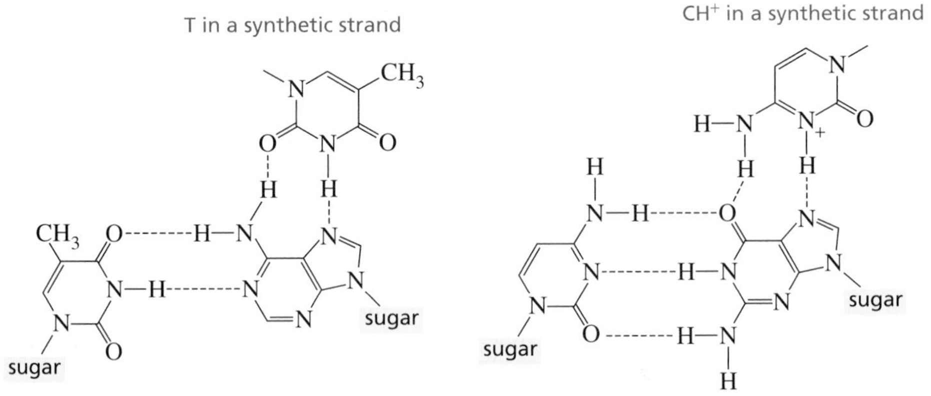

Figure 25.19 ▶
Hoogsteen base pairing: a T in a synthetic strand of oligonucleotides binds to the A of an A—T base pair in double-stranded DNA; a protonated C (CH^+) in the synthetic strand binds to the G of a G—C base pair in DNA.

T in a synthetic strand

CH^+ in a synthetic strand

thymine–adenine base pair

cytosine–guanine base pair

A problem with synthetic oligonucleotides being used to target DNA is that the synthetic strands are susceptible to enzymes, such as restriction endonucleases, that catalyze the hydrolysis of DNA. Consequently, other polymers are being designed that will recognize specific DNA sequences but will not be enzymatically hydrolyzed. A compound that has shown some promise is a polymer of phosphorothioates. The polymer differs from DNA in that the negatively charged oxygen bonded to the phosphorus is replaced by a negatively charged sulfur (Figure 25.20). The polymer binds to both DNA and RNA with complementary base pairing, and it also has been found to bind to certain proteins. Oligonucleotide phosphorothioates consisting of various lengths of deoxycytidine residues have recently been found to be effective in protecting T cells from the HIV virus.

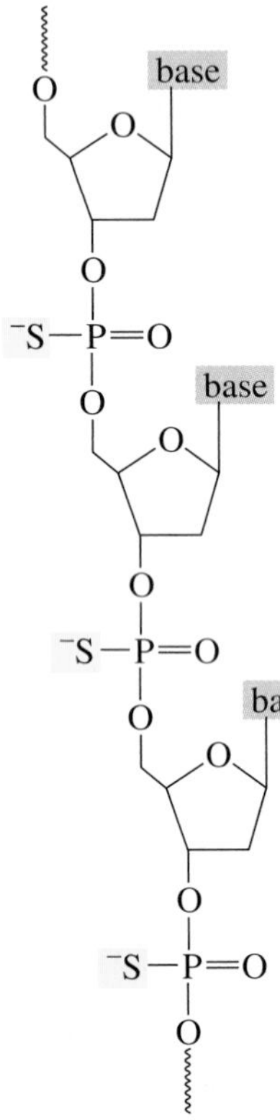

▲ **Figure 25.20**
A synthetic oligonucleotide with negatively charged sulfurs in place of negatively charged oxygens.

PROBLEM 27

5-Bromouracil, a highly mutagenic compound, is used in cancer chemotherapy. When administered to a patient, it is converted to the triphosphate and incorporated into DNA in place of thymine, which it resembles sterically. Why does it cause mutations? (*Hint*: The bromo substituent increases the stability of the enol tautomer.)

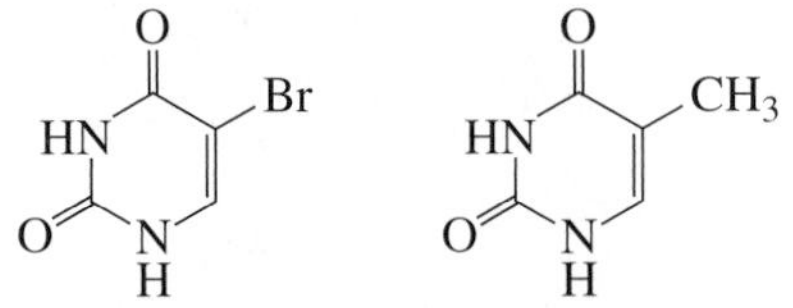

5-bromouracil

thymine

---

gene therapy (page 1093)
genetic code (page 1085)
high-energy bond (page 1072)
Hoogsteen base pairing (page 1099)
human genome (page 1081)
informational strand (page 1082)
in-line displacement mechanism (page 1068)
intron (page 1082)
major groove (page 1079)
minor groove (page 1079)
nucleic acid (page 1075)
nucleoside (page 1065)
nucleotide (page 1065)
oligonucleotide (page 1075)
phosphoanhydride bond (page 1068)
phosphoryl transfer reaction (page 1069)
polynucleotide (page 1075)
primary structure (page 1075)
prokaryotic organism (page 1083)
promoter site (page 1082)
rational drug design (page 1098)
replication (page 1080)
replication fork (page 1081)
restriction endonuclease (page 1091)
restriction fragment (page 1091)
retrovirus (page 1098)
ribonucleic acid (RNA) (page 1063)
ribonucleotide (page 1065)
ribosome (page 1083)
ribozyme (page 1083)
RNA splicing (page 1083)
sedimentation constant (page 1083)
semiconservative replication (page 1081)
sense strand (page 1082)
site-specific recognition (page 1099)
stacking interactions (page 1079)
stop codon (page 1086)
template strand (page 1082)
transcription (page 1082)
translation (page 1086)

PROBLEMS

28. Name the following compounds:

a.

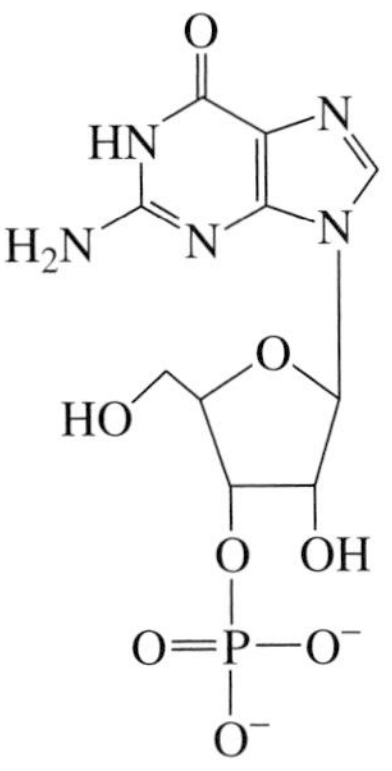

c.

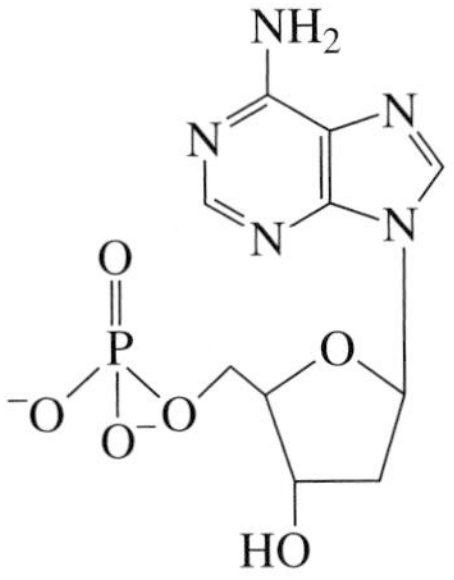

b.

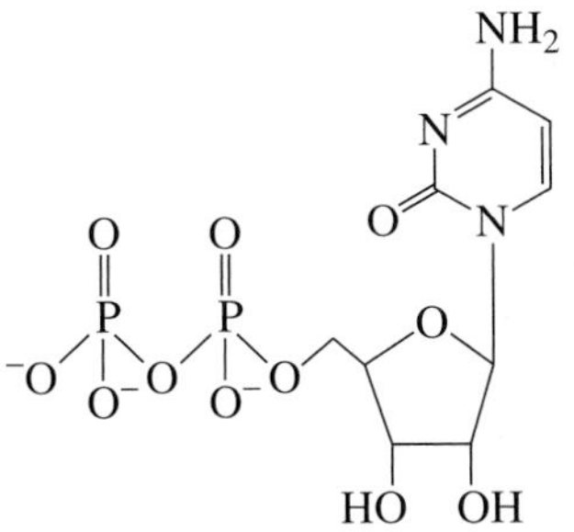

d.

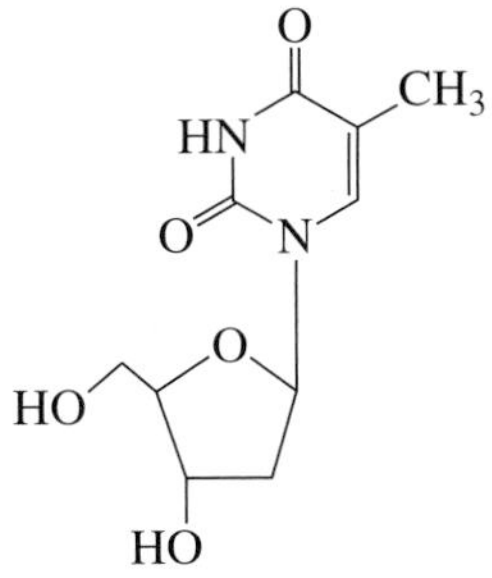

29. What nonapeptide is coded for by the following piece of mRNA?

5′—AAA—GUU—GGC—UAC—CCC—GGA—AUG—GUG—GUC—3′

30. What would be the base sequence of the segment of DNA that is responsible for the biosynthesis of the following hexapeptide?

Gly-Ser-Arg-Val-His-Glu

31. Propose a mechanism for the following reaction:

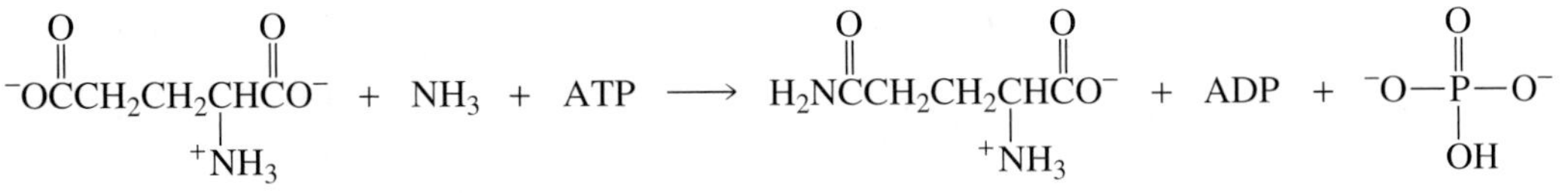

32. Match the codon with the anticodon:

codon	anticodon
AAA	ACC
GCA	CCU
CUU	UUU
AGG	AGG
CCU	UGA
GGU	AAG
UCA	GUC
GAC	UGC

33. Using the single-letter abbreviations for the amino acids in Table 21.1, write the sequence of amino acids in a polypeptide represented by the first four letters in your first name. Do not use any letter twice. (Because not all letters are assigned to amino acids, you might have to use one or two letters in your last name.) Write the sequence of bases in mRNA that would result in the synthesis of that polypeptide. Write the sequence of bases in the sense strand of DNA that would result in formation of the appropriate mRNA.

34. Which of the following pairs of dinucleotides are present in equal amounts in DNA?
a. CC and GG
b. CG and GT
c. CA and TG
d. CG and AT
e. GT and CA
f. TA and AT

35. Why is the codon a triplet rather than a doublet or a quartet?

36. RNAase, the enzyme that catalyzes the hydrolysis of RNA, has two catalytically active histidine residues at its active site. One of the histidine residues is catalytically active in its acidic form and the other is catalytically active in its basic form. Propose a mechanism for RNAase.

37. The amino acid sequences of peptide fragments obtained from a normal protein and from the same protein synthesized by a defective gene were compared. They were found to differ in only one peptide fragment. The primary sequences of the fragments are shown here.

Normal: Gln-Tyr-Gly-Thr-Arg-Tyr-Val

Mutant: Gln-Ser-Glu-Pro-Gly-Thr

a. What is the defect in DNA?
b. It was later determined that the normal peptide fragment is an octapeptide with a C-terminal Val-Leu. What is the C-terminal amino acid of the mutant peptide?

38. Whether the mechanism requiring activation of a carboxylate ion by ATP involves attack of the carboxylate ion on the α-phosphorus or the β-phosphorus of ATP cannot be determined from the reaction products because AMP and pyrophosphate are obtained as products in both mechanisms. The mechanisms, however, can be distinguished by a labeling experiment in which the enzyme, the carboxylate ion, ATP, and radioactively labeled pyrophosphate are incubated, and the ATP is isolated. If the isolated ATP is radioactive, the mechanism involves attack on the α-phosphorus. If it is not radioactive, the mechanism involves attack on the β-phosphorus. Explain these conclusions.

39. What would be the results of the experiment in Problem 38 if radioactive AMP were added to the incubation mixture instead of radioactive pyrophosphate?

40. Which cytosine in the following sense strand of DNA could cause the most damage to the organism if it were deaminated?

5′—A—T—G—T—C—G—C—T—A—A—T—C—3′

41. Sodium nitrite, a common food preservative (page 660), is capable of causing mutations in an acidic environment by converting cytosines to uracils. Explain how this occurs.

42. The first amino acid incorporated into a polypeptide chain during its biosynthesis in prokaryotes is *N*-formylmethionine. Explain the purpose of the formyl group.

Special Topics in Organic Chemistry

You have studied polymers synthesized by biological systems—proteins, carbohydrates, and nucleic acids. **Chapter 26** discusses polymers synthesized by chemists—synthetic polymers. These polymers have physical properties that make them useful in everyday life—in applications such as beverage bottles, clothing, food wrap, and compact discs.

Chapter 27 introduces the reactions of heterocyclic compounds—cyclic compounds in which one of the ring atoms is a heteroatom (an atom other than carbon). Many natural products—such as caffeine, nicotine, chlorophyll, and serotonin—and most drugs are heterocyclic compounds.

Chapter 28 discusses pericyclic reactions—reactions that occur as a result of a cyclic reorganization of electrons. In this chapter you will learn how to use the conservation of orbital symmetry theory to explain the relationships among reactant, product, and reaction conditions in a pericyclic reaction.

Throughout this text you have designed a wide variety of multistep organic syntheses. In **Chapter 29** we will explore in greater depth how organic chemists approach a synthetic problem. A few examples of some complicated syntheses are examined step by step so that you can gain an understanding of the power of organic synthesis and some of the ways in which chemists go about solving difficult problems.

Chapter 30 introduces you to medicinal chemistry. Here you will see how many of our commonly used drugs were discovered, and you will learn about some of the techniques used to develop new drugs.

26

Synthetic Polymers

Super Glue

Probably no group of synthetic compounds is more important to modern life than synthetic polymers. A **polymer** is a large molecule made by linking together repeating units of small molecules called **monomers.** The process of linking them up is called **polymerization.**

$$nM \xrightarrow{\text{polymerization}} -M-M-M-M-M-M-M-M-M-$$

monomer   **polymer**

ethylene monomers   **polyethylene**

Unlike small organic molecules that are of interest because of their chemical properties, these giant molecules are interesting because of their physical properties that make them useful in everyday life. Some synthetic polymers resemble natural substances, but most are materials that are quite different from those found in nature. Such diverse items as photographic film, records, compact discs, rugs, food wrap, artificial joints, Super Glue, toys, beverage bottles, weather stripping, automobile body parts, shoe soles, and condoms are made of synthetic polymers.

Polymers can be divided into two broad groups—**synthetic polymers** and **biopolymers** (natural polymers). Synthetic polymers are synthesized by scientists, whereas biopolymers are synthesized by organisms and are essential for our very existence. Examples of biopolymers are DNA, the storage molecule for genetic information—the molecule that determines whether a fertilized egg becomes a human or a honeybee; RNA and proteins, the molecules by which biochemical transformations are made to occur; and polysaccharides. Their structures and properties are presented in other chapters. In this chapter, we will explore synthetic polymers.

Humans first relied on *natural polymers* for clothing, wrapping themselves in animal skins and furs. Later, they learned to spin natural fibers into thread and to weave the thread into cloth. Now much of our clothing is made of *synthetic polymers* (nylon, polyester, polyacrylonitrile). Many people prefer clothing made of natural polymers (cotton, wool, silk), but it has been estimated that if synthetic polymers were not available, all the arable land in the United States would have to be used for the production of cotton and wool for clothing.

The first plastic—a polymer capable of being molded—was celluloid. Invented in 1856 by Alexander Parke, it was a mixture of nitrocellulose and camphor. Celluloid was used in the manufacture of billiard balls and piano keys, replacing scarce ivory. This discovery provided a reprieve for many elephants but caused some moments of consternation in billiard parlors because nitrocellulose is flammable and explosive. Celluloid was used for motion picture film until it was replaced by cellulose acetate, a less dangerous polymer.

The first synthetic fiber was rayon. In 1865, the French silk industry was threatened by an epidemic that killed many silkworms, highlighting the need for an artificial silk substitute. Louis Chardonnet accidentally discovered the starting material for a synthetic fiber when, while wiping up some spilled nitrocellulose from a table, he noticed long, silk-like strands adhering to both the cloth and the table. "Chardonnet silk" was introduced at the Paris Exposition in 1891. It was called rayon because it was so shiny that it appeared to give off rays of light.

The first synthetic rubber was synthesized by German chemists in 1917. Their efforts were in response to a severe shortage of raw materials—a result of blockading during World War I.

Hermann Staudinger was the first to recognize that the various polymers being produced were not disorderly conglomerates of monomers but rather were made up of chains of monomers joined together. Today, the synthesis of polymers has grown from a process carried out with little chemical understanding to a sophisticated science in which molecules are engineered with predetermined specifications in order to produce new materials tailored to fit human needs. New polymers that make new products possible are constantly being designed. Recent examples include Lycra, a fabric with elastic properties, and Dyneema, the strongest fabric commercially available.

Polymer chemistry is part of the larger discipline of **materials science,** which involves creation of new materials to replace metals, glass, ceramics, wood, cardboard, and paper. Currently there are approximately 30,000 patented polymers in the United States.

Alexander Parke (1813–1890) was born in Birmingham, England. He called the polymer that he invented "pyroxylin," but was unable to market it.

*The inventor **John Wesley Hyatt (1837–1920)** was born in New York. Because a New York firm was offering a prize of $10,000 for a substitute for ivory billiard balls, Hyatt improved the synthesis of pyroxylin. He changed its name to "celluloid" and patented a method for making billiard balls. He, however, did not win the prize.*

Louis-Marie-Hilaire Bernigaud, Comte de Chardonnet (1839–1924) was born in France. In the early stages of his career, he was an assistant to Louis Pasteur. Because the rayon he initially produced was made from nitrocellulose, it was dangerously flammable. Eventually, chemists learned to remove some of the nitro groups after the fiber was formed, which made the fiber much less flammable but not as strong.

Hermann Staudinger (1881–1965), the son of a professor, was born in Germany. He became a professor at the Technical Institute of Karlsruhe and at the University of Freiburg. He received the Nobel Prize in chemistry in 1953 for his contributions to polymer chemistry.

26.1 GENERAL CLASSES OF SYNTHETIC POLYMERS

Synthetic polymers can be divided into two major classes depending on their method of preparation. **Chain-growth polymers** (also known as **addition polymers**) are made by the addition of monomers to the end of a growing chain. The end of the chain is reactive because it is a radical, a cation, or an anion. Polystyrene—used for disposable food containers, insulation, "beanbag" pillows (Styrofoam), and toothbrush handles—is an example of a chain-growth polymer.

3-D Molecules:
Styrene;
Polystyrene

Step-growth polymers (also called **condensation polymers**) are made by combining two molecules while, in most cases, removing a small molecule—generally water or an alcohol. The reacting molecules have reactive functional groups at each end. Unlike chain-growth polymerization, which requires the individual molecules to add to the end of a growing chain, any two reactive molecules can combine in step-growth polymerization. Dacron is an example of a step-growth polymer.

Dacron is the most common of the group of polymers known as **polyesters**—polymers with many ester groups. Polyesters are used for clothing and are responsible for the wrinkle-resistant behavior of many fabrics. Polyester is also used to make the plastic film needed for making magnetic recording tape, which is known by the trade name Mylar. This film is tear-resistant and, when processed, has a tensile strength nearly as great as that of steel. The polymer used to make soft drink bottles is also a polyester.

26.2 CHAIN-GROWTH POLYMERS

The monomers used most commonly in chain-growth polymerization are ethylene and substituted ethylenes. In the chemical industry, monosubstituted ethylenes are known as **alpha olefins.** Polymers formed from ethylene or substituted ethylenes are called **vinyl polymers.** Some of the many vinyl polymers synthesized by chain-growth polymerization are listed in Table 26.1.

When plastics are recycled, the various types must be separated from one another. To aid in the separation, many states require manufacturers to include a recycling symbol on their products to indicate the type of plastic it is. You are probably familiar with these symbols, often found on the bottom of plastic containers, which consist of three arrows around one of seven numbers. The lower the number in the middle of the symbol, the greater the ease with which the material can be recycled. The abbreviation below the symbol indicates the type of polymer from which the container is made: 1 for poly(ethylene terephthalate) (PET), 2 for high-density polyethylene (HDPE), 3 for poly(vinyl chloride) (V), 4 for low-density polyethylene (LDPE), 5 for polypropylene (PP), 6 for polystyrene (PS), and 7 for all other plastics.

Chain-growth polymerization proceeds by one of three possible mechanisms: **radical polymerization, cationic polymerization,** or **anionic polymerization.** Each mechanism has three distinct phases: an *initiation step* that starts the polymerization, *propagation steps* that allow the chain to grow, and *termination steps* that stop

TABLE 26.1   Some Important Chain-Growth Polymers and Their Uses

Monomer	Repeating Unit	Polymer Name	Uses
$CH_2{=}CH_2$	$-CH_2-CH_2-$	polyethylene	film, toys, bottles, plastic bags
$CH_2{=}CH$ $\quad\ \ \|$ $\quad\ \ Cl$	$-CH_2-CH-$ $\qquad\ \|$ $\qquad\ Cl$	poly(vinyl chloride)	"squeeze" bottles, pipe, siding, flooring
$CH_2{=}CH-CH_3$	$-CH_2-CH-$ $\qquad\ \|$ $\qquad\ CH_3$	polypropylene	molded caps, margarine tubs, indoor/outdoor carpeting, upholstery
$CH_2{=}CH$ $\qquad\ $	$-CH_2-CH-$ $\qquad\ $	polystyrene	packaging, toys, clear cups, egg cartons, hot drink cups
$CF_2{=}CF_2$	$-CF_2-CF_2-$	poly(tetrafluoroethylene) Teflon	nonsticking surfaces, liners, cable insulation
$CH_2{=}CH$ $\qquad\ \|$ $\qquad\ C{\equiv}N$	$-CH_2-CH-$ $\qquad\ \|$ $\qquad\ C{\equiv}N$	poly(acrylonitrile) Orlon, Acrilan	rugs, blankets, yarn, apparel, simulated fur
$CH_2{=}C-CH_3$ $\qquad\ \|$ $\qquad\ COCH_3$ $\qquad\ \ \|$ $\qquad\ \ O$	$\quad\quad\ CH_3$ $\qquad\ \|$ $-CH_2-C-$ $\qquad\ \|$ $\qquad\ COCH_3$ $\qquad\ \ \|$ $\qquad\ \ O$	poly(methyl methacrylate) Plexiglas, Lucite	lighting fixtures, signs, solar panels, skylights
$CH_2{=}CH$ $\qquad\ \|$ $\qquad\ OCCH_3$ $\qquad\ \ \|$ $\qquad\ \ O$	$-CH_2-CH-$ $\qquad\ \|$ $\qquad\ OCCH_3$ $\qquad\ \ \|$ $\qquad\ \ O$	poly(vinyl acetate)	latex paints, adhesives

the growth of the chain. We will see that the choice of mechanism depends on the structure of the monomer *and* the initiator used to activate the monomer.

Radical Polymerization

For chain-growth polymerization to occur by a radical mechanism, a radical initiator must be added to the monomer to convert some of the monomer molecules into radicals. The initiator breaks homolytically into radicals and each radical adds to an alkene monomer, converting it into a radical. This radical reacts with another monomer, adding a new subunit that propagates the chain. The radical site is now at the end of the most recent unit added to the end of the chain. This is called the **propagating site.**

chain-initiating steps

$$RO\!-\!OR \xrightarrow{\Delta} 2\ RO\cdot$$

a radical initiator   radicals

$$RO\cdot \ +\ CH_2\!=\!CH \longrightarrow ROCH_2\dot{C}H$$
$$\underset{Z}{|} \qquad\qquad \underset{Z}{|}$$

chain-propagating steps

propagating sites

$$ROCH_2\dot{C}H\ +\ CH_2\!=\!CH \longrightarrow ROCH_2CHCH_2\dot{C}H$$
$$\underset{Z}{|} \qquad \underset{Z}{|} \qquad\qquad \underset{Z}{|}\ \ \underset{Z}{|}$$

$$ROCH_2CHCH_2\dot{C}H\ +\ CH_2\!=\!CH \longrightarrow ROCH_2CHCH_2CHCH_2\dot{C}H \xrightarrow{\text{etc.}}$$
$$\underset{Z}{|}\ \ \underset{Z}{|} \qquad \underset{Z}{|} \qquad\qquad \underset{Z}{|}\ \ \underset{Z}{|}\ \ \underset{Z}{|}$$

This process is repeated over and over. Hundreds or even thousands of alkene monomers can add, one at a time, to the growing chain. Eventually the chain reaction stops because the propagating sites are destroyed. Propagating sites can be destroyed when two chains combine at their propagating sites; when two chains undergo disproportionation, with one chain being oxidized to an alkene and the other being reduced to an alkane; or when a chain reacts with an impurity that consumes the radical.

methods to terminate the chain

chain combination

$$2\ RO\!\!\left[CH_2CH\right]_n\!\!CH_2\dot{C}H \longrightarrow RO\!\!\left[CH_2CH\right]_n\!\!CH_2CHCHCH_2\!\!\left[CHCH_2\right]_n\!\!OR$$

disproportionation

$$2\ RO\!\!\left[CH_2CH\right]_n\!\!CH_2\dot{C}H \longrightarrow RO\!\!\left[CH_2CH\right]_n\!\!CH\!=\!CH\ +\ RO\!\!\left[CH_2CH\right]_n\!\!CH_2CH_2$$

reaction with an impurity

$$RO\!\!\left[CH_2CH\right]_n\!\!CH_2\dot{C}H\ +\ \text{impurity} \longrightarrow RO\!\!\left[CH_2CH\right]_n\!\!CH_2CH\!-\!\text{impurity}$$

Thus, *radical polymerizations* have chain-initiating, chain-propagating, and chain-terminating steps similar to the radical reactions discussed in Sections 3.18 and 8.2.

As long as the polymer has a high molecular weight, the groups at the ends of the polymer are relatively unimportant in determining the physical properties of the polymer and are generally not even specified. It is the rest of the molecule that determines the properties of the polymer.

The molecular weight of the polymer can be controlled by a process known as **chain transfer.** In chain transfer, the growing chain reacts with a molecule XY in a manner that allows X· to terminate the chain, leaving behind Y· to initiate a new chain. XY can be a solvent, a radical initiator, or any molecule with a bond that can be cleaved homolytically.

$$-CH_2 \left[CH_2CH \right]_n CH_2\dot{C}H + XY \longrightarrow -CH_2 \left[CH_2CH \right]_n CH_2CHX + Y·$$
$$\qquad\qquad\quad Z \qquad\quad Z \qquad\qquad\qquad\qquad\quad Z \qquad\quad Z$$

Chain-growth polymerization of monosubstituted ethylenes exhibits a marked preference for **head-to-tail addition,** where the head of one monomer is attached to the tail of another.

| tail | head |

$$CH_2{=}CH$$
$$\qquad\quad Z$$

$$-CH_2CHCH_2CH- \qquad -CH_2CHCHCH_2- \qquad -CHCH_2CH_2CH-$$
$$\quad\; Z \qquad\quad Z \qquad\qquad\quad Z \;\; Z \qquad\qquad\quad Z \qquad\qquad Z$$
head-to-tail   **head-to-head**   **tail-to-tail**

Head-to-tail addition of a substituted ethylene results in a polymer in which every other carbon bears a substituent.

$$CH_2{=}CH \longrightarrow -CH_2CHCH_2CHCH_2CHCH_2CHCH_2CHCH_2CH-$$
$$\qquad\; Cl \qquad\qquad\qquad Cl \quad\;\; Cl \quad\;\; Cl \quad\;\; Cl \quad\;\; Cl \quad\;\; Cl$$
poly(vinyl chloride)

3-D Molecules:
Vinyl chloride;
Poly(vinyl chloride)

Head-to-tail addition is preferred for steric reasons because the propagating site preferentially attacks the less sterically hindered unsubstituted sp^2 carbon of the alkene. Groups that stabilize radicals also favor head-to-tail addition. For example, when Z is a phenyl substituent, the benzene ring stabilizes the radical by electron delocalization, so the propagating site is the carbon that bears the phenyl substituent.

$$-CH_2\dot{C}H \quad\longleftrightarrow\quad -CH_2CH \quad\longleftrightarrow\quad -CH_2CH \quad\longleftrightarrow\quad -CH_2CH \quad\longleftrightarrow\quad -CH_2\dot{C}H$$

In cases where Z is small—which makes steric considerations less important—and is less able to stabilize the growing end of the chain by electron delocalization, some head-to-head addition and some tail-to-tail addition also occur. This has been observed primarily in situations where Z = F. Abnormal addition, however, has never been found to constitute more than 10% of the overall chain.

Monomers that most readily undergo chain-growth polymerization by a radical mechanism are those in which the substituent (Z) is able to stabilize the growing radical species by electron delocalization. Examples of monomers that undergo radical polymerization are shown in Table 26.2.

TABLE 26.2 Examples of Alkenes That Undergo Radical Polymerization

$CH_2\!\!=\!\!CH$ styrene	$CH_2\!\!=\!\!CH$ $\underset{\parallel}{\underset{O}{OCCH_3}}$ vinyl acetate	$CH_2\!\!=\!\!CCH_3$ $\underset{\parallel}{\underset{O}{COCH_3}}$ methyl methacrylate
$CH_2\!\!=\!\!CH$ Cl vinyl chloride	$CH_2\!\!=\!\!CH$ $C\!\!\equiv\!\!N$ acrylonitrile	$CH_2\!\!=\!\!CH$ $CH\!\!=\!\!CH_2$ 1,3-butadiene

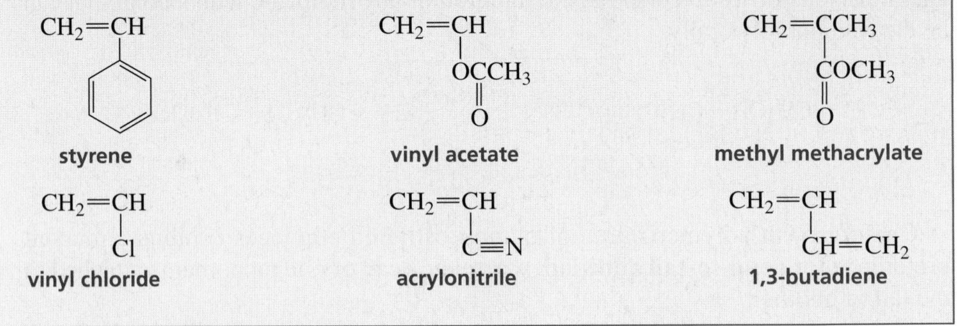

Any compound that readily undergoes homolytic cleavage to form radicals that are sufficiently energetic to convert an alkene into a radical can serve as a radical initiator for radical polymerization. Several radical initiators are shown in Table 26.3.

TABLE 26.3 Some Radical Initiators

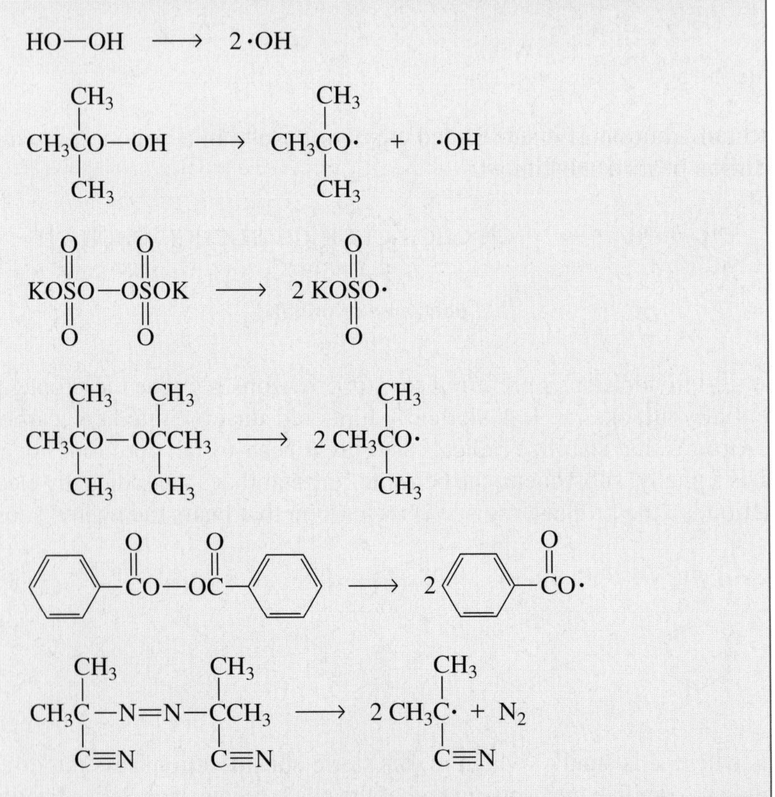

$$HO\!-\!OH \longrightarrow 2\,\cdot OH$$

$$\underset{CH_3}{\overset{CH_3}{CH_3CO\!-\!OH}} \longrightarrow \underset{CH_3}{\overset{CH_3}{CH_3CO\cdot}} + \ \cdot OH$$

$$\underset{\underset{O}{\parallel}}{\overset{\overset{O}{\parallel}}{KOSO\!-\!OSOK}} \longrightarrow 2\ \underset{\underset{O}{\parallel}}{\overset{\overset{O}{\parallel}}{KOSO\cdot}}$$

$$\underset{CH_3\quad CH_3}{\overset{CH_3\quad CH_3}{CH_3CO\!-\!OCCH_3}} \longrightarrow 2\ \underset{CH_3}{\overset{CH_3}{CH_3CO\cdot}}$$

$$\text{Ph}\!-\!\underset{\overset{O}{\parallel}}{C}O\!-\!O\underset{\overset{O}{\parallel}}{C}\!-\!\text{Ph} \longrightarrow 2\ \text{Ph}\!-\!\underset{\overset{O}{\parallel}}{C}O\cdot$$

$$\underset{C\equiv N \quad C\equiv N}{\overset{CH_3 \quad CH_3}{CH_3C\!-\!N\!\!=\!\!N\!-\!CCH_3}} \longrightarrow 2\ \underset{C\equiv N}{\overset{CH_3}{CH_3C\cdot}} + N_2$$

A common feature of all radical initiators is a relatively weak bond that readily undergoes homolytic cleavage. In all but one of the radical initiators shown in Table 26.3, the weak bond is an oxygen–oxygen bond. Two factors enter into the choice of radical initiator for a particular chain-growth polymerization. The first is the desired solubility property of the initiator. For example, potassium persulfate is often used if the initiator needs to be water soluble, whereas an initiator with several carbons is chosen if the initiator must be soluble in a nonpolar solvent. The second factor is the temperature at which the polymerization reaction is to be carried out. For example, a *tert*-butoxy radical is relatively stable, so an initiator that forms a *tert*-butoxy radical is used for polymerizations carried out at relatively high temperatures.

PROBLEM 1◆

What monomer would you use to form each of the following polymers?

a. 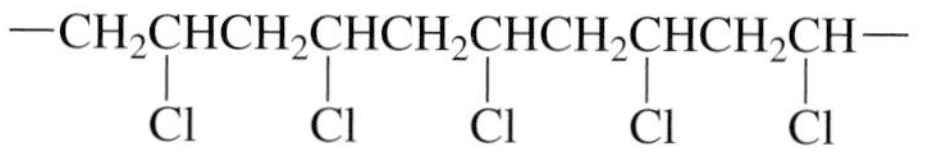 $-CH_2CHCH_2CHCH_2CHCH_2CHCH_2CH-$
with Cl substituents below each CH: Cl, Cl, Cl, Cl, Cl

b. 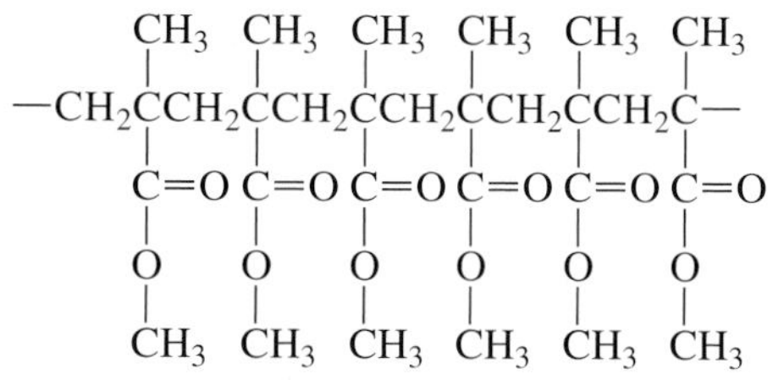
$$-CH_2CCH_2CCH_2CCH_2CCH_2C-$$
with CH_3 groups above each C, and $C=O$ with O and CH_3 below each C

c. $-CF_2CF_2CF_2CF_2CF_2CF_2CF_2CF_2CF_2CF_2-$

PROBLEM 2◆

Which polymer would be more apt to contain abnormal head-to-head linkages, poly(vinyl chloride) or polystyrene?

PROBLEM 3

Draw a segment of polystyrene that contains abnormal head-to-head and tail-to-tail linkages.

PROBLEM 4

Show the mechanism for the formation of a segment of poly(vinyl chloride) containing three units of vinyl chloride and initiated by hydrogen peroxide.

Branching of the Polymer Chain

If the propagating site abstracts a hydrogen atom from a chain, a branch can grow off the chain at that point.

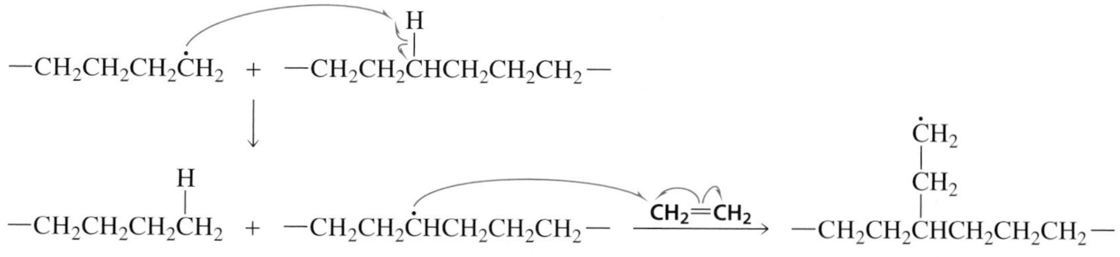

$$-CH_2CH_2CH_2\overset{\cdot}{C}H_2 \quad + \quad -CH_2CH_2\overset{H}{CHCH_2CH_2CH_2-}$$

$$\downarrow$$

$$-CH_2CH_2CH_2\overset{H}{CH_2} \quad + \quad -CH_2CH_2\overset{\cdot}{C}HCH_2CH_2CH_2- \quad \xrightarrow{CH_2=CH_2} \quad -CH_2CH_2CHCH_2CH_2CH_2-$$
with $\overset{\cdot}{C}H_2$ and CH_2 branch

Abstraction of a hydrogen atom from a carbon near the end of a chain leads to short branches, while abstraction of a hydrogen atom from a carbon near the middle of a chain results in long branches. Short branches are more likely to be formed than long ones.

chain with short branches   chain with long branches

Branching greatly affects the physical properties of the polymer. Linear, unbranched chains can pack together more closely than branched chains can. Consequently, linear polyethylene (known as high-density polyethylene) is a relatively hard plastic, used for the production of such things as artificial hip joints, while branched polyethylene (low-density polyethylene) is a much more flexible polymer, used for trash bags and dry cleaning bags.

PROBLEM 5◆

Polyethylene can be used for the production of beach chairs and beach balls. Which of these items is made from more highly branched polyethylene?

PROBLEM 6

Draw a short segment of branched polystyrene that shows the linkages at the branch point.

Cationic Polymerization

In cationic polymerization, the initiator is an electrophile that adds to the alkene, causing it to become a cation. The initiator most often used in cationic polymerization is a Lewis acid, such as BF_3 or $AlCl_3$. The advantage of such an initiator is that it does not have an accompanying nucleophile that could act as a chain terminator, as would be the case with a proton-donating acid such as HCl. The cation formed in the initiation step reacts with a second monomer, forming a new cation that reacts in turn with a third monomer. As each subsequent monomer adds to the chain, the positively charged propagating site always ends up on the last added unit.

chain-initiating step

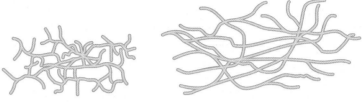

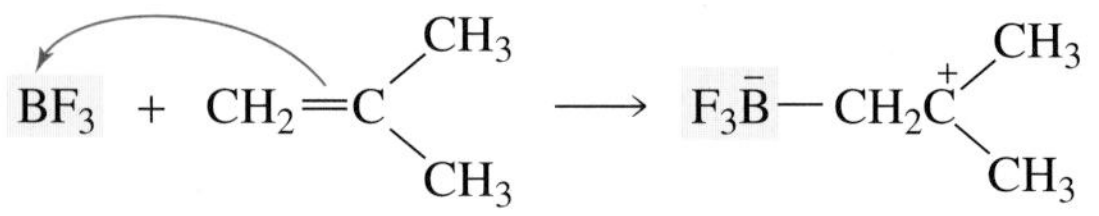

chain-propagating steps

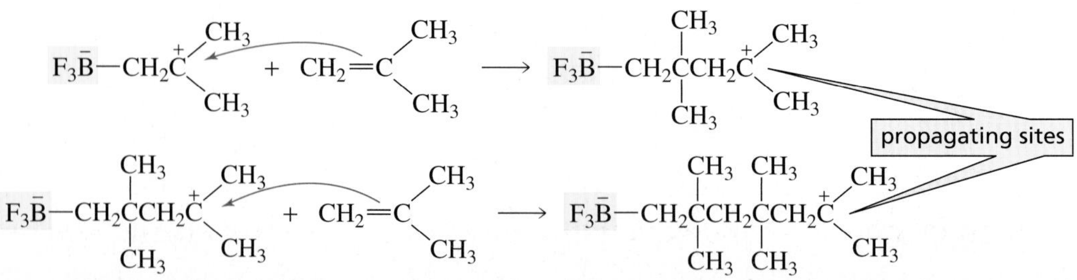

Cationic polymerization can be terminated by loss of a proton or by addition of a nucleophile that reacts with the propagating site. The chain can also be terminated by a chain transfer reaction with the solvent (XY).

chain-terminating steps

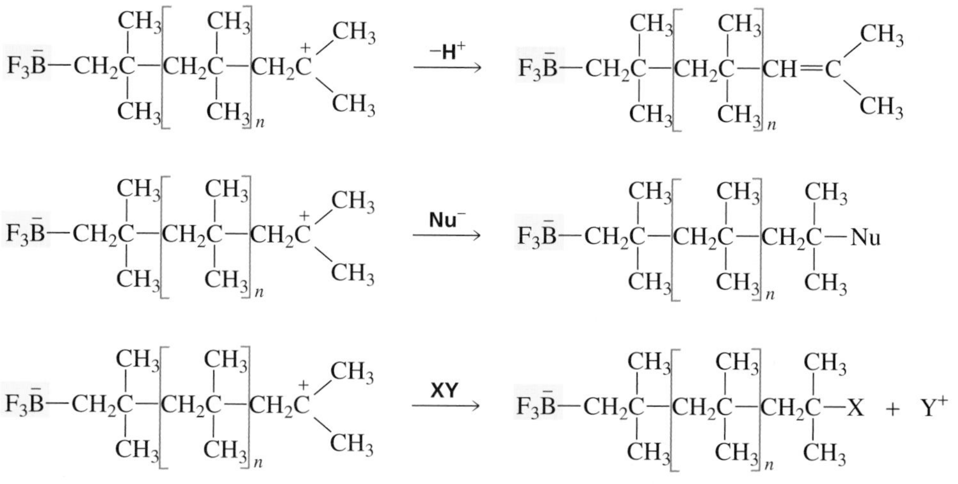

The carbocation intermediates formed during cationic polymerization, like any other carbocations, can undergo rearrangement by either a 1,2-hydride shift or a 1,2-methyl shift if rearrangement leads to a more stable carbocation (Section 3.14). For example, the polymer formed from cationic polymerization of 3-methyl-1-butene contains both unrearranged and rearranged units. The unrearranged propagating site is a secondary carbocation, while the rearranged propagating site—obtained by a 1,2-hydride shift—is a more stable tertiary carbocation. The extent of rearrangement depends on the reaction temperature.

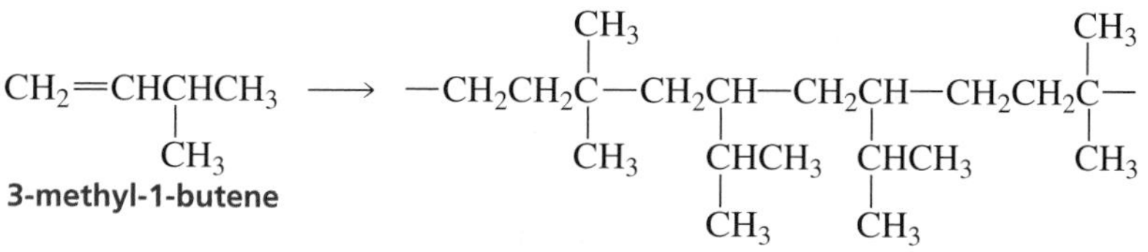

3-methyl-1-butene

unrearranged propagating site   rearranged propagating site

Monomers that are best able to undergo polymerization by a cationic mechanism are those with electron-donating substituents that can stabilize the positive charge at the propagating site. Examples of monomers that undergo cationic polymerization are shown in Table 26.4.

TABLE 26.4   Examples of Alkenes That Undergo Cationic Polymerization

$CH_2{=}CH$ $CH_2{=}CCH_3$ $CH_2{=}CH$ $CH_2{=}CH$

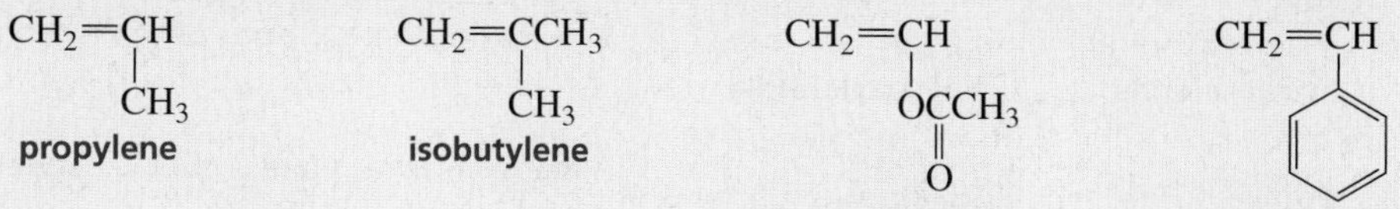

propylene   isobutylene   vinyl acetate   styrene

PROBLEM 7◆

List the following groups of monomers in order of decreasing ability to undergo cationic polymerization:

a. 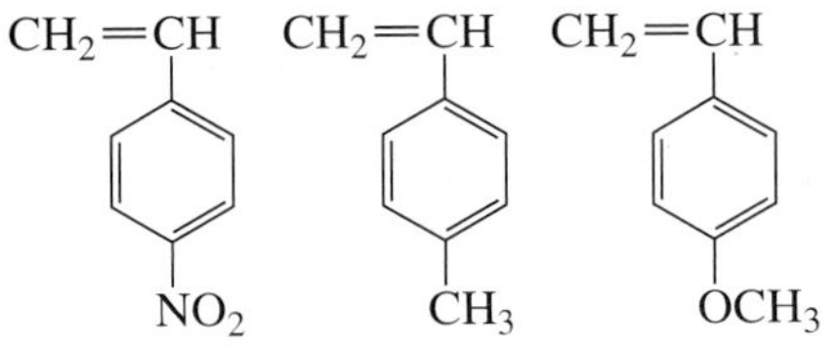 $CH_2\!=\!CH$ $CH_2\!=\!CH$ $CH_2\!=\!CH$

(three benzene rings bearing substituents: NO_2, CH_3, OCH_3)

b. $CH_2\!=\!CHCH_3$ $CH_2\!=\!CHOCCH_3$ $CH_2\!=\!CHCOCH_3$

                                       O   O

c. 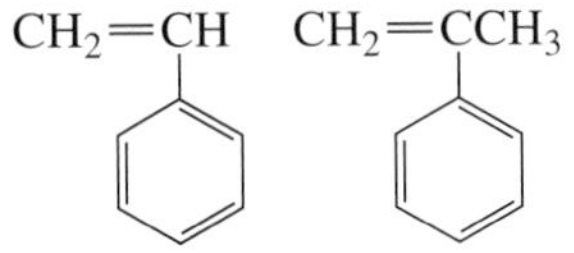 $CH_2\!=\!CH$ $CH_2\!=\!CCH_3$

(two benzene rings)

Anionic Polymerization

In anionic polymerization, the initiator is a nucleophile that reacts with the alkene to form a propagating site that is an anion. Nucleophilic attack on an alkene is not an easy reaction because alkenes are themselves electron rich. Therefore, the initiator must be a very good nucleophile such as sodium amide or butyllithium, and the alkene must contain an electron-withdrawing substituent to decrease its electron density. Some alkenes that undergo polymerization by an anionic mechanism are shown in Table 26.5.

chain-initiating step

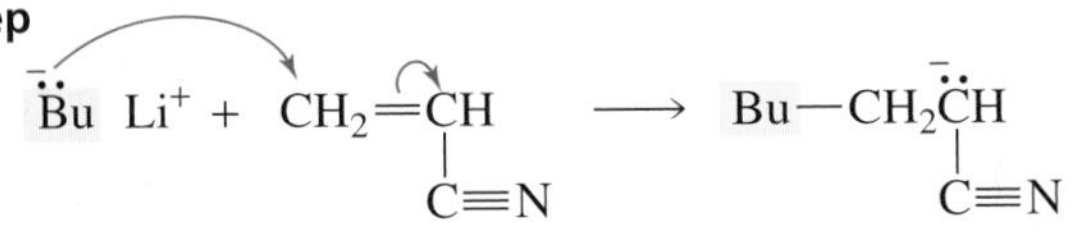

$$\ddot{B}u\ Li^+ + CH_2\!=\!CH \longrightarrow Bu\!-\!CH_2\ddot{C}H$$
$$\quad\quad\quad\quad\quad\quad | \quad\quad\quad\quad\quad\quad\quad |$$
$$\quad\quad\quad\quad\quad\quad C\!\equiv\!N \quad\quad\quad\quad\quad\quad C\!\equiv\!N$$

chain-propagating steps

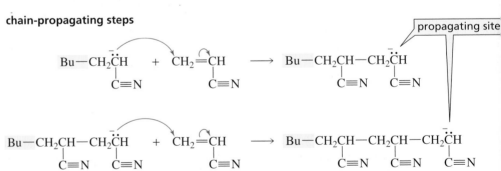

propagating site

TABLE 26.5 Examples of Alkenes That Undergo Anionic Polymerization

$CH_2\!=\!CH$	$CH_2\!=\!CH$	$CH_2\!=\!CCH_3$	$CH_2\!=\!CH$
$\|$	$\|$	$\|$	
Cl	$C\!\equiv\!N$	$COCH_3$	(benzene ring)
		$\|\|$	
		O	
vinyl chloride	**acrylonitrile**	**methyl methacrylate**	**styrene**

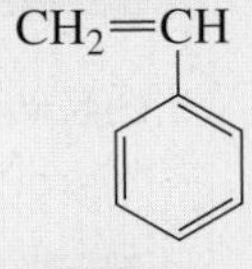

The chain can be terminated by a chain transfer reaction with the solvent or by reaction with an impurity in the reaction mixture. If the solvent cannot donate a proton to terminate the chain, and if all impurities that can react with a carbanion are rigorously excluded, chain propagation will continue until all the monomer has been consumed. At this point the propagating site will still be active, so the polymerization reaction will continue if more monomer is added to the system. Such nonterminated chains are called **living polymers** because the chains remain active until they are "killed." Living polymers usually result from anionic polymerization because the chains cannot be terminated by proton loss from the polymer as they can in cationic polymerization, or by disproportionation or radical recombination as they can in radical polymerization.

Super Glue is a polymer of methyl α-cyanoacrylate. Because the monomer has two electron-withdrawing groups, it requires only a moderately good nucleophile to initiate anionic polymerization. An OH group of cellulose or a nucleophilic group of a protein can act as an initiator. You may well have experienced this reaction if you have ever spilled a drop of Super Glue on your fingers. A nucleophilic group of the protein on the surface of the skin initiates the polymerization reaction with the result that two fingers can become firmly glued together. The ability to form covalent bonds with groups on the surfaces of the objects to be glued together is what gives Super Glue its amazing strength.

3-D Molecules:
Methyl α-cyanoacrylate;
Poly(methyl α-cyano-
acrylate)

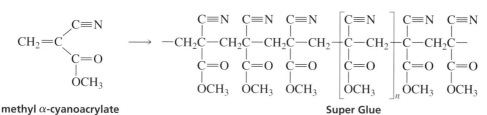

methyl α-cyanoacrylate   **Super Glue**

PROBLEM 8◆

List the following groups of monomers in order of decreasing ability to undergo anionic polymerization:

a. CH$_2$=CH   CH$_2$=CH   CH$_2$=CH

NO$_2$ CH$_3$ OCH$_3$

b. CH$_2$=CHCH$_3$ CH$_2$=CHCl   CH$_2$=CHC≡N

We have seen that the substituent on the alkene determines the best mechanism for chain-growth polymerization. Alkenes with substituents that can stabilize radicals readily undergo radical polymerization; alkenes with electron-donating substituents that can stabilize cations undergo cationic polymerization; and alkenes with electron-withdrawing substituents that can stabilize anions undergo anionic polymerizations.

Some alkenes undergo polymerization by more than one mechanism. For example, styrene can undergo polymerization by radical, cationic, and anionic mechanisms because the phenyl group can stabilize benzylic radicals, benzylic cations, and benzylic anions. The particular mechanism followed for the polymerization of styrene depends on the nature of the initiator chosen to start the polymerization reaction.

Although ethylene and substituted ethylenes are the monomers most commonly used for chain-growth polymerization reactions, other compounds can polymerize as well. For example, epoxides undergo chain-growth polymerization reactions. If the initiator is a nucleophile such as HO^- or RO^-, polymerization occurs by an anionic mechanism.

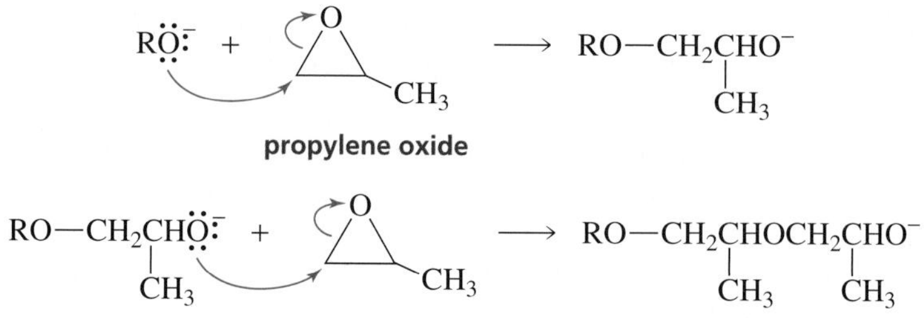

propylene oxide

If the initiator molecule is a Lewis acid or a proton-donating acid, polymerization of epoxides occurs by a cationic mechanism. Polymerization reactions that involve ring-opening reactions, such as the polymerization of propylene oxide, are called **ring-opening polymerizations.**

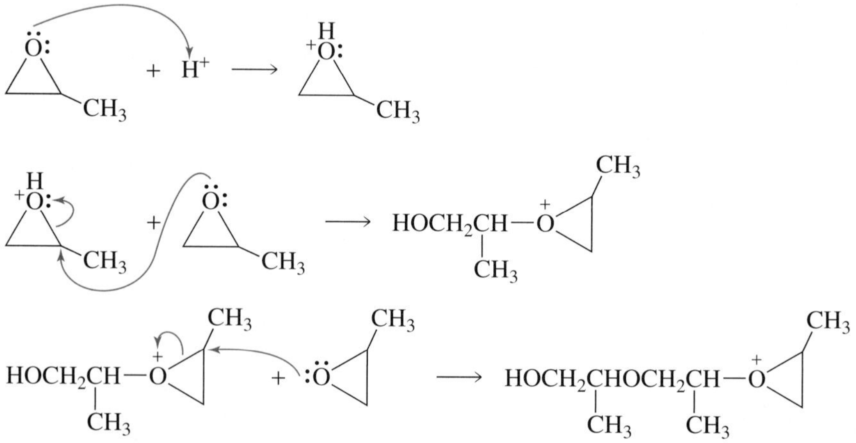

PROBLEM 9

Explain why when propylene oxide undergoes anionic polymerization, nucleophilic attack occurs at the less substituted carbon of the epoxide, but when it undergoes cationic polymerization, nucleophilic attack occurs at the more substituted carbon.

PROBLEM 10

Describe the polymerization of 2,2-dimethyloxirane

a. by an anionic mechanism.   **b.** by a cationic mechanism.

PROBLEM 11◆

What monomer and what type initiator would you use to synthesize each of the following polymers?

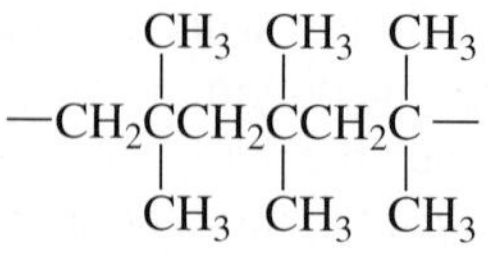

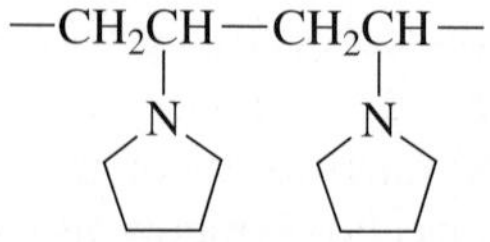

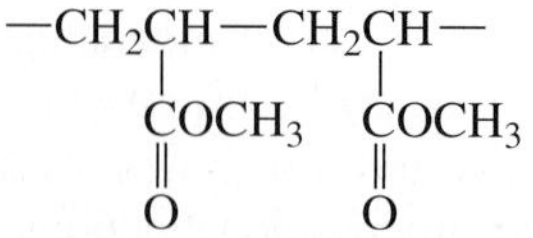

PROBLEM 12◆

Draw a short segment of the polymer formed from cationic polymerization of 3,3-dimethyloxacyclobutane.

3,3-dimethyloxacyclobutane

26.3 STEREOCHEMISTRY OF POLYMERIZATION. ZIEGLER–NATTA CATALYSTS

Polymers formed from monosubstituted ethylenes can exist in three possible configurations: isotactic, syndiotactic, and atactic. An **isotactic polymer** has all the substituents on the same side of the fully extended carbon chain. *Iso* and *taxis* are Greek for "the same" and "order," respectively. In a **syndiotactic polymer** (*syndio* means "alternating"), the substituents regularly alternate on both sides of the carbon chain. The substituents in an **atactic polymer** are randomly oriented.

isotactic configuration (same side)

syndiotactic configuration (both sides)

The configuration of the polymer affects its physical properties. Polymers in the isotactic or syndiotactic configuration are more likely to be crystalline solids because positioning the substituents in a regular order allows for a more regular packing arrangement. Polymers in the atactic configuration are more disordered and cannot pack together as well, so these polymers are softer and less rigid.

The configuration of the polymer depends on the mechanism by which polymerization occurs. In general, radical polymerization leads primarily to branched polymers in the atactic configuration. Cationic polymerization produces polymers with a considerable fraction of the chains in the isotactic or syndiotactic configuration. Anionic polymerization produces polymers with the most stereoregularity. The percentage of chains in the isotactic or syndiotactic configuration increases as the polymerization temperature decreases.

In 1953, Karl Ziegler and Giulio Natta found that the structure of a polymer could be controlled if the growing end of the chain and the incoming monomer were coordinated with an aluminum–titanium initiator. These initiators are now called **Ziegler–Natta catalysts.** Long, unbranched polymers with either the isotactic or the syndiotactic configuration can be prepared using Ziegler–Natta catalysts. Whether the chain is isotactic or syndiotactic depends on the particular Ziegler–Natta catalyst used. These catalysts revolutionized the field of polymer chemistry because they allow the synthesis of stronger and stiffer polymers that have greater resistance to cracking and heat. High-density polyethylene is prepared using a Ziegler–Natta process.

The mechanism of the Ziegler–Natta-catalyzed polymerization of a substituted ethylene is shown in Figure 26.1. The monomer forms a π complex with the metal at an open coordination site and the coordinated alkene is inserted between titanium and the growing polymer chain, thereby extending the polymer chain. Because a new coordination site opens up during insertion, the process can be repeated over and over.

Karl Ziegler (1898–1973), the son of a minister, was born in Germany. He was a professor at the University of Frankfurt and then at the University of Heidelberg.

Karl Ziegler and Giulio Natta did not work together, but each independently developed the catalyst system used in polymerization. They shared the 1963 Nobel Prize in chemistry.

Giulio Natta (1903–1979) was the son of an Italian judge. He was a professor at the Polytechnic Institute in Milan, where he became the director of the Industrial Chemistry Research Center.

▲ Figure 26.1
The mechanism of the Ziegler–Natta-catalyzed polymerization of a substituted ethylene. A monomer forms a π complex with an open coordination site of titanium and then is inserted between titanium and the growing polymer.

Polyacetylene is another polymer prepared by a Ziegler–Natta process. It is a **conducting polymer** because the conjugated double bonds in polyacetylene make it possible to conduct electricity down its backbone after several electrons are removed from or added to the backbone.

$$HC{\equiv}CH \xrightarrow{\text{a Ziegler–Natta catalyst}} -CH{=}CH{-}\!\!\left[\!CH{=}CH\right]_{\!n}\!\!CH{=}CH{-}$$

acetylene   polyacetylene

26.4 POLYMERIZATION OF DIENES. THE MANUFACTURE OF RUBBER

When the bark of a rubber tree is cut, a sticky white liquid oozes out. This is the same liquid found inside the stalks of dandelions and milkweed. The sticky material is latex, a suspension of rubber particles in water. Natural rubber is a polymer of 2-methyl-1,3-butadiene (isoprene) (Section 24.6). On average, a molecule of rubber contains 5000 isoprene units. All the double bonds in natural rubber are cis. Rubber is a waterproof material because it consists of a tangle of hydrocarbon chains that have no affinity for water.

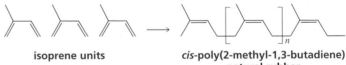

isoprene units   *cis*-poly(2-methyl-1,3-butadiene)
                                       natural rubber

Gutta-percha (from the Malaysian words *getah* meaning "gum" and *percha* meaning "tree") is a naturally occurring isomer of rubber in which all the double bonds are trans. Like rubber, gutta-percha is exuded by certain trees but it is much less common. It is also harder and more brittle than rubber. It is the filling material that dentists use in root canals and the material used for the casing of golf balls.

PROBLEM 13

Draw a short segment of gutta-percha.

By mimicking nature, scientists have learned to make synthetic rubber with properties tailored to meet human needs. These materials are called synthetic rubbers because they have some of the properties of natural rubber, including being waterproof and elastic, but they have some improved properties as well—they are tougher, more flexible, and more durable than natural rubber.

Latex being collected from a rubber tree.

Synthetic rubbers have been made by polymerizing dienes other than isoprene. One synthetic rubber is a polymer of 1,3-butadiene in which all the double bonds are cis. Polymerization is carried out in the presence of a Ziegler–Natta catalyst so the configuration of the double bonds in the polymer can be controlled.

1,3-butadiene monomers → **a Ziegler–Natta catalyst** → *cis*-**poly(1,3-butadiene)** **a synthetic rubber**

Neoprene is a synthetic rubber made by polymerizing 2-chloro-1,3-butadiene in the presence of a Ziegler–Natta catalyst that causes all the double bonds in the polymer to have the trans configuration. Neoprene is used to make shoe soles, tires, hoses, and coated fabrics.

$$CH_2{=}CCH{=}CH_2$$
2-chloro-1,3-butadiene chloroprene → **a Ziegler–Natta catalyst** → neoprene

Charles Goodyear (1800–1860), the son of an inventor of farm implements, was born in Connecticut. He patented the process of vulcanization in 1844. The process was so simple, however, that it could be easily copied, so he spent many years contesting infringements on his patent. In 1852, with Daniel Webster as his lawyer, he obtained the right to the patent.

A problem common to both natural and most synthetic rubber is that the polymers are very soft and sticky. They can be hardened by a process known as *vulcanization*. Charles Goodyear discovered this process while trying to improve the properties of rubber. He accidentally spilled a mixture of rubber and sulfur on a hot stove. To his surprise, the mixture became hard but flexible. He called the heating of rubber with sulfur **vulcanization,** after Vulcan, the Roman god of fire.

Heating rubber with sulfur causes **cross-linking** of separate polymer chains through disulfide bonds (Figure 26.2). Rather than the individual chains just being entangled together, the vulcanized chains are covalently bonded together in one giant molecule. Because the polymer has double bonds, the chains have bends and kinks that prevent them from forming a tightly packed crystalline polymer. When rubber is stretched, the chains straighten out along the direction of the pull. Cross-linking prevents the polymer from being torn when it is stretched, and the cross-links provide a reference framework for the material to return to when the stretching force is removed.

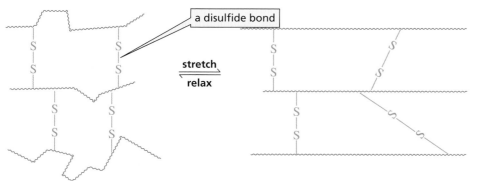

a disulfide bond

stretch / relax

◀ **Figure 26.2**
The rigidity of rubber is increased by cross-linking the polymer chains with disulfide bonds. When rubber is stretched, the randomly coiled chains straighten out and orient themselves along the direction of the stretch.

The physical properties of rubber can be controlled by regulating the amount of sulfur used in the vulcanization process. Rubber made with 1–3% sulfur is soft and stretchy and is used to make rubber bands. Rubber made with 3–10% sulfur is more rigid and is used in the manufacture of tires. Goodyear's name can be found on many tires sold today. The story of rubber is an example of a scientist taking a natural material and finding ways to improve its useful properties.

> **PROBLEM 14**
>
> The polymer formed from a diene such as 1,3-butadiene contains vinyl branches. Propose an anionic polymerization mechanism to account for the formation of these branches.
>
> $$CH_2=CHCH=CH_2 \longrightarrow -CH_2CH=CHCH_2CHCH_2CH_2CH=CHCH_2-$$
> $$ \underset{CH=CH_2}{|}$$

26.5 COPOLYMERS

The polymers we have discussed so far are formed from only one type of monomer and are called **homopolymers.** Often two (or more) different monomers are used to form a polymer. The resulting product is called a **copolymer.** Increasing the number of different monomers used to form the copolymer dramatically increases the number of different copolymers that can be formed. Even if only two kinds of monomers are used, copolymers with very different properties can be prepared by varying the amounts of each monomer. Both chain-growth polymers and step-growth polymers can be copolymers. Many of the synthetic polymers used today are copolymers. Table 26.6 shows some common copolymers and the monomers from which they are synthesized.

TABLE 26.6   Some Examples of Copolymers and Their Uses

Monomer	Copolymer Name	Uses
$CH_2=CH$ ($\overset{|}{Cl}$) **vinyl chloride** + $CH_2=CCl$ ($\overset{|}{Cl}$) **vinylidene chloride**	Saran	film for wrapping food
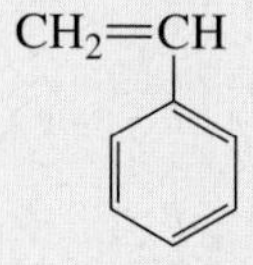 $CH_2=CH$ (phenyl) **styrene** + $CH_2=CH$ ($\overset{|}{C\equiv N}$) **acrylonitrile**	SAN	dishwasher-safe objects, vacuum cleaner parts
$CH_2=CH$ ($\overset{|}{C\equiv N}$) **acrylonitrile** + $CH_2=CH$ ($\overset{|}{CH=CH_2}$) **1,3-butadiene** + $CH_2=CH$ (phenyl) **styrene**	ABS	bumpers, crash helmets, telephones, luggage
$CH_2=CCH_3$ ($\overset{|}{CH_3}$) **isobutylene** + $CH_2=CHC=CH_2$ ($\overset{|}{CH_3}$) **isoprene**	butyl rubber	inner tubes, balls, inflatable sporting goods

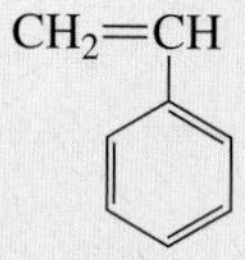

There are four types of copolymers. In an **alternating copolymer,** the two monomers alternate. In a **block copolymer,** there are blocks of each kind of monomer. In a **random copolymer,** the distribution of monomers is random. A **graft copolymer** contains branches derived from one monomer grafted onto a backbone derived from another monomer. These structural differences extend the range of physical properties available to the scientist designing the copolymer.

an alternating copolymer	ABABABABABABABABABABABA
a block copolymer	AAAAABBBBBAAAAABBBBBAAA
a random copolymer	AABABABBABAABBABABBAAAB

a graft copolymer   AAAAAAAAAAAAAAAAAAAAAAA
                    B   B   B
                    B   B   B
                    B   B   B
                    B   B   B
                    B   B   B
                    B   B   B

26.6 STEP-GROWTH POLYMERS

Step-growth polymers are formed by the intermolecular reaction of bifunctional molecules (molecules with two functional groups). When the functional groups react, in most cases a small molecule such as H_2O, alcohol, or HCl is lost. This is why these polymers are also called *condensation polymers*.

There are two types of step-growth polymers. One type is formed by the reaction of a single monomer that possesses two different functional groups, A and B. Functional group A of one monomer reacts with functional group B of another monomer.

$$A—B \quad A—B \longrightarrow A—X—B$$

The other type of step-growth polymer is formed by the reaction of two different bifunctional monomers. One monomer contains two A functional groups and the other monomer contains two B functional groups.

$$A—A \quad B—B \longrightarrow A—X—B$$

The formation of step-growth polymers, unlike the formation of chain-growth polymers, does not involve chain reactions. Any two monomers (or short chains) can react. The progress of a typical step-growth polymerization is shown schematically in Figure 26.3. When the reaction is 50% complete (12 bonds have formed between 25 monomers), the reaction products are primarily dimers and trimers. Even at 75% completion, no long chains have been formed. This means that if step-growth polymerization is to lead to long-chain polymers, very high yields must be achieved.

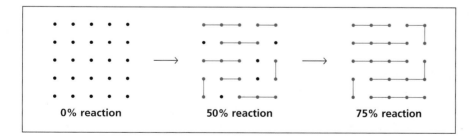

◀ **Figure 26.3**
Progress of a step-growth polymerization.

0% reaction   50% reaction   75% reaction

Polyamides

Nylon is the common name of a synthetic **polyamide.** Nylon 6 is an example of a step-growth polymer formed by a monomer with two different functional groups. The carboxylic acid group of one monomer reacts with the amino group of another monomer, resulting in the formation of amide groups. Structurally, this reaction is similar to the polymerization of α-amino acids to form proteins (Section 21.9).

*Nylon was first synthesized in 1931 by **Wallace Carothers** (1896–1937). He was born in Iowa and received a Ph.D. from the University of Illinois. He taught there and at Harvard before being hired by Du Pont to head their program in basic science. Nylon was introduced to the public in 1939, but its widespread use was delayed until after World War II because all nylon produced during the war was used by the military. Carothers died unaware of the era of synthetic fibers that occurred after the war.*

This particular nylon is called nylon 6 because it is formed from the polymerization of 6-aminohexanoic acid, a compound that contains six carbons.

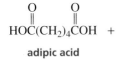

6-aminohexanoic acid

nylon 6
a polyamide

The starting material for nylon 6 is ε-caprolactam. The lactam is opened by hydrolysis.

ε-caprolactam

ε-aminocaproic acid
6-aminohexanoic acid

3-D Molecules:
ε-Caprolactam;
Nylon 6;
Nylon 66

Nylon 66 is an example of a step-growth polymer formed by two different bifunctional monomers—adipic acid and 1,6-hexanediamine. It is called nylon 66 because it is a polyamide formed from a six-carbon diacid and a six-carbon diamine.

adipic acid + **1,6-hexanediamine**

nylon 66

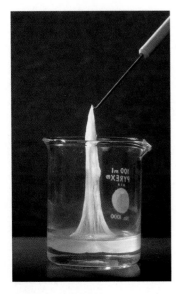

Nylon is pulled from a beaker of adipoyl chloride and 1,6-hexanediamine.

Nylon first found wide use in textiles and carpets. Because it is resistant to stress, it is now used in many other applications, such as mountaineering ropes, for tire cords and fishing line, and as a substitute for metal in bearings and gears. The extended applications of nylon precipitated a search for new "super fibers" with super strength and super heat resistance.

PROBLEM 15◆

a. Draw a short segment of nylon 4.

b. From what lactam is it synthesized?

c. Draw a short segment of nylon 44.

PROBLEM 16

Write the equation that explains what will happen if a scientist working in the laboratory spills sulfuric acid on her nylon 66 stockings.

One new super fiber is Kevlar, a polymer of 1,4-benzenedicarboxylic acid and 1,4-diaminobenzene. Incorporation of aromatic rings into polymers has been found to result in polymers with great physical strength. Aromatic polyamides are known as **aramides.** Kevlar is an aramide with a tensile strength greater than that of steel. Army helmets are made of Kevlar. Kevlar is also used for lightweight bulletproof vests and high-performance skis. Because it is stable at very high temperatures, it is used in the protective clothing worn by firefighters.

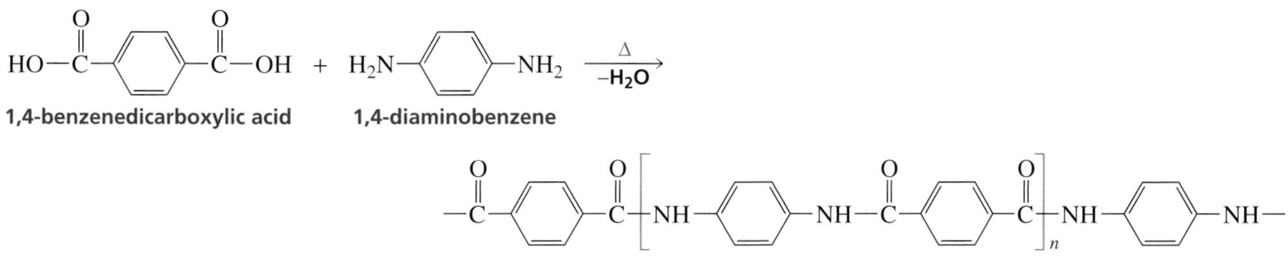

1,4-benzenedicarboxylic acid   **1,4-diaminobenzene**

Kevlar
an aramide

Kevlar owes its strength to the way in which the individual polymer chains interact with each other. The chains are hydrogen bonded, which causes them to form a sheet structure.

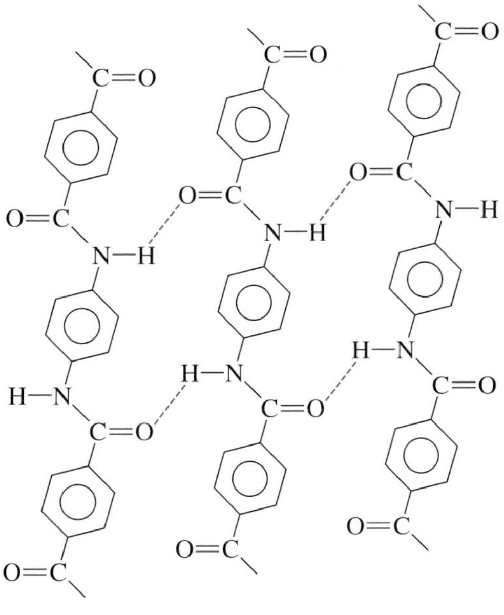

Polyesters

Polyesters are step-growth polymers in which the monomer units are joined together by ester groups. They have found wide commercial use as fibers, plastics, and coatings. The most common polyester is known by the trade name Dacron and is made by the transesterification of dimethyl terephthalate with ethylene glycol. High resilience, durability, and moisture resistance are the properties of this polymer that contribute to its "wash-and-wear" characteristics.

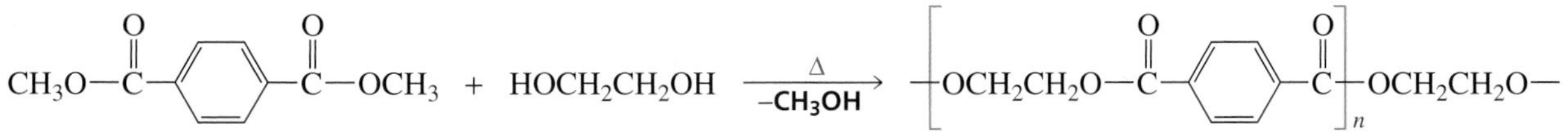

dimethyl terephthalate   **1,2-ethanediol**
                              ethylene glycol

poly(ethylene terephthalate)
Dacron
a polyester

Kodel polyester is formed by the transesterification of dimethyl terephthalate with 1,4-di(hydroxymethyl)cyclohexane. The stiff polyester chain causes the fiber to have a harsh feel that can be softened by blending it with wool or cotton.

CH₃O—C(=O)—⟨benzene⟩—C(=O)—OCH₃ + HOCH₂—⟨cyclohexane⟩—CH₂OH →(Δ, −CH₃OH)

dimethyl terephthalate   **1,4-di(hydroxymethyl)cyclohexane**

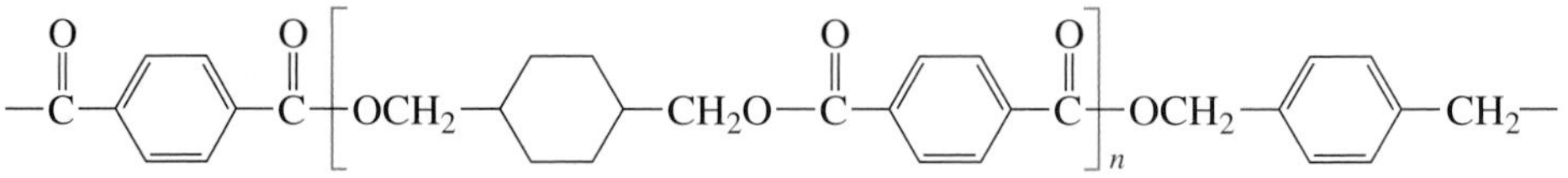

—C(=O)—⟨benzene⟩—C(=O)—[OCH₂—⟨cyclohexane⟩—CH₂O—C(=O)—⟨benzene⟩—C(=O)—]ₙ OCH₂—⟨benzene⟩—CH₂—

Kodel

PROBLEM 17

What happens to polyester slacks if aqueous NaOH is spilled on them?

Polyesters with two ester groups bonded to the same carbon are known as **polycarbonates.** Lexan is produced by the reaction of phosgene with bisphenol A. Lexan is a strong and transparent polymer used for bulletproof windows and traffic-light lenses. In recent years, polycarbonates have become important polymers in the automobile industry as well as in the manufacture of compact discs.

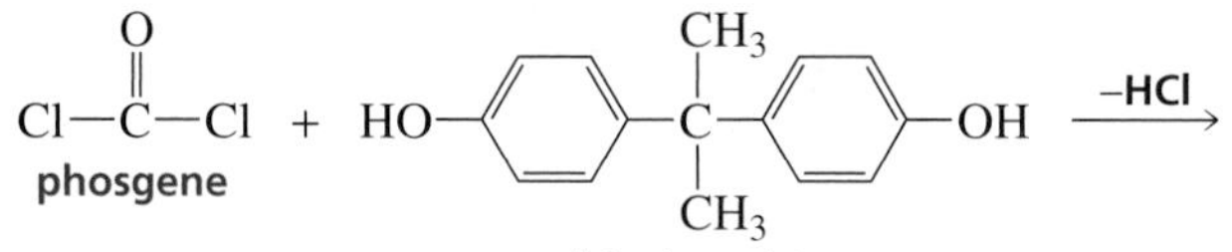

Cl—C(=O)—Cl + HO—⟨benzene⟩—C(CH₃)(CH₃)—⟨benzene⟩—OH →(−HCl)

phosgene   **bisphenol A**

—⟨benzene⟩—C(CH₃)(CH₃)—⟨benzene⟩—O—[C(=O)—O—⟨benzene⟩—C(CH₃)(CH₃)—⟨benzene⟩—O—]ₙ C(=O)—O—⟨benzene⟩—C(CH₃)(CH₃)—⟨benzene⟩—O—C(=O)—

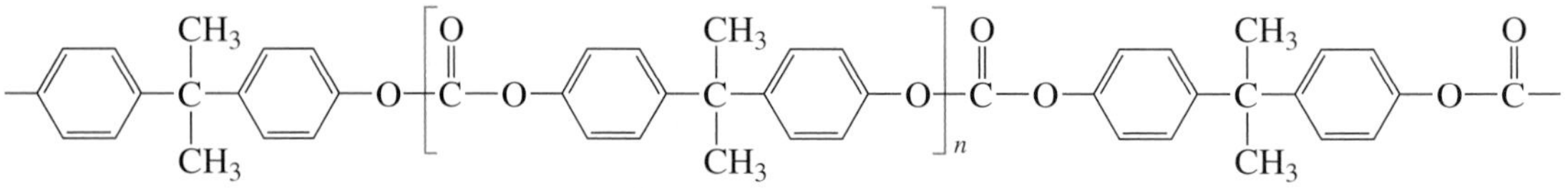

Lexan
a polycarbonate

Epoxy Resins

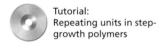
Tutorial:
Repeating units in step-growth polymers

Epoxy resins are the strongest adhesives known. They can adhere to almost any kind of surface and are resistant to solvents and to extremes of temperature. When an epoxy cement is used, a low-molecular-weight *prepolymer* (the most common is a polymer of bisphenol A and epichlorohydrin) is mixed with a *hardener*, a compound that will react with the prepolymer to form a cross-linked polymer.

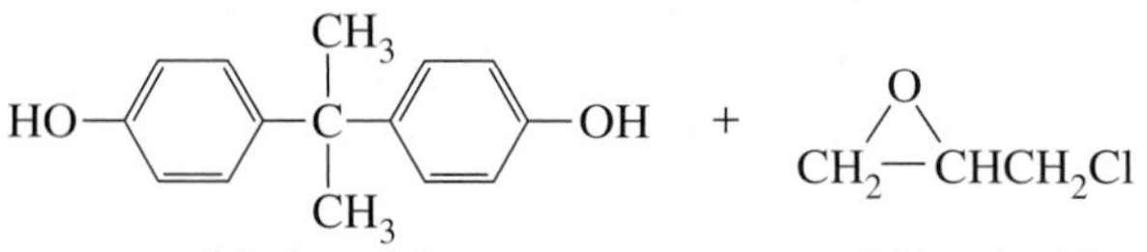

HO—⟨benzene⟩—C(CH₃)(CH₃)—⟨benzene⟩—OH + CH₂—CHCH₂Cl (epoxide)

bisphenol A   **epichlorohydrin**

↓ −HCl

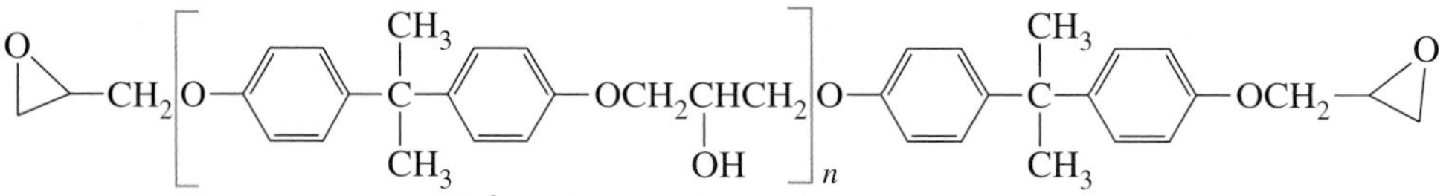

⟨epoxide⟩—CH₂[O—⟨benzene⟩—C(CH₃)(CH₃)—⟨benzene⟩—OCH₂CHCH₂]ₙ O—⟨benzene⟩—C(CH₃)(CH₃)—⟨benzene⟩—OCH₂—⟨epoxide⟩
OH

prepolymer

↓ H₂NCH₂CH₂NHCH₂CH₂NH₂
hardener

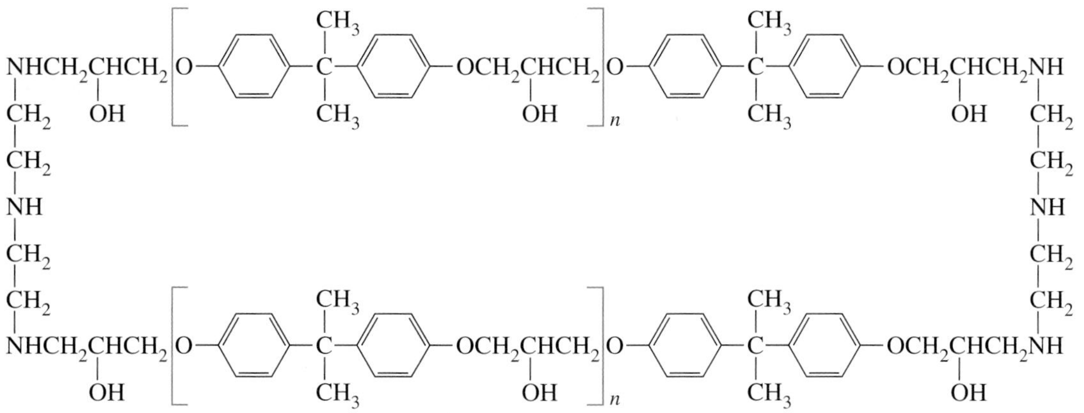

The structure at the top of the page shows:

NHCH$_2$CHCH$_2$ | O—[benzene ring]—C(CH$_3$)(CH$_3$)—[benzene ring]—OCH$_2$CHCH$_2$ | O—[benzene ring]—C(CH$_3$)(CH$_3$)—[benzene ring]—OCH$_2$CHCH$_2$NH (with OH, CH$_2$, CH$_2$, NH, CH$_2$, CH$_2$ chain), continuing to

NHCH$_2$CHCH$_2$ | O—[benzene ring]—C(CH$_3$)(CH$_3$)—[benzene ring]—OCH$_2$CHCH$_2$ | O—[benzene ring]—C(CH$_3$)(CH$_3$)—[benzene ring]—OCH$_2$CHCH$_2$NH

an epoxy resin

PROBLEM 18

a. Propose a mechanism for the formation of the prepolymer formed by bisphenol A and epichlorohydrin.

b. Propose a mechanism for the reaction of the prepolymer with the hardener.

Polyurethanes

A **urethane** (also called a carbamate) is a compound that has an ester group and an amide group bonded to the same carbon. Urethanes can be prepared by treating an isocyanate with an alcohol.

$$RN{=}C{=}O \ + \ ROH \ \longrightarrow \ RNH{-}\overset{\displaystyle O}{\overset{\|}{C}}{-}OR$$

an isocyanate **an alcohol** **a urethane**

Polyurethanes are polymers that contain urethane groups. One of the most common polyurethanes is prepared by the polymerization of toluene-2,6-diisocyanate and ethylene glycol. If the reaction is carried out in the presence of a blowing agent, the product is a polyurethane foam. Blowing agents are gases such as nitrogen or carbon dioxide. Chlorofluorocarbons—low-boiling liquids that vaporize on heating—used to be used but have been banned for environmental reasons (Section 8.9). Polyurethane foams are used for furniture stuffing, carpet backings, and insulation. Notice that polyurethanes prepared from diisocyanates and diols are the only step-growth polymers that we have seen in which a small molecule is *not* lost during polymerization.

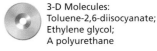

3-D Molecules:
Toluene-2,6-diisocyanate;
Ethylene glycol;
A polyurethane

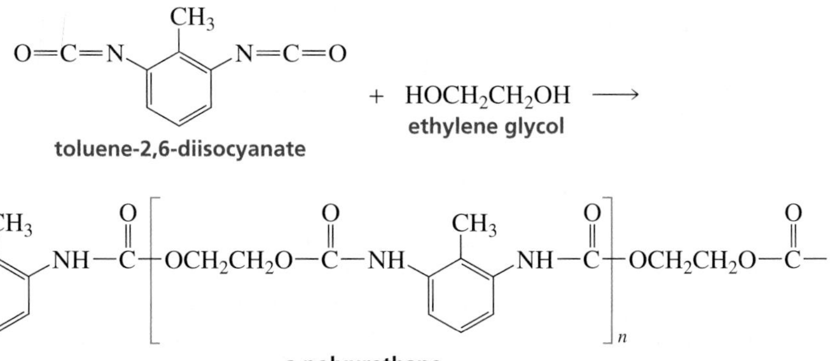

O=C=N—[benzene ring with CH$_3$]—N=C=O + HOCH$_2$CH$_2$OH $\longrightarrow$
                                                              ethylene glycol

toluene-2,6-diisocyanate

$$-\overset{\displaystyle O}{\overset{\|}{C}}{-}NH{-}[\text{ring, } CH_3]{-}NH{-}\overset{\displaystyle O}{\overset{\|}{C}}{-}\left[OCH_2CH_2O{-}\overset{\displaystyle O}{\overset{\|}{C}}{-}NH{-}[\text{ring, } CH_3]{-}NH{-}\overset{\displaystyle O}{\overset{\|}{C}}{-}OCH_2CH_2O{-}\overset{\displaystyle O}{\overset{\|}{C}}{-}\right]_n$$

a polyurethane

One of the most important uses of polyurethanes is in fabrics with elastic properties such as spandex (Lycra). These materials are block copolymers in which some of the polymer segments are polyurethanes, some are polyesters, and some are polyamides. The blocks of polyurethane are soft, amorphous segments that become crystalline on stretching (See Section 26.7). When the tension is released, they revert to the amorphous state.

PROBLEM 19

If a small amount of glycerol is added to the reaction mixture of toluene-2,6-diiso-cyanate and ethylene glycol during the synthesis of polyurethane foam, a much stiffer foam is obtained. Explain.

$$CH_2 \!-\! CH \!-\! CH_2$$
$$|||$$
$$OH \quad OH \quad OH$$
glycerol

26.7 PHYSICAL PROPERTIES OF POLYMERS

The individual chains of a polymer such as polyethylene are held together by van der Waals forces. Because these forces operate only at small distances, they are strongest if the polymer chains can line up in an ordered, closely packed array. The regions of the polymer in which the chains are highly ordered with respect to one another are called **crystallites** (Figure 26.4). Between the crystallites are amorphous, noncrystalline regions in which the chains are randomly oriented. The more crystalline—more ordered—the polymer is, the denser, harder, and more resistant to heat it is (Table 26.7). If the polymer chains possess substituents (as does poly[methyl methacrylate], for example) or have branches that prevent them from packing closely together, the density of the polymer is reduced.

Table 26.7 Properties of Polyethylene as a Function of Crystallinity					
Crystallinity (%)	55	62	70	77	85
Density (g/cm³)	0.92	0.93	0.94	0.95	0.96
Melting point (°C)	109	116	125	130	133

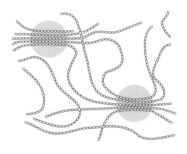

▲ **Figure 26.4**
The circled regions are crystallites, regions where the polymer chains are highly ordered, similar to the ordering found in crystals. Between the circles, the polymer chains are randomly oriented, and these regions are noncrystalline.

Thermoplastic Polymers

Plastics can be classified according to the physical properties imparted to them by the way in which their individual chains are arranged. **Thermoplastic polymers** have both ordered crystalline regions and amorphous, noncrystalline regions. Thermoplastic polymers are hard at room temperature, but soft enough to be molded when heated because the individual chains can slip past one another at elevated temperatures. Thermoplastic polymers are the plastics we encounter most often in our daily lives—in combs, toys, light-switch plates, and telephone casings, for example. They are the plastics that are relatively easily cracked.

Thermosetting Polymers

Very strong and rigid materials can be obtained if the polymer chains are cross-linked. The greater the degree of cross-linking, the more rigid the polymer. Such cross-linked polymers are called **thermosetting polymers.** After they are hardened, they cannot be remelted by heating because the cross-links are covalent bonds, not intermolecular van der Waals forces. Cross-linking reduces the mobility of the polymer chains, causing the polymer to be a relatively brittle material. Because thermosetting polymers do not have the wide range of properties characteristic of thermoplastic polymers, they are less widely used.

Melmac, a highly cross-linked thermosetting polymer of melamine and formaldehyde, is a hard and moisture-resistant material. Because it is a colorless polymer, it can be made into materials with pastel colors. It is used to make lightweight dishes and counter surfaces.

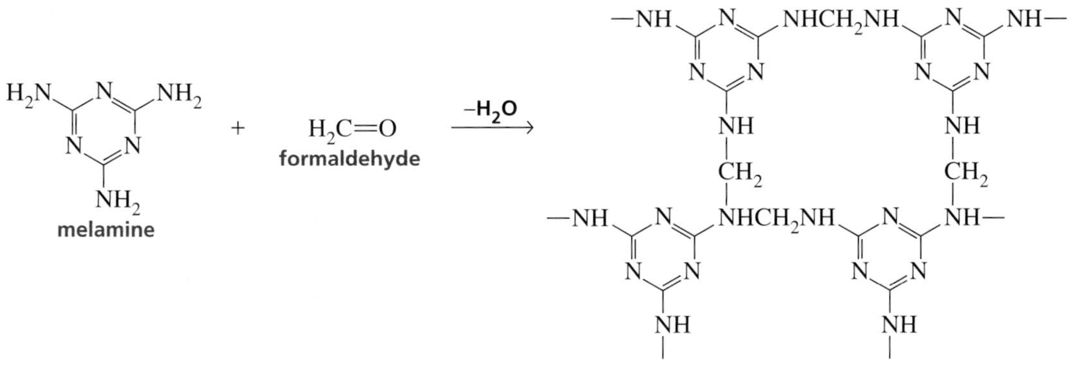

DESIGNING A POLYMER

A polymer used for making dental impressions must be soft enough initially to mold around the teeth but must become hard enough later to maintain a fixed shape. The polymer commonly used for dental impressions contains three-membered aziridine rings that react to cross-link the chains.

Because arizidine rings are not very reactive (Section 27.1), cross-linking occurs relatively slowly so most of the hardening of the polymer does not occur until the polymer is removed from the patient's mouth.

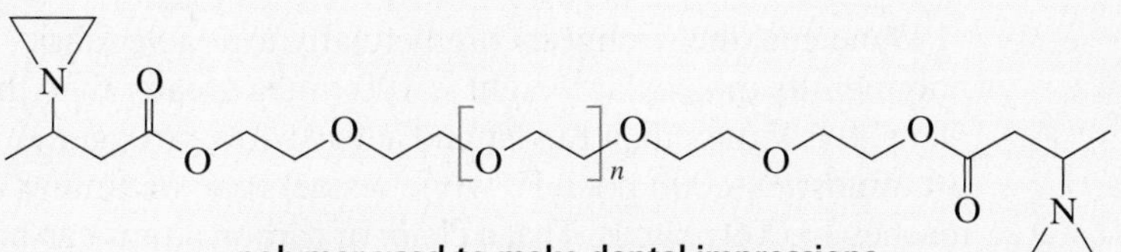

polymer used to make dental impressions

A polymer used for making contact lenses must be sufficiently hydrophilic to allow lubrication of the eye. Therefore, the polymer used for contact lenses has many OH groups.

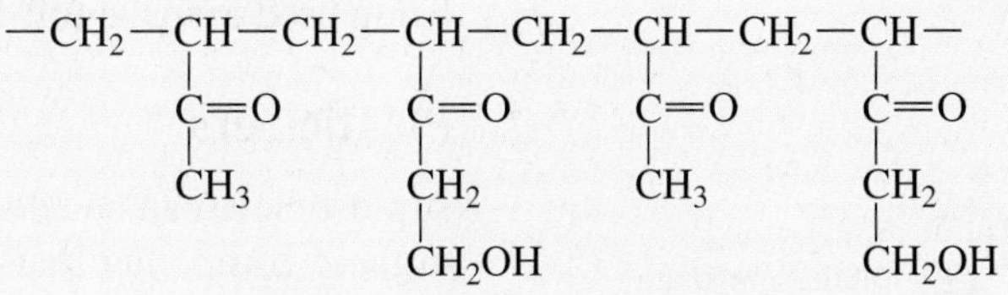

polymer used to make contact lenses

PROBLEM 20

Propose a mechanism for the formation of Melmac.

PROBLEM 21

Bakelite was the first of the thermosetting polymers. It is a highly cross-linked polymer formed from the acid-catalyzed polymerization of phenol and formaldehyde. It is a much

Leo Hendrik Baekeland (1863–1944) discovered Bakelite while looking for a substitute for shellac in his home laboratory. He was born in Belgium and became a professor of chemistry at the University of Ghent. A fellowship brought him to the United States in 1889, and he decided to stay. His hobby was photography, and he invented photographic paper that could be developed under artificial light. He sold the patent to Eastman-Kodak.

darker polymer than Melmac, so the color range of the products made from Bakelite is limited. Propose a structure for Bakelite.

Elastomers

An **elastomer** is a plastic that stretches and then reverts to its original shape. It is a randomly oriented amorphous polymer. It must have some cross-linking so the chains do not slip over one another. When elastomers are stretched, the random chains stretch out, but there are insufficient van der Waals forces to maintain them in that arrangement. When the stretching force is removed, they go back to their random shapes. Rubber is an elastomer.

Oriented Polymers

Polymers that are stronger than steel or that conduct electricity almost as well as copper can be made by taking the polymer chains obtained by conventional polymerization, stretching them out, and putting them back together in a parallel fashion (Figure 26.5). Such polymers are called **oriented polymers.** Converting conventional polymers into oriented polymers has been compared to "uncooking" spaghetti. The conventional polymer is disordered cooked spaghetti, while the oriented polymer is ordered raw spaghetti.

Figure 26.5 ▶
The creation of an oriented polymer.

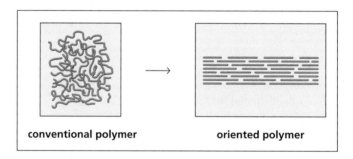

conventional polymer   oriented polymer

Dyneema, the strongest commercially available fabric, is an oriented polyethylene polymer. Its molecular weight is 100 times greater than that of high-density polyethylene. It is lighter than Kevlar and at least 40% stronger. A rope made of Dyneema can lift almost 119,000 pounds, while a steel rope of similar size fails before the weight reaches 13,000 pounds. That a chain of carbon atoms can be stretched and properly oriented to produce a material stronger than steel is somewhat astounding. Dyneema is being used to make full-face crash helmets, protective fencing suits, and hang gliders.

Plasticizers

3-D Molecule:
Dibutyl phthalate

A plasticizer can be added to a polymer to make it more flexible. A **plasticizer** is an organic compound that dissolves in the polymer and allows the polymer chains to slide past one another. Dibutyl phthalate is a commonly used plasticizer. It is added to poly(vinyl chloride)—normally a brittle polymer—to make products such as vinyl raincoats, shower curtains, and garden hoses.

$$\text{dibutyl phthalate}$$

dibutyl phthalate
a plasticizer

An important criterion to consider in choosing a plasticizer is its permanence. Permanence refers to how well the plasticizer remains in the polymer. The "new car smell" appreciated by car owners is the odor of the plasticizer that has vaporized from the vinyl upholstery. When a significant amount of the plasticizer has evaporated, the upholstery becomes brittle and cracks. Phthalates with higher molecular weights and lower vapor pressures than those of dibutyl phthalate are now commonly used for car interiors.

Biodegradable polymers are polymers that can be broken into small segments by enzyme-catalyzed reactions. The enzymes are produced by microorganisms. The carbon–carbon bonds of chain-growth polymers are inert to enzyme-catalyzed reactions, so they are nonbiodegradable unless bonds that *can* be broken by enzymes are inserted into the polymer. Then, when the polymer is buried as waste, microorganisms present in the ground can degrade the polymer. One method of making a polymer biodegradable involves inserting hydrolyzable ester groups into the polymer. For example, if the following acetal is added to an alkene undergoing radical polymerization, ester groups will be inserted into the polymer. These "weak links" are susceptible to enzyme-catalyzed hydrolysis.

26.8 BIODEGRADABLE POLYMERS

Polymer chemistry has evolved into a multibillion-dollar industry. More than 2.5×10^{13} kilograms of synthetic polymers are produced in the United States each year, and we can expect many more new materials to be developed by scientists in the years to come.

KEY TERMS

addition polymer (page 1105)
alpha olefin (page 1106)
alternating copolymer (page 1120)
anionic polymerization (page 1106)
aramide (page 1122)
atactic polymer (page 1117)
biodegradable polymer (page 1129)
biopolymer (page 1104)
block copolymer (page 1120)
cationic polymerization (page 1106)
chain-growth polymer (page 1105)
chain transfer (page 1109)
condensation polymer (page 1106)
conducting polymer (page 1118)
copolymer (page 1120)
cross-linking (page 1119)

crystallites (page 1126)
elastomer (page 1128)
epoxy resin (page 1124)
graft copolymer (page 1120)
head-to-tail addition (page 1109)
homopolymer (page 1120)
isotactic polymer (page 1117)
living polymer (page 1115)
materials science (page 1105)
monomer (page 1104)
oriented polymer (page 1128)
plasticizer (page 1128)
polyamide (page 1121)
polycarbonate (page 1124)
polyester (page 1123)
polymer (page 1104)

polymer chemistry (page 1105)
polymerization (page 1104)
polyurethane (page 1125)
propagating site (page 1108)
radical polymerization (page 1106)
random copolymer (page 1120)
ring-opening polymerization (page 1116)
step-growth polymer (page 1106)
syndiotactic polymer (page 1117)
synthetic polymer (page 1104)
thermoplastic polymer (page 1126)
thermosetting polymer (page 1126)
urethane (page 1125)
vinyl polymer (page 1106)
vulcanization (page 1119)
Ziegler–Natta catalyst (page 1117)

PROBLEMS

22. Draw short segments of the polymers obtained from the following monomers. Indicate whether the polymerization is a chain-growth or a step-growth polymerization.

a. $CH_2{=}CHF$

b. $CH_2{=}CHCO_2H$

c. $HO(CH_2)_5\overset{\displaystyle O}{\overset{\|}{C}}OH$

d. $Cl\overset{\displaystyle O}{\overset{\|}{C}}(CH_2)_5\overset{\displaystyle O}{\overset{\|}{C}}Cl \ + \ H_2N(CH_2)_5NH_2$

e. 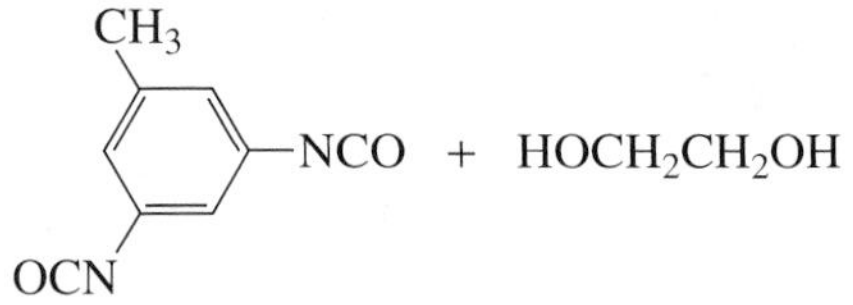 [structure: benzene ring with CH_3 at top, $-NCO$ on right, OCN at lower left] $+ \ HOCH_2CH_2OH$

23. Draw the repeating unit of the step-growth polymer that will be formed from each of the following reactions:

a. 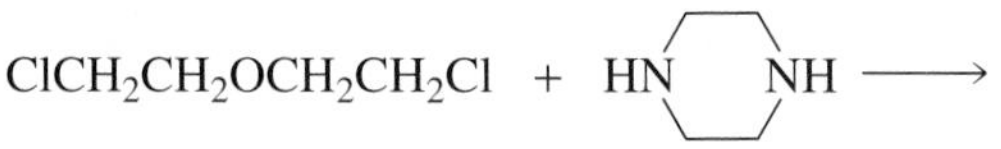 $ClCH_2CH_2OCH_2CH_2Cl \ + \ HN\overset{\frown}{\underset{\smile}{}}NH \longrightarrow$

b. 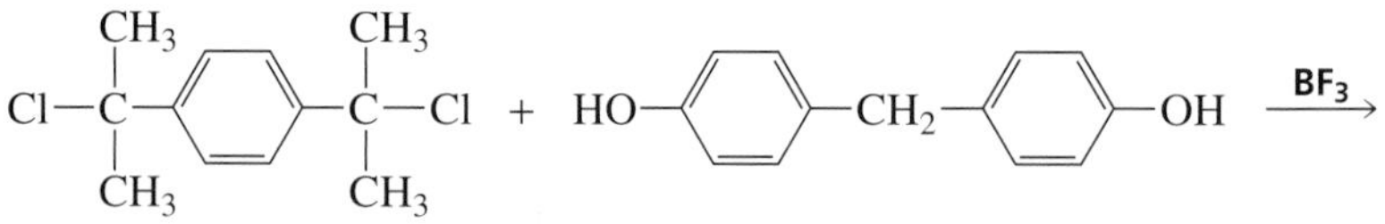 $Cl{-}\overset{\underset{\displaystyle CH_3}{|}}{\underset{\underset{\displaystyle CH_3}{|}}{C}}{-}$[benzene ring]$-\overset{\underset{\displaystyle CH_3}{|}}{\underset{\underset{\displaystyle CH_3}{|}}{C}}{-}Cl \ + \ HO{-}$[benzene ring]$-CH_2-$[benzene ring]$-OH \ \xrightarrow{\ BF_3\ }$

c. $H_2N{-}$[benzene ring]$-OCH_2CH_2CH_2O{-}$[benzene ring]$-NH_2 \ + \ H\overset{\displaystyle O}{\overset{\|}{C}}{-}\overset{\displaystyle O}{\overset{\|}{C}}H \longrightarrow$

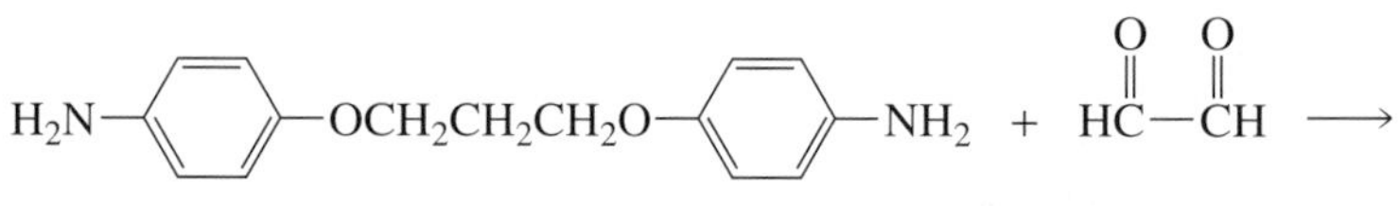

d. 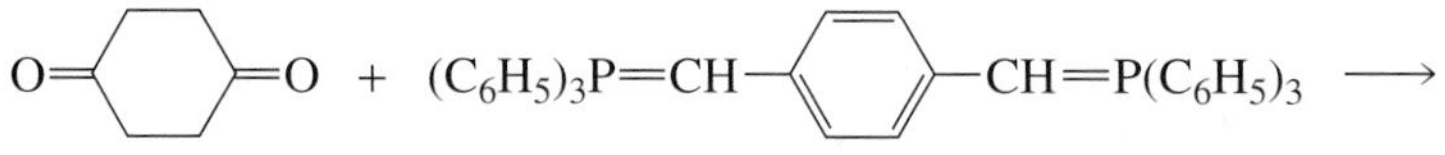 $O{=}$[cyclohexane ring]$={=}O \ + \ (C_6H_5)_3P{=}CH{-}$[benzene ring]$-CH{=}P(C_6H_5)_3 \longrightarrow$

24. Draw the structure of the monomer or monomers used to synthesize the following polymers. Indicate whether each polymer is a chain-growth polymer or a step-growth polymer.

a. $-CH_2CH-$
    $|$
    CH_2CH_3

b. $-CH_2CHO-$
    $|$
    CH_3

c. $-CH_2\overset{\underset{\displaystyle CH_3}{|}}{C}{=}CHCH_2-$

d. $-CH_2CH_2CH_2CH_2\overset{\displaystyle O}{\overset{\|}{C}}O-$

e. 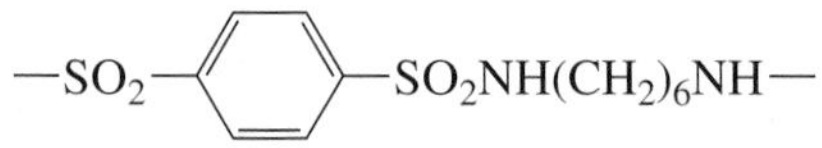 $-SO_2-$[benzene ring]$-SO_2NH(CH_2)_6NH-$

f. 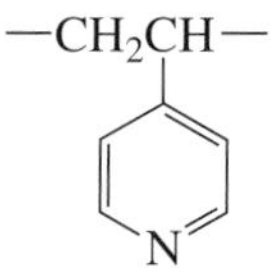 $-CH_2CH-$ [attached to pyridine ring with N]

g. 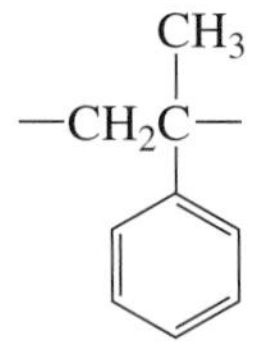 $-CH_2\overset{\underset{\displaystyle CH_3}{|}}{C}{-}$ [attached to benzene ring]

h. 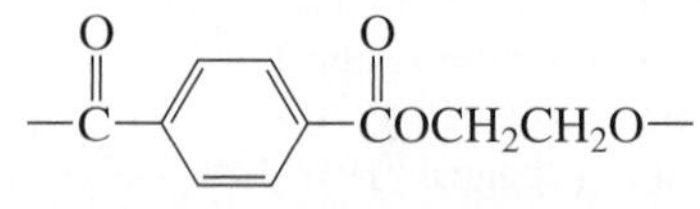 $-\overset{\displaystyle O}{\overset{\|}{C}}{-}$[benzene ring]$-\overset{\displaystyle O}{\overset{\|}{C}}OCH_2CH_2O-$

25. Explain why the configuration of a polymer of isobutylene is not isotactic, syndio-tactic, or atactic.

26. Draw short segments of the polymers obtained from the following compounds under the given reaction conditions:

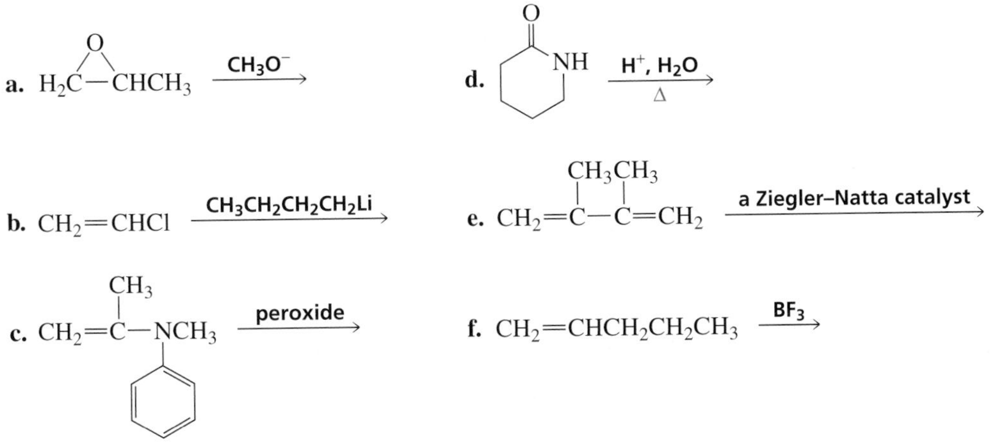

a. H$_2$C—CHCH$_3$ (epoxide) $\xrightarrow{\text{CH}_3\text{O}^-}$

b. CH$_2$=CHCl $\xrightarrow{\text{CH}_3\text{CH}_2\text{CH}_2\text{CH}_2\text{Li}}$

c. CH$_2$=C(CH$_3$)—NCH$_3$ (phenyl) $\xrightarrow{\text{peroxide}}$

d. (lactam with NH) $\xrightarrow[\Delta]{\text{H}^+,\ \text{H}_2\text{O}}$

e. CH$_2$=C(CH$_3$)—C(CH$_3$)=CH$_2$ $\xrightarrow{\text{a Ziegler–Natta catalyst}}$

f. CH$_2$=CHCH$_2$CH$_2$CH$_3$ $\xrightarrow{\text{BF}_3}$

27. Quiana is a synthetic fabric that feels very much like silk.
   a. What monomers are used to synthesize Quiana?
   b. Is Quiana a nylon or a polyester?

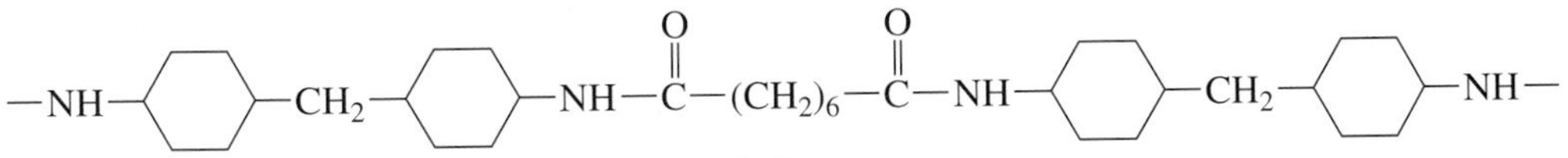

—NH—⟨⟩—CH$_2$—⟨⟩—NH—C(=O)—(CH$_2$)$_6$—C(=O)—NH—⟨⟩—CH$_2$—⟨⟩—NH—

Quiana

28. Explain why a random copolymer is obtained when 3,3-dimethyl-1-butene undergoes cationic polymerization.

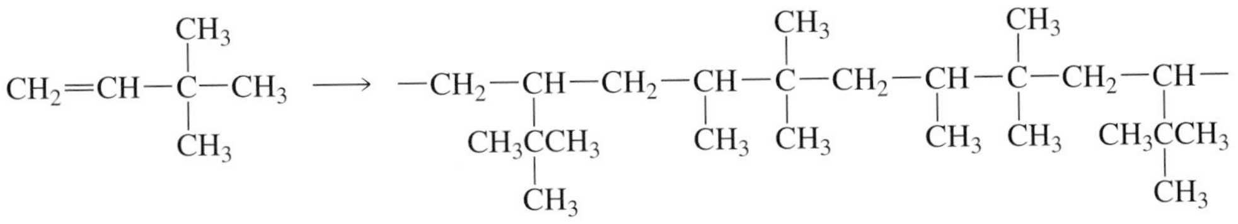

CH$_2$=CH—C(CH$_3$)(CH$_3$)—CH$_3$ ⟶ —CH$_2$—CH(CH$_3$CCH$_3$CH$_3$)—CH$_2$—CH—C(CH$_3$)(CH$_3$)—CH$_2$—CH(CH$_3$)(CH$_3$)—C(CH$_3$)(CH$_3$)—CH$_2$—CH(CH$_3$CCH$_3$CH$_3$)—

29. Polly Propylene starts two polymerization reactions. One flask contains a monomer that polymerizes by a chain-growth mechanism, and the other flask contains a monomer that polymerizes by a step-growth mechanism. When the reactions are terminated and the contents of the flasks analyzed, one flask contains a high-molecular-weight polymer and some monomer but very little material of intermediate molecular weight. The other flask contains mainly material of intermediate molecular weight and very little monomer or high-molecular-weight material. Which flask is which? Explain.

30. Poly(vinyl alcohol) is a polymer used to make fibers and adhesives. It is synthesized by hydrolysis or alcoholysis of the polymer obtained from polymerization of vinyl acetate.
   a. Why is poly(vinyl alcohol) not prepared by polymerizing vinyl alcohol?
   b. Is poly(vinyl acetate) a polyester?

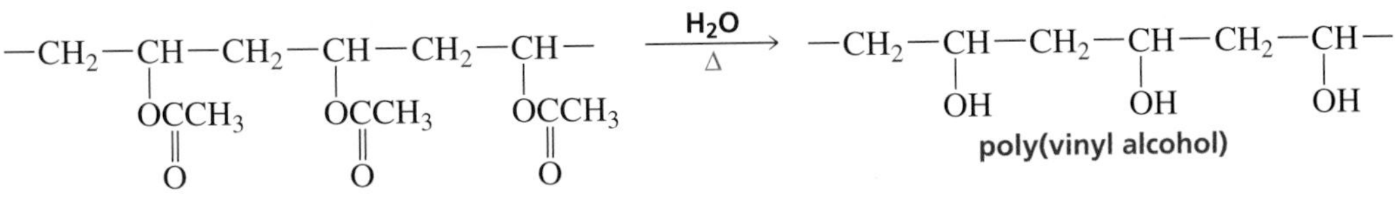

—CH$_2$—CH(OCCH$_3$, O)—CH$_2$—CH(OCCH$_3$, O)—CH$_2$—CH(OCCH$_3$, O)— $\xrightarrow[\Delta]{\text{H}_2\text{O}}$ —CH$_2$—CH(OH)—CH$_2$—CH(OH)—CH$_2$—CH(OH)—

poly(vinyl acetate)   **poly(vinyl alcohol)**

31. Five different repeating units are found in the polymer obtained by cationic polymerization of 4-methyl-1-pentene. Identify these repeating units.

32. If a peroxide is added to styrene, the polymer known as polystyrene is formed. If a small amount of 1,4-divinylbenzene is added to the reaction mixture, a stronger and more rigid polymer is formed. Draw a short section of this more rigid polymer.

$$CH_2=CH-\langle\text{benzene ring}\rangle-CH=CH_2$$

1,4-divinylbenzene

33. A particularly strong and rigid polyester used for electronic parts is marketed under the trade name Glyptal. Glyptal is a polymer of terephthalic acid and glycerol. Draw a segment of the polymer and explain why it is so strong.

34. Draw a short section of the polymer obtained from anionic polymerization of β-propiolactone.

35. Which monomer would give a greater yield of polymer, 5-hydroxypentanoic acid or 6-hydroxyhexanoic acid? Explain your choice.

36. When rubber balls and other objects made of natural rubber are exposed to the air for long periods of time, they turn brittle and crack. This does not happen to objects made of polyethylene. Explain.

37. Why do vinyl raincoats become brittle as they get old, even if they are not exposed to air or to any pollutants?

38. The polymer shown here is synthesized by hydroxide-ion-promoted hydrolysis of an alternating copolymer of *para*-nitrophenyl methacrylate and acrylate.
  a. Propose a mechanism for the formation of the alternating copolymer.
  b. Explain why hydrolysis of the copolymer to form the polymer occurs much more rapidly than hydrolysis of *para*-nitrophenyl acetate.

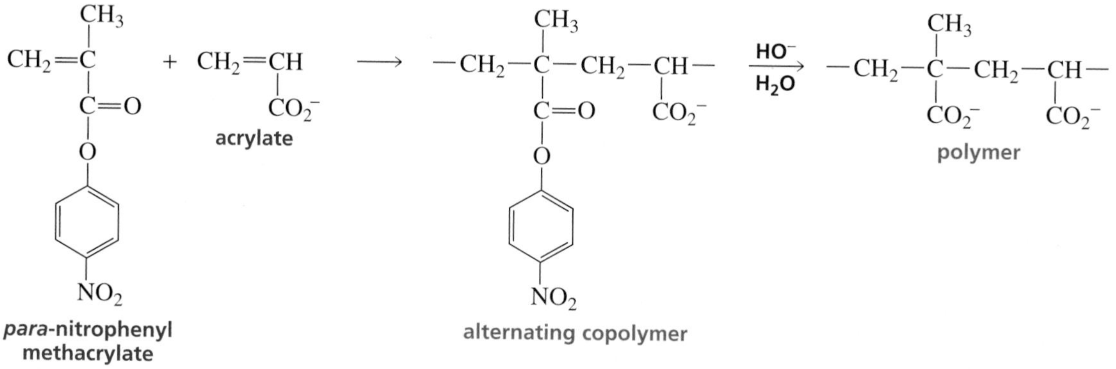

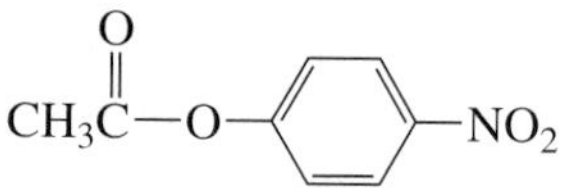

para-nitrophenyl methacrylate + acrylate ⟶ alternating copolymer $\xrightarrow[\text{H}_2\text{O}]{\text{HO}^-}$ polymer

para-nitrophenyl acetate

39. An alternating copolymer of styrene and vinyl acetate can be turned into a graft copolymer by hydrolyzing it and then adding ethylene oxide. Draw the structure of the graft copolymer.

40. How could "head-to-head" poly(vinyl bromide) be synthesized?

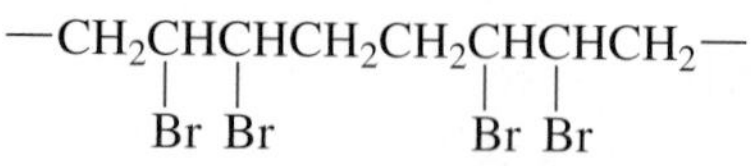

$$-CH_2CHCHCH_2CH_2CHCHCH_2-$$
$$\qquad\;|\;\;|\qquad\qquad\;\;|\;\;|$$
$$\qquad Br\;Br\qquad\qquad\;Br\;Br$$

"head-to-head" poly(vinyl bromide)

27

Heterocyclic Compounds

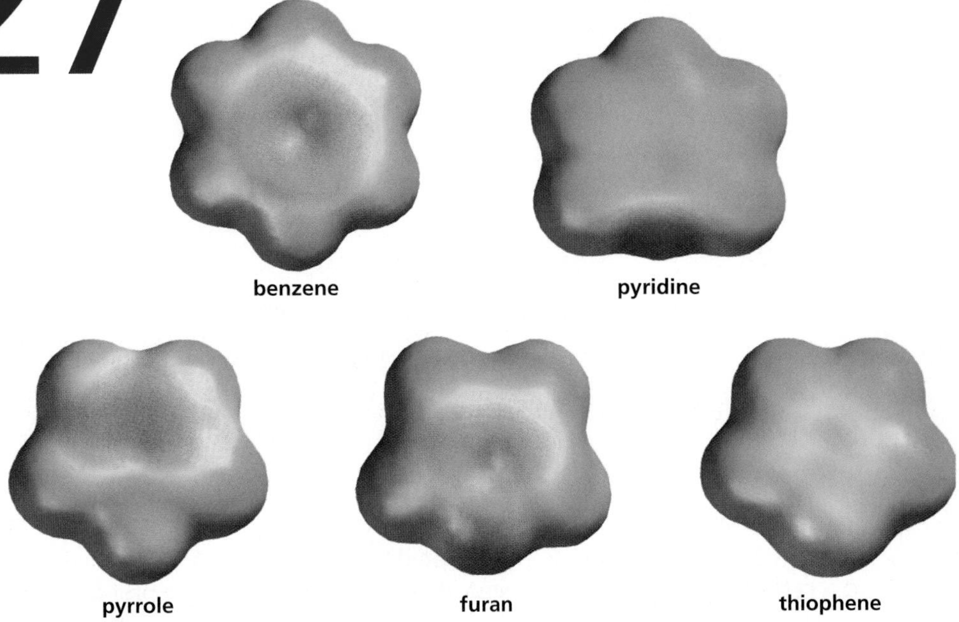

benzene

pyridine

pyrrole

furan

thiophene

Heterocyclic compounds, or **heterocycles,** are cyclic compounds in which one or more of the atoms of the ring are heteroatoms. A **heteroatom** is an atom other than carbon. The name comes from the Greek word *heteros,* which means "different." A variety of atoms, such as N, O, S, Se, P, Si, B, and As, can be incorporated into ring structures. In this chapter, however, we will consider only the most prevalent heterocyclic compounds—the ones that contain the heteroatoms N, O, and S.

Heterocycles make up an exceedingly important class of compounds—more than half of all known organic compounds are heterocycles. Almost all the compounds we know as drugs, most vitamins (Chapter 23), and many other natural products are heterocycles.

A **natural product** is a compound synthesized by a plant or an animal. **Alkaloids** are natural products containing one or more nitrogen heteroatoms that are found in the leaves, bark, roots, or seeds of plants. Examples include caffeine (found in tea leaves, coffee beans, and cola nuts) and nicotine (found in tobacco leaves). Morphine is an alkaloid obtained from opium, the juice derived from a species of poppy. Morphine is 50 times stronger than aspirin as an analgesic, but it is addictive and suppresses respiration. Heroin is a synthetic compound that is made by acetylating morphine (Section 30.3).

Other heterocycles include Valium, a synthetic tranquilizer, and serotonin, a neurotransmitter. Serotonin is responsible for, among other things, the feeling of having had enough to eat. When food is ingested, brain neurons are signaled to release serotonin. A widely used diet drug (actually a combination of two drugs, fenfluramine and phentermine) popularly known as fen/phen works by causing brain neurons to release extra serotonin (Chapter 15, page 627). After finding that many of those who took fenfluramine had abnormal echocardiograms due to heart valve problems, the Food and Drug Administration asked the manufacturer of these diet drugs to withdraw their products. There is some evidence that faulty metabolism of serotonin plays a role in manic depressive psychoses.

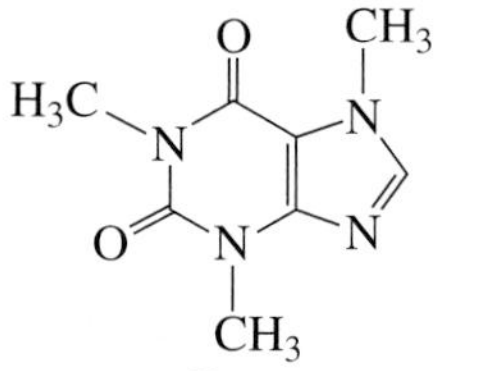

caffeine 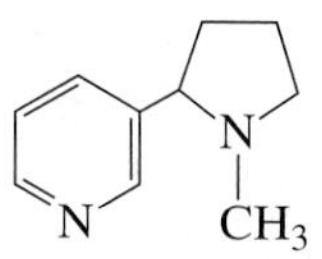   nicotine 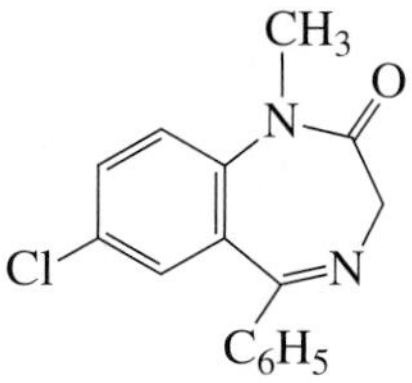   Valium 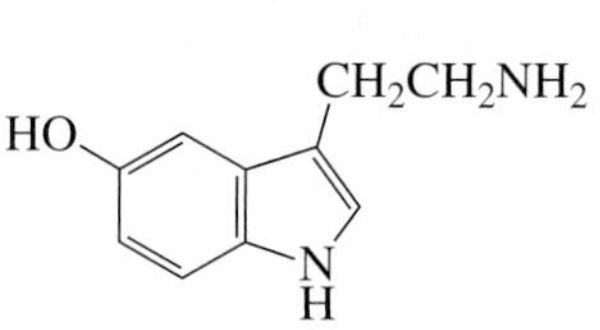   serotonin

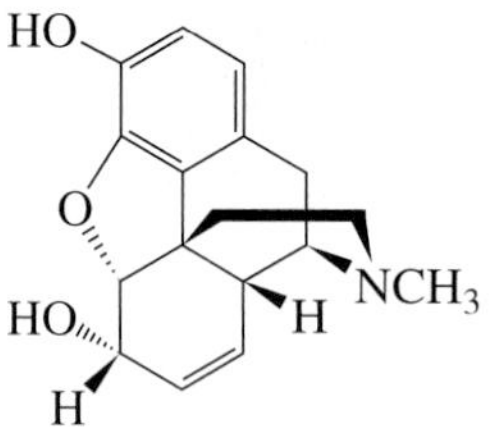

morphine 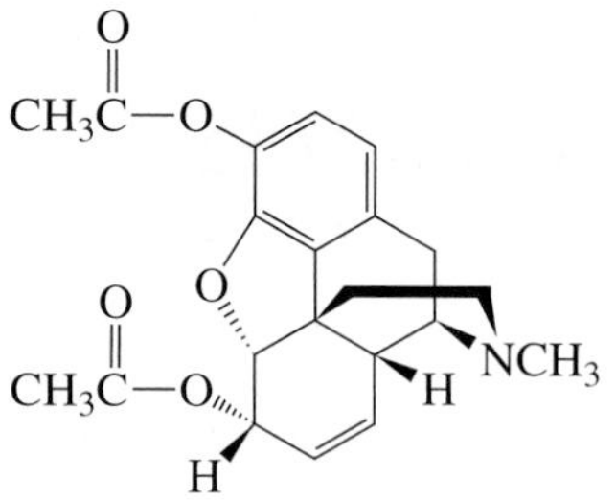   heroin

27.1 SATURATED HETEROCYCLES

Saturated heterocycles do not contain double bonds. A saturated heterocycle can be named as a cycloalkane using a prefix to denote the heteroatom ("aza" for nitrogen, "oxa" for oxygen, or "thia" for sulphur). There are, however, other acceptable names. Some of the more commonly used names are shown here.

oxacyclopropane azacyclopropane thiacyclopropane oxacyclobutane  thiacyclobutane  oxacyclopentane
  oxirane   aziridine   thiirane
ethylene oxide   ethyleneimine   tetrahydrofuran

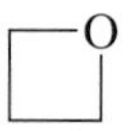

Heterocyclic rings are numbered so that the heteroatoms have the lowest possible numbers.

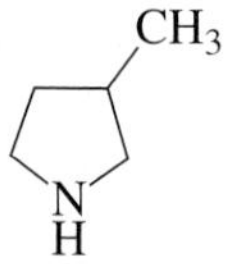

3-methylazacyclopentane 2-methylazacyclohexane N-ethylazacyclopentane
  3-methylpyrrolidine   2-methylpiperidine   N-ethylpyrrolidine

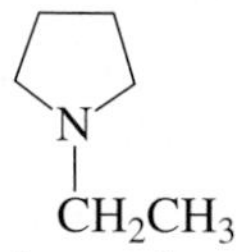

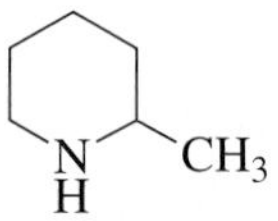

1,3-dithiacyclohexane   2,2-dimethyloxacyclohexane
  1,3-dithiane   2,2-dimethyltetrahydropyran

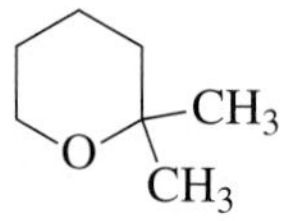

PROBLEM 1◆

Name the following compounds:

a. 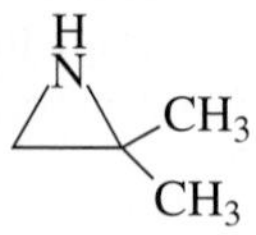   b.

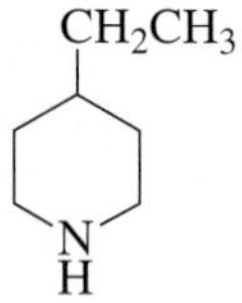

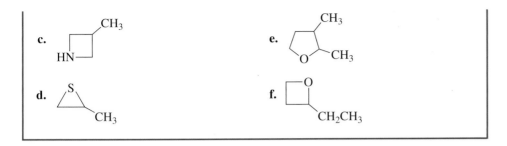

Saturated heterocycles containing five or more atoms have physical and chemical properties typical of acyclic compounds that contain the same heteroatom. For example, tetrahydrofuran, tetrahydropyran, and 1,4-dioxane are typical ethers. Because ethers have little chemical reactivity, these compounds are commonly used as solvents (Section 11.5).

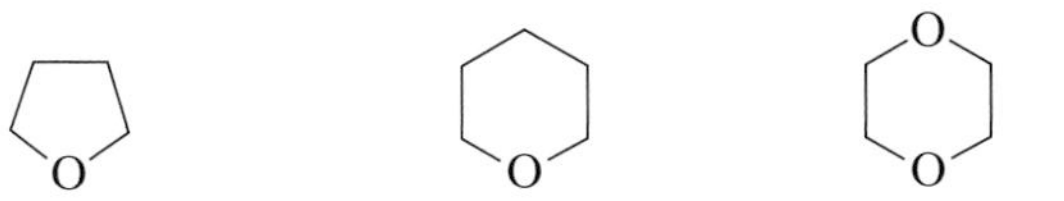

| tetrahydrofuran | tetrahydropyran | 1,4-dioxane |

Pyrrolidine, piperidine, and morpholine are typical secondary amines. *N*-Methylpyrrolidine and quinuclidine are typical tertiary amines. The conjugate acids of these amines have pK_a values expected for ammonium ions.

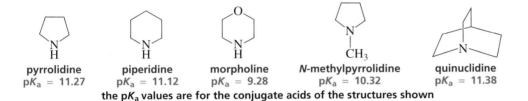

| pyrrolidine | piperidine | morpholine | *N*-methylpyrrolidine | quinuclidine |
| pK_a = 11.27 | pK_a = 11.12 | pK_a = 9.28 | pK_a = 10.32 | pK_a = 11.38 |

the pK_a values are for the conjugate acids of the structures shown

Because of its three-membered ring, the aziridinium ion has a considerably lower pK_a than that of a typical secondary ammonium ion. The C—N—C internal bond angle is smaller than usual, causing the external bond angles to be somewhat larger than usual. The larger external bond angles cause the orbital that nitrogen uses to overlap the orbital of hydrogen to have more *s* character than a typical sp^3 orbital. (Recall from Section 1.14 that the larger the bond angle, the greater the *s* character.) The greater *s* character makes nitrogen more electronegative, which lowers the pK_a (Section 5.9).

3-D Molecules:
Aziridinium ion;
Pyrrolidine;
Piperidine;
Morpholine

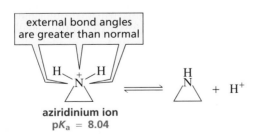

aziridinium ion
pK_a = 8.04

PROBLEM 2◆

Why is the pK_a of the conjugate acid of morpholine significantly lower than the pK_a of the conjugate acid of piperidine?

PROBLEM 3◆

a. Draw the structure of 3-quinuclidinone.

b. What is the approximate pK_a of its conjugate acid?

c. Which has a lower pK_a, the conjugate acid of 3-bromoquinuclidine or the conjugate acid of 3-chloroquinuclidine?

Compounds with saturated five- and six-membered rings undergo the same reactions as their open-chain analogs.

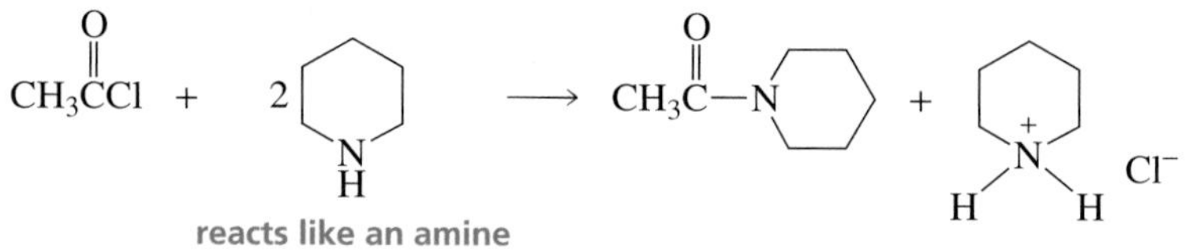

reacts like an amine

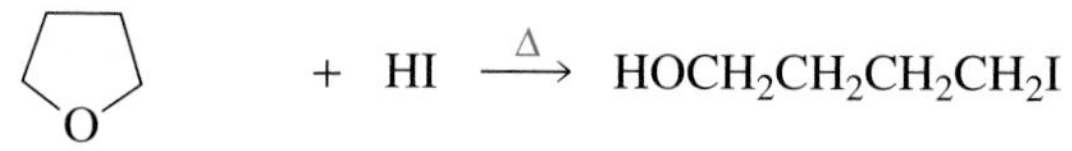

reacts like an ether

PROBLEM 4◆

Give the product of each of the following reactions:

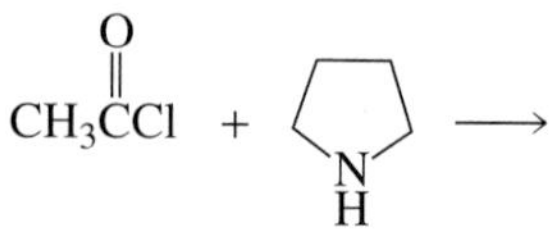

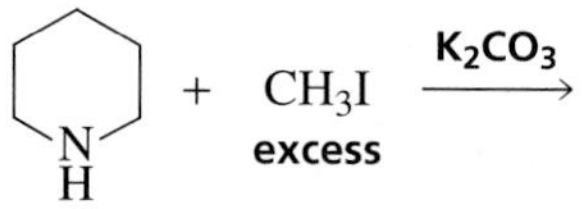

When you learned about the reactions of cyclopropanes (Section 8.7), epoxides (Section 11.6), and arene oxides (Section 11.7), you saw that compounds with three-membered rings do not have the same chemical reactivity as their open-chain analogs. Three-membered ring compounds readily undergo ring-opening reactions because this releases the angle strain and torsional strain inherent in their structures.

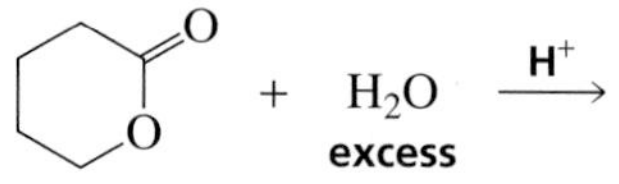

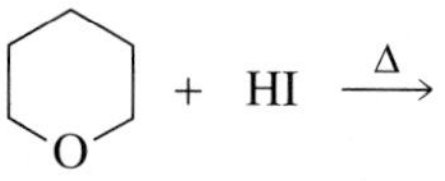

strain energy:	25 kcal/mol	14 kcal/mol	13 kcal/mol	9 kcal/mol
	105 kJ/mol	59 kJ/mol	54 kJ/mol	38 kJ/mol

Epoxides undergo ring-opening reactions under both acidic and basic conditions, but aziridines—three-membered rings that contain a nitrogen heteroatom—only undergo ring-opening reactions under acidic conditions (Sections 11.6 and 22.3). Apparently, relief of strain is not sufficient to make up for the highly basic (poor) leaving group when the nitrogen atom is not protonated.

27.2 UNSATURATED FIVE-MEMBERED-RING HETEROCYCLES

Pyrrole, Furan, and Thiophene

Pyrrole, furan, and **thiophene** are five-membered-ring heterocycles. Each of these compounds has three pairs of delocalized π electrons. Two of the pairs are shown as π bonds, and one pair is shown as a pair of nonbonding electrons on the heteroatom. Furan and thiophene have a second pair of nonbonding electrons that are not part of the π cloud. These electrons are in an sp^2 orbital perpendicular to the *p* orbitals. Pyrrole, furan, and thiophene are aromatic because they are cyclic and planar, every carbon in the ring has a *p* orbital, and the π cloud contains *three* pairs of π electrons (Section 14.1).

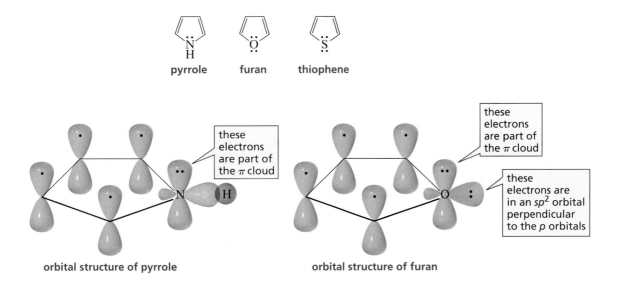

pyrrole furan thiophene

orbital structure of pyrrole

orbital structure of furan

Pyrrole is an extremely weak base because the pair of electrons shown as nonbonding electrons is part of the π cloud. When pyrrole is protonated, it's aromaticity is destroyed. Therefore, the conjugate acid of pyrrole is a very strong acid ($pK_a = -3.8$).

The resonance contributors of pyrrole show that nitrogen donates the electrons depicted in its structure as nonbonding electrons into the five-membered ring.

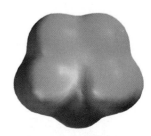

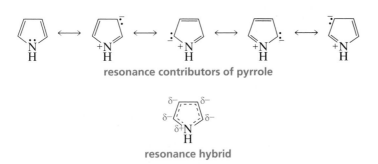

resonance contributors of pyrrole

resonance hybrid

pyrrolidine

pyrrole

Pyrrolidine has a dipole moment of 1.57 D because the nitrogen atom is electron-withdrawing. Pyrrole has a slightly larger dipole moment (1.80 D), but the electrostatic potential maps show the two dipole moments are in opposite directions. This indicates that nitrogen's ability to donate electrons into the ring by resonance more than makes up for its inductive electron withdrawal (Section 15.3).

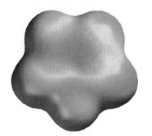

$\mu = 1.57 \text{ D}$ $\mu = 1.80 \text{ D}$

3-D Molecules:
Pyrrole;
Furan;
Thiophene

In Section 6.6 we saw that the more stable and more nearly equivalent the resonance contributors, the greater the resonance energy. The resonance energies of pyrrole, furan, and thiophene are not as great as the resonance energies of benzene and the cyclopentadienyl anion, compounds for which the resonance contributors are all equivalent. Thiophene, with the least electronegative heteroatom, has the greatest resonance energy of these five-membered heterocycles, and furan, with the most electronegative heteroatom, has the smallest resonance energy.

relative resonance energies of some aromatic compounds

Pyrrole, furan, and thiophene are more reactive than benzene toward electrophilic aromatic substitution.

Because pyrrole, furan, and thiophene are aromatic, they undergo electrophilic aromatic substitution reactions. All three compounds are more reactive than benzene toward electrophilic substitution because, as shown for pyrrole, there is increased electron density on each of the carbons due to delocalization of the nonbonding pair of electrons onto the ring carbon atoms. Because the rate-limiting step of an electrophilic substitution reaction is attachment of the electrophile to the aromatic ring, increasing the electron density of the ring carbons increases the rate of electrophilic substitution (Section 15.3).

mechanism for electrophilic aromatic substitution

Pyrrole, furan, and thiophene undergo electrophilic substitution at C-2.

+ Br_2 ⟶ 2-bromofuran + HBr

+ HNO_3 $\xrightarrow{(CH_3CO)_2O}$ 2-methyl-5-nitropyrrole + H_2O

Pyrrole, furan, and thiophene undergo electrophilic substitution preferentially at C-2.

Substitution occurs preferentially at C-2 because the intermediate obtained by attaching a substituent at this position is more stable than the intermediate obtained by attaching a substituent at C-3. Both intermediates have a resonance contributor in which all the atoms (except H) have complete octets. The intermediate resulting from C-2 substitution of pyrrole has *two* additional resonance contributors—each has a positive charge on a relatively stable secondary allylic carbon. The intermediate resulting from C-3 substitution has only *one* additional resonance contributor—it has a positive charge on a secondary carbon, so its predicted stability is less than that of a resonance contributor with a positive charge on a secondary allylic carbon (Figure 27.1).

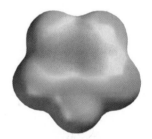

Figure 27.1
Structures of the intermediates that can be formed from the reaction of an electrophile with pyrrole at C-2 and C-3.

If both positions adjacent to the heteroatom are occupied, electrophilic substitution will take place at C-3.

3-bromo-2,5-dimethylfuran

Furan is not as reactive as pyrrole in electrophilic substitution reactions. The oxygen of furan is more electronegative than the nitrogen of pyrrole, so the oxygen is not as effective as nitrogen in donating electrons into the ring. Thiophene is less reactive than furan toward electrophilic substitution because sulfur's p electrons are in a $3p$ orbital, which overlaps less effectively than the $2p$ orbital of nitrogen or oxygen with the $2p$ orbital of carbon. The electrostatic potential maps illustrate the different electron densities of the three rings.

relative reactivity toward electrophilic aromatic substitution

pyrrole   furan   thiophene   benzene

The relative reactivities of the five-membered-ring heterocycles are reflected in the Lewis acid required to catalyze a Friedel–Crafts acylation reaction (Section 14.12). Benzene requires $AlCl_3$, a relatively strong Lewis acid. Thiophene is more reactive than benzene, so it can undergo a Friedel–Crafts reaction using $SnCl_4$, a weaker Lewis acid. An even weaker Lewis acid, BF_3, can be used when the substrate is furan. Pyrrole is so reactive that an anhydride is used instead of a more reactive acyl chloride, and no catalyst is necessary.

phenylethanone

2-acetylthiophene

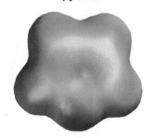

pyrrole

furan

thiophene

The resonance hybrid of pyrrole indicates that there is a partial positive charge on the nitrogen. Therefore, when pyrrole is protonated, it is not protonated on nitrogen—it is protonated on C-2. Remember, a proton is an electrophile and, like other electrophiles, attaches to the position most vulnerable to electrophilic attack.

Pyrrole is unstable in strongly acidic solutions because it polymerizes readily.

Neutral pyrrole is more acidic ($pK_a = \sim 17$) than a saturated neutral amine ($pK_a = \sim 36$) because the nitrogen in pyrrole is sp^2 hybridized and thus is more electronegative than the sp^3 nitrogen of a saturated amine. In addition, because the nitrogen in pyrrole donates electrons into the ring, it has a partial positive charge, which further increases its electron-withdrawing ability and, therefore, the acidity of its proton (Table 27.1).

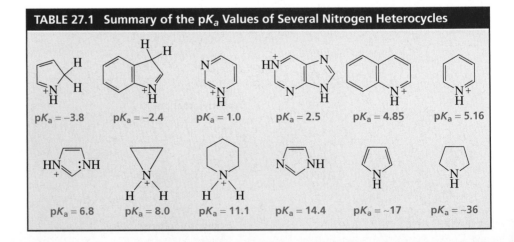

TABLE 27.1 Summary of the pK_a Values of Several Nitrogen Heterocycles

$pK_a = -3.8$	$pK_a = -2.4$	$pK_a = 1.0$	$pK_a = 2.5$	$pK_a = 4.85$	$pK_a = 5.16$
$pK_a = 6.8$	$pK_a = 8.0$	$pK_a = 11.1$	$pK_a = 14.4$	$pK_a = \sim 17$	$pK_a = \sim 36$

PROBLEM 5

When pyrrole is added to a dilute solution of D_2SO_4 in D_2O, 2-deuteriopyrrole is formed. Propose a mechanism to account for the formation of this compound.

PROBLEM 6

Use resonance contributors to explain why pyrrole is protonated on C-2 rather than on nitrogen.

PROBLEM 7

Even though nitrogen is considerably more electronegative than carbon, pyrrole ($pK_a \sim 17$) is less acidic than cyclopentadiene ($pK_a = 15$). Explain.

Indole, Benzofuran, and Benzothiophene

Indole, benzofuran, and benzothiophene contain a five-membered aromatic ring fused to a benzene ring. The rings are numbered in a way that gives the heteroatom the lowest possible number. Indole, benzofuran, and benzothiophene are aromatic because they are cyclic and planar, every carbon in the ring has a p orbital, and the π cloud contains *five* pairs of π electrons (Section 14.1). Notice that the electrons shown as nonbonding electrons in indole are part of the π cloud. Therefore, the conjugate acid of indole, like the conjugate acid of pyrrole, is a strong acid ($pK_a = -2.4$).

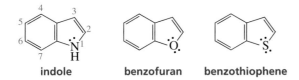

indole   benzofuran   benzothiophene

Electrophilic aromatic substitution in indole, benzofuran, and benzothiophene takes place on the five-membered ring because it is more reactive toward electrophilic aromatic substitution than the benzene ring. Indole undergoes electrophilic aromatic substitution primarily at C-3, benzofuran primarily at C-2, and benzothiophene about equally well at C-2 and C-3.

3-D Molecules:
Indole;
Benzofuran;
Benzothiophene

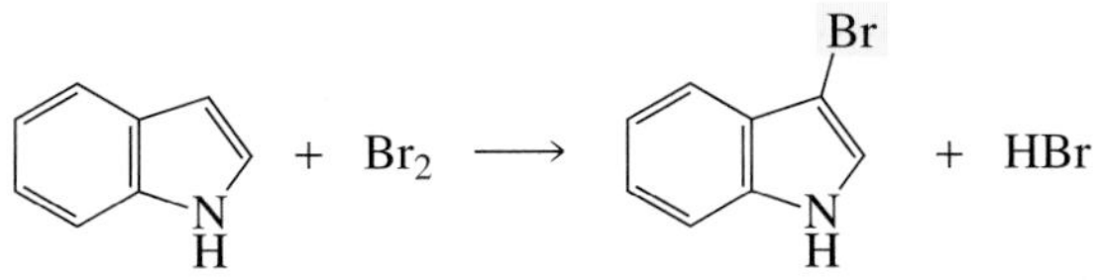

indole   + Br₂ ⟶   3-bromoindole   + HBr

benzofuran + $CH_3\overset{O}{\overset{\|}{C}}O\overset{O}{\overset{\|}{C}}CH_3$ ⟶ 2-acetylbenzofuran $\overset{O}{\overset{\|}{C}}CH_3$ + $CH_3\overset{O}{\overset{\|}{C}}OH$

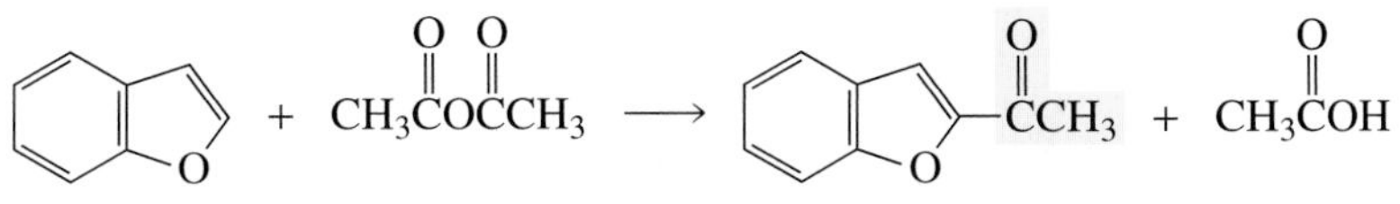

benzothiophene + Br₂ ⟶ 3-bromobenzothiophene + 2-bromobenzothiophene —Br + HBr

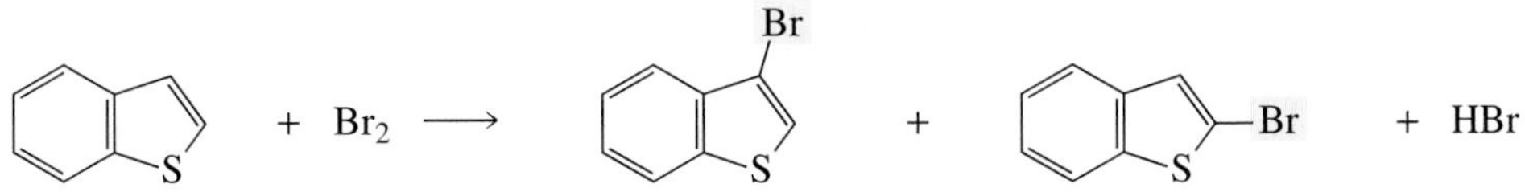

The difference in orientation can be explained by comparing the stabilities of the intermediates obtained as a result of electrophilic attack at the C-2 and C-3 positions (Figure 27.2). Substitution at C-2 gives only one resonance contributor with a relatively stable predicted stability because the six-π-electron system of the benzene ring is disrupted in the other four resonance contributors. Substitution at C-3 also gives a resonance contributor with a relatively stable predicted stability. The predicted stability of its second resonance contributor depends on the nature of the heteroatom. In the case of indole, the second resonance contributor is predicted to be *more* stable than the four contributors in which the aromaticity of the benzene ring has been lost, so indole undergoes electrophilic substitution primarily at C-3. In the case of benzofuran, the second resonance contributor is predicted to be *less* stable than the four contributors with a disrupted benzene ring because oxygen is more electronegative than nitrogen, so oxygen is less able to tolerate a positive charge. Benzofuran, therefore, undergoes electrophilic aromatic substitution primarily at C-2. Benzothiophene undergoes electrophilic aromatic substitution equally well at C-2 and C-3, indicating that the two intermediates have similar stabilities.

▲ **Figure 27.2**
Structures of the intermediates that can be formed from the reaction of an electrophile with indole at C-2 and C-3.

27.3 UNSATURATED SIX-MEMBERED-RING HETEROCYCLES

Pyridine

When one of the carbons of a benzene ring is replaced by a nitrogen, the resulting compound is called **pyridine.**

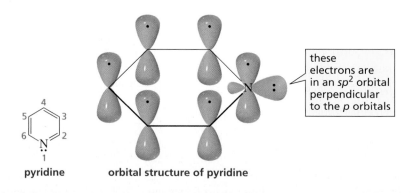

these electrons are in an sp^2 orbital perpendicular to the p orbitals

pyridine   orbital structure of pyridine

Pyridine is a tertiary amine and undergoes reactions characteristic of tertiary amines. For example, pyridine undergoes S_N2 reactions with alkyl halides (Section 9.3), and it reacts with hydrogen peroxide to form an *N*-oxide (Section 11.11).

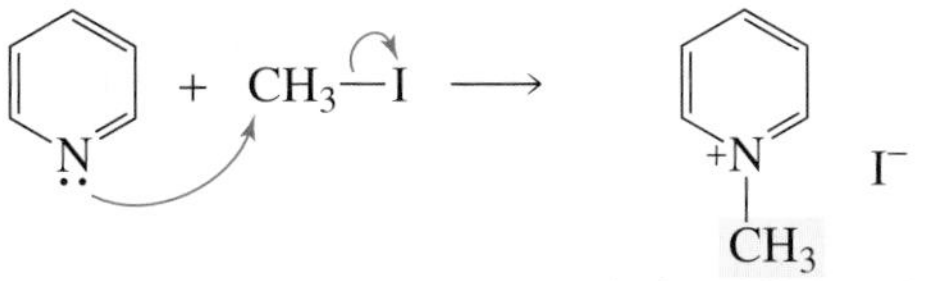

N-methylpyridinium iodide

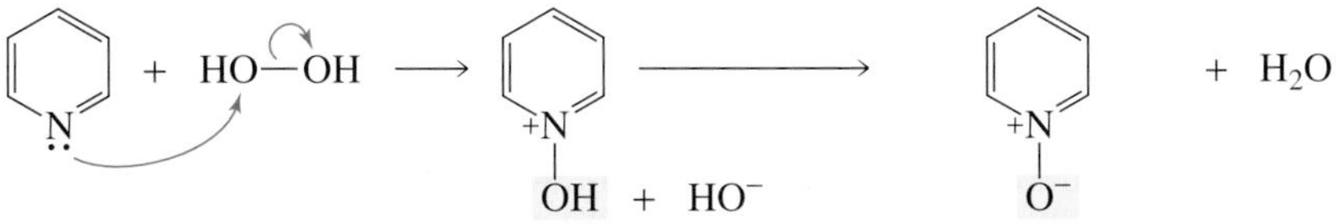

$pK_a = 0.79$ **pyridine-N-oxide**

PROBLEM 8 / SOLVED

Will an amide be formed from the reaction of an acyl chloride with an aqueous solution of pyridine? Explain your answer.

SOLUTION An amide will *not* be formed because the positively charged nitrogen causes pyridine to be an excellent leaving group. Therefore, the final product of the reaction will be a carboxylic acid. (If the final pH of the solution is greater than the pK_a of the carboxylic acid, the carboxylic acid will be predominantly in its basic form.)

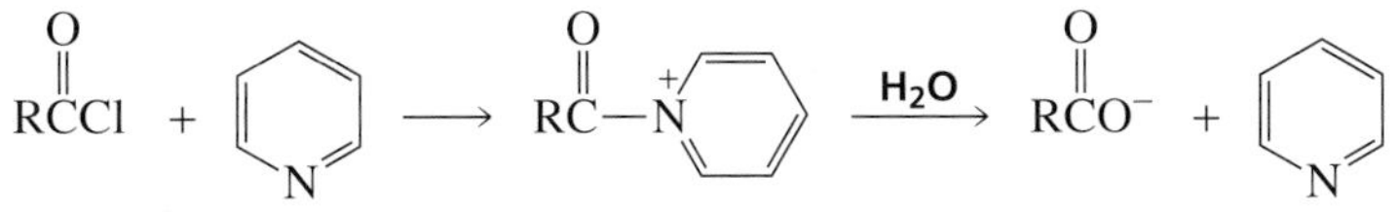

The pyridinium ion is a stronger acid than a typical ammonium ion because the acidic hydrogen of a pyridinium ion is attached to an sp^2 hybridized nitrogen, which is more electronegative than an sp^3 hybridized nitrogen (Section 5.9).

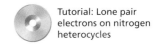

Tutorial: Lone pair electrons on nitrogen heterocycles

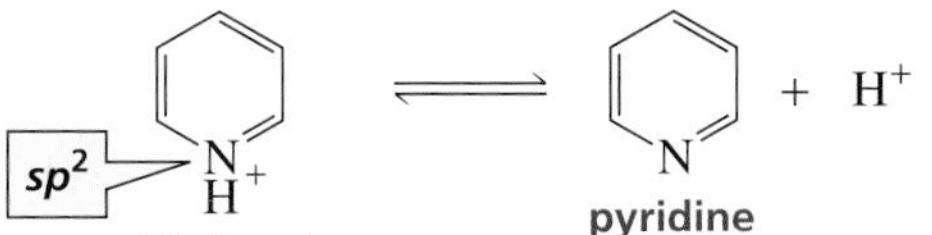

pyridinium ion
$pK_a = 5.16$

pyridine

piperidinium ion
$pK_a = 11.12$

piperidine

Pyridine is aromatic. Like benzene, it has two uncharged resonance contributors. Because of the electron-withdrawing nitrogen, it also has three charged resonance contributors that benzene does not have.

resonance contributors of pyridine

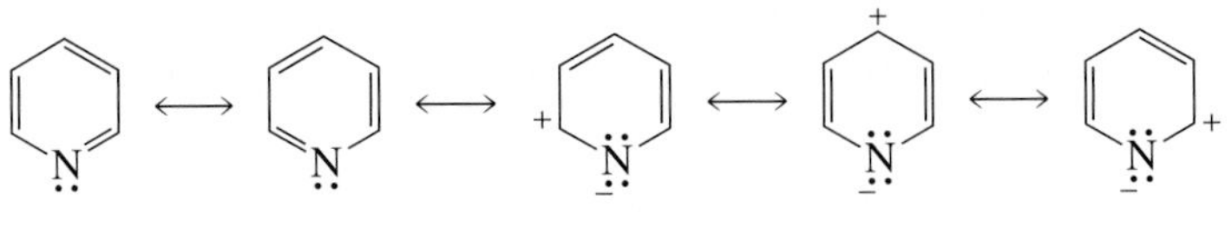

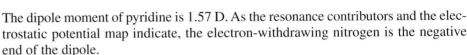

The dipole moment of pyridine is 1.57 D. As the resonance contributors and the electrostatic potential map indicate, the electron-withdrawing nitrogen is the negative end of the dipole.

$$\mu = 1.57 \text{ D}$$

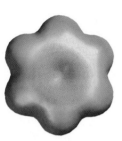

benzene

Because it is aromatic, pyridine (like benzene) undergoes electrophilic aromatic substitution reactions.

mechanism for electrophilic aromatic substitution

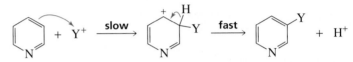

Pyridine's electron-withdrawing nitrogen causes the ring carbons to have significantly less electron density than the ring carbons of benzene. Pyridine, therefore, is less reactive than benzene toward electrophilic aromatic substitution. It is even less reactive than nitrobenzene. (Recall from Section 15.3 that an electron-withdrawing nitro group strongly deactivates a benzene ring toward electrophilic aromatic substitution.)

pyridine

relative reactivity toward electrophilic aromatic substitution

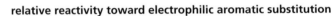

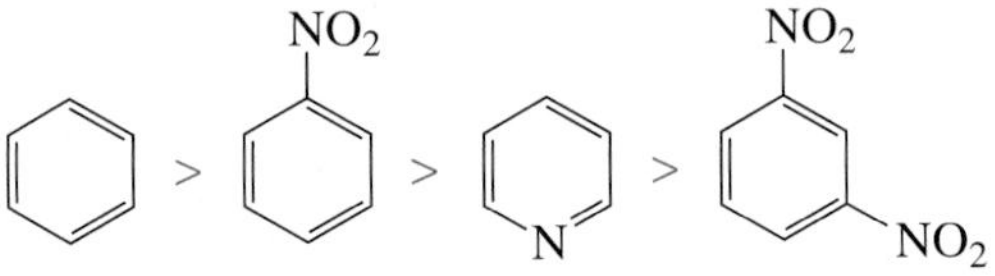

Pyridine undergoes electrophilic aromatic substitution reactions only under vigorous conditions, and the yields of these reactions are often quite low. If the nitrogen becomes protonated under the reaction conditions, the reactivity is further decreased because a positively charged nitrogen is more electron-withdrawing than a neutral nitrogen.

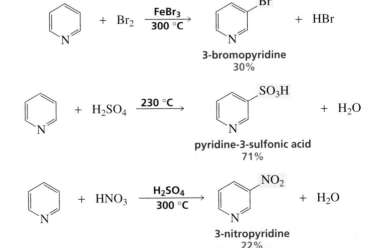

Electrophilic aromatic substitution of pyridine takes place at C-3 because the most stable intermediate is obtained by placing an electrophilic substituent at this position (Figure 27.3). When the substituent is placed at C-2 or C-4, one of the resulting resonance contributors is particularly unstable because its nitrogen atom has an incomplete octet *and* a positive charge.

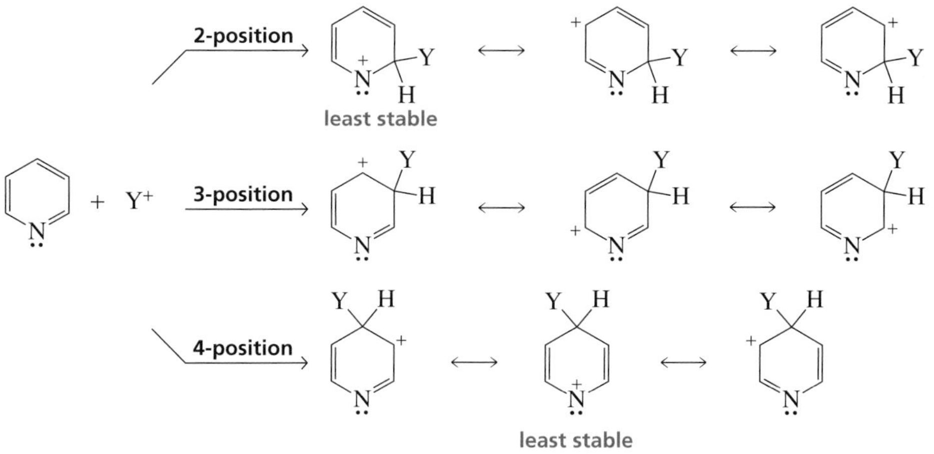

◀ **Figure 27.3**
Structures of the intermediates that can be formed from the reaction of an electrophile with pyridine.

We have seen that highly deactivated benzene rings do not undergo Friedel–Crafts alkylation or acylation reactions. Therefore pyridine, whose reactivity is similar to that of a highly deactivated benzene, does not undergo these reactions.

+ CH₃CH₂Cl $\xrightarrow{\text{AlCl}_3}$ **no electrophilic aromatic substitution reaction**

PROBLEM 9 ◆

Give the product of the following reaction:

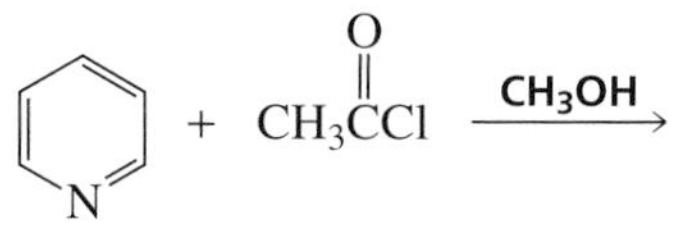

Pyridine-*N*-oxide is considerably more reactive toward electrophilic aromatic substitution than pyridine because the oxygen atom can donate electrons into the ring by resonance. Pyridine-*N*-oxide undergoes electrophilic aromatic substitution primarily at C-4. (Recall from Section 14.10 that the ⁺NO₂ electrophile is generated from the reaction of HNO₃ with H₂SO₄.) Because the *N*-oxide can be converted to pyridine by treatment with PCl₃, electrophilic aromatic substitution of pyridine-*N*-oxide provides a synthetically useful way to introduce an electrophile at the 4-position of pyridine.

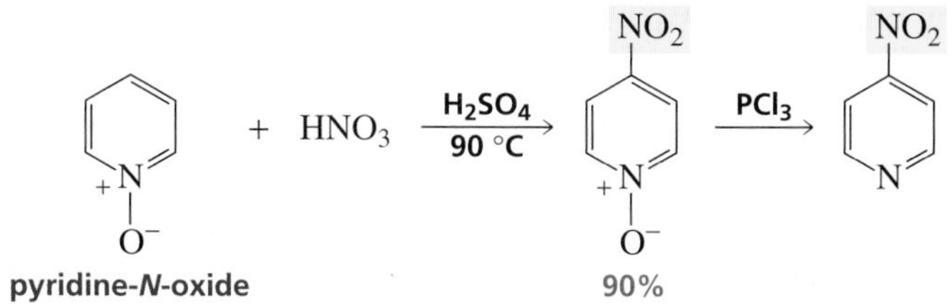

pyridine-*N*-oxide   90%

Pyridine is *more* reactive than benzene toward *nucleophilic* aromatic substitution for the same reason it is *less* reactive than benzene toward *electrophilic* aromatic

Pyridine is *less* reactive than benzene toward electrophilic aromatic substitution and *more* reactive than benzene toward nucleophilic aromatic substitution.

substitution—because the electron-withdrawing nitrogen causes the ring carbons to have significantly less electron density than the ring carbons of benzene.

mechanism for nucleophilic aromatic substitution

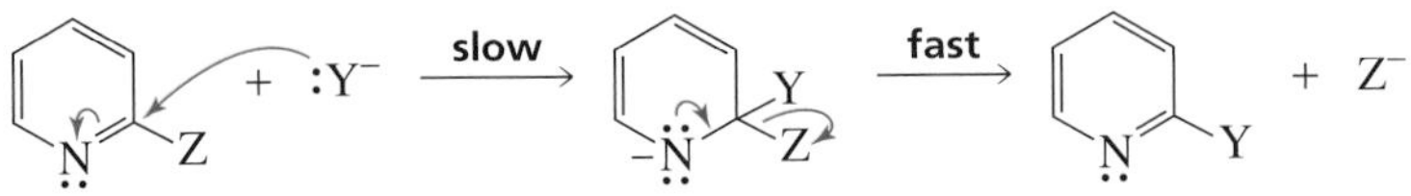

Nucleophilic aromatic substitution of pyridine takes place at C-2 and C-4 because attack at these positions leads to the most stable intermediate. Only when nucleophilic attack occurs at these positions is a resonance contributor obtained that has the greatest electron density on nitrogen, the most electronegative of the ring atoms (Figure 27.4).

Figure 27.4 ▶
Structures of the intermediates that can be formed from the reaction of a nucleophile with pyridine.

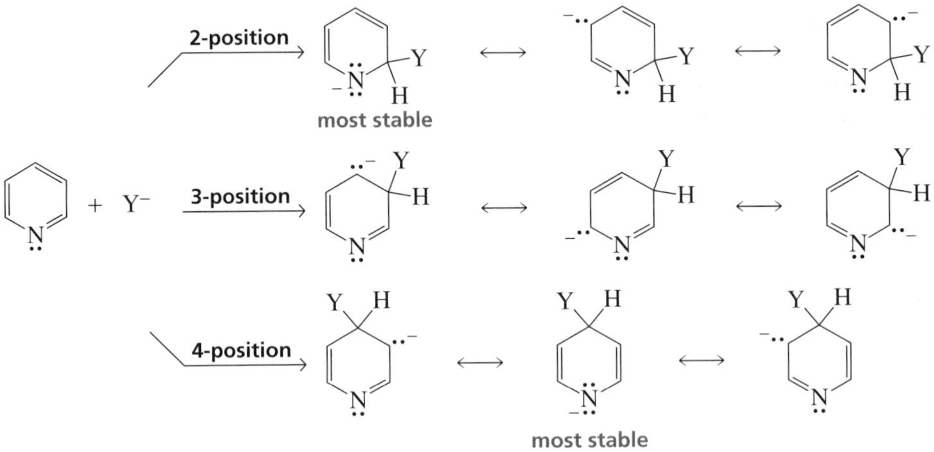

Pyridine undergoes nucleophilic aromatic substitution at C-2 and C-4.

PROBLEM 10

Draw resonance contributors to show why pyridine-*N*-oxide is more reactive than pyridine toward electrophilic aromatic substitution and why it undergoes electrophilic aromatic substitution primarily at C-4.

If the leaving groups at C-2 and C-4 are different, the incoming nucleophile will preferentially substitute for the weaker base (the better leaving group).

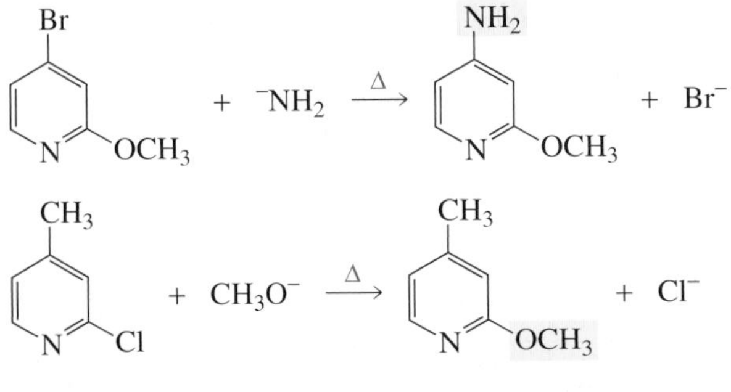

PROBLEM 11

Compare the mechanisms for the following reactions:

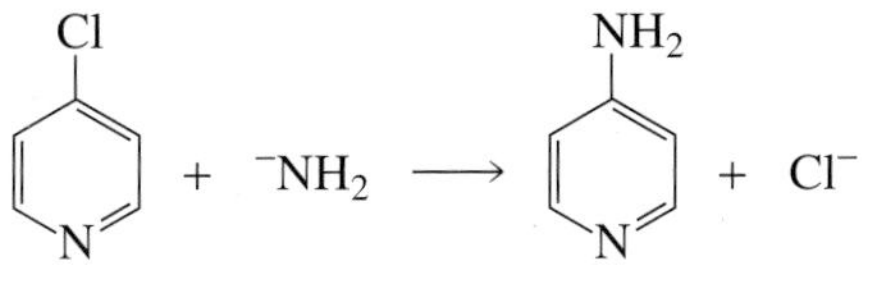

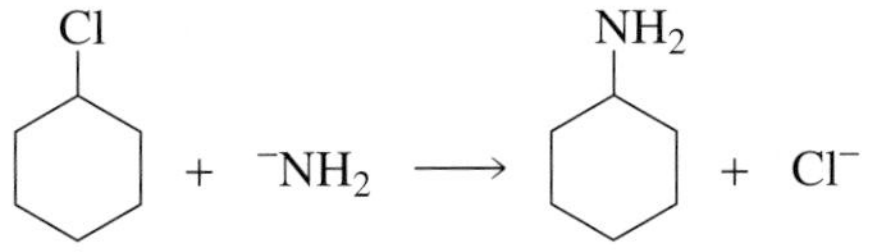

PROBLEM 12

a. Propose a mechanism for the following reaction:

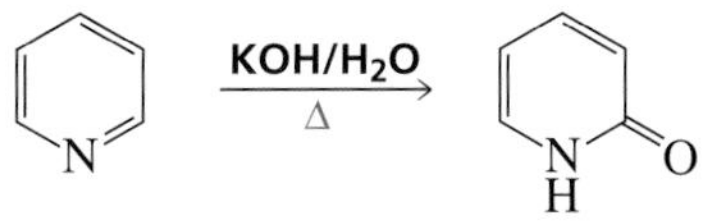

b. What other product is formed?

Substituted pyridines undergo many of the side-chain reactions that substituted benzenes undergo.

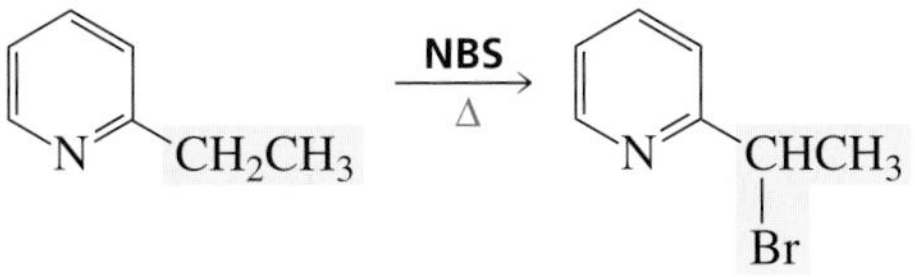

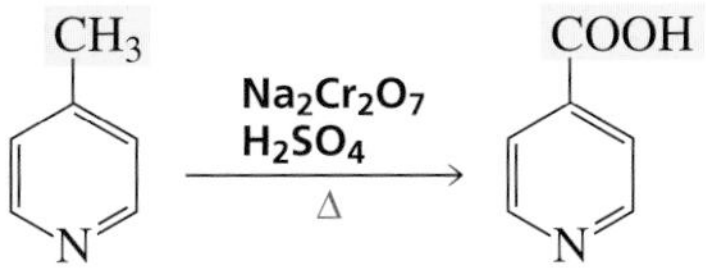

When 2- or 4-aminopyridine is diazotized, α-pyridone or γ-pyridone is formed. The mechanism for the conversion of a primary amino group into a diazonium group is shown in Section 15.12. Apparently the diazonium salt reacts immediately with water to form a hydroxypyridine (Section 15.10). The product of the reaction is a pyridone because the keto form of a hydroxypyridine is more stable than the enol form.

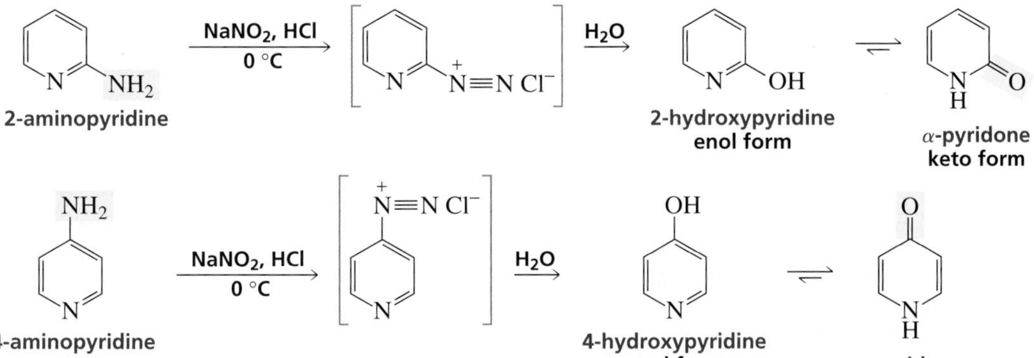

2-aminopyridine — 2-hydroxypyridine enol form — α-pyridone keto form

4-aminopyridine — 4-hydroxypyridine enol form — γ-pyridone keto form

The electron-withdrawing nitrogen causes the α-hydrogens of alkyl groups attached to the 2- and 4-positions of the pyridine ring to have about the same acidity as the α-hydrogens of ketones (Section 19.1).

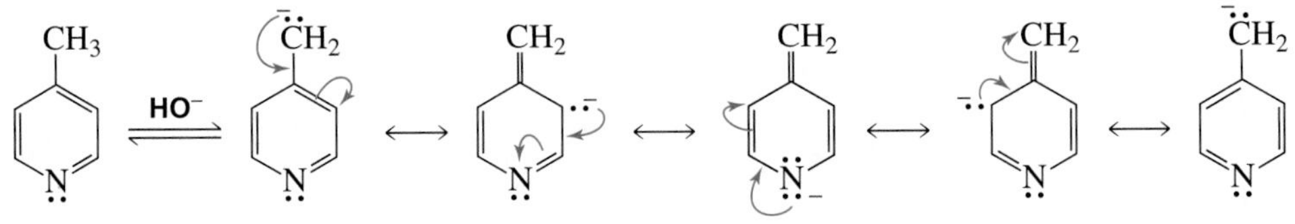

Consequently, they can be removed by base, and the resulting carbanions can react as nucleophiles.

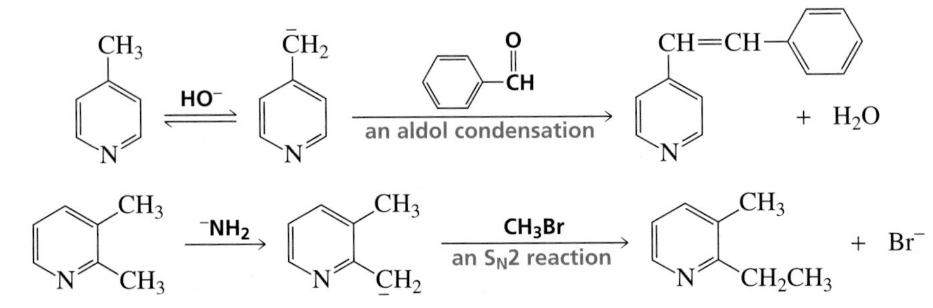

PROBLEM 13

Show how the following compounds could be prepared from pyridine:

a.

b.

c.

d.

PROBLEM 14◆

Rank the following compounds in order of decreasing ease of removing a proton from a methyl group:

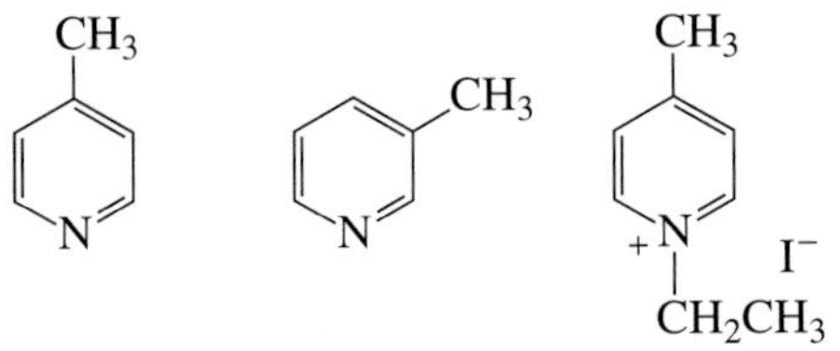

Quinoline and Isoquinoline

Quinoline and isoquinoline are known as *benzopyridines* because they have both a benzene ring and a pyridine ring. Like benzene and pyridine, they are aromatic compounds. The pK_a values of their conjugate acids are similar to the pK_a of the conjugate acid of pyridine. (In order for the carbons in quinoline and isoquinoline to have the same numbers, the nitrogen in isoquinoline is assigned the 2-position, not the lowest possible number.)

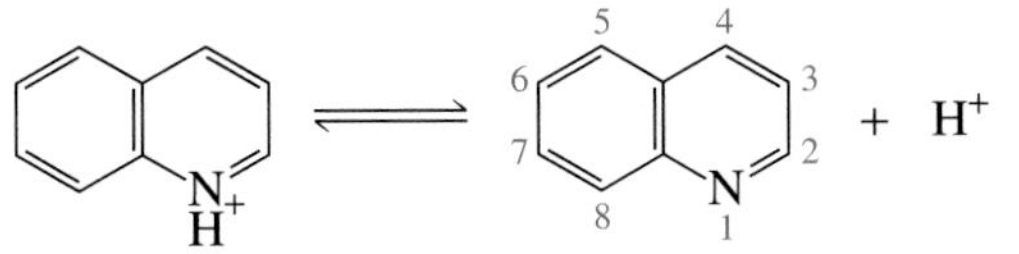

quinoline

pK$_a$ = 4.85

isoquinoline

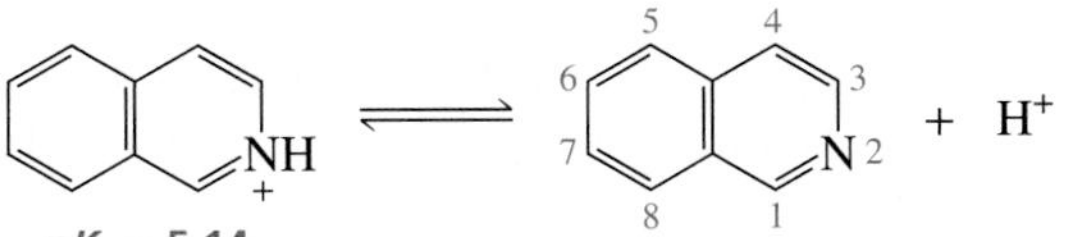

pK$_a$ = 5.14

Quinoline and isoquinoline undergo *electrophilic* aromatic substitution on the *benzene* ring because a benzene ring is more reactive than a pyridine ring toward electrophilic substitution. Substitution takes place primarily at C-5 and C-8, which is reminiscent of naphthalene's preference for electrophilic aromatic substitution at the α-position (Section 15.16).

3-D Molecules:
Quinoline;
Isoquinoline

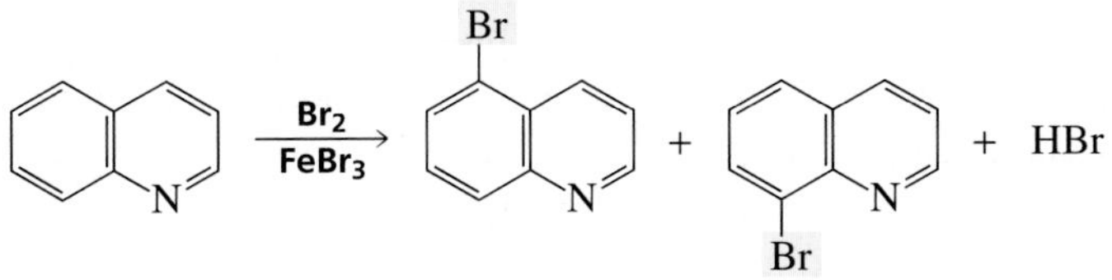

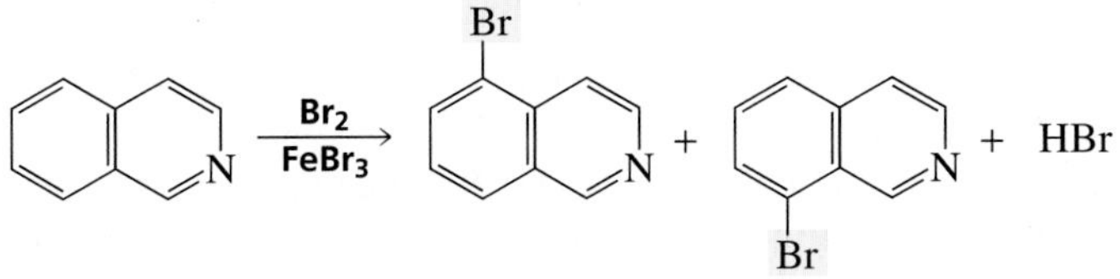

Quinoline and isoquinoline undergo *nucleophilic* aromatic substitution on the *pyridine* ring because a pyridine ring is more reactive than a benzene ring toward nucleophilic substitution. As expected, quinoline undergoes nucleophilic substitution at C-2 and C-4.

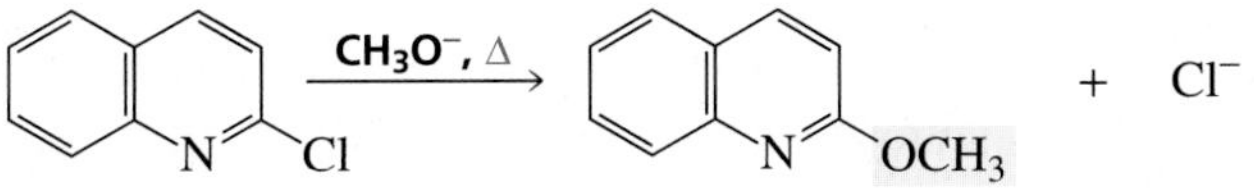

Isoquinoline undergoes nucleophilic aromatic substitution only at C-1.

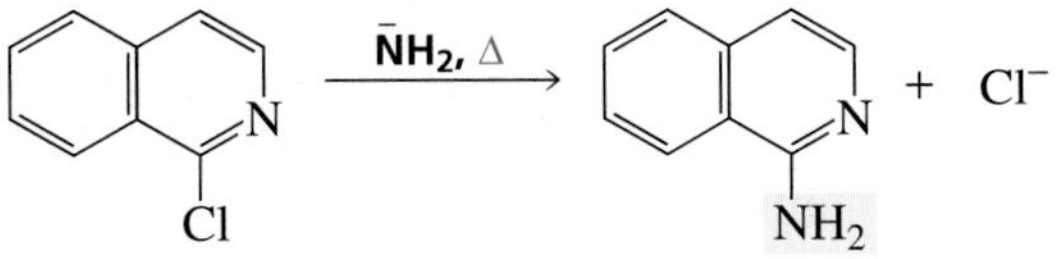

PROBLEM 15

Give the products of the following reactions:

a. 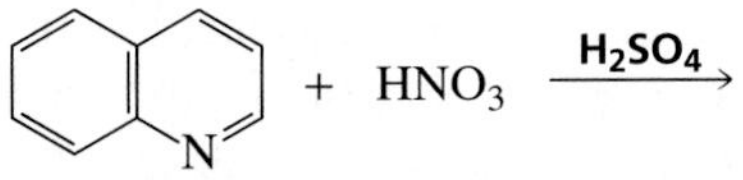 + HNO$_3$ $\xrightarrow{\text{H}_2\text{SO}_4}$

b. 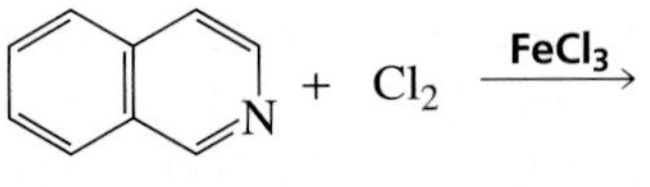 + Cl$_2$ $\xrightarrow{\text{FeCl}_3}$

c. [structure: 4-chloroquinoline] $\xrightarrow{\text{CH}_3\text{CH}_2\text{MgBr}}$ **d.** [structure: 1-bromoisoquinoline] $\xrightarrow{\text{CH}_3\text{CH}_2\text{MgBr}}$

27.4
BIOLOGICALLY
IMPORTANT
HETEROCYCLES

Proteins are naturally occurring polymers of α-amino acids (Chapter 21). Three of the 20 most common naturally occurring amino acids contain heterocyclic rings. Proline contains a pyrrolidine ring, tryptophan contains an indole ring, and histidine contains an imidazole ring.

[structures: proline, tryptophan, histidine]

proline   tryptophan   histidine

We have already seen that a saturated pyrrolidine ring behaves like a typical secondary amine (Section 27.1), and we studied the chemistry of indole (Section 27.2). Imidazole, the heterocyclic ring of histidine, is the first heterocyclic compound we have encountered that has two heteroatoms.

Imidazole

Tutorial: Recognizing common heterocyclic rings in complex molecules

Imidazole is an aromatic compound because it is cyclic and planar, every carbon in the ring has a p orbital, and the π cloud contains *three* pairs of π electrons (Section 14.1). The electrons drawn as lone-pair electrons on N-1 are part of the π cloud because they are in a p orbital, while the lone-pair electrons on N-3 are not part of the π cloud because they are in an sp^2 orbital, perpendicular to the p orbitals. The resonance energy of imidazole is 14 kcal/mol (59 kJ/mol), significantly less than the resonance energy of benzene (36 kcal/mol or 151 kJ/mol).

these electrons are in an sp^2 orbital perpendicular to the p orbitals

these electrons are part of the π cloud

orbital structure of imidazole

resonance contributors of imidazole

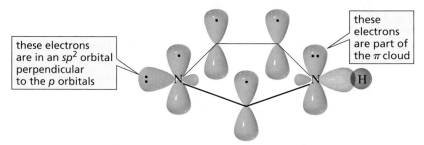

Imidazole can be protonated because the pair of nonbonding electrons in the sp^2 orbital are not part of the π cloud. Because the conjugate acid of imidazole has a pK_a

of 6.8, it exists in both the protonated and unprotonated forms at physiological pH (pH = 7.3). We have seen that this is one of the reasons why histidine, the imidazole-containing amino acid, is an important catalytic component of many enzymes (Section 22.9).

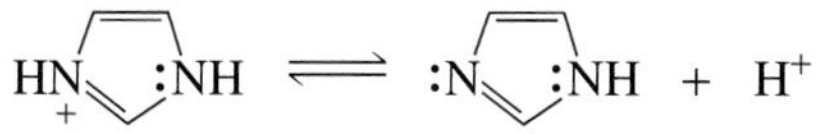

$$HN \overset{+}{\underset{}{}} :NH \;\rightleftharpoons\; :N \quad :NH \;+\; H^+$$

pK$_a$ = 6.8

Because of the second ring nitrogen, neutral imidazole is a stronger acid (pK$_a$ = 14.4) than neutral pyrrole (pK$_a$ ~ 17).

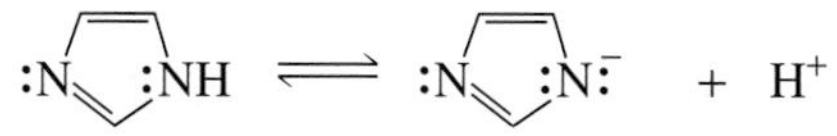

$$:N \quad :NH \;\rightleftharpoons\; :N \quad :N\overset{-}{:} \;+\; H^+$$

pK$_a$ = 14.4

Notice that both protonated imidazole and the imidazole anion have two equivalent resonance contributors. This means that C-4 and C-5 become equivalent when imidazole is either protonated or deprotonated.

protonated imidazole   **imidazole anion**

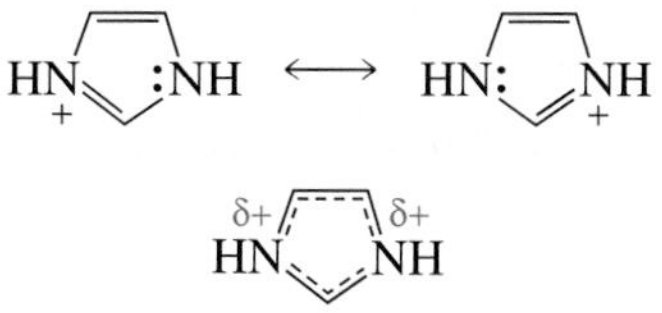

$$HN \overset{+}{\underset{}{}} :NH \;\longleftrightarrow\; HN: \underset{+}{} NH \qquad :N \quad :N\overset{-}{:} \;\longleftrightarrow\; \overset{-}{:}N \quad :N$$

$$\overset{\delta+}{HN} \cdots \overset{\delta+}{NH} \qquad\qquad \overset{\delta-}{:N} \cdots \overset{\delta-}{N:}$$

resonance hybrid   **resonance hybrid**

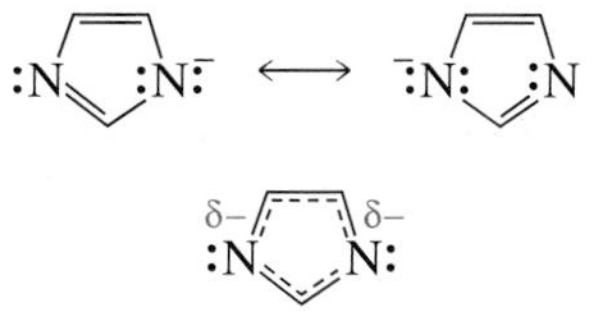

PROBLEM 16

a. Give the major product of the following reaction:

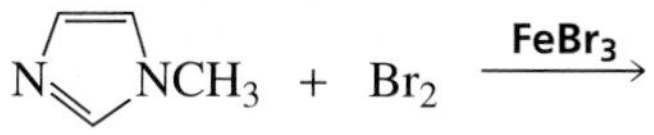

$$N \quad NCH_3 \;+\; Br_2 \;\xrightarrow{\;FeBr_3\;}$$

b. List imidazole, pyrrole, and benzene in order of decreasing reactivity toward electrophilic aromatic substitution.

PROBLEM 17 ◆

Imidazole boils at 257 °C, while N-methylimidazole boils at 199 °C. Explain this difference in boiling points.

PROBLEM 18 ◆

What percent of imidazole will be protonated at physiological pH (pH = 7.3)?

Purine and Pyrimidine

Nucleic acids (DNA and RNA) contain substituted **purines** and substituted **pyrimidines** (Section 25.1). Unsubstituted purine and pyrimidine are not found in nature. Notice that hydroxypurines and hydroxypyrimidines are more stable in the keto form. We have seen that the preference for the keto form is crucial for proper base pairing in DNA (Section 25.7).

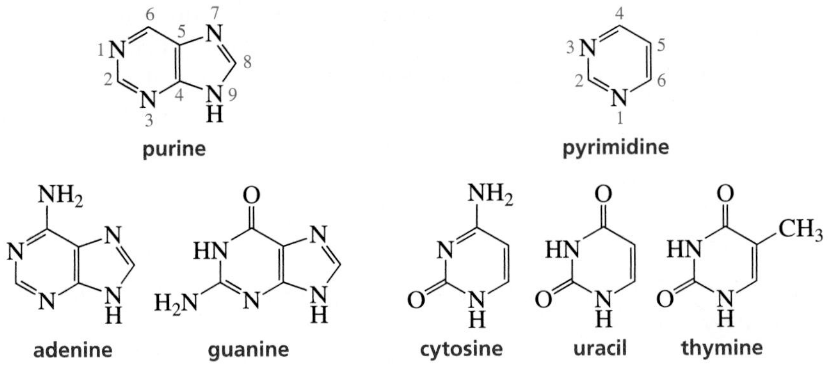

purine

pyrimidine

adenine   guanine   cytosine   uracil   thymine

Substituted purines and pyrimidines were some of the first compounds to be studied by organic chemists. Uric acid was isolated in 1776 and alloxan in 1818.

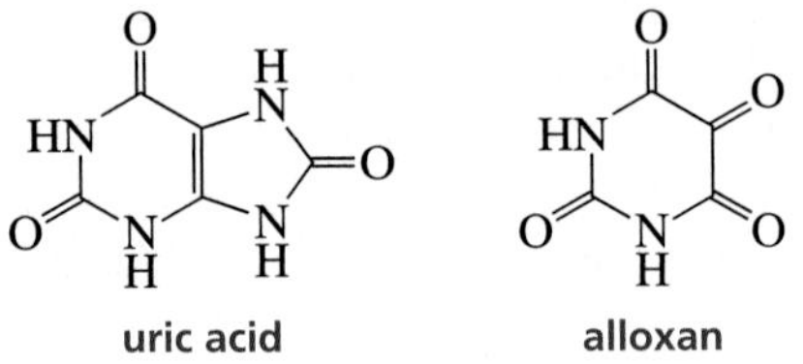

uric acid   alloxan

Because of the second nitrogen, pyrimidine is a much weaker base than pyridine (the conjugate acid is a stronger acid).

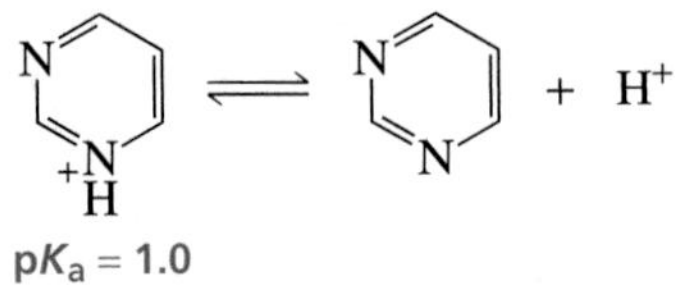

$pK_a = 1.0$

The pK_a values of the N-1 hydrogen in uracil, thymine, and cytosine are 9.5, 9.8, and 12.1, respectively. The relative acidity of this hydrogen means that substituents can easily be placed on N-1.

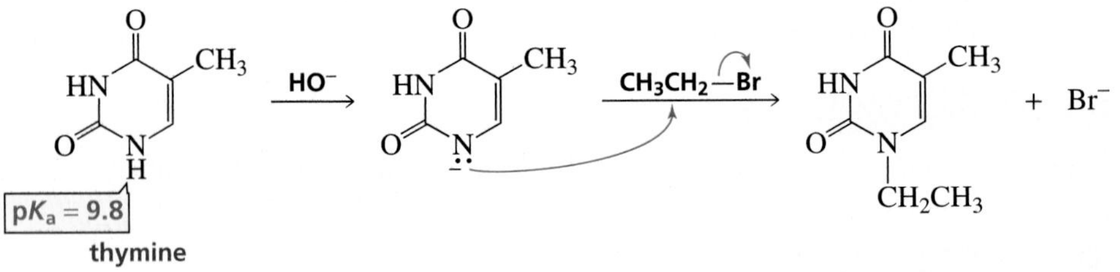

$pK_a = 9.8$

thymine

Because it has two electron-withdrawing heteroatoms, pyrimidine is considerably less reactive than pyridine toward electrophilic aromatic substitution. It can undergo electrophilic aromatic substitution only if there are strongly electron-donating substituents on the ring. Cytosine and uracil undergo electrophilic aromatic substitution reactions because they have electron-donating amino and/or "hydroxy" substituents. The most activated position of an unsubstituted pyrimidine toward electrophilic aromatic substitution is C-5 (it corresponds to C-3 of pyridine), and the electron-donating substituents of cytosine and uracil are located so that they further activate this position. 5-Bromouracil and 5-fluorouracil are used as anticancer agents (Sections 23.8 and 25.17).

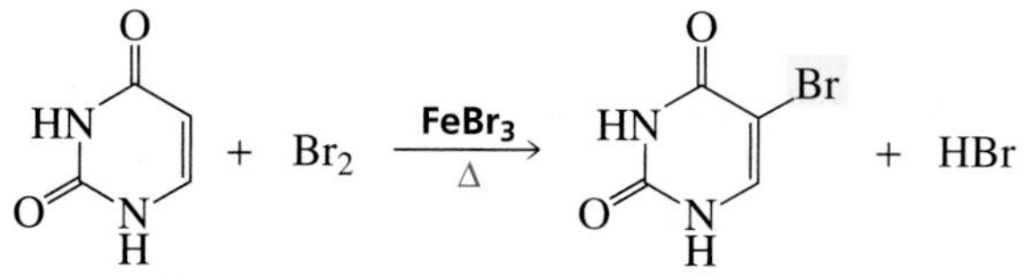

uracil + Br₂ →[FeBr₃][Δ] 5-bromouracil + HBr

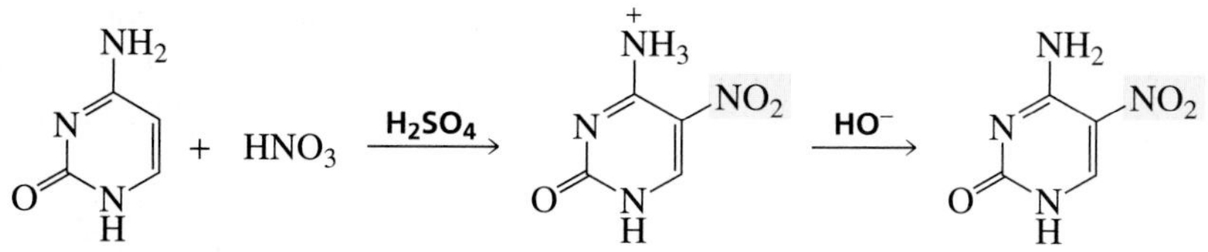

cytosine + HNO₃ →[H₂SO₄] →[HO⁻] 5-nitrocytosine

Pyrimidine is more reactive than pyridine toward nucleophilic aromatic substitution for the same reason that it is less reactive than pyridine toward electrophilic aromatic substitution.

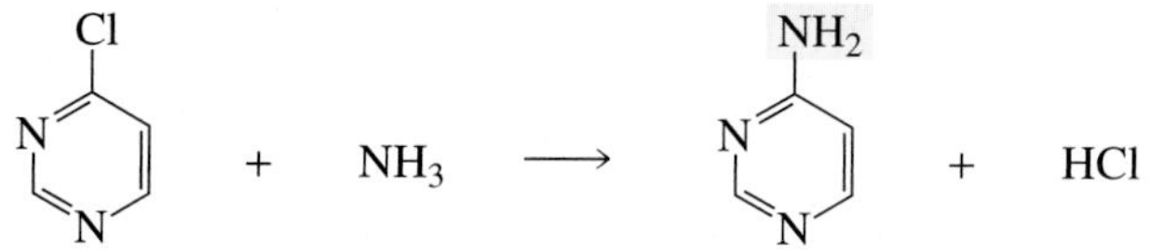

4-chloropyrimidine + NH₃ ⟶ 4-aminopyrimidine + HCl

Purine consists of a pyrimidine ring fused to an imidazole ring, so it has the properties of both ring systems. The electron-donating imidazole ring makes the protonated pyrimidine ring less acidic ($pK_a = 2.5$) than unsubstituted protonated pyrimidine ($pK_a = 1.0$). The electron-withdrawing pyrimidine ring makes the hydrogen on N-9 ($pK_a = 8.9$) more acidic than the corresponding N-1 hydrogen of imidazole ($pK_a = 14.4$).

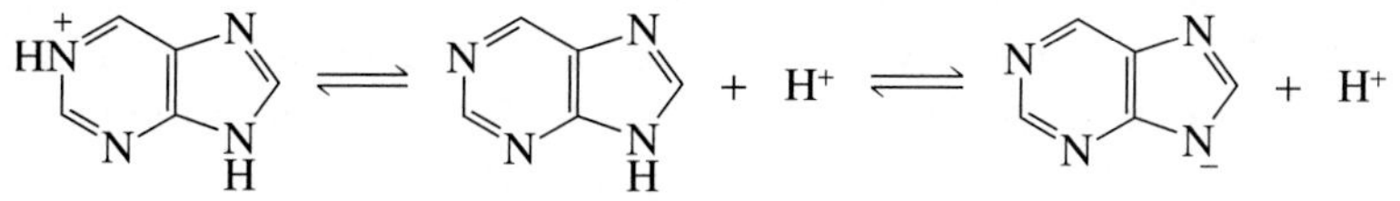

$pK_a = 2.5$ $pK_a = 8.9$

Adenine can be deaminated to give hypoxanthine, and guanine can be deaminated to give xanthine. **Deamination** is loss of ammonia due to reaction with water. Hypoxanthine and xanthine are both enzymatically oxidized to uric acid. Humans form about 600 mg of uric acid each day as a result of the metabolism of no-longer-needed nucleic acids. Uric acid is oxidized to allantoin, which is excreted (Section 16.5).

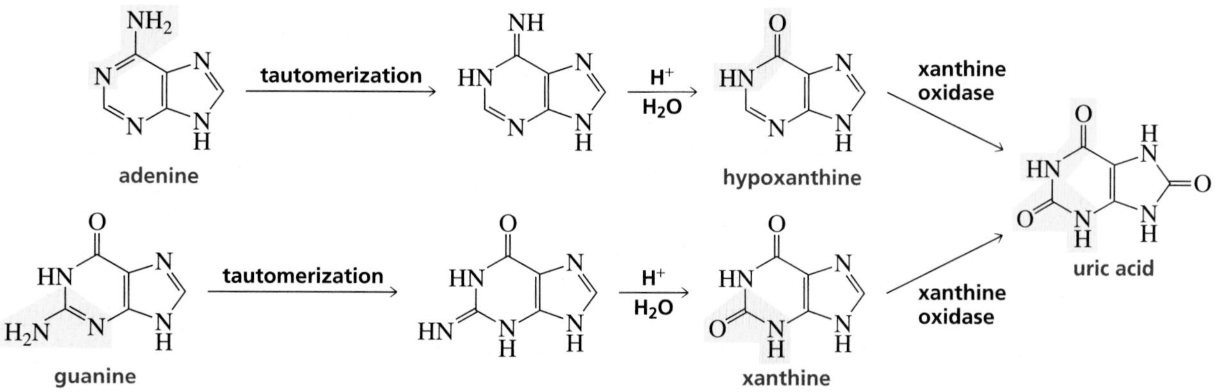

adenine →[tautomerization] →[H⁺][H₂O] hypoxanthine →[xanthine oxidase] uric acid

guanine →[tautomerization] →[H⁺][H₂O] xanthine →[xanthine oxidase]

Porphyrin

Substituted porphyrins are important, naturally occurring heterocyclic compounds. A **porphyrin ring system** consists of four pyrrole rings joined by one-carbon bridges. Heme, which is found in hemoglobin and myoglobin, contains an iron ion (Fe^{2+}) ligated by the four nitrogens of a porphyrin ring system. **Ligation** is the sharing of non-bonding electrons with a metal ion. The porphyrin ring system of heme is known as **protoporphyrin IX.** The ring system plus the iron atom is called **iron protoporphyrin IX.** In hemoglobin or myoglobin, the iron is also ligated to a histidine of the protein component (globin), and its sixth ligand is oxygen or carbon dioxide.

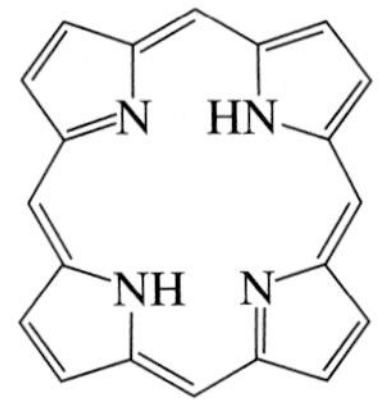

a porphyrin ring system

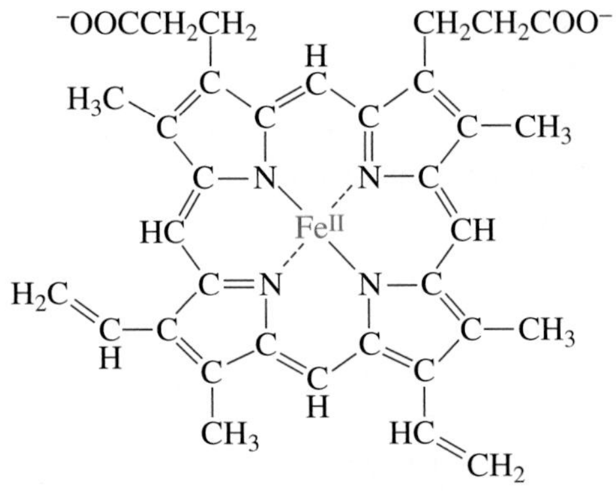

iron protoporphyrin IX
heme

Hemoglobin is responsible for transporting oxygen to cells and carbon dioxide away from cells, while myoglobin is responsible for storing oxygen in cells. Carbon monoxide is about the same size and shape as O_2 but CO binds more tightly than O_2 to Fe^{2+}. Consequently, breathing carbon monoxide can be fatal because it prevents the transport of oxygen in the bloodstream.

The extensive, conjugated system of porphyrin gives blood its characteristic red color. Its high molar absorptivity (about 160,000) allows concentrations as low as 1×10^{-8} M to be detected by UV spectroscopy (Section 12.16).

The biosynthesis of porphyrin involves the formation of porphobilinogen from two molecules of δ-aminolevulinic acid. The precise mechanism for this biosynthesis is as yet unknown. A possible mechanism starts with the formation of an imine between the enzyme that catalyzes the reaction and one of the molecules of δ-aminolevulinic acid. An aldol-type condensation occurs between the imine and a free molecule of δ-aminolevulinic acid. Nucleophilic attack by the amino group on the imine closes the ring. The enzyme is then eliminated, and removal of a proton creates the aromatic ring.

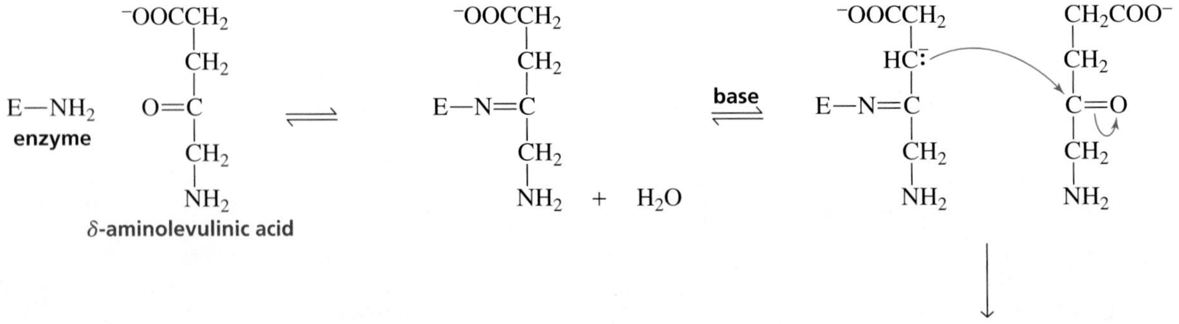

δ-aminolevulinic acid

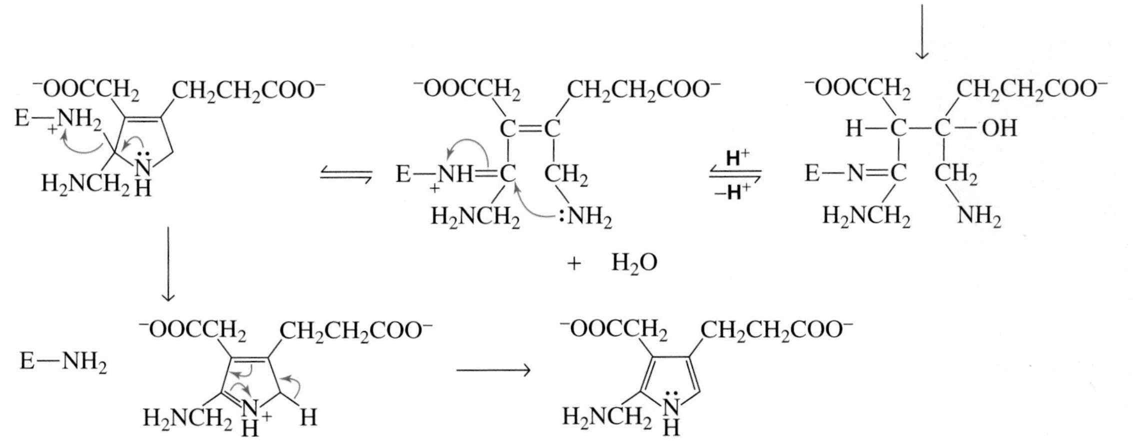

The reaction scheme showing the formation of porphobilinogen:

$$E\text{—}NH_2 + \cdots \rightleftharpoons E\text{—}NH{=}C \cdots CH_2 \quad \underset{-H^+}{\overset{H^+}{\rightleftharpoons}} \quad H\text{—}C\text{—}C\text{—}OH$$

$$+ \; H_2O$$

porphobilinogen

Four porphobilinogen molecules react to form porphyrin.

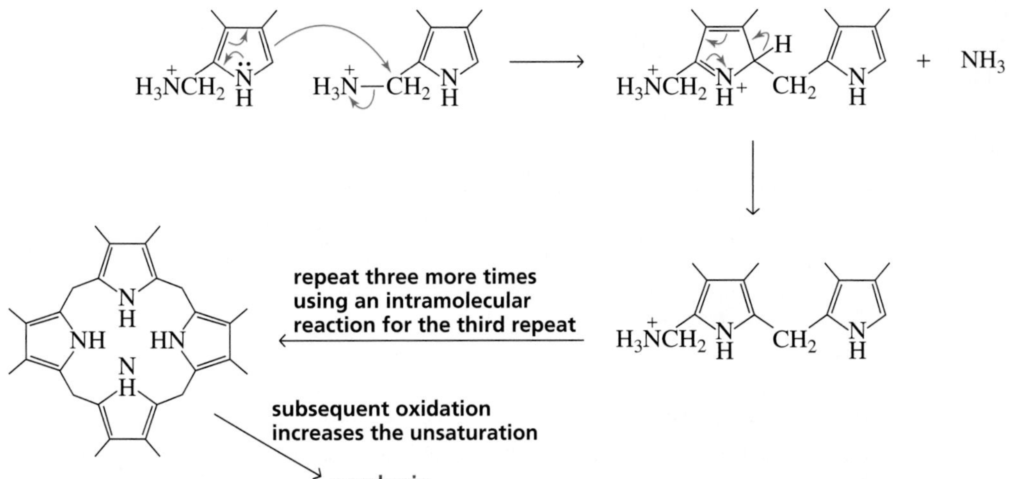

$$\cdots \longrightarrow \cdots + \; NH_3$$

**repeat three more times
using an intramolecular
reaction for the third repeat**

**subsequent oxidation
increases the unsaturation**

porphyrin

The ring system in chlorophyll *a*, the substance responsible for the green color of plants, is similar to porphyrin but contains a cyclopentanone ring, and one of its pyrrole rings is partially reduced. The metal atom in chlorophyll *a* is magnesium (Mg^{2+}).

3-D Molecule:
Chlorophyll *a*

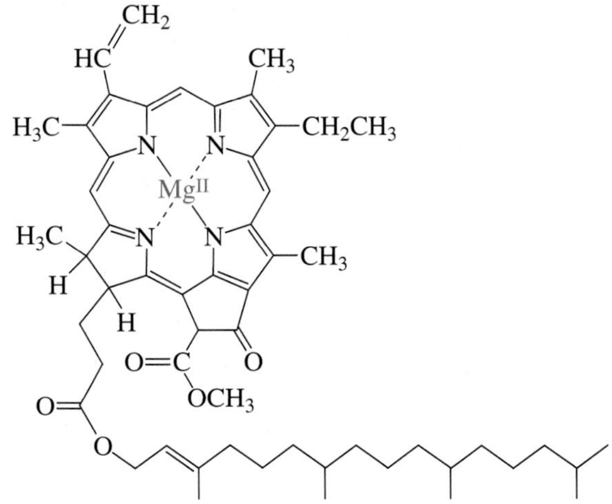

chlorophyll *a*

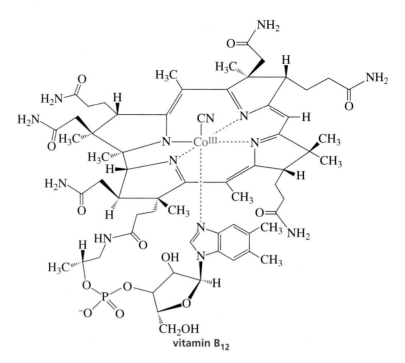

vitamin B₁₂

3-D Molecule:
Vitamin B₁₂

Vitamin B_{12} also has a ring system similar to that of porphyrin, but one of the methine bridges is missing. This is known as a **corrin ring system.** The metal atom in vitamin B_{12} is cobalt (Co^{3+}). The chemistry of vitamin B_{12} is discussed in Section 23.7.

PROBLEM 19◆

Is porphyrin aromatic?

PROBLEM 20

Show how the last two porphobilinogen molecules are incorporated into the porphyrin ring.

PORPHYRIN, BILIRUBIN, AND JAUNDICE

The average human turns over about 6 g of hemoglobin each day. The protein portion (globin) and the iron are reutilized, but the porphyrin ring is broken down. First it is reduced to biliverdin, a green compound, which is subsequently reduced to bilirubin, a yellow compound. If more bilirubin is formed than can be excreted by the liver, bilirubin accumulates in the blood. When the concentration of bilirubin in the blood reaches a certain level, it diffuses into the tissues, causing them to become yellow. This condition is known as jaundice.

SUMMARY OF REACTIONS

1. Electrophilic aromatic substitution reactions

    a. pyrrole, furan, and thiophene (Section 27.2)

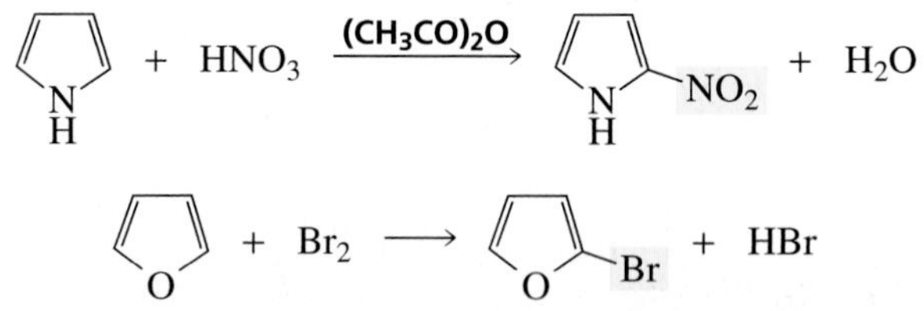

$$\text{pyrrole} + HNO_3 \xrightarrow{(CH_3CO)_2O} \text{2-nitropyrrole} + H_2O$$

$$\text{furan} + Br_2 \longrightarrow \text{2-bromofuran} + HBr$$

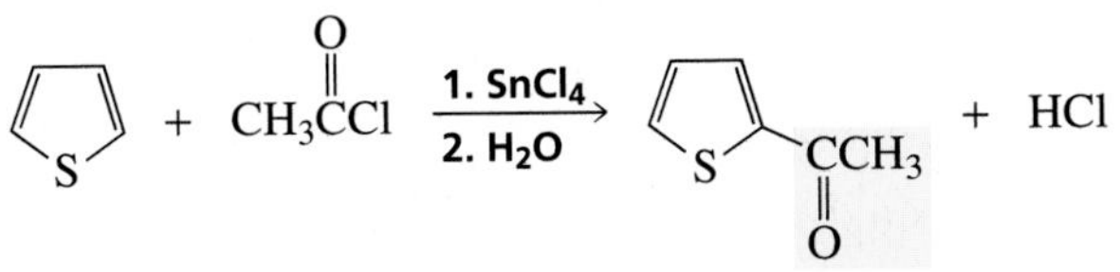

b. pyridine and pyridine-*N*-oxide (Section 27.3)

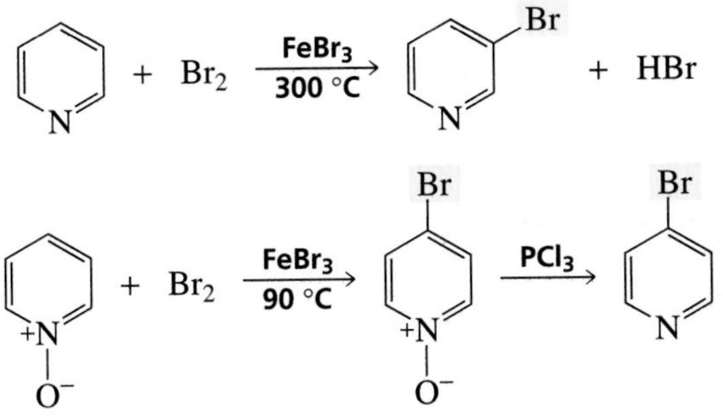

c. indole, benzofuran, and benzothiophene (Section 27.2)

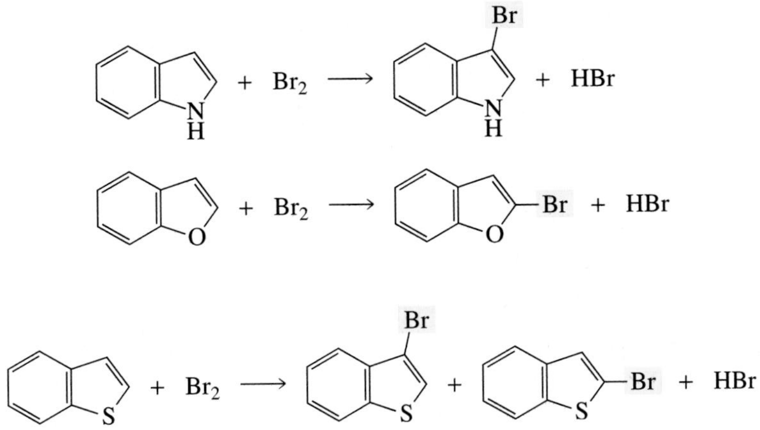

d. quinoline and isoquinoline (Section 27.3)

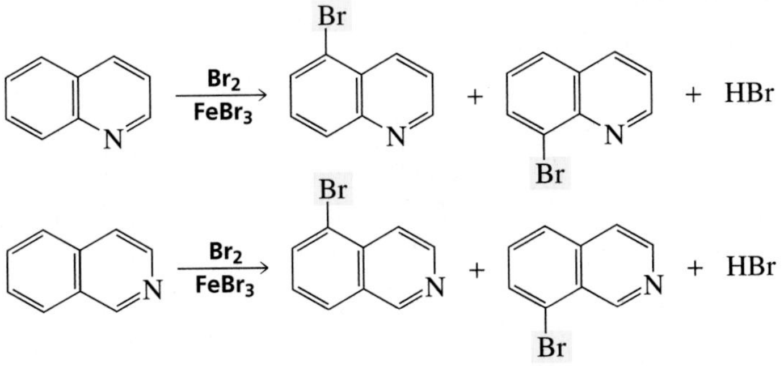

2. Nucleophilic aromatic substitution reactions

    a. pyridine and pyrimidine (Sections 27.3 and 27.4)

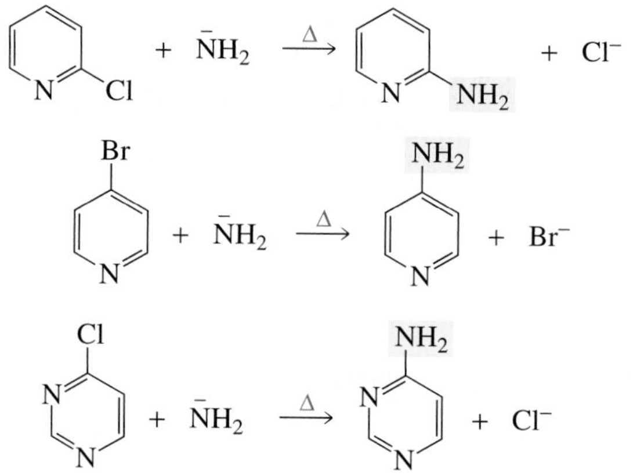

b. quinoline and isoquinoline (Section 27.3)

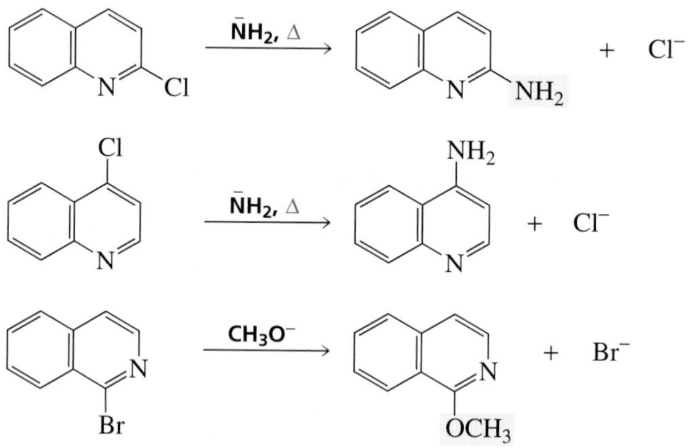

KEY TERMS

alkaloid (page 1133)
corrin ring system (page 1156)
deamination (page 1153)
furan (page 1137)
heteroatom (page 1133)
heterocycle (page 1133)
heterocyclic compound (page 1133)
imidazole (page 1150)
iron protoporphyrin IX (page 1154)
ligation (page 1154)
natural product (page 1133)
porphyrin ring system (page 1154)
protoporphyrin IX (page 1154)
purine (page 1151)
pyridine (page 1142)
pyrimidine (page 1151)
pyrrole (page 1137)
saturated heterocycle (page 1134)
thiophene (page 1137)

PROBLEMS

21. Name the following compounds:

a. 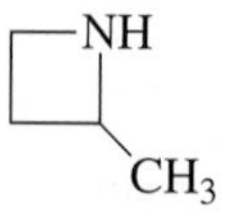

c. 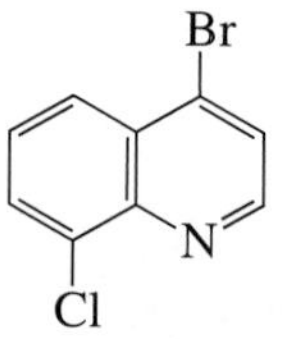

e. 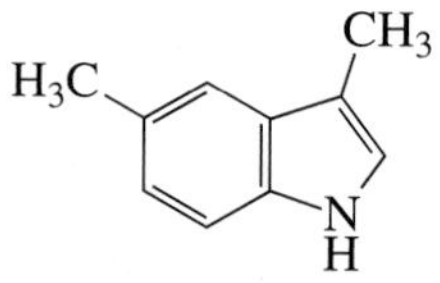

b. 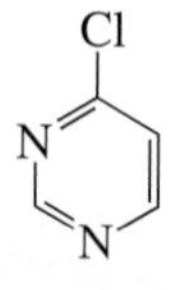

d. 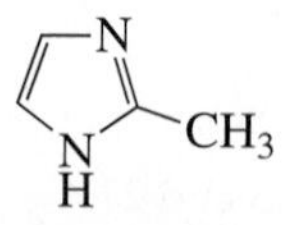

f. 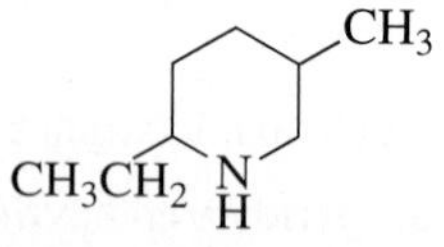

22. Give the product of each of the following reactions:

a. 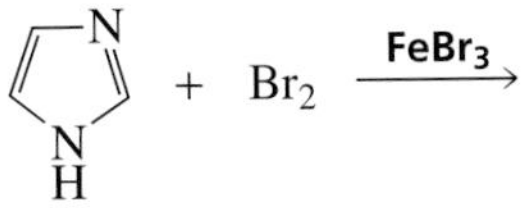 (imidazole) + Br$_2$ $\xrightarrow{\text{FeBr}_3}$

g. 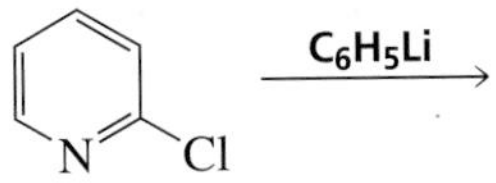 (2-chloropyridine) $\xrightarrow{\text{C}_6\text{H}_5\text{Li}}$

b. (4-bromopyrimidine) + (piperidine) $\longrightarrow$

h. (2-methylfuran) + CH$_3$$\overset{\text{O}}{\overset{\|}{\text{C}}}$Cl $\longrightarrow$

c. (3-methylazetidine, NH) + CH$_3$CH$_2$OH $\xrightarrow{\text{H}^+}$

i. (8-methylquinoline) + Cl$_2$ $\xrightarrow{\text{FeCl}_3}$

d. CH$_3$$\overset{\text{O}}{\overset{\|}{\text{C}}}$Cl + (benzofuran) $\xrightarrow[\text{2. H}_2\text{O}]{\text{1. AlCl}_3}$

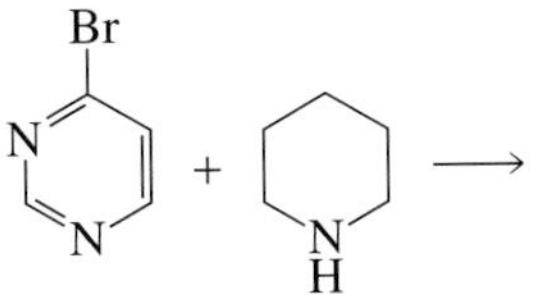

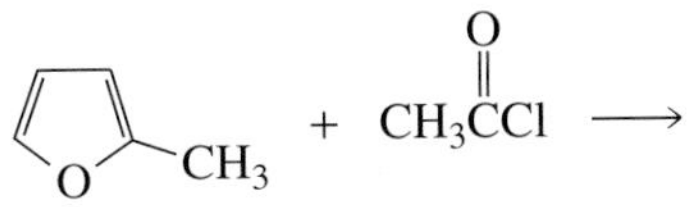

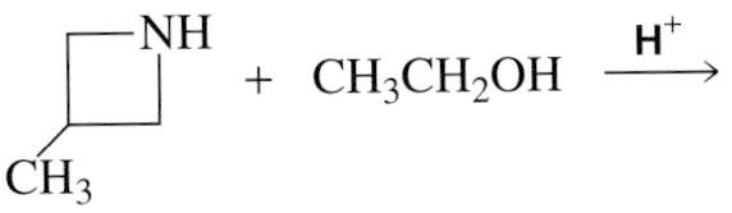

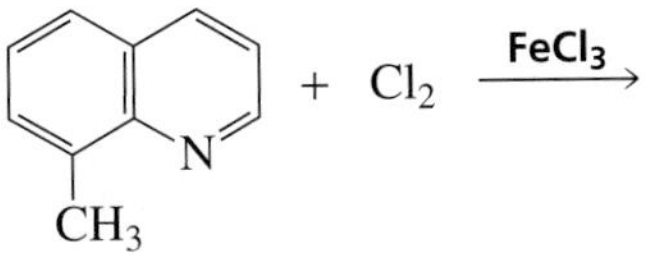

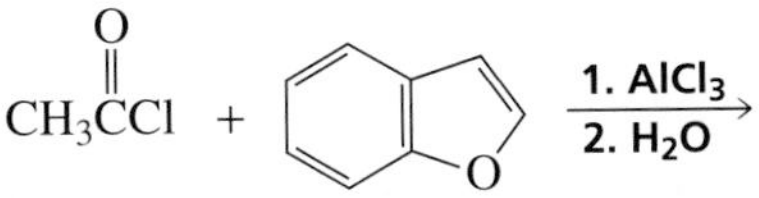

j. 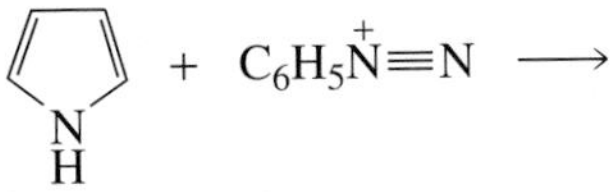 (pyrrole) + C$_6$H$_5\overset{+}{\text{N}}\equiv$N $\longrightarrow$

e. 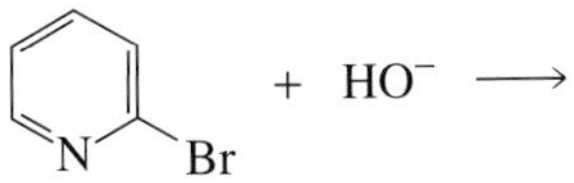 (2-bromopyridine) + HO$^-$ $\longrightarrow$

k. 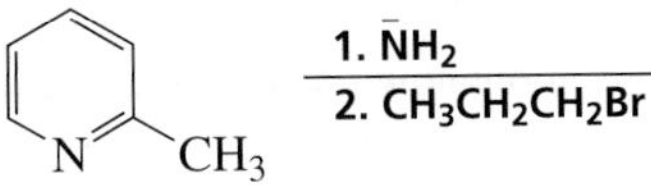 (2-methylpyridine) $\xrightarrow[\text{2. CH}_3\text{CH}_2\text{CH}_2\text{Br}]{\text{1. }\bar{\text{N}}\text{H}_2}$

f. 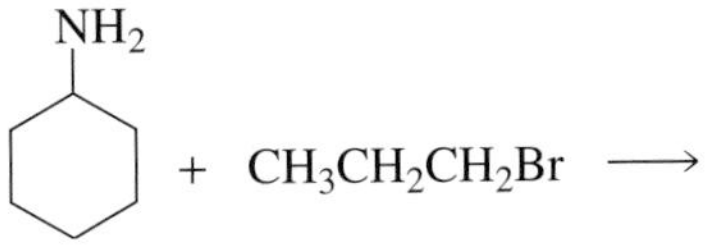 (cyclohexylamine, NH$_2$) + CH$_3$CH$_2$CH$_2$Br $\longrightarrow$

l. 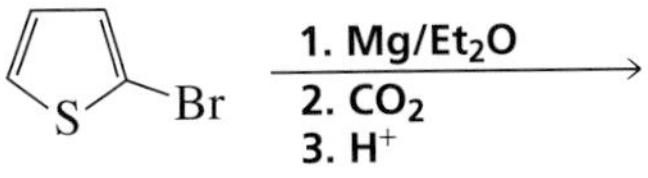 (2-bromothiophene) $\xrightarrow[\text{3. H}^+]{\begin{array}{l}\text{1. Mg/Et}_2\text{O}\\\text{2. CO}_2\end{array}}$

23. List the following compounds in order of decreasing acidity:

      (plus piperidinium structure)

24. Which of the following compounds is easier to decarboxylate?

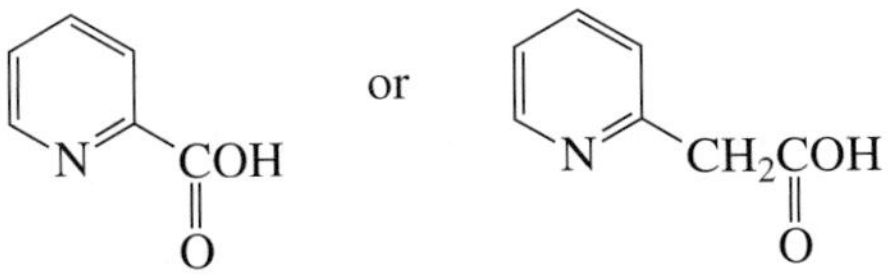

(pyridine-2-COOH) or (pyridine-2-CH$_2$COOH)

25. Rank the following compounds in order of decreasing reactivity in an electrophilic aromatic substitution reaction:

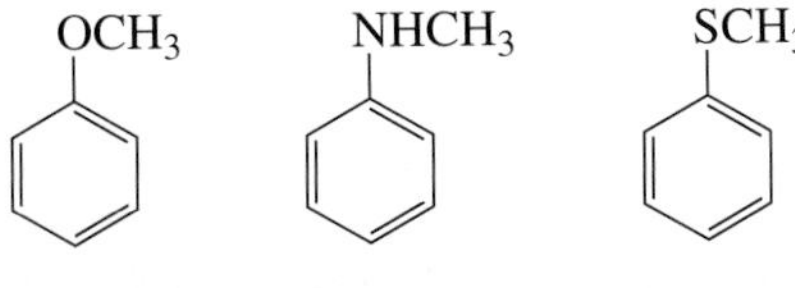

OCH$_3$ (anisole)   NHCH$_3$ (N-methylaniline)   SCH$_3$ (thioanisole)

26. One of the following compounds undergoes electrophilic aromatic substitution predominantly at C-3, and one undergoes electrophilic aromatic substitution predominantly at C-4. Which is which?

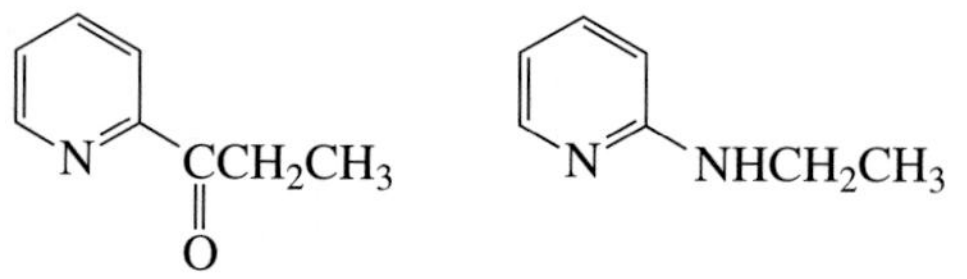

27. Benzene undergoes electrophilic aromatic substitution reactions with aziridines in the presence of a Lewis acid such as $AlCl_3$.
  a. What are the major and minor products of the following reaction?

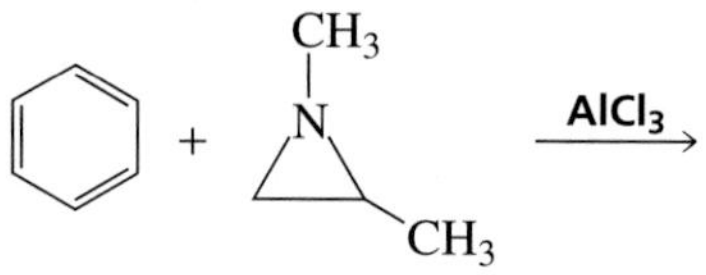

  b. Would you expect epoxides to undergo similar reactions?

28. The dipole moments of furan and tetrahydrofuran are in the same direction. One compound has a dipole moment of 0.70 D, and the other has a dipole moment of 1.73 D. Which is which?

29. Show how the vitamin niacin can be synthesized from nicotine.

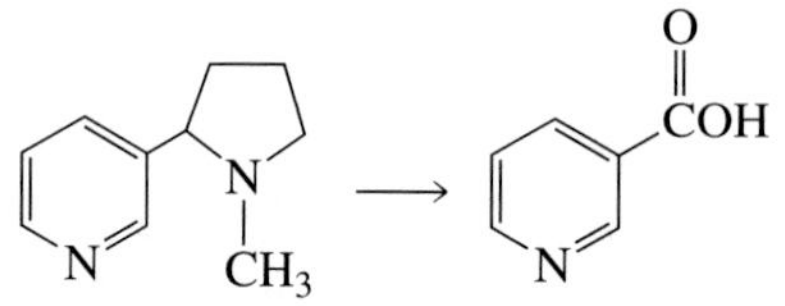

30. Propose a mechanism for the following reaction:

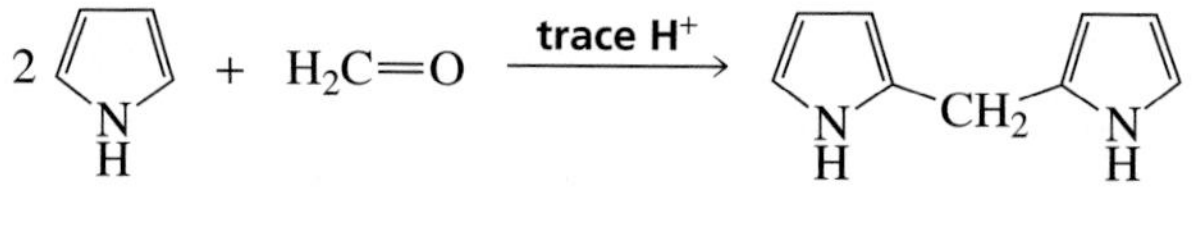

31. The chemical shifts of the C-2 hydrogen in the NMR spectra of pyrrole, pyridine, and pyrrolidine are δ 2.82, δ 6.42, and δ 8.50. Match each chemical shift with its heterocycle.

32. Explain why protonation of aniline has a dramatic effect on its UV spectrum, while protonation of pyridine has only a small effect on its UV spectrum.

33. Explain why pyrrole $(pK_a \sim 17)$ is a much stronger acid than ammonia $(pK_a = 36)$.

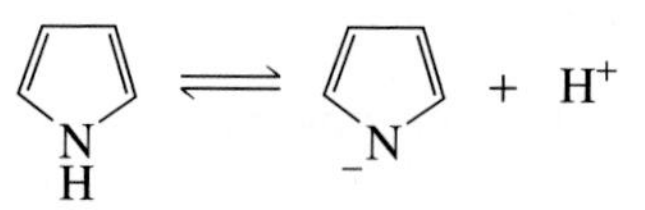

$$NH_3 \rightleftharpoons {}^-NH_2 + H^+$$
$$pK_a = 36$$

34. Quinolines are commonly synthesized by a method known as the *Skraup synthesis.* The Skraup synthesis involves the reaction of aniline with glycerol under acidic conditions. Nitrobenzene is added to the reaction mixture to serve as an oxidizing agent. The first step in the synthesis is the dehydration of glycerol to propenal.

$$CH_2-CH-CH_2 \xrightarrow[\Delta]{H_2SO_4} CH_2{=}CH-CH{=}O + 2 H_2O$$

$$\underset{\text{glycerol}}{\underset{|}{OH} \;\; \underset{|}{OH} \;\; \underset{|}{OH}}$$

a. What product would be obtained if *para*-ethylaniline were used instead of aniline?
b. What product would be obtained if 3-hexen-2-one were used instead of glycerol?
c. What starting materials are needed for the synthesis of 2,7-diethyl-3-methylquinoline?

35. Propose a mechanism for each of the following reactions:

a.

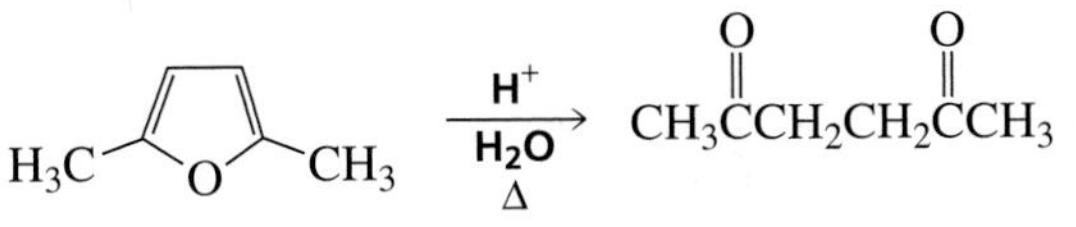

b.

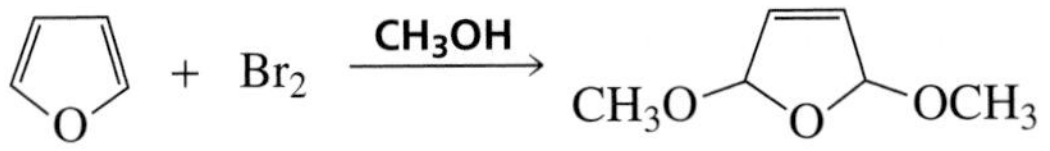

36. Give the major product of each of the following reactions:

a.

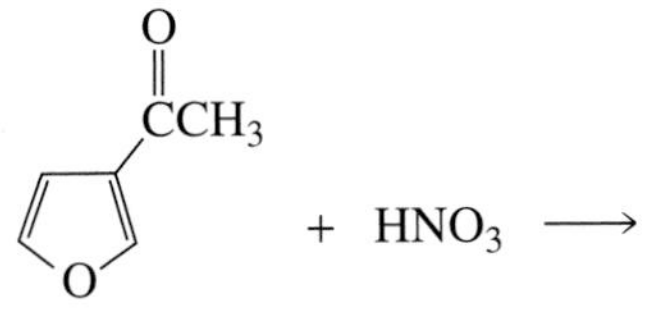

b.

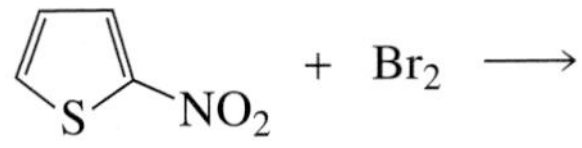

c.

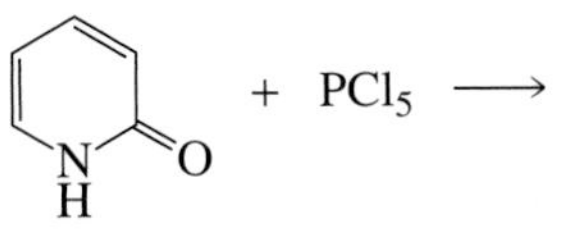

d.

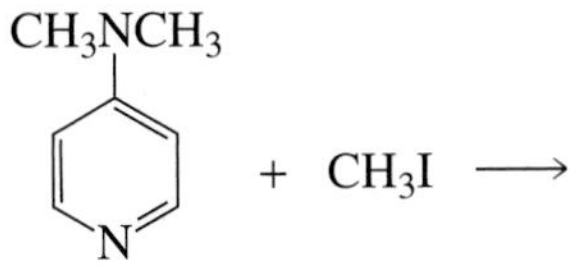

e.

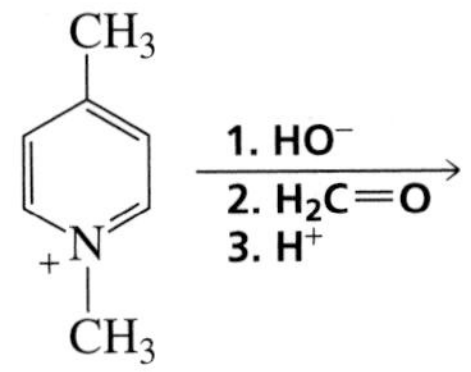

f.

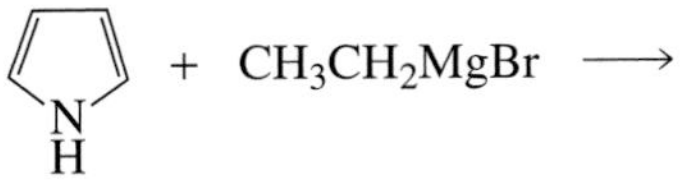

g.

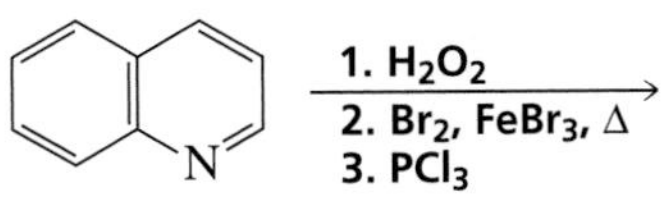

37. Propose a mechanism for the following reaction:

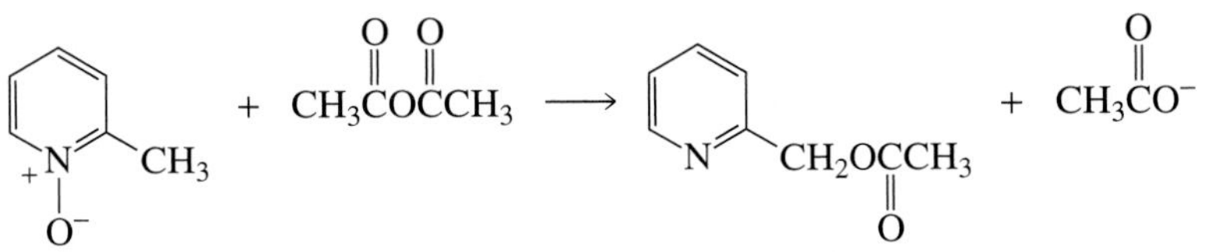

38. Propose a different mechanism than the one shown in Section 27.4 for the biosynthesis of porphobilinogen.

39. Pyrrole reacts with excess *para*-(*N,N*-dimethylamino)benzaldehyde to form a highly colored compound. Give the structure of the colored compound.

40. 2-Phenylindole is prepared from the reaction of acetophenone and phenylhydrazine, a method known as the Fischer indole synthesis. Propose a mechanism for this reaction. (*Hint:* The reactive species is the enamine tautomer of the phenylhydrazone.)

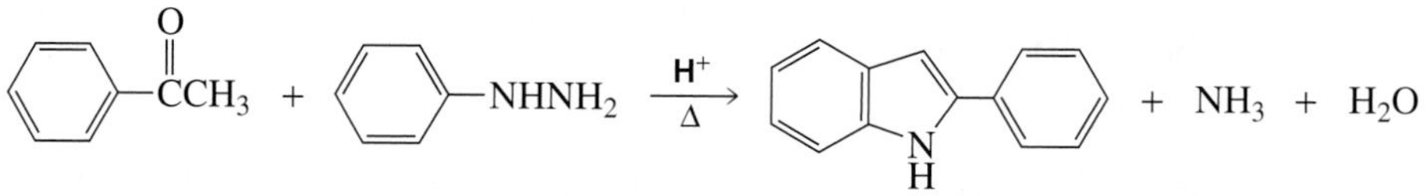

41. What starting materials are required for the synthesis of the following compounds using the Fischer indole synthesis? (*Hint:* See Problem 40.)

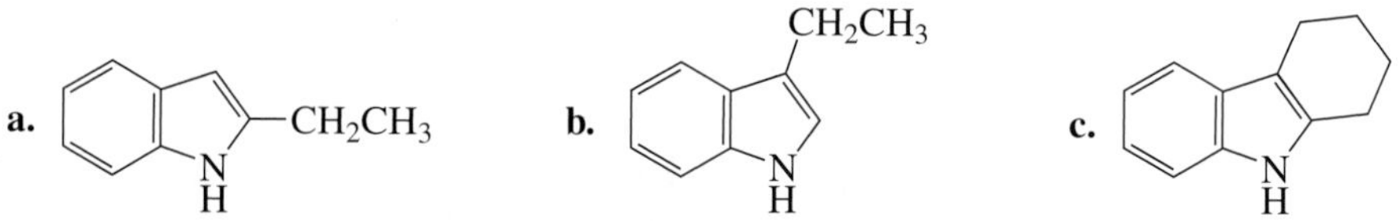

42. Organic chemists work with tetraphenylporphyrins rather than porphyrins because tetraphenylporphyrins are much more resistant to air oxidation. Tetraphenylporphyrin can be prepared by the reaction of benzaldehyde with pyrrole. Propose a mechanism for the formation of the ring system shown here.

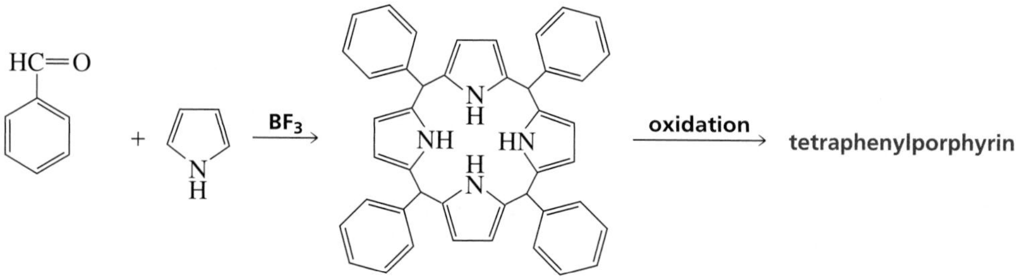

28

Pericyclic Reactions

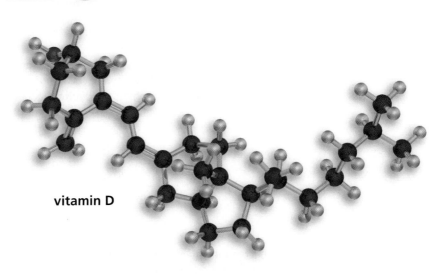

vitamin D

Reactions of organic compounds can be divided into three classes—polar reactions, radical reactions, and pericyclic reactions. The most common and the ones most familiar to you are polar reactions. A **polar reaction** is one in which a nucleophile reacts with an electrophile. Both electrons in the new bond come from the nucleophile.

a polar reaction

$$H:\ddot{\underset{..}{O}}: \quad + \quad \overset{\delta+}{CH_3}-\overset{\delta-}{Br} \quad \longrightarrow \quad CH_3OH \quad + \quad Br^-$$

A **radical reaction** is one in which a new bond is formed using one electron from each of the reactants.

a radical reaction

$$CH_3\dot{C}H_2 \quad + \quad Cl-Cl \quad \longrightarrow \quad CH_3CH_2Cl \quad + \quad \cdot Cl$$

A **pericyclic reaction** occurs when the electrons in the reactant(s) are reorganized in a cyclic manner. In this chapter we will look at the three most common types of pericyclic reactions—electrocyclic reactions, cycloaddition reactions, and sigmatropic rearrangements.

An **electrocyclic reaction** is an intramolecular reaction in which a new σ (sigma) bond is formed between the ends of a conjugated π (pi) system. This reaction is easy to recognize—the product is a cyclic compound that has one *fewer* π bond than the reactant.

**28.1
THREE KINDS OF
PERICYCLIC
REACTIONS**

an electrocyclic reaction

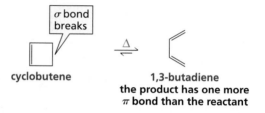

1,3,5-hexatriene ⇌ (Δ) 1,3-cyclohexadiene

new σ bond

the product has one fewer
π bond than the reactant

Electrocyclic reactions are reversible. In the reverse direction, an electrocyclic reaction is one in which a σ bond in a cyclic reactant breaks, forming a conjugated π system that has one *more* π bond than the cyclic reactant.

σ bond breaks

cyclobutene ⇌ (Δ) 1,3-butadiene

the product has one more
π bond than the reactant

In a **cycloaddition reaction,** two different π-bond-containing molecules react to form a cyclic compound. Each of the reactants loses a π bond, and the resulting cyclic product has two new σ bonds. The Diels–Alder reaction is a familiar example of a cycloaddition reaction (Section 7.9).

a cycloaddition reaction

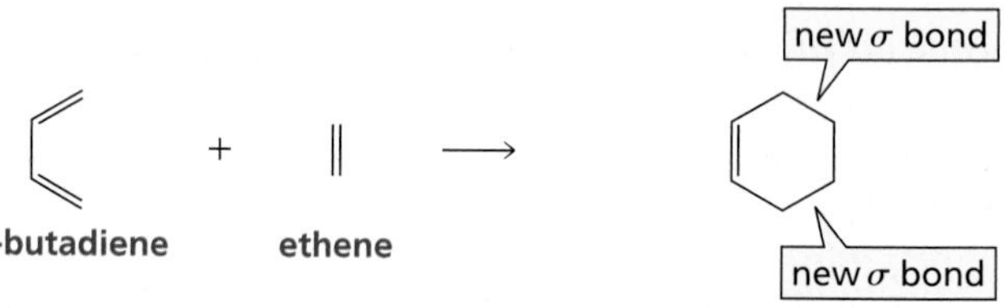

1,3-butadiene + ethene ⟶ cyclohexene

new σ bond

new σ bond

the product has two fewer π
bonds than the sum of the
π bonds in the reactants

In a **sigmatropic rearrangement,** a σ bond is broken in the reactant, a new σ bond is formed in the product, and the π bonds rearrange. The number of π bonds does not change (the reactant and the product have the same number of π bonds). The σ bond that is broken can be in the middle of the π system or at the end of the π system. The π system consists of the double-bonded carbons and the carbons immediately adjacent to them.

sigmatropic rearrangements

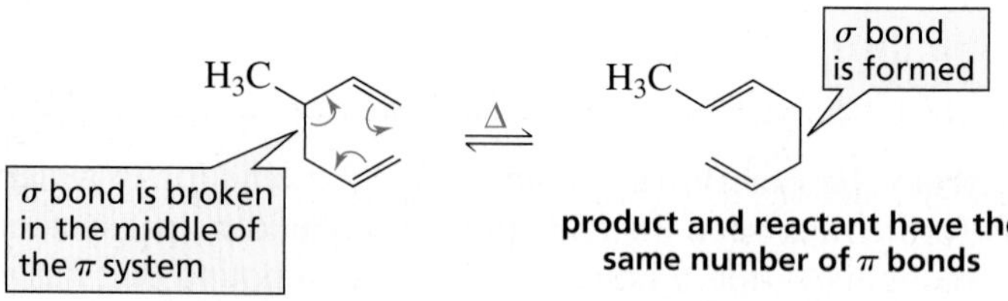

H₃C ⇌ (Δ) H₃C

σ bond
is formed

σ bond is broken
in the middle of
the π system

product and reactant have the
same number of π bonds

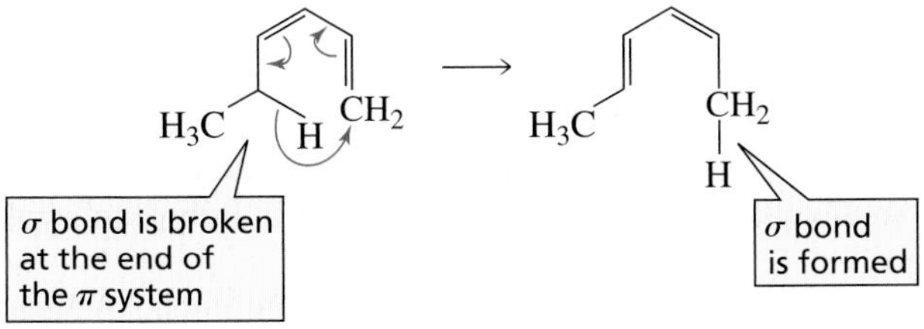

$$H_3C \xrightarrow{H} CH_2 \longrightarrow H_3C \xrightarrow{} CH_2$$

σ bond is broken at the end of the π system

σ bond is formed

Roald Hoffmann and **Kenichi Fukui** shared the 1981 Nobel Prize in chemistry for the conservation of orbital symmetry theory and the frontier orbital theory. **R. B. Woodward** did not receive a share in the prize because he died two years before it was awarded, and Alfred Nobel's will stipulates that the prize cannot be awarded posthumously. Woodward, however, received the 1965 Nobel Prize in chemistry for his work in organic synthesis (page 1221).

Notice that electrocyclic reactions and sigmatropic rearrangements occur within a single π system (they are *intra*molecular reactions), whereas cycloaddition reactions involve the interaction of two different π systems (they are *inter*molecular reactions). The three kinds of pericyclic reactions share the following common features:

a. They are all concerted reactions. This means that all the electron reorganization takes place in a single step. Therefore, there is one transition state and no intermediates.

b. Because the reactions are concerted, they are highly stereoselective.

c. The reactions are generally not affected by catalysts or by a change in solvent.

We will see that the configuration of the product formed in a pericyclic reaction depends on the following conditions:

a. the configuration of the reactant,

b. the number of double bonds in the reactant, and

c. whether the reaction is a thermal reaction or a photochemical reaction.

Roald Hoffmann was born in Poland in 1937 and came to the United States when he was 12. He received a B.S. from Columbia and a Ph.D. from Harvard. When the conservation of orbital symmetry theory was proposed, he and Woodward were both on the faculty at Harvard. Hoffmann is presently a professor of chemistry at Cornell.

A **photochemical reaction** is one that takes place when a reactant absorbs light. A **thermal reaction** takes place *without* the absorption of light. Despite its name, a thermal reaction does not necessarily require more heat than what is available to the reaction at room temperature. Some thermal reactions do require additional heat in order to take place at a reasonable rate, but others readily occur at, or even below, room temperature.

For many years, pericyclic reactions puzzled chemists. Why did some pericyclic reactions take place only under thermal conditions, while others took place only under photochemical conditions, and yet others were successfully carried out under both thermal and photochemical conditions? Another puzzling aspect of pericyclic reactions was the configurations of the products that were formed. After many pericyclic reactions had been investigated, it became apparent that if a pericyclic reaction could take place under both thermal and photochemical conditions, the configuration of the product formed under one set of conditions was different from the configuration of the product formed under the other set of conditions. For example, if the cis isomer was obtained under thermal conditions, the trans isomer was obtained under photochemical conditions and vice versa.

It took two very talented chemists, each bringing their own expertise to the problem, to solve the puzzling behavior of pericyclic reactions. In 1965, R. B. Woodward, an experimentalist, and Roald Hoffmann, a theorist, developed the **conservation of orbital symmetry theory** to explain the relationship among the structure and configuration of the reactant, the conditions (thermal and/or photochemical) under which the reaction takes place, and the configuration of the product. Because the behavior of pericyclic reactions is so precise, it is not surprising that everything about their behavior can be explained by one simple theory. The difficult part was having the insight to arrive at the theory.

Kenichi Fukui (1918-1998) was born in Japan. He was a professor at Kyoto University until 1982 when he became president of the Kyoto Institute of Technology. He was the first Japanese to receive the Nobel Prize in chemistry.

The conservation of orbital symmetry theory states that *in-phase orbitals overlap during the course of a pericyclic reaction*. The conservation of orbital symmetry theory was based on the **frontier orbital theory** put forth by Kenichi Fukui in 1954. Although Fukui's theory was more than 10 years old, it had been overlooked because of its mathematical complexity and Fukui's failure to apply it to stereoselective reactions.

According to the conservation of orbital symmetry theory, whether or not a compound will undergo a pericyclic reaction under particular conditions *and* what product will be formed both depend on molecular orbital symmetry. To understand pericyclic reactions, therefore, we must now spend some time looking at molecular orbital theory. We will then be able to understand how the symmetry of a molecular orbital controls both the conditions under which a pericyclic reaction takes place and the configuration of the product that is formed.

PROBLEM 1◆

Examine the following pericyclic reactions. For each reaction, indicate whether it is an electrocyclic reaction, a cycloaddition reaction, or a sigmatropic rearrangement.

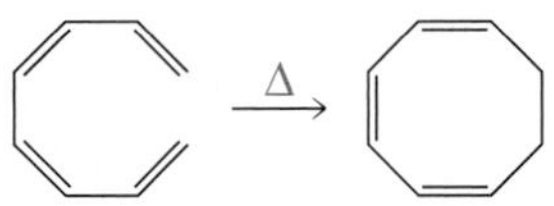

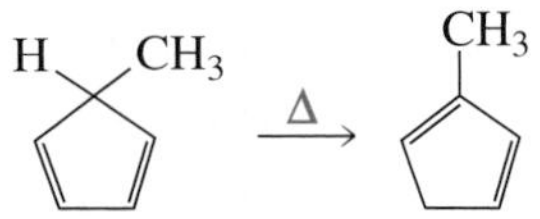

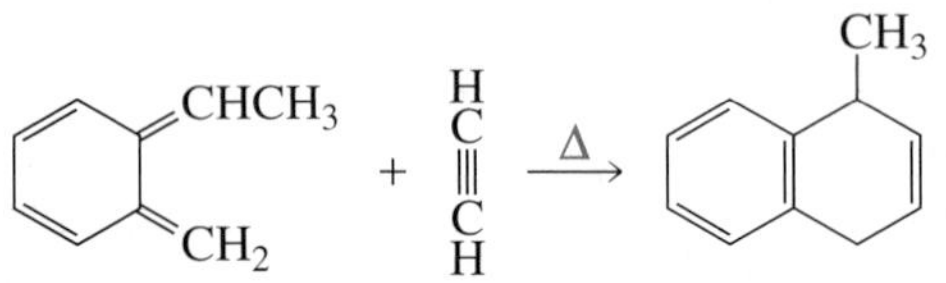

28.2 MOLECULAR ORBITALS AND ORBITAL SYMMETRY

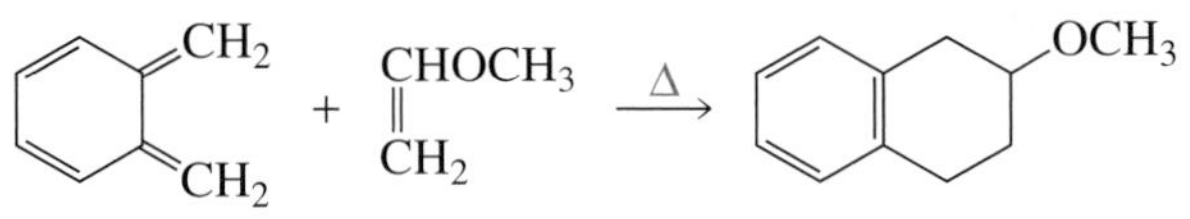

The overlap of *p* atomic orbitals to form π molecular orbitals can be described mathematically using quantum mechanics. The result of the mathematical treatment can be described simply in nonmathematical terms using **molecular orbital (MO) theory.** You were introduced to molecular orbital theory in Sections 1.6, 6.11, and 7.4, where you learned the following:

a. The two lobes of a *p* orbital have opposite phases. When two in-phase atomic orbitals interact, a covalent bond is formed. When two out-of-phase atomic orbitals interact, a node is created between the two nuclei.

b. Electrons fill molecular orbitals according to the same rules that are used when they fill atomic orbitals: An electron goes into the available molecular orbital with the lowest energy, and only two electrons can occupy a particular molecular orbital (Section 1.2).

You should take a few minutes to review these sections.

Because the π-bonding portion of a molecule is perpendicular to the framework of the σ bonds, the π bonds can be treated independently. Each carbon atom that forms a π bond has a *p* orbital, and the *p* orbitals of the carbon atoms combine to produce a π molecular orbital. Thus, a molecular orbital can be described by the **linear combination of atomic orbitals (LCAO).** In a π molecular orbital, each electron that previously occupied a *p* atomic orbital surrounding an individual carbon nucleus now surrounds the entire part of the molecule that is included in the interacting *p* orbitals.

A molecular orbital description of ethene is shown in Figure 28.1. (To show the different phases of the two lobes of a *p* orbital, one phase is represented by a blue lobe and the other phase by a lavender lobe.)[1] Because ethene has one π bond, it has two *p* atomic orbitals that combine to produce two π molecular orbitals. The in-phase interaction of the two *p* atomic orbitals of ethene gives a **bonding π molecular orbital,** designated by ψ_1 (ψ is the Greek letter psi). The bonding molecular orbital is of lower energy than the isolated *p* atomic orbitals. The two *p* atomic orbitals of ethene can also interact out of phase. Interaction of out-of-phase orbitals gives an **antibonding π molecular orbital,** ψ_2, which is of higher energy than the *p* atomic orbitals. The bonding molecular orbital results from additive interaction of the atomic orbitals, while the antibonding molecular orbital results from subtractive interaction. In other words, the interaction of in-phase orbitals holds atoms together, while the interaction of out-of-phase orbitals pushes atoms apart. Because electrons reside in the available molecular orbitals with the lowest energy and two electrons can occupy a molecular orbital, the two π electrons of ethene reside in the bonding π molecular orbital. This molecular orbital picture describes all molecules with one π bond.

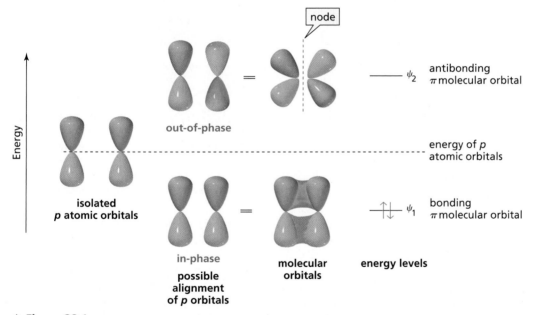

▲ **Figure 28.1**
Interaction of in-phase *p* atomic orbitals gives a bonding π molecular orbital that is lower in energy than the *p* atomic orbitals. Interaction of out-of-phase *p* atomic orbitals gives an antibonding π molecular orbital that is higher in energy than the *p* atomic orbitals.

1,3-Butadiene has two conjugated π bonds, so it has four *p* atomic orbitals (Figure 28.2). Four atomic orbitals can combine linearly in four different ways. Consequently, there are four π molecular orbitals: ψ_1, ψ_2, ψ_3, and ψ_4. Notice that orbitals are conserved: Four atomic orbitals combine to produce four molecular orbitals. Half are bonding molecular orbitals (ψ_1 and ψ_2) and the other half are antibonding molecular orbitals (ψ_3 and ψ_4). Because the four π electrons will reside in the available

> Orbitals are conserved—two atomic orbitals combine to produce two molecular orbitals, four atomic orbitals combine to produce four molecular orbitals, six atomic orbitals combine to produce six molecular orbitals, etc.

[1] Because the different phases of the *p* orbital result from the different mathematical signs ($+$ and $-$) of the wave function of the electron, some chemists represent the different phases by a ($+$) and a ($-$).

molecular orbitals with the lowest energy, there are two electrons in ψ_1 and two electrons in ψ_2. Remember that although the molecular orbitals have different energies, they are all valid and they all coexist. This molecular orbital picture describes all molecules with two conjugated π bonds.

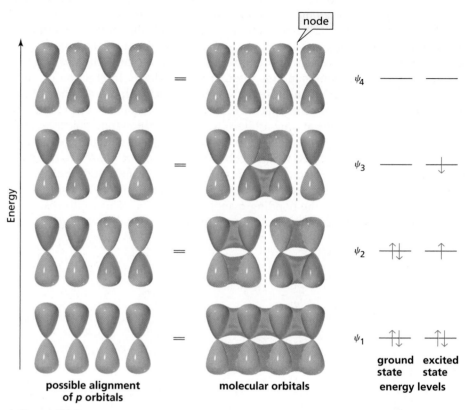

possible alignment of p orbitals   **molecular orbitals**   **ground state excited state energy levels**

▲ **Figure 28.2**
Four *p* atomic orbitals interact to give the four π molecular orbitals of 1,3-butadiene.

Tutorial:
π Molecular orbitals

If you examine the interacting orbitals in Figure 28.2, you will see that in-phase orbitals interact to give a bonding interaction and out-of-phase orbitals interact to create a node. Recall that a node is a place in which there is zero probability of finding an electron (Section 1.5). You will also see that as the energy of the molecular orbital increases, the number of bonding interactions decreases and the number of nodes *between* the nuclei increases. For example, ψ_1 has three bonding interactions and zero nodes between the nuclei, ψ_2 has two bonding interactions and one node between the nuclei, ψ_3 has one bonding interaction and two nodes between the nuclei, and ψ_4 has zero bonding interactions and three nodes between the nuclei. *Notice that a molecular orbital is bonding if the number of bonding interactions is greater than the number of nodes between the nuclei, and a molecular orbital is antibonding if the number of bonding interactions is fewer than the number of nodes between the nuclei.*

The normal electronic configuration of a molecule is known as its **ground state.** In the ground state of 1,3-butadiene, the *highest occupied molecular orbital (HOMO)* is ψ_2, and the *lowest unoccupied molecular orbital (LUMO)* is ψ_3. If a molecule absorbs light of an appropriate wavelength, the light will promote an electron from its ground-state HOMO to its LUMO (from ψ_2 to ψ_3). The molecule is then in an **excited state.** In the excited state, the HOMO is ψ_3 and the LUMO is ψ_4. In a thermal reaction, the reactant is in its ground state; in a photochemical reaction, the reactant is in an excited state.

Some molecular orbitals are symmetric and some are asymmetric, and they are easy to distinguish. Imagine holding a mirror perpendicular to the plane of the paper along the center of the molecular orbital. If the left half of the molecular orbital is a mirror image of the right half, the molecular orbital is symmetric. If it is not, it is asymmetric. In Figure 28.2, ψ_1 and ψ_3 are **symmetric molecular orbitals,** and ψ_2 and ψ_4 are **asymmetric molecular orbitals.** There is an even easier way to determine whether a molecular orbital is symmetric or asymmetric. If the p orbitals at the ends of the molecular orbital are identical (both have blue lobes on the top and lavender lobes on the bottom), the molecular orbital is symmetric. If the two end p orbitals are not identical, the molecular orbital is asymmetric. Notice that as the molecular orbitals increase in energy, they alternate from being symmetric to being asymmetric. *This means that the ground-state HOMO and the excited-state HOMO always have opposite symmetries—one is symmetric and the other is asymmetric.* A molecular orbital description of 1,3,5-hexatriene, a compound with three conjugated double bonds, is shown in Figure 28.3. As a review, examine the figure to see the following:

The ground-state HOMO and the excited-state HOMO have opposite symmetries.

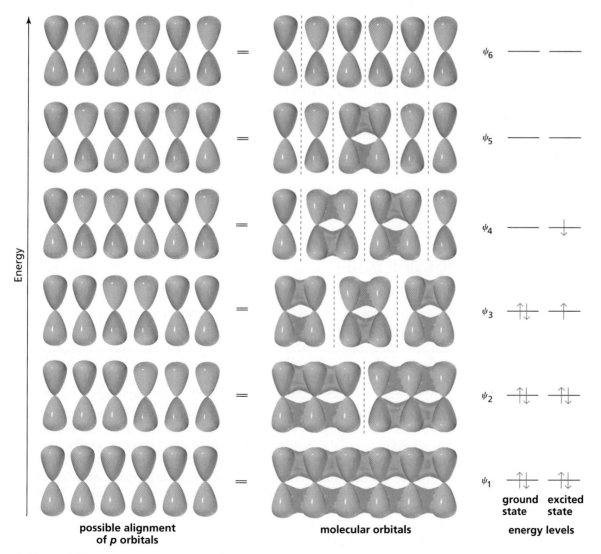

▲ **Figure 28.3**
Six p atomic orbitals interact to give the six π molecular orbitals of 1,3,5-hexatriene.

- the distribution of electrons in the ground and excited states;
- that the number of bonding interactions decreases and the number of nodes increases as the molecular orbitals increase in energy;
- that the molecular orbitals alternate between being symmetric and being asymmetric as the molecular orbitals increase in energy.

Although the chemistry of a compound is determined by all its molecular orbitals, we can learn a great deal about the chemistry of a compound by looking at only the **highest occupied molecular orbital (HOMO)** and the **lowest unoccupied molecular orbital (LUMO).** These two molecular orbitals are known as the **frontier orbitals.** Now we will see that simply by evaluating *one* of the frontier molecular orbitals of the reactant(s) in a pericyclic reaction, we can predict the conditions under which the reaction will occur (thermal or photochemical, or both) and the products that will be formed.

PROBLEM 2◆

Answer the following questions for the π molecular orbitals of 1,3,5-hexatriene:

a. Which are the bonding orbitals and which are the antibonding orbitals?

b. Which orbitals are the HOMO and the LUMO in the ground state?

c. Which orbitals are the HOMO and the LUMO in the excited state?

d. Which orbitals are symmetric and which are asymmetric?

e. What is the relationship between HOMO and LUMO and symmetric and asymmetric orbitals?

PROBLEM 3◆

a. How many π molecular orbitals does 1,3,5,7-octatetraene have?

b. What is the designation of its HOMO (ψ_1, ψ_2, etc.)?

c. How many nodes does its highest energy π molecular orbital have between the nuclei?

PROBLEM 4

Give a molecular orbital description for each of the following:

a. 1,3-pentadiene

b. 1,4-pentadiene

c. 1,3,5-heptatriene

d. 1,3,5,8-nonatetraene

28.3
ELECTROCYCLIC REACTIONS

An *electrocyclic reaction* is an intramolecular reaction in which the rearrangement of π electrons leads to a cyclic product that has one fewer π bond than the reactant. An electrocyclic reaction is completely stereoselective—it preferentially forms one stereoisomer. For example, when (2E,4Z,6E)-octatriene undergoes an electrocyclic reaction under thermal conditions, only the cis product is formed; when (2E,4Z,6Z)-octatriene undergoes an electrocyclic reaction under thermal conditions, only the trans product is formed. Recall that E means the high-priority groups are on opposite sides of the double bond, and Z means the high-priority groups are on the same side of the double bond (Section 3.5).

However, when the reactions are carried out under photochemical conditions, the products have opposite configurations. The compound that forms the cis isomer

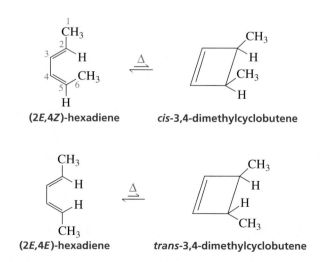

(2E,4Z,6E)-octatriene Δ *cis*-5,6-dimethyl-1,3-cyclohexadiene

(2E,4Z,6Z)-octatriene Δ *trans*-5,6-dimethyl-1,3-cyclohexadiene

under thermal conditions forms the trans isomer under photochemical conditions, and the compound that forms the trans isomer under thermal conditions forms the cis isomer under photochemical conditions.

(2E,4Z,6E)-octatriene *hv* *trans*-5,6-dimethyl-1,3-cyclohexadiene

(2E,4Z,6Z)-octatriene *hv* *cis*-5,6-dimethyl-1,3-cyclohexadiene

Under thermal conditions, (2*E*,4*Z*)-hexadiene cyclizes to *cis*-3,4-dimethyl-cyclobutene, and (2*E*,4*E*)-hexadiene cyclizes to *trans*-3,4-dimethylcyclobutene.

(2E,4Z)-hexadiene Δ *cis*-3,4-dimethylcyclobutene

(2E,4E)-hexadiene Δ *trans*-3,4-dimethylcyclobutene

As we saw with the octatrienes, the configuration of the product changes if the reactions are carried out under photochemical conditions: The trans isomer is obtained from (2E,4Z)-hexadiene instead of the cis isomer; the cis isomer is obtained from (2E,4E)-hexadiene instead of the trans isomer.

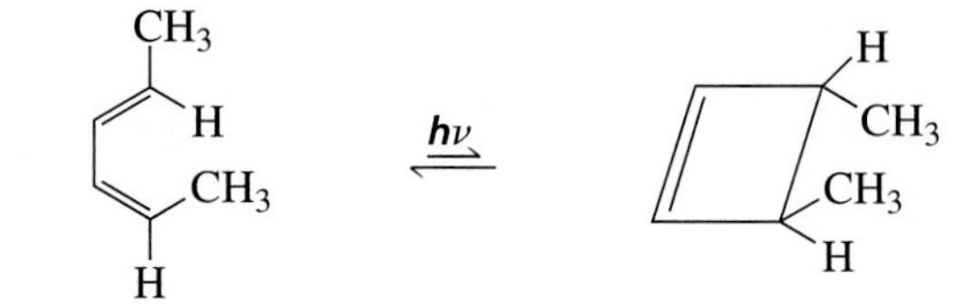

(2E,4Z)-hexadiene $\xrightleftharpoons{h\nu}$ *trans*-3,4-dimethylcyclobutene

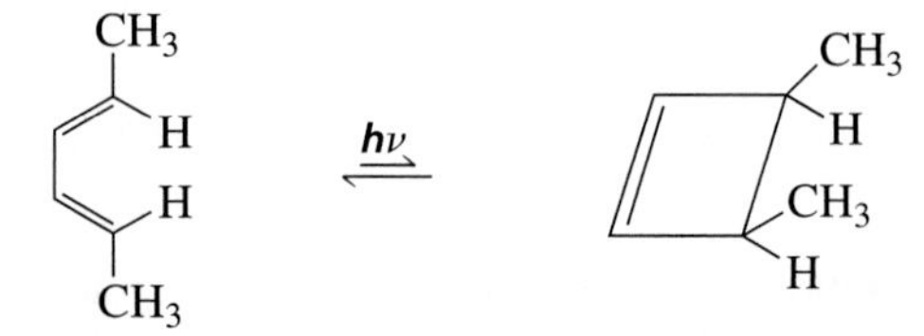

(2E,4E)-hexadiene $\xrightleftharpoons{h\nu}$ *cis*-3,4-dimethylcyclobutene

Tutorial:
Electrocyclic reactions

Electrocyclic reactions are reversible. The cyclic compound is favored for electrocyclic reactions that form six-membered rings, whereas the open-chain compound is favored for electrocyclic reactions that form four-membered rings because of the angle strain and torsional strain associated with four-membered rings (Section 2.11).

Now we will use what we have learned about molecular orbitals to explain the configuration of the products of the previous reactions. We will then be able to predict the configuration of the product of any other electrocyclic reaction.

The product of an electrocyclic reaction results from forming a new σ bond. In order to form this bond, the p orbitals at the ends of the conjugated system must rotate so they can overlap head-to-head (and rehybridize to sp^3). Rotation can occur in two ways. If both orbitals rotate in the same direction (both clockwise or both counterclockwise), **ring closure is conrotatory.**

ring closure is
conrotatory

If the orbitals rotate in opposite directions (one clockwise, the other counterclockwise), **ring closure is disrotatory.**

ring closure is
disrotatory

The mode of ring closure depends on the symmetry of the HOMO of the compound undergoing ring closure. Only the symmetry of the HOMO is important in determining the course of the reaction because this is where the highest energy electrons are. These are the most loosely held electrons and therefore the ones most easily moved during a reaction.

To form the new σ bond, the orbitals must rotate so that in-phase p orbitals overlap because in-phase overlap is a bonding interaction. Out-of-phase overlap would be an antibonding interaction. If the HOMO is symmetric (the end orbitals are identical), rotation will have to be disrotatory to achieve in-phase overlap. In other words, disrotatory ring closure is symmetry-allowed, while conrotatory ring closure is symmetry-forbidden.

HOMO is symmetric

If the HOMO is asymmetric, rotation has to be conrotatory in order to achieve in-phase overlap. In other words, conrotatory ring closure is symmetry-allowed, while disrotatory ring closure is symmetry-forbidden.

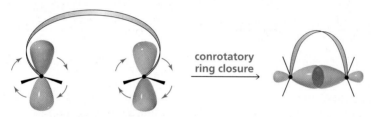

HOMO is asymmetric

Notice that a **symmetry-allowed pathway** is one in which in-phase orbitals overlap; a **symmetry-forbidden pathway** is one in which out-of-phase orbitals would overlap. A symmetry-allowed reaction can take place under relatively mild conditions. If a reaction is symmetry-forbidden, it cannot take place by a concerted pathway. If a symmetry-forbidden reaction takes place at all, it must do so by a nonconcerted mechanism.

Now we are ready to learn why the electrocyclic reactions discussed at the beginning of this section form the indicated products, and why the configuration of the product changes if the reaction is carried out under photochemical conditions.

The ground-state HOMO (ψ_3) of a compound with three conjugated π bonds, such as (2E,4Z,6E)-octatriene, is symmetric (Figure 28.3). This means that ring closure under *thermal conditions* is disrotatory. In disrotatory ring closure of (2E,4Z,6E)-octatriene, the methyl groups are both pushed up (or down) which results in the *cis* product being formed.

> A symmetry-allowed pathway requires in-phase orbital overlap.

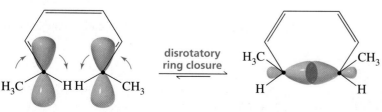

(2E,4Z,6E)-octatriene   *cis*-**5,6-dimethyl-1,3-cyclohexadiene**

In disrotatory ring closure of (2*E*,4*Z*,6*Z*)-octatriene, one methyl group is pushed up and the other is pushed down, which results in the *trans* product being formed.

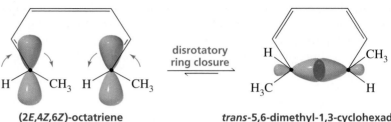

(2E,4Z,6Z)-octatriene   *trans*-**5,6-dimethyl-1,3-cyclohexadiene**

If the reaction takes place under *photochemical conditions,* we must consider the excited-state HOMO rather than the ground-state HOMO. The excited-state HOMO (ψ_4) of a compound with three π bonds is asymmetric. Therefore, under photochemical conditions, (2*E*,4*Z*,6*Z*)-octatriene undergoes conrotatory ring closure, so both methyl groups are pushed down (or up) and the *cis* product is formed.

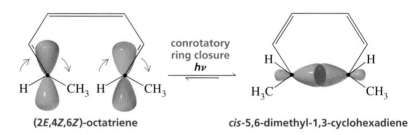

(2E,4Z,6Z)-octatriene   *cis*-**5,6-dimethyl-1,3-cyclohexadiene**

The symmetry of the HOMO of the compound undergoing ring closure controls the stereochemical outcome of an electrocyclic reaction.

We have just seen why the configuration of the product formed under photochemical conditions is the opposite of the configuration of the product formed under thermal conditions: the ground state HOMO is symmetric—so disrotatory ring closure occurs, whereas the excited state HOMO is asymmetric—so conrotatory ring closure occurs. Thus, the stereochemical outcome of an electrocyclic reaction depends on the symmetry of the HOMO of the compound undergoing ring closure.

Now let's see why ring closure of (2*E*,4*Z*)-hexadiene forms *cis*-3,4-dimethylcyclobutene. The compound undergoing ring closure has two conjugated π bonds. The ground-state HOMO of a compound with two conjugated π bonds is asymmetric (Figure 28.2), so ring closure is conrotatory. Conrotatory ring closure of (2*E*,4*Z*)-hexadiene leads to the cis product.

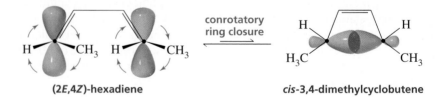

(2E,4Z)-hexadiene   *cis*-**3,4-dimethylcyclobutene**

Similarly, conrotatory ring closure of (2*E*,4*E*)-hexadiene leads to the trans product.

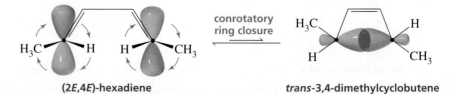

(2E,4E)-hexadiene   *trans*-**3,4-dimethylcyclobutene**

However, if the reaction is carried out under photochemical conditions, the excited-state HOMO of a compound with two conjugated π bonds is symmetric. (Recall that the ground-state HOMO and the excited-state HOMO have opposite symmetries.) So (2E,4Z)-hexadiene will undergo disrotatory ring closure, resulting in the trans product, whereas (2E,4E)-hexadiene will undergo disrotatory ring closure and form the cis product.

We have seen that the ground-state HOMO of a compound with two conjugated double bonds is asymmetric, whereas the ground-state HOMO of a compound with three conjugated double bonds is symmetric. If we examine molecular orbital diagrams for compounds with four, five, six, and more conjugated double bonds, we can conclude that *the ground-state HOMO of a compound with an even number of conjugated double bonds is asymmetric, whereas the ground-state HOMO of a compound with an odd number of conjugated double bonds is symmetric.* Therefore, from the number of conjugated double bonds in a compound, we can immediately tell whether ring closure will be conrotatory (an even number of conjugated double bonds) or disrotatory (an odd number of conjugated double bonds) under thermal conditions. However, if the reaction takes place under photochemical conditions, everything is reversed because the ground-state and excited-state HOMOs have opposite symmetries—if the ground-state HOMO is symmetric, the excited-state HOMO is asymmetric, and vice versa.

> **The ground-state HOMO of a compound with an even number of conjugated double bonds is asymmetric.**

> **The ground-state HOMO of a compound with an odd number of conjugated double bonds is symmetric.**

We have seen that the stereochemistry of an electrocyclic reaction depends on the mode of ring closure, and the mode of ring closure depends on the number of conjugated π bonds in the reactant *and* on whether the reaction is carried out under thermal or photochemical conditions. What we have learned about electrocyclic reactions can be summarized by the **selection rules** listed in Table 28.1. These are also known as the **Woodward–Hoffmann rules** for electrocyclic reactions.

Table 28.1 Woodward–Hoffmann Rules for Electrocyclic Reactions		
Number of conjugated π bonds	**Reaction conditions**	**Allowed mode of ring closure**
Even number	Thermal	Conrotatory
	Photochemical	Disrotatory
Odd number	Thermal	Disrotatory
	Photochemical	Conrotatory

The rules in Table 28.1 are for determining whether a given electrocyclic reaction is "allowed by orbital symmetry." There are also selection rules to determine whether cycloaddition reactions (Table 28.3) and sigmatropic rearrangements (Table 28.4) are "allowed by orbital symmetry." It can be rather burdensome to memorize these rules (and worrisome if they are forgotten during an exam), but all the rules can be summarized by the word "TE-AC." How to use "TE-AC" is explained in Section 28.7.

PROBLEM 5

a. For conjugated systems with two, three, four, five, six, and seven conjugated π bonds, construct quick molecular orbitals (just draw the p orbitals at the ends of the conjugated system as they are drawn on page 1174) to show whether the HOMO is symmetric or asymmetric.

b. Using these drawings, convince yourself that the Woodward-Hoffmann rules in Table 28.1 are valid.

c. Using these drawings, convince yourself that the "TE-AC" shortcut method for learning the information in Table 28.1 is valid (see Section 28.7).

PROBLEM 6◆

a. Under thermal conditions, will ring closure of (2*E*,4*Z*,6*Z*,8*E*)-decatetraene be conrotatory or disrotatory?

b. Will the product have the cis or the trans configuration?

c. Under photochemical conditions, will ring closure be conrotatory or disrotatory?

d. Will the product have the cis or the trans configuration?

The series of reactions in Figure 28.4 illustrates just how easy it is to determine the mode of ring closure and therefore the product of an electrocyclic reaction. The reactant of the first reaction has three conjugated double bonds and is undergoing ring closure under thermal conditions. Ring closure, therefore, is disrotatory (Table 28.1). Disrotatory ring closure of this reactant causes the hydrogens to be cis in the ring-closed product. To determine the relative configurations of the hydrogens, draw them in the reactant and then draw arrows showing disrotatory ring closure (Figure 28.4[a]).

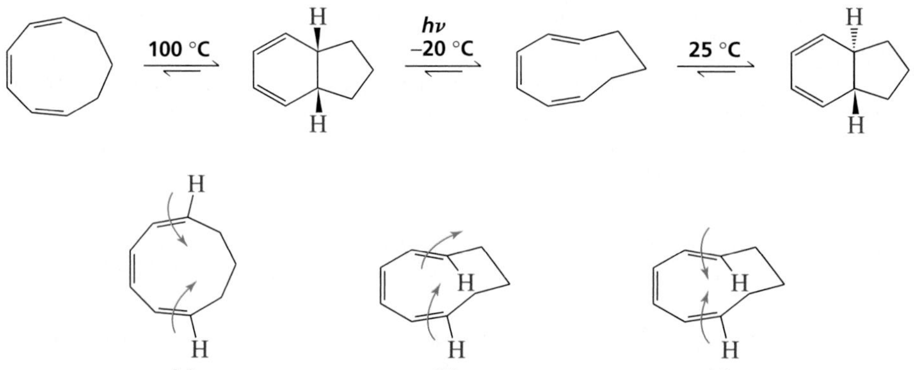

▶ **Figure 28.4**
Determining the stereo-chemistry of the product of an electrocyclic reaction.

(a)   (b)   (c)

The second step is a ring-opening electrocyclic reaction that takes place under photochemical conditions. Because of the principle of microscopic reversibility (Section 14.11), the orbital symmetry rules used for a ring-closure reaction also apply to the reverse ring-opening reaction. The compound undergoing reversible ring closure has three conjugated double bonds. Because the reaction occurs under photochemical conditions, ring opening (and ring closure) is conrotatory. (Notice that the number of conjugated double bonds used to determine the mode of ring opening/closure in reversible electrocyclic reactions is the number in the compound undergoing ring closure.) In order for conrotatory rotation to result in a product with cis hydrogens, the hydrogens in the compound undergoing ring closure must point in the same direction (Figure 28.4[b]).

The third step is a thermal ring closure of a compound with three conjugated double bonds, so ring closure is disrotatory. Drawing the hydrogens and the arrows (Figure 28.4[c]) allows you to determine the configurations of the hydrogens in the ring-closed product.

Notice that in all these electrocyclic reactions, if the bonds to the substituents in the reactant point in *opposite directions* [as in Figure 28.4(a)], the substituents will be cis in the product if ring closure is disrotatory and trans if ring closure is conrotatory. On the other hand, if they point in the *same direction* [as in Figure 28.4(b) or 28.4(c)], they will be trans in the product if ring closure is disrotatory and cis if ring closure is conrotatory (Table 28.2).

Table 28.2 Configuration of the Product of an Electrocyclic Reaction

Substituents in the reactant	Mode of ring closure	Configuration of the product
Point in opposite directions	Disrotatory	cis
	Conrotatory	trans
Point in the same direction	Disrotatory	trans
	Conrotatory	cis

PROBLEM 7◆

Which of the following are correct? Correct any false statements.

a. A conjugated diene with an even number of double bonds undergoes conrotatory ring closure under thermal conditions.

b. A conjugated diene with an asymmetric HOMO undergoes conrotatory ring closure under thermal conditions.

c. A conjugated diene with an odd number of double bonds has a symmetric HOMO.

PROBLEM 8◆

a. Identify the mode of ring closure for each of the following electrocyclic reactions.

b. Are the indicated hydrogens cis or trans?

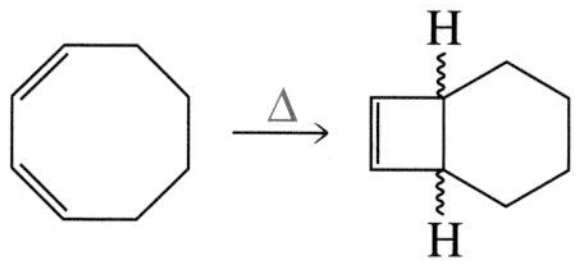

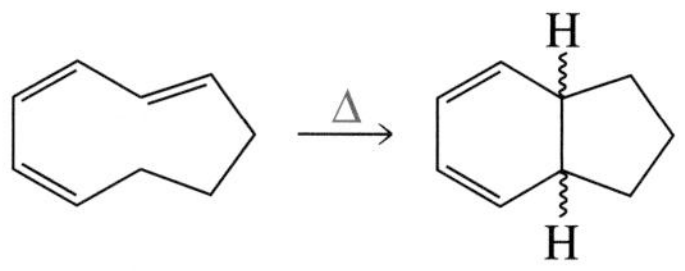

28.4 CYCLOADDITION REACTIONS

In a *cycloaddition reaction,* two different π-bond-containing molecules react to form a cyclic molecule. The Diels–Alder reaction is one of the best known examples of a cycloaddition reaction (Section 7.9). Cycloaddition reactions are classified according to the number of π electrons that interact in the reaction. The Diels–Alder reaction is a [4+2] cycloaddition reaction because one reactant has four interacting π electrons and the other reactant has two interacting π electrons. Only the π electrons participating in electron rearrangement are counted.

[4 + 2] cycloaddition (a Diels–Alder reaction)

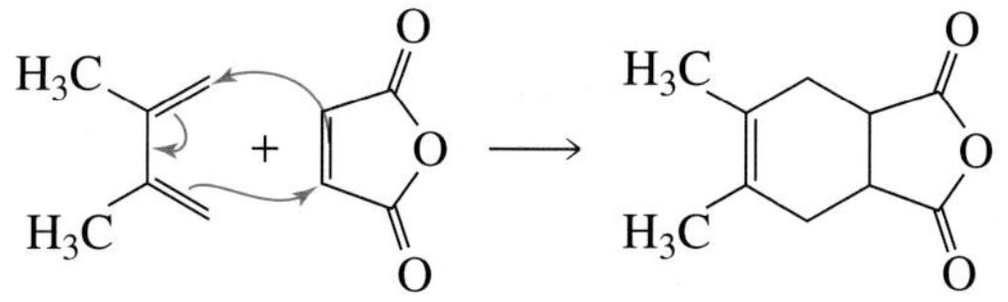

[2 + 2] cycloaddition

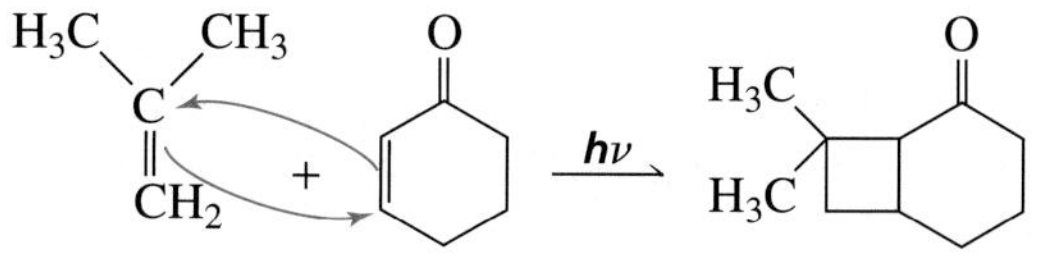

[8 + 2] cycloaddition

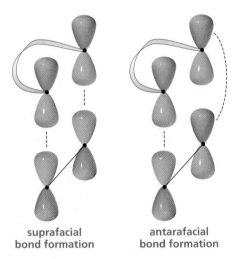

In a cycloaddition reaction, the orbitals of one molecule must overlap with the orbitals of the second molecule. Therefore, the frontier molecular orbitals of both reactants must be evaluated to determine the outcome of the reaction. Because the new σ bonds in the product are formed by donation of electron density from one reactant to the other reactant, we must consider the HOMO of one of the molecules and the LUMO of the other because only an empty orbital can accept electrons. It does not matter which reacting molecule's HOMO is used. It is required only that we use the HOMO of one and the LUMO of the other.

There are two modes of orbital overlap for the simultaneous formation of two σ bonds—suprafacial and antarafacial. **Bond formation is suprafacial** if both σ bonds form on the same side of the π system. **Bond formation is antarafacial** if the two σ bonds form on opposite sides of the π system. Suprafacial bond formation is similar to syn addition, whereas antarafacial bond formation resembles anti addition (Section 4.17).

suprafacial
bond formation

antarafacial
bond formation

A cycloaddition reaction that forms a four-, five-, or six-membered ring must involve suprafacial bond formation. The geometric constraints of these small rings make the antarafacial approach highly unlikely even if it is symmetry-allowed. (Remember that symmetry-allowed means the overlapping orbitals are in phase.) Antarafacial bond formation is more likely in cycloaddition reactions that form larger rings.

Frontier orbital analysis of a [4+2] cycloaddition reaction shows that overlap of in-phase orbitals to form the two new σ bonds requires suprafacial orbital overlap (Figure 28.5). This is true whether we use the HOMO of the alkene (a system with one π bond; Figure 28.1) and the LUMO of the diene (a system with two conjugated π bonds; Figure 28.2), or the LUMO of the alkene and the HOMO of the diene. Now we can understand why Diels–Alder reactions occur with relative ease (Section 7.9).

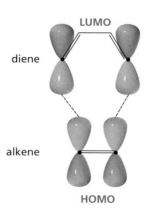

◀ **Figure 28.5**
Frontier molecular orbital analysis of a [4 + 2] cycloaddition reaction. The HOMO of either of the reactants can be used with the LUMO of the other. Both situations require suprafacial overlap for bond formation.

A [2+2] cycloaddition reaction does not occur under thermal conditions but does take place under photochemical conditions.

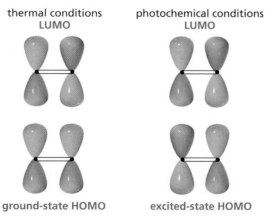

The frontier molecular orbitals in Figure 28.6 show why this is so. Under thermal conditions, suprafacial overlap is not symmetry-allowed (the overlapping orbitals are out of phase). Antarafacial overlap is symmetry-allowed but is not possible because of the small size of the ring. Under photochemical conditions, however, the reaction can take place because the excited-state HOMO has symmetry opposite that of the ground-state HOMO. Therefore, overlap of the excited-state HOMO of one alkene with the LUMO of the second alkene involves symmetry-allowed suprafacial bond formation. Notice in the photochemical reaction that only one of the reactants is in an excited state. Because of the very short lifetimes of excited states, it is unlikely that two reactants in their excited states would find one another to interact. The selection rules for cycloaddition reactions are summarized in Table 28.3.

Tutorial:
Cycloaddition reactions

thermal conditions
LUMO

photochemical conditions
LUMO

ground-state HOMO

excited-state HOMO

◀ **Figure 28.6**
Frontier molecular orbital analysis of a [2 + 2] cycloaddition reaction under thermal and photochemical conditions.

Table 28.3 Woodward–Hoffmann Rules for Cycloaddition Reactions

Sum of the number of π bonds in the reacting systems of both reagents	Reaction conditions	Allowed mode of ring closure
Even number	Thermal	Antarafacial[a]
	Photochemical	Suprafacial
Odd number	Thermal	Suprafacial
	Photochemical	Antarafacial[a]

[a]Although antarafacial ring closure is symmetry-allowed, it can occur only with large rings.

COLD LIGHT

A reverse [2+2] cycloaddition reaction is responsible for the cold light given off by light sticks. A light stick contains a thin glass vial that holds a mixture of sodium hydroxide and hydrogen peroxide suspended in a solution of diphenyloxalate and a dye. When the vial breaks, two nucleophilic acyl substitution reactions occur—recall that phenoxide ion is a relatively good leaving group—that form a compound with an unstable four-membered ring.

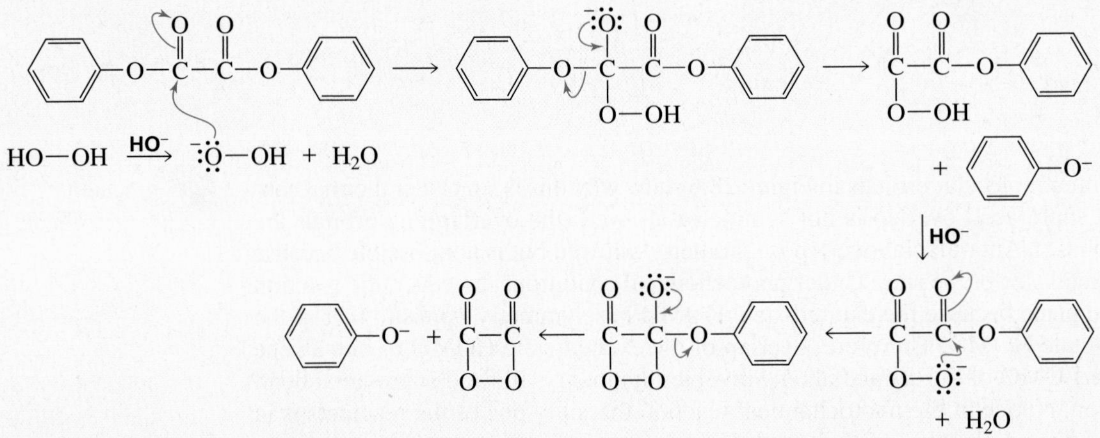

Suprafacial overlap to form a four-membered ring can take place only under photochemical conditions, so one of the reactants must be in an excited state. Therefore, one of the two carbon dioxide molecules formed when the four-membered ring breaks is in an excited state (indicated by an asterisk). When the electron in the excited state drops down to the ground state, a photon of ultraviolet light is released—which is *not* visible to the human eye. However, in the presence of a dye, the excited carbon dioxide molecule can transfer some of its energy to the dye molecule, which causes an electron in the dye to be promoted to an excited state. When the electron of the dye drops down to the ground state, a photon of visible light is released—which *is* visible to the human eye. In Section 28.6 you will see that a similar reaction is responsible for the light given off by fireflies.

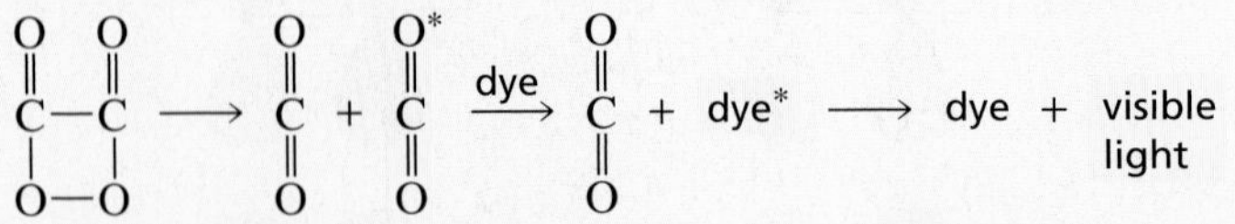

PROBLEM 9

Explain why maleic anhydride reacts rapidly with 1,3-butadiene but does not react at all with ethene under thermal conditions.

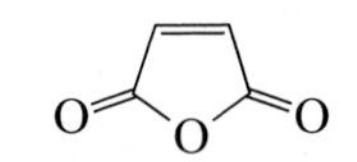

maleic anhydride

PROBLEM 10 / SOLVED

Compare the reaction of 2,4,6-cycloheptatrienone with cyclopentadiene to that with ethene. Why does 2,4,6-cycloheptatrienone use two π electrons in one reaction and four π electrons in the other?

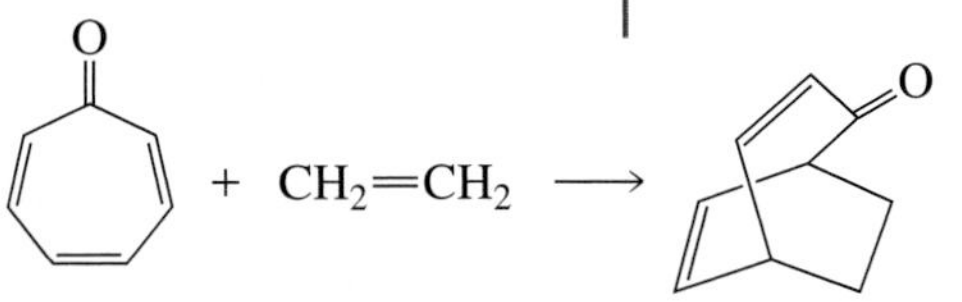

SOLUTION Both reactions are [4+2] cycloaddition reactions. When 2,4,6-cyclohep-tatrienone reacts with cyclopentadiene, it uses two of its π electrons because cyclopen-tadiene is the four-π-electron reactant. When 2,4,6-cycloheptatrienone reacts with ethene, it uses four of its π electrons because ethene is the two-π-electron reactant.

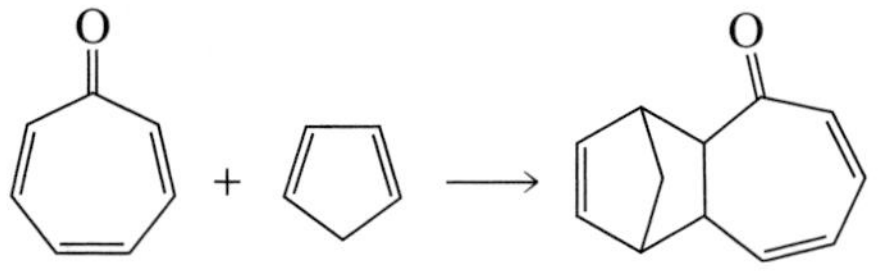

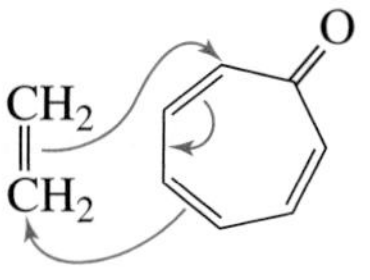

PROBLEM 11◆

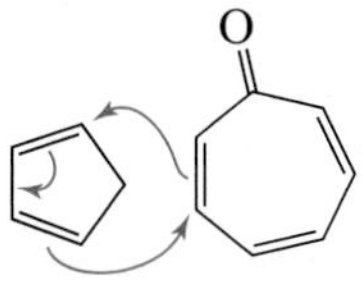

Will a concerted reaction take place between 1,3-butadiene and 2-cyclohexenone in the presence of ultraviolet light?

28.5 SIGMATROPIC REARRANGEMENTS

The last class of concerted pericyclic reactions that we will consider is the group of reactions known as *sigmatropic rearrangements*. In a sigmatropic rearrange-ment, a σ bond in the reactant is broken, a new σ bond is formed, and the π electrons rearrange. The σ bond that breaks is a bond to an allylic carbon. It can be a σ bond between a carbon and a hydrogen, between a carbon and another car-bon, or between a carbon and an oxygen, nitrogen, or sulfur. "Sigmatropic" comes from the Greek word *tropos,* which means "change," so sigmatropic means "sigma-change."

The numbering system used to describe a sigmatropic rearrangement differs from any numbering system you have seen previously. First, mentally break the σ bond in the reactant, and then look at the new σ bond in the product. Count the number of atoms in each of the fragments that connect the broken σ bond and the new σ bond. The two numbers are put in brackets with the smaller number stated first. For example, in the following [2,3] sigmatropic rearrangement, two atoms (N, N) con-nect the old and new σ bonds in one fragment and three atoms (C, C, C) connect the old and new σ bonds in the other fragment.

a [2,3] sigmatropic rearrangement

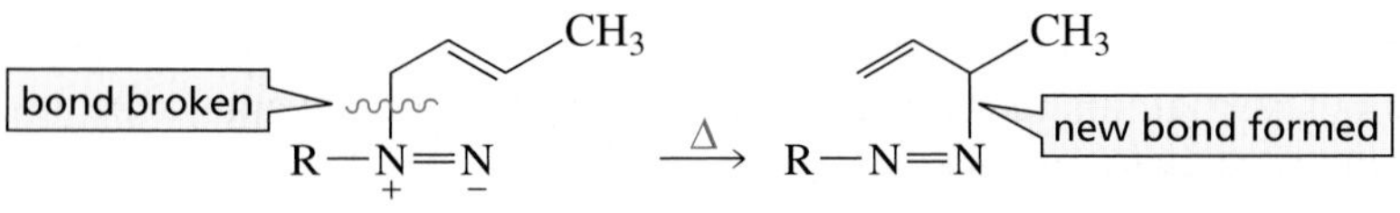

a [1,5] sigmatropic rearrangement

$$CH_3CH-CH=CH-CH=CH_2 \xrightarrow{\Delta} CH_3CH=CH-CH=CH-CH_2$$

bond broken ⟶ (at H) | H   new bond formed

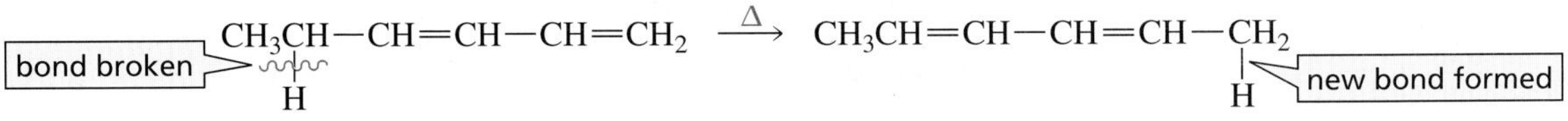

a [1,3] sigmatropic rearrangement

$$CH_3 \qquad\qquad CH_3$$
$$CH_3C-CH=CH_2 \xrightarrow{\Delta} CH_3C=CH-CH_2$$

bond broken ⟶ | CH_3   new bond formed

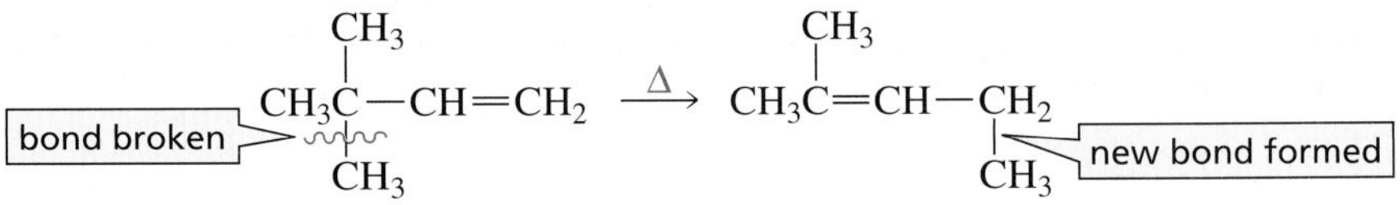

a [3,3] sigmatropic rearrangement

bond broken ⟶ $\xrightarrow{\Delta}$ new bond formed

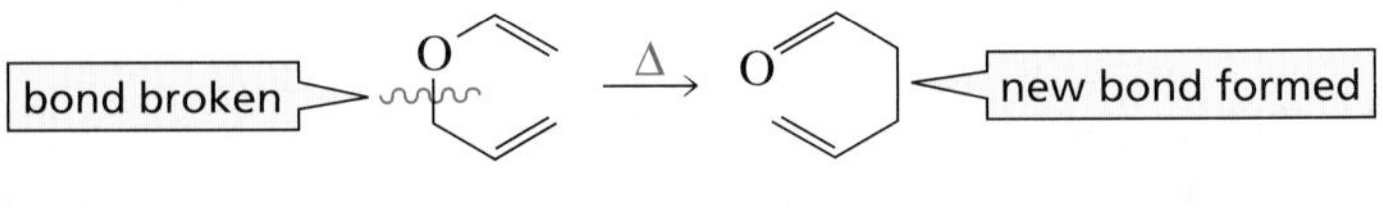

PROBLEM 12

a. Name the kind of sigmatropic rearrangement that occurs in each of the following reactions.

b. Using arrows, show the electron rearrangement that takes place in each of the reactions.

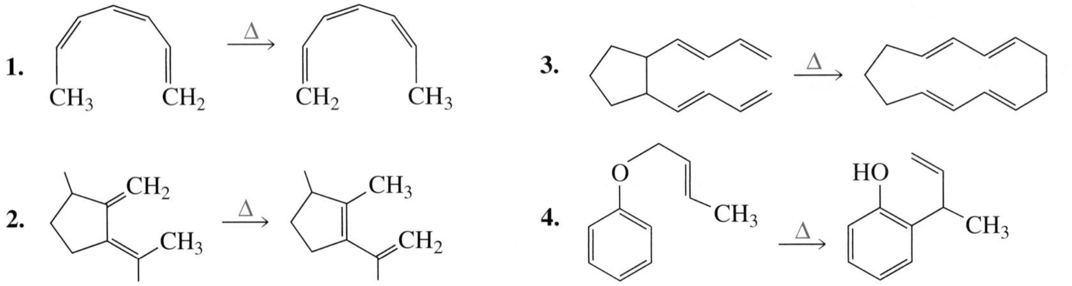

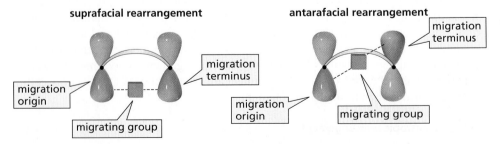

In the transition state of a sigmatropic rearrangement, the group that migrates is partially bonded to the migration origin and partially bonded to the migration terminus. There are two possible modes for rearrangement. If the migrating group remains on the same face of the π system, the **rearrangement is suprafacial.** If the migrating group moves to the opposite face of the π system, the **rearrangement is antarafacial.**

suprafacial rearrangement

migration origin
migration terminus
migrating group

antarafacial rearrangement

migration terminus
migration origin
migrating group

Sigmatropic rearrangements have cyclic transition states. If the transition state has six or fewer atoms in the ring, rearrangement must be suprafacial because of the geometric constraints of small rings.

A [1,3] sigmatropic rearrangement involves a π bond and a pair of σ electrons, or we can say that it involves two pairs of electrons. A [1,5] sigmatropic rearrangement involves two π bonds and a pair of σ electrons (three pairs of electrons), and a [1,7] sigmatropic rearrangement involves four pairs of electrons. We can use the same symmetry rules for sigmatropic rearrangements that we used for cycloaddition reactions if we substitute "number of pairs of electrons" for "sum of the number of π bonds." (Compare Tables 28.3 and 28.4.) *Recall that the ground-state HOMO of a compound with an even number of conjugated double bonds is asymmetric, whereas the ground-state HOMO of a compound with an odd number of conjugated double bonds is symmetric.*

Table 28.4 Woodward–Hoffmann Rules for Sigmatropic Rearrangements		
Number of pairs of electrons in the reacting system	**Reaction conditions**	**Allowed mode of ring closure**
Even number	Thermal	Antarafacial[a]
	Photochemical	Suprafacial
Odd number	Thermal	Suprafacial
	Photochemical	Antarafacial[a]
[a]Although antarafacial ring closure is symmetry-allowed, it can occur only with large rings.		

A **Cope**[2] **rearrangement** is a [3,3] sigmatropic rearrangement of a 1,5-diene. A **Claisen**[3] **rearrangement** is a [3,3] sigmatropic rearrangement of an allyl vinyl ether. Both rearrangements form six-membered-ring transition states. Therefore, the reactions must be able to take place by a suprafacial pathway. Whether or not a suprafacial pathway is symmetry-allowed depends on the number of pairs of electrons involved in the rearrangement (Table 28.4). Because [3,3] sigmatropic rearrangements involve three pairs of electrons, they occur by a suprafacial pathway under thermal conditions. Therefore, both Cope and Claisen rearrangements readily take place under thermal conditions.

Tutorial:
Sigmatropic
rearrangements

a Cope rearrangement

$$\text{(structure with } C_6H_5 \text{ and } CH_3 \xrightarrow{\Delta} \text{product with } C_6H_5 \text{ and } CH_3\text{)}$$

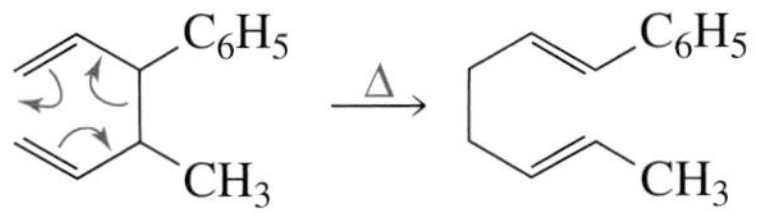

a Claisen rearrangement

$$\text{(allyl vinyl ether with } O \text{ and } CH_3 \xrightarrow{\Delta} \text{product with } O \text{ and } CH_3\text{)}$$

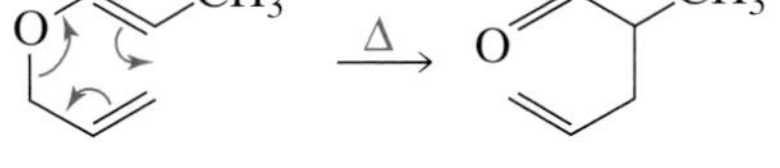

PROBLEM 13◆

a. Give the product of the following reaction:

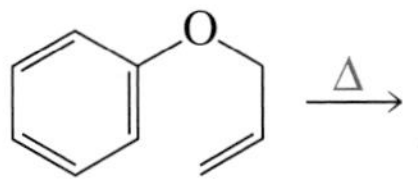

$$\text{(phenyl allyl ether)} \xrightarrow{\Delta}$$

---

[2]Cope also discovered the Cope elimination (page 467).
[3]Claisen also discovered the Claisen condensation (page 848).

b. If the terminal sp^2 carbon of the substituent bonded to the benzene ring is labeled with ^{14}C, where will the label be in the product?

Migration of Hydrogen

When a hydrogen migrates in a sigmatropic rearrangement, the *s* orbital of hydrogen is partially bonded to both the migration origin and the migration terminus in the transition state. Therefore, a [1,3] sigmatropic migration of hydrogen involves a four-membered-ring transition state. Because two pairs of electrons are involved (the HOMO is asymmetric), the selection rules require an antarafacial rearrangement for a 1,3-hydrogen shift under thermal conditions (Table 28.4). Consequently, 1,3-hydrogen shifts do not occur under thermal conditions because the four-membered-ring transition state does not allow the required antarafacial rearrangement.

migration of hydrogen

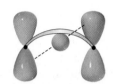

suprafacial rearrangement   antarafacial rearrangement

1,3-Hydrogen shifts can take place if the reaction is carried out under photochemical conditions because the HOMO is symmetric under photochemical conditions, which means that hydrogen can migrate by a suprafacial pathway (Table 28.4).

1,3-hydrogen shifts

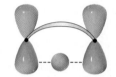

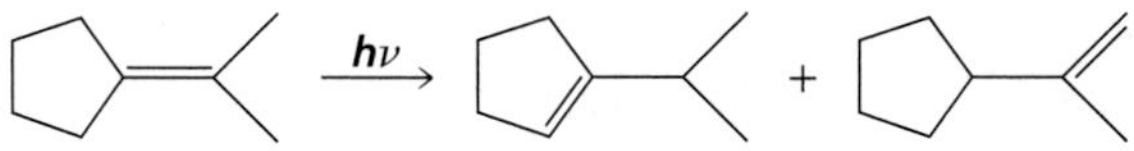

Two products are obtained in the previous reaction because two different allylic hydrogens can undergo a 1,3-hydrogen shift.

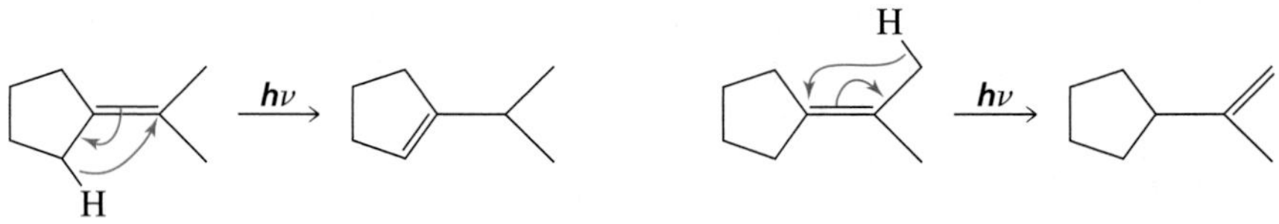

[1,5] Sigmatropic migrations of hydrogen are well known. They involve three pairs of electrons, so they take place by a suprafacial pathway under thermal conditions.

1,5-hydrogen shifts

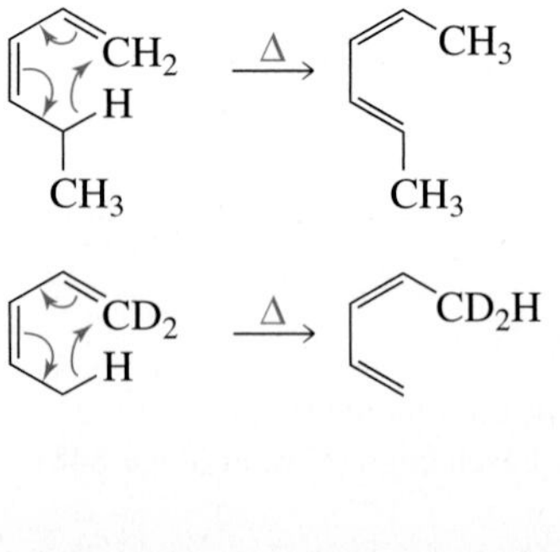

PROBLEM 14

Why was a deuterated compound used in the last example?

PROBLEM 15

Account for the difference in the products obtained under photochemical and thermal conditions.

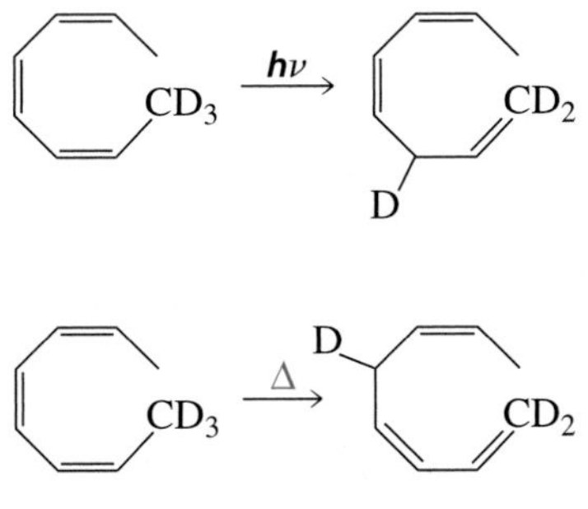

[1,7] Sigmatropic hydrogen migrations involve four pairs of electrons. They can take place under thermal conditions because the eight-membered-ring transition state allows the required antarafacial rearrangement.

1,7-hydrogen shift

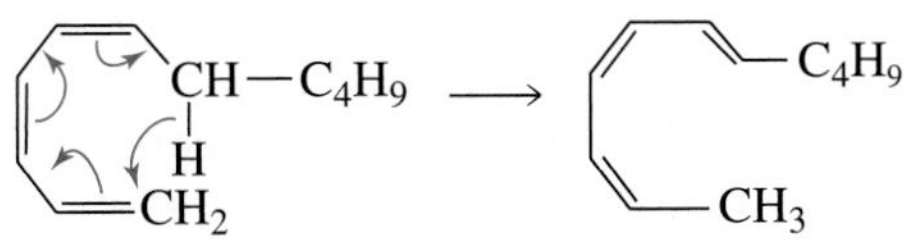

PROBLEM 16 / SOLVED

5-Methyl-1,3-cyclopentadiene rearranges to give a mixture of 5-methyl-1,3-cyclopenta-diene, 1-methyl-1,3-cyclopentadiene, and 2-methyl-1,3-cyclopentadiene. Show how these products are formed.

SOLUTION Notice that both equilibria involve [1,5] sigmatropic rearrangements. Al-though a hydride moves from one carbon to an adjacent carbon, the rearrangements are not called 1,2-shifts because this would not account for all the atoms involved in the re-arranged π electron system.

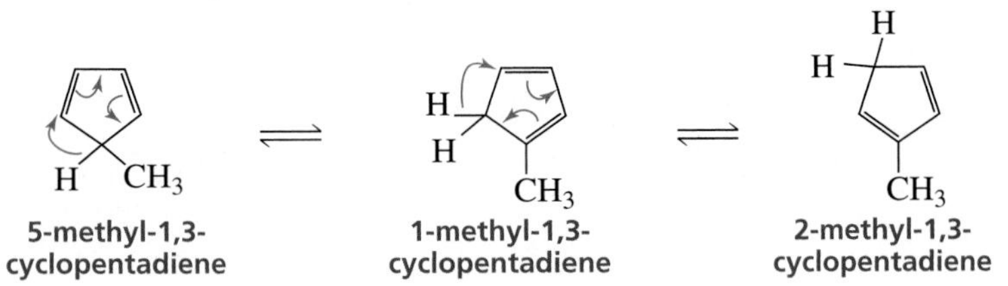

5-methyl-1,3-cyclopentadiene ⇌ 1-methyl-1,3-cyclopentadiene ⇌ 2-methyl-1,3-cyclopentadiene

Migration of Carbon

Unlike hydrogen, which can migrate in only one way because of its spherical *s* or-bital, carbon has two ways to migrate because it has a two-lobed *p* orbital. Carbon can simultaneously interact with the migration origin and the migration terminus using one of its lobes.

carbon migrating with one of its lobes interacting

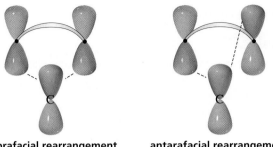

suprafacial rearrangement   **antarafacial rearrangement**

Carbon can also simultaneously interact with the migration source and the migration terminus using both of its lobes.

carbon migrating with both of its lobes interacting

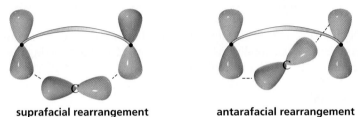

suprafacial rearrangement   **antarafacial rearrangement**

If the reaction requires a suprafacial rearrangement, carbon will migrate using one of its lobes if the HOMO is symmetric, and will migrate using both of its lobes if the HOMO is asymmetric.

When carbon migrates with only one of its p lobes interacting with the migration source and migration terminus, the migrating group retains its configuration because bonding is always to the same lobe. When carbon migrates with both of its p lobes interacting, bonding in the reactant and bonding in the product involve different lobes. Therefore, migration occurs with inversion of configuration.

The following [1,3] sigmatropic rearrangement has a four-membered-ring transition state that requires a suprafacial pathway. The reacting system has two pairs of electrons, so its HOMO is asymmetric. Therefore, the migrating carbon interacts with the migration source and the migration terminus using both of its lobes, so it undergoes inversion of configuration.

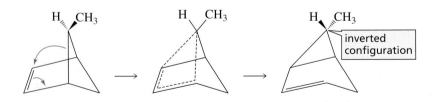

inverted configuration

PROBLEM 17

[1,3] Sigmatropic migrations of hydrogen cannot occur under thermal conditions, but [1,3] sigmatropic migrations of carbon can occur under thermal conditions. Explain.

PROBLEM 18 ◆

a. Will thermal 1,3-migrations of carbon occur with retention or inversion of configuration?

b. Will thermal 1,5-migrations of carbon occur with retention or inversion of configuration?

Biological Cycloaddition Reactions

**28.6
PERICYCLIC
REACTIONS IN
BIOLOGICAL
SYSTEMS**

Exposure to ultraviolet light causes skin cancer. This is one of the reasons for the current concern about the thinning ozone layer. The ozone layer absorbs ultraviolet radiation high in the atmosphere, protecting organisms on the Earth's surface (Section 8.9). One cause of skin cancer is the formation of *thymine dimers*. At any point in DNA where there are two adjacent thymine residues, a [2+2] cycloaddition reaction can occur, resulting in the formation of a thymine dimer. Because [2+2] cycloaddition reactions take place only under photochemical conditions, the reaction takes place only in the presence of ultraviolet light.

two adjacent thymine residues on DNA $\xrightarrow{h\nu}$ **mutation-causing thymine dimer**

Thymine dimers can cause cancer because they interfere with the structural integrity of DNA. Any modification of DNA structure can lead to mutations and possibly to cancer.

Fortunately, there are enzymes that repair damaged DNA. When a repair enzyme recognizes a thymine dimer, it removes it and inserts two individual thymine residues. Repair enzymes, however, are not perfect, and some damage always remains uncorrected. People who do not have the repair enzyme (called DNA photolyase) that removes thymine dimers rarely live beyond the age of 20. Fortunately, this genetic defect is rare.

Fireflies are one of several species that emit light as a result of a reverse [2+2] cycloaddition reaction, similar to the reaction that produces the cold light in light sticks (Section 28.4). Fireflies have an enzyme (luciferase) that catalyzes the reaction between luciferin, ATP, and molecular oxygen that forms a compound with an unstable four-membered ring. When the four-membered ring breaks, an electron in oxyluciferin is in an excited state because suprafacial overlap can occur only under photochemical conditions. When the electron in the excited state drops down to the ground state, a photon of light is released (see the scheme on the next page).

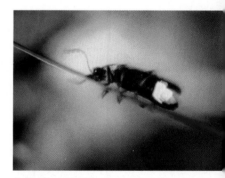

A firefly with its stomach aglow

3-D Molecules:
Thymine dimer;
DNA photolyase

A Biological Reaction Involving an Electrocyclic Reaction and a Sigmatropic Rearrangement

7-Dehydrocholesterol, a steroid found in skin, is converted into vitamin D_3 by two pericyclic reactions. The first is an electrocyclic reaction that opens one of the six-membered rings in the starting material to form provitamin D_3. This reaction occurs under photochemical conditions. The provitamin then undergoes a [1,7]

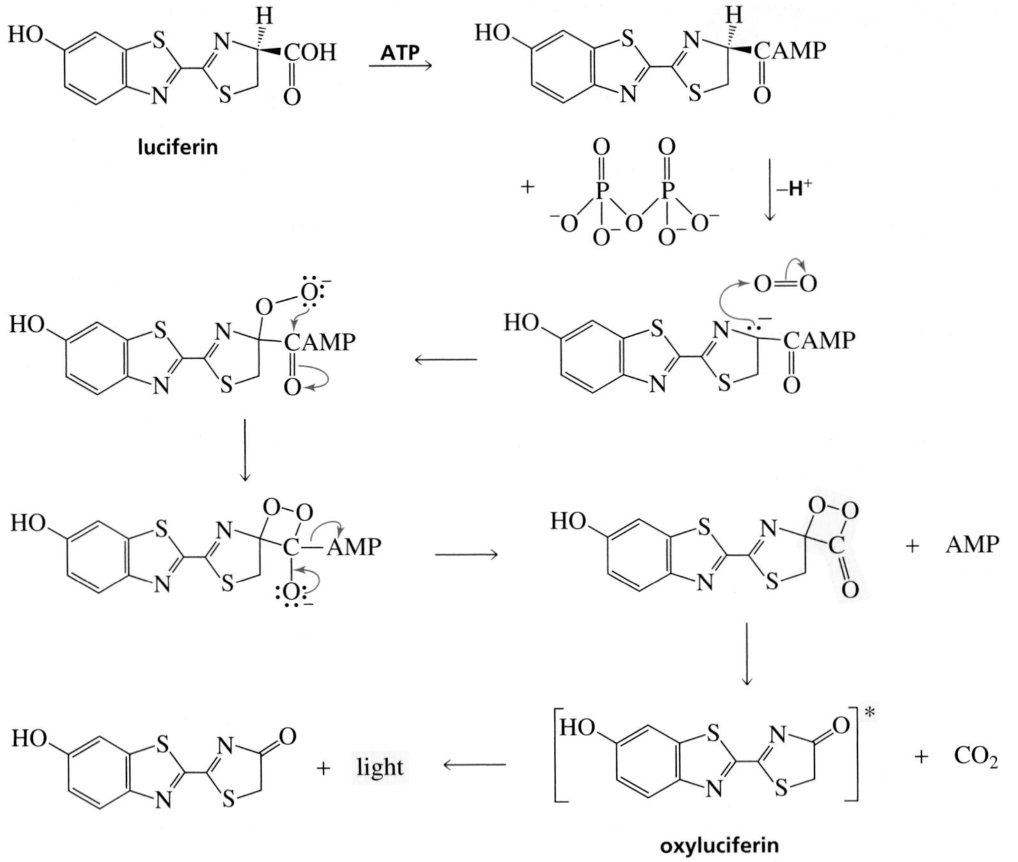

luciferin

oxyluciferin

sigmatropic rearrangement to form vitamin D$_3$. The sigmatropic rearrangement takes place under thermal conditions and is slower than the electrocyclic reaction, so vitamin D$_3$ continues to be synthesized for several days after exposure to sunlight. Vitamin D$_3$ is not the active form of the vitamin. The active form requires two successive hydroxylations of vitamin D$_3$—the first occurs in the liver and the second in the kidneys.

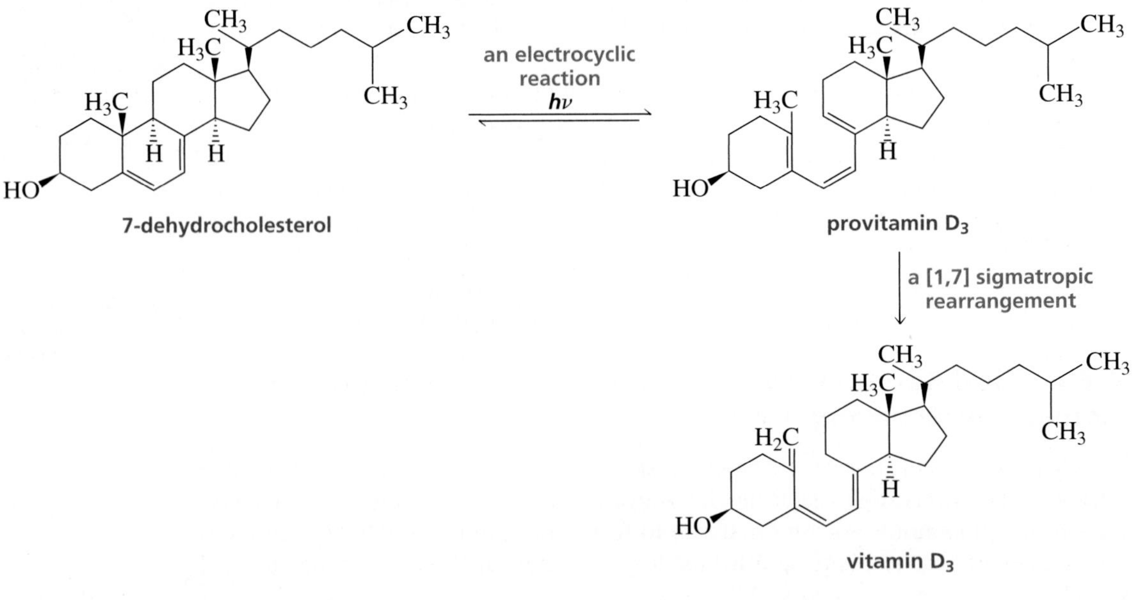

7-dehydrocholesterol

an electrocyclic reaction
$h\nu$

provitamin D$_3$

a [1,7] sigmatropic rearrangement

vitamin D$_3$

A deficiency in vitamin D, which can be prevented by getting enough sun, causes a disease known as rickets. Rickets is characterized by deformed bones and stunted growth. Too much vitamin D is also harmful because it causes the calcification of soft tissues. It is thought that skin pigmentation evolved to protect the skin from the sun's ultraviolet rays in order to prevent the synthesis of too much vitamin D_3. This agrees with the observation that peoples who are indigenous to countries close to the equator have greater skin pigmentation.

PROBLEM 19◆

Does the [1,7] sigmatropic rearrangement that converts provitamin D_3 to vitamin D_3 involve suprafacial or antarafacial rearrangement?

PROBLEM 20

Explain why photochemical ring closure of provitamin D_3 to form 7-dehydrocholesterol results in the hydrogen and methyl substituents being trans to one another.

The selection rules that determine the outcome of electrocyclic reactions, cycloaddition reactions, and sigmatropic rearrangements are summarized in Tables 28.1, 28.3, and 28.4, respectively. This is still a lot to remember. Fortunately, the selection rules for all pericyclic reactions can be summarized in one word: "**TE-AC**."

28.7 SUMMARY OF THE SELECTION RULES FOR PERICYCLIC REACTIONS

- If **TE** describes the reaction (**T**hermal/**E**ven), the outcome *is given* by **AC** (**A**ntarafacial or **C**onrotatory).
- If one of the letters of **TE** is incorrect (it is not **T**hermal/**E**ven but is **T**hermal/**O**dd or **P**hotochemical/**E**ven), the outcome *is not given* by **AC** (the outcome is **S**uprafacial or **D**isrotatory).
- If both of the letters of **TE** are incorrect (**P**hotochemical/**O**dd), the outcome *is given* by **AC** (**A**ntarafacial or **C**onrotatory)—"two negatives make a positive."

KEY TERMS

antarafacial bond formation (page 1178)
antarafacial rearrangement (page 1182)
antibonding π molecular orbital (page 1167)
asymmetric molecular orbital (page 1169)
bonding π molecular orbital (page 1167)
Claisen rearrangement (page 1183)
conrotatory ring closure (page 1172)
conservation of orbital symmetry theory (page 1165)
Cope rearrangement (page 1183)
cycloaddition reaction (page 1164)
disrotatory ring closure (page 1172)

electrocyclic reaction (page 1163)
excited state (page 1168)
frontier orbital analysis (page 1178)
frontier orbitals (page 1170)
frontier orbital theory (page 1166)
ground state (page 1168)
highest occupied molecular orbital (HOMO) (page 1170)
linear combination of atomic orbitals (LCAO) (page 1166)
lowest unoccupied molecular orbital (LUMO) (page 1170)
molecular orbital (MO) theory (page 1166)
pericyclic reaction (page 1163)
photochemical reaction (page 1165)
polar reaction (page 1163)

radical reaction (page 1163)
selection rules (page 1175)
sigmatropic rearrangement (page 1164)
suprafacial bond formation (page 1178)
suprafacial rearrangement (page 1182)
symmetric molecular orbital (page 1169)
symmetry-allowed pathway (page 1173)
symmetry-forbidden pathway (page 1173)
thermal reaction (page 1165)
Woodward–Hoffmann rules (page 1175)

PROBLEMS

21. Give the product of each of the following reactions:

a. 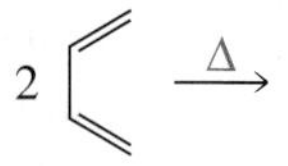

e. 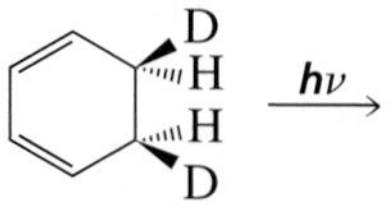

b. 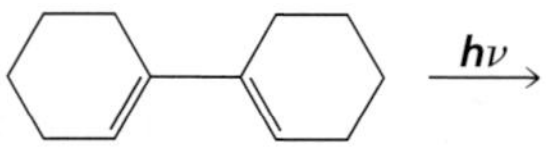

f. 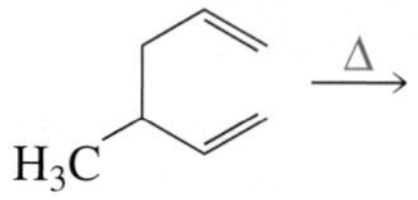

c. 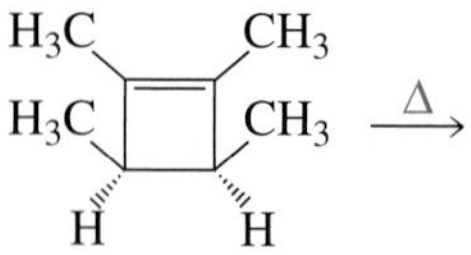

g. 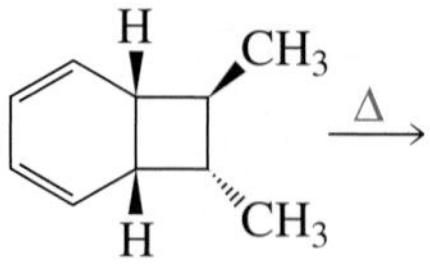

d. 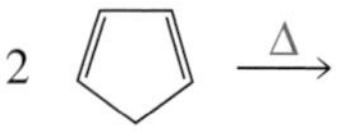

h. 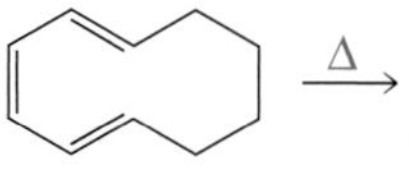

22. Give the product formed when each of the following compounds undergoes an electrocyclic reaction

a. under thermal conditions

b. under photochemical conditions

1. 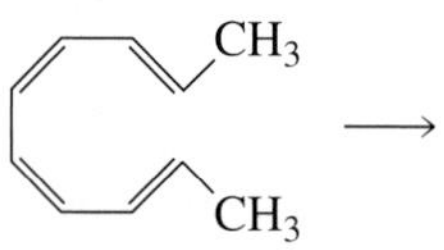

2. 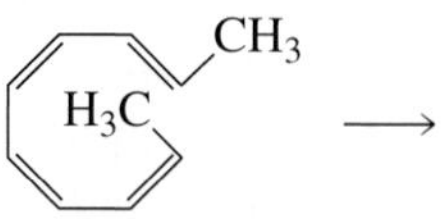

23. Chorismate mutase is an enzyme that catalyzes a pericyclic reaction that forms prephenate, which is subsequently converted into the amino acids phenylalanine and tyrosine. What kind of a pericyclic reaction does chorismate mutase catalyze?

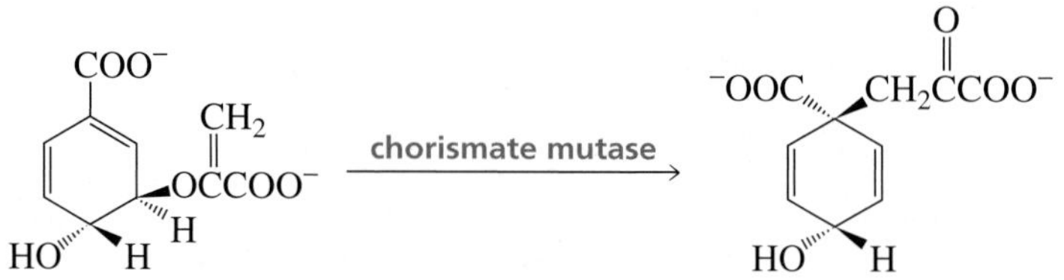

chorismate prephenate

24. Account for the difference in the products of the following reactions:

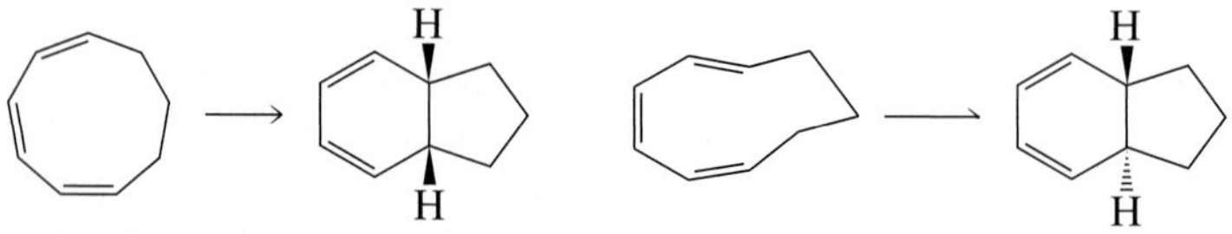

25. Show how norbornane could be prepared from cyclopentadiene.

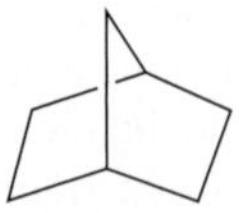

norbornane

26. Give the product of each of the following reactions:

a. 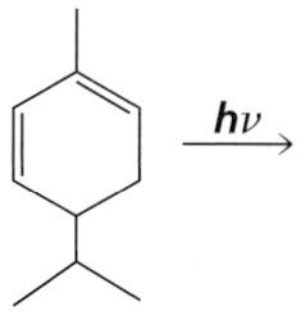 $\xrightarrow{h\nu}$

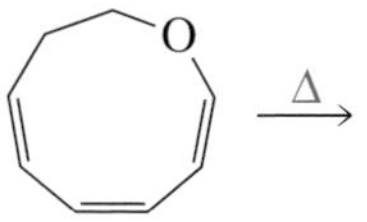

b. 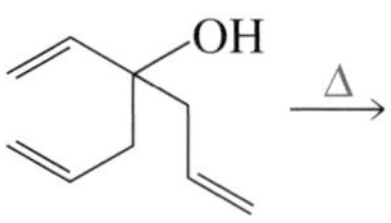 $\xrightarrow{\Delta}$

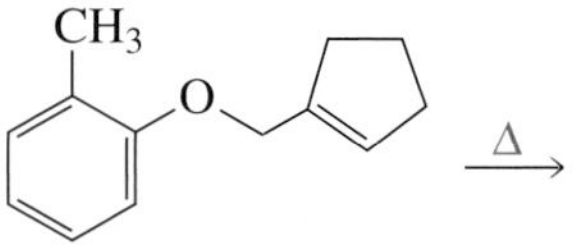

c. 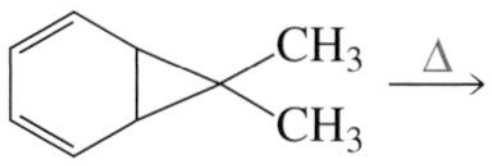 $\xrightarrow{\Delta}$

d. $\xrightarrow{\Delta}$

e. $\xrightarrow{\Delta}$

27. Which is the product of the following [1,3] sigmatropic rearrangement, A or B?

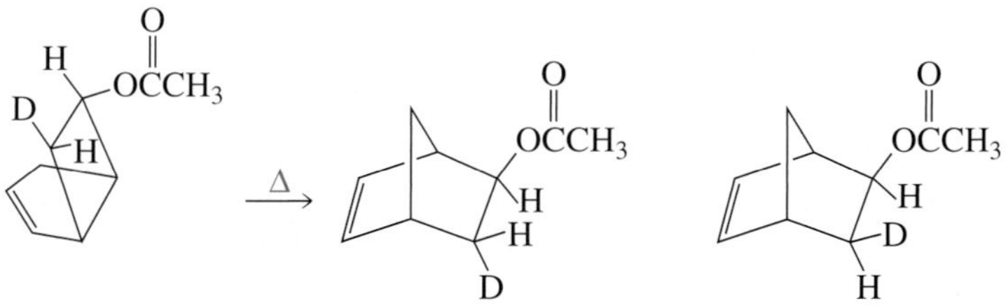

28. Dewar benzene is a highly strained isomer of benzene. In spite of its thermodynamic instability, it is very stable kinetically. It will rearrange to benzene, but only if heated to a very high temperature. Why is it kinetically stable?

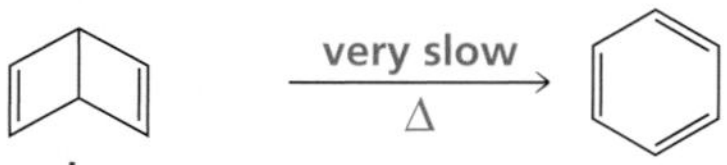

Dewar benzene

29. If the following compounds are heated, one will form one product from a [1,3] sigmatropic rearrangement and the other will form two products from two different [1,3] sigmatropic rearrangements. Give the products of the reactions.

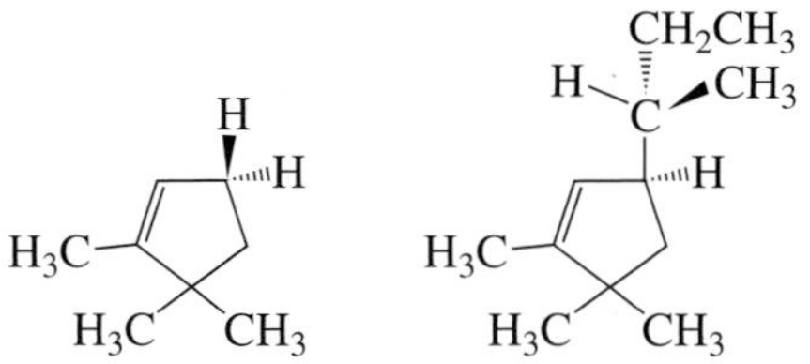

30. When the following compound is heated, a product is formed that shows an infrared absorption band at 1715 cm^{-1}. Draw the structure of the product.

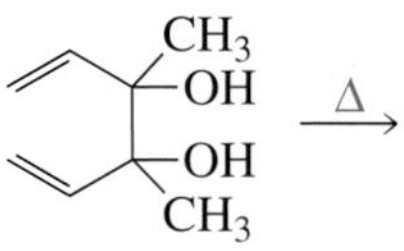

31. Two products are formed in the following [1,7] sigmatropic rearrangement, one due to hydrogen migration and the other to deuterium migration. Show the configuration of the products by replacing A and B with the appropriate substituents (H or D).

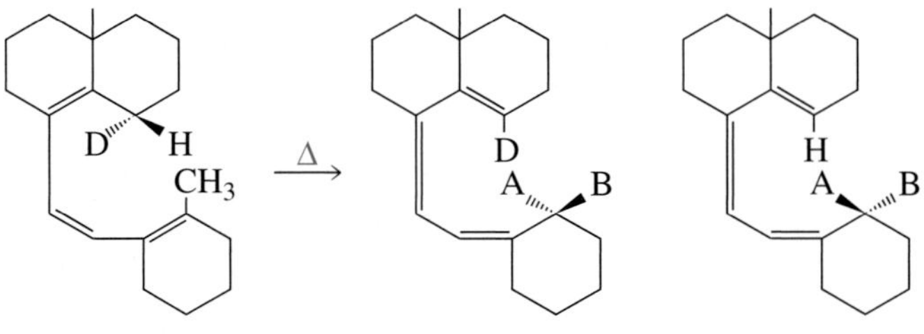

32. a. Propose a mechanism for the following reaction. (*Hint:* An electrocyclic reaction is followed by a Diels–Alder reaction.)
   b. What would be the reaction product if *trans*-2-butene were used instead of ethene?

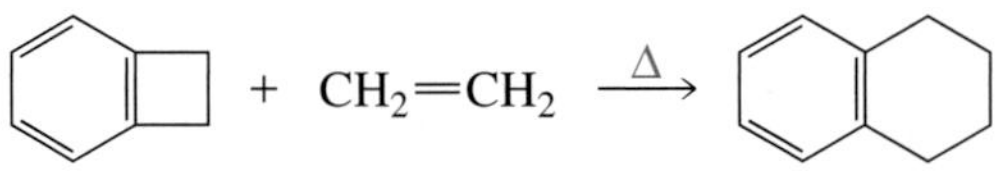

33. Explain why two different products are formed from disrotatory ring closure of (2*E*,4*Z*,6*Z*)-octatriene, but only one product is formed from disrotatory ring closure of (2*E*,4*Z*,6*E*)-octatriene.

34. Give the product of each of the following sigmatropic rearrangements:

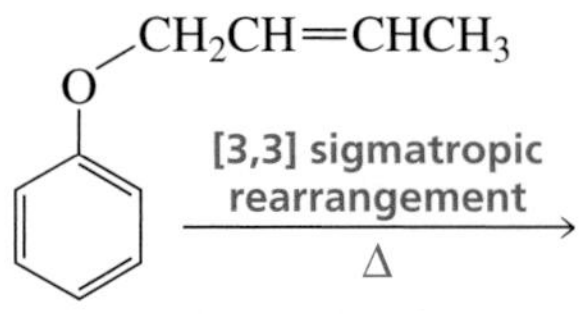

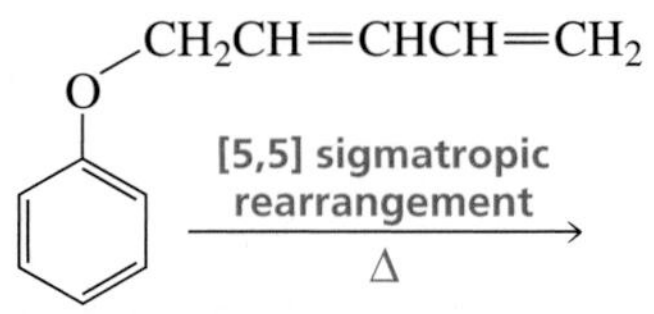

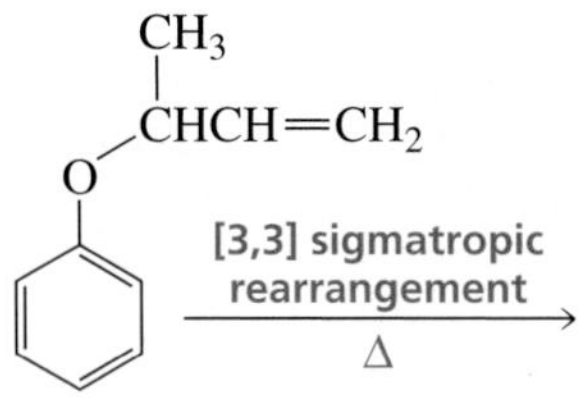

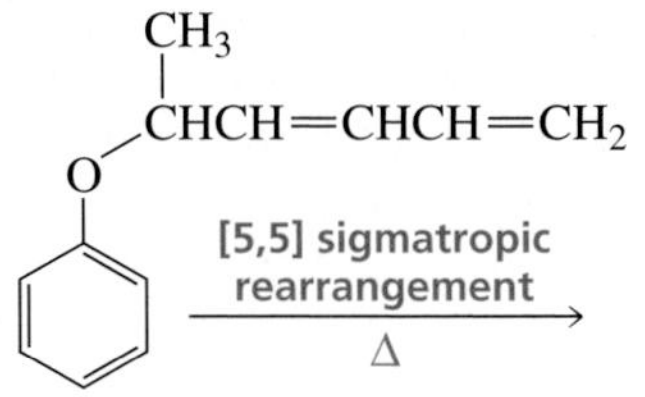

35. *cis*-3,4-Dimethylcyclobutene undergoes thermal ring opening to form the two products shown. One of the products is formed in 99% yield, the other in 1% yield. Which is which?

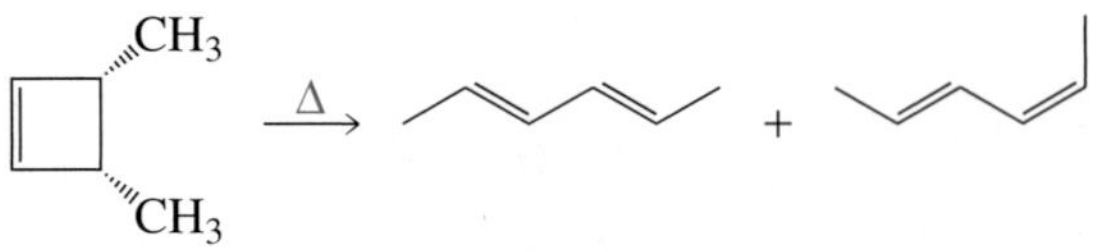

36. If isomer A is heated to about 100 °C, a mixture of isomers A and B is formed. Explain why there is no trace of isomer C or D.

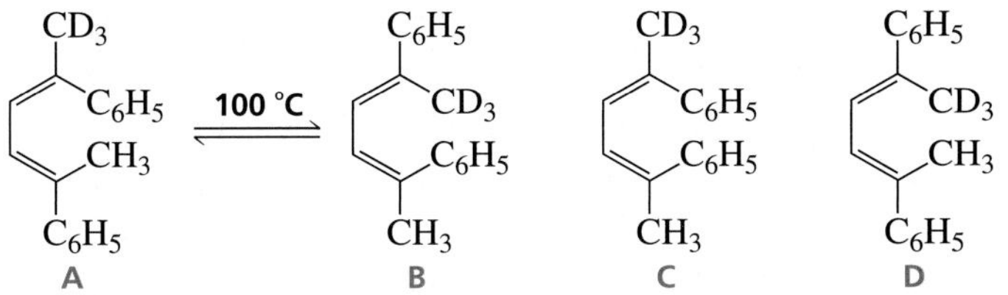

37. Propose a mechanism for the following reaction:

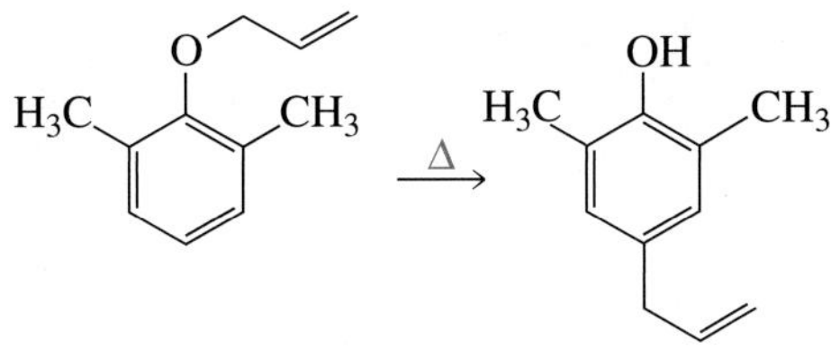

38. Compound A will not undergo a ring-opening reaction under thermal conditions, but compound B will. Explain.

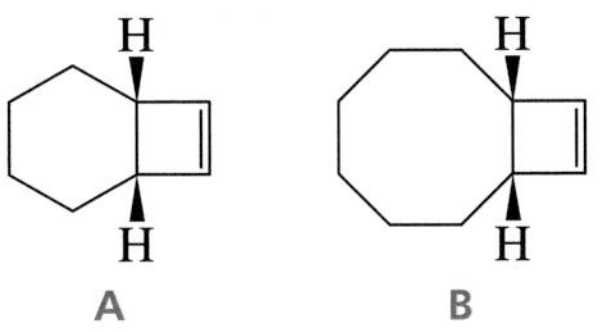

39. Professor Perry C. Click found that heating any one of the following isomers resulted in scrambling of the deuterium to all three positions on the five-membered ring. Propose a mechanism to account for this observation.

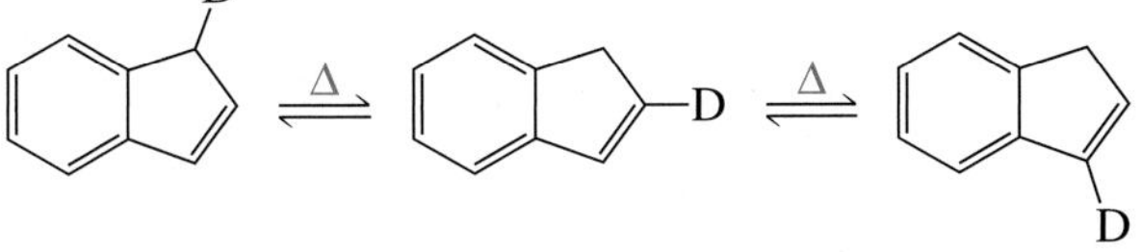

40. How could the following transformation be carried out, using only heat or light?

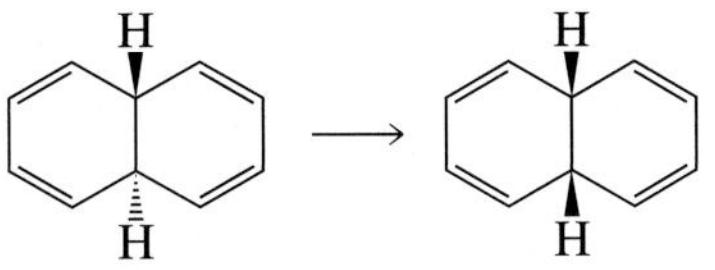

41. Show the steps involved in the following reaction:

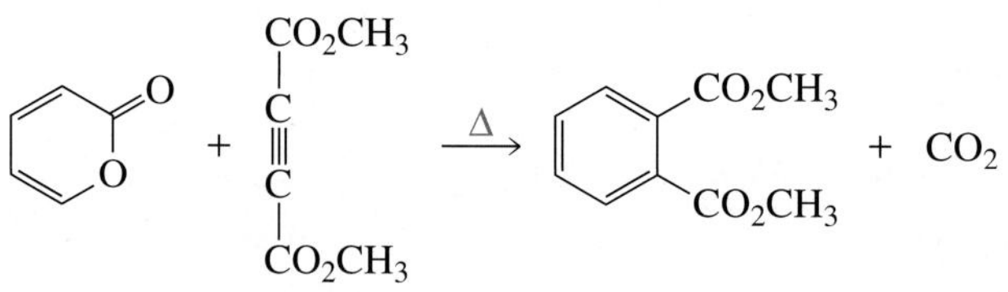

42. Propose a mechanism for the following reaction:

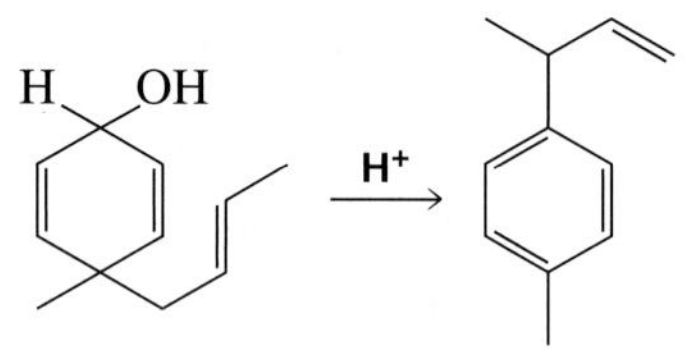

29

More About Multistep Organic Synthesis

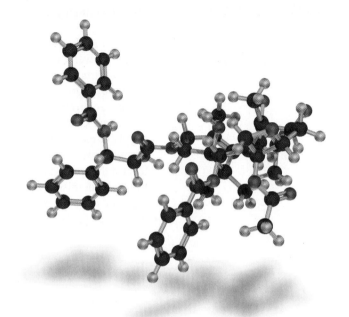

Taxol

Organic synthesis is the preparation of organic compounds from other organic compounds. The word "synthesis" comes from the Greek word *synthesis,* which means "a putting together." Because of the ability of carbon atoms to form rings, chains, and multiple bonds, an almost infinite number of organic compounds can be conceived. Chemists synthesize many hundreds of thousands of them each year. Many organic compounds exist only because chemists can synthesize them, but many are synthesized in nature. Compounds synthesized in nature, such as penicillin G and cocaine, are called **natural products.**

natural products

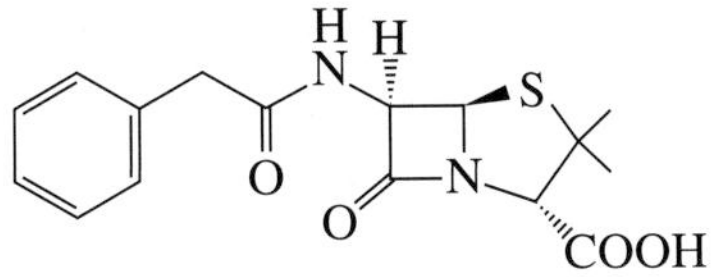

penicillin G

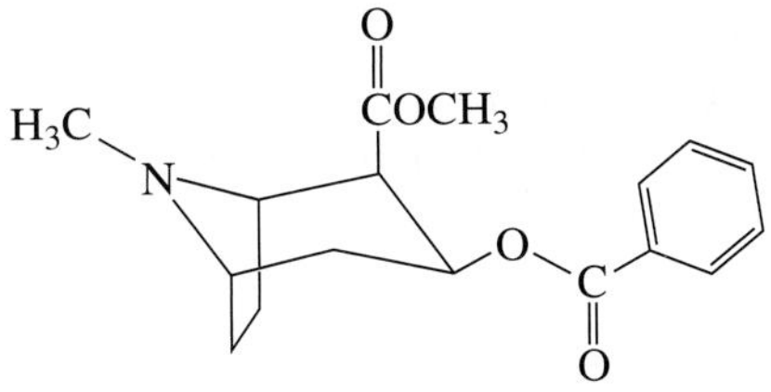

cocaine

You have been introduced to many aspects of organic synthesis (Sections 5.11, 10.12, 15.8, 17.13, 18.9, and 19.20), and you have had the opportunity to design some relatively simple syntheses. In this chapter we will examine in greater depth how organic chemists approach a synthetic problem. We will start by describing some of the basic strategies chemists use when designing syntheses; then we will take a look at how they use these tools to carry out the syntheses of some complicated organic molecules. It is impossible to present all the tools available to the synthetic chemist, but this chapter will give you some idea of the thought processes involved in designing a synthesis and of the great diversity of operations available to the synthetic chemist.

Organic chemists synthesize compounds for many reasons—to study their properties, to answer a variety of chemical questions, because they have unusual shapes or other unusual structural features, or because they have useful properties. One reason chemists synthesize natural products is to provide us with greater supplies of these compounds than nature can produce. For example, Taxol—a compound that has been successful in treating certain kinds of cancer—is extracted from the bark of *Taxus*, the Pacific yew tree. The supply of natural Taxol is limited because yew trees are uncommon and grow very slowly, and because stripping the bark kills the tree. Additionally, *Taxus* forests serve as habitats for the spotted owl, an endangered species, so harvesting the trees would accelerate the owl's demise. Once chemists determined the structure of Taxol, however, efforts could be undertaken to synthesize it in order to make it more widely available as an anticancer drug. A successful synthesis has been accomplished.

Taxol

Once a compound has been synthesized, chemists can study its properties to learn how it works; then they can design and synthesize safer or more potent analogs. For example, chemists have found that the anticancer activity of taxol is substantially reduced if its four ester groups are hydrolyzed. This gives one small clue as to how the molecule functions.

3-D Molecule:
Taxol

Organic chemists can structurally modify a natural product to see if they can produce a related compound with different properties. For example, cocaine is a natural product found in the leaves of the coca plant. In the nineteenth century, cocaine was widely used as a local anesthetic because it deadens nerve endings quite effectively. Its wide use in medicine meant that legal supplies of the drug were readily available for misuse. An analog of the drug was synthesized that had the pain-deadening effects but lacked the psychological side effects (Section 30.3). This compound was called procaine (Novocain). Substituting Novocain for cocaine in medical practice reduced the early abuse of cocaine.

procaine
Novocain

Many syntheses, such as the synthesis of an ester from an acyl chloride, can be done in a single step (Section 16.7).

$$CH_3\overset{O}{\overset{\|}{C}}Cl \ + \ CH_3CH_2OH \ \longrightarrow \ CH_3\overset{O}{\overset{\|}{C}}OCH_2CH_3 \ + \ HCl$$

acetyl chloride   ethanol   ethyl acetate

Some syntheses, however, require many steps. The synthesis of adenine from hydrogen cyanide and ammonia is an example of a **multistep synthesis.** The number of steps required to synthesize a particular compound depends on the starting materials that are available.

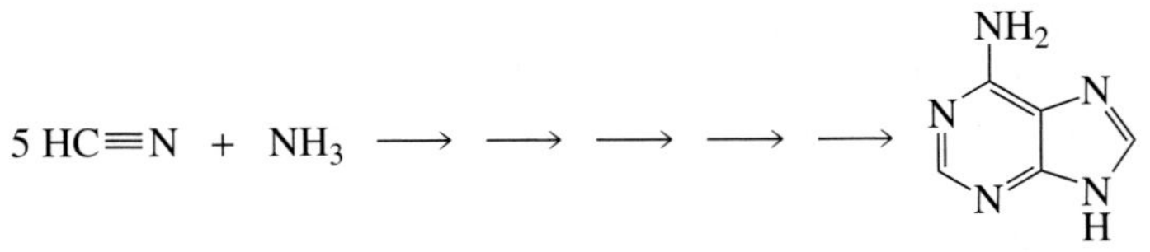

$$5 \text{ HC} \equiv \text{N} + \text{NH}_3 \longrightarrow \longrightarrow \longrightarrow \longrightarrow \longrightarrow$$

adenine

The design of a complicated synthesis requires analysis, insight, and intuition coupled with a sound knowledge of the reactivity of organic compounds. The particular synthetic scheme that results depends on the background and training of the individual designing the synthesis because generally several different approaches can be used. There are often many "correct" answers to a synthesis problem. The only "incorrect" answer is one that fails to produce a pure sample of the desired compound. One method may be superior to others, however, because it leads to a higher yield, requires fewer steps, uses less complicated reactions, requires less expensive starting materials, or avoids toxic materials and flammable solvents (Section 5.11).

Frequently, a chemist plans a synthetic route by first constructing a **synthetic tree.** A synthetic tree is an outline of the available routes that lead to the desired product—the **target molecule (TM)**—from available starting materials. Available starting materials are those compounds that can be obtained from chemical suppliers. For example, the following synthetic tree indicates that four different compounds (A, E, L, and O) can be converted into the target molecule, and that there is more than one way to produce each of the four from available starting materials.

a synthetic tree

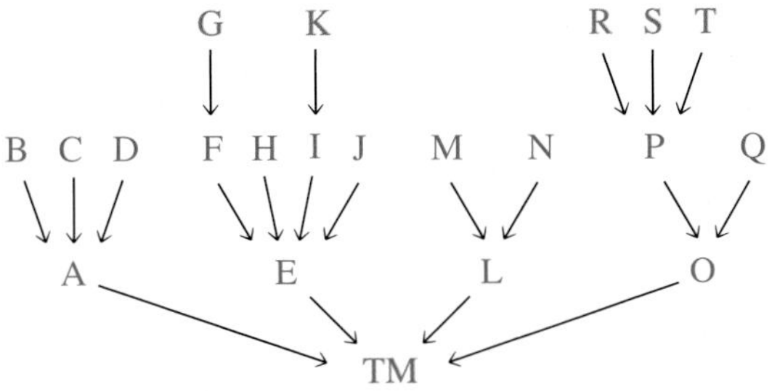

A **linear synthesis** builds a molecule step by step from the starting materials. In a **convergent synthesis,** pieces of the molecule are prepared and then assembled late in the synthesis. A completely convergent synthesis would involve the combination of all the subunits of a molecule in a single step. Because the overall yield of the target molecule decreases as the number of sequential steps that must be performed increases, convergent syntheses are preferred over linear syntheses.

a linear synthesis

$$A \longrightarrow B \longrightarrow C \longrightarrow D \longrightarrow E \longrightarrow F \longrightarrow G$$

a convergent synthesis

$$A \longrightarrow B \longrightarrow C + F \longrightarrow G$$
$$\uparrow$$
$$E$$
$$\uparrow$$
$$D$$

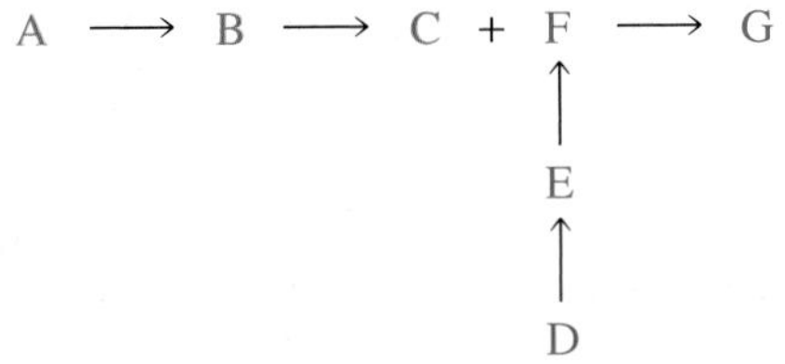

Before starting a synthesis, a chemist carefully examines each of the possible routes to the desired product, aiming for a scheme with the fewest steps, the greatest convergence, the lowest overall cost, the highest yield, and the easiest purification. There are now computer programs that can assist in the design of complicated syntheses. Computer programmers have designed these programs to use "thought processes" similar to those presented in this chapter.

PROBLEM 1 ◆

Starting with A, compare the overall yield of G using a six-step linear synthesis and then using the five-step convergent synthesis just outlined, if each of the reactions employed gives an 80% yield (a relatively high laboratory yield).

Combinatorial organic synthesis is a technique that uses the concept of mass production to prepare a large number of compounds with closely related structures. This technique looks at an organic compound as a group of building blocks connected together. By systematically changing one building block at a time, a **combinatorial library** (a group of structurally related compounds) is obtained. An example of the use of combinatorial organic synthesis by the pharmaceutical industry in an attempt to find new tranquilizers is described in Section 30.11.

SEMISYNTHETIC DRUGS

Chemists have learned to synthesize compounds jointly with nature. For example, in synthesizing the anticancer drug Taxol, the yew tree is responsible for the first part of the synthesis. Chemists extract the drug precursor from the needles of the tree, and the precursor is converted to Taxol in the laboratory. Thus, the precursor is isolated from a renewable resource, while the drug itself could be obtained only by killing the tree.

Sometimes the chemist does the first part of the synthesis, and the synthetic compound is fed to a biological organism that completes the synthesis. Penicillin V, for example, is prepared by feeding 2-phenoxyethanol to a culture of the mold *Penicillium*. The mold oxidizes the alcohol to the carboxylic acid it needs for the side chain of penicillin V.

2-phenoxyethanol → (*Penicillium mold*) → **penicillin V**

29.1 FUNCTIONAL GROUP INTRODUCTION, REMOVAL, AND INTERCONVERSION

The more methods that are available to form a specific functional group, the easier it is to design a synthesis. The different reactions a particular functional group undergoes tend to be related to one another. The different ways a functional group can be synthesized, however, tend to be very different. Therefore, it is a good idea to have a classification scheme that allows easy access to the many available reactions. For this purpose, we have assembled lists of reactions that produce particular functional groups. These lists are found in Appendix IV. The reactions listed there will be familiar to you from earlier chapters.

In designing a synthesis, we often find that the desired product contains more carbons than the available starting material contains. In other words, the carbon skeleton in the reactant is smaller than the carbon skeleton in the product. Such syntheses require reactions that form new carbon–carbon bonds. Appendix V contains a compilation of reactions that form carbon–carbon bonds.

We have seen that a functional group can be *introduced* into an alkane by first halogenating the alkane and then replacing the halogen with a nucleophile (Sections 8.4 and 9.3).

$$CH_3CH_2CH_2CH_3 \xrightarrow[h\nu]{Br_2} CH_3\underset{\underset{Br}{|}}{C}HCH_2CH_3 \xrightarrow{^-Nu} CH_3\underset{\underset{Nu}{|}}{C}HCH_2CH_3 + Br^-$$

We have also seen how a functional group can be *removed* from a molecule.

$$CH_3CH_2CH=CH_2 \xrightarrow{\underset{Pt}{H_2}} CH_3CH_2CH_2CH_3$$

$$CH_3CH_2\overset{\overset{O}{\|}}{C}CH_3 \xrightarrow[HO^-,\,\Delta]{H_2NNH_2} CH_3CH_2CH_2CH_3$$

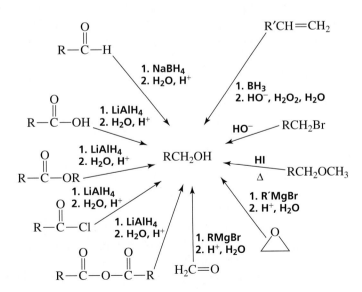

We have explored many ways in which one functional group can be converted into another functional group—reactions known as a **functional group interconversions.** For example, a primary alcohol can be formed from the wide variety of other functional groups shown in Figure 29.1. The method that would actually be used depends on the primary alcohol being synthesized and the available starting materials.

◀ **Figure 29.1**
Functional group interconversions that lead to the synthesis of a primary alcohol.

PROBLEM 2 ◆

Which of the functional group interconversions in Figure 29.1 cannot be used for the synthesis of isobutyl alcohol?

PROBLEM 3 ◆

Which of the functional group interconversions in Figure 29.1 change the carbon skeleton of the starting material?

PROBLEM 4

Without looking at Appendix IV, list as many ways to synthesize butanoic acid in one step from another organic compound as you can recall.

PROBLEM 5

Design a synthesis of 1-butanamine using 1-butene as the starting material.

29.2 MORE ABOUT RETROSYNTHETIC ANALYSIS. DISCONNECTIONS

The synthesis of a complicated molecule from simple starting materials is not always obvious. We have seen that it is often easier to work backward from the desired product to the available starting materials (Section 5.11). This process is called **retrosynthesis** or **retrosynthetic analysis.** In a retrosynthetic analysis, the chemist dissects a molecule into smaller and smaller pieces until readily available starting materials are reached.

retrosynthetic analysis

$$\text{target molecule} \implies Y \implies X \implies W \implies \text{starting materials}$$

There are two kinds of steps in a retrosynthetic analysis: disconnections and functional group interconversions. A **disconnection** involves breaking a bond to carbon to produce two ionic fragments. One fragment is positively charged and one is negatively charged. The fragments of a disconnection are called **synthons.** Synthons are generally not real compounds—they are imaginary constructs. For example, if we consider the retrosynthetic analysis of cyclohexanol, a disconnection gives an α-hydroxycarbocation and a hydride ion. The α-hydroxycarbocation and the hydride ion are synthons.

retrosynthetic analysis

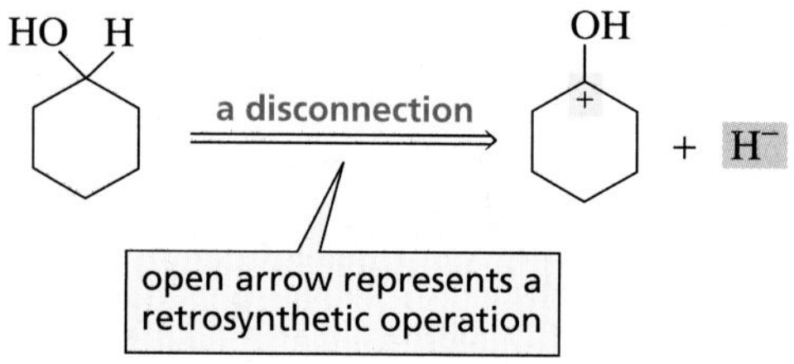

a disconnection

open arrow represents a retrosynthetic operation

A synthetic equivalent is the reagent used as a source of the imaginary synthon.

A **synthetic equivalent** is the reagent actually used as the source of the synthon. Cyclohexanone is the synthetic equivalent for the α-hydroxycarbocation, while sodium borohydride is the synthetic equivalent for hydride ion. So cyclohexanol, the target molecule, can be prepared by treating cyclohexanone with sodium borohydride.

synthesis

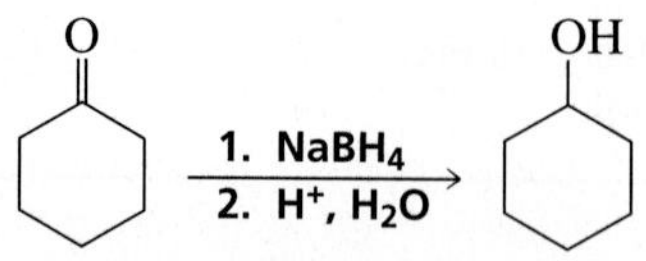

1. NaBH₄
2. H⁺, H₂O

When carrying out a disconnection, you must decide, after breaking the bond, which fragment gets the positive charge and which gets the negative charge. In the retrosynthetic analysis of cyclohexanol, we could have given the positive charge to the hydrogen, and many acids (HCl, HBr, etc.) could have been used as synthetic equivalents for H^+. However, we would have been at a loss to find a synthetic equivalent for an α-hydroxycarbanion. Therefore, when we carry out the disconnection, we assign the positive charge to the carbon and the negative charge to the hydrogen.

Cyclohexanol can also be disconnected by breaking the carbon–oxygen bond instead of the carbon–hydrogen bond, forming a carbocation and hydroxide ion.

retrosynthetic analysis

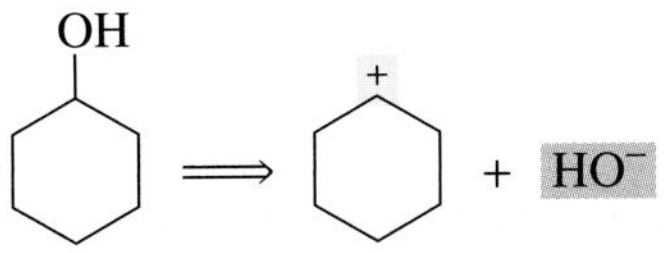

The problem then becomes choosing a synthetic equivalent for the carbocation. A synthetic equivalent for a positively charged synthon needs an electron-withdrawing group at just the right place. Cyclohexyl bromide, with an electron-withdrawing bromine, is a synthetic equivalent for the cyclohexyl carbocation. Cyclohexanol, therefore, can be prepared by treating cyclohexyl bromide with hydroxide ion. This method, however, is not as good as reduction of cyclohexanone because some of the alkyl halide is converted into an alkene, so the overall yield of the target molecule is lower.

synthesis

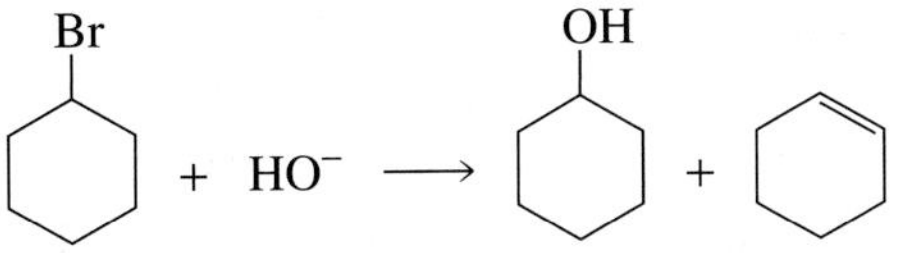

Retrosynthetic analysis shows that 1-methylcyclohexanol can be formed from the reaction of cyclohexanone, the synthetic equivalent for the α-hydroxycarbocation, and methylmagnesium bromide, the synthetic equivalent for the methyl anion (Section 17.4).

retrosynthetic analysis

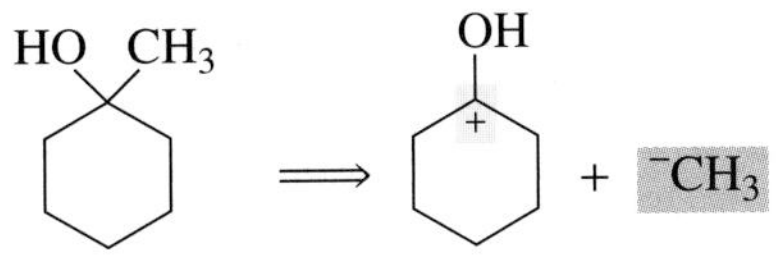

synthesis

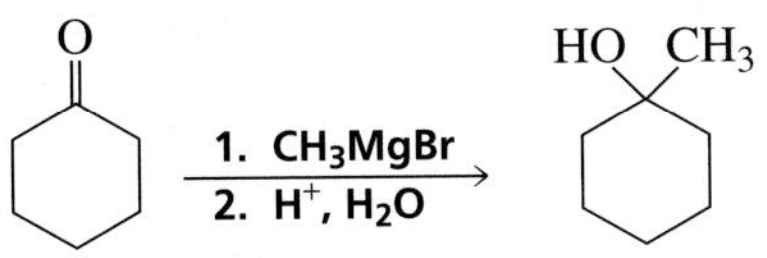

Other disconnections of 1-methylcyclohexanol are possible. For example, one of the ring carbon–carbon bonds could be broken. However, these are not useful disconnections because the synthetic equivalents of these synthons are not easily prepared. A retrosynthetic step must lead to readily obtainable starting materials. If not, how will the starting materials be obtained to make the disconnected compound?

retrosynthetic analysis

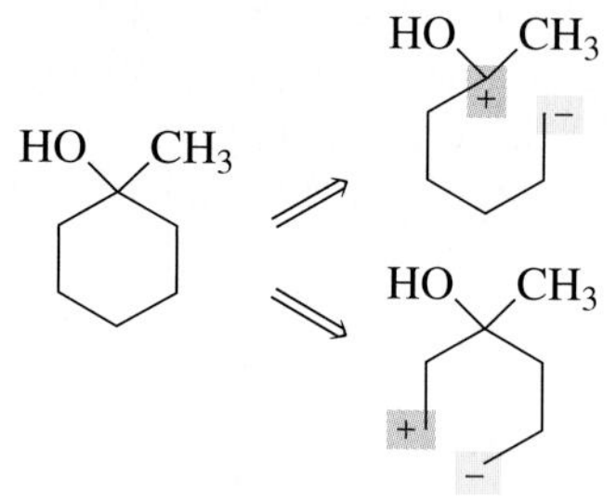

2-Methylcyclohexanone can be disconnected into an α-carbanion and a methyl cation.

retrosynthetic analysis

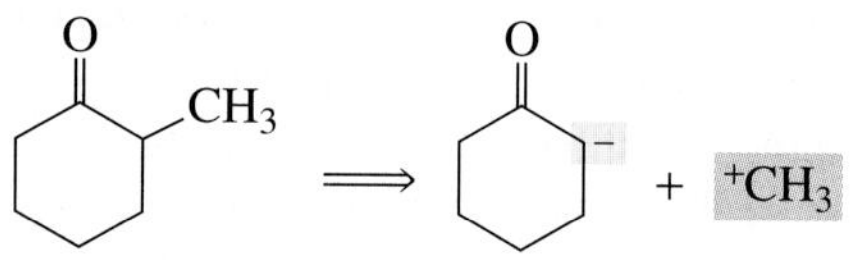

The α-carbanion can be prepared by treating cyclohexanone with LDA (Section 19.7). Methyl iodide is a synthetic equivalent for the methyl cation.

synthesis

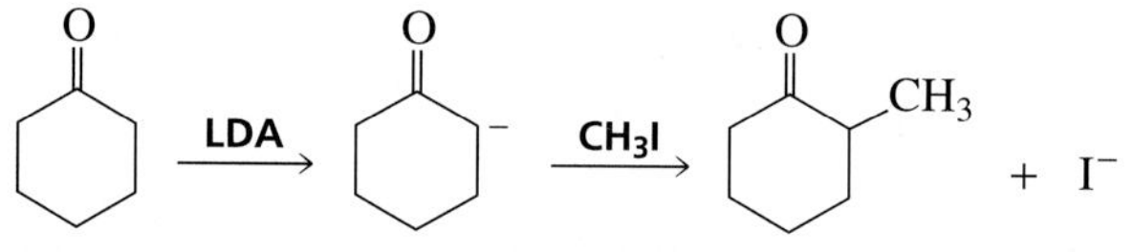

2-Aminocyclohexanone can be disconnected into an electrophilic α-carbocation and a nucleophilic amide anion.

retrosynthetic analysis

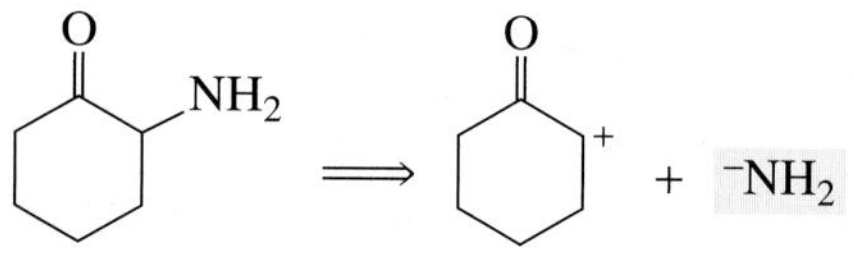

The α-carbon of a ketone is normally nucleophilic because an α-hydrogen can be removed by a strong base to form an α-carbanion, as was done in the synthesis of 2-methylcyclohexanone. To carry out this synthesis, the α-carbon must be electrophilic. In other words, we must reverse its normal polarity. Reversing the normal polarity is called ***umpolung*** (German for "polarity reversal"). The α-carbon can be made electrophilic by substituting an electron-withdrawing group for an α-hydrogen. The α-bromo-substituted ketone is easily produced by treating the ketone with bromine under acidic conditions (Section 19.4).

synthesis

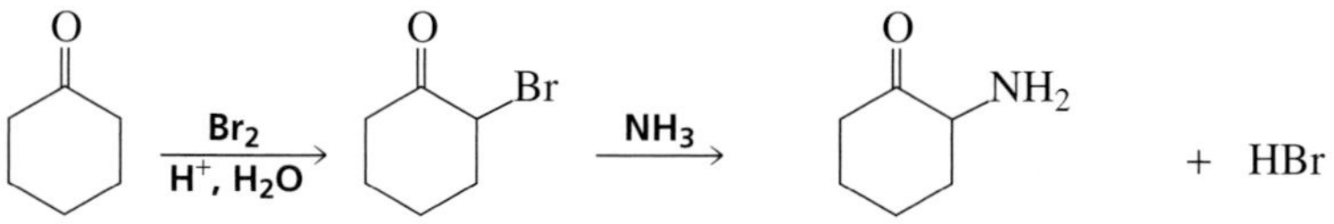

2-aminocyclohexanone

PROBLEM 6

How could you apply the principle of *umpolung* to bromocyclohexane? In other words, how could you convert an ordinarily electrophilic bromocyclohexane into a nucleophile?

PROBLEM 7 ◆

Label each of the following as a normal synthon or an *umpolung* synthon:

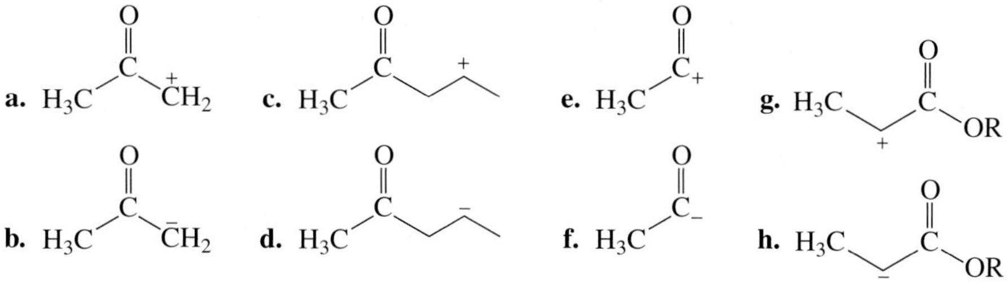

Butanal can be disconnected into either a *formyl cation* and a *propyl anion* or a *propyl cation* and a *formyl anion.* The formyl cation is the normal synthon; the formyl anion is the *umpolung* synthon.

retrosynthetic analysis

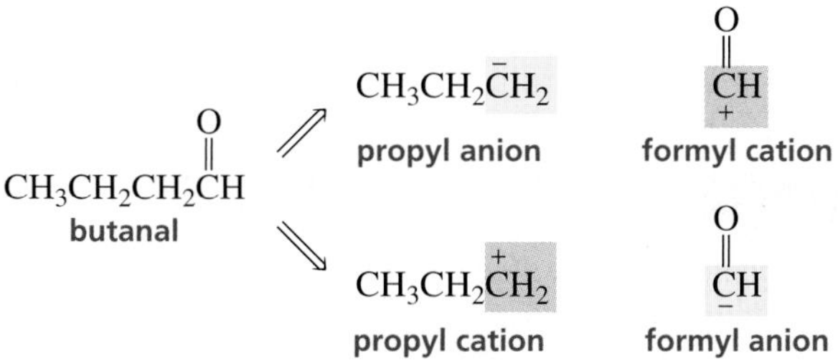

Ethyl formate is the synthetic equivalent for the formyl cation and propylmagnesium bromide is the synthetic equivalent for the propyl anion. Using these reagents for the synthesis of butanal presents a problem because once butanal is formed, it will react with the Grignard reagent, forming a secondary alcohol (Section 17.4). Therefore, we need to look at the other disconnection. It may provide a more promising synthesis because it does not require the use of a Grignard reagent.

Propyl bromide is a synthetic equivalent for the propyl cation, and dithianyllithium is the synthetic equivalent for the formyl anion. Dithianyllithium is easily prepared by treating 1,3-dithiane with a strong base such as butyllithium because the C-2 hydrogens of 1,3-dithiane are relatively acidic. Reaction of dithianyllithium with an alkyl halide forms a thioacetal. The carbon–sulfur bonds are cleaved by mercuric ion in an aqueous acetonitrile solution, forming a hydrate that loses water to form the target molecule.

synthesis

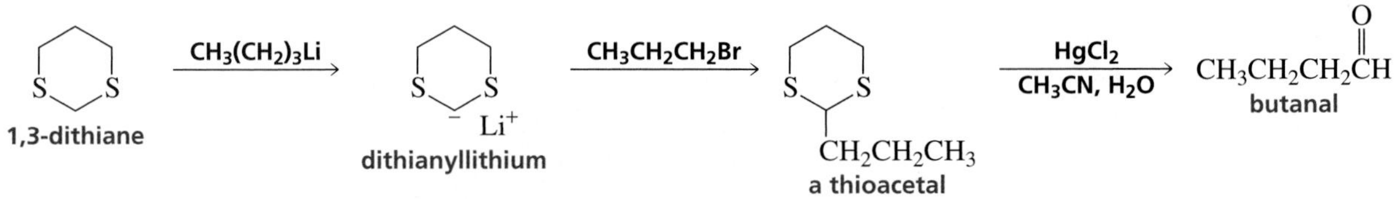

1,3-dithiane → (CH₃(CH₂)₃Li) → dithianyllithium → (CH₃CH₂CH₂Br) → a thioacetal → (HgCl₂, CH₃CN, H₂O) → CH₃CH₂CH₂CH (butanal)

PROBLEM 8 ◆

What change would you make in the preceding synthesis if the desired product were

a. heptanal?   **b.** 2-pentanone?

Because dithianyllithium is a formyl anion synthetic equivalent, it can be used to convert aldehydes and ketones into 1,2-dioxygenated compounds.

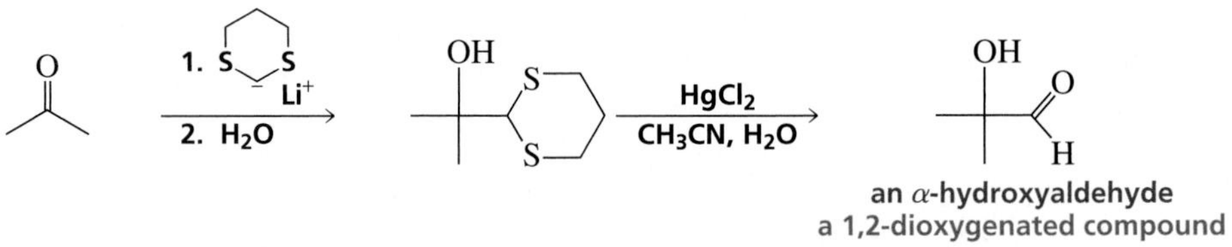

an α-hydroxyaldehyde
a 1,2-dioxygenated compound

1,2-Dioxygenated compounds can also be prepared using cyanide ion rather than dithianyllithium as the nucleophile. When dithianyllithium is the nucleophile, the product is an α-hydroxyaldehyde; when cyanide is the nucleophile, the product is an α-hydroxynitrile that can be converted to an α-hydroxycarboxylic acid. Notice that in both cases the number of carbon atoms in the starting material is increased by one.

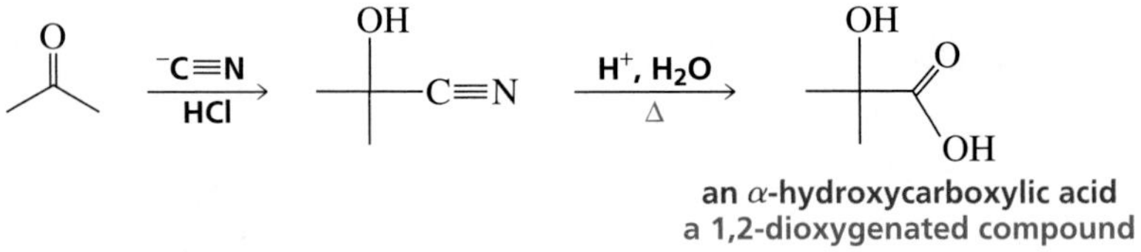

an α-hydroxycarboxylic acid
a 1,2-dioxygenated compound

PROBLEM 9

Using bromocyclohexane as a starting material, how could you synthesize the following compounds?

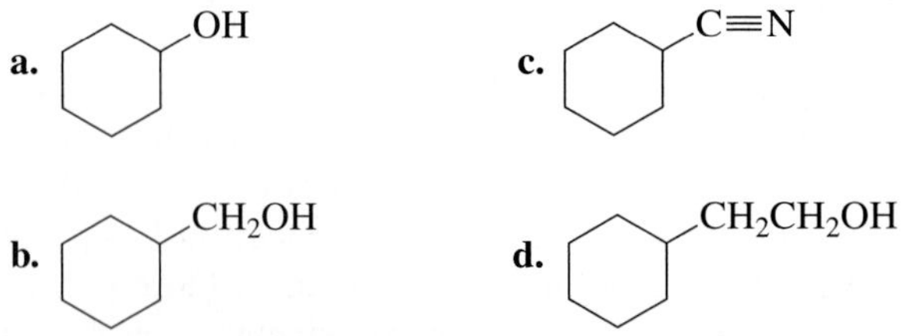

a. (OH)

c. (C≡N)

b. (CH₂OH)

d. (CH₂CH₂OH)

PROBLEM 10

Give a synthetic equivalent for each of the following:

a. $\overset{+}{C}H_2CH$ (with =O)   **b.** $\overset{-}{C}H_2CH$ (with =O)

c. $\overset{O}{\overset{\|}{\underset{+}{C}}}CH_3$

e. $\overset{O}{\overset{\|}{C}}H$

g. $\overset{+}{C}H_2CH_2CH_3$

i. $\overset{+}{C}H_2CH_2OH$

d. $\overset{O}{\overset{\|}{\underset{-}{C}}}CH_3$

f. $\overset{O}{\overset{\|}{\underset{-}{C}}}CH_2CH_3$

h. $\overset{-}{C}H_2CH_2CH_3$

j. $\overset{+}{C}H_2CH_2CH_2OH$

PROBLEM 11 / SOLVED

Describe how the following compounds could be prepared using the given starting materials. Perform retrosynthetic analyses to help you arrive at your answers:

a. (structure) $\Longrightarrow$ (structures) $H_2C=O$

b. (structure) $\Longrightarrow$ (structure)

SOLUTION TO 11a

(reaction scheme with oxidation, Mg, PBr₃, HBr/peroxide, H₂C=O steps)

As we have seen in planning the syntheses of 1,2-dioxygenated compounds, the relative locations of the two functional groups in the target molecule provide valuable hints for designing a synthesis. A 1,3-dioxygenated compound such as 3-hydroxybutanal can be disconnected to give an α-hydroxycarbocation and an α-carbanion. These are both synthons for the same aldehyde. Therefore, a 1,3-dioxygenated compound can be synthesized by means of an aldol addition. Acetaldehyde is used as the starting material for the synthesis of 3-hydroxybutanal. Dehydration of the β-hydroxycarbonyl compound forms 2-butenal, an α,β-unsaturated carbonyl compound.

29.3 RETROSYNTHETIC ANALYSIS OF DIOXYGENATED COMPOUNDS

retrosynthetic analysis

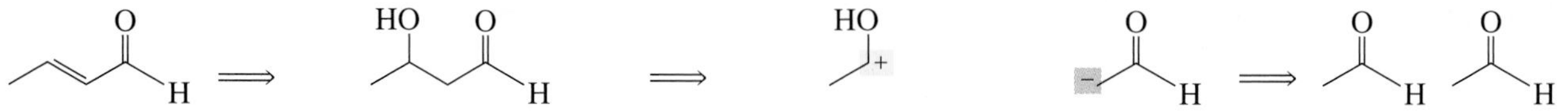

2-butenal ⟹ 3-hydroxybutanal a 1,3-dioxygenated compound ⟹ an α-hydroxycarbocation an α-carbanion ⟹ acetaldehyde

Retrosynthetic analysis shows that 2,5-hexanedione can be used as the starting material for the synthesis of 3-methyl-2-cyclopentenone, an α,β-unsaturated ketone.

retrosynthetic analysis

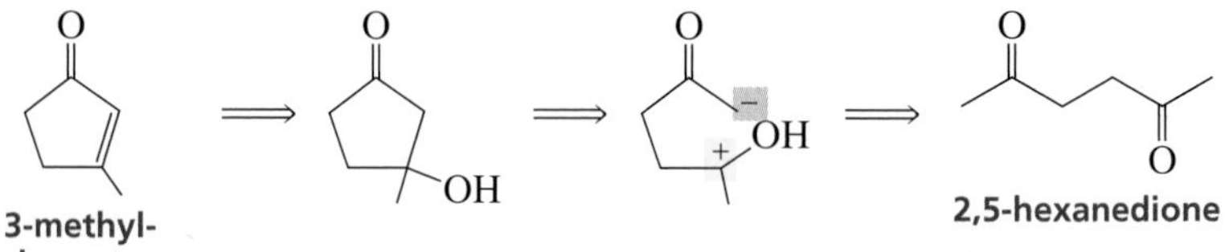

3-methyl-
2-cyclopentenone ⟹ ⟹ ⟹ 2,5-hexanedione

PROBLEM 12

Using retrosynthetic analysis, plan a synthesis of the following compound from the given starting materials:

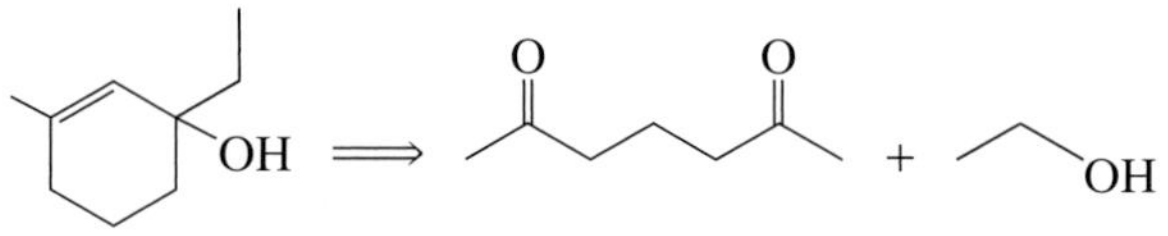

Disconnection of the following 1,4-dioxygenated compound can be done in two different ways. We have seen that the synthetic equivalent for an α-carbocation is an α-brominated derivative (Section 29.2).

retrosynthetic analysis

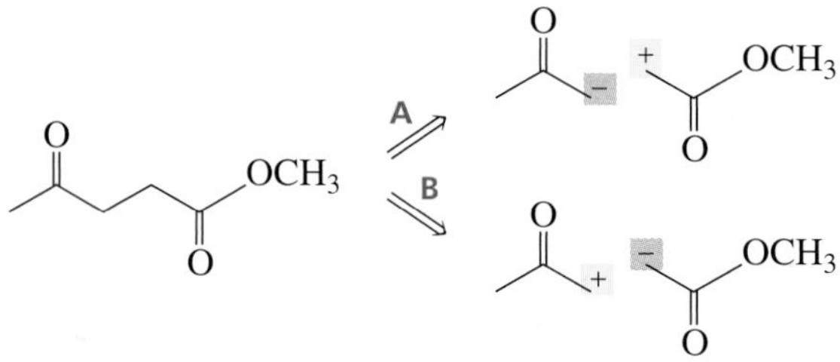

Now we need to determine how to generate the α-carbanion. Normally α-carbanions are generated by using a base to remove an α-hydrogen. However, bromination increases the acidity of the protons bonded to the brominated carbon. Therefore, if a base is used to create the α-carbanion, the base may remove a proton from the α-carbon of the brominated carbonyl compound rather than from the α-carbon of the nonbrominated carbonyl compound.

This problem can be solved if an enamine is used as the synthetic equivalent of the α-carbanion (Section 19.9). Because esters cannot form enamines, disconnection A leads to the best starting materials. The enamine can be prepared by treating the ketone with a secondary amine. After the enamine has reacted with the α-brominated ester, the carbonyl group can be regenerated by hydrolysis under conditions that won't also hydrolyze the ester (Section 17.7).

synthesis

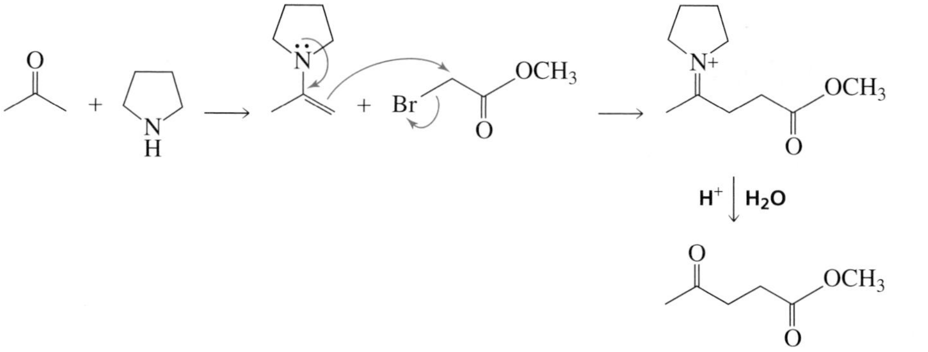

a 1,4-dioxygenated compound

PROBLEM 13

Using a base to generate an α-carbanion can result in the following condensation, known as the **Darzen's condensation.** Propose a mechanism for this reaction.

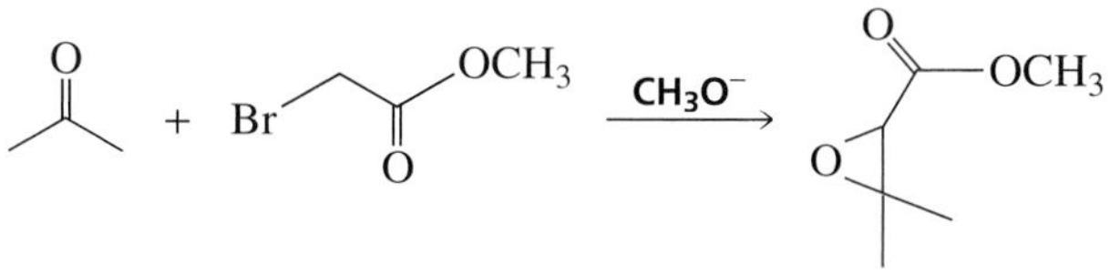

PROBLEM 14

Explain why an α-monobromocarbonyl compound cannot be prepared by treating a carbonyl compound with bromine under basic conditions.

Retrosynthetic analysis indicates that 1,5-dioxygenated compounds can be synthesized by the reaction of an α-carbanion with a carbonyl compound that has a positively charged β-carbon.

retrosynthetic analysis

a 1,5-dioxygenated compound

An α,β-unsaturated carbonyl compound is a synthetic equivalent for the β-carbocation. This suggests that a Michael reaction will form the desired 1,5-dioxygenated product (Section 19.10). The Michael reaction shows how the polar reactivity of a functional group, such as a carbonyl group, can be transmitted through a carbon–carbon double bond. Transmission of reactivity through double bonds is called **vinylogy.**

synthesis

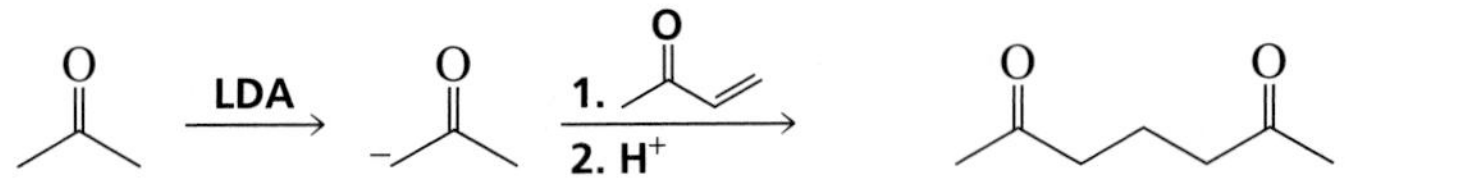

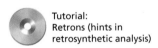

2,6-heptanedione
a 1,5-dioxygenated compound

PROBLEM 15

Describe how the following compound could be synthesized from compounds containing no more than six carbon atoms:

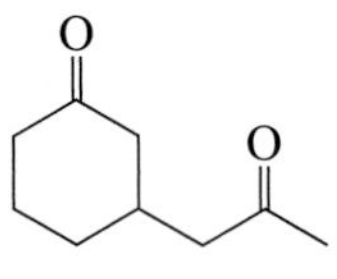

Retrosynthetic analysis shows that a 1,6-dioxygenated compound can be prepared by nucleophilic attack of an α-carbanion on a carbonyl compound with a positively charged γ-carbon.

retrosynthetic analysis

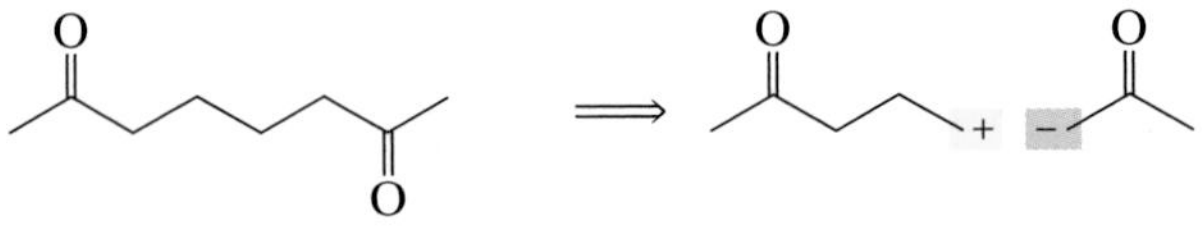

a 1,6-dioxygenated compound

Because there are no good general methods for synthesizing γ-halosubstituted carbonyl compounds, we must consider another route. Recognizing that a 1,6-dioxygenated compound can be prepared by cleaving a cyclohexene provides an easy route to the target compound because a compound with a six-membered ring can be readily prepared by means of a Diels–Alder reaction (Section 7.9). The convergence (assembly of pieces) of the Diels–Alder reaction is one of its most useful features.

Tutorial:
Retrons (hints in
retrosynthetic analysis)

retrosynthetic analysis

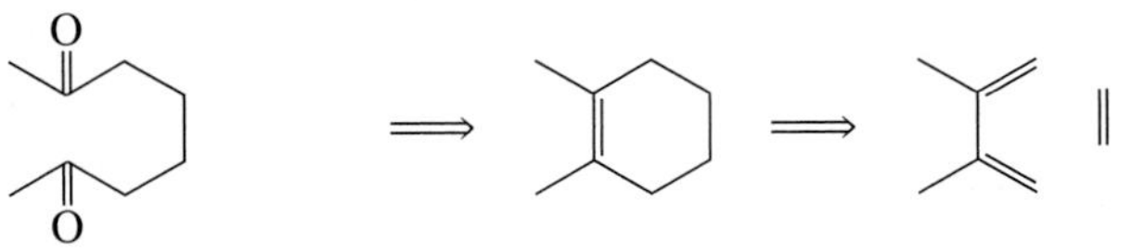

a 1,6-dioxygenated compound

When the target molecule has a six-membered ring, it is quite easy to determine that a Diels–Alder reaction should be employed in the synthetic scheme. However, when the product does not have a six-membered ring, as in the case of the previous synthesis, it is not immediately obvious that a Diels–Alder reaction should be used.

Another synthesis in which the use of a Diels–Alder reaction is not immediately obvious is the following extrusion reaction. In an **extrusion reaction,** a neutral molecule (for example, carbon dioxide, carbon monoxide, nitrogen, or sulfur dioxide) is eliminated. A Diels–Alder reaction leads to a compound that can readily eliminate carbon monoxide in order to form the hexasubstituted benzene ring.

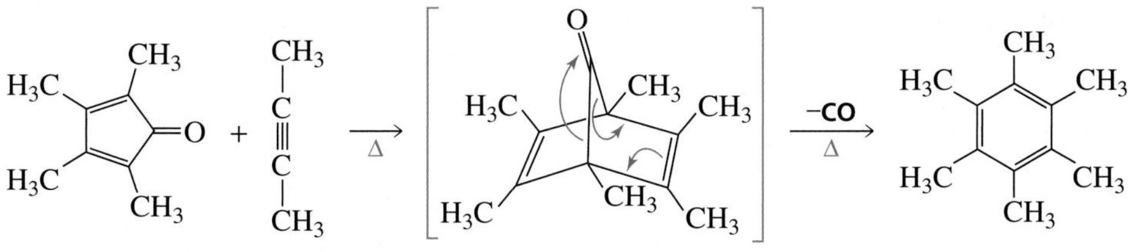

PROBLEM 16

Do a retrosynthetic analysis on each of the following compounds, ending with available starting materials:

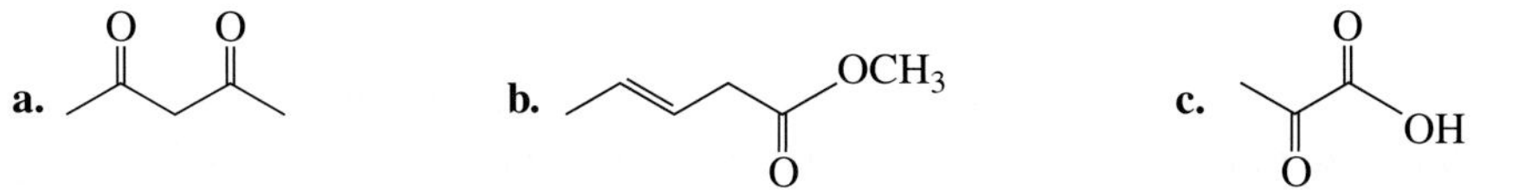

PROBLEM 17 / SOLVED

How could you synthesize the following compounds from starting materials containing no more than four carbons?

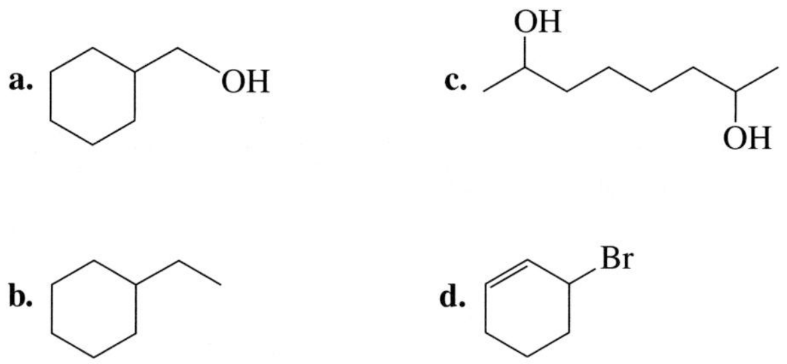

SOLUTION TO 17a   The six-membered ring indicates that the compound can be synthesized by means of a Diels–Alder reaction.

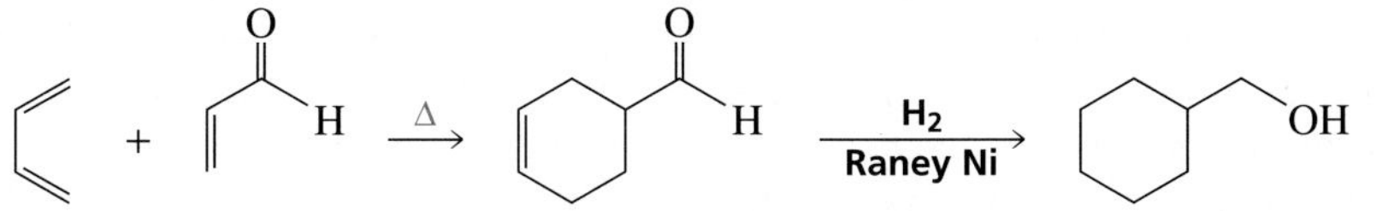

We have seen that the Wittig reaction is used to form carbon–carbon double bonds (Section 17.11). This is the most useful procedure for forming compounds with carbon–carbon double bonds because the placement of the double bond is unambiguous. Retrosynthetically, the reaction is represented as follows:

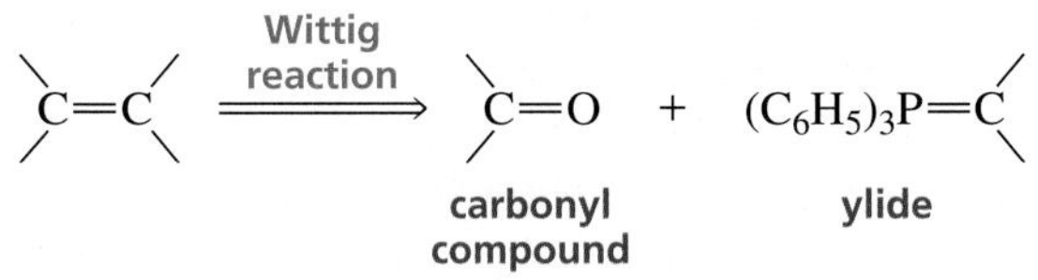

When the target alkene has different substituents bonded to the sp^2 carbons, you must carefully consider which part of the product should be derived from the carbonyl compound and which part from the ylide. For example, the following enone has two disconnections. Disconnection A is preferred because the ylide is stabilized by resonance and the ketone is readily available.

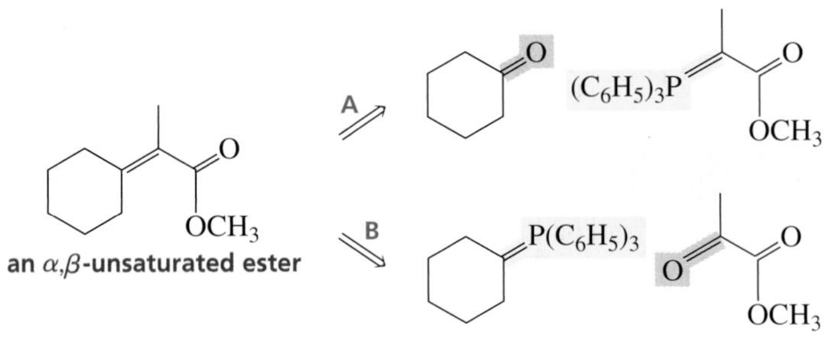

an α,β-unsaturated ester

Catalytic hydrogenation of Wittig products is a useful method for synthesizing complicated alkanes with precisely controlled structures.

PROBLEM 18 ◆

a. What reagents would you use for the synthesis of 3-ethyl-1-pentene using a Wittig reaction?

b. What advantage does this method of synthesis have over a synthesis using an alkyl halide and acetylide ion?

PROBLEM 19

Propose a method for synthesizing the following compound using starting materials containing no more than six carbons and triphenylphosphine:

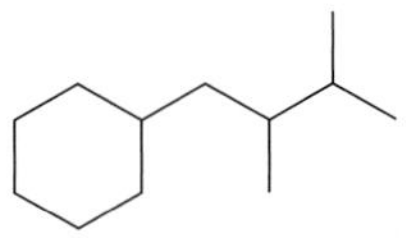

Sometimes when a complicated molecule is broken down into its component parts, the route to its synthesis becomes readily apparent. At first glance, buspirone appears to be difficult to synthesize. However, it can be broken down into readily available starting materials. Recognizing the smaller pieces simplifies the retrosynthetic analysis. With very complicated molecules, this is often the best approach. You can think of this approach as the "divide and conquer" method of analysis. This method leads to convergence, which in turn leads to higher yields.

retrosynthetic analysis

buspirone

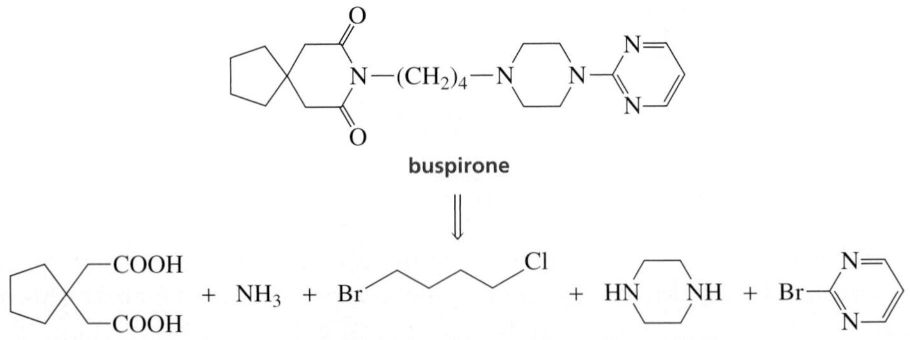

A **protecting group** protects a functional group from a synthetic operation that it would not otherwise survive. We have seen that 1,2- and 1,3-diols can be used to protect the carbonyl groups of aldehydes and ketones from reaction with nucleophiles (Section 17.9). For example, LiAlH₄ can reduce both functional groups of a keto ester. The ketone group can be protected from reduction, however, by being converted into a ketal. After the ester has been reduced, the protecting group can easily be removed by acid-catalyzed hydrolysis. It is critical that protecting groups can be removed under conditions that will not affect other groups in the molecule.

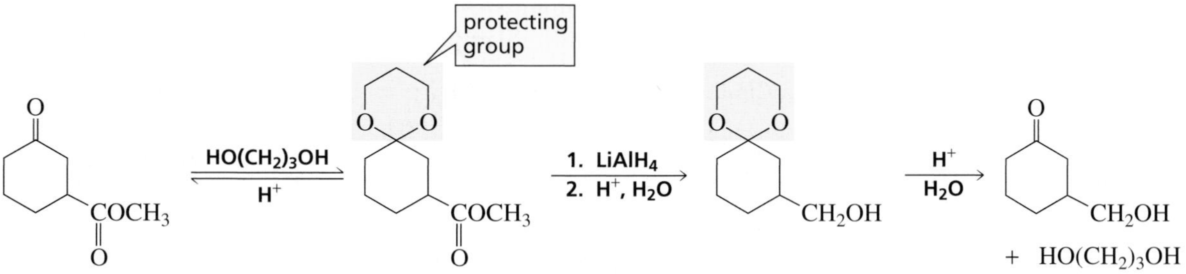

29.4 MORE ABOUT PROTECTING GROUPS

Protecting groups should be used only when absolutely necessary because each time a protecting group is used, it must be attached and then taken off. This adds two linear steps to the synthesis, which decreases the overall yield of the target compound.

The following disconnection indicates that the desired transformation can be carried out by treating a ketone with a Grignard reagent.

retrosynthetic analysis

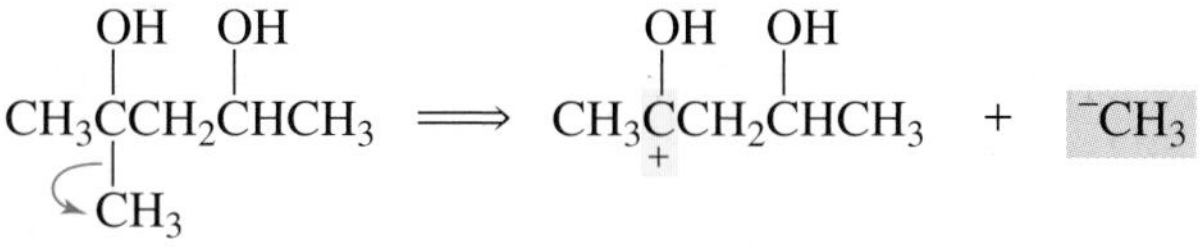

$$\underset{\underset{CH_3}{|}}{CH_3CCH_2CHCH_3} \overset{OH\ \ OH}{\Longrightarrow} \underset{+}{CH_3CCH_2CHCH_3} + \ ^-CH_3$$

However, a problem arises because the labile proton of the OH group in the hydroxy-ketone will react with the Grignard reagent (Section 11.8). Therefore, the OH group must be protected in order to make it immune to the Grignard reagent. There are many protecting groups available for alcohols, but the *tert*-butyldimethylsilyl (TBDMS) group has several advantages. It can be put on an alcohol under mild conditions, it is unaffected by acid and Grignard reagents, and it is easily removed by treatment with tetrabutylammonium fluoride. The affinity of fluoride ion for silicon is due to the unusually large energy of the Si—F bond (135 kcal/mol or 565 kJ/mol).

3-D Molecule: TBDMS protected alcohol

synthesis

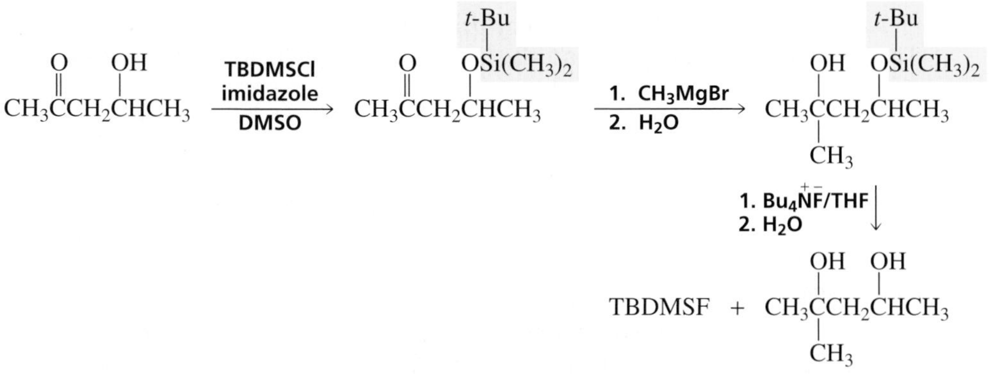

A carboxyl group can be protected by being converted into an oxazoline because oxazolines are not attacked by nucleophiles. The carboxyl group can be regenerated by acid-catalyzed hydrolysis.

$$CH_3CCH_2CH_2COH + HOCH_2CCH_3 \xrightarrow{\Delta} CH_3CCH_2CH_2C \underset{O}{\overset{N}{\diagdown}} \underset{CH_3}{\overset{CH_3}{\diagup}}$$

an oxazoline

$$\downarrow CH_3MgBr$$

$$HOCH_2CCH_3 + CH_3CCH_2CH_2COH \xleftarrow[H_2O]{H^+} CH_3CCH_2CH_2C \underset{O}{\overset{N}{\diagdown}} \underset{CH_3}{\overset{CH_3}{\diagup}}$$

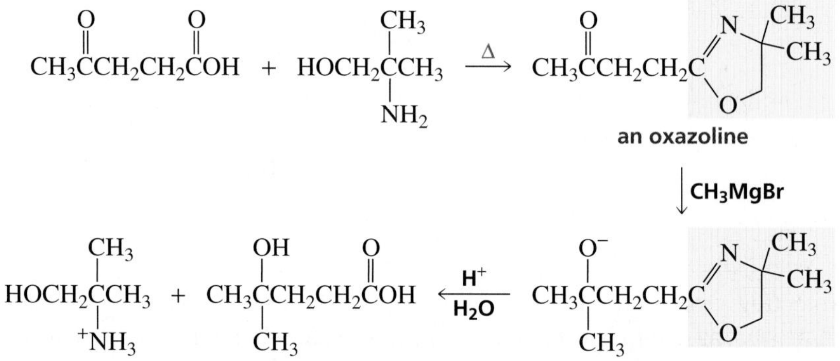

3-D Molecule:
An oxazoline

If only the OH group of the carboxyl group needs to be protected, the carboxylic acid can be converted into an ester. When the protection is no longer needed, hydrolysis will regenerate the carboxylic acid.

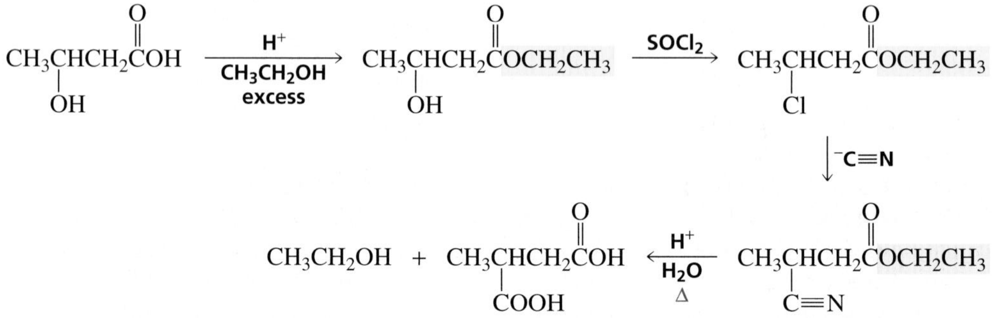

$$CH_3CHCH_2COH \xrightarrow[\substack{CH_3CH_2OH \\ excess}]{H^+} CH_3CHCH_2COCH_2CH_3 \xrightarrow{SOCl_2} CH_3CHCH_2COCH_2CH_3$$

$$\downarrow {}^-C{\equiv}N$$

$$CH_3CH_2OH + CH_3CHCH_2COH \xleftarrow[\substack{H_2O \\ \Delta}]{H^+} CH_3CHCH_2COCH_2CH_3$$

An amino group can be protected by being converted into an amide (Section 15.7). The acetyl group can subsequently be removed by acid-catalyzed hydrolysis.

aniline $\xrightarrow{CH_3CCl}$ acetanilide $\xrightarrow[\substack{H_2SO_4}]{HNO_3}$ p-nitroacetanilide $\xrightarrow[\substack{2.\ HO^-}]{1.\ HCl,\ H_2O,\ \Delta}$ p-nitroaniline $+ CH_3CO^-$

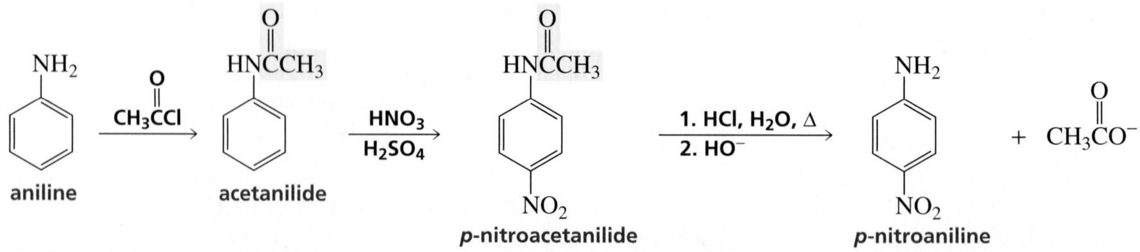

An amino group can also be protected with di-*tert*-butyl dicarbonate. Because this group can be removed under mildly acidic conditions, it is used as a protecting group when the compound also contains a group that would be adversely affected by acid-catalyzed hydrolysis.

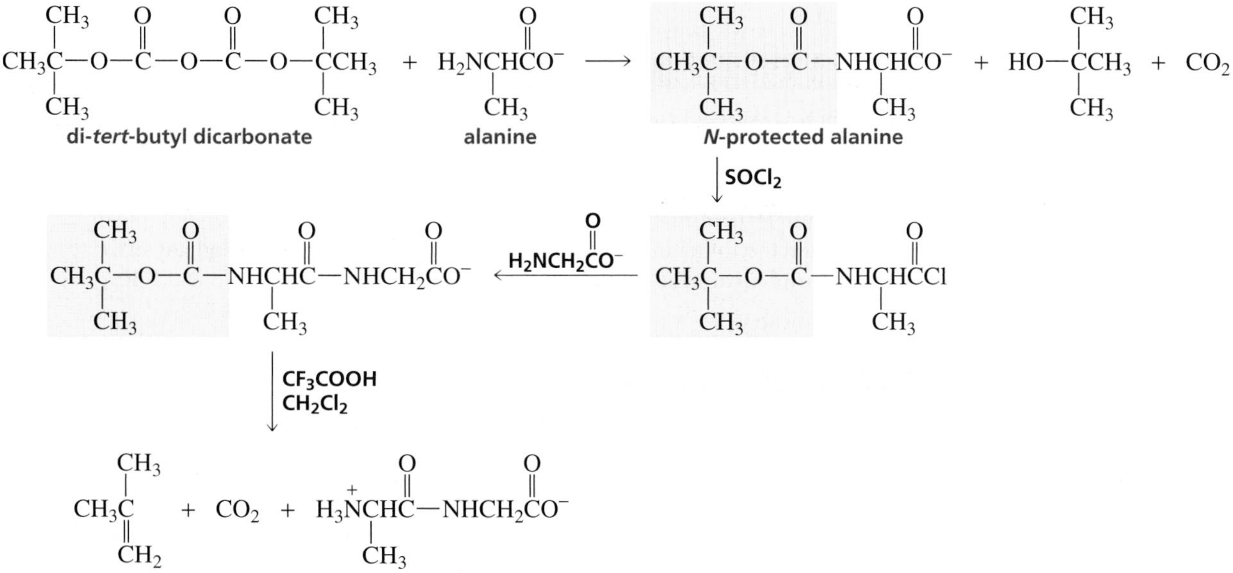

An amino group can also be protected with di-*tert*-butyl dicarbonate reaction scheme showing di-*tert*-butyl dicarbonate + alanine → *N*-protected alanine, followed by SOCl₂, H₂NCH₂CO⁻, and CF₃COOH/CH₂Cl₂ steps.

Another advantage the *tert*-butoxycarbonyl group has as a protecting group is that the products formed when it is removed are gases (isobutylene and carbon dioxide), so they force the reaction to completion as they escape (Section 21.9).

PROBLEM 22

Why is di-*tert*-butyl dicarbonate rather than acetyl chloride used as a protecting group in the preceding reaction?

PROBLEM 23 ◆

What products would be formed from the preceding reaction if alanine's amino group were not protected?

The target molecule of a synthesis may be one of several stereoisomers. The actual number of stereoisomers depends on the number of double bonds and chirality centers in the molecule because each double bond can exist in the *E* or *Z* configuration (Section 3.5), and each chirality center can have an *R* or *S* configuration (Section 4.5). In addition, if the target molecule has rings with a common bond, the rings can be either trans fused or cis fused (Section 2.15). In designing a synthesis, care must be taken to make sure that each double bond, each chirality center, and each ring fusion in the target molecule has the appropriate configuration. If the stereochemistry of the reactions is not controlled, the resulting mixture of stereoisomers may be difficult or even impossible to separate. Organic chemists must therefore consider the stereochemical outcomes of all reactions in planning a synthesis and use highly stereoselective reactions to achieve the desired configurations.

We have seen examples of stereoselective reactions in previous chapters, with the simplest being the S$_N$2 reaction. In an S$_N$2 reaction, the configuration of the chirality

29.5 CONTROL OF STEREOCHEMISTRY

center is inverted during the course of the reaction, so two successive S_N2 reactions at a chirality center lead to retention of configuration. S_N1 reactions are much less useful to the synthetic chemist because they lead to racemic mixtures. Some other stereoselective reactions with which you are familiar are addition of bromine to an alkene (forms anti addition products) and addition of H_2 to an alkene (forms syn addition products).

PROBLEM 24

Explain how you could use S_N2 reactions to synthesize either enantiomer of 2-bromooctane starting with (*S*)-2-octanol. (*Hint:* See Section 11.3.)

PROBLEM 25

How could the following compounds be prepared from 1-butyne and any other organic reagent with fewer than four carbon atoms? (*Hint:* See Sections 5.8 and 5.10.)

a. *cis*-3-hexene   **b.** *trans*-3-heptene

PROBLEM 26 ◆

How could *trans*-2-butene be converted to *cis*-2-butene? (*Hint:* See Section 10.10.)

Some stereoselective reactions are also enantioselective. An **enantioselective reaction** is one that forms an excess of one enantiomer of a pair. Several methods can be used to synthesize an enantiomerically pure target molecule. We will look at the four most commonly used methods.

One method employs an enzyme to catalyze the synthetic transformation. Because enzymes are chiral, enzyme-catalyzed reactions can result in the exclusive formation of one enantiomer (Section 4.18). For example, ketones are enzymatically reduced to alcohols by enzymes known as alcohol dehydrogenases. Whether the *R* or the *S* enantiomer is formed depends on the particular alcohol dehydrogenase used: Alcohol dehydrogenase from the bacterium *Lactobacillus kefir* forms *R* alcohols, while alcohol dehydrogenases from yeast, horse liver, and the bacterium *Thermoanaerobium brocki* form *S* alcohols. These alcohol dehydrogenases use NADPH as the actual reducing agent (Section 23.2). The use of an enzyme is not a universally useful method for controlling the configuration of a product because enzymes require substrates of very specific size and shape (Section 22.8).

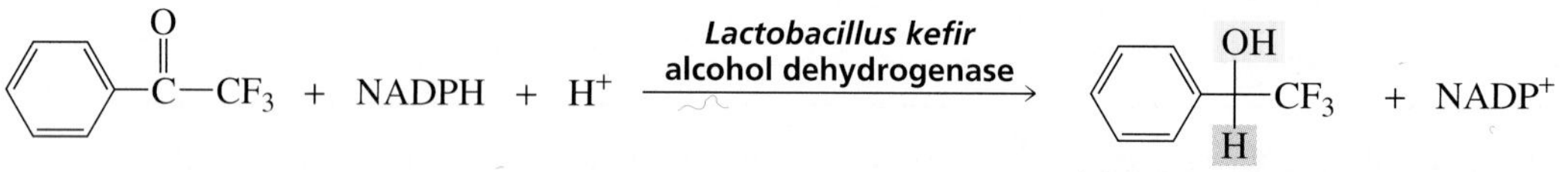

(*R*)-2,2,2-trifluoro-1-phenyl-1-ethanol

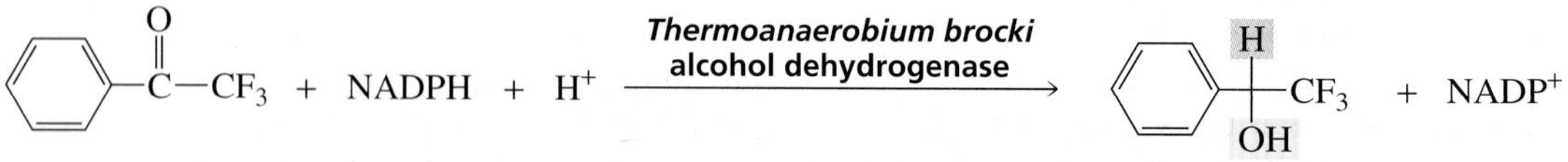

(*S*)-2,2,2-trifluoro-1-phenyl-1-ethanol

A second method used to prepare a single enantiomer employs a chiral auxiliary. A **chiral auxiliary** is an enantiomerically pure compound that, when attached to a reactant, causes the product with the desired configuration to be formed preferentially. When the reaction is over, the chiral auxiliary is removed. An example of a chiral auxiliary is the chiral hydrazine used in the following reaction. The hydrazine reacts with

Tutorial:
A chiral auxiliary

a ketone, forming a chiral hydrazone. The chiral auxiliary blocks one face of the α-carbon from the bulky base (LDA) used to form the α-carbanion (Section 19.7) and also blocks the subsequent approach of the alkyl halide. Consequently, alkylation of the α-carbanion occurs on only one face. When the chiral auxiliary is removed by hydrolysis, only a single enantiomer of the α-alkylated ketone is left behind. In the absence of the auxiliary, the alkylation reaction would form a racemic mixture.

a chiral hydrazine

a chiral hydrazone

1. LDA/THF
2. $CH_3CH_2CH_2I$

new chirality center

A disadvantage of using a chiral auxiliary is that a molar equivalent of the auxiliary must be used. However, it can often be recovered and reused.

A third method used to control the configuration of a target molecule employs an enantiomerically pure starting material. For example, an enantiomerically pure epoxide of an allylic alcohol can be prepared by treating the alcohol with *tert*-butyl hydroperoxide, titanium isopropoxide, and enantiomerically pure diethyl tartrate (DET). The structure of the epoxide depends on the enantiomer of diethyl tartrate used.

t-BuOOH
(isoPrO)₄Ti
(–)DET

t-BuOOH
(isoPrO)₄Ti
(+)DET

an allyl alcohol

This third method was developed by K. Barry Sharpless, who was born in Philadelphia in 1941. He received a Ph.D. in chemistry from Stanford and served as a professor at MIT, Stanford, and again at MIT, where he developed this procedure. He is currently at the Scripps Research Institute in La Jolla, Calif.

An enantiomerically pure epoxide is a useful starting material for a wide variety of optically active compounds because an epoxide can easily be converted into a compound with two adjacent chirality centers. This is because epoxides are very susceptible to attack by nucleophiles. In the following example, an allylic alcohol is converted into an enantiomerically pure epoxide, which is used to form an enantiomerically pure diol.

t-BuOOH
(isoPrO)₄Ti
(–)DET

1. NaH, $C_6H_5CH_2Br$
2. H_2O, H^+

PROBLEM 27

What is the product of the reaction of methylmagnesium bromide with either of the enantiomerically pure epoxides that can be prepared from (E)-3-methyl-2-pentene by the preceding method? Assign R or S configurations to the chirality centers of each product.

PROBLEM 28 ◆

Is the addition of Br_2 to an alkene such as *trans*-2-pentene a stereoselective reaction? Is it a stereospecific reaction? Is it an enantioselective reaction?

A fourth method used to control stereochemistry is to employ a reactant that has a chirality center with the needed configuration. Reactants having chirality centers with the desired configuration are often naturally occurring molecules such as amino acids or carbohydrates. For example, cysteine was used for the synthesis of enantiomerically pure biotin. An advantage to this method is that naturally occurring compounds are often inexpensive, even if they have many chirality centers.

cysteine   several steps →   biotin

PROBLEM 29

Show how the following compound could be synthesized from the given starting materials:

29.6 SELECTED EXAMPLES OF SYNTHESES

Organic chemists prepare thousands of compounds each year. In this section we will look at several interesting syntheses. First we will look at the syntheses of triptycene and cubane, compounds that were synthesized because of their unusual structures; then we will look at the syntheses of lysergic acid and caryophyllene, two natural products. These syntheses illustrate the power of organic synthesis and the ability of chemists to come up with clever solutions to difficult problems. They are presented here to show you examples of brilliantly designed syntheses. It may have taken the scientists who designed them many months (or even years) of false starts and wrong turns before the successful synthesis was achieved, so don't be intimidated by these examples. You certainly would not be expected at this point in your study of organic chemistry to design these kind of syntheses on your own. Instead, try to appreciate how they bring together many of the organic reactions that are familiar to you.

Triptycene was first synthesized in 1942 by P. D. Bartlett. Once it had been synthesized, its cation was found to be considerably less stable than the triphenylmethyl cation that exists in an almost-planar conformation. This allowed chemists to conclude that electron delocalization requires a planar conformation in order to be effective.

Paul D. Bartlett (1907–1997) was born in Michigan. He received a Ph.D. from Harvard University and was a professor of chemistry at Harvard and at Texas Christian University.

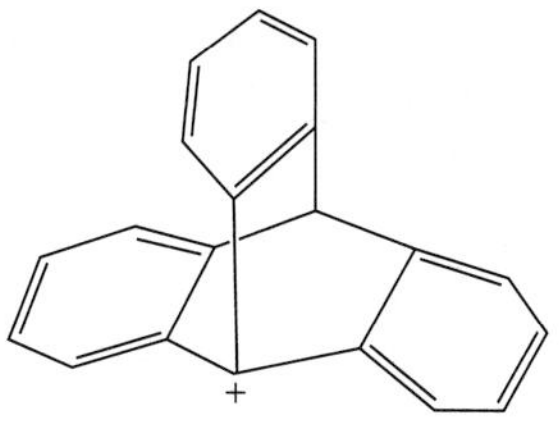

triptycene cation 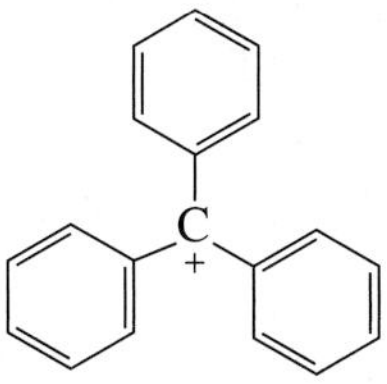   **triphenylmethyl cation**

The only step that forms a carbon–carbon bond in the synthesis of triptycene is a Diels–Alder cycloaddition of anthracene and 1,4-benzoquinone. All the other steps serve only to convert the newly installed ring into a benzene ring. Treatment of the enedione with hydrochloric acid is an acid-catalyzed enolization of a compound whose aromatic enol tautomer is more stable than its keto tautomer. The product is oxidized to a quinone by sodium bromate (Section 18.11). Oxime formation (Section 17.7) followed by reduction with Na₂S forms the diamino-substituted benzene ring. The amino substituents are removed by diazotization followed by treatment with hypophosphorous acid (Section 15.10).

3-D Molecule:
Triptycene

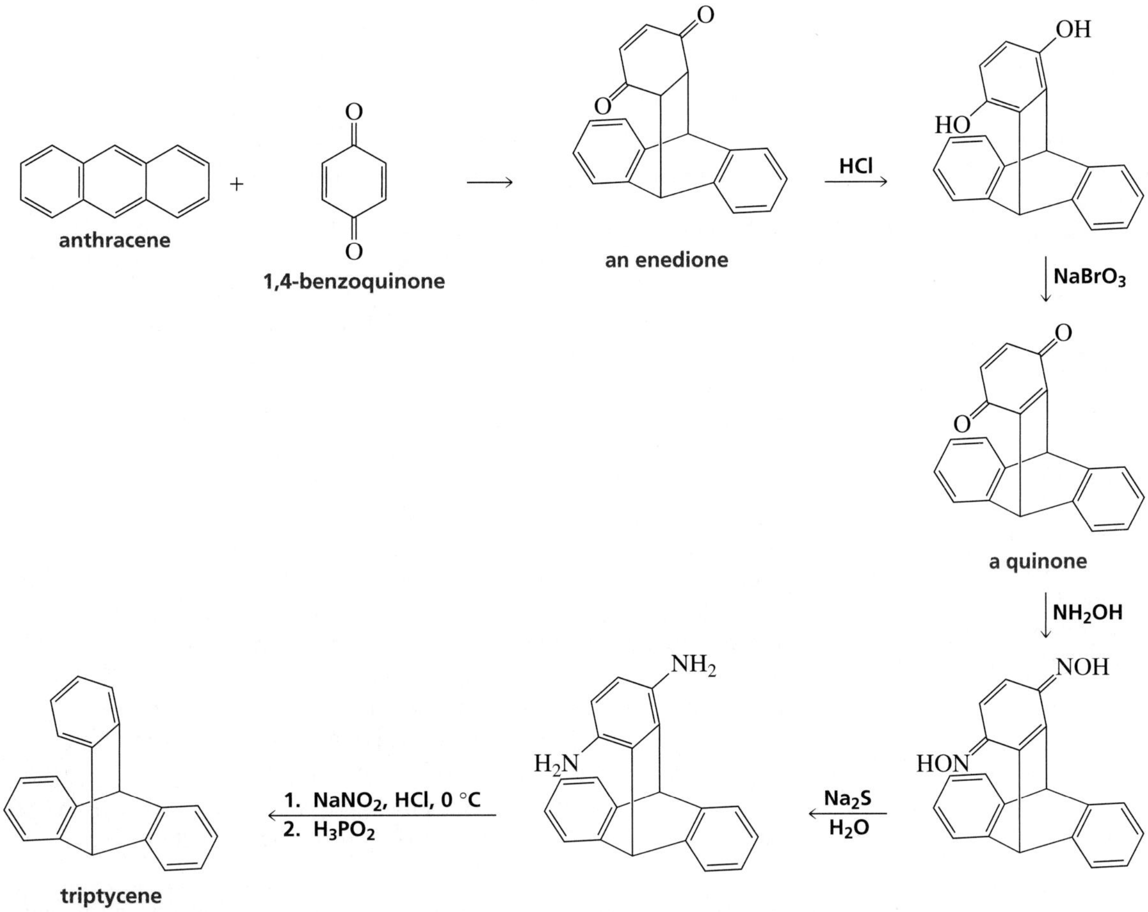

anthracene + **1,4-benzoquinone** → **an enedione** $\xrightarrow{\text{HCl}}$

$\downarrow$ **NaBrO₃**

a quinone

$\downarrow$ **NH₂OH**

$\xleftarrow[\text{H₂O}]{\text{Na₂S}}$

$\xleftarrow[\text{2. H₃PO₂}]{\text{1. NaNO₂, HCl, 0 °C}}$

triptycene

PROBLEM 30

Show the mechanism for the conversion of the enedione to the diol in the synthetic scheme for triptycene.

A more convergent synthesis of triptycene has since been developed. It has been prepared in a single step by a Diels–Alder reaction of anthracene with a diazonium salt of anthranilic acid.

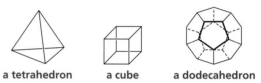

anthranilic acid   anthracene   1. NaNO₂, HCl, 0 °C   2. HO⁻   triptycene

PROBLEM 31

Propose a mechanism for the formation of triptycene from the diazonium salt of anthranilic acid. (*Hint:* A benzyne intermediate is formed; Section 15.14.)

Plato was born in Athens, Greece, in about 428 B.C. Plato and other philosophers of his time believed that world structures could be explained in terms of geometric forms. The Platonic solids are the five "perfect" polyhedra, all faces and angles of which are congruent: the tetrahedron, octahedron, and icosahedron are built of equilateral triangles, the cube is built of squares, and the dodecahedron is built of equilateral pentagons.

Chemists have synthesized three of the Platonic solids—the tetrahedron, the cube, and the dodecahedron.

a tetrahedron   a cube   a dodecahedron

Cubane is the common name for the C_8H_8 hydrocarbon that resembles a cube. The severe strain (angle strain and torsional strain) of this compound was the major problem that had to be overcome in its synthesis. Any approach had to rely on methods that would work with severely strained systems. It was also important to use as few steps as possible, since each step forms a new strained intermediate that has to be stable enough to survive.

The first successful synthesis of cubane was reported by Philip Eaton in 1964. The starting material for the synthesis was the Diels–Alder adduct of 2-bromo-cyclopentadienone. In the first step of the synthesis, one of the ketone groups is protected by being converted into an acetal. The first four-membered ring is formed by a [2+2] cycloaddition reaction under photochemical conditions (Section 28.4). The third step is a **Favorskii ring contraction** of an α-bromoketone (Chapter 19, Problems 65 and 66). The carboxyl group is removed after first being converted into a perester. When the perester is heated, the weak oxygen–oxygen bond breaks homolytically. As a result, CO_2 is eliminated, and the resultant cubyl radical abstracts a hydrogen atom from the solvent. The protecting group is removed by hydrolysis, and a second Favorskii ring contraction completes the synthesis of the cube. The remaining carboxyl group is removed by the same technique used to remove the first carboxyl group. This is an example of an iterative synthesis. An **iterative synthesis** is one in which a reaction sequence is carried out more than once.

Philip Eaton was born in Brooklyn in 1936. He received an A.B. from Princeton and a Ph.D. from Harvard. He was a professor of chemistry at the University of California, Berkeley, and is now at the University of Chicago. Eaton has recently reported the synthesis of octanitrocubane—cubane with a nitro group at each of the corners—which is expected to be a powerful explosive (page 94).

HO(CH₂)₂OH, HCl   hν

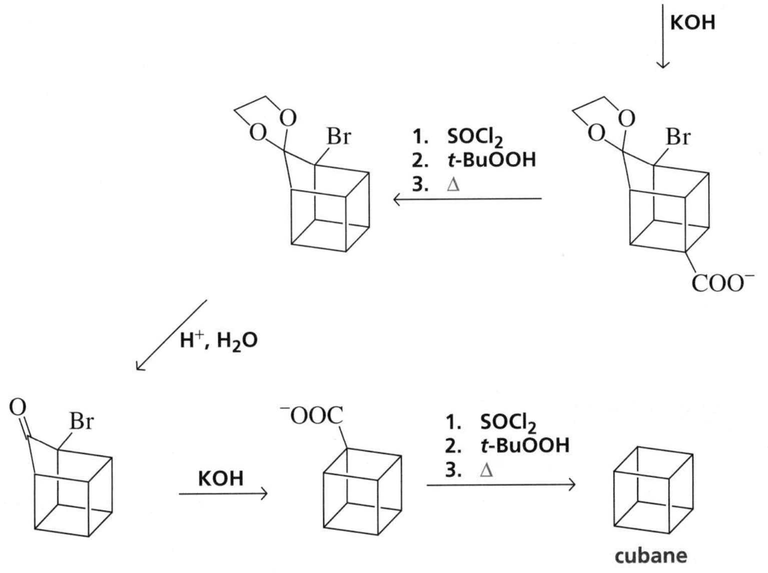

cubane

PROBLEM 32

Write out all the intermediates formed in the synthesis of cubane.

PROBLEM 33

2-Bromo-2,4-cyclopentadienone cannot be isolated at room temperature. Why?

PROBLEM 34

2-Bromo-2,4-cyclopentadienone can be formed by treatment of 2,3,4-tribromocyclopentanone with diethylamine.

a. Show how the compound is formed.

b. Suggest a synthesis for the tribromo compound from 2-cyclopentenone.

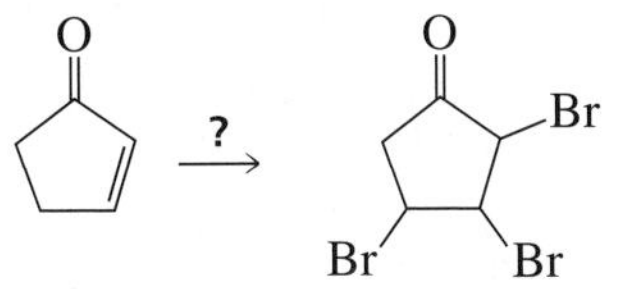

A more convergent synthesis of cubane was later carried out by Rowland Petit. The first step is a Diels–Alder reaction of a cyclobutadiene equivalent and 2,5-dibromo-1,4-benzoquinone. Cyclobutadiene itself cannot be used as the starting material because its antiaromaticity makes it too unstable (Section 14.5). A metal complex of cyclobutadiene is stable enough to be handled easily. The complex is mixed with the dibromosubstituted quinone in the presence of ceric ammonium nitrate, which liberates cyclobutadiene from the metal complex. The second step is a [2+2] cycloaddition reaction under photochemical conditions. The third step involves Favorskii ring contractions of the two α-bromoketones. The carboxyl groups are removed by the same method Eaton employed.

This synthesis was designed as a result of a retrosynthetic analysis that indicated that a pair of Favorskii ring contractions was the best route to cubane. From that point it was relatively easy to decide to use a quinone and cyclobutadiene to get to the intermediate required for the ring contractions.

Notice that in the syntheses of triptycene and cubane, reactions that remove functional groups are important because the target molecule in each case is a hydrocarbon.

Rowland Petit (1927–1981) was born in Australia. He earned two Ph.D.s in chemistry, the first at the University of Adelaide in Australia and the second at the University of London. He was a professor of chemistry at the University of Texas, Austin.

PROBLEM 35

Propose a mechanism for the Favorskii ring contraction used in the synthesis of cubane.

Thousands of natural products have been synthesized by organic chemists. The two natural product syntheses we will discuss use several reactions with which you are familiar. These syntheses should give you an insight into how chemists plan a synthesis based on a sound knowledge of mechanism and structural theory.

Lysergic acid was first synthesized by R. B. Woodward in 1954. The diethylamide of lysergic acid (LSD) is a better-known compound because of its hallucinogenic properties. It is a tribute to Woodward's ability as a synthetic organic chemist that the next synthesis of lysergic acid was not accomplished until 1969.

Lysergic acid is a tetracyclic molecule derived from indole. Because indole is unstable under acidic conditions (Section 27.2), Woodward designed a synthesis that did not form the indole portion of the molecule until the final step. His synthesis began with a dihydroindole carboxylic acid with the nitrogen of the five-membered ring protected by being converted into an amide. This starting material contained two of the required rings. An intramolecular Friedel–Crafts acylation reaction was used to form the third ring. Bromination of the α-carbon followed by an S_N2 substitution of the bromine by an amine and removal of the carbonyl protecting group produced a compound that could form the fourth ring by means of an aldol condensation. The carbonyl group was reduced, and the resulting hydroxyl group was activated by thionyl chloride and replaced by a cyano group. The cyano group and the protecting amide group were then hydrolyzed. The last step is the reverse of hydrogenation of a double bond and was carried out using a typical hydrogenation catalyst in the absence of hydrogen.

3-D Molecule:
Lysergic acid

The reaction scheme across the top of the page shows, from left to right:

lysergic acid (structure with CO₂H group, NCH₃, and HN-containing ring system) ← **1. H₂O, HO⁻, Δ** / **2. Raney Ni** ← (structure with C≡N group, NCH₃, and C₆H₅—C(=O)—N ring system) ← **1. NaBH₄** / **2. SOCl₂** / **3. NaCN** ← (structure with ketone O, NCH₃, and C₆H₅—C(=O)—N ring system) + H₂O

PROBLEM 36

Write out all the intermediates formed in Woodward's synthesis of lysergic acid.

PROBLEM 37

Show the retrosynthetic analysis that gets you to the starting material used by Woodward. For each step, indicate whether carbon–carbon bonds are formed, functional groups are introduced, functional groups are removed, or functional groups are interconverted.

PROBLEM 38

Give the reaction that shows why indole is unstable under acidic conditions.

Caryophyllene is an oil found in cloves. The most important strategic element in its synthesis was the realization that the nine-membered ring could be formed by fragmenting a six-membered ring fused to a five-membered ring. Then the problem was to design the synthesis of the compound that would undergo fragmentation.[1]

The first step in the synthesis is a photochemical [2+2] cycloaddition reaction of 2-methylpropene and 2-cyclohexenone. One α-hydrogen of the resulting ketone is substituted with a methoxycarbonyl group, and the second α-hydrogen on the same carbon is substituted with a methyl group. An acetylide anion is used to add a three-carbon fragment to the ketone carbonyl group. The triple bond of the alkyne is reduced, the acetal is hydrolyzed, and the resulting aldehyde is oxidized to a carboxylic acid that forms a lactone. A Claisen condensation and hydrolysis of the lactone form the 3-oxocarboxylic acid that is decarboxylated to give the cyclic ketone. Reduction with sodium borohydride gives the diol. Because the secondary alcohol is a more reactive nucleophile than the more sterically hindered tertiary alcohol, the desired tosylate is formed. The stage is now set for the key fragmentation reaction. This reaction involves removal of a proton from the OH group with base and expulsion of the tosylate leaving group. Epimerization of the chirality center alpha to the carbonyl group occurs once the ketone has been formed. **Epimerization** is changing the configuration of a chirality center by removing a proton from it and then reprotonating the molecule at the same atom. In the final step of the synthesis, a Wittig reaction is used to form the exocyclic carbon–carbon double bond.

PROBLEM 39 ◆

Why is the first step in the synthesis of caryophyllene carried out under photochemical conditions? (*Hint:* See Section 28.4.)

Robert B. Woodward (1917–1979) *was born in Boston and first became acquainted with chemistry in his home laboratory. He entered MIT at the age of 16 and received a Ph.D. the same year that those who entered MIT with him received their B.A.'s. He went to Harvard as a postdoctoral fellow and remained there for his entire career. Cholesterol, cortisone, strychnine, reserpine (the first tranquilizing drug), chlorophyll, tetracycline, and vitamin B₁₂ are just some of the complicated organic molecules Woodward synthesized. He received the Nobel Prize in chemistry in 1965. A description of his synthesis of lysergic acid can be found in* J. Am. Chem. Soc., *1954, 76, 5256; 1956, 78, 3087.*

R. B. Woodward lecturing at Harvard.

---

[1]The synthesis was designed and executed by E. J. Corey (Chapter 5, page 257) and his students at Harvard University. A description of the synthesis is published in *J. Am. Chem. Soc., 1963, 85, 362; 1964, 86, 485.*

PROBLEM 40

Propose a mechanism for the fragmentation reaction in the synthesis of caryophyllene.

PROBLEM 41

Explain why, in the second step in the synthesis of caryophyllene, an α-proton is removed preferentially from one side of the carbonyl group.

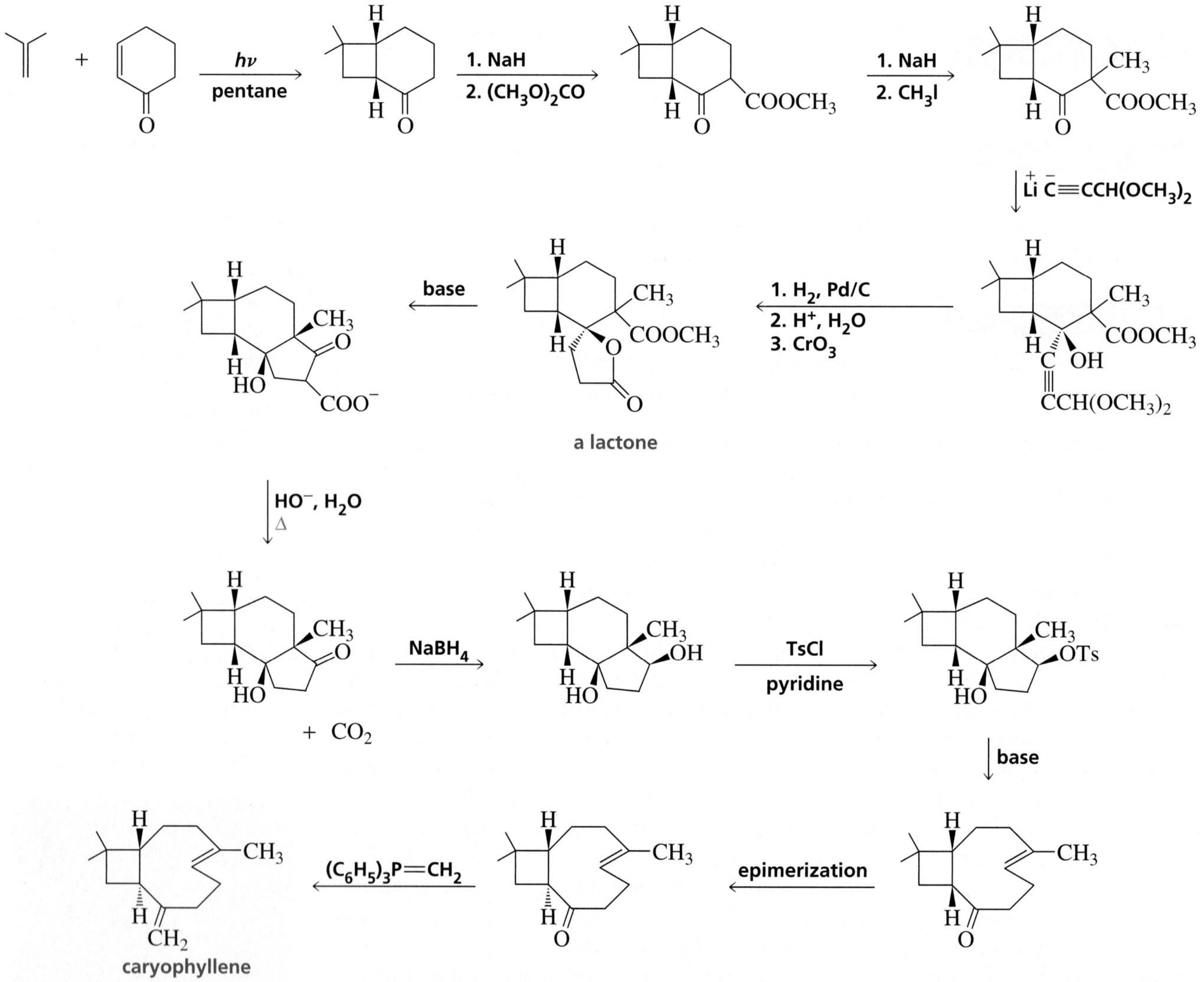

a lactone

caryophyllene

You should now have an appreciation of some of the mental processes involved in designing a synthesis. The most important factor in designing a synthesis is to have good command of organic reactions. The more reactions you know, the better your chances of coming up with a useful synthesis. The guiding factor in planning a synthesis is to keep it as simple as possible. The simpler the synthetic plan, the greater the chance it will be successful.

KEY TERMS

chiral auxiliary (page 1214)
combinatorial library (page 1198)

combinatorial organic synthesis
(page 1198)

convergent synthesis (page 1197)
Darzen's condensation (page 1207)

42. Describe how each of the following compounds could be synthesized using starting materials containing no more than five carbon atoms:
    a. 1-hexanol   d. 2-hexanol
    b. 1-heptanol   e. 2-methyl-2-hexanol
    c. 1-octanol   f. 4-methyl-4-heptanol

43. a. Disconnect the following molecule at bond **a** and give a retrosynthetic analysis for its synthesis.
    b. Disconnect the following molecule at bond **b** and give a retrosynthetic analysis for its synthesis.

$$\text{HO} \quad \text{CH}_2 - \text{CH}=\text{CH}_2$$

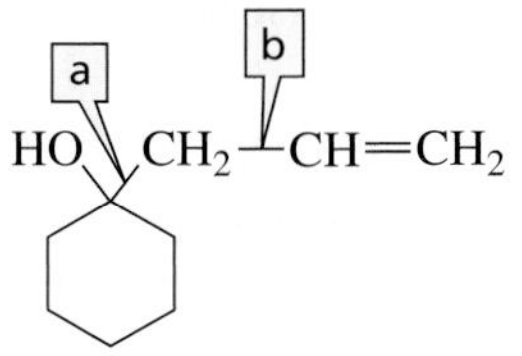

44. Using retrosynthetic analysis, plan a synthesis of the following target molecules using only the given starting materials and any needed inorganic reagents.

target molecules

a.

b.

c.

d.

starting materials

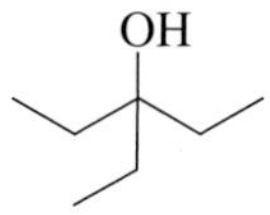

CO_2

CH_3CH_2OH

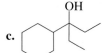—OH

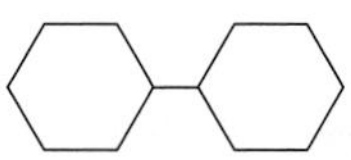

45. What reagents are required to convert 1-methylcyclohexene into the following compounds?

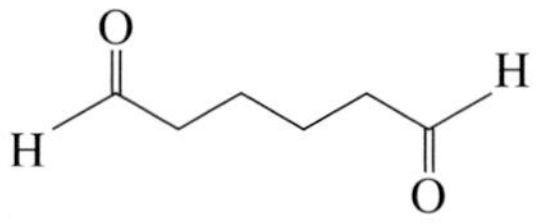

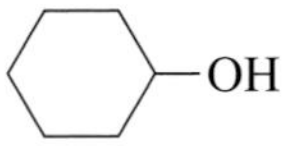

a. 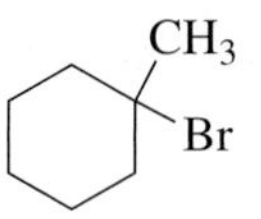   c. 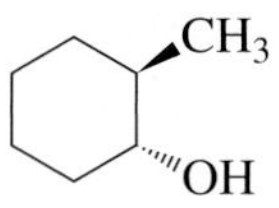   e. 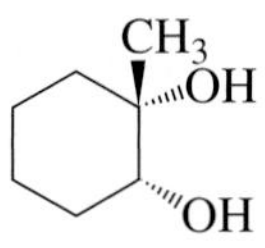   g.

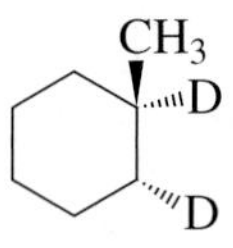

b.   d.   f.   h.

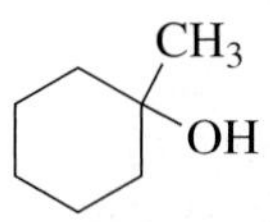

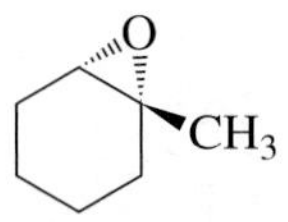

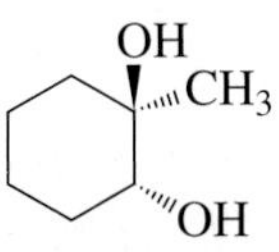

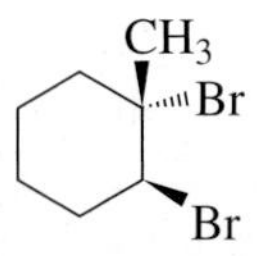

46. Describe a method for the synthesis of each of the following compounds from starting materials containing no more than four carbon atoms:

a.

b.

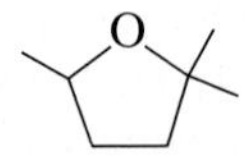

47. Sodium borohydride is used to reduce cyclohexanone to cyclohexanol. Why can't sodium hydride be used?

48. Describe two methods for the synthesis of the following compound:

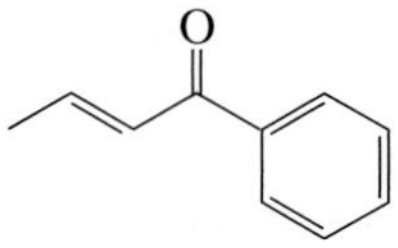

49. Show how each of the following compounds can be prepared from the given starting material. In each case, you will need to use a protecting group.

a. $ClCH_2CH_2CH$ ⟶ $H_2NCH_2CH_2CH_2CH$

b. $CH_3CHCH_2COCH_3$ ⟶ $CH_3CHCH_2CCH_3$
        $\quad\quad\; OH$ $\quad\; OH \quad OH$

c.

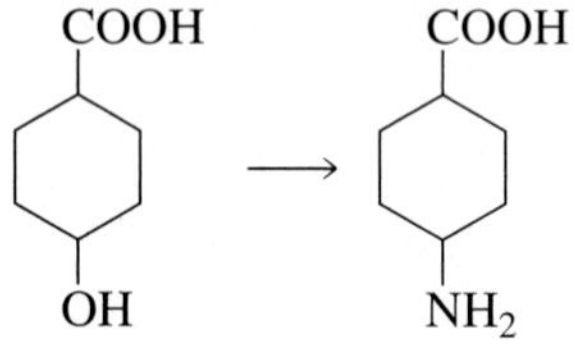

d.

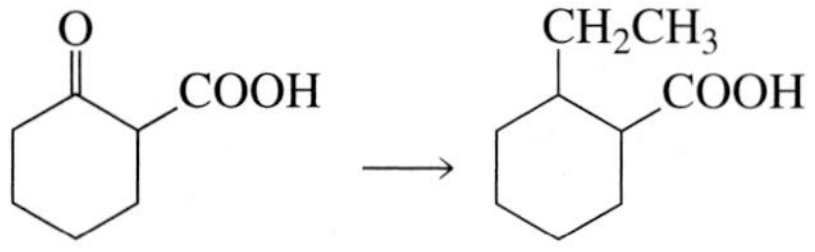

50. Describe how the following compounds could be synthesized from starting materials containing no more than four carbon atoms:

a.

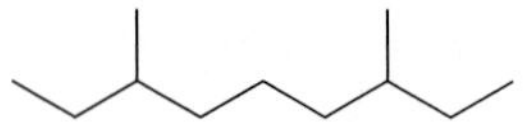

b.

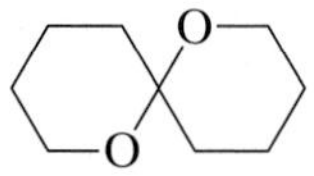

51. Do a retrosynthetic analysis on each of the following compounds, ending with the required starting materials:

a.

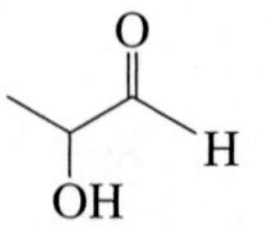

b.

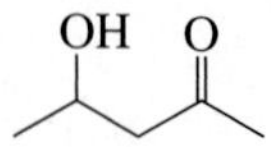

c.

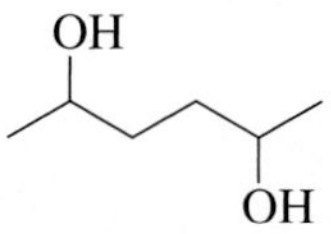

d.

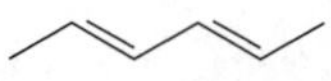

e.

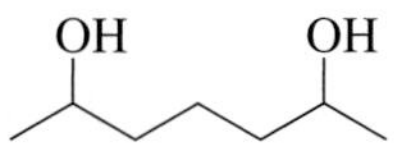

f.

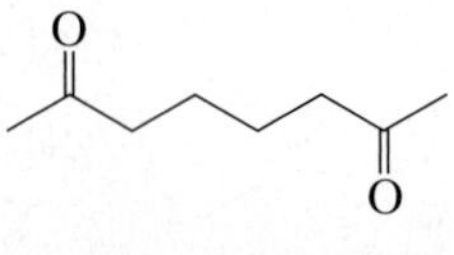

52. Outline syntheses for the following compounds using starting materials containing no more than five carbon atoms:

a.

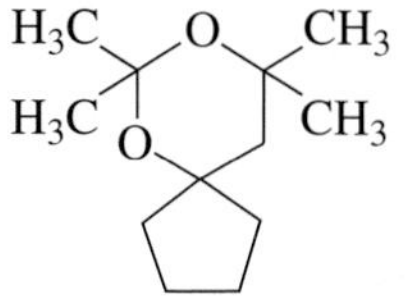

b. 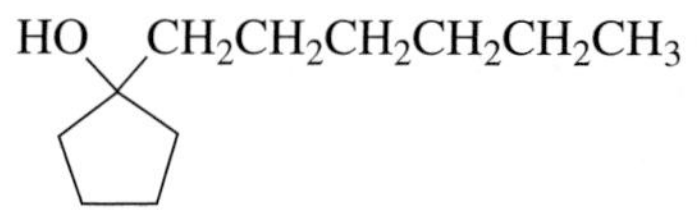 HO CH₂CH₂CH₂CH₂CH₂CH₃

53. Using *para*-benzoquinone (see Problem 55) as one of the starting materials, what diene would you use in order to synthesize the following compounds?

a.

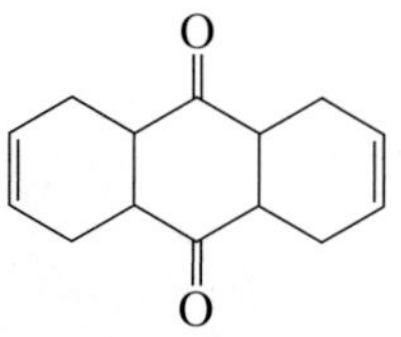

b.

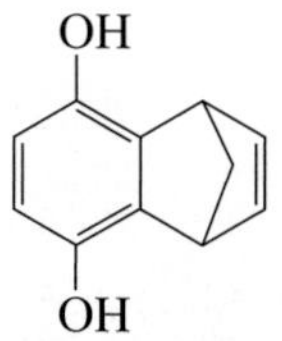

c.

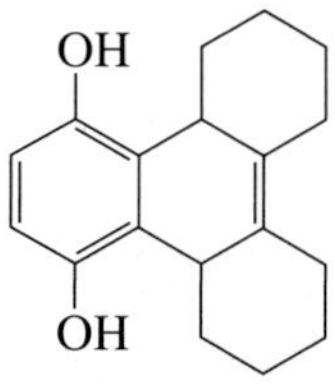

54. Propose a mechanism for each of the following reactions:

a.

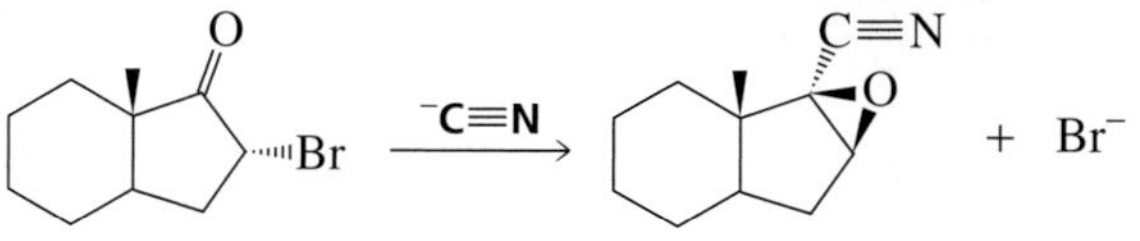

b.

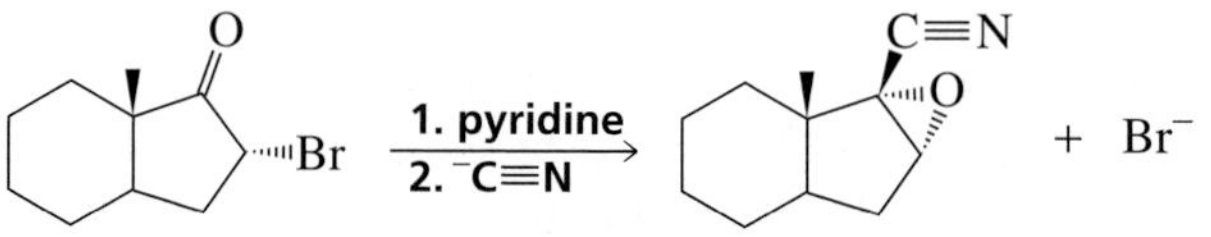

55. The following *para*-quinone derivative can be used to protect OH and NH₂ groups by converting them into esters and amides, respectively. An advantage of this protecting group is that it can be removed by a mild reducing agent. Consequently, removing it will not hurt groups in the molecule that would be affected by acid- or base-catalyzed hydrolysis. Explain why the protecting group can be removed by reduction. (*Hint:* See Section 18.11.)

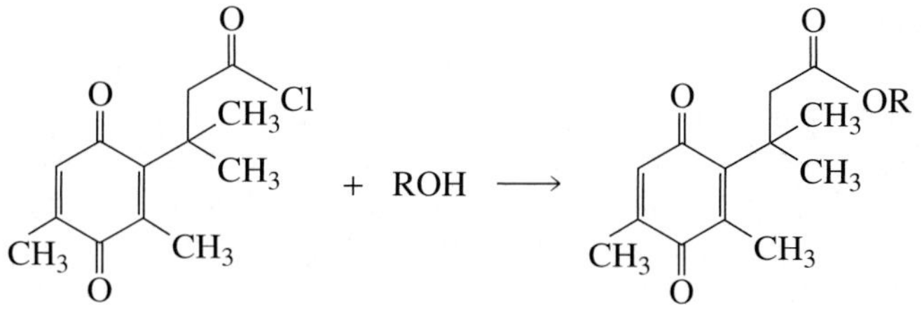

56. Show how each of the following compounds could be prepared from the given starting material:

a.

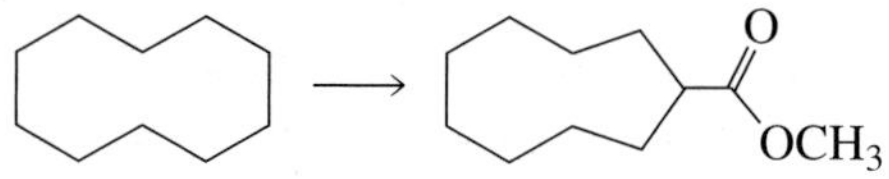

b.

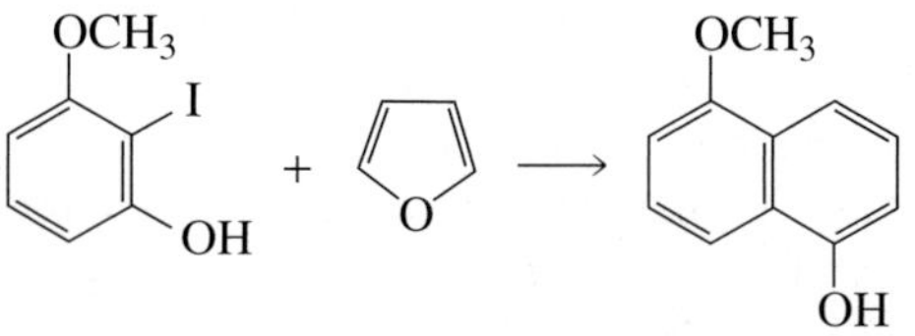

30

The Organic Chemistry of Drugs:

Discovery and Design

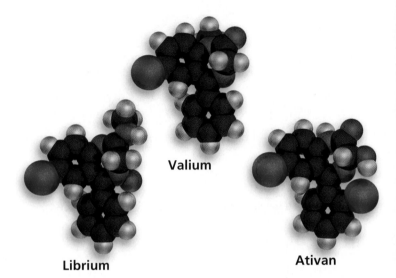

Valium

Librium

Ativan

A **drug** is a compound that interacts with a biological molecule, triggering a physiological effect. Drugs have been used by humans for thousands of years to alleviate pain and illness. By trial and error, people learned which herbs, berries, roots, and bark could be used for medicinal purposes. However, as late as the beginning of the twentieth century, relatively few truly effective medicinal agents were known. Local anesthetics had just been discovered, and there were only two analgesics to relieve major pain. There were no drugs for the dozens of functional, degenerative, neurological, and psychiatric disorders; no hormone therapies; no vitamins; and—most significantly—no effective drug for the cure of any infectious disease. One reason that families had many children was because some of the children were bound to succumb to childhood diseases. Life spans were generally short. In 1900, for example, the average life expectancy in the United States was 46 years for a man and 48 years for a woman. Now there is a drug for treating almost every disease, and this is reflected by current life expectancies: 72 years for a man and 79 years for a woman.

The shelves of a typical modern pharmacy contain almost 2000 preparations, most of which contain a single active ingredient, usually an organic compound. In 1999, 2,821,700,000 prescriptions were dispensed in the United States—the most widely prescribed drugs are listed in Table 30.1. Antibiotics are the most widely prescribed class of drugs in the world. In the developed world, heart drugs are the most widely prescribed class, partly because they are generally taken for the remainder of the patient's life. In recent years prescriptions for psychotropic drugs have decreased as doctors have become more aware of problems associated with addiction, and prescriptions for asthma have increased, reflecting a greater incidence (or awareness) of the disease.

Throughout this book we have encountered many drugs, vitamins, and hormones and, in many cases, we have discussed the mechanism by which each compound produces its physiological effect. Table 30.2 lists some of these compounds, their uses, and where in the text they are discussed. Now we will take a look at how drugs are discovered, how they are named, and some of the techniques currently used by scientists in their search for new drugs.

Table 30.1   The Most Widely Prescribed Drugs in the United States

Proprietary name	Generic name	Structures	Use
Albuterol	albuterol	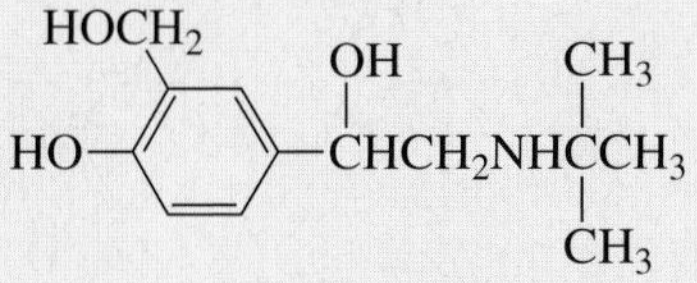	bronchodilator
Norvasc	amlodipine	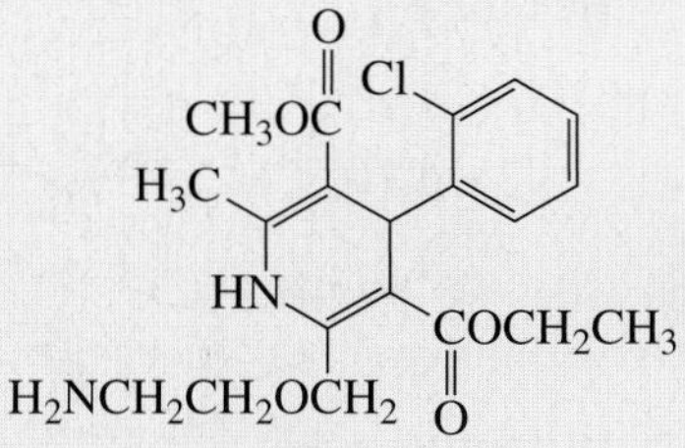	antihypertensive
Claritin	loratadine	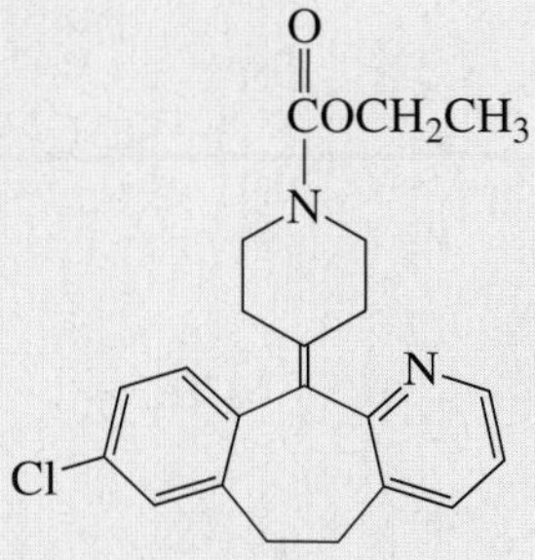	antihistamine
Trimox	amoxicillin	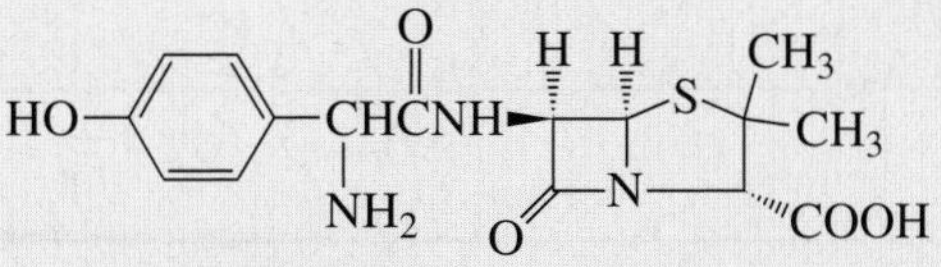	antibiotic
Prozac	fluoxetine	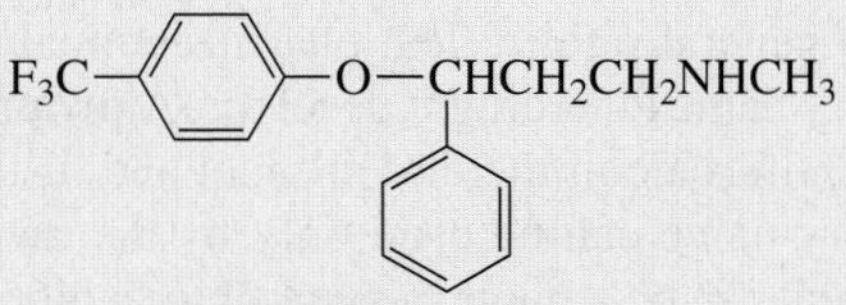	antidepressant
Premarin	conjugated estrogens	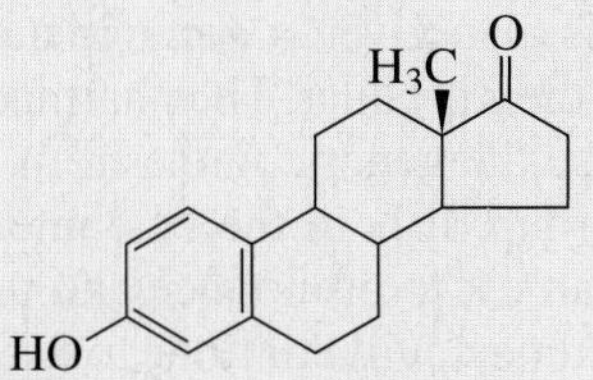	hormone replacement therapy

continues

Table 30.1 Continued			
Proprietary name	**Generic name**		**Use**
Synthroid	levothyroxine	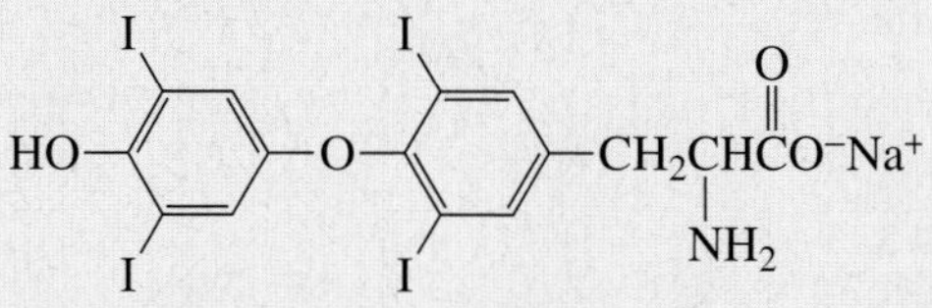	treatment of hypothyroidism
Lipitor	atorvastatin	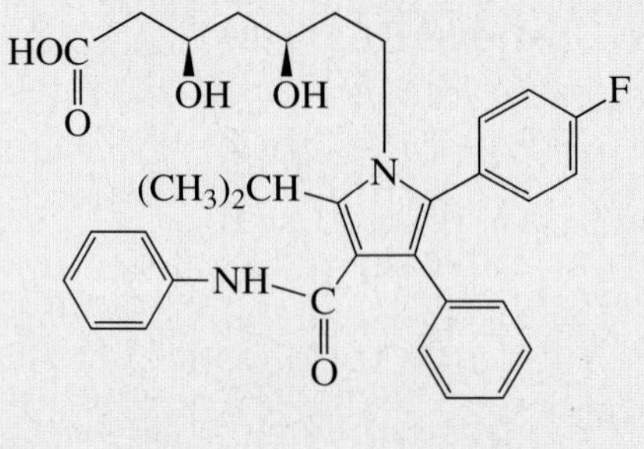	statin (cholesterol-reducing drug)
Prilosec	omeprazole	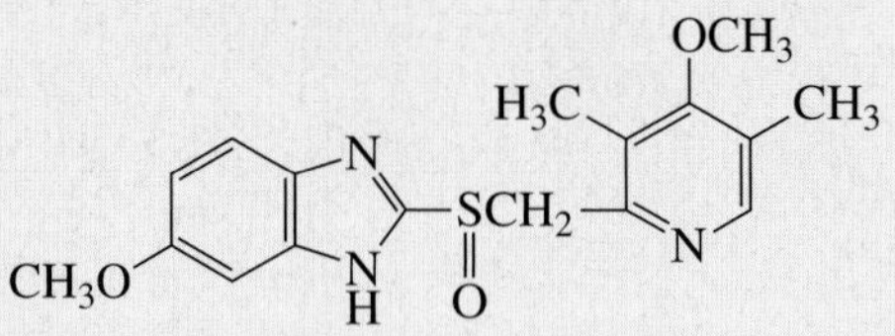	treatment of stomach acid reflux
Hydrocodone with APAP	hydrocodone with APAP	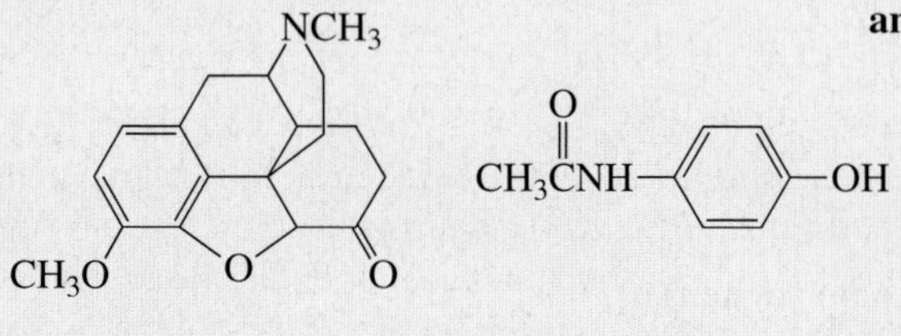	analgesic

30.1 NAMING DRUGS

The most accurate names of drugs are the chemical names that define their structures. However, these names are too long and complicated to appeal to physicians and to the general public. The pharmaceutical company is allowed to choose the proprietary name for a drug it develops. A **proprietary name (trade name** or **brand name)** identifies a commercial product and distinguishes it from other products. A proprietary name can be used only by the owner of the registered trademark—a **trademark** can be a name, a symbol, or a picture. It is in the best interest of the company to choose a name that is easy to remember and pronounce so that when the patent expires, the public will continue to request the drug by its proprietary name.

Each drug is also given a **generic name** that any pharmaceutical company can use to identify their product. The pharmaceutical company that develops a drug is allowed to choose the generic name from a list of 10 possible names provided by an independent group. It is in the best interest of the company to choose the generic name that is hardest to pronounce and the least likely to be remembered, so physicians and consumers will continue to use the familiar proprietary name. Proprietary names must always be capitalized, while generic names are not capitalized.

Table 30.2	Drugs, Hormones, and Vitamins Discussed in Earlier Chapters	
sulfa drugs	the first antibiotics	Section 23.8
tetracycline	broad-spectrum antibiotic	Section 4.8
puromycin	broad-spectrum antibiotic	Section 25.13
nonactin	ionophorous antibiotic	Section 11.9
penicillin	antibiotic	Sections 16.5 and 16.14
gentamicin	antibiotic	Section 20.19
gramicidin S	antibiotic	Section 21.8
aspirin	analgesic, anti-inflammatory agent	Sections 19.9 and 24.5
enkephalins	pain killers	Section 21.8
diethyl ether	anesthetic	Section 11.5
sodium pentothal	sedative hypnotic	Section 11.5
5-fluorouracil	anticancer agent	Section 23.8
methotrexate	anticancer agent	Section 23.8
Taxol	anticancer agent	Chapter 29, introduction
thalidomide	sedative with teratogenic side effects	Section 4.13
AZT*	anti-AIDS agent	Section 25.17
warfarin	anticoagulant	Section 23.9
epinephrine	vasoconstrictor, bronchodilator	Section 9.11
vitamins		Chapter 23, Sections 8.8, 17.11, 20.19, 24.4, 24.7, and 27.4
hormones		Sections 21.8 and 24.9–24.11

*3′-Azido-2′-deoxythymidine

Drug manufacturers are permitted to patent and retain exclusive rights to the drugs they develop. A patent is valid for 20 years. Once the patent expires, other drug companies can market the drug under the generic name or under their own proprietary names, which are called branded generic names. For example, the antibiotic ampicillin is sold as Penbritin by the company that held the original patent. Now that the patent has expired, this drug is sold by other companies as Ampicin, Ampilar, Amplital, Binotal, Nuvapen, Pentrex, Ultrabion, Viccillin, and 30 other branded generic names.

The over-the-counter drugs that line the shelves of drugstores are available without prescription. They are often mixtures containing one or more active ingredients (generic or proprietary drugs) plus sweeteners and inert fillers. For example, the preparations called Advil (Whitehall Laboratories), Motrin (Upjohn), and Nuprin (Bristol-Meyers Squibb) all contain ibuprofen, a mild analgesic and anti-inflammatory drug. Ibuprofen was patented in Britain in 1964 by Boots, Inc., and the U.S. Food and Drug Administration (FDA) approved its use as a nonprescription drug in 1984.

30.2 LEAD COMPOUNDS

The goal of the medicinal chemist is to find compounds that have potent effects on given diseases with minimum side effects. In other words, a drug must react selectively with its target and have a minimal effect on the host cell. A drug must get to the right place in the body, at the right concentration, and at the right time. Therefore, a drug must have the appropriate solubility to allow it to be transported to the target cell. If it is taken orally, it must be insensitive to the acid conditions of the stomach, and it also must resist enzymatic degradation before it reaches its target. Finally, it must eventually be either excreted or degraded to harmless compounds that can be excreted.

Foxglove.

Herbs used by humans since ancient times provided the starting point for the development of our current arsenal of drugs. The active ingredients were isolated from herbs and roots used in traditional medicine. Foxglove, for instance, furnished digitoxin, a cardiac stimulant. The bark of the cinchona tree yielded quinine for relief from malaria. Willow bark contains salicylates used to control fever and pain. The sticky juice of the oriental opium poppy provided morphine for severe pain and codeine for the control of cough. By 1882 there were more than 50 different herbs commonly used to make medicines. Many of these herbs were grown in the gardens of religious establishments used to treat the sick.

Scientists still search the world for plants and berries and the oceans for flora and fauna that might yield new medicinal compounds. Taxol, a compound isolated from the bark of the Pacific yew tree, is a relatively recently recognized anticancer agent (Chapter 29, introduction).

Once a naturally occurring drug is isolated and its structure determined, it can serve as a prototype in a search for other biologically active compounds. The prototype is called a **lead compound.** Analogs of the lead compound are synthesized in order to find one that might have improved therapeutic properties or fewer side effects. The analog may have a different substituent than the lead compound, a branched chain instead of a straight chain, or a different ring system. Changing the structure of the lead compound is called **molecular modification.**

30.3 MOLECULAR MODIFICATION

A classic example of molecular modification is the development of synthetic local anesthetics from cocaine. Cocaine comes from the leaves of *Erythroxylon coca,* a bush native to the highlands of the South American Andes. It is an effective local anesthetic, but it produces disturbing effects on the central nervous system (CNS) ranging from initial euphoria to severe depression. By degrading the cocaine molecule step by step (removing the methoxycarbonyl group and cleaving the seven-membered-ring system), scientists identified the portion of the molecule that carries the local anesthetic activity without the damaging CNS effects. This knowledge gave an improved lead compound, an ester of benzoic acid with the alcohol component of the ester having a terminal tertiary amino group.

Coca leaves.

cocaine
lead compound

improved lead compound

Hundreds of esters were then synthesized, putting substituents on the aromatic ring, changing the alkyl groups bonded to the nitrogen, and changing the length of the connecting alkyl chain. Successful anesthetics obtained as a result of molecular modification were Benzocaine, a topical anesthetic, and procaine, commonly known by the trade name Novocain.

Because the ester group of procaine is hydrolyzed relatively rapidly by serum esterases (enzymes that catalyze ester hydrolysis), procaine has a short half-life. Therefore, compounds with less easily hydrolyzed amide groups were synthesized. In this way lidocaine, one of the most widely used injectable anesthetics, was discovered. The rate at which lidocaine is hydrolyzed is further decreased by the two *ortho*-methyl substituents, which provide steric hindrance to the vulnerable carbonyl group.

Benzocaine

procaine
Novocain

lidocaine
Xylocaine

Later, physicians recognized that the action of an anesthetic administered *in vivo* (in a living organism) could be lengthened considerably if it were administered along with epinephrine. Because epinephrine is a vasoconstrictor, it reduces the blood supply, allowing the drug to remain at its targeted site for a longer period of time.

In screening the structurally modified compounds for biological activity, scientists were surprised to find that replacing the ester linkage of procaine with an amide linkage led to a compound, procainamide hydrochloride, that had activity as a cardiac depressant in addition to its activity as a local anesthetic. It is currently used clinically as an antiarrhythmic.

procainamide hydrochloride

Morphine is the most widely used analgesic for severe pain. Although scientists have learned how to synthesize morphine, all commercial morphine is obtained from opium, the juice obtained from a species of poppy. Morphine occurs in opium to the extent of 10%. Morphine is the standard by which other painkilling medications are measured. Codeine, which has one-tenth the analgesic activity of morphine, profoundly inhibits the cough reflex. Although 3% of opium is codeine, most commercial codeine is obtained by methylating morphine. Acetylating the two OH groups of morphine forms heroin, a drug that has been banned in most countries because it is widely abused. Heroin is less polar than morphine, so it crosses the blood/brain barrier more rapidly, resulting in a more rapid "high." Because acetic anhydride is used to acetylate morphine, both heroin and acetic acid are formed as products. Drug enforcement agencies use dogs trained to recognize the pungent odor of acetic acid.

morphine

codeine

heroin

Molecular modification of codeine led to dextromethorphan, the active ingredient in most cough medicines. Etorphine was synthesized when scientists realized that analgesic potency was related to the ability of the drug to bind hydrophobically to the opiate receptor (Section 30.6). Etorphine is about 2000 times more potent than morphine, but it is not safe enough for use in humans. It has been used to tranquilize elephants and other large animals. Pentazocine is useful in obstetrics because it dulls the pain of labor but does not depress the respiration of the infant as morphine does.

Paul Ehrlich

Paul Ehrlich (1854–1915) *was a German bacteriologist. He received a medical degree from the University of Leipzig and was a professor at the University of Berlin. In 1892 he developed an effective diphtheria antitoxin. For his work on immunity, he received the 1908 Nobel Prize in medicine or physiology, sharing it with Ilya Ilich Mechnikov.*

Ilya Ilich Mechnikov (1845–1916)*, later known as Elié Metchnikoff, was born in the Ukraine. He was the first scientist to recognize that white blood cells were important in resistance to infection. Metchnikoff succeeded Pasteur as director of the Pasteur Institute.*

dextromethorphan   etorphine   pentazocine

30.4 RANDOM SCREENING

The lead compound for the development of most drugs is found by random screening thousands of compounds. A **random screen,** also known as a **blind screen,** is the search for a pharmacologically active compound without any information about what chemical structures might show activity. The first blind screen was carried out by Paul Ehrlich, who was searching for a "magic bullet" against trypanosomes—the microorganisms that cause African sleeping sickness. After testing more than 900 compounds against trypanosomes, Ehrlich tested some of his compounds against other bacteria. Compound 606 (salvarsan) was found to be dramatically effective against the microorganisms (spirochetes) that cause syphilis.

An important part of random screening is recognizing an effective compound. This requires the development of an assay for the desired biological activity. Some assays can be done *in vitro* (in glass)—for example, searching for a compound that will inhibit a particular enzyme. Others are done *in vivo*—for example, searching for a compound that will save a mouse from a lethal dose of a virus. One problem with *in vivo* assays is that drugs can be metabolized differently by different animals. An effective drug in a mouse may be less effective or useless in a human. Another problem is regulation of the dosages of both the virus and the drug. If the dosage of the virus is too high, it might kill the mouse in spite of the presence of a biologically active compound that could save it. If the dosage of the potential drug is too high, the drug might kill the mouse when a lesser dosage would have saved it.

The observation that azo dyes effectively dyed wool fibers (animal protein) gave scientists the idea that azo dyes might selectively bind to bacterial proteins, too. Well over 10,000 dyes were screened *in vitro* in antibacterial tests, but none showed any antibiotic activity. Some scientists argued that the dyes should be screened *in vivo* because what physicians really needed were antibacterial agents that would cure infections in humans and animals, not in test tubes.

The *in vivo* studies were done in mice that had been infected with a bacterial culture. Now the luck of the investigators improved. Several dyes turned out to counteract gram-positive infections. The least toxic of these, Prontosil (a bright red dye), became the first drug to treat bacterial infections.

Prontosil

Gerhard Domagk (1895–1964) was a research scientist at I. G. Farbenindustrie, a German manufacturer of dyes and other chemicals. He carried out studies that showed Prontosil to be an effective antibacterial agent. His daughter, who was dying of a streptococcal infection as a result of cutting her finger, was the first patient to receive and be cured by the drug (1935). Prontosil received wider fame when it was used to save the life of Franklin D. Roosevelt, Jr., son of the U.S. president. Domagk received the Nobel Prize in medicine or physiology in 1939, but Hitler did not allow Germans to accept Nobel Prizes because Carl von Ossietsky, a German who was in a concentration camp, had been awarded the Nobel Prize for peace in 1935. Domagk was eventually able to accept the prize in 1947.

The fact that Prontosil was inactive *in vitro* but active *in vivo* should have suggested that the dye was converted to an active compound by the mammalian organism, but this did not occur to the bacteriologists, who were content to have found a useful antibiotic. When scientists at the Pasteur Institute later investigated Prontosil, they noted that mice given the compound did not excrete a red compound. Urine analysis showed that the mice excreted *para*-acetamidobenzenesulfonamide, a colorless compound. Chemists knew that anilines are acetylated *in vivo*, so they prepared the nonacetylated compound (sulfanilamide). When it was tested in mice infected with streptococcus, all the mice were cured, while untreated control mice died. Sulfanilamide was the first of the sulfa drugs.

$$CH_3\overset{\overset{\displaystyle O}{\|}}{C}-NH-\!\!\!\left\langle \;\;\right\rangle\!\!\!-SO_2NH_2$$

para-acetamidobenzenesulfonamide

$$H_2N-\!\!\!\left\langle \;\;\right\rangle\!\!\!-SO_2NH_2$$

para-aminobenzenesulfonamide
sulfanilamide

We have seen that sulfanilamide acts by inhibiting the bacterial enzyme that incorporates *para*-aminobenzoic acid into folic acid (Section 23.8). Thus, sulfanilamide is a bacteriostatic drug, not a bactericidal drug. A **bacteriostatic drug** inhibits the further growth of the bacteria, whereas a **bactericidal drug** kills the bacteria. Sulfanilamide acts as an inhibitor because the sulfonamide and carboxylic acid groups have similar sizes. Many successful drugs have been designed using similar isosteric replacements (Sections 23.8 and 23.9).

Many drugs have been discovered accidentally. Nitroglycerin, the drug used to relieve the symptoms of angina pectoris, was discovered when workers handling nitroglycerine in the explosives industry experienced severe headaches. Investigation revealed that the headaches were caused by nitroglycerine's ability to produce a marked dilation of blood vessels. The pain associated with an angina attack results from the inability of the blood vessels to supply the heart with adequate blood. Nitroglycerine relieves the pain by dilating cardiac blood vessels.

30.5 SERENDIPITY IN DRUG DEVELOPMENT

$$CH_2-ONO_2$$
$$CH-ONO_2$$
$$CH_2-ONO_2$$
nitroglycerin

The tranquilizer Librium is another drug that was discovered accidentally. Leo Sternbach synthesized a series of quinazoline 3-oxides, but none of them showed any pharmacological activity. One of the compounds was not submitted for testing because it was not the quinazoline 3-oxide he had set out to synthesize. Two years after the project was abandoned, a laboratory worker came across this compound when cleaning up the lab, and Sternbach decided he might as well submit it for testing before it was thrown away. The compound was shown to have tranquilizing properties and, when its structure was investigated, was found be a benzodiazepine 4-oxide. Methylamine, instead of displacing the chloro substituent to form a different quinazoline 3-oxide, had added to the imine group of the six-membered ring, causing this ring to open and reclose to a seven-membered ring. The compound was given the trade name Librium when it was put into clinical use in 1960.

Leo H. Sternbach

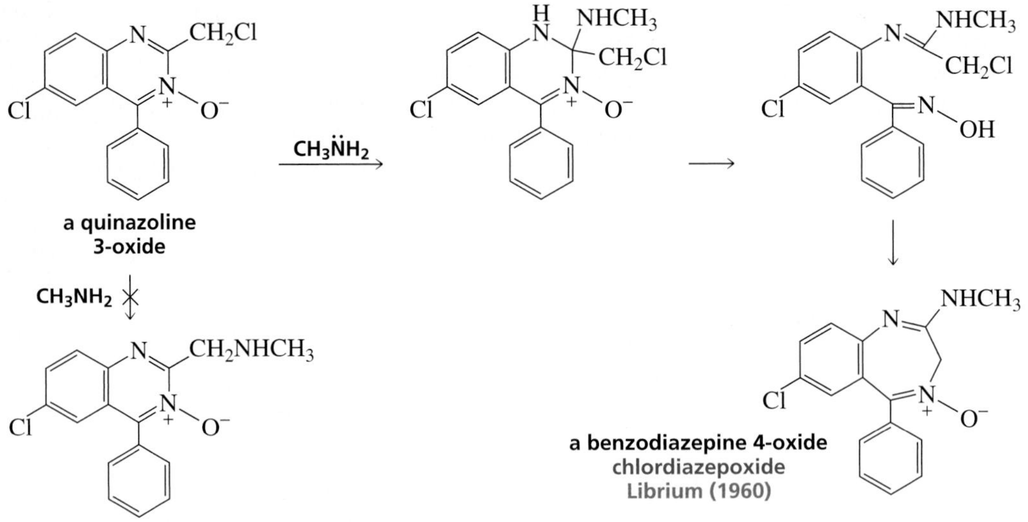

a quinazoline
3-oxide

a benzodiazepine 4-oxide
chlordiazepoxide
Librium (1960)

Librium was structurally modified in an attempt to find other tranquilizers. One successful modification produced Valium, a tranquilizer almost 10 times more potent than Librium. Currently there are 8 benzodiazepines in clinical use as tranquilizers in the United States and some 15 others abroad. Xanax is one of the most widely prescribed medications.

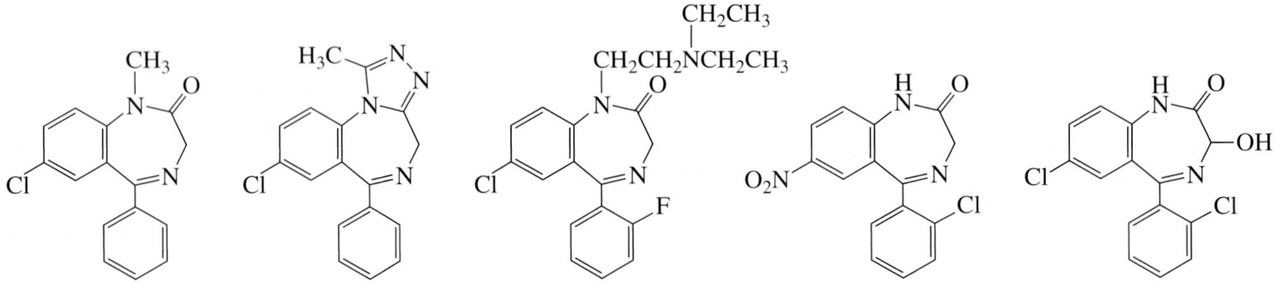

diazepam
Valium (1963)

alprazolam
Xanax (1970)

flurazepam
Dalmane (1970)

clonazepam
Klonopin (1975)

lorazepam
Ativan (1977)

30.6 RECEPTORS

Some drugs exert their physiological effects by interacting with a particular cellular binding site called a **receptor.** Drug receptors are generally glycoproteins or lipoproteins. Some receptors are part of cell membranes, while others are found in the cytoplasm (the material outside the nucleus). A drug interacts with its receptor using the same kinds of bonding interactions—hydrogen bonding, electrostatic attractions, van der Waals interactions—that we have encountered in other examples of molecular recognition (Section 22.8). The most important factor in bringing together a drug and a receptor is a snug fit: The greater the affinity of a drug for its binding site, the higher its potential biological activity.

Knowing something about the molecular basis of drug action allows scientists to design and synthesize compounds that might have the desired biological activity. For example, when excess histamine is produced by the body, it causes the symptoms associated with the common cold and allergic responses. This is thought to be the result of the protonated ethylamino group anchoring the histamine molecule to a negatively charged portion of the histamine receptor.

histamine
histamine receptor

Drugs that have been found to be effective antihistamines bind to the histamine receptor but do not trigger the same response as histamine. Like histamine, these drugs have an amino group that binds to the histamine receptor. The drugs also have bulky groups that keep the histamine molecule from approaching the receptor. Because these compounds interfere with the natural action of histamine, they are called antihistamines.

3-D Molecules:
Histamine;
Diphenhydramine;
Promethazine;
Promazine

antihistamines

diphenhydramine
Benadryl

promethazine
Promine

promazine
Talofen

Acetylcholine is a neurohormone that enhances peristalsis, wakefulness, and memory and is essential for nerve transmission. A deficiency of brain cell receptors that bind acetylcholine (cholinergic receptors) contributes to the characteristic loss of memory in Alzheimer's disease. Cholinergic receptors are structurally similar to those that bind histamine. Therefore, antihistamines and cholinergic agents show overlapping activities. As a result, the antihistamine diphenhydramine has been used to treat insomnia and to combat motion sickness.

acetylcholine
cholinergic receptor

Excess histamine production by the body also causes the hypersecretion of stomach acid by the cells of the stomach lining, leading to the development of ulcers. The antihistamines that block the histamine receptors (thereby preventing the allergic responses associated with excess histamine production) have no effect on HCl production. This fact led scientists to conclude that a second kind of histamine receptor triggers the release of acid into the stomach.

Because 4-methylhistamine was found to cause weak inhibition of HCl secretion, it was used as a lead compound. About 500 molecular modifications were performed over a 10-year period before four clinically useful antiulcer agents were found. Two of these are cimetidine and ranitidine. Notice that steric blocking of the receptor site is not a factor in these compounds. Compared with the antihistamines, the effective antiulcer drugs have more polar imidazole rings, longer side chains, and less basic substituents on the side chains.

Leo H. Sternbach was born in Austria in 1908. In 1918, after World War I and the dissolution of the Austro-Hungarian Empire, Sternbach's father moved to Krakow in the recreated Poland and obtained a concession to open a pharmacy. As a pharmacist's son, Sternbach was accepted at the Jagiellonian University School of Pharmacy, where he received both a Master of Pharmacy and a Ph.D. in chemistry. With discrimination against Jewish scientists growing in Eastern Europe in 1937, Sternbach moved to Switzerland to work with Ružička (p. 1046) at the ETH (the Swiss Federal Institute of Technology). In 1941, Hoffmann–LaRoche brought Sternbach and several other scientists out of Europe. Sternbach became a research chemist at their U. S. headquarters in Nutley, N. J., where he later became director of medicinal chemistry.

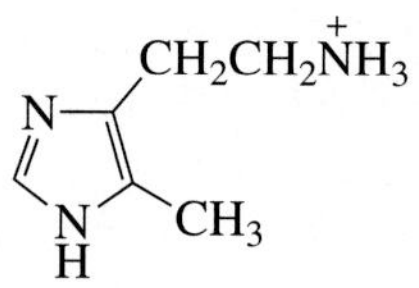

4-methylhistamine

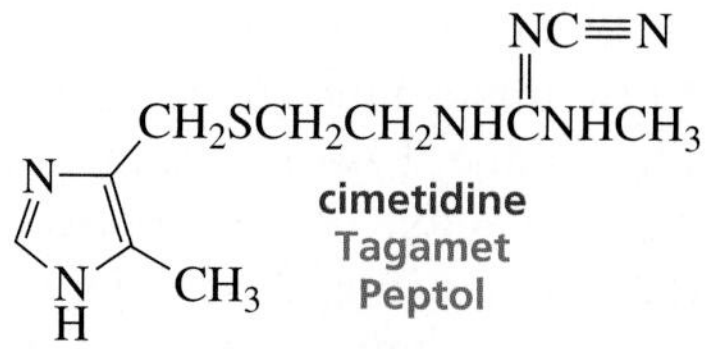

cimetidine
Tagamet
Peptol

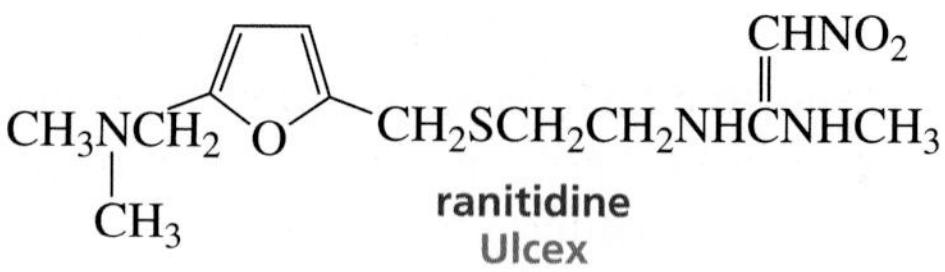

ranitidine
Ulcex
Zantac

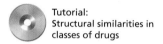
Tutorial:
Structural similarities in classes of drugs

Observations that serotonin was implicated in the generation of migraine attacks led to the development of drugs that bind to serotonin receptors. Sumatriptan, introduced in 1991, relieved not only the pain associated with migraines but also many of migraine's other symptoms—light and sound sensitivity and nausea.

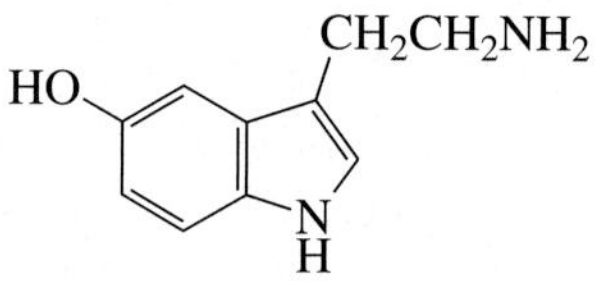

serotonin

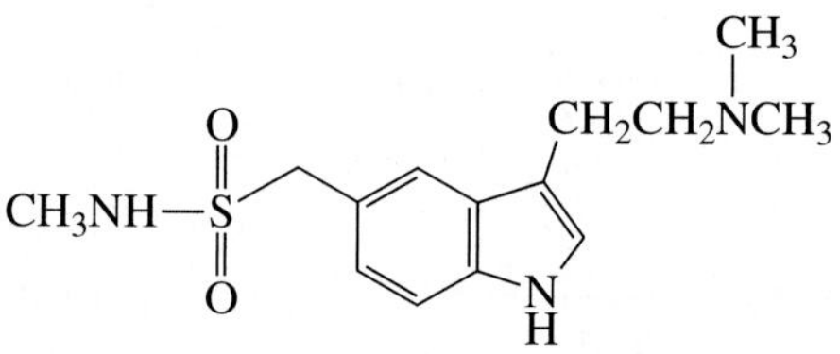

sumatriptan
Imitrex

The success of sumatriptan spurred the search for other antimigraine agents by molecular modification, and three new triptans were introduced in 1997 and 1998. These second generation triptans showed some improvements over sumatriptan—a longer half-life, reduced cardiac side effects, or improved CNS penetration.

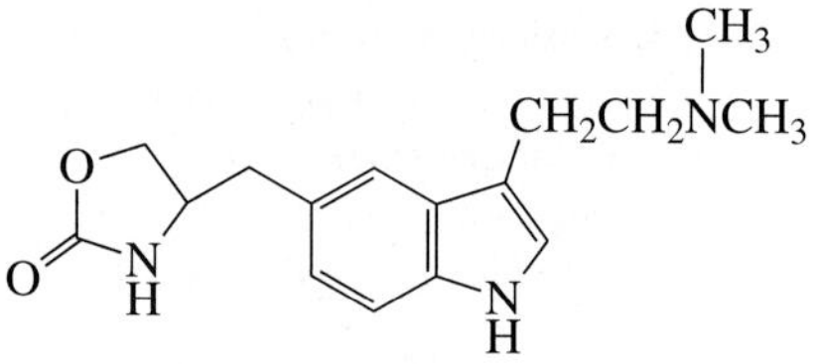

zolmitriptan
Zomig

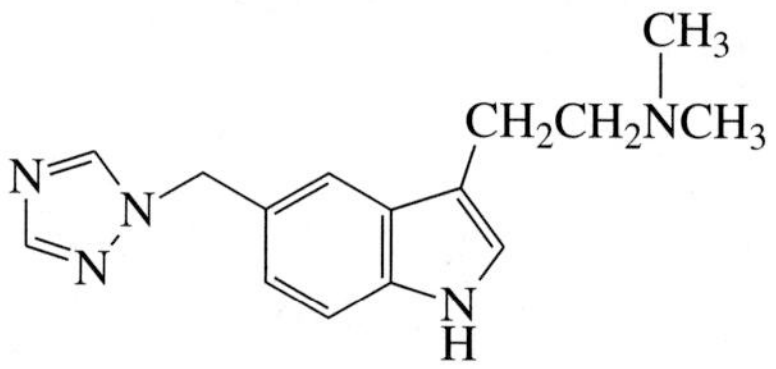

rizatriptan
Maxalt

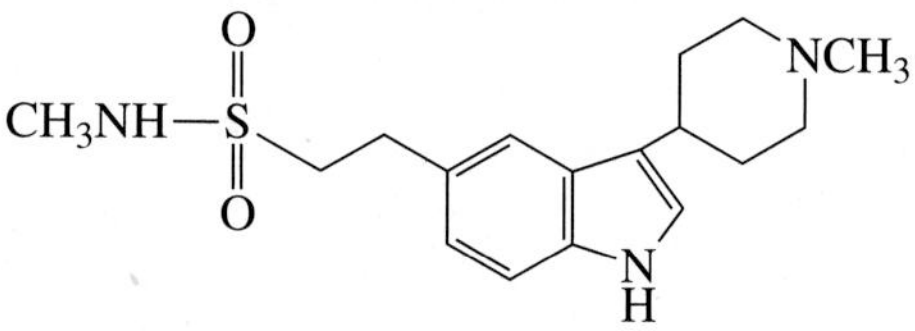

naratriptan
Amerge

It is not unusual when screening modified compounds to find a compound with completely different pharmacological activity than the lead compound. For example, a molecular modification of a sulfonamide, an antibiotic, led to tolbutamide, which is a drug with hypoglycemic activity (Section 23.8).

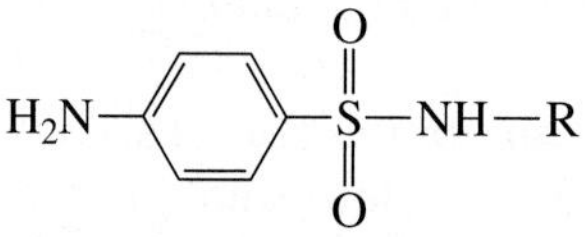

a sulfonamide

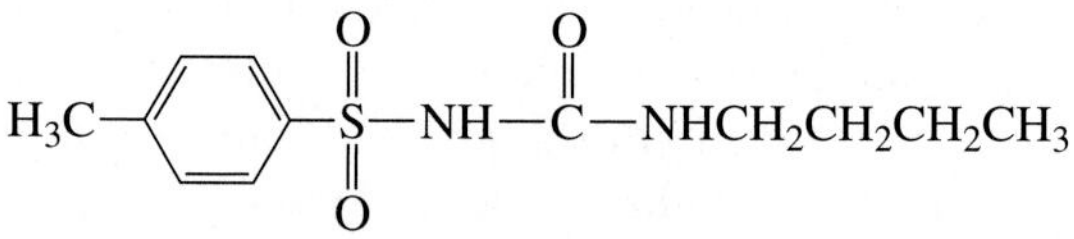

tolbutamide

Molecular modification of promethazine, an antihistamine, led to chlorpromazine. Chlorpromazine did not show any antihistamine activity, but it lowered body temperature. This drug found clinical use in chest surgery, where patients previously had to be cooled down by wrapping them in cold wet sheets. Because wrapping in cold wet sheets had been an old method of calming psychotic patients, a French psychiatrist tried the drug on some of his patients. He found that chlorpromazine was able to suppress psychotic symptoms to the point that his patients assumed almost normal behavioral characteristics. However, they soon developed uncoordinated involuntary movements. After thousands of molecular modifications, thioridazine was found to have the appropriate calming effect with less problematic side effects. It is now in clinical use as an antipsychotic.

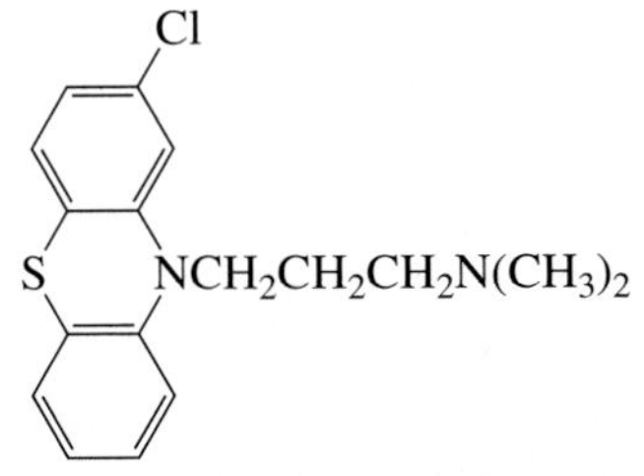

chlorpromazine
Thorazine

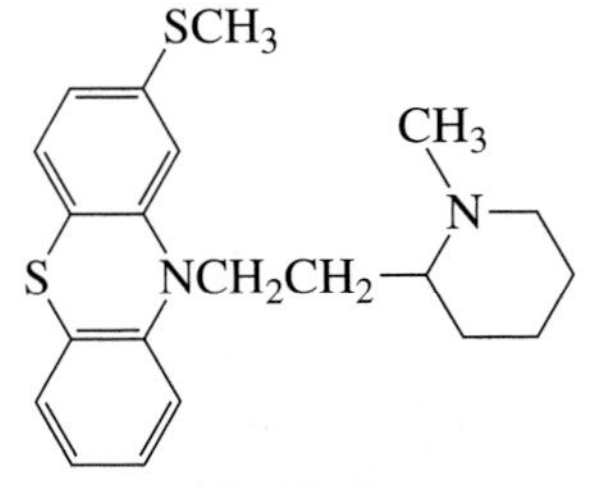

thioridazine
Mellaril

Sometimes a drug initially developed for one purpose is later found to have properties that make it a better drug for a different purpose. Beta-blockers were originally intended to be used to alleviate the pain associated with angina by reducing the amount of work done by the heart. They were later found to have hypotensive (antihypertensive) properties, so now they are used primarily to manage hypertension, a disease quite prevalent in the Western world.

In earlier chapters, we discussed several drugs that act by inhibiting enzymes (Sections 23.8 and 23.9). Penicillin, for example, destroys bacteria by inhibiting the enzyme that synthesizes bacterial cell walls (Section 16.14).

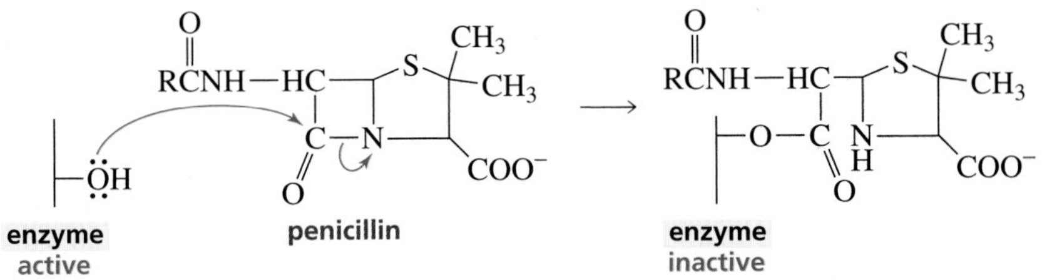

30.7
DRUGS AS ENZYME INHIBITORS

enzyme
active

penicillin

enzyme
inactive

Bacteria develop resistance to penicillin by secreting an enzyme (β-lactamase) that hydrolyzes penicillin before it can interfere with bacterial cell wall synthesis.

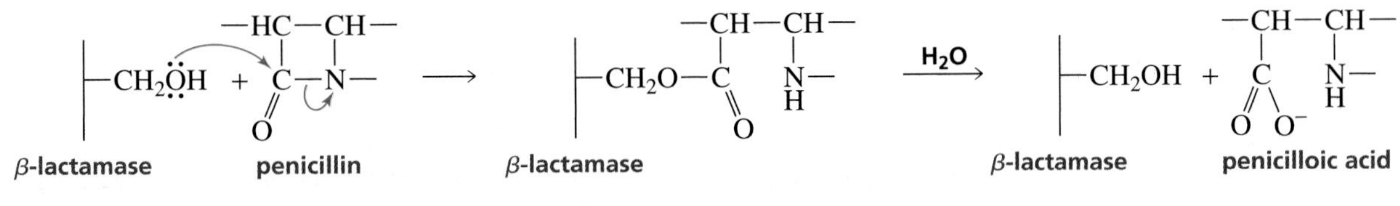

β-**lactamase**   **penicillin** β-**lactamase** β-**lactamase**   **penicilloic acid**

Chemists have developed drugs that inhibit β-lactamase. If such a drug is given to the patient along with penicillin, the antibiotic is not destroyed. This is an example of a drug that has no therapeutic effect itself but acts by protecting a therapeutic drug.

One β-lactamase inhibitor is a sulfone, which is easily prepared from penicillin by oxidizing the sulfur atom with a peroxyacid (Section 18.4).

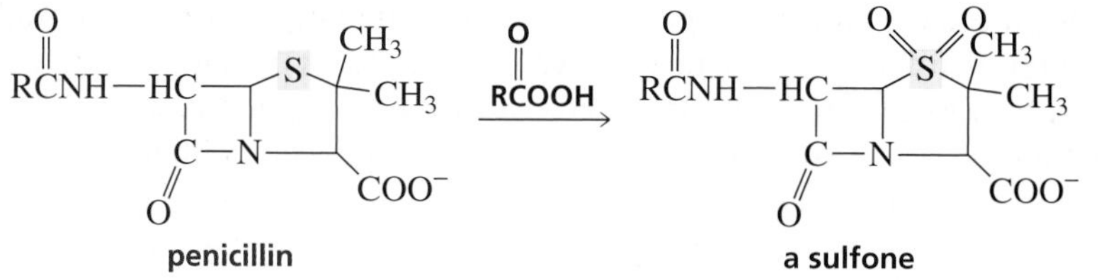

penicillin   a sulfone

Because the sulfone looks like the original antibiotic, β-lactamase accepts it as a substrate, forming an ester as it does with penicillin. If the ester were then hydrolyzed, β-lactamase would be free to react with penicillin. However, the electron-withdrawing sulfone provides an alternative pathway to hydrolysis that forms a stable imine. Because imines are susceptible to nucleophilic attack, an amino group at the active site of β-lactamase reacts with the imine, forming a second covalent bond between the enzyme and the sulfone. This inactivates β-lactamase, thereby wiping out the resistance to penicillin. The sulfone is another example of a mechanism-based **suicide inhibitor** (Section 23.8).

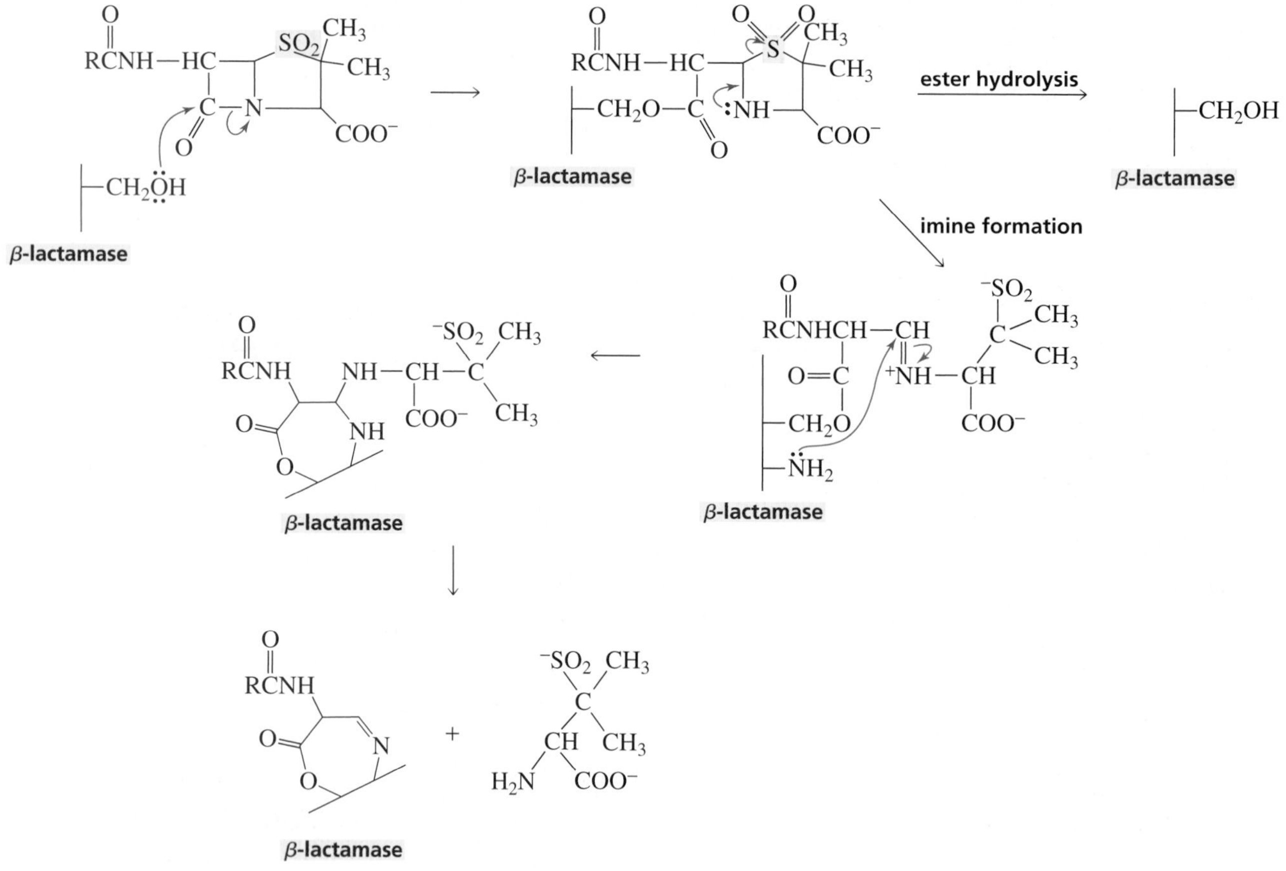

What makes the sulfone such an effective inhibitor is that the reactive imine group does not appear until after the sulfone has been bound to the enzyme that is to be inactivated. In other words, the inhibitor has a specific target. In contrast, if an imine were directly administered to the patient, it would be very nonspecific, reacting with whatever nucleophile it first encountered.

When two drugs are given simultaneously to a patient, their combined effect can be additive, antagonistic, or synergistic. That is, the effect of two drugs used in combination can be equal to, less than, or greater than the sum of the effects obtained when the drugs are administered individually. Administering penicillin and the sulfone in combination results in **drug synergism.** The sulfone inhibits the β-lactamase so penicillin won't be destroyed and can inhibit the enzyme that synthesizes bacterial cell walls.

Another reason to administer two drugs in combination is that if some bacteria are resistant to one of them, the second drug will minimize the chance that the resistant strain will proliferate. For example, two antimicrobial agents, isoniazid and rifampin, are given in combination to treat tuberculosis.

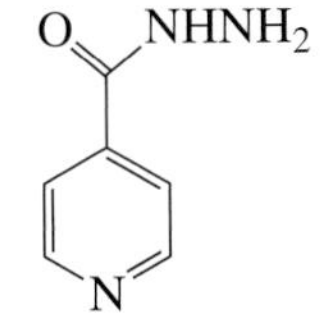

isoniazid
Nydrazid

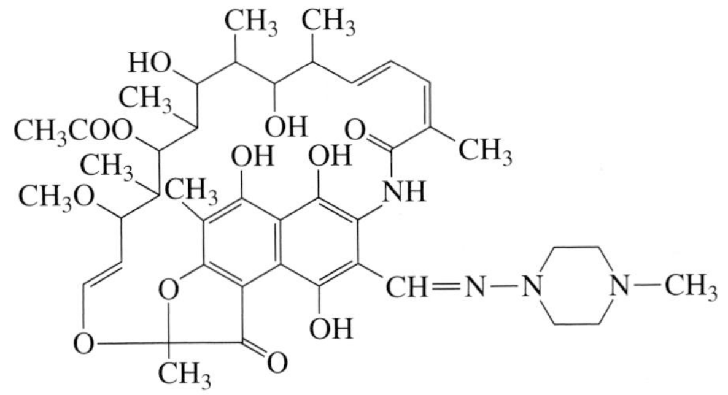

rifampin
Rifadin

Typically it takes a bacterial strain 15 to 20 years to evolve resistance to antibiotics. The fluoroquinolones, the last class of antibiotics to be discovered until very recently, were discovered more than 30 years ago, so **drug resistance** has become an increasingly important problem in medicinal chemistry. More and more bacteria in recent years have become resistant to all antibiotics, even vancomycin—until recently the antibiotic of last resort.

The antibiotic activity of the fluoroquinolones results from their ability to inhibit DNA gyrase, an enzyme required for transcription (Section 25.10). Fortunately, the bacterial and mammalian forms of the enzyme are sufficiently different that the fluoroquinolones inhibit only the bacterial enzyme.

There are many different fluoroquinolones. They all have fluorine substituents, which increase the lipophilicity of the drug to enable it to penetrate into tissues and cells. If either the carboxyl group or the double bond in the 4-pyridinone ring is removed, all activity is lost. By changing the substituents on the piperazine ring, excretion of the drug can be shifted from the liver to the kidney, which is useful to patients with impaired liver function. The substituents on the piperazine ring also affect the half-life of the drug.

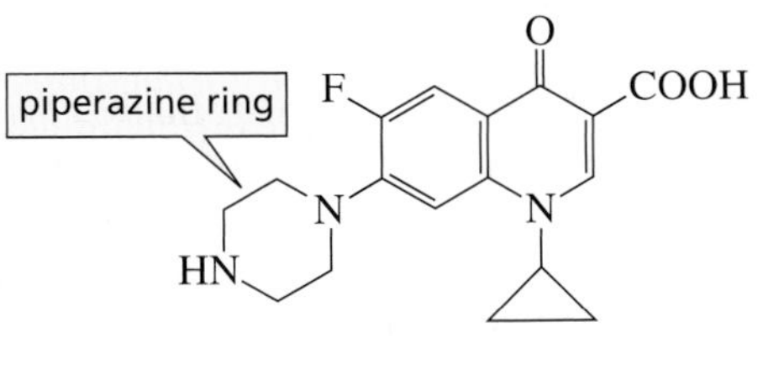

3-D Molecules:
Rifampin;
Isoniazid

ciprofloxacin
Cipro
active against gram-negative bacteria

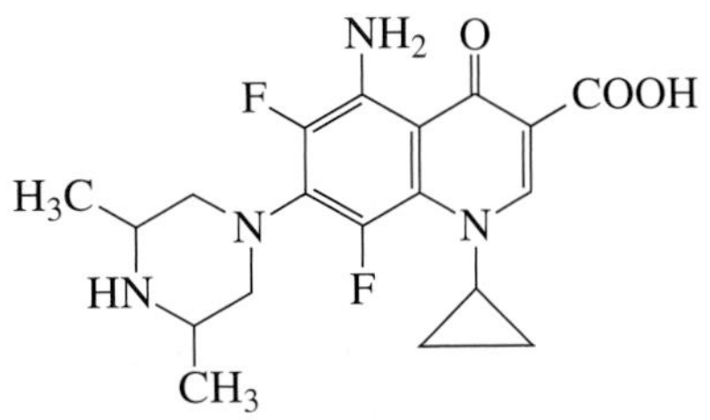

sparfloxacin
Zagam
**active against gram-negative bacteria
and gram-positive bacteria**

The approval of Zyvox by the FDA in April 2000 was met by the medical community with great relief. Zyvox is the first in a new family of antibiotics—the oxazolidinones. In clinical trials Zyvox was found to cure three-fourths of the patients infected with bacteria that had become resistant to all other antibiotics.

linezolid
Zyvox

Zyvox is a synthetic compound. It was designed by scientists to inhibit bacterial growth at a different point than any other antibiotic. Zyvox inhibits the initiation of protein synthesis by preventing the formation of the complex between the first amino-acid-bearing tRNA, mRNA, and the 30S ribosome (Section 25.13). Because of its new mode of activity, resistance is expected to be rare at first and hopefully slow to emerge.

30.8 DESIGNING A SUICIDE SUBSTRATE

It is important for a drug to have a minimum of undesirable side effects. A drug must be administered in sufficient quantity to achieve a therapeutic effect. However, too much of a drug can be lethal. The **therapeutic index** of a drug is the ratio of the lethal dose to the therapeutic dose. The higher the therapeutic index, the greater the margin of safety of the drug.

Penicillin is an effective antibiotic that has a high therapeutic index because it interferes with cell wall synthesis—and bacterial cells have cell walls but human cells do not. What else is characteristic about cell walls that could lead to the design of an antibiotic? We know that enzymes and other proteins are polymers of L-amino acids. Cell walls, however, contain both L-amino acids and D-amino acids. Therefore, if the racemization of naturally occurring L-amino acids to mixtures of L- and D-amino acids could be prevented, D-amino acids would not be available for incorporation into cell walls, and bacterial cell wall synthesis could be stopped.

We have seen that amino acid racemization is catalyzed by an enzyme that requires pyridoxal phosphate as a coenzyme (Section 23.6). What we need, then, is a compound that will inhibit this enzyme. Because the natural substrate for this enzyme is an amino acid, an amino acid analog should be a good potential inhibitor.

The first step in racemization is removal of the α-hydrogen of the amino acid. If the inhibitor has a leaving group on the β-carbon, the electrons left behind when the proton is removed can displace the leaving group instead of being delocalized into the pyridine ring. (Compare the mechanism shown on the following page with that shown for racemization in Section 23.6.) Transimination with the enzyme forms a reactive α,β-unsaturated amino acid that reacts irreversibly with the imine formed by the enzyme and the coenzyme. Because the enzyme is now bound to the coenzyme in an amine linkage rather than in an imine linkage, the enzyme can no longer undergo a transimination reaction with its amino acid substrate. The enzyme has been irreversibly inactivated. This is another example of an inhibitor that does not become chemically active until it is at the active site of the targeted enzyme.

transimination
with E – (CH₂)₄NH₂

$$R— \ = \ ^-O-\overset{\overset{O}{\|}}{\underset{\overset{\|}{O^-}}{P}}-OCH_2—$$

inactivated enzyme

The enormous cost involved in synthesizing and testing thousands of modified compounds in an attempt to find an active drug led scientists to develop a more rational approach to the design of biologically active molecules. They realized that if a physical or chemical property of a series of compounds could be correlated with biological activity, they would know what property of the drug was related to that specific biological activity. Armed with this knowledge, scientists could design compounds that would have a good chance of exhibiting the desired activity. This would be a great improvement over the random approach to molecular modification they had traditionally employed.

The first hint that a physical property of a drug could be related to biological activity was made almost 100 years ago when scientists recognized that chloroform ($CHCl_3$), diethyl ether, cyclopropane, and nitrous oxide (N_2O) were all useful general anesthetics. Clearly, the chemical structures of these diverse compounds could not account for their similar pharmacological effects. Instead, some physical property must explain the similarity of their biological activities.

In the early 1960s, Corwin Hansch postulated that the biological activity of a drug depended on two processes. First, the drug must be able to get from the point where it enters the body to the receptor where it exerts its effect. For example, an anesthetic must be able to cross the aqueous milieu (blood) and penetrate the lipid barrier of nerve cell membranes. Secondly, when a drug reaches its receptor, it must properly interact with it.

Chloroform, diethyl ether, cyclopropane, and nitrous oxide were each put into a mixture of 1-octanol (a nonpolar compound) and water. When the amount of drug dissolving in each of the layers was measured, it turned out that they all had a similar **distribution coefficient** (the ratio of the amount of a compound dissolving in each of the solvents). In other words, the distribution coefficient could be related to biological activity. Compounds with lower distribution coefficients could not penetrate the nonpolar cell membrane. Compounds with greater distribution coefficients (more nonpolar compounds) could not cross the aqueous phase. This meant that the distribution coefficient

30.9 QUANTITATIVE STRUCTURE– ACTIVITY RELATION- SHIPS (QSAR)

Corwin H. Hansch was born in North Dakota in 1918. He received a B.S. from the University of Illinois and a Ph.D. from New York University. He has been a professor of chemistry at Pomona College since 1946.

of a compound could be used to determine whether or not a compound should be tested *in vivo*. This technique of relating a property of a series of compounds to biological activity is known as a **quantitative structure–activity relationship (QSAR).**

Determining the physical property of a drug cannot take the place of *in vivo* testing because how the drug will behave once it reaches a suitable receptor cannot be predicted by the distribution coefficient alone. Nevertheless, QSAR provides a way to shortcut extensive molecular modification.

In the following example, QSAR was useful in determining not only the structure of a potentially active drug but also in determining something about the structure of the receptor site. A series of substituted 2,4-diaminopyrimidines, used as inhibitors of dihydrofolate reductase (Section 23.8), was investigated.

a 2,4-diaminopyrimidine

The potency of the inhibitors could be described by the following equation, where σ and π are substituent parameters:

$$\text{potency} = 0.80\pi - 7.34\sigma - 8.14$$

The σ parameter is a measure of the electron-donating or electron-withdrawing ability of the substituents R and R'. The negative coefficient of σ indicates that potency is increased by electron donation (Chapter 16, Problem 69). The fact that increasing the basicity of the drug increases its potency suggests that the protonated drug is more active than the nonprotonated drug.

The π parameter is a measure of the hydrophobicity of the substituents. Potency was found to be better related to π when the π value for the more hydrophobic of the two substituents was used rather than the sum of the π values for both substituents. This suggests that the receptor has a hydrophobic pocket that can accommodate one but not both of the substituents.

In a search for a new analgesic, the potency of the drug could be described by the following equation, where *HA* indicates whether R is a hydrogen bond acceptor and *B* is a steric factor. Analysis indicated that the vinyl substituted compound should be prepared.

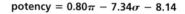

$$\text{potency} = -4.45 - 0.73HA + 6.5B - 1.55(B)^2$$

In addition to solubility and substituent parameters, some of the properties that have been correlated with biological activity are oxidation–reduction potentials, molecular size, interatomic distances between functional groups, degree of ionization, and configuration.

30.10 MOLECULAR MODELING

Because the shape of a molecule determines whether it will be recognized by a receptor and, therefore, whether it will exhibit biological activity, compounds with similar biological activity often have similar structures. Because computers can draw molecular models of compounds on a video display and move them around to assume different conformations, the use of computer **molecular modeling** allows more rational drug design. There are computer programs that allow chemists to scan existing collections of thousands of compounds to find those with appropriate structural and conformational properties.

Any compound that shows promise can be drawn on a computer display along with the three-dimensional image of a receptor site. For example, the binding of distamycin A, a compound with both antibiotic and antiviral activity, to the minor groove of DNA is shown in Figure 30.1.

The fit between the compound and the receptor may suggest modifications that can be made to the compound to result in more favorable binding. In this way, the selection of compounds to be synthesized for the purpose of screening for biological activity will be more rational and will allow pharmacologically active compounds to be more rapidly discovered. This technique will become more valuable as scientists learn more about receptor sites.

The need for large collections of compounds that can be screened for biological activity in the constant search for new drugs has led organic chemists to a synthetic strategy that employs the concept of mass production. This strategy, called **combinatorial organic synthesis,** involves the synthesis of a large group of related compounds (known as a library) by covalently connecting sets of building blocks of various structures. For example, if a compound can be synthesized by connecting three different building blocks, and if each set of building blocks contains 10 interchangeable compounds, 1000 ($10 \times 10 \times 10$) different compounds can be prepared. This approach clearly mimics nature, which uses amino acids and nucleic acids as building blocks to synthesize an enormous number of different proteins and nucleic acids.

The first requirement in combinatorial synthesis is the availability of an assortment of reactive small molecules to be used as building blocks. Because of the ready availability of amino acids, the use of combinatorial synthesis made its first appearance in the creation of peptide libraries. Peptides, however, have limited use as therapeutic agents because they generally cannot be taken orally and are rapidly metabolized. Currently, organic chemists are attempting to create libraries of small organic molecules.

An example of the approach used in combinatorial synthesis is the creation of a library of benzodiazepines. These compounds can be thought of as originating from three different sets of building blocks: a substituted 2-aminobenzophenone, an amino acid, and an alkylating agent.

30.11 COMBINATORIAL ORGANIC SYNTHESIS

▲ **Figure 30.1**
The antibiotic distamycin A (purple atoms) bound in the minor groove of DNA.

The 2-aminobenzophenone is attached to a solid support in a manner that allows it to be readily removed by acid hydrolysis (Figure 30.2). An *N*-protected amino acyl fluoride is then added. After the amide is formed, the protecting group is removed and the seven-membered ring is created as a result of imine formation. A base is

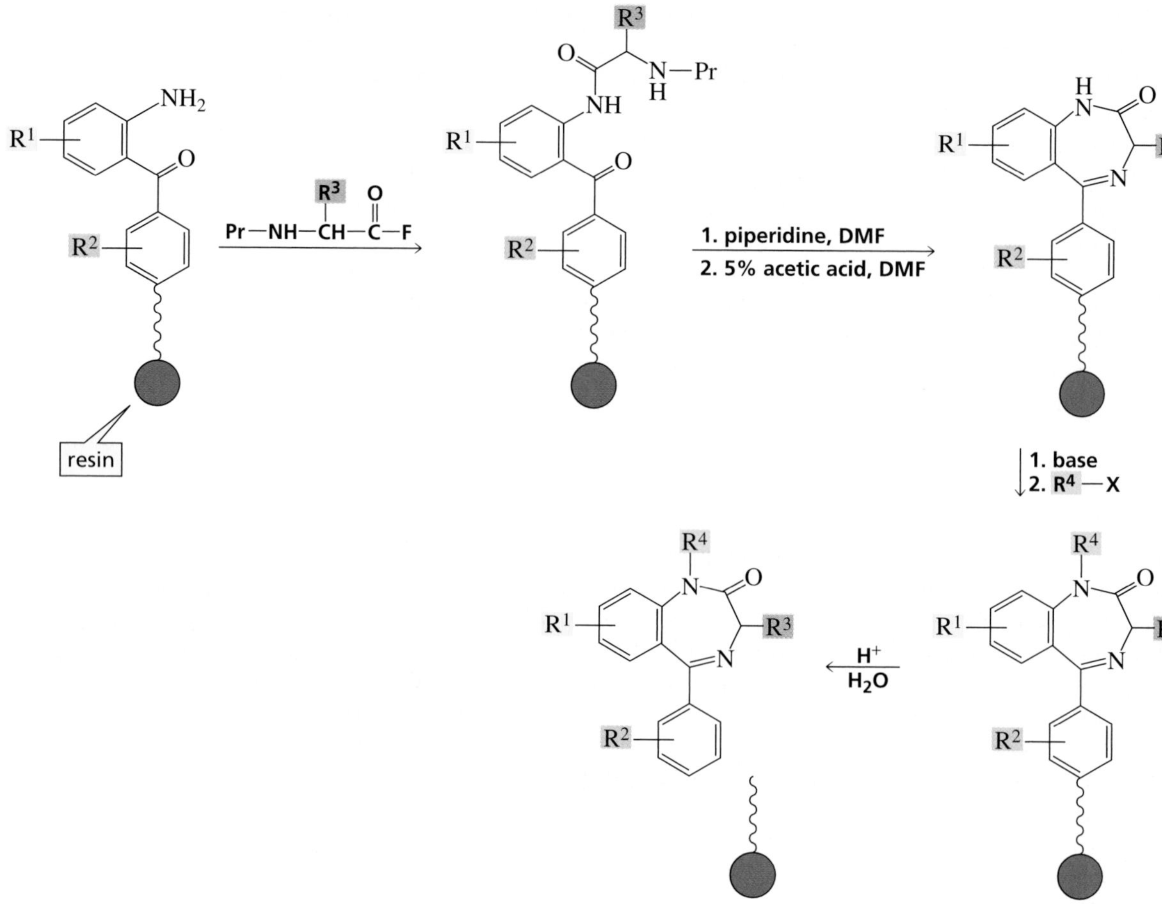

▲ **Figure 30.2**
Combinatorial organic synthesis of benzodiazepines.

added to remove the amide hydrogen, forming a nucleophile that reacts with the added alkylating agent. The final product is then removed from the solid support.

The solid support containing the 2-aminobenzophenone can be divided into several portions, and a different amino acid can then be added to each portion. Each ring-closed product can also be divided into several portions, and a different alkylating agent can be added to each portion. In this way, many different benzodiazepines can be prepared.

Tutorial:
Combinatorial synthesis

30.12
ANTIVIRAL DRUGS

Relatively few clinically useful drugs have been developed for viral infections. This slow progress is due to the nature of viruses and the way they replicate. Viruses are smaller than bacteria. They consist of nucleic acid (either DNA or RNA) surrounded by a coat of protein. A virus penetrates a host cell or merely injects its nucleic acid. In either case, its nucleic acid is transcribed and is integrated into the host genome. Viruses have no metabolism of their own; they rely on the enzymes of the host cell to complete their life cycle.

Most **antiviral drugs** are analogs of nucleosides. They interfere with DNA or RNA synthesis and in this way prevent the virus from replicating.

cytarabine
Cytosar
used against rabies

vidarabine
Vira-A
used against encephalitis and epithelial keratitis

idoxuridine
Herplex
approved for topical ophthalmic use

acyclovir
Aclovir
used against shingles (herpes zoster), chickenpox, and herpes simplex infections

ribavirin
Viramid
a broad-spectrum antiviral agent

By some estimates, the average cost of launching a new drug is about $230 million. This cost has to be amortized quickly by the manufacturer because the starting date of a patent is the date the drug is first discovered. A patent is good for 20 years from the date of the patent application but, because it takes an average of 12 years to market a drug after its initial discovery, the patent protects the discoverer of the drug for an average of only 8 years. In addition, the average lifetime of a drug is only 15 to 20 years. After that time it is generally replaced by a newer and improved drug.

Why does it cost so much to develop a new drug? First of all, the Food and Drug Administration (FDA) has very high standards that must be met before it approves a drug for a particular use. Before the U.S. government became involved with the regulation of drugs, it was not uncommon for charlatans to dispense useless and even harmful medical preparations. Starting in 1906, Congress passed laws governing the manufacture, distribution, and use of drugs. These laws are amended from time to time to reflect changing situations. The present law requires that all new drugs used by physicians must first be thoroughly tested for effectiveness and safety.

An important factor leading to the high price of many drugs is the low success rate in progressing from the initial concept for a drug to an approved product: Only one or two of every hundred compounds tested become lead compounds; out of a hundred structural modifications of a lead compound, only one is worthy of further study; and only 10% of these compounds actually become marketable drugs.

30.13 ECONOMICS OF DRUGS. GOVERNMENTAL REGULATIONS

ORPHAN DRUGS

Because of the high cost associated with developing a drug, drug companies are reluctant to carry out research on drugs for rare diseases. Even if a company were to find a drug, there would be no way to recoup its expenditure because of the limited demand. Drugs that no one wants to develop are called **orphan drugs.** In 1983, the U.S. Congress passed the Orphan Drugs Act, which creates public subsidies to fund research and tax credits for up to 50% of the costs of developing and marketing drugs for diseases or conditions that affect fewer than 200,000 people. In addition, the company that develops the drug has four years of exclusive marketing rights if the drug is nonpatentable. In the 10 years prior to passage of this act, fewer than 10 drugs were developed to treat orphan diseases. Today, they are more than 100, and some 600 others are now in development.

Drugs developed as orphan drugs include AZT (to treat AIDS), Taxol (to treat ovarian cancer), Exosurf Neonatal (to treat respiratory distress syndrome in infants), and Opticrom (to treat corneal swelling).

KEY TERMS

antiviral drug (page 1244)
bactericidal drug (page 1233)
bacteriostatic drug (page 1233)
blind screen (page 1232)
brand name (page 1228)
combinatorial organic
    synthesis (page 1243)
distribution coefficient (page 1241)
drug (page 1226)

drug resistance (page 1239)
drug synergism (page 1239)
generic name (page 1228)
lead compound (page 1230)
molecular modeling (page 1243)
molecular modification (page 1230)
orphan drug (page 1245)
proprietary name (page 1228)

quantitative structure–activity
    relationship (QSAR) (page 1242)
random screen (page 1232)
receptor (page 1234)
suicide inhibitor (page 1238)
therapeutic index (page 1240)
trademark (page 1228)
trade name (page 1228)

PROBLEMS

1. What is the chemical name of each of the following drugs?
   a. benzocaine   **b.** procaine

2. Based on the lead compound for the development of procaine and lidocaine, propose structures for other compounds that you would like to see tested for potential use as anesthetics.

3. Which of the following compounds is more likely to exhibit activity as a tranquilizer?

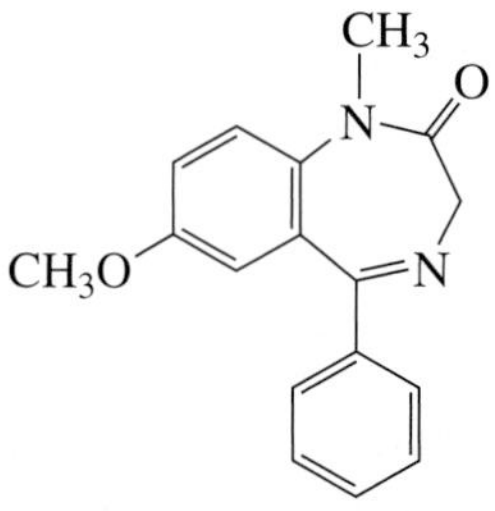

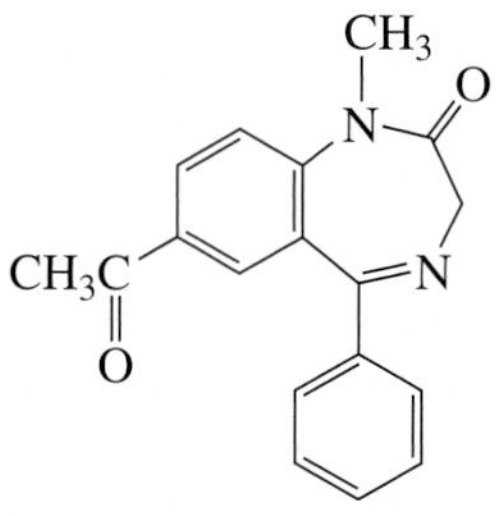

4. Which compound is more likely to be a general anesthetic?

$$CH_3CH_2CH_2OH \quad\quad \text{or} \quad\quad CH_3OCH_2CH_3$$

5. What accounts for the ease of imine formation between β-lactamase and the sulfone antibiotic that counteracts penicillin resistance?

6. For each of the following pairs of compounds, indicate the compound that you would expect to be a more potent inhibitor of dihydrofolate reductase:

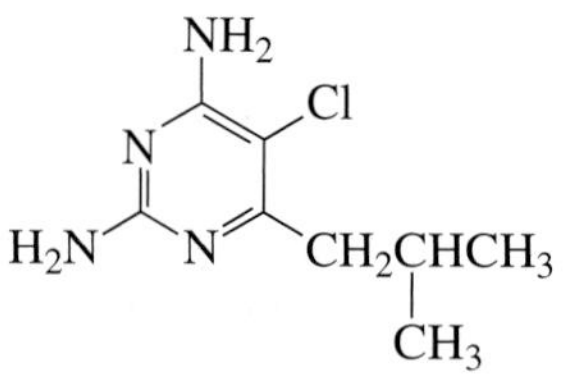

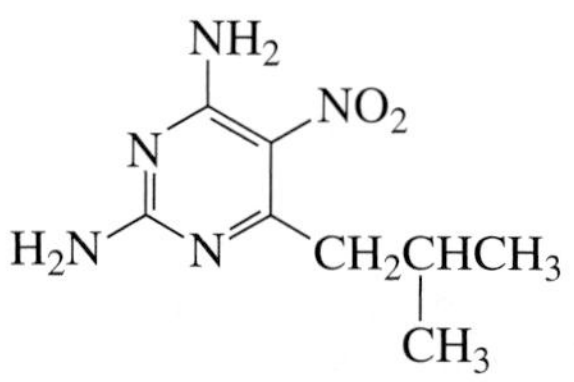

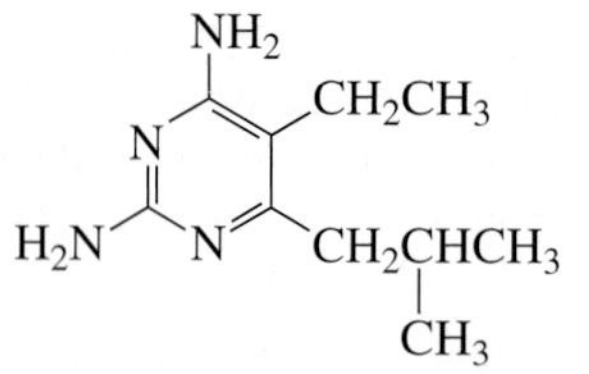

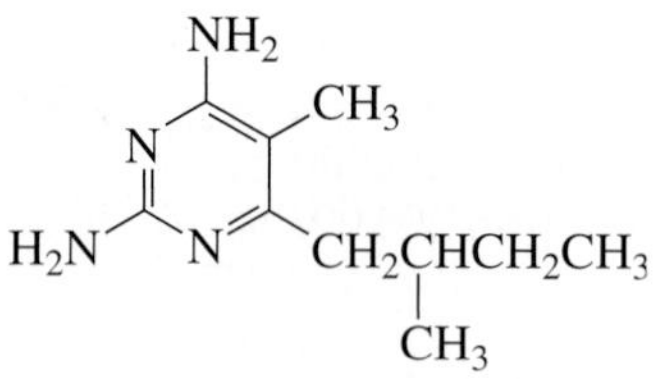

7. The lethal dose of tetrahydrocannabinol in mice is 2.0 g/kg, and the therapeutic dose is 20 mg/kg. The lethal dose of sodium pentothal in mice is 100 mg/kg, and the therapeutic dose is 30 mg/kg. Which is a safer drug?

8. The following compound is a suicide inhibitor of the enzyme that catalyzes amino acid racemization. Propose a mechanism that explains how it irreversibly inactivates the enzyme.

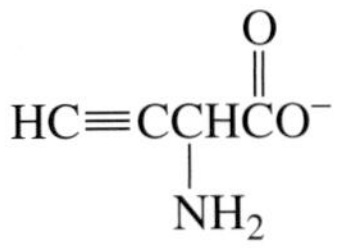

9. Explain how each of the antiviral drugs shown in Section 30.12 differs from the naturally occurring nucleoside that it most closely resembles.

10. Show a mechanism for the formation of a benzodiazepine 4-oxide from the reaction of a quinazoline 3-oxide with methylamine.

11. Show how Valium could be synthesized from benzoyl chloride, *para*-chloroaniline, methyl iodide, and the ethyl ester of glycine.

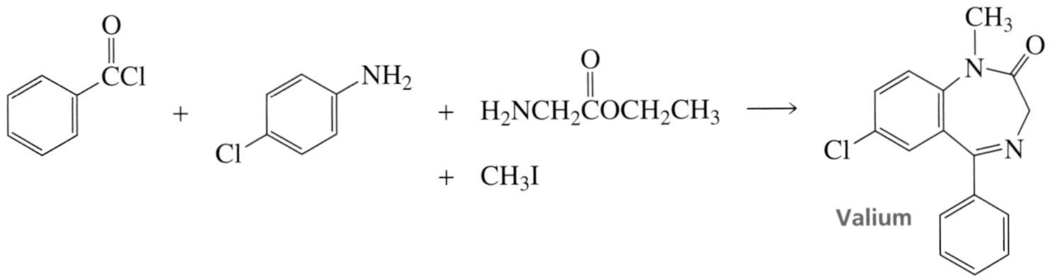

Physical Properties of Organic Compounds

Physical Properties of Alkenes

Name	Structure	mp (°C)	bp (°C)	Density (g/mL)
Ethene	$CH_2\!=\!CH_2$	−169	−104	
Propene	$CH_2\!=\!CHCH_3$	−185	−47	
1-Butene	$CH_2\!=\!CHCH_2CH_3$	−185	−6.3	
1-Pentene	$CH_2\!=\!CH(CH_2)_2CH_3$		30	0.641
1-Hexene	$CH_2\!=\!CH(CH_2)_3CH_3$	−138	64	0.673
1-Heptene	$CH_2\!=\!CH(CH_2)_4CH_3$	−119	94	0.697
1-Octene	$CH_2\!=\!CH(CH_2)_5CH_3$	−101	122	0.715
1-Nonene	$CH_2\!=\!CH(CH_2)_6CH_3$	−81	146	0.730
1-Decene	$CH_2\!=\!CH(CH_2)_7CH_3$	−66	171	0.741
cis-2-Butene	cis-$CH_3CH\!=\!CHCH_3$	−180	37	0.650
trans-2-Butene	trans-$CH_3CH\!=\!CHCH_3$	−140	37	0.649
Methylpropene	$CH_2\!=\!C(CH_3)_2$	−140	−6.9	0.594
cis-2-Pentene	cis-$CH_3CH\!=\!CHCH_2CH_3$	−180	37	0.650
trans-2-Pentene	trans-$CH_3CH\!=\!CHCH_2CH_3$	−140	37	0.649
Cyclohexene		−104	83	0.811

Physical Properties of Alkynes

Name	Structure	mp (°C)	bp (°C)	Density (g/mL)
Ethyne	$HC\!\equiv\!CH$	−82	−84.0	
Propyne	$HC\!\equiv\!CCH_3$	−101.5	−23.2	
1-Butyne	$HC\!\equiv\!CCH_2CH_3$	−122	8.1	
2-Butyne	$CH_3C\!\equiv\!CCH_3$	−24	27	0.694
1-Pentyne	$HC\!\equiv\!C(CH_2)_2CH_3$	−98	39.3	0.695
2-Pentyne	$CH_3C\!\equiv\!CCH_2CH_3$	−101	55.5	0.714
3-Methyl-1-butyne	$HC\!\equiv\!CCH(CH_3)_2$		29	0.665
1-Hexyne	$HC\!\equiv\!C(CH_2)_3CH_3$	−132	71	0.715
2-Hexyne	$CH_3C\!\equiv\!C(CH_2)_2CH_3$	−92	84	0.731
3-Hexyne	$CH_3CH_2C\!\equiv\!CCH_2CH_3$	−101	81	0.725
1-Heptyne	$HC\!\equiv\!C(CH_2)_4CH_3$	−81	100	0.733
1-Octyne	$HC\!\equiv\!C(CH_2)_5CH_3$	−80	127	0.747
1-Nonyne	$HC\!\equiv\!C(CH_2)_6CH_3$	−50	151	0.757
1-Decyne	$HC\!\equiv\!C(CH_2)_7CH_3$	−44	174	0.766

Physical Properties of Cyclic Saturated Alkanes

Name	mp (°C)	bp (°C)	Density (g/mL)
Cyclopropane	−128	−33	
Cyclobutane	−80	−12	
Cyclopentane	−94	50	0.751
Cyclohexane	6.5	81	0.779
Cycloheptane	−12	118	0.811
Cyclooctane	14	149	0.834
Methylcyclopentane	−142	72	0.749
Methylcyclohexane	−126	100	0.769
cis-1,2-Dimethylcyclopentane	−62	99	0.772
trans-1,2-Dimethylcyclopentane	−120	92	0.750

Physical Properties of Ethers

Name	Structure	mp (°C)	bp (°C)	Density (g/mL)
Dimethyl ether	CH_3OCH_3	−141	−24.8	
Diethyl ether	$CH_3CH_2OCH_2CH_3$	−116	34.6	0.706
Dipropyl ether	$CH_3(CH_2)_2O(CH_2)_2CH_3$	−123	88	0.736
Diisopropyl ether	$(CH_3)_2CHOCH(CH_3)_2$	−86	69	0.725
Dibutyl ether	$CH_3(CH_2)_3O(CH_2)_3CH_3$	−98	142	0.764
Divinyl ether	$CH_2{=}CHOCH{=}CH_2$		35	
Diallyl ether	$CH_2{=}CHCH_2OCH_2CH{=}CH_2$		94	0.830
Tetrahydrofuran	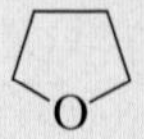	−108	66	0.889
Dioxane	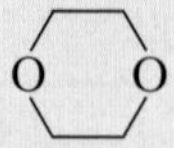	12	101	1.034

Physical Properties of Alcohols

Name	Structure	mp (°C)	bp (°C)	Solubility (g/100 g H_2O at 25 °C)
Methanol	CH_3OH	−97.8	64	∞
Ethanol	CH_3CH_2OH	−114.7	78	∞
1-Propanol	$CH_3(CH_2)_2OH$	−127	97.4	∞
1-Butanol	$CH_3(CH_2)_3OH$	−90	118	7.9
1-Pentanol	$CH_3(CH_2)_4OH$	−78	138	2.3
1-Hexanol	$CH_3(CH_2)_5OH$	−52	157	0.6
1-Heptanol	$CH_3(CH_2)_6OH$	−36	176	0.2
1-Octanol	$CH_3(CH_2)_7OH$	−15	196	0.05
2-Propanol	$CH_3CHOHCH_3$	−89.5	82	∞
2-Butanol	$CH_3CHOHCH_2CH_3$	−115	99.5	12.5
2-Methyl-1-propanol	$(CH_3)_2CHCH_2OH$	−108	108	10.0
2-Methyl-2-propanol	$(CH_3)_3COH$	25.5	83	∞
3-Methyl-1-butanol	$(CH_3)_2CH(CH_2)_2OH$	−117	130	2
2-Methyl-2-butanol	$(CH_3)_2COHCH_2CH_3$	−12	102	12.5
2,2-Dimethyl-1-propanol	$(CH_3)_3CCH_2OH$	55	114	∞
Allyl alcohol	$CH_2{=}CHCH_2OH$	−129	97	∞
Cyclopentanol	C_5H_9OH	−19	140	s. sol.
Cyclohexanol	$C_6H_{11}OH$	24	161	s. sol.
Benzyl alcohol	$C_6H_5CH_2OH$	−15	205	4

Physical Properties of Alkyl Halides

Name	bp (°C)			
	Fluoride	*Chloride*	*Bromide*	*Iodide*
Methyl	−78.4	−24.2	3.6	42.4
Ethyl	−37.7	12.3	38.4	72.3
Propyl	−2.5	46.6	71.0	102.5
Isopropyl	−9.4	34.8	59.4	89.5
Butyl	32.5	78.4	100	130.5
Isobutyl		68.8	90	120
sec-Butyl		68.3	91.2	120.0
tert-Butyl		50.2	73.1	dec.
Pentyl	62.8	108	130	157.0
Hexyl	92	133	154	179

Physical Properties of Amines

Name	Structure	mp (°C)	bp (°C)	Solubility (g/100 g H$_2$O at 25 °C)
Primary Amines				
Methylamine	CH_3NH_2	−93	−6.3	v. sol.
Ethylamine	$CH_3CH_2NH_2$	−81	17	∞
Propylamine	$CH_3(CH_2)_2NH_2$	−83	48	∞
Isopropylamine	$(CH_3)_2CHNH_2$	−95	33	∞
Butylamine	$CH_3(CH_2)_3NH_2$	−49	78	v. sol.
Isobutylamine	$(CH_3)_2CHCH_2NH_2$	−85	68	∞
sec-Butylamine	$CH_3CH_2CH(CH_3)NH_2$	−72	63	∞
tert-Butylamine	$(CH_3)_3CNH_2$	−67	46	∞
Cyclohexylamine	$C_6H_{11}NH_2$	−18	134	s. sol.
Secondary Amines				
Dimethylamine	$(CH_3)_2NH$	−93	7.4	v. sol.
Diethylamine	$(CH_3CH_2)_2NH$	−50	55	10.0
Dipropylamine	$(CH_3CH_2CH_2)_2NH$	−63	110	10.0
Dibutylamine	$(CH_3CH_2CH_2CH_2)_2NH$	−62	159	s. sol.
Tertiary Amines				
Trimethylamine	$(CH_3)_3N$	−115	2.9	91
Triethylamine	$(CH_3CH_2)_3N$	−114	89	14
Tripropylamine	$(CH_3CH_2CH_2)_3N$	−93	157	s. sol.

Physical Properties of Benzene and Substituted Benzenes

Name	Structure	mp (°C)	bp (°C)	Solubility (g/100 g H$_2$O at 25 °C)
Aniline	$C_6H_5NH_2$	−6	184	3.7
Benzene	C_6H_6	5.5	80.1	s. sol.
Benzaldehyde	C_6H_5CHO	−26	178	s. sol.
Benzamide	$C_6H_5CONH_2$	132	290	s. sol.
Benzoic acid	C_6H_5COOH	122	249	0.34
Bromobenzene	C_6H_5Br	−30.8	156	insol.
Chlorobenzene	C_6H_5Cl	−45.6	132	insol.
Nitrobenzene	$C_6H_5NO_2$	5.7	210.8	s. sol.
Phenol	C_6H_5OH	43	182	s. sol.
Styrene	$C_6H_5CH=CH_2$	−30.6	145.2	insol.
Toluene	$C_6H_5CH_3$	−95	110.6	insol.

Physical Properties of Carboxylic Acids

Name	Structure	mp (°C)	bp (°C)	Solubility (g/100 g H$_2$O at 25 °C)
Formic acid	HCOOH	8.4	101	∞
Acetic acid	CH$_3$COOH	16.6	118	∞
Propionic acid	CH$_3$CH$_2$COOH	−21	141	∞
Butanoic acid	CH$_3$(CH$_2$)$_2$COOH	−5	162	∞
Pentanoic acid	CH$_3$(CH$_2$)$_3$COOH	−34	186	4.97
Hexanoic acid	CH$_3$(CH$_2$)$_4$COOH	−4	202	0.97
Heptanoic acid	CH$_3$(CH$_2$)$_5$COOH	−8	223	0.24
Octanoic acid	CH$_3$(CH$_2$)$_6$COOH	17	237	0.068
Nonanoic acid	CH$_3$(CH$_2$)$_7$COOH	15	255	0.026
Decanoic acid	CH$_3$(CH$_2$)$_8$COOH	32	270	0.015

Physical Properties of Dicarboxylic Acids

Name	Structure	mp (°C)	Solubility (g/100 g H$_2$O at 25 °C)
Oxalic acid	HOOCCOOH	189	s
Malonic acid	HOOCCH$_2$COOH	136	v. sol.
Succinic acid	HOOC(CH$_2$)$_2$COOH	185	s. sol.
Glutaric acid	HOOC(CH$_2$)$_3$COOH	98	v. sol.
Adipic acid	HOOC(CH$_2$)$_4$COOH	151	s. sol.
Pimelic acid	HOOC(CH$_2$)$_5$COOH	106	s. sol.
Phthalic acid	1,2-C$_6$H$_4$(COOH)$_2$	231	s. sol.
Maleic acid	*cis*-HOOCCH=CHCOOH	130.5	v. sol.
Fumaric acid	*trans*-HOOCCH=CHCOOH	302	s. sol.

Physical Properties of Acyl Chlorides and Acid Anhydrides

Name	Structure	mp (°C)	bp (°C)
Acetyl chloride	CH$_3$COCl	−112	51
Propionyl chloride	CH$_3$CH$_2$COCl	−94	80
Butyryl chloride	CH$_3$(CH$_2$)$_2$COCl	−89	102
Valeryl chloride	CH$_3$(CH$_2$)$_3$COCl	−110	128
Acetic anhydride	CH$_3$(CO)O(CO)CH$_3$	−73	140
Succinic anhydride	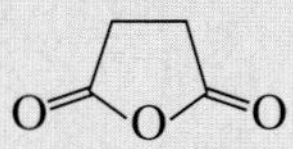		120

Physical Properties of Esters

Name	Structure	mp (°C)	bp (°C)
Methyl formate	$HCOOCH_3$	−100	32
Ethyl formate	$HCOOCH_2CH_3$	−80	54
Methyl acetate	CH_3COOCH_3	−98	57.5
Ethyl acetate	$CH_3COOCH_2CH_3$	−84	77
Propyl acetate	$CH_3COO(CH_2)_2CH_3$	−92	102
Methyl propionate	$CH_3CH_2COOCH_3$	−87.5	80
Ethyl propionate	$CH_3CH_2COOCH_2CH_3$	−74	99
Methyl butyrate	$CH_3CH_2CH_2COOCH_3$	−84.8	102.3
Ethyl butyrate	$CH_3CH_2CH_2COOCH_2CH_3$	−93	121

Physical Properties of Amides

Name	Structure	mp (°C)	bp (°C)
Formamide	$HCONH_2$	3	200 d*
Acetamide	CH_3CONH_2	82	221
Propanamide	$CH_3CH_2CONH_2$	80	213
Butanamide	$CH_3(CH_2)_2CONH_2$	116	216
Pentanamide	$CH_3(CH_2)_3CONH_2$	106	232

*d means the substance decomposes.

Physical Properties of Aldehydes

Name	Structure	mp (°C)	bp (°C)	Solubility (g/100 g H_2O at 25 °C)
Formaldehyde	$HCHO$	−92	−21	v. sol.
Acetaldehyde	CH_3CHO	−121	21	∞
Propionaldehyde	CH_3CH_2CHO	−81	49	16
Butyraldehyde	$CH_3(CH_2)_2CHO$	−96	75	7
Pentanal	$CH_3(CH_2)_3CHO$	−92	103	s. sol.
Hexanal	$CH_3(CH_2)_4CHO$	−56	131	s. sol.
Heptanal	$CH_3(CH_2)_5CHO$	−43	153	0.1
Octanal	$CH_3(CH_2)_6CHO$		171	insol.
Nonanal	$CH_3(CH_2)_7CHO$		192	insol.
Decanal	$CH_3(CH_2)_8CHO$	−5	209	insol.
Benzaldehyde	C_6H_5CHO	−26	178	0.3

Physical Properties of Ketones

Name	Structure	mp (°C)	bp (°C)	Solubility (g/100 g H_2O at 25 °C)
Acetone	CH_3COCH_3	−95	56	∞
2-Butanone	$CH_3COCH_2CH_3$	−86	80	25.6
2-Pentanone	$CH_3CO(CH_2)_2CH_3$	−78	102	5.5
2-Hexanone	$CH_3CO(CH_2)_3CH_3$	−57	127	1.6
2-Heptanone	$CH_3CO(CH_2)_4CH_3$	−36	151	0.4
2-Octanone	$CH_3CO(CH_2)_5CH_3$	−16	173	insol.
2-Nonanone	$CH_3CO(CH_2)_6CH_3$	−7	195	insol.
2-Decanone	$CH_3CO(CH_2)_7CH_3$	14	210	insol.
3-Pentanone	$CH_3CH_2COCH_2CH_3$	−40	102	4.8
3-Hexanone	$CH_3CH_2CO(CH_2)_2CH_3$		123	1.5
3-Heptanone	$CH_3CH_2CO(CH_2)_3CH_3$	−39	149	0.3
Acetophenone	$CH_3COC_6H_5$	19	202	insol.
Propiophenone	$CH_3CH_2COC_6H_5$	18	218	insol.

pKa Values

Compound	pKa	Compound	pKa	Compound	pKa
CH₃C≡N⁺H	−10.1	O₂N—⟨C₆H₄⟩—N⁺H₃	1.0	CH₃—⟨C₆H₄⟩—COH (O)	4.3
HI	−10	pyrimidine (N⁺H)	1.0	CH₃O—⟨C₆H₄⟩—COH (O)	4.5
HBr	−9	Cl₂CHCOH (O)	1.3	⟨C₆H₅⟩—N⁺H₃	4.6
CH₃CH (⁺OH)	−8	HSO₄⁻	2.0	CH₃COH (O)	4.8
CH₃CCH₃ (⁺OH)	−7.3	H₃PO₄	2.1	quinoline (N⁺H)	4.9
HCl	−7	purine (HN⁺)	2.5	CH₃—⟨C₆H₄⟩—N⁺H₃	5.1
CH₃S⁺H (H)	−6.8	FCH₂COH (O)	2.7	pyridine (N⁺H)	5.2
CH₃COCH₃ (⁺OH)	−6.5	ClCH₂COH (O)	2.8	CH₃O—⟨C₆H₄⟩—N⁺H₃	5.3
CH₃COH (⁺OH)	−6.1	BrCH₂COH (O)	2.9	CH₃C=N⁺HCH₃ (CH₃)	5.5
H₂SO₄	−5	ICH₂COH (O)	3.2	CH₃CCH₂CH (O O)	5.9
pyrrole (N⁺H)	−3.8	HF	3.2	HON⁺H₃	6.0
CH₃CH₂OCH₂CH₃ (H⁺)	−3.6	HNO₂	3.4	H₂CO₃	6.4
CH₃CH₂O⁺H (H)	−2.4	O₂N—⟨C₆H₄⟩—COH (O)	3.4	imidazole (HN—N⁺H)	6.8
CH₃O⁺H (H)	−2.5	HCOH (O)	3.8	H₂S	7.0
H₃O⁺	−1.7	Br—⟨C₆H₄⟩—N⁺H₃	3.9	O₂N—⟨C₆H₄⟩—OH	7.1
HNO₃	−1.3	Br—⟨C₆H₄⟩—COH (O)	4.0	H₂PO₄⁻	7.2
CH₃SO₃H	−1.2	pyridine—COH (O)	4.2	⟨C₆H₅⟩—SH	7.8
⟨C₆H₅⟩—SO₃H	−0.60				
CH₃CNH₂ (⁺OH)	0.0				
F₃CCOH (O)	0.2				
Cl₃CCOH (O)	0.64				
pyridine N—OH (N⁺)	0.79				

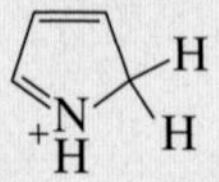

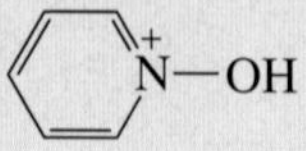

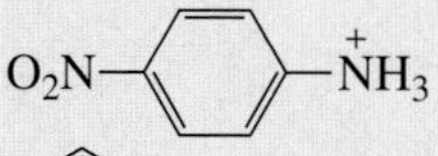

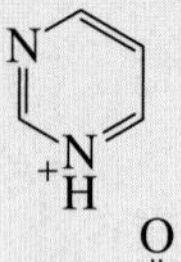

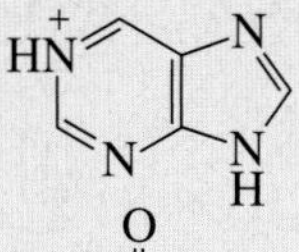

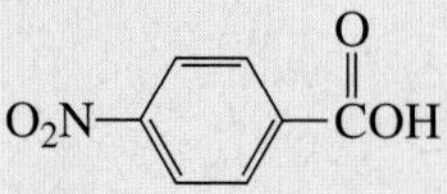

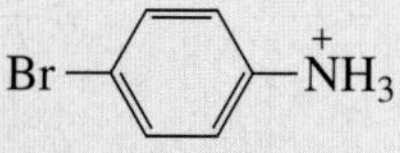

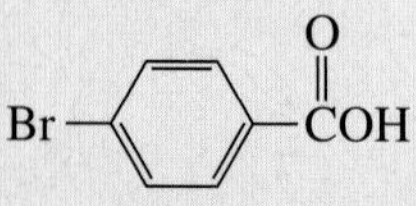

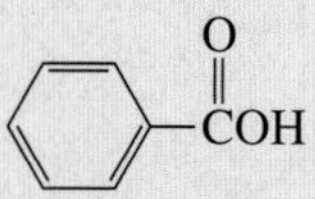

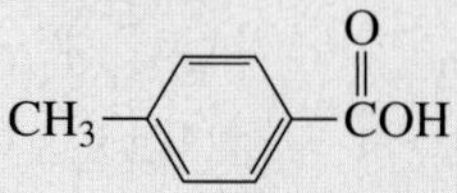

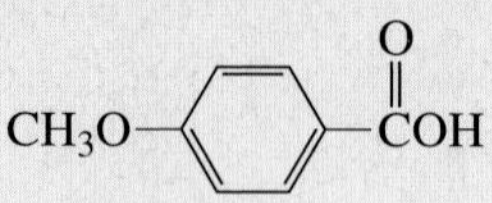

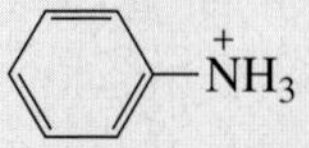

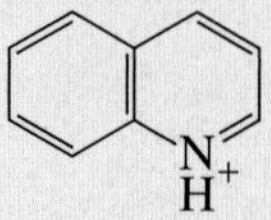

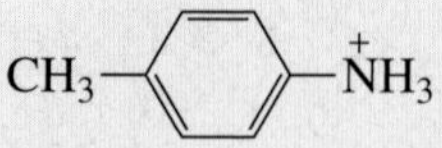

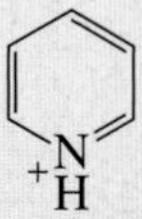

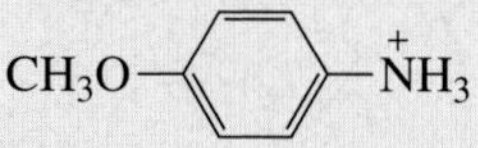

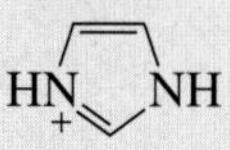

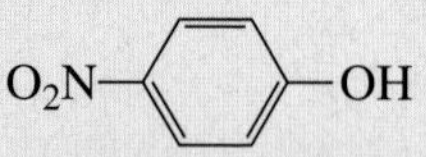

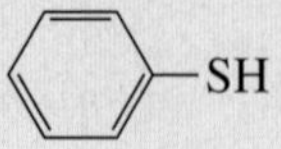

pKa Values (Continued)

Compound	pK_a	Compound	pK_a	Compound	pK_a
aziridinium (ring N with H, H, +)	8.0	cyclohexyl-NH_3^+	10.7	$CH_3CH(=O)$	17
$H_2N\overset{+}{N}H_3$	8.1	$(CH_3)_2\overset{+}{N}H_2$	10.7	$(CH_3)_3COH$	18
CH_3COOH (with O)	8.2	piperidinium (ring N+ H H)	11.1	$CH_3\overset{O}{C}CH_3$	20
$CH_3CH_2NO_2$	8.6	$CH_3CH_2\overset{+}{N}H_3$	11.0	$CH_3\overset{O}{C}OCH_2CH_3$	24.5
$CH_3\overset{O}{C}CH_2\overset{O}{C}CH_3$	8.9	pyrrolidinium (ring N+ H H)	11.3	$HC\equiv CH$	25
purine	8.9	HPO_4^{2-}	12.3	$CH_3C\equiv N$	25
$HC\equiv N$	9.1	CF_3CH_2OH	12.4	$CH_3\overset{O}{C}N(CH_3)_2$	30
morpholinium (ring O, N+ H H)	9.3	$CH_3CH_2O\overset{O}{C}CH_2\overset{O}{C}OCH_2CH_3$	13.3	NH_3	36
$Cl-C_6H_4-OH$	9.4	$HC\equiv CCH_2OH$	13.5	pyrrolidine (N H)	36
$\overset{+}{N}H_4$	9.4	$H_2N\overset{O}{C}NH_2$	13.7	CH_3NH_2	40
$HOCH_2CH_2\overset{+}{N}H_3$	9.5	$CH_3\overset{CH_3}{\underset{CH_3}{\overset{+}{N}}}CH_2CH_2OH$	13.9	$C_6H_5-CH_3$	41
$H_3\overset{+}{N}CH_2\overset{O}{C}O^-$	9.8	imidazole (N NH)	14.4	benzene	43
C_6H_5-OH	10.0	CH_3OH	15.5	$CH_2=CHCH_3$	43
$CH_3-C_6H_4-OH$	10.2	H_2O	15.7	$CH_2=CH_2$	44
HCO_3^-	10.2	CH_3CH_2OH	16.0	cyclopropane	46
CH_3NO_2	10.2	$CH_3\overset{O}{C}NH_2$	16	CH_4	50
$H_2N-C_6H_4-OH$	10.3	$C_6H_5\overset{O}{C}CH_3$	16.0	CH_3CH_3	50
CH_3CH_2SH	10.5	pyrrole (N H)	~17		
$(CH_3)_3\overset{+}{N}H$	10.6				
$CH_3\overset{O}{C}CH_2\overset{O}{C}OCH_2CH_3$	10.7				
$CH_3\overset{+}{N}H_3$	10.7				

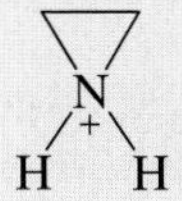

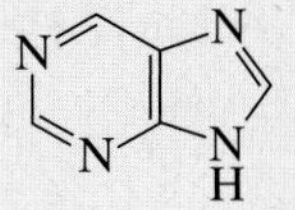

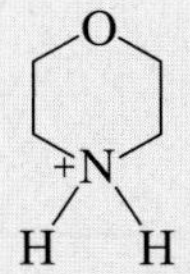

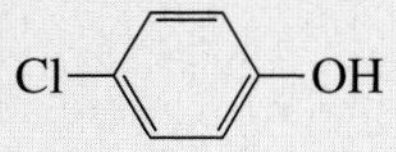

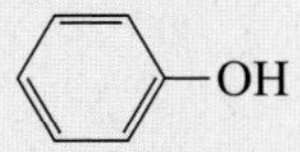

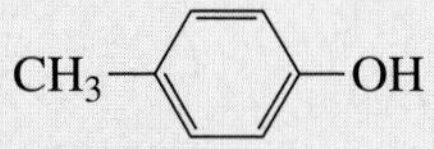

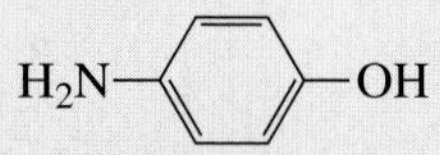

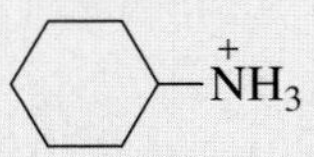

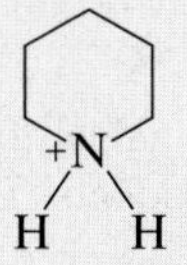

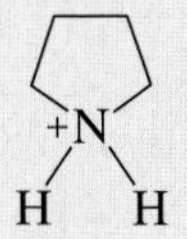

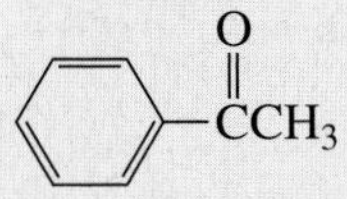

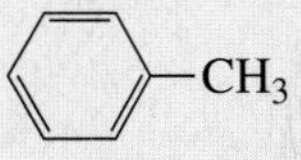

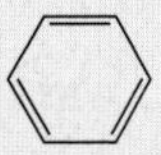

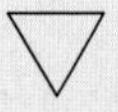

III

Derivations of Rate Laws

A **reaction mechanism** is a detailed analysis of how the chemical bonds (or the electrons) in the reactants rearrange to form the products. The mechanism for a given reaction must obey the observed rate law for the reaction.

A **rate law** tells how the rate of a reaction depends on the concentration of the species involved in the reaction.

First-Order Reaction

$$A \xrightarrow{k_1} \text{products}$$

Change in the concentration of A with respect to time:

$$\frac{-d[A]}{dt} = k_1[A]$$

Let a = the initial concentration of A;
let x = the concentration of A that has reacted up to time t.
Therefore, the concentration of A left at time $t = (a - x)$.
Substitution in the previous equation gives:

$$\frac{-d(a - x)}{dt} = k_1(a - x)$$

$$\frac{-da}{dt} + \frac{dx}{dt} = k_1(a - x)$$

$$0 + \frac{dx}{dt} = k_1(a - x)$$

$$\frac{dx}{(a - x)} = k_1 \, dt$$

Integration of the previous equation gives

$$-\ln(a - x) = k_1 t + \text{constant}$$

At $t = 0$, $x = 0$; therefore,

$$\text{constant} = -\ln a$$

$$-\ln(a - x) = k_1 t - \ln a$$

$$\ln \frac{a}{a - x} = k_1 t$$

$$\ln \frac{a - x}{a} = -k_1 t$$

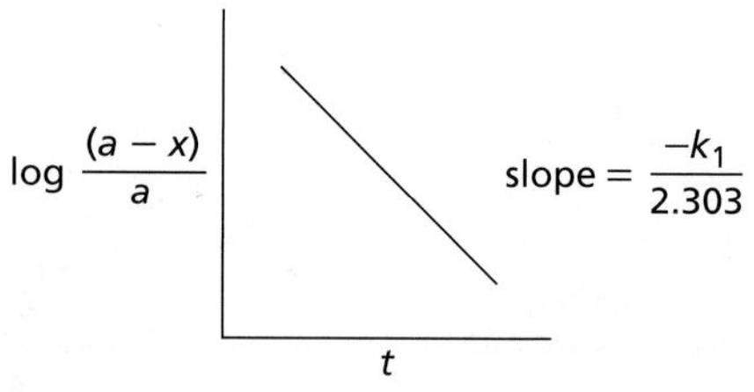

Plot of $\log \frac{(a - x)}{a}$ versus t with slope $= \dfrac{-k_1}{2.303}$

Half-Life of a First-Order Reaction

The **half-life** ($t_{1/2}$) of a reaction is the time it takes for half the reactant to react (or for half the product to form).

$$\ln \frac{a}{(a - x)} = k_1 t$$

At $t_{1/2}$, $x = \dfrac{a}{2}$; therefore,

$$\ln \frac{a}{\left(a - \dfrac{a}{2}\right)} = k_1 t_{1/2}$$

$$\ln \frac{a}{\dfrac{a}{2}} = k_1 t_{1/2}$$

$$\ln 2 = k_1 t_{1/2}$$

$$0.693 = k_1 t_{1/2}$$

$$t_{1/2} = \frac{0.693}{k_1}$$

Notice that the half-life of a first-order reaction is independent of the concentration of the reactant.

Second-Order Reaction

$$A + B \xrightarrow{k_2} \text{products}$$

Change in the concentration of A with respect to time:

$$\frac{-d[A]}{dt} = k_2[A][B]$$

Let $a =$ the initial concentration of A;
let $b =$ the initial concentration of B;
let $x =$ the concentration of A that has reacted at time t.
Therefore, the concentration of A left at time $t = (a - x)$, and the concentration of B left at time $t = (b - x)$.
Substitution gives:

$$\frac{dx}{dt} = k_2 (a - x)(b - x)$$

For the case where $a = b$ (this condition can be arranged experimentally):

$$\frac{dx}{dt} = k_2 (a - x)^2$$

$$\frac{dx}{(a - x)^2} = k_2 \, dt$$

Integrating the equation gives:

$$\frac{1}{(a - x)} = k_2 t + \text{constant}$$

At $t = 0$, $x = 0$; therefore,

$$\text{constant} = \frac{1}{a}$$

$$\frac{1}{(a - x)} - \frac{1}{a} = k_2 t$$

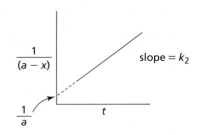

Half-Life of a Second-Order Reaction

$$\frac{1}{(a - x)} - \frac{1}{a} = k_2 t$$

At $t_{1/2}$, $x = \frac{a}{2}$; therefore,

$$\frac{1}{a} = k_2 t_{1/2}$$

$$t_{1/2} = \frac{1}{k_2 a}$$

Pseudo-First-Order Reaction

It is easier to determine a first-order rate constant than a second-order rate constant because the kinetic behavior of a first-order reaction is independent of the initial concentration of the reactant. Therefore, a first-order rate constant can be determined without knowing the initial concentration of the reactant. Determination of a second-order rate constant requires not only that the initial concentration of the reactants be known but also that the initial concentrations of the two reactants be identical in order to simplify the kinetic equation.

However, if the concentration of one of the reactants in a second-order reaction is much greater than the concentration of the other, the reaction can be treated as a first-order reaction. Such a reaction is known as a **pseudo-first-order reaction.**

$$\frac{-d\,[\text{A}]}{dt} = k_2\,[\text{A}]\,[\text{B}]$$

If $[\text{B}] \gg [\text{A}]$,

$$\frac{-d\,[\text{A}]}{dt} = k_2{}'\,[\text{A}]$$

The rate constant obtained for a pseudo-first-order reaction $(k_2{}')$ includes the concentration of B, but k_2 can be determined by carrying out the reaction at several different concentrations of B and determining the slope of a plot of the observed rate versus [B].

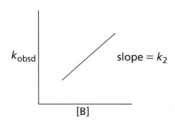

Summary of Methods Used to Synthesize a Particular Functional Group

SYNTHESIS OF ACETALS

1. Acid-catalyzed reaction of an aldehyde or a ketone with 2 equivalents of an alcohol (17.8).

SYNTHESIS OF ACID ANHYDRIDES

1. Reaction of an acyl halide with a carboxylate ion (16.7).
2. Preparation of a cyclic anhydride by heating a dicarboxylic acid (16.19).

SYNTHESIS OF ACYL CHLORIDES OR ACYL BROMIDES

1. Reaction of a carboxylic acid with $SOCl_2$, PCl_3, or PBr_3 (16.18).

SYNTHESIS OF ALCOHOLS

1. Acid-catalyzed hydration of an alkene (3.13).
2. Oxymercuration–demercuration of an alkene (3.16).
3. Hydroboration–oxidation of an alkene (3.17).
4. Reaction of an alkyl halide with HO^- (9.4).
5. Reaction of a Grignard reagent with an epoxide (11.8).
6. Reduction of an aldehyde, ketone, acyl chloride, anhydride, ester, or carboxylic acid (17.5, 17.6, 18.1).
7. Reaction of a Grignard reagent with an aldehyde, ketone, acyl chloride, or ester (17.4, 17.6).
8. Cleavage of an ether with HI or HBr (11.5).

SYNTHESIS OF ALDEHYDES

1. Hydroboration–oxidation of a terminal alkyne with disiamylborane followed by $H_2O_2 + HO^- + H_2O$ (5.7).
2. Oxidation of a primary alcohol with pyridinium chlorochromate (18.2).
3. Swern oxidation of a primary alcohol with dimethyl sulfoxide, oxalyl chloride, and triethylamine (18.2).
4. Rosenmund reduction: catalytic hydrogenation of an acyl chloride (18.1).
5. Reaction of an acyl chloride with lithium tri(*tert*-butoxy)aluminum hydride (18.1).
6. Cleavage of a 1,2-diol with periodic acid (18.6).
7. Ozonolysis of an alkene followed by work-up under reducing conditions (18.7).
8. Reaction of dithianyllithium with an alkyl halide followed by treatment with an aqueous solution of mercuric ion (29.2).

SYNTHESIS OF ALKANES

1. Catalytic hydrogenation of an alkene or an alkyne (3.19, 5.8).
2. Reaction of a Grignard reagent with a source of protons (11.8).
3. Wolff–Kishner or Clemmensen reduction of an aldehyde or a ketone (14.14, 17.7).
4. Reduction of a thioacetal or thioketal with H_2 and Raney nickel (17.10).
5. Reaction of a Gilman reagent with an alkyl halide (11.8).
6. Preparation of a cyclopropane by the reaction of an alkene with a carbene (3.17).

SYNTHESIS OF ALKENES

1. Elimination of hydrogen halide from an alkyl halide (10.1, 10.2, 10.3).
2. Acid-catalyzed dehydration of an alcohol (11.4).
3. Hofmann elimination reaction: elimination of a proton and a tertiary amine from a quaternary ammonium hydroxide (11.11).
4. Exhaustive methylation of an amine followed by a Hofmann elimination reaction (11.11).

5. Hydrogenation of an alkyne with Lindlar's catalyst to form a cis alkene (5.8).
6. Reduction of an alkyne with Na (or Li) and liquid ammonia to form a trans alkene (5.8).
7. Formation of a cyclic alkene using a Diels–Alder reaction (7.9, 28.4).
8. Wittig reaction: reaction of an aldehyde or a ketone with a phosphonium ylide (17.11).

SYNTHESIS OF ALKYL HALIDES
1. Addition of hydrogen halide (HX) to an alkene (3.9).
2. Addition of HBr + peroxide (3.18).
3. Addition of halogen to an alkene (3.15).
4. Addition of hydrogen halide or a halogen to an alkyne (5.5).
5. Radical halogenation of an alkane, an alkene, or an alkyl benzene (8.2, 8.5).
6. Reaction of an alcohol with hydrogen halide, $SOCl_2$, PCl_3, or PBr_3 (11.1, 11.2).
7. Reaction of a sulfonate ester with halide ion (11.3).
8. Cleavage of an ether with HI or HBr (11.5).
9. Halogenation of an α-carbon of an aldehyde, a ketone, or a carboxylic acid (19.4, 19.5).
10. Hunsdiecker reaction: reaction of a carboxylic acid with Br_2 and Ag_2O (19.17).

SYNTHESIS OF ALKYNES
1. Elimination of hydrogen halide from a vinyl halide (10.10).
2. Two successive eliminations of hydrogen halide from a vicinal dihalide or a geminal dihalide (10.10).
3. Reaction of an acetylide ion (formed by removing a proton from a terminal alkyne) with an alkyl halide (5.10).

SYNTHESIS OF AMIDES
1. Reaction of an acyl chloride, an acid anhydride, or an ester with ammonia or with an amine (16.7, 16.8, 16.9).
2. Reaction of a carboxylic acid with ammonia or with an amine and heat (16.12).
3. Reaction of a carboxylic acid and an amine with dicyclohexylcarbodiimide (21.9, 21.10).
4. Reaction of a nitrile with a secondary or tertiary alcohol (p. 731).

SYNTHESIS OF AMINES
1. Reaction of an alkyl halide with NH_3, RNH_2, or R_2NH (9.4).
2. Reaction of an alkyl halide with azide ion followed by reduction of the alkyl azide (16.15).
3. Reduction of an imine, a nitrile, or an amide (18.1).
4. Reductive amination of an aldehyde or a ketone (17.7).
5. Gabriel synthesis of primary amines: reaction of a primary alkyl halide with potassium phthalimide (16.15).
6. Reduction of a nitro compound (15.2).

SYNTHESIS OF AMINO ACIDS
1. Hell–Volhard–Zelinski reaction: halogenation of a carboxylic acid followed by treatment with excess NH_3 (21.6).
2. Reductive amination of an α-keto acid (21.6).

SYNTHESIS OF CARBOXYLIC ACIDS
1. Oxidation of a primary alcohol (18.2).
2. Oxidation of an aldehyde (18.3).
3. Ozonolysis of a monosubstituted alkene or a 1,2-disubstituted alkene followed by work-up under oxidizing conditions (18.7).
4. Ozonolysis of an alkyne (18.8).
5. Oxidation of an alkyl benzene (15.2).
6. Hydrolysis of an acyl halide, acid anhydride, ester, amide, or nitrile (16.7, 16.8, 16.9, 16.13, 16.16).
7. Haloform reaction: reaction of a methyl ketone with excess Br_2 (or Cl_2 or I_2) + HO^- (19.4).
8. Reaction of a Grignard reagent with CO_2 (17.4).

9. Malonic ester synthesis (19.18).

SYNTHESIS OF CYANOHYDRINS
1. Reaction of an aldehyde or a ketone with sodium cyanide and HCl (17.4).

SYNTHESIS OF 1,2-DIOLS
1. Reaction of an epoxide with water (11.6).
2. Reaction of an alkene with osmium tetroxide or potassium permanganate (18.5).

SYNTHESIS OF DISULFIDES
1. Mild oxidation of a thiol (21.7).

SYNTHESIS OF ENAMINES
1. Reaction of an aldehyde or a ketone with a secondary amine (17.7).

SYNTHESIS OF EPOXIDES
1. Reaction of an alkene with a peroxyacid (18.4).
2. Reaction of a halohydrin with hydroxide ion (p. 477).
3. Reaction of an aldehyde or a ketone with a sulfonium ylide (p. 784).

SYNTHESIS OF ESTERS
1. Reaction of an acyl halide or an acid anhydride with an alcohol (16.7, 16.8).
2. Acid-catalyzed reaction of an ester or a carboxylic acid with an alcohol (16.9, 16.12).
3. Reaction of an alkyl halide or a sulfonate ester with a carboxylate ion (9.4, 11.3).
4. Oxidation of a ketone (18.3).
5. Preparation of a methyl ester by the reaction of a carboxylate ion with diazomethane (15.12).

SYNTHESIS OF ETHERS
1. Acid-catalyzed addition of an alcohol to an alkene (3.13).
2. Alkoxymercuration–demercuration of an alkene (3.16).
3. Williamson ether synthesis: reaction of an alkoxide ion with an alkyl halide (10.9).
4. Formation of symmetrical ethers by heating an acidic solution of a primary alcohol (11.4).

SYNTHESIS OF HALOHYDRINS
1. Reaction of an alkene with Br_2 (or Cl_2) and H_2O (3.15).
2. Reaction of an epoxide with a hydrogen halide (11.6).

SYNTHESIS OF IMINES
1. Reaction of an aldehyde or a ketone with a primary amine (17.7).

SYNTHESIS OF KETONES
1. Mercuric acid-catalyzed hydration of an alkyne (5.6).
2. Hydroboration–oxidation of an alkyne (5.7).
3. Oxidation of a secondary alcohol (18.2).
4. Cleavage of a 1,2-diol with periodic acid (18.6).
5. Ozonolysis of an alkene (18.7).
6. Friedel–Crafts acylation of an aromatic ring (14.12).
7. Preparation of a methyl ketone by the acetoacetic ester synthesis (19.19).
8. Reaction of a Gilman reagent with an acyl chloride (11.8).
9. Preparation of a cyclic ketone by the reaction of the next size smaller cyclic ketone with diazomethane (p. 786).

SYNTHESIS OF α,β-UNSATURATED KETONES
1. Elimination from an α-haloketone (19.6).
2. Selenenylation of a ketone followed by oxidative elimination (p. 874).

SYNTHESIS OF NITRILES
1. Reaction of an alkyl halide with cyanide ion (9.4).

SYNTHESIS OF SUBSTITUTED BENZENES

1. Halogenation with Br_2 or Cl_2 and a Lewis acid (14.9).
2. Nitration with $HNO_3 + H_2SO_4$ (14.10).
3. Sulfonation: reaction with H_2SO_4 (14.11).
4. Friedel–Crafts acylation (14.12).
5. Friedel–Crafts alkylation (14.13).
6. Sandmeyer reaction: reaction of an arenediazonium salt with CuBr, CuCl, or CuCN (15.10).
7. Formation of a phenol by reaction of an arenediazonium salt with water (15.10).
8. Formation of an aniline by reaction of a benzyne intermediate with $^-NH_2$ (15.14).

SYNTHESIS OF SULFIDES

1. Reaction of a thiol with an alkyl halide (9.4, 11.10).
2. Catalytic hydrogenation of a disulfide (21.7).

SYNTHESIS OF THIOLS

1. Reaction of an alkyl halide with hydrogen sulfide (9.4).

Summary of Methods Used to Form Carbon–Carbon Bonds

1. Reaction of an acetylide ion with an alkyl halide or a sulfonate ester (5.10, 9.4, 11.3).
2. Diels–Alder reaction (7.9, 28.4).
3. Reaction of a Grignard reagent with an epoxide (11.8).
4. Friedel–Crafts alkylation and acylation (14.12, 14.13, 14.14).
5. Reaction of cyanide ion with an alkyl halide or a sulfonate ester (9.4, 11.3).
6. Reaction of cyanide ion with an aldehyde or a ketone (17.4).
7. Reaction of a Grignard reagent with an aldehyde, a ketone, an ester, an amide, an epoxide, or CO_2 (17.4, 17.6).
8. Reaction of an alkene with a carbene (3.17).
9. Reaction of a lithium dialkylcuprate with an α,β-unsaturated ketone or an α,β-unsaturated aldehyde (17.14).
10. Aldol addition (19.11, 19.12, 19.13).
11. Claisen condensation (19.14, 19.15).
12. Malonic ester synthesis and acetoacetic ester synthesis (19.18, 19.19).
13. Michael addition reaction (19.10).
14. Alkylation of an enamine (19.9).
15. Alkylation of the α-carbon of a carbonyl compound (19.9).

Spectroscopy Tables

Mass Spectrometry

Common Fragment Ions*

m/z	ion	m/z	ion
14	CH_2	46	NO_2
15	CH_3	47	CH_2SH, CH_3S
16	O	48	$CH_3S + H$
17	OH	49	CH_2Cl
18	H_2O, NH_4	51	CHF_2
19	F, H_3O	53	C_4H_5
26	$C \equiv N$	54	$CH_2CH_2C \equiv N$
27	C_2H_3	55	C_4H_7, $CH_2 = CHC = O$
28	C_2H_4, CO, N_2, $CH = NH$	56	C_4H_8
29	C_2H_5, CHO	57	C_4H_9, $C_2H_5C = O$
30	CH_2NH_2, NO		
31	CH_2OH, OCH_3		$\overset{O}{\overset{\|}{}}$
32	O_2 (air)	58	$CH_3CCH_2 + H$, $C_2H_5CHNH_2$, $(CH_3)_2NCH_2$, $C_2H_5NHCH_2$, C_2H_2S
33	SH, CH_2F		
34	H_2S		$\overset{O}{\overset{\|}{}}$
35	Cl	59	$(CH_3)_2COH$, $CH_2OC_2H_5$, $COCH_3$,
36	HCl		$CH_2C = O + H$, CH_3OCHCH_3,
39	C_3H_3		$\underset{NH_2}{}$
40	$CH_2C \equiv N$		
41	C_3H_5, $CH_2C \equiv N + H$, C_2H_2NH		CH_3CHCH_2OH
42	C_3H_6	60	$CH_2COOH + H$, CH_2ONO
43	C_3H_7, $CH_3C = O$, C_2H_5N		
44	$CH_2CH = O + H$, CH_3CHNH_2, CO_2, $NH_2C = O$, $(CH_3)_2N$		
45	CH_3CHOH, CH_2CH_2OH, CH_2OCH_3, COOH, $CH_3CHO + H$		

*All of these ions have a single positive charge.

Mass Spectrometry

Common Fragment Lost

Molecular ion minus	Fragment lost	Molecular Ion minus	Fragment lost
1	H		
15	CH_3	43	C_3H_7, $CH_3\overset{O}{\overset{\|}{C}}$, $CH_2{=}CHO$, HCNO, CH_3 + $CH_2{=}CH_2$
17	HO		
18	H_2O	44	$CH_2{=}CHOH$, CO_2, N_2O, $CONH_2$, $NHCH_2CH_3$
19	F		
20	HF	45	CH_3CHOH, CH_3CH_2O, CO_2H, $CH_3CH_2NH_2$
26	$CH{\equiv}CH$, $C{\equiv}N$		
27	$CH_2{=}CH$, $HC{\equiv}N$	46	H_2O + $CH_2{=}CH_2$, CH_3CH_2OH, NO_2
28	$CH_2{=}CH_2$, CO, (HCN + H)	47	CH_3S
29	CH_3CH_2, CHO	48	CH_3SH, SO, O_3
30	NH_2CH_2, CH_2O, NO	49	CH_2Cl
31	OCH_3, CH_2OH, CH_3NH_2	51	CHF_2
32	CH_3OH, S	52	C_4H_4, C_2N_2
33	HS, (CH_3 and H_2O)	53	C_4H_5
34	H_2S	54	$CH_2{=}CHCH{=}CH_2$
35	Cl	55	$CH_2{=}CHCHCH_3$
36	HCl, 2 H_2O	56	$CH_2{=}CHCH_2CH_3$, $CH_3CH{=}CHCH_3$
37	HCl + H	57	C_4H_9
38	C_3H_2, C_2N, F_2	58	NCS, NO + CO, CH_3COCH_3
39	C_3H_3, HC_2N	59	$CH_3O\overset{O}{\overset{\|}{C}}$, $CH_3\overset{O}{\overset{\|}{C}}NH_2$
40	$CH_3C{\equiv}CH$		
41	$CH_2{=}CHCH_2$	60	C_3H_7OH
42	$CH_2{=}CHCH_3$, $CH_2{=}C{=}O$, $\overset{CH_2}{CH_2{-}CH_2}$, NCO		

¹H NMR CHEMICAL SHIFTS

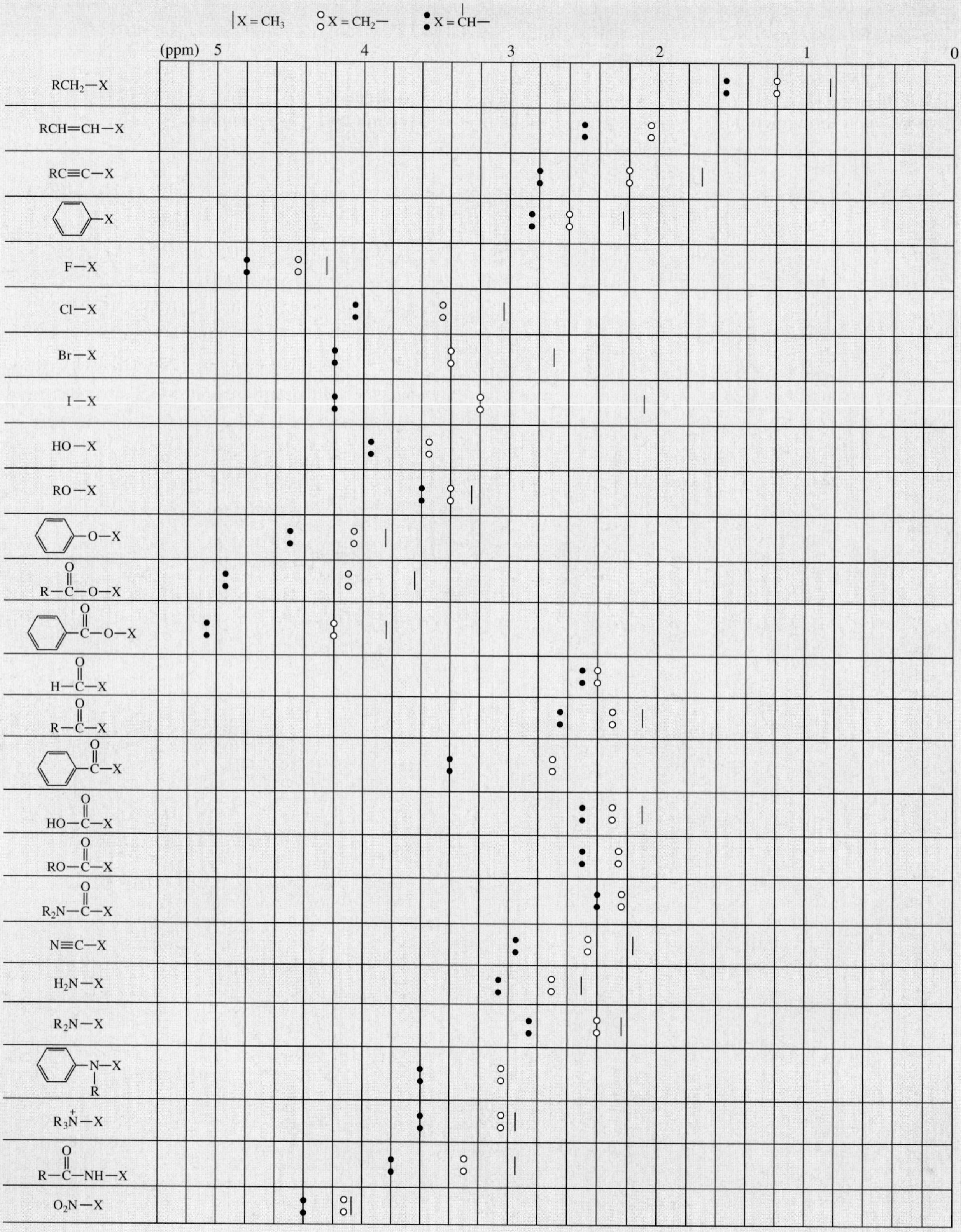

| X = CH₃ | X = CH₂— | X = CH— |

(ppm) 5   4   3   2   1   0

RCH₂—X

RCH=CH—X

RC≡C—X

⬡—X

F—X

Cl—X

Br—X

I—X

HO—X

RO—X

⬡—O—X

$R-\overset{O}{\overset{\|}{C}}-O-X$

$⬡-\overset{O}{\overset{\|}{C}}-O-X$

$H-\overset{O}{\overset{\|}{C}}-X$

$R-\overset{O}{\overset{\|}{C}}-X$

$⬡-\overset{O}{\overset{\|}{C}}-X$

$HO-\overset{O}{\overset{\|}{C}}-X$

$RO-\overset{O}{\overset{\|}{C}}-X$

$R_2N-\overset{O}{\overset{\|}{C}}-X$

N≡C—X

H₂N—X

R₂N—X

⬡—N—X
    |
    R

$R_3\overset{+}{N}-X$

$R-\overset{O}{\overset{\|}{C}}-NH-X$

O₂N—X

Characteristic Infrared Group Frequencies (S = strong, M = medium, W = weak). (Courtesy of N.B. Colthup, Stamford Research Laboratories, American Cyanamid Company, and the editor of the Journal of the Optical Society.) Overtone bands are marked 2ν.

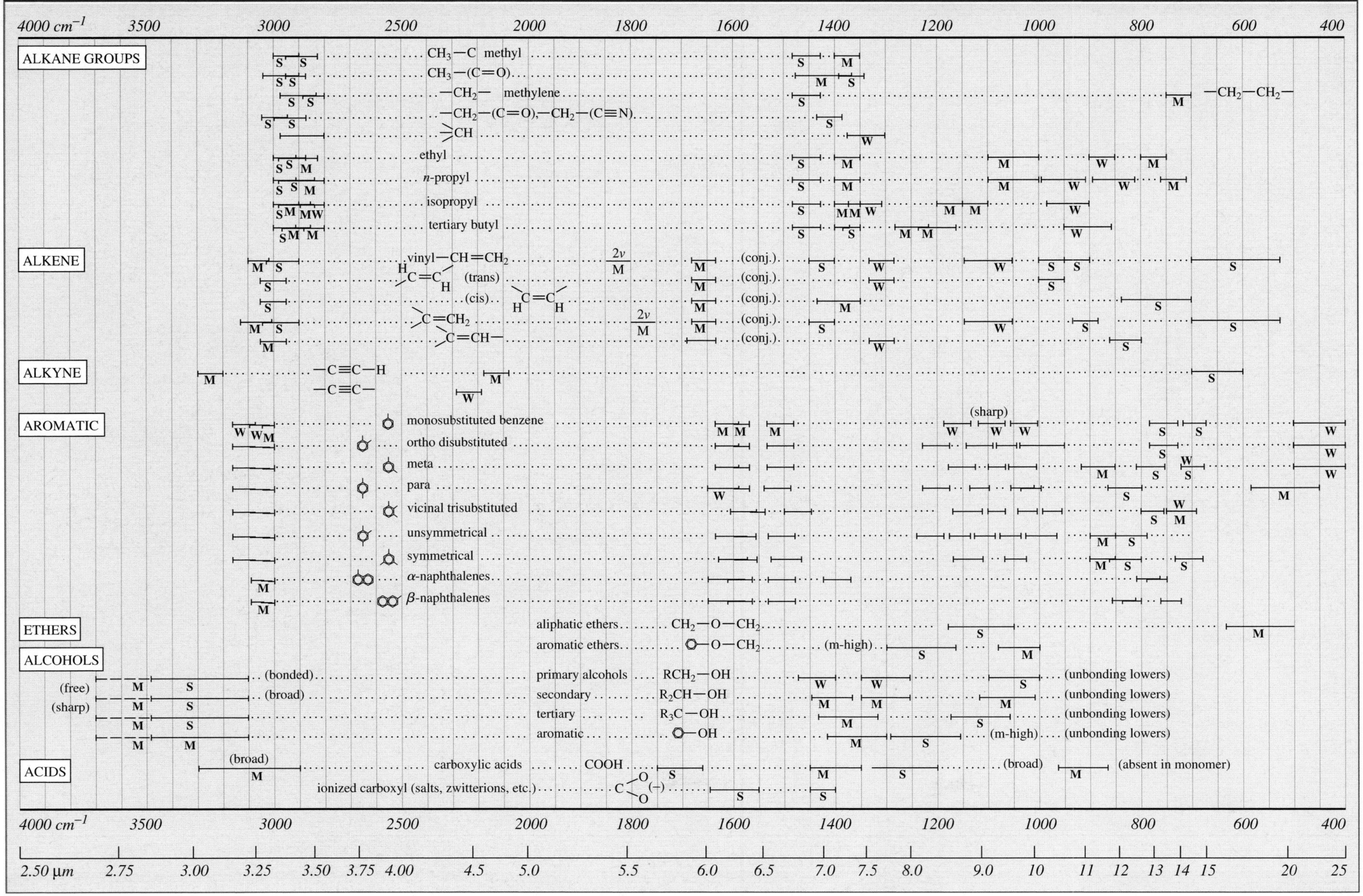

ALKANE GROUPS

CH₃—C methyl
CH₃—(C=O).
—CH₂— methylene
—CH₂—(C=O), —CH₂—(C≡N).
≡CH
ethyl
n-propyl
isopropyl
tertiary butyl

ALKENE

vinyl—CH=CH₂
H C=C H (trans)
C=C H (cis)
C=CH₂
C=CH
—CH₂—CH₂—

ALKYNE

—C≡C—H
—C≡C—

AROMATIC

monosubstituted benzene
ortho disubstituted
meta
para
vicinal trisubstituted
unsymmetrical
symmetrical
α-naphthalenes.
β-naphthalenes.

ETHERS

aliphatic ethers.   CH₂—O—CH₂
aromatic ethers.   —O—CH₂

ALCOHOLS

primary alcohols   RCH₂—OH   (unbonding lowers)
secondary   R₂CH—OH   (unbonding lowers)
tertiary   R₃C—OH   (unbonding lowers)
aromatic   —OH   (unbonding lowers)
(free)
(sharp)
(bonded)
(broad)

ACIDS

carboxylic acids   COOH   C=O (+) (absent in monomer)
ionized carboxyl (salts, zwitterions, etc.).

(conj.)   (m-high)   (m-high)   (broad)   (sharp)

4000 cm⁻¹   3500   3000   2500   2000   1800   1600   1400   1200   1000   800   600   400

2.50 μm   2.75   3.00   3.25   3.50   3.75   4.00   4.5   5.0   5.5   6.0   6.5   7.0   7.5   8.0   9.0   10   11   12   13   14   15   20   25

Characteristic Infrared Group Frequencies (Continued)

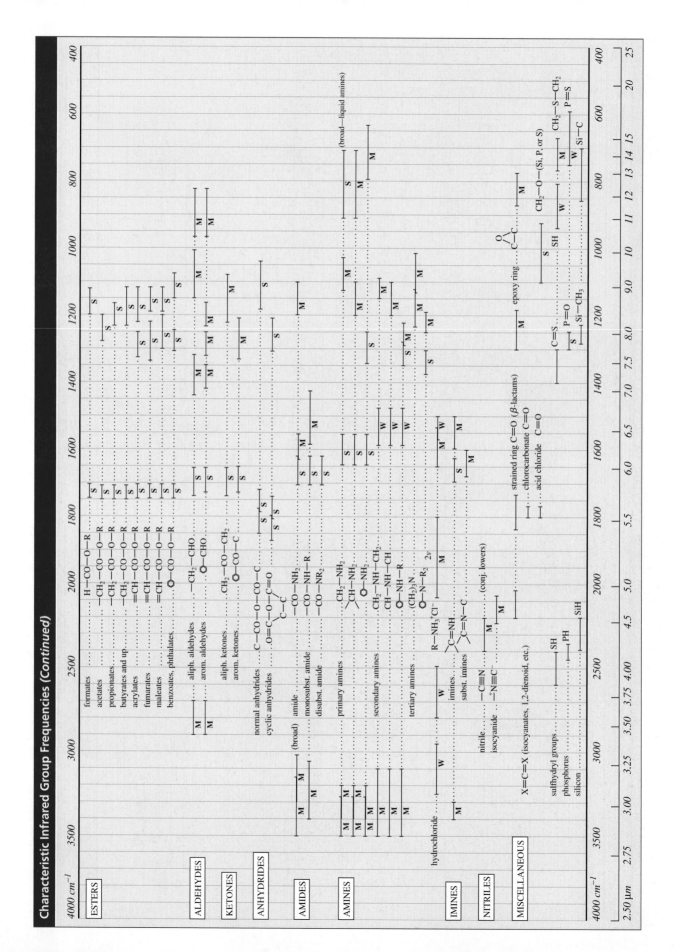

Characteristic Infrared Group Frequencies *(Continued)*

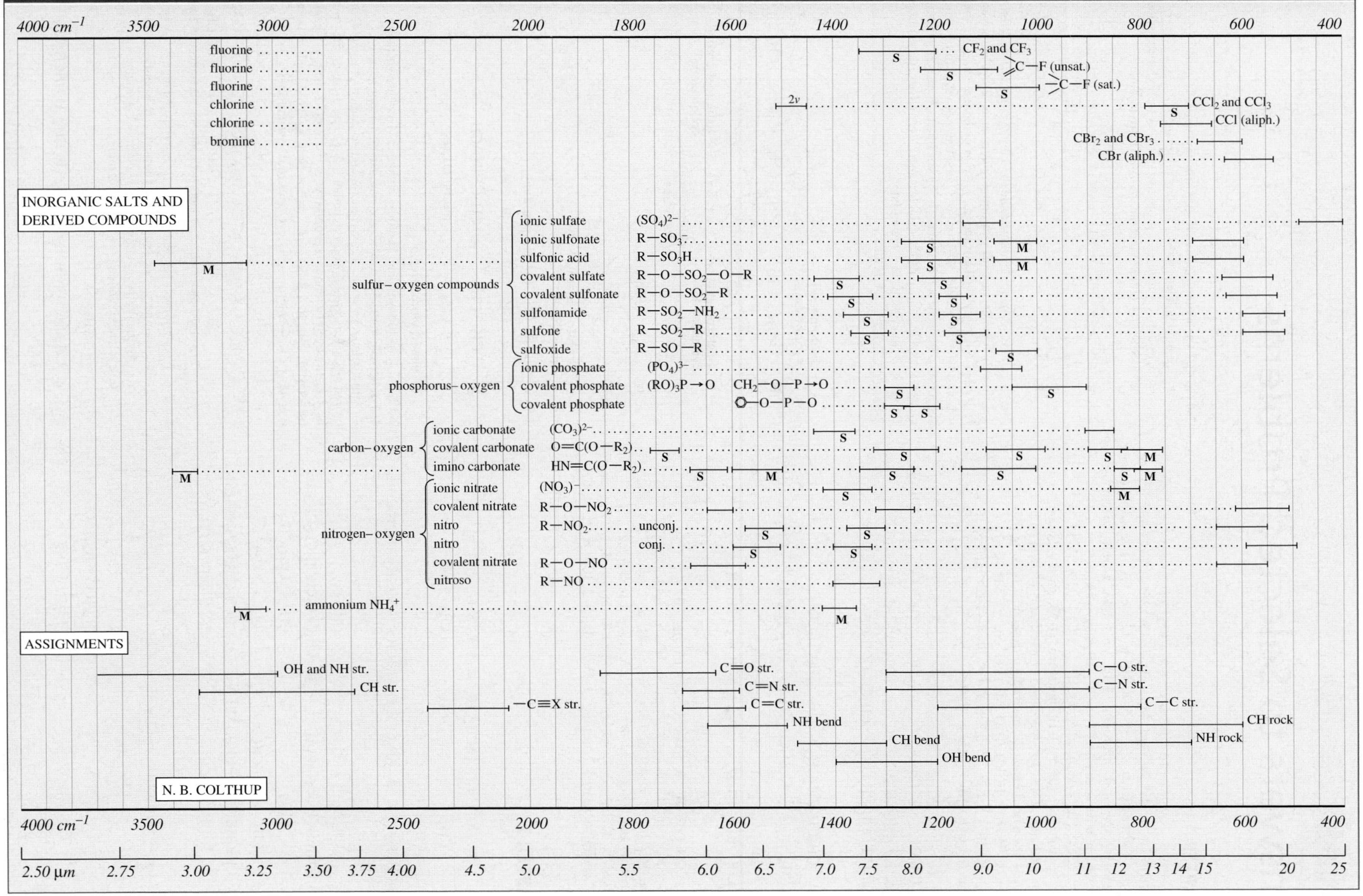

4000 cm⁻¹ 3500 3000 2500 2000 1800 1600 1400 1200 1000 800 600 400
fluorine
fluorine
fluorine
chlorine
chlorine
bromine
CF₂ and CF₃
C—F (unsat.)
C—F (sat.)
2ν
CCl₂ and CCl₃
CCl (aliph.)
CBr₂ and CBr₃
CBr (aliph.)
INORGANIC SALTS AND DERIVED COMPOUNDS
sulfur–oxygen compounds
ionic sulfate (SO₄)²⁻
ionic sulfonate R—SO₃⁻
sulfonic acid R—SO₃H
covalent sulfate R—O—SO₂—O—R
covalent sulfonate R—O—SO₂—R
sulfonamide R—SO₂—NH₂
sulfone R—SO₂—R
sulfoxide R—SO—R
phosphorus–oxygen
ionic phosphate (PO₄)³⁻
covalent phosphate (RO)₃P→O CH₂—O—P→O
covalent phosphate —O—P—O
carbon–oxygen
ionic carbonate (CO₃)²⁻
covalent carbonate O=C(O—R₂)
imino carbonate HN=C(O—R₂)
nitrogen–oxygen
ionic nitrate (NO₃)⁻
covalent nitrate R—O—NO₂
nitro R—NO₂ unconj.
nitro conj.
covalent nitrate R—O—NO
nitroso R—NO
ammonium NH₄⁺
S
M
ASSIGNMENTS
OH and NH str.
CH str.
—C≡X str.
C=O str.
C=N str.
C=C str.
NH bend
CH bend
OH bend
C—O str.
C—N str.
C—C str.
CH rock
NH rock
N. B. COLTHUP
4000 cm⁻¹ 3500 3000 2500 2000 1800 1600 1400 1200 1000 800 600 400
2.50 μm 2.75 3.00 3.25 3.50 3.75 4.00 4.5 5.0 5.5 6.0 6.5 7.0 7.5 8.0 9.0 10 11 12 13 14 15 20 25

Answers to Selected Problems

CHAPTER 1

1-1. 8 + 8, 8 + 9, 8 + 10   **1-2.** $4s$ **1-3.** Cl $1s^2 2s^2 2p^6 3s^2 3p^5$;   Br $1s^2 2s^2 2p^6 3s^2 3p^6 4s^2 3d^{10} 4p^5$;   I $1s^2 2s^2 2p^6 3s^2 3p^6 4s^2 3d^{10} 4p^6 5s^2 4d^{10} 5p^5$ **1-4. a.** KCl **b.** Cl_2 **1-5. a.** HF **b.** LiH **c.** H_2 **1-8. a.** oxygen **b.** oxygen **c.** oxygen **d.** hydrogen   **1-12.** yes   **1-13. a.** relative lengths: $Br_2 > Cl_2$; relative strengths: $Cl_2 > Br_2$ **b.** relative lengths: HBr > HCl > HF; relative strengths: HF > HCl > HBr   **1-15.** σ **1-16.** $> 104°5$ $< 109°5$ **1-17.** hydrogen   **1-18.** most = water, least = methane   **1-19.** $sp^2 = 33.3\%$ s character, $sp^3 = 25\%$ s character—the more s character, the stronger the bond   **1-23. a, d, e, h**   **1-24. a. 1.** $^+NH_4$ **2.** HCl **3.** H_2O **4.** H_3O^+ **b. 1.** $^-NH_2$ **2.** Br^- **3.** NO_3^- **4.** HO^- **1-26. a.** 5.2 **b.** 3.4×10^{-3} **1-27.** 8.16×10^{-8} **1-29. a.** CH_3COO^- **b.** $^-NH_2$ **c.** H_2O

1-31. $CH_3NH^- > CH_3O^- > CH_3NH_2 > CH_3\overset{\overset{\displaystyle O}{\|}}{C}O^- > CH_3OH$

1-32. a. 2.0×10^5 **b.** 3.2×10^{-7} **c.** 1.0×10^{-5} **d.** 4.0×10^{-13}

1-33. a. $CH_3OCH_2CH_2OH$ **c.** $CH_3CH_2OCH_2CH_2OH$

b. $CH_3CH_2CH_2\overset{+}{O}H_2$ **d.** $CH_3CH_2\overset{\overset{\displaystyle O}{\|}}{C}OH$

1-34. $\underset{\underset{\displaystyle F}{|}}{CH_3CHOH} > \underset{\underset{\displaystyle F}{|}}{CH_2CH_2OH} > \underset{\underset{\displaystyle Cl}{|}}{CH_2CH_2OH} > CH_3CH_2OH$

1-36. a. F^- **b.** I^- **1-37. a.** oxygen **b.** H_2S **c.** CH_3SH

1-38. a. $CH_3C\overset{+}{\equiv}NH$ **b.** CH_3CH_3 **c.** $F_3C\overset{\overset{\displaystyle O}{\|}}{C}OH$ **d.** sp^2 **e.** $sp > sp^2 > sp^3$ **f.** $sp > sp^2 > sp^3$ **g.** HNO_3 **1-40.** 10.4   **1-41. a.** 10.4 **b.** 2.7 **c.** 4.9 **d.** 7.3 **e.** 9.3   **1-42. a. 1.** neutral **2.** neutral **3.** 1/2 neutral and 1/2 charged **4.** charged **5.** charged **6.** charged **7.** charged **b. 1.** charged **2.** charged **3.** charged **4.** charged **5.** 1/2 charged and 1/2 neutral **6.** neutral **7.** neutral   **1-43. a. 1.** 4.9 **2.** 10.7 **b. 1.** > 6.9 **2.** < 8.7

CHAPTER 2

2-1. a. $\underset{\underset{\displaystyle CH_3}{|}}{CH_3CHOH}$ **c.** $\underset{\underset{\displaystyle CH_3}{|}}{CH_3CH_2CHI}$ **e.** $\underset{\underset{\displaystyle CH_3}{|}}{CH_3\overset{\overset{\displaystyle CH_3}{|}}{C}NH_2}$

b. $\underset{\underset{\displaystyle CH_3}{|}}{CH_3CHCH_2CH_2F}$ **d.** $\underset{\underset{\displaystyle CH_3}{|}}{CH_3\overset{\overset{\displaystyle CH_3}{|}}{C}CH_2Cl}$ **f.** $\underset{\underset{\displaystyle CH_3}{|}}{CH_3CHCH_2CH_2CH_2CH_2Br}$

2-2. a. $\underset{\underset{\displaystyle CH_3}{|}}{CH_3CHCHCH_2CH_3}$ **d.** $\underset{\underset{\displaystyle CH_3\ \ CH_2CH_2CH_3}{|\quad\quad|}}{CH_3\overset{\overset{\displaystyle CH_3}{|}}{C}CH_2CHCH_2CH_2CH_3}$

b. $\underset{\underset{\displaystyle CH(CH_3)_2}{|}}{CH_3\overset{\overset{\displaystyle CH_3}{|}}{C}HCH_2\overset{\overset{\displaystyle CH_3}{|}}{C}\text{—}\overset{\overset{\displaystyle CH_3}{|}}{C}HCH_2CH_3}$ **e.** $\underset{\underset{\displaystyle CH_2CH(CH_3)_2}{|}}{CH_3\overset{\overset{\displaystyle CH_3}{|}}{C}HCH_2\overset{\overset{\displaystyle CH_3}{|}}{C}HCHCH_2CH_3}$

c. $\underset{\underset{\displaystyle CH_2CH_3}{|}}{CH_3CH_2CH_2\overset{\overset{\displaystyle CH_2CH_3}{|}}{C}CH_2CH_2CH_2CH_2CH_3}$

f. $\underset{\underset{\displaystyle (CH_3)_2CCH_3}{|}}{CH_3CH_2CH_2CHCH_2CH_2CH_3}$

2-4. a. 2,2,4-trimethylhexane **b.** 2,2-dimethylbutane **c.** 2,5-dimethylheptane **d.** 3,3-diethylhexane **e.** 3,3-diethyl-4-methyl-5-propyloctane **f.** 3-methyl-4-propylheptane **g.** 5-ethyl-4,4-dimethyloctane **h.** 4-isopropyloctane

2-5. a. **c.** **b.** **d.**

2-6. a. 1-ethyl-2-methylcyclopentane **b.** ethylcyclobutane **c.** 4-ethyl-1,2-dimethylcyclohexane **d.** 3,6-dimethyldecane **e.** 2-cyclopropylpentane **f.** 1-ethyl-3-isobutylcyclohexane **g.** 5-isopropylnonane **h.** 1-*sec*-butyl-4-isopropylcyclohexane   **2-7. a.** *sec*-butyl chloride, 2-chlorobutane, **secondary b.** isoheptyl chloride, 1-chloro-5-methylhexane, **primary c.** cyclohexyl bromide, bromocyclohexane, **secondary d.** isopropyl fluoride, 2-fluoropropane, **secondary**

2-8. a.

chloromethyl-cyclohexane

b.

1-chloro-2-methyl-cyclohexane   1-chloro-3-methyl-cyclohexane   1-chloro-4-methyl-cyclohexane

c.

1-chloro-1-methyl-cyclohexane

2-9. a. $CH_3CH_2CH_2CH_2CH_3$ **pentane**

b. $\underset{\underset{\displaystyle CH_3}{|}}{CH_3\overset{\overset{\displaystyle CH_3}{|}}{C}CH_3}$ **c.** $\underset{\underset{\displaystyle CH_3}{|}}{CH_3CHCH_2CH_3}$ **methylbutane**

dimethylpropane

2-10. a. 1. methoxyethane **2.** ethoxyethane **3.** 4-methoxyoctane **4.** 1-propoxybutane **5.** 2-isopropoxypentane **6.** 1-isopropoxy-3-methylbutane **b.** no **c. 1.** ethyl methyl ether **2.** diethyl ether **4.** butyl propyl ether **5.** isopentyl isopropyl ether   **2-12. a.** 1-pentanol, **primary b.** 4-methylcyclohexanol, **secondary c.** 5-chloro-2-methyl-2-pentanol, **tertiary d.** 5-methyl-3-hexanol, **secondary e.** 2,6-dimethyl-4-octanol, **secondary f.** 4-chloro-3-ethylcyclohexanol, **secondary**

2-13. a. $\underset{\underset{\displaystyle OH}{|}}{CH_3\overset{\overset{\displaystyle CH_3}{|}}{C}CH_2CH_2CH_3}$ **b.** $\underset{\underset{\displaystyle OH}{|}}{CH_3CH_2\overset{\overset{\displaystyle CH_3}{|}}{C}CH_2CH_3}$ **c.** $\underset{\underset{\displaystyle OH\ CH_3}{|\quad\ |}}{CH_3\overset{\overset{\displaystyle CH_3}{|}}{C}\text{—}CHCH_3}$

   2-methyl-2-pentanol   3-methyl-3-pentanol   2,3-dimethyl-2-butanol

2-14. a. hexylamine, 1-hexanamine, **b.** butylpropylamine, *N*-propyl-1-butanamine **c.** *sec*-butylisobutylamine, *N*-isobutyl-2-butanamine **d.** ethylmethylpropylamine, *N,N*-diethyl-1-propanamine **e.** cyclohexylamine, cyclohexanamine

2-15. a. $\underset{\underset{\displaystyle CH_3}{|}}{CH_3CHCH_2NHCH_2CH_2CH_3}$ **b.** $CH_3CH_2NHCH_2CH_3$

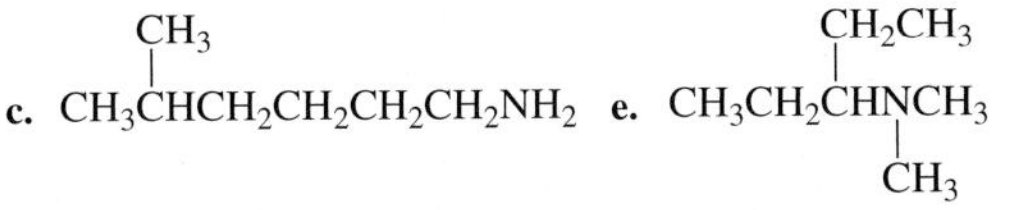

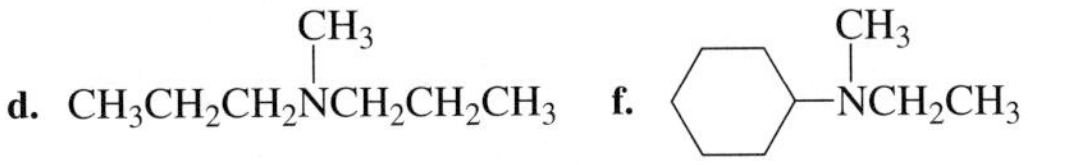

2-16. a. 6-methyl-1-heptanamine, isooctylamine, primary **b.** 3-methyl-*N*-propyl-1-butanamine, isopentylpropylamine, secondary **c.** *N*-ethyl-*N*-methylethanamine, diethylmethylamine, tertiary **d.** 2,5-dimethyl-cyclohexanamine, no common name, primary **2-17. a.** 104.5° **b.** 107.3° **c.** 104.5° **d.** 109.5° **2-18. a.** 1, 4, 5 **b.** 1, 2, 4, 5, 6

2-21. 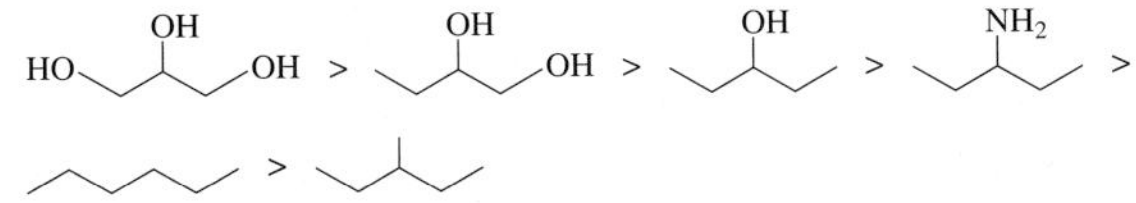

2-22. a. $CH_3CH_2CH_2CH_2CH_2CH_2Br > CH_3CH_2CH_2CH_2CH_2Br >$ $CH_3CH_2CH_2CH_2Br$

b. 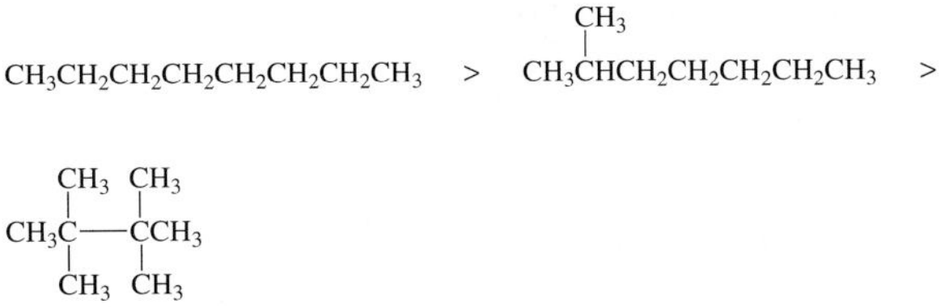

c. $CH_3CH_2CH_2CH_2CH_2OH > CH_3CH_2CH_2CH_2OH >$ $CH_3CH_2CH_2CH_2Cl > CH_3CH_2CH_2CH_2CH_3$

2-23. a. $HOCH_2CH_2CH_2OH > CH_3CH_2CH_2OH >$ $CH_3CH_2CH_2CH_2OH > CH_3CH_2CH_2CH_2Cl$

b. 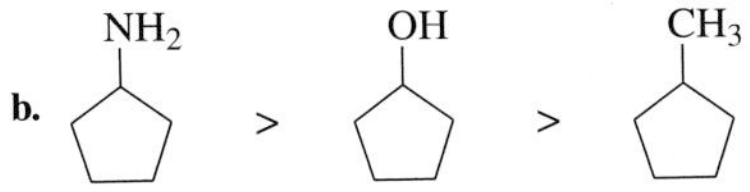

2-24. ethanol

2-26. a. 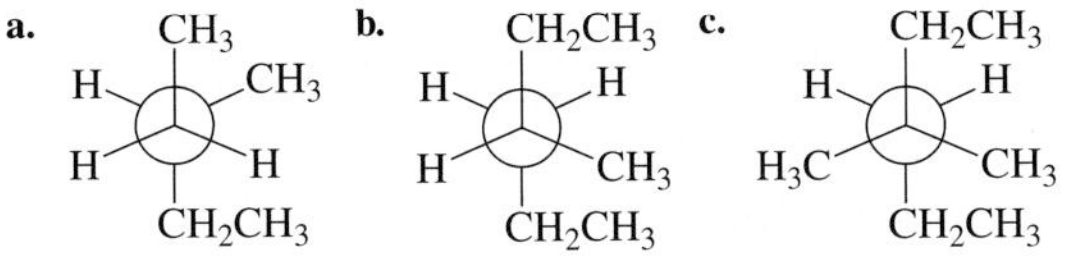

2-27. a. 135° **b.** 140° **2-31. a.** cis **b.** cis **c.** cis **d.** trans **e.** trans **f.** trans **2-33. a.** one equatorial and one axial **b.** both equatorial and both axial **c.** both equatorial and both axial **d.** one equatorial and one axial **e.** one equatorial and one axial **f.** both equatorial and both axial

CHAPTER 3

3-1. a. C_5H_8 **b.** C_4H_6 **c.** $C_{10}H_{16}$ **d.** C_8H_{10} **3-2. a.** 3 **b.** 4 **c.** 1 **d.** 3

3-4. a. 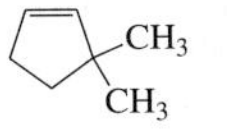 **b.** 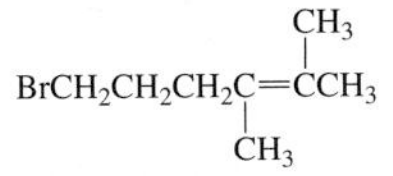 **c.** $CH_3CH_2OCH{=}CH_2$

d. $CH_2{=}CHCH_2OH$

3-5. a. 4-methyl-2-pentene **b.** 2-chloro-3,4-dimethyl-3-hexene **c.** 1-bromocyclopentene **d.** 1-bromo-4-methyl-3-hexene **e.** 1,5-dimethyl-cyclohexene **f.** 1-butoxy-1-propene

3-6. a. 1 and 3

b. 1. 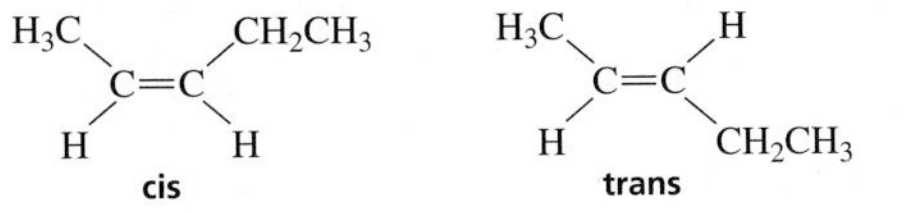

3. 

3-8. nucleophiles: H^- CH_3O^- $CH_3C{\equiv}CH$ NH_3; electrophiles: $AlCl_3$ $CH_3\overset{+}{C}HCH_3$ **3-10. a.** all (except H) **b.** *tert*-butyl **c.** *tert*-butyl **d.** −1.7 kcal/mol or −7.1 kJ/mol **3-13. a. 1.** $\Delta G° = -15$; $K_{eq} = 6.2 \times 10^{10}$ **2.** $\Delta G° = -16$; $K_{eq} = 1.9 \times 10^8$ **b.** the greater the temperature, the more negative the $\Delta G°$ **c.** the greater the temperature, the smaller the K_{eq} **3-14. a.** −19 kcal/mol **b.** −37 kcal/mol **c.** exothermic **d.** exergonic **3-15. a.** a and b **b.** b **c.** c **3-18.** decreasing; increasing **3-19. a.** 1×10^{-7} M s^{-1} **b.** decrease the rate **c.** none **3-22. a.** 1 **b.** 2 **c.** k_2 **d.** k_{-1} **e.** k_{-1} **f.** B→C **g.** C→B

3-23. a. 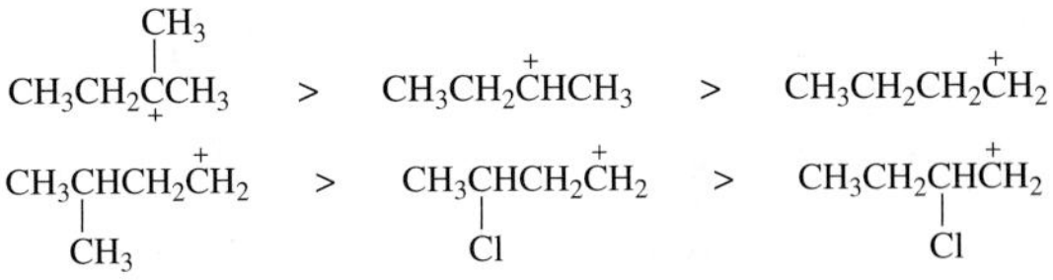

3-24. a. none **b.** ethyl cation **3-25.** 2-methylpropene

3-26. a. 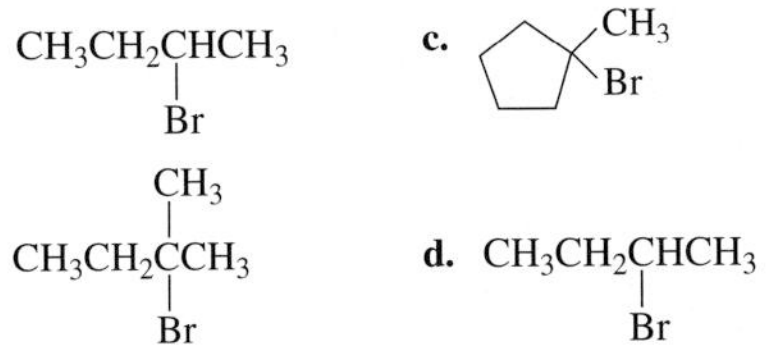

3-27. a. 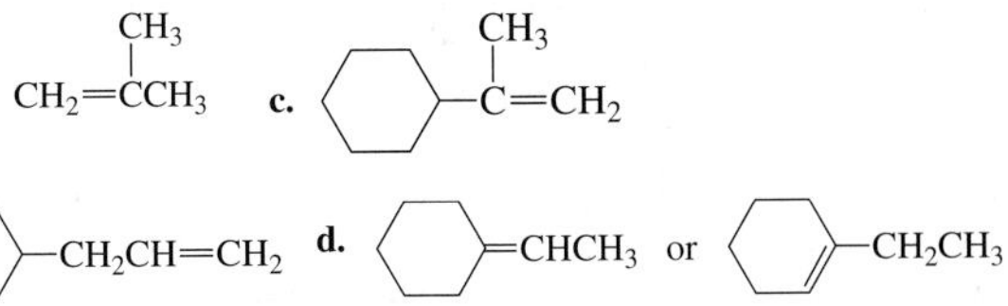

3-28. 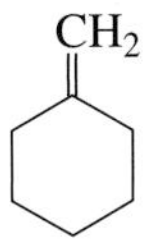 **3-29.** 16

3-35. a. 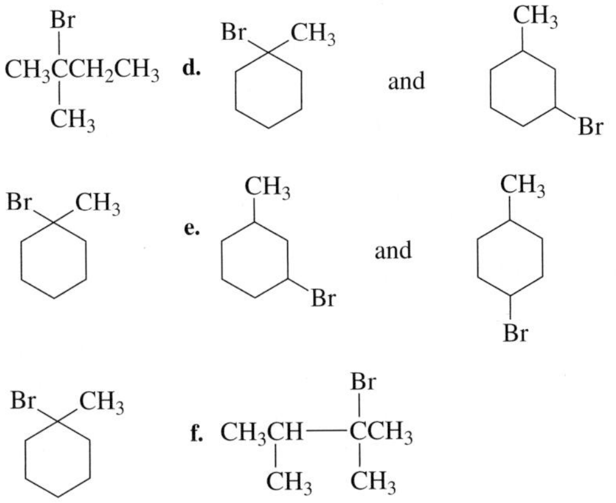

3-37. a. 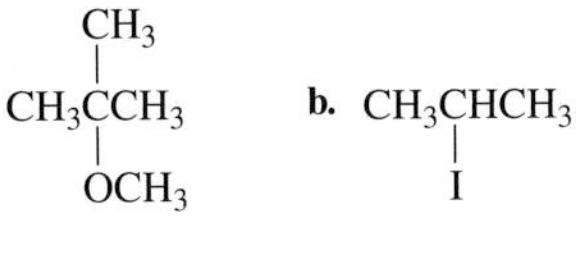

c. $CH_3CH_2CHCH_3$ **d.** $CH_3CH_2CHCH_3$
        $|$ $|$
        OH OCH_3

3-39. $CH_3CH_2CHCH_2I$ **3-42.** 2/3 mole
              $|$
              Cl

3-43. a. $CH_3CHCHCH_3$ **b.**
                $|$
                OH

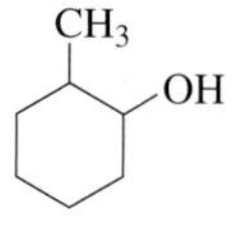

3-45. A

CHAPTER 4

4-1. a. $CH_3CH_2CH_2OH$ CH_3CHOH $CH_3CH_2OCH_3$ **b.** 7
                                   $|$
                                   CH_3

4-2. a.

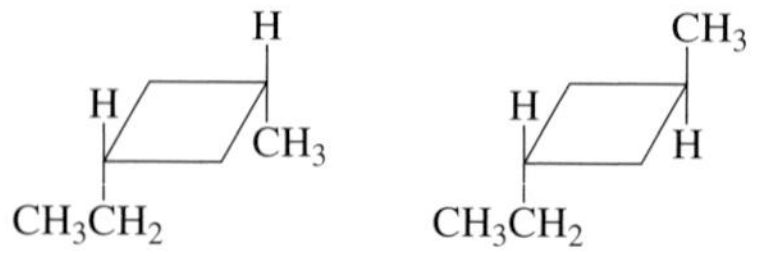

b.

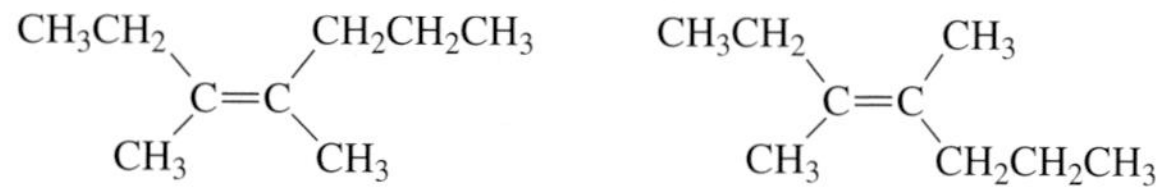

c. 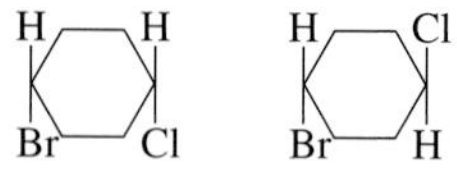 **d.**

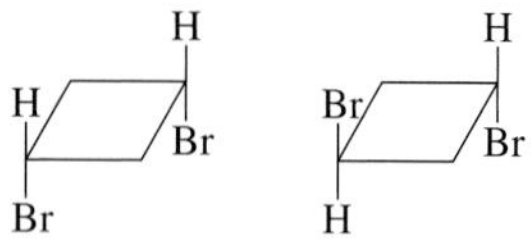

4-3. a, c, and f **4-4.** a, c, and f **4-6. a.** P, F, J, L, G, R, Q, N, Z **b.** T, M, O, A, U, V, C, D, E, H, I, X, Y **4-8. a.** R **b.** R **c.** R **d.** R
4-9. a. enantiomers **b.** enantiomers **c.** enantiomers **d.** enantiomers
4-10.

	1	3	2	4
a.	$-CH_2OH$	$-CH_3$	$-CH_2CH_2OH$	$-H$

	2	1	4	3
b.	$-CH=O$	$-OH$	$-CH_3$	$-CH_2OH$

	2	3	1	4
c.	$-CH(CH_3)_2$	$-CH_2CH_2Br$	$-Cl$	$-CH_2CH_2CH_2Br$

	2	3	1	4
d.	$-CH=CH_2$	$-CH_2CH_3$	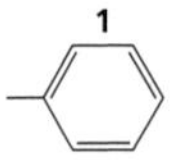	$-CH_3$

4-11. a. S **b.** R **c.** S **d.** S **4-12.** 268° **4-13. a.** levorotatory
b. dextrorotatory **4-14. a.** −24° **b.** 0° **4-15. a.** +79° **b.** 0° **c.** −79°
4-16. 58% **4-18. a.** enantiomers **b.** identical **c.** diastereomers
4-20. a. 8 **b.** $2^8 = 256$ **c.** 1
4-24.

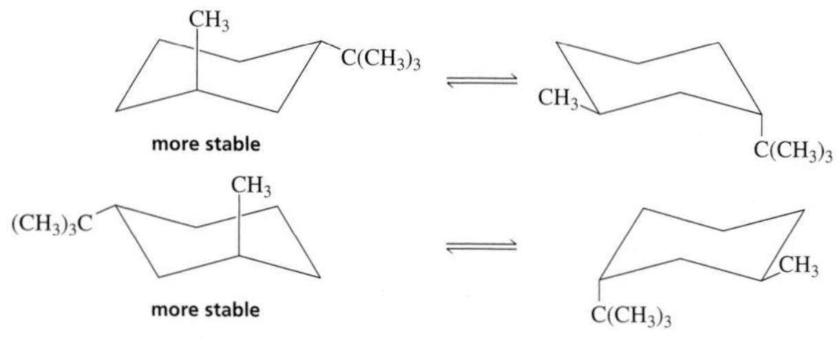

4-25. 1-chloro-1-methylcyclooctane, *cis*-1-chloro-5-methylcyclooctane, *trans*-1-chloro-5-methylcyclooctane **4-27.** b, d, and f

4-31. a. 4 **b.** **4-32. a.** (2R,3R)-2,3-dichloropentane

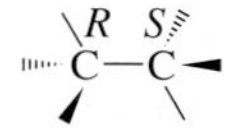

b. (2R,3R)-2-bromo-3-chloropentane **c.** (1r, 3s)-1,3-cyclopentanediol
d. (3R,4S)-3-chloro-4-methylhexane **4-34.** R **4-35. a.** R **b.** R **c.** S
4-36. a. diastereotopic **b.** homotopic **c.** enantiotopic **d.** homotopic
e. diastereotopic **f.** diasteriotopic **4-38. a.** no **b.** no **c.** no **d.** yes **e.** no
f. no

4-41. a. **c.**

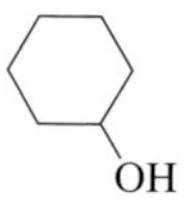

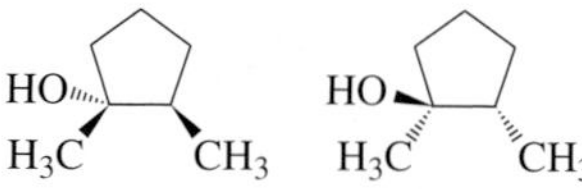

b. 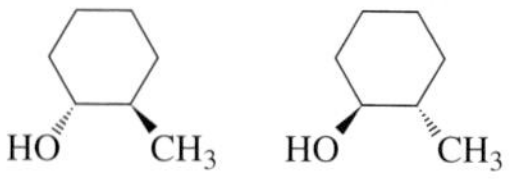 **d.**

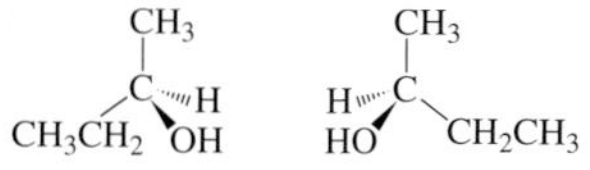

4-44. a.

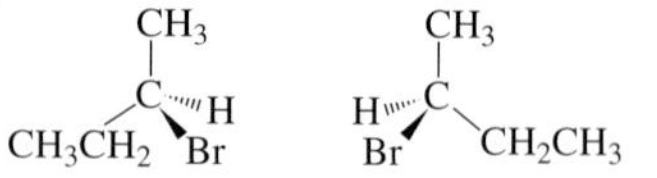

b.

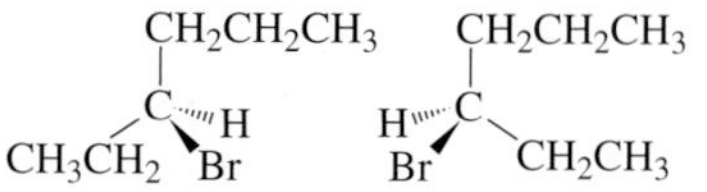

c.

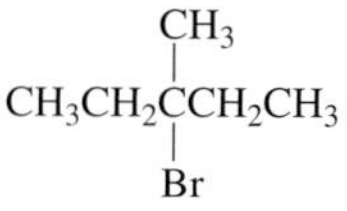

d.

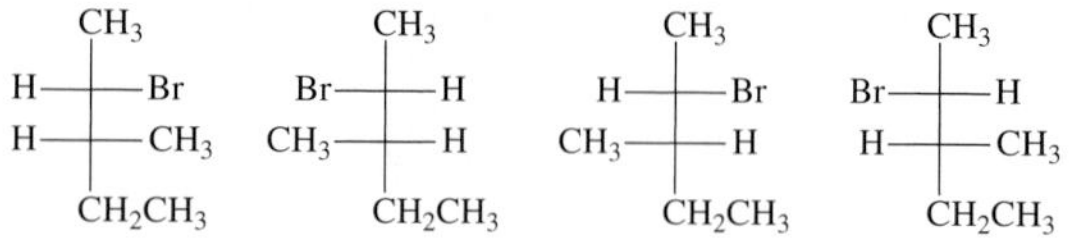

e.

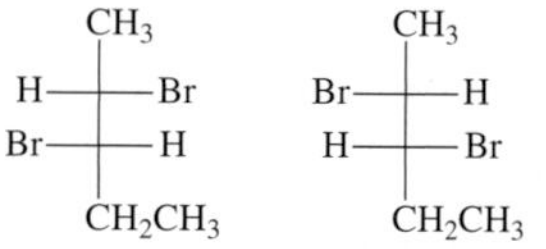

f.

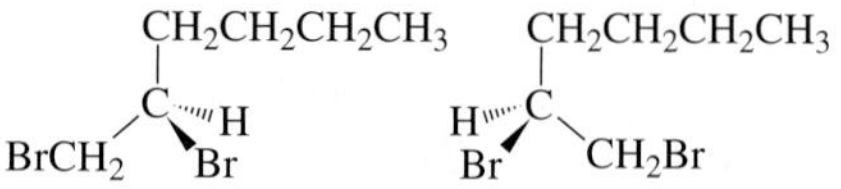

4-47. a. 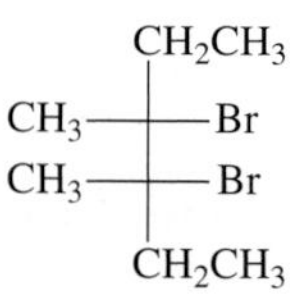 **b.**

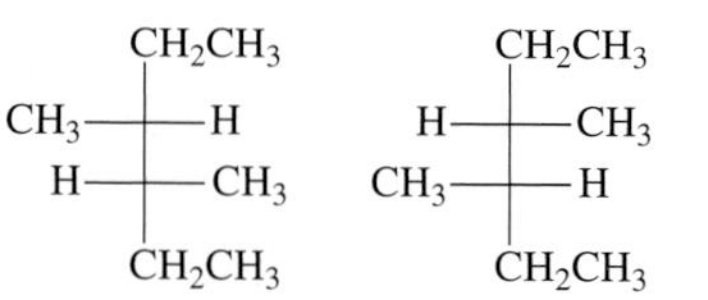

c.

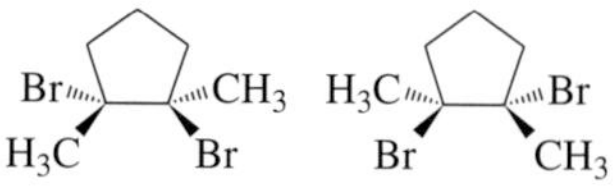

d.

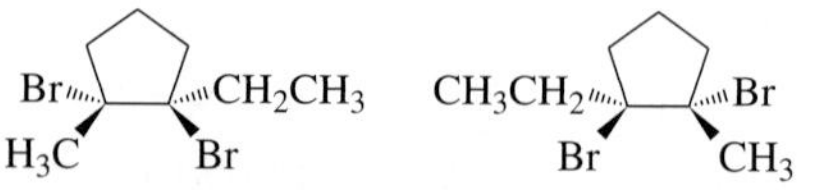

e. 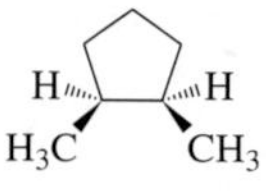 **f.**

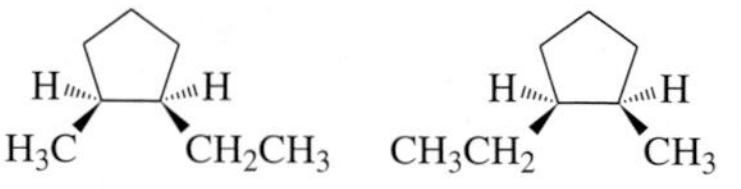

CHAPTER 5

5-1. $C_{14}H_{20}$ **5-2. a.** $ClCH_2CH_2C{\equiv}CCH_2CH_3$ **b.**

c. $CH_3CHC \equiv CH$ **d.** $HC \equiv CCH_2Cl$ **e.** $HC \equiv CCH_2CCH_3$
       | |
      CH_3 CH_3
                                                                 |
                                                                CH_3

f. $CH_3C \equiv CCH_3$ **5-3. a.** 5-bromo-2-pentyne **b.** 6-bromo-2-chloro-4-octyne **c.** 1-methoxy-2-pentyne **d.** 3-ethyl-1-hexyne

5-4.

$HC \equiv CCH_2CH_2CH_2CH_3$ $CH_3C \equiv CCH_2CH_2CH_3$ $CH_3CH_2C \equiv CCH_2CH_3$
       1-hexyne   **2-hexyne**   **3-hexyne**
    butylacetylene   methylpropylacetylene   diethylacetylene

$HC \equiv CCHCH_2CH_3$ $HC \equiv CCH_2CHCH_3$ $CH_3CHC \equiv CCH_3$ $CH_3CC \equiv CH$
         | | | |
        CH_3 CH_3 CH_3 CH_3
                                                                                                   |
                                                                                                  CH_3
  3-methyl-1-pentyne   **4-methyl-1-pentyne**   **4-methyl-2-pentyne**   **3,3-dimethyl-1-butyne**
   sec-butylacetylene   isobutylacetylene   isopropylmethylacetylene   *tert*-butylacetylene

5-6. If the less stable reactant has the more stable transition state, or if the difference in the stabilities of the reactants is greater than the difference in the stabilities of the transition states.

5-7. a. $CH_2 = CCH_2CH_3$ **d.** $HC = CCH_2CH_3$
                 | | |
                Br Br Br

b. $CH_3CCH_2CH_3$ **e.** $ClCHCHCH_2CH_3$
       | | |
      Br Cl Cl
       |
      Br

c. $BrCH = CHCH_2CH_3$

5-8. a. $H_3C \quad Br$ **b.** $H_3C \quad CH_3$ $H_3C \quad Br$
           $C=C$ $C=C$ $C=C$
        $Br \quad CH_3$ $H \quad Br$ $H \quad CH_3$

5-9. $CH_3CH_2\overset{O}{\overset{\|}{C}}CH_2CH_2CH_3$ and $CH_3CH_2CH_2\overset{O}{\overset{\|}{C}}CH_2CH_2CH_3$

5-10. a. $CH_3C \equiv CH$ **b.** $CH_3CH_2C \equiv CCH_2CH_3$ **c.** $HC \equiv C$—⬡

5-11. a. $CH_2 = CCH_3$
                 |
                OH

b. $CH_3CH = CCH_2CH_2CH_3$ and $CH_3CH_2C = CHCH_2CH_3$
              | |
             OH OH

c. $CH_2 = C$—⬡   and CH_3C—⬡
             | |
            OH OH

5-12. a. 1. $CH_3CH_2\overset{O}{\overset{\|}{C}}CH_3$ **b. 1.** $CH_3CH_2\overset{O}{\overset{\|}{C}}CH_3$

2. $CH_3CH_2CH_2\overset{O}{\overset{\|}{C}}H$ **2.** $CH_3CH_2\overset{O}{\overset{\|}{C}}CH_3$

c. 1. $CH_3CH_2CH_2\overset{O}{\overset{\|}{C}}CH_3$ and $CH_3CH_2\overset{O}{\overset{\|}{C}}CH_2CH_3$

2. $CH_3CH_2CH_2\overset{O}{\overset{\|}{C}}CH_3$ and $CH_3CH_2\overset{O}{\overset{\|}{C}}CH_2CH_3$

5-13. acetylene

5-14. a. $CH_3CH_2CH_2C \equiv CH$ or $CH_3CH_2C \equiv CCH_3 \xrightarrow{H_2 \atop Pt}$

b. $CH_3C \equiv CCH_3 \xrightarrow[\text{catalyst}]{H_2 \atop \text{Lindlar's}}$ **c.** $CH_3CH_2C \equiv CCH_3 \xrightarrow{Na \atop NH_3}$

d. $CH_3CH_2CH_2CH_2C \equiv CH \xrightarrow[\text{catalyst}]{H_2 \atop \text{Lindlar's}}$ or $\xrightarrow{Na \atop NH_3}$

5-15. 25   **5-16. a.** $CH_3\overset{+}{C}H_2$ **b.** $H_2C = \overset{+}{C}H$

CHAPTER 6

6-1. a. 1. 3 **2.** 2 **b. 1.** 5 **2.** 5   **6-3. a.** all are the same length
b. 2/3 of a negative charge   **6-6. a.** $CH_3\overset{+}{N}HCH_2$

b. ⬡—$\overset{\text{\Large<}}{}$ $^+$ **c.** $CH_3\overset{+}{O}CH_2$ **d.** ⬡(with $\overset{+}{C}HCH_3$)

6-9. a. $CH_3CH = CHOH$ **b.** $CH_3CH = CHOH$ **6-10. a.** ethylamine
b. ethoxide ion **c.** ethoxide ion

6-11.

⬡—COOH   >   ⬡—OH   >   ⬡—$CH_2\overset{+}{N}H_3$ >   ⬡—CH_2OH

6-12. a. $\psi_1 = 3;\ \psi_2 = 2$ **b.** $\psi_1 = 5;\ \psi_2 = 4$

CHAPTER 7

7-1. a. 1,5-cyclooctadiene **b.** 1-hepten-4-yne **c.** 4-methyl-1,4-hexadiene
d. 2,4-dimethyl-4-hexen-1-ol **e.** 1,6-dimethyl-1,3-cyclohexadiene
f. 3-butyn-1-ol **g.** 1,3,5-heptatriene **h.** 5-vinyl-5-octen-1-yne

7-2. a.

$H_3C \quad H$ $H_3C \quad H$
   $C=C$ $C=C$
$H_3C \quad$ $H_3C \quad CH_3$
      $C=C$ $C=C$
   $H \quad CH_3$ $H \quad H$
(E)-2-methyl-2,4-hexadiene   **(Z)-2-methyl-2,4-hexadiene**

b.

$H_3C \quad H$ $H_3C \quad H$
   $C=C$ $C=C$
 $H \quad$ $H \quad CH_2CH_3$
    $C=C$ $C=C$
 $H \quad CH_2CH_3$ $H \quad H$
(2E,4E)-2,4-heptadiene   **(2E,4Z)-2,4-heptadiene**

$H \quad H$ $H \quad CH_2CH_3$
   $C=C$ $C=C$
$H_3C \quad$ $H_3C \quad H$
     $C=C$ $C=C$
  $H \quad CH_2CH_3$ $H \quad H$
(2Z,4E)-2,4-heptadiene   **(2Z,4Z)-2,4-heptadiene**

c.

$H \quad H$ $H \quad H$
  $C=C$ $C=C$
$H \quad$ $H \quad CH_3$
    $C=C$ $C=C$
 $H \quad CH_3$ $H \quad H$
(E)-1,3-pentadiene   **(Z)-1,3-pentadiene**

7-3.

$\quad CH_3 \quad\quad CH_3$
$CH_3C = CHCH = CCH_3$ > $CH_3CH = CHCH = CHCH_3$ >
2,5-dimethyl-2,4-hexadiene   **2,4-hexadiene**

$CH_3CH = CHCH = CH_2$ > $CH_2 = CHCH_2CH = CH_2$
1,3-pentadiene   **1,4-pentadiene**

7-4. ψ_3 has three nodes, ψ_4 has four nodes

7-5.

CH_3	CH_3	CH_3	
H—Br	H—Br	Br—H	$CH_2 = CHCH_2CH_2CCH_3$
CH_2	CH_2	CH_2	$\quad\quad\quad\quad\quad\quad\quad$	Cl
CH_2	CH_2	CH_2	
H—Br	Br—H	H—Br	**no other isomers are possible**
CH_3	CH_3	CH_3	

7-7.

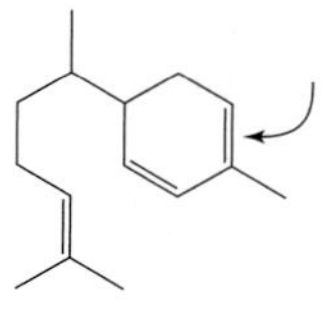

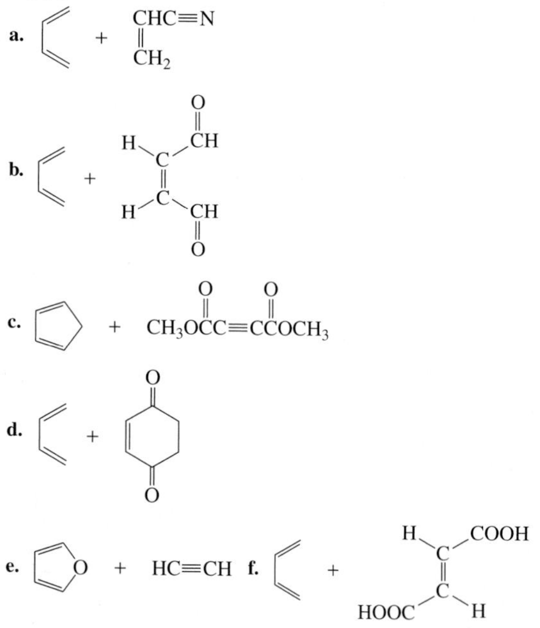

7-18. a. exo-2-chlorobicyclo[2.2.1]heptane **b.** 3-methylbicyclo[4.4.0]decane **c.** bicyclo[2.2.2]octane **d.** *cis*-6,7-dimethylbicyclo[3.2.0]heptane

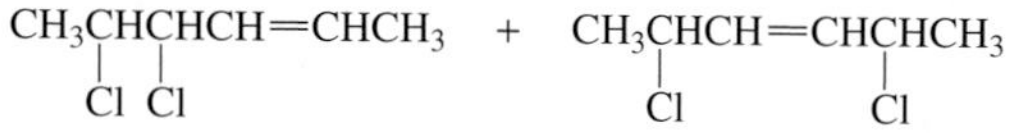

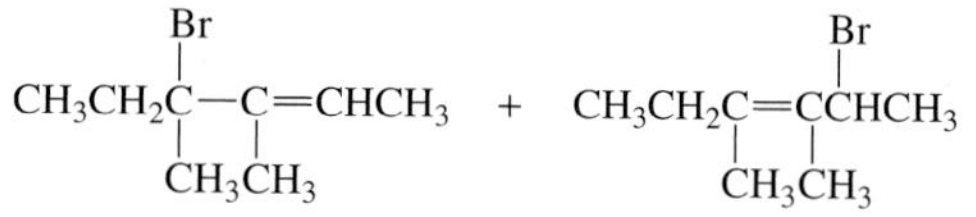

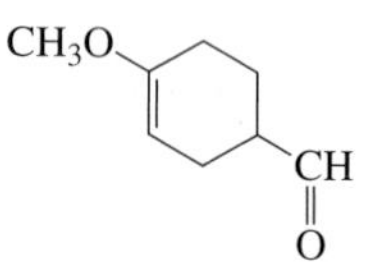

CHAPTER 8

8-5. a. 3 **b.** 3 **c.** 1 **d.** 5 **e.** 2 **f.** 1 **g.** 5 **h.** 4

8-6.

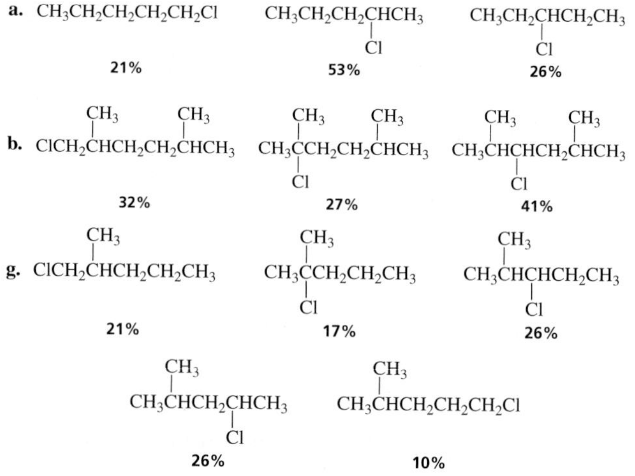

a. $CH_3CH_2CH_2CH_2CH_2Cl$ $CH_3CH_2CH_2CHCH_3$ with Cl $CH_3CH_2CHCH_2CH_3$ with Cl

21%   53%   26%

b.   ClCH₂CHCH₂CH₂CHCH₃ ; CH₃CCH₂CH₂CHCH₃ (Cl) ; CH₃CHCHCH₂CHCH₃ (Cl)

32%   27%   41%

g.   ClCH₂CHCH₂CH₂CH₃ ; CH₃CCH₂CH₂CH₃ (Cl) ; CH₃CHCHCH₂CH₃ (Cl)

21%   17%   26%

CH₃CHCH₂CHCH₃ (Cl) ; CH₃CHCH₂CH₂CH₂Cl

26%   10%

8-7. a.   CH₃CH₂CHCH₃ (Br)   CH₃CH₂CH₂CH₂Br

$4 \times 82 = 328$ $6 \times 1 = 6$

$$\frac{328}{328+6} = \frac{328}{334} = 0.98 = 98\%$$

b.   CH₃CCH₂CH₂CCH₃ (Br)(CH₃)

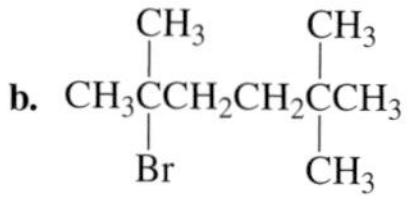

$1 \times 1600 = 1600$

other products $\begin{cases} 9 \times 1 = 9 \\ 2 \times 82 = 164 \\ 2 \times 82 = 164 \\ 6 \times 1 = 6 \end{cases}$ 1600 ... 1943

$$\frac{1600}{1943} = 0.82 = 82\%$$

8-8. 99.4% (vs. 36%)

8-9.

a.   CH₃CH₂CH₂CH₂CH₂Br   CH₃CH₂CHCH₂CH₃ (Br)   CH₃CH₂CH₂CHCH₃ (Br)

1%   33%   66%

b.   BrCH₂CHCH₂CH₂CHCH₃   CH₃CCH₂CH₂CHCH₃ (Br)   CH₃CHCHCH₂CHCH₃ (Br)

0.3%   90.4%   9.3%

g.   BrCH₂CHCH₂CH₂CH₃   CH₃CCH₂CH₂CH₃ (Br)   CH₃CHCHCH₂CH₃ (Br)

0.3%   82.6%   8.5%

CH₃CHCH₂CHCH₃ (Br)   CH₃CH₂CH₂CH₂CH₂CH₂Br

8.5%   0.2%

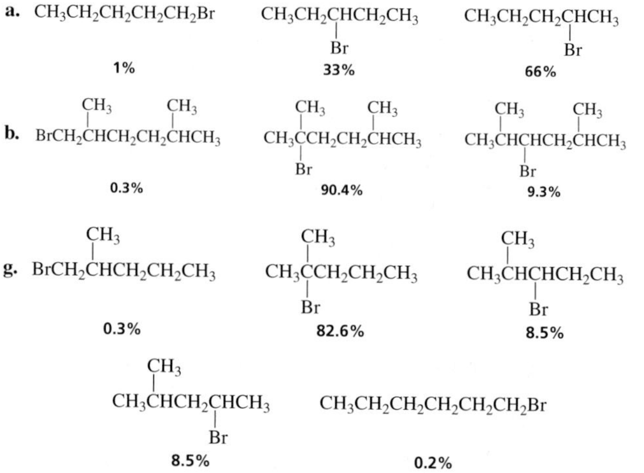

8-10. a. chlorination **b.** bromination **c.** no preference

8-13. a.

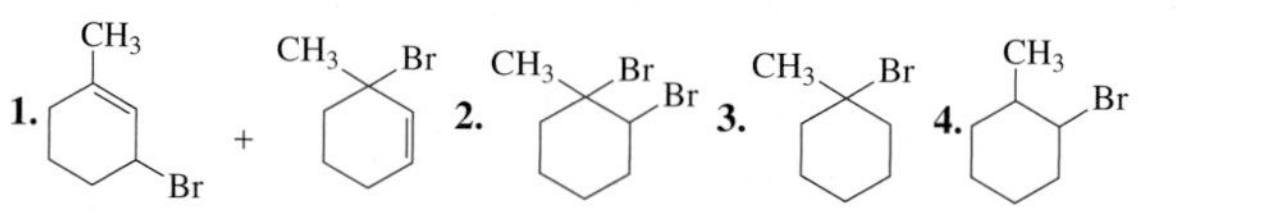

1.   **2.**   **3.**   **4.**

7-9. a.   CH₃CHCHCH=CHCH₃ (Cl)(Cl)   +   CH₃CHCH=CHCHCH₃ (Cl)(Cl)

b.   CH₃CH₂C—C=CHCH₃ (Br)(CH₃)(CH₃)   +   CH₃CH₂C=CCHCH₃ (Br)(CH₃)(CH₃)

c.   CH₃CHCH=CCH₂CH₃ (Br)(CH₃)   +   CH₃CH=CHCCH₂CH₃ (Br)(CH₃)

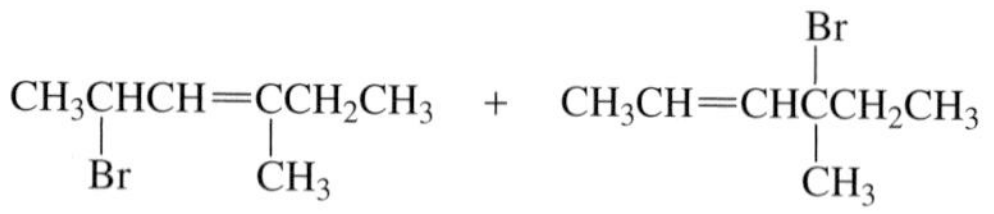

d.

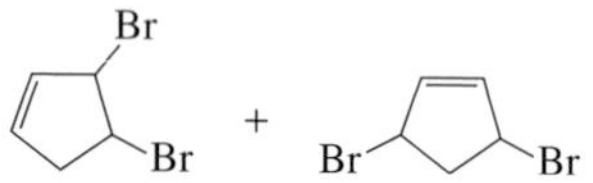

7-10. a. 3 **b.** 4-bromo-2-pentene

7-12. a.   **c.**

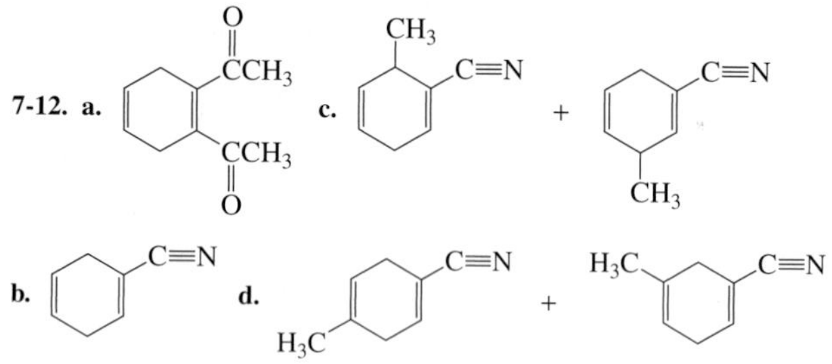

b.   **d.**

7-14.

7-15. a and **d**

7-17. a.

b.

c.

d.

e.   **f.**

b. 1. (structures: CH₃-substituted cyclohexene with H, Br; CH₃ cyclohexene with Br, H; Br, CH₃ cyclohexene; H₃C, Br cyclohexene) **2.** (two cyclohexane structures with H, Br / Br, CH₃ and Br, H / CH₃, Br)

3. (H₃C, Br cyclohexane) **4.** (four cyclohexane structures: H, H / Br, CH₃; H, H / CH₃, Br; Br, H / H, CH₃; H, Br / CH₃, H)

9-25. a. decrease **b.** decrease **c.** increase

9-26. a. $CH_3Br + HO^- \longrightarrow CH_3OH + Br^-$

b. $CH_3I + HO^- \longrightarrow CH_3OH + I^-$

c. $CH_3Br + NH_3 \longrightarrow CH_3\overset{+}{N}H_3 + Br^-$

d. $CH_3Br + HO^- \xrightarrow{\text{DMSO}} CH_3OH + Br^-$

e. $CH_3Br + NH_3 \xrightarrow{\text{EtOH}} CH_3\overset{+}{N}H_3 + Br^-$

9-28. dimethyl sulfoxide **9-29.** HO^- in 50% water/50% ethanol

CHAPTER 9

9-1. decrease **9-2.**

$CH_3CH_2CH_2CH_2CH_2Br > CH_3CHCH_2CH_2Br > CH_3CH_2CHCH_2Br > CH_3CH_2CBr$ (with CH₃ substituents)

9-4. a. RO^- **b.** RS^- **9-6. a.** aprotic **b.** aprotic **c.** protic **d.** aprotic

9-7. a. $CH_3CH_2Br + HO^-$ **d.** $CH_3Br + I^-$

b. $CH_3CHCH_2Br + HO^-$ (CH₃ substituent) **e.** $BrCH_2CH_2CH_2CH_2NH_2$

c. $CH_3CH_2Cl + CH_3S^-$

9-11. a. $CH_3CH_2OCH_3$ **b.** $CH_3CH_2N_3$ **c.** $CH_3CH_2\overset{+}{N}(CH_3)_3\ Br^-$

9-13.

$CH_3CBr > CH_3CHBr > CH_3CH_2CH_2Br > CH_3Br$ (with CH₃ substituents)

9-14.

$CH_3CH_2CH_2CH_3 > CH_3CHCH_2CH_2CH_3 > CH_3CHCH_2CH_2CH_3 >$ (with Br, Br, Cl substituents respectively)

$ClCH_2CH_2CH_2CH_2CH_3$

9-15. a, b, c

9-16.

$CH_2=CHCCH_3$ with $O-C=O-CH_3$ (kinetic product) and $CH_2CH=CCH_3$ with $O-C=O-CH_3$ (thermodynamic product)

kinetic product   **thermodynamic product**

9-17. *trans*-4-methylcyclohexanol **9-19. a** and **b**

9-20. a. $CH_3CH_2CH_2I$ **e.** (benzene)–CH_2CH_2Br

b. CH_3OCH_2Cl **f.** (benzene)–CH_2Br

c. CH_3CH_2CHBr (CH₃ substituent) **g.** $CH_3CH=CHCHCH_3$ (Br substituent)

d. $CH_3CH_2CHCH_2Br$ (CH₃ substituent)

9-21. a. $CH_3CH_2CH_2I$ **e.** (benzene)–CH_2CHCH_3 (Br substituent)

b. CH_3OCH_2Cl **f.** (benzene)–CH_2Br

c. equally reactive **g.** $CH_3CH=CHCHCH_3$ (Br substituent)

d. $CH_3CH_2CH_2CHBr$ (CH₃ substituent)

CHAPTER 10

10-2. a. $CH_3CH=CHCH_3$ **d.** $CH_3CHCH=CHCH_3$ (CH₃ substituent)

b. $CH_2=CHCH_2CH_3$ **e.** (cyclohexene)

c. $CH_3C=CHCH_2CH_3$ (CH₃ substituent) **f.** $CH_3CH=CHCH=CH_2$

10-3. a. $CH_3CH_2CHCH_2CH_3$ (Br substituent) **c.** $CH_3CHCHCH_2CH_3$ (CH₃ and Br substituents)

b. (benzene)–$CH_2CHCH_2CH_3$ (Br substituent) **d.** (cycloheptene with Br substituent)

10-4.

$CH_3C=CCH_2CH_3 > CH_3CHC=CHCH_3 > CH_3CHCH=CCH_2CH_3$ (with CH₃, CH₃, CH₂ and CH₃ substituents respectively)

10-7. a. E2 $CH_3CH=CHCH_3$ **d.** E2 $CH_3C=CH_2$ (CH₃ substituent)

b. E1 $CH_3CH=CHCH_3$ **e.** E1 $CH_3C=CCH_3$ (CH₃ substituents)

c. E1 $CH_3C=CH_2$ (CH₃ substituent) **f.** E2 $CH_3CCH=CH_2$ (CH₃ substituents)

10-8. a. 96% **b.** 1.2%

10-9. a. 1. $CH_3CH_2CH=CCH_3$ (CH₃ substituent) **2.** CH_3CH_2 and H / H and (benzene) on C=C

3. CH_3CH_2 and H / H and $CH=CH_2$ on C=C

b. no

10-10. a. H_3C and H / C=C with (benzene) groups **b.** (benzene) and H / C=C with H_3C and (benzene) groups

10-15. ~1 **10-16.** it will increase **10-18. a. 1.** no reaction **2.** no reaction **3.** substitution and elimination **4.** substitution and elimination **b. 1.** primarily substitution **2.** substitution and elimination **3.** substitution and elimination **4.** elimination

10-20. a. $CH_3CH=CH_2$ **b.** $CH_3CH_2CH=CH_2$

10-22.

$$CH_3\underset{\underset{CH_3}{|}}{\overset{\overset{CH_3}{|}}{C}}{-}OH \;+\; CH_3\underset{\underset{CH_3}{|}}{\overset{\overset{CH_3}{|}}{C}}{-}OCH_2CH_3 \;+\; CH_3\overset{\overset{CH_3}{|}}{C}{=}CH_2$$

10-25. a. HO$\diagup\!\diagdown\!\diagup\!\diagdown$Br **c.** HO$\diagup\!\diagdown\!\diagup\!\diagdown\!\diagup$Br

b. HO$\diagup\!\diagdown\!\diagup$Br

CHAPTER 11

11-1. $3° > 1° > 2°$

11-2. a. $CH_3CH_2\underset{\underset{Br}{|}}{CH}CH_3$ **c.** $CH_3\underset{\underset{Br}{|}}{\overset{\overset{CH_3}{|}}{C}}{-}\underset{\underset{CH_3}{|}}{CH}CH_3$

b. [cyclopentane with CH₃ and Cl]

d. [cyclohexane with CH₃, CH₂CH₃, Cl]

11-3. a. [structures W and X]

b. [structures Y and Z]

11-6. [cyclopentane with Br and CH₃]

11-10. No, F^- is too poor a nucleophile.

11-12. a. [epoxide]—$CH_2CH_2CH_3$ **c.** [tetramethyl epoxide]

b. [bicyclic epoxide] **d.** [epoxide]—CH_2CH_3

11-13. 2

11-14. a. $HOCH_2\underset{\underset{OCH_3}{|}}{\overset{\overset{CH_3}{|}}{C}}CH_3$ **c.** $HO\underset{\underset{CH_3}{|}}{\overset{\overset{OCH_3}{|}}{CH}}{-}\underset{\underset{CH_3}{|}}{C}CH_3$

b. $CH_3OCH_2\underset{\underset{CH_3}{|}}{\overset{\overset{OH}{|}}{C}}CH_3$ **d.** $CH_3O\underset{\underset{CH_3}{|}}{\overset{\overset{OH}{|}}{CH}}{-}\underset{\underset{CH_3}{|}}{C}CH_3$

11-15. noncyclic ether

11-17. [phenol with H and CH₃] and [phenol with D, OH, and CH₃]

11-18. The products are the same.

11-22. a. $CH_3CH_2CH_2CH_2CH_2OH$ **c.** [cyclohexane]—CH_2CH_2OH

b. [benzene]—$CH_2CH_2CH_2OH$

11-25. They all occur. **11-26.** $(CH_3)_4Si$

11-31. $CH_3C=CH_2 \;+\; \underset{\underset{CH_3}{|}}{\overset{\overset{CH_3}{|}}{N}}CH_2CH_2CH_3$

11-32. a. $CH_3CH=CH_2 \;+\; \underset{\underset{CH_3}{|}}{\overset{\overset{CH_3}{|}}{N}}CH_3$ **c.** [methylenecyclohexane] $+\; N(CH_3)_3$

b. $CH_2=CHCH\underset{\underset{}{}}{\overset{\overset{CH_3}{|}}{}}CH_2CH_2\overset{\overset{CH_3}{|}}{N}CH_3$ **d.** $CH_3N\underset{}{}CH_2\overset{\overset{CH_3}{|}}{CH}CH_2CH=CH_2$

11-35. carbanion-like

11-36.

a. $CH_3\underset{\underset{OH}{|}}{N} \;+\; CH_2=CHCH_3$ **c.** $CH_2=CH_2 \;+\; \underset{\underset{OH}{|}\;\;\underset{CH_3}{}}{N}CH_2\overset{}{CH}CH_3$

b. $CH_3\underset{\underset{[phenyl]}{|}}{\overset{\overset{OH}{|}}{N}} \;+\; CH_2=CHCH_3$ **d.** $CH_2=CHCH_2CH_2CH_2CH_2\overset{\overset{}{}}{N}CH_3$, OH

CHAPTER 12

12-2. $m/z = 57$ **12-5.** 8 **12-6.** C_6H_{14} **12-8. a.** 2-methoxy-2-methylpropane **b.** 2-methoxybutane **c.** 1-methoxybutane

12-9. $CH_2=\overset{+}{\underset{\cdot\cdot}{O}}H$

12-10. a. 2-pentanone **b.** 3-pentanone **12-12. a.** 3-pentanol **b.** 2-pentanol
12-13. a. $2000\ cm^{-1}$ **b.** $8\ \mu m$ **c.** $2\ \mu m$ **12-14. a.** $2500\ cm^{-1}$ **b.** $50\ \mu m$

12-15. a. 1. $C=O$ **2.** $C-H$ stretch **3.** $C\equiv C$
b. 1. $C-Cl$ **2.** $C-O$

12-16. a. carbon-nitrogen stretch of an amide **b.** carbon-oxygen stretch of phenol **c.** carbon-oxygen double bond stretch of a ketone **d.** carbon-oxygen stretch **12-17.** sp^3

12-18. a. [lactone] > [cyclohexanone] > [lactam NH]

b. [structures] > [structures] > [structures]

c. $H{-}\overset{\overset{O}{\|}}{C}{-}H \;>\; CH_3{-}\overset{\overset{O}{\|}}{C}{-}H \;>\; CH_3{-}\overset{\overset{O}{\|}}{C}{-}CH_3$

12-19. ethanol dissolved in carbon disulfide **12-21.** ketone **12-24.** CO_2, 2-butyne, H_2, Cl_2, ethene **12-25.** *trans*-2-hexene **12-26.** methyl vinyl ketone **12-27. a.** UV light **b.** visible light **12-28.** $4.1 \times 10^{-5}\ M$

12-29. a. [structure] > [structure with CH₃] > [structure with CH₃]

b. C₆H₅—CH=CH—C₆H₅ > (biphenyl) > C₆H₅—CH=CH₂ > (benzene)

CHAPTER 13

13-1. 43 MHz **13-2. a.** 8.46 T **b.** 11.75 T **13-3. a.** 4 **b.** 4 **c.** 3
d. 3 **e.** 2 **f.** 3 **g.** 1 **h.** 5 **i.** 3 **j.** 3 **k.** 3 **l.** 4 **m.** 2 **n.** 1 **o.** 3
13-6. a. 0.4 ppm **b.** 0.4 ppm **c.** 40 Hz **13-7. a.** 1.5 ppm **b.** 30 hertz
13-8. to the right of the TMS signal

13-9. a. CH₃CH₂CH₂Cl (most/least) **b.** CH₃CH₂COCH₃ (most/least) **c.** CH₃CHCHBr Br Br (most/least)

13-10. a. CH₃CH₂CHCH₃ Cl **d.** CH₃CH₂CH=CH₂

b. CH₃CH₂CHCH₃ Cl **e.** CH₃CH₂CH=O **g.** CH₃CHCCH₂CH₃ CH₃ =O

c. CH₃CH₂CH₂Cl **f.** CH₃CHOCH₃ CH₃ **h.** CH₃CHCHBr Br Br

13-11.

a. CH₃CH₂CH₂Cl (a b c) **e.** CH₃CHCH₂OCH₃ CH₃ (a b d c) **i.** CH₃CH₂CHCH₂CH₃ OCH₃ (a b d c)

b. CH₃CHCHCH₃ Cl CH₃ (a d b c) **f.** CH₃CH₂CH=O (a b c) **j.** CH₃CH₂CH₂OCHCH₃ CH₃ (a c d e b)

c. CH₃CH₂CH₂COCH₃ (a b c =O d) **g.** CH₃CH₂CH₂CCH₃ =O (a b d c)

d. CH₃CHCHBr Br Br (a b c) **h.** ClCH₂CH₂CH₂Cl (b a)

13-12. The compounds have different integration ratios: 2 : 9, 1 : 3 and 2 : 1
13-14.

CH₃—C₆H₄—CH₃

13-15. 9.25 ppm = hydrogens that protrude out; −2.88 ppm = hydrogens
that point in **13-17.** first spectrum = 1-iodopropane
13-20. a. propyl benzene **b.** 3-pentanone **c.** ethyl benzoate

13-23. a. CH₃OCH₂CH₂OCH₃ **c.** HC(=O)—C₆H₄—CH(=O)

b. CH₃CCH₂CH₂CCH₃ (=O =O)
or
CH₃OCH₂C≡CCH₂OCH₃
or
(dioxane ring with two CH₃)

13-25. CH₃O—C₆H₄—CH₃ **13-28.** pure ethanol

13-30. propanamide
13-31. a. 1. 3 **2.** 3 **3.** 2 **4.** 3 **5.** 4 **6.** 3 **7.** 2 **8.** 3 **9.** 4 **10.** 3

b. 1. CH₃CH₂CH₂Br **4.** CH₃COCH₃ (CH₃ / CH₃) **7.** (H)(H)C=C(Br)(H)

2. CH₃CH₂OCH₃ **5.** CH₃CH₂COCH₃ (=O) **8.** CH₃CHCH (CH₃) =O

3. CH₃CHCH₃ Br **6.** (H₃C)(H₃C)C=C(H)(H) **9.** (chlorobenzene ring with H's)

10. CH₃CCH₂CH₂CCH₃ (=O =O)

13-34.

a. CH₃CH₂CH₂CH₂CH₂CCH₂CH₂CH₂CH₂CH₃ (=O) **c.** CH₃CHCHCl₂ Cl

b. Br—C₆H₄—CH₂CH₃ **d.** CH₃CH₂CH=CHCH₂CH₃

CHAPTER 14

14-1. a. 4 **b.** It will be aromatic if it is cyclic, planar, and every carbon in
the ring has a p orbital. **14-2.** b, c, e, g **14-5.** quinoline = sp^2;
indole = p; imidazole = 1 is sp^2 and 1 is p; purine = 2 are sp^2 and 1 is
p; pyrimidine = both are sp^2 **14-7.** Cyclopentadiene has the lower pK_a.
14-9. a. Cyclopropane has the lower pK_a. **b.** 3-bromocyclopropene
14-10. a **14-12.** 1 bonding, 2 nonbonding, and 0 antibonding; 2 are in the
bonding π MO and 1 is in each of the two nonbonding MOs
14-15. Hydrated FeBr₃ cannot activate Br₂ for nucleophilic attack by
accepting electrons from it because it has accepted a pair of electrons from
water. **14-18. a.** ethylbenzene **b.** isopropylbenzene **c.** *sec*-butylbenzene
d. *tert*-pentylbenzene **e.** *tert*-butylbenzene **f.** 3-phenylpropene

CHAPTER 15

15-1. a. CH₃—(ring)—NH₂ **b.** CH₃—(ring)—OH **c.** CH₃—(ring)—CH₃

15-2. a. (ring with Cl, Cl) **d.** (ring with NO₂, Br, I) **g.** (ring with CH₃, Cl) **j.** (ring with C≡N, Cl)

b. (ring with OH, Br) **e.** CH₃CHCH₂CH₂CH₂CH₃ (phenyl) **h.** (ring with HC=O, NO₂, O₂N)

c. (ring with NH₂, NO₂) **f.** CH₃CH₂CHCH₂CH₃ CH₂ (phenyl) **i.** (ring with CH₃, CH₃)

15-3. a. 1,3,5-tribromobenzene **b.** 3-nitrophenol **c.** *para*-bromotoluene **d.** 1,2-dichlorobenzene

15-4. a.

b. CH$_2$OCH$_3$ **c.** COOH / COOH **d.** CH$_2$CH$_2$NH$_2$

15-6. a. donates electrons by resonance and withdraws electrons inductively **b.** donates electrons inductively **c.** withdraws electrons by resonance and withdraws electrons inductively **d.** donates electrons by resonance and withdraws electrons inductively **e.** donates electrons by resonance and withdraws electrons inductively **f.** withdraws electrons inductively **15-7. a.** phenol > toluene > benzene > bromobenzene > nitrobenzene **b.** toluene > chloromethylbenzene > dichloromethyl-benzene > difluoromethylbenzene

15-10. a. CH$_2$CH$_2$CH$_3$ / NO$_2$ + CH$_2$CH$_2$CH$_3$ / NO$_2$ **d.** SO$_3$H / NO$_2$

b. Br / NO$_2$ + Br / NO$_2$

e.

c. CH / NO$_2$ **f.** C≡N / NO$_2$

15-11. They are all meta directors.

15-12.

a. ClCH$_2$COH **c.** H$_3$NCH$_2$COH **e.** HCOH **g.** FCH$_2$COH

b. O$_2$NCH$_2$COH **d.** HOCCH$_2$COH **f.** COOH **h.** COOH / Cl

15-15. a. no reaction **c.** no reaction

b. NH$_2$ / Br, Br / Br **d.** CH$_3$ / CH$_3$ + CH$_3$ / CH$_3$

15-17. a. OCH$_3$ / NO$_2$ / F **b.** COOH / COOH / Cl **c.** COOH / Br / Cl

15-21. a. no **b.** phenol

15-26. CH$_3$CHCH$_3$ / Cl + CH$_3$CHCH$_3$ / OH + CH$_3$CH=CH$_2$

15-29.
a. 1-chloro-2,4-dinitrobenzene > *p*-chloronitrobenzene > chlorobenzene
b. chlorobenzene > *p*-chloronitrobenzene > 1-chloro-2,4-dinitrochlorobenzene

CHAPTER 16

16-2. a. butanenitrile, propyl cyanide **b.** ethanoic propanoic anhydride, acetic propionic anhydride **c.** potassium butanoate **d.** pentanoyl chloride, valeryl chloride **e.** isobutyl butanoate, isobutyl butyrate **f.** *N,N*-dimethylhexanamide **g.** γ-butyrolactam **h.** cyclopentanecarboxylic acid
16-3. C—O bond in an alcohol; no double-bond character
16-4. The shortest bond has the highest frequency.

a. CH$_3$—C(=O)—O—CH$_3$ **b.** CH$_3$—C(=O)—O—CH$_3$

1 = longest / 3 = shortest   1 = highest frequency / 3 = lowest frequency

16-5. the carbonyl oxygen

16-7. a. no reaction **b.** CH$_3$—C(=O)—O—C(=O)—CH$_3$ **c.** no reaction

16-8. true
16-12. a. CH$_3$CH$_2$CH$_2$OH **c.** (CH$_3$)$_2$NH **e.** HO—C$_6$H$_4$—NO$_2$
b. CH$_3$CH$_2$NH$_2$ **d.** CH$_3$CO$^-$ **f.** H$_2$O

16-20. a. CH$_3$CH$_2$CH$_2$O$^-$ **b.** H$^+$ would make the amine unreactive by protonating it. HO$^-$ would form the hydrolysis product. RO$^-$ would form an ester. **16-21. a.** the alcohol **b.** the carboxylic acid
16-24.

CH$_3$CNH—C$_6$H$_4$—NO$_2$ > CH$_3$CNH—C$_6$H$_4$(NO$_2$) > CH$_3$CNH—C$_6$H$_5$ > CH$_3$CNH—C$_6$H$_{11}$

16-25. a. pentyl bromide **b.** isohexyl bromide **c.** benzyl bromide **d.** cyclohexyl bromide

16-27. a. CH$_3$CH$_2$CH$_2$Br **c.** cyclohexyl—Br
b. CH$_3$CHCH$_2$Br / CH$_3$ **d.** BrCH$_2$CH$_2$Br

16-30. a. (CH$_3$CH$_2$)$_2$NCN(CH$_2$CH$_3$)$_2$ **d.** no reaction
b. CH$_3$COCCH$_2$COCCH$_3$ **e.** HOCOH ⇌ CO$_2$ + H$_2$O
c. H$_2$NCNHCH$_3$ **f.** CH$_3$CH$_2$—(lactone ring)

CHAPTER 17

17-2. a. 3-methylpentanal, β-methylvaleraldehyde **b.** 4-heptanone, dipropyl ketone **c.** 2-methyl-4-heptanone, isobutyl propyl ketone **d.** 4-phenylbutanal, γ-phenylbutyraldehyde **e.** 4-ethylhexanal, γ-ethylcaproaldehyde **f.** 1-hepten-3-one, butyl vinyl ketone
17-3. a. 6-hydroxy-3-heptanone **b.** 2-oxocyclohexylmethanenitrile **c.** 3-formylpentanenitrile **17-4. a.** 2-heptanone **b.** *para*-nitroacetophenone
17-5. a. two; (*R*)-3-methyl-3-hexanol and (*S*)-3-methyl-3-hexanol **b.** one; 2-methyl-2-pentanol

17-6.

$$CH_3\overset{O}{\overset{\|}{C}}CH_2CH_3 \ + \ CH_3CH_2CH_2MgBr$$

$$CH_3CH_2\overset{O}{\overset{\|}{C}}CH_2CH_2CH_3 \ + \ CH_3MgBr$$

17-7. Grignard reagent + formaldehyde

17-12. a. $CH_3\underset{CH_3}{\overset{|}{C}}HCH_2OH$

c. ⬡—CH_2OH

b. ⬡—OH

d. ⬡—$\underset{OH}{\overset{|}{C}}HCH_3$

17-13. an alkane and a carboxylate ion

17-15. 2-propanol and 3-pentanol

17-16. a. ⬡—$\overset{O}{\overset{\|}{C}}NHCH_3$

c. $CH_3\overset{O}{\overset{\|}{C}}NHCH_2CH_3$

b. $CH_3\overset{O}{\overset{\|}{C}}NH_2$

d. $CH_3\overset{O}{\overset{\|}{C}}\underset{CH_2CH_3}{\overset{CH_2CH_3}{N}}$

17-20.

a. ⬠=NCH_2CH_3 + H_2O

c. ⬡—$\underset{CH_3}{\overset{|}{C}}=N(CH_2)_5CH_3$ + H_2O

b. ⬠—$\underset{CH_2CH_3}{\overset{CH_2CH_3}{N}}$ + H_2O

d. ⬡—$\underset{CH_3}{\overset{|}{C}}=N$—⬡ + H_2O

17-24. the ketone with the nitro substituents

17-28.

a. ⬡—$\underset{CH_3}{\overset{OH}{\overset{|}{C}}}CH_2CH_3$ **S**

c. $CH_3\underset{CH_3}{\overset{OH}{\overset{|}{C}}}CH_2CH_2CH_3$

compound does not have a chirality center

b. $CH_3\underset{H}{\overset{OH}{\overset{|}{C}}}$⬡ **R**

d. $CH_3CH_2CH_2\underset{CH_3}{\overset{OH}{\overset{|}{C}}}CH_2CH_3$ **S**

17-30. a. (bicyclic structure with C≡N and =O)

c. $CH_3C=CHC\underset{CH_3}{\overset{OH}{\overset{|}{C}}}CH_3$ (with CH_3)

b. (bicyclic structure with OH)

d. $CH_3\underset{CH_3}{\overset{CH_3}{\overset{|}{C}}}-CH_2\overset{O}{\overset{\|}{C}}CH_3$

CHAPTER 18

18-1. a. reduction **b.** neither **c.** oxidation **d.** oxidation **e.** reduction **f.** neither

18-2. a. $CH_3CH_2CH_2CH_2CH_2OH$
b. $CH_3CH_2CH_2CH_2NH_2$
c. $CH_3CH_2CH_2\underset{H}{\overset{}{C}}=\underset{H}{\overset{CH_3}{C}}$
d. ⬡—OH
e. no reaction
f. CH_3CH_2OH
g. $CH_3\overset{O}{\overset{\|}{C}}H$
h. ⬡—$NHCH_3$

18-4.

a. ⬡—CH_2NH_2
c. $CH_3CH_2\underset{}{\overset{OH}{\overset{|}{C}}}HCH_2CH_3$
e. $CH_2CH_2CH_2NHCH_2CH_3$

b. ⬡—CH_2OH
d. ⬡—CH_2OH + CH_3CH_2OH
f. $CH_3CH_2CH_2CH_2OH$

18-7. a. 1. $CH_3CH_2\overset{O}{\overset{\|}{C}}CH_2CH_3$ **3.** no reaction **5.** ⬡=O

2. $CH_3CH_2CH_2CH_2\overset{O}{\overset{\|}{C}}OH$ **4.** $CH_3\overset{O}{\overset{\|}{C}}CH_2\overset{O}{\overset{\|}{C}}CH_2CH_3$ **6.** $HO\overset{O}{\overset{\|}{C}}CH_2CH_2\overset{O}{\overset{\|}{C}}OH$

b. 1. $CH_3CH_2\overset{O}{\overset{\|}{C}}CH_2CH_3$ **3.** no reaction **5.** ⬡=O

2. $CH_3CH_2CH_2CH_2\overset{O}{\overset{\|}{C}}H$ **4.** $CH_3\overset{O}{\overset{\|}{C}}CH_2\overset{O}{\overset{\|}{C}}CH_2CH_3$ **6.** $H\overset{O}{\overset{\|}{C}}CH_2CH_2\overset{O}{\overset{\|}{C}}H$

18-11. a. cyclohexene **b.** 1-butene **c.** *trans*-2-pentene **d.** *cis*-2-pentene

18-15. a. $CH_3\underset{OH}{\overset{CH_3}{\overset{|}{C}}}\underset{OH}{\overset{|}{C}}HCH_2CH_3$ **b.** ⬡ (with CH_2OH and OH)

18-17. (bicyclohexylidene structure)

18-18. a. (cyclohexane with OH, OH, C(CH₃)₃) **b.** (cyclohexane with OH, OH, C(CH₃)₃)

18-21. a. 2,3-dimethyl-2-butene **b.** *cis*-4-octene and *trans*-4-octene
18-22. whether the compound has the cis or the trans configuration

18-24. a. ⬡—C≡CH **b.** $CH_3CH_2C≡CCH_2CH_2CH_2C≡CCH_2CH_3$

18-26. a.

$$\overset{Br}{\underset{}{\diagup}} \xrightarrow[HBr/\Delta]{HO^-} \overset{OH}{\underset{}{\diagup}} \xrightarrow[\substack{1.\ NaBH_4 \\ 2.\ H^+,H_2O}]{H_2CrO_4} \overset{O}{\underset{}{\diagup}}$$

b. cyclohexene
$\xrightarrow[\substack{1.\ OsO_4 \\ 2.\ H_2O,\ NaHSO_3 \\ or \\ 1.\ KMnO_4,\ HO^- \\ 2.\ H_2O}]{}$ (cyclohexane with OH, OH)

$\xrightarrow[\substack{1.\ RCOOH \\ 2.\ HO^-}]{}$ (cyclohexane with OH, OH)

CHAPTER 19

19-2. a. $CH_3\overset{O}{\overset{\|}{C}}CH_2C≡N$ a β-keto nitrile **b.** $CH_3O\overset{O}{\overset{\|}{C}}CH_2\overset{O}{\overset{\|}{C}}OCH_3$ a β-diester

19-3. a. $CH_3\overset{O}{\overset{\|}{C}}H$ > HC≡CH > $CH_2=CH_2$ > CH_3CH_3

b. $CH_3\overset{O}{\overset{\|}{C}}CH_2\overset{O}{\overset{\|}{C}}CH_3$ > $CH_3\overset{O}{\overset{\|}{C}}CH_2\overset{O}{\overset{\|}{C}}OCH_3$ > $CH_3O\overset{O}{\overset{\|}{C}}CH_2\overset{O}{\overset{\|}{C}}OCH_3$ > $CH_3\overset{O}{\overset{\|}{C}}CH_3$

c. > ... > ...

19-5. a. $CH_3CH=CCH_2CH_3$ (OH) **b.** [structure with OH, C=CH$_2$] **c.** [cyclohexenol]

d. [structure] and [structure] — **more stable**

e. $CH_3CH_2C=CHCCH_2CH_3$ (OH, O) — **more stable** — and $CH_3CH=CCH_2CCH_2CH_3$ (OH, O)

f. [structure] $-CH=CCH_3$ (OH) — **more stable** — and [structure] $-CH_2C=CH_2$ (OH)

19-7. The rate-determining step must be removal of the proton from the α-carbon of the ketone.

19-11. [cyclohexanone with D substituents]

19-13.

a. [cyclohexanone] →(1. LDA/THF, 2. ICH$_2$CH=CH$_2$) → [product with CH$_2$CH=CH$_2$]

b. $CH_3CH_2CCH_2CH_3$ →(1. LDA/THF, 2. [benzyl]—CH$_2$Br) → [product] $-CH_2CHCCH_2CH_3$ with CH$_3$

19-15. a. [cyclohexenone], $CH_3CCH_2CCOCH_3$, CH_3O^-

b. $CH_3CCH=CH_2$, $CH_3CH_2OCCH_2COCH_2CH_3$, $CH_3CH_2O^-$

19-17. a. $CH_3CH_2CH_2CH$ (O) **c.** [cyclohexyl] $-CH_2CH$ (O)

b. CH_3CCH_3 (O) **d.** $CH_3CH_2CCH_2CH_3$ (O)

19-23.

a. $CH_3CH_2CH_2CCHCOCH_3$ with CH_2 / CH_3 **b.** $CH_3CHCH_2CCHCOCH_2CH_3$ with CH_3 / $CHCH_3$ / CH_3

19-28. a. [cyclobutane with OH, CH$_3$, CCH$_3$(O)] **b.** [cyclooctane with OH, CH$_3$, O]

19-31. a. **c.** [structure with OH, CH (O)]

b. [HO bicyclic structure] **d.** [structure with OH, CCH$_3$(O)]

19-33. **a** and **d**; **b** doesn't have a carboxyl group. **19-34. a.** methyl bromide **b.** methyl bromide (twice) **c.** benzyl bromide **d.** propyl bromide **e.** isobutyl bromide **f.** propyl bromide and methyl bromide

19-36. a. [cyclohexyl] $-COH$ (O) **b.** $HOCCH_2CH_2CH_2CH_2CH_2CH_2COH$ (O...O)

19-37. a. ethyl bromide **b.** pentyl bromide **c.** benzyl bromide
19-40. 7 **19-41. a.** 3 **b.** 7

CHAPTER 20

20-1. D-Ribose is an aldopentose. D-Sedoheptulose is a ketoheptose. D-Mannose is an aldohexose. **20-3. a.** L-glyceraldehyde **b.** L-glyceraldehyde **c.** D-glyceraldehyde **20-4. a.** enantiomers **b.** diastereomers
20-5. a. D-ribose **b.** L-talose **c.** L-allose **20-6. a.** (2R,3S,4R,5R)-2,3,4,5,6-pentahydroxyhexanal **b.** (2S,3R,4S,5S)-2,3,4,5,6-pentahydroxyhexanal
20-7. D-psicose **20-8. a.** A ketoheptose has four chirality centers (2^4 = 16 stereoisomers). **b.** An aldoheptose has five chirality centers (2^5 = 32 stereoisomers). **c.** A ketotriose has no chirality centers; therefore, it has no stereoisomers. **20-9. a.** D-iditol **b.** D-iditol and D-gulitol
20-10. a. 1. D-altrose 2. L-galactose **b.** 1. D-tagatose 2. D-fructose
20-12. D-tagatose, D-galactose, and D-talose **20-13. a.** L-gulose **b.** L-gularic acid **c.** D-allose and L-allose, D-altrose and L-altrose, L-altrose and L-talose, D-galactose and L-galactose **20-14. a.** D-arabinose and D-ribulose **b.** D-allose and D-psicose **c.** L-gulose and L-sorbose **d.** D-talose and D-tagatose
20-15. D-gulose and D-idose **20-16. a.** D-gulose and D-idose **b.** L-xylose and L-lyxose **20-17. a.** D-glucose and D-mannose **b.** D-erythrose and D-threose **c.** L-allose and L-altrose **20-19. A** = D-glucose **B** = D-mannose **C** = D-arabinose **D** = D-erythrose **20-23. a.** the OH group at C-4 **b.** the OH group at C-2, C-3, and C-4 **c.** the OH group at C-3 and C-1 **20-25.** A protonated amine is not a nucleophile. **20-30. a.** They cannot receive type A, B, or AB blood because these have sugar components that type O blood does not have. **b.** They cannot give blood to those with type A, B, or O blood because AB blood has sugar components that these other blood types do not have.

CHAPTER 21

21-2. a. (R)-alanine **b.** (R)-aspartate **c.** The α-carbon of all the D-amino acids except cysteine has the R-configuration. **21-3.** Ile, Thr
21-4. because of the electron-withdrawing ammonium group

21-6. a. $HOCCH_2CH_2CHCOH$ with $^+NH_3$ **c.** $^-OCCH_2CH_2CHCO^-$ with $^+NH_3$

b. $HOCCH_2CH_2CHCO^-$ with $^+NH_3$ **d.** $^-OCCH_2CH_2CHCO^-$ with NH_2

21-9. a. 5.43 **b.** 10.76 **c.** 5.68 **21-10. a.** Asp **b.** Arg **c.** Asp **d.** Met
21-12. Leucine and isoleucine have similar polarities and pI values, so they show up as one spot. **21-15.** His > Val > Ser > Asp

21-18. a. L-Ala, L-Asp, L-Glu **b.** L-Ala and D-Ala, L-Asp and D-Asp, L-Glu and D-Glu **21-20. a.** $2^8 = 25,600,000,000$ **b.** 20^{100} **21-21.** the bonds on either side of the α-carbon **21-26. a.** 5.8% **b.** 4.4% **21-29.** Gly-Arg-Trp-Ala-Glu-Leu-Met-Pro-Val-Asp **21-31.** Edman's reagent would release two amino acids in approximately equal amounts. **21-33.** Leu-Tyr-Lys-Arg-Met-Phe-Arg-Ser **21-34.** 110 Å in an α-helix and 260 Å in a straight chain **21-35.** nonpolar groups on the outside and polar groups on the inside **21-36. a.** cigar-shaped protein **b.** subunit of a hexamer

CHAPTER 22

22-1. $\Delta H^\dagger$, E_a, $\Delta S^\dagger$, $\Delta G^\dagger$, k_{rate} **22-5.** Hydroxide ion acts as a nucleophile. **22-8.** 2.5×10^6 **22-11.** close to one **22-12. a.** 4

b.

22-14. The nitro groups cause the phenoxide ion to be a better leaving group than the carboxylate ion. **22-16.** Ser-Ala-Phe **22-18.** Arginine forms a direct hydrogen bond, lysine forms an indirect hydrogen bond. **22-20.** NAM **22-21.** The acid denatures the enzyme. **22-24.** Putting a substituent on cysteine could interfere with binding or catalysis of the substrate.

CHAPTER 23

23-1. It increases the susceptibility of the carbonyl carbon to nucleophilic attack, increases the nucleophilicity of water, and stabilizes the negative charge on the transition state. **23-2.** pyrophosphate **23-3. a.** 7 **b.** 3 isolated from 2 others **23-12. a.** pyridoxal phosphate and α-ketoglutarate **b.** **c.** S-adenosylmethionine

23-13. alanine and aspartate
23-19.

$$CH_3CH_2CSCoA \xrightarrow[\text{biotin}]{E} CH_3CHCSCoA \xrightarrow[\text{coenzyme } B_{12}]{E} CH_2CH_2CSCoA$$

23-20. the methylene group of N^5,N^{10}-methylene-THF

CHAPTER 24

24-2. hexanoic acid, 3 malonic acids, and glutaric acid **24-3.** no **24-4.** glyceryl tripalmitate **24-6.** Integral proteins will have a higher percentage of nonpolar amino acids. **24-7.** The bacteria could synthesize phosphoacylglycerols with more saturated fatty acids. **24-12.** The two halves are synthesized in a head-to-tail fashion and are joined together in a tail-to-tail linkage. **24-20.** cis fused; trans fused **24-21.** β-substituent **24-24.** Two are axial substituents and one is an equatorial substituent.

CHAPTER 25

25-4. phosphocreatine
25-11. a. 3′——C—C—T—G—T—T—A—G—A—C—G——5′
b. guanine **25-12. a.** 25 Å **b.** 34 Å **c.** 46 Å **25-14.** 10^{10}Å
25-17. Met-Asp-Pro-Val-Ile-Lys-His **25-18.** Met-Asp-Pro-Leu-Leu-Asn
25-20. 5′——G-C-A-T-G-G-A-C-C-C-C-G-T-T-A-T-T-A-A-A-C-A-C——3′

25-22.

hypoxanthine   xanthine

25-24. a

CHAPTER 26

26-1. a. $CH_2=CHCl$ **b.** $CH_2=CCH_3$ **c.** $CF_2=CF_2$

26-2. poly(vinyl chloride) **26-5.** beach balls

26-7. a.

b. $CH_2=CHOCCH_3$ > $CH_2=CHCH_3$ > $CH_2=CHCOCH_3$
c. $CH_2=CCH_3$ $CH_2=CH$

26-8. a.

b. $CH_2=CHC\equiv N$ > $CH_2=CHCl$ > $CH_2=CHCH_3$

26-11. a. $CH_2=CCH_3 + BF_3$ **c.** + CH_3O^-

b. $CH_2=CH + BF_3$ **d.** $CH_2=CH + BuLi$

26-12.

26-15. a.

b.

c.

CHAPTER 27

27-1. a. 2,2-dimethylaziridine **b.** 4-ethylpiperidine **c.** 3-methylazacyclobutane **d.** 2-methylthiacyclopropane **e.** 2,3-dimethyltetrahydrofuran **f.** 2-ethyloxacyclobutane **27-2.** The electron-withdrawing oxygen stabilizes the conjugate base.
27-3. a. **b.** pK_a ~8 **c.** 3-chloroquinuclidine

27-4. a. $CH_3\overset{\overset{O}{\|}}{C}-N\langle\rangle$

c. $HOCH_2CH_2CH_2CH_2\overset{\overset{O}{\|}}{C}OH$

b. $HOCH_2CH_2CH_2CH_2CH_2I$ **d.** (piperidinium ring) $\overset{+}{N}$ with H_3C CH_3 $\quad I^-$

27-9. $CH_3\overset{\overset{O}{\|}}{C}OCH_3$

27-14.

(4-methyl-N-ethylpyridinium iodide) $\overset{+}{N}$ CH_2CH_3 I^- $>$ (4-methylpyridine) N $>$ (3-methylpyridine with CH_3) N

27-17. Imidazole forms intramolecular hydrogen bonds that *N*-methylimidazole cannot form. **27-18.** 24% **27-19.** yes

CHAPTER 28

28-1. a. electrocyclic reaction **b.** sigmatropic rearrangement **c.** cycloaddition reaction **d.** cycloaddition reaction **28-2. a.** bonding orbitals = ψ_1,ψ_2,ψ_3; antibonding orbitals = ψ_4,ψ_5,ψ_6; **b.** ground state HOMO = ψ_3; ground state LUMO = ψ_4 **c.** excited state HOMO = ψ_4; excited state LUMO = ψ_5 **d.** symmetric orbitals = ψ_1,ψ_3,ψ_5; asymmetric orbitals = ψ_2,ψ_4,ψ_6 **e.** The HOMO and LUMO have opposite symmetries. **28-3. a.** 8 **b.** ψ_4 **c.** 8 **28-6. a.** conrotatory **b.** trans **c.** disrotatory **d.** cis **28-7. a.** correct **b.** correct **c.** correct **28-8. 1. a.** conrotatory **b.** trans **2. a.** disrotatory **b.** cis **28-11.** yes **28-13.**

28-18. a. inversion **b.** retention **28-19.** antarafacial

CHAPTER 29

29-1. The linear synthesis results in a 21% yield; the five-step convergent synthesis results in a 41% yield. **29-2.** The reaction of a Grignard reagent with ethylene oxide. **29-3.** The reaction of a Grignard reagent with ethylene oxide or with formaldehyde. **29-7. a.** *umpolung* **b.** normal **c.** normal **d.** *umpolung* **e.** normal **f.** *umpolung* **g.** *umpolung* **h.** normal **29-8. a.** Use 1-bromohexane instead of 1-bromopropane. **b.** Use 2-methyl-1,3-dithiane instead of 1,3-dithiane.

29-18. a. $CH_3CH_2\overset{\overset{\overset{O}{\|}}{}}{\underset{CH_2CH_3}{C}}HCH$ $+$ $CH_2{=}P(C_6H_5)_3$

b. Substitution will not compete with elimination. **29-20.** It will become an alkane as a result of protonation by the OH group.

29-21. $CH_3\underset{COOH}{CH}CH_2\overset{\overset{O}{\|}}{C}COOH$

29-23. Both Ala-Gly and Ala-Ala would be formed.

29-26.

29-28. It is a stereoselective reaction; it is a stereospecific reaction; it is not an enantioselective reaction. **29-39.** Only under photochemical conditions can the required suprafacial mode of ring closure occur.

Glossary

absolute configuration   the three-dimensional structure of a chiral compound. The configuration is designated by *R* or *S*.

absorption band   a peak in a spectrum that occurs as a result of absorption of energy.

acetal $\begin{array}{c} \text{OR} \\ | \\ \text{RCH} \\ | \\ \text{OR} \end{array}$

acetoacetic ester synthesis   synthesis of a methyl ketone using ethyl acetoacetate as the starting material.

achiral (optically inactive)   an achiral molecule has a conformation with a superimposable mirror image.

acid   a substance that donates a proton

acid anhydride   R—C(=O)—O—C(=O)—R

acid–base reaction   a reaction in which an acid donates a proton to a base or accepts a share in a base's electrons.

acid catalyst   a catalyst that increases the rate of a reaction by donating a proton.

acid-catalyzed reaction   a reaction catalyzed by an acid.

acid dissociation constant   a measure of the degree to which an acid dissociates in solution.

activating substituent   a substituent that increases the reactivity of an aromatic ring. Electron-donating substituents activate aromatic rings toward electrophilic attack, and electron-withdrawing substituents activate aromatic rings toward nucleophilic attack.

active site   a pocket or cleft in an enzyme where the substrate is bound.

acyclic   noncyclic.

acyl adenylate   a carboxylic acid derivative with AMP as the leaving group.

acyl–enzyme intermediate   an intermediate formed when an amino acid residue of an enzyme is acetylated.

acyl group   a carbonyl group bonded to an alkyl group or to an aryl group.

acyl halide   R—C(=O)—Cl

acyl phosphate   a carboxylic acid derivative with a phosphate leaving group.

acyl pyrophosphate   a carboxylic acid derivative with a pyrophosphate leaving group.

1,2-addition (direct addition)   addition to the 1- and 2-positions of a conjugated system.

1,4-addition (conjugate addition)   addition to the 1- and 4-positions of a conjugated system.

addition polymer (chain-growth polymer)   a polymer made by adding monomers to the growing end of a chain.

addition reaction   a reaction in which atoms or groups are added to the reactant.

adrenal cortical steroids   glucocorticoids and mineralocorticoids.

alcohol   a compound with an OH group in place of one of the hydrogens of an alkane; (ROH).

alcoholysis   reaction with an alcohol.

aldaric acid   a dicarboxylic acid with an OH group bonded to each carbon. Obtained by oxidizing the aldehyde and primary alcohol groups of an aldose.

aldehyde

alditol   a compound with an OH group bonded to each carbon. Obtained by reducing an aldose or a ketose.

aldol addition   a reaction between two molecules of an aldehyde (or two molecules of a ketone) that connects the α-carbon of one with the carbonyl carbon of the other.

aldol condensation   an aldol addition followed by elimination of water.

aldonic acid   a carboxylic acid with an OH group bonded to each carbon. Obtained by oxidizing the aldehyde group of an aldose.

aldose   a polyhydroxyaldehyde.

aliphatic compound   a nonaromatic organic compound.

alkaloid   a natural product with one or more nitrogen heteroatoms found in the leaves, bark, or seeds of plants.

alkane   a hydrocarbon that contains only single bonds.

alkene   a hydrocarbon that contains a double bond.

alkoxymercuration   addition of alcohol to an alkene using a mercuric salt of a carboxylic acid as a catalyst.

alkylation reaction   a reaction that adds an alkyl group to a reactant.

alkyl halide   a compound with a halogen in place of one of the hydrogens of an alkane.

alkyl substituent (alkyl group)   formed by removing a hydrogen from an alkane.

alkyl tosylate   an ester of *para*-toluenesulfonic acid.

alkyne   a hydrocarbon that contains a triple bond.

allene   a compound with two adjacent double bonds.

allyl group CH_2=$CHCH_2$—

allylic carbon   an sp^3 carbon adjacent to a vinylic carbon.

allylic cation   a compound with a positive charge on an allylic carbon.

alpha olefin   a monosubstituted olefin.

alternating copolymer   a copolymer in which two monomers alternate.

ambident nucleophile   a nucleophile with two nucleophilic sites.

amide   R—C(=O)—NH_2, R—C(=O)—NHR, R—C(=O)—NR_2

amine   a compound with a nitrogen in place of one of the hydrogens of an alkane; (RNH_2, R_2NH, R_3N).

amine inversion   the configuration of an sp^3 hybridized nitrogen with a nonbonding pair of electrons rapidly turns inside out.

amino acid   an α-aminocarboxylic acid. Naturally occurring amino acids have the L configuration.

amino acid analyzer   an instrument that automates the ion-exchange separation of amino acids.

amino acid residue   a monomeric unit of a peptide or protein.

aminolysis   reaction with an amine.

amino sugar   a sugar in which one of the OH groups is replaced by an NH_2 group.

amphoteric compound   a compound that can behave either as an acid or as a base.

anabolic steroids   steroids that aid in the development of muscle.

anabolism   reactions living organisms carry out in order to synthesize complex molecules from simple precursor molecules.

anchimeric assistance (intramolecular catalysis) catalysis in which the catalyst that facilitates the reaction is part of the molecule undergoing reaction.

androgens male sex hormones.

angle strain the strain introduced into a molecule as a result of its bond angles being distorted from their ideal values.

angstrom unit of length 100; picometers $= 10^{-8}$ cm $= 1$ angstrom

angular methyl group a methyl substituent at the 10- or 13-position of a steroid ring system.

anion-exchange resin a positively charged resin used in ion-exchange chromatography.

anionic polymerization chain-growth polymerization in which the initiator is a nucleophile; the propagation site therefore is an anion.

annulation reaction a ring-forming reaction.

annulene a monocyclic hydrocarbon with alternating double and single bonds.

anomeric carbon the carbon in a cyclic sugar that is the carbonyl carbon in the open-chain form.

anomeric effect the preference for the axial position shown by certain substituents bonded to the anomeric carbon of a six-membered ring sugar.

anomers two cyclic sugars that differ in configuration only at the carbon that is the carbonyl carbon in the open-chain form.

antarafacial bond formation formation of two σ bonds on opposites sides of the π system.

antarafacial rearrangement rearrangement in which the migrating group moves to the opposite face of the π system.

anti addition an addition reaction in which the two added substituents add to opposite sides of the molecule.

antiaromatic a cyclic and planar compound with an uninterrupted ring of p orbital-bearing atoms containing an even number of pairs of π electrons.

antibiotic a compound that interferes with the growth of a microorganism.

antibodies compounds that recognize foreign particles in the body.

antibonding molecular orbital a molecular orbital that results when two atomic orbitals with opposite signs interact. Electrons in an antibonding orbital decrease bond strength.

anticodon the three bases at the bottom of the middle loop in a tRNA.

anti conformer the most stable of the staggered conformers.

anti elimination an elimination reaction in which the two substituents eliminated are removed from opposite sides of the molecule.

antigene agent a polymer designed to bind to DNA at a particular site.

antigens compounds that can generate a response from the immune system.

anti-periplanar parallel substituents on opposite sides of a molecule.

antisense agent a polymer designed to bind to mRNA at a particular site.

antisense strand (template strand) the strand in DNA that is read during transcription.

antiviral drug a drug that interferes with DNA or RNA synthesis in order to prevent a virus from replicating.

apoenzyme an enzyme without its cofactor.

applied magnetic field the externally applied magnetic field.

aprotic solvent a solvent that does not have a hydrogen bonded to an oxygen or to a nitrogen.

aramide an aromatic polyamide.

arene oxide an aromatic compound that has had one of its double bonds converted to an epoxide.

aromatic a cyclic and planar compound with an uninterrupted ring of p orbital-bearing atoms containing an odd number of pairs of π electrons.

Arrhenius equation relates the rate constant of a reaction to the energy of activation and to the temperature at which the reaction is carried out ($k = Ae^{-E_a/RT}$).

aryl group a benzene or a substituted-benzene group.

asymmetrical ether an ether with two different substituents bonded to the oxygen.

asymmetric molecular orbital a molecular orbital in which the left (or top) half is not a mirror of the right (or bottom) half.

atactic polymer a polymer in which the substituents are randomly oriented on the extended carbon chain.

atomic number the number of protons (or electrons) that the neutral atom has.

atomic orbital an orbital associated with an atom.

atomic weight the average mass of the atoms in the naturally occurring element.

aufbau principle states that an electron will always go into the available orbital with the lowest energy.

automated solid-phase peptide synthesis an automated technique that synthesizes a peptide while its C-terminal amino acid is attached to a solid support.

autoradiograph the exposed photographic plate obtained in autoradiography.

autoradiography a technique used to determine the base sequence of DNA.

auxochrome a substituent that, when attached to a chromophore, alters the λ_{max} and intensity of absorption of UV/Vis radiation.

axial bond a bond of the chair conformation of cyclohexane that is perpendicular to the plane in which the chair is drawn (an up–down bond).

aziridine a three-membered ring compound in which one of the ring atoms is a nitrogen.

azo linkage a $-N=N-$ bond.

back side attack nucleophilic attack on the side of the carbon opposite to the side bonded to the leaving group.

bactericidal drug a drug that kills bacteria.

bacteriostatic drug a drug that inhibits the further growth of bacteria.

Baeyer–Villiger oxidation oxidation of aldehydes or ketones with H_2O_2 to form carboxylic acids or esters, respectively.

banana bond the σ bonds in small rings that are weaker as a result of overlapping at an angle rather than overlapping head-on.

base¹ a substance that accepts a proton.

base² a heterocyclic compound (a purine or a pyrimidine) in DNA and RNA.

base catalyst a catalyst that increases the rate of a reaction by removing a proton.

base peak the peak with the greatest abundance in a mass spectrum.

basicity describes the tendency of a compound to share its electrons with a proton.

Beer–Lambert law relationship among the absorbance of UV/Vis light, the concentration of the sample, the length of the light path, and the molar absorptivity ($A = cl\epsilon$).

bending vibration a vibration that does not occur along the line of the bond. It results in changing bond angles.

benzoyl group a benzene ring bonded to a carbonyl group.

benzyl group

benzylic carbon an sp^3 hybridized carbon bonded to a benzene ring.

benzylic cation a compound with a positive charge on a benzylic carbon.

benzyne intermediate a compound with a triple bond in place of one of the double bonds of benzene.

bicyclic compound a compound that contains two rings that share at least one carbon.

bifunctional molecule a molecule with two functional groups.

bile acids steroids that act as emulsifying agents so that water-insoluble compounds can be digested.

bimolecular reaction (second-order reaction) a reaction whose rate depends on the concentration of two reactants.

biochemistry (biological chemistry) the chemistry of biological systems.

biodegradable polymer a polymer that can be broken into small segments by an enzyme-catalyzed reaction.

bioorganic compound   an organic compound that is found in biological systems.

biopolymer   a polymer that is synthesized in nature.

biosynthesis   synthesis in a biological system.

biotin   the coenzyme required by enzymes that catalyze carboxylation of a carbon adjacent to an ester or a keto group.

Birch reduction   the partial reduction of benzene to 1,4-cyclohexadiene.

blind screen (random screen)   the search for a pharmacologically active compound without any information about what chemical structures might show activity.

block copolymer   a copolymer in which there are regions (blocks) of each kind of monomer.

blue shift   a shift to a shorter wavelength.

boat conformation   the conformation of cyclohexane that roughly resembles a boat.

boiling point   the temperature at which vapor pressure = atmospheric pressure.

bonding molecular orbital   a molecular orbital that results when two in-phase atomic orbitals interact. Electrons in a bonding orbital increase bond strength.

bond length   the internuclear distance between two atoms at minimum energy (maximum stability).

brand name (proprietary name, trade name)   identifies a commercial product and distinguishes it from other products. It can be used only by the owner of the registered trademark.

bridged bicyclic compound   a bicyclic compound in which rings share two nonadjacent carbons.

Brønsted acid   a substance that donates a proton.

Brønsted base   a substance that accepts a proton.

buffer   an acid and its conjugate base.

carbanion   a compound containing a negatively charged carbon.

carbene   a species with a carbon that has a nonbonded pair of electrons and an empty orbital (H_2C:)

carbocation   a compound containing a positively charged carbon.

carbocation rearrangement   the rearrangement of a carbocation to a more stable carbocation.

carbohydrate   a sugar, a saccharide. Naturally occurring carbohydrates have the D configuration.

α-carbon   a carbon adjacent to a carbonyl carbon.

carbon acid   a compound that contains a carbon that is bonded to a relatively acidic hydrogen.

carbonyl addition (direct addition)   nucleophilic addition to the carbonyl carbon.

carbonyl carbon   the carbon of a carbonyl group.

carbonyl compound   a compound that contains a carbonyl group.

carbonyl group   a carbon doubly bonded to an oxygen.

carbonyl oxygen   the oxygen of a carbonyl group.

carboxyl group   COOH

$$\text{carboxylic acid}\quad R-\overset{\overset{\displaystyle O}{\|}}{C}-OH$$

carboxylic acid derivative   a compound that is hydrolyzed to a carboxylic acid.

carboxyl oxygen   the single bonded oxygen of a carboxlic acid or an ester.

carotenoid   a class of compounds (a tetraterpene) responsible for the red and orange colors of fruits, vegetables, and fall leaves.

catabolism   reactions living organisms carry out in order to break down complex molecules into simple molecules and energy.

catalyst   a species that increases the rate at which a reaction occurs without being consumed in the reaction. Because it does not change the equilibrium constant of the reaction, it does not change the amount of product that is formed.

catalytic antibody   a compound that facilitates a reaction by forcing the conformation of the substrate in the direction of the transition state.

catalytic hydrogenation   the addition of hydrogen to a double or a triple bond with the aid of a metal catalyst.

cation-exchange resin   a negatively charged resin used in ion-exchange chromatography.

cationic polymerization   chain-growth polymerization where the initiator is an electrophile; the propagation site therefore is a cation.

cephalin   a phosphoacylglycerol in which the second OH group of phosphate has formed an ester with ethanolamine.

cerebroside   a sphingolipid in which the terminal OH group of sphingosine is bonded to a sugar residue.

chain-growth polymer (addition polymer)   a polymer made by adding monomers to the growing end of a chain.

chain transfer   a growing polymer chain reacts with a molecule XY in a manner that allows X to terminate the chain, leaving behind Y to initiate a new chain.

chair conformation   the conformation of cyclohexane that roughly resembles a chair. It is the most stable conformation of cyclohexane.

chemical exchange   the transfer of a proton from one molecule to another.

chemically equivalent protons   protons with the same connectivity relationship to the rest of the molecule.

chemical shift   the location of a signal in an NMR spectrum. It is measured downfield from a reference compound (most often TMS).

chiral (optically active)   a chiral molecule has a nonsuperimposable mirror image.

chiral auxiliary   an enantiomerically pure compound that, when attached to a reactant, causes a product with particular configuration to be formed.

chirality center   a tetrahedral atom bonded to four different groups.

cholesterol   a steroid that is the precursor of all other animal steroids.

chromatography   a separation technique in which the mixture to be separated is dissolved in a solvent and the solvent passed through a column packed with an absorbent stationary phase.

chromophore   the part of a molecule responsible for a UV or visible spectrum.

cine substitution   substitution at the carbon adjacent to the carbon that was bonded to the leaving group.

cis-fused   two cyclohexane rings fused together such that, if the second ring were considered to be two substituents of the first ring, one substituent would be in an axial position and the other would be in an equatorial position.

cis isomer   the isomer with identical substituents on the same side of the double bond.

cis-trans isomers   geometric isomers.

citric acid cycle (see Krebs cycle)   a series of reactions that converts the actyl group of acetyl-CoA into two molecules of CO_2.

Claisen condensation   a reaction between two molecules of an ester that connects the α-carbon of one with the carbonyl carbon of the other and eliminates an alkoxide ion.

Claisen rearrangement   a [3,3] sigmatropic rearrangement of an allyl vinyl ether.

α-cleavage   homolytic cleavage of an alpha substituent.

Clemmensen reduction   a reaction that reduces the carbonyl group of a ketone to a methylene group using Zn(Hg)/HCl.

codon   a sequence of three bases in mRNA that specifies the amino acid to be incorporated into a protein.

coenzyme   a cofactor that is an organic molecule.

coenzyme A   a thiol used by biological organisms to form thioesters.

coenzyme B_{12}   the coenzyme required by enzymes that catalyze certain rearrangement reactions.

cofactor   an organic molecule or a metal ion that certain enzymes need in order to catalyze a reaction.

coil conformation (loop conformation)   that part of a protein that is highly ordered but not in an α-helix or a β-pleated sheet.

combination band   occurs at the sum of two fundamental absorption frequencies ($v_1 + v_2$).

combinatorial library a group of structurally related compounds.

combinatorial organic synthesis the synthesis of a library of compounds by covalently connecting sets of building blocks of varying structure.

common intermediate an intermediate that two compounds have in common.

common name nonsystematic nomenclature.

competitive inhibitor a compound that inhibits an enzyme by competing with the substrate for binding at the active site.

complete racemization formation of a pair of enantiomers in equal amounts.

complex carbohydrate a carbohydrate containing two or more sugar molecules linked together.

concerted reaction a reaction in which all the bond-making and bond-breaking processes occur in one step.

condensation polymer (step-growth polymer) a polymer made by combining two molecules while removing a small molecule (usually water or an alcohol).

condensation reaction a reaction combining two molecules while removing a small molecule (usually water or an alcohol).

conducting polymer a polymer that can conduct electricity.

configuration the three-dimensional structure of a particular atom in a compound. The configuration is designated by *R* or *S*.

configurational isomers stereoisomers that cannot interconvert unless a covalent bond is broken. Cis-trans isomers and optical isomers are configurational isomers.

conformation the three-dimensional shape of a molecule at a given instant that can change as a result of rotations about σ bonds.

conformational analysis the investigation of the various conformations of a compound and their relative stabilities.

conformers (conformational isomers) different conformations of a molecule.

conjugate acid a compound accepts a proton to form its conjugate acid.

conjugate addition 1,4-addition.

conjugate base a compound loses a proton to form its conjugate base.

conjugated double bonds double bonds separated by one single bond.

conrotatory ring closure achieves head-to-head overlap of *p* orbitals by rotating the orbitals in the same direction.

conservation of orbital symmetry theory a theory that explains the relationship between the structure and configuration of the reactant, the conditions under which a pericyclic reaction takes place, and the configuration of the product.

constitutional isomers (structural isomers) molecules that have the same molecular formula but differ in the way the atoms are connected.

contributing resonance structure (resonance contributor, resonance structure) a structure with localized electrons that approximates the true structure of a compound with delocalized electrons.

convergent synthesis a synthesis in which pieces of the target compound are individually prepared and then assembled.

Cope elimination reaction elimination of a proton and a hydroxyl amine from an amine oxide.

Cope rearrangement a [3,3] sigmatropic rearrangement of a 1,5-diene.

copolymer a polymer formed using two or more different monomers.

corrin ring system a porphyrin ring system without one of the methine bridges.

COSY spectrum a 2-D NMR spectrum that shows coupling between sets of protons.

coupled protons protons that split each other. Coupled protons have the same coupling constant.

coupling constant the distance (in hertz) between two adjacent peaks of a split NMR signal.

coupling reaction a reaction that joins two alkyl groups.

covalent bond a bond created as a result of sharing electrons.

covalent catalysis (nucleophilic catalysis) catalysis that occurs as a result of a nucleophile forming a covalent bond with one of the reactants.

Cram's rule the rule used to determine the major product of a carbonyl addition reaction in a compound with a chirality center adjacent to the carbonyl group.

cross-conjugation nonlinear conjugation.

crossed (mixed) aldol addition an aldol addition in which two different carbonyl compounds are used.

cross-linking connecting polymer chains by intermolecular bond formation.

crown ether a cyclic molecule that contains several ether linkages.

crown–guest complex the complex formed when a crown ether binds a substrate.

cryptand a three-dimensional polycyclic compound that binds a substrate by encompassing it.

cryptate the complex formed when a cryptand binds a substrate.

crystallites regions of a polymer in which the chains are highly ordered.

C-terminal amino acid the terminal amino acid of a peptide (or protein) that has a free carboxyl group.

cumulated double bonds double bonds that are adjacent to one other.

Curtius rearrangement conversion of an acyl chloride into a primary amine using azide ion ($^-N_3$).

cyanohydrin
$$\overset{\displaystyle OH}{\underset{\displaystyle C\equiv N}{\underset{|}{\overset{|}{RCR(H)}}}}$$

cycloaddition reaction a reaction in which two π-bond-containing molecules react to form a cyclic compound.

[4 + 2] cycloaddition reaction a cycloaddition reaction in which 4 π electrons come from one reactant and 2 π electrons come from the other reactant.

cycloalkane an alkane with its carbon chain arranged in a closed ring.

deactivating substituent a substituent that decreases the reactivity of an aromatic ring. Electron-withdrawing substituents deactivate aromatic rings toward electrophilic attack, and electron-donating substituents deactivate aromatic rings toward nucleophilic attack.

deamination loss of ammonia.

decarboxylation loss of carbon dioxide.

degenerate orbitals orbitals that have the same energy.

dehydration loss of water.

dehydrogenase an enzyme that carries out an oxidation reaction by removing hydrogen from the substrate.

dehydrohalogenation elimination of a proton and a halide ion.

delocalization energy (resonance energy) the extra stability a compound achieves as a result of having delocalized electrons.

delocalized electrons electrons that are shared by more than two atoms.

denaturation destruction of the highly organized tertiary structure of a protein.

deoxygenation removal of an oxygen from a reactant.

deoxyribonucleic acid (DNA) a polymer of deoxyribonucleotides.

deoxyribonucleotide a nucleotide in which the sugar component is D-2′-deoxyribose.

deoxy sugar a sugar in which one of the OH groups has been replaced by an H.

DEPT ^{13}C NMR spectrum a series of four spectra that distinguishes among $-CH_3$, $-CH_2$, and $-CH$ groups.

depurination elimination of a purine ring.

detergent a salt of a sulfonic acid.

deuterium kinetic isotope effect ratio of the rate constant obtained for a compound containing hydrogen and the rate constant obtained for an identical compound in which one or more of the hydrogens have been replaced by deuterium.

dextrorotatory the enantiomer that rotates polarized light in a clockwise direction.

diastereomer a configurational stereoisomer that is not an enantiomer.

diastereotopic hydrogens two hydrogens bonded to a carbon that when replaced in turn with a deuterium results in a pair of diastereomers.

1,3-diaxial interaction the interaction between an axial substituent and the other two axial substituents on the same side of the cyclohexane ring.

diazonium ion $Ar\overset{+}{N}\equiv N$ or $R\overset{+}{N}\equiv N$.

diazonium salt   a diazonium ion and an anion ($Ar\overset{+}{N}\equiv N\ X^-$).

Dieckmann condensation   an intramolecular Claisen condensation.

dielectric constant   a measure of how well a solvent can insulate opposite charges from one another.

Diels–Alder reaction   a $[4\ +\ 2]$ cycloaddition reaction.

diene   a hydrocarbon with two double bonds.

dienophile   an alkene that reacts with a diene in a Diels–Alder reaction.

β-diketone   a ketone with a second carbonyl group at the β-position.

dimer   a molecule formed by the joining together of two identical molecules.

dinucleotide   two nucleotides linked by phosphodiester bonds.

dipeptide   two amino acids linked by an amide bond.

dipole–dipole interaction   an interaction between the dipole of one molecule and the dipole of another.

dipole moment (μ)   a measure of the separation of charge in a bond or in a molecule.

direct addition   1,2-addition.

direct displacement mechanism   a reaction in which the nucleophile displaces the leaving group in a single step.

direct substitution   substitution at the carbon that was bonded to the leaving group.

disaccharide   a compound containing two sugar molecules linked together.

disconnection   breaking a bond to carbon to give a simpler species.

disproportionation   transfer of a hydrogen atom by a radical to another radical with the result that an alkane and an alkene are formed.

disrotatory ring closure   achieves head-to-head overlap of p orbitals by rotating the orbitals in opposite directions.

dissociation energy   the amount of energy required to break a bond, or the amount of energy released when a bond is formed.

dissolving metal reduction   a reduction using sodium or lithium metal dissolved in liquid ammonia.

distribution coefficient   the ratio of the amount of a compound dissolving in each of two solvents in contact with each other.

disulfide bridge   a disulfide (—S—S—) bond in a peptide or protein.

DNA (deoxyribonucleic acid)   a polymer of deoxyribonucleotides.

double bond   a sigma bond and a pi bond between two atoms.

doublet   an NMR signal split into two peaks.

doublet of doublets   an NMR signal split into four peaks of approximately equal height. Caused by splitting a signal into a doublet by one hydrogen and into another doublet by another (nonequivalent) hydrogen.

drug   a compound that reacts with a biological molecule, triggering a physiological effect.

drug resistance   biological resistance to a particular drug.

drug synergism   when the effect of two drugs used in combination is greater than the sum of the effects obtained when administered individually.

eclipsed conformation   a conformation in which the bonds on adjacent carbons are aligned as viewed looking down the carbon–carbon bond.

E conformation   the conformation of a carboxylic acid or carboxylic acid derivative in which the carbonyl oxygen and the substituent bonded to the carboxyl oxygen or nitrogen are on opposite sides of the single bond.

Edman's reagent   phenyl isothiocyanate. A reagent used to determine the N-terminal amino acid of a polypeptide.

effective magnetic field   the magnetic field that a proton "senses" through the surrounding cloud of electrons.

effective molarity   the concentration of the reagent that would be required in an intermolecular reaction for it to have the same rate as an intramolecular reaction.

E isomer   the isomer with the high-priority groups on opposite sides of the double bond.

elastomer   a polymer that can stretch and then revert back to its original shape.

electrocyclic reaction   a reaction in which a π bond in the reactant is lost so that a cyclic compound with a new σ bond can be formed.

electromagnetic radiation   radiant energy that displays wave properties.

electron affinity   the energy given off when an atom acquires an electron.

electronegative element   an element that readily acquires an electron.

electronegativity   tendency of an atom to pull electrons toward itself.

electronic transition   promotion of an electron from its HOMO to its LUMO.

electron sink   site to which electrons can be delocalized.

electrophile   an electron-deficient atom or molecule.

electrophilic addition reaction   an addition reaction in which the first species that adds to the reactant is an electrophile.

electrophilic aromatic substitution   a reaction in which an electrophile substitutes for a hydrogen of an aromatic ring.

electrophilic catalysis   catalysis in which the species that facilitates the reaction is an electrophile.

electrophoresis   a technique that separates amino acids on the basis of their pI values.

electropositive element   an element that readily loses an electron.

electrostatic attraction   attractive force between opposite charges.

electrostatic catalysis   stabilization of a charge by an opposite charge.

elemental analysis   a determination of the relative proportions of the elements present in a compound.

α-elimination   removal of two atoms or groups from the same carbon.

β-elimination   removal of two atoms or groups from adjacent carbons.

elimination reaction   a reaction that involves the elimination of atoms (or molecules) from the reactant.

empirical formula   the relative numbers of the different kinds of atoms in a molecule.

enamine   an α,β-unsaturated tertiary amine.

enantiomerically pure   containing only one enantiomer.

enantiomeric excess (optical purity)   how much excess of one enantiomer is present in a mixture of a pair of enantiomers.

enantiomers   nonsuperimposable mirror-image molecules.

enantioselective reaction   a reaction that forms an excess of one enantiomer.

enantiotopic hydrogens   two hydrogens bonded to a carbon that is bonded to two other groups that are nonidentical.

endergonic reaction   a reaction with a positive $\Delta G°$.

endo   a substituent is endo if it is closer to the longer or more unsaturated bridge.

endopeptidase   an enzyme that hydrolyzes a peptide bond that is not at the end of a peptide chain.

endothermic reaction   a reaction with a positive $\Delta H°$.

enkephalins   pentapeptides synthesized by the body to control pain.

enolization   keto–enol interconversion.

enthalpy   the heat given off ($-\Delta H°$) or the heat absorbed ($+\Delta H°$) during the course of a reaction.

entropy   a measure of the freedom of motion in a system.

enzyme   a protein that is a catalyst.

epimerization   changing the configuration of a chirality center by removing a proton from it and then reprotonating the molecule at the same site.

epimers   monosaccharides that differ in configuration at only one carbon.

epoxidation   formation of an epoxide.

epoxide (oxirane)   an ether in which the oxygen is incorporated into a three-membered ring.

epoxy resin   formed by mixing a low-molecular-weight prepolymer with a compound that forms a cross-linked polymer.

equatorial bond   a bond of the chair conformer of cyclohexane that juts out from the ring in approximately the same plane that contains the chair.

equilibrium constant   the ratio of products to reactants at equilibrium or the ratio of the rate constants for the forward and reverse reactions.

equilibrium control   thermodynamic control.

E1 reaction   a first-order elimination reaction.

E2 reaction   a second-order elimination reaction.

erythro enantiomers   the pair of enantiomers with similar groups on the same side when drawn in a Fischer projection.

essential amino acid   an amino acid that humans must obtain from their diet because they cannot synthesize it or cannot synthesize it in adequate amounts.

essential oils   fragrances and flavorings isolated from plants that do not leave residues when they evaporate. Most are terpenes.

ester

$$R-\overset{\overset{\displaystyle O}{\|}}{C}-OR$$

estrogens   female sex hormones.

ether   a compound containing an oxygen bonded to two carbons (ROR).

eukaryote   a unicellular or multicellular body whose cell or cells contain a nucleus.

excited-state electronic configuration   the electronic configuration that results when an electron in the ground-state electronic configuration has been moved to a higher-energy orbital.

exergonic reaction   a reaction with a negative $\Delta G°$.

exhaustive methylation   reaction of an amine with excess methyl iodide resulting in the formation of a quaternary ammonium iodide.

exo   a substituent is exo if it closer to the shorter or more saturated bridge.

exon   a stretch of bases in DNA that are a portion of a gene.

exopeptidase   an enzyme that hydrolyzes a peptide bond at the end of a peptide chain.

exothermic reaction   a reaction with a negative $\Delta H°$.

experimental energy of activation ($E_a = \Delta H^{\ddagger} - RT$)   a measure of the approximate energy barrier to a reaction. (It is approximate because it does not contain an entropy component.)

extrusion reaction   a reaction in which a neutral molecule (for example, CO_2, CO, or N_2) is eliminated from a molecule.

fat   a triester of glycerol that exists as a solid at room temperature.

fatty acid   a long-chain carboxylic acid.

fibrous protein   a water-insoluble protein in which the polypeptide chains are arranged in bundles.

fingerprint region   the right-hand third of an IR spectrum where the absorption bands are characteristic of the compound as a whole.

first-order rate constant   the rate constant of a first-order reaction.

first-order reaction (unimolecular reaction)   a reaction whose rate depends on the concentration of one reactant.

Fischer esterification reaction   the reaction of a carboxylic acid with alcohol in the presence of an acid catalyst to form an ester.

Fischer projection   a method of representing the spatial arrangement of groups bonded to a chirality center. The chirality center is the point of intersection of two perpendicular lines; the horizontal lines represent bonds that project out of the plane of the paper toward the viewer, and the vertical lines represent bonds that point back from the plane of the paper away from the viewer.

flagpole hydrogens (transannular hydrogens)   the two hydrogens in the boat conformation of cyclohexane that are closest to each other.

flavin adenine dinucleotide (FAD)   a coenzyme required in certain oxidation reactions. It is reduced to $FADH_2$, which can act as a reducing agent in another reaction.

flavin mononucleotide (FMN)   a coenzyme required in certain oxidation reactions. It is reduced to $FMNH_2$, which can act as a reducing agent in another reaction.

formal charge   the number of valence electrons − (the number of nonbonding electrons +1/2 the number of bonding electrons).

Fourier transform NMR   a technique in which all the nuclei are excited simultaneously by an rf pulse, their relaxation monitored, and the data mathematically converted to a spectrum.

free energy of activation ($\Delta G^{\ddagger}$)   the true energy barrier to a reaction.

free induction decay   relaxation of excited nuclei.

frequency   the velocity of a wave divided by its wavelength (in units of cycles/s).

Friedel–Crafts acylation   an electrophilic substitution reaction that puts an acyl group on a benzene ring.

Friedel–Crafts alkylation   an electrophilic substitution reaction that puts an alkyl group on a benzene ring.

frontier orbital analysis   determining the outcome of a pericyclic reaction using frontier orbitals.

frontier orbitals   the HOMO and the LUMO.

frontier orbital theory   a theory that, like the conservation of orbital symmetry theory, explains the relationships among reactant, product, and reaction conditions in a pericyclic reaction.

functional group   the center of reactivity in a molecule.

functional group interconversion   conversion of one functional group into another functional group.

functional group region   the left-hand two-thirds of an IR spectrum where most functional groups show absorption bands.

furanose   a five-membered ring sugar.

furanoside   a five-membered ring glycoside.

fused bicyclic compound   a bicyclic compound in which the rings share two adjacent carbons.

Gabriel synthesis   conversion of an alkyl halide into a primary amine using phthalimide as a starting material.

gauche   X and Y are gauche to each other in this Newman projection.

gauche conformer   a staggered conformer in which the largest substituents are gauche to each other.

gauche interaction   the interaction between two atoms or groups that are gauche to each other.

***gem*-dialkyl effect**   two alkyl groups on a carbon whose effect is to increase the probability that the molecule will be in the proper conformation for ring closure.

***gem*-diol (hydrate)**   a compound with two OH groups on the same carbon.

geminal coupling   the mutual splitting of two nonidentical protons bonded to the same carbon.

geminal dihalide   a compound with two halogen atoms bonded to the same carbon.

gene   a segment of DNA.

general-acid catalysis   catalysis in which a proton is transferred to the reactant during the slow step of the reaction.

general-base catalysis   catalysis in which a proton is removed from the reactant during the slow step of the reaction.

generic name   a non-commercially restricted name for a drug.

gene therapy   a technique that inserts a synthetic gene into the DNA of an organism defective in that gene.

genetic code   the amino acid specified by each three-base sequence of mRNA.

geometric isomers   cis-trans (or E,Z) isomers.

Gibbs standard free energy change ($\Delta G°$)   the difference between the free energy content of the products and the free energy content of the reactants at equilibrium under standard conditions (1 M, 25 °C, 1 atm).

Gilman reagent   an organocuprate, prepared from the reaction of an organolithium reagent with cuprous iodide, used to replace a halogen with an alkyl group.

globular protein   a water-soluble protein that tends to have a roughly spherical shape.

gluconeogenesis   the synthesis of D-glucose from pyruvate.

glycol   a compound containing two or more OH groups.

glycolysis (glycolytic cycle)   the sequence of reactions that convert D-glucose into two molecules of pyruvate.

glycoprotein   a protein that is covalently bonded to a polysaccharide.

glycoside   the acetal of a sugar.

glycosidic bond   the bond between the anomeric carbon and the alcohol in a glycoside.

α-1,4′-glycosidic linkage a glycosidic linkage between the C-1 oxygen of one sugar and the C-4 of a second sugar with the oxygen atom of the glycosidic linkage in the axial position.

β-1,4′-glycosidic linkage a glycosidic linkage between the C-1 oxygen of one sugar and the C-4 of a second sugar with the oxygen atom of the glycosidic linkage in the equatorial position.

graft copolymer a copolymer that contains branches of a polymer of one monomer grafted onto the backbone of a polymer made from another monomer.

Grignard reagent the compound that results when magnesium is inserted between the carbon and halogen of an alkyl halide (RMgBr, RMgCl).

ground-state electronic configuration a description of which orbitals the electrons of an atom or molecule occupy when they are all in their lowest energy orbitals.

half-chair conformation the least stable conformation of cyclohexane.

haloform reaction the reaction of a halogen and HO⁻ with a methyl ketone.

halogenation reaction with halogen (Br_2, Cl_2).

halohydrin an organic molecule that contains a halogen and an OH group on adjacent carbons.

Hammond postulate states that the transition state will be more similar in structure to the species (reactants or products) that it is closer to energetically.

Haworth projection a way to show the structure of a sugar; the five- and six-membered rings are represented as being flat.

head-to-tail addition the head of one molecule is added to the tail of another molecule.

heat of combustion the amount of heat given off when a carbon-containing compound reacts completely with O_2 to form CO_2 and H_2O.

heat of formation the heat given off when a compound is formed from its elements under standard conditions.

heat of hydrogenation the heat ($-\Delta H°$) released in a hydrogenation reaction.

Heisenberg uncertainty principle states that both the precise location and the momentum of an atomic particle cannot be simultaneously determined.

α-helix the backbone of a polypeptide coiled in a right-handed spiral with hydrogen bonding occurring within the helix.

Hell–Volhard–Zelinski (HVZ) reaction heating a carboxylic acid with Br_2 + P in order to convert it into an α-bromocarboxylic acid.

$$
\text{hemiacetal} \quad \underset{\overset{|}{OR}}{\overset{\overset{OH}{|}}{RCH}}
$$

$$
\text{hemiketal} \quad \underset{\overset{|}{OR}}{\overset{\overset{OH}{|}}{RCR}}
$$

Henderson–Hasselbalch equation $pK_a = pH + \log[HA]/[A^-]$

heptose a monosaccharide with 7 carbons.

HETCOR spectrum a 2-D NMR spectrum that shows coupling between protons and the carbons to which they are attached.

heteroatom an atom other than carbon and hydrogen.

heterocyclic compound (heterocycle) a cyclic compound in which one or more of the atoms of the ring are heteroatoms.

heterogeneous catalyst a catalyst that is insoluble in the reaction mixture.

heterolytic bond cleavage (heterolysis) breaking a bond with the result that both bonding electrons stay with one of the atoms.

hexose a monosaccharide with six carbons.

high-energy bond a bond that releases a great deal of energy when it is broken.

highest occupied molecular orbital (HOMO) the molecular orbital of highest energy that contains an electron.

high-resolution NMR spectroscopy NMR spectroscopy that uses a spectrometer with a high operating frequency.

Hofmann degradation exhaustive methylation of an amine, followed by reaction with Ag_2O, followed by heating to achieve a Hofmann elimination reaction.

Hofmann elimination (anti-Zaitsev elimination) a hydrogen is removed from the β-carbon bonded to the most hydrogens.

Hofmann elimination reaction elimination of a proton and a tertiary amine from a quaternary ammonium hydroxide.

Hofmann rearrangement conversion of an amide into an amine using Br_2/HO^-.

holoenzyme an enzyme plus its cofactor.

homogeneous catalyst a catalyst that is soluble in the reaction mixture.

homolog a member of a homologous series.

homologous series a family of compounds in which each member differs from the next by one methylene group.

homolytic bond cleavage (homolysis) breaking a bond with the result that each of the atoms gets one of the bonding electrons.

homopolymer a polymer that contains only one kind of monomer.

homotopic hydrogens two hydrogens bonded to a carbon, which is bonded to two other groups that are identical.

Hoogsteen base pairing the pairing between a base in a synthetic strand of DNA and a base pair in double-stranded DNA.

Hooke's law an equation that describes the motion of a vibrating spring.

hormone an organic compound synthesized in a gland and delivered by the bloodstream to its target tissue.

Hückel's rule states that for a compound to be aromatic its cloud of electrons must contain $(4n + 2)$ π electrons, where n is an integer. This is the same as saying it must contain an odd number of pairs of π electrons.

human genome the total DNA of a human cell.

Hund's rule states that when there are degenerate orbitals, an electron will occupy an empty orbital before it will pair up with another electron.

Hunsdiecker reaction conversion of a carboxylic acid into an alkyl halide by heating a heavy metal salt of the carboxylic acid with bromine or iodine.

hybrid orbital an orbital formed by mixing (hybridizing) orbitals.

$$
\text{hydrate (\textit{gem}-diol)} \quad \underset{\overset{|}{OH}}{\overset{\overset{OH}{|}}{RCR(H)}}
$$

hydrated water has been added to a compound.

hydration addition of water to a compound.

hydrazone $R_2C{=}NNH_2$

hydride ion a negatively charged hydrogen.

1,2-hydride shift the movement of a hydride ion from one carbon to an adjacent carbon.

hydroboration–oxidation the addition of borane to an alkene or an alkyne followed by reaction with hydrogen peroxide and hydroxide ion.

hydrocarbon a compound that contains only carbon and hydrogen.

α-hydrogen usually a hydrogen bonded to the carbon adjacent to a carbonyl carbon.

hydrogenation addition of hydrogen.

hydrogen bond an unusually strong dipole–dipole attraction (5 kcal/mol) between a hydrogen bonded to O, N, or F and the nonbonding electrons of an O, N, or F of another molecule.

hydrogen ion (proton) a positively charged hydrogen.

hydrolysis reaction with water.

hydrophobic interactions interactions between nonpolar groups. They increase stability by decreasing the amount of structured water (increasing entropy).

hyperconjugation delocalization of electrons by overlap of carbon–hydrogen or carbon–carbon σ bonds with an empty p orbital.

imine $R_2C{=}NR$

inclusion compound a compound that specifically binds a metal ion or an organic molecule.

induced dipole–induced dipole interaction an interaction between a temporary dipole in one molecule and the dipole the temporary dipole induces in another molecule.

induced-fit model a model that describes the specificity of an enzyme for its substrate: The shape of the active site does not become completely complementary to the shape of the substrate until after the enzyme binds the substrate.

inductive electron donation donation of electrons through σ bonds.

inductive electron withdrawal withdrawal of electrons through a σ bond.

inflection point the midpoint of the flattened-out region of a titration curve.

informational strand (sense strand) the strand in DNA that is not read during transcription; it has the same sequence of bases as the synthesized mRNA strand (with a U, T difference).

infrared radiation electromagnetic radiation familiar to us as heat.

infrared spectroscopy uses infrared energy to provide a knowledge of the functional groups in a compound.

infrared (IR) spectrum a plot of percent transmission versus wave number (or wavelength) of infrared radiation.

initiation step the step in which radicals are created, or the step in which the radical needed for the first propagation step is created.

in-line displacement mechanism nucleophilic attack on a phosphorus concerted with breaking a phosphoanhydride bond.

interchain disulfide bridge a disulfide bridge between two cysteine residues in different peptide chains.

intermediate a species formed during a reaction that is not the final product of the reaction.

intermolecular reaction a reaction that takes place between two molecules.

internal alkyne an alkyne with the triple bond not at the end of the carbon chain.

intimate ion pair the covalent bond that joined the cation and the anion has broken, but the cation and anion are still next to each other.

intrachain disulfide bridge a disulfide bridge between two cysteine residues in the same peptide chain.

intramolecular catalysis (anchimeric assistance) catalysis in which the catalyst that facilitates the reaction is part of the molecule undergoing reaction.

intramolecular reaction a reaction that takes place within a molecule.

intron a stretch of bases in DNA that contain no genetic information.

inversion of configuration turning the configuration of the carbon inside-out like an umbrella in a windstorm so that the resulting product has a configuration opposite to that of the reactant.

iodoform test addition of I_2/HO^- to a methyl ketone forms a yellow precipitate of triiodomethane.

ion–dipole interaction the interaction between an ion and the dipole of a molecule.

ion-exchange chromatography a technique that uses a column packed with an insoluble resin to separate compounds on the basis of their charges and polarities.

ionic bond a bond formed through the attraction of two ions of opposite charges.

ionization energy the energy required to remove an electron from an atom.

ionophore a compound that binds metal ions tightly.

iron protoporphyrin IX the porphyrin ring system of heme plus an iron atom.

isoelectric point (pI) the pH at which there is no net charge on an amino acid.

isolated double bonds double bonds separated by more than one single bond.

isomers nonidentical compounds with the same molecular formula.

isoprene rule head-to-tail linkage of isoprene units.

isopropyl split a split in the IR absorption band attributable to a methyl group. It is characteristic of an isopropyl group.

isotactic polymer a polymer in which all the substituents are on the same side of the fully extended carbon chain.

isotopes atoms with the same number of protons but different numbers of neutrons.

iterative synthesis a synthesis in which a reaction sequence is carried out more than once.

IUPAC nomenclature systematic nomenclature of chemical compounds.

Kekulé structure a model that represents the bonds between atoms as lines.

ketal

keto–enol tautomerism (keto–enol interconversion) interconversion of keto and enol tautomers.

keto–enol tautomers a ketone and its isomeric α,β-unsaturated alcohol.

β-keto ester an ester with a second carbonyl group at the β-position.

ketone

ketose a polyhydroxyketone.

Kiliani–Fischer synthesis a method used to increase the number of carbons in an aldose by one, resulting in the formation of a pair of C-2 epimers.

kinetic control when a reaction is under kinetic control, the relative amounts of the products depend on the rates at which they are formed.

kinetic isotope effect a comparison of the rate of reaction of a compound with the rate of reaction of an identical compound in which one of the atoms has been replaced by an isotope.

kinetic product the product that is formed the fastest.

kinetic resolution separation of enantiomers based on the difference in their rate of reaction with an enzyme.

kinetics the field of chemistry that deals with the rates of chemical reactions.

kinetic stability chemical reactivity, indicated by $\Delta G^{\ddagger}$. If $\Delta G^{\ddagger}$ is large, the compound is kinetically stable (not very reactive). If $\Delta G^{\ddagger}$ is small, the compound is kinetically unstable (very reactive).

Kolbe–Schmitt carboxylation reaction a reaction that uses CO_2 to carboxylate phenol.

Krebs cycle (citric acid cycle, tricarboxylic acid cycle, TCA cycle) a series of reactions that converts the acetyl group of acetyl-CoA into two molecules of CO_2.

lactam a cyclic amide.

lactone a cyclic ester.

λ_{max} the wavelength at which there is maximum UV/Vis absorbance.

lead compound the prototype in a search for other biologically active compounds.

leaning when a line drawn over the outside peaks of an NMR signal points in the direction of the signal given by the protons that cause the splitting.

leaving group the group that is displaced in a nucleophilic substitution reaction.

Le Châtelier's principle states that if an equilibrium is disturbed, the components of the equilibrium will adjust in a way that will offset the disturbance.

lecithin a phosphoacylglycerol in which the second OH group of phosphate has formed an ester with choline.

levorotatory the enantiomer that rotates polarized light in a counterclockwise direction.

Lewis acid a substance that accepts an electron pair.

Lewis base a substance that donates an electron pair.

Lewis structure a model that represents the bonds between atoms as lines or dots and the valence electrons as dots.

ligation   sharing of nonbonding electrons with a metal.

linear combination of atomic orbitals (LCAO)   the combination of atomic orbitals to produce a molecular orbital.

linear conjugation   the atoms in the conjugated system are in a linear arrangement.

linear synthesis   a synthesis that builds a molecule step by step from starting materials.

lipid   a water-insoluble compound found in a living system.

lipid bilayer   two layers of phosphoacylglycerols arranged so that their polar heads are on the outside and their nonpolar fatty acid chains are on the inside.

lipoate   a coenzyme required in certain oxidation reactions.

living polymer   a nonterminated chain-growth polymer that remains active. This means that the polymerization reaction can continue upon addition of more monomer.

localized electrons   electrons that are restricted to a particular locality.

lock-and-key model   a model that describes the specificity of an enzyme for its substrate: The substrate fits the enzyme like a key fits a lock.

London forces   Induced dipole–induced dipole interactions.

lone-pair electrons (nonbonding electrons)   valence electrons not used in bonding.

long-range coupling   splitting of a proton by a proton more than three σ bonds away.

loop conformation (coil conformation)   that part of a protein that is highly ordered but not in an α-helix or β-pleated sheet.

lowest unoccupied molecular orbital (LUMO)   the molecular orbital of lowest energy that does not contain an electron.

Lucas test   a test that determines whether an alcohol is primary, secondary, or tertiary.

magnetic anisotropy   the term used to describe the greater freedom of a π electron cloud to move in response to a magnetic field as a consequence of its greater polarizability compared with σ electrons.

magnetic resonance imaging (MRI)   NMR used in medicine. The difference in the way water is bound in different tissues produces the signal variation between organs as well as between healthy and diseased states.

magnetogyric ratio   a property (measured in rad $T^{-1}s^{-1}$) that depends on the magnetic properties of a particular kind of nucleus.

major groove   the wider and deeper of the two alternating grooves in DNA.

malonic ester synthesis   synthesis of a carboxylic acid using diethyl malonate as the starting material.

Markovnikov's rule   the actual rule is: "When a hydrogen halide adds to an asymmetrical alkene, the addition occurs such that the halogen attaches itself to the sp^2 carbon of the alkene bearing the lowest number of hydrogen atoms." A more useful version is: The electrophile adds to the sp^2 carbon that is bonded to the greater number of hydrogens.

mass number   the number of protons plus the number of neutrons in an atom.

mass spectrometry   provides a knowledge of the molecular weight, molecular formula, and certain structural features of a compound.

mass spectrum   a plot of the relative abundance of the positively charged fragments produced in a mass spectrometer versus their m/z values.

materials science   the science of creating new materials to be used in place of known materials such as metal, glass, wood, cardboard, and paper.

Maxam–Gilbert sequencing   a technique used to sequence restriction fragments.

McLafferty rearrangement   rearrangement of the molecular ion of a ketone. The bond between the α- and β-carbons breaks, and a γ-hydrogen migrates to the oxygen.

mechanism-based inhibitor (suicide inhibitor)   a compound that inactivates an enzyme by undergoing part of its normal catalytic mechanism.

mechanism of a reaction   a description of the step-by-step process by which reactants are changed into products.

melting point   the temperature at which a solid becomes a liquid.

membrane   the material that surrounds the cell in order to isolate its contents.

mercaptan (thiol)   the sulfur analog of an alcohol (RSH).

meso compound   a compound that contains chirality centers and a plane of symmetry.

metabolism   reactions living organisms carry out in order to obtain the energy they need and to synthesize the compounds they require.

meta-directing substituent   a substituent that directs an incoming substituent meta to an existing substituent.

metal-activated enzyme   an enzyme that has a loosely bound metal ion.

metal-ion catalysis   catalysis in which the species that facilitates the reaction is a metal ion.

metalloenzyme   an enzyme that has a tightly bound metal ion.

methine hydrogen   a tertiary hydrogen.

methylene group   a CH_2 group.

1,2-methyl shift   the movement of a methyl group with its bonding electrons from one carbon to an adjacent carbon.

micelle   a spherical aggregation of molecules each with a long hydrophobic tail and a polar head, arranged so that the polar head points to the outside of the sphere.

Michael reaction   the addition of an α-carbanion to the β-carbon of an α,β-unsaturated carbonyl compound.

minor groove   the narrower and more shallow of the two alternating grooves in DNA.

mixed (crossed) aldol addition   an aldol addition in which two different carbonyl compounds are used.

mixed anhydride   an acid anhydride with two different R groups.

$$R-\overset{\overset{O}{\|}}{C}-O-\overset{\overset{O}{\|}}{C}-R'$$

mixed Claisen condensation   a Claisen condensation in which two different esters are used.

mixed triacylglycerol   a triacylglycerol in which the fatty acid components are different.

molar absorptivity   the absorbance obtained from a 1.00-M solution in a cell with a 1.00-cm light path.

molecular ion (parent ion)   peak in the mass spectrum with the greatest m/z.

molecular modeling   computer-assisted design of a compound with particular structural characteristics.

molecular modification   changing the structure of a lead compound.

molecular orbital   an orbital associated with a molecule.

molecular orbital theory   describes a model in which the electrons occupy orbitals as they do in atoms but with the orbitals extending over the entire molecule.

molecular recognition   the recognition of one molecule by another as a result of specific interactions; for example, the specificity of an enzyme for its substrate.

molozonide   an unstable intermediate containing a five-membered ring with three oxygens in a row that is formed from the reaction of an alkene with ozone.

monomer   a repeating unit in a polymer.

monosaccharide (simple carbohydrate)   a single sugar molecule.

monoterpene   a terpene that contains 10 carbons.

MRI scanner   an NMR spectrometer used in medicine for whole-body NMR.

multiplet   an NMR signal split into more than 7 peaks.

multiplicity   the number of peaks in an NMR signal.

multistep synthesis   preparation of a compound by a route that requires several steps.

mutarotation   a slow change in optical rotation to an equilibrium value.

$N + 1$ rule   an ^1H NMR signal for a hydrogen with N equivalent hydrogens bonded to an adjacent carbon is split into $N + 1$ peaks. A ^{13}C NMR signal for a carbon bonded to N hydrogens is split into $N + 1$ peaks.

N-glycoside   a glycoside with a nitrogen instead of an oxygen at the glycosidic linkage.

N-terminal amino acid   the terminal amino acid of a peptide (or protein) that has a free amino group.

natural abundance atomic weight the average mass of the atoms in the naturally occurring element.

natural product a product synthesized in nature.

neurotransmitter a compound that transmits nerve impulses.

nicotinamide adenine dinucleotide (NAD⁺) a coenzyme required in certain oxidation reactions. It is reduced to NADH, which can act as a reducing agent in another reaction.

nicotinamide adenine dinucleotide phosphate (NADP⁺) a coenzyme required in certain oxidation reactions. It is reduced to NADPH, which can act as a reducing agent in another reaction.

NIH shift the 1,2-hydride shift of a carbocation (obtained from an arene oxide) that leads to an enone.

nitration substitution of a nitro group (NO_2) for a hydrogen of a benzene ring.

nitrile a compound that contains a carbon–nitrogen triple bond ($RC{\equiv}N$).

nitrosamine (*N*-nitroso compound) $R_2NN{=}O$.

NMR spectroscopy the absorption of electromagnetic radiation to determine the structural features of an organic compound. In the case of ¹H NMR spectroscopy, it determines the carbon–hydrogen framework.

node that part of an orbital in which there is zero probability of finding an electron.

nominal mass mass rounded to the nearest whole number.

nonbonding electrons (lone-pair electrons) valence electrons not used in bonding.

nonbonding molecular orbital the *p* orbitals are too far apart to overlap significantly, so the molecular orbital that results neither favors nor disfavors bonding.

nonpolar covalent bond a bond formed between two atoms that share the bonding electrons equally.

nonreducing sugar a sugar that cannot be oxidized by reagents such as Ag^+ and Cu^+. Nonreducing sugars are not in equilibrium with the open-chain aldose or ketose.

normal alkane (straight-chain alkane) an alkane in which the carbons form a contiguous chain with no branches.

nucleic acid the two kinds of nucleic acid are DNA and RNA.

nucleophile an electron-rich atom or molecule.

nucleophilic acyl substitution reaction a reaction in which a group bonded to an acyl or aryl group is substituted by another group.

nucleophilic addition–elimination–nucleophilic addition reaction a nucleophilic addition reaction that is followed by an elimination reaction that is followed by a nucleophilic addition reaction. Acetal formation is an example: An alcohol adds to the carbonyl carbon, water is eliminated, and a second molecule of alcohol adds to the dehydrated product.

nucleophilic addition–elimination reaction a nucleophilic addition reaction that is followed by an elimination reaction. Imine formation is an example: An amine adds to the carbonyl carbon, and water is eliminated.

nucleophilic addition reaction a reaction that involves the addition of a nucleophile to a reagent.

nucleophilic aromatic substitution a reaction in which a nucleophile substitutes for a substituent of an aromatic ring.

nucleophilic catalysis (covalent catalysis) catalysis that occurs as a result of a nucleophile forming a covalent bond with one of the reactants.

nucleophilic catalyst a catalyst that increases the rate of a reaction by acting as a nucleophile.

nucleophilicity a measure of how readily an atom or molecule with a pair of nonbonding electrons attacks an atom.

nucleophilic substitution reaction a reaction in which a nucleophile substitutes for an atom or group.

nucleoside a heterocyclic base (a purine or a pyrimidine) bonded to the anomeric carbon of a sugar (D-ribose or D-2′-deoxyribose).

nucleotide a heterocycle attached in the β-position to a phosphorylated ribose.

observed rotation the amount of rotation observed in a polarimeter.

octet rule states that an atom will give up, accept, or share electrons in order to achieve a filled shell. Because a filled second shell contains eight electrons, this is known as the octet rule.

off-resonance decoupling the mode in ¹³C NMR spectroscopy in which spin–spin splitting occurs between carbons and the hydrogens attached to them.

oil a triester of glycerol that exists as a liquid at room temperature.

olefin an alkene.

oligomer a protein with more than one peptide chain.

oligonucleotide three to ten nucleotides linked by phosphodiester bonds.

oligopeptide three to ten amino acids linked by amide bonds.

oligosaccharide three to ten sugar molecules linked by glycosidic bonds.

open-chain compound an acyclic compound.

operating frequency the frequency at which an NMR spectrometer operates.

optical isomers stereoisomers that contain chirality centers.

optically active rotates the plane of polarized light.

optically inactive does not rotate the plane of polarized light.

optical purity (enantiomeric excess) how much excess of one enantiomer is present in a mixture of a pair of enantiomers.

orbital the volume of space around the nucleus in which an electron is most likely to be found.

orbital hybridization mixing of orbitals.

organic compound a compound that contains carbon.

organic synthesis preparation of organic compounds from other organic compounds.

organometallic compound a compound containing a carbon–metal bond.

oriented polymer a polymer obtained by stretching out polymer chains and putting them back together in a parallel fashion.

orphan drugs drugs for diseases or conditions that affect fewer than 200,000 people.

ortho/para-directing substituent a substituent that directs an incoming substituent ortho and para to an existing substituent.

osazone the product obtained by treating an aldose or a ketose with excess phenylhydrazine. An osazone contains two imine bonds.

overtone band an absorption that occurs at a multiple of the fundamental absorption frequency $(2v_l, 3v_l)$.

oxidation loss of electrons by an atom or molecule.

oxidation–reduction reaction (redox reaction) a reaction that involves the transfer of electrons from one species to another.

oxidative cleavage an oxidation reaction that cuts the reactant into two or more pieces.

oxime $R_2C{=}NOH$

oxirane (epoxide) an ether in which the oxygen is incorporated into a three-membered ring.

oxonium ion a compound with a positively charged oxygen.

oxyanion a compound with a negatively charged oxygen.

oxymercuration addition of water using a mercuric salt of a carboxylic acid as a catalyst.

ozonide the five-membered ring compound formed as a result of rearrangement of a molozonide.

ozonolysis reaction of a carbon–carbon double or triple bond with ozone.

packing the fitting of individual molecules into a frozen crystal lattice.

paraffin an alkane.

parent hydrocarbon the longest continuous carbon chain in a molecule.

parent ion (molecular ion) peak in the mass spectrum with the greatest m/z.

partial hydrolysis a technique that hydrolyzes only some of the peptide bonds in a polypeptide.

partial racemization formation of a pair of enantiomers in unequal amounts.

Pauli exclusion principle states that no more than two electrons can occupy an orbital and that the two electrons must have opposite spin.

pentose a monosaccharide with five carbons.

peptide polymer of amino acids linked together by amide bonds. A peptide contains fewer amino acid residues than a protein.

peptide bond the amide bond that links the amino acids in a peptide or protein.

peptide nucleic acid (PNA) a polymer containing both an amino acid and a base designed to bind to specific residues on DNA or mRNA.

pericyclic reaction a concerted reaction that takes place as the result of a cyclic rearrangement of electrons.

peroxy acid a carboxylic acid with an OOH group instead of an OH group.

perspective formula a method of representing the spatial arrangement of groups bonded to a chirality center. Two bonds are drawn in the plane of the paper; a solid wedge is used to depict a bond that projects out of the plane of the paper toward the viewer, and a hatched wedge is used to represent a bond that projects back from the plane of the paper away from the viewer.

pH the pH scale is used to describe the acidity of a solution ($pH = -\log [H^+]$).

pH-activity profile a plot of the activity of an enzyme as a function of the pH of the reaction mixture.

phase transfer catalysis catalysis of a reaction by providing a way to bring a polar reagent into a nonpolar phase so that the reaction between a polar and a nonpolar compound can occur.

phase transfer catalyst a compound that carries a polar reagent into a nonpolar phase.

phenone $\overset{\overset{\text{O}}{\|}}{C_6H_5CR}$

phenyl group C_6H_5-

phenylhydrazone $R_2C{=}NNHC_6H_5$

pheromone a compound secreted by an animal that stimulates a physiological or behavioral response from a member of the same species.

phosphatidic acid a phosphoacylglycerol in which only one of the OH groups of phosphate is in an ester linkage.

phosphoacylglycerol (phosphoglyceride) a compound formed when two OH groups of glycerol form esters with fatty acids and the terminal OH group forms a phosphate ester.

phosphoanhydride bond the bond holding two phosphoric acid molecules together.

phospholipid a lipid that contains a phosphate group.

phosphoryl transfer reaction the transfer of a phosphate group from one compound to another.

photochemical reaction a reaction that takes place when a reactant absorbs light.

photosynthesis the synthesis of glucose and O_2 from CO_2 and H_2O.

pH-rate profile a plot of the observed rate for a reaction as a function of the pH of the reaction mixture.

pi (π) bond a bond formed as a result of side-to-side overlap of p orbitals.

pi-complex a complex formed between an electrophile and a triple bond.

pinacol rearrangement rearrangement of a vicinal diol.

pK_a describes the tendency of a compound to lose a proton ($pK_a = -\log K_a$, where K_a is the acid dissociation constant).

plane of symmetry an imaginary plane that bisects a molecule into mirror images.

plasticizer an organic molecule that dissolves in a polymer and allows the polymer chains to slide by each other.

β-pleated sheet the backbone of a polypeptide that is extended in a zigzag structure with hydrogen bonding between neighboring chains.

polar covalent bond a bond formed by the unequal sharing of electrons.

polarimeter an instrument that measures the rotation of polarized light.

polarizability an indication of the ease with which the electron cloud of an atom can be distorted.

polarized light light that oscillates only in one plane.

polar reaction the reaction between a nucleophile and an electrophile.

polyamide a polymer in which the monomers are amides.

polycarbonate a step-growth polymer in which the dicarboxylic acid is carbonic acid.

polyene a compound that has several double bonds.

polyester a polymer in which the monomers are esters.

polymer a large molecule made by linking monomers together.

polymer chemistry the field of chemistry that deals with synthetic polymers; part of the larger discipline known as materials science.

polymerization the process of linking up monomers to form a polymer.

polynucleotide many nucleotides linked by phosphodiester bonds.

polypeptide many amino acids linked by amide bonds.

polysaccharide a compound containing more than ten sugar molecules linked together.

polyunsaturated fatty acid a fatty acid with more than one double bond.

polyurethane a polymer in which the monomers are urethanes.

porphyrin ring system consists of four pyrrole rings joined by one-carbon bridges.

primary alcohol an alcohol in which the OH group is bonded to a primary carbon.

primary alkyl halide an alkyl halide in which the halogen is bonded to a primary carbon.

primary alkyl radical a radical with the unpaired electron on a primary carbon.

primary amine an amine with one alkyl group bonded to the nitrogen.

primary carbocation a carbocation with the positive charge on a primary carbon.

primary carbon a carbon bonded to only one other carbon.

primary hydrogen a hydrogen bonded to a primary carbon.

primary structure (of a nucleic acid) the sequence of bases in the nucleic acid.

primary structure (of a protein) the sequence of amino acids in the protein.

principle of microscopic reversibility states that the mechanism for a reaction in the forward direction has the same intermediates and the same rate-determining step as the mechanism for the reaction in the reverse direction.

prochiral carbonyl carbon a carbonyl carbon that will become a chirality center if it is attacked by a group unlike any of the groups already bonded to it.

prochirality center a carbon bonded to two hydrogens that will become a chirality center if one of the hydrogens is replaced by deuterium.

prokaryote a unicellular body without a nucleus.

promoter site a short sequence of bases at the beginning of a gene.

propagating site the reactive end of a chain-growth polymer.

propagation step in the first of a pair of propagation steps, a radical (or an electrophile or a nucleophile) reacts to produce another radical (or an electrophile or a nucleophile) that reacts in the second propagation step to produce the radical (or the electrophile or the nucleophile) that was the reactant in the first propagation step.

proprietary name (trade name, brand name) identifies a commercial product and distinguishes it from other products.

pro-R-hydrogen replacing this hydrogen with deuterium creates a chirality center with the R configuration.

pro-S-hydrogen replacing this hydrogen with deuterium creates a chirality center with the S configuration.

prostacyclin a lipid, derived from arachidonic acid, that dilates blood vessels and inhibits platelet aggregation.

prosthetic group a tightly bound coenzyme.

protecting group a reagent that protects a functional group from a synthetic operation that it would otherwise not survive.

protein a polymer containing 40 to 4000 amino acids linked by amide bonds.

protic solvent a solvent that has a hydrogen bonded to an oxygen or a nitrogen.

proton a positively charged hydrogen (hydrogen ion); a positively charged particle in an atomic nucleus.

proton-decoupled ^{13}C NMR spectrum a ^{13}C NMR spectrum in which all the signals appear as singlets because there is no coupling between the ^{13}C nucleus and its bonded hydrogens.

proton transfer reaction a reaction in which a proton is transferred from an acid to a base.

protoporphyrin IX the porphyrin ring system of heme.

pseudo-first-order reaction a second-order reaction in which the concentration of one of the reactants is much greater than the other, which allows the reaction to be treated as a first-order reaction.

pyranose a six-membered ring sugar.

pyranoside a six-membered ring glycoside.

pyridoxal phosphate the coenzyme required by enzymes that catalyze certain transformations of amino acids.

quantitative structure–activity relationship (QSAR) the relation between a particular property of a series of compounds and their biological activity.

quantum numbers numbers arising from the quantum mechanical treatment of an atom that describe the properties of the electrons in the atom.

quartet an NMR signal split into four peaks.

quaternary ammonium ion an ion containing a nitrogen bonded to four alkyl groups (R_4N^+).

quaternary ammonium salt a quaternary ammonium ion and an anion ($R_4N^+X^-$).

quaternary structure a description of the way the individual polypeptide chains of a protein are arranged with respect to each other.

racemic mixture (racemate, racemic modification) a mixture of equal amounts of a pair of enantiomers.

radical an atom or molecule with an unpaired electron.

radical addition reaction an addition reaction in which the first species that adds is a radical.

radical anion a species with a negative charge and an unpaired electron.

radical cation a species with a positive charge and an unpaired electron.

radical chain reaction a reaction in which radicals are formed and react in repeating propagating steps.

radical inhibitor a compound that traps radicals.

radical initiator a compound that creates radicals.

radical polymerization chain-growth polymerization in which the initiator is a radical; the propagation site therefore is a radical.

radical reaction a reaction in which a new bond is formed using one electron from one reagent and one electron from another reagent.

radical substitution reaction a substitution reaction that has a radical intermediate.

random coil the conformation of a totally denatured protein.

random copolymer a copolymer with a random distribution of monomers.

random screen (blind screen) the search for a pharmacologically active compound without any information about what chemical structures might show activity.

rate constant a measure of how easy it is to reach the transition state of a reaction (to get over the energy barrier to the reaction).

rate-determining step (rate-limiting step) the step in a reaction that has the transition state with the highest energy.

rational drug design designing drugs with a particular structure to achieve a specific purpose.

R configuration after assigning relative priorities to the four groups bonded to a chirality center, if the lowest priority group is on a vertical axis in a Fischer projection (or pointing away from the viewer in a perspective formula), an arrow drawn from the highest priority group to the next-highest priority group goes in a clockwise direction.

reaction coordinate diagram describes the energy changes that take place during the course of a reaction.

reactivity–selectivity principle states that the greater the reactivity of a species, the less selective it will be.

receptor site the site at which a drug binds in order to exert its physiological effect.

redox reaction (oxidation–reduction reaction) a reaction that involves the transfer of electrons from one species to another.

red shift a shift to a longer wavelength.

reducing sugar a sugar that can be oxidized by reagents such as Ag^+ or Br_2. Reducing sugars are in equilibrium with the open-chain aldose or ketose.

reduction gain of electrons by an atom or molecule.

reductive amination the reaction of an aldehyde or a ketone with ammonia or with a primary amine in the presence of a reducing agent (H_2/Raney Ni).

reference compound a compound added to the sample whose NMR spectrum is to be taken. The positions of the signals in the NMR spectrum are measured from the position of the signal given by the reference compound.

regioselective reaction a reaction that leads to the preferential formation of one constitutional isomer over another.

relative configuration the configuration of a compound relative to the configuration of another compound.

relative rate obtained by dividing the actual rate constant by the rate constant of the slowest reaction in the group being compared.

replication the synthesis of identical copies of DNA.

replication fork the position on DNA at which replication begins.

resolution of a racemic mixture separation of a racemic mixture into the individual enantiomers.

resonance a compound with delocalized electrons is said to have resonance.

resonance contributor (resonance structure, contributing resonance structure) a structure with localized electrons that approximates the true structure of a compound with delocalized electrons.

resonance electron donation donation of electrons through p orbital overlap with neighboring π bonds.

resonance electron withdrawal withdrawal of electrons through p orbital overlap with neighboring π bonds.

resonance energy (delocalization energy) the extra stability associated with a compound as a result of its having delocalized electrons.

resonance hybrid the actual structure of a compound with delocalized electrons; it is represented by two or more structures with localized electrons.

resonances NMR absorption signals.

restriction endonuclease an enzyme that cleaves DNA at a specific base sequence.

restriction fragment a fragment that is formed when DNA is cleaved by a restriction endonuclease.

retrosynthesis (retrosynthetic analysis) working backward (on paper) from the target molecule to available starting materials.

retrovirus a virus whose genetic information is stored in its RNA.

rf radiation radiation in the radiofrequency region of the electromagnetic spectrum.

ribonucleic acid (RNA) a polymer of ribonucleotides.

ribonucleotide a nucleotide in which the sugar component is D-ribose.

ribosome a particle composed of about 40% protein and 60% RNA on which protein biosynthesis takes place.

ribozyme an RNA molecule that acts as a catalyst.

ring current the movement of π electrons around an aromatic benzene ring.

ring-expansion rearrangement rearrangement of a carbocation in which the positively charged carbon is bonded to a cyclic compound and as a result of rearrangement the size of the ring increases by one carbon.

ring-flip (chair–chair interconversion) the conversion of the chair conformer of cyclohexane into the other chair conformer. Bonds that are axial in one chair conformer are equatorial in the other chair conformer.

ring-opening polymerization a chain-growth polymerization that involves opening the ring of the monomer.

Ritter reaction reaction of a nitrile with a secondary or tertiary alcohol to form a secondary amide.

RNA (ribonucleic acid) a polymer of ribonucleotides.

RNA splicing the step in RNA processing that cuts out nonsense bases and splices informational pieces together.

Robinson annulation a Michael reaction followed by an intramolecular aldol condensation.

Rosenmund reduction reduction of an acyl chloride to an aldehyde using H_2 and a deactivated palladium catalyst.

Ruff degradation a method used to shorten an aldose by one carbon.

Sandmeyer reaction the reaction of an aryl diazonium salt with a cuprous salt.

saponification hydrolysis of an ester (such as a fat) under basic conditions.

saturated hydrocarbon a hydrocarbon that is completely saturated with hydrogen (contains no double or triple bonds).

Schiemann reaction the reaction of an arenediazonium salt with HBF_4.

Schiff base $R_2C=NR$

s-cis conformation the conformation in which two double bonds are on the same side of a single bond.

S configuration after assigning relative priorities to the four groups bonded to a chirality center, if the lowest priority group is on a vertical axis in a Fischer projection (or pointing away from the viewer in a perspective formula), an arrow drawn from the highest priority group to the next-highest priority group goes in a counterclockwise direction.

secondary alcohol an alcohol in which the OH group is bonded to a secondary carbon.

secondary alkyl halide an alkyl halide in which the halogen is bonded to a secondary carbon.

secondary alkyl radical a radical with the unpaired electron on a secondary carbon.

secondary amine an amine with two alkyl groups bonded to the nitrogen.

secondary carbocation a carbocation with the positive charge on a secondary carbon.

secondary carbon a carbon bonded to two other carbons.

secondary hydrogen a hydrogen bonded to a secondary carbon.

secondary structure a description of the conformation of the backbone of a protein.

second-order rate constant the rate constant of a second-order reaction.

second-order reaction (bimolecular reaction) a reaction whose rate depends on the concentration of two reactants.

sedimentation constant designates where a species sediments in an ultracentrifuge.

selection rules the rules that determine the outcome of a pericyclic reaction.

selenenylation reaction conversion of an α-bromoketone into an α,β-unsaturated ketone via formation of a selenoxide.

semicarbazone $R_2C=NNHCNH_2$ (with O double-bonded to C)

semiconservative replication the mode of replication that results in a daughter molecule of DNA having one of the original DNA strands plus a newly synthesized strand.

sense strand (informational strand) the strand in DNA that is not read during transcription; it has the same sequence of bases as the synthesized mRNA strand (with a U, T difference).

separated charges a positive and a negative charge that can be neutralized by the movement of electrons.

sesquiterpene a terpene that contains 15 carbons.

shielding phenomenon caused by electron donation to the environment of a proton. The electrons shield the proton from the full effect of the applied magnetic field. The more a proton is shielded, the farther to the right its signal appears in an NMR spectrum.

sigma (σ) bond a bond with a cylindrically symmetrical distribution of electrons.

sigmatropic rearrangement a reaction in which a σ bond is broken in the reactant, a new σ bond is formed in the product, and the π bonds rearrange.

Simmons–Smith reaction formation of a cyclopropane using $CH_2I_2 + Zn(Cu)$.

simple carbohydrate (monosaccharide) a single sugar molecule.

simple triacylglycerol a triacylglycerol in which the fatty acid components are the same.

single bond a σ bond.

singlet an unsplit NMR signal.

site-specific mutagenesis a technique that substitutes one amino acid of a protein for another.

site-specific recognition recognition of a particular site on DNA.

skeletal structure shows the carbon–carbon bonds as lines and does not show the carbon–hydrogen bonds.

S_NAr reaction a nucleophilic aromatic substitution reaction.

S_N1 reaction a unimolecular nucleophilic substitution reaction.

S_N2 reaction a bimolecular nucleophilic substitution reaction.

soap a sodium or potassium salt of a fatty acid.

solid-phase synthesis a technique in which one end of the compound being synthesized is covalently attached to a solid support.

solvation the interaction between a solvent and another molecule (or ion).

solvent-separated ion pair the cation and anion are separated by a solvent molecule.

solvolysis reaction with the solvent.

specific-acid catalysis catalysis in which the proton is fully transferred to the reactant before the slow step of the reaction.

specific-base catalysis catalysis in which the proton is completely removed from the reactant before the slow step of the reaction.

specific rotation the amount of rotation that will be caused by a compound with a concentration of 1.0 g/mL in a sample tube 1.0 dm long.

spectroscopy study of the interaction of matter and electromagnetic radiation.

sphingolipid a lipid that contains sphingosine.

sphingomyelin a sphingolipid in which the terminal OH group of sphingosine is bonded to phosphocholine or phosphoethanolamine.

spin-coupled ^{13}C NMR spectrum a ^{13}C NMR spectrum in which each signal for a carbon is split by the hydrogens bonded to that carbon.

spin coupling the atom that gives rise to an NMR signal is coupled to the rest of the molecule.

spin decoupling the atom that gives rise to an NMR signal is decoupled from the rest of the molecule.

spin–spin coupling the splitting of a signal in an NMR spectrum described by the $N + 1$ rule.

α-spin state nuclei in this spin state have their magnetic moments oriented in the same direction as the applied magnetic field.

β-spin state nuclei in this spin state have their magnetic moments oriented opposite to the direction of the applied magnetic field.

spirocyclic compound a bicyclic compound in which the rings share one carbon.

splitting diagram a diagram that describes the splitting of a set of protons.

squalene a triterpene that is a precursor of steroid molecules.

stacking interactions van der Waals interactions between the mutually induced dipoles of adjacent pairs of bases in DNA.

staggered conformation a conformation in which the bonds on one carbon bisect the bond angle on the adjacent carbon when viewed looking down the carbon–carbon bond.

step-growth polymer (condensation polymer) a polymer made by combining two molecules while removing a small molecule (usually water or an alcohol).

stereochemistry the field of chemistry that deals with the structures of molecules in three dimensions.

stereoelectronic effects the combination of steric effects and electronic effects.

stereogenic center (chirality center) a tetrahedral atom bonded to four different substituents.

stereoisomers isomers that differ in the way the atoms are arranged in space.

stereoselective reaction a reaction that leads to the preferential formation of one stereoisomer over another.

stereospecific reaction a reaction in which the reactant can exist as stereoisomers and each stereoisomeric reactant leads to a different stereoisomeric product or set of products.

steric effects effects due to the fact that groups occupy a certain volume of space.

steric hindrance refers to bulky groups at the site of a reaction that make it difficult for the reactants to approach each other.

steric strain (van der Waals strain, van der Waals repulsion) the repulsion between the electron cloud of an atom or group of atoms and the electron cloud of another atom or group of atoms.

steroid a class of compounds that contains a steroid ring system.

stop codon a codon that says "stop protein synthesis here."

Stork enamine reaction uses an enamine as a nucleophile in a Michael reaction.

straight-chain alkane (normal alkane) an alkane in which the carbons form a contiguous chain with no branches.

s-trans conformation a conformation in which two double bonds are on opposite sides of a single bond.

Strecker synthesis a method used to synthesize an amino acid: An aldehyde reacts with NH_3, forming an imine that is attacked by cyanide ion. Hydrolysis of the product gives an amino acid.

stretching frequency the frequency at which a stretching vibration occurs.

stretching vibration a vibration occurring along the line of the bond.

structural isomers (constitutional isomers) molecules that have the same molecular formula but differ in the way the atoms are connected.

structural protein a protein that gives strength to a biological structure.

α-substituent a substituent on the opposite side of a steroid ring system as the angular methyl groups.

β-substituent a substituent on the same side of a steroid ring system as the angular methyl groups.

α-substitution reaction a reaction that puts a substituent on an α-carbon in place of an α-hydrogen.

substrate the reactant of an enzyme-catalyzed reaction.

subunit an individual chain of an oligomer.

suicide inhibitor (mechanism-based inhibitor) a compound that inactivates an enzyme by undergoing part of its normal catalytic mechanism.

sulfide (thioether) the sulfur analog of an ether (RSR).

sulfonate ester the ester of a sulfonic acid (RSO_2OR).

sulfonation substitution of a hydrogen of a benzene ring by a sulfonic acid group ($-SO_3H$).

suprafacial bond formation formation of two σ bonds on the same side of the π system.

suprafacial rearrangement rearrangement in which the migrating group remains on the same face of the π system.

symmetrical anhydride an acid anhydride with identical R groups.

$$R-\overset{\overset{\displaystyle O}{\|}}{C}-O-\overset{\overset{\displaystyle O}{\|}}{C}-R$$

symmetrical ether an ether with two identical substituents bonded to the oxygen.

symmetric molecular orbital a molecular orbital in which the left half is a mirror image of the right half.

symmetry-allowed pathway a pathway that leads to overlap of in-phase orbitals.

symmetry-forbidden pathway a pathway that leads to overlap of out-of-phase orbitals.

syn addition an addition reaction in which the two added substituents add to the same side of the molecule.

syndiotactic polymer a polymer in which the substituents regularly alternate on both sides of the fully extended carbon chain.

syn elimination an elimination reaction in which the two substituents eliminated are removed from the same side of the molecule.

syn-periplanar parallel substituents on the same side of a molecule.

synthetic equivalent the reagent actually used as the source of a synthon.

synthetic polymer a polymer that is not synthesized in nature.

synthetic tree an outline of the available routes to get to the desired product from available starting materials.

synthon a fragment of a disconnection.

systematic nomenclature nomenclature based on structure.

target molecule the desired end product of a synthesis.

tautomerism interconversion of tautomers.

tautomers isomers that differ in the location of a double bond and a hydrogen.

template strand (antisense strand) the strand in DNA that is read during transcription.

terminal alkyne an alkyne with the triple bond at the end of the carbon chain.

termination step when two radicals combine to produce a molecule in which all the electrons are paired.

terpene a lipid, isolated from a plant, that contains carbon atoms in multiples of five.

terpenoid a terpene that contains oxygen.

tertiary alcohol an alcohol in which the OH group is bonded to a tertiary carbon.

tertiary alkyl halide an alkyl halide in which the halogen is bonded to a tertiary carbon.

tertiary alkyl radical a radical with the unpaired electron on a tertiary carbon.

tertiary amine an amine with three alkyl groups bonded to the nitrogen.

tertiary carbocation a carbocation with the positive charge on a tertiary carbon.

tertiary carbon a carbon bonded to three other carbons.

tertiary hydrogen a hydrogen bonded to a tertiary carbon.

tertiary structure a description of the three-dimensional arrangement of all the atoms in a protein.

tetraene a hydrocarbon with four double bonds.

tetrahedral bond angle the bond angle (109.5°) formed by adjacent bonds of an sp^3 hybridized carbon.

tetrahedral carbon an sp^3 hybridized carbon; a carbon that forms covalent bonds using four sp^3 hybridized orbitals.

tetrahedral intermediate the intermediate formed in a nucleophilic acyl substitution reaction.

tetrahydrofolate (THF) the coenzyme required by enzymes that catalyze reactions that donate a group containing a single carbon to their substrates.

tetraterpene a terpene that contains 40 carbons.

tetrose a monosaccharide with four carbons.

therapeutic index the ratio of the lethal dose of a drug to the therapeutic dose.

thermal cracking using heat to break a molecule apart.

thermal reaction a reaction that takes place without the reactant having to absorb light.

thermodynamic control when a reaction is under thermodynamic control, the relative amounts of the products depend on their stabilities.

thermodynamic product the most stable product.

thermodynamics the field of chemistry that describes the properties of a system at equilibrium.

thermodynamic stability is indicated by $\Delta G°$. If $\Delta G°$ is negative, the products are more stable than the reactants. If $\Delta G°$ is positive, the reactants are more stable than the products.

thermoplastic polymer a polymer that has both ordered crystalline regions and amorphous noncrystalline regions.

thermosetting polymer cross-linked polymers that, after they are hardened, cannot be remelted by heating.

thiamine pyrophosphate (TPP) the coenzyme required by enzymes that catalyze a reaction that transfers a two-carbon fragment to a substrate.

thiirane a three-membered ring compound in which one of the ring atoms is a sulfur.

thin-layer chromatography a technique that separates compounds on the basis of their polarity.

thioester the sulfur analog of an ester.

thioether (sulfide) the sulfur analog of an ether (RSR).

thiol (mercaptan) the sulfur analog of an alcohol (RSH).

threo enantiomers the pair of enantiomers with similar groups on opposite sides when drawn in a Fischer projection.

titration curve a plot of pH versus added equivalents of hydroxide ion.

Tollens test an aldehyde can be identified by observation of the formation of a silver mirror in the presence of Tollens' reagent (Ag_2O/NH_3).

torsional strain the repulsion felt by the bonding electrons of one substituent as they pass close to the bonding electrons of another substituent.

trademark a registered name, symbol, or picture.

trade name (proprietary name, brand name) identifies a commercial product and distinguishes it from other products.

transamination a reaction in which an amino group is transferred from one compound to another.

transannular hydrogens (flagpole hydrogens) the two hydrogens in the boat conformation of cyclohexane that are closest to each other.

transcription the synthesis of mRNA from a DNA blueprint.

transesterification reaction the reaction of an ester with an alcohol to form a different ester.

trans-fused two cyclohexane rings fused together such that if the second ring were considered to be two substituents of the first ring, both substituents would be in equatorial positions.

transimination the reaction of a primary amine with an imine to form a new imine and a primary amine derived from the original imine.

trans isomer the isomer with identical substituents on opposite sides of the double bond.

transition state the highest point on a hill in a reaction coordinate diagram. In the transition state, bonds in the reactant that will break are partially broken and bonds in the product that will form are partially formed.

transition state analog a compound that is structurally similar to the transition state of an enzyme-catalyzed reaction.

translation the synthesis of a protein from an mRNA blueprint.

transmetallation metal exchange.

triacylglycerol the compound formed when the three OH groups of glycerol are esterified with fatty acids.

triene a hydrocarbon with three double bonds.

trigonal planar carbon an sp^2 hybridized carbon.

triose a monosaccharide with three carbons.

tripeptide three amino acids linked by amide bonds.

triple bond a σ bond plus two π bonds.

triplet an NMR signal split into three peaks.

triterpene a terpene that contains 30 carbons.

twist-boat conformation (skew-boat conformation) a conformation of cyclohexane.

ultraviolet light electromagnetic radiation with wavelengths ranging from 180 to 400 nm.

umpolung reversing the normal polarity of a functional group.

unimolecular reaction (first-order reaction) a reaction whose rate depends on the concentration of one reactant.

unsaturated hydrocarbon a hydrocarbon that contains one or more double or triple bonds.

urethane a compound with a carbonyl group that is both an amide and an ester.

UV/Vis spectroscopy the absorption of electromagnetic radiation in the ultraviolet and visible regions to determine information about conjugated systems.

valence electron an electron in an unfilled shell.

valence shell electron pair repulsion (VSEPR) model combines the concept of atomic orbitals with the concept of shared electron pairs and the minimization of electron pair repulsion.

van der Waals forces (London forces) induced dipole–induced dipole interactions.

van der Waals radius a measure of the effective size of an atom or group. A repulsive force occurs (van der Waals repulsion) if two atoms approach each other at a distance less than the sum of their van der Waals radii.

vector sum takes into account both the magnitudes and the directions of the bond dipoles.

vicinal dihalide a compound with halogens bonded to adjacent carbons.

vicinal diol (vicinal glycol) a compound with OH groups bonded to adjacent carbons.

vinyl group CH_2=CH—

vinylic carbon a carbon in a carbon–carbon double bond.

vinylic cation a compound with a positive charge on a vinylic carbon.

vinylic radical a compound with an unpaired electron on a vinylic carbon.

vinylogy transmission of reactivity through double bonds.

vinyl polymer a polymer in which the monomers are ethylene or a substituted ethylene.

visible light electromagnetic radiation with wavelengths ranging from 400 to 780 nm.

vitamin a substance needed in small amounts for normal body function that the body cannot synthesize or cannot synthesize in adequate amounts.

vitamin KH_2 the coenzyme required by the enzyme that catalyzes the carboxylation of glutamate side chains.

vulcanization increasing the flexibility of rubber by heating it with sulfur.

wave equation an equation that describes the behavior of each electron in an atom or a molecule.

wave functions a series of solutions of a wave equation.

wavelength distance from any point on one wave to the corresponding point on the next wave (usually in units of μm or nm).

wavenumber the number of waves in 1 cm.

wax an ester formed from a long-chain carboxylic acid and a long-chain alcohol.

wedge-and-dash structure a method of representing the spatial arrangement of groups. Wedges are used to represent bonds that point out of the plane of the paper toward the viewer, and dashed lines are used to represent bonds that point back from the plane of the paper away from the viewer.

Williamson ether synthesis formation of an ether from the reaction of an alkoxide ion with an alkyl halide.

Wittig reaction the reaction of an aldehyde or a ketone with a phosphonium ylide, resulting in formation of an alkene.

Wolff–Kishner reduction a reaction that reduces the carbonyl group of a ketone to a methylene group using NH_2NH_2/HO^-.

Woodward–Fieser rules allow the calculation of the λ_{max} of the $\pi \longrightarrow \pi^*$ transition for compounds with four or fewer conjugated double bonds.

Woodward–Hoffmann rules a series of selection rules for pericyclic reactions.

ylide a compound with opposite charges on adjacent, covalently bonded atoms with complete octets.

Zaitsev's rule the more stable alkene product is obtained by removing a proton from the β-carbon that is bonded to the fewest hydrogens.

Z conformation the conformation of a carboxylic acid or carboxylic acid derivative in which the carbonyl oxygen and the substituent bonded to the carboxyl oxygen or nitrogen are on the same side of the single bond.

Ziegler–Natta catalyst an aluminum–titanium initiator that controls the stereochemistry of a polymer.

Z isomer the isomer with the high-priority groups on the same side of the double bond.

zwitterion a compound with a negative charge and a positive charge on nonadjacent atoms.

Photo Credits

Periodic Table of the Elements

Main groups

Main groups

Transition metals

1 1A	2 2A	3 3B	4 4B	5 5B	6 6B	7 7B	8 8B	9 8B	10	11 1B	12 2B	13 3A	14 4A	15 5A	16 6A	17 7A	18 8A
1 H 1.00794																	2 He 4.00260
3 Li 6.941	4 Be 9.01218											5 B 10.81	6 C 12.011	7 N 14.0067	8 O 15.9994	9 F 18.998403	10 Ne 20.1797
11 Na 22.98977	12 Mg 24.305											13 Al 26.98154	14 Si 28.0855	15 P 30.97376	16 S 32.066	17 Cl 35.453	18 Ar 39.948
19 K 39.0983	20 Ca 40.078	21 Sc 44.9559	22 Ti 47.88	23 V 50.9415	24 Cr 51.996	25 Mn 54.9380	26 Fe 55.847	27 Co 58.9332	28 Ni 58.69	29 Cu 63.546	30 Zn 65.39	31 Ga 69.72	32 Ge 72.61	33 As 74.9216	34 Se 78.96	35 Br 79.904	36 Kr 83.80
37 Rb 85.4678	38 Sr 87.62	39 Y 88.9059	40 Zr 91.224	41 Nb 92.9064	42 Mo 95.94	43 Tc (98)	44 Ru 101.07	45 Rh 102.9055	46 Pd 106.42	47 Ag 107.8682	48 Cd 112.41	49 In 114.82	50 Sn 118.710	51 Sb 121.757	52 Te 127.60	53 I 126.9045	54 Xe 131.29
55 Cs 132.9054	56 Ba 137.33	57 *La 138.9055	72 Hf 178.49	73 Ta 180.9479	74 W 183.85	75 Re 186.207	76 Os 190.2	77 Ir 192.22	78 Pt 195.08	79 Au 196.9665	80 Hg 200.59	81 Tl 204.383	82 Pb 207.2	83 Bi 208.9804	84 Po (209)	85 At (210)	86 Rn (222)
87 Fr (223)	88 Ra 226.0254	89 †Ac 227.0278	104 Rf (261)	105 Db (262)	106 Sg (266)	107 Bh (264)	108 Hs (269)	109 Mt (268)	110 (271)	111 (272)	112 (277)		114 (289)		116 (289)		118 (293)

*Lanthanide series

58 Ce 140.12	59 Pr 140.9077	60 Nd 144.24	61 Pm (145)	62 Sm 150.36	63 Eu 151.96	64 Gd 157.25	65 Tb 158.9254	66 Dy 162.50	67 Ho 164.9304	68 Er 167.26	69 Tm 168.9342	70 Yb 173.04	71 Lu 174.967

†Actinide series

90 Th 232.0381	91 Pa 231.0359	92 U 238.0289	93 Np 237.048	94 Pu (244)	95 Am (243)	96 Cm (247)	97 Bk (247)	98 Cf (251)	99 Es (252)	100 Fm (257)	101 Md (258)	102 No (259)	103 Lr (262)

Masses in parentheses apply to mass number of the longest-lived isotope.

s block elements

p block elements

d block elements

f block elements

Common Functional Groups

Alkane $\quad RCH_3$

Alkene
$$\underset{\text{internal}}{\text{C}=\text{C}} \quad \underset{\text{terminal}}{\text{C}=\text{CH}_2}$$

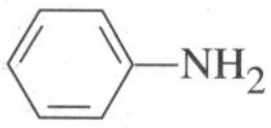

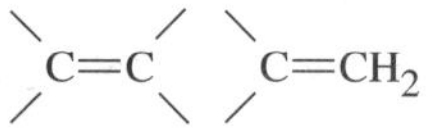

Alkyne $\quad \underset{\text{internal}}{RC \equiv CR} \quad \underset{\text{terminal}}{RC \equiv CH}$

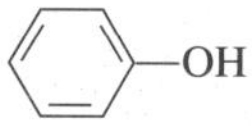

Nitrile $\quad RC \equiv N$

Ether $\quad R-O-R$

Thiol $\quad RCH_2-SH$

Sulfide $\quad R-S-R$

Disulfide $\quad R-S-S-R$

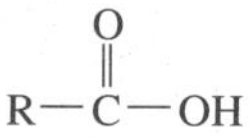

Epoxide

Aniline

Phenol

Carboxylic acid $\quad R-\overset{\overset{\displaystyle O}{\|}}{C}-OH$

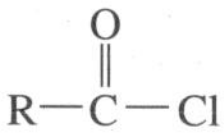

Acyl chloride $\quad R-\overset{\overset{\displaystyle O}{\|}}{C}-Cl$

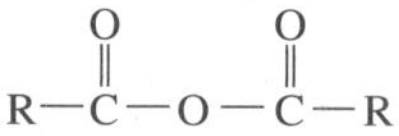

Acid anhydride $\quad R-\overset{\overset{\displaystyle O}{\|}}{C}-O-\overset{\overset{\displaystyle O}{\|}}{C}-R$

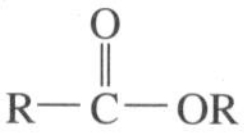

Ester $\quad R-\overset{\overset{\displaystyle O}{\|}}{C}-OR$

Amide $\quad R-\overset{\overset{\displaystyle O}{\|}}{C}-NH_2 \quad -NHR \quad -NR_2$

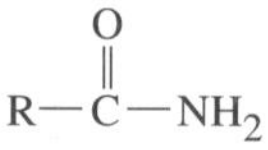

Aldehyde $\quad R-\overset{\overset{\displaystyle O}{\|}}{C}-H$

Ketone $\quad R-\overset{\overset{\displaystyle O}{\|}}{C}-R$

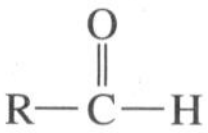

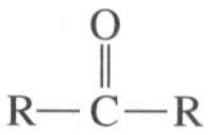

	primary	secondary	tertiary
Alkyl halide	$R-CH_2-X$ X = F, Cl, Br, or I	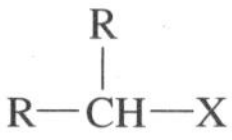 $R-\overset{\overset{\displaystyle R}{\|}}{C}H-X$	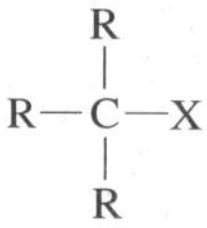 $R-\overset{\overset{\displaystyle R}{\|}}{\underset{\underset{\displaystyle R}{\|}}{C}}-X$
Alcohol	$R-CH_2-OH$	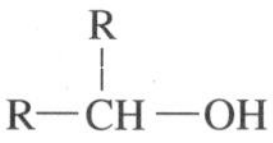 $R-\overset{\overset{\displaystyle R}{\|}}{C}H-OH$	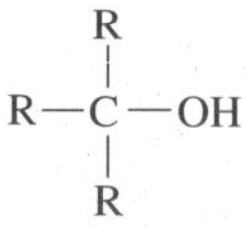 $R-\overset{\overset{\displaystyle R}{\|}}{\underset{\underset{\displaystyle R}{\|}}{C}}-OH$
Amine	$R-NH_2$	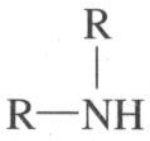 $R-\overset{\overset{\displaystyle R}{\|}}{N}H$	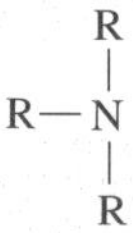 $R-\overset{\overset{\displaystyle R}{\|}}{\underset{\underset{\displaystyle R}{\|}}{N}}$